INFINITE SERIES

$$\sum_{n=m}^{\infty} cr^n = \frac{cr^m}{1-r} \quad \text{(geometric series)}$$

$$e^x = \sum_{n=0}^{\infty} \frac{x^n}{n!}$$

$$\sin x = \sum_{n=0}^{\infty} \frac{(-1)^n}{(2n+1)!} x^{2n+1}$$

$$\cos x = \sum_{n=0}^{\infty} \frac{(-1)^n}{(2n)!} x^{2n}$$

$$(1+x)^s = \sum_{n=0}^{\infty} \binom{s}{n} x^n \quad \text{for } -1 < x < 1 \quad \text{(binomial series)}$$

$$f(x) = \sum_{n=0}^{\infty} \frac{f^{(n)}(a)}{n!} (x-a)^n \quad \text{(Taylor series)}$$

$$r_n(x) = \frac{f^{(n+1)}(t_x)}{(n+1)!} (x-a)^{n+1} \quad \text{(nth Taylor remainder)}$$

GEOMETRIC FORMULAS

 area: $\dfrac{1}{2} bh$

 area: bh

 area: πr^2
circumference: $2\pi r$

 area: $\dfrac{\theta}{2} r^2$

 volume: abc
surface area: $2ab + 2ac + 2bc$

 volume: $\frac{4}{3}\pi r^3$
surface area: $4\pi r^2$

 volume: $\pi a^2 h$
surface area: $2\pi ah$

 volume: $\frac{1}{3}\pi a^2 h$
surface area: $\pi a \sqrt{a^2 + h^2}$

$$|PQ| = \sqrt{(x_2 - x_1)^2 + (y_2 - y_1)^2} \quad \text{(distance formula)}$$

$$y - y_1 = m(x - x_1) \quad \text{(point-slope equation of a line)}$$

$$y = mx + b \quad \text{(slope-intercept equation of a line)}$$

ALGEBRAIC FORMULAS

$$|a + b| \leq |a| + |b| \qquad -|a| \leq a \leq |a|$$

$$ax^2 + bx + c = 0 \quad \text{if } x = \frac{-b + \sqrt{b^2 - 4ac}}{2a} \quad \text{or} \quad x = \frac{-b - \sqrt{b^2 - 4ac}}{2a} \quad \text{(quadratic formula)}$$

$$n! = n(n - 1)(n - 2) \cdots (3)(2)(1)$$

$$\binom{s}{n} = \frac{s(s - 1)(s - 2) \cdots (s - (n - 1))}{n!} \quad \text{for } n \geq 1 \quad \text{(binomial coefficient)}$$

TRIGONOMETRIC FORMULAS

$$\sin(x \pm y) = \sin x \cos y \pm \cos x \sin y \qquad \sin^2 x + \cos^2 x = 1 \qquad \sec^2 x = 1 + \tan^2 x$$

$$\cos(x \pm y) = \cos x \cos y \mp \sin x \sin y \qquad \sin(-x) = -\sin x \qquad \cos(-x) = \cos x$$

$$\tan(x \pm y) = \frac{\tan x \pm \tan y}{1 \mp \tan x \tan y} \qquad \sin 2x = 2 \sin x \cos x \qquad \cos 2x = \cos^2 x - \sin^2 x$$

$$\sin^2 x = \frac{1 - \cos 2x}{2} \qquad \cos^2 x = \frac{1 + \cos 2x}{2}$$

EXPONENTIAL AND LOGARITHMIC FORMULAS

$$a^r = e^{r \ln a} \qquad \log_a bc = \log_a b + \log_a c$$

$$a^{b+c} = a^b a^c \qquad \log_a b^r = r \log_a b$$

$$(a^b)^c = a^{bc} \qquad \log_a x = \frac{\ln x}{\ln a}$$

$$a^{-b} = \frac{1}{a^b}$$

VECTOR FORMULAS

$$|PQ| = \sqrt{(x_1 - x_0)^2 + (y_1 - y_0)^2 + (z_1 - z_0)^2} \quad \text{(distance formula)}$$

$$\| a_1 \mathbf{i} + a_2 \mathbf{j} + a_3 \mathbf{k} \| = \sqrt{a_1^2 + a_2^2 + a_3^2}$$

$$\mathbf{a} \cdot \mathbf{b} = a_1 b_1 + a_2 b_2 + a_3 b_3 = \| \mathbf{a} \| \| \mathbf{b} \| \cos \theta$$

$$\mathbf{a} \times \mathbf{b} = (a_2 b_3 - a_3 b_2)\mathbf{i} + (a_3 b_1 - a_1 b_3)\mathbf{j} + (a_1 b_2 - a_2 b_1)\mathbf{k}$$

$$\| \mathbf{a} \times \mathbf{b} \| = \| \mathbf{a} \| \| \mathbf{b} \| \sin \theta$$

$$\mathbf{v} = \frac{d\mathbf{r}}{dt} \qquad \mathbf{a} = \frac{d\mathbf{v}}{dt}$$

$$\| \mathbf{v} \| = \frac{ds}{dt} = \sqrt{\left(\frac{dx}{dt}\right)^2 + \left(\frac{dy}{dt}\right)^2 + \left(\frac{dz}{dt}\right)^2}$$

$$\mathbf{T}(t) = \frac{\mathbf{r}'(t)}{\| \mathbf{r}'(t) \|} = \frac{d\mathbf{r}/dt}{\| d\mathbf{r}/dt \|}$$

$$\mathbf{N}(t) = \frac{\mathbf{T}'(t)}{\| \mathbf{T}'(t) \|} = \frac{d\mathbf{T}/dt}{\| d\mathbf{T}/dt \|}$$

$$\kappa(t) = \frac{\| \mathbf{T}'(t) \|}{\| \mathbf{r}'(t) \|} = \frac{\| \mathbf{v} \times \mathbf{a} \|}{\| \mathbf{v} \|^3}$$

$$f_x(x, y, z) = \frac{\partial f}{\partial x} = \lim_{h \to 0} \frac{f(x + h, y, z) - f(x, y, z)}{h}$$

$$f_y(x, y, z) = \frac{\partial f}{\partial y} = \lim_{h \to 0} \frac{f(x, y + h, z) - f(x, y, z)}{h}$$

$$f_z(x, y, z) = \frac{\partial f}{\partial z} = \lim_{h \to 0} \frac{f(x, y, z + h) - f(x, y, z)}{h}$$

$$z = f(x_0, y_0) + f_x(x_0, y_0)(x - x_0) + f_y(x_0, y_0)(y - y_0)$$

(equation of the tangent plane of f at $(x_0, y_0, f(x_0, y_0))$)

$$\operatorname{grad} f(x, y, z) = \nabla f(x, y, z) = \frac{\partial f}{\partial x}(x, y, z)\mathbf{i} + \frac{\partial f}{\partial y}(x, y, z)\mathbf{j} + \frac{\partial f}{\partial z}(x, y, z)\mathbf{k}$$

$$\operatorname{div} \mathbf{F}(x, y, z) = \nabla \cdot \mathbf{F}(x, y, z) = \frac{\partial M}{\partial x}(x, y, z) + \frac{\partial N}{\partial y}(x, y, z) + \frac{\partial P}{\partial z}(x, y, z)$$

$$\operatorname{curl} \mathbf{F}(x, y, z) = \nabla \times \mathbf{F}(x, y, z) = \left(\frac{\partial P}{\partial y} - \frac{\partial N}{\partial z}\right)\mathbf{i} + \left(\frac{\partial M}{\partial z} - \frac{\partial P}{\partial x}\right)\mathbf{j} + \left(\frac{\partial N}{\partial x} - \frac{\partial M}{\partial y}\right)\mathbf{k}$$

arc length: $L = \int_a^b \sqrt{\left(\frac{dx}{dt}\right)^2 + \left(\frac{dy}{dt}\right)^2 + \left(\frac{dz}{dt}\right)^2} \, dt$ area: $A = \iint_R 1 \, dA$

surface area: $S = \iint_R \sqrt{[f_x(x, y)]^2 + [f_y(x, y)]^2 + 1} \, dA$ volume: $V = \iiint_D 1 \, dV$

$$\int_C f(x, y, z) \, ds = \int_a^b f(x(t), y(t), z(t)) \left\|\frac{d\mathbf{r}}{dt}\right\| \, dt \qquad \int_C \mathbf{F} \cdot d\mathbf{r} = \int_a^b \mathbf{F}(x(t), y(t), z(t)) \cdot \frac{d\mathbf{r}}{dt} \, dt$$

$$\iint_\Sigma g(x, y, z) \, dS = \iint_R g(x, y, f(x, y)) \sqrt{[f_x(x, y)]^2 + [f_y(x, y)]^2 + 1} \, dA$$

$$\iint_\Sigma \mathbf{F} \cdot \mathbf{n} \, dS = \pm \iint_R [M(x, y, f(x, y))f_x(x, y) + N(x, y, f(x, y))f_y(x, y) - P(x, y, f(x, y))] \, dA$$

$$\int_C \operatorname{grad} f \cdot d\mathbf{r} = f(x_1, y_1, z_1) - f(x_0, y_0, z_0) \quad \text{(Fundamental Theorem of Line Integrals)}$$

$$\int_C M(x, y) \, dx + N(x, y) \, dy = \iint_R \left(\frac{\partial N}{\partial x} - \frac{\partial M}{\partial y}\right) dA \quad \text{(Green's Theorem)}$$

$$\int_C \mathbf{F} \cdot d\mathbf{r} = \iint_\Sigma (\operatorname{curl} \mathbf{F}) \cdot \mathbf{n} \, dS \quad \text{(Stokes's Theorem)}$$

$$\iint_\Sigma \mathbf{F} \cdot \mathbf{n} \, dS = \iiint_D \operatorname{div} \mathbf{F}(x, y, z) \, dV \quad \text{(Divergence Theorem)}$$

CALCULUS
WITH ANALYTIC GEOMETRY

Sixth Edition

Robert Ellis
University of Maryland

Denny Gulick
University of Maryland

Australia · Canada · Mexico · Singapore · Spain · United Kingdom · United States

CALCULUS WITH ANALYTIC GEOMETRY
Ellis / Gulick

Executive Editors:
Michele Baird, Maureen Staudt &
Michael Stranz

Project Development Manager:
Linda de Stefano

Marketing Coordinators:
Lindsay Annett and Sara Mercurio

Production/Manufacturing Supervisor:
Donna M. Brown

Pre-Media Services Supervisor:
Dan Plofchan

Rights and Permissions Specialists:
Kalina Hintz and Bahman Naraghi

Cover Image
Getty Images*

The Adaptable Courseware Program
consists of products and additions to
existing Thomson products that are
produced from camera-ready copy.
Peer review, class testing, and
accuracy are primarily the responsibility
of the author(s).

ISBN-13: 978-0-7593-1379-2
ISBN-10: 0-7593-1379-2

International Divisions List

Asia (Including India):
Thomson Learning
(a division of Thomson Asia Pte Ltd)
5 Shenton Way #01-01
UIC Building
Singapore 068808
Tel: (65) 6410-1200
Fax: (65) 6410-1208

Australia/New Zealand:
Thomson Learning Australia
102 Dodds Street
Southbank, Victoria 3006
Australia

Latin America:
Thomson Learning
Seneca 53
Colonia Polano
11560 Mexico, D.F., Mexico
Tel (525) 281-2906
Fax (525) 281-2656

Canada:
Thomson Nelson
1120 Birchmount Road
Toronto, Ontario
Canada M1K 5G4
Tel (416) 752-9100
Fax (416) 752-8102

UK/Europe/Middle East/Africa:
Thomson Learning
High Holborn House
50-51 Bedford Row
London, WC1R 4LS
United Kingdom
Tel 44 (020) 7067-2500
Fax 44 (020) 7067-2600

Spain (Includes Portugal):
Thomson Paraninfo
Calle Magallanes 25
28015 Madrid
España
Tel 34 (0)91 446-3350
Fax 34 (0)91 445-6218

PREFACE

Like its predecessors, *Calculus* contains all the topics that normally constitute a course in calculus of one and several variables. It is suitable for sequences taught in three semesters or in four or five quarters. In the three-semester case, the first semester would usually include the introductory chapter (Chapter 1), the three chapters on limits and derivatives (Chapters 2 to 4), and the initial chapter on integrals (Chapter 5). The second semester would then include the rest of the discussion of integration (Chapters 6 to 8) and some combination of the chapters on sequences and series (Chapter 9), polar coordinates and conic sections (Chapter 10), and the introduction to vectors and vector-valued functions (Chapters 11 and 12). The third semester would include the remainder of those chapters, along with the material on calculus of several variables (Chapters 13 and 14) and Chapter 15, which contains the theorems of Green and Stokes as well as the Divergence Theorem.

New to this Edition

Noteworthy features that are new to this edition include:

Many passages, especially the introductory passages, have been rewritten to be more accessible to the reader and to motivate the main ideas under discussion.

At the end of nearly every section's exercise set throughout Chapters 2-12 and in many sections in the remainder of the text there are mini-projects that are designed to expand on the concepts put forth in the section.

Quadratic approximation of functions has been added to Section 3.8 to complement linear approximation and better prepare the reader for Taylor polynomials. The chapter on applications of the integral (Chapter 6) now appears before the chapter on techniques of integration (Chapter 8). This change reflects the general availability of software packages such as Mathematica, MATLAB, Maple, and Derive that perform symbolic integration. One technique—substitution in integrals—is still presented in Section 5.6, before the applications of the integral; the other standard techniques have been retained and appear in Chapter 8. Nevertheless, with a little care in selecting exercises, Chapter 8 could easily be presented before Chapter 6 if desirable.

An initial discussion of parametrized curves has been placed in Chapter 6 so that the applications of length and area can involve such curves.

A new section (Section 14.9) on parametrized surfaces expands the class of surfaces whose areas can be computed and to which one can apply Stokes's and the Divergence Theorems in Chapter 15. New examples and exercises have been added in both Chapter 14 and Chapter 15 to reflect this change.

New exercises, both applied and mathematical, appear through the text.

Organization

Although we have been careful in selecting the order in which the topics appear in this edition, there is flexibility in the choice of topics and the order in which they are introduced. Chapter 1 (which includes an introduction to the trigonometric, logarithmic, and exponential functions) is preliminary and can be covered quickly if the reader's preparation is sufficient. With a little care, techniques of integration (Chapter 8) can be discussed before applications of the integral (Chapter 6). In addition, sequences and series (Chapter 9) can be studied any time after Chapter 8, and conic sections (Sections 10.3 to 10.5) can be considered any time after Chapter 4.

Pedagogical Features

The pedagogical features that have made the earlier editions so helpful to both students and instructors have been retained or improved for this edition.

Whenever possible, we use geometric and intuitive motivation to introduce concepts and results so that readers may readily absorb the definitions and theorems that follow.

The concepts are supported by numerous examples, with graphical and numerical emphasis where appropriate.

The topical development, in which we employ numerous worked examples and approximately 1000 illustrations, aims for clarity and precision without overburdening the reader with formalism. In keeping with this goal, we have placed the more difficult proofs in the Appendix and have retained in the body of the text the more illuminating and illustrative proofs from first-year calculus. In the chapters on calculus of several variables we have proved selected theorems that aid comprehension of the material.

Exercises appear both at the ends of the sections and, for review, at the end of each chapter. Each set begins with a full complement of routine exercises (graphical, numerical, and analytical) to provide practice in using the ideas and methods presented in the text. These are followed by applied problems and by other exercises of a more challenging nature. The especially difficult exercises are identified with an asterisk.

Exercises requiring the use of a calculator or computer are indicated by the symbol ▦ . One or two projects that require more thought appear at the end of most sections from Chapter 2 on.

Topics for Discussion appear in the Review sections of Chapters 2 to 15. They are ideal for discussion in class and for writing assignments.

Chapters 3 to 15 all end with a collection of Cumulative Review Exercises, which are intended to reinforce the main ideas of the previous chapters.

In the interest of accuracy, every exercise has been completely worked by each of the authors. Answers to odd-numbered exercises (except those requiring longer explanations) appear at the back of the book.

Throughout the book, statements of definitions, theorems, lemmas, and corollaries, as well as important formulas, are highlighted with tints for easy identification. Numbering is consecutive throughout each chapter for definitions and theorems, and consecutive within each section for examples and formulas. We use the symbol ■ to signal the end of a proof and ❏ for the end of the solution to an example.

Lists of Key Terms and Expressions, Key Formulas, and Key Theorems appear at the end of each chapter before the Review Exercises.

On the endpapers we have assembled important formulas and results to facilitate reference.

Pronunciation of difficult terms and names is shown in footnotes on the pages where they first appear.

Solutions Manuals

The **Student Solutions Manual**, by the authors, is available for purchase. It contains worked-out solutions for all odd-numbered exercises in the text.

The **Instructor's Solutions Manual**, by the authors, is available free of charge and in two volumes to instructors who adopt this text. It contains worked-out solutions for all exercises and a separate answer section for all exercises. (Answers to odd-numbered exercises are given at the back of the book.)

Acknowledgements

We are grateful to many people who have helped us in a variety of ways as we prepared the various editions of this book. Our thanks go to these reviewers:

Linda J. S. Allen, *Texas Tech University*
Robert M. Anderson, *Boise State University*
James Angelos, *Central Michigan University*
Michael Bleicher, *Clark Atlanta University*
Gary A. Bogar, *Montana State University*
Jack Ceder, *University of California–Santa Barbara*
Richard M. Davitt, *University of Louisville*
Frank Glaser de Lugo, *California State Polytechnic Institute–Pomona*
Benny Evans, *Oklahoma State University*
W. E. Fitzgibbon, *University of Houston*
Martin Flashman, *Humboldt State University*
Herbert A. Gindler, *San Diego State University*
Stuart Goldenberg, *California Polytechnic State University*
Dorian Goldfeld, *Columbia University*
John Gosselin, *University of Georgia*
Kamel N. Haddad, *California State University–Bakersfield*
Judykay Hartzell, *University of Nebraska–Omaha*
Chung-Wu Ho, *Southern Illinois University–Edwardsville*
Arnold J. Insel, *Illinois State University*
David Johnson, *Lehigh University*
Ronald Knill, *Tulane University*
Cecilia Knoll, *Florida Institute of Technology*
Jack Lamoreaux, *Brigham Young University*
M. Paul Latiolais, *Portland State University*
Daniel McCallum, *University of Arkansas–Little Rock*
Giles Maloof, *Boise State University*
Maurice Monahan, *South Dakota State University*
Robert Myers, *Oklahoma State University*
Thomas Roe, *South Dakota State University*
David Ryeburn, *Simon Fraser University*
Laurence Small, *Los Angeles Pierce College*
Kirby C. Smith, *Texas A&M University*
Hugo S. H. Sun, *California State University–Fresno*
Stephen Willard, *University of Alberta*

Many thanks are also due these survey respondents:

John Akeroyd, *University of Arkansas*
Maria Calzada, *Loyola University*
Todd Cochrane, *Kansas State University*
Don Curlovic, *University of South Carolina–Sumter*
Tim Feeman, *Villanova University*
Norman J. Finizio, *University of Rhode Island*
Michael B. Gregory, *University of North Dakota*

Boris Hasselblatt, *Tufts University*
Lee Johnson, *Virginia Polytechnic Institute and State University*
Roger Johnson, *Georgia Institute of Technology*
Sidney Kolpas, *Glendale College*
Peter Kuhfittig, *Milwaukee School of Engineering*
David Minda, *University of Cincinnati*
S. James Taylor, *University of Virginia*
Sergey Yuzvinsky, *University of Oregon*

    Many of our colleagues at the University of Maryland have made contributions to the original writing of this book and to the revisions; we wish to express our appreciation to

Jeffrey Adams	Henry King
Stuart Antman	Umberto Neri
Kenneth Berg	John Osborn
Michael Boyle	Jonathan Rosenberg
Jeff Cooper	Jason Schultz
Jerome Dancis	C. Robert Warner
Paul Green	Peter Wolfe
Frances Gulick	Scott Wolpert
John Horváth	Mishael Zedek

    We are also grateful for comments and suggestions from

Peter M. Gibson, *University of Alabama–Huntsville*
Gregory Grant, *University of Maryland*
Jonathan Sandow, *Yeshiva College*
Steven Schonefeld, *Tri-State University*
William V. Thayer, *St. Louis Community College–Meramec*

    Finally we want to thank Jon Fuller and Andy Gates for their assistance in bringing this edition to publication.

Robert Ellis
Denny Gulick
March 2003

TO THE READER

When you begin to study calculus, you will find that you have encountered many of its concepts and techniques before. Calculus makes extensive use of plane geometry and algebra, two branches of mathematics with which you are already familiar. However, added to these is a third ingredient, which may be new to you: the notion of limit and of limiting processes. From the idea of limit arise the two principal concepts that form the nucleus of calculus; these are the derivative and the integral.

The derivative can be thought of as a rate of change, and this interpretation has many applications. For example, we may use the derivative to find the velocity of an object, such as a rocket, or to determine the maximum and minimum values of a function. In fact, the derivative provides so much information about the behavior of functions that it greatly simplifies graphing them. Because of its broad applicability, the derivative is as important in such disciplines as physics, engineering, economics, and biology as it is in pure mathematics.

The definition of the integral is motivated by the familiar notion of area. Although the methods of plane geometry enable us to calculate the areas of polygons, they do not provide ways of finding the areas of plane regions whose boundaries are curves other than circles. By means of the integral we can find the areas of many such regions. We will also use it to calculate volumes, centers of gravity, lengths of curves, work, and hydrostatic force.

The derivative and the integral have found many diverse uses. The following list, taken from the examples and exercises in this book, illustrates the variety of fields in which these powerful concepts are employed.

Application	Section
Marginal cost and marginal revenue	3.1, 3.3, 5.4
Weight of an astronaut	3.2
Angular velocity and acceleration of a pendulum	3.4, 3.5
Windpipe pressure during a cough	4.1
Dating of a bone or lunar rock sample	4.4
Atmospheric pressure	4.4
Population prediction	4.4, 9.2
Amount of an anesthetic needed during an operation	4.4
Cost of insulating an attic floor	4.6
Blood resistance in vascular branching	4.6
Surface area of a cell in a beehive	4.6
Probable distance of an electron from the center of an atom	4.6
Escape velocity from the earth's gravitational field	4.8, 12.7
Terminal speed of a falling object	4.8, 7.4, 7.8
Blood alcohol levels	Chapter 4 Review
Buffon's needle problem	5.4
Rate of flow of blood in an artery	5.5
Probability of failure in certain systems	5.6
Equity of distribution of income	5.8
Volume of the great Pyramid of Cheops	6.1
Length of a (hanging) telephone line	6.4
Energy required to empty water from a swimming pool	6.4
Center of gravity	6.5, 14.7
Force of water on a dam	6.6
Work done on a piston	Chapter 6 Review
Calibration of a water clock	Chapter 6 Review
Contour of the Gateway Arch in St. Louis	7.4
Shape of a hanging cable	7.4, 7.8
Banking of curves	7.5
Intensity of thermal radiation	7.6, 8.7
Spread of a disease	8.4
Chemical reaction rate	8.4
Thermonuclear reactions	8.4
Present sale value	8.7
Mean life of an atom	8.7
Location of an electron in an atom	8.7
Pareto's law of distribution of income	8.7
Harmonics of a stringed musical instrument	9.2
Compound and continuous interest	9.2
Multiplier effect in economics	9.4
Period of a pendulum	9.8
Shape of a suspended bridge cable	9.10, 10.3
Light-gathering power of a telescope	Chapter 9 Review
Location of the source of a sound	10.3
Orbit of Mars	10.3
Corner mirror on the moon	11.2
Motion of a point on the rim of a tire	12.1

The concepts basic to calculus can be traced, in uncrystallized form, to the time of the ancient Greeks. However, it was only in the sixteenth and early seventeenth centuries that mathematicians developed refined techniques for determining tangents to curves and areas of plane regions. These mathematicians and their ingenious techniques set the stage for Isaac Newton (1642–1727) and Gottfried Leibniz (1646–1716), who are usually credited with the "invention" of calculus because they codified the techniques of calculus and put them into a general setting; moreover, they recognized the importance of the fact that finding derivatives and finding integrals are inverse processes.

During the next 150 years calculus matured bit by bit, and by the middle of the nineteenth century it had become, mathematically, much as we know it today. Thus the definitions and theorems presented in this book were all known a century ago. What is newer is the great diversity of applications, with which we will try to acquaint you throughout the book.

Robert Ellis • Denny Gulick

CONTENTS

13   PARTIAL DERIVATIVES   813

15   CALCULUS OF VECTOR FIELDS   991

14   MULTIPLE INTEGRALS   897

APPENDIX   A-1

INDEX OF SYMBOLS   I-1

INDEX   I-3

The San Andreas Fault in
California.
(James Balog/Black Star)

1 FUNCTIONS

The development of mathematics parallels the human endeavor to understand our physical environment. Numbers were created for counting and measurement, inequalities were introduced to compare sizes, and functions were invented to express the dependence of one physical or geometric quantity on another. Calculus, in turn, was primarily an outgrowth of an attempt to understand the rate at which a variable quantity changes. Thus calculus is first and foremost a study of functions.

This chapter serves as a review of basic concepts related to functions. It begins with real numbers, equations, and inequalities. Functions and their graphs are then introduced. Several general classes of functions are discussed, including polynomial, rational, trigonometric, exponential, and logarithmic functions. These functions appear frequently in science and technology. For example, polynomial functions can describe the motion of a falling object, and logarithmic functions are used in order to measure the magnitude of an earthquake.

If you are already familiar with most of the definitions and concepts in the chapter, we suggest that you read Chapter 1 quickly and proceed to Chapter 2.

1.1   THE REAL NUMBERS

Real numbers, their properties, and their relationships are basic to calculus. Therefore we begin with a description of some important properties of real numbers.

Types of Real Numbers and the Real Number Line

The best-known real numbers are the **integers:**

$$0, \pm 1, \pm 2, \pm 3, \ldots$$

From the integers we derive the **rational numbers.** These are the real numbers that can be written in the form p/q, where p and q are integers and $q \neq 0$. Thus $\frac{48}{37}$, -17, and 1.41 (which is equal to $\frac{141}{100}$) are rational numbers. Any real number that is not rational is called an **irrational number.** Examples of irrational numbers are π and $\sqrt{2}$. (See Exercise 88 at the end of this section for a proof that $\sqrt{2}$ is irrational.)

There is an order $<$ on the real numbers. If $a \neq b$, then either a is less than b or a is greater than b, that is, $a < b$ or $a > b$. For example, $5 < 7$ and $-1 > -2$. If a is less than or equal to b, we write $a \leq b$. If a is greater than or equal to b, we write $a \geq b$. For example, $x^2 \geq 0$ for any real number x. We say that a is **positive** if $a > 0$ and **negative** if $a < 0$. If $a \geq 0$, we say that a is **nonnegative.**

The real numbers can be represented as points on a horizontal line in such a way that if $a < b$, then the point on the line corresponding to the number a lies to the left of the point on the line corresponding to the number b (Figure 1.1).

The equal sign = was first used by Robert Recorde in his algebra book *The Whetstone of Witte,* published in 1557. He chose the two bars because "Noe 2 thynges can be moare equalle" than two parallel bars. It is interesting that the symbols < and > were first introduced nearly a century later, by the mathematician Thomas Harriot.

FIGURE 1.1   The real line.

Such a line is called the **real number line** or **real line.** We think of the real numbers as points on the real line, and vice versa. Thus we say that the negative numbers lie to the left of 0 and the positive numbers lie to the right of 0.

In this book the numbers we consider are almost exclusively real numbers. For that reason we frequently refer to real numbers simply as numbers.

Intervals

Certain sets of real numbers, called **intervals,** appear with great frequency in calculus. They can be grouped into nine categories:

Kind of interval	Notation	Description
Open interval	(a, b)	all x such that $a < x < b$
Closed interval	$[a, b]$	all x such that $a \leq x \leq b$
Half-open interval	$(a, b]$	all x such that $a < x \leq b$
Half-open interval	$[a, b)$	all x such that $a \leq x < b$
Open interval	(a, ∞)	all x such that $a < x$
Open interval	$(-\infty, a)$	all x such that $x < a$
Closed interval	$[a, \infty)$	all x such that $a \leq x$
Closed interval	$(-\infty, a]$	all x such that $x \leq a$
The real line	$(-\infty, \infty)$	all real numbers

Intervals of the form (a, b), $[a, b]$, $(a, b]$, and $[a, b)$ are **bounded intervals,** and a and b are the **endpoints** of each of these intervals. Figure 1.2 shows the four types of bounded intervals. Intervals of the form (a, ∞), $(-\infty, a)$, $[a, \infty)$, $(-\infty, a]$, and $(-\infty, \infty)$ are **unbounded intervals,** and a is the **endpoint** of each of the first four of these intervals. A number that is in an interval but is not an endpoint of the interval is called an **interior point** of the interval.

> **Caution:** The symbols ∞ and $-\infty$ used above are called "infinity" and "minus infinity," respectively. They do not represent numbers.

Notice that (b, b), $(b, b]$, and $[b, b)$ contain no numbers. More generally, (a, b), $(a, b]$, and $[a, b)$ contain no numbers if $b \leq a$. Whenever we write (a, b), $(a, b]$, or $[a, b)$, we make the implicit assumption that $a < b$. Likewise, we write $[a, b]$ only when $a \leq b$. Notice that $[b, b]$ contains the single number b.

Inequalities and Their Properties

Statements such as $a < b$, $a \leq b$, $a > b$, and $a \geq b$ are called **inequalities.** We list several basic laws for inequalities. In what follows, a, b, c, and d are assumed to be real numbers.

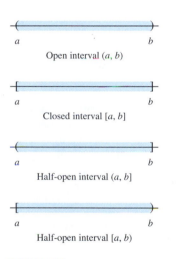

Open interval (a, b)

Closed interval $[a, b]$

Half-open interval $(a, b]$

Half-open interval $[a, b)$

Trichotomy: Either $a < b$, or $a > b$, or $a = b$,
and only one of these holds for any given a and b. (1)

Transitivity: If $a < b$ and $b < c$, then $a < c$. (2)

Additivity: If $a < b$ and $c < d$, then $a + c < b + d$. (3)

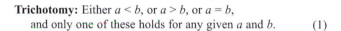

Positive multiplicativity: If $a < b$ and $c > 0$, then $ac < bc$. (4)

Negative multiplicativity: If $a < b$ and $c < 0$, then $ac > bc$. (5)

Replacing $<$ by $\leq$ and $>$ by $\geq$ in laws (2) – (5) yields four new laws for inequalities, which we will also find useful.

The word "trichotomy" in (1) means a threefold division. The trichotomy law states that any two numbers a and b are related in exactly one of the three ways listed in (1). For example, given the two numbers 3.1416 and π, we have either $3.1416 < \pi$, $3.1416 = \pi$, or $3.1416 > \pi$. (The last is actually correct.)

For simplicity of notation, two inequalities are sometimes combined. For example, if $a \leq b$ and $b \leq c$, then we can write $a \leq b \leq c$.

> **Caution:** The multiplication laws, (4) and (5), must be carefully observed. To illustrate their use, we will present several examples. In each the problem is to solve an inequality, which means to find all real numbers that satisfy the inequality.

EXAMPLE 1 Solve the inequality $1/x < 3$.

Solution First we observe that 0 cannot be a solution because division by 0 is undefined. Next we multiply both sides of the inequality by x in order to eliminate x from the denominator on the left side. For positive x, (4) yields $1 < 3x$, or $\frac{1}{3} < x$. Thus the numbers in $\left(\frac{1}{3}, \infty\right)$ constitute one part of the solution of the given inequality. For negative x, (5) yields $1 > 3x$, or $\frac{1}{3} > x$. Since the last inequality is satisfied by all $x < 0$, a second part of the solution consists of all x in $(-\infty, 0)$. Therefore the complete solution consists of all numbers in the interval $(-\infty, 0)$ and all numbers in the interval $(\frac{1}{3}, \infty)$. ❏

When the solution of an inequality forms only one interval, we will write only that interval as the solution. However, if the solution of an inequality consists of more than one interval, we will refer to the solution as the **union** of these intervals. Thus the solution of the inequality $1/x < 3$ in Example 1 consists of the union of the intervals $(-\infty, 0)$ and $(\frac{1}{3}, \infty)$.

EXAMPLE 2   Solve the inequality $-1 < -2x + 3 \le 2$.

Solution   First subtracting 3 throughout, we find that

$$-4 < -2x \le -1$$

Then dividing throughout by -2 and reversing the inequality signs, we obtain

$$2 > x \ge \frac{1}{2}$$

Thus the solution consists of all numbers x satisfying $\frac{1}{2} \le x < 2$, that is, the interval $[\frac{1}{2}, 2)$. ❏

In solving most inequalities, the goal is to find values of x for which a certain expression in x is positive (or negative). One must be careful to observe the negative multiplicativity rule (5) when multiplying or dividing by negative numbers. The rules imply that xy is positive if x and y are both positive or both negative, and xy is negative if one of x and y is negative while the other is positive. The same rules apply to x/y, since $1/y$ and y have the same sign. In general, a product of several non-zero numbers is negative if the product has an odd number of negative factors. Otherwise the product is positive.

EXAMPLE 3   Solve the inequality

$$\frac{x^2 - 4x + 3}{x + 2} > 0$$

Solution   First we factor the numerator to obtain $x^2 - 4x + 3 = (x - 1)(x - 3)$. Therefore the given inequality is equivalent to

$$\frac{(x - 1)(x - 3)}{x + 2} > 0$$

Next we draw a diagram (Figure 1.3) that shows the signs of the factors $x - 1$, $x - 3$, and $x + 2$ appearing in the inequality. Using these results we deduce the values of x for which $(x - 1)(x - 3)/(x + 2) > 0$. From Figure 1.3 we conclude that the solution of the given inequality is the union of the intervals $(-2, 1)$ and $(3, \infty)$. ❏

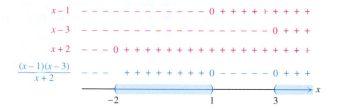

FIGURE 1.3

If we had wished to solve the inequality $(x^2 - 4x + 3)/(x + 2) \geq 0$, we would have used the same diagram, but at the end we would have selected the union of those intervals on which $(x^2 - 4x + 3)/(x + 2)$ is nonnegative, namely, $(-2, 1]$ and $[3, \infty)$.

In Section 2.5 we will discuss a second method of solving inequalities that uses results from calculus.

Absolute Value

The **distance** between a and b on the real line is either $a - b$ or $b - a$, whichever is nonnegative (Figure 1.4). Likewise, the distance between 0 and b is either $b - 0 = b$ or $0 - b = -b$, whichever is nonnegative. The distance between b and 0 is the basis for the definition of the absolute value of b.

FIGURE 1.4

DEFINITION 1.1

The absolute value of any real number b is b if $b \geq 0$ and is $-b$ if $b < 0$. The absolute value of b is denoted $|b|$. Thus

$$|b| = \begin{cases} -b & \text{for } b < 0 \\ b & \text{for } b \geq 0 \end{cases}$$

For example, $|6| = 6$, $|0| = 0$, $|-5| = -(-5) = 5$, and $|8 - 17| = |-9| = -(-9) = 9$. Notice that $|b|$ is the greater of b and $-b$, whichever is nonnegative. Thus

$$|b| \geq 0$$

Geometrically, $|b|$ is the distance between 0 and b. More generally, $|a - b|$ is the distance between the numbers a and b.

We will use the following properties of absolute value:

$$|-a| = |a| \quad \text{and} \quad |a - b| = |b - a| \tag{6}$$

$$|ab| = |a| |b| \quad \text{and} \quad |b^2| = |b|^2 \tag{7}$$

$$-|b| \le b \le |b| \tag{8}$$

$$|a + b| \le |a| + |b| \tag{9}$$

$$|a - b| \ge ||a| - |b|| \tag{10}$$

Except for (9) and (10), these properties follow directly from Definition 1.1. We will verify (9) and leave (10) as an exercise. To verify (9), we first use (8):

$$-a \le |a| \quad \text{and} \quad a \le |a|$$

$$-b \le |b| \quad \text{and} \quad b \le |b|$$

Adding these inequalities vertically yields

$$-(a + b) = -a - b \le |a| + |b| \quad \text{and} \quad a + b \le |a| + |b|$$

Since $|a + b|$ is either $a + b$ or $-(a + b)$, it follows that

$$|a + b| \le |a| + |b|$$

which is (9).

Next we show that if $b > 0$, then

$$|x| < b \quad \text{if and only if} \quad -b < x < b \tag{11}$$

To prove (11) we notice that $|x| < b$ means that

$$\text{if } x \ge 0, \quad \text{then } x < b$$

and

$$\text{if } x < 0, \quad \text{then } -x < b, \quad \text{or equivalently, } -b < x$$

From (11) we see that the solution of the inequality $|x| < b$ is the open interval $(-b, b)$.

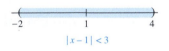

FIGURE 1.5(a)

EXAMPLE 4   Solve the inequality $|x - 1| < 3$.

Solution   By (11) with x replaced by $x - 1$ and b replaced by 3, the inequality $|x - 1| < 3$ is equivalent to

$$-3 < x - 1 < 3$$

or equivalently, $-2 < x < 4$. Thus the solution is $(-2, 4)$.   ❏

Geometrically, $|x - 1| < 3$ means that the distance between x and 1 is less than 3 (Figure 1.5(a)). More generally, $|x - a| < d$ means the distance between x and a

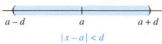

FIGURE 1.5(b)

FIGURE 1.6

Approximate Numerical Values

is less than d. Thus $|x - a| < d$ if and only if x lies in the interval $(a - d, a + d)$ (Figure 1.5(b)). Algebraically,

$$|x - a| < d \quad \text{if and only if} \quad a - d < x < a + d$$

EXAMPLE 5   Find all values of x such that $0 < |x - a| < d$, where a is any number and d is any positive number.

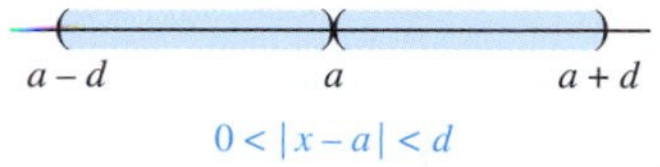

Solution   The double inequality $0 < |x - a| < d$ means that

$$0 < |x - a| \quad \text{and} \quad |x - a| < d$$

From $0 < |x - a|$ we know that $x \neq a$. By our comments above, the values of x satisfying $|x - a| < d$ lie in the interval $(a - d, a + d)$. These two observations give the complete solution, which is the union of the intervals $(a - d, a)$ and $(a, a + d)$ (Figure 1.6).   ❑

In many respects mathematics is a very precise subject. For example, if the legs of a right triangle are 1 meter long, then the hypotenuse is $\sqrt{2}$ meters long; similarly, the circumference of a circle of radius 1 is 2π. However, for numerical computations, calculators and computers can only use numbers with decimal representations of limited length. This means using decimal approximations for $\sqrt{2}$ and π. When we want to signify that a number b is an approximation for another number a, we will write a ≈ b. The symbol ≈ is to be read "is approximately equal to."

Today's calculators display somewhere between 10 and 15 digits, whereas computers usually display anywhere from 6 to 20 digits. There are two common methods of determining, say, a 10-digit approximation for π, the first 20 digits of whose decimal expansion are known to be

$$3.1415926535897932384$$

One way merely drops all digits after the tenth digit, which yields

$$\pi \approx 3.141592653$$

This procedure is called **chopping** or **truncating.** Another procedure is called **rounding,** in which the tenth digit is rounded up by 1 in the event that the eleventh digit is 5 or greater. For π the eleventh digit is 5, so the tenth digit (which is 3) is rounded up to 4. We obtain the 10-digit approximation

$$\pi \approx 3.141592654$$

Many calculators and computers use rounding when exhibiting decimal expansions of numbers.

Because calculators and computers associate finite decimal expansions with numbers, it might not be surprising that complicated expressions involving such numbers can lead to inaccurate answers. To help minimize this possibility, some calculators display up to 12 digits (plus a 3-digit exponent), but hold in storage 14 digits. For example, the TI-85 displays

$$2.62373843983 \text{ E}178$$

for 7.8^{200}, although it holds

$$2.6237384398267 \text{ E}178$$

in storage, out of view.

On computers, Mathematica normally carries a machine precision of 16 digits and displays only a few digits unless asked for more. Thus for 7.8^{200} it might display

$$2.62374 \times 10^{178}$$

However, when asked for, say, 30 digits, it would give its best:

$$2.623738439825925986 \times 10^{178}$$

which as you will notice does not contain 30 digits!

Despite all their accuracy, calculators and computers sometimes produce erroneous results or fail to provide answers. Two problems that can occur are overflow and underflow.

1. Overflow: If an operation produces a number that is too large for the instrument to store, it will stop and display an error message. For example, try to calculate 1000^{1000} on your calculator or computer, and see what happens.

2. Underflow: If an operation produces a number that is too small to store, then the number will be treated as 0. This may yield an absurd answer or an error message. For example, try to calculate $(0.001)^{1000}$ or $1/(0.001)^{1000}$.

Another problem that can occur in using calculators or computers involves subtracting two nearly equal numbers. For example, suppose that $x = 12{,}345.2$ and $y = 12{,}345.1$, both accurate to 5 digits. Then $x - y = 0.1$ is accurate to at most 1 digit. Thus there is a loss of 4 significant digits (from 5 to 1).

We can frequently rearrange the formula we are calculating in order to minimize overflow, underflow, and difference errors. Thus we would likely improve the instrument's answer if we calculate

$$\frac{1000^{1000}}{999^{1000}} \qquad \text{as} \qquad \left(\frac{1000}{999}\right)^{1000}$$

$$\frac{237^{1000}}{237^{998}} \qquad \text{as} \qquad 237^{1000-998} = 237^2$$

$$\frac{(0.012)^{1000}}{(0.011)^{1000}} \qquad \text{as} \qquad \left(\frac{0.012}{0.011}\right)^{1000}$$

$$\frac{1}{\sqrt{25{,}000} - \sqrt{24{,}999}} \qquad \text{as} \qquad \sqrt{25{,}000} + \sqrt{24{,}999}$$

Until now we have not talked about how many digits are reasonable for a given problem. The answer to this question varies, depending upon circumstance. For example, in measuring a length of wooden fencing, estimates accurate to a tenth of an inch or to a millimeter should certainly suffice. By contrast, in the preparation of computer chips or in the analysis of wave lengths, an error of a single micron (that is, a millionth of a meter) might well make a difference.

If we make calculations with information we are given, the accuracy of the result will depend on the accuracy of the given values. Thus we will need to be careful in the interpretation of our numerical answers. In the remainder of the book we will generally adopt the following convention.

1. If the numbers that enter a calculation are considered to be accurate and the calculation is performed on a calculator or computer, we will give the instrument's output.
2. If the numbers entering a calculation are known (or expected) to be accurate to, say, two decimal places, then we will give the answer only to two decimal places.

This concludes our discussion of real numbers, inequalities, and absolute values. The concepts and rules we have given will play an important part in our study of calculus.

EXERCISES 1.1

In Exercises 1–4 determine whether $a < b$ or $a > b$.

1. $a = \frac{4}{9}$, $b = \frac{7}{16}$ **2.** $a = -\frac{1}{7}$, $b = -0.142857$

3. $a = \pi^2$, $b = 9.8$ **4.** $a = (3.2)^2$, $b = 10$

5. Use the fact that $(\sqrt{2})^2 = 2$ to determine whether $\sqrt{2} < 1.41$, $\sqrt{2} = 1.41$, or $\sqrt{2} > 1.41$.

6. Use the fact that $(\sqrt{11})^2 = 11$ to determine whether $\sqrt{11} < 3.3$, $\sqrt{11} = 3.3$, or $\sqrt{11} > 3.3$.

In Exercises 7–14 state whether the interval is open, half-open, or closed and whether it is bounded or unbounded. Then sketch the interval on the real line.

7. $[-4, 5]$ **8.** $(-2, -1)$ **9.** $(-\infty, 3)$

10. $[\frac{3}{2}, \frac{5}{2})$ **11.** $[0, \infty)$ **12.** $(5, 7)$

13. $(-\infty, -1]$ **14.** $[-\frac{1}{2}, \frac{1}{2}]$

In Exercises 15–18 write the union of the two intervals as a single interval.

15. $(-3, 2)$ and $[1, 4)$ **16.** $(-\infty, 0]$ and $[0, 3)$

17. $(1, 3)$ and $(2, \infty)$ **18.** $(-\infty, \frac{1}{2}]$ and $(0, \infty)$

In Exercises 19–38 solve the inequality.

19. $-6x - 2 > 5$ **20.** $4 - 3x \geq 7$

21. $-1 \leq 2x - 3 < 4$ **22.** $-0.1 < 3x + 4 < 0.1$

23. $(x - 1)(x + 1/2) \geq 0$ **24.** $(x - 1)(x - 2)(x - 3) \leq 0$

25. $x(x - 2/3)(x + 1/3) < 0$ **26.** $\dfrac{x}{(x - 1)(x + 2)} > 0$

27. $\dfrac{(2x - 1)^2}{(x + 1)(x + 3)} > 0$ **28.** $\dfrac{(2x - 3)(4x + 1)}{x - 2} \leq 0$

29. $4x^3 - 6x^2 \leq 0$ **30.** $3x^2 - 2x - 1 \geq 0$

31. $8x - \dfrac{1}{x^2} > 0$ **32.** $8x + \dfrac{1}{x^2} < 0$

33. $\dfrac{4x(x^2 - 6)}{x^2 - 4} < 0$ **34.** $\dfrac{2x(x^2 - 3)}{(x^2 + 1)^3} \geq 0$

35. $\dfrac{t^2 + t - 2}{(t^2 - 1)^3} \geq 0$ **36.** $\dfrac{t^2 - 2t - 3}{t^2 - 8t + 15} > 0$

37. $\dfrac{2 - x}{\sqrt{9 - 6x}} > 0$ **38.** $\dfrac{2x^2 - 1}{(1 - x^2)^{1/2}} < 0$

In Exercises 39–42 solve the inequality.

39. $\dfrac{1}{x + 1} > \dfrac{3}{2}$ (*Hint:* Write the inequality as $1/(x + 1) - 3/2 > 0$. Then rewrite the left side as a single fraction.)

40. $\dfrac{1}{3 - x} < -2$ **41.** $\dfrac{x + 1}{x - 1} \leq \dfrac{1}{2}$ **42.** $\dfrac{2 - 5x}{3 - 4x} \geq -2$

In Exercises 43–46 evaluate the expression.

43. $-|-3|$ **44.** $|-\sqrt{2}|^2$

45. $|-5| + |5|$ **46.** $|-5| - |5|$

In Exercises 47–58 solve the equation.

47. $|x| = 1$ **48.** $|x| = \pi$

49. $|x - 1| = 2$ **50.** $|2x - \frac{1}{2}| = \frac{1}{2}$

51. $|6x + 5| = 0$ **52.** $|3 - 4x| = 2$

53. $|x| = |x|^2$ **54.** $|x| = |1 - x|$

55. $|x + 1|^2 + 3|x + 1| - 4 = 0$

56. $|x - 2|^2 - |x - 2| = 6$

57. $|x + 4| = |x - 4|$

58. $|x - 1| = |2x + 1|$

In Exercises 59–70 solve the inequality.

59. $|x - 2| < 1$ **60.** $|x - 4| < 0.1$

61. $|x + 1| < 0.01$ **62.** $|x + \frac{1}{2}| \leq 2$

63. $|x + 3| \geq 3$ **64.** $|x - 0.3| > 1.5$

65. $|2x + 1| \geq 1$ **66.** $|3x - 5| \leq 2$

67. $|2x - \frac{1}{3}| > \frac{2}{3}$ **68.** $0 < |x - 1| < 0.5$

69. $-1 < |4 - 2x| < 1$ **70.** $|x - a| \leq d$

 In Exercises 71–73 modify the expression, and then find its approximate value by calculator or computer.

71. $\dfrac{69^{800}}{59^{800}}$ **72.** $\dfrac{221^{907}}{221^{897}}$

73. $\dfrac{(0.123)^{9000}}{(0.125)^{9000}}$

 74. a. Show that $1/(\sqrt{25{,}000} - \sqrt{24{,}998}) = \frac{1}{2}(\sqrt{25{,}000} + \sqrt{24{,}998})$.

   b. Compare the value your calculator gives for $1/(\sqrt{25{,}000} - \sqrt{24{,}998})$ with the value it gives for $\frac{1}{2}(\sqrt{25{,}000} + \sqrt{24{,}998})$. If they are different, which do you think is the more accurate?

75. Find all numbers x the sum of whose distances from 12 and from 13 exceeds 4. Draw a figure to illustrate your solution.

76. Find all numbers x with the property that the distance from x to 2 is less than twice the distance from x to 3.

77. If $x^2 \leq 25$, is it necessarily true that $x \leq 5$? Explain.

78. If $x^3 > 125$, is it necessarily true that $x > 5$? Explain.

79. Is $1/x < x$ for all nonzero x? Explain.

80. a. Show that $x < x^2$ for $x < 0$ or $x > 1$.

   b. Show that $x^2 < x$ for $0 < x < 1$.

81. Use the definition of absolute value to prove the following.

   a. $|ab| = |a||b|$

   b. $-|b| \leq b \leq |b|$

   c. $|a - b| = |b - a|$

82. a. Use property (9) of absolute value to prove that

$$|a - b| \geq |a| - |b|$$

   for all real numbers a and b. (*Hint:* Show that $|c| \geq |c + b| - |b|$ for all c, and replace c by $a - b$.)

   b. Use (6) and part (a) to prove that for all real numbers a and b,

$$|a - b| \geq |b| - |a|$$

   c. Use parts (a) and (b) to prove that

$$|a - b| \geq ||a| - |b||$$

   for all real numbers a and b.

*83. Show that $|a + b| = |a| + |b|$ if and only if $ab \geq 0$ (which means that $a = 0$, $b = 0$, or a and b have the same sign).

84. Prove that if $a < b$, then $a < (a + b)/2 < b$. How is $(a + b)/2$ related to a and b on the real line? The number $(a + b)/2$ is called the **arithmetic mean** of a and b.

*85. Prove that if $0 < a < b$, then $a < \sqrt{ab} < (a + b)/2$. The number $\sqrt{ab}$ is called the **geometric mean** of a and b. (*Hint:* $(\sqrt{b/2} - \sqrt{a/2})^2 > 0$.)

*86. Let $0 < a < b$, and let h be defined by

$$\frac{1}{h} = \frac{1}{2}\left(\frac{1}{a} + \frac{1}{b}\right)$$

Show that $a < h < b$. The number h is called the **harmonic mean** of a and b.

*87. Let $0 < a < b$. Show that

$$\sqrt{b} - \sqrt{a} < \sqrt{b - a}$$

*88. Prove that $\sqrt{2}$ is irrational. (*Hint:* Assume that $\sqrt{2} = p/q$, where p and q are integers such that at most one of them is divisible by 2. It can be shown that a square integer is divisible by 2 only if it is also divisible by 4. Use this fact to show first that p is divisible by 2 and then that q is also divisible by 2. This contradicts the assumption.)

*89. Prove that $\sqrt{3}$ is irrational. (*Hint:* Use the method of Exercise 88.)

90. A rectangle R has length x and width y.

   a. Write an inequality that expresses the condition that the area of R is less than 10.

   b. Write an inequality that expresses the condition that the perimeter of R is at least 47.

91. Show that a square has the largest area of all rectangles having a given perimeter. (*Hint:* Let P be the perimeter. Show that if a rectangle of perimeter P has adjacent sides a and b, with $0 < a \leq b$, then the area of the rectangle is ab and that of the square is $[(a + b)/2]^2$. Now use Exercise 85.)

92. Show that if a square and a circle have equal perimeters, then the circle has a larger area than the square. (*Hint:* Show first that a circle of perimeter P has area $P^2/4\pi$.)

93. Using the results of Exercises 91 and 92, show that a circle has an area larger than any rectangle of equal perimeter.

1.2   POINTS AND LINES IN THE PLANE

In calculus we often encounter lines and curves lying in a given plane. Using the correspondence between numbers and the points on a line, we will identify points in the plane with pairs of numbers. This will enable us to describe curves in the plane by means of equations.

The Plane

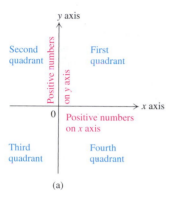

We construct a coordinate system for the plane by drawing two real number lines that are perpendicular to one another and that intersect at their zero points. We call their point of intersection the **origin,** denoted 0. It is customary to have one of the lines horizontal, with the positive numbers located to the right of 0, and to call it the **x axis.** The other line is usually called the **y axis,** with the positive numbers lying above 0 (Figure 1.7(a)). These axes divide the plane into four quadrants, as shown in Figure 1.7(a), and are called the **coordinate axes.**

Now let P be any point in the plane. Draw lines l_1 and l_2 through P perpendicular to the two axes (Figure 1.7(b)). Then the number corresponding to the point on the x axis that lies on l_1 is the **x coordinate** of P, and the number corresponding to the point on the y axis that lies on l_2 is the **y coordinate** of P. Call these numbers a and b, respectively. Then we associate P with the ordered pair (a, b) of numbers. In this way every point in the plane is associated with one and only one ordered pair of numbers, and every ordered pair of real numbers is associated with one and only one point in the plane. Consequently we identify points in the plane with ordered pairs of real numbers. If P is identified with (a, b), we sometimes write $P(a, b)$ for P. Notice that the origin, which is the point whose x and y coordinates are both 0, is not the same as the number 0. Figure 1.8 shows several points in the plane, along with their coordinates.

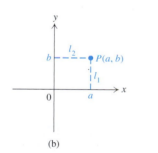

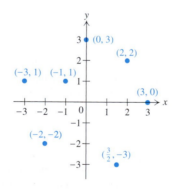

FIGURE 1.7   A coordinate system for the plane.

FIGURE 1.8   The coordinates of some points in the plane.

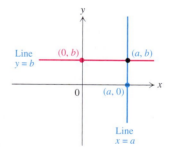

FIGURE 1.9   Lines perpendicular to the axes.

Observe that the x axis consists of all points of the form $(x, 0)$. Similarly, the y axis consists of all points of the form $(0, y)$. The line perpendicular to the x axis and passing through a given point (a, b) contains all points of the form (a, y), and we describe this line by the equation $x = a$. This line crosses the x axis at the point $(a, 0)$ (Figure 1.9). Analogously, the line perpendicular to the y axis and passing

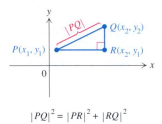

$$|PQ|^2 = |PR|^2 + |RQ|^2$$

FIGURE 1.10

through the point (a, b) contains all points of the form (x, b), and its equation is $y = b$. This line crosses the y axis at the point $(0, b)$ (Figure 1.9). Moreover, notice that

$| a |$ is the distance between the point (a, b) and the y axis

and

$| b |$ is the distance between the point (a, b) and the x axis

Two points in the plane are the same if and only if they have the same x coordinates and the same y coordinates. Thus (a, b) and (c, d) are the same point if and only if $a = c$ and $b = d$.

The Distance Between Two Points

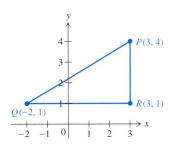

FIGURE 1.11

To find the distance between any two points $P(x_1, y_1)$ and $Q(x_2, y_2)$ in the plane, we use the Pythagorean Theorem. The point $R(x_2, y_1)$ in Figure 1.10 has the same x coordinate as Q and the same y coordinate as P. Therefore triangle PQR has a right angle at R. The distance $| PQ |$ between P and Q is given by the formula

Distance Formula
$$|PQ| = \sqrt{(x_2 - x_1)^2 + (y_2 - y_1)^2}$$

which is an immediate consequence of the following form of the Pythagorean Theorem:

$$| PQ |^2 = (x_2 - x_1)^2 + (y_2 - y_1)^2$$

EXAMPLE 1   Let $P = (3, 4)$ and $Q = (-2, 1)$. Find the distance between P and Q (Figure 1.11).

Solution   By the distance formula,

$$|PQ| = \sqrt{(-2 - 3)^2 + (1 - 4)^2} = \sqrt{25 + 9} = \sqrt{34} \quad \square$$

Lines in the Plane

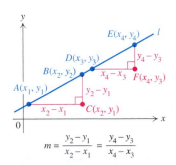

$$m = \frac{y_2 - y_1}{x_2 - x_1} = \frac{y_4 - y_3}{x_4 - x_3}$$

FIGURE 1.12

Identification of points in the plane with ordered pairs of real numbers allows us to use algebra to describe geometric objects such as lines. When we speak of an **equation of a line *l*,** we mean an equation such as

$$y = 2x + 5$$

having the property that a point P lies on l if and only if the equation is satisfied by the coordinates of P. We will give two methods of obtaining an equation of a line. Since any vertical line is perpendicular to the x axis and hence has an equation of the form $x = x_1$ for some number x_1, we will only consider equations of nonvertical lines in the following discussion.

Let l be any nonvertical line, and let (x_1, y_1) and (x_2, y_2) be any two distinct points on l (Figure 1.12). Also let

$$m = \frac{y_2 - y_1}{x_2 - x_1} \tag{1}$$

It follows from similar triangles that the fraction in (1) is independent of the points (x_1, y_1) and (x_2, y_2) we choose on the line. We call the fraction in (1) the **slope** of the line *l*. For any point (x, y) on *l* different from (x_1, y_1),

$$m = \frac{y - y_1}{x - x_1} \tag{2}$$

Multiplying the equation in (2) by $x - x_1$, we obtain

Point-Slope Equation

$$y - y_1 = m \, (x - x_1) \tag{3}$$

which also holds when $x = x_1$ and $y = y_1$. A point-slope equation of a line is determined by the slope *m* and any given point (x_1, y_1) on the line.

EXAMPLE 2   Find a point-slope equation of the line passing through (2, 1) with the given slope, and sketch the line.
a. slope 0   b. slope $\frac{1}{2}$ c. slope –3

Solution   In each case we use (3), with $x_1 = 2$ and $y_1 = 1$. For (a) we have $m = 0$, so that (3) becomes

$$y - 1 = 0(x - 2) = 0$$

The line is horizontal; it is sketched in Figure 1.13. For (b) we have $m = \frac{1}{2}$, so that (3) becomes

$$y - 1 = \frac{1}{2}(x - 2)$$

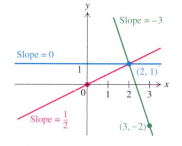

FIGURE 1.13

To sketch the line, we need a second point on it. Now if we let $x = 0$ in the equation, we obtain

$$y - 1 = \frac{1}{2}(0 - 2) = -1$$

so that $y = 0$. Thus the line passes through both (0, 0) and (2, 1). It is also sketched in Figure 1.13. For part (c), $m = -3$, so that (3) becomes

$$y - 1 = -3(x - 2)$$

Since the slope is –3, or $\frac{-3}{1}$, moving 1 unit to the right and 3 units downward gives a second point (3, –2) on the line. The line also appears in Figure 1.13.   ❑

Figure 1.13 shows that the line with slope 0 is horizontal. More generally, from (3) we see that any line having slope 0 is horizontal. Such a line has an equation of the form

$$y - y_1 = 0, \quad \text{or} \quad y = y_1$$

When we draw a line with a given point-slope equation $y - y_1 = m \, (x - x_1)$, we start from the point (x_1, y_1) and draw a line with slope *m*. For example, in part (c) above, the slope was –3, so an increase of 1 unit in the value of *x* on the line caused a change of –3 units in the value of *y*. More generally, if the slope of a given line is *m*, then a change of 1 unit in the value of *x* (say, from x_1 to $x_1 + 1$) causes a change of m units in the value of y (from y_1 to $y_1 + m$) (Figure 1.14). Consequently if (x_1, y_1) is on the line, so is $(x_1 + 1, y_1 + m)$.

Observe that if $m > 0$, then a line with slope m slants upward from the left to right, and if $m < 0$, then the line slants downward from left to right. The larger $|m|$ is, the steeper the line is (Figure 1.15).

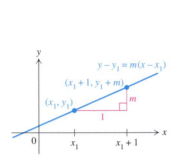

FIGURE 1.14   **FIGURE 1.15**

Caution: The slope of a vertical line is undefined.

To describe a second type of equation for a line l, recall that we have assumed from the outset that l is *not* vertical. Thus l must cross the y axis. The y coordinate of the point of intersection of l and the y axis is called the **y intercept** of l. Rewriting (3), we have

$$y = mx + y_1 - mx_1$$

Now let $b = y_1 - mx_1$. Then the equation becomes

> **Slope-Intercept Equation**
>
> $$y = mx + b \tag{4}$$

If $x = 0$, then $y = b$. Thus b is the y coordinate of the point of intersection of l and the y axis. As you can see, the slope-intercept equation of the line l is determined by the slope m and the y intercept b of the line.

EXAMPLE 3   Find the slope-intercept equation of the line with slope 3 and y intercept -1. Sketch the line.

Solution   By (4) the equation is

$$y = 3x - 1$$

Since $(0, -1)$ is a point on the line and the slope is 3, another point on the line is $(0 + 1, -1 + 3)$, or $(1, 2)$. This enables us to sketch the line (Figure 1.16).   ❑

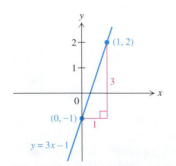

FIGURE 1.16

Finally, we mention that either of these types of equations may be used for a given line. The choice is a matter of convenience.

Parallel and Perpendicular Lines

We can determine when two lines are parallel or perpendicular to each other by considering their slopes.

First we discuss parallel lines. Let l_1 and l_2 be two nonvertical lines whose equations are

$$y = m_1 x + b_1 \quad \text{and} \quad y = m_2 x + b_2$$

respectively. Recall from geometry that two lines are *not* parallel if and only if they have precisely one point in common. This fact permits us to characterize parallel (nonvertical) lines.

THEOREM   1.2

Let l_1 and l_2 be nonvertical lines with slopes m_1 and m_2. Then l_1 and l_2 are parallel if and only if $m_1 = m_2$.

Proof If $m_1 = m_2$, then either l_1 and l_2 do not intersect, in which case they are parallel, or they intersect at a point (x_0, y_0), in which case

$$m_1 x_0 + b_1 = y_0 = m_2 x_0 + b_2$$

Since $m_1 = m_2$, we have $b_1 = b_2$. Thus since the slope-intercept equation uniquely determines a line, l_1 and l_2 are identical and therefore parallel. Hence, if $m_1 = m_2$, then l_1 and l_2 are parallel.

Now suppose $m_1 \neq m_2$. Observe that (x, y) is on both lines if and only if

$$y = m_1 x + b_1 = m_2 x + b_2$$

Since $m_1 \neq m_2$, the equation $m_1 x + b_1 = m_2 x + b_2$ has the unique solution

$$x_0 = \frac{b_2 - b_1}{m_1 - m_2}$$

Thus $(x_0, m_1 x_0 + b_1)$ is a point on both lines, and no other point lies on both lines. Hence if $m_1 \neq m_2$, then the lines are not parallel. This means that if the lines are parallel, then $m_1 = m_2$.   ■

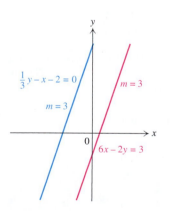

FIGURE 1.17   Parallel nonvertical lines.

EXAMPLE 4   Show that the lines l_1 and l_2 described by the equations

$$6x - 2y = 3 \quad \text{and} \quad \frac{1}{3} y - x - 2 = 0$$

are parallel.

Solution We rewrite the equations for l_1 and l_2 as slope-intercept equations, obtaining

$$y = 3x - \frac{3}{2} \quad \text{and} \quad y = 3x + 6$$

respectively. Therefore $m_1 = m_2 = 3$, so that by Theorem 1.2 the lines l_1 and l_2 are parallel (Figure 1.17).   ❑

EXAMPLE 5 Show that the lines l_1 and l_2 described by the equations

$$x + 2y = 1 \quad \text{and} \quad x - y = 2$$

are not parallel. Find their point of intersection.

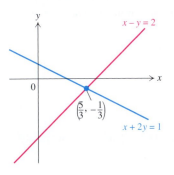

FIGURE 1.18

Solution The equations $x + 2y = 1$ and $x - y = 2$ are equivalent to $y = -\frac{1}{2}x + \frac{1}{2}$ and $y = x - 2 = 1 \cdot x - 2$, respectively, which are in slope-intercept form. Therefore $m_1 = -\frac{1}{2}$ and $m_2 = 1$. Hence by Theorem 1.2 the lines l_1 and l_2 are not parallel. To find their point of intersection, we solve the equations

$$y = -\frac{1}{2}x + \frac{1}{2} \quad \text{and} \quad y = x - 2$$

for x and y. We obtain

$$-\frac{1}{2}x + \frac{1}{2} = y = x - 2, \quad \text{so that} \quad \frac{3}{2}x = \frac{5}{2}$$

Thus $x = \frac{5}{3}$ and hence $y = \frac{5}{3} - 2 = -\frac{1}{3}$. Therefore $(\frac{5}{3}, -\frac{1}{3})$ is the point of intersection of l_1 and l_2 (Figure 1.18). ❑

Now we present a characterization of perpendicular (nonvertical) lines.

THEOREM 1.3

Let l_1 and l_2 be nonvertical lines with slopes m_1 and m_2. Then l_1 and l_2 are perpendicular if and only if $m_1 m_2 = -1$.

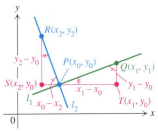

(a)

Proof Suppose l_1 is perpendicular to l_2, and l_1 and l_2 intersect at $P(x_0, y_0)$. Let $Q(x_1, y_1)$ and $R(x_2, y_2)$ be points distinct from P on the lines l_1 and l_2, respectively (Figure 1.19(a)). Since l_1 is perpendicular to l_2, triangle *PRS* and triangle *QPT* are similar. Thus

$$m_1 = \frac{y_1 - y_0}{x_1 - x_0} = \frac{x_0 - x_2}{y_2 - y_0} = \frac{1}{\dfrac{y_2 - y_0}{x_0 - x_2}} = -\frac{1}{\dfrac{y_2 - y_0}{x_2 - x_0}} = -\frac{1}{m_2}$$

Therefore $m_1 m_2 = -1$. Conversely, suppose $m_1 m_2 = -1$. Let l_3 be perpendicular to l_1, and let l_3 have slope m_3, so that $m_1 m_3 = -1$. This implies that $m_2 = m_3$. Then by Theorem 1.2, l_2 and l_3 are parallel, so l_1 is perpendicular to l_2. ∎

EXAMPLE 6 Show that the lines l_1 and l_2 with equations

$$2x - 8y = 3 \quad \text{and} \quad y = 5 - 4x$$

are perpendicular.

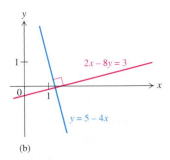

(b)

FIGURE 1.19 Perpendicular nonvertival lines.

Solution The slopes of l_1 and l_2 are

$$m_1 = \frac{2}{8} = \frac{1}{4} \quad \text{and} \quad m_2 = -4$$

Since $m_1 m_2 = -1$, l_1 and l_2 are perpendicular by Theorem 1.3 (Figure 1.19(b)). ❑

Slope is one of the central concepts of calculus, and we will return to it in Chapters 2–4.

EXERCISES 1.2

1. Draw a set of coordinate axes and plot the following points.
   a. $(2, 1)$ **b.** $(-1, 3)$ **c.** $(4, 0)$
   d. $(0, -\frac{3}{2})$ **e.** $(1, -1)$ **f.** $(-2, -2)$
   g. $(0, 0)$ **h.** $(3, -\frac{1}{3})$

2. Let (a, b) be any point in the second quadrant. Describe the locations of the following points.
   a. $(-a, b)$ **b.** $(a, -b)$ **c.** $(-a, -b)$

In Exercises 3–12 determine the distance between the given points.

3. $(3, 0)$ and $(-2, 0)$ 4. $(0, 0)$ and $(3, 4)$
5. $(2, 1)$ and $(6, -3)$ 6. $(-1, -3)$ and $(-2, 2)$
7. $(6, 5)$ and $(-3, -4)$
8. $(\sqrt{6}, \sqrt{3})$ and $(3\sqrt{6}, -\sqrt{3})$
9. $(\sqrt{2}, 1)$ and $(\sqrt{3}, 2)$ 10. (a, b) and (b, a)
11. (a, a) and (b, b)
12. $(a + e, b + e)$ and $(c + e, d + e)$

In Exercises 13–20 find an equation of the line described. Then sketch the line.

13. The line through $(2, -1)$ with slope 3
14. The line through $(-3, -2)$ with slope 1
15. The line through $(\frac{1}{2}, \frac{1}{2})$ with slope -1
16. The line through $(0, \pi)$ with slope 0
17. The line with slope -1 and y intercept 0
18. The line with slope $\frac{1}{2}$ and y intercept -1
19. The line with slope 3 and y intercept -3
20. The line with slope -2 and y intercept $\frac{5}{2}$

In Exercises 21–26 determine the slope m and y intercept b of the line with the given equation. Then sketch the line.

21. $y = -x$ 22. $y = 2x - 3$
23. $y - 1 = 2(x + 3)$ 24. $x - \frac{1}{2}y = 2$
25. $2x + y - 4 = 0$ 26. $\frac{2}{3}y - \frac{1}{3}x = 2$

In Exercises 27–36 decide which pairs of lines are parallel, which are perpendicular, and which are neither. For any pair that is not parallel, find the point of intersection.

27. $y - 2x = 3$ and $y + \frac{1}{2}x = -1$
28. $y = 3 - 2x$ and $3x + \frac{3}{2}y - 4 = 0$

29. $x - y = -1$ and $x = y$
30. $x + y = -1$ and $x = y$
31. $2x + 3y = -1$ and $2y + 2 = 3(x - 1)$
32. $2x + 3y = -1$ and $3x + 2y = 2$
33. $x = 2$ and $x = -5$
34. $x = -1$ and $y = 4$
35. $3y + 6x = 1$ and $y - 3 = -2x$
36. $x - 2y = 8$ and $2x - y = -8$

In Exercises 37–42 find an equation of the line that is parallel to the given line l and passes through the given point P.

37. $l : y = 3x - 1; P = (2, -1)$
38. $l : y = -\frac{1}{2}x + 4; P = (-1, 0)$
39. $l : x + y = 1; P = (0, 0)$
40. $l : 3y + 2x = 5; P = (-1, -3)$
41. $l : -3y + 2x = 8; P = (2, 1)$
42. $l : 5x - 2y - 1 = 0; P = (3, 3)$

In Exercises 43–48 find an equation of the line that is perpendicular to the given line l and passes through the given point P.

43. $l : y = 2x + 1; P = (-1, -3)$
44. $l : y = -\frac{1}{3}x - 2; P = (0, 0)$
45. $l : 2x + 3y - 6 = 0; P = (2, 3)$
46. $l : 3x - y = 0; P = (1, 3)$
47. $l : y - 1 = 2(x - 3); P = (4, -5)$
48. $l : y + 4 = -\frac{3}{5}(x - \frac{1}{2}); P = (-1, \frac{1}{2})$

49. Let l be the line that contains two given points, (x_1, y_1) and (x_2, y_2), with $x_1 \neq x_2$. Show that an equation of l is

$$y - y_1 = \frac{y_2 - y_1}{x_2 - x_1}(x - x_1)$$

(This equation is called a **two-point equation** of l.)

In Exercises 50–52 find a two-point equation of the given line.

50. The line containing $(3, 4)$ and $(1, 3)$
51. The line containing $(-2, 4)$ and $(-1, 3)$
52. The line containing $(-\frac{3}{2}, -\frac{1}{2})$ and $(\frac{1}{2}, 2)$

In Exercises 53–62 sketch the region in the plane satisfying the given conditions.

53. $x > 0$ 54. $y \leq 0$

55. $y > x$ (Hint: Consider first the line $y = x$.)

56. $x < -y$

57. $x < 0$ and $y < 0$ 58. $x > 0$ and $y < 0$

59. $x < 2$ and $y > 4$ 60. $x \leq 3$ and $y \leq 2$

61. $x \geq -1$ and $y \geq \frac{1}{2}$ 62. $x > 2$ and $y < 1$

63. **a.** Describe the region consisting of all (x, y) for which $(x, y) = (x, -y)$.

   b. Describe the region consisting of all (x, y) for which $(x, y) = (-x, y)$.

64. Suppose two vertices of a rectangle R are $(2, 5)$ and $(7, 1)$, and the sides of R are parallel to the coordinate axes. Determine the other vertices of R.

65. Suppose the sides of a square S are 4 units long and are parallel to the coordinate axes. If $(-3, 3)$ is the vertex of S closest to the origin, find the other vertices of S.

66. Let $(2, 1)$, $(-3, -2)$, and (a, b) form a triangle. Show that the collection of points (a, b) for which the triangle is isosceles, and for which (a, b) is the vertex common to the two sides of equal length, is a line (with one point deleted). Find an equation of that line.

67. Show that the triangle with vertices $(-1, 2)$, $(\sqrt{3} - 1, 3)$, and $(-1, 4)$ is equilateral.

68. Show that the midpoints of the sides of any rectangle are the vertices of a rhombus (a quadrilateral with all sides of equal length). (*Hint:* Let the vertices of the rectangle be $(0, 0)$, $(a, 0)$, $(0, b)$, and (a, b).)

*69. Show that in any triangle the sum of the squares of the lengths of the medians (the line segments joining the vertices to the midpoints of the opposite sides) is equal to three fourths the sum of the squares of the lengths of the sides. (*Hint:* Pick the vertices of the triangle judiciously.)

*70. Show that the sum of the squares of the lengths of the sides of a parallelogram is equal to the sum of the squares of the lengths of the diagonals.

Applications

71. A ramp connects the ground to the loading platform of a warehouse. If the platform is 5 feet high and the ramp has a slope of 0.28, how far from the wall of the warehouse is the base of the ramp?

72. A bookcase is 7 feet tall, 4 feet wide, and 1 foot deep. What is the minimum vertical distance that is necessary in order to raise the bookcase from a horizontal position on the floor to a vertical position against the wall?

1.3   FUNCTIONS

Scientists and mathematicians often consider correspondences between two sets of numbers. For example, each temperature in degrees Celsius corresponds to a temperature in degrees Fahrenheit. There is also a correspondence between the radius of a circle and the circle's area. In mathematics such correspondences are called functions.

DEFINITION   1.4

A **function** consists of a domain and a rule. The **domain** is a set of real numbers. The **rule** assigns to each number in the domain one and only one number.

Functions are normally denoted f, g, or h, and the numbers in the domain are usually denoted by x, t, a, b, or c. The value assigned by a function f to a member x of its domain is written $f(x)$ and is read "f of x" or "the value of f at x." The collection of values (numbers) $f(x)$ that a given function assigns to the members of its domain is called the **range** of f.

We can think of a function as a machine that takes the members of the domain and applies the rule to each to produce the members of the range. As Figure 1.20 suggests, any member x of the domain goes into the machine, is acted on by the function f, and is transformed into a member $f(x)$ of the range.

FIGURE 1.20

Caution: We stress two key points in the definition:

1. A function must make an assignment to *each* number in the domain. For example, if the domain of a function f is $(-\infty, \infty)$, then f must make an assignment to each real number.

2. A function can assign *only one* number to any given number in the domain. Thus, for example, f cannot assign both 5 and 6 to a single number in its domain.

Examples of Functions

We give several examples of functions. Let f be the function whose domain consists of all real numbers and whose rule assigns to any real x the number $x^2 - 1$. Then we write

$$f(x) = x^2 - 1 \quad \text{for all } x$$

Next, if the domain of a function g consists of all real numbers except 3 and if g assigns $(x-1)/(x-3)$ to each such number x, then g is described by

$$g(x) = \frac{x-1}{x-3} \quad \text{for } x \neq 3$$

Now consider the function h that relates temperature in degrees Celsius with temperature in degrees Fahrenheit. The domain of h consists of all numbers greater than or equal to -273.15 (absolute zero) and assigns to each such number x the number $\frac{9}{5}x + 32$. Thus

$$h(x) = \frac{9}{5}x + 32 \quad \text{for } x \geq -273.15$$

When the rule of a function is described by one formula or equation, we normally specify the numbers in the domain after the rule. If the domain consists of all real numbers for which the formula or equation is meaningful, then we may omit mention of the domain. Thus we may write

$$f(x) = x^2 - 1 \quad \text{and} \quad g(x) = \frac{x-1}{x-3}$$

without specifying the domains of f and g. However, it is necessary to give the domain of h if

$$h(x) = \frac{9}{5}x + 32 \quad \text{for } x \geq -273.15$$

because $\frac{9}{5}x + 32$ makes sense even for numbers less than -273.15.

When there can be no misinterpretation, we will sometimes let the rule of a function stand for the function itself. For instance, we can replace f and g defined above by $x^2 - 1$ and $(x-1)/(x-3)$, respectively.

Sometimes two or more formulas may be needed to define a function. For example,

$$f(x) = \begin{cases} 1 & \text{for } x < 0 \\ x & \text{for } 0 \leq x \leq 2 \\ x^2 & \text{for } 2 < x < 3 \\ x^3 & \text{for } x \geq 3 \end{cases}$$

defines a single function.

Our next example concerns a formula for motion that comes from physics and will appear frequently throughout the remainder of the book. If an object is subject only to the force of gravity, and if at time $t = 0$ the height of the object is h_0 meters above the ground and the velocity is v_0 meters per second, then the object's height $h(t)$ in meters above the ground at time t (seconds) is given approximately by

$$h(t) = -4.9t^2 + v_0 t + h_0 \qquad (1)$$

If height were measured in feet and the velocity in feet per second, then the formula in (1) would become

$$h(t) = -16t^2 + v_0 t + h_0 \qquad (2)$$

Conversion Table

1 inch = 2.54 centimeters

1 foot = 0.3048 meters

1 meter ≈ 39.37 inches ≈ 3.281 feet

1 mile per hour ≈ 1.467 feet per second ≈ 0.4470 meters per second

In either case, h_0 is the **initial height** and v_0 is the **initial velocity** of the object. We emphasize that the formulas in (1) and (2) are valid only as long as the object is subject only to the force of gravity and is unaffected by other forces such as air resistance. We will adhere to this assumption unless otherwise noted. Implicit in (1) and (2) is the fact that if the object is moving upward at $t = 0$, then $v_0 > 0$, whereas if the object is moving downward at $t = 0$, then $v_0 < 0$.

According to sports announcers, the fastest pitchers in the major leagues throw the baseball upwards of 95 miles per hour, which is approximately 140 feet per second. Keeping the units in feet, we will see how long it would take for a baseball to hit the ground if it is hurled by a pitcher from one of the world's great skyscrapers.

EXAMPLE 1   Located in Paris, the Eiffel Tower is 984 feet tall. Suppose a pitcher standing on the tower 832 feet above the ground throws a baseball straight down with an initial speed of 144 feet per second. Ignoring air resistance, find a formula for the height of the ball until it hits the ground, and determine how long it is in the air.

Solution   We measure time so that $t = 0$ at the instant the ball is thrown. By assumption, $h_0 = 832$ and $v_0 = -144$. By (2) the height of the ball is given by

$$h(t) = -16t^2 - 144t + 832 \qquad (3)$$

until the ball strikes the ground. To determine how long it takes for the ball to reach the ground, we find the values of t for which $h(t) = 0$:

$$-16(t^2 + 9t - 52) = 0$$
$$-16(t - 4)(t + 13) = 0$$
$$t = 4 \quad \text{or} \quad t = -13$$

Since the ball is thrown at time $t = 0$, it follows that it hits the ground at time $t = 4$. Thus it takes 4 seconds for the ball to hit the ground. Consequently $h(t) = -16t^2 - 144t + 832$ for $0 \leq t \leq 4$.   ❑

EXAMPLE 2   Substance A undergoes a chemical reaction in which molecules of A are transformed into molecules of substance B. Suppose the initial amount of A is 3, and the rate $f(x)$ at which x grams of A are turned into B is proportional to the product of x and $3 - x$. Express $f(x)$ in terms of x.

Solution   Since $f(x)$ is proportional to $x(3 - x)$, there is a number $c \neq 0$ such that

$$f(x) = cx(3 - x)$$

There are initially 3 grams of substance A, so this relation holds only for $0 \leq x \leq 3$. Consequently

$$f(x) = cx(3 - x) \quad \text{for } 0 \leq x \leq 3 \quad \square$$

We now discuss some general classes of functions that will be helpful to us later.

Polynomial and Rational Functions

First we consider the polynomial functions, which are especially amenable to the methods of calculus. Examples of polynomial functions are

$$f(x) = 2x^3 - 4x - 1 \qquad f(x) = \frac{1}{5}x$$

$$f(x) = \pi x^2 + 13 \qquad f(x) = 17$$

In general, a **polynomial** (or **polynomial function**) is any function f whose rule can be expressed in the form

$$f(x) = c_n x^n + c_{n-1} x^{n-1} + \cdots + c_1 x + c_0$$

where $c_n, c_{n-1}, \ldots, c_1,$ and c_0 are real numbers and n is a nonnegative integer (called the **degree** of the polynomial if $c_n \neq 0$). Zero-degree polynomials have the form

$$f(x) = c_0 \quad \text{with } c_0 \neq 0$$

and are called **constant functions.** Thus $f(x) = 17$ provides an example of a constant function. By convention, no degree is assigned to the constant polynomial 0. First-degree polynomials are of the form

$$f(x) = c_1 x + c_0 \quad \text{with } c_1 \neq 0$$

and are called **linear functions.** The particular linear function defined by $f(x) = x$ is called the **identity function.**

Quotients of polynomials form a second class of functions, called **rational functions.** Examples are

$$f(x) = \frac{1}{x} \qquad\qquad f(x) = x^2 + \sqrt{3}x$$

$$f(x) = \frac{x^4 + 3x^3 + 2x - \pi}{3x^3 - 4x + 1} \qquad f(x) = \frac{x^2 + 3x - 2}{x - 3}$$

Notice that a polynomial is a rational function whose denominator is the constant function 1. For example, the polynomial $f(x) = x^2 + \sqrt{3}x$ can be rewritten

$$f(x) = x^2 + \sqrt{3}x = \frac{x^2 + \sqrt{3}x}{1}$$

EXAMPLE 3   Let f be the function defined by

$$f(x) = \frac{x^2 + x - 2}{x^2 + 5x - 6}$$

Find the domain of f.

Solution   Since $x^2 + 5x - 6 = (x - 1)(x + 6)$, the denominator is 0 for $x = 1$ and $x = -6$. Thus the domain of f consists of all numbers except 1 and -6.   ❑

> **Caution:**   It is also possible to factor the numerator in Example 3, which yields $x^2 + x - 2 = (x - 1)(x + 2)$. Thus if $x \neq 1$ and $x \neq -6$, then
>
> $$f(x) = \frac{(x - 1)(x + 2)}{(x - 1)(x + 6)} = \frac{x + 2}{x + 6}$$
>
> Although the expression $(x + 2)/(x + 6)$ is meaningful for $x = 1$, the number 1 is not in the domain of f. Thus the domain of a function must be determined from the original description of the function.

> **Power functions** constitute a special class of rational functions. They have the form
>
> $$f(x) = x^n$$
>
> where n is an integer. Examples are x^2, x^5, x^{-1}, and x^{-3}. Recall that
>
> $$x^{-1} = \frac{1}{x}, \quad x^{-2} = \frac{1}{x^2}, \quad x^{-3} = \frac{1}{x^3}, \quad \ldots, \quad \text{and} \quad x^{-n} = \frac{1}{x^n}$$
>
> for any nonzero integer n. The domain of x^n consists of all real numbers if $n > 0$. If $n < 0$, then the domain contains all real numbers except 0, since division by 0 is not defined. We define $x^0 = 1$ for $x \neq 0$.

Root Functions

We begin with the square root function. By definition the square root function assigns to each nonnegative number x the nonnegative number y such that $y^2 = x$. We denote y by $\sqrt{x}$ or $x^{1/2}$. We emphasize that $\sqrt{x}$ is defined only for $x \geq 0$ and that $\sqrt{x} \geq 0$ for all $x \geq 0$. Consequently we may write $\sqrt{14}$, $\sqrt{\frac{1}{2}}$, and $\sqrt{0}$, but $\sqrt{-5}$ has no meaning, since there is no real number whose square is -5. Moreover, $\sqrt{4} = 2$, not ± 2. In Chapter 2 we will prove that a square root does in fact exist for each $x \geq 0$. Thus we can define the **square root function** by

$$f(x) = \sqrt{x} \quad \text{for } x \geq 0$$

Since $|x|^2 = x^2$ by the definition of absolute value, there is an intimate relation between square root and absolute value:

$$|x| = \sqrt{x^2}$$

Next we define the **cube root function.** It assigns to any real number x the unique number y such that $y^3 = x$. We denote y by $\sqrt[3]{x}$ or $x^{1/3}$. In contrast to the square root function, the cube root function has as its domain all real numbers, including negative numbers. For example, $\sqrt[3]{-1} = -1$, $\sqrt[3]{-8} = -2$, and $\sqrt[3]{-\frac{1}{27}} = -\frac{1}{3}$.

More generally, we can define the *n*th root function for any positive integer *n*. If *n* is odd, then for any real number *x* the *n*th root $\sqrt[n]{x}$ is the number *y* such that $y^n = x$. (Later we will show that for every *x* there is a unique number *y* with this property.) If *n* is even, then for any *nonnegative* number *x* the *n*th root $\sqrt[n]{x}$ is the *nonnegative* number *y* such that $y^n = x$. Although $(-y)^n = x$ also when *n* is even, we exclude negative values for $\sqrt[n]{x}$ when *n* is even. Thus $(-2)^4 = 2^4 = 16$, but $\sqrt[4]{16} = 2$. In this way $\sqrt[n]{x}$ is unique, whether *n* is odd or even, so that the *n*th root function really is a function.

Another notation for $\sqrt[n]{x}$ is $x^{1/n}$. Therefore

$$\sqrt[n]{x} = x^{1/n}$$

Observe that

$$\sqrt[n]{x} \text{ is defined } \begin{cases} \text{for any real number } x \text{ if } n \text{ is odd} \\ \text{for any nonnegative number } x \text{ if } n \text{ is even} \end{cases}$$

and

$$\sqrt[n]{x} = x^{1/n} = y \quad \text{if and only if } y^n = x \text{ (with } y \geq 0 \text{ if } n \text{ is even)}$$

It follows that

$$\sqrt[5]{-32} = -2, \quad \sqrt[3]{\frac{27}{8}} = \frac{3}{2}, \quad (16)^{1/4} = 2, \text{ and } \sqrt[6]{(-2)^6} = 2$$

In contrast, the expressions $\sqrt[4]{-1}$ and $\sqrt[6]{-\frac{43}{55}}$ have no meaning as real numbers, since -1 and $-\frac{43}{55}$ are negative and hence not in the domains of the respective root functions. For every positive integer *n* we also have

$$\sqrt[n]{1} = 1 \quad \text{and} \quad \sqrt[n]{0} = 0$$

Equality of Functions

We say that two functions *f* and *g* are **equal,** or the same, if *f* and *g* have the same domain and $f(x) = g(x)$ for each *x* in the common domain. Thus if

$$f(x) = x^2$$

and

$$g(x) = x^2 \quad \text{for } x \geq 1$$

then *f* and *g* are distinct functions because their domains are different. But if

$$f(x) = x^2 \qquad\qquad\qquad \text{for } x \geq -10$$
$$g(x) = (x-1)^2 + 2x - 1 \qquad \text{for } x \geq -10$$
$$h(y) = y^2 \qquad\qquad\qquad \text{for } y \geq -10$$

then *f, g,* and *h* are the same function, because their domains are identical and their rules all assign the same number to each number in the domain. To summarize: If two functions have the same domain and assign the same value to each number in their domain, then the two functions are equal.

Historical Comment

The notion of function was the result of a long development of mathematical thought. Mathematicians from antiquity to the Middle Ages had at best a vague idea of the concept of function. As calculus developed during the latter part of the seventeenth century, the need to make the notion of function precise gradually became apparent. The word "function," which derives from the Latin word for "perform," seems to have been used first by Gottfried Leibniz* (1646–1716).

The Swiss mathematician Leonhard Euler † (1707–1783) was the first to adopt the expression $f(x)$ for the value of function. He systematically classified many collections of functions. Nevertheless, it was not until the middle of the nineteenth century that the formulation given in Definition 1.4 emerged. This was the work of P. G. Lejeune-Dirichlet ‡ (1805–1859). Although there are more formal, set-theoretic versions of the definition of function, the form in Definition 1.4 will suffice for us.

* **Leibniz:** Pronounced *"Libe* -nits."
† **Euler:** Pronounced *"Oi* -ler."
‡ **Dirichlet:** Pronounced "Di-ri- *shlay.* "

EXERCISES 1.3

In Exercises 1–14 find the numerical value of the function at the given values of a.

1. $f(x) = \sqrt{3}$; $a = \sqrt{5}, \pi$

2. $f(x) = 2x^2 - 3$; $a = 1, -2$

3. $f(x) = 1 - x + x^3$; $a = 0, -1$

4. $f(x) = 1/x$; $a = 2, \frac{1}{2}$

5. $g(x) = 1/(2x^2)$; $a = \sqrt{2}$

6. $g(x) = \sqrt{x}$; $a = 4, \frac{1}{25}$

7. $g(t) = \sqrt[3]{t}$; $a = 27, -\frac{1}{8}$

8. $g(t) = |2 - t|$; $a = 6$

9. $f(x) = \dfrac{x - 1}{x^2 + 4}$; $a = 2$

10. $f(x) = \dfrac{3x^2 - 4x - 1}{2x^2 + 5x - 3}$; $a = -1$

11. $f(x) = \dfrac{-2}{169} x^2 + \dfrac{4}{13} x + 3$; $a = 3.2, 25.5$

12. $f(x) = \sqrt{\dfrac{192{,}000}{x} - 6}$; $a = 10{,}000, 21{,}729$

13. $g(x) = \dfrac{100.24 \, x^6}{(0.24)x^6 - 1}$; $a = 0.5, -7.31$

14. $g(x) = 4.5 \, x^{1/2} - x^{3/2}$; $a = 3, 1.64$

In Exercises 15–36 find the domain of the function.

15. $f(x) = x^3 - 4x + 1$

16. $f(x) = x^6 - \sqrt{2}x^3 - \pi$

17. $k(x) = 1 + x^3$ for $-2 \le x \le 8$

18. $f(x) = 2x - 3x^5$ for $x < 4$

19. $f(x) = \sqrt{x + 2}$ 20. $f(x) = \sqrt{2 - 3x}$

21. $f(x) = \sqrt{x(x - 1)}$ 22. $f(t) = \sqrt{4 - 9t^2}$

23. $f(t) = \sqrt{3 - \dfrac{1}{t^2}}$ 24. $f(t) = \dfrac{t}{\sqrt{t + 5}}$

25. $f(t) = \sqrt[3]{1 - t^2}$ 26. $g(x) = (x - 6)^{1/4}$

27. $g(x) = \dfrac{2}{x - 1}$ 28. $g(x) = \dfrac{3x - 1}{x - 3}$

29. $g(w) = \dfrac{2w - 8}{w^2 - 16}$ 30. $g(w) = \dfrac{w - 1}{w^2 - w - 6}$

31. $k(x) = \dfrac{2x - 3}{x^2 + 4}$ 32. $k(x) = \dfrac{1}{x + 1} - \dfrac{2}{x - 1}$

33. $f(x) = \begin{cases} 2x \text{ for } -4 \le x \le -1 \\ 3 \text{ for } 0 < x < 6 \end{cases}$

34. $f(x) = \begin{cases} x^2 + 1 \text{ for } x \le 2 \\ x^2 - 1 \text{ for } x > \sqrt{5} \end{cases}$

*35. $f(x) = \sqrt{1 - \sqrt{9 - x^2}}$

*36. $f(x) = \sqrt{4 - \sqrt{1 + 9x^2}}$

In Exercises 37–42 determine the range of the function.

37. $f(x) = -1$ **38.** $f(x) = 3x - 2$

39. $f(x) = 3x - 2$ for $x < 4$ **40.** $f(x) = \sqrt{1 - x^2}$

41. $f(x) = \dfrac{1}{x - 1}$ ***42.** $f(x) = \dfrac{x^2 - 1}{x^2 + 1}$

43. Determine which of the following define a function. Explain your reason for any that do not define a function.

  a. The domain consists of the number -2, which is assigned the number π.

  b. The domain consists of the number -2, which is assigned the numbers -2 and π.

  c. $f(x) = \pm \sqrt{x}$

  d. $f(x) = \pm \sqrt{x^2 + 1}$

  e. $g(x) = \begin{cases} x - 1 & \text{for } x < 0 \\ 12x - 6 & \text{for } x > 0 \end{cases}$

  f. $g(x) = \begin{cases} 2 - 4x & \text{for } x < 0 \\ x^2 & \text{for } x > 1 \end{cases}$

  g. $g(x) = \begin{cases} 4x + 1 & \text{for } x \leq 2 \\ 2x^3 - 7 & \text{for } x \geq 2 \end{cases}$

  h. $g(x) = \begin{cases} 2 - 3x^3 & \text{for } x \leq 1 \\ 3x^4 - 3 & \text{for } x \geq 1 \end{cases}$

  i. $f(t) = \begin{cases} t^2 & \text{for } t \text{ rational} \\ t & \text{for } t \text{ irrational} \end{cases}$

  *j. $f(t) = \begin{cases} t^2 & \text{for } t^2 \text{ rational} \\ t & \text{for } t \text{ irrational} \end{cases}$

44. In each of the following, determine whether f and g are the same.

  a. $f(x) = 1 - x^2$; $g(x) = 1 - x^2$ for $-1 < x < 1$

  b. $f(x) = \sqrt{x}$ for $x \geq 0$; $g(x) = \sqrt{x}$

  c. $f(x) = \sqrt{x^2}$; $g(x) = |x|$

  d. $f(x) = \dfrac{x^3 - 4x}{x^3 - 4x}$; $g(x) = 1$

  e. $f(x) = \dfrac{x - 1}{x^2 - 1}$; $g(x) = \dfrac{1}{x + 1}$

  f. $f(x) = \dfrac{x^2 - 5x + 6}{x + 2}$; $g(x) = x - 3$ for $x \neq -2$

45. Which of the following functions are the same?

  a. $f_1(x) = \sqrt{1 - 6x + 9x^2}$

  b. $f_2(x) = 1 - 3x$

  c. $f_3(t) = 1 - 3t$

  d. $f_4(w) = 1 - 3w$ for $w \geq 0$

  e. $f_5(t) = |1 - 3t|$

  f. $f_6(x) = \dfrac{(1 - 3x)^2}{1 - 3x}$

46. Let
$$f(x) = \sqrt{x^2 + 1} - 1 \quad \text{and} \quad g(x) = \frac{x^2}{1 + \sqrt{x^2 + 1}}$$

  a. Find the domains of f and g.

  b. Show that $f = g$.

47. Let
$$f(x) = x - \sqrt{x^2 - 1} \quad \text{and} \quad g(x) = \frac{1}{x + \sqrt{x^2 - 1}}$$

  a. Find the domains of f and g.

  b. Show that $f = g$.

48. Find a formula for the function f that assigns to each x greater than -1 the number obtained by squaring x, then subtracting $2x$, and finally adding $\sqrt{2}$.

49. Find a formula for the function f that assigns to each nonnegative x the number obtained by dividing x by 5, then taking the cube root of the quotient, and finally multiplying the result by the product of $\frac{1}{2}$ and x^2.

50. The volume V of a rectangular box with square base is 60 cubic centimeters. Express the length l (in centimeters) of a vertical side as a function of the length s of a side of the base.

51. Find a formula for the function A that expresses the area of an equilateral triangle in terms of the length of one of its sides.

Applications

52. Determine the height of the ball in Example 1 after it has traveled for

  a. 2 seconds

  b. 3 seconds

   c. 3.9 seconds

 53. a. Transform the formula derived in Example 1 into one corresponding to (1), in which distance is measured in meters.

  b. Using part (a), show that the baseball hits the ground in (approximately) 4 seconds.

 54. In order to discover the height above ground of a ten–story apartment, Pat drops a ball from its balcony. If the ball hits the grass below after 2.5 seconds, determine the height (in meters) from which the ball was dropped.

55. Suppose that the ball in Exercise 54 was thrown from the same location but hit the grass after 2 seconds. Determine the speed with which the ball was thrown.

56. A wrench is dropped from Chicago's Sears Tower, some 1350 feet above ground (100 feet from the top).

  a. Express the height of the wrench as a function of time while it is falling.

b. Find the distance the wrench travels during the first 3 seconds.

c. Determine how long it takes for the wrench to hit the ground.

57. A ball is thrown downward from the roof of a building 30 meters high, with an initial speed of 5 meters per second.

   a. Find the approximate height of the ball after 1/2 second and after 1 second.

   b. Approximately how long does it take for the ball to reach a window 10 meters above the ground?

58. A tank has the form of a right circular cylinder with hemispherical ends. Its volume V is 100 cubic meters.

   a. Find the length L of the cylinder in terms of the radius of the hemispheres.

    b. Find the length L (to the nearest centimeter) if the radius of the hemispheres is 2 meters.

    c. How much longer (to the nearest centimeter) would the cylindrical portion of the tank need to be if the radius of the hemispheres were 1 meter instead of 2 meters?

59. The period T of a pendulum of length L that swings under the influence only of gravity is given approximately by $T = 2\pi\sqrt{L/g}$, where $g = 9.8$ meters per second squared.

   a. Write L as a function of T.

    b. The length of the Foucault pendulum at the Smithsonian Institution is approximately 21.8 meters. Determine its period.

60. The Tee-rific Company produces 100,000 golf tees daily and sells them for 5¢ apiece. Assume that the total cost of producing one tee is 2¢. Find the company's profit P (in cents) in terms of the number of working days.

61. Starting at noon, A flies 2400 miles from New York to San Francisco at a velocity of 400 miles per hour. B starts the same trip at 2:00 P.M. the same day with a velocity of 800 miles per hour. Express the distance D between A and B at any instant between noon and 5:00 P.M. in terms of the time in hours elapsed after noon.

62. Two cars depart from the same location at the same time. One travels north at 40 miles per hour and the other travels east at 50 miles per hour. Find a formula for the function D that expresses in terms of t the distance between the cars t hours after departure.

63. In the study of the response to acetylcholine by a frog's heart, the formula

$$R(x) = \frac{x}{c + dx}$$

arises, where x denotes the concentration of the drug and c and d are positive constants.

   a. Find $R(0)$ and $R(2)$. What is the physical significance of $R(0)$?

   b. Find a formula that expresses the concentration x in terms of $R(x)$.

1.4 GRAPHS

The most concise way of describing a function is by means of formulas like

$$f(x) = \frac{x^2}{x + 1}$$

In calculus, however, it is often useful and instructive to sketch a picture of a function. In fact, historically one of the major achievements of calculus was that it enabled one to draw accurate pictures of many functions.

Graph of a Function

The pictorial representation of a function is called a graph.

DEFINITION 1.5   Let f be a function. Then the set of all points $(x, f(x))$ such that x is in the domain of f is called the **graph** of f.

In many cases, by studying the formula that defines a function we can easily draw a satisfactory sketch of its graph.

EXAMPLE 1   Let $f(x) = \frac{9}{5}x + 32$. Sketch the graph of f.

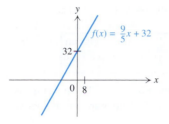

FIGURE 1.21

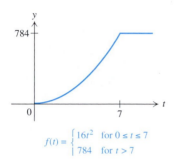

FIGURE 1.22

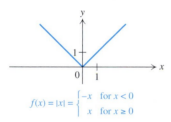

FIGURE 1.23   The absolute value function.

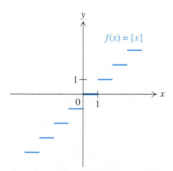

FIGURE 1.24   The greatest integer function

Solution   By Definition 1.5, the graph of *f* consists of the set of all points (x, y) in the plane for which $y = \frac{9}{5}x + 32$. This is a line with slope $\frac{9}{5}$ and *y* intercept 32. The graph is shown in Figure 1.21.   ❑

If *f* is any linear function, then $f(x) = mx + b$. As we saw in Section 1.2, the graph of *f* is a line with slope *m* and *y* intercept *b*.

EXAMPLE 2   Sketch the graph of the function defined by

$$f(t) = \begin{cases} 16t^2 & \text{for } 0 \le t \le 7 \\ 784 & \text{for } t > 7 \end{cases}$$

Solution   We first tabulate several points $(t, f(t))$ on the graph of *f*:

t	0	$\frac{1}{2}$	1	2	3	4	5	6	7	10
$f(t)$	0	4	16	64	144	256	400	576	784	784

Since the values of $f(t)$ shown here are much larger than those of *t*, we use a smaller scale on the *y* axis than on the *t* axis. Plotting the points in the table and connecting them with a curve, we obtain the graph shown in Figure 1.22.   ❑

EXAMPLE 3   Sketch the graph of the **absolute value function:**

$$f(x) = |x|$$

Solution   By the definition of absolute value we have

$$f(x) = |x| = \begin{cases} -x & \text{for } x < 0 \\ x & \text{for } x \ge 0 \end{cases}$$

Therefore the part of the graph to the right of the *y* axis coincides with the line $y = x$, and the part to the left of the *y* axis coincides with the line $y = -x$. Thus the graph is as shown in Figure 1.23.   ❑

EXAMPLE 4   The **greatest integer,** or **staircase, function** is defined by

$$f(x) = [x]$$

where the symbol $[x]$ denotes the greatest integer not larger than *x*. Sketch the graph of *f*.

Solution   The values of *f* are integers. Moreover, if *n* is any integer and if $n \le x < n + 1$, then $[x] = n$. Therefore $f(x) = n$ for all *x* in $[n, n + 1)$, where *n* is any integer. The graph of *f* is sketched in Figure 1.24.   ❑

EXAMPLE 5   Let $f(x) = 1/x$. Sketch the graph of *f*.

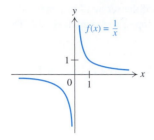

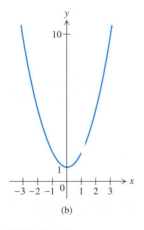

FIGURE 1.25

Solution   We determine several points $(x, f(x))$ on the graph of f:

x	$\frac{1}{4}$	$\frac{1}{2}$	1	2	4	$-\frac{1}{4}$	$-\frac{1}{2}$	-1	-2	-4
$f(x)$	4	2	1	$\frac{1}{2}$	$\frac{1}{4}$	-4	-2	-1	$-\frac{1}{2}$	$-\frac{1}{4}$

Note that for $x > 0$, $1/x$ decreases as x increases. For $x < 0$, $1/x$ increases as x decreases. Note also that $x = 0$ is not in the domain of f. The graph of f is shown in Figure 1.25. ❑

EXAMPLE 6   Let

$$f(x) = x^2 + 1 \quad \text{and} \quad g(x) = \frac{x^3 - x^2 + x - 1}{x - 1}$$

Sketch the graphs of f and g.

Solution   First we assemble the following table of points $(x, f(x))$ on the graph of f:

x	-2	-1	$-\frac{1}{2}$	0	$\frac{1}{2}$	1	2
$f(x)$	5	2	$\frac{5}{4}$	1	$\frac{5}{4}$	2	5

Notice that $x^2 + 1$ increases as x increases and is positive, or as x decreases and x is negative. Filling in between the points with a smooth curve, we obtain the graph in Figure 1.26(a). Next we turn to the function g. Although the formula for g looks complicated, it can be simplified:

$$g(x) = \frac{x^3 - x^2 + x - 1}{x - 1} = \frac{(x^2 + 1)(x - 1)}{x - 1} = x^2 + 1 \quad \text{for } x \neq 1$$

It follows that $f(x) = g(x)$ whenever $x \neq 1$. Moreover, 1 is in the domain of f but is not in the domain of g. Therefore the graph of g, which is displayed in Figure 1.26(b), is the same as the graph of f except that we have left a hole in the graph of g above the point 1 on the x axis in order to emphasize that 1 is not in the domain of g. ❑

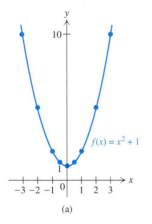

FIGURE 1.26

In Chapter 3 we will encounter many functions whose rules can be simplified as we did for g in Example 6.

In previous examples we found several points on the graph of a function and used this information to sketch the graph of the function. However, this method of analyzing graphs of functions will be replaced later by very general and powerful methods of calculus that allow us to draw graphs without plotting numerous points.

Until now we have started with a formula for a function and drawn its graph. Next we will determine conditions under which a curve in the plane is the graph of a function. Since a function can assign only one value to each number in its domain, a curve can represent the graph of a function provided that no more than one point of the curve lies on any given vertical line (Figure 1.27(a)). In that case, we can describe the function as follows. The domain of the function consists of all values of *x* that are the first coordinates of points on the curve, and the rule assigns to each such *x* the *unique* value *y* such that (*x, y*) is on the curve. Thus if each vertical line contains no more than one point on the curve, each *x* is assigned a unique value *f* (*x*), as required by the definition of a function. The curve shown in Figure 1.27(b) cannot be the graph of a function, since there are vertical lines containing more than one point on the curve.

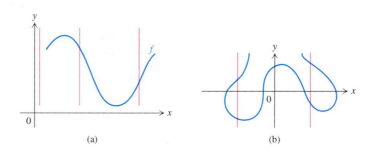

(a)   (b)

FIGURE 1.27   The curve in (a) is the graph of a function; the curve in (b) is not.

Finally, we observe that two functions are equal precisely when their graphs are identical.

Graphing Functions by Calculator

Today high-speed computers and graphics calculators can produce accurate graphs of many functions by plotting large numbers of points. These instruments normally graph such a function for *x* and *y* values in certain specified intervals, which we will call, respectively, the **x window** and the **y window.** The standard, or default, windows might be [−10, 10]. If $f(x) = x^{1/3}$ and we use the standard windows, then the graph of *f* would be as in Figure 1.28(a). If we wished to analyze the graph closer to the origin, we might choose the *x* window to be [−2, 2] and the *y* window to be [−2, 2], as in Figure 1.28(b).

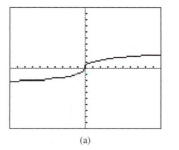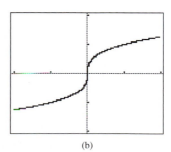

(a)   (b)

FIGURE 1.28

There are several ways in which the graph drawn on a computer or calculator may be misleading.

1. It may lead us to think that certain numbers are in the domain of f when they are not. For example, let

$$f(x) = \frac{x^2 + 4x + 4}{x + 2}$$

Then $f(x) = x + 2$ except for $x = -2$, which is not in the domain of f. However, the graph, as it appears on a graphics calculator, seems to have no breaks (Figure 1.29(a)).

2. It may ignore subtle but important features of the graph. For example, let $f(x) = x^5 - x^4$. If the windows are standard, then it appears as though the graph of f lies along the x axis on the interval for $0 \le x \le 1$ (Figure 1.29(b)). To show that this is not true, we need only to take the windows to be, say, $[-1, 1]$.

3. A single window may not provide all the details one needs in order to adequately understand the graph. For example, consider the function defined by

$$f(x) = \frac{x^3}{x^2 - 0.5}$$

If we use a small x window like $[-3, 3]$, then we obtain information about the part of the graph near the origin but learn nothing about the graph far away from the origin (see Figure 1.29(c)). By contrast, if we use a larger x window like $[-10, 10]$, then the part of the graph near the origin is obscured (see Figure 1.29(d)).

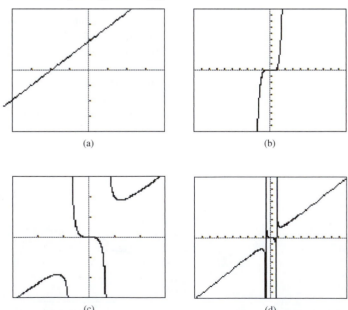

(a)

(b)

(c)

(d)

FIGURE 1.29

In conclusion, computers and graphics calculators can assist in understanding the graph of a function, although we must be careful in the interpretation of such graphs. Nevertheless, one of the triumphs of calculus is the possibility of analyzing the graph of a function without necessarily resorting to a computer or calculator and without plotting numerous points on the graph.

Dependent and Independent Variables

We generally use the expression (x, y) to denote a point in the plane. We can describe the graph of a function f as the set of all points (x, y) such that x is in the domain of f and $y = f(x)$. For this reason we sometimes specify a function by writing y in terms of x. For instance, we may write

$$y = \frac{9}{5}x + 32 \quad \text{instead of} \quad f(x) = \frac{9}{5}x + 32$$

$$y = x^2 \quad \text{instead of} \quad f(x) = x^2$$

$$y = \frac{1}{x} \quad \text{instead of} \quad f(x) = \frac{1}{x}$$

We call x an **independent variable.** Since the value of y depends on that of x, y is a **dependent variable.** We will sometimes say that y is a variable depending on x. We will use both notations, $f(x)$ and y, for the value assigned to a number x in the domain of a function. Also, we frequently will use other, more suggestive letters in place of x and y. For example, rather than writing $y = \frac{9}{5}x + 32$ for the conversion from degrees Celsius to degrees Fahrenheit, we would likely write $F = \frac{9}{5}C + 32$.

Graph of an Equation

There are many equations in x and y that are not of the form $y = f(x)$. Examples of such equations are

$$x^2 + y^2 = 4 \quad \text{and} \quad x = y^2$$

The **graph of an equation** in x and y is the collection of points (x, y) in the plane that satisfy the equation.

EXAMPLE 7   Sketch the graph of the equation $x^2 + y^2 = 4$.

Solution   The graph consists of all points (x, y) that satisfy $x^2 + y^2 = 4$. Recall that the distance r between any point (x, y) in the plane and the origin, $(0, 0)$, is given by

$$r = \sqrt{(x - 0)^2 + (y - 0)^2} = \sqrt{x^2 + y^2}$$

Hence any point (x, y) at a distance r from the origin satisfies the equation

$$x^2 + y^2 = r^2$$

and so the collection of points satisfying $x^2 + y^2 = 4$ consists of those points whose distance from the origin is 2. Thus the graph of $x^2 + y^2 = 4$ is a circle of radius 2 centered at the origin (Figure 1.30).   ❑

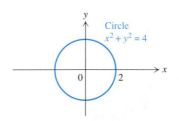

FIGURE 1.30

EXAMPLE 8   Sketch the graph of the equation $x = y^2$.

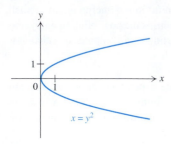

FIGURE 1.31

Solution   We first find a few points satisfying the equation:

x	0	$\frac{1}{2}$	1	2	4
y	0	$\pm\sqrt{2}/2$	± 1	$\pm\sqrt{2}$	± 2

Since $x = y^2$ and $y^2 \geq 0$ for any y, x cannot be negative. For each positive value of x there are two values of y, namely, $\sqrt{x}$ and $-\sqrt{x}$. As x increases, so must $|y|$. The graph is sketched in Figure 1.31. ❑

EXAMPLE 9   Sketch the graph of the equation $x^2 + y^4 = -1$.

Solution   Since $x^2 \geq 0$ and $y^4 \geq 0$ for any x and y, there are no numbers x and y satisfying the equation. Consequently there are no points on the graph. ❑

We will continue to sketch and study graphs in the remainder of this book.

EXERCISES 1.4

In Exercises 1–24 sketch the graph of the function.

1. $f(x) = \frac{1}{2}x + 1$

2. $f(x) = 1 - 3x$ for $-1 \leq x \leq 2$

3. $f(x) = x^2$ **4.** $f(x) = x^2 + 2$

5. $f(x) = x^2 - 1$ for $-2 \leq x \leq 2$

6. $f(x) = x^3$ **7.** $y = \sqrt{x}$

8. $y = \sqrt{4 - x^2}$ (*Hint:* See Example 7.)

9. $y = \sqrt{1 - x^2}$ **10.** $y = \dfrac{x^2 - 16}{x - 4}$ **11.** $y = \dfrac{1 - x^2}{x + 1}$

12. $g(x) = |x|$ for $-2 \leq x \leq 3$ **13.** $g(x) = x|x|$

14. $g(x) = |x + 1|$ (*Hint:* Consider the cases $x + 1 \geq 0$ and $x + 1 < 0$ separately.)

15. $g(x) = |x - 2|$ **16.** $f(t) = |t| + t$

17. $f(t) = |t| - t$ **18.** $f(t) = \dfrac{|t|}{t}$

19. $f(x) = -[x]$ **20.** $f(x) = [-x]$

21. $f(x) = x - [x]$

22. $f(x) = \begin{cases} x^2 \text{ for } 0 \leq x \leq 2 \\ 4 \text{ for } x > 2 \end{cases}$

23. $f(x) = \begin{cases} x \text{ for } x < 0 \\ 2x \text{ for } x \geq 0 \end{cases}$ **24.** $f(x) = \begin{cases} x^2 \text{ for } x < 0 \\ -x \text{ for } x \geq 0 \end{cases}$

In Exercises 25–27 sketch the graph of the function.

25. The function f defined by

$$f(x) = \begin{cases} 3 & \text{for } x \neq 1 \\ 5 & \text{for } x = 1 \end{cases}$$

A function whose graph resembles that of f can be called a "hiccup" function.

26. The **sign**, or **signum**, function, defined by

$$f(x) = \begin{cases} -1 & \text{for } x < 0 \\ 0 & \text{for } x = 0 \\ 1 & \text{for } x > 0 \end{cases}$$

27. The **diving board** function, defined by

$$f(x) = \begin{cases} 0 & \text{for } x < 0 \\ 1 & \text{for } x \geq 0 \end{cases}$$

In Exercises 28–33 have a computer or graphics calculator plot the graph of the function on the given x window. Alter the y window as needed in order to be able to discern the major features of the graph. By reading off the y coordinate of the lowest point on the graph, approximate the smallest value assumed by the function on the given interval. It may be helpful to adjust the window.

28. $f(x) = 2x^2 - 10x + 13$; $[0, 5]$

29. $f(x) = x^7 - x^5$; $[-1, 1]$

30. $f(x) = x^2(x - 5)^2$; $[-3, 7]$

31. $f(x) = \dfrac{4x^3 - x + 1}{x^4 + 1}$; $[-3, 3]$

32. $f(x) = \dfrac{x^6 - x^4 + x^2 + 1}{x^4 + 1}$; $[-10, 10]$

33. $f(x) = 210x^4 - 107x^3 + 18x^2 - x$; $[0, 0.5]$

In Exercises 34–46 sketch the graph of the equation. In each case determine whether the graph is that of a function.

34. $x^2 + y^2 = 1$ **35.** $x^2 + y^2 = 9$

36. $x^2 + y^2 = 9$ for $x \geq 0$ **37.** $x^2 + y^2 = 4$ for $y \leq 0$

38. $x^2 + y^2 = 0$ **39.** $x = \frac{1}{2} y^2$

40. $y = x^2$ for $x \leq 0$ **41.** $x^2 = y^2$

42. $y^2 = x^3$ **43.** $xy = 0$

44. $|x| = |y|$ **45.** $|x| + |y| = 1$

46. $|x| - |2y| = 1$

47. Let $f(x) = x^3$.

   a. On the same screen, plot the graphs of $y = f(x)$, $y = f(x-1)$, and $y = f(x-2)$.

   b. On the same screen, plot the graphs of $y = f(x)$, $y = f(x+1)$, and $y = f(x+2)$.

   c. On the basis of parts (a) and (b), how do you think one obtains the graph of $y = f(x+c)$ from the graph of f? Consider the two cases $c > 0$ and $c < 0$ separately.

48. Which of the graphs in Figure 1.32 are graphs of functions?

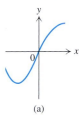 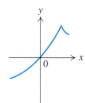 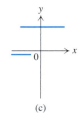

     (a)   (b)   (c)

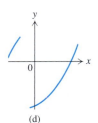 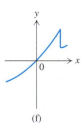 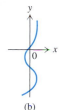

     (d)   (e)   (f)

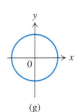 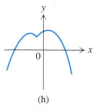

     (g)   (h)

FIGURE 1.32   Graphs for Exercise 48.

49. Let f be the function with domain $[-4, 4]$ whose graph is sketched in Figure 1.33. From the graph, determine the solution (if any) of

   a. $f(x) = 0$ **b.** $f(x) > 0$ **c.** $f(x) < 0$ **d.** $f(x) = -10$

   e. $f(x) > -4$

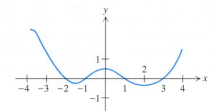

FIGURE 1.33   Graph for Exercise 49.

50. Can a horizontal line pass through more than one point on the graph of a function? Explain.

51. Let $f(x) = (x-1)^p (x-r)^q$. Find possible integer values of p, q, and r so that the graph in Figure 1.34(a) could be the graph of f.

52. Let $f(x) = (x+1)^p (x-1)^q (x-2)^r$. Find possible integer values of p, q, and r so that the graph in Figure 1.34(b) could be the graph of f.

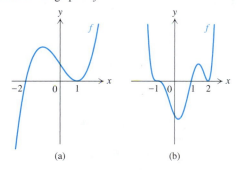

     (a)   (b)

FIGURE 1.34   (a) Graph for Exercise 51.   (b) Graph for Exercise 52.

Applications

53. The graphs of possible revenue, cost and profit functions are shown in Figure 1.35. Decide which is which.

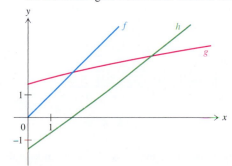

FIGURE 1.35   Graph for Exercise 53.

54. You are to start a new company that manufactures compact discs. In Figure 1.36 the horizontal axis represents the number of CD's produced, and the vertical axis represents the corresponding profit. Assuming that it is possible to make so many CD's that they cannot all be sold, determine which of the graphs is the most likely profit curve.

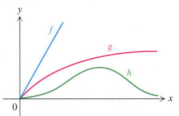

FIGURE 1.36   Graphs for Exercise 54.

55. Suppose you walk along a straight path from your dormitory to the mathematics building, which is 300 meters away. After 100 meters you pass the Student Union. Assuming that you walk at the rate of 1 meter per second, sketch the graph of your distance from the Student Union as a function of time t, with $t = 0$ corresponding to the instant you leave your dormitory.

56. The height of a ball that is dropped into the sand from a height of h_0 meters is given in Figure 1.37.
   a. Find an approximate value for h_0.
   b. Why is the graph horizontal for $t \geq t^*$?
   c. Suppose a superball were dropped from the same height onto a hardwood floor. What might the corresponding graph of its height look like?

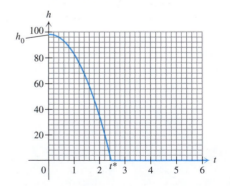

FIGURE 1.37   Graph for Exercise 56.

57. On the weekend the Budge Agency rents Buicks at the rate of \$30 plus 1¢ per mile, whereas the Econ Agency charges \$25 plus 2¢ per mile.

   a. On the same coordinate system sketch the graphs of the cost of renting cars from the two agencies, as functions of the number of miles driven.
   b. What does the point of intersection of the two graphs imply about which agency you should choose to rent from?

58. a. Let f be a function, and let a be any nonzero number in the domain of f. Find an expression for the slope of the line joining $(0, 0)$ and $(a, f(a))$ (Figure 1.38(a)).
   b. For the function whose graph appears in Figure 1.38(b), determine which of the numbers $f(2)/2$ and $f(5)/5$ is larger.

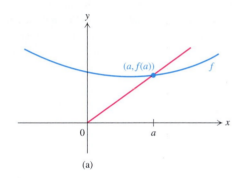

(a)

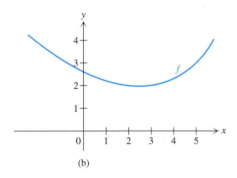

(b)

FIGURE 1.38   Graph for Exercise 58.

59. The gravitational force F that the earth exerts on a unit mass depends on whether the mass is inside or outside the earth. Let R denote the radius of the earth and r the distance of the unit mass from the center of the earth. Then F is given by

$$F = \begin{cases} \dfrac{GMr}{R^3} & \text{for } 0 < r < R \\[2mm] \dfrac{GM}{r^2} & \text{for } r \geq R \end{cases}$$

where G denotes the universal gravitational constant and M the mass of the earth. Sketch the graph of F as a function of r. (For the graph include $[0, 2R]$ on the horizontal axis.)

*60. Postage for domestic first class letters is 37¢ for the first ounce or part ounce and 23¢ for each additional ounce or part ounce up to 14 ounces. Let $P(x)$ denote the postage for a letter weighing x ounces, and assume that the domain of P is $(0, 14)$. Express $P(x)$ in terms of the greatest integer function.

1.5   AIDS TO GRAPHING

There are certain aids to graphing that do not depend on calculus. Among these are the location of coordinate intercepts, symmetry properties, and translations.

Intercepts

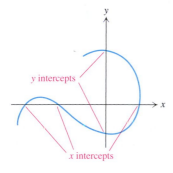

FIGURE 1.39

The intercepts of a graph determine the points where the graph meets the axes. We use the term **x intercept** for the x coordinate of any point where the graph meets the x axis. Similarly, the **y intercept** is the y coordinate of any point where the graph meets the y axis (Figure 1.39). (This definition of y intercept coincides with the original definition given in Section 1.2 for lines.) In theory it is very simple to locate the x and y intercepts from a given equation. More specifically, to find the x intercepts, set $y = 0$ in the equation and solve for x. To find the y intercepts, set $x = 0$ in the equation and solve for y.

EXAMPLE 1   Find the x and y intercepts of the graph of the equation

$$y = x^2 - 2x - 3$$

Solution   To find the y intercepts, we let $x = 0$ and solve for y. This gives us -3 as the y intercept. To find the x intercepts, we let $y = 0$ and solve for x:

$$0 = x^2 - 2x - 3 = (x - 3)(x + 1)$$

It follows that -1 and 3 are the two x intercepts of the graph (Figure 1.40). ❑

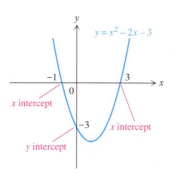

FIGURE 1.40

The graph of a function f has a y intercept only if 0 is in the domain of f, and in that case the y intercept is $f(0)$. Since $f(0)$ is unique, there can be at most one y intercept of the graph of a function. For example, if

$$f(x) = 2x^3 - 4x^2 + \sqrt{2}$$

then the y intercept is $\sqrt{2}$, since $f(0) = \sqrt{2}$. However, if

$$g(x) = \sqrt{x - 1}$$

then there is no y intercept because 0 is not in the domain of g.

In contrast, the graph of a function f may have many x intercepts. The graph of f has an x intercept at x if $f(x) = 0$. Such a value of x is called a **zero,** or **root,** of f. In the event that $f(x) = ax^2 + bx + c$ with $a \neq 0$, then any zeros of f are given by the **quadratic formula:**

$$x = \frac{-b - \sqrt{b^2 - 4ac}}{2a} \quad \text{and} \quad x = \frac{-b + \sqrt{b^2 - 4ac}}{2a} \tag{1}$$

(see Exercise 40). It follows that if the **discriminant** $b^2 - 4ac$ is positive, then the function has two zeros and hence the graph has two x intercepts. For example, let $f(x) = x^2 - 2x - 2$. By the quadratic formula, the zeros of f are

$$\frac{-(-2) - \sqrt{(-2)^2 - 4(1)(-2)}}{2(1)} \quad \text{and} \quad \frac{-(-2) + \sqrt{(-2)^2 - 4(1)(-2)}}{2(1)}$$

which equal $1 - \sqrt{3}$ and $1 + \sqrt{3}$ (Figure 1.41(a)). However, if the discriminant is negative, then the function has no zeros and thus no x intercepts. Thus if $g(x) = x^2 - 2x + 2$, then g has no zeros because the discriminant $(-2)^2 - 4(1)(2)$ is negative (Figure 1.41(b)). Finally, we observe that if the discriminant is 0, then the function has only one zero. For example, if $h(x) = x^2 - 2x + 1 = (x - 1)^2$, then h has exactly one zero, namely 1 (Figure 1.41(c)).

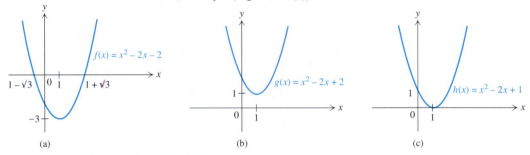

FIGURE 1.41

Although there are complicated formulas for zeros of third- and fourth-degree polynomials, there is no formula that automatically produces zeros of polynomials of degree five or higher. However, even though we might not be able to find the precise values of zeros for such a function, many computer algebra systems and many calculators are able to give very good estimates for such zeros. For example, if $f(x) = x^5 + x^3 + x + 1$, then we obtain -0.6368829168 for an approximate value of the lone zero of f. The graph in Figure 1.42 reinforces this assertion. There are also methods in calculus that will help us to approximate zeros, and hence x intercepts, of graphs of functions.

A function need not have intercepts, as Example 2 illustrates.

EXAMPLE 2   Let $f(x) = 1/x$. Find the intercepts of the graph of f.

Solution   Since 0 is not in the domain of f, there is no y intercept. Moreover, since there is no x such that $0 = 1/x$, there is no x intercept. The graph of f, which is shown in Figure 1.43, has no intercepts at all.   ❑

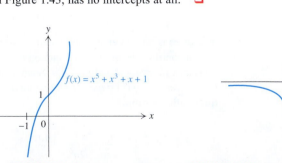

FIGURE 1.42   **FIGURE 1.43**   A graph with no intercepts.

Symmetry

The Taj Mahal, at Agra, India (*David Sutherland/Tony Stone Images*)

Symmetry is a form of balance. For example, part of the beauty of the Taj Mahal lies in the symmetry visible in the building and its surroundings. In mathematics we are interested in symmetry in graphs, especially symmetry with respect to the *x* and *y* axes and the origin. There are very simple criteria for determining the existence of such symmetry.

Types of symmetry	Conditions for symmetry
Symmetry with respect to the *x* axis	$(x, -y)$ is on the graph whenever (x, y) is
Symmetry with respect to the *y* axis	$(-x, y)$ is on the graph whenever (x, y) is
Symmetry with respect to the origin	$(-x, -y)$ is on the graph whenever (x, y) is

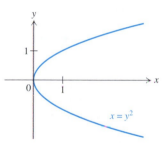

FIGURE 1.44   Symmetry with respect to the *x* axis.

For symmetry with respect to the *x* axis, the points above and below the *x* axis can be considered reflections of each other in the *x* axis, which we regard as a mirror (Figure 1.44).

EXAMPLE 3   Show that the graph of the equation $x = y^2$ is symmetric with respect to the *x* axis.

Solution   We must show that if (x, y) is on the graph, then $(x, -y)$ is also on the graph. Thus we must show that

$$\text{if } x = y^2, \quad \text{then } x = (-y)^2$$

But this is clear, since

$$(-y)^2 = y^2$$

Figure 1.45 supports our proof.   ❏

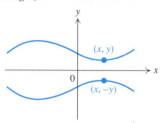

FIGURE 1.45   The graph is symmetric with respect to the *x* axis.

If only *even* powers of *y* appear in an equation, then the graph of the equation is automatically symmetric with respect to the *x* axis, since for any integer *n*, $(-y)^{2n} = y^{2n}$. By this test, the graphs of the following equations are all symmetric with respect to the *x* axis:

$$x^3y^4 - xy^8 + y^2 \; = \; 13$$

$$x^3 + y^4 \; = \; y^6 - 1$$

$$x^4y^{-4} + x^5y^2 + x^6 \; = \; 9$$

For symmetry with respect to the *y* axis, the points to the left and to the right of the *y* axis can be considered reflections of each other in the *y* axis, which acts as a mirror (Figure 1.46).

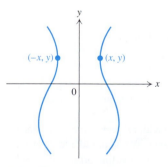

FIGURE 1.46   Symmetry with respect to the *y* axis.

EXAMPLE 4   Show that the graph of the equation $y = x^2$ is symmetric with respect to the *y* axis.

Solution   We must show that

$$\text{if } y = x^2, \quad \text{then } y = (-x)^2$$

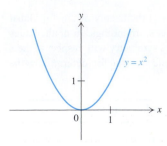

FIGURE 1.47   The graph is symmetric with respect to the *y* axis.

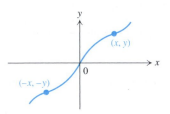

FIGURE 1.48   Symmetry with respect to the origin.

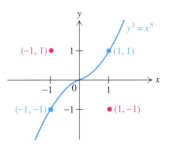

FIGURE 1.49   The graph is symmetric with respect to the origin.

Translations

But this is clear because

$$(-x)^2 = x^2$$

Therefore the graph is symmetric with respect to the *y* axis (Figure 1.47).   ❏

It is noteworthy that the graph of a nonzero function cannot be symmetric with respect to the *x* axis because two distinct points on the graph cannot be on the same vertical line. However, many functions have graphs that are symmetric with respect to the *y* axis. Such functions are called **even functions.** In order for a function to be even, $-x$ must be in the domain of *f* whenever *x* is, and the relation $f(-x) = f(x)$ must hold. Any rational function in which only even powers of *x* occur is even. Thus the following functions are even:

$$f(x) = 3x^4 - 2x^2 + 5 \quad \text{and} \quad g(x) = \frac{x^2 - 1}{x^2 + 1}$$

For symmetry with respect to the origin, the origin itself acts as a mirror (Figure 1.48).

EXAMPLE 5   Show that the graph of the equation $y^3 = x^5$ is symmetric with respect to the origin but not with respect to either axis.

Solution   Since $(-y)^3 = (-x)^5$ whenever $y^3 = x^5$, the graph is symmetric with respect to the origin. Next we observe that $(1, 1)$ is on the graph, but neither $(1, -1)$ nor $(-1, 1)$ is on it. Therefore the graph is not symmetric with respect to either axis (Figure 1.49).   ❏

A function whose graph is symmetric with respect to the origin is called an **odd function.** In order for a function *f* to be odd, $-x$ must be in the domain of *f* whenever *x* is, and the relation $f(-x) = -f(x)$ must hold. Some examples of odd functions are given by

$$f(x) = x^3 \quad \text{and} \quad g(x) = x(x^2 + 1)$$

Knowledge of symmetry properties reduces the work of sketching graphs, because we can use information about part of the graph to draw the remaining parts.

Graphing an equation is frequently made easier by changing from one set of axes to another. We will call the original set of axes the *x* and *y* axes and the new set the *X* and *Y* axes. We assume that the *X* and *Y* axes are obtained by moving the *x* and *y* axes in such a way that each is parallel to its original position (Figure 1.50). Under this condition we say that the *X* and *Y* axes are obtained from the *x* and *y* axes by a **translation.**

Any point *P* in the plane has coordinates (x, y) with respect to the *x* and *y* axes, and coordinates (X, Y) with respect to the *X* and *Y* axes. We would like to determine how these two sets of coordinates are related. Suppose the origin of the *XY* coordinate system has coordinates (a, b) in the *xy* coordinate system (Figure 1.50). By inspection we see that

$$x = X + a \quad \text{and} \quad y = Y + b$$

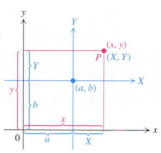

FIGURE 1.50   Translation of axes.

Solving for X and Y, we obtain

$$X = x - a \quad \text{and} \quad Y = y - b \tag{2}$$

In particular, horizontal and vertical translations are achieved as follows:

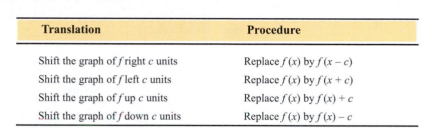

Translation	Procedure
Shift the graph of f right c units	Replace $f(x)$ by $f(x - c)$
Shift the graph of f left c units	Replace $f(x)$ by $f(x + c)$
Shift the graph of f up c units	Replace $f(x)$ by $f(x) + c$
Shift the graph of f down c units	Replace $f(x)$ by $f(x) - c$

EXAMPLE 6   Sketch the graphs of

   a. $g(x) = |x - 2|$ b. $h(x) = |x + 3|$ c. $k(x) = |x| - 1$

Solution   Let $f(x) = |x|$, whose graph is shown in Figure 1.51(a). Using the table above, we obtain the shifted graphs in Figure 1.51(b) – (d).   ❑

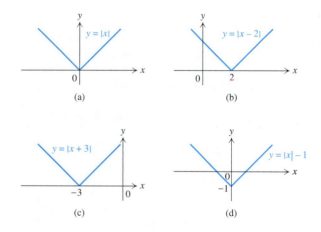

FIGURE 1.51

Examples 7 and 8 involve both a vertical and a horizontal translation.

EXAMPLE 7   Sketch the graph of the equation

$$y - 1 = (x - 2)^2$$

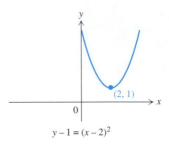

$$y - 1 = (x - 2)^2$$

FIGURE 1.52   The graph has been translated two units to the right and one unit upward.

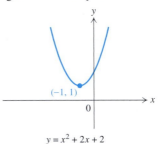

$$y = x^2 + 2x + 2$$

FIGURE 1.53   The graph has been translated one unit to the left and one unit upward.

Completing the Square

Solution   The given equation can be rewritten as

$$y = (x - 2)^2 + 1$$

We know the graph of $y = x^2$. According to the table above, the effect of having $x - 2$ in the equation rather than x is to shift the graph of $y = x^2$ to the right 2 units. Analogously, the effect of the $+ 1$ term is to shift the graph of $(x - 2)^2$ up 1 unit. Therefore the graph of $y = (x - 2)^2 + 1$, and hence the graph of the given equation, is as shown in Figure 1.52.   ❏

EXAMPLE 8   Sketch the graph of the equation

$$y = x^2 + 2x + 2$$

Solution   We rewrite the right side as follows:

$$x^2 + 2x + 2 = (x^2 + 2x + 1) + 1 = (x + 1)^2 + 1$$

Thus the original equation becomes

$$y = (x + 1)^2 + 1$$

According to the table above, the effect of having $x + 1$ in the equation rather than x is to shift the graph of $y = x^2$ to the left 1 unit. Similarly, the effect of having the $+ 1$ term is to shift the graph of $y = (x + 1)^2$ up 1 unit. The graph is shown in Figure 1.53.   ❏

In Example 8 we replaced $x^2 + 2x + 2$ by $(x + 1)^2 + 1$, a step known as **completing the square.** This can be performed for any expression of the form $ax^2 + bx + c$, where $a \neq 0$. First we rewrite the quadratic expression $ax^2 + bx + c$ in the following way by factoring out the leading coefficient a :

$$ax^2 + bx + c = a\left(x^2 + \frac{b}{a}x\right) + c \tag{3}$$

Then we take $b/2a$, which is one half the coefficient of x in the parentheses, and square it. The result is added to and subtracted from the expression in the parentheses. Thus the right side of (3) becomes

$$a\left[x^2 + \frac{b}{a}x + \left(\frac{b}{2a}\right)^2 - \left(\frac{b}{2a}\right)^2\right] + c$$

Since the first three terms in the brackets constitute the square $(x + b/2a)^2$, we separate them from the remainder of the expression, obtaining

$$a\left[x^2 + \frac{b}{a}x + \left(\frac{b}{2a}\right)^2\right] - a\left(\frac{b}{2a}\right)^2 + c = a\left(x + \frac{b}{2a}\right)^2 - \frac{b^2}{4a} + c$$

Letting $X = x + b/2a$, we have

$$ax^2 + bx + c = aX^2 - \frac{b^2}{4a} + c \tag{4}$$

The point of completing the square is that there is no X term in the right side of (4). This simplifies graphing the equation. Some other examples of completing the square are

$$x^2 - 2x = (x-1)^2 - 1$$

$$3x^2 - x + 1 = 3\left(x - \frac{1}{6}\right)^2 + \frac{11}{12}$$

$$-2x^2 + x + 2 = -2\left(x - \frac{1}{4}\right)^2 + \frac{17}{8}$$

In Example 9 it will be necessary to complete the square twice.

EXAMPLE 9   Show that the graph of

$$x^2 - 4x + y^2 + 6y = -4$$

is a circle.

Solution   We complete the square for terms in x and for terms in y:

$$x^2 - 4x = x^2 - 4x + 4 - 4 = (x-2)^2 - 4$$

$$y^2 + 6y = y^2 + 6y + 9 - 9 = (y+3)^2 - 9$$

Then we substitute these expressions into the original equation, obtaining

$$(x-2)^2 - 4 + (y+3)^2 - 9 = -4 \quad \text{or} \quad (x-2)^2 + (y+3)^2 = 9$$

Setting

$$X = x - 2 \quad \text{and} \quad Y = y + 3$$

we obtain

$$X^2 + Y^2 = 9$$

The last is an equation of a circle of radius 3 centered at the origin of the XY coordinate system. Since that origin has coordinates $(2, -3)$ in the xy system, the graph of the original equation is as shown in Figure 1.54.   ❑

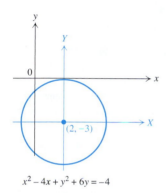

$$x^2 - 4x + y^2 + 6y = -4$$

FIGURE 1.54   The axes have been translated two units to the right and three units downward.

Observe that it is the presence of the first-degree terms $-4x$ and $6y$ that made it difficult to graph the original equation in Example 9. Completing the square and translating gives an equation in X and Y with no first-degree terms, which is easier to graph.

EXERCISES 1.5

In Exercises 1–16 determine all intercepts of the graph of the equation. Then decide whether the graph is symmetric with respect to the x axis, the y axis, or the origin.

1. $x = 3y^2 - 2$

2. $4x^2 + y^2 = 12$

3. $x^2 - y^2 = 1$

4. $x^2 = y^{15} - y^9$

5. $x^4 = 3y^3$

6. $x^4 = 3y^3 + 4$

7. $x^2 y^4 - 2x^4 = 1$

8. $2y^2 x - 4x^2 = xy^4$

9. $y = x - \dfrac{1}{x}$

10. $y = \dfrac{x}{1 + x^2}$

11. $y = [\, x \,]$

12. $|\, x - 3 \,| = |\, y + 5 \,|$

13. $y = \sqrt{9 - x^2}$ **14.** $y^2 = x\sqrt{1 + x^2}$

15. $\sqrt{x} + \sqrt{y} = 1$ **16.** $y^2 = \dfrac{x^2 + 1}{x^2 - 1}$

In Exercises 17–26 sketch the graph. List the intercepts and describe the symmetry (if any) of the graph.

17. $y = \frac{1}{3}x$ **18.** $2x = -y^2$

19. $y = x^2 - 3$ **20.** $|x| = 2$

21. $|y| = 1$ **22.** $4x^2 + 4y^2 = 9$

23. $x = \sqrt{4 - y^2}$ **24.** $y = \sqrt{25 - x^2}$

25. $|y| = |3x|$ **26.** $|2y - 3| = |x + 4|$

In Exercises 27–38 sketch the graph of the given equation with the help of a suitable translation. Show both the x and y axes and the X and Y axes.

27. $(x - 1)^2 + (y - 3)^2 = 4$ **28.** $(x + 2)^2 + (y + 4)^2 = \frac{1}{4}$

29. $x^2 - 2x + y^2 = 3$ **30.** $x^2 + y^2 + 4y = -1$

31. $x^2 + y - 3 = 0$ **32.** $x^2 - 4x + y = 5$

33. $x^2 + y^2 + 4x - 6y + 13 = 0$

34. $x^2 - 6x + y^2 - y = -9$

35. $y + 4 = \dfrac{1}{x + 2}$ **36.** $y = |x - 4|$

37. $x - 2 = |y - 2|$ **38.** $y = \sqrt{x + 3}$

39. Determine which of the following functions are even, which are odd, and which are neither.

  a. $f(x) = -x$ **b.** $f(x) = 5x^2 - 3$

  c. $f(x) = x^3 + 1$ **d.** $f(x) = (x - 2)^2$

  e. $f(x) = (x^2 + 3)^3$ **f.** $y = x(x^2 + 1)^2$

  g. $y = \dfrac{x}{x^2 + 4}$ **h.** $y = |x|$

  i. $y = \dfrac{|x|}{x}$

40. Let $f(x) = ax^2 + bx + c$, where $a \neq 0$. Show that any real zeros of f are given by (1). (*Hint:* Prove that $ax^2 + bx + c = 0$ if and only if

$$\left(x + \frac{b}{2a}\right)^2 = \frac{b^2 - 4ac}{4a^2}$$

Then solve for x. Note that such a (real) zero exists only if $b^2 - 4ac \geq 0$. There are two zeros if $b^2 - 4ac > 0$, whereas there is only one zero if $b^2 - 4ac = 0$.)

In Exercises 41–44 use the result of Exercise 40 to determine the zeros, if any, of the function.

41. $f(x) = x^2 - 3x + 1$ **42.** $f(x) = 3x^2 + 2x - 1$

43. $g(t) = 2t^2 + 7t + 7$ **44.** $g(t) = 8t^2 - 8t + 2$

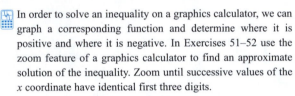

In Exercises 45–46 approximate all zeros of the function to the nearest hundredth.

45. $f(x) = -4.9x^2 + 5.1x + 1.2$

46. $f(x) = \sqrt{2}\,x^2 + \pi x + 1$

47. Let $f(x) = ax^2 + bx + c$ and $g(x) = ax^2 + bx - c$, with $a \neq 0$. Prove that if f has no real zeros, then g has two real zeros.

48. Let $f(x) = ax^2 + bx + c$ with $a \neq 0$. Suppose f has two zeros, z_1 and z_2. Express $z_1 + z_2$ in terms of a, b, and c.

In Exercises 49–50 use the zoom feature of a graphics calculator to approximate the coordinates of the points of intersection of f and g. Zoom until successive values of the x coordinate have identical first three digits.

49. $f(x) = x^4 - 1$, $g(x) = x^3 + 1$

50. $f(x) = 1 - 1/x$, $g(x) = x^2$

In order to solve an inequality on a graphics calculator, we can graph a corresponding function and determine where it is positive and where it is negative. In Exercises 51–52 use the zoom feature of a graphics calculator to find an approximate solution of the inequality. Zoom until successive values of the x coordinate have identical first three digits.

51. $x^3 + 1 \geq -x - 2$ (*Hint:* Let $f(x) = x^3 + x + 3$, and determine where $f(x) \geq 0$.)

52. $4x^3 + 4x < 3$

53. a. Let $f(x) = x^2$ and $g(x) = f(x + 3)$. Using a suitable translation, sketch the graph of g.

  b. Let $f(x) = |x|$ and $g(x) = f(x - 2)$. Sketch the graph of g.

54. Let f be a function, and let $g(x) = f(x) - 3$. What is the relationship between the graphs of f and g?

55. Suppose that f is a function and that $g(x) = f(x) + d$ for all x in the domain of f, where d is a constant. What is the relationship between the graphs of f and g?

56. Let $f(x) = |x - 1| + |x + 1|$.

  a. Determine whether the graph of f is symmetric with respect to either axis or the origin.

  b. Find alternative expressions for $f(x)$ in the three cases $x < -1$, $-1 \leq x \leq 1$, and $x > 1$, and use this information to sketch the graph of f.

57. Suppose the graph of an equation is symmetric with respect to both axes. Prove that it is symmetric with respect to the origin. Is the converse true?

58. Suppose the graph of an equation is symmetric with respect to the y axis and the origin. Is it necessarily symmetric with respect to the x axis? Explain.

59. Suppose $c > 0$. If $f(c - x) = f(c + x)$ for all x, what property of symmetry does the graph of f have?

60. Show that the points twice as far from the point $(2, -3)$ as from the point $(-1, 0)$ form a circle, and find the center and radius of that circle.

Applications

61. Assume that the graph of f in Figure 1.55 represents a wave in the ocean at time $t = 0$. Assume that after t seconds the wave is represented by the graph of f_t, where

$$f_t(x) = f(x - t)$$

Sketch the graphs of f_1, f_2, and f_4. As time increases, is the wave moving to the left or to the right?

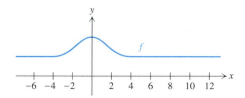

FIGURE 1.55   Graph for Exercise 61.

1.6   COMBINING FUNCTIONS

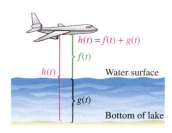

FIGURE 1.56   The function h is the sum of the functions f and g.

Many functions arise as combinations of other functions. For example, suppose an airplane is flying over a large lake. Then at any time t, the height $h(t)$ of the airplane above the bottom of the lake is the sum of the height $f(t)$ of the airplane above the surface of the lake and the depth $g(t)$ of the lake directly below the airplane (Figure 1.56). In other words,

$$h(t) = f(t) + g(t)$$

Certain properties of the sum h of f and g can be determined directly from the properties of f and g. Similarly, certain properties of the difference, the product, and the quotient of two functions can be determined directly from the properties of those functions.

Sums, Differences, Products, and Quotients of Functions

Let f and g be functions. We define the **sum** $f + g$, the **difference** $f - g$, and the **product** fg to be the functions whose domains consist of all numbers in the domains of *both* f and g and whose rules are given by

$$(f + g)(x) = f(x) + g(x)$$

$$(f - g)(x) = f(x) - g(x)$$

$$(fg)(x) = f(x)\, g(x)$$

In each case the domain is the expected one, consisting of those values of x for which both $f(x)$ and $g(x)$ are defined. Because division by 0 is excluded, we give the definition of the quotient of two functions separately. The **quotient** f / g is the function whose domain consists of all numbers x in the domains of both f and g for which $g(x) \neq 0$ and whose rule is given by

$$\left(\frac{f}{g}\right)(x) = \frac{f(x)}{g(x)}$$

EXAMPLE 1   Let $f(x) = 1/x$ and $g(x) = \sqrt{x}$. Find the domain and rule of $f + g$.

Solution   The domain of f consists of all nonzero numbers, and the domain of g consists of all nonnegative numbers. The only numbers in both domains are the positive numbers, which constitute the domain of $f + g$. For the rule we have

$$(f + g)(x) = f(x) + g(x) = \frac{1}{x} + \sqrt{x} \quad \text{for } x > 0 \quad \square$$

EXAMPLE 2   Let $f(x) = \sqrt{4 - x^2}$ and $g(x) = \sqrt{x - 1}$. Find the domain and rule of fg.

Solution   The domain of f is the interval $[-2, 2]$, and the domain of g is the interval $[1, \infty)$. Therefore the domain of fg is the set of numbers in both $[-2, 2]$ and $[1, \infty)$. This is just the interval $[1, 2]$. The rule is given by

$$(fg)(x) = f(x) \, g(x) \quad = \quad \sqrt{4 - x^2} \, \sqrt{x - 1}$$

$$= \quad \sqrt{(4 - x^2)(x - 1)} \quad \text{for } 1 \leq x \leq 2 \quad \square$$

Caution: Example 2 illustrates a surprising fact about combinations of functions. Although we found that the domain of fg is the interval $[1, 2]$, the expression $\sqrt{(4 - x^2)(x - 1)}$ is also meaningful for x in $(-\infty, -2]$. This is true because

$$(4 - x^2)(x - 1) \geq 0 \quad \text{for } x \leq -2$$

Hence we must be careful to determine the domain of fg from the separate domains of f and g rather than from the rule for fg. Similar comments hold for the domains of $f + g$, $f - g$, and f/g.

EXAMPLE 3   Let $f(x) = x + 5$ and $g(x) = 4x^2 - 1$. Find the domain and rule of f/g.

Solution   Since the domains of f and g are all real numbers and since $g(x) = 0$ for $x = -1/2$ and for $x = 1/2$, it follows that the domain of f/g consists of all real numbers except $-1/2$ and $1/2$. The rule of f/g is given by

$$\left(\frac{f}{g}\right)(x) = \frac{f(x)}{g(x)} = \frac{x + 5}{4x^2 - 1} \quad \text{for } x \neq \frac{1}{2} \text{ and } x \neq -\frac{1}{2} \quad \square$$

We can add or multiply more than two functions. For example, if f, g, and h are functions, then for all x common to the domains of f, g, and h we have

$$(f + g + h)(x) = f(x) + g(x) + h(x) \quad \text{and} \quad (fgh)(x) = f(x) \, g(x) \, h(x)$$

In a similar way we can add or multiply more than three functions.

A special case of the product occurs when one of the functions is a constant function:

$$g(x) = c \quad \text{for all } x$$

For any function f, the domain of the product cf is the same as the domain of f. For example,

$$(2f)(x) = 2f(x) \quad \text{and} \quad (\pi f)(x) = \pi f(x) \quad \text{for all } x \text{ in the domain of } f$$

These very simple products occur often in calculus.

Let f and g be functions and c a nonzero number such that

$$g = cf$$

Then we say that the values of g are **proportional** to the values of f. For instance, if $g(r)$ denotes the area of a circle of radius r and if $f(r) = r^2$, then $g(r)$ is proportional to $f(r)$, or $g(r) = cf(r)$, where $c = \pi$. In other words, the area of a circle is proportional to the square of the radius.

Finally, any polynomial

$$g(x) = c_n x^n + c_{n-1} x^{n-1} + \cdots + c_1 x + c_0$$

can be considered as a combination of constant functions and the identity function $f(x) = x$. Indeed

$$g(x) = c_n (f(x))^n + c_{n-1}(f(x))^{n-1} + \cdots + c_1 f(x) + c_0$$

so that

$$g = c_n f^n + c_{n-1} f^{n-1} + \cdots + c_1 f + c_0$$

This fact makes polynomials particularly amenable to the methods of calculus.

Composition of Functions

Another way of combining functions that occurs frequently in calculus is illustrated by the following example. The sound from an explosion extends in every direction, filling out an expanding spherical region. The volume V of the spherical region depends on the radius r of the sphere, which in turn depends on the time t elapsed since the explosion. The volume V is given by

$$V(r) = \frac{4}{3} \pi r^3$$

The rate at which the radius r grows depends on the medium and corresponds to the speed of sound in the medium. For instance, in air held at 20° C the speed of sound is 343 meters per second, so that

$$r(t) = 343t$$

Thus the volume V is a function of r, which in turn is a function of time t. Therefore the function h that represents the volume V in terms of time t is given by

$$h(t) = V(r(t)) = \frac{4}{3} \pi (343t)^3$$

In general, if f and g are two functions, then the **composition** $g \circ f$ of f and g is defined as the function whose rule is

$$(g \circ f)(x) = g(f(x)) \tag{1}$$

FIGURE 1.57

and whose domain consists of all numbers x in the domain of f for which the number $f(x)$ is in the domain of g. (This is just the set of all numbers x for which the right side of (1) is defined.) The expression $g \circ f$ is read "g of f," "g composed with f," or "g circle f." The function $g \circ f$ is the result of performing f and then performing g. If we think of a function as a machine that manufactures numbers in its range out of numbers in its domain, then the composite function $g \circ f$ can be represented as in Figure 1.57.

EXAMPLE 4   Let $f(x) = \sqrt{x - 1}$ and $g(x) = 1/x$. Determine the functions $g \circ f$ and $f \circ g$, and then find $g(f(5))$ and $f(g(\frac{1}{4}))$.

Solution   The rule of $g \circ f$ is given by

$$g(f(x)) = g(\sqrt{x - 1}) = \frac{1}{\sqrt{x - 1}}$$

The domain of f is $[1, \infty)$, so the domain of $g \circ f$ consists of those numbers x in $[1, \infty)$ for which $\sqrt{x - 1}$ is in the domain of g, that is, for which $\sqrt{x - 1} \neq 0$. Since this excludes $x = 1$, the domain of $g \circ f$ is the interval $(1, \infty)$.

The rule for $f \circ g$ is given by

$$f(g(x)) = f\left(\frac{1}{x}\right) = \sqrt{\frac{1}{x} - 1}$$

The domain of g is the set of nonzero numbers, so the domain of $f \circ g$ consists of those nonzero numbers x such that $1/x$ is in the domain $[1, \infty)$ of f. Since $1/x$ is in $[1, \infty)$ whenever x is in $(0, 1]$, the domain of $f \circ g$ is $(0, 1]$.

Finally, we calculate that

$$g(f(5)) = \frac{1}{\sqrt{5 - 1}} = \frac{1}{2} \quad \text{and} \quad f\left(g\left(\frac{1}{4}\right)\right) = \sqrt{\frac{1}{\frac{1}{4}} - 1} = \sqrt{3} \quad \square$$

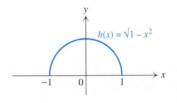

FIGURE 1.58

Example 4 shows that the functions $g \circ f$ and $f \circ g$ may not be equal. In fact, they are usually unequal. In Example 4 not only are the rules for $g \circ f$ and $f \circ g$ different, but the domains of $g \circ f$ and $f \circ g$ do not even have any numbers in common. The difference between $g \circ f$ and $f \circ g$ is much the same as the difference between first taking an examination and then seeing the correct answers, and first seeing the correct answers and then taking the examination.

In calculus it is important to be able to determine whether a given function h is the composite of two simpler functions because if that is the case then properties of h can often be obtained from corresponding properties of the simpler functions. In the next example we will consider the function h that assigns to the x coordinate of a point on the unit semicircle its y coordinate (Figure 1.58); we will decompose h into two simpler functions, f and g.

EXAMPLE 5   Let $h(x) = \sqrt{1 - x^2}$. Write h as the composite of two functions f and g.

Solution   Notice that $h(x) = \sqrt{f(x)}$, where $f(x) = 1 - x^2$. Thus if we let $g(x) = \sqrt{x}$, then

$$h(x) = \sqrt{1 - x^2} = \sqrt{f(x)} = g(f(x))$$

Consequently $h = g \circ f$. (There are other ways of expressing h as the composite of two functions, but the way we have chosen is a natural and convenient one.)   ❑

We now use composition of functions to define a function that will be useful later. This is the function $x^{m/n}$, where m and n are integers having no common integer factors other than -1 and 1 and where $n > 0$. Let

$$f(x) = x^m \quad \text{and} \quad g(x) = x^{1/n}$$

Then we define $x^{m/n}$ by

$$x^{m/n} = (g \circ f)(x) = (x^m)^{1/n} \quad \text{domain:} \quad \begin{cases} \text{all } x \text{ if } m > 0, n \text{ odd} \\ \text{all } x \geq 0 \text{ if } m > 0, n \text{ even} \\ \text{all } x \neq 0 \text{ if } m < 0, n \text{ odd} \\ \text{all } x > 0 \text{ if } m < 0, n \text{ even} \end{cases}$$

Thus the function x^r has been defined for every rational number r. In Figure 1.59 we have sketched the graphs of four power functions of the form $x^{m/n}$ in order to illustrate the restrictions on domains for various powers.

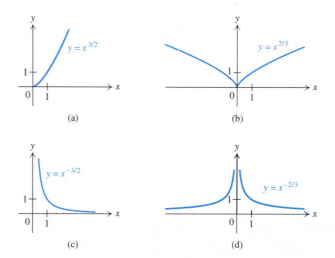

FIGURE 1.59

To see where a function that involves a rational power might arise, let x denote the ratio of the minimum to the maximum volume of a cylinder in an internal combustion engine. Under certain conditions, the efficiency of the engine is given by

$$h(x) = (1 - x)^{0.4}$$

Composites of functions also play an important role in the subject of "chaos," which, in addition to mathematics, has recently attracted much attention in such diverse fields as meteorology and medicine. Consider any function f whose range is contained in its domain. Then the numbers

$$f(x), f(f(x)), f(f(f(x))), \ldots$$

are called the **iterates** of f at x. In particular,

$$f(x) = \text{the first iterate of } f \text{ at } x$$
$$f(f(x)) = \text{the second iterate of } f \text{ at } x$$

and so on.

For example, if $f(x) = x^2 - 1$, then the first three iterates of f at -1 are

$$f(-1) = 0$$
$$f(f(-1)) = f(0) = -1$$
$$f(f(f(-1))) = f(-1) = 0$$

It is apparent that the successive iterates of f at -1 oscillate between -1 and 0. In contrast, if $g(x) = x^2 + 1$, then the first three iterates of g at -1 are

$$g(-1) = 2$$
$$g(g(-1)) = g(2) = 5$$
$$g(g(g(-1))) = g(5) = 26$$

Evidently successive iterates of -1 for g increase without bound.

These examples give only a hint as to the variety and complexity of behavior for iterates of even very simple functions.

EXERCISES 1.6

In Exercises 1–8 let $f(x) = 2x^2 + x - 4$ and $g(x) = 3 - x^2$. Find the specified values.

1. $(f + g)(-1)$ **2.** $(f - g)(2)$ **3.** $(fg)(\frac{1}{2})$

4. $(f/g)(-3)$ **5.** $f(g(1))$ **6.** $g(f(0))$

7. $\dfrac{f(x) - f(2)}{x - 2}$ **8.** $\dfrac{g(a) - g(-1)}{a + 1}$

In Exercises 9–14 let $f(x) = \dfrac{x - 1}{x^2 + 1}$ and $g(x) = x^{1/4}$. Find the specified values.

9. $(f + g)(16)$ **10.** $(f - g)(1)$ **11.** $(fg)(9)$

12. $(f/g)(\frac{1}{4})$ **13.** $f(g(1))$ **14.** $g(f(1))$

In Exercises 15–18 find the domains and rules of $f + g$, fg, and f/g.

15. $f(x) = \dfrac{2}{x - 1}$ and $g(x) = x - 1$

16. $f(x) = \dfrac{x + 2}{x - 3}$ and $g(x) = \dfrac{x + 3}{x^2 - 4}$

17. $f(t) = t^{3/4}$ and $g(t) = t^2 + 3$

18. $f(t) = \sqrt{1 - t^2}$ and $g(t) = \sqrt{2 + t - t^2}$

In Exercises 19–26 find the domain and rule of $g \circ f$ and $f \circ g$.

19. $f(x) = 1 - x$ and $g(x) = 2x + 5$

20. $f(x) = x^2 + 2x + 3$ and $g(x) = x - 1$

21. $f(x) = x^2$ and $g(x) = \sqrt{x}$

22. $f(x) = x^6$ and $g(x) = x^{3/4}$

23. $f(x) = \sqrt{x}$ and $g(x) = x^2 - 5x + 6$

24. $f(x) = \dfrac{1}{x}$ and $g(x) = x^2 - 3x - 10$

25. $f(x) = \dfrac{1}{x - 1}$ and $g(x) = \dfrac{1}{x + 1}$

26. $f(x) = \sqrt{x^2 + 3}$ and $g(x) = \sqrt{x^2 - 4}$

In Exercises 27–34 write h as the composite $g \circ f$ of two functions f and g (neither of which is equal to h).

27. $h(x) = \sqrt{x - 3}$

28. $h(x) = (1 - x^2)^{3/2}$

29. $h(x) = (3x^2 - 5\sqrt{x})^{1/3}$

30. $h(x) = \left(x + \dfrac{1}{x} \right)^{5/2}$

31. $h(x) = \dfrac{1}{(x + 3)^2 + 1}$

32. $h(x) = \dfrac{1}{(x^3 - 2x^2)^5}$

33. $h(x) = \sqrt{\sqrt{x} - 1}$

34. $h(x) = \sqrt{[x] - 1}$

35. Find g if $f(x) = |x|$ and $(f + g)(x) = |x| - |x - 2|$.

36. Find g if $f(x) = (x^2 - 4)/(x + 3)$ and $(fg)(x) = 1$, for $x \neq 2$, -2, and -3.

37. Suppose f is defined on $[0, 4]$ and $g(x) = f(x + 3)$. What is the domain of g?

38. Suppose f is defined on $[a, b]$ and $g(x) = f(x + c)$ for a fixed c. What is the domain of g?

39. Figure 1.60 includes color graphs of four functions: f, g, $f + g$, and fg. Identify each of the functions with a corresponding colored graph.

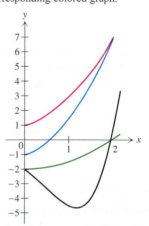

FIGURE 1.60   Graphs for Exercise 39.

40. Use Figure 1.61 to find the approximate value of d such that the highest point on the graph of $g + d$ is the same height as the highest point on the graph of f.

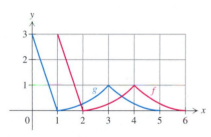

FIGURE 1.61   Graphs for Exercise 40.

41. Consider the graphs of f and g, which appear in Figure 1.62. Let $h(x) = g(x + c)$, where c is a fixed number. Find an approximate value of c such that

    a. $h = f$ **b.** $h(1) = 2$

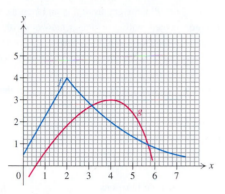

FIGURE 1.62   Graphs for Exercise 41.

42. The graph of f appears in Figure 1.63. Sketch the graphs of g and h, where

    a. $g(x) = f(2x)$ **b.** $h(x) = f(x/2)$

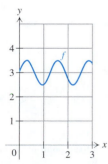

FIGURE 1.63   Graphs for Exercise 42.

43. Let $f(x) = 1 + \sqrt{x + 1}$, $g(x) = 1 + \sqrt{2 - x}$, and $h(x) = f(x) + g(x)$. Plot the graphs of f, g, and h on the same calculator screen with viewing windows $[-10, 10]$. Explain why the graph of h is so short.

44. For which functions f is there a function g such that $f = g^2$?

45. For which functions f is there a function g such that $f = g^3$?

46. For which functions f is there a function g such that $f = 1/g$?

47. For which functions f is there a function g such that

$$f = \sqrt{1 + g}$$

48. Let f and g be even functions.
  a. Show that $f + g$ is an even function.
  b. Show that fg is an even function.

49. Let f be even and g odd. Show that fg is an odd function.

50. Let $f(x) = 1/x$. Show that $f(f(x)) = x$ for all $x \neq 0$.

51. Let a be a real number and $f(x) = a - x$. Show that $f(f(x)) = x$ for all x.

52. Let $f(x) = 1/(1 - x)$. Show that $f(f(f(x))) = x$ for all x different from 0 and 1.

53. Let $f(x) = ax + b$, where a and b are constants, and let p be any real number. Show that if

$$g(x) = f(x + p) - f(x)$$

then g is a constant function.

54. Let c be any number, and define a function f by $f(x) = cx$. Show that $f(x + y) = f(x) + f(y)$ for all numbers x and y.

55. Let f be a function, and let

$$g(x) = \frac{1}{2}\left[f(x) + f(-x)\right] \text{ and } h(x) = \frac{1}{2}\left[f(x) - f(-x)\right]$$

  a. Show that g is an even function.
  b. Show that h is an odd function.
  c. Show that $f = g + h$. (Thus every function can be written as the sum of an even and an odd function.)

56. Let

$$T(x) = \begin{cases} 2x & \text{for } 0 \leq x \leq \dfrac{1}{2} \\ 2(1 - x) & \text{for } \dfrac{1}{2} < x \leq 1 \end{cases}$$

Find a value of x in $(0, 1)$ such that $T(x) = x$. (Such a value of x is called a **fixed point** of the function T.)

In Exercises 57–61 let $f(x) = cx - cx^2$, where c is a constant. By calculating a large number (up to 100, if necessary) of iterates, make a reasonable guess for the behavior of the nth iterate of f at 0.3 for large values of n.

57. 0.5   **58.** 2.5   **59.** 3.2   **60.** 3.5   **61.** 3.83

62. Let $f(x) = 4x(1 - x)$. Calculate the first 100 iterates of f at 0.3, and see if you observe any pattern or repetitions in the iterates.

Applications

63. The revenue function R for a certain product is given by

$$R(x) = 5x^2 - \frac{x^4}{10}$$

The cost function C is given by

$$C(x) = 4x^2 - 24x + 38$$

The profit function P is defined as the difference $R - C$. Find the equation that describes P. Then find $P(1)$ and $P(2)$, and show that it is possible to lose money and also possible to make a profit.

64. Recall that the volume $V(r)$ of a spherical balloon of radius r is given by the formula

$$V(r) = \frac{4}{3}\pi r^3 \quad \text{for } r \geq 0$$

Suppose the radius is given by $r(t) = 3\sqrt{t}$. Write a formula for the volume in terms of t.

65. A sphere with surface area s has a radius $r(s)$ given by

$$r(s) = \frac{1}{2}\sqrt{\frac{s}{\pi}}$$

  a. Using the formula in Exercise 64, find a formula for the volume of a sphere in terms of its surface area.
  b. Determine the volume corresponding to a surface area of 6.

66. According to Newton's Law of Gravitation, if two bodies are a distance r apart, then the gravitational force $F(r)$ exerted by one body on the other is given by

$$F(r) = \frac{k}{r^2} \quad \text{for } r > 0$$

where k is a positive constant. Suppose that as a function of time t, the distance between the two bodies is given by

$$r(t) = 4000\left(\frac{1 + t}{1 + t^2}\right) \quad \text{for } t \geq 0$$

Find a formula for the force in terms of time.

1.7   TRIGONOMETRIC FUNCTIONS

You are doubtless already acquainted with the trigonometric functions. These functions arise in geometry, but they are also applicable to the study of sound, the motion of a pendulum, and many other phenomena involving rotation or oscillation. For this reason, they are important in the study of calculus.

Radian Measure

We will first discuss the measurement of angles. Recall that there are 360 degrees in a circle (Figure 1.64). However, a more convenient unit of angle measurement for calculus is the **radian,** which is chosen so that a circle contains 2π radians. Thus 2π radians = 360 degrees, so that 1 radian is equal to $360/2\pi$ degrees, or $180/\pi$ degrees (approximately 57.2958 degrees). It is easy to convert radians to degrees or degrees to radians:

If $f(x)$ is the number of degrees in x radians, then $f(x) = \dfrac{180}{\pi}\, x$.

If $g(x)$ is the number of radians in x degrees, then $g(x) = \dfrac{\pi}{180}\, x$.

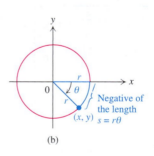

FIGURE 1.64   A circle contains 360 degrees and 2π radians.

The following is a table of conversions for some common angles.

Degrees	0°	30°	45°	60°	90°	120°	135°	150°	180°	270°	360°
Radians	0	$\dfrac{\pi}{6}$	$\dfrac{\pi}{4}$	$\dfrac{\pi}{3}$	$\dfrac{\pi}{2}$	$\dfrac{2\pi}{3}$	$\dfrac{3\pi}{4}$	$\dfrac{5\pi}{6}$	π	$\dfrac{3\pi}{2}$	2π

In this book we will use radians to measure angles except where otherwise indicated.

Let us draw a circle centered at the origin, with arbitrary radius $r > 0$. Next, we place an angle of θ radians so that its vertex is at the origin and its initial side is on the positive x axis. The angle opens counterclockwise if $\theta \geq 0$ and clockwise if $\theta < 0$ (Figure 1.65(a) and (b)). We allow θ to be greater than 2π. For example, an angle of 3π radians can be obtained by rotating a line through one full revolution (2π radians) and an extra half-revolution. Thus an angle of 3π radians has the same initial and terminal sides as an angle of π radians. The same is true of $-\frac{7}{3}\pi$ and $-\frac{1}{3}\pi$ radians, since $-\frac{7}{3}\pi = -\frac{1}{3}\pi - 2\pi$. Since the circle contains 2π radians and its circumference is $2\pi r$, an angle of 1 radian intercepts an arc of length r on the circle. Thus if $\theta > 0$, an angle of θ radians intercepts an arc of length θr on the circle. Calling this arc length s, we have

$$s = r\theta \tag{1}$$

(a)

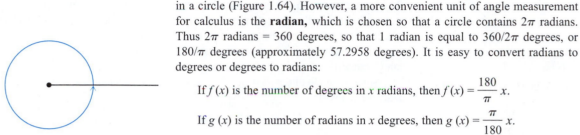

(b)

FIGURE 1.65

For $\theta < 0$ formula (1) holds if we think of s as the negative of the length of the arc intercepted on the circle (Figure 1.65(b)).

Definitions of the Trigonometric Functions

We define the sine and cosine functions with reference to a circle of radius r centered at the origin and an angle of θ radians, positioned as shown in Figure 1.66.

The terminal side of the angle intersects the circle at a unique point (x, y). We define the **sine function** and **cosine function** by

$$\sin \theta = \frac{y}{r} \quad \text{and} \quad \cos \theta = \frac{x}{r} \tag{2}$$

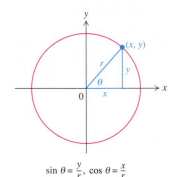

$\sin \theta = \frac{y}{r}, \cos \theta = \frac{x}{r}$

FIGURE 1.66

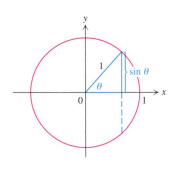

FIGURE 1.67

The properties of similar triangles imply that $\sin \theta$ and $\cos \theta$ depend only on θ, not on the value of r. If $r = 1$, then $\sin \theta = y$, which is the length of a halfchord* (Figure 1.67). The domains of the sine and cosine functions are $(-\infty, \infty)$. Notice that in the expression $\sin \theta$, θ represents a *number*. Thus we write $\sin 1$ instead of $\sin$ (1 radian).

Since an angle of θ radians and one of $\theta + 2\pi$ radians have the same terminal side, we can write

$$\sin \theta = \sin (\theta + 2\pi) \quad \text{and} \quad \cos \theta = \cos (\theta + 2\pi)$$

Thus the sine and cosine functions are periodic; they both have period 2π. Consequently, for any integer n and any number θ,

$$\sin \theta = \sin (\theta + 2n\pi) \quad \text{and} \quad \cos \theta = \cos (\theta + 2n\pi) \tag{3}$$

For a circle of radius r centered at the origin, the distance between the origin and a point (x, y) on the circle is r. Using the distance formula, we have

$$x^2 + y^2 = r^2 \tag{4}$$

Substituting for x and y from (2) gives us

$$r^2 \cos^2 \theta + r^2 \sin^2 \theta = r^2$$

which yields the famous Pythagorean Identity

$$\sin^2 \theta + \cos^2 \theta = 1 \tag{5}$$

The table below lists some values of the sine and the cosine. We leave verification of these values to you.

θ	0	$\dfrac{\pi}{6}$	$\dfrac{\pi}{4}$	$\dfrac{\pi}{3}$	$\dfrac{\pi}{2}$	$\dfrac{2\pi}{3}$	$\dfrac{3\pi}{4}$	$\dfrac{5\pi}{6}$	π
$\sin \theta$	0	$\dfrac{1}{2}$	$\dfrac{\sqrt{2}}{2}$	$\dfrac{\sqrt{3}}{2}$	1	$\dfrac{\sqrt{3}}{2}$	$\dfrac{\sqrt{2}}{2}$	$\dfrac{1}{2}$	0
$\cos \theta$	1	$\dfrac{\sqrt{3}}{2}$	$\dfrac{\sqrt{2}}{2}$	$\dfrac{1}{2}$	0	$-\dfrac{1}{2}$	$-\dfrac{\sqrt{2}}{2}$	$-\dfrac{\sqrt{3}}{2}$	-1

*This fact seems to have influenced the etymology of the word "sine." In the fifth century A.D. the Indian mathematician Aryabbatta called the sine of an angle by the name "ardhajya," meaning "halfchord." Soon the word condensed to "jya," or "chord." When the Arabs translated the word into their written language, they contracted the word to "jb." Later "jb" somehow became "jaib," which already meant "curve" or "bosom." Around the twelfth century, when mathematics was translated into Latin, the Arabic "jaib" was translated into "sinus," the Latin equivalent for "bosom," from which our word "sine" comes.

Since we normally use x to represent numbers in the domain of a function, we will usually follow that convention for the sine and cosine functions and replace θ by x. Thus (5) becomes

$$\sin^2 x + \cos^2 x = 1 \tag{6}$$

By plotting a few points on the graphs of the sine and cosine functions and using (3), we obtain the graphs of these functions, shown in Figure 1.68(a) and (b).

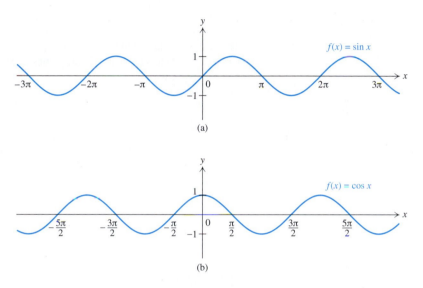

(a)

(b)

FIGURE 1.68   The graphs of the sine and cosine functions.

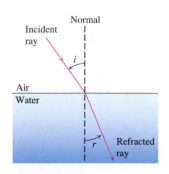

FIGURE 1.69

A classic formula that employs the sine function is Snell's Law, which relates the angle at which a light ray of a single color (that is, a monochromatic ray) in air is bent when it enters another medium such as water. If i is the angle of incidence and r is the angle of refraction, as in Figure 1.69, then Snell's Law says that

$$\sin i = \mu \, \sin r \tag{7}$$

where μ is the **index of refraction** of the water. The index of refraction depends only on the color and is defined by

$$\mu = \frac{\text{speed of the ray in air}}{\text{speed of the ray in water}}$$

For the visible spectrum, μ varies from approximately 1.330 for red to 1.342 for violet. The variation in the indices for different colors yields different angles of refraction for various colors. It is this variation that gives rise to rainbows in the sky. We will discuss various features of the rainbow in more detail in Section 13.3.

There are four other basic trigonometric functions that are defined in terms of the sine and the cosine. These are the **tangent,** the **cotangent,** the **secant,** and the

cosecant functions. Their definitions are

$$\tan x = \frac{\sin x}{\cos x} \quad \text{for } x \neq n\pi + \frac{\pi}{2}, n \text{ any integer}$$

$$\cot x = \frac{\cos x}{\sin x} \quad \text{for } x \neq n\pi, n \text{ any integer}$$

$$\sec x = \frac{1}{\cos x} \quad \text{for } x \neq n\pi + \frac{\pi}{2}, n \text{ any integer}$$

$$\csc x = \frac{1}{\sin x} \quad \text{for } x \neq n\pi, n \text{ any integer}$$

The values of these functions can be quickly computed from the corresponding values of the sine and cosine. Their graphs appear in Figure 1.70(a) – (d).

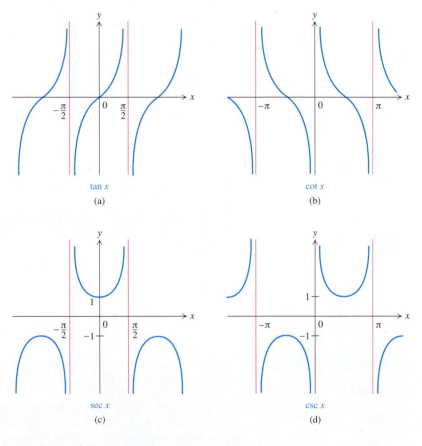

FIGURE 1.70 The graphs of the tangent, cotangent, secant, and cosecant functions.

Observe that if θ is the angle between the positive x axis and any nonvertical line l (Figure 1.71), then

$$\tan \theta = \text{slope of } l \qquad (8)$$

This fact relates slopes of lines and the trigonometric functions.

Trigonometric Identities

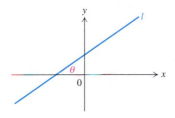

There are many equations, called **trigonometric identities,** that describe relationships among the various trigonometric functions. The most important ones are

$$
\begin{aligned}
\sin^2 x + \cos^2 x &= 1 && (9) \\
\sin (x \pm y) &= \sin x \cos y \pm \cos x \sin y && (10) \\
\cos (x \pm y) &= \cos x \cos y \mp \sin x \sin y && (11) \\
\sin (-x) &= -\sin x && (12) \\
\cos (-x) &= \cos x && (13)
\end{aligned}
$$

Other trigonometric identities can be derived from these. Below we list several of the more useful ones, with n denoting an arbitrary integer.

FIGURE 1.71   The slope of a nonvertical line is the tangent of the angle θ.

$$\sec^2 x = 1 + \tan^2 x \qquad\qquad \sin 2x = 2 \sin x \cos x$$

$$\csc^2 x = 1 + \cot^2 x \qquad\qquad \cos 2x = \cos^2 x - \sin^2 x$$

$$\sin (x + 2n\pi) = \sin x \qquad\qquad \sin^2 x = \frac{1 - \cos 2x}{2}$$

$$\cos (x + 2n\pi) = \cos x \qquad\qquad \cos^2 x = \frac{1 + \cos 2x}{2}$$

$$\sin\left(\frac{\pi}{2} - x\right) = \sin\left(\frac{\pi}{2} + x\right) = \cos x \qquad \tan (x \pm y) = \frac{\tan x \pm \tan y}{1 \mp \tan x \tan y}$$

$$\cos\left(\frac{\pi}{2} - x\right) = -\cos\left(\frac{\pi}{2} + x\right) = \sin x$$

In addition to these identities we add one important set of inequalities that we will use frequently:

$$-1 \le \sin x \le 1 \quad \text{and} \quad -1 \le \cos x \le 1$$

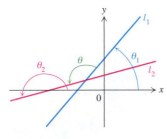

FIGURE 1.72

You can justify these inequalities either from the definitions of $\sin x$ and $\cos x$ or from the graphs in Figure 1.68(a) and (b).

Suppose l_1 and l_2 are two intersecting lines. Then we define the **angle from l_1 to l_2** to be the smallest nonnegative angle θ through which l_1 must be rotated counterclockwise about the point of intersection in order to coincide with l_2 (see θ in Figure 1.72). Thus $0 \le \theta < \pi$. By using trigonometric identities, we can express θ in terms of the slopes of l_1 and l_2.

THEOREM 1.6

Let l_1 and l_2 be two nonvertical lines that are not perpendicular, with slopes m_1 and m_2, respectively. The tangent of the angle θ from l_1 to l_2 is given by

$$\tan \theta = \frac{m_2 - m_1}{1 + m_1 m_2}$$

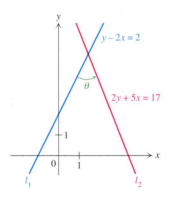

FIGURE 1.73

Proof Let θ_1 and θ_2 be angles with initial sides along the positive x axis and terminal sides along l_1 and l_2, respectively, chosen so that $0 \leq \theta_1 < \pi$ and $\theta_2 \geq \theta_1$ (Figure 1.72). Then $\theta = \theta_2 - \theta_1$. Since $m_1 = \tan \theta_1$ and $m_2 = \tan \theta_2$,* and since $m_1 m_2 \neq -1$ (why?), we have

$$\tan \theta = \tan (\theta_2 - \theta_1) = \frac{\tan \theta_2 - \tan \theta_1}{1 + \tan \theta_1 \tan \theta_2} = \frac{m_2 - m_1}{1 + m_1 m_2} \quad \blacksquare$$

EXAMPLE 1 Let the equations of l_1 and l_2 be

$$y - 2x = 2 \quad \text{and} \quad 2y + 5x = 17$$

Find the tangent of the angle θ from l_1 to l_2 (Figure 1.73).

Solution From the equations of l_1 and l_2 we find that $m_1 = 2$ and $m_2 = -\frac{5}{2}$. Therefore

$$\tan \theta = \frac{m_2 - m_1}{1 + m_1 m_2} = \frac{-\frac{5}{2} - 2}{1 + 2(-\frac{5}{2})} = \frac{-\frac{9}{2}}{-4} = \frac{9}{8} \quad \square$$

The trigonometric functions have many important applications. They will also be valuable to us in illustrating numerous features of calculus.

EXERCISES 1.7

1. Convert the following degree measures to radians.
   a. $210°$ **b.** $-405°$ **c.** $1°$
2. Convert the following radian measures to degrees.
   a. $\dfrac{\pi}{8}$ **b.** $-\dfrac{3\pi}{10}$ **c.** $\dfrac{13\pi}{6}$
3. Find the following values.
   a. $\sin \dfrac{11\pi}{6}$ **b.** $\sin\left(-\dfrac{2\pi}{3}\right)$ **c.** $\cos \dfrac{5\pi}{4}$
   d. $\cos\left(-\dfrac{7\pi}{6}\right)$ **e.** $\tan \dfrac{4\pi}{3}$ **f.** $\tan\left(-\dfrac{\pi}{4}\right)$
   g. $\cot \dfrac{\pi}{6}$ **h.** $\cot\left(-\dfrac{17\pi}{3}\right)$ **i.** $\sec 3\pi$
   j. $\sec\left(-\dfrac{\pi}{3}\right)$ **k.** $\csc \dfrac{\pi}{2}$ **l.** $\csc\left(-\dfrac{5\pi}{3}\right)$
4. For each of the following intervals, state which of the six trigonometric functions have positive values throughout the interval.
   a. $\left(0, \dfrac{\pi}{2}\right)$ **b.** $\left(\dfrac{\pi}{2}, \pi\right)$ **c.** $\left(\pi, \dfrac{3\pi}{2}\right)$
   d. $\left(\dfrac{3\pi}{2}, 2\pi\right)$ **e.** $(3\pi, 4\pi)$

*If $\theta_2 \geq \pi$ (as in Figure 1.72), then $\tan \theta_2 = \tan (\theta_2 - \pi) = m_2$.

In Exercises 5–6 find the values of the remaining four trigonometric functions under the given conditions.

5. $\sin x = \frac{4}{5}$ and $\cos x = -\frac{3}{5}$

6. $\cos x = \frac{1}{3}$ and $\tan x = 2\sqrt{2}$

In Exercises 7–10 solve the equation for x in $[0, 2\pi)$.

7. $\sin x = -\frac{1}{2}$ **8.** $\tan x = \sqrt{3}$

9. $\sin 2x = \sin x$ **10.** $\cos 2x = \cos x$

In Exercises 11–15 solve the inequality for x in $[0, 2\pi)$.

11. $\sin x > -\frac{1}{2}$ **12.** $\cos x \le 0$

13. $\tan x \ge 1$ **14.** $\sin x \le \cos x$

15. $\cot x \ge \tan x$

16. Decide whether each of the following functions is even or odd.

a. $\sin x$ **b.** $\cos x$ **c.** $\tan x$

d. $\cot x$ **e.** $\sec x$ **f.** $\csc x$

In Exercises 17–24 sketch the graph of the function. Indicate any intercepts and symmetry, and determine whether the function is even, odd, or neither.

17. $\cos(\pi - x)$ **18.** $\sin\left(\frac{\pi}{4} - x\right)$

19. $\tan\left(x + \frac{\pi}{2}\right)$ **20.** $|\cot x|$

21. $\sec(2\pi - x)$ **22.** $2 - \csc x$

23. $\sin 2x$ **24.** $\cos \frac{x - \pi}{2}$

In Exercises 25–26 use the identities given in this section to compute the given value.

25. $\sin \frac{7\pi}{12}$ $\left(Hint: \frac{7\pi}{12} = \frac{\pi}{3} + \frac{\pi}{4}\right)$ **26.** $\cos \frac{5\pi}{12}$

27. Solve the equation $2\sin^2 x + \sin x - 1 = 0$.

28. Solve the equation $4\cos^2 x - 4\sqrt{3}\cos x + 3 = 0$.

29. Using (9), prove that $1 + \tan^2 x = \sec^2 x$.

30. Using (9), prove that $1 + \cot^2 x = \csc^2 x$.

31. Using identities (10)–(13), prove that the following identities hold:

a. $\sin(\pi - x) = \sin x$ **b.** $\sin\left(\frac{3\pi}{2} - x\right) = -\cos x$

c. $\cos(\pi - x) = -\cos x$ **d.** $\cos\left(\frac{3\pi}{2} - x\right) = -\sin x$

32. a. Using (9), show that $\cos x = \sqrt{1 - \sin^2 x}$ for $0 \le x \le \pi/2$ and for $3\pi/2 \le x \le 2\pi$. Show also that $\cos x = -\sqrt{1 - \sin^2 x}$ for $\pi/2 \le x \le 3\pi/2$.

b. Using (9), show that $\sin x = \sqrt{1 - \cos^2 x}$ for $0 \le x \le \pi$. Show also that $\sin x = -\sqrt{1 - \cos^2 x}$ for $\pi \le x \le 2\pi$.

In Exercises 33–34 find the tangent of the angle θ from the first line to the second line.

33. $y = 4x - 2$; $3y = -2x + 7$

34. $y = 3x + 9$; $4y - 11x = 6$

 35. For a function f, if there is a smallest positive number a such that $f(x + a) = f(x)$ for all x in the domain of f, then a is called the **period** of f. Plot each of the following functions, and from the graph guess the period. Then prove that your guess is correct.

a. $\tan x$ **b.** $\sin 3x$

c. $|\sin x|$ **d.** $\cos\left(-2x + \frac{\pi}{2}\right)$

In Exercises 36–37 prove the identities pertaining to the right triangle shown in Figure 1.74.

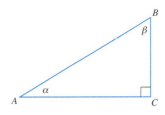

FIGURE 1.74   Figure for Exercises 36–37.

36. $\cos 2\alpha + \cos 2\beta = 0$ **37.** $\cos(\alpha - \beta) = \sin 2\alpha$

38. Let r be any rational number and let $f(x) = \sin x + \sin rx$. Show that f is periodic. (*Hint:* Let $r = m/n$, where m and n are integers.)

 39. For each of the given values of a, calculate the iterates

$$a, \cos a, \cos(\cos a), \ldots$$

until the first three displayed digits do not change. From your results, make a conjecture about the behavior of the iterates for any real number a.

a. $a = 1$ **b.** $a = 12$ **c.** $a = 100$ **d.** $a = -5$

Applications

 40. Through how many complete revolutions does a bicycle wheel with radius 1 foot turn when the bicycle travels 1 mile?

41. A beacon is located 2 miles from a straight coastline, as depicted in Figure 1.75. Express the distance between the beacon and the illuminated point x on the coastline as a function of the angle θ in Figure 1.75.

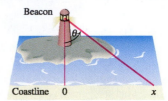

FIGURE 1.75   Figure for Exercise 41.

42. Prove that the sine of an angle inscribed in a circle of unit diameter is the length of the chord of the subtended arc. (*Hint:* First assume that one side of the angle is a diameter and use the fact that the resulting triangle is a right triangle (Figure 1.76). Then use the fact that all inscribed angles with the same subtended arc are equal.)

FIGURE 1.76   Figure for Exercise 42.

43. A ladybug is standing on the floor, and an 8-foot giant walks toward her. When the giant is 8 feet away, how big an angle does the giant occupy in the ladybug's field of vision? What happens as the giant comes closer? (This suggests why it is sometimes better to sit far back in a theater with empty rows directly in front of you than to sit in the second row directly behind someone.)

44. A rectangular sheet of metal 18 feet wide and 30 feet long is bent as in Figure 1.77 to form a trough. Find a formula for the function that expresses the volume V of the trough in terms of the angle θ.

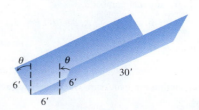

FIGURE 1.77   Figure for Exercise 44.

*45. a. Show that the perimeter $p_n(r)$ of a regular polygon of n sides inscribed in a circle of radius r is given by

$$p_n(r) = 2nr \sin \frac{\pi}{n}$$

 b. Using the result of part (a), find the radius of the smallest circle that can circumscribe the Pentagon building, each of whose outer walls is 921 feet long.

46. A sack of sand weighing 50 pounds is being dragged by a person whose arm makes an angle of x radians with the ground. If the coefficient of friction is $\mu > 0$, then the minimum force F necessary to drag the sack is given by

$$F(x) = \frac{50\mu}{\mu \sin x + \cos x}$$

 a. If the arm makes an angle of $\pi/4$ radians with the ground, what is the minimum force necessary to drag the load?

 b. If the arm makes an angle of $\pi/3$ radians with the ground, what is the minimum force necessary to drag the load?

 c. What happens if the arm makes an angle of $\pi/2$ radians with the ground?

47. Suppose a ball of mass m is attached to a string of length L and is rotated in a vertical plane with enough velocity v so that the string remains taut (Figure 1.78). Then the tension T in the string, which depends on the angle θ that the string makes with the downward vertical, is given by

$$T = m\left(\frac{v^2}{L} + g \cos \theta\right) \quad \text{for } 0 \le \theta < 2\pi \quad (14)$$

where g is the (negative) acceleration due to gravity.

FIGURE 1.78   Figure for Exercise 47.

 a. From your intuition, at which point in the path of the ball would the tension be greatest, and at which would it be least?

 b. From (14), find the value of θ at which T is greatest and the value at which T is least. Do these values agree with your intuition?

48. In a certain industrial area the amount of sulfur dioxide pollutant released into the atmosphere due to burning fossil fuels varies according to the season. Suppose that we wish to model the amount A of pollutant (in tons) released into the atmosphere at time t (in weeks) by means of the formula

$$A = 1 + b \cos \frac{\pi}{26} t$$

What are the reasonable values of b ?

49. A single respiratory cycle includes one inhalation and one exhalation. During one respiratory cycle of a certain person at rest, the rate of flow R (in liters per second) of air into a person's lungs at time t (in seconds) is given by

$$R = 0.5 \sin \frac{2\pi}{5} t$$

a. How long does it take to complete one respiratory cycle?

b. How many respiratory cycles are completed in one minute?

c. Graph one complete cycle, starting at time $t = 0$.

d. Interpret the meaning of *positive* and *negative* values of R.

 e. To the nearest hundredth, find R when $t = 3$ seconds.

1.8   EXPONENTIAL AND LOGARITHMIC FUNCTIONS

Exponential and logarithmic functions appear in many applications of mathematics and the sciences. Since they are discussed in precalculus books and are available on calculators, you may already be familiar with them. In this section we will define these functions by means of their graphical properties. In Chapters 5 and 6 we will return to a discussion of the exponential and logarithmic functions.

Exponential Functions

In Section 1.6 we defined $a^{m/n}$, where m and n are integers with $n \neq 0$, and a is positive. In this section we will define a^x, where x is *any* real number and $a > 0$.

Before we define a^x, let us consider a specific example. Taking $a = 3$ and $x = \sqrt{2}$, we will try to determine a reasonable value for $3^{\sqrt{2}}$. Because the decimal expansion of $\sqrt{2}$ begins 1.414213562, we compute by calculator that

$$3^{1.4} \approx 4.655536722 \qquad 3^{1.414} \approx 4.727695035$$

$$3^{1.41421} \approx 4.728785881 \qquad 3^{1.4142135} \approx 4.728804064$$

$$3^{1.414213562} \approx 4.728804386$$

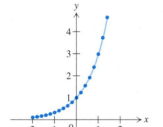

FIGURE 1.79   For x rational, the points $(x, 3^x)$ lie on a smooth curve.

From these calculations it appears that the decimal expansion of $3^{\sqrt{2}}$ should begin 4.728804.... By using more and more digits in the decimal expansion of $\sqrt{2}$, we could find more and more digits in the decimal expansion of $3^{\sqrt{2}}$. An analogous approximation could be made with decimal expansions if $\sqrt{2}$ were replaced by any other irrational number. The situation can be described geometrically as follows. For rational values of x, the points $(x, 3^x)$ lie on a smooth curve (Figure 1.79). Consequently, we define $3^{\sqrt{2}}$ to be the y coordinate of the point that lies on the curve directly above the point $(\sqrt{2}, 0)$.

More generally, if a is any positive number with $a \neq 1$, then for rational values of x the points (x, a^x) lie on a unique smooth curve (Figure 1.80). Thus we define

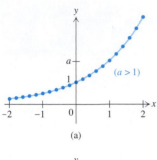

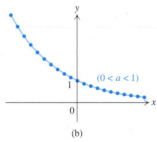

FIGURE 1.80   For x rational, the points (x, a^x) lie on a smooth curve.

a^x for any fixed $a > 0$ and any real number x as follows:

> a^x = the y coordinate of the point that lies above the point $(x, 0)$ and lies on the smooth curve that contains all points of the form (r, a^r), where r is a rational number.

We call the function f defined by $f(x) = a^x$ the **exponential function with base a**. In the expression a^x, the number a is the **base**, and x is the **exponent** or **power**. No matter what the values of a and x are (as long as $a > 0$), a^x is the "limiting value" of a^r, where r is rational and r gets closer and closer to x.

The domain of a^x is the set of all real numbers, and the range is the set of all positive real numbers unless $a = 1$, in which case the range is the single number 1. Moreover, for any positive number a, the y intercept of a^x is 1, and there is no x intercept. As the graph of a^x suggests,

$$\text{for } a > 1: \quad \text{if } x < z \quad \text{then } a^x < a^z \tag{1}$$

and

$$\text{for } 0 < a < 1: \quad \text{if } x < z \quad \text{then } a^x > a^z \tag{2}$$

It follows from (1) and (2) that if $a > 0$ and $a \neq 1$, then $a^x \neq a^z$ whenever $x \neq z$. Equivalently,

$$a^x = a^z \quad \text{if and only if} \quad x = z \tag{3}$$

Suppose, for example, that we wish to solve the equation $2^x = 2^{5/x}$. From (3) we deduce that $x = 5/x$, which implies that $x^2 = 5$. Consequently $x = \sqrt{5}$ or $x = -\sqrt{5}$.

The various laws of exponents that apply to a^x for rational values of x remain valid for arbitrary real values of x:

Laws of Exponents

Let a and b be fixed positive numbers. For any numbers x and y we have

i. $a^{x+y} = a^x a^y$ v. $a^{-x} = \dfrac{1}{a^x} = \left(\dfrac{1}{a}\right)^x$

ii. $a^{xy} = (a^x)^y = (a^y)^x$ vi. $\left(\dfrac{a}{b}\right)^x = \dfrac{a^x}{b^x}$

iii. $(ab)^x = a^x b^x$ vii. $a^0 = 1$

iv. $a^1 = a$

We can use the various laws in order to simplify expressions involving exponents. For example, using Laws (ii), (i), and (v), respectively, we find that

$$\frac{2^{\sqrt{3}} \, 4^{\sqrt{3}}}{2^{3\sqrt{3}-1}} = \frac{2^{\sqrt{3}}(2^2)^{\sqrt{3}}}{2^{3\sqrt{3}-1}} \overset{(ii)}{=} \frac{2^{\sqrt{3}} 2^{2\sqrt{3}}}{2^{3\sqrt{3}-1}} \overset{(i)}{=} \frac{2^{3\sqrt{3}}}{2^{3\sqrt{3}} 2^{-1}} \overset{(v)}{=} 2$$

From Law (v) of Exponents,

$$a^{-x} = \left(\frac{1}{a}\right)^x$$

which means, for example, that the graph of a^{-x} is the same as the graph of $(1/a)^x$.

Of all the possible bases a for an exponential function, one outshines the others in importance. It is the one that was first called e by the Swiss mathematician Leonhard Euler. This number e is an irrational number whose decimal expansion begins

$$e = 2.71828182845904523536 \ldots$$

The function e^x is called the **natural exponential function,** or simply the **exponential function.** Since $2 < e < 3$, it follows that the graph of e^x lies between the graphs of 2^x and 3^x (Figure 1.81).

The natural exponential function appears with great frequency in both mathematics and applications. Therefore it is useful to be able to approximate numbers of the form e^x on a calculator or computer. The method of obtaining e^x on a calculator varies, depending on the instrument. But the basic idea is the same. For example, to evaluate $e^{0.287}$ on some calculators, we would press the exponential key, then the decimal, 2, 8, 7, and ENTER keys in that order. Displayed on the calculator would be the approximate value 1.332424213, meaning that

$$e^{0.287} \approx 1.332424213$$

If we use the powerful mathematical software package Mathematica, then we must input Exp[.287] in order to evaluate $e^{0.287}$.

Although we will devote Section 4.4 to applications of the natural exponential function, we give one example here. By the **Bouguer-Lambert Law,** which is fundamental to the study of photometry, if a beam of light with intensity c at the surface of a lake travels vertically downward, then at a depth of x meters the intensity $f(x)$ of the beam is given by the formula

$$f(x) = ce^{-1.4x} \tag{4}$$

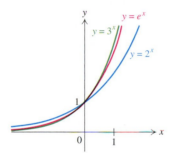

FIGURE 1.81

where the constant 1.4 is the so-called **absorption coefficient** of pure water.

EXAMPLE 1   Use a calculator to find the ratio of the intensity 10 meters below the surface of a lake to the intensity at the surface itself.

Solution   The ratio we seek is $f(10)/f(0)$, where f is defined by (4). By calculator we find that

$$\frac{f(10)}{f(0)} = \frac{ce^{-1.4(10)}}{ce^0} = e^{-14} \approx 8.315287191 \times 10^{-7} \quad \square$$

From the value obtained in the example, it is apparent why most flora cannot survive at depths greater than 10 meters.

Logarithmic Functions

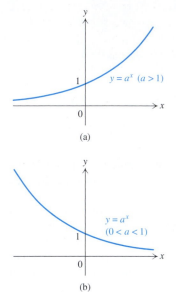

(a)

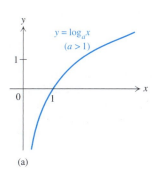

(b)

FIGURE 1.82

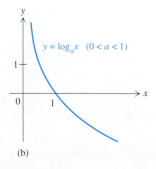

(a)

(b)

FIGURE 1.83

Let $a > 0$ with $a \neq 1$. Notice that the graph of the exponential function to the base a rises as x increases if $a > 1$ (Figure 1.82(a)), and falls as x increases if $0 < a < 1$ (Figure 1.82(b)). This suggests that for every positive number x there is one and only one number y such that $a^y = x$. That number y is called the logarithm of x to the base a and is denoted by $\log_a x$. Thus $\log_a x$ is the number y such that $a^y = x$. It follows that

$$\log_a x = y \quad \text{if and only if} \quad a^y = x \tag{5}$$

For example, $\log_2 8$ equals the unique number y such that $2^y = 8$. Therefore $\log_2 8 = 3$. In addition, $\log_a 1 = 0$ since $a^0 = 1$. The function that associates $\log_a x$ to any given positive x is the **logarithm function to the base a.**

The domain of $\log_a x$ consists of all positive numbers, and the range includes all real numbers. The graph of $\log_a x$ appears as in Figure 1.83, sloping upward if $a > 1$ and sloping downward if $0 < a < 1$. In other words,

$$\text{for } a > 1: \quad \text{if } x < z \quad \text{then } \log_a x < \log_a z$$

$$\text{for } 0 < a < 1: \quad \text{if } x < z \quad \text{then } \log_a x > \log_a z$$

It follows that

$$\log_a x = \log_a z \quad \text{if and only if} \quad x = z$$

There are Laws of Logarithms reminiscent of the Laws of Exponents:

Laws of Logarithms
Let a be a fixed positive number with $a \neq 1$. For any positive numbers x and y we have

i. $\log_a (xy) = \log_a x + \log_a y$	iv. $\log_a (x^c) = c \log_a x$ for any number c
ii. $\log_a \dfrac{1}{x} = -\log_a x$	v. $\log_a 1 = 0$
iii. $\log_a \dfrac{x}{y} = \log_a x - \log_a y$	vi. $\log_a a = 1$

Besides the Laws of Logarithms there are two additional formulas relating $\log_a x$ and a^x. For one of them we substitute a^y for x in the first equation of (5) and obtain

$$\log_a (a^y) = y \quad \text{for all } y \tag{6}$$

For the other one we substitute $\log_a x$ for y in the second equation of (5), obtaining

$$a^{\log_a x} = x \quad \text{for } x > 0 \tag{7}$$

For example,

$$\log_{10} 100 = \log_{10} (10^2) = 2 \quad \text{and} \quad \log_e (e^3) = 3$$

and

$$e^{\log_e 2} = 2 \quad \text{and} \quad 10^{\log_{10} 23} = 23$$

From (5) it follows that the functions a^x and $\log_a x$ are inverses of one another. Because the two functions are inverses, if (x, y) is on the graph of a^x, then (y, x) is on the graph of $\log_a x$. In other words the graphs of a^x and $\log_a x$ are symmetric with respect to the line $y = x$. You can check that this is true by comparing Figures 1.80 and 1.83. We will devote Chapter 6 to inverse functions.

The base of a logarithmic function can be any positive number except 1, which means, for example, that $\sqrt{2}$ and $1/\pi$ are perfectly reasonable bases. However, from the viewpoint of mathematical theory and practical applications the base is almost always greater than 1, and moreover, there are two bases that are far more widely used than any other: e and 10. In calculus and in other branches of mathematics the logarithm to the base e plays a special role and is called the **natural logarithmic function.** In calculus books its value at x is normally denoted by $\ln x$, so we write

$$\ln x \quad \text{instead of } \log_e x$$

(The expression "ln" stands for the Latin phrase "logarithmus naturalis.") In terms of the natural exponential and natural logarithmic functions, (6) and (7) become

$$\ln (e^x) = x \quad \text{and} \quad e^{\ln x} = x \tag{8}$$

In numerical calculation the logarithm to the base 10 has historically been important because of decimal notation and is called the **common logarithmic function.** Its value at x is frequently denoted by $\log x$, so that we would write

$$\log x \quad \text{instead of } \log_{10} x$$

The graphs of the natural and the common logarithmic functions are shown in Figure 1.84.

> **Caution:**   There is considerable variation in the notation used to denote the natural and common logarithmic functions. In advanced mathematics books and in many computer programs the natural logarithmic function is denoted by "log." However, on most calculators, "log" is used for the common logarithm, and "ln" denotes the natural logarithm.

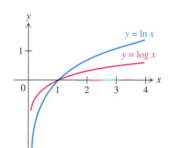

FIGURE 1.84

Change of Base Formula

It is sometimes necessary, or convenient, to convert from logarithms in one base to logarithms in another base. Since calculators normally have keys only for natural logarithms and common logarithms, this conversion may even be necessary when one wishes to use a calculator to make computations involving logarithms to other bases. For that reason we will now derive a conversion formula for logarithms.

Suppose that a and b are positive numbers different from 1, and we wish to express $\log_b x$ in terms of $\log_a x$. Let

$$r = \log_b x \tag{9}$$

which by (5) is equivalent to $x = b^r$. Since $a \neq 1$ by hypothesis, we can take logarithms to the base a of both sides of the equation $x = b^r$ to deduce that

$$\log_a x = \log_a (b^r)$$

By (iv) in the Laws of Logarithms, $\log_a (b^r) = r \log_a b$, so that

$$\log_a x = r \log_a b \qquad (10)$$

Since $b \neq 1$ by hypothesis, it follows that $\log_a b \neq 0$. Therefore we may divide both sides of (10) by $\log_a b$. This yields

$$r = \frac{\log_a x}{\log_a b}$$

Finally, using (9) we substitute $\log_b x$ for r to obtain the following conversion formula, which is customarily called the Change of Base Formula:

Change of Base Formula

$$\log_b x = \frac{\log_a x}{\log_a b} \quad \text{for } x > 0$$

In particular, if $a = e$, we derive the formula for converting to natural logarithms:

$$\log_b x = \frac{\ln x}{\ln b} \qquad (11)$$

In addition, if $x = a$ in the Change of Base Formula, we obtain

$$\log_b a = \frac{1}{\log_a b}$$

a formula that is occasionally of use.

John Napier (1550–1617)
The invention of logarithms is generally credited to Napier, a Scottish nobleman for whom mathematics was a hobby. His invention arose through his desire to associate ratios of pairs of numbers in a geometric progression with ratios of pairs of numbers in an arithmetic progression. Thus the name *logarithmus,* which means "ratio number" and has been anglicized to "logarithm." What resulted from Napier's creation was an association between quotients and differences, namely Law of Logarithms (iii).

EXAMPLE 2   Use a calculator and the Change of Base Formula to approximate $\log_2 3$.

Solution   Since most calculators have keys for both natural and common logarithms, we can use the Change of Base Formula with either $a = e$ or $a = 10$. We obtain the same answer either way:

$$\log_2 3 = \frac{\ln 3}{\ln 2} = \frac{\log 3}{\log 2} \approx 1.584962501 \quad \square$$

At this point we observe that in the same way that $\log_a x$ can be given in terms of the natural logarithm, a^x can be given in terms of the natural exponential function. To be more precise, we recall the second equation in (8), with a replacing x:

$$a = e^{\ln a}$$

From (ii) in the Laws of Exponents, we obtain the following formula:

$$a^x = (e^{\ln a})^x = e^{x \ln a} \qquad (12)$$

Formula (12) shows that any exponential function a^x can be expressed in terms of the natural exponential function—a fact that will be useful later, especially in Section 7.6.

Earthquakes

Near the epicenter of the 1989 Loma Prieta earthquake (*Jonathan Turk*)

As an application of logarithms to the base 10, we discuss the magnitude of an earthquake. Of the many formulas proposed for converting seismographic readings into a reasonable scale of the magnitude of an earthquake, the most commonly used is the Richter, or Gutenberg-Richter, scale devised by the American geologists Beno Gutenberg and Charles F. Richter during the 1930s. In order to describe measurement on the Richter scale, we first introduce a reference, or zero-level, earthquake, which by definition is any earthquake whose largest seismic wave would measure 0.001 millimeter (1 micron) on a standard seismograph located 100 kilometers from the epicenter of the earthquake. The **magnitude** M of a given earthquake is defined by the formula

$$M = \log \frac{x}{a} = \log x - \log a \qquad (13)$$

where x is the amplitude measured on a standard seismograph of the largest seismic wave of the given earthquake and a is the amplitude on the same seismograph of the largest seismic wave of a zero-level earthquake with the same epicenter. The fraction x/a is called the **intensity** of the earthquake. Values of a for various distances from the epicenter have been tabulated, so when an earthquake occurs, a person needs only to measure the number x and know the distance to the epicenter in order to be able to assess the magnitude of the earthquake.

Suppose that at a given location, the intensity of an earthquake is x/a. If a second earthquake located at the same spot is 10 times as intense as the first, then the magnitudes M_1 and M_2 of the two earthquakes are given by

$$M_1 = \log \frac{x}{a} \quad \text{and} \quad M_2 = \log \frac{10x}{a}$$

Therefore

$$M_2 = \log \frac{10x}{a} = \log 10 + \log \frac{x}{a} = 1 + \log \frac{x}{a} = 1 + M_1$$

Thus multiplying the intensity of an earthquake by 10 corresponds to an increase of 1 in the magnitude of the earthquake.

Next we compare the intensities of two famous earthquakes.

EXAMPLE 3 The great San Francisco earthquake of 1906 had a magnitude of 8.4. On October 17, 1989, the devastating Loma Prieta earthquake near San Francisco registered a magnitude of 7.1. How many times greater was the intensity of the 1906 earthquake than the intensity of the 1989 earthquake?

Solution Let x_1/a and x_2/a denote the intensities of the 1906 and 1989 earthquakes, respectively, and let M_1 and M_2 denote the corresponding magnitudes. By (13),

$$M_1 = \log \frac{x_1}{a} \quad \text{and} \quad M_2 = \log \frac{x_2}{a}$$

Therefore

$$M_1 - M_2 = \log \frac{x_1}{a} - \log \frac{x_2}{a} = \log \frac{x_1/a}{x_2/a}$$

so that

$$\frac{x_1/a}{x_2/a} = 10^{M_1 - M_2} = 10^{8.4 - 7.1} = 10^{1.3} \approx 19.95262315$$

Consequently, the 1906 earthquake was approximately 20 times as intense as the 1989 earthquake. ☐

The physical damage caused by an earthquake increases with magnitude, but fortunately the frequency of earthquakes decreases with magnitude. The following table indicates damage and frequency in relation to magnitude.

Magnitude	Result near the epicenter	Approximate number per year
≥ 8	near total damage	0.2
7.0–7.9	serious damage to buildings	14
6.0–6.9	moderate damage to buildings	185
5.0–5.9	slight damage to buildings	1,000
4.0–4.9	felt by most people	2,800
3.0–3.9	felt by some people	26,000
2.0–2.9	not felt but recorded	800,000

Theoretically, the magnitude of an earthquake can be any real number. In practice, however, a standard seismograph can record only those earthquakes whose magnitudes exceed 2. On the other end of the scale, no recorded magnitude has ever exceeded 9.0, and only half a dozen times has one been as high as 8.5.

The photograph at the beginning of the chapter portrays a small portion of the San Andreas fault. This 750-mile-long fracture in the earth's crust separates the Pacific and North American plates and runs roughly from Los Angeles past San Francisco in a northwesterly direction. Movement along the fault averages approximately 2 inches per year and has caused many severe earthquakes in the past, including both earthquakes discussed in Example 3.

EXERCISES 1.8

In Exercises 1–12 simplify the expression.

1. $\ln e^3$ **2.** $\ln \sqrt{e}$ **3.** $e^{\ln 3x}$

4. $e^{-4 \ln x}$ **5.** $\ln (e^{\ln e})$ **6.** $\ln |\ln (1/e)|$

7. $7^{\log_7 2x}$ **8.** $\log_4 4^x$ **9.** $\log_9 3$

10. $\log_2 \frac{1}{4}$ **11.** $\log_{1/4} 2^x$ **12.** $e^{x - 2 \ln x}$

In Exercises 13–16 use a calculator to find the approximate value.

13. $e^{-1.24}$ **14.** $e^{-1.24 \times 10^{-4}}$ **15.** $2^{7/2}$ **16.** $10^{-5/3}$

In Exercises 17–20 calculate the logarithm by using (11).

17. $\log_3 5$ **18.** $\log_{1/2} \frac{1}{3}$ **19.** $\log_\pi e$ **20.** $\log_{\sqrt{2}} \sqrt{\pi}$

In Exercises 21–22 find the error.

21. $(x \ln x)^2 \stackrel{?}{=} x^2 \ln x^2 \stackrel{?}{=} 2x^2 \ln x$

22. $\dfrac{e^3}{e^{\sqrt{2}} e^{\sqrt{2}}} \stackrel{?}{=} \dfrac{e^3}{e^{\sqrt{2} \cdot \sqrt{2}}} \stackrel{?}{=} \dfrac{e^3}{e^2} \stackrel{?}{=} e$

In Exercises 23–24 determine whether f is an even function, an odd function, or neither.

23. $f(x) = e^x /(e^{2x} + 1)$

24. $f(x) = \ln (e^{3x} + 1)$

In Exercises 25–28 sketch the graph of f.

25. $f(x) = e^{2 + x}$

26. $f(x) = 2^{2 - x}$

27. $f(x) = \ln (x + 1)$

28. $f(x) = \ln (ex)$

In Exercises 29–34 determine at which points the graphs of the given pair of functions intersect.

29. $f(x) = e^x$ and $g(x) = e^{1 - x}$ (*Hint:* Take natural logarithms.)

30. $f(x) = e^x$ and $g(x) = e^{-x^2}$

31. $f(x) = e^{3x}$ and $g(x) = 3e^x$

32. $f(x) = 5e^{-2x}$ and $g(x) = 3e^x$

33. $f(x) = \log_3 x$ and $g(x) = \log_2 x$

34. $f(x) = 3^x$ and $g(x) = 2^{(x^2)}$ (*Hint:* Take natural logarithms.)

35. Solve $\ln x + \ln (3x - 1) = 0$ for x.

36. Let $f(x) = \ln (x + \sqrt{x^2 - 9}) + \ln (x - \sqrt{x^2 - 9})$ for $x \ge 3$. Show that f is a constant function, and find the constant.

37. Show that $\ln (x + \sqrt{x^2 - 1}) = -\ln (x - \sqrt{x^2 - 1})$ for $x \ge 1$.

38. Let a and b be distinct positive numbers different from 1. Show that $\log_a x \ne \log_b x$ for $x \ne 1$.

39. Use the Laws of Exponents to prove that

$$a^{b + c + d} = a^b \, a^c \, a^d \text{ for all } b, c, \text{ and } d \text{ and all } a > 0$$

40. Let $a > 0$. Using the Change of Base Formula, show that $\log_{1/a} x = -\log_a x$ for $x > 0$.

 41. Let $f(x) = \ln (4x) - \ln x^3 + \ln x^2$. Plot f on a graphics calculator, and use properties of logarithms to explain the appearance of the graph.

 42. Let $f(x) = e^x$ and $g(x) = c \ln x$. Use a graphics calculator to determine an approximate value of c such that the graphs of f and g touch, but do not cross, each other.

43. a. Let $f(x) = ax + b$. Show that $f(x + 1) - f(x) = a$.

    b. Let $g(x) = ba^x$, where a is positive and $b \ne 0$. Show that $g(x + 1)/g(x) = a$.

 In Exercises 44–46 let $f(x) = ce^x$ for all x. For the given constant c, determine whether the iterates of f at 0 approach a single number. If they do, approximate that number.

44. 1   **45.** $\frac{1}{5}$ **46.** 0.4

Applications

47. If an earthquake has magnitude 2, find the amplitude of its largest wave 100 kilometers from the epicenter.

48. Suppose the amplitude of the maximal seismic wave is doubled. By how much is the magnitude of the earthquake increased?

49. The magnitude of the Good Friday Alaskan earthquake of March 28, 1964, is given sometimes as 8.4 and sometimes as 8.5. What is the ratio of the maximum amplitude of an earthquake of magnitude 8.4 to that of an earthquake of magnitude 8.5?

50. The strongest earthquakes ever recorded occurred off the coast of Ecuador and Colombia in 1906, and in Japan in 1933. Each had a magnitude of 8.9. Find the ratio of the amplitude of the largest wave of such a quake to the corresponding amplitude of a zero-level quake.

51. During the 1950's, scientists devised an experimental formula relating the energy E (in ergs) of an earthquake or explosion to the Richter scale magnitude M of the occurrence. The formula that arose is

$$\log E = 11.4 + 1.5M$$

During the Gulf War of 1991, the United Nations forces used explosives amounting to 90 kilotons. Using the fact that a kiloton of explosives releases approximately 10^{20} ergs of energy, determine the magnitude M of an earthquake that would release the same amount of energy.

Suppose the frequency of a particular sound wave is 1000 hertz (cycles per second). Let I_0 be the intensity of the sound wave at the threshold of audibility, approximately 10^{-16} watts per square centimeter. If x denotes the intensity of the sound wave, then the **noise level** $L(x)$ of the sound wave is given by

$$L(x) = 10 \log \frac{x}{I_0}$$

The units for $L(x)$ are called decibels, in honor of Alexander Graham Bell. Exercises 52–56 concern noise level.

52. The noise level of a whisper is about 30 decibels, and that of ordinary conversation is around 50 decibels. Determine the ratio of the intensity of a whisper to that of conversation.

53. Determine the number of decibels that corresponds to each of the following intensities.

a. 10^{-12} (threshold of hearing)

b. 10^{-11} (rustling leaves)

c. 10^{-2} (power mower)

d. 10 (jackhammer)

54. What is the difference in the noise levels of two sounds, one of which is 1000 times as intense as the other?

55. What is the ratio of the intensity of a given sound to that of one that is 100 decibels higher?

56. The human ear can just barely distinguish between two sounds if one is 0.6 decibels higher than the other. What is the ratio of the intensity of one sound to that of another sound that is lower than the first and is just barely distinguishable from the first sound?

 57. Halley's Law states that the barometric pressure $p(t)$ in inches of mercury at t miles above sea level is given by

$$p(t) \approx 29.92 \, e^{-0.2t} \quad \text{for } t \geq 0$$

Find the barometric pressure

a. at sea level

b. 5 miles above sea level

c. 10 miles above sea level

 58. Suppose that a living organism died t years ago. The number t frequently can be assessed by carbon-14 dating. If p percent of the original C^{14} in the organism is now present, then t is given approximately by

$$t = -\frac{\ln(p/100)}{0.000124}$$

Find the approximate age of each of the following objects with the given value of p.

a. mammal tusk, where $p = 1$

b. wooden post, where $p = 60$

 59. When an X-ray beam passes through an object (Figure 1.85), the intensity I (energy per unit time per unit area) of the beam at a distance x meters from the point of entry is given by

$$I = I_0 e^{-\sigma x}$$

where I_0 is the intensity at entry and σ is the linear absorption coefficient (in meters^{-1}) of the material of the object. The number σ depends on the material and the wavelength of the X-rays. For aluminum, $\sigma = 5.4 \times 10^2$ for X-rays with wavelength equal to 5×10^{-11} meters, and $\sigma = 4.1 \times 10^3$ for X-rays with wavelength equal to 10^{-10} meters. Suppose two X-ray beams with wavelengths 5×10^{-11} and 10^{-10} meters pass through a sheet of aluminum that is 2 millimeters thick. If the beams have the same intensity at entry, find the ratio of the intensity of the emerging 5×10^{-11}-meter beam to that of the emerging 10^{-10}-meter beam.

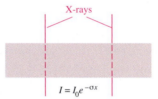

FIGURE 1.85   Figure for Exercise 59.

REVIEW
Key Terms and Expressions

Absolute value

Function

Domain, rule, and range of a function

Polynomial function; rational function

Graph of a function; graph of an equation

Independent variable; dependent variable

x intercept; y intercept

Symmetry with respect to the x axis; with respect to the y axis; with respect to the origin

Even function; odd function

Translation of axes

Sum, product, and quotient of two functions

Composition of two functions

Radian

Trigonometric function

Exponential function

Logarithmic function

Key Formulas

$|ab| = |a||b|$

$|a + b| \leq |a| + |b|$

$|a - b| \geq |a| - |b|$

$\sec^2 x = 1 + \tan^2 x$

$\csc^2 x = 1 + \cot^2 x$

$\log_b x = \dfrac{\log_a x}{\log_a b}$

$\sin 2x = 2 \sin x \cos x$

$\cos 2x = \cos^2 x - \sin^2 x$

$m_1 = m_2$ for parallel lines

$m_1 m_2 = -1$ for perpendicular lines

$\sin^2 x + \cos^2 x = 1$

$a^{x+y} = a^x a^y$

$(a^{xy}) = (a^x)^y = (a^y)^x$

$\log_a (xy) = \log_a x + \log_a y$

$\log_a (x^c) = c \log_a x$

$\log_a (a^x) = x$

$a^{\log_a x} = x$

Review Exercises

In Exercises 1–6 solve for *x*.

1. $\dfrac{2x + 3}{(x + 4)^3} > 0$

2. $\dfrac{x}{x - 1} + \dfrac{x + 2}{x - 4} \leq 0$

3. $\dfrac{2}{3 - x} \geq 4$

4. $|x + 1| = |2x - 3|$

5. $|4 - 6x| < \frac{1}{2}$

6. $\left|\dfrac{x + 1}{x - 1}\right| \leq 1$

7. Find the value of $(|a| + a)/2$ for
  a. $a \geq 0$ **b.** $a < 0$

8. Show that for all *x* and *y* we have $2xy \leq x^2 + y^2$.

In Exercises 9–11 find an equation of the line described. Then sketch the line.

9. The line through $(1, -3)$ and $(0, 1)$

10. The line through $(-2, 2)$ with slope $-\frac{1}{2}$

11. The line with slope $-\frac{1}{3}$ and *y* intercept 6

12. Find the distance between the points $(-4, -1)$ and $(-2, 1/2)$.

13. Determine whether the lines $y = 2x - 3$ and $x + 2y = 5$ are parallel, perpendicular, or neither.

14. Find the point of intersection of the lines $x + 3y = 4$ and $2x - 3y = 5$.

15. Let *l* be the line with equation $2x + 4y = 1$, and let $P = (1, -2)$.

  a. Find an equation of the line parallel to *l* and passing through *P*.

  b. Find an equation of the line perpendicular to *l* and passing through *P*.

16. Show that the lines $2x + 3y = 1$, $2x + 3y = 6$, $3x - 2y = 2$, and $3x - 2y = 6$ form a rectangle.

17. Show that the lines $y = x$, $y = -x + 2$, $y = -x + 10$, and $y = x + 8$ form a square.

18. Are the functions x^2 and $\sqrt{x^4}$ equal? Explain.

In Exercises 19–22 find the domain of the function.

19. $f(x) = \dfrac{x^2 + 3x - 4}{x(x^2 + 4x + 3)}$

20. $f(t) = \sqrt{t^3 - 4t}$

21. $f(x) = \ln \dfrac{e^x}{e^x - 1}$

22. $f(x) = \dfrac{\ln(1 - x)}{\ln x}$

In Exercises 23–26 sketch the graph of the function.

23. $f(x) = 2|x - 3|$

24. $f(x) = \sqrt{2 - x^2}$

25. $g(x) = \begin{cases} x^3 & \text{for } x \leq 1 \\ -|x| & \text{for } x > 1 \end{cases}$

26. $y = \dfrac{1}{x + 2}$

In Exercises 27–30 sketch the graph of the equation. List the intercepts and describe the symmetry (if any) of the graph.

27. $x + y = 1$

28. $x + y^2 = 1$

29. $2x^2 + 2y^2 = 6$

30. $y = \frac{1}{2}(\sin x + |\sin x|)$

In Exercises 31–36 sketch the graph of the function. Indicate any intercepts and symmetry, and determine whether the function is even, odd, or neither.

31. $f(x) = 1 - \sin x$

32. $f(x) = \cos |x|$

33. $f(x) = \tan(x - \pi/4)$

34. $f(x) = \sin(\pi x /2)$

35. $f(x) = \ln(1 - |x|)$

36. $f(x) = e^{x+3}$

In Exercises 37–40 determine whether the function is even, odd, or neither.

37. $f(x) = x^2 - \cos x$

38. $f(x) = x^2 + \sin x$

39. $f(x) = e^x + e^{-x}$

40. $f(x) = e^x - e^{-x}$

In Exercises 41–44 sketch the graph of the given equation with the help of a suitable translation. Show both the x and y axes and the X and Y axes.

41. $x^2 + 6x + y + 4 = 0$ **42.** $x^2 - 4x + y^2 = 5$

43. $x - 2 = \dfrac{1}{y + 3}$ **44.** $x^2 - x + y^2 - 4y = \frac{19}{4}$

45. Let $f(x) = |x - 2| + |x - 3|$ and $g(x) = |x - 2| - |x - 3|$.

  a. Find formulas for f and g that do not contain absolute value signs.

  b. Sketch the graphs of f and g.

46. Prove that $|x| = \sqrt{x^2}$.

47. Let

$$f(x) = \frac{x + 2}{x^2 - 4x + 3} \quad \text{and} \quad g(x) = \frac{x + 1}{x^2 - 2x - 3}$$

  Find the domains and rules of $f - g$ and f/g.

48. Find the domains and rules of $g \circ f$ and $f \circ g$, where

$$f(x) = \frac{x^2 - 1}{2x} \quad \text{and} \quad g(x) = \sqrt{x + 1}$$

49. Let $f(x) = \sqrt{x + 1}$, $g(x) = \sqrt{x + 2}$, and $h(x) = \sqrt{(x + 1)(x + 2)}$.

  a. Find the domains of f, g, and h.

  b. Find the domain of fg, and compare it with the domain of h.

50. Let $f(x) = \sqrt{1 - x^2}$. Show that $f(f(x)) = x$ for $0 \le x \le 1$.

51. Let $f(x) = x/(x - 1)$.

  a. Show that $f(f(x)) = x$ for $x \ne 1$.

  b. Use part (a) to show that $f(f(f(f(x)))) = x$ for $x \ne 1$.

52. Let

$$f(x) = x + \sqrt{x^2 + 1} \quad \text{and} \quad g(x) = x + \sqrt{x^2 - 1}$$

Show that

$$g(\sqrt{x^2 + 1}) = f(x) \quad \text{for } x \ge 0$$

$$f(\sqrt{x^2 - 1}) = g(x) \quad \text{for } x \ge 1$$

53. Let

$$f(x) = \frac{ax + b}{cx - a}$$

Show that if $a^2 + bc \ne 0$, then $f(f(x)) = x$ for all x in the domain of f.

54. Let $h(x) = \dfrac{1}{\sqrt{x^2 + 4}}$. Write h as the composite of two functions f and g (neither of which is equal to h).

55. Find the values of the remaining four trigonometric functions if $\sin x = -\frac{2}{3}$ and $\tan x = -2\sqrt{5}/5$.

56. Solve the inequality $\cos x < -\frac{1}{2}$ for x in $[0, 2\pi)$.

57. Solve the inequality $|\sin x| \ge |\cos x|$ for x in $[0, 2\pi)$.

58. **a.** For $a = \frac{1}{2}$, 2, and 3 show that the graphs of $\log_{\sqrt{a}} \sqrt{x}$ and $\log_a x$ are identical. (*Hint:* To graph $\log_a x$, use the formula $\log_a x = (\ln x)/(\ln a)$.)

  b. Prove that $\log_{\sqrt{a}} \sqrt{x} = \log_a x$ for any positive number a with $a \ne 1$.)

*59. In each of the following parts, find numbers a and b such that the equation holds for all x.

  a. $\sin x + \sqrt{3} \cos x = a \sin(x + b)$

  b. $\sin x + \sqrt{3} \cos x = a \cos(x + b)$

In Exercises 60–62 solve the inequality.

60. $2 \sin x \ge \sin(2x)$ for x in $[0, 2\pi]$ **61.** $2e^x \ge e^{2x}$

62. $2 \ln x \ge \ln(2x)$

63. Identify each graph in Figure 1.86(a)–(b) with one of the following functions.

  a. $f(x) = 1 - x^2$ **b.** $g(x) = e^{2x}$

  c. $h(x) = 1 + \ln(x + 1)$ **d.** $k(x) = 1 + \sin x$

  e. $F(x) = \dfrac{16}{(x + 4)^2}$ **f.** $G(x) = \dfrac{2x^2 + 1}{x^2 + 1}$

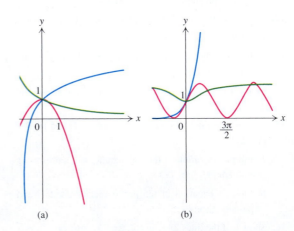

(a)   (b)

FIGURE 1.86   Graphs for Exercise 63.

64. Given the partial graph of f in Figure 1.87, complete the graph of f, assuming that

   a. f is an even function

   b. f is an odd function

   c. the graph of f is symmetric with respect to the y axis

   d. the graph of f is symmetric with respect to the origin

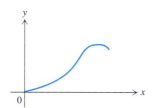

FIGURE 1.87   Graph for Exercise 64.

65. Prove that $\sin x \le \tan x \le 2 \sin x$ for $0 \le x \le \pi/3$. (*Hint:* What are the largest and the smallest values of $\cos x$ for $0 \le x \le \pi/3$?)

 66. Approximate $1/(1 - \cos (10^{-10}))$. (*Hint:* Use (9) of Section 1.7.)

67. Show that if a square and an equilateral triangle have the same perimeter, then the square has the larger area. Is this true for an arbitrary equilateral triangle and arbitrary rectangle of equal perimeter? Explain.

68. A line with positive slope m contains the point $(-4, 0)$. This line, the negative x axis, and the positive y axis form a triangle. Find a formula for the function A that expresses the area of the triangle in terms of m.

69. Three vertices of a parallelogram are located at $(0, 1)$, $(2, 0)$, and $(3, 2)$. Determine the three possible locations of the fourth vertex.

Applications

70. A ball was thrown twice from a balcony. The first time it was thrown straight up and the second time it was thrown straight down, but each time the initial speed was the same. Assume that on the second time it took 2 fewer seconds for the ball to hit the ground. What was the initial speed in meters per second?

71. Suppose that you drive from A to D, passing through either B or C (Figure 1.88). If you travel at a constant speed of 60 kilometers per hour and the trip lasts 1 hour, do you pass through B ? Explain.

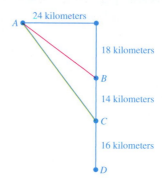

FIGURE 1.88   Figure for Exercise 71.

72. The angular displacement $f(t)$ of a pendulum bob is given by

$$f(t) = a \cos 2\pi \omega t$$

where ω is the frequency and a the maximum displacement. Draw the graph of this function for $\omega = 1$ and $a = 2$.

73. Consider a circular racetrack with a radius of 600 feet. Suppose a straight, narrow path connects two points A and B on the track that are diametrically opposite one another. If a horse is 800 feet around the track from point A, how far from the path is the horse?

74. A motorist drives down a road toward the intersection on a heavily traveled boulevard (Figure 1.89). A house stands 40 feet from the center of the road and 60 feet from the center of the boulevard. Assuming that the motorist drives down the center of the road, determine how far the motorist will be from the center of the intersection when the angle θ shown in Figure 1.89 is $\pi/6$.

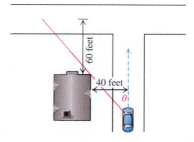

FIGURE 1.89   Figure for Exercise 74.

75. The population of Bismarck, North Dakota, increased from 34,703 in 1970 to 44,485 in 1980. Letting $t = 0$ in 1970 and $t = 10$ in 1980, predict what the population P of Bismarck was in the year 2000 if

 a. P grew in a linear fashion, that is, $P(t) = at + b$, where a and b are constants

 b. P grew exponentially, that is, $P(t) = ae^{bt}$, where a and b are constants

76. The formula

$$\log N = 8.73 - 1.15 \, M$$

has been used in order to approximate the average number N of earthquakes per year with a magnitude of between $M - 0.05$ and $M + 0.05$ (that is, within 0.05 of M) for $M \geq 7$.

 a. Determine the magnitude M for which $N = 1$.

 b. Determine the average number of earthquakes per year with magnitude between 7.45 and 8.05. (*Hint:* Take $M = 7.5, 7.6, \ldots, 8.0$ in succession.)

Tower Falls, in Yellowstone
National Park.
(E. Cooper/Superstock)

2 LIMITS AND CONTINUITY

In Chapter 2 we study the notions of limit and continuity. These concepts are fundamental to the main subjects of calculus: the derivative and the integral.

Chapter 2 opens with a discussion of two seemingly unrelated topics—the line tangent to the graph of a function and the velocity of a moving object. It turns out that these two problems have similar mathematical formulations that lead to the definition of limit. After defining limits and giving rules for evaluating them, we turn to properties of continuous functions, and end the chapter with a discussion of approximating zeros of functions.

2.1   TANGENT LINES AND VELOCITY

We devote Section 2.1 to the two enormously important concepts of tangent lines and velocity, which we will show have a surprising relationship to one another.

Tangent Lines

Since the time of ancient Greece, mathematicians have been interested in tangent lines. Early mathematicians studied tangents to simple curves such as circles and spirals (Figure 2.1). Euclid, who was the most prominent of them, conceived of a

73

line tangent to a circle as a line that touches the circle at exactly one point. Yet the idea of a tangent line "touching" a curve does not lend itself well to drawing a tangent line, nor does it give a procedure for deriving an equation of a tangent line, which is important in calculus.

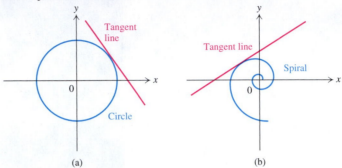

FIGURE 2.1

The line tangent to a circle at a given point P can be described in another way. If Q is any other point on the circle, then the line joining P and Q is called a **secant line** (Figure 2.2(a)). It turns out that the line tangent to the circle at P is the limit of secant lines through P, in the sense that the slope of the secant line approaches the slope of the tangent line as Q approaches P (Figure 2.2(b)). This property of the tangent line provides a method for finding an equation of the tangent line, even when the curve to which it is tangent is not a circle.

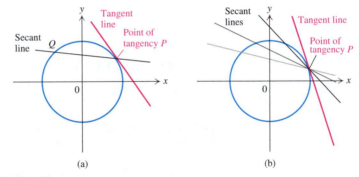

FIGURE 2.2

Before we give the general definition of a line tangent to the graph of a function, we first consider the specific function $f(x) = x^2$ and select $(1, f(1)) = (1, 1)$ to be the point P at which we wish to define the tangent line L (Figure 2.3(a)). Recall that a line is determined by its slope and any point on the line. Since we know that $(1, 1)$ is to lie on the tangent line L, we only need to specify the slope of the tangent in order to determine L. To find the slope, choose any point $(x, f(x))$ on the graph of f that is *not* $(1, 1)$. Then the slope of the secant line through $(1, 1)$ and $(x, f(x))$ is

$$\frac{f(x) - f(1)}{x - 1} = \frac{x^2 - 1}{x - 1} = \frac{(x - 1)(x + 1)}{x - 1} = x + 1 \qquad (1)$$

As x approaches 1, the slope $x + 1$ of the secant line approaches 2, and the secant line seems to slant more and more the way the graph of f does near $(1, 1)$ (Figure

2.3(b)). We say that the "limit" of $(f(x) - f(1))/(x - 1)$ is 2 as x approaches 1, and write it symbolically as

$$\lim_{x \to 1} \frac{f(x) - f(1)}{x - 1} = 2$$

More generally, if f is any function, then the slope of the secant line through $(a, f(a))$ and any other point $(x, f(x))$ on the graph of f is

$$\frac{f(x) - f(a)}{x - a}$$

(Figure 2.4(a)). In case the numbers $(f(x) - f(a))/(x - a)$ approach a limiting value as x approaches a (Figure 2.4(b)), we will define the line tangent to the graph of f at $(a, f(a))$ to be the line through $(a, f(a))$ whose slope is that limit, written as

$$\lim_{x \to a} \frac{f(x) - f(a)}{x - a} \tag{2}$$

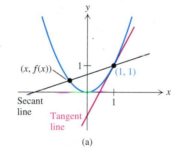 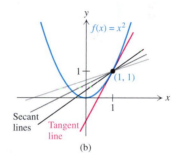

(a)   (b)

FIGURE 2.3

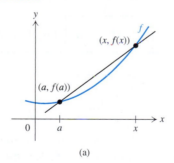 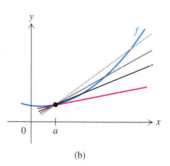

(a)   (b)

FIGURE 2.4

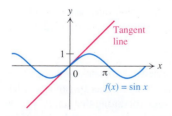

FIGURE 2.5

It is not always as simple to find the limit by manipulating the expression $(f(x) - f(a))/(x - a)$ as it was in (1). For example, suppose $f(x) = \sin x$, and consider the line tangent to the graph of f at $(0, 0)$ (Figure 2.5). By (2) the tangent line in question has slope

$$\lim_{x \to 0} \frac{\sin x - \sin 0}{x - 0} = \lim_{x \to 0} \frac{\sin x}{x} \tag{3}$$

Here there is no way of canceling terms in the numerator and denominator. Since $\sin x$ approaches 0 as x approaches 0, the quotient $(\sin x)/x$ might appear to

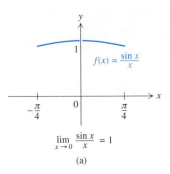

$$\lim_{x \to 0} \frac{\sin x}{x} = 1$$

(a)

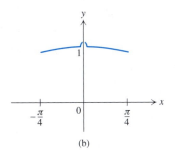

(b)

FIGURE 2.6

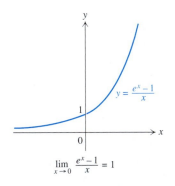

$$\lim_{x \to 0} \frac{e^x - 1}{x} = 1$$

FIGURE 2.7

Velocity

approach 0/0. But 0/0 is undefined, so if the limit in (3) exists, we must find it by a different technique. Having no easy way of rewriting $(\sin x)/x$ to obtain the limit, we use a calculator to find the values of $(\sin x)/x$ for certain values of x close to 0:

x	$\pm\ 0.1$	$\pm\ 0.01$	$\pm\ 0.001$	$\pm\ 0.0001$
$\dfrac{\sin x}{x}$	.9983341665	.9999833334	.9999998333	.9999999983

From the table one might guess that $(\sin x)/x$ approaches 1 as x approaches 0, that is,

$$\lim_{x \to 0} \frac{\sin x}{x} = 1 \qquad (4)$$

Furthermore, the computer-drawn graph of $(\sin x)/x$ in Figure 2.6(a) also suggests that the limit should be 1. However, our limiting value of 1 is only a guess. It is more in doubt than the other limits we have found, since we calculated $(\sin x)/x$ for only a few values of x. The true graph could possibly be the one shown in Figure 2.6(b). This would suggest a different limit. With the definition of limit that we will give in Section 2.2 it will be possible to conclude with certainty that $\lim_{x \to 0} (\sin x)/x$ exists and equals 1.

In a similar way, if $f(x) = e^x$, then for the slope of the line tangent to the graph of f at (0, 1), we are led to the limit

$$\lim_{x \to 0} \frac{f(x) - f(0)}{x - 0} = \lim_{x \to 0} \frac{e^x - 1}{x}$$

Calculating $(e^x - 1)/x$ for a few values of x near 0, we assemble the following table:

x	-0.1	0.01	-0.001	0.0001
$\dfrac{e^x - 1}{x}$	0.9516258196	1.005016708	0.9995001666	1.000050002

The table suggests that

$$\lim_{x \to 0} \frac{e^x - 1}{x} = 1 \qquad (5)$$

and Figure 2.7 reinforces this suggestion. The limit in (5) was obtained by calculating values of $(e^x - 1)/x$ for x near 0; this, however, does not constitute a *proof* of (5). Nevertheless we will assume that (5) is valid until Section 7.2, where it will be proved.

The limit in (2) also appears in the study of velocity. Suppose a spacecraft is launched from earth and travels vertically upward. If the rocket travels 36 miles during a 2-minute time interval, then the average velocity during that interval is 18 miles per minute. In general, during any time interval the average velocity of the rocket is defined by

$$\text{average velocity} = \frac{\text{distance traveled}}{\text{elapsed time}}$$

Launch of Discovery. (NASA)

Instead of finding average velocity we would like to be able to calculate the velocity at a particular moment (sometimes called instantaneous velocity). This is the number that can be read from a speedometer. One way to calculate the velocity is to consider it as the limit of average velocities. In particular, let $f(t)$ be the height (in miles) of the rocket t minutes after launch. If t is a little larger than 2, then the distance traveled in the interval from 2 to t is $f(t) - f(2)$, and the elapsed time is $t - 2$. Consequently the average velocity during the time interval from 2 to t is given by

$$\frac{f(t) - f(2)}{t - 2}$$

The closer t is to 2, the closer we would expect the average velocity to be to the velocity at 2. It is therefore natural to define the velocity of the rocket at time 2 to be the limit of $(f(t) - f(2))/(t - 2)$ as t approaches 2, which in limit notation we would write as

$$\lim_{t \to 2} \frac{f(t) - f(2)}{t - 2}$$

More generally, the velocity $v(t_0)$ at time t_0 of an object traveling in a straight line with position $f(t)$ at time t is given by

$$v(t_0) = \lim_{t \to t_0} \frac{f(t) - f(t_0)}{t - t_0} \tag{6}$$

To illustrate (6), suppose a spacecraft is headed vertically upward, and that $f(t)$ is its height (in miles) t minutes after launch, with

$$f(t) = t^2 \quad \text{for } 0 \le t \le 2$$

Then the velocity $v(1)$ of the spacecraft at time $t = 1$ is given by

$$v(1) = \lim_{t \to 1} \frac{f(t) - f(1)}{t - 1} = \lim_{t \to 1} \frac{t^2 - 1}{t - 1} = \lim_{t \to 1} (t + 1) = 2$$

Therefore $v(1) = 2$ (miles per minute).

In Section 2.2 we will make a precise definition of limit which will include the limits for the slope of a tangent line and velocity that we have presented informally in this section.

EXERCISES 2.1

In Exercises 1–8 guess the value of the limit.

1. $\displaystyle\lim_{x \to -1} (x + 4)$

2. $\displaystyle\lim_{x \to 5} (-2x + 7)$

3. $\displaystyle\lim_{h \to 0} (2h + h^2)$

4. $\displaystyle\lim_{h \to 0} \left(1 - \frac{h^2}{2}\right)$

5. $\displaystyle\lim_{x \to 2} \frac{2x - 5}{4x + 3}$

6. $\displaystyle\lim_{x \to 1/2} \frac{3x - 2}{4x - 1}$

7. $\displaystyle\lim_{x \to 3} \frac{\sqrt{x + 1}}{2x - 1}$

8. $\displaystyle\lim_{x \to -1/2} \frac{1}{2}\sqrt{\frac{1}{x} + 6}$

In Exercises 9–18 first simplify the given expression and then guess the value of the limit.

9. $\lim\limits_{x \to -2} \dfrac{x^2 - 4}{x + 2}$

10. $\lim\limits_{x \to 1} \dfrac{x^2 + 4x - 5}{x - 1}$

11. $\lim\limits_{x \to -5} \dfrac{x^2 + 4x - 5}{x + 5}$

12. $\lim\limits_{x \to \pi} \dfrac{x^2 - 2x + 1}{(x - 1)^2}$

13. $\lim\limits_{x \to 1} \dfrac{x^3 - 1}{x - 1}$

14. $\lim\limits_{x \to -2} \dfrac{x^3 + 8}{x + 2}$

15. $\lim\limits_{x \to 0} 3(\sin^2 x + \cos^2 x)$

16. $\lim\limits_{x \to \pi} \dfrac{\sin 2x}{\sin x \cos x}$

17. $\lim\limits_{x \to -1} -\dfrac{|x|}{x}$

18. $\lim\limits_{x \to -6} \dfrac{x^2}{|x|}$

In Exercises 19–27 evaluate the function at 0.1, 0.01, and 0.001, and at −0.1, −0.01, and −0.001. Then guess the value of $\lim\limits_{x \to 0} f(x)$.

19. $f(x) = \dfrac{\tan x}{x}$

20. $f(x) = \dfrac{\sin 2x}{x}$

21. $f(x) = \dfrac{1 - \cos x}{x}$

22. $f(x) = \dfrac{1 - \cos x}{x^2}$

23. $f(x) = \dfrac{\sin 3x}{\sin 5x}$

24. $f(x) = \dfrac{x^3}{x - \sin x}$

25. $f(x) = \csc x - \cot x$

26. $f(x) = \dfrac{\ln(1 + x)}{x}$

27. $f(x) = (e^{3x} - 1)/x$

28. Suppose $\lim\limits_{x \to 1} f(x) = 2$. Which of the graphs in Figure 2.8 could be the graph of f?

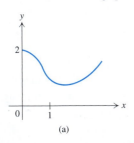

(a)

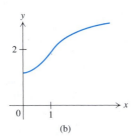

(b)

(c)

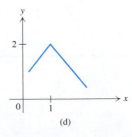

(d)

FIGURE 2.8 Graphs for Exercise 28.

29. Find the value of $\lim\limits_{x \to 1.2} f(x)$ that is suggested by the graph of f in Figure 2.9.

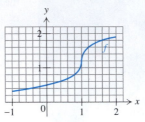

FIGURE 2.9 Graph for Exercise 29.

30. Let $f(x) = x - x^3$. Use Figure 2.10 to determine the approximate slope of the line tangent to the graph of f at $(a, f(a))$.

   a. $a = 0$ **b.** $a = 0.6$

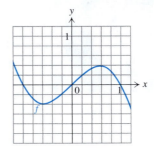

FIGURE 2.10 Graph for Exercise 30.

In Exercises 31–34 plot the graph of f on $(-1, 1)$. From the graph, decide whether $\lim\limits_{x \to 0} f(x)$ exists, and find the limit if it does. (*Hint:* You may need to zoom in toward the origin in order to make your decision.)

31. $f(x) = x \sin \dfrac{1}{x}$

32. $f(x) = \sin 1/x$

33. $f(x) = \sin e^{-1/x}$

34. $f(x) = e^{-1/x^2}$

35. a. Plot the graph of $(\ln x)/(x - 1)$ for $0.5 \le x \le 1.5$ with $x \neq 1$.

    b. From the graph obtained in part (a), guess the value of

$$\lim_{x \to 1} \frac{\ln x}{x - 1}$$

36. Consider Figure 2.11, which displays the graph of a function with $\lim\limits_{x \to 2} f(x) = 4$. From the graph, estimate how close x must be to 2 in order to ensure that $f(x)$ is within 0.5 of 4.

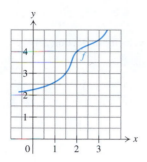

FIGURE 2.11 Graph for Exercise 36.

37. For the function f whose graph appears in Figure 2.12, determine which is larger,

$$\frac{f(2) - f(1)}{2 - 1} \quad \text{or} \quad \frac{f(5) - f(2)}{5 - 2}$$

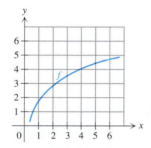

FIGURE 2.12 Graph for Exercise 37.

38. Suppose $\lim_{x \to a} f(x) = -2$ and $\lim_{x \to a} g(x) = 3$.

a. Guess the value of $\lim_{x \to a} [f(x)g(x)]$.

b. Guess the value of $\lim_{x \to a} \dfrac{f(x)}{g(x)}$.

In Exercises 39–42 simplify the quotient $(f(x) - f(a))/(x - a)$, and then guess the slope of the line tangent to the graph of f at $(a, f(a))$.

39. $f(x) = x^2$; $a = 3$ **40.** $f(x) = x^2 + 2x$; $a = -2$
41. $f(x) = 2x^2 + 1$; $a = 4$ **42.** $f(x) = x^3 + 1$; $a = -2$

In Exercises 43–46 evaluate $(f(x) - f(2))/(x - 2)$ for $x = 1.9$, 1.99, and 1.999, and at 2.1, 2.01, and 2.001. Then guess the slope of the line tangent to the graph of f at $(2, f(2))$.

43. $f(x) = \ln x$ **44.** $f(x) = e^{-x}$

45. $f(x) = \tan\left(\dfrac{\pi}{2} x\right)$ **46.** $f(x) = \dfrac{x^2 - 4}{x^2 + 1}$

Applications

47. Suppose the height of a spacecraft at time t is given by $f(t) = t^2$ for $0 \leq t \leq 2$. Find the average velocity of the

spacecraft during the time interval between 1/2 and t (for $t \neq 1/2$), and then find its velocity at time 1/2.

48. Suppose the height of a spacecraft at time t is given by $f(t) = 2t^2 + 1$ for $0 \leq t \leq 3$. Find an expression for the spacecraft's average velocity during the time interval between 2 and t (for $t \neq 2$), and then find its velocity at time 2.

49. Suppose you would like to estimate the velocity, in miles per hour, of a car at time $t = 50$ minutes, and you have the following recorded data:

t (minutes)	48	49	50	51	52
distances (miles)	27.1	27.8	28.4	29	29.2

a. Which data would you ignore and which data would you use?

b. What is your estimate of the velocity at $t = 50$?

50. According to the theory of thermodynamics, the total energy of a system in thermodynamic equilibrium must remain as small as possible. As a result, when the volume is decreased through compression, physical characteristics of the solid may change. For example, the well-known semi-conductor silicon has a diamond structure at ambient pressure; however, under sufficient pressure this structure is transformed into the metallic tin structure called β-tin. Graphs of the energy for the diamond structure and the β-tin structure appear in Figure 2.13. The slope of the dashed line that is tangent to the two curves represents the pressure at which the transformation from one structure to the other takes place. Approximate that slope.

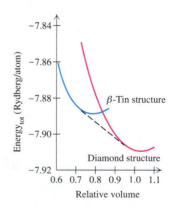

FIGURE 2.13 The total energy curves for the diamond and β-tin phases of silicon plotted against relative volume. The dashed line is the common tangent of the energy curves for the two phases. The slope of this line gives the transformation pressure.

2.2 DEFINITION OF LIMIT

In our discussion of tangent lines and velocity in Section 2.1 we were led to limits of the form

$$\lim_{x \to a} \frac{f(x) - f(a)}{x - a} \tag{1}$$

Some examples, such as

$$\lim_{x \to 1} \frac{x^2 - 1}{x - 1}$$

which was discussed in Section 2.1, are easy to evaluate. However, other limits in the form of (1), such as

$$\lim_{x \to 0} \frac{\sin x - \sin 0}{x - 0} = \lim_{x \to 0} \frac{\sin x}{x}$$

which we also discussed in Section 2.1, are not so accessible. Moreover, we will encounter many diverse limits that don't involve fractions. As a result, we need a way of defining limits that allows us to evaluate limits not only arising from tangents and velocity but in many other arenas. Thus our present goal is to make precise what we mean by

$$\lim_{x \to a} f(x) = L \tag{2}$$

for a given function of f and numbers a and L. We will assume that f is defined for all numbers in some interval about a except possibly a itself. The reason we don't require $f(a)$ to be defined is that the function appearing in the limit may well not be defined at a. For example, as we noted in Section 2.1,

$$\lim_{x \to 1} \frac{x^2 - 1}{x - 1} \quad \text{exists and equals 2}$$

even though the function $(x^2 - 1) / (x - 1)$ is not defined at 1.

Intuitively, we interpret the statement in (2) to mean:

> We can make $f(x)$ as close to L as we like if we take
> x close enough to a (but different from a) $\tag{3}$

Let us explore the idea in (3) with the limit

$$\lim_{x \to 1} (2x + 7) = 9$$

In the notation of (2), $f(x) = 2x + 7$, $a = 1$, and $L = 9$, so that (3) becomes:

> We can make $2x + 7$ as close to 9 as we like if we
> take x close enough to 1 (but different from 1)

For example, suppose that we wish to make sure that $2x + 7$ is within 0.1 of 9. How close would we need to take x to 1 in order to be successful? In other words, how close must x be to 1 in order that $2x + 7$ be within 0.1 of 9?

To answer this question, let us notice that the distance between $2x + 7$ and 9 is $|(2x + 7) - 9| = |2x - 2|$. Thus we need to show that

we can make $|2x - 2| < 0.1$ if we take

x close enough to 1 (but $x \neq 1$)

Now the distance between x and 1 is $|x - 1|$, and the statement in (3) implies that $|x - 1|$ cannot be 0. Hence we must have $|x - 1| > 0$. Therefore we must find a positive number, which we will denote by the Greek letter δ (delta), such that

$$\text{if } 0 < |x - 1| < \delta, \quad \text{then} \quad |2x - 2| < 0.1 \tag{4}$$

But it is clear that

$$\text{if } 0 < |x - 1| < \delta, \quad \text{then } |2x - 2| = 2|x - 1| < 2\delta \tag{5}$$

Comparing (4) and (5), we see that if $2\delta = 0.1$, that is, if we take $\delta = 0.05$, then (5) will be the same as (4), and hence we will have achieved our goal.

In an analogous fashion, suppose we wish to have the distance between $2x + 7$ and 9 be less than 0.01. Then (4) would be replaced by the following statement:

$$\text{if } 0 < |x - 1| < \delta, \quad \text{then} \quad |2x - 2| < 0.01 \tag{6}$$

Comparing (6) and (5), we observe that if $2\delta = 0.01$, that is, if we take $\delta = 0.005$, then (5) will be the same as (6), so once again, we will have achieved our goal.

In fact, what we have just performed for 0.1 and 0.01 could be done for any positive number. To convince you of this, we will let the Greek letter ε (epsilon) denote any positive number. Suppose that we wish to make $2x + 7$ within a distance ε of 9. Then (4) would be replaced by

$$\text{if } 0 < |x - 1| < \delta, \quad \text{then} \quad |2x - 2| < \varepsilon \tag{7}$$

Comparing (7) and (5), we see that if we have $2\delta = \varepsilon$, that is, if we take $\delta = \varepsilon/2$, then (7) will be valid. Consequently our goal will have been achieved.

From our discussion we have substantial evidence that we could make $2x + 7$ as close to 9 as we like (that is, within any positive distance ε) if we take x sufficiently close to 1 (that is, within a distance $\delta = \varepsilon/2$). In mathematical notation, this means that for any positive number ε there is a positive number δ such that

$$\text{if } 0 < |x - 1| < \delta, \quad \text{then} \quad |(2x + 7) - 9| < \varepsilon$$

This leads us to the following definition of limit, which is a mathematically precise version of the statement in (3).

DEFINITION 2.1

DEFINITION OF LIMIT

Let f be a function defined at each point of some open interval containing a, except possibly at a itself. Then a number L is the **limit of $f(x)$ as x approaches a** (or is the **limit of f at a**) if for every number $\varepsilon > 0$ there is a number $\delta > 0$ such that

$$\text{if } 0 < |x - a| < \delta, \quad \text{then} \quad |f(x) - L| < \varepsilon \tag{8}$$

If L is the limit of $f(x)$ as x approaches a, then we write

$$\lim_{x \to a} f(x) = L$$

If such an L can be found, then we say that the **limit of f at a exists,** or that f **has a limit at a,** or that $\lim_{x \to a} f(x)$ exists.

It follows from Definition 2.1 that a function can have at most one limit L at a. (For a proof of this see the Appendix.) Thus

> The limit of f at a is unique.

We comment that if $\lim_{x \to a} f(x) = L$, then the smaller ε is, the smaller δ normally will need to be in order that (8) be satisfied. This fact is illustrated in Figure 2.14 (a) and (b): Any value of δ that works for $\varepsilon = 1/2$ also works for $\varepsilon = 1$. In any case, according to Definition 2.1, no matter how small a number ε is, there must be a corresponding value of δ such that (8) holds. Thus we usually think of ε as representing a very small number.

Although Definition 2.1 is not needed in order to convince ourselves that limits like $\lim_{x \to 1} (2x + 7)$ exist, and to figure out the values of their limits, there are other limits that are not so easy to find. In fact, before the advent of a formal definition

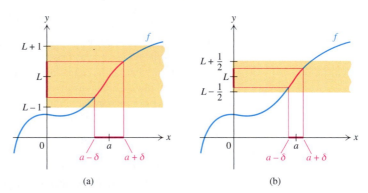

(a)   (b)

FIGURE 2.14

of limit in the early 19th century, mathematicians and scientists used their intuitive ideas of limits to derive many fascinating results. However, their intuition sometimes failed them and led them to invalid conclusions. As a result, it is essential for modern mathematics that we have a precise definition of limit.

To show how Definition 2.1 can be used to prove that limits exist, we will use it in Examples 1 and 2 to verify two basic limits.

EXAMPLE 1   Suppose a and c are any numbers. Show that

$$\lim_{x \to a} c = c \tag{9}$$

Solution   By $\lim_{x \to a} c$ we mean $\lim_{x \to a} f(x)$, where $f(x) = c$ for all x. Let $\varepsilon > 0$. We must find a number $\delta > 0$ such that

$$\text{if } 0 < |x - a| < \delta, \quad \text{then } |f(x) - c| < \varepsilon$$

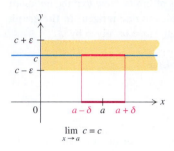

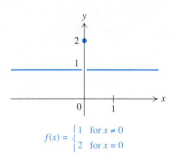

$$f(x) = \begin{cases} 1 & \text{for } x \neq 0 \\ 2 & \text{for } x = 0 \end{cases}$$

FIGURE 2.15

FIGURE 2.16   "Hiccup" function.

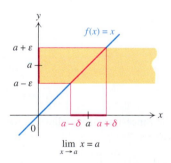

$$\lim_{x \to a} x = a$$

FIGURE 2.17

But for *any* positive number δ,

$$\text{if } 0 < |x - a| < \delta, \quad \text{then } |f(x) - c| = |c - c| = 0 < \varepsilon$$

(See Figure 2.15.) Hence in this case we may choose δ to be any positive number. ❏

From Example 1 it follows that

$$\lim_{x \to 3} 1 = 1, \quad \lim_{x \to -\sqrt{2}} \frac{\pi}{17} = \frac{\pi}{17}, \quad \text{and} \quad \lim_{x \to -2} (-\pi) = -\pi$$

Since the limit of a function f at a does not depend on the value (if any) of f at a, it also follows from Example 1 that if f is the "hiccup" function defined by

$$f(x) = \begin{cases} 1 & \text{for } x \neq 0 \\ 2 & \text{for } x = 0 \end{cases}$$

then

$$\lim_{x \to 0} f(x) = \lim_{x \to 0} 1 = 1$$

(See Figure 2.16.)

EXAMPLE 2   Suppose that a is any number. Show that

$$\lim_{x \to a} x = a \tag{10}$$

Solution   Here $f(x) = x$ for all x. Let $\varepsilon > 0$. We must find a number $\delta > 0$ such that

$$\text{if } 0 < |x - a| < \delta, \quad \text{then } |x - a| < \varepsilon$$

Here we can choose δ to be ε because

$$\text{if } 0 < |x - a| < \delta = \varepsilon, \quad \text{then } |x - a| < \varepsilon$$

(See Figure 2.17.)   ❏

From Example 2 we see that

$$\lim_{x \to 4\pi} x = 4\pi \quad \text{and} \quad \lim_{x \to -\sqrt{2}} x = -\sqrt{2}$$

A slight alteration in the solution would show that for any fixed numbers a, b, and m,

$$\lim_{x \to a} (mx + b) = ma + b \tag{11}$$

$$\lim_{x \to a} |x| = |a| \tag{12}$$

(see Exercises 29 and 30). It follows, for example, that

$$\lim_{x \to 3} (-4x + 5) = -4 \cdot 3 + 5 = -7$$

and

$$\lim_{x \to -3} |x| = |-3| = 3$$

Tangent Lines

In Section 2.1 we discussed tangent lines in terms of limits. Now that the notion of limit has been defined in Definition 2.1, we define the **line tangent to the graph of f at $(a, f(a))$** to be the line through $(a, f(a))$ with slope m_a given by

> **Slope of Tangent Line**
>
> $$m_a = \lim_{x \to a} \frac{f(x) - f(a)}{x - a}$$

(13)

provided that the limit exists. In that case, an equation of the line tangent to the graph of f at $(a, f(a))$ is as follows:

> **Equation of Tangent Line**
>
> $$y - f(a) = m_a(x - a), \quad \text{or} \quad y = f(a) + m_a(x - a)$$

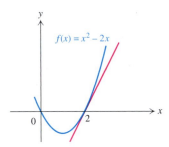

$f(x) = x^2 - 2x$

FIGURE 2.18

EXAMPLE 3   Let $f(x) = x^2 - 2x$. Find an equation of the line tangent to the graph of f at $(2, 0)$ (Figure 2.18).

Solution   By (13) the slope of the line tangent at $(2, 0)$ is

$$\lim_{x \to 2} \frac{f(x) - f(2)}{x - 2}$$

provided that the limit exists. However

$$\lim_{x \to 2} \frac{f(x) - f(2)}{x - 2} = \lim_{x \to 2} \frac{(x^2 - 2x) - 0}{x - 2} = \lim_{x \to 2} \frac{x(x - 2)}{x - 2} = \lim_{x \to 2} x$$

By Example 2, $\lim_{x \to 2} x = 2$. Therefore

$$\lim_{x \to 2} \frac{f(x) - f(2)}{x - 2} = 2$$

so that the slope of the line tangent at $(2, 0)$ is 2. Since the tangent line passes through $(2, 0)$, an equation of the desired tangent line is

$$y - 0 = 2(x - 2), \quad \text{or simply } y = 2(x - 2) \qquad \square$$

Suppose the graph of the function f in Example 3 is plotted on a graphics calculator, and we repeatedly zoom in on the graph near the point $(2, 0)$. As Figure 2.19(a)–(c) shows, eventually the graph looks like a straight line—the line tangent to the graph of f at $(2, 0)$. By contrast, let $g(x) = |x|$, and repeatedly zoom in on the graph of g near $(0, 0)$. No matter how large the magnification, the graph near $(0, 0)$ always has the same v shape (Figure 2.20). The fact that the graph of g near $(0, 0)$ does not resemble a straight line suggests that there is no line tangent at $(0, 0)$. Thus a graphics calculator can help us decide whether there is a line tangent to the graph of a function at a given point.

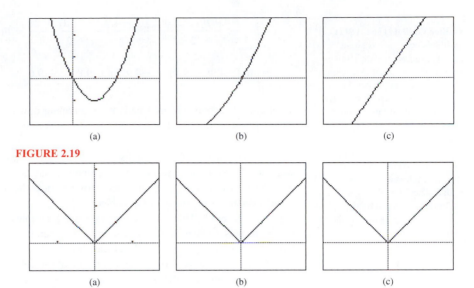

FIGURE 2.19

(a)   (b)   (c)

(a)   (b)   (c)

FIGURE 2.20

Velocity

Now consider an object traveling in a straight line with position $f(t)$ at time t. From our discussion of velocity in Section 2.1, we define the **velocity** $v(t_0)$ at time t_0 by the formula

$$v(t_0) = \lim_{t \to t_0} \frac{f(t) - f(t_0)}{t - t_0} \tag{14}$$

provided that the limit exists.

If we have a reasonable formula for the position $f(t)$ of an object at time t, then we can use (14) to find the velocity $v(t_0)$ at a given time t_0. This is illustrated in the next example. Recall from (1) in Section 1.3 that if an object is subject only to the force of gravity, and if distances are measured in meters and time in seconds, then the object's height $h(t)$ at time t is given by

$$h(t) = -4.9t^2 + v_0 t + h_0 \tag{15}$$

where h_0 is the initial height and v_0 is the initial velocity.

EXAMPLE 4   Galileo is reported to have dropped two iron balls from an upper balcony of the Leaning Tower of Pisa, approximately 49 meters above ground. If we neglect air resistance, determine how fast the iron balls would have been traveling after 2 seconds in flight.

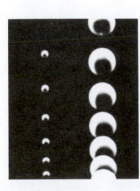

A multiflash photo showing two objects of different sizes and weights dropped at the same moment. (*Courtesy of Educational Development Center, Newton, MA*)

Solution   By our assumptions, $v_0 = 0$ since the iron balls began their descent with zero velocity, and $h_0 = 49$ because the balls were dropped from 49 meters above ground. Thus (15) becomes $h(t) = -4.9t^2 + 49$. Then by (14) with $t_0 = 2$, and by (11),

$$v(2) = \lim_{t \to 2} \frac{h(t) - h(2)}{t - 2} = \lim_{t \to 2} \frac{(-4.9t^2 + 49) - (-4.9(2^2) + 49)}{t - 2}$$

$$= \lim_{t \to 2} \frac{-4.9(t^2 - 2^2)}{t - 2} = \lim_{t \to 2} \frac{-4.9(t - 2)(t + 2)}{t - 2} = \lim_{t \to 2} \left[-4.9(t + 2) \right]$$

$$= \lim_{t \to 2} (-4.9t - 9.8) = -4.9(2) - 9.8 = -19.6$$

Thus after 2 seconds the iron balls were traveling downward at the rate of 19.6 meters per second. ❏

The fact that both iron balls—one weighing 5 kilograms and the other weighing 1/2 kilogram—would hit the ground simultaneously supported Galileo's hypothesis that the velocity of an object moving under the sole influence of gravity is independent of the weight of the object.

The answer of −19.6 (meters per second) in Example 4 is negative because the height decreases when the object descends. In general, an object traveling upward has a positive velocity, and an object traveling downward has a negative velocity. When we are interested only in the magnitude of the velocity, we talk of the speed of an object. **Speed** is defined as the absolute value of velocity. Thus an object moving with a velocity of −10 meters per second has a speed of | −10 | meters per second, or 10 meters per second. After 2 seconds the speed of the balls in Example 4 is therefore 19.6 meters per second.

Equivalent Forms of the Limit   There are two equations that are equivalent to $\lim_{x \to a} f(x) = L$:

$$\lim_{x \to a} f(x) = L \quad \text{if and only if} \quad \lim_{x \to a} (f(x) - L) = 0 \qquad (16)$$

and

$$\lim_{x \to a} f(x) = L \quad \text{if and only if} \quad \lim_{h \to 0} f(a + h) = L \qquad (17)$$

Thus the following three statements are equivalent:

$$\lim_{x \to -1} x^2 = 1, \quad \lim_{x \to -1} (x^2 - 1) = 0, \quad \text{and} \quad \lim_{h \to 0} (-1 + h)^2 = 1$$

The equivalent forms of the limit statement given in (16) and (17) will be important when we evaluate limits of exponential and trigonometric functions in Section 2.3. They can be proved by referring to Definition 2.1 (see Exercises 32–33).

You may wonder under what conditions a limit does not exist. Three such conditions are illustrated in Figure 2.21(a)–(c). In Figure 2.21(a) there is a jump in the graph at 0; in Figure 2.21(b), $| f(x) |$ becomes arbitrarily large as x approaches 0; finally, in Figure 2.21(c), $f(x)$ oscillates wildly as x approaches 0. In each case, $\lim_{x \to 0} f(x)$ does not exist.

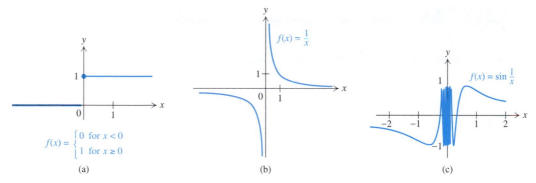

FIGURE 2.21   The graph of three functions that do not have a limit at 0

Historical Commentary

The notion of limit is rather complicated. In fact, mathematicians talked about limits for centuries before they were able to define the concept clearly. Even the ancient Greeks had some feeling for limits; for example, Archimedes found an approximation of the value of 2π as the "limit" of the perimeters of regular polygons inscribed in a circle of radius 1 by letting the number of sides grow without bound. Through the Middle Ages and into the Renaissance, mathematicians used various types of limits to compute areas. The seventeenth–century giants of calculus, Isaac Newton (1642–1727) and Gottfried Leibniz (1646–1716), had a good intuitive understanding of limits and even computed very complicated limits. However, neither they nor anyone before them actually defined the concept.

The merely intuitive quality of the idea of limit hampered progress in the development of calculus for a century after Newton and Leibniz. In 1754 the French mathematician Jean-le-Rond d'Alembert* (1717–1783) suggested that the logical basis of calculus would eventually reside in the concept of limit. In time the great French mathematician Augustin-Louis Cauchy† (1789–1857) molded calculus into a form much like the discipline we have today. However, the definition contained in his 1821 treatise *Cours d'Analyse* still relied on the intuitive notion of a variable approaching a fixed value:

> When the successive values attributed to a variable approach indefinitely close to a fixed value so as to end by differing from it by as little as one wishes, this latter value is called the limit of all the others.††

By today's standards this definition is imprecise. Nevertheless, it was a large step toward the formulation proposed in the 1860's by the great German mathematician Karl Weierstrass§ (1815–1897). It is his definition of limit that we presented in Definition 2.1 and that is accepted by mathematicians.

  * **d'Alembert:** Pronounced "dah-lem-*bair*."
  † **Cauchy:** Pronounced "*Co*-shee."
  †† Cauchy, Augustin-Louis, *Oeuvres Complètes,* 2nd series, vol. 3 (Paris, 1882, 1932), p. 19.
  § **Weierstrass:** Pronounced "*Vire*-shtrahss."

EXERCISES 2.2

In Exercises 1–10 use the results of this section to evaluate the given limit.

1. $\lim\limits_{x \to 2} (-5)$

2. $\lim\limits_{x \to -\sqrt{3}} \dfrac{\sqrt{2}}{3}$

3. $\lim\limits_{x \to 1/2} x$

4. $\lim\limits_{x \to \pi} x$

5. $\lim\limits_{x \to -1} (2x + 5)$

6. $\lim\limits_{x \to 0} \left(-4x - \dfrac{1}{3}\right)$

7. $\lim\limits_{x \to 3/2} |x|$

8. $\lim\limits_{x \to -0.1} |x|$

9. $\lim\limits_{x \to 0} f(x)$, where $f(x) = \begin{cases} 2x & \text{for } x \neq 0 \\ \frac{1}{3} & \text{for } x = 0 \end{cases}$

10. $\lim\limits_{x \to 0} f(x)$, where $f(x) = \begin{cases} -\frac{3}{2}x + \frac{1}{4} & \text{for } x \neq \frac{1}{2} \\ 0 & \text{for } x = \frac{1}{2} \end{cases}$

In Exercises 11–14 use the definition of limit to verify the given limit.

11. $\lim\limits_{x \to 0} (4x + 7) = 7$

12. $\lim\limits_{x \to 3} (-2x + 5) = -1$

13. $\lim\limits_{x \to 0} (x^2 + 5) = 5$

*14. $\lim\limits_{x \to 1} \dfrac{2x}{x^2 + 1} = 1$

In Exercises 15–20 find an equation of the line l tangent to the graph of f at the given point.

15. $f(x) = \pi$; $(3, \pi)$

16. $f(x) = -\frac{1}{2}x + 2$; $(8, -2)$

17. $f(x) = 4x^2 + \frac{1}{2}$; $(\frac{1}{2}, \frac{3}{2})$

18. $f(x) = -\frac{1}{2}x^2 + 3$; $(2, 1)$

19. $f(x) = x^2 + 3x$; $(1, 4)$

20. $f(x) = -x^2 + 4x$; $(-1, -5)$

In Exercises 21–25 $f(t)$ is the position at time t of an object moving along the x axis. Find the velocity of the object at time t_0.

21. $f(t) = 2t + 1$; $t_0 = 2$

22. $f(t) = -3t + \frac{1}{5}$; $t_0 = 0$

23. $f(t) = -16t^2$; $t_0 = 4$

24. $f(t) = -16t^2 + 100$; $t_0 = 1$

25. $f(t) = -16t^2 + 8t + 54$; $t_0 = 2$

26. Which of the graphs in Figure 2.22 appear to be graphs of functions with a limit at 2?

27. Can the graph of a function f have the same tangent line at $(-3, f(-3))$ and at $(2, f(2))$? Give an example showing that it can happen, or explain why it cannot happen.

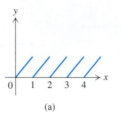

(a)

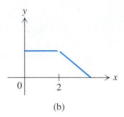

(b)

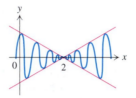

(c)

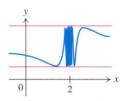

Infinitely many oscillations near 2

(d)

FIGURE 2.22 Graphs for Exercise 26.

28. In each of the following, approximate the limit with the help of a calculator or computer. Then tell how your answer is related to a tangent line of a suitable function.

a. $\lim\limits_{x \to 1} \dfrac{e - e^x}{x - 1}$

b. $\lim\limits_{x \to 0} \dfrac{\sin(x^3) - x}{x}$

c. $\lim\limits_{x \to 1} \dfrac{2 \ln x}{x - 1}$

29. Let a, b, and m be any real numbers. Show that

$$\lim\limits_{x \to a} (mx + b) = ma + b$$

(*Hint:* Consider the cases $m = 0$ and $m \neq 0$ separately. For the case $m \neq 0$, let $\delta = \varepsilon / |m|$.)

30. Show that $\lim\limits_{x \to a} |x| = |a|$. (*Hint:* $\big||x| - |a|\big| \leq |x - a|$.)

31. Determine how small a positive number x must be in order that x plus the square of x is less than

a. 0.11

b. any preassigned positive number ε

32. Show that the statements $\lim\limits_{x \to a} f(x) = L$ and $\lim\limits_{x \to a} (f(x) - L) = 0$ are equivalent.

33. Show that the statements $\lim\limits_{x \to a} f(x) = L$ and $\lim\limits_{h \to 0} f(a + h) = L$ are equivalent.

34. Let f be defined at each point of some open interval containing a, except possibly at a itself. Consider the statement

For every $\delta > 0$ there is an $\varepsilon > 0$ such that if $0 < |x - a| < \delta$, then $|f(x) - L| < \varepsilon$.

Is this equivalent to the statement "$\lim_{x \to a} f(x) = L$"? Explain.

Applications

35. Find the velocity of the balls in Example 4 after they dropped for 1 second.

36. Suppose we try to calculate the velocity of the balls in Example 4 after 4 seconds. Why is the velocity *not* −39.2 meters per second?

37. Suppose Galileo had thrown the balls downward with a speed of 6 meters per second.
  a. Determine the velocity of the balls after 2 seconds in flight.
  b. Determine the speed of the balls after 2 seconds in flight.

38. Suppose two stones are thrown simultaneously from a bridge 20 meters above a river, one vertically upward with initial velocity v_0, and the other vertically downward with initial velocity $-v_0$. Let $V_1(t)$ be the velocity of the first stone at any time t until it hits the river, and $V_2(t)$ the velocity of the second stone at any time t until it hits the river.
  a. Find the difference $D(t) = V_1(t) - V_2(t)$ between the velocity of the two stones until one of them hits the river.
  b. Determine the value of v_0 such that $D(t) = 8$ meters per second.

39. Suppose a ball is thrown from a balcony 40 meters above ground, and assume that after 1 second the velocity of the ball is −9.8 meters per second.
  a. Determine the ball's initial velocity.
  b. Would the answer to part (a) be the same if the balcony were higher? Explain.

Projects

1. Consider the functions f, g, and h, defined by

$$f(x) = \begin{cases} \sin(1/x) & \text{for } x \neq 0 \\ 0 & x = 0 \end{cases}$$

$$g(x) = \begin{cases} x \sin(1/x) & \text{for } x \neq 0 \\ 0 & x = 0 \end{cases}$$

$$h(x) = \begin{cases} x^2 \sin(1/x) & \text{for } x \neq 0 \\ 0 & x = 0 \end{cases}$$

Plot the graphs of the functions. Then tell in a sentence how the behavior of f, g, and h differ as x approaches 0. Next, zoom toward the point $(0, 0)$, and determine which, if any, of the functions should have a tangent line at the point $(0, 0)$. Using the information you have gained, and perhaps additional graphs, conjecture for which positive rational numbers r the function k defined by

$$k(x) = \begin{cases} x^r \sin(1/x) & \text{for } x \neq 0 \\ 0 & x = 0 \end{cases}$$

has a tangent line at the point $(0, 0)$.

2. Suppose that $\lim_{x \to a} f(x)$ exists, and f has a nonhorizontal tangent line at $(a, f(a))$. Then there is a way of determining a reasonable approximate value for δ that works for a given small value of ε. Here is what to do. Zoom in on the graph near $(a, f(a))$. Then pick $(a, f(a))$ and a second point near $(a, f(a))$, and find the slope of the line that the two points determine. Use that information to find a number c such that $\delta = c\varepsilon$ for small values of ε. For each function described below, assume that $\varepsilon = 0.01$, and find an approximate value of δ.
  a. $f(x) = \sin 2x$; $a = 0$
  b. $f(x) = x^2 + 3x + 1$; $a = 1$
  c. $f(x) = 2x^2 - 5$; $a = 2$

2.3   LIMIT RULES AND EXAMPLES

Limits appear, either explicitly or implicitly, throughout calculus. Fortunately it is not necessary to apply the definition of limit every time a limit is to be evaluated. The reason is that most of the functions that appear in calculus are combinations (such as sums, products, quotients, and composites) of simpler functions, and there are rules for finding limits of combinations of functions whose limits are already known.

For example, the function $|x| + 3$ is the sum of the functions $|x|$ and 3, and we know how to find the limits of $|x|$ and 3 from (12) and (9) in Section 2.2. Similarly, the function x^2 is the product $x \cdot x$, and we know the limit of x at any number a by (10) in Section 2.2. Once we know the rules for finding limits of combinations, we will be able to calculate the limits of $|x| + 3$ and x^2 with ease.

The following theorem gives five basic rules for finding limits of combinations of functions. It is proved in the Appendix.

THEOREM 2.2
Limit Theorem

If $\lim_{x \to a} f(x)$ and $\lim_{x \to a} g(x)$ exist, then $\lim_{x \to a} [f(x) + g(x)]$, $\lim_{x \to a} cf(x)$, $\lim_{x \to a} [f(x) + g(x)]$, and $\lim_{x \to a} [f(x)g(x)]$ exist, and

Sum Rule: $\lim_{x \to a}[f(x) + g(x)] = \lim_{x \to a} f(x) + \lim_{x \to a} g(x)$

Constant Multiple Rule: $\lim_{x \to a} cf(x) = c \lim_{x \to a} f(x)$ for any constant c

Difference Rule: $\lim_{x \to a}[f(x) - g(x)] = \lim_{x \to a} f(x) - \lim_{x \to a} g(x)$

Product Rule: $\lim_{x \to a}[f(x)g(x)] = \lim_{x \to a} f(x) \lim_{x \to a} g(x)$

If $\lim_{x \to a} f(x)$ and $\lim_{x \to a} g(x)$ exist and $\lim_{x \to a} g(x) \neq 0$, then $\lim_{x \to a}(f(x) / g(x))$ exists, and

Quotient Rule: $\lim_{x \to a} \dfrac{f(x)}{g(x)} = \dfrac{\lim_{x \to a} f(x)}{\lim_{x \to a} g(x)}$

EXAMPLE 1   Evaluate the following limits.

a. $\lim_{x \to -2} (|x| + 3)$ **b.** $\lim_{x \to -2} x^2$ **c.** $\lim_{x \to -2} \dfrac{x^2}{|x| + 3}$

Solution   **a.** We consider $|x| + 3$ as the sum of $|x|$ and 3, and use the Sum Rule. Since $\lim_{x \to -2} |x| = |-2| = 2$ and $\lim_{x \to -2} 3 = 3$, we find that

$$\lim_{x \to -2} (|x| + 3) = \lim_{x \to -2} |x| + \lim_{x \to -2} 3 = 2 + 3 = 5$$

b.   We consider x^2 as the product $x \cdot x$ and use the Product Rule. Since $\lim_{x \to -2} x = -2$, it follows that

$$\lim_{x \to -2} x^2 = \lim_{x \to -2} (x \cdot x) = (\lim_{x \to -2} x)(\lim_{x \to -2} x) = (-2)(-2) = 4$$

c.   Observe that $x^2/(|x| + 3)$ is the quotient of x^2 and $|x| + 3$, whose limits at -2 we know from parts (a) and (b). We conclude from the Quotient Rule that

$$\lim_{x \to -2} \frac{x^2}{|x| + 3} = \frac{\lim_{x \to -2} x^2}{\lim_{x \to -2} (|x| + 3)} = \frac{4}{5} \qquad \square$$

We often encounter sums and products of more than two functions. The results are analogous to those in the Sum and Product Rules: If $\lim_{x \to a} f_1(x)$, $\lim_{x \to a} f_2(x)$, . . . , $\lim_{x \to a} f_n(x)$ exist, then

$$\lim_{x \to a} [f_1(x) + f_2(x) + \cdots + f_n(x)] = \lim_{x \to a} f_1(x) + \lim_{x \to a} f_2(x) + \cdots + \lim_{x \to a} f_n(x) \quad (1)$$

and

$$\lim_{x \to a} [f_1(x)f_2(x)\cdots f_n(x)] = \left[\lim_{x \to a} f_1(x)\right]\left[\lim_{x \to a} f_2(x)\right]\cdots\left[\lim_{x \to a} f_n(x)\right] \quad (2)$$

One consequence of (2) is that

$$\lim_{x \to a} x^n = \lim_{x \to a} \overbrace{(x \cdot x \cdots x)}^{n \text{ factors}} = \overbrace{(\lim_{x \to a} x)(\lim_{x \to a} x) \cdots (\lim_{x \to a} x)}^{n \text{ factors}} = \overbrace{a \cdot a \cdots a}^{n \text{ factors}} = a^n$$

so that

$$\lim_{x \to a} x^n = a^n \quad \text{for any positive integer } n \quad (3)$$

By means of the formulas obtained thus far, we can find the limit of *any* polynomial at *any* number. In fact, if $c_n, c_{n-1}, \ldots, c_1$, and c_0 are arbitrary real numbers and n is an arbitrary nonnegative integer, then by (1) through (3),

$$\lim_{x \to a} (c_n x^n + c_{n-1} x^{n-1} + \cdots + c_1 x + c_0)$$

$$= c_n \lim_{x \to a} x^n + c_{n-1} \lim_{x \to a} x^{n-1} + \cdots + c_1 \lim_{x \to a} x + \lim_{x \to a} c_0$$

$$= c_n a^n + c_{n-1} a^{n-1} + \cdots + c_1 a + c_0$$

It follows that to find the limit of any polynomial f at any point a, we only need to evaluate $f(a)$. In other words,

Limit of a Polynomial

$$\lim_{x \to a} f(x) = f(a)$$

(4)

For example,

$$\lim_{x \to 2} (4x^3 - 6x^2 - 9x) = 4(2)^3 - 6(2)^2 - 9(2) = 32 - 24 - 18 = -10$$

Since a rational function can always be expressed as the quotient of two polynomial functions f and g, (4) and the Quotient Rule together tell us that

Limit of a Rational Function

$$\lim_{x \to a} \frac{f(x)}{g(x)} = \frac{f(a)}{g(a)}$$

(5)

provided that $g(a) \neq 0$. Therefore to find the limit of any rational function at a point a in its domain, we need only evaluate the function at a. In particular,

$$\lim_{x \to -1} \frac{x^3 + 3x + 1}{x^2 - 3\sqrt{5}\, x} = \frac{(-1)^3 + 3(-1) + 1}{(-1)^2 - 3\sqrt{5}(-1)} = \frac{-3}{1 + 3\sqrt{5}}$$

We mention that (5) has a particularly simple form if $f(x) = 1$ and $g(x) = x^n$:

$$\lim_{x \to a} \frac{1}{x^n} = \frac{1}{a^n} \quad \text{for } a \neq 0$$

Although the Quotient Rule does not guarantee the existence of $\lim_{x \to a} f(x)/g(x)$ when $\lim_{x \to a} g(x) = 0$, sometimes it is still possible to evaluate such limits. That was the case with the limits that appeared in the solutions of Examples 3 and 4 in Section 2.2. The next example provides a slightly more complex illustration of this.

EXAMPLE 2   Find $\displaystyle\lim_{x \to -2} \frac{x^3 + 2x^2 - x - 2}{x^2 - 4}$.

Solution   Since $\lim_{x \to -2} (x^2 - 4) = 0$, we cannot apply the Quotient Rule to this function in its original form. However, since $x^3 + 2x^2 - x - 2 = (x + 2)(x^2 - 1)$ and $x^2 - 4 = (x + 2)(x - 2)$, we have

$$\lim_{x \to -2} \frac{x^3 + 2x^2 - x - 2}{x^2 - 4} = \lim_{x \to -2} \frac{(x + 2)(x^2 - 1)}{(x + 2)(x - 2)} = \lim_{x \to -2} \frac{x^2 - 1}{x - 2}$$

$$= \frac{(-2)^2 - 1}{-2 - 2} = -\frac{3}{4}$$

where the last limit to appear was evaluated by (5).   ❏

The limit rules can also help in showing that a limit does not exist. Before we show that the limit of $1/x$ does not exist as x approaches 0, notice from Figure 2.23 that $1/x$ becomes larger and larger in absolute value as x approaches 0, which suggests that $1/x$ does not approach any *number L* as x tends to 0.

EXAMPLE 3   Show that $\lim_{x \to 0} 1/x$ does not exist.

Solution   Suppose that $\lim_{x \to 0} 1/x$ exists, and let $L = \lim_{x \to 0} 1/x$. Since $1 = x(1/x)$, the Product Rule and Example 2 of Section 2.2 would then imply that

$$1 = \lim_{x \to 0} 1 = \lim_{x \to 0} \left(x \cdot \frac{1}{x} \right) = \left(\lim_{x \to 0} x \right)\left(\lim_{x \to 0} \frac{1}{x} \right) = 0 \cdot L = 0$$

which is obviously false. Therefore $\lim_{x \to 0} 1/x$ cannot exist (Figure 2.23).   ❏

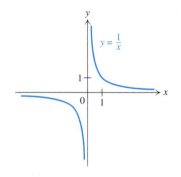

FIGURE 2.23

Some Prominent Limits

There are several limits that will appear frequently in the remainder of the book. First, if r is any fixed rational number, then

$$\lim_{x \to a} x^r = a^r \tag{6}$$

which is valid for any nonzero number a in the domain of x^r and is even valid for $a = 0$ if $r = m/n$, where m and n are positive integers, with n odd. This will be a consequence of results to be proved in Section 7.3, so until then we will assume it.

From (6) we know that

$$\lim_{x \to -32} x^{1/5} = (-32)^{1/5} = -2 \quad \text{and} \quad \lim_{x \to 9} x^{3/2} = 9^{3/2} = 27$$

Since $\sqrt[n]{x} = x^{1/n}$, it follows from (6) that

$$\lim_{x \to a} \sqrt[n]{x} = \sqrt[n]{a} \quad \begin{cases} \text{for all } a \text{ if } n \text{ is odd} \\ \text{for all } a > 0 \text{ if } n \text{ is even} \end{cases} \tag{7}$$

In particular,

$$\lim_{x \to a} \sqrt{x} = \sqrt{a} \quad \text{for } a > 0 \tag{8}$$

For example,

$$\lim_{x \to 1/4} \sqrt{x} = \sqrt{\frac{1}{4}} = \frac{1}{2}$$

EXAMPLE 4 Evaluate $\lim_{x \to 9} \dfrac{x(\sqrt{x} - 3)}{x - 9}$.

Solution Since $\lim_{x \to 9} (x - 9) = 0$, the Quotient Rule cannot be applied directly. However, if we factor the denominator, then we can cancel terms and then use (8) and the Quotient Rule:

$$\lim_{x \to 9} \frac{x(\sqrt{x} - 3)}{x - 9} = \lim_{x \to 9} \frac{x(\sqrt{x} - 3)}{x(\sqrt{x} - 3)(\sqrt{x} + 3)}$$

$$= \lim_{x \to 9} \frac{x}{\sqrt{x} + 3} = \frac{9}{\sqrt{9} + 3} = \frac{3}{2} \qquad \square$$

Next we mention that for any real number a,

$$\lim_{x \to a} e^x = e^a \tag{9}$$

This result will follow immediately from our discussion in Section 7.2. For now let us observe that if we just assume that

$$\lim_{h \to 0} e^h = e^0 = 1 \tag{10}$$

then (9) is valid for any real number a. Indeed, using the Law of Exponents, the Constant Multiple Rule, and (10), we obtain

$$\lim_{h \to 0} e^{a+h} = \lim_{h \to 0} (e^a e^h) = e^a \lim_{h \to 0} e^h \overset{(10)}{=} e^a \cdot 1 = e^a$$

By (17) in Section 2.2 this implies (9).

One can also show that

$$\lim_{x \to a} \ln x = \ln a \quad \text{for } a > 0 \tag{11}$$

a formula that we will assume. (See Exercise 32 in Section 2.5.)

Now we calculate two limits that involve the sine and cosine functions, and which will be used later.

Example 5 Show that $\lim_{x\to 0} \sin x = 0$ and $\lim_{x\to 0} \cos x = 1$.

Solution   Let ε be any positive number less than $\pi/2$. First we notice from Figure 2.24 that for $0 < x < \pi/2$:

$$0 < \underbrace{\text{length of segment } BC}_{\sin x} < \underbrace{\text{length of segment } AC}_{|AC|} < \underbrace{\text{length of arc } AC}_{x}$$

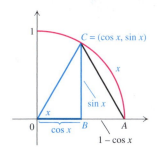

FIGURE 2.24

Therefore if $0 < x < \pi/2$, then $0 < \sin x < x$. Next we notice that if $-\pi/2 < x < 0$, then $0 < -x < \pi/2$, so that by the preceeding inequalities, $0 < \sin (-x) = -\sin x < -x$, and thus $x < \sin x < 0$. It follows that if $-\pi/2 < x < 0$ or if $0 < x < \pi/2$, then $| \sin x | < | x |$. Letting $\delta = \varepsilon$, we find that

$$\text{if}\quad 0 < | x - 0 | < \delta, \quad \text{then}\quad | \sin x - 0 | = | \sin x | < | x | < \delta = \varepsilon$$

Consequently, $\lim_{x\to 0} \sin x = 0$.

To prove that $\lim_{x\to 0} \cos x = 1$, we observe from Figure 2.24 that

$$| \cos x - 1 | = | AB | < | AC | < x \quad \text{if}\quad 0 < x < \pi/2$$

Since $\cos(-x) = \cos x$, it follows that

$$\text{if}\quad 0 < | x | < \pi/2, \quad \text{then}\quad | \cos x - 1 | < | x |$$

As with the sine function above, we let $\delta = \varepsilon$, and conclude that

$$\text{if}\quad 0 < | x - 0 | < \delta, \quad \text{then}\quad | \cos x - 1 | < | x | < \varepsilon$$

This means that $\lim_{x\to 0} \cos x = 1$, as we wish to show.   ❑

Our next goal is to show that $\lim_{x\to a} \sin x = \sin a$ and $\lim_{x\to a} \cos x = \cos a$ for any number a. Because

$$\lim_{x\to a} f\,(x) = L \qquad \text{if and only if}\qquad \lim_{h\to 0} f(a + h) = L$$

(see (17) in Section 2.2), we can instead show that

$$\lim_{h\to 0} \sin (a + h) = \sin a \qquad \text{and}\qquad \lim_{h\to 0} \cos (a + h) = \cos a$$

EXAMPLE 6   Show that for any number a,

$$\lim_{x\to a} \sin x = \sin a \quad \text{and}\quad \lim_{x\to a} \cos x = \cos a \tag{12}$$

Solution   Let a be a fixed number. To prove that $\lim_{h\to 0} \sin (a + h) = \sin a$ and hence that $\lim_{x\to a} \sin x = \sin a$, we use the trigonometric identity

$$\sin (a + h) = \sin a \cos h + \sin h \cos a \tag{13}$$

Since a is fixed, $\sin a$ and $\cos a$ are constants. Applying the Sum and Constant Multiple Rules to (13) and using Example 5, we find that

$$\lim_{h \to 0} \sin (a + h) = \lim_{h \to 0} (\sin a \cos h + \sin h \cos a)$$
$$= \sin a \lim_{h \to 0} \cos h + \cos a \lim_{h \to 0} \sin h$$
$$= (\sin a) \cdot 1 + (\cos a) \cdot 0$$
$$= \sin a$$

For the proof that $\lim_{h \to 0} \cos (a + h) = \cos a$ we use the trigonometric identity

$$\cos (a + h) = \cos a \cos h - \sin a \sin h$$

and in a similar way conclude that

$$\lim_{h \to 0} \cos (a + h) = \lim_{h \to 0} (\cos a \cos h - \sin a \sin h)$$
$$= \cos a \lim_{h \to 0} \cos h - \sin a \lim_{h \to 0} \sin h$$
$$= (\cos a) \cdot 1 - (\sin a) \cdot 0$$
$$= \cos a \quad \blacksquare$$

From Example 6 it follows that the sine and cosine functions have the property that the limit at any number a is the value of the function at a. The other trigonometric functions have the same property, as can be verified from (12) by using the limit rules. For example,

$$\lim_{x \to a} \tan x = \lim_{x \to a} \frac{\sin x}{\cos x} = \frac{\lim_{x \to a} \sin x}{\lim_{x \to a} \cos x} = \frac{\sin a}{\cos a} = \tan a \tag{14}$$

for any number a in the domain of the tangent function.

In summary, we can now find the limits of many kinds of functions.

polynomials   rational functions   basic trigonometric functions
e^x $\ln x$ x^n for any integer n

More precisely, if f represents any of these functions, then f has a limit at any point a in its domain, and the limit is $f(a)$. (The same is true for x^r, where r is any rational number, with some stipulations on a.) A function f that has the property that

$$\lim_{x \to a} f(x) = f(a)$$

is said to be **continuous** at a. Functions that are continuous at each number in their domains are of great importance in calculus, and will be the focus of Section 2.5.

The Squeezing Theorem

The limit rules presented thus far are effective for finding limits of many common functions. However, they don't apply in the evaluation of limits like

$$\lim_{x \to 0} \frac{\sin x}{x} \tag{15}$$

As we mentioned in Section 2.1, this limit yields the slope of the line tangent to the graph of the sine function at the point (0, 0) (Figure 2.5). In Section 2.1 we used a

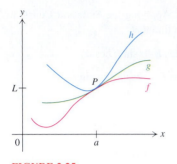

FIGURE 2.25

few calculations with values of x near 0 in order to conjecture that the limit in (15) is 1. Although the limit rules studied so far don't yield definitive information about the limit, and we cannot just evaluate the function $(\sin x)/x$ at 0 because it is not defined at 0, nevertheless we will be able to prove that the limit is 1 with help of the Squeezing Theorem.

In effect, the Squeezing Theorem says that if the graphs of f and h converge at a point P in the plane (Figure 2.25) and if the graph of g is "squeezed" between the graphs of f and h, then the graph of g converges with the graphs of f and h at P. This result, which is sometimes called the Pinching Theorem, will be proved in the Appendix.

THEOREM   2.3
Squeezing Theorem

Assume that $f(x) \le g(x) \le h(x)$ for all x in some open interval about a except possibly a itself. If $\lim_{x \to a} f(x) = \lim_{x \to a} h(x) = L$, then $\lim_{x \to a} g(x)$ exists and $\lim_{x \to a} g(x) = L$.

The Squeezing Theorem is tailor-made for the next examples.

EXAMPLE 7   Show that

$$\lim_{x \to 0} \frac{\sin x}{x} = 1 \tag{16}$$

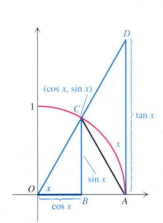

FIGURE 2.26

Solution   Using Figure 2.26, we obtain the following equations, which are valid for $0 < x < \pi/2$:

$$\text{area of triangle } OAC = \frac{1}{2}|OA|\,|BC| = \frac{1}{2} \cdot 1 \cdot \sin x = \frac{\sin x}{2}$$

$$\text{area of sector } OAC = \frac{x}{2\pi}\,(\text{area of circle}) = \frac{x}{2\pi}\,\pi = \frac{x}{2}$$

$$\text{area of triangle } OAD = \frac{1}{2}|OA|\,|AD| = \frac{1}{2} \cdot 1 \cdot \tan x = \frac{1}{2} \cdot \frac{\sin x}{\cos x}$$

It is geometrically clear that

$$\text{area of triangle } OAC \le \text{area of sector } OAC \le \text{area of triangle } OAD$$

so that

$$\frac{\sin x}{2} \le \frac{x}{2} \le \frac{1}{2} \cdot \frac{\sin x}{\cos x} \quad \text{for } 0 < x < \frac{\pi}{2}$$

Multiplying both sides of the first inequality by $2/x$ and multiplying both sides of the second inequality by $(2 \cos x)/x$, we obtain

$$\frac{\sin x}{x} \le 1 \quad \text{and} \quad \cos x \le \frac{\sin x}{x} \tag{17}$$

Combining the inequalities in (17) and using the fact that

$$\cos(-x) = \cos x \qquad \text{and} \qquad \frac{\sin(-x)}{-x} = \frac{\sin x}{x}$$

we obtain

$$\cos x \leq \frac{\sin x}{x} \leq 1 \qquad \text{for } 0 < |x| < \frac{\pi}{2}$$

Since $\lim_{x\to 0} \cos x = 1 = \lim_{x\to 0} 1$, it follows from the Squeezing Theorem that

$$\lim_{x\to 0} \frac{\sin x}{x} = 1 \qquad\qquad \Box$$

In the next example we present a companion to the limit appearing in Example 7.

EXAMPLE 8   Show that

$$\lim_{x\to 0} \frac{\cos x - 1}{x} = 0$$

Solution   Notice that $\lim_{x\to 0} x = 0$, so we cannot apply the Quotient Rule directly. However,

$$\lim_{x\to 0} \frac{\cos x - 1}{x} = \lim_{x\to 0} \left(\frac{\cos x - 1}{x}\right)\left(\frac{\cos x + 1}{\cos x + 1}\right) = \lim_{x\to 0} \frac{\cos^2 x - 1}{x(\cos x + 1)}$$

$$= \lim_{x\to 0} \frac{-\sin^2 x}{x(\cos x + 1)} = \lim_{x\to 0} \left(\frac{\sin x}{x}\right)\left(\frac{-\sin x}{\cos x + 1}\right)$$

Now, by Example 7,

$$\lim_{x\to 0} \frac{\sin x}{x} = 1$$

Furthermore

$$\lim_{x\to 0} \frac{-\sin x}{\cos x + 1} = \frac{0}{1 + 1} = 0$$

by the Sum and the Quotient Rules. Thus the Product Rule tells us that

$$\lim_{x\to 0} \frac{\cos x - 1}{x} = \lim_{x\to 0} \frac{\sin x}{x} \lim_{x\to 0} \frac{-\sin x}{\cos x + 1} = 1 \cdot 0 = 0 \qquad \Box$$

The results of Examples 7 and 8 can be used to evaluate other limits, as Example 9 illustrates.

EXAMPLE 9   Evaluate $\lim\limits_{x \to 0} \dfrac{\sin x}{x^{2/3}}$.

Solution   From Example 7 we know that

$$\lim_{x \to 0} \frac{\sin x}{x} = 1$$

To use this result, we rewrite $(\sin x)/x^{2/3}$:

$$\frac{\sin x}{x^{2/3}} = \left(\frac{\sin x}{x} \right) x^{1/3}$$

Since $\lim_{x \to 0} x^{1/3} = 0$ by (7), it follows from the Product Rule that

$$\lim_{x \to 0} \frac{\sin x}{x^{2/3}} = \lim_{x \to 0} \left[\left(\frac{\sin x}{x} \right) x^{1/3} \right] = \lim_{x \to 0} \frac{\sin x}{x} \lim_{x \to 0} x^{1/3} = 1 \cdot 0 = 0 \qquad \square$$

The Substitution Rule

The final limit rule in this section is normally called the Substitution Rule, and will allow us to evaluate limits of composite functions such as

$$\lim_{x \to 3} \sqrt{25 - x^2} \quad \text{and} \quad \lim_{x \to 0} \frac{\sin 2x}{x}$$

As you would expect, the Substitution Rule involves a substitution. Let us see informally how the rule works. Consider the limit

$$\lim_{x \to 3} \sqrt{25 - x^2}$$

First notice that $\sqrt{25 - x^2}$ is the composite $(g \circ f)(x) = (g(f(x))$, where $f(x) = 25 - x^2$ and $g(x) = \sqrt{x}$. First we substitute y for $f(x)$, that is, we let $y = 25 - x^2$. Next we find that if x approaches 3, then $25 - x^2$ approaches $25 - 9 = 16$, that is, y approaches 16. Therefore

$$\lim_{x \to 3} \sqrt{25 - x^2} \quad \text{is equivalent to} \quad \lim_{y \to 16} \sqrt{y}$$

By (8) we know that $\lim_{y \to 16} \sqrt{y} = \sqrt{16} = 4$. Consequently

$$\lim_{x \to 3} \sqrt{25 - x^2} = \lim_{y \to 16} \sqrt{y} = 4$$

In general, to evaluate $\lim_{x \to a} (g(f(x)))$, we may substitute y for $f(x)$ and apply the following result, which is precisely stated and proved in the Appendix:

> **Substitution Rule**
>
> $$\lim_{x \to a} g(f(x)) = \lim_{y \to c} g(y), \quad \text{where} \quad c = \lim_{x \to a} f(x)$$

For the Substitution Rule to be applicable with a given limit, it is necessary to assume that $\lim_{y \to c} g(y)$ exists. It is also necessary that $f(x) \neq c$ for all x in some open interval about a except possibly at $x = a$. This latter assumption, which is a technical detail, will be satisfied in all the examples we encounter.

When we use the Substitution Rule to find $\lim_{x \to a} g(f(x))$, we first substitute y for $f(x)$, then determine c by the formula $\lim_{x \to a} y = \lim_{x \to a} f(x) = c$, and finally compute $\lim_{y \to c} g(y)$. Frequently the process is straightforward.

EXAMPLE 10   Evaluate $\lim_{x \to -1} e^{(x^4)}$.

Solution   Let $y = x^4$, and notice that

$$\lim_{x \to -1} y = \lim_{x \to -1} x^4 = (-1)^4 = 1$$

By the Substitution Rule and (9),

$$\lim_{x \to -1} e^{(x^4)} = \lim_{y \to 1} e^y = e^1 = e \qquad \square$$

With experience you often will be able to apply the Substitution Rule mentally, without having to write down all the steps, as we did in Example 10. That will be especially true if f is continuous at a and g is continuous at $f(a)$. In that case, the Substitution Rule reduces to

$$\lim_{x \to a} g(f(x)) = g(f(a)) \tag{18}$$

In particular, (18) applies to the limit in Example 10, because x^4 is continuous at -1 and e^x is continuous at 1. Thus

$$\lim_{x \to -1} e^{(x^4)} = e^1 = e$$

More generally, since e^x is continuous at each number, one can show that if $\lim_{x \to a} f(x)$ exists, then

$$\lim_{x \to a} e^{f(x)} = e^{\lim_{x \to a} f(x)} \tag{19}$$

We will use this formula from time to time.

In contrast to Example 10, we cannot use (18) to evaluate the limit in the next example because we cannot substitute 0 for x in the expression $(\sin 2x)/x$.

EXAMPLE 11   Evaluate $\lim_{x \to 0} \dfrac{\sin 2x}{x}$.

Solution   Because of the appearance of $2x$ in the numerator, we substitute $y = 2x$ and notice that

$$\lim_{x \to 0} y = \lim_{x \to 0} 2x = 0$$

From the fact that $x = y/2$, the Substitution and Constant Multiple Rules, and (16), we conclude that

$$\lim_{x \to 0} \frac{\sin 2x}{x} = \lim_{y \to 0} \frac{\sin y}{y/2} = \lim_{y \to 0} 2\,\frac{\sin y}{y} = 2 \lim_{y \to 0} \frac{\sin y}{y} = 2 \cdot 1 = 2 \qquad \square$$

EXERCISES 2.3

In Exercises 1–24 use the results of this section to evaluate the limit.

1. $\lim_{x \to 4} (3x^2 - 5\sqrt{x} - 6|x|)$ **2.** $\lim_{x \to \sqrt{2}} (x^2 + 5)(\sqrt{2}x + 1)$

3. $\lim_{y \to 64} (\sqrt[3]{y} + \sqrt{y})^2$ **4.** $\lim_{t \to 0} \dfrac{2t^{1/3} - 4}{-3t^{1/3} + 5}$

5. $\lim_{x \to e} (\ln x)/x$ **6.** $\lim_{x \to -\pi/3} 3x^2 \cos x$

7. $\lim_{x \to 0} e^x \cos x$ **8.** $\lim_{x \to e} \dfrac{\ln x}{e^{-x}}$

9. $\lim_{y \to 2\pi/3} \dfrac{\pi \sin y \cos y}{y}$ **10.** $\lim_{y \to 0} \dfrac{\pi \sin y \cos y}{y}$

11. $\lim_{x \to \sqrt{5}} (9 - x^2)^{-5/2}$ **12.** $\lim_{t \to 3\pi/2} \sin\left(\dfrac{\pi}{2} \sin t\right)$

13. $\lim_{x \to \pi/6} e^{\sin x}$ **14.** $\lim_{x \to e} \dfrac{\ln (\ln x)}{\ln x}$

15. $\lim_{x \to 0} \ln\left(\dfrac{e^x - 1}{x}\right)$ **16.** $\lim_{x \to 0} \dfrac{e^{2x} - 1}{x}$

17. $\lim_{x \to 0} \dfrac{\cos 4x - 1}{x}$ **18.** $\lim_{t \to 0} \dfrac{\cos^2 t - 1}{t}$

19. $\lim_{t \to 0} \dfrac{\cos t - 1}{\sqrt[3]{t}}$ **20.** $\lim_{y \to 0} y \cot y$

21. $\lim_{x \to 0} \dfrac{\sin x}{\sin 2x}$ **22.** $\lim_{x \to 0} \dfrac{\sin^2 x}{1 - \cos x}$

23. $\lim_{x \to \pi} \dfrac{\tan^2 x}{1 + \sec x}$ **24.** $\lim_{x \to 0} \dfrac{1 - \cos 3x}{\sin 3x}$

In Exercises 25–36 reduce the expression and then evaluate the limit.

25. $\lim_{x \to -1} \dfrac{x^2 - 1}{x + 1}$ **26.** $\lim_{x \to 2} \dfrac{x^2 - 4}{x^3 - 8}$

27. $\lim_{x \to -2} \dfrac{x^4 - 16}{4 - x^2}$ **28.** $\lim_{x \to -4} \dfrac{x^2 - 16}{|x| - 4}$

29. $\lim_{x \to 3} \dfrac{x^2 - x - 6}{x^3 - 3x^2 + x - 3}$ **30.** $\lim_{x \to 100} \dfrac{x - 100}{\sqrt{x} - 10}$

31. $\lim_{y \to 2} \dfrac{\sqrt{y} - \sqrt{2}}{y^2 - 2y}$ **32.** $\lim_{y \to 1/2} \dfrac{6y - 3}{y(1 - 2y)}$

33. $\lim_{x \to -2} \left(\dfrac{x^2}{x + 2} - \dfrac{4}{x + 2}\right)$ **34.** $\lim_{x \to 0} \dfrac{1 + 1/x}{2 + 1/x}$

35. $\lim_{x \to 0} \dfrac{e^{2x} - 1}{e^x - 1}$ **36.** $\lim_{x \to 0} \dfrac{e^{3x} - 1}{e^x - 1}$

In Exercises 37–40 use the Squeezing Theorem to evaluate the limit.

37. $\lim_{x \to 0} x \cos \dfrac{1}{x^2}$

38. $\lim_{x \to 1} (x - 1)^4 \sin\left(\tan \dfrac{\pi x}{2}\right)$

39. $\lim_{x \to 1} (\ln x) \sin \dfrac{1}{x - 1} \cos x$

40. $\lim_{x \to 1} \dfrac{(\ln x)^2}{1 + e^{1/(x-1)}}$

41. Suppose that $x^4 \le f(x) \le 2x^4$ for all x in $[-1, 1]$. Find $\lim_{x \to 0} f(x)/x^2$.

42. a. Draw the graph of a function f that satisfies the condition $1 + x < f(x) < 1 + x + x^2$ for $0 < |x| < 1$.
   b. Find $\lim_{x \to 0} f(x)$.

In Exercises 43–50 find an equation of the line l tangent to the graph of f at the given point.

43. $f(x) = x^6 - 1$; $(1, 0)$ **44.** $f(x) = \dfrac{1}{x}$; $\left(2, \dfrac{1}{2}\right)$

45. $f(x) = \dfrac{1}{x + 3}$; $\left(-1, \dfrac{1}{2}\right)$ **46.** $f(x) = \sqrt{x}$; $(16, 4)$

47. $f(x) = \cos x$; $(0, 1)$ **48.** $f(x) = \sin 4x$; $(0, 0)$

49. $f(x) = \sqrt{4 - x^2}$; $(1, \sqrt{3})$

50. $f(x) = \sqrt{x + x^2}$; $(1, \sqrt{2})$

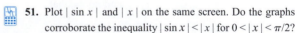   **51.** Plot $|\sin x|$ and $|x|$ on the same screen. Do the graphs corroborate the inequality $|\sin x| < |x|$ for $0 < |x| < \pi/2$?

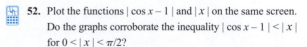   **52.** Plot the functions $|\cos x - 1|$ and $|x|$ on the same screen. Do the graphs corroborate the inequality $|\cos x - 1| < |x|$ for $0 < |x| < \pi/2$?

In Exercises 53–56 determine whether f is continuous at a.

53. $f(x) = x^2 - 4x + 3$; $a = 2$

54. $f(x) = \sqrt{x}\,(x^2 + 4)$; $a = 4$

55. $f(x) = \dfrac{x^4 + x^2 - 2}{x^2 - 1}$; $a = 0$

56. $f(x) = \dfrac{1}{2x - 5x^3}$; $a = -1$

57. Figure 2.27 shows the graphs of f and g, each of which has a limit at 2.

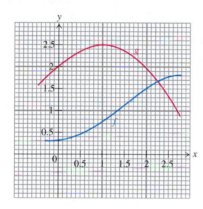

FIGURE 2.27   Figure for Exercise 57

Determine the following limits:

a. $\displaystyle \lim_{x \to 2} [f(x) + g(x)]$

b. $\displaystyle \lim_{x \to 2} [f(x)g(x)]$

c. $\displaystyle \lim_{x \to 2} (x - 4)g(x)$

58. Let $f(x) = x^2 - 2x$. Show that the line $y = 8x - 25$ is tangent to the graph of f. What is the point of tangency?

59. Verify that the following pairs of functions have a common tangent line at the indicated point.

a. $f(x) = x^2$ and $g(x) = x^3$ at $(0, 0)$

b. $f(x) = x^2 + 1$ and $g(x) = -x^2 + 1$ at $(0, 1)$

60. Let $f(x) = 1/x$.

a. For $a \neq 0$, prove that an equation of the line tangent to the graph of f at $(a, 1/a)$ is

$$y - \frac{1}{a} = \frac{-1}{a^2}(x - a)$$

b. Show that the area A of the triangle formed by the

coordinate axes and the tangent line in part (a) is independent of the number a.

61. Suppose $\lim_{x \to -1} f(x) = 4$ and $\lim_{x \to -1} [f(x) - g(x)] = 6$. Show that $\lim_{x \to -1} g(x)$ exists and find $\lim_{x \to -1} g(x)$.

62. Suppose $\lim_{x \to \sqrt{2}} f(x) = 3$ and $\lim_{x \to \sqrt{2}} (fg)(x) = -\sqrt{2}$. Show that $\lim_{x \to \sqrt{2}} g(x)$ exists and find $\lim_{x \to \sqrt{2}} g(x)$.

63. Let $f(x) = 1/x$ and $g(x) = -1/x$. Show that $\lim_{x \to 0} [f(x) + g(x)]$ exists although neither $\lim_{x \to 0} f(x)$ nor $\lim_{x \to 0} g(x)$ exists. (This example shows that the limit of a sum $f + g$ can exist even though it cannot be calculated by applying the Sum Rule to $f + g$.)

64. Suppose $\lim_{x \to a} f(x)$ exists and $\lim_{x \to a} [f(x) + g(x)]$ does not exist. Prove that $\lim_{x \to a} g(x)$ does not exist.

65. a. Give an example of a function f defined on $[-1, 1]$ for which $\lim_{x \to 0} x f(x)$ does not exist.

   b. Give an example of a function f defined on $[-1, 1]$ for which $\lim_{x \to 0} x f(x) = 100$.

  *c. Give a general condition on a function f under which $\lim_{x \to 0} x f(x) = 0$.

66. Suppose that $\lim_{x \to a} f(x) = 0$ and $|g(x)| \leq M$ for a fixed M and for all $x \neq a$. Prove that $\lim_{x \to a} f(x)g(x) = 0$. (*Hint:* Notice that $-M|f(x)| \leq f(x)g(x) \leq M|f(x)|$.)

67. Using Exercise 66, evaluate

$$\lim_{x \to 0} \frac{x \sin^2 (1/x)}{1 + \sin^2(1/x)}$$

68. a. Suppose that $\lim_{x \to a} g(x) = 0$. Show that in order for $\lim_{x \to a} f(x)/g(x)$ to exist, it must be true that $\lim_{x \to a} f(x) = 0$. (*Hint:* Note that $f = (f/g)g$.)

   b. Can $\lim_{x \to a} f(x) = 0$ and $\lim_{x \to a} g(x) = 0$, and yet $\lim_{x \to a} f(x)/g(x)$ fail to exist? Explain why it could not happen, or give an example to show that it can happen.

   c. Use part (a) to prove that the following limits do not exist.

   (i) $\displaystyle \lim_{x \to 0} \frac{x^2 + 1}{x(x + 2)}$ **(ii)** $\displaystyle \lim_{x \to 1} \frac{\sqrt{x}}{x - 1}$

   (iii) $\displaystyle \lim_{x \to 0} \frac{\cos x}{x}$

69. Suppose $\lim_{x \to a} f(x) = L$ and $\lim_{x \to a} g(x) = M$. Determine what relationship (if any) must exist between L and M if

a. $f(x) \leq g(x)$ for all $x \neq a$

b. $f(x) < g(x)$ for all $x \neq a$

70. Suppose that $f(x) \geq 0$ for all x in the domain of f and $\lim_{x \to a} [f(x)]^2 = L > 0$. Prove that $\lim_{x \to a} f(x) = \sqrt{L}$. (*Hint:* Use the Substitution Rule.)

71. Using the Substitution Rule, prove that if $\lim_{x \to 0} f(x) = L$, then $\lim_{x \to 0} f(x^n) = L$ for any positive integer n.

72. Using Exercise 71, show that $\lim_{x \to 0} \sin x^n = 0$ for any positive integer n.

73. a. Find the error in the following argument:

$$\lim_{x \to 0} \left[x \left(\frac{1}{x^2 + x} \right) \right] \overset{?}{=} \lim_{x \to 0} x \lim_{x \to 0} \frac{1}{x^2 + x}$$

$$\overset{?}{=} 0 \lim_{x \to 0} \frac{1}{x^2 + x} \overset{?}{=} 0$$

b. Find $\lim_{x \to 0} \left[x \left(\dfrac{1}{x^2 + x} \right) \right]$.

74. A bee's cell in a hive is a regular hexagonal prism open at the front, with a trihedral vertex at the back (the right end in Figure 2.28). It can be shown that the surface area of a cell with vertex angle θ is given by

$$S(\theta) = 6ab + \frac{3}{2} b^2 \left(-\cot \theta + \frac{\sqrt{3}}{\sin \theta} \right) \quad \text{for } 0 < \theta < \frac{\pi}{2}$$

where a and b are positive constants. Show that for any θ in $(0, \pi/2)$, S is continuous at θ.

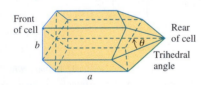

FIGURE 2.28   Figure for Exercise 74.

75. The kinetic energy of a body with mass m is given by

$$K(t) = \frac{1}{2} m [v(t)]^2$$

where $v(t)$ is the velocity of the body at time t. Suppose $v(t) = 50/(1 + t^2)$ for all $t > 0$. Show for any positive t that K is continuous at t.

Project

1. Consider the ellipse

$$\frac{x^2}{p^2} + \frac{y^2}{q^2} = 1, \text{ where } p > 0 \text{ and } q > 0$$

and let $0 < a < p$ (Figure 2.29).

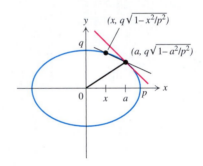

FIGURE 2.29   Figure for Project 1.

a. Show that the point $(a, q\sqrt{1 - a^2/p^2})$ is on the ellipse.

b. For x sufficiently close to a, find the slope of the secant line through $(a, q\sqrt{1 - a^2/p^2})$ and $(x, q\sqrt{1 - x^2/p^2})$.

c. Use the information of (b) to find the slope m_a of the line tangent to the ellipse at $(a, q\sqrt{1 - a^2/p^2})$.

d. Find the slope M_a of the radial line joining $(0, 0)$ to the point $(a, q\sqrt{1 - a^2/p^2})$, and then the slope L_a of the line perpendicular to the radial line.

e. Show that $L_a = m_a$ if and only if $p = q$, that is, if and only if the ellipse is a circle.

2.4   ONE-SIDED AND INFINITE LIMITS

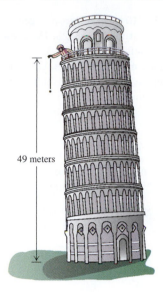

In Example 4 of Section 2.2 we considered the velocity of an iron ball dropped by Galileo from the Leaning Tower of Pisa (Figure 2.30). Until the ball hit the ground its position $h(t)$ at time t was given by

$$h(t) = -4.9t^2 + 49 = -4.9\,(t^2 - 10)$$

Now we would like to find the velocity of the ball as it struck the ground. Since $h(\sqrt{10}) = 0$, the ball hit the ground $\sqrt{10}$ seconds after it was dropped. After that the ball remained on the ground, so obviously the position of the ball for $t > \sqrt{10}$ will not help us calculate its velocity on impact. However, for $t < \sqrt{10}$ the average velocity of the ball during the time interval $[t, \sqrt{10}]$ was

$$\frac{h(\sqrt{10}) - h(t)}{\sqrt{10} - t}, \qquad \text{which equals} \qquad \frac{h(t) - h(\sqrt{10})}{t - \sqrt{10}} \qquad (1)$$

49 meters

If we keep $t < \sqrt{10}$ and let t approach $\sqrt{10}$, the average velocity in (1) should approach the velocity of the ball as it struck the ground. Thus the velocity should be a kind of "half limit," or "left-hand limit" (because t is required to be less than and hence to the left of $\sqrt{10}$ as it approaches $\sqrt{10}$). In defining left-hand limits, we will simply replace the inequality $0 < |x - a| < \delta$ in Definition 2.1 by the inequality $a - \delta < x < a$, which requires x to be to the left of a.

FIGURE 2.30

DEFINITION   2.4

Let f be defined on some open interval (c, a). A number L is the **limit of $f(x)$ as x approaches a from the left** (or the **left-hand limit of f at a**) if for every $\varepsilon > 0$ there is a number $\delta > 0$ such that

$$\text{if } a - \delta < x < a, \quad \text{then } |f(x) - L| < \varepsilon$$

In this case we write

$$\lim_{x \to a^-} f(x) = L$$

and say that the **left-hand limit of f at a exists** (Figure 2.31).

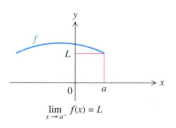

$$\lim_{x \to a^-} f(x) = L$$

FIGURE 2.31

Right-hand limits are treated in a completely analogous way. Suppose that f is defined on some interval (a, c) to the right of a. A number L is the **limit of $f(x)$ as x approaches a from the right** (or the **right-hand limit of f at a**) if for every $\varepsilon > 0$ there is a number $\delta > 0$ such that

$$\text{if } a < x < a + \delta \qquad \text{then } |f(x) - L| < \varepsilon$$

In this case we write

$$\lim_{x \to a^+} f(x) = L$$

and say that the **right-hand limit of f at a exists** (Figure 2.32).

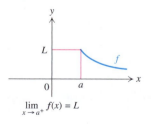

$$\lim_{x \to a^+} f(x) = L$$

FIGURE 2.32

It follows from the preceding definitions that for the height function h described at the beginning of the section,

$$\lim_{t \to \sqrt{10}^-} \frac{h(t) - h(\sqrt{10})}{t - \sqrt{10}} = \lim_{t \to \sqrt{10}^-} \frac{-4.9(t^2 - 10)}{t - \sqrt{10}}$$

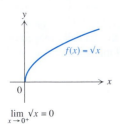

$f(x) = \sqrt{x}$

$$\lim_{x \to 0^+} \sqrt{x} = 0$$

FIGURE 2.33

$$= \lim_{t \to \sqrt{10}^-} -4.9(t + \sqrt{10}) = -9.8\sqrt{10}$$

so that as the ball hit the ground its velocity was $-9.8\sqrt{10} \approx -31$ meters per second.

One can also show that $\lim_{x \to 0^+} \sqrt{x} = 0$, but that neither $\lim_{x \to 0^-} \sqrt{x}$ nor $\lim_{x \to 0} \sqrt{x}$ exists (because $\sqrt{x}$ is not defined to the left of 0 (Figure 2.33)).

Right-hand limits and left-hand limits are called **one-sided limits.** Ordinary limits are called **two-sided limits.** All the rules given in Section 2.3 for finding limits of combinations of functions remain valid for one-sided limits.

EXAMPLE 1   Find $\lim_{x \to 0^+} \dfrac{x^{3/2} - 3x + 2}{x - 2\sqrt{x} + 1}$.

Solution   Since $x^{3/2} = (\sqrt{x})^3$ and $\lim_{x \to 0^+} \sqrt{x} = 0$, it follows from the version of the Product Rule for one-sided limits that $\lim_{x \to 0^+} x^{3/2} = 0$. Consequently

$$\lim_{x \to 0^+} \frac{x^{3/2} - 3x + 2}{x - 2\sqrt{x} + 1} = \frac{0 - 0 + 2}{0 - 0 + 1} = 2 \qquad \square$$

Care must be used in applying the Substitution Rule to one-sided limits.

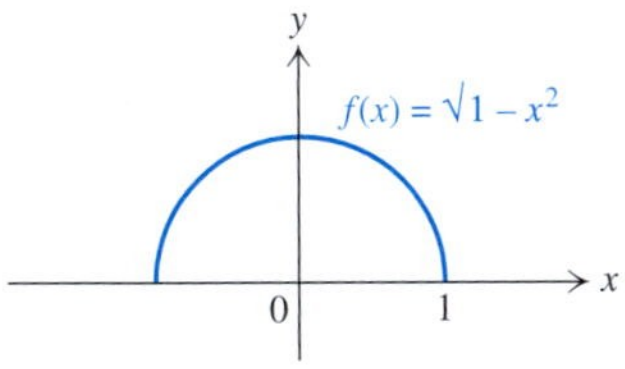

$f(x) = \sqrt{1 - x^2}$

FIGURE 2.34

EXAMPLE 2   Find $\lim_{x \to 1^-} \sqrt{1 - x^2}$.

Solution   Let y be the expression $1 - x^2$. Since $1 - x^2$ approaches 0 from the right as x approaches 1 from the left, it follows that y approaches 0 from the right as x approaches 1 from the left, so we have

$$\lim_{x \to 1^-} \sqrt{1 - x^2} = \lim_{y \to 0^+} \sqrt{y} = 0$$

(Figure 2.34). $\square$

Although $\lim_{x \to 1^-} \sqrt{1 - x^2}$ exists by Example 2, $\lim_{x \to 1^+} \sqrt{1 - x^2}$ does not exist because $1 - x^2$ is negative whenever x lies to the right of 1, and the square root of a negative number is not defined (as a real number).

The next theorem, which is proved in the Appendix, reveals the relationship between the existence of a two-sided limit at a and the one-sided limits at a.

THEOREM 2.5   Let f be defined on an open interval about a, except possibly at a itself. Then $\lim_{x \to a} f(x)$ exists if and only if both one-sided limits, $\lim_{x \to a^+} f(x)$ and $\lim_{x \to a^-} f(x)$, exist and

$$\lim_{x \to a^+} f(x) = \lim_{x \to a^-} f(x)$$

In that case,

$$\lim_{x \to a} f(x) = \lim_{x \to a^+} f(x) = \lim_{x \to a^-} f(x)$$

When a function is defined by a formula involving more than one equation, one-sided limits are often useful in showing that the function has a limit at a given point.

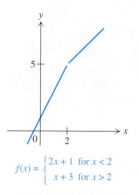

$$f(x) = \begin{cases} 2x+1 & \text{for } x < 2 \\ x+3 & \text{for } x > 2 \end{cases}$$

FIGURE 2.35

EXAMPLE 3   Let

$$f(x) = \begin{cases} 2x+1 & \text{for } x < 2 \\ x+3 & \text{for } x > 2 \end{cases}$$

Find $\lim_{x \to 2} f(x)$.

Solution   Since $\lim_{x \to 2} (2x+1) = 5$ and $\lim_{x \to 2} (x+3) = 5$, Theorem 2.5 assures us that

$$\lim_{x \to 2^-} f(x) = \lim_{x \to 2^-} (2x+1) = 5 \quad \text{and} \quad \lim_{x \to 2^+} f(x) = \lim_{x \to 2^+} (x+3) = 5$$

Since both one-sided limits exist and are equal, a second application of Theorem 2.5 now yields

$$\lim_{x \to 2} f(x) = 5$$

The function is graphed in Figure 2.35.   ❑

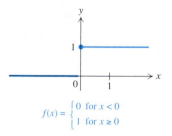

$$f(x) = \begin{cases} 0 & \text{for } x < 0 \\ 1 & \text{for } x \geq 0 \end{cases}$$

FIGURE 2.36   Diving board function.

EXAMPLE 4   Let f be the diving board function, defined by

$$f(x) = \begin{cases} 0 & \text{for } x < 0 \\ 1 & \text{for } x \geq 0 \end{cases}$$

Find the left-hand and right-hand limits of f at 0. Show that f has no two-sided limit at 0.

Solution   Since $\lim_{x \to 0^+} 1 = 1$, we know that $\lim_{x \to 0^+} f(x) = 1$ (see Figure 2.36). Similarly, since $\lim_{x \to 0^-} 0 = 0$, we know that $\lim_{x \to 0^-} f(x) = 0$. Thus both one-sided limits exist, and they are unequal. Therefore f has no two-sided limit at 0 by Theorem 2.5.   ❑

Infinite Limits and Vertical Asymptotes

If $\lim_{x \to a^+} f(x)$ does not exist, it may happen that as x approaches a from the right, the value of $f(x)$ becomes indefinitely large or becomes negative and indefinitely large in absolute value (Figure 2.37(a) and (b)). The value of $f(x)$ may behave similarly when the left-hand limit at a does not exist (Figure 2.37(c) and (d)). We introduce a special terminology to cover such cases.

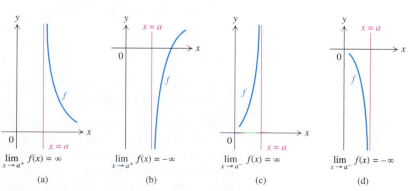

$$\lim_{x \to a^+} f(x) = \infty \qquad \lim_{x \to a^+} f(x) = -\infty \qquad \lim_{x \to a^-} f(x) = \infty \qquad \lim_{x \to a^-} f(x) = -\infty$$

(a)   (b)   (c)   (d)

FIGURE 2.37

DEFINITION 2.6

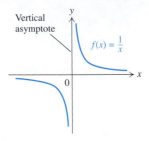

FIGURE 2.38

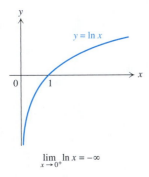

$$\lim_{x \to 0^+} \ln x = -\infty$$

FIGURE 2.39

Let f be defined on some open interval (a, c).

a. If for every number N there is some $\delta > 0$ such that

$$\text{if } a < x < a + \delta, \quad \text{then } f(x) > N$$

then we say that the **limit of $f(x)$ as x approaches a from the right** is ∞. In that case we write $\lim_{x \to a^+} f(x) = \infty$ (Figure 2.37(a)).

b. If for every number N there is some $\delta > 0$ such that

$$\text{if } a < x < a + \delta, \quad \text{then } f(x) < N$$

then we say that the **limit of $f(x)$ as x approaches a from the right is $-\infty$.** In that case we write $\lim_{x \to a^+} f(x) = -\infty$ (Figure 2.37(b)).

c. In either case (a) or (b) the vertical line $x = a$ is called a **vertical asymptote of the graph of f,** and we say that f has an **infinite right-hand limit at a.**

There are analogous definitions for the limits

$$\lim_{x \to a^-} f(x) = \infty \quad \text{and} \quad \lim_{x \to a^-} f(x) = -\infty$$

Definition 2.6 implies that $\lim_{x \to 0^+} 1/x = \infty$ and that $\lim_{x \to 0^-} 1/x = -\infty$ (Figure 2.38). This means, in particular, that the line $x = 0$ is a vertical asymptote of the graph of $1/x$. Similarly, if a is any number, then

$$\lim_{x \to a^+} \frac{1}{x - a} = \infty \quad \text{and} \quad \lim_{x \to a^-} \frac{1}{x - a} = -\infty$$

One can also show that

$$\lim_{x \to 0^+} \ln x = -\infty$$

(Figure 2.39).

Now suppose that

$$\lim_{x \to a^+} f(x) = \lim_{x \to a^-} f(x) = \infty$$

Then we write $\lim_{x \to a} f(x) = \infty$ and say that the **limit of $f(x)$ as x approaches a is** ∞ and that **f has an infinite limit at a** (Figure 2.40(a)). Similar comments hold if ∞ is replaced by $-\infty$ (Figure 2.40(b)). One can show that if $f(x) = 1/x^2$, then $\lim_{x \to 0} f(x) = \infty$ (Figure 2.40(c)), so that $1/x^2$ has an infinite (two-sided) limit at 0. By contrast, if $g(x) = 1/x$, then because $\lim_{x \to 0^-} 1/x = -\infty$ and $\lim_{x \to 0^+} 1/x = \infty$, the two one-sided limits are distinct, so that g does not have an infinite (two-sided) limit at 0.

Caution: If f has an infinite limit at a, then f does not have a limit at a in the sense of Definition 2.1.

The line $x = a$ is a vertical asymptote whenever $\lim_{x \to a^+} f(x) = \infty$ or $-\infty$, and whenever $\lim_{x \to a^-} f(x) = \infty$ or $-\infty$. Pictorially, if $x = a$ is a vertical asymptote of the graph of f, then the graph of f is nearly vertical near the line $x = a$, the points $(x, f(x))$ on the graph of f getting arbitrarily far from the x axis and approaching the line $x = a$ as x approaches a from one or possibly both sides (see Figures 2.37 and 2.40(a) and (b)).

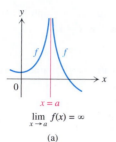

$$\lim_{x \to a} f(x) = \infty$$

(a)

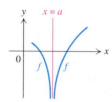

$$\lim_{x \to a} f(x) = -\infty$$

(b)

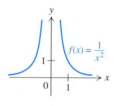

$$\lim_{x \to 0} \frac{1}{x^2} = \infty$$

FIGURE 2.40

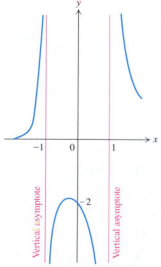

Vertical asymptote   Vertical asymptote

FIGURE 2.41

EXAMPLE 5   Let

$$f(x) = \frac{x + 2}{x^2 - 1}$$

Find all vertical asymptotes of the graph of f.

Solution   If a is any number other than 1 or -1, then $a^2 - 1 \neq 0$, so that

$$\lim_{x \to a} \frac{x + 2}{x^2 - 1} = \frac{a + 2}{a^2 - 1}$$

which is a number. Hence the only possible vertical asymptotes are the lines $x = 1$ and $x = -1$. Since

$$\frac{x + 2}{x^2 - 1} = \frac{x + 2}{x + 1} \cdot \frac{1}{x - 1}$$

it follows that

$$\lim_{x \to 1^+} \frac{x + 2}{x^2 - 1} = \lim_{x \to 1^+} \left(\frac{x + 2}{x + 1} \cdot \frac{1}{x - 1} \right) = \infty$$

and

$$\lim_{x \to -1^-} \frac{x + 2}{x^2 - 1} = \lim_{x \to -1^-} \left(\frac{x + 2}{x - 1} \cdot \frac{1}{x + 1} \right) = \infty$$

Thus $x = 1$ and $x = -1$ are indeed vertical asymptotes of the graph of f. Alternatively we could show that $\lim_{x \to 1^-} f(x) = -\infty$ and $\lim_{x \to -1^+} f(x) = -\infty$, which imply that $x = 1$ and $x = -1$ are vertical asymptotes. The graph of f appears in Figure 2.41.   ❑

In general, if f is any rational function with $f(x) = p(x)/q(x)$ (where p and q are polynomials), then the line $x = a$ is a vertical asymptote of the graph of f if $p(a) \neq 0$ and $q(a) = 0$. For instance, if

$$f(x) = \frac{(x - 2)(x + 3)}{x(x - \pi)(x - 4)(x + 1)^2}$$

then the graph of f has vertical asymptotes $x = 0$, $x = \pi$, $x = 4$, and $x = -1$.

Caution: It is not always possible to determine the vertical asymptotes of the graph of a rational function by merely finding out where the denominator vanishes. For example, if

$$f(x) = \frac{x + 1}{x^2 - 1}$$

then the denominator equals 0 for $x = -1$ as well as $x = 1$. However,

$$\frac{x + 1}{x^2 - 1} = \frac{x + 1}{(x + 1)(x - 1)} = \frac{1}{x - 1}$$

so $\lim_{x \to -1} f(x) = -1/2$. Consequently $x = -1$ is *not* a vertical asymptote of the graph of f, and the only vertical asymptote is the line $x = 1$.

Vertical Tangent Lines

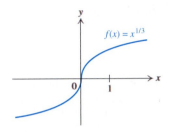

$f(x) = x^{1/3}$

FIGURE 2.42   Vertical tangent line at $(0, 0)$.

Let $f(x) = x^{1/3}$. From Figure 2.42 it would appear that the vertical line $x = 0$ is tangent to the graph of f at $(0, 0)$. However, since

$$\frac{f(x) - f(0)}{x - 0} = \frac{x^{1/3} - 0}{x - 0} = \frac{1}{x^{2/3}} \tag{2}$$

and $1/x^{2/3}$ becomes arbitrarily large as x tends to 0, it follows that

$$\lim_{x \to 0} \frac{f(x) - f(0)}{x - 0}$$

does not exist (in the sense of Definition 2.1), so that there is no line tangent to the graph of f according to the definition of tangent line given in Section 2.2. Nevertheless, the ratio in (2) has a two-sided infinite limit at 0:

$$\lim_{x \to 0} \frac{f(x) - f(0)}{x - 0} = \lim_{x \to 0} \frac{1}{x^{2/3}} = \infty$$

This leads us to make the following definition.

DEFINITION 2.7   Suppose f is continuous at a. If

$$\lim_{x \to a} \frac{f(x) - f(a)}{x - a} = \infty \quad \text{or} \quad \lim_{x \to a} \frac{f(x) - f(a)}{x - a} = -\infty$$

then we say that the graph of f has a **vertical tangent line at** $(a, f(a))$. In that case the vertical line $x = a$ is called the **line tangent to the graph of f at a.**

By Definition 2.7 the graph of the function $x^{1/3}$ has a vertical tangent at $(0, 0)$. The same is true of $x^{1/n}$, for any positive odd integer n (see Exercise 67).

We summarize the possible behavior of $f(x)$ as x approaches a number a.

1. $\lim_{x \to a} f(x)$ exists. In this case it is a unique number.

2. $\lim_{x \to a} f(x)$ does not exist, but it can be assigned either $-\infty$ or ∞. In this case it is called an infinite limit.

3. $\lim_{x \to a} f(x)$ has no meaning.

EXERCISES 2.4

In Exercises 1–8 determine the one–sided limit.

1. $\lim\limits_{x \to -2^+} (x^3 + 3x - 5)$

2. $\lim\limits_{x \to \pi/3^-} \cos(x + \pi/6)$

3. $\lim\limits_{x \to 2^-} \dfrac{x^2 - 4}{x - 2}$

4. $\lim\limits_{x \to 2^+} \dfrac{x^2 - 4}{x - 2}$

5. $\lim\limits_{x \to 1^+} \dfrac{x^2 + 3x - 4}{x^2 - 1}$

6. $\lim\limits_{x \to -2^-} \dfrac{x^2 - 3x - 10}{x^2 - 9}$

7. $\lim\limits_{t \to 5^+} \dfrac{|t - 5|}{5 - t}$

8. $\lim\limits_{t \to -3^+} \sqrt{t + 3}$

In Exercises 9–20 determine the infinite limit.

9. $\lim\limits_{x \to 0^-} 4/x^3$

10. $\lim\limits_{x \to 0^+} 4/x^3$

11. $\lim\limits_{x \to 0} -1/x^2$

12. $\lim\limits_{x \to 0^+} \dfrac{2}{x^{1/4}}$

13. $\lim\limits_{y \to -1^-} \dfrac{\pi}{y + 1}$

14. $\lim\limits_{z \to 3} \dfrac{2}{(z - 3)^2}$

15. $\lim\limits_{z \to \pi/2^-} \tan z$

16. $\lim\limits_{z \to -\pi/2^+} \sec z$

17. $\lim_{x \to 1^-} \ln (1 - |x|)$

18. $\lim_{x \to 1^+} \dfrac{\ln (x - 1)}{\ln (x^2 + 1)}$

19. $\lim_{x \to 0^+} \dfrac{1}{1 - e^{\sqrt{x}}}$

20. $\lim_{x \to 0^-} \ln (\sec x - 1)$

In Exercises 21–46 decide which of the given one-sided or two-sided limits exist as numbers, which as ∞, which as $-\infty$, and which do not exist. Where the limit is a number, evaluate it.

21. $\lim_{x \to 5^+} \dfrac{1}{x\sqrt{x - 5}}$

22. $\lim_{x \to 1^+} (\sqrt{x^2 - x} + x)$

23. $\lim_{x \to 0^+} \sqrt{x} \cos 2x$

24. $\lim_{x \to 0^-} \sqrt{-x^2 \sin 3x}$

25. $\lim_{x \to 0} \sqrt{\dfrac{1}{x^2}}$

26. $\lim_{x \to 0} \dfrac{x}{|x|}$

27. $\lim_{x \to -1/2^-} \dfrac{4x - 7}{x + \frac{1}{2}}$

28. $\lim_{x \to -1^+} \dfrac{x^2 + 5x + 4}{x + 1}$

29. $\lim_{x \to 2^-} \dfrac{x + 1}{2(x^2 - 4)}$

30. $\lim_{x \to 1^+} \dfrac{x^2 - 3x + 2}{x^2 - 2x + 1}$

31. $\lim_{y \to 3^-} \dfrac{-1}{\sqrt{3 - y}}$

32. $\lim_{y \to -3^+} \dfrac{\sqrt{9 - y^2}}{y + 3}$

33. $\lim_{y \to 0^-} (5 + \sqrt{1 + y^2})/\sqrt{y}$

34. $\lim_{t \to \pi/2} \sec^2 t$

*35. $\lim_{t \to 0} \dfrac{1 - \cos t}{t^2}$

36. $\lim_{x \to 0^-} \dfrac{\sqrt{1 + x} - \sqrt{1 - x}}{x}$

37. $\lim_{x \to 3^-} \left(\dfrac{1}{x - 3} - \dfrac{6}{x^2 - 9} \right)$

38. $\lim_{h \to 0^+} \left(\dfrac{1}{h} - \dfrac{1}{\sqrt{h}} \right)$

39. $\lim_{x \to 0^+} e^{2 + \ln x}$

40. $\lim_{x \to 0^+} \sqrt{e^x - 1}$

41. $\lim_{x \to 0^-} \sqrt{e^x - 1}$

42. $\lim_{x \to 1^-} \ln (\ln x)$

43. $\lim_{x \to 1^+} \ln (\ln x)$

44. $\lim_{x \to 0^+} \dfrac{1}{(\sin x) \ln (1 - x)}$

45. $\lim_{x \to 0} f(x)$, where $f(x) = \begin{cases} 2x - 4 \text{ for } x < 0 \\ -(x + 2)^2 \text{ for } x \geq 0 \end{cases}$

46. $\lim_{x \to -2} f(x)$, where $f(x) = \begin{cases} -1 + 4x \text{ for } x < -2 \\ -9 \text{ for } x > -2 \end{cases}$

In Exercises 47–58 find all vertical asymptotes (if any) of the graph of f.

47. $f(x) = \dfrac{x + 2}{x - 3}$

48. $f(x) = \dfrac{x^2 - 1}{x^2 - 4}$

49. $f(x) = \dfrac{\sin x}{x(x^2 - 1)}$

50. $f(x) = \dfrac{x^2 - 4x - 12}{x^2 - x - 6}$

51. $f(x) = \dfrac{x^2 + 2x - 15}{x^3 + 7x^2 + 10x}$

52. $f(x) = \dfrac{x + 1}{|x - 1|}$

53. $f(x) = \dfrac{x + 1/x}{x^4 + 1}$

54. $f(x) = \sqrt{\dfrac{x}{x - 1}}$

55. $f(x) = \dfrac{\sin x}{x}$

56. $f(x) = \dfrac{\cos x}{x}$

57. $f(x) = \tan x$

58. $f(x) = \csc x$

In Exercises 59–61 show that f has a vertical tangent line at the given point. Find an equation of the tangent line l.

59. $f(x) = x^{1/5}$; $(0, 0)$

60. $f(x) = (x + 1)^{1/3}$; $(-1, 0)$

61. $f(x) = 1 - 5x^{3/5}$; $(0, 1)$

62. For each of the six graphs shown in Figure 2.43, decide whether the two-sided limit, the right-hand limit, or the left-hand limit at 2 exists, or whether none of these limits exists. If the two-sided limit does not exist, determine whether there is an infinite (two-sided or one-sided) limit at 2.

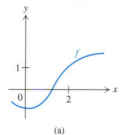

(a)

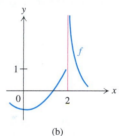

(b)

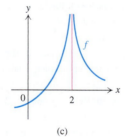

(c)

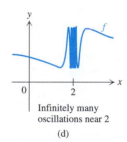

Infinitely many oscillations near 2

(d)

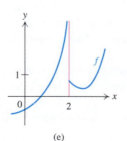

(e)

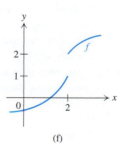

(f)

FIGURE 2.43   Graphs for Exercise 62.

63. Draw the graph of a function with domain [0, 4] and having the following properties:

    (i) $f(0) = -1$, $f(2) = 2$, and $f(4) = 1$

    (ii) $\lim_{x \to 1^-} f(x) = 1$

    (iii) $\lim_{x \to 1^+} f(x) = 3$

    (iv) $\lim_{x \to 4^-} f(x) = 0$

64. Let $f(x) = \dfrac{4x + |x|}{5x - 3|x|}$. Determine which of the following exist and evaluate those that do exist.

    a. $\lim_{x \to 0^+} f(x)$ **b.** $\lim_{x \to 0^-} f(x)$ **c.** $\lim_{x \to 0} f(x)$

65. Let $[x]$ denote the greatest integer less than or equal to x. Show that for each integer n,

    $$\lim_{x \to n^-} [x] = n - 1 \quad \text{and} \quad \lim_{x \to n^+} [x] = n$$

66. Suppose $\lim_{x \to a} f(x) = \infty$ and $\lim_{x \to a} g(x) = \infty$. Show by an example that we cannot conclude that

    $$\lim_{x \to a} [f(x) - g(x)] = 0$$

67. Let $f(x) = x^{1/n}$, where n is a positive odd integer. Show that the graph of f has a vertical tangent at (0, 0).

Applications

68. Suppose a rock is dropped into a well 49 meters deep.
    a. Determine the velocity of the rock when it hits the bottom of the well.
    b. If the bottom of the well is lined with thick sand so that the rock does not bounce, what is the velocity of the rock just after it hits the bottom of the well?

 69. Tower Falls in Yellowstone Park, pictured at the beginning of the chapter, is approximately 40 meters tall. Suppose that before the water droplets reach the top of the falls, they are descending (vertically) at a rate of –2 meters per second.
    a. Assuming no air resistance, determine the velocity of the droplets when they splash at the bottom of the falls.
    b. Guess the height $h_{1/2}$ at which the velocity of a droplet is half its "terminal" velocity. Then compute $h_{1/2}$ exactly.

 70. On a hardwood floor, a superball can bounce to 90% of the initial height. Suppose that a superball is dropped from 2 meters.
    a. Determine the velocity of the superball when it hits the floor.
    ***b.** Determine the velocity of the superball as it bounces up from the floor.

c. Why are the answers to parts (a) and (b) not negatives of one another?

71. Suppose a disk of radius r is perpendicular to the x axis (Figure 2.44) and has a uniform charge per unit area. Then at any point x along the x axis that is not on the disk, the electric field E of the disk is given by the equation

$$E(x) = c\left(\frac{x}{|x|} - \frac{x}{(x^2 + r^2)^{1/2}}\right)$$

where c is a nonzero constant. Determine whether $\lim_{x \to 0} E(x)$ exists.

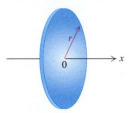

FIGURE 2.44   Figure for Exercise 71.

Projects

1. Consider the functions

$$f(x) = \left(1 + \frac{1}{x}\right)^x, \quad g(x) = \left(1 + \frac{1}{2x}\right)^x, \quad \text{and} \quad h(x) = \left(1 + \frac{1}{x}\right)^{3x}$$

    a. By sketching the graphs of f, g, and h, and zooming in toward 0, guess what the one-sided limits as x approaches 0 from the right should be. (*Hint:* They should be related to the number e.)
    b. Let a and b be positive numbers, and let

    $$k(x) = \left(1 + \frac{1}{ax}\right)^{bx}$$

    Use the information from (a) to guess $\lim_{x \to 0^+} k(x)$.

2. Suppose you throw a ball vertically at a height of 2 meters above ground, with initial velocity v_0.
    a. Find the length of time t_0 until the ball hits the ground.
    b. Let v_h be hitting velocity, that is, the velocity at the moment that the ball hits the ground. Express v_h in terms of v_0.
    c. Determine v_0 such that $v_h = -4$ meters per second.
    d. Let u_h be hitting velocity when the initial velocity is $-v_0$. Compare u_h and v_h.
    e. Suppose that the ball is dropped (that is, $v_0 = 0$), and after it bounces once, it rises to 80 percent of its original height. Find the hitting velocities at the moment the ball hits the ground the first time and the second time.

2.5   CONTINUITY, THE INTERMEDIATE VALUE THEOREM, AND THE BISECTION METHOD

In Section 2.3 we briefly mentioned that many common functions (like polynomials and trigonometric functions) are continuous at various points in their domains. In this section we will talk much more about continuity, and will, among other things, be able to show that every positive number has a square root. We will also discuss a method of approximating a zero of a function whose zeros are not obvious (such as $f(x) = x^5 + x^3 + x^2 - 1$).

DEFINITION   2.8

A function f is **continuous** at a number a in its domain if

$$\lim_{x \to a} f(x) = f(a)$$

A function f is **discontinuous** at a number a in its domain if f is not continuous at a.

The advantage of a function f continuous at a is that the limit of f at a can be found by just plugging in a for x in $f(x)$. This is true for polynomial, rational and basic trigonometric functions, as well as the natural exponential and logarithmic functions and the various root functions. By contrast, if f is *not* continuous at a, then the procedure of plugging in is *not* valid, as is evident with

$$\lim_{x \to 0} \frac{\sin x}{x}$$

After all, $(\sin x)/x$ is not even defined at 0.

Geometrically, the graph of a function that is continuous at a cannot have a break at the point $(a, f(a))$. In Figure 2.45 (on the next page), f is continuous at a, whereas g and h are discontinuous at a.

We emphasize that the definition of continuity includes three conditions:

    a. f is defined at a,

    b. $\lim_{x \to a} f(x)$ exists,

    c. $\lim_{x \to a} f(x)$ equals $f(a)$.

As we observed in Section 2.3, polynomial and rational functions, basic trigonometric functions, the natural exponential function, and the natural logarithmic function are continuous at every point of their domains. Thus a function like $f(x) = (x^2 - 3x + 2)/(x^2 + 5x - 6)$ is continuous at each x except $x = 1$ and $x = -6$, since at those numbers the denominator is zero.

Results on the continuity of combinations of functions follow immediately from the corresponding results for limits.

THEOREM 2.9

Suppose f and g are continuous at a, and let c be any number. Then $f + g$, cf, and fg are continuous at a. If $g(a) \neq 0$, then f / g is continuous at a.

As a result of Theorem 2.9, a function such as $f(x) = x^{3/2} - 2x + 1$ is continuous at every positive number because each of the functions $x^{3/2}$ and $-2x + 1$ is continuous at such numbers.

Turning to composite functions, we suppose that f is continuous at a, and that g is continuous at $f(a)$. Then it follows from the continuity of g, the Substitution Rule, and the continuity of f that

$$\lim_{x \to a} g(f(x)) = g(\lim_{x \to a} f(x)) = g(f(a))$$

Thus $g \circ f$ is continuous at a. Therefore we have the following result.

THEOREM 2.10   If f is continuous at a and g is continuous at $f(a)$, then $g \circ f$ is continuous at a.

To illustrate Theorem 2.10, we note that if $g(x) = \sin(e^x)$, then g is continuous at every number because $\sin x$ and e^x are continuous at each number. In addition, many functions encountered in applications are continuous wherever they are defined, as the following examples illustrate:

1. The volume of a sphere as a function of its radius:

$$V(r) = \frac{4}{3}\pi r^3 \quad \text{for } r > 0$$

2. The height (in seconds) of an object subject only to the earth's gravitational force, with initial velocity v_0, and initial height h_0:

$$h(t) = -4.9t^2 + v_0 t + h_0$$

3. The period of a pendulum as a function of its length:

$$T(L) = 2\pi \sqrt{L/g} \quad \text{for } L > 0$$

4. The concentration of a drug as a function of time (in minutes after injection):

$$C(t) = c(e^{-bt} - e^{-at}) \quad \text{for } t > 0, \text{ where } a, b, \text{ and } c \text{ are positive constants}$$

5. The cost (in dollars) of producing calculators as a function of the number produced:

$$C(x) = 200(x + 100)^{0.9} \quad \text{for } 1000 < x < 1,000,000$$

Sometimes functions are idealized so as to be continuous and thus easier to work with. This occurs especially in business and economics. For instance, the cost function mentioned above might actually be defined only for integer values of x. But for purposes of analysis, the domain of the function may be considered to be the entire interval $(1000, 1,000,000)$.

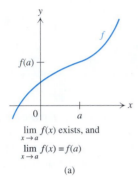

$\lim_{x \to a} f(x)$ exists, and

$\lim_{x \to a} f(x) = f(a)$

(a)

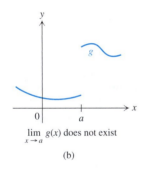

$\lim_{x \to a} g(x)$ does not exist

(b)

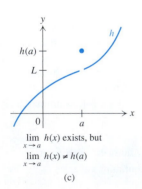

$\lim_{x \to a} h(x)$ exists, but

$\lim_{x \to a} h(x) \neq h(a)$

(c)

FIGURE 2.45

One-Sided Continuity

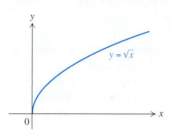

FIGURE 2.46   The square root function.

In (8) of Section 2.3 we stated that

$$\lim_{x \to a} \sqrt{x} = \sqrt{a} \qquad \text{for any } a > 0$$

This means that the square root function is continuous at every positive number (Figure 2.46). By contrast, $\sqrt{x}$ is technically not continuous at 0 because $\sqrt{x}$ is not defined for $x < 0$ and hence $\lim_{x \to 0} \sqrt{x}$ does not exist as a two-sided limit. However,

$$\lim_{x \to 0^+} \sqrt{x} = 0 = \sqrt{0}$$

We express this fact by saying that the square root function is continuous from the right at 0.

DEFINITION   2.11

A function f is **continuous from the right** at a point a in its domain if

$$\lim_{x \to a^+} f(x) = f(a)$$

A function f is **continuous from the left** at a point a in its domain if

$$\lim_{x \to a^-} f(x) = f(a)$$

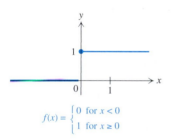

$$f(x) = \begin{cases} 0 & \text{for } x < 0 \\ 1 & \text{for } x \geq 0 \end{cases}$$

FIGURE 2.47   The diving board function.

Since one-sided limits are as easy to determine as two-sided limits, we can ascertain the one-sided continuity of most of the functions we know. In particular, we observe that for the diving board function f in Figure 2.47,

$$\lim_{x \to 0^+} f(x) = 1 = f(0) \quad \text{and} \quad \lim_{x \to 0^-} f(x) = 0 \neq f(0)$$

It follows that f is continuous from the right at 0 but is not continuous from the left at 0.

We remark that a function is continuous at a if and only if it is both continuous from the right and continuous from the left at a. This follows directly from Theorem 2.5.

EXAMPLE 1   Let

$$f(x) = \begin{cases} x^2 & \text{for } x \leq 1 \\ x & \text{for } x > 1 \end{cases}$$

Show that f is continuous at 1.

Solution   Since

$$\lim_{x \to 1^-} f(x) = \lim_{x \to 1^-} x^2 = 1 = f(1)$$

and

$$\lim_{x \to 1^+} f(x) = \lim_{x \to 1^+} x = 1 = f(1)$$

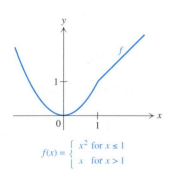

$$f(x) = \begin{cases} x^2 & \text{for } x \leq 1 \\ x & \text{for } x > 1 \end{cases}$$

FIGURE 2.48

f is continuous from the left and from the right at 1. Therefore f is continuous at 1 (see Figure 2.48).   ❑

In calculus, many of the most important theorems involve functions that are continuous on given intervals in their domains. It is therefore important that we know precisely what this means.

DEFINITION 2.12

> **a.** A function is **continuous on an open interval** (***a, b***), or simply **continuous on** (***a, b***), if it is continuous at every point in (*a, b*).
> **b.** A function is **continuous on a closed interval** **[*a, b*],** or simply **continuous on [*a, b*],** if it is continuous at every point in (*a, b*) and is also continuous from the right at *a* and continuous from the left at *b*.

There are analogous definitions of continuity on a half-open interval or on an interval of the form (a, ∞), $[a, \infty)$, $(-\infty, a)$, $(-\infty, a]$, or $(-\infty, \infty)$.

In addition to many other functions, the polynomial functions, the sine and cosine functions, and the natural exponential functions are continuous on every open interval, including the entire real line $(-\infty, \infty)$. Other functions are continuous on more limited open intervals:

$$\tan x \text{ is continuous on } (-\pi/2, \pi/2), \text{ on } (\pi/2, 3\pi/2), \text{ and so forth}$$

$$\frac{x}{(x + 1)(x - 2)} \text{ is continuous on } (-\infty, -1), \text{ on } (-1, 2), \text{ and on } (2, \infty)$$

Functions that are continuous on closed intervals will be of special interest to us in the remainder of this book. Of course, polynomial functions, the sine and cosine functions, and the natural exponential functions are continuous on any given closed interval. The function $\sqrt{x}$, which is not defined for $x < 0$, is continuous on $[0, \infty)$ because it is continuous at each point in $(0, \infty)$ and is continuous from the right at 0.

EXAMPLE 2   Let $f(x) = \sqrt{4 - x^2}$. Show that f is continuous on $[-2, 2]$.

Solution   Notice that $\sqrt{4 - x^2}$ is the composite of $4 - x^2$ (which is continuous at every real number) and $\sqrt{x}$ (which is continuous at every $x > 0$). Since $4 - x^2 > 0$ for $-2 < x < 2$, it follows from Theorem 2.10 that f is continuous at every x in $(-2, 2)$. Next, we substitute $y = 4 - x^2$ and thereby obtain

$$\lim_{x \to -2^+} f(x) = \lim_{x \to -2^+} \sqrt{4 - x^2} = \lim_{y \to 0^+} \sqrt{y} = 0 = f(-2)$$

$$\lim_{x \to 2^-} f(x) = \lim_{x \to 2^-} \sqrt{4 - x^2} = \lim_{y \to 0^+} \sqrt{y} = 0 = f(2)$$

Therefore f is continuous from the right at -2 and from the left at 2. Consequently f is continuous on $[-2, 2]$ (see Figure 2.49).   ❑

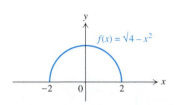

FIGURE 2.49

A service toss during the U.S. Open tennis tournament. (*Eric Miller/The Gamma Liaison Network*)

Many functions appearing in applications are defined and continuous on closed intervals, as the following list shows:

1. Degrees Celsius as a function of degrees Fahrenheit:

$$C(x) = \frac{5}{9}(x - 32) \quad \text{for } x \geq -459.67$$

2. The speed of a tennis ball tossed up for a serve, as a function of time:

$$v(t) = 16 - 32t \quad \text{for } 0 \leq t \leq \frac{1}{2}$$

3. Magnitude of an earthquake as a function of the amplitude of the largest seismic wave:

$$M(x) = \log\frac{x}{a} \quad \text{for } x \geq a$$

where a is the amplitude of the largest seismic wave for a corresponding zero-level earthquake.

The Intermediate Value Theorem

The Intermediate Value Theorem is an important result about functions that are continuous on closed intervals. As we will see, it can be used in order to help locate zeros of continuous functions and to show that every positive number has a square root. The theorem, which is proved in the Appendix, asserts that the range of a function that is continuous on a closed interval cannot skip any intermediate values. For example, if a function f is continuous throughout the interval [2, 6], and if $f(2) = 1$ and $f(6) = 4$, then every number between 1 and 4 must be in the range of f (see Figure 2.50).

THEOREM 2.13
Intermediate Value Theorem

Suppose f is continuous on a closed interval $[a, b]$. Let p be any number between $f(a)$ and $f(b)$, so that $f(a) \leq p \leq f(b)$ or $f(b) \leq p \leq f(a)$. Then there exists a number c in $[a, b]$ such that $f(c) = p$.

Pictorially, the Intermediate Value Theorem asserts that the graph of a continuous function cannot jump over a horizontal line (Figure 2.51(a)). In physical terms, you might think of the Intermediate Value Theorem as saying that you cannot cross an infinitely long river without either getting wet or jumping over it (Figure 2.51(b)).

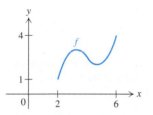

FIGURE 2.50

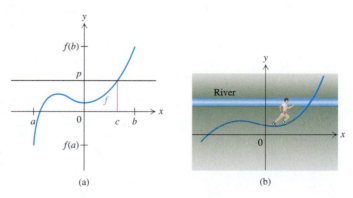

(a)   (b)

FIGURE 2.51

The Intermediate Value Theorem implies that the domain of the square root function consists of *all* nonnegative numbers, as asserted in Chapter 1 when the function was defined. To prove the assertion, we select any nonnegative number p and show that there is a number c such that $c \geq 0$ and $c^2 = p$. To that end, we let $f(x) = x^2$ for $x \geq 0$ and observe that f is continuous and

$$f(0) = 0 \leq p \leq p^2 + 2p + 1 = (p + 1)^2 = f(p + 1)$$

Thus $f(0) \leq p \leq f(p + 1)$, so the Intermediate Value Theorem says that there is a number c in $[0, p + 1]$ such that $f(c) = p$, or equivalently, $c^2 = p$. Thus p has a square root. Since p was an arbitrarily chosen nonnegative number, it follows that the square root is defined for every nonnegative number. Similar comments apply to the other root functions.

In the Intermediate Value Theorem the hypothesis that f is continuous is crucial. For example, if

$$f(x) = \begin{cases} 0 & \text{for } -2 \leq x < 0 \\ 1 & \text{for } 0 \leq x \leq 2 \end{cases}$$

then the range of f does not contain all numbers between 0 and 1; in fact, it contains none of them. Of course, f is not continuous on $[-2, 2]$ because it is not continuous at 0, and this is why its graph can "skip" the values between 0 and 1.

Positive and Negative Values of Functions

The consequences of the Intermediate Value Theorem that we will discuss in this section involve numbers d for a function f such that $f(d) = 0$. Recall that such a number is called a zero, or root, of f. Zeros of functions of the form $f(x) = ax^2 + bx + c$, where $a \neq 0$, can be located by means of the quadratic formula (see Section 1.5), but for higher-degree polynomials and functions in general there is no simple formula from which we can determine a zero. This is unfortunate because if the function f happens to be a profit function, then the zeros of f would represent the "break-even" points at which there is neither profit nor loss, and it might be desirable to know these points.

Now let f be continuous on an interval I. If f has both positive and negative values on I, then the Intermediate Value Theorem implies that $f(x) = 0$ for some x in I, that is, f has a zero in I. Equivalently, if f has *no* zero in I, then either $f(x) > 0$ for all x in I or $f(x) < 0$ for all x in I. This fact yields a procedure for discovering the intervals on which a continuous function f is positive and those on which f is negative, provided that we can determine the zeros of f.

EXAMPLE 3   Let $f(x) = (x + 1)^2 (x - 2)(x - 3)$. Determine the intervals on which f is positive and those on which f is negative.

Solution   Evidently the zeros of f are -1, 2, and 3, so we will determine the sign of $f(x)$ on each of the intervals $(-\infty, -1)$, $(-1, 2)$, $(2, 3)$, and $(3, \infty)$ by evaluating f at a convenient number c in each interval:

Interval	c in interval	$f(c)$	Sign of $f(x)$ on interval
$(-\infty, -1)$	-2	20	$+$
$(-1, 2)$	0	6	$+$
$(2, 3)$	$\dfrac{5}{2}$	$-\dfrac{49}{16}$	$-$
$(3, \infty)$	4	50	$+$

Consequently f is positive on $(-\infty, -1)$, $(-1, 2)$, and $(3, \infty)$, and is negative on $(2, 3)$.   ❑

The method used in Example 3 applies to a function such as a rational function, even if it is not defined on the whole real line $(-\infty, \infty)$.

EXAMPLE 4   Let

$$f(x) = \frac{x(x-2)(x^2 + 2x + 4)}{x^3 + 1}$$

Determine the intervals on which f is positive and those on which f is negative.

Solution   Since $x^3 + 1 = 0$ if $x = -1$, the domain of f is the union of $(-\infty, -1)$ and $(-1, \infty)$. Next we observe that

$$x^2 + 2x + 4 = (x^2 + 2x + 1) + 3 = (x + 1)^2 + 3 \; > \; 0 \quad \text{for all } x$$

so that $f(x) = 0$ only if $x = 0$ or $x = 2$. Therefore we will determine the sign of $f(x)$ on each of the intervals $(-\infty, -1)$, $(-1, 0)$, $(0, 2)$, and $(2, \infty)$:

Interval	c in interval	$f(c)$	Sign of $f(x)$ on interval
$(-\infty, -1)$	-2	$-\dfrac{32}{7}$	$-$
$(-1, 0)$	$-\dfrac{1}{2}$	$\dfrac{65}{14}$	$+$
$(0, 2)$	1	$-\dfrac{7}{2}$	$-$
$(2, \infty)$	3	$\dfrac{57}{28}$	$+$

Consequently f is positive on $(-1, 0)$ and $(2, \infty)$, and is negative on $(-\infty, -1)$ and $(0, 2)$.   ❑

Finding the intervals on which a function f is positive (or negative) is equivalent to solving the inequality $f(x) > 0$ (or $f(x) < 0$), so one can consider Example 4 as requesting us to determine the solutions of the inequalities

$$\frac{x(x-2)(x^2 + 2x + 4)}{x^3 + 1} > 0 \quad \text{and} \quad \frac{x(x-2)(x^2 + 2x + 4)}{x^3 + 1} < 0$$

Thus in addition to the method of solving inequalities discussed in Section 1.1 we now have a second method, which uses the Intermediate Value Theorem.

The Bisection Method

Suppose a function g represents the profit from the sale of grain and is defined by

$$g(x) = x^5 + x^3 + x^2 - 1 \qquad \text{for } x \text{ in } [0, 1]$$

where $g(x)$ is in hundreds of dollars and x is measured in tons. Since g is continuous on $[0, 1]$, and since $g(0) = -1$ and $g(1) = 2$, the Intermediate Value Theorem assures us that there is a number d in $[0, 1]$ such that $g(d) = 0$. The number d represents the break–even point at which there is neither profit nor loss. Although the

Intermediate Value Theorem implies the existence of d, it does not tell us the exact value of d.

In trying to find d we could attempt to factor $g(x)$. But that would be futile, so finding the precise solution of the equation $g(x) = 0$ seems impossible. Thus we must be satisfied with approximating d. One method of doing so is called the bisection method.

The bisection method is based on the ancient Roman proverb "divide and conquer," and applies to any function f that is continuous on an interval $[a, b]$ and has the property that $f(a)$ and $f(b)$ have different signs (so either $f(a) < 0$ and $f(b) > 0$ or $f(a) > 0$ and $f(b) < 0$).

Because f is continuous on $[a, b]$, and $f(a)$ and $f(b)$ have different signs, the Intermediate Value Theorem implies that f has a zero in $[a, b]$. Let c be the midpoint of $[a, b]$. If $f(c) = 0$, then c is a zero of f in $[a, b]$, and the search stops. However, if $f(c) \neq 0$, then

either $f(a)$ and $f(c)$ have different signs, so let $I_1 = [a, c]$

or $f(c)$ and $f(b)$ have different signs, so let $I_1 = [c, b]$

Whichever half of $[a, b]$ becomes I_1, we know by the Intermediate Value Theorem that I_1 contains a zero of f, since $f(x)$ has different signs at the endpoints of I_1. Continuing in the same way, we obtain successively smaller subintervals $I_2, I_3, \ldots$, each containing a zero of f. Because the process involves bisecting intervals, it is called the **bisection method** for approximating a zero of f.

The process could be continued indefinitely, so we must have a way of knowing when to stop, that is, when we have found a number c^* sufficiently close to a zero d of f. The bisection method yields successive subintervals of $[a, b]$, each containing d and each half as long as its immediate predecessor. Letting the midpoint of the nth subinterval I_n be c_n, we find that after n steps in the bisection method,

$$|c_n - d| < \frac{b - a}{2^n}$$

Thus if we wish to perform the bisection method until c_n is within, say, 10^{-2} of the zero d, we need only to determine n such that

$$\frac{b - a}{2^n} < 10^{-2}$$

For example, if $a = 0$ and $b = 1$, then this means that

$$\frac{1}{2^n} < \frac{1}{100}, \quad \text{or equivalently}, \quad 2^n > 100$$

This occurs if $n \geq 7$.

In applying the bisection method, we perform a sequence of steps, called an **algorithm,** and stop when we achieve the desired accuracy. We call the desired accuracy the allowable **tolerance.** The algorithm, which could easily be programmed into a computer or calculator, can be described as follows:

INPUT: A continuous function f, numbers a and b such that $f(a)$ and $f(b)$ have opposite signs, and an allowable error, or tolerance, ε.

OUTPUT: A number c that approximates a zero of f with error less than ε.

STEP 1: Let the interval I be $[a, b]$.

STEP 2: Let c be the midpoint of I.

STEP 3: Compute the length L of I. If $L \leq 2\varepsilon$, then c is an approximate zero of f with error $< \varepsilon$, so STOP. Otherwise CONTINUE.

STEP 4: Compute $f(c)$. If $f(c) = 0$, then c is a zero of f, so STOP. Otherwise CONTINUE.

STEP 5: If $f(a)$ and $f(c)$ have different signs, replace I by $[a, c]$. Otherwise replace I by $[c, b]$.

STEP 6: Return to Step 2, using the new interval I.

Eventually the process stops (with either Step 3 or Step 4). When the process does stop, the value of c approximates a zero of f with error less than ε.

EXAMPLE 5 Let $f(x) = x^5 + x^3 + x^2 - 1$. Use the bisection method to find a number in $[0, 1]$ that approximates a zero of f with an error less than $1/16$.

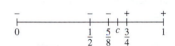

FIGURE 2.52

Solution Since $f(0) = -1$ and $f(1) = 2$, we can begin by letting $a = 0$ and $b = 1$. Using the algorithm, we assemble the following table and the accompanying graph (Figure 2.52):

Interval	Length	Midpoint c	$f(c)$
$[0, 1]$	1	$\dfrac{1}{2}$	≈ -0.59
$\left[\dfrac{1}{2}, 1\right]$	$\dfrac{1}{2}$	$\dfrac{3}{4}$	≈ 0.22
$\left[\dfrac{1}{2}, \dfrac{3}{4}\right]$	$\dfrac{1}{4}$	$\dfrac{5}{8}$	≈ -0.27
$\left[\dfrac{5}{8}, \dfrac{3}{4}\right]$	$\dfrac{1}{8}$		

Since the length of $\left[\frac{5}{8}, \frac{3}{4}\right]$ is $\frac{1}{8}$, and neither $\frac{5}{8}$ nor $\frac{3}{4}$ is a zero of f, the midpoint $\frac{11}{16}$ of $\left[\frac{5}{8}, \frac{3}{4}\right]$ is less than $\frac{1}{16}$ from a zero of f. ☐

Caution: If $f(x) \geq 0$ (or if $f(x) \leq 0$) for *all* x in some open interval containing a zero d of f, then the bisection method cannot be used in order to approximate d. For example, if

$$f(x) = \frac{(x-1)^2}{x^2 + 3}$$

then since $f(x) \geq 0$ for all x, we cannot use the bisection method to approximate the zero of f, which is evidently 1.

If $f(a)$ and $f(b)$ have opposite signs, and if f is continuous on $[a, b]$, then the bisection method is applicable and guarantees that we can find an approximate zero of f with any desired accuracy. However, the method is slow, since the successive intervals, and hence the successive error estimates, are only halved at each step (which in computer terms amounts to achieving one extra "bit" of accuracy per step). Nevertheless, the bisection method is employed in computer calculations of zeros of functions, usually in conjunction with other algorithms that we will present in Section 3.8.

Appoximating Zeros by Calculator

Most graphics calculators have a zoom feature that serves as a substitute for the bisection method. To approximate a zero of a function f, we would first graph the function on the calculator in order to determine the general location of the desired zero of f. Then we would zoom in on the graph near that zero to obtain a better approximation to the zero. By repeatedly zooming in on the graph and reading off the x coordinate of the point on the graph that appears to be nearest the x axis, we would obtain better and better estimates for the zero of f.

For example, if we were to use the zoom feature on our calculator to approximate the zero in $[0, 1]$ for the function f in Example 5, we would fairly quickly find that 0.699 appears to approximate the zero of f to 3 places. This estimate is superior to the estimate $\frac{11}{16} = 0.6875$, which we found in Example 5 by the bisection method.

Finally, we observe that whereas the Intermediate Value Theorem can guarantee the existence of a zero for a function, the bisection method helps in approximating the zero.

EXERCISES 2.5

In Exercises 1–4 determine whether f is continuous at a.

1. $f(x) = \sqrt{x} \cos x \,; a = \dfrac{\pi}{4}$

2. $f(x) = e^x \ln x \,; a = 1$

3. $f(x) = \begin{cases} \dfrac{\sin x}{x} & \text{for } x \neq 0 \\ 1 & \text{for } x = 0 \end{cases} \quad a = 0$

4. $f(x) = \begin{cases} \dfrac{\sin 5x}{x} & \text{for } x \neq 0 \\ 1 & \text{for } x = 0 \end{cases} \quad a = 0$

5. Which of the graphs in Figure 2.53 represent functions that are continuous at 2?

6. For which of the following functions can we define $f(a)$ so as to make f continuous at a? For those we can, find the value of $f(a)$ that makes f continuous at a.

a. $f(x) = \dfrac{x^2 - 9}{x - 3} \,; a = 3$

b. $f(x) = \dfrac{x^2 + 5x + 4}{x - 1} \,; a = 1$

c. $f(x) = \dfrac{e^x - 1}{x} \,; a = 0$

d. $f(x) = \sin \dfrac{1}{x} \,; a = 0$

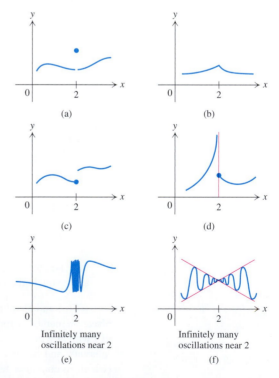

(a)

(b)

(c)

(d)

Infinitely many oscillations near 2

(e)

Infinitely many oscillations near 2

(f)

FIGURE 2.53 Graphs for Exercise 5.

In Exercises 7–14 determine whether f is continuous or discontinuous at a. If f is discontinuous at a, determine whether f is continuous from the right at a, continuous from the left at a, or neither.

7. $f(x) = \sqrt{x-2}; a = 2$

8. $f(x) = \ln(x^2 + 1); a = 0$

9. $f(x) = \sqrt{1 - e^x}; a = 0$

10. $f(x) = \tan\sqrt{x} - \pi/2; a = \pi/2$

11. $f(x) = \begin{cases} \dfrac{|x-4|}{x-4} & \text{for } x \neq 4 \\ e^{4-x} & \text{for } x = 4 \end{cases}$ $a = 0, 4$

12. $f(x) = \begin{cases} -1 & \text{for } x < 0 \\ 0 & \text{for } x = 0 \\ 1 & \text{for } x > 0 \end{cases}$ $a = 0$

13. $f(t) = t^2\sqrt{t^2 - t^4}; a = 0, 1$

14. $f(t) = \sqrt{\sqrt{t+2} - t}; a = 2$

15. For which of the following functions f can we redefine $f(a)$ so as to make f continuous at a?

a. $f(x) = \begin{cases} x & \text{for } x \neq 3 \\ 2 & \text{for } x = 3 \end{cases}$ $a = 3$

b. $f(x) = \begin{cases} 2x - 3 & \text{for } x < 1 \\ 0 & \text{for } x = 1 \\ 3x - 2 & \text{for } x > 1 \end{cases}$ $a = 1$

In Exercises 16–31 explain why f is continuous on the given interval or intervals

16. $f(x) = \dfrac{1-x}{1+x^2}; (-\infty, \infty)$

17. $f(x) = \dfrac{\sin x}{x}; (-\infty, 0); (0, \infty)$

18. $f(x) = \begin{cases} (\sin x)/x & \text{for } x \neq 0 \\ 1 & \text{for } x = 0 \end{cases}; (-\infty, \infty)$

19. $f(x) = \sqrt{x + 3}; [-3, \infty)$

20. $f(x) = \sqrt{x^2 - x^4}; [-1, 1]$

21. $f(x) = \dfrac{1}{\sqrt{x^2 - x^4}}; (0, 1)$

22. $f(x) = \sqrt{16x^4 - x^2}; (-\infty, -\frac{1}{4}], [\frac{1}{4}, \infty)$

23. $f(x) = \sqrt{\dfrac{1-x}{3x-2}}; (\frac{2}{3}, 1]$

24. $f(x) = \dfrac{\sqrt{x-1}}{\sqrt{2-x}}; [1, 2)$

25. $f(x) = \tan x; (-\pi/2, \pi/2)$

26. $f(x) = \tan\left(\dfrac{1}{x^2 + 1}\right); (-\infty, \infty)$

27. $f(x) = \cot(1/x); (1/\pi, \infty)$

28. $f(x) = \sec\sqrt{x}; [0, \pi^2/4)$

29. $f(t) = \sqrt{e^t}; (-\infty, \infty)$

30. $f(t) = \ln(\ln t); (1, \infty)$

31. $f(t) = \ln(1 - e^t); (-\infty, 0)$

32. a. Plot the graph of $\ln x$ on the interval $(0.5, 1.5)$. Does the graph suggest that $\lim_{x \to 1} \ln x = 0$?

b. Using the formula

$$\ln x = \ln\frac{x}{a} + \ln a$$

along with the limit in part (a) and the Substitution Rule, prove that

$$\lim_{x \to a} \ln x = \ln a \qquad \text{for } a > 0$$

that is, prove that $\ln x$ is a continuous function.

In Exercises 33–42 show that the equation has at least one solution.

33. $x^4 - x - 1 = 0$ (*Hint:* Let $f(x) = x^4 - x - 1$, and apply the Intermediate Value Theorem to f on $[-1, 1]$.)

34. $x^3 - x - 5 = 0$ **35.** $x^2 + \dfrac{1}{x} = 1$

36. $x^2 + \dfrac{1}{2x} = 2$ **37.** $x^3 + x^2 + x - 2 = 0$

38. $x^{7/3} + x^{5/3} - 1 = 0$ **39.** $\cos x = x$

40. $\sin x = 1 - x$ **41.** $e^{-x} = x$

42. $x \ln x = 1$

In Exercises 43–50 solve the inequality for x.

43. $(2-x)^2(40 - 8x) > 0$

44. $(x+1)(x-2)(2x - \frac{1}{4}) \geq 0$

45. $(x^4 + x)(x + 3) \leq 0$

46. $\dfrac{x(x-4)}{2(x-2)^2} < 0$

47. $\dfrac{(x-1)(x-3)}{(2x+1)(2x-1)} \geq 0$

48. $\dfrac{4x^2 - 1}{(x^2 - 1)^3} > 0$

49. $\dfrac{-2(x-1)(x^2 + 2x + 4)}{27 - x^3} < 0$

50. $\dfrac{3x(x^3 - 8)}{(x^3 + 1)^3} \leq 0$

In Exercises 51–54 use the bisection method to approximate a zero of f with an error less than $\frac{1}{16}$.

51. $f(x) = x^2 - 2$ (Your answer is an approximation to $\sqrt{2}$.)
52. $f(x) = x^2 - 7$
53. $f(x) = x^3 - 3$
54. $f(x) = x^3 + x - 3$

In Exercises 55–58 use the bisection method to find an approximation of the given number with an error less than $\frac{1}{16}$.

55. $\sqrt{5}$ **56.** $\sqrt{12}$ **57.** $\sqrt{0.7}$ **58.** $\sqrt[3]{10}$

In Exercises 59–62 use the zoom feature to approximate all zeros of f. Continue until the first three digits of successive approximations are identical.

59. $f(x) = \cos x - x$
60. $f(x) = \tan x - 2x$ for $0 < x < \pi/2$
61. $f(x) = e^{-x} - x$
62. $f(x) = 2 - x + \ln x$
63. Show that every real number has a cube root. (*Hint:* Follow the ideas in the text concerning the existence of square roots.)
64. Show that every real number is the tangent of a number in $(-\pi/2, \pi/2)$.
65. Use the Intermediate Value Theorem to show that among all circles with radius no larger than 10 centimeters, there is one whose area is 200 square centimeters.
66. Use the Intermediate Value Theorem to show that among all right circular cylinders of height 10 meters and radius of the base not exceeding 1 meter, there is one whose volume is 25 cubic meters.
67. For which positive integers n is there a number c such that $c = c^n + 1$?
68. We say that the function f "changes sign at a" if there is a number $\delta > 0$ such that $f(x) < 0$ for $a - \delta < x < a$ and $f(x) > 0$ for $a < x < a + \delta$, or similarly, if $f(x) > 0$ for $a - \delta < x < a$ and $f(x) < 0$ for $a < x < a + \delta$. Use the Intermediate Value Theorem to prove that a polynomial function of positive degree n can change signs at most at n numbers. Then give an example to show that a polynomial function f of degree n may change signs fewer than n times. (*Hint:* Find such an example when $n = 3$, so that f is a cubic polynomial.)
69. Which of the following functions do you think are continuous, and which do you think have discontinuities? Explain your reasoning.
  a. The circumference of a circle as a function of its radius.

b. Your height as a function of time.
c. The required postage of a letter as a function of its weight.
d. The speed of an airplane as a function of time during a normal flight.
e. The velocity of a bouncing ball.
70. If highway A always runs northeast and highway B always runs southeast, must they meet one another? Explain.
  a. Would it be possible for highways A and B not to meet each other? Explain.
  b. If neither highway is longer than 3500 miles, must they meet one another? Explain.
71. During a cough, the rate of flow (volume per unit time) of air that passes through a human windpipe is sometimes modeled by

$$F(r) = k(r_0 - r)r^4 \quad \text{for} \quad \frac{1}{2}r_0 < r < r_0$$

where r_0 is the normal radius of the windpipe and r is the contracted radius during the cough. Show that F is continuous at every r in the interval $(\frac{1}{2}r_0, r_0)$.
72. The **Gompertz growth curve**, which appears in population studies, is the graph of the function f defined by

$$f(x) = ae^{(-be^{-cx})} \quad \text{for } x > 0$$

where a, b, and c are positive constants. Show that f is continuous at every $x > 0$.

Applications

73. A cylinder of height h is inscribed in a sphere of radius r (Figure 2.54).
  a. Show that the volume V of the cylinder is given by

$$V(h) = \pi r^2 h - \frac{\pi}{4}h^3 \quad \text{for } 0 \le h \le 2r$$

  b. Show that V is continuous on the interval $[0, 2r]$.

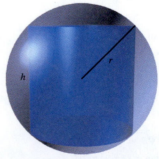

FIGURE 2.54 Figure for Exercise 73. A cylinder of height h is inscribed in a sphere of radius r.

74. The air pressure $p(x)$ (in atmospheres) at an altitude of x meters above sea level is given by

$$p(x) = e^{-1.25 \times 10^{-4}x} \quad \text{for } x \geq 0$$

Show that p is continuous on $[0, \infty)$. (This shows why there is no fixed altitude at which a mountaineer must don a gas mask.)

75. If we neglect air resistance, then the range of a ball shot at an angle θ with respect to the x axis and with initial velocity v_0 is given by

$$R(\theta) = \left(\frac{v_0^2}{g}\right) \sin 2\theta \quad \text{for } 0 \leq \theta \leq \frac{\pi}{2}$$

Show that R is continuous on $[0, \pi/2]$.

76. A ring of radius a carries a uniform electric charge Q. The electric field intensity $E(x)$ at any point x along the axis of the ring is given by

$$E(x) = \frac{Qx}{(x^2 + a^2)^{3/2}}$$

Show that E is continuous on the whole real line.

77. The gravitational force F that the earth exerts on a unit mass located a distance r from the center of the earth is given by

$$F = \begin{cases} \dfrac{GMr}{R^3} & \text{for } r < R \\ \dfrac{GM}{r^2} & \text{for } r \geq R \end{cases}$$

where R is the radius of the earth, M is its mass, and G is the universal gravitational constant. Is the force F a continuous function of the distance r on $(0, \infty)$? Explain your answer.

Project

1. a. Let $f(x) = \sin 1/x$ for $x \neq 0$ and $f(0) = 0$. Show that f is continuous except at 0.

  b. Let $g(x) = x \sin 1/x$ for $x \neq 0$ and $g(0) = 0$. Show that g is continuous at each real number (including 0).

  c. Let $h(x) = x \sin 1/x$ for $x \neq 0$ and $h(0) = 1$. Show that h is continuous at every number except 0.

  d. Can you use the information in (c) as a starting-off point to define a function k that is discontinuous only at the numbers 0 and 1/2, or is discontinuous only at a finite number of numbers in $[0, 1]$?

REVIEW

Key Terms and Expressions

Limit
Right-hand limit
Left-hand limit
Tangent line

Velocity
Vertical asymptote
Continuity
Continuity on a closed interval

Key Formulas

$$\lim_{x \to a} [f(x) + g(x)] = \lim_{x \to a} f(x) + \lim_{x \to a} g(x)$$

$$\lim_{x \to a} [cf(x)] = c \lim_{x \to a} f(x)$$

$$\lim_{x \to a} [f(x)g(x)] = \lim_{x \to a} f(x) \lim_{x \to a} g(x)$$

$$\lim_{x \to a} \frac{f(x)}{g(x)} = \frac{\lim_{x \to a} f(x)}{\lim_{x \to a} g(x)} \text{ if } \lim_{x \to a} g(x) \neq 0$$

$$\lim_{x \to a} g(f(x)) = \lim_{y \to c} g(y), \text{ where } c = \lim_{x \to a} f(x)$$

Key Theorems

Sum Rule
Constant Multiple Rule
Product Rule
Quotient Rule

Substitution Rule
Squeezing Theorem
Intermediate Value Theorem

Review Exercises

In Exercises 1–10 evaluate the limit.

1. $\lim\limits_{x \to 2} \dfrac{-4x + 3}{x^2 - 1}$

2. $\lim\limits_{x \to 1/2} \dfrac{2x - 1}{8x^2 - 4x}$

3. $\lim\limits_{x \to 1} x\sqrt{x + 3}$

4. $\lim\limits_{x \to 4} \dfrac{\sqrt{x} - 2}{4 - x}$

5. $\lim\limits_{x \to \pi/4} \sec^3\left(x + \dfrac{3\pi}{4}\right)$

6. $\lim\limits_{x \to 0} \dfrac{\tan^2 x}{x}$

7. $\lim\limits_{v \to 0} \dfrac{5 \sin 6v}{4v}$

8. $\lim\limits_{v \to 0} \dfrac{\sin^6 v}{6v^6}$

9. $\lim\limits_{x \to 0} \dfrac{1 - e^x}{1 + e^x}$

10. $\lim\limits_{x \to -2} \ln (5 - x^2)^2$

In Exercises 11–20 determine whether the limit exists as a number, as ∞, or as $-\infty$. If the limit exists, evaluate it.

11. $\lim\limits_{x \to -10} \dfrac{4x}{(x + 10)^2}$

12. $\lim\limits_{x \to 4} \dfrac{2x - 8\sqrt{x} + 8}{\sqrt{x} - 2}$

13. $\lim\limits_{x \to 0^-} \sqrt{9 + \sqrt{-x}}$

14. $\lim\limits_{x \to 5} \dfrac{x^2 - 2x - 15}{x^2 - x - 20}$

15. $\lim\limits_{w \to 0} \dfrac{\sqrt{2 + 3w} - \sqrt{2 - 3w}}{w}$

16. $\lim\limits_{w \to 0^+} \dfrac{\cos\sqrt{w}}{\sin w}$

17. $\lim\limits_{w \to 0^+} \cos \dfrac{1}{w}$

18. $\lim\limits_{x \to 0^+} \dfrac{\sin x}{\sqrt{x}}$

19. $\lim\limits_{x \to 0^+} \dfrac{e^x}{\ln x}$

20. $\lim\limits_{x \to 1^+} \dfrac{e^x}{\ln x}$

In Exercises 21–30 find all vertical asymptotes (if any) of the graph of f.

21. $f(x) = \dfrac{3 - 2|x|}{x - 4}$

22. $f(x) = \dfrac{(x + 3)^2}{4 - 9x^2}$

23. $f(x) = \dfrac{x^2 + 3x - 4}{x^2 - 5x - 14}$

24. $f(x) = \dfrac{x^2 + x - 2}{x + 2}$

25. $f(x) = \dfrac{(x - 2)^2}{x^2 - 4}$

26. $f(x) = \dfrac{\tan x}{x}$

27. $f(x) = e^{-1/x}$

28. $f(x) = \ln (x^2 - 2x - 3)$

29. $f(x) = \dfrac{\ln x}{e^x}$

*30. $f(x) = \dfrac{x}{e^x - 1}$

In Exercises 31–34 determine whether f is continuous or discontinuous at a. If f is discontinuous at a, determine whether f is continuous from the right at a, continuous from the left at a, or neither.

31. $f(x) = x^2 - 13;\ a = -\sqrt{13}$

*32. $f(x) = \sqrt{\sqrt{x + 3} - 2x};\ a = 1$

33. $f(x) = \begin{cases} 3x - 4 & \text{for } x < 2 \\ 3x + 4 & \text{for } x \geq 2 \end{cases} \quad a = 2$

34. $f(x) = \begin{cases} \sqrt{2 - x} & \text{for } x < 1 \\ 1 & \text{for } x = 1 \quad a = 1 \\ 2x^2 - x & \text{for } x > 1 \end{cases}$

In Exercises 35–40 explain why f is continuous on the given interval.

35. $f(x) = \dfrac{x^2 + 6x - 16}{x - 2};\ (2, \infty)$

36. $f(x) = \sqrt{5 - 2x};\ (-\infty, \frac{5}{2}]$

37. $f(x) = \sqrt{\dfrac{3x - \sqrt{2}}{4 - x}};\ \left[\dfrac{\sqrt{2}}{3}, 4\right)$

38. $f(x) = \sec \dfrac{1}{1 + x^2};\ (-\infty, \infty)$

39. $f(x) = \sqrt{e^{(x^2)}};\ (-\infty, \infty)$

40. $f(x) = \ln (1 - e^x);\ (-\infty, 0)$

41. Let

$$f(x) = \dfrac{x^2 + 6x - 16}{x - 2}$$

Find the value for $f(2)$ that would make f continuous at 2.

42. Suppose the graph of f is as drawn in Figure 2.55, with domain [0, 4.5].

   a. At which integers does f have a left-hand limit?

   b. At which integers does f have a right-hand limit?

   c. At which integers does f have a limit?

   d. At which integers is f continuous from the left?

   e. At which integers is f continuous from the right?

   f. At which integers is f continuous?

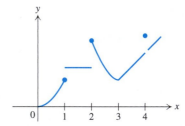

FIGURE 2.55   Graph for Exercise 42.

43. Consider the graphs of f and g in Figure 2.56. Evaluate the following limits, assuming $\lim\limits_{x \to 0^+} g(x) = \infty$.

   a. $\lim\limits_{x \to 0^+} \dfrac{f(x)}{g(x)}$

   b. $\lim\limits_{x \to a} \dfrac{f(x)}{g(x)}$

   c. $\lim\limits_{x \to b^-} \dfrac{f(x)}{g(x)}$

   d. $\lim\limits_{x \to b^+} \dfrac{f(x)}{g(x)}$

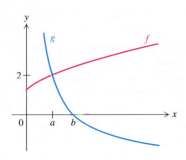

FIGURE 2.56   Graph for Exercise 43.

44. Consider the functions f, g, and h, whose graphs appear in Figure 2.57. In each of the following, determine whether the limit exists as a number, is ∞ or $-\infty$, or does not exist.

a. $\lim\limits_{x\to 2} (f(x) + g(x))$ **b.** $\lim\limits_{x\to 2} (f(x) - g(x))$

c. $\lim\limits_{x\to 2} f(x)g(x)$ **d.** $\lim\limits_{x\to 2} (f(x) + h(x))$

e. $\lim\limits_{x\to 2} f(x)h(x)$ **f.** $\lim\limits_{x\to 2} \dfrac{f(x)}{h(x)}$

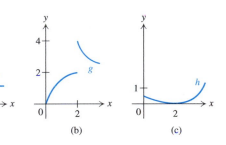

(a)   (b)   (c)

FIGURE 2.57   Graphs for Exercise 44.

In Exercises 45–46 solve the inequality for x.

45. $\dfrac{(2x - 3)(4 - x)}{(1 + x)^2} < 0$ **46.** $\dfrac{x + 4}{(x^2 - 4)(1 + x^2)} \geq 0$

In Exercises 47–48 show that the equation has at least one solution in the given interval. Then use the bisection method to approximate the solution with an error less than $\frac{1}{16}$.

47. $x^3 + 3x - 3 = 0$; $[-1, 1]$ **48.** $\sqrt{x} + 1 = x^2$; $[1, 2]$

In Exercises 49–50 use the zoom feature to approximate all solutions of the given equation. Continue until the first three digits of successive approximations are identical.

49. $x + \sin x = \cos x$ **50.** $e^{-x^2/2} = 0.3x$

*51. Let a and b be positive constants. Find

$$\lim_{x\to 0} \frac{e^{-ax} - e^{-bx}}{x}$$

52. Show that every number in $[-1, 1]$ is the sine of some number in $[-\pi/2, \pi/2]$.

53. What is wrong with the following statement? "The limit of the product of two functions is the product of the limits of the two functions."

54. a. Show that if $\lim\limits_{x\to a} f(x) = L$, then

$$\lim_{x\to -a} f(-x) = L$$

b. Suppose that f is either an even or an odd function. Using part (a), show that if f is continuous at a, then f is continuous at $-a$.

55. a. Show that if f is continuous at a, then so is $|f|$.

b. Show that the converse of part (a) is false by letting

$$f(x) = \begin{cases} -1 & \text{for } x \leq 2 \\ 1 & \text{for } x > 2 \end{cases}$$

56. a. Show that if $\lim\limits_{x\to a} f(x) = \infty$, then $\lim\limits_{x\to a} 1/f(x) = 0$.

b. Show by means of an example that the converse of part (a) is not true. (*Hint:* Choose a function that has both positive and negative values on every open interval about 0.)

57. Verify that the following pairs of functions have parallel tangent lines at the indicated points.

a. $f(x) = 2\sin x$ at $(0, 0)$ and $g(x) = -1/x^2$ at $(1, -1)$

b. $f(x) = x^3$ at $(-1, -1)$ and $g(x) = \frac{3}{8}x^2$ at $(4, 6)$

c. $f(x) = x^2 + 5x + 1$ at $(-4, -3)$ and $g(x) = -3x - 7$ at $(3, -16)$

58. Verify that the following pairs of functions have perpendicular tangent lines at the indicated points.

a. $f(x) = \sin x$ at $(0, 0)$ and $g(x) = 1/x$ at $(1, 1)$

b. $f(x) = \frac{1}{2}\sin x$ at $(0, 0)$ and $g(x) = 1/x^2$ at $(1, 1)$

c. $f(x) = x - 10$ at $(6, -4)$ and $g(x) = -x + 3$ at $(-5, 8)$

59. Let $f(x) = \sqrt{x}$. Show that the line $4y = x + 4$ is tangent to the graph of f. What is the point of tangency?

60. Suppose a thimble is dropped into a well 15 meters deep. What is its velocity after 1.5 seconds? What is its speed at that moment?

61. The drip from a faucet takes $\frac{1}{4}$ second to reach the basin.

a. How high is the faucet?

b. With what speed does the drip hit the basin?

Topics for Discussion

1. Explain in your own words how velocity and slopes of tangent lines are related.

2. State in your own words the definition of limit. Why do we pick the ε first, and then the δ?

3. Consider a function f that is defined on an open interval containing 0, with $f(0) = 0$. Suppose that no matter how much we zoom in on $(0, 0)$, the graph appears wobbly. Can f have a limit at 0? Can f have a line tangent at $(0, 0)$? Include examples to support your answers.

4. Consider the following statement:

   The limit of the sum of two functions is the sum of the limits of the two functions.

   Give a counterexample to it, and then correct the statement.

5. Describe the Squeezing Theorem in your own words, and draw a picture to accompany it.

6. What is the relationship between two-sided and one-sided limits? Under what conditions would you use one-sided limits to find a two-sided limit?

7. How are the Intermediate Value Theorem and the bisection method related to the existence of and search for a zero of a given function?

Derivatives can help describe the motion of this surfer in Hawaii.
(T. Nakamura/Superstock)

3 DERIVATIVES

In Chapter 2 we used limits of the form

$$\lim_{x \to a} \frac{f(x) - f(a)}{x - a} \tag{1}$$

to define the slope of a tangent line and the velocity of an object moving in a straight line. However, the limit in (1) also appears in many other contexts, such as marginal revenue and marginal cost in economics, acceleration in physics, and reaction rates in chemistry. Because of the various interpretations of the limit in (1), we will isolate it as an abstract mathematical entity, called the derivative, and study its properties in Chapters 3 and 4.

3.1 THE DERIVATIVE

We are now ready to give the definition of one of the two central concepts of calculus: the derivative.

DEFINITION 3.1
Definition of Derivative

Let a be a number in the domain of a function f. If

$$\lim_{x \to a} \frac{f(x) - f(a)}{x - a}$$

exists, we call this limit the **derivative of f at a** and denote it by $f'(a)$, so that

$$f'(a) = \lim_{x \to a} \frac{f(x) - f(a)}{x - a} \tag{2}$$

If the limit in (2) exists, we say that **f has a derivative at a,** that f is **differentiable at a,** or that **$f'(a)$ exists.**

The three bold-faced phrases appearing in the last sentence of Definition 3.1 are all in common use, and we will use them interchangeably. The expression $f'(a)$ is read "the derivative of f at a" or "f prime of a."

By definition,

$f'(a)$ is the slope of the line tangent to the graph of f at $(a, f(a))$.

Therefore an equation of the line tangent to the graph of f at $(a, f(a))$ is

$$y - f(a) = f'(a)(x - a), \quad \text{or equivalently,} \quad y = f(a) + f'(a)(x - a)$$

EXAMPLE 1 Let $f(x) = \frac{1}{4}x^2 + 1$. Find $f'(-1)$ and $f'(3)$, and draw the lines tangent to the graph of f at the corresponding points.

Solution Using (2), we obtain

$$f'(-1) = \lim_{x \to -1} \frac{(\frac{1}{4}x^2 + 1) - [\frac{1}{4}(-1)^2 + 1]}{x - (-1)} = \lim_{x \to -1} \frac{\frac{1}{4}x^2 - \frac{1}{4}}{x + 1} = \lim_{x \to -1} \frac{\frac{1}{4}(x^2 - 1)}{x + 1}$$

$$= \lim_{x \to -1} \frac{\frac{1}{4}(x + 1)(x - 1)}{x + 1} = \lim_{x \to -1} \frac{1}{4}(x - 1) = -\frac{1}{2}$$

We also obtain

$$f'(3) = \lim_{x \to 3} \frac{(\frac{1}{4}x^2 + 1) - [\frac{1}{4}(3^2) + 1]}{x - 3} = \lim_{x \to 3} \frac{\frac{1}{4}x^2 - \frac{9}{4}}{x - 3} = \lim_{x \to 3} \frac{\frac{1}{4}(x^2 - 9)}{x - 3}$$

$$= \lim_{x \to 3} \frac{\frac{1}{4}(x - 3)(x + 3)}{x - 3} = \lim_{x \to 3} \frac{1}{4}(x + 3) = \frac{3}{2}$$

The lines tangent to the graph at the corresponding points are shown in Figure 3.1. ❑

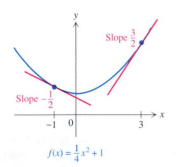

FIGURE 3.1 Tangent lines.

Because $f'(a)$ represents the slope of the line tangent to the graph of f at $(a, f(a))$, we sometimes call $f'(a)$ the **slope** of f at a. Observe that if $|f'(a)|$ is large, then near $(a, f(a))$ the graph of f will be close to a steep tangent line. By contrast, if $|f'(a)|$ is small, then near $(a, f(a))$ the graph will be close to a gently sloping tangent line (Figure 3.2).

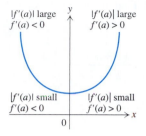

$|f'(a)|$ large
$f'(a) < 0$

y

$|f'(a)|$ large
$f'(a) > 0$

$|f'(a)|$ small
$f'(a) < 0$

$|f'(a)|$ small
$f'(a) > 0$

x

0

FIGURE 3.2

At this point let us observe that if $f'(a)$ exists, then the letter x in (2) can be replaced by other letters. For example, if t represents time, we might write

$$f'(a) = \lim_{t \to a} \frac{f(t) - f(a)}{t - a} \qquad (3)$$

In particular, suppose an object moves along a straight line on which a coordinate system has been set up, and let $f(t)$ denote the position of the object at time t. In Section 2.2 we defined the velocity $v(t_0)$ of the object at time t_0 by the formula

$$v(t_0) = \lim_{t \to t_0} \frac{f(t) - f(t_0)}{t - t_0}$$

In terms of derivatives, this is expressed as

$$v(t_0) = f'(t_0)$$

Thus $v(t_0)$ is the derivative of the position function at time t_0.

Marginal Cost and Marginal Revenue

The notion of derivative is also ideally suited to the fundamental economic concepts of marginal cost and marginal revenue. Suppose a company produces a product such as paint or honey. If little is produced, then the cost per unit of producing additional amounts of the product might be quite high (because the production process is inefficient). As mass production is introduced, the cost per unit of producing additional amounts would probably decline.

To be more specific, let $C(x)$ denote the cost of producing x units of a given product, and consider the cost per unit of producing additional amounts of the product when the production level is a units. If $x > a$, then $C(x) - C(a)$ is the additional cost required to produce the additional quantity $x - a$ of the product. Therefore

$$\frac{C(x) - C(a)}{x - a} = \text{average cost per unit of increasing production from } a \text{ to } x \text{ units}$$

Since

$$\frac{C(x) - C(a)}{x - a} = \frac{C(a) - C(x)}{a - x}$$

the ratio $(C(x) - C(a))/(x - a)$ represents the corresponding average cost when $x < a$. If the ratio approaches a limit as x approaches a, we call that limit the **marginal cost** $m_C(a)$ of producing additional amounts when the production level is a. Thus

$$m_C(a) = \lim_{x \to a} \frac{C(x) - C(a)}{x - a} = C'(a) \qquad (4)$$

For instance, if $C(x)$ is the cost in hundreds of dollars of producing x tons of clover honey, then the marginal cost $m_C(2)$ is by definition the cost per ton of producing more honey when the production level is at 2 tons. Intuitively, we would expect this to be very nearly the same as the cost per ton of producing the final ounce of a total of 2 tons of honey.

EXAMPLE 2   Let $C(x) = 1 + 8x - x^2$ for $0 \le x \le 3$. Find the marginal cost at 2.

Solution   Using (4), we find that

$$m_C(2) = \lim_{x \to 2} \frac{C(x) - C(2)}{x - 2} = \lim_{x \to 2} \frac{(1 + 8x - x^2) - (1 + 8 \cdot 2 - 2^2)}{x - 2}$$

$$= \lim_{x \to 2} \frac{-x^2 + 8x - 12}{x - 2} = \lim_{x \to 2} \frac{-(x - 2)(x - 6)}{x - 2} = \lim_{x \to 2} -(x - 6) = 4 \quad \square$$

Similarly, let $R(x)$ denote the revenue (money received) when the production level is x. Then the **marginal revenue** $m_R(a)$ from producing additional amounts when the production level is a is given by

$$m_R(a) = \lim_{x \to a} \frac{R(x) - R(a)}{x - a} = R'(a) \tag{5}$$

In each of these interpretations—velocity, marginal cost, and marginal revenue—the derivative represents a "rate of change." In fact, if f is any function differentiable at a point a, then we can think of $f'(a)$ as the **(instantaneous) rate of change of f at a.** Correspondingly, $[f(x) - f(a)]/(x - a)$ is called the **average rate of change of f** on the interval whose endpoints are x and a.

There is a simple relationship between differentiability and continuity, which is stated in the following theorem.

THEOREM   3.2

If f is differentiable at a, then f is continuous at a, that is

$$\lim_{x \to a} f(x) = f(a)$$

Proof   To show that f is continuous at a, it suffices to show that

$$\lim_{x \to a} f(x) = f(a), \quad \text{or equivalently,} \quad \lim_{x \to a} [f(x) - f(a)] = 0$$

Notice that $\lim_{x \to a} [f(x) - f(a)]/(x - a)$ exists by hypothesis. Hence by the Product Rule we have

$$\lim_{x \to a} [f(x) - f(a)] = \lim_{x \to a} \left[\frac{f(x) - f(a)}{x - a} \cdot (x - a) \right]$$

$$= \lim_{x \to a} \frac{f(x) - f(a)}{x - a} \cdot \lim_{x \to a} (x - a)$$

$$= f'(a) \cdot 0 = 0 \quad \blacksquare$$

The converse of Theorem 3.2 is false: A function can be continuous but not differentiable at a given point. For example, if $f(x) = |x|$, then f is continuous at 0; however, f is not differentiable at 0, as we prove now.

EXAMPLE 3   Let $f(x) = |x|$. Show that $f'(0)$ does not exist.

Solution   We find that

$$\lim_{x \to 0^+} \frac{f(x) - f(0)}{x - 0} = \lim_{x \to 0^+} \frac{|x| - 0}{x - 0} = \lim_{x \to 0^+} \frac{x - 0}{x - 0} = 1$$

and

$$\lim_{x \to 0^-} \frac{f(x) - f(0)}{x - 0} = \lim_{x \to 0^-} \frac{|x| - 0}{x - 0} = \lim_{x \to 0^-} \frac{-x - 0}{x - 0} = -1$$

Since the two one-sided limits are different, we conclude from Theorem 2.5 that

$$\lim_{x \to 0} \frac{f(x) - f(0)}{x - 0}$$

does not exist, which is equivalent to $f'(0)$ not existing.   ❑

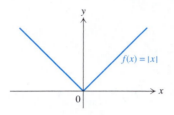

FIGURE 3.3   There is no tangent line at the origin.

Since $|x|$ does not have a derivative at 0, it follows that the graph of $|x|$ does not have a tangent line at $(0, 0)$. Notice that the graph of f is bent, or pointed, at $(0, 0)$ (Figure 3.3). More generally, if the graph of a function f is bent, or pointed, at $(a, f(a))$, then f is not differentiable at a, and there is no line tangent to the graph at $(a, f(a))$. Thus differentiability of f at a is associated with smoothness of the graph at $(a, f(a))$.

In the remainder of this chapter we will discover that many functions, including the ones we normally encounter in applications, are differentiable at all, or almost all, points in their domains. By Theorem 3.2, this means that such functions are also continuous at those points. The fact that differentiability implies continuity will be employed many times throughout the remainder of the book.

The Derivative as a Function

Nearly every function we will encounter is differentiable at all numbers, or all but finitely many numbers, in its domain. The function f' whose domain is the collection of numbers at which f is differentiable and whose value at any such number x is given by

$$f'(x) = \lim_{t \to x} \frac{f(t) - f(x)}{t - x} \tag{6}$$

is called the **derivative of** f.

Notice that in (6), x represents any number at which f is differentiable. However, when the limit on the right side of (6) is evaluated, t is the variable (which may or may not represent time) and x is regarded as a constant.

EXAMPLE 4   Let $f(x) = x^2$. Show that $f'(x) = 2x$ for all x.

Solution   It follows from (6) that for all x,

$$f'(x) = \lim_{t \to x} \frac{f(t) - f(x)}{t - x} = \lim_{t \to x} \frac{t^2 - x^2}{t - x} = \lim_{t \to x} \frac{(t - x)(t + x)}{t - x}$$

$$= \lim_{t \to x} (t + x) = 2x \quad ❑$$

EXAMPLE 5   Let $f(x) = \sqrt{x}$. Show that $f'(x) = \dfrac{1}{2\sqrt{x}}$ for $x > 0$.

Solution   From (6) we find that for $x > 0$,

$$
\lim_{t \to x} \frac{f(t) - f(x)}{t - x} = \lim_{t \to x} \frac{\sqrt{t} - \sqrt{x}}{t - x}
$$

$$
= \lim_{t \to x} \left(\frac{\sqrt{t} - \sqrt{x}}{t - x} \cdot \frac{\sqrt{t} + \sqrt{x}}{\sqrt{t} + \sqrt{x}} \right)
$$

$$
= \lim_{t \to x} \frac{t - x}{(t - x)(\sqrt{t} + \sqrt{x})}
$$

$$
= \lim_{t \to x} \frac{1}{\sqrt{t} + \sqrt{x}} = \frac{1}{\sqrt{x} + \sqrt{x}} = \frac{1}{2\sqrt{x}}
$$

Therefore

$$
f'(x) = \frac{1}{2\sqrt{x}} \quad \text{for } x > 0 \qquad \square
$$

Many functions we encounter in calculus are derivatives. Indeed, from our comments earlier in the section we find that

> velocity is the derivative of the position function: $v(t) = f'(t)$
>
> marginal cost is the derivative of the cost function: $m_C(x) = C'(x)$
>
> marginal revenue is the derivative of the revenue function: $m_R(x) = R'(x)$

Other Notations for the Derivative

The expression $f'(x)$ is neither the only notation for the derivative nor the oldest. Some of the other notations for derivatives are

$$
\dot{u}, \quad \frac{dy}{dx}, \quad \frac{d}{dx} f(x), \quad Df(x), \quad D_x f
$$

We will outline the origins of these notations.

Newton developed his calculus in the middle 1660s. He often thought of functions in terms of motion—a reasonable approach, since he was the foremost physicist of the seventeenth century. He referred to his functions as "fluents," and he called his derivatives "fluxions." Later on, he let u denote a function and $\dot{u}$ the derivative of u. The "dot" notation is still in frequent use in certain branches of science.

At about the same time, Leibniz thought of his functions as relations between variables such as x and y. Motivated by the small triangles that appeared when he attempted to find tangents to curves (Figure 3.4), he adopted the notation $\Delta y / \Delta x$ in connection with the derivative. Here Δy and Δx signify small changes in y and in x, respectively. Today we would write

$$
\lim_{\Delta x \to 0} \frac{\Delta y}{\Delta x}
$$

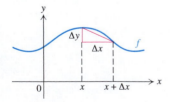

FIGURE 3.4

emphasizing the limiting process. Later, thinking of the derivative as the quotient of "infinitesimally" small differences in the variables y and x, Leibniz used dy/dx for the derivative of a variable y that depends on x. Thus if $y = x^2$, then using the fact that the derivative of the function x^2 is $2x$ (see Example 4), Leibniz would write

$$\frac{dy}{dx} = 2x \tag{7}$$

British mathematicians adopted the dot notation of Newton, while Continental mathematicians stuck to the dy/dx notation of Leibniz. The fact that the British and Continental mathematicians shunned each other's notation was only one symptom of a schism that had developed between British and Continental mathematicians over the "invention" of calculus. As a result of this conflict, mathematical communication across the English Channel dwindled, ultimately slowing mathematical progress in England. Finally, at the start of the nineteenth century, the Analytic Society at Cambridge University was founded for the expressed purpose of advancing "the principles of pure d-ism as opposed to the dot-age of the university." Thereafter the Leibniz notation permeated English mathematics, and mathematical communication across the Channel grew.

A variant form of the Leibniz form dy/dx is

$$\frac{d}{dx} f(x) \tag{8}$$

which combines the function and variable notations. An advantage of (8) is that it allows us to include the value of the function in the expression for the derivative. For example, instead of (7) we can write

$$\frac{d}{dx} (x^2) = 2x$$

This keeps the formula for the function to be differentiated in front of us. For these reasons we will frequently employ the Leibniz notation for the derivative, especially in examples.

> **Caution:** Note carefully that the expression dy/dx is merely the derivative of y as a function of x. Although it has the appearance of a fraction, it is to be regarded as a single entity. We have not assigned separate meanings to dx and dy. Moreover, to indicate the derivative of y at a particular number such as 3 in this notation, we must write something more, such as
>
> $$\frac{dy}{dx}(3), \quad \text{or} \quad \left.\frac{dy}{dx}\right|_{x=3}$$
>
> where the vertical line signifies that the derivative is to be evaluated at the number 3.

In addition to Newton's and Leibniz's notations for the derivative, another notation was initiated by the French-Italian mathematician Joseph Louis Lagrange* (1736–1813). He designated the derivative of a function f at x by the expression

*__Lagrange:__ Pronounced "La-*grahnj*."

$f'(x)$, and called f' the function "derived from f." When a variable y is used for f, the derivative f' is sometimes denoted by y'. Because of the simplicity and utility of the form $f'(x)$, it has been very popular during the past 200 years, and we will use it with regularity in the text.

Finally, we mention that the derivative is sometimes signified by the letter D, so that $f'(x)$ would be denoted by $Df(x)$ or $D_x f(x)$.

In Table 3.1 we list five common ways of expressing derivatives and show the relationships between them.

TABLE   3.1
Common Notations for Derivatives

Function	Derivative as a function	Derivative at a point
$f(x) = x^2$	$f'(x) = 2x$	$f'(4) = 8$
$y = x^2$	$\dfrac{dy}{dx} = 2x,$ or $\dfrac{d}{dx}(x^2) = 2x$	$\dfrac{dy}{dx}\bigg|_{x=4} = 8$
$f(x) = x^2$	$\dfrac{d}{dx}f(x) = 2x,$ or $\dfrac{d}{dx}(x^2) = 2x$	$\dfrac{d}{dx}f(x)\bigg|_{x=4} = 8$
$u = t^2$	$\dot{u} = 2t$	$\dot{u}(4) = 8$ or $\dot{u}\big|_{t=4} = 8$
$f(x) = x^2$	$Df(x) = 2x$	$Df(4) = 8$

EXERCISES 3.1

In Exercises 1–12 use Definition 3.1 to find $f'(a)$ for the given value of a.

1. $f(x) = 5; a = 4$ **2.** $f(x) = -4x + 7; a = 0$
3. $f(x) = 2x + 3; a = 1$ **4.** $f(x) = x^2 - 2; a = -1$
5. $f(x) = x^3; a = 0$ **6.** $f(x) = 1/x; a = -2$
7. $f(x) = 2x - 3/x; a = 1$ **8.** $f(x) = 4x^2 + 1/x^2; a = -1$
9. $f(x) = 1/\sqrt{x}; a = 4$ **10.** $f(x) = |x|; a = -\sqrt{2}$

11. $f(x) = \begin{cases} x^2 \text{ for } x < 2 \\ 4x - 4 \text{ for } x \geq 2 \end{cases}$ $a = 2$

12. $f(x) = \begin{cases} \sqrt{x} \text{ for } x \leq \frac{1}{16} \\ 2x + \frac{1}{8} \text{ for } x > \frac{1}{16} \end{cases}$ $a = \frac{1}{16}$

In Exercises 13–20 use (6) to find the derivative of the function.

13. $f(x) = -\pi$ **14.** $f(x) = 3x - 7$
15. $f(x) = -5x^2$ **16.** $f(x) = -5x^2 + x$
17. $g(x) = x^3$ **18.** $g(x) = 1/x$

19. $k(x) = \dfrac{1}{x^2} - \sqrt{7}$

20. $k(x) = x^{1/3}$
(*Hint:* $t - x = (t^{1/3} - x^{1/3})(t^{2/3} + t^{1/3}x^{1/3} + x^{2/3})$.)

In Exercises 21–24 find dy/dx.

21. $y = \frac{7}{3}$ **22.** $y = 1 - \frac{1}{2}x$
23. $y = 3x^2 + 1$ **24.** $y = 1/x^3$

In Exercises 25–28 find $\dfrac{dy}{dx}\bigg|_{x=2}$.

25. $y = 0.25$ **26.** $y = -5x + 9$
27. $y = x^2 - 3$ **28.** $y = -1/x$

In Exercises 29–36 use Definition 3.1 to determine whether the given function has a derivative at a. Where applicable, find the derivative at a.

29. $f(x) = x^{1/3}$; $a = 0$ **30.** $f(x) = x^{7/3}$; $a = 0$
31. $f(x) = |x| - x$; $a = 0$ **32.** $g(x) = |x - 3|$; $a = 3$
33. $g(x) = |x + 3|$; $a = 3$ **34.** $g(x) = 1/x^2$; $a = 0$

35. $k(x) = \begin{cases} -x^2 + 4x & \text{for } x < 0 \\ x^2 - 1 & \text{for } x \geq 0 \end{cases}$ $a = 0$

36. $k(x) = \begin{cases} 3x^2 + 4x & \text{for } x < 0 \\ x^2 + 4x & \text{for } x \geq 0 \end{cases}$ $a = 0$

In Exercises 37–40 find an equation of the line tangent to the graph of f at the given point.

37. $f(x) = x^2$; $(-2, 4)$ **38.** $f(x) = 1/x$; $(-3, -\frac{1}{3})$
39. $f(x) = \sqrt{x}$; $(4, 2)$ **40.** $f(x) = \sin x$; $(0, 0)$

41. Let f be the function whose graph appears in Figure 3.5. Estimate $f'(0)$, $f'(1)$, $f'(2)$, and $f'(3)$.

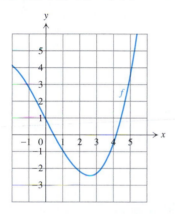

FIGURE 3.5   Graph for Exercise 41.

42. Let f be the function whose graph appears in Figure 3.6. Determine graphically which is larger, the average rate of change of f on $[1, 3]$ or the average rate of change of f on $[2, 5]$.

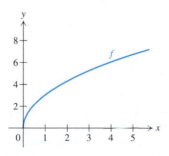

FIGURE 3.6   Graph for Exercise 42.

In Exercises 43–48 plot the graph of f for x in the window $[-10, 10]$. From the graph, determine the intervals on which f is decreasing and those on which f is increasing.

43. $f(x) = x^2 - 4x + 1$ **44.** $f(x) = x^3 - 3x$
45. $f(x) = 0.6x^4 - 0.8x^3 - 2.4x^2$ **46.** $f(x) = x + \sin x$
47. $f(x) = e^x - 2$ **48.** $f(x) = \ln(x^2 + x)$

In Exercises 49–50 let

$$m_a = \frac{f(a + 0.1) - f(a)}{0.1}$$

Find the values of m_a and $f'(a)$, and then compute $|f'(a) - m_a|$.

49. $f(x) = x^2$; $a = 2$ **50.** $f(x) = \frac{1}{3}x^3$; $a = -1$

In Exercises 51–54 plot the graph of the function

$$\frac{f(x + 0.1) - f(x)}{0.1}$$

From the resulting graphs, guess a formula for f'.

51. $f(x) = \frac{1}{2}x^2$ for $-1 \leq x \leq 1$
52. $f(x) = \sin x$ for $0 \leq x \leq 2\pi$
53. $f(x) = \cos x$ for $0 \leq x \leq 2\pi$
54. $f(x) = e^x$ for $0 \leq x \leq 2$

If $f'(a)$ exists, then by zooming in on the point $(a, f(a))$, we may find that the graph of f looks essentially like a straight line L_a in a neighborhood of the point. The slope of L_a is (approximately) $f'(a)$. In Exercises 55–58 take two points on the graph of f that are near $(a, f(a))$, and get an approximate value of the slope of L_a and hence of $f'(a)$.

55. $f(x) = x^3 - 5x$; $a = -2$ **56.** $f(x) = \sin x^2$; $a = 3$
57. $f(x) = e^x$; $a = \frac{1}{2}$ **58.** $f(x) = \ln x$; $a = 1.3$

59. Each of the following limits is of the form

$$\lim_{x \to a} \frac{f(x) - f(a)}{x - a}$$

for an appropriate function f and an appropriate point a. In each case find a formula for f, and calculate $f'(a)$.

a. $\lim_{x \to 2} \dfrac{x^4 - 16}{x - 2}$ **b.** $\lim_{x \to 2} (x^3 + 2x^2 + 4x + 8)$

60. Assume that $f'(a)$ exists. Use (17) of Section 2.2 to prove that

$$f'(a) = \lim_{h \to 0} \frac{f(a + h) - f(a)}{h}$$

61. Assume that $f'(a)$ exists. Express

$$\lim_{h \to 0} \frac{f(a - h) - f(a)}{h}$$

in terms of $f'(a)$. (*Hint:* Show that

$$\lim_{h \to 0} \frac{f(a - h) - f(a)}{h} = -\lim_{h \to 0} \frac{f(a - h) - f(a)}{-h}$$

and then use the Substitution Rule and Exercise 60.)

*62. Assume that $f'(a)$ exists. Express

$$\lim_{h \to 0} \frac{f(a + h) - f(a - h)}{h}$$

in terms of $f'(a)$. (*Hint:* Note that $f(a + h) - f(a - h) = f(a + h) - f(a) + f(a) - f(a - h)$. Then use the limit rules and Exercise 61.)

63. a. Suppose f is an even function. If $f'(a) = 2$, find $f'(-a)$. (*Hint:* Draw the graph of such a function f.)
b. Suppose f is an odd function. If $f'(a) = 2$, find $f'(-a)$. (*Hint:* Draw the graph of such a function f.)

64. a. Suppose f and g are functions such that $f = g$ on an open interval containing a. If $f'(a)$ exists, show that $g'(a)$ exists and that $f'(a) = g'(a)$.
b. Let $g(x) = |x|$. Find $g'(x)$ for $x > 0$, and then for $x < 0$.

65. Suppose f and g are functions such that $f(a) = g(a)$, and assume that $f'(a)$ and $g'(a)$ exist. Does it follow that $f'(a) = g'(a)$? Explain.

66. Suppose $y = 8x - 23$ is an equation of the line tangent to the graph of a function f at $(5, 17)$. Find $f'(5)$.

67. Suppose f is differentiable at a, and let

$$g(x) = \begin{cases} \dfrac{f(x) - f(a)}{x - a} & \text{if } x \neq a \\ f'(a) & \text{if } x = a \end{cases}$$

Show that g is continuous at a. (*Hint:* Use the definitions of continuity and $f'(a)$.)

In Exercises 68–69 plot the graph of f and repeatedly zoom in on the origin. Then tell why it is impossible to define f at 0 in such a way that $f'(0)$ exists.

68. $f(x) = \sin(1/x)$ **69.** $f(x) = x \sin(1/x)$
70. a. Let $f(x) = x^2$. For any value of r, find an equation of the line tangent to the graph of f at (r, r^2).
*b. Determine all points (p, q) in the plane that lie on some line tangent to the graph of f. (*Hint:* Draw the

graph of f, and after some experimentation, guess the answer and justify it.)

Applications

71. Suppose that while waiting for a big wave, a surfer is being propelled toward the beach in such a way that the distance (in meters) traveled is given by $f(t) = 2t^2$ for $0 \leq t \leq 1$, where t is in seconds. Find the velocity of the surfer at $t = \frac{1}{2}$.

72. Suppose the cost of preparing x barrels of clover honey for sale is $C(x)$ dollars, where

$$C(x) = 400x - (0.1)x^2 \quad \text{for } 0 \leq x \leq 1000$$

Find the marginal cost at 40.

73. Suppose a cost function C is given by $C(x) = 10{,}000 + 3/x$ for $1 \leq x \leq 100$. Find the marginal cost at 50.

74. Suppose the revenue resulting from the sale of x barrels of clover honey is $R(x)$ dollars, where

$$R(x) = 450x^{1/2} \quad \text{for } x \geq 0$$

Find the marginal revenue at 16.

75. Suppose it costs $C(x)$ thousand dollars to produce x thousand gallons of antifreeze, where

$$C(x) = 3 + 12x - 2x^2 \quad \text{for } 0 \leq x \leq 3$$

a. Find the marginal cost at 1.
b. Find the value of a for which $m_C(a) = \frac{1}{2}C(1)$.

76. In Exercise 75 suppose the cost of purchasing the ethylene glycol, a major ingredient of antifreeze, is increased by \$500 per thousand gallons. Find the marginal cost at 1, and compare your answer with the answer to part (a) of Exercise 75.

77. In Exercise 75 suppose the rent of the buildings in which the antifreeze is produced is raised \$1000 per year. Find the marginal cost at 1, and compare your answer with the answer to part (a) of Exercise 75.

78. Two ships leave port at the same time. One travels north at 15 knots (that is, 15 nautical miles per hour), and the other west at 20 knots. Show that the distance between the ships increases at a constant rate, and determine the rate of increase.

79. Two toy boats start simultaneously from virtually the same point in a lake and travel in straight lines at an angle of 60° with each other (Figure 3.7).

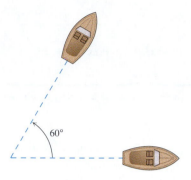

a. If the speed of each is 2 meters per minute, at what rate does the distance between them increase? (*Hint:* Use the Law of Cosines.)

b. If the distance between them increases at 3 meters a minute and both boats have the same speed, determine their speed.

FIGURE 3.7 Figure for Exercise 79.

3.2   DIFFERENTIABLE FUNCTIONS

Like Section 3.1, this section is devoted to a discussion of basic ideas related to the derivative. We begin by finding the derivatives of several functions, each of which is differentiable at every number in its domain.

DEFINITION   3.3   If f is differentiable at each number in its domain, then f is a **differentiable function.**

EXAMPLE 1   Let $f(x) = c$, where c is a constant. Show that $f'(x) = 0$ for all x.

Solution   By the definition of the derivative we find that

$$f'(x) = \lim_{t \to x} \frac{f(t) - f(x)}{t - x} = \lim_{t \to x} \frac{c - c}{t - x} = \lim_{t \to x} 0 = 0 \quad \text{for all } x \quad \square$$

EXAMPLE 2   Let $f(x) = x$. Show that $f'(x) = 1$ for all x.

Solution   Again by the definition of the derivative we find that

$$f'(x) = \lim_{t \to x} \frac{f(t) - f(x)}{t - x} = \lim_{t \to x} \frac{t - x}{t - x} = \lim_{t \to x} 1 = 1 \quad \text{for all } x \quad \square$$

We have just now found the derivatives of the functions c and x, and in Section 3.1 we found the derivative of the function x^2. With a little more effort we can find the derivative of x^n, where n is any positive integer. Let $f(x) = x^n$. Since

$$t^n - x^n = (t - x)(t^{n-1} + t^{n-2} x + t^{n-3} x^2 + \cdots + tx^{n-2} + x^{n-1})$$

we have

$$f'(x) = \lim_{t \to x} \frac{t^n - x^n}{t - x}$$

$$= \lim_{t \to x} (t^{n-1} + t^{n-2}\,x + t^{n-3}\,x^2 + \cdots + tx^{n-2} + x^{n-1})$$

$$\overbrace{\;= x^{n-1} + x^{n-2} \cdot x + x^{n-3} \cdot x^2 + \cdots + x \cdot x^{n-2} + x^{n-1}\;}^{n \text{ terms}}$$

$$= nx^{n-1}$$

We have thus shown that constant functions and x^n are differentiable functions. In the Leibniz notation the formula for the derivative of x^n is

$$\frac{d}{dx}(x^n) = nx^{n-1} \quad \text{for any integer } n > 0 \tag{1}$$

For example, if $f(x) = x^{17}$, then $f'(x) = 17x^{16}$.

Recall from Section 3.1 that the derivative is sometimes called the rate of change; in particular, this terminology is used when the function is to be interpreted geometrically or physically. If the function is given by expressing a variable y in terms of a variable x, we call the derivative dy/dx the **(instantaneous) rate of change of y with respect to x.** In the next example we will use (1) to find a rate of change.

EXAMPLE 3   Find the rate of change of the volume of a cube with respect to its side length.

Solution   The volume V of a cube with side length s is given by

$$V = s^3$$

We are to find dV/ds. From (1) with $n = 3$ it follows that

$$\frac{dV}{ds} = 3s^2$$

Thus the rate of change of the volume with respect to the side length is $3s^2$, that is, 3 times the area of a side.   ❑

We have not yet determined the derivative of the sine or cosine function. In order to achieve this goal, we will use an alternative form of (6) from Section 3.1, obtained by replacing t with $x + h$:

$$f'(x) = \lim_{h \to 0} \frac{f(x + h) - f(x)}{h} \tag{2}$$

The Derivatives of sin x and cos x

In our computation of the derivatives of the sine and cosine functions we will use the limits derived in Examples 7 and 8 of Section 2.3:

$$\lim_{h \to 0} \frac{\sin h}{h} = 1 \quad \text{and} \quad \lim_{h \to 0} \frac{\cos h - 1}{h} = 0$$

For the derivative of the sine function, we let $f(x) = \sin x$ and use (2) along with the above limits. We obtain

$$f'(x) = \lim_{h \to 0} \frac{\sin(x + h) - \sin x}{h}$$

$$\overset{\underset{\text{Section 1.7}}{(10)\text{ in}}}{=} \lim_{h \to 0} \frac{(\sin x \cos h + \cos x \sin h) - \sin x}{h}$$

$$= \lim_{h \to 0} \left(\frac{\sin x(\cos h - 1)}{h} + \frac{\cos x \sin h}{h} \right)$$

$$= \sin x \lim_{h \to 0} \frac{\cos h - 1}{h} + \cos x \lim_{h \to 0} \frac{\sin h}{h}$$

$$= (\sin x) \cdot 0 + (\cos x) \cdot 1 = \cos x \quad \text{for all } x$$

Consequently $f'(x) = \cos x$, or equivalently,

$$\frac{d}{dx} \sin x = \cos x \quad \text{for all } x \tag{3}$$

This establishes that the sine function is differentiable, and its derivative is the cosine function.

For the derivative of the cosine function, we let $f(x) = \cos x$ and find that

$$f'(x) = \lim_{h \to 0} \frac{\cos(x + h) - \cos x}{h}$$

$$\overset{\underset{\text{Section 1.7}}{(11)\text{ in}}}{=} \lim_{h \to 0} \frac{(\cos x \cos h - \sin x \sin h) - \cos x}{h}$$

$$= \lim_{h \to 0} \left(\frac{\cos x(\cos h - 1)}{h} - \frac{\sin x \sin h}{h} \right)$$

$$= \cos x \lim_{h \to 0} \frac{\cos h - 1}{h} - \sin x \lim_{h \to 0} \frac{\sin h}{h}$$

$$= (\cos x) \cdot 0 - (\sin x) \cdot 1 = -\sin x \quad \text{for all } x$$

Consequently $f'(x) = -\sin x$, or equivalently,

$$\frac{d}{dx} \cos x = -\sin x \quad \text{for all} \tag{4}$$

Therefore the cosine function is differentiable, and its derivative is the negative of the sine function.

Caution: The formulas in (3) and (4) for the derivatives of the sine and cosine functions are valid only when the variable is given in terms of radian measure. If the sine and cosine are given in terms of degree measure, the derivatives must be modified, as Exercise 71 of Section 3.4 confirms.

In order to visualize the derivatives of functions such as $\sin x$ and $\cos x$ on a graphics calculator, let us recall from (2) that if h is small enough, then

$$\frac{f(x + h) - f(x)}{h}$$

is close to $f'(x)$. If we let $f(x) = \sin x$ and $h = 0.1$, then the quotient becomes

$$\frac{\sin(x + 0.1) - \sin x}{0.1} \tag{5}$$

Plotting (5) on a graphics calculator produces the graph in Figure 3.8(a), which is a facsimile of $\cos x$ and supports (3). Analogously, if $f(x) = \cos x$, then the corresponding plot (Figure 3.8(b)) yields a facsimile of $-\sin x$, which supports (4). It is important to be aware that the graph of $(f(x + h) - f(x))/h$ does not always look like the graph of f'. However, for many of the functions commonly encountered in calculus, it will produce a reasonable copy of f', provided that h is small.

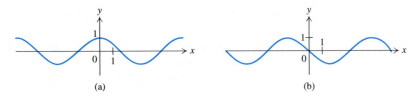

(a)   (b)

FIGURE 3.8   (a) Approximate derivative of $\sin x$. (b) Approximate derivative of $\cos x$.

The Derivative of e^x

As with $\sin x$ and $\cos x$, we can use (2) in order to find the derivative of the natural exponential function. To that end, let $f(x) = e^x$. By (5) of Section 2.1, with h replacing x,

$$\lim_{h \to 0} \frac{f(h) - f(0)}{h} = \lim_{h \to 0} \frac{e^h - 1}{h} = 1$$

Thus $f'(0) = 1$. To obtain the derivative of f at any real number x, we notice that

$$\lim_{h \to 0} \frac{f(x + h) - f(x)}{h} = \lim_{h \to 0} \frac{e^{x+h} - e^x}{h} = \lim_{h \to 0} \frac{e^x e^h - e^x}{h}$$

$$= e^x \lim_{h \to 0} \frac{e^h - 1}{h} = e^x \cdot 1 = e^x$$

Therefore $f'(x) = e^x$, or equivalently,

$$\frac{d}{dx} e^x = e^x \tag{6}$$

In other words, the natural exponential function is its own derivative:

$$f'(x) = f(x) \quad \text{for all } x$$

We emphasize that the derivation of (6) relied on formula (5) of Section 2.1, which was based on numerical calculation rather than proof. In Section 7.2 we will return to (6).

The functions c, x^n, $\sin x$, $\cos x$, and e^x, which we have discussed in this section, are differentiable. However, there are functions that are not differentiable. Indeed, we showed in Example 3 of Section 3.1 that the function $|x|$ is not differentiable at 0; this means that $|x|$ is not a differentiable function. (Nevertheless, since $|x| = x$ for $x > 0$ and $|x| = -x$ for $x < 0$, it follows that $|x|$ is differentiable at all $x \neq 0$.)

Differentiability on Intervals   Although $|x|$ is not a differentiable function, it is differentiable at every number in $(-\infty, 0)$ and every number in $(0, \infty)$. If I is an open interval, then we say that a function f is **differentiable on I** if f is differentiable at each point of I. As a result, $|x|$ is differentiable on $(-\infty, 0)$ and on $(0, \infty)$. In Chapter 4 we will be especially interested in functions that are differentiable on open intervals.

If I is a closed interval $[a, b]$ with $a < b$, then we say that f is **differentiable on I** if f is differentiable on (a, b) and if the one-sided limits

$$\lim_{t \to a^+} \frac{f(t) - f(a)}{t - a} \quad \text{and} \quad \lim_{t \to b^-} \frac{f(t) - f(b)}{t - b} \tag{7}$$

both exist. Of course, if f is differentiable at a or at b, then the corresponding one–sided limit in (7) exists. However, there are functions for which the one–sided limits in (7) exist even though f is not differentiable at b. For example, if $f(t) = |t|$, then as we saw in Example 3 of Section 3.1,

$$\lim_{t \to 0^+} \frac{|t| - 0}{t - 0} = 1 \quad \text{and} \quad \lim_{t \to 0^-} \frac{|t| - 0}{t - 0} = -1$$

Therefore the absolute value function is differentiable on every interval of the form $[-c, 0]$ or $[0, c]$, where $c > 0$, despite the fact that the function is *not* differentiable at 0.

Differentiability on a closed unbounded interval can be defined similarly. We say that a function f is differentiable on $[a, \infty)$ if f is differentiable at every number in (a, ∞) and if the first one-sided limit in (7) exists. Likewise, f is differentiable on $(-\infty, b]$ if f is differentiable at every number in $(-\infty, b)$ and if the second one-sided limit in (7) exists. For example, the absolute value function is differentiable on $(-\infty, 0]$ and on $[0, \infty)$.

Next we study the differentiability of the square root function.

EXAMPLE 4   Let $f(x) = \sqrt{x}$ (see Figure 3.9). Show that f is differentiable on $(0, \infty)$ but not on $[0, \infty)$.

Solution   By Example 5 of Section 3.1, $f'(x) = 1/(2\sqrt{x})$ for $x > 0$. Consequently f is differentiable on $(0, \infty)$. For $x = 0$ we have

$$\lim_{t \to 0^+} \frac{f(t) - f(0)}{t - 0} = \lim_{t \to 0^+} \frac{\sqrt{t}}{t} = \lim_{t \to 0^+} \frac{1}{\sqrt{t}} = \infty$$

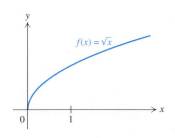

FIGURE 3.9

Since the right-hand limit at 0 does not exist, we conclude that f is not differentiable on $[0, \infty)$.   ❑

Increasing and Decreasing Functions

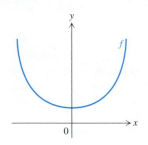

FIGURE 3.10

Notice in Figure 3.10 that to the left of the y axis the graph of f falls, whereas to the right of the axis the graph of f rises. We say that f is decreasing on $(-\infty, 0]$ and is increasing on $[0, \infty)$. More generally, if f is a function defined on an interval I, then we say that

f is **increasing on I** if $f(x) < f(z)$ whenever x and z are in I and $x < z$
f is **decreasing on I** if $f(x) > f(z)$ whenever x and z are in I and $x < z$

(see Figure 3.11).

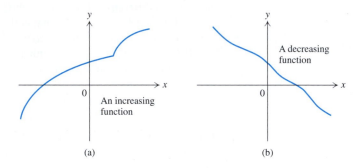

(a)   (b)

FIGURE 3.11

In order to relate these notions to the derivative, observe that if f is increasing on an open interval I, then $z - x$ and $f(z) - f(x)$ have the same sign, so that

$$\frac{f(z) - f(x)}{z - x} > 0 \quad \text{for all } x \text{ and } z \text{ in } I$$

If in addition f is differentiable on I, then for any x in I,

$$f'(x) = \lim_{z \to x} \frac{f(z) - f(x)}{z - x} \geq 0$$

The conclusion is that if f is differentiable and increasing on I, then $f'(x) \geq 0$ for every x in I. Analogously, if f is differentiable and decreasing on I, then $f'(x) \leq 0$ for every x in I.

In Chapter 4 we will be able to prove a near converse: Let I be an open interval. If $f'(x) > 0$ for all x in I, then f is increasing on I; similarly, if $f'(x) < 0$ for all x in I, then f is decreasing on I.

EXAMPLE 5   Let $f(x) = x^3 - 12x$. Use the observations above to show that f is increasing on $(-\infty, -2)$ and $(2, \infty)$ and is decreasing on $(-2, 2)$.

Solution   By (2), for any number x we have

$$f'(x) = \lim_{h \to 0} \frac{[(x + h)^3 - 12(x + h)] - (x^3 - 12x)}{h} = \lim_{h \to 0} \frac{3x^2h + 3xh^2 + h^3 - 12h}{h}$$

$$= \lim_{h \to 0} (3x^2 + 3xh + h^2 - 12) = 3x^2 - 12$$

Thus

$$f'(x) = 3x^2 - 12 = 3(x + 2)(x - 2)$$

Since $3(x + 2)(x - 2) > 0$ whenever x is in $(-\infty, -2)$ or $(2, \infty)$, f is increasing on $(-\infty, -2)$ and $(2, \infty)$. Similarly, $3(x + 2)(x - 2) < 0$ whenever x is in $(-2, 2)$ so that f is decreasing on $(-2, 2)$.   ❏

Intervals on which a function is increasing or decreasing are very important in the study of the graph of a function, which is a major topic of Chapter 4.

EXERCISES 3.2

In Exercises 1–10 use the results of this section to find the derivative of the given function at the given numbers.

1. $f(x) = -2$; $a = 1$ **2.** $f(x) = \sqrt{3}$; $a = -3$

3. $f(x) = x^2$; $a = \frac{3}{2}$, 0 **4.** $f(x) = x^3$; $a = -\frac{1}{4}$

5. $f(x) = x^4$; $a = \sqrt[3]{2}$ **6.** $f(x) = x^5$; $a = -2$

7. $f(x) = x^{10}$; $a = 1$ **8.** $f(t) = \sin t$; $a = \pi/4$, $\pi/3$

9. $f(t) = \cos t$; $a = 0$, $-\pi/3$ **10.** $f(t) = e^t$; $t = 1$, $\ln 3$

In Exercises 11–26 use either (6) of Section 3.1 or (2) of this section to find the derivative of the given function.

11. $f(x) = -2x - 1$ **12.** $f(x) = 1 - x^2$

13. $f(x) = x^5$ **14.** $f(x) = x^6$

15. $f(x) = x/(x + 1)$ **16.** $f(x) = (2x - 1)/(x + 3)$

17. $f(x) = (x^2 - 1)/(x^2 + 1)$ **18.** $y = 5 \sin x$

19. $y = -3 \cos x$ **20.** $y = \cos x - \sin x$

21. $y = x^{2/3}$

(*Hint:* $\dfrac{t^{2/3} - x^{2/3}}{t - x} = \dfrac{(t^{1/3} - x^{1/3})(t^{1/3} + x^{1/3})}{(t^{1/3} - x^{1/3})(t^{2/3} + t^{1/3} x^{1/3} + x^{2/3})}$.)

22. $y = x^{3/2}$

(*Hint:* $\dfrac{t^{3/2} - x^{3/2}}{t - x} = \dfrac{(t^{1/2} - x^{1/2})(t + t^{1/2} x^{1/2} + x)}{(t^{1/2} - x^{1/2})(t^{1/2} + x^{1/2})}$.)

*23. $y = \sqrt{x - 1}$ (*Hint:* Rationalize the numerator of the fraction that arises.)

*24. $y = \sin 2x$

*25. $f(x) = e^{2x}$ (*Hint:* $e^{2(x+h)} = e^{2x} e^{2h}$.)

*26. $f(x) = e^{3x}$

In Exercises 27–32 show that f is differentiable on the given interval.

27. $f(x) = x^2 + x$; $(-\infty, \infty)$

28. $f(x) = 2x^3 - \sqrt{x}$; $(0, \infty)$

29. $f(x) = 1/(4 - x)$; $(4, \infty)$

30. $f(x) = 1/(4 - x^2)$; $(-\infty, -2)$

31. $f(x) = |x - 1|$; $[1, \infty)$

32. $f(x) = |2 + 3x|$; $(-\infty, -\frac{2}{3}]$

33. Find a in each of the following cases.

   a. $f(x) = -2x^2$; $f'(a) = 12$

   b. $f(x) = 3x + x^2$; $f'(a) = 13$

   c. $f(x) = 1/x$; $f'(a) = -\frac{1}{9}$ (There are two possible values for a.)

   d. $f(x) = \sin x$; $f'(a) = \sqrt{3}/2$ (There are infinitely many possible values for a.)

34. Corroborate the fact that e^x is its own derivative by plotting e^x and $(e^{x+0.1} - e^x)/0.1$ together.

*35. Plot $\frac{1}{2}x^{-1/2}$ and $((x + 0.1)^{1/2} - x^{1/2})/0.1$ together, with x window $[0, 1]$ and y window $[0, 3]$. Why are the graphs not virtually coincident?

36. Suppose that a function f is differentiable on an open interval I. Show that if f is decreasing on I, then $f'(x) \leq 0$ for all x in I.

37. Suppose that $f'(x) = (x + 1)/(x - 1)$ for x in the domain of f. Which of the graphs in Figure 3.12 is the graph of f?

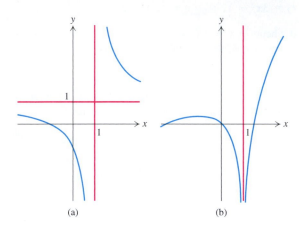

(a)   (b)

FIGURE 3.12   Graphs for Exercise 37.

38. Figure 3.13 includes the graph of f and three candidates for f'. Tell which is the graph of f', and then tell why the others are impostors.

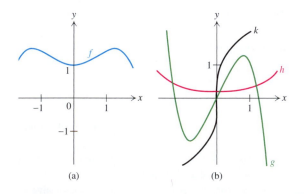

(a)   (b)

FIGURE 3.13   Graphs for Exercise 38.

39. Match the graphs in Figure 3.14(a) with the graphs of their derivatives in Figure 3.14(b).

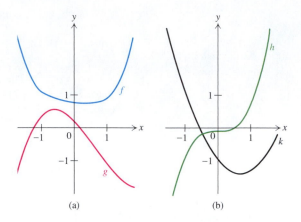

(a)   (b)

FIGURE 3.14   Graphs for Exercise 39.

40. Figure 3.15 displays the graphs of three functions f, g and h. Determine which of the three functions is the derivative of one of the other two and explain your answer.

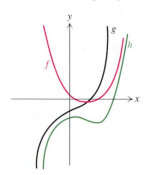

FIGURE 3.15   Graphs for Exercise 40.

41. Assume that the graph of a function f has a tangent line at $(a, f(a))$. The line **normal** to the graph of f at $(a, f(a))$ is the line through $(a, f(a))$ perpendicular to the tangent line. Find equations of the lines normal to the graphs of the following functions at the indicated points on their graphs.

a. $f(x) = x^2$; $(-1, 1)$ **b.** $f(x) = 2x - 3$; $(-1, -5)$
c. $f(x) = \sin x$; $(0, 0)$

***42.** Let $f(x) = x^2$. Show that every normal to the graph of f except the normal passing through the origin intersects the graph of f twice.

Applications

43. Suppose an object moving along the x axis has x coordinate $f(t)$ at time t given by

$$f(t) = -3 \sin t$$

Find a formula for the velocity $v(t)$ at any time t, and in particular find $v(\pi/6)$.

44. Newton's Universal Law of Gravitation states that when an astronaut is a distance r from the center of the earth, the astronaut's weight W is given by

$$W = \frac{GMm}{r^2}$$

where M is the mass of the earth, m is the mass of the astronaut, and G is the universal gravitational constant.
   a. Find a formula for the rate of change of the weight of the astronaut with respect to the distance r.
   b. Show that this rate of change is negative.
   c. What does the result of (b) mean physically?

45. The gravitational force F that the earth exerts on a unit mass depends on the distance r between the mass and the center of the earth. It is given by

$$F = \begin{cases} GMr/R^3 & \text{for } 0 < r < R \\ GM/r^2 & \text{for } r \ge R \end{cases}$$

where R is the radius of the earth, M is its mass, and G is the universal gravitational constant.
   a. Is F differentiable on $(0, \infty)$? Explain your answer.
   b. Let a, b, and c be positive constants, and let

$$f(r) = \begin{cases} ar/b & \text{for } 0 < r < b \\ c/r^2 & \text{for } r \ge b \end{cases}$$

Assuming that a and b are fixed, is there a positive constant c such that f is differentiable on $(0, \infty)$? Explain.

46. Suppose a revenue function R is given by

$$R(x) = \begin{cases} 4x & \text{for } 0 \le x \le 1 \\ 6x - x^2 - 1 & \text{for } 1 < x \le 3 \end{cases}$$

   a. Show that there is a marginal revenue at $x = 1$, and calculate it.
   b. Using part (a), show that the revenue function is differentiable on $[0, 3]$.

47. Let $R(x)$ be the revenue in dollars resulting from the harvest of x thousand bushels of wheat, and suppose

$$R(x) = \begin{cases} 1432x & \text{for } 0 < x < 6 \\ 8592 & \text{for } x \ge 6 \end{cases}$$

Show that R is differentiable on $(0, 6)$ and on $(6, \infty)$, but is not differentiable at 6. (If the government were to set a ceiling on the amount of wheat a single farmer can sell, then the function described above could represent the farmer's revenue.)

48. Show that the rate of change of the circumference of a circle with respect to the radius is constant.

49. Find the rate of change of the area of a circle with respect to its radius.

50. Find the rate of change of the area of a square with respect to the length of a side.

51. a. Find the rate of change of the area of an equilateral triangle with respect to the length of a side.
   b. Let $A(x)$ denote the area of the triangle when the length of a side is x. Find a value of x for which $A'(x) = A(x)$.

52. a. Find the rate of change of the volume of a sphere with respect to the radius.
   b. What relationship does the rate of change bear to the surface area of the sphere?

53. Suppose a student drives to college for a calculus examination, and suppose that the graph shown in Figure 3.16 represents miles from home as a function of time (in minutes). Use the graph to answer the following questions.
   a. After how many minutes does the student turn around to return home for a forgotten calculator?
   b. How long does the student stop at a red light?
   c. During which period does the student drive at a constant speed?

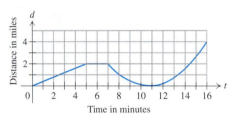

FIGURE 3.16   Travel from home to campus.

54. When radioactive iodine I^{128} disintegrates, the rate of change of the amount $A(t)$ present at time t is proportional to the amount at that time. The constant of proportionality is -0.028.
   a. Write an equation that relates $A(t)$ and the rate of change in $A(t)$.
   b. Suppose $A(1) = 3$. Find $A'(1)$.

Project

1. This project is designed to show that if $f'(a) > 0$ then although we may imagine that f must be increasing in a neighborhood of a, in fact that is not necessarily true.

   a. First let

$$f_1(x) = \begin{cases} \sin\frac{1}{x} & \text{for } x \neq 0 \\ 0 & \text{for } x = 0 \end{cases}$$

   With the help of a graphics calculator, with window $|x| \leq 1$ and $|y| \leq 1$, determine whether f_1 is increasing, or decreasing, in some open interval about 0. Explain why f_1 is, or is not, continuous or differentiable at 0.

   b. Second, let

$$f_2(x) = \begin{cases} x\sin\frac{1}{x} & \text{for } x \neq 0 \\ 0 & \text{for } x = 0 \end{cases}$$

   With the help of a graphics calculator, with window $|x| \leq 0.5$ and $|y| \leq 0.2$, determine whether f_2 is increasing, or is decreasing, in some open interval about 0. Is f_2 continuous at 0? Is f_2 differentiable at 0? Explain your answers.

   c. Third, let

$$f_3(x) = \begin{cases} x^2\sin\frac{1}{x} & \text{for } x \neq 0 \\ 0 & \text{for } x = 0 \end{cases}$$

   Using a graphics calculator, with window $|x| \leq 0.5$ and $|y| \leq 0.2$, show that $f_3'(0) = 0$, and that f_3 is sometimes increasing and sometimes decreasing in any open interval about 0.

   d. If it were true that $f_3'(0) > 0$, then we would have succeeded in finding a function with positive derivative at a number a that is sometimes increasing and sometimes decreasing in each open interval about a. However, we can succeed if, in effect, we rotate the graph of f_3 so it tilts upward. To do that, you are asked to find a constant $c > 0$ such that if

$$f(x) = \begin{cases} cx + x^2\sin\frac{1}{x} & \text{for } x \neq 0 \\ 0 & \text{for } x = 0 \end{cases}$$

   then $f'(0) > 0$, and moreover, the graph of f appears on your graphics calculator to be sometimes increasing and sometimes decreasing on every open interval containing 0.

3.3   DERIVATIVES OF COMBINATIONS OF FUNCTIONS

Since there are limit rules for sums, differences, products, and quotients of functions, it is natural to ask whether there are corresponding rules for derivatives. There are such rules, but some of the formulas for the derivatives of combinations of functions are quite different from their counterparts for limits. The desirability of such rules can be shown by means of the following example. Recall from (1) in Section 1.3 that the height $h(t)$ of a falling object at any time t is given by

$$h(t) = -4.9t^2 + v_0 t + h_0$$

This is a fairly complicated formula, and finding the derivative $h'(t)$ directly from the definition would be tedious. But if we knew how to find the derivative of a combination of functions from the derivatives of the individual functions, then obtaining $h'(t)$ would be much simpler.

Derivative of a Sum

For the sum of two differentiable functions we have the following result.

THEOREM   3.4

The Sum Rule

If f and g are differentiable at a, then $f + g$ is also differentiable at a, and

$$(f + g)'(a) = f'(a) + g'(a)$$

Proof   Using the limit rules, we find that

$$(f + g)'(a) = \lim_{t \to a} \frac{(f + g)(t) - (f + g)(a)}{t - a}$$

$$= \lim_{t \to a} \left(\frac{f(t) - f(a)}{t - a} + \frac{g(t) - g(a)}{t - a} \right)$$

$$= \lim_{t \to a} \frac{f(t) - f(a)}{t - a} + \lim_{t \to a} \frac{g(t) - g(a)}{t - a}$$

$$= f'(a) + g'(a) \quad \blacksquare$$

Theorem 3.4 says that the derivative of the sum of two functions at a given number is equal to the sum of the derivatives, provided that the two functions are differentiable at that number. Alternatively, the formula in Theorem 3.4 can be rewritten as

$$(f + g)'(x) = f'(x) + g'(x)$$

for all x at which both f and g are differentiable. This formula is reminiscent of the formula for the limit of a sum of two functions.

EXAMPLE 1   Let $k(x) = x + \sin x$. Find a formula for $k'(x)$, and then compute $k'(\pi/4)$.

Solution   Let $f(x) = x$ and $g(x) = \sin x$, so that $k = f + g$. By Example 2 and (3) of Section 3.2,

$$f'(x) = 1 \quad \text{and} \quad g'(x) = \cos x$$

Therefore Theorem 3.4 implies that

$$k'(x) = f'(x) + g'(x) = 1 + \cos x \quad \text{for all } x$$

Letting $x = \pi/4$, we conclude that

$$k'\left(\frac{\pi}{4}\right) = 1 + \cos \frac{\pi}{4} = 1 + \frac{\sqrt{2}}{2} \qquad \square$$

The sum rule for derivatives can also be stated in the Leibniz notation. If u and v are variables that depend on x, then

$$\frac{d}{dx}(u + v) = \frac{du}{dx} + \frac{dv}{dx}$$

EXAMPLE 2   Find $\dfrac{d}{dx}(\sin x + \cos x)$.

Solution   By (3) and (4) of Section 3.2,

$$\frac{d}{dx} \sin x = \cos x \quad \text{and} \quad \frac{d}{dx} \cos x = -\sin x$$

Consequently

$$\frac{d}{dx}(\sin x + \cos x) = \frac{d}{dx}\sin x + \frac{d}{dx}\cos x = \cos x - \sin x \qquad \square$$

With practice you should be able to find the derivatives of sums without writing intermediate steps. Thus for Example 2 one would normally just write

$$\frac{d}{dx}(\sin x + \cos x) = \cos x - \sin x$$

Caution: Note that $(f + g)'(a)$ may exist even when $f'(a)$ or $g'(a)$ fails to exist (see Exercise 61). Theorem 3.4 *does not* say that if $(f + g)'(a)$ exists, then $f'(a)$ and $g'(a)$ exist. Rather, it says that if $f'(a)$ and $g'(a)$ both exist, then $(f + g)'(a)$ exists.

We can readily extend Theorem 3.4 to the sum of more than two functions. Indeed, if $f_1, f_2, \ldots, f_n$ are each differentiable at a, so is $f_1 + f_2 + \cdots + f_n$, and

$$(f_1 + f_2 + \cdots + f_n)'(a) = f_1'(a) + f_2'(a) + \cdots + f_n'(a)$$

Hence if $f(x) = x^2 + x - 12 + \sin x$, then by separating f into the functions $f_1(x) = x^2$, $f_2(x) = x, f_3(x) = -12$, and $f_4(x) = \sin x$ and differentiating each of these four functions, we deduce that

$$f'(x) = 2x + 1 + \cos x$$

Derivative of a Constant Multiple of a Function

For a constant multiple of a differentiable function we have the following result.

THEOREM 3.5

The Constant Multiple Rule

If f is differentiable at a, then for any number c the function cf is differentiable at a, and

$$(cf)'(a) = cf'(a)$$

Proof From the definition of differentiability at a we have

$$(cf)'(a) = \lim_{t \to a}\frac{(cf)(t) - (cf)(a)}{t - a} = \lim_{t \to a}\frac{cf(t) - cf(a)}{t - a}$$

$$= c\lim_{t \to a}\frac{f(t) - f(a)}{t - a} = cf'(a) \quad \blacksquare$$

The formula in Theorem 3.5 can be rewritten as

$$(cf)'(x) = cf'(x)$$

for all x at which f is differentiable. It reminds us of the formula for the limit of a constant times a function.

EXAMPLE 3   Let $k(x) = 4\sqrt{x}$. Find $k'(x)$.

Solution   If $f(x) = \sqrt{x}$, then $k(x) = 4f(x)$. Since $f'(x) = 1/(2\sqrt{x})$ by Example 5 of Section 3.1, it follows from Theorem 3.5 that

$$k'(x) = 4f'(x) = 4\left(\frac{1}{2\sqrt{x}}\right) = \frac{2}{\sqrt{x}} \quad \square$$

In the Leibniz notation, if u is a variable depending on x and if c is any number, then

$$\frac{d}{dx}(cu) = c\frac{du}{dx}$$

Thus for the derivative of the function in Example 3 we would write

$$\frac{d}{dx}(4\sqrt{x}) = \frac{2}{\sqrt{x}}$$

The addition and constant multiple rules for derivatives, along with (1) in Section 3.2, imply that all polynomials are differentiable functions. They yield the following formula for the derivative of any polynomial:

$$\frac{d}{dx}(c_n x^n + c_{n-1} x^{n-1} + \cdots + c_1 x + c_0)$$
$$= nc_n x^{n-1} + (n-1)c_{n-1} x^{n-2} + \cdots + 2c_2 x + c_1 \tag{1}$$

EXAMPLE 4   Find $\left.\dfrac{d}{dx}(3x^8 - \sqrt{2}x^5 + \tfrac{3}{2}x^3 + 20x + 1)\right|_{x=-1}$.

Solution   First we use (1) to find the derivative of the function at an arbitrary x:

$$\frac{d}{dx}\left(3x^8 - \sqrt{2}x^5 + \frac{3}{2}x^3 + 20x + 1\right) = 3(8x^7) - \sqrt{2}(5x^4) + \frac{3}{2}(3x^2) + 20$$

$$= 24x^7 - 5\sqrt{2}x^4 + \frac{9}{2}x^2 + 20$$

Then we evaluate the derivative at $x = -1$:

$$\left.\frac{d}{dx}\left(3x^8 - \sqrt{2}x^5 + \frac{3}{2}x^3 + 20x + 1\right)\right|_{x=-1} = 24(-1)^7 - 5\sqrt{2}(-1)^4 + \frac{9}{2}(-1)^2 + 20$$

$$= -24 - 5\sqrt{2} + \frac{9}{2} + 20$$

$$= \frac{1}{2} - 5\sqrt{2} \quad \square$$

The formula in (1) plays a significant role when we wish to find the velocity of an object that is moving vertically above the ground and is subject only to the force of gravity. Suppose the initial height of the object is h_0 and its initial velocity is v_0. Then as we observed at the beginning of the section, the object's height $h(t)$ in meters above the ground after t seconds is given by the formula

$$h(t) = -4.9t^2 + v_0 t + h_0 \qquad (2)$$

until the object hits the ground. By (1), it follows that

$$v(t) = h'(t) = -9.8t + v_0$$

so that the velocity at time t depends on the initial velocity but not the initial position. If length is measured in feet, then $h(t) = -16t^2 + v_0 t + h_0$, so that $v(t) = -32t + v_0$.

Old Faithful, in Yellowstone National Park. *(J. Warden/Super-stock)*

EXAMPLE 5   At Old Faithful, the most famous geyser in the world, a powerful stream of boiling water shoots up from the ground. It takes many of the droplets in the stream approximately 2.5 seconds to reach their zenith. Ignoring air resistance, determine the initial velocity and maximum height of such a droplet.

Solution   First we want to determine the initial velocity, v_0, in the equation

$$v(t) = -9.8t + v_0$$

However, when the droplet is at its highest point, it is not moving up or down, so its velocity is 0. By hypothesis this occurs after 2.5 seconds. Therefore

$$0 = v(2.5) = -9.8(2.5) + v_0, \quad \text{so that } v_0 = 24.5$$

Consequently the initial velocity is approximately 24.5 meters per second. Since the initial height $h_0 = 0$, the height $h(t)$ after t seconds is given by

$$h(t) = -4.9t^2 + 24.5t$$

The maximum height occurs when $t = 2.5$. As a result, to calculate the maximum height we need to calculate $h(2.5)$:

$$h(2.5) = -4.9(2.5)^2 + 24.5(2.5) = 30.625$$

It follows that the maximum height of the droplet is 30.625 meters (nearly 100 feet) above the ground.   ❑

As a second consequence of the addition and constant multiple theorems for derivatives, we obtain the difference rule for derivatives:

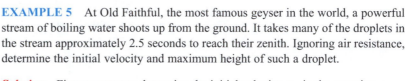

$$(f - g)'(x) = f'(x) - g'(x) \qquad (3)$$

for all x at which f and g are differentiable.

Marginal Profit

An important application of the derivative of the difference of two functions occurs in economics. Recall that if R and C are the revenue and cost functions, then the marginal revenue m_R and the marginal cost m_C are given by

$$m_R(x) = R'(x) \quad \text{and} \quad m_C(x) = C'(x)$$

for all x at which R and C are differentiable. Now let P be the profit function, defined by

$$P(x) = R(x) - C(x)$$

for all x in the domains of both R and C. The **marginal profit** m_P is defined by

$$m_P(x) = P'(x)$$

Then (3) tells us that

$$m_P(x) = R'(x) - C'(x) = m_R(x) - m_C(x)$$

for all x at which m_R and m_C are defined. Thus marginal profit equals marginal revenue minus marginal cost.

EXAMPLE 6   Suppose the revenue and cost functions are given by

$$R(x) = 3x \quad \text{and} \quad C(x) = 4\sqrt{x} \quad \text{for } x > 0$$

Find a formula for the marginal profit function.

Solution   From (1) we know that $m_R(x) = 3$, and from Example 3 we know that $m_C(x) = 2/\sqrt{x}$. We conclude that

$$m_P(x) = m_R(x) - m_C(x) = 3 - 2/\sqrt{x} \quad \square$$

Derivative of a Product

As we have seen, the derivative of a sum of functions is the sum of the derivatives, and the derivative of a difference of functions is the difference of the derivatives. By analogy it is tempting to assume that the derivative of a product of functions is the product of the derivatives. But if that were the case, we would conclude that

$$\frac{d}{dx}(x^2) = \frac{d}{dx}(x \cdot x) \overset{?}{=} \frac{d}{dx}(x)\frac{d}{dx}(x) = 1 \cdot 1 = 1$$

However, that is incorrect because

$$\frac{d}{dx}(x^2) = 2x$$

The correct formula for the derivative of a product, discovered by Leibniz, is often called the Leibniz Rule.

THEOREM   3.6
The Product Rule

If f and g are differentiable at a, then fg is also differentiable at a, and

$$(fg)'(a) = f'(a)g(a) + f(a)g'(a)$$

Proof   We will employ the device of writing $f(t)g(t) - f(a)g(a)$ in what may appear to be a curious way, but one that will allow us to use the hypothesis that f and g are differentiable at a. You can check that

$$\frac{f(t)g(t) - f(a)g(a)}{t - a} = \frac{[f(t) - f(a)]g(t) + f(a)[g(t) - g(a)]}{t - a}$$

It also follows from Theorem 3.2 and the fact that g is differentiable at a that

$$\lim_{t \to a} g(t) = g(a)$$

Using this information and several limit rules, we deduce that

$$(fg)'(a) = \lim_{t \to a} \frac{f(t)g(t) - f(a)g(a)}{t - a}$$

$$= \lim_{t \to a} \left[\frac{f(t) - f(a)}{t - a} g(t) + f(a)\frac{g(t) - g(a)}{t - a} \right]$$

$$= \lim_{t \to a} \frac{f(t) - f(a)}{t - a} \lim_{t \to a} g(t) + f(a)\lim_{t \to a} \frac{g(t) - g(a)}{t - a}$$

$$= f'(a)g(a) + f(a)g'(a) \quad \blacksquare$$

The formula in Theorem 3.6 can be rewritten as

$$(fg)'(x) = f'(x)g(x) + f(x)g'(x)$$

for all x at which both f and g are differentiable.

EXAMPLE 7   Let $k(x) = x \sin x$. Find $k'(x)$.

Solution   If $f(x) = x$ and $g(x) = \sin x$, then $k = fg$. Since

$$f'(x) = 1 \quad \text{and} \quad g'(x) = \cos x$$

we have

$$k'(x) = f'(x)g(x) + f(x)g'(x) = \sin x + x \cos x \qquad \square$$

If u and v are variables depending on x, then in the Leibniz notation,

$$\frac{d}{dx}(uv) = \frac{du}{dx}v + u\frac{dv}{dx} \tag{4}$$

EXAMPLE 8   Find $\dfrac{d}{dx}(x^2 e^x)$.

Solution   Since

$$\frac{d}{dx}(x^2) = 2x \quad \text{and} \quad \frac{d}{dx}e^x = e^x$$

we apply (4) and find that

$$\frac{d}{dx}\left(x^2\,e^x\right) = \left[\frac{d}{dx}\left(x^2\right)\right]e^x + x^2\left[\frac{d}{dx}\,e^x\right] = 2x\,e^x + x^2\,e^x \qquad \square$$

We can extend Theorem 3.6 to the derivative of the product of more than two functions. However, the larger the number of functions, the more complicated the formula becomes. With three functions f, g, and h, the formula is

$$(fgh)'(x) = f'(x)g(x)h(x) + f(x)g'(x)h(x) + f(x)g(x)h'(x) \tag{5}$$

EXAMPLE 9   Let $k(x) = x^2 \sin x \cos x$. Find $k'(x)$.

Solution   Let $f(x) = x^2$, $g(x) = \sin x$, and $h(x) = \cos x$. Then

$$f'(x) = 2x, \quad g'(x) = \cos x, \quad \text{and} \quad h'(x) = -\sin x$$

Since $k = fgh$, we use (5) to conclude that

$$k'(x) = 2x \sin x \cos x + x^2 \cos^2 x - x^2 \sin^2 x \qquad \square$$

Derivative of a Quotient

We now investigate the derivative of the quotient of two functions.

THEOREM   3.7

The Quotient Rule

If f and g are differentiable at a and $g(a) \neq 0$, then f/g is also differentiable at a, and

$$\left(\frac{f}{g}\right)'(a) = \frac{f'(a)g(a) - f(a)g'(a)}{[g(a)]^2}$$

Proof   Since $g'(a)$ exists by hypothesis, it follows from Theorem 3.2 that g is continuous at a, so that $\lim_{t \to a} g(t) = g(a)$. Because $g(a) \neq 0$ by hypothesis, it follows that $g(t) \neq 0$ for all t in some open interval about a. Therefore f/g is defined throughout some open interval about a, and the following limits exist:

$$\left(\frac{f}{g}\right)'(a) = \lim_{t \to a} \frac{\dfrac{f(t)}{g(t)} - \dfrac{f(a)}{g(a)}}{t - a} = \lim_{t \to a} \frac{f(t)g(a) - f(a)g(t)}{(t - a)g(a)g(t)}$$

$$\overset{=\,0}{}$$

$$= \lim_{t \to a} \frac{f(t)g(a) - f(a)g(a) + f(a)g(a) - f(a)g(t)}{(t - a)g(a)g(t)}$$

$$= \lim_{t \to a} \frac{f(t)g(a) - f(a)g(a)}{(t - a)g(a)g(t)} + \lim_{t \to a} \frac{f(a)g(a) - f(a)g(t)}{(t - a)g(a)g(t)}$$

$$= \lim_{t \to a} \left(\frac{f(t) - f(a)}{t - a} \cdot \frac{g(a)}{g(a)g(t)}\right) - \lim_{t \to a} \left(\frac{f(a)}{g(a)g(t)} \cdot \frac{g(t) - g(a)}{t - a}\right)$$

$$= f'(a)\frac{g(a)}{[g(a)]^2} - \frac{f(a)}{[g(a)]^2}g'(a) = \frac{f'(a)g(a) - f(a)g'(a)}{[g(a)]^2} \qquad \blacksquare$$

The formula in Theorem 3.7 can be rewritten as

$$\left(\frac{f}{g}\right)'(x) = \frac{f'(x)g(x) - f(x)g'(x)}{[g(x)]^2}$$

for all x at which both f and g are differentiable and $g(x) \neq 0$.

Every rational function is the quotient of two polynomials. Since polynomials are differentiable at every real number, the quotient rule just cited shows that every rational function is differentiable at every number in its domain.

EXAMPLE 10   Let $k(x) = \dfrac{9x^7}{x^2 + 1}$. Find $k'(x)$.

Solution   If $f(x) = 9x^7$ and $g(x) = x^2 + 1$, then $k = f/g, f'(x) = 63x^6$, and $g'(x) = 2x$. Therefore

$$k'(x) = \frac{f'(x)g(x) - f(x)g'(x)}{[g(x)]^2} = \frac{63x^6(x^2 + 1) - (9x^7)(2x)}{(x^2 + 1)^2}$$

$$= \frac{45x^8 + 63x^6}{(x^2 + 1)^2} \qquad \square$$

If u and v are variables depending on x, then in the Leibniz notation,

$$\frac{d}{dx}\left(\frac{u}{v}\right) = \frac{\dfrac{du}{dx}v - u\dfrac{dv}{dx}}{v^2} = \frac{v\dfrac{du}{dx} - u\dfrac{dv}{dx}}{v^2}$$

EXAMPLE 11   Show that $\dfrac{d}{dx}\tan x = \sec^2 x$.

Solution   Since $\tan x = (\sin x)/\cos x$, we conclude that

$$\frac{d}{dx}\tan x = \frac{d}{dx}\frac{\sin x}{\cos x} = \frac{\cos x \dfrac{d}{dx}\sin x - \sin x \dfrac{d}{dx}\cos x}{\cos^2 x}$$

$$= \frac{\cos x \cos x - \sin x(-\sin x)}{\cos^2 x} = \frac{\cos^2 x + \sin^2 x}{\cos^2 x}$$

$$= \frac{1}{\cos^2 x} = \sec^2 x \quad \square$$

In exactly the same way you can show that

$$\frac{d}{dx}\cot x = -\csc^2 x$$

The formula for the derivative of a quotient becomes more concise when $f(x) = 1$ for all x. In this case the formula is

$$\left(\frac{1}{g}\right)'(x) = \frac{-g'(x)}{[g(x)]^2} \tag{6}$$

or in the Leibniz notation, with $u = 1$,

$$\frac{d}{dx}\left(\frac{1}{v}\right) = \frac{-1}{v^2}\frac{dv}{dx}$$

EXAMPLE 12   Show that $\dfrac{d}{dx}\sec x = \sec x \tan x$.

Solution   From (6) we obtain

$$\frac{d}{dx}\sec x = \frac{d}{dx}\left(\frac{1}{\cos x}\right) = \frac{-(-\sin x)}{\cos^2 x} = \frac{1}{\cos x}\cdot\frac{\sin x}{\cos x} = \sec x \tan x \quad \square$$

Similarly,

$$\frac{d}{dx}\csc x = -\csc x \cot x$$

EXAMPLE 13   Show that $\dfrac{d}{dx}(x^{-n}) = -nx^{-n-1}$, for any positive integer n.

Solution   Combining (6) of this section and (1) of Section 3.2, we find that

$$\frac{d}{dx}(x^{-n}) = \frac{d}{dx}\left(\frac{1}{x^n}\right) = \frac{-nx^{n-1}}{(x^n)^2} = \frac{-nx^{n-1}}{x^{2n}} = -nx^{n-1-2n} = -nx^{-n-1} \quad \square$$

For example,

$$\frac{d}{dx}\left(\frac{1}{x}\right) = -\frac{1}{x^2} \quad \text{and} \quad \frac{d}{dx}(x^{-17}) = -17x^{-18}$$

The result in Example 13 is useful for many applications. For example, Boyle's Law states that if P represents pressure of a gas and V the volume of the solid in which the gas is located, then P and V are related by the equation $PV = c$, where c is a constant. Equivalently,

$$V = \frac{c}{P}$$

If we wish to determine the rate of change in the volume with respect to pressure, then by Example 13 and the constant multiple rule we find that

$$\frac{dV}{dP} = \frac{-c}{P^2}$$

Analogously Newton's Law of Gravitation says that the force $F(r)$ of attraction between two bodies of mass M and m that are separated by a distance r is given by $F(r) = GMm/r^2$. To find the rate of change of the attractive force F with respect to r we use Example 13 and the constant multiple rule to find that

$$F'(r) = \frac{-2GMm}{r^3}$$

Example 13, combined with formula (1) of Section 3.2, yields

$$\frac{d}{dx}(x^n) = nx^{n-1} \quad \text{for any nonzero integer } n \tag{7}$$

If we assume that $0 \cdot x^{-1} = 0$, then (7) holds even when $n = 0$. Using the results of the next section, we will be able to show that (7) remains valid when n is replaced by any nonzero rational number.

The existence of differentiation rules has also made it possible to program computers to find the derivatives of even quite complicated functions. Programs such as Derive, Maple, and Mathematica can accept as input a formula for a function and then print out a formula for the derivative of the function. This is referred to as **symbolic differentiation.**

EXERCISES 3.3

In Exercises 1–32 find the derivative of the given function.

1. $f(x) = -4x^3$

2. $f(x) = x^5 - x^8$

3. $f(x) = 4x^4 + 3x^3 + 2x^2 + x$

4. $f(t) = 7t^{-5}$

5. $f(t) = -\dfrac{4}{t^9}$

6. $f(t) = -3t^2 + \dfrac{3}{t^6}$

7. $g(x) = (2x - 3)(x + 5)$

8. $g(x) = (x - 2x^2)^2$

9. $g(x) = \left(1 + \dfrac{1}{x}\right)\left(2 - \dfrac{1}{x}\right)$

10. $g(x) = \left(x - \dfrac{1}{x}\right)\left(x^2 - \dfrac{1}{x^2}\right)$

11. $g(x) = -4x^{-3} + 2\cos x$

12. $g(x) = 3\sin x + 5\cos x$

13. $f(z) = -2z^3 + 4\sec z$

14. $f(z) = z^{-11} + \pi\tan z$

15. $f(z) = z^2 \sin z$

16. $f(x) = \sin x \cos x$

17. $f(x) = \sin^2 x$

18. $f(x) = (2x + 1)(x - \tan x)$

19. $f(x) = \dfrac{2x + 3}{4x - 1}$

20. $f(x) = \dfrac{-x^2 + 3}{x^2 + 9}$

21. $f(t) = \dfrac{t + 2}{t^2 + 4t + 4}$

22. $f(t) = \dfrac{t^2 - 5t + 4}{t^2 + t - 20}$

23. $f(t) = \dfrac{t^2 + 5t + 4}{t^2 + t - 20}$

24. $f(t) = \dfrac{-2t}{\sin t}$

25. $f(x) = \cot x$ (*Hint:* $\cot x = (\cos x)/\sin x$.)

26. $f(x) = \csc x$

27. $f(y) = \sqrt{y}\,\sec y$

28. $g(y) = \dfrac{\csc y}{y^2}$

29. $f(x) = e^x - 1/e^x$

30. $f(x) = e^x \sin x$

31. $f(t) = \dfrac{2t}{e^t}$

32. $f(t) = \dfrac{e^t}{1 + e^t}$

In Exercises 33–43, find dy/dx.

33. $y = (3x + 1)(2x^2 - 5x)$

34. $y = x + 1/x$

35. $y = x^2 + \dfrac{1}{x^2}$

36. $y = \dfrac{x^2 + x + 1}{x^2 - x + 1}$

37. $y = (x^3 - 1)/(x^4 + 1)$

38. $y = 4x \sec x$

39. $y = \csc x \sec x$

40. $y = \dfrac{x^2 + \sqrt{x}}{\sin x \cos x}$

41. $y = (x \sin x)/(x^2 + 1)$

42. $y = e^x(1 - \cos x)$

43. $y = \dfrac{2x \sin x}{e^x}$

44. Use trigonometric identities to find dy/dx in each case.

    a. $y = \sin 2x$ **b.** $y = \sin(-x)$

    c. $y = \cos^2(x/2)$

In Exercises 45–50 find $f'(a)$.

45. $f(x) = -7/x^9$; $a = 1$

46. $f(x) = (3x - \sin x)(x^2 + \cos x)$; $a = 0$

47. $f(x) = (3x^2 - 4x)/\pi$; $a = -2$

48. $f(x) = (2x^2 - 1)/(x^2 + 1)$; $a = \sqrt{3}$

49. $f(x) = xe^x$; $a = 1$

50. $f(x) = (\sec x)/x^2$; $a = \pi$

In Exercises 51–54 find an equation of the line tangent to the graph of f at the given point.

51. $f(x) = x^2 - 3x - 4$; $(2, -6)$

52. $f(x) = (x + 1)/(x - 1)$; $(3, 2)$

53. $f(x) = \sin x - \cos x$; $(\pi/2, 1)$

54. $f(x) = 2e^x$; $(0, 2)$

55. Let $f(x) = 2x^3 - 9x^2 + 12x + 1$. Find the points on the graph of f at which the tangent line is horizontal.

56. Let $f(x) = 3x^2 - 2x + 1$. We wish to derive an equation of the line tangent to the graph of f at $(1, 2)$. Describe the error in the following argument, and correct the erroneous part:

Since $f'(x) = 6x - 2$ yields the slope of the tangent line, an equation of the tangent line is given by
$$y - 2 = (6x - 2)(x - 1)$$

57. a. Let $f(x) = 1 + x^2$ and $g(x) = -1 - x^2$. Determine the number of lines that are tangent simultaneously to the graphs of f and g. Find the points of tangency.

   b. Let $f(x) = a + x^2$ and $g(x) = -b - x^2$, where $a > 0$ and $b > 0$. Determine the number of lines that are tangent simultaneously to the graphs of f and g. Find the points of tangency.

58. Suppose the lines $y = x + 1$ and $y = 3x - 1$ are tangent to the graph of a function f at the points $(0, 1)$ and $(1, 2)$, respectively. Show that f cannot be a polynomial function with degree 2 or less.

59. Let $f(x) = \cos x$, and let

$$g_1(x) = 1, \quad g_2(x) = 1 - \frac{x^2}{2!} \quad \text{and} \quad g_3(x) = 1 - \frac{x^2}{2!} + \frac{x^4}{4!}$$

where $2! = 2 \cdot 1$, $4! = 4 \cdot 3 \cdot 2 \cdot 1$, and in general, $n! = n(n-1)(n-2) \cdots 3 \cdot 2 \cdot 1$, and is read n **factorial.**

   a. Plot the graphs of f, g_1, g_2, and g_3. Notice that the graph of g_2 is closer to the graph of $\cos x$ than the graph of g_1 is, and the graph of g_3 is closer to the graph of $\cos x$ than the graph of g_2 is.

   b. Using g_1, g_2, and g_3 as a guide, guess a polynomial of degree 14 whose graph is essentially the graph of $\cos x$ on the interval $[-2\pi, 2\pi]$.

60. Using the fact that
$$\frac{d}{dx} \cos x = -\sin x$$

along with the result of Exercise 59, guess a polynomial h of degree 13 whose graph is essentially the graph of $\sin x$ on the interval $[-2\pi, 2\pi]$.

61. Let $f(x) = |x|$ and $g(x) = -|x|$. Find a simple formula for $f + g$. Then show that $(f + g)'(x)$ exists for all x, whereas $f'(x)$ and $g'(x)$ exist only for all nonzero x.

62. Prove that if $f'(a)$ and $(f + g)'(a)$ exist, then $g'(a)$ exists.

63. Suppose $g'(a)$ and $(fg)'(a)$ exist and $g(a) \neq 0$. Prove that $f'(a)$ exists.

64. a. Suppose that $f'(a)$, $g'(a)$, and $h'(a)$ exist. Prove that $(f + g + h)'(a)$ exists and that
$$(f + g + h)'(a) = f'(a) + g'(a) + h'(a)$$

   by using the idea of the proof of Theorem 3.4.

   b. Prove that if $f'(a)$, $g'(a)$, and $h'(a)$ exist, then
$$(fgh)'(a) = f'(a)g(a)h(a) + f(a)g'(a)h(a)$$
$$+ f(a)g(a)h'(a)$$

   (*Hint:* First, think of fgh as $(fg)h$ and apply Theorem 3.6 to the two functions fg and h. Then rewrite $(fg)'(a)$ whenever it appears, with the help of Theorem 3.6.)

65. Suppose f is differentiable at a, g is continuous at a, and $f(a) = 0$. Prove that fg is differentiable at a, and find a simple formula for $(fg)'(a)$. (*Hint:* First show that
$$\frac{f(x)g(x) - f(a)g(a)}{x - a} = \frac{f(x) - f(a)}{x - a} g(x)$$

66. Suppose that $f'(a)$ and $g'(a)$ exist, and that $f(a) \neq 0$ and $g(a) \neq 0$.

   a. Let $h = fg$. Show that $\dfrac{h'(a)}{h(a)} = \dfrac{f'(a)}{f(a)} + \dfrac{g'(a)}{g(a)}$.

   b. Let $k = f/g$. Show that $\dfrac{k'(a)}{k(a)} = \dfrac{f'(a)}{f(a)} - \dfrac{g'(a)}{g(a)}$.

67. Prove the quotient rule (Theorem 3.7) by assuming (6) and then by treating f/g as $f(1/g)$ and using the product rule (Theorem 3.6).

Applications

68. Two balls are thrown at the same time, from the same height and in the vertical direction, and with the same initial speed of 10 meters per second. The first ball is thrown upward and the second is thrown downward. Compare the velocities of the two balls.

69. Assume that a ball is shot upward from 8 feet above the ground with initial velocity of 128 feet per second. Find the velocity of the ball when it returns to the 8-foot level.

70. Suppose that a 5-foot-long chandelier falls from a 35-foot ceiling and that you are 6 feet tall and standing directly under the chandelier (Figure 3.17).

  a. How long do you have to get out of the way?

  b. If you do not duck or get out of the way, how fast will the chandelier be traveling when it hits your head?

  c. If you were only 5 feet tall, how much greater would the speed of the chandelier be when it hit your head?

FIGURE 3.17   The falling chandelier for Exercise 70.

71. Suppose you are asked to design a connecting road that joins a highway and a warehouse, as depicted in Figure 3.18(a).

  a. Show that there is exactly 1 polynomial of degree 2 that has the added property that the resulting connecting road and highway represent a function that is differentiable at −1.

  b. Again you are asked to design a connecting road, but this time it is supposed to join two highways, as depicted in Figure 3.18(b). Let f denote the resulting function. Show that it is impossible for f to coincide with a polynomial of degree 2 on $[-1, 0]$ and be differentiable at both −1 and 0.

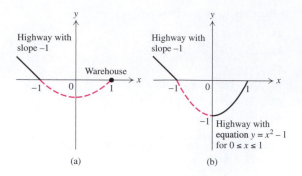

FIGURE 3.18   Figures for Exercise 71.

72. The center of gravity of a long jumper traverses a parabolic path during a jump. Suppose the height $h(x)$ in feet of the jumper's center of gravity during a 26-foot jump satisfies

$$h(x) = \frac{-2}{169}x^2 + \frac{4}{13}x + 3 \quad \text{for } 0 \le x \le 26$$

where x is the distance in feet along the ground (Figure 3.19). Show that h is differentiable on $[0, 26]$ and find a formula for $h'(x)$.

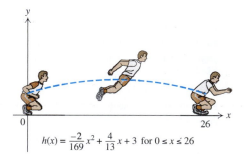

FIGURE 3.19   A long jumper.

73. Suppose you are dragging a sack of sand weighing 50 pounds. If your arm makes an angle of x radians with the ground and if the coefficient of friction is $\mu > 0$, then the minimum force $F(x)$ necessary to drag the sack is given by

$$F(x) = \frac{50\mu}{\mu \sin x + \cos x} \quad \text{for } 0 \le x \le \frac{\pi}{2}$$

Show that F is differentiable on $[0, \pi/2]$ and find a formula for $F'(x)$.

74. The response $R(x)$ to a given concentration x of acetylcholine on a frog's heart is given by

$$R(x) = \frac{x}{a + bx} \quad \text{for } 0 \le x \le 1$$

where a and b are fixed positive constants. Show that R is differentiable on $[0, 1]$ and find a formula for $R'(x)$.

75. The formula

$$f(x) = \frac{100kx^n}{1 + kx^n} \quad \text{for } x > 0$$

where n is a positive integer and k is a constant, appears in the study of the saturation of hemoglobin with oxygen. Find a formula for $f'(x)$.

76. Suppose a yogurt firm finds that its revenue and cost functions are given by

$$R(x) = 15x^{1/2} - x^{3/2} \quad \text{and} \quad C(x) = 3x^{1/2} + 4$$

respectively, for $0.5 \le x \le 5$. Here x is measured in thousands of gallons, and $R(x)$ and $C(x)$ are measured in hundreds of dollars.

   a. Find a formula for the marginal profit m_P, and calculate $m_P(1)$.

   b. Show that $m_P(4) = 0$.

77. Suppose the cost and revenue functions of a gingerbread manufacturing firm are described by

$$C(x) = \frac{x + 3}{\sqrt{x} + 1} \quad \text{and} \quad R(x) = \sqrt{x} \quad \text{for } 1 \le x \le 15$$

Find a value of x for which the profit is 0, and show that for no value of x is the marginal profit $m_P(x) = 0$.

Project

1. Suppose that you are asked to design a connecting road joining two highways, one of which has slope -1 and begins at $(-1, 0)$ and the other of which has slope 1 and begins at $(0, r)$, as depicted in Figure 3.20. However, it is necessary that the connecting road be represented by the graph of the polynomial f, and have smooth joints with the highways, in the sense that $f'(-1) = -1$ and $f'(0) = 1$.

   a. Show that there is such a function of the form $f(x) = ax^2 + bx + c$ if and only if $r = 0$. In that case, draw the graph of f.

   b. Show that for every number r, there is such a function of the form $f(x) = ax^3 + bx^2 + cx + d$. Is the road unique, that is, is there only one cubic polynomial that satisfies the requirements? Draw the graph of f.

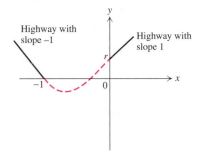

FIGURE 3.20 Figure for the Project.

3.4   THE CHAIN RULE

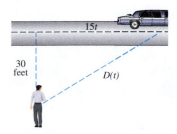

FIGURE 3.21

Suppose you are standing 30 feet away from the center of Pennsylvania Avenue, down which a Presidential motorcade is passing at 15 feet per second (just under 10 miles per hour). Suppose also that for best results in videotaping the Presidential limousine, the distance between you and the limousine must not change faster than 12 feet per second. Is the rate of change of the distance within that range 5 seconds after the limousine passes in front of you?

    Setting out to answer the question, we notice that t seconds after the limousine passes directly in front of you, the limousine has traveled $15t$ feet, so that the distance $D(t)$ between you and the limousine is given by

$$D(t) = \sqrt{(30)^2 + (15t)^2} = 15\sqrt{4 + t^2}$$

(See Figure 3.21). The question then becomes: Is $D'(5) \le 12$? Unfortunately, although we know the derivative of the square root and the function $4 + t^2$, we don't yet have a rule for finding the derivative of the composite function $\sqrt{4 + t^2}$. In this section we will derive such a rule, called the Chain Rule. It will be used repeatedly throughout the text.

Let f be differentiable at a and g differentiable at $f(a)$. In order to get an idea of what $(g \circ f)'(a)$ should be, we assume that $f(x) \neq f(a)$ for all x near a. Then by using the Product Rule for limits, as well as the Substitution Rule with $y = f(x)$, we conclude that

$$\lim_{x \to a} \frac{(g \circ f)(x) - (g \circ f)(a)}{x - a} = \lim_{x \to a} \left(\frac{g(f(x)) - g(f(a))}{f(x) - f(a)} \cdot \frac{f(x) - f(a)}{x - a} \right)$$

$$= \lim_{x \to a} \frac{g(f(x)) - g(f(a))}{f(x) - f(a)} \lim_{x \to a} \frac{f(x) - f(a)}{x - a}$$

$$= \lim_{y \to f(a)} \frac{g(y) - g(f(a))}{y - f(a)} \lim_{x \to a} \frac{f(x) - f(a)}{x - a}$$

$$= g'(f(a))f'(a)$$

Thus $(g \circ f)'(a) = g'(f(a))f'(a)$. This formula, which is the substance of the following theorem, is valid even if it is not true that $f(x) \neq f(a)$ for all x near a. The complete proof (which is more difficult than the argument just presented) is given in the Appendix.

THEOREM   3.8

The Chain Rule

Let f be differentiable at a, and let g be differentiable at $f(a)$. Then $g \circ f$ is differentiable at a, and

$$(g \circ f)'(a) = g'(f(a))f'(a) \tag{1}$$

The Chain Rule can be interpreted as saying that

$$(g \circ f)'(x) = g'(f(x))f'(x)$$

for all x such that f is differentiable at x and g is differentiable at $f(x)$. The formula in (1) is often written in the form

$$\frac{d}{dx} g(f(x)) = g'(f(x))f'(x) \tag{2}$$

EXAMPLE 1   Let $k(x) = 15\sqrt{4 + x^2}$. Find $k'(x)$.

Solution   Let $f(x) = 4 + x^2$ and $g(x) = 15\sqrt{x}$. Since

$$f'(x) = 2x \quad \text{and} \quad g'(x) = \frac{15}{2\sqrt{x}}$$

we conclude that

$$k'(x) = g'(f(x))f'(x) = 15 \frac{1}{2\sqrt{4 + x^2}} (2x) = \frac{15x}{\sqrt{4 + x^2}} \qquad \square$$

Since the function k in Example 1 is equivalent to the distance function D in the Presidential motorcade vignette, we are now able to answer the question raised with respect to videotaping the limousine: Letting $x = 5$, we find that

$$D'(5) = k'(5) = \frac{15 \cdot 5}{\sqrt{4 + 5^2}} \approx 13.9$$

Consequently 5 seconds after the Presidential limousine passes in front of you, the distance between you and the limousine is changing at approximately 13.9 feet per second, which is well above the assumed 12 feet per second limit for best results.

EXAMPLE 2   Let $k(x) = e^{cx}$, where c is a constant. Show that $k'(x) = ce^{cx}$.

Solution   Let $f(x) = cx$ and $g(x) = e^x$. Then $k = g \circ f$. Since

$$f'(x) = c \quad \text{and} \quad g'(x) = e^x$$

we conclude that

$$k'(x) = g'(f(x))f'(x) = (e^{f(x)})(c) = ce^{cx} \qquad \square$$

Using Example 2, we can find the derivative of a^x for any positive value of a. To show this, recall from (12) in Section 1.8 that

$$a^x = e^{x \ln a}$$

If we let $c = \ln a$ in Example 2, then we obtain

$$\frac{d}{dx} a^x = (\ln a) \, e^{(\ln a) x}$$

so that

$$\frac{d}{dx} a^x = (\ln a) \, a^x \tag{3}$$

For example,

$$\frac{d}{dx} 2^x = (\ln 2) \, 2^x$$

More generally, if f is any differentiable function, we find by the Chain Rule that

$$\frac{d}{dx} e^{f(x)} = e^{f(x)} f'(x) = f'(x) \, e^{f(x)} \tag{4}$$

Thus

$$\frac{d}{dx} e^{\sin x} = e^{\sin x} \frac{d}{dx} (\sin x) = e^{\sin x} \cos x$$

In order to use the Chain Rule to differentiate a given composite function, we must regard the given function as the composite $g \circ f$ of two functions f and g. Thus we need to decide which functions are to be f and g. Since

$$(g \circ f)(x) = g(f(x))$$

we call f the "inner" function and g the "outer" function. For example, with the function $(x - 2x^3)^{-11}$ we would choose $f(x) = x - 2x^3$ and $g(x) = x^{-11}$. Analogously, for the function $\cos x^5$ we would choose $f(x) = x^5$ and $g(x) = \cos x$. In practice the choices of f and g are usually made mentally and not written down. Finally, we mention that in finding $(g \circ f)'(x) = g'(f(x))f'(x)$ we first compute $g'(x)$ and replace x by $f(x)$ to obtain $g'(f(x))$, and then multiply the result by $f'(x)$.

EXAMPLE 3   Find the derivatives of the following functions.
a. $k(x) = (x - 2x^3)^{-11}$ **b.** $y = \cos x^5$

Solution   **a.** Using the comments just above, we find that

$$k'(x) = [-11(x - 2x^3)^{-12}](1 - 6x^2) = (-11 + 66x^2)(x - 2x^3)^{-12}$$

b. Here we have

$$\frac{dy}{dx} = [-\sin x^5](5x^4) = -5x^4 \sin x^5$$

or equivalently,

$$\frac{d}{dx} [\cos x^5] = [-\sin x^5](5x^4) = -5x^4 \sin x^5 \qquad \square$$

The Chain Rule assumes a very suggestive form in the Leibniz notation. Suppose the functions f and g in the Chain Rule are already given, and let

$$u = f(x) \quad \text{and} \quad y = g(u)$$

Then $y = g(f(x))$, $du/dx = f'(x)$, and $dy/du = g'(u)$. Therefore by (2),

$$\frac{dy}{dx} = \frac{d}{dx} g(f(x)) = g'(f(x))f'(x) = g'(u)f'(x) = \frac{dy}{du} \frac{du}{dx}$$

or more concisely,

$$\frac{dy}{dx} = \frac{dy}{du} \frac{du}{dx} \tag{5}$$

Formula (5) is the Chain Rule in the Leibniz notation.

Caution: Notice that if dy/du and du/dx were quotients rather than expressions for derivatives, we could cancel the du's and make (5) into an identity. But we stress that *du* has not been defined as an entity, and consequently it is not legitimate to cancel the *du*'s. Nonetheless, the resemblance between (5) and an algebraic identity makes it easy to remember.

EXAMPLE 4   Let $y = \sin^8 x$. Find *dy/dx*.

Solution   We let $u = \sin x$. Then $y = \sin^8 x = u^8$, so it follows from (5) that

$$\frac{dy}{dx} = \frac{dy}{du}\frac{du}{dx} = (8u^7)(\cos x) = 8 \sin^7 x \cos x \qquad \square$$

In (5) there are two derivatives of *y*: *dy/dx* and *dy/du*. It is important to realize that these stand for derivatives of different functions. In fact, by comparing (5) with (2) we see that

$$\frac{dy}{dx} \quad \text{represents} \quad \frac{d}{dx} g(f(x))$$

whereas

$$\frac{dy}{du} \quad \text{represents} \quad g'(u) = g'(f(x)), \text{ that is, } g' \text{ evaluated at } f(x).$$

Thus when we use variables, it is important to specify the variable with respect to which we differentiate.

Formula (5) is especially useful when *y* is not given explicitly in terms of *x* but is given in terms of an intermediate variable, as in the next example.

EXAMPLE 5   Suppose the radius *r* of a balloon varies with respect to time according to the equation $r = 1 + 1/t^2$. Find the rate of change of the balloon's volume with respect to time.

Solution   If *V* denotes the volume, then $V = \frac{4}{3}\pi r^3$, while by assumption $r = 1 + 1/t^2$. Therefore (5) tells us that

$$\frac{dV}{dt} = \frac{dV}{dr}\frac{dr}{dt} = 4\pi r^2\left(\frac{-2}{t^3}\right) = \frac{-8\pi}{t^3}\left(1 + \frac{1}{t^2}\right)^2 = \frac{-8\pi}{t^7}(t^2 + 1)^2 \qquad \square$$

We have not yet found a formula for the derivative of ln *x*, nor have we even shown that ln *x* has a derivative. However, assuming that it is differentiable, we will now use the Chain Rule to find a formula for its derivative.

Recall from Section 1.8 that $e^{\ln x} = x$, so that

$$\frac{d}{dx}(e^{\ln x}) = \frac{d}{dx}(x) = 1 \tag{6}$$

By the Chain Rule (or by (4) with $f(x) = \ln x$) we find that

$$\frac{d}{dx}(e^{\ln x}) = (e^{\ln x})\frac{d}{dx}(\ln x) = x\frac{d}{dx}(\ln x) \tag{7}$$

Combining (6) and (7), we deduce that

$$x \frac{d}{dx} (\ln x) = 1$$

Dividing by x yields

$$\frac{d}{dx} (\ln x) = \frac{1}{x} \tag{8}$$

Thus we have the derivative of $\ln x$. We will return to the differentiability of $\ln x$ in Section 5.7.

From (7) of Section 3.3 we know that

$$\frac{d}{dx} (x^n) = nx^{n-1} \quad \text{for any nonzero integer } n$$

Now let n be replaced by any nonzero real number r. In order to prove that

$$\frac{d}{dx} (x^r) = rx^{r-1}$$

we observe that $x^r = e^{r \ln x}$, and use the Chain Rule and (8):

$$\frac{d}{dx} (x^r) = \frac{d}{dx} e^{r \ln x} = e^{r \ln x} \frac{d}{dx} (r \ln x) = e^{r \ln x} \left(\frac{r}{x} \right) = x^r \left(\frac{r}{x} \right) = rx^{r-1}$$

More simply,

$$\frac{d}{dx} (x^r) = rx^{r-1} \text{ for all } r \neq 0 \tag{9}$$

EXAMPLE 6   Find $\dfrac{d}{dx} (x^{-2/5})$.

Solution   Applying (9), we find that

$$\frac{d}{dx} (x^{-2/5}) = -\frac{2}{5} x^{(-2/5)-1} = -\frac{2}{5} x^{-7/5} \quad \square$$

EXAMPLE 7   Let $k(x) = (3x^2 + 1)^{7/4}$. Find $k'(x)$.

Solution   Using (9) and the Chain Rule, we find that

$$k'(x) = \frac{7}{4} (3x^2 + 1)^{(7/4)-1}(6x) = \frac{21}{2} x(3x^2 + 1)^{3/4} \quad \square$$

EXAMPLE 8   Suppose y is a differentiable function of x. Express the derivative (with respect to x) of each of the following in terms of x, y, and dy/dx.
a. y^3 **b.** $\sin y$ **c.** $2x^3 y^4$

Solution **a.** It might appear that the derivative of y^3 with respect to x would be $3y^2$, but this is false because we must differentiate y^3 with respect to x, not with respect to y. To differentiate correctly, we use the Chain Rule in the form

$$\frac{dv}{dx} = \frac{dv}{dy}\frac{dy}{dx} \tag{10}$$

where $v = y^3$. We obtain

$$\frac{d}{dx}(y^3) = \left(\frac{d}{dy}(y^3)\right)\frac{dy}{dx} = 3y^2\frac{dy}{dx}$$

b. Using (10) with $v = \sin y$, we have

$$\frac{d}{dx}(\sin y) = \left(\frac{d}{dy}(\sin y)\right)\frac{dy}{dx} = (\cos y)\frac{dy}{dx}$$

c. Since $2x^3y^4$ is the product of the functions $2x^3$ and y^4, we take the derivative of the product, using (10) with $v = y^4$ when we differentiate y^4 with respect to x:

$$\frac{d}{dx}(2x^3y^4) = \left[\frac{d}{dx}(2x^3)\right]y^4 + 2x^3\frac{d}{dx}(y^4) = 6x^2y^4 + 2x^3\left(4y^3\frac{dy}{dx}\right)$$

$$= 6x^2y^4 + 8x^3y^3\frac{dy}{dx} \quad \square$$

The Compound Chain Rule

The Chain Rule can be carried a step further. Let

$$k(x) = (h \circ g \circ f)(x) = h(g(f(x)))$$

and let f be differentiable at x, g differentiable at $f(x)$, and h differentiable at $g(f(x))$. Since

$$k(x) = (h \circ g \circ f)(x) = h((g \circ f)(x))$$

a first application of the Chain Rule yields

$$k'(x) = h'(g(f(x)))(g \circ f)'(x)$$

A second application of the Chain Rule, to the term on the right, yields

$$k'(x) = h'(g(f(x)))g'(f(x))f'(x) \tag{11}$$

The derivative of k is thus obtained in a chainlike fashion. In the formula, the derivative of h at the number $g(f(x))$ appears first, then the derivative of g at the number $f(x)$, and finally the derivative of f at the number x.

EXAMPLE 9 Let $k(x) = \cos^3 4x$. Find $k'(x)$, and calculate $k'(\pi/6)$.

Solution Since $k(x) = (\cos 4x)^3$, we let $f(x) = 4x$, $g(x) = \cos x$, and $h(x) = x^3$. Then $k(x) = h(g(f(x)))$. By (11) we find that

$$k'(x) = (3\cos^2 4x)(-\sin 4x)(4) = -12\cos^2 4x \sin 4x$$

In particular,

$$k'\left(\frac{\pi}{6}\right) = -12 \cos^2 \frac{2\pi}{3} \sin \frac{2\pi}{3}$$

$$= -12\left(-\frac{1}{2}\right)^2\left(\frac{\sqrt{3}}{2}\right) = -\frac{3}{2}\sqrt{3} \qquad \square$$

In Leibniz notation, (11) becomes

$$\frac{du}{dx} = \frac{du}{dv}\frac{dv}{dy}\frac{dy}{dx} \tag{12}$$

where $y = f(x)$, $v = g(y)$, and $u = h(v)$.

EXAMPLE 10   Suppose y is a differentiable function of x. Find $\dfrac{d}{dx}(\sin^3 y)$ in terms of y and dy/dx.

Solution   If we let $v = \sin y$ and $u = v^3$, then

$$\sin^3 y = (\sin y)^3 = v^3 = u$$

so that by (12),

$$\frac{d}{dx}(\sin^3 y) = \frac{du}{dx} = \frac{du}{dv}\frac{dv}{dy}\frac{dy}{dx} = (3v^2)(\cos y)\frac{dy}{dx} = (3\sin^2 y \cos y)\frac{dy}{dx} \qquad \square$$

With a little practice you will probably find it easier to apply the Chain Rule from the outside inward, without introducing intermediate variables. For Example 10 one could write

$$\frac{d}{dx}(\sin^3 y) = (3\sin^2 y)(\cos y)\frac{dy}{dx}$$

The Chain Rule can be applied to even longer composites. The procedure is always the same: Differentiate from the outside inward, and multiply the resulting derivatives (evaluated at the appropriate numbers). For example,

$$\frac{d}{dx}[\sin(\cos(\tan^5 x))] = [\cos(\cos(\tan^5 x))][-\sin(\tan^5 x)](5\tan^4 x)(\sec^2 x)$$

Summary of Differentiation Rules

We have now presented all the basic rules of differentiation. Using them, you will be able to differentiate all sorts of functions, including very complicated ones. To have the rules readily available, we list them here:

1. $(f + g)'(x) = f'(x) + g'(x)$	$\dfrac{d}{dx}(u + v) = \dfrac{du}{dx} + \dfrac{dv}{dx}$
2. $(cf)'(x) = cf'(x)$	$\dfrac{d}{dx}(cu) = c\dfrac{du}{dx}$

$$3. \quad (fg)'(x) = f'(x)g(x) + f(x)g'(x) \qquad \frac{d}{dx}(uv) = \frac{du}{dx}v + u\frac{dv}{dx}$$

$$4. \quad \left(\frac{f}{g}\right)'(x) = \frac{f'(x)g(x) - f(x)g'(x)}{[g(x)]^2} \qquad \frac{d}{dx}\left(\frac{u}{v}\right) = \frac{v\frac{du}{dx} - u\frac{dv}{dx}}{v^2}$$

$$5. \quad (g \circ f)'(x) = \frac{d}{dx}g(f(x)) = g'(f(x))f'(x) \qquad \frac{dy}{dx} = \frac{dy}{du}\frac{du}{dx}$$

There are still some functions whose derivatives cannot be computed via these rules. However, the derivative of such a function can sometimes be computed directly from the definition. For instance, if $f(x) = x|x|$, then we cannot apply any of the rules to obtain $f'(0)$ because $|x|$ is not differentiable at 0. Nevertheless, using the definition of the derivative, we find that

$$f'(0) = \lim_{x \to 0} \frac{f(x) - f(0)}{x - 0} = \lim_{x \to 0} \frac{x|x| - 0}{x - 0} = \lim_{x \to 0} |x| = 0$$

But the great majority of the differentiable functions you will encounter can be differentiated by rules 1–5.

EXERCISES 3.4

In Exercises 1–38 find the derivative of the function.

1. $f(x) = x^{9/4}$

2. $f(x) = x^{2/3} - 7x^{-1/3}$

3. $f(x) = (1 - 3x)^{3/2}$

4. $f(x) = (4 - 3x^2)^{400}$

5. $f(x) = x\sqrt{2 - 7x^2}$

6. $f(x) = \sqrt{4x} - \sqrt{x}$

7. $f(t) = \sin 5t$

8. $f(t) = 3 \cos \pi t^2$

9. $f(t) = \sin^4 t + \cos^4 t$

10. $g(x) = \tan^6 x$

11. $g(x) = \sqrt[3]{1 - \sin x}$

12. $g(x) = (\tan x - \cot x)^{-1/3}$

13. $f(x) = \cos(\sin x)$

14. $f(x) = \tan(\csc x)$

15. $f(x) = \left(\frac{x - 1}{x + 1}\right)^3$

16. $f(x) = (\sqrt{x^2 - 1})/x$

17. $f(x) = \dfrac{1}{x\sqrt{5 - 2x}}$

18. $f(x) = \dfrac{(x^2 + 1)^2}{(x^4 + 1)^4}$

19. $f(x) = x \cos(1/x)$

20. $f(x) = x^2 \sec(1/x^3)$

21. $g(z) = \sqrt{2z - (2z)^{1/3}}$

22. $g(z) = [z^7 + (z^2 - 1)^5]^{-2}$

23. $g(z) = \cos^2(3z^6)$

24. $g(z) = \sqrt{1 + \sin^2 z}$

25. $f(x) = \cos(1 + \tan 2x)$

26. $f(x) = \cot^2(2\sqrt{3x + 1})$

27. $f(x) = e^{(x^5)}$

28. $f(x) = e^{-\sqrt{x}}$

29. $f(t) = \tan e^{3t}$

30. $f(t) = e^{-t} \sin at$

31. $g(t) = \ln(t^2 + 1)$

32. $g(t) = \ln(\ln t)$

33. $g(t) = (\ln t)^2$

34. $g(t) = \ln(\sin e^t)$

35. $f(x) = 3^{5x-7}$

36. $f(x) = 6x/(1 + 6^x)$

37. $f(x) = \log_3(x^2 + 4)$

38. $f(x) = x \log_2 x$

In Exercises 39–50 find dy/dx.

39. $y = 3x^{-2/3}$

40. $y = \left(2x + \dfrac{1}{x}\right)^{-6}$

41. $y = -x\sqrt{1 + 3x^2}$

42. $y = \dfrac{1}{(x^8 + 1)^{12} + 1}$

43. $y = \left(\dfrac{1}{x \sin x}\right)^{2/3}$

44. $y = \sin\sqrt{3x + 1}$

45. $y = \tan^3 \frac{1}{2}x$

46. $y = \csc(1 - 3x)^2$

47. $y = e^{\cos x}$

48. $y = e^{-1/x}$

49. $y = a^x \cos bx$

50. $y = a^x \ln x$

In Exercises 51–60 suppose that y is a differentiable function of x. Express the derivative of the given function with respect to x in terms of x, y, and dy/dx.

51. y^5

52. $y^{-2/3}$

53. $2/y$

54. $\cos y^2$

55. $\sin\sqrt{y}$

56. $\sec\sqrt{y^2 - 1}$

57. x^3y^2

58. $1/(x^2 - xy + y^3)$

59. $\sqrt{x^2 + y^2}$

60. $\cos x^5y^3$

61. Let $f(x) = \ln(x + \sqrt{x^2 - 1})$. Show that $f'(x) = 1/\sqrt{x^2 - 1}$.

62. Let $f(x) = \ln \dfrac{x + \sqrt{x^2 - 1}}{x - \sqrt{x^2 - 1}}$. Use properties of logarithms and Exercise 61 to find a simple formula for $f'(x)$.

In Exercises 63–68 find an equation of the line l tangent to the graph of f at the given point.

63. $f(x) = 1/(x + 1)^2$; $(0, 1)$

64. $f(x) = (1 + x^{1/3})^{2/3}$; $(-8, 1)$

65. $f(x) = -2 \cos 3x$; $(\pi/3, 2)$

66. $f(x) = \sin \sqrt{x}$; $(\pi^2, 0)$

67. $f(x) = 2e^{-3x}$; $(0, 2)$

68. $f(x) = \ln (\sin x)$; $\left(\dfrac{\pi}{6}, -\ln 2 \right)$

69. Plot $[\ln (x + 0.1) - \ln x]/0.1$ with both windows equal to $[0, 5]$. Is the graph you obtain consistent with (8)? Explain.

70. Assume that f and g are differentiable functions, and suppose that the values of $f, f', g,$ and g' at $-3, 0, 1,$ and 2 are prescribed as in the table below:

x	$f(x)$	$f'(x)$	$g(x)$	$g'(x)$
-3	2	4	6	11
0	1	2	-3	-7
1	16	-3	4	13
2	-3	2	0	$\sqrt{2}$

Determine $(f \circ g)'(x)$ and $(g \circ f)'(x)$ for as many values of x as the table allows.

71. From Section 1.7 we know that x degrees represents $(\pi/180) x$ radians. Let s and c represent the sine and cosine functions in terms of degrees. This means that

$$s(x) = \sin \left(\frac{\pi}{180} x \right) \quad \text{and} \quad c(x) = \cos \left(\frac{\pi}{180} x \right)$$

a. Express the derivatives $s'(x)$ and $c'(x)$ in terms of $s(x)$ and $c(x)$.

b. Plot the graphs of s and c. (From the graphs you might well understand why radian measure is more satisfactory than degree measure in calculus.)

72. Suppose that E is a function such that $E'(x) = E(x)$ for all real numbers x. (The function e^x is such a function.)

a. Let c be a number. Use the Chain Rule to find

$$\frac{d}{dx} E(cx)$$

b. Let $g(x) = [E(cx)]/e^{cx}$. Find $g'(x)$. (In Section 4.4 the result of (b) will be used in order to find a formula for E.)

73. a. Let f be differentiable, and let $g(x) = f(-x)$. Use the Chain Rule to find $g'(x)$ in terms of $f'(x)$.

b. Prove that the derivative of an even function is an odd function.

c. Prove that the derivative of an odd function is an even function.

Applications

74. The angular displacement $\theta (t)$ of a pendulum bob at time t is given by

$$\theta (t) = a \cos 2\pi\omega t$$

where ω is the frequency and a is the maximum displacement (Figure 3.22). Find the rate of change of the angular displacement as a function of time. (The rate of change is called the **angular velocity** of the bob.)

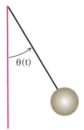

FIGURE 3.22   Angular displacement of a pendulum bob.

75. Suppose the amount $F(t)$ of water (in tons) that flows through a dam from midnight until t hours after midnight is controlled so that

$$f(t) = \frac{336,000}{\pi} \left(1 + \cos \frac{\pi t}{24} \right) \quad \text{for } 0 \le t \le 24$$

Show that F is differentiable on $[0, 24]$, and find a formula for the rate of flow of water (in tons per hour).

76. If we ignore air resistance, then the range $R (\theta)$ of a baseball hit at an angle θ with respect to the x axis and with initial velocity v_0 is given by

$$R(\theta) = \frac{v_0^2}{g} \sin 2\theta \quad \text{for } 0 \le \theta \le \frac{\pi}{2}$$

where g is the acceleration due to gravity.

a. If $v_0 = 30$ (meters per second) and $g = 9.8$ (meters per second per second), calculate $R'(\pi/4)$.

b. Determine those values of θ for which $R'(\theta) > 0$.

77. Under certain conditions the percentage efficiency of an internal combustion engine is given by

$$E = 100\left(1 - \frac{v}{V}\right)^{0.4}$$

where V and v are, respectively, the maximum and minimum volumes of air in each cylinder.

a. If V is kept constant, find the derivative of E with respect to v.

b. If v is kept constant, find the derivative of E with respect to V.

78. If two bodies are a distance r apart, then the gravitational force $F(r)$ exerted by one body on the other is given by

$$F(r) = \frac{k}{r^2} \quad \text{for } r > 0$$

where k is a positive constant. Suppose that as a function of time, the distance between the two bodies is given by

$$r(t) = 64 + 48t - 16t^2 \quad \text{for } 0 < t < 3$$

a. Find the rate of change of the force with respect to time.

b. Show that $(F \circ r)'(1) = -(F \circ r)'(2)$.

79. Suppose a rocket is launched vertically with a maximum velocity of v_0 miles per second, which is reached moments after takeoff. Suppose also that the velocity $v(r)$ in miles per second when the rocket is a distance of r miles from the center of the earth is given by the formula

$$v(r) = \sqrt{\frac{192{,}000}{r} + v_0^2 - 48} \quad \text{for } r \geq 4000$$

a. Find a formula for the rate of change in velocity with respect to r.

b. If $v_0 = 8$ miles per second, what is the rate of change of the rocket's velocity with respect to the distance from the center of the earth when that distance is 24,000 miles?

80. The weight $W(t)$ of a tree limb is a function of the age t of the limb. It can be approximated by the formula

$$W(t) = kt^r \quad \text{for } t > 0$$

where r is a rational number between 0 and 1 and k is a positive constant. For $t > 0$, the **specific rate of growth** is defined to be $W'(t)/W(t)$. Find a formula for the specific rate of growth.

81. Recall that the volume V of a spherical balloon is related to the radius r of the balloon by the formula

$$V = \frac{4}{3}\pi r^3$$

Suppose the radius is increasing at the constant rate of 10 inches per minute. Using the Chain Rule, find the rate of change of V with respect to time.

82. The surface area S of the balloon mentioned in Exercise 81 is related to the balloon's radius by the formula

$$S = 4\pi r^2$$

Using the Chain Rule, find the rate of change of the volume with respect to the surface area. (*Hint:* $r = (S/(4\pi))^{1/2}$.)

83. If A, x, and h denote the area, length of side, and altitude of an equilateral triangle, respectively, then they are related by the formulas

$$A = \frac{\sqrt{3}}{4}x^2 \quad \text{and} \quad x = \frac{2\sqrt{3}}{3}h$$

Using the Chain Rule, find the rate of change of the area with respect to the altitude, and determine the rate of change when $h = \sqrt{3}$.

84. Cider is poured into a cylindrical vat 4 feet in diameter and 5 feet tall. After t seconds the cider level is $\frac{1}{3}t$ feet above the base of the vat. Show that the rate of change of the volume with respect to time is constant.

85. The **demand** for a product gives the quantity $D(x)$ that can be sold when the price of one unit of the product is x. The demand almost always has a negative derivative.

a. Explain why the demand should have a negative derivative.

b. Let

$$D(x) = \sqrt{3 - 2x} \quad \text{for} \quad 0 < x < \tfrac{3}{2}$$

Show that $D'(x) < 0$ for $0 < x < \tfrac{3}{2}$.

86. Suppose the research department of the Bulb Company determines that the demand for bulbs is given by

$$D(x) = 1000\left(\frac{6}{x - 16}\right)^{1/3} \quad \text{for } 17 < x < 37$$

where x is in cents.

a. Find $D'(22)$.

b. Show that $D'(x) < 0$ for all x in $(17, 37)$.

Projects

1. This problem harks back to the project at the end of Section 3.2, which involved showing that there is a function f such that $f'(0) > 0$ but with the peculiar property that on the calculator or computer screen the graph of f appears *not* to be increasing throughout any open interval containing 0. The function that had such properties was given by

$$f(x) = \begin{cases} cx + x^2 \sin \frac{1}{x} & \text{for } x \neq 0 \\ 0 & \text{for } x = 0 \end{cases}$$

for some positive constant c. With differentiation rules at our disposal, we can now calculate $f'(x)$ for all nonzero values of x.

   a. Find $f'(0)$, and show that $f'(0) > 0$. Then by using the derivative $f'(x)$ for x near 0, show that f is not increasing throughout any neighborhood of 0 if $c < 1$.

   b. Suppose that g is a polynomial, and assume that $g'(a) > 0$. Prove that there is some interval about a on which $g'(x) > 0$. (*Hint:* You will need to use the fact that g' is continuous.)

2. A number p is a **fixed point** of f provided that $f(p) = p$, that is, the graph of f and the line $y = x$ meet at the point $(p, f(p))$. For example, if $f(x) = \frac{1}{2} \sin x$, then 0 is a fixed point because $f(0) = \frac{1}{2} \sin 0 = 0$. Also, if $f(x) = x^3 - 6$, then 2 is a fixed point because $f(2) = 2^3 - 6 = 2$. It can be shown that if p is fixed point of f and if $|f'(p)| < 1$, then $f(x)$ is closer to p than x is, for all x in some open interval about p. In that case, p is an **attracting fixed point.** Analogously, if $|f'(p)| > 1$, then $f(x)$ is farther away from p for x close to p. In that case, p is a **repelling fixed point.** For f above, 0 is an attracting fixed point, and for g above, 2 is a repelling fixed point.

   Next, suppose $f(p) = q$ and $f(q) = p$, where $p \neq q$. Then p is a **period-2 point** for f. For example, if $f(x) = 1/x$, then $f(-2) = -1/2$ and $f(-1/2) = -2$, so that $-1/2$ is a period-2 point for f. Notice that $(f \circ f)(p) = f(f(p)) = f(q) = p$, so that p is a fixed point for $f \circ f$. Therefore if p is a period-2 point of f, and if $|(f \circ f)'(p)| < 1$, then $(f \circ f)(x)$ is nearer to p than x is, for all x in some open interval about p. In that case, p is an **attracting period-2 point** for f. There is an analogous definition for repelling period-2 point. When is $|(f \circ f)'(p)| < 1$? By the Chain Rule,

$$(f \circ f)'(x) = f'(f(x)) f'(x)$$

Letting $x = p$ and noting that $f(p) = q$, we find that

$$(f \circ f)'(p) = f'(f(p))f'(p) = f'(q)f'(p) = f'(p)f'(q)$$

Consequently,

$$|(f \circ f)'(p)| < 1 \quad \text{if and only if} \quad |f'(p)||f'(q)| < 1$$

Similarly,

$$|(f \circ f)'(p)| > 1 \quad \text{if and only if} \quad |f'(p)||f'(q)| > 1$$

Now we are ready for the exercises.

   a. Show that 0 is an attracting point for $0.5 \sin x$, and that 2 is a repelling fixed point for $x^3 - 6$.

   b. Let $f(x) = -\frac{1}{2} x^2 - x + \frac{1}{2}$. Show that 1 is a period-2 point for f, and determine whether it is attracting or repelling.

   c. Let $f(x) = 3.2x - 3.2x^2$. Use a calculator to find an approximate period-2 point for f, and then determine whether it is attracting or repelling.

   d. Suppose $f(p) = q$, $f(q) = r$, and $f(r) = p$. Then p is a period-3 point for f. Find a formula for $(f \circ f \circ f)'(x)$, and use it to find a simple formula for $(f \circ f \circ f)'(p)$ analogous to the one we derived above.

   e. Let

$$T(x) = \begin{cases} 2x & \text{for } 0 \leq x \leq 1/2 \\ 2 - 2x & \text{for } 1/2 < x \leq 1 \end{cases}$$

The function T is called the **tent function** because its graph looks like a tent (Figure 3.23). Show that 2/3 is a fixed point, 2/5 is a period-2 point, and 2/7 is a period-3 point. Are they attracting, or are they repelling? Can you find other fixed points, period-2 points, and period-3 points for T?

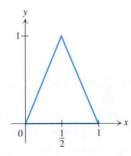

FIGURE 3.23 The graph of the tent function for Project 2.

3.5   HIGHER DERIVATIVES

An advertisement claims that a new car can accelerate from 0 miles per hour to 60 miles per hour in just 6.3 seconds. It is clear from that assertion that acceleration is related to the rate of change of velocity. In this section we will define the "derivative of the derivative," or equivalently the second derivative of a function, and then show that acceleration is the second derivative of the position function.

If f is a function, then f' is the function that assigns the number $f'(x)$ to each x at which f is differentiable. Since f' is a function, we can carry the process a step further and define $f''(a)$ by the formula

$$f''(a) = (f')'(a) = \lim_{x \to a} \frac{f'(x) - f'(a)}{x - a}$$

whenever this limit exists. We call $f''(a)$ the **second derivative of f at a.** It is often read "f double prime of a." Correspondingly, $f'(a)$ is often called the **first derivative of f at a.** Since $f''(a)$ is merely the derivative of f' at a, finding second derivatives is no more difficult than finding first derivatives.

EXAMPLE 1   Let $f(x) = \sin x$. Find a formula for $f''(x)$.

Solution   Since $f'(x) = \cos x$, it follows that $f''(x) = -\sin x$.   ❏

EXAMPLE 2   Let $f(x) = 3x^{1/2}$. Find a formula for $f''(x)$.

Solution   We find that

$$f'(x) = 3\left(\frac{1}{2} x^{-1/2}\right) = \frac{3}{2} x^{-1/2}$$

$$f''(x) = \frac{3}{2}\left(-\frac{1}{2} x^{-3/2}\right) = -\frac{3}{4} x^{-3/2}$$ ❏

For any positive integer $n \geq 3$ we can define the **nth derivative $f^{(n)}(a)$ of f at a** by letting $f^{(n-1)}$ denote the $(n - 1)$st derivative and then letting

$$f^{(n)}(a) = (f^{(n-1)})'(a) = \lim_{x \to a} \frac{f^{(n-1)}(x) - f^{(n-1)}(a)}{x - a}$$

if the limit exists. The second derivative, the third derivative, and so on are called **higher derivatives,** to distinguish them from the first derivative. We say that f is **twice differentiable** if $f''(x)$ exists for all x in the domain of f, and f is **n times differentiable** if $f^{(n)}(x)$ exists for all x in the domain of f.

EXAMPLE 3   Let $f(x) = \cos x$. Show that $f^{(4)}(x) = f(x)$ for all x.

Solution   By taking four successive derivatives we deduce that

$$f'(x) = -\sin x \qquad f^{(3)}(x) = \sin x$$

$$f''(x) = -\cos x \qquad f^{(4)}(x) = \cos x = f(x)$$ ❏

EXAMPLE 4 Let $f(x) = x^5 - 3x^4 + 2x - 1$. Find all higher derivatives of f.

Solution We obtain

$$f'(x) = 5x^4 - 12x^3 + 2 \qquad f^{(4)}(x) = 120x - 72$$

$$f''(x) = 20x^3 - 36x^2 \qquad f^{(5)}(x) = 120$$

$$f^{(3)}(x) = 60x^2 - 72x \qquad f^{(6)}(x) = 0$$

It is apparent also that $f^{(n)}(x) = 0$ for all $n > 6$. ❑

Notice that for the polynomial f in Example 4, the degree is 5 and $f^{(n)}(x) = 0$ for $n \geq 6$. More generally, the $(n + 1)$st and all higher derivatives of any polynomial of degree n are equal to 0. This means that if we differentiate any given polynomial enough times, we will eventually obtain a derivative that is 0.

EXAMPLE 5 Let $f(x) = e^{cx}$. Find a formula for the nth derivative of f.

Solution By the Chain Rule,

$$f'(x) = ce^{cx} \qquad f''(x) = c^2 e^{cx} \qquad f^{(3)}(x) = c^3 e^{cx}$$

In general, for any positive integer n,

$$f^{(n)}(x) = c^n e^{cx} \qquad ❑$$

In the Leibniz notation the second, third, and fourth derivatives are written

$$\frac{d^2 y}{dx^2}, \quad \frac{d^3 y}{dx^3}, \quad \text{and} \quad \frac{d^4 y}{dx^4}$$

They are read "d squared y, dx squared," and so on.

EXAMPLE 6 Let $y = x \sin x$. Using the Leibniz notation, find the first three derivatives of y.

Solution With the help of the product rule we take successive derivatives and obtain

$$\frac{dy}{dx} = \sin x + x \cos x$$

$$\frac{d^2 y}{dx^2} = \cos x + \cos x - x \sin x = 2 \cos x - x \sin x$$

$$\frac{d^3 y}{dx^3} = -2 \sin x - \sin x - x \cos x = -3 \sin x - x \cos x \qquad ❑$$

Of the higher derivatives, the second derivative is the most frequently used. Since the derivative represents the slope of a function, the second derivative represents the rate of change of the slope of the function. This aspect of the second derivative will play a significant role when we analyze graphs of functions in Chapter 4.

Acceleration

Just as the first derivative can be identified with velocity, the second derivative can be identified with acceleration. In conversation we refer to the acceleration of a car or the acceleration of an object due to gravity. In order to define acceleration we will first define average acceleration. Our discussion will parallel the discussion of velocity in Chapter 2.

Let the motion of an object be in a straight line, and let $v(t)$ be the velocity of the object at time t, as defined in Section 2.2. The **average acceleration** during a time interval $[t_0, t]$ is defined to be

$$\frac{\text{difference in velocity}}{\text{time elapsed}} = \frac{(\text{velocity at time } t) - (\text{velocity at time } t_0)}{t - t_0}$$

$$= \frac{v(t) - v(t_0)}{t - t_0}$$

It seems reasonable to assume that the acceleration at time t_0 should be close to the average acceleration during a time interval having t_0 as one of its endpoints, provided that the length of the interval is small enough. Therefore we define the **acceleration** $a(t_0)$ of the object at time t_0 by the formula

$$a(t_0) = \lim_{t \to t_0} \frac{v(t) - v(t_0)}{t - t_0} = v'(t_0)$$

provided that this limit exists.

If $f(t)$ denotes the position of an object at time t and if the first and second derivatives of f exist at t, then

$$v(t) = f'(t) \quad \text{and} \quad a(t) = v'(t) = f''(t)$$

Hence acceleration is the first derivative of the velocity function and the second derivative of the position function.

Suppose an object moves in a vertical direction and is subject only to the influence of gravity. By (2) in Section 3.3 its height in meters above the ground at time t (in seconds) is given by the formula

$$h(t) = -4.9t^2 + v_0 t + h_0 \tag{1}$$

where h_0 is the initial height and v_0 is the initial velocity. It is a simple matter to calculate the acceleration of the object.

EXAMPLE 7 Find the acceleration of an object whose position is described by (1).

Solution Differentiating, we have

$$v(t) = f'(t) = -9.8t + v_0 \quad \text{and} \quad a(t) = v'(t) = -9.8 \quad \square$$

A consequence of Example 7 is the fact that any object moving under the sole influence of gravity accelerates at a constant rate, namely -9.8 meters per second per second.

EXERCISES 3.5

In Exercises 1–20 find $f''(x)$.

1. $f(x) = 5x - 3$ 　　　　**2.** $f(x) = x^3 + 3x + 2$

3. $f(x) = -12x^5 + \frac{1}{2}x^4 - \sqrt{1 - x}$

4. $f(x) = (x + 1)/(x - 1)$

5. $f(x) = 2/(1 - 4x)^2$ 　　**6.** $f(x) = 1/\sqrt{x}$

7. $f(x) = ax^{-n}$ 　　　　**8.** $f(x) = 2x^2 - 4000/x$

9. $f(x) = 1/(x^3 - 1)$ 　　**10.** $f(x) = \pi x^{5/2} + (\cos x)/x$

11. $f(x) = \tan x$ 　　　　**12.** $f(x) = \sec x$

13. $f(x) = (x^2 + \sin x)^3$ 　**14.** $f(x) = x \cot (-4x)$

15. $f(x) = \sqrt{1 + \sin x}$ 　**16.** $f(x) = \sqrt{1 + \sqrt{x}}$

17. $f(x) = e^{1/x}$ 　　　　**18.** $f(x) = e^{\sin x}$

19. $f(x) = \ln (1 + x^2)$ 　**20.** $f(x) = (\ln x)/x$

In Exercises 21–34 find d^2y/dx^2.

21. $y = x^{3/2}$ 　　　　　**22.** $y = \frac{17}{4} - \frac{4}{9}x^2 - 7x^6$

23. $y = (x^4 - \tan x)^3$ 　**24.** $y = (1 - x^2)^{3/2}$

25 $y = ax^2 + bx + c$ 　**26.** $y = \sqrt{1 + x^4}$

27. $y = 1/(3 - x)$ 　　　**28.** $y = x/(x^2 - 1)$

29. $y = \csc x$ 　　　　　**30.** $y = \sin x + \cos x$

31. $y = e^x \sin x$ 　　　　**32.** $y = e^{(e^x)}$

33. $y = x^2 \ln x$ 　　　　**34.** $y = \ln (e^{2x} - 1)$

In Exercises 35–43 find the third derivative of the function.

35. $f(x) = -4x^2 + 5$ 　　**36.** $f(x) = x^8 - 4x^6 + 3x^4 - 2x^2$

37. $f(x) = \sin x^2$ 　　　**38.** $f(x) = x \cos x$

39. $f(x) = 1/x$ 　　　　**40.** $f(x) = 1/(3x^{1/2})$

41. $f(x) = 3x/(4x + 5)$ 　**42.** $f(x) = \ln x$

43. $f(x) = e^{(x^2)}$

In Exercises 44–50 find d^3y/dx^3.

44. $y = 3x^2$ 　　　　　**45.** $y = x^{7/2} - 2x^{5/2}$

46. $y = 1/(35x^{3/2})$ 　　**47.** $y = \csc x$

48. $y = 2/(1 - x)^2$ 　　**49.** $y = ax^3 + bx^2 + cx + d$

50. $y = ax^4 + bx^3 + cx^2 + dx + e$

In Exercises 51–56 find the fourth derivative of the function.

51. $f(x) = 3x^8 + \frac{3}{4}x^6 - 4x^{3/4} + 2x^{-1}$

52. $f(x) = \sin x - \cos x$ 　**53.** $f(x) = \sin \pi x$

54. $f(x) = ax^4 + bx^3 + cx^2 + dx + e$

55. $f(x) = e^{-\sqrt{2}\,x}$ 　　　**56.** $f(x) = x \ln x$

In Exercises 57–60 find the velocity and acceleration of an object moving along the x axis and having the given position function.

57. $f(t) = -16t^2 + 3t + 4$ 　**58.** $f(t) = -16t^2 - \frac{1}{2}t + 100$

59. $f(t) = 2 \sin t - 3 \cos t$ 　**60.** $f(t) = 3 - 1/t^2$

61. Show that the $(n + 1)$st derivative of any polynomial of degree n is 0. What is the $(n + 2)$nd derivative of such a polynomial?

62. Let an object move along the x axis with a position function given by $f(t) = t^2 + 25/(t^2 + 1)$ for $t \geq 0$.

 a. Find the velocity and acceleration of the object for any time $t > 0$.

 b. At what time does the object change direction?

 c. For what values of t is the velocity negative?

In Exercises 63–68 find a formula for the nth derivative of f, for $n \geq 1$.

63. $f(x) = e^x$ 　　　　　**64.** $f(x) = e^{3x}$

65. $f(x) = x e^{-x}$ 　　　　**66.** $f(x) = 1/x$

67. $f(x) = 1/(1 - x)$ 　　**68.** $f(x) = \ln x$

69. Figure 3.24 exhibits the graphs of f, f' and f'', labeled as g, h, and k. Determine which of g, h, and k is f and which is f'. Justify your answer.

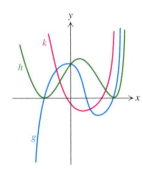

FIGURE 3.24 Graph for Exercise 69.

*70. Verify the following statements.

 a. If $f(x) = \sin x$ and n is a nonnegative integer, then

$$f^{(2n)}(x) = (-1)^n \sin x$$

 and

$$f^{(2n+1)}(x) = (-1)^n \cos x$$

 b. If $f(x) = \cos x$ and n is a nonnegative integer, then

$$f^{(2n)}(x) = (-1)^n \cos x$$

 and

$$f^{(2n+1)}(x) = (-1)^{n+1} \sin x$$

71. Suppose the function f has the property that $f'(x) = x^2 f(x)$.

a. Find $f''(x)$ in terms of $f(x)$ and x.

b. If $f(3) = \pi$, find $f''(3)$.

72. Let $f(x) = x^3 + ax^2 + bx + c$, where a, b, and c are constants. Determine the constants in such a way that $f(0) = f'(1) = f''(2) = f^{(3)}(3)$.

73. Let $h = fg$. Find $h''(x)$ in terms of f and g and their derivatives.

74. Let $h = g \circ f$. Find $h''(x)$ in terms of f and g and their derivatives.

Applications

75. The angular displacement $f(t)$ of a pendulum bob at time t is given by

$$f(t) = a \cos 2\pi\omega t$$

where ω is the frequency and a is the maximum displacement. The first and second derivatives of the angular displacement are the angular velocity and the angular acceleration, respectively, of the bob. Find the angular acceleration of the bob.

76. The **Gompertz growth curve**, which appears in population studies, is the graph of the function f defined by

$$f(x) = ae^{(-be^{-cx})}$$

Find $f''(x)$.

77. The distance s traveled by a moving particle is given by $s = ae^{kt} + be^{-kt}$, where a, b, and k are positive constants and t denotes time. Show that the acceleration of the particle is proportional to the distance traveled.

*78. Suppose you are asked to design a connector joining two highways, as shown in Figure 3.25.

a. Find the polynomial of minimal degree that yields a function differentiable at both -1 and 0. (Thus the road is smooth.)

b. If it were necessary to make the resulting function twice differentiable at both -1 and 0, what is the minimal degree that the polynomial would need to have?

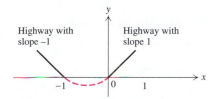

FIGURE 3.25   Graph for Exercise 78.

79. Let $f(x) = a/(1 + be^{-kax})$ for $x \geq 0$, where a, b, and k are positive constants. The graph of f is a **logistic curve,** and it has played a significant role in population ecology during the past century and a half.

a. Show that $f''(x) = \dfrac{a^3bk^2e^{-kax}(-1 + be^{-kax})}{(1 + be^{-kax})^3}$.

b. Use the formula in (a) to determine the value of c such that $f''(c) = 0$.

c. Find $f(c)$ in terms of a.

*80. A baseball player chasing a fly ball runs in a straight line toward the right field fence. Set up a coordinate system, with feet as units, such that the y axis represents the fence and the player runs along the negative x axis toward the origin. Suppose the player's velocity in feet per second is

$$v(x) = \frac{1}{60}x^2 + \frac{11}{10}x + 25 \quad \text{for } -30 \leq x \leq 0$$

when the player is located at x. What is the acceleration when the player is 1 foot from the fence? (*Hint:* Use the Chain Rule.)

Project

1. As you know, $-1 \leq \sin x \leq 1$ for all x, so the graph of $\sin x$ lies in a horizontal strip in the xy plane. By contrast, the graph of each nonconstant polynomial cannot be constrained to a horizontal strip (Figure 3.26). Nevertheless, it is possible to find polynomials whose graphs look like the graph of $\sin x$ for x not too far away from 0. This project seeks out such polynomials. Let $f(x) = \sin x$.

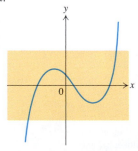

FIGURE 3.26   The graph of a nonconstant polynomial.

a. We begin with $p_1(x) = x$. Show that $p_1(0) = f(0)$ and $p_1'(0) = f'(0)$. Draw the graphs of f and p_1 on your graphics calculator and determine an interval I_1 about 0 over which the values of p_1 and f appear to be within 0.1 of each other.

b. Next, let $p_3(x) = x - x^3 / (3 \cdot 2)$. Show that

$$p_3(0) = f(0), \; p_3'(0) = f'(0),$$
$$p_3''(0) = f''(0), \; p_3^{(3)}(0) = f^{(3)}(0)$$

Again, draw the graphs of f and p_3 on your graphics calculator and determine an interval I_3 about 0 over which the values of p_3 and f appear to be within 0.1 of each other. Is I_3 larger than I_1?

c. Now, let $p_5(x) = x - x^3 / (3 \cdot 2) + x^5 / (5 \cdot 4 \cdot 3 \cdot 2)$. Show that

$$p_5(0) = f(0), \quad p_5'(0) = f'(0), \quad p_5''(0) = f''(0),$$

$$p_5^{(3)}(0) = f^{(3)}0, \quad p_5^{(4)}(0) = f^{(4)}(0), \quad p_5^{(5)}(0) = f^{(5)}(0)$$

Draw the graphs of f and p_5 on your graphics calculator and determine an interval I_5 about 0 over which the values of p_5 and f appear to be within 0.1 of each other. Is I_5 larger than I_3?

d. Following the pattern set by p_1, p_3, and p_5, write down a formula for p_7, and analyse it in the same way we did for the other polynomials. Moreover, one can continue with $p_9, p_{11},\ldots$.Doing so yields polynomials of ever higher degrees that approximate $\sin x$ in ever longer intervals about 0. The study of polynomials that approximate functions like $\sin x$ or $\cos x$ or e^x are very important, and will be a focus of study in Chapter 9.

3.6   IMPLICIT DIFFERENTIATION

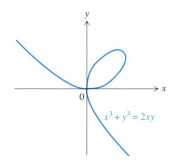

FIGURE 3.27

René Descartes (1596–1650)
The folium of Descartes is named after the Frenchman René Descartes. As a teenager Descartes tired of studying and joined the army in order to see the world. Later he became a renowned philosopher and mathematician. The Cartesian plane is named after him. The problem of finding formulas for lines tangent to the folium of Descartes puzzled mathematicians during the sixteenth century.

The differentiable functions we have encountered so far can be described by equations in which y is expressed in terms of x, as in

$$y = \frac{9x^2}{x^2 + 1} \quad \text{and} \quad y = \sin x^3$$

Such equations are said to **define y explicitly.** However, suppose y is a differentiable function of x, and instead of having a formula for y in terms of x, we are given an equation such as

$$x^3 + y^3 = 2xy \tag{1}$$

whose graph (Figure 3.27) is called a **folium of Descartes,** after the French mathematician René Descartes. Notice that the folium is actually the graph of no fewer than 3 functions (Figure 3.27).Descartes was the first to obtain formulas for tangents to figures such as the folium. The goal of this section is to be able to find derivatives of smooth curves that are not necessarily graphs of functions.

Because y does not appear alone on one side of the equation in (1), we say that the equation **defines y implicitly in terms of x.** This section focuses on a procedure, called implicit differentiation, for determining the derivative of y.

Let y be a differentiable function that is defined implicitly in terms of x. For simplicity, suppose that y is defined implicitly by the equation

$$x^2 + y^2 = 9 \tag{2}$$

Assume that we want an equation of the line that is tangent to the graph of y at a point on its graph. In order to find an equation of the tangent line from the given equation, we need to be able to find dy/dx. Our usual methods do not apply because y is not given explicitly in terms of x. An alternative method of finding dy/dx when *x* and *y* are related by an equation involves the following four steps:

1. Notice that each side of the equation is a differentiable function of *x* because by assumption *y* is a differentiable function of *x*. (For the equation in (2), this means that $x^2 + y^2$, as well as 9, are differentiable functions of *x*.)
2. Take the derivatives of both sides of the given equation separately.
3. Equate the derivatives of the two sides. (The derivatives of the two sides are equal because the two original sides are equal.)
4. Solve for the derivative dy/dx.

The method just outlined is called **implicit differentiation.**

We now perform implicit differentiation to find dy/dx for the equation in (2).

EXAMPLE 1   Suppose *y* is a differentiable function of *x* that satisfies $x^2 + y^2 = 9$. Use implicit differentiation to find a formula for dy/dx.

Solution   Following the procedure just outlined, we differentiate both sides of the equation $x^2 + y^2 = 9$ and equate the derivatives:

$$\frac{d}{dx}(x^2 + y^2) = \frac{d}{dx}(9)$$

$$\frac{d}{dx}(x^2) + \frac{d}{dx}(y^2) = 0$$

$$2x + 2y\frac{dy}{dx} = 0$$

Solving for dy/dx, we conclude that

$$\frac{dy}{dx} = -\frac{2x}{2y} = -\frac{x}{y}, \quad \text{provided that } y \neq 0 \qquad \square$$

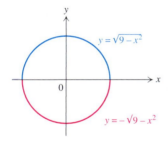

FIGURE 3.28

The equation $x^2 + y^2 = 9$, whose graph is a circle, defines two differentiable functions $y = \sqrt{9 - x^2}$ and $y = -\sqrt{9 - x^2}$ on (−3, 3) (see Figure 3.28). The graph of the first function is the top semicircle, and the graph of the second function is the lower semicircle. If we differentiated either function, we would conclude once again that $dy/dx = -x/y$. Indeed

$$\text{if } y = \sqrt{9 - x^2}, \quad \text{then } \frac{dy}{dx} = \frac{-x}{\sqrt{9 - x^2}} = -\frac{x}{y}$$

$$\text{if } y = -\sqrt{9 - x^2}, \quad \text{then } \frac{dy}{dx} = \frac{x}{\sqrt{9 - x^2}} = -\frac{x}{y}$$

Nevertheless, there are many equations in *x* and *y* that cannot be solved for *y* in terms of *x* (such as the equation in Example 3 below). In such cases implicit differentiation is the relevant method of finding derivatives.

In Example 1 the derivative dy/dx is expressed in terms of both *x* and *y*. This will usually be the case when we differentiate an equation implicitly. Moreover, it is usually not possible to find the value of dy/dx at a particular value of *x* unless the corresponding value of *y* is known. For example, corresponding to *x* = −2 there are two points on the circle $x^2 + y^2 = 9$, so that in order to specify a unique point on the circle one must specify the value of y as well as the value of x.

A line l is **tangent to the graph of an equation** at a point (a, b) provided that l is tangent at (a, b) to the graph of a function y that satisfies the equation. With this definition, we are ready to find an equation of a line tangent to the graph of the circle mentioned in Example 1.

EXAMPLE 2   Find an equation of the line that is tangent to the circle $x^2 + y^2 = 9$ at the point $(-2, \sqrt{5})$.

Solution   The point $(-2, \sqrt{5})$ lies on the top half of the circle, which is the graph of a differentiable function y (see Figure 3.28 again). By Example 1, $dy/dx = -x/y$, so that

$$\left.\frac{dy}{dx}\right|_{(-2,\sqrt{5})} = -\frac{-2}{\sqrt{5}} = \frac{2}{\sqrt{5}}$$

Therefore the slope of the tangent line we seek is $2/\sqrt{5}$, so an equation of the line is

$$y - \sqrt{5} = \frac{2}{\sqrt{5}}(x + 2) \qquad \square$$

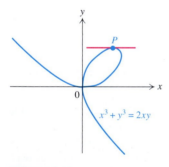

FIGURE 3.29   At P the tangent line is horizontal.

EXAMPLE 3   The folium of Descartes is the graph of

$$x^3 + y^3 = 2xy \tag{3}$$

Find a point P in the first quadrant that is on the folium and has the property that the tangent line at P is horizontal (Figure 3.29).

Solution   Following the procedure already discussed, we differentiate both sides of (3) to obtain

$$\frac{d}{dx}(x^3) + \frac{d}{dx}(y^3) = \frac{d}{dx}(2xy)$$

$$3x^2 + 3y^2\frac{dy}{dx} = 2y + 2x\frac{dy}{dx}$$

Then we solve for dy/dx:

$$\frac{dy}{dx} = \frac{2y - 3x^2}{3y^2 - 2x}$$

The line tangent to the folium of Descartes at $P = (a, b)$ is horizontal provided that $dy/dx = 0$ at P. But this means that $2b - 3a^2 = 0$, that is, $b = 3a^2/2$. To determine the coordinates of P, we substitute a for x and $3a^2/2$ for y in (3), and then solve for a:

$$a^3 + \left(\frac{3}{2}a^2\right)^3 = 2a\left(\frac{3}{2}a^2\right)$$

$$a^3 + \frac{27}{8}a^6 = 3a^3$$

$$27a^6 = 16a^3$$

Since P is to be in the first quadrant by hypothesis, it follows that $a \neq 0$, so that $27a^3 = 16$. Thus $a = \sqrt[3]{16/27} = 2^{4/3}/3$. Therefore

$$b = \frac{3a^2}{2} = \frac{3(2^{8/3})}{2(9)} = \frac{2^{5/3}}{3}$$

We conclude that the tangent line is horizontal at the point $P = (2^{4/3}/3, 2^{5/3}/3)$. ❑

If y is a function defined implicitly in terms of x, and if d^2y/dx^2 exists, we can differentiate dy/dx to obtain d^2y/dx^2.

EXAMPLE 4 Consider the circle $x^2 + y^2 = 9$. Find d^2y/dx^2 by implicit differentiation.

Solution From Example 1 we know that

$$\frac{dy}{dx} = \frac{-x}{y} \quad \text{for } y \neq 0 \tag{4}$$

Taking the derivative of each side of (4) with respect to x, then substituting $-x/y$ for dy/dx, and finally using the fact that $x^2 + y^2 = 9$, we deduce that

$$\frac{d^2y}{dx^2} = \frac{(-1)y - (-x)\dfrac{dy}{dx}}{y^2} = \frac{-y + x\left(-\dfrac{x}{y}\right)}{y^2} = \frac{-(y^2 + x^2)}{y^3} = \frac{-9}{y^3} \quad ❑$$

If y is differentiable and defined implicitly in terms of x, then it is generally possible to solve for d^2y/dx^2 in terms of y and x, as we did in Example 4. The principle followed in finding d^2y/dx^2 is the same as the principle followed in finding dy/dx.

In our final example assume that an object travels around the circle $x^2 + y^2 = 9$, so that both x and y are functions of time, represented by t.

EXAMPLE 5 Assume that x and y are differentiable functions of t and satisfy the equation $x^2 + y^2 = 9$. Find dy/dt in terms of x, y, and dx/dt.

Solution Differentiating both sides of the given equation implicitly with respect to t, we obtain

$$2x\frac{dx}{dt} + 2y\frac{dy}{dt} = 0$$

Therefore

$$2y\frac{dy}{dt} = -2x\frac{dx}{dt}$$

from which we conclude that

$$\frac{dy}{dt} = -\frac{x}{y}\frac{dx}{dt}$$

provided that $y \neq 0$. ❑

If we desired the value of dy/dt at an instant when $x = 1$ and $y = -2\sqrt{2}$, and if $dx/dt = 5$ at that instant, then we would substitute these values into the equation

$$\frac{dy}{dt} = -\frac{x}{y}\frac{dx}{dt}$$

to obtain

$$\frac{dy}{dt} = -\frac{1}{-2\sqrt{2}}(5) = \frac{5}{4}\sqrt{2}$$

How can we determine whether there is a differentiable function y that satisfies an equation in x and y? Although the answer to this question is not simple, such a function often does exist. In this section and in Section 3.7 we will assume without proof that all equations appearing in the text and in the exercises do in fact define one or more differentiable functions y (either implicitly or explicitly).

EXERCISES 3.6

In Exercises 1–18 use implicit differentiation to find the derivative of y with respect to x.

1. $3y^2 = 2x^4$

2. $y^2 = x^3/(2 - x)$

3. $y^2 + y = (1 + x)/(1 - x)$

4. $x^2 = y^2/(y^2 - 1)$

5. $\sec y - \tan x = 0$

6. $\sqrt{x} + (1/\sqrt{y}) = 2$

7. $\dfrac{\sin y}{y^2 + 1} = 3x$

8. $x = \dfrac{1 - \sqrt{y}}{1 + \sqrt{y}}$

9. $x^2 + x^2y^2 + y^3 = 3$

10. $x^2 = \dfrac{x - y^2}{x + y}$

11. $x^2 + y^2 = \dfrac{y^2}{x^2}$

12. $\sqrt{1 + xy} = \dfrac{x}{y} + \dfrac{y}{x}$

13. $\sqrt{xy} + \sqrt{x + 2y} = 4$

14. $2xy = (x^2 + y^3)^{3/2}$

15. $xe^y = y + x^2$

16. $e^{xy} = 2x + y$

17. $y + \sqrt{y}\ln x = x^2 + y^2$

18. $\ln(x - y) = xy$

In Exercises 19–30 use implicit differentiation to find the derivative of y with respect to x at the given point.

19. $x^2 + y^2 = y$; $(0, 1)$

20. $x^2 - y^2 = 1$; $(\sqrt{3}, \sqrt{2})$

21. $xy = 2$; $(-2, -1)$

22. $x^4 + xy^3 = 0$; $(-1, 1)$

23. $x^3 + 2xy = 5$; $(1, 2)$

24. $x^2 + 3xy + 2y^2 = 6$; $(-1, -1)$

25. $x^2 + x/y = -2$; $(1, -\frac{1}{3})$

26. $x/(x + 2y) = 1 - y$; $(1, 0)$

27. $(\sqrt{x} + 1)(\sqrt{y} + 2) = 8$; $(1, 4)$

28. $\sin x = \cos y$; $(\pi/6, \pi/3)$

29. $2e^{x^2y} = x$; $(2, 0)$

30. $\ln(x^2 - 3y) = x - y - 1$; $(2, 1)$

In Exercises 31–34 find an equation of the line l tangent to the graph of the equation at the given point.

31. $xy^2 = 18$; $(2, -3)$

32. $x^2 + y^2 = 3y$; $(-\sqrt{2}, 2)$

33. $\sin(x + y) = 2x$; $(0, \pi)$

34. $y^2 = x^3/(2 - x)$; $(1, 1)$

In Exercises 35–38 use implicit differentiation to find d^2y/dx^2.

35. $x^2 - y^4 = 6$

36. $2xy^2 = 4$

37. $x^2\sin 2y = 1$

38. $xe^{2y} = y$

In Exercises 39–44 assume that x and y are differentiable functions of t. Find dy/dt in terms of x, y, and dx/dt.

39. $y^2 - x^2 = 4$

40. $x^2 + y^3 = x$

41. $x\sin y = 2$

42. $x^4y^2 = y$

43. $y = \cos xy^2$

44. $y + \ln x + e^y = 1$

45. Each of the following equations implicitly describes a single function y that can also be given explicitly. Find dy/dx by implicit differentiation. Then solve the equation explicitly for y. Finally, differentiate again to check the implicit differentiation.

    a. $y^3 = x^2$ **b.** $8/y = x^2 + 4$ **c.** $y^3 = x^2/(x^2 - 1)$

46. Find the line tangent to the graph of $x^3 + y^3 = 3xy$ at $(\frac{3}{2}, \frac{3}{2})$. Show that the normal at $(\frac{3}{2}, \frac{3}{2})$ passes through the origin.

47. Find a point Q in the first quadrant at which the tangent to the folium of Descartes (given in (3)) has slope -1.

48. The **angle** from a curve C_1 to a curve C_2 at a point of intersection (x_0, y_0) is defined as the angle θ from the line l_1 tangent to C_1 at (x_0, y_0) to the line l_2 tangent to C_2 at (x_0, y_0). The tangent of the angle is given by

$$\tan\theta = \frac{m_2 - m_1}{1 + m_1m_2}$$

where m_1 is the slope of l_1 and m_2 is the slope of l_2. Find the angles from the curve $x^2 + y^2 = 1$ to the curve $(x - 1)^2 + y^2 = 1$ at the two points of intersection.

49. Let l be the line tangent to the astroid $x^{2/3} + y^{2/3} = 4$ (Figure 3.30) at $(2\sqrt{2}, 2\sqrt{2})$. Find the area of the triangle formed by l and the coordinate axes.

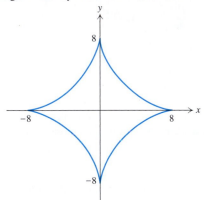

FIGURE 3.30   The astroid $x^{2/3} + y^{2/3} = 4$.

50. Consider the astroid $x^{2/3} + y^{2/3} = 4$ (Figure 3.30).

   a. Find equations of the four lines that are tangent to the astroid and whose slopes are either 1 or –1.

   b. Determine the area A of the square formed by the tangents found in part (a).

*51. Consider the lemniscate $(x^2 + y^2)^2 = x^2 - y^2$, shown in Figure 3.31. Find the points on the graph at which the tangents are horizontal. (*Hint:* There are four such points. In finding them you will need to use the equation of the lemniscate twice.)

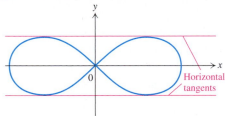

FIGURE 3.31   Lemniscate $(x^2 + y^2)^2 = x^2 - y^2$.

Project

1. A marble rests as far down as it can inside a glass that is set on a flat table. The object of this project is to determine the point at which the marble and the glass touch each other.

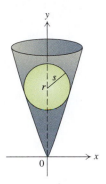

FIGURE 3.32   Figure for the project.

   a. Suppose the glass is conical (as in Figure 3.32). First convince yourself that the marble cannot touch the vertex of the glass. Then determine how far above the table the marble touches the glass. (*Hint:* The cross section of the marble in the xy plane has boundary that is the circle $(y - r)^2 + x^2 = s^2$, with center $(0, r)$ and radius s. Likewise, the cross section of the cone in the xy plane has boundary consisting of the two lines $y = cx$ and $y = -cx$. Here r, s, and c are positive constants. Because of symmetry, we will only be concerned with the portions of the circle $(y - r)^2 + x^2 = s^2$ and the line $y = cx$ in the first quadrant. The goal is to determine the point (a, b) at which the line and circle are tangent to one another. Then b is the desired number.)

   b. Suppose the glass is parabolic, that is, its cross section in the xy plane has the form $y = cx^2$ with $c > 0$. Show that it is impossible to find a number c such that the marble rests at the bottom of the glass.

3.7   RELATED RATES

When a spherical balloon is inflated, the radius r and the volume V of the balloon are functions of time t. Even if we do not know formulas for r and V as functions of t, at least we know that r and V are related by the equation

$$V = \frac{4}{3}\pi r^3 \tag{1}$$

Using the Chain Rule to differentiate V with respect to t, we find that

$$\frac{dV}{dt} = \frac{dV}{dr}\frac{dr}{dt} = 4\pi r^2 \frac{dr}{dt} \tag{2}$$

The rates dV/dt and dr/dt are related by (2). Therefore we say that dV/dt and dr/dt are **related rates.** If we know, for example, the values of r and dV/dt at a specific time t_0, then we can find the value of dr/dt at t_0 by solving equation (2) for dr/dt.

EXAMPLE 1   Suppose a spherical balloon is inflated at the rate of 10 cubic centimeters per minute. How fast is the radius of the balloon increasing when the radius is 5 centimeters?

Solution   By assumption, the volume V of the balloon is increasing at 10 cubic centimeters per minute. Thus we know that $dV/dt = 10$. We wish to find

$$\frac{dr}{dt}\bigg|_{t=t_0}$$

where t_0 is the instant at which $r = 5$. The formula for the volume is given in (1), and by (2),

$$\frac{dV}{dt} = 4\pi r^2 \frac{dr}{dt}$$

Substituting 10 for dV/dt and 5 for r, we conclude that

$$10 = 4\pi(5)^2 \frac{dr}{dt}\bigg|_{t=t_0}$$

so that

$$\frac{dr}{dt}\bigg|_{t=t_0} = \frac{10}{4\pi(5)^2} = \frac{1}{10\pi}$$

Therefore when the radius is 5 centimeters, the radius is increasing at the rate of $1/(10\pi)$ centimeter per minute.   ❑

Observe that in Example 1 we did not need to know the time t_0 at which $r = 5$. All we needed to know were the values of r and dV/dt at t_0.

Example 1 is a classic related rates problem: We are given one or more rates of change (dV/dt in Example 1), and are asked to find another rate of change (dr/dt in Example 1). The general procedure for solving such a problem consists of the following steps:

1.   Identify and label the different variables. Include the variable whose rate is to be evaluated and those whose rates are given. It may be helpful to sketch a drawing at this stage.

2. Find an equation relating the variables whose rates of change are known and the variable whose rate of change is desired. This may involve similar triangles, the Pythagorean Theorem, or trigonometric identities.

3. Differentiate both sides of the equation, either by using the Chain Rule or by differentiating implicitly, in order to find a relation between the rates of change.

4. Solve for the desired rate of change by using the given values of the variables and their rates.

We followed this procedure in the solution of Example 1, and will continue to follow it in the remaining examples of the section.

EXAMPLE 2   Suppose that the bigger the balloon in Example 1 becomes, the slower it is inflated. Specifically, suppose that when the volume V is greater than 10 cubic centimeters, the balloon is inflated at the rate of $8/V$ cubic centimeters per minute. How fast is the radius of the balloon increasing when the radius is 2 centimeters?

Solution   As before, let r denote the radius of the balloon. Our goal is to find

$$\left. \frac{dr}{dt} \right|_{t=t_0}$$

where t_0 is the instant at which $r = 2$. On the one hand, (2) implies that

$$\frac{dV}{dt} = 4\pi r^2 \frac{dr}{dt} \tag{3}$$

On the other hand, we have assumed that $dV/dt = 8/V$ whenever $V > 10$. Therefore

$$\frac{dV}{dt} = \frac{8}{V} = \frac{8}{\frac{4}{3}\pi r^3} = \frac{6}{\pi r^3} \tag{4}$$

Equating the expressions given in (3) and (4) for dV/dt, we deduce that

$$4\pi r^2 \frac{dr}{dt} = \frac{6}{\pi r^3}$$

or equivalently,

$$\frac{dr}{dt} = \frac{6}{\pi r^3} \frac{1}{4\pi r^2} = \frac{3}{2\pi^2 r^5}$$

For the time t_0 at which $r = 2$ we have

$$\left. \frac{dr}{dt} \right|_{t=t_0} = \frac{3}{2\pi^2 2^5} = \frac{3}{64\pi^2}$$

Consequently, when the radius is 2 centimeters, the radius is increasing at the rate of $3/(64\pi^2)$ centimeters per minute.   ❏

In the next example we will use a diagram to help discover an equation relating the variables.

EXAMPLE 3 One end of a 13-foot ladder is on the floor, and the other end rests on a vertical wall (Figure 3.33). If the bottom end is drawn away from the wall at 3 feet per second, how fast is the top of the ladder sliding down the wall when the bottom of the ladder is 5 feet from the wall?

FIGURE 3.33 Sliding ladder.

Solution At any given instant, let y be the height of the top of the ladder above the floor, and let x be the distance between the base of the wall and the bottom of the ladder, as in Figure 3.33. By hypothesis, $dx/dt = 3$, and we must find the value of

$$\left. \frac{dy}{dt} \right|_{t=t_0}$$

where t_0 is the time at which $x = 5$. By the Pythagorean Theorem, x and y are related by the equation

$$x^2 + y^2 = 13^2 = 169 \qquad (5)$$

Differentiating the left and right sides of (5) implicitly with respect to t, we obtain

$$2x \frac{dx}{dt} + 2y \frac{dy}{dt} = 0$$

which we solve for dy/dt:

$$2y \frac{dy}{dt} = -2x \frac{dx}{dt}$$

$$\frac{dy}{dt} = -\frac{x}{y} \frac{dx}{dt} \qquad (6)$$

Now at the time t_0 at which the base of the ladder is 5 feet from the wall, we have $x = 5$. Therefore by (5),

$$y^2 = 169 - 5^2 = 144$$

and thus $y = 12$ (Figure 3.34). Substituting $x = 5$, $y = 12$, and $dx/dt = 3$ into (6), we conclude that

$$\left. \frac{dy}{dt} \right|_{t = t_0} = -\frac{5}{12}(3) = -\frac{5}{4}$$

Thus when the bottom of the ladder is 5 feet from the wall, the top is sliding down at the rate of 5/4 feet per second.   ❑

FIGURE 3.34   The moment the bottom of the ladder is 5 feet from the wall.

Caution: In the preceding example, if we had used the diagram in Figure 3.34 (which represents the situation only at time t_0) in order to try to find the rate at which the top of the ladder slipped, we would not have been able to obtain a relationship between the rates dx/dt and dy/dt. In particular, x does not appear in Figure 3.34. Thus it is essential to draw a figure that represents the situation at *any* instant, rather than at a particular instant.

Consider once again the ladder in Example 3. Suppose that we desire to determine how fast the angle of elevation θ of the ladder changes at the instant when the bottom of the ladder is 5 feet from the wall (see Figure 3.33 again). This is equivalent to seeking $d\theta/dt$ when $x = 5$. First we find that θ and y are related by the equation $\sin \theta = y/13$. Differentiating the equation implicitly with respect to t, we deduce that

$$(\cos \theta) \frac{d\theta}{dt} = \frac{dy/dt}{13}$$

At the moment when the bottom of the ladder is 5 feet from the wall, $\cos \theta = \frac{5}{13}$ and $dy/dt = -\frac{5}{4}$ (by Example 3). We conclude that

$$\frac{5}{13}\frac{d\theta}{dt} = \frac{-5/4}{13} = -\frac{5}{52}, \quad \text{so that} \quad \frac{d\theta}{dt} = -\frac{1}{4}$$

Consequently the angle decreases at the rate of $\frac{1}{4}$ radians per second at the moment in question.

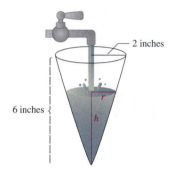

FIGURE 3.35   Filling a paper cup.

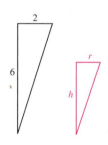

FIGURE 3.36

EXAMPLE 4   Water is poured into a conical paper cup at the rate of $\frac{2}{3}$ cubic inches per second. If the cup is 6 inches tall and the top of the cup has a radius of 2 inches, how fast does the water level rise when the water is 4 inches deep?

Solution   At any time t let h be the height of the water, V the volume of the water, and r the radius of the top surface of the water (Figure 3.35). Recall that the volume V of a cone is given by

$$V = \frac{1}{3}\pi r^2 h$$

Since the water enters the cup at the rate of $\frac{2}{3}$ cubic inches per second, we have $dV/dt = \frac{2}{3}$. We wish to determine the value of

$$\frac{dh}{dt}\Big|_{t=t_0}$$

where t_0 is the time at which $h = 4$. Using the similar triangles in Figure 3.36, which we have obtained from Figure 3.35, we find that

$$\frac{r}{h} = \frac{2}{6}$$

so that $r = h/3$. As a result we can express V in terms of h alone:

$$V = \frac{1}{3}\pi\left(\frac{h}{3}\right)^2 h = \frac{\pi h^3}{27}$$

Differentiating with respect to t, we find that

$$\frac{dV}{dt} = \frac{dV}{dh}\frac{dh}{dt} = \frac{\pi h^2}{9}\frac{dh}{dt}$$

so that

$$\frac{dh}{dt} = \frac{9}{\pi h^2}\frac{dV}{dt} \tag{7}$$

At time t_0 we have $h = 4$ and $dV/dt = \frac{2}{3}$. Substituting these values into (7) yields

$$\frac{dh}{dt}\Big|_{t=t_0} = \frac{9}{\pi(4)^2}\cdot\frac{2}{3} = \frac{3}{8\pi}$$

Thus the water level is rising at the rate of $3/(8\pi)$ inches per second when the water is 4 inches deep.   ❑

EXAMPLE 5   Pat walks at the rate of 5 feet per second toward a street light whose lamp is 20 feet above the base of the light. If Pat is 6 feet tall, determine the rate of change of the length of Pat's shadow at the moment Pat is 24 feet from the base of the lamppost.

Solution   At any given time, let x be the distance between Pat and the lamppost and y the length of Pat's shadow (Figure 3.37). We wish to find

$$\left.\frac{dy}{dt}\right|_{t=t_0}$$

for the value of t_0 at which $x = 24$. By the similar triangles appearing in Figure 3.38,

$$\frac{x+y}{20} = \frac{y}{6}$$

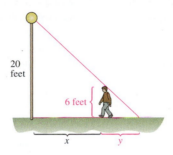

FIGURE 3.37 Pat walking toward a street light.   **FIGURE 3.38**

so that

$$6x + 6y = 20y, \quad \text{or equivalently,} \quad y = \frac{3}{7}x$$

Differentiating with respect to t, we obtain

$$\frac{dy}{dt} = \frac{3}{7}\frac{dx}{dt}$$

Since Pat walks at the rate of 5 feet per second toward the lamppost, it follows that $dx/dt = -5$, so that

$$\frac{dy}{dt} = \frac{3}{7}(-5) = -\frac{15}{7}$$

Consequently Pat's shadow shrinks at the rate of $\frac{15}{7}$ feet per second at the moment in question.   ❑

Notice that in the solution we did not need to use the fact that at some moment Pat was 24 feet from the base of the lamppost. This information turned out to be extraneous to the solution.

EXERCISES 3.7

1. Suppose the radius of a spherical balloon is shrinking at $\frac{1}{2}$ centimeter per minute. How fast is the volume decreasing when the radius is 4 centimeters?

2. Suppose a snowball remains spherical while it melts, with the radius shrinking at 1 inch per hour. How fast is the volume of the snowball decreasing when the radius is 2 inches?

3. Suppose the volume of the snowball in Exercise 2 shrinks at the rate of $2/V$ (cubic inches per hour), so that $dV/dt = -2/V$. How fast is the radius changing when the radius is $\frac{1}{2}$ inch?

4. A spherical balloon is inflated at the rate of 3 cubic centimeters per minute. How fast is the radius of the balloon increasing when the radius is 6 centimeters?

5. Suppose a spherical balloon grows in such a way that after t seconds, $V = 4\sqrt{t}$ (cubic centimeters). How fast is the radius changing after 64 seconds?

6. A spherical balloon is losing air at the rate of 2 cubic centimeters per minute. How fast is the radius of the balloon shrinking when the radius is 8 centimeters?

7. Water leaking onto a floor creates a circular pool with an area that increases at the rate of 3 square centimeters per minute. How fast is the radius of the pool increasing when the radius is 10 centimeters?

8. A point moves around the circle $x^2 + y^2 = 9$. When the point is at $(-\sqrt{3}, \sqrt{6})$, its x coordinate is increasing at the rate of 20 units per second. How fast is its y coordinate changing at that instant?

9. Suppose the top of the ladder in Example 3 is being pushed up the wall at the rate of 1 foot per second. How fast is the base of the ladder approaching the wall when it is 3 feet from the wall?

10. A ladder 15 feet long leans against a vertical wall. Suppose that when the bottom of the ladder is x feet from the wall, the bottom is being pushed toward the wall at the rate of $\frac{1}{2}x$ feet per second. How fast is the top of the ladder rising at the moment the bottom is 5 feet from the wall?

11. A board 5 feet long slides down a wall. At the instant the bottom end is 4 feet from the wall, the other end is moving down the wall at the rate of 2 feet per second. At that moment,
    a. how fast is the bottom end sliding along the ground?
    b. how fast is the area of the region between the board, ground, and wall changing?

12. Suppose the water in Example 4 is poured in at the rate of $\frac{3}{2}$ cubic inches per second. How fast is the water level rising when the water is 2 inches deep?

13. A water trough is 12 feet long, and its vertical cross section is an equilateral triangle with sides 2 feet long. Water is pumped into the trough at a rate of 3 cubic feet per minute. How fast is the water level rising when the depth of the water is $\frac{1}{2}$ foot?

14. A beacon on a lighthouse 1 mile from shore revolves at the rate of 10π radians per minute. Assuming that the shoreline is straight, calculate the speed at which the spotlight is sweeping across the shoreline as it lights up the sand 2 miles from the lighthouse. (*Hint:* In Figure 3.39, x is the coordinate of the point on the shore at which the light shines. Thus dx/dt is the speed of the image of the spotlight moving across the shoreline, and $d\theta/dt = 10\pi$.)

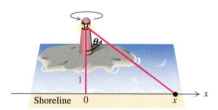

FIGURE 3.39   Figure for Exercise 14.

15. Boyle's Law states that if the temperature of a gas remains constant, then the pressure p and the volume V of the gas satisfy the equation $pV = c$, where c is a constant. If the volume is decreasing at the rate of 10 cubic centimeters per second, how fast is the pressure increasing when the pressure is 100 pounds per square centimeter and the volume is 20 cubic centimeters?

16. If a sample of air expands with no loss or gain in heat (that is, adiabatically), then the pressure p and volume V of the sample are related by the equation

$$pV^{1.4} = c$$

where c is a positive constant. Assume that at a certain instant the volume is 30 cubic centimeters and the pressure is 3×10^6 dynes per square centimeter. Assume also that the volume is increasing at the rate of 2 cubic centimeters per second. At what rate is the pressure changing at that moment?

17. As a low energy helium nucleus moves toward an atom of the heavy element tin, it undergoes Rutherford scattering, which makes it move in a hyperbolic orbit with respect to the tin atom (Figure 3.40). Assume that the hyperbolic orbit is described by $x^2 - y^2 = c^2$. Assume that at the moment the helium nucleus is at the point $(2c, c\sqrt{3})$ we have $dy/dt = 2$. Find dx/dt at that moment.

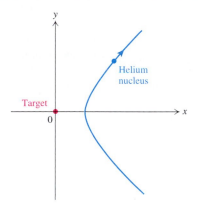

FIGURE 3.40   Figure for Exercise 17.

18. The escape velocity of a star with a radius of r kilometers and mass of M kilograms is given by

$$v_{esc} = \sqrt{\frac{2GM}{r}}$$

where $G = 6.67 \times 10^{-11}$. Suppose that an old star with mass 4×10^{30} kilograms is in the process of collapsing and becoming an exceedingly dense neutron star. Assume that at the instant the star has radius 45,000 kilometers, its radius is decreasing at the rate of 3×10^{-6} kilometers per second. How fast is its escape velocity increasing at that instant?

19. A person is pushing a box up the ramp in Figure 3.41 at the rate of 3 feet per second. How fast is the box rising?

FIGURE 3.41   Figure for Exercise 19.

20. A rope is attached to the bow of a sailboat coming in for the evening. Assume that the rope is drawn in over a pulley 5 feet higher than the bow at the rate of 2 feet per second, as shown in Figure 3.42. How fast is the boat docking when the length of rope from bow to pulley is 13 feet?

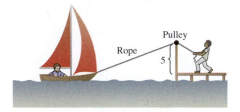

FIGURE 3.42   Figure for Exercise 20.

21. Suppose the rope in Exercise 20 is pulled so that the boat docks at a constant rate of 2 feet per second. How fast is the rope being pulled in when the boat is 12 feet from the dock?

22. As in Exercise 20 assume that the boat is pulled in by a rope attached to the bow passing through a pulley 5 feet above the bow. Assume also that the distance between the bow and the dock decreases as the cube root of the distance; that is, if the distance at time t is y feet, then $dy/dt = -y^{1/3}$ (feet per second). How fast is the length of the rope shrinking when the bow is 8 feet from the dock?

23. A Flying Tiger is making a nose dive along a parabolic path having the equation $y = \frac{1}{100}x^2 + 1$, where x and y are measured in feet. Assume that the sun is directly above the y axis, that the ground is the x axis, and that the distance from the plane to the ground is decreasing at the constant rate of 100 feet per second. How fast is the shadow of the plane moving along the ground when the plane is 2501 feet above the earth's surface? Assume that the sun's rays are vertical.

24. The tortoise and the hare are having their famous race, each moving along a straight line. The tortoise, moving at a constant rate of 10 feet per minute, is 4 feet from the finish line when the hare wakes up 5001 feet from the finish line and darts off after the tortoise. Let x be the distance from the tortoise to the finish line, and suppose the distance y from the hare to the finish line is given by

$$y = 5001 - 2500\sqrt{4 - x}$$

a. How fast is the hare moving when the tortoise is 3 feet from the finish line?

b. Who wins? By how many feet?

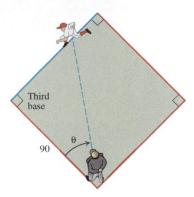

Third base

90

θ

25. A baseball diamond is a square with sides 90 feet long. Suppose a baseball player is advancing from second to third base at the rate of 24 feet per second, and an umpire is standing on home plate. Let θ be the angle between the third baseline and the line of sight from the umpire to the runner (Figure 3.43). How fast is θ changing when the runner is 30 feet from third base?

26. Maple and Main Streets are straight and perpendicular to each other. A stationary police car is located on Main Street $\frac{1}{4}$ mile from the intersection of the two streets. A sports car on Maple Street approaches the intersection at the rate of 40 miles per hour. How fast is the distance between the two cars decreasing when the sports car is $\frac{1}{8}$ mile from the intersection?

27. Suppose in Exercise 26 that the sports car approaches the intersection in such a way that the distance between the sports car and the police car decreases at 30 miles per hour. How far from the intersection would the sports car be at the moment when it is traveling 50 miles per hour?

28. A spotlight is on the ground 100 feet from a building that has vertical sides. A person 6 feet tall starts at the spotlight and walks directly toward the building at a rate of 5 feet per second.

a. How fast is the top of the person's shadow moving down the building when the person is 50 feet away from it?

b. How fast is the top of the shadow moving when the person is 25 feet away?

29. A kite 100 feet above the ground is being blown away from the person holding its string. It moves in a direction parallel to the ground and at the rate of 10 feet per second. At what rate must the string be let out when the length of string already let out is 200 feet?

30. A helicopter flies parallel to the ground at an altitude of $\frac{1}{2}$ kilometer and at a speed of 2 kilometers per minute. If the helicopter flies along a straight line that passes directly over the White House, at what rate is the distance between the helicopter and the White House changing 1 minute after the helicopter flies over the White House?

31. When a rocket is 2 kilometers high, it is moving vertically upward at a speed of 300 kilometers per hour. At that instant, how fast is the angle of elevation of the rocket increasing, as seen by an observer on the ground 5 kilometers from the launching pad?

32. The largest of all ferris wheels was built by the engineer G. W. Gale Ferris for the World's Columbian Exposition in Chicago in 1893. It was 125 feet in radius and could hold 2160 people at a time. Suppose it revolved at 2 radians per minute. How fast would a passenger rise when the passenger was 75 feet higher than the center of the ferris wheel and was rising?

A smaller version of Ferris's wheel. *(Doug Armand/Tony Stone Images)*

33. A street light 16 feet high casts a shadow on the ground from a ball that is falling toward a point on the ground 15 feet from the base of the street light (See Figure 3.44). Suppose that when the ball is 6 feet high, it is falling at the rate of 10 feet per second. How fast is the shadow of the ball moving along the ground at that instant?

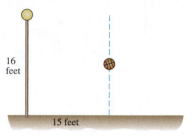

FIGURE 3.44   Figure for Exercise 33.

*34. Water is released from a conical tank 50 inches tall and 30 inches in radius, and falls into a rectangular tank whose base has an area of 400 square inches (Figure 3.45). The rate of release is controlled so that when the height of the water in the conical tank is x inches, the height is decreasing at the rate of $50 - x$ inches per minute. How fast is the water level in the rectangular tank rising when the height of the water in the conical tank is 10 inches? (*Hint:* The total amount of water in the two tanks is constant.)

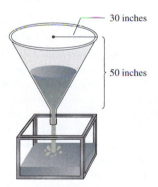

30 inches

50 inches

FIGURE 3.45   Figure for Exercise 34.

35. A helicopter 3000 feet high is moving horizontally at the rate of 100 feet per second. It flies directly over a searchlight that rotates so as to always illuminate the helicopter. At how many radians per second is the searchlight rotating when the distance between the helicopter and searchlight is 5000 feet?

36. Suppose a deer is standing 20 feet from a highway on which a car is traveling at a constant rate of v feet per

second. Let θ be the angle made by the highway and the line of sight from a passenger to the deer (Figure 3.46). Show that

$$\frac{d\theta}{dt} = \frac{20v}{400 + x^2}$$

(Notice that for x close to 0, $d\theta/dt$ is approximately $v/20$, and thus for the passenger to keep the deer in focus, the passenger's eyes must rotate at the approximate rate of $v/20$ radians per second. This suggests why at large velocities it may be impossible to keep a stationary object near the highway in focus.)

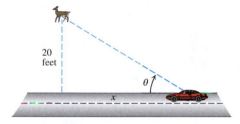

20 feet

θ

x

FIGURE 3.46   Spotting a deer.

*37. A 10-foot-square sign of negligible thickness revolves about a vertical axis through its center at a rate of 10 revolutions per minute. An observer far away sees it as a rectangle of variable width. How fast is the width changing when the sign appears to be 6 feet wide and is increasing in width? (*Hint:* View the sign from above, and consider the angle it makes with a line pointing toward the observer.)

38. At night a patrol boat approaches a point on shore along the curve $y = -\frac{1}{2}x^3$, as indicated in Figure 3.47. If the boat moves along the curve so that $dx/dt = -x$, and if its spotlight is pointed straight ahead, determine how fast the illuminated spot on the shore moves when $x = -2$. (*Hint:* You will need to find the x intercept of the line tangent to the curve $y = -\frac{1}{2}x^3$.)

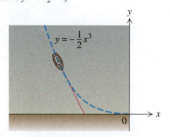

$y = -\frac{1}{2}x^3$

FIGURE 3.47   A patrol boat.

39. A deer 5 feet long and 6 feet tall, whose rump is 4 feet above ground as in Figure 3.48, approaches a street light with lamp 20 feet above ground. If the deer proceeds at 3 feet per second, how fast is the rear of the shadow moving when the front of the deer is

a. 48 feet from the street light?

b. 24 feet from the street light?

(*Hint:* In parts (a) and (b), determine which yields the shadow, the head or the rump of the deer.)

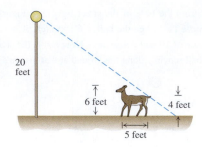

FIGURE 3.48 A deer shadow.

3.8 POLYNOMIAL APPROXIMATION AND THE NEWTON-RAPHSON METHOD

One of the many triumphs of the derivative is that it allows us to approximate values of differentiable functions. In fact, calculators and computers use methods related to the derivative in order to provide approximate values of such functions as the trigonometric and exponential functions.

In this section we will study approximation of differentiable functions by lines and parabolas, and then will complete the chapter by discussing a method called the Newton-Raphson method for approximating the zeros of a differentiable function.

Linear Approximation and Parabolic Approximation

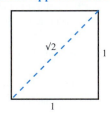

FIGURE 3.49

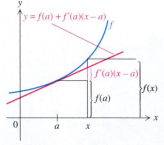

FIGURE 3.50 Linear approximation for $f(x)$.

We all know that the diagonal of a unit square has length $\sqrt{2}$ (Figure 3.49). But what is the value of $\sqrt{2}$ in decimal form? Often 1.414 or 1.41421 are given to represent $\sqrt{2}$; calculators and computers give better estimates, such as 1.414213562. However, even the ancient Greek mathematicians recognized that $\sqrt{2}$ is irrational, so that we can never write down all the digits in the decimal expansion of $\sqrt{2}$. This raises the question of how calculators and computers are programmed so as to provide approximate values of irrational numbers such as $\sqrt{2}$, or e, or π.

The method that we will describe now uses judiciously chosen polynomials in order to approximate values of functions that are not so accessible. In this section we will keep the polynomials simple—having degree 2 or less—and will nevertheless be able to achieve very reasonable approximations with little effort.

We proceed by considering a function f that is differentiable at a given number a. By the way it was defined, the line tangent to the graph of f at $(a, f(a))$ has the property that for values of x near a, $f(x)$ is close to the corresponding y-value on the tangent line, which is $f(a) + f'(a)(x - a)$. We write this relationship as

$$f(x) \approx f(a) + f'(a)(x - a) \quad \text{for} \quad x \text{ near } a \tag{1}$$

Because $(x, f(x))$ is close to a corresponding point on the line tangent to the graph of f, this type of approximation of $f(x)$ is called a **tangent line approximation,** or more simply, a **linear approximation** (See Figure 3.50). Notice that in order for the approximation in (1) to be helpful, it is necessary that $f(a)$ and $f'(a)$ be easily computed.

Now we will apply these ideas to estimate $\sqrt{2}$.

EXAMPLE 1   Use (1) to find a linear approximation of $\sqrt{2}$.

Solution   First we notice that $\sqrt{2}$ is a square root, so we consider f to be the square root function: $f(x) = \sqrt{x}$. We need to find a value of a so that $f(a)$ and $f'(a)$ can be easily computed, and so that 2 is close to a. Since $\sqrt{1.96} = 1.4$ and 1.96 is near 2, we can let $a = 1.96$ and $x = 2$, so that indeed x is near a. Next we observe that $f'(x) = 1/(2\sqrt{x})$, so that

$$f(x) = f(1.96) = \sqrt{1.96} = 1.4 \quad \text{and} \quad f'(a) = \frac{1}{2\sqrt{1.96}} = \frac{1}{2.8}$$

Having set the stage, we are ready to use (1):

$$\sqrt{2} = f(2) \approx f(1.96) + f'(1.96)(2 - 1.96) = 1.4 + \frac{1}{2.8}(0.04) \approx 1.414285714 \quad \square$$

As we mentioned before, the value of $\sqrt{2}$ produced on a calculator is 1.414213562, so the approximation 1.414285714 deduced in Example 1, is accurate to 4 decimal places—pretty good for such little effort!

Because we can only expect $f(a) + f'(a)(x - a)$ to be a good estimate for $f(x)$ when x is close to a, we often rewrite (1) by replacing x with $a + h$. With this substitution, (1) becomes

$$f(a + h) \approx f(a) + f'(a)h \tag{2}$$

Notice that x is close to a if and only if h is close to 0, so in using (2) we consider only values of h close to 0.

EXAMPLE 2   Use (2) to approximate $\sin \frac{7\pi}{36}$.

Solution   In this case we will let $f(x) = \sin x$. Since $7\pi/36$ is close to $\pi/6$, and since $f(\pi/6)$ and $f'(\pi/6)$ are easy to evaluate, we are ready to apply (2) with $a = \pi/6$ and $h = 7\pi/36 - \pi/6 = \pi/36$:

$$\sin \frac{7\pi}{36} \approx \sin \frac{\pi}{6} + \left(\cos \frac{\pi}{6}\right)\frac{\pi}{36} = \frac{1}{2} + \frac{\sqrt{3}}{2}\frac{\pi}{36} \approx 0.5755749735 \quad \square$$

A natural question at this point is the following: How good is the approximation in Example 2? In order to give an answer to this question, let us define the **error** that occurs when we substitute $f(a) + f'(a)h$ for $f(a + h)$:

$$\text{error} = |f(a + h) - [f(a) + f'(a)h]|$$

It can be shown that if there exists a positive number M such that

$$|f''(x)| \leq M \text{ for all } x \text{ between } a \text{ and } a + h$$

then

$$\text{error} \leq \frac{1}{2} Mh^2 \tag{3}$$

Although (3) does not tell us the exact error, (3) does tell us that the error can be no larger than $\frac{1}{2} Mh^2$. We call $\frac{1}{2} Mh^2$ an **error bound** for the error.

To find an error bound that arises when we use the linear approximation to estimate $\sin \frac{7\pi}{36}$, as we did in Example 2, we once again let $f(x) = \sin x$, $a = \pi/6$, and $h = \pi/36$. Then $f'(x) = \cos x$ and $f''(x) = -\sin x$. Since

$$|f''(x)| = |\sin x| \leq 1$$

we may use $M = 1$ in (3) to deduce that

$$\text{error} \leq \frac{1}{2} Mh^2 = \frac{1}{2} \cdot 1 \cdot \left(\frac{\pi}{36}\right)^2 < 0.004$$

Consequently if we use 0.5755749732 from Example 2 as an estimate of $\sin \frac{7\pi}{36}$, then the error will be less than 0.004. You can check whether the value for $\sin \frac{7\pi}{36}$ given by your calculator agrees.

An important feature of the linear approximation is that the smaller h is, the smaller we expect the error to be in approximating $f(a + h)$ by $f(a) + f'(a)h$. If we replace a in (2) by any number x in the domain of f, then we have

$$f(x + h) \approx f(x) + f'(x)h \tag{4}$$

which is equivalent to

$$f(x + h) - f(x) \approx f'(x)h \tag{5}$$

The number $f'(x)h$ appearing on the right side of (4) and (5) is the adjustment that needs to be made to $f(x)$ in order to achieve the tangent line approximation of $f(x + h)$. The number $f'(x)h$ is usually called the **differential of f (at x with increment h),** and is denoted by df. Thus

$$df = f'(x)h$$

Of course, df depends on both x and h, although these variables do not appear in the expression df. The relationship between the differential and $f(x + h) - f(x)$ is illustrated in Figure 3.51.

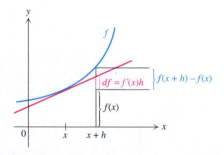

FIGURE 3.51   Differential versus first difference at x.

If $g(x) = x$, then we denote dg by dx. Since $g'(x) = 1$, it follows that

$$dx = dg = g'(x)h = h$$

Therefore we may replace the number h by dx when writing the differential df of a function f:

$$df = f'(x)dx \tag{6}$$

The use of differentials shortens certain manipulations that will appear in Chapters 5–8. If u is a variable depending on x, then by (6),

$$du = u'(x)dx = \left(\frac{du}{dx}\right)dx \tag{7}$$

For example, if $u = x^2$, then

$$du = \left(\frac{du}{dx}\right)dx = \left(\frac{d}{dx}(x^2)\right)dx = 2x\,dx$$

In the same way, if $v = x \sin x$, then

$$du = \left(\frac{du}{dx}\right)dx = \left(\frac{d}{dx}(x \sin x)\right)dx = (\sin x + x \cos x)\,dx$$

For the tangent line approximation, we assume that f is differentiable at a, and approximate the values of $f(x)$ for x near a by values along the tangent line. If we let the function p_1 be given by

$$p_1(x) = f(a) + f'(a)(x - a)$$

then it is $p_1(x)$ that approximates $f(x)$ for x near a. The index 1 in p_1 signifies that p_1 is a polynomial of degree no more than 1. (Of course, the degree of p_1 is 0 in the special case that $f'(a) = 0$.)

Parabolic Approximation

Graphs of functions are normally curves, rather than straight lines. Thus it makes sense to conjecture that we might be able to get better approximations for values of $f(x)$ for x near a by using polynomials of degree 2, rather than polynomials of degree 1 (or 0 in special cases). What polynomial p_2 of degree 2 should we use for the best possible approximations?

To address this issue, we will assume that $f'(a)$ and $f''(a)$ both exist. Since the tangent approximation p_1 has the property that $p_1(a) = f(a)$ and $p_1'(a) = f'(a)$, we are led to ask that p_2 have not only the same value and derivative at a as f has, but also the same second derivative at a. You can check that if p_2 is given by

$$p_2(x) = f(a) + f'(a)(x - a) + \frac{1}{2}f''(a)(x - a)^2 \tag{8}$$

then $p_2(a) = f(a)$, $p_2'(a) = f'(a)$, and $p_2''(a) = f''(a)$ (See Exercise 55). Because p_2 is a polynomial of degree 2 (or less in special cases), and the graph of each polynomial of degree 2 is a parabola, we can say that p_2 represents a

parabolic (or quadratic) approximation of *f* near *a*.

We now return to $\sqrt{2}$, and estimate its value by a parabolic approximation.

EXAMPLE 3   Use (8) to find a parabolic approximation of $\sqrt{2}$.

Solution   As in Example 1, we let $f(x) = \sqrt{x}$, $a = 1.96$, and $x = 2$. We find that

$$f'(a) = \frac{1}{2\sqrt{x}} \quad \text{and} \quad f''(x) = -\frac{1}{4x^{3/2}}$$

Therefore

$$f(1.96) = \sqrt{1.96} = 1.4, \ f'(1.96) = \frac{1}{2\sqrt{1.96}} = \frac{1}{2.8},$$

and

$$f''(1.96) = -\frac{1}{4(1.96)^{3/2}} = -\frac{1}{4(1.4)^3}$$

and hence

$$\sqrt{2} \approx p_2(2) = f(1.96) + f'(1.96)(2 - 1.96) + \frac{1}{2}f''(1.96)(2 - 1.96)^2$$

$$= 1.4 + \frac{1}{2.8}(0.04) + \frac{1}{2}\left(-\frac{1}{4(1.4)^3}\right)(0.04)^2$$

$$\approx 1.414212828 \quad \blacksquare$$

How does the parabolic approximation of $\sqrt{2}$ obtained in Example 3 compare with the linear approximation obtained in Example 1? Let us see:

actual value, to 9 decimal places: 1.414213562

value by linear approximation: 1.414285714

value by parabolic approximation: 1.414212828

As you can see, the extra effort in finding and using the second derivative resulted in better accuracy with the parabolic approximation.

Now we will apply the linear and quadratic approximations to a problem involving the cooling of a liquid, and see how the quadratic approximation is better than the linear approximation.

Suppose a cup of boiling water (at 100° Celsius) is left in a room with a constant temperature of 20° Celsius. Then Newton's Law of Cooling says that the temperature $T(t)$ after t minutes is given (approximately) by the formula

$$T(t) = 20 + 80e^{-kt} \tag{9}$$

where k is a positive constant. Let us assume that the constant $k = .3$, so that T becomes

$$T(t) = 20 + 80e^{-.3t}$$

Notice that $T(0) = 100$.

EXAMPLE 4   For $t = 0$, 0.5, 1, 2, and 4, compare the actual values of the temperature from (9) with the approximate values when using the linear and the quadratic approximations with $a = 0$ and $f = T$.

Solution   First we find p_1 and p_2. For p_1 we have

$$T'(t) = -24e^{-.3t}, \quad \text{so that} \quad T'(0) = -24, \quad \text{and thus} \quad p_1(t) = 100 - 24t$$

For p_2 we have

$$T''(t) = 7.2e^{-.3t}, \quad \text{so that} \quad T''(0) = 7.2$$

and thus

$$p_2(t) = 100 - 24t + \frac{7.2}{2} t^2 = 100 - 24t + 3.6t^2$$

Now we are ready to compare the values of $T(t)$, $p_1(t)$, and $p_2(t)$ when $t = 0$, 0.5, 1, 2, and 4:

	$t = 0$	$t = .5$	$t = 1$	$t = 2$	$t = 4$
T	100	88.86	79	63	44
p_1	100	88	76	52	4
p_2	100	88.9	79.6	66.4	61.6

Evidently, the linear approximation is pretty good after half a minute, reasonable after 1 minute, not so good after 2 minutes, and awful after 4 minutes. By contrast, the quadratic approximation is much better through the 2nd minute, though the error increases markedly by the 4th minute   ❑

For the temperature of cooling water, if we would use cubic approximation or higher degree polynomial approximations, we would find increasingly improved approximate values as time increases. Higher degree polynomial approximations will be studied in detail in Chapter 9.

We should also observe that, as you can see, the temperature drops the fastest at first, and then more gently as time passes. Eventually the temperature will level out as it comes to the room temperature of 20° C.

As occurred with the examples above, generally a parabolic approximation will be a better estimate of $f(x)$ than a linear approximation, for x near a. The polynomials p_1 and p_2 are well-known and are called **Taylor polynomials** after the English mathematician Brook Taylor. We will discuss Taylor polynomials in more depth in Chapter 9.

Finally, we note that in physics and engineering, linear approximation is often used in order to give an approximate solution to a problem. For example, a pendulum, such as is in a pendulum clock, has the property that if it is released from its maximum angle θ_0, then for any time $t \geq 0$, the angle θ that the pendulum rod makes with the vertical (see Figure 3.52) must satisfy the equation

FIGURE 3.52

$$\frac{d^2\theta}{dt^2} = -c^2 \sin\theta \tag{10}$$

where c depends on the length of the rod. Now there is no way to find a simple formula for $\theta(t)$ in (10). However, the linear approximation to the function $\sin\theta$ for θ near 0 is θ, that is, for small values of θ, we have $\sin\theta \approx \theta$. Therefore physicists often write

$$\frac{d^2\theta}{dt^2} = -c^2\theta \tag{11}$$

instead of (10). You can check that if $\theta(t) = \theta_0 \cos(ct)$, then $\theta(t)$ satisfies (11).

The Newton-Raphson Method

Suppose that we wish to find the point at which the graph of $\cos x$ crosses the line $y = x$. To find it, we need to locate a number d such that $\cos d = d$, that is, we need to solve the equation $\cos x - x = 0$. The bisection method in Section 2.5 gives a way of approximating solutions of equations. Now we will describe a second, much more efficient, method called the Newton-Raphson method. It is named for two mathematicians, Isaac Newton and Joseph Raphson.

Let d be a zero of a function f. We would like to approximate the numerical value of d. Assume that f is differentiable on an open interval I containing d. We begin the Newton-Raphson method by choosing a number c_1 in I that we believe is close to d (Figure 3.53). There are two possibilities:

1. If $f(c_1) = 0$, then c_1 is a zero of f, and we have completed our search.
2. If $f(c_1) \neq 0$, then provided that the line l tangent to the graph of f at $(c_1, f(c_1))$ is not horizontal, we let c_2 be the x intercept of l (Figure 3.53), and replace c_1 by c_2.

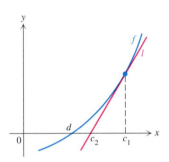

FIGURE 3.53   First step in the Newton-Raphson method.

If possibility 2 occurs, then ideally c_2 will be closer to d and hence will be a better approximation of d than c_1 is. (This is indeed the case in Figure 3.53.)

For c_2 there are again two possibilities: $f(c_2) = 0$ or $f(c_2) \neq 0$. If $f(c_2) = 0$, then c_2 is a zero of f and the process stops. However, if $f(c_2) \neq 0$ and the line tangent to the graph of f at $(c_2, f(c_2))$ is not horizontal, then we apply the same procedure to c_2 to obtain c_3 (Figure 3.54).

For the process of selecting $c_1, c_2, c_3, \ldots$ to be successful, we need to find a formula for c_2 in terms of c_1, c_3 in terms of c_2, and so on. To that end, we recall that l is tangent to the graph of f at $(c_1, f(c_1))$, so has slope $f'(c_1)$. Therefore an equation of l is

$$y = f(c_1) + f'(c_1)(x - c_1)$$

Since c_2 is the x intercept of l, it follows that c_2 satisfies

$$0 = f(c_1) + f'(c_1)(c_2 - c_1)$$

Solving for c_2, we find that

$$c_2 - c_1 = -\frac{f(c_1)}{f'(c_1)}$$

and thus

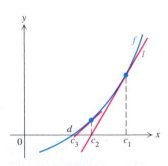

FIGURE 3.54   Second step in the Newton-Raphson method.

$$c_2 = c_1 - \frac{f(c_1)}{f'(c_1)}$$

Repeating the analysis but with c_1 replaced by c_2 and c_2 by c_3, we find that

$$c_3 = c_2 - \frac{f(c_2)}{f'(c_2)}$$

This procedure may be continued. After we have obtained the approximation c_n, the next approximation c_{n+1} is given by the formula

$$c_{n+1} = c_n - \frac{f(c_n)}{f'(c_n)} \tag{12}$$

The method of approximating a zero of f by means of (12) is called the **Newton-Raphson method.**

Ideally we would like to stop the process when an approximate zero c_n and the zero d of f are within an acceptable tolerance ε of each other, that is, when

$$|c_n - d| < \varepsilon$$

But since we are trying to find the value of d, and hence don't know it, a reasonable alternative is to stop the process when successive approximations c_n and c_{n+1} are within ε of each other, that is, when

$$|c_{n+1} - c_n| < \varepsilon \tag{13}$$

This means, for example, that if we wish to approximate a zero to four decimal places (that is, $\varepsilon = 10^{-4}$), then we would carry out the Newton-Raphson method until we obtain two successive values c_n and c_{n+1} that agree to four decimal places. Then c_{n+1} would be the desired approximate zero.

Now by (12),

$$c_{n+1} - c_n = -\frac{f(c_n)}{f'(c_n)}$$

so that (13) is equivalent to

$$\left| \frac{f(c_n)}{f'(c_n)} \right| < \varepsilon \tag{14}$$

The algorithm we will use in finding an approximate zero for a given function uses the criterion in (14) to determine when to stop.

INPUT: A differentiable function f, an initial value c, and an allowable error ε.

OUTPUT: A number c that is within ε of the preceding approximation.

STEP 1: Compute $f(c)$. If $f(c) = 0$, then c is a zero of f, so STOP. Otherwise CONTINUE.

STEP 2: Compute $f'(c)$. If $f'(c) = 0$, then STOP because the method cannot be continued. Otherwise CONTINUE.

The Newton-Raphson Method
In his study of the orbits of the planets, Johannes Kepler (1571–1630) encountered an equation equivalent to
$$x + c \sin x - a = 0$$
where c and a are constants. He found an approximate solution to the equation by a delicate iteration process. Some 40 years later, Newton proposed a powerful alternative method of approximating zeros that involved the derivative, which was later systematized by Joseph Raphson (1648–1715).

STEP 3: Compute $f(c)/f'(c)$. If $|f(c)/f'(c)| < \varepsilon$, then $c - (f(c)/f'(c))$ is the desired approximate zero with error ε, so STOP. Otherwise CONTINUE.

STEP 4: Replace c by $c - (f(c)/f'(c))$.

STEP 5: Repeat Steps 1–4 with the new number found in STEP 4, continuing until the process stops with STEP 1, 2, or 3.

One can facilitate the Newton-Raphson method on a calculator by letting

$$g(x) = x - \frac{f(x)}{f'(x)} \tag{15}$$

If we select an initial value c_1, then

$$c_2 = g(c_1), \qquad c_3 = g(c_2), \qquad c_4 = g(c_3), \qquad \text{and so on}$$

Thus the successive approximations $c_2, c_3, c_4,\ldots$, are the iterates of c_1 for g.

On a calculator it is reasonable to terminate the process when two successive iterates appearing on the screen are identical. If the calculator displays k digits to the right of the decimal, then such a pair of iterates will be less than 10^{-k} apart.

EXAMPLE 5   Let $f(x) = \cos x - x$. (See Figure 3.55.) Use the Newton-Raphson method to approximate a zero of f. Continue the process until successive approximations obtained by calculator are identical.

Solution   First we notice that f is a continuous function, and that $f(0) = \cos 0 - 0 = 1$ and $f(\pi/2) = \cos \pi/2 - \pi/2 = -\pi/2$. Therefore the Intermediate Value Theorem asserts that f has a zero in the interval $(0, \pi/2)$. To start the Newton-Raphson method, we can let $c_1 = 1$, which is in the interval $(0, \pi/2)$. Since $f'(x) = -\sin x - 1$, the function g in (15) is given by

$$g(x) = x - \frac{\cos x - x}{-\sin x - 1}$$

Using a calculator, we obtain the following sequence:

$$c_1 = 1$$
$$c_2 = g(c_1) \approx 0.7503638678$$
$$c_3 = g(c_2) \approx 0.7391128909$$
$$c_4 = g(c_3) \approx 0.7390851334$$
$$c_5 = g(c_4) \approx 0.7390851332$$
$$c_6 = g(c_5) \approx 0.7390851332$$

Since $c_6 = c_5$, accurate to 10 digits, we take 0.7390851332 as the desired approximation to a zero of f—which is also the number d for which $\cos d = d$. ❑

The Newton-Raphson method works very effectively in approximating square or cube roots, as the next example shows.

EXAMPLE 6   Approximate $\sqrt{2}$ by using the Newton-Raphson method until successive approximations obtained by calculator are identical.

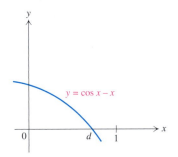

$y = \cos x - x$

FIGURE 3.55

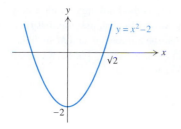

FIGURE 3.56

Solution Notice that if $f(x) = x^2 - 2$, then $f(\sqrt{2}) = 0$ (Figure 3.56). Thus we seek an approximate zero of f. Since 1.96 is close to 2 and $\sqrt{1.96} = 1.4$, we let $c_1 = 1.96$. Now $f'(x) = 2x$, so the function g in (15) is given by

$$g(x) = x - \frac{x^2 - 2}{2x}$$

By calculator we find that

$$c_1 = 1.96$$
$$c_2 = g(c_1) \approx 1.490204082$$
$$c_3 = g(c_2) \approx 1.416151068$$
$$c_4 = g(c_3) \approx 1.414214888$$
$$c_5 = g(c_4) \approx 1.414213562$$
$$c_6 = g(c_5) \approx 1.414213562$$

We conclude that $\sqrt{2} \approx 1.414213562$. ❑

Notice that the approximation obtained in Example 6 by the Newton-Raphson method is far superior to the approximations obtained in Examples 1 and 3 by means of linear and parabolic approximation.

If we had selected $c_1 = 1$, in Example 6, then we would have obtained the same conclusion after 5 calculations as before. In fact, the function g has the property that whatever positive value c_1 we pick, the iterates $c_2, c_3, \ldots$ approach 1.414213562 with great rapidity. To test this assertion out you might let $c_1 = 0.01$ or $c_1 = 1000$.

How carefully must the initial value of c be chosen in order for the Newton-Raphson method to provide the desired approximate zero in an acceptable number of steps? The answer depends on the function. For example, in Example 6 if we had let $c_1 = 10$ (which is not very close to a zero of f), then it would have taken only two more approximations before stopping. However, an unfortunate choice of c_1 may even lead to the failure of the Newton-Raphson method to work at all. Indeed, successive values of c_i might lie outside the domain of f, fail to approach a zero of f, or approach the wrong zero (see Figure 3.57 and Exercises 51–54).

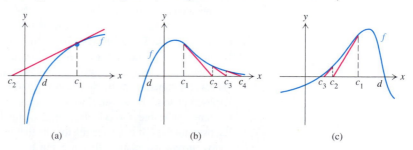

(a)   (b)   (c)

FIGURE 3.57 (a) c_2 lies outside the domain. (b) The values of c_i fail to converge to a zero of f. (c) The values of c_i converge to the wrong zero of f.

It is apparent from the preceding comments that the Newton-Raphson method does not always work, even if the function has a zero in a given interval. However, under fairly general conditions the Newton-Raphson method not only yields a very good approximation, but also produces it expeditiously. Indeed, assume that f has a zero d in the interval $[a, b]$, and that there are positive numbers m and M such that $|f'(x)| \geq m$ and $|f''(x)| \leq M$ for all x in $[a, b]$. This means that $|f'|$ is not too small and $|f''|$ is not too large on $[a, b]$. Then it can be shown that

$$|d - c_{n+1}| < \frac{M}{2m} |d - c_n|^2$$

If we let $|d - c_n| = e_n$, which is the error of the nth approximation c_n, we conclude that

$$e_{n+1} < \frac{M}{2m} e_n^2 \tag{16}$$

To illustrate the preceding comments, let us return to the approximation of $\sqrt{2}$ that appeared in Example 6. Let $f(x) = \sqrt{x}$ and $d = \sqrt{2}$. We pick $a = 1$ and $b = 4$ because $1 \leq \sqrt{2} \leq 4$, and f' and f'' are easy to evaluate at 1 and 4. For $1 \leq x \leq 4$,

$$|f'(x)| = \frac{1}{2\sqrt{x}} \geq \frac{1}{2\sqrt{4}} = \frac{1}{4}$$

and

$$|f''(x)| = \left|-\frac{1}{4x^{3/2}}\right| \leq \frac{1}{4 \cdot 1} = \frac{1}{4}$$

Therefore we can take $m = M = \frac{1}{4}$, so that $M/(2m) = 1/2 < 1$. In this case, we deduce from (16) that $e_{n+1} < e_n^2$. This means that if $e_n < 10^{-k}$, then $e_{n+1} < 1 \cdot e_n^2 < 10^{-2k}$. For example, if $e_n < 10^{-4}$, then $e_{n+1} < 10^{-8}$. Thus the error bound decreases very rapidly with each step. This is in stark contrast with the performance of the bisection method, for which each step merely halves the error bound.

Despite the speed with which convergence can be obtained with the Newton-Raphson method, it requires an evaluation of the derivative of the function in question. For many simple functions this poses no problem. However, for more complicated functions the formula in the Newton-Raphson method can become a real challenge. For example, suppose that $f(x) = (e^x \sin x)/(x^2 + 1)$, and try writing out the formula for g in (15). To circumvent this problem, a variation of the Newton-Raphson method uses secant lines instead of tangent lines, and is called the **secant method** (see the project for this section). It gives convergence that is in general far superior to the bisection method but a little slower than the Newton-Raphson method. However, since it does not rely on derivatives, the secant method is more applicable for calculator and computer calculations than the Newton-Raphson method. It is interesting to note that in approximating a zero of a function, many current calculators first use the bisection method in order to get into a reasonable range of the zero, and then switch to the secant method for a quick zoom toward the zero.

EXERCISES 3.8

In Exercises 1–14 approximate the given number by using a linear approximation.

1. $\sqrt{101}$

2. $\sqrt{99.5}$

3. $\sqrt[3]{29}$

4. $\sqrt[4]{17}$

5. $(28)^{4/3}$

6. $1/[1 + (1.0175)^2]$

7. $\cos 2\pi/13$

8. $\cot 11\pi/36$

9. $\sec 4\pi/17$

10. $\tan 99\pi/100$

11. $e^{0.1}$

12. $e^{-0.124}$

13. $\ln 1.1$

14. $\ln 0.95$

In Exercises 15–18 compute df for the given values of a and h.

15. $f(x) = \sqrt{x}$; $a = 4$, $h = 0.2$

16. $f(x) = \sqrt[3]{x}$; $a = 64$, $h = -0.1$

17. $f(x) = \sqrt{1 + x^3}$; $a = 2$, $h = 0.01$

18. $f(x) = \sqrt{1 + x^3}$; $a = 2$, $h = -0.001$

In Exercises 19–24 find the differential by using (6) or (7).

19. $f(x) = 5x^3 + 2$

20. $f(x) = \sin x^2$

21. $f(x) = \sin(\cos x)$

22. $u = x\sqrt{x-1}$

23. $u = \sqrt{1 + x^4}$

24. $u = (x^2 + 3)/(x^3 - 4)$

In Exercises 25–27 use (3) to find an upper bound for the error introduced by using the linear approximation to estimate the given number.

25. $\cos 2\pi/13$; $a = \pi/6$

26. $\sqrt{101}$; $a = 100$

27. $\sqrt[3]{28}$; $a = 27$

In Exercises 28–33 approximate the given number by using a parabolic approximation.

28. $\sqrt{101}$

29. $\sqrt[3]{29}$

30. $(28)^{4/3}$

31. $\cos \dfrac{2\pi}{13}$

32. $e^{0.1}$

33. $\ln 1.1$

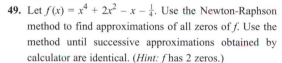
In Exercises 34–43 use the Newton-Raphson method to find an approximate solution of the given equation in the given interval. Use the method until successive approximations obtained by calculator are identical.

34. $x^3 + x = 1$; $(-\infty, \infty)$ (*Hint:* Let $f(x) = x^3 + x - 1$.)

35. $x^3 - 3x - 1 = 0$; $[1, 2]$ (This equation is related to the study of inscribing a nine-sided polygon in a circle.)

36. $2x^3 - 5x - 3 = 0$; $[1, 2]$

37. $x^3 - 2x - 5 = 0$; $[0, 3]$ (Newton solved this equation himself.)

38. $x^2 + 4x^6 = 2$; $[0, 1]$

39. $2x^3 - 5x - 3 = 0$; $(-1, 0)$

40. $x^4 + \sin x = 0$; $[-2, -1/2]$

41. $\tan x = x$; $(\pi/2, 3\pi/2)$

42. $\ln x = x - 3$; $[1, 10]$

43. $e^{-x} = x$; $[0, 1]$

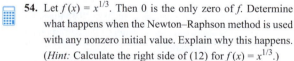
In Exercises 44–47 use the Newton–Raphson method to find an approximate value of the given root. Use the method until successive approximations obtained by calculator are identical.

44. $\sqrt{0.2}$ **45.** $\sqrt{15}$ **46.** $\sqrt[4]{13}$ **47.** $\sqrt[3]{9}$

48. Let $f(x) = x^4 + 2x^3 - x - 1$. Use the Newton-Raphson method to find approximations of all zeros of f. Use the method until successive approximations obtained by calculator are identical.

49. Let $f(x) = x^4 + 2x^2 - x - \frac{1}{4}$. Use the Newton-Raphson method to find approximations of all zeros of f. Use the method until successive approximations obtained by calculator are identical. (*Hint:* f has 2 zeros.)

50. Suppose that in approximating a zero of f by the Newton-Raphson method we find that $f(c_n) = 0$ and $f'(c_n) \neq 0$ for some n. What does this imply about $c_{n+1}, c_{n+2}, \dots$?

51. Let $f(x) = x^4 + x^2 + 8x - 1$, which has a zero in the interval $[-2, 0]$. Determine what happens when the Newton-Raphson method is used with the initial value of c equal to -1.

52. Let $f(x) = \sqrt{x} - \frac{1}{2}$. In trying to use the Newton-Raphson method to find a zero of f, determine what goes wrong when we choose

  a. $c_1 = 1$ **b.** $c_1 = 4$

53. Let $f(x) = x^3 - x$, and suppose we attempt to approximate a zero of f in $(0, 1)$. Determine what happens when the Newton-Raphson method is used with the initial value of c equal to $1/\sqrt{5}$.

54. Let $f(x) = x^{1/3}$. Then 0 is the only zero of f. Determine what happens when the Newton–Raphson method is used with any nonzero initial value. Explain why this happens. (*Hint:* Calculate the right side of (12) for $f(x) = x^{1/3}$.)

55. Suppose f has a second derivative at a, and let

$$p_2(x) = f(a) + f'(a)(x - a) + \tfrac{1}{2}f''(a)(x - a)^2$$

Prove that

$$p_2(a) = f(a), p_2'(a) = f'(a) \quad \text{and} \, p_2''(a) = f''(a)$$

56. Let $g(x) = x - \dfrac{f(x)}{f'(x)}$, as in (15), and assume that $f''(x)$ exists. Show that

$$g'(x) = \frac{f(x) f''(x)}{[f'(x)]^2}$$

Applications

57. The outside surface of a hemispherical dome of radius 20 feet is to be given a coat of paint, $\frac{1}{100}$ inch thick. Use differentials to approximate the volume of paint needed for the job. (*Hint:* Approximate the change in the volume of a hemisphere when the radius increases from 20 feet to $20\frac{1}{1200}$ feet.)

58. Approximate the volume of material in a spherical ball with inner radius 5 inches and outer radius 5.137 inches.

59. One of the most striking notions of Einstein's Theory of Relativity is his time dilation concept. If a person P_1 moves with velocity v with respect to an observer P_2, then a watch carried by P_1 appears to be running more slowly to P_2 than to P_1 by a (dilation) factor of

$$\frac{1}{\sqrt{1 - v^2/c^2}}$$

Here c is the velocity of light, approximately 186,000 miles per second.

   a. If $v = c/2$, what is the dilation factor?

   b. If $v = c/3600$, which is 186,000 miles per *hour*, estimate the dilation factor. (*Hint:* Consider the function $f(x) = 1/\sqrt{1 - x}$ near 0.)

60. Suppose that pressure is measured in atmospheres and volume in liters. If the temperature of 1 mole of an ideal gas is held constant at 0° Celsius, then the pressure p and volume V of the gas are related by the equation

$$p = \frac{22.414}{V}$$

Suppose the volume changes from 20 to 20.35 liters. Use a parabolic approximation to estimate the corresponding change in the pressure.

61. Let $P(t)$ represent the population of woodpeckers in a certain region at time t, and suppose the graph of the derivative P' is as in Figure 3.58. Assume that at time $t = 0$ there are 500 woodpeckers.

   a. Use (2), with P and 0 replacing f and a, to approximate $P(1)$.

   b. Use (2), along with the result of part (a), to approximate $P(2)$.

   c. Continuing in the same manner as in (a) and (b), approximate $P(3)$, $P(4)$, $P(5)$, $P(6)$, and $P(7)$.

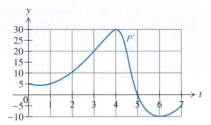

FIGURE 3.58 Population of woodpeckers.

62. A 6-inch-square sheet of metal is cut as in Figure 3.59 and the edges are bent to form a pan. The value of x for which the volume is maximized must satisfy

$$3x^3 + 12x^2 + 10x - 6 = 0$$

Use the Newton-Raphson method to approximate this value of x. Continue until successive approximations obtained by calculator are identical.

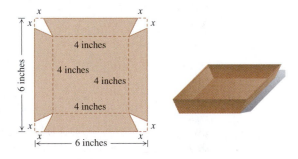

FIGURE 3.59 Forming a pan from a square sheet of metal.

Project

1. If the derivative of f is complicated, it may not be convenient to use the Newton-Raphson method. An alternative is the secant method, which uses two seeds, c_1 and c_2. Then instead of using the line tangent at $(c_n, f(c_n))$ to find c_{n+1}, the secant method uses the secant line that joins the points $(c_{n-1}, f(c_{n-1}))$ and $(c_n, f(c_n))$ (Figure 3.60). The formula that arises is

$$c_{n+1} = c_n - \frac{[f(c_n)](c_n - c_{n-1})}{f(c_n) - f(c_{n-1})}$$

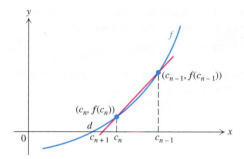

FIGURE 3.60 The secant method.

a. Prove that c_{n+1} is the x intercept of the secant line from which it is obtained.

b. For the two functions below (whose derivatives are not complicated), use the secant method to approximate a zero of f until successive calculator approximations are identical.

   i. $f(x) = x^3 - x - 5$

   ii. $f(x) = \sec x - 3$ for $0 \le x < \pi/2$

REVIEW

Key Terms and Expressions

Derivative

Function differentiable at a point

Differentiable function;
   function differentiable on an interval

Second derivative; higher derivatives

Chain Rule

Implicit differentiation

Acceleration

Linear approximation

Differential

Parabolic approximation

Newton-Raphson method

Key Formulas

$(f + g)'(x) = f'(x) + g'(x)$

$(cf)'(x) = cf'(x)$

$(fg)'(x) = f'(x)g(x) + f(x)g'(x)$

$\left(\dfrac{f}{g}\right)'(x) = \dfrac{f'(x)g(x) - f(x)g'(x)}{(g(x))^2}$

$(g \circ f)'(x) = [g'(f(x))]\,f'(x)$

$\dfrac{d}{dx}(x^r) = rx^{r-1}$

$\dfrac{d}{dx}(u + v) = \dfrac{du}{dx} + \dfrac{dv}{dx}$

$\dfrac{d}{dx}(cu) = c\dfrac{du}{dx}$

$\dfrac{d}{dx}(uv) = v\dfrac{du}{dx} + u\dfrac{dv}{dx}$

$\dfrac{d}{dx}\left(\dfrac{u}{v}\right) = \dfrac{v\dfrac{du}{dx} - u\dfrac{dv}{dx}}{v^2}$

$\dfrac{dy}{dx} = \dfrac{dy}{du}\dfrac{du}{dx}$

$df = f'(x)\,dx$

Review Exercises

In Exercises 1–10 find the derivative of the given function at x.

1. $f(x) = -4x^3 + 2/x^2$ **2.** $f(x) = \sqrt{x}(x^2 - 3)^{4/7}$

3. $g(x) = x/(2x - 1)^2$ **4.** $g(x) = 1/(4 - x^2)^{3/2}$

5. $f(t) = \cos t \sin 2t$ **6.** $f(t) = t^2 \sin \dfrac{1}{t}$

7. $f(t) = 5t \tan t + 3 \sec 3t$ **8.** $f(t) = \tan(\sin t^2)$

9. $f(t) = e^{2t} \ln(3 + e^t)$ **10.** $f(t) = (\ln t)/(1 - e^t)$

In Exercises 11–18 find dy/dx.

11. $y = 4x^3 - \sqrt{3}\,x + 2/(5x)$ **12.** $y = x^2 \tan^2 x$

13. $y = \dfrac{\sin x}{1 - \sec x}$

14. $y = (3x - 5)^{5/9}$

15. $y = x^3 \sqrt{x^2 - 4}$

16. $y = \dfrac{x^2 - x + 1}{x^2 + x + 1}$

17. $y = xe^x - 5e^{-x}$ **18.** $y = (1 + \ln x)/(1 - \ln x)$

In Exercises 19–26 find an equation of the line l tangent to the graph of f at the given point.

19. $f(x) = 3x^3 - 2x^2 + 4$; $(1, 5)$

20. $f(x) = \dfrac{2x - 1}{5x + 2}$; $(0, -1/2)$

21. $f(x) = x \cos \sqrt{2}\,x$; $(0, 0)$

22. $f(x) = \sin x - 3 \cos 2x$; $(\pi/6, -1)$

23. $f(x) = x\sqrt{x - 1}$; $(5, 10)$

24. $f(x) = \sqrt{1 + 3e^x}$; $(0, 2)$

25. $f(x) = \begin{cases} 2 \sin x & \text{for } x < 0 \\ 3x^2 + 2x & \text{for } x \geq 0 \end{cases}$ $(0, 0)$

26. $f(x) = \begin{cases} x^2 \sin (1/x) & \text{for } x \neq 0 \\ 0 & \text{for } x = 0 \end{cases}$ $(0, 0)$

In Exercises 27–32 find the second derivative of f.

27. $f(x) = \frac{1}{4} x^{12} - 6x^6 + 12$ **28.** $f(x) = \cos (3 - x)$

29. $f(t) = (t^2 + 9)^{3/2}$ **30.** $f(t) = \dfrac{2t + 1}{2t - 1}$

31. $f(x) = x^2 + x \ln x$ **32.** $f(x) = x \sin x$

In Exercises 33–38 find dy/dx by implicit differentiation.

33. $3y^3 - 4x^2 y + xy = -5$ **34.** $x^2 + y^2 = x^2/y^2$

35. $y(\sqrt{x} + 1) = x$ **36.** $xe^y + ye^x = 1$

37. $y^3 + \sin xy^2 = 3/2$ **38.** $x \tan x^2 y = y^2 + 1$

In Exercises 39–40 use implicit differentiation to find the derivative at the indicated point.

39. $2x^3 - \sin 4y = x^2 y + 2$; $(1, 0)$

40. $x^2 - xy^2 + y^3 = 13$; $(-1, 2)$

In Exercises 41–42 assume that x and y are differentiable functions of t. Find dy/dt in terms of x, y, and dx/dt.

41. $xy = 3$ **42.** $y = \sin xy^2$

In Exercises 43–46 find df in terms of dx.

43. $f(x) = x^2 \cos x$ **44.** $f(x) = \dfrac{(3x - 1)^{2/3}}{x}$

45. $f(x) = (x - e^x)^5$ **46.** $f(x) = \dfrac{\tan x^2}{x}$

In Exercises 47–48 approximate the given number by using first a linear and then a parabolic approximation.

47. $1 + \sqrt{10}$ **48.** $\sec 0.26\pi$

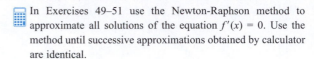 In Exercises 49–51 use the Newton-Raphson method to approximate all solutions of the equation $f'(x) = 0$. Use the method until successive approximations obtained by calculator are identical.

49. $f(x) = x^2 + 4 \cos x$ **50.** $f(x) = e^x - 4x^2$

51. $f(x) = (x - 1) \ln x - x$

52. Let $f(x) = x^2 + 2x - 3$ and $g(x) = x^2 - \frac{9}{4}x + \frac{5}{4}$. Show that the lines tangent to the graphs of f and g at the point of their intersection are perpendicular to one another.

53. Let $f(x) = x^2 + 1$ and $g(x) = x^2 - \cos (\pi/(x^2 + 1))$. Show that the lines tangent to the graphs of f and g at their point of intersection are identical.

54. Let $f(x) = x^2 + 3x$. Find $f'(3)$ from the definition of the derivative.

55. Suppose f is differentiable at a point $a > 0$. Find

$$\lim_{x \to a} \frac{f(x) - f(a)}{x^{1/2} - a^{1/2}}$$

in terms of $f'(a)$. (*Hint:* Eliminate the fractional exponents from the denominator.)

56. Let

$$f(x) = \begin{cases} cx (2 - x) & \text{for } x < 0 \\ \sin x & \text{for } x \geq 0 \end{cases}$$

a. Show that f is continuous at 0 for every value of c.

b. Find the unique value of c for which f is differentiable at 0.

57. Let f be differentiable at 0 and let $g(x) = f(x^2)$. Show that $g'(0) = 0$.

58. Let $y = a \sin x + b \cos x$, where a and b are constant. Show that $d^2y/dx^2 + y = 0$.

Applications

59. A triangle is formed by the coordinate axes and the line that passes through the point $(4, 0)$ and is tangent to the graph of $y = 1/x$. Find the area of the triangle.

60. The volume V of a right circular cone of height h and radius r is given by $V = \pi r^2 h/3$.

a. Find the rate of change of the volume with respect to the radius, assuming that the height is constant.

b. Find the rate of change of the volume with respect to the height, assuming that the radius is constant.

c. Find the rate of change of the height with respect to the volume, assuming that the radius is constant.

61. Suppose water flows into a hemispherical pool in such a way that the volume of water in the pool increases at a constant rate with respect to time. Which of the graphs in Figure 3.61 could be the graph of the rate of change of the depth of water in the pool as a function of time?

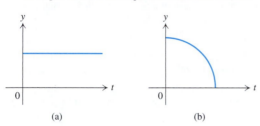

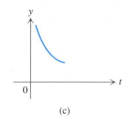 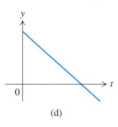

(a)   (b)

(c)   (d)

FIGURE 3.61   Graphs for Exercise 61.

62. Water is poured into a conical paper cup so that the height increases at the constant rate of 1 inch per second. If the cup is 6 inches tall and its top has a radius of 2 inches, how fast is the volume of water in the cup increasing when the height is 3 inches?

63. A person 6 feet tall walks at a rate of 150 feet per minute toward a light tower whose searchlight is located 40 feet above the ground. Show that the length of the person's shadow shrinks at a constant rate.

64. A whisperjet travels east from San Francisco toward St. Louis at 500 miles per hour, and a turbojet travels north from New Orleans toward St. Louis at 600 miles per hour. Find the rate of change of the distance between them when the jets are 300 miles apart, and the whisperjet is 100 miles from St. Louis.

65. Suppose it costs $C(x)$ thousand dollars to manufacture x thousand gallons of house paint, where

$$C(x) = 5 + 3x + 2\sqrt{x} \quad \text{for } 0 \le x \le 2$$

Suppose also that the revenue accruing from selling x thousand gallons of the paint is $R(x)$ thousand dollars, where

$$R(x) = 10\sqrt{x}$$

Find the value of a for which $m_C(a) = m_R(a)$.

66. A cubical cardboard box measures 3 feet on an outer edge. If the cardboard is $\frac{1}{2}$ inch thick, use differentials to approximate the volume of sand that can be poured into the box.

Topics for Discussion

1. What is the relationship between differentiability and continuity of a function? Does one imply the other? Explain your answer, giving appropriate examples.

2. Describe the relationship between tangent lines and derivatives. Can a function f have a tangent line through $(a, f(a))$ without having a derivative at a? Explain your answer.

3. The derivative of a function is sometimes described as the "instantaneous rate of change" of the function. Describe in your own words what this means, and relate it to velocity.

4. Describe how the derivative can be approximated on a calculator.

5. What is implicit differentiation, and under what conditions would you be apt to use it?

6. What are "related rates"? Describe some ways in which we can obtain a relation between two rates.

7. Explain the Newton–Raphson method for approximating zeros of a function. Which is usually more efficient in approximating such a zero, the Newton-Raphson method or the bisection method? Give an example for which the bisection method works but the Newton-Raphson method does not.

Cumulative Review, Chapters 1–2

In Exercises 1–4 solve the inequality.

1. $\dfrac{x(x^2 - 3)}{(1 - x)^3} > 0$ **2.** $\dfrac{6t^2 - 6t + 2}{t^2 - t} \le 0$

3. $\left| \dfrac{1}{|x|} - 3 \right| < 2$

4. $2 \sin x - 1 \le 0$ for x in $[0, 2\pi)$

5. Let $f(x) = 6x^2 - x - 2$. Determine the values of x for which $f(x) > 0$.

6. Show that $(1 + \sin x + \cos x)^2 - 2(1 + \sin x + \cos x) = \sin 2x$.

7. Let $f(x) = \sqrt{\sqrt{\sqrt{x^2 - 1} - x}}$. Find the domain of f.

8. Let $f(x) = \dfrac{x^2}{x - 1}$ and $g(x) = \dfrac{1}{4 - x}$.

    a. Find the domain of $f \circ g$.

    b. Find a formula for $(f \circ g)(x)$.

In Exercises 9–12 determine whether the limit exists as a number, as ∞, or as $-\infty$. If the limit exists, evaluate it.

9. $\displaystyle\lim_{x \to 2} \frac{x^2 - 3x + 2}{x^2 - 5x + 6}$

10. $\displaystyle\lim_{x \to 0} \frac{\sqrt{1 + xe^x} - \sqrt{1 - xe^x}}{x}$

11. $\displaystyle\lim_{x \to 0} \frac{|x|^3 - x^2}{x^3 + x^2}$

12. $\displaystyle\lim_{x \to 0^+} \frac{\sqrt{1 + x^2} - \sqrt{1 - x^2}}{x^3}$

13. Let $f(x) = \sqrt{2x^2 - 4}$. Find $\displaystyle\lim_{x \to 2} \frac{f(x) - f(2)}{x - 2}$.

14. Let

$$f(x) = \begin{cases} x + 1 & \text{for } x < -1 \\ (x + 1)^2 & \text{for } x \geq -1 \end{cases}$$

Is f continuous at -1? Explain why or why not.

15. Sketch the graph of a function f with all of the following properties.

    a. The domain of f is $[0, 4]$.

    b. f is continuous on $(0, 1)$, $[1, 3]$, and $(3, 4]$.

    c. f is not continuous at 1 or at 3, and is not continuous from the right at 0.

16. Use the bisection method to approximate $\sqrt[4]{15}$ with an error of at most $\frac{1}{16}$.

17. Let a be a positive number. Define a function f by $f(x) = \frac{1}{2}(x + a/x)$. Show that $f(\sqrt{a}) = \sqrt{a}$.

A winding road through an autumn landscape.
(*Pete Turner/The Image Bank*)

4 APPLICATIONS OF THE DERIVATIVE

Chapter 4 is really the high point of the study of the derivative, because it concentrates on applications of the derivative. Through new derivative-related concepts we will be able to analyze functions and their graphs. We will also be able to apply the derivative to problems in such widely varying areas as engineering and the physical, biological, and social sciences. Section 4.4, in particular, is devoted to applications related to exponential growth and decay, and Section 4.6 contains many diverse physical applications.

4.1 MAXIMUM AND MINIMUM VALUES

Suppose the profit function for an oil company is given by

$$f(x) = x - x^3 \quad \text{for } 0 \le x \le 1 \tag{1}$$

where x is the amount of oil produced per year in millions of gallons, $f(x)$ is the profit per year in millions of dollars, and the domain is restricted to [0, 1] because

209

of limitations on plant space and personnel. Naturally, the company wishes to make the largest profit possible, which would correspond to a value of x such that $f(x)$ is the maximum value of f on $[0, 1]$. Does such a value of x exist?

Before we answer this question, let us define precisely what we mean by a maximum value, or a minimum value, of a function.

DEFINITION   4.1

a. A function f **has a maximum value on a set** I if there is a number d in I such that $f(x) \leq f(d)$ for all x in I (Figure 4.1). We call $f(d)$ the **maximum value of f on I.**

b. A function f **has a minimum value on a set** I if there is a number c in I such that $f(x) \geq f(c)$ for all x in I (Figure 4.1). We call $f(c)$ the **minimum value of f on I.**

c. A value of f that is either a maximum value or a minimum value on I is called an **extreme value of f on I.**

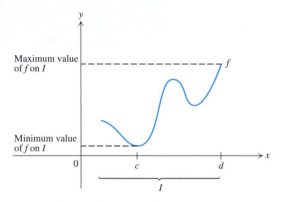

FIGURE 4.1

If the set I is the complete domain of f, and if f has a maximum value on I, then this maximum value is called the **maximum value** (or sometimes the **global maximum value**) **of f.** Similarly, when it exists, the minimum value of f on its domain is the **minimum value** (or sometimes the **global minimum value**) **of f.**

From the definitions, it is evident that

> $f(d)$ is the maximum value of f if and only if $(d, f(d))$ is at least as high as any point on the graph of f.

> $f(c)$ is the minimum value of f if and only if $(c, f(c))$ is at least as low as any point on the graph of f.

Both cases are illustrated in Figure 4.1.

A function *f* may or may not have extreme values on a set *I*, depending on *f* and on *I*. The following simple examples illustrate four of the possibilities:

1. If $f(x) = x$, then on [0, 1] the function *f* has the maximum value of 1 and the minimum value of 0 (Figure 4.2(a)).
2. If $f(x) = \tan x$, then on $(-\pi/2, \pi/2)$ the function *f* has neither a maximum value nor a minimum value (Figure 4.2(b)).
3. If $f(x) = |x|$ for $-1 \le x < 0$ and $0 < x \le 1$, and if $f(0) = 1$, then on $[-1, 1]$ the function has the maximum value of 1 but has no minimum value because *f* does not assume the value 0 (Figure 4.2(c)).
4. If $f(x) = x$ for $-\infty < x < \infty$, then on $(-\infty, \infty)$ the function has neither a maximum value nor a minimum value (Figure 4.2(d)).

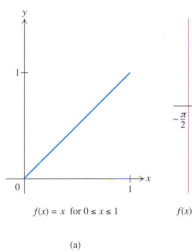

$f(x) = x$ for $0 \le x \le 1$

(a)

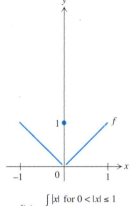

$f(x) = \tan x$ for $-\dfrac{\pi}{2} < x < \dfrac{\pi}{2}$

(b)

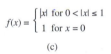

$f(x) = \begin{cases} |x| & \text{for } 0 < |x| \le 1 \\ 1 & \text{for } x = 0 \end{cases}$

(c)

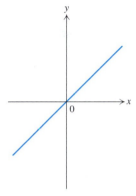

$f(x) = x$ for $-\infty < x < \infty$

(d)

FIGURE 4.2

Notice that in the second example the domain of *f* is an open interval, whereas in the third the function is not continuous, and finally in the fourth the domain is unbounded. The following theorem, which is proved in the Appendix, shows that if a function does not have any of these liabilities, so that it is continuous on a closed, bounded interval, then it must have extreme values.

THEOREM   4.2
Maximum-Minimum
Theorem

Let *f* be continuous on a closed, bounded interval [*a*, *b*]. Then *f* has a maximum and a minimum value on [*a*, *b*].

The function given in (1) by

$$f(x) = x - x^3 \quad \text{for } 0 \le x \le 1$$

is continuous on the closed interval [0, 1]. Therefore by the Maximum-Minimum Theorem, *f* has a maximum value on [0, 1]. But the theorem does not tell us where in [0, 1] the maximum value occurs, nor does it tell us how to find the maximum value. Even an accurate graph of *f* or computer estimates will not give us the exact maximum value. It is the next theorem that provides a method for finding this value.

THEOREM   4.3

Suppose c is an interior point of an interval I, and $f(c)$ is an extreme value of f on I. If $f'(c)$ exists, then $f'(c) = 0$.

Proof   The statement of the theorem is equivalent to the assertion that if c is any number interior to I such that $f'(c)$ exists and is not equal to 0, then $f(c)$ is *not* an extreme value of f on I. Therefore we assume that $f'(c) \neq 0$. Consider first the case in which $f'(c) > 0$. Since

$$f'(c) = \lim_{x \to c} \frac{f(x) - f(c)}{x - c} > 0$$

we have

$$\frac{f(x) - f(c)}{x - c} > 0$$

for all $x \neq c$ in some open interval about c. For such x,

$$\text{if } x > c, \text{ then } f(x) - f(c) = \overbrace{(x - c)}^{\text{positive}} \left(\overbrace{\frac{f(x) - f(c)}{x - c}}^{\text{positive}} \right) > 0$$

Therefore $f(x) > f(c)$, so that f does not have a maximum value at c. In the same way,

$$\text{if } x < c, \text{ then } f(x) - f(c) = \overbrace{(x - c)}^{\text{negative}} \left(\overbrace{\frac{f(x) - f(c)}{x - c}}^{\text{positive}} \right) < 0$$

Thus $f(x) < f(c)$, so that f does not have a minimum value at c. Hence if $f'(c) > 0$, then f has neither a maximum value nor a minimum value at c. The case in which $f'(c) < 0$ is treated analogously, so we leave it as an exercise. ∎

Let f be defined on an interval I, and suppose that f assumes an extreme value at an interior point c of I. Then Theorem 4.3 implies that either $f'(c) = 0$ or f is not differentiable at c. We will call such a number a **critical number** of f. Thus

c is a critical number of f if $f'(c) = 0$ or f is not differentiable at c

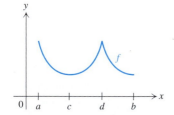

FIGURE 4.3   The extreme values of f on $[a, b]$ occur at endpoints (a or b) or critical numbers (c or d).

In the terminology of critical numbers, Theorem 4.3 implies that a function f can assume an extreme value on an interval I only at a critical number in I or at an endpoint of the interval (Figure 4.3).

If I is a closed, bounded interval $[a, b]$, then the Maximum-Minimum Theorem guarantees that any function f that is continuous on $[a, b]$ has a maximum value and a minimum value on $[a, b]$. Therefore the preceding discussion provides the following general scheme for determining the extreme values of a continuous function f on $[a, b]$:

Finding Extreme Values on [a, b]
Compute the values of *f* at all critical numbers in (*a, b*) and at the endpoints *a* and *b*. The largest of those values is the maximum value of *f* on [*a, b*]; the smallest of those values is the minimum value of *f* on [*a, b*].

Most of the functions we will encounter are differentiable at all numbers in their domains. For such functions the critical numbers are just the numbers at which the derivative is 0. If *f* does not have very many critical numbers in (*a, b*), we can apply the method just described, provided that we can evaluate *f* at the critical numbers.

EXAMPLE 1   Let $f(x) = x - x^3$. Find the extreme values of *f* on [0, 1], and determine at which numbers in [0, 1] they occur.

Solution   As we remarked earlier, *f* has extreme values on [0, 1] because it is continuous on [0, 1]. Since *f* is differentiable, the critical numbers of *f* are the values of *x* for which $f'(x) = 0$. But

$$f'(x) = 1 - 3x^2$$

so that $f'(x) = 0$ if $1 - 3x^2 = 0$, that is, if $x = -\frac{1}{3}\sqrt{3}$ or $x = \frac{1}{3}\sqrt{3}$. Since $-\frac{1}{3}\sqrt{3}$ is not in the interval [0, 1], we conclude that an extreme value of *f* on [0, 1] can occur only at one of the endpoints 0 and 1 or at the critical number $\frac{1}{3}\sqrt{3}$ in (0, 1). To decide which of these give the extreme values, we compute the corresponding values of *f*:

$$f(0) = 0, \quad f\left(\frac{1}{3}\sqrt{3}\right) = \frac{1}{3}\sqrt{3} - \left(\frac{1}{3}\sqrt{3}\right)^3 = \frac{2}{9}\sqrt{3}, \quad \text{and} \quad f(1) = 0$$

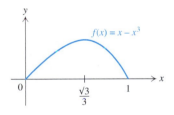

$f(x) = x - x^3$

FIGURE 4.4

Consequently the minimum value of *f* on [0, 1] is 0, and it occurs at 0 and 1. The maximum value of *f* on [0, 1] is $\frac{2}{9}\sqrt{3}$, and it occurs at $\frac{1}{3}\sqrt{3}$ (see Figure 4.4).   ❑

By Example 1, the profit function introduced at the beginning of the section has a maximum value of $\frac{2}{9}\sqrt{3}$, which would not have been so easy to guess.

Caution: It is possible for a function *f* to have a critical number in (*a, b*) that does *not* correspond to an extreme value of *f* on [*a, b*]. Indeed, let $f(x) = 2x^3$ for *x* in [−1, 1]. Now 0 is a critical number of *f* because $f'(0) = 0$. However, $f(0)$ is not an extreme value because the minimum and maximum values of *f* on [−1, 1] are −2 and 2, respectively (Figure 4.5).

Sometimes the extreme values of a function on an interval [*a, b*] occur at the endpoints. For example, this occurs for any continuous function that has no critical number in (*a, b*) (such as $f(x) = x$ on [0, 1]).

Finally we mention that a function *f* may have an extreme value at a critical number *c* for which $f'(c)$ does not exist. In particular, if $f(x) = x^{2/3}$, then $f'(0)$ does not exist. The reason is that

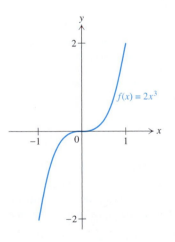

FIGURE 4.5   There is no extreme value at the critical number 0.

$$\frac{f(x) - f(0)}{x - 0} = \frac{x^{2/3} - 0}{x - 0} = \frac{1}{x^{1/3}}$$

$f(x) = 2x^3$

and since $\lim_{x \to 0} 1/x^{1/3}$ does not exist, neither does

$$\lim_{x \to 0} \frac{f(x) - f(0)}{x - 0}$$

Thus $f'(0)$ does not exist, so 0 is a critical number of f. However, since $f(x) \geq 0$ for all x and $f(0) = 0$, f has a minimum value at 0 (Figure 4.6).

Applications

Now we apply our method of determining extreme values on closed, bounded intervals to physical problems.

EXAMPLE 2   A landowner wishes to use 2000 meters of fencing to enclose a rectangular region. Suppose one side of the property lies along a stream and thus does not need to be fenced in. What should the lengths of the sides be in order to maximize the area?

Solution   Any rectangular region the landowner could enclose must have a length x and a width y (Figure 4.7). Therefore the area A we are to maximize is given by xy. We need to express the area in terms of one variable, say x. Since the length of fencing that is available is 2000 meters, we have

$$x + 2y = 2000$$

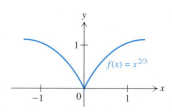

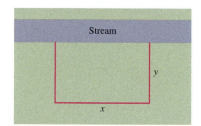

FIGURE 4.6

FIGURE 4.7   Rectangular region with 2000 meters of fencing.

Therefore $2y = 2000 - x$, so that $y = 1000 - x/2$. As a result, A can be written as a function of x alone:

$$A(x) = xy = x\left(1000 - \frac{x}{2}\right) = 1000x - \frac{1}{2}x^2$$

Since x must be nonnegative, it follows that $0 \leq x \leq 2000$. Thus the problem has been reduced to finding the maximum value of A on $[0, 2000]$. Notice that $A'(x) = 1000 - x$, so that $A'(x) = 0$ only when $x = 1000$. As a result, A can have its maximum value on $[0, 2000]$ only at the critical number 1000 or at one of the endpoints, 0 and 2000. However,

$$A(0) = 0, \quad A(1000) = 1000(1000) - \frac{1}{2}(1000)^2 = 500{,}000, \quad \text{and} \quad A(2000) = 0$$

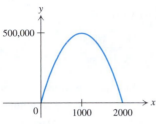

$$A(x) = 1000x - \frac{x^2}{2} \quad \text{for } 0 \le x \le 2000$$

FIGURE 4.8

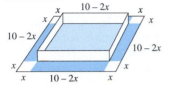

FIGURE 4.9

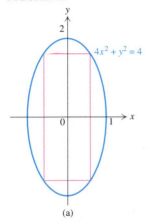

(a)

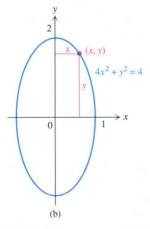

(b)

FIGURE 4.10

Thus the maximum value of A occurs for $x = 1000$ (Figure 4.8). Since $y = 1000 - x/2$, it follows that if $x = 1000$, then $y = 1000 - 500 = 500$. Consequently the fence should enclose a region 1000 meters long and 500 meters wide. ❑

If a house were placed along one side, how would that affect the dimensions of the region of maximum area? See Exercise 44.

EXAMPLE 3 A metal box (without a top) is to be constructed from a square sheet of metal 10 inches on a side by first cutting square pieces of the same size from the corners of the sheet and then folding up the sides (Figure 4.9). Find the dimensions of the box with the largest volume that can be so constructed.

Solution Let x be the length (in inches) of the sides of the squares that are cut out (Figure 4.9). Since the original sheet of metal is 10 inches on a side, we have $0 \le x \le 5$. The resulting box has a height of x inches and a base that is $10 - 2x$ inches on a side. This means that the volume V of the box is given by

$$V = x(10 - 2x)^2 = 4x^3 - 40x^2 + 100x \quad \text{for } 0 \le x \le 5$$

Thus the problem reduces to finding the maximum value of V on $[0, 5]$. To determine the critical numbers of V in $(0, 5)$, we first compute $V'(x)$:

$$V'(x) = 12x^2 - 80x + 100 = 4(3x^2 - 20x + 25) = 4(3x - 5)(x - 5)$$

Therefore $V'(x) = 0$ for $x = \frac{5}{3}$ or $x = 5$. Thus the only critical number in $(0, 5)$ is $\frac{5}{3}$. Since

$$V(0) = 0, \quad V\left(\frac{5}{3}\right) = \frac{5}{3}\left(\frac{20}{3}\right)^2 = \frac{2000}{27}, \quad \text{and} \quad V(5) = 0$$

it follows that the maximum value of V occurs for $x = \frac{5}{3}$. The corresponding value of $10 - 2x$ is $10 - 2(\frac{5}{3}) = \frac{20}{3}$. Consequently the box with the maximum volume has a height of $\frac{5}{3}$ inches and a square base $\frac{20}{3}$ inches on a side. ❑

EXAMPLE 4 A rectangle is to be inscribed in the ellipse $4x^2 + y^2 = 4$ in such a way that its sides are parallel to the axes (Figure 4.10(a)). Find the lengths of the sides of the rectangle with maximum area.

Solution Let x denote half the length of the horizontal side, and let y denote half the length of the vertical side of the rectangle (Figure 4.10(b)). Then the area A of the rectangle is given by

$$A = (2x)(2y) = 4xy$$

Our goal is to find values of x and y that yield the maximum possible value of A. Since the upper right corner of the rectangle is the point (x, y), and since it is on the ellipse, we know that $4x^2 + y^2 = 4$. Therefore

$$y^2 = 4 - 4x^2, \quad \text{so that} \quad y = \sqrt{4 - 4x^2} = 2\sqrt{1 - x^2}$$

Thus

$$A = 4xy = 8x\sqrt{1 - x^2} \quad \text{for } 0 \le x \le 1$$

To find the maximum value of A, we first take the derivative of A:

$$A'(x) = 8\sqrt{1 - x^2} - \frac{8x^2}{\sqrt{1 - x^2}} = \frac{8 - 16x^2}{\sqrt{1 - x^2}}$$

As a result, $A'(x) = 0$ for $16x^2 = 8$, that is, $x = 1/\sqrt{2}$. Thus the only critical number of A in $(0, 1)$ is $1/\sqrt{2}$. Since $A(0) = 0 = A(1)$ and since $A(1/\sqrt{2}) = 4$, it follows that the maximum value of A occurs for $x = 1/\sqrt{2}$. We conclude that the rectangle with the largest area occurs when the horizontal side has length $2x = \sqrt{2}$ and the vertical side has length $2y = 4\sqrt{1 - (1/\sqrt{2})^2} = 2\sqrt{2}$.   ◻

We will return to applications of extreme values in Section 4.6, where we will consider extreme values of functions defined on arbitrary intervals, not just closed, bounded intervals.

EXERCISES 4.1

In Exercises 1–18 find all critical numbers (if any) of the given function.

1. $f(x) = x^2 + 4x + 6$ **2.** $f(x) = 4x^3 - 6x^2 - 9x$

3. $f(x) = 3x^4 + 4x^3 - 12x^2 + 1$

4. $f(x) = x^5 - 5x^3 + 10x - 3$

5. $g(x) = x + 1/x$ **6.** $g(x) = x - 4/x^2$

7. $k(t) = t/\sqrt{t^2 + 1}$ **8.** $k(t) = 2t - t^{2/3}$

9. $f(x) = \sin x$ **10.** $f(x) = \cos \sqrt{x}$

11. $f(x) = x + \sin x$ **12.** $f(x) = \tan x$

13. $f(z) = |z - 2|$ **14.** $f(z) = z|z + 3|$

15. $f(x) = x^2 e^x$ **16.** $f(x) = 1/(e^x - 1)$

17. $f(x) = x \ln x$ **18.** $f(x) = \ln(2x + e^{-x})$

In Exercises 19–34 find all extreme values (if any) of the given function on the given interval. Determine at which numbers in the interval these values occur.

19. $f(x) = x^2 - x$; $[0, 2]$ **20.** $g(x) = 1/x$; $(0, 3]$

21. $f(t) = -1/(2t)$; $(0, \infty)$ **22.** $f(t) = t^2 - 4/t$; $[1, 3)$

23. $k(z) = \sqrt{1 + z^2}$; $[-2, 3]$

24. $k(z) = \sqrt{1 + z}$; $(3, 8]$

25. $f(x) = \sqrt{|x|}$; $(-1, 2)$ **26.** $f(x) = x^{2/3}$; $[-8, 8]$

27. $f(x) = -\sin \sqrt[3]{x}$; $[-\pi^3/27, \pi^3/8]$

28. $f(x) = \cos \pi x$; $(\frac{1}{3}, 1]$ **29.** $f(x) = \tan x/2$; $(-\pi/2, \pi/6)$

30. $f(x) = \csc 3x$; $[\pi/18, \pi/4]$

31. $f(x) = xe^x$; $[-2, 0]$ **32.** $f(x) = e^x - e^{2x}$; $[0, 1]$

33. $f(x) = x - 2\ln x$; $[\frac{1}{2}, 2]$

34. $f(x) = \ln \dfrac{x}{x^2 + 1}$; $[\frac{1}{2}, 3]$

In Exercises 35–36 use the Newton-Raphson method to approximate all the critical numbers of f in the given interval. To obtain your initial guess, plot the graph of f. Then continue until successive iterations obtained by the calculator are identical.

35. $f(x) = \frac{1}{4}x^4 + x^3 - x - 1$ for $-3 \le x \le 1$

36. $f(x) = x \cos x$ for $-5 \le x \le 5$

In Exercises 37–38 use the method for Exercises 35–36 to approximate all critical numbers of f in the given interval, and use this information to determine the (approximate) extreme values of f on the given interval.

37. $f(x) = x^2 + e^x$; $[-1, 1]$

38. $f(x) = \frac{1}{4}x^4 + x^3 + 2x + 6$; $[-4, 0]$

39. Suppose c and d are not both 0, and let

$$f(x) = \frac{ax + b}{cx + d}$$

Show that f has no critical numbers unless $ad - bc = 0$, in which case f is a constant function.

40. Assume that f is defined on I and that $g = -f$. Prove that $f(x_0)$ is the maximum value of f on I if and only if $g(x_0)$ is the minimum value of g on I.

41. Prove Theorem 4.3 for the case $f'(c) < 0$. (All that is required is to change certain inequalities in the proof for the case $f'(c) > 0$.)

42. In the proof of Theorem 4.3, where do we use the fact that c is an interior point of I?

*43. Prove **Darboux's Theorem:** Let f be differentiable on $[a, b]$, and let $f'(a) < m < f'(b)$. Then there exists some number c in (a, b) such that $f'(c) = m$. (*Hint:* Let $g(x) = f(x) - mx$ for $a \leq x \leq b$. Show that $g'(a) < 0$ and $g'(b) > 0$, and conclude from this that g does not assume its minimum value on $[a, b]$ at either a or b. Now apply the Maximum-Minimum Theorem and Theorem 4.3 to the function g.)

Applications

44. In Example 2, suppose a house 30 meters long will be located on the boundary of the field, so that 30 meters of the boundary require no fencing.

  a. If the house is placed along one side, then determine the dimensions of the field that maximize its area.

  b. If the house is to be placed diagonally (at a 45° angle) in one corner of the field, as in Figure 4.11, then determine the dimensions of the field that maximize its area.

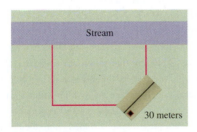

FIGURE 4.11   Figure for Exercise 44.

45. You wish to fence in a rectangular garden in your 100×90 foot back yard, with one side of the garden against the house (see Figure 4.12)

  a. Suppose you have 20 feet of available fencing for the garden. Find the dimensions of the garden that yields the largest possible area A of the garden.

  b. Suppose that instead of 20 feet of fencing there are 80 feet of fencing. Find the area A_1 of the largest possible garden.

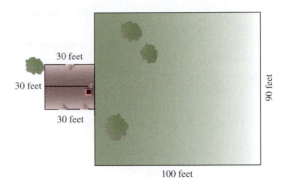

FIGURE 4.12   Figure for Exercise 45.

46. Suppose $R(x)$, $C(x)$, and $P(x)$ denote the revenue, cost, and profit resulting from the manufacture and sale of x units of an item. Recall that

$$P(x) = R(x) - C(x) \quad \text{for } x \geq 0$$

Assume that it is possible to make a maximum profit by manufacturing x_0 units of the item. Show that if R and C are differentiable and $x_0 > 0$, then $R'(x_0) = C'(x_0)$ (that is, the marginal revenue at x_0 equals the marginal cost at x_0).

47. A mass connected to a spring moves along the x axis so that its x coordinate at time t is given by

$$x(t) = \sin 2t + \sqrt{3} \cos 2t$$

What is the maximum distance of the mass from the origin?

48. Suppose the velocity of an object moving along a straight line is given by

$$v(t) = 5(9 + 3t - 2t^2) \quad \text{for } 0 < t \leq 4$$

At what time is the position maximized?

49. At a distance x from one end of a beam, the bending moment is given by

$$M = \frac{1}{2} wx(L - x) \quad \text{for } 0 \leq x \leq L$$

where w is the uniform weight density of the beam. Show that M is largest at the midpoint of the beam.

50. If we neglect air resistance, then the range of a ball (or any projectile) shot at an angle θ with respect to the x axis and with an initial velocity v_0 is given by

$$R(\theta) = \frac{v_0^2}{g} \sin 2\theta \quad \text{for } 0 \leq \theta \leq \frac{\pi}{2}$$

where g is the acceleration due to gravity (9.8 meters per second per second).

a. Show that the maximum range is attained when $\theta = \pi/4$.

b. If $v_0 = 30$ meters per second and the aim is to snuff out a smoldering cigarette lying on the ground 50 meters away, at what angle should the ball be hit?

c. The maximum height reached by the ball is

$$y_{\max} = \frac{v_0^2 \sin^2\theta}{2g}$$

Why would it be a bad idea to hit the ball so that $y_{\max}$ is maximized?

51. A company plans to invest \$50,000 for the next 4 years, initially buying oil stocks. If it seems profitable to do so, the oil stocks will be sold before the 4-year period has lapsed, and the revenue from the sale of the stocks will be placed in tax-free municipal bonds. According to the company's analysis, if the oil stocks are sold after t years, then the net profit $P(t)$ in dollars for the 4-year period is given by

$$P(t) = 2(20 - t)^3 t \quad \text{for } 0 \leq t \leq 4$$

Determine whether the company should switch from oil stocks to municipal bonds, and if so, after what period of time.

52. According to one model of coughing, the flow F (volume per unit time) of air through the windpipe during a cough is a function of the radius r of the windpipe, given by

$$F = k(r_0 - r)r^4 \quad \text{for } \frac{1}{2}r_0 \leq r \leq r_0$$

where k is a positive constant and r_0 is the normal (noncoughing) radius.

a. Find the value of r that maximizes the flow F.

b. According to the same model, the velocity v of air through the windpipe during a cough is given by

$$v = \frac{k}{\pi}(r_0 - r)r^2 \quad \text{for } \frac{1}{2}r_0 \leq r \leq r_0$$

Find the value of r that maximizes the velocity v.

c. During a cough the windpipe is constricted. According to parts (a) and (b), is that likely to assist or hinder the cough?

53. The Spice-of-Life Company is preparing to create shipping crates. The company wishes the volume of each crate to be 6 cubic feet, with the crate's base to be square

between 1 and 2 feet on a side. Assume that the material for the bottom costs \$5, the sides \$2, and the top \$1 per square foot. Find the dimensions that yield the minimum cost.

54. A cable TV company wishes to place an amplifier station at a point on a street and run wires from the station to three houses. One house is adjacent to the street, and two are 50 feet from the street, as in Figure 4.13. Where should the station be located in order to minimize the total length of wire required to service all three houses?

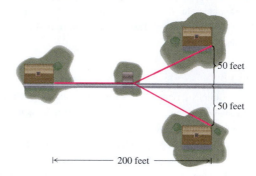

FIGURE 4.13 Figure for Exercise 54.

55. A landowner wishes to use 3 miles of fencing to enclose an isosceles triangular region of as large an area as possible. What should be the lengths of the sides of the triangle?

56. A wire of length L is cut into two pieces. One piece is bent to form a square, and the other is bent to form a circle. Determine the minimum possible value for the sum A of the areas of the square and the circle. If the wire is actually cut, is there a maximum value of A?

57. A rectangular sheet of metal 80 centimeters wide and 1000 centimeters long is folded along the center to form a triangular trough (Figure 4.14). Two extra pieces of metal are attached to the ends of the trough, and the trough is then filled with water.

a. How deep should the trough be to maximize its capacity?

b. What is the maximum capacity of the trough?

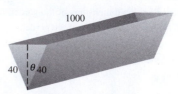

FIGURE 4.14 Figure for Exercise 57.

58. A rectangle is inscribed in a semicircle of radius r with one side lying on the diameter of the semicircle. Find the maximum possible area of the rectangle.

59. An isosceles triangle has base 6 and height 12. Find the maximum possible area of a rectangle that can be placed inside the triangle with one side on the base of the triangle.

60. An isosceles triangle is inscribed in a circle of radius r. Find the maximum possible area of the triangle.

61. A cylinder is inscribed in a cone of height H and base radius R (Figure 4.15). Determine the largest possible volume for the cylinder.

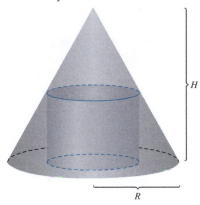

FIGURE 4.15   A cylinder inscribed in a cone. See Exercise 61.

Project

1. Consider nonnegative numbers x and y such that $x + y = 1$, and let m and n be positive integers. We want to explore properties of the maximum value P of the product $x^m y^n$. Such numbers appear in statistics, as we will see below.

   a. Let $m = 1 = n$. Find the largest possible product P of xy, and the corresponding values of x and y.

   b. Let $m = n \neq 1$. Find the largest possible product P of $x^m y^n$, and the corresponding values of x and y. Does the answer surprise you? Explain why or why not.

   c. Now assume that $m \neq n$. Again find the largest possible product P of $x^m y^n$, as well as the corresponding values of x and y. What can you say about the behavior of P if m is fixed and n increases without bound?

   d. Suppose we wish to estimate the probability p of rolling a 3 with a loaded die. We roll the die n times and obtain m 3s in a particular order. The probability of this is known to be $p^m (1 - p)^{n - m}$. The **maximum likelihood estimate** of p based on the n rolls is the value of x that maximizes $x^m (1 - x)^{n - m}$ on $[0, 1]$. Show that the maximum likelihood estimate of p is m/n.

4.2   THE MEAN VALUE THEOREM

Michel Rolle (1652–1719)

Early in his mathematical career Rolle was a vocal critic of calculus and even described the subject as "a collection of ingenious fallacies." Later in life he moderated his views and was able to see the merit of the subject.

The main theorem in Section 4.2 is the Mean Value Theorem, which not only has powerful consequences for derivatives and integrals, but also enters into solutions of applied problems. For example, we will use the Mean Value Theorem to help determine whether a person getting off a toll road has been speeding. (See Example 2.)

We begin with a special case of the Mean Value Theorem called Rolle's Theorem, which is named after the seventeenth-century French mathematician Michel Rolle.*

---

* **Rolle:** Pronounced "Role."

THEOREM   4.4
Rolle's Theorem

Let f be continuous on $[a, b]$ and differentiable on (a, b). If $f(a) = f(b)$, then there is a number c in (a, b) such that $f'(c) = 0$.

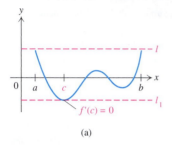

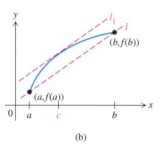

(a)

(b)

FIGURE 4.16

Proof   If f is constant, then its derivative is 0, so that $f'(c) = 0$ for each c in (a, b). If f is not constant, then its maximum and minimum values (which exist by the Maximum-Minimum Theorem) are distinct. Since $f(a) = f(b)$, either the maximum or the minimum must occur at a number c in (a, b). By hypothesis, f is differentiable at c, so $f'(c) = 0$ by Theorem 4.3.   ■

Notice that the condition $f(a) = f(b)$ cannot be eliminated from Rolle's Theorem. For example, if $f(x) = x$, then $f'(x) = 1$ for all x in any interval (a, b), so that $f'(c) \neq 0$ for all c in (a, b).

Rolle's Theorem implies that at some point $(c, f(c))$ on the graph of f, the slope of the tangent line l_1 is 0 (Figure 4.16(a)). But the line l joining $(a, f(a))$ to $(b, f(b))$ is horizontal because $f(a) = f(b)$. Thus l_1 and l are parallel.

Is there a comparable theorem when $f(a) \neq f(b)$? In other words, must there be a number c in (a, b) such that the line l_1 tangent to the graph of f at $(c, f(c))$ is parallel to the line l joining $(a, f(a))$ and $(b, f(b))$ (Figure 4.16(b))? Since two nonvertical lines are parallel only if their slopes are equal, we can rephrase the question as follows: Does there necessarily exist a number c in (a, b) such that

$$\overbrace{f'(c)}^{\substack{\text{slope of} \\ \text{the graph} \\ \text{of } f \text{ at } c}} = \overbrace{\frac{f(b) - f(a)}{b - a}}^{\text{slope of } l}$$

holds? The Mean Value Theorem supplies the answer.

THEOREM   4.5
Mean Value Theorem

Let f be continuous on $[a, b]$ and differentiable on (a, b). Then there is a number c in (a, b) such that

$$f'(c) = \frac{f(b) - f(a)}{b - a} \qquad (1)$$

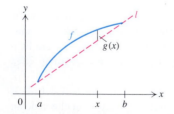

FIGURE 4.17

Proof   We introduce an auxiliary function g that allows us to simplify the proof by using Rolle's Theorem. The function g is defined by

$$g(x) = f(x) - \left[f(a) + \frac{f(b) - f(a)}{b - a}(x - a) \right] \quad \text{for } a \leq x \leq b$$

(See Figure 4.17.) Now g is continuous on $[a, b]$ and differentiable on (a, b), since g is a simple combination of f, constant functions, and a linear function. Substituting first a for x and then b for x in the equation above, we find that

$$g(a) = g(b) = 0$$

so that by Rolle's Theorem there is a number c in (a, b) such that $g'(c) = 0$. However,

$$g'(x) = f'(x) - \frac{f(b) - f(a)}{b - a} \quad \text{for } a < x < b$$

and thus

$$0 = g'(c) = f'(c) - \frac{f(b) - f(a)}{b - a}$$

Solving for $f'(c)$, we obtain

$$f'(c) = \frac{f(b) - f(a)}{b - a} \quad \blacksquare$$

The following example illustrates the Mean Value Theorem for a particular function f and a particular interval $[a, b]$.

EXAMPLE 1   Let $f(x) = \frac{1}{3}x^3 + 2x$. Find a number c in $(0, 3)$ such that

$$f'(c) = \frac{f(3) - f(0)}{3 - 0}$$

Solution   Since

$$\frac{f(3) - f(0)}{3 - 0} = \frac{15 - 0}{3 - 0} = 5$$

we seek a number c in $(0, 3)$ such that $f'(c) = 5$. But

$$f'(x) = x^2 + 2$$

so that c must satisfy

$$c^2 + 2 = 5$$

Therefore $c^2 = 3$, and since c must be in $(0, 3)$, we conclude that $c = \sqrt{3}$.   ❑

When the renowned physicist André Marie Ampère first stated the Mean Value Theorem around 200 years ago, the terms "average" and "mean" were synonymous. If $f(t)$ denotes the position of an object on the x axis at time t, then the average (or mean) velocity during the interval $[a, b]$ is

$$\frac{f(b) - f(a)}{b - a}$$

Thus by the Mean Value Theorem the mean velocity during an interval $[a, b]$ is equal to the velocity $f'(c)$ at some instant c in (a, b).

EXAMPLE 2   Pat enters the Pennsylvania Turnpike at the Valley Forge toll station, and receives a ticket with the time 11 A.M. stamped on it. At noon Pat exits the turnpike at the Harrisburg East toll station, some 79 miles east of Valley Forge, and goes on. A few moments later a police officer gives Pat a ticket for speeding on the turnpike, which has a speed limit of 55 miles per hour. Show that Pat had in fact been speeding somewhere on the pike.

Solution Let $f(t)$ denote the distance that Pat has traveled during the t hours since 11 A.M. Notice that by assumption,

$$\frac{f(1) - f(0)}{1 - 0} = \frac{79 - 0}{1 - 0} = 79$$

It is natural to suppose that f is a differentiable function. Then the Mean Value Theorem implies that $f'(c) = 79$ for some value of c in $(0, 1)$. Since $f'(c)$ is the velocity of the car at time c, we conclude that at that instant the velocity was 79 miles per hour, and indeed Pat was speeding. ❑

EXERCISES 4.2

In Exercises 1–10 find all numbers c in the interval (a, b) for which the line tangent to the graph of f is parallel to the line joining $(a, f(a))$ and $(b, f(b))$.

1. $f(x) = x^2 - 6x$; $a = 0$, $b = 4$
2. $f(x) = x - 3x^2$; $a = -1$, $b = 3$
3. $f(x) = x^3 - 6x$; $a = -2$, $b = 0$
4. $f(x) = x^3 - 6x$; $a = -2$, $b = 2$
5. $f(x) = x^3 + 4$; $a = -2$, $b = 1$
6. $f(x) = x^3 - 2$; $a = -3$, $b = 3$
7. $f(x) = x^3 - 3x^2 + 3x + 1$; $a = -2$, $b = 2$
8. $f(x) = -3 + \sqrt{x}$; $a = 0$, $b = 1$
9. $f(x) = 1 + x^{1/3}$; $a = 1$, $b = 8$
10. $f(x) = 3\left(x + \dfrac{1}{x}\right)$; $a = \frac{1}{3}$, $b = 3$

In Exercises 11–12 first find the slope m of the line joining $(a, f(a))$ and $(b, f(b))$. Then use the Newton-Raphson method to estimate the values of c for which $f'(c) = m$. Continue the process until successive iterations obtained by the calculator are identical.

11. $f(x) = x^2 - 2/x$; $a = 1$, $b = 2$
12. $f(x) = \sin(\pi x/2)$; $a = -1$, $b = 1$
13. Let $f(x) = Ax^2 + Bx + C$, where A, B, and C are constants with $A \neq 0$.
    a. Show that for any interval $[a, b]$, the number c guaranteed by the Mean Value Theorem is the midpoint of $[a, b]$.
    b. What does this mean geometrically?
14. Let $f(x) = |x|$. Show that $f(-2) = f(2)$, but there is no number c in $(-2, 2)$ such that $f'(c) = 0$. Does this result contradict Rolle's Theorem? Explain.

15. Suppose $|f'(x)| \leq M$ for $a \leq x \leq b$. Using the Mean Value Theorem, prove that $|f(b) - f(a)| \leq M(b - a)$, so that $f(a) - M(b - a) \leq f(b) \leq f(a) + M(b - a)$.

In Exercises 16–18 use the result of Exercise 15 to determine lower and upper bounds for the given number.

16. $\sqrt{101}$ (*Hint:* Let $f(x) = \sqrt{x}$, $a = 100$, and $b = 101$, and then find bounds for $f(101) = \sqrt{101}$.)
17. $28^{2/3}$ \qquad 18. $33^{1/5}$
19. Use the result of Exercise 15 to estimate how far 1.7 can be from the true value of $\sqrt{3}$. (*Hint:* Take $a = 2.89$, $b = 3$.)
20. Use the result of Exercise 15 to estimate how far 2.2 can be from the true value of $\sqrt{5}$.
21. Use Rolle's Theorem to show that there is a solution of the equation $\tan x = 1 - x$ in $(0, 1)$. (*Hint:* Let $f(x) = (x - 1)\sin x$, and find $f(0)$, $f(1)$, and $f'(x)$.)
22. Use Rolle's Theorem to show that there is a solution of the equation $\cot x = x$ in $(0, \pi/2)$. (*Hint:* Let $f(x) = x \cos x$ for x in $[0, \pi/2]$.)
23. Use the Mean Value Theorem to show that if $\pi/2 < x < \pi$ or $\pi < x < 3\pi/2$ and if x approximates π, then $x + \sin x$ is a better approximation for π, that is

$$|x + \sin x - \pi| < |x - \pi|$$

24. Let $x = 3.14$. Use the result of Exercise 23 to find an approximation to π that is better than 3.14. (The actual value of π is 3.14159265358979, accurate to 15 digits. How good an approximation did your calculator produce?)

25. Recall that a number a is a fixed point of a function f if $f(a) = a$.
  a. Prove that if $f'(x) \neq 1$ for every real number x, then f has at most one fixed point.
  b. Let $f(x) = \sin \frac{1}{2}x$. Using part (a), prove that 0 is the only fixed point of f.

26. a. Let g be continuous on $[a, b]$ and differentiable on (a, b). Assume that $g(x) = 0$ for two values of x in (a, b). Show that there is at least one value c in (a, b) such that $g'(c) = 0$.
  b. Let $g(x) = x^4 - 20x^3 - 25x^2 - x + 1$. Use (a) to show that there is some c in $[-1, 1]$ such that $4c^3 - 60c^2 - 50c - 1 = 0$.

27. Suppose f is differentiable at every number in $[-1, 1]$, and $f'(-1) = 1 = f'(1)$. Assume in addition that $f(-1) = 0 = f(1)$.
  a. Explain why f has at least one zero in $(-1, 1)$.
  b. Explain why f' must have at least two zeros in $(-1, 1)$.

*28. Let f be continuous on $[a, b]$ and differentiable on (a, b), and assume that f is not linear. Show that there are numbers s and t in (a, b) such that

$$f'(s) > \frac{f(b) - f(a)}{b - a} \quad \text{and} \quad f'(t) < \frac{f(b) - f(a)}{b - a}$$

(*Hint:* Consider a point $(r, f(r))$ on the graph of f but not on the line joining $(a, f(a))$ and $(b, f(b))$. Use the Mean Value Theorem on $[a, r]$ and then on $[r, b]$.)

*29. Let $f(x) = x^m (x - 1)^n$, where m and n are positive integers. Show that the number c guaranteed by Rolle's Theorem is unique and that it divides $[0, 1]$ into segments whose lengths have ratio m/n.

*30. Suppose f' assumes a value m at most n times. Use the Mean Value Theorem to show that any line with slope m intersects the graph of f at most $n + 1$ times.

Applications

 31. A racing car accelerated from rest, traveling 2400 feet in 12 seconds. Must the car have been traveling at least 130 miles per hour at some moment during that time interval? Explain. (*Hint:* Convert to miles and hours.)

32. In the aftermath of a car accident it is concluded that one driver slowed to a halt in 9 seconds while skidding 400 feet. If the speed limit was 30 miles per hour, can it be proved that the driver had been speeding? (*Hint:* 30 miles per hour is equal to 44 feet per second.)

*33. A cable of a suspension bridge hangs in the shape of a parabola (which is the graph of a second-degree polynomial). Consider a cable that joins two points at $(0, 2)$ and $(2, 1)$, and passes through the point $(1, 0)$, and let l be the line joining $(0, 2)$ and $(2, 1)$ (Figure 4.18). Find the point on the cable whose vertical distance from l is greatest.

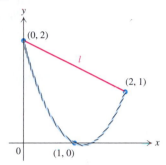

FIGURE 4.18   Graph for Exercise 33.

Project

1. The idea for this project came from Rolle's Theorem.
  a. Let f be continuous on $[a, b]$ and differentiable on (a, b). Assume that $f(x) = 0$ for 3 values of x in (a, b). Show that there are at least 2 values of x in (a, b) such that $f'(x) = 0$.
  b. Let $g(x) = \sin x$ for x in $[0, b]$. Determine the smallest positive value of b such that the conditions in (a) are satisfied. Are there 2 values at which $g'(x) = 0$? Explain your answer after sketching the graph of g.
  c. Define a function g that is continuous on $[a, b]$ and differentiable on (a, b), and such that $g(x) = 0$ for 3 values of x in (a, b) but $g'(x) = 0$ for exactly 4 values of x in (a, b).
  d. Let f be continuous on $[a, b]$ and differentiable on (a, b). Assume that $f(x) = 0$ for n values of x in (a, b), where n is any positive integer greater than 2. Determine the minimum number m of values of x in (a, b) such that $f'(x) = 0$. Give reasons for your answer.
  e. Suppose that $f'(a) = a_7 x^7 + a_6 x^6 + \cdots + a_1 x + a_0$. What is the minimum number of critical numbers of f? Explain your answer.

4.3 CONSEQUENCES OF THE MEAN VALUE THEOREM

Suppose a Metroliner begins moving along a straight path after stopping at Penn Station, and during the next 4 minutes its velocity is recorded from the speedometer. From that information, how could we tell how far the train has traveled from Penn Station by the end of the 4-minute interval? More generally, could we determine how far the Metroliner has traveled at any specific time during the 4-minute interval? The answer to both questions is yes, provided that the velocity function is reasonable. The answer will result from Theorem 4.6, which tells us that a differentiable function is (nearly) determined by its derivative. We will also discuss a second result, which provides a criterion for determining whether a function is increasing or decreasing on a given interval. These results will play a key role in the applications that will appear in Section 4.6.

Antiderivatives

If f is a function defined on an interval I, then any continuous function F on I such that $f'(x) = f(x)$ for each interior point of I is called an **antiderivative** of f (since f is the derivative of F). Thus on any given interval, x^3 is an antiderivative of $3x^2$ and $\sin x$ is an antiderivative of $\cos x$.

Our investigation of antiderivatives begins with the following theorem, which implies that if the graphs of two functions have identical slopes at each number in an interval, then the functions differ by a constant on that interval (Figure 4.19).

THEOREM   4.6

> **a.** Let f be continuous on an interval I. If $f'(x)$ exists and equals 0 for each interior point x of I, then f is constant on I.
>
> **b.** Let f and g be continuous on an interval I. If $f'(x)$ and $g'(x)$ exist and are equal for each interior point x of I, then $f - g$ is constant on I. In other words, there is a constant C such that $f(x) = g(x) + C$ for all x in I.

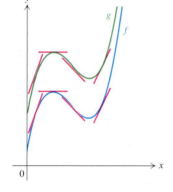

FIGURE 4.19

Proof   To prove (a), let x and z be arbitrary numbers in I with $x < z$. By the Mean Value Theorem there is a number c in the interval (x, z) such that

$$\frac{f(z) - f(x)}{z - x} = f'(c) \tag{1}$$

By assumption, $f'(c) = 0$, and thus (1) reduces to

$$f(z) - f(x) = 0$$

Therefore $f(z) = f(x)$. It follows that f assigns the same value to any two points in I, so f is constant on I. To prove (b), notice that

$$(f - g)'(x) = f'(x) - g'(x) = 0$$

so $f - g$ satisfies the conditions of part (a). Consequently $f - g$ is constant on I.   ■

By part (b) of Theorem 4.6, once we know a single antiderivative F of a given function, all other antiderivatives can be ascertained by adding constants to F. It follows that on any interval I the only antiderivatives of $3x^2$ are functions of the form $x^3 + C$, and the only antiderivatives of $\cos x$ are functions of the form $\sin x + C$.

EXAMPLE 1 Find the antiderivatives of the following functions.
a. $x^2 - 4x$ **b.** $\sec^2 x$

Solution **a.** Since $\frac{1}{3}x^3$ is an antiderivative of x^2, and $-2x^2$ is an antiderivative of $-4x$, it follows that $\frac{1}{3}x^3 - 2x^2$ is an antiderivative of $x^2 - 4x$. Thus the antiderivatives of $x^2 - 4x$ have the form $\frac{1}{3}x^3 - 2x^2 + C$, where C is any constant.
b. One antiderivative of $\sec^2 x$ is $\tan x$, so the antiderivatives of $\sec^2 x$ have the form $\tan x + C$, where C is any constant. ❏

There is precisely one antiderivative that has a specified value at a prescribed point, as the next example illustrates.

EXAMPLE 2 Let f be such that $f'(x) = \cos x$ and $f(\pi/2) = -1$. Determine the function f.

Solution Since f and $\sin x$ are both antiderivatives of $\cos x$, by Theorem 4.6(b) there is a constant C such that

$$f(x) = \sin x + C$$

for the appropriate constant C. To determine C, we use the assumption that $f(\pi/2) = -1$, which yields

$$-1 = f\left(\frac{\pi}{2}\right) = \sin \frac{\pi}{2} + C = 1 + C$$

Therefore $C = -2$, and hence

$$f(x) = \sin x - 2 \qquad ❏$$

Examples 1 and 2 illustrate two features of functions that have antiderivatives on an interval I. If f has an antiderivative F, then the following hold:

> f has infinitely many antiderivatives $F + C$, where C is a constant.
>
> f has a unique antiderivative with a specified value at a prescribed number.

These features will be basic in the study of integration (Chapters 5–8).

The notion of antiderivative plays an important role in the study of motion. Recall that the velocity function v of an object moving along a straight line is the derivative of the position function f. This means that f is an antiderivative of v. Similarly, since the acceleration function a is the derivative of the velocity function v, it follows that v is an antiderivative of a. We will use this information to derive the basic law of motion

$$h(t) = -4.9t^2 + v_0 t + h_0 \tag{2}$$

which was stated in (1) of Section 1.3, and which describes the height of an object near the earth, under the assumption that the object is influenced only by the force of gravity.

For the derivation of (2), we will assume that (as experiments have shown) the acceleration of the object is constant, given by $a(t) = -9.8$ for all t in an interval I.

By the observations above, acceleration is the derivative of velocity. Therefore

$$\frac{dv}{dt} = a = -9.8$$

In other words, v is an antiderivative of -9.8. Theorem 4.6(b) then implies that

$$v(t) = -9.8t + C_0 \qquad (3)$$

for some appropriate constant C_0. If v_0 denotes the velocity at time 0, then it follows by substituting 0 for t in (3) that $C_0 = v_0$. Consequently

$$v(t) = -9.8t + v_0 \qquad (4)$$

Because the height h is an antiderivative of v and $-4.9t^2 + v_0 t$ is an antiderivative of $-9.8t + v_0$ (which equals v), Theorem 4.6(b) implies that

$$h(t) = -4.9t^2 + v_0 t + C_1$$

where C_1 is a suitable constant. If h_0 denotes the height at time 0, then it follows by substituting 0 for t that $C_1 = h_0$. As a result

$$h(t) = -4.9t^2 + v_0 t + h_0$$

which is (2). Consequently we have proved that if $a(t) = -9.8$ for all t in an interval I, then the height of the object is given by (2).

Returning to the Metroliner discussion at the outset of the section, we can use the information on antiderivatives to find out how far the train has traveled since leaving Penn Station, provided that the velocity $v(t)$ at time t is represented by a simple function. For example, suppose that the velocity $v(t)$ is given in miles per minute by

$$v(t) = \frac{3}{16}t^2 - \frac{1}{32}t^3, \quad \text{for } 0 \le t \le 4$$

where t is the time after departure. We will measure distance from Penn Station, so that $f(0) = 0$. Notice first that $v(t) = t^2(6 - t)/32$, so that when $0 \le t \le 4$ the velocity is nonnegative, which means that the train is moving forward. Next we find that $t^3/16$ is an antiderivative of $3t^2/16$, and $-t^4/128$ is an antiderivative of $-t^3/32$. As a result, an antiderivative of v has the form

$$f(t) = \frac{t^3}{16} - \frac{t^4}{128} + C, \quad \text{for } 0 \le t \le 4$$

for some constant C. Substituting 0 for t, we find that $f(0) = C$, so that $C = 0$. It follows that the distance the train has traveled after t minutes is

$$f(t) = \frac{t^3}{16} - \frac{t^4}{128}, \quad \text{for } 0 \le t \le 4$$

In particular, when $t = 4$, that is, after 4 minutes, we have $f(4) = 4 - 2 = 2$, so that the Metroliner has traveled 2 miles during the initial 4 minutes after leaving Penn Station.

The conclusion is that if we know the function representing the velocity of an object moving in a straight line, then provided we are able to determine an antiderivative of the function, we can deduce the position of the object relative to its starting point.

Increasing and Decreasing Functions

Let f be defined on an interval I. Recall from Section 3.2 that

f is increasing on I if $f(x) < f(z)$ whenever x and z are in I and $x < z$

f is decreasing on I if $f(x) > f(z)$ whenever x and z are in I and $x < z$

Graphically, a function is increasing on I if its graph slopes upward to the right (Figure 4.20(a)). It is decreasing on I if its graph slopes downward to the right (Figure 4.20(b)). For example, if $f(x) = x^2$, then f is decreasing on $(-\infty, 0]$ and increasing on $[0, \infty)$ (Figure 4.21(a)). If $f(x) = x^3$, then f is increasing on $(-\infty, \infty)$ (Figure 4.21(b)).

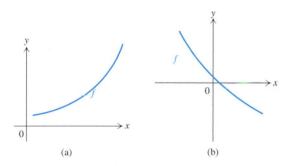

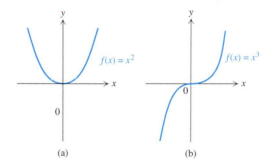

FIGURE 4.20   (a) The graph of an increasing function. (b) The graph of a decreasing function.

FIGURE 4.21   (a) f is decreasing on $(-\infty, 0]$ and increasing on $[0, \infty)$. (b) f is increasing on $(-\infty, \infty)$.

The next theorem provides a criterion that helps in determining whether a function is increasing or decreasing on a given interval.

THEOREM   4.7

Let f be continuous on an interval I and differentiable at each interior point of I.

a. If $f'(x) > 0$ at each interior point of I, then f is increasing on I. Moreover, f is increasing on I if $f'(x) > 0$ except for a finite number of points x in I.

b. If $f'(x) < 0$ at each interior point of I, then f is decreasing on I. Moreover, f is decreasing on I if $f'(x) < 0$ except for a finite number of points x in I.

Proof   We will prove only (a); the proof of (b) is analogous. Assume that $f'(x) > 0$ at each interior point of I, and let x and z be arbitrary numbers in I, with $x < z$. By assumption f is continuous on $[x, z]$ and differentiable on (x, z). Thus we can apply the Mean Value Theorem to f in order to find a number c in (x, z) such that

$$f'(c) = \frac{f(z) - f(x)}{z - x}$$

Since $f'(c) > 0$ and $z - x > 0$, this means that

$$f(z) - f(x) = f'(c)(z - x) > 0$$

Therefore $f(z) > f(x)$, and thus f is increasing on I.

Next, assume that $f'(x) > 0$ for all x in I except a single number c. By the preceding argument f is increasing on the subinterval I_1 of I with right endpoint c and on the subinterval I_2 of I with left endpoint c. If x and z are in I and $x < c < z$, then $f(x) < f(c)$ because f is increasing on I_1; similarly $f(c) < f(z)$ because f is increasing on I_2. It follows that $f(x) < f(z)$, so that f is increasing on I. The same reasoning applies to show that if $f'(x) > 0$ except at a finite number of points x in I, then f is increasing on I.   ■

If f is continuous on a closed interval $[a, b]$ and $f'(x) > 0$ for $a < x < b$, then it is an immediate consequence of Theorem 4.7 that f is increasing on $[a, b]$. Similar comments apply for decreasing functions and for intervals of infinite length.

The next four examples illustrate the use of Theorem 4.7.

EXAMPLE 3   Let $f(x) = 2x^3 + 3x^2 - 12x - 3$. On which intervals is f increasing and on which is it decreasing?

Solution   First we find that

$$f'(x) = 6x^2 + 6x - 12 = 6(x + 2)(x - 1)$$

Next we assemble Figure 4.22 to determine the sign of $f'(x)$.

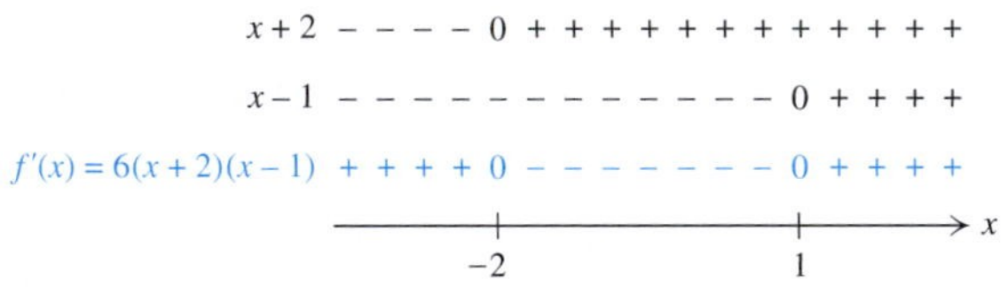

$$x + 2 \quad - \; - \; - \; - \; 0 \; + + + + + + + + + + + + +$$
$$x - 1 \quad - \; - \; - \; - \; - \; - \; - \; - \; - \; - \; - \; 0 \; + + + +$$
$$f'(x) = 6(x + 2)(x - 1) \; + + + + \; 0 \; - \; - \; - \; - \; - \; - \; - \; 0 \; + + + +$$

$$\xrightarrow{\qquad \qquad \qquad -2 \qquad \qquad \qquad 1 \qquad \qquad} x$$

FIGURE 4.22

It follows that $f'(x) > 0$ for x in $(-\infty, -2)$ and $(1, \infty)$, and $f'(x) < 0$ for x in $(-2, 1)$. Therefore Theorem 4.7 implies that f is increasing on $(-\infty, -2]$ and $[1, \infty)$ and is decreasing on $[-2, 1]$ (Figure 4.23).   ❑

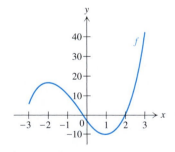

FIGURE 4.23 f is increasing on $(-\infty, -2]$ and $[1, \infty)$ and is decreasing on $[-2, 1]$.

EXAMPLE 4   Let $f(x) = \frac{1}{3}x^3 - x^2 + x - 5$. Show that f is increasing on $(-\infty, \infty)$ (Figure 4.24).

Solution   First notice that

$$f'(x) = x^2 - 2x + 1 = (x - 1)^2$$

Since $f'(x) > 0$ for all x except $x = 1$ where $f'(x) = 0$, it follows from Theorem 4.7(a) that f is increasing on the entire real line $(-\infty, \infty)$.   ❑

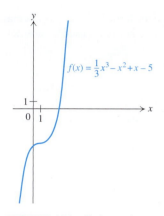

$f(x) = \frac{1}{3}x^3 - x^2 + x - 5$

FIGURE 4.24   *f* is increasing on $(-\infty, \infty)$.

EXAMPLE 5   Let $f(x) = x^2 e^x$. Determine the intervals on which *f* is increasing and those on which it is decreasing.

Solution   We begin by taking the derivative of *f*:

$$f'(x) = 2xe^x + x^2 e^x = x(2 + x)e^x$$

Again we assemble a table (Figure 4.25) to determine the sign of $f'(x)$:

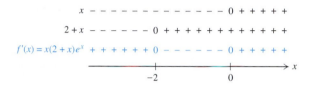

FIGURE 4.25

(Since $e^x > 0$ for all *x*, the table needs no separate entry for e^x.) It follows from the table that *f* is increasing on $(-\infty, -2]$ and $[0, \infty)$, and is decreasing on $[-2, 0]$. The graph in Figure 4.26 confirms these assertions.   ❑

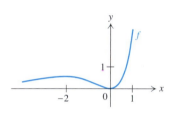

FIGURE 4.26

Caution: Theorem 4.7 says that if $f'(x) > 0$ ($f'(x) < 0$) for all *x* in an interval *I*, then *f* is increasing (decreasing) on *I*. The fact that *I* is an *interval* is critical. In the following example, *f* has a negative derivative throughout its domain, but is *not* decreasing on its domain—which is not an interval.

EXAMPLE 6   Let $f(x) = 1/x$. Show that $f'(x) < 0$ for all *x* in the domain of *f*. Then determine whether *f* is decreasing on its domain.

Solution   Since $f'(x) = -1/x^2$, it follows that $f'(x) < 0$ for all *x* in the domain of *f*. However, *f* is *not* decreasing on its domain because for each *x* in $(-\infty, 0)$ and each *z* in $(0, \infty)$, the inequality $f(x) < f(z)$ holds (see Figure 4.27). However, *f* is decreasing on each of the intervals $(-\infty, 0)$ and $(0, \infty)$ separately.   ❑

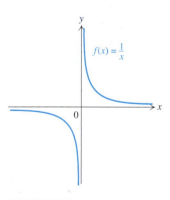

$f(x) = \frac{1}{x}$

FIGURE 4.27

Theorem 4.7 can be effectively employed in order to derive inequalities. In particular, if *f* is continuous on $[a, \infty]$ and $f'(x) > 0$ for all *x* in (a, ∞), then it follows from Theorem 4.7(a) that *f* is increasing on $[a, \infty)$. Thus $f(x) > f(a)$ for all $x > a$.

EXAMPLE 7   Show that the following inequalities hold for all $x > 0$.
a. $e^x > 1$ **b.** $e^x > 1 + x$

Solution   To prove (a), we let $f(x) = e^x$. Because $f(0) = 1$ and $f'(x) = e^x > 0$, the observation preceding the example implies that $f(x) > f(0) = 1$ for all $x > 0$. It follows that $e^x > 1$, which proves (a).

To prove (b) we first notice that $e^x > 1$ is equivalent to $e^x - 1 > 0$. Now let $g(x) = e^x - x$. Then $g(0) = 1$. In addition, $g'(x) = e^x - 1$, so by (a) we know that $g'(x) > 0$ for all $x > 0$. By the observation preceding the example, this means that $g(x) > g(0) = 1$ for all $x > 0$. This is equivalent to $e^x - x > 1$, or rather, $e^x > 1 + x$.   ❑

The notions of increasing and decreasing functions furnish the key to finding extreme values of functions defined on intervals that are not closed or are not bounded. Section 4.6 is devoted to that topic.

EXERCISES 4.3

In Exercises 1–12 find all antiderivatives of the given function.

1. 0

2. -2

3. $3x$

4. $-6x + 5$

5. $-x^2$

6. $4x^2 + 6x - 1$

7. $\sin x$

8. $\csc^2 \pi x$

9. $\sin x \cos x$

10. x^n for $n \neq -1$

11. e^x

12. e^{-x}

In Exercises 13–20 determine the function f satisfying the given conditions.

13. $f'(x) = -2, f(0) = 0$

14. $f'(x) = \frac{1}{2}x, f(\frac{1}{2}) = -1$

15. $f'(x) = x^2, f(0) = -5$

16. $f'(x) = -\frac{3}{2}x^2, f(-1) = -\frac{1}{2}$

17. $f'(x) = \cos x, f(\pi/3) = 1$

18. $f'(x) = \sec \dfrac{x}{2} \tan \dfrac{x}{2}, f(\pi/2) = 2$

19. $f'(x) = e^x, f(0) = 10$

20. $f'(x) = \dfrac{1}{x}, f(e) = -3$

In Exercises 21–28 determine all functions f satisfying the given conditions.

21. $f''(x) = 0$ (*Hint:* Use Theorem 4.6 twice.)

22. $f''(x) = 0, f'(-2) = 1$

23. $f''(x) = 0, f'(0) = -1, f(0) = 2$

24. $f''(x) = 0, f'(2) = 3, f(-1) = 1$

25. $f''(x) = \sin x, f'(\pi) = -2, f(0) = 4$

26. $f^{(3)}(x) = 0$ (*Hint:* Use Theorem 4.6 three times.)

27. $f^{(4)}(x) = 0$

28. $f^{(n)}(x) = 0$ for any positive integer n

In Exercises 29–50 find the intervals on which the given function is increasing and those on which it is decreasing.

29. $f(x) = x^2 + x + 1$

30. $f(x) = x^3 - 12x + 4$

31. $f(x) = x^3 - x^2 + x - 1$

32. $f(x) = 4x^3 - 6x^2 - 9x$

33. $f(x) = x^4 - 2x^3 + 1$

34. $f(x) = x^5 + x^3 + 2x - 1$

35. $f(x) = x^5 + x^3 - 2x + 1$

36. $f(x) = x|x|$

37. $g(x) = \sqrt{16 - x^2}$

38. $g(x) = \sqrt{9x^2 - 4}$

39. $g(x) = 1/(x + 3)$

40. $g(x) = (x - 2)/(x - 1)$

41. $k(x) = 1/(x^2 + 1)$

42. $k(x) = 1/(x^2 - 4)$

43. $f(t) = \tan t$

44. $f(t) = \sin t$

45. $f(t) = 2 \cos t - t$

46. $f(t) = \sin t + \cos t$

47. $f(x) = xe^x$

48. $f(x) = e^x - 3x$

49. $f(x) = x - \ln x$

50. $f(x) = (\ln x)/x$

In Exercises 51–54 determine on which intervals f is increasing and on which intervals f is decreasing.

51.

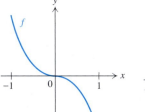

52.

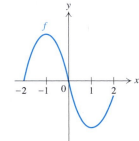

53.

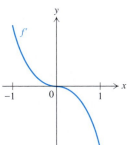

54.

In Exercises 55–58 show that the given inequality holds for the given values of x.

55. $x^4 - 4x > -3$ for $x > 1$

56. $4x^2 + 1/x > 5$ for $x > 1$

57. $\frac{1}{4}x + 1/x > 1$ for $x > 2$

58. $1/x + \tan(1/x) < 1 + \pi/4$ for $x > 4/\pi$

59. a. In Example 7 we showed that $e^x > 1 + x$ for all $x > 0$. Use this result to show that $e^x > 1 + x + \frac{1}{2}x^2$ for $x > 0$.

   b. Using the result of part (a), show that $e^x > 1 + x + \frac{1}{2}x^2 + \frac{1}{6}x^3$ for $x > 0$.

    c. Plot e^x and $1 + x + \frac{1}{2}x^2 + \frac{1}{6}x^3$ in order to determine whether the inequality in (b) is corroborated.

60. a. Let $f(x) = x - \sin x$. Show that f is increasing.

    b. Use part (a) to show that $\sin x > x$ on $(-\infty, 0)$ and $\sin x < x$ on $(0, \infty)$.

61. Let f and g be continuous on $[a, \infty)$ and differentiable on (a, ∞). Assume that $f(a) \geq g(a)$ and $f'(x) > g'(x)$ for all $x > a$. Prove that $f(x) > g(x)$ for any $x > a$.

62. Show that $\tan x > x$ for all x in $(0, \pi/2)$.

63. Suppose n is a positive integer greater than 1. Prove that $(1 + x)^n > 1 + nx$ for $x > 0$.

64. Using Exercises 60(b) and 61, prove that $\cos x > 1 - x^2/2$ for all $x > 0$.

65. Using 61 and 64, show that $\sin x > x - x^3/6$ for all $x > 0$.

66. a. Show that there is exactly one value of x for which $\cos x = 2x$. (*Hint:* Let $f(x) = \cos x - 2x$, and determine how many zeros f could have.)

    b. Use the Newton-Raphson method to approximate the value of x for which $\cos x = 2x$. Continue until successive iterations obtained by the calculator are identical.

67. Consider the function f that is continuous on $[-1, 1]$, such that $f(0) = 0$ and

$$f'(x) = \begin{cases} 3x^2 & \text{for } -1 < x < 0 \\ 2e^x & \text{for } 0 < x < 1 \end{cases}$$

Write down a formula for f on $[-1, 1]$, making sure that your choice for f is continuous at 0. Then sketch the graph of f. (*Hint:* The rule of f will have 2 parts, just as f' does. You will need to join the parts carefully, so that f will be continuous at 0.)

68. Let $f(x) = x^3 + ax^2 + bx + c$, where a, b, and c are constants.

    a. Show that f is increasing on $(-\infty, \infty)$ if $a^2 \leq 3b$.

    b. Assume that $a^2 > 3b$. Show that f is increasing on

$$(-\infty, (-a - \sqrt{a^2 - 3b})/3]$$

    and on

$$[(-a + \sqrt{a^2 - 3b})/3, \infty)$$

    and is decreasing on

$$[(-a - \sqrt{a^2 - 3b})/3, (-a + \sqrt{a^2 - 3b})/3]$$

69. Figure 4.28 shows the graph of a function g. Assume that g is the derivative of a function f such that $f(0) = -2$. Draw a rough sketch of the graph of f.

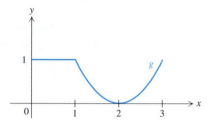

FIGURE 4.28 Graph for Exercise 69.

70. The graph of f' appears in Figure 4.29.

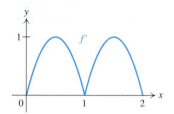

FIGURE 4.29 Graph for Exercise 70.

    a. Draw a rough sketch of the graph of f''.

    b. Do you have enough information to sketch the graph of f? If so, sketch it. If not, supply necessary information and then sketch it.

71. Let F and G be antiderivatives of f and g, respectively, on an interval I.

    a. Prove that $F + G$ is an antiderivative of $f + g$ on I.

    b. Prove that cF is an antiderivative of cf on I for any constant c.

    c. Is FG necessarily an antiderivative of fg? Explain.

    d. Is F/G necessarily an antiderivative of f/g? Explain.

72. Use Theorem 4.6(a) to prove that only the constant functions satisfy the inequality

$$|f(x) - f(y)| \leq |x - y|^2$$

for all x and y. Is this true for any integer exponents of $|x - y|$ other than 2? Explain your answer.

*73. Suppose f is differentiable on $(-\infty, \infty)$, and for each x and y we have $f(x + y) = f(x) + f(y)$. Show that there is a fixed number c such that $f(x) = cx$ for all x. (*Hint:* Using (2) of Section 3.2, show that $f'(x) = f'(0)$ for all x, and let $c = f'(0)$.)

Applications

74. Show that if a particle has zero velocity for any given period of time, then it stands still during that period.

75. The rate of a certain autocatalytic reaction is given by

$$v(x) = kx(a - x) \quad \text{for } 0 \leq x \leq a$$

where a is the original amount of the substance, x is the amount of the substance produced by the reaction, and k is positive. Determine where the rate of reaction is increasing and where it is decreasing.

76. A weight W is suspended from a ceiling by three strings of length L so that the points of attachment form an equilateral triangle of side S (Figure 4.30). If the weights of the various strings are negligible, then the tension T in the three strings attached to the ceiling is given by

$$T = \frac{W}{3\sqrt{1 - S^2/3L^2}}$$

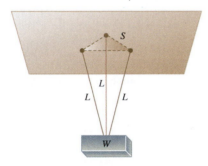

FIGURE 4.30   Figure for Exercise 76.

a. Show that for fixed values of S and L, T is an increasing function of W.

b. Show that for fixed values of W and L, T is an increasing function of S.

c. Show that for fixed values of S and W, T is a decreasing function of L.

77. An electrical network consists of a resistor, a capacitor, and an inductor connected in series with a generator producing a voltage V that oscillates between positive and negative according to the equation $V = V_0 \sin \omega t$, where V_0 is the maximum voltage and ω is the angular frequency with $\omega > 0$. The steady-state current also oscillates, with the maximum current i_{max} given by the formula

$$i_{max} = \frac{V_0}{\sqrt{\left(\dfrac{1}{\omega C} - \omega L\right)^2 + R^2}}$$

where L, R, and C are positive constants. On what interval is i_{max} an increasing function of ω?

78. According to Hooke's Law, the force $f(x)$ exerted by a spring that has been expanded a distance x is given by $f(x) = -kx$, where k is a positive constant called the **spring constant.** Find a function U such that $f(x) = -dU/dx$. (Such a function U is called a **potential energy function** for the force f. If $U(0) = 0$, then $U(x)$ represents the amount of energy stored in the spring when it is expanded a distance x.)

79. Suppose a child picks up a rope with length l and mass m, and whirls it overhead with angular velocity ω. The resulting centrifugal force creates tension T in the rope that varies from point to point, depending on the distance r of the point from the stationary end of the rope grabbed by the child's hand. The laws of physics imply that $dT/dr = -m\omega^2 r/l$, and that $T = 0$ for $r = l$. Express T as a function of r.

*80. On the moon, the acceleration due to gravity is approximately 1.6 meters per second per second. Assume that a person jumps with the same initial velocity on the moon as on earth. If the person can high jump 2 meters on earth, how high would that jump be on the moon?

Project

1. Recall that the natural exponential function e^x has the following three properties: $e^0 = 1$, e^x is differentiable, and e^x satisfies the famous law

$$e^{a+b} = e^a e^b \qquad \text{Law of Exponents}$$

In this exercise we will show that if f is any positive function such that $f(0) = 1, f'(0) = 1$, and

$$f(a + b) = f(a)f(b) \quad \text{for all } a \text{ and } b \qquad (5)$$

then $f(x) = e^x$.

a. Let x be any number, and use (5) to show that $(f(x + h) - (f(x))/h = f(x)(f(h) - f(0))/h$

b. Use (a) and the fact that $f'(0)$ exists and equals 1 to show that $f'(x) = f(x)f'(0) = f(x)$ for all x.

c. Use (b) to show that $\frac{d}{dx}\ln f(x) = 1$. Then use Theorem 4.6(b) to prove that $\ln f(x) = x + c$ for some constant c. (*Hint:* What is $\frac{d}{dx}\ln f(x)$?)

d. Use (c) to show that $f(x) = e^{x+c}$.

e. Use (d) and the Law of Exponents, as well as the hypothesis that $f(0) = 1$, to conclude that $f(x) = e^x$ for all x.

4.4   EXPONENTIAL GROWTH AND DECAY

In this section one of the consequences of the Mean Value Theorem will assist in the analysis of functions representing population growth and radioactive decay.

For an application related to population growth, we observe that certain strains of algae and bacteria reproduce at a rate proportional to the number present at any given time. This means that if $f(t)$ represents the population at time t, so that $f'(t)$ represents the rate of growth of the population at time t, then $f(t)$ and $f'(t)$ are proportional, so that there is a positive constant k such that

$$f'(t) = kf(t) \qquad (1)$$

In Example 1 we will consider a specific case.

Turning to an application involving radioactive decay, we mention that carbon has three isotopes that occur in nature: C^{12} and C^{13}, which are stable, and C^{14}, which is unstable and radioactive. In the atmosphere the loss of C^{14} atoms through radioactive decay is offset by the creation of new C^{14} atoms by cosmic radiation. In living plants the loss of C^{14} atoms is offset by the intake of C^{14} during photosynthesis, and in living animals the C^{14} is replenished through consumption of vegetation. It turns out that the ratio between the three isotopes of carbon is essentially constant, both in the atmosphere and in living organisms. However, in a dead organism the ratio varies with time, with the number of C^{14} atoms decreasing at a rate proportional to the amount present at any given instant. Consequently if $f(t)$ denotes the amount of C^{14} in, say, a bone at time t, then f satisfies (1) for an appropriate negative constant k. This is the basis for estimating age by "carbon 14 dating," as we will see in Example 2.

Thus the equation in (1) arises in the description of two quite different physical phenomena. In fact, it arises in the solutions of many diverse physical and mathematical problems. It would therefore be of interest to determine all functions that satisfy (1). Of course, the constant function 0 satisfies (1), but so do other functions, such as e^{kt}. Indeed, if $f(t) = e^{kt}$, then the Chain Rule implies that

$$f'(t) = \frac{d}{dt}(e^{kt}) = ke^{kt} = kf(t)$$

Now let us determine the form of all functions f satisfying (1).

Assume that f is a continuous function on $[0, \infty)$ satisfying (1) on $(0, \infty)$, and let

$$g(t) = e^{-kt}f(t) \qquad (2)$$

where k is the constant appearing in (1). Then differentiating and using (1), we find that

$$g'(t) = -ke^{-kt}f(t) + e^{-kt}f'(t) \overset{(1)}{=} -ke^{-kt}f(t) + e^{-kt}kf(t) = 0$$

for all $t > 0$. Then by Theorem 4.6, g is a constant function, so that for some number C we have

$$g(t) = C \quad \text{for } t \geq 0$$

Therefore from (2) we see that

$$f(t) = g(t)e^{kt} = Ce^{kt} \quad \text{for } t \ge 0 \tag{3}$$

Thus any function f satisfying (1) must be a constant multiple of e^{kt}. Finally, observe that

$$f(0) = Ce^{k \cdot 0} = C$$

so we can substitute $f(0)$ for C in (3). This yields

$$f(t) = f(0)e^{kt} \quad \text{for } t \ge 0$$

We summarize our result in a theorem.

THEOREM   4.8

Suppose f is continuous on $[0, \infty)$ and

$$f'(t) = kf(t) \quad \text{for } t > 0$$

Then

$$f(t) = f(0)e^{kt} \quad \text{for } t \ge 0 \tag{4}$$

If a function f satisfies (4), then we say that f **grows exponentially** if $k > 0$ and that f **decays exponentially** if $k < 0$. For future reference we notice that by the same analysis, if f satisfies (1) for all real numbers t, then f also satisfies (4) for all real numbers t.

EXAMPLE 1   It is known that a certain kind of algae in the Dead Sea can double in population every 2 days. Assuming that the population of algae grows exponentially, beginning now with a population of 1,000,000, determine what the population will be after one week.

Solution   Let $f(t)$ denote the number of algae t days from now, for $t \ge 0$. We wish to find $f(7)$. By hypothesis, $f(0) = 1,000,000$, and combined with (4) this means that

$$f(t) = f(0)e^{kt} = 1,000,000e^{kt} \quad \text{for } t \ge 0 \tag{5}$$

In order to find $f(7)$, we first determine k. By using (5) and the hypothesis that $f(2) = 2,000,000$ we deduce that

$$2,000,000 = f(2) = 1,000,000e^{2k}$$

Thus

$$2 = e^{2k}$$

so that by taking logarithms of both sides, we have

$$\ln 2 = 2k$$

and therefore

$$k = \frac{1}{2} \ln 2$$

Consequently from (5),

$$f(t) = 1,000,000 e^{(\ln 2)t/2} \tag{6}$$

The right side of (6) can be simplified by noting that

$$e^{(\ln 2)t/2} = (e^{\ln 2})^{t/2} = 2^{t/2}$$

which means that

$$f(t) = 1,000,000 (2^{t/2}) \quad \text{for } t \geq 0$$

To find $f(7)$, we use the last equation with $t = 7$ and find that

$$f(7) = 1,000,000 (2^{7/2}) \approx 11,313,708 \qquad \square$$

The time it takes for a population to double is called its **doubling time.** Thus the doubling time for the algae in Example 1 is 2 days. More generally, if a population grows exponentially with doubling time d, then the solution of Example 1 can be modified to show that at time t the population is given by

$$f(t) = f(0) 2^{t/d} \quad \text{for } t \geq 0$$

(see Exercise 6).

The **half-life** of a radioactive substance is the length of time it takes for half of a given amount of the substance to disintegrate through radiation (Figure 4.31). The half-life of C^{14} is approximately 5730 years. This means that 5730 years after an organism dies, it should contain half as much C^{14} as it did when it was alive.

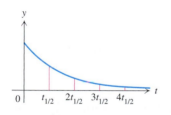

FIGURE 4.31   The half-life is the time $t_{1/2}$ it takes for half of a substance to disintegrate. After two half-lives one quarter will remain, and so forth.

EXAMPLE 2   Let $f(t)$ denote the amount of C^{14} present in an organism t years after its death. Determine the constant k in equation (4) and the percentage of C^{14} that should remain 1000 years after the death of the organism.

Solution   If t denotes the number of years after death, then

$$f(t) = f(0) e^{kt} \quad \text{for } t \geq 0$$

The hypothesis that the half-life of C^{14} is 5730 years means that $f(5730) = 1/2 f(0)$, so that

$$\frac{1}{2} = \frac{f(5730)}{f(0)} = e^{5730k}$$

Taking logarithms, we obtain

$$5730k = \ln \frac{1}{2} = -\ln 2$$

so that

$$k = \frac{-\ln 2}{5730}$$

Therefore

$$f(t) = f(0)e^{-(\ln 2)t/5730} \quad \text{for } t \geq 0 \tag{7}$$

To determine the percentage of C^{14} left after 1000 years, we need only calculate $f(1000)/f(0)$:

$$\frac{f(1000)}{f(0)} = e^{-(\ln 2)1000/5730} \approx 0.886$$

Thus approximately 88.6 percent remains after a millennium.   ❑

EXAMPLE 3   Suppose that of the original amount of C^{14} in a human bone uncovered in Kenya only $\frac{1}{10}$ remains today. How long ago did death occur?

Solution   If t denotes the number of years from death until now, then we know by hypothesis that

$$f(t) = \frac{1}{10} f(0)$$

The Iceman

In 1991 the exceptionally well-preserved body of a prehistoric person, now known as the Iceman, was discovered high in the Italian Alps as the result of a glacier meltback. Carbon dating fixed the age of the Iceman and the copper ax found with him as 5300 years, suggesting that the Copper Age occurred earlier than previously believed.

and by (7) that

$$f(t) = f(0)e^{-(\ln 2)t/5730}$$

Consequently

$$e^{-(\ln 2)t/5730} = \frac{f(t)}{f(0)} = \frac{1}{10}$$

Taking logarithms, we find that

$$-\frac{\ln 2}{5730} t = \ln \frac{1}{10} = -\ln 10$$

Solving for t, we obtain

$$t = \frac{\ln 10}{\ln 2} (5730) \approx 19{,}000$$

Thus death occurred approximately 19,000 years ago.   ❑

Carbon dating is effective only for dates between approximately 1500 A.D. and 50,000 B.C. For objects significantly older, such as million-year-old fossils and billion-year-old lunar rocks, the potassium-argon and rubidium-strontium dating methods can be used (see Exercises 25–27).

Finally, we consider the air pressure $f(t)$ at an altitude of t meters above sea level. The higher the altitude is, the lower the air pressure is. Moreover, it turns out that to a high degree of accuracy, the rate at which the air pressure decreases with increasing altitude is proportional to the pressure. Consequently it is reasonable to assume that f satisfies (4), so that there is a negative constant k such that

$$f(t) = f(0)e^{kt} \quad \text{for } t \geq 0$$

If t is measured in meters and $f(t)$ in "atmospheres," with $f(0) = 1$ (atmosphere), then experimental results indicate that

$$k \approx -1.25 \times 10^{-4}$$

so that we take

$$f(t) = e^{-1.25 \times 10^{-4}t} \quad \text{for } t \geq 0 \tag{8}$$

EXAMPLE 4   The atmospheric pressure outside a jet passenger plane is 0.28 atmospheres. Determine the altitude of the plane.

Solution   If t denotes the altitude of the plane in meters above sea level, then by hypothesis, $f(t) = 0.28$. To find t, we use (8):

$$0.28 = f(t) = e^{-1.25 \times 10^{-4}t}$$

Taking logarithms of both sides, we obtain

$$\ln 0.28 = -1.25 \times 10^{-4}t$$

so that

$$t = \frac{\ln 0.28}{-1.25 \times 10^{-4}} \approx 10{,}200 \text{ (meters)}$$

Consequently the plane is approximately 10,200 meters above sea level.   ❑

An altitude of 10,200 meters is approximately 33,500 feet, a typical altitude of a plane in transcontinental flight. Since at that altitude the air pressure is about $\frac{1}{4}$ the air pressure at sea level, it is not surprising that the pressure inside the plane would be regulated.

As we have seen, problems concerning exponential growth or decay involve four quantities: $t, k, f(0)$, and $f(t)$. If we know any three of them (or if we know the ratio $f(t)/f(0)$ and either t or k), then we can determine the remaining quantity by using (4). The problem normally revolves about finding that remaining quantity.

The functions arising in our examples concerning exponential growth and decay are determined from experimental data in idealized conditions. As a result, we should expect the formulas we obtain to be only approximately accurate. Moreover, we frequently idealize the functions themselves. After all, the number of bacteria or C^{14} atoms is always an integer, and yet we use a continuous function to represent the number of bacteria at various times. However, these idealizations lead us to very valuable information about different sorts of natural phenomena, and this is the utility of the theory of exponential growth and decay.

In Leibniz notation the equation in (1) becomes

$$\frac{dy}{dx} = ky \tag{9}$$

Equations such as (9), in which y and its derivatives appear, are called differential equations. Finding all solutions to such an equation, as we did in Theorem 4.8, is called "solving the equation." We will return to this topic in Chapter 7.

EXERCISES 4.4

1. How long would it take for the algae mentioned in Example 1 to
   a. quadruple in number?   b. triple in number?

2. Suppose that a colony of bacteria is growing exponentially. If 12 hours are required for the number of bacteria to grow from 4000 to 6000, find the doubling time.

3. Experiment has shown that under ideal conditions and a constant temperature of 28.5°C, the population of a certain type of flour beetle doubles in 6 days and 20 hours. Suppose that there are now 1500 such beetles. How long ago were there 1200?

4. If the population of the world is not unduly affected by war, famine, or new technology, then it is reasonable to assume that the population will grow (at least for a long period of time) at an exponential rate. Using this assumption and the census figures which show that

   the world population in 1962 was 3,150,000,000

   the world population in 1978 was 4,238,000,000

   determine what the world population should have been in 2000.

5. Suppose the populations of two countries are growing exponentially. Suppose also that one country has a population of 50,000,000 and a doubling time of 20 years, whereas the other has a population of 20,000,000 and a doubling time of 10 years. How long will it be until the two countries have the same population?

6. Suppose the population $f(t)$ of a given species grows exponentially, so that $f(t) = f(0)e^{kt}$ for some positive constant k.
   a. Show that the population doubles during any time

interval of duration $(\ln 2)/k$. Thus $(\ln 2)/k$ is the doubling time d.
   b. Show that $f(t) = f(0)2^{t/d}$.

7. Suppose the amount $f(t)$ of a radioactive substance decays exponentially, so that $f(t) = f(0)e^{kt}$ for some negative constant k.
   a. Show that the amount decreases by half in any time interval of duration $-(\ln 2)/k$. Thus $-(\ln 2)/k$ is the half-life h.
   b. Show that $f(t) = f(0)(\frac{1}{2})^{t/h}$.

8. Suppose an unknown radioactive substance produces 4000 counts per minute on a Geiger counter at a certain time, and only 500 counts per minute 4 days later. Assuming that the amount of radioactive substance is proportional to the number of counts per minute, determine the half-life of the radioactive substance.

9. Suppose you have a cache of radium, whose half-life is approximately 1590 years. How long would you have to wait for one tenth of it to disappear?

10. Find the approximate percentage of C^{14} still remaining in the 5300-year-old Iceman mentioned in the text.

11. The so-called "Pittsburgh man," unearthed near the town of Arella, Pennsylvania, shows that civilization existed there from around 13,000 B.C. to 12,300 B.C. Calculate the difference between the percentage of C^{14} lost prior to 2000 A.D. from a bone of someone who died in 12,300 B.C. and that from someone who died in 13,000 B.C.

12. Richard Leakey's "1470" skull, found in Kenya, is reputed to be 1,800,000 years old. Show that the percentage of C^{14} remaining now would be negligible, and hence that in this instance dating by means of C^{14} would be meaningless.

13. Iodine 131, which has been used for treating cancer of the thyroid gland, is also used to detect leaks in water pipes. It has a half-life of 8.14 days. Suppose a water company wishes to use 100 milligrams of iodine 131 to search for a leak and must take delivery 2 days before using it. How much iodine 131 should be purchased?

14. Suppose a small quantity of radon gas, which has a half-life of 3.8 days, is accidentally released into the air in a laboratory. If the resulting radiation level is 50% above the "safe" level, how long should the laboratory remain vacated?

15. Cabins in jet passenger planes are often pressurized to a pressure equivalent to that at 1600 meters above sea level. Use (8) to determine the ratio of the cabin pressure to the atmospheric pressure at sea level.

16. Mountain climbers normally wear oxygen masks when they are higher than 7000 meters above sea level. Determine the air pressure at 7000 meters.

A climber on Mt. Everest, which is approximately 8850 meters tall. (*Art Wolfe/Allstock*)

17. Halley's Law states that the barometric pressure $p(t)$ in inches of mercury at t miles above sea level is given by

$$p(t) \approx 29.92 e^{-0.2t} \quad \text{for } t \geq 0$$

Find the barometric pressure

a. at sea level   b. 5 miles above sea level

c. 10 miles above sea level

18. Assume that the air pressure $p(x)$ in pounds per square foot at x feet above sea level is given by

$$p(x) \approx 2140 e^{-0.000035x} \quad \text{for } x \geq 0$$

and that an airplane is losing altitude at the rate of 20 miles per hour. At what rate is the air pressure just outside the plane increasing when the plane is 2 miles above sea level?

19. For an operation a dog is anesthetized with sodium pentobarbitol, which is eliminated exponentially from the blood stream. Assume that of any sodium pentobarbitol in the blood stream, half is eliminated in 5 hours. Assume also that to anesthetize a dog, 20 milligrams of sodium pentobarbitol are required for each kilogram of body mass. What single dose of sodium pentobarbitol would be required to anesthetize a 10-kilogram dog for half an hour?

20. If a sum of S dollars is invested at p percent interest and interest is compounded continuously, then the amount $A(t)$ of money accumulated after t years is given by

$$A(t) = S e^{pt/100} \quad \text{for } t \geq 0$$

In terms of S, how much money will there be after 10 years, if the interest rate is 6 percent?

21. Using the equation of Exercise 20, determine what interest rate would be required in order to double a sum of money within 10 years.

22. If one of your grandparents had put $100 into a savings account 75 years ago at 4 percent interest compounded continuously, how much would be in the account now? Suppose Thomas Jefferson had done the same 200 years ago, but at 3 percent interest. How much would be in the account now?

23. The **Unimolecular Reaction Theory** states that if a substance is dissolved in a large container of solvent, then the rate of reaction is proportional to the amount of the remaining substance. Suppose a sugar cube 1 cubic inch in volume is dropped into a jug of iced tea. If there is $\frac{3}{4}$ of a cubic inch of the cube left after 1 minute, then when is there $\frac{1}{2}$ cubic inch left?

24. When an electric capacitor discharges electricity through resistance, the charge on the capacitor decreases at a rate proportional to the amount of charge. If the charge was 5×10^{-2} coulombs 4 seconds ago and is 10^{-3} coulombs now, what was the charge 1 second ago?

25. Suppose a "parent" substance decays exponentially into a "daughter" substance so that the amount $P(t)$ of the parent remaining after t years is given by

$$P(t) = P(0)e^{-\lambda t}$$

The amount $D(t)$ of the daughter substance is given by

$$D(t) = P(0) - P(t)$$

Show that

$$t = \frac{1}{\lambda} \ln \left(\frac{D(t)}{P(t)} + 1 \right)$$

(This formula can be used to determine the age of an object; see Exercise 26.)

26. For the decay of rubidium (Rb^{87}) into strontium (Sr^{87}) the value of λ in Exercise 25 is 1.39×10^{-11}.

    a. Biotite taken from a sample of granite in the Grand Canyon contained 202 parts per million of rubidium and 3.96 parts per million of strontium.* Determine the age of the sample of granite.

    b. Some lunar breccias, formed by meteors and collected during the Apollo 16 mission, were determined by the rubidium-strontium method to be approximately 4.53 billion years old. Assuming that the breccias contained no strontium when they were formed, determine the present ratio of strontium to rubidium in the breccias.

27. Potassium (K^{40}) decays into two substances, calcium (Ca^{40}) and argon (Ar^{40}). By measuring the amounts of potassium and argon it is possible to determine the age of an object from which they are extracted. The formula, which is more complicated than the one in Exercise 25, is

$$t = (1.885)10^9 \ln \left(9.068 \frac{D(t)}{P(t)} + 1 \right)$$

where $D(t)$ is the amount of argon and $P(t)$ the amount of potassium at time t (in years).

a. A sample of basalt taken from lava in the Grand Canyon contained 1.95×10^{-12} moles per gram of argon and 2.885×10^{-8} moles per gram of potassium.* Determine the age of the basalt.

b. A rock from the lunar plains collected during the Apollo 16 mission was determined by the potassium-argon method to be approximately 4.19 billion years old. Assuming that the rock contained no argon when it was formed, determine the present ratio of the amount of argon to the amount of potassium in the lunar rock.

A lunar breccia collected by the Apollo 16 mission. (NASA)

4.5   THE FIRST AND SECOND DERIVATIVE TESTS

Suppose we have a function like $f(x) = 4x^3 + 9x^2 - 12x + 3$. By calculation we find that $f'(x) = 0$ if $x = -2$ or $x = 1/2$, which means that the graph of f has horizontal tangent lines at the points $(-2, f(-2))$ and $(\frac{1}{2}, f(\frac{1}{2}))$. However, without some extra knowledge, we find it difficult to figure out whether the graph might have a "hilltop" or a "valley bottom" when $x = -2$, and similarly when $x = 1/2$. (We will address this issue in Example 1.) Yet hilltops and valley bottoms, which are obvious in Figure 4.32(a) and (b), can be very important in the analysis of the graph of a function. Because of their local, or relative, nature, we call the values of f at such points by the name of "relative extreme values."

*Nations, Dale, *The Record of Geological Time: A Vicarious Trip* (New York: McGraw-Hill, 1975), p. 48.

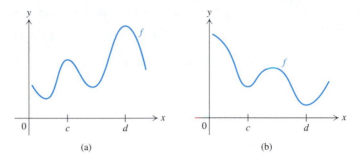

FIGURE 4.32 (a) *f* has a relative maximum at *c* and at *d*. (b) *f* has a relative minimum at *c* and at *d*.

DEFINITION 4.9

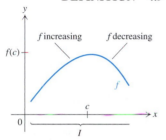

FIGURE 4.33 $f(c)$ is the maximum value of *f* on *I*.

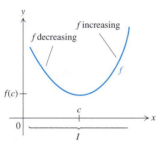

FIGURE 4.34 $f(c)$ is the minimum value of *f* on *I*.

The First Derivative Test

A function *f* has a **relative maximum value** (respectively, a **relative minimum value**) at *c* if $f(c)$ is the maximum value (respectively, the minimum value) of *f* on an open interval containing *c*. A value that is either a relative maximum value or relative minimum value is called a **relative extreme value.**
Sometimes relative extreme values are called **local extreme values**.

The hilltops and valley bottoms on the graph of a differentiable function *f* correspond to the relative extreme values of *f*. By Theorem 4.3,

the relative extreme values occur only at critical numbers of *f*

Thus in the search for relative extreme values of a function *f*, we normally locate all critical numbers. Then it remains to decide which of the critical numbers, if any, yield relative extreme values of *f*. There are two tests, the First Derivative Test and the Second Derivative Test, which help in this decision. Both tests are based on the following observations.

1. If a continuous function *f* is increasing on the portion of an interval *I* to the left of *c* and decreasing on the portion to the right of *c*, then $f(c)$ is the maximum value of *f* on *I* (Figure 4.33). Likewise, if *f* is decreasing on the portion of *I* to the left of *c* and increasing on the portion to the right of *c*, then $f(c)$ is the minimum value of *f* on *I* (Figure 4.34).

2. By Theorem 4.7 a continuous function *f* is increasing on an interval *I* if $f'(x) > 0$ for all interior points of *I*; similarly *f* is decreasing on an interval *I* if $f'(x) < 0$ for all interior points of *I*.

Before we state and prove the First Derivative Test, we introduce the following terminology. We say that the derivative f' **changes from positive to negative at *c*** if there exists some number $\delta > 0$ such that $f'(x) > 0$ for all *x* in $(c - \delta, c)$ and $f'(x) < 0$ for all *x* in $(c, c + \delta)$. The definition of f' **changes from negative to positive at *c*** results from replacing $f'(x) > 0$ by $f'(x) < 0$, and *vice versa*.

EXAMPLE 1 Let $f(x) = 4x^3 + 9x^2 - 12x + 3$. Determine where f' changes from positive to negative and where it changes from negative to positive.

Solution First we find the derivative of f:

$$f'(x) = 12x^2 + 18x - 12 = 6(2x^2 + 3x - 2) = 6(2x - 1)(x + 2)$$

To determine where f' changes sign, we assemble the chart in Figure 4.35.

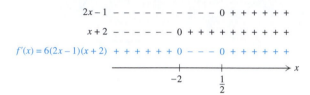

FIGURE 4.35

Consequently f' changes from positive to negative at -2 and from negative to positive at $\frac{1}{2}$. ❏

Now we are ready to state and prove the First Derivative Test.

THEOREM 4.10
First Derivative Test

Let f be differentiable on an open interval about the number c except possibly at c, where f is continuous.

  a. If f' changes from positive to negative at c, then f has a relative maximum value at c.

  b. If f' changes from negative to positive at c, then f has a relative minimum value at c.

Proof Because the proofs of the two parts are so similar, we prove only part (a). Since f' changes from positive to negative at c, there is a number $\delta > 0$ such that $f'(x) > 0$ for all x in $(c - \delta, c)$ and $f'(x) < 0$ for all x in $(c, c + \delta)$. By Theorem 4.7, f is increasing on $[c - \delta, c]$ and decreasing on $[c, c + \delta]$. Therefore $f(c)$ is the maximum value of f on $[c - \delta, c + \delta]$, and so f has a relative maximum value at c. ■

EXAMPLE 2 Let $f(x) = 4x^3 + 9x^2 - 12x + 3$. Show that f has a relative maximum value at -2 and a relative minimum value at $\frac{1}{2}$.

Solution By Example 1 we know that f' changes from positive to negative at -2. Thus the First Derivative Test implies that f has a relative maximum value at -2. Similarly, f' changes from negative to positive at $\frac{1}{2}$, so that f has a relative minimum value at $\frac{1}{2}$. ❏

Together, the First Derivative Test and Theorem 4.7 can help us sketch the graph of a continuous function f. The procedure is as follows. We compute the derivative of f and examine it:

  1. From Theorem 4.7, if $f'(x) > 0$ for all x interior to an interval, then f is increasing on that interval, whereas if $f'(x) < 0$ for all x interior to an interval, then f is decreasing on that interval (Figure 4.36).

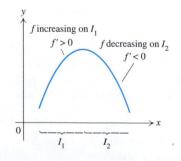

FIGURE 4.36

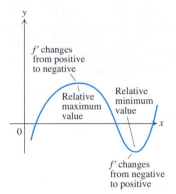

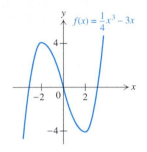

2. From the First Derivative Test, if f' changes from positive to negative at c, then f has a relative maximum value at c; if f' changes from negative to positive at c, then f has a relative minimum value at c (Figure 4.37).

EXAMPLE 3   Let $f(x) = \frac{1}{4}x^3 - 3x$. Find the relative extreme values of f, and then sketch the graph of f.

Solution   The derivative is given by

$$f'(x) = \frac{3}{4}x^2 - 3 = \frac{3}{4}(x^2 - 4) = \frac{3}{4}(x + 2)(x - 2)$$

The pertinent information about the sign of $f'(x)$ is summarized in Figure 4.38.

$$
\begin{array}{l}
x + 2 \quad - - - - - - \; 0 + + + + + + + + + + + \\
x - 2 \quad - - - - - - - - - - \; 0 + + + + + + \\
f'(x) = \frac{3}{4}(x+2)(x-2) \; + + + + + + \; 0 - - - \; 0 + + + + + + \\
\end{array}
$$

FIGURE 4.38

We conclude that f is increasing on $(-\infty, -2]$ and on $[2, \infty)$ and is decreasing on $[-2, 2]$. Furthermore, f' changes from positive to negative at -2 and from negative to positive at 2. Therefore by the First Derivative Test, $f(-2) = 4$ is a relative maximum value and $f(2) = -4$ is a relative minimum value of f. We can now make a fairly accurate sketch of the graph of f, as shown in Figure 4.39.   ❑

EXAMPLE 4   Let $f(x) = (x - 1)^2(x - 3)^2$. Sketch the graph of f.

Solution   First we find the derivative of f:

$$
\begin{aligned}
f'(x) &= 2(x - 1)(x - 3)^2 + 2(x - 1)^2(x - 3) \\
&= 2(x - 1)(x - 3)[(x - 3) + (x - 1)] \\
&= 4(x - 1)(x - 2)(x - 3)
\end{aligned}
$$

Then we prepare Figure 4.40 concerning the sign of $f'(x)$.

$$
\begin{array}{l}
x - 1 \quad - - - - \; 0 + + + + + + + + + + + + + \\
x - 2 \quad - - - - - - - - \; 0 + + + + + + + + \\
x - 3 \quad - - - - - - - - - - - - \; 0 + + + + \\
f'(x) = 4(x-1)(x-2)(x-3) \; - - - - \; 0 + + + \; 0 - - - \; 0 + + + + \\
\end{array}
$$

FIGURE 4.40

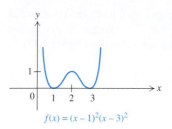

$$f(x) = (x-1)^2(x-3)^2$$

FIGURE 4.41

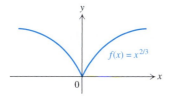

$$f(x) = x^{2/3}$$

FIGURE 4.42

The Second Derivative Test

From Figure 4.40 we find that f is increasing on the intervals $[1, 2]$ and $[3, \infty)$ and is decreasing on the intervals $(-\infty, 1]$ and $[2, 3]$. We also find that f' changes from negative to positive at 1 and at 3 and from positive to negative at 2. As a result, the First Derivative Test implies that $f(1) = 0$ and $f(3) = 0$ are relative minimum values and that $f(2) = 1$ is a relative maximum value. This information enables us to sketch the graph of f (Figure 4.41).   ❑

Sometimes we can conclude from the First Derivative Test that a relative extreme value exists at a number c even if $f'(c)$ does not exist. For example, if $f(x) = x^{2/3}$, then

$$f'(x) = \frac{2}{3}x^{-1/3} \quad \text{for } x \neq 0$$

Consequently $f'(x) < 0$ for $x < 0$, and $f'(x) > 0$ for $x > 0$. Thus f' changes from negative to positive at 0. Since f is continuous on $(-\infty, \infty)$, the First Derivative Test tells us that $f(0) = 0$ is a relative minimum value of f (Figure 4.42). Nevertheless, $f'(0)$ does not exist.

Now we present our second result on the location of relative extreme values. The result is called the Second Derivative Test because of the importance of the second derivative in the conditions of the theorem.

THEOREM   4.11
Second Derivative Test

> Assume that $f'(c) = 0$ and that $f''(c)$ exists.
>
> **a.** If $f''(c) < 0$, then $f(c)$ is a relative maximum value of f.
>
> **b.** If $f''(c) > 0$, then $f(c)$ is a relative minimum value of f.
>
> If $f''(c) = 0$, then from this test alone we cannot draw any conclusions about a relative extreme value of f at c.

Proof   We prove (a). By hypothesis,

$$f''(c) = \lim_{x \to c} \frac{f'(x) - f'(c)}{x - c} < 0$$

Since $f'(c) = 0$ by hypothesis, it follows that for all $x \neq c$ in some interval $(c - \delta, c + \delta)$,

$$\frac{f'(x)}{x - c} = \frac{f'(x) - f'(c)}{x - c} < 0$$

If $c - \delta < x < c$, then $x - c < 0$, so that $f'(x) > 0$. If $c < x < c + \delta$, then $x - c > 0$, so that $f'(x) < 0$. This means that f' changes from positive to negative at c. We conclude by the First Derivative Test that f has a relative maximum value at c. The proof of (b) is analogous to the proof of (a).

If both $f'(c) = 0$ *and* $f''(c) = 0$, it is possible for f to have a relative minimum value at c, or a relative maximum value at c, or neither (Figure 4.43(a)–(c)), so we cannot conclude anything about a possible relative extreme value of f at c.   ■

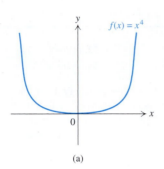

(a)

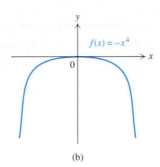

(b)

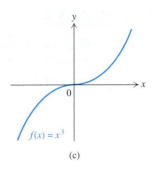

(c)

FIGURE 4.43

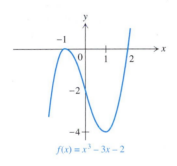

$f(x) = x^3 - 3x - 2$

FIGURE 4.44

EXAMPLE 5   Let $f(x) = x^3 - 3x - 2$. Using the Second Derivative Test, find the relative extreme values of f.

Solution   By differentiation we obtain

$$f'(x) = 3x^2 - 3 = 3(x - 1)(x + 1) \quad \text{and} \quad f''(x) = 6x$$

Therefore $f'(x) = 0$ when $x = -1$ or $x = 1$. Since

$$f''(-1) = -6 < 0 \quad \text{and} \quad f''(1) = 6 > 0$$

we know from the Second Derivative Test that $f(-1) = 0$ is a relative maximum value of f, whereas $f(1) = -4$ is a relative minimum value of f. These are the only relative extreme values of f. (The graph of f is shown in Figure 4.44.)   ❑

The Second Derivative Test does not always apply, as the following example shows.

EXAMPLE 6   Let $f(x) = 3x^4 - 4x^3$. Find the relative extreme values of f, and sketch the graph of f.

Solution   Differentiating, we have

$$f'(x) = 12x^3 - 12x^2 = 12x^2(x - 1) \tag{1}$$

and

$$f''(x) = 36x^2 - 24x$$

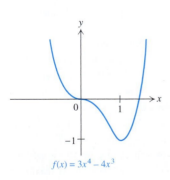

$f(x) = 3x^4 - 4x^3$

FIGURE 4.45

Consequently $f'(x) = 0$ for $x = 0$ or $x = 1$. Since $f''(1) = 12$, the Second Derivative Test tells us that $f(1) = -1$ is a relative minimum value of f. However, since $f''(0) = 0$, the Second Derivative Test cannot be applied to f at 0. But we observe from (1) that $f'(x) < 0$ for all x in $(-\infty, 1)$ except 0, so that f is decreasing on $(-\infty, 1)$. Therefore f cannot have an extreme value at 0. The information we now have leads us to draw the graph in Figure 4.45.   ❑

In Section 4.7 we will return to the function in Example 6 and study its graph further.

EXERCISES 4.5

In Exercises 1–8 determine the values of c at which f' changes from positive to negative, or from negative to positive.

1. $f(x) = x^2 + 6x - 11$ **2.** $f(x) = x^3 - x^2 - x + 2$
3. $f(x) = 2x^4 - 4x^2 + 3$ **4.** $f(x) = x/(x^3 - 2)$
5. $f(t) = \dfrac{t^2 - t + 1}{t^2 + t + 1}$ **6.** $f(t) = \dfrac{1}{\sqrt{t - t^2}}$
7. $f(t) = \sin t + \frac{1}{2}t$ **8.** $f(t) = \sin t - \cos t$

In Exercises 9–24 use the First Derivative Test to determine the relative extreme values (if any) of the function.

9. $f(x) = -3x^2 + 3x + 7$ **10.** $f(x) = x^3 - 12x + 2$
11. $f(x) = x^3 + 3x^2 + 4$ **12.** $f(x) = x^4 - 8x^2 + 1$
13. $g(x) = 4x^2 - 1/x$ **14.** $g(x) = 1/(x^2 + 1)$
15. $f(x) = x/(16 + x^3)$ **16.** $f(x) = \sqrt{|x|} + 1$
17. $f(x) = x\sqrt{1 - x^2}$ **18.** $f(x) = \dfrac{x^2 + x + 1}{x^2 - x + 1}$
19. $k(x) = \cos x + \frac{1}{2}x$ **20.** $k(x) = \sin x - \dfrac{\sqrt{3}}{2}x$
21. $k(x) = \sin\left(\dfrac{x^2}{1 + x^2}\right)$ **22.** $k(x) = \dfrac{\cos x}{1 + \sin x}$
23. $f(x) = x^2 e^{-x}$ **24.** $f(x) = \ln x - \ln(x^2 + 1)$

In Exercises 25–36 use the Second Derivative Test to determine the relative extreme values (if any) of the function.

25. $f(x) = -4x^2 + 3x - 1$ **26.** $f(x) = x^3 + 6x^2 + 9$
27. $f(x) = x^3 - 3x^2 - 24x + 1$ **28.** $f(x) = x^4 + \frac{1}{2}x$
29. $f(x) = 3x^4 - 4x^3 - \frac{9}{2}x^2 + \frac{1}{2}$ **30.** $g(x) = (x^2 + 2)^6$
31. $f(t) = t^2 + 1 + 1/t$ **32.** $f(t) = t^3 - 48/t^2$
33. $f(t) = \sin t + \cos t$ **34.** $f(t) = t + \cos 2t$
35. $f(t) = e^t - e^{-t}$ **36.** $f(t) = t^2 - 8 \ln t$

In Exercises 37–48 use the First Derivative Test or the Second Derivative Test to determine the relative extreme values, if any, of the function. Then sketch the graph of the function.

37. $f(x) = x^2 + 8x + 12$ **38.** $f(x) = x^3 - 3x$
39. $f(x) = x^3 + 3x$ **40.** $f(x) = 2x^3 - 3x^2 - 12x + 5$
41. $f(x) = x^4 + 4x$ **42.** $f(x) = x^5 - 5x$
43. $f(x) = (x^2 - 1)^2$ **44.** $f(x) = x^2(x + 3)^2$
45. $f(x) = (x - 2)^2(x + 1)^2$ **46.** $f(x) = \sqrt{x - x^2}$
47. $f(x) = e^x - x$ **48.** $f(x) = \ln(1 + x) + \ln(1 - x)$

In Exercises 49–54 find a point c such that $f'(c) = f''(c) = 0$. Using the derivative of f, show that f does not have a relative extreme value at c.

49. $f(x) = x^5$ **50.** $f(x) = x^{7/3}$

51. $f(x) = x^5 - x^3$ **52.** $f(x) = (x - 2)^3$
53. $f(x) = x^3 e^x$ **54.** $f(x) = x + \sin x$

In Exercises 55–58 plot the graph of f. Using the zoom feature of the calculator, approximate to within 0.1 all values of c such that $f(c)$ is a relative extreme value, and identify each as a relative maximum value or a relative minimum value.

55. $f(x) = x^5 - x^4 + x^2 - 1$
56. $f(x) = (10x + 2)/(x^4 + 1)$
57. $f(x) = 3 \sin x - x$ for $-2\pi \le x \le 2\pi$
58. $f(x) = e^x + \sin x$ for $-6 \le x \le 3$
59. Prove Theorem 4.11(b).
60. Let $g(x) = (x - 3)^2 (x - 5)^2$. Using only the graph of $(x - 1)^2 (x - 3)^2$ in Figure 4.41 (no calculator or derivative allowed), sketch the graph of g.
*61. Find a function f such that $f'(0) = 0$ and $f''(0)$ does not exist, and such that
  a. $f(0)$ is a relative minimum value
  b. $f(0)$ is not a relative extreme value

Projects

1. If $f'(c) = 0$, then we have two tests—the First Derivative Test and the Second Derivative Test—to help decide whether $f(c)$ is a relative extreme value. The question is: Which test should we use? The answer is that it depends on the function (e.g., how complicated f'' is, if it exists), and on our feeling. Yet the two tests are not equivalent. In fact, whenever the Second Derivative Test applies, so does the First Derivative Test. However, the converse is not valid in general. This project is devoted to a comparison of the two tests.

  a. Show that if we can deduce that $f(c)$ is a relative maximum value by the Second Derivative Test, then we can deduce it from the First Derivative Test.
  b. Give an example of a function f for which the First Derivative Test shows that $f(0)$ is a relative maximum value, but for which $f''(0) = 0$, so that the Second Derivative Test cannot be applied.
  c. Give an example of a function g for which the First Derivative Test shows that $g(0)$ is a relative maximum value, but for which $g''(0)$ does not exist, so that the Second Derivative Test cannot be applied.

2. Let r be a constant such that $0 < r \le 4$. If we define

$$f(x) = rx(1 - x) = rx - rx^2 \quad \text{for } 0 \le x \le 1$$

then $f(0) = 0 = f(1)$, and the graph of f has one hump (Figure 4.46). However, the composite function

$$(f \circ f)(x) = f(f(x)) = r f(x)(1 - f(x))$$
$$= r^2 x(1-x)(1 - rx + rx^2)$$

can have either no relative minimum value or a single relative minimum value in the open interval $(0, 1)$, depending on the value of r. This project will focus on the values of r for which $f \circ f$ has a relative minimum value in $(0, 1)$.

a. Show that $f'(1/2) = 0$ for any r.

b. Use the Chain Rule to prove that $(f \circ f)'(1/2) = 0$.

c. Write out $(f \circ f)$ as a fourth degree polynomial, and take the derivative $(f \circ f)'(x)$. Of course the derivative is a third-degree polynomial.

d. The Factor Theorem states that if g is a polynomial function of degree 3 and if $g(s) = 0$, then $g(x) = (x - s)h(x)$, where h is a degree 2 polynomial. Since we know that $(f \circ f)'(1/2) = 0$, it follows that

$$(f \circ f)'(x) = \left(x - \frac{1}{2}\right)(ax^2 + bx + c) = \left(x - \frac{1}{2}\right) h(x)$$

for appropriate constants a, b, and c. Find the values of a, b, and c.

e. Show that if $0 < r \le 2$, then $ax^2 + bx + c < 0$ for all x in $[0, 1]$, so that the only root of $(f \circ f)'$ is 1/2. Then use the First Derivative Test to show that f has a relative maximum value at 1/2. *Remark:* Since $(f \circ f)'$ has only one zero, and it corresponds to a relative maximum value of $f \circ f$, there is no relative minimum value for $f \circ f$ on $(0, 1)$.

f. Show that if $2 < r \le 4$, then $h(x) = 0$ for two values of x in $[0, 1]$. By using the Second Derivative Test, show that $(f \circ f)$ has a relative minimum value, and find that value.

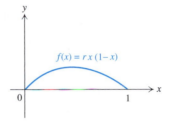

FIGURE 4.46

4.6   EXTREME VALUES ON AN ARBITRARY INTERVAL

The Maximum-Minimum Theorem applies to closed, bounded intervals. If the interval I is not closed or is not bounded, then a function f that is continuous on I does not automatically have a maximum (or a minimum) value on I. However, if f has a single critical number in I, then as we will see, another test may apply to yield an extreme value for f on I. Functions with a single critical number in a given interval arise in many applications, such as the determination of the thickness of insulation that minimizes the total cost of insulating and heating one's home. In this section we will analyze that application, as well as others that also are modeled by functions with a single critical number.

Throughout the section we will assume that the function f is continuous on an interval I. The procedure for finding extreme values of f on I rests on the following observation:

> If c is the only critical number of f in the interior of I and if $f(c)$ is a relative maximum (minimum) value, then $f(c)$ is the maximum (minimum) value of f on I.   (1)

Figure 4.47(a) and (b) supports this observation.

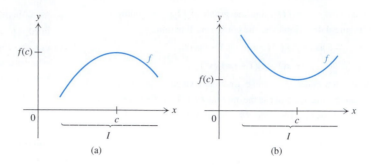

FIGURE 4.47   (a) A relative maximum is the maximum if there is only one critical point. (b) A relative minimum is the minimum if there is only one critical point.

Armed with (1) and the First and Second Derivative Tests, we have a method for determining extreme values of many functions f that are continuous on an arbitrary interval I:

> **1.** Find a number c that is the only critical number of f in I.
>
> **2.** Use the First or Second Derivative Test (if applicable) to verify that $f(c)$ is a relative maximum (or minimum) value.
>
> **3.** Conclude from (1) that $f(c)$ is the maximum (or minimum) value of f on I.

EXAMPLE 1   Assume that we wish to make a right cylindrical can with top and bottom, and with a capacity of 2000 cubic centimeters (that is, 2 liters). Find the dimensions that minimize the surface area.

FIGURE 4.48   A cylindrical can with a volume of 2000 cubic centimeters.

Solution   Let r and h denote the radius and height of the can (Figure 4.48), respectively. Let S denote the surface area of the can (including top and bottom), so that

$$
S = \overbrace{\pi r^2}^{\substack{\text{area of} \\ \text{top}}} + \overbrace{\pi r^2}^{\substack{\text{area of} \\ \text{bottom}}} + \overbrace{2\pi rh}^{\substack{\text{area of} \\ \text{vertical portion}}} = 2\pi r^2 + 2\pi rh \qquad (2)
$$

The goal is to minimize S. In order to obtain S as a function of one variable, say r, we will express h in terms of r. Let V represent the volume of the can, so that $V = \pi r^2 h$. By hypothesis, $V = 2000$, so that $2000 = \pi r^2 h$ and thus

$$
h = \frac{2000}{\pi r^2} \qquad (3)
$$

Substituting for h in (2) yields

$$
S = 2\pi r^2 + 2\pi r\left(\frac{2000}{\pi r^2}\right) = 2\pi r^2 + \frac{4000}{r} \quad \text{for } r > 0
$$

To find the minimum value of S as a function of r, we first differentiate S:

$$
S'(r) = 4\pi r - \frac{4000}{r^2}
$$

It follows that $S'(r) = 0$ if (and only if)

$$4\pi r^3 = 4000, \quad \text{or equivalently,} \quad r = \frac{10}{\sqrt[3]{\pi}}$$

Thus $10/\sqrt[3]{\pi}$ is the only critical number of S as a function of r. Notice that

$$S''(r) = 4\pi + \frac{8000}{r^3} > 0 \quad \text{for all } r > 0$$

By the Second Derivative Test, $S(10/\sqrt[3]{\pi})$ is a relative minimum value of S. Since S has only one critical number on $(0, \infty)$, it follows from (1) that $S(10/\sqrt[3]{\pi})$ is the minimum value of S. By (3) the corresponding value of h is given by

$$h = \frac{2000}{\pi(10/\sqrt[3]{\pi})^2} = \frac{20}{\sqrt[3]{\pi}}$$

As a result, the surface area of the can is minimum if the radius is $10/\sqrt[3]{\pi}$ centimeters and the height is $20/\sqrt[3]{\pi}$ centimeters. ❑

EXAMPLE 2 At noon a sailboat is 20 kilometers south of a freighter. The sailboat is traveling east at 20 kilometers per hour, and the freighter is traveling south at 40 kilometers per hour. If visibility is 10 kilometers, could the people on the two ships ever see each other?

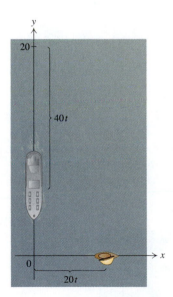

FIGURE 4.49

Solution   We measure time in hours, and let $t = 0$ correspond to noon. Then at any time $t \geq 0$, the sailboat has traveled $20t$ kilometers and the freighter has traveled $40t$ kilometers. From Figure 4.49 we see that at time t the distance D between the two ships is given by

$$D = \sqrt{(20t)^2 + (20 - 40t)^2} \quad \text{for } t \geq 0$$

The goal is to determine whether $D \leq 10$ for any such value of t, and to accomplish this, we will find the minimum value of D. But notice that the minimum value of D occurs for the same value of t as the minimum value of D^2, the square of the distance. Moreover, the derivative of D^2 is easier to obtain than the derivative of D. So we let $E = D^2$, which means that for $t \geq 0$,

$$E = (20t)^2 + (20 - 40t)^2 = 400[t^2 + (1 - 2t)^2] = 400(1 - 4t + 5t^2)$$

Then

$$E'(t) = 400(-4 + 10t) = 800(5t - 2)$$

Therefore $E'(t) = 0$ if

$$5t - 2 = 0, \quad \text{or equivalently,} \quad t = \frac{2}{5}$$

Since $E'(t) < 0$ for $0 < t < \frac{2}{5}$ and $E'(t) > 0$ for $t > \frac{2}{5}$, it follows from the First Derivative Test that $E(\frac{2}{5})$ is a relative minimum value. Thus it is the minimum value of E since $\frac{2}{5}$ is the only critical number. Consequently $D(\frac{2}{5})$ is the minimum value of D. Next,

$$D\left(\frac{2}{5}\right) = \sqrt{\left(20 \cdot \frac{2}{5}\right)^2 + \left(20 - 40 \cdot \frac{2}{5}\right)^2} = \sqrt{8^2 + 4^2} = \sqrt{80} < 10$$

so we conclude that the people on the two ships could see each other. In particular, they could see each other $\frac{2}{5}$ of an hour after noon, that is, at 12:24 P.M. ❏

EXAMPLE 3 According to the theory of quantum mechanics, the probability function for an electron in the ground state of a hydrogen atom to be located at a distance r from the center of the nucleus is given by

$$P(r) = \frac{4r^2}{a^3} e^{-2r/a} \quad \text{for } r > 0 \tag{4}$$

Here a is a constant that is approximately 5.29×10^{-11}, and r is measured in meters. Find the most probable distance of the electron from the center.

Solution Mathematically we are asked to find the value of r that maximizes P. From (4) we find that

$$P'(r) = \frac{8r}{a^3} e^{-2r/a} + \frac{4r^2}{a^3} e^{-2r/a}\left(-\frac{2}{a}\right) = \frac{8r}{a^3}\left(1 - \frac{r}{a}\right) e^{-2r/a}$$

Thus $P'(r) = 0$ only for $r = a$, so that a is the only critical number of P. Since $P'(r) > 0$ for $r < a$ and $P'(r) < 0$ for $r > a$, it follows from the First Derivative Test that P has a relative maximum value at a. We conclude from (1) that P is maximum for $r = a$. (The graph of P is shown in Figure 4.50.) ❏

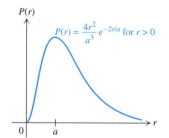

$P(r)$

$P(r) = \dfrac{4r^2}{a^3} e^{-2r/a}$ for $r > 0$

FIGURE 4.50

EXAMPLE 4 A forest ranger is in a forest 2 miles from a straight road. A car is located 5 miles down the road (Figure 4.51). If the forest ranger can walk 3 miles per hour in the forest and 4 miles per hour along the road, toward what point on the road should the ranger walk in order to minimize the time needed to walk to the car?

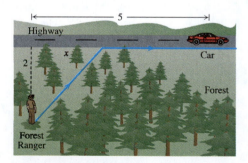

FIGURE 4.51

Solution Let x be the distance shown in Figure 4.51. Then by the Pythagorean Theorem the ranger walks $\sqrt{x^2 + 4}$ miles in the forest. The ranger also walks $5 - x$ miles along the road. Using the formula

$$\text{time} = \frac{\text{distance}}{\text{speed}}$$

twice (once for the forest walk and once for the road walk), we find that the total time T the ranger walks is given by

$$T = \overbrace{\frac{\sqrt{x^2 + 4}}{3}}^{\text{time in forest}} + \overbrace{\frac{5 - x}{4}}^{\text{time on road}} \quad \text{for } 0 \le x \le 5$$

Since

$$T'(x) = \frac{2x}{2 \cdot 3\sqrt{x^2 + 4}} - \frac{1}{4} = \frac{x}{3\sqrt{x^2 + 4}} - \frac{1}{4}$$

it follows that $T'(x) = 0$ if

$$3\sqrt{x^2 + 4} = 4x$$

Since x must be nonnegative, when we solve for x we obtain

$$9(x^2 + 4) = 16x^2$$

$$7x^2 = 36$$

$$x = \frac{6}{7}\sqrt{7}$$

Thus $\frac{6}{7}\sqrt{7}$ is the only critical number of T. Since

$$T''(x) = \frac{3\sqrt{x^2 + 4} - x(3x/\sqrt{x^2 + 4})}{9(x^2 + 4)} = \frac{3(x^2 + 4) - 3x^2}{9(x^2 + 4)^{3/2}} = \frac{4}{3(x^2 + 4)^{3/2}}$$

we find that $T''(\frac{6}{7}\sqrt{7}) > 0$. By the Second Derivative Test, $T(\frac{6}{7}\sqrt{7})$ is a relative minimum value. Thus by (1), $T(\frac{6}{7}\sqrt{7})$ is the minimum value of T since $\frac{6}{7}\sqrt{7}$ is the only critical number. Therefore the total walking time is minimized if the ranger walks toward the point $\frac{6}{7}\sqrt{7}$ miles (approximately 2.3 miles) down the road.   ❑

Since the domain of T in Example 4 is really $[0, 5]$, we could have solved the problem by finding the critical number $\frac{6}{7}\sqrt{7}$ of T on $[0, 5]$ and then using the procedure described in Section 4.1: computing $T(0)$, $T(\frac{6}{7}\sqrt{7})$, and $T(5)$ and determining the smallest of these three values. The answer would have been the same—that $T(\frac{6}{7}\sqrt{7})$ is the minimum value of T on $[0, 5]$. Thus when the interval I is closed and bounded, we can sometimes use more than one method in determining extreme values.

In our final applied problem, we will study the financial effect of placing insulation on a bare attic floor.

Let $C_i(x)$ be the purchase price in dollars of insulation x inches thick. We make the reasonable assumption that

$$C_i(x) = ax \quad \text{for } x > 0$$

where a is a positive constant. If $C_h(x)$ denotes the cost per year of heating the home with this insulation, then we assume further that

$$C_h(x) = \frac{b}{x} \quad \text{for } x > 0$$

where b is also a positive constant.* Over a 10-year period, the total cost $C(x)$ to the homeowner will be

$$C(x) = C_i(x) + 10C_h(x) = ax + \frac{10b}{x} \quad \text{for } x > 0$$

The value of x that yields the minimum cost $C(x)$ over a 10-year period represents the ideal amount of insulation the homeowner should buy. With this in mind we state and solve the following example, in which we assume that the attic floor has an area of 1000 square feet, $a = 50$, and $b = 295$, which would be reasonable in a region like Washington, D.C.

EXAMPLE 5   Find the value of x that minimizes the cost $C(x)$, where

$$C(x) = 50x + \frac{2950}{x} \quad \text{for } x > 0$$

Solution   Differentiating, we obtain

$$C'(x) = 50 - \frac{2950}{x^2} \quad \text{and} \quad C''(x) = \frac{5900}{x^3}$$

It follows that $C'(x) = 0$ only for $x = \sqrt{\frac{2950}{50}} = \sqrt{59}$, so that $\sqrt{59}$ is the only critical number of C. Since $C''(x) > 0$ for $x > 0$, it follows from the Second Derivative Test that $C(\sqrt{59})$ is a relative minimum value of C. We conclude from (1) that the minimum cost $C(x)$ occurs when $x = \sqrt{59}$ (inches), that is, for approximately 7.7 inches of insulation.   ❏

Our idealized formula in the insulation example does not take into account all possible variables. For instance, it does not reflect installation costs or the fact that some heat is lost through the walls and floor. Nevertheless, the problem as stated and solved is representative of fairly common conditions.

---

* We will tell how to compute a and b for a particular case. First, $a = qA$, where q is the cost in dollars per square foot of insulation 1 inch thick and A is the area of the attic floor in square feet. Second, $b = dcAp$, where d is the number of degree days in the area, c is the conductivity of the insulation, and p is the price per British thermal unit of fuel. If the price in dollars per therm on the gas bill is p_0, then

$$p = \frac{3}{2} p_0 \left(\frac{1}{100,000} \right)$$

In Washington, D.C., values for 1990 were approximately as follows: $d = 4200$, $c = 6.5$, $p_0 = 0.57$, and $q = .05$. If the attic has 1000 square feet of floor, then a is approximately 50 and b is approximately 295. You should be able to find out the number of degree days and the price p_0 in your area from your utility company.

General Procedure for Applied Extreme Value Problems

There is a general procedure we have used in solving the applied problems in this section. Below we list the major features of the procedure as a guide for you in solving applied problems involving extreme values.

1. After reading the problem carefully, choose a letter for the quantity to be maximized or minimized, and choose auxiliary variables for the other quantities appearing in the problem.

2. Express the quantity to be maximized or minimized in terms of the auxiliary variables. A diagram is often useful.

3. Choose one auxiliary variable, say x, to serve as master variable, and use the information given in the problem to express all other auxiliary variables in terms of x. Again a diagram may be helpful.

4. Use the results of the preceding steps to express the given quantity to be maximized or minimized as a function, say f, of x alone. Note the values of x that are pertinent to the problem.

5. Determine the critical numbers of f.

6. Use the theory of this chapter to find the desired maximum or minimum value of f. This usually involves applying (1) in conjunction with the First Derivative Test or the Second Derivative Test. If the domain of f is the closed interval $[a, b]$, this can be accomplished alternatively by evaluating f at the critical numbers and at a and b.

It Pays to Check!

According to a rumor attributed to the great physicist Richard Feynman, an Air Force scientist once proposed an exotic wing shape to maximize the range of the airplane. Unfortunately, he forgot to check that his solution really yielded the maximum range. The outcome was that his design gave the minimum range, not the maximum range.

EXERCISES 4.6

1. In an autocatalytic chemical reaction a substance A is converted into a substance B in such a manner that

$$\frac{dx}{dt} = kx(a - x)$$

where x is the concentration of substance B at time t, a is the initial concentration of substance A, and k is a positive constant. Determine the value of x at which the rate dx/dt of the reaction is maximum.

2. If $C(x)$ is the cost of manufacturing an amount x of a given product and p is the price per unit amount, then the profit $P(x)$ obtained by selling an amount x is

$$P(x) = px - C(x)$$

(Notice that there is a loss if $P(x)$ is negative.)

a. If $C(x) = cx$ and $c < p$, is there a maximum profit?

b. If $C(x) = (x - 1)^2 + 2$, find the maximum profit.

3. Consider the circuit shown in Figure 4.52, consisting of a battery having a constant source voltage E, constant internal resistance r, and a variable external resistance R. When current flows through the circuit, the power P dissipated in the external resistance is given by

$$P = \frac{E^2 R}{(R + r)^2}$$

Assume that E and r are positive constants. Show that the largest power dissipation occurs when $R = r$.

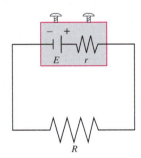

FIGURE 4.52   Figure for Exercise 3.

4. The power output of an electric generator is VI, where V is the constant terminal voltage and I is the variable current. The power loss is $P + I^2R$, where P is constant and I^2R is the power lost to heat through the internal resistance R. The efficiency E of the generator is given by

$$E = \frac{\text{power output}}{\text{power input}} = \frac{VI}{VI + P + I^2R}$$

Assume that V, P, and R are positive constants. Find the current for which the efficiency is maximum.

5. For an electron in the $2p$ state of an excited hydrogen atom, the probability function for the electron to be located at a distance r from the atom's center is given by

$$P(r) = \frac{\pi r^4}{6a^5} e^{-r/a} \quad \text{for } r > 0$$

Find the most probable distance of the electron from the center of the atom.

6. According to one model, the time rate R at which a tumor grows is given by

$$R = Ax \ln \frac{B}{x} \quad \text{for } 0 < x < B \qquad (5)$$

where A and B are positive constants, and x is the radius of the tumor.

   a. Use a graphics calculator to view the graph of R for the case $A = B = 5$. From the graph, do you think that (5) is a reasonable model for tumor growth? Explain your answer.

   b. Find the radius of the tumor at which the tumor is growing most rapidly with respect to time.

7. In a transatlantic cable let x be the ratio of the radius of the core to the thickness of the covering. It can be shown that the speed s of a signal passing through the cable is given by

$$s = -kx^2 \ln x \quad \text{for } 0 < x < 1$$

where k is a positive constant. What value of x will maximize s?

8. The concentration of a drug in the blood stream t seconds after injection into a muscle is given by

$$y = c(e^{-bt} - e^{-at}) \quad \text{for } t \geq 0$$

where a, b, and c are positive constants with $a > b$.

   a. Find the time at which the concentration is maximum.

   b. Find the time at which the concentration of the drug in the blood stream is decreasing most rapidly.

9. Find the two positive numbers whose sum is 18 and whose product is as large as possible.

10. Find the two real numbers whose difference is 16 and whose product is as small as possible.

11. A crate open at the top has vertical sides, a square bottom, and a volume of 4 cubic meters. If the crate has the least possible surface area, find its dimensions.

12. Suppose the crate in Exercise 11 has a top. Find the dimensions of the crate with minimum surface area.

13. An outdoor track is to be created in the shape of a rectangle with semicircles at two opposite ends. If the perimeter of the track is 440 yards, find the dimensions of the track for which the area of rectangular portion is maximized.

14. A Norman window is a window in the shape of a rectangle with a semicircle attached at the top (Figure 4.53). Assuming that the perimeter of the window is 12 feet, find the dimensions that allow the maximum amount of light to enter.

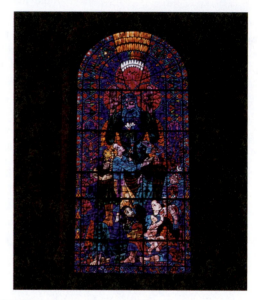

FIGURE 4.53   Figure for Exercise 14. Canterbury Cathedral, England. (*E. Carle/Superstock*)

15. Suppose a window has the shape of a rectangle with an equilateral triangle attached at the top. Assuming that the perimeter of the window is 12 feet, find the dimensions that allow the maximum amount of light to enter.

16. At 3 P.M. an oil tanker traveling west in the ocean at 15 kilometers per hour passes the same point as a luxury liner that arrived at the same spot at 2 P.M. while traveling north at 25 kilometers per hour. At what time were the ships closest together?

17. A horse breeder plans to set aside a rectangular region of 1 square kilometer for horses and wishes to build a wooden fence to enclose the region. Since one side of the region will run along a well-traveled highway, the breeder decides to make that side more attractive, using wood that costs three times as much per meter as the

wood for the other sides. What dimensions will minimize the cost of the fence?

18. A manufacturer wishes to produce rectangular containers with square bottoms and tops, each container having a capacity of 250 cubic inches. If the material used for the top and the bottom costs twice as much per square inch as the material for the sides, what dimensions will minimize the cost?

In Exercise 19 we present a mathematical problem that arises in two completely different settings (see Exercises 20 and 21).

19. Let p, q, and r be positive constants with $q < r$, and let

$$f(\theta) = p - q \cot \theta + \frac{r}{\sin \theta} \quad \text{for } 0 < \theta < \frac{\pi}{2}$$

Show that f has a minimum value on $(0, \pi/2)$ at the value of θ for which $\cos \theta = q/r$.

20. This problem derives from the biological study of vascular branching. Assume that a major blood vessel A leads away from the heart (P in Figure 4.54) and that in order for the heart to feed an organ at R, there must be an auxiliary artery somewhere between P and Q. The resistance $\mathcal{R}$ of the blood as it flows along the path PSR is given by

$$\mathcal{R}(\theta) = k\frac{a - b \cot \theta}{r_1^4} + \frac{kb}{r_2^4 \sin \theta} \quad \text{for } 0 < \theta < \frac{\pi}{2}$$

where k, a, b, r_1, and r_2 are positive constants with $r_1 > r_2$ (see Figure 4.54). Where should the contact at S be made to produce the least resistance? (*Hint*: Using the result of Exercise 19, find the cosine of the angle θ for which $\mathcal{R}(\theta)$ is minimized.)

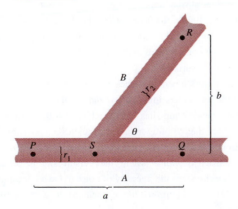

FIGURE 4.54   Figure for Exercise 20.

21. A bee's cell in a hive is a regular hexagonal prism open at the front, with a trihedral apex at the back (the top in Figure 4.55). It can be shown that the surface area of a cell with apex θ is given by

$$S(\theta) = 6ab + \frac{3}{2}b^2\left(-\cot \theta + \frac{\sqrt{3}}{\sin \theta}\right) \quad \text{for } 0 < \theta < \frac{\pi}{2}$$

where a and b are positive constants. Show that the surface area is minimized if $\cos \theta = 1/\sqrt{3}$, so that $\theta \approx 54.7°$. (*Hint*: Use the result of Exercise 19.) Experiments have shown that bee cells have an average angle within $2'$ (less than one tenth of one degree) of $54.7°$.

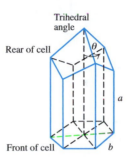

FIGURE 4.55   Figure for Exercise 21.

22. A ring of radius a carries a uniform electric charge Q. The electric field intensity at any point x along the axis of the ring is given by

$$E(x) = \frac{Qx}{(x^2 + a^2)^{3/2}}$$

At what point on the axis is the electric field the greatest?

23. If the electric charge is uniformly distributed throughout a circular cylinder (such as a telephone wire) of radius a, then at any point whose distance from the axis of the cylinder is r, the electric field intensity is given by

$$E(r) = \begin{cases} cr & \text{for } 0 \leq r \leq a \\ ca^2/r & \text{for } r > a \end{cases}$$

where c is a positive constant.

a. Show that $E(r)$ is maximum for $r = a$.

b. Is E differentiable at a? Explain your answer.

24. Find the points on the line $y = 2x - 4$ that are closest to the point $(1, 3)$.

25. Find the points on the parabola $y = x^2 + 2x$ that are closest to the point $(-1, 0)$.

26. Of all the triangles that pass through the point $(1, 1)$ and have two sides lying on the coordinate axes, one has the

smallest area. Determine the lengths of its sides.

27. Consider Figure 4.56, consisting of the parabola $y = x^2$ and an arbitrary point $(0, p)$ on the y axis.

   a. For which values of p is the origin the only point on the parabola that is closest to $(0, p)$?

   b. For which values of p are there two points on the parabola that are closest to $(0, p)$?

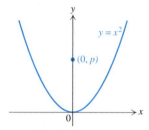

FIGURE 4.56   Graph for Exercise 27.

28. A 12-foot wire is cut into 12 pieces, which are soldered together to form a rectangular frame whose base is twice as long as it is wide (as in Figure 4.57). The frame is then covered with paper.

FIGURE 4.57   Figure for Exercise 28.

   a. How should the wire be cut if the volume of the frame is to be maximized?

   b. How should the wire be cut if the total surface area of the paper is to be maximized?

29. Toward what point on the road should the ranger in Example 4 walk in order to minimize the travel time to the car if the car is located

   a. 10 miles down the road?

   b. $\frac{1}{2}$ mile down the road?

   c. an arbitrary number c of miles down the road?

30. It is known that homing pigeons fly faster over land than over water. Assume that they fly 10 meters per second over land but only 8 meters per second over water. If a pigeon is located at the edge of a straight river 500 meters

wide and must fly to its nest, located 1300 meters away on the opposite side of the river (Figure 4.58), what path would minimize its flying time?

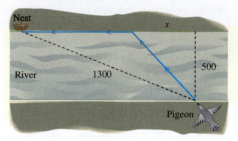

FIGURE 4.58   Figure for Exercise 30.

31. A rectangular printed page is to have margins 2 inches wide at the top and the bottom and margins 1 inch wide on each of the two sides. If the page is to have 35 square inches of printing, determine the minimum possible area of the page itself.

32. A cylindrical can with bottom but no top has volume V. Find the radius of the can with the smallest possible surface area.

33. A cylinder is inscribed in a sphere with radius R. Find the height of the cylinder with the maximum possible volume.

34. Find the radius of the cone with given volume V and minimum surface area. (*Hint:* The surface area S of a cone with radius r and height h is given by $S = \pi r \sqrt{r^2 + h^2}$.)

35. Find the length of the longest thin, rigid pipe that can be carried from one 10-foot-wide corridor to a similar corridor at right angles to the first. Assume that the pipe has negligible diameter. (*Hint:* Find the length of the shortest line that touches the inside corner of the hallways and extends to the two walls. Use an angle as an auxiliary variable.)

36. After work a person wishes to sit in a large park along a path 300 meters long. At the ends of the path there are two construction sites, one of which is 8 times as noisy as the other. In order to have the quietest repose, how far from the quieter site should the person sit? (*Hint:* The intensity of noise where the person sits is directly proportional to the intensity of noise at the source and is inversely proportional to the square of the distance from the source.)

37. A real estate firm can borrow money at 5% interest per year and can lend the money out. If the amount of money it can lend is inversely proportional to the square of the interest rate at which it lends, what interest rate would maximize the firm's profit per year? (*Hint:* Let x be the loan interest rate. Notice that the profit is the product of the amount borrowed by the firm and the difference between the interest rates at which it lends and borrows.)

38. A company has a daily fixed cost of $5000. If the company produces x units daily, then the daily cost in dollars for labor and materials is $3x$. The daily cost of equipment maintenance is $x^2/2{,}500{,}000$. What daily production minimizes the total daily cost per unit of production? (*Hint:* The cost per unit is the total cost $C(x)$ divided by x.)

39. A company sells 1000 units of a certain product annually, with no seasonal fluctuations in demand. It always reorders the same number x of units, stocks unsold units until no more remain, and then reorders again. If it costs b dollars to stock one unit for 1 year and there is a fixed cost of c dollars each time the company reorders, how many units should be reordered each time to minimize the total annual cost of reordering and stocking? (*Hint:* The company will have an average inventory of $x/2$ units and must reorder $1000/x$ times per year. Find the annual stocking and reordering costs and minimize their sum.)

*40. A farmer wishes to employ tomato pickers to harvest 62,500 tomatoes. Each picker can harvest 625 tomatoes per hour and is paid $6 per hour. In addition, the farmer must pay a supervisor $10 per hour and pay the union $10 for each picker employed.

   a. How many pickers should the farmer employ to minimize the cost of harvesting the tomatoes?

   b. What is the minimum cost to the farmer?

*41. Suppose light travels from point A in a first medium (such as air) to point B in a second medium (such as water). Assume the velocity of the light in the first medium is u, and in the second medium is v. According to Fermat's Principle, the light travels the path requiring the least time. Using Fermat's Principle and Figure 4.59,

show that the light is bent at the boundary of the two mediums according to Snell's Law:

$$\frac{\sin\theta}{u} = \frac{\sin\phi}{v}$$

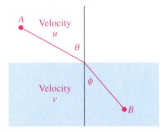

FIGURE 4.59   Snell's Law:

$$\frac{\sin\theta}{u} = \frac{\sin\phi}{v}$$

See Exercise 42.

Project

1. Most post offices in the United States have the following limit on the size of a parcel that can be mailed by parcel post: The sum of the length of its longest side and its girth (the largest perimeter of a cross-section perpendicular to the longest side) can be no more than 108 inches.

   a. Find the dimensions of the rectangular parallelepiped with a square base having the largest volume that can be mailed. (There are two cases to be considered, depending on which side is longest.)

   b. Find the dimensions of the right circular cylinder having the largest volume that can be mailed. (Again, there are two cases to consider.)

   c. Find the dimensions of the cube having the largest volume that can be mailed.

   d. Show that it is possible for a parcel to be mailable and yet have a larger volume than a parcel that is not mailable. (*Hint:* Examine your solutions to (a)–(c).)

4.7   CONCAVITY AND INFLECTION POINTS

Notice that in Figure 4.60(a), if we slide the tangent line along the graph from left to right, the slope of the tangent line increases. Consequently f' seems to be increasing. In a similar way, the slope of the tangent line in Figure 4.60(b) decreases as the tangent slides from left to right, and as a result f' appears to decrease. Knowing those intervals on which f' increases and those on which f'

decreases facilitates graphing the function *f*. These concepts will help us analyze the logistic curve, which appears in the study of population growth.

The observations in the preceding paragraph lead us to make the following definition.

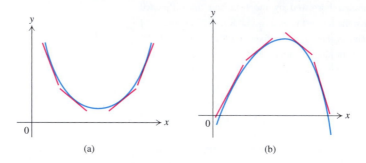

(a)   (b)

FIGURE 4.60   (a) The slope of the tangent line increases. (b) The slope of the tangent line decreases.

DEFINITION   4.12

Let *f* be differentiable on an open interval *I*. The graph of *f* is **concave upward on *I*** provided that f' is increasing on *I*. The graph of *f* is **concave downward on *I*** provided that f' is decreasing on *I*.

By Theorem 4.7, if $f''(x) > 0$ for all *x* in *I*, then f' is increasing on *I*, so that the graph of *f* is concave upward on *I*. Analogously, if $f''(x) < 0$ for all *x* in *I*, then f' is decreasing on *I*, so that the graph of *f* is concave downward on *I*. These facts constitute the proof of the following theorem.

THEOREM 4.13

Assume that f'' exists on an open interval *I*.
   a. If $f''(x) > 0$ for all *x* in *I*, then the graph of *f* is concave upward on *I*.
   b. If $f''(x) < 0$ for all *x* in *I*, then the graph of *f* is concave downward on *I*.

By using Theorem 4.13, we see that the graph in Figure 4.61 represents $f(x) = x^2$ and that the graph in Figure 4.62 does not. This is true because $f''(x) = 2$ for all *x*, so that the graph of *f* is concave upward on $(-\infty, \infty)$. More generally, if $f(x) = ax^2 + bx + c$, where $a \neq 0$, then $f''(x) = 2a$. Thus the graph of *f* is either concave upward on $(-\infty, \infty)$ or concave downward on $(-\infty, \infty)$. The graph of such a function is called a **parabola,** one of the basic types of conic sections, which we will study in Section 10.3.

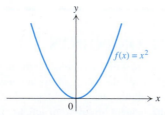

FIGURE 4.61

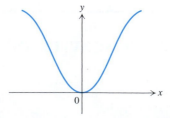

FIGURE 4.62   Not the graph of $y = x^2$.

Next we look at third- and fourth-degree polynomials.

EXAMPLE 1   Let $f(x) = 3x^4 - 4x^3$. Find the intervals on which the graph of f is concave upward and those on which it is concave downward. Then sketch the graph of f.

Solution   From Example 6 of Section 4.5 we know that

$$f'(x) = 12x^3 - 12x^2 = 12x^2(x - 1)$$

$$f''(x) = 36x^2 - 24x = 12x(3x - 2)$$

and that $f(1) = -1$ is a relative minimum value of f. Now we determine the sign of $f''(x)$ from Figure 4.63.

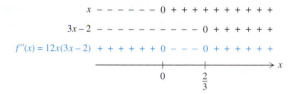

FIGURE 4.63

Using the sign of $f''(x)$ along with Theorem 4.13, we deduce that the graph of f is concave upward on $(-\infty, 0)$ and on $(\frac{2}{3}, \infty)$ and is concave downward on $(0, \frac{2}{3})$. From this information we conclude that the graph of f is as shown in Figure 4.64.   ❑

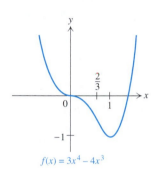

$f(x) = 3x^4 - 4x^3$

FIGURE 4.64

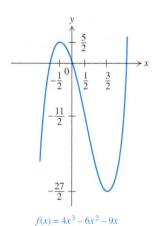

$f(x) = 4x^3 - 6x^2 - 9x$

FIGURE 4.65

Inflection Points

EXAMPLE 2   Let $f(x) = 4x^3 - 6x^2 - 9x$. Find the intervals on which the graph of f is concave upward and those on which it is concave downward. Then sketch the graph of f.

Solution   Differentiating f, we have

$$f'(x) = 12x^2 - 12x - 9 = 3(4x^2 - 4x - 3)$$

$$= 3(2x - 3)(2x + 1) = 12\left(x - \frac{3}{2}\right)\left(x + \frac{1}{2}\right)$$

Next we find that

$$f''(x) = 24x - 12 = 24\left(x - \frac{1}{2}\right)$$

By Theorem 4.13 the graph of f is concave downward on $(-\infty, 1/2)$ and concave upward on $(\frac{1}{2}, \infty)$. From the first derivative we know that the critical numbers are $-\frac{1}{2}$ and $\frac{3}{2}$. Since $f''(-\frac{1}{2}) = -24$ and $f''(\frac{3}{2}) = 24$, it follows from the Second Derivative Test that $f(-\frac{1}{2}) = \frac{5}{2}$ is a relative maximum value of f and $f(\frac{3}{2}) = -\frac{27}{2}$ is a relative minimum value of f. We can now draw the graph of f (Figure 4.65).   ❑

When an epidemic (like a flu epidemic) strikes a city, the disease spreads slowly at first, because there are so few infected people to spread the disease. Then as time increases, the disease spreads faster because there are more infected people to

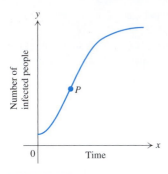

FIGURE 4.66

spread the disease. At some moment, however, the rate at which the disease is spreading begins to decrease and approach 0, because there are so few susceptible uninfected people remaining. The curve in Figure 4.66, which plots the number of infected people against time, models the spread of the disease and is related to the logistic curve, which we will discuss in Example 6.

If the curve in Figure 4.66 is the graph of a function *f*, what we have just observed is equivalent to saying that to the left of the special point *P* the graph of *f* is concave upward and to the right of *P* the graph is concave downward. Knowing points on a graph at which the concavity changes in this way facilitates drawing the graph. This leads us to the following definition.

DEFINITION   4.14   Assume that there is a (possibly vertical) line tangent to the graph of *f* at $(c, f(c))$. Then $(c, f(c))$ is an **inflection point** of the graph of *f* if there is a number $\delta > 0$ such that the graph of *f* is concave upward on $(c - \delta, c)$ and concave downward on $(c, c + \delta)$ (or vice versa).

Figure 4.67 illustrates the relationship between concavity and inflection points. Observe that an inflection point is a point on the graph of the function in question.

Assume that f'' exists and is continuous on an interval containing *c*. Assume also that the graph of *f* has an inflection point at $(c, f(c))$, so there is a change of concavity at $(c, f(c))$. Because of the continuity of f'' it is possible to show that $f''(c) = 0$ (see Exercise 51). Thus we are led to the following method of finding inflection points for many functions:

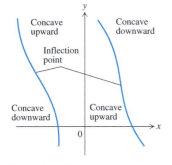

FIGURE 4.67

> **How to Find Inflection Points**
> 1. Find the values of *c* for which $f''(c) = 0$.
> 2. For each value of *c* found in step 1, determine whether f'' changes sign at *c*.
> 3. If f'' changes sign at *c*, the point $(c, f(c))$ is an inflection point.

EXAMPLE 3   Let $f(x) = x^4 - 6x^2 + 8x + 10$. Find the inflection points of the graph of *f*.

Solution   The derivatives of *f* are

$$f'(x) = 4x^3 - 12x + 8$$

$$f''(x) = 12x^2 - 12 = 12(x + 1)(x - 1)$$

Figure 4.68 shows the sign of $f''(x)$.

FIGURE 4.68

Observe that $f''(1) = f''(-1) = 0$ and that f'' changes sign at both 1 and -1. It follows that $(1, f(1)) = (1, 13)$ and $(-1, f(-1)) = (-1, -3)$ are inflection points.   ❑

EXAMPLE 4   Let $f(x) = \sin x$ for $-2\pi \leq x \leq 2\pi$. Find the inflection points of the graph of f, discuss its concavity, and sketch it.

Solution   The derivatives are

$$f'(x) = \cos x \quad \text{and} \quad f''(x) = -\sin x$$

Now $f''(x) = 0$ for $x = -\pi$, 0, and π. Since f'' changes sign at each of these values of x, it follows that $(-\pi, 0)$, $(0, 0)$, and $(\pi, 0)$ are inflection points. Moreover, the graph of f is concave downward on $(-2\pi, -\pi)$ and $(0, \pi)$ and concave upward on $(-\pi, 0)$ and $(\pi, 2\pi)$. The graph of the sine function appeared in Figure 1.68(a); we show it again in Figure 4.69, restricted in domain to the interval $[-2\pi, 2\pi]$.   ❑

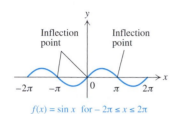

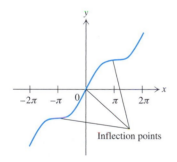

$f(x) = \sin x$ for $-2\pi \leq x \leq 2\pi$

FIGURE 4.69

Notice that the graph of the sine function in Figure 4.69 has an inflection point wherever it crosses the x axis. Since the sine function (without restricted domain) is periodic with period 2π, its graph therefore has an inflection point wherever it crosses the x axis, that is, at $(n\pi, 0)$ for any integer n. This shows that the graph of a function can have infinitely many inflection points.

EXAMPLE 5   Let $g(x) = x + \sin x$. Find the inflection points of the graph of g, discuss its concavity, and sketch it.

Solution   Differentiating g, we have

$$g'(x) = 1 + \cos x \quad \text{and} \quad g''(x) = -\sin x$$

If $f(x) = \sin x$, then $g''(x) = f''(x)$, so g'' changes sign exactly where f'' does. Hence the discussion of the sine function just after Example 4 tells us that $(n\pi, g(n\pi)) = (n\pi, n\pi)$ is an inflection point of the graph of g for any integer n. We also deduce that the graph of g is concave downward on each interval $(2n\pi, (2n + 1)\pi)$ and concave upward on each interval $((2n + 1)\pi, (2n + 2)\pi)$. Does g have any relative extreme values? Note that $g'(x) = 0$ only if $\cos x = -1$, which means $x = (2n + 1)\pi$ for some integer n. Moreover, $g'(x) > 0$ unless $x = (2n + 1)\pi$. It follows from Theorem 4.7 that g is increasing on each bounded closed interval and hence on $(-\infty, \infty)$. Thus g has *no* relative extreme values. The graph of g is drawn in Figure 4.70.   ❑

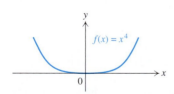

Inflection points

FIGURE 4.70

Caution: The graph of f need not have an inflection point at $(c, f(c))$ simply because $f''(c) = 0$. Indeed, if $f(x) = x^4$, then $f''(x) = 12x^2 \geq 0$ for all x, so f'' does not change sign at all, and in particular, it does not change sign at 0. Consequently $(0, 0)$ is not an inflection point, even though $f''(0) = 0$ (Figure 4.71).

$f(x) = x^4$

FIGURE 4.71   No inflection points.

Inflection points are interesting in applications as well as in graphing functions. Consider the following example.

EXAMPLE 6   Consider $f(x) = \dfrac{a}{1 + be^{-kax}}$ for $x \geq 0$, where a and k are positive numbers and $b > 1$. Find the inflection point of the graph of f and show that its y coordinate is $a/2$.

Solution   Taking derivatives, we find that

$$f'(x) = \frac{-a(be^{-kax})(-ka)}{(1 + be^{-kax})^2} = a^2bk\frac{e^{-kax}}{(1 + be^{-kax})^2}$$

and

$$f''(x) = a^2bk\frac{-ake^{-kax}(1 + be^{-kax})^2 - (e^{-kax})2(1 + be^{-kax})(-abke^{-kax})}{(1 + be^{-kax})^4}$$

$$= a^2bk\frac{ake^{-kax}(-1 + be^{-kax})}{(1 + be^{-kax})^3}$$

It follows that $f''(x)$ changes from positive to negative at the number x such that

$$be^{-kax} = 1, \qquad \text{or equivalently,} \qquad b = e^{kax}$$

Let x_0 be this value. Then $\ln b = kax_0$, so that

$$x_0 = \frac{\ln b}{ka}$$

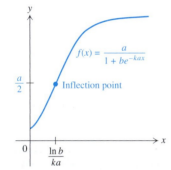

$f(x) = \dfrac{a}{1 + be^{-kax}}$

Inflection point

FIGURE 4.72   The logistic curve.

Consequently there is an inflection point at $(x_0, f(x_0))$. Substituting $(\ln b)/ka$ for x in the equation for f, we obtain

$$f\left(\frac{\ln b}{ka}\right) = \frac{a}{1 + be^{-ka[(\ln b)/ka]}} = \frac{a}{1 + b(1/b)} = \frac{a}{2}$$

Therefore the y coordinate of the inflection point is $a/2$. The graph of f is shown in Figure 4.72.   ❑

The function f in Example 6 was introduced in 1838 by the Belgian mathematician P. F. Verhulst in order to model population growth of paramecia. The graph of f is known as the **logistic curve.** Notice that since $f'(x) > 0$ and $f''(x) > 0$ for $0 < x < x_0$, it follows that if x represents time, then the population grows at an ever-increasing rate for $0 < x < x_0$. By contrast, if $x > x_0$, then $f'(x) > 0$ and $f''(x) < 0$, so that the population continues to grow but at an ever-decreasing rate. Thus x_0 represents the time when the population is growing the fastest. It could correspond to the moment when space begins to be a factor in inhibiting population growth.

One could also model an epidemic of, say, influenza, in a city with a logistic curve, where the carrying capacity denotes the number of people in the city who are susceptible. Health officials would be very much interested in the point at which the rate of infection begins decreasing. That point corresponds to the inflection point of the logistic curve.

Curves that are concave upward are common in economics texts. For example, the curve that portrays interest rates as a function of the amount of money people want to save or invest is concave upward and is represented in Figure 4.73.

Finally we include Figure 4.74 to illustrate and compare several concepts related to concavity and inflection points.

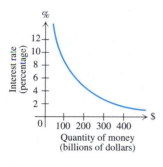

FIGURE 4.73

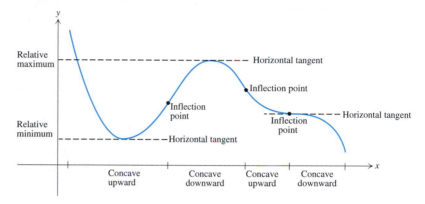

FIGURE 4.74

EXERCISES 4.7

In Exercises 1–10 find the intervals on which the graph of the function is concave upward and those on which it is concave downward.

1. $f(x) = -\frac{3}{2}x^2 + x$

2. $f(x) = x^2 + 2x + 4$

3. $f(x) = x^3 - 6x^2 + 12x - 4$

4. $f(x) = x^4 - 6x^2 + 8$

5. $g(x) = \dfrac{x}{x^2 + 1}$

6. $g(x) = x\sqrt{x-1}$

7. $g(x) = xe^x$

8. $g(x) = x^2 e^{-x}$

9. $f(x) = x \ln x$

10. $f(x) = x^2 + \ln x$

In Exercises 11–20 find the intervals on which the graph of the function is concave upward and those on which it is concave downward. Then sketch the graph of the function.

11. $f(x) = x^3 + 8$

12. $f(x) = x^3 - 6x^2 + 9x + 2$

13. $g(x) = x^4 - 4x$

14. $g(x) = 3x^5 - 5x^3$

15. $f(x) = x + \dfrac{1}{x}$

16. $f(x) = x\sqrt{x^2 - 4}$

17. $f(x) = \sin 2x$

18. $f(x) = \cos \frac{1}{2}x - 1$

19. $f(x) = \sec x$

20. $f(x) = \csc \frac{1}{4}x$

In Exercises 21–32 find all inflection points (if any) of the graph of the function. Then sketch the graph of the function.

21. $f(x) = (x + 2)^3$

22. $f(x) = x^3 + 3$

23. $f(x) = x^3 + 3x^2 - 9x - 2$

24. $f(x) = x^3 - \frac{3}{2}x^2 - 6x$

25. $g(x) = 3x^4 + 4x^3$

26. $g(x) = x^4 - 2x^3$

27. $g(x) = x^9 - 3x^3$

28. $g(x) = x^{1/3}$

29. $g(x) = \frac{2}{3}x^{2/3} - \frac{3}{5}x^{5/3}$

30. $g(x) = x\sqrt{1 - x^2}$

31. $f(t) = \tan t$

*32. $f(t) = t - \cos t$

In Exercises 33–35 plot the graph of f'', and then use the Newton-Raphson method to approximate all values of c for which $(c, f(c))$ is an inflection point. Continue until the output of the calculator does not change.

33. $f(x) = x^7 - x^5 + x^3 + 3x^2$

34. $f(x) = \sin x^2$ for $-\pi/2 \le x \le \pi/2$

35. $f(x) = e^x - \sin x$ for $-e \le x < \infty$

36. Use Figure 4.75 to determine the intervals on which the graph is concave upward and the intervals on which the graph of f is concave downward. Assume that the curve in the figure is the graph of

a. f **b.** f' **c.** f''

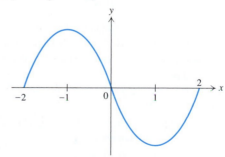

FIGURE 4.75   Graph for Exercise 36.

37. Suppose the curve in Figure 4.76 is the graph of f'. Determine at which of the numbers a, b, c, and d we could have the following:

a. f has its largest value

b. f has its smallest value

c. f'' has its largest value

d. f'' has its smallest value

e. the graph of f has an inflection point

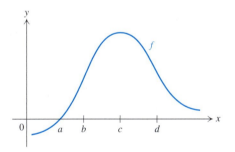

FIGURE 4.76   Graph for Exercise 37.

38. Suppose the curve in Figure 4.77 is the graph of f. Determine at which of the numbers a, b, c, and d we have the following:

a. f' has its largest value

b. f' has its smallest value

c. the graph of f has an inflection point

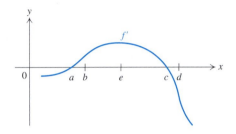

FIGURE 4.77   Graph for Exercise 38.

39. Consider the function f in Figure 4.78.

a. On which intervals (with integer endpoints) does f'' appear to be negative, and on which does it appear to be positive?

b. For which points $(x, f(x))$ with integer values of x does it appear that f has an inflection point?

c. Suppose that $g' = f$. For which integer values of x is $(x, g(x))$ an inflection point of g?

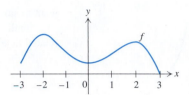

FIGURE 4.78   Graph for Exercise 39.

40. In each of the following, draw the graph of a continuous function f having the given properties.

a. f is increasing and its graph is concave upward on $(-\infty, 0)$, and f is decreasing and its graph is concave downward on $(0, \infty)$.

b. f is decreasing and its graph is concave upward on $(-\infty, 1)$, f is increasing and its graph is concave upward on $(1, 2)$, and f is decreasing and its graph is concave upward on $(2, \infty)$.

c. f is decreasing and its graph is concave upward on $(-\infty, 1)$, f is increasing and its graph is concave upward on $(1, 2)$, and f is increasing and its graph is concave downward on $(2, \infty)$.

d. f is decreasing and its graph is concave downward on $(-\infty, 0)$, f is increasing and its graph is concave downward on $(0, 1)$, f is increasing and its graph is concave upward on $(1, 5)$, and f is decreasing and its graph is concave downward on $(5, \infty)$.

41. Let $f(x) = 1/(1 + x^2)$ and $g(x) = e^{-x^2/2}$. Plot the graphs of f and g on $[-2, 2]$, and determine which graph is more concave near the point $(0, 1)$. Then justify your answer by comparing $f''(0)$ and $g''(0)$.

42. Recall that a function f is even if $f(-x) = f(x)$ for all x, and f is odd if $f(-x) = -f(x)$ for all x.

a. If f is even and its graph is concave upward on $(0, \infty)$, what is the concavity on $(-\infty, 0)$?

b. If f is odd and its graph is concave upward on $(0, \infty)$, what is the concavity on $(-\infty, 0)$?

43. How does the concavity of the graph of a function f compare with the concavity of the graph of $-f$?

44. Show by giving an example that the graph of the function fg need not be concave upward on an open interval I even if the graph of f is concave upward on I and the graph of g is concave upward on I.

45. Let $f(x) = x^{1/3}$. Show that $(0, 0)$ is an inflection point of the graph of f, although neither $f'(0)$ nor $f''(0)$ exists. (Thus it is not absolutely necessary for either the first or the second derivative to exist in order to have an inflection point.)

46. Let $f(x) = x^5 - cx^3$, where c is a constant. Show that the graph of f has an inflection point at $(0, 0)$.

 47. Let

$$g(x) = \frac{1}{20} x^5 + \frac{1}{4} x^4 + \frac{1}{2} cx^2$$

Use your graphics calculator (or computer) to graph g'' for $c = -5$, $c = -4$, $c = -1$, $c = 0$, and $c = 1$. What can you tell about the possible number of inflection points for the graph of g?

48. Let n be a positive integer and $f(x) = x^n$. Show that the graph of f has at most one inflection point. Determine those values of n for which the inflection point exists, and find the inflection point.

49. It is known that a polynomial of degree k can have at most k real zeros. Use this fact to determine the maximum number of inflection points of the graph of a polynomial of degree n, where $n \geq 2$.

*50. Let $f(x) = \dfrac{\sin x}{x}$ for $x > 0$.

  a. Show that $f''(x) = g(x)/x^3$, where

  $$g(x) = -x^2 \sin x - 2x \cos x + 2 \sin x$$

  b. Use Theorem 4.7 to show that $g(x) < 0$ for $0 < x < \pi/2$. Use this result to prove that the graph of f is concave downward on $(0, \pi/2)$.

  c. Sketch the graph of f. (*Hint:* Make use of (b) and the fact that as x increases, the graph oscillates between the graphs of $y = 1/x$ and $y = -1/x$. Also use the fact that the graph of f touches the graph of $y = 1/x$ where $\sin x = 1$ and touches the graph of $y = -1/x$ where $\sin x = -1$. Note also that $f'(\pi/2) = -4/\pi^2$ and $f'(x) \neq 0$ for all x in $(0, \pi/2)$, since $\tan x > x$. Thus f decreases on $(0, \pi/2)$.)

51. Show that if f'' exists and is continuous on an open interval containing c and if f has an inflection point at $(c, f(c))$, then $f''(c) = 0$.

52. Let $h(t)$ denote the height at time t of an object moving in a vertical direction above the ground. Assume that the object is subject only to the force of gravity.

  a. Show that the graph of h is concave downward.

  b. In terms of the behavior of the velocity v, tell why the result of (a) is true.

Project

1. In this project we analyze the concavity and potential number of inflection points for polynomials up to degree 5.

  a. Let $f(x) = ax^2 + bx + c$, with $a \neq 0$. Show that f has one extreme value, and its graph *cannot* have any inflection points.

  b. Let $f(x) = ax^3 + bx^2 + cx + d$, with $a \neq 0$. Show that f always has one inflection point. Then write down an example in which f has 2 relative extreme values, and a second example in which f has *no* relative extreme values.

  c. Let $f(x) = ax^4 + bx^3 + cx^2 + dx + r$, with $a \neq 0$. Determine the possible number of relative extreme values, and the possible number of inflection points. Does f necessarily have any extreme values? Explain your answer.

  d. Let $f(x) = ax^5 + bx^4 + cx^3 + dx^2 + rx + s$, with $a \neq 0$. Determine the possible number of relative extreme values, and the possible number of inflection points. Give examples of as many varieties as you can.

  e. From the information you have assembled in (a)–(d), suppose you have a polynomial of degree n. What can you say about the possible number of relative extreme values and the possible number of inflection points if

  i. n is an even integer?

  ii. n is an odd integer?

4.8 LIMITS AT INFINITY

Voyager 2 was launched with the idea that it would fly past the outer planets of our solar system, leave the solar system and, in essence, fly infinitely far away. One question faced by scientists was how fast the spacecraft would have to travel (shortly after lift-off) in order to achieve this goal. In order to answer this question we will consider the limit of $f(x)$ as x is positive and becomes arbitrarily large, that

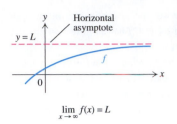

$$\lim_{x \to \infty} f(x) = L$$

FIGURE 4.79

is, as x approaches ∞. Analogously, we will consider the limit of $f(x)$ as x is negative and $|x|$ becomes arbitrarily large, that is, as x approaches $-\infty$. A knowledge of such limits will aid us in sketching the parts of the graph of f that are far from the y axis and can be important in applications (see Example 4).

We can think of the limit of $f(x)$ as x approaches ∞ as a kind of left-hand limit because x approaches ∞ from the left (Figure 4.79). However, we must find a way of expressing mathematically how "near" a number x is to ∞. To do so, we notice that the larger x is, the nearer it is to ∞. Analogous statements can be made for x approaching $-\infty$. These observations suggest the following definition.

DEFINITION 4.15

a. Let f be defined on an interval (a, ∞). A number L is the **limit of $f(x)$ as x approaches** ∞ if for every $\varepsilon > 0$ there is a number M such that

$$\text{if } x > M, \text{ then } |f(x) - L| < \varepsilon$$

In this case we write

$$\lim_{x \to \infty} f(x) = L$$

We say that the **limit of $f(x)$ exists as x approaches** ∞, or that f **has a limit at** ∞.

b. Let f be defined on an interval $(-\infty, a)$. A number L is the **limit of $f(x)$ as x approaches** $-\infty$ if for every $\varepsilon > 0$ there is a number M such that

$$\text{if } x < M, \text{ then } |f(x) - L| < \varepsilon$$

In this case we write

$$\lim_{x \to -\infty} f(x) = L$$

We say that the **limit of $f(x)$ exists as x approaches** $-\infty$ or that f **has a limit at** $-\infty$.

c. If either $\lim_{x \to \infty} f(x) = L$ or $\lim_{x \to -\infty} f(x) = L$, then we call the horizontal line $y = L$ a **horizontal asymptote of the graph of f** (Figure 4.79).

The number M in Definition 4.15 corresponds to the number δ in all other definitions of limits given so far. We think of x as "close to" ∞ when $x > M$ and M is a large number, just as we say that x is "close to" a when $a - \delta < x < a$ and δ is a small positive number. If $\lim_{x \to \infty} f(x) = L$ (or if $\lim_{x \to -\infty} f(x) = L$), then the point $(x, f(x))$ approaches the asymptote $y = L$ as x becomes arbitrarily large positively (or negatively).

From Definition 4.15 it follows that

$$\lim_{x \to \infty} \frac{1}{x} = 0 \quad \text{and} \quad \lim_{x \to -\infty} \frac{1}{x} = 0$$

(Figure 4.80).

The basic limit theorems remain true for limits at ∞ or $-\infty$. Thus the limit as x approaches ∞ or $-\infty$ is unique when it exists. Furthermore, if $\lim_{x \to \infty} f(x)$ and $\lim_{x \to \infty} g(x)$ exist, we have

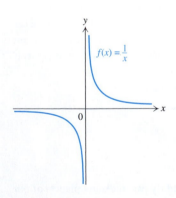

$f(x) = \frac{1}{x}$

FIGURE 4.80

$$\lim_{x \to \infty} [f(x) + g(x)] = \lim_{x \to \infty} f(x) + \lim_{x \to \infty} g(x)$$

$$\lim_{x \to \infty} f(x)g(x) = \lim_{x \to \infty} f(x) \lim_{x \to \infty} g(x)$$

There are corresponding formulas for limits at $-\infty$.

From the formulas for the limits of a product we conclude that if n is any positive integer, then

$$\lim_{x\to\infty}\frac{1}{x^n}=0 \quad\text{and}\quad \lim_{x\to-\infty}\frac{1}{x^n}=0 \tag{1}$$

Indeed,

$$\lim_{x\to\infty}\frac{1}{x^n}=\lim_{x\to\infty}\overbrace{\left(\frac{1}{x}\cdot\frac{1}{x}\cdots\frac{1}{x}\right)}^{n\text{ factors}}=\overbrace{\left(\lim_{x\to\infty}\frac{1}{x}\right)\left(\lim_{x\to\infty}\frac{1}{x}\right)\cdots\left(\lim_{x\to\infty}\frac{1}{x}\right)}^{n\text{ factors}}$$

$$=0\cdot 0\cdots 0=0$$

The limit at $-\infty$ follows analogously.

EXAMPLE 1   Find the following limits.

a. $\displaystyle\lim_{x\to\infty}\frac{x^2+2x-5}{2x^2-6x-1}$ **b.** $\displaystyle\lim_{x\to-\infty}\frac{3x+1}{4-x^3}$

Solution   **a.** We divide both numerator and denominator by x^2, which is the highest power of x in the denominator, and then use (1) along with the limit theorems. We obtain

$$\lim_{x\to\infty}\frac{x^2+2x-5}{2x^2-6x-1}=\lim_{x\to\infty}\frac{\dfrac{x^2}{x^2}+\dfrac{2x}{x^2}-\dfrac{5}{x^2}}{\dfrac{2x^2}{x^2}-\dfrac{6x}{x^2}-\dfrac{1}{x^2}}=\lim_{x\to\infty}\frac{1+\dfrac{2}{x}-\dfrac{5}{x^2}}{2-\dfrac{6}{x}-\dfrac{1}{x^2}}$$

$$=\frac{1+0-0}{2-0-0}=\frac{1}{2}$$

b. As in part (a), we divide by the largest power of x appearing in the denominator, this time x^3:

$$\lim_{x\to-\infty}\frac{3x+1}{4-x^3}=\lim_{x\to-\infty}\frac{\dfrac{3x}{x^3}+\dfrac{1}{x^3}}{\dfrac{4}{x^3}-\dfrac{x^3}{x^3}}=\lim_{x\to-\infty}\frac{\dfrac{3}{x^2}+\dfrac{1}{x^3}}{\dfrac{4}{x^3}-1}=\frac{0-0}{0-1}=0 \qquad \square$$

EXAMPLE 2   Let $a>0$. Show that $\lim_{x\to\infty}(\sqrt{x^2-a^2}-x)=0$.

Solution   First we rewrite the expression in the limit:

$$\sqrt{x^2-a^2}-x=(\sqrt{x^2-a^2}-x)\frac{\sqrt{x^2-a^2}+x}{\sqrt{x^2-a^2}+x}$$

$$=\frac{x^2-a^2-x^2}{\sqrt{x^2-a^2}+x}=\frac{-a^2}{\sqrt{x^2-a^2}+x}$$

$$=\frac{-a^2/x}{\sqrt{1-(a^2/x^2)}+1}$$

Since

$$\lim_{x \to \infty} \frac{-a^2/x}{\sqrt{1 - (a^2/x^2)} + 1} = \frac{0}{\sqrt{1 - 0} + 1} = 0$$

it follows that

$$\lim_{x \to \infty} (\sqrt{x^2 - a^2} - x) = 0 \qquad \square$$

Pictorially Example 2 tells us that far to the right of the *y* axis the points on the graphs of $y = \sqrt{x^2 - a^2}$, and $y = x$ are arbitrarily close together. (See the project for this section.)

Analogously, the line $y = a$ is a horizontal asymptote of a function f if far to the right or to the left of the *y* axis the points of the graphs f and $y = a$ are arbitrarily close together. In Example 3 we will determine the asymptotes of a graph.

EXAMPLE 3 Let

$$f(x) = \frac{2x + 1}{3x - 1}$$

Sketch the graph of f.

Solution   We first find the derivatives of f:

$$f'(x) = \frac{2(3x - 1) - 3(2x + 1)}{(3x - 1)^2} = \frac{-5}{(3x - 1)^2}$$

$$f''(x) = \frac{30}{(3x - 1)^3}$$

Consequently f is decreasing on $(-\infty, \frac{1}{3})$ and on $(\frac{1}{3}, \infty)$, and its graph is concave downward on $(-\infty, \frac{1}{3})$ and concave upward on $(\frac{1}{3}, \infty)$. For the asymptotes we observe that

$$\lim_{x \to 1/3^+} f(x) = \lim_{x \to 1/3^+} \frac{2x + 1}{3x - 1} = \infty$$

$$\lim_{x \to 1/3^-} f(x) = \lim_{x \to 1/3^-} \frac{2x + 1}{3x - 1} = -\infty$$

$$\lim_{x \to \infty} f(x) = \lim_{x \to -\infty} f(x) = \frac{2}{3}$$

Therefore the line $x = \frac{1}{3}$ is a vertical asymptote, and the line $y = \frac{2}{3}$ is a horizontal asymptote. This information enables us to sketch the graph (Figure 4.81). $\square$

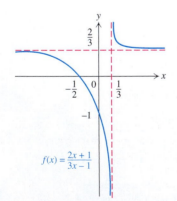

$$f(x) = \frac{2x + 1}{3x - 1}$$

FIGURE 4.81

Since limits are unique, the graph of a function cannot have more than two horizontal asymptotes, one arising from $\lim_{x \to \infty} f(x)$ and one from $\lim_{x \to -\infty} f(x)$. (See Exercises 48 and 49 for examples in which two distinct horizontal asymptotes actually arise.) Thus for a given function, there may be 0, 1, or 2 horizontal asymptotes. If p and q are polynomial functions and f the rational function defined by $f(x) = p(x)/q(x)$, then we can be more specific about possible horizontal asymptotes of the graph of f:

degree of p < degree of q: one horizontal asymptote, $y = 0$

degree of p = degree of q: one horizontal asymptote, $y = c$, with $c \neq 0$

degree of p > degree of q: no horizontal asymptote

Escape Velocity

Let us suppose that a spacecraft such as Voyager 2, which was launched September 20, 1977, escapes the earth's gravity and drifts infinitely far from earth. Then the velocity v of the spacecraft remains positive as a function of the distance r from the center of the earth. In the absence of any additional rocket boost, the velocity decreases as the distance r increases. The smallest initial velocity the spacecraft can have and still escape the earth's gravity is called the **escape velocity** of the earth. We will denote it by v_{esc}. In mathematical terms, v_{esc} is the unique initial velocity for which

$$\lim_{r \to \infty} v(r) = 0 \qquad (2)$$

In fact, if $\lim_{r \to \infty} v(r) > 0$, then a smaller initial velocity would allow the spacecraft to escape. Our goal is to determine the value of v_{esc}, under the assumption that the spacecraft is subject only to the earth's gravitational force.

Let the mass and radius of the earth be M and R, respectively, and let m be the mass of the spacecraft. The Law of Conservation of Energy implies that the total energy (comprised of the kinetic energy and the potential energy) of the spacecraft must remain constant if the craft is subject only to gravitational force. Thus there is a constant c such that

$$\overbrace{\frac{1}{2} m[v(r)]^2}^{\text{kinetic energy}} + \overbrace{\frac{-GmM}{r}}^{\text{potential energy}} = c \quad \text{for all } r \geq R \qquad (3)$$

Now we are prepared to find a formula for the escape velocity of a spacecraft.

EXAMPLE 4 Use (2) and (3) to determine the escape velocity v_{esc} of a spacecraft leaving the earth's surface.

Solution First we notice that $\lim_{r \to \infty} GmM/r = 0$. Next, by (2) we have

$$\lim_{r \to \infty} \frac{1}{2} m[v(r)]^2 = 0$$

Using this information, along with (3), we deduce that

$$c = \lim_{r \to \infty} \left\{ \frac{1}{2} m[v(r)]^2 - \frac{GmM}{r} \right\} = \lim_{r \to \infty} \frac{1}{2} m[v(r)]^2 - \lim_{r \to \infty} \frac{GmM}{r} = 0 - 0 = 0$$

Thus $c = 0$. Setting $c = 0$ in (3) and solving for $v(r)$, we find that

$$v(r) = \sqrt{\frac{2GM}{r}} \quad \text{for } r \geq R$$

The Voyager spacecraft, an example of an interplanetary robot craft. Voyager 2, launched in 1977, has explored all four jovian planets and their ring and satellite systems.

Since the spacecraft has velocity v_{esc} when $r = R$, we conclude that

$$v_{\text{esc}} = \sqrt{\frac{2GM}{R}} \qquad \square \tag{4}$$

It follows from (4) that the escape velocity of the earth depends on the mass and radius of the earth. However, it does *not* depend on the mass of the spacecraft. With the accepted values of

$$G = 6.67 \times 10^{-11} \ (\text{N} \cdot \text{m}^2/\text{kg}^2), \quad M = 5.98 \times 10^{24} \ (\text{kg}), \quad \text{and} \quad R = 6.37 \times 10^6 \ (\text{m})$$

we find that $v_{\text{esc}} \approx 11.2$ kilometers per second, or approximately 25,000 miles per hour.

Actually the preceeding analysis completely ignored the sun. The escape velocity of approximately 25,000 miles per hour that we obtained is only sufficient to allow a spacecraft to escape the earth's gravity and travel out into the solar system. The reality is that the sun is massive enough, with a mass $M_s \approx 2 \cdot 10^{30}$ kilograms, to demonstrably affect the velocity required for escape of a spacecraft into deep space—even though the distance r_s between the sun and the earth is approximately 93,000,000 miles (or more precisely, 1.5×10^{11} meters).

In order to take the sun's gravity into account, we would need to add the potential energy of the sun to the left side of (3), yielding

$$\frac{1}{2} m [v(r)]^2 + \frac{-GmM}{r} + \frac{-GmM_s}{r_s} = c \quad \text{for all} \quad r \geq R$$

From calculations like those in the solution of Example 4, and again letting $r = R$, we deduce a modified form of (4) for the velocity required to escape from the solar system, which we will denote by V_{esc}:

$$V_{\text{esc}} = \sqrt{\frac{2GM}{6.37 \times 10^6} + \frac{2GM_s}{1.5 \times 10^{11}}}$$

Using the approximate values for G, M, and M_s given above, we find that $V_{\text{esc}} \approx$ 43.6 kilometers per second, or about 97,000 miles per hour. So you see that the effect of the sun's gravity is powerful.

Infinite Limits at Infinity

We conclude this section by examining a type of limit that combines infinite limits with limits at ∞.

DEFINITION   4.16

Let f be defined on an interval (a, ∞). If for any real number N there is some number M such that

$$\text{if } x > M, \quad \text{then } f(x) > N$$

we say that the **limit of $f(x)$ as x approaches ∞ is ∞** and write

$$\lim_{x \to \infty} f(x) = \infty$$

(see Figure 4.82). Alternatively, we say that f **has an infinite limit at ∞.**

The definitions of

$$\lim_{x \to \infty} f(x) = -\infty, \quad \lim_{x \to -\infty} f(x) = \infty, \quad \text{and} \quad \lim_{x \to -\infty} f(x) = -\infty$$

are completely analogous.

Using Definition 4.16, one can show that for any positive integer n,

$$\lim_{x \to \infty} x^n = \infty$$

Furthermore,

$$\lim_{x \to -\infty} x^n = \begin{cases} \infty & \text{for } n \text{ even} \\ -\infty & \text{for } n \text{ odd} \end{cases}$$

It follows that if p is a polynomial of degree at least 1, then

$$\lim_{x \to \infty} p(x) = \infty \quad \text{or} \quad \lim_{x \to \infty} p(x) = -\infty$$

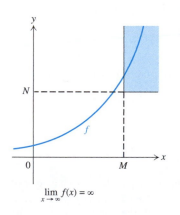

$$\lim_{x \to \infty} f(x) = \infty$$

FIGURE 4.82

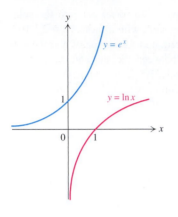

FIGURE 4.83

and

$$\lim_{x \to -\infty} p(x) = \infty \quad \text{or} \quad \lim_{x \to -\infty} p(x) = -\infty$$

Which infinite limits occur depends on the coefficient of the term in p with the highest power. For example,

$$\lim_{x \to \infty} (x - x^2) = -\infty \quad \text{and} \quad \lim_{x \to -\infty} (x - x^2) = -\infty$$

whereas

$$\lim_{x \to \infty} (x^3 - 2x + 1) = \infty \quad \text{and} \quad \lim_{x \to -\infty} (x^3 - 2x + 1) = -\infty$$

Now we turn to limits at ∞ associated with the logarithm and exponential functions. From the graphs of e^x and $\ln x$ (Figure 4.83), it is clear that

$$\lim_{x \to \infty} e^x = \infty \tag{5}$$

$$\lim_{x \to -\infty} e^x = 0 \tag{6}$$

$$\lim_{x \to \infty} \ln x = \infty \tag{7}$$

The limits in (5) and (7) can be verified by observing that the exponential and logarithm functions are increasing and noting that for any positive integer n,

$$e^{\ln n} = n \quad \text{and} \quad \ln e^n = n$$

The limit in (6) follows in a similar way from the fact that for any positive integer n,

$$e^{-\ln n} = \frac{1}{n}$$

Finally we comment on the relevance of horizontal asymptotes to the logistic curve, which was defined in Section 4.7 to be the graph of the function

$$f(x) = \frac{a}{1 + be^{-kax}} \quad \text{for } x \geq 0$$

where a, b, and k are positive constants with $b > 1$. From (6) we deduce that

$$\lim_{x \to \infty} e^{-kax} = 0$$

As a result,

$$\lim_{x \to \infty} f(x) = \lim_{x \to \infty} \frac{a}{1 + be^{-kax}} = a$$

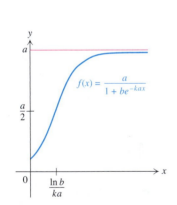

FIGURE 4.84

(see Figure 4.84). If we consider $f(x)$ as representing the population of a colony of bacteria at time x, then the number a to which the population tends with time is called the **carrying capacity** of the medium in which the population lives. The carrying capacity is the theoretical upper limit of the population. In Exercise 58 we study the population of paramecia in a saline solution, as analyzed in 1934 by the biologist G. F. Gause.

EXERCISES 4.8

In Exercises 1–22 find the given limit.

1. $\lim\limits_{x\to\infty} 2/(x-3)$ **2.** $\lim\limits_{x\to\infty} 4/(2-x)$

3. $\lim\limits_{x\to\infty} \dfrac{x}{3x+2}$ **4.** $\lim\limits_{x\to\infty} \dfrac{4x^2}{\sqrt{2}\,x-3}$

5. $\lim\limits_{x\to\infty} \dfrac{2x^2+x-1}{x^2-x+4}$ **6.** $\lim\limits_{t\to\infty} \dfrac{(t-1)(2t+1)}{(3t-2)(t+4)}$

7. $\lim\limits_{t\to\infty} \dfrac{t}{t^{1/2}+2t^{-1/2}}$ **8.** $\lim\limits_{x\to\infty} \dfrac{\sin x}{x}$

9. $\lim\limits_{x\to-\infty} \dfrac{\cos x}{\sqrt{x^2-1}}$ **10.** $\lim\limits_{x\to-\infty} \dfrac{x-\frac{1}{2}}{\frac{1}{2}x+1}$

11. $\lim\limits_{x\to-\infty} \dfrac{x^2}{4x^3-9}$ **12.** $\lim\limits_{x\to-\infty} \dfrac{2-3x-4x^2}{3x^2+6x+10}$

13. $\lim\limits_{x\to\infty} (x-\sqrt{4x^2-1})$ **14.** $\lim\limits_{x\to\infty} \left(3\sqrt{\dfrac{x^2}{4}-1}-\dfrac{3}{2}x\right)$

15. $\lim\limits_{x\to\infty} \tan\dfrac{1}{x}$ ***16.** $\lim\limits_{x\to-\infty} x\tan\dfrac{1}{x}$

17. $\lim\limits_{x\to\infty} e^{-x}$ **18.** $\lim\limits_{x\to-\infty} (x-1)\sin\dfrac{1}{x-1}$

19. $\lim\limits_{x\to-\infty} e^{-1/x}$ **20.** $\lim\limits_{x\to\infty} \ln(1/x)$

21. $\lim\limits_{x\to\infty} \dfrac{1}{\ln x}$ **22.** $\lim\limits_{x\to\infty} e^x \ln x$

In Exercises 23–36 find the horizontal asymptote of the graph of the function. Then sketch the graph of the function.

23. $f(x)=1/(x-2)$ **24.** $f(x)=-1/(x+3)$

25. $f(x)=3x/(2x-4)$ **26.** $f(x)=4/(1-x)^2$

27. $f(x)=\dfrac{2x(x+3)}{9-x^2}$ **28.** $f(x)=\dfrac{\sqrt{x+2}}{(x+2)^2}$

29. $f(x)=\dfrac{x+2}{x-1}$ **30.** $f(x)=\dfrac{2x-3}{4-6x}$

31. $f(x)=\sqrt{\dfrac{3-x}{4-x}}$ **32.** $f(x)=\sqrt{\dfrac{1-3x}{-2-x}}$

33. $f(x)=1/(x^2-4)$ **34.** $f(x)=-2/(x^2-9)$

35. $f(x)=\ln(1+e^x)$ **36.** $f(x)=\dfrac{e^x-e^{-x}}{e^x+e^{-x}}$

In Exercises 37–42 find all vertical and horizontal asymptotes of the graph of f. You may wish to use a graphics calculator to assist you.

37. $f(x)=\dfrac{x^2-20x+50}{\sqrt{2x^4-512}}$ **38.** $f(x)=(\tan x)/x$

39. $f(x)=\dfrac{\ln(1-x)}{\ln(1+x)}$ **40.** $f(x)=\dfrac{\ln(1+e^x)}{\ln(2+e^{3x})}$

41. $f(x)=xe^{1/x}$ **42.** $f(x)=\dfrac{x^{4/3}+x^{1/3}-2}{x^{4/3}-16}$

It follows from the Substitution Rule that

$$\lim_{x\to\infty} f(x) = \lim_{x\to0^+} f(1/x) \quad \text{and} \quad \lim_{x\to-\infty} f(x) = \lim_{x\to0^-} f(1/x)$$

In Exercises 43–46 use these formulas to evaluate the limit.

43. $\lim\limits_{x\to\infty} \dfrac{2x^2+1}{3x^2-5}$ **44.** $\lim\limits_{x\to-\infty} \dfrac{4x^3-9x^2}{-7x^3+17}$

45. $\lim\limits_{x\to\infty} \dfrac{\sqrt{1+x^2}}{x}$ **46.** $\lim\limits_{x\to-\infty} \dfrac{\sqrt{1+x^2}}{x}$

47. Consider the graph of the function f in Figure 4.85. For each of the following, find either the value of the limit (as a number, ∞, or $-\infty$), or the asymptote(s), or write "undefined."

  a. $\lim\limits_{x\to-\infty} f(x)$ **b.** $\lim\limits_{x\to\infty} f(x)$

  c. $\lim\limits_{x\to-2^-} f(x)$ **d.** $\lim\limits_{x\to-2^+} f(x)$

  e. $\lim\limits_{x\to-2} f(x)$ **f.** $\lim\limits_{x\to0^-} f(x)$

  g. $\lim\limits_{x\to0^+} f(x)$ **h.** $\lim\limits_{x\to0} f(x)$

  i. vertical asymptote(s)

  j. horizontal asymptote(s)

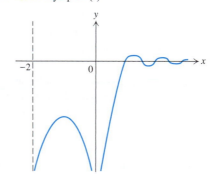

FIGURE 4.85   Graph for Exercise 47.

48. Let

$$f(x) = \dfrac{x|x|}{x^2+1}$$

Show that the graph of f has two horizontal asymptotes, and determine them.

49. Let

$$f(x) = \dfrac{\sqrt{2+4x^2}}{x}$$

Show that the graph of f has two horizontal asymptotes, and determine them.

50. Suppose $\lim_{x \to -\infty} f(x)/g(x) = 1$ and $\lim_{x \to -\infty} g(x) = \infty$. Show that $\lim_{x \to -\infty} f(x) = \infty$.

51. Suppose there is a number M such that $f(x) \geq 0$ for all $x \geq M$, and assume that $\lim_{x \to \infty} [f(x)]^2 = L$. Show that $\lim_{x \to \infty} f(x) = \sqrt{L}$.

52. Let $f(x) = (x - \pi) \cos \pi x$.
    a. Try to find $\lim_{x \to \infty} f(x)$ by letting $x = 10^2$, 10^3, and 10^4.
    b. Does $\lim_{x \to \infty} f(x)$ exist? Support your answer.

Applications

53. In this exercise you will find an alternative formula for the escape velocity given in (4). The alternative formula uses only the acceleration g due to gravity on earth and the radius R of the earth.
    a. Use Newton's Law of Gravitation and Newton's Second Law of Motion to deduce that for any object on earth (such as a spaceship) with mass m,
    $$mg = \frac{GMm}{R^2}$$
    where M is as before the mass of the earth.
    b. From (4) and part (a), show that
    $$v_{\text{esc}} = \sqrt{2gR}$$
    c. Use (b) to calculate v_{esc}. Does your result agree with the value of 11.2 kilometers per second obtained from (4)?

54. Suppose a rocket is launched vertically upward and its velocity $v(r)$ in miles per second at a distance of r miles from the center of the earth is given by the formula
    $$v(r) = \sqrt{\frac{192{,}000}{r} + v_0^2 - 48} \quad \text{for } r \geq 4000$$
    where v_0 is constant and represents the velocity of the rocket at burnout.
    a. If $v_0 = 8$ (miles per second), what is the limit of the velocity as the distance grows without bound?
    b. Find the number $v_0 > 0$ for which $\lim_{r \to \infty} v(r) = 0$.

55. According to one model, the velocity of a falling parachutist is given by
    $$v(t) = v^*(1 - e^{-at})$$
    where t denotes time, and v^* and a are positive constants.
    a. Find $\lim_{t \to \infty} v(t)$.
    b. What is the physical interpretation of the result in part (a)?

56. Suppose a rocket is launched from the surface of the earth. If we disregard the effect of the sun, the work required to propel the rocket from the surface (3960 miles from the center of the earth) to x miles above the surface is given by the formula
    $$W(x) = GMm\left(\frac{1}{3960} - \frac{1}{x}\right)$$
    where G, M, and m are constants. Find the work required to send the rocket from here to the end of the universe. Leave your answer in terms of G, M, and m.

57. According to Weiss's Law of Excitation of Tissue, the strength S of an electric current is related to the time t the current takes to excite tissue by the formula
    $$S(t) = \frac{a}{t} + b \quad \text{for } t > 0$$
    where a and b are constants. Then the limit $\lim_{t \to \infty} S(t)$ is the threshold strength of current below which the tissue will never be excited. Find $\lim_{t \to \infty} S(t)$.

58. In 1934 the biologist G. F. Gause placed 20 paramecia in 5 cubic centimeters of a saline solution with a constant amount of food and measured their growth on a daily basis. He found that the population $f(t)$ at any time t (in days) was approximately
    $$f(t) = \frac{4490}{1 + e^{5.4094 - 1.0235t}} \quad \text{for } t \geq 0$$
    Determine the carrying capacity of the medium.

59. Suppose the current $I(t)$ flowing in an electrical circuit at time t is given by
    $$I(t) = \frac{100}{1 + t^2} + 3 \sin \frac{30t}{\pi} \quad \text{for } t \geq 0$$
    Show that
    $$\lim_{t \to \infty}\left(I(t) - 3 \sin \frac{30t}{\pi}\right) = 0$$
    Thus for large values of t, $I(t)$ is very nearly equal to $3 \sin (30t/\pi)$. The expression $3 \sin (30t/\pi)$ is called the steady-state current, and the expression $100/(1 + t^2)$ is the transient current (since it is significant only for small values of t).

60. Suppose an object is released and moves through a viscous fluid that tends to resist the motion of the object. Then the velocity v increases with time, and may approach a **terminal velocity** v_T, which depends on the mass of the object and the viscosity of the fluid. If the

fluid's resistance is proportional to the object's velocity, then as a function of time t the velocity is given by

$$v = v_T(1 - e^{-gt/v_T})$$

where $g = 9.8$ (meters per second per second) is the acceleration due to gravity.

a. Find $\lim_{t \to \infty} v(t)$.

b. If the object is a tiny fog droplet (which is frequently on the order of 5×10^{-6} meters in radius), and is falling in the sky near earth, then a reasonable value for the terminal velocity is 2.7×10^{-2} meters per second (which is equivalent to 1 meter every 37 seconds). If the fog droplet begins falling at time $t = 0$, determine how long it takes for the velocity of the droplet to reach half of the terminal velocity.

Project

1. The feature of a horizontal asymptote for a function f is that the graph of f snuggles up to the asymptote as x gets very large positively or negatively. However, the graph of f can snuggle up to an inclined line as well. For instance, in Example 2 we showed that $\lim_{x \to \infty}(\sqrt{x^2 - a^2} - x) = 0$. A direct consequence is that the graph of $\sqrt{x^2 - a^2}$ approaches the line $y = x$ as x increases without bound (Figure 4.86). More generally, we will consider a nonhorizontal line $y = mx + b$ which the graph of f approaches for x far away from 0. Intuitively, this means that when x is large, $f(x)$ is close to $mx + b$. More precisely, we will say that a nonhorizontal line $y = mx + b$ is a **slant asymptote** for f if

$$\lim_{x \to \infty} [f(x) - (mx + b)] = 0$$

or

$$\lim_{x \to -\infty} [f(x) - (mx + b)] = 0$$

This project concerns slant asymptotes.

a. Let $R(x) = \dfrac{2x^5 + 3x}{-x^4 + \pi}$. Find the slant asymptote.

b. Let P be a polynomial of degree m, Q a polynomial of degree n, and $R(x) = P(x)/Q(x)$ for all x for which $Q(x) \neq 0$. Under what conditions on P and Q does R have a slant asymptote?

c. Let $f(x) = \sqrt{64 x^2 + 1}$. Find the slant asymptote.

d. Let $f(x) = 3 + \sqrt{64 x^2 + 1}$. Find the slant asymptote.

e. Let $f(x) = 2x - (\sin x)/x$. Determine whether or not f has a slant asymptote, and if so, find it.

f. Let $f(x) = \ln x$. Determine whether or not f has a slant asymptote, and if so, find it.

g. Can a function f have two slant asymptotes? Either explain why it cannot, or give an example of a function f that has two slant asymptotes.

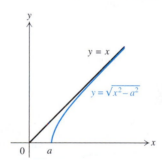

FIGURE 4.86   Graph for the project.

4.9   GRAPHING

As we have seen throughout Chapter 4, a knowledge of derivatives helps greatly in sketching the graphs of functions. In this section we collect and illustrate the methods we have encountered for sketching graphs. The most important ones are listed in Table 4.1.

TABLE 4.1

Important Properties of the Graph of f

Property	Test
f has y intercept c	$f(0) = c$
f has x intercept c	$f(c) = 0$
Graph of f is symmetric with respect to the $\begin{cases} y \text{ axis} \\ \text{origin} \end{cases}$	$f(-x) = f(x)$ $f(-x) = -f(x)$
f has a relative maximum value at c	$\begin{cases} f'(c) = 0 \text{ and } f' \text{ changes from positive to negative at } c \\ f'(c) = 0 \text{ and } f''(c) < 0 \end{cases}$
f has a relative minimum value at c	$\begin{cases} f'(c) = 0 \text{ and } f' \text{ changes from negative to positive at } c \\ f'(c) = 0 \text{ and } f''(c) > 0 \end{cases}$
f is increasing on an interval I	$f'(x) > 0$ for all except finitely many x in I
f is decreasing on an interval I	$f'(x) < 0$ for all except finitely many x in I
Graph of f is concave upward on an open interval I	$f''(x) > 0$ for all x in I
Graph of f is concave downward on an open interval I	$f''(x) < 0$ for all x in I
$(c, f(c))$ is an inflection point of the graph of f	f'' changes sign at c (and usually $f''(c) = 0$)
f has a vertical asymptote $x = c$	$\lim_{x \to c^+} f(x) = \pm\infty$ or $\lim_{x \to c^-} f(x) = \pm\infty$
f has a horizontal asymptote $y = d$	$\lim_{x \to \infty} f(x) = d$ or $\lim_{x \to -\infty} f(x) = d$

EXAMPLE 1 Let

$$f(x) = \frac{2}{1 + x^2}$$

Sketch the graph of f, noting all relevant properties listed in Table 4.1.

Solution Since $f(0) = 2$, the y intercept is 2. However, there are no x intercepts because $f(x) > 0$ for all x. The fact that $f(-x) = f(x)$ implies that the graph of f is symmetric with respect to the y axis. For the derivatives of f we have

$$f'(x) = \frac{-4x}{(1 + x^2)^2}$$

and

$$f''(x) = \frac{-4(1 + x^2)^2 + 4x(2)(1 + x^2)(2x)}{(1 + x^2)^4} = \frac{4(3x^2 - 1)}{(1 + x^2)^3}$$

Since $f'(x) > 0$ for $x < 0$ and $f'(x) < 0$ for $x > 0$, it follows that f is increasing on $(-\infty, 0]$ and decreasing on $[0, \infty)$, so that $f(0) = 2$ is the maximum value of f. Next we display the sign of $f''(x)$ in Figure 4.87.

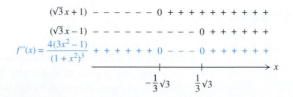

FIGURE 4.87

From the figure we see that the graph of f is concave upward on $(-\infty, -\frac{1}{3}\sqrt{3})$ and on $(\frac{1}{3}\sqrt{3}, \infty)$ and is concave downward on $(-\frac{1}{3}\sqrt{3}, \frac{1}{3}\sqrt{3})$, with inflection points at $(-\frac{1}{3}\sqrt{3}, \frac{3}{2})$ and $(\frac{1}{3}\sqrt{3}, \frac{3}{2})$. Finally, we notice that

$$\lim_{x \to \infty} \frac{2}{1 + x^2} = \lim_{x \to -\infty} \frac{2}{1 + x^2} = 0$$

which means that the x axis is a horizontal asymptote of the graph of f. We are now ready to sketch the graph of f, shown in Figure 4.88. ❑

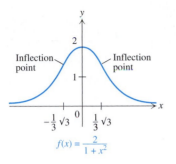

$$f(x) = \frac{2}{1 + x^2}$$

FIGURE 4.88

EXAMPLE 2 Let

$$f(x) = \frac{x}{(x + 1)^2}$$

Sketch the graph of f, noting all relevant properties listed in Table 4.1.

Solution First, $f(x) = 0$ precisely when $x = 0$, so that the x intercept and y intercept are both 0. Second, there is no symmetry with respect to either the y axis or the origin, because

$$f(-x) = \frac{-x}{(-x + 1)^2}$$

and consequently $f(-x)$ is different from $f(x)$ and from $-f(x)$. Next we calculate the first and second derivatives of f:

$$f'(x) = \frac{(x + 1)^2 - 2x(x + 1)}{(x + 1)^4} = \frac{1 - x}{(x + 1)^3}$$

and

$$f''(x) = \frac{(-1)(x + 1)^3 - (1 - x)3(x + 1)^2}{(x + 1)^6} = \frac{2(x - 2)}{(x + 1)^4}$$

The signs of $f'(x)$ and $f''(x)$ are shown in Figure 4.89.

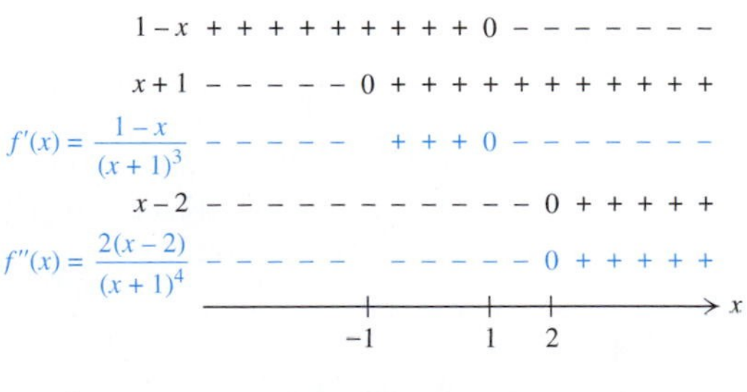

$$
\begin{array}{l}
1 - x \quad + + + + + + + + + 0 - - - - - - - \\
x + 1 \quad - - - - - 0 + + + + + + + + + + + \\
f'(x) = \dfrac{1 - x}{(x + 1)^3} \quad - - - - - \quad + + + 0 - - - - - - - \\
x - 2 \quad - - - - - - - - - - - 0 + + + + + \\
f''(x) = \dfrac{2(x - 2)}{(x + 1)^4} \quad - - - - - \quad - - - - - 0 + + + + + \\
\end{array}
$$

FIGURE 4.89

From the sign of $f'(x)$ it follows that f is increasing on $(-1, 1]$ and is decreasing on $(-\infty, -1)$ and on $[1, \infty)$. Thus $f(1) = \frac{1}{4}$ is a relative maximum value of f. From the sign of $f''(x)$ we deduce that the graph of f is concave downward on $(-\infty, -1)$

and on $(-1, 2)$ and is concave upward on $(2, \infty)$, with an inflection point at $(2, \frac{2}{9})$. Finally, we determine the asymptotes. Since

$$\lim_{x \to \infty} \frac{x}{(x+1)^2} = \lim_{x \to -\infty} \frac{x}{(x+1)^2} = 0$$

the line $y = 0$ is a horizontal asymptote. Since

$$\lim_{x \to -1} \frac{x}{(x+1)^2} = -\infty$$

the line $x = -1$ is a vertical asymptote. Now we are ready to sketch the graph of f, shown in Figure 4.90. ❑

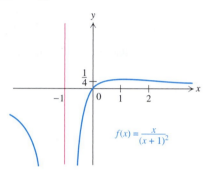

FIGURE 4.90

We have found it valuable to graph a function in order to understand its various characteristics. If two functions are graphed on the same coordinate plane, their graphs may form the boundary of a plane region (Figure 4.91). In Section 5.8 we will be interested in regions of this type. To draw such a region, we need only draw the graphs of the two functions and note where the graphs cross each other. We illustrate this procedure in Example 3.

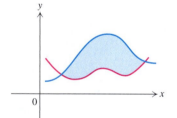

FIGURE 4.91

EXAMPLE 3 Sketch the graphs of f and g, where

$$f(x) = \frac{2}{1 + x^2} \quad \text{and} \quad g(x) = x^2$$

Shade the region they enclose.

Solution We graphed f in Example 1, and we know the graph of g from Figure 4.61. To find where the two graphs intersect, we solve the equation $f(x) = g(x)$ for x, that is, we solve

$$\frac{2}{1 + x^2} = x^2$$

We find that

$$x^4 + x^2 - 2 = 0, \quad \text{or} \quad (x^2 + 2)(x^2 - 1) = 0$$

Thus $x = 1$ or $x = -1$. Therefore the graphs of f and g intersect at $(-1, 1)$ and $(1, 1)$. The region is shown in Figure 4.92. ❑

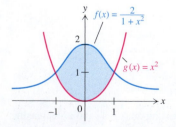

FIGURE 4.92

Our final example concerns the function

$$f(t) = c(e^{-bt} - e^{-at}) \quad \text{for } t \geq 0 \tag{1}$$

that has been employed to model the concentration of a drug in the blood at any time after the drug has been injected into a patient's muscle. The constant c corresponds to the amount of the drug that is injected into the muscle, a represents the rate at which the drug is absorbed into the blood stream, and b expresses the rate at which it is eliminated from the blood stream. The numbers a and b vary with the type of drug used, but in order for the concentration f to be positive, we must have $a > b > 0$.

In the solution we will use this observation about a differentiable function:

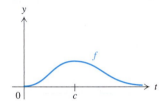

If $f(0) = 0$, $f(t) > 0$ for all $t > 0$, $\lim_{t \to \infty} f(t) = 0$, and $f'(c) = 0$ for exactly one value of c, then $f(c)$ is the maximum value of f on $[0, \infty)$.

Figure 4.93 illustrates the situation.

FIGURE 4.93

EXAMPLE 4   Sketch the graph of f described in (1).

Solution   First we observe that $f(0) = 0$, and $f(t) > 0$ for $t > 0$ because $a > b$. Thus the x and y intercepts are 0. Moreover, since a and b are positive,

$$\lim_{t \to \infty} f(t) = \lim_{t \to \infty} c(e^{-bt} - e^{-at}) = c(0 - 0) = 0$$

Differentiating f, we find that

$$f'(t) = c(ae^{-at} - be^{-bt}) \quad \text{and} \quad f''(t) = c(b^2 e^{-bt} - a^2 e^{-at}) \tag{2}$$

Therefore $f'(t) = 0$ if and only if

$$c(ae^{-at} - be^{-bt}) = 0$$

Since $c > 0$, this is equivalent to

$$be^{-bt} = ae^{-at}$$

Taking logarithms and solving for t, we find that

$$\ln b - bt = \ln a - at$$

which means that

$$(a - b)t = \ln a - \ln b$$

Consequently $f'(t) = 0$ if and only if $t = t_0$, where

$$t_0 = \frac{\ln a - \ln b}{a - b}$$

From the remark preceding the example, we conclude that $f(t_0)$ is the maximum value of f. From the second derivative in (2) you can verify that $f''(2t_0) = 0$ and that there is a change in the sign of f'' from negative to positive at $2t_0$. It follows that the graph of f has an inflection point at $(2t_0, f(2t_0))$. The graph is as shown in Figure 4.94. ❑

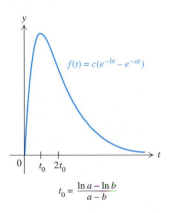

$$t_0 = \frac{\ln a - \ln b}{a - b}$$

FIGURE 4.94

The graph of f in Figure 4.94 has properties that we would expect from a model for the concentration of a drug.

1. The initial concentration in the blood stream is 0 at the time of injection.
2. The concentration in the blood stream rather quickly reaches a maximum value and then gradually declines, eventually leaving only trace amounts of the drug in the blood.

The inflection point of the graph of f corresponds to the time $2t_0$ at which the *rate of change* of the concentration stops decreasing and starts increasing.

EXERCISES 4.9

In Exercises 1–40 sketch the graph of the given function, noting all relevant properties listed in Table 4.1.

1. $f(x) = x^3 + 3x + 2$
2. $f(x) = x^3 - 8x^2 + 16x - 3$
3. $f(x) = x^4 + 8x^3 + 36x^2 - 3$ 4. $f(x) = x^4 - 6x^2$
5. $g(x) = x + 4/x$ 6. $g(x) = 2 + 1/x^2$
7. $g(x) = \dfrac{1}{x} - \dfrac{1}{x^2}$ 8. $g(x) = \dfrac{-x}{x+1}$
9. $k(x) = \dfrac{1+x}{1-x}$ 10. $k(x) = \dfrac{x+2}{x-3}$
11. $k(x) = 4x/(1 + 4x^2)$ 12. $k(x) = x^2/(1 + x^2)$
13. $f(t) = 1/(t^2 - 1)$ 14. $f(t) = t/(t^2 - 1)$
15. $f(t) = t^2/(t^2 - 1)$ 16. $f(t) = 1/(t^2 - t)$
17. $f(z) = 2z^2/(z + 1)^2$ 18. $f(z) = \dfrac{-z^2}{(2-z)^2}$
19. $f(x) = x^3/(1 - x^2)$ 20. $f(x) = \dfrac{(x+1)^2}{x-1}$
21. $f(x) = 1/(x^3 - 1)$ 22. $f(x) = x\sqrt{1 + x}$
23. $f(x) = \sqrt{x-1} + \sqrt{x+1}$
24. $f(x) = (1 - x^2)^{3/2}$ 25. $f(x) = x^2/\sqrt{1 - x^2}$
26. $f(x) = \dfrac{1}{\sqrt{x}} + \sqrt{x}$ 27. $f(x) = 4x - 3x^{4/3}$
28. $f(x) = (x + 2)(x - 1)^{1/3}$ 29. $g(x) = |\sin x|$
30. $g(x) = \sin x + \cos x$
31. $g(x) = \sqrt{3}\,\sin x + \cos x$
32. $g(x) = \dfrac{x}{2} + \sin x$ 33. $g(t) = \sin^2 t$
34. $g(t) = \tan^2 t$ 35. $f(x) = e^x/(1 + e^x)$
36. $f(x) = e^{-1/x}$ 37. $f(x) = \ln(e^x + e^{-x})$
38. $f(x) = \ln(1 + e^x)$ 39. $f(x) = \ln(1 + \ln x)$
40. $f(x) = \ln(1 - \ln x)$

In Exercises 41–44 use the formulas for the function and its first and second derivatives as an aid in sketching the graph of the given function. Note all relevant properties listed in Table 4.1.

41. $f(x) = x\sqrt{x^2 - 1}$, $f'(x) = \dfrac{2x^2 - 1}{(x^2 - 1)^{1/2}}$, and

$$f''(x) = \dfrac{x(2x^2 - 3)}{(x^2 - 1)^{3/2}}$$

42. $f(x) = x\sqrt{1 + x^2}$, $f'(x) = \dfrac{1 + 2x^2}{(1 + x^2)^{1/2}}$, and

$$f''(x) = \dfrac{x(3 + 2x^2)}{(1 + x^2)^{3/2}}$$

43. $f(x) = \dfrac{x}{\sqrt{1 - x}}$, $f'(x) = \dfrac{2 - x}{2(1 - x)^{3/2}}$, and

$$f''(x) = \dfrac{4 - x}{4(1 - x)^{5/2}}$$

44. $f(x) = \dfrac{\sqrt{x}}{1 + \sqrt{x}}$, $f'(x) = \dfrac{x^{-1/2}}{2(1 + x^{1/2})^2}$, and

$$f''(x) = \dfrac{-(x^{-3/2} + 3x^{-1})}{4(1 + x^{1/2})x^3}$$

In Exercises 45–47 sketch the graph of the function, using the Newton-Raphson method where necessary to find approximate zeros, critical points, and inflection points.

45. $f(x) = x^4 + 4x^3 + 4x^2 - 2$
46. $f(x) = 2x^4 + x^3 + x$
47. $f(x) = (x - 1)/[x(x + 1)]$

In Exercises 48–50 plot the graph of f and note all asymptotes. Then use the zoom features of the calculator to approximate the relative extreme values.

48. $f(x) = x^9 - x^5 + x^4 - x^3$

49. $f(x) = \dfrac{1}{x} + \dfrac{1}{x-1} + \dfrac{1}{x-2}$

50. $f(x) = \dfrac{1}{x} - \dfrac{1}{x-1} + \dfrac{1}{x-2}$

In Exercises 51–56 graph each pair of functions. Shade the region(s) the graphs enclose.

51. $f(x) = x^2$, $g(x) = x$

52. $f(x) = x^2 + 4$ and $g(x) = 12 - x^2$

53. $f(x) = x^3 + x$ and $g(x) = 3x^2 - x$ (*Hint:* The region has two parts.)

54. $f(x) = x^3 + x^2 + 1$ and $g(x) = x^3 + x + 1$

55. $g(x) = 2x/\sqrt{1 + x^2}$ and $k(x) = x/\sqrt{1 - x^2}$ (*Hint:* The region has two parts.)

56. $g(t) = t - \cos t$ for $-\pi/4 \le t \le 7\pi/4$ and $k(t) = t + \sin t$ for $-\pi/4 \le t \le 7\pi/4$

REVIEW

Key Terms and Expressions

Maximum value; minimum value
Critical point
Increasing function; decreasing function
Relative maximum value; relative minimum value
Concave upward; concave downward
Inflection point

Limit at infinity
Horizontal asymptote
Infinite limit at infinity
Parabola
Exponential growth
Radioactive decay

Key Theorems

Maximum-Minimum Theorem
Rolle's Theorem
Mean Value Theorem

First Derivative Test
Second Derivative Test

Review Exercises

In Exercises 1–2 find the critical numbers of the given function.

1. $f(x) = x^2 \sqrt{2 - x}$ **2.** $f(x) = \cos x^{1/3}$

In Exercises 3–6 find the extreme values of f on the given interval. Determine at which numbers in the interval they are assumed.

3. $f(x) = x^2 + x + 1$; $[-2, 2]$

4. $f(x) = \dfrac{x}{3 + x^2}$; $[-4, 4]$

5. $f(x) = x - \sqrt{1 - x^2}$; $[-1, 1]$

6. $f(x) = x^{2/3} - x$; $\left[-\frac{1}{8}, \frac{1}{8}\right]$

7. Let

$$f(x) = \begin{cases} x + 1 & \text{for } x \le 0 \\ x - 1 & \text{for } x > 0 \end{cases}$$

  a. Show that $f(-1) = f(1) = 0$ but that there is no number c in $(-1, 1)$ such that $f'(c) = 0$.
  b. Why does this not contradict Rolle's Theorem?

8. Let f be a function that is continuous on $[0, 2]$ and differentiable on $(0, 2)$. Suppose that $f(0) < f(2)$ but that f is *not* increasing on $[0, 2]$. Does f' necessarily take on both positive and negative values on $(0, 2)$?

In Exercises 9–12 find all functions satisfying the given conditions.

9. $f'(x) = x^2 - \sin x$

10. $f'(x) = 5, f(0) = -3$

11. $f''(x) = x^2 - 4$

12. $f''(x) = 0, f(0) = 0, f(1) = -1$

In Exercises 13–16 determine the intervals on which f is increasing and those on which f is decreasing.

13. $f(x) = \frac{1}{3}x^3 - x^2 + x - 2$

14. $f(x) = x^{1/3} - x$

15. $f(x) = \sin x - \frac{1}{8} \tan x$

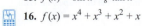 **16.** $f(x) = x^4 + x^3 + x^2 + x$

17. Use Theorem 4.7 to prove that $\sqrt{x + 3} \geq \sqrt{3} + x/4$ for $0 \leq x \leq 1$.

*18. Show that $\sin x \geq 2x/\pi$ for $0 \leq x \leq \pi/2$.

In Exercises 19–22 use the First Derivative Test or the Second Derivative Test to determine the relative extreme values of the function.

19. $f(x) = 3x^4 - 10x^3 + 6x^2 + 3$

20. $f(x) = \dfrac{x - 1}{x^2 + 3}$

21. $f(x) = (x + 1)^2(x - 2)^4$

22. $f(x) = 2\sqrt{x + 1} - \sqrt{x - 1}$

In Exercises 23–26 determine the intervals on which the graph of f is concave upward and the intervals on which the graph is concave downward.

23. $f(x) = \frac{1}{2}x^4 + x^3 - 6x^2$ **24.** $f(x) = \sqrt{x} + 1/x$

25. $f(x) = \dfrac{1}{1 + x^4}$ **26.** $f(x) = \sin x + \frac{1}{4}\sin 2x$

In Exercises 27–28 let f be a function with the given derivative. Find the intervals on which f is increasing and those on which it is decreasing, as well as the intervals on which the graph of f is concave upward and those on which it is concave downward.

27. $f'(x) = (x - 1)^3(x + 1)$ **28.** $f'(x) = xe^x$

In Exercises 29–38 sketch the graph of the function, indicating all relevant properties listed in Table 4.1 of Section 4.9.

29. $f(x) = x^3 - 6x - 1$ **30.** $f(x) = x^4 + 2x^3 + 1$

31. $f(x) = \dfrac{1}{x^3} - \dfrac{1}{x}$ **32.** $f(x) = \frac{1}{4}x - \sqrt{x}$

33. $k(x) = \dfrac{x^2 + 1}{x^2 - 4}$ **34.** $f(x) = \cos^2 x$

35. $f(x) = e^{-2x} - e^{-3x}$ **36.** $f(x) = \ln(e^x - e^{-x})$

37. $f(x) = \ln(x^2 - 4)$ **38.** $f(x) = \ln(4 - x^2)$

In Exercises 39–40 sketch the graphs of each pair of functions. Shade the regions they enclose.

39. $f(x) = x^2 + 4x$; $g(x) = x - 2$

40. $f(x) = -8/(7 + 5x)$; $g(x) = 8/(x^2 - 1)$

41. Let $f(x) = x^m(x - 1)^n$, where m and n are integers greater than or equal to 2. Determine the values of m and n for which the relative extreme values of f are $f(0), f(\frac{1}{4})$, and $f(1)$.

42. A trapezoid is to have three sides of equal length L. Find the length of the fourth side that yields the trapezoid with maximum area.

43. Find the maximum area of a rectangle in the first quadrant with one vertex at the origin and the opposite vertex on the graph of $y = e^{-x}$ (Figure 4.95).

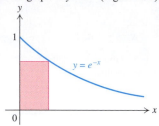

FIGURE 4.95   Graph for Exercise 43.

44. Find the point on the graph of $y = x^{1/2}$ that is closest to the point $(4, 0)$.

Applications

45. Uranium 238 is a radioactive isotope. Of any given amount, 85.719 percent remains after 1 billion years.
  a. Compute the half-life of uranium 238.
  b. If some uranium 238 was collected in 1945, what percentage of it would remain after 10 million years?

46. Suppose that the value of a bond grows with an interest rate of r percent compounded continuously. If the value of the account is B after t years, then the **present value** of the account is $Be^{-rt/100}$. Assuming an annual interest rate of 8%, is it a good deal for you to pay \$3000 for a bond that is registered to be worth \$5000 in 6 years?

47. Suppose the distance $D(v)$ a car can travel on one tank of gas at a velocity of v miles per hour is given by

$$D(v) = \frac{\sqrt{3}}{48}(80v^{3/2} - v^{5/2}) \quad \text{for } 0 \leq v \leq 75$$

What velocity maximizes D (and hence minimizes fuel consumption per mile)?

48. A certain type of vitamin capsule is to have the shape of a cylinder of height h with hemispheres of radius r on each end of the cylinder. For a given amount S of surface area, find the value of r that maximizes the volume of the capsule.

49. A toolshed with a square base and a flat roof is to have a volume of 800 cubic feet. If the floor costs \$6 per square foot, the roof \$2 per square foot, and the sides \$5 per square foot, determine the dimensions of the most economical shed.

50. An entrepreneur makes and sells perfume in spherical bottles that cost $10 + 60\pi r^2$ cents to make, where r is the radius of the bottle in centimeters. Suppose the revenue on each cubic centimeter of perfume is 15 cents and the largest bottle that can be made has a radius of 5 centimeters.

a. What radius will maximize the profit on each bottle of perfume?

b. What radius will minimize the profit on each bottle?

51. An airline company offers a round-trip group flight from New York to London. If x people sign up for the flight, the cost of each ticket is to be $1000 - 2x$ dollars. Find the maximum revenue the airline company can receive from the sale of tickets for the flight.

52. A swimming pool whose bottom has area $5000/\pi$ square feet is to be built in the form of a rectangle with semicircles attached at two opposite ends of the rectangle. Give the dimensions of the pool having a minimum perimeter.

53. The time dilation factor in Einstein's Theory of Relativity is a function of the velocity v of a moving object and is given by

$$f(v) = \frac{1}{\sqrt{1 - v^2/c^2}} \quad \text{for } 0 < v < c$$

where c is the velocity of light. Sketch the graph of f.

54. A carpenter wishes to illuminate a certain point P on the floor with a lamp 5 feet away, which may be raised to any level between 0 and 8 feet above the floor (Figure 4.96). How high above the floor should the carpenter raise the lamp so that the intensity of light at P will be a maximum? (*Hint:* The intensity of light at P is proportional to the cosine of the angle θ that the incident light makes with respect to the vertical, and it is inversely proportional to the square of the distance r from P to the light source.)

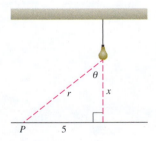

FIGURE 4.96 Figure for Exercise 54.

55. A farmer plans to fence off three rectangular grazing fields along a straight river by erecting one fence parallel to the river and four fences perpendicular to the river. If the total area of the three plots is to be 6400 square feet, what length of fence parallel to the river will minimize the total length of fence required?

56. A small private oil field has a maximum daily yield of 100 barrels. The owner estimates that the daily profit in dollars from a daily production of x barrels is

$$P(x) = 100x - 5x^2 \quad \text{for } 0 \le x \le 100$$

a. What daily profit or loss will result from maximum daily production?

b. What daily production will result in maximum daily profit, and what will that daily profit be?

57. If $C(x)$ denotes the cost of producing x units of a certain product, then by our discussion in Section 3.1, $C'(x)$ is the marginal cost at x, and the **average cost** $C_a(x)$ of the units is given by

$$C_a(x) = \frac{C(x)}{x}$$

A reasonable production strategy is to try to minimize the average cost C_a.

a. Assuming that C is differentiable on $(0, \infty)$, show that $C_a'(x_0) = 0$ at the production level x_0 for which $C_a(x_0) = C'(x_0)$.

b. Assume that $C_a''(x_0) > 0$. Show that $C_a(x_0)$ is the minimum value of C_a. (This result yields a basic principle of economics: Average cost is minimized at the production level for which average cost equals marginal cost. It also implies that the line from the origin to $(x_0, C(x_0))$ is tangent to the graph of C (see Figure 4.97).

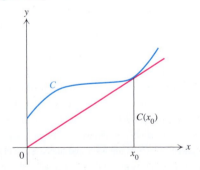

FIGURE 4.97 Figure for Exercise 57.

58. A projectile is fired with an initial velocity and height of v_0 and h_0, respectively, and at an angle of θ with respect to the horizontal. In terms of the horizontal distance, its height h is given by the formula

$$h(x) = -\frac{g}{2v_0^2}(\sec^2\theta)x^2 + (\tan\theta)x + h_0 \quad \text{for } 0 \le x \le x_0$$

where g is the acceleration due to gravity, and x_0 denotes the range of the projectile, that is, the total horizontal distance that the projectile travels before hitting the ground.

a. Find the maximum height of the projectile.

b. Find a formula for the range of the projectile in terms of v_0, h_0, and θ. Show that the formula you obtain is consistent with the formula given in Exercise 50 of Section 4.1 under the stipulation that $h_0 = 0$.

59. Exercise 30 in Section 4.6 concerns a pigeon flying toward its nest, first over water and then over land. Suppose the width of the river is a meters and the nest is located k meters downstream (Figure 4.98). Assume also that the pigeon flies v meters per second over water and w meters per second over land, with $v < w$.

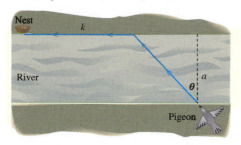

FIGURE 4.98   Figure for Exercise 59.

a. Let θ denote the angle shown in the figure. Express the total flight time T as a function of θ.

b. Show that if $T'(\theta_0) = 0$, then $\sin\theta_0 = v/w$.

c. Two pigeons, Jo and Mo, have a race. Suppose that Jo flies at 20 meters per second over water and 24 meters per second over land, whereas Mo flies at 18 meters per second over water and 26 meters per second over land. Jo and Mo start from the same point at the same time, 3000 meters up the river and 1000 meters across the river. Which pigeon arrives first at the nest?

60. A power company wishes to install a temporary power line. The purchase price of the wire is proportional to the cross-sectional area x of the wire, and the cost due to power loss while the wire is in use is proportional to the reciprocal of x. Thus there are constants a and b such that the total cost $C(x)$ is given by $C(x) = ax + b/x$. Determine the value of the cross-sectional area that minimizes the cost. Show that for such a cross-sectional area the purchase price is equal to the cost due to power loss.

61. During the sixteenth and seventeenth centuries, wine merchants in Linz, Austria, calculated the price of a barrel of wine by first inserting a measuring rod as far as possible into the taphole located in the middle of the side of the barrel (Figure 4.99), and then charging the customer according to the length l of rod inside the barrel. Assuming that the barrel is a circular cylinder, calculate the dimensions of the barrel with largest volume for a given length l of rod.

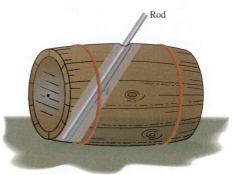

FIGURE 4.99   Figure for Exercise 61.

62. A real estate firm borrows money and then lends the money to its customers at 18 percent interest per year. The amount of money it can borrow is proportional to the square of the interest rate it pays for the money. At what interest rate should the firm borrow money in order to maximize its profit per year from lending?

63. Two houses are being built, one 50 feet from the street and the other 75 feet from the street, as in Figure 4.100. At what point on the edge of the street should a telephone pole be located to minimize the sum of the distances to the two houses?

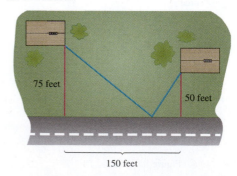

FIGURE 4.100   Figure for Exercise 63.

*64. A cylindrical container with water H feet deep has a spout h feet below the water line. Torricelli's Theorem states that the speed of the water as it flows through the spout is $\sqrt{2gh}$. Determine the value of h that maximizes the distance R as shown in Figure 4.101. Then find the maximum possible value of R. (*Hint:* The distance R equals the product of the velocity $\sqrt{2gh}$ and time t_0, that is, $R = \sqrt{2gh}\, t_0$, where t_0 is the time it takes for a droplet to fall from the spout to the base level of the container. Determine the value of t_0 by using (2) in Section 1.3.)

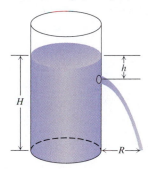

FIGURE 4.101   Figure for Exercise 64.

65. A driver involved in a car accident was given two alcohol tests. The first occurred 2 hours after the accident, and the alcohol content was fixed at 0.07%; the second occurred 4 hours after the accident, with an alcohol content of 0.05%. Assuming that the alcohol in a person's system decreases exponentially with time, and that driving with an alcohol content greater than 0.10% constitutes driving while intoxicated, determine whether the driver was intoxicated at the time of the accident.

Topics for Discussion

1. What methods do we have for finding an extreme value of a function on an interval?
2. Tell what Rolle's Theorem and the Mean Value Theorem say geometrically. What are the major consequences of the Mean Value Theorem in Chapter 4?
3. What is the difference between a maximum value and a relative maximum value? If f has a relative maximum value at c, must $f'(c) = 0$? Explain your answer.
4. Explain how the first and second derivatives help in sketching the graph of a function.
5. In your own words, explain concavity and inflection points.
6. What are vertical and horizontal asymptotes, and how can one find them?
7. Suppose that the degrees of polynomials p and q are m and n, respectively. Under what conditions does p/q have a horizontal asymptote? a vertical asymptote?

Cumulative Review, Chapters 1–3

In Exercises 1–4 solve the inequality.

1. $18x(4x^6 - 1) > 0$
2. $\|\,x\,| - 3| \le 4$
3. $\sqrt{2} \le \sqrt{1 + x^4} < 4$
4. $\sin x \ge 2\cos^2 x - 1$
5. Let $f(x) = \dfrac{1}{\sqrt{x + 3}}$ and $g(x) = \dfrac{1}{2x - 1}$.
   a. Find the domain of $f \circ g$.
   b. Find a formula for $(f \circ g)(x)$.
6. Let $h(x) = 1 + \sin(3x^2 - 2)$. Find functions f and g, neither of which equals h, such that $h = f \circ g$.

In Exercises 7–10 determine whether the limit exists as a number, as ∞, or as $-\infty$. If the limit exists, evaluate it.

7. $\displaystyle\lim_{x \to 4^+} \frac{x(x + 4)}{16 - x^2}$
8. $\displaystyle\lim_{x \to -3^-} \frac{3|x| + 9}{x^2 - 9}$
9. $\displaystyle\lim_{x \to 0^+} \frac{\sin 2x - 2\sqrt{x}\,\sin x + 4x^2}{x}$
10. $\displaystyle\lim_{x \to 0^+} \ln(1 - \ln 2x)$
11. Let $f(x) = (x - 4)^2 + 1$. Find the point (a, b) on the graph of f at which the tangent line is perpendicular to the line $y - 2x = 8$.
12. Let
$$f(x) = \begin{cases} (x - 1)^2 & \text{for } x < 1 \\ x^3 - x^2 & \text{for } x \ge 1 \end{cases}$$
   a. Determine whether f is continuous at 1.
   b. Determine whether f is differentiable at 1.
13. Let $f(x) = 2/x^{1/2}$. Use the definition of the derivative to find $f'(4)$.

In Exercises 14–15 find the derivative of f.

14. $f(x) = \sin(\cos x^3)$
15. $f(x) = \dfrac{1}{x - e^{\tan x}}$
16. Let $f(x) = \ln\left(\dfrac{-4}{1 - e^x}\right)$.
   a. Find the domain of f.   b. Find the derivative of f.
   c. Find $\displaystyle\lim_{x \to 0^+} f(x)$.
17. Show that
$$\frac{d}{dx}\left[3(2x + 1)^{5/2} - 5(2x + 1)^{3/2}\right] = 30x(2x + 1)^{1/2}$$
18. Suppose
$$f(t) = \frac{t}{t^2 + 1}$$
represents the position of an object at any time $t \ge 0$.
   a. At what time will the direction of motion of the object change?
   b. What is the acceleration of the object at that time?

19. Suppose y is a differentiable function such that $xy^2 + 3y - 4x = 17$. Find dy/dx.

20. Suppose a point moves along the right branch of the hyperbola $x^2 - 2y^2 = 1$ in such a way that its y coordinate decreases at the rate of 3 units per second. How fast is its x coordinate increasing when the point is located at $(\sqrt{3}, -1)$?

21. The area of a circular hole in the ground increases at the rate of $2\pi\sqrt{r}$ feet per minute, where r is the radius of the hole in feet. How large is the area of the hole when its radius is increasing at the rate of 2 feet per minute?

 22. Let $f(x) = x^4 - x^3 + x - 2$. Use the Newton-Raphson method to approximate a zero of f. Continue until successive iterations obtained by the calculator are identical.

23. According to one tall tale, during half-time at a football game a baton twirler lofted the baton straight up with such a speed that it fell back to earth one week later, during the next game's halftime. How many feet high would the baton have gone?

A hang-glider high over the ocean.
(*Tony Stone Images*)

5   THE INTEGRAL

Since the time of the ancient Greeks, mathematicians have attempted to calculate areas of plane regions. The most basic plane region is the rectangle, whose area is the product of its base and its height (Figure 5.1(a)). The ancient Greeks used Euclidean geometry to deduce the areas of parallelograms and triangles. They also knew how to compute the area of any polygon by partitioning it into triangles (Figure 5.1(b)).

However, with regions having curved boundaries, a different approach is needed. For example, can you guess the area of the region in either Figure 5.1(c) or (d)? The region in Figure 5.1(c) occurs in the famous eighteenth-century problem called "Buffon's needle problem;" we will find its area in Section 5.4. The region in Figure 5.1(d) occurs in the study of the distribution of a nation's income; we will discuss it in Section 5.8.

It was Archimedes* who made the first notable advance in calculating areas of regions with curved boundaries. He made an ingenious use of the "method of exhaustion," based on an idea of Eudoxus (ca. 408–355 B.C.). With this method, larger and larger polygons of known area are inscribed in such a region so that it would eventually be "exhausted." The area of the region was then the "limit" of

*Archimedes: Pronounced "Ar-ki-*mee*-deez."

287

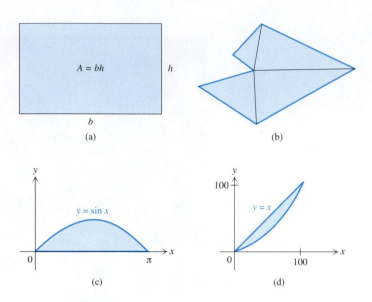

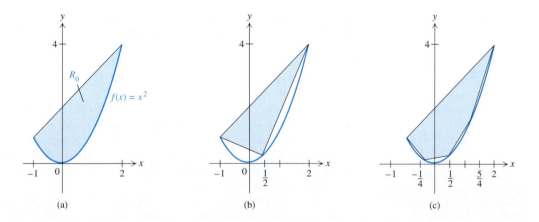

FIGURE 5.1

the areas of the inscribed figures. (However, Archimedes had no formal notion of limit.) For example, to find the area of a parabolic region such as R_0 in Figure 5.2(a), Archimedes inscribed an increasing number of triangles, as shown in Figure 5.2(b) and (c). From his calculations he concluded that the area of the region R_0 should be $\frac{4}{3}$ of the area of the single inscribed triangle in Figure 5.2(b).* Since it is possible to show that this triangle has area $\frac{27}{8}$, we conclude from Archimedes' result that the area of R_0 is $\frac{9}{2}$. The definition of area that we will give will also imply that the area of R_0 is $\frac{9}{2}$. The fact that the modern definition of area stems from

FIGURE 5.2

*For further details of his procedure, see Example 5 of Section 9.4.

Archimedes' method of exhaustion is a tribute to his genius.

In this chapter we will use the problem of computing area to motivate the notion of definite integral. The definition of the integral will also be motivated by the problem of finding the distance traveled by a moving object whose velocity is known. Later we will show that the integral occurs naturally in numerous other settings, including lengths of curves and surface area, moments and centers of gravity, and various concepts in electricity and magnetism.

5.1   PREPARATION FOR THE DEFINITE INTEGRAL

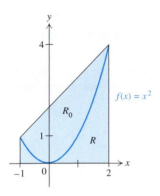

FIGURE 5.3

As a preview of the method of finding area, we consider the region R bounded by the parabola $f(x) = x^2$, the x axis, and the lines $x = -1$ and $x = 2$ (Figure 5.3). Since the regions R and R_0 shown in Figure 5.3 together comprise a trapezoid whose area is $\frac{15}{2}$, we can find the area of R_0 by calculating the area of R and subtracting the result from $\frac{15}{2}$.

Suppose we inscribe rectangles in the region R, as shown in Figure 5.4(a) and (b). Then the sum of the areas of the rectangles is less than the area of R. Similarly, if we circumscribe rectangles about R, as in Figure 5.5(a) and (b), then the sum of the areas of the rectangles is greater than the area of R. Of course, the area of each rectangle is the product of its base and its height.

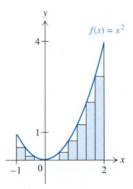 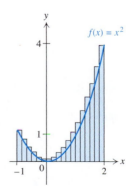

FIGURE 5.4

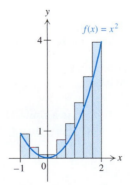

FIGURE 5.5

The crucial observation to make about this process is that as the bases of the rectangles become smaller and smaller, the sum of the areas of the rectangles appears to approach the area of R. This suggests that the area of R should be defined as the limit (in a sense to be clarified later) of the sum of the areas of inscribed or circumscribed rectangles. Our definition of area will be based on this idea.

Our assertions thus far about the area of R have rested on three basic properties we expect area to possess:

The Rectangle Property: The area of a rectangle is the product of its base and height.

The Addition Property: The area of a region composed of several smaller regions that overlap in at most a line segment is the sum of the areas of the smaller regions.

The Comparison Property: The area of a region that contains a second region is at least as large as the area of the second region.

You should understand where each of these properties was employed in the preceding discussion. They will play a major role in the definition we will give of area.

Partitions

Now let us consider any region R bounded by the graph of a nonnegative function f that is continuous on an interval $[a, b]$, by the x axis, and by the lines $x = a$ and $x = b$, where $a < b$ (Figure 5.6(a)). We call R **the region between the graph of f and the x axis on $[a, b]$.** Using the three basic properties of area listed above, we set out to define the area of the region R.

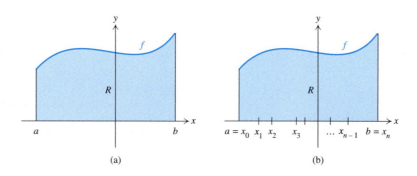

(a)   (b)

FIGURE 5.6 The region between the graph of f and the x axis on $[a, b]$.

For any positive integer n we divide $[a, b]$ into subintervals by introducing points of subdivision $x_0, x_1, \ldots, x_n$ (Figure 5.6(b)).

DEFINITION 5.1

A **partition** of $[a, b]$ is a finite set P of points $x_0, x_1, \ldots, x_n$ such that $a = x_0 < x_1 < \cdots < x_n = b$. We describe P by writing

$$P = \{x_0, x_1, \ldots, x_n\}$$

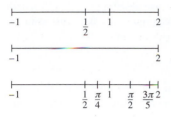

FIGURE 5.7

By definition, any partition of $[a, b]$ must contain both a and b. Except for a and b, the number of points and their placement in $[a, b]$ are arbitrary. For example, $\{-1, \frac{1}{2}, 1, 2\}$ is a partition of $[-1, 2]$, as are $\{-1, 2\}$ and $\{-1, \frac{1}{2}, \pi/4, 1, \pi/2, 3\pi/5, 2\}$ (Figure 5.7). However, $\{\frac{1}{4}, \frac{2}{5}, \frac{1}{2}, \frac{7}{8}, 2\}$ is not a partition of $[-1, 2]$ because it does not include -1.

The n subintervals into which a partition $P = \{x_0, x_1, \ldots, x_n\}$ divides $[a, b]$ are $[x_0, x_1], [x_1, x_2], \ldots, [x_{n-1}, x_n]$. Their lengths are $x_1 - x_0, x_2 - x_1, \ldots, x_n - x_{n-1}$, respectively. We denote the length $x_k - x_{k-1}$ of the kth subinterval $[x_{k-1}, x_k]$ by Δx_k. Thus

$$\Delta x_k = x_k - x_{k-1}$$

In particular, for the partition $\{-1, \frac{1}{2}, 1, 2\}$ of $[-1, 2]$, we have

$$x_0 = -1, \qquad x_1 = \frac{1}{2}, \qquad x_2 = 1, \qquad x_3 = 2$$

and

$$\Delta x_1 = \frac{1}{2} - (-1) = \frac{3}{2}, \qquad \Delta x_2 = 1 - \frac{1}{2} = \frac{1}{2},$$

$$\Delta x_3 = 2 - 1 = 1$$

The length $b - a$ of $[a, b]$ can be written in terms of the lengths $\Delta x_1, \Delta x_2, \ldots, \Delta x_n$ of the subintervals:

$$b - a = \Delta x_1 + \Delta x_2 + \cdots + \Delta x_n$$

Lower and Upper Sums

Having chosen a partition P of $[a, b]$, we inscribe over each subinterval derived from P the largest rectangle that lies inside the region R, as was done in Figure 5.4(a) and (b). Since we are assuming that f is continuous on $[a, b]$, we know from the Maximum-Minimum Theorem (Section 4.1) that for each k between 1 and n there exists a smallest value m_k of f on the kth subinterval $[x_{k-1}, x_k]$. If we choose m_k as the height of the kth rectangle R_k, then R_k will be the largest (tallest) rectangle that can be inscribed in R over $[x_{k-1}, x_k]$ (Figure 5.8). Doing this for each subinterval, we create n inscribed rectangles $R_1, R_2, \ldots, R_n$, all lying inside the region R.

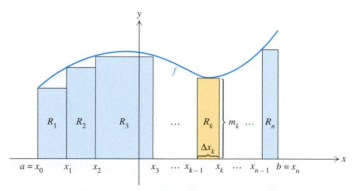

FIGURE 5.8

For each k between 1 and n the rectangle R_k has base $[x_{k-1}, x_k]$ with length Δx_k and has height m_k. Hence the area of R_k is the product $m_k \, \Delta x_k$. Just as in the case of the parabolic region in Figure 5.4(a) and (b), the sum

$$m_1 \, \Delta x_1 + m_2 \, \Delta x_2 + \cdots + m_n \, \Delta x_n$$

of the areas of all the rectangles should be no larger than the area of R. We denote this sum $L_f(P)$ and call it the **lower sum** of f associated with the partition P. Thus

$$L_f(P) = m_1 \, \Delta x_1 + m_2 \, \Delta x_2 + \cdots + m_n \, \Delta x_n \qquad (1)$$

No matter how we define the area of R, this area must be *at least as large* as the lower sum $L_f(P)$ associated with *any* partition P of $[a, b]$.

EXAMPLE 1   Let $f(x) = x^2$ for $-1 \le x \le 2$. Find $L_f(P)$ and $L_f(P')$ for the partitions

$$P = \left\{ -1, \frac{1}{2}, 1, 2, \right\} \quad \text{and} \quad P' = \left\{ -1, -\frac{1}{2}, \frac{1}{2}, 1, 2 \right\}$$

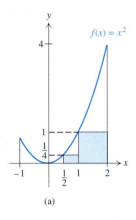

(a)

Solution   The subintervals associated with the partition P are $[-1, \frac{1}{2}]$, $[\frac{1}{2}, 1]$, and $[1, 2]$. Using the graph in Figure 5.9(a), we find that the minimum values of f on the three subintervals are given by

$$m_1 = f(0) = 0, \quad m_2 = f\left(\frac{1}{2}\right) = \frac{1}{4}, \quad m_3 = f(1) = 1$$

Since

$$\Delta x_1 = \frac{1}{2} - (-1) = \frac{3}{2}, \qquad \Delta x_2 = 1 - \frac{1}{2} = \frac{1}{2}, \quad \text{and} \quad \Delta x_3 = 2 - 1 = 1$$

it follows from (1) that

$$L_f(P) = m_1 \cdot \Delta x_1 + m_2 \cdot \Delta x_2 + m_3 \cdot \Delta x_3 = 0 \cdot \frac{3}{2} + \frac{1}{4} \cdot \frac{1}{2} + 1 \cdot 1 = \frac{9}{8}$$

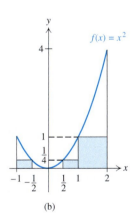

(b)

FIGURE 5.9

In a similar way, the four subintervals associated with the partition P' are $[-1, -\frac{1}{2}]$, $[-\frac{1}{2}, \frac{1}{2}]$, $[\frac{1}{2}, 1]$, and $[1, 2]$. Using the graph in Figure 5.9(b), we find that the minimum values of f on the four subintervals are given by

$$m_1 = f\left(-\frac{1}{2}\right) = \frac{1}{4}, \quad m_2 = f(0) = 0, \quad m_3 = f\left(\frac{1}{2}\right) = \frac{1}{4}, \text{ and } m_4 = f(1) = 1$$

Moreover,

$$\Delta x_1 = -\frac{1}{2} - (-1) = \frac{1}{2}, \quad \Delta x_2 = \frac{1}{2} - \left(-\frac{1}{2}\right) = 1, \quad \Delta x_3 = 1 - \frac{1}{2} = \frac{1}{2}, \quad \text{and } \Delta x_4 = 2 - 1 = 1$$

Consequently from (1) we conclude that

$$L_f(P') = m_1 \cdot \Delta x_1 + m_2 \cdot \Delta x_2 + m_3 \cdot \Delta x_3 + m_4 \cdot \Delta x_4 = \frac{1}{4} \cdot \frac{1}{2} + 0 \cdot 1 + \frac{1}{4} \cdot \frac{1}{2} + 1 \cdot 1 = \frac{5}{4} \quad \square$$

In Example 1 we computed two lower sums: $L_f(P) = \frac{9}{8}$ and $L_f(P') = \frac{5}{4}$. The larger of these is $\frac{5}{4}$. Since the area A of the region R between the graph of f and the x axis on $[-1, 2]$ must be at least as large as *any* lower sum, it follows that $A \geq \frac{5}{4}$.

We also mention that on each subinterval, the minimum value of f can be at the left endpoint, the right endpoint, or somewhere in the middle. For example, if we consider P', then m_3 is the value of f at the left endpoint of the subinterval, whereas m_1 is the value of f at the right endpoint of the subinterval, and m_2 is the value of f at the midpoint of the subinterval. The common feature is that for each k, m_k is the minimum value of f on the corresponding subinterval of $[-1, 2]$.

By a procedure similar to inscribing rectangles to compute a lower sum, we can also circumscribe rectangles and compute an upper sum. Let $P = \{x_0, x_1, \ldots, x_n\}$ be a given partition of $[a, b]$, and let f be continuous and nonnegative on $[a, b]$. Then the Maximum-Minimum Theorem implies that for each k between 1 and n there exists a largest value M_k of f on the kth subinterval $[x_{k-1}, x_k]$ (Figure 5.10). Consequently if we let M_k be the height of the kth rectangle R_k, then R_k will be the smallest possible rectangle circumscribing the appropriate portion of R (Figure 5.10). The area of R_k is $M_k \, \Delta x_k$, and the sum

$$M_1 \, \Delta x_1 + M_2 \, \Delta x_2 + \cdots + M_n \, \Delta x_n$$

of the areas of the circumscribed rectangles should be no smaller than the area of R. We denote this sum $U_f(P)$ and call it the **upper sum** of f associated with the partition P. Thus

$$U_f(P) = M_1 \, \Delta x_1 + M_2 \, \Delta x_2 + \cdots + M_n \, \Delta x_n \qquad (2)$$

No matter how we define the area of R, this area must be *no larger* than $U_f(P)$ for *any* partition P of $[a, b]$.

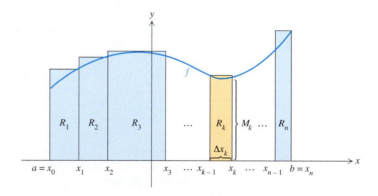

FIGURE 5.10

Now we revisit the function f and partitions P and P' in Example 1, and calculate the corresponding upper sums.

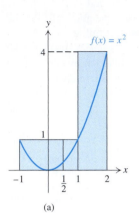

(a)

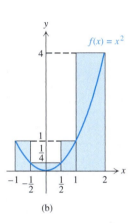

(b)

FIGURE 5.11

EXAMPLE 2 Let $f(x) = x^2$ for $-1 \le x \le 2$. Find $U_f(P)$ and $U_f(P')$ for the partitions

$$P = \left\{-1, \frac{1}{2}, 1, 2\right\} \quad \text{and} \quad P' = \left\{-1, -\frac{1}{2}, \frac{1}{2}, 1, 2\right\}$$

Solution As in Example 1, the three subintervals associated with the partition P are $\left[-1, \frac{1}{2}\right]$, $\left[\frac{1}{2}, 1\right]$, and $[1, 2]$, and thus

$$\Delta x_1 = \frac{3}{2}, \quad \Delta x_2 = \frac{1}{2}, \quad \text{and} \quad \Delta x_3 = 1$$

Using the graph in Figure 5.11(a), we find that the maximum values of f on the three subintervals are given by

$$M_1 = f(-1) = 1, \quad M_2 = f(1) = 1, \quad \text{and} \quad M_3 = f(2) = 4$$

It follows from (2) that

$$U_f(P) = M_1 \cdot \Delta x_1 + M_2 \cdot \Delta x_2 + M_3 \cdot \Delta x_3 = 1 \cdot \frac{3}{2} + 1 \cdot \frac{1}{2} + 4 \cdot 1 = 6$$

In a similar way, the four subintervals associated with the partition P' are $\left[-1, -\frac{1}{2}\right]$, $\left[-\frac{1}{2}, \frac{1}{2}\right]$, $\left[\frac{1}{2}, 1\right]$, and $[1, 2]$, so we have

$$\Delta x_1 = \frac{1}{2}, \quad \Delta x_2 = 1, \quad \Delta x_3 = \frac{1}{2}, \quad \text{and} \quad \Delta x_4 = 1$$

Using the graph of f in Figure 5.11(b), we find the maximum values of f on the four subintervals:

$$M_1 = f(-1) = 1, \quad M_2 = f\left(-\frac{1}{2}\right) = f\left(\frac{1}{2}\right) = \frac{1}{4}, \quad M_3 = f(1) = 1, \quad \text{and} \quad M_4 = f(2) = 4$$

We conclude from (2) that

$$U_f(P') = M_1 \cdot \Delta x_1 + M_2 \cdot \Delta x_2 + M_3 \cdot \Delta x_3 + M_4 \cdot \Delta x_4 = 1 \cdot \frac{1}{2} + \frac{1}{4} \cdot 1 + 1 \cdot \frac{1}{2} + 4 \cdot 1 = \frac{21}{4} \; \square$$

In contrast to the lower sums, where the finer partition P' resulted in a larger lower sum than the partition P did, here P' resulted in a smaller upper sum: $U_f(P) = 6$, whereas $U_f(P') = \frac{21}{4}$. Since the area A of the region R between the graph of f and the x axis on $[-1, 2]$ must be no larger than any upper sum, and no smaller than any lower sum, we conclude that

$$\frac{5}{4} \le A \le \frac{21}{4}$$

More generally, if f is any continuous nonnegative function on $[a, b]$ and P any partition of $[a, b]$, then

$$L_f(P) \le \text{area of } R \le U_f(P) \tag{3}$$

Later we will be able to find the exact value of A. However, what we can say at this time is that the smaller the subintervals of $[a, b]$ are, the closer the lower sum and the upper sum appear to be (see Figures 5.4 and 5.5); in fact, as the lengths of the subintervals shrink toward 0, the lower sum and the upper sum approach a single number which we will call the area of R. That we will do in Section 5.2.

Riemann sums

George Friedrich Bernhard Riemann (1826–1866)
Riemann had a profound influence on future mathematics as he formalized theories in many areas of mathematics, especially integration. His results were so far-reaching that mathematicians of his time were not able to fully comprehend their significance. Unfortunately he died at the age of 39 from tuberculosis.

Lower sums and upper sums, which use the minimum and maximum values of the function on the subintervals defined by a partition P, are important because the area A of R is guaranteed to lie somewhere between them (as stated in (3)). However, we can generally get a better approximation of the area A by using values of f that lie *between* the minimum and maximum values of f on the subintervals defined by P.

To understand this, compare Figure 5.12(a) with Figures 5.8 and 5.10. The height of each rectangle in Figure 5.12(a) is intermediate to the corresponding inscribed rectangle in Figure 5.8 and the corresponding circumscribed rectangle in Figure 5.10. This is shown for a representative subinterval $[x_{k-1}, x_k]$ in Figure 5.12(b), where the height of the rectangle is $f(t_k)$ for an appropriately placed t_k. Since

$$\underbrace{m_k}_{\substack{\text{minimum value of}\\ f \text{ on } [x_{k-1}, x_k]}} \leq \underbrace{f(t_k)}_{\substack{\text{arbitrary value of } f\\ \text{on } [x_{k-1}, x_k]}} \leq \underbrace{M_k}_{\substack{\text{maximum value of}\\ f \text{ on } [x_{k-1}, x_k]}} \quad \text{for} \quad k = 1, 2, \ldots, n \qquad (4)$$

it follows that the sum

$$f(t_1)\,\Delta x_1 + f(t_2)\,\Delta x_2 + \cdots + f(t_n)\,\Delta x_n$$

lies between the lower sum $L_f(P)$ and the upper sum $U_f(P)$. Therefore

$$L_f(P) \leq f(t_1)\,\Delta x_1 + f(t_2)\,\Delta x_2 + \cdots + f(t_n)\,\Delta x_n \leq U_f(P) \qquad (5)$$

The sum in (5) is called a Riemann* sum. Since $L_f(P)$ and $U_f(P)$ approximate the area A, so does the Riemann sum.

The sum in (5) makes sense whether f is nonnegative or not. Therefore in the definition of Riemann sum we will assume only that f is continuous on $[a,b]$.

DEFINITION 5.2

Let f be continuous on $[a, b]$, and let $P = \{x_0, x_1, \ldots, x_n\}$ be any partition of $[a, b]$. For each k between 1 and n, let t_k be an arbitrary number in $[x_{k-1}, x_k]$. Then the sum

$$f(t_1)\,\Delta x_1 + f(t_2)\,\Delta x_2 + \cdots + f(t_n)\,\Delta x_n$$

is called a **Riemann sum** for f on $[a, b]$ and is denoted $\sum_{k=1}^{n} f(t_k)\,\Delta x_k$ (Σ is the Greek letter sigma). Thus

$$\sum_{k=1}^{n} f(t_k)\,\Delta x_k = f(t_1)\,\Delta x_1 + f(t_2)\,\Delta x_2 + \cdots + f(t_n)\,\Delta x_n \qquad (6)$$

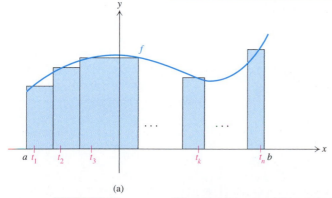

(a)

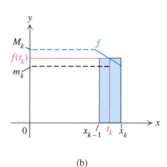

(b)

FIGURE 5.12   *Riemann:** Pronounced *"Ree-mahn."*

We mentioned that (4) and (5) hold if f is a nonnegative function. Actually, even if f is *not* nonnegative, both (4) and (5) hold for exactly the same reasons. Thus every Riemann sum for a partition P lies between the lower sum $L_f(P)$ and the upper sum $U_f(P)$.

For any partition $P = \{x_0, x_1, x_2, \ldots, x_n\}$, the points $t_1, t_2, \ldots, t_n$ may be chosen in any way we wish. If we choose them so that $f(t_1), f(t_2), \ldots, f(t_n)$ are the minimum (or maximum) values of f on the subintervals, then the Riemann sum becomes the lower sum (or upper sum). As we mentioned above, such choices for the t_k's usually will not give the best approximations for a given partition. Three simple choices that may provide better approximations are the left endpoints, the right endpoints, and the midpoints, of the subintervals. They yield, in turn, the following:

left sum: t_k is the left endpoint of $[x_{k-1}, x_k]$

right sum: t_k is the right endpoint of $[x_{k-1}, x_k]$

midpoint sum: t_k is the midpoint of $[x_{k-1}, x_k]$

EXAMPLE 3 Let $f(x) = x^2$ and let $[-1, 2]$ be partitioned by $P = \{-1, 0, 1, \frac{3}{2}, 2\}$. Find the three Riemann sums listed above.

Solution For each of the Riemann sums, the t_k are shown in Figure 5.13. Using that information, we find the following values:

left sum: $\displaystyle\sum_{k=1}^{4} f(t_k)\,\Delta x_k = \overset{1}{\overbrace{f(-1)}} \cdot 1 + \overset{0}{\overbrace{f(0)}} \cdot 1 + \overset{1}{\overbrace{f(1)}} \cdot \frac{1}{2} + \overset{9/4}{\overbrace{f\left(\frac{3}{2}\right)}} \cdot \frac{1}{2} = \frac{21}{8}$

right sum: $\displaystyle\sum_{k=1}^{4} f(t_k)\,\Delta x_k = \overset{0}{\overbrace{f(0)}} \cdot 1 + \overset{1}{\overbrace{f(1)}} \cdot 1 + \overset{9/4}{\overbrace{f\left(\frac{3}{2}\right)}} \cdot \frac{1}{2} + \overset{4}{\overbrace{f(2)}} \cdot \frac{1}{2} = \frac{33}{8}$

midpoint sum: $\displaystyle\sum_{k=1}^{4} f(t_k)\,\Delta x_k = \overset{1/4}{\overbrace{f\left(-\frac{1}{2}\right)}} \cdot 1 + \overset{1/4}{\overbrace{f\left(\frac{1}{2}\right)}} \cdot 1 + \overset{25/16}{\overbrace{f\left(\frac{5}{4}\right)}} \cdot \frac{1}{2} + \overset{49/16}{\overbrace{f\left(\frac{7}{4}\right)}} \cdot \frac{1}{2}$

$= \dfrac{45}{16}$ □

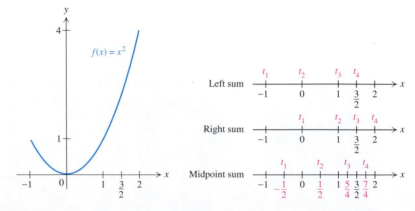

FIGURE 5.13

When the number n of subintervals of $[a, b]$ is large, a calculator or computer becomes very useful, and computations are simplified if the n subintervals have the same length, namely $(b - a)/n$. In this case,

$$\sum_{k=1}^{n} f(t_k)\, \Delta x_k = \frac{b - a}{n}\, [f(t_1) + f(t_2) + \cdots + f(t_n)] \tag{7}$$

We also observe that the lower and upper sums can be evaluated only after finding various minimum and maximum values of f. By contrast, the left, right, and midpoint sums do not depend on such minima and maxima, and hence are generally easier to evaluate.

EXAMPLE 4   Let $f(x) = 1/x$ for $1 \leq x \leq 2$, and let $[1, 2]$ be subdivided into 10 subintervals of equal length. Find the corresponding left sum, right sum, and midpoint sum of f on $[1, 2]$.

Solution   We use (7) with $a = 1$, $b = 2$, and $n = 10$, first letting t_k be the left endpoint of $[x_{k-1}, x_k]$, then the right endpoint, and finally the midpoint, for $k = 1, 2, \ldots, 10$. By calculator we find that

$$\text{left sum:} \quad \frac{2-1}{10}\left[\frac{1}{1} + \frac{1}{1.1} + \frac{1}{1.2} + \cdots + \frac{1}{1.9}\right] \approx 0.7187714032$$

$$\text{right sum:} \quad \frac{2-1}{10}\left[\frac{1}{1.1} + \frac{1}{1.2} + \cdots + \frac{1}{1.9} + \frac{1}{2}\right] \approx 0.6687714032$$

$$\text{midpoint sum:} \quad \frac{2-1}{10}\left[\frac{1}{1.05} + \frac{1}{1.15} + \cdots + \frac{1}{1.95}\right] \approx 0.6928353604$$

Thus we have completed the solution.   ❏

If we had let $n = 100$, then we would have obtained the following results:

Method	Riemann sum
left sum	0.6956534305
right sum	0.6906534305
midpoint sum	0.6931440556

Distance from Velocity

Another context in which Riemann sums appear naturally is that of finding distance from velocity. In Sections 2.2 and 3.1 we defined the velocity of an object moving in a straight line as the derivative of the position function. Now we turn the table around, by assuming that we know the velocity of an object over a period of time, and asking how we can determine the distance D traveled during that time interval.

To answer this question, let us assume that an object is moving in a straight line, and that its velocity $v(t)$ at any time t represents a continuous, nonnegative function on the time interval $[a, b]$. If the velocity is constant during the interval, say $v(t) = c$ for $a \leq t \leq b$, then the distance D is given by

$$D = \text{velocity} \times \text{time} = c(b - a) \qquad (8)$$

However, suppose that the velocity is not constant. Let $P = \{t_0, t_1, t_2, \ldots, t_n\}$ be a partition of $[a, b]$. Then the distance D that the object travels during the time interval from $t = a$ to $t = b$ is the sum of the distances traveled during each of the time subintervals. If the subintervals are small enough then during such a subinterval the velocity is nearly constant. This means that if we choose any point t_1^* in the first subinterval $[t_0, t_1]$, then as in (8), the distance traveled during that time subinterval is approximately

$$\text{velocity} \times \text{time} = v(t_1^*)\,(t_1 - t_0) = v(t_1^*)\,\Delta t_1$$

The same reasoning holds for each of the other time subintervals. Consequently, if we choose points $t_1^*, t_2^*, \ldots, t_n^*$ in the successive time subintervals $[t_0, t_1]$, $[t_1, t_2]$, $\ldots, [t_{n-1}, t_n]$, then we conclude that the total distance D traveled during the time interval $[a, b]$ is given (approximately) by the formula

$$D \approx v(t_1^*)\Delta t_1 + v(t_2^*)\Delta t_2 + \cdots + v(t_n^*)\Delta t_n$$

Observe that the sum on the right is a Riemann sum for v on $[a, b]$.

EXAMPLE 5   In October of 2000 the Current Eliminator V drag car set a world's record by accelerating from rest (i.e., zero velocity) to an astounding 137 miles per hour in (approximately) 8.8 seconds. Suppose that readings of the speedometer in the car were registered every 0.8 seconds, with the results appearing in the table, where time is in seconds and velocity is in miles per hour:

time	0	0.8	1.6	2.4	3.2	4.0	4.8	5.6	6.4	7.2	8.0	8.8
velocity	0	22	50	76	96	112	124	130	133	135	136	137

Approximate the distance the drag car traveled during the 8.8 seconds by considering a partition of the 8.8 seconds with 11 subintervals of equal length, and evaluating the corresponding left sum and right sum.

Solution   Before finding appropriate Riemann sums that approximate the distance that the car traveled during the 8.8 seconds, we need to convert so that the units of time (seconds) and of velocity (miles per hour) are compatible. Because all time intervals in the table are 0.8 seconds, it is easiest to convert the time to hours, with the length of each of the 11 subintervals equaling $(0.8)/3600$ hours. This means that in the Riemann sum, $\Delta t_k = 0.8/3600$ for $k = 1, 2, \ldots 11$. Taking the left endpoint of the kth subinterval as t_k, we obtain the following calculation for the left sum:

$$\sum_{k=1}^{n} f(t_k)\,\Delta t_k = \left(\frac{0.8}{3600}\right)(0 + 22 + 50 + 76 + 96 + 112 + 124 + 130 + 133 + 135 + 136) \approx .2253$$

Taking the right endpoint of the kth subinterval as t_k, we find that the corresponding right sum is given by

$$\sum_{k=1}^{n} f(t_k)\,\Delta t_k = \left(\frac{0.8}{3600}\right)(22 + 50 + 76 + 96 + 112 + 124 + 130 + 133 + 135 + 136 + 137) \approx .2558$$

Thus the left sum with 11 equal subintervals is approximately 0.2253, and the corresponding right sum is approximately 0.2558.   ❏

The left and right sums in Example 5 are quite close together. In reality the Current Eliminator V drag car traveled 0.25 mile in 8.8 seconds.

You can imagine that the smaller we take the subintervals, the better we would expect the Riemann sums to approximate the distance D traveled by the drag car. Indeed, as the subintervals become infinitely small, the Riemann sums approach a single number, just as they did for the area of a region R. In the next section we will give meaning—called the definite integral—to the limit of Riemann sums as the subintervals shrink to nothing. Then both area and distance will be special examples of the definite integral.

EXERCISES 5.1

In Exercises 1–3 compute $L_f(P)$ and $U_f(P)$ for the indicated partition.

In Exercises 4–6 compute the left sum and the right sum for the indicated partition.

1.

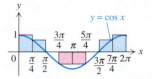

2.

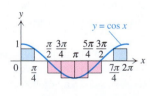

3.

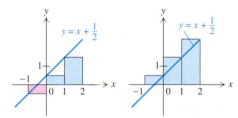

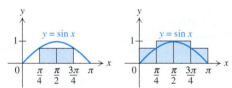

4.

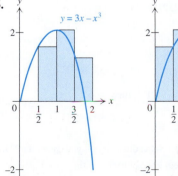

5.

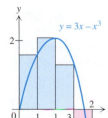

6.

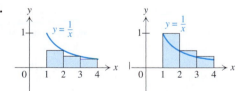

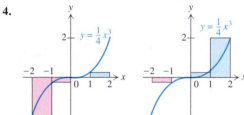

In Exercises 7–12 compute $L_f(P)$ and $U_f(P)$.

7. $f(x) = x + 2$; $P = \{-1, -\frac{1}{2}, 0, \frac{1}{2}, 1, \frac{3}{2}, 2\}$

8. $f(x) = x^2$; $P = \{-1, -\frac{1}{2}, 0, \frac{1}{2}, 1, \frac{3}{2}, 2\}$

9. $f(x) = -1/x$; $P = \{-4, -3, -2, -1\}$

10. $f(x) = \sin x$; $P = \{0, \pi/6, \pi/4, \pi/3, \pi/2\}$

11. $f(x) = \sin x + \cos x$; $P = \{0, \pi/4, \pi/2\}$

12. $f(x) = x^4 - 2x^2$; $P = \{-2, -1, 0, 1, 2\}$

In Exercises 13–16 use a calculator to approximate $L_f(P)$ and $U_f(P)$.

13. $f(x) = \sqrt{x}$; $P = \{0, \frac{1}{9}, \frac{2}{9}, \ldots, \frac{8}{9}, 1\}$

14. $f(x) = \sqrt{1 - x^2}$; $P = \{-1, -\frac{3}{4}, -\frac{1}{2}, -\frac{1}{4}, 0, \frac{1}{4}, \frac{1}{2}, \frac{3}{4}, 1\}$

15. $f(x) = \ln x$; $P = \{0.5, 0.75, 1, 1.25, 1.5, 1.75, 2\}$

16. $f(x) = e^{-2x} - 1$; $P = \{-1, -0.8, -0.6, -0.4, -0.2, 0, 0.2, 0.4, 0.6, 0.8, 1\}$

In Exercises 17–20 compute the left sum, right sum, and midpoint sum for the given function and partition.

17. $f(x) = 2x - 3$; $P = \{-2, -1, 0, 1\}$

18. $f(x) = 2x^2 - 1$; $P = \{-1, 0, \frac{1}{2}, 1\}$

19. $f(x) = 1/x^2$; $P = \{1, \frac{3}{2}, 2, \frac{5}{2}, 3\}$

20. $f(x) = \sin x$; $P = \{0, \pi/2, \pi, 2\pi\}$

In Exercises 21–24 use a calculator to compute the left sum, right sum, and midpoint sum for the given function and interval, using a partition having the indicated number of subintervals of the same length.

21. $f(x) = \sqrt{4 + x^2}$; $[-1, 1]$; $n = 10$

22. $f(x) = \sin x$; $[0, \pi]$; $n = 50$

23. $f(x) = \ln x$; $[2, 4]$; $n = 20$

24. $f(x) = e^{2x}$; $[0, 3]$; $n = 100$

In Exercises 25–27 approximate the area A of the region between the graph of f and the x axis on the given interval by using the indicated Riemann sum and a partition having the indicated number of subintervals of the same length.

25. $f(x) = 1/x^2$; $[1, 3]$; left sum; $n = 4$

26. $f(x) = |x|$; $[-2, 2]$; midpoint sum; $n = 8$

27. $f(x) = \tan x$; $[0, \pi/3]$; upper sum; $n = 50$

28. Why is it impossible to find a function f and a partition P such that $L_f(P) = 3$ and $U_f(P) = 2$?

29. Let P be a partition of $[0, 1]$, and let

$$f(x) = \begin{cases} 1 & \text{for } x = 0 \\ \dfrac{1}{x} & \text{for } 0 < x \le 1 \end{cases}$$

Try to find $U_f(P)$. What difficulty do you encounter? Does the same problem arise in trying to find $L_f(P)$?

30. Assume that f and g are continuous and nonnegative on $[a, b]$ and that $f(x) \le g(x)$ for $a \le x \le b$. Show that for any partition P of $[a, b]$ the inequalities

$$L_f(P) \le L_g(P) \quad \text{and} \quad U_f(P) \le U_g(P)$$

hold.

31. Suppose that f is increasing on $[a, b]$ and $P = \{x_0, x_1, x_2, \ldots, x_n\}$ is a partition of $[a, b]$ such that

$$\Delta x_k = \frac{b - a}{n} \text{ for } 1 \le k \le n$$

Show that

$$U_f(P) - L_f(P) = [f(b) - f(a)]\left(\frac{b - a}{n}\right)$$

32. Suppose f is continuous on $[1, 3]$ and has values appearing in the following table:

x	1	1.5	1.7	2.1	2.5	3
$f(x)$	2	1	.5	.2	0	.1

Approximate the area A between the graph of f and the x axis on $[1, 3]$ by using the information given in the table and an appropriate

a. left sum   **b.** right sum

Application

33. Suppose a campus bookstore projects that the profits (in thousands of dollars) per month for the 4-month period from September 1 to December 31 will be given by the function

$$P(x) = \frac{10^4 x}{1 + x^2} \quad \text{for } 0 \le x \le 4$$

a. Using the lower sum corresponding to monthly increments in time, approximate the projected profit.

b. Using the lower sum corresponding to semimonthly increments in time, approximate the projected profit. Explain why your answer is larger than the answer to part (a).

34. A skydiver drops from an airplane. At the end of each of the first six seconds the diver's speed (in meters per second) is checked, and reads as follows:

time	1	2	3	4	5	6
speed	20	37	45	50	53	55

Use appropriate left and right sums to approximate the distance the diver falls during the six–second period. How much do they differ by?

Projects

1. Let f be an arbitrary function continuous on $[a, b]$. Tell what you can about f if f satisfies the given condition. (*Hint:* A sketch of the graph of f and appropriate rectangles for P may help.)

   a. $L_f(P)$ = left sum for P, for every partition P of $[a, b]$
   b. $L_f(P)$ = right sum for P, for every partition P of $[a, b]$
   c. $L_f(P)$ = midpoint sum for P, for every partition P of $[a, b]$
   d. $L_f(P) = U_f(P)$, for some partition P of $[a, b]$
   e. left sum for P = right sum for P for every partition P of $[a, b]$
   f. left sum for P = right sum for P, for every partition P of $[a, b]$ whose subintervals have equal length

2. This project takes us back to the days of Archimedes, when he used polygons inscribed in a circle in order to approximate the area of the circle. Let us consider the unit circle, centered at the origin. As you know, its area is π (≈ 3.14159265358979). For convenience let us assume that each inscribed polygon is regular (the sides have equal length) and has one vertex at the point $(1, 0)$ (Figure 5.14 (a)–(c)).

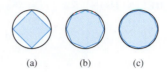

(a)   (b)   (c)

FIGURE 5.14   Figure for Project 2

   a. Find the area A_4 of the inscribed square.
   b. Find the area A_6 of the inscribed hexagon. (*Hint:* The hexagon is composed of 6 equilateral triangles.)
   c Find the area A_8 of the inscribed octagon. (*Hint:* If two vertices of a component triangle of the octagon are the origin and $(1, 0)$, then what is the third vertex?)
   d. For any positive integer n, find the area A_n of the inscribed polygon of n sides.
   e. Find the minimum number n such that the area A_n is greater than 3.
   f. Find the minimum number n such that the area A_n is greater than 3.14.

5.2   THE DEFINITE INTEGRAL

In Section 5.1 we introduced the Riemann sum

$$f(t_1)\,\Delta x_1 + f(t_2)\,\Delta x_2 + \cdots + f(t_n)\,\Delta x_n$$

for any function f continuous (but not necessarily nonnegative) on $[a, b]$ and for a partition P. We mentioned that as the subintervals of a partition of $[a, b]$ shrink, the Riemann sums approach a specific number. In fact, it is possible to show (and we do so in the Appendix) that there is a unique number I such that if a partition has small enough subintervals, then any Riemann sum with respect to the partition is as close to I as we wish. The following theorem states this result in mathematical terms:

THEOREM 5.3

For any function f that is continuous on $[a, b]$, there is a number I with the following property:

For any $\varepsilon > 0$, there is a number $\delta > 0$ such that if $P = \{x_0, x_1, \ldots, x_n\}$ is any partition of $[a, b]$ each of whose subintervals has length less than δ, and if $x_{k-1} \le t_k \le x_k$ for each k between 1 and n, then the associated Riemann sum $\sum_{k=1}^{n} f(t_k)\,\Delta x_k$ satisfies

$$\left| I - \sum_{k=1}^{n} f(t_k)\,\Delta x_k \right| < \varepsilon \qquad (1)$$

The unique number I that the Riemann sums approach is called the definite integral of f from a to b.

DEFINITION 5.4   Let f be continuous on $[a, b]$. The **definite integral of f from a to b** is the unique number I which the Riemann sums approach (see (1) above). This number is denoted by

$$\int_a^b f(x)\, dx$$

The symbol $\int$ is called an **integral sign,** the numbers a and b are called the **limits of integration,** and the function f appearing in the integral is called the **integrand.**

We sometimes express (1) by saying that "the integral $\int_a^b f(x)\, dx$ is the limit of the Riemann sums as the partitions become finer and finer." More succinctly, we write

$$\int_a^b f(x)\, dx = \lim_{\|P\| \to 0} \sum_{k=1}^n f(t_k)\Delta x_k \qquad (2)$$

where $\|P\|$ denotes the largest of the lengths of the subintervals associated with P and is called the **norm** of P.

The definite integral $\int_a^b f(x)\, dx$ is a number depending only on f, a, and b. Sometimes a is called the **lower limit** and b the **upper limit** of the integral. The variable x appearing in the integral is a "dummy variable"; it may be replaced by any other variable, such as t or u, not already in use. This means that

$$\int_a^b f(x)\, dx = \int_a^b f(t)\, dt = \int_a^b f(u)\, du$$

For the specific case in which $f(x) = x^2$, $a = 0$, and $b = 3$, we have

$$\int_0^3 x^2\, dx = \int_0^3 t^2\, dt = \int_0^3 u^2\, du$$

Caution: For the present we will attach no particular meaning to the expression dx in $\int_a^b f(x)\, dx$, although it arose originally from the differential. This expression dx will play a role later when we develop methods for computing definite integrals.

We now formally define area in terms of the definite integral.

DEFINITION 5.5   Let f be continuous and nonnegative on $[a, b]$, and let R be the region between the graph of f and the x axis on $[a, b]$ (see Figure 5.15). The **area** of R is defined to be

$$\int_a^b f(x)\, dx$$

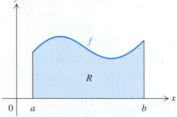

FIGURE 5.15   The region between the graph of f and the x axis on $[a, b]$.

History of $\int$

The integral sign was first used by Leibniz on October 29, 1675. The symbol is derived from the first letter of the Latin word "summa," which means sum.

Occasionally it is possible to calculate $\int_a^b f(x)\, dx$ from the definition of the integral by calculating Riemann sums directly.

EXAMPLE 1   Let $f(x) = c$ for $a \le x \le b$. Show that $\int_a^b c\, dx = c(b - a)$.

Solution   Suppose that $P = \{x_0, x_1, \ldots, x_n\}$ is any partition of $[a, b]$. Because f assumes only the value c, it follows that no matter which numbers $t_1, t_2, \ldots, t_n$ we choose in order to compute the corresponding Riemann sum for f and P, we obtain

$$\sum_{k=1}^n f(t_k)\, \Delta x_k = c\Delta x_1 + c\Delta x_2 + \cdots + c\Delta x_n = c(\Delta x_1 + \Delta x_2 + \cdots + \Delta x_n) = c(b - a)$$

Since every Riemann sum equals $c(b - a)$, it follows from (2) that

$$\int_a^b c\, dx = \lim_{\|P\| \to 0} \sum_{k=1}^n f(t_k)\Delta x_k = c(b - a) \qquad \Box$$

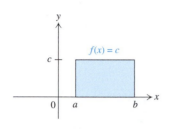

FIGURE 5.16   Area $= c(b - a)$.

From the formula in Example 1 we find that

$$\int_2^5 4\, dx = 4(5 - 2) = 12 \quad \text{and} \quad \int_{-1}^1 \pi\, dx = \pi\, [1 - (-1)] = 2\,\pi$$

We also notice that the result of Example 1 is geometrically understandable. After all, if $c > 0$, then the integral $\int_a^b c\, dx$ represents the area $c(b - a)$ of the rectangle in Figure 5.16.

EXAMPLE 2   Show that

$$\int_a^b x\, dx = \frac{1}{2}(b^2 - a^2)$$

Solution   We will find a number I such that the midpoint sum of f for any partition of $[a, b]$ actually equals I. To that end, let $P = \{x_0, x_1, \ldots, x_n\}$ be any partition of $[a, b]$. For $1 \le k \le n$ we choose

$$t_k = \frac{1}{2}(x_k + x_{k-1})$$

which is the midpoint of $[x_{k-1}, x_k]$. Then the resulting midpoint sum is

$$\sum_{k=1}^n f(t_k)\Delta x_k = \frac{1}{2}(x_1 + x_0)\Delta x_1 + \frac{1}{2}(x_2 + x_1)\Delta x_2 + \cdots + \frac{1}{2}(x_n + x_{n-1})\Delta x_n$$

$$= \frac{1}{2}(x_1 + x_0)(x_1 - x_0) + \frac{1}{2}(x_2 + x_1)(x_2 - x_1) + \cdots + \frac{1}{2}(x_n + x_{n-1})(x_n - x_{n-1})$$

$$= \frac{1}{2}(x_1^2 - x_0^2) + \frac{1}{2}(x_2^2 - x_1^2) + \frac{1}{2}(x_3^2 - x_2^2) + \frac{1}{2}(x_4^2 - x_3^2) + \cdots + \frac{1}{2}(x_n^2 - x_{n-1}^2)$$

$$= \frac{1}{2}x_1^2 - \frac{1}{2}x_0^2 + \frac{1}{2}x_2^2 - \frac{1}{2}x_1^2 + \frac{1}{2}x_3^2 - \frac{1}{2}x_2^2 + \frac{1}{2}x_4^2 - \frac{1}{2}x_3^2 + \cdots + \frac{1}{2}x_n^2 - \frac{1}{2}x_{n-1}^2$$

$$= \frac{1}{2}x_n^2 - \frac{1}{2}x_0^2 = \frac{1}{2}(b^2 - a^2)$$

Thus, no matter which partition of $[a, b]$ we pick, if we choose t_k as the midpoint of the kth subinterval for each k, then the associated midpoint sum is precisely $\frac{1}{2}(b^2 - a^2)$. We conclude that

$$\int_a^b x\, dx = \frac{1}{2}(b^2 - a^2) \quad \square$$

For example, if $a = 3$ and $b = 7$, we have

$$\int_3^7 x\, dx = \frac{1}{2}(7^2 - 3^2) = 20$$

From this and Definition 5.5 it follows that the area of the trapezoidal region shown in Figure 5.17(a) is 20. In contrast, if $a = -3$ and $b = 1$, we have

$$\int_{-3}^1 x\, dx = \frac{1}{2}[1^2 - (-3)^2] = -4$$

which clearly is not the area of the region shaded in Figure 5.17(b) because area is never negative.

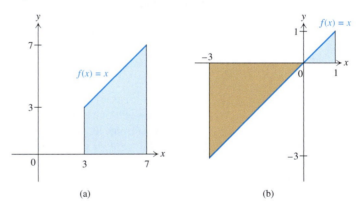

(a)   (b)

FIGURE 5.17

Only slight modifications are necessary in order to prove that for any constant c,

$$\int_a^b c\, x\, dx = \frac{c}{2}(b^2 - a^2) = c\int_a^b x\, dx \tag{3}$$

This will also follow directly from Theorem 5.17 in Section 5.5.

One can use much the same general technique as in the solution of Example 2 to show that for any constant c,

$$\int_a^b c\, x^2\, dx = \frac{c}{3}(b^3 - a^3) \tag{4}$$

(see Exercise 35 for the case $c = 1$). In particular, if $a = -1$, $b = 2$, and $c = 1$, then (4) tells us that

$$\int_{-1}^{2} x^2 \, dx = \frac{1}{3} (2^3 - (-1)^3) = 3$$

This implies that the area of the parabolic region R in Figure 5.3 is 3. Since the area of the trapezoid composed of R and R_0 is $\frac{15}{2}$, it follows that the area of the region R_0 is $\frac{15}{2} - 3 = \frac{9}{2}$. This is compatible with the results obtained by Archimedes more than 2000 years ago.

Evaluating integrals such as $\int_a^b x^2 \, dx$ by using Riemann sums is tedious. To find an integral such as $\int_0^\pi \sin x \, dx$ by this method would be even more difficult. Fortunately, we will derive a method (in Section 5.4) for computing definite integrals that does not involve Riemann sums.

How Accurate Are Riemann Sums?

We know by Definition 5.4 that Riemann sums for a continuous function f on an interval $[a, b]$ approximate the integral $\int_a^b f(x) \, dx$, and we know from Theorem 5.3 that the approximation can be as good as we wish by choosing a partition with small enough subintervals. But it is reasonable to ask how small the subintervals in a partition P must be in order that any Riemann sum associated with P will be guaranteed to approximate the exact value of the integral with a specified accuracy. To that end, we define the "error" incurred in using a Riemann sum to approximate the integral. Because the left, right, and midpoint Riemann sums are the most easily calculated Riemann sums, we will focus on errors for those Riemann sums.

Assume that f is continuous on a closed interval $[a, b]$, and that the partition P subdivides $[a, b]$ into n subintervals of equal length $(b - a)/n$. The ***n*th left sum error** is defined by

$$E_n^L = \left| \int_a^b f(x) \, dx - \text{the corresponding left sum} \right|$$

Similarly, the ***n*th right sum error** is defined by

$$E_n^R = \left| \int_a^b f(x) \, dx - \text{the corresponding right sum} \right|$$

and the ***n*th midpoint sum error** is defined by

$$E_n^M = \left| \int_a^b f(x) \, dx - \text{the corresponding midpoint sum} \right|$$

Later we will show that

$$\int_1^2 \frac{1}{x} \, dx = \ln 2 \approx 0.6931471806$$

rounded to 10 digits. It then follows from the calculations in Example 4 in Section 5.1 and the table following it that

$$E_{10}^L \approx 0.0256242226 \qquad E_{100}^L \approx 0.0025062499$$

$$E_{10}^{R} \approx 0.0243757774 \qquad E_{100}^{R} \approx 0.0024937501$$

$$E_{10}^{M} \approx 0.0003118202 \qquad E_{100}^{M} \approx 0.0000031250$$

Notice that for each of the three types of Riemann sums, the error decreases as n increases, and the midpoint error is significantly less than the left or right errors. However, it would be really valuable to have an idea of how small the error is in those cases when we do not know the exact value of the integral. The faster the error decreases as n increases, the more efficient the method will be. Indeed, keeping n small reduces calculation time because calculating the Riemann sum $\sum_{k=1}^{n} f(t_k) \Delta x_k$ involves computing and adding n terms.

Formulas that provide upper bounds for the left, right, and midpoint sum errors appear in Exercises 38–39. However, the numerical approximations (the Trapezoidal Rule and Simpson's Rule) to be discussed in Section 8.6 will in general be far more efficient estimates of the integral and will in general entail much smaller errors.

Concluding Remark

We end this section with two comments. First, the integral $\int_a^b f(x) \, dx$ can be defined for some functions that are not continuous on $[a, b]$. Such functions actually appear in many applications, especially in engineering. In general, we say that a function f defined on $[a, b]$ is **integrable** on $[a, b]$ if there exists a number I with the property that the Riemann sums tend to I, as in Theorem 5.3. In Section 5.3 we will give examples of integrable functions that are not continuous on a closed interval.

Second, we can only expect a single Riemann sum to provide an *approximate* value for an integral, not the *exact* value. In Section 5.4 we will derive a method for evaluating integrals exactly. Not only will that method not involve computing Riemann sums, but also, it will provide a powerful and surprising relationship between integrals and derivatives.

EXERCISES 5.2

In Exercises 1–4 approximate $\int_a^b f(x) \, dx$ by computing $L_f(P)$ and $U_f(P)$.

1. $\int_{-1}^{3} 2x \, dx$; $P = \{-1, 0, 1, 2, 3\}$
2. $\int_{-1}^{3} |x| \, dx$; $P = \{-1, -\frac{1}{2}, 0, \frac{1}{2}, 1, \frac{3}{2}, 2, \frac{5}{2}, 3\}$
3. $\int_{-\pi/4}^{\pi/4} 3 \sin x \, dx$; $P = \{-\pi/4, 0, \pi/4\}$
4. $\int_0^2 |x - 1| \, dx$; $P = \{0, \frac{1}{2}, 1, \frac{3}{2}, 2\}$

In Exercises 5–8 approximate the integral by Riemann sums with the indicated partitions, first using the left sum, then the right sum, and finally the midpoint sum.

5. $\int_1^3 (x^2 - x) \, dx$; $P = \{1, 2, 3\}$
6. $\int_0^4 (x - 3) \, dx$; $P = \{0, 1, 2, 3, 4\}$
7. $\int_{-\pi}^{0} \cos x \, dx$; $P = \{-\pi, -2\pi/3, -\pi/3, 0\}$
8. $\int_1^5 1/x \, dx$; $P = \{1, 2, 3, 4, 5\}$

In Exercises 9–13 approximate the integral by the given type of Riemann sum, using a partition having the indicated number of subintervals of the same length.

9. $\int_0^{\pi} \sin x \, dx$; midpoint sum; $n = 10$
10. $\int_1^2 \dfrac{1}{1 + x^2} \, dx$; left sum; $n = 10$
11. $\int_{1.1}^{1.2} \ln(1 + e^x) \, dx$; right sum; $n = 20$
12. $\int_2^3 e^{-x} \, dx$; upper sum; $n = 20$
13. $\int_{-1}^{1} e^{x^2} \, dx$; lower sum; $n = 20$

In Exercises 14–18 compute the definite integral by using the results of this section.

14. $\int_{-2}^{3} 4\, dx$ **15.** $\int_{-3}^{3} x\, dx$ **16.** $\int_{-2.7}^{2.9} -x\, dx$

17. $\int_{-5}^{0} -3x^2\, dx$ **18.** $\int_{-1}^{4} \pi x^2\, dx$

In Exercises 19–20 approximate the area A of the region between the graph of f and the x axis on $[a, b]$ by using the left sum with the indicated partition.

19. $f(x) = x/(x + 1)$, $a = 0$, $b = 2$, $P = \{0, \frac{1}{2}, 1, 2\}$

 20. $f(x) = x^2$, $a = 1$, $b = 3$, P divides $[1, 3]$ into 10 subintervals of equal length.

In Exercises 21–24 find the area A of the region between the graph of f and the x axis on the given interval.

21. $f(x) = \frac{5}{2}$; $[-2, 3]$
22. $f(x) = \sqrt{3}$; $[\sqrt{3}, 3\sqrt{3}]$
23. $f(x) = x$; $[1, 4]$
24. $f(x) = -x$; $[-3, -1]$

25. Let f and g be continuous on $[a, b]$, and suppose that $f(x) \le g(x)$ for $a \le x \le b$. Show that

$$\int_a^b f(x)\, dx \le \int_a^b g(x)\, dx$$

In Exercises 26–30, use Exercise 25 to verify the given inequalities.

26. $\displaystyle\int_0^1 x^6\, dx \le \int_0^1 x\, dx$ **27.** $\displaystyle\int_1^2 x\, dx \le \int_1^2 x^6\, dx$

28. $\displaystyle\int_1^2 \frac{1}{x^6}\, dx \le \int_1^2 \frac{1}{x}\, dx$ **29.** $\displaystyle\int_0^{\pi/2} \sin x\, dx \le \int_0^{\pi/2} x\, dx$

30. $\displaystyle\int_0^{\pi/4} \sin x\, dx \le \int_0^{\pi/4} \cos x\, dx$

*31. Using Examples 1 and 2 as guides, prove that

$$\int_a^b (x + 4)\, dx = \frac{1}{2}(b^2 - a^2) + 4(b - a)$$

32. Using the method of Example 2, show that for any constant c and any numbers a and b with $a < b$, we have

$$\int_a^b cx\, dx = \frac{c}{2}(b^2 - a^2)$$

33. Using the result of Exercise 32, evaluate the following integrals.

a. $\displaystyle\int_1^3 4x\, dx$ **b.** $\displaystyle\int_2^6 -\frac{1}{2}x\, dx$

c. $\displaystyle\int_{-5}^{5} \sqrt{3}x\, dx$ **d.** $\displaystyle\int_{-1/\pi}^{0} \pi x\, dx$

34. Suppose $f(x) = mx + c$ for $a \le x \le b$. Let P be any partition that divides $[a, b]$ into equal subintervals. Show that the midpoint sum gives the exact value of $\int_a^b f(x)\, dx$.

*35. Let $0 \le a < b$. Show that

$$\int_a^b x^2\, dx = \frac{1}{3}(b^3 - a^3)$$

(*Hint:* Use the fact that if $x_{k-1} < x_k$, then

$$x_{k-1}^2 \le \frac{1}{3}(x_k^2 + x_k x_{k-1} + x_{k-1}^2) < x_k^2$$

and then let $t_k = (x_k^2 + x_k x_{k-1} + x_{k-1}^2)/3$. Then use the method of Example 2.)

36. Eventually we will show that for any numbers a and b with $a < b$, we have

$$\int_a^b x^3\, dx = \frac{1}{4}(b^4 - a^4)$$

Use this formula to evaluate the following integrals.

a. $\displaystyle\int_0^2 x^3\, dx$ **b.** $\displaystyle\int_{-1}^{1} x^3\, dx$

37. Using the results of Examples 1 and 2, together with Exercises 35 and 36, guess a formula for

$$\int_a^b x^n\, dx$$

that would hold for any positive integer n.

38. Suppose f' is continuous on $[a, b]$. It can be shown that if $K \ge$ the maximum value of $|f'|$ on $[a, b]$, then

$$E_n^L \le \frac{K}{2n}(b - a)^2 \quad \text{and} \quad E_n^R \le \frac{K}{2n}(b - a)^2$$

The right sides of the inequalities are **error bounds** for the left sum and right sum, respectively. Notice that they approach zero as n increases without bound. Find error bounds for the following.

a. $\displaystyle\int_1^2 \frac{1}{x}\, dx$ **b.** $\displaystyle\int_0^3 \sin x\, dx$ **c.** $\displaystyle\int_1^2 e^{1/x}\, dx$

39. Suppose f'' is continuous on $[a, b]$. It can be shown that if $K \ge$ the maximum value of $|f''|$ on $[a, b]$, then

$$E_n^M \le \frac{K}{24n^2}(b - a)^3$$

The right side of the inequality is an **error bound** for the midpoint sum. Notice that it approaches zero as n increases without bound. Find error bounds for the following.

a. $\displaystyle\int_1^2 \frac{1}{x}\, dx$ **b.** $\displaystyle\int_0^3 \sin x\, dx$ **c.** $\displaystyle\int_0^1 e^{-x^2}\, dx$

40. Let f be any odd function, and let P be any partition that divides $[-a, a]$ into subintervals of equal length. Describe a simple method of finding a Riemann sum for $\int_{-a}^{a} f(x)\, dx$ so that

$$\sum_{k=1}^{n} f(t_k)\, \Delta x_k = 0$$

(*Hint:* Recall that $f(-x) = -f(x)$ for $-a \le x \le a$.)

*41. Let a and b be arbitrary real numbers, with $a < b$. Show that $\int_a^b e^x dx = e^b - e^a$. (*Hint:* By applying the Mean Value Theorem to each subinterval $[x_{k-1}, x_k]$ associated with a partition P, and using the fact that e^x is an increasing function of x, show that

$$e^{x_{k-1}} < \frac{e^{x_k} - e^{x_{k-1}}}{x_k - x_{k-1}} < e^{x_k}$$

Use this information to show that $L_f(P) < e^b - e^a < U_f(P)$ for every partition P.)

 42. The area A of the unit disk centered at the origin is given by

$$A = 2 \int_{-1}^{1} \sqrt{1 - x^2}\, dx$$

Let P divide the interval $[-1, 1]$ into 100 subintervals of equal length. Approximate A by the corresponding

i. left sum   **ii.** right sum   **iii.** midpoint sum

 43. The area A of the ellipse pictured in Figure 5.18 is given by $A = 2(b/a) \int_{-a}^{a} \sqrt{a^2 - x^2}\, dx$.

a. Let $a = 3$ and $b = 2$. Approximate A by the midpoint sum, where the partition P has 100 subintervals of equal length.

b. By further experimentation if needed, conjecture a simple formula for A that involves a, b, and π, but does not involve an integral.

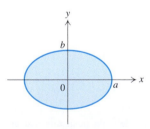

FIGURE 5.18   Graph for Exercise 43.

44. The function

$$f(x) = \frac{1}{\sqrt{2\pi}} e^{-x^2/2} \quad \text{for } x \text{ real}$$

is the **standard normal density** (Figure 5.19).

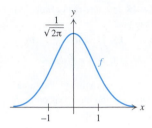

FIGURE 5.19   Graph for Exercise 44.

a. Use the midpoint sum with 100 subintervals to show that

$$\int_{-1}^{1} \frac{1}{\sqrt{2\pi}} e^{-x^2/2}\, dx \approx 0.68$$

(This means that approximately 68% of the area under the graph of f lies between $x = -1$ and $x = 1$, that is, it lies within one standard deviation of 0.)

b. Use the midpoint sum with 100 subintervals to show that

$$\int_{-2}^{2} \frac{1}{\sqrt{2\pi}} e^{-x^2/2}\, dx \approx 0.95$$

Applications

45. Suppose one wishes to approximate the area A of the lake depicted in Figure 5.20. Using the measurements shown, which were taken 100 feet apart, approximate the area, using a midpoint sum with three subintervals.

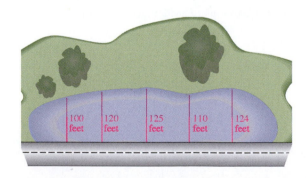

FIGURE 5.20   Graph for Exercise 45.

Project

1. In this project we show that sometimes the integral of a function f can be defined (that is, the Riemann sums approach a unique number I as the subintervals of partitions shrink) even if f is not continuous.

a. Let $f(x) = 1$ for all x in $[0, 1]$ except $x = \frac{1}{2}$, and $f(\frac{1}{2}) = 0$. Show that I exists, and find its value.

b. Let $g(x) = 1$ for all x in $[0, 1]$ except for a finite collection of numbers $m_1, m_2, \ldots, m_k$ in $(0, 1)$, and $g(m_j) = 0$ for $j = 1, 2, \ldots, k$. Show that I exists, and find its value.

c. Let $h(x) = 1$ for all rational numbers x in $[0, 1]$, and $h(x) = 0$ for all irrational numbers x in $[0, 1]$. Show that there is *no* number I that all Riemann sums approach. Thus the integral $\int_0^1 h(x)\, dx$ cannot be defined by Riemann sums. (*Hint:* Every interval $[c, d]$ with $c < d$ contains both rational and irrational numbers.)

5.3   SPECIAL PROPERTIES OF THE DEFINITE INTEGRAL

In this section we return to the three basic properties of area from which our definition of integral was derived, and we present them in terms of integrals. The theorems of this section will be used repeatedly throughout the remainder of the book.

Before reformulating the three basic properties in terms of integrals, we recall that in Definition 5.4 we defined $\int_a^b f(x)\, dx$ under the stipulation that $a < b$. For theoretical purposes and for later applications it will be convenient to give meaning to $\int_a^a f(x)\, dx$ and $\int_b^a f(x)\, dx$ when $a < b$.

DEFINITION   5.6

> Let f be continuous on $[a, b]$. Then
> $$\int_a^a f(x)\, dx = 0 \quad \text{and} \quad \int_b^a f(x)\, dx = -\int_a^b f(x)\, dx$$

The definition $\int_a^a f(x)\, dx = 0$ is consistent with our expectation that a line segment has area 0 (Figure 5.21).

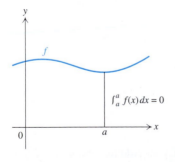

FIGURE 5.21   The area of a line segment is 0.

EXAMPLE 1   Evaluate $\displaystyle\int_4^1 x^2\, dx$.

Solution   By Definition 5.6 we have
$$\int_4^1 x^2\, dx = -\int_1^4 x^2\, dx$$
and by (4) in Section 5.2 we know that
$$\int_1^4 x^2\, dx = \frac{1}{3}(4^3 - 1^3) = 21$$
Therefore we conclude that
$$\int_4^1 x^2\, dx = -21 \qquad \square$$

Integral Forms of the Three Basic Properties

The first property to appear is an integral form of the Rectangle Property, which is related to Example 1 of Section 5.2.

THEOREM 5.7
Rectangle Property

For any numbers a, b, and c,

$$\int_a^b c\,dx = c(b - a)$$

Proof   If $a < b$, then the result follows directly from Example 1 of Section 5.2. If $a = b$, then by Definition 5.6,

$$\int_a^b c\,dx = \int_a^a c\,dx = 0 = c(b - a)$$

Finally, if $a > b$, then by combining Definition 5.6 and Example 1 of Section 5.2, we obtain

$$\int_a^b c\,dx = -\int_b^a c\,dx = -c(a - b) = c(b - a) \quad ■$$

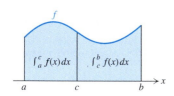

For the Addition Property we give a version that in special cases can be interpreted geometrically as follows: The area of a region composed of two smaller regions overlapping only in a line segment is the sum of the areas of the two regions (Figure 5.22).

FIGURE 5.22

Since the proof is technical, we have placed it in the Appendix.

THEOREM 5.8
Addition Property

Let f be continuous on an interval containing a, b, and c. Then

$$\int_a^b f(x)\,dx = \int_a^c f(x)\,dx + \int_c^b f(x)\,dx$$

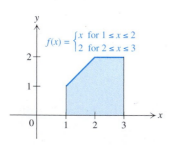

FIGURE 5.23

EXAMPLE 2   Evaluate $\int_1^3 f(x)\,dx$, where

$$f(x) = \begin{cases} x & \text{for } 1 \le x \le 2 \\ 2 & \text{for } 2 \le x \le 3 \end{cases}$$

(See Figure 5.23.)

Solution   Notice that f is continuous on $[1, 3]$. By the Addition Property,

$$\int_1^3 f(x)\,dx = \int_1^2 f(x)\,dx + \int_2^3 f(x)\,dx$$

Now by Examples 1 and 2 of Section 5.2,

$$\int_1^2 f(x)\,dx = \int_1^2 x\,dx = \frac{1}{2}(2^2 - 1^2) = \frac{3}{2}$$

and

$$\int_2^3 f(x)\,dx = \int_2^3 2\,dx = 2(3-2) = 2$$

We conclude that

$$\int_1^3 f(x)\,dx = \int_1^2 f(x)\,dx + \int_2^3 f(x)\,dx = \frac{3}{2} + 2 = \frac{7}{2} \qquad \square$$

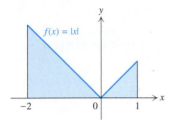

FIGURE 5.24

EXAMPLE 3   Evaluate $\int_1^{-2} |x|\,dx$ (see Figure 5.24).

Solution   With the help of the Addition Property and Definition 5.6, we find that

$$\int_1^{-2} |x|\,dx = \int_1^0 |x|\,dx + \int_0^{-2} |x|\,dx = \int_1^0 x\,dx + \int_0^{-2} -x\,dx$$

$$= -\int_0^1 x\,dx + \int_{-2}^0 x\,dx = -\frac{1}{2}(1^2 - 0^2) + \frac{1}{2}[0^2 - (-2)^2]$$

$$= -\frac{1}{2} - 2 = -\frac{5}{2} \qquad \square$$

It turns out that the Addition Property is valid not only for functions that are continuous on $[a, b]$ but also functions that are continuous on $[a, b]$ except at a finite number of points in $[a, b]$, provided that at such points of discontinuity the function has both left and right limits. Such functions are said to be **piecewise continuous** because they are composed of continuous "pieces." For example, consider

$$f(x) = \begin{cases} 2 & \text{if } -1 \le x \le 0 \\ x & \text{if } 0 < x \le 1 \end{cases}$$

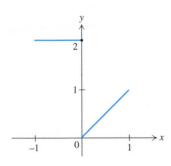

FIGURE 5.25

(See Figure 5.25.) Then f consists of two separate parts, one part defined on $[-1, 0]$ and the other part defined on $(0, 1]$. Thus f is piecewise continuous on $[-1, 1]$.

It is possible to prove that if f is piecewise continuous on $[a, b]$ and is continuous except at $\{c_1, c_2, \ldots, c_n\}$ in (a, b), then $\int_a^b f(x)\,dx$ exists, and that

$$\int_a^b f(x)\,dx = \int_a^{c_1} f(x)\,dx + \int_{c_1}^{c_2} f(x)\,dx + \cdots + \int_{c_n}^b f(x)\,dx$$

where each of the integrals on the right is evaluated by considering the continuous extension of f to the interval of integration.

For example, if

$$f(x) = \begin{cases} 2 & \text{if } -1 \le x \le 0 \\ x & \text{if } 0 < x \le 1 \end{cases}$$

then

$$\int_{-1}^1 f(x)\,dx = \int_{-1}^0 f(x)\,dx + \int_0^1 f(x)\,dx = \int_{-1}^0 2\,dx + \int_0^1 x\,dx = 2 + \frac{1}{2} = \frac{5}{2}$$

The third property of area is the Comparison Property. We will need only a special case of the Comparison Property. In geometric terms it implies that the area of any region is at least as large as that of any inscribed rectangle and no larger than that of any circumscribed rectangle (Figure 5.26).

THEOREM 5.9
Comparison Property

Let f be continuous on $[a, b]$, and suppose $m \le f(x) \le M$ for all x in $[a, b]$. Then

$$m(b - a) \le \int_a^b f(x)\, dx \le M(b - a)$$

Proof Let $P = \{x_0, x_1, \ldots, x_n\}$ be any partition of $[a, b]$, and let t_k be in $[x_{k-1}, x_k]$ for each k. Since $m \le f(x) \le M$ for all x in $[a, b]$, it follows that necessarily $m \le f(t_k) \le M$ for all $k = 1, 2, \ldots, n$. Therefore since

$$\sum_{k=1}^n \Delta x_k = b - a$$

we have

$$m(b - a) = \sum_{k=1}^n m\Delta x_k \le \sum_{k=1}^n f(t_k)\Delta x_k \le \sum_{k=1}^n M\Delta x_k = M(b - a)$$

The inequality above is valid for *any* Riemann sum, and hence is valid for the limit of the Riemann sums, which is the integral $\int_a^b f(x)\, dx$. Consequently

$$m(b - a) \le \int_a^b f(x)\, dx \le M(b - a) \quad \blacksquare$$

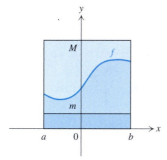

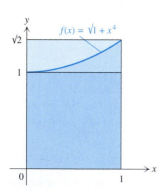

FIGURE 5.26

FIGURE 5.27

The Comparison Property is frequently used to estimate the value of an integral that cannot be computed exactly or easily. A number less than or equal to a given integral is called a **lower bound** for the integral, and a number greater than or equal to the integral is an **upper bound** for the integral.

EXAMPLE 4   Using the Comparison Property, find lower and upper bounds for

$$\int_0^1 \sqrt{1 + x^4}\, dx$$

Solution   If $0 \le x \le 1$, then $1 \le 1 + x^4 \le 2$, so that

$$1 \le \sqrt{1 + x^4} \le \sqrt{2} \quad \text{for } 0 \le x \le 1$$

(See Figure 5.27.) By the Comparison Property it follows that

$$1(1 - 0) \le \int_0^1 \sqrt{1 + x^4}\, dx \le \sqrt{2}(1 - 0)$$

so that

$$1 \le \int_0^1 \sqrt{1 + x^4}\, dx \le \sqrt{2}$$

Thus 1 is a lower bound and $\sqrt{2}$ is an upper bound for the integral.   ❑

In Example 4 of Section 5.5 we will obtain a better, that is, smaller upper bound for $\int_0^1 \sqrt{1 + x^4}\, dx$.

Consequences of the Comparison Property

Our first consequence of the Comparison Property tells us that the integral of a nonnegative function is nonnegative.

COROLLARY 5.10

Let f be nonnegative and continuous on $[a, b]$. Then

$$\int_a^b f(x)\, dx \geq 0$$

Proof   By hypothesis, $f(x) \geq 0$ for $a \leq x \leq b$. Therefore it is permissible to let $m = 0$ in Theorem 5.9. This implies that

$$\int_a^b f(x)\, dx \geq 0(b - a) = 0 \quad \blacksquare$$

From Corollary 5.10 it follows that the area of any region covered by Definition 5.5 is a nonnegative number. A second consequence of the Comparison Property is an integral form of the Mean Value Theorem.

THEOREM 5.11
Mean Value Theorem for Integrals

Let f be continuous on $[a, b]$. Then there is a number c in $[a, b]$ such that

$$\int_a^b f(x)\, dx = f(c)(b - a)$$

Proof   If $a = b$, then the result is obvious. Thus assume that $a < b$, and let m and M be the minimum and the maximum values of f on $[a, b]$. By the Comparison Property we know that

$$m(b - a) \leq \int_a^b f(x)\, dx \leq M(b - a)$$

Since $a < b$, and hence $b - a > 0$, this means that

$$m \leq \frac{\int_a^b f(x)\, dx}{b - a} \leq M$$

Since f is continuous on $[a, b]$, the Intermediate Value Theorem asserts that there is a number c in $[a, b]$ such that

$$\frac{\int_a^b f(x)\, dx}{b - a} = f(c)$$

that is,

$$\int_a^b f(x)\, dx = f(c)(b - a) \quad \blacksquare$$

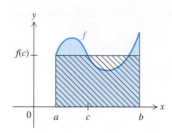

FIGURE 5.28

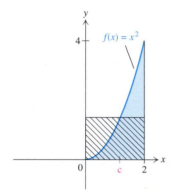

FIGURE 5.29

It follows from the Mean Value Theorem for Integrals that if f is nonnegative and continuous on $[a, b]$, then the area of the region bounded above by the graph of f is the same as the area of a rectangle whose height is $f(c)$ for some properly chosen c in $[a, b]$ (Figure 5.28).

EXAMPLE 5   Let $f(x) = x^2$, and let A be the area of the region R between the graph of f and the x axis on $[0, 2]$. Find the value of c such that A is the same as the area of the rectangle with base $[0, 2]$ and height $f(c)$ (Figure 5.29).

Solution   By (4) of Section 5.2,

$$A = \int_0^2 x^2 \, dx = \frac{1}{3}(2^3 - 0^3) = \frac{8}{3}$$

Thus the rectangle must have area $\frac{8}{3}$ and the base must have length 2. Therefore the height $f(c)$ of the rectangle must be $\frac{4}{3}$. Consequently $c^2 = f(c) = \frac{4}{3}$, so that $c = \sqrt{4/3}$.   ❑

The value

$$\frac{1}{b - a} \int_a^b f(x) \, dx \tag{1}$$

that arises from the Mean Value Theorem for Integrals is called the **mean** (or **average**) **value** of f on the interval $[a, b]$. When the interval is obvious, the mean or average value of f is denoted f_{av}.

EXAMPLE 6   Suppose that a ball drops into a hole 4 feet deep. Find the mean height during its fall.

Solution   Measured in terms of feet, the height $h(t)$ of the ball after t seconds is given by $h(t) = -16t^2$ until the ball hits the bottom of the hole, which occurs after $\frac{1}{2}$ second. Why? It follows that the mean value h_{av} of the height is given by

$$h_{av} = \frac{1}{1/2 - 0} \int_0^{1/2} (-16t^2) \, dt = 2 \int_0^{1/2} (-16t^2) \, dt$$

By (4) in Section 5.2,

$$\int_0^{1/2} (-16t^2) \, dt = -\frac{16}{3}\left[\left(\frac{1}{2}\right)^3 - 0^3\right] = -\frac{2}{3}$$

Therefore

$$h_{av} = 2\left(-\frac{2}{3}\right) = -\frac{4}{3}$$

We conclude that the mean height of the ball is $-\frac{4}{3}$ feet.   ❑

The reason that the mean height of the ball in Example 6 is not –2 feet is that the ball drops more slowly at the beginning of its journey and more swiftly at the end.

Later we will see that if f represents velocity, then the mean value of velocity on $[a, b]$ is the same as the average velocity defined in Section 2.1. Corresponding statements apply to mean cost and mean revenue.

EXERCISES 5.3

In Exercises 1–4 use the Rectangle Property to evaluate the integral.

1. $\displaystyle\int_3^5 7\,dx$ **2.** $\displaystyle\int_{-1}^2 -3\,dx$

3. $\displaystyle\int_2^{-1} -10\,du$ **4.** $\displaystyle\int_1^{-1} 5\,dx$

In Exercises 5–8 evaluate the integrals to corroborate the Addition Property for integrals. Use the results of Section 5.2 in your calculations.

5. $\displaystyle\int_0^1 x\,dx + \int_1^2 x\,dx = \int_0^2 x\,dx$

6. $\displaystyle\int_3^4 x^2\,dx + \int_4^3 x^2\,dx = \int_3^3 x^2\,dx$

7. $\displaystyle\int_1^0 y^2\,dy + \int_0^2 y^2\,dy = \int_1^2 y^2\,dy$

8. $\displaystyle\int_{-2}^{-3} -y\,dy + \int_{-3}^{-6} -y\,dy = \int_{-2}^{-6} -y\,dy$

In Exercises 9–12 let f be a continuous function on $(-\infty, \infty)$. Use the Addition Property to find the values of a and b that make the equation true.

9. $\displaystyle\int_0^2 f(x)\,dx + \int_3^0 f(x)\,dx = \int_a^b f(x)\,dx$

10. $\displaystyle\int_{1/2}^{-1/2} f(x)\,dx + \int_{-1}^{1/2} f(x)\,dx = \int_a^b f(x)\,dx$

11. $\displaystyle\int_a^b f(t)\,dt - \int_5^3 f(t)\,dt = \int_3^1 f(t)\,dt$

12. $\displaystyle\int_\pi^{2\pi} f(t)\,dt - \int_a^b f(t)\,dt = \int_{3\pi}^{2\pi} f(t)\,dt$

In Exercises 13–16 find the maximum and minimum values of the given function on the given interval. Then use the Comparison Property to find upper and lower bounds for the area of the region between the graph of the function and the x axis on the given interval.

13. $f(x) = 1/x$; [2, 3]   **14.** $g(x) = \sin x$; [$\pi/4$, $\pi/2$]
15. $g(x) = \cos x$; [$\pi/4$, $\pi/3$]   **16.** $h(t) = \tan t$; [0, $\pi/3$]

In Exercises 17–20 find the mean value of f on the given interval.

17. $f(x) = x$; [0, 1]   **18.** $f(x) = x$; [−2, 2]
19. $f(x) = x^2$; [−1, 1]   **20.** $f(x) = |x|$; [−2, 3]

21. Let $f(x) = x$ for $a \le x \le b$. Show that the mean value of f on [a, b] is $(a + b)/2$.

*22. Let $0 \le a < b$ and let $f(x) = x^2$.
  a. Show that the mean value of f on [a, b] is
  $$\tfrac{1}{3}(a^2 + ab + b^2)$$
  b. Show that there is a number c in [a, b] such that
  $$c^2 = \frac{1}{3}(a^2 + ab + b^2)$$

23. Let
$$f(x) = \begin{cases} -x & \text{for } x < 0 \\ x^2 & \text{for } x \ge 0 \end{cases}$$
Find the area A of the region between the graph of f and the x axis on [−1, 1].

24. Let
$$f(x) = \begin{cases} 1 & \text{for } 0 \le x \le 1 \\ x^2 & \text{for } x > 1 \end{cases}$$
Find the area A of the region between the graph of f and the x axis on [0, 4].

25. Consider the graph of f in Figure 5.30.

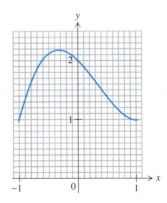

FIGURE 5.30   Graph for Exercise 25.

  a. By the graph alone, predict the mean value of f on [−1, 1].
  b. Use the left sum with 10 subintervals to approximate $\int_{-1}^1 f(x)\,dx$ and then the mean value of f. (The actual mean value of f on [−1, 1] is $\tfrac{5}{3}$.)

26. Let $f(x) = \sin x^2$ for $0 \le x \le \sqrt{\pi}$.
  a. Plot the graph of f.

b. Predict the mean value f_{av} of f on $[0, \sqrt{\pi}]$. Call your prediction A.

c. Use the midpoint sum with $n = 100$ in order to find an approximate value B of f_{av}.

d. Compare the values A and B.

27. Prove the Addition Property for the cases $c < a < b$ and $b < a < c$.

28. Show that the Comparison Property is a consequence of the Mean Value Theorem for Integrals.

Applications

29. If T represents the temperature at time t, then the mean temperature during an interval of length b is $\frac{1}{b} \int_0^b T(t)\, dt$. In degrees Fahrenheit the average temperature for each of the months of the year in New York City is as follows:

Jan	Feb	Mar	Apr	May	Jun	Jul	Aug	Sep	Oct	Nov	Dec
32	33	41	53	62	71	77	75	68	58	47	36

First prove, using the Addition Property, that the mean temperature during the year is the average of the mean temperatures during the twelve months. Then use the values in the table to find the mean temperature in New York during the year.

30. Suppose that the temperature T is taken regularly during a 24-hour period. By definition the average of n successive readings is given by

$$\frac{1}{n} \sum_{k=1}^n T\left(\frac{24k}{n}\right)$$

By using an appropriate Riemann sum, show that if n is large then the average temperature over n readings approximates the mean temperature over the same period.

Project

1. This project focuses on the integral of a function f that is continuous on $[a, b]$ and whose values are positive except possibly at the endpoints a and b.

a. Show that $\int_{\pi/4}^{\pi/2} \sin x\, dx > 0$.

b. Show that $\int_0^{\pi/2} \sin x\, dx > 0$ (*Hint:* Use the Addition Property with $a = 0$, $b = \pi/2$, and $c = \pi/4$, and then apply the Comparison Property to each of the resulting integrals.)

c. Suppose f is continuous on $[a, b]$. Assume that $f(x) \geq 0$ for all x in $[a, b]$ and $f(x) > 0$ for at least one value of x in $[a, b]$. Show that $\int_a^b f(x)\, dx > 0$.

d. Let f be a continuous and nonnegative on $[a, b]$, and assume that $\int_a^b f(x)\, dx = 0$. Show that $f(x) = 0$ for *all* x in $[a, b]$.

5.4   THE FUNDAMENTAL THEOREM OF CALCULUS

The purpose of this section is to develop a general method for evaluating $\int_a^b f(x)\, dx$ that does not necessitate computing various sums. The method will allow us to evaluate many (but not all) of the integrals that arise in applications.

We begin by letting I be any interval and c any number in I. Suppose f is continuous on I. Then for each x in I, f is necessarily continuous on the closed interval whose endpoints are c and x. Consequently we can associate with any such x the number $\int_c^x f(t)\, dt$ in order to obtain a function G that is defined by

$$G(x) = \int_c^x f(t)\, dt \quad \text{for } x \text{ in } I \tag{1}$$

Here we have used t inside the integral, rather than x, because x appears as a limit of integration.

For example, if $f(x) = x$, $I = (-\infty, \infty)$, and $c = 1$, then by (3) in Section 5.2,

$$G(0) = \int_1^0 t\, dt = -\int_0^1 t\, dt = -\frac{1}{2}(1^2 - 0^2) = -\frac{1}{2}$$

$$G(1) = \int_1^1 t \, dt = 0$$

$$G(2) = \int_1^2 t \, dt = \frac{1}{2}(2^2 - 1^2) = \frac{3}{2}$$

In fact, for any x,

$$G(x) = \int_1^x t \, dt = \frac{1}{2}(x^2 - 1)$$

Notice that in this example, $G'(x) = x = f(x)$ for all x. We will prove more generally that if f is *any* function that is continuous on an interval I, and if G is defined as in (1), then $G'(x) = f(x)$ for all x in I, so that G is an antiderivative of f on I.

To see geometrically why we can expect to have $G' = f$, let us suppose that f is continuous and nonnegative on I, and let x and z be any numbers in I with $x < z$, as in Figure 5.31. Since $f \geq 0$, it follows from the definition of G in (1) that

$G(z)$ = area of the region between the graph of f and the x axis on $[c, z]$

$G(x)$ = area of the region between the graph of f and the x axis on $[c, x]$

Therefore $G(z) - G(x)$ is the area of the entire shaded region in Figure 5.31, and if z is close to x, this area appears to be close to the area $f(x)(z - x)$ of the darkly shaded rectangle in the figure. We deduce that if $z > x$ and z is close to x, then $G(z) - G(x)$ should be close to $f(x)(z - x)$, and hence that

$$\frac{G(z) - G(x)}{z - x} \quad \text{should be close to } f(x)$$

A similar analysis would show that this approximation also holds if $z < x$ and z is close to x. We conclude intuitively that

$$\lim_{z \to x} \frac{G(z) - G(x)}{z - x} = f(x)$$

that is,

$$G'(x) = f(x)$$

We now state the result in the preceding paragraph formally and prove it.

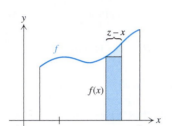

FIGURE 5.31

THEOREM 5.12

Let f be continuous on an interval I (containing more than one point) and let c be any point in I. Define G by the equation

$$G(x) = \int_c^x f(t) \, dt \quad \text{for all } x \text{ in } I$$

Then G is differentiable on I, and

$$G'(x) = f(x) \quad \text{for all } x \text{ in } I \tag{2}$$

In particular, f has an antiderivative on I.

Proof   Let x be any number interior to I. We must show that

$$\lim_{z \to x} \frac{G(z) - G(x)}{z - x} = f(x)$$

To that end, let z be any number in I with $z \neq x$. The Addition Property tells us that

$$G(z) - G(x) = \int_c^z f(t)\, dt - \int_c^x f(t)\, dt = \int_x^z f(t)\, dt$$

If ε is any positive number, then because f is assumed to be continuous at x, there is a $\delta > 0$ such that if $|t - x| < \delta$, then $|f(t) - f(x)| < \varepsilon$, that is, $f(x) - \varepsilon < f(t) < f(x) + \varepsilon$. Now suppose that $0 < |z - x| < \delta$. By the Comparison Property, we find that if $x < z$, then

$$(f(x) - \varepsilon)(z - x) \leq \int_x^z f(t)\, dt = G(z) - G(x) \leq (f(x) + \varepsilon)(z - x)$$

and if $x > z$, then

$$(f(x) - \varepsilon)(x - z) \leq \int_z^x f(t)\, dt = G(x) - G(z) \leq (f(x) + \varepsilon)(x - z)$$

If we divide through the first set of inequalities by $z - x$, and the second set by $x - z$, then each set of inequalities reduces to

$$f(x) - \varepsilon \leq \frac{G(z) - G(x)}{z - x} \leq f(x) + \varepsilon, \text{ that is, } \left| \frac{G(z) - G(x)}{z - x} - f(x) \right| \leq \varepsilon$$

Therefore

$$G'(x) = \lim_{z \to x} \frac{G(z) - G(x)}{z - x} = f(x)$$

If x is an endpoint of I, then $G'(x)$ is just the appropriate one-sided limit, and the foregoing argument can be altered to apply.   ■

EXAMPLE 1   Let $G(x) = \displaystyle\int_1^x (1/t^2)\, dt$ for $x > 0$. Find $G'(x)$.

Solution   By Theorem 5.12, with $I = (0, \infty)$ and $c = 1$, we have

$$G'(x) = \frac{1}{x^2} \quad \text{for } x > 0 \qquad \square$$

EXAMPLE 2   Let $G(x) = \displaystyle\int_0^x t \sin t^3\, dt$ for all x. Find $G'(x)$.

Solution   Again by Theorem 5.12,

$$G'(x) = x \sin x^3 \qquad \square$$

At this point let us recall from Theorem 4.6 that any two antiderivatives of a function f differ by a constant. For example, the antiderivatives of $2x$ all have the form $x^2 + C$, where C is a constant. For easy reference we list in Table 5.1 a few of the basic functions, along with their antiderivatives. To verify any entry in the column of antiderivatives, simply differentiate it and observe that its derivative is the same as the corresponding entry in the left column.

TABLE 5.1

Table of Antiderivatives

Function	Antiderivative
c (c a constant)	$cx + C$ (C any constant)
x	$\dfrac{1}{2} x^2 + C$
x^2	$\dfrac{1}{3} x^3 + C$
$px + q$ (p and q constants)	$\dfrac{1}{2} px^2 + qx + C$
x^r ($r \neq -1$)	$\dfrac{1}{r+1} x^{r+1} + C$
$\dfrac{1}{x}$ ($x > 0$)	$\ln x + C$
$\sin x$	$-\cos x + C$
$\cos x$	$\sin x + C$
e^x	$e^x + C$

Now we are ready for the most important theorem in calculus. It shows that the notions of derivative and integral are intimately related, and also provides an effective method of evaluating numerous integrals.

THEOREM 5.13
Fundamental Theorem
of Calculus

Let f be continuous on $[a, b]$.

a. The function G defined by

$$G(x) = \int_a^x f(t)\, dt, \quad \text{for} \quad x \quad in \quad [a, b]$$

is an antiderivative of f on $[a, b]$.

b. If F is any antiderivative of f on $[a, b]$, then

$$\int_a^b f(x)\, dx = F(b) - F(a)$$

Proof   Part (a) follows immediately from Theorem 5.12. To prove (b), we let G be the function defined in (a). Then

$$G(a) = \int_a^a f(t)\, dt = 0 \quad \text{and} \quad G(b) = \int_a^b f(t)\, dt$$

Now if F is any antiderivative of f, then by Theorem 4.6 we know that $F = G + C$

for some constant C. Consequently

$$\int_a^b f(t)\,dt = G(b) = G(b) - G(a) = [F(b) - C] - [F(a) - C]$$

$$= F(b) - F(a) \quad \blacksquare$$

EXAMPLE 3   Evaluate $\int_0^2 x^2\,dx$.

Solution   From Table 5.1 we know that if $F(x) = \frac{1}{3}x^3$ then F is an antiderivative of x^2, so that by the Fundamental Theorem,

$$\int_0^2 x^2\,dx = F(2) - F(0) = \frac{1}{3} \cdot 2^3 - \frac{1}{3} \cdot 0^3 = \frac{8}{3} \quad \square$$

It is usually simplest to dispense with the symbol F when evaluating integrals. Instead we write, for example,

$$\int_0^2 x^2\,dx = \frac{x^3}{3}\bigg|_0^2 = \frac{1}{3} \cdot 2^3 - \frac{1}{3} \cdot 0^3 = \frac{8}{3}$$

where the expression $\big|_0^2$ indicates the numbers at which the antiderivative is to be evaluated, in this case 0 and 2.

EXAMPLE 4   Evaluate $\int_1^4 x^{1/2}\,dx$.

Solution   Since $\frac{2}{3}x^{3/2}$ is an antiderivative of $x^{1/2}$, the Fundamental Theorem asserts that

$$\int_1^4 x^{1/2}\,dx = \frac{2}{3}x^{3/2}\bigg|_1^4 = \frac{2}{3}(4^{3/2}) - \frac{2}{3}(1^{3/2}) = \frac{16}{3} - \frac{2}{3} = \frac{14}{3} \quad \square$$

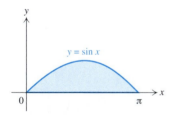

$y = \sin x$

FIGURE 5.32   The area under one arch of the sine curve is 2.

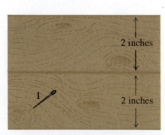

2 inches

2 inches

FIGURE 5.33   Buffon's needle problem.

EXAMPLE 5   Evaluate $\int_0^\pi \sin x\,dx$.

Solution   Because $-\cos x$ is an antiderivative of $\sin x$, we know from the Fundamental Theorem that

$$\int_0^\pi \sin x\,dx = -\cos x\bigg|_0^\pi = -\cos \pi - (-\cos 0) = -(-1) - (-1) = 2 \quad \square$$

From Example 5 it follows that the area of the region in Figure 5.32 is 2. Without the Fundamental Theorem this result would have been very hard to obtain. The integral in Example 5 arises in the famous eighteenth-century problem known as **Buffon's needle problem.** In this problem a 1-inch-long needle is dropped onto a hardwood floor whose boards are 2 inches wide. The question is: What is the probability that the needle will come to rest lying across two boards (Figure 5.33)?

See the project for this section for the details.

EXAMPLE 6   Evaluate $\displaystyle\int_{-1}^{1} (1 + 2x)\, dx$.

Solution   Using Table 5.1 or some trial and error, we find that $x + x^2$ is an antiderivative of $1 + 2x$. Thus

$$\int_{-1}^{1} (1 + 2x)\, dx = (x + x^2)\Big|_{-1}^{1} = (1 + 1^2) - [-1 + (-1)^2] = 2 \qquad \square$$

EXAMPLE 7   Evaluate $\displaystyle\int_{-1}^{1} e^x\, dx$.

Solution   As indicated in Table 5.1, e^x is an antiderivative of e^x. Therefore

$$\int_{-1}^{1} e^x\, dx = e^x\,\Big|_{-1}^{1} = e^1 - e^{-1} = e - 1/e \qquad \square$$

We can extend the Fundamental Theorem of Calculus to the case in which the lower limit of integration is greater than the upper limit.

COROLLARY 5.14

Let f be continuous on $[a, b]$. Then for any antiderivative F of f,
$$\int_{b}^{a} f(t)\, dt = F(a) - F(b)$$

Proof   By the Fundamental Theorem,

$$\int_{a}^{b} f(t)\, dt = F(b) - F(a)$$

Therefore by Definition 5.6,

$$\int_{b}^{a} f(t)\, dt = -\int_{a}^{b} f(t)\, dt = -[F(b) - F(a)] = F(a) - F(b) \qquad \blacksquare$$

As a consequence of the Fundamental Theorem and Corollary 5.14, once we know an antiderivative F of f, we can compute $\int_{a}^{b} f(t)\, dt$ by the formula

$$\int_{a}^{b} f(t)\, dt = F(b) - F(a) \tag{3}$$

whether $a < b$, $a > b$, or $a = b$.

EXAMPLE 8   Find $\displaystyle\int_{2}^{1} 2t^3\, dt$.

Solution   Since $\dfrac{d}{dt}\left(\dfrac{1}{2}t^4\right) = 2t^3$, we deduce from (3) that

$$\int_{2}^{1} 2t^3 \, dt = \frac{1}{2} t^4 \Big|_{2}^{1} = \frac{1}{2} \cdot 1^4 - \frac{1}{2} \cdot 2^4 = \frac{1}{2} - 8 = -\frac{15}{2} \qquad \square$$

Formula (2) can be restated as

$$\frac{d}{dx} \int_{c}^{x} f(t) \, dt = f(x) \tag{4}$$

Using the Fundamental Theorem and the Chain Rule we can evaluate

$$\frac{d}{dx} \int_{g(x)}^{h(x)} f(t) \, dt$$

where g and h are differentiable functions of x, and where f is continuous between $g(x)$ and $h(x)$ for all appropriate values of x. Indeed, let F be any antiderivative of f. By (3) with $g(x)$ replacing a and $h(x)$ replacing b, we have

$$\int_{g(x)}^{h(x)} f(t) \, dt = F(h(x)) - F(g(x))$$

Therefore using the Chain Rule and the fact that $F' = f$, we find that

$$\frac{d}{dx} \int_{g(x)}^{h(x)} f(t) \, dt = \frac{d}{dx} [(F(h(x)) - F(g(x)))] = F'(h(x)) \, h'(x) - F'(g(x)) \, g'(x)$$

$$= f(h(x)) \, h'(x) - f(g(x)) \, g'(x)$$

We conclude that

$$\frac{d}{dx} \int_{g(x)}^{h(x)} f(t) \, dt = f(h(x)) \, h'(x) - f(g(x)) \, g'(x) \tag{5}$$

EXAMPLE 9   Let $K(x) = \int_{x^2}^{x^3} \sin t^7 \, dt$. Find $K'(x)$.

Solution   From (5) with $f(x) = \sin x^7$, $g(x) = x^2$, and $h(x) = x^3$, we deduce that

$$K'(x) = \frac{d}{dx} \int_{x^2}^{x^3} \sin t^7 \, dt = [\sin (x^3)^7](3x^2) - [\sin (x^2)^7](2x) = 3x^2 \sin (x^{21}) - 2x \sin (x^{14}) \qquad \square$$

Differentiation and Integration as Inverse Processes

According to (4), if we start with a continuous function f, integrate it to obtain $\int_{c}^{x} f(t) \, dt$, and then differentiate, the result is the original function f. Thus the differentiation has nullified the integration. On the other hand, if we start with a function F having a continuous derivative, first differentiate, and then integrate, we obtain $\int_{c}^{x} F'(t) \, dt$. But by the Fundamental Theorem of Calculus,

$$\int_{c}^{x} F'(t) \, dt = F(x) - F(c) \tag{6}$$

so we obtain the original function F altered by at most a constant. This time the integration has essentially nullified the differentiation. Thus the two basic processes of calculus, differentiation and integration, are inverses of each other.

Furthermore, whenever we know the derivative F' of a function F, (6) gives us an integration formula. For example, we know already that

$$\frac{d}{dx}\tan x = \sec^2 x$$

Therefore (6) tells us that

$$\int_{\pi/4}^{x} \sec^2 t \, dt = \tan x - \tan \frac{\pi}{4} = \tan x - 1$$

provided that x is in the interval $(-\pi/2, \pi/2)$.

Formula (6) has numerous applications. For example, in economics the marginal revenue function m_R is by definition the derivative of the total revenue function R. Thus by (6),

$$R(x) - R(c) = \int_{c}^{x} R'(t) \, dt = \int_{c}^{x} m_R(t) \, dt \tag{7}$$

Likewise, the marginal cost function m_C is by definition the derivative of the total cost function C. Consequently by (6),

$$C(x) - C(c) = \int_{c}^{x} C'(t) \, dt = \int_{c}^{x} m_C(t) \, dt \tag{8}$$

In physics, the velocity of a particle moving along a straight line is the derivative of the position function. If we use t for the independent variable representing time, f for the position, v for velocity, and u for the variable of integration, we obtain, by suitable substitution in (6),

$$f(t) - f(t_0) = \int_{t_0}^{t} v(u) \, du \tag{9}$$

In (9) the number t_0 is arbitrary, and it plays the same role as c does in (6). In applications t_0 is usually a special instant of time. When t_0 is the moment at which motion begins, it is called the **initial time.**

The acceleration a of a particle is the derivative of the velocity v. Hence we obtain

$$v(t) - v(t_0) = \int_{t_0}^{t} a(u) \, du \tag{10}$$

Near the surface of the earth, the acceleration due to gravity is essentially constant, approximately -9.8 meters per second per second. Assuming that an object is under the sole influence of gravity, we can derive the formula

$$h(t) = -4.9t^2 + v_0 t + h_0 \tag{11}$$

for the height of the object at time t. Formula (11) was first presented (without proof) in Section 1.3.

EXAMPLE 10 Suppose an object experiences a constant acceleration of -9.8 meters per second per second. Assume that at time $t = 0$ its initial height is h_0 and initial velocity is v_0. Show that the height $h(t)$ of the object at any time $t > 0$ is given by (11).

Solution Using (10) with $t_0 = 0$ and $a(u) = -9.8$, we find that

$$v(t) - v_0 = v(t) - v(0) = \int_0^t -9.8 \, du = -9.8u \Big|_0^t = -9.8t$$

Thus

$$v(t) = v_0 - 9.8t$$

From this equation, and from (9) with $t_0 = 0$ and f replaced by h, we find that

$$h(t) - h_0 = h(t) - h(0) = \int_0^t (v_0 - 9.8u) \, du = (v_0 u - 4.9u^2) \Big|_0^t = v_0 t - 4.9t^2$$

Therefore $h(t) = -4.9t^2 + v_0 t + h_0$, which is (11). ◻

Because differentiation and integration arose from apparently unrelated problems (such as tangents and areas), it was only after mathematicians had worked for centuries with concepts related to derivatives and integrals separately that Isaac Barrow, who was Newton's teacher, discovered and proved the Fundamental Theorem. His proof was completely geometric, and his terminology far different from ours. Beginning with the work of Newton and Leibniz, the theorem grew in importance, eventually becoming the cornerstone for the study of integration.

EXERCISES 5.4

In Exercises 1–10 find the derivative of each function.

1. $F(x) = \displaystyle\int_0^x t(1 + t^3)^{29} \, dt$

2. $F(x) = \displaystyle\int_3^x \frac{1}{(t + t^3)^{16}} \, dt$

3. $F(y) = \displaystyle\int_y^2 \frac{1}{t^3} \, dt$

4. $F(t) = \displaystyle\int_t^0 x \sin x \, dx$ **5.** $F(x) = \displaystyle\int_0^{x^2} t \sin t \, dt$

6. $F(x) = \displaystyle\int_0^{-x} e^{(t^2)} \, dt$

7. $G(y) = \displaystyle\int_y^{y^2} (1 + t^2)^{1/2} \, dt$

8. $F(x) = \displaystyle\int_{x^2}^{x^3} (1 + t^2)^{1/2} \, dt$

9. $F(x) = \dfrac{d}{dx} \displaystyle\int_0^{4x} (1 + t^2)^{4/5} \, dt$

10. $G(y) = \dfrac{d}{dy} \displaystyle\int_{\sin y}^{2y} \cos t \, dt$

In Exercises 11–40 use (3) to evaluate the integral.

11. $\displaystyle\int_0^1 4 \, dx$ **12.** $\displaystyle\int_1^{12} 0 \, dx$

13. $\displaystyle\int_1^3 -y \, dy$ **14.** $\displaystyle\int_5^2 -4t \, dt$

15. $\displaystyle\int_1^{-3} 3u \, du$ **16.** $\displaystyle\int_{-b}^b x^5 \, dx$, b a constant

17. $\int_0^1 x^{100}\,dx$

18. $\int_0^2 u^{1/2}\,du$

19. $\int_{-1}^1 u^{1/3}\,du$

20. $\int_{16}^2 x^{5/4}\,dx$

21. $\int_1^4 x^{-7/9}\,dx$

22. $\int_0^1 x^{12/5}\,dx$

23. $\int_{-1.5}^{2\pi} (5-x)\,dx$

24. $\int_0^3 \left(\frac{1}{2}x - 4\right) dx$

25. $\int_{-4}^{-1} (5x + 14)\,dx$

26. $\int_0^{\pi/6} \cos x\,dx$

27. $\int_{-\pi}^{\pi/3} \cos x\,dx$

28. $\int_{\pi/3}^{\pi/4} \sin t\,dt$

29. $\int_{\pi/3}^{-\pi/4} \sin t\,dt$

30. $\int_2^3 \frac{1}{x^3}\,dx$

31. $\int_1^2 \frac{1}{y^4}\,dy$

32. $\int_{-1}^{-2} \left(x - \frac{5}{x^3}\right) dx$

33. $\int_2^4 \frac{1}{x}\,dx$

34. $\int_1^e \frac{2}{x}\,dx$

35. $\int_0^2 e^x\,dx$

36. $\int_1^{\ln 3} e^x\,dx$

37. $\int_{\pi/6}^{\pi/2} \csc^2 t\,dt$

38. $\int_0^{\pi/4} \sec x \tan x\,dx$

39. $\int_0^{\pi/2} \left(\frac{d}{dx} \sin^5 x\right) dx$

40. $\int_{-1}^1 \left(\frac{d}{dx} \sqrt{1 + x^4}\right) dx$

In Exercises 41–52 compute the area A of the region between the graph of f and the x axis on the given interval.

41. $f(x) = x^4$; $[-1, 1]$

42. $f(x) = 1/x^2$; $[-2, -1]$

43. $f(x) = \sin x$; $[0, 2\pi/3]$

44. $f(x) = \cos x$; $[-\pi/2, \pi/3]$

45. $f(x) = x^{1/2}$; $[1, 4]$

46. $f(x) = x^{1/3}$; $[1, 8]$

47. $f(x) = \sec^2 x$; $[0, \pi/4]$

48. $f(x) = \csc x \cot x$; $[\pi/4, \pi/2]$

49. $f(x) = 1/x$; $[1/e, 1]$

50. $f(x) = 2 + e^x$; $[0, 4]$

*51. $f(x) = \cos^2 x$; $[0, \pi/2]$

   (*Hint:* What is the derivative of $x/2 + (\sin 2x)/4$?)

52. $f(x) = \sec x \tan^3 x$; $[0, \pi/3]$

   (*Hint:* What is the derivative of $(\sec^3 x)/3 - \sec x$?)

53. **a.** Show that if n is an odd positive integer, then

$$\int_{-1}^1 x^n\,dx = 0.$$

   b. Show that if n is an even nonnegative integer, then

$$\int_{-1}^1 x^n\,dx = \frac{2}{n + 1}.$$

*54. Suppose $f(1) = 10$ and the graph of the derivative f' is as in Figure 5.34. Find $f(3)$.

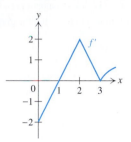

FIGURE 5.34   Graph for Exercise 54

55. Evaluate $\int_0^x f(t)\,dt$ for each of the following functions. By differentiating the resulting function, verify formula (4) of this section.

   a. $f(x) = x$ **b.** $f(x) = -2x^2$
   c. $f(x) = -\sin x$ **d.** $f(x) = 10x^4$

56. In each of the following, verify formula (6) by first differentiating F, then integrating F' from a to x, and finally comparing that result with $f(x)$.

   a. $F(x) = x + 2$; $a = 1$ **b.** $F(x) = x^3$; $a = 1$
   c. $F(x) = x^4$; $a = -1$

57. Find the number I satisfying

$$(x_0^2 + 4x_0)\,\Delta x_1 + \cdots + (x_{n-1}^2 + 4x_{n-1})\,\Delta x_n \le I$$
$$\le (x_1^2 + 4x_1)\,\Delta x_1 + \cdots + (x_n^2 + 4x_n)\,\Delta x_n$$

for every partition $P = \{x_0, x_1, \ldots, x_n\}$ of $[1, 2]$.

58. Find the number I satisfying

$$(\cos x_1 - \sin x_1)\,\Delta x_1 + \cdots + (\cos x_n - \sin x_n)\,\Delta x_n \le I$$
$$\le (\cos x_0 - \sin x_0)\,\Delta x_1 + \cdots + (\cos x_{n-1} - \sin x_{n-1})\,\Delta x_n$$

for every partition $P = \{x_0, x_1, \ldots, x_n\}$ of $[0, \pi/2]$.

Applications

59. Suppose the velocity of a car, which starts from the origin at $t = 0$ and moves along the x axis, is given by

$$v(t) = 10t - t^2 \quad \text{for } 0 \le t \le 10$$

Find the position of the car

   a. at any time t, with $0 \le t \le 10$.
   b. when its acceleration is 0.

60. The velocity of a bob moving along the x axis on a spring varies with time according to the equation

$$v(t) = 2 \sin t + 3 \cos t$$

At $t = 0$ the position of the bob is 1. Express the position of the bob as a function of time.

61. The flow of water through a dam is controlled so that the rate $F'(t)$ of flow in tons per hour is given by the equation

$$F'(t) = 14,000 \sin \frac{\pi t}{24} \quad \text{for } 0 \le t \le 24$$

How many tons of water flow through the dam per day? (*Hint:* Use formula (6) and the fact that

$$\frac{d}{dt}\left(-\frac{336,000}{\pi} \cos \frac{\pi t}{24}\right) = F'(t)$$

for $0 \le t \le 24$.)

62. Suppose the velocity of a person walking up and down Main Street is as shown in Figure 5.35.

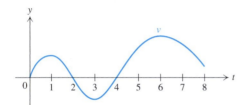

FIGURE 5.35 Graph for Exercise 62.

a. At what time(s) did the person change directions?
b. At what time(s) was the person farthest from the starting position?
c. At what time(s) was the acceleration equal to zero?

63. Suppose the revenue from the first pound of soap sold is $4 and the marginal revenue is given by

$$m_R(x) = 4 - 0.02x \quad \text{for } 1 \le x \le 400$$

a. What is the revenue accruing from a sale of 30 pounds of soap? (*Hint:* Use formula (7).)
b. What is the revenue at the point at which the marginal revenue is 0?

64. Assume that the cost of producing the first two pounds of soap is $10.98 and that the marginal cost is given by

$$m_C(x) = 3 - 0.1x \quad \text{for } 0 \le x \le 30$$

Find the total cost involved in producing 30 pounds of soap. (*Hint:* Use formula (8).)

65. To make a balloon rise faster, the balloonist drops a sandbag from the bottom of the balloon.
a. If the sandbag is ejected at an elevation of 528 feet and takes 6 seconds to hit the ground, what was the speed

of the balloon at the time of ejection?
b. If the balloon rises at 4 feet per second and is 992 feet above the ground when the sandbag is ejected, how long will it take the sandbag to hit the ground?

A hot-air balloon. (*Larry Brownstein/Rainbow*)

66. A train is cruising at 60 miles per hour when suddenly the engineer notices a cow on the track ahead of the train. The engineer applies the brakes, causing a constant deceleration in the train. Two minutes later the train grinds to a halt, barely touching the cow, which is too frightened to move. How far back was the train when the brakes were applied?

67. Radium disintegrates at a variable rate. Let $R(t)$ be the rate at which radium disintegrates at time t. Express the total amount lost between times t_1 and t_2 as an integral.

68. a. For $a \le t \le b$ let $v(t)$ be the velocity of an object at time t. Using the Fundamental Theorem, show that the mean value

$$\frac{\int_a^b v(t)\, dt}{b - a}$$

of the velocity is the average velocity of the object as defined in Section 2.1.
b. Suppose a ball thrown from the top of a cliff has the velocity

$$v(t) = -20 - 32t \quad \text{for } 0 \le t \le 3$$

Find the ball's average velocity during its flight.

69. a. Let $C(x)$ be the cost of producing x units of a product for $a \le x \le b$. Using the Fundamental Theorem, show that the mean value

$$\frac{\int_a^b m_C(x)\, dx}{b - a}$$

of the cost is the average cost between the ath and bth units produced, as defined in Section 3.1.

b. Suppose an umbrella manufacturing company has a marginal cost function given by

$$m_C(x) = \frac{1}{x^{1/2}} \quad \text{for } 1 \le x \le 4$$

where x represents thousands of umbrellas produced and $m_C(x)$ is measured in thousands of dollars. Find the mean cost between the one-thousandth umbrella produced and the four-thousandth umbrella produced.

70. Suppose the voltage in an electrical circuit varies with time according to the formula

$$V(t) = 110 \sin t \quad 0 \le t \le \pi$$

Find the mean voltage in the circuit during the given time interval.

71. A cylindrical tank 100 feet high and 100 feet in diameter is full of water. The work W (in foot-pounds) required to pump all the water out of the tank is given by

$$W = (2500\pi)(62.5) \int_0^{100} (100 - y) \, dy$$

Compute the work required.

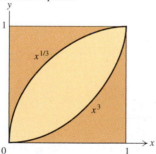

FIGURE 5.36   Graph for Exercise 72.

72. A tile manufacturing company plans to produce enamel tiles with the design and color scheme shown in Figure 5.36. Will more peach enamel be required than cream?

73. A landing strip for an airport is to be built so as to accommodate airplanes landing at 150 miles per hour and decelerating at the rate of 10 miles per hour per second. How long should the landing strip be if it is required to be 60% longer than the stopping distance of the planes?

74. Suppose a carton of chocolate bars slides down an inclined plane with an initial velocity of v_0. The force F acting on the carton is given by

$$F = mg \sin \theta - \mu mg \cos \theta$$

where m is the mass of the carton, g is the acceleration due to gravity, θ is the angle between the plane and the horizontal, and μ is the coefficient of friction between the carton and the inclined plane. Show that the distance s the carton travels in t seconds is

$$s = \frac{g}{2} (\sin \theta - \mu \cos \theta) \, t^2 + v_0 t$$

(*Hint:* Use Newton's Second Law of Motion, $F = ma$, and the ideas of Example 10.)

75. The **linear momentum** of an object is the product of its mass and velocity. Newton's Second Law of Motion is sometimes expressed in the form

$$F = \frac{dp}{dt} \tag{12}$$

where F is the force and p is the linear momentum, both expressed as functions of time t. When a force acts on an object during a time interval $[t_1, t_2]$, as when a baseball is hit by a bat, the change $p(t_2) - p(t_1)$ in the linear momentum of the object is called the **impulse** of the force.

a. Use (12) to express the impulse between t_1 and t_2 as an integral.

b. A ball with mass 0.1 kilogram falls vertically and hits the floor with a speed of 5 meters per second. It remains in contact with the floor 10^{-3} seconds, and rebounds with a speed of 4.5 meters per second. First find the impulse of the force exerted on the ball by the floor, and then use it and part (a) to determine the average force exerted on the ball by the floor during the time of contact. (*Note:* A force of 1 Newton equals 1 kilogram meter per second per second.)

Project

1. In a famous eighteenth-century problem known as **Buffon's needle problem,** a needle 1 inch long is dropped onto a hardwood floor whose boards are 2 inches wide. We are asked to determine the probability that the needle will come to rest lying across two boards. As in Figure 5.37(a), we let P be the southernmost point of the needle (or the left-hand endpoint of the needle if the needle lies horizontally). Let y denote the distance from P to the next crack northward, so that $0 \le y < 2$. Let

θ denote the positive angle the needle makes with the horizontal to the right of P, with $0 \le \theta < \pi$.

a. Convince yourself that the needle crosses the crack *only* when the pair (θ, y) satisfies $y \le \sin\theta$, that is, when (θ, y) lies in the darker shaded region in Figure 5.37(b).

b. Convince yourself that the total set of possibilities for the needle can be identified with the rectangular region with $0 \le y < 2$ and $0 \le \theta < \pi$.

c. The proportion of times that the needle crosses the crack is the fraction

$$\frac{\text{area of the darker shaded region}}{\text{area of rectangle}}$$

It is reasonable to take this proportion to be the probability sought. Calculate this probability, that is, proportion.

d. Suppose that the needle is 2 inches long (instead of 1 inch long). Draw the counterpart to Figure 5.37(b), and find the probability that a dropped needle would come to rest lying across two boards.

e. Suppose that the needle is 2 inches long and the boards 5 inches wide. Then find the corresponding probability.

f. What happens to the picture and the probability if the needle is longer than 2 inches, and the boards are 2 inches wide?

Buffon's needle problem

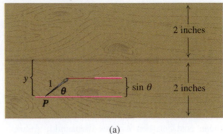

(a)

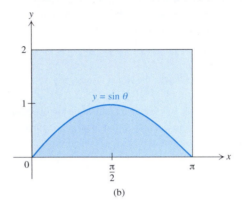

(b)

FIGURE 5.37   Figures for the project.

5.5   INDEFINITE INTEGRALS AND INTEGRATION RULES

When we evaluate a definite integral $\int_a^b f(x)\,dx$ using the Fundamental Theorem of Calculus, the basic problem is to find an antiderivative of f. In this section we present some elementary rules that will help us find antiderivatives of combinations of functions. Recall that if F is an antiderivative of f, then for any constant C the function $F + C$ is also an antiderivative of f, since

$$(F + C)' = F' + C' = f$$

Consequently every function f that has an antiderivative has infinitely many of them—one for each choice of C. By Theorem 4.6 the functions of the form $F + C$, where C is an arbitrary constant, are the *only* antiderivatives of f. As a result, mathematicians group the antiderivatives of f together and call them the indefinite integral of f.

DEFINITION   5.15   Let f have an antiderivative on an interval I. The collection of antiderivatives of f is called the **indefinite integral** of f on I and is denoted by $\int f(x)\,dx$.

Mathematicians use the notation

$$\int f(x)\, dx = F(x) + C$$

when finding indefinite integrals. For example,

$$\int x^2\, dx = \frac{1}{3}x^3 + C$$

This equation means that $\frac{1}{3}x^3$ is a function whose derivative is x^2 and that the only other functions having derivative x^2 are of the form $\frac{1}{3}x^3 + C$, where C is a constant.

In the notation of indefinite integrals, Table 5.1 in Section 5.4 yields Table 5.2. In the table c, C, p, q, and r are constants.

TABLE 5.2

Some Common Indefinite Integrals

$\int c\, dx = cx + C$	$\int \frac{1}{x}\, dx = \ln x + C$
$\int x\, dx = \frac{1}{2}x^2 + C$	$\int \sin x\, dx = -\cos x + C$
$\int x^2\, dx = \frac{1}{3}x^3 + C$	$\int \cos x\, dx = \sin x + C$
$\int (px + q)\, dx = \frac{1}{2}px^2 + qx + C$	$\int e^x\, dx = e^x + C$
$\int x^r\, dx = \frac{1}{r+1}x^{r+1} + C \ (r \neq -1)$	

We now show how to obtain the indefinite integrals of certain combinations of functions from the indefinite integrals of the individual functions.

THEOREM 5.16

If f and g have antiderivatives on an interval I, then

$$\int [f(x) + g(x)]\, dx = \int f(x)\, dx + \int g(x)\, dx$$

Proof This formula means that the sum of an antiderivative of f and an antiderivative of g is an antiderivative of $f + g$. To prove this, let F and G be antiderivatives of f and g, respectively. Then

$$(F + G)' = F' + G' = f + g$$

Hence $F + G$ is an antiderivative of $f + g$. ∎

If f and g are continuous on $[a, b]$, then by the Fundamental Theorem we have the corresponding equation for definite integrals:

$$\int_a^b [f(x) + g(x)] \, dx = \int_a^b f(x) \, dx + \int_a^b g(x) \, dx \qquad (1)$$

Moreover, Theorem 5.16 can be extended to the sum of more than two functions (see Exercise 63).

THEOREM 5.17   If f has an antiderivative on an interval I, and c is a real number, then

$$\int cf(x) \, dx = c \int f(x) \, dx$$

Proof   This formula means that c times an antiderivative of f is an antiderivative of cf. But if F is an antiderivative of f, then

$$(cF)' = c(F') = cf$$

so that cF is an antiderivative of cf. ■

If f is continuous on $[a, b]$, then the Fundamental Theorem of Calculus implies that the corresponding formula for definite integrals holds:

$$\int_a^b cf(x) \, dx = c \int_a^b f(x) \, dx \qquad (2)$$

From Theorems 5.16 and 5.17 we immediately have the following corollary, whose proof is left as an exercise (see Exercise 62).

COROLLARY 5.18   If f and g have antiderivatives on an interval I, then

$$\int [f(x) - g(x)] \, dx = \int f(x) \, dx - \int g(x) \, dx$$

If f and g are continuous on $[a, b]$, then the Fundamental Theorem yields

$$\int_a^b [f(x) - g(x)] \, dx = \int_a^b f(x) \, dx - \int_a^b g(x) \, dx \qquad (3)$$

EXAMPLE 1   Evaluate the indefinite integral $\int (5x - 3 \cos x) \, dx$.

Solution   Using first Corollary 5.18 and then Theorem 5.17, we obtain

$$\int (5x - 3 \cos x) \, dx = \int 5x \, dx - \int 3 \cos x \, dx$$

$$= 5 \int x \, dx - 3 \int \cos x \, dx = 5\left(\frac{1}{2}x^2\right) - 3 \sin x + C$$

$$= \frac{5}{2}x^2 - 3 \sin x + C \qquad □$$

EXAMPLE 2   Evaluate the definite integral $\int_0^1 (6x^2 + 5e^x)\, dx$.

Solution   First method: We break the integral into its components and then integrate. Using (1) and (2), we obtain

$$\int_0^1 (6x^2 + 5e^x)\, dx \overset{(1)}{=} \int_0^1 6x^2\, dx + \int_0^1 5e^x\, dx$$

$$\overset{(2)}{=} 6 \int_0^1 x^2\, dx + 5 \int_0^1 e^x\, dx$$

$$= 6 \left(\frac{1}{3} x^3 \right) \Big|_0^1 + 5(e^x) \Big|_0^1$$

$$= 6 \left(\frac{1}{3} - 0 \right) + 5(e - 1) = 5e - 3$$

Second method: Here we first find an antiderivative of $6x^2 + 5e^x$ and then evaluate it at the limits of integration, which are 0 and 1:

$$\int_0^1 (6x^2 + 5e^x)\, dx = (2x^3 + 5e^x) \Big|_0^1 = (2 + 5e) - 5 = 5e - 3 \qquad \square$$

Repeated applications of Theorems 5.16 and 5.17 yield the indefinite integral of a general polynomial:

$$\int \left[c_n x^n + c_{n-1} x^{n-1} + \cdots + c_1 x + c_0 \right] dx$$

$$= \frac{c_n}{n+1} x^{n+1} + \frac{c_{n-1}}{n} x^n + \cdots + \frac{c_1}{2} x^2 + c_0 x + C$$

Thus to find the integral of a polynomial

$$c_n x^n + c_{n-1} x^{n-1} + \cdots + c_1 x + c_0$$

just integrate the terms $c_n x^n$, $c_{n-1} x^{n-1}$, . . . , $c_1 x$ and c_0 one by one and add the results.

EXAMPLE 3   Evaluate $\int (x^4 - 3x^2 + 4x - 2)\, dx$.

Solution   Integrating the terms one by one, we find that

$$\int (x^4 - 3x^2 + 4x - 2)\, dx = \frac{1}{5} x^5 - 3 \left(\frac{1}{3} x^3 \right) + 4 \left(\frac{1}{2} x^2 \right) - 2x + C$$

$$= \frac{1}{5} x^5 - x^3 + 2x^2 - 2x + C \quad \square$$

Formula (3) yields a simple proof of the following result, which we could also have proved, with more difficulty, in Section 5.3.

COROLLARY 5.19
General Comparison
Property

Let f and g be continuous on $[a, b]$, with $g(x) \leq f(x)$ for $a \leq x \leq b$. Then

$$\int_a^b g(x) \, dx \leq \int_a^b f(x) \, dx$$

Proof From the hypotheses it follows that $f(x) - g(x) \geq 0$ for $a \leq x \leq b$. Then by (3) and Corollary 5.10 with $f - g$ replacing f,

$$\int_a^b f(x) \, dx - \int_a^b g(x) \, dx \stackrel{(3)}{=} \int_a^b [f(x) - g(x)] \, dx \stackrel{\text{Corollary 5.10}}{\geq} 0$$

Consequently

$$\int_a^b g(x) \, dx \leq \int_a^b f(x) \, dx \quad \blacksquare$$

In effect, Corollary 5.19 says that the larger the function, the larger the definite integral. Of course if $b < a$, then $\int_a^b g(x) \, dx \geq \int_a^b f(x) \, dx$ (see Exercise 61).

Another consequence of Corollary 5.19 is that if $h(x) \leq g(x) \leq f(x)$ for $a \leq x \leq b$, then

$$\int_a^b h(x) \, dx \leq \int_a^b g(x) \, dx \leq \int_a^b f(x) \, dx \tag{4}$$

These inequalities can lead to good lower and upper bounds for integrals that are difficult or impossible to find directly.

EXAMPLE 4   Show that $1 \leq \int_0^1 \sqrt{1 + x^4} \, dx \leq \frac{4}{3}$.

Solution We have

$$1 \leq 1 + x^4 \leq 1 + 2x^2 + x^4 = (1 + x^2)^2$$

which implies that

$$1 \leq \sqrt{1 + x^4} \leq \sqrt{(1 + x^2)^2} = 1 + x^2$$

From (4) we conclude that

$$1 = \int_0^1 1 \, dx \leq \int_0^1 \sqrt{1 + x^4} \, dx \leq \int_0^1 (1 + x^2) \, dx = \left(x + \frac{x^3}{3} \right) \Big|_0^1 = \frac{4}{3} \quad \square$$

Since $\frac{4}{3} < \sqrt{2}$, Example 4 yields a smaller upper bound for the integral $\int_0^1 \sqrt{1 + x^4}\, dx$ than the one found in Example 4 of Section 5.3. An even smaller upper bound will be obtained in Section 8.6.

Since $-|f(x)| \le f(x) \le |f(x)|$ for $a \le x \le b$, we obtain as a special case of (4) the inequalities

$$-\int_a^b |f(x)|\, dx = \int_a^b -|f(x)|\, dx \le \int_a^b f(x)\, dx \le \int_a^b |f(x)|\, dx$$

Therefore, by the definition of the absolute value,

$$\left| \int_a^b f(x)\, dx \right| \le \int_a^b |f(x)|\, dx \qquad (5)$$

EXERCISES 5.5

In Exercises 1–14 evaluate the indefinite integral.

1. $\displaystyle\int (2x - 7)\, dx$

2. $\displaystyle\int (2x^2 - 7x^3 + 4x^4)\, dx$

3. $\displaystyle\int (2x^{1/3} - 3x^{3/4} + x^{2/5})\, dx$ **4.** $\displaystyle\int (x^{3/2} + 4x^{1/2} - \pi)\, dx$

5. $\displaystyle\int \left(t^5 - \frac{1}{t^4} \right) dt$

6. $\displaystyle\int \left(\sqrt{y} + \frac{1}{\sqrt{y}} \right) dy$

7. $\displaystyle\int (2 \cos x - 5x)\, dx$

8. $\displaystyle\int (\theta^2 + \sec^2 \theta)\, d\theta$

9. $\displaystyle\int (3 \csc^2 x - x)\, dx$

10. $\displaystyle\int \left(\frac{1}{y^3} - \frac{1}{y} + 2y \right) dy$

11. $\displaystyle\int (2t + 1)^2\, dt$ (*Hint:* Expand the binomial.)

12. $\displaystyle\int \left(t + \frac{1}{t} \right)^2 dt$

13. $\displaystyle\int \left(1 + \frac{1}{x} \right)^2 dx$

14. $\displaystyle\int (\sqrt{x} - 3e^x)\, dx$

In Exercises 15–34 evaluate the definite integral.

15. $\displaystyle\int_{-1}^2 (3x - 4)\, dx$

16. $\displaystyle\int_1^4 \left(\sqrt{x} + \frac{1}{2\sqrt{x}} \right)^2 dx$

17. $\displaystyle\int_{\pi/4}^{\pi/2} (-7 \sin x + 3 \cos x)\, dx$

18. $\displaystyle\int_0^\pi (\sin x - 8x^2)\, dx$

19. $\displaystyle\int_{-\pi/4}^{-\pi/2} \left(3x - \frac{1}{x^2} + \sin x \right) dx$

20. $\displaystyle\int_{-1/4}^{1/4} (4t - 3)^2\, dt$

21. $\displaystyle\int_{\pi/3}^{\pi/4} (3 \sec^2 \theta + 4 \csc^2 \theta)\, d\theta$

22. $\displaystyle\int_0^1 (3 - 2e^t)\, dt$

23. $\displaystyle\int_1^{1/3} (3t + 2)^3\, dt$

24. $\displaystyle\int_{-\pi/4}^0 (\sec \theta)(\tan \theta + \sec \theta)\, d\theta$

(*Hint:* First multiply out the product.)

25. $\displaystyle\int_{\pi/2}^\pi \left(\pi \sin x - 2x + \frac{5}{x^2} + 2\pi \right) dx$

26. $\displaystyle\int_1^0 x(2x + 5)\, dx$

27. $\displaystyle\int_{-1}^1 (2x + 5)(2x - 5)\, dx$

28. $\displaystyle\int_2^1 (x + 3)^2(x + 1)\, dx$ **29.** $\displaystyle\int_4^7 |x - 5|\, dx$

30. $\displaystyle\int_2^1 \left(4x - \frac{1}{x} \right) dx$ **31.** $\displaystyle\int_0^\pi (\sin x - 2e^x)\, dx$

32. $\displaystyle\int_{1/2}^2 \frac{x - 1}{x}\, dx$

33. $\displaystyle\int_4^6 f(x)\, dx$, where

$$f(x) = \begin{cases} 2x & \text{for } 4 \le x \le 5 \\ 20 - 2x & \text{for } 5 < x \le 6 \end{cases}$$

34. $\displaystyle\int_0^{\pi/2} f(x)\,dx$, where

$$f(x) = \begin{cases} \sec^2 x & \text{for } 0 \le x \le \pi/4 \\ \csc^2 x & \text{for } \pi/4 < x \le \pi/2 \end{cases}$$

In Exercises 35–42 differentiate the function F. Then give F in terms of an indefinite integral. For example, if $F(x) = \cos x^2$, then since $d(\cos x^2)/dx = -2x \sin x^2$, we obtain

$$\int -2x \sin x^2\,dx = \cos x^2 + C$$

35. $F(x) = (1 + x^2)^{10}$ **36.** $F(x) = \frac{1}{2}(1+x)^2$
37. $F(x) = x \sin x - \cos x$ **38.** $F(x) = \tan(2x+1)$
39. $F(x) = 3 \sin^7 x$ **40.** $F(x) = x \sin x \cos x$
41. $F(x) = e^{(x^2)} - e^{-x}$ **42.** $F(x) = x \ln x - x$

In Exercises 43–50 find the area A of the region between the graph of f and the x axis on the given interval.

43. $f(x) = 3x^2 + 4;\ [-1, 1]$
44. $f(x) = \frac{1}{2}x^3 + 3x;\ [1, 2]$
45. $f(x) = 3\sqrt{x} - 1/\sqrt{x};\ [1, 4]$
46. $f(x) = 8x^{1/3} - x^{-1/3};\ [1, 8]$
47. $f(x) = 2 \sin x + 3 \cos x;\ [\pi/4, \pi/2]$
48. $f(x) = |x + 1|;\ [-2, 0]$
49. $f(x) = 2x - 4/x;\ [2, 4]$
50. $f(x) = \frac{1}{2}e^x - \frac{1}{3}x;\ [0, 3]$
51. Use the inequalities $0 \le \sin x \le x$ for $0 \le x \le 1$ to show that

a. $0 \le \displaystyle\int_0^1 \sin x^2\,dx \le \frac{1}{3}$

b. $0 \le \displaystyle\int_0^{\pi/6} \sin^{3/2} x\,dx \le \frac{2}{5}(\pi/6)^{5/2}$

52. Use the inequalities $0 \le \sin x \le x$ for $0 \le x \le \frac{1}{2}$ to show that

$$0 \le \int_0^{1/2} x \sin x\,dx \le \frac{1}{24}$$

53. Let $f(x) = 1 - x$. Show that $|\int_0^2 f(x)\,dx| < \int_0^2 |f(x)|\,dx$. (This inequality shows that the inequality sign in (5) cannot in general be replaced by an equals sign.)
54. Let f be continuous on $[a, b]$, and let $|f(x)| \le M$ for $a \le x \le b$. Prove that

$$\left| \int_a^b f(x)\,dx \right| \le M(b - a)$$

55. Use Exercise 54 to find an upper bound for

$$\left| \int_{-\pi/3}^{-\pi/4} \tan x\,dx \right|$$

56. Show that

$$\lim_{\varepsilon \to 0^+} \frac{1}{\varepsilon} \int_0^\varepsilon x \sin x\,dx = 0$$

(*Hint:* $0 \le \sin x \le x$ for $x \ge 0$.)

*57. Suppose f has a bounded derivative on $[a, b]$, so that there is a number M such that $|f'(x)| \le M$ for $a \le x \le b$. Assume that $f(a) = 0$.

a. Using the Mean Value Theorem, show that

$$|f(x)| \le M(x - a) \quad \text{for } a \le x \le b$$

b. Using the result of (a) and formula (5), show that

$$\left| \int_a^b f(x)\,dx \right| \le \frac{M}{2}(b - a)^2$$

58. Using Exercise 57(b), find upper bounds for the absolute values of the following integrals.

a. $\displaystyle\int_1^{\sqrt{5}} (x^2 - 1)^{3/2}\,dx$

*b. $\int_0^{\pi/6} \sin^{3/2} x\,dx$ (*Hint:* Show first that $d(\sin^{3/2} x)/dx \le \frac{3}{4}\sqrt{2}$ for $0 \le x \le \pi/6$.)

*59. Show that for every polynomial f of degree 3 or less,

$$\int_{-1}^1 f(x)\,dx = f\left(-\frac{1}{\sqrt{3}}\right) + f\left(\frac{1}{\sqrt{3}}\right) \tag{6}$$

Formula (6) is known as the two-point **Gauss quadrature formula** for $[-1, 1]$. For functions other than polynomials of degree 3 or less, the right side of (6) yields an approximation of the left side.

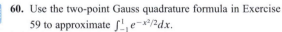 **60.** Use the two-point Gauss quadrature formula in Exercise 59 to approximate $\int_{-1}^1 e^{-x^2/2}\,dx$.

61. Suppose that $b < a$ and that f and g are continuous, with $g(x) \le f(x)$ for $b \le x \le a$. Use Corollary 5.19 to prove that

$$\int_a^b g(x)\,dx \ge \int_a^b f(x)\,dx$$

62. Prove Corollary 5.18.
63. Prove that if $f_1, f_2, \ldots, f_n$ have antiderivatives on an interval I, then

$$\int [f_1(x) + f_2(x) + \cdots + f_n(x)] \, dx$$

$$= \int f_1(x) \, dx + \int f_2(x) \, dx + \cdots + \int f_n(x) \, dx$$

*64. Find an upper bound for the area $\pi/4$ of the quarter of the unit circle in the first quadrant by calculating the area of the region formed by the tangent to the circle $x^2 + y^2 = 1$ at $(\sqrt{3}/2, \frac{1}{2})$, the x and y axes, and the line $x = 1$.

*65. **a.** Use the method of Exercise 64 to find an upper bound for the area $\pi/2$ of the quarter of the ellipse

$$\frac{x^2}{4} + y^2 = 1$$

   in the first quadrant. (*Hint:* Consider the tangent line at $(\sqrt{2}, 1/\sqrt{2})$.)

   b. Find a lower bound for the area $\pi/2$ by constructing a triangle inside the ellipse.

66. Let a, b, c, and d be positive numbers with $a < b$. The region R bounded by the graph of a function of the form $f(x) = cx + d$, the x axis, and the lines $x = a$ and $x = b$ is a right trapezoid. Show that the area A of R is given by

$$\left(\frac{f(a) + f(b)}{2} \right) (b - a)$$

*67. Let f and g be continuous on $[a, b]$. Show that

$$\left(\int_a^b f(x)g(x) \, dx \right)^2 \le \int_a^b [f(x)]^2 \, dx \int_a^b [g(x)]^2 \, dx$$

(*Hint:* Let $p(r) = \int_a^b [f(x) + rg(x)]^2 \, dx$ for all real numbers r. Show that $p(r)$ has the form $Ar^2 + Br + C$, where A, B, and C are constants. Then show that $p(r) \ge 0$ for all r and

deduce that $B^2 - 4AC \le 0$, which yields the desired inequality.)

Application

68. Let r denote the radius of a cylindrical artery of length l and let x denote the distance of a given blood cell from the center of a cross section of the artery. The volume V per unit time of the flow of blood through the artery is given by the formula

$$V = \int_0^r \frac{k}{l} x(r^2 - x^2) \, dx$$

where k is a constant depending on the difference in pressure at the two ends of the artery and on the viscosity of the blood. Calculate V.

Project

1. In each part of this project, assume that f is differentiable on $[0, \infty)$, with $f(0) = 0$ but $f(x) \ne 0$ for $x \ne 0$.

   a. Assume that $\int_0^x f(t) \, dt = (f(x))^2$ for all x. First show that $f(x) = 2f(x)f'(x)$. Then show that f is unique, and find a formula for $f(x)$.

   b. Assume that $\int_0^x f(t) \, dt = (f(x))^3$ for all x. First show that $f(x) = 3[f(x)]^2 f'(x)$. Then show that f is unique, and find a formula for $f(x)$. (*Hint:* What is the derivative of $\frac{1}{2} f^2$?)

   c. Assume that $\int_0^x f(t) \, dt = (f(x))^4$ for all x. First show that $f(x) = 4[f(x)]^3 f'(x)$. Then show that f is unique, and find a formula for $f(x)$.

5.6   INTEGRATION BY SUBSTITUTION

In the preceding section we transformed addition and constant multiple theorems for derivatives into corresponding theorems for integrals (see Theorems 5.16 and 5.17). Presently we will transform the Chain Rule in the form of

$$\frac{d}{dx} G(f(x)) = G'(f(x))f'(x)$$

into a theorem for integrals. The result we will obtain will be as useful in integration as the Chain Rule is in differentiation, and it will allow us to express many integrals, such as

$$\int x \sqrt{2x + 1} \, dx \quad \text{and} \quad \int \sin 3x \cos 3x \, dx$$

in terms of functions familiar to us. The latter integral appears in the study of electric power. (See Exercise 66.)

THEOREM 5.20

Let f and g be functions, with both $g \circ f$ and f' continuous on an interval I. If G is an antiderivative of g, then

$$\int g(f(x))f'(x)\, dx = G(f(x)) + C \qquad\qquad (1)$$

Proof Since G is an antiderivative of g, we have $G'(x) = g(x)$. Therefore the Chain Rule implies that

$$\frac{d}{dx} G(f(x)) = G'(f(x))f'(x) = g(f(x))f'(x)$$

In terms of indefinite integrals this becomes

$$\int g(f(x))f'(x)\, dx = G(f(x)) + C$$

which is (1). ∎

In the process of applying (1) it is usually convenient to substitute u for $f(x)$ and du for $f'(x)\, dx$. Thus we obtain

$$\int g(\overbrace{f(x)}^{u})\ \overbrace{f'(x)\, dx}^{du} = \int g(u)\, du = G(u) + C = G(f(x)) + C$$

which shows clearly the integration of g to obtain G. For this reason, evaluating an integral by means of (1) is called **integration by substitution.** We illustrate the method in the examples that follow.

EXAMPLE 1   Find $\displaystyle\int 3x^2(x^3 + 5)^9\, dx$.

Solution   We let

$$u = x^3 + 5, \quad \text{so that} \quad du = 3x^2\, dx$$

Then

$$\int 3x^2(x^3 + 5)^9\, dx = \int \overbrace{(x^3 + 5)^9}^{u^9}\ \overbrace{(3x^2)\, dx}^{du} = \int u^9\, du = \frac{1}{10} u^{10} + C$$

$$= \frac{1}{10}(x^3 + 5)^{10} + C \qquad \square$$

Notice that after the integration was performed we resubstituted $x^3 + 5$ for u, so the answer would be expressed in terms of the original variable x. The variable u is only a temporary convenience.

We observe that it would also be possible to find $\int 3x^2(x^3 + 5)^9\, dx$ by expanding the polynomial $3x^2(x^3 + 5)^9$ and integrating term by term. However, integrating by substitution is much more efficient.

EXAMPLE 2   Find $\displaystyle\int \sin^4 x \cos x \, dx$.

Solution   We let

$$u = \sin x, \quad \text{so that} \quad du = \cos x \, dx$$

Then

$$\int \sin^4 x \cos x \, dx = \int \overbrace{(\sin x)^4}^{u^4} \overbrace{\cos x \, dx}^{du} = \int u^4 \, du = \frac{1}{5} u^5 + C = \frac{1}{5} \sin^5 x + C \qquad \square$$

The substitution of u for $x^3 + 5$ worked well in Example 1 because $du = 3x^2 \, dx$ and because $3x^2 \, dx$ appeared in the original integral. Similarly, the substitution of u for $\sin x$ worked well in Example 2 because $du = \cos x \, dx$ and because $\cos x \, dx$ appeared in the original integral. The method of substitution can still be applied if merely a constant multiple of du appears in the original integral.

EXAMPLE 3   Find $\displaystyle\int \frac{1}{2} \cos 2x \, dx$.

Solution   We let

$$u = 2x, \quad \text{so that} \quad du = 2 \, dx, \quad \text{and thus} \quad dx = \tfrac{1}{2} \, du$$

Then

$$\int \frac{1}{2} \cos 2x \, dx = \int \frac{1}{2} (\cos \overbrace{2x}^{u}) \overbrace{dx}^{\frac{1}{2} du} = \int \frac{1}{2} (\cos u) \frac{1}{2} \, du$$

$$= \frac{1}{4} \int \cos u \, du = \frac{1}{4} \sin u + C = \frac{1}{4} \sin 2x + C \qquad \square$$

We can use the result of Example 3 to evaluate $\int \cos^2 x \, dx$. Indeed, by using the trigonometric identity

$$\cos^2 x = \frac{1}{2} + \frac{1}{2} \cos 2x$$

along with Example 3 and the integration rules of Section 5.5, we find that

$$\int \cos^2 x \, dx = \int \left(\frac{1}{2} + \frac{1}{2} \cos 2x \right) dx = \int \frac{1}{2} \, dx + \int \frac{1}{2} \cos 2x \, dx$$

$$= \frac{1}{2} x + \frac{1}{4} \sin 2x + C$$

Therefore

$$\int \cos^2 x \, dx = \frac{1}{2} x + \frac{1}{4} \sin 2x + C \qquad (2)$$

By a similar argument,

$$\int \sin^2 x \, dx = \frac{1}{2} x - \frac{1}{4} \sin 2x + C \qquad (3)$$

(See Exercise 59.) The integrals $\int \sin^2 x \, dx$ and $\int \cos^2 x \, dx$ will occur from time to time throughout the remainder of this book.

EXAMPLE 4   Find $\displaystyle\int e^x \sqrt{1 - e^x} \, dx$.

Solution   We let

$$u = 1 - e^x, \quad \text{so that} \quad du = -e^x \, dx$$

Then

$$\int e^x \sqrt{1 - e^x} \, dx = \int \overbrace{\sqrt{1 - e^x}}^{\sqrt{u}} \overbrace{e^x \, dx}^{(-1)\, du} = \int \sqrt{u} \, (-1) \, du = -\int u^{1/2} \, du$$

$$= -\frac{2}{3} u^{3/2} + C = -\frac{2}{3} (1 - e^x)^{3/2} + C \qquad \square$$

Occasionally it is convenient to solve for x (or some expression involving x) in terms of u in order to complete the substitution in the original integral.

EXAMPLE 5   Find $\displaystyle\int x\sqrt{2x + 1} \, dx$.

Solution   To simplify the expression $\sqrt{2x + 1}$, we let

$$u = 2x + 1, \quad \text{so that} \quad du = 2 \, dx$$

Then

$$\int x\sqrt{2x + 1} \, dx = \int x \overbrace{\sqrt{2x + 1}}^{u^{1/2}} \overbrace{dx}^{\frac{1}{2}\, du}$$

Thus we still need to find x in terms of u. From the equation $u = 2x + 1$ we deduce that

$$x = \frac{1}{2} (u - 1)$$

Therefore

$$\int x\sqrt{2x + 1} \, dx = \int \overbrace{x}^{\frac{1}{2}(u-1)} \overbrace{\sqrt{2x + 1}}^{u^{1/2}} \overbrace{dx}^{\frac{1}{2}\, du}$$

$$= \int \frac{1}{2}(u-1)u^{1/2} \cdot \frac{1}{2}\,du = \frac{1}{4}\int (u^{3/2} - u^{1/2})\,du$$

$$= \frac{1}{4}\left(\frac{2}{5}u^{5/2} - \frac{2}{3}u^{3/2}\right) + C$$

$$= \frac{1}{10}(2x+1)^{5/2} - \frac{1}{6}(2x+1)^{3/2} + C \quad \square$$

Frequently there is more than one substitution that will work. For instance, in Example 5 we could have let $u = \sqrt{2x+1}$. Then

$$du = \frac{1}{\sqrt{2x+1}}\,dx, \quad \text{so that} \quad u\,du = dx$$

Solving for x in the equation $u = \sqrt{2x+1}$, we obtain $x = \frac{1}{2}(u^2 - 1)$. Thus

$$\int x\sqrt{2x+1}\,dx = \int \overbrace{x}^{\frac{1}{2}(u^2-1)}\ \overbrace{\sqrt{2x+1}}^{u}\ \overbrace{dx}^{u\,du}$$

$$= \int \frac{1}{2}(u^2 - 1)u^2\,du$$

$$= \int \left(\frac{1}{2}u^4 - \frac{1}{2}u^2\right)du = \frac{1}{10}u^5 - \frac{1}{6}u^3 + C$$

$$= \frac{1}{10}(2x+1)^{5/2} - \frac{1}{6}(2x+1)^{3/2} + C$$

Even though we used a different substitution, the final answer remains the same.

EXAMPLE 6   Find $\displaystyle\int x^5\sqrt{x^2-1}\,dx$.

Solution   In order to simplify the integrand, we let

$$u = \sqrt{x^2-1}, \quad \text{so that} \quad du = \frac{x}{\sqrt{x^2-1}}\,dx, \quad \text{and thus} \quad u\,du = x\,dx$$

Then we factor out an x from x^5 so that $x\,dx$, which equals $u\,du$, appears in the integral:

$$\int x^5\sqrt{x^2-1}\,dx = \int x^4\ \overbrace{\sqrt{x^2-1}}^{u}\ \overbrace{x\,dx}^{u\,du}$$

Now we need to write x^4 in terms of u:

$$u = \sqrt{x^2-1}, \quad \text{so} \quad u^2 = x^2 - 1$$

Thus

$$x^2 = u^2 + 1, \quad \text{so} \quad x^4 = (u^2+1)^2$$

Therefore

$$\int x^5 \sqrt{x^2 - 1} \, dx = \int \overbrace{x^4}^{(u^2+1)^2} \overbrace{\sqrt{x^2-1}}^{u} \overbrace{x \, dx}^{u \, du} = \int (u^2 + 1)^2 \, u \cdot u \, du$$

$$= \int (u^6 + 2u^4 + u^2) \, du = \frac{1}{7} u^7 + \frac{2}{5} u^5 + \frac{1}{3} u^3 + C$$

$$= \frac{1}{7} \left(\sqrt{x^2 - 1} \right)^7 + \frac{2}{5} \left(\sqrt{x^2 - 1} \right)^5 + \frac{1}{3} \left(\sqrt{x^2 - 1} \right)^3 + C \; \square$$

Substitution with Definite Integrals

Suppose we wish to evaluate a definite integral of the form $\int_a^b g(f(x))f'(x) \, dx$. Using (1) and the Fundamental Theorem of Calculus, we find that

$$\int_a^b g(f(x))f'(x) \, dx = G(f(x)) \, \bigg|_a^b = G(f(b)) - G(f(a)) \qquad (4)$$

However, since G is an antiderivative of g, we also have

$$\int_{f(a)}^{f(b)} g(u) \, du = G(u) \, \bigg|_{f(a)}^{f(b)} = G(f(b)) - G(f(a)) \qquad (5)$$

From (4) and (5) it follows that

$$\int_a^b g(f(x))f'(x) \, dx = \int_{f(a)}^{f(b)} g(u) \, du \qquad (6)$$

Thus we now have two methods of evaluating a definite integral of the form $\int_a^b g(f(x))f'(x) \, dx$. One way is to find the indefinite integral $\int g(f(x))f'(x) \, dx$ by substitution, and then evaluate it between the limits a and b, as in (4). The other way is to use (6), which involves using substitution, but with the limits of integration changed before we integrate. Formula (6) is called the **change of variable formula** in integration.

We will illustrate both methods of evaluating a definite integral by substitution in our final example.

EXAMPLE 7   Evaluate $\displaystyle\int_0^1 \frac{x^5}{(x^6 + 1)^3} \, dx$.

Solution   For the first method we find the indefinite integral $\int x^5/(x^6 + 1)^3 \, dx$ and then evaluate it between 0 and 1. To achieve this, we let

$$u = x^6 + 1, \quad \text{so that} \quad du = 6x^5 \, dx$$

Then

$$\int \frac{x^5}{(x^6 + 1)^3} \, dx = \int \overbrace{\frac{1}{(x^6 + 1)^3}}^{1/u^3} \overbrace{x^5 \, dx}^{\frac{1}{6} \, du} = \int \frac{1}{u^3} \cdot \frac{1}{6} \, du$$

$$= \frac{1}{6} \left(-\frac{1}{2} \cdot \frac{1}{u^2} \right) + C = -\frac{1}{12} \cdot \frac{1}{(x^6 + 1)^2} + C$$

Therefore

$$\int_0^1 \frac{x^5}{(x^6 + 1)^3}\, dx = -\frac{1}{12} \cdot \frac{1}{(x^6 + 1)^2}\Big|_0^1 = -\frac{1}{12} \cdot \frac{1}{4} - \left(-\frac{1}{12}\right) = \frac{1}{16}$$

For the second method we make the same substitution as before but accompany it with a change in limits of integration. Since $u = x^6 + 1$, it follows that

$$\text{if } x = 0 \quad \text{then } u = 1, \quad \text{and} \quad \text{if } x = 1 \quad \text{then } u = 2$$

Consequently

$$\int_0^1 \frac{x^5}{(x^6 + 1)^3}\, dx = \int_1^2 \frac{1}{u^3} \cdot \frac{1}{6}\, du = \frac{1}{6}\left(-\frac{1}{2} \cdot \frac{1}{u^2}\right)\Big|_1^2$$

$$= \frac{1}{6}\left(-\frac{1}{8}\right) - \frac{1}{6}\left(-\frac{1}{2}\right) = \frac{1}{16} \qquad \square$$

EXERCISES 5.6

In Exercises 1–18 evaluate the integral by making the indicated substitution.

1. $\displaystyle\int \sqrt{4x - 5}\, dx; u = 4x - 5$

2. $\displaystyle\int (1 - 5x^2)^{2/3}(10x)\, dx; u = 1 - 5x^2$

3. $\displaystyle\int \cos \pi x\, dx; u = \pi x$

4. $\displaystyle\int 3 \sin (-2x)\, dx; u = -2x$

5. $\displaystyle\int x \cos x^2\, dx; u = x^2$

6. $\displaystyle\int \sin^3 t \cos t\, dt; u = \sin t$

7. $\displaystyle\int \cos^{-4} t \sin t\, dt; u = \cos t$

8. $\displaystyle\int \frac{2t - 3}{(t^2 - 3t + 1)^2}\, dt; u = t^2 - 3t + 1$

9. $\displaystyle\int \frac{2t - 3}{(t^2 - 3t + 1)^{7/2}}\, dt; u = t^2 - 3t + 1$

10. $\displaystyle\int x\sqrt{x - 1}\, dx; v = x - 1$

11. $\displaystyle\int (x - 1)\sqrt{x + 1}\, dx; v = x + 1$

12. $\displaystyle\int x^2 \sqrt{x + 3}\, dx; v = x + 3$

13. $\displaystyle\int \sec x \tan x \sqrt{3 + \sec x}\, dx; u = 3 + \sec x$

14. $\displaystyle\int \frac{\sqrt[3]{x}}{(\sqrt[3]{x} + 1)^5}\, dx; u = \sqrt[3]{x} + 1$

15. $\displaystyle\int x\, e^{(x^2)}\, dx; u = x^2$

16. $\displaystyle\int \frac{e^x}{1 + e^x}\, dx; u = 1 + e^x$

17. $\displaystyle\int \frac{x}{x^2 + 1}\, dx; u = x^2 + 1$

*18. $\displaystyle\int \frac{1}{x + \sqrt{x}}\, dx; u = \sqrt{x} + 1$

In Exercises 19–44 evaluate the integral.

19. $\displaystyle\int 3x^2(x^3 + 1)^{12}\, dx$ **20.** $\displaystyle\int x^3(2 - 5x^4)^7\, dx$

21. $\displaystyle\int \sqrt{x}\, (4 + x^{3/2})\, dx$ **22.** $\displaystyle\int \left(1 - \frac{1}{x^2}\right)\left(x + \frac{1}{x}\right)^{-3}\, dx$

23. $\displaystyle\int \sqrt{3x + 7}\, dx$ **24.** $\displaystyle\int \sqrt{4 - 2x}\, dx$

25. $\displaystyle\int (1 + 4x)\sqrt{1 + 2x + 4x^2}\, dx$

26. $\displaystyle\int \cos 7x \, dx$

27. $\displaystyle\int_{-1}^{3} \sin \pi x \, dx$

28. $\displaystyle\int_{0}^{1} t^9 \sin t^{10} \, dt$

29. $\displaystyle\int \sin^6 t \cos t \, dt$

30. $\displaystyle\int \cos^{-3} t \sin t \, dt$

31. $\displaystyle\int \sqrt{\sin 2z} \cos 2z \, dz$

32. $\displaystyle\int \sin 3z \sqrt{1 - \cos 3z} \, dz$

33. $\displaystyle\int_{0}^{\pi/4} \frac{\sin z}{\cos^2 z} \, dz$

34. $\displaystyle\int_{\pi/2}^{\pi/6} \frac{\cos z}{\sin^3 z} \, dz$

35. $\displaystyle\int \frac{1}{\sqrt{z}} \sec^2 \sqrt{z} \, dz$

36. $\displaystyle\int \frac{1}{z^2} \csc^2 \frac{1}{z} \, dz$

37. $\displaystyle\int w \left(\sqrt{w^2 + 1} + \frac{1}{\sqrt{w^2 + 1}} \right) dw$

38. $\displaystyle\int_{4}^{1} \frac{\sqrt{1 + \sqrt{x}}}{\sqrt{x}} \, dx$

39. $\displaystyle\int_{1}^{8} x^{-2/3} \sqrt{1 + 4x^{1/3}} \, dx$

40. $\displaystyle\int_{-1}^{0} w(\sqrt{1 - w^2} + \sin \pi w^2) \, dw$

41. $\displaystyle\int e^{2x} \sin (1 + e^{2x}) \, dx$

42. $\displaystyle\int \frac{e^{\sqrt{x}}}{\sqrt{x}} \, dx$

43. $\displaystyle\int \frac{x^3}{1 + x^4} \, dx$

44. $\displaystyle\int \frac{(\ln x)^2}{x} \, dx$

In Exercises 45–52 evaluate the integral.

45. $\displaystyle\int x \sqrt{x + 2} \, dx$

46. $\displaystyle\int \frac{x}{\sqrt{x + 3}} \, dx$

47. $\displaystyle\int_{1}^{3} 4x \sqrt{6 - 2x} \, dx$

48. $\displaystyle\int x^2 \sqrt{x + 4} \, dx$

49. $\displaystyle\int t^2 \sqrt{1 - 8t} \, dt$

50. $\displaystyle\int e^{3t} \sqrt{1 + e^t} \, dt$

51. $\displaystyle\int_{-1}^{2} \frac{t^2}{\sqrt{t + 2}} \, dt$

52. $\displaystyle\int_{0}^{1} \frac{\sqrt{x}}{\sqrt{1 + \sqrt{x}}} \, dx$

In Exercises 53–58 find the area A of the region between the graph of f and the x axis on the given interval.

53. $f(x) = \sqrt{x + 1}$; $[0, 3]$

54. $f(x) = \sin \pi x$; $[0, 1]$

55. $f(x) = \dfrac{x}{(x^2 + 1)^2}$; $[1, 2]$

56. $f(x) = x \sqrt{x^2 - 9}$; $[3, 5]$

57. $f(x) = \dfrac{1}{x^2} \left(1 + \dfrac{1}{x}\right)^{1/2}$; $\left[\dfrac{1}{8}, \dfrac{1}{3}\right]$

58. $f(x) = -x^{1/3} (1 + x^{4/3})^{1/3}$; $[-1, 0]$

59. By using a trigonometric identity, show that

$$\int \sin^2 x \, dx = \frac{1}{2} x - \frac{1}{4} \sin 2x + C$$

60. Let a be any number and k any integer.
  a. Verify that $\int_{a}^{a+k\pi} \sin^2 x \, dx = k\pi/2$.
  b. Verify that $\int_{a}^{a+k\pi} \cos^2 x \, dx = k\pi/2$.

61. a. By making the substitution $u = 1 - x$, show that

$$\int_{0}^{1} x^n (1 - x)^m \, dx = \int_{0}^{1} x^m (1 - x)^n \, dx$$

for any nonnegative integers m and n.
  b. Use part (a) to evaluate

$$\int_{0}^{1} x^2 (1 - x)^{10} \, dx$$

62. Let $a > 0$, and let f be continuous on $[0, a]$.
  a. Making the substitution $u = a - x$, show that

$$\int_{0}^{a} \frac{f(x)}{f(x) + f(a - x)} \, dx = \int_{0}^{a} \frac{f(a - u)}{f(u) + f(a - u)} \, du$$

  b. Use part (a) to show that

$$\int_{0}^{a} \frac{f(x)}{f(x) + f(a - x)} \, dx = \frac{a}{2}$$

(The answer is independent of f!)
  c. Use part (b) to evaluate

$$\int_{0}^{1} \frac{x^4}{x^4 + (1 - x)^4} \, dx$$

63. Let f be continuous, and suppose $\int f(x) \, dx = F(x) + C$.
  a. Prove that $\int f(ax + b) \, dx = [F(ax + b)]/a + C$ for any $a \neq 0$ and any b.
  b. Use (a) to evaluate $\int \sin (ax + b) \, dx$.
  c. Use (a) to evaluate $\int (ax + b)^n \, dx$, for any integer $n \neq 0, -1$.

64. Show that $\int_{ca}^{cb} (1/x) \, dx = \int_{a}^{b} (1/x) \, dx$ for any positive numbers a, b, and c. (*Hint:* Let $u = x/c$.)

65. Find the fallacy in the following argument: Since $(1 + x^2)^{-1} > 0$ we have $\int_{-1}^{1} (1 + x^2)^{-1} \, dx > 0$. However, by substituting $u = 1/x$ we obtain

$$\int_{-1}^{1} (1 + x^2)^{-1} \, dx = \int_{-1}^{1} \left(1 + \frac{1}{u^2}\right)^{-1} \left(-\frac{1}{u^2}\right) du$$

$$= -\int_{-1}^{1} (1 + u^2)^{-1} \, du$$

$$= -\int_{-1}^{1} (1 + x^2)^{-1} \, dx$$

which implies that $\int_{-1}^{1} (1 + x^2)^{-1} \, dx = 0$.

Applications

66. In an electric circuit, the **electric power** P produced by the voltage V and current I is given by $P = VI$. Frequently both voltage and current are sinusoidal as functions of time. Assume that V_0 and I_0 are constants. Find the mean power

$$\frac{1}{\pi} \int_0^{\pi} V(t)\, I(t)\, dt$$

during the time interval $[0, \pi]$, assuming that $V(t) = V_0 \sin 3t$ and

a. $I(t) = I_0 \sin 3t$ **b.** $I(t) = I_0 \cos 3t$

67. According to the theory of quantum mechanics, an electron moving along the x axis does not have a definite position at any particular instant, but instead is represented by a wave function ψ. The probability of finding the electron in an interval $[a, b]$ is

$$\int_a^b [\psi(x)]^2\, dx$$

For an electron confined to the interval $[0, 1]$ with impenetrable barriers at 0 and 1, the wave function is given by $\psi(x) = \sqrt{2} \sin(2\pi x)$.

a. Find the value of $\int_0^{1/4} 2 \sin^2(2\pi x)\, dx$, which is the probability of finding the electron in the interval $[0, 1/4]$.

b. Show that

$$\int_0^1 |\psi(x)|^2\, dx = 1$$

and explain the physical significance of this result.

68. The probability of "breakdowns" of certain systems, such as automobile accidents at a busy intersection, cars arriving at a toll booth, lifetime of a battery, or earthquakes, can be modeled by the **exponential density function** f, given by

$$f(t) = \frac{1}{\lambda} e^{-t/\lambda} \quad \text{for } t \ge 0$$

where λ is a positive constant associated with the system. The probability $P(t)$ that such a breakdown will occur during the time interval $[0, t]$ is given by

$$P(t) = \int_0^t f(s)\, ds \quad \text{for } t \ge 0$$

a. Show that $\lim_{t \to \infty} P(t) = 1$.

b. Assume that the formula for P holds for the lifetime of car batteries, with $\lambda = 2$. Find the probability that a randomly selected car battery will last at most 1 year.

c. Using $\lambda = 2$ in part (b), find the value of t^* such that $P(t^*) = 0.5$.

69. Suppose an object with mass m_1 is located at a on the x axis, and another object with mass m_2 is to the left of the first object on the x axis. Newton's Law of Gravitation implies that the work W required to move the second object from $x_2 < a$ to $x_1 < a$ is given by

$$W = \int_{x_1}^{x_2} \frac{Gm_1 m_2}{(x - a)^2}\, dx$$

where G is a constant. Show that

$$W = \frac{Gm_1 m_2 (x_2 - x_1)}{(x_1 - a)(x_2 - a)}$$

Project

1. Let f be continuous on the interval $[-a, a]$. This project concerns functions that are even (that is, $f(-x) = f(x)$ for $-a \le x \le a$) and those that are odd (that is, $f(-x) = -f(x)$ for $-a \le x \le a$). We start with two supporting results, in (a) and (b).

a. Prove that

$$\int_{-a}^0 f(x)\, dx = \int_0^a f(-x)\, dx$$

b. Prove that

$$\int_{-a}^a f(x)\, dx = \int_0^a [f(x) + f(-x)]\, dx$$

c. Show that if f is odd, then $\int_{-a}^a f(x)\, dx = 0$.

d. Show that if f is even, then $\int_{-a}^a f(x)\, dx = 2 \int_0^a f(x)\, dx$.

e. Suppose that $\int_{-c}^c f(x)\, dx = 0$ for all c such that $0 \le c \le a$. Does this imply that f is an odd function on $[-a, a]$? Explain your answer.

5.7 THE LOGARITHM

Until now our discussion of the natural logarithm has been based on an informal definition of the logarithm. Having introduced integrals, we are now in a position to give a formal definition of the natural logarithm and derive several of its properties, including three that we have exploited in earlier chapters:

$$\text{its value at 1:} \quad \ln 1 = 0 \tag{1}$$

$$\text{its derivative:} \quad \frac{d}{dx}\ln x = \frac{1}{x} \tag{2}$$

$$\text{the law of logarithms:} \quad \ln bc = \ln b + \ln c \tag{3}$$

To prepare for the definition of the natural logarithm, let $f(x) = 1/x$. Then f is continuous on $(0, \infty)$, so we may define a function G by the formula

$$G(x) = \int_1^x \frac{1}{t}\,dt \quad \text{for all } x > 0 \tag{4}$$

Then

$$G(1) = \int_1^1 \frac{1}{t}\,dt = 0 \tag{5}$$

By Theorem 5.12, G is differentiable on $(0, \infty)$, and

$$\frac{dG}{dx} = \frac{1}{x} \tag{6}$$

Comparing (1) with (5) and (2) with (6), we see that G and $\ln x$ have the same derivative and also the same value at 1. However, Theorem 4.6 implies that there is exactly one function defined on $(0, \infty)$ with a given derivative and a given value at 1. Thus it is reasonable to formally define the **natural logarithm function** by

$$\ln x = \int_1^x \frac{1}{t}\,dt \tag{7}$$

For $x > 1$, $\ln x$ may be considered as the area of the region in Figure 5.38.

It is immediate from (7) that

$$\ln 1 = 0 \quad \text{and} \quad \frac{d}{dx}\ln x = \frac{1}{x} \tag{8}$$

That the well-known Law of Logarithms is valid for $\ln x$ will be proved in Theorem 5.21.

The indefinite integral form of (7) is

$$\int \frac{1}{x}\,dx = \ln x + C \tag{9}$$

As a result, we have a formula for $\int t^r\,dt$ when $r = -1$. This means that finally we have found a formula for the indefinite integral $\int t^r\,dt$ for every real number r.

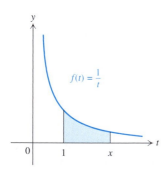

$f(t) = \frac{1}{t}$

FIGURE 5.38

From (8) we know that ln x is an increasing function (because $x > 0$). Thus

$$x < z \quad \text{if and only if} \quad \ln x < \ln z \tag{10}$$

Moreover,

$$\frac{d^2}{dx^2} \ln x = \frac{d}{dx}\left(\frac{1}{x}\right) = -\frac{1}{x^2}$$

so that the graph of ln x is concave downward. One can also prove that

$$\lim_{x \to 0^+} \ln x = -\infty \quad \text{and} \quad \lim_{x \to \infty} \ln x = \infty$$

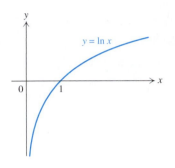

FIGURE 5.39

(See Exercise 50.) Using these facts, along with the equation ln $1 = 0$, we deduce that the graph of ln x has the shape in Figure 5.39.

Since ln $1 = 0$ and $\lim_{x \to \infty} \ln x = \infty$, the Intermediate Value Theorem implies that there is a (unique) number e such that ln $e = 1$. It can be shown that e is irrational and that

$$e = 2.71828182845904523536\ldots$$

To prove that the expansion is nonrepeating is not easy. However, using the Addition and Comparison Properties of the integral, we can show that $e < 3$.

EXAMPLE 1   Show that $e < 3$.

Solution   By (10), the inequality $e < 3$ is equivalent to ln $e <$ ln 3, that is, $1 <$ ln 3. Thus we need only show that ln $3 > 1$. First we notice that $1/t$ is a decreasing function on $[1, \infty)$, so that the minimum value of $1/t$ on any subinterval $[r, s]$ of $[1, \infty)$ is $1/s$. Thus from the Addition and Comparison Properties of the integral,

$$\ln 3 = \int_1^{5/4} \frac{1}{t}\,dt + \int_{5/4}^{6/4} \frac{1}{t}\,dt + \int_{6/4}^{7/4} \frac{1}{t}\,dt + \cdots + \int_{11/4}^{3} \frac{1}{t}\,dt$$

$$\geq \int_1^{5/4} \frac{1}{5/4}\,dt + \int_{5/4}^{6/4} \frac{1}{6/4}\,dt + \int_{6/4}^{7/4} \frac{1}{7/4} + \cdots + \int_{11/4}^{3} \frac{1}{3}\,dt$$

$$= \frac{1}{4}\left(\frac{4}{5} + \frac{4}{6} + \frac{4}{7} + \cdots + \frac{4}{12}\right)$$

$$\approx 1.019877345$$

Therefore ln $3 > 1$, which is what we wished to prove.   ❏

Since ln x is an antiderivative of $1/x$, we can easily evaluate any integral of the form $\int_a^b 1/x\,dx$ in terms of the natural logarithm, whenever a and b are positive.

EXAMPLE 2   Evaluate $\displaystyle\int_2^6 \frac{1}{x}\,dx$ in terms of logarithms.

Solution   By the preceding comment,

$$\int_2^6 \frac{1}{x}\, dx = (\ln x)\ \Big|_2^6 = \ln 6 - \ln 2 \qquad \square$$

Next we will state and prove the Law of Logarithms.

THEOREM 5.21
Law of Logarithms

For all $b > 0$ and $c > 0$,

$$\ln bc = \ln b + \ln c$$

Proof Fix $b > 0$. For any $x > 0$ let

$$g(x) = \ln bx$$

The Chain Rule yields

$$g'(x) = \left(\frac{1}{bx}\right) b = \frac{1}{x}$$

Therefore g and $\ln x$ have the same derivative. By Theorem 4.6 they differ by a constant C, that is,

$$\ln bx = \ln x + C$$

Substituting $x = 1$ in this equation and noting that $\ln 1 = 0$, we obtain

$$\ln b = \ln 1 + C = C$$

As a result,

$$\ln bx = \ln x + \ln b$$

When $x = c$, this is equivalent to the equation that was to be verified. ■

The following properties of logarithms can be proved from the Law of Logarithms:

$$\ln b^r = r \ln b \qquad \text{for } b > 0 \tag{11}$$

$$\ln \frac{1}{b} = -\ln b \qquad \text{for } b > 0 \tag{12}$$

$$\ln \frac{b}{c} = \ln b - \ln c \quad \text{for } b, c > 0 \tag{13}$$

The formulas can also be proved by using derivatives (see Exercises 47–49).

Using (13), we can simplify the answer obtained in Example 2. Indeed (13) implies that

$$\ln 6 - \ln 2 = \ln \frac{6}{2} = \ln 3$$

Thus

$$\int_2^6 \frac{1}{x}\, dx = \ln 3$$

If f is any positive differentiable function, then we can apply the Chain Rule to conclude that

$$\frac{d}{dx} \ln f(x) = \frac{1}{f(x)} f'(x) = \frac{f'(x)}{f(x)} \tag{14}$$

EXAMPLE 3   Find

$$\frac{d}{dx} \ln (x^2 + x)^{1/3}$$

Solution   By using first (11) and then (14) we find that

$$\frac{d}{dx} \ln (x^2 + x)^{1/3} \overset{(11)}{=} \frac{d}{dx} \left[\frac{1}{3} \ln(x^2 + x) \right] \overset{(14)}{=} \frac{1}{3} \cdot \frac{1}{x^2 + x} \cdot \frac{d}{dx} (x^2 + x)$$

$$= \frac{1}{3} \cdot \frac{1}{x^2 + x} (2x + 1) = \frac{2x + 1}{3(x^2 + x)} \qquad \square$$

Since the domain of the function $\ln x$ is $(0, \infty)$, the domain of $\ln (-x)$ is $(-\infty, 0)$. Hence by the Chain Rule we have

$$\frac{d}{dx} \ln (-x) = \frac{1}{-x} (-1) = \frac{1}{x}$$

Thus the function $\ln |x|$, which is equal to $\ln (-x)$ on $(-\infty, 0)$, and equal to $\ln x$ on $(0, \infty)$, is an antiderivative of $1/x$ on $(-\infty, 0)$ and on $(0, \infty)$. This gives us the formula

$$\int \frac{1}{x} dx = \ln |x| + C \tag{15}$$

EXAMPLE 4   Express $\int_{-8}^{-7} \frac{1}{x} dx$ in terms of logarithms.

Solution   From (15) we deduce that

$$\int_{-8}^{-7} \frac{1}{x} dx = \ln |x| \, \Big|_{-8}^{-7} = \ln |-7| - \ln |-8| = \ln 7 - \ln 8 = \ln \frac{7}{8} \qquad \square$$

Next we will evaluate

$$\int \frac{f'(x)}{f(x)} dx$$

by substitution. To accomplish that, we let

$$u = f(x), \quad \text{so that} \quad du = f'(x) \, dx$$

Then by (15),

$$\int \frac{f'(x)}{f(x)} dx = \int \overset{1/u}{\overbrace{\frac{1}{f(x)}}} \overset{du}{\overbrace{f'(x) \, dx}} = \int \frac{1}{u} du \overset{(15)}{=} \ln |u| + C = \ln |f(x)| + C$$

which yields the formula

$$\int \frac{f'(x)}{f(x)}\, dx = \ln |f(x)| + C \tag{16}$$

It is not necessary to memorize (16) if you remember that when you see an integral of the form in (16), you should think about substituting $u = f(x)$.

EXAMPLE 5   Find $\displaystyle\int \frac{x^4}{x^5 + 1}\, dx$.

Solution   Let

$$u = x^5 + 1, \quad \text{so that} \quad du = 5x^4\, dx$$

Then

$$\int \frac{x^4}{x^5 + 1}\, dx = \int \overbrace{\frac{1}{x^5 + 1}}^{1/u} \overbrace{x^4\, dx}^{\frac{1}{5}\, du} = \int \frac{1}{u} \cdot \frac{1}{5}\, du$$

$$= \frac{1}{5} \ln |u| + C = \frac{1}{5} \ln |x^5 + 1| + C \qquad \square$$

Formula (16) can be used to find the integrals of the four remaining basic trigonometric functions that we have not yet found: $\tan x$, $\sec x$, $\cot x$, and $\csc x$.

To find $\int \tan x\, dx$, we first recall that

$$\tan x = \frac{\sin x}{\cos x}$$

Since the numerator is essentially the derivative of the denominator, we can apply (16) by letting

$$u = \cos x, \quad \text{so that} \quad du = -\sin x\, dx$$

We find that

$$\int \tan x\, dx = \int \overbrace{\frac{1}{\cos x}}^{1/u} \overbrace{\sin x\, dx}^{(-1)\, du} = \int \frac{1}{u} (-1)\, du = -\ln |u| + C = -\ln |\cos x| + C$$

Therefore

$$\int \tan x\, dx = -\ln |\cos x| + C$$

The evaluation of $\int \cot x\, dx$ is analogous. However, the evaluations of $\int \sec x\, dx$ and $\int \csc x\, dx$ are not so transparent. For $\int \sec x\, dx$ recall that

$$\frac{d}{dx} \tan x = \sec^2 x \quad \text{and} \quad \frac{d}{dx} \sec x = \sec x \tan x$$

Thus

$$\frac{d}{dx}(\sec x + \tan x) = \sec x \tan x + \sec^2 x = (\sec x)(\sec x + \tan x)$$

Since

$$\int \sec x \, dx = \int \frac{(\sec x)(\sec x + \tan x)}{\sec x + \tan x} \, dx$$

the calculations above tell us that the numerator of the latter integral is the derivative of the denominator, so we can once again use (16). We let

$$u = \sec x + \tan x, \quad \text{so that} \quad du = (\sec x)(\sec x + \tan x) \, dx$$

Then

$$\int \sec x \, dx = \int \overbrace{\frac{1}{\sec x + \tan x}}^{1/u} \overbrace{[\sec x(\tan x + \sec x)]}^{du} \, dx$$

$$= \int \frac{1}{u} \, du = \ln|u| + C = \ln|\sec x + \tan x| + C$$

Consequently

$$\int \sec x \, dx = \ln|\sec x + \tan x| + C \tag{17}$$

The evaluation of $\int \csc x \, dx$ is similar (see Exercise 36).

Logarithmic Differentiation

The formula in (16) can be useful in the differentiation of complicated functions involving products, quotients, and powers. Indeed, from (16),

$$\frac{d}{dx} \ln|f(x)| = \frac{f'(x)}{f(x)}$$

so that

$$f'(x) = f(x) \frac{d}{dx} \ln|f(x)| \tag{18}$$

or in the Leibniz notation,

$$\frac{dy}{dx} = y \frac{d}{dx} \ln|y| \tag{19}$$

The idea is to first use the properties of logarithms to rewrite $\ln|f(x)|$ as a combination of logarithms of the component functions of f. For example, let

$$f(x) = \frac{(x^3 - 1)^8 (2x + \sin x)^{1/3}}{(x^2 + 1)^{1/2}}$$

Then

$$\ln|f(x)| = \ln|x^3 - 1|^8 + \ln|2x + \sin x|^{1/3} - \ln(x^2 + 1)^{1/2}$$

$$= 8 \ln |x^3 - 1| + \frac{1}{3} \ln |2x + \sin x| - \frac{1}{2} \ln (x^2 + 1)$$

Then we differentiate the right–hand side and use (18) to obtain $f'(x)$. This method of finding $f'(x)$ is called **logarithmic differentiation,** and is illustrated in the next example.

EXAMPLE 6 Let

$$f(x) = \frac{(x^3 - 1)^8 (2x + \sin x)^{1/3}}{(x^2 + 1)^{1/2}}$$

Use logarithmic differentiation to find $f'(x)$.

Solution By the comments preceding the example,

$$\ln |f(x)| = 8 \ln |x^3 - 1| + \frac{1}{3} \ln |2x + \sin x| - \frac{1}{2} \ln (x^2 + 1)$$

Next we differentiate $\ln |f(x)|$ to obtain

$$\frac{d}{dx} \ln |f(x)| = \frac{24x^2}{x^3 - 1} + \frac{1}{3} \frac{2 + \cos x}{2x + \sin x} - \frac{x}{x^2 + 1}$$

Now (18) yields

$$f'(x) = \frac{(x^3 - 1)^8 (2x + \sin x)^{1/3}}{(x^2 + 1)^{1/2}} \left(\frac{24x^2}{x^3 - 1} + \frac{2 + \cos x}{6x + 3 \sin x} - \frac{x}{x^2 + 1} \right) \qquad \square$$

We will return to the study of logarithms in Section 7.3, where we will establish relationships between the natural logarithm and other logarithms.

EXERCISES 5.7

In Exercises 1–4 evaluate the integral. Express your answer in terms of logarithms.

1. $\displaystyle\int_2^8 \frac{1}{x}\,dx$ **2.** $\displaystyle\int_{1/9}^{1/4} \frac{-1}{3x}\,dx$

3. $\displaystyle\int_{-4}^{-12} \frac{2}{t}\,dt$ **4.** $\displaystyle\int_{-1/16}^{-1/8} \frac{1}{t}\,dt$

In Exercises 5–12 find the domain and the derivative of the function.

5. $f(x) = \ln (x + 1)$ **6.** $k(t) = \ln (t^2 + 4)^3$

7. $f(x) = \ln \sqrt{\dfrac{x - 3}{x - 2}}$ **8.** $f(x) = \dfrac{\ln x}{x - 1}$

9. $f(t) = \sin (\ln t)$ **10.** $g(u) = \ln (\sin u)$

11. $f(x) = \ln (\ln x)$ **12.** $f(x) = \ln (x + \sqrt{x^2 - 1})$

In Exercises 13–14 find dy/dx by implicit differentiation.

13. $x \ln (y^2 + x) = 1 + 5y$ **14.** $y \ln \dfrac{y}{x} = \sin y^2$

In Exercises 15–18 find the domain, intercepts, relative extreme values, inflection points, concavity, and asymptotes for the given function. Then draw its graph.

15. $f(x) = \ln |x|$ **16.** $f(x) = \ln (x - 2)$

17. $f(x) = \ln (1 + x^2)$ **18.** $f(x) = (\ln x)^2$

 19. Let $f(x) = x^{-2} - \ln x$. Use the Newton-Raphson method to approximate a zero of f. Continue until successive iterations obtained by calculator are identical.

 20. Let $f(x) = x^2 + x \ln x$. Use the Newton-Raphson method to approximate a relative extreme value of f. Continue until successive iterations obtained by calculator are identical.

In Exercises 21–35 evaluate the integral.

21. $\displaystyle\int \frac{1}{x-1}\,dx$ **22.** $\displaystyle\int \frac{2}{1-4x}\,dx$

23. $\displaystyle\int \frac{x}{x^2+4}\,dx$ **24.** $\displaystyle\int \frac{x^2}{1-x^3}\,dx$

25. $\displaystyle\int_0^{\pi/3} \frac{\sin x}{1-3\cos x}\,dx$ **26.** $\displaystyle\int \frac{1}{x(1+\ln x)}\,dx$

27. $\displaystyle\int_{-1}^{0} \frac{x+2}{x^2+4x-1}\,dx$ **28.** $\displaystyle\int_1^4 \frac{1}{\sqrt{x}\,(1+\sqrt{x})}\,dx$

29. $\displaystyle\int \frac{\ln z}{z}\,dz$ **30.** $\displaystyle\int \frac{(\ln z)^5}{z}\,dz$

31. $\displaystyle\int \frac{\ln(\ln t)}{t\ln t}\,dt$ **32.** $\displaystyle\int \frac{\tan\sqrt{t}}{\sqrt{t}}\,dt$

33. $\displaystyle\int \cot t\,dt$

34. $\displaystyle\int \frac{1}{1+x^{1/3}}\,dx$ (*Hint:* Substitute $u = 1 + x^{1/3}$.)

*35. $\displaystyle\int \frac{x}{1+x\tan x}\,dx$ (*Hint:* Substitute $u = x\sin x + \cos x$.)

*36. Evaluate $\int \csc x\,dx$. (*Hint:* Pattern the solution after the evaluation of $\int \sec x\,dx$ in the text.)

In Exercises 37–39 find the area A of the region between the graph of f and the x axis on the given interval.

37. $f(x) = \dfrac{1}{x}$; $[e, e^2]$ **38.** $f(x) = \dfrac{x}{2-x^2}$; $[-2, -\sqrt{3}]$

*39. $f(x) = \dfrac{\sin^3 x}{\cos x}$; $\left[\dfrac{\pi}{4}, \dfrac{\pi}{3}\right]$

 40. a. Adapt the ideas used in the solution of Example 1 in order to show that $\int_1^{2.9} 1/t\,dt > 1$. (*Hint:* Find a suitable partition P of $[1, 2.9]$, and use the Comparison Property.)
  b. Use part (a) to prove that $e < 2.9$.

In Exercises 41–46 use logarithmic differentiation to find the derivative of the given function.

41. $f(x) = (x+1)^{1/5}(2x+3)^2(7-4x)^{-1/2}$

42. $f(x) = (1+\cos x)^{2/3}(x^2+x+1)^{4/5}x^{1.1}$

43. $y = \sqrt[3]{\dfrac{(x+3)^2(2x-1)}{(4x+5)^4}}$ **44.** $y = \dfrac{\sqrt{x^2+1}\,\sin^3 x}{x^2\sqrt{2x^2+1}}$

45. $y = \dfrac{x^{3/2}\,e^{-x^2}}{1-e^x}$ **46.** $y = \dfrac{x^2\ln x}{(2x+1)^{3/2}\cos x}$

47. Prove that $\ln b^r = r\ln b$ for $b > 0$ and r rational by showing that the functions $\ln x^r$ and $r\ln x$ have the same derivative and the same value at 1.

48. Prove that $\ln(1/b) = -\ln b$ for $b > 0$
  a. by using (11).
  b. by showing that the functions $\ln(1/x)$ and $-\ln x$ have

the same derivative and the same value at 1.

49. Prove that $\ln(b/c) = \ln b - \ln c$ for $b, c > 0$
  a. by using (12) and the Law of Logarithms.
  b. by showing that the functions $\ln(x/c)$ and $\ln x - \ln c$ have the same derivative and the same value at c.

*50. Use Exercises 47–48 to show that
$$\lim_{x\to\infty}\ln x = \infty \quad \text{and} \quad \lim_{x\to 0^+}\ln x = -\infty$$

51. Using lower and upper sums, show that for each integer $n \ge 2$,
$$\frac{1}{2}+\frac{1}{3}+\cdots+\frac{1}{n} < \ln n < 1 + \frac{1}{2}+\cdots+\frac{1}{n-1}$$

 52. Exercise 50 and the inequality in Exercise 51 together show that if n is large enough, then the sum
$$1 + \frac{1}{2}+\frac{1}{3}+\frac{1}{4}+\cdots+\frac{1}{n}$$

is as large as we please. Use this information along with Exercise 51 to find a positive integer n such that the sum is larger than
  a. 20   **b.** 100

53. a. Use the inequality $1/t \le 1/\sqrt{t}$ for $t \ge 1$ to show that
$$\ln x = \int_1^x \frac{1}{t}\,dt \le 2(\sqrt{x}-1) \quad \text{for } x \ge 1$$
  b. Use part (a) to show that
$$\lim_{x\to\infty}\frac{\ln x}{x} = 0$$
  c. Use part (b) and (12) to show that $\lim_{x\to 0^+} x\ln x = 0$.

54. Sketch the graph of $x\ln x$. Use Exercise 53(c) and the fact that $\ln x = -1$ for $x \approx 0.37$.

55. Recall that $\ln e = 1$, and let
$$f(x) = \frac{\ln x}{x}$$
  a. Show that f is increasing on $(0, e]$ and decreasing on $[e, \infty)$.
  b. Find the maximum value of f on $(0, \infty)$.
  c. Sketch the graph of f. Use the second derivative of f, Exercise 53, and the fact that $\ln x = 3/2$ for $x \approx 4.5$.

56. Use Exercise 57(b) of Section 5.5 to find an upper bound for the integral $\int_1^2 \ln x\,dx$. (The exact value is $(2\ln 2) - 1$.)

57. When a quantity of gas expands from an initial volume V_1 to a final volume V_2, the amount of work W done by the gas during the expansion is given by

$$W = \int_{V_1}^{V_2} P \, dV \qquad (20)$$

where P is the pressure expressed as a function of the volume V. During an expansion in which the temperature remains constant, P is related to the volume by means of Boyle's Law:

$$P = \frac{c}{V} \qquad (21)$$

where c is a constant. Using (20) and (21), obtain a formula for W that involves logarithms.

Applications

58. A beanbag factory has a marginal revenue function

$$m_R(x) = \frac{2}{x+1}$$

where x denotes thousands of beanbags sold and $m_R(x)$ denotes dollars received per beanbag.

a. Determine the total revenue function R. (*Hint:* Use (7) of Section 5.4 and the fact that $R(0) = 0$. What is the derivative of $\ln(x+1)$?)

b. Demonstrate that R is a reasonable total revenue function by showing that R is increasing and concave downward on the interval $(0, \infty)$.

 59. The equation

$$\ln w = 4.4974 + 3.135 \ln s$$

has been used to relate the weight w (in kilograms) to the sitting height s (in meters) of people. Using this equation, find the weight of a person whose sitting height is

a. 1 meter   **b.** $\frac{1}{2}$ meter

 60. Let y represent the weight in ounces of a baby and x the age in months. It has been conjectured that y and x are related by the equation

$$\ln y - \ln(341.5 - y) = c(x - 1.66) \quad \text{for } 0 \leq x \leq 9$$

where c is a positive constant.

a. Show that dy/dx, the rate of weight increase with respect to age of the baby, is a positive function.

b. Find the age x_0 at which dy/dx is maximum, and find the corresponding weight. According to the equation, it is at this age that the baby gains weight fastest.

Projects

1. This project relates approximate values of the natural logarithm to parabolic approximation as discussed in Section 3.8.

a. Find an equation of the line tangent to the curve $y = 1/t$ at $(1, 1)$.

b. Show that for small $h > 0$, the area below the tangent line on $[1, 1+h]$ is $h - h^2/2$ (which is approximately $\int_1^{1+h} (1/t) \, dt$).

c. Show that for $h < 0$ with h near 0, the area below the tangent line on $[1+h, 1]$ is $-(h - h^2/2)$. (which is approximately $\int_{1+h}^1 (1/t) \, dt$).

d. Together (b) and (c) yield

$$\ln(1 + h) \approx h - \frac{1}{2}h^2$$

when $|h|$ is small. Show that this formula is precisely the parabolic approximation formula for $\ln x$ with h near 0 (see Section 3.8).

2. In this project we will show that under mild conditions, a "law of logarithms" kind of condition yields the natural logarithm function. To that end, suppose f is defined on $(0, \infty)$, is differentiable at 1, and satisfies

$$f(xy) = f(x) + f(y) \qquad (22)$$

for all x and y in $(0, \infty)$.

a. Prove that $f(1) = 0$.

b. Show that

$$f\left(\frac{y}{x}\right) = f(y) - f(x)$$

for all x and y in $(0, \infty)$. (Hint: $y = x(y/x)$.)

c. Deduce from (a) and (b) that if $0 < |h| < x$, then

$$\frac{f(x+h) - f(x)}{h} = \frac{f(\frac{x+h}{x})}{h} = \frac{1}{x} \frac{f(1 + h/x) - f(1)}{h/x}$$

d. From (c) and the hypothesis that f is differentiable at 1, conclude that f' exists on $(0, \infty)$ and that

$$f'(x) = \frac{1}{x} f'(1) \quad \text{for } x > 0$$

It follows from (a) and (d) and the definition of $\ln x$ that if f is a differentiable function satisfying (22), then

$$f(x) = f'(1) \ln x \quad \text{for } x > 0$$

Finally, if we assume that $f'(1) = 1$, then $f(x) = \ln x$ for $x > 0$. Thus with mild extra conditions, the "law of logarithms" type of formula in (22) yields the natural logarithm function.

5.8   ANOTHER LOOK AT AREA

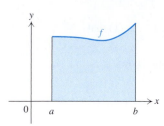

FIGURE 5.40

In Section 5.2 we defined the area of a region of the type shown in Figure 5.40 to be $\int_a^b f(x)\,dx$. However, this definition does not apply to regions like that shown in Figure 5.41(a) (which might, for instance, represent the surface of a lake). Using further properties that we naturally expect area to have, we will now extend the earlier definition of area to include regions whose lower boundaries are not necessarily horizontal.

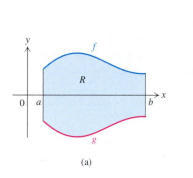

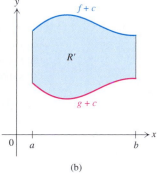

(a)   (b)

FIGURE 5.41

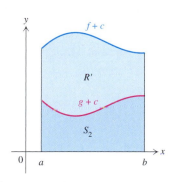

FIGURE 5.42

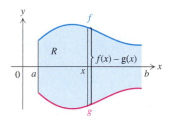

FIGURE 5.43

Let f and g be continuous on an interval $[a, b]$, and assume that $f(x) \geq g(x)$ for $a \leq x \leq b$. We will define the area of the region R that is bounded above by the graph of f, below by the graph of g, and on the sides by the lines $x = a$ and $x = b$ (Figure 5.41(a)). We call R the **region between the graphs of f and g on $[a, b]$.**

Since the area of a region should not be affected by shifting the region vertically, the area A of the region R in Figure 5.41(a) should be the same as the area of the region R' in Figure 5.41(b), which lies above the x axis and is bounded by the graphs of $f + c$ and $g + c$ for an appropriate constant c. However, in Figure 5.42 the area of the region S_1 (which is bounded above by the graph of $f + c$ and below by the x axis) should be the sum of the areas of R' and S_2. Now by Definition 5.5, we have the formulas for the areas of S_1 and S_2. Therefore we would expect that

$$A = \overbrace{\int_a^b [f(x) + c]\,dx}^{\text{area of } S_1} - \overbrace{\int_a^b [g(x) + c]\,dx}^{\text{area of } S_2} = \int_a^b [f(x) - g(x)]\,dx$$

Thus we are led to the following formal definition of the area A:

DEFINITION   5.22   Let f and g be continuous on $[a, b]$, with $f(x) \geq g(x)$ for $a \leq x \leq b$. The area A of the region R between the graphs of f and g on $[a, b]$ is given by

$$A = \int_a^b [f(x) - g(x)]\,dx \qquad (1)$$

Notice that for $a \leq x \leq b$ the integrand $f(x) - g(x)$ represents the height of R at x (Figure 5.43).

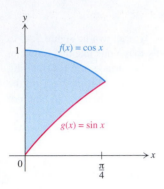

FIGURE 5.44

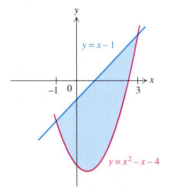

FIGURE 5.45

EXAMPLE 1   Let $f(x) = \cos x$ and $g(x) = \sin x$. Find the area A of the region between the graphs of f and g on $[0, \pi/4]$ (Figure 5.44).

Solution   Since $\cos x \geq \sin x$ on $[0, \pi/4]$, it follows from (1) that

$$A = \int_0^{\pi/4} [f(x) - g(x)]\, dx = \int_0^{\pi/4} (\cos x - \sin x)\, dx$$

$$= (\sin x + \cos x)\Big|_0^{\pi/4} = \left(\frac{1}{2}\sqrt{2} + \frac{1}{2}\sqrt{2}\right) - (0 + 1) = \sqrt{2} - 1 \qquad \square$$

If we seek the area of the region between two graphs that cross exactly twice but we are not given the interval over which to integrate, we first determine where the graphs cross and which graph lies above the other, and then integrate over the corresponding (bounded) interval.

EXAMPLE 2   Find the area A of the region between the graphs of $y = x^2 - x - 4$ and $y = x - 1$ (Figure 5.45).

Solution   First we determine the x coordinates of the points at which the two curves intersect:

$$x^2 - x - 4 = x - 1$$

$$x^2 - 2x - 3 = 0$$

$$(x + 1)(x - 3) = 0$$

$$x = -1 \qquad \text{or} \qquad x = 3$$

Thus the region whose area we seek lies between the graphs of $y = x^2 - x - 4$ and $y = x - 1$ on $[-1, 3]$. Since

$$x - 1 \geq x^2 - x - 4 \quad \text{for } -1 \leq x \leq 3$$

(see Figure 5.45), it follows that the height of the region at x is $[(x - 1) - (x^2 - x - 4)]$, so by (1),

$$A = \int_{-1}^{3} [(x - 1) - (x^2 - x - 4)]\, dx = \int_{-1}^{3} (-x^2 + 2x + 3)\, dx$$

$$= \left(-\frac{1}{3}x^3 + x^2 + 3x\right)\Big|_{-1}^{3} = (-9 + 9 + 9) - \left(\frac{1}{3} + 1 - 3\right)$$

$$= \frac{32}{3} \qquad \square$$

Now suppose the graphs of f and g cross at one or more points in (a, b). Then the region R between the graphs of f and g on $[a, b]$ is composed of several regions R_1, R_2, ..., each of the type whose area we have already defined (Figure 5.46). We naturally define the total area A of R to be the sum of the areas of those regions. In order to calculate A, that is, the area of the region between the graphs of f and g on $[a, b]$, we first determine those subintervals on which $f - g \geq 0$ and those on which $f - g \leq 0$. Then we integrate over those subintervals separately. Example 3 illustrates this technique.

EXAMPLE 3   Let $f(x) = \sin x$ and $g(x) = \cos x$. Find the area A of the region between the graphs of f and g on $[0, 2\pi]$ (Figure 5.47).

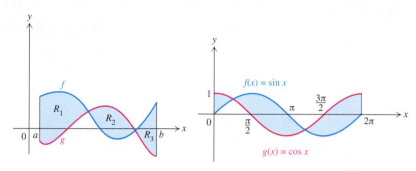

FIGURE 5.46   **FIGURE 5.47**

Solution   In order to determine where $\sin x \geq \cos x$ and where $\sin x \leq \cos x$, we first find the values of x in $[0, 2\pi]$ for which $\sin x = \cos x$, which is equivalent to solving $\tan x = 1$ for x. But $\tan x = 1$ for $x = \pi/4$ and $x = 5\pi/4$. From this you can verify that $\sin x \geq \cos x$ on $[\pi/4, 5\pi/4]$ and $\cos x \geq \sin x$ on $[0, \pi/4]$ and $[5\pi/4, 2\pi]$. Therefore

$$A = \int_0^{\pi/4} (\cos x - \sin x)\, dx + \int_{\pi/4}^{5\pi/4} (\sin x - \cos x)\, dx + \int_{5\pi/4}^{2\pi} (\cos x - \sin x)\, dx$$

$$= (\sin x + \cos x)\Big|_0^{\pi/4} + (-\cos x - \sin x)\Big|_{\pi/4}^{5\pi/4} + (\sin x + \cos x)\Big|_{5\pi/4}^{2\pi}$$

$$= (\sqrt{2} - 1) + (2\sqrt{2}) + (1 + \sqrt{2}) = 4\sqrt{2} \qquad \square$$

The region in the next example may at first appear not to be of the form covered by our discussion, but in fact it is.

EXAMPLE 4   Find the area A of the region R between the parabola $x = \frac{1}{2}y^2$ and the line $y = 2x - 2$ (Figure 5.48).

Solution   First we determine the x coordinates of the points at which the parabola and the line intersect:

$$x = \frac{1}{2}y^2 = \frac{1}{2}(2x - 2)^2 = 2x^2 - 4x + 2$$

which means that

$$2x^2 - 5x + 2 = 0$$

or

$$(2x - 1)(x - 2) = 0$$

Consequently $x = \frac{1}{2}$ or $x = 2$. Next, observe from Figure 5.48 that R may be broken up into two parts: the part between $x = 0$ and $x = \frac{1}{2}$ on the x axis and the part between

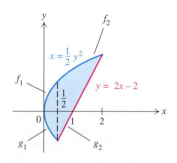

FIGURE 5.48

$x = \frac{1}{2}$ and $x = 2$ on the x axis. The part of R between $x = 0$ and $x = \frac{1}{2}$ lies between the graphs of f_1 and g_1, where

$$f_1(x) = \sqrt{2x} \quad \text{and} \quad g_1(x) = -\sqrt{2x}$$

The part of R between $x = \frac{1}{2}$ and $x = 2$ lies between the graphs of f_2 and g_2, where

$$f_2(x) = \sqrt{2x} \quad \text{and} \quad g_2(x) = 2x - 2$$

Therefore

$$A = \int_0^{1/2} (f_1(x) - g_1(x))\, dx + \int_{1/2}^2 (f_2(x) - g_2(x))\, dx$$

$$= \int_0^{1/2} \left[\sqrt{2x} - (-\sqrt{2x}) \right] dx + \int_{1/2}^2 \left[\sqrt{2x} - (2x - 2) \right] dx$$

$$= \int_0^{1/2} 2\sqrt{2x}\, dx + \int_{1/2}^2 (\sqrt{2x} - 2x + 2)\, dx$$

$$= \frac{4\sqrt{2}}{3} x^{3/2} \Big|_0^{1/2} + \left(\frac{2\sqrt{2}}{3} x^{3/2} - x^2 + 2x \right) \Big|_{1/2}^2$$

$$= \left(\frac{2}{3} - 0 \right) + \left[\left(\frac{8}{3} - 4 + 4 \right) - \left(\frac{1}{3} - \frac{1}{4} + 1 \right) \right] = \frac{9}{4} \quad \square$$

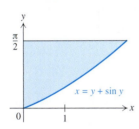

FIGURE 5.49

Reversing the Roles of x and y

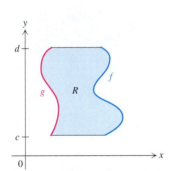

FIGURE 5.50

Instead of considering a region R as the region between the graphs of two functions of x, it is sometimes convenient to consider R as the region between the graphs of two functions of y (Figure 5.49). Then the area is computed by integrating along the y axis, instead of along the x axis.

EXAMPLE 5   Let R be the region between the y axis and the graph of the equation $x = y + \sin y$ on $[0, \pi/2]$ (Figure 5.50). Find the area A of R.

Solution   Since $y + \sin y \geq 0$ for $0 \leq y \leq \pi/2$, it follows that

$$A = \int_0^{\pi/2} (y + \sin y)\, dy = \left(\frac{1}{2} y^2 - \cos y \right) \Big|_0^{\pi/2}$$

$$= \frac{1}{2} \left(\frac{\pi}{2} \right)^2 - (-1) = \frac{\pi^2}{8} + 1 \quad \square$$

In Example 5 we have no way of describing y in terms of x; therefore it would be impossible to use (1) to determine the area of the region.

If a region can be described both in terms of x and in terms of y, the area is the same, whether we integrate with respect to x or with respect to y. To support this claim, we now return to the region described in Example 4 and find its area by integrating along the y axis.

EXAMPLE 6   Find the area A of the region R between the parabola $x = \frac{1}{2} y^2$ and the line $y = 2x - 2$ (Figure 5.51).

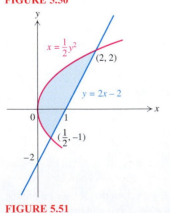

FIGURE 5.51

Solution   First we determine the y coordinates of points at which the parabola and the line intersect:

$$\frac{1}{2} y^2 = x = \frac{1}{2}(y + 2)$$

$$y^2 - y - 2 = 0$$

$$(y + 1)(y - 2) = 0$$

$$y = -1 \quad \text{or} \quad y = 2$$

Since $\frac{1}{2}(y + 2) \geq \frac{1}{2} y^2$ for $-1 \leq y \leq 2$, it follows that

$$A = \int_{-1}^{2} \left[\frac{1}{2}(y + 2) - \frac{1}{2} y^2 \right] dy = \frac{1}{2} \int_{-1}^{2} (y + 2 - y^2)\, dy$$

$$= \frac{1}{2} \left(\frac{1}{2} y^2 + 2y - \frac{1}{3} y^3 \right) \Big|_{-1}^{2}$$

$$= \frac{1}{2} \left[\left(2 + 4 - \frac{8}{3} \right) - \left(\frac{1}{2} - 2 + \frac{1}{3} \right) \right] = \frac{9}{4}$$

(the same value for the area we found in Example 4).   ❑

Cavalieri's Principle

We continue this section with a discussion of Cavalieri's Principle. Let f and g be continuous on an interval $[a, b]$, and let A be the area of the region between the graphs of f and g. Using the fact that

$$|f(x) - g(x)| = \begin{cases} f(x) - g(x) & \text{if } f(x) \geq g(x) \\ g(x) - f(x) & \text{if } g(x) \geq f(x) \end{cases}$$

we obtain the formula

$$A = \int_{a}^{b} |f(x) - g(x)|\, dx \tag{2}$$

which holds regardless of the relationship between f and g.

Now suppose f is any continuous function defined on $[a, b]$, and let $g = f + c$, where c is a fixed positive constant. Then the area of the region between the graphs of f and g on $[a, b]$ is

$$\int_{a}^{b} |f(x) - g(x)|\, dx = \int_{a}^{b} c\, dx = c(b - a)$$

In other words, if the distance between the lower and upper boundaries of a region is constant, then the region's area is the product of its length and height, just as in the case of rectangles (Figure 5.52(a)).

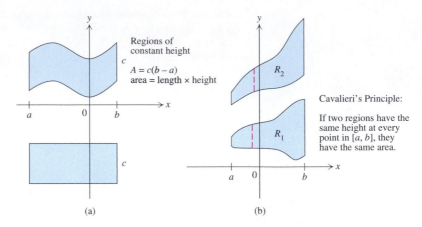

FIGURE 5.52

The result just obtained is a special case of Cavalieri's Principle. This principle was stated in 1635 by the Milanese mathematician Bonaventura Cavalieri, who used it to help derive formulas for areas of plane regions. In effect Cavalieri's Principle says that if R_1 and R_2 are two regions like those in Figure 5.52(b), with side boundaries $x = a$ and $x = b$, and if the height of R_1 (measured perpendicular to the x axis) is the same as the height of R_2 at every point in $[a, b]$ (Figure 5.52(b)), then R_1 and R_2 have the same area. Cavalieri came to this conclusion several decades before Newton and Leibniz introduced the general concept of integral. However, by using (2), we can easily prove Cavalieri's Principle for the case in which the upper boundaries of R_1 and R_2 are the graphs of continuous functions f_1 and f_2 and the lower boundaries are the graphs of continuous functions g_1 and g_2. In that case the fact that R_1 and R_2 have identical height at every point in $[a, b]$ means that

$$|f_1(x) - g_1(x)| = |f_2(x) - g_2(x)| \quad \text{for } a \le x \le b$$

Consequently

$$\int_a^b |f_1(x) - g_1(x)| \, dx = \int_a^b |f_2(x) - g_2(x)| \, dx$$

which means, by (2), that the area of R_1 and the area of R_2 are the same.

Other Interpretations of Area

The area of the region between the graphs of f and g on an interval $[a, b]$ has many interpretations, depending on the particular meaning attached to f and g.

For example, if v represents the velocity of an object during the time period $[a, b]$ and if $v(t) \ge 0$ for $a \le t \le b$, then the net distance the object has traveled during the period is given by $\int_a^b v(t) \, dt$ (see Figure 5.53(a)). More generally, suppose v_1 and v_2 denote the velocities of two objects that start from the same position at the same time and suppose $v_2(t) \ge v_1(t) \ge 0$ for $a \le t \le b$, as in Figure 5.53(b). Then the area of the shaded region represents the distance between the objects at the end of that time interval.

Areas of regions such as the one shown in Figure 5.53(c) play a role in the economic study of the distribution of national income. In the figure, R is bounded above by the line $y = x$ and below by the so-called **Lorenz curve** $y = L(x)$. For any x

between 0 and 100, $L(x)$ is the percentage of the national income owned by the lowest x percentage of families in the country. Values for income shares in various years have been surprisingly stable, which means that over the years L has changed very little. If the income were distributed equally, then the Lorenz curve would be the line $y = x$. Thus the shaded region in Figure 5.53(c) represents the inequity in income distribution. The larger the area is, the larger the inequality in income distribution.

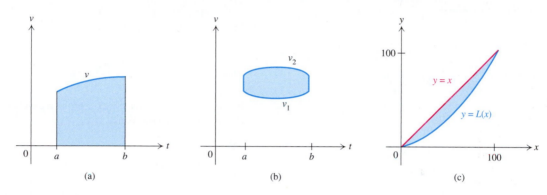

FIGURE 5.53   (a) The area under the velocity curve represents distance.   (b) The area between the velocity curves represents distance.   (c) A Lorenz diagram.

EXERCISES 5.8

In Exercises 1–6 find the area A of the shaded region.

1.

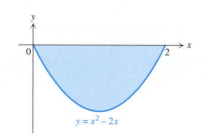

$y = x^2 - 2x$

2.

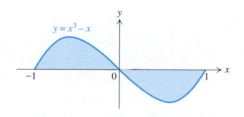

$y = x^3 - x$

3.

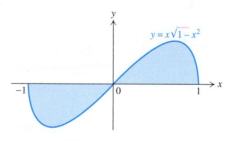

$y = x\sqrt{1 - x^2}$

4.

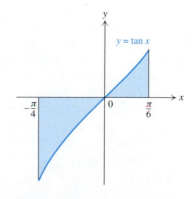

$y = \tan x$

5.

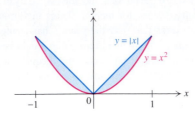

6.

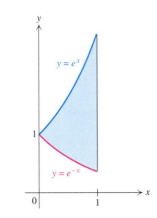

In Exercises 7–12 find the area A of the region between the graph of f and the x axis on the given interval.

7. $f(x) = x^2 + 2x$; $[-1, 3]$

8. $f(x) = \cos x - \sin x$; $[0, \pi/3]$

9. $f(x) = \dfrac{x}{\sqrt{1 + x^2}}$; $[-1, \sqrt{7}]$

10. $f(x) = \dfrac{x}{x^2 - 1}$; $[-\tfrac{1}{2}, \tfrac{1}{3}]$

11. $f(x) = \dfrac{\ln x}{x}$; $[\tfrac{1}{2}, 2]$

12. $f(x) = e^{-x} - 1$; $[0, 3]$

In Exercises 13–24 find the area A of the region between the graphs of the functions on the given interval.

13. $f(x) = x^2$, $g(x) = x^3$; $[-2, 1]$

14. $f(x) = 1/x$, $g(x) = 1/x^2$; $[\tfrac{1}{2}, 2]$

15. $g(x) = x^2 + 4x$, $k(x) = x - 2$; $[-3, 0]$

16. $g(x) = 2x^3 + 2x^2$, $k(x) = 2x^3 - 2x$; $[-4, 2]$

17. $f(x) = \sec^2 x$, $g(x) = \sec x \tan x$; $[-\pi/3, \pi/6]$

18. $f(x) = \sin 2x$, $g(x) = 2 \cot x$; $[\pi/3, 2\pi/3]$

19. $g(x) = \sin^2 x$, $k(x) = \tan x$; $[-\pi/4, \pi/4]$

20. $g(x) = x^3 + 3x + \cos^2 x$, $k(x) = 4x^2 - \sin^2 x + 1$; $[-1, 2]$

21. $f(x) = x\sqrt{2x + 3}$, $g(x) = x^2$; $[-1, 3]$

22. $f(x) = x(x^2 + 1)^5$, $g(x) = x^2 (x^3 + 1)^5$; $[-1, 1]$

23. $f(x) = e^{2x}$, $g(x) = e^x$; $[-1, 1]$

24. $f(x) = e^x$, $g(x) = 1/e^x$; $[-1, 2]$

In Exercises 25–30 the graphs of f and g enclose a region. Determine the area A of that region.

25. $f(x) = x^3$, $g(x) = x^{1/3}$

26. $f(x) = x^2 + 3$, $g(x) = 12 - x^2$

27. $f(x) = x^2 + 1$, $g(x) = 2x + 9$

28. $f(x) = x^3 + x$, $g(x) = 3x^2 - x$

29. $f(x) = x^3 + 1$, $g(x) = (x + 1)^2$

30. $f(x) = 2 - \sqrt{x}$, $g(x) = \dfrac{\sqrt{x} + 1}{2\sqrt{x}}$

In Exercises 31–34 find the area A of the region between the graphs of the given equations.

31. $y^2 = 6x$ and $x^2 = 6y$

32. $y^2 = x$ and $y = x - 2$

33. $y^2 = 2x - 5$ and $y = x - 4$

34. $y^2 = 3x$ and $y = x^2 - 2x$ (*Hint:* The curves intersect at the points $(0, 0)$ and $(3, 3)$.)

In Exercises 35–36 the graphs of the three equations enclose a region. Determine the area A of that region.

35. $y = x + 2$, $y = -3x + 6$, $y = (2 - x)/3$

36. $y = \tfrac{3}{2}x$ for $x \geq 0$; $y = -\tfrac{3}{2}x$ for $x \leq 0$; $y = -x^2 - \tfrac{3}{2}x + 4$

In Exercises 37–39 find the area A of the region between the graphs of the given equations.

37. $x = y^2 - y$ and $x = y - y^2$

38. $x = 0$ and $x = \cos y$ for $-\pi/2 \leq y \leq 3\pi/2$

39. $x = y^2$ and $x = 6 - y - y^2$

40. Calculate the area A_b of the region shaded in Figure 5.54. Then find the limit of the area as b tends to ∞.

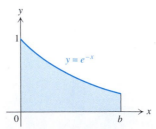

FIGURE 5.54 Graph for Exercise 40.

41. For any positive number a, let R_a and S_a denote the regions colored light blue and dark blue, respectively, in Figure 5.55, and let r_a be the ratio of the area of R_a to the area of S_a. Show that r_a is independent of a.

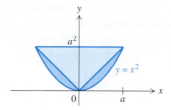

FIGURE 5.55   Graph for Exercise 41.

42. Let f, g, and h be continuous on $[a, b]$. Using (2), show that the area A of the region between the graphs of f and g on $[a, b]$ is the same as the area of the region between the graphs of $f + h$ and $g + h$ on $[a, b]$.

Application

43. Suppose the revenue per unit time and the cost per unit time of an athletic equipment company are given, respectively, by

$$f(t) = \sqrt{t} + 3 \quad \text{and} \quad g(t) = t^{1/3} + 2 \quad \text{for } 0 \le t \le 4$$

where t is in months and $f(t)$ and $g(t)$ are in thousands of dollars per month. Determine if the company is able to earn a profit of \$5000 during the four-month period.

Project

1. Let f be differentiable on $[0, \infty)$ and $f(x) > 0$ for $x > 0$. In this project we will compare the area $A(x)$ under the graph of f on $[0, x]$ with the area $B(x)$ of the rectangle with vertices $(0, 0)$, $(x, 0)$, $(x, f(x))$, and $(0, f(x))$ (Figure 5.56).

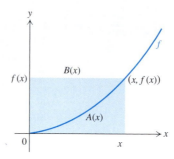

FIGURE 5.56   Figure for the project.

a. Let $c > 0$ and $r > 1$, and let $f(x) = cx^{r-1}$ for $x \ge 0$. Show that for each $x > 0$, $B(x)/A(x)$ depends only on r (and not on either x or c).

b. This part demonstrates that the functions f described in (a) are the only ones for which $B(x)/A(x)$ does not depend on x. To that end, suppose that g is differentiable on $[0, \infty)$ and $g(x) > 0$ for $x > 0$. Then $A(x) = \int_0^x g(t)\, dt$ and $B(x) = xg(x)$. Suppose $B(x)/A(x) = r$, or equivalently, $B(x) = rA(x)$, for all $x > 0$.

i.   By taking derivatives of both sides of the equation $B(x) = rA(x)$, show that $g(x) + xg'(x) = rg(x)$, or equivalently, $g'(x)/g(x) = (r-1)/x$ for all $x > 0$.

ii.   Integrate the two sides of the last equation in (i), and deduce that there is a constant c such that $\ln g(x) = \ln cx^{r-1}$. (*Hint:* Let $\ln c = C$, where C is the constant appearing from indefinite integration, and use the Law of Logarithms.)

iii.   Conclude that $g(x) = cx^{r-1}$ for all $x \ge 0$.

5.9   WHO INVENTED CALCULUS?

It is often said that Newton and Leibniz invented calculus. From the vantage point of our knowledge of the derivative and the integral—as well as of the Fundamental Theorem of Calculus, which unites them—we may ask why these two great mathematicians are credited with founding calculus.

Long before Newton and Leibniz, mathematicians knew how to calculate tangents to various curves. Using horizontal tangents, they produced a general procedure for determining maxima and minima of curves. As regards areas, even Archimedes was able to find areas of regions bounded by a few of the more common curves. By the early seventeenth century, the foremost mathematicians had developed very refined ways of evaluating areas and computing what we now call integrals.

Even the Fundamental Theorem did not originate with Newton or Leibniz. Although he did not have our concept of the derivative and the integral, Isaac Barrow, Newton's teacher at Cambridge, stated and proved a geometric equivalent of the Fundamental Theorem. The proof can be found in his *Lectiones Opticae et Geometricae.*

What was left of sufficient importance to justify calling Newton and Leibniz the inventors of calculus? There remained the need for a systematic method of differentiating and integrating general classes of functions as well as a need for a symbolism that would lend itself to formal, rather than geometrically intuitive, proofs. Newton and Leibniz recognized that the Fundamental Theorem could be used to systematically find integrals of functions they encountered. They also discovered rules for the derivatives and the integrals of combinations of functions (see Sections 3.3 and 5.5). Newton conceived his ideas first, during the middle 1660s, whereas Leibniz began his work during the early 1670s. However, Leibniz published his ideas first, and his symbolism and notation were far superior to those of Newton. In fact, Leibniz's notation survived and is in common use today.

Although we do not minimize the profound implications of Newton's and Leibniz's work, their ideas did rely heavily on previous results. Great as the achievements of Newton and Leibniz were, their ideas constituted only a link between the earlier conception of calculus and the modern approach to the subject. The following passage by L. C. Karpinski is pertinent.

> There is a strong temptation on the part of professional mathematicians and scientists to seek always to ascribe great discoveries and inventions to single individuals. Such ascription serves a didactic end in centering attention upon certain fundamental aspects of the subjects, much as the history of events is conveniently divided into epochs for purposes of exposition. There is in such attributions and division, however, the serious danger that too great a significance will be attached to them. Rarely— perhaps never—is a single mathematician or scientist entitled to receive the full credit for an "innovation," nor does any one age deserve to be called the "renaissance" of an aspect of culture. Back of any discovery or invention there is invariably to be found an evolutionary development of ideas making its geniture possible. The history of the calculus furnishes a remarkably apt illustration of this fact.*

REVIEW

Key Terms and Expressions

The Rectangle Property

The Addition Property

The Comparison Property

Partition

Lower sum; upper sum

Riemann sum

Right sum; left sum; midpoint sum

Definite integral

Integrand; limit of integration; integral sign

Area

Average (or mean) value of a function on an interval

Indefinite integral

Method of substitution

Change of variables

Natural logarithm

* Karpinski, L. C., "Is There Progress in Mathematical Discovery and Did the Greeks Have Analytic Geometry?" *Isis 27* (1937), 46–52.

Key Formulas

$$\int_a^b c \, dx = c(b-a)$$

$$\int_a^b f(x) \, dx = \int_a^c f(x) \, dx + \int_c^b f(x) \, dx$$

$$\int_a^b f(x) \, dx = F(b) - F(a)$$

If $h(x) \le g(x) \le f(x)$ for $a \le x \le b$, then

$$\int_a^b h(x) \, dx \le \int_a^b g(x) \, dx \le \int_a^b f(x) \, dx$$

$$\int_a^b g(f(x))f'(x) \, dx = \int_{f(a)}^{f(b)} g(u) \, du$$

$$\int cf(x) \, dx = c \int f(x) \, dx$$

$$\int [f(x) + g(x)] \, dx = \int f(x) \, dx + \int g(x) \, dx$$

If $f(x) \ge g(x)$ on $[a, b]$, then $A = \displaystyle\int_a^b [f(x) - g(x)] \, dx$

$$\ln x = \int_1^x \frac{1}{t} \, dt \qquad \text{and} \qquad \frac{d}{dx} \ln x = \frac{1}{x}$$

$$\ln bc = \ln b + \ln c$$

$$\ln b^r = r \ln b$$

Key Theorems

Fundamental Theorem of Calculus

Mean Value Theorem for Integrals
Law of Logarithms

Review Exercises

1. Let $f(x) = x \sin x$ and $P = \{0, \pi/3, 2\pi/3, \pi, 3\pi/2\}$. Compute the left sum, right sum, and midpoint sum.

2. Let $f(x) = \sqrt{\sin x}$. Compute the midpoint sum for f corresponding to the partition of $[0, 2]$ into 20 intervals of equal length.

In Exercises 3–12 find the indefinite integral.

3. $\displaystyle\int (x^{3/5} - 8x^{5/3}) \, dx$ 4. $\displaystyle\int (3 \cos x - 2 \sin x) \, dx$

5. $\displaystyle\int (x^3 - 3x + 2 - 2/x) \, dx$ 6. $\displaystyle\int (4 - x)^9 \, dx$

7. $\displaystyle\int \frac{1 + \sqrt{x+1}}{\sqrt{x+1}} \, dx$ 8. $\displaystyle\int \frac{1}{x^2} e^{1/x} \, dx$

9. $\displaystyle\int \cos^3 3t \sin 3t \, dt$ 10. $\displaystyle\int \frac{\tan(\ln t)}{t} \, dt$

11. $\displaystyle\int \sqrt{1 + \sqrt{x}} \, dx$ *12. $\displaystyle\int \frac{\tan^2 x}{x - \tan x} \, dx$

In Exercises 13–24 evaluate the definite integral.

13. $\displaystyle\int_{-1}^{-2} \left(\frac{5}{3}x^{2/3} - \frac{4}{x^3}\right) dx$ 14. $\displaystyle\int_1^2 \frac{x^2 + 2x + 3}{x} \, dx$

15. $\displaystyle\int_0^2 (x^2 + 3)(x^3 + 9x + 1)^{1/3} \, dx$

16. $\displaystyle\int_0^\pi (\sqrt{x} - 3 \sin x) \, dx$ 17. $\displaystyle\int_{-8}^{-2} \frac{-1}{5u} \, du$

18. $\displaystyle\int_{-3}^2 (u + |u|) \, du$ 19. $\displaystyle\int_0^1 \frac{1 + e^x}{x + e^x} \, dx$

20. $\displaystyle\int_0^1 t^5 \sqrt{1 - t^2} \, dt$ 21. $\displaystyle\int_{-\pi/4}^{\pi/2} \frac{\cos t}{1 + \sin t} \, dt$

22. $\displaystyle\int_2^5 \frac{x}{\sqrt{x-1}} \, dx$ 23. $\displaystyle\int_{1/26}^{1/7} \frac{1}{x^2} \left(\frac{x+1}{x}\right)^{1/3} dx$

24. $\displaystyle\int_{-1}^{\pi/2} f(x) \, dx$, where $f(x) = \begin{cases} x^3 - 2x^2 & \text{for } -1 \le x < 0 \\ \sin x & \text{for } x \ge 0 \end{cases}$

In Exercises 25–28 find the area A of the region between the graph of f and the x axis on the given interval.

25. $f(x) = \dfrac{7}{4} x^2 \sqrt{x} + \dfrac{1}{\sqrt{x}}$; $[2, 4]$

26. $f(x) = x + 2 \sin x$; $[-\pi/2, \pi]$

*27. $f(x) = \dfrac{e^x - e^{2x}}{1 + e^x}$; $[1, \ln 3]$

28. $f(x) = \begin{cases} x - 3 & \text{for } x \le 3 \\ x^2 - 9 & \text{for } x > 3 \end{cases}$; $[1, 4]$

In Exercises 29–31 find the area A of the region between the graphs of the given functions.

29. $f(x) = 2x^5 + 5x^4$; $g(x) = 2x^5 + 20x^2$

30. $y = \dfrac{x^2 + 2}{\sqrt{x+1}}$; $y = \dfrac{3x + 2}{\sqrt{x+1}}$

31. $x = 2y^3 + y^2 + 5y - 7$; $x = y^3 + 4y^2 + 3y - 7$

32. Let $f(x) = x + 2 \sin x$. Find the average value of f on the interval $[0, \pi]$.

In Exercises 33–34 sketch the graph of f, giving all pertinent information.

33. $f(x) = \ln(x^2 - 4)$ 34. $f(x) = \ln(2 - x - x^2)$

In Exercises 35–42 find the derivative of the given function.

35. $F(x) = \displaystyle\int_0^x t\sqrt{1 + t^5}\, dt$ **36.** $G(y) = \displaystyle\int_{2y}^{\sin y} t \sin^2 t\, dt$

37. $F(x) = \displaystyle\int_1^{\ln x} \frac{1}{t}\, dt$ **38.** $f(x) = \ln(\tan x + \sec x)$

39. $f(x) = \ln x + \ln 1/x$ **40.** $f(x) = \ln(\ln(\ln x))$

41. $f(x) = \dfrac{(4 - \cos x)^{1/3}\sqrt{2x - 5}}{\sqrt[3]{x + 5}}$

42. $f(x) = \sqrt[3]{\dfrac{(3x + 7)(x^2 - 1)^2}{x^2 e^{2x}}}$

In Exercises 43–46 determine which of the integrals can be readily evaluated by means of a substitution. For any that can, evaluate the integral.

43. a. $\displaystyle\int \sqrt{x^2 + 6}\, dx$ **b.** $\displaystyle\int x\sqrt{x^2 + 6}\, dx$

   c. $\displaystyle\int x^{1/2}\sqrt{x^2 + 6}\, dx$

44. a. $\displaystyle\int \sin\sqrt{x}\, dx$ **b.** $\displaystyle\int \sqrt{x}\sin\sqrt{x}\, dx$

   c. $\displaystyle\int \frac{1}{\sqrt{x}}\sin\sqrt{x}\, dx$

45. a. $\displaystyle\int \ln(x + 1)\, dx$ **b.** $\displaystyle\int (x + 1)\ln(x + 1)\, dx$

   c. $\displaystyle\int \frac{\ln(x + 1)}{x + 1}\, dx$

46. a. $\displaystyle\int \sin(e^{2x})\, dx$ **b.** $\displaystyle\int e^x \sin(e^{2x})\, dx$

   c. $\displaystyle\int e^{2x} \sin(e^{2x})\, dx$

47. Using the fact that $x^4 \le 1 + x^4 \le 2x^4$ for $x \ge 1$, show that

$$\frac{26}{3} \le \int_1^3 \sqrt{1 + x^4}\, dx \le \frac{26}{3}\sqrt{2}$$

48. Using the fact that

$$1 - \frac{1}{2}x^2 \le \cos x \le 1 - \frac{1}{2}x^2 + \frac{1}{24}x^4 \quad \text{for } 0 \le x \le 1$$

find lower and upper bounds for $\displaystyle\int_0^1 \cos\sqrt{x}\, dx$.

49. a. Prove that $\int_1^x (1/t)\, dt \le \int_1^x 1\, dt$ for $x \ge 1$.
   b. Using part (a), prove that $0 \le \ln x \le x - 1$ for $x \ge 1$.
   c. Using part (b), find upper and lower bounds for $\int_1^2 x \ln x\, dx$.

50. a. Let r be a rational number greater than -1. Using the fact that $1/t \le t^r$ for $t \ge 1$, show that

$$\ln x \le \frac{1}{r + 1}(x^{r+1} - 1) \quad \text{for } x \ge 1$$

   b. Let s be a rational number less than -1. Using ideas similar to those in part (a), show that

$$\ln x \ge \frac{1}{s + 1}(x^{s+1} - 1) \quad \text{for } x \ge 1$$

   c. In (i) and (ii), plot the graphs of $\ln x$ and g on the same screen, using $[0, 10]$ and $[-2, 3]$ for the x window and y window, respectively. Are the relative positions of the graphs consistent with the inequalities in parts (a) and (b)?
      i. $g(x) = 10(x^{0.1} - 1)$ **ii.** $g(x) = -10(x^{-0.1} - 1)$

51. Suppose that f is a linear function defined on the interval $[a, b]$. Show that the midpoint sum equals the average of the left and right sums.

*52. Suppose the graph of f is concave upward. Determine whether the midpoint sum is greater than, or less than, the average of the left and right sums. Explain your answer geometrically.

53. Let $G(x) = \int_c^x f(t)\, dt$, where f is a function that is continuous on an open interval I and c is in I.
   a. What property of f would imply that G is increasing on I?
   b. What are the critical numbers of G in I?
   c. What property of f would imply that the graph of G is concave upward on I?

*54. The following problem appeared in Goursat's *Cours d'Analyse,* one of the most influential early calculus books. Suppose $p(x)$ is a polynomial of degree 7, such that $(x - 1)^4$ is a factor of $p(x) + 1$ and $(x + 1)^4$ is a factor of $p(x) - 1$. Find $p(x)$. (*Hint:* The given information implies that $(x - 1)^3$ and $(x + 1)^3$ are factors of $p'(x)$, so that $p'(x) = B(x - 1)^3(x + 1)^3$ for some constant B.)

Applications

55. A car moving initially at 44 feet per second (that is, 30 miles per hour) decelerates at the constant rate of 4 feet per second per second. Determine how far the car travels before coming to a stop.

56. From noon to 2 P.M. the temperature T increases with time t in such a way that

$$\frac{dT}{dt} = t^2 + 2t \quad \text{for } 0 \le t \le 2$$

If $T(0) = 60$, find $T(2)$.

57. Suppose a circuit has constant resistance R and variable current i given by

$$i = 110 \sin 120\pi t$$

The rate of heat production in the circuit is $i^2 R$. Find the average rate of heat production during the time interval $[0, \frac{1}{60}]$. (*Hint:* See the definition of average value in Section 5.3.)

58. Some psychologists believe that a numerical measure of a child's ability to learn during the first 4 years of life is approximately described by the function

$$f(x) = \frac{5}{3x \ln x - 5x + 10} \quad \text{for } \frac{1}{2} \leq x \leq 4$$

where x is the age in years. At what age can such a child learn best?

59. Suppose a rocket moves in outer space (where gravitational forces are negligible) with initial speed v_0 and initial mass m_0. If gas is ejected from the rocket with speed u_0, then the Law of Conservation of Momentum from physics implies that the speed v of the rocket increases as the mass m decreases in such a way that

$$\frac{dv}{dm} = -\frac{u_0}{m}$$

   a. Show that $v = v_0 + u_0 \ln (m_0/m)$.

   b. For chemical fuels, 3000 meters per second would be a large value for u_0. Assuming that $u_0 = 3000$, determine the ratio m/m_0 that would result in a change of velocity of 3×10^7 meters per second, which is one tenth the speed of light. (From the fact that the answer is essentially zero, it should be clear that chemical fuels are not suitable for interstellar travel.)

60. As a meteor approaches earth, its weight (the earth's gravitational force on the meteor) varies with the distance r from the center of the earth. The weight is given by $w = k/r^2$, where k is a constant. Let r_1 and r_2 be two distances, with $0 < r_1 < r_2$, and let w_1 and w_2 be the corresponding weights of the meteor. Show that the average weight of the meteor over the interval $[r_1, r_2]$ is $\sqrt{w_1 w_2}$.

61. The specific heat H of a solid like nickel, fluorspar, or diamond is the amount of heat per unit mass needed in order to raise the temperature of the mass 1° Kelvin (or equivalently, Celsius) under constant pressure. It has been found that at low temperatures H depends on the temperature T in degrees Kelvin according to the formula

$$H = 1945 \left(\frac{T}{T_D}\right)^3$$

where T_D is the so-called Debye temperature, which is a specific temperature characteristic to a given solid. The increase in internal energy per mole as the temperature of the crystal is increased from T_1 to T_2 degrees Kelvin is given by $\int_{T_1}^{T_2} H \, dt$. Using the value $T_D = 474$ for fluorspar, determine the increase in energy per mole required to raise the temperature of a chunk of fluorspar from 20K to 50K. Your answer will be in terms of joules per mole.

Topics for Discussion

1. What are the geometric properties of area that are implied by the definite integral of a nonnegative function?

2. Explain the difference between a definite integral and an indefinite integral.

3. Give reasons why the Fundamental Theorem of Calculus is important.

4. Tell in your own words what the phrase "differentiation and integration are inverse processes" means.

5. Give an example of functions f and g with $\int f(x)g(x) \, dx \neq \int f(x) \, dx \int g(x) \, dx$. Explain why this form of a product rule for integration fails for most functions.

6. Riemann sums are designed to approximate definite integrals. Suppose f is continuous on $[a, b]$, and P is a partition of $[a, b]$. Is there always a Riemann sum corresponding to P that equals the integral $\int_a^b f(x) \, dx$ exactly? Explain your answer.

7. Suppose f is continuous on $[a, b]$, and let P be a partition of $[a, b]$ that has n subintervals of the same length. By Theorem 5.3, as n increases, the corresponding Riemann sums approach $\int_a^b f(x) \, dx$. When these Riemann sums are actually computed and n is sufficiently large, might the value obtained by a calculator or computer diverge from the value of the integral because of round-off error? (*Hint:* Since a calculator can only store, say, 16 digits, what might happen if $n > 10^{16}$?)

Cumulative Review, Chapters 1–4

1. Solve the inequality $-\dfrac{(2 + \sin x)}{(x - 3)^2}\left(\dfrac{2 - x}{4 - x}\right)^{-1} < 0$.

2. Let $f(x) = \dfrac{2x + 1}{3x + 2}$ and $g(x) = \dfrac{2x - 1}{2 - 3x}$.

   a. Find the domains of $f \circ g$ and $g \circ f$.

   b. Show that $(f \circ g)(x) = (g \circ f)(x)$ for all x in the domains of both f and g.

   c. Are $f \circ g$ and $g \circ f$ equal? Explain why or why not.

In Exercises 3–5 find the limit.

3. $\displaystyle\lim_{x\to\pi/2^-}\frac{\cos x}{\sin x-1}$ **4.** $\displaystyle\lim_{x\to\infty}\frac{2x^3+3x^2-2}{-5x^3+x-9}$

5. $\displaystyle\lim_{x\to\pi/4}\frac{\tan x-1}{x-\pi/4}$

6. Let
$$f(x)=\sqrt{\frac{x^2(1+x)}{1-x}}\quad\text{for }-1\le x<1$$

Determine whether or not f is

a. continuous at 0   **b.** differentiable at 0

7. Let $f(x)=e^{1/(x^2+1)}$. Find $f'(x)$.

8. Find the slope of the line tangent to the circle $x^2+2x+y^2-4y=0$ at the point on the circle lying on the positive y axis.

9. Let $f(x)=2x^3-9x^2+14x-6$. Show that f is an increasing function.

 10. Show that the equation $x^{15}+x^9+100x-9=0$ has exactly one real solution. Then use the Newton-Raphson method to approximate the solution.

11. Let
$$f(x)=cx-\frac{1}{x^3}-\frac{1}{2}\quad\text{for some constant }c\ne0$$

Determine the value of c for which the equations $f(x)=0$ and $f'(x)=0$ have the same root, and find that root.

12. Suppose that f is differentiable and $f'(x)=2f(x)+1$ for all x. Show that $f''(x)$ exists for all x, and express $f''(x)$ in terms of $f(x)$.

13. Sketch the graph of a function f defined on $[-2,\infty)$ with all the following properties.
  a. f'' is continuous on $(-2,\infty)$.
  b. f is increasing on $(-2,-1)$ and its graph is concave downward on $(-2,-1)$.
  c. f is decreasing on $(-1,0)$ and its graph is concave downward on $(-1,0)$.
  d. f is decreasing on $(0,1)$ and its graph is concave upward on $(0,1)$.
  e. The graph of f has an inflection point at the point $(2,3)$.
  f. $y=4$ is a horizontal asymptote of the graph of f.

In Exercises 14–15 sketch the graph of f, noting all relevant properties.

14. $f(x)=\dfrac{(x+2)^2}{x-1}$ **15.** $f(x)=1-\dfrac{1}{x}+\dfrac{1}{x^2}$

16. Suppose a spherical balloon is being deflated in such a way that its surface area decreases at the rate of 1 square inch per second. How fast is the radius of the balloon decreasing when the surface area in square inches equals its volume in cubic inches?

17. A flower pot is accidentally knocked off a ledge on the top floor of an apartment building. When it is 128 feet from the ground, the pot is falling at the rate of 32 feet per second.
  a. What is the velocity of the pot 1 second later?
  b. How long does it take for the pot to fall the last 128 feet to the ground?

18. Suppose the acceleration a of an object is related to its velocity v in such a way that $a^3=7v^2+v$. How fast is the acceleration increasing with respect to time when the velocity is 1?

19. A certain toll road averages 24,000 cars per day when the toll is 3 dollars per car. A survey concludes that 300 fewer cars would use the toll road for each 5 cent increase in the toll. What toll would maximize the total revenue?

20. Suppose one leg of a given right triangle is 6 inches long and the other leg is 2 inches long. Determine the dimensions of the rectangle with maximum area that can be inscribed in the triangle, assuming that two sides of the rectangle lie along the legs of the triangle.

21. A radar detector submerged in water tracks an approaching airplane (Figure 5.57). When the axis of the monitor makes an angle of $\pi/4$ radians with respect to the vertical, the monitor is rotating at the rate of 1/10 radian per second. How fast is the angle of incidence θ of the airplane changing at that instant? (*Hint:* Use Snell's Law, which is (7) in Section 1.7, with $\mu=1.33$.)

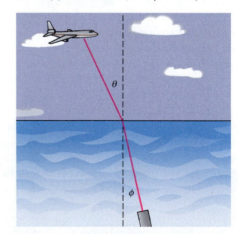

FIGURE 5.57 Figure for Exercise 21.

Fireworks in Orange County, California. (*S. Maimon/Superstock*)

6

APPLICATIONS OF THE INTEGRAL

Chapter 6 is devoted to applications of the integral. The mathematical applications will include volume, length of a curve, and surface area. Most of these applications involve a curve that is the graph of a function. However, some involve curves (circles, for example) that are not graphs of functions but are defined by means of a parameter t. Such curves will be introduced in Section 6.7. The physical applications will include the computation of the work done by a force, moments of a plane region, and hydrostatic force on a plane surface. In order to present the various applications in a unified manner we will base our presentations on the fact that if f is a function continuous on an interval $[a, b]$, then $\int_a^b f(x)\, dx$ is, by definition, the limit of the Riemann sums for f on $[a, b]$ as the lengths of the subintervals derived from the partitions of $[a, b]$ approach 0.

We will describe briefly here our procedure for introducing the applications of the integral. For each application our goal will be to find a formula for a quantity I (such as the volume of a solid region or the work done by a force). We will proceed in the following way.

1. Using properties that we expect the quantity to have, we will find a function f continuous on an interval $[a, b]$ with the property that for each partition P of $[a, b]$ there are numbers $t_1, t_2 \ldots, t_n$ in the n subintervals derived from P such that I should be approximately equal to the Riemann sum

$$\sum_{k=1}^{n} f(t_k) \, \Delta x_k$$

(Recall that $\Delta x_k = x_k - x_{k-1}$, which is the length of the kth subinterval of P.) In each case it will be reasonable to expect that $\sum_{k=1}^{n} f(t_k) \, \Delta x_k$ should approach I as the length of the largest subinterval of P tends to 0. As explained in Section 5.2, this idea is expressed by writing

$$I = \lim_{\|P\| \to 0} \sum_{k=1}^{n} f(t_k) \, \Delta x_k$$

2. We will conclude that

$$I = \int_{a}^{b} f(x) \, dx$$

In each case step 1 is the crucial one and is the justification for the formula in step 2.

6.1   VOLUME

In our first application of the integral we will derive formulas for volumes of solid regions in space. The formulas will enable us to find the volumes of such well-known solids as spheres, cylinders, pyramids, and cones, as well as the volumes of many other solids. In Chapter 14 we will give a general definition of volume from which one could obtain all the formulas that will appear in this section.

The Cross-Sectional Method   Consider a solid region D having the following description. For every x in an interval $[a, b]$, the plane perpendicular to the x axis at x intersects D in a plane region having area $A(x)$ (Figure 6.1(a)). Our goal is to define the volume of D.

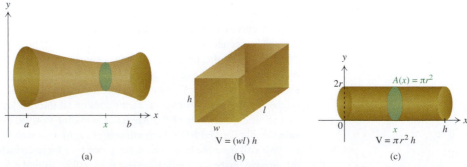

(a)   (b)   (c)

FIGURE 6.1   (a) A solid with cross-sectional area $A(x)$ at x. (b) A parallelepiped with rectangular cross sections. (c) A right circular cylinder. The cross sections are circular.

If the cross-sectional area is constant, that is,

$$A(x) = A_0 \quad \text{for } a \le x \le b$$

then we define the volume to be $A_0(b - a)$, the product of the constant cross-sectional area A_0 and the length of the interval $[a, b]$. This is consistent, for example, with the formula for the volume of a rectangular parallelepiped (Figure 6.1(b)).

For another example in which the cross-sectional area is constant, consider a right circular cylinder of radius r and height h (Figure 6.1(c)). Here,

$$A(x) = \pi r^2 \quad \text{for } 0 \le x \le h$$

Consequently

$$V = \pi r^2 h$$

Now suppose the cross-sectional area A of the solid region is continuous, but not necessarily constant (Figure 6.2(a)). Let $P = \{x_0, x_1, \ldots, x_n\}$ be a partition of $[a, b]$. The planes perpendicular to the x axis at $x_0, x_1, \ldots, x_n$ decompose D into n "slices," just as a bread machine slices a loaf of bread. We will approximate the volume of each slice of D. For each integer k between 1 and n, let ΔV_k be the volume of the kth slice, that is, the portion of D between x_{k-1} and x_k (Figure 6.2(b)). If the width Δx_k of the kth slice is small and t_k is an arbitrary number in $[x_{k-1}, x_k]$, then ΔV_k is approximately equal to the product of the cross-sectional area $A(t_k)$ and thickness Δx_k. Thus

$$\Delta V_k \approx (\text{cross-sectional area}) \times (\text{thickness}) = A(t_k)\, \Delta x_k$$

Since the volume V of D is the sum of $\Delta V_1, \Delta V_2, \ldots, \Delta V_n$, it follows that V should be approximately

$$\sum_{k=1}^{n} \overbrace{A(t_k)}^{\substack{\text{cross-}\\ \text{sectional}\\ \text{area}}} \overbrace{\Delta x_k}^{\text{thickness}}$$

which is a Riemann sum for A on $[a, b]$. Therefore it is reasonable to expect that

$$V = \lim_{\|P\| \to 0} \sum_{k=1}^{n} A(t_k)\, \Delta x_k = \int_a^b A(x)\, dx$$

Thus we are led to the following definition.

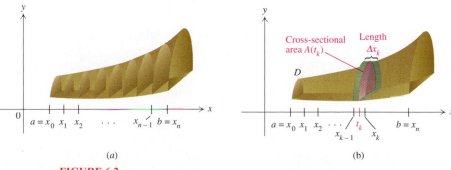

(a)   (b)

FIGURE 6.2

DEFINITION 6.1

Suppose a solid region D has cross-sectional area $A(x)$ for $a \leq x \leq b$, where A is continuous on $[a, b]$. Then the **volume** V of D is defined by

$$V = \int_a^b A(x)\, dx \tag{1}$$

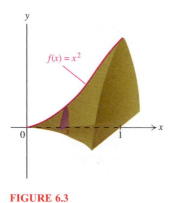

FIGURE 6.3

EXAMPLE 1   Find the volume V of the solid D in Figure 6.3, whose cross section at x is semicircular with radius x^2, for $0 \leq x \leq 1$.

Solution   Recall that the area of a semicircular region with radius r is $\frac{1}{2}\pi r^2$. Thus the cross-sectional area $A(x)$ at x is given by

$$A(x) = \frac{1}{2}\,\pi(x^2)^2 = \frac{1}{2}\,\pi x^4$$

Therefore we conclude from (1) that

$$V = \int_0^1 A(x)\, dx = \int_0^1 \frac{1}{2}\,\pi x^4\, dx = \frac{1}{10}\,\pi x^5\,\Big|_0^1 = \frac{\pi}{10} \quad \square$$

Reversing the Roles of x and y   We can just as well reverse the roles of x and y and integrate with respect to y by the corresponding formula

$$V = \int_c^d A(y)\, dy \tag{2}$$

We will use (2) to calculate the volume of the pyramid in Figure 6.4(a).

EXAMPLE 2   Suppose a pyramid is 4 units tall and has a square base, 3 units on a side (Figure 6.4(a)). Find the volume V of the pyramid.

Solution   Let us place the y axis along the axis of the pyramid, the origin lying at the base of the pyramid, as in Figure 6.4(b). Notice that the pyramid extends from 0 to 4 on the y axis, so the limits of integration will be 0 and 4. Next we observe that the cross section at any y in $[0, 4]$ is a square. From the similar triangles in Figure 6.4(c) we see that the side length $s(y)$ of the square at y satisfies

$$\frac{s(y)}{3} = \frac{4 - y}{4}, \quad \text{that is,} \quad s(y) = \frac{3}{4}\,(4 - y)$$

Therefore the cross-sectional area at y is given by

$$A(y) = [s(y)]^2 = \frac{9}{16}\,(16 - 8y + y^2)$$

It follows from (2) that

$$V = \int_0^4 \frac{9}{16}\,(16 - 8y + y^2)\, dy = \frac{9}{16}\left(16y - 4y^2 + \frac{1}{3}\,y^3\right)\Big|_0^4$$

$$= \frac{9}{16}\left(64 - 64 + \frac{64}{3}\right) = 12 \quad \square$$

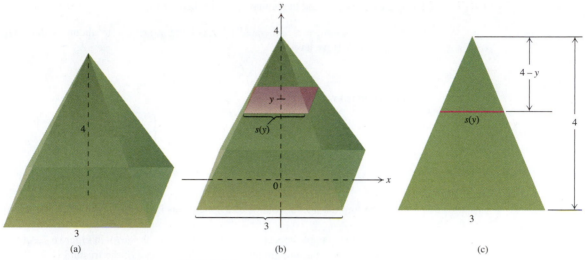

FIGURE 6.4 (a) A pyramid. (b) The cross sections are squares. (c) The similar triangles provide a formula for $s(y)$.

The Great Pyramid of Cheops.
(*Will & Deni McIntyre/Photo Researchers*)

More generally, if the pyramid has height h and a square base with side length a, then by substituting h for 4 and a for 3 in Example 2, we would find that

$$V = \frac{1}{3} a^2 h$$

(see the project). For example, the Great Pyramid of Cheops was originally (approximately) 482 feet tall and 754 feet square at the base. Therefore its volume was approximately $\frac{1}{3}(754)^2(482)$ cubic feet, which is just over 91,000,000 cubic feet.

The Disc Method

When the graph of a continuous, nonnegative function f on an interval $[a, b]$ is revolved about the x axis, it generates a solid region having circular cross sections, that is, cross sections that are circular discs (Figure 6.5). Since the radius of the cross section at x is $f(x)$, it follows that

$$A(x) = \pi[f(x)]^2$$

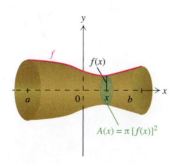

FIGURE 6.5 A solid of revolution has circular cross sections.

Thus from (1) we obtain a formula for the volume V of the solid that is generated:

$$V = \int_a^b \pi[f(x)]^2 \, dx \tag{3}$$

Because the cross sections are discs (see Figure 6.5), the use of (3) to compute volume is called the **disc method.**

The formulas for the volumes of many well-known figures follow readily from equation (3).

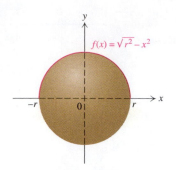

FIGURE 6.6   A sphere is a solid of revolution.

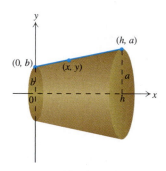

FIGURE 6.7   A frustum of a cone is a solid of revolution.

The Apollo 17 Command Module. (NASA/JSC)

The Washer Method

EXAMPLE 3   Find the volume V of a sphere of radius r.

Solution   A sphere is generated by revolving a semicircle about its diameter (Figure 6.6). If we let

$$f(x) = \sqrt{r^2 - x^2} \quad \text{for } -r \le x \le r$$

then from (3) we obtain

$$V = \int_{-r}^{r} \pi \, (\sqrt{r^2 - x^2})^2 \, dx = \pi \int_{-r}^{r} (r^2 - x^2) \, dx$$

$$= \pi \left(r^2 x - \frac{1}{3} x^3 \right) \Big|_{-r}^{r} = \frac{4}{3} \pi r^3 \quad \square$$

The formula for the volume of a sphere was known to Archimedes about 250 B.C.

Our next example concerns a frustum of a cone, as shown in Figure 6.7. A two-point equation of the straight line along the upper edge of the frustum is

$$\frac{y - b}{x - 0} = \frac{a - b}{h - 0}$$

or

$$y = b + \frac{a - b}{h} x \tag{4}$$

The frustum is generated by revolving the graph of this equation on $[a, b]$ about the x axis.

EXAMPLE 4   Find the volume V of the frustum of the cone shown in Figure 6.7.

Solution   Since y in (4) represents the radius of the cross section of the frustum at x, it follows from (3) that

$$V = \int_{0}^{h} \pi \left(b + \frac{a - b}{h} x \right)^2 dx$$

$$= \frac{\pi h}{3(a - b)} \left(b + \frac{a - b}{h} x \right)^3 \Big|_{0}^{h} = \frac{\pi h}{3(a - b)} (a^3 - b^3)$$

$$= \frac{\pi h}{3} (a^2 + ab + b^2) \quad \square$$

If $b = 0$ in Example 4, then the resulting region is a complete cone, and the volume reduces to the well-known formula

$$V = \frac{1}{3} \pi a^2 h$$

Next, we present a formula for the volume of the solid region generated by revolving a more general plane region about the x axis. Let f and g be functions such that

$$0 \le g(x) \le f(x) \quad \text{for } a \le x \le b$$

Then the plane region between the graph of f and the x axis on $[a, b]$ is composed of the region between the graphs of f and g on $[a, b]$ and the region between the graph of g and the x axis on $[a, b]$. Therefore the volume of the solid generated by revolving the entire region about the x axis should be equal to the sum of the volumes of the solids generated by revolving the two component regions about the x axis (Figure 6.8(a)). Accordingly, the volume V of the solid generated by revolving the region between the graphs of f and g on $[a, b]$ is given by

$$V = \int_a^b \pi[(f(x))^2 - (g(x))^2]\, dx \qquad (5)$$

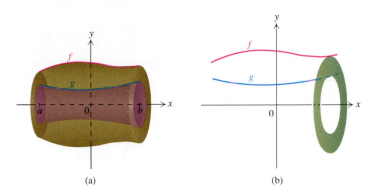

(a)   (b)

FIGURE 6.8   (a) A more general solid of revolution. (b) The cross sections resemble washers.

When g is different from 0, the method of finding volumes by (5) is sometimes called the **washer method** because the cross sections resemble washers (Figure 6.8(b)).

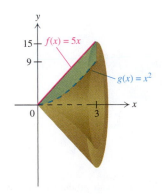

FIGURE 6.9

EXAMPLE 5   Let $f(x) = 5x$ and $g(x) = x^2$, and let R be the region between the graphs of f and g on $[0, 3]$. Find the volume V of the solid obtained by revolving R about the x axis (Figure 6.9).

Solution   Since $5x \geq x^2$ for $0 \leq x \leq 3$, (5) implies that

$$V = \int_0^3 \pi(25x^2 - x^4)\, dx = \pi \left(\frac{25x^3}{3} - \frac{x^5}{5} \right) \Bigg|_0^3$$

$$= \pi \left(225 - \frac{243}{5} \right) = \frac{882\pi}{5} \qquad \square$$

The Shell Method

Now we turn to a method of finding volumes that employs thin shells rather than cross-sections.

To begin, let us determine the volume V of a cylindrical shell obtained by revolving a rectangle about the y axis (Figure 6.10). Suppose the rectangle is bounded by the x axis, the line $y = c$, and the lines $x = a$ and $x = b$, where $b \geq a \geq 0$ and $c \geq 0$; then since the volume of the cylindrical shell is the difference of the

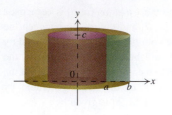

FIGURE 6.10 The volume of the cylindrical shell is equal to $\pi c(b^2 - a^2)$.

volumes of the outer and the inner cylinders, it follows that

$$V = \overbrace{\pi b^2 c}^{\substack{\text{volume of} \\ \text{outer cylinder}}} - \overbrace{\pi a^2 c}^{\substack{\text{volume of} \\ \text{inner cylinder}}} = \pi c(b^2 - a^2)$$

If we replace a by x_{k-1} and b by x_k, then we obtain

$$V = \pi c(x_k^2 - x_{k-1}^2) = \pi c(x_k + x_{k-1})(x_k - x_{k-1}) \qquad (6)$$

This formula will be useful when partitions are introduced.

Now let f and g be continuous functions on $[a, b]$, with $a \geq 0$, and such that

$$g(x) \leq f(x) \quad \text{for } a \leq x \leq b$$

We wish to define the volume V of the solid region in Figure 6.11(a), obtained by revolving about the y axis the region R between the graphs of f and g on $[a, b]$. Let $P = \{x_0, x_1, \ldots, x_n\}$ be any partition of $[a, b]$. For each k between 1 and n, let t_k be the midpoint $(x_k + x_{k-1})/2$ of the subinterval $[x_{k-1}, x_k]$. If Δx_k is small, the volume ΔV_k of the portion of the solid between the revolved lines $x = x_{k-1}$ and $x = x_k$ is approximately equal to the volume of the corresponding cylindrical shell with height $f(t_k) - g(t_k)$ (Figure 6.11(b)). By (6), with $f(t_k) - g(t_k)$ replacing c, this means that

$$\Delta V_k \approx \pi(f(t_k) - g(t_k))(x_k + x_{k-1})(x_k - x_{k-1}) = 2\pi t_k \, (f(t_k) - g(t_k)) \, \Delta x_k$$

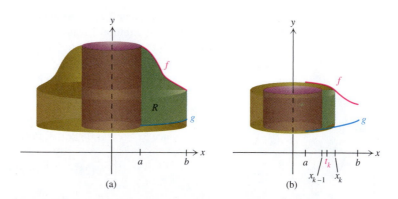

(a)   (b)

FIGURE 6.11 (a) The region R is revolved around the y axis to produce a solid of revolution. (b) A cylindrical shell is produced by revolving the rectangle around the y axis.

Therefore the volume V of the solid, which is the sum of $\Delta V_1, \Delta V_2, \ldots, \Delta V_n$, should be approximately

$$\sum_{k=1}^{n} 2\pi t_k \, (f(t_k) - g(t_k)) \, \Delta x_k$$

which is a Riemann sum for $2\pi x(f - g)$ on $[a, b]$. Therefore it is reasonable to expect that

$$V = \lim_{\|P\| \to 0} \sum_{k=1}^{n} 2\pi t_k(f(t_k) - g(t_k)) \, \Delta x_k = \int_a^b 2\pi x(f(x) - g(x)) \, dx$$

Thus we are led to the following formula for volume:

$$V = \int_a^b 2\pi x(f(x) - g(x)) \, dx \tag{7}$$

The emphasis on cylindrical shells justifies the name **shell method** for this way of computing volumes (see Figure 6.11(b)).

EXAMPLE 6   Let $f(x) = (x - 2)^2 + 1$ for $2 \le x \le 3$, and let R be the region between the graph of f and the x axis on $[2, 3]$. Find the volume V of the solid obtained by revolving R about the y axis (Figure 6.12).

Solution   Since the height of the shell at any x in $[2, 3]$ is $(x - 2)^2 + 1$, (7) implies that

$$V = \int_2^3 2\pi x[(x - 2)^2 + 1] \, dx = 2\pi \int_2^3 (x^3 - 4x^2 + 5x) \, dx$$

$$= 2\pi \left(\frac{1}{4} x^4 - \frac{4}{3} x^3 + \frac{5}{2} x^2 \right) \bigg|_2^3$$

$$= 2\pi \left[\left(\frac{81}{4} - 36 + \frac{45}{2} \right) - \left(4 - \frac{32}{3} + 10 \right) \right] = \frac{41}{6} \pi \quad \square$$

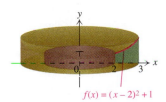

$f(x) = (x - 2)^2 + 1$

FIGURE 6.12

We will use (7) to find the volume of a solid rather like a single-scoop ice cream cone.

EXAMPLE 7   Let $f(x) = 3 - x^2$ and $g(x) = 3x - 1$, and let R be the region between the graphs of f and g on $[0, 1]$. Find the volume V of the solid generated by revolving R about the y axis (Figure 6.13).

Solution   Since $3 - x^2 \ge 3x - 1$ for $0 \le x \le 1$, it follows that the height of the solid at any x in $[0, 3]$ is $(3 - x^2) - (3x - 1)$. Thus by (7),

$$V = \int_0^1 2\pi x[(3 - x^2) - (3x - 1)] \, dx$$

$$= 2\pi \int_0^1 (-x^3 - 3x^2 + 4x) \, dx = 2\pi \left(-\frac{1}{4} x^4 - x^3 + 2x^2 \right) \bigg|_0^1$$

$$= \frac{3\pi}{2} \quad \square$$

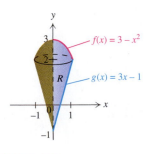

$f(x) = 3 - x^2$

$g(x) = 3x - 1$

R

FIGURE 6.13

The volumes of certain solids of revolution can be evaluated either by the washer method or by the shell method. As you would expect, the result is the same, whichever method is used. We support this claim with the following example.

EXAMPLE 8   Let R be the region between the graphs of the equations $y = x^2$ and $y = 2x$ on $[0, 2]$. Find the volume V of the solid generated by revolving R about the x axis (Figure 6.14) by

a. the washer method   **b.** the shell method

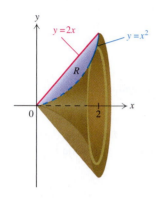

FIGURE 6.14

Solution   **a.** For the washer method we integrate with respect to x, using (5). We obtain

$$V = \int_0^2 \pi[(2x)^2 - (x^2)^2]\, dx = \pi \int_0^2 (4x^2 - x^4)\, dx$$

$$= \pi \left(\frac{4}{3} x^3 - \frac{1}{5} x^5 \right)\bigg|_0^2 = \pi \left(\frac{32}{3} - \frac{32}{5} \right)$$

$$= \frac{64\pi}{15}$$

b.   For the shell method we integrate with respect to y, using (7) with x replaced by y. In order to integrate with respect to y we need to consider the two equations $y = 2x$ and $y = x^2$ as defining functions of y. We obtain the equations

$$x = \tfrac{1}{2} y \quad \text{and} \quad x = \sqrt{y}$$

To determine the limits of integration, we find the y coordinates of the points at which the line $x = \tfrac{1}{2} y$ and the parabola $x = \sqrt{y}$ intersect:

$$\tfrac{1}{2} y = x = \sqrt{y}$$

$$\tfrac{1}{4} y^2 = y$$

$$y^2 - 4y = 0$$

$$y = 0 \quad \text{or} \quad y = 4$$

Since $\sqrt{y} \geq \tfrac{1}{2} y$ for $0 \leq y \leq 4$, we conclude from (7) that

$$V = \int_0^4 2\pi y \left(\sqrt{y} - \frac{1}{2} y \right) dy = 2\pi \int_0^4 \left(y^{3/2} - \frac{1}{2} y^2 \right) dy$$

$$= 2\pi \left(\frac{2}{5} y^{5/2} - \frac{1}{6} y^3 \right)\bigg|_0^4 = 2\pi \left(\frac{64}{5} - \frac{64}{6} \right)$$

$$= \frac{64\pi}{15} \quad \square$$

EXERCISES 6.1

In Exercises 1–8 let R be the region between the graph of the given function and the x axis on the given interval. Find the volume V of the solid obtained by revolving R about the x axis.

1. $f(x) = x^{3/2}$; $[0, 1]$ **2.** $f(x) = 1 + x^2$; $[-1, 2]$
3. $f(x) = \sqrt{\cos x}$; $[0, \pi/6]$
4. $f(x) = \sqrt{1 + \sin^2 x}$; $[-\pi/2, \pi/2]$
5. $f(x) = \sec x$; $[-\pi/4, 0]$ **6.** $f(x) = (\ln x)/\sqrt{x}$; $[1, e]$
7. $f(x) = x(x^3 + 1)^{1/4}$; $[1, 2]$ **8.** $f(x) = \sqrt{x}(1 - x)^{1/4}$; $[0, 1]$

In Exercises 9–10 let R be the region between the graph of f and the y axis on the given interval. Find the volume V of the solid obtained by revolving R about the y axis.

9. $f(y) = \sqrt{1 + y^3}$; $[1, 2]$
10. $f(y) = \sqrt{\sin y \cos y} \,(1 + \cos^2 y)^{1/3}$; $[0, \pi/2]$

In Exercises 11–14 let R be the region between the graphs of f and g on the given interval. Find the volume V of the solid obtained by revolving R about the x axis.

11. $f(x) = \sqrt{x + 1}$, $g(x) = \sqrt{x - 1}$; $[1, 3]$
12. $f(x) = x + 1$, $g(x) = x - 1$; $[1, 4]$
13. $f(x) = \cos x + \sin x$, $g(x) = \cos x - \sin x$; $[0, \pi/4]$
*14. $f(x) = 2x - x^2$, $g(x) = x^2 - 2x$; $[0, 1]$ (*Hint:* The washer method will not work. Why?)

In Exercises 15–18 find the volume V of the solid generated by revolving about the x axis the region between the graphs of the given equations.

15. $y = \frac{1}{2}x^2 + 3$ and $y = 12 - \frac{1}{2}x^2$
16. $y = x^{1/2}$ and $y = 2x^{1/4}$
17. $y = 5x$ and $y = x^2 + 2x + 2$
*18. $y = x^3 + 2$ and $y = x^2 + 2x + 2$

In Exercises 19–22 let R be the region between the graph of the function and the x axis on the given interval. Use the shell method to find the volume V of the solid generated by revolving R about the y axis.

19. $f(x) = (x - 1)^2$; $[0, 2]$
20. $f(x) = \sin x^2$; $[\sqrt{\pi/2}, \sqrt{\pi}]$
21. $g(x) = \dfrac{1}{\sqrt{1 - x^2}}$; $[0, \sqrt{3}/2]$
*22. $g(x) = \sqrt{1 + \sqrt{x}}$; $[0, 4]$

In Exercises 23–24 let R be the region between the graph of f and the y axis on the given interval. By interchanging the roles of x and y in (7), find the volume V of the solid generated by revolving R about the x axis.

23. $f(y) = \dfrac{\ln y}{y^2}$; $[1, 2]$ **24.** $f(y) = y^2 \sqrt{1 + y^4}$; $[0, 1]$

In Exercises 25–26 let R be the region between the graphs of f and g on the given interval. Use the shell method to find the volume V of the solid obtained by revolving R about the y axis.

25. $f(x) = 1$, $g(x) = x - 2$; $[1, 3]$
26. $f(x) = \sqrt{1 - x^2}$, $g(x) = -2 + \sqrt{1 + x^2}$; $[0, 1]$

In Exercises 27–28 let R be the region between the graphs of f and g on the given interval. Use the shell method to find the volume V of the solid obtained by revolving R about the x axis.

27. $f(y) = y^2 + 1$, $g(y) = y\sqrt{1 + y^3}$; $[0, 1]$
28. $f(y) = \dfrac{1}{(y + 2)^2}$, $g(y) = \dfrac{1}{y + 2}$; $[0, 2]$ (*Hint:* To evaluate the integral, make the substitution $u = y + 2$.)

In Exercises 29–30 find the volume V of the solid generated by revolving the region between the graphs of the equations about the y axis.

29. $y = 2x$ and $y = x^2$ **30.** $y^2 = x$ and $y = 3x$

In Exercises 31–36 find the volume V of the solid with the given information about its cross-sections.

31. The base of the solid is an isosceles right triangle whose legs L_1 and L_2 are each 4 units long. Any cross section perpendicular to L_1 is semicircular (see Figure 6.15).

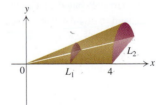

FIGURE 6.15 The solid for Exercise 31.

32. The base of the solid is a square centered at the origin, with sides of length 6 and parallel to the axes. The area of each cross section perpendicular to the edge of the base equals the distance from the cross section to the origin.

33. The solid has a circular base with radius 1, and the cross sections perpendicular to a fixed diameter of the base are squares. (*Hint:* Center the base at the origin.)

34. The base of a solid is a circle with radius 2, and the cross sections perpendicular to a fixed diameter of the base are equilateral triangles.

35. The base is an equilateral triangle each side of which has length 10. The cross sections perpendicular to a given altitude of the triangle are squares.

36. The base is an equilateral triangle with altitude 10. The cross sections perpendicular to a given altitude of the triangle are semicircles.

37. Suppose f is continuous on $[a, b]$ and the graph of f lies above the line $y = c$. Write down a formula for the volume V of the solid obtained by revolving about the line $y = c$ the region between the graph of f and the line $y = c$ on $[a, b]$.

38. Use the result of Exercise 37 to find the volume V of the solid obtained by revolving about the line $y = -1$ the region between the graph of the equation $y = \sqrt{x + 1}$ and the line $y = -1$ on the interval $[0, 1]$.

39. Find the volume V of the solid obtained by revolving about the line $y = 1$ the region between the graph of the equation $y = e^{-2x}$ and the x axis on the interval $[0, 1]$. (*Hint:* Pattern your solution after Exercise 37.)

40. Suppose f and g are continuous on $[a, b]$, and let c be such that $c \leq g(x) \leq f(x)$ for $a \leq x \leq b$. Write down a formula for the volume V of the solid obtained by revolving about the line $y = c$ the region between the graphs of f and g on $[a, b]$.

41. Use the result of Exercise 40 to find the volume V of the solid obtained by revolving about the line $y = 1$ the region between the graphs of $y = x^2 - x + 1$ and $y = 2x^2 - 4x + 3$.

42. Use the result of Exercise 40 to find the volume V of the solid obtained by revolving about the line $y = 2$ the region between the graphs of $y = x + 1$ and $y = x^2 - 2x + 2$.

43. Suppose the Great Pyramid of Cheops had been built with equilateral triangular cross sections instead of square cross sections but had the same height of 482 feet and base 754 feet on a side. What percentage of the original volume would have resulted?

44. Let a sphere of radius r be sliced at a distance h from its center. Show that the volume V of the smaller piece cut off is given by

$$V = \frac{\pi(r - h)^2}{3}(2r + h)$$

45. Let f be continuous and nonnegative on $[a, \infty)$. If

$$\lim_{b \to \infty} \int_a^b \pi[f(x)]^2 \, dx$$

converges, then we say that the volume of the solid obtained by revolving the graph of f about the x axis is finite. Otherwise, we say that the volume is infinite. Determine whether the graph of each of the following functions generates a solid with finite volume when the graph is revolved about the x axis. For any that is finite, find the volume V.
   a. $f(x) = 1/x$ for $x \geq 1$
   b. $f(x) = 1/x^2$ for $x \geq 1$
   c. $f(x) = 1/x^{1/2}$ for $x \geq 1$

46. Cavalieri's Principle for volume states that if two solids have the same cross-sectional area at each x between a and b, then the two solids have the same volume. Prove Cavalieri's Principle for volume by using (1).

47. Let f be continuous and nonnegative on $[a, b]$, and assume that $c \leq a$. Let R be the region between the graph of f and the x axis on $[a, b]$. Find a formula for the volume V of the solid obtained by revolving R about the line $x = c$.

48. Use the result of Exercise 47 to find the volume V of the solid obtained by revolving about the line $x = -1$ the region between the graph of $y = x^4$ and the x axis on $[0, 1]$.

49. Find the volume V of the solid obtained by revolving about the line $x = 2$ the region between the graph of $y = x^2$ and the x axis on $[2, 3]$. (*Hint:* Pattern your solution after Exercise 47.)

50. Let f and g be continuous on $[a, b]$, with $g(x) \leq f(x)$ for $a \leq x \leq b$. Let R be the region between the graphs of f and g on $[a, b]$, and assume that $c \leq a$. Find a formula for the volume V of the solid obtained by revolving R about the line $x = c$.

51. Use the result of Exercise 50 to find the volume V of the solid obtained by revolving about the line $x = -1$ the region between the graphs of $y = 2x$ and $y = -2x^2 + 4x$.

52. Use the result of Exercise 50 to find the volume V of the solid obtained by revolving about the line $x = -5$ the region between the graphs of $y = x^2 + 4$ and $y = 2x^2 + x + 2$.

Applications

53. As viewed from above, a swimming pool has the shape of the ellipse

$$\frac{x^2}{400} + \frac{y^2}{100} = 1$$

The cross sections of the pool perpendicular to the ground and parallel to the y axis are squares. If the units are feet, determine the volume V of the pool.

54. Find the volume V of the solid generated when the ellipse

$$\frac{x^2}{a^2} + \frac{y^2}{b^2} = 1$$

is revolved about the x axis. With $a = 100$ and $b = 25$, your answer could be the volume of gas needed to fill a dirigible having these dimensions.

55. The ivory stones used in the Oriental game of *go* have approximately the shape of the solid generated by revolving a certain ellipse about the x axis. If the length of a *go* stone is 2 centimeters and its height is 1 centimeter, what is its volume? (*Hint:* Let $a = \frac{1}{2}$ and $b = 1$ in Exercise 54.)

56. Suppose a ring is obtained by revolving about the x axis the region between the curve $y = 4 - x^2$ and the line $y = 1$ for $-\sqrt{3} \leq x \leq \sqrt{3}$. Determine the volume V of the ring.

57. A soda glass has the shape of the surface generated by revolving the graph of $y = 6x^2$ for $0 \leq x \leq 1$ about the y axis. Soda is extracted from the glass through a straw at the rate of $\frac{1}{2}$ cubic inch per second. How fast is the depth of soda decreasing when the depth is $\frac{3}{2}$ inches?

58. The Washington Monument has the shape of an obelisk. Its sides slant gradually inward as they rise to the base of

The Washington Monument. (*Rob Crandall/Stock Boston*)

the small pyramid at the top. The base of the monument is square, 16.80 meters on a side. At the base of the small pyramid, 152.49 meters above ground, the walls are 10.50 meters on a side. Finally, the small pyramid is 16.79 meters tall. Determine the total volume V of the monument.

59. A wooden golf tee has approximately the dimensions of the solid obtained by revolving about the x axis the region bounded by the graphs of f and g, where

$$f(x) = \begin{cases} \frac{1}{2}x & \text{for } 0 \leq x \leq \frac{1}{2} \\ \frac{1}{4} & \text{for } \frac{1}{2} \leq x \leq \frac{7}{2} \\ \frac{1}{4}[1 + (x - \frac{7}{2})^2] & \text{for } \frac{7}{2} \leq x \leq \frac{9}{2} \\ \frac{1}{2} & \text{for } \frac{9}{2} \leq x \leq 5 \end{cases}$$

and

$$g(x) = \begin{cases} 0 & \text{for } 0 \leq x \leq \frac{9}{2} \\ x - \frac{9}{2} & \text{for } \frac{9}{2} \leq x \leq 5 \end{cases}$$

(Figure 6.16). Here x and $g(x)$ are measured in centimeters. Determine how much wood goes into a golf tee.

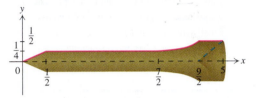

FIGURE 6.16   Figure for Exercise 59.

60. In order to create a ring, a hole with radius 1 centimeter is drilled through the center of a sphere of radius 2 centimeters. Find the volume V removed.

61. A wok is in the shape of a solid obtained by revolving about the y axis the curve shown in Figure 6.17. Assuming that the units are centimeters, find the capacity of the wok.

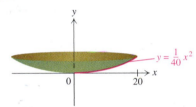

FIGURE 6.17   The wok for Exercise 61.

Project

1. This project concerns solids D in the shape of a "pyramid" such as that in Figure 6.18, whose cross section at x perpendicular to the x axis is a regular polygon (or is a circle), for $0 \leq x \leq h$, where h is a positive number.

a. Let $A(x) = $ the area of the cross section perpendicular to the x axis at x, and let A be the area of the largest cross section of D. Find a formula for $A(x)$ in terms of x, h, and A.

b. Find a formula for the volume V of D in terms of h and A.

c. Suppose that the cross sections are square (as for the Great Pyramid of Cheops), and that $A = a^2$. Use (b) to show that $V = \frac{1}{3} a^2 h$ (which conforms to the result of Example 2).

d. Suppose that D is a circular cone, with base (largest) radius equal to a. Show that $V = \frac{\pi}{3}a^2 h$, a formula known to the ancient Greeks.

e. Suppose that D is an equilateral triangular pyramid, with largest triangle having side length a. Find a simple formula for the volume V of D, and show that the volume is smaller than the volume in (c).

f. Finally, suppose that D is a hexagonal pyramid, with largest hexagon having side length a. Find a simple formula for the volume V of D.

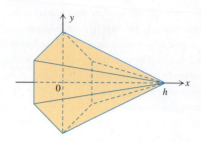

FIGURE 6.18   Figure for the project.

6.2   LENGTH OF A CURVE

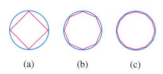

(a)   (b)   (c)

FIGURE 6.19   The circumference of a circle is approximated by the perimeters of inscribed polygons.

The ancient Greeks estimated the circumference of a circle by inscribing a polygon of n sides and then computing the perimeter of the polygon. They surmised that the larger n was, the better the perimeter of the polygon approximated the actual circumference of the circle (Figure 6.19). In this section we will use this basic idea to define and compute the lengths of many curves.

Each curve that we consider will be the graph of a function f with a continuous derivative on a closed interval $[a, b]$. If f is linear, that is, if the graph of f is a line segment, then the length L of the graph is the distance between $(a, f(a))$ and $(b, f(b))$, so that

$$L = \sqrt{(b - a)^2 + [f(b) - f(a)]^2}$$

(Figure 6.20(a)).

When f is not necessarily linear, we let $P = \{x_0, x_1, \ldots, x_n\}$ be any partition of $[a, b]$, and approximate the graph of f by a polygonal line l whose vertices are $(x_0, f(x_0)), (x_1, f(x_1)), \ldots, (x_n, f(x_n))$ (Figure 6.20(b)). Let ΔL_k be the length of the portion of the graph of f joining $(x_{k-1}, f(x_{k-1}))$ and $(x_k, f(x_k))$. If Δx_k is small, ΔL_k is approximately equal to the length of the line segment joining $(x_{k-1}, f(x_{k-1}))$ and $(x_k, f(x_k))$. In other words,

$$\Delta L_k \approx \sqrt{(x_k - x_{k-1})^2 + [f(x_k) - f(x_{k-1})]^2} \qquad (1)$$

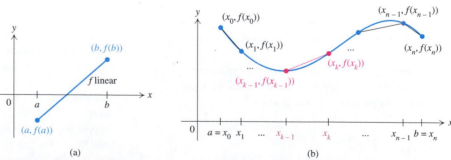

(a)   (b)

FIGURE 6.20   (a) The length of a line segment is the distance between its endpoints. (b) The length of a curve is approximated by the length of a polygonal line.

The Mean Value Theorem, applied to f on the interval $[x_{k-1}, x_k]$, implies that

$$f(x_k) - f(x_{k-1}) = f'(t_k)(x_k - x_{k-1})$$

for some t_k in (x_{k-1}, x_k). Therefore (1) can be rewritten

$$\Delta L_k \approx \sqrt{(x_k - x_{k-1})^2 + [f'(t_k)(x_k - x_{k-1})]^2} = \sqrt{1 + [f'(t_k)]^2}\,(x_k - x_{k-1})$$

Therefore the total length L of the graph of f, which is the sum of the lengths ΔL_1, $\Delta L_2, \ldots, \Delta L_n$, should be approximately

$$\sum_{k=1}^{n} \sqrt{1 + [f'(t_k)]^2}\,\Delta x_k$$

itself a Riemann sum for $\sqrt{1 + (f')^2}$ on $[a, b]$. Therefore it is reasonable to expect that

$$L = \lim_{\|P\| \to 0} \sum_{k=1}^{n} \sqrt{1 + [f'(t_k)]^2}\,\Delta x_k = \int_a^b \sqrt{1 + [f'(x)]^2}\,dx$$

This leads us to make the following definition.

DEFINITION 6.2

Let f have a continuous derivative on $[a, b]$. Then the **length** L of the graph of f on $[a, b]$ is defined by

$$L = \int_a^b \sqrt{1 + [f'(x)]^2}\,dx \qquad (2)$$

In Definition 6.2 the derivative of f is assumed to be continuous so that the integrand $\sqrt{1 + [f'(x)]^2}$ will be continuous, and hence the integral will be defined by Definition 5.4.

EXAMPLE 1   Let $f(x) = x^{3/2} + 1$ for $0 \leq x \leq \frac{4}{3}$. Find the length L of the graph of f (Figure 6.21).

Solution   Since $f'(x) = \frac{3}{2} x^{1/2}$, we know from (2) that

$$L = \int_0^{4/3} \sqrt{1 + \left(\frac{3}{2} x^{1/2}\right)^2}\,dx = \int_0^{4/3} \sqrt{1 + \frac{9}{4} x}\,dx$$

To evaluate the integral, we let

$$u = 1 + \frac{9}{4} x, \quad \text{so that} \quad du = \frac{9}{4}\,dx$$

If $x = 0$ then $u = 1$, and if $x = \frac{4}{3}$ then $u = 1 + \frac{9}{4} \cdot \frac{4}{3} = 4$. Therefore

$$L = \int_0^{4/3} \overbrace{\sqrt{1 + \frac{9}{4} x}}^{\sqrt{u}}\ \overbrace{dx}^{\frac{4}{9} du} = \int_1^4 \sqrt{u}\,\frac{4}{9}\,du = \frac{4}{9}\left(\frac{2}{3} u^{3/2}\right)\Big|_1^4 = \frac{56}{27} \qquad \square$$

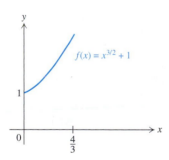

$f(x) = x^{3/2} + 1$

FIGURE 6.21

* van Heuraet: Pronounced "von-*Hoe*-r$\bar{a}$t."

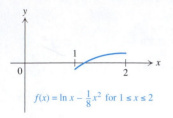

FIGURE 6.22

EXAMPLE 2 Let

$$f(x) = \ln x - \frac{1}{8} x^2 \quad \text{for } 1 \leq x \leq 2$$

Find the length L of the graph of f (Figure 6.22).

Solution Since $f'(x) = 1/x - x/4$, it follows from (2) that

$$L = \int_1^2 \sqrt{1 + \left(\frac{1}{x} - \frac{x}{4} \right)^2} \, dx = \int_1^2 \sqrt{1 + \left(\frac{1}{x^2} - \frac{1}{2} + \frac{x^2}{16} \right)} \, dx$$

$$= \int_1^2 \sqrt{\frac{1}{x^2} + \frac{1}{2} + \frac{x^2}{16}} \, dx = \int_1^2 \sqrt{\left(\frac{1}{x} + \frac{x}{4} \right)^2} \, dx$$

$$= \int_1^2 \left(\frac{1}{x} + \frac{x}{4} \right) dx = \left(\ln x + \frac{x^2}{8} \right) \Big|_1^2$$

$$= \ln 2 + \frac{3}{8} \qquad \square$$

Notice in the solution of Example 2 that by adding 1 to $(f'(x))^2$ we obtained the perfect square $(1/x + x/4)^2$ inside the radical sign. This enabled us to eliminate the radical sign so that we could perform the integration.

When the integration cannot readily be performed, Riemann sums help us find approximate values for the length of a graph.

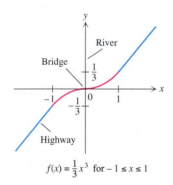

FIGURE 6.23

EXAMPLE 3 A highway runs northeast across a river, which is represented by the y axis in Figure 6.23. To achieve a crossing at right angles, a curve is introduced into the highway. If

$$f(x) = \frac{1}{3} x^3 \quad \text{for } -1 \leq x \leq 1$$

then this curve is represented by the graph of f. Suppose the center of the curve is to be painted with a continuous white stripe. By using the left sum with $n = 100$, find the approximate length of the stripe.

Solution Since $f'(x) = x^2$, we know from (2) that

$$L = \int_{-1}^1 \sqrt{1 + (x^2)^2} \, dx = \int_{-1}^1 \sqrt{1 + x^4} \, dx$$

We cannot find the exact value of the integral. However, by calculating the left sum with $n = 100$, we find that

$$L = \int_{-1}^1 \sqrt{1 + x^4} \, dx \approx 2.178953107 \qquad \square$$

EXERCISES 6.2

In Exercises 1–11 find the length L of the graph of the given function.

1. $f(x) = 2x + 3$ for $1 \le x \le 5$
2. $f(x) = 2/3x^{3/2}$ for $1 \le x \le 4$
3. $f(x) = x^2 - \frac{1}{8} \ln x$ for $2 \le x \le 3$
4. $g(x) = x^3 + \dfrac{1}{12x}$ for $1 \le x \le 3$
5. $k(x) = x^4 + \dfrac{1}{32x^2}$ for $1 \le x \le 2$
6. $g(x) = \ln(x^2 - 1)$ for $2 \le x \le 5$
7. $f(x) = \ln(1 + x^2) - \dfrac{1}{8}\left(\dfrac{x^2}{2} + \ln x\right)$ for $1 \le x \le 2$
8. $f(x) = \ln(1 + x^3) + \dfrac{1}{12}\left(\dfrac{1}{x} - \dfrac{x^2}{2}\right)$ for $1 \le x \le 2$
9. $f(x) = -\frac{1}{4}\sin x + \ln(\sec x + \tan x)$ for $\pi/4 \le x \le \pi/3$
10. $f(x) = \frac{1}{2}(e^x + e^{-x})$ for $0 \le x \le \ln 2$
*11. $f(x) = \tan x - \frac{1}{8}\left(x + \frac{1}{2}\sin 2x\right)$ for $0 \le x \le \pi/4$

In Exercises 12–17 let f be a function defined on the given interval and having the indicated derivative f'. (Such a function f exists by the Fundamental Theorem of Calculus.) Find the length L of the graph of f.

12. $f'(x) = \sqrt{x^2 - 1}$; $[2, 3]$ 13. $f'(x) = \tan x$; $[0, \pi/4]$
14. $f'(x) = \sqrt{\tan^2 x - 1}$; $[\frac{2}{3}\pi, \frac{3}{4}\pi]$
15. $f'(x) = \sqrt{x^n - 1}$; $[2, 4]$; n a positive integer
16. $f'(x) = \sqrt{\sqrt{x} - 1}$; $[25, 100]$
17. $f'(x) = \sqrt{x^{1/n} - 1}$; $[1, 2^n]$; n a positive integer
18. **a.** Let $f(x) = x^{1+1/(2n)}$, for $a \le x \le b$. Show that the length L of the graph of f is given by

$$L = \int_a^b \sqrt{1 + (2n + 1)^2 \, x^{1/n}/(2n)^2} \; dx$$

   *b. Show that if $u = \sqrt{1 + (2n + 1)^2 \, x^{1/n}/(2n)^2}$, then the integral in part (a) becomes an integral of a polynomial, and hence can (at least in theory) be evaluated.

   c. Using the substitution in part (b), evaluate the integral in part (a) when $n = 1$, $a = 0$, and $b = 1$.

19. Let $r > 0$. The graph of the equation

$$x^{2/3} + y^{2/3} = r^{2/3}$$

is called an **astroid** (Figure 6.24).

   a. For the portion of the astroid in the first quadrant, express y as a function of x.

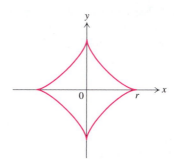

FIGURE 6.24   The astroid $x^{2/3} + y^{2/3} = r^{2/3}$ for Exercise 19.

   b. Let $\epsilon > 0$. Find the length L_ϵ of the portion of the astroid in the first quadrant for $\epsilon \le x \le r$. Then find the limit $L = \lim_{\epsilon \to 0+} L_\epsilon$, and thereby determine the length of one-fourth of the astroid and hence the length of the whole astroid. (Later we will have an easier way to find the length of the astroid. See Exercise 10 in Section 6.8.)

Application

*20. An electric wire connecting a telephone pole to a house hangs in the shape of a catenary $y = \frac{c}{2}(e^{x/c} + e^{-x/c})$, where the units are feet (Figure 6.25). Find the length of the wire. (*Hint:* First find the value of c.)

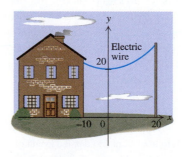

FIGURE 6.25   Figure for Exercise 20.

Projects

1. This project is parallel to Project 2 in Section 5.1, where we found approximations to the area of the unit circle. We all know that the circumference of the unit circle centered at the origin is 2π. In order to approximate the circumference, Archimedes inscribed polygons, as in

Figures 6.26 (a)–(c), and found the lengths of those polygons.

a. Find the length L_4 of the inscribed square.

b. Find the length L_6 of the inscribed hexagon. (*Hint:* What are the coordinates of a vertex that is neighbor to (1, 0)?)

c. Find the length L_8 of the inscribed octagon. (*Hint:* What are the coordinates of a vertex that is neighbor to (1, 0)?)

d. Find a formula for the length L_n of the inscribed polygon of n sides, and then determine the minimum number n such that the length L_n is greater than 6.

e. Find the minimum number n such that the length L_n of the inscribed polygon of n sides is greater than 6.28.

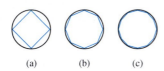

(a)   (b)   (c)

FIGURE 6.26   Figure for Project 1

2. In this project we look at the length of curves that have an infinite number of wobbles.

a. Let $f(x) = \sin(\pi/x)$ for $0 < x \le 1$ (Figure 6.27 (a))

   i. Show that $f(\frac{1}{n}) = 0$ and $|f(\frac{2}{2n+1})| = 1$ for all $n \ge 1$. (Thus $|f(2/3)| = |f(2/5)| = |f(2/7)| = \cdots = 1$).

   ii. Let L_2 = the length of the graph of f for $1/2 \le x \le 1$. Using the fact that $1/2 < 2/3 < 1$ and the fact that a straight line is the shortest distance between two points in the plane, show that $L_2 > 2$.

   iii. Let L_3 = the length of the graph of f for $1/3 \le x \le 1$. Using the fact that $1/3 < 2/5 < 1/2 < 2/3 < 1$, show that $L_3 > 4$.

   iv. Let L_n = the length of the graph of f for $1/n \le x \le 1$. Use the ideas of (ii) and (iii) to show that $L_n > 2n - 2$.

   Note: It follows from (iv) that the length of the graph of f for $0 < x \le 1$ could not be finite, even if it were defined.

b. Let

$$g(x) = \begin{cases} x \sin(\pi/x) & \text{for } 0 < x \le 1 \\ 0 & \text{for } x = 0 \end{cases}$$

   (See Figure 6.27 (b)).

   i. Show that $g(\frac{1}{n}) = 0$ and $|g(\frac{2}{2n+1})| = \frac{2}{2n+1}$, for all $n \ge 1$.

   ii. Let M_2 = the length of the graph of g for $1/2 \le x \le 1$. Show that $M_2 > 2(2/3) = 4/3$.

   iii. Let M_3 = the length of the graph of g for $1/3 \le x \le 1$. Show that $M_3 > 2(2/3) + 2(2/5) = 4/3 + 4/5$.

   iv. Let M_n = the length of the graph of g for $1/n \le x \le 1$. Show that

$$M_n > 2\left(\frac{2}{3}\right) + 2\left(\frac{2}{5}\right) + 2\left(\frac{2}{7}\right) + \cdots$$

$$+ 2\left(\frac{2}{2n-1}\right) > 1 + \frac{1}{2} + \frac{1}{3} + \frac{1}{4} + \cdots + \frac{1}{n-1}$$

   v. Show that $1 + \frac{1}{2} + \frac{1}{3} + \frac{1}{4} + \cdots + \frac{1}{n-1}$ is a left sum (and upper sum also) for $\int_1^n 1/x \, dx$.

   vi. Using the fact that $\int_1^n 1/x \, dx = \ln n$, as well as your knowledge about $\lim_{x \to \infty} \ln x$, show that the graph of g would have infinite length (if it were defined), despite the fact that g is continuous on [0, 1]. This shows that we cannot define the length of the graph of every continuous function.

c. Let

$$h(x) = \begin{cases} x^3 \sin(\pi/x) & \text{for } x \ne 0 \\ 0 & \text{for } x = 0 \end{cases}$$

   (See Figure 6.27(c)).

   i. Prove that h is not only continuous but also differentiable at 0 (as well as $x \ne 0$).

   ii. From Definition 6.2 we know that the length L of the portion of the graph of h for $0 \le x \le 1$ is defined and finite. First estimate the value of L by using Riemann sums with partitions having a large number of subintervals. Then use a technique similar to that in part (b) and see if the estimates you get by this technique remains finite as n increases without bound.

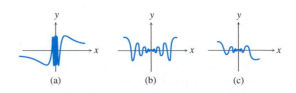

(a)   (b)   (c)

FIGURE 6.27   Figure for Project 2

6.3 AREA OF A SURFACE

(a)

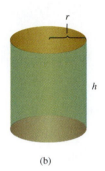

(b)

FIGURE 6.28 (a) The surface area is $6s^2$. (b) The surface area is $2\pi rh$.

A higher-dimensional version of the length of a curve is the area of a surface. It has been known since antiquity that the surface area S of a cube of side s is given by

$$S = 6s^2$$

(Figure 6.28(a)), and the surface area S of a cylinder of radius r and height h is given by

$$S = 2\pi rh$$

(Figure 6.28(b)). However, our analysis of the areas of other surfaces will be based on the surface area of a frustum of a cone. If the frustum has slant height l and radii r_1 and r_2, as in Figure 6.29(a), then the surface area S is given by

$$S = 2\pi \left(\frac{r_1 + r_2}{2} \right) l = \pi(r_1 + r_2)l \tag{1}$$

We will prove (1) in Section 14.3. It can also be derived from the formula for the surface area of a cone (see Exercise 8).

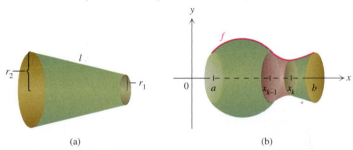

(a) (b)

FIGURE 6.29 (a) The frustum of a cone. (b) The surface area of the band is approximated by the area of a frustum of a cone.

More generally, suppose f is defined on $[a, b]$, and $f(x) \geq 0$ for $a \leq x \leq b$. We will derive a formula for the area S of the surface obtained by revolving the graph of f about the x axis (Figure 6.29(b)). In order to be able to use the length of the graph of f on $[a, b]$ in our calculations of surface area, we will also assume that f is continuously differentiable on $[a, b]$. Now let $P = \{x_0, x_1, \ldots, x_n\}$ be a partition of $[a, b]$, and let ΔS_k be the area of the portion of the surface between the revolved lines $x = x_{k-1}$ and $x = x_k$. If Δx_k is small, the surface area S_k is approximately equal to the surface area of the frustum of the corresponding cone, that is, the frustum whose slant height is equal to the length of the line between $(x_{k-1}, f(x_{k-1}))$ and $(x_k, f(x_k))$, and the radii of whose ends are $f(x_{k-1})$ and $f(x_k)$ (see Figure 6.29(b)). From (1) this means that

$$\Delta S_k \approx \pi[f(x_{k-1}) + f(x_k)] \sqrt{(x_k - x_{k-1})^2 + [f(x_k) - f(x_{k-1})]^2} \tag{2}$$

As in our discussion of lengths of curves in Section 6.2, the Mean Value Theorem can be applied to show that for some t_k in $[x_{k-1}, x_k]$,

$$\sqrt{(x_k - x_{k-1})^2 + [f(x_k) - f(x_{k-1})]^2} = \sqrt{1 + [f'(t_k)]^2} \, \Delta x_k \tag{3}$$

Since Δx_k is assumed to be small, x_{k-1} and x_k are close together, with t_k between them. Since f' and hence f are continuous on $[x_{k-1}, x_k]$, it follows that $f(x_{k-1}) + f(x_k)$

should be approximately $f(t_k) + f(t_k)$, that is, $2f(t_k)$. So substituting from (3) into (2), we deduce that

$$\Delta S_k \approx \pi[2f(t_k)]\sqrt{1 + [f'(t_k)]^2}\, \Delta x_k$$

Consequently the area S of the complete surface, which equals the sum of ΔS_1, $\Delta S_2, \ldots, \Delta S_n$, should be approximately

$$\sum_{k=1}^{n} 2\pi f(t_k)\sqrt{1 + [f'(t_k)]^2}\, \Delta x_k$$

which is a Riemann sum for $2\pi f\sqrt{1 + (f')^2}$ on $[a, b]$. Therefore we are led to define the surface area S by

$$S = \lim_{\|P\|\to 0} \sum_{k=1}^{n} 2\pi f(t_k)\sqrt{1 + [f'(t_k)]^2}\, \Delta x_k = \int_a^b 2\pi f(x)\sqrt{1 + [f'(x)]^2}\, dx$$

DEFINITION 6.3

Let f be nonnegative and continuously differentiable on $[a, b]$. The **area** of the surface obtained by revolving the graph of f about the x axis is defined by

$$S = \int_a^b 2\pi f(x)\sqrt{1 + [f'(x)]^2}\, dx \qquad (4)$$

EXAMPLE 1   Let $f(x) = x^3$ for $0 \le x \le 1$. Find the area S of the surface obtained by revolving the graph of f about the x axis (Figure 6.30).

Solution   Since $f'(x) = 3x^2$, it follows from (4) that

$$S = \int_0^1 2\pi x^3 \sqrt{1 + (3x^2)^2}\, dx = 2\pi \int_0^1 x^3 \sqrt{1 + 9x^4}\, dx$$

To evaluate the integral, we let

$$u = 1 + 9x^4, \quad \text{so that} \quad du = 36x^3\, dx$$

Now if $x = 0$ then $u = 1$, and if $x = 1$ then $u = 10$. Therefore

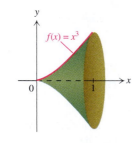

FIGURE 6.30

$$S = 2\pi \int_0^1 x^3 \sqrt{1 + 9x^4}\, dx = 2\pi \int_0^1 \overbrace{\sqrt{1 + 9x^4}}^{\sqrt{u}}\ \overbrace{x^3\, dx}^{\frac{1}{36}du}$$

$$= 2\pi \int_1^{10} \sqrt{u}\, \frac{1}{36}\, du = \frac{\pi}{18}\left(\frac{2}{3} u^{3/2}\right)\Big|_1^{10}$$

$$= \frac{\pi}{27}(10^{3/2} - 1) \qquad \square$$

EXAMPLE 2   Let $f(x) = \sqrt{1 - x^2}$ for $0 \le x \le \frac{1}{2}$. Find the area S of the portion of the sphere obtained by revolving the graph of f about the x axis (Figure 6.31).

Solution   Since $f'(x) = -x/\sqrt{1 - x^2}$, it follows from (4) that

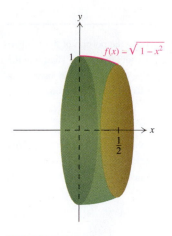

FIGURE 6.31   A portion of a sphere.

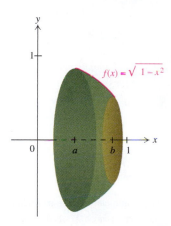

FIGURE 6.32   A portion of a sphere.

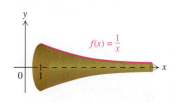

FIGURE 6.33   Gabriel's horn.

$$S = \int_0^{1/2} 2\pi \sqrt{1 - x^2} \sqrt{1 + \left(\frac{-x}{\sqrt{1 - x^2}}\right)^2} \, dx$$

$$= 2\pi \int_0^{1/2} \sqrt{1 - x^2} \sqrt{1 + \frac{x^2}{1 - x^2}} \, dx$$

$$= 2\pi \int_0^{1/2} \sqrt{1 - x^2} \sqrt{\frac{1}{1 - x^2}} \, dx = 2\pi \int_0^{1/2} 1 \, dx = 2\pi x \Big|_0^{1/2} = \pi \quad \square$$

Had we considered the portion of the sphere for which $a \leq x \leq b$, where $-1 < a < b < 1$ (Figure 6.32), similar calculations would show that

$$S = 2\pi(b - a)$$

Thus, rather surprisingly, the surface area depends only on the length of the interval $[a, b]$ and not on its location within $(-1, 1)$ (see Exercise 10).

We end this section with an analyis of the volume and the surface area of a figure that looks like a long horn.

EXAMPLE 3   Let $f(x) = 1/x$, and let $b > 1$. Denote by R_b the region between the graph of f and the x axis on $[1, b]$, and let D_b be the solid generated by revolving R_b about the x axis. (Figure 6.33)

a.   Letting V_b denote the volume of D_b, show that $\lim_{b \to \infty} V_b = \pi$.

b.   Letting S_b denote the surface area of D_b, show that $\lim_{b \to \infty} S_b = \infty$.

Solution   **a.** We use (3) of Section 6.1, with the limits of integration replaced by 1 and b:

$$V_b = \int_1^b \pi \frac{1}{x^2} \, dx = \pi \left(-\frac{1}{x}\right)\Big|_1^b = \pi \left(1 - \frac{1}{b}\right)$$

Evidently $\lim_{b \to \infty} V_b = \pi$.

b. We use (4) of this section, again with the limits of integration replaced by 1 and b:

$$S_b = \int_1^b 2\pi \frac{1}{x} \sqrt{1 + \left(-\frac{1}{x^2}\right)^2} \, dx = \int_1^b 2\pi \frac{1}{x} \sqrt{1 + \frac{1}{x^4}} \, dx$$

Since

$$\frac{2\pi}{x} \sqrt{1 + \frac{1}{x^4}} \geq \frac{2\pi}{x} \quad \text{for } 1 \leq x \leq b$$

it follows from the General Comparison Property (Corollary 5.19) that

$$S_b \geq \int_1^b \frac{2\pi}{x} \, dx = 2\pi \ln b$$

Because $\lim_{b \to \infty} 2\pi \ln b = \infty$, we conclude that

$$\lim_{b \to \infty} S_b = \infty \quad \square$$

The infinitely long "horn" that results when b approaches ∞ is called **Gabriel's horn.** A surprising consequence of the results of Example 3 is the fact that one could fill the interior of Gabriel's horn with a finite amount of paint (namely, π cubic units of paint), whereas it would take an infinite amount of paint to cover the surface of Gabriel's horn!

EXERCISES 6.3

In Exercises 1–7 find the area S of the surface generated by revolving about the x axis the graph of f on the given interval.

1. $f(x) = \sqrt{4 - x^2}$; $\left[-\frac{1}{2}, \frac{3}{2}\right]$

2. $f(x) = \frac{1}{3}x^3$; $[0, \sqrt{2}]$

3. $f(x) = \sqrt{x}$; $[2, 6]$

4. $f(x) = 2\sqrt{1 - x}$; $[-1, 0]$

5. $f(x) = \frac{1}{2}(e^x + x^{-x})$; $[0, 1]$

6. $f(x) = \frac{1}{2}x^3 + \frac{1}{6x}$; $[1/\sqrt{2}, 1]$

7. $f(x) = \frac{1}{4}x^4 + \frac{1}{8x^2}$; $[1, \sqrt{2}]$

8. Derive (1) from the formula $S = \pi r l$ for the surface area of a cone with slant height l and radius r. (*Hint:* The surface area of the frustum is the difference of the surface areas of two cones, one with radius r_2 and one with radius r_1.)

9. Let $0 < a < r$. The portion of the sphere of radius r obtained by revolving the graph of

$$y = \sqrt{r^2 - x^2} \quad \text{for } -a \leq x \leq a$$

about the x axis is called a **zone** of the sphere.

a. Determine the surface area S of the zone.

b. Check your result in part (a) by letting $a = r$ and seeing whether you obtain the surface area of the sphere.

10. Assume that $[a, b]$ is a closed interval contained in the open interval $(-1, 1)$. Let $f(x) = \sqrt{1 - x^2}$, and let S be the area of the surface obtained by revolving the graph of f on $[a, b]$ about the x axis (refer to Figure 6.32). Show that

$$S = 2\pi(b - a)$$

(Thus the surface area depends only on the width of the interval $[a, b]$ and not on its location within $(-1, 1)$.)

11. The graph of the equation $x^{2/3} + y^{2/3} = 1$ is an astroid (Figure 6.34). Let $0 < \epsilon < 1$. Find the area S_ϵ of the

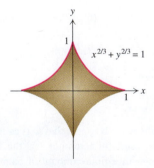

FIGURE 6.34 The surface for Exercise 11.

surface obtained by revolving the portion of the top half of the astroid for $\epsilon \leq x \leq 1$ about the x axis. Then find the limit $S = \lim_{\epsilon \to 0+} S_\epsilon$. (Later we will have an easier way of finding the area S. See Exercise 10 in Section 6.8.)

Applications

12. A yo-yo is made from a sphere by removing a slice from its center, as in Figure 6.35. Suppose the yo-yo is cut from a sphere of radius 3 centimeters in such a way that its width is

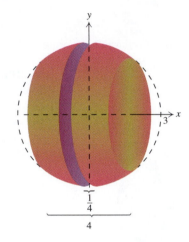

FIGURE 6.35 The yo-yo for Exercise 12.

4 centimeters, with a slit $\frac{1}{4}$ centimeter wide in its middle (Figure 6.35). Find the surface area S of the outer surfaces of the yo-yo.

13. A wok is in the shape of a solid obtained by revolving about the y axis the curve shown in Figure 6.36. Assume that the units are centimeters. Find the interior surface area of the wok. (*Hint:* Use (4) with the roles of x and y reversed.)

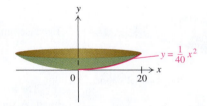

FIGURE 6.36 The wok for Exercise 13.

Project

1. With the results on Gabriel's horn under our belt, we will consider another hornlike object. Let $g(x) = 1/x^2$ for $x \geq 1$, and for $b > 1$ let D_b denote the solid obtained by revolving the graph of g on $[1, b]$ about the x axis.

    a. Find the volume V_b of D_b, and find $\lim_{b \to \infty} V_b$.

    b. Prove that the surface area S_b of D_b satisfies $S_b \leq 15$ for all $b > 1$. (*Hint:* First find a reasonable positive number M such that $\sqrt{1 + (g'(x))^2} \leq M$ for all $x \geq 1$.)

    c. Can you find a number $S < 14$ such that $S_b \leq S$ for all $b > 1$?

    d. Suppose that $g_r(x) = 1/x^r$ for $x \geq 1$, where $r \leq 1$. Let D_{rb} denote the solid obtained by revolving the graph of g_r on $[1, b]$ about the x axis, and S_{rb} the surface area of D_{rb}. On the one hand, we know that the surface area S_{1b} of Gabriel's horn approaches ∞ as b approaches ∞. On the other hand, we know from part (b) that the surface area S_{2b} of D_{2b} is no larger than 15 as b increases without bound. Is the surface area S_{rb} of D_{rb} bounded as a function of b, for each r with $r > 1$? Explain your answer.

6.4   WORK

Next we consider the physical concept of work. To introduce this concept, let us imagine a person pushing a wheelbarrow across a yard, from point a to point b. If a constant force c is exerted all the way across the yard, then the amount of work done is defined to be $c(b - a)$, that is, the force times the distance traveled by the wheelbarrow (Figure 6.37). (The product $c(b - a)$ should remind you of the area of a rectangle.) This is a reasonable definition, since we would expect the work done to increase if the person pushes the wheelbarrow farther or with a greater force.

How should the work be defined when the force is variable? The integral will enable us to do this. In order to relate integrals to work, we make certain assumptions.

1. The object on which the force acts moves in a straight line, so that we may think of the object as moving along the x axis from a point a to a point b.

2. At each point x between a and b a certain force $F(x)$ is exerted on the object.

3. The function F is continuous on $[a, b]$.

Furthermore, we adopt the convention that force is positive if exerted in the direction of the positive x axis and negative if exerted in the direction of the negative axis.

As we have seen, if the force F exerted on an object moving from a to b is constant, say $F(x) = c$, then the work W done is $W = c(b - a)$. When F is not necessarily constant, we let $P = \{x_0, x_1, \ldots, x_n\}$ be any partition of $[a, b]$, and for each k between 1 and n we let t_k be an arbitrary point in the subinterval $[x_{k-1}, x_k]$. If Δx_k is small, then as the object moves from x_{k-1} to x_k, the amount of work ΔW_k done by the force on the object is approximately $F(t_k) \Delta x_k$. It is also reasonable to expect the work W done by the force when the object moves from a to b to be the sum of $\Delta W_1, \Delta W_2, \ldots, \Delta W_n$, each one of which represents the work done on the object as it travels over one of the successive subintervals of $[a, b]$. Therefore W should be approximately equal to

$$\sum_{k=1}^{n} \overset{\text{force}}{F(t_k)} \, \overset{\text{distance}}{\Delta x_k}$$

work = force × distance = $c(b - a)$

FIGURE 6.37

which is a Riemann sum for F on $[a, b]$. Consequently W should be defined by

$$W = \lim_{\|P\| \to 0} \sum_{k=1}^{n} F(t_k)\,\Delta x_k = \int_a^b F(x)\,dx$$

DEFINITION 6.4   Suppose an object moves from a to b under the influence of a force F, where F is a continuous function on $[a, b]$. Then the **work** W done by the force on the object is defined by

$$W = \int_a^b F(x)\,dx \qquad (1)$$

It is implicit in Definition 6.4 that work is positive if the force acts in the direction of motion of the object and negative if the force opposes the motion of the object.

There are several kinds of units for measuring work.

1. The set of units that has by now gained wide acceptance worldwide is the International System (abbreviated SI), established by an international committee in 1960. It is based on the metric system, with kilograms, meters, and seconds as the standard units of mass, length, and time, respectively. In the SI system, force is measured in newtons and work in joules (newton-meters).
2. In the cgs (that is, centimeter-gram-second) system, force is measured in terms of dynes, and work in ergs (dyne-centimeters).
3. In the British system, the basic units are slugs, feet, and seconds. In this system, force is given in terms of pounds, and work in foot-pounds.

In our first example we return to the wheelbarrow and compute the work done on it as it moves across the yard, under the condition that it leaks all the while.

EXAMPLE 1   Suppose we push a leaking wheelbarrow 100 meters and (because of the leak) exert a decreasing force in newtons given by

$$F(x) = 60\left(1 - \frac{x^2}{20{,}000}\right) \quad \text{for } 0 \le x \le 100$$

Find the work W done by the force on the wheelbarrow.

Solution   Since the force at any x in $[0, 100]$ is $60(1 - x^2/20{,}000)$, it follows from (1) that

$$W = \int_0^{100} 60\left(1 - \frac{x^2}{20{,}000}\right)dx = 60\left(x - \frac{x^3}{60{,}000}\right)\Bigg|_0^{100}$$

$$= 5000 \text{ (joules)} \qquad \square$$

Hooke's Law

As another example we consider the work done when a spring is stretched. **Hooke's Law** (stated in 1676 by the English geometer Robert Hooke) says that the elastic force $G(x)$ exerted by a spring which has been extended x units beyond its natural length is proportional to x (Figure 6.38). The force is negative, since it opposes the expansion of the spring and hence is exerted in the direction of the negative x axis (Figure 6.38). Thus there is a positive constant k such that

$$G(x) = -kx$$

and it is known that this formula is fairly reliable for reasonably small values of x.

The force $F(x)$ needed in order to hold a spring extended x units beyond its natural length is opposite to $G(x)$ and equal in magnitude, which means that

$$F(x) = -G(x) = kx$$

It follows from (1) that the work W required in order to stretch a spring from a units extended to b units is given by the formula

$$W = \int_a^b F(x)\,dx = \int_a^b kx\,dx \tag{2}$$

Since the force that produces the work opposes the force exerted by the spring, we say that W is the work done *against* the force of the spring.

EXAMPLE 2   Suppose the work required to stretch a certain spring 1 centimeter beyond its natural length is 10^5 ergs. Find the work W required to stretch it from 1 centimeter beyond its natural length to 3 centimeters beyond.

Solution   Our assumption implies that

$$10^5 = \int_0^1 kx\,dx \tag{3}$$

and we desire the value of W, given by

$$W = \int_1^3 kx\,dx$$

From (3) we can determine k:

$$10^5 = \int_0^1 kx\,dx = \frac{1}{2}kx^2 \Big|_0^1 = \frac{1}{2}k$$

so that $k = 2 \times 10^5$. Therefore

$$W = \int_1^3 kx\,dx = \int_1^3 (2 \times 10^5)\,x\,dx = 10^5 x^2 \Big|_1^3 = 8 \times 10^5 \ \text{(ergs)} \qquad \square$$

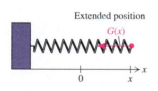

Natural position

Extended position

$G(x)$

FIGURE 6.38   Hooke's Law: The restoring force $G(x)$ is proportional to the displacement x.

Work Required to Empty a Tank

In the next illustration we will find a formula for the work required to pump water from a tank. We will use the properties of work that led us to Definition 6.4, along with one additional property, which we will prove in Chapter 15: The *same* amount of work is required to move an object against the force of gravity from a point P to a point Q, no matter what path from P to Q the object travels. Thus we may assume

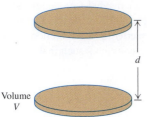

FIGURE 6.39

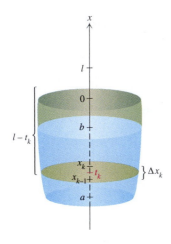

FIGURE 6.40

that the object moves along a vertical line, and in that case we must exert a force at least as great as the weight of the object to balance the force of gravity.

We wish to compute the amount of work done in pumping water out of a tank and up to a certain level. Suppose first that we lift a solid region of water d feet, (so that *each* particle of water rises d feet), and suppose the region has volume V cubic feet (Figure 6.39). By definition the work W done by the lifting force is the product of the weight of the water and the distance d the water is moved. Since the water weighs 62.5 pounds per cubic foot, the weight of the water is $62.5V$, so that

$$W = (\text{weight}) \times (\text{distance}) = 62.5Vd$$

(If the liquid in the tank were not water, the number 62.5 would be replaced by the weight of 1 cubic foot of the other liquid.)

Now suppose the x axis is chosen to be vertical, with the positive numbers above the origin, placed at a convenient reference point for measuring depth, such as the top of the tank. Assume that the water to be pumped extends from a to b on the x axis (Figure 6.40), and let l be the level to which the water is to be pumped. Furthermore, for each x in $[a, b]$ let $A(x)$ be the cross-sectional area of the tank at x, and assume that A is continuous on $[a, b]$. Let $P = \{x_0, x_1, \ldots, x_n\}$ be any partition of $[a, b]$. For each k between 1 and n, let t_k be any number in $[x_{k-1}, x_k]$. If Δx_k is small, the volume of the water between the levels x_{k-1} and x_k is approximately $A(t_k)\,\Delta x_k$. Each particle of water in that portion must travel a distance of approximately $l - t_k$ in order to reach the level l. Therefore our earlier discussion implies that the work ΔW_k required to lift the water contained between x_{k-1} and x_k is approximately the product of the weight $62.5\,A(t_k)\,\Delta x_k$ and the distance $l - t_k$ traveled, that is, approximately

$$62.5\,A(t_k)\,\Delta x_k\,(l - t_k)$$

As a result, the work W required to lift all the water between the levels $x = a$ and $x = b$, which is the sum of $\Delta W_1, \Delta W_2, \ldots, \Delta W_n$, should be approximately

$$\sum_{k=1}^{n} 62.5\,A(t_k)\,\Delta x_k\,(l - t_k)$$

which is a Riemann sum for $62.5\,A(x)\,(l - x)$. Therefore the work W done in lifting the water is given by

$$W = \lim_{\|P\| \to 0} \sum_{k=1}^{n} 62.5(l - t_k)A(t_k)\,\Delta x_k = \int_{a}^{b} 62.5(l - x)A(x)\,dx \quad (4)$$

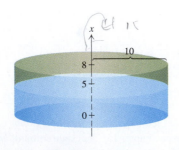

FIGURE 6.41

EXAMPLE 3   A swimming pool has the shape of a right circular cylinder with radius 10 feet and depth 8 feet. Assume that the pool contains water to a depth of 5 feet. Find the work W required to pump all but 1 foot of water to the top of the pool.

Solution   We place the origin at the bottom of the pool (Figure 6.41). Then $l = 8$, so a particle of water x feet from the bottom is to be raised $8 - x$ feet, and moreover, the water to be pumped out extends from 1 to 5 on the x axis. Furthermore, the cross-sectional area $A(x)$ is constant, with

$$A(x) = \pi(10)^2 = 100\pi$$

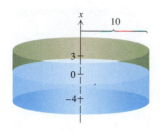

FIGURE 6.42

Thus it follows from (4) that

$$W = \int_1^5 62.5(8 - x)(100\pi) \, dx = 6250\pi \int_1^5 (8 - x) \, dx$$

$$= 6250\pi \left(8x - \frac{1}{2}x^2 \right) \Big|_1^5 = 125,000\pi \text{ (foot-pounds)} \quad \square$$

Placement of the origin in pumping problems is a matter of convenience. Were we to place the origin at water level in Example 3 (Figure 6.42), we would still have $A(x) = 100\pi$. However, l would be 3, a particle of water corresponding to x would be raised $3 - x$ feet, and the water to be pumped would extend from -4 to 0 on the x axis. Therefore we would find that

$$W = \int_{-4}^0 62.5(3 - x)(100\pi) \, dx = 6250\pi \int_{-4}^0 (3 - x) \, dx$$

$$= 6250\pi \left(3x - \frac{1}{2}x^2 \right) \Big|_{-4}^0 = 125,000\pi \text{ (foot-pounds)}$$

as we found in Example 3.

EXAMPLE 4   A hemispherical tank with radius 10 feet is filled with water (Figure 6.43). Find the amount of work W required to pump all the water in the tank to 6 feet above the top of the tank.

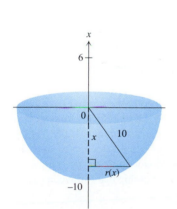

FIGURE 6.43   Hemispherical tank.

Solution   This time we place the origin at the top of the tank. Then $l = 6$, a particle of water corresponding to x is to be raised $6 - x$ feet, and the water to be pumped out extends from -10 to 0 on the x axis. From Figure 6.43 we see that the radius $r(x)$ of the cross section at x satisfies the equation $[r(x)]^2 = 100 - x^2$. Therefore

$$A(x) = \pi[r(x)]^2 = \pi(100 - x^2)$$

Thus it follows from (4) that

$$W = \int_{-10}^0 62.5(6 - x)\pi(100 - x^2) \, dx$$

$$= (62.5)\pi \int_{-10}^0 (600 - 100x - 6x^2 + x^3) \, dx$$

$$= 62.5\pi \left(600x - 50x^2 - 2x^3 + \frac{1}{4}x^4 \right) \Big|_{-10}^0$$

$$= 62.5\pi(6000 + 5000 - 2000 - 2500)$$

$$= 406,250\pi \text{ (foot-pounds)} \quad \square$$

Work vs Kinetic Energy

Suppose an object such as a ball of mass m is moving along the x axis under a force $F(x)$ and with an acceleration $a(x)$. By Newton's Second Law of Motion, the force and acceleration are related by the formula

$$F(x) = ma(x)$$

If the object moves from point x_0 to x_1, then since acceleration is the derivative (with respect to time) of the velocity, we find that

$$F(x) = ma(x) = m\frac{dv}{dt} \tag{5}$$

Using the Chain Rule and observing that $v = dx/dt$, we obtain

$$m\frac{dv}{dt} = m\frac{dv}{dx}\frac{dx}{dt} = m\frac{dv}{dx}v \tag{6}$$

The formulas in (5) and (6) together yield

$$F(x) = mv\frac{dv}{dx} \tag{7}$$

From (7) and the substitution of $v = v(x)$ and $dv = \dfrac{dv}{dx}\,dx$, we find that

$$\int_{x_0}^{x_1} F(x)\,dx = \int_{x_0}^{x_1}\left(mv\frac{dv}{dx}\right)dx = \int_{v_0}^{v_1} mv\,dv = \frac{1}{2}mv_1^2 - \frac{1}{2}mv_0^2$$

Thus the work W done by the force F on the ball as it travels from x_0 to x_1 is given by

$$W = \int_{x_0}^{x_1} F(x)\,dx = \frac{1}{2}mv_1^2 - \frac{1}{2}mv_0^2 \tag{8}$$

The quantity $\frac{1}{2}mv^2$ is called the **kinetic energy** of the ball. Therefore (8) expresses the fact that the work done by the force on the ball as the ball travels from x_0 to x_1 is equal to the change in the kinetic energy of the object. It is remarkable that the work done by the force depends only on the initial and final velocities of the ball.

Now that we are acquainted with the notions of work and kinetic energy, we can derive the Law of Conservation of Energy for a spaceship that is traveling away from the earth.

Let

R = the radius of the earth G = the universal gravitational constant

M = the mass of the earth m = the mass of the spaceship

Newton's Law of Gravitation asserts that the gravitational force F on the craft is given by

$$F = -\frac{GMm}{r^2}$$

(The minus sign appears in this formula because the force is directed toward the earth.) As a result, the work $W(b)$ done by the gravitational force as the craft travels from the surface of the earth to a distance b from the center of the earth is given by

$$W(b) = \int_R^b -\frac{GMm}{r^2}\,dr = \frac{GMm}{r}\bigg|_R^b = GMm\left(\frac{1}{b} - \frac{1}{R}\right) \tag{9}$$

The formula in (8) implies that the work is equal to the change in the kinetic energy of the craft:

$$W(b) = \int_R^b F(r)\, dr = \frac{1}{2} m[v(b)]^2 - \frac{1}{2} m[v(R)]^2 \tag{10}$$

Together (9) and (10) imply that

$$GMm\left(\frac{1}{b} - \frac{1}{R}\right) = \frac{1}{2} m[v(b)]^2 - \frac{1}{2} m[v(R)]^2 \tag{11}$$

When the spacecraft is a distance b from the center of the earth, the **potential energy** of the craft is defined to be $-GMm/b$. Thus we can rewrite (11) in the following form:

$$\underset{\text{kinetic energy}}{\underbrace{\frac{1}{2} m[v(b)]^2}} + \underset{\text{potential energy}}{\underbrace{\left(-GMm\frac{1}{b}\right)}} = \frac{1}{2} m[v(R)]^2 - GMm\frac{1}{R} \tag{12}$$

Notice that the right-hand side of (12) is constant as the spacecraft recedes from earth. Thus we have proved the famous **Law of Conservation of Energy:**

> The sum of the kinetic energy and potential
> energy is constant (for all values of b).

The formula in (11) also yields a formula for the escape velocity (which we discussed initially in Section 4.8). If a spacecraft has precisely the escape velocity, then $\lim_{b \to \infty} v(b) = 0$, so that taking the limits of the expression in (11) as b approaches ∞ yields

$$\frac{GMm}{R} = \frac{1}{2} m[v(R)]^2, \quad \text{or equivalently,} \quad v(R) = \sqrt{2GM/R}$$

Consequently the initial velocity shortly after liftoff must be $\sqrt{2GM/R}$. This is the same as the formula for the escape velocity derived in Section 4.8.

EXERCISES 6.4

1. Determine the work W done on the wheelbarrow of Example 1 if it is pushed only 60 meters.

2. An elevator in the Empire State Building weighs 1600 pounds. Find the work W required to raise the elevator from ground level to the 102nd story, some 1200 feet above ground level.

3. A 10-pound bag of groceries is to be carried up a flight of stairs 8 feet tall. Find the work W done on the bag.

4. Suppose a 120-pound person carries the bag in Exercise 3 up the stairs. Find the total work W done by the person in walking up the stairs with the bag.

5. A sailboat is stationary in the middle of a lake until a strong gust of wind blows it along a straight line. Suppose the force in newtons exerted on the sails by the wind when the boat is x kilometers from its starting point is

$$F(x) = 10^4 \sin x \quad \text{for } 0 \le x \le \pi$$

Find the work W done on the sails by the gust of wind.

6. Suppose a person pushes a 1/2-centimeter-long thumbtack into a bulletin board and the force (in dynes) exerted when the depth of the thumbtack in the bulletin board is x centimeters is given by

$$F(x) = 10{,}000(1 + 2x)^2 \quad \text{for } 0 \leq x \leq \frac{1}{2}$$

Find the work W done in pushing the thumbtack all the way into the board.

7. A bottle of wine has a cork 5 centimeters long. A person uncorking the bottle exerts a force to overcome the force of friction between the cork and the bottle. Suppose the applied force in dynes is given by

$$F(x) = 2 \times 10^6 (5 - x) \quad \text{for } 0 \leq x \leq 5$$

where x represents the length in centimeters of the cork extending from the bottle. Determine the work W done in removing the cork.

8. When a certain spring is expanded 10 centimeters from its natural position and held fixed, the force necessary to hold it is 4×10^6 dynes. Find the work W required to stretch the spring an additional 10 centimeters.

9. Find the work W required to stretch the spring described in Exercise 8 from 20 to 30 centimeters beyond its natural length.

10. If 6×10^7 ergs of work are required to compress a spring 4 centimeters from its natural length, find the work W necessary to compress the spring an additional 4 centimeters. (*Hint:* Hooke's Law is also valid for compressing springs.)

11. If 6×10^7 ergs of work are required to compress a spring from its natural length of 10 centimeters to a length of 5 centimeters, find the work W necessary to stretch the spring from its natural length to a length of 12 centimeters.

12. Find the work W necessary to pump all the water out of the swimming pool in Example 3.

13. Find the work W required to pump the water out of the swimming pool in Figure 6.44. Assume that the pool is initially full, and that all the water is pumped to a level 1 foot above the top of the pool.

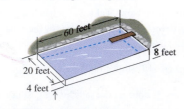

FIGURE 6.44   The swimming pool for Exercise 13.

14. A tank has the shape of the surface generated by revolving the parabolic segment $y = \frac{1}{2}x^2$ for $0 \leq x \leq 4$

about the y axis. If the tank is full of a fluid weighing 80 pounds per cubic foot, find the work W required to pump the contents of the tank to a level 4 feet above the top of the tank. (*Hint:* Integrate along the y axis.)

15. A tank in the shape of an inverted cone 12 feet tall and 3 feet in radius is full of water. Calculate the work W required to pump all the water over the edge of the tank.

16. Suppose that in Exercise 15 just half the water is pumped out the top edge of the tank and the remaining water is pumped to a level 3 feet above the top of the tank. Compute the total work W done.

17. Suppose a large gasoline tank has the shape of a half-cylinder 8 feet in diameter and 10 feet long (Figure 6.45). If the tank is full, find the work W necessary to pump all the gasoline to the top of the tank. Assume the gasoline weighs 42 pounds per cubic foot.

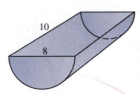

FIGURE 6.45   Figure for Exercise 17.

18. In each of the following, find the work W done by the force F graphed in the figure.

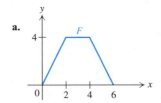

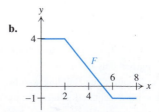

19. Suppose a spring is extended 6 centimeters from its natural length and then compressed to its natural length. How much work is done?

20. A car has run out of gas, and is being pushed on a level road toward a gas station. The force used as the car proceeds toward the station 200 meters away is described

in Figure 6.46. Approximate the work W done in moving the car to the station by estimating the area under the graph of F in the figure

a. by counting appropriate rectangles.

b. by calculating the left sum with $n = 10$.

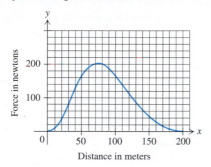

FIGURE 6.46   Graph for Exercise 20.

21. Suppose an object moves a distance of 20 centimeters along the x axis, and that the force (in dynes) acting on the object is measured at intervals of 2 centimeters, with the following results (in order):

1.21,   2.90,   3.01,   3.52,   3.41,   3.19,   2.78,   2.76, 2.83,   2.90,   2.84

By calculating the right sum, approximate the amount of work W done by the force.

22. A stationary proton located at the origin of the x axis exerts an attractive force on an electron located at a point x on the negative x axis. The force is given by

$$F(x) = \frac{a}{x^2}$$

where $a = 2.3 \times 10^{-28}$, x is measured in meters, and $F(x)$ is measured in newtons. Determine the work W done on the electron by this force when the electron moves from $x = -10^{-9}$ to $x = -10^{-10}$.

23. A ball weighing 0.2 pound is thrown vertically upward. Its height after t seconds is given by

$$h(t) = 6 + 8t - 16t^2$$

until the ball strikes the ground again.

a. Find the work W done on the ball by gravity while the ball descends from its maximum height to the ground.

b. Find the work W done on the ball on its descent if it is caught 6 feet above the ground.

24. A rocket lifts off the pad at Cape Canaveral. According to Newton's Law of Gravitation, the force of gravity on the rocket is given by

$$F(x) = \frac{-GMm}{x^2}$$

where M is the mass of the earth, m is the mass of the rocket, G is a universal constant, and x is the distance (in miles) between the rocket and the center of the earth. Take the radius of the earth to be 4000 miles, so that $x \geq 4000$ miles.

a. Find the work W done against gravity when the rocket rises 1000 miles. Express your answer in terms of G, M, and m.

b. Find the work W done against gravity when the rocket rises b miles.

c. Find the limit of the work found in (b) as b approaches ∞, and determine whether it is possible, with a finite amount of work, to send the rocket arbitrarily far away.

25. A bucket of cement weighing 200 pounds is hoisted by means of a windlass from the ground to the tenth story of an office building, 80 feet above the ground.

a. If the weight of the rope used is negligible, find the work W required to make the lift.

b. Assume that a chain weighing 1 pound per foot is used in (a), instead of the lightweight rope. Find the work W required to make the lift. (*Hint:* As the bucket is raised, the length of chain that must be lifted decreases.)

26. A container is lifted vertically at the rate of 2 feet per second. Water is leaking out of the container at the rate of $\frac{1}{2}$ pound per second. If the initial weight of water and container is 20 pounds, find the work W done in raising the container 10 feet. (*Hint:* For $0 \leq x \leq 10$, determine the total weight of water and container when the container has been raised x feet.)

27. A bucket containing water is raised vertically at the rate of 2 feet per second. Water is leaking out of the container at the rate of $\frac{1}{2}$ pound per second. If the bucket weighs 1 pound and initially contains 20 pounds of water, determine the amount of work W required to raise the bucket until it is empty.

*28. A building demolisher consists of a 2000-pound ball attached to a crane by a 100-foot chain weighing 3 pounds per foot. At night the chain is wound up and the ball is secured to a point 100 feet high. Find the work W done by gravity on the ball and the chain when the ball is lowered from its nighttime position to its daytime position at ground level.

29. The Great Pyramid of Cheops had square cross sections and was (approximately) 482 feet high and 754 feet square at the base. If the rock used to build the pyramid weighed 150 pounds per cubic foot, find the work W required to lift the rock into place as the pyramid was built.

*30. A cylindrical tank has a radius of 3 feet and height of 10 feet. Water is to be pumped from a lake into the base of the tank, which is 20 feet above the lake. Assuming that the lake is so large that its water level does not change during the pumping, write an integral for the work W required to fill the tank.

*31. A thin steel plate in the shape of an isosceles triangle with base 6 feet and height 4 feet is lying flat on the ground (Figure 6.47). Suppose the weight (in pounds) of any part of the plate equals its area (in square feet). How much work is required in order to raise the plate to a vertical position, assuming the base remains in contact with the ground? (*Hint:* Let the x axis be perpendicular to the base of the triangle, with the origin on the base. Consider any partition P of $[0, 4]$, and approximate the work by a Riemann sum.)

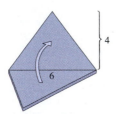

FIGURE 6.47   Figure for Exercise 31.

Project

1. A 5-pound bucket initially containing 20 pounds of water is hoisted by chain from the ground to the top of a bungee jumping tower, 120 feet above ground. Suppose that the basket is leaking water at a steady rate of 0.2 pounds per second, that the bucket is raised at the rate of 3 feet per second, and that the chain weighs 4 ounces per foot of length.

a. Find the number of seconds it takes for the bottom of the bucket to rise x feet from the ground (with $x \le 120$).

b. How much water has leaked from the bucket by the time that the bucket has risen x feet?

c. Find the work W required to raise the bucket to the top of the tower.

d. Suppose that instead of leaking 0.2 pounds per second the bucket leaks at a steady rate of 0.6 pounds per second until there is no more water in the bucket. Determine the work W_0 required to raise the bucket from the ground to the top of the tower. Is your answer different from your answer to part (c)? Explain why, or why not.

e. In addition to the hypotheses (including 0.6 pound per second leak), suppose that the chain is tapered in a linear fashion, so it weighs 4 ounces per foot at the bottom and 2 ounces per foot at the top. Find the work W_1 required to raise the bucket from the ground to the top of the tower.

6.5   MOMENTS AND CENTER OF GRAVITY

Children playing on a seesaw quickly learn that a heavier child has more effect on the rotation of the seesaw than does a lighter child and that a lighter child can balance a heavier one by moving farther away from the axis of rotation (Figure 6.48). In this section we define a quantity called "moment," which measures the tendency of a mass to produce rotation. We will use moments to define a point called the "center of gravity" of a set of points in the plane.

Moments of Point Masses

Let us first consider the idealized situation in which an object of positive mass m is concentrated at a point (x, y) in the plane. Such an object is called a **point mass.** The **moment** of the point mass **about the y axis** is defined to be mx; we may think of mx as a measure of the tendency of the point mass to rotate about the y axis

(Figure 6.49). The larger x or m is, the larger the magnitude of the moment. Thus our definition of moment is consistent with the observation that it is easier to rotate a seesaw about its axis the heavier or farther from the axis one is.

FIGURE 6.48

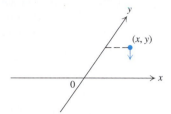

FIGURE 6.49   Moment = mx.

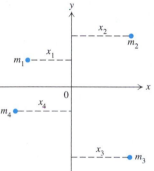

FIGURE 6.50

FIGURE 6.51

Next, suppose there are several point masses whose masses are $m_1, m_2, \ldots, m_n$, located at the respective points $(x_1, y_1), (x_2, y_2), \ldots, (x_n, y_n)$ in the plane (Figure 6.50). (In the event that these point masses lie on a line, we could associate them with n small children sitting on a seesaw; see Figure 6.51). We define the **moment** M_y of the collection of point masses **about the y axis** to be the sum of the moments of the individual point masses about the y axis:

$$M_y = m_1 x_1 + m_2 x_2 + \cdots + m_n x_n \tag{1}$$

If we think of M_y as a measure of the tendency of the collection of masses to produce a rotation about the y axis, then there will be *no* tendency for rotation if $M_y = 0$. In this case the collection of point masses is said to be in **equilibrium** with respect to the y axis. When several children are placed on a seesaw so that equilibrium exists, they can easily rotate the seesaw by pushing against the ground with their feet.

Analogously, we can define the **moment** M_x of the point masses $m_1, m_2, \ldots, m_n$ **about the x axis** by setting

$$M_x = m_1 y_1 + m_2 y_2 + \cdots + m_n y_n \tag{2}$$

Then the point masses are in equilibrium with respect to rotation about the x axis if $M_x = 0$.

Now let $m = m_1 + m_2 + \cdots + m_n$ be the combined mass of the point masses just considered, and let us seek a point $(\bar{x}, \bar{y})$ with the property that if we place a point mass with mass m at $(\bar{x}, \bar{y})$, then its moments about the x and y axes will be M_x and M_y, respectively. The moment of the single point mass about the y axis is $m\bar{x}$, and its moment about the x axis is $m\bar{y}$; hence by (1) and (2) we need $\bar{x}$ and $\bar{y}$ such that

$$m\bar{x} = M_y = m_1 x_1 + m_2 x_2 + \cdots + m_n x_n$$

and

$$m\bar{y} = M_x = m_1 y_1 + m_2 y_2 + \cdots + m_n y_n$$

These conditions on x and y tell us that

$$\bar{x} = \frac{m_1 x_1 + m_2 x_2 + \cdots + m_n x_n}{m} = \frac{M_y}{m}$$

$$\bar{y} = \frac{m_1 y_1 + m_2 y_2 + \cdots + m_n y_n}{m} = \frac{M_x}{m} \tag{3}$$

The point $(\bar{x}, \bar{y})$ is called the **center of gravity, center of mass,** or **centroid,** of the given collection of point masses.

EXAMPLE 1   Find the two moments and the center of gravity of objects with masses 2, 4, 5, and 7 located at the points (1, 2), (–3, 1), (–1, –2), and (0, 3), respectively.

Solution   By (1) and (2), the moments are given by

$$M_y = 2(1) + 4(-3) + 5(-1) + 7(0) = -15$$

and

$$M_x = 2(2) + 4(1) + 5(-2) + 7(3) = 19$$

Since $m = 2 + 4 + 5 + 7 = 18$, it follows from (3) that

$$\bar{x} = \frac{M_y}{m} = \frac{-15}{18} = \frac{-5}{6} \quad \text{and} \quad \bar{y} = \frac{M_x}{m} = \frac{19}{18}$$

Therefore the center of gravity of the collection is the point $(-\frac{5}{6}, \frac{19}{18})$ (Figure 6.52).   ❑

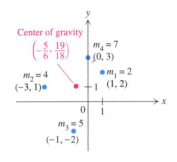

FIGURE 6.52

We mention an interesting feature of the center of mass. Suppose a fireworks rocket is shot into the air and suddenly explodes into numerous fragments. Before the explosion the firework is essentially a point mass and follows an approximately parabolic path. Since the explosion is internal to the firework and independent of the gravitational force, the motion of the center of mass is unaffected by the explosion and thus continues along the same (essentially) parabolic path.

Moments of Plane Regions about the *y* Axis

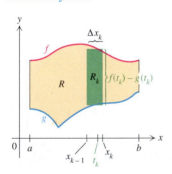

FIGURE 6.53

We assume from now on that mass is distributed uniformly throughout a plane region R instead of being concentrated at several points. This might be the case if R represented a thin metal plate, or **lamina.** We assume that R is the region between the graphs of two continuous functions f and g on an interval $[a, b]$, where

$$g(x) \le f(x) \quad \text{for } a \le x \le b$$

(Figure 6.53). We also assume that the mass of any subregion of R is equal to the subregion's area.

Let $P = \{x_0, x_1, \ldots, x_n\}$ be any partition of $[a, b]$. For k between 1 and n, let t_k be an arbitrary number in the subinterval $[x_{k-1}, x_k]$, and let R_k be the portion of R lying between the lines $x = x_{k-1}$ and $x = x_k$. If Δx_k is small, the area ΔA_k of R_k is approximately $[f(t_k) - g(t_k)] \Delta x_k$ (Figure 6.53). Moreover, the moment ΔM_k of R_k about the y axis should be approximately equal to the moment that would result if the entire mass of R_k were concentrated on the line $x = t_k$, that is,

$$\Delta M_k \approx (\text{distance to } y \text{ axis}) \times (\text{area}) = t_k \Delta A_k = t_k[f(t_k) - g(t_k)] \Delta x_k$$

Therefore the moment M_y of the whole region R, which is the sum of the moments $\Delta M_1, \Delta M_2, \ldots, \Delta M_n$, should be approximately

$$\sum_{k=1}^{n} \overbrace{t_k}^{\substack{\text{distance} \\ \text{to } y \text{ axis}}} \overbrace{[f(t_k) - g(t_k)]}^{\text{height}} \overbrace{\Delta x_k}^{\text{width}}$$

which is a Riemann sum for $x(f - g)$ on $[a, b]$. Hence we are led to define the moment M_y of R about the y axis by

$$M_y = \lim_{\|P\| \to 0} \sum_{k=1}^{n} t_k[f(t_k) - g(t_k)]\,\Delta x_k = \int_a^b x[f(x) - g(x)]\,dx \qquad (4)$$

A noteworthy special case occurs when $g = 0$, for then (4) reduces to

$$M_y = \int_a^b xf(x)\,dx$$

The integral on the right-hand side of this equation makes sense regardless of geometric or physical considerations, and it is normally called the **moment of f** (in contrast to the moment of a region). Moments of functions play a central role in statistics and probability.

EXAMPLE 2   Find the moment about the y axis of the rectangle R bounded by the lines $x = a$, $x = b$, $y = 0$, and $y = c$.

Solution   Here we have $f(x) = c$ and $g(x) = 0$ (Figure 6.54). Therefore the height of the rectangle at any x in $[a, b]$ is c, so

$$M_y = \int_a^b x(c - 0)\,dx = \left.\frac{c}{2}x^2\right|_a^b = \frac{c(b^2 - a^2)}{2} = c(b - a)\left(\frac{a + b}{2}\right) \qquad \square$$

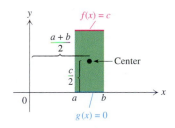

$f(x) = c$

$\dfrac{a+b}{2}$

Center

$\dfrac{c}{2}$

$g(x) = 0$

FIGURE 6.54

Notice that $c(b - a)$ is the area of R and $(a + b)/2$ is the distance from the center of R to the y axis. Notice also that this is the same moment that would result from having the entire mass concentrated at a point on the line $x = (a + b)/2$, midway between the lines $x = a$ and $x = b$.

Moments of Plane Regions about the x Axis

We now turn to the moment of a region about the x axis. First, consider the region R between the graph of a nonnegative continuous function f and the x axis on $[a, b]$. Assume temporarily that f is constant, say

$$f(x) = c \quad \text{for } a \leq x \leq b$$

Then R is a rectangle (see Figure 6.54). We know by analogy with Example 2 that the moment M_x of R about the x axis should be defined by

$$M_x = \text{area of } R \times \text{distance from center of } R \text{ to } x \text{ axis}$$

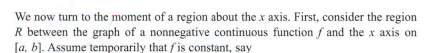

$$= [c(b - a)]\left(\frac{c}{2}\right) = \frac{1}{2}c^2(b - a) \qquad (5)$$

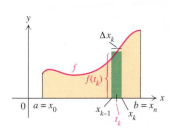

Δx_k

f

$f(t_k)$

$a = x_0$ x_{k-1} t_k x_k $b = x_n$

FIGURE 6.55

If f is not necessarily constant, then as before we let $P = \{x_0, x_1, \ldots, x_n\}$ be any partition of $[a, b]$. For each k between 1 and n, we let R_k be the portion of R lying between the lines $x = x_{k-1}$ and $x = x_k$. If Δx_k is small and t_k is an arbitrary number in the subinterval $[x_{k-1}, x_k]$, then the moment ΔM_k of R_k about the x axis should be approximately equal to the moment about the x axis of a rectangle with height $f(t_k)$ and base $[x_{k-1}, x_k]$ along the x axis (Figure 6.55). By (5) this means that

$$\Delta M_k \approx \frac{1}{2}[f(t_k)]^2(x_k - x_{k-1}) = \frac{1}{2}[f(t_k)]^2\,\Delta x_k$$

Since the moment M_x of R about the x axis should be the sum of ΔM_1, ΔM_2, . . . , ΔM_n, it follows that M_x should be approximately

$$\sum_{k=1}^{n} \frac{1}{2} [f(t_k)]^2 \, \Delta x_k$$

and this is a Riemann sum for $\frac{1}{2} f^2$ on $[a, b]$. Accordingly, we are led to define M_x by

$$M_x = \lim_{\|P\| \to 0} \sum_{k=1}^{n} \frac{1}{2} [f(t_k)]^2 \, \Delta x_k = \int_a^b \frac{1}{2} [f(x)]^2 \, dx$$

We could expand this argument to show that the moment about the x axis of the region between the graphs of two continuous functions f and g on an interval $[a, b]$ should be defined by

$$M_x = \int_a^b \frac{1}{2} \{[f(x)]^2 - [g(x)]^2\} \, dx$$

We incorporate our definitions of the moments of a region R about the two coordinate axes into a single definition that also includes the definition of the center of gravity of R.

DEFINITION 6.5

Let f and g be continuous on $[a, b]$, with

$$g(x) \le f(x) \quad \text{for } a \le x \le b$$

and let R be the region between the graphs of f and g on $[a, b]$. Then the **moment M_x of R about the x axis** is given by

$$M_x = \int_a^b \frac{1}{2} \{[f(x)]^2 - [g(x)]^2\} \, dx$$

and the **moment M_y of R about the y axis** is given by

$$M_y = \int_a^b x[f(x) - g(x)] \, dx$$

If R has positive area A, then the **center of gravity** (or **center of mass,** or **centroid**) of R is the point $(\bar{x}, \bar{y})$ defined by

$$\bar{x} = \frac{M_y}{A} \quad \text{and} \quad \bar{y} = \frac{M_x}{A}$$

EXAMPLE 3   Let R be the semicircular region bounded by the y axis and the graphs of f and g, where

$$f(x) = \sqrt{r^2 - x^2} \quad \text{and} \quad g(x) = -\sqrt{r^2 - x^2} \quad \text{for } 0 \le x \le r$$

(Figure 6.56). Find the moments and the center of gravity of R.

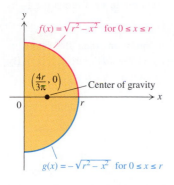

$f(x) = \sqrt{r^2 - x^2}$ for $0 \leq x \leq r$

$\left(\dfrac{4r}{3\pi}, 0\right)$ — Center of gravity

$g(x) = -\sqrt{r^2 - x^2}$ for $0 \leq x \leq r$

FIGURE 6.56

Solution   Definition 6.5 tells us directly that

$$M_x = \int_0^r \frac{1}{2}\left[(r^2 - x^2) - (r^2 - x^2)\right] dx = \frac{1}{2}\int_0^r 0\ dx = 0$$

and

$$M_y = \int_0^r x\left[\sqrt{r^2 - x^2} - (-\sqrt{r^2 - x^2})\right] dx = 2\int_0^r x\sqrt{r^2 - x^2}\ dx$$

$$= \frac{-2}{3}(r^2 - x^2)^{3/2}\Big|_0^r = \frac{2}{3}r^3$$

Since the area A of R is $\frac{1}{2}\pi r^2$, we have

$$\overline{x} = \frac{M_y}{A} = \frac{\frac{2}{3}r^3}{\frac{1}{2}\pi r^2} = \frac{4r}{3\pi}$$

and

$$\overline{y} = \frac{M_x}{A} = \frac{0}{\frac{1}{2}\pi r^2} = 0$$

Consequently the center of gravity $(4r/3\pi, 0)$ lies on the x axis, approximately $0.42r$ from the origin (Figure 6.56).   ❑

We observe that if a plate occupying the region R in Example 3 is placed on a pin located at the center of gravity of R, then in theory the plate would be in balance, that is, the plate would not tilt in any direction.

Notice that in Example 3, R was symmetric with respect to the x axis, and the center of gravity turned out to lie on the x axis. More generally, it is possible to show the following:

1.   If R is symmetric with respect to the line $x = c$, then the center of gravity of R lies on the line $x = c$; that is, $\overline{x} = c$ (Figure 6.57(a)).

2.   If R is symmetric with respect to the line $y = d$, then the center of gravity of R lies on the line $y = d$; that is, $\overline{y} = d$ (Figure 6.57(b)).

Therefore symmetry can reduce the effort required to locate the center of gravity of a region.

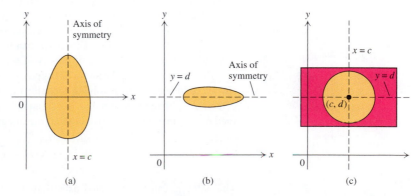

FIGURE 6.57   (a) Center of gravity lies on the axis of symmetry. (b) Center of gravity lies on the axis of symmetry. (c) Center of gravity of circle or rectangle lies at the center.

EXAMPLE 4   Show that the center of gravity of a circle or a rectangle is its center.

Solution   If (c, d) is the center of the region in question, then the region is obviously symmetric with respect to the lines $x = c$ and $y = d$ (Figure 6.57(c)). By the observation above, the center of gravity of the region is (c, d).   ❑

The Theorem of Pappus and Guldin

Pappus of Alexandria (ca. 300 A.D.)

Pappus's great work was his *Mathematical Collection,* which was devoted to geometry and its history. In it he cited no fewer than 30 mathematicians from earlier ages. The preface to Book 3 is addressed to a female teacher of geometry by the name of Pandrosian. This suggests that at that time women played a significant role in Alexandrian mathematics.

Assume that f and g are continuous on an interval $[a, b]$, with $a \geq 0$ and

$$g(x) \leq f(x) \quad \text{for } a \leq x \leq b$$

Let R be the region between the graphs of f and g on $[a, b]$, with area A. If we compare the formula in Definition 6.5 for the moment M_y about the y axis with formula (7) in Section 6.1 for the volume V of the solid obtained by revolving R about the y axis, we see that

$$V = 2\pi M_y$$

But since $M_y = \overline{x} A$ we may also rewrite this volume as

$$V = 2\pi \overline{x} A$$

Notice that $\overline{x}$ is the distance from the center of gravity of R to the y axis, and the y axis is by assumption the axis of revolution for R. A corresponding result holds when R is revolved about the x axis. These results were first anticipated by the mathematician Pappus of Alexandria about 300 A.D., but they are often credited as well to the Swiss mathematician Paul Guldin, who lived some 1300 years later.

THEOREM 6.6
Theorem of Pappus and Guldin

Let R be a plane region lying completely to one side of a line l. Then the volume V of the solid region generated by revolving R about l is given by

$$V = 2\pi b A$$

where A is the area of R and b is the distance from the center of gravity of R to the line l.

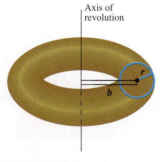

Axis of revolution

FIGURE 6.58   A torus.

Theorem 6.6 implies that the volume of the solid generated by revolving a plane region about a line outside the region is equal to the product of the area of the plane region and the distance the center of gravity travels as the region is revolved once about the line. This result can be used to compute the volume of a solid when the area and center of gravity of the plane region are known.

Before we apply the Theorem of Pappus and Guldin, let us consider a doughnut-shaped solid obtained by revolving about the y axis the circle $(x - b)^2 + y^2 = r^2$, whose center is $(b, 0)$ and whose radius is r (Figure 6.58), with $b \geq r$. If we used the shell method, (7) in Section 6.1 would yield the integral

$$V = \int_{b-r}^{b+r} 4\pi x \sqrt{r^2 - (x - b)^2} \, dx$$

This integral can be evaluated exactly, but only with the help of an integration method to appear in Section 8.3. By contrast, the Theorem of Pappus and Guldin yields a simple solution to the problem.

EXAMPLE 5   Let $b \geq r > 0$. Consider the solid obtained by revolving a circle of radius r about a line lying a distance b from the center of the circle (Figure 6.58). Show that the volume V of the solid is given by

$$V = 2\pi^2 r^2 b$$

Solution   Since the center of gravity of the circle is its center $(b, 0)$, the Theorem of Pappus and Guldin implies that

$$V = 2\pi b(\pi r^2) = 2\pi^2 r^2 b \quad \square$$

We note that mathematicians call a solid of the variety discussed in Example 5 a **torus**. It plays a leading role in certain geometry topics.

EXERCISES 6.5

1. Two children weighing 15 and 20 kilograms are sitting on opposite sides of a seesaw, both 2 meters from the axis of revolution. Where on the seesaw should a 10-kilogram toddler sit in order to achieve equilibrium?

2. Suppose the 10-kilogram toddler in Exercise 1 sits down on the seesaw first, 1 meter to the left of the axis of revolution. If the two larger children wish to sit on opposite sides of the axis, equidistant from it, find the location of these children that will ensure equilibrium.

3. Determine whether the center of mass of the system consisting of the earth and moon lies inside or outside the earth. Assume that the radius of the earth is 6.37×10^3 km, the mass of the earth is 5.98×10^{24} kg, the mass of the moon is 7.35×10^{22} kg, and the distance between the centers of the earth and the moon is 3.84×10^5 km. When computing the center of mass, consider the earth and the moon as point masses.

4. A mobile consists of three weights attached to a square piece of cardboard of negligible weight, to be suspended from the ceiling by a string. Assume that the weights have mass 50, 30, and 20 grams, respectively. Suppose that a coordinate system with origin at the center of the square has been set up and that the points at which the weights are attached to the square are $(3, -1)$, $(4, 2)$, and $(-1, 1)$, respectively in this coordinate system. Determine the center of gravity of the mobile; that is, find the point on the cardboard at which the string must be attached for the mobile to be balanced.

In Exercises 5–11 calculate the center of gravity of the region R between the graphs of f and g on the given interval.

5. $f(x) = x$, $g(x) = -2$; $[0, 2]$
6. $f(x) = 2x - 1$, $g(x) = x - 2$; $[2, 5]$
7. $f(x) = 2 - x$, $g(x) = -(2 - x)$; $[0, 2]$
8. $f(x) = 3x$, $g(x) = x^2$; $[0, 1]$
9. $f(x) = (x + 1)^2$, $g(x) = (x - 1)^2$; $[1, 2]$
10. $f(x) = x + 1$, $g(x) = \sqrt{x + 1}$; $[0, 3]$
11. $f(x) = \dfrac{1}{\sqrt{x - 1}}$, $g(x) = \dfrac{1}{\sqrt{x + 1}}$; $[2, 5]$
12. Find the center of gravity of the region between the graphs of f and g, where $f(x) = 11 - x^2$ and $g(x) = x^2 + 3$.
13. Find the center of gravity of the region between the graphs of f and g, where $f(x) = 2 - x^2$ and $g(x) = |x|$.
14. Find the center of gravity of the region between the graphs of $y^2 = 2x - 5$ and $y = x - 4$.
15. Find the center of gravity of the region between the graphs of $y = x + 2$, $y = -3x + 6$, and $y = (2 - x)/3$.

In Exercises 16–20 use the symmetry of the region R to determine the center of gravity of R.

16. R is bounded by the parallelogram with vertices at $(0, 0)$, $(0, 2)$, $(1, 1)$, and $(1, 3)$.
17. R is bounded by the hexagon with vertices at $(0, 0)$, $(0, 6)$, $(1, 1)$, $(1, 5)$, $(-1, 1)$, and $(-1, 5)$.
18. R is bounded by the ellipse

$$\frac{(x - 3)^2}{a^2} + \frac{(y + 1)^2}{b^2} = 1$$

19. R is bounded by the astroid $x^{2/3} + y^{2/3} = 1$.
20. R is bounded by the curve $y = 1/x$ and the lines $y = 1 + x$ and $y = -1 + x$.

Exercises 21–24 involve some regions that are important in engineering and architecture. Find the center of gravity of each region.

21. The region is the triangle shown in Figure 6.59(a). (You should recognize the center of gravity as a point familiar from geometry.)

22. The region is the **parabolic sector** shown in Figure 6.59(b).

23. The region is the **parabolic spandrel** shown in Figure 6.59(c).

24. The region is the fingerlike shape shown in Figure 6.59(d).

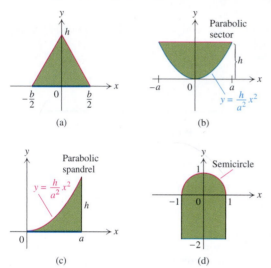

(a) (b)

(c) (d)

FIGURE 6.59 Figures for Exercises 21–24.

25. Use the Theorem of Pappus and Guldin to obtain an alternate derivation of the center of gravity of the semicircular region R in Example 3.

26. Use the Theorem of Pappus and Guldin and the formula for the volume of a cone to find the center of gravity of a right triangle with legs of length a and c.

27. Let a square $ABCD$ have side length 2 and diagonal AC. Use the Theorem of Pappus and Guldin to find the volume V of the solid obtained by revolving the square about the line that passes through B and is parallel to AC.

28. An equilateral trapezoid has sides with length 8, 4, $2\sqrt{2}$, and $2\sqrt{2}$. Use the Theorem of Pappus and Guldin to find the center of gravity of the trapezoid.

29. Suppose f is an even function defined on $[-a, a]$, and let R be the region between the graph of f and the x axis on $[-a, a]$. Show that the moment of R about the y axis is 0 and hence that the center of gravity of R lies on the y axis.

30. Suppose f is an odd function defined on $[-a, a]$ and is nonnegative on $[0, a]$. Let R be the region between the

graph of f and the x axis on $[-a, a]$. Show that the center of gravity of R is the origin.

31. Let R be a plane region with center of gravity $P = (\bar{x}, \bar{y})$. Suppose R is contained in a circle with center at P. Show that the volume of the solid that results from revolving R about a line tangent to the circle is the same for any such line.

32. Sometimes a region R can be divided into regions R_1, R_2, . . . , R_n, each of whose center of gravity is easily computed. Then the center of gravity $(\bar{x}, \bar{y})$ of R may be determined from the centers of gravity $(\bar{x}_1, \bar{y}_1)$, $(\bar{x}_2, \bar{y}_2)$, , . . . , $(\bar{x}_n, \bar{y}_n)$ of R_1, R_2, . . . , R_n by using the fact that the moment of R about either the x axis or the y axis is the sum of the moments of R_1, R_2, . . . , R_n about that axis. This yields

$$\bar{x} \times \text{area of } R = (\bar{x}_1 \times \text{area of } R_1) + (\bar{x}_2 \times \text{area of } R_2)$$
$$+ \cdots + (\bar{x}_n \times \text{area of } R_n)$$

and

$$\bar{y} \times \text{area of } R = (\bar{y}_1 \times \text{area of } R_1) + (\bar{y}_2 \times \text{area of } R_2)$$
$$+ \cdots + (\bar{y}_n \times \text{area of } R_n)$$

Using these equations, determine the centers of gravity of the regions shown in

a. Figure 6.60(a). (*Hint:* The center of gravity lies outside the region.)

b. Figure 6.60(b).

c. Figure 6.60(c).

d. Figure 6.60(d).

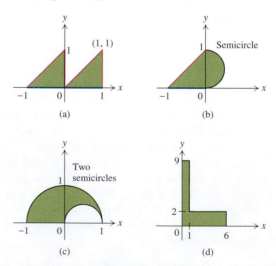

(a) (b)

(c) (d)

FIGURE 6.60 Figures for Exercise 32.

Project

1. Assume that c is a positive number. Let $f_c(x) = x^2 + c$ for $-1 \leq x \leq 1$, and let R_c be the region between the graph of f_c and the x axis on $[-1, 1]$.

   a. Draw the graph of R_c for a very small positive value of c, and indicate where you think that the center of gravity of R_c should be.

   b. Draw the graph of R_2, and indicate where you think that the center of gravity of R_2 should be.

   c. Find the center of gravity of R_c for an arbitrary positive number c.

   d. Find the number c^* such that the center of gravity of R_{c^*} lies on the boundary of R_{c^*}, and show that if $c > c^*$, then the center of gravity of R_c lies inside R_c.

6.6   HYDROSTATIC FORCE

We devote this section to hydrostatic force, the force exerted by stationary water on a surface such as a plate, a wall, or a dam.

Hydrostatic Pressure

A diver under water experiences pressure due to the weight of the water above. Moreover, the deeper the diver descends, the greater the pressure is. Let us analyze this phenomenon and define it formally.

Consider a horizontal plate A square feet in area at a depth of x feet below the surface of the water (Figure 6.61). The water directly above the plate exerts a force F equal to its weight on the plate. Since the volume of the water directly above the plate is xA cubic feet and water weighs 62.5 pounds per cubic foot, the force F is given by the formula

$$F = \underbrace{62.5}_{\substack{\text{weight per} \\ \text{cubic foot}}} \quad \underbrace{xA}_{\substack{\text{volume in} \\ \text{cubic feet}}}$$

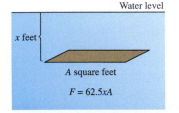

Water level

x feet

A square feet

$F = 62.5xA$

FIGURE 6.61

Dividing by A, we find that

$$\frac{F}{A} = 62.5x \tag{1}$$

The quantity F/A is called the **hydrostatic pressure** on the plate and is measured in pounds per square foot.

The hydrostatic pressure on a horizontal plate depends only on the depth of the plate, not on the area of the plate. As a result, we talk about the hydrostatic pressure at a point under water, meaning the hydrostatic pressure on any horizontal plate containing that point. Hydrostatic pressure is exerted not only downward but in all directions. In fact, Pascal's Principle states that the hydrostatic pressure at a point in water is exerted equally in all directions (Figure 6.62). If the point is x feet below the surface of the water, then the magnitude of the pressure at that point is $62.5x$. Therefore, the deeper the plate is submerged, the greater the hydrostatic pressure on it is—a fact well known to divers.

A scuba diver in the West Indies.
(*E. Manewal/Superstock*)

Water level

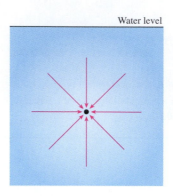

FIGURE 6.62   Pressure exerted equally from all directions.

We are now ready to analyze the hydrostatic force on a vertical plate or surface that is either totally or partially submerged in water.

Hydrostatic Force on a Vertical Plate

The Glen Canyon Dam.
(*Superstock*)

Archimedes

One of the most famous stories about Archimedes concerns his discovery that a body in a fluid is buoyed up by a force equal to the weight of the displaced fluid. To determine whether a "gold" crown made for King Hieron was pure gold, Archimedes placed the crown and an equal weight of gold on a balance system and immersed them into the water of the public baths. The crown rose and the gold fell, proving that the crown was not pure gold. Archimedes was so excited by this discovery that he jumped out of the baths, forgot to clothe himself, and ran home through the streets of Syracuse shouting "Eureka, Eureka!"

The analysis we make of the hydrostatic force on a vertical plate applies equally well to a vertical wall or dam in water. You may think of any of these in the following discussion.

First we place the x axis vertically, with the positive direction upward. Let $x = c$ represent the water level, and let the portion of the plate submerged in water extend from a to b on the x axis (Figure 6.63(a) and (b)). Notice that in Figure 6.63(a) the top of the plate is higher than the water level, so that $c = b$, whereas in Figure 6.63(b) the top of the plate is lower than the water level, so that $c > b$. Next we let $w(x)$ be the width of the plate at x and assume that w is continuous on $[a, b]$.

Let $P = \{x_0, x_1, \ldots, x_n\}$ be any partition of $[a, b]$. For each k between 1 and n, let t_k be an arbitrary number in the subinterval $[x_{k-1}, x_k]$. If Δx_k is small, the area of the portion S_k of the plate lying between x_{k-1} and x_k on the x axis is approximately $w(t_k) \Delta x_k$ (Figure 6.63(b)). In addition, from (1) we know that the pressure at any point on S_k is approximately $(62.5)(c - t_k)$. Therefore, the force ΔF_k on S_k is approximately the product of the pressure $(62.5)(c - t_k)$ on S_k and its area $w(t_k) \Delta x_k$. Thus

$$\Delta F_k \approx (\text{pressure}) \times (\text{area}) \approx (62.5)(c - t_k)w(t_k) \Delta x_k$$

Since the force F on the plate is the sum of the forces $\Delta F_1, \Delta F_2, \ldots, \Delta F_n$, it follows that F should be approximately

$$\sum_{k=1}^{n} \overbrace{(62.5)(c - t_k)}^{\text{pressure}} \overbrace{w(t_k) \Delta x_k}^{\text{area}}$$

and this is a Riemann sum for the function $(62.5)(c - x)w$ on $[a, b]$. Accordingly, the force F should be defined by

$$F = \lim_{\|P\| \to 0} \sum_{k=1}^{n} (62.5)(c - t_k)w(t_k) \Delta x_k = \int_{a}^{b} (62.5)(c - x)w(x)\, dx$$

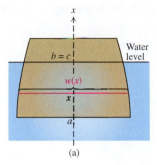
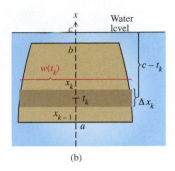

FIGURE 6.63

DEFINITION 6.7   Assume that a vertical plate is submerged or partially submerged in water with the water level located at $x = c$. Let the submerged portion of the plate extend from a to b on the x axis. Let $w(x)$ be the width of the plate for $a \le x \le b$, and assume that w is continuous on $[a, b]$. Then the **hydrostatic force** F on the plate due to the water is given by

$$F = \int_a^b 62.5(c - x)w(x) \, dx \qquad (2)$$

If water is replaced by some other fluid, such as gasoline or alcohol, then the number 62.5 must be replaced by the weight of 1 cubic foot of that fluid.

EXAMPLE 1   An isosceles triangular plate 3 feet tall and 1.5 feet wide is located on a vertical wall of a swimming pool, the bottom vertex 4 feet below water level (Figure 6.64). Find the hydrostatic force F on the plate.

Solution   For convenience we place the origin at the bottom vertex of the plate. Then the plate extends from $x = 0$ to $x = 3$, and the depth at a given x is $4 - x$ (Figure 6.64). Moreover, since the width of the triangle is $\frac{1}{2}$ the height, it follows from similar triangles that at any x the width $w(x)$ is given by

$$w(x) = \frac{1}{2}x$$

Consequently (2) implies that the force F is given by

$$F = \int_0^3 (62.5)(4 - x)\left(\frac{1}{2}x\right) dx = 31.25 \int_0^3 (4x - x^2) \, dx$$

$$= 31.25 \left(2x^2 - \frac{x^3}{3} \right)\Big|_0^3 = (31.25)(9) = 281.25 \text{ (pounds)} \qquad \square$$

Had the triangle been placed with the vertex at the top, 1 foot below water level, then we could place the origin again at the bottom of the plate (Figure 6.65). Once again the plate would extend from $x = 0$ to $x = 3$ and the depth at a given x

would be $4 - x$. But now at any x the width $w(x)$ would, by similar triangles, satisfy

$$\frac{w(x)}{1.5} = \frac{3 - x}{3} \quad \text{so that} \quad w(x) = \frac{3 - x}{2}$$

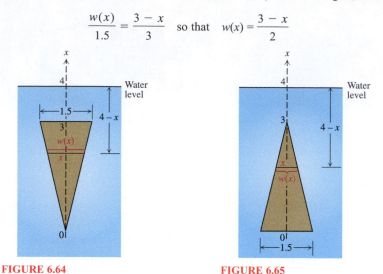

FIGURE 6.64   **FIGURE 6.65**

(Figure 6.65). Therefore

$$F = \int_0^3 (62.5)(4 - x)\left(\frac{3 - x}{2}\right) dx = 31.25 \int_0^3 (12 - 7x + x^2)\, dx$$

$$= 31.25\left(12x - \frac{7}{2}x^2 + \frac{x^3}{3}\right)\Bigg|_0^3 = (31.25)(13.5) = 421.875 \text{ (pounds)}$$

EXERCISES 6.6

1. Suppose the top edge of the plate in Example 1 were at water level. Compute the hydrostatic force F on the plate.

2. Suppose a rectangular dam is 1000 feet wide and 100 feet high. Compute the hydrostatic force F on the dam when the water level is
   a. 100 feet above the bottom.
   b. 50 feet above the bottom.

3. Suppose the end of a water trough has the shape of an equilateral triangle (pointed down) with side length 2 feet. Compute the hydrostatic force F on the end of the trough when the water level is 1 foot above the bottom. (*Hint:* Let the origin be at water level.)

4. A flat circular floodlight on a vertical wall of a swimming pool has a radius of 1 foot, and its highest point is 3 feet below the water level. Compute the hydrostatic force F on the light.

5. A pool surrounding a water fountain has flat vertical sides that contain lights having the shape of an equilateral triangle with sides of length 2 feet. The arrangement of the lights is shown in Figure 6.66. Assuming that the water level is 3 feet above the bottoms of the triangles, compute the hydrostatic force F on each type of light. (*Hint:* Let the origin be at water level.)

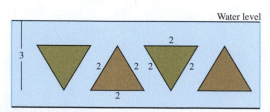

FIGURE 6.66   Figure for Exercise 5.

6. A cubical block of marble 1 foot on a side rests at the bottom of a swimming pool 6 feet deep. Determine the total hydrostatic force F on the five sides touching the water.

7. A cubical block of marble 1 foot on each side is submerged in water but does not touch bottom. Suppose two of its faces are parallel to the water surface. Show that the difference between the hydrostatic force exerted on the bottom face and the hydrostatic force on the top face is equal to 62.5 pounds. (This difference is called the **buoyant force.** Archimedes's Principle states that any body submerged in water is buoyed up by a force equal to the weight of the water it displaces.)

*8. a. Using (2), show that the hydrostatic force F equals $62.5hA$, where A is the area of the submerged portion of the plate and h is the depth of the water at the center of gravity of the submerged portion.

   b. Using the results of part (a) and Example 3 of Section 6.5, calculate the force on a vertical semicircular dam with radius 100 feet when the water level is at the top of the dam (Figure 6.67).

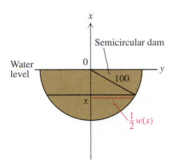

FIGURE 6.67   Figure for Exercise 8.

9. Suppose that in our analysis of hydrostatic pressure we replace the assumption that the plate is vertical by the assumption that it makes an angle θ with the vertical (Figure 6.68). Show that the hydrostatic force F on the portion between x_{k-1} and x_k would be

$$62.5(c - t_k)[(\sec \theta)w(t_k)\,\Delta x_k]$$

and that the hydrostatic force F would be given by

$$F = \int_a^b 62.5 \sec \theta\,(c - x)w(x)\,dx$$

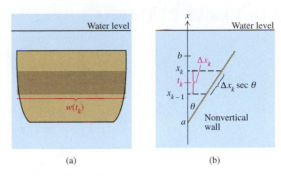

FIGURE 6.68   Figures for Exercise 9. (a) Front view of wall. (b) Side view of wall.

10. Suppose a rectangular earth-filled dam 1000 feet wide makes an angle of $\pi/3$ with the vertical. Using the result of Exercise 9, compute the hydrostatic force F on the dam when the water level is 50 feet above the bottom.

11. Consider the swimming pool in Figure 6.69. Assuming the pool is full of water, find the hydrostatic force on
    a. the smaller end of the pool.
    b. the larger end of the pool.
    c. each of the two sides of the pool.
    d. the bottom of the pool.

FIGURE 6.69   The swimming pool for Exercise 11.

12. Using the method employed to derive formula (2), show that the hydrostatic force F on the sides of a cylindrical tank filled with water and having a radius of 20 feet and a height of 30 feet is given by

$$F = \int_0^{30} (62.5)40\pi x\,dx$$

13. In an offshore search for oil in water 100 feet deep, drilling is done through a cylindrical pipe having an exterior diameter of 3 feet. Using a formula like the one obtained in Exercise 12, compute the hydrostatic force F on the exterior of the pipe.

6.7 PARAMETRIZED CURVES

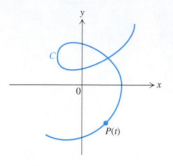

FIGURE 6.70 A parametrized curve.

Until now virtually every curve we have discussed has been the graph of a function. We complete this chapter with a discussion of more general planar curves called parametrized curves.

To begin, suppose a tiny bug travels on a piece of paper that represents a portion of the plane. At each time t, the bug occupies a point $P(t)$, and as t varies, the bug traces out a curve C (Figure 6.70). We can describe C by two functions f and g, defined on an interval I, that represent the coordinates of the points on C:

$$x = f(t) \quad \text{and} \quad y = g(t) \quad \text{for } t \text{ in } I \tag{1}$$

We say that C is **parametrized** by the equations in (1). The equations are called **parametric equations** for C, and t is called a **parameter.** As a result, the point $P(t)$ on C is given by

$$P(t) = (f(t), g(t)) \quad \text{for } t \text{ in } I$$

Frequently there is a geometrical or physical interpretation of the parameter such as an angle or time.

EXAMPLE 1 Sketch the curve C parametrized by

$$x = 1 - 2t \quad \text{and} \quad y = -3 + 4t \quad \text{for all } t$$

and indicate the direction $P(t)$ moves along C as t increases.

Solution We first find a single equation relating x and y that does not involve t. To that end, we solve for t in the first parametric equation to obtain $t = \frac{1}{2} - \frac{1}{2}x$. Next we substitute for t in the second parametric equation:

$$y = -3 + 4\left(\frac{1}{2} - \frac{1}{2}x\right)$$

$$y = -1 - 2x$$

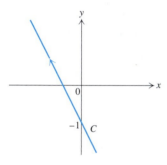

FIGURE 6.71

Thus every point (x, y) on C lies on the straight line $y = -1 - 2x$, which has slope -2 and y intercept -1 (Figure 6.71). Notice from the parametric equations that $P(0) = (1, -3)$ and $P(1) = (-1, 1)$. Therefore as t increases, $P(t)$ moves upward and to the left on C, as Figure 6.71 indicates. ❑

More generally, any set of parametric equations of the form

$$x = a + bt \quad \text{and} \quad y = c + dt \quad \text{with } b \neq 0 \text{ or } d \neq 0 \tag{2}$$

represents a straight line. In the sequel we will be interested in parametrizations of not only lines but also line segments.

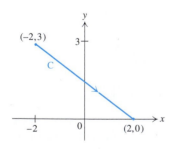

FIGURE 6.72

EXAMPLE 2 Find a parametrization for a line segment C starting at the point $(-2, 3)$ and ending at $(2, 0)$ (Figure 6.72).

Solution Suppose we let $t = 0$ at the starting point $(-2, 3)$, and let $t = 1$ at the ending point $(2, 0)$. Then a parametrization for C has the form

$$x = -2 + at \quad \text{and} \quad y = 3 + bt \quad \text{for} \quad 0 \le t \le 1$$

When $t = 0$, we have $(x, y) = (-2, 3)$. In order that $(x, y) = (2, 0)$ when $t = 1$, a and b must be chosen so that

$$2 = -2 + a(1) = -2 + a \quad \text{and} \quad 0 = 3 + b(1) = 3 + b$$

Consequently $a = 4$ and $b = -3$. It follows that

$$x = -2 + 4t \quad \text{and} \quad y = 3 - 3t \quad \text{for} \quad 0 \le t \le 1$$

serves as the desired parametrization.   ❏

We remark that there are infinitely many parametrizations for any given curve. Even for line segments like that in Example 2, there are infinitely many, among them

$$x = -3 + t \quad \text{and} \quad y = 3.75 - .75t \quad \text{for} \quad 1 \le t \le 5$$

for example. Nevertheless, some parametrizations for a given finite curve C are likely to be easier to use than others.

EXAMPLE 3   Let $r > 0$. Describe the curve C with parametric equations

$$x = r \cos t \quad \text{and} \quad y = r \sin t \quad \text{for } 0 \le t \le 2\pi$$

Solution   Let $P(t)$ be a point on C, so that $P(t) = (r \cos t, r \sin t)$. We will show that for $0 \le t \le 2\pi$, $P(t)$ is on the circle $x^2 + y^2 = r^2$. To that end, we square both sides of each of the given equations and add:

$$x^2 + y^2 = (r \cos t)^2 + (r \sin t)^2 = r^2(\cos^2 t + \sin^2 t) = r^2$$

Therefore the points satisfying the given equations also satisfy the equation $x^2 + y^2 = r^2$, and thus the points lie on the designated circle. As t increases, the point $P(t)$ traverses the circle, with $P(0) = (r, 0)$, $P(\pi/2) = (0, r)$, $P(\pi) = (-r, 0)$, $P(3\pi/2) = (0, -r)$, and finally, $P(2\pi) = (r, 0)$. It follows that C is the circle $x^2 + y^2 = r^2$, traversed exactly once and in the counterclockwise direction (Figure 6.73).   ❏

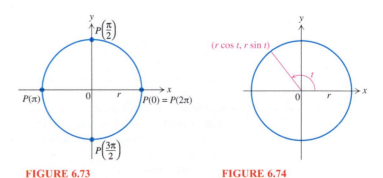

FIGURE 6.73   **FIGURE 6.74**

The parameter t in Example 3 may be interpreted as the angle from the positive x axis to the line segment joining the origin to the point (x, y) on the circle (Figure 6.74). If we imagine a bug starting at $(r, 0)$ and moving around the circle, then t

would represent the angle (in radians) through which the line segment has rotated about the center of the circle when the bug is at the point $(r \cos t, r \sin t)$.

The parametric equations

$$x = r \cos t \quad \text{and} \quad y = r \sin t \quad \text{for } 0 \leq t \leq 4\pi$$

represent the same circle as the one in Example 3, but this time the circle is traced out twice in the counterclockwise direction, once for $0 \leq t \leq 2\pi$ and the second time for $2\pi \leq t \leq 4\pi$.

Next, suppose an object traverses a circle of radius r, starting at time $t = 0$ on the positive x axis. Then the **angular velocity** of the object is the rate of change of the central angle θ with respect to time (Figure 6.75). Let us assume that the angular velocity is constant, and denote it by ω (the Greek letter omega). Then $\theta = \omega t$. It follows that for t in an appropriate time interval I, the object's path can be described parametrically by

$$x = r \cos \omega t \quad \text{and} \quad y = r \sin \omega t \quad \text{for } t \text{ in } I \tag{3}$$

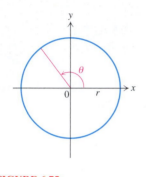

FIGURE 6.75

Since

$$\cos \omega(t + 2\pi/\omega) = \cos (\omega t + 2\pi) = \cos \omega t$$

and

$$\sin \omega(t + 2\pi/\omega) = \sin (\omega t + 2\pi) = \sin \omega t$$

a consequence of (3) is that the object takes $2\pi/\omega$ units of time to complete one revolution of the circle.

Perhaps the most famous curve described parametrically is the **cycloid**, which is traced out by a point on a circle of radius r as the circle rolls along a line (Figure 6.76). In Section 12.1 we will show that the cycloid is parametrized by

$$x = r(t - \sin t) \quad \text{and} \quad y = r(1 - \cos t) \tag{4}$$

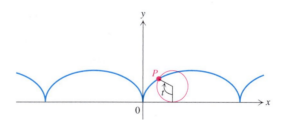

FIGURE 6.76   A cycloid is traced out by P as the circle rolls along the x axis.

where t represents the central angle through which the circle has rotated (see Figure 6.76 again). Although it is tempting to think that a point on a circle might trace out semicircles, this is not the case. Evidently the highest point on the cycloid occurs when y has the largest value, that is, when $\cos t = -1$. This happens when $t = \pi + 2n\pi$ for any integer n. Moreover, the cycloid has a cusp whenever $y = 0$, that is, when $t = n\pi$ for any integer n.

We also observe that if g is any function, then the curve C parametrized by

$$x = t \quad \text{and} \quad y = g(t)$$

is the graph of g. Thus the graph of any function can be parametrized.

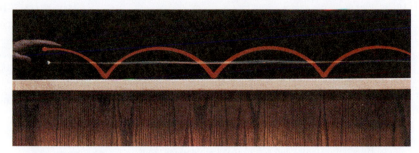

The path traced out by a point on the rim of a rolling wheel is a cycloid. For this photograph, taken with camera shutter held open, one light was attached to the rim of the wheel and another light to its center. (*Courtesy of Henry Leap and Jim Lehman*)

EXERCISES 6.7

In Exercises 1–12 first find an equation relating x and y, when possible. Then sketch the curve C whose parametric equations are given, and indicate the direction $P(t)$ moves as t increases.

1. $x = 2 \cos t$ and $y = 2 \sin t$ for $0 \le t \le \pi/2$
2. $x = 3 \sin t$ and $y = 3 \cos t$ for $-\pi/2 \le t \le \pi/2$
3. $x = 2 - \cos t$ and $y = -1 - \sin t$ for $0 \le t \le 2\pi$
4. $x = -1 + \frac{3}{2} \sin t$ and $y = \frac{1}{2} - \frac{3}{2} \cos t$ for $-\pi \le t \le 3\pi$
5. $x = -2 + 3t$ and $y = 2 - 3t$ for all t
6. $x = 5 - t$ and $y = -4$ for $t \ge 0$
7. $x = 3$ and $y = -1 - t$ for $0 \le t \le 1$
8. $x = t$ and $y = \sqrt{1 - t^2}$ for $-1 \le t \le 1$
9. $x = t^3$ and $y = t^2$ for all t
10. $x = e^t$ and $y = e^{-t}$ for all t
11. $x = e^{-t}$ and $y = e^{3t}$ for all t
12. $x = e^t$, $y = \frac{1}{2}(e^t + e^{-t})$ for all t

 13. For each pair of curves described parametrically, determine if there is a value of t in $[0, 2\pi]$ for which the corresponding points on the two curves coincide. If so, give the value(s) of t. (*Hint:* Plot the curves sequentially.)
   a. $x = \cos t$ and $y = \sin t$; $x = \sin t$ and $y = \cos t$
   b. $x = \cos t$ and $y = \sin t$; $x = -\sin t$ and $y = \cos t$
   c. $x = 1 + \cos t$ and $y = \sin t$; $x = \sin t$ and $y = \cos t$
14. Let $a, b > 0$, and consider the ellipse parametrized by
   $$x = a \cos t \quad \text{and} \quad y = b \sin t \quad \text{for } 0 \le t \le 2\pi$$
   Find a representation of the ellipse in rectangular coordinates.
15. Consider the Folium of Descartes, parametrized by
   $$x = \frac{3t}{1 + t^3} \quad \text{and} \quad y = \frac{3t^2}{1 + t^3} \quad \text{for all } t \ne -1$$

   a. Find an equation in rectangular coordinates for the folium. (*Hint:* Compute $x^3 + y^3$.)

 b. Using a graphics calculator, determine which portion of the folium is traced out when t increases without bound, and when t increases toward -1.

 16. Let $b > 0$. Consider the curve $C_{\pi b}$ parametrized on $[0, b]$ by
   $$x = \cos t + \cos(\pi t) \quad \text{and} \quad y = \sin t - \sin(\pi t)$$

   a. What can you say about the total curve as b increases without bound?
   b. Replace π by 2 or 3, or any other positive integer, and see how the graph is altered. How can you account for the change?

 17. Using a graphics calculator or computer, plot the graph C_n of
   $$x = \sin t \quad \text{and} \quad y = \sin(nt) \quad \text{for } 0 \le t \le 2\pi$$
   where $n = 2, 4, 6$, and 8.

   a. Determine the behavior of C_n as n increases.
   *b. Why are the outer loops thinner than the inner loops? (Compare dy/dx for $t = 0$ with dy/dx for t near $\pi/2$ or $3\pi/2$.)

 18. Let $r > 0$, and consider the curve defined parametrically by
   $$x = \cos^r t \quad \text{and} \quad y = \sin^r t \quad \text{for all } 0 \le t \le 2\pi$$
   Determine a rational value of r such that the whole graph appearing on the screen

   a. lies only in the first quadrant. Explain why this happens.
   b. appears to consist of portions of the x and y axes only, even when you zoom in. Explain why this happens.

19. a. Plot the curve parametrized by

$$x = \frac{2t}{1 + t^2} \quad \text{and} \quad y = \frac{1 - t^2}{1 + t^2} \quad \text{for all } t$$

    b. By squaring both x and y, find an equation of the graph in rectangular coordinates.

 20. a. Plot the curve parametrized by

$$x = \frac{t^2 + 1}{t^2 - 1} \quad \text{and} \quad y = \frac{2t}{t^2 - 1} \quad \text{for all } t$$

    b. By squaring both x and y, find an equation of the graph in rectangular coordinates.

21. Write down three sets of parametric equations whose combined graph is the capital letter B.

Project

1. A **Lissajous*** figure is a curve that has a parametrization of the form

$$x = \cos(mt) \quad \text{and} \quad y = \sin(nt) \quad \text{for} \quad 0 \le t \le 2\pi$$

where m and n are positive numbers (normally integers). These figures are named for the French physicist Jules Lissajous (1822–1880). Let L_{mn} denote the Lissajous figure for the given pair m and n.

    a. Suppose $m = n$. Prove that L_{mn} is a circle (which is a degenerate Lissajous figure).

    b. Suppose m and n are even, and that n divides m. Prove that the origin is not on L_{mn}. (In addition to this case, one can show that any time n is odd, the origin is not on L_{mn}.)

    c. Let $m = 4$ and $n = 6$. Prove that the origin is not on L_{mn}.

    d. Let $m = 3$ and $n = 6$. Prove that the origin is on L_{mn}.

 e. Let m and n be distinct even integers such that n is even and m divides n. By experimenting with your calculator or computer, determine how many loops are in L_{mn}.

    f. Consider the figure $L_{3\pi}$, where $m = 3$ and $n = \pi$. (Strictly speaking, $L_{3\pi}$ is not a Lissajous figure. Why?) What happens when you trace out the figure for $0 \le t \le b$ and let b increase without bound? Is a point on the figure ever repeated? Explain why or why not.

6.8   LENGTH OF A CURVE GIVEN PARAMETRICALLY

In Section 6.2 we derived the integral for the length of the graph of a function on an interval $[a, b]$. Now we will find the lengths of parametrized curves, and in particular we will determine the length of one arch of the cycloid.

    Suppose that a curve C is given parametrically by

$$x = f(t) \quad \text{and} \quad y = g(t) \quad \text{for } a \le t \le b$$

where f and g have continuous derivatives on $[a, b]$. Our derivation of the formula for the length L of C will parallel the derivation of the formula given in Definition 6.2. Let $P = \{t_0, t_1, \ldots, t_n\}$ be any partition of $[a, b]$, and for $1 \le k \le n$ let $(x_k, y_k) = (f(t_k), g(t_k))$ be the corresponding point on C (Figure 6.77). Let ΔL_k be the length of the portion of the curve joining (x_{k-1}, y_{k-1}) and (x_k, y_k). If Δx_k is small, ΔL_k is approximately equal to the length of the line segment joining (x_{k-1}, y_{k-1}) and (x_k, y_k). In other words,

$$\Delta L_k \approx \sqrt{(x_k - x_{k-1})^2 + (y_k - y_{k-1})^2}$$

$$= \sqrt{[f(t_k) - f(t_{k-1})]^2 + [g(t_k) - g(t_{k-1})]^2}$$

*__Lissajous:__ Pronounced "lissajoo."

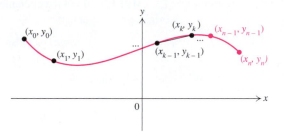

FIGURE 6.77

By the Mean Value Theorem there are numbers t'_k and t''_k in $[t_{k-1}, t_k]$ such that

$$f(t_k) - f(t_{k-1}) = f'(t'_k)\,\Delta t_k \quad \text{and} \quad g(t_k) - g(t_{k-1}) = g'(t''_k)\,\Delta t_k$$

Therefore the total length L of the graph of the curve, which is the sum of the lengths $\Delta L_1, \Delta L_2, \ldots, \Delta L_n$, should be approximately

$$\sum_{k=1}^{n} \sqrt{[f'(t'_k)]^2 + [g'(t''_k)]^2}\,\Delta t_k$$

Although this sum is not a Riemann sum (because t'_k and t''_k may be distinct numbers), it is closely related to a Riemann sum, and moreover, it can be shown that

$$\lim_{\|P\| \to 0} \sum_{k=1}^{n} \sqrt{[f'(t'_k)]^2 + [g'(t''_k)]^2}\,\Delta t_k = \int_a^b \sqrt{[f'(t)]^2 + [g'(t)]^2}\,dt$$

Consequently it is reasonable to define the length L of the curve C by the formula

$$L = \int_a^b \sqrt{[f'(t)]^2 + [g'(t)]^2}\,dt \tag{1}$$

If $x = f(t)$ and $y = g(t)$, then in Leibniz notation (1) becomes

$$L = \int_a^b \sqrt{\left(\frac{dx}{dt}\right)^2 + \left(\frac{dy}{dt}\right)^2}\,dt \tag{2}$$

Notice that if C is parametrized by

$$x = t \quad \text{and} \quad y = g(t) \quad \text{for } a \le t \le b$$

then (1) becomes

$$L = \int_a^b \sqrt{1 + [g'(t)]^2}\,dt$$

which is equivalent to (2) of Section 6.2 but with g substituted for f. Thus (1) is a more general formula for length than (2) of Section 6.2.

It is possible to prove that the formula in (1) yields the same value for the length of a curve C if f and g are replaced by any other pair of continuously differentiable functions that parametrize C. We express this by saying that the length of a curve is **independent of parametrization.**

EXAMPLE 1   Find the circumference L of the circle of radius r centered at the origin, given parametrically by

$$x = r \cos t \quad \text{and} \quad y = r \sin t \quad \text{for } 0 \le t \le 2\pi$$

Solution   Since

$$\frac{dx}{dt} = -r \sin t \quad \text{and} \quad \frac{dy}{dt} = r \cos t$$

it follows from (2) that

$$L = \int_0^{2\pi} \sqrt{(-r \sin t)^2 + (r \cos t)^2} \, dt = \int_0^{2\pi} \sqrt{r^2 \sin^2 t + r^2 \cos^2 t} \, dt$$

$$= \int_0^{2\pi} \sqrt{r^2} \, dt = r \int_0^{2\pi} 1 \, dt = 2\pi r$$

Thus the circumference of the circle of radius r is $2\pi r$, a fact familiar from plane geometry.   ❑

In 1658 the British architect and mathematician Christopher Wren calculated the length of an arch of a cycloid, as we do in the next example.

EXAMPLE 2   Find the length L of one arch of the cycloid given parametrically by

$$x = r(t - \sin t) \quad \text{and} \quad y = r(1 - \cos t) \quad \text{for } 0 \le t \le 2\pi$$

(Figure 6.78).

Solution   First we notice that

$$\frac{dx}{dt} = r - r \cos t \quad \text{and} \quad \frac{dy}{dt} = r \sin t$$

Therefore by (2) we have

$$L = \int_0^{2\pi} \sqrt{(r - r \cos t)^2 + (r \sin t)^2} \, dt$$

$$= \int_0^{2\pi} \sqrt{r^2 - 2r^2 \cos t + r^2 \cos^2 t + r^2 \sin^2 t} \, dt$$

$$= \int_0^{2\pi} \sqrt{2r^2 - 2r^2 \cos t} \, dt$$

$$= r \int_0^{2\pi} \sqrt{2(1 - \cos t)} \, dt$$

Sir Christopher Wren (1632–1723)
At the time of London's Great Fire, Wren was professor of astronomy and taught geometry at Oxford University. His works encompassed mechanics, optics, celestial mechanics, and mathematical physics as well as mathematics. After the Great Fire he played a leading role in the rebuilding of St. Paul's Cathedral and more than fifty other churches and public buildings. As a result, Wren's fame as an architect far exceeds his reputation as a mathematician.

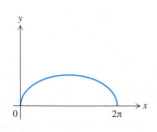

FIGURE 6.78

By the half-angle formula for sin ($t/2$),

$$\frac{1 - \cos t}{2} = \sin^2 \frac{t}{2}, \quad \text{so that} \quad \sqrt{2(1 - \cos t)} = \sqrt{4 \sin^2 \frac{t}{2}}$$

Since $\sin t/2 \geq 0$ for $0 \leq t \leq 2\pi$, we conclude that

$$L = r \int_0^{2\pi} \sqrt{4 \sin^2 \frac{t}{2}} \, dt = 2r \int_0^{2\pi} \sin \frac{t}{2} \, dt = -4r \cos \frac{t}{2} \Big|_0^{2\pi} = 8r \quad \square$$

Motion along a Parametrized Curve

Suppose an object moves along a curve C parametrized by

$$x = f(t) \quad \text{and} \quad y = g(t) \quad \text{for } a \leq t \leq b \tag{3}$$

For simplicity we will assume that C does not cross itself, so that the object never returns a second time to any point on C. Since t is the parameter in (3), we will use T to represent time, with $T = 0$ corresponding to the value a of the parameter t when the object is at the initial point of C. For any time $T > 0$ let L denote the length of the portion of C from the initial point of C to the point on C at which the object is located at time T.

If C is a segment of the x axis and an object is moving in the positive direction along C, then dL/dT is the rate of change of the distance traveled by the object. In analogy with straight line motion, we define the **speed** of the object traversing the curve C by

$$v = \frac{dL}{dT} \tag{4}$$

The parameter t and time T are related to one another, each corresponding to a unique point on the curve C. Notice that when L is considered as a function of t, it represents the length of the portion of C from the initial point of C to the point on C corresponding to t. Therefore it follows from (2) and Theorem 5.12 that

$$\frac{dL}{dt} = \sqrt{\left(\frac{dx}{dt}\right)^2 + \left(\frac{dy}{dt}\right)^2} \tag{5}$$

Before we give an example that uses (5), let us recall from (3) of Section 6.7 that the path of an object traversing a circle of radius r with a constant angular velocity ω can be parametrized with respect to time T by

$$x = r \cos \omega T \quad \text{and} \quad y = r \sin \omega T$$

Under these conditions we can use (5) to derive a very simple relation between the speed v defined in (4) and the angular velocity ω of the object.

EXAMPLE 3   Suppose

$$x = r \cos \omega T \quad \text{and} \quad y = r \sin \omega T$$

where r and ω are positive constants. Show that $v = r\omega$.

Solution   Since $dx/dT = -r\omega \sin \omega T$ and $dy/dT = r\omega \cos \omega T$, (5) implies that

$$\frac{dL}{dT} = \sqrt{(-r\omega \sin \omega T)^2 + (r\omega \cos \omega T)^2} = r\omega$$

Along with (4), this tells us that

$$v = \frac{dL}{dT} = r\omega$$

Consequently the speed of the object equals the radius of the circle multiplied by the angular velocity. ❑

Special Properties of the Cycloid

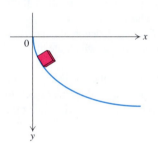

FIGURE 6.79   A slide in the shape of an inverted cycloid.

Cycloid
Galileo was one of the first people to recognize the significance of the cycloid. It played an important role in the development of calculus as mathematicians sought to calculate the length and center of gravity of an arch, and the area under an arch. It also serves as the answer to the famous tautochrone and brachistochrone problems described in this section. Because of attention and controversy caused by the cycloid during the seventeenth century, it was sometimes called the "apple of discord."

Suppose an inverted cycloid represents a frictionless slide, and a smooth ice cube glides down the cycloid under the sole influence of gravity (Figure 6.79). Let us determine the time that is required for the ice cube to slide to the bottom of the cycloid.

To make the solution as simple as possible, let the positive y axis be pointed downward, so that the portion of the cycloid under consideration is given by

$$x = r(t - \sin t) \quad \text{and} \quad y = r(1 - \cos t) \quad \text{for } 0 \le t \le \pi \tag{6}$$

Notice that the ice cube begins its journey at the origin, where $t = 0$, and reaches the bottom of the slide, where $t = \pi$.

Considering time T as a function of the parameter t, and letting T^* denote the time it takes for the ice cube to glide down to the bottom of the slide, we find that

$$T^* = T(\pi) - T(0)$$

By the Fundamental Theorem of Calculus,

$$T^* = T(\pi) - T(0) = \int_0^\pi \frac{dT}{dt} \, dt \tag{7}$$

We will compute T^* by means of this integral.

It follows from principles of physics that the speed v of the ice cube is related to the y coordinate of the ice cube by the equation

$$v = \sqrt{2gy} \tag{8}$$

where g denotes the gravitational constant. By the Chain Rule,

$$\frac{dL}{dt} = \frac{dL}{dT} \frac{dT}{dt} \tag{9}$$

Using first (9), then (4) and (5), and finally (8), we deduce that

$$\frac{dT}{dt} \overset{(9)}{=} \frac{dL/dt}{dL/dT} \overset{(4)}{\underset{(5)}{=}} \frac{\sqrt{(dx/dt)^2 + (dy/dt)^2}}{v} \overset{(8)}{=} \frac{\sqrt{(dx/dt)^2 + (dy/dt)^2}}{\sqrt{2gy}} \tag{10}$$

Next we use the parametrization of the cycloid in (6). Taking derivatives of x and y, we find that

$$\frac{dx}{dt} = r(1 - \cos t) \quad \text{and} \quad \frac{dy}{dt} = r \sin t$$

If we substitute for y, dx/dt, and dy/dt, we obtain

$$\frac{\sqrt{(dx/dt)^2 + (dy/dt)^2}}{\sqrt{2gy}} = \frac{\sqrt{r^2(1 - \cos t)^2 + r^2 \sin^2 t}}{\sqrt{2gr(1 - \cos t)}} = \frac{\sqrt{r^2(2 - 2\cos t)}}{\sqrt{2gr(1 - \cos t)}} = \sqrt{\frac{r}{g}}$$

Therefore by (10),

$$\frac{dT}{dt} = \sqrt{\frac{r}{g}}$$

Consequently (7) yields

$$T^* = \int_0^\pi \sqrt{\frac{r}{g}}\, dt = \pi \sqrt{\frac{r}{g}}$$

Brachistochrone and Tautochrone Problems
The brachistochrone problem, posed by Johann Bernoulli in 1696, is one of the most celebrated in mathematics. It is said that when Newton received Bernoulli's letter containing the problem as a challenge, Newton spent 12 straight hours (4 p.m. to 4 a.m.) producing a solution. The tautochrone problem was first solved by the Dutch mathematician Christiaan Huygens (1629–1695), who later applied the tautochrone feature in his construction of pendulum clocks.

FIGURE 6.80   The cycloid is a tautochrone.

Thus the time it takes the ice cube to roll from the origin to the base of the cycloid depends only on gravity and the radius of the cycloid—provided that no forces other than gravity influence the motion of the ice cube.

A question that attracted the attention of scientists in the eighteenth century was this: How does the time it takes the ice cube to reach the bottom of the slide depend on the initial position of the ice cube on the slide? What may be surprising is that even if the ice cube begins its journey part way down the slide, it takes exactly the same amount of time to reach the bottom. In other words, if ice cubes start simultaneously at the points A, B, and C in Figure 6.80, they will reach the bottom of the slide at the same instant. To make this assertion seem plausible, notice that the higher on the slide an ice cube starts, the farther the ice cube must travel, but also the more steeply and quickly it begins its descent. One says that the cycloid is a **tautochrone,** the word coming from the Greek words "tauto," meaning "the same," and "chrone," meaning "time."

The inverted cycloid has another remarkable quality. If an ice cube is to descend from any point P to a point Q that is lower than P but not directly under P and is subject only to the influence of gravity, then it turns out that the path that gets the ice cube most quickly to Q is a path along a portion of an inverted cycloid. Thus the cycloid is a **brachistochrone,** whose meaning in Greek is "shortest time."

Surface Area

Let C be a curve given parametrically by

$$x = f(t) \quad \text{and} \quad y = g(t) \quad \text{for } a \le t \le b$$

where g is assumed to be nonnegative on $[a, b]$. For simplicity we assume that each point of C corresponds to only one value of t in $[a, b]$. When C is revolved about the x axis, a surface is generated (Figure 6.81). By a variation of the discussion leading to the formula for surface area in (4) of Section 6.3, we obtain the following formula for the surface area S of the surface generated by C:

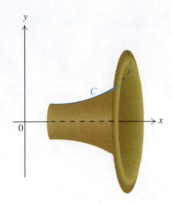

FIGURE 6.81

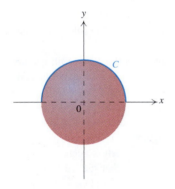

FIGURE 6.82

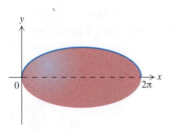

FIGURE 6.83

$$S = \int_a^b 2\pi g(t)\sqrt{[f'(t)]^2 + [g'(t)]^2}\,dt \qquad (11)$$

In Leibniz notation, (11) becomes

$$S = \int_a^b 2\pi y\sqrt{\left(\frac{dx}{dt}\right)^2 + \left(\frac{dy}{dt}\right)^2}\,dt \qquad (12)$$

EXAMPLE 4   Show that the surface area S of a sphere of radius r is $4\pi r^2$.

Solution   Suppose the center of the sphere is at the origin, so that if C is the semicircle defined by

$$x = r\cos t \quad\text{and}\quad y = r\sin t \quad\text{for } 0 \le t \le \pi$$

then the sphere is the surface obtained by revolving C about the x axis (Figure 6.82). It follows from (12) that

$$S = \int_0^\pi 2\pi r\sin t\sqrt{(-r\sin t)^2 + (r\cos t)^2}\,dt$$

$$= 2\pi \int_0^\pi r\sin t\sqrt{r^2\sin^2 t + r^2\cos^2 t}\,dt$$

$$= 2\pi \int_0^\pi r^2\sin t\,dt = 2\pi r^2(-\cos t)\Big|_0^\pi = 4\pi r^2 \quad \square$$

EXAMPLE 5   Suppose that one arch of the cycloid given parametrically by

$$x = r(t - \sin t) \quad\text{and}\quad y = r(1 - \cos t) \quad\text{for } 0 \le t \le 2\pi$$

is revolved around the x axis. Find the area S of the surface generated (Figure 6.83).

Solution   From (11) or (12) and the half-angle formula $\sin^2 t/2 = (1 - \cos t)/2$ we find that

$$S = \int_0^{2\pi} 2\pi r(1 - \cos t)\sqrt{[r(1 - \cos t)]^2 + (r\sin t)^2}\,dt$$

$$= \int_0^{2\pi} 2\pi r(1 - \cos t)\sqrt{r^2(1 - 2\cos t + \cos^2 t + \sin^2 t)}\,dt$$

$$= \int_0^{2\pi} \pi r^2\, 2(1 - \cos t)\sqrt{2(1 - \cos t)}\,dt$$

$$= \pi r^2 \int_0^{2\pi} [2(1 - \cos t)]^{3/2}\,dt = \pi r^2 \int_0^{2\pi} \left(4\sin^2\frac{t}{2}\right)^{3/2}\,dt$$

$$= 8\pi r^2 \int_0^{2\pi} \sin^3\frac{t}{2}\,dt = 8\pi r^2 \int_0^{2\pi}\left(1 - \cos^2\frac{t}{2}\right)\sin\frac{t}{2}\,dt$$

Next we let

$$u = \cos \frac{t}{2}, \quad \text{so that} \quad du = -\frac{1}{2} \sin \frac{t}{2} \, dt$$

Now if $t = 0$ then $u = 1$, and if $t = 2\pi$ then $u = -1$. Therefore

$$S = 8\pi r^2 \int_0^{2\pi} \overbrace{\left(1 - \cos^2 \frac{t}{2}\right)}^{1 - u^2} \overbrace{\sin \frac{t}{2} \, dt}^{-2 \, du}$$

$$= 8\pi r^2 \int_1^{-1} (1 - u^2)(-2) \, du = -16\pi r^2 \left(u - \frac{1}{3} u^3\right)\Big|_1^{-1}$$

$$= -16\pi r^2 \left[\left(-1 + \frac{1}{3}\right) - \left(1 - \frac{1}{3}\right)\right] = \frac{64\pi}{3} r^2 \quad \square$$

EXERCISES 6.8

In Exercises 1–3 find the length L of the curve described parametrically.

1. $x = 1 - t^2$ and $y = 1 + t^3$ for $0 \leq t \leq 1$

2. $x = e^t \sin t$ and $y = e^t \cos t$ for $0 \leq t \leq \pi$

3. $x = \sin t - t \cos t$ and $y = t \sin t + \cos t$ for $0 \leq t \leq \pi/2$

In Exercises 4–6 find the area S of the surface generated by revolving about the x axis the curve with the given parametric representation.

4. $x = \frac{1}{2} t^2$ and $y = t$ for $\sqrt{3} \leq t \leq 2\sqrt{2}$

5. $x = \frac{2}{3}(1 - t)^{3/2}$ and $y = \frac{2}{3}(1 + t)^{3/2}$ for $-\frac{3}{4} \leq t \leq 0$

6. $x = \sin^2 t$ and $y = \sin t \cos t$ for $0 \leq t \leq \pi/2$

7. Let $0 \leq \alpha < \beta \leq \pi$ and $r > 1$, and consider the circular arc C parametrized by

$$x = r \cos t \quad \text{and} \quad y = r \sin t \quad \text{for } \alpha \leq t \leq \beta$$

  a. Show, that the area S of the surface obtained by revolving C about the x axis does not depend only on the difference $\beta - \alpha$.

  b. Does the result of part (a) contradict the assertion that follows Example 2 of Section 6.3? Explain why or why not.

8. Suppose an object moves in an elliptical orbit parametrized by

$$x = 2 \cos t \quad \text{and} \quad y = 3 \sin t$$

where t represents time.

  a. Use (4) and (5) to find a formula for the velocity.

  b. Determine the points on the ellipse at which the object is moving fastest and the points at which it is moving slowest.

9. Let $a, b > 0$, and consider the ellipse parametrized by

$$x = a \cos t \quad \text{and} \quad y = b \sin t \quad \text{for } 0 \leq t \leq 2\pi$$

  a. Find an integral that represents the circumference C_{ab} of the ellipse. (The integral is called an **elliptic integral.**)

  b. Use the formula obtained in (a) to find C_{ab} when $a = b > 0$.

  c. Use the result of (a) to determine $\lim_{b \to 0^+} C_{ab}$.

10. Let $r > 0$. The equations

$$x = r \cos^3 t \quad \text{and} \quad y = r \sin^3 t \quad \text{for } 0 \leq t \leq 2\pi$$

parametrize an astroid (see Figure 6.84).

  a. Find the length L of the astroid. (*Hint:* Remember that square roots are nonnegative.)

  b. Show that the astroid is alternatively the graph of

$$x^{2/3} + y^{2/3} = r^{2/3}$$

  c. Find the area S of the surface generated by revolving C about the x axis. (*Hint:* S is equal to twice the surface area of the part for which $0 \leq t \leq \pi/2$.)

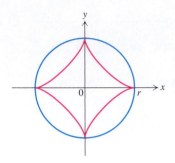

FIGURE 6.84 An astroid.

11. Let C be the curve parametrized by
$$x = \cos^n t \quad \text{and} \quad y = \sin^n t \quad \text{for } 0 \le t \le \pi/2$$
where n is an integer with $n \ge 2$. Let S be the surface area of the surface generated by revolving C about the x axis. Determine the positive integers n for which we can readily carry out the integration given in (12) to obtain the surface area S.

*12. Using (7) and (10), find the time T_{st} it takes for an ice cube to slide on the straight line from the origin to the bottom point $P = (\pi r, 2r)$ of the cycloid, and show that $T_{st} > T^*$.

Project

1. **a. Cornu's spiral** (Figure 6.85), which arises in the study of diffraction, is the curve parametrized by

$$x = x(t) = \int_0^t \cos \frac{\pi s^2}{2} \, ds$$

and

$$y = y(t) = \int_0^t \sin \frac{\pi s^2}{2} \, ds \quad \text{for all } t$$

Show that $x(-t) = -x(t)$ and $y(-t) = -y(t)$. How is this fact reflected in Figure 6.85?

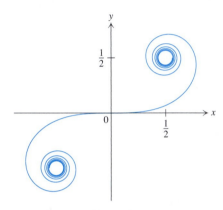

FIGURE 6.85 Cornu's spiral for the project.

b. Let a and b be any real numbers with $a < b$. Find the length L of the part of Cornu's spiral that is traced out for $a \le t \le b$.

REVIEW

Key Terms

Cross-sectional method
Length of a curve
Area of a surface
Work
Moment
Center of gravity

Center of mass
Centroid
Hydrostatic force
Parametric equations
Parameter

Key Theorem

Theorem of Pappus and Guldin

Formulas for Applications of the Integral

Type	Formula
Cross-sectional volume	$V = \displaystyle\int_a^b A(x)\,dx$
Volume of revolution about x axis (disc method)	$V = \begin{cases} \displaystyle\int_a^b \pi[f(x)]^2\,dx \\ \displaystyle\int_a^b \pi\{[f(x)]^2 - [g(x)]^2\}\,dx \end{cases}$
Volume of revolution about y axis (shell method)	$V = \displaystyle\int_a^b 2\pi x[f(x) - g(x)]\,dx$
Length of a curve	$L = \displaystyle\int_a^b \sqrt{1 + [f'(x)]^2}\,dx$
Area of a surface	$S = \displaystyle\int_a^b 2\pi f(x)\sqrt{1 + [f'(x)]^2}\,dx$
Work	$W = \displaystyle\int_a^b f(x)\,dx$
Work to pump water from a tank	$W = \displaystyle\int_a^b 62.5(l - x)A(x)\,dx$
Moment about x axis	$M_x = \displaystyle\int_a^b \frac{1}{2}\{[f(x)]^2 - [g(x)]^2\}\,dx$
Moment about y axis	$M_y = \displaystyle\int_a^b x[f(x) - g(x)]\,dx$
Center of gravity	$(\bar{x}, \bar{y})\colon \bar{x} = \dfrac{M_y}{A}$ and $\bar{y} = \dfrac{M_x}{A}$
Volume of revolution about an axis	$V = 2\pi bA$
Hydrostatic force	$F = \displaystyle\int_a^b 62.5(c - x)w(x)\,dx$
Length of a curve (parametric)	$L = \displaystyle\int_a^b \sqrt{[f'(t)]^2 + [g'(t)]^2}\,dt$ $= \displaystyle\int_a^b \sqrt{\left(\dfrac{dx}{dt}\right)^2 + \left(\dfrac{dy}{dt}\right)^2}\,dt$
Area of a surface (parametric)	$S = \displaystyle\int_a^b 2\pi g(t)\sqrt{[f'(t)]^2 + [g'(t)]^2}\,dt$ $= \displaystyle\int_a^b 2\pi y\sqrt{\left(\dfrac{dx}{dt}\right)^2 + \left(\dfrac{dy}{dt}\right)^2}\,dt$

Review Exercises

In Exercises 1–2 find the volume V of the solid obtained by revolving about the x axis the region between the graphs of f and g on the given interval.

1. $f(x) = 1 + \sqrt{x}$, $g(x) = e^{-x}$; $[0, 1]$

2. $f(x) = \cos x$, $g(x) = \sec^2 x$; $[-\pi/6, \pi/4]$

In Exercises 3–4 find the volume V of the solid obtained by revolving about the y axis the region between the graphs of f and g on the given interval.

3. $f(x) = e^{(x^2)}$, $g(x) = e^{-x^2}$; $[0, 1]$

4. $f(x) = \sqrt{1 + x}$, $g(x) = (\ln x)/x^2$; $[1, 3]$

5. Let $f(x) = x^3$, and let R be the region between the graph of f and the x axis on $[0, 2]$. Then R is called a **cubical spandrel.**

   a. Using the shell method, find the volume V of the solid obtained by revolving R about the y axis.

   b. Find the center of gravity of R.

   c. Rework part (a) using the Theorem of Pappus and Guldin.

6. The graph of $f(x) = x^2$ divides the rectangle bounded by the lines $x = 0$, $x = 2$, $y = 0$, and $y = 4$ into a lower region A and an upper region B. Compare the volumes of the solids obtained by revolving A and B

   a. about the x axis.

   b. about the y axis.

7. Suppose that c is a fixed positive number. Let $f(x) = x + c/x$ and $g(x) = x$, and let R be the region between the graphs of f and g on $[1, 3]$.

   a. Find the volume V_1 of the solid obtained by revolving R about the y axis.

   b. Find the volume V_2 of the solid obtained by revolving R about the x axis.

   c. Use the results of (a) and (b) to show that there is no positive value of c for which $V_1 = V_2$.

8. One way of constructing a ring is to remove a cylinder from the center of a sphere. If the resulting ring is $\frac{1}{4}$ inch wide, show that the volume of the ring is $\pi/384$ cubic inches, regardless of the radius r of the sphere from which the ring is made (which could even be as big as the sun). (*Hint:* Revolve the region R depicted in Figure 6.86 about the x axis, and compute the volume V of the solid generated.)

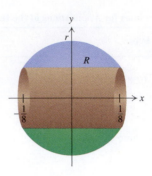

FIGURE 6.86   A ring is formed when R is revolved about the x axis. See Exercise 8.

9. The horizontal cross sections of a cone 10 feet high are regular hexagons. The hexagon at the base has sides 2 feet long. Find the volume V of the cone.

FIGURE 6.87   Figure for Exercise 10.

10. A wedge is cut out of a circular cylinder of cheese with a radius a by first making a cut halfway through the cheese perpendicular to the axis of the cylinder and then making another cut halfway through the cheese at an angle θ with respect to the first cut (Figure 6.87). Find the volume V of cheese removed. (*Hint:* The cross sections of the wedge perpendicular to the first cut are rectangles.)

11. Find the volume V of the wedge in Figure 6.88, each of whose horizontal cross sections is a rectangle.

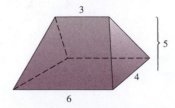

FIGURE 6.88   The wedge for Exercise 11.

In Exercises 12–13 find the length L of the graph of f.

12. $f(x) = \ln(\sin x)$ for $\pi/6 \le x \le 5\pi/6$

13. $f(x) = e^x + \frac{1}{4}e^{-x}$ for $0 \le x \le 1$

 14. By calculating a midpoint sum with $n = 10$, estimate the length of the graph of f, where $f(x) = x^{1/2}$ for $1 \le x \le 2$.

15. Let $f(x) = e^x + \frac{1}{4}e^{-x}$ for $0 \le x \le 1$. Find the area S of the surface obtained by revolving the graph of f about the x axis.

16. A spherical gasoline tank 40 feet in diameter contains gasoline to a depth of 5 feet in the middle. How many cubic feet of gasoline are in the tank?

17. Suppose 4 foot-pounds of work are required in order to stretch a spring from a natural length of $\frac{1}{2}$ foot to a length of 1 foot. Find the work W necessary to stretch the spring from a length of $\frac{3}{2}$ feet to a length of 2 feet.

18. A cylindrical well 20 feet deep and 3 feet in radius is dug. Assuming that the soil weighs 150 pounds per cubic foot, calculate the work W required to raise the soil to ground level.

19. An underground tank containing 144π cubic feet of water has the shape of a right circular cone with its vertex at the bottom. If the radius of the cone is 10 feet at the top and the height is 20 feet, find the work W required to pump all the water to a point 16 feet above the top of the tank.

20. Let $f(x) = 1 + 2x - x^2$ and $g(x) = x^2 - 2x + 1$, and let R be the region between the graphs of f and g on $[0, 2]$. Find the center of gravity of R.

21. Find the center of gravity of the region bounded by the graphs of f and g, where $f(x) = x$ and $g(x) = x^2$.

22. Find the center of gravity of the region bounded by the graphs of the equations $y^2 = 3x$ and $y = x^2 - 2x$. (*Hint:* The curves meet for $x = 0$ and $x = 3$.)

23. **a.** Find the center of gravity of the region R in Exercise 7.
b. Rework Exercise 7 using the result of part (a), along with the Theorem of Pappus and Guldin.

24. Let R be the region between the graphs of $y = (x - 1)^2$ and $y - 2 = -(x - 1)^2$. Using the Theorem of Pappus and Guldin, show that the volume of the solid obtained by revolving R about the x axis equals the volume of the solid obtained by revolving R about the y axis.

In Exercises 25–26 sketch the graph of the curve represented parametrically.

25. $x = 2t + 1$ and $y = 4 - 6t$

26. $x = 3 \cos t$ and $y = 3 \sin t$ for $-\pi/2 \le t \le \pi/4$

In Exercises 27–28 find the length L of the curve given parametrically.

27. $x = \frac{1}{4}(e^{2t} - e^{-2t}) - t$ and $y = e^t + e^{-t}$ for $0 \le t \le 1$

28. $x = \frac{1}{3}\cos^3 t$ and $y = \frac{1}{2}\sin^2 t$ for $0 \le t \le \pi/2$.

In Exercises 29–30 find the area S of the surface obtained by revolving about the x axis the curve with the given parametric equations.

29. $x = \pi + \frac{1}{3}t^3$ and $y = \frac{1}{2}t^2$ for $1 \le t \le \sqrt{3}$

30. $x = e^t \cos t$ and $y = e^t \sin t$ for $0 \le t \le \pi/2$

Applications

31. A chamber in the shape of a right circular cylinder contains a gas, which can be compressed or expanded by a piston (Figure 6.89). If the chamber is maintained at a constant temperature, then according to Boyle's Law the pressure p and the volume V of the gas are related by
$$pV = c$$
where c is a constant. The force F on the piston is the product of its area A and the pressure in the gas (that is, $F = pA$).
a. Show that the force F exerted on the piston by the gas is c/x when the piston is x units from the left end of the cylinder.
b. Compute the work W done on the piston by the gas when the piston moves from $x = 1$ to $x = 2$.

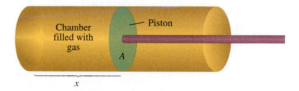

FIGURE 6.89   Figure for Exercise 31.

32. A vertical dam 100 feet tall has the shape of a trapezoid 200 feet wide at the top and 100 feet wide at the bottom. Find the hydrostatic force F on the dam
a. if the water is 50 feet deep.
b. if the water is 100 feet deep.

33. Suppose a cylindrical barrel with radius 1 foot and height 3 feet is full of olive oil weighing 57.3 pounds per cubic foot.
a. Calculate the force on the bottom of the barrel when the barrel is upright (Figure 6.90(a)).
b. Calculate the force on each end of the barrel when the barrel lies on its side (Figure 6.90(b)).

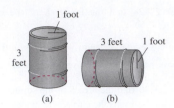

FIGURE 6.90 Figure for Exercise 33.

*34. A major tenet of the theory of relativity is that the mass m of an object depends on its velocity v according to the formula

$$m = \frac{m_0}{\sqrt{1 - v^2/c^2}}$$

where m_0 is the "rest mass" corresponding to $v = 0$, and c is the speed of light. In the theory of relativity,

$$F = ma = m\frac{dv}{dt} \quad \text{is replaced by} \quad F = \frac{d}{dt}(mv)$$

(When v is small compared to c, m is essentially constant and thus $F \approx ma$.)

a. Using the Chain Rule, show that

$$F = v\frac{d}{dx}(mv)$$

b. Show that

$$\frac{d}{dv}(mv) = \frac{m_0}{(1 - v^2/c^2)^{3/2}}$$

c. Using (a), the substitution v for $v(x)$, and then applying (b), show that the work W done as an object moves from point x_0 to x_1 is given by

$$W = \int_{x_0}^{x_1} F(x)\,dx = \int_{v_0}^{v_1} \frac{m_0 v}{(1 - v^2/c^2)^{3/2}}\,dv$$

d. Evaluate the integral in part (c) to obtain

$$W = \frac{m_0 c^2}{\sqrt{1 - v_1^2/c^2}} - \frac{m_0 c^2}{\sqrt{1 - v_0^2/c^2}}$$

(In Section 9.10 we will conclude that if v_0 and v_1 are small compared to c, then $W \approx \frac{1}{2}m_0 v_1^2 - \frac{1}{2}m_0 v_0^2$, which is the classical result we obtained in (8) of Section 6.4.)

35. A water clock is a glass water container with a hole in the bottom through which water can trickle. Such a clock is calibrated for measuring time by placing markings on the exterior of the container corresponding to water levels at equally spaced times (Figure 6.91).

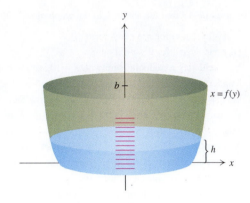

FIGURE 6.91 Figure for Exercise 35.

a. Let f be a function of y that is continuous on an interval $[0, b]$, and suppose the container has the shape of the surface obtained by revolving the graph of f on $[0, b]$ about the y axis. If V denotes the volume of water and h the height of the water level above the bottom, then h depends on time t, and V depends on h and t. Show that

$$\frac{dV}{dt} = \pi[f(h)]^2\frac{dh}{dt}$$

b. Suppose A is the area of the hole in the bottom of the container. If the hole is very small, then it is known from physics that the rate dV/dt depends on the height h according to the equation

$$\frac{dV}{dt} = cA\sqrt{h}$$

where c is a (negative) constant. Find a formula for f if dh/dt is to be a constant k. (In such a case, the markings on the exterior would be equally spaced.)

*36. Newton's Law of Gravitation states that the gravitational force F between two point masses m and M separated by a distance r is given by

$$F = \frac{GmM}{r^2}$$

Suppose that the point mass M is replaced by a uniform bar of mass M and length L, as in Figure 6.92. Show that the gravitational force becomes

$$F = \frac{GmM}{r(r + L)}$$

(*Hint:* Consider any partition P of $[0, L]$, and approximate F by a Riemann sum for P.)

FIGURE 6.92   Figure for Exercise 36.

*37. Consider a star whose radius is R. Suppose the mass density $D(r)$ at any given point in the star depends only on the distance r between the point and the center of the star (Figure 6.93). Show that the mass M of the star is given by

$$M = \int_0^R 4\pi r^2 D(r)\, dr$$

(*Hint:* Partition the star into spherical shells with masses $\Delta M_1, \Delta M_2, \ldots, \Delta M_n$. Note that each ΔM_k is approximately mass density times volume.)

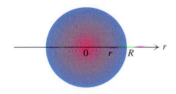

FIGURE 6.93   The star for Exercise 37.

Topics for Discussion

1. Describe the procedure for showing that a geometric or physical quantity equals the value of a definite integral.
2. Suppose you misuse (5) in Section 6.1 as you attempt to find the volume V of a solid, and obtain a negative answer B. Describe conditions under which $V = |B|$, and conditions under which $V \neq |B|$.
3. Suppose a solid D is generated by revolving a region R about the x axis. When would it be convenient to find the volume of D by means of
   a. the disc method (integrating with respect to x)?
   b. the shell method (integrating with respect to y)?
4. Suppose f is continuously differentiable on $[a, b]$, and let L denote the length of the graph of f on $[a, b]$. In defining L, suppose that each of the polygonal line segments that were employed were replaced by a horizontal and a vertical line segment (see (1) of Section 6.2). Could the same value of L be obtained in all cases? You might see what would occur if f were linear but not constant.

5. An analogue of Cavalieri's Principle for lengths of curves is the following: Suppose f and g are continuously differentiable on $[a, b]$, and have the same slope at each x in the interval. Then the graphs of f and g have the same lengths. Under what conditions on f and g would this apply?
6. Suppose a region R in the plane is symmetric with respect to the x axis or the y axis or the origin. Explain how these characteristics relate to the moments and center of gravity of R. Can the center of gravity of R be the origin without R being symmetric with respect to either axis? Give an example or explain why not.
7. Define a parametrized curve. Under what conditions is it possible to consider a parametrized curve as the graph of a function? Write down parametric equations for general lines and for circles.
8. Derive the formula for the length of the graph of a function from the length for a parametrized curve.

Cumulative Review, Chapters 1–5

In Exercises 1–4 find the limit.

1. $\displaystyle\lim_{x \to 3^+} \frac{\sqrt{x^2 - 9}}{x - 3}$ 2. $\displaystyle\lim_{x \to 0} \frac{\sec x - 1}{x}$

3. $\displaystyle\lim_{x \to 0} \sin x \sin \frac{1}{x}$ 4. $\displaystyle\lim_{x \to \infty} \frac{2x^2 - \sin x}{4 - 3x^2}$

5. Let $f(x) = \ln \left| \dfrac{1 - x}{1 + x} \right|$.
   a. Find the domain of f.
   b. Show that f is an odd function.
   c. Are there any vertical asymptotes to the graph of f?

6. Let $f(x) = \dfrac{1}{2} \ln \left| \dfrac{1 - \cos x}{1 + \cos x} \right|$. Show that $f'(x) = \csc x$.
7. Let $f(x) = (x - 1) \cos x$. Find a formula for $f^{(24)}(x)$.
8. Find an equation of the line tangent to the graph of the equation $xy^3 - x^2 = y^2 + xy + 5$ at the point $(3, 2)$.
9. A shadow in the shape of an equilateral triangle is growing 9 square inches per minute. At what rate does the height of the triangle grow when the area is $\sqrt{3}$ square inches?

In Exercises 10–11 sketch the graph of f, indicating all pertinent information.

10. $f(x) = -x^4 + 3x^2 - 2$ 11. $f(x) = x^5 - \dfrac{10}{3}x^3 + 5x$
12. Suppose a shoreline has the shape of the parabola $y = \frac{1}{5}x^2$, where x and y are measured in miles, and that a fog light located at $(0, 2)$ revolves at the rate of $\frac{1}{2}$ radian per

second. How fast does the x coordinate of the point of illumination on the shoreline change at the instant the point $(1, \frac{1}{5})$ is illuminated?

13. Suppose the velocity of an object is given by $v(t) = \frac{1}{4}t^2 + \sin t$ for $0 \leq t \leq \pi/2$. For what value of t in $[0, \pi/2]$ is the acceleration maximal?

14. A cylindrical tank with a volume of 40π cubic meters is to be built. Materials for the sides cost \$10 per square meter, and for the top and bottom the materials cost \$25 per square meter. Find the dimensions that will minimize the cost.

15. Find the point(s) on the graph of $y = 2/(1 + x^2)$ that are closest to the origin.

16. Suppose the acceleration of an object is given by

$$a(t) = \frac{1}{t(\ln t + 2)} \quad \text{for } t \geq 1$$

If $v(1) = 5$, find $v(e)$.

17. Let $g(x) = \int_x^{x+\pi} \sin^{2/3} t \, dt$. Show that g is a constant function.

In Exercises 18–21 evaluate the integral.

18. $\displaystyle \int \frac{1}{\sqrt{2x - 3}} \, dx$ 19. $\displaystyle \int_0^\pi \frac{\sin x}{2 + \cos x} \, dx$

20. $\displaystyle \int x(1 + \sqrt{x})^{1/3} \, dx$ 21. $\displaystyle \int \frac{1}{t} (\ln t)^{5/3} \, dt$

22. **a.** Determine, if possible, a positive integer n such that $\int_1^n 1/x^{.9} \, dx > 1000$.

   b. Determine, if possible, a positive integer n such that $\int_1^n 1/x^{1.1} \, dx > 1000$.

Mt. Rainier reflected in Benck Lake. (*H. Richard Johnson/Tony Stone Images*)

7 INVERSE FUNCTIONS, L'HÔPITAL'S RULE, AND DIFFERENTIAL EQUATIONS

Any function *f* associates with each number *x* in its domain a unique number *y* in its range. Is it possible to reverse this procedure and associate *x* with the number *y*? Although it is not always possible to do so and still obtain a function, in some cases this can be done. The resulting function is called the inverse of *f*. By applying this procedure to certain familiar functions we obtain many new and useful functions.

We begin this chapter by discussing general properties of inverses. Then we formally define the natural exponential function (as the inverse of the natural logarithmic function, which itself was formally defined in Section 5.7), general exponential functions, hyperbolic functions, and inverse trigonometric functions. Section 7.6 is devoted to l'Hôpital's Rule for evaluating limits of quotients whose denominators tend to 0. The chapter concludes with a brief introduction to differential equations.

431

7.1 INVERSE FUNCTIONS

Certain pairs of functions are "opposites" of one another. For example, the function f representing the conversion from degrees Celsius to degrees Fahrenheit is opposite to the function g representing the conversion from degrees Fahrenheit to degrees Celsius. In mathematical terms, how are f and g opposite to each other? Notice that the equation $f(x) = y$ means that a temperature of x degrees Celsius is the same as a temperature of y degrees Fahrenheit. This is equivalent to saying that a temperature of y degrees Fahrenheit is the same as a temperature of x degrees Celsius, which means that $g(y) = x$. Thus $f(x) = y$ is equivalent to $g(y) = x$. We use the term "inverse" to describe such a relationship between two functions.

DEFINITION 7.1

Let f be a function. Then f **has an inverse** provided that there is a function g such that the domain of g is the range of f and such that

$$f(x) = y \quad \text{if and only if} \quad g(y) = x \tag{1}$$

for all x in the domain of f and all y in the range of f. In this case, g is the **inverse** of f, and is designated by f^{-1}. Thus

$$f(x) = y \quad \text{if and only if} \quad f^{-1}(y) = x \tag{2}$$

for all x in the domain of f and all y in the range of f.

If f has an inverse f^{-1}, then by Definition 7.1, f^{-1} is a function. This means that for any y in the domain of f^{-1} there can be only one number x such that $f(x) = y$. Consequently f^{-1} is not only a function, but its values are uniquely determined by (2).

Many functions have inverses. For a simple example, let $f(x) = x^3$ and $g(y) = x^{1/3}$. Since

$$x^3 = y \quad \text{if and only if} \quad y^{1/3} = x$$

it follows that

$$f(x) = y \quad \text{if and only if} \quad g(y) = x$$

Therefore Definition 7.1 implies that g is the inverse of f, which means that $g = f^{-1}$.

Caution: The function f^{-1} is almost always different from the function $1/f$. For example, if $f(x) = x^3$, then

$$f^{-1}(x) = x^{1/3}, \quad \text{whereas} \quad \left(\frac{1}{f}\right)(x) = \frac{1}{x^3}$$

In particular,

$$f^{-1}(8) = 8^{1/3} = 2, \quad \text{whereas} \quad \left(\frac{1}{f}\right)(8) = \frac{1}{8^3} = \frac{1}{512}$$

The difference between f^{-1} and $1/f$ is similar to the difference between walking backwards and walking upside down on one's hands.

Often the clearest way to describe a function is to write a formula for its rule by one or more procedures. If we know that a function f has an inverse, we can sometimes generate a formula for the inverse by the following method:

> **1.** Write $y = f(x)$.
>
> **2.** Solve for x in terms of y.
>
> **3.** Write $f^{-1}(y)$ for x in step 2.

In order to see how the method works, let us first recall that if x represents a temperature in degrees Celsius, then the temperature is $\frac{9}{5}x + 32$ in degrees Fahrenheit. Thus if

$$f(x) = \frac{9}{5}x + 32 \quad \text{for } x \geq -273.15$$

then f is the Celsius-to-Fahrenheit function. (Recall that –273.15 degrees Celsius corresponds to absolute zero, the lowest possible temperature.) We will find a formula for f^{-1}, which is the Fahrenheit-to-Celsius function.

EXAMPLE 1   Let $f(x) = \frac{9}{5}x + 32$ for $x \geq -273.15$. Find a formula for the inverse of f, and determine the domain of f^{-1}.

Solution   Following the steps listed above, we obtain

$$y = \frac{9}{5}x + 32$$

$$\frac{9}{5}x = y - 32$$

$$x = \frac{5}{9}(y - 32)$$

Thus

$$f^{-1}(y) = \frac{5}{9}(y - 32)$$

Because we customarily use x as the independent variable, we replace y by x to obtain

$$f^{-1}(x) = \frac{5}{9}(x - 32)$$

Since f is continuous and increasing, and since $f(-273.15) = -459.67$, the range of f is $[-459.67, \infty)$. Therefore the domain of f^{-1} is $[-459.67, \infty)$. Consequently f^{-1} is completely defined by the formula

$$f^{-1}(x) = \frac{5}{9}(x - 32) \quad \text{for } x \geq -459.67 \quad \square$$

EXAMPLE 2   Let $f(x) = 8x^3 - 1$. Write a formula for the inverse of f.

Solution   Again following the steps listed above, we find that

$$y = 8x^3 - 1$$

$$\frac{y + 1}{8} = x^3$$

$$x = \left(\frac{y + 1}{8}\right)^{1/3} = \frac{1}{2}(y + 1)^{1/3}$$

Thus

$$f^{-1}(y) = \frac{1}{2}(y + 1)^{1/3}$$

Converting to the variable x, we have

$$f^{-1}(x) = \frac{1}{2}(x + 1)^{1/3} \qquad \square$$

It is not always possible to find a simple formula for an inverse. Indeed, suppose $f(x) = x^7 + 8x^3 + 4x - 2$ for all x. As we will see a bit later, f has an inverse; however, there is no simple formula for f^{-1}.

Properties of Inverses

First, observe from Definition 7.1 that the domain of f^{-1} is the range of f, whereas the range of f^{-1} is the domain of f. From this observation and from (2) we can derive three elementary relationships between f and f^{-1}, which we group together in a theorem.

THEOREM 7.2

Let f have an inverse. Then f and f^{-1} have the following properties:

a. f^{-1} has an inverse, and $(f^{-1})^{-1} = f$

b. $f^{-1}(f(x)) = x$ for all x in the domain of f

c. $f(f^{-1}(y)) = y$ for all y in the range of f

Proof   To prove (a), recall first that the domain and range of f are the range and domain, respectively, of f^{-1}. Furthermore, (2) is equivalent to

$$f^{-1}(y) = x \quad \text{if and only if} \quad f(x) = y$$

for all y in the domain of f^{-1} and all x in the range of f^{-1}. Hence f^{-1} has an inverse, which is evidently f. To prove (b), simply substitute $f(x)$ for y in the equation $f^{-1}(y) = x$ in (2). Similarly, substitution of $f^{-1}(y)$ for x in the equation $f(x) = y$ in (2) establishes (c).   ■

We would like to have a simple criterion that tells us whether a given function has an inverse. If f has an inverse, then for each y in the range of f there is a single x such that $f(x) = y$; equivalently, two distinct values of x in the domain of f are associated with two distinct values of y in the range of f. Conversely, if any two distinct values of x in the domain of f are assigned two distinct values in the range of f, then for any y in the range of f we can let $g(y) = x$ for the unique x such that $f(x) = y$, thereby defining a function g that satisfies (1). Thus g is the inverse of f. In summary, we have the following criterion for the existence of an inverse of a function.

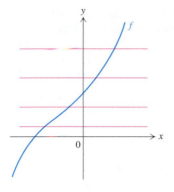

FIGURE 7.1 *f* has an inverse.

A function *f* has an inverse if and only if for any two numbers x_1 and x_2 in the domain of *f*,

$$\text{if } x_1 \neq x_2, \quad \text{then} \quad f(x_1) \neq f(x_2) \tag{3}$$

A function that satisfies (3) is said to be **one-to-one.** Pictorially the condition in (3) means that a horizontal line can intersect the graph of *f* at most once (Figure 7.1). Thus if $f(x) = x^2$, then *f* does not have an inverse because the horizontal line $y = 1$ intersects the graph of *f* twice (Figure 7.2).

Although (3) can frequently help in deciding that a function does not have an inverse, it is not so easy to use (3) to determine that a function has an inverse. For that purpose one usually applies the following theorem.

THEOREM 7.3   Every increasing function and every decreasing function has an inverse.

Proof   Let *f* be increasing, and suppose that x_1 and x_2 are in the domain of *f* with $x_1 \neq x_2$. Then either $x_1 < x_2$, in which case $f(x_1) < f(x_2)$ (because *f* is increasing), or $x_2 < x_1$, in which case $f(x_2) < f(x_1)$ (again because *f* is increasing). In either case, $f(x_1) \neq f(x_2)$. Therefore *f* satisfies the criterion in (3) and hence has an inverse. The proof for decreasing functions is similar. ■

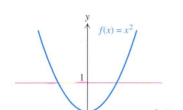

FIGURE 7.2 *f* does not have an inverse.

Suppose the domain of a function *f* is an interval *I*, and *f* is continuous on *I*. Then Theorem 7.3, along with Theorem 4.7 in Section 4.3, yields the following criterion that can be effective in showing that *f* has an inverse:

If $f'(x) > 0$ for every interior point *x* of *I* (or if $f'(x) < 0$ for every interior point of *I*), with the possible exception of a finite number of *x*'s for which $f'(x) = 0$, then *f* has an inverse.

Thus we can conclude in each case below that *f* has an inverse:

$$f(x) = x^7 + 8x^3 + 4x - 2 \qquad f'(x) = 7x^6 + 24x^2 + 4 > 0 \tag{4}$$

$$f(x) = 8x^3 - 1 \qquad\qquad f'(x) = 24x^2 \geq 0 \tag{5}$$
$$f'(x) = 0 \quad \text{only if} \quad x = 0$$

$$f(x) = \sin x \qquad\qquad f'(x) = \cos x > 0 \tag{6}$$
$$\text{for } -\pi/2 \leq x \leq \pi/2 \qquad \text{for } -\pi/2 < x < \pi/2$$

$$f(x) = \ln x \qquad\qquad f'(x) = \frac{1}{x} > 0 \quad \text{for } x > 0 \tag{7}$$

We say that *f* **has an inverse on an interval** *I* provided that when the domain of *f* is restricted to *I*, the new function has an inverse. By Theorem 7.3 a function *f* has an inverse on an interval *I* if *f* is either increasing or decreasing on *I*.

EXAMPLE 3   Let $f(x) = x^2$. Show that *f* has an inverse on $[0, \infty)$ and also has an inverse on $(-\infty, 0]$.

Solution   Since $f'(x) = 2x$, we have $f'(x) > 0$ for $x < 0$, and $f'(x) > 0$ for $x > 0$. Thus f is increasing on $[0, \infty)$ and is decreasing on $(-\infty, 0]$. We conclude that f has an inverse on $[0, \infty)$ and an inverse on $(-\infty, 0]$.   ❏

We observe that for the function $f(x) = x^2$ appearing in Example 3, the inverse on $[0, \infty)$ is the usual square root function.

Graphs of Inverses

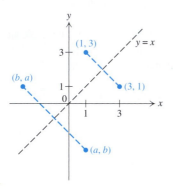

FIGURE 7.3

To find a method of graphing the inverse of f, we begin by noting that if $(1, 3)$ is on the graph of f, then $f(1) = 3$, so that $f^{-1}(3) = 1$. This means that $(3, 1)$ is on the graph of f^{-1}. In general (a, b) is on the graph of f if and only if (b, a) is on the graph of f^{-1}. But $(1, 3)$ and $(3, 1)$ are symmetric with respect to the line $y = x$, as are (a, b) and (b, a) (Figure 7.3). Thus the graph of f^{-1} is obtained by simply reflecting the graph of f through the line $y = x$.

EXAMPLE 4   For each function f, sketch the graphs of f and f^{-1} on the same coordinate system.

     a. $f(x) = 8x^3 - 1$ **b.** $f(x) = \sin x$ for $-\pi/2 \le x \le \pi/2$

Solution   In each case the graph of f^{-1} is obtained by reflecting the graph of f through the line $y = x$. The graphs appear in Figure 7.4(a) and (b).   ❏

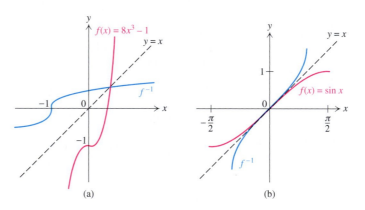

FIGURE 7.4

Continuity and Differentiability of Inverse Functions

When a function f has an inverse, it is natural to ask which properties of f are shared by f^{-1}. In calculus two properties naturally come to mind: continuity and differentiability. In this subsection we will show that the inverse of a continuous function is continuous and, under mild restrictions, the inverse of a differentiable function is differentiable.

     The next theorem concerns the continuity of inverses. Its proof appears in the Appendix.

THEOREM 7.4

Let f be continuous on an interval I, and let the values assigned by f to the points in I form the interval J. If f has an inverse, then f^{-1} is continuous on J.

Before stating the theorem on differentiability of inverses, we temporarily assume that the inverse f^{-1} of a function f is differentiable and analyze the derivative of f^{-1}. To begin with, we recall that the graph of f^{-1} can be obtained from the graph of f by reflecting the graph of f through the line $y = x$. As Figure 7.5 suggests, we would expect f^{-1} to increase (or decrease) rapidly at $f(a)$ if f increases (or decreases) slowly at a, and vice versa. This means that $|(f^{-1})'(f(a))|$ should be large if $|f'(a)|$ is small, and small if $|f'(a)|$ is large. Moreover, if $f'(a) = 0$, then it appears that f^{-1} is not differentiable at $f(a)$ because the line tangent to the graph of f^{-1} is vertical. Now we are ready to state and prove the theorem.

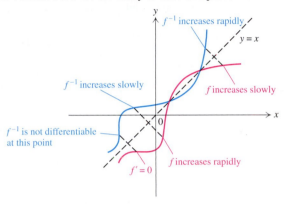

FIGURE 7.5

THEOREM 7.5
Suppose that f has an inverse and is continuous on an open interval I containing a. Assume also that $f'(a)$ exists, $f'(a) \neq 0$, and $f(a) = c$. Then $(f^{-1})'(c)$ exists, and

$$(f^{-1})'(c) = \frac{1}{f'(a)} \qquad (8)$$

Proof   Using the fact that $f^{-1}(c) = a$ and the definition of the derivative, we find that

$$(f^{-1})'(c) = \lim_{y \to c} \frac{f^{-1}(y) - f^{-1}(c)}{y - c} = \lim_{y \to c} \frac{f^{-1}(y) - a}{f(f^{-1}(y)) - f(a)} \qquad (9)$$

provided that the latter limit exists. We will simultaneously show that it does exist and find its value. First notice that f^{-1} is continuous at c by Theorem 7.4. Therefore

$$\lim_{y \to c} f^{-1}(y) = f^{-1}(c) = a$$

so that if $x = f^{-1}(y)$, then x approaches a as y approaches c. Moreover, the fact that f^{-1} has an inverse and $f^{-1}(c) = a$ implies that $f^{-1}(y) \neq a$ for $y \neq c$. Consequently (9) and the Substitution Rule for Limits (with x substituting for $f^{-1}(y)$) imply that

$$(f^{-1})'(c) = \lim_{y \to c} \frac{f^{-1}(y) - a}{f(f^{-1}(y)) - f(a)} = \lim_{x \to a} \frac{x - a}{f(x) - f(a)}$$

$$= \frac{1}{\displaystyle\lim_{x \to a} \frac{f(x) - f(a)}{x - a}} = \frac{1}{f'(a)} \qquad \blacksquare$$

Caution: Notice from (8) that $(f^{-1})'(c)$ is obtained by finding the derivative of f at a, not at c.

EXAMPLE 5   Let $f(x) = 8x^3 - 1$.

a. Find $(f^{-1})'(7)$ by using the fact that $f^{-1}(x) = \frac{1}{2}(x + 1)^{1/3}$.

b. Find $(f^{-1})'(7)$ by using (8) and the fact that $f(1) = 7$.

Solution   **a.** The formula for f^{-1} appeared in the solution of Example 2. Differentiating f^{-1}, we find that

$$(f^{-1})'(x) = \frac{1}{6}(x + 1)^{-2/3}$$

so that

$$(f^{-1})'(7) = \frac{1}{6}(7 + 1)^{-2/3} = \frac{1}{6} \cdot \frac{1}{8^{2/3}} = \frac{1}{6} \cdot \frac{1}{4} = \frac{1}{24}$$

b. Since $f(1) = 7$ and $f'(x) = 24x^2$, (8) implies that

$$(f^{-1})'(7) = \frac{1}{f'(1)} = \frac{1}{24(1^2)} = \frac{1}{24}$$

This is the same answer that we found in part (a).   ❑

Two comments are in order. First, in part (a) of Example 5 we were able to find $(f^{-1})'(7)$ by means of the formula for $f^{-1}(x)$. Since frequently no formula for $f^{-1}(x)$ is known, this method of finding the derivative of the inverse has limited utility.

Second, before we could apply (8) to calculate $(f^{-1})'(7)$ in part (b), it was essential to know that $f(1) = 7$. In general, when we set out to evaluate $(f^{-1})'(c)$ by means of (8), we must first determine the value of a for which $f(a) = c$, that is, we must solve the equation $f(x) = c$ for x. Unfortunately, this is often difficult or impossible. Nevertheless, it is sometimes possible to solve the equation $f(x) = c$ by trial and error. For example, if $f(x) = x^7 + 8x^3 + 4x - 2$, then we can readily ascertain that a solution of $f(x) = -2$ is $x = 0$. We will use this observation in the solution of Example 6.

EXAMPLE 6   Let $f(x) = x^7 + 8x^3 + 4x - 2$. Find $(f^{-1})'(-2)$.

Solution   In order to use (8), we must first find the value of a for which $f(a) = -2$. But $f(0) = -2$, so $a = 0$. Since $f'(x) = 7x^6 + 24x^2 + 4$, it follows that $f'(0) = 4$. Thus we conclude from (8) that

$$(f^{-1})'(-2) = \frac{1}{f'(0)} = \frac{1}{4} \quad ❑$$

If we had desired the number $(f^{-1})'(11)$ for the function f in Example 6, we would first have needed to determine the value of a for which $f(a) = 11$. By checking the values of f at a few numbers, we would have found that $f(1) = 11$. Then (8) would have yielded

$$(f^{-1})'(11) = \frac{1}{f'(1)} = \frac{1}{35}$$

By contrast, if we had wished to find the number $f'(10)$, that would be more difficult because there is no readily available number a such that $f(a) = 10$, although by the Newton-Raphson method or some other calculator or computer technique we could find an approximate value of a.

Formula (8) takes a very simple form in the Leibniz notation. Suppose $y = f(x)$ and f has an inverse, so that $x = f^{-1}(y)$. Then dy/dx is the derivative of f, whereas dx/dy is the derivative of f^{-1}. Using this notation, we may rewrite formula (8) in the easily remembered form

$$\frac{dx}{dy} = \frac{1}{\dfrac{dy}{dx}} \tag{10}$$

EXAMPLE 7   Let $y = x^5 + 2x$. Find $\dfrac{dx}{dy}$ and $\dfrac{dx}{dy}\bigg|_{y=-3}$.

Solution   The function y has an inverse because its derivative, $5x^4 + 2$, is a positive function. Using (10), we deduce that

$$\frac{dx}{dy} = \frac{1}{\dfrac{dy}{dx}} = \frac{1}{5x^4 + 2} \tag{11}$$

Since $y = x^5 + 2x$, it follows that $y = -3$ for $x = -1$. We conclude from (11) that

$$\frac{dx}{dy}\bigg|_{y=-3} = \frac{1}{5(-1)^4 + 2} = \frac{1}{7} \quad \square$$

As the solution of Example 7 illustrates, when we wish to evaluate dx/dy at a given value c of y by using (10), we must evaluate the right side of (10) at the value a of x for which $y = c$.

EXERCISES 7.1

In Exercises 1–14 determine whether the given function has an inverse. If an inverse exists, give the domain and range of the inverse and graph the function and its inverse.

1. $f(x) = x^5$

2. $f(x) = 5x^7 + 4x^3$

3. $f(x) = -x^8$

4. $f(x) = 4\sqrt[5]{x}$

5. $f(t) = \sqrt{4 - t}$

6. $f(t) = \sqrt{1 - t^2}$

7. $f(x) = x + |x|$

8. $f(x) = x + \sin x$

9. $f(x) = x - \sin x$

10. $f(x) = x^2 + \sin x$

11. $f(z) = \tan z$

12. $f(z) = \tan z$ for $-\pi/2 < z < \pi/2$

13. $g(t) = \ln(3 - t)$

14. $g(t) = \ln t^2$

In Exercises 15–18 plot the graph of f. From the graph determine whether f appears to have an inverse.

15. $f(x) = x^4 + 2x^3 + 2x^2 + x$

16. $f(x) = 2x^3 + 2x^2 + x$

17. $f(x) = \dfrac{x^3}{1 + x^2}$

18. $f(x) = \dfrac{x^3 - x}{1 + x^2}$

19. Determine which of the functions h, j, and k is the inverse of f, and which is the inverse of g.

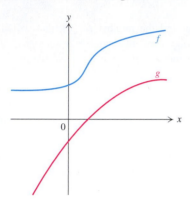

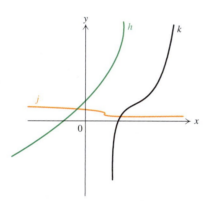

20. Let f be the function whose graph is shown in Figure 7.6. Make a rough sketch of the graphs of $1/f$ and f^{-1} on the same coordinate system.

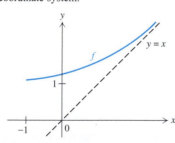

FIGURE 7.6   Graph for Exercise 20.

21. Let f be a periodic function; that is, suppose there is a positive number a such that

$$f(a + x) = f(x)$$

for all x in the domain of f. Show that f does not have an inverse.

In Exercises 22–24 use Exercise 21 to show that the given function does not have an inverse.

22. $f(x) = \sin x - \cos x$

23. $f(x) = 2 \tan 3x - 3 \cos 4x + 17$

24. $f(x) = x - [x]$

In Exercises 25–30 find a formula for the inverse of the function.

25. $f(x) = -4x^3 - 1$ **26.** $f(x) = -2x^5 + \dfrac{9}{4}$

27. $g(x) = \sqrt{1 + x}$ **28.** $g(t) = \sqrt{3 - 2t}$

29. $k(t) = \dfrac{t - 1}{t + 1}$ **30.** $k(t) = \dfrac{3t + 5}{t - 4}$

In Exercises 31–40 find an interval on which f has an inverse. (*Hint:* Find an interval on which $f' > 0$ or on which $f' < 0$.)

31. $f(x) = x^2 - 4$ **32.** $f(x) = x^2 - 3x + 2$

33. $f(x) = x^3 - 5x + 1$ **34.** $f(x) = x^3 + 5x + 1$

35. $f(x) = \dfrac{1}{1 + x^2}$ **36.** $f(x) = \dfrac{x}{1 + x^2}$

37. $f(x) = \cos x$ **38.** $f(x) = \tan x$

39. $f(x) = \sin^2 x$ **40.** $f(x) = \sec x$

In Exercises 41–48 use (8) to calculate $(f^{-1})'(c)$. (*Hint:* Find a by inspection.)

41. $f(x) = x^3 + 7$; $c = 6$ **42.** $f(x) = 5x^5 + 4x^3$; $c = 9$

43. $f(x) = x + \sin x$; $c = 0$ **44.** $f(x) = x + \sqrt{x}$; $c = 2$

45. $f(x) = 4 \ln x$; $c = 0$

46. $f(x) = \tan x$ for $-\pi/2 < x < \pi/2$; $c = \sqrt{3}$

47. $f(t) = 3t - (1/t^3)$ for $t < 0$; $c = -2$

*48. $f(t) = t \ln t$; $c = 2e^2$

In Exercises 49–54 find dx/dy.

49. $y = x^9 + 7x$ **50.** $y = x - 2/x$ for $x < 0$

51. $y = \ln (x^3 + 1)$ **52.** $y = x + \cos x$

53. $y = \sin x$ for $-\pi/2 < x < \pi/2$

54. $y = \tan x$ for $-\pi/2 < x < \pi/2$

55. Suppose the graph of the derivative of a function f is as shown in Figure 7.7. On which subintervals of $[1, 9]$ does f have an inverse?

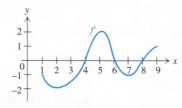

FIGURE 7.7   Figure for Exercise 55.

56. Suppose f has a continuous derivative on the interval $[0, 6]$. Assume also that f' is increasing on $[0, 4]$, f' is decreasing on $[4, 6]$, and

$$f'(0) = -1, \quad f'(3) = 0, \quad f'(4) = 2, \quad \text{and} \quad f'(5) = 0$$

On which subintervals of $[0, 6]$ does f have an inverse?

57. Let

$$f(x) = \int_0^x \sqrt{1 + t^4} \, dt \quad \text{for all } x$$

a. Show that f has an inverse.

b. Let $c = f(1)$. Find $(f^{-1})'(c)$. (*Hint:* Do not attempt to evaluate $f(1)$.)

58. Let

$$f(x) = \int_0^{x^3} \sin^6(t^2) \, dt \quad \text{for all } x$$

a. Show that f has an inverse.

b. Let $c = f(\sqrt[6]{\pi/6})$. Find $(f^{-1})'(c)$.

59. a. Can a polynomial of even degree have an inverse? Explain.

b. Can a polynomial of odd degree have an inverse? Explain.

60. Assume that f has an inverse.

a. Suppose the graph of f lies in the first quadrant. In which quadrant does the graph of f^{-1} lie?

b. Suppose the graph of f lies in the second quadrant. In which quadrant does the graph of f^{-1} lie?

61. Show that if f and g both have inverses, then $g \circ f$ has an inverse and

$$(g \circ f)^{-1} = f^{-1} \circ g^{-1}$$

62. Using (7), along with Exercises 1, 29, and 61, show that the following functions have inverses.

a. $k(x) = \left(\dfrac{x - 1}{x + 1}\right)^5$

b. $k(x) = \ln\left(\dfrac{x - 1}{x + 1}\right)$

c. $k(x) = \dfrac{\ln x - 1}{\ln x + 1}$

63. Assume that g has an inverse, and let $f(x) = -x$. Using Exercise 61, show that

$$(g \circ f)^{-1}(x) = -g^{-1}(x)$$

for all x in the domain of g^{-1}.

64. a. Show that the inverse of an increasing function is increasing.

b. Show that the inverse of a decreasing function is decreasing.

65. Assume that f has an inverse, and let a be a fixed number. Let

$$g(x) = f(x + a)$$

for all x such that $x + a$ is in the domain of f. Show that g has an inverse and that $g^{-1}(x) = f^{-1}(x) - a$.

66. Assume that f has an inverse, and let a be a fixed number different from 0. Let

$$g(x) = f(ax)$$

for all x such that ax is in the domain of f. Show that g has an inverse and that $g^{-1}(x) = f^{-1}(x)/a$.

67. Let $y = x^3$. Then

$$\frac{dy}{dx} = 3x^2 \quad \text{and} \quad \frac{dx}{dy} = \frac{1}{3y^{2/3}}$$

Show that equation (10) holds for this function if we evaluate dy/dx at $x = 2$ and dx/dy at $y = 8$ but that it does not hold if we evaluate dy/dx at $x = 2$ and dx/dy at $y = 2$.

68. Let f be a function with an inverse, and suppose $f'(a) = 0$. Show that f^{-1} is not differentiable at $f(a)$. (*Hint:* Prove by contradiction, using the Chain Rule and differentiating the equation $f^{-1}(f(x)) = x$ implicitly.)

69. Using Exercise 68, show that the following functions are not differentiable at the given value of c.

a. f^{-1}, where $f(x) = x + \sin x$; $c = \pi$

b. f^{-1}, where $f(x) = x^5 + x^3 - 4$; $c = -4$

70. Let $0 \le a < b$, and let f be nonnegative, increasing, and continuous on $[a, b]$, so that f^{-1} exists. Let A_1 and A_2 be the areas of the regions R_1 and R_2, respectively, in Figure 7.8.

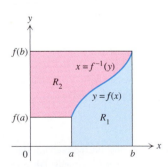

FIGURE 7.8 Figure for Exercise 70.

a. Show that $A_1 + A_2 = bf(b) - af(a)$.

b. Use (a) to prove that

$$\int_a^b f(x) \, dx = bf(b) - af(a) - \int_{f(a)}^{f(b)} f^{-1}(y) \, dy$$

71. Use Exercise 70(b) to evaluate $\int_0^1 [(x-1)^{1/3} + 1]^{1/2}\, dx$.

72. a. Let $f(x) = -\sqrt{x}$ for $x > 0$. Show that the graphs of f and f^{-1} are concave upward on their respective domains.

b. Let $f(x) = \sqrt{x}$ for $x > 0$. Show that the graph of f is concave downward on $(0, \infty)$, whereas the graph of f^{-1} is concave upward on $(0, \infty)$.

*73. Assume that f has an inverse, that $f''(f^{-1}(x))$ exists, and that $f'(f^{-1}(x)) \neq 0$. Show that

$$(f{-}1)''(x) = \frac{-f''(f^{-1}(x))}{[f'(f^{-1}(x))]^3}$$

74. Let the domain of f be an open interval I, and assume that f^{-1} exists. Suppose f'' exists on I and $f'(x) \neq 0$ and $f''(x) \neq 0$ for all x in I.

a. Suppose f is increasing on I. Show that the graph of f^{-1} is concave upward on its domain if the graph of f is concave downward on I, and the graph of f^{-1} is concave downward on its domain if the graph of f is concave upward on I. (*Hint:* Use Exercise 73.)

b. Suppose f is decreasing on I. Show that the graph of f^{-1} is concave upward on its domain if the graph of f is concave upward on I, and the graph of f^{-1} is concave downward on its domain if the graph of f is concave downward on I.

c. Suppose that a is in I and that the graph of f has an inflection point at $(a, f(a))$. What can you say about an inflection point for the graph of f^{-1}?

75. Let $f(x) = x^{2/3}$ for $0 \leq x \leq 8$. Find the length L of the graph of f. (*Hint:* Since $f'(0)$ does not exist, the formula in Section 6.2 for length does not apply. However, L equals the length of the graph of f^{-1}.)

*76. Suppose a person is traversing a path that has the shape of the graph of $y = x^{2/3}$ from the point $(0, 0)$ to the point $(1, 1)$. Find the halfway point on the curve. (*Hint:* Use the idea of the solution of Exercise 75.)

Applications

77. Let f be the function representing the conversion from inches to centimeters, and let g be the function representing the conversion from centimeters to inches.

Then

$$f(x) = 2.54x \quad \text{for } x \geq 0$$

and

$$g(x) = \frac{1}{2.54}x \quad \text{for } x \geq 0$$

Show that f and g are inverses of one another.

78. Let

$$f(v) = \frac{m_0}{\sqrt{1 - v^2/c^2}} \quad \text{for } v \geq 0$$

and

$$g(m) = c\sqrt{1 - m_0^2/m^2} \quad \text{for } m \geq m_0$$

where m_0 and c are constants. Show that g is the inverse of f. (The functions f and g arise in the theory of relativity. If c is the speed of light in a vacuum and m_0 is the rest mass of a particle, then $f(v)$ is the mass of the particle as it moves with velocity v, and $g(m)$ is the velocity of the particle when it has mass m.)

Project

1. In this project we will consider functions that are their own inverses, that is, that satisfy $f = f^{-1}$. To be more precise, f and f^{-1} have the same domains, and $f^{-1}(x) = f(x)$ for all x in the domain of f and of f^{-1}.

a. Show that the following functions are their own inverses.

i. $f(x) = x$

ii. $f(x) = -x$

iii. $f(x) = a/x$ for $x > 0$, where a is any fixed nonzero number

iv. $f(x) = \sqrt{r^2 - x^2}$ for $0 \leq x \leq r$, where r is any fixed positive number

v. $f(x) = -\sqrt{r^2 - x^2}$ for $-r \leq x \leq 0$ where r is any fixed positive number

b. What can you say about the graphs of functions that are their own inverses?

c. Find another function g that is its own inverse.

d. Show that if h is its own inverse and if h is also increasing, then $h(x) = x$.

7.2 THE NATURAL EXPONENTIAL FUNCTION

The natural exponential function is one of the most important functions in mathematics and has many applications, as we have already seen in Section 4.4 on exponential growth and decay.

In Section 1.8 we defined the natural exponential function informally. Now that we have a formal definition of the natural logarithm function, it is possible to define e^x formally, thereby eliminating the assumptions we have made about e^x. Moreover, from the definition to be given in this section, we will be able to prove, without resorting to numerical calculation, that e^x is differentiable and its derivative is e^x.

Recall from Section 5.7 that the natural logarithm is formally defined by

$$\ln x = \int_1^x \frac{1}{t}\, dt \quad \text{for } x > 0$$

This implies that

$$\frac{d}{dx} \ln x = \frac{1}{x} > 0 \quad \text{for all } x > 0$$

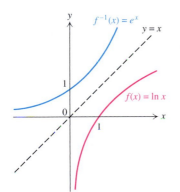

so that $\ln x$ is increasing on $(0, \infty)$ and hence has an inverse. We call the inverse the **natural exponential function** and denote its value at x by e^x. Since the domain of e^x is the range of $\ln x$, and vice versa, we conclude that

$$y = e^x \quad \text{if and only if} \quad x = \ln y \quad \text{for any } x \text{ and for } y > 0$$

In particular, since $\ln 1 = 0$, it follows that $e^0 = 1$. Furthermore, Theorem 7.2 implies that

$$e^{\ln x} = x \quad \text{for } x > 0 \tag{1}$$

$$\ln e^x = x \quad \text{for all } x \tag{2}$$

FIGURE 7.9

By reflecting the graph of $\ln x$ through the line $y = x$, we immediately obtain the graph of e^x (Figure 7.9). Since

$$\lim_{x \to \infty} \ln x = \infty \quad \text{and} \quad \lim_{x \to 0^+} \ln x = -\infty$$

we conclude that

$$\lim_{x \to \infty} e^x = \infty \quad \text{and} \quad \lim_{x \to -\infty} e^x = 0 \tag{3}$$

These limits are corroborated by the graph of e^x in Figure 7.9.

The Law of Exponents can be derived from the Law of Logarithms.

THEOREM 7.6
Law of Exponents

For all numbers b and c,

$$e^{b+c} = e^b e^c \tag{4}$$

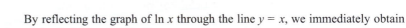

Proof From (2) we know that

$$\ln(e^{b+c}) = b + c$$

The Law of Logarithms and further applications of (2) yield

$$\ln(e^b e^c) = \ln e^b + \ln e^c = b + c$$

Therefore

$$\ln(e^{b+c}) = \ln(e^b e^c)$$

Since $\ln x = \ln y$ only if $x = y$, it follows that

$$e^{b+c} = e^b e^c \qquad \blacksquare$$

If $c = -b$, then (4) becomes

$$e^{b-b} = e^b e^{-b}$$

Since $e^{b-b} = e^0 = 1$, it follows that $1 = e^b e^{-b}$, so that

$$e^{-b} = \frac{1}{e^b} \tag{5}$$

Next we turn to the derivative of e^x. Since the natural logarithm function is differentiable, (10) of Section 7.1 implies that if $y = e^x$, then

$$\frac{d}{dx} e^x = \frac{1}{\dfrac{d}{dy} \ln y} = \frac{1}{\dfrac{1}{y}} = y = e^x$$

Therefore

$$\frac{d}{dx} e^x = e^x \tag{6}$$

In other words, the natural exponential function *is its own derivative.*

Using (6) and the Chain Rule, we can differentiate many functions formed from the exponential function. In particular, for any differentiable function f,

$$\frac{d}{dx} e^{f(x)} = f'(x) e^{f(x)} \tag{7}$$

For example

$$\frac{d}{dx} e^{x \sin x} = (\sin x + x \cos x) e^{x \sin x}$$

The integral counterparts of the formulas in (6) and (7) are

$$\int e^x \, dx = e^x + C$$

and

$$\int f'(x) e^{f(x)} \, dx = e^{f(x)} + C \tag{8}$$

The formula in (8) can also be verified by substituting $u = f(x)$.

EXAMPLE 1   Evaluate

$$\int_0^\pi (\sin x) e^{\cos x}\, dx$$

Solution   Let

$$u = \cos x, \quad \text{so that} \quad du = -\sin x\, dx$$

For the change in the limits of integration we find that

$$\text{if } x = 0 \quad \text{then } u = 1, \quad \text{and} \quad \text{if } x = \pi \quad \text{then } u = -1$$

Consequently

$$\int_0^\pi (\sin x) e^{\cos x}\, dx = \int_0^\pi \overbrace{e^{\cos x}}^{e^u} \overbrace{\sin x\, dx}^{-du} = -\int_1^{-1} e^u\, du$$

$$= -e^u \Big|_1^{-1} = -e^{-1} - (-e^1)$$

$$= e - e^{-1} \qquad \square$$

Normal Density Functions

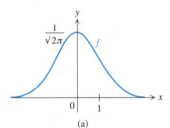

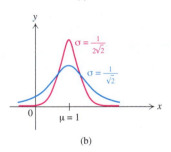

FIGURE 7.10   (a) The normal density function

$$f(x) = \frac{1}{\sigma \sqrt{2\pi}}\, e^{-(x-\mu)^2/2\sigma^2}$$

(b) Two normal density functions with the same value of μ, but different values of σ.

One of the most famous curves in the literature is the graph of the function f defined by

$$f(x) = \frac{1}{\sqrt{2\pi}}\, e^{-x^2/2}$$

and called the **standard normal density.** You have undoubtedly seen the bell-shaped graph (Figure 7.10(a)) in connection with data that are said to be "normally distributed," like test scores or heights of people.

The symmetry, the maximum value, and the horizontal asymptote for f are obvious from the graph. What may not be so obvious is the location of the points of inflection.

EXAMPLE 2   Find the points of inflection of the graph of the standard normal density f.

Solution   By the Chain Rule,

$$f'(x) = \frac{1}{\sqrt{2\pi}}\, e^{-x^2/2}\left(-\frac{2x}{2}\right) = \frac{-x}{\sqrt{2\pi}}\, e^{-x^2/2}$$

and

$$f''(x) = -\frac{1}{\sqrt{2\pi}}\, e^{-x^2/2} - \frac{x}{\sqrt{2\pi}}\, e^{-x^2/2}\left(-\frac{2x}{2}\right) = \frac{1}{\sqrt{2\pi}}\,(x^2 - 1)e^{-x^2/2}$$

Thus f'' changes from positive to negative at -1 and from negative to positive at 1, so that the inflection points are

$$\left(-1, \frac{1}{\sqrt{2\pi}}\, e^{-1/2}\right) \quad \text{and} \quad \left(1, \frac{1}{\sqrt{2\pi}\, e^{-1/2}}\right) \qquad \square$$

More generally, a **normal density** function is defined by

$$f(x) = \frac{1}{\sigma\sqrt{2\pi}} e^{-(x-\mu)^2/2\sigma^2}$$

where μ and σ are constants that together determine the complexion of the graph of f (Figure 7.10(b)):

1. The graph of f is symmetric with respect to the vertical line $x = \mu$.

2. The maximum value of f is $f(\mu) = \dfrac{1}{\sigma\sqrt{2\pi}}$ (see Exercise 43).

3. The graph of f has inflection points

$$\left(\mu - \sigma, \frac{1}{\sigma\sqrt{2\pi}} e^{-1/2}\right) \quad \text{and} \quad \left(\mu + \sigma, \frac{1}{\sigma\sqrt{2\pi}} e^{-1/2}\right)$$

(see Exercise 44).

In Section 8.7 we will return to the normal density function and relate it to the study of probability.

EXERCISES 7.2

In Exercises 1–4 find the derivative of the function.

1. $f(x) = \ln(e^x + e^{-x})$ **2.** $f(x) = (x+1)e^{-x}$

3. $y = \dfrac{e^x + 1}{e^x - 1}$ **4.** $y = e^{2x} \ln x$

In Exercises 5–6 find dy/dx by implicit differentiation.

5. $x^2 e^y = \ln xy$ **6.** $e^{xy} = y^2 - x$

7. Let $f(x) = e^x \sin x$.

   a. Find $f^{(8)}(x)$.   **b.** Guess the 80th derivative of f.

8. Let $f(x) = xe^x$. Find a formula for the nth derivative of f, where n is any positive integer.

9. Let $y = xe^{2x}$. Show that $x\dfrac{dy}{dx} = y(1 + 2x)$.

10. Let $y = xe^{-x^2/2}$. Show that $x\dfrac{dy}{dx} = y(1 - x^2)$.

In Exercises 11–12 sketch the graph of the function, noting any relative extreme values, concavity, inflection points, and asymptotes.

11. $f(x) = e^x + e^{-x}$ **12.** $f(x) = \dfrac{e^x}{1 + e^x}$

In Exercises 13–23 find the integral.

13. $\displaystyle\int e^{ex}\, dx$ **14.** $\displaystyle\int e^{\sqrt{2}\,x+3}\, dx$

15. $\displaystyle\int_0^{\pi/3} \frac{e^{\tan y}}{\cos^2 y}\, dy$ **16.** $\displaystyle\int_{-1}^{1} \frac{e^y - e^{-y}}{e^y + e^{-y}}\, dy$

17. $\displaystyle\int \frac{e^{2t}}{\sqrt{e^{2t} - 4}}\, dt$ **18.** $\displaystyle\int e^t\, e^{(e^t)}\, dt$

19. $\displaystyle\int \frac{e^{-t}\ln(1 + e^{-t})}{1 + e^{-t}}\, dt$ **20.** $\displaystyle\int \sqrt{e^x}\, dx$

21. $\displaystyle\int \frac{1}{1 + e^{-x}}\, dx$ ***22.** $\displaystyle\int \frac{1}{1 + e^x}\, dx$

*23. $\displaystyle\int e^{(x - e^x)}\, dx$

24. Show that

$$\int_0^1 e^{(1-t)a}\, e^{tb}\, dt = \frac{e^b - e^a}{b - a} \quad \text{for } a \neq b$$

25. Find the point at which the line $y = -4x - 7$ is tangent to the graph of $y = e^{x^2 - 4}$.

*26. Let $f(x) = e^x$ and $g(x) = \ln x$. Show that the graphs of f and g do not intersect.

In Exercises 27–28 find a formula for the inverse of the function.

27. $f(x) = \dfrac{e^x - 1}{e^x + 1}$ **28.** $f(x) = \dfrac{3e^x - 2}{e^x + 4}$

In Exercises 29–30 use the Newton-Raphson method to approximate the smallest solution of the equation. Continue until successive approximations obtained by calculator are identical.

29. $e^{-x} = 2 - x$ **30.** $e^x = 3x$

31. Let $f(x) = \frac{1}{2}(e^x + e^{-x})$ for $x \geq 0$ and $g(x) = \ln(x + \sqrt{x^2 - 1})$ for $x \geq 1$. Plot the graphs of f and g on the same screen. From the plot, do you think that g could be the inverse of f? (We will return to this question in the project for Section 7.4.)

32. Find the value of c such that $\int_0^c e^{-x}\, dx = \frac{1}{2}$.

33. Find the area A of the region bounded by the graphs of $y = e^{2x}$ and $y = e^{-2x}$ and the line $x = 1/2$.

34. Find the area A of the region in the first quadrant bounded by the curves $y = 3e^x$ and $y = 2 + e^{2x}$.

35. Let $f(x) = e^{(x^2)}$ and $g(x) = e^{-x^2}$. Find the volume V of the solid obtained by revolving about the y axis the region between the graphs of f and g on $[0, 1]$.

36. Find the volume V of the solid obtained by revolving about the line $y = 1$ the region between the graph of the equation $y = e^{-2x}$ and the x axis on the interval $[0, 1]$. (*Hint:* Use Exercise 37 of Section 6.1.)

37. Find the length L of the curve described parametrically by $x = e^t \sin t$ and $y = e^t \cos t$ for $0 \leq t \leq \pi$.

38. Let $f(x) = e^x + \frac{1}{4}e^{-x}$ for $0 \leq x \leq 1$.
 a. Determine the length L of the graph of f.
 b. Determine the surface area S of the surface obtained by revolving the graph of f about the x axis.

39. Evaluate $\int_1^e \ln x\, dx$ by using the formula

$$\int_a^b f(x)\, dx = bf(b) - af(a) - \int_{f(a)}^{f(b)} f^{-1}(y)\, dy$$

appearing in Exercise 70 of Section 7.1.

40. Let R be a positive number. Show that

$$\int_0^{\pi/2} e^{-R \sin x}\, dx < \frac{\pi}{2R}$$

which is an inequality known as **Jordan's inequality.** (*Hint:* By Review Exercise 18 in Chapter 4, $\sin x \geq 2x/\pi$ for $0 \leq x \leq \pi/2$.)

41. Let f be a function satisfying $f(0) = 1$ and $f(a + b) = f(a)f(b)$ for all real numbers a and b. Suppose that f is differentiable at 0, with $f'(0) = 1$.
 a. Using the definition of derivative, show that f is differentiable on $(-\infty, \infty)$, and that $f'(x) = f(x)$ for all x. (*Hint:* Use (2) of Section 3.2.)
 b. Show that

$$\frac{d}{dx}(e^{-x} f(x)) = 0 \quad \text{for all } x$$

 c. Use part (b), along with Theorem 4.6, to show that $f(x) = e^x$ for all x.

42. This exercise is designed to give an alternative proof of the Law of Exponents (Theorem 7.6).
 a. Show that for any real number b,

$$\frac{d}{dx}(e^{-x} e^{b+x}) = 0$$

 b. Using (a) and Theorem 4.6, show that $e^{-x} e^{b+x} = e^b$ for all x.
 c. Use (b) to prove that $e^{-x} = 1/e^x$.
 d. Use (b) and (c) to prove that $e^{b+c} = e^b e^c$.

Exercises 43–44 refer to a normal density f with mean μ and standard deviation σ.

43. Show that the maximum value of f is $f(\mu) = 1/(\sigma \sqrt{2\pi})$.

44. Determine the inflection points of the graph of f.

Applications

45. Let the path C of a particle in motion be given parametrically by

$$x = f(t) \quad \text{and} \quad y = g(t) \quad \text{for all } t$$

where $f'(t) = 2f(t)$ and $g'(t) = 6g(t)$ for all t.
 a. Solve the equations for f and g, and then write y as a function of x.
 b. Suppose $f(0) = 2$ and $g(0) = 3$. Determine the curve C, and sketch it.

46. The **Gompertz growth curve,** which appears in population studies, is the graph of the function f defined by

$$g(x) = ae^{(-be^{-cx})} \quad \text{for } x \geq 0$$

where a, b, and c are positive constants with $b > 1$. Show that the inflection point of the graph of g is $((\ln b)/c, a/e)$.

47. If an electrical circuit has a capacitor with a capacitance C, a battery with voltage V, and a resistor with resistance R connected in series, then the charge $Q(t)$ on the capacitor at time $t \geq 0$ is governed by the equation

$$Q(t) = VC + ce^{-t/RC}$$

where c is a suitable constant.
 a. If $Q(0) = 0$, find c.
 b. Assuming that C, V, and R are positive and that $Q(0) = 0$, sketch the graph of Q.
 c. At what time t is the capacitor 90 percent charged? (In other words, for what t is $Q(t) = 0.9\, VC$?)

Project

1. This project is the counterpart for the natural exponential function of the project in Section 5.7 in which we showed that with mild additional conditions, if a function satisfies the "law of logarithms," then it is guaranteed to be the natural logarithm function.

   Let f be a function satisfying $f(0) = 1$ and $f(a + b) = f(a)f(b)$ for all real numbers a and b. Assume also that f is differentiable at 0, and let $k = f'(0)$.

   a. Show that $f(x) > 0$ for all x.

   b. Using given properties of f and general derivative formula (2) in Section 3.2, show that f is differentiable on $(-\infty, \infty)$ and in fact $f'(x) = kf(x)$ for all x.

   c. Show that

   $$\frac{d}{dx}(e^{-x}f(x)) = 0 \quad \text{for all } x$$

   d. Use part (c), along with Theorem 4.6, to show that $f(x) = e^{kx}$ for all x.

   e. Instead of assuming that $f(0) = 1$, suppose we assume that $f(0) = r$ for an arbitrary value of r. How would the resulting function be different from the natural exponential function?

7.3   GENERAL EXPONENTIAL AND LOGARITHMIC FUNCTIONS

In Section 7.2 we formally defined the natural exponential function e^x. In this section we will use the natural exponential function to formally define a^r, where a is any positive number and r is any real number. As we will illustrate in the exercises, functions of the form a^x appear in physical applications, from the dimensions of fractal sets to diffusion processes.

In defining a^r we would want to make certain that the exponential law

$$(u^v)^w = u^{vw} \tag{1}$$

is valid. This requirement actually determines a unique choice for the value of a^r. Let us see why. Recall from (1) of Section 7.2 that $a = e^{\ln a}$ for all $a > 0$. Therefore if (1) is to hold, then we must have

$$a^r = (e^{\ln a})^r = e^{r \ln a}$$

Consequently we define

$$a^r = e^{r \ln a} \quad \text{for any real number } r \tag{2}$$

Notice that the expression $e^{r \ln a}$ makes sense for every positive number a and every real number r. In the expression a^r, the number a is the **base** and r is the **exponent** (or **power**).

From the definition of a^r, we observe that

$$a^0 = e^{0 \ln a} = e^0 = 1$$

and

$$a^1 = e^{1 \ln a} = e^{\ln a} = a$$

Thus

$$a^0 = 1 \quad \text{and} \quad a^1 = a \tag{3}$$

Next we take the natural logarithms of both sides of the equation in (2). We find that

$$\ln a^r = \ln e^{r \ln a}$$

By (2) of Section 7.2,

$$\ln e^{r \ln a} = r \ln a$$

so we conclude that

$$\ln a^r = r \ln a \qquad (4)$$

From the Law of Exponents (Theorem 7.6), along with the definition of a^r, we can prove very general laws of exponents.

THEOREM 7.7
General Laws of Exponents

Let $a > 0$, and let b and c be any numbers. Then the following formulas hold:

  a. $a^{b+c} = a^b a^c$

  b. $a^{-b} = 1/a^b$

  c. $(a^b)^c = a^{bc}$

Proof   To prove (a), we use the definition of a^r, along with the Law of Exponents:

$$a^{b+c} = e^{[(b+c)\ln a]} = e^{(b \ln a + c \ln a)} = e^{b \ln a} e^{c \ln a} = a^b a^c$$

To prove (b), we take $c = -b$ in (a) and refer to (3) to obtain

$$1 \overset{(3)}{=} a^0 = a^{b-b} = a^b a^{-b}$$

Dividing by a^b (which is never 0 because e^x is never 0), we conclude that

$$a^{-b} = \frac{1}{a^b}$$

To prove (c), we use (4), obtaining

$$(a^b)^c = e^{c \ln (a^b)} \overset{(4)}{=} e^{bc \ln a} = a^{bc} \qquad \blacksquare$$

From the General Laws of Exponents it follows, for example, that

$$\pi^2 \pi^{-4} = \pi^{2-4} = \pi^{-2}$$

However, the laws apply even when the exponents are not rational. For instance,

$$\pi^{\sqrt{2}} \pi^{\sqrt{5}} = \pi^{\sqrt{2}+\sqrt{5}}$$

Exponential and Power Functions

In (2), if the base a stays fixed and r varies, we obtain the exponential function a^x defined by

$$a^x = e^{x \ln a} \qquad (5)$$

The derivative of the function a^x is obtained by differentiating $e^{x \ln a}$ by means of the Chain Rule:

$$\frac{d}{dx} a^x = \frac{d}{dx} (e^{x \ln a}) = (e^{x \ln a})(\ln a) = (\ln a)a^x$$

or in more condensed form,

$$\frac{d}{dx} a^x = (\ln a)a^x \tag{6}$$

Thus the derivative of a^x is merely a constant multiple of a^x. Moreover, since $a^x > 0$ for all x and $\ln a > 0$ only for $a > 1$, (6) implies that a^x is increasing if $a > 1$ and is decreasing if $0 < a < 1$. It also follows from (6) that

$$\frac{d^2}{dx^2} a^x = (\ln a)^2 \, a^x$$

so that if $a \neq 1$, the graph of a^x is concave upward on $(-\infty, \infty)$. Finally, formula (5) above, along with (3) of Section 7.2, implies that

$$\text{if } a > 1, \quad \text{then } \lim_{x \to \infty} a^x = \infty \quad \text{and} \quad \lim_{x \to -\infty} a^x = 0$$

whereas

$$\text{if } 0 < a < 1, \quad \text{then } \lim_{x \to \infty} a^x = 0 \quad \text{and} \quad \lim_{x \to -\infty} a^x = \infty$$

With this information we can draw the graph of a^x. As we notice in Figure 7.11, the appearance of the graph of a^x depends on the value of a.

Although the natural exponential function is by far the most prevalent of the exponential functions occurring in applications, exponential functions with other bases do appear. (See Exercise 43.)

If in (2) the base varies and the exponent r is fixed, we obtain the **power function** x^r defined by

$$x^r = e^{r \ln x} \tag{7}$$

The derivative of x^r is obtained by using the Chain Rule:

$$\frac{d}{dx} x^r = \frac{d}{dx} (e^{r \ln x}) = e^{r \ln x} \left(\frac{r}{x}\right) = \frac{r}{x} x^r = rx^{r-1}$$

or more simply,

$$\frac{d}{dx} x^r = rx^{r-1} \tag{8}$$

This formula is familiar from (9) of Section 3.4. It follows from (8) that

$$\frac{d^2}{dx^2} x^r = \frac{d}{dx} (rx^{r-1}) = r(r-1)x^{r-2}$$

By using the derivatives of x^r, we can draw the graph of x^r. As you might well expect, the appearance of the graph depends on the value of r, as shown in Figure 7.12.

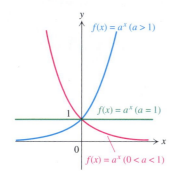

FIGURE 7.11

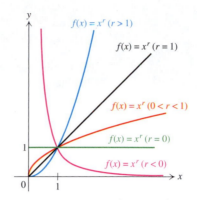

FIGURE 7.12

It is also possible for *both* the base and the exponent in a^r to vary; in that case, we obtain a function of the form $f(x)^{g(x)}$. Then by (2),

$$f(x)^{g(x)} = e^{g(x)\ln f(x)} \tag{9}$$

Any time we must perform an operation (such as differentiation) on a function of the form $f(x)^{g(x)}$, we use (9) to write the function with base e.

EXAMPLE 1   Let $f(x) = (2x)^{\sin x}$. Find $f'(x)$.

Solution   By (9) we know that

$$f(x) = (2x)^{\sin x} = e^{\sin x \ln 2x}$$

Using the Chain Rule, we find that

$$f'(x) = e^{\sin x \ln 2x} \left[\cos x \ln 2x + \frac{(\sin x)2}{2x} \right]$$

$$= (2x)^{\sin x} \left(\cos x \ln 2x + \frac{\sin x}{x} \right) \qquad \Box$$

Now we turn to integrals of a^x and x^r. If we divide both sides of (6) by $\ln a$, we obtain

$$\frac{d}{dx} \left(\frac{1}{\ln a} a^x \right) = a^x$$

In integral form this becomes

$$\int a^x \, dx = \frac{1}{\ln a} a^x + C \tag{10}$$

EXAMPLE 2   Evaluate $\displaystyle\int_0^3 3^x \, dx$.

Solution   By (10) we have

$$\int_0^3 3^x \, dx = \frac{1}{\ln 3} \, 3^x \Big|_0^3 = \frac{1}{\ln 3} \, 3^3 - \frac{1}{\ln 3} \, 3^0 = \frac{27}{\ln 3} - \frac{1}{\ln 3} = \frac{26}{\ln 3} \qquad \square$$

The formula for the integral of x^r is obtained from (8) by first replacing r by $r + 1$ to obtain

$$\frac{d}{dx}(x^{r+1}) = (r+1)x^r$$

and then dividing both sides by $r + 1$, which produces

$$\frac{d}{dx}\left(\frac{1}{r+1} \, x^{r+1}\right) = x^r \quad \text{for } r \neq -1$$

In integral form this becomes

$$\int x^r \, dx = \frac{1}{r+1} \, x^{r+1} + C \quad \text{for } r \neq -1 \tag{11}$$

EXAMPLE 3   Evaluate $\displaystyle\int_1^2 x^{\sqrt{3}} \, dx$.

Solution   By (11),

$$\int_1^2 x^{\sqrt{3}} \, dx = \frac{1}{\sqrt{3}+1} \, x^{\sqrt{3}+1} \Big|_1^2 = \frac{1}{\sqrt{3}+1}(2^{\sqrt{3}+1}) - \frac{1}{\sqrt{3}+1}(1^{\sqrt{3}+1})$$

$$= \frac{2(2^{\sqrt{3}}) - 1}{\sqrt{3}+1} \qquad \square$$

Logarithms to Different Bases   In the preceding subsection we observed that a^x is increasing for $a > 1$ and decreasing for $0 < a < 1$. Therefore, provided that $a > 0$ and $a \neq 1$, there is an inverse for a^x, which we denote by $\log_a x$ and call the **logarithm to the base a.** When $a = e$, we obtain the natural logarithm function, which we write as before as $\ln x$; thus

$$\ln x = \log_e x$$

For the special case $a = 10$, $\log_{10} x$ is called the **common logarithm of x** and is usually written $\log x$.

For logarithms to the base a, parts (b) and (c) of Theorem 7.2 become

$$a^{\log_a x} = x \tag{12}$$

and

$$\log_a a^x = x \tag{13}$$

We can show that the logarithm to any base is a constant multiple of the natural logarithm.

THEOREM 7.8

For any $a > 0$ such that $a \neq 1$ we have

$$\log_a x = \frac{\ln x}{\ln a} \quad \text{for all } x > 0 \tag{14}$$

Proof   The simplest way to verify (14) is to take natural logarithms of both sides of (12) and use (4) with $\log_a x$ replacing r:

$$\ln x = \ln (a^{\log_a x}) = (\log_a x)(\ln a)$$

Since $a \neq 1$ by assumption, it follows that $\ln a \neq 0$, so that

$$\log_a x = \frac{\ln x}{\ln a} \quad \blacksquare$$

From (14) we can easily find the derivative of $\log_a x$:

$$\frac{d}{dx} \log_a x = \frac{d}{dx} \frac{\ln x}{\ln a} = \frac{1}{\ln a} \frac{d}{dx} \ln x = \frac{1}{x \ln a}$$

Formula (14) also makes it possible to calculate $\log_a x$ once we know the natural logarithms of x and a. Since most calculators have keys for natural logarithms, the conversion formula is very useful.

EXAMPLE 4   Calculate $\log_2 3$ by using natural logarithms.

Solution   By (14),

$$\log_2 3 = \frac{\ln 3}{\ln 2}$$

By calculator we find that

$$\log_2 3 \approx 1.584962501 \quad \square$$

Another consequence of (14) is the General Law of Logarithms.

THEOREM 7.9
General Law of Logarithms

If a, b, and c are positive, then

$$\log_a bc = \log_a b + \log_a c$$

Proof   By virtue of (14) and the original Law of Logarithms (Theorem 5.21), we find that

$$\log_a bc \overset{(14)}{=} \frac{\ln bc}{\ln a} = \frac{\ln b + \ln c}{\ln a} = \frac{\ln b}{\ln a} + \frac{\ln c}{\ln a} \overset{(14)}{=} \log_a b + \log_a c \quad \blacksquare$$

From (4) and (14) it follows that for any r and any positive a and b,

$$\boxed{\log_a b^r = r \log_a b}$$

Finally, (14) enables us to determine the general shape of the graph of $\log_a x$ from the graph of $\ln x$ because $\log_a x$ is just a constant multiple of $\ln x$. Figure 7.13 shows the graphs for $a > 1$ and for $0 < a < 1$.

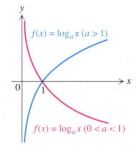

$f(x) = \log_a x \, (a > 1)$

$f(x) = \log_a x \, (0 < a < 1)$

FIGURE 7.13

Fractals

Numbers such as $\log_2 3$ appear in the study of geometric fractals. Consider the set G obtained by deleting the central triangle from a given triangle (Figure 7.14(a)), and then performing the same operation infinitely often on the resulting triangles (Figure 7.14(b) and (c)). The set G is called a **Sierpinski gasket,** after the Polish mathematician Waclaw Sierpinski (1882–1969).

(a)   (b)   (c)

FIGURE 7.14   The construction of a Sierpinski gasket.

More generally, let a set S^* in the plane or in space be created in the following way:

1. Subdivide a given set S into n nonoverlapping congruent subsets, each of which is a reduced copy of S. Let r denote the ratio of the dimension of the subsets to that of S.

2. Retain m of the subsets $S_1, S_2, \ldots, S_m$.

3. Divide each of the subsets $S_1, S_2, \ldots, S_m$ into n congruent subsets, and retain m of the subsets of each S_k.

4. Continue the process indefinitely to obtain S^*.

> **Origin of "Fractal"**
>
> The term "fractal" was introduced by the contemporary mathematician Benoit Mandelbrot for sets that have dimensions that are fractional.

The set S^* has the property that no matter how much we zoom in on a portion of it, the same shapes appear in ever smaller form. For this reason the set is called **self-similar,** and is often referred to as a **fractal.** The **fractal dimension** of S^* is defined to be $\log_{1/r} m$. In the case of the Sierpinski gasket, the original subset is subdivided into four subtriangles, each side of which is half as long as the corresponding side of the original triangle. Thus $r = 1/2$. Since three are retained, $m = 3$. Consequently the fractal dimension of S^* is $\log_2 3 \approx 1.58$.

EXERCISES 7.3

In Exercises 1–10 find the derivative of the function.

1. $g(x) = 2^{-x}$

2. $g(x) = \log_{10} x$

3. $y = t^t$

4. $y = t^{\sin t}$

5. $y = t^{2/t}$

6. $y = \left(1 + \dfrac{1}{t}\right)^t$

7. $f(x) = (\cos x)^{\cos x}$

8. $f(x) = (\ln x)^{\ln x}$

9. $f(x) = (2x)^{\sqrt{2}}$

10. $f(x) = (\cos x)^{\pi x^2}$

In Exercises 11–12 sketch the graph of the function, noting all pertinent information.

11. $f(x) = 2^x - 2^{-x}$

12. $f(x) = \log_2(1 + x^2)$

In Exercises 13–20 find the integral.

13. $\displaystyle\int 2^x \, dx$

14. $\displaystyle\int_1^2 10^x \, dx$

15. $\displaystyle\int_{-2}^0 3^{-x} \, dx$

16. $\displaystyle\int_0^1 4^{3x} \, dx$

17. $\displaystyle\int x \cdot 5^{-x^2} \, dx$

18. $\displaystyle\int (\log_2 x)/x \, dx$

19. $\displaystyle\int x^{2\pi}\,dx$ **20.** $\displaystyle\int (\sin x)^e \cos x\,dx$

21. Find the area A of the region between the graph of $y = x \cdot 2^{(x^2)}$ and the x axis on $[1, 2]$.

22. Find the area A of the region bounded by the graph of $y = 2^{-x} - \frac{1}{4}$ and the two axes.

23. Find the area A of the region between the graphs of $y = 2^x$ and $y = 3^x$ on $[0, 2]$.

24. Find the area A of the region bounded by the line $x = 16$ and the graphs of

$$y = \frac{3}{x} \quad \text{and} \quad y = \frac{\log_2 x}{x}$$

In Exercises 25–28 calculate the logarithm by using Theorem 7.8.

25. $\log_3 5$ **26.** $\log_{1/2} \frac{1}{3}$

27. $\log_\pi e$ **28.** $\log_{\sqrt{2}} \sqrt{\pi}$

29. Find the volume V of the solid generated by revolving about the x axis the region between the graph of $y = (\log_3 x)/\sqrt{x}$ and the x axis on $[1, 3]$.

30. Find the volume V of the solid generated by revolving about the line $y = -1$ the region between the graph of $y = 2^x$ and the x axis on $[0, 1]$.

31. By Exercise 55 of Section 5.7, $(\ln x)/x$ is increasing on $(0, e]$ and decreasing on $[e, \infty)$. Using this fact, prove that $a^b > b^a$ for $0 < b < a \le e$ and for $e \le a < b$.

32. Use Exercise 31 to show that
a. $\pi^e < e^\pi$
b. $2^{\sqrt{x}} > x$ for $0 < x < 4$

33. Let $f(x) = x \log_a x$. Determine the values of a for which the graph of f is concave upward on $(0, \infty)$ and the values of a for which the graph is concave downward on $(0, \infty)$.

34. Let $f(x) = (2^x + 1)/(2^x - 3)$. Find a formula for $f^{-1}(x)$.

35. Find the value of b such that the graph of $\log_2 (bx)$ is exactly 1.5 units above the graph of $\log_2 x$.

36. Use the General Law of Exponents to prove that
$$a^{b+c+d} = a^b a^c a^d \quad \text{for all } b, c, \text{ and } d \text{ and all } a > 0$$

37. a. Using Theorem 7.8, prove the change of base formula

$$\log_b x = \log_b a \log_a x$$

where a, b, and x are positive.
b. Using (a), find a formula that converts logarithms to the base 7 into logarithms to the base 4.

38. Let $r \neq 0$. Show that the functions x^r and $x^{1/r}$, defined for $x > 0$, are inverses of one another. (Thus even when r is irrational we might call $x^{1/r}$ the rth root of x.)

39. Let a be a positive number.

a. Show that there is exactly one square with one vertex at the origin and opposite vertex on the graph of $y = a/x^a$ for $x > 0$.
b. Find the area A of the square. Your answer should be a function of a.

Applications

40. Consider the set C obtained by deleting $(\frac{1}{3}, \frac{2}{3})$ from the interval $[0, 1]$, then deleting the open middle third from each remaining closed interval, and continuing the process indefinitely (Figure 7.15). The set C is the **Cantor set,** named after the Danish mathematician Georg Cantor (1845–1918), who was one of the fathers of modern set theory. Find the fractal dimension of C.

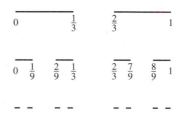

FIGURE 7.15 The construction of the Cantor set.

41. Consider the **Sierpinski carpet** S, obtained by subdividing the unit square into 9 congruent subsquares and deleting the middle one, then performing the same procedure on each of the remaining squares, and continuing the process indefinitely (Figure 7.16). Find the fractal dimension of S.

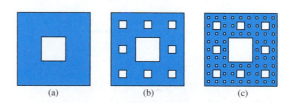

(a) (b) (c)

FIGURE 7.16 The contructions of the Sierpinski carpet.

42. Start with a set S in the plane (such as a square or triangle). Create other interesting fractal shapes S^* by the procedure described in the text. (Either use a computer, or draw the first few stages in the construction.)

43. Consider the diffusion of a new process into an already existing market. If $t = 0$ represents the date of introduction of the new process, then according to one model the cumulative level of diffusion y of the new process at any time $t > 0$ is given by

$$y = ka^{(b^t)}$$

where a, b, and k are constants that lie in the interval $(0, 1)$.

a. Find dy/dt, and show that it is positive for all $t > 0$.

b. What does the result of (a) imply about changes in the level of diffusion of the new process?

c. Find $\lim_{t \to \infty} y(t)$.

44. The radioactive isotope Polonium-214 has a half-life of 0.00014 seconds. Suppose 1 kilogram of the polonium appears in your sink.

a. Find the values of a and b such that the amount in grams after t seconds is given by $f(t) = a\left(\frac{1}{2}\right)^{bt}$.

b. After a $\frac{1}{100}$ second blink, would a gram of polonium remain? Explain your answer.

Project

1. Logarithms were invented by John Napier, who considered two particles, P and Q, traveling along paths described in Figure 7.17, according to the formulas

$$P'(t) = 10^7 - P(t) \quad \text{for } t > 0$$

$$Q(t) = 10^7 t \quad \text{for } t > 0$$

In this way P travels slower and slower on its finite path as it approaches the point a distance 10^7 from the origin, while Q travels at a constant rate on its infinitely long path. Napier defined his logarithmic function, which we denote by the symbol N, by the equation

$$N(10^7 - P(t)) = Q(t) = 10^7 t \quad \text{for } t > 0 \qquad (15)$$

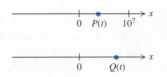

FIGURE 7.17   Graphs for the project.

In this project, you will analyze the function N.

a. By taking derivatives of both sides of (15) with respect to t, show that

$$[N'(10^7 - P(t))][P(t) - 10^7] = 10^7 \quad \text{for } t > 0 \qquad (16)$$

b. Letting $x = 10^7 - P(t)$ in (16), show that

$$N'(x) = \frac{-10^7}{x} \quad \text{for } 0 < x < 10^7$$

and therefore that

$$N(x) = -10^7 \ln x + C \quad \text{for } 0 < x \le 10^7$$

c. Using the fact that $N(10^7) = 0$, prove that

$$N(x) = -10^7 \ln x + 10^7 \ln 10^7 \quad \text{for } 0 < x \le 10^7$$

d. Let $b = e^{-1/10^7}$. (Try calculating this on your calculator!) Show that

$$-10^7 = \log_b e$$

With the help of Exercise 37(a) conclude that

$$-10^7 \ln x = (\log_b e) \ln x = \log_b x$$

e. Using (c) and (d), show that

$$N(x) = \log_b x + C \quad \text{for } 0 < x \le 10^7$$

where $C = 10^7 \ln 10^7$. This proves that N is actually the logarithm to the base b, shifted by C.

7.4   HYPERBOLIC FUNCTIONS

In this section we briefly describe the hyperbolic functions, which are defined in terms of the exponential function and have applications in engineering.

The Hyperbolic Functions

The two most important hyperbolic functions are defined as follows:

$$\sinh x = \frac{e^x - e^{-x}}{2} \quad \text{and} \quad \cosh x = \frac{e^x + e^{-x}}{2}$$

We read $\sinh x$ as "hyperbolic sine of x"; mathematicians often give it the shortened pronunciation "shin x" or "sinch x." Analogously, cosh x is read "hyperbolic cosine of x" or, for short, "cosh x."

Notice that sinh x is an odd function with sinh $0 = 0$, and that cosh x is an even function with cosh $0 = 1$. Since $0 < e^x < 1$ for $x < 0$, and $e^x > 1$ for $x > 0$, it follows that

$$\sinh x < 0 \quad \text{for } x < 0 \quad \text{and} \quad \sinh x > 0 \quad \text{for } x > 0 \tag{1}$$

$$\cosh x > 0 \quad \text{for all } x \tag{2}$$

The derivatives of sinh x and cosh x are given by

$$\frac{d}{dx} \sinh x = \frac{e^x + e^{-x}}{2} = \cosh x \tag{3}$$

$$\frac{d}{dx} \cosh x = \frac{e^x - e^{-x}}{2} = \sinh x \tag{4}$$

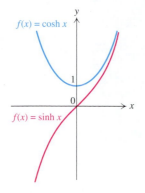

FIGURE 7.18

From (2) and (3) we conclude that sinh x is an increasing function, and from (1) and (4) we conclude that cosh x is decreasing on $(-\infty, 0]$ and increasing on $[0, \infty)$, so that cosh x has a minimum value at 0. Evidently

$$\frac{d^2}{dx^2} \sinh x = \sinh x \quad \text{and} \quad \frac{d^2}{dx^2} \cosh x = \cosh x$$

The graphs of these hyperbolic functions are shown in Figure 7.18.

Direct calculation shows that

$$\cosh^2 x - \sinh^2 x = \frac{e^{2x} + 2 + e^{-2x}}{4} - \frac{e^{2x} - 2 + e^{-2x}}{4} = 1$$

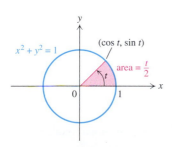

FIGURE 7.19

Thus

$$\boxed{\cosh^2 x - \sinh^2 x = 1} \tag{5}$$

the fundamental identity relating sinh x and cosh x. Formula (5) is very reminiscent of the fundamental trigonometric identity $\sin^2 x + \cos^2 x = 1$, and formulas (3) and (4) remind us of the formulas for the derivatives of the sine and the cosine. Whereas the point $(\cos t, \sin t)$ lies on the unit circle (Figure 7.19), formula (5) shows that $(\cosh t, \sinh t)$ lies on the hyperbola $x^2 - y^2 = 1$ (Figure 7.20). The unit circle and the hyperbola are related in yet another way. The region shaded in Figure 7.19 has area $t/2$, and the region shaded in Figure 7.20 also has area $t/2$ (see Exercise 47).

If c is a nonzero constant, then the graph of the function

$$y = c \cosh \frac{x}{c}$$

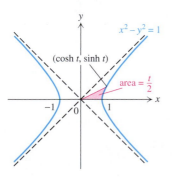

FIGURE 7.20

is called a **catenary,** from the Latin word *catena,* meaning chain. A flexible, inelastic chain of uniform density suspended at its two ends will take the shape of an arc of a catenary when it is subjected only to the influence of gravity (Figure 7.21). Thus a cable attached to telephone poles hangs in this shape.

Gateway Arch to the West
It turns out that certain structures are strongest when built in the shape of a catenary. With this in mind, the famous Finnish architect Eero Saarinen designed the "Gateway Arch to the West" in St. Louis in a shape closely related to a catenary (Figure 7.22).

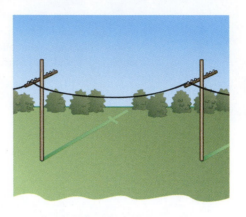

FIGURE 7.21 The cable hangs in the shape of a catenary.

Continuing the analogy of the hyperbolic functions with trigonometric functions, we define the other four hyperbolic functions in terms of $\sinh x$ and $\cosh x$:

$$\tanh x = \frac{\sinh x}{\cosh x} \qquad \operatorname{sech} x = \frac{1}{\cosh x}$$

$$\coth x = \frac{\cosh x}{\sinh x} \qquad \operatorname{csch} x = \frac{1}{\sinh x}$$

FIGURE 7.22 St. Louis Arch.
(*Bachmann/The Image Works*)

(see Figure 7.23(a)–(d)). The derivatives of these hyperbolic functions follow readily from the derivatives of $\sinh x$ and $\cosh x$:

$$\frac{d}{dx} \tanh x = \operatorname{sech}^2 x \qquad \frac{d}{dx} \operatorname{sech} x = -\operatorname{sech} x \tanh x$$

$$\frac{d}{dx} \coth x = -\operatorname{csch}^2 x \qquad \frac{d}{dx} \operatorname{csch} x = -\operatorname{csch} x \coth x$$

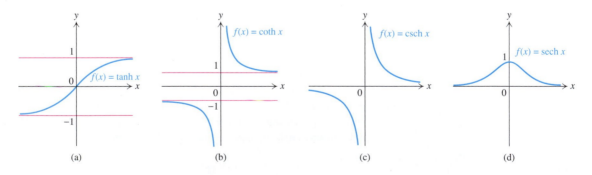

(a)   (b)   (c)   (d)

FIGURE 7.23

(see Exercises 17–20). The six hyperbolic functions are related by many identities, called **hyperbolic identities.** We list a few of them:

$$\tanh^2 x + \operatorname{sech}^2 x = 1 \qquad\qquad \coth^2 x - \operatorname{csch}^2 x = 1$$

$$\sinh(-x) = -\sinh x \qquad\qquad \cosh(-x) = \cosh x$$

$$\sinh(x \pm y) = \sinh x \cosh y \pm \cosh x \sinh y$$

$$\cosh(x \pm y) = \cosh x \cosh y \pm \sinh x \sinh y$$

$$\tanh(x \pm y) = \frac{\tanh x \ \pm \ \tanh y}{1 \ \pm \ \tanh x \tanh y}$$

These identities, which are easily verified from the definitions of the hyperbolic functions, are very similar to the corresponding trigonometric identities listed in Section 1.7. However, in one way the hyperbolic functions are very different from trigonometric functions. Whereas all the trigonometric functions are periodic, *none* of the hyperbolic functions is periodic. The reason is that the hyperbolic functions are defined from the exponential function, which is certainly not periodic.

The Inverse Hyperbolic Sine Function

As you might expect, the hyperbolic functions have inverses, at least when their domains are suitably restricted. Because we will have little need for them later on, we study only the inverse hyperbolic sine function, to give you an idea of how to analyze the inverse hyperbolic functions.

Since $\sinh x$ is an increasing function, and $(-\infty, \infty)$ is its domain and range, it follows that $\sinh x$ has an inverse with the same domain and range. We denote the value of the inverse at x by $\sinh^{-1} x$:

$$\sinh^{-1} x = y \quad \text{if and only if} \quad \sinh y = x \quad \text{for all } x \text{ and all } y$$

In order to derive a formula for $\sinh^{-1} x$ in terms of x, we note that

$$\sinh y = \frac{e^y - e^{-y}}{2} = \frac{e^{2y} - 1}{2e^y}$$

Hence if $x = \sinh y$, then

$$2e^y x = e^{2y} - 1, \quad \text{or} \quad (e^y)^2 - 2xe^y - 1 = 0$$

This equation is a quadratic equation in e^y, that is, has the form $z^2 + bz + c = 0$, where $z = e^y$, $b = -2x$, and $c = -1$. We use the quadratic formula to solve for e^y:

$$e^y = \frac{2x + \sqrt{4x^2 + 4}}{2} = x + \sqrt{x^2 + 1} \tag{6}$$

We can find y by taking natural logarithms in (6). This yields

$$y = \ln\left(x + \sqrt{x^2 + 1}\right)$$

Thus our formula for $\sinh^{-1} x$ is

$$\sinh^{-1} x = \ln\left(x + \sqrt{x^2 + 1}\right) \quad \text{for all } x \tag{7}$$

Differentiating (7), we obtain

$$\frac{d}{dx} \sinh^{-1} x = \frac{1 + \dfrac{2x}{2\sqrt{x^2 + 1}}}{x + \sqrt{x^2 + 1}} = \frac{1}{\sqrt{x^2 + 1}} \qquad (8)$$

The associated indefinite integral is

$$\int \frac{1}{\sqrt{1 + x^2}}\, dx = \sinh^{-1} x + C \qquad (9)$$

Formula (9) expands the class of functions that we can integrate.

EXAMPLE 1   Evaluate $\displaystyle\int_{-1}^{3} \frac{1}{\sqrt{1 + x^2}}\, dx$.

Solution   From (9) and (7) we obtain

$$\int_{-1}^{3} \frac{1}{\sqrt{1 + x^2}}\, dx \overset{(9)}{=} \sinh^{-1} x \Big|_{-1}^{3} = \sinh^{-1} 3 - \sinh^{-1}(-1)$$

$$\overset{(7)}{=} \ln(3 + \sqrt{10}) - \ln(-1 + \sqrt{2})$$

$$= \ln \frac{3 + \sqrt{10}}{-1 + \sqrt{2}} \qquad \square$$

EXERCISES 7.4

In Exercises 1–12 find the numerical value of the expression.

1. $\sinh 0$ **2.** $\cosh 0$ **3.** $\tanh 0$

4. $\tanh 1$ **5.** $\coth(-1)$ **6.** $\sinh(\ln 2)$

7. $\sinh(\ln 3)$ **8.** $\cosh(\ln 3)$ **9.** $\coth(\ln 4)$

10. $\operatorname{csch}(\ln \pi^2)$ **11.** $\operatorname{sech}(\ln \sqrt{2})$ **12.** $\sinh^{-1} \frac{4}{3}$

In Exercises 13–16 simplify the expression.

13. $\sinh(\ln x)$ **14.** $\cosh(\ln x)$

15. $\tanh(\ln x)$ **16.** $\sinh^{-1} \dfrac{1 - x^2}{2x}$

In Exercises 17–20 establish the formula.

17. $(d/dx) \tanh x = \operatorname{sech}^2 x$ **18.** $(d/dx) \coth x = -\operatorname{csch}^2 x$

19. $(d/dx) \operatorname{sech} x = -\operatorname{sech} x \tanh x$

20. $(d/dx) \operatorname{csch} x = -\operatorname{csch} x \coth x$

In Exercises 21–26 differentiate the function.

21. $f(x) = \operatorname{sech} \sqrt{x}$ **22.** $f(x) = \cosh \sqrt{1 - x^2}$

23. $f(x) = \sinh^2 \sqrt{1 - x^2}$ **24.** $f(x) = e^{\operatorname{csch} x}$

25. $f(x) = \cosh(\tan e^{2x})$ **26.** $f(x) = \sinh^{-1}(-3x^2)$

27. Let $y = \dfrac{\sinh^{-1} x}{\sqrt{1 + x^2}}$. Show that $(1 + x^2)\dfrac{dy}{dx} + xy = 1$.

In Exercises 28–34 find the integral.

28. $\displaystyle\int \operatorname{csch}^2 x\, dx$ **29.** $\displaystyle\int \operatorname{sech}^2 x\, dx$

30. $\displaystyle\int \tanh x\, dx$ **31.** $\displaystyle\int 2^x \sinh 2^x\, dx$

32. $\displaystyle\int e^x \sinh x\, dx$ **33.** $\displaystyle\int_{5}^{10} \frac{1}{\sqrt{x^2 + 1}}\, dx$

34. $\displaystyle\int_{0}^{1} \frac{x}{\sqrt{1 + x^4}}\, dx$ (*Hint:* Substitute $u = x^2$.)

35. Find the area A of the region between the catenary $y = 4 \cosh x/4$ and the x axis on $[-4, 4]$.

36. Let $f(x) = \cosh x$ for $0 \le x \le \ln 2$. Find the length L of the graph of f.

37. Suppose we substitute sinh x for cosh x in Exercise 36. Why is it more difficult to find the corresponding length? (Do not attempt to evaluate the integral.)

38. Let $f(x) = \cosh x$. Find the surface area S of the surface generated by revolving about the x axis the graph of f on $[0, 1]$.

39. Let $f(x) = 1/\sqrt{1 + x^2}$. Find the center of gravity of the region between the graph of f and the x axis on $[0, 1]$.

40. Show that $\cosh x \geq 1$ for all x by using the fact that $z + 1/z \geq 2$ for all $z > 0$.

41. Show that
$$\cosh x > \frac{e^{|x|}}{2} \quad \text{for all } x$$

42. a. Prove that $e^x = \cosh x + \sinh x$ for all x.
   b. Prove that $e^{-x} = \cosh x - \sinh x$ for all x.

43. a. Prove that $(\cosh x + \sinh x)^n = \cosh nx + \sinh nx$ for all x. (*Hint:* Use Exercise 42(a).)
   b. Prove that $(\cosh x - \sinh x)^n = \cosh nx - \sinh nx$ for all x. (*Hint:* Use Exercise 42(b).)

44. a. Prove that $\sinh 2x = 2 \sinh x \cosh x$.
   b. Prove that $\cosh 2x = \cosh^2 x + \sinh^2 x = 2 \sinh^2 x + 1$

45. Let $f(x) = \frac{1}{2}(1 + \tanh x)$. Show that $f(x) = 1/(1 + e^{-2x})$. (Thus the graph of f is a logistic curve, as described in Example 6 of Section 4.7.)

*46. It can be shown that for $t > 0$,
$$\int_1^{\cosh t} \sqrt{x^2 - 1}\, dx = \int_0^t \sinh^2 u\, du$$
Assuming this fact and using Exercise 44(b), prove that for $t > 0$,
$$\int_1^{\cosh t} \sqrt{x^2 - 1}\, dx = \frac{\sinh 2t}{4} - \frac{t}{2}$$

*47. With the help of Exercises 44(a) and 46 show that the shaded region in Figure 7.20 has area $t/2$.

Applications

48. A falling skydiver experiences a "drag force" due to air resistance. The drag force increases with the magnitude of the velocity v of the diver and is approximately equal to kv^2, where k is a constant. As a result, v is given approximately by
$$v(t) = -\sqrt{\frac{mg}{k}} \tanh\left(\sqrt{\frac{kg}{m}}\, t\right)$$
where m is the mass of the skydiver and g is the acceleration due to gravity. Find $\lim_{t \to \infty} v(t)$. The speed of the skydiver can never exceed the magnitude of this

limit, which is called the **terminal velocity.**

49. The shape of the Gateway Arch to the West is based on the curve
$$y = 694 - 69 \cosh \frac{x}{100}$$
which is the red curve located inside the arch in Figure 7.24, and is given in terms of feet.

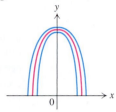

FIGURE 7.24 Figure for Exercise 49.

   a. Find the maximum value of y. (This is a little less than the height of the arch.)
   b. Find the distance between the two x intercepts of the curve. (This is a little less than the outer width at the base of the arch.)

Project

1. It is clear from the discussion of $\cosh x$ that it does not have an inverse, but that if we restrict the domain to $[0, \infty)$, then the resulting function does have an inverse, which we will denote by $\cosh^{-1} x$.

   a. Find the domain and the range of $\cosh^{-1} x$.
   b. Prove that $\cosh^{-1} x = \ln(x + \sqrt{x^2 - 1})$. In so doing, explain where you use the fact that the restricted domain of $\cosh x$ is $[0, \infty)$.
   c. Use part (b) to prove that the following formulas are valid.
      i. $\sinh^{-1} \sqrt{x^2 - 1} = \cosh^{-1} x \quad$ for $\quad x \geq 1$
      ii. $\cosh^{-1} \sqrt{x^2 + 1} = \sinh^{-1} x \quad$ for $\quad x \geq 0$
   d. Suppose we restricted the domain of $\cosh x$ to $(-\infty, 0]$. How would that change the formula for $\cosh^{-1} x$?
   e. Let
$$f(x) = c \cosh \frac{x}{c} = c\left(\frac{e^{x/c} + e^{-x/c}}{2}\right) \quad \text{for } x \geq 0$$
where c is a positive constant. Show that
$$f^{-1}(x) = c \ln\left(\frac{x + \sqrt{x^2 - c^2}}{c}\right) \quad \text{for} \quad x \geq c$$
   f. Does $\tanh x$ have an inverse? If it does, then find the domain and range of the inverse, and a rule for it. If it does not, then why not?

7.5   THE INVERSE TRIGONOMETRIC FUNCTIONS

In this section we give special names to six inverse functions, one for each of the basic trigonometric functions. The inverse trigonometric functions are used in evaluating integrals as well as in other applications.

For example, did you ever wonder how the decimal expansion of π, which begins

$$3.14159265358979...$$

is computed? Although Archimedes and other ancient mathematicians were able to derive rough approximations to π, now one can obtain the decimal expansion to a billion places. One of the best methods for estimating π during the past century has been through use of the inverse tangent function, which we introduce in this section. (We will need to wait until Section 9.8 to provide formulas using the inverse tangent function that give good estimates of π.) Not only are inverse trigonometric functions used in finding many kinds of integrals, but they also have many applications in life. In one application we will show that they can be used in order to determine how close, or how far away, one should sit in a movie house so that the screen will subtend the largest possible angle with respect to your eye.

The Inverse Sine Function

Because the sine function is periodic and hence is not one-to-one, the sine function does not have an inverse. However, if we restrict the domain of the sine function to $[-\pi/2, \pi/2]$, then the resulting function is increasing (see Figure 7.25(a)). Hence the restricted function has an inverse, which is called the **inverse sine function** (or the **arcsine function**). There are two common notations for the value of the inverse sine at x:

$$\sin^{-1} x \quad \text{and} \quad \arcsin x$$

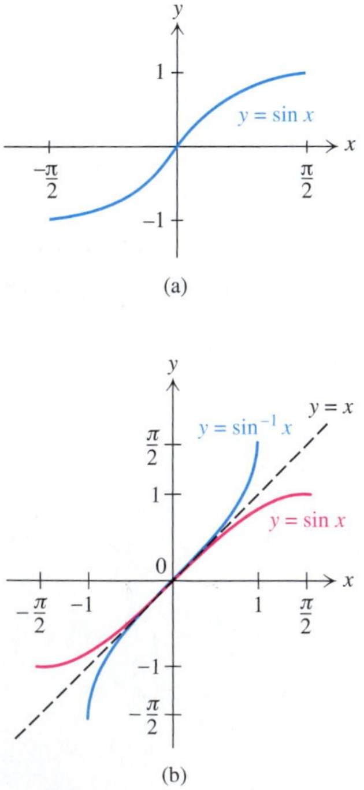

The former has gained prominence on calculators; in Mathematica the latter is used in the form of ArcSin $[x]$. In the discussion that follows we will primarily use the notation $\sin^{-1}x$ because it emphasizes the fact that the function is inverse to the (restricted) sine function. We caution again that, as with any inverse function, $\sin^{-1} x$ is not the same as $1/\sin x$.

The domain of $\sin^{-1}$ is $[-1, 1]$, and its range is $[-\pi/2, \pi/2]$. As a consequence,

$$\sin^{-1} x = y \quad \text{if and only if} \quad \sin y = x$$
$$\text{for } -1 \leq x \leq 1 \quad \text{and} \quad -\pi/2 \leq y \leq \pi/2$$

Thus if $-1 \leq x \leq 1$, then $\sin^{-1} x$ is the number y between $-\pi/2$ and $\pi/2$ whose sine is x. For example,

$$\sin^{-1}(-1) = -\frac{\pi}{2}, \quad \sin^{-1} 0 = 0, \quad \sin^{-1} \frac{1}{2} = \frac{\pi}{6}, \quad \sin^{-1} \frac{\sqrt{2}}{2} = \frac{\pi}{4}$$

The graph of the inverse sine function appears in Figure 7.25(b). We also have the equations

$$\sin^{-1}(\sin x) = x \quad \text{for } -\pi/2 \leq x \leq \pi/2$$
$$\sin(\sin^{-1} x) = x \quad \text{for } -1 \leq x \leq 1$$

(a)

(b)

FIGURE 7.25

Caution: Although $\sin^{-1}(\sin x)$ is defined for all real numbers, the equation $\sin^{-1}(\sin x) = x$ is valid only for x in $[-\pi/2,\ \pi/2]$ because that interval is the range of the inverse sine function. Thus care must be exercised when we evaluate $\sin^{-1}(\sin x)$ with x outside $[-\pi/2,\ \pi/2]$.

EXAMPLE 1   Evaluate $\sin^{-1}\left(\sin \dfrac{5\pi}{6}\right)$.

Solution   First we use the fact that $\sin 5\pi/6 = 1/2$, so that

$$\sin^{-1}\left(\sin \frac{5\pi}{6}\right) = \sin^{-1}\frac{1}{2}$$

Since $\sin \pi/6 = 1/2$, it follows that $\sin^{-1} 1/2 = \pi/6$, so that

$$\sin^{-1}\left(\sin \frac{5\pi}{6}\right) = \frac{\pi}{6} \qquad \square$$

Sometimes an expression involving both trigonometric functions and their inverses occurs. It may be possible to simplify such an expression by drawing and analyzing a judiciously chosen right triangle.

EXAMPLE 2   Simplify the expression $\sec (\sin^{-1} \sqrt{x})$.

Solution   We will evaluate $\sec (\sin^{-1} \sqrt{x})$ by evaluating $\sec y$ for the value of y in $(-\pi/2,\ \pi/2)$ such that $\sin^{-1} \sqrt{x} = y$, that is, $\sin y = \sqrt{x}$. Since $\sin y = \sqrt{x} \geq 0$, it follows that $0 \leq y < \pi/2$. We draw the triangle in Figure 7.26 with acute angle y, hypotenuse with length 1, and opposite side with length $\sqrt{x}$ (to reflect the fact that $\sin y = \sqrt{x}$). By the Pythagorean Theorem, the remaining side has length $\sqrt{1 - (\sqrt{x})^2} = \sqrt{1 - x}$. From the triangle we find that $\sec y = 1/\sqrt{1 - x}$. $\square$

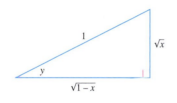

FIGURE 7.26

Next we derive a formula for the derivative of the inverse sine function. Toward that end, observe that any x in $[-1,\ 1]$ can be written as $\sin y$ for an appropriate y with $-\pi/2 \leq y \leq \pi/2$. For such a value of y, $\cos y \geq 0$, so that

$$\cos y = \sqrt{1 - \sin^2 y} = \sqrt{1 - x^2}$$

Therefore by (10) of Section 7.1 we find that

$$\frac{d}{dx} \sin^{-1} x = \frac{1}{\dfrac{d}{dy} \sin y} = \frac{1}{\cos y} = \frac{1}{\sqrt{1 - x^2}}$$

Consequently

$$\frac{d}{dx} \sin^{-1} x = \frac{1}{\sqrt{1 - x^2}} \qquad \text{for } -1 < x < 1 \qquad (1)$$

EXAMPLE 3   Let $f(x) = \sin^{-1} x^4$. Find $f'(x)$.

Solution   Using (1), along with the Chain Rule, we find that

$$f'(x) = \frac{1}{\sqrt{1 - (x^4)^2}} (4x^3) = \frac{4x^3}{\sqrt{1 - x^8}} \quad \square$$

The integral formula corresponding to (1) is

$$\int \frac{1}{\sqrt{1 - x^2}} \, dx = \sin^{-1} x + C \tag{2}$$

We have thus expanded the collection of integrals that we can evaluate.

EXAMPLE 4   Find $\displaystyle\int \frac{1}{\sqrt{9 - x^2}} \, dx$.

Solution   We wish the denominator to have the form $\sqrt{1 - u^2}$ for a suitable u, so we first prepare the integral by factoring out the 9 in the square root:

$$\int \frac{1}{\sqrt{9 - x^2}} \, dx = \int \frac{1}{\sqrt{9(1 - x^2/9)}} \, dx = \frac{1}{3} \int \frac{1}{\sqrt{1 - (x/3)^2}} \, dx$$

Next we make the substitution

$$u = \frac{x}{3}, \quad \text{so that} \quad du = \frac{1}{3} \, dx$$

Therefore by (2),

$$\int \frac{1}{\sqrt{9 - x^2}} \, dx = \frac{1}{3} \int \overbrace{\frac{1}{\sqrt{1 - (x/3)^2}}}^{1/\sqrt{1 - u^2}} \overbrace{dx}^{3 \, du} = \frac{1}{3} \int \frac{1}{\sqrt{1 - u^2}} 3 \, du$$

$$\overset{(2)}{=} \sin^{-1} u + C = \sin^{-1} \frac{x}{3} + C \quad \blacksquare$$

By the same method as used in Example 4 you can show that

$$\int \frac{1}{\sqrt{a^2 - x^2}} \, dx = \sin^{-1} \frac{x}{a} + C \quad \text{for } a > 0 \tag{3}$$

A second way of proving (3) is to differentiate the right-hand side of (3) with the help of the Chain Rule and (1):

$$\frac{d}{dx} \sin^{-1} \frac{x}{a} = \frac{1}{\sqrt{1 - (x/a)^2}} \left(\frac{1}{a}\right) = \frac{1}{\sqrt{a^2 - x^2}}$$

The result is exactly the integrand in (3).

EXAMPLE 5   Find $\displaystyle\int_3^{2 + \sqrt{2}} \frac{1}{\sqrt{4x - x^2}} \, dx$.

Solution   We first evaluate the indefinite integral. In order to make the denominator have the form $\sqrt{a^2 - u^2}$ for suitable a and u, we first rewrite the denominator by completing the square under the radical:

$$\int \frac{1}{\sqrt{4x - x^2}}\, dx = \int \frac{1}{\sqrt{4 - (4 - 4x + x^2)}}\, dx = \int \frac{1}{\sqrt{4 - (x - 2)^2}}\, dx$$

Now we make the substitution

$$u = x - 2, \quad \text{so that} \quad du = dx$$

and then use (3) with $a = 2$:

$$\int \underbrace{\frac{1}{\sqrt{4 - (x - 2)^2}}}_{1/\sqrt{4 - u^2}}\, \overset{du}{\overbrace{dx}} = \int \frac{1}{\sqrt{4 - u^2}}\, du \overset{(3)}{=} \sin^{-1}\frac{u}{2} + C = \sin^{-1}\frac{x - 2}{2} + C$$

Thus

$$\int \frac{1}{\sqrt{4x - x^2}}\, dx = \int \frac{1}{\sqrt{4 - (x - 2)^2}}\, dx = \sin^{-1}\frac{x - 2}{2} + C$$

Therefore

$$\int_3^{2 + \sqrt{2}} \frac{1}{\sqrt{4x - x^2}}\, dx = \left(\sin^{-1}\frac{x - 2}{2}\right)\Bigg|_3^{2 + \sqrt{2}} = \sin^{-1}\frac{\sqrt{2}}{2} - \sin^{-1}\frac{1}{2}$$

$$= \frac{\pi}{4} - \frac{\pi}{6} = \frac{\pi}{12} \quad \square$$

In conclusion, we mention that although it is customary to choose the interval $[-\pi/2, \pi/2]$ as the range of the inverse sine function, we could equally well have used $[\pi/2, 3\pi/2]$, or any other interval on which the sine function has an inverse.

The Inverse Tangent Function

To define an inverse for the tangent function, we restrict the tangent function to $(-\pi/2, \pi/2)$. The resulting inverse function is called the **inverse tangent function** (or the **arctangent function**). Its domain is $(-\infty, \infty)$, and its range is $(-\pi/2, \pi/2)$. We usually write its value at x as $\tan^{-1} x$ or $\arctan x$. As a consequence,

$$\tan^{-1} x = y \quad \text{if and only if} \quad \tan y = x$$
$$\text{for any } x \text{ and for } -\pi/2 < y < \pi/2$$

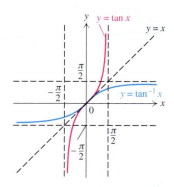

FIGURE 7.27

Thus for any x, $\tan^{-1} x$ is the number y between $-\pi/2$ and $\pi/2$ whose tangent is x. For instance,

$$\tan^{-1} 0 = 0, \quad \tan^{-1} 1 = \frac{\pi}{4}, \quad \text{and} \quad \tan^{-1}(-\sqrt{3}) = -\frac{\pi}{3}$$

The graph of the inverse tangent function is shown in Figure 7.27. We also have the relations

$$\tan^{-1}(\tan x) = x \quad \text{for } -\pi/2 < x < \pi/2$$
$$\tan(\tan^{-1} x) = x \quad \text{for all } x$$

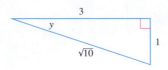

FIGURE 7.28

EXAMPLE 6   Evaluate $\csc\left(\tan^{-1}\left(-\tfrac{1}{3}\right)\right)$.

Solution   We will evaluate $\csc\left(\tan^{-1}\left(-\tfrac{1}{3}\right)\right)$ by evaluating $\csc y$ for the value of y in $(-\pi/2,\ \pi/2)$ such that $\tan^{-1}\left(-\tfrac{1}{3}\right) = y$, that is, $\tan y = -\tfrac{1}{3}$. Since $\tan y < 0$, it follows that $-\pi/2 < y < 0$, so that $\csc y < 0$. By the Pythagorean Theorem, the hypotenuse of the triangle in Figure 7.28 has length $\sqrt{10}$. Therefore $\csc y = -\sqrt{10}$. As a result,

$$\csc\left(\tan^{-1}\left(-\tfrac{1}{3}\right)\right) = \csc y = -\sqrt{10} \qquad \square$$

If $x = \tan y$, then

$$\sec^2 y = \tan^2 y + 1 = x^2 + 1$$

Therefore by (10) in Section 7.1 we infer that

$$\frac{d}{dx}\tan^{-1} x = \frac{1}{\dfrac{d}{dy}\tan y} = \frac{1}{\sec^2 y} = \frac{1}{x^2 + 1}$$

In other words,

$$\frac{d}{dx}\tan^{-1} x = \frac{1}{x^2 + 1} \tag{4}$$

EXAMPLE 7   Find $\dfrac{d}{dx}\tan^{-1} e^{3x}$.

Solution   By (4), along with the Chain Rule,

$$\frac{d}{dx}\tan^{-1} e^{3x} = \frac{1}{(e^{3x})^2 + 1}\,(e^{3x}\cdot 3) = \frac{3e^{3x}}{e^{6x} + 1} \qquad \square$$

The indefinite integral version of (4) is

$$\int \frac{1}{x^2 + 1}\,dx = \tan^{-1} x + C \tag{5}$$

EXAMPLE 8   Find $\displaystyle\int \frac{1}{x^2 + 16}\,dx$.

Solution   We wish the denominator to have the form $u^2 + 1$ for suitable u, so we factor out 16:

$$\int \frac{1}{x^2 + 16}\,dx = \int \frac{1}{16\left(\dfrac{x^2}{16} + 1\right)}\,dx = \frac{1}{16}\int \frac{1}{(x/4)^2 + 1}\,dx$$

Then we let

$$u = \frac{x}{4}, \quad \text{so that} \quad du = \frac{1}{4} dx$$

Therefore by (5),

$$\int \frac{1}{x^2 + 16} dx = \frac{1}{16} \int \overbrace{\frac{1}{(x/4)^2 + 1}}^{1/(u^2+1)} \overset{4\,du}{\overbrace{dx}} = \frac{1}{16} \int \frac{1}{u^2 + 1} 4 \, du$$

$$= \frac{1}{4} \tan^{-1} u + C = \frac{1}{4} \tan^{-1} \frac{x}{4} + C \qquad \square$$

In general, if $a > 0$, then

$$\int \frac{1}{x^2 + a^2} dx = \frac{1}{a} \tan^{-1} \frac{x}{a} + C \tag{6}$$

This can be shown by letting $u = x/a$ and proceeding exactly as in the solution of Example 8, or by differentiating both sides of (6).

EXAMPLE 9   Find $\displaystyle\int \frac{1}{2x^2 + 2x + 1} dx$.

Solution   We wish the denominator to have the form $u^2 + a^2$ for suitable u and a, so we factor out 2 and then complete the square:

$$\int \frac{1}{2x^2 + 2x + 1} dx = \frac{1}{2} \int \frac{1}{x^2 + x + 1/2} dx = \frac{1}{2} \int \frac{1}{(x + 1/2)^2 + (1/2)^2} dx$$

Now we make the substitution

$$u = x + \frac{1}{2}, \quad \text{so that} \quad du = dx$$

and conclude from (6), with $a = 1/2$, that

$$\int \frac{1}{2x^2 + 2x + 1} dx = \frac{1}{2} \int \frac{1}{u^2 + (1/2)^2} du$$

$$= \frac{1/2}{1/2} \tan^{-1} \frac{u}{1/2} + C = \tan^{-1} 2\left(x + \frac{1}{2}\right) + C$$

$$= \tan^{-1}(2x + 1) + C \qquad \square$$

We conclude this subsection with an application of the inverse tangent function.

EXAMPLE 10   In a certain movie theater, the bottom edge of the 15-foot-tall screen at the front is 5 feet above eye level. Suppose we wish to sit so that the angle θ in Figure 7.29 is as large as possible. How far should we sit from the screen?

Solution   Let x denote the distance in feet from our eyes to the front of the theater. We seek the maximum value of θ, regarded as a function of x. If ϕ is as shown in Figure 7.29, then

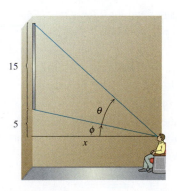

FIGURE 7.29

$$\tan\left(\theta(x) + \phi(x)\right) = \frac{20}{x}, \quad \text{so that} \quad \theta(x) + \phi(x) = \tan^{-1}\frac{20}{x}$$

and

$$\tan\phi(x) = \frac{5}{x}, \quad \text{so that} \quad \phi(x) = \tan^{-1}\frac{5}{x}$$

Therefore

$$\theta(x) = [\theta(x) + \phi(x)] - \phi(x) = \tan^{-1}\frac{20}{x} - \tan^{-1}\frac{5}{x} \quad \text{for } x > 0$$

Consequently by the Chain Rule,

$$\theta'(x) = \frac{-20/x^2}{1 + (20/x)^2} - \frac{-5/x^2}{1 + (5/x)^2}$$

$$= \frac{-20}{x^2 + 400} + \frac{5}{x^2 + 25} = \frac{-15(x^2 - 100)}{(x^2 + 400)(x^2 + 25)}$$

Since $x > 0$, this implies that $\theta'(x) = 0$ for $x = 10$. Since $\theta'(x) > 0$ for $0 < x < 10$ and $\theta'(x) < 0$ for $x > 10$, it follows from (1) in Section 4.6 that θ achieves its maximum value for $x = 10$. Therefore the largest view of the screen is attained when we are seated 10 feet away from the front.   ❑

It may seem strange that the largest view is attained if we sit only 10 feet from the front. In fact at that distance the screen occupies too large an angle for our eyes. From the standpoint of viewing pleasure, the optimal distance is substantially larger than the distance calculated above.

The Remaining Inverse Trigonometric Functions

Now we turn to the inverses of the remaining trigonometric functions. We define the inverse cosine, inverse cotangent, inverse secant, and inverse cosecant functions by the following formulas:

$$\cos^{-1} x = y \quad \text{if and only if} \quad \cos y = x \quad \text{for } -1 \le x \le 1 \text{ and for } 0 \le y \le \pi$$

$$\cot^{-1} x = y \quad \text{if and only if} \quad \cot y = x \quad \text{for any } x \text{ and for } 0 < y < \pi$$

$$\sec^{-1} x = y \quad \text{if and only if} \quad \sec y = x \quad \text{for } |x| \ge 1 \text{ and for } 0 \le y < \pi/2$$
$$\text{or } \pi \le y < 3\pi/2$$

$$\csc^{-1} x = y \quad \text{if and only if} \quad \csc y = x \quad \text{for } |x| \ge 1 \text{ and for } 0 < y \le \pi/2$$
$$\text{or } \pi < y \le 3\pi/2$$

The graphs of the four functions are in Figure 7.30(a)–(d).

The same process that we used in finding the derivatives of the inverse sine and inverse tangent functions yields the derivatives of the inverse functions just defined:

$$\frac{d}{dx}\cos^{-1} x = \frac{-1}{\sqrt{1 - x^2}} \qquad \frac{d}{dx}\sec^{-1} x = \frac{1}{x\sqrt{x^2 - 1}}$$

$$\frac{d}{dx}\cot^{-1} x = \frac{-1}{x^2 + 1} \qquad \frac{d}{dx}\csc^{-1} x = \frac{-1}{x\sqrt{x^2 - 1}}$$

Corresponding integration formulas are as follows (where we assume that $a > 0$):

$$\int \frac{1}{\sqrt{a^2 - x^2}}\, dx = -\cos^{-1}\frac{x}{a} + C \tag{7}$$

$$\int \frac{1}{x^2 + a^2}\, dx = -\frac{1}{a}\cot^{-1}\frac{x}{a} + C \tag{8}$$

$$\int \frac{1}{x\sqrt{x^2 - a^2}}\, dx = \frac{1}{a}\sec^{-1}\frac{x}{a} + C \tag{9}$$

$$\int \frac{1}{x\sqrt{x^2 - a^2}}\, dx = -\frac{1}{a}\csc^{-1}\frac{x}{a} + C \tag{10}$$

Notice that we can use either the inverse sine or the inverse cosine function to evaluate the integral in (3) and (7). From now on we will use the inverse sine function for this purpose. Similar comments hold for the inverse tangent and inverse cotangent functions in (6) and (8), and for the inverse secant and inverse cosecant functions in (9) and (10). For uniformity we will use only the inverse tangent and inverse secant functions hereafter.

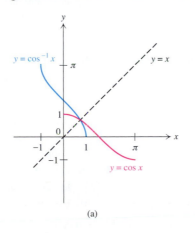

(a)

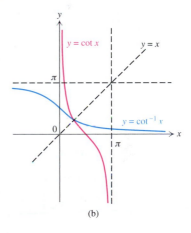

(b)

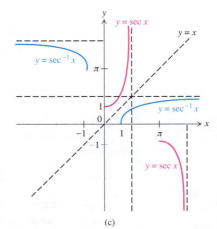

(c)

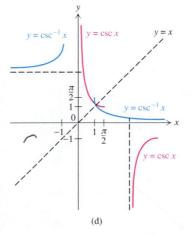

(d)

FIGURE 7.30

EXERCISES 7.5

In Exercises 1–10 find the numerical value of the expression.

1. $\sin^{-1} \sqrt{3}/2$
2. $\sin^{-1} 1$
3. $\cos^{-1} \sqrt{2}/2$
4. $\cos^{-1} \left(-\frac{1}{2}\right)$
5. $\tan^{-1} \left(-1/\sqrt{3}\right)$
6. $\tan^{-1} \sqrt{3}$
7. $\cot^{-1} \sqrt{3}$
8. $\cot^{-1} (-1)$
9. $\sec^{-1} \left(-\sqrt{2}\right)$
10. $\csc^{-1} 2\sqrt{3}/3$

In Exercises 11–18 find the numerical value of the expression.

11. $\sin \left(\sin^{-1} \left(-\frac{1}{2}\right)\right)$
12. $\sin \left(\cos^{-1} \sqrt{2}/2\right)$
13. $\tan \left(\sec^{-1} \sqrt{2}\right)$
14. $\sec \left(\cos^{-1} \sqrt{3}/2\right)$
15. $\csc \left(\cot^{-1} \left(-\sqrt{3}\right)\right)$
16. $\tan^{-1} (\tan 0)$
17. $\sin^{-1} (\cos \pi/6)$
18. $\cot^{-1} (\tan \pi/3)$

In Exercises 19–26 simplify the expression.

19. $\cos \left(\sin^{-1} x\right)$
20. $\sin \left(\sec^{-1} x\right)$
21. $\sec \left(\tan^{-1} x\right)$
22. $\tan \left(\csc^{-1} x/2\right)$
23. $\cos \left(\cot^{-1} x^2\right)$
24. $\tan \left(\cos^{-1} \sqrt{1 - x}\right)$
*25. $\cos \left(2 \sin^{-1} x\right)$
*26. $\sin \left(2 \sin^{-1} x\right)$

In Exercises 27–32 find the derivative of f.

27. $f(x) = \cos^{-1} (-3x)$
28. $f(x) = x \sin^{-1} (x^2)$
29. $f(t) = \tan^{-1} \sqrt{t}$
30. $f(x) = \tan^{-1} \left(\dfrac{x + 1}{x - 1}\right)$
31. $f(x) = \cot^{-1} \sqrt{1 - x^2}$
32. $f(u) = \sec^{-1} (-u)$

In Exercises 33–52 evaluate the indefinite integral.

33. $\displaystyle\int \frac{1}{x^2 + 16}\, dx$
34. $\displaystyle\int \frac{1}{t^2 + 6}\, dt$
35. $\displaystyle\int \frac{1}{9x^2 + 16}\, dx$
36. $\displaystyle\int \frac{1}{x^2 + 4x + 7}\, dx$
37. $\displaystyle\int \frac{1}{2x^2 + 4x + 6}\, dx$
38. $\displaystyle\int \frac{1}{t^2 - 3t + 3}\, dt$
39. $\displaystyle\int \frac{1}{\sqrt{9 - 4x^2}}\, dx$
40. $\displaystyle\int \frac{1}{\sqrt{4 - 9x^2}}\, dx$
41. $\displaystyle\int \frac{1}{x\sqrt{x^2 - 25}}\, dx$
42. $\displaystyle\int \frac{1}{x\sqrt{4x^4 - 25}}\, dx$
43. $\displaystyle\int \frac{x^3}{\sqrt{1 - x^8}}\, dx$
44. $\displaystyle\int \frac{1}{x \ln x \sqrt{(\ln x)^2 - 1}}\, dx$
45. $\displaystyle\int \frac{e^{-x}}{1 + e^{-2x}}\, dx$
46. $\displaystyle\int \frac{e^{3x}}{\sqrt{1 - e^{6x}}}\, dx$
47. $\displaystyle\int \frac{\tan^{-1} (2x)}{1 + 4x^2}\, dx$
48. $\displaystyle\int \frac{1}{x[1 + (\ln x)^2]}\, dx$
49. $\displaystyle\int \frac{\cos t}{9 + \sin^2 t}\, dt$
50. $\displaystyle\int \frac{\sec^2 x}{\sqrt{4 - \tan^2 x}}\, dx$

51. $\displaystyle\int \frac{\cos 4x}{\sin 4x \sqrt{16 \sin^2 4x - 4}}\, dx$
52. $\displaystyle\int \frac{x^{n - 1}}{1 + x^{2n}}\, dx$, n a positive integer

In Exercises 53–55 evaluate the definite integral.

53. $\displaystyle\int_0^2 \frac{1}{\sqrt{16 - x^2}}\, dx$
54. $\displaystyle\int_{-2}^{2\sqrt{3} - 2} \frac{1}{u^2 + 4u + 8}\, du$
55. $\displaystyle\int_{4\sqrt{3}/3}^4 \frac{1}{x\sqrt{x^2 - 4}}\, dx$

 56. Plot the graph of $y = \sin^{-1} (\sin x)$ with windows $-6 \leq x \leq 6$ and $-1.6 \leq y \leq 1.6$. Explain the appearance of the graph.

57. Find the area A of the region bounded by the y axis and the graphs of
$$y = \frac{1}{x^2 - 2x + 4} \quad \text{and} \quad y = \frac{1}{3}$$

58. Let $f(x) = 1/(1 - x^2)^{1/4}$, and let R be the region between the graph of f and the x axis on $[0, 1/2]$. Find the volume V of the solid obtained by revolving R about the x axis.

59. Let $f(y) = 1/\sqrt{1 - y^4}$, and let R be the region between the graph of f and the y axis on $[0, \frac{1}{2}\sqrt{2}]$. Find the volume V of the solid obtained by revolving R about the x axis.

60. Let $f(x) = \frac{1}{3} x^3 + x - \frac{1}{4} \tan^{-1} x$ for $0 \leq x \leq 1$. Find the length L of the graph of f.

61. Evaluate $\int_0^1 \sin^{-1} x\, dx$ by using the formula
$$\int_a^b f(x)\, dx = bf(b) - af(a) - \int_{f(a)}^{f(b)} f^{-1}(y)\, dy$$
appearing in Exercise 70 of Section 7.1.

*62. Let a, $b > 0$. By making the substitution $u = \tan x$, evaluate
$$\int \frac{1}{a^2 \sin^2 x + b^2 \cos^2 x}\, dx$$

63. Prove that the graph of the inverse tangent function has an inflection point at $x = 0$.

*64. Prove that
$$\sin^{-1} x + \sin^{-1} y = \sin^{-1} \left(x\sqrt{1 - y^2} + y\sqrt{1 - x^2}\right)$$
provided that the value of the expression on the left-hand side lies in $[-\pi/2, \pi/2]$. (*Hint:* Use the formula for $\sin (a + b)$.)

65. Prove that

$$\tan^{-1}\frac{x}{\sqrt{1-x^2}} = \sin^{-1}x \quad \text{for} \ -1 < x < 1$$

(*Hint:* You can prove this by differentiating both sides. You might convince yourself that the formula is correct by drawing a right triangle whose sides have length x, $\sqrt{1-x^2}$, and 1.)

*66. Prove that

$$\tan^{-1}x + \tan^{-1}y = \tan^{-1}\!\left(\frac{x+y}{1-xy}\right) \quad \text{for} \ \ xy \neq 1$$

provided that the value of the expression on the left-hand side lies in $(-\pi/2,\ \pi/2)$.

67. Using Exercise 66, verify the following relations involving the number $\pi/4$. These can be used to estimate π. (We will use (d) for this purpose in Chapter 9.)
   a. $\tan^{-1}\frac{1}{2} + \tan^{-1}\frac{1}{3} = \pi/4$
   b. $2\tan^{-1}\frac{1}{3} + \tan^{-1}\frac{1}{7} = \pi/4$
   c. $\tan^{-1}\frac{120}{119} - \tan^{-1}\frac{1}{239} = \pi/4$
   *d. $4\tan^{-1}\frac{1}{5} - \tan^{-1}\frac{1}{239} = \pi/4$ (*Hint:* Use Exercise 66 twice to find $4\tan^{-1}\frac{1}{5}$, and then use part (c).)

68. Using Exercise 66, find a number c such that $\tan^{-1}\frac{1}{4} + \tan^{-1}c = \pi/4$.

69. Let $f(x) = \tan^{-1}x + \tan^{-1}(1/x)$. Show that f is constant on each of the intervals $(-\infty, 0)$ and $(0, \infty)$. Find the constants.

*70. Suppose a, b, and c are numbers satisfying $bc = 1 + a^2$. Show that

$$\tan^{-1}\frac{1}{a+b} + \tan^{-1}\frac{1}{a+c} = \tan^{-1}\frac{1}{a} \qquad (11)$$

provided that a, $a + b$, and $a + c$ are not 0 and that the value of the expression on the left-hand side lies in $(-\pi/2,\ \pi/2)$. This formula was discovered by C. L. Dodgson (Lewis Carroll). The equation in Exercise 67(a) results from (11) if $a = b = 1$ and $c = 2$.

*71. Show that the following identities are valid:
   a. $\sin^{-1}\!\left(\frac{x}{3} - 1\right) = \frac{\pi}{2} - 2\sin^{-1}\sqrt{1 - \frac{x}{6}}$
   b. $\sin^{-1}\!\left(\frac{x}{3} - 1\right) = 2\sin^{-1}\frac{\sqrt{x}}{\sqrt{6}} - \frac{\pi}{2}$

*72. Let $f(x) = \tan^{-1}\!\left(\frac{x+1}{x-1}\right)$.

   a. Show that $f'(x) = \dfrac{-1}{x^2 + 1}$ for $x \neq 1$.
   b. Find a formula for f''.
   c. Show that $\lim_{x\to 1^+} f(x) = \pi/2$ and $\lim_{x\to 1^-} f(x) = -\pi/2$.
   d. Sketch the graph of f.

   e. Show that there is no constant C such that $f(x) = -\tan^{-1}x + C$ for *all* $x \neq 1$, although the functions f and $-\tan^{-1}x$ have the same derivative when $x \neq 1$.
   f. Find constants C_1 and C_2 such that
$$f(x) = -\tan^{-1}x + C_1 \quad \text{for } x < 1$$
$$f(x) = -\tan^{-1}x + C_2 \quad \text{for } x > 1$$

73. a. Use Theorem 4.6 to show that there is a constant c such that
$$\sin^{-1}x + \cos^{-1}x = c \quad \text{for } -1 \leq x \leq 1$$
   b. Determine the value of c in part (a).

Applications

74. Suppose a large oil painting of water lilies by Claude Monet is on display at a gallery. In order to accommodate a large number of viewers at one time, the painting is placed with its bottom and top edges 7 and 12 feet above the floor, respectively (Figure 7.31). Assume that a spectator's eyes are 5 feet above the floor, and that the spectator has the best view when the angle θ shown in the illustration is largest. Determine which distance x from the wall will allow the best view of the water lilies.

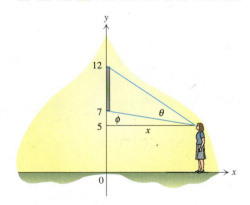

FIGURE 7.31   Figure for Exercise 74.

75. Suppose that the painting in Exercise 74 is to be hung on the wall in a way that provides the best view to observers standing precisely 10 feet from the wall. Assuming that an observer's eyes are 5 feet from the floor, how high should the picture be hung? Would the answer be the same for an observer standing 15 feet from the wall? Explain your answer.

76. An observer watches a rocket blast off 20,000 feet away. At the moment the rocket appears to make an angle of $\pi/4$ radians with the horizon (Figure 7.32), the observer calculates that the angle of elevation of the rocket is changing at the rate of $\pi/60$ radians per second and that the rocket seems to move vertically. How fast is the rocket then traveling?

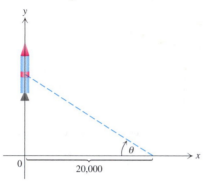

FIGURE 7.32   Figure for Exercise 76.

 77. Suppose two apartment buildings are 20 meters and 15 meters tall, and are 30 meters apart. Because they need as much sunlight as possible, tomato plants are to be planted so as to encounter the maximum sunlight per day (Figure 7.33). Assuming that the sun passes directly overhead, determine

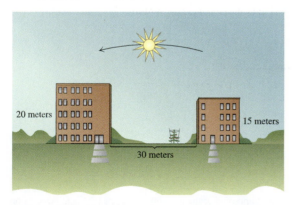

FIGURE 7.33   Figure for Exercise 77.

    a. the distance from the smaller apartment building at which the tomatoes should be planted.

    b. the total length of time per day the plants will be in the sunlight.

 78. The **rated speed** v_R of a banked curve on a road is the maximum speed a car can attain on the curve without skidding outward, under the assumption that there is no friction between the road and the tires (under icy road conditions, for example). The rated speed is given by

$$v^2 = \rho g \tan \theta$$

where $g = 9.8$ (meters per second per second) is the acceleration due to gravity, ρ is the radius of curvature of the curve, and θ is the banking angle (Figure 7.34).

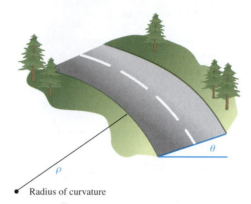

    Radius of curvature

FIGURE 7.34   Figure for Exercise 78.

    a. Express the angle θ in terms of v, ρ, and g.

    b. If a curve is to have a rated speed of 18 meters per second (approximately 40 mph) and a radius of curvature of 60 meters, at what angle should it be banked?

    c. Suppose the radius of curvature is a constant 100 meters, but that the banking angle is variable. Suppose also that a professional stunt driver rounds the curve, accelerating or decelerating as necessary to keep the car at the maximum safe speed. If at a certain instant the driver's speed is 18 meters per second and is decreasing at the rate of 1 meter per second, how fast is the banking angle changing at that instant?

Project

1. A hiker on one rim of a canyon watches as a boulder falls from the rim of the opposite wall of the canyon, 800 feet away. Let θ denote the angle of depression of the hiker's line of sight, as shown in Figure 7.35.

    a. Using the fact that the vertical distance $s(t)$ the boulder falls in t seconds is $16t^2$ feet, find θ in terms of t.

    b. Find a formula for $d\theta/dt$ until the boulder hits the bottom of the canyon. (Leave your answer as a function of t alone.) Explain why the derivative is positive.

c. Find the angle θ at which the boulder appears to be falling the fastest, that is, when $d\theta/dt$ is maximized.

d. Suppose that the bottom of the canyon is 1024 feet below the top, and that the rim of the opposite wall of the canyon is a distance D from the hiker. How large would D need to be so that the angle θ would appear to be increasing during the boulder's entire fall?

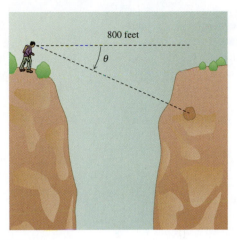

FIGURE 7.35   Figure for the project.

7.6   L'HÔPITAL'S RULE

In Chapter 2 we stated the Quotient Theorem for Limits:

$$\lim_{x \to a} \frac{f(x)}{g(x)} = \frac{\lim_{x \to a} f(x)}{\lim_{x \to a} g(x)}$$

provided that the limits on the right side exist *and* $\lim_{x \to a} g(x) \neq 0$. However, there are many examples of quotient functions f/g that have limits at a even though $\lim_{x \to a} g(x) = 0$. A nontrivial example is $(\sin x)/x$, with $a = 0$:

$$\lim_{x \to 0} \frac{\sin x}{x} = 1$$

More generally, if f is differentiable at a, then

$$\lim_{x \to a} \frac{f(x) - f(a)}{x - a} = f'(a)$$

even though $\lim_{x \to a} (x - a) = 0$. Nevertheless, except for such examples, we have thus far no systematic method for evaluating limits of quotients in which the denominator approaches 0. Examples are

$$\lim_{x \to \pi/2^+} \frac{\cos x}{\sin x - 1} \quad \text{and} \quad \lim_{x \to 0} \frac{e^x - x - 1}{x^2} \qquad (1)$$

In this section we will describe a method for evaluating limits such as those in (1). More particularly, we will be interested in limits of the form

$$\lim_{x \to *} f(x) \qquad (2)$$

where we use $*$ to represent a, a^+, a^-, ∞, or $-\infty$. The method to be described is called **l'Hôpital's Rule,**† and it is named after the seventeenth- and eighteenth-century French mathematician the Marquis de l'Hôpital.

One version of l'Hôpital's Rule is the following:

† **L'Hôpital:** Pronounced "*Lo*-pi-tal."

The Marquis de l'Hôpital (1661–1704)

L'Hôpital's Rule should actually be called "Bernoulli's Rule" because it appears in correspondence from Johann Bernoulli to l'Hôpital. But l'Hôpital and Bernoulli had made an agreement under which l'Hôpital paid Bernoulli a monthly fee for solutions to certain problems, with the understanding that Bernoulli would tell no one of the arrangement. L'Hôpital's Rule first appeared publicly in l'Hôpital's 1696 calculus book. It was only recently discovered that the rule, its proof, and relevant examples all appeared in a 1694 letter from Bernoulli to l'Hôpital.

L'Hôpital's Rule

Suppose that $\lim_{x \to *} f(x)$ and $\lim_{x \to *} g(x)$ are both 0, and assume that $g'(x) \neq 0$ for x near $*$. Then

$$\lim_{x \to *} \frac{f(x)}{g(x)} = \lim_{x \to *} \frac{f'(x)}{g'(x)}$$

provided that the latter limit exists (as a number, or as ∞ or $-\infty$).

In the Appendix, this version of l'Hôpital's Rule is stated precisely and proved by means of a generalization of the Mean Value Theorem. However, for the special case in which $*$ is a, $f(a) = 0 = g(a)$, both f' and g' are continuous at a, and $g'(a) \neq 0$, there is a simple proof that relies only on the Quotient Rule for Limits:

$$\lim_{x \to a} \frac{f(x)}{g(x)} = \lim_{x \to a} \frac{f(x) - f(a)}{g(x) - g(a)} = \lim_{x \to a} \frac{\dfrac{f(x) - f(a)}{x - a}}{\dfrac{g(x) - g(a)}{x - a}}$$

$$= \frac{\lim_{x \to a} \dfrac{f(x) - f(a)}{x - a}}{\lim_{x \to a} \dfrac{g(x) - g(a)}{x - a}} = \frac{f'(a)}{g'(a)} = \frac{\lim_{x \to a} f'(x)}{\lim_{x \to a} g'(x)} = \lim_{x \to a} \frac{f'(x)}{g'(x)}$$

The Indeterminate Form 0/0

If $\lim_{x \to *} f(x) = 0 = \lim_{x \to *} g(x)$, then we say that $\lim_{x \to *} f(x)/g(x)$ has the **indeterminate form 0/0.** It is called "indeterminate" because without additional analysis we cannot determine whether the limit exists, or if it does, what its value is. For example,

$$\lim_{x \to 0} \frac{x^2}{x} = 0, \quad \lim_{x \to 0} \frac{x}{x^3} = \infty, \quad \text{and} \quad \lim_{x \to 0} \frac{\sin 2x}{x} = 2$$

although in each case the limit of the numerator and the limit of the denominator are 0.

Normally when we use l'Hôpital's Rule, the differentiability of f and g is obvious, as is a suitable interval on which $g'(x) \neq 0$. Therefore in the examples that follow we will not refer to these two hypotheses of l'Hôpital's Rule.

EXAMPLE 1   Find $\lim_{x \to 0} \dfrac{\sin 4x}{\sin 3x}$.

Solution   Notice first that

$$\lim_{x \to 0} \sin 4x = 0 = \lim_{x \to 0} \sin 3x$$

As a result we can apply l'Hôpital's Rule, which yields

$$\lim_{x \to 0} \frac{\sin 4x}{\sin 3x} = \lim_{x \to 0} \frac{4 \cos 4x}{3 \cos 3x} = \frac{4 \cdot 1}{3 \cdot 1} = \frac{4}{3} \qquad \square$$

EXAMPLE 2   Evaluate $\displaystyle\lim_{x \to \pi/2^-} \frac{\cos x}{\sin x - 1}$.

Solution   Since

$$\lim_{x \to \pi/2^-} \cos x = 0 = \lim_{x \to \pi/2^-} (\sin x - 1)$$

we can apply l'Hôpital's Rule to obtain

$$\lim_{x \to \pi/2^-} \frac{\cos x}{\sin x - 1} = \lim_{x \to \pi/2^-} \frac{-\sin x}{\cos x} = \lim_{x \to \pi/2^-} (-\tan x) = -\infty \qquad \square$$

EXAMPLE 3   Find $\displaystyle\lim_{x \to \infty} \frac{(\pi/2) - \tan^{-1} x}{1/x}$.

Solution   Since $\displaystyle\lim_{x \to \infty} \tan^{-1} x = \pi/2$, we deduce that

$$\lim_{x \to \infty} \left(\frac{\pi}{2} - \tan^{-1} x \right) = 0 = \lim_{x \to \infty} \frac{1}{x}$$

Therefore l'Hôpital's Rule applies to give us

$$\lim_{x \to \infty} \frac{(\pi/2) - \tan^{-1} x}{1/x} = \lim_{x \to \infty} \frac{-1/(1 + x^2)}{-1/x^2} = \lim_{x \to \infty} \frac{x^2}{1 + x^2} = 1 \qquad \square$$

In the next example we will determine the limit by applying l'Hôpital's Rule twice in succession.

EXAMPLE 4   Find $\displaystyle\lim_{x \to 0} \frac{e^x - x - 1}{x^2}$.

Solution   Since

$$\lim_{x \to 0} (e^x - x - 1) = 0 = \lim_{x \to 0} x^2$$

l'Hôpital's Rule tells us that

$$\lim_{x \to 0} \frac{e^x - x - 1}{x^2} = \lim_{x \to 0} \frac{e^x - 1}{2x}$$

provided that the latter limit exists. However,

$$\lim_{x \to 0} (e^x - 1) = 0 = \lim_{x \to 0} 2x$$

so that a second application of l'Hôpital's Rule yields

$$\lim_{x \to 0} \frac{e^x - 1}{2x} = \lim_{x \to 0} \frac{e^x}{2} = \frac{1}{2}$$

We conclude that

$$\lim_{x\to 0}\frac{e^x - x - 1}{x^2} = \lim_{x\to 0}\frac{e^x - 1}{2x} = \frac{1}{2} \qquad \square$$

The Indeterminate Form ∞/∞ Suppose that $\lim_{x\to *} f(x) = \infty$ or $-\infty$, and that $\lim_{x\to *} g(x) = \infty$ or $-\infty$. Then we say that $\lim_{x\to *} f(x)/g(x)$ has the indeterminate form ∞/∞. L'Hôpital's Rule is valid in this case:

$$\lim_{x\to *}\frac{f(x)}{g(x)} = \lim_{x\to *}\frac{f'(x)}{g'(x)}$$

provided that the latter limit exists (as a number, or as ∞ or $-\infty$).

In the next example we determine the relationship between e^x and x^2 as x grows without bound.

EXAMPLE 5   Find $\lim_{x\to\infty}\dfrac{e^x}{x^2}$.

Solution   Evidently

$$\lim_{x\to\infty} e^x = \infty = \lim_{x\to\infty} x^2$$

so that by l'Hôpital's Rule we have

$$\lim_{x\to\infty}\frac{e^x}{x^2} = \lim_{x\to\infty}\frac{e^x}{2x}$$

Since

$$\lim_{x\to\infty} e^x = \infty = \lim_{x\to\infty} 2x$$

another application of l'Hôpital's Rule is necessary. We find that

$$\lim_{x\to\infty}\frac{e^x}{2x} = \lim_{x\to\infty}\frac{e^x}{2} = \infty$$

Consequently

$$\lim_{x\to\infty}\frac{e^x}{x^2} = \infty \qquad \square$$

In a similar way you can show that

$$\lim_{x\to\infty}\frac{e^x}{x^n} = \infty$$

for any positive integer n. We sometimes express this by saying that as x approaches ∞, e^x goes to infinity faster than any power of x.

EXAMPLE 6   Find $\lim_{x\to 0^+}\dfrac{1/x}{e^{1/x^2}}$.

Solution   Since

$$\lim_{x \to 0^+} \frac{1}{x} = \infty = \lim_{x \to 0^+} e^{1/x^2}$$

it follows from l'Hôpital's Rule that

$$\lim_{x \to 0^+} \frac{1/x}{e^{1/x^2}} = \lim_{x \to 0^+} \frac{-1/x^2}{e^{1/x^2}(-2/x^3)}$$

$$= \lim_{x \to 0^+} \frac{xe^{-1/x^2}}{2} = \frac{0 \cdot 0}{2} = 0 \qquad □$$

Let us apply the limit just found. Suppose

$$f(x) = \begin{cases} e^{-1/x^2} & \text{for } x \neq 0 \\ 0 & \text{for } x = 0 \end{cases}$$

and suppose we wish to determine $f'(0)$. By definition,

$$f'(0) = \lim_{x \to 0} \frac{f(x) - f(0)}{x - 0} = \lim_{x \to 0} \frac{f(x) - 0}{x - 0} = \lim_{x \to 0} \frac{e^{-1/x^2}}{x}$$

if the limit exists. It is tempting to use l'Hôpital's Rule on the fraction $e^{-1/x^2}/x$, but unfortunately a straight application of the rule does not help. However, we get results by transforming both numerator and denominator to obtain

$$\lim_{x \to 0} \frac{e^{-1/x^2}}{x} = \lim_{x \to 0} \frac{1/x}{e^{1/x^2}}$$

From Example 6 we know that

$$\lim_{x \to 0^+} \frac{1/x}{e^{1/x^2}} = 0$$

and by a similar calculation,

$$\lim_{x \to 0^-} \frac{1/x}{e^{1/x^2}} = 0$$

Therefore

$$f'(0) = \lim_{x \to 0} \frac{e^{-1/x^2}}{x} = 0$$

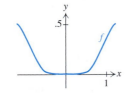

FIGURE 7.36

It is possible to show that for every positive integer n, $f^{(n)}(0) = 0$. In other words, the graph of f is as flat as possible at the origin (see Figure 7.36).

Other Indeterminate Forms

There are many other indeterminate forms. We list a few of them below, and give an example of each:

Indeterminate form	Example
$0 \cdot \infty$	$\displaystyle\lim_{x \to 0^+} (x \ln x)$
0^0	$\displaystyle\lim_{x \to 0^+} x^x$
1^∞	$\displaystyle\lim_{x \to \infty} \left(1 + \frac{1}{x}\right)^x$
∞^0	$\displaystyle\lim_{x \to \infty} x^{1/x}$
$\infty - \infty$	$\displaystyle\lim_{x \to 0^+} (\csc x - \cot x)$

When we prepare to find limits such as those just described, our first goal will be to rewrite the given limit in a way that enables us to use l'Hôpital's Rule.

EXAMPLE 7   Find $\displaystyle\lim_{x \to 0^+} x \ln x$.

Solution   Since

$$\lim_{x \to 0^+} x = 0 \quad \text{and} \quad \lim_{x \to 0^+} \ln x = -\infty$$

the given limit is of the form $0 \cdot \infty$ (more precisely, $0 \cdot (-\infty)$). However, we can transform it into the indeterminate form ∞/∞ by rewriting it as

$$\lim_{x \to 0^+} \frac{\ln x}{1/x}$$

because

$$\lim_{x \to 0^+} \ln x = -\infty \quad \text{and} \quad \lim_{x \to 0^+} \frac{1}{x} = \infty$$

By l'Hôpital's Rule we conclude that

$$\lim_{x \to 0^+} \frac{\ln x}{1/x} = \lim_{x \to 0^+} \frac{1/x}{-1/x^2} = \lim_{x \to 0^+} (-x) = 0$$

so that

$$\lim_{x \to 0^+} x \ln x = 0 \qquad \square$$

In the next example we apply l'Hôpital's Rule to the logarithm of an expression in order to evaluate the limit of the expression.

EXAMPLE 8   Find $\displaystyle\lim_{x \to 0^+} x^x$.

Solution   The limit has the indeterminate form 0^0. By (9) in Section 7.3, $x^x = e^{x \ln x}$, and consequently

$$\lim_{x \to 0^+} x^x = \lim_{x \to 0^+} e^{x \ln x}$$

Since the exponential function is continuous, it follows that

$$\lim_{x \to 0^+} e^{x \ln x} = e^{\lim_{x \to 0^+} (x \ln x)}$$

if the limit on the right side exists. But by Example 7,

$$\lim_{x \to 0^+} x \ln x = 0$$

Consequently

$$\lim_{x \to 0^+} x^x = \lim_{x \to 0^+} e^{x \ln x} = e^0 = 1 \qquad \square$$

Let $f(x) = x^x$. We will use the information from Example 8 to help sketch the graph of f. Notice that the domain of f is $(0, \infty)$. Since $x^x = e^{x \ln x} > 0$, f has no intercepts. Differentiating x^x, we obtain

$$\frac{d}{dx}(x^x) = \frac{d}{dx}(e^{x \ln x}) = e^{x \ln x}(\ln x + 1) = x^x(\ln x + 1) \qquad (3)$$

Thus $f'(x) = 0$ if $\ln x + 1 = 0$, that is, if $\ln x = -1$, or $x = 1/e$. We use (3) to obtain the second derivative of x^x:

$$\frac{d^2}{dx^2}(x^x) = x^x(\ln x + 1)^2 + x^x\left(\frac{1}{x}\right)$$

$$= x^x\left[(\ln x + 1)^2 + \frac{1}{x}\right] > 0 \quad \text{for all } x > 0$$

It follows that $f(1/e) = (1/e)^{1/e} \approx 0.69$ is the absolute minimum value of f, and the graph is concave upward on $(0, \infty)$. The only major feature of the graph we need to attend to is its appearance near the y axis. This is determined in part by $\lim_{x \to 0^+} x^x$, which is 1 by Example 8. We now have the information needed to sketch the graph, which appears in Figure 7.37.

EXAMPLE 9   Show that $\displaystyle\lim_{x \to \infty}\left(1 + \frac{1}{x}\right)^x = e$.

Solution   The limit has the indeterminant form 1^∞. As in Example 8, we first find the limit of the logarithm of the expression on the left. This means finding a number b such that

$$\lim_{x \to \infty} \ln\left(1 + \frac{1}{x}\right)^x = b$$

Our answer for the original limit will then be e^b. We find that

$$\lim_{x \to \infty} \ln\left(1 + \frac{1}{x}\right)^x = \lim_{x \to \infty} x \ln\left(1 + \frac{1}{x}\right) = \lim_{x \to \infty} \frac{\ln(1 + 1/x)}{1/x}$$

This expression is now prepared for l'Hôpital's Rule because

$$\lim_{x \to \infty} \ln\left(1 + \frac{1}{x}\right) = 0 = \lim_{x \to \infty} \frac{1}{x}$$

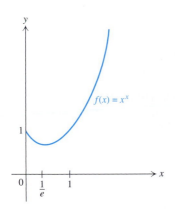

$f(x) = x^x$

FIGURE 7.37

As a result,

$$\lim_{x \to \infty} \frac{\ln(1 + 1/x)}{1/x} = \lim_{x \to \infty} \frac{\frac{1}{1 + 1/x}\left(-\frac{1}{x^2}\right)}{-1/x^2} = \lim_{x \to \infty} \frac{1}{1 + 1/x} = \frac{1}{1 + 0} = 1$$

Thus $b = 1$, so that

$$\lim_{x \to \infty} \left(1 + \frac{1}{x}\right)^x = e^1 = e \quad \square$$

The number e could be defined by means of the limit in Example 9, as is done in some textbooks.

Thermal Radiation

All bodies emit thermal radiation, which is produced when charged particles near the surface of the body are accelerated by thermal agitation. There is a whole range of possible accelerations, with a corresponding spectrum of wavelengths for the radiation. The intensity I of the radiation varies with the wavelength λ. We will discuss thermal radiation relating to a blackbody, which is an ideal body that absorbs all radiation that impinges on the body. A classic model for blackbody radiation was provided by the **Rayleigh-Jeans Law:**

$$I_0(\lambda) = \frac{2\pi c k T}{\lambda^4} \tag{4}$$

where c is the speed of light in a vacuum, k is the Boltzman constant, and T is the temperature. Although the Rayleigh-Jeans Law, which was proposed over 100 years ago, agrees with experiment for large values of λ, it is unreasonable for small values of λ because it implies that an infinite amount of energy could be radiated.

This problem was resolved by Max Planck, who proposed the formula

$$I_1(\lambda) = \frac{2\pi h c^2}{\lambda^5(e^{hc/\lambda k T} - 1)} \tag{5}$$

The constant h is now known as Planck's constant. The formula in (5) eliminated the problem for small values of λ. Additionally, $I_1(\lambda)$ is approximately $I_0(\lambda)$ when λ is large, as the following example confirms.

EXAMPLE 10   Show that $\lim_{\lambda \to \infty} I_1(\lambda)/I_0(\lambda) = 1$.

Solution   By simplifying $I_1(\lambda)/I_0(\lambda)$, we find that

$$\lim_{\lambda \to \infty} \frac{I_1(\lambda)}{I_0(\lambda)} = \lim_{\lambda \to \infty} \frac{ch/\lambda}{kT(e^{hc/\lambda k T} - 1)}$$

Since

$$\lim_{\lambda \to \infty} \frac{ch}{\lambda} = 0 = \lim_{\lambda \to \infty} kT(e^{hc/\lambda k T} - 1)$$

l'Hôpital's Rule applies, and yields

$$\lim_{\lambda \to \infty} \frac{ch/\lambda}{kT(e^{hc/\lambda k T} - 1)} = \lim_{\lambda \to \infty} \frac{-ch/\lambda^2}{kTe^{hc/\lambda k T}(-hc/\lambda^2 kT)} = \lim_{\lambda \to \infty} \frac{1}{e^{hc/\lambda k T}} = 1 \quad \square$$

Max Planck (1858–1947)

Planck's derivation of $I_1(\lambda)$ was based on his hypothesis that molecules absorb and radiate energy in discrete units called **quanta.** This idea eventually led to the development of quantum mechanics during the 1920s. Planck's fundamental research has been given highest honors. In 1918 he received the Nobel Prize in physics, and currently the most prestigious research institutes in Germany are named after him.

EXERCISES 7.6

In Exercises 1–50 use l'Hôpital's Rule to find the limit.

1. $\displaystyle\lim_{x\to a}\frac{x^{16}-a^{16}}{x-a}$

2. $\displaystyle\lim_{x\to 0}\frac{\sin x}{x}$

3. $\displaystyle\lim_{x\to 0}\frac{\cos x - 1}{x}$

4. $\displaystyle\lim_{x\to \pi/2^-}\frac{\cos x}{\sin x - 1}$

5. $\displaystyle\lim_{x\to 0^+}\frac{\sin 8\sqrt{x}}{\sin 5\sqrt{x}}$

6. $\displaystyle\lim_{x\to 1}\frac{\ln x}{x-1}$

7. $\displaystyle\lim_{x\to \infty}\frac{e^{1/x}-1}{1/x}$

8. $\displaystyle\lim_{x\to \infty}\frac{\tan 1/x}{1/x}$

9. $\displaystyle\lim_{x\to \pi/2^-}\frac{\tan x}{\sec x + 1}$

10. $\displaystyle\lim_{x\to 0^+}\frac{x^{-1/2}}{\ln x}$

11. $\displaystyle\lim_{x\to \infty}\frac{x}{\ln x}$

12. $\displaystyle\lim_{x\to \infty}\frac{x}{e^{2x}}$

13. $\displaystyle\lim_{x\to \infty}\frac{\ln(1+x)}{\ln x}$

14. $\displaystyle\lim_{x\to \infty}\frac{\ln x}{\ln(\ln x)}$

15. $\displaystyle\lim_{x\to 0}\frac{1-\cos x}{\sin x}$

16. $\displaystyle\lim_{x\to 0}\frac{\tan 4x}{\tan 2x}$

17. $\displaystyle\lim_{x\to \pi/2^-}\frac{\tan 4x}{\tan 2x}$

18. $\displaystyle\lim_{x\to 0^-}\frac{\tan x}{x^2}$

19. $\displaystyle\lim_{x\to 0}\frac{1-\cos 2x}{1-\cos 3x}$

20. $\displaystyle\lim_{x\to 0^+}\frac{1-\cos\sqrt{x}}{\sin x}$

21. $\displaystyle\lim_{x\to 0}\frac{\sqrt{1+x}-\sqrt{1-x}}{x}$

22. $\displaystyle\lim_{x\to 0^+}\frac{e^{1/x}}{\ln x}$

23. $\displaystyle\lim_{x\to 0^+}\frac{\sin x}{e^{\sqrt{x}}-1}$

24. $\displaystyle\lim_{x\to 0}\frac{e^{(x^2)}-1}{e^x - 1}$

25. $\displaystyle\lim_{x\to 0}\frac{5^x - 3^x}{x}$

26. $\displaystyle\lim_{x\to 1}\frac{4^x - 3^x - 1}{x-1}$

27. $\displaystyle\lim_{x\to 0}\frac{\sin^{-1}x}{x}$

28. $\displaystyle\lim_{x\to 1^-}\frac{(\pi/2)-\sin^{-1}x}{\sqrt{1-x^2}}$

29. $\displaystyle\lim_{x\to 0}\frac{x^3}{x-\sin x}$

30. $\displaystyle\lim_{x\to 1}\frac{\ln x - x + 1}{x^3 - 3x + 2}$

31. $\displaystyle\lim_{x\to 0}\frac{\tanh x - \sinh x}{x^2}$

32. $\displaystyle\lim_{x\to \pi/2}(2x-\pi)\sec x$

33. $\displaystyle\lim_{x\to 0}(\csc x - \cot x)$

34. $\displaystyle\lim_{x\to \pi/2}(\pi^2 - 4x^2)\tan x$

35. $\displaystyle\lim_{x\to 0^+}\sin x\ln(\sin x)$

36. $\displaystyle\lim_{x\to 0^+}x^{\sin x}$

37. $\displaystyle\lim_{x\to 0^+}\left(\ln\frac{1}{x}\right)^x$

38. $\displaystyle\lim_{x\to 1/2^-}(\tan \pi x)^{1-2x}$

39. $\displaystyle\lim_{x\to \pi/2^-}\frac{\ln(\cos x)}{\tan x}$

40. $\displaystyle\lim_{x\to \infty}\frac{e^x}{x^3}$

41. $\displaystyle\lim_{x\to \infty}x\sin\frac{1}{x}$

42. $\displaystyle\lim_{x\to \infty}\frac{\ln x}{x^2}$

43. $\displaystyle\lim_{x\to \infty}\frac{\ln(x^2+1)}{\ln x}$

44. $\displaystyle\lim_{x\to \infty}\frac{\log_4 x}{x}$

45. $\displaystyle\lim_{x\to \infty}\frac{e^{(e^x)}}{e^x}$

46. $\displaystyle\lim_{x\to \infty}\left(1-\frac{1}{x}\right)^x$

47. $\displaystyle\lim_{x\to \infty}\left(1+\frac{1}{x^2}\right)^x$

48. $\displaystyle\lim_{x\to \infty}\frac{1}{x(\pi/2 - \tan^{-1}x)}$

49. $\displaystyle\lim_{x\to \infty}x^{-1/2}\ln(\ln x)$

50. $\displaystyle\lim_{x\to \infty}x^2\left(1-x\sin\frac{1}{x}\right)$

The limits appearing in Exercises 51–52 appear as the first two illustrations of l'Hôpital's Rule in l'Hôpital's original text. Find these limits, assuming that $a > 0$.

51. $\displaystyle\lim_{x\to a}\frac{a^2 - ax}{a-\sqrt{ax}}$

52. $\displaystyle\lim_{x\to a}\frac{\sqrt{2a^3 x - x^4}-a\sqrt[3]{a^2 x}}{a-\sqrt[4]{ax^3}}$

In Exercises 53–54 sketch the graph of the function, noting any relative extreme values and concavity. Use l'Hôpital's Rule to determine the horizontal asymptotes, if any, and to determine the behavior of the function near 0.

53. $f(x) = xe^{-x}$

54. $f(x) = x\ln x$

55. Let

$$f(x) = \begin{cases} x(\ln x)^2 & \text{for } x > 0 \\ 0 & \text{for } x = 0 \end{cases}$$

Show that $\lim_{x\to 0^+} f(x) = f(0)$, which means that f is continuous from the right at 0.

56. Let

$$f(x) = \begin{cases} \dfrac{\cos x}{x^2 - \pi^2/4} & \text{for } x \neq -\dfrac{\pi}{2} \text{ and } \dfrac{\pi}{2} \\ -\dfrac{1}{\pi} & \text{for } x = -\dfrac{\pi}{2} \text{ or } \dfrac{\pi}{2} \end{cases}$$

a. Show that f is continuous at $-\pi/2$ and $\pi/2$.

b. Show that f is differentiable at $-\pi/2$ and $\pi/2$.

57. a. Why is the following "application" of l'Hôpital's Rule invalid?

$$\lim_{x\to \pi/2}\frac{\sin x}{x} = \lim_{x\to \pi/2}\frac{\cos x}{1} = \frac{0}{1} = 0$$

b. Evaluate $\displaystyle\lim_{x\to \pi/2}\frac{\sin x}{x}$.

58. Try to use l'Hôpital's Rule to find $\lim_{x\to \infty} x/\sqrt{x^2+1}$, and see what happens. Then determine the limit by a different method.

59. Try to evaluate $\lim_{x\to 0} e^{-1/x^2}/x$ by applying l'Hôpital's Rule directly to the fraction $e^{-1/x^2}/x$, and see what happens.

60. Show that $\lim_{x \to 0^+} x^{(x^x)} = 0$.

61. Show that $\lim_{x \to 0^+} (x^x)^x = 1$.

62. Find the value of c such that

$$\lim_{x \to \infty} \left(\frac{x}{x + c} \right)^x = e^3$$

*63. Find

$$\lim_{x \to 0} \frac{\displaystyle\int_0^x (x - t) \sin (t^2) \, dt}{\ln(1 + x^4)}$$

64. Let $f(x) = c \cosh(x/c)$, where c is a positive constant. Find

$$\lim_{x \to \infty} \frac{f^{-1}(x)}{c \ln 2x}$$

(*Hint:* Use the formula for $f^{-1}(x)$ in the project of Section 7.4, along with l'Hôpital's Rule.)

Applications

65. A right triangle T in the first quadrant has legs on the axes and hypotenuse tangent to the curve $y = e^{-x}$.

    a. Show that for any $\varepsilon > 0$ the area of T is less than ε if the base is sufficiently large.

    b. Show that the area of T is maximal if the base of T has length 2.

66. This exercise shows that Planck's function I_1 tends to 0 at 0 and at ∞.

    a. Use Example 10 and (4) to show that $\lim_{\lambda \to \infty} I_1(\lambda) = 0$.

    b. Let $a = hc/kT$. Show that $\lim_{\lambda \to \infty} e^{a\lambda}/\lambda^5 = \infty$.

    c. Use (b) to show that $\lim_{\lambda \to 0^+} \lambda^5(e^{a/\lambda} - 1) = \infty$.

    d. Use (c) to show that $\lim_{\lambda \to 0^+} I_1(\lambda) = 0$.

Project

1. Let f and g be defined on $(0, \infty)$. We say that f is of **higher order of magnitude** than g, and that g has **lower order of magnitude** than f, if $\lim_{x \to \infty} f(x)/g(x) = \infty$. In this case we write $f >> g$, or equivalently, $g << f$. In addition, we say that f and g have **equal order of magnitude** if $\lim_{x \to \infty} f(x)/g(x) = L$ for an appropriate positive number L, and we signify this by writing $f == g$. This project compares orders of magnitude for a variety of functions all of which have infinite limits at infinity.

Consider the following 9 functions:

$$x^x, \ x^k, \ x^{1/m}, \ (\ln x)^n, \ e^x, \ \sinh x, \ \cosh x, \ a^x, \text{ and } b^x$$

where k, m, and n are positive integers, and where a and b are any numbers such that $1 < a < e$ and $e < b$.

    a. Rank the 9 functions from least to highest order of magnitude, indicating any that are of equal order or magnitude.

    b. Find a function f that has a higher order of magnitude than any in part (a).

    c. Find a positive function g that has a lower order of magnitude than any in part (a).

7.7 INTRODUCTION TO DIFFERENTIAL EQUATIONS

In Section 4.4 we showed that if k is any fixed number, then every function f that satisfies the equation

$$f'(t) = kf(t) \tag{1}$$

has the form $f(t) = ce^{kt}$, for an appropriate constant c. Such a function represents exponential growth if $k > 0$ and exponential decay if $k < 0$. In the Leibniz notation, (1) is written

$$\frac{dy}{dt} = ky$$

where y denotes the function f. Equations that involve an unknown function and its derivatives are called differential equations. The final two sections of Chapter 7 are devoted to an introduction to differential equations. Differential equations are of fundamental importance in applications, not only in mathematics but in virtually all sciences and engineering. In Section 7.7 we will introduce the terminology relevant to differential equations. Then in Section 7.8 we will discuss two important methods of solving certain types of differential equations, and will present several applications.

Basic Notions in Differential Equations

An equation that involves an unknown function y, various derivatives of y, and possibly other known functions is called a differential equation. Some examples of differential equations are

$$\frac{dy}{dt} = -9.8t \tag{2}$$

$$\frac{dy}{dt} + 2y = 100 \tag{3}$$

$$\frac{d^2y}{dt^2} + 6\frac{dy}{dt} + 9y = \cos t \tag{4}$$

$$\frac{d^2y}{dx^2} + \sin y = 0 \tag{5}$$

Notice that the equations are written in the Leibniz notation, which is customary. The independent variable is usually t when the variable denotes time; otherwise the variable is normally x.

Differential equations can describe many phenomena. For example, equations (2)–(5) can represent, respectively, the motion of a falling body, the change in the size of a population, the flow of current in an electric circuit, and the motion of a pendulum.

The **order** of a differential equation is the order of the highest-order derivative of y appearing in the equation. Equations (2) and (3) are first-order differential equations; (4) and (5) are of second order. We will be concerned primarily with first-order equations.

If y is a function that satisfies a given differential equation, we say that y is a **solution** of the differential equation.

EXAMPLE 1   Show that if $y = -4.9t^2$, then y is a solution of

$$\frac{dy}{dt} = -9.8t$$

Solution   Differentiation of y yields immediately that $dy/dt = -9.8t$, so y is a solution of the given differential equation.   ❑

The equation in Example 1 represents the velocity of a falling object, moving vertically under the sole influence of gravity.

EXAMPLE 2   Let $y = 50 - 30e^{-2t}$. Show that y is a solution of

$$\frac{dy}{dt} + 2y = 100$$

Solution   Notice that $dy/dt = 60e^{-2t}$, so that

$$\frac{dy}{dt} + 2y = 60e^{-2t} + 2\,(50 - 30e^{-2t}) = 100 \quad \square$$

The differential equation appearing in Example 2 could arise in the study of population growth. For example, suppose a population grows but has a fixed upper bound, due to limited food and space. If the upper bound of the population y is B, then $y < B$ by hypothesis. As $B - y$ approaches 0, in all likelihood the population growth levels off, so that dy/dt approaches 0 also. The simplest relationship between dy/dt and $B - y$ is that they are proportional to one another, that is,

$$\frac{dy}{dt} = k(B - y), \quad \text{or equivalently,} \quad \frac{dy}{dt} + ky = kB$$

for some constant k. In the case of Example 2, $k = 2$ and $kB = 100$, so $B = 50$.

EXAMPLE 3   Let $y = \dfrac{4}{50} \cos t + \dfrac{3}{50} \sin t$. Show that y is a solution of

$$\frac{d^2y}{dt^2} + 6\frac{dy}{dt} + 9y = \cos t$$

Solution   First we observe that

$$\frac{dy}{dt} = -\frac{4}{50} \sin t + \frac{3}{50} \cos t$$

$$\frac{d^2y}{dt^2} = -\frac{4}{50} \cos t - \frac{3}{50} \sin t$$

Therefore

$$\frac{d^2y}{dt^2} + 6\frac{dy}{dt} + 9y = \left(-\frac{4}{50}\cos t - \frac{3}{50}\sin t\right) + 6\left(-\frac{4}{50}\sin t + \frac{3}{50}\cos t\right)$$

$$+ 9\left(\frac{4}{50}\cos t + \frac{3}{50}\sin t\right)$$

$$= \cos t$$

Consequently y satisfies the given equation. $\square$

Suppose an electric circuit consists of a voltage source, a capacitance of $\frac{1}{9}$ farad, and a resistance of 6 ohms. Let y represent the charge (in coulombs) on the capacitor. If the impressed voltage oscillates with time and is $\cos t$ volts at time t, then it can be shown that y satisfies the differential equation appearing in Example 3.

Not all solutions of differential equations can be expressed in the form $y = f(t)$. For example, the equation

$$t^2 + ty^3 + \frac{1}{2}e^{-2y} = 0 \tag{6}$$

yields a solution of the differential equation

$$2t + y^3 + (3ty^2 - e^{-2y})\frac{dy}{dt} = 0 \tag{7}$$

in the sense that there is a function y defined implicitly by (6), and we can obtain the differential equation (7) by differentiating both sides of equation (6) implicitly. Such a solution is called an **implicit solution.** In contrast, a solution in the form $y = f(t)$ is called an **explicit solution.** The solutions given in Examples 1 to 3 are explicit solutions.

How many solutions can a differential equation have? For Example 1, the function $y = -4.9t^2$ is a solution of the differential equation $dy/dt = -9.8t$. Theorem 4.6 then tells us that every solution of the differential equation has the form

$$y = -4.9t^2 + C \tag{8}$$

For that reason, we say that equation (8) is the general solution of $dy/dt = -9.8t$. Similarly, if C, C_1, and C_2 are arbitrary constants, then it can be shown that the functions defined by

$$y = Ce^{-2t} + 50 \tag{9}$$

and

$$y = \frac{4}{50} \cos t + \frac{3}{50} \sin t + C_1 e^{-3t} + C_2 te^{-3t} \tag{10}$$

are general solutions of (3) and (4), respectively, in the sense that any solution of the differential equation in (3) or (4) can be derived from (9) and (10) by choosing the constant C, or the constants C_1 and C_2 appropriately. As you can see, differential equations frequently have infinitely many solutions, depending on one or more constants. We will call a solution, whether explicit or implicit, the **general solution** if every solution can be obtained from it by specifying the constants suitably.

A solution in which all of the constants (like C, C_1, and C_2 in (8)–(10)) are specific numbers is a **particular solution.** The solutions we found in Examples 1 to 3 are particular solutions. To obtain a particular solution from a general solution, we often require that the solution satisfy **initial** (or **boundary**) **conditions,** which specify the value of the solution or certain of its derivatives at specific numbers in the domain. Thus the solution of (2) such that $y(1) = 3$ has the initial condition $y(1) = 3$. Similarly, the solution of (3) for which $y(0) = 2$ and $y'(0) = -4$ has the initial conditions $y(0) = 2$ and $y'(0) = -4$. The problem of finding a particular solution of a differential equation that satisfies one or more initial conditions is called an **initial value problem.** In this book when we ask for a particular solution to a first-order differential equation satisfying the initial condition $y(a) = b$, there will be only one such solution.

EXAMPLE 4   Let $y = Ce^{-2t} + 50$. Find the value of C for which y is the particular solution of

$$\frac{dy}{dt} + 2y = 100$$

such that $y(0) = 10$.

Solution   By (9), $y = Ce^{-2t} + 50$ is a general solution of the differential equation. Thus we need only find the constant C for which $y(0) = 10$. We have

$$10 = y(0) = Ce^{-2(0)} + 50$$

so that $-40 = C$. Consequently the particular solution for which $y(0) = 10$ is

$$y = -40e^{-2t} + 50 \qquad \square$$

Direction Fields

As we mentioned above, for a first-order differential equation like

$$\frac{dy}{dx} = x + 2y \tag{11}$$

there is a unique solution that corresponds to any initial condition. What is perhaps surprising is that we can observe the general behavior of solutions to (11) without actually solving (11), but rather by drawing a picture that is related to the solutions of the equation.

To explain how a picture can describe the solutions of a differential equation, we first notice that if y is the particular solution of (11) that satisfies the initial condition $y(x_0) = y_0$, then the graph of y passes through the point (x_0, y_0). At that point the slope of y is $x_0 + 2y_0$, since

$$\frac{dy}{dx}\bigg|_{(x_0,\, y_0)} = x_0 + 2y_0$$

For example, the slope of the solution that passes through $(-1, 2)$ is $-1 + (2)(2) = 3$. We can indicate this fact by drawing a small line segment that passes through the point $(-1, 2)$ and has slope 3 (Figure 7.38). We will call such a line segment a **slope segment.** In general, the slope segment through a given point (x_0, y_0) is a linear approximation to the solution satisfying $y(x_0) = y_0$. (You might wish to refer to Section 3.8 for a discussion of linear approximations.) Taking a significant number of well-scattered points in the plane and drawing corresponding slope segments, we create a picture called the **direction field** (or **slope field**) of the differential equation. The direction field for the differential equation in (11) is indicated in Figure 7.39(a).

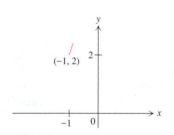

FIGURE 7.38 The slope segment at $(-1, 2)$ for the differential equation $\dfrac{dy}{dx} = x + 2y.$

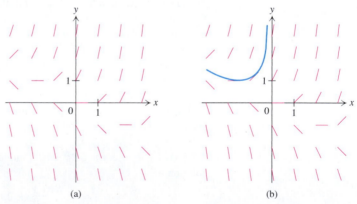

(a)   (b)

FIGURE 7.39 Part of the slope field of the differential equation $\dfrac{dy}{dx} = x + 2y.$

From the direction field of a differential equation, we can connect associated slope segments as in Figure 7.39(b) to obtain a smooth curve that (approximately) represents a solution. The curve that results is a **solution curve,** or **integral curve,** and corresponds to the solution that passes through any point on it. The solution curves for a differential equation yield important information about the solutions of the equation—even if we might not be able to find a reasonable formula for the solutions of the differential equation. For example, from Figure 7.40(a) we deduce that the solutions of the differential equation die out as x increases without bound, whereas Figure 7.40(b) suggests that the solutions are unbounded as x increases without bound.

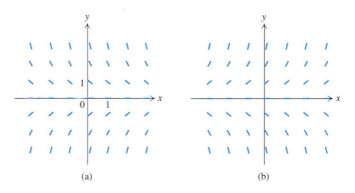

(a)   (b)

FIGURE 7.40

Now we have completed the preliminary discussion of differential equations. In the next section we will discuss the method of solving two basic kinds of differential equations.

EXERCISES 7.7

In Exercises 1–8 verify that the function y satisfies the given differential equation.

1. $\dfrac{dy}{dt} = e^{3t};\ y = \dfrac{1}{3}e^{3t}$

2. $\dfrac{dy}{dt} = 2t + 3y;\ y = 5e^{3t} - \dfrac{2}{3}t - \dfrac{2}{9}$

3. $2\dfrac{dy}{dt} - y^2 = 1;\ y = \tan t + \sec t$ for $0 < t < \pi/2$

4. $t\left(\dfrac{dy}{dt}\right)^2 - y\dfrac{dy}{dt} + 1 = 0;\ y = 3 + \dfrac{t}{3}$

5. $\dfrac{d^2y}{dx^2} + 4y = 0;\ y = \sin 2x - \cos 2x$

6. $\dfrac{d^2y}{dx^2} + y = 2e^{-x};\ y = e^{-x} + \sin x$

7. $\dfrac{d^2y}{dx^2} + 4\dfrac{dy}{dx} + 4y = 0;\ y = xe^{-2x}$

8. $\dfrac{d^3y}{dx^3} + 64y = 0;\ y = e^{-4x}$

In Exercises 9–12 verify that y is the particular solution that satisfies the initial conditions.

9. $\dfrac{dy}{dt} + 5y = -4e^{-3t};\ y(0) = -2;\ y = -2e^{-3t}$

10. $\left(\dfrac{dy}{dt}\right)^2 = 4(y + 1);\ y(0) = y(2) = 0;\ y = t^2 - 2t$

11. $x\dfrac{dy}{dx} - y = x^2\sqrt{1 + x^4};\ y(0) = y'(0) = 0;$

$y = x\displaystyle\int_0^x \sqrt{1 + t^4}\,dt$

12. $\dfrac{d^2y}{dx^2} - 2\dfrac{dy}{dx} + 2y = 0;\ y(0) = 0,\ y'(0) = 1;$

$y = e^x \sin x$

In Exercises 13–18 plot the direction field of the differential equation.

13. $\dfrac{dy}{dx} = 2x$

14. $\dfrac{dy}{dx} = \dfrac{1}{x}$

15. $\dfrac{dy}{dx} = -\dfrac{x}{y}$ **16.** $\dfrac{dy}{dx} = -4xy$

17. $\dfrac{dy}{dx} = \dfrac{x}{y}$ **18.** $\dfrac{dy}{dx} = x - y$

19. Show that if $y = \sqrt{r^2 - x^2}$, the graph of which is a semicircle, then y satisfies the differential equation

$$y\,\frac{dy}{dx} + x = 0$$

20. Show that if $y = c \cosh x/c$, the graph of which is a catenary, then y satisfies the differential equation

$$\frac{d^2y}{dx^2} = \frac{1}{c}\sqrt{1 + \left(\frac{dy}{dx}\right)^2}$$

21. Find the value of the constant c such that $y = x^2$ is a solution of

$$c\left(\frac{dy}{dx}\right)^2 - x\,\frac{dy}{dx} + y = 0$$

22. Consider the differential equation $\dfrac{d^2y}{dx^2} + y = 0$.

    a. Show that $\sin x$ and $\cos x$ are solutions.

    b. Show that $C_1 \sin x + C_2 \cos x$ is a solution, for all constants C_1 and C_2.

    c. Show that $\sin(x + \pi/3)$ is a solution, either directly or by showing that it can be written in the form $C_1 \sin x + C_2 \cos x$ and then using part (b).

23. Suppose that g is a differentiable function, and consider the differential equation $dy/dx = g(y)$. What can you conclude about the graphs of the solutions of the differential equation if

    a. $g(y) > 0$ for all $y > 0$

    b. $g'(y) > 0$ and $g(y) > 0$ for all $y > 0$

24. **a.** Write down a differential equation that has *no* solution.

    b. Write down a differential equation that has *exactly one* solution.

25. Consider the differential equation $(dy/dx)^2 = c$, where c is a real number. Find all values of c for which there is a solution of the differential equation.

Applications

26. In order for a cell to grow, it must receive nutrients through its surface. Notice that the surface area is in square units and the volume is in cubic units, and that the volume should be directly related to the weight of the cell. In one model of the growth in weight of the cell, the rate of change of the weight is proportional to the surface area. Thus if $W(t)$ denotes the weight of the cell at time t, then it is plausible that for some positive constant c,

$$\frac{dW}{dt} = cW^{2/3}$$

Find a positive value of k such that $W = (\tfrac{1}{3}ct + 1)^k$ is a solution.

27. Suppose glucose is injected at a constant rate a into one's blood stream, and let C denote the concentration of glucose in the blood stream at any moment. Assume that the glucose is eliminated at a constant rate b. Then one model of the concentration is that C satisfies the differential equation

$$\frac{dC}{dt} + aC = b$$

Determine the value of c (in terms of a and b) such that $C = c(1 - e^{-at})$ is a solution of the differential equation.

28. Suppose an object of mass m is close to the earth's surface and is falling toward earth with air resistance proportional to velocity. It follows from Newton's Second Law of Motion that the position of the object is given approximately by the differential equation

$$\frac{d^2y}{dt^2} - \frac{p}{m}\frac{dy}{dt} = -g$$

where p is a constant. Show that if $y = mgt/p + C_1 + C_2\,e^{pt/m}$, then y is a solution to the differential equation.

7.8   METHODS OF SOLVING DIFFERENTIAL EQUATIONS

Now that we have introduced the basic terminology of differential equations, we are ready to analyze two groups of differential equations: separable differential equations and linear first-order differential equations.

Separable Differential Equations

The easiest kind of differential equation to solve is a **separable differential equation,** which is one that can be written in the form

$$P(x) + Q(y)\frac{dy}{dx} = 0 \tag{1}$$

or in the more symmetric differential notation,

$$P(x)\,dx + Q(y)\,dy = 0, \quad \text{or} \quad Q(y)\,dy = -P(x)\,dx \tag{2}$$

(Notice that the x's and y's in (2) are separated from one another, hence the name "separable differential equation.")

A special type of separable differential equation occurs if $Q = -1$. In that case (1) becomes

$$P(x) - \frac{dy}{dx} = 0, \quad \text{or} \quad \frac{dy}{dx} = P(x)$$

This is an **integrable differential equation,** and as we know from the definition of indefinite integrals, the general solution is

$$y = \int P(x)\,dx$$

For example,

$$\frac{dy}{dx} = \sec^2 x$$

has the general solution

$$y = \int \sec^2 x\,dx = \tan x + C$$

In order to find a general solution of (1) when Q is not necessarily -1, we integrate both sides of (1) with respect to x. This yields

$$\int P(x)\,dx + \int \left[Q(y(x))\frac{dy}{dx} \right] dx = C$$

A substitution in the second integral yields

$$\int P(x)\,dx + \int Q(y)\,dy = C \tag{3}$$

This is also equivalent to

$$\int Q(y)\,dy = - \int P(x)\,dx \tag{4}$$

When we carry out the integration in either (3) or (4) we obtain the general (possibly implicit) solution of (1).

EXAMPLE 1   Let k be a constant. Find the general solution of the separable differential equation

$$\frac{1}{y}\,dy = \frac{k}{x}\,dx$$

for which $x > 0$ and $y > 0$.

Solution   Integrating both sides of the given equation, or equivalently, applying (4) with $Q(y) = 1/y$ and $P(x) = -k/x$, and noting that $x > 0$ and $y > 0$ by hypothesis, we obtain

$$\int \frac{1}{y}\,dy = \int \frac{k}{x}\,dx, \quad \text{so that} \quad \ln y = k \ln x + C_1$$

where C_1 is the constant of integration. It follows that

$$y = e^{\ln y} = e^{k \ln x + C_1} = e^{C_1} e^{k \ln x} = e^{C_1} x^k$$

Consequently if $C = e^{C_1}$, then the general solution of the differential equation is

$$y = Cx^k \quad \square$$

To see where the equation in Example 1 arises, we turn to growth in humans or organs. If we let x denote the size of a growing object, then the **specific (or relative) growth rate** at x is defined to be

$$\frac{1}{x}\frac{dx}{dt}, \quad \text{or equivalently,} \quad \frac{dx/dt}{x}$$

which is the ratio of the rate of change of x to x itself. The larger x is, the larger dx/dt needs to be to retain the same specific growth rate. Thus a baby's specific growth rate is apt to be larger than that of an adolescent.

Let x and y denote the sizes of two organs in the same living organism (like a young person). Empirical data support the claim that frequently the specific growth rates at x and y are approximately proportional to one another. This means that there is a constant k such that

$$\frac{1}{y}\frac{dy}{dt} = \frac{k}{x}\frac{dx}{dt} \tag{5}$$

Now (5) is not in the form appearing in Example 1. However, (5) is equivalent to

$$\frac{dy/dt}{dx/dt} = \frac{ky}{x}$$

By the Chain Rule,

$$\frac{dy}{dt} = \frac{dy}{dx}\frac{dx}{dt}, \quad \text{so that} \quad \frac{dy}{dx} = \frac{dy/dt}{dx/dt}$$

Consequently (5) can be rewritten as

$$\frac{dy}{dx} = \frac{ky}{x}$$

Separating the variables, we obtain

$$\frac{1}{y}\,dy = \frac{k}{x}\,dx \tag{6}$$

This is the equation appearing in Example 1. We conclude that

$$y = Cx^k \text{ for an appropriate constant } C \tag{7}$$

The equation in (7) is called an **allometric relation.**

To illustrate an allometric relation, we refer to a study by D. R. Platt,* in which the weight W and length L of western hognose snakes were measured. He showed that to a great degree of accuracy they satisfied the equation

$$W = 446\,L^{2.99}$$

* University of Kansas Publications, *Museum of Natural History* 18, No. 4 (1969), pp. 253–420.

where the lengths ranged from less than 0.3 meter to over 0.6 meter.

If data for x and y are given, we can tell if they (approximately) satisfy the equation $y = Cx^k$ for appropriate values of C and k. Indeed, if $y = Cx^k$, then $\ln y = \ln C + k \ln x$, so that when we plot $\ln x$ versus $\ln y$, we obtain a straight line with slope k and y intercept $\ln C$. For the hognose snake, the line is given by

$$\ln W = \ln 446 + 2.99 \ln L$$

with slope 2.99 and y intercept $\ln 446$.

A second example of a separable differential equation arises from the velocity of a spaceship as a function of the distance from the earth. Let v denote the velocity of a spaceship at an altitude of x kilometers above the surface of the earth. We will assume that the spaceship is under the influence of a gravitational force that varies with altitude; however, we will disregard air resistance. Using Newton's Law of Gravitation and his Second Law of Motion, one can show that under these conditions the velocity v satisfies the differential equation

$$v \frac{dv}{dx} = -\frac{gR^2}{(R + x)^2} \tag{8}$$

where R denotes the radius of the earth and g is the magnitude of the gravitational acceleration at the earth's surface.

EXAMPLE 2   Find the general solution of (8), and the particular solution for which the initial velocity (at the earth's surface) is 12 kilometers per second (that is, 43,200 kilometers per hour).

Solution   The differential equation in (8) is equivalent to

$$v \, dv = -\frac{gR^2}{(R + x)^2} \, dx$$

which is separable. This is equivalent to

$$\int v \, dv = \int -\frac{gR^2}{(R + x)^2} \, dx$$

which by integration yields the general solution

$$\frac{1}{2} v^2 = \frac{gR^2}{R + x} + C \tag{9}$$

For the particular solution we substitute $x = 0$ and $v = 12$ in (9) to obtain

$$\frac{1}{2} (12)^2 = \frac{gR^2}{R + 0} + C = gR + C$$

Thus

$$C = \frac{1}{2} (12)^2 - gR = 72 - gR$$

so that the particular solution is given implicitly by

$$\frac{1}{2} v^2 = \frac{gR^2}{R + x} + 72 - gR = 72 - \frac{gRx}{R + x}$$

If we assume that v is positive, then

$$v = \sqrt{144 - \frac{2gRx}{R + x}} \qquad \square$$

Notice that as the distance x above the earth's surface increases without bound, the velocity v approaches the limit

$$\lim_{x \to \infty} v(x) = \lim_{x \to \infty} \sqrt{144 - \frac{2gRx}{R + x}} = \sqrt{144 - 2gR}$$

Since the gravitational constant g is approximately 9.8×10^{-3} kilometers per second per second, and the earth's radius R is approximately 6.37×10^3 kilometers,

$$\lim_{x \to \infty} v(x) \approx \sqrt{144 - 2(9.8 \times 10^{-3})(6.37 \times 10^3)} \approx 4.4$$

Thus as the spaceship recedes from the earth, its velocity approaches a limiting, or terminal, velocity of 4.4 kilometers per second.

Linear First-Order Differential Equations

A **linear first-order differential equation** has the form

$$\frac{dy}{dx} + P(x)y = Q(x) \qquad (10)$$

where P and Q are continuous. The equation

$$f'(x) = kf(x)$$

which in the notation of differential equations is

$$\frac{dy}{dx} = ky$$

is the special case of (10) for which $P(x) = -k$ and $Q(x) = 0$. We studied this equation in Section 4.4 in connection with exponential growth and decay. There we found solutions defined on $[0, \infty)$.

The solution of (10) is obtained by letting S be an antiderivative of P, so that $S'(x) = P(x)$, and then multiplying both sides of (10) by $e^{S(x)}$. This yields

$$e^{S(x)} \frac{dy}{dx} + e^{S(x)} P(x)y = e^{S(x)} Q(x)$$

Now observe that

$$\frac{d}{dx}(ye^{S(x)}) = \frac{dy}{dx} e^{S(x)} + ye^{S(x)} S'(x)$$

$$= e^{S(x)} \frac{dy}{dx} + e^{S(x)} P(x)y$$

Therefore

$$\frac{d}{dx}(ye^{S(x)}) = e^{S(x)} Q(x) \qquad (11)$$

Integrating both sides of this equation, we obtain

$$ye^{S(x)} = \int e^{S(x)} Q(x) \, dx$$

so that

$$y = e^{-S(x)} \int e^{S(x)} Q(x)\, dx, \quad \text{where } S(x) = \int P(x)\, dx \tag{12}$$

The expression $e^{S(x)}$, by which we multiplied (10) in order to obtain (11), is called an **integrating factor** for the differential equation because the left side of (11) is easily integrated.

EXAMPLE 3   Let a and b be constants, with $a \neq 0$. Find the general solution of $dy/dx + ay = b$.

Solution   This equation has the form of (10) with $P(x) = a$ and $Q(x) = b$. Since ax is an antiderivative of P, $S(x) = ax$, so (12) implies that

$$y = e^{-ax} \int e^{ax} b\, dx = e^{-ax}\left(\frac{b}{a} e^{ax} + C\right) = \frac{b}{a} + Ce^{-ax} \quad \square$$

Under certain ideal laboratory conditions, the surface temperature y of an object changes in time at a rate proportional to the difference between the temperature of the object and that of the surrounding medium, which we will assume is constantly y_0. This gives rise to the differential equation

$$\frac{dy}{dt} = k(y - y_0), \quad \text{or equivalently,} \quad \frac{dy}{dt} - ky = -ky_0$$

where k is a negative constant. This equation, known as **Newton's Law of Cooling,** is equivalent to the equation in Example 3 if $a = -k$ and $b = -ky_0$. Consequently the general solution is

$$y = \frac{-ky_0}{-k} + Ce^{kt} = y_0 + Ce^{kt}$$

Since $k < 0$, we find that

$$\lim_{t \to \infty} y(t) = \lim_{t \to \infty} (y_0 + Ce^{kt}) = y_0$$

so that the temperature of the object approaches the "ambient" temperature of the medium as time passes. The formula for y has been used by investigators to determine the length of time between death and discovery of the corpse (see Exercise 32).

Next we derive a formula for the velocity of an object that moves through a viscous medium. When a cherry sinks to the bottom of a glass of lime juice, or a marble sinks in a can of motor oil, it experiences a resistive force called **drag** that is frequently proportional to the velocity v. Thus the resistive force is approximately bv, for some constant b. Since the velocity is in the downward direction and the drag force is in the upward direction, it follows that $b < 0$. Now suppose that the only forces acting on the object are the gravitational force and a drag force proportional to the velocity, and choose the positive direction to be upward. Then at any time t the total force on the object is $-mg + bv$, where m is the mass of the object and g is the gravitational constant. However, by Newton's Second Law of Motion, the total force equals ma. Since $ma = m\, dv/dt$, we obtain the differential equation

$$-mg + bv = m\frac{dv}{dt} \tag{13}$$

Unlike the drag force on small, slow-moving objects, the drag force on large, fast-moving objects like skydivers is approximately proportional to the square of the velocity. (*G. Savage/Vandtstadt/ Photo Researchers*)

or equivalently,

$$\frac{dv}{dt} - \frac{bv}{m} = -g \tag{14}$$

Equation (14) has the form of (10) with $P(t) = -b/m$ and $Q(t) = -g$. Since an antiderivative of $-b/m$ is $-bt/m$, it follows from (12) that the general solution of (14) is given by

$$v = e^{bt/m} \int e^{-bt/m}(-g)\, dt = e^{bt/m}\left(\frac{mg}{b} e^{-bt/m} + C\right)$$

$$= \frac{mg}{b} + Ce^{bt/m}$$

Now if $v = 0$ when $t = 0$, then $0 = mg/b + C$, so that $C = -mg/b$ and thus

$$v = \frac{mg}{b}(1 - e^{bt/m})$$

Since $b < 0$ and $m > 0$, it follows that $m/b < 0$, so that $\lim_{t\to\infty} e^{bt/m} = 0$. We conclude that if the object is allowed to fall forever, then its **limiting velocity,** or **terminal velocity,** would be mg/b.

Electric Circuits

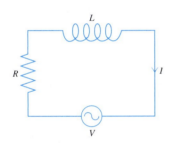

FIGURE 7.41

Assume that a closed electric circuit contains a resistance R (in ohms), an inductance L (in henrys), and an applied voltage V (in volts) (Figure 7.41). Let I be the electric current (in amperes) flowing through the circuit. You may think of the resistance as tending to impede the flow of electricity, the inductance as tending to keep the electricity flowing at a constant rate, and the voltage as tending to increase or decrease the flow of electricity, depending on the direction attached to the voltage. Kirchhoff's Second Law says that the applied voltage is equal to the sum of the voltage drops in the rest of the circuit. This law gives rise to the differential equation

$$L\frac{dI}{dt} + RI = V(t)$$

In the following example we will assume that the inductance and resistance are constant and that the voltage varies with time.

Gustav Kirchhoff (1824–1887)
Kirchhoff was a German physicist famous for his work in electricity and spectroscopy.

EXAMPLE 4   Let L and R be nonzero constants. Find the general solution of $L(dI/dt) + RI = e^{-t}$.

Solution   The equation is equivalent to

$$\frac{dI}{dt} + \frac{R}{L} I = \frac{1}{L} e^{-t}$$

which conforms to (10) if $P(t) = R/L$ and $Q(t) = (1/L)\, e^{-t}$. Since $(R/L)t$ is an antiderivative of P, $S(t) = (R/L)t$, so we conclude from (12) that

$$I = e^{-(R/L)t} \int e^{(R/L)t} \frac{1}{L} e^{-t}\, dt$$

$$= \frac{1}{L} e^{-(R/L)t} \int e^{(R-L)t/L}\, dt$$

Thus

$$I = \frac{1}{L} e^{-(R/L)t} \left(\frac{L}{R-L} e^{(R-L)t/L} + C \right)$$

$$= \frac{1}{R-L} e^{-t} + \frac{C}{L} e^{-(R/L)t} \qquad \square$$

EXERCISES 7.8

In Exercises 1–6 find the general solution of the separable differential equation.

1. $\dfrac{dy}{dx} = \dfrac{x}{y}$

2. $\dfrac{2y}{y^2+1} \dfrac{dy}{dx} = \dfrac{1}{x^2}$

3. $(1+x^2)\, dy = (1+y^2)\, dx$

4. $dy = \dfrac{y}{1+y}\, dx$

5. $\dfrac{1+e^x}{1-e^{-y}}\, dy + e^{x+y}\, dx = 0$

6. $\dfrac{dy}{dx} = x\sqrt{1-y^2}$

In Exercises 7–10 find the particular solution of the separable differential equation that satisfies the initial condition.

7. $e^{-2y}\, dy = (x-2)\, dx\,;\, y(0) = 0$

8. $(\ln y)^2 \dfrac{dy}{dx} = x^2 y\,;\, y(2) = 1$

9. $y^2 x \dfrac{dy}{dx} - x + 1 = 0\,;\, y(1) = 3$

10. $\sqrt{x^2+1}\, \dfrac{dy}{dx} = \dfrac{x}{y}\,;\, y(\sqrt{3}) = 2$

In Exercises 11–18 find the general solution of the linear differential equation.

11. $\dfrac{dy}{dx} + \dfrac{1}{x^2} y = 0$

12. $\dfrac{dy}{dx} - (\sinh x)\, y = 0$

13. $\dfrac{dy}{dx} + 2y = 4$

14. $\dfrac{dy}{dx} - ay = f(x)$, where f is continuous

15. $\dfrac{dy}{dx} - y = \dfrac{1}{1-e^{-x}}$

16. $\dfrac{dy}{dx} + y \cos x = \cos x$

17. $\dfrac{dy}{dt} + y \tan t = \tan t$ for $-\pi/2 < t < \pi/2$

18. $\dfrac{dy}{dt} - y \tan t = e^{\sin t}$ for $-\pi/2 < t < \pi/2$

In Exercises 19–22 find the particular solution of the linear differential equation that satisfies the initial condition.

19. $\dfrac{dy}{dx} + 5y = -4e^{-3x}\,;\, y(0) = -4$

20. $x \dfrac{dy}{dx} - 4y = x\,;\, y(1) = 1$

21. $(\cos x) \dfrac{dy}{dx} + y = 1$ for $0 < x < \pi/2\,;\, y(\pi/4) = 2$

22. $\dfrac{dy}{dx} - 2xy = x\,;\, y(0) = 0$

23. A differential equation of the form $dy/dx = f(y/x)$ is called **homogeneous.**

   a. Let $v = y/x$. Show that

   $$\frac{dy}{dx} = x \frac{dv}{dx} + v \qquad (15)$$

   b. Use (15) to show that a homogeneous differential equation $dy/dx = f(y/x)$ can be rewritten in the form

   $$\frac{1}{f(v) - v}\, dv = \frac{1}{x}\, dx \qquad (16)$$

24. Show that the differential equation

   $$\frac{dy}{dx} = \frac{x+y}{x}$$

   is homogeneous. Then find the general solution by using (16).

25. Find the general solution of the differential equation

   $$\frac{y^2}{x^2+y^2} + \frac{2xy}{x^2+y^2} \frac{dy}{dx} = 0$$

   which is not separable but which can be made separable by multiplying each side by a suitable expression in x and y.

26. Find the general solution of the differential equation

   $$L \frac{dI}{dt} + RI = 2$$

   where L and R are nonzero constants.

Applications

27. Recall that the limiting velocity of an object falling with a drag force proportional to velocity is mg/b.

   a. Let $v = mg/b$. Use (13) to show that $dv/dt = 0$.

   b. Give a physical explanation of why dv/dt should be zero at the limiting velocity.

28. Let L denote the length of a pendulum and T its period of oscillation. If the angle through which the pendulum swings is small (say less than 10°) and if the pendulum is assumed to be subject to no forces other than gravity, then T and L very nearly satisfy the differential equation

$$\frac{dT}{dL} = \frac{T}{2L}$$

   Show that the period T is proportional to the square root of the length L. (Experiment has shown that $T \approx 2\pi\sqrt{L/g}$.)

29. When an object is dropped, its speed v depends on the distance s fallen. If air resistance is negligible, s satisfies the equation

$$v = \sqrt{2gs} \qquad (17)$$

   where g is the acceleration due to gravity. Since $v = ds/dt$, (17) can be written as the differential equation

$$\frac{ds}{dt} = \sqrt{2gs} \qquad (18)$$

   a. Find the particular solution of (18) for which $s(0) = 0$.

   b. Show that the formula you obtain in (a) agrees with (1) and (2) in Section 1.3.

30. The radius r and the volume V of a spherical drop of water are related by the formula

$$V = \frac{4}{3}\pi r^3$$

   Show that the specific growth rate of the volume equals three times the specific growth rate of the radius.

31. Data relating heat production to weight of various warm-blooded animals have been assembled. The following table itemizes a few of the statistics:

	Pigeon	Hen	Dog	Human	Cow
Mean daily heat production (in calories)	25	110	460	1400	6000
Mass (in kilograms)	0.44	3	19	80	520

   Determine values of C and k so that these data satisfy to

a high degree the allometric relation in (7). Let x denote heat production and y denote mass.

32. Suppose a corpse is found at noon, and at that moment has a temperature of 87° F. One-half hour later the corpse has a temperature of 83° F. Assuming that normal body temperature is 98.6° F and the air temperature is constantly 75° F, determine at what time death occurred. (*Hint:* Let $t = 0$ at noon.)

33. An inflexible hanging cable is fastened at two points A and B (Figure 7.42). If the cable is subject only to the gravitational force and the tension at A and B, then the curve it forms is the graph of a solution f of

$$f''(x) = c\{1 + [f'(x)]^2\}^{1/2}$$

where c is a constant depending on the mass per unit length of the cable and on the tension at A and B. If we let $y = f'$, then the differential equation becomes

$$\frac{dy}{dx} = c(1 + y^2)^{1/2} \qquad (19)$$

FIGURE 7.42   The hanging cable for Exercise 33.

   a. Use (9) of Section 7.4 to show that $y = \sinh(cx + C)$ is the general solution of (19).

   b. Use part (a) to show that

$$f(x) = \frac{1}{c}\cosh(cx + C) + D$$

   (If $C = 0 = D$, then the cable hangs in the shape of a catenary, as asserted in Section 7.4.)

34. Consider a rocket in free space, being propelled away from the earth by the thrust of the exhaust gases escaping through the tail of the rocket. Assume that the velocity v_e of the exhaust relative to the rocket is constant, and denote the total mass of the rocket and its decreasing supply of fuel by M. Then it can be shown from the Law of Conservation of Momentum that

$$M\frac{dv}{dM} = -v_e \qquad (20)$$

where v denotes the velocity of the rocket relative to the earth. Let $M = M_0$ and $v = v_0$ at the instant the rockets are turned on. Find the solution of (20) for which $v = v_0$ when $M = M_0$.

Project

1. A natural extension of a linear first-order differential equation is an equation of the form

$$\frac{dy}{dx} = q_1(x) + q_2(x)y + q_3(x)y^2 \qquad (21)$$

where q_1, q_2, and q_3 are continuous. Such a differential equation is called a **Riccati equation** after the Venetian savant Jacopo Francesco Riccati (1676–1754), who was instrumental in introducing Newton's work to Italy.

Notice that if $q_3 = 0$, then (21) is a linear first-order differential equation and can be solved by the methods of this section. However, if $q_3 \neq 0$, then Riccati's equation cannot always be solved by elementary means. Nevertheless, if one solution of (21) is known, then the following method will allow us to find others by solving an associated linear first-order differential equation.

a. Suppose y is a nonzero solution of (21), and let c be a nonzero constant such that $c \neq 1$. Show that in general cy cannot be a solution of (21).

b. Suppose y_1 is a solution of (21). Show that the substitution $y = y_1 + 1/v$ leads to the linear first-order equation

$$\frac{dv}{dx} = -(q_2 + 2q_3 y_1)v - q_3 \qquad (22)$$

(*Hint:* After finding dy/dx in terms of dy_1/dx, v, and dv/dx, use the fact that y_1 is a solution of the equation in (21).)

c. Show that $y_1 = e^x$ is a solution of the Riccati equation

$$\frac{dy}{dx} = \frac{1}{3}e^x + \frac{1}{3}y + \frac{1}{3}e^x y^2$$

d. Using the substitution in (b) with $y_1 = e^x$, find another solution of the equation in (c). (*Hint:* For simplicity, take the constant of integration to be 0 when solving (22).)

REVIEW

Key Terms and Expressions

Inverse function
Exponential function
Exponent; base
Hyperbolic function

Differential equation
General solution
Particular solution
Initial value problem

Key Formulas

$$f^{-1}(f(x)) = x \quad \text{and} \quad f(f^{-1}(x)) = x$$

$$\frac{dx}{dy} = \frac{1}{dy/dx}$$

$$(f^{-1})'(c) = \frac{1}{f'(a)}, \quad \text{where } c = f(a)$$

$$\int \frac{1}{\sqrt{1-x^2}}\,dx = \sin^{-1} x + C$$

$$\text{and} \quad \frac{d}{dx}\sin^{-1} x = \frac{1}{\sqrt{1-x^2}}$$

$$\int \frac{1}{x^2+1}\,dx = \tan^{-1} x + C$$

$$\text{and} \quad \frac{d}{dx}\tan^{-1} x = \frac{1}{x^2+1}$$

$$\int \frac{1}{x\sqrt{x^2-1}}\,dx = \sec^{-1} x + C$$

$$\text{and} \quad \frac{d}{dx}\sec^{-1} x = \frac{1}{x\sqrt{x^2-1}}$$

$$\int e^x\,dx = e^x + C \quad \text{and} \quad \frac{d}{dx}e^x = e^x$$

$$a^r = e^{r \ln a} \quad \text{for } a > 0$$

$$a^{b+c} = a^b a^c, \quad a^{-b} = \frac{1}{a^b}, \quad \text{and} \quad (a^b)^c = a^{bc}$$

$$\log_a bc = \log_a b + \log_a c$$

$$\log_a b^r = r \log_a b$$

$$\log_a x = \frac{\ln x}{\ln a}$$

$$\cosh^2 x - \sinh^2 x = 1$$

$$\frac{d}{dx}\sinh x = \cosh x$$

$$\frac{d}{dx}\cosh x = \sinh x$$

Key Theorems

General Laws of Exponents
General Law of Logarithms

l'Hôpital's Rule

Review Exercises

In Exercises 1–6 determine whether the function has an inverse.

1. $f(x) = 3x^3 + 5x^5 - 10$ **2.** $f(x) = \dfrac{x^2 - 1}{x^2 + 1}$

3. $f(x) = 1 - \dfrac{1}{x}$ **4.** $f(x) = x - \dfrac{1}{x}$

5. $g(x) = \sin 2x + \cos x$ **6.** $g(x) = x + \cos x$

In Exercises 7–8 find a formula for f^{-1}.

7. $f(x) = \dfrac{3x - 2}{-x + 1}$ **8.** $f(x) = \dfrac{x^3}{4 - x^3}$

9. Let f be an even function on $(-a, a)$. Show that f does not have an inverse.

10. Let
$$f(x) = \frac{e^x}{1 + e^x}$$

   a. Show that f has an inverse.
   b. Find $(f^{-1})'(1/2)$.

In Exercises 11–18 find dy/dx.

11. $y = \ln(1 + 2^x)$ **12.** $y = x^{\cos x}$
13. $y = \tan^{-1}(\sinh x)$ **14.** $y = x^{(2^x)}$
15. $y = \sin^{-1}(1 - x^2)^{1/3}$ **16.** $y = \dfrac{\sinh x}{x}$
17. $y = \log_4(\tan^{-1} x^2)$ **18.** $y \sinh^{-1} x + e^y = y^5$

In Exercises 19–34 evaluate the integral.

19. $\displaystyle\int \frac{e^x}{\sqrt{1 + e^x}}\, dx$ **20.** $\displaystyle\int \frac{e^x}{\sqrt{1 - e^{2x}}}\, dx$

21. $\displaystyle\int \frac{e^x}{\sqrt{1 + e^{2x}}}\, dx$ **22.** $\displaystyle\int \frac{1}{\sqrt{e^{2x} - 1}}\, dx$

23. $\displaystyle\int \frac{e^x}{e^x + e^{-x}}\, dx$ **24.** $\displaystyle\int e^x \cosh x \, dx$

25. $\displaystyle\int_0^1 x^2\, 5^{-x^3}\, dx$ **26.** $\displaystyle\int \frac{\sin x}{\sqrt{1 - 4\cos^2 x}}\, dx$

27. $\displaystyle\int \frac{3}{1 + 4t^2}\, dt$ **28.** $\displaystyle\int \frac{t^4}{t^{10} + 1}\, dt$

29. $\displaystyle\int_{-5/4}^{5/4} \frac{1}{\sqrt{25 - 4t^2}}\, dt$ **30.** $\displaystyle\int \frac{e^{1 + \ln x}}{\sqrt{1 + x^2}}\, dx$

31. $\displaystyle\int \operatorname{sech} x \, dx$ **32.** $\displaystyle\int \frac{\sqrt{x}}{\sqrt{1 + x^3}}\, dx$

33. $\displaystyle\int \frac{x}{x^4 + 4x^2 + 10}\, dx$ **34.** $\displaystyle\int \frac{1}{x\sqrt{1 - (\ln x)^2}}\, dx$

In Exercises 35–42 evaluate each limit by l'Hôpital's Rule.

35. $\displaystyle\lim_{t \to 0} \frac{\sinh at}{t}$ **36.** $\displaystyle\lim_{x \to 0} \frac{\tan x - x}{\sin x - x}$

37. $\displaystyle\lim_{x \to 0^+} \frac{\ln x}{\ln(\sin x)}$ **38.** $\displaystyle\lim_{x \to 0^+} \frac{e^{1/x}}{1 - \cot x}$

39. $\displaystyle\lim_{x \to 0^+} \left(\frac{1}{x} - \frac{1}{\tan^{-1} x} \right)$ **40.** $\displaystyle\lim_{x \to \infty} x^{1/2} \sin \frac{1}{x}$

41. $\displaystyle\lim_{x \to \infty} \left(\frac{x + 1}{x - 1} \right)^x$ **42.** $\displaystyle\lim_{x \to 1^+} (\ln x)^{x - 1}$

In Exercises 43–46 find the general solution of the differential equation.

43. $e^{(y^2)}\, dx + x^2 y \, dy = 0$

44. $x\sqrt{1 - y^2} + y\sqrt{1 - x^2}\, \dfrac{dy}{dx} = 0$

45. $\dfrac{dy}{dx} + \dfrac{2}{x} y = x^2 + 6$ for $x > 0$

46. $\dfrac{dy}{dx} - y \cot x = \csc x$ for $0 < x < \pi$

In Exercises 47–50 find the particular solution that satisfies the given initial condition.

47. $2xy \, dx = (y + 1)\, dy;\ y(0) = 1$
48. $y(1 + x^2)\, dy + (y^2 + 1)\, dx = 0;\ y(0) = \sqrt{3}$
49. $\dfrac{dy}{dx} - 2y = 3;\ y(0) = 2$
50. $\dfrac{dy}{dx} + y \tan x = \sec x$ for $-\pi/2 < x < \pi/2;\ y(0) = \pi/4$

51. Let $f(x) = (2x)^x$. Sketch the graph of f, indicating any relative extreme values, concavity, and asymptotes.

52. Let $f(x) = \dfrac{x + a}{x + b}$ and let $a \neq b$. Find the value of b for which $f = f^{-1}$.

53. Use (10) of Section 7.1 and a hyperbolic identity to derive the formula
$$\frac{d}{dx} \sinh^{-1} x = \frac{1}{\sqrt{1 + x^2}}$$

54. Verify the following two equations:
$$\cosh(2u) - 1 = \frac{1}{2}(e^u - e^{-u})^2$$
$$\cosh(2u) + 1 = \frac{1}{2}(e^u + e^{-u})^2$$

55. a. Use the inequality $e^x > 1 + x$ for all $x \neq 0$ to show that

$$e^x < \frac{1}{1-x} \quad \text{for } x < 1$$

b. Show that

$$\ln \frac{x+1}{x} < \frac{1}{x} < \ln \frac{x}{x-1} \quad \text{for } x > 1$$

(*Hint:* Replace x by $1/x$ in part (a).)

 c. Plot the graphs of

$$\ln\left(\frac{x+1}{x}\right) - \frac{1}{x} \quad \text{and} \quad \ln\left(\frac{x}{x-1}\right) - \frac{1}{x}$$

on the same screen, with x window [1, 10] and y window [–2, 2]. Do the graphs corroborate the result in part (b)?

56. Let

$$f(x) = \left(1 + \frac{1}{x}\right)^x$$

Show that f is increasing on $(0, \infty)$. (*Hint:* Using Exercise 55(b), show that $\ln f(x)$ is increasing.)

 57. Let

$$f(x) = (\sin x)^x \quad \text{and} \quad g(x) = (\sin x)^{\sin x}$$

a. Plot the graphs of f and g for $0 \leq x \leq 3$, and conjecture $\lim_{x\to 0^+} [f(x)/g(x)]$.

*b. Compute $\lim_{x\to 0^+} f(x)/g(x)$.

 58. Let

$$f(x) = 2^x \quad \text{and} \quad g(x) = f(0) + f'(0)x + \frac{1}{2}f''(0)x^2$$

a. Plot $f(x)$ and $g(x)$ for $-1 \leq x \leq 2$ and $0 \leq y \leq 2$ to see that $f(x)$ and $g(x)$ are nearly equal for x near 0.

b. To discover how closely g approximates f near 0, find the largest positive integer n such that

$$\lim_{x\to 0} \frac{f(x) - g(x)}{x^n} = 0$$

59. Let $f(x) = x + \sin x$ and $g(x) = x$. Show that $\lim_{x\to\infty} [f(x)/g(x)]$ exists but $\lim_{x\to\infty} [f'(x)/g'(x)]$ does not exist.

60. Evaluate

$$\lim_{x\to\infty} \left(\frac{2x+1}{2x-1}\right)^x$$

After evaluating the limit, calculate $[(2x + 1)/(2x - 1)]^x$ for $x = 2$ and compare it with the limit.

61. Let

$$f(x) = \frac{\sqrt{x}}{x-1}e^{\sqrt{x}} \quad \text{and} \quad g(x) = \frac{1}{\sqrt{x}(x-1)}e^{\sqrt{x}}$$

Find the area A of the region between the graphs of f and g on [4, 9]. (*Hint:* To evaluate the integral, substitute $u = \sqrt{x}$.)

62. Let

$$f(x) = \frac{x}{\sqrt{x^2 - 1}} \quad \text{and} \quad g(x) = \frac{\sqrt{x^2 - 1}}{x}$$

Find the area A of the region between the graphs of f and g on $[\sqrt{2}, 2]$.

Applications

63. A rectangle with base on the x axis has its two upper vertices on the graph of $y = e^{-x^2}$. Show that if these vertices are inflection points of the graph, then the rectangle has the maximum possible area.

64. Let y be the population of a given species. Assume that the death rate is proportional to y, and that the birth rate is proportional to the product of the number of males and the number of females (assumed to be equal). Then

$$\frac{dy}{dt} = ay^2 - by$$

where a and b are positive constants. Show that if $0 < y(t) < b/a$ for all $t > 0$, then the population decreases forever.

65. The magnitude F of the gravitational force on a particle of mass m located at a distance R from the center of a slender homogeneous rod with mass M and length L is given by

$$F = \frac{GmM}{L} \int_{-L/2}^{L/2} \frac{1}{x^2 + R^2} \, dx$$

where G is a gravitational constant.

a. Evaluate the integral and hence F.

b. Suppose that L approaches 0 and R is fixed. Show that F approaches GmM/R^2.

66. Suppose an ice cube melts at a rate proportional to its surface area.

a. Let V denote the volume of the ice cube. Find the number $r > 0$ such that $dV/dt = kV^r$ for some negative constant k.

b. Show that $V = (\frac{1}{3}kt + C)^3$ is a solution of the differential equation.

c. If the ice cube begins to melt at time $t = 0$, what is the physical significance of C?

d. Express the length of time it takes the ice cube to melt completely in terms of k and C.

e. Use the Chain Rule to find a differential equation that relates the side length s of the cube to its rate of change ds/dt.

Topics for Discussion

1. Suppose f has an inverse. Tell what the following properties of f imply about f^{-1}: $(a, f(a))$ is an inflection point, $f'(x) > 0$ for all x in the domain of f, $\lim_{x \to a} f(x) = \infty$, and $f'(a) = 0$.

2. If f has an inverse, does it follow that f' also has an inverse? Explain why or why not.

3. From your knowledge of the derivatives of $\log_a x$ and a^x, why do you think that $\ln x$ and e^x are called "natural"?

4. Discuss what it means for a limit to be of indeterminate form. Tell why 0^∞ and 1^0 are not in the list of indeterminate forms discussed in the text. Should 2^∞ be in the list? Explain why or why not.

5. Discuss what it means for e^x to approach ∞ faster than x^n for any positive integer n as x approaches ∞. Compare the rates at which e^x and x^{-n} approach 0 as x approaches $-\infty$. Also compare the rates at which $-\ln x$ and x^{-n} approach ∞ as x approaches 0 from the right.

6. Define in your own words what constitutes a differential equation, a general solution of a differential equation, and an initial value problem.

7. Explain why the process of finding an indefinite integral is equivalent to solving a particular type of a differential equation.

Cumulative Review, Chapters 1–6

In Exercises 1–2 find the limit.

1. $\displaystyle \lim_{x \to \sqrt{2}} \frac{2\sqrt{2} - 2x}{8 - 4x^2}$ 2. $\displaystyle \lim_{x \to \infty} e^{-x}(1 + \sin^2(e^x))$

In Exercises 3–4 find $f'(x)$.

3. $f(x) = \dfrac{1 + e^{-x}}{1 + e^x}$ 4. $f(x) = \displaystyle\int_x^{2x+1} e^{(t^2)}\, dt$

5. Let $f(x) = \ln x - \frac{1}{8} x^2$. Show that

$$\sqrt{1 + (f'(x))^2} = \frac{1}{x} + \frac{x}{4}$$

6. Consider the graph of the equation $2x^{3/2} - 2y^{3/2} = 3xy^{1/2}$.
   a. Show that if (a, b) is on the graph, then $a \geq b \geq 0$.
   b. Show that the slope of any line tangent to the graph is positive.

7. Let $f(x) = \dfrac{x - 4}{2x + 6}$. Show that $f(x) > \frac{3}{20}$ for all $x > 7$ by

   a. solving the inequality $\dfrac{x - 4}{2x + 6} > \dfrac{3}{20}$.

b. showing that f is increasing on $[7, \infty)$ and then using this fact.

In Exercises 8–9 sketch the graph of f, indicating all pertinent information.

8. $f(x) = \dfrac{\cos x}{1 - \sin x}$ 9. $f(x) = e^{(1 - e^x)}$

10. Find a positive value of c that makes the parabolas $y = x^2 + 3$ and $x = cy^2$ have the same tangent line at a point, and find the point of tangency.

11. A circular racetrack has a radius of 200 feet. Pat sits in the best seat in the stands, located at a point on the edge of the track. At a certain instant a greyhound is on the track, 200 feet from Pat. The dog is moving around the track with a speed of 50 feet per second. How fast is the distance between the greyhound and Pat changing at that instant? (*Hint:* Use the Law of Cosines.)

12. A rectangular window is to be placed inside a parabolic arch whose height and base are each 36 inches. Determine the largest possible area A of the window.

In Exercises 13–14 evaluate the integral.

13. $\displaystyle\int \frac{x}{4 + x^2}\, dx$ 14. $\displaystyle\int_0^{\pi/3} \frac{1 + \tan^2 x}{\sec x}\, dx$

15. Suppose the velocity of a car (in miles per hour) is given by $v(t) = 40 + 40/(4 + t^2)$ for $0 \leq t \leq 2$, where t is in hours. How far does the car travel in the two hours?

16. Let
$$f(x) = x^2 + 2x \text{ and } g(x) = \frac{4}{4 + x} - 1$$
Find the area A of the region between the graphs of f and g on $[0, 1]$.

17. Find the area A of the region that lies in the first quadrant below the graph of $(y^2/x^4) - x^3 = 8$ and between the lines $x = 1$ and $x = 2$.

18. Find the value of c for which
$$\int_{1/2}^2 2/x^2\, dx = \int_{1/2}^2 (x^2 + c)\, dx$$

19. Let R be the region between the graphs of $y = 4 - x^2$ and $y = 1 - (|x| - 1)^2$.
   a. Determine the volume V of the double-holed solid obtained by revolving R about the x axis.
   b. Determine the center of gravity of R.

20. Suppose a pyramidal water tank with square base 20 feet on a side is 30 feet tall, and rests on its base. If the tank is full of water, determine the work W required to pump all but the bottom three feet of water to a point two feet above the top of the tank.

A ten-mark bill honoring Gauss and including a normal density curve. (*David Grossman*)

8 TECHNIQUES OF INTEGRATION

We have encountered numerous applications of definite integrals, especially in Chapter 6, which is devoted to applications of the integral. The main issue in evaluating an integral $\int_a^b f(x)\,dx$ involves finding an antiderivative of f. For those definite integrals appearing in Chapters 5–7 we have been able to find the needed antiderivatives. However, there are other integrals, such as

$$\int_0^\pi e^{-x} \cos x\,dx \quad \text{and} \quad \int_{-5/2}^{5/2} \sqrt{25 - 4x^2}\,dx$$

which we are as yet not able to evaluate directly because the needed antiderivatives are not easily found. The first integral appears in the study of electrical circuits (see Section 8.1), whereas the second integral yields the area of an ellipse (see Section 8.3). With the techniques of integration that we discuss in the present chapter, you will be able to evaluate those integrals and many others of interest in mathematics and in applications.

In Section 5.6 we discussed one technique: the substitution method. Although it is the most widely used aid to integration, there are others in common use. In Sections 8.1 to 8.4 we describe other prominent techniques of integration, including integration by parts, by trigonometric substitution, and by partial fractions. Together these techniques will greatly increase the collection of functions we can integrate.

Sections 8.5 and 8.6 offer other alternatives in the evaluation of integrals. Section 8.5 concerns symbolic integration by computer. These methods are most effective for finding integrals that cannot be easily evaluated. Section 8.6 is dedicated to the computation of approximate values of definite integrals whose numerical values are difficult or impossible to calculate. We conclude the chapter with a discussion of what are called improper integrals.

501

Below we summarize the basic integrals that have so far appeared.

$$\int x^r \, dx = \frac{1}{r+1} x^{r+1} + C \qquad \int \frac{1}{x} \, dx = \ln|x| + C$$

$$\int \sin x \, dx = -\cos x + C \qquad \int \tan x \, dx = -\ln|\cos x| + C$$

$$\int \cos x \, dx = \sin x + C \qquad \int \sec x \, dx = \ln|\sec x + \tan x| + C$$

$$\int \cot x \, dx = \ln|\sin x| + C \qquad \int \csc x \, dx = -\ln|\csc x + \cot x| + C$$

$$\int e^x \, dx = e^x + C \qquad \int a^x \, dx = \frac{1}{\ln a} a^x + C$$

$$\int \frac{1}{\sqrt{a^2 - x^2}} \, dx = \sin^{-1} \frac{x}{a} + C \qquad \int \frac{1}{x^2 + a^2} \, dx = \frac{1}{a} \tan^{-1} \frac{x}{a} + C$$

8.1   INTEGRATION BY PARTS

In Section 5.5 we converted the addition and constant multiple theorems for derivatives into results about integration, and in Section 5.6 we converted the Chain Rule into the method of integration by substitution. In this section we will recast the product theorem for differentiation in terms of integrals. The resulting integration theorem will allow us to evaluate many integrals, such as

$$\int \ln x \, dx \quad \text{and} \quad \int e^{-x} \cos x \, dx$$

which we have not yet evaluated.

Let F and G be differentiable functions. From the product rule we know that

$$(FG)'(x) = F'(x)G(x) + F(x)G'(x)$$

or by rearranging,

$$F(x)G'(x) = (FG)'(x) - F'(x)G(x) \tag{1}$$

If F' and G' are continuous, then we can restate (1) in the language of indefinite integrals as

$$\int F(x)G'(x) \, dx = \int [(FG)'(x) - F'(x)G(x)] \, dx$$

$$= \int (FG)'(x) \, dx - \int F'(x)G(x) \, dx$$

or equivalently,

$$\int F(x)G'(x)\,dx = F(x)G(x) - \int F'(x)G(x)\,dx \qquad (2)$$

Now suppose we wish to evaluate $\int f(x)\,dx$ but cannot readily do so. If f can be rewritten as the product FG', then (2) tells us that

$$\int f(x)\,dx = \int F(x)G'(x)\,dx = F(x)G(x) - \int F'(x)G(x)\,dx \qquad (3)$$

If in addition $\int F'(x)G(x)\,dx$ can be readily evaluated, then $\int f(x)\,dx$ can be evaluated by means of (3). This method of evaluating $\int f(x)\,dx$, by "splitting" the integrand f into two parts F and G', is known as **integration by parts.** We summarize this result and its definite integral version as a theorem.

THEOREM 8.1
Integration by Parts

Let F and G be differentiable on $[a, b]$, and assume that F' and G' are continuous on $[a, b]$. Then

$$\int F(x)G'(x)\,dx = F(x)G(x) - \int F'(x)G(x)\,dx \qquad (4)$$

and

$$\int_a^b F(x)G'(x)\,dx = F(x)G(x)\Big|_a^b - \int_a^b F'(x)G(x)\,dx \qquad (5)$$

Formula (4) can be simplified by formally substituting $u = F(x)$ and $v = G(x)$, so that $du = F'(x)\,dx$ and $dv = G'(x)\,dx$. We obtain

$$\int u\,dv = uv - \int v\,du$$

which is shorter and easier to remember than (4). In this notation (3) becomes

$$\int f(x)\,dx = \int u\,dv = uv - \int v\,du \qquad (6)$$

To apply (6), we must choose u and dv so that $f(x)\,dx = u\,dv$. In particular, $f(x)$ is to be split into two parts, one becoming u and the other joining with dx to become dv. In order to express $\int u\,dv$ as $uv - \int v\,du$, we must be able to determine du from u and v from dv. As a result, u and dv should be chosen so that

u is easy to differentiate.

If $dv = g(x)\,dx$, then an antiderivative of g can be easily found.

$\int v\,du$ is no more difficult (and hopefully easier) to integrate than $\int u\,dv$ is.

Let us see how all this works in specific examples.

EXAMPLE 1 Find $\int x \cos x\,dx$.

Solution The integrand $x \cos x$ can naturally be split into the two parts x and $\cos x$. We let

$$u = x \quad \text{and} \quad dv = \cos x\,dx$$

Then $$du = dx \quad \text{and} \quad v = \sin x$$

Consequently integration by parts yields

$$\int \overset{u}{\overbrace{x}} \overset{dv}{\overbrace{\cos x \, dx}} = \overset{u}{\overbrace{x}} \overset{v}{\overbrace{\sin x}} - \int \overset{v}{\overbrace{\sin x}} \overset{du}{\overbrace{dx}} = x \sin x + \cos x + C \quad \square$$

Integrals like the one in Example 1 appear in the analysis of the motion of a vibrating string, such as a cello or guitar string plucked at its center (see Exercise 67).

As we said earlier, it is important that the integral $\int v \, du$ in (6) be no more difficult to integrate than $\int u \, dv$. For example, in calculating $\int x \cos x \, dx$, suppose we had let $u = \cos x$ and $dv = x \, dx$. Then we could write $du = -\sin x \, dx$ and $v = \frac{1}{2} x^2$, but the integration would not be made any easier, because in that case,

$$\int x \cos x \, dx = \overset{v}{\overbrace{\frac{1}{2} x^2}} \overset{u}{\overbrace{\cos x}} - \int \overset{v}{\overbrace{\frac{1}{2} x^2}} \overset{du}{\overbrace{(-\sin x) \, dx}}$$

and the second integral is harder to evaluate than the first.

Observe also that in Example 1 we could have let v be any function of the form $\sin x + C$, where C is a constant. However, it is only necessary to choose one antiderivative for v. After a single v has been chosen, the integral $\int v \, du$ must be evaluated. After evaluating this integral, we include an arbitrary constant C.

EXAMPLE 2   Find $\displaystyle\int 2xe^{3x} \, dx.$

Solution   Since $2xe^{3x}$ splits into $2x$ and e^{3x}, we let

$$u = 2x \quad \text{and} \quad dv = e^{3x} \, dx$$

Then

$$du = 2 \, dx \quad \text{and} \quad v = \frac{1}{3} e^{3x}$$

Therefore

$$\int \overset{u}{\overbrace{2x}} \overset{dv}{\overbrace{e^{3x} \, dx}} = \overset{u}{\overbrace{2x}} \overset{v}{\overbrace{\left(\frac{1}{3} e^{3x}\right)}} - \int \overset{v}{\overbrace{\frac{1}{3} e^{3x}}} \overset{du}{\overbrace{2 \, dx}} = \frac{2}{3} xe^{3x} - \int \frac{2}{3} e^{3x} dx$$

$$= \frac{2}{3} xe^{3x} - \frac{2}{9} e^{3x} + C \quad \square$$

EXAMPLE 3   Find $\displaystyle\int \ln x \, dx.$

Solution   This time we write $\ln x$ as $(\ln x) \cdot 1$, so that $\ln x$ is the product of $\ln x$ and 1. This simple maneuver prepares the integral for integration by parts. We let

$$u = \ln x \quad \text{and} \quad dv = 1 \, dx = dx$$

Then
$$du = \frac{1}{x}\, dx \quad \text{and} \quad v = x$$

Therefore

$$\int \overbrace{\ln x}^{u}\,\overbrace{dx}^{dv} = \overbrace{(\ln x)}^{u}\,\overbrace{x}^{v} - \int \overbrace{x}^{v}\cdot\underbrace{\frac{1}{x}}_{du}\, dx = x\ln x - \int 1\,dx$$

$$= x\ln x - x + C \qquad \square$$

Notice that in the solution of Example 3, we let $u = \ln x$ and $dv = dx$, which led to the simpler integral $\int v\,du = \int x(1/x)\,dx$. A corresponding choice of u can be effective if the integrand involves an inverse trigonometric function, as you will discover in the exercises.

In the next example we will need two successive applications of integration by parts in order to evaluate the integral.

EXAMPLE 4   Find $\displaystyle\int_0^1 x^2 e^{-x}\,dx$.

Solution   We start by letting

$$u = x^2 \quad \text{and} \quad dv = e^{-x}\,dx$$

Then
$$du = 2x\,dx \quad \text{and} \quad v = -e^{-x}$$

Therefore

$$\int_0^1 \overbrace{x^2}^{u}\,\overbrace{e^{-x}\,dx}^{dv} = \overbrace{x^2}^{u}\overbrace{(-e^{-x})}^{v}\,\bigg|_0^1 - \int_0^1 \overbrace{(-e^{-x})}^{v}\,\overbrace{2x\,dx}^{du} = -e^{-1} + \int_0^1 2xe^{-x}\,dx \qquad (7)$$

Next we perform integration by parts on $\int_0^1 2xe^{-x}\,dx$, with

$$u = 2x \quad \text{and} \quad dv = e^{-x}\,dx$$

Then
$$du = 2\,dx \quad \text{and} \quad v = -e^{-x}$$

We find that

$$\int_0^1 \overbrace{2x}^{u}\,\overbrace{e^{-x}\,dx}^{dv} = \overbrace{(2x)}^{u}\,\overbrace{(-e^{-x})}^{v}\,\bigg|_0^1 - \int_0^1 \overbrace{(-e^{-x})}^{v}\overbrace{2\,dx}^{du} = -2e^{-1} + \int_0^1 2e^{-x}\,dx \qquad (8)$$

The last integral in (8) can be evaluated directly, and in fact,

$$\int_0^1 2e^{-x}\,dx = -2e^{-x}\,\bigg|_0^1 = -2e^{-1} + 2 \qquad (9)$$

Now we combine the results of (7)–(9) to conclude that

$$\int_0^1 x^2 e^{-x}\,dx = -e^{-1} + (-2e^{-1}) + (-2e^{-1} + 2) = -5e^{-1} + 2 \qquad \square$$

In the following example we will see that two successive integrations by parts do not quite complete the solution.

EXAMPLE 5   Find $\displaystyle\int e^{-x}\cos x\,dx$.

Solution   Let

$$u = e^{-x} \quad\text{and}\quad dv = \cos x\,dx$$

Then

$$du = -e^{-x}\,dx \quad\text{and}\quad v = \sin x$$

Therefore

$$\int \overbrace{e^{-x}}^{u}\,\overbrace{\cos x\,dx}^{dv} = \overbrace{e^{-x}}^{u}\,\overbrace{\sin x}^{v} - \int \overbrace{(\sin x)}^{v}\,\overbrace{(-e^{-x})\,dx}^{du}$$

$$= e^{-x}\sin x + \int e^{-x}\sin x\,dx$$

Next we apply a second integration by parts to $\int e^{-x}\sin x\,dx$ by letting

$$u = e^{-x} \quad\text{and}\quad dv = \sin x\,dx$$

Then

$$du = -e^{-x}\,dx \quad\text{and}\quad v = -\cos x$$

Thus

$$\int \overbrace{e^{-x}}^{u}\,\overbrace{\sin x\,dx}^{dv} = \overbrace{e^{-x}}^{u}\,\overbrace{(-\cos x)}^{v} - \int \overbrace{(-\cos x)}^{v}\,\overbrace{(-e^{-x})\,dx}^{du}$$

$$= -e^{-x}\cos x - \int e^{-x}\cos x\,dx$$

Combining the results of the two integrations by parts, we obtain

$$\int e^{-x}\cos x\,dx = e^{-x}\sin x + \int e^{-x}\sin x\,dx$$

$$= e^{-x}\sin x - e^{-x}\cos x - \int e^{-x}\cos x\,dx$$

Notice that the original integral has once again appeared—in the last integral. To complete the solution, we add $\int e^{-x}\cos x\,dx$ to both sides of the equation and obtain

$$2\int e^{-x}\cos x\,dx = e^{-x}\sin x - e^{-x}\cos x + C_1$$

Consequently

$$\int e^{-x}\cos x\,dx = \frac{1}{2}(e^{-x}\sin x - e^{-x}\cos x) + C \quad\square$$

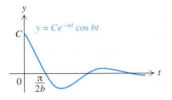

FIGURE 8.1

If we had applied integration by parts the second time with $u = \sin x$ and $dv = e^{-x}\, dx$ (instead of keeping $u = e^{-x}$), then the second integration would not have led to a solution of the original integral (see Exercise 53).

Integrals such as the one in Example 5 arise in the study of oscillatory motion. For example, suppose a spring moves back and forth in a container filled with a viscous fluid such as oil, which acts as a damping force on the spring. If there are no forces at work on the spring other than the damping force and gravity, then the amount the spring is stretched or compressed from its natural length as it oscillates may be given as a function of time t by

$$y = Ce^{-at} \cos bt$$

where a, b, and C are suitable nonzero constants (Figure 8.1). Positive values of y indicate that the spring is stretched, whereas negative values of y indicate that the spring is compressed. The average of the stretching and compressing over a time period $[0, T]$ is then $(1/T) \int_0^T Ce^{-at} \cos bt\, dt$.

In conclusion we remark that integration by parts is effective with integrals involving a polynomial and an exponential, a logarithmic, or a trigonometric function. More specifically, integration by parts is especially well adapted to integrals of the form

$$\int (\text{polynomial}) \sin ax\, dx, \quad \int (\text{polynomial}) \cos ax\, dx,$$

$$\int (\text{polynomial})\, e^{ax}\, dx, \quad \int (\text{polynomial}) \ln x\, dx$$

In all except $\int (\text{polynomial}) \ln x\, dx$, the most effective choice of u is the polynomial, since the derivatives of a polynomial are simpler than the polynomial (and the antiderivatives of $\sin ax$, $\cos ax$, and e^{ax} are easy to find). For $\int (\text{polynomial}) \ln x\, dx$ the choice of $u = \ln x$ is effective, and in fact, $\int v\, du$ becomes the integral of a polynomial.

Reduction Formulas for
$\int \sin^n x\, dx$ **and** $\int \cos^n x\, dx$

Integrals such as $\int \sin^n x\, dx$ and $\int \cos^n x\, dx$, for $n \geq 2$, appear occasionally in applications such as the study of binary stars (see Exercise 66). These integrals can be evaluated by means of integration by parts.

First we will discuss a formula for $\int \sin^n x\, dx$, for $n \geq 2$. In order to use integration by parts, we notice that $\sin^n x = (\sin^{n-1} x)(\sin x)$, and we let

$$u = \sin^{n-1} x \quad \text{and} \quad dv = \sin x\, dx$$

Then $\qquad du = (n-1)\sin^{n-2} x \cos x\, dx \quad \text{and} \quad v = -\cos x$

Therefore

$$\int \sin^n x\, dx = \int \overset{u}{\overbrace{\sin^{n-1} x}}\ \overset{dv}{\overbrace{\sin x\, dx}}$$

$$= \overset{u}{\overbrace{(\sin^{n-1} x)}}\ \overset{v}{\overbrace{(-\cos x)}} - \int \overset{v}{\overbrace{(-\cos x)}}\ \overset{du}{\overbrace{(n-1)\sin^{n-2} x \cos x\, dx}}$$

$$= -\sin^{n-1} x \cos x + (n-1) \int \sin^{n-2} x \cos^2 x \, dx$$

$$= -\sin^{n-1} x \cos x + (n-1) \int \sin^{n-2} x (1 - \sin^2 x) \, dx$$

Thus

$$\int \sin^n x \, dx = -\sin^{n-1} x \cos x + (n-1) \int \sin^{n-2} x \, dx - (n-1) \int \sin^n x \, dx$$

Combining the terms that contain $\int \sin^n x \, dx$ yields

$$n \int \sin^n x \, dx = -\sin^{n-1} x \cos x + (n-1) \int \sin^{n-2} x \, dx$$

so that we have the following formula for $\int \sin^n x \, dx$:

$$\int \sin^n x \, dx = -\frac{1}{n} \sin^{n-1} x \cos x + \frac{n-1}{n} \int \sin^{n-2} x \, dx \qquad (10)$$

The formula that arises for $\int \cos^n x \, dx$ is established in exactly the same way:

$$\int \cos^n x \, dx = \frac{1}{n} \cos^{n-1} x \sin x + \frac{n-1}{n} \int \cos^{n-2} x \, dx \qquad (11)$$

The formulas in (10) and (11) are called **reduction formulas,** because the exponent is reduced (from n to $n-2$). By using the formulas repeatedly, we can reduce the power of the sine or the cosine to 0 or 1. We mention that if $n - 2 = 0$, then by convention $\sin^{n-2} x$ and $\cos^{n-2} x$ are both 1.

Using (10) with $n = 2$, we find that

$$\int \sin^2 x \, dx = -\frac{1}{2} \sin x \cos x + \frac{1}{2} \int 1 \, dx = -\frac{1}{2} \sin x \cos x + \frac{1}{2} x + C$$

Since $-\frac{1}{2} \sin x \cos x = -\frac{1}{4} \sin 2x$, it follows that

$$\int \sin^2 x \, dx = -\frac{1}{4} \sin 2x + \frac{1}{2} x + C$$

which is equivalent to (3) of Section 5.6, and which we obtained by integrating by substitution. Similarly, if $n = 2$ then (11) yields

$$\int \cos^2 x \, dx = \frac{1}{4} \sin 2x + \frac{1}{2} x + C$$

which is equivalent to (2) of Section 5.6.

In Exercises 45 and 46 you are asked to verify that

$$\int \sin^4 x \, dx = -\frac{1}{4} \sin^3 x \cos x - \frac{3}{8} \sin x \cos x + \frac{3}{8} x + C \qquad (12)$$

$$\int \cos^4 x \, dx = \frac{1}{4} \cos^3 x \sin x + \frac{3}{8} \cos x \sin x + \frac{3}{8} x + C \qquad (13)$$

EXAMPLE 6   Find $\displaystyle\int \cos^5 x \, dx$.

Solution   Letting $n = 5$ in the reduction formula for $\int \cos^n x \, dx$, we obtain

$$\int \cos^5 x \, dx = \frac{1}{5} \cos^4 x \sin x + \frac{4}{5} \int \cos^3 x \, dx$$

A second application of the reduction formula, to $\int \cos^3 x \, dx$, yields

$$\int \cos^3 x \, dx = \frac{1}{3} \cos^2 x \sin x + \frac{2}{3} \int \cos x \, dx$$

$$= \frac{1}{3} \cos^2 x \sin x + \frac{2}{3} \sin x + C_1$$

Consequently

$$\int \cos^5 x \, dx = \frac{1}{5} \cos^4 x \sin x + \frac{4}{5} \left(\frac{1}{3} \cos^2 x \sin x + \frac{2}{3} \sin x + C_1 \right)$$

$$= \frac{1}{5} \cos^4 x \sin x + \frac{4}{15} \cos^2 x \sin x + \frac{8}{15} \sin x + C \quad \square$$

For the definite integral $\int_0^{\pi/6} \cos^5 x \, dx$, we would have

$$\int_0^{\pi/6} \cos^5 x \, dx = \left(\frac{1}{5} \cos^4 x \sin x + \frac{4}{15} \cos^2 x \sin x + \frac{8}{15} \sin x \right) \Bigg|_0^{\pi/6}$$

$$= \frac{1}{5} \left(\frac{\sqrt{3}}{2} \right)^4 \frac{1}{2} + \frac{4}{15} \left(\frac{\sqrt{3}}{2} \right)^2 \frac{1}{2} + \frac{8}{15} \left(\frac{1}{2} \right)$$

$$= \frac{9}{160} + \frac{12}{120} + \frac{8}{30} = \frac{203}{480}$$

EXERCISES 8.1

In Exercises 1–28 find the indefinite integral.

1. $\displaystyle\int x \sin x \, dx$

2. $\displaystyle\int x \sec^2 x \, dx$

3. $\displaystyle\int x \ln x \, dx$

4. $\displaystyle\int x \ln x^2 \, dx$

5. $\displaystyle\int (\ln x)^2 \, dx$

6. $\displaystyle\int x^2 \ln x \, dx$

7. $\displaystyle\int x^3 \ln x \, dx$

8. $\displaystyle\int x e^{-x} \, dx$

9. $\displaystyle\int x^2 e^{4x} \, dx$

10. $\displaystyle\int x^2 \sin x \, dx$

11. $\displaystyle\int x^3 \cos x \, dx$

12. $\displaystyle\int e^x \sin x \, dx$

13. $\displaystyle\int e^{3x} \cos 3x \, dx$

14. $\displaystyle\int \frac{\sin x}{e^x} \, dx$

15. $\displaystyle\int t \cdot 2^t \, dt$

16. $\displaystyle\int t \cdot 3^{-t} \, dt$

17. $\displaystyle\int t^2 \cdot 4^t \, dt$

18. $\displaystyle\int \log_6 x \, dx$

19. $\displaystyle\int t \sinh t \, dt$

20. $\displaystyle\int t^2 \cosh t \, dt$

21. $\displaystyle\int \tan^{-1} x \, dx$

22. $\displaystyle\int \sin^{-1} 3x \, dx$

23. $\displaystyle\int \cos^{-1}(-7x) \, dx$

24. $\displaystyle\int x^n \ln x \, dx$, where n is a positive integer

25. $\displaystyle\int x^n \ln x^m \, dx$, where m and n are positive integers

26. $\int \sin (\ln x) \, dx$ (*Hint:* Let $u = \sin (\ln x)$, and integrate by parts twice.)

27. $\int \cos (\ln x) \, dx$

28. $\int x \ln (x + 1) \, dx$ (*Hint:* Take $u = \ln (x + 1)$, $dv = x \, dx$, and $v = (x^2 - 1)/2$.)

In Exercises 29–34 evaluate the definite integral.

29. $\int_0^1 xe^{5x} \, dx$ **30.** $\int_0^{\pi/2} (x + x \sin x) \, dx$

31. $\int_0^{\pi} t^2 \cos t \, dt$ **32.** $\int_1^4 x^{3/2} \ln x \, dx$

33. $\int_{-\pi/3}^{\pi/4} x \sec^2 x \, dx$ **34.** $\int_1^b x^2 (\ln x)^2 \, dx \ (b > 0)$

In Exercises 35–42 first make a substitution and then use integration by parts to evaluate the integral.

35. $\int_0^1 \ln (x + 1) \, dx$ **36.** $\int e^x \ln (1 + e^x) \, dx$

37. $\int x \sin ax \, dx$ **38.** $\int e^{6x} \cos e^{3x} \, dx$

39. $\int \sin x \tan^{-1}(\cos x) \, dx$ **40.** $\int \frac{1}{x} \sin^{-1}(\ln x) \, dx$

41. $\int \cos \sqrt{t} \, dt$ **42.** $\int \frac{(\ln t)^2}{t^2} \, dt$

In Exercises 43–44 evaluate the integral with the help of the reduction formulas in (10) and (11).

43. $\int_0^{\pi/2} \cos^3 \frac{x}{2} \, dx$ **44.** $\int \sin^5 x \, dx$

45. Verify equation (12).   **46.** Verify equation (13).

In Exercises 47–50 use integration by parts to establish the reduction formula.

47. $\int \cos^n x \, dx = \frac{1}{n} \cos^{n-1} x \sin x + \frac{n-1}{n} \int \cos^{n-2} x \, dx$, where n is an integer greater than or equal to 2

48. $\int (\ln x)^n \, dx = x (\ln x)^n - n \int (\ln x)^{n-1} \, dx$, where n is a positive integer (*Hint:* Let $u = (\ln x)^n$.)

49. $\int \frac{1}{(x^2 + a^2)^{n+1}} \, dx = \frac{x}{2na^2(x^2 + a^2)^n}$
$$+ \frac{2n - 1}{2na^2} \int \frac{1}{(x^2 + a^2)^n} \, dx$$

(*Hint:* Use integration by parts on $\int \frac{1}{(x^2 + a^2)^n} \, dx$ with $u = \frac{1}{(x^2 + a^2)^n}$ and $v = x$.)

*50. $\int \sec^n x \, dx = \frac{\sec^{n-2} x \tan x}{n - 1} + \frac{n - 2}{n - 1} \int \sec^{n-2} x \, dx$

where n is an integer greater than or equal to 2 (*Hint:* Let $u = \sec^{n-2} x$ and $v = \tan x$. After integrating by parts, you will have an integral of the form $\int g(x) \tan^2 x \, dx$. Write this as $\int g(x)(\sec^2 x - 1) \, dx = \int g(x) \sec^2 x \, dx - \int g(x) \, dx$, and solve algebraically for the original integral.)

In Exercises 51–52 use the results of Exercises 48 and 50 to evaluate the integral.

51. $\int (\ln x)^3 \, dx$ **52.** $\int_{\pi/3}^{\pi/4} \sec^5 x \, dx$

53. In the solution of Example 5, suppose for the second application of integration by parts we let $u = \sin x$ and $dv = e^{-x} \, dx$. Show that integration then yields the equation $\int e^{-x} \cos x \, dx = \int e^{-x} \cos x \, dx$. (In effect this means that the second integration by parts nullifies the first integration.)

54. Verify the following formulas, where $a \neq 0$, $b \neq 0$.

a. $\int e^{ax} \sin bx \, dx = \frac{e^{ax}}{a^2 + b^2} (a \sin bx - b \cos bx) + C$

b. $\int e^{ax} \cos bx \, dx = \frac{e^{ax}}{a^2 + b^2} (a \cos bx + b \sin bx) + C$

In Exercises 55–58 find the area A of the region between the graph of f and the x axis on the given interval.

55. $f(x) = \ln x$; $[1, 2]$ **56.** $f(x) = x \ln x$; $[1/e, e]$
57. $f(x) = x \sin x$; $[0, \pi/2]$ **58.** $f(x) = \tan^{-1} x$; $[0, 1]$
59. Find the area A of the region bounded by the graphs of $y = 3x \ln x$ and $y = x^2 \ln x$.

In Exercises 60–61 let V be the region between the graph of f and the x axis on the given interval. Find the volume V of the solid obtained by revolving R about the x axis.

60. $f(x) = \sqrt{x} \, e^x$; $[0, 1]$ **61.** $f(x) = \sin^{3/2} x$; $[0, \pi]$
62. Let $f(x) = x^2 - \frac{1}{8} \ln x$. Find the surface area S of the surface generated by revolving about the x axis the graph of f on $[1, 2]$.

In Exercises 63–64 calculate the center of gravity of the region R between the graphs of f and g on the given interval.

63. $f(x) = \cos x$, $g(x) = \sin x$; $[0, \pi/4]$
64. $f(x) = 1 + \ln x$, $g(x) = 1 - \ln x$; $[1, 2]$

Applications

65. Let f be a continuous nonnegative function on $[a, b]$. Then the **moment** about the y axis of the region between the graph of f and the x axis on $[a, b]$ is defined as $\int_a^b xf(x)\,dx$. Find the moments about the y axis arising from the following functions.

a. $f(x) = \cos x$ for $0 \le x \le \pi/2$

b. $f(x) = \cos x$ for $-\pi/2 \le x \le \pi/2$

c. $f(x) = \ln x$ for $1 \le x \le 2$

Moments are important in physics, engineering, and statistics. We will return to them in Section 8.6.

66. Certain binary stars are believed to have identical masses, which we denote by m. Spectroscopic measurements (based on the Doppler shift) yield an "observed mass" m_0. The true mass m is then estimated by means of the formula

$$m = \frac{1}{c}\,m_0, \quad \text{where } c = \int_0^{\pi/2} \sin^4 x\,dx$$

Determine the number c.

An x-ray image of the binary star system ARLac in the constellation Lacerta. (*Max-Planck-Institut für Physik und Astrophysik/Science Photo Library/Photo Researchers*)

67. Suppose a taut string of length L is pulled at its center, and let H be the maximum displacement at the center (Figure 8.2). Assume that

$f(x)$ is the vertical displacement at x for $-\dfrac{L}{2} \le x \le \dfrac{L}{2}$

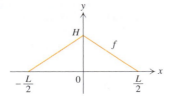

FIGURE 8.2 Figure for Exercise 67.

In the analysis of the string's motion after it is released, the integral

$$\int_{-L/2}^{L/2} f(x)\,\cos \frac{\pi}{L}\,x\,dx$$

appears.

a. Find a formula for $f(x)$ for $-L/2 \le x \le L/2$, assuming that f is linear on $[-L/2, 0]$ and $[0, L/2]$.

b. Using the function derived in part (a), evaluate the resulting integral. (*Hint:* Notice that the integrand is an even function.)

68. If a closed electric circuit has resistance R, inductance L, and applied voltage V, then as we saw in Section 7.8, the current I flowing through the circuit satisfies the differential equation

$$L\frac{dI}{dt} + RI = V(t)$$

Assuming that $V(t) = \sin t$, find the general solution of the differential equation.

Project

1. a. Assume that f has a continuous second derivative on $[a, b]$. Use integration by parts to show that

$$f(b) = f(a) + f'(a)(b - a) + \int_a^b f''(x)(b - x)\,dx$$

b. Assume that f has a continuous third derivative on $[a, b]$. Use integration by parts to show that

$$f(b) = f(a) + f'(a)(b - a) +$$

$$\frac{f''(a)}{2}(b - a)^2 + \frac{1}{2}\int_a^b f^{(3)}(x)(b - x)^2\,dx$$

The formula in part (a) should remind you of the linear approximation formula in Section 3.8 (with an appended term!), and the formula in (b) should remind you of the quadratic approximation formula in Section 3.8 (also with an appended term).

c. Find the first three derivatives of $\tan x$ and then evaluate

$$\int_0^{\pi/4} (\sec^2 x \tan x)\left(x - \frac{\pi}{4}\right)dx \quad \text{and}$$

$$\int_0^{\pi/4} (6\sec^4 x - 4\sec^2 x)\left(x - \frac{\pi}{4}\right)^2 dx$$

8.2   TRIGONOMETRIC INTEGRALS

Integrals such as

$$\int \sin^5 x \cos^4 x \, dx, \quad \int \tan^3 x \sec^3 x \, dx, \quad \text{and} \quad \int \sin 5x \cos 3x \, dx$$

are called **trigonometric integrals** because their integrands are combinations of trigonometric functions. This section is devoted to trigonometric integrals, especially those in which the integrands are composed of powers of the basic trigonometric functions. Trigonometric integrals play a critical role in the evaluation of the integrals that we will study in Section 8.3.

Integrals of the Form
∫ sin^m x cos^n x dx

We have already encountered integrals of the form $\int \sin^m x \cos^n x \, dx$. Indeed, if $n = 0$, then the integral becomes $\int \sin^m x \, dx$; likewise, if $m = 0$, then the integral becomes $\int \cos^n x \, dx$; both were analyzed in Section 8.1 by means of reduction formulas.

Next, if $n = 1$, then $\int \sin^m x \cos^n x \, dx$ becomes $\int \sin^m x \cos x \, dx$, which is evaluated by substituting $u = \sin x$ (see Example 2 of Section 5.6). Similarly, if $m = 1$, then we obtain $\int \sin x \cos^n x \, dx$, which is evaluated by substituting $u = \cos x$.

Now we consider those integrals of the form $\int \sin^m x \cos^n x \, dx$ in which m and n are positive integers with $m \geq 2$ and $n \geq 2$. There are two cases to consider: when m or n is odd (or both are odd), and when m and n are both even.

If n is odd, as in $\int \sin^2 x \cos^3 x \, dx$, we have the following procedure:

> **1.** Factor out $\cos x$.
> **2.** Write the rest of the integrand in terms of $\sin x$, using the Pythagorean Identity $\cos^2 x = 1 - \sin^2 x$.
> **3.** Substitute $u = \sin x$, so that $du = \cos x \, dx$.

EXAMPLE 1   Find $\displaystyle\int \sin^2 x \cos^3 x \, dx$.

Solution   Since $\cos x$ appears to the power 3, which is odd, we factor out $\cos x$ and write the rest of the integral in terms of $\sin x$:

$$\int \sin^2 x \cos^3 x \, dx = \int \sin^2 x \cos^2 x \cos x \, dx$$

$$= \int (\sin^2 x)(1 - \sin^2 x) \cos x \, dx$$

Now we evaluate the integral by substituting

$$u = \sin x, \quad \text{so that} \quad du = \cos x \, dx$$

We obtain

$$\int \sin^2 x \cos^3 x \, dx = \int \overbrace{(\sin^2 x)}^{u^2} \overbrace{(1 - \sin^2 x)}^{1-u^2} \overbrace{\cos x \, dx}^{du}$$

$$= \int u^2 (1 - u^2) \, du = \int (u^2 - u^4) \, du = \frac{1}{3} u^3 - \frac{1}{5} u^5 + C$$

$$= \frac{1}{3} \sin^3 x - \frac{1}{5} \sin^5 x + C \quad \square$$

If m is odd in $\int \sin^m x \cos^n x \, dx$, we follow the same general procedure, but in this case we factor out $\sin x$ and write the rest of the integrand in terms of $\cos x$, using the Pythagorean Identity $\sin^2 x = 1 - \cos^2 x$.

EXAMPLE 2   Find $\displaystyle\int \sin^5 x \cos^4 x \, dx$.

Solution   Using the suggestion made above, we factor out $\sin x$ and rearrange as follows:

$$\int \sin^5 x \cos^4 x \, dx = \int \sin^4 x \sin x \cos^4 x \, dx$$

$$= \int (\sin^2 x)^2 \cos^4 x \sin x \, dx$$

$$= \int (1 - \cos^2 x)^2 \cos^4 x \sin x \, dx$$

Next we substitute

$$u = \cos x, \quad \text{so that} \quad du = -\sin x \, dx$$

Therefore

$$\int \sin^5 x \cos^4 x \, dx = \int \overbrace{(1 - \cos^2 x)^2}^{(1-u^2)^2} \overbrace{\cos^4 x}^{u^4} \overbrace{\sin x \, dx}^{-du}$$

$$= -\int (1 - u^2)^2 \, u^4 \, du = -\int (u^4 - 2u^6 + u^8) \, du$$

$$= -\frac{1}{5} u^5 + \frac{2}{7} u^7 - \frac{1}{9} u^9 + C$$

$$= -\frac{1}{5} \cos^5 x + \frac{2}{7} \cos^7 x - \frac{1}{9} \cos^9 x + C \quad \square$$

If m and n are both even in the integral $\int \sin^m x \cos^n x \, dx$, the evaluation is more complicated. Three trigonometric identities help reduce the exponents m and n:

$$\sin x \cos x = \frac{1}{2} \sin 2x \tag{1}$$

$$\sin^2 x = \frac{1 - \cos 2x}{2} \tag{2}$$

$$\cos^2 x = \frac{1 + \cos 2x}{2} \tag{3}$$

In addition, the following two formulas, derived in Section 5.6, will be helpful:

$$\int \sin^2 x \, dx = \frac{1}{2} x - \frac{1}{4} \sin 2x + C \tag{4}$$

and

$$\int \cos^2 x \, dx = \frac{1}{2} x + \frac{1}{4} \sin 2x + C \tag{5}$$

EXAMPLE 3   Find $\int \sin^2 x \cos^4 x \, dx$.

Solution   Using (1) and (3) together, we find that

$$\int \sin^2 x \cos^4 x \, dx = \int (\sin^2 x \cos^2 x) \cos^2 x \, dx$$

$$= \int (\sin x \cos x)^2 \cos^2 x \, dx$$

$$\overset{(1),\,(3)}{=} \int \left(\frac{1}{2} \sin 2x \right)^2 \left(\frac{1 + \cos 2x}{2} \right) dx$$

$$= \frac{1}{8} \int (\sin^2 2x)(1 + \cos 2x) \, dx$$

$$= \frac{1}{8} \int \sin^2 2x \, dx + \frac{1}{8} \int \sin^2 2x \cos 2x \, dx$$

For the first integral on the right we let

$$u = 2x, \quad \text{so that} \quad du = 2 \, dx$$

and for the second integral we let

$$v = \sin 2x, \quad \text{so that} \quad dv = 2 \cos 2x \, dx$$

Then with the help of (4) we find that

$$\int \sin^2 x \cos^4 x \, dx = \frac{1}{8} \int \underbrace{\sin^2 \overbrace{2x}^{u} \, \overbrace{dx}^{\frac{1}{2}du}}_{} + \frac{1}{8} \int \underbrace{\overbrace{\sin^2 2x}^{v^2} \overbrace{\cos 2x \, dx}^{\frac{1}{2}dv}}_{}$$

$$= \frac{1}{8} \int (\sin^2 u) \frac{1}{2} \, du + \frac{1}{8} \int v^2 \cdot \frac{1}{2} \, dv$$

$$\overset{(4)}{=} \frac{1}{16}\left(\frac{1}{2}u - \frac{1}{4}\sin 2u\right) + \frac{1}{16}\left(\frac{1}{3}v^3\right) + C$$

$$= \frac{1}{16}\left(x - \frac{1}{4}\sin 4x\right) + \frac{1}{48}\sin^3 2x + C \quad \square$$

An alternative way to evaluate $\int \sin^m x \cos^n x \, dx$ when m and n are even is to use the identity $\sin^2 x + \cos^2 x = 1$, and thereby transform the integral into integrals of the form $\int \sin^k x \, dx$ or of the form $\int \cos^k x \, dx$, which can be evaluated by the reduction formulas (10) and (11) of Section 8.1. Thus for the integral $\int \sin^4 x \cos^6 x \, dx$ we would have

$$\int \sin^4 x \cos^6 x \, dx = \int (\sin^2 x)^2 \cos^6 x \, dx$$

$$= \int (1 - \cos^2 x)^2 \cos^6 x \, dx$$

$$= \int (\cos^6 x - 2\cos^8 x + \cos^{10} x) \, dx$$

$$= \int \cos^6 x \, dx - 2\int \cos^8 x \, dx + \int \cos^{10} x \, dx$$

and to evaluate the integrals on the right we would use (11) of Section 8.1.

Table 8.1 summarizes our analysis of $\int \sin^m x \cos^n x \, dx$.

TABLE 8.1

Evaluation of $\int \sin^m x \cos^n x \, dx$ for $m \geq 0$ and $n \geq 0$

	Method	Useful identities
n odd	substitute $u = \sin x$	$\cos^2 x = 1 - \sin^2 x$
m odd	substitute $u = \cos x$	$\sin^2 x = 1 - \cos^2 x$
m and n even	reduce to smaller powers of m or n	$\begin{cases} \sin x \cos x = \dfrac{1}{2}\sin 2x \\[2mm] \sin^2 x = \dfrac{1 - \cos 2x}{2} \\[2mm] \cos^2 x = \dfrac{1 + \cos 2x}{2} \end{cases}$

Integrals of the Form
$\int \tan^m x \sec^n x \, dx$

For many nonnegative integer values of m and n, integrals of the form $\int \tan^m x \sec^n x \, dx$ are handled by substitution. Again, the procedure we use to evaluate the integral depends on whether m and n are even or odd.

If n is even and $n > 0$, as in $\int \tan^3 x \sec^4 x \, dx$, the procedure is as follows:

1. Factor out $\sec^2 x$.
2. Write the rest of the integrand in terms of $\tan x$, using the identity $\sec^2 x = 1 + \tan^2 x$.
3. Substitute $u = \tan x$, so that $du = \sec^2 x \, dx$.

EXAMPLE 4   Find $\displaystyle\int \tan^3 x \sec^4 x \, dx$.

Solution   Since $\sec x$ appears to the power 4, which is even, we factor out $\sec^2 x$ and write the rest of the integrand in terms of $\tan x$:

$$\int \tan^3 x \sec^4 x \, dx = \int \tan^3 x \sec^2 x \sec^2 x \, dx$$

$$= \int \tan^3 x \, (1 + \tan^2 x) \sec^2 x \, dx$$

Now we make the substitution

$$u = \tan x, \quad \text{so that} \quad du = \sec^2 x \, dx$$

It follows that

$$\int \tan^3 x \sec^4 x \, dx = \int \overbrace{\tan^3 x}^{u^3} \overbrace{(1 + \tan^2 x)}^{1 + u^2} \overbrace{\sec^2 x \, dx}^{du}$$

$$= \int u^3 \, (1 + u^2) \, du = \int (u^3 + u^5) \, du$$

$$= \frac{1}{4} u^4 + \frac{1}{6} u^6 + C$$

$$= \frac{1}{4} \tan^4 x + \frac{1}{6} \tan^6 x + C \quad \square$$

Next let us assume that m is odd and $n > 0$ in $\int \tan^m x \sec^n x \, dx$, as in $\int \tan^3 x \sec^3 x \, dx$. In this case we continue as follows:

1. Factor out $\sec x \tan x$.
2. Write the rest of the integrand in terms of $\sec x$, using the identity $\tan^2 x = \sec^2 x - 1$.
3. Substitute $u = \sec x$, so that $du = \sec x \tan x \, dx$.

EXAMPLE 5   Find $\displaystyle\int \tan^3 x \sec^3 x \, dx$.

Solution   Since tan x appears to the power 3, which is odd, we first factor out sec x tan x and then rewrite the rest of the integrand in terms of sec x:

$$\int \tan^3 x \sec^3 x \, dx = \int (\tan^2 x \sec^2 x) \sec x \tan x \, dx$$

$$= \int [(\sec^2 x - 1) \sec^2 x] \sec x \tan x \, dx$$

At this point we substitute

$$u = \sec x, \quad \text{so that} \quad du = \sec x \tan x \, dx$$

We obtain

$$\int \tan^3 x \sec^3 x \, dx = \int [\overbrace{(\sec^2 x - 1)}^{u^2-1}\ \overbrace{\sec^2 x}^{u^2}]\ \overbrace{\sec x \tan x \, dx}^{du}$$

$$= \int (u^2 - 1)u^2 \, du = \int (u^4 - u^2) \, du$$

$$= \frac{1}{5} u^5 - \frac{1}{3} u^3 + C$$

$$= \frac{1}{5} \sec^5 x - \frac{1}{3} \sec^3 x + C \quad \square$$

As Examples 4 and 5 illustrate, if n is even or m odd, we can readily evaluate $\int \tan^m x \sec^n x \, dx$. Now let us consider the case in which n is odd and m is even. By using the identity $\tan^2 x = \sec^2 x - 1$ we can reduce the problem to finding integrals of the form $\int \sec^n x \, dx$, where n is odd. However, these integrals are not so easy to evaluate, even when $n = 1$ or $n = 3$. If $n = 1$, then $\int \sec^n x \, dx$ becomes $\int \sec x \, dx$, which by (17) of Section 5.7 is given by

$$\int \sec x \, dx = \ln|\sec x + \tan x| + C \tag{6}$$

The integration of $\int \sec^3 x \, dx$ is more involved. Using integration by parts, we let

$$u = \sec x \quad \text{and} \quad dv = \sec^2 x \, dx$$

Then
$$du = \sec x \tan x \, dx \quad \text{and} \quad v = \tan x$$
Therefore

$$\int \sec^3 x \, dx = \int \overbrace{\sec x}^{u}\ \overbrace{\sec^2 x \, dx}^{dv} = \overbrace{\sec x}^{u}\ \overbrace{\tan x}^{v} - \int \overbrace{\tan x}^{v}\ \overbrace{\sec x \tan x \, dx}^{du}$$

$$= \sec x \tan x - \int \sec x \tan^2 x \, dx$$

$$= \sec x \tan x - \int (\sec x)(\sec^2 x - 1)\, dx$$

$$= \sec x \tan x - \int \sec^3 x\, dx + \int \sec x\, dx$$

By combining both occurrences of $\int \sec^3 x\, dx$, we obtain

$$\int \sec^3 x\, dx = \frac{1}{2} \sec x \tan x + \frac{1}{2} \int \sec x\, dx$$

so that by (6),

$$\int \sec^3 x\, dx = \frac{1}{2} \sec x \tan x + \frac{1}{2} \ln |\sec x + \tan x| + C \qquad (7)$$

The integral $\int \sec^3 x\, dx$ arises many times in applications, including lengths of curves and areas of surfaces (see Exercises 64 and 65).

EXAMPLE 6   Find $\int \tan^2 x \sec x\, dx$.

Solution   Since m is even and n is odd, we use the identity $\tan^2 x = \sec^2 x - 1$ to reduce the integrand to terms involving powers of $\sec x$ and then use (6) and (7):

$$\int \tan^2 x \sec x\, dx = \int (\sec^2 x - 1)(\sec x)\, dx = \int (\sec^3 x - \sec x)\, dx$$

$$= \int \sec^3 x\, dx - \int \sec x\, dx$$

$$= \frac{1}{2} \sec x \tan x + \frac{1}{2} \ln |\sec x + \tan x|$$

$$- \ln |\sec x + \tan x| + C$$

$$= \frac{1}{2} \sec x \tan x - \frac{1}{2} \ln |\sec x + \tan x| + C \quad \square$$

Our various ways of evaluating integrals of the form $\int \tan^m x \sec^n x\, dx$ are summarized in Table 8.2.

TABLE 8.2
Evaluation of $\int \tan^m x \sec^n x\, dx$ for $m \geq 0$ and $n > 0$

	Method	Useful identity
n even	substitute $u = \tan x$	$\sec^2 x = 1 + \tan^2 x$
m odd	substitute $u = \sec x$	$\tan^2 x = \sec^2 x - 1$
m even, n odd	reduce to powers of $\sec x$ alone	$\tan^2 x = \sec^2 x - 1$

Since $\csc^2 x = 1 + \cot^2 x$, the same techniques allow us to find integrals of the form

$$\int \cot^m x \csc^n x \, dx$$

(see Exercises 28–33).

Conversion to Sine and Cosine

Trigonometric integrals that are not of the forms discussed so far can often be simplified by expressing the integrands in terms of sines and cosines and then using the methods already discussed.

EXAMPLE 7   Find $\displaystyle\int \cos^2 x \tan^5 x \, dx$.

Solution   We write $\tan^5 x$ in terms of sines and cosines and combine like terms:

$$\int \cos^2 x \tan^5 x \, dx = \int \cos^2 x \, \frac{\sin^5 x}{\cos^5 x} \, dx = \int \frac{\sin^4 x}{\cos^3 x} \sin x \, dx$$

$$= \int \frac{(1 - \cos^2 x)^2}{\cos^3 x} \sin x \, dx$$

Substituting

$$u = \cos x, \quad \text{so that} \quad du = -\sin x \, dx$$

gives us

$$\int \cos^2 x \tan^5 x \, dx = \int \frac{\overbrace{(1 - \cos^2 x)^2}^{(1-u^2)^2/u^3}}{\cos^3 x} \overbrace{\sin x \, dx}^{-du}$$

$$= -\int \frac{(1 - u^2)^2}{u^3} \, du = -\int \frac{1 - 2u^2 + u^4}{u^3} \, du$$

$$= \int \left(-\frac{1}{u^3} + \frac{2}{u} - u \right) du$$

$$= \frac{1}{2u^2} + 2 \ln |u| - \frac{u^2}{2} + C$$

$$= \frac{1}{2 \cos^2 x} + 2 \ln |\cos x| - \frac{\cos^2 x}{2} + C \quad \square$$

Integrals of the Form $\int \sin ax \cos bx \, dx$

Evaluating such integrals depends on the trigonometric identity

$$\sin x \cos y = \frac{1}{2} \sin (x - y) + \frac{1}{2} \sin (x + y)$$

With the appropriate replacements, this identity becomes

$$\sin ax \cos bx = \frac{1}{2} \sin (a - b)x + \frac{1}{2} \sin (a + b)x \tag{8}$$

Notice that $\frac{1}{2} \sin (a - b)x$ and $\frac{1}{2} \sin (a + b)x$ are easy to integrate by substitution.

EXAMPLE 8   Find $\displaystyle\int \sin 5x \cos 3x \, dx$.

Solution   Using (8) with $a = 5$ and $b = 3$, we find that

$$\int \sin 5x \cos 3x \, dx = \int \left(\frac{1}{2} \sin 2x + \frac{1}{2} \sin 8x \right) dx$$

$$= -\frac{1}{4} \cos 2x - \frac{1}{16} \cos 8x + C \quad \square$$

Integrals of the form

$$\int \sin ax \sin bx \, dx \quad \text{and} \quad \int \cos ax \cos bx \, dx$$

can be found by similar techniques (see Exercises 50–53).

EXERCISES 8.2

In Exercises 1–58 evaluate the integral.

1. $\displaystyle\int \sin^3 x \cos^2 x \, dx$ **2.** $\displaystyle\int \sin^3 x \cos^3 x \, dx$

3. $\displaystyle\int \sin^3 3x \cos 3x \, dx$ **4.** $\displaystyle\int_0^{\pi/2} \sin^2 t \cos^5 t \, dt$

5. $\displaystyle\int \frac{1}{x^2} \sin^5 \frac{1}{x} \cos^2 \frac{1}{x} \, dx$ **6.** $\displaystyle\int \sin^8 6x \cos^3 6x \, dx$

7. $\displaystyle\int \sin^2 y \cos^2 y \, dy$ **8.** $\displaystyle\int_0^{\pi/2} \sin^4 x \cos^2 x \, dx$

9. $\displaystyle\int \sin^4 x \cos^4 x \, dx$ **10.** $\displaystyle\int \sin^6 w \cos^4 w \, dw$

11. $\displaystyle\int \sin^{-10} x \cos^3 x \, dx$ **12.** $\displaystyle\int \sin^{-17} x \cos^5 x \, dx$

13. $\displaystyle\int (1 + \sin^2 x)(1 + \cos^2 x) \, dx$

14. $\displaystyle\int \sin^5 x \cos^{1/2} x \, dx$ **15.** $\displaystyle\int_0^{\pi/4} \frac{\sin^3 x}{\cos^2 x} \, dx$

16. $\displaystyle\int \frac{\cos^3 x}{\sin^{5/2} x} \, dx$ **17.** $\displaystyle\int \tan^5 x \sec^2 x \, dx$

18. $\displaystyle\int \tan^3 5x \sec^2 5x \, dx$ **19.** $\displaystyle\int_0^{\pi/4} \tan^5 t \sec^4 t \, dt$

20. $\displaystyle\int x \tan^3 x^2 \sec^4 x^2 \, dx$ **21.** $\displaystyle\int_{5\pi/4}^{4\pi/3} \tan^3 x \sec x \, dx$

22. $\displaystyle\int \tan x \sec^3 x \, dx$ **23.** $\displaystyle\int \frac{1}{\sqrt{x}} \tan^3 \sqrt{x} \sec^3 \sqrt{x} \, dx$

24. $\displaystyle\int \tan^4 (1 - y) \sec^4 (1 - y) \, dy$

25. $\displaystyle\int \tan^3 x \sec^4 x \, dx$ **26.** $\displaystyle\int_0^{\pi/3} \tan x \sec^{3/2} x \, dx$

27. $\displaystyle\int \tan x \sec^5 x \, dx$ **28.** $\displaystyle\int \csc^3 x \, dx$

29. $\displaystyle\int \cot^3 x \csc^2 x \, dx$ **30.** $\displaystyle\int \cot^3 s \csc^4 s \, ds$

31. $\displaystyle\int_{\pi/4}^{\pi/2} \cot^3 x \csc^3 x \, dx$ **32.** $\displaystyle\int_{\pi/3}^{\pi/4} \cot x \csc^3 x \, dx$

33. $\displaystyle\int \cot x \csc^{-2} x \, dx$ **34.** $\displaystyle\int \frac{\cot t}{\csc^3 t} \, dt$

35. $\displaystyle\int \frac{\tan x}{\cos^3 x} \, dx$ **36.** $\displaystyle\int \tan^3 x \csc^2 x \, dx$

37. $\displaystyle\int \frac{\tan^2 x}{\sec^5 x} \, dx$ **38.** $\displaystyle\int \sin^5 w \cot^3 w \, dw$

39. $\displaystyle\int \frac{\tan x}{\sec^2 x}\, dx$ **40.** $\displaystyle\int \frac{\tan^5 x}{\sec^4 x}\, dx$

41. $\displaystyle\int \tan^2 x\, dx$ (*Hint:* $\tan^2 x = \sec^2 x - 1$.)

42. $\displaystyle\int \tan^3 x\, dx$ **43.** $\displaystyle\int \tan^4 x\, dx$

44. $\displaystyle\int \cot^5 x\, dx$ **45.** $\displaystyle\int \sin 2x \cos 3x\, dx$

46. $\displaystyle\int_0^{2\pi/3} \sin x \cos 2x\, dx$ **47.** $\displaystyle\int \sin(-4x)\cos(-2x)\, dx$

48. $\displaystyle\int \sin 3x \cos \tfrac{1}{2}x\, dx$ **49.** $\displaystyle\int \sin \tfrac{1}{2}x \cos \tfrac{2}{3}x\, dx$

50. $\displaystyle\int \sin ax \sin bx\, dx$ (*Hint:* Use the identity $\sin x \sin y =$

$\tfrac{1}{2}\cos(x-y) - \tfrac{1}{2}\cos(x+y)$.)

51. $\displaystyle\int \sin 2x \sin 3x\, dx$

52. $\displaystyle\int \cos ax \cos bx\, dx$ (*Hint:* Use the identity $\cos x \cos y =$

$\tfrac{1}{2}\cos(x+y) + \tfrac{1}{2}\cos(x-y)$.)

53. $\displaystyle\int \cos 5x \cos(-3x)\, dx$

54. $\displaystyle\int \frac{1}{1+\sin x}\, dx$ (*Hint:* Multiply the integrand by
$(1-\sin x)/(1-\sin x)$.)

55. $\displaystyle\int_{\pi/4}^{\pi/2} \frac{1}{1+\cos x}\, dx$ ***56.** $\displaystyle\int \frac{1}{(1+\sin x)^2}\, dx$

57. $\displaystyle\int \frac{1+\cos x}{\sin x}\, dx$ **58.** $\displaystyle\int \frac{1+\sin x}{\cos x}\, dx$

59. Let $n > 2$. Establish the following reduction formula for
$\int \tan^n x\, dx$:

$$\int \tan^n x\, dx - \frac{1}{n-1}\tan^{n-1} x - \int \tan^{n-2} x\, dx$$

*60. Evaluate

$$\int \frac{\sin x - 5\cos x}{\sin x + \cos x}\, dx$$

by finding numbers a and b such that

$$\sin x - 5\cos x = a(\sin x + \cos x) + b(\cos x - \sin x)$$

In Exercises 61–63 find the area A of the region between the graph of f and the x axis on the given interval.

61. $f(x) = \sin^2 x \cos^3 x$; $[0, \pi/2]$
62. $f(x) = \sec^4 x$; $[-\pi/3, \pi/4]$
63. $f(x) = \tan^3 x \sec x$; $[\pi/4, \pi/3]$
64. Find the area A of the portion of the region between the graphs of $y = \tan^3 x$ and $y = \tfrac{1}{4}\tan x \sec^4 x$ that lies between the lines $x = 0$ and $x = \pi/3$.
65. Let C be described parametrically by $x = \ln(\sec t + \tan t)$ and $y = \sec t$ for $0 \le t \le \pi/4$. Find the surface area S of the surface generated by revolving C about the x axis.

Project

1. Let m and n be positive integers. Prove the following.

a. $\displaystyle\int_{-\pi}^{\pi} \sin mx \cos nx\, dx = 0$

Select a choice of m and n, with $m \ne n$, $m > 1$, and $n > 1$. Look at the graph of $y = \sin mx \cos nx$, and see whether it appears as though the integral *should* be 0.

b. $\displaystyle\int_{-\pi}^{\pi} \cos mx \cos nx\, dx = \begin{cases} 0 & \text{if } m \ne n \\ \pi & \text{if } m = n \end{cases}$

c. $\displaystyle\int_{-\pi}^{\pi} \sin mx \sin nx\, dx = \begin{cases} 0 & \text{if } m \ne n \\ \pi & \text{if } m = n \end{cases}$

8.3   TRIGONOMETRIC SUBSTITUTIONS

In Section 5.6 we introduced integration by substitution. The substitutions were always of the form $u = g(x)$. As we will see, it is also possible to make substitutions of the form $x = g(u)$.

In order to prepare for this kind of substitution, we consider the substitution

$$u = \sin^{-1} x, \quad \text{so that} \quad du = \frac{1}{\sqrt{1-x^2}}\, dx \qquad (1)$$

To simplify the equations in (1), we first notice that $u = \sin^{-1} x$ is equivalent to $x = \sin u$, and by the definition of the inverse sine function, $-\pi/2 \leq u \leq \pi/2$. Since $\cos u \geq 0$ for $-\pi/2 \leq u \leq \pi/2$, we deduce that

$$\sqrt{1 - x^2} = \sqrt{1 - \sin^2 u} = \cos u$$

and thus the second equation in (1) can be rewritten

$$du = \frac{1}{\cos u} \, dx, \quad \text{or equivalently,} \quad dx = \cos u \, du$$

Consequently instead of making the substitution appearing in (1), we can equally well substitute

$$x = \sin u, \quad \text{so that} \quad dx = \cos u \, du \tag{2}$$

where $-\pi/2 \leq u \leq \pi/2$.

More generally, in order to evaluate an integral $\int f(x) \, dx$, we can make a substitution of the form

$$x = g(u), \quad \text{so that} \quad dx = g'(u) \, du \tag{3}$$

thereby obtaining the integral formula

$$\int f(x) \, dx = \int f(g(u)) g'(u) \, du \tag{4}$$

The corresponding definite integral formula is

$$\int_{g(a)}^{g(b)} f(x) \, dx = \int_{a}^{b} f(g(u)) g'(u) \, du \tag{5}$$

Usually the function g in (3) involves trigonometric functions (as when $x = \sin u$ in (2)); in such cases the substitution in (3) is called a **trigonometric substitution.** Trigonometric substitutions are especially valuable when the integral contains square roots of the form $\sqrt{a^2 - x^2}$, $\sqrt{x^2 + a^2}$, or $\sqrt{x^2 - a^2}$, where $a \geq 0$.

Integrals Containing $\sqrt{a^2 - x^2}$

If we let $x = a \sin u$, with $a > 0$ and $-\pi/2 \leq u \leq \pi/2$, then $a \cos u \geq 0$, so that

$$\sqrt{a^2 - x^2} = \sqrt{a^2 - a^2 \sin^2 u} = \sqrt{a^2(1 - \sin^2 u)} = \sqrt{a^2 \cos^2 u} = a \cos u$$

Thus if an integral contains $\sqrt{a^2 - x^2}$, we can eliminate the square root by substituting

$$x = a \sin u, \quad \text{so that} \quad dx = a \cos u \, du$$

EXAMPLE 1   Find $\displaystyle\int \frac{1}{x^2 \sqrt{16 - x^2}} \, dx$.

Solution   Because $\sqrt{16 - x^2} = \sqrt{4^2 - x^2}$, we substitute

$$x = 4 \sin u, \quad \text{so that} \quad dx = 4 \cos u \, du$$

Then

$$\int \frac{1}{x^2\sqrt{16-x^2}}\,dx = \int \frac{1}{16\sin^2 u\sqrt{16-16\sin^2 u}}\,(4\cos u)\,du$$

$$= \int \frac{1}{(16\sin^2 u)\,4\sqrt{1-\sin^2 u}}\,(4\cos u)\,du$$

$$= \frac{1}{16}\int \frac{\cos u}{\sin^2 u\,\cos u}\,du = \frac{1}{16}\int \csc^2 u\,du$$

$$= -\frac{1}{16}\cot u + C$$

In order to write the answer in terms of the original variable x, we recall that $x = 4\sin u$, so that $\sin u = x/4$. This leads us to draw the triangle in Figure 8.3, in which $x = 4\sin u$. From the triangle we see that

$$\cot u = \frac{\sqrt{16-x^2}}{x}$$

Therefore

$$\int \frac{1}{x^2\sqrt{16-x^2}}\,dx = -\frac{1}{16}\cot u + C = -\frac{\sqrt{16-x^2}}{16x} + C \quad \square$$

FIGURE 8.3

Had the integral in Example 1 been

$$\int \frac{x^2}{\sqrt{16-x^2}}\,dx$$

then the same substitution $x = 4\sin u$, along with (4) of Section 8.2, would have yielded

$$\int \frac{x^2}{\sqrt{16-x^2}}\,dx = \int \frac{16\sin^2 u}{\sqrt{16-16\sin^2 u}}\,(4\cos u)\,du = \int \frac{16\sin^2 u}{4\cos u}\,(4\cos u)\,du$$

$$= 16\int \sin^2 u\,du = 8u - 4\sin 2u + C$$

As always, we must write the answer in terms of the original variable x. Using the double angle formula and referring to Figure 8.3 again, we find that

$$\sin 2u = 2\sin u\cos u = 2\left(\frac{x}{4}\right)\frac{\sqrt{16-x^2}}{4} = \frac{x}{8}\sqrt{16-x^2}$$

Since $x = 4\sin u$, it follows that $u = \sin^{-1}(x/4)$, and thus

$$\int \frac{x^2}{\sqrt{16-x^2}}\,dx = 8u - 4\sin 2u + C = 8\sin^{-1}\frac{x}{4} - \frac{x}{2}\sqrt{16-x^2} + C$$

EXAMPLE 2   Evaluate $\displaystyle\int_{-5/2}^{5/2}\sqrt{25-4x^2}\,dx$.

Solution Because $\sqrt{25 - 4x^2} = \sqrt{5^2 - (2x)^2}$, we are led to substitute

$$2x = 5 \sin u, \quad \text{so that} \quad x = \frac{5}{2} \sin u, \quad \text{and thus} \quad dx = \frac{5}{2} \cos u \, du$$

For the limits of integration we notice that

$$\text{if } x = -\frac{5}{2} \text{ then } u = -\frac{\pi}{2}, \quad \text{and} \quad \text{if } x = \frac{5}{2} \text{ then } u = \frac{\pi}{2}$$

Therefore with the help of (5) of Section 8.2, we conclude that

$$\int_{-5/2}^{5/2} \sqrt{25 - 4x^2} \, dx = \int_{-5/2}^{5/2} \sqrt{5^2 - (2x)^2} \, dx$$

$$= \int_{-\pi/2}^{\pi/2} \sqrt{5^2 - 5^2 \sin^2 u} \left(\frac{5}{2} \cos u\right) du$$

$$= \int_{-\pi/2}^{\pi/2} 5\sqrt{1 - \sin^2 u} \left(\frac{5}{2} \cos u\right) du = \frac{25}{2} \int_{-\pi/2}^{\pi/2} \cos^2 u \, du$$

$$= \frac{25}{2} \left(\frac{1}{2} u + \frac{1}{4} \sin 2u\right)\Big|_{-\pi/2}^{\pi/2}$$

$$= \frac{25}{2} \left(\frac{\pi}{4} - \left(-\frac{\pi}{4}\right)\right) = \frac{25}{4} \pi \quad \square$$

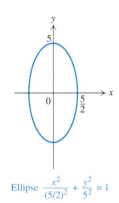

Ellipse $\dfrac{x^2}{(5/2)^2} + \dfrac{y^2}{5^2} = 1$

FIGURE 8.4

We can use Example 2 to find the area A of the region enclosed by the ellipse

$$\frac{x^2}{(5/2)^2} + \frac{y^2}{5^2} = 1$$

(Figure 8.4). This area is twice the area of the region between the graph of $y = \sqrt{25 - 4x^2}$ and the x axis on the interval $[-5/2, 5/2]$. Using the result of Example 2, we find that

$$A = 2 \int_{-5/2}^{5/2} \sqrt{25 - 4x^2} \, dx = 2 \left(\frac{25\pi}{4}\right) = \frac{25\pi}{2}$$

More generally, it is possible to show that the area of the region enclosed by the ellipse

$$\frac{x^2}{a^2} + \frac{y^2}{b^2} = 1 \quad \text{with } a > 0 \text{ and } b > 0$$

is πab (see Exercise 57). This result will be employed in the study of planetary orbits in Section 12.7.

Integrals Containing
$\sqrt{x^2 + a^2}$

If we let $x = a \tan u$, with $a > 0$ and $-\pi/2 < u < \pi/2$, then $a \sec u > 0$, so that

$$\sqrt{x^2 + a^2} = \sqrt{a^2 \tan^2 u + a^2} = \sqrt{a^2(\tan^2 u + 1)} = \sqrt{a^2 \sec^2 u} = a \sec u$$

Thus if an integral contains $\sqrt{x^2 + a^2}$, we can eliminate the square root by substituting

$$x = a \tan u, \quad \text{so that} \quad dx = a \sec^2 u \, du$$

EXAMPLE 3   Find $\displaystyle\int \frac{1}{x^2\sqrt{x^2 + 1}} \, dx.$

Solution   We substitute

$$x = \tan u, \quad \text{so that} \quad dx = \sec^2 u \, du$$

Therefore

$$\int \frac{1}{x^2\sqrt{x^2 + 1}} \, dx = \int \frac{1}{\tan^2 u \sqrt{\tan^2 u + 1}} \, (\sec^2 u) \, du$$

$$= \int \frac{1}{\tan^2 u \sec u} \, (\sec^2 u) \, du = \int \frac{\sec u}{\tan^2 u} \, du$$

$$= \int \frac{\cos u}{\sin^2 u} \, du = -\frac{1}{\sin u} + C$$

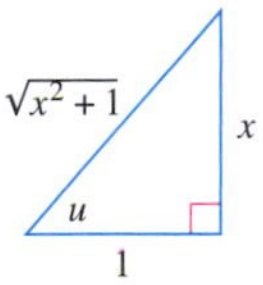

To give the answer in terms of x, we use the triangle in Figure 8.5, with $x = \tan u$. We find that

$$\int \frac{1}{x^2\sqrt{x^2 + 1}} \, dx = -\frac{1}{\sin u} + C = -\frac{\sqrt{x^2 + 1}}{x} + C \quad \square$$

FIGURE 8.5

EXAMPLE 4   Find $\displaystyle\int \frac{1}{\sqrt{4 + 16x^2}} \, dx.$

Solution   Because

$$\sqrt{4 + 16x^2} = \sqrt{4(1 + 4x^2)} = 2\sqrt{1 + (2x)^2}$$

we substitute

$$2x = \tan u, \quad \text{so that} \quad x = \frac{1}{2} \tan u, \quad \text{and thus} \quad dx = \frac{1}{2} \sec^2 u \, du$$

Then

$$\int \frac{1}{\sqrt{4 + 16x^2}} \, dx = \int \frac{1}{2\sqrt{1 + (2x)^2}} \, dx = \int \frac{1}{2\sqrt{1 + \tan^2 u}} \left(\frac{1}{2} \sec^2 u\right) du$$

$$= \int \frac{1}{2 \sec u} \left(\frac{1}{2} \sec^2 u\right) du$$

$$= \frac{1}{4} \int \sec u \, du$$

Using (6) of Section 8.2 for $\int \sec u \, du$, and then using the fact that $\tan u = 2x$, so that $\sec u = \sqrt{\tan^2 u + 1} = \sqrt{4x^2 + 1}$, we conclude that

$$\int \frac{1}{\sqrt{4 + 16x^2}} \, dx = \frac{1}{4} \int \sec u \, du = \frac{1}{4} \ln |\sec u + \tan u| + C$$

$$= \frac{1}{4} \ln |\sqrt{4x^2 + 1} + 2x| + C \quad \square$$

Integrals Containing $\sqrt{x^2 - a^2}$

If we let $x = a \sec u$, with $0 \le u < \pi/2$ or $\pi \le u < 3\pi/2$, then $a \tan u \ge 0$, so that

$$\sqrt{x^2 - a^2} = \sqrt{a^2 \sec^2 u - a^2} = \sqrt{a^2 (\sec^2 u - 1)} = \sqrt{a^2 \tan^2 u} = a \tan u$$

Thus if an integral contains $\sqrt{x^2 - a^2}$, then we can eliminate the square root by substituting

$$x = a \sec u, \qquad \text{so that} \qquad dx = a \sec u \tan u \, du$$

EXAMPLE 5   Find $\displaystyle\int_{-6}^{-3} \frac{\sqrt{x^2 - 9}}{x} \, dx$.

Solution   Since $\sqrt{x^2 - 9} = \sqrt{x^2 - 3^2}$, we let

$$x = 3 \sec u, \qquad \text{so that} \qquad dx = 3 \sec u \tan u \, du$$

For the limits of integration we observe that

$$\text{if } x = -6 \quad \text{then } u = \frac{4\pi}{3}, \quad \text{and} \quad \text{if } x = -3 \quad \text{then } u = \pi$$

Thus if x is in the interval $[-6, -3]$, it follows that u is in $[\pi, \frac{4}{3}\pi]$. For such values of u we have $\tan u \ge 0$, so that

$$\sqrt{x^2 - 9} = \sqrt{9 \sec^2 u - 9} = \sqrt{9 \tan^2 u} = 3 \tan u$$

Therefore

$$\int_{-6}^{-3} \frac{\sqrt{x^2 - 9}}{x} \, dx = \int_{4\pi/3}^{\pi} \frac{\sqrt{9 \sec^2 u - 9}}{3 \sec u} (3 \sec u \tan u) \, du$$

$$= \int_{4\pi/3}^{\pi} \frac{3 \tan u}{3 \sec u} (3 \sec u \tan u) \, du$$

$$= 3 \int_{4\pi/3}^{\pi} \tan^2 u \, du = 3 \int_{4\pi/3}^{\pi} (\sec^2 u - 1) \, du$$

$$= 3(\tan u - u) \Big|_{4\pi/3}^{\pi} = 3(\tan \pi - \pi) - 3 \left(\tan \frac{4\pi}{3} - \frac{4\pi}{3} \right)$$

$$= \pi - 3\sqrt{3} \quad \square$$

Table 8.3 summarizes the trigonometric substitutions that we have discussed.

TABLE 8.3

Trigonometric Substitutions

Expression in integrand	Substitution
$\sqrt{a^2 - x^2}$	$x = a \sin u$, with $-\pi/2 \le u \le \pi/2$ $dx = a \cos u \, du$
$\sqrt{x^2 + a^2}$	$x = a \tan u$, with $-\pi/2 < u < \pi/2$ $dx = a \sec^2 u \, du$
$\sqrt{x^2 - a^2}$	$x = a \sec u$, with $0 \le u < \pi/2$ or $\pi \le u < 3\pi/2$ $dx = a \sec u \tan u \, du$

Integrals Containing
$\sqrt{bx^2 + cx + d}$

By completing the square in $bx^2 + cx + d$ we can express $\sqrt{bx^2 + cx + d}$ in terms of $\sqrt{a^2 - x^2}$, $\sqrt{x^2 + a^2}$, or $\sqrt{x^2 - a^2}$, for suitable $a > 0$. Then a trigonometric substitution eliminates the square root as before.

EXAMPLE 6   Find $\displaystyle\int \frac{1}{\sqrt{9x^2 + 6x + 2}} \, dx$.

Solution   First we complete the square in the denominator:

$$\sqrt{9x^2 + 6x + 2} = \sqrt{(3x + 1)^2 + 1}$$

Then we let

$$3x + 1 = \tan u, \quad \text{so that} \quad 3 \, dx = \sec^2 u \, du, \quad \text{and thus} \quad dx = \frac{1}{3} \sec^2 u \, du$$

and notice that $\sqrt{(3x + 1)^2 + 1} = \sqrt{\tan^2 u + 1} = \sec u$. Using (6) of Section 8.2, we find that

$$\int \frac{1}{\sqrt{9x^2 + 6x + 2}} \, dx = \int \frac{1}{\sqrt{(3x + 1)^2 + 1}} \, dx = \int \frac{1}{\sec u} \left(\frac{1}{3} \sec^2 u\right) du$$

$$= \frac{1}{3} \int \sec u \, du = \frac{1}{3} \ln | \sec u + \tan u | + C$$

Referring to our formulas for $\tan u$ and $\sec u$, we conclude that

$$\int \frac{1}{\sqrt{9x^2 + 6x + 2}} \, dx = \frac{1}{3} \ln |\sqrt{(3x + 1)^2 + 1} + 3x + 1| + C$$

$$= \frac{1}{3} \ln |\sqrt{9x^2 + 6x + 2} + 3x + 1| + C \quad \square$$

EXERCISES 8.3

In Exercises 1–45 evaluate the integral.

1. $\displaystyle\int_0^{1/2} \sqrt{1 - 4x^2}\, dx$ **2.** $\displaystyle\int_0^4 \sqrt{16 - x^2}\, dx$

3. $\displaystyle\int_{-2}^2 \sqrt{1 - \frac{x^2}{4}}\, dx$ **4.** $\displaystyle\int \sqrt{1 + (2x - 1)^2}\, dx$

5. $\displaystyle\int \frac{1}{(3 - x^2)^{3/2}}\, dx$ **6.** $\displaystyle\int_{4\sqrt{2}}^8 \frac{1}{(t^2 - 16)^{3/2}}\, dt$

7. $\displaystyle\int \frac{1}{(x^2 + 1)^{3/2}}\, dx$ **8.** $\displaystyle\int \frac{1}{(2x^2 + 1)^{3/2}}\, dx$

9. $\displaystyle\int_0^1 \frac{1}{(3x^2 + 2)^{5/2}}\, dx$ **10.** $\displaystyle\int \frac{x^2}{(x^2 - 2)^{5/2}}\, dx$

11. $\displaystyle\int_2^{2\sqrt{2}} \frac{\sqrt{x^2 - 4}}{x}\, dx$ **12.** $\displaystyle\int \frac{1}{(y^2 - 4)^{3/2}}\, dy$

13. $\displaystyle\int \frac{1}{(9 + t^2)^2}\, dt$ **14.** $\displaystyle\int_1^{\sqrt{2}} \frac{1}{\sqrt{2x^2 - 1}}\, dx$

15. $\displaystyle\int_0^{5/4} \frac{1}{\sqrt{25 - 4x^2}}\, dx$ **16.** $\displaystyle\int \frac{x^2}{\sqrt{1 - x^2}}\, dx$

17. $\displaystyle\int \frac{1}{\sqrt{4x^2 + 4x + 2}}\, dx$ **18.** $\displaystyle\int_1^{\sqrt{3}} \frac{1}{\sqrt{1 + x^2}}\, dx$

19. $\displaystyle\int \frac{1}{(1 - 2w^2)^{5/2}}\, dw$

20. $\displaystyle\int_{\sqrt{2}-2}^{(2\sqrt{3}/3)-2} \frac{1}{(t^2 + 4t + 3)^{3/2}}\, dt$

21. $\displaystyle\int_0^1 \frac{1}{2x^2 - 2x + 1}\, dx$ ***22.** $\displaystyle\int \sqrt{z^2 - 9}\, dz$

23. $\displaystyle\int \sqrt{x - x^2}\, dx$ **24.** $\displaystyle\int \frac{\sqrt{25 - z^2}}{z^2}\, dz$

25. $\displaystyle\int \frac{x^2}{\sqrt{9x^2 - 1}}\, dx$ **26.** $\displaystyle\int_{-4/\sqrt{3}}^{-2\sqrt{2}} \frac{1}{x\sqrt{x^2 - 4}}\, dx$

27. $\displaystyle\int \frac{1}{x\sqrt{x^2 + 4}}\, dx$ **28.** $\displaystyle\int \frac{1}{x^2\sqrt{9 - 4x^2}}\, dx$

29. $\displaystyle\int \frac{1}{x^2\sqrt{4x^2 - 9}}\, dx$ **30.** $\displaystyle\int \frac{1}{w^2\sqrt{4w^2 + 9}}\, dw$

31. $\displaystyle\int \frac{x^2}{\sqrt{1 + x^2}}\, dx$ **32.** $\displaystyle\int \frac{x^2}{(1 + x^2)^{3/2}}\, dx$

33. $\displaystyle\int \frac{1}{(9x^2 - 4)^{5/2}}\, dx$ **34.** $\displaystyle\int \frac{\sqrt{4 + x^2}}{x^3}\, dx$

35. $\displaystyle\int \sqrt{4 + x^2}\, dx$ **36.** $\displaystyle\int \frac{\sqrt{x^2 + 1}}{x^4}\, dx$

37. $\displaystyle\int_{3\sqrt{2}}^6 \frac{1}{x^4\sqrt{x^2 - 9}}\, dx$ **38.** $\displaystyle\int \frac{2x - 3}{\sqrt{4x - x^2 - 3}}\, dx$

39. $\displaystyle\int \frac{x}{\sqrt{2x^2 + 12x + 19}}\, dx$

40. $\displaystyle\int \frac{1}{(w^2 + 2w + 5)^{3/2}}\, dw$

*41. $\displaystyle\int \sqrt{x^2 + 6x + 5}\, dx$ **42.** $\displaystyle\int \frac{e^{3w}}{\sqrt{1 - e^{2w}}}\, dw$

43. $\displaystyle\int e^w \sqrt{1 + e^{2w}}\, dw$

44. $\displaystyle\int_{\sqrt{2}}^2 \sec^{-1} x\, dx$

45. $\displaystyle\int x \sin^{-1} x\, dx$ (*Hint:* Use integration by parts.)

46. a. Evaluate $\displaystyle\int \left(\frac{1 - x}{1 + x}\right)^{1/2} dx$ by letting $x = \cos 2u$ and recalling that $1 + \cos 2u = 2 \cos^2 u$ and $1 - \cos 2u = 2 \sin^2 u$.

    *b. Evaluate $\displaystyle\int_1^{\cosh 2} \left(\frac{x - 1}{x + 1}\right)^{1/2} dx$ by letting $x = \cosh 2u$.

In Exercises 47–49 find the area A of the region between the graph of f and the x axis on the given interval.

47. $f(x) = \sqrt{1 - x^2}$; $[0, 1]$

48. $f(x) = \dfrac{x^2}{\sqrt{1 - x^2}}$; $[-1/2, 1/2]$

49. $f(x) = \sqrt{9 + x^2}$; $[0, 3]$

50. Let $f(x) = \sin x$ for $0 \le x \le \pi$, and let D be the solid generated by revolving the graph of f around the x axis.
    a. Sketch D.
    b. Find the surface area S of D.

51. The curve C parametrized by

$$x = \sqrt{2}\, \cos t \quad \text{and} \quad y = 2 \sin t \quad \text{for } 0 \le t \le \pi$$

is half an ellipse, and the surface obtained by revolving C around the x axis is called an **ellipsoid**. Determine the surface area S of the ellipsoid.

52. Let $f(x) = \sqrt{1 - x^2}$ and $g(x) = -(1 + x)$. Calculate the center of gravity of the region R between the graphs of f and g on $[0, 1]$.

*53. Let $f(x) = x^2 - x$ for $0 \le x \le 1$. Find the length L of the graph of f.

*54. Let $f(x) = x^2$ for $0 \le x \le 1/2$. Find the length L of the graph of f.

Applications

55. Suppose a washer for a faucet has the shape of the solid generated by revolving about the x axis the figure bounded by the y axis, the lines $x = 1$ and $y = \frac{1}{2}$, and the curve

$$y = 1 + \sqrt{1 - x^2} \quad \text{for} \quad 0 \le x \le 1$$

Determine the volume V of the washer.

56. Suppose a large gasoline tank has the shape of a half-cylinder 4 feet in radius and 10 feet long (Figure 8.6). If the tank is full of gasoline weighing 42 pounds per cubic foot, find the work W necessary to pump all the gasoline through a spout at the top.

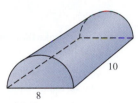

10

8

FIGURE 8.6   Figure for Exercise 56.

57. a. Show that the region enclosed by the ellipse

$$\frac{x^2}{a^2} + \frac{y^2}{b^2} = 1$$

has area πab.

 b. The Ellipse in Washington, D.C., which is located next to the White House, is approximately 1500 feet long and 1280 feet wide. Use the result of part (a) to determine how many square feet of grass should be purchased in order to sod the entire Ellipse. (*Hint:* Assume that an equation of the Ellipse is $x^2/750^2 + y^2/640^2 = 1$.)

The Ellipse in Washington, D.C. (*Everett C. Johnson/Folio Inc.*)

58. a. In order to determine the prices of various sizes of doughnuts, a supermarket manager would like to have a formula for the volume of a doughnut (Figure 8.7). The manager finds the formula

$$V = \int_{R-r}^{R+r} 4\pi x \sqrt{r^2 - (x - R)^2}\; dx \quad \text{where } r < R$$

in a calculus book but does not know how to evaluate the integral. Make the substitution $x = R + r \sin u$ for $-\pi/2 \le u \le \pi/2$, and then evaluate the integral to obtain the formula $V = 2\pi^2 Rr^2$.

b. Which doughnut should cost more, one with $R = 4$ and $r = 2$, or one with $R = 6$ and $r = 1$?

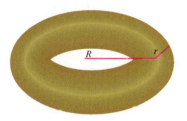

R   r

FIGURE 8.7   Doughnut.

Project

1. a. Let m and n be positive integers, with n an odd integer. Determine for which m and n it is feasible to use trigonometric substitution and then apply methods discussed in Section 8.2 in order to evaluate

$$\int \frac{x^m}{(1 - x^2)^{n/2}}\; dx$$

b. Let m and n be positive integers.

i. Determine for which m and n it is feasible to use one or more types of substitution to convert the integral to a trigonometric integral, and then apply methods discussed in Section 8.2 in order to evaluate

$$\int \frac{x^m}{\sqrt{1 - x^n}}\; dx$$

ii. Determine values of m and n such that $m \ge 2$ and $n \ge 2$, and such that a substitution of the form $u = g(x)$ is the easiest substitution to use in order to evaluate the integral in (i).

8.4 PARTIAL FRACTIONS

In this section we set up a three-part procedure that, in principle, could be used to integrate any rational function. We will use that procedure to find integrals such as

$$\int \frac{2x + 3}{x^3 + 2x^2 + x}\, dx \quad \text{and} \quad \int \frac{x^2 + 2x + 7}{x^3 + x^2 - 2}\, dx$$

The formula

$$\int \frac{1}{x^2 + a^2}\, dx = \frac{1}{a} \tan^{-1} \frac{x}{a} + C \quad \text{where } a \neq 0 \tag{1}$$

derived in Section 7.5, will come into play.

Preparing Rational Functions for Integration

Before integrating any rational function, we normally must make three preparations, which we list as (i), (ii), and (iii).

> (i) When applicable, divide the numerator of the rational function by its denominator, to ensure that the degree of the numerator of any resulting remainder is *less* than the degree of the denominator.

EXAMPLE 1   Perform operation (i) on $\dfrac{2x^3}{x^2 + 3}$.

Solution   Since the degree of the numerator is greater than the degree of the denominator, we perform operation (i), dividing as if the numerator and denominator were integers:

$$
\begin{array}{r}
2x \\
x^2 + 3 \overline{\smash{)}2x^3 } \\
-\underline{(2x^3 + 6x)} \\
-6x
\end{array}
$$

Therefore

$$\frac{2x^3}{x^2 + 3} = 2x - \frac{6x}{x^2 + 3}$$

Notice that the degree of the numerator in the remainder is 1, while the degree of the denominator is 2.   ❑

It follows from Example 1 that

$$\int \frac{2x^3}{x^2 + 3}\, dx = \int 2x\, dx - \int \frac{6x}{x^2 + 3}\, dx$$

Since the first integral on the right side of the equation is easy to find, the problem of finding the integral on the left is reduced to finding the second integral on the right.

EXAMPLE 2   Perform operation (i) on $\dfrac{x^2 - 1}{x^2 + 1}$.

Solution   In this case the degree of both numerator and denominator is 2, and since the degree of the numerator is *not less than* the degree of the denominator, we must divide:

$$x^2 + 1 \overline{\smash{\big)}\,x^2 - 1} \quad \begin{array}{c} 1 \\ \\ \end{array}$$
$$\begin{array}{c} x^2 + 1 \\ \hline -2 \end{array}$$

Therefore

$$\frac{x^2 - 1}{x^2 + 1} = 1 - \frac{2}{x^2 + 1}$$

and the right-hand side complies with (i).   ❑

In the case of

$$\frac{x^2 + 3}{2x^3 + x^2 + 1}$$

operation (i) is not required, since the degree of the numerator is already less than the degree of the denominator.

After (i) has been performed, operation (ii) is to be performed on the resulting remainder, which is itself a rational function.

> (ii) Factor the numerator and the denominator of the remainder into linear and quadratic factors, that is, into expressions of the form
>
> $$\text{constants,} \quad (ax + b)^r, \quad \text{and} \quad (ax^2 + bx + c)^s$$
>
> where r and s are positive integers. Factor quadratic expressions into real factors wherever possible. Finally, reduce the fraction if any common factors appear in both the numerator and the denominator.

EXAMPLE 3   Perform operation (ii) on $\dfrac{2x + 4}{x^2 + 3x + 2}$.

Solution   First, we factor the numerator and denominator:

$$2x + 4 = 2(x + 2)$$

$$x^2 + 3x + 2 = (x + 2)(x + 1)$$

Therefore

$$\frac{2x + 4}{x^2 + 3x + 2} = \frac{2(x + 2)}{(x + 2)(x + 1)} = \frac{2}{x + 1} \quad ❑$$

EXAMPLE 4   Perform operation (ii) on $\dfrac{2x + 3}{x^3 + 2x^2 + x}$.

Solution   Here we find that

$$x^3 + 2x^2 + x = x(x^2 + 2x + 1) = x(x + 1)^2$$

As a result,

$$\frac{2x + 3}{x^3 + 2x^2 + x} = \frac{2x + 3}{x(x + 1)^2} \qquad \square$$

In Examples 3 and 4 the factorization was easy. In more difficult cases the following result, which is often called the Factor Theorem, can simplify finding factors of a polynomial $P(x)$.

THEOREM 8.2
Factor Theorem

The expression $x - a$ is a factor of $P(x)$ if and only if $P(a) = 0$.

Proof   If $x - a$ is a factor of $P(x)$, then there is a polynomial function R such that

$$P(x) = (x - a)R(x)$$

But then

$$P(a) = (a - a)R(a) = 0$$

To prove the converse, let $P(x)$ be given by

$$P(x) = c_n x^n + c_{n-1} x^{n-1} + \cdots + c_1 x + c_0$$

and assume that $P(a) = 0$. Then

$$P(a) = c_n a^n + c_{n-1} a^{n-1} + \cdots + c_1 a + c_0 = 0$$

so that

$$P(x) = P(x) - P(a)$$

$$= c_n(x^n - a^n) + c_{n-1}(x^{n-1} - a^{n-1}) + \cdots + c_1(x - a) \qquad (2)$$

But for any integer $k \geq 2$, we know that

$$x^k - a^k = (x - a)(x^{k-1} + x^{k-2} a + \cdots + xa^{k-2} + a^{k-1})$$

Therefore $x - a$ is a factor of each of the summands appearing on the right side of (2). Thus $x - a$ is a factor of $P(x)$.   ■

EXAMPLE 5   Perform (ii) on $\dfrac{x^2 + 2x + 7}{x^3 + x^2 - 2}$.

Solution   By the quadratic formula, the numerator has no real roots, and hence cannot be factored. To factor the denominator, notice first that

$$x^3 + x^2 - 2 = 0 \quad \text{for } x = 1$$

By the Factor Theorem, $x - 1$ must therefore be a factor of $x^3 + x^2 - 2$. Carrying out the division of $x^3 + x^2 - 2$ by $x - 1$ gives us

$$x^3 + x^2 - 2 = (x - 1)(x^2 + 2x + 2)$$

Again the quadratic formula tells us that $x^2 + 2x + 2$ has no real roots, so cannot be factored. Thus we have

$$\frac{x^2 + 2x + 7}{x^3 + x^2 - 2} = \frac{x^2 + 2x + 7}{(x - 1)(x^2 + 2x + 2)} \tag{3}$$

as the desired way of rewriting the original function.   ❑

After performing operations (i) and (ii) we obtain a rational function g such that the denominator of $g(x)$ is a product of constants, factors of the form $(ax + b)^r$, and factors of the form $(ax^2 + bx + c)^s$ (where $ax^2 + bx + c$ cannot be factored). The final step in the preparation involves equating $g(x)$ with a sum of terms arising from the factors appearing in the denominator of $g(x)$. For every factor $(ax + b)^r$ appearing in the denominator of $g(x)$ we include an expression of the form

$$\frac{A_1}{ax + b} + \frac{A_2}{(ax + b)^2} + \cdots + \frac{A_r}{(ax + b)^r} \tag{4}$$

where $A_1, A_2, \ldots, A_r$ are constants to be determined. For every factor of the form $(ax^2 + bx + c)^s$ we include an expression of the form

$$\frac{B_1 x + C_1}{ax^2 + bx + c} + \frac{B_2 x + C_2}{(ax^2 + bx + c)^2} + \cdots + \frac{B_s x + C_s}{(ax^2 + bx + c)^s} \tag{5}$$

where again the numbers $B_1, B_2, \ldots, B_s$ and $C_1, C_2, \ldots, C_s$ are constants to be determined. For instance,

$$\frac{2x + 3}{x(x + 1)^2} = \frac{A}{x} + \frac{B}{x + 1} + \frac{C}{(x + 1)^2} \tag{6}$$

and

$$\frac{x^2 + 2x + 7}{(x - 1)(x^2 + 2x + 2)} = \frac{A}{x - 1} + \frac{Bx + C}{x^2 + 2x + 2} \tag{7}$$

where the constants A, B, and C must be determined.

We are now ready to describe the third and final preparation.

> (iii) Rewrite the transformed rational function $g(x)$ as a sum of expressions of the types appearing in (4) and (5).

Because of the fractions that arise from step (iii), this method of transforming rational functions is known as the method of **partial fractions.**

Examples of Integration by Partial Fractions

In the final three examples of the section we will perform (i)–(iii) as needed, and then we will integrate the resulting integrals. The first integral is the simplest; it occurs in studies involving chemical reactions and population growth.

EXAMPLE 6   Evaluate $\displaystyle\int \frac{1}{x^2 - 8x + 15} \, dx$.

Solution   Since

$$x^2 - 8x + 15 = (x - 5)(x - 3)$$

it follows that

$$\frac{1}{x^2 - 8x + 15} = \frac{A}{x - 5} + \frac{B}{x - 3} \tag{8}$$

where A and B are constants. To determine A and B, we clear fractions by multiplying both sides of the equation by $(x - 5)(x - 3)$. The result is

$$1 = A(x - 3) + B(x - 5) \tag{9}$$

The equation in (8) and hence the equation in (9) hold for all x except 3 and 5. However, since both sides of (9) are polynomials, this means that (9) must actually hold for all x, including 3 and 5. For $x = 3$, (9) becomes

$$1 = A(0) + B(-2), \quad \text{so that} \quad B = -\frac{1}{2}$$

For $x = 5$, (9) becomes

$$1 = A(2) + B(0), \quad \text{so that} \quad A = \frac{1}{2}$$

(We chose to substitute $x = 3$ and $x = 5$ in (9) because for each of these values only one term on the right side of (9) is *not* zero.) Substituting $A = \frac{1}{2}$ and $B = -\frac{1}{2}$ in (8), we obtain

$$\frac{1}{x^2 - 8x + 15} = \frac{1}{2} \cdot \frac{1}{x - 5} - \frac{1}{2} \cdot \frac{1}{x - 3}$$

Therefore

$$\int \frac{1}{x^2 - 8x + 15} \, dx = \int \frac{1}{2} \cdot \frac{1}{x - 5} \, dx + \int \frac{-1}{2} \cdot \frac{1}{x - 3} \, dx$$

$$= \frac{1}{2} \ln |x - 5| - \frac{1}{2} \ln |x - 3| + C_1$$

$$= \frac{1}{2} \ln \left| \frac{x - 5}{x - 3} \right| + C_1$$

Thus we have evaluated the integral under consideration.   ❑

In certain chemical reactions, molecules of two different substances react and thereby produce molecules of a third substance. These are called **bimolecular**

A red precipitate is formed when solutions of sodium hydroxide and iron nitrate are mixed. (*Charles D. Winters*)

reactions. We denote by a and b the initial concentrations of the original substances and by y the concentration of the third substance. For some such reactions the rate dy/dt at which the reaction occurs is given by

$$\frac{dy}{dt} = r(a - y)(b - y) \tag{10}$$

where r is a suitable positive constant. If the units are chosen appropriately, and if $a = 5$, $b = 3$, and $r = 1$, then (10) becomes

$$\frac{dy}{dt} = (5 - y)(3 - y) = y^2 - 8y + 15$$

This is a separable differential equation (see Section 7.8), whose general solution can be written in the form

$$t = \int \frac{1}{y^2 - 8y + 15} \, dy \tag{11}$$

provided that y is never 3 or 5. Combining (11) with the result of Example 6, we find that

$$t = \frac{1}{2} \ln \left| \frac{y - 5}{y - 3} \right| + C_1$$

Therefore

$$\ln \left| \frac{y - 5}{y - 3} \right| = 2t - 2C_1$$

so that

$$\left| \frac{y - 5}{y - 3} \right| = C \, e^{2t} \tag{12}$$

where $C = e^{-2C_1}$.

If at time $t = 0$ the mixture contains none of the resulting substance, then $y = 0$ at that time. Substituting $t = 0$ and $y = 0$ in (12) yields $C = \frac{5}{3}$. Since y is assumed to be differentiable and hence continuous, and since y is never 3, it follows from (12) that y remains less than 3. Thus we may remove the absolute value signs in (12) and solve for y, obtaining

$$y = \frac{15e^{2t} - 15}{5e^{2t} - 3}$$

Since

$$\lim_{t \to \infty} \frac{15e^{2t} - 15}{5e^{2t} - 3} = \lim_{t \to \infty} \frac{15 - 15e^{-2t}}{5 - 3e^{-2t}} = 3$$

we conclude that as time increases, the amount of the third substance approaches 3 units.

EXAMPLE 7   Evaluate $\displaystyle\int \frac{x^2 + 2x + 7}{x^3 + x^2 - 2}\, dx.$

Solution   To achieve (ii) and (iii), we combine (3) and (7):

$$\frac{x^2 + 2x + 7}{x^3 + x^2 - 2} = \frac{A}{x - 1} + \frac{Bx + C}{x^2 + 2x + 2} \tag{13}$$

Next, to clear fractions in the right-hand sum of (13), we multiply through by $(x - 1)\,(x^2 + 2x + 2)$ and obtain

$$x^2 + 2x + 7 = A(x^2 + 2x + 2) + (Bx + C)(x - 1)$$

This equation must hold for all x, so in particular it must hold for $x = 1$. For $x = 1$ we get

$$10 = A(5), \quad \text{so that} \quad A = 2$$

Therefore

$$x^2 + 2x + 7 = 2(x^2 + 2x + 2) + (Bx + C)(x - 1)$$

Next we combine like powers of x:

$$x^2 + 2x + 7 = (2 + B)x^2 + (4 + C - B)x + (4 - C) \tag{14}$$

This can be true for all values of x only if the coefficients of like powers of x on both sides of (14) are equal. Thus

$$1 = 2 + B, \quad \text{so that} \quad B = -1$$

$$7 = 4 - C, \quad \text{so that} \quad C = -3$$

(Since $4 + C - B = 4 - 3 + 1 = 2$, the coefficients of x on both sides of (14) are equal as well.) As a result, we can rewrite (13) as follows:

$$\frac{x^2 + 2x + 7}{x^3 + x^2 - 2} = \frac{2}{x - 1} - \frac{x + 3}{x^2 + 2x + 2}$$

Thus

$$\int \frac{x^2 + 2x + 7}{x^3 + x^2 - 2}\, dx = \int \frac{2}{x - 1}\, dx - \int \frac{x + 3}{x^2 + 2x + 2}\, dx$$

To evaluate the right-hand integral, we first complete the square in the denominator to obtain

$$x^2 + 2x + 2 = (x + 1)^2 + 1$$

and then substitute

$$u = x + 1, \quad \text{so that} \quad du = dx \quad \text{and} \quad x + 3 = u + 2$$

Therefore

$$\int \frac{x+3}{x^2+2x+2}\,dx = \int \frac{x+3}{(x+1)^2+1}\,dx = \int \frac{u+2}{u^2+1}\,du$$

$$= \int \frac{u}{u^2+1}\,du + 2\int \frac{1}{u^2+1}\,du$$

$$= \frac{1}{2}\int \frac{2u}{u^2+1}\,du + 2\int \frac{1}{u^2+1}\,du$$

$$= \frac{1}{2}\ln(u^2+1) + 2\tan^{-1}u + C_1$$

$$= \frac{1}{2}\ln(x^2+2x+2) + 2\tan^{-1}(x+1) + C_1$$

Consequently

$$\int \frac{x^2+2x+7}{x^3+x^2-2}\,dx = \int \frac{2}{x-1}\,dx - \int \frac{x+3}{x^2+2x+2}\,dx = 2\ln|x-1|$$

$$-\frac{1}{2}\ln(x^2+2x+2) - 2\tan^{-1}(x+1) + C_2 \quad \square$$

Finally, we evaluate an integral in which the quadratic factor x^2+1 in the denominator is squared.

EXAMPLE 8   Evaluate $\displaystyle\int \frac{1}{x(x^2+1)^2}\,dx$.

Solution   To execute operation (iii), we write

$$\frac{1}{x(x^2+1)^2} = \frac{A}{x} + \frac{Bx+C}{x^2+1} + \frac{Dx+E}{(x^2+1)^2}$$

Next we clear fractions:

$$1 = A(x^2+1)^2 + (Bx+C)x(x^2+1) + (Dx+E)x \qquad (15)$$

For $x=0$ this equation becomes

$$1 = A(1) + C(0) + E(0), \quad \text{so that} \quad A=1$$

After we replace A by 1, multiply out, and combine like powers, equation (15) becomes

$$1 = (1+B)x^4 + Cx^3 + (2+B+D)x^2 + (C+E)x + 1$$

Equating coefficients of like powers of x, we find that

$$1+B=0, \quad \text{so that} \quad B=-1$$

$$C=0$$

$$2+B+D=0, \quad \text{so that} \quad D=-1$$

$$C+E=0, \quad \text{so that} \quad E=0$$

Therefore

$$\frac{1}{x(x^2 + 1)^2} = \frac{1}{x} - \frac{x}{x^2 + 1} - \frac{x}{(x^2 + 1)^2}$$

so that

$$\int \frac{1}{x(x^2 + 1)^2} \, dx = \int \frac{1}{x} \, dx - \int \frac{x}{x^2 + 1} \, dx - \int \frac{x}{(x^2 + 1)^2} \, dx$$

Substituting

$$u = x^2 + 1, \quad \text{so that} \quad du = 2x \, dx$$

we find that

$$-\int \frac{x}{x^2 + 1} \, dx - \int \frac{x}{(x^2 + 1)^2} \, dx = -\int \frac{1}{u} \cdot \frac{1}{2} \, du - \int \frac{1}{u^2} \cdot \frac{1}{2} \, du$$

$$= -\frac{1}{2} \ln|u| + \frac{1}{2} \cdot \frac{1}{u} + C_1$$

$$= -\frac{1}{2} \ln(x^2 + 1) + \frac{1}{2} \cdot \frac{1}{x^2 + 1} + C_1$$

Consequently

$$\int \frac{1}{x(x^2 + 1)^2} \, dx = \int \frac{1}{x} \, dx - \int \frac{x}{x^2 + 1} \, dx - \int \frac{x}{(x^2 + 1)^2} \, dx$$

$$= \ln|x| - \frac{1}{2} \ln(x^2 + 1) + \frac{1}{2(x^2 + 1)} + C_2 \quad \square$$

When we evaluate the integral of a rational function by partial fractions, the result can always, at least in principle, be given in terms of rational functions, logarithms, and inverse tangents. You can check that this is true with Examples 6 to 8.

EXERCISES 8.4

In Exercises 1–29 find the integral.

1. $\displaystyle\int \frac{x}{x + 1} \, dx$

2. $\displaystyle\int \frac{x^2}{x^2 + 1} \, dx$

3. $\displaystyle\int \frac{x^2}{x^2 - 1} \, dx$

4. $\displaystyle\int \frac{t^2 - 1}{t^2 + 1} \, dt$

5. $\displaystyle\int \frac{x^2 + 4}{x(x - 1)^2} \, dx$

6. $\displaystyle\int \frac{2x^3 + x^2 + 12}{x^2 - 4} \, dx$

7. $\displaystyle\int_3^4 \frac{5}{(x - 2)(x + 3)} \, dx$

8. $\displaystyle\int \frac{5x}{(x - 2)(x + 3)} \, dx$

9. $\displaystyle\int \frac{3t}{t^2 - 8t + 15} \, dt$

10. $\displaystyle\int \frac{2}{x^2 - x - 6} \, dx$

11. $\displaystyle\int_{-1}^0 \frac{x^2 + x + 1}{x^2 + 1} \, dx$

12. $\displaystyle\int \frac{2x + 5}{x^2 + 3x + 2} \, dx$

13. $\displaystyle\int \frac{x^2 + x + 1}{x^2 - 1} \, dx$

14. $\displaystyle\int \frac{x^2 + 2x + 1}{x^2 - 1} \, dx$

15. $\displaystyle\int_0^1 \frac{u - 1}{u^2 + u + 1} \, du$

16. $\displaystyle\int \frac{2x}{(x + 1)^2} \, dx$

17. $\displaystyle\int \frac{3x}{x^2 - 4x + 4} \, dx$

18. $\displaystyle\int \frac{4x}{x^3 + 6x^2 + 11x + 6} \, dx$

19. $\displaystyle\int \frac{-x}{x^3 - 3x^2 + 2x}\,dx$ **20.** $\displaystyle\int \frac{x+1}{x^3 - 2x^2 + x}\,dx$

21. $\displaystyle\int \frac{u^3}{(u+1)^2}\,du$ **22.** $\displaystyle\int_0^1 \frac{x^3}{(x+1)^3}\,dx$

23. $\displaystyle\int \frac{1}{(1-x^2)^2}\,dx$

$\left(\text{Hint: } \dfrac{1}{(1-x^2)^2} = \left(\dfrac{1}{2(x+1)} - \dfrac{1}{2(x-1)}\right)^2.\right)$

24. $\displaystyle\int \frac{1}{x^4 + 1}\,dx$

(*Hint:* $x^4 + 1 = (x^2 + \sqrt{2}x + 1)(x^2 - \sqrt{2}x + 1)$.)

25. $\displaystyle\int \frac{x}{(x+1)^2(x-2)}\,dx$ **26.** $\displaystyle\int \frac{1}{(x-1)(x^2+1)^2}\,dx$

27. $\displaystyle\int \frac{-x^3 + x^2 + x + 3}{(x+1)(x^2+1)^2}\,dx$ **28.** $\displaystyle\int \frac{1}{(x^2+1)(x-2)}\,dx$

29. $\displaystyle\int \frac{x^2 - 1}{x^3 + 3x + 4}\,dx$

(*Hint:* Find a root of the denominator.)

In Exercises 30–36 find the integral by means of the indicated substitution.

30. $\displaystyle\int \frac{\sqrt{x}+1}{x+1}\,dx;\ u = \sqrt{x}$

31. $\displaystyle\int \frac{1}{x\sqrt{x+1}}\,dx;\ u = \sqrt{x+1}$

32. $\displaystyle\int_0^{\pi/4} \tan^3 x\,dx;\ u = \tan x$

33. $\displaystyle\int \frac{\sqrt{x}}{1 + \sqrt[3]{x}}\,dx;\ u = \sqrt[6]{x}$

34. $\displaystyle\int \frac{1}{\sqrt[4]{x}\,(1 + \sqrt{x})}\,dx;\ u = \sqrt[4]{x}$

*35. $\displaystyle\int_{-5/3}^{-1} \sqrt{\frac{x+1}{x-1}}\,dx;\ u = \sqrt{\frac{x+1}{x-1}}$

*36. $\displaystyle\int_{-1/9}^{-1/2} \frac{1}{x}\left(\frac{x}{x+1}\right)^{1/3}\,dx;\ u = \left(\frac{x}{x+1}\right)^{1/3}$

In Exercises 37–43 evaluate the integral by first using substitution or integration by parts and then using partial fractions.

37. $\displaystyle\int \frac{\sin^2 x \cos x}{\sin^2 x + 1}\,dx$ **38.** $\displaystyle\int \frac{e^x}{1 - e^{2x}}\,dx$

39. $\displaystyle\int \frac{e^x}{1 - e^{3x}}\,dx$ **40.** $\displaystyle\int \frac{e^x}{e^{2x} + 3e^x + 2}\,dx$

41. $\displaystyle\int x \tan^{-1} x\,dx$ **42.** $\displaystyle\int x^3 \tan^{-1} x\,dx$

43. $\displaystyle\int \ln(x^2 + 1)\,dx$

44. Evaluate

$$\int \frac{x^2}{(x^2-4)^2}\,dx$$

$\left(\text{Hint: } \dfrac{x^2}{(x^2-4)^2} = \dfrac{1}{4}\left(\dfrac{1}{x-2} + \dfrac{1}{x+2}\right)^2.\right)$

45. Evaluate

$$\int \frac{\sqrt{x+4}}{x^2}\,dx.$$

(*Hint:* Let $u = \sqrt{x+4}$ and use Exercise 44.)

46. Let r be a real number different from 0 and -1. Evaluate

$$\int \frac{1}{x - x^{r+1}}\,dx$$

by first substituting $u = x^r$ and then using partial fractions.

47. Use the result of Exercise 46 to evaluate

$$\int \frac{1}{x - x^{11}}\,dx$$

48. The reduction formula

$$\int \frac{1}{(x^2 + a^2)^{n+1}}\,dx = \frac{x}{2na^2(x^2 + a^2)^n}$$
$$+ \frac{2n-1}{2na^2} \int \frac{1}{(x^2 + a^2)^n}\,dx$$

which appeared in Exercise 49 of Section 8.1, helps in the evaluation of certain rational functions. Use partial fractions, along with the above formula, to evaluate the integral

$$\int \frac{x^2 - x + 1}{x(x^2 + 1)^2}\,dx$$

In Exercises 49–50 find the area A of the region between the graph of f and the x axis on the given interval.

49. $f(x) = \dfrac{x^3}{x^2 + 1}$; $[0, 3]$

50. $f(x) = \dfrac{4x^2}{(x^2 - 1)(x^2 + 1)}$; $\left[0, \dfrac{\sqrt{3}}{3}\right]$

51. Find the area A of the region between the graphs of

$$y = \frac{x^2}{(x-2)(x^2+1)} \quad \text{and} \quad y = \frac{1}{x-3}$$

52. Let $f(x) = \dfrac{1}{2x^3 - x^4}$, and let R be the region between the graph of f and the x axis on $[1/2, 1]$. Find the volume V of the solid generated by revolving R about the y axis.

53. Let $f(y) = \dfrac{1}{(y + 2)^2}$, and $g(y) = \dfrac{1}{y + 2}$, and let R be the region between the graphs of f and g on $[0, 2]$. Find the volume V of the solid obtained by revolving R about the x axis.

54. Let $f(x) = \dfrac{x}{x + 2}$, and let R be the region between the graph of f and the x axis on $[0, 2]$.

    a. Find the center of gravity of R.

    b. Use (a) and the Theorem of Pappus and Guldin to determine the volume V of the solid generated by revolving R about the x axis.

55. Let $g(x) = \ln (x^2 - 1)$ for $2 \le x \le 5$. Find the length L of the graph of f.

56. Find the length L of the curve described parametrically by $x = \sin^{-1} t$ and $y = \frac{1}{2} \ln (1 - t^2)$ for $0 \le t \le \frac{1}{2}$.

Applications

57. A disease is running through the town of Lavender. Suppose that y denotes the proportion of the population consisting of those people who are infectious, and $1 - y$ is the proportion of the population that is susceptible. If there is free movement between those who are infectious and those who are susceptible, then the number of contacts between the infectious and the susceptible is proportional to $y(1 - y)$, so that the rate of spread of the disease is given by

$$\frac{dy}{dt} = ay(1 - y)$$

where a is a positive constant. This differential equation has the general solution

$$t = \int \frac{1}{ay(1 - y)}\, dy$$

Use partial fractions to evaluate the integral.

58. The equation

$$\frac{dy}{dt} = r(a - y)(b - y)$$

which appeared in (10), describes the rate at which a bimolecular chemical reaction may occur. It follows that

$$t = \int \frac{1}{r(a - y)(b - y)}\, dy \qquad (16)$$

    a. Assuming that $a \neq b$, use partial fractions to evaluate the integral in (16). Show that

$$\left| \frac{y - a}{y - b} \right| = C_1 e^{(a - b)rt}$$

where C_1 is a constant.

    b. Assume that $b < a$. If the values of y are to lie in (b, a), let $C = C_1$; likewise, if the values of y are to lie in $(-\infty, b)$ or (a, ∞), let $C = -C_1$. Under these conditions, conclude that

$$y = \frac{a + bC\, e^{(a - b)rt}}{1 + C\, e^{(a - b)rt}}$$

59. The differential equation

$$\frac{dy}{dt} = r(a - y)(b - y)(c - y)$$

where a, b, c, and r are positive constants, arises in the study of thermonuclear reactions. Assume that a, b, and c are distinct. Then the general solution of the differential equation is

$$t = \int \frac{1}{r(a - y)(b - y)(c - y)}\, dy$$

    a. Use partial fractions to integrate the right-hand side, and then show that

$$D\, e^{(a - c)(b - a)(c - b)rt} = (a - y)^{c - b}\, (b - y)^{a - c}(c - y)^{b - a}$$

where D is a constant.

    b. Suppose that $y = 0$ when $t = 0$. Evaluate the corresponding constant D.

Project

1. The substitution $u = \tan (x/2)$, which is equivalent to the substitution $x = 2 \tan^{-1} u$, may appear to be a strange substitution, but the results from using it can be very good. The project involves this substitution and its consequences.

a. Show that $dx = \dfrac{2}{1 + u^2}\, du$.

b. Use an appropriately labeled triangle to show that

$$\sin (x/2) = \frac{u}{\sqrt{1 + u^2}} \quad \text{and} \quad \cos (x/2) = \frac{1}{\sqrt{1 + u^2}}$$

c. Show that $\sin x = \dfrac{2u}{1 + u^2}$ and $\cos x = \dfrac{1 - u^2}{1 + u^2}$.

d. Use the information above to find $\displaystyle\int \frac{1}{2 + \sin x}\, dx$.

e. Find $\displaystyle\int \frac{1}{3 - \cos x}\, dx$.

8.5   SYMBOLIC INTEGRATION

In recent years several computer programs, called computer algebra systems, or CAS, have been developed that perform symbolic (that is, indefinite) integration of a large collection of functions. Among the more prominent programs are Derive, Maple, Mathematica, and MATLAB. These programs generally are capable of finding all the integrals that we have evaluated in this book, and many more.

Once you understand the format of a particular program, integration can be relatively effortless. Below we present several integrals, along with the responses given by Mathematica.

Integral	Mathematica response
1. $\displaystyle\int_0^3 x^2\sqrt{9 - x^2}\, dx$	$\dfrac{81\ \text{Pi}}{16}$
2. $\displaystyle\int x^2\sqrt{1 - 9x^2}\, dx$	$\text{Sqrt}\,[1 - 9x^2]\left(\dfrac{-x}{72} + \dfrac{x^3}{4}\right) + \dfrac{\text{ArcSin}\,[3x]}{216}$
3. $\displaystyle\int e^{-2x} \cos 3x\, dx$	$\dfrac{-2\,\text{Cos}\,[3x]}{13E^{2x}} + \dfrac{3\,\text{Sin}\,[3x]}{13\ E^{2x}}$
4. $\displaystyle\int \dfrac{x^3 + 2x^2 + x + 1}{2x^3 - x^2 - 5x - 2}\, dx$	$\dfrac{x}{2} + \dfrac{19\,\text{Log}\,[-2 + x]}{15} + \dfrac{\text{Log}\,[1 + x]}{3} - \dfrac{7\,\text{Log}\,[1 + 2x]}{20}$
5. $\displaystyle\int \dfrac{1}{x^6 + 1}\, dx$	$(-2\,\text{ArcTan}\,[\text{Sqrt}[3] - 2x] + 4\,\text{ArcTan}\,[x] +$ $2\,\text{ArcTan}\,[\text{Sqrt}[3] + 2x] - \text{Sqrt}[3]\,\text{Log}\,[1 - \text{Sqrt}[3]x + x^2]$ $+ \text{Sqrt}[3]\,\text{Log}[1 + \text{Sqrt}[3]x + x^2])/12$
6. $\displaystyle\int \dfrac{x^8}{a^2 + x^2}\, dx$	$-(a^6x) + \dfrac{a^4x^3}{3} - \dfrac{a^2x^5}{5} + \dfrac{x^7}{7} + a^7\,\text{ArcTan}\left[\dfrac{x}{a}\right]$
7. $\displaystyle\int \sin (x^x)\, dx$	$\text{Integrate}\,[\text{Sin}[x^x], x]$

A few comments are in order. First, the nomenclature is a little different from that of our text: Pi denotes π, Sqrt denotes the square root, ArcSin and ArcTan denote $\sin^{-1}$ and $\tan^{-1}$, E denotes the natural exponential, and Log denotes the natural logarithm. Except for the omission of the additive constant C, the answers given by Mathematica for integrals 1 to 3 are in agreement with the answers derived in Examples 1 to 3. For integrals 4 to 5 Mathematica pulls no surprises; it can perform much more complicated partial fractions than we would want to do by hand. Integral 6 involves an integrand with the parameter a. Finally, for integral 7 there is no answer using elementary functions; after trying to find such an answer and failing in its mission, Mathematica responds with the original integral.

EXERCISES 8.5

In Exercises 1–18 use a table of integrals or a computer algebra system to evaluate the given integral.

1. $\displaystyle\int \sqrt{x^2 + 9}\, dx$

2. $\displaystyle\int \frac{1}{(16 - x^2)^{3/2}}\, dx$

3. $\displaystyle\int_0^1 e^{5x} \sin \frac{1}{2} x\, dx$

4. $\displaystyle\int_1^2 \frac{1}{x(3x - 2)}\, dx$

5. $\displaystyle\int \frac{1}{4x^2 - 9}\, dx$

6. $\displaystyle\int \frac{1}{x\sqrt{2x - 4x^2}}\, dx$

7. $\displaystyle\int \frac{\sqrt{10x - \frac{1}{4}x^2}}{x}\, dx$

8. $\displaystyle\int \frac{\sqrt{\frac{1}{9}x^2 - 5}}{x^2}\, dx$

9. $\displaystyle\int \frac{e^{\sqrt{x}} \sinh 2\sqrt{x}}{\sqrt{x}}\, dx$

10. $\displaystyle\int_{\pi/6}^{\pi/2} \frac{\cos x}{\sin^2 x} \sqrt{1 + \sin^2 x}\, dx$

11. $\displaystyle\int_0^1 e^{2x} \cos e^x\, dx$

12. $\displaystyle\int \frac{x^3}{3 - 2x^2}\, dx$

13. $\displaystyle\int \sin(\ln x)\, dx$

14. $\displaystyle\int \frac{\sqrt{x}}{(2 - 3\sqrt{x})^2}\, dx$

15. $\displaystyle\int \sqrt{2\sqrt{x} - x}\, dx$

16. $\displaystyle\int \frac{1}{x \ln x\sqrt{-4 + \ln x}}\, dx$

17. $\displaystyle\int x\sqrt{\frac{2 + x}{2 - x}}\, dx$

18. $\displaystyle\int \sqrt{\frac{x}{x^3 + 1}}\, dx$

In Exercises 19–22 use a computer algebra system to determine the computer's response to the integral.

19. $\displaystyle\int x^4\sqrt{a^2 + x^2}\, dx$

20. $\displaystyle\int \frac{x^2 + a^2}{x^4 + a^4}\, dx$

21. $\displaystyle\int \sqrt{x^3 + 1}\, dx$

22. $\displaystyle\int \sin x^3\, dx$

8.6   THE TRAPEZOIDAL RULE AND SIMPSON'S RULE

Despite the various methods that exist for calculating integrals—those we have described as well as many we have not—it is frequently difficult or impossible to express an indefinite integral in terms of familiar functions and thereby compute the exact numerical values of corresponding definite integrals. For example, we are unable to evaluate exactly the integral

$$\int_0^1 \sqrt{1 + x^4}\, dx \tag{1}$$

Such an integral turns up in the analysis of highways and railroad tracks (see Example 3 in Section 6.2). Although it is impossible to find the exact numerical value of (1), there are numerical methods for approximating its value. One such method, approximation by Riemann sums (including left sums, right sums, and midpoint sums), was already discussed in Section 5.1.

When we approximate the integral $\int_a^b f(x)\, dx$ of a nonnegative function f by a Riemann sum, we are, in effect, replacing the segment of the graph of f on each

FIGURE 8.8   Riemann sum: Horizontal lines replace portions of the graph.

subinterval by a suitable horizontal line (Figure 8.8); the Riemann sum then equals the sum of the areas of the corresponding rectangles. If we allow the segments of the graph of f to be replaced instead by nonhorizontal lines or parabolas, then with nearly the same effort, we can generally approximate $\int_a^b f(x)\,dx$ with much greater precision than with Riemann sums. It is to these more refined types of approximation that we now turn. For purposes of exposition, we will assume that f is continuous and nonnegative throughout $[a, b]$. Despite this restriction, the formulas we will derive remain valid for functions that are not necessarily nonnegative.

The Trapezoidal Rule

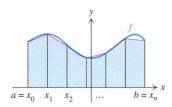

FIGURE 8.9   Trapezoidal Rule: Possibly nonhorizontal lines replace portions of the graph.

To describe the first type of approximation, we take a partition $P = \{x_0, x_1, \ldots, x_n\}$ that divides the interval $[a, b]$ into n subintervals of equal length $(b - a)/n$. We join each pair of points $(x_{k-1}, f(x_{k-1}))$ and $(x_k, f(x_k))$ on the graph of f by a straight line, thereby creating trapezoidal regions (Figure 8.9). Since we assumed that f is nonnegative, the formula for the area of the kth trapezoid is

$$\left(\frac{f(x_{k-1}) + f(x_k)}{2}\right)(x_k - x_{k-1}) = \frac{b - a}{2n}\left[f(x_{k-1}) + f(x_k)\right]$$

(see Exercise 66 of Section 5.5). Using this formula for each of the subintervals, we obtain the sum

$$\frac{b - a}{2n}\{[f(x_0) + f(x_1)] + [f(x_1) + f(x_2)] + \cdots + [f(x_{n-1}) + f(x_n)]\}$$

as an approximation of $\int_a^b f(x)\,dx$. In other words,

$$\int_a^b f(x)\,dx \approx \frac{b - a}{2n}\big[f(x_0) + 2f(x_1) + 2f(x_2) + \cdots$$
$$+ 2f(x_{n-1}) + f(x_n)\big]$$

(2)

Because of the way it was derived, (2) is called the **Trapezoidal Rule.**

EXAMPLE 1   Use the Trapezoidal Rule with $n = 10$ to compute an approximate value of $\int_1^2 1/x\,dx$.

Solution   Using the formula in (2) with $f(x) = 1/x$, $a = 1$, $b = 2$, and $n = 10$, we find that

$$\int_1^2 \frac{1}{x}\,dx \approx \frac{1}{20}\left[f(1) + 2f\left(\frac{11}{10}\right) + 2f\left(\frac{12}{10}\right) + \cdots + 2f\left(\frac{19}{10}\right) + f(2)\right]$$

$$= \frac{1}{20}\left(1 + \frac{20}{11} + \frac{20}{12} + \cdots + \frac{20}{19} + \frac{1}{2}\right)$$

$$\approx 0.6937714032 \quad \square$$

The actual value of $\int_1^2 1/x\,dx$ is $\ln 2$, which equals 0.6931471806 (rounded to 10 digits).

The value given by the Trapezoidal Rule approximates, but normally is not equal to, the value of the corresponding definite integral. In order to measure the difference between the value of a definite integral and the approximation by the Trapezoidal Rule, we define the ***n*th Trapezoidal Rule error** E_n^T by

$$E_n^T = \left| \int_a^b f(x)\, dx - \frac{b-a}{2n} [f(x_0) + 2f(x_1) + \cdots + 2f(x_{n-1}) + f(x_n)] \right|$$

Using the approximate value 0.6931471806 for ln 2, we find from Example 1 that

$$E_{10}^T \approx |\, 0.6931471806 - 0.6937714032\,| = 0.0006242226 = 6.242226 \times 10^{-4}.$$

There is an upper bound for E_n^T if f has a continuous second derivative on $[a, b]$. Let

$$K_T = \text{the maximum of } |f''(x)| \quad \text{for } a \le x \le b$$

Then it can be proved that the error E_n^T incurred in using the Trapezoidal Rule to approximate $\int_a^b f(x)\, dx$ satisfies

$$E_n^T \le \frac{K_T}{12n^2} (b-a)^3 \tag{3}$$

Notice that as n increases, the right side of (3) shrinks in proportion to $1/n^2$.

EXAMPLE 2   Determine a value of n that guarantees an error of no more than 10^{-3} in the approximation by the Trapezoidal Rule of $\ln 2 = \int_1^2 1/x\, dx$.

Solution   To use (3), we let $a = 1$, $b = 2$, and $f(x) = 1/x$. We find that $f''(x) = 2/x^3$, so that

$$|f''(x)| = \left| \frac{2}{x^3} \right| \le 2 \quad \text{for } 1 \le x \le 2$$

Thus $K_T = 2$, so by (3) we conclude that

$$E_n^T \le \frac{2(2-1)^3}{12n^2} = \frac{2}{12n^2} = \frac{1}{6n^2}$$

To ensure that $E_n^T \le 10^{-3}$, we choose n so that $1/(6n^2) \le 10^{-3}$, that is, $n^2 \ge 10^3/6 = 166\frac{2}{3}$. Therefore if $n \ge 13$, then $E_n^T \le 10^{-3}$.   ❑

We emphasize that the Trapezoidal Rule error E_n^T is rarely exactly equal to the right side of (3), but rather is bounded above by it. Also if K_T cannot be easily computed, then in (3) we can substitute any larger number for K_T.

For Example 2 we found that $E_{13}^T \le 10^{-3}$. Thus if $T = $ the value of the Trapezoidal Rule for $n = 13$, then

$$\left| \int_1^2 \frac{1}{x}\, dx - T \right| \le 10^{-3}$$

or equivalently,

$$T - 10^{-3} < \int_1^2 \frac{1}{x}\, dx < T + 10^{-3} \tag{4}$$

But by the Trapezoidal Rule applied to $\int_1^2 \frac{1}{x}\,dx$ with $n = 13$,

$$T = \frac{1}{26}\left[1 + 2\left(\frac{13}{14}\right) + 2\left(\frac{13}{15}\right) + \cdots + 2\left(\frac{13}{25}\right) + \frac{1}{2}\right] \approx 0.6935167303$$

(accurate to 10 places). Consequently $\int_1^2 1/x\ dx$ must be within 10^{-3} of 0.6935167303.

Simpson's Rule

Next we substitute parabolas for the straight lines used in the Trapezoidal Rule (Figure 8.10). We begin by taking a partition $P = \{x_0, x_1, \ldots, x_n\}$, but now in addition to assuming that the partition divides the interval $[a, b]$ into n subintervals of equal length

$$h = \frac{b - a}{n} \tag{5}$$

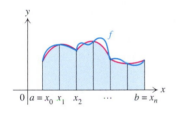

FIGURE 8.10

we will assume that n is an *even* number. You can verify that the parabola passing through three consecutive points $(x_0, f(x_0))$, $(x_1, f(x_1))$, and $(x_2, f(x_2))$ on the graph of f is defined by

$$p(x) = f(x_0) + \frac{f(x_1) - f(x_0)}{h}(x - x_0)$$

$$+ \frac{f(x_0) - 2f(x_1) + f(x_2)}{2h^2}(x - x_0)(x - x_1) \tag{6}$$

Since p is only a second-degree polynomial in the variable x, it is of course possible to find

$$\int_{x_0}^{x_2} p(x)\,dx$$

Using (6), we conclude that

$$\int_{x_0}^{x_2} p(x)\,dx = \frac{h}{3}[f(x_0) + 4f(x_1) + f(x_2)]$$

(see Exercise 23). Since f is by assumption nonnegative, the integral just evaluated represents the area of the region between the graph of p and the x axis on $[x_0, x_2]$.

The same is done for the subintervals $[x_2, x_4]$, $[x_4, x_6]$, $\ldots$, $[x_{n-2}, x_n]$; each of the indexes 2, 4, 6, $\ldots$, $n - 2$, n must be even, which explains why we required n to be even. The sum of the areas under the parabolas so obtained serves as an approximation to $\int_a^b f(x)\,dx$. That sum can be shown to be the right-hand member of the formula

$$\int_a^b f(x)\,dx \approx \frac{b - a}{3n}[f(x_0) + 4f(x_1) + 2f(x_2) + 4f(x_3) + \cdots$$
$$+ 2f(x_{n-2}) + 4f(x_{n-1}) + f(x_n)] \tag{7}$$

This formula is widely known as **Simpson's Rule,** after the Englishman Thomas Simpson. We emphasize that it holds for all functions that are continuous on $[a, b]$, whether or not they are nonnegative. The beauty of the formula is that it is essentially as easy to use as the Trapezoidal Rule or any Riemann sum, and yet it generally produces far more accurate estimates than do the other kinds of approximation presented. In fact, variants of Simpson's Rule are actually used by computers to approximate definite integrals.

To illustrate the accuracy of Simpson's Rule, we return to the approximation of $\int_1^2 1/x \, dx$.

EXAMPLE 3   Compute an approximate value for $\int_1^2 1/x \, dx$ by using Simpson's Rule with $n = 10$.

Solution   By Simpson's Rule,

$$\int_1^2 \frac{1}{x} \, dx \approx \frac{1}{3(10)} \left[f(1) + 4f\left(\frac{11}{10}\right) + 2f\left(\frac{12}{10}\right) + 4f\left(\frac{13}{10}\right) + \right.$$

$$\left. \cdots + 2f\left(\frac{18}{10}\right) + 4f\left(\frac{19}{10}\right) + f(2) \right]$$

$$= \frac{1}{30}\left(1 + \frac{40}{11} + \frac{20}{12} + \frac{40}{13} + \frac{20}{14} + \frac{40}{15} + \frac{20}{16} + \frac{40}{17} + \frac{20}{18} + \frac{40}{19} + \frac{1}{2}\right)$$

$$\approx 0.6931502307 \quad \square$$

Since $\ln 2 \approx 0.6931471806$, we find that the estimate of $\int_1^2 1/x \, dx$ given by Simpson's Rule is within $0.00001 = 10^{-5}$ of the actual value. Notice that the estimate is much more accurate than the one obtained in Example 1 by the Trapezoidal Rule.

The ***n*th Simpson's Rule error** E_n^S incurred in using Simpson's Rule to approximate the integral is defined by

$$E_n^S = \left| \int_a^b f(x) \, dx - \frac{b - a}{3n} [f(x_0) + 4f(x_1) + 2f(x_2) + \cdots + 4f(x_{n-1}) + f(x_n)] \right|$$

Using the approximate value 0.6931471806 for ln 2, along with the result of Example 3, we find that

$$E_{10}^S \approx |\, 0.6931471806 - 0.6931502307 \,| = 3.0501 \times 10^{-6}$$

Recalling that $E_{10}^T \approx 6.242226 \times 10^{-4}$, we see that for $n = 10$, Simpson's Rule gives a much more precise value for $\int_1^2 1/x \, dx$ than does the Trapezoidal Rule.

An upper bound for E_n^S is obtained from the fourth derivative of f on $[a, b]$. Let us assume that f has a continuous fourth derivative on $[a, b]$, and let

$$K_S = \text{the maximum of } |f^{(4)}(x)| \quad \text{for } a \leq x \leq b$$

Then the error E_n^S incurred in using Simpson's Rule to approximate $\int_a^b f(x) \, dx$ satisfies

$$E_n^S \leq \frac{K_S}{180n^4} (b - a)^5 \qquad (8)$$

As n increases, the right side of E_n^S in (8) shrinks in proportion to $1/n^4$. As a result, approximation by Simpson's Rule becomes very accurate as n becomes large.

EXAMPLE 4   Determine a value of n that guarantees an error of no more than 10^{-6} in the approximation by Simpson's Rule of $\int_1^2 1/x\, dx$.

Solution   We use (8) with $a = 1$, $b = 2$, and $f(x) = 1/x$. We find that $f^{(4)}(x) = 24/x^5$, so that

$$|f^{(4)}(x)| = \left|\frac{24}{x^5}\right| \le 24 \quad \text{for } 1 \le x \le 2$$

Thus $K_S \le 24$, so by (8) we conclude that

$$E_n^S \le \frac{24(2-1)^5}{180n^4} = \frac{24}{180n^4} = \frac{2}{15n^4}$$

To be certain that $E_n^S \le 10^{-6}$, we choose n so that

$$\frac{2}{15n^4} \le 10^{-6}, \quad \text{that is,} \quad n \ge \left(\frac{2}{15} \times 10^6\right)^{1/4} \approx 19.10885584$$

Thus it suffices to let $n = 20$ in order that the Simpson Rule error bound for $\int_1^2 1/x\, dx$ is less than 10^{-6}.   ❑

From Example 4 we see that with $n = 20$ we can guarantee an approximation of $\int_1^2 1/x\, dx$ to within 10^{-6}. However, in order to achieve this guarantee we needed to be able to find an upper bound of $|f^{(4)}(x)|$ for all x in $[1, 2]$.

EXAMPLE 5   Compute an approximate value of $\int_0^1 \sqrt{1 + x^4}\, dx$ by using Simpson's Rule with $n = 10$.

Solution   Simpson's Rule yields

$$\int_0^1 \sqrt{1 + x^4}\, dx \approx \frac{1}{30}\left[f(0) + 4f(0.1) + 2f(0.2) + \cdots + 4f(0.9) + f(1)\right]$$

$$\approx 1.089429384 \quad ❑$$

Finding an upper bound for E_n^S in Example 5 involves first calculating $f^{(4)}$. By means of a computer algebra system we obtain

$$f^{(4)}(x) = \frac{60x^8 - 168x^4 + 12}{(1 + x^4)^{7/2}}$$

Then it is necessary to find a reasonable upper bound for $|f^{(4)}(x)|$ for x in $[0, 1]$. This is not so simple. However, by utilizing a computer algebra system such as Mathematica, MATLAB, Derive, or Maple, we find that the maximum value K_S of $|f^{(4)}(x)|$ on $[0, 1]$ satisfies

$$14 < K_S < 15$$

Therefore

$$E_{10}^S \le \frac{15}{180(10^4)}(1-0)^5 < 10^{-5}$$

Along with the result of Example 5, this means that the value of $\int_0^1 \sqrt{1+x^4}\,dx$ must be within 10^{-5} of 1.089429384.

As we conclude this section, we present the error estimates of the Trapezoidal Rule and Simpson's Rule alongside each other:

$$E_n^T \le \frac{K_T}{12n^2}(b-a)^3 \quad \text{and} \quad E_n^S \le \frac{K_S}{180n^4}(b-a)^5$$

where K_T and K_S are the maximum values of $|f^{(2)}(x)|$ and $|f^{(4)}(x)|$ for x in $[a, b]$, respectively. In order to get an idea of the speed with which the error bounds shrink as n increases, we notice that if $m = 10n$, then

$$\frac{K_T}{12\,m^2}(b-a)^3 = \frac{K_T}{12(10n)^2}(b-a)^3 = \frac{K_T}{100(12n^2)}(b-a)^3$$

Thus when n is increased by a factor of 10,

the error bound E_n^T is reduced by a factor of 100

Similarly,

the error bound E_n^S is reduced by a factor of 10,000

From the observations of the preceding two paragraphs, you can see why Simpson's Rule is so often used when approximating integrals whose exact values cannot be determined.

EXERCISES 8.6

In Exercises 1–4 approximate the given integral by each of the Trapezoidal and Simpson's Rules, using the indicated number of subintervals.

1. $\int_1^3 \frac{1}{x}\,dx; n = 6$ **2.** $\int_0^1 \sin(x^2)\,dx; n = 10$

3. $\int_0^1 e^{-x^2}\,dx; n = 10$ **4.** $\int_0^1 e^{-x^2}\,dx; n = 40$

In Exercises 5–8 use either (3) or (8) to find an upper bound for the error in approximating the given integral by the indicated method and number of subintervals. In some cases it may be convenient to use a reasonable upper bound for K_T or K_S, rather than the exact value.

5. $\int_1^3 \frac{1}{x}\,dx;$ Trapezoidal Rule; $n = 10$

6. $\int_0^1 \sqrt{1+x^2}\,dx;$ Trapezoidal Rule; $n = 10$

7. $\int_{-1}^2 \frac{1}{3+x}\,dx;$ Simpson's Rule; $n = 10$

8. $\int_1^4 x \ln x\,dx;$ Simpson's Rule; $n = 20$

In Exercises 9–10 use a computer algebra system, along with (8), to find an upper bound for the error in approximating the given integral by Simpson's Rule with $n = 10$.

9. $\int_1^2 \sqrt{1+x^2}\,dx$ **10.** $\int_2^4 \frac{1}{1+x^3}\,dx$

In Exercises 11–12 approximate the area A of the region between the graph of f and the x axis on the given interval by using Simpson's Rule with $n = 10$.

11. $f(x) = \dfrac{1}{1 + \cos x}$; $[-\pi/3, \pi/3]$

12. $f(x) = \dfrac{\pi \cos x}{x}$; $[\pi/2, 3\pi/2]$

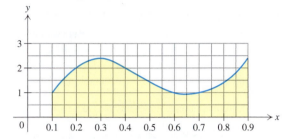

FIGURE 8.11   Graph for Exercise 13.

13. Approximate the area A of the region shown in Figure 8.11, by using Simpson's Rule with
a. $n = 4$ **b.** $n = 8$

 14. a. Use Simpson's Rule with $n = 100$ to approximate
$$\int_{1}^{b} \frac{1}{x} \sqrt{1 + 1/x^4} \, dx$$
for the given values of b.
i. 10   **ii.** 100   **iii.** 1000
b. Does it appear as though the values of the integral approach a limit as b increases without bound? Explain.

15. It is known that $\pi = \int_{0}^{1} 4/(1 + x^2) \, dx$.
a. Approximate π by using
i. the Trapezoidal Rule with $n = 10$.
ii. Simpson's Rule with $n = 10$.
b. Using the fact that $\pi \approx 3.141592654$, estimate the errors that arise from the approximations of π in (a).

16. Approximate $\int_{0}^{\pi/2} \sqrt{1 + \cos x} \, dx$ by using Simpson's Rule with $n = 2$. Calculate the exact value of the integral by using the relation
$$\sqrt{1 + \cos x} = \sqrt{2} \cos \frac{x}{2}$$
and estimate the error that arises from the approximation.

17. Find the smallest positive integer n that is guaranteed by (8) to yield an error of no more than 10^{-4} in the approximation of $\ln 8$ by Simpson's Rule by considering $\ln 8$ as
a. $\int_{1}^{8} \frac{1}{x} \, dx$ **b.** $3 \int_{1}^{2} \frac{1}{x} \, dx$
(Thus making algebraic changes that shorten the interval of integration can drastically reduce the computations required in using Simpson's Rule to approximate definite integrals.)

18. Prove that the Trapezoidal Rule gives the exact value for $\int_{a}^{b} f(x) \, dx$ if f is linear. (*Hint:* Use (3).)

19. a. Prove that Simpson's Rule gives the exact value for $\int_{a}^{b} f(x) \, dx$ if f is any polynomial of degree at most 3. (*Hint:* Use (8).)
b. Let $f(x) = x^3 - 2x^2 + 3x - 1$ for $-2 \le x \le 1$. By letting $n = 10$, corroborate the statement that Simpson's Rule gives the exact value of $\int_{-2}^{1} g(x) \, dx$ whenever g is a polynomial of degree less than or equal to 3.

20. Let $f(x) = x^4$ for $0 \le x \le 1$, and let $n = 2$. Show that
$$E_n^S = \frac{K_s (b - a)^5}{180n^4}$$
so that the error bound in (8) can actually be achieved. (A similar remark is valid for (3), where the function $f(x) = x^2$ for $0 \le x \le 1$ suffices.)

21. Assume that $f(x) \ge 0$ for $a \le x \le b$, and let the Trapezoidal Rule give the value T_n for the approximation of $\int_{a}^{b} f(x) \, dx$.
a. Show that if the graph of f is concave upward on $[a, b]$, then
$$T_n \ge \int_{a}^{b} f(x) \, dx$$
(*Hint:* Draw a picture, and note the relationship of the trapezoids to the region under the graph of f.)
b. Suppose that the graph of f is concave upward on (a, b). Will the Trapezoidal Rule approximation increase or decrease if n is doubled?
c. Would your answer change if the graph of f is concave downward? Explain why or why not.

22. Prove that the sum appearing in the Trapezoidal Rule for $\int_{a}^{b} f(x) \, dx$ is a genuine Riemann sum for f on $[a, b]$. (*Hint:* Use the Intermediate Value Theorem on each subinterval $[x_{k-1}, x_k]$ associated with the partition $P = \{x_0, x_1, \ldots, x_n\}$ to obtain a number c_k in $[x_{k-1}, x_k]$ such that
$$\frac{f(x_{k-1}) + f(x_k)}{2} = f(c_k)$$

*23. Let p be the second-degree polynomial defined by (6). Show that
$$\int_{x_0}^{x_2} p(x) \, dx = \frac{h}{3} [f(x_0) + 4f(x_1) + f(x_2)]$$

Applications

24. A car travels for 2 hours on an interstate highway without stopping. The driver records the car's speed in miles per hour every quarter hour, as indicated in the following table:

Time	0	$\frac{1}{4}$	$\frac{1}{2}$	$\frac{3}{4}$	1	$\frac{5}{4}$	$\frac{3}{2}$	$\frac{7}{4}$	2
Velocity	20	50	56	54	52	49	54	52	30

Express the distance the car traveled as a definite integral and use Simpson's Rule to approximate the distance.

25. On a windy day the following wind speeds in miles per hour are recorded in 2-hour intervals:

Time	8 a.m.	10 a.m.	noon	2 p.m.	4 p.m.	6 p.m.	8 p.m.
Speed	20	24	18	28	21	15	10

Use Simpson's Rule with an appropriate value of n to approximate the mean wind speed during the 12-hour period.

26. During a blizzard the following rates of snow accumulation (in inches per hour) were recorded:

Time	1 p.m.	2 p.m.	3 p.m.	4 p.m.	5 p.m.	6 p.m.	7 p.m.
Rate	0.3	1.3	2.7	3.1	2.9	1.5	0.2

   a. Use the Trapezoidal Rule with an appropriate n in order to approximate the total accumulation of snow.

   b. Use Simpson's Rule with an appropriate n in order to approximate the total accumulation of snow. How much does this answer differ from that in part (a)?

   c. Use (b) to determine the mean rate of accumulation during the 6-hour period.

27. Suppose the rate (in persons per day) at which an epidemic of influenza spreads through a city is as shown in Figure 8.12. Use Simpson's Rule to approximate the total number of flu cases in the city.

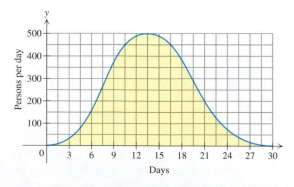

FIGURE 8.12   Graph for Exercise 27.

28. Use Simpson's Rule with $n = 50$ to approximate the length L of a bridge cable in the shape of the curve $y = 16 - \frac{1}{16} x^2$ for $-16 \le x \le 16$.

29. The equations

$$x = 2 \cos t \quad \text{and} \quad y = 3 \sin t \text{ for } \quad 0 \le t \le 2\pi$$

parametrize an ellipse. Use Simpson's Rule with $n = 30$ to estimate the circumference L of the ellipse.

30. An egg has the shape of the ellipse generated by revolving about the x axis the half ellipse parametrized by

$$x = 2 \cos t \quad \text{and} \quad y = 3 \sin t \text{ for } \quad 0 \le t \le \pi$$

Use Simpson's Rule with $n = 20$ to estimate the surface area S of the egg.

31. Use Simpson's Rule to approximate the area of the oil spill shown in Figure 8.13. The measurements were taken 1 mile apart.

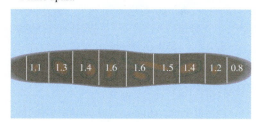

FIGURE 8.13   Graph for Exercise 31.

32. The state of Idaho is depicted in Figure 8.14. Use Simpson's Rule with $n = 6$ to approximate the area of Idaho. (The area of Idaho is in reality 82,412 square miles.)

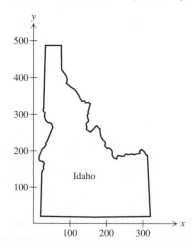

FIGURE 8.14   Graph for Exercise 32.

33. From sales records collected over several years the Java Coffee Corporation prepared the graph in Figure 8.15 for its marginal revenue. Use Simpson's Rule to approximate the revenue the corporation would receive from exporting 500,000 tons of coffee.

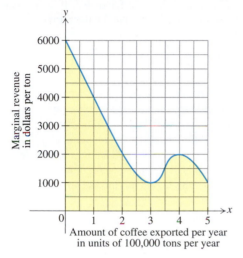

FIGURE 8.15 Graph for Exercise 33.

Project

1. Let f be continuous on $[a, b]$, and let $[a, b]$ be partitioned into n equal subintervals, where n is an *even* integer. This project focuses on relationships between the various approximations of $\int_a^b f(x)\, dx$ we have discussed:

L_n = left sum
M_n = midpoint sum
R_n = right sum
T_n = Trapezoidal Rule
S_n = Simpson's Rule

a. Show that $T_n = \frac{1}{2}(L_n + R_n)$. (Thus T_n is the average of the left and right sums.)

b. Show that $S_{2n} = \frac{1}{3} T_n + \frac{2}{3} M_n$.

c. Find positive constants u, v, and w such that $S_{2n} = uL_n + vM_n + wR_n$.

d. Suppose that f is increasing, and the graph of f is concave upward on $[a, b]$. Order L_n, M_n, R_n, and T_n from smallest to largest.

e. Suppose that $a > 0$, and that f is a continuous odd function on $[-a, a]$. Which of L_n, M_n, R_n, T_n, and S_n are necessarily equal to zero?

8.7 IMPROPER INTEGRALS

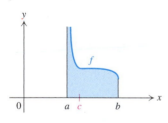

FIGURE 8.16

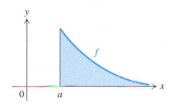

FIGURE 8.17

So far we have defined $\int_a^b f(x)\, dx$ only for f continuous on $[a, b]$. It follows from the Maximum-Minimum Theorem that such a function f is bounded on $[a, b]$ in the sense that for some number M, $|f(x)| \leq M$ for all x in $[a, b]$. More generally, if I is any interval, then we say that f is **bounded on I** if there is a constant M such that

$$|f(x)| \leq M \quad \text{for all } x \text{ in } I$$

A function not bounded on a given interval I inside its domain is said to be **unbounded on I.** In this section we will consider the geometric question of whether an area can be defined for the region under the graph of a nonnegative function that is unbounded on a bounded interval (Figure 8.16). We would also like to know whether it is possible to define the area of the region under the graph of a nonnegative function on an unbounded interval (Figure 8.17). Although it may seem that the areas of such regions must be infinite, we will find that in certain cases they should be defined to be finite. To accomplish this we will define the definite integral when either the integrand or the interval of integration is unbounded. Such integrals are called **improper integrals,** as opposed to integrals of continuous functions over bounded intervals, which are called **proper integrals.** (Of course, improper integrals are as proper an object of study as proper integrals, but this is the usual and traditional terminology.)

To simplify the discussion, we say that f is **unbounded near** c if f is unbounded either on every open interval of the form (c, x) or on every open interval of the form (x, c). For example, $1/x$ and $1/\sqrt{x}$ are unbounded near 0.

We now consider a function f that is continuous at every point in $(a, b]$ and unbounded near a (see Figure 8.16). By assumption f is continuous on the interval $[c, b]$ for any c in (a, b), so that $\int_c^b f(x)\,dx$ is defined for such c. If the one-sided limit $\lim_{c \to a^+} \int_c^b f(x)\,dx$ exists, then we define $\int_a^b f(x)\,dx$ by the formula

$$\int_a^b f(x)\,dx = \lim_{c \to a^+} \int_c^b f(x)\,dx$$

and say that the integral $\int_a^b f(x)\,dx$ **converges.** Otherwise we say that $\int_a^b f(x)\,dx$ **diverges** and do not assign any number to the integral. If f is nonnegative on $(a, b]$, then as c approaches a from the right, the integrals $\int_c^b f(x)\,dx$ represent areas of larger and larger regions (Figure 8.18(a) and (b)). In addition, if $\int_a^b f(x)\,dx$ converges, the **area** of the region between the graph of f and the x axis on $[a, b]$ is defined to be $\int_a^b f(x)\,dx$. By contrast, if f is nonnegative on $(a, b]$ and if $\int_a^b f(x)\,dx$ diverges, then we say that the corresponding region has infinite area and write

$$\int_a^b f(x)\,dx = \infty$$

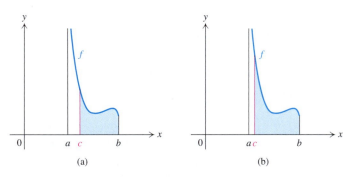

(a)   (b)

FIGURE 8.18

EXAMPLE 1   Show that $\displaystyle\int_0^1 \frac{1}{\sqrt{x}}\,dx$ converges and compute its value.

Solution   The integral is of the kind just discussed, because $1/\sqrt{x}$ is continuous at every point in $(0, 1]$ and is unbounded near 0. Thus

$$\int_0^1 \frac{1}{\sqrt{x}}\,dx = \lim_{c \to 0^+} \int_c^1 \frac{1}{\sqrt{x}}\,dx$$

provided that the limit exists. Now for $0 < c < 1$ we have

$$\int_c^1 \frac{1}{\sqrt{x}}\,dx = 2\sqrt{x}\,\Big|_c^1 = 2(1 - \sqrt{c})$$

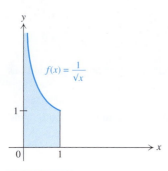

FIGURE 8.19

and

$$\lim_{c \to 0^+} 2(1 - \sqrt{c}) = 2$$

Consequently $\int_0^1 1/\sqrt{x}\, dx$ converges, and

$$\int_0^1 \frac{1}{\sqrt{x}}\, dx = 2$$

This means that the region shaded in Figure 8.19 has area 2. ❑

EXAMPLE 2   Show that $\displaystyle\int_0^1 \frac{1}{x}\, dx$ diverges.

Solution   The function $1/x$ is continuous on $(0, 1]$ and unbounded near 0. For $0 < c < 1$ we have

$$\int_c^1 \frac{1}{x}\, dx = \ln x \Big|_c^1 = \ln 1 - \ln c = -\ln c$$

But

$$\lim_{c \to 0^+} (-\ln c) = \infty$$

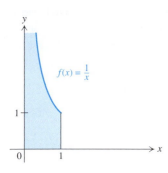

FIGURE 8.20

Therefore the integral $\int_0^1 1/x\, dx$ diverges, which means that the region shaded in Figure 8.20 has infinite area. ❑

If f is continuous at every point of $[a, b)$ and is unbounded near b, the integral $\int_c^a f(x)\, dx$ is defined for all c in (a, b). If

$$\lim_{c \to b^-} \int_a^c f(x)\, dx$$

exists, then we define $\int_a^b f(x)\, dx$ to be that limit and say that the integral $\int_a^b f(x)\, dx$ **converges.** Otherwise, we say that $\int_a^b f(x)\, dx$ **diverges.** If f is nonnegative, then $\int_a^b f(x)\, dx$ represents the area of the corresponding region and is finite or infinite depending on the convergence or divergence of $\int_a^b f(x)\, dx$.

EXAMPLE 3   Determine whether $\displaystyle\int_0^{\pi/2} \tan x\, dx$ converges.

Solution   The integrand is continuous on $[0, \pi/2)$ and is unbounded near $\pi/2$. For $0 < c < \pi/2$ we have

$$\int_0^c \tan x\, dx = -\ln (\cos x) \Big|_0^c = -\ln (\cos c) + \ln (\cos 0) = -\ln (\cos c)$$

Since

$$\lim_{c \to \pi/2^-} \cos c = 0$$

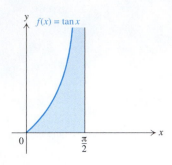

FIGURE 8.21

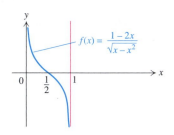

FIGURE 8.22

it follows that

$$\lim_{c \to \pi/2^-} \left[-\ln (\cos c) \right] = \infty$$

Consequently the integral $\int_0^{\pi/2} \tan x \, dx$ diverges, and the corresponding region has infinite area (Figure 8.21). □

If f is continuous at every point of (a, b) and is unbounded near both a and b, we say that $\int_a^b f(x) \, dx$ **converges** if for some point d in (a, b) both the integrals $\int_a^d f(x) \, dx$ and $\int_d^b f(x) \, dx$ converge. Otherwise, we say that $\int_a^b f(x) \, dx$ **diverges.**

EXAMPLE 4   Determine whether $\displaystyle\int_0^1 \frac{1 - 2x}{\sqrt{x - x^2}} \, dx$ converges.

Solution   The integrand is unbounded near both the endpoints 0 and 1 and is continuous on $(0, 1)$ (Figure 8.22). Consequently the integral is of the type under consideration. If we let $d = \frac{1}{2}$, then we need to analyze the convergence of

$$\int_0^{1/2} \frac{1 - 2x}{\sqrt{x - x^2}} \, dx \quad \text{and} \quad \int_{1/2}^1 \frac{1 - 2x}{\sqrt{x - x^2}} \, dx$$

For $0 < c < \frac{1}{2}$ we have

$$\int_c^{1/2} \frac{1 - 2x}{\sqrt{x - x^2}} \, dx = 2\sqrt{x - x^2} \, \Big|_c^{1/2} = 2\left(\sqrt{\frac{1}{4}} - \sqrt{c - c^2} \right) = 1 - 2\sqrt{c - c^2}$$

Since

$$\lim_{c \to 0^+} \sqrt{c - c^2} = 0$$

this implies that

$$\int_0^{1/2} \frac{1 - 2x}{\sqrt{x - x^2}} \, dx = \lim_{c \to 0^+} \int_c^{1/2} \frac{1 - 2x}{\sqrt{x - x^2}} \, dx = \lim_{c \to 0^+} \left(1 - 2\sqrt{c - c^2}\right) = 1$$

A similar computation shows that the second improper integral also converges and that

$$\int_{1/2}^1 \frac{1 - 2x}{\sqrt{x - x^2}} \, dx = -1$$

Therefore the original integral converges, and

$$\int_0^1 \frac{1 - 2x}{\sqrt{x - x^2}} \, dx = \int_0^{1/2} \frac{1 - 2x}{\sqrt{x - x^2}} \, dx + \int_{1/2}^1 \frac{1 - 2x}{\sqrt{x - x^2}} \, dx = 1 - 1 = 0 □$$

The last type of unbounded function to be considered in this subsection consists of functions that are continuous at every point of $[a, b]$ except at a point d in (a, b), near which f is unbounded. We say that $\int_a^b f(x) \, dx$ **converges** if both the integrals $\int_a^d f(x) \, dx$ and $\int_d^b f(x) \, dx$ converge, and otherwise we say that $\int_a^b f(x) \, dx$ **diverges.** For example

$$\int_{-1}^{2} \frac{1}{x^{1/3}} \, dx$$

converges (see Exercise 18) and

$$\int_{-1}^{2} \left(\frac{1}{x} + \frac{1}{x^2} \right) dx$$

diverges (see Exercise 19). Again, the interpretation of $\int_{a}^{b} f(x) \, dx$ as the area of a region applies if f is nonnegative.

Integrals over Unbounded Intervals

Integrals of the form $\int_{a}^{\infty} f(x) \, dx$ and $\int_{-\infty}^{b} f(x) \, dx$ are also called improper integrals. If f is continuous on $[a, \infty)$, then $\int_{a}^{b} f(x) \, dx$ is a proper integral for any $b \geq a$. This fact allows us to say that $\int_{a}^{\infty} f(x) \, dx$ **converges** if $\lim_{b \to \infty} \int_{a}^{b} f(x) \, dx$ exists. In that event,

$$\int_{a}^{\infty} f(x) \, dx = \lim_{b \to \infty} \int_{a}^{b} f(x) \, dx$$

As before, the integral **diverges** otherwise. The improper integral $\int_{-\infty}^{b} f(x) \, dx$ is handled in an analogous way. When $f \geq 0$, the connection between these integrals and areas holds as before.

EXAMPLE 5   Show that the shaded region in Figure 8.23 has finite area, and compute the area.

Solution   We must evaluate $\int_{1}^{\infty} e^{-x} \, dx$, which is equal to $\lim_{b \to \infty} \int_{1}^{b} e^{-x} \, dx$. For $b \geq 1$ we have

$$\int_{1}^{b} e^{-x} \, dx = -e^{-x} \Big|_{1}^{b} = -e^{-b} + e^{-1}$$

Since $\lim_{b \to \infty} -e^{-b} = 0$, the improper integral converges, and

$$\int_{1}^{\infty} e^{-x} \, dx = \lim_{b \to \infty} \int_{1}^{b} e^{-x} \, dx = \lim_{b \to \infty} (-e^{-b} + e^{-1}) = 0 + e^{-1} = e^{-1}$$

By our definition this is the area of the region in Figure 8.23.   ❑

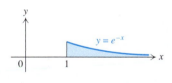

FIGURE 8.23

It is possible to apply improper integrals to velocities and distances. Recall that the distance function is an antiderivative of the velocity function.

EXAMPLE 6   Suppose a projectile is launched from earth at time $t = 1$ with a velocity $v(t) = 1/t$ for $t \geq 1$. Show that the projectile is eventually as far away from earth as one can imagine.

Solution   Let $b > 1$. Since the distance traveled between time $t = 1$ and time $t = b$ is $\int_{1}^{b} v(t) \, dt$, we must first compute $\int_{1}^{b} v(t) \, dt$. But

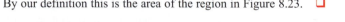

$$\int_{1}^{b} v(t) \, dt = \int_{1}^{b} \frac{1}{t} \, dt = \ln b - \ln 1 = \ln b$$

Then since $\lim_{b \to \infty} \ln b = \infty$, we know that $\int_1^\infty 1/t \, dt$ diverges and that as time passes the projectile goes farther and farther from earth—farther than any preassigned distance. ❑

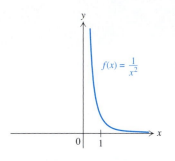

$f(x) = \dfrac{1}{x^2}$

FIGURE 8.24

Integrals can be improper in more than one way. For example, the integrand in $\int_0^\infty 1/x^2 \, dx$ is unbounded near 0, and the interval of integration is also unbounded (see Figure 8.24). In such a case we say that the integral converges if for some $d > 0$, both $\int_0^d 1/x^2 \, dx$ and $\int_d^\infty 1/x^2 \, dx$ (each of which is improper in only one way) converge. It can be proved that if $\int_0^d f(x) \, dx$ and $\int_d^\infty f(x) \, dx$ both converge for a single d in $(0, \infty)$, then the integrals converge for *all* d in $(0, \infty)$. Therefore to show that $\int_0^\infty f(x) \, dx$ diverges, it is enough to find *one* d in $(0, \infty)$ for which $\int_0^d f(x) \, dx$ or $\int_d^\infty f(x) \, dx$ diverges.

EXAMPLE 7   Show that $\displaystyle\int_0^\infty \frac{1}{x^2} \, dx$ diverges.

Solution   Let $d = 1$. We will show that $\int_0^1 1/x^2 \, dx$, and hence $\int_0^\infty 1/x^2 \, dx$, diverges. To determine that $\int_0^1 1/x^2 \, dx$ diverges we compute $\int_c^1 1/x^2 \, dx$ for any c in $(0, 1)$. Since

$$\int_c^1 \frac{1}{x^2} \, dx = -\frac{1}{x}\bigg|_c^1 = \frac{1}{c} - 1$$

and since $\lim_{c \to 0+} (1/c - 1) = \infty$, we know that $\int_0^1 1/x^2 \, dx$ diverges. This implies that $\int_0^\infty 1/x^2 \, dx$ diverges. ❑

We say that $\int_{-\infty}^\infty f(x) \, dx$ **converges** if both $\int_{-\infty}^d f(x) \, dx$ and $\int_d^\infty f(x) \, dx$ converge, for some number d. In that case,

$$\int_{-\infty}^\infty f(x) \, dx = \int_{-\infty}^d f(x) \, dx + \int_d^\infty f(x) \, dx$$

EXAMPLE 8   Determine whether $\displaystyle\int_{-\infty}^\infty \frac{1}{x^2 + 1} \, dx$ converges.

Solution   The integrand is continuous on $(-\infty, \infty)$. To show that the integral converges, we will show that

$$\int_{-\infty}^0 \frac{1}{x^2 + 1} \, dx \quad \text{and} \quad \int_0^\infty \frac{1}{x^2 + 1} \, dx$$

both converge. First we have

$$\int_{-\infty}^0 \frac{1}{x^2 + 1} \, dx = \lim_{a \to -\infty} \int_a^0 \frac{1}{x^2 + 1} \, dx = \lim_{a \to -\infty} \tan^{-1} x \bigg|_a^0$$

$$= \lim_{a \to -\infty} (\tan^{-1} 0 - \tan^{-1} a) = \frac{\pi}{2}$$

A similar calculation shows that

$$\int_0^\infty \frac{1}{x^2 + 1}\, dx = \frac{\pi}{2}$$

Therefore the given improper integral converges, and moreover,

$$\int_{-\infty}^\infty \frac{1}{x^2 + 1}\, dx = \int_{-\infty}^0 \frac{1}{x^2 + 1}\, dx + \int_0^\infty \frac{1}{x^2 + 1}\, dx = \frac{\pi}{2} + \frac{\pi}{2} = \pi \quad \square$$

Sometimes it is more important to know whether an improper integral converges than it is to know the value of the integral (if it converges). The following theorem, which is proved in more advanced texts, assists us in ascertaining the convergence or divergence of improper integrals of the form $\int_a^\infty f(x)\, dx$.

THEOREM 8.3
Comparison Property

Assume that f is continuous on $[a, \infty)$, and suppose that $0 \le f(x) \le g(x)$ for $a \le x < \infty$.

a.   If $\int_a^\infty g(x)\, dx$ converges, then so does $\int_a^\infty f(x)\, dx$.

b.   If $\int_a^\infty f(x)\, dx$ diverges, then so does $\int_a^\infty g(x)\, dx$.

Geometrically, Theorem 8.3 implies that if the graph of f lies below the graph of g and there is finite area between the graph of g and the x axis on $[a, \infty)$, then there is finite area between the graph of f and the x axis on $[a, \infty)$ (Figure 8.25).

For an application of the Comparison Property, recall from the solution of Example 6 that $\int_1^\infty 1/x\, dx$ diverges. If p is any real number such that $p < 1$, then $x \ge x^p$ whenever $x \ge 1$, so that $1/x^p \ge 1/x$ whenever $x \ge 1$. Therefore it follows from the Comparison Property that $\int_1^\infty 1/x^p\, dx$ diverges whenever $p \le 1$. In fact it is possible to show that

$$\int_1^\infty \frac{1}{x^p}\, dx \begin{cases} \text{diverges whenever } p \le 1 \\ \text{converges whenever } p > 1 \end{cases}$$

(see Exercise 62). We will use this information in Chapter 9.

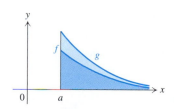

FIGURE 8.25

Normal Probability Distributions

In Section 7.2 we defined a normal density function by the formula

$$f(x) = \frac{1}{\sigma\sqrt{2\pi}} e^{-(x-\mu)^2/2\sigma^2}$$

where μ and σ are constants that together determine the complexion of the graph of f (Figure 8.26). The graph of f is symmetric with respect to the line $x = \mu$. The number μ is called the **mean** of the distribution. The number σ tells us how pointed the graph of f is, and is called the **standard deviation** of the distribution.

The function F that is defined on $(-\infty, \infty)$ by

$$F(b) = \int_{-\infty}^b \frac{1}{\sigma\sqrt{2\pi}} e^{-(x-\mu)^2/2\sigma^2}\, dx$$

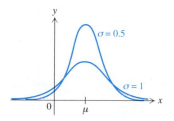

FIGURE 8.26   Two normal density functions with the same mean but different standard deviations.

is called a **normal probability distribution.** Because the integrand is positive for all x, in order to prove that the integral defining $F(b)$ converges for all numbers b, it is only necessary to show that

$$\int_{-\infty}^{\infty} \frac{1}{\sigma\sqrt{2\pi}} e^{-(x-\mu)^2/2\sigma^2}\, dx \quad \text{converges} \tag{1}$$

Karl Friedrich Gauss (1777–1855)
The normal density, which is often called the "Gaussian distribution," is featured on German 10-mark bills, along with a picture of Gauss. Gauss was a remarkable child prodigy; moreover, many regard him as one of the three greatest mathematicians of all time, along with Archimedes and Newton. In addition to absolutely fundamental results in mathematics, Gauss made notable contributions to astronomy and electricity (see Section 15.6). A perfectionist, he strove to make each of his works complete, concise, and polished. In that vein he adopted as his seal a tree with few fruit, and bearing his motto: *Pauca sed matura* (Few, but ripe).

Although (1) is true for every value of μ and σ, we will demonstrate that it is true in the special case for which $\mu = 0$ and $\sigma = 1$.

EXAMPLE 9 Show that $\displaystyle\int_{-\infty}^{\infty} \frac{1}{\sqrt{2\pi}} e^{-x^2/2}\, dx$ converges.

Solution Evidently the given integral converges if and only if $\int_{-\infty}^{\infty} e^{-x^2/2}\, dx$ converges. Since the graph of $e^{-x^2/2}$ is symmetric with respect to the y axis, it suffices to show that $\int_0^{\infty} e^{-x^2/2}\, dx$ converges. To that end observe that

$$x^2 - 2x + 1 = (x-1)^2 \geq 0, \qquad \text{so that} \qquad -\frac{x^2}{2} \leq \frac{1}{2} - x$$

Since the natural exponential function is increasing,

$$0 \leq e^{-x^2/2} \leq e^{(1/2)-x} = e^{1/2} e^{-x}$$

Now

$$\lim_{b\to\infty} \int_0^b e^{1/2} e^{-x}\, dx = \lim_{b\to\infty} \left(-e^{1/2} e^{-x}\right)\Big|_0^b = \lim_{b\to\infty} e^{1/2}\left(-e^{-b} + 1\right) = e^{1/2}$$

Since $\int_0^{\infty} e^{(1/2)-x}\, dx$ converges, it follows from Theorem 8.3 that $\int_0^{\infty} e^{-x^2/2}\, dx$ also converges. This is what we desired to show. ❑

In Section 14.2 we will be able to prove that

$$\int_{-\infty}^{\infty} \frac{1}{\sqrt{2\pi}} e^{-x^2/2}\, dx = 1$$

which means that the area of the complete region bounded above by the standard normal density and below by the x axis is 1 (Figure 8.27).

Many random quantities, such as heights or weights of people or the velocity of a molecule in a gas, are (essentially) normally distributed. In mathematical terms this means that there are suitable constants μ and σ with the following property:

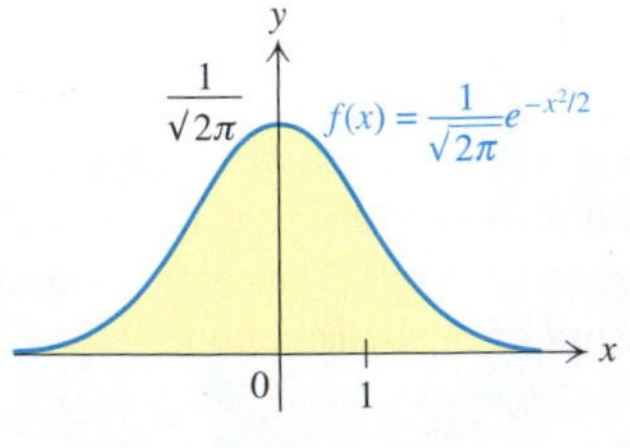

FIGURE 8.27 The area of the region between the standard normal density function and the x axis is 1.

For any real numbers a and b with $a < b$, let R_{ab} denote the fraction of quantities whose values are between a and b. Then

$$R_{ab} = \int_a^b \frac{1}{\sigma\sqrt{2\pi}} e^{-(x-\mu)^2/2\sigma^2}\, dx$$

Notice that R_{ab} is the area under the graph of $e^{-(x-\mu)^2/2\sigma^2}/\sigma\sqrt{2\pi}$ and between the lines $x = a$ and $x = b$ (Figure 8.28). In particular, if $\mu = 0$, $\sigma = 1$, $a = -1$, and $b = 1$, then we find (by computer or calculator) that

$$\int_{-1}^{1} \frac{1}{\sqrt{2\pi}}\, e^{-x^2/2}\, dx \approx 0.68$$

This implies that about 68% of the region under the graph of the standard normal density lies between the lines $x = -1$ and $x = 1$. It can be shown that, more generally, about 68% of the region under the graph of any normal density function lies between the lines $x = \mu - \sigma$ and $x = \mu + \sigma$, that is, within 1 standard deviation of the mean (Figure 8.29). Thus if data are normally distributed, then we can expect that approximately 68% of such data would lie within 1 standard deviation of the mean. For the percentage of data that are expected to be within 2 standard deviations of the mean, see Exercise 80.

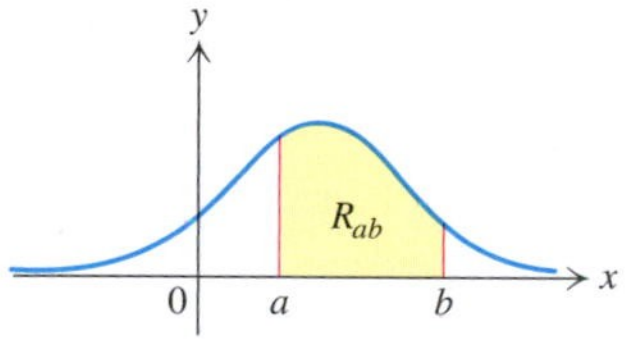

FIGURE 8.28 R_{ab} is the area between the lines $x = a$ and $x = b$.

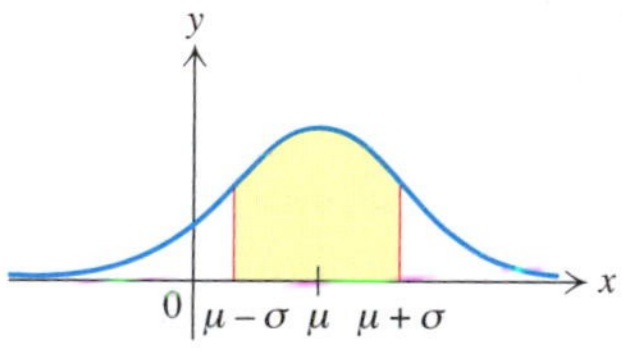

FIGURE 8.29   Approximately 68% of the area lies within 1 standard deviation of the mean.

Suppose the mean of the heights of adults in a town is 66 inches and the standard deviation is 2 inches, and assume that the heights of adults are (approximately) normally distributed. Because the mean is 66, we know that 50% of the adults are no larger than 66 inches tall. However, what percentage would be between 64 and 68 inches tall? Since the standard deviation is assumed to be 2, it follows from the preceding paragraph that the answer is approximately 68%. Moreover, approximately half of that group, or 34%, have heights between 66 and 68 inches, so we can conclude that approximately 84% ($= 50\% + 34\%$) are no larger than 68 inches tall.

EXERCISES 8.7

In Exercises 1–24 determine whether the improper integral converges. If it does, determine the value of the integral.

1. $\displaystyle\int_{0}^{1} \frac{1}{x^{0.9}}\, dx$

2. $\displaystyle\int_{0}^{1} \frac{1}{x^{1.1}}\, dx$

3. $\displaystyle\int_{3}^{4} \frac{1}{(t-4)^2}\, dt$

4. $\displaystyle\int_{1}^{5} \frac{1}{(x-1)^{1.5}}\, dx$

5. $\displaystyle\int_{3}^{4} \frac{1}{\sqrt[3]{x-3}}\, dx$

6. $\displaystyle\int_{0}^{2} \frac{x}{\sqrt[3]{x^2-2}}\, dx$

7. $\displaystyle\int_{0}^{\pi/2} \sec^2 \theta\, d\theta$

8. $\displaystyle\int_{-\pi/2}^{\pi/4} \tan x\, dx$

9. $\displaystyle\int_{0}^{\pi} \frac{\sin x}{\sqrt{1+\cos x}}\, dx$

10. $\displaystyle\int_{0}^{2} \frac{\ln x}{x}\, dx$

11. $\int_1^2 \dfrac{1}{w \ln w}\, dw$

12. $\int_1^2 \dfrac{1}{w(\ln w)^{1/2}}\, dw$

13. $\int_0^1 \dfrac{3x^2 - 1}{\sqrt[3]{x^3 - x}}\, dx$

14. $\int_0^1 \dfrac{3x^2 - 1}{x^3 - x}\, dx$

15. $\int_0^\pi \csc^2 x\, dx$

16. $\int_{-\pi/2}^{\pi/2} \sec \theta\, d\theta$

17. $\int_0^1 \dfrac{1}{e^t - e^{-t}}\, dt$

18. $\int_{-1}^2 \dfrac{1}{x^{1/3}}\, dx$

19. $\int_{-1}^2 \left(\dfrac{1}{x} + \dfrac{1}{x^2}\right) dx$

20. $\int_0^2 \dfrac{1}{(x-1)^{4/3}}\, dx$

21. $\int_0^2 \dfrac{1}{(x-1)^{1/3}}\, dx$

22. $\int_{-\pi/2}^{\pi/2} \dfrac{1}{x^2} \sin \dfrac{1}{x}\, dx$

23. $\int_0^1 \dfrac{1}{\sqrt{1 - t^2}}\, dt$

24. $\int_0^1 \dfrac{e^t}{\sqrt{e^t - 1}}\, dt$

In Exercises 25–60 determine whether the improper integral converges. If it does, determine the value of the integral.

25. $\int_0^\infty \dfrac{1}{x}\, dx$

26. $\int_1^\infty \dfrac{1}{x^{1.1}}\, dx$

27. $\int_0^\infty \dfrac{1}{(2 + x)^\pi}\, dx$

28. $\int_{-\infty}^0 \sqrt{4 - x}\, dx$

29. $\int_0^\infty \sin y\, dy$

30. $\int_0^\infty \cos x\, dx$

31. $\int_0^\infty \dfrac{1}{(1 + x)^3}\, dx$

32. $\int_{-\infty}^0 \dfrac{1}{1 - w}\, dw$

33. $\int_0^\infty \dfrac{x}{1 + x^2}\, dx$

34. $\int_0^\infty \dfrac{x}{(1 + x^2)^4}\, dx$

35. $\int_3^\infty \ln x\, dx$

36. $\int_{-\infty}^{-1} \ln(-x)\, dx$

37. $\int_2^\infty \dfrac{1}{x(\ln x)^3}\, dx$

38. $\int_2^\infty \dfrac{1}{x \ln x}\, dx$

39. $\int_{-\infty}^0 \dfrac{1}{(x + 3)^2}\, dx$

40. $\int_{2/\pi}^\infty \dfrac{1}{x^2} \sin \dfrac{1}{x}\, dx$

41. $\int_1^\infty \dfrac{1}{\sqrt{x}(\sqrt{x} + 1)}\, dx$

42. $\int_{-\infty}^0 e^{4x}\, dx$

43. $\int_0^\infty e^{4x}\, dx$

44. $\int_0^\infty \dfrac{1}{x^2} e^{-x}\, dx$

45. $\int_0^\infty x e^{-x}\, dx$

46. $\int_0^\infty e^x \sin x\, dx$

47. $\int_1^\infty \dfrac{1}{\sqrt{x^2 - 1}}\, dx$

48. $\int_{-\infty}^{-2} \dfrac{1}{x^2 + 4}\, dx$

49. $\int_1^\infty \dfrac{1}{t\sqrt{t^2 - 1}}\, dt$

50. $\int_0^\infty \dfrac{1}{t^2 - 2t + 1}\, dt$

51. $\int_{-2}^\infty \dfrac{1}{t^2 + 4t + 8}\, dt$

52. $\int_0^\infty \dfrac{x \cos x - \sin x}{x^2}\, dx$ $\left(\text{Hint: Compute } \dfrac{d}{dx}\left(\dfrac{\sin x}{x}\right).\right)$

53. $\int_{-\infty}^\infty x\, dx$

54. $\int_{-\infty}^\infty x \sin x^2\, dx$

55. $\int_{-\infty}^\infty x \sin x\, dx$

56. $\int_{-\infty}^\infty \dfrac{x^3}{x^4 + 1}\, dx$

57. $\int_{-\infty}^\infty \dfrac{x^3}{(x^4 + 1)^2}\, dx$

58. $\int_{-\infty}^\infty x e^{-x^2}\, dx$

59. $\int_{-\infty}^\infty \dfrac{1}{x^2 - 6x + 10}\, dx$

60. $\int_{-\infty}^\infty \dfrac{x \sin x + 2 \cos x - 2}{x^3}\, dx$

(*Hint:* Differentiate $(1 - \cos x)/x^2$, and use the fact that $\lim_{x \to 0} (1 - \cos x)/x^2 = 1/2$.)

61. a. Show that

$$\frac{1}{x(x + 1)} = \frac{1}{x} - \frac{1}{x + 1}$$

b. Does the equation

$$\int_1^\infty \frac{1}{x(x + 1)}\, dx \overset{?}{=} \int_1^\infty \frac{1}{x}\, dx - \int_1^\infty \frac{1}{x + 1}\, dx$$

hold? Explain why or why not.

62. a. Show that $\int_0^1 1/x^p\, dx$ converges if $p < 1$ and diverges otherwise.

b. Show that $\int_1^\infty 1/x^p\, dx$ converges if $p > 1$ and diverges otherwise.

c. Conclude from (a) and (b) that $\int_0^\infty 1/x^p\, dx$ diverges for every p.

In Exercises 63–68 decide whether the region between the graph of the integrand and the x axis on the interval of integration has finite area. If it does, calculate the area.

63. $\int_{-\infty}^0 \dfrac{1}{(x - 3)^2}\, dx$

64. $\int_{-\infty}^3 \dfrac{1}{(x - 3)^2}\, dx$

65. $\int_2^\infty \dfrac{\ln x}{x}\, dx$

66. $\int_0^1 \dfrac{\ln x}{-x}\, dx$

67. $\int_2^\infty \dfrac{1}{\sqrt{x + 1}}\, dx$

68. $\int_{-1}^1 \dfrac{1}{\sqrt{x + 1}}\, dx$

In Exercises 69–74 use the Comparison Property to determine whether the integral converges.

69. $\displaystyle\int_1^\infty \frac{1}{1+x^4}\,dx$ **70.** $\displaystyle\int_0^\infty \frac{1}{\sqrt{2+\sin x}}\,dx$

71. $\displaystyle\int_1^\infty \frac{\sin^2 x}{\sqrt{1+x^3}}\,dx$

72. $\displaystyle\int_1^\infty \frac{\ln x}{x^3}\,dx$ (*Hint:* $\ln x < x$ for $x \geq 1$.)

73. $\displaystyle\int_1^\infty \frac{1}{x}\sqrt{1+\frac{1}{x^4}}\,dx$ ***74.** $\displaystyle\int_1^\infty \frac{\sin^2 x}{x}\,dx$

75. a. Suppose f and g are continuous on $[a,\infty)$. Show that if $\int_a^\infty f(x)\,dx$ and $\int_a^\infty g(x)\,dx$ converge, then $\int_a^\infty (f(x)+g(x))\,dx$ converges.

    b. Show that if $\int_a^\infty f(x)\,dx$ converges, then $\int_a^\infty cf(x)\,dx$ converges for any number c.

    c. Show by a simple example that $\int_a^\infty (f(x)+g(x))\,dx$ can converge without $\int_a^\infty f(x)\,dx$ or $\int_a^\infty g(x)\,dx$ converging.

76. Let

$$I_n = \int_0^\infty x^n e^{-x}\,dx$$

for any nonnegative integer n. Using integration by parts, show that

$$I_n = nI_{n-1}$$

for $n \geq 1$. From this show that

$$I_n = n(n-1)(n-2)\cdots 2\cdot 1$$

77. Suppose that $\int_a^\infty f(t)\,dt$ converges, and let $x > a$.

    a. Show that $\int_a^\infty f(t)\,dt = \int_a^x f(t)\,dt + \int_x^\infty f(t)\,dt$.

    b. Use (a) to show that $\dfrac{d}{dx}\displaystyle\int_x^\infty f(t)\,dt = -f(x)$.

78. Let $a > 0$, and let f be continuous on $[a,\infty)$. Show that the improper integrals

$$\int_a^\infty f(x)\,dx \quad \text{and} \quad \int_0^{1/a} \frac{f(1/t)}{t^2}\,dt$$

either both converge or both diverge.

79. a. Show that $\displaystyle\int_1^\infty \frac{1}{x^{3/2}+1}\,dx$ converges.

    b. By finding the inverse of $1/(x^{3/2}+1)$, show that

$$\int_0^{1/2} \left(\frac{1}{x}-1\right)^{2/3} dx = \int_1^\infty \frac{1}{x^{3/2}+1}\,dx$$

(See Figure 8.30.)

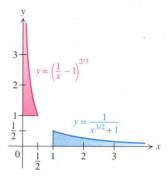

FIGURE 8.30 Graphs for Exercise 79.

 80. Assume that data are normally distributed, with $\mu = 0$ and $\sigma = 1$. Use Simpson's Rule with $n = 100$ in order to approximate the percentage of data that should lie within

    a. 2 standard deviations of the mean.

    b. 3 standard deviations of the mean.

 81. This exercise is designed to support the assertion that

$$\int_{-\infty}^\infty e^{-x^2/2}/\sqrt{2\pi}\,dx = 1$$

    a. Use Simpson's Rule with $n = 100$ to approximate

$$\int_{-5}^5 e^{-x^2/2}/\sqrt{2\pi}\,dx$$

    b. Find the smallest positive integer b such that when you use Simpson's Rule with $n = 100$, the value your computer or calculator gives for $\int_{-b}^b e^{-x^2/2}/\sqrt{2\pi}\,dx$ is 1.

82. Let μ and σ be arbitrary numbers. Use integration by substitution to show that for any numbers r and s,

$$\int_{\mu+r\sigma}^{\mu+s\sigma} \frac{1}{\sigma\sqrt{2\pi}}\,e^{-(x-\mu)^2/2\sigma^2}\,dx = \int_r^s \frac{1}{\sqrt{2\pi}}\,e^{-x^2/2}\,dx$$

(This result shows that, in particular, the percentage of normally distributed data that lie within n standard deviations of the mean is the same percentage as if $\mu = 0$ and $\sigma = 1$.)

Applications

83. In the statistical analysis of the length of time that it takes to repair a machine, it has been suggested that the mean time of repair is $\int_0^\infty ct\,e^{-ct}\,dt$, for an appropriate positive constant c and unit of time t. Evaluate the integral.

84. Economists calculate the **present sale value** P of land that can be rented for R dollars annually by the formula

$$P = \int_0^\infty Re^{-rt/100} \, dt$$

where r is the prevailing interest rate and where $e^{-rt/100}$ is called the **discounting factor.** Show that

$$P = \frac{100R}{r}$$

85. If A is the amount of radioactive substance at time 0, then the amount $f(t)$ remaining at any time $t > 0$ (measured in years) is given by

$$f(t) = Ae^{kt}$$

where k is a negative constant. The **mean life** M of an atom in the substance is given by

$$M = -\frac{1}{A} \int_0^\infty tkf(t) \, dt$$

a. Using $k = -1.24 \times 10^{-4}$, find the mean life of a C^{14} atom.

b. Using $k = -4.36 \times 10^{-4}$, find the mean life of a radium atom.

86. The function that indicates the probability that an electron in the ground state of hydrogen is located at a distance r from the nucleus is given by

$$P(r) = \frac{4}{a_0^3} r^2 \, e^{-2r/a_0}$$

where $a_0 = 5.29 \times 10^{-11}$ and is called the **first Bohr radius.** The probability P that an electron is located outside the first Bohr radius is given by

$$P = \int_{a_0}^\infty P(r) \, dr$$

a. Calculate P.

b. Show that $\int_0^\infty P(r) \, dr = 1$, and explain the physical significance of this result.

87. Let s denote the annual subsistence level of people in the United States. For $x \geq s$ let $G(x)$ be the number of people with annual income between s and x, and assume that there is a continuous function g such that $G(x) = \int_s^x g(t) \, dt$ for all $x \geq s$. Let P be the total population of the United States, and assume that every individual's income is at least at subsistence level. Pareto's Law of distribution of income in a population asserts that under very general conditions,

$$P - G(x) = \int_x^\infty g(t) \, dt \approx cx^{-1.5} \quad \text{for } x \geq s$$

where c is a constant. Assuming that

$$\int_x^\infty g(t) \, dt = cx^{-1.5} \quad \text{for } x \geq s$$

and using Exercise 77, determine the function g. (Data collected in England, Germany, Ireland, Italy, Peru, and the United States confirm Pareto's Law to a remarkable degree.)

88. Suppose the heights of adults in a town are normally distributed, with a mean of 66 inches and standard deviation of 3 inches. Use Simpson's Rule with $n = 100$ to approximate the percentage of adults that

a. are between 70 and 71 inches tall.

b. are taller than 70 inches. (*Hint:* First find the percentage between 66 and 70 inches tall.)

.89. Suppose the mean weight of 200 babies born in the Easy-Birth Hospital is 7.5 pounds, with a standard deviation of 1 pound.

a. Use Simpson's Rule with $n = 100$ to approximate the number of babies that should have weighed between 8 pounds and 12 pounds at birth.

b. Determine how many babies should have weighed under 6.5 pounds at birth.

90. Show that $\int_0^\infty I_0(\lambda) \, d\lambda$ diverges, where $I_0(\lambda)$ is the intensity of blackbody radiation with wave length λ, and is given by the Rayleigh-Jeans Law as

$$I_0(\lambda) = \frac{2\pi ckT}{\lambda^4}$$

where c is the speed of light in a vacuum, k is the Boltzman constant, and T is the temperature. (The fact that the integral diverged was a physical impossibility, and led Max Planck to propose a different formula.)

Projects

1. a. Consider the rational function $P(x)/Q(x)$ in reduced form, and let the degrees of P and Q be m and n, respectively. Under what conditions on m and n can we be sure that $\int_a^\infty P(x)/Q(x) \, dx$ converges for some lower limit of integration a?

b. Using part (a), find conditions on P and Q such that $\int_0^\infty P(x)/Q(x) \, dx$ converges.

c. Describe a continuous function f defined on $[0, \infty)$ such that f is *unbounded*, $\lim_{x \to \infty} f(x)$ does *not* exist, but such that $\int_0^\infty f(x)\, dx$ *does* converge. Don't try to find a formula for f, just a description of f. (*Hint:* The graph of f should have plenty of narrow spikes!)

2. a. Show that $\int_0^1 \sin(1/x)\, dx$ converges.

b. How would you get an estimate for the value of $\int_0^1 \sin(1/x)\, dx$ that has an error of less than 10^{-3}? (*Hint:* You need to integrate from ε to 1, for an appropriately small value of ε.)

c. Let f be bounded and continuous on $(0, 1]$. Show that $\int_0^1 f(x)\, dx$ converges.

REVIEW

Key Terms and Expressions

Integration by parts
Trapezoidal Rule
Simpson's Rule

Proper integral; improper integral
Convergent improper integral; divergent improper integral

Key Formulas

$$\int u\, dv = uv - \int v\, du$$

$$\int_{g(a)}^{g(b)} f(x)\, dx = \int_a^b f(g(u))g'(u)\, du, \quad \text{where } x = g(u)$$

$$\int_a^b f(x)\, dx \approx \frac{b - a}{2n} [f(x_0) + 2f(x_1) + 2f(x_2) + \cdots + 2f(x_{n-1}) + f(x_n)]$$

$$\int_a^b f(x)\, dx \approx \frac{b - a}{3n} [f(x_0) + 4f(x_1) + 2f(x_2) + 4f(x_3) + \cdots + 2f(x_{n-2}) + 4f(x_{n-1}) + f(x_n)], \quad \text{where } n \text{ is even}$$

Review Exercises

In Exercises 1–24 evaluate the indefinite integral.

1. $\displaystyle\int \ln(x^2 + 9)\, dx$ **2.** $\displaystyle\int (\ln x)^4\, dx$

3. $\displaystyle\int x \csc^2 x\, dx$ **4.** $\displaystyle\int x^3 e^{3x}\, dx$

5. $\displaystyle\int x \cosh x\, dx$ **6.** $\displaystyle\int x^2 \sin^{-1} x\, dx$

7. $\displaystyle\int x \cos^2 x\, dx$ **8.** $\displaystyle\int \cos^2 x \sin^4 x\, dx$

9. $\displaystyle\int x^2 \sin x^3 \cos x^3\, dx$ **10.** $\displaystyle\int \sec^4 x\, dx$

11. $\displaystyle\int \tan^5 x\, dx$ **12.** $\displaystyle\int \frac{\tan^3 x}{\sec^5 x}\, dx$

13. $\displaystyle\int x^3 \cos x^2\, dx$ **14.** $\displaystyle\int \sqrt{e^t - 1}\, dt$

15. $\displaystyle\int t^2 \sqrt{1 - 3t}\, dt$ **16.** $\displaystyle\int \frac{t^7}{(1 - t^4)^3}\, dt$

17. $\displaystyle\int \frac{\cos x}{1 + \cos x}\, dx$

$\left(\textit{Hint: Multiply the integrand by } \dfrac{1 - \cos x}{1 - \cos x}.\right)$

18. $\displaystyle\int \frac{x^3}{\sqrt{x^8 - 1}}\, dx$ **19.** $\displaystyle\int \frac{x^2}{(x^2 + 2x + 10)^{5/2}}\, dx$

20. $\displaystyle\int \frac{1}{x^3 - 1}\, dx$ **21.** $\displaystyle\int \frac{x^4}{(x^2 + 1)^2}\, dx$

22. $\displaystyle\int \frac{1}{x(x^2 + x + 1)}\, dx$ **23.** $\displaystyle\int \frac{x}{x^2 + 3x - 18}\, dx$

24. $\displaystyle\int \frac{\sqrt{2x + 1}}{x + 1}\, dx$ (*Hint:* Substitute $u = \sqrt{2x + 1}$ and then use partial fractions.)

In Exercises 25–52 determine whether the integral is proper or improper. If the integral is improper, determine whether it converges. If it is either a proper integral or a convergent improper integral, evaluate the integral.

25. $\displaystyle\int_{-2}^{1} \frac{1}{3x + 4}\, dx$ **26.** $\displaystyle\int_{\pi}^{3\pi/2} \frac{\cos x}{1 + \sin x}\, dx$

27. $\displaystyle\int_{0}^{\pi/4} x \sec x \tan x\, dx$ **28.** $\displaystyle\int_{0}^{1} \frac{x}{x + 2}\, dx$

29. $\displaystyle\int_{1}^{4} \frac{1}{1 + \sqrt{x}}\, dx$ **30.** $\displaystyle\int_{1}^{5} \frac{x}{\sqrt{x - 1}}\, dx$

31. $\displaystyle\int_{0}^{\pi/4} \sin^4 x \cos^3 x\, dx$ **32.** $\displaystyle\int_{0}^{\pi/3} \tan^5 x \sec x\, dx$

33. $\displaystyle\int_{0}^{\pi/4} (\tan^3 x + \tan^5 x)\, dx$ **34.** $\displaystyle\int_{0}^{1} \frac{x^3}{\sqrt{1 - x^2}}\, dx$

35. $\displaystyle\int_{1}^{\sqrt{2}} \frac{\sqrt{x^2 - 1}}{x^2}\, dx$ **36.** $\displaystyle\int_{0}^{1} \frac{x^2}{1 + x^6}\, dx$

37. $\displaystyle\int_{0}^{\sqrt[3]{2}} \frac{\sqrt{x}}{\sqrt{1 - x^3}}\, dx$ (*Hint:* Substitute $u = x^{3/2}$.)

38. $\displaystyle\int_{0}^{1} x^5\sqrt{1 - x^2}\, dx$ **39.** $\displaystyle\int_{0}^{\sqrt{3}} \sqrt{x^2 + 1}\, dx$

40. $\displaystyle\int_{-1}^{1} \frac{x}{x^2 + 2x + 5}\, dx$ **41.** $\displaystyle\int_{-5}^{0} \frac{x}{x^2 + 4x - 5}\, dx$

42. $\displaystyle\int_{-1}^{1} \frac{5x^3}{x^2 + x - 6}\, dx$

43. $\displaystyle\int_{0}^{\pi/2} \frac{1}{1 - \sin x}\, dx$

$\left(\text{\textit{Hint:} Multiply the integrand by } \dfrac{1 + \sin x}{1 + \sin x}.\right)$

44. $\displaystyle\int_{0}^{\pi/4} \frac{1}{1 + \tan x}\, dx$ (*Hint:* Substitute $u = 1 + \tan x$ and

then use partial fractions.)

45. $\displaystyle\int_{0}^{1} x \ln x\, dx$ **46.** $\displaystyle\int_{0}^{\infty} x \ln x\, dx$

47. $\displaystyle\int_{1}^{\infty} \frac{1}{x (\ln x)^2}\, dx$ **48.** $\displaystyle\int_{3}^{\infty} \frac{1}{1 + \ln x}\, dx$

49. $\displaystyle\int_{0}^{\infty} e^{-x} \cos x\, dx$ **50.** $\displaystyle\int_{1}^{\infty} x^2 e^{-x^3}\, dx$

51. $\displaystyle\int_{1}^{\infty} \frac{1}{x(x^2 + 1)}\, dx$ **52.** $\displaystyle\int_{0}^{\infty} \frac{1}{x(x^2 + 4)}\, dx$

53. Use integration by parts to evaluate $\displaystyle\int \frac{\ln x}{x}\, dx$.

54. a. Using the substitution $x = \sin u$, show that

$$\int_{0}^{1} x^m (1 - x^2)^n\, dx = \int_{0}^{\pi/2} \sin^m u \cos^{2n + 1} u\, du$$

where m and n are nonnegative integers.

b. Use part (a) to evaluate

$$\int_{0}^{1} x^3 (1 - x^2)^{10}\, dx$$

55. a. By making the substitution $x = u/(1 - u)$ show that

$$\int_{0}^{b} \frac{x^{m-1}}{(1 + x)^{m+n}}\, dx = \int_{0}^{b/(1 + b)} u^{m-1}(1 - u)^{n-1}\, du$$

for any positive integers m and n and any positive number b.

b. Use part (a) to show that

$$\int_{0}^{\infty} \frac{x^3}{(1 + x)^5}\, dx = \int_{0}^{1} u^3\, du = \frac{1}{4}$$

56. a. Show that the integral

$$I_n = \int_{0}^{1} (1 - x^2)^n\, dx$$

satisfies the reduction formula

$$I_n = \frac{2n}{2n + 1} I_{n-1}$$

for any positive integer n. (*Hint:* Write $(1 - x^2)^n = (1 - x^2)^{n-1} - x^2(1 - x^2)^{n-1}$, express I_n as a sum of two integrals, and evaluate the second integral by parts.)

b. Use part (a) to evaluate $\int_{0}^{1} (1 - x^2)^4\, dx$.

*57. Use the trigonometric identities

$$\tan \frac{x}{2} = \frac{\sin x}{1 + \cos x} \quad \text{and} \quad \sec^2 \frac{x}{2} = \frac{2}{1 + \cos x}$$

to find

$$\int \frac{\tan\left(\dfrac{\pi}{4} + \dfrac{x}{2}\right)}{\sec^2 \dfrac{x}{2}}\, dx$$

58. Approximate $\int_{0}^{1} \sqrt{1 - x^2}\, dx$ (which has the value $\pi/4$) by using

a. the Trapezoidal Rule with $n = 10$.

b. Simpson's Rule with $n = 10$.

59. Approximate $\int_0^2 \sqrt{2x - x^2}\, dx$ by using

  a. the Trapezoidal Rule with $n = 10$.

  b. Simpson's Rule with $n = 10$.

60. Determine values of n that ensure errors of less than 10^{-4} when $\int_2^{2.5} \sqrt{x^2 - 1}\, dx$ is approximated by the Trapezoidal Rule and by Simpson's Rule.

 61. Using a calculator, approximate the integral in Exercise 60 with an error less than 10^{-4}

  a. by the Trapezoidal Rule.

  b. by Simpson's Rule.

In Exercises 62–63 find the area A of the region between the graph of f and the x axis on the given interval.

62. $f(x) = \cos^3 x$; $[0, \pi/2]$

63. $f(x) = \sqrt{9 - x^2}$; $[-3, 0]$

64. Let $f(x) = \sqrt{1 + x^5}$. Use Simpson's Rule with $n = 10$ to approximate the area of the region between the graph of f and the x axis on $[0, 4]$.

In Exercises 65–69 determine whether the area of the region between the graph of f and the x axis on the given interval is finite or infinite. If the area is finite, determine its numerical value.

65. $f(x) = \dfrac{\cos x}{1 + \sin x}$; $[\pi, 3\pi/2)$

66. $f(x) = \dfrac{\cos x}{\sqrt{1 + \sin x}}$; $[\pi, 3\pi/2)$

67. $f(x) = \dfrac{x^3}{2 + x^4}$; $(-\infty, \infty)$

68. $f(x) = \dfrac{x^3}{\sqrt{2 + x^4}}$; $(-\infty, \infty)$

69. $f(x) = \dfrac{x^3}{(2 + x^4)^2}$; $(-\infty, \infty)$

Applications

70. a. Suppose two electrons are separated by a distance a. If it were possible to bring the two electrons together, the amount of work required to overcome the electrostatic repulsion of the two electrons would be $\int_0^a k/r^2\, dr$, where k is a positive constant. Is it possible to bring the electrons together? Explain.

  b. Suppose an electron and a proton are separated by a distance a. If it were possible to take the electron arbitrarily far away from the proton, the amount of work required to overcome the electrostatic attraction between the two particles would be $\int_a^\infty k/r^2\, dr$, where k is the same constant as in part (a). In theory, is it possible to do so? Explain.

*71. The average velocity $\bar{v}$ of a gas molecule with mass m equals the ratio A/B, where

$$A = \int_0^\infty v^3\, e^{-mv^2/2kT}\, dv \quad \text{and} \quad B = \int_0^\infty v^2\, e^{-mv^2/2kT}\, dv$$

where T is the temperature and k is Boltzmann's constant. Show that

$$\bar{v} = \sqrt{\frac{8kT}{\pi m}}$$

(*Hint:* To evaluate A and B use integration by parts; then for B use the fact that

$$\int_0^\infty e^{-x^2/2}\, dx = \sqrt{\frac{\pi}{2}}$$

Topics for Discussion

1. Give a plausible reason for the term "integration by parts."

2. The integration by parts formula is based on the product rule for differentiation. Obtain a similar formula based on the quotient rule for differentiation. Show that your formula reduces to a special case of the integration by parts formula.

3. Of the integration techniques discussed in Sections 8.1 to 8.4, which two are the most likely to be used together? Explain your answer.

4. Explain why Simpson's Rule tends to be more accurate than the Trapezoidal Rule.

5. Using the Trapezoidal Rule and Simpson's Rule as a guide, describe a rule that you think in theory would be more accurate than each of these rules.

6. What is the difference between proper and improper integrals? Describe the various ways in which an integral can be improper.

Cumulative Review, Chapters 1–7

In Exercises 1–2 evaluate the limit.

1. $\displaystyle\lim_{x \to 0} \frac{\sin x - \sin(\sin x)}{x^3}$ **2.** $\displaystyle\lim_{x \to \infty} \left(1 - \frac{3}{x}\right)^x$

3. Find the value(s) of c for which $\displaystyle\lim_{x \to \infty} (\sqrt{cx + 1} - \sqrt{x})$ exists.

In Exercises 4–5 find $f'(x)$.

4. $f(x) = \sqrt{\dfrac{x + 1}{3x - 2}}$ **5.** $f(x) = \dfrac{e^{2x}}{e^x + 1}$

6. Let $f(x) = \dfrac{1}{1-x}$. Find a formula for $f^{(5)}(x)$.

7. Suppose that the velocity and acceleration of a given particle are positive and are related by the equation $a^2 - a = 2v^2 - 6v$. Show that $v = 3a$ at the instant that $da/dt = 6\, dv/dt$.

8. A car going 60 feet per second passes a car going 50 feet per second along a straight highway. The distance between the drivers is 10 feet when they are alongside one another. Let θ denote the angle made by one edge of the highway and the line joining the two drivers. How fast is θ changing when $\theta = \pi/3$?

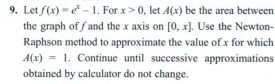

9. Let $f(x) = e^x - 1$. For $x > 0$, let $A(x)$ be the area between the graph of f and the x axis on $[0, x]$. Use the Newton-Raphson method to approximate the value of x for which $A(x) = 1$. Continue until successive approximations obtained by calculator do not change.

In Exercises 10–11 sketch the graph of f, indicating all pertinent information.

10. $f(x) = x^{2/3}(x - 2)^{1/3}$ 11. $f(x) = \dfrac{x}{(x-3)^2}$

12. Find the largest possible area A of an isosceles trapezoid inscribed in a circle of radius a with the diameter of the circle serving as the largest side of the trapezoid.

13. Of all right triangles with hypotenuse 1, which has the smallest ratio of circumference to area? (*Hint:* Express the ratio in terms of one of the angles of the triangle.)

14. A radioactive substance has a half-life of 24 hours. How long will it take for 99% of a given amount to disintegrate?

In Exercises 15–16 evaluate the integral.

15. $\displaystyle\int \dfrac{2x + 4}{x^2 + 4x - 5}\, dx$ 16. $\displaystyle\int \dfrac{(1 + \sqrt{\sin x})}{\sqrt{\sin x}} \cos x\, dx$

17. Consider the region R bounded by the curves $y = e^{2x}$ and $y = e^{-x}$ and the line $x = \frac{1}{2}$. Find the area A of R.

18. The base of a solid is semicircular with diameter 2. The cross sections perpendicular to the diameter of the base are also semicircular. Find the volume V of the solid.

19. Let $f(x) = \frac{3}{2} x^{2/3}$ for $-27 \leq x \leq -1$. Find the length L of the graph of f.

20. Let C be parametrized by $x = \frac{1}{3} t^3$ and $y = \frac{1}{5} t^5$ for $0 \leq t \leq 1$. Determine the surface area S of the surface generated by revolving C about the x axis.

21. Show that if f is continuous on $(-\infty, \infty)$, then for any $a, b,$ and c we have
$$\int_a^b f(cx)\, dx = \frac{1}{c} \int_{ca}^{cb} f(x)\, dx$$

22. Suppose that f has the property that
$$f(t) = \int_0^t f(s)\, dx \quad \text{for all} \quad t \geq 0$$
Show that $f(t) = 0$ for all $t \geq 0$.

An endless sequence of railroad ties. (*Frank La Bua/Liaison International*)

9 SEQUENCES AND SERIES

From very early times people have estimated the numerical values of mathematical expressions such as π and $\sqrt{2}$. In ancient clay tablets there are lists of approximate square roots. Later, Hipparchus and Ptolemy compiled tables of trigonometric values for use in astronomy. More recently, tables of trigonometric, logarithmic, and exponential values were indispensable to mathematicians and scientists and frequently appeared as appendixes to mathematics and science books. The French mathematician Pierre-Simon Laplace (1749–1827) even asserted that the invention of logarithms had doubled the life of astronomers.

Today tables of values have been supplanted by calculators and computers. However, we might inquire how calculators and computers derive trigonometric, logarithmic, and exponential values. One method is to use the values of closely related polynomials because their values are relatively easy to calculate. The focus of Chapter 9 is the approximation of functions by polynomials.

The ideas central to approximation of functions by polynomials are introduced in Section 9.1. They lead to the notion of sequence in Section 9.2 and then to series in Section 9.4. Finally, in Sections 9.8 to 9.10 we use series to approximate functions by polynomials.

567

9.1 POLYNOMIAL APPROXIMATION

In preceding chapters we have found approximate values of numbers such as e, π, and $\ln 2$. Through methods of the present chapter we will be able (at least theoretically) to estimate these numbers, as well as the values of many functions, with any prescribed degree of accuracy.

Suppose we wish to approximate a value $f(x)$ of a function f and we know the value $f(a)$ at a nearby number a. Because values of polynomials are easy to compute, we will use the value $p(x)$ of a polynomial p to approximate $f(x)$. The polynomial to be used will be manufactured from higher derivatives of f at a, so we will need to assume that f has such higher derivatives at a. The discussion in this first section of Chapter 9 is preliminary, serving as a platform for the rest of the chapter. For simplicity we will assume that $a = 0$ in this section; in Section 9.9 we will allow a to be any real number, and will also discuss the accuracy of the approximations.

Throughout the discussion in this section we will assume that a given function f has derivatives of all orders at 0, that is, we will assume that $f^{(n)}(0)$ exists for $n \geq 1$. For every positive integer n, our goal is the following: Among *all* polynomials of degree at most n, we wish to find the polynomial that "best" approximates f near 0. As we proceed, it will become clear what "best" means.

Evidently the constant polynomial p_0 that best approximates $f(x)$ for x near 0 is defined by $p_0(x) = f(0)$ (Figure 9.1). As we showed in Section 3.8, the linear polynomial p_1 that best approximates $f(x)$ for x near 0 is

$$p_1(x) = f(0) + f'(0)x$$

Notice that $p_1(0) = f(0)$, $p_1'(0) = f'(0)$, and the graph of p_1 is the line tangent to the graph of f at $(0, f(0))$ (Figure 9.2). In addition, in Section 3.8 we found the quadratic polynomial p_2 that best approximates $f(x)$ for x near 0 is

$$p_2(x) = f(0) + f'(0)x + \frac{f''(0)}{2}x^2$$

By its definition, p_2 has the property that $p_2(0) = f(0)$, $p_2'(0) = f'(0)$, and $p_2''(0) = f''(0)$. Figure 9.3 includes the graphs of f, as well as p_0, p_1, and p_2, and shows that p_2 is a better approximation to f near 0 than either p_0 or p_1 is.

To illustrate these polynomials, consider $f(x) = e^{3x}$. Since $f'(x) = 3e^{3x}$ and $f''(x) = 9e^{3x}$, we have $f(0) = 1$, $f'(0) = 3$, and $f''(0) = 9$. Therefore

$$p_0(x) = 1, \quad p_1(x) = 1 + f'(0)x = 1 + 3x, \quad p_2(x) = 1 + f'(0)x + \frac{f''(0)}{2}x^2 = 1 + 3x + \frac{9}{2}x^2$$

As you can check on your calculator or computer, for x near 0, $p_1(x)$ is a better approximation of $f(x)$ than $p_0(x)$ is, and $p_2(x)$ is a better approximation of $f(x)$ than either $p_0(x)$ or $p_1(x)$ is.

To simplify the formula for the higher-degree polynomials that will approximate f near 0, we define the **factorial** of any nonnegative integer n as follows:

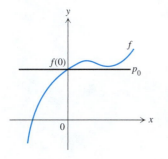

FIGURE 9.1

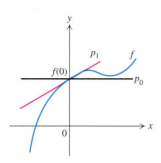

FIGURE 9.2

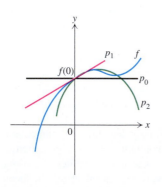

FIGURE 9.3

$$n! = n(n-1)(n-2)(n-3) \cdots 3 \cdot 2 \cdot 1 \quad \text{for } n \geq 1$$
$$0! = 1$$

Thus

$$1! = 1 \qquad\qquad 4! = 4 \cdot 3 \cdot 2 \cdot 1 = 24$$

$$2! = 2 \cdot 1 = 2 \qquad\qquad 5! = 5 \cdot 4 \cdot 3 \cdot 2 \cdot 1 = 120$$

$$3! = 3 \cdot 2 \cdot 1 = 6 \qquad\qquad 6! = 6 \cdot 5 \cdot 4 \cdot 3 \cdot 2 \cdot 1 = 720$$

An often-used property of factorials that follows directly from the definition is

$$(n+1)! = (n+1)n! \quad \text{for } n \geq 0$$

Now we define a polynomial p_n of degree (at most) n by the formula

$$p_n(x) = f(0) + f'(0)x + \frac{f''(0)}{2!}x^2 + \frac{f^{(3)}(0)}{3!}x^3 + \cdots + \frac{f^{(n)}(0)}{n!}x^n \quad (1)$$

The polynomial p_n is called the **nth Taylor polynomial** of f about 0, after the English mathematician Brook Taylor. The functions p_n and f have the same value at 0, as well as the same derivatives at 0, up through the nth derivative. The larger n is, the more derivatives of p_n coincide with those of f at 0. It is therefore plausible that as n gets larger, p_n becomes a better approximation to f near 0.

In our first example, we will derive the nth Taylor polynomial p_n for the function defined by $f(x) = e^x$.

Brook Taylor (1685–1731)
Taylor polynomials appeared in 1715 in Taylor's treatise *The Method of Increments*. Taylor used his polynomials to approximate solutions of equations. He became secretary of the Royal Society of London, but resigned at age 34 to pursue his writing interests. His work on perspective has found modern application in aerial photography.

EXAMPLE 1   Let $f(x) = e^x$. Show that the nth Taylor polynomial p_n of f is given by

$$p_n(x) = 1 + x + \frac{x^2}{2!} + \frac{x^3}{3!} + \cdots + \frac{x^n}{n!}$$

Then write down formulas for p_1, p_2, and p_5.

Solution   By (1),

$$p_n(x) = f(0) + f'(0)x + \frac{f''(0)}{2!}x^2 + \frac{f^{(3)}(0)}{3!}x^3 + \cdots + \frac{f^{(n)}(0)}{n!}x^n$$

where $f(0), f'(0), f''(0), \ldots, f^{(n)}(0)$ need to be determined. Now for each $k \leq n$, $f^{(k)}(x) = e^x$, so that $f^{(k)}(0) = e^0 = 1$ for each k. Therefore

$$p_n(x) = 1 + 1 \cdot x + \frac{1}{2!}x^2 + \frac{1}{3!}x^3 + \frac{1}{4!}x^4 + \cdots + \frac{1}{n!}x^n$$

$$= 1 + x + \frac{1}{2}x^2 + \frac{1}{6}x^3 + \frac{1}{24}x^4 + \cdots + \frac{1}{n!}x^n$$

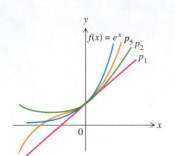

FIGURE 9.4

Thus we have a formula for p_n. Letting $n = 1, 2$, and 5, we obtain the three desired Taylor polynomials:

$$p_1(x) = 1 + x$$

$$p_2(x) = 1 + x + \frac{1}{2}x^2$$

$$p_5(x) = 1 + x + \frac{1}{2}x^2 + \frac{1}{6}x^3 + \frac{1}{24}x^4 + \frac{1}{120}x^5$$

In Figure 9.4 we include the graph of e^x, along with the graphs of p_1, p_2, and p_5.   ❑

If $f(x) = e^x$, then since $f(1) = e^1 = e$, it follows that we can use $p_n(1)$ to approximate e. For $n = 10$ we obtain

$$p_{10}(1) = 1 + 1 + \frac{1}{2!} + \frac{1}{3!} + \cdots + \frac{1}{10!} \approx 2.718281801$$

Since the value of e is 2.718281828 (accurate to 10 digits), we conclude that $|p_{10}(1) - e| < 3 \times 10^{-8}$. If we used a larger value of n, we would find that $p_n(1)$ is a better estimate of e. Later in the chapter we will find that if n is sufficiently large, then $p_n(1)$ approximates e with any prescribed degree of accuracy.

The next example involves $\ln x$, and will give us a method for estimating $\ln 2$.

EXAMPLE 2   Let $f(x) = \ln(1 + x)$. Find a formula for the nth Taylor polynomial of f, and then calculate $p_n(1)$.

Solution   First we calculate the derivatives of f:

$$f(x) = \ln(1 + x) \qquad\qquad f(0) = 0$$

$$f'(x) = \frac{1}{1 + x} \qquad\qquad f'(0) = 1$$

$$f''(x) = \frac{-1}{(1 + x)^2} \qquad\qquad f''(0) = -1$$

$$f^{(3)}(x) = \frac{(-1)^2 2!}{(1 + x)^3} \qquad\qquad f^{(3)}(0) = 2!$$

$$f^{(4)}(x) = \frac{(-1)^3 3!}{(1 + x)^4} \qquad\qquad f^{(4)}(0) = -(3!)$$

In general, for $k \geq 1$,

$$f^{(k)}(x) = \frac{(-1)^{k-1}(k-1)!}{(1 + x)^k}, \qquad f^{(k)}(0) = (-1)^{k-1}(k-1)!$$

Consequently

$$p_n(x) = f(0) + f'(0)x + \frac{f''(0)}{2!}x^2 + \frac{f^{(3)}(0)}{3!}x^3 + \cdots + \frac{f^{(n)}(0)}{n!}x^n$$

$$= 0 + (1)x + \frac{-1}{2!}x^2 + \frac{2!}{3!}x^3 + \cdots + \frac{(-1)^{n-1}(n-1)!}{n!}x^n$$

$$= x - \frac{x^2}{2} + \frac{x^3}{3} - \frac{x^4}{4} + \cdots + \frac{(-1)^{n-1}}{n}x^n$$

We conclude that

$$p_n(1) = 1 - \frac{1}{2} + \frac{1}{3} - \frac{1}{4} + \cdots + (-1)^{n-1}\frac{1}{n} \qquad \square$$

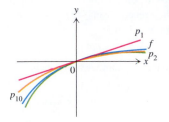

FIGURE 9.5

If we let $n = 10$ and $n = 20$, then we find that

$$p_{10}(1) = 1 - \frac{1}{2} + \frac{1}{3} - \frac{1}{4} + \frac{1}{5} - \frac{1}{6} + \frac{1}{7} - \frac{1}{8} + \frac{1}{9} - \frac{1}{10} \approx 0.6456349206$$

$$p_{20}(1) = 1 - \frac{1}{2} + \frac{1}{3} + \cdots + \frac{1}{19} - \frac{1}{20} \approx 0.6687714032$$

Since $f(1) = \ln(1 + 1) = \ln 2$, and since the value of $\ln 2$ is 0.6931471806 (accurate to 10 digits), we conclude that $p_{20}(1)$ approximates $\ln 2$ with an error less than 0.025. From the discussion it becomes apparent that we may be able to give an approximation of $\ln 2$ to any desired degree of accuracy by computing the sum $1 - \frac{1}{2} + \frac{1}{3} - \frac{1}{4} + \cdots + (-1)^{n-1}/n$ for sufficiently large n. Figure 9.5 shows the graphs of $\ln(1 + x)$, p_1, p_2, and p_{10}.

By the next example it will follow that Taylor polynomials do not always give good estimates for their corresponding functions.

EXAMPLE 3   Let $f(x) = 1/(1 - x)$. Find a formula for the nth Taylor polynomial of f, and compute $p_n(2)$.

Solution   The derivatives of f are

$$f(x) = \frac{1}{1 - x} \qquad\qquad f(0) = 1$$

$$f'(x) = \frac{1}{(1 - x)^2} \qquad\qquad f'(0) = 1$$

$$f''(x) = \frac{2!}{(1 - x)^3} \qquad\qquad f''(0) = 2!$$

$$f^{(3)}(x) = \frac{3!}{(1 - x)^4} \qquad\qquad f^{(3)}(0) = 3!$$

In general,

$$f^{(k)}(x) = \frac{k!}{(1 - x)^{k+1}}, \qquad f^{(k)}(0) = k!$$

As a result,

$$p_n(x) = f(0) + f'(0)x + \frac{f''(0)}{2!}x^2 + \frac{f^{(3)}(0)}{3!}x^3 + \cdots + \frac{f^{(n)}(0)}{n!}x^n$$

$$= 1 + x + x^2 + x^3 + \cdots + x^n$$

In particular,

$$p_n(2) = 1 + 2 + 2^2 + 2^3 + \cdots + 2^n \qquad \square$$

We observe that although $f(2) = 1/(1 - 2) = -1$, the numbers $p_n(2)$ increase without bound as n increases. Therefore in the case of the function in Example 3, the numbers $p_n(2)$ do not approach $f(2)$; in fact they recede from $f(2)$.

Let us recapitulate our conclusions from Examples 1 to 3:

For $f(x) = e^x$: $p_1(1), p_2(1), p_3(1), \ldots$ approach the number $e = f(1)$.

For $f(x) = \ln(1 + x)$: $p_1(1), p_2(1), p_3(1), \ldots$ approach the number $\ln 2 = f(1)$.

For $f(x) = 1/(1 - x)$: $p_1(2), p_2(2), p_3(2), \ldots$ increase without bound

and do not approach $f(2)$.

In each case we have produced a sequence $p_0(a), p_1(a), p_2(a), \ldots,$ of numbers as candidates for an approximation of $f(a)$. Determination of conditions under which a sequence of numbers approaches a specific number is the focus of Sections 9.2 and 9.3. In addition, we notice that $p_n(x)$ is the sum of the $n + 1$ terms $f(0), f'(0)x,$ $f''(0)x^2/2!, \ldots, f^{(n)}(0)x^n/n!$; therefore as n increases, $p_n(x)$ is the sum of an ever-increasing number of terms. We are thus led to consider sums of infinitely many terms. These will be called infinite series, and will be defined in Section 9.4.

EXERCISES 9.1

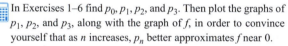 In Exercises 1–6 find $p_0, p_1, p_2,$ and p_3. Then plot the graphs of $p_1, p_2,$ and p_3, along with the graph of f, in order to convince yourself that as n increases, p_n better approximates f near 0.

1. $f(x) = \sin x$ **2.** $f(x) = \cos x$

3. $f(x) = e^{-2x}$ **4.** $f(x) = \ln(2 + x)$

5. $f(x) = \sin(2x)$ **6.** $f(x) = (1 - x)^{1/2}$

In Exercises 7–16 find a formula for an arbitrary Taylor polynomial of f.

7. $f(x) = x^2 - x - 2$ $(n \geq 2)$

8. $f(x) = x^5 + x^3 + 4$ $(n \geq 5)$

9. $f(x) = \dfrac{1}{1 + x}$ **10.** $f(x) = \dfrac{1}{1 + 2x}$

11. $f(x) = e^{-x}$ **12.** $f(x) = e^{3x}$

13. $f(x) = \cosh x$ **14.** $f(x) = \cos x$

15. $f(x) = \sin x$

16. $f(x) = \ln \dfrac{1 + x}{1 - x}$

     (*Hint:* $\ln \dfrac{1 + x}{1 - x} = \ln(1 + x) - \ln(1 - x)$.)

17. Consider the function f defined by

$$f(x) = \begin{cases} e^{-1/x^2} & \text{for } x \neq 0 \\ 0 & \text{for } x = 0 \end{cases}$$

Plot the graph of f. Then by zooming in on the region near 0, guess the Taylor polynomial p_2.

In Exercises 18–25 find the nth Taylor polynomial of f for the given values of n.

18. $f(x) = \sin x^2$; $n = 3$ **19.** $f(x) = e^{-(x^2)}$; $n = 3$

20. $f(x) = \sin^{-1} x$; $n = 2$ **21.** $f(x) = \ln(\cos x)$; $n = 2$

22. $f(x) = \tan x$; $n = 3$ and $n = 4$

23. $f(x) = \sec x$; $n = 2$ and $n = 3$

24. $f(x) = \begin{cases} \dfrac{\sin x}{x} & \text{for } x \neq 0 \\ 1 & \text{for } x = 0 \end{cases}$ $n = 2$

*25. $f(x) = \begin{cases} e^{-1/x^2} & \text{for } x \neq 0 \\ 0 & \text{for } x = 0 \end{cases}$ $n = 2$

26. Find the third Taylor polynomial of $\tan^{-1} x$, and use it to approximate $\pi/4$.

27. Let $f(x) = \sqrt{1 + x}$.

   a. Find the second Taylor polynomial of f.

   b. Use the solution to part (a) to approximate $\sqrt{2}$.

   c. Use the solution to part (a) to approximate $\sqrt{1.1}$.

28. Assume that $f^{(3)}(0)$ exists. Verify that the third Taylor polynomial of f, defined by (1), has the same value and the same first, second, and third derivatives at 0 as f does.

9.2   SEQUENCES

The functions appearing thus far have been defined on intervals of real numbers. This section and the next are devoted to functions of a different type—those with domains consisting only of integers. Examples are provided by the population of the United States taken by the US Census Bureau every decade, or the profit of IBM at the end of each tax year, or the water level in the Great Salt Lake each day at noon. Functions such as these are called sequences.

DEFINITION 9.1   A **sequence** is a function whose domain is the collection of all integers greater than or equal to a given integer m (usually 0 or 1).

If a is a sequence and n is an integer in its domain, then the value of a at n is usually denoted by a_n, rather than $a(n)$. The sequence itself is then written as $\{a_n\}_{n=m}^{\infty}$. It is possible to specify a sequence by giving a formula for a_n. For example, we can write

$$a_n = \frac{1}{n} \quad \text{for } n \geq 1$$

for the sequence that assigns to each positive integer its reciprocal. For this sequence,

$$a_1 = \frac{1}{1} = 1, \quad a_2 = \frac{1}{2}, \quad a_3 = \frac{1}{3}, \quad \text{and so on}$$

The sequence is usually written $\{1/n\}_{n=1}^{\infty}$.

The numbers a_n in the sequence $\{a\}_{n=m}^{\infty}$ are called the **terms** of the sequence. The subscript n is an **index,** and m is the **initial index.** Sometimes we identify a sequence by writing out the first few terms, with the understanding that the remaining terms follow an obvious pattern. For example, we might write

$$1, \frac{1}{2}, \frac{1}{3}, \frac{1}{4}, \cdots \quad \text{for the sequence} \left\{ \frac{1}{n} \right\}_{n=1}^{\infty}$$

Convergent Sequences

For any sequence $\{a_n\}_{n=m}^{\infty}$ we may want to know the behavior of a_n as n increases without bound. For example, if $a_n = 1/n$, then

$$a_1 = 1, \quad a_2 = \frac{1}{2}, \quad a_3 = \frac{1}{3}, \quad a_4 = \frac{1}{4}, \quad a_5 = \frac{1}{5}, \ldots$$

and it appears that a_n is as close to 0 as we wish if we take n sufficiently large. However, for

$$a_n = \left(1 + \frac{0.05}{n} \right)^n$$

it might not be so clear whether a_n approaches some specific number as n increases without bound. But suppose a_n represents a bank balance after 1 year with interest compounded n times per year. Then the behavior of a_n for large n would be of interest to an investor or a bank director. Now we state precisely what it means for the numbers a_n in a sequence to approach a fixed value as n increases without bound.

DEFINITION 9.2   Let $\{a_n\}_{n=m}^{\infty}$ be a sequence. A number L is the **limit** of $\{a_n\}_{n=m}^{\infty}$ if for every $\varepsilon > 0$ there is an integer N such that

$$\text{if } n \geq N, \quad \text{then } |a_n - L| < \varepsilon \tag{1}$$

In this case we write

$$\lim_{n \to \infty} a_n = L$$

If such a number L exists, we say that $\{a_n\}_{n=m}^{\infty}$ **converges** (or **converges to L**), or that $\lim_{n \to \infty} a_n$ **exists.** If such a number L does not exist, we say that $\{a_n\}_{n=m}^{\infty}$ **diverges** or that $\lim_{n \to \infty} a_n$ **does not exist.**

As with limits of functions, the limit of a sequence is unique if it exists.

EXAMPLE 1   Let c be any number, and let $a_n = c$ for $n \geq 1$. Show that $\lim_{n \to \infty} a_n = c$.

Solution   For any $\varepsilon > 0$, let $N = 1$. Observe that

$$\text{if } n \geq N, \quad \text{then } |a_n - c| = |c - c| = 0 < \varepsilon$$

Consequently $\lim_{n \to \infty} a_n = c$ (Figure 9.6).   ❑

We express the result of Example 1 by writing $\lim_{n \to \infty} c = c$. Specifically, $\lim_{n \to \infty} 1 = 1$ and $\lim_{n \to \infty} (-5) = -5$.

EXAMPLE 2   Show that $\{1/n\}_{n=1}^{\infty}$ converges and that $\lim_{n \to \infty} 1/n = 0$.

Solution   For any $\varepsilon > 0$, let N be an integer greater than $1/\varepsilon$. It follows that

$$\text{if } n \geq N, \quad \text{then } \left| \frac{1}{n} - 0 \right| = \frac{1}{n} \leq \frac{1}{N} < \varepsilon$$

This shows that $\lim_{n \to \infty} 1/n = 0$ (Figure 9.7).   ❑

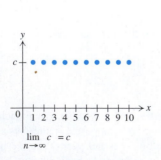

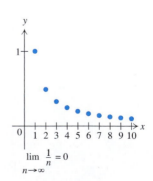

FIGURE 9.6   **FIGURE 9.7**

The sequence $\{1/n\}_{n=1}^{\infty}$ is known as the **harmonic sequence;** notice that $1/n$ is the harmonic mean of $1/(n-1)$ and $1/(n+1)$ (see Exercise 86 of Section 1.1). There is also a musical interpretation for the harmonic sequence. The fundamental tone of a guitar or violin string of length L is obtained by vibrating the whole string. The first harmonic (or overtone) is obtained by lightly touching the vibrating string at its midpoint. The second harmonic (or overtone) is obtained by lightly touching the vibrating string at a point one third of the way up the string. In general, the $(n-1)$st harmonic is obtained by lightly touching the vibrating string at a point $1/n$ of the way up the string. If L is taken to be 1, then the lengths that determine the fundamental and the harmonics form the harmonic sequence $\{1/n\}_{n=1}^{\infty}$.

EXAMPLE 3   Show that $\{(-1)^n\}_{n=1}^{\infty}$ diverges.

Solution   We have

$$(-1)^n = \begin{cases} 1 & \text{for } n \text{ even} \\ -1 & \text{for } n \text{ odd} \end{cases}$$

A vibrating guitar string. (From *Sounds of Music* by Charles Taylor, BBC London)

so the graph is as shown in Figure 9.8. If $L \geq 0$, then for odd values of n we have

$$| (-1)^n - L | = |-1 - L| \geq 1$$

But if $L < 0$, then for even values of n we have

$$| (-1)^n - L | = | 1 - L | \geq 1$$

Consequently for any ε such that $0 < \varepsilon \leq 1$, there is no number L satisfying (1). Therefore the given sequence diverges.   ❑

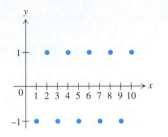

FIGURE 9.8

The sequence in Example 3 diverges because the numbers in the sequence oscillate between 1 and −1 (and hence do not approach a limit). A sequence can also diverge because the numbers in the sequence grow arbitrarily large in absolute value. An example of such a sequence is $\{n^2\}_{n=0}^{\infty}$. To describe such sequences we make the following definition.

DEFINITION 9.3   Let $\{a_n\}_{n=m}^{\infty}$ be a sequence. If for every number M there is an integer N such that

$$\text{if } n \geq N, \quad \text{then } a_n > M \tag{2}$$

we say that $\{a_n\}_{n=m}^{\infty}$ **diverges to** ∞, and we write

$$\lim_{n \to \infty} a_n = \infty$$

Similarly, if for every number M there is an integer N such that

$$\text{if } n \geq N, \quad \text{then } a_n < M$$

we say that $\{a_n\}_{n=m}^{\infty}$ **diverges to** $-\infty$, and we write

$$\lim_{n \to \infty} a_n = -\infty$$

We could apply Definition 9.3 to prove that $\lim_{n \to \infty} n^2 = \infty$. But referring to Definition 9.3 to show that a sequence diverges, or referring to Definition 9.2 to show that a sequence such as those in Examples 1 to 3 converges or diverges, is tedious. One way to avoid constantly using Definitions 9.2 and 9.3 arises from the fact that the definition of $\lim_{n \to \infty} a_n = L$ is very reminiscent of the definition of $\lim_{x \to \infty} f(x) = L$ (see Definition 4.15), and similarly, that the definition of $\lim_{n \to \infty} a_n = \infty$ (or $-\infty$) is reminiscent of the definition of $\lim_{x \to \infty} f(x) = \infty$ (or $-\infty$). These observations lead us to the following theorem.

THEOREM 9.4   Let $\{a_n\}_{n=m}^{\infty}$ be a sequence, L a number, and f a function defined on $[m, \infty)$ such that $f(n) = a_n$ for $n \geq m$.

a.   If $\lim_{x \to \infty} f(x) = L$, then $\{a_n\}_{n=m}^{\infty}$ converges and $\lim_{n \to \infty} a_n = L$.

b.   If $\lim_{x \to \infty} f(x) = \infty$ (or $\lim_{x \to \infty} f(x) = -\infty$), then $\{a_n\}_{n=m}^{\infty}$ diverges, and $\lim_{n \to \infty} a_n = \infty$ (or $\lim_{n \to \infty} a_n = -\infty$).

Thus under the conditions of (a) or (b),

$$\lim_{n \to \infty} a_n = \lim_{x \to \infty} f(x)$$

Proof   To prove (a) we assume that $\lim_{x\to\infty} f(x) = L$, and let $\varepsilon > 0$. Then there is some integer N such that

$$\text{if } x \geq N, \quad \text{then } |f(x) - L| < \varepsilon$$

This implies that

$$\text{if } n \geq N, \quad \text{then } |a_n - L| = |f(n) - L| < \varepsilon$$

so that by (1), $\lim_{n\to\infty} a_n = L$. This proves (a). For (b), we assume that $\lim_{x\to\infty} f(x) = \infty$. Then for any M there is an integer N such that

$$\text{if } x \geq N, \quad \text{then } f(x) > M$$

so that

$$\text{if } n \geq N, \quad \text{then } a_n > M$$

But by (2), this means that $\lim_{n\to\infty} a_n = \infty$. Similarly, $\lim_{x\to\infty} f(x) = -\infty$ implies that $\lim_{n\to\infty} a_n = -\infty$. ■

FIGURE 9.9

The result of part (a) of Theorem 9.4 is illustrated in Figure 9.9.

EXAMPLE 4   Show that $\lim_{n\to\infty} n^2 = \infty$.

Solution   Let

$$f(x) = x^2 \quad \text{for } x \geq 1$$

Then $f(n) = n^2$ for $n \geq 1$. Since $\lim_{x\to\infty} x^2 = \infty$, we conclude from Theorem 9.4 that

$$\lim_{n\to\infty} n^2 = \infty \qquad \square$$

EXAMPLE 5   Show that

$$\lim_{n\to\infty} \left(1 + \frac{1}{n}\right)^n = e$$

Solution   If we let

$$f(x) = \left(1 + \frac{1}{x}\right)^x \quad \text{for } x \geq 1$$

then $f(n) = (1 + 1/n)^n$ for $n \geq 1$. By Example 9 of Section 7.6 we know that $\lim_{x\to\infty} f(x) = e$. Consequently Theorem 9.4 yields

$$\lim_{n\to\infty} \left(1 + \frac{1}{n}\right)^n = e \qquad \square$$

The limit in Example 5 is sometimes used to define the number e. By letting $g(x) = (1 + 0.05/x)^x$ and using the same ideas as in Example 5, we can deduce that $\lim_{x \to \infty} g(x) = e^{0.05}$. Thus

$$\lim_{n \to \infty} \left(1 + \frac{0.05}{n}\right)^n = e^{0.05} \approx 1.05127 \qquad (3)$$

(see Exercise 25). This implies that the sequence in (1) converges to $e^{0.05}$. The limit in (3) is the basis of compounding interest "continuously." If a bank offers 5 percent interest compounded continuously, then (3) tells us that during one year $1000 grows to approximately $1051.27.

EXAMPLE 6   Show that

$$\lim_{n \to \infty} \frac{1}{n^r} = 0 \quad \text{for } r > 0$$

Solution   Let

$$f(x) = \frac{1}{x^r} \quad \text{for } x \geq 1$$

Then $f(n) = 1/n^r$ for $n \geq 1$, and by our analysis of power functions in Section 7.3, we know that $\lim_{x \to \infty} f(x) = 0$. As a result, Theorem 9.4 implies that

$$\lim_{n \to \infty} \frac{1}{n^r} = 0 \qquad \square$$

It follows from Example 6 that the sequences $\{1/n^{1/2}\}_{n=1}^{\infty}$ and $\{1/n^3\}_{n=1}^{\infty}$ converge to 0.

EXAMPLE 7   Let r be any number. Show that

$$\lim_{n \to \infty} r^n = \begin{cases} 0 & \text{for } |r| < 1 \\ 1 & \text{for } r = 1 \end{cases}$$

and that $\{r^n\}_{n=1}^{\infty}$ diverges for all other values of r.

Solution   First we consider nonnegative values of r. Let

$$f(x) = r^x \quad \text{for } x \geq 1$$

so that $f(n) = r^n$ for $n \geq 1$. It follows from our analysis of exponential functions in Section 7.3 that

$$\lim_{x \to \infty} r^x = \begin{cases} 0 & \text{for } 0 \leq r < 1 \\ 1 & \text{for } r = 1 \\ \infty & \text{for } r > 1 \end{cases}$$

(Figure 9.10). By Theorem 9.4 this means that

$$\lim_{n \to \infty} r^n = \begin{cases} 0 & \text{for } 0 \leq r < 1 \\ 1 & \text{for } r = 1 \\ \infty & \text{for } r > 1 \end{cases} \qquad (4)$$

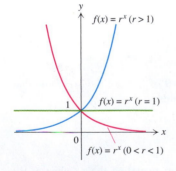

FIGURE 9.10

Thus $\{r^n\}_{n=1}^{\infty}$ diverges for $r > 1$ and converges for $0 \leq r \leq 1$. Next we consider negative values of r. If $r = -1$, then $\{r^n\}_{n=1}^{\infty}$ becomes $\{(-1)^n\}_{n=1}^{\infty}$, which diverges by Example 3. If $r \neq -1$, then since $|r^n| = |r|^n$, we know from (4) that

$$\lim_{n \to \infty} |r^n| = \lim_{n \to \infty} |r|^n = \begin{cases} 0 & \text{for } -1 < r < 0 \\ \infty & \text{for } r < -1 \end{cases}$$

It follows that $\lim_{n \to \infty} r^n = 0$ when $-1 < r < 0$ and that $\lim_{n \to \infty} r^n$ does not exist when $r < -1$. This completes the solution of the example. ❑

From Example 7 we conclude that

$$\left\{ \left(\frac{1}{2} \right)^n \right\}_{n=1}^{\infty}, \text{ which is the sequence } \frac{1}{2}, \frac{1}{4}, \frac{1}{8}, \ldots, \text{ converges to } 0$$

$$\{1^n\}_{n=1}^{\infty}, \text{ which is the sequence } 1, 1, 1, \ldots, \text{ converges to } 1$$

$$\{1.1^n\}_{n=1}^{\infty}, \text{ which is the sequence } 1.1, 1.1^2, 1.1^3, \ldots, \text{ diverges}$$

For any number r, the sequence $\{r^n\}_{n=1}^{\infty}$ in Example 7 is called a geometric sequence. More generally, for any nonzero constant c, the sequence $\{cr^n\}_{n=m}^{\infty}$ is called a **geometric sequence.** The first term of the sequence, cr^m, is the **initial term.** Since

$$\frac{cr^{n+1}}{cr^n} = r$$

for any $n \geq m$, we call r the **common ratio** of the sequence.

EXAMPLE 8 Show that

$$\lim_{n \to \infty} \sqrt[n]{c} = 1 \quad \text{for } c > 0$$

Solution Notice that

$$\sqrt[n]{c} = c^{1/n} = e^{(1/n) \ln c}$$

Thus we let

$$f(x) = e^{(1/x) \ln c} \quad \text{for } x \geq 1$$

so f is continuous and $f(n) = e^{(1/n) \ln c} = \sqrt[n]{c}$ for $n \geq 1$. Since $\lim_{x \to \infty} (1/x) \ln c = 0$, we have

$$\lim_{x \to \infty} f(x) = \lim_{x \to \infty} e^{(1/x) \ln c} = e^0 = 1$$

It follows from Theorem 9.4 that $\lim_{n \to \infty} \sqrt[n]{c} = 1$. ❑

EXAMPLE 9 Show that

$$\lim_{n \to \infty} \sqrt[n]{n} = 1$$

Solution   Imitating the solution of Example 8, we notice that

$$\sqrt[n]{n} = n^{1/n} = e^{(1/n)\ln n}$$

Thus we let

$$f(x) = e^{(1/x)\ln x} \quad \text{for } x \geq 1$$

so that f is continuous and $f(n) = e^{(1/n)\ln n} = \sqrt[n]{n}$ for $n \geq 1$. Since

$$\lim_{x \to \infty} x = \infty = \lim_{x \to \infty} \ln x$$

l'Hôpital's Rule implies that

$$\lim_{x \to \infty} \frac{\ln x}{x} = \lim_{x \to \infty} \frac{1/x}{1} = \lim_{x \to \infty} \frac{1}{x} = 0$$

and thus

$$\lim_{x \to \infty} f(x) = \lim_{x \to \infty} e^{(1/x)\ln x} = e^0 = 1$$

From Theorem 9.4 we conclude that $\lim_{n \to \infty} \sqrt[n]{n} = 1$.   ❑

Population Dynamics

Suppose that the population of a certain species (such as black bears or Canada geese) is considered as a function of the year. Then we obtain a sequence $\{P_n\}_{n=0}^{\infty}$, where P_n denotes the population at the end of year n, and where $n = 0$ corresponds to some conveniently chosen initial year (such as 1900). When a sequence $\{P_n\}_{n=0}^{\infty}$ denotes population, the study of the sequence and its limit $\lim_{n \to \infty} P_n$ is called **population dynamics.** As they try to encourage or restrict growth of certain species, conservationists are naturally interested in the population dynamics of those species.

At this point we will discuss two formulas that could be used to predict the population of a particular species in the future. In each case the population P_{n+1} after $n + 1$ units of time will be a function of the population P_n after n units of time.

Our first formula says that to obtain P_{n+1} we add to P_n a certain fraction $r_0 P_n$ of P_n, where r_0 is independent of n:

$$P_{n+1} = P_n + r_0 P_n \quad \text{for } n \geq 0 \tag{5}$$

The constant r_0 is called the **specific** (or **relative**) **rate of growth** of the population, because

$$r_0 = \frac{P_{n+1} - P_n}{P_n} = \frac{\text{change in population}}{\text{population}}$$

For example, if the population of black bears in a certain region increases from 10,000 to 10,300 in one year, then the rate of growth r_0 would be

$$\frac{10{,}300 - 10{,}000}{10{,}000} = 0.03$$

which corresponds to the 3 percent increase in population during the year. Similarly, if the population decreased to 9,700, then r_0 would equal -0.03, corresponding to the 3 percent decrease in population during the year.

If we let $r = 1 + r_0$, then (5) becomes

$$P_{n+1} = r P_n \quad \text{for } n \geq 0 \tag{6}$$

Notice that if $r > 1$, then the population is growing, whereas if $0 < r < 1$, then the population is shrinking. Finally, if $r = 1$, then the population is constant. Starting with $n = 0$ in (6) and letting n increase, we obtain successively the equations

$$P_1 = rP_0$$

$$P_2 = rP_1 = r^2P_0$$

$$P_3 = rP_2 = r^3P_0$$

$$P_4 = rP_3 = r^4P_0$$

It follows that in general,

$$P_n = r^nP_0 \quad \text{for } n \geq 0 \tag{7}$$

Thus $\{P_n\}_{n=0}^{\infty}$ is a geometric sequence, with common ratio r and initial term P_0. When it is used in the context of population dynamics, the sequence is sometimes called a **Malthus sequence,** after the English professor of history and political economy Thomas Malthus (1766–1834), whose famous essay on population growth had a remarkable influence on Charles Darwin's theory of evolution.

It follows from Example 7 that if $0 \leq r < 1$, then $\lim_{n \to \infty} P_n = 0$ and hence the population would die out. In contrast, if $r > 1$, then $\lim_{n \to \infty} P_n = \infty$, so the population would grow without bound.

EXAMPLE 10   Suppose that on June 1 there are 100 gnats per acre of land, and let P_n be the gnat population n days after June 1. Assume that the specific rate of growth of the gnat population is 0.3. Determine the number of gnats per acre by August 1.

Solution   By hypothesis, $P_0 = 100$. Moreover, we are assuming that $r_0 = 0.3$, so that $r = 1.3$. Since there are 61 days between June 1 and August 1, we are asked to find the population P_{61} of gnats after 61 days, which by (7) is given by

$$P_{61} = r^{61}P_0 = (1.3)^{61}\,100 \approx 892{,}000{,}000 \qquad \square$$

Eight hundred million gnats per acre is an unimaginable number of gnats; it is also unrealistic because of limitations of natural resources (such as food) and because of predators and disease. However, Example 10 helps us understand that although (7) could perhaps be reasonable if the population were dying out, it cannot long be reasonable for a population that is growing. As a result, (7) has only limited value in the study of population dynamics. Nevertheless, before turning to the next formula for population, we observe that if in (7) we substitute $f(n)$ for P_n, e^k for r, and $f(0)$ for P_0, then (7) becomes

$$f(n) = f(0)(e^k)^n = f(0)e^{kn}$$

which is what the exponential growth and decay formula (4) of Section 4.4 reduces to when $t = n$. Because the variable in (7) is confined to the nonnegative integers, (7) is called a "discrete analogue" of the exponential growth and decay formula.

A second formula for population incorporates not only the population P_n after n units of time, but also the existence of a number M called the carrying capacity, which has the property that if the population should ever exceed M, then the population would have to decrease. The formula we have in mind is

$$P_{n+1} = P_n + \frac{r(M - P_n)}{M} P_n \qquad (8)$$

where M and r are constants (r is called the **growth parameter** of the population). Formula (8) was devised by the nineteenth-century Belgian mathematician P. F. Verhulst for studying human populations. Because of this, the sequence $\{P_n\}_{n=0}^{\infty}$ is sometimes called a **Verhulst sequence.** Notice from (8) that if $P_n > M$, then $P_{n+1} < P_n$, so that the population indeed decreases when it exceeds M; similarly, if $P_n < M$, then $P_{n+1} > P_n$ (see Exercise 47). Observe also from (8) that the specific rate of growth $(P_{n+1} - P_n)/P_n$ is given by

$$\frac{P_{n+1} - P_n}{P_n} = \left(\frac{M - P_n}{M}\right) r$$

Although r is fixed in the equation, the specific rate of growth is in general not constant because P_n varies with n; moreover, the closer P_n is to the carrying capacity M, the smaller in absolute value the growth rate is.

In (8) there are four quantities: P_n, P_{n+1}, r, and M. Any time we know three of the quantities, we can determine the possible values for the fourth.

EXAMPLE 11   The 1900 and 1910 census figures for the United States were given as follows:

$$1900: 76{,}212{,}168 \qquad 1910: 92{,}228{,}496$$

In order to use the given data to predict the population in 1920, let us assume, as was actually done around the turn of the century, that the carrying capacity of the United States would be approximately 197,000,000. From this information, predict the population in 1920.

Solution   If P_0 and P_1 denote the census figures for 1900 and 1910, respectively, then our goal is to calculate P_2. First we will find r. For ease of calculation let us round off all numbers to the nearest hundred thousand, so that

$$P_0 = 76.2 \times 10^6 \quad P_1 = 92.2 \times 10^6 \quad M = 197 \times 10^6$$

If we let $n = 0$ in (8) and solve for r, we obtain

$$r = \frac{M(P_1 - P_0)}{P_0(M - P_0)} \approx \frac{(197 \times 10^6)(92.2 \times 10^6 - 76.2 \times 10^6)}{(76.2 \times 10^6)(197 \times 10^6 - 76.2 \times 10^6)} \approx 0.342$$

It then follows from (8), with $n = 1$, that

$$P_2 = P_1 + \frac{r(M - P_1)}{M} P_1$$

$$\approx 92.2 \times 10^6 + \frac{(0.342)(197 \times 10^6 - 92.2 \times 10^6)}{197 \times 10^6} (92.2 \times 10^6)$$

$$\approx 109 \times 10^6$$

Therefore the predicted population of the United States in 1920 would be approximately 109 million.   ❑

The actual census figure for 1920 was 106,021,537, which is approximately 2.7 percent below the predicted figure given in the solution of the example. If we were to use the same value of r in calculating the predicted census figure for 1930 from the actual figures for 1920, we would find an excellent correlation between the predicted and actual data (see Exercise 50).

Suppose a population is changing according to the Verhulst equation (8). Since the population increases when it is less than the carrying capacity M and decreases when it is greater than M, it is tempting to conjecture that in time the population would approach M. Indeed, this is the case when the growth parameter r is not too large. However, to see what can happen for larger r, let us fix $M = 1$ and assume that the initial population $P_0 = 0.15$ (say, in millions). With these values of M and P_0 and five selected values of r, we use (8) to calculate successive values of P_n and observe their behavior:

$r = 1.5$ P_n eventually approaches 1

$r = 2.2$ P_n eventually bounces between 0.746 and 1.163

$r = 2.5$ P_n eventually bounces between 1.158, 0.701, 1.224, and 0.536

$r = 2.54$ P_n eventually bounces between eight numbers

$r = 2.55$ P_n eventually bounces between sixteen numbers

(see Figure 9.11). This type of oscillation continues as r increases, until it reaches a certain critical number that is approximately 2.570, at which point P_n no longer oscillates between a finite set of numbers, nor does it approach a limit. Instead it jumps so wildly that the word "chaotic" is often used to describe its behavior. Such wild behavior can in fact occur with populations of certain insects under special conditions. Gypsy moth populations in the Northeast seem to exhibit such chaotic behavior year by year—a kind of feast or famine. We mention also that sequences displaying chaotic behavior have occurred in the analysis of, for example, weather prediction, turbulent flow, laser physics, and kinetics of chemical reactions.

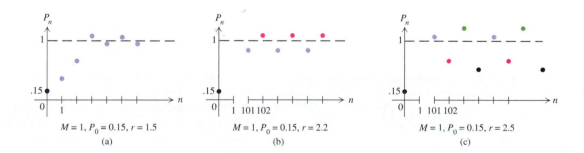

FIGURE 9.11   (a) P_n converges to 1. (b) P_n oscillates between 2 points. (c) P_n oscillates between 4 points.

EXERCISES 9.2

In Exercises 1–4 write the initial four terms of the sequence.

1. $\left\{\dfrac{1}{n}\right\}_{n=3}^{\infty}$

2. $\left\{\dfrac{1}{3^n}\right\}_{n=0}^{\infty}$

3. $\left\{\dfrac{k-1}{k+1}\right\}_{k=1}^{\infty}$

4. $\left\{k-\dfrac{1}{k}\right\}_{k=5}^{\infty}$

In Exercises 5–12 use Definition 9.2 or 9.3 to verify the equation.

5. $\lim\limits_{n\to\infty}(-2)=-2$

6. $\lim\limits_{n\to\infty}\dfrac{1}{n^2}=0$

7. $\lim\limits_{n\to\infty}\dfrac{3n+1}{n}=3$

8. $\lim\limits_{n\to\infty}\dfrac{2n-1}{n+1}=2$

9. $\lim\limits_{n\to\infty}\sqrt{n}=\infty$

10. $\lim\limits_{n\to\infty}(-2n^3)=-\infty$

11. $\lim\limits_{k\to\infty}e^k=\infty$

12. $\lim\limits_{k\to\infty}\dfrac{k^2+1}{k}=\infty$

In Exercises 13–32 evaluate the limit as a number, ∞, or $-\infty$.

13. $\lim\limits_{n\to\infty}\left(\pi+\dfrac{1}{n}\right)$

14. $\lim\limits_{n\to\infty}(\pi-n)$

15. $\lim\limits_{j\to\infty}(0.8)^j$

16. $\lim\limits_{j\to\infty}\dfrac{3^j}{2^j}$

17. $\lim\limits_{n\to\infty}e^{-n}$

18. $\lim\limits_{n\to\infty}e^{1/n}$

19. $\lim\limits_{n\to\infty}\dfrac{n+3}{n^2-2}$

20. $\lim\limits_{n\to\infty}\dfrac{5n^2+1}{4-3n^2}$

21. $\lim\limits_{n\to\infty}\dfrac{2n^2-4}{-n-5}$

22. $\lim\limits_{n\to\infty}\cos\dfrac{\pi}{n}$

23. $\lim\limits_{n\to\infty}n\sin\dfrac{\pi}{n}$

24. $\lim\limits_{n\to\infty}\ln\dfrac{1}{n}$

25. $\lim\limits_{n\to\infty}\left(1+\dfrac{0.05}{n}\right)^n$

26. $\lim\limits_{k\to\infty}\left(1+\dfrac{1}{3k}\right)^k$

27. $\lim\limits_{k\to\infty}(1+k)^{1/(2k)}$

28. $\lim\limits_{k\to\infty}\tan^{-1}k$

29. $\lim\limits_{k\to\infty}\sin^{-1}\left(\dfrac{1}{\sqrt{2}}\cos\dfrac{1}{k}\right)$

30. $\lim\limits_{k\to\infty}\sqrt[k]{2k}$

31. $\lim\limits_{n\to\infty}\displaystyle\int_{-1/n}^{1/n}e^x\,dx$

32. $\lim\limits_{n\to\infty}\displaystyle\int_{1+1/n}^{2-1/n}\dfrac{1}{x}\,dx$

In Exercises 33–38 determine whether the sequence converges or diverges. If it converges, find its limit.

33. $\{-4n\}_{n=-2}^{\infty}$

34. $\left\{\dfrac{n-1}{n}\right\}_{n=1}^{\infty}$

35. $\left\{\dfrac{1}{n^2-1}\right\}_{n=2}^{\infty}$

36. $\{2^n\}_{n=1}^{\infty}$

37. $\left\{\left(-\dfrac{1}{3}\right)^n\right\}_{n=1}^{\infty}$

38. $\left\{\dfrac{1}{n}-n\right\}_{n=2}^{\infty}$

In Exercises 39–42 find the smallest positive integer n for which $|a_n|<\varepsilon$.

39. $a_n=(0.25)^n;\ \varepsilon=10^{-5}$

40. $a_n=(0.9)^n;\ \varepsilon=10^{-5}$

41. $a_n=1-\sqrt[n]{1.2};\ \varepsilon=10^{-3}$

42. $a_n=1-\sqrt[n]{n};\ \varepsilon=2\times10^{-2}$

In Theorem 9.4 we have $\lim_{n\to\infty}a_n=\lim_{x\to\infty}f(x)=\lim_{x\to0^+}f(1/x)$. In Exercises 43–44 find an appropriate function f corresponding to the sequence $\{a_n\}_{n=1}^{\infty}$, and plot $f(1/x)$. Then guess $\lim_{n\to\infty}a_n$.

43. $a_n=n\sinh e^{-n}$

44. $a_n=n^{10}\ln(1+e^{-n})$

45. Prove that a convergent sequence has a unique limit.

*46. Prove that if $\lim_{n\to\infty}a_{2n}=\lim_{n\to\infty}a_{2n+1}$, then $\lim_{n\to\infty}a_n$ exists and

$$\lim_{n\to\infty}a_n=\lim_{n\to\infty}a_{2n}=\lim_{n\to\infty}a_{2n+1}$$

47. Show that in a Verhulst sequence given by (8), if $P_n>M$, then $P_{n+1}<P_n$, and if $P_n<M$, then $P_{n+1}>P_n$.

48. Suppose that the specific growth rate of the gnat population in Example 10 was only 0.2. Determine the number of gnats per acre on August 1.

49. Using the census figures given in Example 11, along with the formula in (7) for a Malthus sequence, predict the population for 1920. Is the prediction better when we use a Malthus sequence than when we use a Verhulst sequence?

Applications

50. The actual census figures for 1910 and 1920 are as follows:

1910: 92,228,496   1920: 106,021,537

a. Using the data above and the value of r determined in Example 11, predict the population in 1930 by means of (8). (The actual figure for 1930 is 123,202,624.)

b. Determine a new value of r by using (8) with the 1910 and 1920 census figures given in this exercise. Then determine the population in 1930 by means of (8). Is this value closer to the figure for 1930 than the value found in part (a)?

51. If \$1000 is deposited in a savings account at an interest rate of r percent per year, then the number of dollars (principal plus interest) in the account after 1 year is $1000(1+0.01r)$. Write a formula for the sequence that gives the amount of money in the account after n years for any positive integer n.

52. Suppose a foundation deposits $1000 into a building fund that pays 5 percent interest compounded annually, and that at the end of each year thereafter the foundation adds $1000 to the account.

  a. Let a_n denote the amount in the account at the end of the nth year (after the $1000 has been added). Express a_{n+1} in terms of a_n.

  b. Show that $a_n = 1000\,(1 + 1.05 + 1.05^2 + \cdots + 1.05^n)$.

  c. Show that $a_n = 20{,}000\,(1.05^{n+1} - 1)$.

53. Suppose P dollars are deposited in a savings account at an interest rate of r percent per year, compounded n times a year.

  a. Show that the amount of money in the account after 1 year is given by

  $$R_n = P\left(1 + \frac{0.01r}{n}\right)^n$$

  dollars (*Hint:* The amount after $1/n$ years is $P(1 + 0.01r/n)$.)

  b. Find

  $$R = \lim_{n \to \infty} P\left(1 + \frac{0.01r}{n}\right)^n$$

  This is the amount in the account after 1 year if the interest is compounded "continuously."

  c. If $P = 1000$ and $r = 5$, find the difference between the amounts after 1 year if the interest is compounded continuously and if it is compounded quarterly.

54. The **Sierpinski carpet** is constructed as follows: Begin with a square region R with sides of length 1. Divide R into 9 subsquares of equal area, and remove the interior but not the boundary of the middle square (Figure 9.12(a)). For each of the remaining subsquares, perform the same operation, leaving a region consisting of 64 smaller squares (Figure 9.12(b)). Let R_n be the region that remains after performing the same operation n times (Figure 9.12(c)). The Sierpinski carpet S consists of all the points in R that are not removed by any of the operations; in other words, S consists of all points that are in R_n for every $n \geq 1$.

  a. Find the area A_n of R_n for any given $n \geq 1$.

  b. Show that $\lim_{n \to \infty} A_n = 0$. (Since S is contained in each R_n, this result implies that the Sierpinski carpet cannot have positive area.)

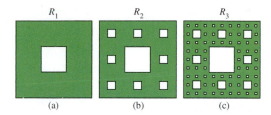

FIGURE 9.12   The construction of the Sierpinski carpet in Exercise 54.

Project

1. This project relates certain Riemann sums with special sequences.

  a. Let

  $$a_n = \frac{1}{n^2} + \frac{2}{n^2} + \frac{3}{n^2} + \cdots + \frac{n}{n^2}, \text{ for } n \geq 1$$

  Show that a_n is a Riemann sum for $\int_0^1 x\,dx$ for each $n \geq 1$. Then find $\lim_{n \to \infty} a_n$.

  b. Let

  $$b_n = \frac{1^2}{n^3} + \frac{2^2}{n^3} + \frac{3^2}{n^3} + \cdots + \frac{n^2}{n^3}, \text{ for } n \geq 1$$

  Show that b_n is a Riemann sum for $\int_0^1 x^2\,dx$ for each $n \geq 1$. Then find $\lim_{n \to \infty} b_n$.

  c. Let k be any positive integer with $k \geq 1$, and let

  $$c_n = \frac{1^k}{n^{k+1}} + \frac{2^k}{n^{k+1}} + \frac{3^k}{n^{k+1}} + \cdots + \frac{n^k}{n^{k+1}}, \text{ for } n \geq 1$$

  Find a suitable integral $\int_0^1 f(x)\,dx$ such that c_n is a Riemann sum for it, for $n \geq 1$. Then find $\lim_{n \to \infty} c_n$.

  d. Suppose that the sequence in part (c) is altered to be

  $$d_n = \frac{2^k}{n^{k+1}} + \frac{3^k}{n^{k+1}} + \cdots + \frac{(n-1)^k}{n^{k+1}}, \text{ for } n \geq 1$$

  Find a suitable integral $\int_0^1 g(x)\,dx$ such that d_n is a Riemann sum for it, for $n \geq 1$. Then find $\lim_{n \to \infty} d_n$.

9.3   CONVERGENCE PROPERTIES OF SEQUENCES

In this section we continue our analysis of the convergence and divergence of sequences. Since sequences are functions, we may add, subtract, multiply, and divide sequences just as we do functions. Rules for computing the limits of combinations of sequences are analogous to the rules for limits of combinations of functions. We present these rules now.

Let $\{a_n\}_{n=m}^{\infty}$ and $\{b_n\}_{n=m}^{\infty}$ be convergent sequences. Then the sum $\{a_n + b_n\}_{n=m}^{\infty}$, any scalar multiple $\{ca_n\}_{n=m}^{\infty}$, the product $\{a_nb_n\}_{n=m}^{\infty}$, and (provided $\lim_{n\to\infty} b_n \neq 0$) the quotient $\{a_n/b_n\}_{n=m}^{\infty}$ all converge, with

$$\lim_{n\to\infty} (a_n + b_n) = \lim_{n\to\infty} a_n + \lim_{n\to\infty} b_n \tag{1}$$

$$\lim_{n\to\infty} ca_n = c \lim_{n\to\infty} a_n \tag{2}$$

$$\lim_{n\to\infty} a_nb_n = \lim_{n\to\infty} a_n \lim_{n\to\infty} b_n \tag{3}$$

$$\lim_{n\to\infty} \frac{a_n}{b_n} = \frac{\displaystyle\lim_{n\to\infty} a_n}{\displaystyle\lim_{n\to\infty} b_n} \tag{4}$$

EXAMPLE 1   Find $\displaystyle\lim_{n\to\infty} \frac{2n - 1}{3n + 5}$.

Solution   We divide the numerator and the denominator by n and then use (1)–(4), along with Examples 1 and 2 of Section 9.2. This yields

$$\lim_{n\to\infty} \frac{2n - 1}{3n + 5} = \lim_{n\to\infty} \frac{2 - 1/n}{3 + 5/n} = \frac{\displaystyle\lim_{n\to\infty} 2 - \lim_{n\to\infty} 1/n}{\displaystyle\lim_{n\to\infty} 3 + 5 \lim_{n\to\infty} 1/n} = \frac{2 - 0}{3 + 0} = \frac{2}{3} \quad \square$$

The version of the Squeezing Theorem for sequences is as follows:

If $\lim_{n\to\infty} a_n = \lim_{n\to\infty} b_n$, and if $\{c_n\}_{n=m}^{\infty}$ is any sequence such that $a_n \leq c_n \leq b_n$ for $n \geq m$, then $\{c_n\}_{n=m}^{\infty}$ converges, and moreover,

$$\lim_{n\to\infty} c_n = \lim_{n\to\infty} a_n = \lim_{n\to\infty} b_n \tag{5}$$

EXAMPLE 2   Show that $\displaystyle\lim_{n\to\infty} \frac{\sin^2 n}{n} = 0$.

Solution   We observe that $0 \leq \sin^2 n \leq 1$, so that

$$0 \leq \frac{\sin^2 n}{n} \leq \frac{1}{n}$$

Since $\lim_{n\to\infty} 0 = 0$ and $\lim_{n\to\infty} 1/n = 0$, it follows from the Squeezing Theorem for sequences that

$$\lim_{n\to\infty} \frac{\sin^2 n}{n} = 0 \quad \square$$

Notice that

$$\{a_n\}_{n=1}^{\infty} \text{ is the sequence } a_1, a_2, a_3, a_4, \ldots$$

Similarly,

$$\{a_{n+1}\}_{n=1}^{\infty} \text{ is the sequence } a_2, a_3, a_4, a_5, \ldots$$

which is the same as the sequence $\{a_n\}_{n=2}^{\infty}$. Since $\{a_n\}_{n=1}^{\infty}$ and $\{a_{n+1}\}_{n=1}^{\infty}$ have the same terms except for a_1 in the first sequence, it follows that $\{a_{n+1}\}_{n=1}^{\infty}$ has a limit if and only if $\{a_n\}_{n=1}^{\infty}$ does, and in that case,

$$\lim_{n \to \infty} a_{n+1} = \lim_{n \to \infty} a_n \tag{6}$$

Thus the sequences

$$1, \frac{1}{2}, \frac{1}{3}, \frac{1}{4}, \frac{1}{5}, \ldots \quad \text{and} \quad \frac{1}{2}, \frac{1}{3}, \frac{1}{4}, \frac{1}{5}, \ldots$$

have the same limit—namely, 0. More generally, if $\{a_n\}_{n=m}^{\infty}$ converges and if k is any integer, then the sequence $\{a_{n+k}\}_{n=m}^{\infty}$ converges, and moreover,

$$\lim_{n \to \infty} a_{n+k} = \lim_{n \to \infty} a_n \tag{7}$$

Consequently in determining whether a sequence converges or diverges, we may ignore the first few terms and apply any of our results to the portion of the sequence that remains.

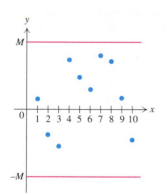

FIGURE 9.13   A bounded sequence.

Bounded Sequences

In analogy with boundedness for a function (see Section 8.7), we say that a sequence $\{a_n\}_{n=m}^{\infty}$ is **bounded** if there is a number M such that $|a_n| \leq M$ for every $n \geq m$ (Figure 9.13). Otherwise, we say that the sequence is **unbounded.** Since $|1/n| \leq \frac{1}{2}$ for $n \geq 2$, and since $|(-1)^n| \leq 1$ for $n \geq 1$, the sequences $\{1/n\}_{n=2}^{\infty}$ and $\{(-1)^n\}_{n=1}^{\infty}$ are bounded. The sequence $\{n^2\}_{n=1}^{\infty}$ is unbounded because no matter what number M we consider, we can always find a value of n such that $n^2 > M$. In the next theorem we will prove that any sequence $\{a_n\}_{n=m}^{\infty}$ that has a limit L is automatically bounded. The basic idea of the proof is that since the numbers a_n approach L as n increases, it follows that from a certain index (say N) onward, a_n is within 1 of L. Thus $\{a_n\}_{n=N}^{\infty}$ is bounded. Since $\{a_n\}_{n=m}^{\infty}$ differs from $\{a_n\}_{n=N}^{\infty}$ only in containing the additional terms $a_m, a_{m+1}, \ldots, a_{N-1}$, it too must be bounded.

THEOREM 9.5

a. If $\{a_n\}_{n=m}^{\infty}$ converges, then $\{a_n\}_{n=m}^{\infty}$ is bounded.

b. If $\{a_n\}_{n=m}^{\infty}$ is unbounded, then $\{a_n\}_{n=m}^{\infty}$ diverges.

Proof   To prove (a), suppose $\lim_{n \to \infty} a_n = L$, which implies that there is an N such that

$$\text{if } n \geq N, \quad \text{then } |a_n - L| < 1$$

Therefore

$$\text{if } n \geq N, \quad \text{then } |a_n| = |a_n - L + L| \leq |a_n - L| + |L| < 1 + |L|$$

Let M be a number larger than $|a_m|, |a_{m+1}|, \ldots, |a_{N-1}|$, and $1 + |L|$. Then $|a_n| \leq M$ for $n \geq m$, and thus the sequence is bounded. Since part (b) is logically equivalent to (a), the proof of the theorem is complete. ∎

Theorem 9.5 tells us immediately that the unbounded sequences

$$\{2^n\}_{n=0}^{\infty} \quad \text{and} \quad \{\ln n\}_{n=2}^{\infty}$$

diverge.

Caution: Theorem 9.5 does *not* imply that all bounded sequences converge, and indeed that is not the case. For example, $\{(-1)^n\}_{n=1}^{\infty}$ is bounded, but it diverges, as we proved in Example 3 of Section 9.2.

We say that the sequence $\{a_n\}_{n=m}^{\infty}$ is **increasing** if as a function it is increasing. This is equivalent to saying that $a_n < a_{n+1}$ for each $n > m$ (Figure 9.14(a)). Similarly, $\{a_n\}_{n=m}^{\infty}$ is **decreasing** if $a_n > a_{n+1}$ for each $n > m$ (Figure 9.14(b)). Thus of the sequences

$$\left\{ 1 - \frac{1}{n} \right\}_{n=1}^{\infty}, \quad \left\{ 1 + 1 + \frac{1}{2!} + \frac{1}{3!} + \cdots + \frac{1}{n!} \right\}_{n=1}^{\infty}, \quad \text{and} \quad \left\{ \sin \frac{1}{n} \right\}_{n=1}^{\infty}$$

the first and second are increasing, and the third is decreasing. As with other types of functions, a sequence need not be increasing or decreasing. For example, $\{(-1)^n\}_{n=m}^{\infty}$ is neither increasing nor decreasing, because it oscillates between 1 and -1.

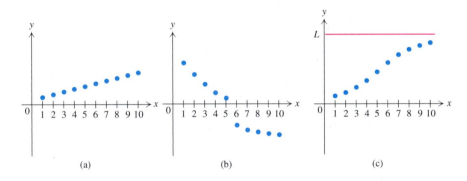

FIGURE 9.14   (a) An increasing sequence. (b) A decreasing sequence. (c) Every bounded increasing sequence has a limit.

Although a bounded sequence may diverge, a bounded sequence that is either increasing or decreasing must converge. The proof of this assertion is given in the Appendix, but Figure 9.14(c) suggests why we might expect it to be true.

THEOREM 9.6   A bounded sequence $\{a_n\}_{n=m}^{\infty}$ that is either increasing or decreasing converges.

One can show that if, for instance, $\{a_n\}_{n=1}^{\infty}$ is increasing and bounded, then $\lim_{n\to\infty} a_n = L$, where L is the smallest number larger than or equal to every term in the sequence. Thus for the bounded, increasing sequence $\{(n-1)/n\}_{n=1}^{\infty}$, the limit 1 is the smallest number greater than $(n-1)/n$ for all positive integers n.

Although Theorem 9.6 will be of theoretical interest to us (especially in Section 9.7), it can be used in showing that certain sequences converge.

EXAMPLE 3   Let $\{a_n\}_{n=0}^{\infty}$ be the sequence defined by

$$a_0 = 0 \quad \text{and} \quad a_{n+1} = a_n^2 + \frac{1}{4} \quad \text{for all } n \geq 0$$

Show that $\{a_n\}_{n=0}^{\infty}$ converges.

Solution   By Theorem 9.6 it suffices to prove that the sequence is bounded and increasing. In order to prove that the sequence is bounded, we need only show that $|a_n| < \frac{1}{2}$ for $n \geq 0$. Since $a_0 = 0$, we have $|a_0| < \frac{1}{2}$. Moreover, if $|a_n| < \frac{1}{2}$, then

$$|a_{n+1}| = a_n^2 + \frac{1}{4} < \left(\frac{1}{2}\right)^2 + \frac{1}{4} = \frac{1}{2}$$

Thus $|a_{n+1}| < \frac{1}{2}$. By induction we conclude that the given sequence is bounded. To prove that it is increasing, we must show that for each $n \geq 0$, we have $a_n < a_{n+1}$, which means that

$$a_n < a_n^2 + \frac{1}{4} \tag{8}$$

But (8) is equivalent to

$$a_n^2 - a_n + \frac{1}{4} > 0, \quad \text{that is,} \quad \left(a_n - \frac{1}{2}\right)^2 > 0$$

which is valid since $a_n \neq \frac{1}{2}$ from the first part. Consequently the given sequence is increasing. Theorem 9.6 then implies that the bounded increasing sequence converges.   ❏

Now that we know the sequence in Example 3 converges, we can find its limit. Let $L = \lim_{n \to \infty} a_n$. Then $L = \lim_{n \to \infty} a_{n+1}$ also. Therefore

$$L = \lim_{n \to \infty} a_{n+1} = \lim_{n \to \infty}\left(a_n^2 + \frac{1}{4}\right) = (\lim_{n \to \infty} a_n)^2 + \frac{1}{4} = L^2 + \frac{1}{4}$$

It follows that

$$L^2 - L + \frac{1}{4} = 0, \text{ or } \left(L - \frac{1}{2}\right)^2 = 0$$

so that $L = \frac{1}{2}$. Consequently $\lim_{n \to \infty} a_n = \frac{1}{2}$.

We conclude by mentioning that in Example 3 we specified a_0 and expressed a_{n+1} in terms of a_n. More generally, if we specify a_0 and express a_{n+1} in terms of $a_0, a_1, a_2, \ldots, a_n$, we say that $\{a_n\}_{n=0}^{\infty}$ is **defined recursively.** Thus the sequence given in Example 3 is defined recursively, as are the sequences in Exercises 32–35 and 38–39. For an example in which a_{n+1} is expressed in terms of a_{n-1} and a_n, see the Project, which concerns the famous Fibonacci sequence.

We mention also that the sequence comprising the Newton-Raphson method is a recursively defined sequence (see Section 3.8). To be more specific, let f be differentiable on an open interval containing a zero d, and let c_1 be an initial guess for d. Then the sequence defined by

$$c_{n+1} = c_n - \frac{f(c_n)}{f'(c_n)} \qquad \text{for } n \geq 1$$

constitutes the algorithm for the Newton-Raphson method, and is recursively defined.

EXERCISES 9.3

In Exercises 1–6 determine whether the sequence is bounded or unbounded.

1. $\left\{1 + \dfrac{2}{n}\right\}_{n=1}^{\infty}$

2. $\{e^{1/n}\}_{n=1}^{\infty}$

3. $\left\{n + \dfrac{1}{2n}\right\}_{n=1}^{\infty}$

4. $\{\cos n^2\}_{n=0}^{\infty}$

5. $\{\cosh k\}_{k=10}^{\infty}$

6. $\{\ln (1 + e^k)\}_{k=1}^{\infty}$

In Exercises 7–30 find the limit.

7. $\displaystyle\lim_{n \to \infty} \left(2 + \frac{1}{n}\right)$

8. $\displaystyle\lim_{n \to \infty} \left(4 - \frac{2}{n}\right)$

9. $\displaystyle\lim_{n \to \infty} \left(\frac{1}{n} - \frac{1}{n+1}\right)$

10. $\displaystyle\lim_{n \to \infty} \left(1 - \frac{1}{n}\right)\left(1 + \frac{1}{n^2}\right)$

11. $\displaystyle\lim_{n \to \infty} \frac{2^{n+1}}{5^{n+2}}$

12. $\displaystyle\lim_{n \to \infty} \frac{2^{n-1} + 3}{3^{n+2}}$

13. $\displaystyle\lim_{k \to \infty} \frac{k+1}{k}$

14. $\displaystyle\lim_{k \to \infty} \frac{2k-1}{k^2-7}$

15. $\displaystyle\lim_{n \to \infty} \frac{4n^3 - 5}{6n^3 + 3}$

16. $\displaystyle\lim_{n \to \infty} \sqrt[n]{5}\left(\frac{n-1}{n+1}\right)$

17. $\displaystyle\lim_{n \to \infty} \sqrt[n]{3n}$

18. $\displaystyle\lim_{n \to \infty} \frac{\sqrt{n^2 - 1}}{2n}$

19. $\displaystyle\lim_{n \to \infty} \frac{\sqrt{n} + 1}{\sqrt{n} - 1}$

20. $\displaystyle\lim_{n \to \infty} (\sqrt{n+1} - \sqrt{n})$

21. $\displaystyle\lim_{n \to \infty} \sqrt{n} (\sqrt{n+1} - \sqrt{n})$

22. $\displaystyle\lim_{n \to \infty} \frac{1 - (1 + 1/n)^2}{1 - (1 + 1/n)}$

23. $\displaystyle\lim_{n \to \infty} \frac{(-1)^n}{n}$

24. $\displaystyle\lim_{n \to \infty} \frac{1}{n^2(n^2 + 1)}$

25. $\displaystyle\lim_{n \to \infty} \frac{n}{e^n}$

26. $\displaystyle\lim_{n \to \infty} \frac{\sin n}{n}$

27. $\displaystyle\lim_{n \to \infty} \frac{\ln (1 + 1/n)}{n}$

28. $\displaystyle\lim_{n \to \infty} n \sin^2 \frac{1}{n}$

29. $\displaystyle\lim_{n \to \infty} \left(n + \frac{1}{n}\right)^{1/n}$

30. $\displaystyle\lim_{n \to \infty} \sqrt[n]{\ln n}$ *(Hint: $1 \leq \ln n \leq n$ for $n \geq 3$.)*

31. From Section 7.4 we know that

$$\tanh x = \frac{\sinh x}{\cosh x} = \frac{e^x - e^{-x}}{e^x + e^{-x}} = \frac{1 - e^{-2x}}{1 + e^{-2x}}$$

Using this formula, prove that $\lim_{n \to \infty} \tanh n = 1$.

32. Let $\{a_n\}_{n=0}^{\infty}$ be defined recursively by

$$a_0 = 2 \quad \text{and} \quad a_{n+1} = \sin a_n \quad \text{for all } n \geq 0$$

It can be shown that the sequence is bounded and decreasing and hence has a limit L. Evaluate the first 100 terms, and then conjecture the value of L.

33. Let $\{a_n\}_{n=0}^{\infty}$ be defined recursively by

$$a_0 = 2 \quad \text{and} \quad a_{n+1} = \cos a_n \quad \text{for all } n \geq 0$$

It can be shown that the sequence has a limit L. Evaluate the first 50 terms, and then conjecture the value of L.

*34. Let f be a continuous function whose range is contained in its domain. Assume that x_0 is in the domain of f, and define $\{x_n\}_{n=1}^{\infty}$ recursively by the formula

$$x_{n+1} = f(x_n) \quad \text{for } n \geq 0$$

a. Suppose $\lim_{n \to \infty} x_n = L$. Show that $f(L) = L$, so L is a fixed point of f.

b. Let $f(x) = x^2 + \frac{3}{4}$ and $x_0 = 0$. Use part (a) to show that $\{x_n\}_{n=0}^{\infty}$ diverges.

c. Let $f(x) = x^2 + \frac{1}{8}$ and $x_0 = 0$. Assuming that $\lim_{n \to \infty} x_n$ exists, use part (a) to find it.

*35. Let $a_1 = 2$, and for each $n \geq 1$, let $a_{n+1} = 2 - 1/a_n$. Use Theorem 9.6 to show that $\{a_n\}_{n=1}^{\infty}$ converges, and find its limit by using Exercise 34.

36. Consider the interval $[0, 1]$. First we remove the middle third, then the middle third of the two remaining line segments, and so on (see Figure 9.15). Let L_n denote the total length of the remaining segments after n steps. Find $\lim_{n \to \infty} L_n$.

FIGURE 9.15   Figure for Exercise 36.

37. a. Assume that $\lim_{n\to\infty} a_n = 0$ and that $\{b_n\}_{n=m}^{\infty}$ is bounded. Show that $\lim_{n\to\infty} a_n b_n = 0$. (*Hint:* Suppose that $|b_n| \le M$ for all n, and show that for any $\varepsilon > 0$, $|a_n|$ is eventually less than ε/M.)

b. Let $\{c_n\}_{n=1}^{\infty}$ be a sequence. Using part (a), show that if $\lim_{n\to\infty} |c_n| = 0$, then $\lim_{n\to\infty} c_n = 0$. (*Hint:* Let $a_n = |c_n|$, and take b_n to be 1 or -1, according to whether $c_n \ge 0$ or $c_n < 0$.)

c. Using part (a), verify the following limits.

i. $\lim\limits_{n\to\infty} \dfrac{\sin n}{n} = 0$

ii. $\lim\limits_{n\to\infty} \dfrac{1}{n^2} \ln\left(1 + \dfrac{(-1)^n}{n}\right) = 0$

iii. $\lim\limits_{n\to\infty} \dfrac{2 + (-1)^n}{e^n} = 0$

*iv. $\lim\limits_{n\to\infty} \dfrac{2n + (-1)^n}{e^{2n}} = 0$

*38. Let $\{a_n\}_{n=1}^{\infty}$ be the sequence $\sqrt{2}, \sqrt{2 + \sqrt{2}}, \sqrt{2 + \sqrt{2 + \sqrt{2}}}, \ldots$, where in general,

$$a_{n+1} = \sqrt{2 + a_n}$$

a. Show that $\{a_n\}_{n=1}^{\infty}$ is a bounded increasing sequence. (*Hint:* Show that if $a_n \le 2$, then $a_n \le a_{n+1} \le 2$.)

b. Using (a) and Theorem 9.6, show that $\{a_n\}_{n=1}^{\infty}$ converges to a number r.

c. Using (b) and the fact that

$$a_{n+1}^2 = 2 + a_n$$

show that $r = 2$.

*39. Let $\{a_n\}_{n=1}^{\infty}$ be the sequence $\sqrt{2}, (\sqrt{2})^{\sqrt{2}}, (\sqrt{2})^{(\sqrt{2}^{\sqrt{2}})}, \ldots$, where in general,

$$a_{n+1} = (\sqrt{2})^{a_n}$$

a. Show that $\{a_n\}_{n=1}^{\infty}$ converges to a number $L \le 2$. (*Hint:* Show that $a_n < 2$ for all n, and use Exercise 55(a) of Section 5.7 to show that $\{a_n\}_{n=1}^{\infty}$ is increasing.)

b. Show that $L = 2$. (*Hint:* Show that $L = (\sqrt{2})^L$, and use Exercise 55(a) of Section 5.7.)

*40. Let $a_n = 1 + \frac{1}{2} + \cdots + 1/n - \ln n$ for $n \ge 1$.

a. Using the definition of the natural logarithm, show that

$$\ln(n+1) - \ln n \ge \frac{1}{n+1} \qquad \text{for } n \ge 1$$

b. Using (a), show that the sequence $\{a_n\}_{n=1}^{\infty}$ is decreasing.

c. Using the fact that

$$\int_1^n \frac{1}{t}\, dt = \int_1^2 \frac{1}{t}\, dt + \int_2^3 \frac{1}{t}\, dt + \cdots + \int_{n-1}^n \frac{1}{t}\, dt$$

for $n \ge 2$ show that

$$\ln n \le 1 + \frac{1}{2} + \cdots + \frac{1}{n-1}$$

and hence that $a_n \ge 0$ for all n.

d. Show that $\{a_n\}_{n=1}^{\infty}$ converges.

The limit c to which the sequence converges is called the **Euler-Mascheroni constant** and has been calculated to be approximately 0.577216 (accurate to six places).

Project

1. In this project we examine the **Fibonacci sequence,** named after the thirteenth-century Italian mathematician known as Fibonacci but whose real name was Leonardo de Pisa.

Suppose a rabbit colony begins with one pair of adult rabbits. Assume that every pair of adult rabbits produces two offspring every month—one male and one female— and that rabbits become adults at the age of 2 months and live forever (a bit far-fetched!). The problem that Fibonacci posed is to find how many pairs of adult rabbits there are after n months. If we let a_n be the number of pairs of adult rabbits after n months have passed, then we have $a_1 = 1$, $a_2 = 1$, and

$$a_{n+1} = a_n + a_{n-1} \qquad \text{for } n \ge 2 \qquad (9)$$

The sequence $\{a_n\}_{n=1}^{\infty}$ is called the **Fibonacci sequence.** Its first few terms are

$$1, 1, 2, 3, 5, 8, 13$$

a. Find the next 5 terms of the sequence, and show that $\lim_{n\to\infty} a_n = \infty$. (This should be no surprise.)

b. Let

$$b_n = \frac{a_{n+1}}{a_n} \qquad \text{for } n \ge 1.$$

Find the first 10 terms of the sequence $\{b_n\}_{n=1}^{\infty}$.

c. Show that for all $n \ge 1$,

$$0 < b_n \le 2 \quad \text{and} \quad b_{n+1} = 1 + \frac{1}{b_n}$$

d. Show that $\{b_n\}_{n=1}^{\infty}$ is *neither* increasing nor decreasing (so we cannot use Theorem 9.6 in order to conclude that $\lim_{n\to\infty} b_n$ exists).

e. It turns out that $\lim_{n\to\infty} b_n$ does exist. Let $\lim_{n\to\infty} b_n = b$. Use (c) to show that $b = \dfrac{1 + \sqrt{5}}{2}$, which is the golden section.

The result of part (e) is that after many years the number of pairs of adult rabbits in the colony would increase by about 62% every month.

9.4   INFINITE SERIES

Of course, we know how to add a finite collection of numbers. But suppose we wish to add the infinite collection of numbers $\frac{1}{2}, \frac{1}{4}, \frac{1}{8}, \frac{1}{16}, \ldots$. If we begin adding them in the order of their appearance, we obtain successively the sums $\frac{1}{2}, \frac{3}{4}, \frac{7}{8}, \frac{15}{16}, \ldots$, which seem to approach 1. Thus it is reasonable to define the sum

$$\frac{1}{2} + \frac{1}{4} + \frac{1}{8} + \frac{1}{16} + \cdots$$

to be 1.

In general, to determine whether it is possible to define the sum of the numbers in a sequence $\{a_n\}_{n=1}^{\infty}$, we simply begin adding the numbers in the order of their indexes ($a_1, a_2, a_3, \ldots$) and ascertain whether the resulting sums approach a limit. The sums obtained, which are

$$s_1 = a_1$$
$$s_2 = a_1 + a_2$$
$$s_3 = a_1 + a_2 + a_3$$
$$\cdot$$
$$\cdot$$
$$\cdot$$

are called partial sums. If the sequence $\{s_n\}_{n=1}^{\infty}$ of partial sums has a limit, we call that limit the sum and denote it $\sum_{n=1}^{\infty} a_n$. But whether the partial sums approach a limit or not, we use the expressions $\sum_{n=1}^{\infty} a_n$ and $a_1 + a_2 + a_3 + \cdots$ to indicate the numbers to be added. Each such expression is called a series. Examples of series are

$$\sum_{n=1}^{\infty} \frac{1}{2^n} = \frac{1}{2} + \frac{1}{4} + \frac{1}{8} + \frac{1}{16} + \cdots$$

and

$$\sum_{n=1}^{\infty} (-1)^n = -1 + 1 - 1 + 1 - 1 + \cdots$$

So far we have discussed partial sums of the sequence $\{a_n\}_{n=1}^{\infty}$ and the sum of the series $\sum_{n=1}^{\infty} a_n$. Corresponding concepts for $\{a_n\}_{n=m}^{\infty}$ (where the initial index m need not be 1) are defined analogously, and appear in Definition 9.7.

DEFINITION 9.7   Let $\{a_n\}_{n=m}^{\infty}$ be a sequence. If $\lim_{j \to \infty} (a_m + a_{m+1} + \cdots + a_{m+j-1})$ exists (as a number), then we say that $\sum_{n=m}^{\infty} a_n$ **converges,** and we call $\lim_{j \to \infty} (a_m + a_{m+1} + \cdots + a_{m+j-1})$ the **sum of the series.** In that case we use the expression $\sum_{n=m}^{\infty} a_n$ to denote the sum of the series, so that

$$\sum_{n=m}^{\infty} a_n = \lim_{j \to \infty} (a_m + a_{m+1} + \cdots + a_{m+j-1})$$

If the limit does not exist, then we say that the series $\sum_{n=m}^{\infty} a_n$ **diverges.**

The numbers $a_m, a_{m+1}, \ldots$ are the **terms** of the series, and the sum

$$s_j = a_m + a_{m+1} + \cdots + a_{m+j-1}$$

of the first j terms is called the **jth partial sum** of the series $\sum_{n=m}^{\infty} a_n$.

Almost all series we will consider are of the form $\sum_{n=1}^{\infty} a_n$ or $\sum_{n=0}^{\infty} a_n$. For $\sum_{n=1}^{\infty} a_n$ the jth partial sum is

$$s_j = a_1 + a_2 + \cdots + a_j$$

and for $\sum_{n=0}^{\infty} a_n$ the jth partial sum is

$$s_j = a_0 + a_1 + \cdots + a_{j-1}$$

Since the limit of a sequence is unique, it follows from Definition 9.7 that the sum of a convergent series is unique. We will use this fact as we solve examples below.

EXAMPLE 1   Show that $\displaystyle\sum_{n=1}^{\infty} \frac{1}{2^n} = \frac{1}{2} + \frac{1}{2^2} + \frac{1}{2^3} + \cdots = 1$.

Solution   Notice that the first few partial sums are

$$s_1 = \frac{1}{2} = 1 - \frac{1}{2} \qquad s_2 = \frac{1}{2} + \frac{1}{2^2} = \frac{3}{4} = 1 - \frac{1}{2^2}$$

$$s_3 = \frac{1}{2} + \frac{1}{2^2} + \frac{1}{2^3} = \frac{7}{8} = 1 - \frac{1}{2^3}$$

In general, for any positive integer j,

$$s_j = \frac{1}{2} + \frac{1}{2^2} + \cdots + \frac{1}{2^j}$$

Thus

$$2s_j = 1 + \frac{1}{2} + \frac{1}{2^2} + \cdots + \frac{1}{2^{j-1}} = 1 + \left(\frac{1}{2} + \frac{1}{2^2} + \cdots + \frac{1}{2^{j-1}} + \frac{1}{2^j} \right) - \frac{1}{2^j}$$

$$= 1 + s_j - \frac{1}{2^j}$$

Therefore $s_j = 1 - 1/2^j$, so that

$$\lim_{j \to \infty} s_j = \lim_{j \to \infty} \left(1 - \frac{1}{2^j} \right) = 1$$

Consequently

$$\sum_{n=1}^{\infty} \frac{1}{2^n} = \lim_{j \to \infty} s_j = 1 \qquad \square$$

EXAMPLE 2   Show that $\displaystyle\sum_{n=0}^{\infty} 2^n = 1 + 2 + 2^2 + 2^3 + \cdots$ diverges.

Solution   For any positive integer j,

$$s_j = 1 + 2 + 2^2 + 2^3 + \cdots + 2^{j-1} \geq 2^{j-1}$$

Evidently

$$\lim_{j \to \infty} s_j = \lim_{j \to \infty} 2^{j-1} = \infty$$

so that $\sum_{n=0}^{\infty} 2^n$ diverges.   ❑

EXAMPLE 3   Show that the series

$$\sum_{n=1}^{\infty} \frac{1}{n(n+1)} = \frac{1}{1 \cdot 2} + \frac{1}{2 \cdot 3} + \frac{1}{3 \cdot 4} + \cdots$$

converges, and find its sum.

Solution   By partial fractions,

$$\frac{1}{n(n+1)} = \frac{1}{n} - \frac{1}{n+1} \qquad \text{for } n \geq 1$$

and hence

$$s_j = \frac{1}{1 \cdot 2} + \frac{1}{2 \cdot 3} + \frac{1}{3 \cdot 4} + \cdots + \frac{1}{(j-1)j} + \frac{1}{j(j+1)}$$

$$= \left(1 - \frac{1}{2}\right) + \left(\frac{1}{2} - \frac{1}{3}\right) + \left(\frac{1}{3} - \frac{1}{4}\right) + \cdots + \left(\frac{1}{j-1} - \frac{1}{j}\right) + \left(\frac{1}{j} - \frac{1}{j+1}\right) \quad (1)$$

Thus

$$s_j = 1 - \frac{1}{j+1}$$

The right side of (1) could be simplified because adjacent pairs of numbers cancel each other. Since

$$\lim_{j \to \infty} s_j = \lim_{j \to \infty} \left(1 - \frac{1}{j+1}\right) = 1$$

we have simultaneously proved that the given series converges and found the sum of the series:

$$\sum_{n=1}^{\infty} \frac{1}{n(n+1)} = \lim_{j \to \infty} s_j = 1 \qquad ❑$$

The series $\sum_{n=1}^{\infty} 1/n(n+1)$ is called a **telescoping series,** because when we write the partial sums as in (1), all but the first and last term cancel.

EXAMPLE 4   Show that the **harmonic series**

$$\sum_{n=1}^{\infty} \frac{1}{n} = 1 + \frac{1}{2} + \frac{1}{3} + \cdots$$

diverges.

Solution   We will show that

$$s_{2^j} \geq 1 + j\left(\frac{1}{2}\right) \quad \text{for each } j \tag{2}$$

and thus that the sequence of partial sums is unbounded and must diverge. To prove (2), we observe that

$$s_{2^1} = s_2 = 1 + \frac{1}{2}$$

$$s_{2^2} = s_4 = 1 + \frac{1}{2} + \frac{1}{3} + \frac{1}{4} \geq 1 + \frac{1}{2} + \overbrace{\frac{1}{4} + \frac{1}{4}}^{1/2} = 1 + \frac{2}{2}$$

$$s_{2^3} = s_8 = 1 + \frac{1}{2} + \frac{1}{3} + \frac{1}{4} + \frac{1}{5} + \frac{1}{6} + \frac{1}{7} + \frac{1}{8}$$

$$\geq 1 + \frac{1}{2} + \overbrace{\frac{1}{4} + \frac{1}{4}}^{1/2} + \overbrace{\frac{1}{8} + \frac{1}{8} + \frac{1}{8} + \frac{1}{8}}^{1/2} \geq 1 + \frac{3}{2}$$

In general we arrange the terms making up s_{2^j} into several groups and then substitute smaller values for the terms so that each group has sum $\frac{1}{2}$:

$$s_{2^j} = 1 + \frac{1}{2} + \left(\frac{1}{3} + \frac{1}{4}\right) + \left(\frac{1}{5} + \frac{1}{6} + \frac{1}{7} + \frac{1}{8}\right) + \cdots + \left(\frac{1}{2^{j-1}+1} + \cdots + \frac{1}{2^j}\right)$$

$$\geq 1 + \frac{1}{2} + \overbrace{\frac{1}{4} + \frac{1}{4}}^{1/2} + \overbrace{\frac{1}{8} + \frac{1}{8} + \frac{1}{8} + \frac{1}{8}}^{1/2} + \cdots + \overbrace{\frac{1}{2^j} + \cdots + \frac{1}{2^j}}^{1/2}$$

$$= 1 + j\left(\frac{1}{2}\right)$$

Since

$$\lim_{j \to \infty} \left[1 + j\left(\frac{1}{2}\right)\right] = \infty$$

it follows that the sequence $\{s_j\}_{j=1}^{\infty}$ of partial sums is unbounded, as we wished to prove; consequently $\sum_{n=1}^{\infty} 1/n$ diverges. ❑

As we have just seen, the harmonic series $\sum_{n=1}^{\infty} 1/n$ diverges even though its terms converge to 0. The divergence of the harmonic series does not imply that the jth partial sum increases rapidly as j increases. On the contrary, s_j grows very slowly. Computer calculations have shown that

$$s_j \geq 20 \quad \text{only if} \quad j \geq 272{,}400{,}600$$

and

$$s_j \geq 100 \quad \text{only if} \quad j \geq 1.5 \times 10^{43} \text{ (approximately)}$$

Caution: There is an important lesson to be learned from the preceding remarks. Namely, computer (or calculator) calculations may suggest incorrect conclusions. For example, if a computer could make one trillion calculations per second (which supercomputers have only recently achieved), then it would take the computer over

$$10{,}000{,}000{,}000{,}000{,}000{,}000{,}000{,}000 \text{ centuries}$$

to compute the partial sum $s_{1.5 \times 10^{43}}$ of the harmonic series, which we mentioned above is approximately 100. The obvious guess from results of the calculator or computer would be that the harmonic series converges, which by Example 4 we know is not true. Thus, even though calculations frequently provide helpful insight into problems, they may also be inconclusive or even misleading. After drawing a tentative conclusion from calculation, we should strive to support the conclusion with careful analysis.

Comparison of $\sum_{n=m_1}^{\infty} a_n$ **and** $\sum_{n=m_2}^{\infty} a_n$

Let us compare the partial sums of the two series $\sum_{n=2}^{\infty} 1/n^2$ and $\sum_{n=5}^{\infty} 1/n^2$. Notice that for any positive integer $j \geq 5$,

$$\overbrace{\frac{1}{2^2} + \frac{1}{3^2} + \cdots + \frac{1}{j^2}}^{\text{partial sum of } \sum_{n=2}^{\infty} 1/n^2} = \left(\frac{1}{2^2} + \frac{1}{3^2} + \frac{1}{4^2} \right) + \overbrace{\left(\frac{1}{5^2} + \frac{1}{6^2} + \cdots + \frac{1}{j^2} \right)}^{\text{partial sum of } \sum_{n=5}^{\infty} 1/n^2} \quad (3)$$

From (3) it follows that if $\sum_{n=2}^{\infty} 1/n^2$ converges, then so does $\sum_{n=5}^{\infty} 1/n^2$, and vice versa. Moreover, if the two series converge, their sums are related by

$$\sum_{n=2}^{\infty} \frac{1}{n^2} = \left(\frac{1}{2^2} + \frac{1}{3^2} + \frac{1}{4^2} \right) + \sum_{n=5}^{\infty} \frac{1}{n^2}$$

A similar argument would show more generally that if $m_2 > m_1$, then

$$\sum_{n=m_1}^{\infty} a_n \quad \text{converges if and only if} \quad \sum_{n=m_2}^{\infty} a_n \text{ converges} \quad (4)$$

In addition, if the two series converge, then

$$\sum_{n=m_1}^{\infty} a_n = (a_{m_1} + a_{m_1+1} + \cdots + a_{m_2-1}) + \sum_{n=m_2}^{\infty} a_n \quad (5)$$

Thus the index m at which the summation of the series $\sum_{n=m}^{\infty} a_n$ begins is irrelevant to the convergence of the series—although, as (5) indicates, the actual sum of the series *is* affected by the index m. For that reason, although most of our theorems will be stated for series of the form $\sum_{n=1}^{\infty} a_n$ (with initial index 1), all these theorems can be applied to a series $\sum_{n=m}^{\infty} a_n$ with arbitrary initial index m.

A Divergence Test

Our first theorem about series presents a condition that must be met in order for a series to converge.

THEOREM 9.8   If $\sum_{n=1}^{\infty} a_n$ converges, then $\lim_{n \to \infty} a_n = 0$.

Proof   Notice that for $n \geq 1$,

$$a_n = (a_1 + a_2 + \cdots + a_n) - (a_1 + a_2 + \cdots + a_{n-1}) = s_n - s_{n-1} \qquad (6)$$

Since $\sum_{n=1}^{\infty} a_n$ converges by assumption, we know that $\lim_{n \to \infty} s_n$ exists. Therefore by (7) of Section 9.3, $\lim_{n \to \infty} s_{n-1}$ also exists, and

$$\lim_{n \to \infty} s_{n-1} = \lim_{n \to \infty} s_n$$

Now we conclude from (6) that

$$\lim_{n \to \infty} a_n = \lim_{n \to \infty} (s_n - s_{n-1}) = \lim_{n \to \infty} s_n - \lim_{n \to \infty} s_{n-1} = 0 \qquad \blacksquare$$

From Theorem 9.8 we immediately obtain a criterion for a series to diverge.

COROLLARY   9.9
nth Term Test for Divergence

If $\lim_{n \to \infty} a_n$ is not 0 (or does not exist), then $\sum_{n=1}^{\infty} a_n$ diverges.

Proof   The statement in the corollary is logically equivalent to the statement of Theorem 9.8.   ■

Caution: The converse of Corollary 9.9 is false. There are series $\sum_{n=1}^{\infty} a_n$ that diverge even though $\lim_{n \to \infty} a_n = 0$. The series $\sum_{n=1}^{\infty} 1/n$ is such an example, because it diverges, yet $\lim_{n \to \infty} 1/n = 0$.

Corollary 9.9 tells us immediately that the following series diverge:

$$\sum_{n=1}^{\infty} 61, \quad \sum_{n=1}^{\infty} \left(1 + \frac{1}{n} \right), \quad \sum_{n=-4}^{\infty} n, \quad \sum_{n=0}^{\infty} 2^{(n^2)}, \quad \sum_{n=0}^{\infty} (-1)^n, \quad \sum_{n=1}^{\infty} n \sin \frac{1}{n}$$

Geometric Series

A **geometric series** is a series of the form $\sum_{n=m}^{\infty} cr^n$, where r and c are constants and $c \neq 0$. The convergence of geometric series depends entirely on the choice of r, as we see next.

THEOREM 9.10
Geometric Series Theorem

Let r be any number, and let $c \neq 0$ and $m \geq 0$. Then the geometric series $\sum_{n=m}^{\infty} cr^n$ converges if and only if $|r| < 1$. For $|r| < 1$,

$$\sum_{n=m}^{\infty} cr^n = \frac{cr^m}{1 - r} \qquad (7)$$

Proof   We consider the cases $|r| \geq 1$ and $|r| < 1$ separately. If $|r| \geq 1$, then $|cr^n| \geq |c|$ for all $n \geq m$, so the terms do not approach 0; thus by Corollary 9.9 the series $\sum_{n=m}^{\infty} cr^n$ diverges. If $|r| < 1$, then we use the identity

$$(1 - r)(1 + r + r^2 + \cdots + r^{j-1}) = 1 - r^j$$

which implies that

$$s_j = cr^m + cr^{m+1} + \cdots + cr^{m+j-1} = cr^m(1 + r + r^2 + \cdots + r^{j-1})$$

$$= cr^m \left(\frac{1 - r^j}{1 - r} \right)$$

Since $\lim_{j \to \infty} r^j = 0$ by Example 7 of Section 9.2, it follows that

$$\lim_{j \to \infty} s_j = \frac{cr^m}{1 - r} \lim_{j \to \infty} (1 - r^j) = \frac{cr^m}{1 - r} \quad \blacksquare$$

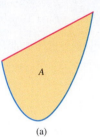

(a)

From the proof of the theorem we see that the jth partial sum is given by

$$\sum_{n=m}^{m+j-1} cr^n = cr^m \left(\frac{1 - r^j}{1 - r} \right)$$

In particular,

$$\sum_{n=0}^{j-1} cr^n = \frac{c(1 - r^j)}{1 - r} \quad \text{and} \quad \sum_{n=1}^{j} cr^n = \frac{c(r - r^{j+1})}{1 - r}$$

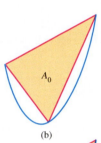

(b)

The number r is called the **ratio** of the geometric series. By the Geometric Series Theorem, the sum of a convergent geometric series is equal to the first term, cr^m, divided by $1 - r$.

EXAMPLE 5   Show that $\displaystyle\sum_{n=0}^{\infty} \left(\frac{1}{4} \right)^n = \frac{4}{3}$.

(c)

Solution   The given series is a geometric series with ratio $r = \frac{1}{4}$; thus the Geometric Series Theorem asserts that the series converges. Since the first term is $\left(\frac{1}{4} \right)^0 = 1$, it follows that

$$\sum_{n=0}^{\infty} \left(\frac{1}{4} \right)^n = \frac{1}{1 - \frac{1}{4}} = \frac{4}{3} \quad \square$$

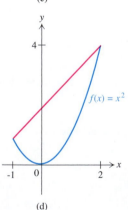

(d)

The series in Example 5 is related to Archimedes' computation of area. When he computed the area of a parabolic region such as that drawn in Figure 9.16(a), he began by inscribing an initial triangle with area A_0 (Figure 9.16(b)). Then he inscribed smaller and smaller triangles in the outer portions of the region, as in Figure 9.16(c). He calculated the sum of the areas of the triangles as

$$A = A_0 \left(1 + \frac{1}{4} + \frac{1}{4^2} + \frac{1}{4^3} + \cdots \right) = A_0 \sum_{n=0}^{\infty} \left(\frac{1}{4} \right)^n$$

By Example 5 we know that the sum of the series is $\frac{4}{3}$, so that

$$A = \tfrac{4}{3} A_0$$

In particular, for the parabolic region described in Figure 9.16(d), it can be shown that $A_0 = \frac{27}{8}$, and thus $A = \frac{4}{3} \cdot \frac{27}{8} = \frac{9}{2}$. This corroborates our results in Section 5.2 (see the comments after formula (4) in Section 5.2). Although Archimedes did not use series in the strict sense, his work involved the germ of this concept.

Geometric series are involved in the study of the concentration of a drug in an individual's system. After a dose c_0 of insulin is injected into a person's system, the concentration of insulin in the system decays exponentially with time. Thus for a suitable constant b, the concentration after t hours is given by $c_0 e^{-bt}$. Now assume that a dose c_0 is injected every t_0 hours. Then just before the $(j + 1)$st injection, the residual concentration c_j is the sum of the residual concentrations of all the preceding j injections, and is given by

$$c_j = \overbrace{c_0 e^{-jbt_0}}^{\substack{\text{residual from} \\ \text{1st injection}}} + \overbrace{c_0 e^{-(j-1)bt_0}}^{\substack{\text{residual from} \\ \text{2nd injection}}} + \cdots + \overbrace{c_0 e^{-bt_0}}^{\substack{\text{residual from} \\ j\text{th injection}}}$$

$$= c_0 e^{-bt_0} + c_0 e^{-2bt_0} + \cdots + c_0 e^{-jbt_0}$$

$$= \sum_{n=1}^{j} c_0 e^{-nbt_0}$$

Thus the limiting pre-injection concentration of insulin in the system is the sum of an infinite series. We will calculate its value in Example 6.

EXAMPLE 6 Evaluate $\displaystyle\sum_{n=1}^{\infty} c_0 e^{-nbt_0}$.

Solution Since $e^{-nbt_0} = (e^{-bt_0})^n$, the given series is a geometric series with ratio $r = e^{-bt_0}$ and first term $c_0 e^{-bt_0}$. It follows that

$$\sum_{n=1}^{\infty} c_0 e^{-nbt_0} = \frac{c_0 e^{-bt_0}}{1 - e^{-bt_0}} \qquad \square$$

Of course the Geometric Series Theorem also tells us that not all geometric series converge. In fact, $\sum_{n=6}^{\infty} 3(5.4)^n$ diverges, since $r = 5.4 > 1$.

There is an intimate relationship between geometric series and repeating decimals. For example the decimal $0.33333\ldots$ is defined by

$$0.33333\ldots = 3\left(\frac{1}{10}\right) + 3\left(\frac{1}{10}\right)^2 + 3\left(\frac{1}{10}\right)^3 + \cdots = \sum_{n=1}^{\infty} 3\left(\frac{1}{10}\right)^n$$

Similarly

$$0.454545\ldots = 45\left(\frac{1}{100}\right) + 45\left(\frac{1}{100}\right)^2 + 45\left(\frac{1}{100}\right)^3 + \cdots$$

$$= \sum_{n=1}^{\infty} 45\left(\frac{1}{100}\right)^n$$

Geometric series allow us to express any repeating decimal as a fraction (and hence as a rational number). For example,

$$0.2700000\ldots = \frac{27}{100}$$

$$0.33333\ldots = \sum_{n=1}^{\infty} 3\left(\frac{1}{10}\right)^n = \frac{3(1/10)}{1 - 1/10} = \frac{1}{3}$$

$$0.45454545 \ldots = \sum_{n=1}^{\infty} 45 \left(\frac{1}{100} \right)^n = \frac{45(1/100)}{1 - 1/100} = \frac{45}{99} = \frac{5}{11}$$

$$0.316454545 \ldots = \frac{316}{1000} + \frac{1}{1000}(0.454545 \ldots)$$

$$= \frac{316}{1000} + \frac{1}{1000} \cdot \frac{5}{11} = \frac{3481}{11,000}$$

Since this procedure will transform any repeating decimal into a fraction, every repeating decimal is a rational number. (The converse is also true: Every rational number has a repeating decimal expansion.) However, the procedure just demonstrated does not work if the decimal is not repeating. For instance, the number $e = 2.71828 \ldots$, which is not a repeating decimal, can be written in the form of a series as

$$2 + 7 \left(\frac{1}{10} \right) + 1 \left(\frac{1}{10} \right)^2 + 8 \left(\frac{1}{10} \right)^3 + 2 \left(\frac{1}{10} \right)^4 + 8 \left(\frac{1}{10} \right)^5 + \cdots$$

The series converges because its partial sums are increasing and bounded (by, say, 3). In fact, the series converges to e. However, the series is *not* a geometric series.

Combinations of Series

If $\sum_{n=1}^{\infty} a_n$ and $\sum_{n=1}^{\infty} b_n$ are two series, we can add them term by term; we can also multiply all the terms of either series by a single number c. These operations generate two new series, $\sum_{n=1}^{\infty} (a_n + b_n)$ and $\sum_{n=1}^{\infty} ca_n$.

THEOREM 9.11

a. If $\sum_{n=1}^{\infty} a_n$ and $\sum_{n=1}^{\infty} b_n$ converge, then $\sum_{n=1}^{\infty} (a_n + b_n)$ also converges, and

$$\sum_{n=1}^{\infty} (a_n + b_n) = \sum_{n=1}^{\infty} a_n + \sum_{n=1}^{\infty} b_n$$

b. If $\sum_{n=1}^{\infty} a_n$ converges and if c is any number, then $\sum_{n=1}^{\infty} ca_n$ also converges, and

$$\sum_{n=1}^{\infty} ca_n = c \sum_{n=1}^{\infty} a_n$$

Proof   To prove (a), let s_j, s_j', and s_j'' be the jth partial sums of $\sum_{n=1}^{\infty} (a_n + b_n)$, $\sum_{n=1}^{\infty} a_n$, and $\sum_{n=1}^{\infty} b_n$, respectively. Then

$$s_j = [(a_1 + b_1) + (a_2 + b_2) + \cdots + (a_j + b_j)]$$

$$= (a_1 + a_2 + \cdots + a_j) + (b_1 + b_2 + \cdots + b_j)$$

$$= s_j' + s_j''$$

As a result,

$$\sum_{n=1}^{\infty} (a_n + b_n) = \lim_{j \to \infty} s_j = \lim_{j \to \infty} (s_j' + s_j'') = \lim_{j \to \infty} s_j' + \lim_{j \to \infty} s_j'' = \sum_{n=1}^{\infty} a_n + \sum_{n=1}^{\infty} b_n$$

The proof of (b) uses the same ideas, so we leave it as an exercise. ■

The result of Theorem 9.11(a) can be extended to the sum of three or more series. We also remark that by combining parts (a) and (b) of Theorem 9.11, we can conclude that if $\sum_{n=1}^{\infty} a_n$ and $\sum_{n=1}^{\infty} b_n$ converge, then $\sum_{n=1}^{\infty} (a_n - b_n)$ converges and

$$\sum_{n=1}^{\infty} (a_n - b_n) = \sum_{n=1}^{\infty} a_n - \sum_{n=1}^{\infty} b_n \tag{8}$$

EXAMPLE 7   Show that the series $\sum_{n=1}^{\infty} \left(\dfrac{4}{2^n} - \dfrac{2}{n(n+1)} \right)$ converges, and find its sum.

Solution   The Geometric Series Theorem implies that

$$\sum_{n=1}^{\infty} \frac{4}{2^n} = \frac{4(\frac{1}{2})}{1 - \frac{1}{2}} = 4$$

From Example 3 and Theorem 9.11(b) we know that

$$\sum_{n=1}^{\infty} \frac{2}{n(n+1)} = 2 \sum_{n=1}^{\infty} \frac{1}{n(n+1)} = 2$$

Consequently (8) assures us that

$$\sum_{n=1}^{\infty} \left(\frac{4}{2^n} - \frac{2}{n(n+1)} \right) = 4 - 2 = 2 \qquad \square$$

Changing Indexes

It is always possible to regard $\sum_{n=m}^{\infty} a_n$ as a series $\sum_{n=0}^{\infty} b_n$ with initial index 0. We accomplish this by letting

$$b_0 = a_m$$

$$b_1 = a_{m+1}$$

$$.$$
$$.$$
$$.$$

$$b_n = a_{m+n}$$

Clearly, the *j*th partial sum of $\sum_{n=0}^{\infty} b_n$ coincides with the *j*th partial sum of $\sum_{n=m}^{\infty} a_n$, for both are the sum of the same *j* numbers. Therefore $\sum_{n=m}^{\infty} a_n$ and $\sum_{n=0}^{\infty} b_n$ are essentially the same series. In particular they have the same sum whenever they converge. Since $b_n = a_{m+n}$ for $n \geq 0$, we may also write

$$\sum_{n=m}^{\infty} a_n = \sum_{n=0}^{\infty} a_{m+n} \tag{9}$$

The change of the initial index from *m* to 0 has been offset by replacing a_n by a_{m+n}. For example,

$$\sum_{n=3}^{\infty} \frac{1}{n!} = \sum_{n=0}^{\infty} \frac{1}{(n+3)!}$$

Caution: It is critical to distinguish between the series $\sum_{n=1}^{\infty} a_n$ and the sequence $\{a_n\}_{n=1}^{\infty}$. If you neglect to maintain this distinction, you are making an error comparable to confusing a book with the pages inside the book.

EXERCISES 9.4

In Exercises 1–5 compute the fourth partial sum of each series.

1. $\sum_{n=1}^{\infty} 1$ **2.** $\sum_{n=1}^{\infty} n$ **3.** $\sum_{n=0}^{\infty} \left(\frac{1}{3}\right)^n$

4. $\sum_{n=1}^{\infty} (-1)^n$ **5.** $\sum_{n=2}^{\infty} \frac{(-1)^n}{n}$

In Exercises 6–14 determine whether the given series must diverge because its terms do not converge to 0.

6. $\sum_{n=1}^{\infty} \left(\frac{-1}{7}\right)^n$ **7.** $\sum_{n=1}^{\infty} \left(1 + \frac{1}{n}\right)$

8. $\sum_{n=0}^{\infty} \frac{n^2}{n+1}$ **9.** $\sum_{n=2}^{\infty} (-1)^n \frac{1}{n^2}$

10. $\sum_{n=1}^{\infty} \sin n\pi$ **11.** $\sum_{n=1}^{\infty} \sin\left(\frac{\pi}{2} - \frac{1}{n}\right)$

12. $\sum_{n=2}^{\infty} \tan\left(\frac{\pi}{2} - \frac{1}{n}\right)$ **13.** $\sum_{n=1}^{\infty} n \sin \frac{1}{n}$

14. $\sum_{n=1}^{\infty} \left(1 + \frac{1}{n}\right) \ln\left(1 + \frac{1}{n}\right)$

In Exercises 15–22 find a formula for the partial sums of the series. For each series, determine whether the partial sums have a limit. If so, find the sum of the series.

15. $\sum_{n=1}^{\infty} 1$ **16.** $\sum_{n=1}^{\infty} \left(\frac{1}{4}\right)^n$

17. $\sum_{n=0}^{\infty} (-1)^n$ **18.** $\sum_{n=3}^{\infty} \frac{1}{n(n-1)}$

19. $\sum_{n=1}^{\infty} \left(\frac{1}{n+1} - \frac{1}{n+2}\right)$ **20.** $\sum_{n=1}^{\infty} \left(\frac{1}{n^3} - \frac{1}{(n+1)^3}\right)$

21. $\sum_{n=1}^{\infty} [n^3 - (n+1)^3]$

*22. $\sum_{n=1}^{\infty} \frac{1}{n(n+1)(n+2)}$ $\left(\text{Hint: } \frac{1}{n(n+1)(n+2)} = \frac{1}{2}\left(\frac{1}{n} - \frac{1}{n+1}\right) - \frac{1}{2}\left(\frac{1}{n+1} - \frac{1}{n+2}\right)\right)$

23. Let $\{a_n\}_{n=1}^{\infty}$ be a sequence. Show that the nth partial sum of the series $\sum_{n=1}^{\infty} (a_n - a_{n+1})$ is $a_1 - a_{n+1}$. Conclude that this series converges if and only if $\lim_{n \to \infty} a_n$ exists,

in which case the sum is $a_1 - \lim_{n \to \infty} a_n$. Series of the form just described (for example, those in Exercises 19–21) are telescoping series.

24. Let $\{a_n\}_{n=1}^{\infty}$ be the Fibonacci sequence (see Project 1 of Section 9.3). Use the fact that $a_{n-1} = a_{n+1} - a_n$ and the idea of Exercise 23 to show that $a_1 + a_2 + \cdots + a_n = a_{n+2} - 1$.

In Exercises 25–34 determine whether or not the series converges, and if so, find its sum.

25. $\sum_{n=1}^{\infty} 5\left(\frac{4}{7}\right)^n$ **26.** $\sum_{n=1}^{\infty} \left(\frac{7}{4}\right)^n$

27. $\sum_{n=0}^{\infty} (-1)^n (0.3)^n$ **28.** $\sum_{n=2}^{\infty} (0.33)^n$

29. $\sum_{n=1}^{\infty} 5\left(\frac{1}{2}\right)^{n+1}$ **30.** $\sum_{n=0}^{\infty} \frac{5^n}{7^{n+1}}$

31. $\sum_{n=1}^{\infty} \frac{3^{n+3}}{5^{n-1}}$ **32.** $\sum_{n=0}^{\infty} \frac{2^n + 5^n}{2^n 5^n}$

33. $\sum_{n=1}^{\infty} (-1)^n \frac{3^n}{2^{n+2}}$ **34.** $\sum_{n=0}^{\infty} (-3)\left(\frac{2}{3}\right)^{2n}$

In Exercises 35–39 express the given series $\sum_{n=1}^{\infty} a_n$ in the form $c + \sum_{n=4}^{\infty} a_n$.

35. $\frac{1}{7} + \frac{1}{7^2} + \frac{1}{7^3} + \cdots$ **36.** $1 + \frac{2}{3} + \frac{2^2}{3^2} + \frac{2^3}{3^3} + \cdots$

37. $\sum_{n=1}^{\infty} \frac{1}{n^2 + 1}$ **38.** $\sum_{n=1}^{\infty} (-1)^n \frac{1}{n}$

39. $\sum_{n=1}^{\infty} \frac{1}{n^2}$

In Exercises 40–47 express the repeating decimal as a fraction.

40. $0.6666666\ldots$ **41.** $0.72727272\ldots$

42. $0.024242424\ldots$ **43.** $0.232232232\ldots$

44. $0.453232232232\ldots$ **45.** $27.56123123123\ldots$

46. $0.00649649649649\ldots$ **47.** $0.86400000\ldots$

48. a. It is known that $\sum_{n=1}^{\infty} 1/n^2 = \pi^2/6$. Use this to find the sum of $\sum_{n=3}^{\infty} 1/n^2$.
   b. It is known that $\sum_{n=1}^{\infty} 1/n^4 = \pi^4/90$. Use this to find the sum of $\sum_{n=1}^{\infty} 1/[\pi^4(n+1)^4]$.

49. It is known that $\sum_{n=1}^{\infty} (-1)^{n+1}/n = \ln 2$. Use this to find the sum of $\sum_{n=4}^{\infty} (-1)^{n+1}/n$.

50. Show that $\sum_{n=0}^{\infty} (-1)^n r^{2n} = 1/(1+r^2)$ if $-1 < r < 1$.

51. a. Determine the smallest value of j such that $\sum_{n=1}^{j} 1/n \geq 3$.

  b. Use a calculator to determine the smallest value of j such that $\sum_{n=1}^{j} 1/n \geq 4$.

52. Prove Theorem 9.11(b).

53. The Swiss mathematician Leonhard Euler used ideas expressed in the Geometric Series Theorem to deduce that if $r > 0$, then

$$\sum_{n=0}^{\infty} \left(\frac{1}{r}\right)^n = \frac{1}{1 - 1/r} = \frac{-r}{1-r} \quad \text{and} \quad \sum_{n=1}^{\infty} r^n = \frac{r}{1-r}$$

Then he concluded that

$$\left(\cdots + \frac{1}{r^4} + \frac{1}{r^3} + \frac{1}{r^2} + \frac{1}{r} + 1\right) + (r + r^2 + \cdots)$$

$$= \sum_{n=0}^{\infty} \left(\frac{1}{r}\right)^n + \sum_{n=1}^{\infty} r^n = \frac{-r}{1-r} + \frac{r}{1-r} = 0$$

Since all terms in the series on the left side of the equation are positive, this is absurd. Why is this argument invalid?

54. Find the fallacy in the following argument: Let

$$a = 1 + 2 + 4 + 8 + \cdots$$

Then

$$2a = 2 + 4 + 8 + 16 + \cdots$$

$$= (1 + 2 + 4 + 8 + \cdots) - 1 = a - 1$$

Thus $a = -1$.

Applications

55. This exercise refers to the discussion just preceding Example 6. Find a formula for the concentration of insulin in the system just after the jth injection. By summing the corresponding infinite series, find the limiting post-injection concentration in the system.

56. A ball is dropped from a height of 1 meter onto a smooth surface. On each bounce the ball rises to 60 percent of the height it reached on the previous bounce. Find the total distance the ball travels.

57. Find the total time T the ball in Exercise 56 is in motion. (*Hint:* Use (1) in Section 1.3 and express T as the sum of a geometric series.)

58. Let r be a positive number with $0 < r < 1$. Suppose that when a lawn is first treated with a certain herbicide, the fraction of weeds killed is r. Assume that with each succeeding treatment,

the additional amount killed =
r (amount killed by the preceding treatment)

  a. Show that if $r = \frac{1}{3}$, then at least half of the weeds will always remain.

  b. Show that if $r = \frac{1}{2}$, then the lawn can be made as nearly weed-free as one might wish.

  c. Let $0 < s < 1$. Find a value of r in $(0, 1)$ so that the lawn can be made to have as close to a proportion s of its original weed count as one might wish.

*59. A person earns $2500 a month and during each month spends a certain fraction p (with $0 < p < 1$) of the total of all money not spent in previous months and the additional $2500 earned during that month. Find a formula for the wealth w_n of such a person after n months, and show that the wealth has a limit as n approaches ∞.

60. In normal times a nation spends about 90 percent of its income. This means that out of a given $100 income, approximately $90 would be spent firsthand, then about 90 percent of the $90 already spent would be spent secondhand, and so on. Thus the same $100 is actually worth many times the initial $100 to the general economy. (The effect of the process of spending and respending a certain amount of money is called the **multiplier effect.**) Under the assumption of a 90 percent reutilization of any given income, how much is $100 really worth to the economy?

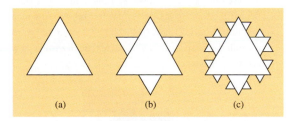

(a)   (b)   (c)

FIGURE 9.17 The construction of a von Koch snowflake. See Exercise 61.

61. A von Koch snowflake is generated as follows. Begin with an equilateral triangle T of side length 1 (Figure 9.17(a)). At the middle of each side of T place an equilateral triangle whose sides have length $\frac{1}{3}$ to obtain a Star of David (Figure 9.17(b)). Then place an equilateral triangle on the middle third of each side of the star (Figure 9.17(c)), and repeat this process ad infinitum to obtain what is called the **von Koch snowflake,** after the Swedish mathematician Helge von Koch, who described it in 1904.

a. By summing an appropriate series, show that the area A of the von Koch snowflake is finite, and find A.

b. By summing an appropriate series, show that the boundary of A, which is called the **von Koch curve**, has infinite length.

62. a. Two trains, each traveling 15 miles per hour, approach each other on a straight track. When the trains are 1 mile apart, a bee begins flying back and forth between the trains at 30 miles per hour. Express the distance the bee travels before the trains collide as an infinite series, and find its sum.

b. Find a simple solution of the bee problem without using series. (*Hint:* Determine how long the bee flies.) (It is said that a similar problem was posed to the great twentieth-century mathematician John von Neumann (1903–1957), who solved it almost instantly in his head. When the poser of the problem suggested that by the quickness of his response, he must have solved the problem the simple way, von Neumann replied that he had actually solved the problem by summing a series.)

*63. One of Zeno's paradoxes purports to prove that Achilles, who runs 10 times faster than the tortoise, cannot overtake the tortoise, who has a 100-yard lead. The argument runs as follows: While Achilles runs the 100 yards, the tortoise runs an additional 10 yards. While Achilles runs that 10 yards, the tortoise runs one additional yard. While Achilles runs that yard, the tortoise runs $\frac{1}{10}$ yard, and so on. Thus Achilles is always behind the tortoise and never catches up. By summing two infinite series, show that Achilles does in fact catch up with the tortoise, and at the same time determine how many yards it takes him to do it.

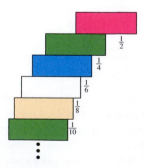

FIGURE 9.18   The blocks for Exercise 64.

*64. An indeterminately large number of identical blocks 1 unit long are stacked on top of each other. Show that it is possible for the top block to protrude as far from the bottom block as we wish without the blocks toppling (Figure 9.18). (*Hint:* The center of gravity of the top block must lie over the second block; the center of gravity of the top two blocks must lie over the third block, and so on. Thus the top block can protrude up to $\frac{1}{2}$ unit from the end of the second block, the second block can protrude up to $\frac{1}{4}$ unit from the end of the third block, the third block can protrude up to $\frac{1}{6}$ unit from the end of the fourth block, and so on. Assuming that the center of gravity of the first $(n-1)$ blocks lies over the end of the nth block, show that the nth block can protrude up to $1/2n$ units from the end of the $(n+1)$st block.)

*65. An ant crawls at the rate of 1 foot per minute along a rubber band, which can be stretched uniformly. Suppose the rubber band is initially 1 yard long and is stretched an additional yard at the end of each minute.

a. If the ant begins at one end of the band, does it ever reach the other end? If so, how long does it take the ant? (*Hint:* Use Exercise 51(a).)

b. Suppose the band is originally 20 feet long and is stretched an additional 20 feet at the end of each minute. How long would it take the ant to reach the end? (*Hint:* Use the data following the solution of Example 4.)

(Adapted from Martin Gardner, "Mathematical Games," *Scientific American* (March 1975) p. 112.)

Project

1. This project is devoted to Abel's Partial Summation Formula, which is an analogue of the integration by parts formula that applies to sums of finite sequences.

Let m and n be integers, and let $\{a_k\}_{m+1}^{n+1}$ and $\{b_k\}_m^n$ be finite sequences. Abel's Partial Summation Formula states that

$$\sum_{k=m+1}^{n} a_k (b_k - b_{k-1})$$

$$= a_{n+1} b_n - a_{m+1} b_m - \sum_{k=m+1}^{n} b_k (a_{k+1} - a_k)$$

a. Prove that

$$-\sum_{k=m+1}^{n} a_k b_{k-1}$$

$$= a_{n+1} b_n - a_{m+1} b_m - \sum_{k=m+1}^{n} b_k a_{k+1}$$

b. Use (a) to prove Abel's Partial Summation Formula.

c. Use Abel's Formula, along with the formulas immediately following the Geometric Series

Theorem, to evaluate $\sum_{k=1}^{n} k 2^k$.

Why do we say that Abel's Formula is an analogue of the integration by parts formula? The sums should remind us of Riemann sums. Indeed, if for $k = m$, $m+1, \ldots , n$ we think of

a_k as $a(k)$, b_k as $b(k)$, $b_k - b_{k-1}$ as db, $a_{k+1} - a_k$ as da

where a and b are continuous functions, and if we substitute integrals for finite sums, then Abel's Formula has a resemblance to the integration by parts formula

$$\int_{m}^{n} a\, db = a(n)b(n) - a(m)b(m) - \int_{m}^{n} b\, da$$

9.5   POSITIVE SERIES: THE INTEGRAL TEST AND THE COMPARISON TESTS

Thus far we have proved that a series converges by actually computing its value. However, for most convergent series the exact sum is difficult or impossible to find. This is true even of such series as $\sum_{n=1}^{\infty} 1/n^2$ or $\sum_{n=1}^{\infty} 1/n^3$. Yet in such cases it may suffice to know at least that the series converges. In Sections 9.5 to 9.7 we will formulate tests for determining the convergence or divergence of series. For the present we will restrict our attention to **positive series,** that is, to series whose terms are positive. For simplicity we assume that the initial index is 1. The partial sums $\{s_j\}_{j=1}^{\infty}$ of a positive series $\sum_{n=1}^{\infty} a_n$ form an increasing sequence:

$$s_j = a_1 + a_2 + \cdots + a_j < a_1 + a_2 + \cdots + a_j + a_{j+1} = s_{j+1} \quad \text{for } j \geq 1$$

Consequently if $\{s_j\}_{j=1}^{\infty}$ is bounded, then $\lim_{j\to\infty} s_j$ exists (by Theorem 9.6), so $\sum_{n=1}^{\infty} a_n$ converges. By contrast, if $\{s_j\}_{j=1}^{\infty}$ is unbounded, then $\lim_{j\to\infty} s_j$ cannot exist (by Theorem 9.5(b)), so $\sum_{n=1}^{\infty} a_n$ diverges.

In this section we will discuss two types of convergence tests, one that compares a given positive series with an improper integral and one that compares a given positive series with another series.

The Integral Test

The test that compares a positive series with an improper integral is the Integral Test. We will assume that $\{a_n\}_{n=1}^{\infty}$ is not only positive but also decreasing, so that for every positive integer n,

$$a_n > a_{n+1} > 0$$

The Integral Test is based on the observation that if f is a continuous, decreasing function defined on the interval $[1, \infty)$ such that $f(n) = a_n$ for all $n \geq 1$, then for each n,

$$a_{n+1} \leq f(x) \leq a_n \quad \text{for all } x \text{ in } [n, n+1]$$

Thus

$$\int_{n}^{n+1} a_{n+1}\, dx \leq \int_{n}^{n+1} f(x)\, dx \leq \int_{n}^{n+1} a_n\, dx$$

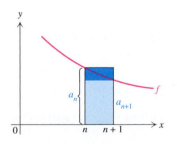

(Figure 9.19). Carrying out the integration, we conclude that

$$a_{n+1} \leq \int_{n}^{n+1} f(x)\, dx \leq a_n \tag{1}$$

FIGURE 9.19

Now we are ready for the test.

THEOREM 9.12
Integral Test

Let $\{a_n\}_{n=1}^{\infty}$ be a positive decreasing sequence, and let f be a continuous decreasing function defined on $[1, \infty)$ such that $f(n) = a_n$ for $n \geq 1$. Then the series $\sum_{n=1}^{\infty} a_n$ converges if and only if the improper integral $\int_1^{\infty} f(x)\, dx$ converges.

Proof Let j be any positive integer. Then by (1) with $n = 1, \ldots, j$, we have

$$0 < a_2 + a_3 + \cdots + a_j = \int_1^2 a_2\, dx + \int_2^3 a_3\, dx + \cdots + \int_{j-1}^{j} a_j\, dx$$

$$\leq \int_1^2 f(x)\, dx + \int_2^3 f(x)\, dx + \cdots + \int_{j-1}^{j} f(x)\, dx$$

$$\leq \int_1^j f(x)\, dx \quad \text{for } j \geq 2 \tag{2}$$

(see Figure 9.20(a)). Similarly,

$$\int_1^j f(x)\, dx \leq a_1 + a_2 + \cdots + a_{j-1} \quad \text{for } j \geq 2 \tag{3}$$

(Figure 9.20(b)). If the improper integral $\int_1^{\infty} f(x)\, dx$ converges, then from (2) we know that the partial sums of $\sum_{n=2}^{\infty} a_n$ are bounded above, so $\sum_{n=2}^{\infty} a_n$, and hence $\sum_{n=1}^{\infty} a_n$, converge by Theorem 9.6. Similarly, if $\sum_{n=1}^{\infty} a_n$ converges, then from (3) we conclude that $\int_1^{\infty} f(x)\, dx$ converges. ■

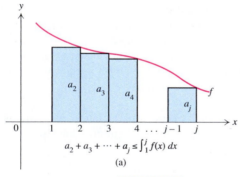

$$a_2 + a_3 + \cdots + a_j \leq \int_1^j f(x)\, dx$$

(a)

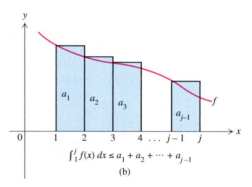

$$\int_1^j f(x)\, dx \leq a_1 + a_2 + \cdots + a_{j-1}$$

(b)

FIGURE 9.20

It follows from (2) and (3) in the proof of the Integral Test that if either $\sum_{n=1}^{\infty} a_n$ or $\int_1^{\infty} f(x)\, dx$ converges, then so does the other, and moreover,

$$\sum_{n=2}^{\infty} a_n \leq \int_1^{\infty} f(x)\, dx \leq \sum_{n=1}^{\infty} a_n$$

The Integral Test is tailor-made for ***p* series,** which are (positive) series of the form $\sum_{n=1}^{\infty} 1/n^p$, where p is a fixed number.

EXAMPLE 1   Show that

$$\sum_{n=1}^{\infty} \frac{1}{n^p} \text{ converges if and only if } p > 1$$

Solution   If $p \leq 0$, the terms $1/n^p$ do not tend to 0 as n increases, so by the nth Term Test (Corollary 9.9) the series $\sum_{n=1}^{\infty} 1/n^p$ diverges. Thus from here on in the proof, we assume that $p > 0$. For $p = 1$ the series is the harmonic series, which we know diverges. So we assume that $p \neq 1$. The ideal function f to use in applying the Integral Test is defined by

$$f(x) = \frac{1}{x^p} \quad \text{for } x \geq 1$$

since $f(n) = 1/n^p$ for $n \geq 1$ and since f is continuous and decreasing on $[1, \infty)$. To determine whether $\int_1^{\infty} f(x)\, dx$ converges, we must ascertain whether

$$\lim_{b \to \infty} \int_1^b \frac{1}{x^p}\, dx$$

exists. However, since $p \neq 1$, we find that

$$\int_1^b \frac{1}{x^p}\, dx = \frac{1}{(-p+1)x^{p-1}}\Big|_1^b = \frac{1}{(-p+1)}\left(\frac{1}{b^{p-1}} - 1\right)$$

Since $\lim_{b \to \infty} (1/b^{p-1})$ exists if $p > 1$ and does not exist if $0 < p < 1$, it follows that $\int_1^{\infty} f(x)\, dx$ converges if $p > 1$ and diverges if $0 < p < 1$. Consequently the Integral Test tells us that $\sum_{n=1}^{\infty} 1/n^p$ converges if $p > 1$ and diverges if $0 < p < 1$.   ❏

From Example 1 we know that $\sum_{n=1}^{\infty} 1/n^{1.001}$ converges and $\sum_{n=1}^{\infty} 1/n^{0.999}$ diverges. We also know that both $\sum_{n=1}^{\infty} 1/n^2$ and $\sum_{n=1}^{\infty} 1/n^3$ converge. More generally, $\sum_{n=1}^{\infty} 1/n^p$ converges for any integer $p > 1$. However, the result of Example 1 does not give us the sum of any of these series. In fact, the value of $\sum_{n=1}^{\infty} 1/n^2$ baffled mathematicians until the middle of the eighteenth century, when Leonhard Euler proved that

$$\sum_{n=1}^{\infty} \frac{1}{n^2} = \frac{\pi^2}{6}$$

Today the sum of the p series $\sum_{n=1}^{\infty} 1/n^p$ is known for every positive even integer p but not for a single odd integer greater than 1. Thus, despite the fact that $\sum_{n=1}^{\infty} 1/n^3$ is known to converge, its sum is unknown (although it has recently been proved that the sum is irrational).

Even though we know that $\sum_{n=1}^{\infty} 1/n^2 = \pi^2/6$, it is not apparent how rapidly the partial sums of the series converge to $\pi^2/6$, or equivalently, how good an estimate of the sum of the series a given partial sum such as $\sum_{n=1}^{100} 1/n^2$ is. To address this question we define the truncation error for partial sums of an arbitrary convergent series. For any convergent series $\sum_{n=1}^{\infty} a_n$ we define the ***jth truncation error*** E_j by the formula

$$E_j = \left| \sum_{n=1}^{\infty} a_n - \sum_{n=1}^{j} a_n \right| = \left| \sum_{n=j+1}^{\infty} a_n \right|$$

If $\{a_n\}_{n=1}^{\infty}$ is a positive sequence, then the jth truncation error reduces to

$$E_j = \sum_{n=j+1}^{\infty} a_n = a_{j+1} + a_{j+2} + \cdots$$

and equals the sum of the terms beyond a_j. The term "truncation" refers to the fact that to obtain the partial sum $\sum_{n=1}^{j} a_n$ one cuts off, or truncates, the infinite sum and dispenses with the terms after the jth term. For example, the 8th truncation error of $\sum_{n=1}^{\infty} 1/n^2$ is

$$\sum_{n=9}^{\infty} \frac{1}{n^2} = \frac{1}{9^2} + \frac{1}{10^2} + \cdots$$

Let us assume, as in the Integral Test, that f is continuous and decreasing with $f(n) = a_n$ for $n \geq 1$. By (1), $a_{n+1} \leq \int_{n}^{n+1} f(x)\, dx$ for all $n \geq 1$. Applying this inequality with $n = j, j + 1, \ldots$, we obtain

$$\sum_{n=j+1}^{\infty} a_n = a_{j+1} + a_{j+2} + \cdots \leq \int_{j}^{j+1} f(x)\, dx + \int_{j+1}^{j+2} f(x)\, dx + \cdots \quad (4)$$

By an extension of the Addition Property for Integrals,

$$\int_{j}^{j+1} f(x)\, dx + \int_{j+1}^{j+2} f(x)\, dx + \cdots = \int_{j}^{\infty} f(x)\, dx \quad (5)$$

Combining (4) and (5) yields

$$\sum_{n=j+1}^{\infty} a_n \leq \int_{j}^{\infty} f(x)\, dx \quad (6)$$

Similarly, by using the inequality $\int_{n}^{n+1} f(x)\, dx \leq a_n$ in (1), we obtain

$$\int_{j}^{\infty} f(x)\, dx \leq \sum_{n=j}^{\infty} a_n, \quad \text{or equivalently,} \quad \int_{j+1}^{\infty} f(x)\, dx \leq \sum_{n=j+1}^{\infty} a_n \quad (7)$$

Together (6) and (7) imply that

$$\int_{j+1}^{\infty} f(x)\, dx \leq \sum_{n=j+1}^{\infty} a_n \leq \int_{j}^{\infty} f(x)\, dx$$

Therefore by the definition of the jth truncation error,

$$\int_{j+1}^{\infty} f(x)\, dx \leq E_j \leq \int_{j}^{\infty} f(x)\, dx \quad (8)$$

The inequalities in (8) can give us a very reasonable estimate for the truncation error, provided f is a positive decreasing function and we are able to evaluate the two related improper integrals.

EXAMPLE 2   Estimate E_{100} for the series $\sum_{n=1}^{\infty} 1/n^2$.

Solution   Let $f(x) = 1/x^2$, so that f is a continuous, decreasing function such that $f(n) = 1/n^2$ for $n \geq 1$. To apply (8) we calculate $\int_{101}^{\infty} 1/x^2\, dx$ and $\int_{100}^{\infty} 1/x^2\, dx$. For any positive integer j,

$$\int_{j}^{\infty} \frac{1}{x^2}\, dx = \lim_{b \to \infty} \int_{j}^{b} \frac{1}{x^2}\, dx = \lim_{b \to \infty} \left(-\frac{1}{x} \right) \Big|_{j}^{b} = \lim_{b \to \infty} \left(-\frac{1}{b} + \frac{1}{j} \right) = \frac{1}{j}$$

Therefore

$$\frac{1}{101} = \int_{101}^{\infty} \frac{1}{x^2} \, dx \le E_{100} \le \int_{100}^{\infty} \frac{1}{x^2} \, dx = \frac{1}{100}$$

Consequently

$$\frac{1}{101} \le E_{100} \le \frac{1}{100} \qquad \square$$

A consequence of Example 2 is that the partial sum $\sum_{n=1}^{j} 1/n^2$ is within 0.01 of $\sum_{n=1}^{\infty} 1/n^2$ if $j \ge 100$. An analogous calculation would show that for any $j > 1$ and any $p > 1$, the jth truncation error E_j for $\sum_{n=1}^{\infty} 1/n^p$ satisfies the inequalities

$$\frac{1}{(p-1)(j+1)^{p-1}} \le E_j \le \frac{1}{(p-1)j^{p-1}} \tag{9}$$

(see Exercise 34). For example, if $p = \frac{3}{2}$, then (9) implies that

$$E_{10,000} < \frac{1}{(1/2)(10,000)^{1/2}} = 0.02 < E_{9999}$$

This means that only after summing the first 10,000 terms of $\sum_{n=1}^{\infty} 1/n^{3/2}$ is the truncation error less than 0.02. We conclude that the series converges very slowly indeed. Nevertheless there are other series that converge much more slowly!

We may sometimes define the function f for the Integral Test on an interval different from $[1, \infty)$. In the next example we will use the Integral Test to show the divergence of the series $\sum_{n=2}^{\infty} 1/(n \ln n)$, and in so doing it will be helpful to define f on $[2, \infty)$.

EXAMPLE 3   Show that $\displaystyle\sum_{n=2}^{\infty} \frac{1}{n \ln n}$ diverges.

Solution   In this case we let

$$f(x) = \frac{1}{x \ln x} \quad \text{for } x \ge 2$$

Then f is continuous and decreasing on $[2, \infty)$, and

$$f(n) = \frac{1}{n \ln n} \quad \text{for } n \ge 2$$

Next we observe that

$$\int_{2}^{\infty} f(x) \, dx = \lim_{b \to \infty} \int_{2}^{b} \frac{1}{x \ln x} \, dx = \lim_{b \to \infty} \left(\ln (\ln x) \Big|_{2}^{b} \right)$$

$$= \lim_{b \to \infty} [\ln (\ln b) - \ln (\ln 2)] = \infty$$

This means that $\int_{2}^{\infty} f(x) \, dx$ diverges. By the Integral Test, the given series also diverges. $\square$

The Integral Test is most effective when the function f to be used is easily integrated, as was the case in Examples 1 and 3.

The Comparison Tests

The next convergence test involves the comparison of two positive series.

THEOREM 9.13
Comparison Test

a. If $\sum_{n=1}^{\infty} b_n$ converges and $0 < a_n \le b_n$ for all $n \ge 1$, then $\sum_{n=1}^{\infty} a_n$ converges, and $\sum_{n=1}^{\infty} a_n \le \sum_{n=1}^{\infty} b_n$.

b. If $\sum_{n=1}^{\infty} b_n$ diverges and $0 < b_n \le a_n$ for all $n \ge 1$, then $\sum_{n=1}^{\infty} a_n$ diverges.

Proof   To prove (a), we first let s_j and s'_j be the jth partial sums of $\sum_{n=1}^{\infty} a_n$ and $\sum_{n=1}^{\infty} b_n$, respectively. Since $\sum_{n=1}^{\infty} b_n$ converges by hypothesis, $\lim_{j \to \infty} s'_j$ exists, and we denote this limit by L. Since $0 < a_n \le b_n$ for all n, it follows that $\{s_j\}_{j=1}^{\infty}$ and $\{s'_j\}_{j=1}^{\infty}$ are increasing and that

$$0 < s_j \le s'_j \le L \quad \text{for } j \ge 1 \tag{10}$$

But this implies that $\{s_j\}_{j=1}^{\infty}$ is a bounded, increasing sequence, so by Theorem 9.6 we know that $\lim_{j \to \infty} s_j$ exists. Consequently $\sum_{n=1}^{\infty} a_n$ converges. Moreover, from (10) it follows that $\sum_{n=1}^{\infty} a_n = \lim_{j \to \infty} s_j \le L = \sum_{n=1}^{\infty} b_n$. This proves (a). To prove (b), we need only observe that if $\sum_{n=1}^{\infty} a_n$ were to converge, then by part (a) the series $\sum_{n=1}^{\infty} b_n$ would necessarily converge, contrary to the hypothesis of (b). Therefore $\sum_{n=1}^{\infty} a_n$ must diverge. ∎

We use the Comparison Test to deduce that a given positive series converges by comparing it with a known convergent series having larger terms. To show that a given positive series diverges, we compare it with a known divergent series having smaller positive terms.

EXAMPLE 4   Show that $\displaystyle\sum_{n=1}^{\infty} \frac{1}{2^n + 1}$ converges.

Solution   By the Geometric Series Theorem we know that the geometric series $\sum_{n=1}^{\infty} 1/2^n$ converges. Since

$$\frac{1}{2^n + 1} \le \frac{1}{2^n} \quad \text{for } n \ge 1$$

it follows from the Comparison Test that $\sum_{n=1}^{\infty} 1/(2^n + 1)$ converges.   ❑

EXAMPLE 5   Show that $\displaystyle\sum_{n=1}^{\infty} \frac{1}{3\sqrt{n} - 1}$ diverges.

Solution   Observe that

$$\frac{1}{3\sqrt{n} - 1} \ge \frac{1}{3\sqrt{n}} \quad \text{for } n \ge 1$$

Since $\sum_{n=1}^{\infty} 1/\sqrt{n}$ is a p series with $p = \frac{1}{2}$, it diverges. Therefore $\sum_{n=1}^{\infty} 1/(3\sqrt{n})$ also diverges. We conclude from the Comparison Test that $\sum_{n=1}^{\infty} 1/(3\sqrt{n} - 1)$ diverges.   ❑

EXAMPLE 6 Show that $\sum_{n=1}^{\infty} \dfrac{1}{2^n - 1}$ converges.

Solution It might be tempting to compare the given series with the convergent series $\sum_{n=1}^{\infty} 1/2^n$. However,

$$\frac{1}{2^n - 1} \geq \frac{1}{2^n} \quad \text{for } n \geq 1$$

and thus it is impossible to determine the convergence or divergence of the given series by comparing it with the series $\sum_{n=1}^{\infty} 1/2^n$. But since $2^n - 1 \geq 2^{n-1}$ for $n \geq 1$,

$$\frac{1}{2^n - 1} \leq \frac{1}{2^{n-1}} \quad \text{for } n \geq 1 \tag{11}$$

and since $\sum_{n=1}^{\infty} 1/2^{n-1}$ is a convergent geometric series, the Comparison Test implies that the given series converges as well.   ❑

In general, when we try to determine whether a given series $\sum_{n=1}^{\infty} a_n$ converges or diverges by using the Comparison Test, we select a series $\sum_{n=1}^{\infty} b_n$ and find a nonzero number c such that either $\sum_{n=1}^{\infty} b_n$ converges and $a_n \leq cb_n$ for all $n \geq 1$, or $\sum_{n=1}^{\infty} b_n$ diverges and $a_n \geq cb_n$ for all $n \geq 1$. However, finding such a number c is sometimes rather difficult. In such instances the following test may be easier to apply.

THEOREM 9.14
Limit Comparison Test

Let $\sum_{n=1}^{\infty} a_n$ and $\sum_{n=1}^{\infty} b_n$ be positive series. Suppose $\lim_{n\to\infty} a_n/b_n = L$, where L is a positive number.

a. If $\sum_{n=1}^{\infty} b_n$ converges, then $\sum_{n=1}^{\infty} a_n$ converges.

b. If $\sum_{n=1}^{\infty} b_n$ diverges, then $\sum_{n=1}^{\infty} a_n$ diverges.

Proof Since $\lim_{n\to\infty} a_n/b_n = L$, there is an integer N such that

$$\frac{1}{2} L \leq \frac{a_n}{b_n} \leq \frac{3}{2} L \quad \text{for } n \geq N$$

Consequently $a_n \leq \frac{3}{2}Lb_n$ and $a_n \geq \frac{1}{2}Lb_n$ for $n \geq N$. From these inequalities and the Comparison Test it follows that if $\sum_{n=1}^{\infty} b_n$ converges, then so in turn do $\sum_{n=N}^{\infty} \frac{3}{2}Lb_n$, $\sum_{n=N}^{\infty} a_n$, and $\sum_{n=1}^{\infty} a_n$. Likewise, if $\sum_{n=1}^{\infty} b_n$ diverges, then so in turn do $\sum_{n=N}^{\infty} \frac{1}{2}LB_n$, $\sum_{n=N}^{\infty} a_n$, and $\sum_{n=1}^{\infty} a_n$.   ■

EXAMPLE 7 Show that $\sum_{n=1}^{\infty} \dfrac{4n - 3}{n^3 - 5n - 7}$ converges.

Solution The terms of the series are rational functions of n. To find a series with which to compare the given series, we disregard all but the highest powers of n appearing in the numerator and denominator, obtaining

$$\frac{4n}{n^3} = \frac{4}{n^2}$$

Therefore we will compare the given series with $\sum_{n=1}^{\infty} 4/n^2$. Since $\sum_{n=1}^{\infty} 1/n^2$ is a convergent p series (with $p = 2$), $\sum_{n=1}^{\infty} 4/n^2$ converges. Because

$$\lim_{n \to \infty} \frac{(4n - 3)/(n^3 - 5n - 7)}{4/n^2} = \lim_{n \to \infty} \frac{4n^3 - 3n^2}{4n^3 - 20n - 28}$$

$$= \lim_{n \to \infty} \frac{4 - 3/n}{4 - 20/n^2 - 28/n^3} = 1$$

part (a) of the Limit Comparison Test implies that the given series converges. ❏

EXAMPLE 8 Show that $\displaystyle\sum_{n=1}^{\infty} \frac{1}{\sqrt[3]{8n^2 - 5n}}$ diverges.

Solution We disregard all but the highest power of n in the denominator, obtaining

$$\frac{1}{\sqrt[3]{8n^2}} = \frac{1}{2n^{2/3}}$$

Accordingly, we compare the given series with $\sum_{n=1}^{\infty} 1/2n^{2/3}$. Since $\sum_{n=1}^{\infty} 1/n^{2/3}$ is a divergent p series (with $p = \frac{2}{3}$), $\sum_{n=1}^{\infty} 1/2n^{2/3}$ diverges. Because

$$\lim_{n \to \infty} \frac{1/\sqrt[3]{8n^2 - 5n}}{1/\sqrt[3]{8n^2}} = \lim_{n \to \infty} \sqrt[3]{\frac{8n^2}{8n^2 - 5n}} = \lim_{n \to \infty} \sqrt[3]{\frac{8}{8 - 5/n}} = 1$$

part (b) of the Limit Comparison Test implies that the given series diverges. ❏

EXERCISES 9.5

In Exercises 1–26 use the Comparison Test, the Limit Comparison Test, or the Integral Test to determine whether the series converges or diverges.

1. $\displaystyle\sum_{n=1}^{\infty} \frac{1}{(n + 1)^2}$

2. $\displaystyle\sum_{n=1}^{\infty} \frac{1}{\sqrt{n}}$

3. $\displaystyle\sum_{n=1}^{\infty} \frac{1}{\sqrt{n^3 + 1}}$

4. $\displaystyle\sum_{n=2}^{\infty} \frac{1}{\sqrt{n^2 - 1}}$

5. $\displaystyle\sum_{n=1}^{\infty} \frac{1}{\sqrt{n^2 + 1}}$

6. $\displaystyle\sum_{n=1}^{\infty} \frac{1}{n + \sqrt{n}}$

7. $\displaystyle\sum_{n=1}^{\infty} \frac{1}{e^{(n^2)}}$ $\left(\text{\emph{Hint:} Compare with } \displaystyle\sum_{n=1}^{\infty} (1/e)^n.\right)$

8. $\displaystyle\sum_{n=2}^{\infty} \frac{1}{n\sqrt{\ln n}}$

9. $\displaystyle\sum_{n=3}^{\infty} \frac{1}{(n - 1)(n - 2)}$

10. $\displaystyle\sum_{n=2}^{\infty} \frac{n}{n^3 - n - 1}$

11. $\displaystyle\sum_{n=2}^{\infty} \frac{n^2 - 1}{n^3 - n - 1}$

12. $\displaystyle\sum_{n=3}^{\infty} \frac{3 + \cos n}{n^2 - 4}$

13. $\displaystyle\sum_{n=1}^{\infty} \frac{n}{7^n}$

14. $\displaystyle\sum_{n=1}^{\infty} \frac{\sqrt{n}}{n^2 + 3}$

15. $\displaystyle\sum_{n=4}^{\infty} \frac{\sqrt{n}}{n^2 - 3}$

16. $\displaystyle\sum_{n=1}^{\infty} \frac{\cos^2 n}{n^{3/2}}$

17. $\displaystyle\sum_{n=1}^{\infty} \frac{n}{(n^3 + 1)^{3/7}}$

18. $\displaystyle\sum_{n=1}^{\infty} \frac{1}{\sqrt[3]{n^2 + 1}}$

19. $\displaystyle\sum_{n=2}^{\infty} \frac{1}{n\sqrt{n^2 - 1}}$

20. $\displaystyle\sum_{n=1}^{\infty} ne^{(-n^2)}$

21. $\displaystyle\sum_{n=1}^{\infty} \frac{\tan^{-1} n}{n^2 + 1}$

22. $\displaystyle\sum_{n=2}^{\infty} \frac{1}{n(\ln n)^2}$

23. $\displaystyle\sum_{n=1}^{\infty} \frac{\ln n}{n^2}$ (*Hint:* $\ln n \le 2\sqrt{n}$ for $n \ge 1$ by Exercise 53 of Section 5.7.)

24. $\displaystyle\sum_{n=1}^{\infty} \frac{1}{1 + 2 + 3 + \cdots + n}$

25. $\displaystyle\sum_{n=1}^{\infty} \sin \frac{1}{n}$

26. $\displaystyle\sum_{n=1}^{\infty} \frac{1}{n\sqrt[n]{n}}$

In Exercises 27–30 find a value of j for which the jth truncation error E_j is less than or equal to 0.02.

27. $\displaystyle\sum_{n=1}^{\infty} \frac{1}{n^3}$

28. $\displaystyle\sum_{n=2}^{\infty} \frac{1}{n^{4/3}}$

29. $\displaystyle\sum_{n=1}^{\infty} \frac{1}{n^{8/7}}$

30. $\displaystyle\sum_{n=2}^{\infty} \frac{1}{n(\ln n)^2}$

31. a. Use (6) and (7) to show that for any $j > 1$,

$$\sum_{n=1}^{j-1} a_n + \sum_{n=j+1}^{\infty} a_n \le \sum_{n=1}^{j-1} a_n + \int_j^{\infty} f(x)\, dx \le \sum_{n=1}^{\infty} a_n$$

b. Use part (a) to show that

$$0 \le \sum_{n=1}^{\infty} a_n - \left(\sum_{n=1}^{j-1} a_n + \int_j^{\infty} f(x)\, dx \right) \le a_j$$

(Thus $\sum_{n=1}^{j-1} a_n + \int_j^{\infty} f(x)\, dx$ approximates $\sum_{n=1}^{\infty} a_n$ with an error at most a_j.)

 32. Use Exercise 31 to approximate $\sum_{n=1}^{\infty} 1/n^4$ with an error at most 10^{-4}.

33. Determine those values of p for which the series $\sum_{n=2}^{\infty} 1/[n\,(\ln n)^p]$ converges.

34. Let $p > 1$. Show that for any integer j, the jth truncation error E_j for $\sum_{n=1}^{\infty} 1/n^p$ satisfies the inequalities

$$\frac{1}{(p-1)(j+1)^{p-1}} \le E_j \le \frac{1}{(p-1)j^{p-1}}$$

35. Prove that if $\sum_{n=1}^{\infty} a_n$ is a convergent positive series, then so is $\sum_{n=1}^{\infty} a_n^2$. (*Hint:* Show that $a_n^2 \le a_n$ for sufficiently large n.)

36. Show that if $\sum_{n=1}^{\infty} a_n$ and $\sum_{n=1}^{\infty} b_n$ are convergent positive series, then $\sum_{n=1}^{\infty} a_n b_n$ converges. (*Hint:* Use the fact that $b_n \le 1$ for sufficiently large values of n.)

*37. Using (2) and (3) from the proof of the Integral Test, find

$$\lim_{j \to \infty} \left(\frac{\sum_{n=1}^{j} 1/n}{\ln j} \right)$$

Applications

38. A rocket is launched from earth. On its journey it consumes one fourth of its fuel during the first 100 miles, one ninth of its initial fuel during the second 100 miles, and in general, $1/(n+1)^2$ of its initial fuel during the nth 100 miles. Does the rocket ever use up all its fuel? (*Hint:* $\sum_{n=1}^{\infty} 1/n^2 = \pi^2/6$.)

39. Imagine an infinitely tall tower with circular floors spaced 1 meter apart (Figure 9.21). Suppose that for $n \ge 2$, the radius of the nth floor is $10/(n \ln n)$.

  a. Is the total circumference of all the floors finite? Explain why or why not.

  b. Is the total area of all the floors finite? Explain why or why not.

Projects

1. In this project we will show that there is always a positive series that diverges more slowly than any given series.

  a. Show that $\displaystyle\sum_{n=2}^{\infty} \frac{1}{n \ln n}$ diverges.

  b. Show that $\displaystyle\sum_{n=3}^{\infty} \frac{1}{n \ln n \ln (\ln n)}$ diverges.

  c. Show that $\dfrac{1}{n} \ge \dfrac{1}{n \ln n} \ge \dfrac{1}{n \ln n \ln n \ln (\ln n)}$ for $n \ge 16$. Use the ideas from this set of inequalities to find a series that diverges even more slowly than the series in (b).

  d. Suppose that $\sum_{n=1}^{\infty} a_n$ is a positive series that diverges. Show that

  $$\sum_{n=1}^{\infty} \frac{a_n}{1 + a_n}$$

  is a divergent series with smaller terms. (Thus there is no "smallest" divergent series.)

  e. Use (d) to find a series that diverges more slowly than the series in (b). (Likely this series is quite different from the series you created in (c)!)

2. In this project we will continue looking at series that involve logarithms.

  a. Show that $\displaystyle\sum_{n=2}^{\infty} \frac{1}{(\ln n)^n}$ converges.

  b. Show that $\displaystyle\sum_{n=2}^{\infty} \frac{1}{(\ln n)^{\ln n}}$ converges. (*Hint:* Make an appropriate substitution and then show that the improper integral converges.)

  c. Let k be any positive integer. Show that

  $$\sum_{n=2}^{\infty} \frac{1}{(\ln n)^k}$$

  diverges. (*Hint:* Compare with

  $$\sum_{n=2}^{\infty} 1/n.)$$

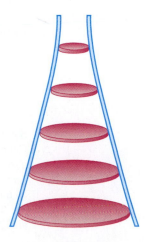

FIGURE 9.21   Figure for Exercise 39.

9.6   POSITIVE SERIES: THE RATIO TEST AND THE ROOT TEST

In this section we present two more convergence tests for positive series: the Ratio Test and the Root Test. These are intrinsic tests in that they involve only the terms of the series being tested; it is not necessary to manufacture another series, an improper integral, or anything else against which to compare the given series. In this sense the Ratio Test and the Root Test are easier to apply than are the Comparison Test and the Integral Test. However, even though no comparison is required in the application of the Ratio Test and the Root Test, in proving the validity of these tests we will compare the given series with a geometric series. Thus both the Ratio Test and the Root Test ultimately depend on the Comparison Test (as well as an understanding of geometric series).

The Ratio Test

The more commonly used of the two tests is the Ratio Test.

THEOREM 9.15
Ratio Test

Let $\sum_{n=1}^{\infty} a_n$ be a positive series. Assume that

$$\lim_{n \to \infty} \frac{a_{n+1}}{a_n} = r \quad (\text{possibly } \infty)$$

a. If $0 \le r < 1$, then $\sum_{n=1}^{\infty} a_n$ converges.

b. If $r > 1$, then $\sum_{n=1}^{\infty} a_n$ diverges.

If $r = 1$, then from this test alone we cannot draw any conclusion about the convergence or divergence of $\sum_{n=1}^{\infty} a_n$.

Proof   First we prove (a). We assume that $0 \le r < 1$, and let s be any number such that $r < s < 1$. Since

$$\lim_{n \to \infty} \frac{a_{n+1}}{a_n} = r \quad \text{and} \quad r < s$$

there is an integer N such that for $n \ge N$ we have $a_{n+1}/a_n \le s$, or equivalently, $a_{n+1} \le a_n s$. By letting $n = N$ we obtain $a_{N+1} \le a_N s$, and then by letting $n = N + 1$ we obtain

$$a_{N+2} \le a_{N+1} s \le (a_N s)s = a_N s^2$$

In general, for any positive integer n we find that

$$0 < a_{N+n} \le a_{N+n-1} s \le a_{N+n-2} s^2 \le \cdots \le a_{N+1} s^{n-1} \le a_N s^n \qquad (1)$$

Since $0 < s < 1$, the geometric series $\sum_{n=0}^{\infty} a_N s^n$ converges. From (1) and the Comparison Test we conclude that $\sum_{n=0}^{\infty} a_{N+n}$, which is the same series as $\sum_{n=N}^{\infty} a_n$, also converges. From the convergence of $\sum_{n=N}^{\infty} a_n$ we conclude that $\sum_{n=1}^{\infty} a_n$ converges. This proves (a). The proof of (b) is analogous, but with $r > s > 1$ and with all the inequalities in the proof of (a) reversed.

To verify the final statement of the theorem, we recall that $\sum_{n=1}^{\infty} 1/n$ diverges and $\sum_{n=1}^{\infty} 1/n^2$ converges, and compute the corresponding ratios:

$$\sum_{n=1}^{\infty} \frac{1}{n} \text{ diverges} \quad \text{and} \quad \lim_{n \to \infty} \frac{1/(n+1)}{1/n} = \lim_{n \to \infty} \frac{n}{n+1} = \lim_{n \to \infty} \frac{1}{1 + 1/n} = 1$$

$$\sum_{n=1}^{\infty} \frac{1}{n^2} \text{ converges} \quad \text{and} \quad \lim_{n \to \infty} \frac{1/(n+1)^2}{1/n^2} = \lim_{n \to \infty} \frac{n^2}{(n+1)^2} = \lim_{n \to \infty} \frac{1}{(1+1/n)^2} = 1$$

Therefore if $\lim_{n \to \infty} a_{n+1}/a_n = 1$, it is impossible to determine the convergence or divergence of the series merely by the Ratio Test.   ■

The Ratio Test furnishes a quick proof that $\sum_{n=0}^{\infty} 1/n!$ converges.

EXAMPLE 1   Use the Ratio Test to show that $\sum_{n=0}^{\infty} 1/n!$ converges.

Solution   We find that

$$r = \lim_{n \to \infty} \frac{a_{n+1}}{a_n} = \lim_{n \to \infty} \frac{1/(n+1)!}{1/n!} = \lim_{n \to \infty} \frac{n!}{(n+1)!} = \lim_{n \to \infty} \frac{1}{n+1} = 0$$

Since $r = 0 < 1$, the series converges.   ❑

In a similar way, you can show that if c is *any* positive number, then $\sum_{n=0}^{\infty} c^n/n!$ converges (Exercise 24).

EXAMPLE 2   Show that $\sum_{n=1}^{\infty} \frac{2^n}{n^2}$ diverges.

Solution   A routine computation yields

$$r = \lim_{n \to \infty} \frac{a_{n+1}}{a_n} = \lim_{n \to \infty} \frac{2^{n+1}/(n+1)^2}{2^n/n^2} = \lim_{n \to \infty} 2\left(\frac{n}{n+1}\right)^2 = 2$$

Since $r > 1$, the series diverges.   ❑

EXAMPLE 3   Show that $\sum_{n=1}^{\infty} \frac{n!}{n^n}$ converges.

Solution   We find that

$$r = \lim_{n \to \infty} \frac{a_{n+1}}{a_n} = \lim_{n \to \infty} \frac{(n+1)!/(n+1)^{n+1}}{n!/n^n} = \lim_{n \to \infty} \frac{(n+1)!}{n!} \cdot \frac{n^n}{(n+1)^{n+1}}$$

$$= \lim_{n \to \infty} (n+1) \frac{n^n}{(n+1)^{n+1}} = \lim_{n \to \infty} \frac{n^n}{(n+1)^n}$$

$$= \lim_{n \to \infty} \frac{1}{[(n+1)/n]^n} = \frac{1}{\lim_{n \to \infty} (1+1/n)^n} = \frac{1}{e}$$

where the last equality results from Example 5 of Section 9.2. Since $r = 1/e < 1$, the series converges.   ❑

Caution: Do not be misled by the solution to Example 3. We have *not* shown that $\sum_{n=1}^{\infty} n!/n^n$ converges to $1/e$. We have merely shown that the ratios of adjacent terms in the series converge to $1/e$. Incidentally, mathematicians have no idea what the sum of the series is.

In case $\lim_{n\to\infty} a_{n+1}/a_n$ fails to exist, no conclusion can be drawn from the Ratio Test (see Exercise 28).

The Root Test

A useful companion of the Ratio Test is the Root Test.

THEOREM 9.16
Root Test

Let $\sum_{n=1}^{\infty} a_n$ be a positive series, and assume that

$$\lim_{n\to\infty} \sqrt[n]{a_n} = r \quad \text{(possibly } \infty)$$

a. If $0 \le r < 1$, then $\sum_{n=1}^{\infty} a_n$ converges.

b. If $r > 1$, then $\sum_{n=1}^{\infty} a_n$ diverges.

If $r = 1$, then from this test alone we cannot draw any conclusion about the convergence or divergence of $\sum_{n=1}^{\infty} a_n$.

Proof   Assume first that $0 \le r < 1$, and let s be any number such that $r < s < 1$. Since

$$\lim_{n\to\infty} \sqrt[n]{a_n} = r \quad \text{and} \quad r < s$$

there is an integer N such that for $n \ge N$ we have $\sqrt[n]{a_n} \le s$, or equivalently, $a_n \le s^n$. Since $s < 1$, the geometric series $\sum_{n=N}^{\infty} s^n$ converges. Then by the Comparison Test $\sum_{n=N}^{\infty} a_n$ converges, and hence $\sum_{n=1}^{\infty} a_n$ also converges. Thus (a) is proved. The case in which $r > 1$ is proved analogously.

To verify the final statement of the theorem, we recall that $\sum_{n=1}^{\infty} 1/n$ diverges and $\sum_{n=1}^{\infty} 1/n^2$ converges, and compute corresponding roots (with the help of Example 9 of Section 9.2):

$$\sum_{n=1}^{\infty} \frac{1}{n} \text{ diverges} \quad \text{and} \quad \lim_{n\to\infty} \sqrt[n]{\frac{1}{n}} = \lim_{n\to\infty} \frac{1}{\sqrt[n]{n}} = 1$$

$$\sum_{n=1}^{\infty} \frac{1}{n^2} \text{ converges} \quad \text{and} \quad \lim_{n\to\infty} \sqrt[n]{\frac{1}{n^2}} = \lim_{n\to\infty} \frac{1}{(\sqrt[n]{n})^2} = \lim_{n\to\infty} \frac{1}{\sqrt[n]{n}} \lim_{n\to\infty} \frac{1}{\sqrt[n]{n}} = 1$$

Therefore if $\lim_{n\to\infty} \sqrt[n]{a_n} = 1$, it is impossible to draw any conclusion from the Root Test about the convergence of $\sum_{n=1}^{\infty} a_n$.   ∎

EXAMPLE 4   Show that $\sum_{n=1}^{\infty} \left(\frac{\ln n}{1000}\right)^n$ diverges.

Solution   A simple computation shows that

$$r = \lim_{n\to\infty} \left[\left(\frac{\ln n}{1000}\right)^n\right]^{1/n} = \lim_{n\to\infty} \frac{\ln n}{1000} = \infty$$

Therefore the series diverges by the Root Test.   ❑

EXAMPLE 5   Show that $\displaystyle\sum_{n=1}^{\infty} \frac{n}{2^n}$ converges.

Solution   Taking the nth roots of the terms of the series and using Example 9 of Section 9.2, we find that

$$r = \lim_{n\to\infty} \sqrt[n]{\frac{n}{2^n}} = \lim_{n\to\infty} \frac{\sqrt[n]{n}}{2} = \frac{1}{2}$$

Thus the Root Test implies that the series converges.   ❑

Once again, it is impossible to draw any conclusion from the Root Test if $\lim_{n\to\infty} \sqrt[n]{a_n}$ does not exist (see Exercise 29).

The Ratio Test is likely to be effective when factorials or powers appear in the terms of the series, whereas the Root Test is likely to be effective when powers (and not factorials) appear in the terms of the series. This is one reason why the Ratio Test is more frequently used than the Root Test.

EXERCISES 9.6

In Exercises 1–23 determine whether the series converges or diverges. In some cases you may need to use tests other than the Ratio and Root Tests.

1. $\displaystyle\sum_{n=0}^{\infty} \frac{n!}{2^n}$ 2. $\displaystyle\sum_{n=1}^{\infty} \frac{n}{10^n}$ 3. $\displaystyle\sum_{n=1}^{\infty} \frac{n!3^n}{10^n}$

4. $\displaystyle\sum_{n=1}^{\infty} \frac{n!}{2^{(n^2)}}$ 5. $\displaystyle\sum_{n=1}^{\infty} \left(\frac{n}{2n+5}\right)^n$ 6. $\displaystyle\sum_{n=1}^{\infty} \frac{n!}{(2n)!}$

7. $\displaystyle\sum_{n=1}^{\infty} \frac{(2n)!}{(n!)^2}$ 8. $\displaystyle\sum_{n=2}^{\infty} \frac{2^{2n-2}}{(2n-2)!}$ 9. $\displaystyle\sum_{n=0}^{\infty} n^{100} e^{-n}$

10. $\displaystyle\sum_{n=1}^{\infty} \frac{(1.1)^n}{n^7}$ 11. $\displaystyle\sum_{n=1}^{\infty} \frac{n^{1.7}}{(1.7)^n}$ 12. $\displaystyle\sum_{n=1}^{\infty} \frac{\ln n}{e^n}$

13. $\displaystyle\sum_{n=0}^{\infty} \frac{n!}{e^n}$ 14. $\displaystyle\sum_{n=1}^{\infty} n\left(\frac{\pi}{4}\right)^n$ 15. $\displaystyle\sum_{n=2}^{\infty} \frac{1}{(\ln n)^n}$

16. $\displaystyle\sum_{n=1}^{\infty} \frac{n+5}{n^3}$ 17. $\displaystyle\sum_{n=1}^{\infty} \frac{1 \cdot 3 \cdot 5 \cdots (2n-1)}{2 \cdot 4 \cdot 6 \cdots (2n)}$

18. $\displaystyle\sum_{n=1}^{\infty} \frac{1 \cdot 3 \cdot 5 \cdots (2n+1)}{2 \cdot 5 \cdot 8 \cdots (3n+2)}$

19. $\displaystyle\sum_{n=1}^{\infty} \frac{(2n)!}{n!(2n)^n}$ (*Hint:* Use the ideas in the solution of Example 3.)

20. $\displaystyle\sum_{n=2}^{\infty} \left(\frac{n!}{n^n}\right)^n$ (*Hint:* $\dfrac{n!}{n^n} \le \dfrac{1}{2}$ for $n \ge 2$.)

*21. $\displaystyle\sum_{n=1}^{\infty} \frac{\sin 1/n!}{\cos 1/n!}$ *22. $\displaystyle\sum_{n=1}^{\infty} \left(\sum_{k=1}^{n} \frac{1}{k}\right)^n$

*23. $\displaystyle\sum_{n=1}^{\infty} a_n$, where $a_n = \begin{cases} 0 & \text{for } n \text{ even} \\ \left(\dfrac{n}{2n+1}\right)^n & \text{for } n \text{ odd} \end{cases}$

24. Let c be any positive number. Show that $\sum_{n=0}^{\infty} c^n/n!$ converges.

*25. **a.** Let $\{a_n\}_{n=1}^{\infty}$ be defined by

$$a_n = \begin{cases} 1/n^n & \text{if } n \text{ is odd} \\ 1/(2n)^{2n} & \text{if } n \text{ is even} \end{cases}$$

Show that the Ratio Test is inconclusive.

b. Show that the series converges by virtue of the Root Test. (It can be shown that if the Ratio Test shows that a series converges or diverges, then so does the Root Test; however, the converse is not true, as this exercise confirms.)

*26. **a.** Show that if $n \ge 3$, then $n/3^n < 1/2^n$.

b. Use part (a) to find an upper bound for the 20th truncation error E_{20} of $\sum_{n=1}^{\infty} n/3^n$.

*27. **a.** Show that if $n \ge 20$, then $\dfrac{2^n}{n!} < \left(\dfrac{2}{21}\right)^{n-20}$.

b. Use part (a) to find an upper bound for the 20th truncation error E_{20} of $\sum_{n=1}^{\infty} 2^n/n!$.

28. **a.** Let $a_{2n} = 1/n^2$ and $a_{2n+1} = 1/(2n+1)^2$. Show that $\lim_{n\to\infty} a_{n+1}/a_n$ does not exist but $\sum_{n=1}^{\infty} a_n$ converges.

b. Construct a positive series $\sum_{n=1}^{\infty} a_n$ such that $\lim_{n\to\infty} a_{n+1}/a_n$ does not exist but $\sum_{n=1}^{\infty} a_n$ diverges.

29. a. Let $\sum_{n=1}^{\infty} a_n$ be the series in Exercise 23. Show that $\lim_{n\to\infty} \sqrt[n]{a_n}$ does not exist but $\sum_{n=1}^{\infty} a_n$ converges.

 b. Construct a positive series $\sum_{n=1}^{\infty} a_n$ such that $\lim_{n\to\infty} \sqrt[n]{a_n}$ does not exist but $\sum_{n=1}^{\infty} a_n$ diverges.

Project

1. In this project we show that if the Ratio Test works for a given positive series $\sum_{n=1}^{\infty} a_n$, then the Root Test works for the same series. (Exercise 25 shows that the converse is not true in general: the Ratio Test may fail but the Root Test succeeds for a given series.)

a. Show that if the positive series $\sum_{n=1}^{\infty} a_n$ converges by the Ratio Test, then it converges by the Root Test. (*Hint:* Use ideas from (1).)

b. Show that if the positive series $\sum_{n=1}^{\infty} a_n$ diverges by the Ratio test, then it diverges by the Root Test. (*Hint:* Use the ideas in (a) but for $r > 1$.)

c. Suppose that $\lim_{n\to\infty} \sqrt[n]{a_n} = 1$. Show that if $\lim_{n\to\infty} a_{n+1}/a_n = L$, then $L = 1$. (Thus if the Root Test fails because the nth roots tend to 1, then the Ratio Test fails also.)

9.7 ALTERNATING SERIES AND ABSOLUTE CONVERGENCE

Now that we have established criteria for the convergence and divergence of positive series, we are ready to study series involving both positive and negative numbers.

Alternating Series

If the terms in a series are alternately positive and negative, we call the series an **alternating series.** For example, the series

$$\sum_{n=1}^{\infty} (-1)^{n+1} 3^n = 3 - 9 + 27 - 81 + \cdots$$

and

$$\sum_{n=1}^{\infty} (-1)^n \frac{1}{(2n)!} = -\frac{1}{2} + \frac{1}{24} - \frac{1}{720} + \cdots$$

are alternating series. There is a simple convergence test credited to Leibniz that often applies to such a series.

THEOREM 9.17
Alternating Series Test

Let $\{a_n\}_{n=1}^{\infty}$ be a decreasing sequence of positive numbers with $\lim_{n\to\infty} a_n = 0$. Then the alternating series $\sum_{n=1}^{\infty} (-1)^{n+1} a_n$ converges. Furthermore, the jth truncation error satisfies the inequality

$$E_j < a_{j+1} \tag{1}$$

The same results hold for $\sum_{n=1}^{\infty} (-1)^n a_n$.

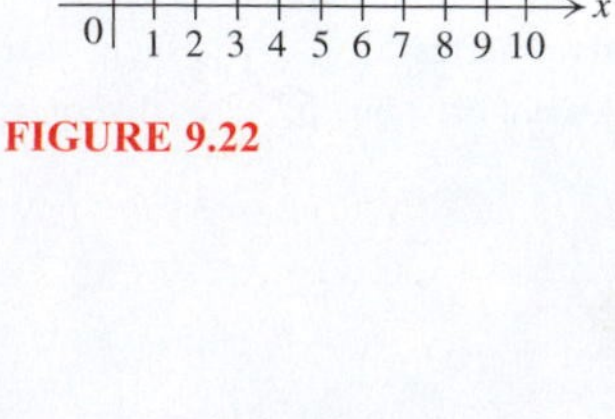

Proof We will prove that $\sum_{n=1}^{\infty} (-1)^{n+1} a_n$ converges and that (1) holds for this series. Because $\sum_{n=1}^{\infty} (-1)^n a_n = \sum_{n=1}^{\infty} -(-1)^{n+1} a_n$, it will then follow that $\sum_{n=1}^{\infty} (-1)^n a_n$ converges and that (1) holds for $\sum_{n=1}^{\infty} (-1)^n a_n$. For the proof we let $\{s_j\}_{j=1}^{\infty}$ be the sequence of partial sums of the series $\sum_{n=1}^{\infty} (-1)^{n+1} a_n$, so that

$$s_j = a_1 - a_2 + a_3 - a_4 + \cdots + (-1)^{j+1} a_j$$

We will first show that $\{s_{2j}\}_{j=1}^{\infty}$ is bounded and increasing (Figure 9.22). Then we can apply Theorem 9.6 to conclude that the sequence $\{s_{2j}\}_{j=1}^{\infty}$ converges. We begin

FIGURE 9.22

by observing that

$$s_{2j+2} - s_{2j} = (a_1 - a_2 + \cdots + a_{2j-1} - a_{2j} + a_{2j+1} - a_{2j+2})$$
$$- (a_1 - a_2 + \cdots + a_{2j-1} - a_{2j}) \tag{2}$$
$$= a_{2j+1} - a_{2j+2} > 0$$

since $\{a_n\}_{n=1}^{\infty}$ is decreasing. Thus the sequence $\{s_{2j}\}_{j=1}^{\infty}$ is increasing. It can be shown similarly that the sequence $\{s_{2j+1}\}_{j=1}^{\infty}$ is decreasing. Next we notice that

$$s_{2j} = a_1 - \overbrace{a_2 + a_3}^{<0} - \overbrace{a_4 + a_5}^{<0} - \cdots - \overbrace{a_{2j-2} + a_{2j-1}}^{<0} - a_{2j} \tag{3}$$

Now (2) and (3) together imply that $s_2 < s_{2j} < a_1$, so that $\{s_{2j}\}_{j=1}^{\infty}$ is bounded. From Theorem 9.6 we conclude that $\{s_{2j}\}_{j=1}^{\infty}$ converges. Next, notice that $s_{2j+1} = s_{2j} + a_{2j+1}$. Since $\lim_{j \to \infty} a_{2j+1} = 0$, we deduce that

$$\lim_{j \to \infty} s_{2j+1} = \lim_{j \to \infty} (s_{2j} + a_{2j+1}) = \lim_{j \to \infty} s_{2j} + \lim_{j \to \infty} a_{2j+1} = \lim_{j \to \infty} s_{2j}$$

As a result, $\lim_{j \to \infty} s_j$ exists. But that is equivalent to the convergence of the series $\sum_{n=1}^{\infty} (-1)^{n+1} a_n$.

Finally, to verify the inequality in (1), let s denote the sum of the series $\sum_{n=1}^{\infty} (-1)^{n+1} a_n$. Recall that $\{s_{2j}\}_{j=1}^{\infty}$ is an increasing sequence with limit s, so that

$$0 < s - s_{2j} \tag{4}$$

Similarly, $\{s_{2j+1}\}_{j=0}^{\infty}$ is a decreasing sequence with limit s, so that

$$s - s_{2j} < s_{2j+1} - s_{2j} \tag{5}$$

Combining (4) and (5) yields

$$0 < s - s_{2j} < s_{2j+1} - s_{2j} = a_{2j+1} \tag{6}$$

Similar computations yield

$$0 < s_{2j+1} - s < s_{2j+1} - s_{2j+2} = a_{2j+2} \tag{7}$$

Since $|s - s_{2j}| = E_{2j}$ and $|s_{2j+1} - s| = E_{2j+1}$, we conclude from (6) and (7) that $E_j < a_{j+1}$ for each positive integer j. Thus we have proved (1), which completes the proof of the theorem. ■

It is usually simple to ascertain whether an alternating series satisfies the hypotheses of the Alternating Series Test. If the hypotheses are satisfied, we can conclude that the series converges.

EXAMPLE 1   Show that the **alternating harmonic series**

$$\sum_{n=1}^{\infty} (-1)^{n-1} \frac{1}{n} = \sum_{n=1}^{\infty} (-1)^{n+1} \frac{1}{n} = 1 - \frac{1}{2} + \frac{1}{3} - \frac{1}{4} + \cdots$$

converges.

Solution   We have $a_n = 1/n$, so that $\{a_n\}_{n=1}^{\infty}$ is a decreasing, positive sequence such that $\lim_{n\to\infty} a_n = 0$. Therefore the alternating harmonic series satisfies the conditions of the Alternating Series Test and consequently converges.   ❏

EXAMPLE 2   Show that $\displaystyle\sum_{n=2}^{\infty} \frac{(-1)^n}{\ln n}$ converges.

Solution   Since $\{1/\ln n\}_{n=2}^{\infty}$ is decreasing and positive with $\lim_{n\to\infty} 1/\ln n = 0$, the series satisfies the hypotheses of the Alternating Series Test. Consequently the series converges.   ❏

In our next example we will use (1) to approximate the sum of an alternating series.

EXAMPLE 3   Find an approximation of the series $\displaystyle\sum_{n=1}^{\infty} (-1)^{n+1} \frac{1}{n^3}$ with an error less than 0.001.

Solution   A partial sum s_j will approximate the sum s of the series with an error less than 0.001 if $E_j < 0.001$, which by (1) occurs if $a_{j+1} < 0.001$. Since $a_{j+1} = 1/(j+1)^3$, our goal is to find a value of j for which $1/(j+1)^3 < 0.001$. By taking $j = 10$ we find that

$$\frac{1}{(j+1)^3} = \frac{1}{11^3} < 0.001$$

and the desired approximation is

$$\sum_{n=1}^{\infty} (-1)^{n+1} \frac{1}{n^3} \approx \frac{1}{1^3} - \frac{1}{2^3} + \frac{1}{3^3} - \frac{1}{4^3} + \frac{1}{5^3} - \frac{1}{6^3} + \frac{1}{7^3} - \frac{1}{8^3} + \frac{1}{9^3} - \frac{1}{10^3}$$

$$\approx 0.9011164764 \qquad ❏$$

Absolute and Conditional Convergence

Since the convergence tests studied so far do not apply directly to a series $\sum_{n=1}^{\infty} a_n$ that is neither positive nor alternating, the normal procedure with such a series is to study the convergence of $\sum_{n=1}^{\infty} |a_n|$. If the terms a_n are not 0, then the latter series is positive and thus amenable to our convergence tests, and as we now prove, convergence of $\sum_{n=1}^{\infty} |a_n|$ implies convergence of $\sum_{n=1}^{\infty} a_n$.

THEOREM 9.18   If $\sum_{n=1}^{\infty} |a_n|$ converges, then $\sum_{n=1}^{\infty} a_n$ converges.

Proof   We may assume that each $a_n \neq 0$ because zero terms do not affect convergence of a series. Since

$$0 < a_n + |a_n| \leq |a_n| + |a_n| = 2|a_n|$$

and since $\sum_{n=1}^{\infty} |a_n|$ and hence $\sum_{n=1}^{\infty} 2|a_n|$ converge by hypothesis, it follows from the Comparison Test that $\sum_{n=1}^{\infty} (a_n + |a_n|)$ converges. Thus the comment preceding (8) of Section 9.4 implies that

$$\sum_{n=1}^{\infty} a_n = \sum_{n=1}^{\infty} [(a_n + |a_n|) - |a_n|]$$

converges.   ■

Using Theorem 9.18, we can conclude that $\sum_{n=1}^{\infty} a_n$ converges by merely observing that the associated series $\sum_{n=1}^{\infty} |a_n|$ converges.

EXAMPLE 4   Show that the series $\displaystyle\sum_{n=1}^{\infty} \frac{\sin n}{n^3}$ converges.

Solution   By calculating a few values of $\sin n$, we find that the series is neither positive nor alternating. Therefore none of the earlier tests applies directly to it. However, since

$$\left| \frac{\sin n}{n^3} \right| \leq \frac{1}{n^3} \quad \text{for } n \geq 1$$

and since $\sum_{n=1}^{\infty} 1/n^3$ converges because it is a p series with $p = 3$, we know by the Comparison Test that $\sum_{n=1}^{\infty} |(\sin n)/n^3|$ converges. Consequently Theorem 9.18 assures us that the given series also converges.   □

> **Caution:**   It is wrong to conclude that the convergence of a series $\sum_{n=1}^{\infty} a_n$ implies the convergence of the series $\sum_{n=1}^{\infty} |a_n|$. A simple example is provided by the alternating harmonic series $\sum_{n=1}^{\infty} (-1)^{n+1} (1/n)$. Its associated positive series is the harmonic series $\sum_{n=1}^{\infty} 1/n$, which we know diverges. Thus when a series $\sum_{n=1}^{\infty} a_n$ converges, the series $\sum_{n=1}^{\infty} |a_n|$ may or may not converge.

In order to distinguish between those convergent series $\sum_{n=1}^{\infty} a_n$ for which $\sum_{n=1}^{\infty} |a_n|$ converges and those for which $\sum_{n=1}^{\infty} |a_n|$ diverges, we make the following definition.

DEFINITION 9.19   Let $\sum_{n=1}^{\infty} a_n$ be a convergent series. If $\sum_{n=1}^{\infty} |a_n|$ converges, we say that the series $\sum_{n=1}^{\infty} a_n$ **converges absolutely.** If $\sum_{n=1}^{\infty} |a_n|$ diverges, we say that the series $\sum_{n=1}^{\infty} a_n$ **converges conditionally.**

In this terminology we say that the series $\sum_{n=1}^{\infty} (\sin n)/n^3$ converges absolutely, whereas the series $\sum_{n=1}^{\infty} (-1)^{n+1} (1/n)$ converges conditionally. Of course, all convergent positive series converge absolutely.

By combining Theorem 9.18 with our tests for positive series, we obtain convergence tests that apply to any series, positive or not.

THEOREM 9.20
Generalized Convergence Tests

Let $\sum_{n=1}^{\infty} a_n$ be a series and $\sum_{n=1}^{\infty} b_n$ a positive series.

a. *Generalized Comparison Test.* If $|a_n| \leq b_n$ for $n \geq 1$, and if $\sum_{n=1}^{\infty} b_n$ converges, then $\sum_{n=1}^{\infty} a_n$ converges (absolutely).

b. *Generalized Limit Comparison Test.* If $\lim_{n\to\infty} |a_n/b_n| = L$, where L is a nonnegative number, and if $\sum_{n=1}^{\infty} b_n$ converges, then $\sum_{n=1}^{\infty} a_n$ converges (absolutely).

c. *Generalized Ratio Test.* Suppose that $a_n \neq 0$ for $n \geq 1$ and that

$$\lim_{n\to\infty} \left| \frac{a_{n+1}}{a_n} \right| = r \quad \text{(possibly } \infty\text{)}$$

If $r < 1$, then $\sum_{n=1}^{\infty} a_n$ converges (absolutely). If $r > 1$, then $\sum_{n=1}^{\infty} a_n$ diverges. If $r = 1$, then from this test alone we cannot draw any conclusion about the convergence of the series.

d. *Generalized Root Test.* Suppose that

$$\lim_{n\to\infty} \sqrt[n]{|a_n|} = r \quad \text{(possibly } \infty\text{)}$$

If $r < 1$, then $\sum_{n=1}^{\infty} a_n$ converges (absolutely). If $r > 1$, then $\sum_{n=1}^{\infty} a_n$ diverges. If $r = 1$, then from this test alone we cannot draw any conclusion about the convergence of the series.

Proof   To prove (a) and (b) (with $L > 0$) and the first parts of (c) and (d), we simply apply Theorems 9.13–9.16 to $\sum_{n=1}^{\infty} |a_n|$ and then conclude from Theorem 9.18 that $\sum_{n=1}^{\infty} a_n$ converges. The case $L = 0$ in (b) must be proved separately. To prove the second parts of (c) and (d), we observe that in each case, $\{a_n\}_{n=1}^{\infty}$ does not converge to 0, and consequently by Corollary 9.9 the series $\sum_{n=1}^{\infty} a_n$ diverges.   ∎

EXAMPLE 5   Show that

$$\sum_{n=1}^{\infty} \frac{x^n}{n} = x + \frac{x^2}{2} + \frac{x^3}{3} + \frac{x^4}{4} + \cdots$$

converges absolutely for $|x| < 1$, converges conditionally for $x = -1$, and diverges for $x = 1$ and for $|x| > 1$.

Solution   If $x = 0$, the series obviously converges. If $x \neq 0$, then

$$\lim_{n\to\infty} \left| \frac{x^{n+1}/(n+1)}{x^n/n} \right| = \lim_{n\to\infty} \left| \frac{n}{n+1} x \right| = |x|$$

Therefore the Generalized Ratio Test implies that the given series converges for $|x| < 1$ and diverges for $|x| > 1$. For $x = 1$ the series becomes the harmonic series $\sum_{n=1}^{\infty} 1/n$, which diverges. For $x = -1$ the series becomes

$$\sum_{n=1}^{\infty} \frac{(-1)^n}{n}$$

This is the negative of the alternating harmonic series and consequently converges. Since

$$\sum_{n=1}^{\infty} \left| \frac{(-1)^n}{n} \right| = \sum_{n=1}^{\infty} \frac{1}{n}$$

which diverges, we conclude that the given series converges conditionally for $x = -1$.   ❏

EXAMPLE 6   Show that

$$\sum_{n=0}^{\infty} \frac{(-1)^n}{2n+1} x^{2n+1} = x - \frac{x^3}{3} + \frac{x^5}{5} - \frac{x^7}{7} + \cdots$$

converges absolutely for $|x| < 1$, converges conditionally for $|x| = 1$, and diverges for $|x| > 1$.

Solution   If $x = 0$, the series converges. If $x \neq 0$, then we have

$$\lim_{n \to \infty} \left| \frac{\dfrac{(-1)^{n+1}}{2(n+1)+1} x^{2(n+1)+1}}{\dfrac{(-1)^n}{2n+1} x^{2n+1}} \right| = \lim_{n \to \infty} \left| \frac{2n+1}{2n+3} x^2 \right| = |x^2|$$

Consequently the Generalized Ratio Test implies that the series converges absolutely for $|x| < 1$ and diverges for $|x| > 1$. It remains to consider the cases in which $|x| = 1$. For $x = -1$ the series becomes

$$\sum_{n=0}^{\infty} \frac{-(-1)^n}{2n+1} = \sum_{n=0}^{\infty} \frac{(-1)^{n+1}}{2n+1}$$

and this converges by the Alternating Series Test. For $x = 1$ the series reduces to

$$\sum_{n=0}^{\infty} \frac{(-1)^n}{2n+1}$$

which also converges by the Alternating Series Test. It is easy to show by using the Integral Test or the Limit Comparison Test that

$$\sum_{n=0}^{\infty} \frac{1}{2n+1}$$

diverges. Hence the given series converges conditionally for $|x| = 1$.   ❏

If $\{a_n\}_{n=1}^{\infty}$ is a sequence and if

$$\lim_{n \to \infty} \left| \frac{a_{n+1}}{a_n} \right| = r < 1 \quad \text{or} \quad \lim_{n \to \infty} \sqrt[n]{|a_n|} = r < 1$$

then we know from the Generalized Ratio and Root Tests that the series $\sum_{n=1}^{\infty} a_n$ converges. It then follows from Theorem 9.8 that

$$\lim_{n \to \infty} a_n = 0$$

This result is useful in itself, and since we will refer to it later, we state it as a corollary to the Generalized Convergence Tests.

COROLLARY   9.21   Let $\{a_n\}_{n=1}^{\infty}$ be a sequence. If

$$\lim_{n \to \infty} \left| \frac{a_{n+1}}{a_n} \right| = r < 1 \quad \text{or} \quad \lim_{n \to \infty} \sqrt[n]{|a_n|} = r < 1$$

then
$$\lim_{n \to \infty} a_n = 0$$

EXAMPLE 7   Show that $\lim_{n \to \infty} x^n/n! = 0$ for all x.

Solution   If $x = 0$, then clearly the limit is 0. If $x \neq 0$, let

$$a_n = \frac{x^n}{n!}$$

Then

$$r = \lim_{n \to \infty} \left| \frac{a_{n+1}}{a_n} \right| = \lim_{n \to \infty} \left| \frac{x^{n+1}/(n+1)!}{x^n/n!} \right|$$

$$= \lim_{n \to \infty} \left| x \frac{n!}{(n+1)!} \right| = |x| \lim_{n \to \infty} \frac{1}{n+1} = 0$$

Since $r < 1$, the result follows from Corollary 9.21.   ❑

It is also possible to prove that if

$$\lim_{n \to \infty} \left| \frac{a_{n+1}}{a_n} \right| = r > 1 \quad \text{or} \quad \lim_{n \to \infty} \sqrt[n]{|a_n|} = r > 1$$

then $\lim_{n \to \infty} |a_n| = \infty$. Thus not only does the series $\sum_{n=1}^{\infty} a_n$ diverge, but the nth term a_n does not even tend to 0.

This section completes our study of tests for convergence and divergence of series. Table 9.1 on page 655 summarizes the tests.

EXERCISES 9.7

In Exercises 1–14 determine whether the series converges or diverges.

1. $\displaystyle\sum_{n=1}^{\infty} (-1)^{n+1} \frac{1}{2n+1}$

2. $\displaystyle\sum_{n=2}^{\infty} (-1)^n \frac{1}{\ln n}$

3. $\displaystyle\sum_{n=1}^{\infty} (-1)^n \frac{2n+1}{5n+1}$

4. $\displaystyle\sum_{n=3}^{\infty} \frac{\cos n\pi}{\sqrt{n}}$

5. $\displaystyle\sum_{n=1}^{\infty} (-1)^n \frac{n+2}{n^2+3n+5}$

6. $\displaystyle\sum_{n=2}^{\infty} (-1)^{n+1} \frac{n+1}{4n}$

7. $\displaystyle\sum_{n=1}^{\infty} (-1)^n \frac{\ln n}{n}$

8. $\displaystyle\sum_{n=1}^{\infty} (-1)^n \frac{(\ln n)^2}{n}$

9. $\displaystyle\sum_{n=1}^{\infty} (-1)^n \frac{(\ln n)^p}{n}$ where p is any positive integer

10. $\displaystyle\sum_{n=1}^{\infty} (-1)^{n+1} \frac{\sqrt{n}}{2n+1}$

11. $\displaystyle\sum_{n=1}^{\infty} (-1)^{n+1} \frac{n!}{100^n}$

12. $\displaystyle\sum_{n=1}^{\infty} (-1)^n \cot\left(\frac{\pi}{2} - \frac{1}{n}\right)$

13. $\displaystyle\sum_{n=1}^{\infty} (-1)^{n+1} \frac{n^2}{2n+1}$

14. $\displaystyle\sum_{n=1}^{\infty} (-1)^{n+1} \frac{1}{n^{1/10}}$

15. For which positive values of p does the series $\sum_{n=1}^{\infty} (-1)^n (1/n^p)$ converge?

In Exercises 16–19 use (1) to find an upper bound for the 10th truncation error E_{10}.

16. $\displaystyle\sum_{n=1}^{\infty} (-1)^{n+1} \frac{1}{n}$ **17.** $\displaystyle\sum_{n=1}^{\infty} (-1)^n \frac{1}{n^2}$

18. $\displaystyle\sum_{n=1}^{\infty} (-1)^{n+1} \frac{1}{n!}$ **19.** $\displaystyle\sum_{n=1}^{\infty} (-1)^n \frac{1}{\sqrt{n+4}}$

In Exercises 20–22 approximate the sum of the given series with an error less than 0.001.

20. $\displaystyle\sum_{n=1}^{\infty} (-1)^n \frac{8}{10^n + 1}$ **21.** $\displaystyle\sum_{n=1}^{\infty} (-1)^{n+1} \frac{1}{n^3}$

22. $\displaystyle\sum_{n=2}^{\infty} (-1)^{n+1} \frac{1}{1 + n + 6n^2}$

In Exercises 23–33 determine which series diverge, which converge conditionally, and which converge absolutely.

23. $\displaystyle\sum_{n=1}^{\infty} (-1)^{n+1} \frac{1}{3n + 4}$ **24.** $\displaystyle\sum_{n=1}^{\infty} n\left(\frac{4}{5}\right)^n$

25. $\displaystyle\sum_{n=1}^{\infty} (-1)^n \frac{n^n}{n!}$ **26.** $\displaystyle\sum_{n=3}^{\infty} (-1)^{n+1} \frac{1}{n(n-2)}$

27. $\displaystyle\sum_{n=1}^{\infty} (-1)^{n+1} \frac{1}{n^{1/n}}$ **28.** $\displaystyle\sum_{n=2}^{\infty} (-1)^n \frac{1}{n(\ln n)}$

29. $\displaystyle\sum_{n=2}^{\infty} (-1)^{n+1} \frac{1}{n(\ln n)^2}$ **30.** $\displaystyle\sum_{n=1}^{\infty} \frac{\sin n}{n^2 + 1}$

31. $\displaystyle\sum_{n=2}^{\infty} (-1)^{n+1} \frac{1}{(\ln n)^n}$ **32.** $\displaystyle\sum_{n=2}^{\infty} (-1)^{n+1} \frac{1}{(\ln n)^{1/n}}$

33. $\displaystyle\sum_{n=1}^{\infty} (-1)^n \frac{1 \cdot 3 \cdot 5 \cdots (2n + 1)}{2 \cdot 5 \cdot 8 \cdots (3n + 2)}$

In Exercises 34–37 use Corollary 9.21 to verify the given limit.

34. $\displaystyle\lim_{n\to\infty} \frac{n!}{n^n} = 0$ **35.** $\displaystyle\lim_{n\to\infty} \frac{(n+1)^2}{n!} = 0$

36. $\displaystyle\lim_{n\to\infty} \frac{x^{2n}}{n^n} = 0$ for all x **37.** $\displaystyle\lim_{n\to\infty} \frac{x^{2n}}{n} = 0$ for $|x| < 1$

38. Show that the series $\displaystyle\sum_{n=1}^{\infty} \frac{n!}{n^n} x^n$ converges for $|x| < e$.

39. Suppose $\sum_{n=1}^{\infty} a_n$ converges and is not necessarily positive. Give an example to show that $\sum_{n=1}^{\infty} a_n^2$ need not converge. (*Hint:* Let $\sum_{n=1}^{\infty} a_n$ be an alternating series whose terms approach 0 very slowly.)

*40. If $\sum_{n=1}^{\infty} a_n$ is absolutely convergent, must $\sum_{n=1}^{\infty} (a_n + a_{n+1})$ be absolutely convergent? Explain your answer.

*41. Let $a \neq 0$, and assume that $\lim_{n\to\infty} a_n = a$ and $a_n \neq 0$ for all n. Show that $\sum_{n=1}^{\infty} |a_{n+1} - a_n|$ converges if and only if

$$\sum_{n=1}^{\infty} \left| \frac{1}{a_{n+1}} - \frac{1}{a_n} \right|$$

converges.

Project

1. The key element in the proof of the Alternating Series Test is that successive partial sums $\{s_j\}_{j=1}^{\infty}$ of an alternating series with sum L alternate: greater than L, then less than L, then greater than L, etc., while approaching L as j approaches ∞. For example, consider the alternating harmonic series $\sum_{n=1}^{\infty} (-1^{n+1})/n$, which we know converges and can be shown to converge to $\ln 2$ (approximately 0.693). For the initial partial sums we have

$s_1 = 1 > \ln 2$

$s_2 = 1 - \dfrac{1}{2} = 0.5 < \ln 2$

$s_3 = 1 - \dfrac{1}{2} + \dfrac{1}{3} \approx 0.833 > \ln 2$

$s_4 = 1 - \dfrac{1}{2} + \dfrac{1}{3} - \dfrac{1}{4} \approx 0.583 < \ln 2$

$s_5 = 1 - \dfrac{1}{2} + \dfrac{1}{3} - \dfrac{1}{4} + \dfrac{1}{5} \approx 0.783 > \ln 2$

As you see, the successive partial sums are alternating around $\ln 2$, at the same time closing in on the number $\ln 2$.

Using the fact that the series obtained from the positive terms of the alternating harmonic series, $1 + \frac{1}{3} + \frac{1}{5} + \cdots$, diverges to ∞, and similarly, the series obtained from the negative terms of the alternating harmonic series, $-\frac{1}{2} - \frac{1}{4} - \frac{1}{6} - \cdots$, diverges to $-\infty$, we can rearrange, or reorder, the terms in order to have a new series that converges to another number. For example, let us see how we do that as we create a new set $\{t_j\}_{j=1}^{\infty}$ of partial sums that approaches the number 1.2:

$t_1 = 1 + \dfrac{1}{3} \approx 1.333 > 1.2$

$t_2 = 1 + \dfrac{1}{3} - \dfrac{1}{2} \approx 0.833 < 1.2$

$t_3 = 1 + \dfrac{1}{3} - \dfrac{1}{2} + \dfrac{1}{5} + \dfrac{1}{7} + \dfrac{1}{9} \approx 1.287 > 1.2$

$t_4 = 1 + \dfrac{1}{3} - \dfrac{1}{2} + \dfrac{1}{5} + \dfrac{1}{7} + \dfrac{1}{9} - \dfrac{1}{4} \approx 1.037 < 1.2$

$t_5 = 1 + \dfrac{1}{3} - \dfrac{1}{2} + \dfrac{1}{5} + \dfrac{1}{7} + \dfrac{1}{9} - \dfrac{1}{4} + \dfrac{1}{11} + \dfrac{1}{13} \approx 1.205 > 1.2$

In creating a new partial sum t_j above, we have always added positive terms until the sum is > 1.2, then always included negative terms until the new sum is < 1.2. If we continue the process in this way, the sequence $\{t_j\}_{j=1}^{\infty}$ will have 1.2 as a limit.

a. Find t_6 and t_7.

b. Suppose that we wish to reorder the terms of $\sum_{n=1}^{\infty}(-1^{n+1})/n$ so as to have the limit 1. Find the first 5 terms of partial sums $\{u_j\}_{j=1}^{\infty}$.

It turns out that any desired number, even a number like π, can be the sum of a suitable reordering of the terms of $\sum_{n=1}^{\infty}(-1^{n+1})/n$. You might try approximating π.

9.8   POWER SERIES

In Examples 5 and 6 of Section 9.7 we determined those values of x for which the series

$$\sum_{n=1}^{\infty} \frac{x^n}{n} \quad \text{and} \quad \sum_{n=0}^{\infty} \frac{(-1)^n}{2n+1} x^{2n+1}$$

converge. Such series are called power series.

DEFINITION 9.22   A series of the form $\sum_{n=0}^{\infty} c_n x^n$ is called a **power series.**

If $c_0 = 0$ we usually write the power series as $\sum_{n=1}^{\infty} c_n x^n$, and if $c_1 = 0$ as well, we normally write the series as $\sum_{n=2}^{\infty} c_n x^n$. For the sake of uniformity in writing power series, we will always assume that $x^0 = 1$ for all x, including $x = 0$, so that

$$\sum_{n=0}^{\infty} c_n x^n = c_0 + c_1 x + c_2 x^2 + c_3 x^3 + \cdots$$

Thus power series are just like extended polynomials, with an infinite number of terms. If $c_n = 0$ for all $n > N$, then the power series is just a polynomial of degree at most N. For instance,

$$2 - 3x + 4x^2 = \sum_{n=0}^{\infty} c_n x^n$$

where

$$c_0 = 2, \quad c_1 = -3, \quad c_2 = 4, \quad \text{and} \quad c_n = 0 \quad \text{for } n \geq 3$$

We can evaluate any polynomial at any real number. It is natural to ask for which real numbers a given power series $\sum_{n=0}^{\infty} c_n x^n$ converges. Thus we define a function whose domain is the collection of those values of x for which the power series converges. The value of the function at any number x in the domain is the sum $\sum_{n=0}^{\infty} c_n x^n$ of the series for x. Notice that every power series automatically converges for $x = 0$, which means that the domain of the associated function contains at least the number 0. The domain may consist of 0 alone, as is illustrated by the following example.

EXAMPLE 1   Show that

$$\sum_{n=0}^{\infty} n! x^n = 1 + x + 2x^2 + 6x^3 + 24x^4 + \cdots$$

converges only for $x = 0$.

Solution If $x \neq 0$, then

$$\lim_{n \to \infty} \left| \frac{(n+1)! x^{n+1}}{n! x^n} \right| = \lim_{n \to \infty} |(n+1)x| = |x| \lim_{n \to \infty} (n+1) = \infty$$

Thus the Generalized Ratio Test implies that the series diverges. ❑

In contrast, a power series can converge for all numbers x.

EXAMPLE 2 Show that

$$\sum_{n=0}^{\infty} \frac{x^n}{n!} = 1 + x + \frac{1}{2} x^2 + \frac{1}{6} x^3 + \frac{1}{24} x^4 + \cdots$$

converges for every number x.

Solution If $x \neq 0$, then

$$\lim_{n \to \infty} \left| \frac{x^{n+1}/(n+1)!}{x^n/n!} \right| = \lim_{n \to \infty} \left| x \frac{n!}{(n+1)!} \right| = |x| \lim_{n \to \infty} \frac{1}{n+1} = 0$$

Thus the Generalized Ratio Test implies that the series converges for all $x \neq 0$, and hence the series converges for all x. ❑

Between the two extremes given in Examples 1 and 2 are power series that converge for some nonzero values of x and diverge for other values of x. One of the most prominent such series is $\sum_{n=0}^{\infty} x^n$, which for each nonzero value of x is a geometric series with ratio x. Thus we know from the Geometric Series Theorem that the series diverges for $|x| \geq 1$ and converges for $|x| < 1$, with

$$\sum_{n=0}^{\infty} x^n = \frac{1}{1-x} \tag{1}$$

Notice in particular that this series converges for all x in the open interval $(-1, 1)$ about 0. In general, if there is a nonzero value of x for which a power series $\sum_{n=0}^{\infty} c_n x^n$ converges, then the power series converges for all values of x in some open interval about 0, as we now prove.

LEMMA 9.23

Let s be any nonzero real number.
a. If $\sum_{n=0}^{\infty} c_n x^n$ converges for $x = s$, then $\sum_{n=0}^{\infty} c_n x^n$ converges absolutely for $|x| < |s|$.
b. If $\sum_{n=0}^{\infty} c_n x^n$ diverges for $x = s$, then $\sum_{n=0}^{\infty} c_n x^n$ diverges for $|x| > |s|$.

Proof To prove (a), let $|x| < |s|$, and rewrite $\sum_{n=0}^{\infty} c_n x^n$ as follows:

$$\sum_{n=0}^{\infty} c_n x^n = \sum_{n=0}^{\infty} c_n s^n \left(\frac{x}{s} \right)^n$$

Suppose $\sum_{n=0}^{\infty} c_n s^n$ converges. Then by Theorem 9.8, $\lim_{n \to \infty} c_n s^n = 0$. Consequently there is a positive integer N such that $|c_n s^n| \leq 1$ for all $n \geq N$. This means that

$$|c_n x^n| = |c_n s^n| \left| \frac{x}{s} \right|^n \leq \left| \frac{x}{s} \right|^n \quad \text{for } n \geq N$$

Because $|x| < |s|$, which means that $|x/s| < 1$, we know that the geometric series $\sum_{n=N}^{\infty} |x/s|^n$ converges absolutely. By the Generalized Comparison Test it follows that $\sum_{n=N}^{\infty} c_n x^n$ converges absolutely, and hence $\sum_{n=0}^{\infty} c_n x^n$ also converges absolutely. This proves (a). To prove (b), we assume that $\sum_{n=0}^{\infty} c_n s^n$ diverges, and we only need to observe that if $\sum_{n=0}^{\infty} c_n x^n$ were to converge, then by reversing the roles of s and x in part (a), we would find that $\sum_{n=0}^{\infty} c_n s^n$ would converge after all, which contradicts our assumption. Therefore $\sum_{n=0}^{\infty} c_n x^n$ diverges. ■

Lemma 9.23 provides much more information about the convergence of a power series than a first reading might reveal. We saw in Examples 1 and 2 that a power series $\sum_{n=0}^{\infty} c_n x^n$ might converge only for $x = 0$ and that it might converge for all x; otherwise, the series must converge for some nonzero value of x and diverge for some other nonzero value of x. Lemma 9.23 implies that in this case there is a number $R > 0$ such that the series converges for $|x| < R$ and diverges for $|x| > R$; this result is fundamental to the study of power series.

THEOREM 9.24   Let $\sum_{n=0}^{\infty} c_n x^n$ be a power series. Then exactly one of the following conditions holds.

a. $\sum_{n=0}^{\infty} c_n x^n$ converges only for $x = 0$.

b. $\sum_{n=0}^{\infty} c_n x^n$ converges for all x.

c.   There is a number $R > 0$ such that $\sum_{n=0}^{\infty} c_n x^n$ converges for $|x| < R$ and diverges for $|x| > R$.

We call the number R in part (c) of Theorem 9.24 the **radius of convergence** of $\sum_{n=0}^{\infty} c_n x^n$. If $\sum_{n=0}^{\infty} c_n x^n$ satisfies (a), then we let $R = 0$. If $\sum_{n=0}^{\infty} c_n x^n$ satisfies (b), then we let $R = \infty$. Therefore *every* power series has a radius of convergence R, which is either a nonnegative number or ∞. The collection of values of x for which $\sum_{n=0}^{\infty} c_n x^n$ converges is called the **interval of convergence** of $\sum_{n=0}^{\infty} c_n x^n$. From conditions (a)–(c) it is apparent that the interval of convergence takes one and only one of the following forms:

$$[0, 0], \quad (-R, R), \quad [-R, R), \quad (-R, R], \quad [-R, R], \quad (-\infty, \infty)$$

In Table 9.1 we verify that each of these types of intervals actually arises as the interval of convergence of a suitable power series.

TABLE   9.1
Examples of Intervals of Convergence

Interval	Example	Reference
$[0, 0]$	$\sum_{n=0}^{\infty} n! x^n$	Example 1
$(-1, 1)$	$\sum_{n=0}^{\infty} x^n$	Theorem 9.10 and the comment after Example 2
$[-1, 1)$	$\sum_{n=1}^{\infty} \dfrac{x^n}{n}$	Example 5 of Section 9.7
$(-1, 1]$	$\sum_{n=1}^{\infty} \dfrac{(-x)^n}{n}$	The preceding with x replaced by $-x$
$[-1, 1]$	$\sum_{n=0}^{\infty} \dfrac{(-1)^n}{2n + 1} x^{2n+1}$	Example 6 of Section 9.7
$(-\infty, \infty)$	$\sum_{n=0}^{\infty} \dfrac{x^n}{n!}$	Example 2

The radius of convergence R need not be 0, 1, or ∞. For example, the power series

$$\sum_{n=0}^{\infty} \frac{x^n}{2^n} = \sum_{n=0}^{\infty} \left(\frac{x}{2}\right)^n$$

converges for all x in $(-2, 2)$, because for each x it may be regarded as a geometric series with ratio $x/2$. Thus $R = 2$. In fact, a similar argument shows that any positive number R is the radius of convergence of some power series.

The next example shows the power of Theorem 9.24.

EXAMPLE 3   Determine the interval of convergence of the power series

$$\sum_{n=1}^{\infty} \frac{x^n}{n^{1/2}}$$

Solution   For $x = -1$ the series becomes $\sum_{n=1}^{\infty} (-1)^n(1/n^{1/2})$, which converges by the Alternating Series Test. For $x = 1$ the series becomes $\sum_{n=1}^{\infty} 1/n^{1/2}$, which diverges because it is a p series with $p = \frac{1}{2}$. Since the original series converges for $x = -1$ and diverges for $x = 1$, Theorem 9.24 assures us that the only possible value for the radius of convergence is $R = 1$. Consequently the interval of convergence is $[-1, 1)$.   ❑

To find the interval of convergence, the usual procedure is first to use the Generalized Ratio Test or the Generalized Root Test to ascertain the radius of convergence R, and then to test the endpoints $x = R$ and $x = -R$ separately for convergence. We illustrate this procedure now.

EXAMPLE 4   Determine the interval of convergence of the power series $\sum_{n=1}^{\infty} nx^n$.

Solution   For $x \neq 0$ we obtain

$$\lim_{n \to \infty} \left| \frac{(n + 1)x^{n+1}}{nx^n} \right| = \lim_{n \to \infty} \left| \frac{n + 1}{n} x \right| = |x|$$

Therefore by the Generalized Ratio Test we know that the series converges for $|x| < 1$ and diverges for $|x| > 1$. By Theorem 9.24 the radius of convergence R is 1. But since we can draw no conclusions from the Generalized Ratio Test when $|x| = 1$, we must test the two values 1 and -1 separately. For those values of x we obtain the series $\sum_{n=1}^{\infty} n$ and $\sum_{n=1}^{\infty} (-1)^n n$, both of which diverge, since the terms in the series fail to converge to 0. It follows that the interval of convergence is $(-1, 1)$.   ❑

Differentiation of Power Series

Since $\sum_{n=0}^{\infty} c_n x^n$ is a function, we may ask whether its derivative exists. Perhaps surprisingly, a power series with a nonzero radius of convergence is *always* differentiable; moreover, the derivative is obtained from $\sum_{n=0}^{\infty} c_n x^n$ by differentiating term by term, the way we differentiate polynomials.

THEOREM 9.25

Differentiation Theorem for Power Series

Let $\sum_{n=0}^{\infty} c_n x^n$ be a power series with radius of convergence $R > 0$. Then $\sum_{n=1}^{\infty} n c_n x^{n-1}$ has the same radius of convergence, and

$$\frac{d}{dx}\left(\sum_{n=0}^{\infty} c_n x^n\right) = \sum_{n=1}^{\infty} n c_n x^{n-1} = \sum_{n=1}^{\infty} \frac{d}{dx}(c_n x^n) \quad \text{for } |x| < R$$

The proof of the Differentiation Theorem is complicated, and we defer it to the Appendix. Notice, however, that the initial index of

$$\sum_{n=1}^{\infty} n c_n x^{n-1} = \sum_{n=1}^{\infty} \frac{d}{dx}(c_n x^n)$$

is 1, not 0, since the derivative of $c_0 x^0$ is 0.

Caution: Although the Differentiation Theorem says that the radii of convergence of the series $\sum_{n=0}^{\infty} c_n x^n$ and $\sum_{n=1}^{\infty} n c_n x^{n-1}$ are the same, it does not imply that the *intervals* of convergence are the same. On the contrary, the interval of convergence of $\sum_{n=1}^{\infty} x^n/n$ is $[-1, 1)$, whereas the interval of convergence of its derivative $\sum_{n=1}^{\infty} x^{n-1}$ is $(-1, 1)$.

EXAMPLE 5   Show that

$$\frac{d}{dx}\left(\sum_{n=0}^{\infty} \frac{x^n}{n!}\right) = \sum_{n=0}^{\infty} \frac{x^n}{n!} \tag{2}$$

Solution   We know from Example 2 that the series $\sum_{n=0}^{\infty} x^n/n!$ converges for all x. The Differentiation Theorem tells us that $\sum_{n=1}^{\infty} n x^{n-1}/n!$ converges as well and that

$$\frac{d}{dx}\left(\sum_{n=0}^{\infty} \frac{x^n}{n!}\right) = \sum_{n=1}^{\infty} \frac{n x^{n-1}}{n!} = \sum_{n=1}^{\infty} \frac{x^{n-1}}{(n-1)!} = \sum_{n=0}^{\infty} \frac{x^n}{n!} \qquad \square$$

By Example 5 we have found a power series, namely $\sum_{n=0}^{\infty} x^n/n!$, that is its own derivative. This is reminiscent of the function e^x. In fact, we will show now that $e^x = \sum_{n=0}^{\infty} x^n/n!$ for all x. To that end, let

$$f(x) = \sum_{n=0}^{\infty} \frac{x^n}{n!} \quad \text{for all } x$$

Then (2) may be rewritten

$$f'(x) = f(x) \quad \text{for all } x$$

Since $f(0) = \sum_{n=0}^{\infty} 0^n/n! = 1$, we conclude from Theorem 4.8 with $k = 1$ that

$$f(x) = e^x \quad \text{for all } x$$

Therefore

$$e^x = \sum_{n=0}^{\infty} \frac{x^n}{n!} = 1 + x + \frac{x^2}{2!} + \frac{x^3}{3!} + \cdots \tag{3}$$

In particular, if $x = 1$ in (3), we obtain the well-known formula for e:

$$e = \sum_{n=0}^{\infty} \frac{1}{n!} = 1 + 1 + \frac{1}{2!} + \frac{1}{3!} + \frac{1}{4!} + \cdots$$

The Differentiation Theorem states that a power series $\sum_{n=0}^{\infty} c_n x^n$ with a nonzero radius of convergence can be differentiated once. However, because the derivative is itself a power series with the same radius of convergence, the derivative also may be differentiated, and thus the original power series can be differentiated twice. By repeating this process, we conclude that a power series with radius of convergence $R > 0$ has derivatives of all orders on $(-R, R)$. The values of the derivatives of the series at 0 are closely related to the numbers $c_0, c_1, c_2, \ldots$, as we see in the next theorem. For convenience, we will denote $f(0)$ by $f^{(0)}(0)$.

THEOREM 9.26   Suppose a power series $\sum_{n=0}^{\infty} c_n x^n$ has radius of convergence $R > 0$. Let

$$f(x) = \sum_{n=0}^{\infty} c_n x^n \quad \text{for } -R < x < R \tag{4}$$

Then f has derivatives of all orders on $(-R, R)$, and

$$f^{(n)}(0) = n!c_n \quad \text{for } n \geq 0 \tag{5}$$

Consequently

$$f(x) = \sum_{n=0}^{\infty} \frac{f^{(n)}(0)}{n!} x^n \quad \text{for } -R < x < R \tag{6}$$

Proof   From the discussion preceding this theorem we know that f has derivatives of all orders on $(-R, R)$. By substituting $x = 0$ into (4) we obtain

$$f^{(0)}(0) = f(0) = c_0 = 0!c_0$$

Next we differentiate both sides of (4), using the Differentiation Theorem:

$$f'(x) = \sum_{n=1}^{\infty} nc_n x^{n-1} \tag{7}$$

By again substituting $x = 0$, we obtain

$$f'(0) = c_1 = 1!c_1$$

Differentiation of both sides of (7) yields

$$f''(x) = \sum_{n=2}^{\infty} n(n-1)c_n x^{n-2}$$

Substituting $x = 0$ once more, we find that

$$f''(0) = 2(1)c_2 = 2!c_2$$

In the same way, performing n differentiations on both sides of (4) and then substituting $x = 0$ into the result yields (5). Finally, if we substitute for c_n from (5) into (4), we obtain (6).   ■

COROLLARY   9.27

Let $R > 0$, and suppose $\sum_{n=0}^{\infty} c_n x^n$ and $\sum_{n=0}^{\infty} b_n x^n$ are power series that converge for $-R < x < R$. If

$$\sum_{n=0}^{\infty} c_n x^n = \sum_{n=0}^{\infty} b_n x^n \quad \text{for } -R < x < R$$

then $c_n = b_n$ for each $n \geq 0$.

Proof   Let $f(x) = \sum_{n=0}^{\infty} c_n x^n = \sum_{n=0}^{\infty} b_n x^n$ for $-R < x < R$. Then Theorem 9.26 implies that

$$c_n = \frac{f^{(n)}(0)}{n!} = b_n \qquad \blacksquare$$

In particular, Corollary 9.27 tells us that two polynomials

$$c_n x^n + c_{n-1} x^{n-1} + \cdots + c_1 x + c_0$$

and

$$b_m x^m + b_{m-1} x^{m-1} + \cdots + b_1 x + b_0$$

which are of course power series having all but a finite number of their coefficients 0, are the same function if and only if

$$m = n \quad \text{and} \quad b_j = c_j \quad \text{for } 1 \leq j \leq n$$

Now we are able to use power series to great advantage. For example, we can substitute any expression for x and still have a valid formula, provided that the expression lies in the interval of convergence of the given power series. Below we will apply this process to the two power series

$$e^x = \sum_{n=0}^{\infty} \frac{x^n}{n!} \quad \text{and} \quad \frac{1}{1-x} = \sum_{n=0}^{\infty} x^n$$

Substitution	Formula
$-x$ for x	$e^{-x} = \displaystyle\sum_{n=0}^{\infty} \frac{(-x)^n}{n!} = \sum_{n=0}^{\infty} \frac{(-1)^n}{n!} x^n$
x^2 for x	$e^{(x^2)} = \displaystyle\sum_{n=0}^{\infty} \frac{1}{n!} (x^2)^n = \sum_{n=0}^{\infty} \frac{1}{n!} x^{2n}$
$-x$ for x	$\dfrac{1}{1+x} = \displaystyle\sum_{n=1}^{\infty} (-1)^n x^n$
$-x^2$ for x	$\dfrac{1}{1+x^2} = \displaystyle\sum_{n=1}^{\infty} (-1)^n x^{2n}$

In addition, we can manipulate power series to obtain new power series. For example,

$$e^x - 1 = \left(\sum_{n=0}^{\infty} \frac{x^n}{n!} \right) - 1 = \left(1 + x + \frac{x^2}{2!} + \frac{x^3}{3!} + \cdots \right) - 1 = \sum_{n=1}^{\infty} \frac{x^n}{n!}$$

and

$$\frac{e^x - 1}{x} = \frac{1}{x} \sum_{n=1}^{\infty} \frac{x^n}{n!} = \sum_{n=1}^{\infty} \frac{x^{n-1}}{n!} = \sum_{n=0}^{\infty} \frac{x^n}{(n+1)!}$$

Integration of Power Series

The next theorem states that we can carry out integration of a power series term by term, just as we do differentiation. We defer the proof of this theorem to the Appendix.

THEOREM 9.28
Integration Theorem for
Power Series

Let $\sum_{n=0}^{\infty} c_n x^n$ be a power series with radius of convergence $R > 0$. Then $\sum_{n=0}^{\infty} (c_n/(n+1))x^{n+1}$ has the same radius of convergence, and

$$\int_0^x \left(\sum_{n=0}^{\infty} c_n t^n \right) dt = \sum_{n=0}^{\infty} \left(\int_0^x c_n t^n \, dt \right) = \sum_{n=0}^{\infty} \frac{c_n}{n+1} x^{n+1} \quad \text{for } |x| < R$$

The Integration Theorem is an invaluable tool for expressing many well-known functions as power series.

EXAMPLE 6 Show that

$$\ln (1 + x) = \sum_{n=0}^{\infty} \frac{(-1)^n}{n+1} x^{n+1} = \sum_{n=1}^{\infty} \frac{(-1)^{n-1}}{n} x^n \quad \text{for } |x| < 1 \quad (8)$$

Solution From (1) we know that

$$\frac{1}{1-x} = \sum_{n=0}^{\infty} x^n \quad \text{for } |x| < 1$$

Replacing x by $-t$ in this equation, we obtain

$$\frac{1}{1+t} = \sum_{n=0}^{\infty} (-1)^n t^n \quad \text{for } |t| < 1 \quad (9)$$

Using (9) and the Integration Theorem, we find that

$$\ln (1 + x) = \int_0^x \frac{1}{1+t} \, dt = \int_0^x \left(\sum_{n=0}^{\infty} (-1)^n t^n \right) dt$$

$$= \sum_{n=0}^{\infty} \frac{(-1)^n}{n+1} x^{n+1} \quad \text{for } |x| < 1 \quad \square$$

We can write the result of Example 6 in expanded form as follows:

$$\ln (1 + x) = x - \frac{x^2}{2} + \frac{x^3}{3} - \frac{x^4}{4} + \cdots$$

The formula holds not only for $|x| < 1$, as Example 6 shows, but also for $x = 1$. The power series expansion is sometimes known as **Mercator's series,** after the Danish mathematician Nicolaus Mercator (about 1620–1687). From Mercator's series one can derive the power series for $\ln [(1 + x)/(1 - x)]$, which has been used in the past to obtain quick and accurate estimates for logarithms.

Using (9) and the Integration Theorem, we can express $\tan^{-1} x$ as a power series.

EXAMPLE 7   Show that

$$\tan^{-1} x = \sum_{n=0}^{\infty} \frac{(-1)^n}{2n+1} x^{2n+1} \quad \text{for } |x| < 1 \tag{10}$$

Solution   If $|t| < 1$, then $|t^2| < 1$. Therefore we may substitute t^2 for t in (9) and obtain

$$\frac{1}{1+t^2} = \sum_{n=0}^{\infty} (-1)^n t^{2n} \quad \text{for } |t| < 1$$

Then the Integration Theorem yields

$$\tan^{-1} x = \int_0^x \frac{1}{1+t^2}\, dt = \int_0^x \left(\sum_{n=0}^{\infty} (-1)^n t^{2n} \right) dt$$

$$= \sum_{n=0}^{\infty} \frac{(-1)^n}{2n+1} x^{2n+1} \quad \text{for } |x| < 1 \quad \square$$

In expanded form the power series for $\tan^{-1} x$ becomes

$$\tan^{-1} x = x - \frac{x^3}{3} + \frac{x^5}{5} - \frac{x^7}{7} + \cdots$$

Gregory's Series
In 1671 this power series was derived by the Scottish mathematician James Gregory (1638–1675). However, two centuries earlier it appeared in the work *Yuktibhā,* in the Indian language Malayalam. The result was attributed to the mathematician Madhava (1340–1425), who lived near Cochin, close to the southernmost tip of India. Incidentally, Gregory not only obtained the series for $\tan^{-1} x$, but many other familiar series. He was one of the first mathematicians to distinguish between convergent and divergent series.

This is called **Gregory's series,** after the Scottish mathematician James Gregory, who achieved (along with, but independently from, Newton) most of the outstanding early results on series.

Although it is not so easy to prove, Gregory's series for $\tan^{-1} x$ also holds for $x = 1$. This value of x yields the following expression for $\pi/4$:

$$\frac{\pi}{4} = \tan^{-1} 1 = \sum_{n=0}^{\infty} (-1)^n \frac{1}{2n+1} = 1 - \frac{1}{3} + \frac{1}{5} - \frac{1}{7} + \cdots$$

This series, which was independently discovered by Leibniz and is often called the Leibniz series, is an alternating series that converges *very* slowly. It is impractical to use it to estimate $\pi/4$, because a large number of terms would be needed to ensure even moderate accuracy. In fact, to guarantee that the truncation error E_j is less than 0.0001 would require that $j \geq 5000$! However, we can use the power series expansion for $\tan^{-1} x$ along with the equation

$$\frac{\pi}{4} = 4 \tan^{-1} \frac{1}{5} - \tan^{-1} \frac{1}{239} \tag{11}$$

which appeared in Exercise 67 of Section 7.5, to obtain a very good approximation of $\pi/4$ with few calculations.

EXAMPLE 8   Using (11), approximate $\pi/4$ with an error less than 0.0001. Use the approximation obtained to estimate π with an error less than 0.0004.

Solution   If we approximate $4 \tan^{-1} \frac{1}{5}$ and $\tan^{-1} \frac{1}{239}$, each with a truncation error less than 0.00005, then the desired approximation of $\pi/4$ will have a truncation error less than 0.0001. The power series for $\tan^{-1} x$ given in (10) satisfies the conditions of the Alternating Series Test for $0 \le x < 1$, so we use (1) of Section 9.7 to find an upper bound for the truncation error introduced by using the first few terms of the series for $4 \tan^{-1} \frac{1}{5}$ and for $\tan^{-1} \frac{1}{239}$. In approximating $4 \tan^{-1} \frac{1}{5}$, we need n large enough to ensure that

$$4 \left| \frac{(-1)^{n+1}}{2(n+1)+1} \left(\frac{1}{5} \right)^{2(n+1)+1} \right| = 4 \left(\frac{1}{2n+3} \right) \left(\frac{1}{5^{2n+3}} \right) < 0.00005$$

which you can check happens if $n \ge 2$. Therefore

$$4 \tan^{-1} \frac{1}{5} \approx 4 \left[\frac{1}{5} - \frac{1}{3} \left(\frac{1}{5} \right)^3 + \frac{1}{5} \left(\frac{1}{5} \right)^5 \right] \approx 0.7895893333$$

with an error less than 0.00005. For $\tan^{-1} 1/239$, we need n large enough so that

$$\left| \frac{(-1)^{n+1}}{2(n+1)+1} \left(\frac{1}{239} \right)^{2(n+1)+1} \right| = \frac{1}{2n+3} \left(\frac{1}{239^{2n+3}} \right) < 0.00005$$

which happens if $n \ge 0$. Consequently

$$\tan^{-1} \frac{1}{239} \approx \frac{1}{239} \approx 0.0041841004$$

with an error less than 0.00005. Therefore by (11),

$$\frac{\pi}{4} = 4 \tan^{-1} \frac{1}{5} - \tan^{-1} \frac{1}{239} \approx 0.7895893333 - 0.0041841004$$

$$= 0.7854052329 \qquad (12)$$

with an error less than $0.00005 + 0.00005 = 0.0001$. Finally, multiplication of the numbers in (12) by 4 yields

$$\pi \approx 3.141620932$$

with an error less than 0.0004.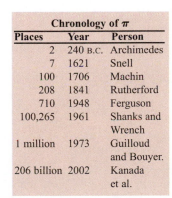

To emphasize the accuracy that can be obtained by means of (11), we mention that in 1706 John Machin employed (11) to calculate the first 100 digits in the decimal expansion of π. More recently formula (11) has been used to compute an approximation of π that is accurate to over 16,000 places. The related formula

$$\pi = 24 \tan^{-1} \frac{1}{8} + 8 \tan^{-1} \frac{1}{57} + 4 \tan^{-1} \frac{1}{239}$$

was used in 1961 by D. Shanks and J. W. Wrench, Jr., to calculate the first 100,000 digits in the decimal expansion of π. It took a computer over 8 hours to perform the task. Recently a high-speed computer has calculated π to over 200 billion places.

In 1996 David Bailey, Peter Borwein and Simon Plouffe discovered the formula

$$\pi = \sum_{k=0}^{\infty} \frac{1}{16^k} \left[\frac{4}{8k+1} - \frac{2}{8k+4} - \frac{1}{8k+5} - \frac{1}{8k+6} \right]$$

This formula can be used to calculate particular binary or hexadecimal digits of π without calculating the preceding digits. For example, the trillionth binary digit of π is 1 and the trillionth hexadecimal digit is 8.

Chronology of π		
Places	**Year**	**Person**
2	240 B.C.	Archimedes
7	1621	Snell
100	1706	Machin
208	1841	Rutherford
710	1948	Ferguson
100,265	1961	Shanks and Wrench
1 million	1973	Guilloud and Bouyer.
206 billion	2002	Kanada et al.

EXERCISES 9.8

In Exercises 1–20 find the interval of convergence of the given series.

1. $\displaystyle\sum_{n=1}^{\infty} \frac{x^n}{n^2}$

2. $\displaystyle\sum_{n=0}^{\infty} 2^n x^n$

3. $\displaystyle\sum_{n=1}^{\infty} \frac{1}{\sqrt{n}\, 3^n} x^n$

4. $\displaystyle\sum_{n=0}^{\infty} \frac{n}{4^n} x^n$

5. $\displaystyle\sum_{n=0}^{\infty} \frac{(-1)^n}{n+1} x^{2n}$

6. $\displaystyle\sum_{n=1}^{\infty} \frac{2}{3^{n+1} n^2} x^n$

7. $\displaystyle\sum_{n=1}^{\infty} \frac{(-1)^n}{2n-1} x^{n+1}$

8. $\displaystyle\sum_{n=1}^{\infty} \frac{n-1}{n^{2n}} x^n$

9. $\displaystyle\sum_{n=1}^{\infty} \frac{(-1)^n}{n^n} x^n$

10. $\displaystyle\sum_{n=1}^{\infty} \frac{2^n}{n^n} x^n$

11. $\displaystyle\sum_{n=0}^{\infty} \frac{n!}{(2n)!} x^n$

12. $\displaystyle\sum_{n=0}^{\infty} \frac{4^{n+1}}{\pi^{n+2}} x^{n+3}$

13. $\displaystyle\sum_{n=1}^{\infty} \frac{2^n}{n(3^{n+2})} x^{n+1}$

14. $\displaystyle\sum_{n=3}^{\infty} \frac{1}{n^3-4} x^n$

15. $\displaystyle\sum_{n=2}^{\infty} (\ln n) x^n$

16. $\displaystyle\sum_{n=2}^{\infty} \frac{\ln n}{n} x^n$

17. $\displaystyle\sum_{n=2}^{\infty} \frac{\ln n}{n^2} x^n$

18. $\displaystyle\sum_{n=2}^{\infty} \frac{1}{\ln n} x^n$

19. $\displaystyle\sum_{n=0}^{\infty} x^{(n^2)}$

20. $\displaystyle\sum_{n=1}^{\infty} x^{n!}$

In Exercises 21–25 find the radius of convergence of the given series.

21. $\displaystyle\sum_{n=1}^{\infty} \frac{n!}{n^n} x^n$

22. $\displaystyle\sum_{n=1}^{\infty} \frac{n^n}{n!} x^n$

23. $\displaystyle\sum_{n=1}^{\infty} \frac{1^2 \cdot 3^2 \cdot 5^2 \cdots (2n-1)^2}{2^2 \cdot 4^2 \cdot 6^2 \cdots (2n)^2} x^{2n}$

24. $\displaystyle\sum_{n=2}^{\infty} (-1)^{n+1} \frac{1 \cdot 3 \cdot 5 \cdots (2n-3)}{2^n n!} x^n$

25. $\displaystyle\sum_{n=1}^{\infty} \frac{1 \cdot 3 \cdot 5 \cdots (2n-1)}{2^n[1 \cdot 4 \cdot 7 \cdots (3n-2)]} x^n$

In Exercises 26–29, let $f(x)$ be the sum of the series. Find $f'(x)$ and $\int_0^x f(t)\, dt$.

26. $\displaystyle\sum_{n=0}^{\infty} \frac{(-1)^n}{n+1} x^n$

27. $\displaystyle\sum_{n=1}^{\infty} (n+1) x^n$

28. $\displaystyle\sum_{n=0}^{\infty} \frac{1}{n^2+1} x^{n+1}$

29. $\displaystyle\sum_{n=1}^{\infty} \frac{5}{n} x^{(n^2)}$

In Exercises 30–37 approximate the value of the integral with an error less than the given error, first using the Integration

Theorem to express the integral as an infinite series and then approximating the infinite series by an appropriate partial sum.

30. $\displaystyle\int_0^1 \cos \sqrt{x}\, dx;\ 10^{-3}$ $\left(\textit{Hint: } \cos x = \displaystyle\sum_{n=0}^{\infty} \frac{(-1)^n}{(2n)!} x^{2n}.\right)$

31. $\displaystyle\int_0^2 \sin x^2\, dx;\ 10^{-2}$ $\left(\textit{Hint: } \sin x = \displaystyle\sum_{n=0}^{\infty} \frac{(-1)^n}{(2n+1)!} x^{2n+1}.\right)$

32. $\displaystyle\int_0^1 \cos x^2\, dx;\ 10^{-7}$

33. $\displaystyle\int_0^1 \frac{1-e^{-x}}{x}\, dx;\ 10^{-3}$

34. $\displaystyle\int_0^{1/5} \tan^{-1} x\, dx;\ 10^{-5}$

35. $\displaystyle\int_0^{1/2} \frac{x^2}{1+x}\, dx;\ 10^{-3}$

36. $\displaystyle\int_0^1 \frac{x^3}{2+x}\, dx;\ 10^{-3}$

*37. $\displaystyle\int_{-1}^0 e^{x^2}\, dx;\ 10^{-3}$

38. Using the power series expansion for e^x given in (3), show that

a. $\cosh x = \displaystyle\sum_{n=0}^{\infty} \frac{1}{(2n)!} x^{2n}$ for all x

b. $\sinh x = \displaystyle\sum_{n=0}^{\infty} \frac{1}{(2n+1)!} x^{2n+1}$ for all x

39. Using formula (1) and the Differentiation Theorem, show that
$$\sum_{n=1}^{\infty} nx^n = \frac{x}{(1-x)^2}.$$

40. Show that both
$$\sum_{n=0}^{\infty} \frac{(-1)^n}{(2n)!} x^{2n} \quad \text{and} \quad \sum_{n=0}^{\infty} \frac{(-1)^n}{(2n+1)!} x^{2n+1}$$
converge for all x.

41. Use the power series expansion for $(e^x-1)/x$ to verify that
$$\lim_{x\to 0} \frac{e^x-1}{x} = 1$$

42. Find a power series expansion for $(e^x-1-x)/x^2$ and use it to evaluate
$$\lim_{x\to 0} \frac{e^x-1-x}{x^2}$$

43. Using (8), evaluate
$$\lim_{x\to 0} \frac{\ln(1+x)}{x}$$

44. Express $\ln[(1+x)/(1-x)]$ as a power series. (*Hint:* $\ln[(1+x)/(1-x)] = \ln(1+x) - \ln(1-x)$.)

45. a. Show that $\ln \dfrac{1}{1-x} = \displaystyle\sum_{n=1}^{\infty} \frac{x^n}{n}$.

b. Using (a), show that $\ln 2 = \sum\limits_{n=1}^{\infty} \dfrac{1}{n2^n}$.

c. Using the fact that

$$\sum_{n=N}^{\infty} \frac{1}{n2^n} \le \sum_{n=N}^{\infty} \frac{1}{N2^n}$$

estimate $\ln 2$ with an error less than 0.01.

46. Approximate $\tan^{-1} \frac{1}{2}$ with an error less than 0.001.

47. Suppose that in Example 8 we had used $n = 3$ rather than $n = 2$ in the estimation of $4 \tan^{-1} \frac{1}{5}$. Find an upper bound for the error introduced for the estimate of $\pi/4$.

48. a. Show that

$$\frac{\tan^{-1} t}{t} = \sum_{n=0}^{\infty} \frac{(-1)^n}{2n+1} t^{2n}$$

$$= 1 - \frac{t^2}{3} + \frac{t^4}{5} - \frac{t^6}{7} + \cdots \quad \text{for } 0 < |t| < 1$$

b. Using part (a), conclude that

$$\lim_{t \to 0} (\tan^{-1} t)/t = 1$$

49. a. Using (1), with x replaced by $-t^4$, show that

$$\frac{t^2}{1 + t^4} = \sum_{n=0}^{\infty} (-1)^n t^{4n+2} \quad \text{for } -1 < t < 1$$

b. Using part (a), express $\int_0^{1/2} t^2/(1 + t^4)\, dt$ as the sum of a power series.

50. The **error function**, which is defined by

$$\text{erf}(x) = \frac{2}{\sqrt{\pi}} \int_0^x e^{-t^2}\, dt \quad \text{for all } x$$

is prominent in statistics. Estimate $\text{erf}(1)$ with an error less than 0.01.

51. a. Using (3), find a power series expansion for xe^x.

b. Show that

$$\sum_{n=0}^{\infty} \frac{1}{n!(n+2)} = 1$$

(*Hint:* Evaluate $\int_0^1 xe^x\, dx$ and $\int_0^1 (\sum_{n=0}^{\infty} c_n x^n)\, dx$, where $\sum_{n=0}^{\infty} c_n x^n$ is the power series obtained in (a).)

52. Let $R > 0$ be arbitrary. Find an example of a power series $\sum_{n=0}^{\infty} c_n x^n$ whose radius of convergence is R.

53. Use (11) and the Gregory series for $\tan^{-1} x$ in order to approximate π to within 10^{-10}. (*Hint:* Pattern your solution after the solution of Example 8.)

54. Show that if $\lim_{n \to 0} |c_{n+1}/c_n| = L \ne 0$, then the radius of convergence of $\sum_{n=0}^{\infty} c_n x^n$ is $1/L$.

55. Let $f(x) = \sum_{n=0}^{\infty} c_n x^n$ have a nonzero radius of convergence.

a. Using Corollary 9.27, prove that if f is an even

function, then $c_n = 0$ for all odd integers n. (*Hint:* Find the series expansion for $f(-x)$, and equate it to the series expansion for $f(x)$.)

b. Prove that if f is an odd function, then $c_n = 0$ for all even integers n.

Applications

56. A pendulum L meters long swings back and forth, with maximum angle of deflection ϕ. Let $x = \sin \phi/2$, and let $g = 32$ (feet per second per second) be the acceleration due to gravity. Then the period of oscillation T of the pendulum is given by the formula

The Foucault Pendulum in the Smithsonian Institution in Washington, D.C. (Smithsonian Institution)

$$T(x) = 2\pi \sqrt{\frac{L}{g}} \left(1 + \sum_{n=1}^{\infty} \frac{1^2 \cdot 3^2 \cdot 5^2 \cdots (2n-1)^2}{2^2 \cdot 4^2 \cdot 6^2 \cdots (2n)^2} x^{2n} \right)$$

The series on the right converges for $|x| < 1$ (see Exercise 23). Notice that

$$\left| \frac{T(x)}{2\pi \sqrt{L/g}} - 1 \right| = \sum_{n=1}^{\infty} \frac{1^2 \cdot 3^2 \cdot 5^2 \cdots (2n-1)^2}{2^2 \cdot 4^2 \cdot 6^2 \cdots (2n)^2} x^{2n} \quad (13)$$

Usually ϕ is very small in pendulum clocks, say, $0 < \phi \le \pi/12$, and consequently

$$0 < x^2 = \sin^2 \phi/2 < 0.018$$

Show that if $x^2 < 0.018$, then the series in (13) is bounded by 0.005. (*Hint:* The coefficient of x^{2n} in the series is less than or equal to $\frac{1}{4}$ for all n.) It is because of (13) that the period of a pendulum of length L is often assumed to be $2\pi \sqrt{L/g}$, rather than $T(x)$ for the correct value of x. We also deduce from (13) that as the maximum angle of deflection of the pendulum shrinks slightly while the clock runs down, the period $T(x)$ stays practically the same, and thus the clock keeps good time.

*57. The probability of getting a 7 or 11 on a roll of two dice is $\frac{8}{36} = \frac{2}{9}$ and the probability of getting some other sum is $1 - \frac{2}{9} = \frac{7}{9}$. This implies that if the dice are rolled repeatedly, then for any integer $n \geq 1$, the probability of rolling a 7 or 11 for the first time on the nth roll is $\left(\frac{7}{9}\right)^{n-1} \frac{2}{9}$. The expected number of rolls required to roll a 7 or 11 the first time is therefore

$$\sum_{n=1}^{\infty} n \left(\frac{7}{9}\right)^{n-1} \frac{2}{9}$$

Find the sum N of the series.

Project

1. a. Let

$$f(x) = \sum_{n=0}^{\infty} \frac{(-1)^n}{(2n)!} x^{2n}$$

and $g(x) = \displaystyle\sum_{n=0}^{\infty} \frac{(-1)^n}{(2n+1)!} x^{2n+1}$ for all x

i. Show that $f'(x) = -g(x)$ and $g'(x) = f(x)$. For all x.
ii. Show that $f''(x) = -f(x)$ and $g''(x) = -g(x)$. For all x.
iii. What functions do you know that satisfy the properties of (i) and (ii)?

b. Let

$$h(x) = \sum_{n=0}^{\infty} \frac{1}{(2n)!} x^{2n}$$

and $k(x) = \displaystyle\sum_{n=0}^{\infty} \frac{1}{(2n+1)!} x^{2n+1}$ for all x

i. Show that $h'(x) = k(x)$ and $k'(x) = h(x)$ for all x.
ii. Show that $h''(x) = h(x)$ and $k''(x) = k(x)$ for all x.
iii. What functions do you know that satisfy the properties of (i) and (ii)?

9.9 TAYLOR SERIES

In the preceding sections we showed that

$$\frac{1}{1-x} = \sum_{n=0}^{\infty} x^n \quad \text{for } -1 < x < 1 \tag{1}$$

$$e^x = \sum_{n=0}^{\infty} \frac{x^n}{n!} \quad \text{for all } x \tag{2}$$

$$\ln(1+x) = \sum_{n=0}^{\infty} \frac{(-1)^n}{n+1} x^{n+1} \quad \text{for } -1 < x < 1 \tag{3}$$

$$\tan^{-1} x = \sum_{n=0}^{\infty} \frac{(-1)^n}{2n+1} x^{2n+1} \quad \text{for } -1 < x < 1 \tag{4}$$

In each case we say that we have a power series representation of the given function. More generally, if f is a function, if I is an open interval containing 0, and if

$$f(x) = \sum_{n=0}^{\infty} c_n x^n \quad \text{for } x \text{ in } I$$

then we say that we have a **power series representation of f on I.** The main advantage of having a power series representation of a function f on I is that the value of f at any point in I is the sum of a convergent series and hence can be approximated by its partial sums. Since the partial sums of a power series are polynomials, values of functions having power series representations are easily approximated and are therefore tractable (with the help of computers, in some cases). In this section we will study functions that have power series representations, and in so doing we will complete the discussion of the approximation of functions by polynomials that we initiated in Section 9.1.

For a given function f we wish to determine whether there is a power series $\sum_{n=0}^{\infty} c_n x^n$ and an open interval I containing 0 such that

$$f(x) = \sum_{n=0}^{\infty} c_n x^n \quad \text{for } x \text{ in } I \tag{5}$$

We can already eliminate many functions such as $|x|$ from consideration, because $|x|$ is not differentiable at 0 and Theorem 9.26 asserts that if f has such a power series representation, then f must have derivatives of all orders on I. Moreover, formula (6) in Theorem 9.26 tells us that if f has a representation in the form of (5), then $c_n = f^{(n)}(0)/n!$. Thus a function can have only one power series representation, and (5) can be rewritten

$$f(x) = \sum_{n=0}^{\infty} \frac{f^{(n)}(0)}{n!} x^n \quad \text{for } x \text{ in } I \tag{6}$$

DEFINITION 9.29   Suppose that f has derivatives of all orders at 0. Then the **Taylor series of f** is the power series

$$\sum_{n=0}^{\infty} \frac{f^{(n)}(0)}{n!} x^n$$

The Taylor series in Definition 9.29 is often referred to as a **Maclaurin series,** after the Scottish mathematician Colin Maclaurin (1698–1746).

From the discussion above we know that once we find a power series representation of a function, that power series must be the Taylor series for the function. It is therefore apparent from (1)–(4) that the series appearing in those formulas are the Taylor series of $1/(1-x)$, e^x, $\ln(1+x)$, and $\tan^{-1} x$, respectively. Moreover, if f is an arbitrary polynomial, say

$$f(x) = c_0 + c_1 x + c_2 x^2 + \cdots + c_n x^n$$

then f is a power series (with $c_j = 0$ for $j > n$). Consequently the polynomial is its own Taylor series.

Now we will derive the Taylor series for the sine function.

EXAMPLE 1   Show that the Taylor series for $\sin x$ is

$$x - \frac{x^3}{3!} + \frac{x^5}{5!} - \frac{x^7}{7!} + \cdots = \sum_{n=0}^{\infty} \frac{(-1)^n}{(2n+1)!} x^{2n+1} \quad \text{for all } x$$

Solution   Let $f(x) = \sin x$. The derivatives of f repeat themselves in groups of four:

$$f(x) = \sin x, \qquad f'(x) = \cos x, \qquad f''(x) = -\sin x, \qquad f^{(3)}(x) = -\cos x,$$

$$f^{(4)}(x) = \sin x, \qquad f^{(5)}(x) = \cos x, \qquad f^{(6)}(x) = -\sin x, \qquad f^{(7)}(x) = -\cos x$$

and so on. As you see, the derivatives of even order involve $\sin x$, whereas the derivatives of odd order involve $\cos x$. More precisely, for each nonnegative integer k,

$$f^{(2k)}(x) = (-1)^k \sin x \quad \text{and} \quad f^{(2k+1)}(x) = (-1)^k \cos x \qquad (7)$$

Therefore

$$f^{(2k)}(0) = (-1)^k \sin 0 = 0 \quad \text{and} \quad f^{(2k+1)}(0) = (-1)^k \cos 0 = (-1)^k$$

In other words,

$$f(0) = 0, f'(0) = 1, f''(0) = 0, f^{(3)}(0) = -1, f^{(4)}(0) = 0, f^{(5)}(0) = 1, \dots$$

Thus for $\sin x$ the Taylor series

$$f(0) + f'(0)x + \frac{f''(0)}{2!}x^2 + \frac{f^{(3)}(0)}{3!}x^3 + \frac{f^{(4)}(0)}{4!}x^4 + \frac{f^{(5)}(0)}{5!}x^5 + \cdots$$

is given by

$$0 + 1 \cdot x + \frac{0}{2!}x^2 + \frac{-1}{3!}x^3 + \frac{0}{4!}x^4 + \frac{1}{5!}x^5 + \cdots$$

Since the coefficients of all even powers of x in the series on the right are 0, we suppress them and write the Taylor series of $\sin x$ as

$$x - \frac{x^3}{3!} + \frac{x^5}{5!} - \frac{x^7}{7!} + \cdots, \quad \text{or more succinctly,} \quad \sum_{n=0}^{\infty} \frac{(-1)^n}{(2n+1)!} x^{2n+1}$$

This completes the solution.   ❑

By an analysis similar to that in Example 1, it is possible to show that for every real number x, the Taylor series for $\cos x$ is

$$1 - \frac{x^2}{2!} + \frac{x^4}{4!} - \frac{x^6}{6!} + \cdots, \quad \text{or equivalently,} \quad \sum_{n=0}^{\infty} \frac{(-1)^n}{(2n)!} x^{2n}$$

However, taken by themselves, these results about the Taylor series of $\sin x$ and $\cos x$ are not very exciting, because we have not yet shown that

$$\sin x = \sum_{n=0}^{\infty} \frac{(-1)^n}{(2n+1)!} x^{2n+1} \quad \text{and} \quad \cos x = \sum_{n=0}^{\infty} \frac{(-1)^n}{(2n)!} x^{2n}$$

In Example 2 we will prove that the first equation is valid for all x. Exercise 35 shows the corresponding result for the second equation.

Despite the fact that the Taylor series for $\sin x$, $\cos x$, e^x, $\ln(1 + x)$, $\tan^{-1} x$, and polynomials converge to the respective functions, it is *not* true that the Taylor series of each function that is infinitely differentiable at 0 converges to the function on an open interval containing 0. An example is given by

$$f(x) = \begin{cases} e^{-1/x^2} & \text{for } x \neq 0 \\ 0 & \text{for } x = 0 \end{cases}$$

By the formula defining f we know that $f(0) = 0$, and an application of l'Hôpital's Rule shows that $f'(0) = 0$ (see the remark following Example 6 of Section 7.6). With effort it is possible to show that f has derivatives of all orders at 0 and that

$$f^{(n)}(0) = 0 \quad \text{for all } n \geq 0$$

As a result, the Taylor series of f is the 0 function. However, since $f(x) = e^{-1/x^2} \neq 0$ for all $x \neq 0$, it follows that the Taylor series of f does not converge to $f(x)$ for any $x \neq 0$.

Because a Taylor series does not always converge to the corresponding function on some open interval about 0, it is important to have conditions under which the Taylor series of a function f does converge on a suitable open interval about 0. Recall from Section 9.1 that the nth Taylor polynomial is given by

$$p_n(x) = f(0) + f'(0)x + \frac{f''(0)}{2!}x^2 + \cdots + \frac{f^{(n)}(0)}{n!}x^n \tag{8}$$

Observe that $p_n(x)$ is the $(n + 1)$st partial sum of the Taylor series of f. In order to measure how well $p_n(x)$ approximates $f(x)$, we define the ***n*th Taylor remainder r_n** of f by the formula

$$r_n(x) = f(x) - p_n(x) \tag{9}$$

The smaller $|r_n(x)|$ is, the better $p_n(x)$ approximates $f(x)$.

A theorem due to Brook Taylor provides us with a means of estimating the nth Taylor remainder, and hence the accuracy involved in replacing $f(x)$ by the value of the nth Taylor polynomial at x.

THEOREM 9.30
Taylor's Theorem

Let n be a nonnegative integer, and suppose $f^{(n+1)}(x)$ exists for each x in an open interval I containing 0. For each $x \neq 0$ in I, there is a number t_x strictly between 0 and x such that

$$f(x) = f(0) + f'(0)x + \cdots + \frac{f^{(n)}(0)}{n!}x^n + \frac{f^{(n+1)}(t_x)}{(n+1)!}x^{n+1} \tag{10}$$

By (8)–(10),

$$r_n(x) = \frac{f^{(n+1)}(t_x)}{(n+1)!}x^{n+1} \tag{11}$$

Equation (10) is known as **Taylor's Formula,** and equation (11) is known as the **Lagrange Remainder Formula.**

Later in the section we will state a slightly more general theorem. Notice that if $n = 0$, then formula (10) in the theorem becomes

$$f(x) = f(0) + f'(t_x)x$$

or equivalently,

$$f(x) - f(0) = f'(t_x)(x - 0)$$

This equation is equivalent to the formula in the Mean Value Theorem (with a, b, and c replaced, respectively, by 0, x, and t_x).

Since $f(x) = p_n(x) + r_n(x)$, it follows that for any x,

$$f(x) = \sum_{n=0}^{\infty} \frac{f^{(n)}(0)}{n!}x^n \quad \text{if and only if} \quad \lim_{n \to \infty} r_n(x) = 0 \tag{12}$$

As a result, we can prove that the Taylor series of f converges to $f(x)$ by verifying that $\lim_{n \to \infty} r_n(x) = 0$. For this we normally try to use the Lagrange Remainder Formula in (11). Unfortunately, the expression $f^{(n+1)}(t_x)$ in (11) usually cannot be evaluated exactly. Indeed, it might be difficult to find a formula for the $(n + 1)$st derivative $f^{(n+1)}(x)$. More importantly, all we know about t_x is that it is a number between 0 and x, so its exact value is usually unknown. Nevertheless, if we can obtain an inequality that bounds $|f^{(n+1)}(t_x)|$ and involves only x and n (but not t_x), then it may be possible to show that $\lim_{n \to \infty} r_n(x) = 0$. This is illustrated in the next example, in which $|f^{(n+1)}(t_x)| \leq 1$ for all x and all n.

EXAMPLE 2   Show that for every real number x, $\sin x$ is the sum of its Taylor series, that is,

$$\sin x = \sum_{n=0}^{\infty} \frac{(-1)^n}{(2n + 1)!} x^{2n+1}$$

Solution   By (7) with n replacing k,

$$|f^{(2n)}(x)| = |(-1)^n \sin x| \leq 1 \quad \text{and} \quad |f^{(2n+1)}(x)| = |(-1)^n \cos x| \leq 1$$

Therefore

$$|f^{(n+1)}(t_x)| \leq 1 \quad \text{for all } x \text{ and all } n \geq 0$$

Thus by (11),

$$|r_n(x)| = \left| \frac{f^{(n+1)}(t_x)}{(n + 1)!} x^{n+1} \right| \leq \frac{|x|^{n+1}}{(n + 1)!} \tag{13}$$

Example 7 of Section 9.7 then yields

$$\lim_{n \to \infty} \frac{|x|^{n+1}}{(n + 1)!} = \lim_{n \to \infty} \frac{|x|^n}{n!} = 0 \quad \text{for all } x$$

Therefore $\lim_{n \to \infty} r_n(x) = 0$ for all x. Consequently (12) tell us that the Taylor series of the sine function converges to $\sin x$ for every x.   ❑

Figure 9.23 includes the graph of $\sin x$, along with its Taylor polynomials p_1, p_5, p_9, and p_{19}. As the figure suggests, the graphs of p_n and f are virtually indistinguishable on larger and larger intervals about 0 as n increases.

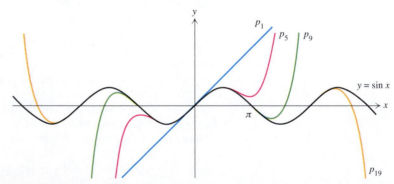

FIGURE 9.23

By contrast, the Taylor series of $1/(1 - x)$, which is $\sum_{n=0}^{\infty} x^n$, converges only on $(-1, 1)$. Consequently outside the interval $(-1, 1)$ we would not expect that the Taylor polynomials would give a good approximation to $1/(1-x)$ (see Figure 9.24).

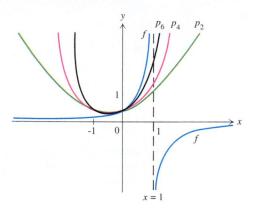

FIGURE 9.24

Numerical Approximation

Suppose we wish to find an integer n such that for any x in $[-\pi, \pi]$, $p_n(x)$ approximates $\sin x$ to within 10^{-9}. By (13),

$$| r_n(x) | \leq \frac{\pi^{n+1}}{(n + 1)!} \quad \text{for } | x | \leq \pi$$

Therefore we need only find a positive integer n such that

$$\frac{\pi^{n+1}}{(n + 1)!} < 10^{-9}$$

By trial and error with a calculator we find that this occurs if $n \geq 20$. Consequently the 20th Taylor polynomial, which is the same as the 19th Taylor polynomial, namely

$$x - \frac{x^3}{3!} + \frac{x^5}{5!} - \frac{x^7}{7!} + \frac{x^9}{9!} - \frac{x^{11}}{11!} + \frac{x^{13}}{13!} - \frac{x^{15}}{15!} + \frac{x^{17}}{17!} - \frac{x^{19}}{19!}$$

differs from $\sin x$ by less than 10^{-9} for *every* x in $[-\pi, \pi]$. You can understand why the graphs of p_{19} and $\sin x$ seem to coincide on the interval $[-\pi, \pi]$ in Figure 9.23.

Similarly the remainder estimate given in (13) is valid for $\cos x$. Accordingly, for any x in $[-\pi, \pi]$ it is possible to estimate $\cos x$ to within 10^{-9} by using the 20th Taylor polynomial of $\cos x$.

EXAMPLE 3   Approximate the number e with an error less than 0.001.

Solution   Let $f(x) = e^x$. Then $f(1) = e^1 = e$, so we want to approximate $f(1)$. If we use $p_n(1)$ as the approximation, then the error will be $| f(1) - p_n(1) | = | r_n(1) |$. Consequently if we find an n such that $| r_n(1) | < 0.001$, then $p_n(1)$ will be the desired approximation. Now by Taylor's Theorem we know that there is a number t_1 between 0 and 1 such that

$$r_n(1) = \frac{f^{(n+1)}(t_1)}{(n+1)!} 1^{n+1}$$

Since $f^{(n+1)}(t_1) = e^{t_1}$ and $e < 3$ (as we observed in Section 5.7), it follows that

$$e^{t_1} < e^1 = e < 3$$

so that

$$r_n(1) = \frac{e^{t_1}}{(n+1)!} < \frac{3}{(n+1)!} \tag{14}$$

By computing the value of $3/(n+1)!$ for $n = 1, 2, \ldots, 6$, we discover that $|r_n(1)| < 0.001$ if $n \geq 6$. Thus $p_6(1)$ is the desired approximation to e. By Example 1 of Section 9.1,

$$p_n(x) = \sum_{k=0}^{n} \frac{1}{k!} x^k$$

Therefore

$$p_6(1) = 1 + 1 + \frac{1}{2!} + \frac{1}{3!} + \frac{1}{4!} + \frac{1}{5!} + \frac{1}{6!} = \frac{1957}{720} \approx 2.718055556$$

We conclude that 2.718055556 approximates e with an error less than 0.001.   ❑

From (14) we find that

$$\left| e - \left(1 + 1 + \frac{1}{2!} + \frac{1}{3!} + \cdots + \frac{1}{n!} \right) \right| = |e - p_n(1)| = |r_n(1)| < \frac{3}{(n+1)!}$$

Since $3/(n+1)!$ gets small as n gets large, we could make the difference between e and $p_n(1)$ as small as we please by taking n large enough.

Since $\sqrt{e} = e^{1/2}$, we could also approximate $\sqrt{e}$ by using the same Taylor polynomials of the function e^x. But for $\sqrt{e}$ we would utilize the numbers $p_n(\frac{1}{2})$ instead of using $p_n(1)$.

EXAMPLE 4   Approximate ln 2 with an error less than 0.1.

Solution   Let $f(x) = \ln(1 + x)$. As in Example 3, we will approximate $f(1)$ by $p_n(1)$ for an appropriate value of n. By the formula for $f^{(k)}(x)$ in Example 2 in Section 9.1, with k replaced by $n + 1$, we have

$$f^{(n+1)}(x) = \frac{(-1)^n n!}{(1+x)^{n+1}}$$

so that by (11), with $x = 1$, there is a number t_1 strictly between 0 and 1 such that

$$r_n(1) = \frac{f^{(n+1)}(t_1)}{(n+1)!} = \frac{(-1)^n n!}{(1+t_1)^{n+1}} \cdot \frac{1}{(n+1)!} = \frac{(-1)^n}{(1+t_1)^{n+1}(n+1)}$$

Since $t_1 > 0$, we have

$$| r_n(1) | < \frac{1}{n+1} \tag{15}$$

so that if $n \geq 9$, then $| r_n(1) | < 0.1$. Thus $p_9(1)$ approximates $\ln 2$ with an error less than 0.1. By Example 2 of Section 9.1,

$$p_n(x) = \sum_{k=1}^{n} \frac{(-1)^{k-1}}{k} x^k$$

Therefore

$$p_9(1) = 1 - \frac{1}{2} + \frac{1}{3} - \frac{1}{4} + \frac{1}{5} - \frac{1}{6} + \frac{1}{7} - \frac{1}{8} + \frac{1}{9} \approx 0.7456349206$$

and this is the desired approximation. ❏

From (15) we find that

$$\left| \ln 2 - \left(1 - \frac{1}{2} + - \cdots + (-1)^{n-1} \frac{1}{n} \right) \right| = |\ln 2 - p_n(1)| = |r_n(1)| < \frac{1}{n+1}$$

Since $1/(n+1)$ gets small as n becomes large, we could make the difference between $\ln 2$ and $p_n(1)$ as small as we please by taking n large enough.

Taylor Series about an Arbitrary Point

If a function f is to have a Taylor series

$$\sum_{n=0}^{\infty} \frac{f^{(n)}(0)}{n!} x^n \tag{16}$$

then f must have derivatives at 0 and hence must be defined in an open interval about 0. Since $\ln x$ is not defined in an open interval about 0, $\ln x$ does not have a Taylor series of the form given in (16) for $f(x) = \ln x$. However, in (3) we presented the Taylor series for $\ln(1+x)$:

$$\ln(1+x) = \sum_{n=0}^{\infty} \frac{(-1)^n}{n+1} x^{n+1} \quad \text{for } -1 < x < 1 \tag{17}$$

Now if $0 < x < 2$, then $-1 < x - 1 < 1$, so that by (17) we find that

$$\ln x = \ln(1 + (x-1)) = \sum_{n=0}^{\infty} \frac{(-1)^n}{n+1} (x-1)^{n+1} \quad \text{for } 0 < x < 2 \tag{18}$$

The series in (18) is also a power series, but it contains powers of $x - 1$ rather than powers of x, and its interval of convergence is centered at 1 rather than at 0.

In general, we are interested in the possibility of expressing a function as a power series

$$\sum_{n=0}^{\infty} c_n(x-a)^n \tag{19}$$

in powers of $x - a$, where a can be any fixed number. All the results we obtained for ordinary power series have their counterparts for power series of the type given in (19). In particular, if f has derivatives of all orders at a, then we call

$$\sum_{n=0}^{\infty} \frac{f^{(n)}(a)}{n!} (x - a)^n \tag{20}$$

the **Taylor series of f about the number a**. The **nth Taylor polynomial p_n** of f about a is defined by

$$p_n(x) = f(a) + f'(a)(x - a) + \frac{f''(a)}{2!} (x - a)^2 + \cdots + \frac{f^{(n)}(a)}{n!} (x - a)^n$$

and the **nth Taylor remainder r_n** of f about a is defined by

$$r_n(x) = f(x) - p_n(x)$$

The next theorem is called Taylor's Theorem. It is a generalization of Theorem 9.30 and is proved in the Appendix.

THEOREM 9.31
Taylor's Theorem

Let n be a nonnegative integer, and suppose $f^{(n+1)}(x)$ exists for each x in an open interval I containing a. For each $x \neq a$ in I, there is a number t_x strictly between a and x such that

$$f(x) = f(a) + f'(a)(x - a) + \cdots + \frac{f^{(n)}(a)}{n!} (x - a)^n + \frac{f^{(n+1)}(t_x)}{(n + 1)!} (x - a)^{n+1} \tag{21}$$

Thus
$$r_n(x) = \frac{f^{(n+1)}(t_x)}{(n + 1)!} (x - a)^{n+1} \tag{22}$$

Equation (21) is known as **Taylor's Formula** and (22) is known as the **Lagrange Remainder Formula**.

EXAMPLE 5   Express the polynomial

$$f(x) = 2x^3 - 9x^2 + 11x - 1 \tag{23}$$

as a polynomial in $x - 2$.

Solution   We wish to write $f(x)$ in the form of (20) with $a = 2$, and to do so we must compute the derivatives of f at 2:

$$f(x) = 2x^3 - 9x^2 + 11x - 1 \qquad\qquad f(2) = 1$$

$$f'(x) = 6x^2 - 18x + 11 \qquad\qquad f'(2) = -1$$

$$f''(x) = 12x - 18 \qquad\qquad f''(2) = 6$$

$$f^{(3)}(x) = 12 \qquad\qquad f^{(3)}(2) = 12$$

$$f^{(n)}(x) = 0 \quad \text{for } n \geq 4 \qquad\qquad f^{(n)}(2) = 0 \quad \text{for } n \geq 4$$

Therefore

$$f(x) = 1 + (-1)(x-2) + \frac{6}{2!}(x-2)^2 + \frac{12}{3!}(x-2)^3$$

$$= 1 - (x-2) + 3(x-2)^2 + 2(x-2)^3 \quad \square$$

Although the form of the polynomial just obtained looks quite different from the polynomial given in (23), both polynomials represent the same function, as you can verify by expanding $(x-2)^2$ and $(x-2)^3$ in powers of x.

Sometimes we can derive a Taylor series about a number a different from 0 by using one or more known Taylor series about 0. The next example illustrates this point.

EXAMPLE 6   Find the Taylor series of $\sin x$ about $\pi/6$, and show that it converges to $\sin x$ for all x.

Solution   Let $f(x) = \sin x$. We could derive the Taylor series of $\sin x$ in powers of $x - \pi/6$ by finding $f^{(n)}(\pi/6)$ for $n = 0, 1, 2, \ldots$. However, observe that

$$\sin x = \sin\left(x - \frac{\pi}{6} + \frac{\pi}{6}\right)$$

By the sum formula,

$$\sin\left(\left(x - \frac{\pi}{6}\right) + \frac{\pi}{6}\right) = \sin\left(x - \frac{\pi}{6}\right)\cos\frac{\pi}{6} + \cos\left(x - \frac{\pi}{6}\right)\sin\frac{\pi}{6}$$

$$= \frac{\sqrt{3}}{2}\sin\left(x - \frac{\pi}{6}\right) + \frac{1}{2}\cos\left(x - \frac{\pi}{6}\right)$$

Since we already know the Taylor series for $\sin x$ and for $\cos x$, we can substitute $x - \pi/6$ for x in their series to obtain

$$\sin x = \frac{\sqrt{3}}{2}\sum_{n=0}^{\infty}(-1)^n\frac{(x - \pi/6)^{2n+1}}{(2n+1)!} + \frac{1}{2}\sum_{n=0}^{\infty}(-1)^n\frac{(x - \pi/6)^{2n}}{(2n)!}$$

$$= \frac{1}{2} + \frac{\sqrt{3}}{2}\left(x - \frac{\pi}{6}\right) - \frac{1}{4}\left(x - \frac{\pi}{6}\right)^2 - \frac{\sqrt{3}}{12}\left(x - \frac{\pi}{6}\right)^3 + \cdots$$

This must be the Taylor series of $\sin x$ about $\pi/6$ because $\sin x$ has a unique representation as a power series in powers of $x - \pi/6$. $\square$

EXAMPLE 7   Using the Taylor series obtained in Example 6, approximate $\sin 7\pi/36$ with an error less than 10^{-5}.

Solution   Since $7\pi/36$ is close to $\pi/6$, the desired approximation of $\sin(7\pi/36)$ will be the value $p_n(7\pi/36)$ of a suitable Taylor polynomial p_n of $\sin x$ about $\pi/6$, which can be obtained from Example 6. We will find an integer n such that $|r_n(7\pi/36)| < 10^{-5}$. To that end we recall from (22) that

$$r_n\left(\frac{7\pi}{36}\right) = \frac{f^{(n+1)}(t_{7\pi/36})}{(n+1)!}\left(\frac{7\pi}{36} - \frac{\pi}{6}\right)^{n+1}$$

where $f(x) = \sin x$ and $t_{7\pi/36}$ lies between $\pi/6$ and $7\pi/36$. We know from (7) that $|f^{(k)}(x)| \leq 1$ for all x. Thus

$$\left|r_n\left(\frac{7\pi}{36}\right)\right| \leq \frac{1}{(n+1)!}\left(\frac{\pi}{36}\right)^{n+1}$$

Now

$$\frac{1}{(n+1)!}\left(\frac{\pi}{36}\right)^{n+1} < 10^{-5} \quad \text{if } n \geq 3$$

Therefore the third Taylor polynomial p_3 furnishes the desired approximation:

$$p_3\left(\frac{7\pi}{36}\right) = \frac{1}{2} + \frac{\sqrt{3}}{2}\left(\frac{\pi}{36}\right) - \frac{1}{4}\left(\frac{\pi}{36}\right)^2 - \frac{\sqrt{3}}{12}\left(\frac{\pi}{36}\right)^3 \approx 0.5735751919 \qquad \square$$

If we had estimated $\sin(7\pi/36)$ by using Taylor polynomials about 0, we would have needed more terms to achieve the accuracy obtained in Example 7 (see Exercise 36). In general, one obtains a better approximation for $f(x)$ by considering the Taylor polynomials about a point a nearer to x.

In this section we have described Taylor series of several kinds of functions: polynomials, e^x, $\sin x$, $\cos x$, $\ln(1+x)$, $\tan^{-1} x$, and others related to these functions. But we caution you that in practice it is usually *very* difficult to write Taylor series for most other functions, unless they are rather closely related to those functions whose Taylor series we already know. As a result, in the exercises we will concentrate on Taylor series of the functions just mentioned and their relatives.

Historical Comment on Series

Power series, which eventually came to be called Taylor series, contributed greatly to the growth of calculus. These series allowed Newton and Gregory and their successors to analyze properties of functions with a single theory and to approximate values of functions easily; as a result, Taylor series became immensely important to mathematicians such as Euler and Lagrange during the eighteenth century. In fact, Lagrange isolated Taylor series as the notion fundamental to calculus. He maintained that all one had to do in order to understand a continuous function f was to find the derivatives $f^{(n)}(a)$ at a given number a of one's choice. Since he, like all other mathematicians at that time, had no clear concept of convergence, he did not realize that not every continuous function has a Taylor series. In time, mathematicians found that Taylor series did not provide the sole key to understanding continuous functions. Moreover, there appeared other kinds of series, such as Fourier series, which are basic to the study of waves. As we have seen, Taylor series can be used to provide estimates of such numbers as e, $\ln 2$, and π; yet the most economical estimates of irrational numbers often come from other series and from other considerations. In spite of the limitations of Taylor series, they have had a major impact on the development of calculus, and they continue to be of theoretical interest.

EXERCISES 9.9

In Exercises 1–6 find the Taylor series of f about a. Do not be concerned with whether the series converges to the given function.

1. $f(x) = 4x^2 - 2x + 1$; $a = -3$

2. $f(x) = 5x^3 + 4x^2 + 3x + 2$; $a = 2$

3. $f(x) = e^x$; $a = 2$ **4.** $f(x) = 1/x$; $a = -1$

5. $f(x) = \ln x$; $a = 2$ **6.** $f(x) = \int_0^x \ln(1 + t)\, dt$; $a = 0$

In Exercises 7–22 find the Taylor series of the given function about a. Use the series already obtained in the text or in previous exercises.

7. $f(x) = \sin 2x$; $a = 0$ **8.** $f(x) = \cos x^2$; $a = 0$

9. $f(x) = \ln 3x$; $a = 1$

10. $g(x) = \ln \dfrac{1 + x}{1 - x}$; $a = 0$ (*Hint:* $\ln \dfrac{1 + x}{1 - x} = \ln(1 + x) - \ln(1 - x)$.)

11. $f(x) = x \ln(1 + x^2)$; $a = 0$

12. $g(x) = \sinh x$; $a = 0$ **13.** $k(x) = 2^x$; $a = 0$

14. $f(x) = 10^x$; $a = 0$ **15.** $f(x) = \dfrac{x - 1}{x + 1}$; $a = 1$

16. $f(x) = \sin^2 x$; $a = 0$ (*Hint:* $\sin^2 x = \frac{1}{2}(1 - \cos 2x)$.)

17. $f(x) = \cos^2 x$; $a = 0$ **18.** $f(x) = \dfrac{x - 1}{1 + x^3}$; $a = 0$

19. $f(x) = \begin{cases} \dfrac{\sin x}{x} & \text{for } x \neq 0 \\ 1 & \text{for } x = 0 \end{cases}$ $a = 0$

20. $f(x) = \begin{cases} \dfrac{(\sin x) - x}{x^3} & \text{for } x \neq 0 \\ -\frac{1}{6} & \text{for } x = 0 \end{cases}$ $a = 0$

21. $f(x) = \sin x$; $a = \pi/3$ **22.** $f(x) = \cos x$; $a = \pi/6$

In Exercises 23–30 use a Taylor polynomial and the Lagrange Remainder Formula or (1) in Section 9.7 to approximate the given number with an error less than the value shown.

23. $e^{1/2}$; 0.001   **24.** $e^{-1/2}$; 0.0001

25. $\ln 1.1$; 0.00001   **26.** $\sin \pi/10$; 0.001

27. $\cos(-\pi/7)$; 0.001   **28.** $\tan^{-1}\frac{1}{2}$; 0.001

29. $\ln 2$; 0.01 (*Hint:* Use Exercise 10.)

30. $\ln \frac{3}{2}$; 0.001 (*Hint:* Use Exercise 10.)

In Exercises 31–33 find a positive integer n such that the plot of p_n seems to coincide with the plot of f on a calculator when you use the given windows.

31. $f(x) = e^{-x}$; x window: $[-2, 2]$; y window: $[-8, 8]$

32. $f(x) = \sin x$; x window: $[-3, 3]$; y window: $[-1.5, 1.5]$

33. $f(x) = \cos x$; x window: $[-4, 4]$; y window: $[-1.5, 1.5]$

34. Let $f(x) = \tan^{-1} x$.

     a. Find a positive integer n such that $|p_n(x) - f(x)| < 0.001$ for all x in $[-\frac{1}{2}, \frac{1}{2}]$.

     b. Explain what happens when you try to find a positive integer n for which $|p_n(x) - f(x)| < 0.001$ for all x in $[-2, 2]$. Does the graph of p_n agree with your findings?

35. **a.** Show that the Taylor series of $\cos x$ is $\sum_{n=0}^{\infty} \dfrac{(-1)^n}{(2n)!} x^{2n}$.

     b. Using the ideas in the solution of Example 2, show that the Taylor series of $\cos x$ converges to $\cos x$ for all x.

36. Find an integer n such that the remainder $r_n(7\pi/36)$ of the Taylor series of $\sin x$ about 0 is less than 10^{-5}.

37. Find the Taylor series about 0 of

$$\frac{-3x + 2}{2x^2 - 3x + 1}$$

by using the partial fraction decomposition of the rational function.

*38. Let $f(x) = \tan x$. Using the fact that $f(0) = 0$ and $f'(x) = 1 + [f(x)]^2$, find the sum of the first six terms in the Taylor series of f about 0.

*39. Let $f(x) = e^{(x^2)} \int_0^x e^{-(t^2)}\, dt$.

     a. Show that $f(0) = 0$ and $f'(x) = 2xf(x) + 1$.

     b. Find the Taylor series of f about 0. (*Hint:* Note that $f^{(n)}(x) = 2(n - 1)f^{(n-2)}(x) + 2xf^{(n-1)}(x)$ for $n \geq 2$.)

*40. Show that if there is a number M such that $|f^{(n)}(x)| \leq M$ for all x and for all $n \geq 0$, then the Taylor series of f about any given point a converges to $f(x)$ for all x.

Applications

41. Let f be the position function of a car that is stationary for $t \leq 0$ and is in motion for $t > 0$. Show that f does not have a power series representation on any open interval containing 0.

42. The integral $\displaystyle\int_0^1 \sin \frac{\pi x^2}{2}\, dx$, which is called a **Fresnel integral,** appears in optics in the study of diffraction. Use a Taylor series and the Integration Theorem to approximate the value of the integral with an error less than 10^{-4}.

Projects

1. This project returns to the Fibonacci sequence, which we discussed in the project in Section 9.3. Recall that the Fibonacci sequence begins

$$a_0 = 1, \, a_1 = 1, \, a_2 = 2, \, a_3 = 3, \, a_4 = 5, \, a_5 = 8, \, a_6 = 13$$

and in general, $a_{n+1} = a_n + a_{n-1}$. The inequality $a_{n+1}/a_n < 2$ holds for all n (see the Project in 9.3).

   Now let

$$f(x) = \sum_{n=0}^{\infty} a_n x^n = 1 + x + 2x^2 + 3x^3 + 5x^4 + 8x^5 + \cdots$$

   a. Show that the Taylor series for f converges for $|x| < 1/2$.

   b. Let the power series $\sum_{n=0}^{\infty} b_n x^n$ be defined as the Taylor series of the function $f(x)(1 - x - x^2)$:

$$f(x)(1 - x - x^2) = \sum_{n=0}^{\infty} b_n x^n$$

   Show that $b_0 = 1$ and $b_1 = 0$, and that $b_n = a_n - a_{n-1} - a_{n-2} = 0$ for all $n \geq 2$.

   c. Use (b) to show that $f(x) = -1/(x^2 + x - 1)$ for $|x| < 1/2$.

   d. Let the roots of $x^2 + x - 1 = 0$ be c_1 and c_2. Use partial fractions to write

$$\frac{-1}{x^2 + x - 1} = \frac{A}{x - c_1} + \frac{B}{x - c_2}$$

   for appropriate constants A and B.

   e. Use (c) and (d) to find a single power series about 0 for $f(x)$.

   f. Show that $1/c_1 = (1 + \sqrt{5})/2$ and $1/c_2 = (1 - \sqrt{5})/2$. Using the fact that the Taylor series for $f(x)$ is unique, show that

$$a_n = \frac{1}{\sqrt{5}} \left[\left(\frac{1 + \sqrt{5}}{2} \right)^n - \left(\frac{1 - \sqrt{5}}{2} \right)^n \right] \quad \text{for all } n$$

   Thus we now have an explicit formula for the numbers of the Fibonacci sequence!

2. The main goal of this project is to show that e is irrational. To that end, we first recall that if $f(x) = e^x$, then $e = f(1)$, and

$$f(1) = 1 + 1 + \frac{1}{2!} + \frac{1}{3!} + \cdots + \frac{1}{n!} + r_n(1)$$

   where $r_n(1)$ is the nth Taylor remainder. From (14) we know that $r_n(1) < 3/(n + 1)!$.

   a. Suppose that e is rational, which means that there are integers a and b such that $e = a/b$. Find an integer $n \geq 2$ (depending on b) such that $n!a/b$ is an integer.

   b. Let $n \geq 2$. Show that

$$n! \left[1 + 1 + \frac{1}{2!} + \frac{1}{3!} + \cdots + \frac{1}{n!} + r_n(1) \right]$$

   cannot be an integer.

   c. Using (a) and (b), derive a contradiction. (Thus e cannot be rational. Consequently e is irrational.)

   d. Show that $\sin 1$ is irrational.

9.10   BINOMIAL SERIES

We conclude Chapter 9 by investigating Taylor series about 0 of functions given by

$$f(x) = (1 + x)^s$$

where s is any fixed number. These functions are called **binomial functions,** because they arise from the binomial, or two-term, expression $1 + x$. They had enormous influence on the early growth of calculus.

   We begin by letting $s = \frac{1}{2}$, so that

$$f(x) = (1 + x)^{1/2}$$

In order to derive the Taylor series of f, we first find the derivatives of f:

$$f(x) = (1 + x)^{1/2} \qquad\qquad\qquad\qquad f(0) = 1$$

$$f'(x) = \frac{1}{2}\,(1 + x)^{-1/2} \qquad\qquad\qquad\qquad f'(0) = \frac{1}{2}$$

$$f''(x) = \left(\frac{1}{2}\right)\left(\frac{-1}{2}\right)(1 + x)^{-3/2} \qquad\qquad f''(0) = -\frac{1}{4}$$

$$f^{(3)}(x) = \left(\frac{1}{2}\right)\left(\frac{-1}{2}\right)\left(\frac{-3}{2}\right)(1 + x)^{-5/2} \qquad f^{(3)}(0) = \frac{3}{8}$$

$$f^{(4)}(x) = \left(\frac{1}{2}\right)\left(\frac{-1}{2}\right)\left(\frac{-3}{2}\right)\left(\frac{-5}{2}\right)(1 + x)^{-7/2} \qquad f^{(4)}(0) = -\frac{15}{16}$$

and in general,

$$f^{(n)}(x) = \left(\frac{1}{2}\right)\left(\frac{-1}{2}\right)\left(\frac{-3}{2}\right)\cdots\left(\frac{-2n + 3}{2}\right)(1 + x)^{(-2n + 1)/2} \quad \text{for } n \geq 2$$

$$f^{(n)}(0) = (-1)^{n + 1}\,\frac{1 \cdot 3 \cdot 5 \cdots (2n - 3)}{2^n} \quad \text{for } n \geq 2$$

Consequently the Taylor series of f is

$$1 + \frac{1}{2}x + \sum_{n=2}^{\infty} (-1)^{n + 1}\,\frac{1 \cdot 3 \cdot 5 \cdots (2n - 3)}{2^n n!}\,x^n \tag{1}$$

which can also be written

$$1 + \frac{1}{2}x + \sum_{n=2}^{\infty} (-1)^{n + 1}\,\frac{1 \cdot 3 \cdot 5 \cdots (2n - 3)}{2 \cdot 4 \cdot 6 \cdots (2n)}\,x^n$$

By the Ratio Test you can show that the radius of convergence of the series in (1) is 1 (see Exercise 24 of Section 9.8).

More generally, let s be any number and let $f(x) = (1 + x)^s$. Then we define the **binomial coefficient** $\displaystyle\binom{s}{n}$ by the formulas

$$\binom{s}{0} = 1 \quad \text{and} \quad \binom{s}{n} = \frac{s(s - 1)(s - 2) \cdots (s - (n - 1))}{n!} \quad \text{for } n \geq 1$$

In particular,

$$\binom{s}{1} = s \quad \text{and} \quad \binom{s}{2} = \frac{s(s - 1)}{2}$$

Moreover, if s is a positive integer, then

$$\binom{s}{n} = \frac{s!}{n!(s - n)!} \quad \text{for } 0 \leq n \leq s$$

Using binomial coefficients, we find that the Taylor series of f can be written

$$\sum_{n=0}^{\infty} \binom{s}{n} x^n$$

This series is called a **binomial series.** It is possible (although not easy) to show that

$$(1 + x)^s = \sum_{n=0}^{\infty} \binom{s}{n} x^n \quad \text{for} \begin{cases} -1 < x < 1 \text{ if } s \le -1 \\ -1 < x \le 1 \text{ if } -1 < s < 0 \\ -1 \le x \le 1 \text{ if } s > 0 \text{ and } s \text{ is not an integer} \\ \text{all } x \text{ if } s \text{ is a nonnegative integer} \end{cases} \quad (2)$$

It follows from (2), for example, that

$$(1 + x)^{1/2} = \sum_{n=0}^{\infty} \binom{1/2}{n} x^n$$

$$= 1 + \frac{1}{2} x + \frac{1}{2!}\left(\frac{1}{2}\right)\left(\frac{-1}{2}\right) x^2 + \frac{1}{3!}\left(\frac{1}{2}\right)\left(\frac{-1}{2}\right)\left(\frac{-3}{2}\right) x^3 + \cdots \quad (3)$$

$$(1 + x)^{1/3} = \sum_{n=0}^{\infty} \binom{1/3}{n} x^n$$

$$= 1 + \frac{1}{3} x + \frac{1}{2!}\left(\frac{1}{3}\right)\left(\frac{-2}{3}\right) x^2 + \frac{1}{3!}\left(\frac{1}{3}\right)\left(\frac{-2}{3}\right)\left(\frac{-5}{3}\right) x^3 + \cdots \quad (4)$$

Newton and Mercator discovered the binomial series independently during the mid-1660s. At that time Mercator had already gained recognition as a mathematician, but the binomial series was among Newton's initial successes, achieved while he was a student at Cambridge University. Thereafter binomial series played a key role in Newton's method of differentiating and integrating, and his facility with series led him to profound insights into the subject we now call calculus.

Before turning to approximate values of numbers, we will estimate the error $r_N(x)$ introduced by taking the Nth Taylor polynomial of $(1 + x)^s$, for certain values of s. If $|s| \le 1$ and $n \ge 1$, then

$$\left| \binom{s}{n} \right| \le \left| \binom{s}{n-1} \right|$$

as you can verify by writing out the binomial coefficients involved and comparing them (see Exercise 19). Then the Nth Taylor remainder $r_N(x)$ satisfies

$$| r_N(x) | \le \sum_{n=N+1}^{\infty} \left| \binom{s}{n} \right| |x|^n \le \sum_{n=N+1}^{\infty} \left| \binom{s}{1} \right| |x|^n = |s| \frac{|x|^{N+1}}{1 - |x|} \quad (5)$$

This estimate of the remainder allows us to approximate various roots of numbers by means of the Taylor polynomials of binomial series.

EXAMPLE 1   Find an approximate value of $\sqrt[3]{28}$ with an error less than 0.0001.

Solution   We begin by writing $\sqrt[3]{28}$ as

$$b\sqrt[3]{1 + x} = b(1 + x)^{1/3}$$

where b is a convenient number and $|x|$ is small:

$$\sqrt[3]{28} = \sqrt[3]{27 + 1} = \sqrt[3]{27}\sqrt[3]{1 + \tfrac{1}{27}} = 3\sqrt[3]{1 + \tfrac{1}{27}}$$

This means that if we find an approximation A to $\sqrt[3]{1 + \tfrac{1}{27}}$ with an error less than, say, 0.00003, then by taking $3A$ for $\sqrt[3]{28}$ we will introduce an error of no more than $3(0.00003) < 0.0001$. From (5) with $s = \tfrac{1}{3}$ and $x = \tfrac{1}{27}$ we obtain

$$|r_N(\tfrac{1}{27})| \le \frac{1}{3} \frac{(1/27)^{N+1}}{26/27} = \frac{1}{78(27)^N}$$

so that $|r_N(\tfrac{1}{27})| < 0.00003$ if $N \ge 2$. Consequently the desired approximation is obtained by taking the first three terms of the series in (4), with $x = \tfrac{1}{27}$:

$$\sqrt[3]{28} = 3\sqrt[3]{1 + \frac{1}{27}} \approx 3\left[1 + \left(\frac{1}{3}\right)\left(\frac{1}{27}\right) + \left(\frac{1}{2!}\right)\left(\frac{1}{3}\right)\left(\frac{-2}{3}\right)\left(\frac{1}{27}\right)^2\right]$$

$$\approx 3.03657979 \quad \square$$

If instead we had desired an estimate of $\sqrt[6]{60}$, we would have rewritten it

$$\sqrt[6]{60} = \sqrt[6]{64\left(1 - \frac{4}{64}\right)} = \sqrt[6]{64}\,\sqrt[6]{1 - \frac{1}{16}} = 2\sqrt[6]{1 - \frac{1}{16}}$$

In this case $\sqrt[6]{60}$ has the form $b\sqrt[6]{1 + x}$, where $b = 2$ and $x = -\tfrac{1}{16}$.

Recall that we are not able to obtain an exact numerical value for $\int_0^1 \sqrt{1 + x^4}\,dx$. However, in Chapter 5 we approximated its value by Riemann sums, and in Example 5 of Section 8.6 we approximated its value by Simpson's Rule. Now that binomial series are available, we can write $\sqrt{1 + x^4}$ as a binomial series and then use the Integration Theorem to once again approximate the integral. As Exercise 21 shows, this method appears to be less efficient in giving good estimates of the integral than Simpson's Rule is.

EXERCISES 9.10

In Exercises 1–6 use a binomial series to approximate the number with an error less than 0.001.

1. $\sqrt{1.05}$ **2.** $\sqrt[3]{9}$ **3.** $\sqrt[4]{83}$

4. $\sqrt[5]{35}$ **5.** $\sqrt[6]{65}$ **6.** $29^{2/3}$

In Exercises 7–14 find the Taylor series of f about a, and write out the first four terms of the series.

7. $f(x) = \dfrac{1}{\sqrt{1 + x}}$; $a = 0$ **8.** $f(x) = (1 + x^2)^{1/3}$; $a = 0$

9. $f(x) = (1 + x)^{-8/5}$; $a = 0$ **10.** $f(x) = \dfrac{1}{\sqrt{1 - x^2}}$; $a = 0$

11. $f(x) = \dfrac{x}{\sqrt{1 - x^2}}$; $a = 0$ **12.** $f(x) = (1 - x^2)^{5/2}$; $a = 0$

13. $f(x) = \sqrt{1 - (x + 1)^2}$; $a = -1$

14. $f(x) = \sqrt{2x - x^2}$; $a = 1$

15. Use (3) with x replaced by $-x^2$, along with the Integration Theorem, to show that

$$\sum_{n=0}^{\infty} \binom{1/2}{n} \frac{(-1)^n}{2n + 1} = \frac{\pi}{4}$$

16. Using the Integration Theorem and the series derived in Exercise 10, show that for $-1 < x < 1$,

$$\sin^{-1} x = x + \frac{1}{2 \cdot 3}x^3 + \frac{1 \cdot 3}{2 \cdot 4 \cdot 5}x^5 + \frac{1 \cdot 3 \cdot 5}{2 \cdot 4 \cdot 6 \cdot 7}x^7 + \cdots$$

$$= x + \sum_{n=1}^{\infty} \frac{1 \cdot 3 \cdot 5 \cdots (2n - 1)}{[2 \cdot 4 \cdot 6 \cdots (2n)](2n + 1)}x^{2n+1}$$

17. Newton found the following integrals by first obtaining power series representations for the integrands and then integrating the power series term by term. In each part below, use Newton's ideas to carry out the integration, and then determine the radius of convergence of the resulting power series. Take $a > 0$.

a. $\displaystyle\int_0^x \sqrt{a^2 + t^2}\,dt$ **b.** $\displaystyle\int_0^x \sqrt{a^2 - t^2}\,dt$

18. a. Find the numerical values of $\dbinom{6}{2}$ and $\dbinom{1/2}{4}$.

b. Show that

$$\sum_{n=0}^{s} \binom{s}{n} = 2^s$$

for any positive integer s. (*Hint:* Use (2).)

c. Show that

$$\sum_{n=0}^{s} \binom{s}{n} (-1)^n = 0$$

for any positive integer s. (*Hint:* Use (2).)

19. Prove that if $|s| \le 1$ and $n \ge 1$, then

$$\left| \binom{s}{n} \right| \le \left| \binom{s}{n-1} \right|$$

20. The first partial sum $1 + x/2$ of the binomial series for $(1 + x)^{1/2}$ is often used as an approximation of $(1 + x)^{1/2}$. Using the Lagrange form of the remainder $r_1(x)$, show that the error introduced is at most $(0.172)x^2$, provided that $|x| < 0.19$.

21. a. Use a binomial series and the Integration Theorem to show that

$$\int_0^1 \sqrt{1 + x^4}\, dx = \sum_{n=0}^{\infty} \binom{1/2}{n} \frac{1}{4n + 1}$$

b. Find the minimum positive integer j such that the jth partial sum of the series in (a) is guaranteed by the Alternating Series Test to be within 10^{-3} of the exact value of the integral. Then find the exact value. (The result, along with the result of Example 5 of Section 8.6, indicates that use of the binomial series is not as efficient in approximating the integral in (a) as Simpson's Rule is.)

Applications

22. When a particle of mass m moves with a high velocity v, the theory of relativity implies that its kinetic energy is given by

$$K = mc^2 \left(\frac{1}{\sqrt{1 - v^2/c^2}} - 1 \right)$$

where c is the speed of light. Using (2), show that when the ratio v/c is small, then K is approximately equal to the usual "Newtonian" kinetic energy $\frac{1}{2}mv^2$. (Thus the relativistic kinetic energy reduces to the Newtonian kinetic energy when the velocity is small.)

23. A tunnel 200 miles long connects two points on the earth's surface. Assuming that the earth's radius is 4000 miles and using Exercise 20, approximate the maximum depth of the tunnel.

24. Suppose a cable hangs from the point A to the origin, as in Figure 9.25, and has a uniformly distributed load of p pounds per foot. Suppose q is the tension force at 0. Assume furthermore that the x coordinate of A is x. Then the cable traces out the parabolic curve given by

$$f(t) = \frac{pt^2}{2q} \quad \text{for } 0 \le t \le x$$

so that in particular,

$$A = \left(x, \frac{px^2}{2q} \right)$$

a. Show that the length L of the cable is given by

$$L = \int_0^x \sqrt{1 + (f'(t))^2}\, dt$$

$$= \sum_{n=0}^{\infty} \binom{1/2}{n} \frac{1}{2n + 1} \frac{p^{2n}}{q^{2n}} x^{2n+1}$$

b. Determine the radius of convergence of the series given in (a).

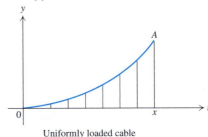

Uniformly loaded cable

FIGURE 9.25 Figure for Exercise 24.

25. Suppose a suspension bridge has span a and sag b (Figure 9.26). Then it can be shown that the length L of the supporting cables is given by

$$L = 2 \int_0^{a/2} \left(1 + \frac{64b^2}{a^4} x^2 \right)^{1/2} dx$$

Using the binomial series and the Integration Theorem, show that L is given approximately by

$$L \approx a \left[1 + \frac{8}{3} \left(\frac{b^2}{a^2} \right) - \frac{32}{5} \left(\frac{b^4}{a^4} \right) \right]$$

and use this formula to approximate the length of a cable on a bridge having a span of 500 feet and a sag of 40 feet.

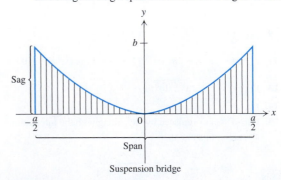

Suspension bridge

FIGURE 9.26 Figure for Exercise 25.

Summary of Convergence and Divergence Tests for Series

Test	Statement of test	Comments
*n*th Term Test	If $\lim_{n \to \infty} a_n \neq 0$, then $\sum_{n=1}^{\infty} a_n$ diverges.	It can show divergence but not convergence.
Comparison Test	If $\|a_n\| \leq b_n$ for $n \geq 1$ and if $\sum_{n=1}^{\infty} b_n$ converges, then $\sum_{n=1}^{\infty} a_n$ converges (absolutely). If $a_n \geq b_n \geq 0$ for $n \geq 1$ and if $\sum_{n=1}^{\infty} b_n$ diverges, then $\sum_{n=1}^{\infty} a_n$ diverges.	Usually $\sum_{n=1}^{\infty} b_n$ is a series whose convergence or divergence is well known.
Limit Comparison Test	Let $\lim_{n \to \infty} \|a_n\|/b_n = L$, where L and b_n are positive numbers. If $\sum_{n=1}^{\infty} b_n$ converges, then $\sum_{n=1}^{\infty} a_n$ converges (absolutely). If $a_n \geq 0$ and $\sum_{n=1}^{\infty} b_n$ diverges, then $\sum_{n=1}^{\infty} a_n$ diverges.	Usually $\sum_{n=1}^{\infty} b_n$ is a series whose convergence or divergence is well known.
Integral Test	Suppose $f(n) = a_n$ for $n \geq 1$ and f is continuous, decreasing, and positive on $[1, \infty)$. Then $\sum_{n=1}^{\infty} a_n$ converges if and only if $\int_1^{\infty} f(x)\, dx$ converges.	The test is most effective if f is easy to integrate.
Ratio Test	Let $\lim_{n \to \infty} \left\| \dfrac{a_{n+1}}{a_n} \right\| = r$ (possibly ∞). If $r < 1$, then $\sum_{n=1}^{\infty} a_n$ converges (absolutely). If $r > 1$, then $\sum_{n=1}^{\infty} a_n$ diverges.	The test is most effective if a_n involves factorials or *n*th powers. If $r = 1$, the test yields no conclusion.
Root Test	Let $\lim_{n \to \infty} \sqrt[n]{\|a_n\|} = r$ (possibly ∞). If $r < 1$, then $\sum_{n=1}^{\infty} a_n$ converges (absolutely). If $r > 1$, then $\sum_{n=1}^{\infty} a_n$ diverges.	The test is most effective if a_n involves *n*th powers. If $r = 1$, the test yields no conclusion.
Alternating Series Test	If $\{a_n\}_{n=1}^{\infty}$ is decreasing and positive and if $\lim_{n \to \infty} a_n = 0$, then $\sum_{n=1}^{\infty} (-1)^n a_n$ and $\sum_{n=1}^{\infty} (-1)^{n+1} a_n$ both converge.	

REVIEW

Key Terms and Expressions

Sequence
Convergent sequence; divergent sequence
Increasing sequence; decreasing sequence
Bounded sequence
Series; sum of a series; convergent series;
   divergent series
Partial sum
Geometric series

Positive series
p series
Absolute convergence; conditional convergence
Power series
Radius of convergence; interval of convergence
Taylor series; nth Taylor polynomial;
   nth Taylor remainder
Binomial series; binomial coefficient

Key Formulas

$$\sum_{n=m}^{\infty} cr^n = \frac{cr^m}{1-r}$$

$$e^x = \sum_{n=0}^{\infty} \frac{x^n}{n!}$$

$$\sin x = \sum_{n=0}^{\infty} \frac{(-1)^n}{(2n+1)!} x^{2n+1}$$

$$\cos x = \sum_{n=0}^{\infty} \frac{(-1)^n}{(2n)!} x^{2n}$$

$$\ln(1+x) = \sum_{n=0}^{\infty} \frac{(-1)^n}{n+1} x^{n+1}$$

$$\frac{1}{1-x} = \sum_{n=0}^{\infty} x^n$$

$$\tan^{-1} x = \sum_{n=0}^{\infty} \frac{(-1)^n}{2n+1} x^{2n+1}$$

$$(1+x)^s = \sum_{n=0}^{\infty} \binom{s}{n} x^n$$

$$f(x) = \sum_{n=0}^{\infty} \frac{f^{(n)}(a)}{n!} (x-a)^n$$

$$r_n(x) = \frac{f^{(n+1)}(t_x)}{(n+1)!} (x-a)^{n+1}$$

Key Theorems

Differentiation Theorem for Power Series
Integration Theorem for Power Series
Taylor's Theorem

Review Exercises

In Exercises 1–6 find the limit.

1. $\displaystyle\lim_{n\to\infty} \frac{n^2 - \sqrt{n}}{4 - n^2}$

2. $\displaystyle\lim_{n\to\infty} \frac{\sqrt{n}}{\ln n}$

3. $\displaystyle\lim_{n\to\infty} \left(1 + \frac{e}{n}\right)^n$

4. $\displaystyle\lim_{k\to\infty} \frac{(2k)!}{2^k k^k}$

5. $\displaystyle\lim_{n\to\infty} (\sqrt{n^2 + n} - \sqrt{n^2 - n})$

6. $\displaystyle\lim_{n\to\infty} \left(\frac{1^3}{n^4} + \frac{2^3}{n^4} + \cdots + \frac{n^3}{n^4}\right)$

   (*Hint:* The sum is a Riemann sum.)

In Exercises 7–16 determine whether the series converges or diverges.

7. $\displaystyle\sum_{n=1}^{\infty} \frac{1}{(n + n\sqrt{n})}$

8. $\displaystyle\sum_{n=1}^{\infty} \sin^2 \frac{1}{n}$

9. $\displaystyle\sum_{n=1}^{\infty} \frac{\sqrt{n}}{n^2 + n + 1}$

10. $\displaystyle\sum_{n=1}^{\infty} n^2 e^{-n/2}$

11. $\displaystyle\sum_{n=2}^{\infty} \frac{6^n}{n^2 (\ln n)^2}$

12. $\displaystyle\sum_{n=2}^{\infty} \frac{(\ln n)^4}{n}$

13. $\displaystyle\sum_{n=4}^{\infty} (-1)^n \frac{\sqrt{n}}{n-3}$

14. $\displaystyle\sum_{n=1}^{\infty} \left(\frac{n^n}{n!}\right)^n$

15. $\displaystyle\sum_{n=1}^{\infty} \frac{\sqrt{n^2+1}-\sqrt{n^2-1}}{n}$

16. $\displaystyle\sum_{n=0}^{\infty} \frac{(2n)!}{2^n n!}$

17. Express the repeating decimal $27.1318318318\ldots$ as a fraction.

18. Suppose $\sum_{n=1}^{\infty} a_n = 5$ and let $b_n = a_1 + a_2 + \cdots + a_n$ for $n \geq 1$.

   a. What is $\lim_{n\to\infty} a_n$?

   b. What is $\lim_{n\to\infty} b_n$?

19. Suppose $\{a_n\}_{n=1}^{\infty}$ is a sequence with $0 < a_n < 1/n$.

   a. Does $\lim_{n\to\infty} a_n$ exist? Explain your answer.

   b. With the information we have about $\{a_n\}_{n=1}^{\infty}$, is it possible to determine whether $\sum_{n=1}^{\infty} a_n$ converges? Explain your answer.

20. Suppose $\{b_n\}_{n=1}^{\infty}$ is a sequence with $1 < b_n < 1 + 1/n$.

   a. What is $\lim_{n\to\infty} b_n$?

   b. With the information we have about $\{b_n\}_{n=1}^{\infty}$, is it possible to determine whether $\sum_{n=1}^{\infty} b_n$ converges?

*21. Let $0 < a_1 < b_1$. Let the sequences $\{a_n\}_{n=1}^{\infty}$ and $\{b_n\}_{n=1}^{\infty}$ be defined by the equations

$$a_{n+1} = \sqrt{a_n b_n} \quad \text{and} \quad b_{n+1} = \frac{1}{2}(a_n + b_n)$$

   a. Show that $a_n < b_n$ for all n. (*Hint:* Use induction.)

   b. Show that $\{a_n\}_{n=1}^{\infty}$ is increasing and $\{b_n\}_{n=1}^{\infty}$ is decreasing.

   c. Show that $\{a_n\}_{n=1}^{\infty}$ converges to a limit L, and $\{b_n\}_{n=1}^{\infty}$ converges to a limit M.

   d. By using the formula $b_{n+1} = \frac{1}{2}(a_n + b_n)$, show that $L = M$. (The number L is called the **arithmetic-geometric mean** of a_1 and b_1.)

   e. Let $a_1 = 1$ and $b_1 = 2$. Approximate L to within 0.01.

22. Using the fact that

$$\ln 2 = \sum_{n=1}^{\infty} (-1)^{n+1} \frac{1}{n} = 1 - \frac{1}{2} + \frac{1}{3} - \frac{1}{4} + \cdots$$

and grouping adjacent pairs of terms, show that

$$\ln 2 = \sum_{n=0}^{\infty} \frac{1}{(2n+1)(2n+2)}$$

$$= \frac{1}{1 \cdot 2} + \frac{1}{3 \cdot 4} + \frac{1}{5 \cdot 6} + \cdots$$

23. Using the fact that

$$\sum_{n=0}^{\infty} (-1)^n \frac{1}{2n+1} = \pi/4$$

determine the sum of

$$\sum_{n=5}^{\infty} (-1)^n \frac{1}{2n+1}$$

24. Show that

$$\sum_{n=0}^{\infty} (-1)^n \frac{2n+3}{(n+1)(n+2)} = 1.$$

$$\left(\text{Hint: } \frac{2n+3}{(n+1)(n+2)} = \frac{1}{n+1} + \frac{1}{n+2}.\right)$$

25. Show that

$$(1-x) \sum_{n=1}^{\infty} n x^n = \frac{x}{1-x} \quad \text{for } |x| < 1$$

26. What is wrong with the following argument? Since

$$\sum_{n=1}^{\infty} x^n = \frac{x}{1-x}$$

if we let $x = -1$, then we obtain

$$\sum_{n=1}^{\infty} (-1)^n = \frac{-1}{2}$$

*27. a. Let $\{a_n\}_{n=1}^{\infty}$ and $\{b_n\}_{n=1}^{\infty}$ be two convergent sequences such that $a_n \leq b_n$ for each n. Show that

$$\lim_{n\to\infty} a_n \leq \lim_{n\to\infty} b_n$$

   b. Prove that if $\sum_{n=1}^{\infty} a_n$ converges absolutely, then

$$\left| \sum_{n=1}^{\infty} a_n \right| \leq \sum_{n=1}^{\infty} |a_n|$$

   (*Hint:* Apply part (a) to the partial sums of $\sum_{n=1}^{\infty} a_n$ and $\sum_{n=1}^{\infty} |a_n|$, and then to the partial sums of $\sum_{n=1}^{\infty} (-a_n)$ and $\sum_{n=1}^{\infty} |a_n|$, and combine the results.)

28. Find the interval of convergence of

$$\sum_{n=2}^{\infty} (-1)^n \frac{(\ln n)^2}{n^2} x^n$$

29. Find the interval of convergence of

$$\sum_{n=0}^{\infty} \frac{3^n}{5^{2n}} x^{3n}$$

30. Find the radius of convergence of

$$\sum_{n=0}^{\infty} \frac{(n!)^2}{(2n)!} x^{2n}$$

31. Find the radius of convergence of

$$\sum_{n=1}^{\infty} \frac{n^{2n}}{(2n)!} x^n$$

32. Find the third Taylor polynomial of $\sec x$ about $\pi/6$.

33. Let $f(x) = \sqrt{1 + x^4}$. Find the second Taylor polynomial of f about 0.

34. Let $f(x) = x^6 - 3x^4 + 2x - 1$.

    a. Find the fifth Taylor polynomial of f about 0.

    b. Find the fourth Taylor polynomial of f about -1.

    c. Find the Taylor series of f about -1.

In Exercises 35–37 find the Taylor series of f about a.

35. $f(x) = \sin x$; $a = \pi/4$

36. $f(x) = \cos 2x$; $a = -\pi/6$

37. $f(x) = \dfrac{x - 1}{x + 1}$; $a = 0$

In Exercises 38–40 use Taylor polynomials to approximate the number with an error less than 0.001.

38. $\sqrt{95}$ **39.** $\sqrt[4]{17}$ **40.** $e^{-1/3}$

41. a. Using the Taylor series for $\tan^{-1} x$, evaluate

$$\lim_{x \to 0} \frac{\tan^{-1} x - x}{x^3}$$

    b. Evaluate the limit in (a) without using Taylor series.

42. Suppose a square is inscribed in a circle of radius 1, and thereafter circles and squares are alternately inscribed in one another, as suggested by Figure 9.27. Determine the area A of the shaded region that results.

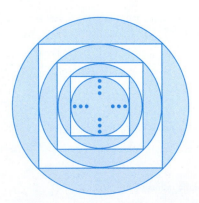

FIGURE 9.27 Figure for Exercise 42.

43. Suppose $\sum_{n=1}^{\infty} a_n$ is a series such that the jth partial sum $s_j = 2 - 2/j$, for each positive integer j.

    a. Find $\sum_{n=1}^{25} a_n$.

b. Determine whether the series converges or diverges, and if it converges, what its sum is.

c. Find $\lim_{n \to \infty} a_n$.

d. Find a formula for a_n in terms of n.

44. The **Bessel function J_0 of order zero** is defined by

$$J_0(x) = \sum_{n=0}^{\infty} (-1)^n \frac{1}{4^n (n!)^2} x^{2n}$$

    a. Show that the series converges for all x.

    b. Show that $xJ_0''(x) + J_0'(x) + xJ_0(x) = 0$ for all x.

45. Using (1) of Section 9.8 with x replaced by t^2, along with the Integration Theorem for Power Series and the formula

$$\tanh^{-1} x = \int_0^x \frac{1}{1 - t^2}\, dt \quad \text{for } -1 < x < 1$$

show that

$$\tanh^{-1} x = \sum_{n=0}^{\infty} \frac{1}{2n + 1} x^{2n + 1}$$

46. Using (9) of Section 7.4, a binomial series, and the Integration Theorem for Power Series, show that

$$\sinh^{-1} x = \sum_{n=0}^{\infty} \binom{-1/2}{n} \frac{x^{2n + 1}}{2n + 1} = x - \frac{1}{2}\left(\frac{1}{3}\right) x^3$$

$$+ \frac{1 \cdot 3}{2 \cdot 4}\left(\frac{1}{5}\right) x^5 - \frac{1 \cdot 3 \cdot 5}{2 \cdot 4 \cdot 6}\left(\frac{1}{7}\right) x^7 + \cdots$$

$$= x + \sum_{n=1}^{\infty} (-1)^n \frac{1 \cdot 3 \cdot 5 \cdots (2n - 1)}{2 \cdot 4 \cdot 6 \cdots (2n)} \frac{1}{2n + 1} x^{2n + 1}$$

47. Estimate $\int_0^{1/4} 1/(1 + x^{3/2})\, dx$ with an error less than 0.001 by using the Taylor series for $1/(1 - x)$. (*Hint:* Replace x by $-x^{3/2}$ in the Taylor series.)

Application

48. The **light-gathering power** of a reflecting telescope, which refers to the amount of light that a given light source can reflect to the mirror focus, is proportional to the area of the parabolic reflecting surface. Suppose that the surface is obtained by rotating part of the graph of $y = \sqrt{4cx}$ about the x axis, where c is a constant related to the shape of the mirror, and let d be the diameter of the mirror, as in Figure 9.28.

    a. Show that the area S of the reflecting surface of the telescope is given by

$$S = \frac{8\pi}{3} c^2 \left[\left(\frac{d^2}{16c^2} + 1 \right)^{3/2} - 1 \right]$$

b. For many reflecting telescopes, $d^2/16c^2$ is rather small. Using the first two terms of the binomial series for $(1 + x)^{3/2}$, show that the surface area is approximately $\pi d^2/4$, which is an expression found in many astronomy books.

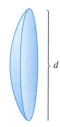

FIGURE 9.28   Figure for Exercise 48.

Topics for Discussion

1. In your own words, explain the difference between a sequence and a series. How are sequences and series related to each other?

2. Explain how results about positive series are used in order to obtain results about series that are not positive.

3. The sum of series may be easy or hard to estimate numerically. Discuss varieties of convergent series for which it is easy, and provide an example for which it is hard. Discuss the role of the truncation error in approximations of sums of convergent series.

4. What is an alternating series? What special results are there about the convergence of an alternating series and the error bounds of such a series?

5. What are the radius of convergence and the interval of convergence of a power series? What effects, if any, do differentiation and integration have on the radius and the interval of convergence of a given series?

6. How can one find out if the Taylor polynomials of a given function f provide a good approximation to the values of f near 0? Give an example of a function whose Taylor polynomials give a good approximation near 0, and one whose Taylor polynomials do not give a good approximation near 0.

7. Discuss ways in which polynomial approximation of functions can be applied.

Cumulative Review, Chapters 1–8

In Exercises 1–2 evaluate the limit.

1. $\displaystyle\lim_{x \to 0^-} \frac{\cos x + \frac{1}{2}x^2 - 1}{x^5}$ 2. $\displaystyle\lim_{x \to \infty} \frac{\ln (1 + \sin^2 (e^x))}{\sqrt{x}}$

3. Let $f(x) = 3^x/x^3$. Find $\lim_{x \to 0^+} f(x)$.

4. Find $\dfrac{d}{dx} (\tan^{-1} \sqrt{x - 1})$.

5. Let $f(x) = \ln (\ln (3x - 4))$.
   a. Find the domain of f.
   b. Find $f'(x)$.

6. Let $f(x) = x^5 - 3x^3$. For what value(s) of x in $[-1, 1]$ is the slope of the line tangent to the graph of f at $(x, f(x))$ greatest?

7. Let $e^x + e^{2x} - \frac{1}{3}y^3 + y = 10$. Suppose there is a point (a, b) on the graph of the equation at which $\dfrac{dy}{dx} (1 - y^2) = -6$. Find the value of a.

8. Let $f(x) = x^2 e^{-x}$. Sketch the graph of f, indicating all pertinent information.

9. A stone dropped into a still pond sends out a circular ripple whose enclosed area increases at a constant rate of $\frac{1}{2}$ square foot per second. How rapidly is the radius of the ripple increasing when the radius is 1 foot?

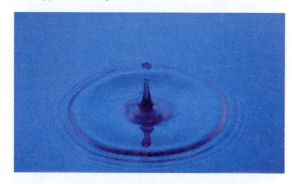

Ripple effect due to waterdrop. (*Yoav Levy/Phototake, Inc.*)

10. A circular cylinder is to be generated by revolving a rectangle with a given perimeter p around one of its sides. Find the largest possible volume of the circular cylinder.

In Exercises 11–13 evaluate the integral.

11. $\displaystyle\int \frac{\cos x}{e^{3x}} dx$ 12. $\displaystyle\int_{\pi/3}^{\pi/2} \sin x \cos^2 \frac{x}{2} dx$

13. $\displaystyle\int_{2\sqrt{2}}^{4} \frac{\sqrt{16 - x^2}}{x^3}\,dx$

14. Find the area A of the portion of the region between the graphs of

$$y = \left(\frac{x}{x - 1}\right)^2 \quad \text{and} \quad y = \left(\frac{x}{x + 2}\right)^2$$

that lies in the second quadrant.

In Exercises 15–16 determine whether the improper integral converges. If it converges, evaluate the integral.

15. $\displaystyle\int_{-\infty}^{\infty} \frac{e^x}{(1 + e^x)^2}\,dx$ **16.** $\displaystyle\int_{1/e}^{e} \frac{1}{x(\ln x)^3}\,dx$

17. Let R be the region between the graphs of $y = 1 + e^x$ and $y = 1 - e^{-x}$ on $[0, \ln 2]$. Find the center of gravity of R.

18. Let D be the solid whose base lies on the xy plane and is bounded by the graphs of $y = 2x^2$ and $y = 3 - x^2$, and whose cross sections perpendicular to the x axis are square. Find the volume V of D.

19. The vertical cross sections of a vat have the shape of the graph of $y = x^2$ for $-2 \leq x \leq 2$. The vat has a horizontal length of 10 feet and is full of water. Find the work W required to pump the water to a level 4 feet above the top of the vat.

The San Francisco Bay Bridge
(*Matt Lambert/Tony Stone Images*)

10 POLAR COORDINATES AND CONIC SECTIONS

FIGURE 10.1   The Mandelbrot set.

Until now we have associated points in the plane only with their rectangular coordinates x and y. In this vein, we have discussed three ways of representing a curve:

a.   as the graph of a function, such as $y = x^2$

b.   as the graph of an equation, such as $x^2 + y^2 = 1$

c.   as the graph of parametric equations, such as $x = t - \sin t$ and $y = 1 - \cos t$

However, there is another coordinate system, called the polar coordinate system, that gives us a good way to describe many curves in the plane. Some curves, such as a circle, can be easily described by both rectangular and polar coordinates. Other curves, such as heart-like cardioids, can readily be described only in polar coordinates. A cardioid appears as the large black body in the middle of the so-called Mandelbrot set shown in Figure 10.1. The Mandelbrot set has become very popular with the advent of computer graphics. (See Project 2 in Section 10.1 for a discussion of the Mandelbrot set.)

In Sections 10.1 and 10.2 we discuss the most common figures described by polar coordinates. We also provide formulas for lengths of curves described in

661

polar coordinates and for areas associated with such curves. The remainder of the chapter—Sections 10.3 to 10.5—is devoted to conic sections: the parabola, the ellipse, and the hyperbola. They play an integral role in studies of flight, orbits of planets, and acoustics.

10.1   THE POLAR COORDINATE SYSTEM

Polar Coordinates
Credit for introducing polar coordinates is generally given to Jakob Bernoulli (1654–1705), elder brother of Johann Bernoulli. Both distinguished mathematicians, they were often bitter rivals.

We begin with a Cartesian coordinate system in the plane. For any point P other than the origin, let r be the distance between P and the origin, and θ an angle having its initial side on the positive x axis and its terminal side on the line segment joining P and the origin (Figure 10.2(a)). Then the pair (r, θ) is called a **set of polar coordinates** for the point P. For each P there are infinitely many possible choices of θ, any two differing by a multiple of 2π. For convenience we also let $(-r, \theta + \pi)$ be a set of polar coordinates for the point P whenever (r, θ) is. (You may think of $(-r, \theta + \pi)$ for $r > 0$ as corresponding to a point reached by moving a distance r in the direction opposite that of the angle $\theta + \pi$ (Figure 10.2(b)).) If (r, θ) is one set of polar coordinates for P, then any other set will be of the form

$$(r, \theta + 2n\pi) \quad \text{or} \quad (-r, \theta + (2n + 1)\pi) \quad \text{for } n \text{ any integer} \tag{1}$$

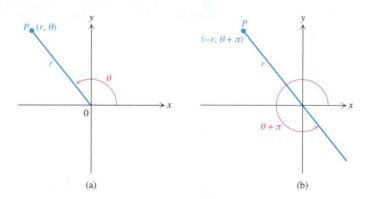

FIGURE 10.2

For example, the point that lies 1 unit from the origin on the positive y axis has polar coordinates $(1, \pi/2)$; by virtue of (1) it also has, among others, polar coordinates $(1, 5\pi/2)$ and $(-1, \pi/2 + \pi) = (-1, 3\pi/2)$ (Figure 10.3). But despite the unlimited number of sets of polar coordinates for any point P other than the origin, we stress that P has *only one* set of polar coordinates (r, θ) such that $r > 0$ and $0 \leq \theta < 2\pi$. Finally, we assign the origin (sometimes called the **pole**) the polar coordinates $(0, \theta)$, where θ may be any number. Figure 10.4 shows several points in the plane, along with some of their polar coordinates.

Since we use the expression (a, b) to denote both Cartesian and polar coordinates, we will explicitly state which kind of coordinates are involved where there is danger of confusion.

Although most maps use Cartesian coordinates, maps in polar coordinates do exist. In addition, the dance a bee performs to communicate the location of a source of food seems to be related to polar coordinates. The orientation of the bee's body

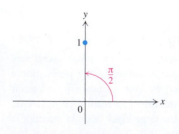

FIGURE 10.3   Polar coordinates $(1, \frac{\pi}{2})$ or $(1, \frac{5\pi}{2})$ or $(-1, \frac{3\pi}{2})$.

locates the direction of the food, and the intensity of the dance indicates the distance to the source (Figure 10.5).

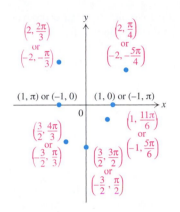

FIGURE 10.4   Polar coordinates of some points.

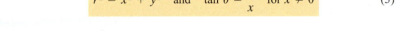

FIGURE 10.5   A bee's dance indicates the location of a food source.

Conversion between Cartesian and Polar Coordinates

Every point in the plane has both Cartesian and polar coordinates. We will now see that it is possible to convert from each type of coordinates to the other. Suppose a point P in the plane has polar coordinates (r, θ) and Cartesian coordinates (x, y). Then from the definition of the sine and the cosine we deduce that

$$x = r \cos \theta \quad \text{and} \quad y = r \sin \theta \tag{2}$$

for all values of r and θ (Figure 10.6). From (2) it follows that x and y are uniquely determined by r and θ and we also obtain formulas for r^2 and $\tan \theta$:

$$r^2 = x^2 + y^2 \quad \text{and} \quad \tan \theta = \frac{y}{x} \quad \text{for } x \neq 0 \tag{3}$$

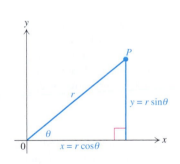

FIGURE 10.6

EXAMPLE 1   Find the Cartesian coordinates of the point P having polar coordinates $(3, 23\pi/6)$.

Solution   From (2) we obtain

$$x = 3 \cos \frac{23\pi}{6} = 3 \cos \left(\frac{-\pi}{6} \right) = 3 \cos \frac{\pi}{6} = \frac{3\sqrt{3}}{2}$$

and

$$y = 3 \sin \frac{23\pi}{6} = 3 \sin \left(\frac{-\pi}{6} \right) = -3 \sin \frac{\pi}{6} = -\frac{3}{2}$$

Therefore $(3\sqrt{3}/2, -\frac{3}{2})$ are the Cartesian coordinates of P.   ❑

Although we cannot determine r and θ uniquely from (x, y) merely by applying (3), it is possible to determine all sets of polar coordinates for a given point (x, y) in Cartesian coordinates.

EXAMPLE 2   Find all sets of polar coordinates for the point P having Cartesian coordinates $(-5, 5\sqrt{3})$.

Solution   First we find the polar coordinates (r, θ) for P such that $r > 0$ and $0 \leq \theta < 2\pi$. From (3) we know that

$$r^2 = (-5)^2 + (5\sqrt{3})^2 = 25 + 75 = 100$$

Therefore $r = 10$. We know also by (3) that

$$\tan \theta = \frac{5\sqrt{3}}{-5} = -\sqrt{3} \tag{4}$$

From (4) and the fact that $(-5, 5\sqrt{3})$ lies in the second quadrant, it follows that $\theta = 2\pi/3$. Thus one set of polar coordinates is $(10, 2\pi/3)$. Consequently by (1) any set of polar coordinates of P must be of the form

$$\left(10, \frac{2\pi}{3} + 2n\pi\right) \quad \text{or} \quad \left(-10, \frac{5\pi}{3} + 2n\pi\right), \text{ where } n \text{ is an integer} \quad \square$$

Caution: Suppose that in the equations

$$r^2 = x^2 + y^2 \quad \text{and} \quad \tan \theta = \frac{y}{x}$$

we let $x = -5$ and $y = 5\sqrt{3}$. The equations become $r^2 = 100$ and $\tan \theta = -\sqrt{3}$. Notice that $r = 10$ and $\theta = -\pi/3$ satisfy these equations. Nevertheless $(10, -\pi/3)$ is not a set of polar coordinates for the point $(-5, 5\sqrt{3})$. The reason is simple: The original point $(-5, 5\sqrt{3})$ lies in the second quadrant, but the point with polar coordinates $(10, -\pi/3)$ lies in the fourth quadrant. Therefore when we convert from Cartesian to polar coordinates, it is not enough merely to choose r and θ to satisfy (3). We must be sure as well that the point with polar coordinates (r, θ) lies in the correct quadrant.

Polar Equations and Graphs

Just as we can graph an equation involving the Cartesian coordinates x and y, we can also graph an equation involving the polar coordinates r and θ. The **polar graph** of such an equation is defined to be the collection of all points in the plane having a set of polar coordinates (r, θ) that satisfies the equation. A **polar equation** of a collection of points in the plane is an equation in r and θ whose polar graph is the given collection of points.

Our next two examples involve polar equations of circles.

EXAMPLE 3   Find a polar equation of the circle $x^2 + y^2 = a^2$, where $a > 0$ (Figure 10.7(a)).

Solution   Using (3), we find immediately that $r^2 = a^2$, or more simply, $r = a$. $\square$

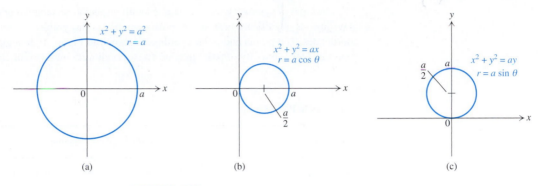

(a)   (b)   (c)

FIGURE 10.7

EXAMPLE 4   Find a polar equation of the circle $x^2 + y^2 = ax$ (Figure 10.7(b)).

Solution   Using (2) and (3), we find that $r^2 = ar \cos \theta$, or $r = a \cos \theta$.   ❑

A similar argument shows that a polar equation of the circle $x^2 + y^2 = ay$ is

$$r = a \sin \theta$$

(Figure 10.7(c)).

Maria Agnesi* (1718–1799)
The daughter of a mathematics professor at the University of Bologna, she published at age 9 an essay promoting education for women. In 1748 her *Foundations of Analysis for the Use of Italian Youth* appeared. It was a two-volume work including calculus, and was one of the first advanced mathematical texts not written in Latin. The **witch of Agnesi,** a curve discussed by Agnesi (see Exercise 56) and derived by means of the circle $r = 2 \sin \theta$, was named a witch through a mistranslation of the Latin "versoria," which really meant "turning."

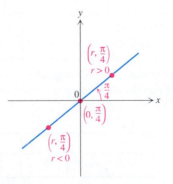

*Agnesi: Pronounced "An-$y\bar{a}$-$z\bar{e}$."

EXAMPLE 5   Find a polar equation of the line passing through the origin and making an angle of $\pi/4$ with respect to the positive x axis (Figure 10.8).

Solution   From the definition of polar coordinates we find that any point on the line has polar coordinates $(r, \pi/4)$, where r may be either positive, negative, or 0. Consequently a polar equation of the line is $\theta = \pi/4$.   ❑

More generally, a polar equation of the line passing through the origin and making an angle θ_0 with the positive x axis is given by $\theta = \theta_0$.

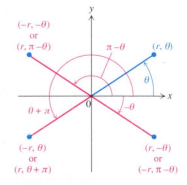

FIGURE 10.8   Polar equation: $\theta = \frac{\pi}{4}$.   **FIGURE 10.9**

The types of symmetry we have studied with respect to rectangular coordinates play just as significant a role in polar graphs as in graphs in rectangular coordinates. The points and polar coordinates shown in Figure 10.9 suggest some conditions for symmetry of polar graphs. These conditions are listed in Table 10.1.

TABLE 10.1

Symmetry in Polar Graphs

Symmetry	Conditions implying symmetry
With respect to the x axis	If (r, θ) satisfies the equation, so does $(r, -\theta)$ or $(-r, \pi - \theta)$
With respect to the y axis	If (r, θ) satisfies the equation, so does $(-r, -\theta)$ or $(r, \pi - \theta)$
With respect to the origin	If (r, θ) satisfies the equation, so does $(-r, \theta)$ or $(r, \theta + \pi)$

However, it is possible for a graph to be symmetric with respect to the origin or one of the axes without satisfying any of the corresponding conditions (see Exercises 44 and 45).

These criteria for symmetry facilitate the drawing of graphs. In the examples that follow we will frequently associate the name of a familiar graph with the equation that defines it. For instance, we say "the circle $r = 2 \cos \theta$" instead of using the longer phrase "the circle whose polar equation is $r = 2 \cos \theta$."

Another aspect of a function f that can assist us in sketching the polar graph of f is related to periodicity. Suppose f has period 2π, that is, f satisfies the condition

$$f(\theta + 2\pi) = f(\theta) \quad \text{for any } \theta \tag{5}$$

Then we can obtain the polar graph of f by letting θ vary from 0 to 2π (or any other interval of length 2π). The reason is that by (5), $(f(\theta), \theta)$ and $(f(\theta + 2n\pi), \theta + 2n\pi)$ are polar coordinates of the same point in the plane, for any integer n.

EXAMPLE 6 Sketch the cardioid $r = 1 + \sin \theta$.

Solution First we observe that $\sin \theta$ has period 2π, so that by the discussion before the example, we only need θ to vary over an interval of 2π at most. Next, we use the fact that $\sin (\pi - \theta) = \sin \theta$, so that (r, θ) satisfies the given equation if $(r, \pi - \theta)$ also does. Therefore by Table 10.1, the cardioid is symmetric with respect to the y axis. From these two observations we need only to let θ vary from $-\pi/2$ to $\pi/2$ in order to sketch half of the graph. Symmetry will yield the remaining half of the graph. From the chart below we obtain the special points appearing in Figure 10.10.

θ	$-\pi/2$	$-\pi/6$	0	$\pi/6$	$\pi/2$
$r = 1 + \sin \theta$	0	$\frac{1}{2}$	1	$\frac{3}{2}$	2

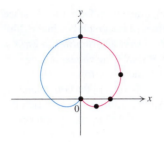

FIGURE 10.10   Cardioid $r = 1 + \sin \theta$.

Next we notice that as θ increases from $-\pi/2$ to $\pi/2$, r increases from 0 to 2. Thus that part of the cardioid to the right of the y axis appears in Figure 10.10 as the portion of the curve in red. Using the symmetry with respect to the y axis, we obtain the complete graph, whose appearance justifies the name "cardioid," which means "heart-shaped."   ❑

Observe that the equation $r = 1 + \sin \theta$ has the same polar graph as the equation $r^2 = r + r \sin \theta$. (The only problem occurs for $r = 0$, and both graphs include the origin, which corresponds to $r = 0$.) However, by (2) and (3) the polar graph of the polar equation $r^2 = r + r \sin \theta$ is the same as the graph of

$$x^2 + y^2 = \sqrt{x^2 + y^2} + y \qquad (5)$$

It is clearly easier to graph the cardioid from its polar equation $r = 1 + \sin \theta$ than from (5).

EXAMPLE 7   Sketch the limaçon* $r = \frac{1}{2} + \cos \theta$.

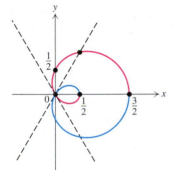

FIGURE 10.11   Limaçon $r = \frac{1}{2} + \cos \theta$.

Solution   Since $\cos(-\theta) = \cos \theta$, it follows from Table 1 that the limaçon is symmetric with respect to the x axis. From the symmetry and the fact that $\cos \theta$ has period 2π, it follows that we need consider only those values of θ corresponding to points above the x axis, that is, between 0 and π; then we can use symmetry to complete the graph. From the chart below we determine the special points in Figure 10.11.

θ	0	$\pi/3$	$\pi/2$	$2\pi/3$	π
$r = \frac{1}{2} + \cos \theta$	$\frac{3}{2}$	1	$\frac{1}{2}$	0	$-\frac{1}{2}$

Now we notice that as θ increases from 0 to $2\pi/3$, r decreases from $\frac{3}{2}$ to 0. Similarly, as θ increases from $2\pi/3$ to π, r decreases from 0 to $-\frac{1}{2}$; since r is negative, the portion of the graph corresponding to $2\pi/3 \leq \theta \leq \pi$ lies in the fourth quadrant instead of the second quadrant. With this information we sketch the part of the graph corresponding to $0 \leq \theta \leq \pi$ (colored red in Figure 10.11) and then use symmetry to sketch the remainder of the graph (colored blue in Figure 10.11). Does the appearance of the graph justify the name "limaçon," which means "snail"?   ❑

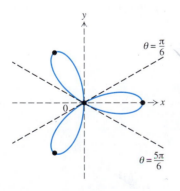

FIGURE 10.12   Three-leaved rose $r = \cos 3\theta$.

EXAMPLE 8   Sketch the three-leaved rose $r = \cos 3\theta$.

Solution   Since $\cos 3(\theta + 2\pi) = \cos 3\theta$, we need only regard θ such that $0 \leq \theta \leq 2\pi$. Next, since $\cos(-3\theta) = \cos 3\theta$, the polar graph is symmetric with respect to the x axis. Therefore we will graph the equation $r = \cos 3\theta$ for $0 \leq \theta \leq \pi$ and then use symmetry to complete the graph. The chart below allows us to plot the special points appearing in Figure 10.12.

θ	0	$\pi/3$	$\pi/2$	$2\pi/3$	π
$r = \cos 3\theta$	1	-1	0	1	-1

* **Limaçon:** Pronounced "*lee*-ma-sohn."

Next we notice that as θ increases from 0 to $\pi/3$, r decreases from 1 to -1; since r is negative for $\pi/6 < \theta \leq \pi/3$, the corresponding part of the graph lies in the third quadrant instead of the first quadrant. Similarly, as θ increases from $\pi/3$ to $2\pi/3$, r increases from -1 to 1; since r is negative for $\pi/3 \leq \theta < \pi/2$, the corresponding part of the graph lies in the third quadrant. Finally, as θ increases from $2\pi/3$ to π, r decreases again from 1 to -1; since r is negative for $5\pi/6 < \theta \leq \pi$, the corresponding part of the graph lies in the fourth quadrant. We use this information to sketch the portion of the graph of $r = \cos 3\theta$ corresponding to $0 \leq \theta \leq \pi$, which appears in Figure 10.12. Because of symmetry with respect to the x axis, the remainder of the graph is obtained by reflection through the x axis, which duplicates the curve we already have. ❏

EXAMPLE 9   Sketch the lemniscate $r^2 = \cos 2\theta$.

Solution   From Table 10.1 it follows that the lemniscate is symmetric with respect to both coordinate axes and the origin. Thus if we sketch the part of the lemniscate in the first quadrant, then we can obtain the complete graph by reflecting that part into the other three quadrants. Because $\cos 2\theta$ is periodic with period π, all points in the first quadrant that lie on the lemniscate can be obtained by letting θ vary from 0 to $\pi/2$ and then plotting the corresponding points (r, θ). As θ varies from 0 to $\pi/4$, the values of r for which $r^2 = \cos 2\theta$ decrease from 1 to 0, yielding the red curve in Figure 10.13. Next, if θ varies from $\pi/4$ to $\pi/2$, then $\cos 2\theta < 0$. Since $r^2 \geq 0$ for all r, we conclude that there is no point (r, θ) on the lemniscate for $\pi/4 < \theta < \pi/2$. Now we are ready to use the information we have gained and the various symmetries in order to complete the sketch of the lemniscate in Figure 10.13. You can see why the curve is named after the Latin word for ribbon. ❏

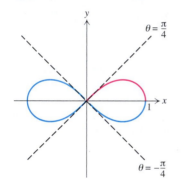

FIGURE 10.13   The lemniscate $r^2 = \cos 2\theta$.

Functions described by polar coordinates can also be described by parametric equations. In particular, if $r = f(\theta)$, then substituting $f(\theta)$ for r in (2) yields the parametric equations

$$x = f(\theta) \cos \theta \quad \text{and} \quad y = f(\theta) \sin \theta \tag{6}$$

Then the graph of the polar equation $r = f(\theta)$ is the graph of the curve described parametrically by (6). Thus one can graph a polar equation on a graphics calculator using either the polar or the parametric mode.

Dictionary of Polar Graphs

Name	Polar equation	Graph	Name	Polar equation	Graph
Circle	$r = a \cos \theta$ with $a > 0$	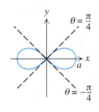	Convex limaçon	$r = a + b \cos \theta$ with $0 < 2b \le a$	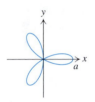
Lemniscate	$r^2 = a^2 \cos 2\theta$		3-leaved rose	$r = a \cos 3\theta$ with $a > 0$	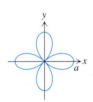
Cardioid	$r = a\,(1 + \cos \theta)$ with $a > 0$		4-leaved rose	$r = a \cos 2\theta$	
Looped limaçon	$r = a + b \cos \theta$ with $0 < a < b$	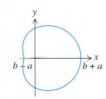	Spiral of Archimedes	$r = a\theta$	
Dimpled limaçon	$r = a + b \cos \theta$ with $0 < b < a < 2b$				

EXERCISES 10.1

1. Find the Cartesian coordinates of the points having the given polar coordinates.

   a. $(3, \pi/4)$ **b.** $(-2, -\pi/6)$
   c. $(3, 7\pi/3)$ **d.** $(5, 0)$
   e. $(-2, \pi/2)$ **f.** $(-2, 3\pi/2)$
   g. $(4, 3\pi/4)$ **h.** $(0, 6\pi/7)$
   i. $(-1, 23\pi/3)$ **j.** $(-1, -23\pi/3)$
   k. $(1, 3\pi/2)$ **l.** $(3, -5\pi/6)$

2. Find all sets of polar coordinates for the points having the given Cartesian coordinates.

   a. $(3, 3)$ **b.** $(4, -4)$
   c. $(0, 5)$ **d.** $(-4, 0)$
   e. $(3, 3\sqrt{3})$ **f.** $(-\frac{1}{3}, \sqrt{3}/3)$
   g. $(-3, \sqrt{3})$ **h.** $(-2\sqrt{3}, 2)$
   i. $(0, 0)$ **j.** $(-5\sqrt{3}, -5)$

In Exercises 3–11 write the equation in polar coordinates. Express the answer in the form $r = f(\theta)$ wherever possible.

3. $2x + 3y = 4$ 4. $y^2 = 4x$
5. $x^2 + 9y^2 = 1$ 6. $9x^2 + y^2 = 4y$
7. $(x^2 + y^2)^2 = x^2 - y^2$ 8. $x^2 + y^2 = 2x$
9. $x^2 + y^2 = 4y$ 10. $y^2 = \dfrac{x^3}{2 - x}$
11. $y^2 = \dfrac{x^2(3 - x)}{1 + x}$

In Exercises 12–18 write the polar equation as an equation in Cartesian coordinates.

12. $r = 5$ 13. $r = 3 \cos \theta$
14. $\tan \theta = 6$ 15. $\cot \theta = 3$
16. $r \cot \theta = 3$ 17. $r = \sin 2\theta$
18. $r = 2 \sin \theta \tan \theta$

19. Show that the polar graph of the equation $r = 1 + \cos \theta$ is the same as the polar graph of $r^2 = r + r \cos \theta$.

20. Show that the polar graph of the equation $r = 2 + \cos \theta$ is not the same as the polar graph of $r^2 = 2r + r \cos \theta$.

In Exercises 21–36 sketch the polar graph of the equation. Each graph has a familiar form. It may be convenient to convert the equation to rectangular coordinates.

21. $r = 5$ 22. $r = -2$
23. $r = 0$ 24. $\theta = 3\pi/2$
25. $\theta = -7\pi/6$ 26. $|\theta| = \pi/3$
27. $r \sin \theta = 5$ 28. $r = 2 \sin \theta$
29. $r = -\frac{3}{2} \cos \theta$ 30. $r \sin (\theta - \pi/2) = 3$
31. $r \cos (\theta - \pi/3) = 2$ 32. $r \cos (\theta + \pi/4) = -2$

33. $r = 2 \cot \theta \csc \theta$ 34. $r = -3 \tan \theta \sec \theta$

35. $r(\sin \theta + \cos \theta) = 1$ 36. $r = \dfrac{2}{3 \cos \theta - 2 \sin \theta}$

In Exercises 37–52 sketch the polar graph of the given equation. Note any symmetries.

37. $r = 1 - \cos \theta$ 38. $r = 3(1 - \sin \theta)$
39. $r = 5(1 + \sin \theta)$ 40. $r = 3 \sin 2\theta$
41. $r = -4 \sin 3\theta$ 42. $r = -\sin 4\theta$
43. $r = 2 \cos 6\theta$

44. $r = \cos \dfrac{\theta}{2}$ (*Hint:* Not all symmetry can be deduced from Table 10.1.)

45. $r = \sin \dfrac{\theta}{2}$ (*Hint:* Not all symmetry can be deduced from Table 10.1.)

46. $r^2 = \sin \theta$ 47. $r^2 = 25 \cos \theta$
48. $r^2 = 9 \sin 2\theta$ 49. $r = 1 + 2 \sin \theta$
50. $r = 3 \tan \theta$ (kappa curve) (*Hint:* Find $\lim_{\theta \to \pi/2^-} r \cos \theta$.)
51. $r = e^{\theta/3}$ (spiral)
52. $r = \sin \theta + \cos \theta$ (*Hint:* Transform the equation into Cartesian coordinates first.)

53. For each of the following equations, plot the polar graph, and then prove the symmetry that you observe.
    a. $r = (\sin 3\theta)(\sin \theta)$ **b.** $r = (\sin 3\theta)(\cos \theta)$
    c. $r = 1 + \cos 2\theta + \cos 3\theta$

54. Consider the polar graph of the equation $r = 1 + 2 \sin n\theta$. Plot the graph for several positive integers n, and determine when small petals lie inside larger ones.

55. **a.** Plot the graph of $r = 1 + \sin n\theta$ for $n = 2, 3, 4$, and 5.
    b. How would you describe the graph of $r = 1 + \sin n\theta$ for any integer $n \geq 2$?

*56. The witch of Agnesi, referred to in this section, is defined as follows: For any number θ with $0 < \theta < \pi$, consider the line that emanates from the origin and makes an angle of θ radians with respect to the positive x axis. It intersects the circle $x^2 + (y - 1)^2 = 1$ at a point $A = (x_0, y)$ and intersects the line $y = 2$ at a point $B = (x, 2)$ (Figure 10.14). Let $P(x, y)$ be the point on the same horizontal line as A and on the same vertical line as B. As θ varies from 0 to π, $P(x, y)$ traces out the witch of Agnesi.
    a. Using the equation $r = 2 \sin \theta$ for the circle, show that the witch is given parametrically by $x = 2 \cot \theta$ and $y = 2 \sin^2 \theta$ for $0 < \theta < \pi$.
    b. Eliminate θ from the equations in part (a) and show

that the witch is the graph of $y = 8/(x^2 + 4)$.

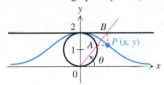

FIGURE 10.14   The witch of Agnesi for Exercise 56.

57. In each of the following, show that both of the given equations have the same graph.
    a. $r = 3(\cos\theta + 1)$ and $r = 3(\cos\theta - 1)$ (*Hint:* Show that if (r, θ) satisfies one equation, then $(-r, \theta + \pi)$ satisfies the other.)
    b. $r = 2(\sin\theta + 1)$ and $r = 2(\sin\theta - 1)$
    c. $r = \theta$ and $r = \theta - 2\pi$

58. Show that the polar graphs of the equations $r = \frac{1}{2}(1 + \cos\theta)$ and $r^2 = -\cos\theta$ intersect at $(1, 0)$, even though no single set of polar coordinates for $(1, 0)$ satisfies both equations.

*59. Find a polar equation of the collection of points the product of whose distances from the points $(1, 0)$ and $(-1, 0)$ is 1.

Projects

1. This project revolves around the number and the configuration of petals that comprise the polar graph of $r = \sin s\theta$ for various positive numbers s. With the help of a calculator, formulate conjectures for the following:
    a. Determine the number of petals in the graph when s is an
      i. even integer   **ii.** odd integer
    b. Let $s = p/q$, where p and q are positive integers without common factors, and with $q > 1$. Determine the number of petals in the graph. Make sure that θ varies over a large enough interval $[0, \alpha]$ so that the graph encloses all the petals available to $r = \sin s\theta$. How does the minimum such α relate to the values of p and q?
    c. Let s be irrational, for example, $s = \pi$. What is the resulting graph when θ varies over $[0, \alpha]$ and α becomes large?

2. In this project we give a glimpse into the creation of the Mandelbrot set (see Figure 10.1 again), named after the Polish-born Benoit Mandelbrot, a present-day mathematician who first brought the set to prominence with his computer renditions of it in the late 1970s. Because the Mandelbrot set can only be described by using complex numbers, we will first give a very brief introduction to complex numbers and operations on them.

We begin by introducing the number i, with the property that $i^2 = -1$, or equivalently, $i = \sqrt{-1}$. The number i is *not* a real number; it is a solution to the equation $x^2 + 1 = 0$.

If a and b are real numbers, then we define $a + bi$, or equivalently, $a + ib$, to be a **complex number.** Thus $3 - 2i$ and $\sqrt{5}i$ are complex numbers. The **real part** of $a + bi$ is a, and the **imaginary part** of $a + bi$ is b. We notice that the real number a can be written as a complex number $a + 0i$; it follows that real numbers can be considered as special complex numbers. The number bi is called a **pure imaginary number,** because its real part is 0.

We can identify $a + bi$ in a natural way with the points in the plane by letting a be the x coordinate and b the y coordinate (Figure 10.15). In this way the real numbers correspond to the points on the x axis, called the **real axis,** and the pure imaginary numbers constitute the y axis, called the **imaginary axis** (Figure 10.15). When the plane is regarded as the collection of complex numbers, it is called the **complex plane.**

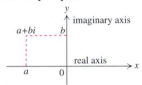

FIGURE 10.15   The complex plane.

The **modulus** $|c|$ of a complex number c is its distance from the origin. In other words, if $c = a + bi$, then

$$|c| = \sqrt{a^2 + b^2}$$

As with the ordinary absolute value, the modulus satisfies the "triangle inequality": $|c + d| \leq |c| + |d|$. We will use this fact in the analysis of the Mandelbrot set.

A set A in the complex plane is said to be **bounded** if there is positive number r such that $|c| \leq r$ for all c in A. Otherwise said, A is bounded if it is contained in a disk whose center is the origin.

The sum and difference of two complex numbers $a + bi$ and $u + vi$ are defined as follows:

sum: $(a + bi) + (u + vi) = (a + u) + (b + v)i$

difference: $(a + bi) - (u + vi) = (a - u) + (b - v)i$

In other words, we add two complex numbers by adding their real parts and their imaginary parts separately, and subtract them in a similar way. Thus

$$(4 - 2i) + (-6 - 3i) = (4 - 6) + (-2 - 3)i = -2 - 5i$$

For the **product** of $a + bi$ and $u + vi$ we multiply as though we had polynomials, obtaining

$$(a + bi)(u + vi) = au + (av + bu)i + bvi^2$$

Then we use the fact that $i^2 = -1$ to obtain

$$(a + bi)(u + vi) = (au - bv) + (av + bu)i$$

For example,

$$(-1 - 4i)(2 - 3i) = [(-1)(2) - (-4)(-3)] + [(-1)(-3)$$
$$+ (-4)(2)]i = -14 - 5i$$

Of importance is the square of a complex number. If $c = a + bi$, then

$$c^2 = (a + bi)(a + bi) = (a^2 - b^2) + 2abi$$

Now we are ready to define the functions that yield the Mandelbrot set. For any complex number c, let g_c be defined by

$$g_c(z) = z^2 + c \quad \text{for all complex numbers } z$$

Thus if $c = \frac{1}{2} - i$, then

$$g_c(1 + 3i) = (1 + 3i)^2 + \left(\frac{1}{2} - i\right)$$

$$= (1 - 9) + 6i + \left(\frac{1}{2} - i\right) = -\frac{15}{2} + 5i$$

Our focus will be on the iterates of 0 for g_c, that is, the (complex) numbers

$$0, \quad g_c(0), \quad g_c(g_c(0)), \quad g_c(g_c(g_c(0))), \quad \text{etc.}$$

where c is allowed to be any complex number. It is surprising that for various values of c, the iterates of c for g_c can have vastly different behaviors. To get just a little idea, let us calculate the first six iterates of 0 for several values of c:

$c = -2$: $0, -2, 2, 2, 2, 2, \ldots$

$c = i$: $0, i, -1 + i, -i, -1 + i, -i, \ldots$

$c = 0.25$: $0, 0.25, 0.3125, 0.347656 \cdots, 0.370863 \cdots,$
$0.387540 \cdots$

$c = 1$: $0, 1, 2, 5, 26, 677$

$c = 1 + i$: $0, 1 + i, 1 + 3i, -7 + 7i, 1 - 97i, -9407 - 193i$

As you see, for $c = -2$, the iterates are eventually 2; for $c = i$ they oscillate between $-i$ and $-1 + i$; for $c = 0.25$ the iterates seem to be inching upward, and for $c = 1$ and $c = 1 + i$ the iterates appear to be unbounded.

The fact that for some values of c the iterates of 0 for g_c are evidently bounded and for other values of c the iterates of 0 for g_c are clearly unbounded leads us to make the following definition. We say that c is in the **Mandelbrot set** M if the iterates of 0 for g_c are bounded. By the definition and the catalogue of iterates above we see that -2 and i are in M, 0.25 appears to be in M, and 1 and $1 + i$ definitely appear *not* to be in M.

Which complex numbers are in M, and which are not? That is a very, very difficult question to answer in general. In fact, the Mandelbrot set is extremely complicated, with a large cardioid (yes, a genuine cardioid) called the "body" in the center, a circular "head" to the left, and then other little disks and tiny blobs and filaments extending out in every direction from the body and the head. For properties of M and methods for drawing M on the computer, you might want to visit the Internet or consult the book *Chaotic Dynamical Systems* by Robert Devaney.

Now it is time to explore the numbers in M.

a. Compute the first six iterates of 0 for g_c and then conjecture whether the given complex number c is in M.

   i. $c = -\dfrac{3}{4}$

   ii. $c = \dfrac{1}{8} + \dfrac{3}{4}i$

   iii. $c = -2 + i$

   iv. $c = \dfrac{1}{10}(1 + i)$

b. Show that M is symmetric about the real axis, that is, show that if $c = a + bi$ is in M, then its **conjugate** $\bar{c} = a - bi$ is in M. (*Hint:* First show that if c and d are complex, then $\overline{c + d} = \bar{c} + \bar{d}$ and $\overline{cd} = \bar{c}\,\bar{d}$.)

c. Prove that if $|c| \leq \frac{1}{4}$, then c is in M. (*Hint:* Let c_n denote the nth iterate of 0 for g_c. Assuming that $|c_n| \leq \frac{1}{2}$, show that $|c_{n+1}| \leq \frac{1}{2}$. Then you can conclude by induction that the iterates of c are bounded in modulus by $\frac{1}{2}$.)

d. This part is devoted to showing that if $|c| > 2$, then c is not in M. To that end, assume that $|c| > 2$, and let $r = |c| - 1$.

   i. Show that $r > 1$, and that if $|z| > |c|$, then $|g_c(z)| = |z^2 + c| \geq r|z|$. (*Hint:* Use the fact that $|z^2 + c| \geq |z|^2 - |c|$.)

   ii. Show that the nth iterate c_n of 0 for g_c has the property that $|c_n| \geq r^{n-1}|c|$.

   iii. Using (i) and (ii), along with convergence properties of $\{r^n\}_{n=1}^{\infty}$, show that c is not in M.

10.2   LENGTH AND AREA IN POLAR COORDINATES

As we saw in Section 10.1, many curves such as circles and cardioids are easier to describe in polar coordinates than in rectangular coordinates. In most cases it is easier to use polar coordinates in order to compute the lengths of such curves and the areas of corresponding regions.

Length in Polar Coordinates

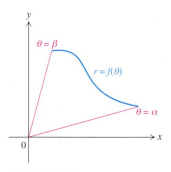

FIGURE 10.16

Consider a nonnegative function f defined on $[\alpha, \beta]$, with $0 \le \beta - \alpha \le 2\pi$. The polar graph of f is the set of points (x, y) with polar coordinates (r, θ) satisfying $r = f(\theta)$ and $\alpha \le \theta \le \beta$ (Figure 10.16). As observed in (6) of Section 10.1, the polar graph of f has the parametric representation

$$x = f(\theta)\cos\theta \quad \text{and} \quad y = f(\theta)\sin\theta \quad \text{for } \alpha \le \theta \le \beta \tag{1}$$

If f' is continuous on $[\alpha, \beta]$, then we may apply (2) of Section 6.8 to the parametric equations in (1) to find the length L of C:

$$L = \int_\alpha^\beta \sqrt{\left(\frac{dx}{d\theta}\right)^2 + \left(\frac{dy}{d\theta}\right)^2}\, d\theta$$

$$= \int_\alpha^\beta \sqrt{[f'(\theta)\cos\theta - f(\theta)\sin\theta]^2 + [f'(\theta)\sin\theta + f(\theta)\cos\theta]^2}\, d\theta$$

$$= \int_\alpha^\beta \sqrt{(f'(\theta))^2(\cos^2\theta + \sin^2\theta) + (f(\theta))^2(\sin^2\theta + \cos^2\theta)}\, d\theta$$

$$= \int_\alpha^\beta \sqrt{(f'(\theta))^2 + (f(\theta))^2}\, d\theta$$

Therefore the length L of C is given by

$$L = \int_\alpha^\beta \sqrt{(f'(\theta))^2 + (f(\theta))^2}\, d\theta \tag{2}$$

In Leibniz notation (2) becomes

$$L = \int_\alpha^\beta \sqrt{r^2 + \left(\frac{dr}{d\theta}\right)^2}\, d\theta$$

EXAMPLE 1   Use (2) to determine the circumference of the circle $r = 5$.

Solution   If $r = 5$, then $f(\theta) = 5$ for $0 \le \theta \le 2\pi$. Then (2) becomes

$$L = \int_0^{2\pi} \sqrt{0^2 + 5^2}\, d\theta = \int_0^{2\pi} 5\, d\theta = 10\pi \quad \square$$

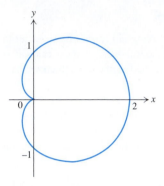

FIGURE 10.17 Cardioid $r = 1 + \cos \theta$.

EXAMPLE 2 Let C be the cardioid $r = 1 + \cos \theta$ for $0 \le \theta \le 2\pi$ (Figure 10.17). Find the length L of C.

Solution Because of the symmetry of C with respect to the x axis, the length L of C is twice the length of the top half of C. Thus we let $f(\theta) = 1 + \cos \theta$ for $0 \le \theta \le \pi$. It follows from (2) that

$$L = 2 \int_0^\pi \sqrt{(-\sin \theta)^2 + (1 + \cos \theta)^2} \, d\theta = 2 \int_0^\pi \sqrt{2 + 2 \cos \theta} \, d\theta$$

since C is symmetric with respect to the x axis. By the half-angle formula for $\cos \theta/2$,

$$\frac{1 + \cos \theta}{2} = \cos^2 \frac{\theta}{2}$$

Since $\cos \theta/2 \ge 0$ for $0 \le \theta \le \pi$, we find that

$$\sqrt{2 + 2 \cos \theta} = \sqrt{4 \left(\frac{1 + \cos \theta}{2} \right)} = \sqrt{4 \cos^2 \frac{\theta}{2}} = 2 \cos \frac{\theta}{2}$$

We conclude that

$$L = 2 \int_0^\pi \sqrt{2 + 2 \cos \theta} \, d\theta = 4 \int_0^\pi \cos (\theta/2) \, d\theta = 8 \sin (\theta/2) \Big|_0^\pi = 8 \quad \square$$

Area in Polar Coordinates

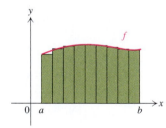

FIGURE 10.18

When we prepared to find the area of a region in rectangular coordinates (in Chapter 5), we divided the region into vertical strips, approximated the area of each such strip by the area of a rectangle, and added the estimates (Figure 10.18). To find the area of a region in polar coordinates (Figure 10.19(a)) we will divide the region into "pie-shaped" sectors (Figure 10.19(b)), approximate the areas of the sectors, and find their sum.

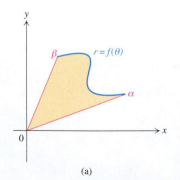

(a)

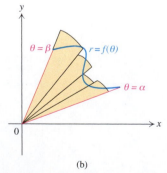

(b)

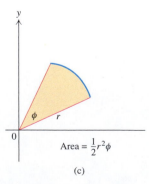

(c)

FIGURE 10.19

We begin by determining the area A of a sector S with angle ϕ and radius r (Figure 10.19(c)). Notice that the area A is given by

$$A = \frac{\overbrace{\phi}^{\text{fraction of circle}}}{(2\pi)} \cdot \overbrace{\pi r^2}^{\text{area of circle}} = \frac{1}{2} r^2 \phi \tag{3}$$

Now consider a continuous, nonnegative function f defined on $[\alpha, \beta]$, with $0 \leq \beta - \alpha \leq 2\pi$, and let R be the region consisting of all points in the plane having polar coordinates that satisfy

$$0 \leq r \leq f(\theta) \quad \text{and} \quad \alpha \leq \theta \leq \beta$$

Our objective is to define the area A of R. Let $P = \{\theta_0, \theta_1, \ldots, \theta_n\}$ be any partition of $[\alpha, \beta]$, and for each k between 1 and n, let $\Delta\theta_k = \theta_k - \theta_{k-1}$. If t_k is an arbitrary number in the interval $[\theta_{k-1}, \theta_k]$, and if $\Delta\theta_k$ is small, then the area ΔA_k of the region R_k between the lines $\theta = \theta_{k-1}$ and $\theta = \theta_k$ is approximately equal to the area of a circular sector of angle $\Delta\theta_k$ and radius $f(t_k)$ (Figure 10.20). Thus by (3) with r replaced by $f(t_k)$ and ϕ replaced by $\Delta\theta_k$, ΔA_k is approximately $\frac{1}{2} [f(t_k)]^2 \Delta\theta_k$. Since the area A of R is the sum of the areas $\Delta A_1, \Delta A_2, \ldots, \Delta A_n$, it follows that A should be approximately equal to

$$\sum_{k=1}^{n} \frac{1}{2} [f(t_k)]^2 \Delta\theta_k$$

and this is a Riemann sum for $\frac{1}{2} f^2$ on $[\alpha, \beta]$. Therefore the area A should be given by the formula

$$A = \lim_{\|P\| \to 0} \sum_{k=1}^{n} \frac{1}{2} [f(t_k)]^2 \Delta\theta_k = \int_{\alpha}^{\beta} \frac{1}{2} [f(\theta)]^2 \, d\theta$$

Thus the area A of the region consisting of all points in the plane having polar coordinates (r, θ) that satisfy $0 \leq r \leq f(\theta)$ and $\alpha \leq \theta \leq \beta$ is given by

$$A = \int_{\alpha}^{\beta} \frac{1}{2} [f(\theta)]^2 \, d\theta \tag{4}$$

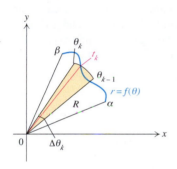

FIGURE 10.20

EXAMPLE 3   Let R be the region enclosed by the cardioid $r = 1 + \cos\theta$ (Figure 10.21). Find the area A of R.

Solution   Let $f(\theta) = 1 + \cos\theta$. If we let $\alpha = 0$ and $\beta = 2\pi$, then by (4), along with (2) of Section 5.6, we have

$$A = \int_0^{2\pi} \frac{1}{2} [f(\theta)]^2 \, d\theta = \frac{1}{2} \int_0^{2\pi} (1 + \cos\theta)^2 \, d\theta$$

$$= \frac{1}{2} \int_0^{2\pi} (1 + 2\cos\theta + \cos^2\theta) \, d\theta$$

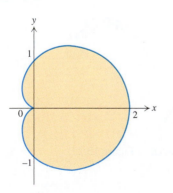

FIGURE 10.21   Cardioid $r = 1 + \cos\theta$.

$$= \frac{1}{2}\left(\theta + 2\sin\theta + \frac{\theta}{2} + \frac{1}{4}\sin 2\theta\right)\Bigg|_0^{2\pi}$$

$$= \frac{3\pi}{2} \quad \square$$

Finding the area of the cardioid in rectangular coordinates would be a formidable task, for it would involve finding the area of the region enclosed by the graph of $x^2 + y^2 = \sqrt{x^2 + y^2} + x$.

Now we consider more general regions. Let f and g be defined on $[\alpha, \beta]$, where $0 \le \beta - \alpha \le 2\pi$, and suppose

$$0 \le g(\theta) \le f(\theta) \quad \text{for } \alpha \le \theta \le \beta$$

Then the area of the region determined by the graph of f between the lines $\theta = \alpha$ and $\theta = \beta$ is the sum of the areas of the darkly colored region determined by the graph of g and the lightly colored region between the graphs of f and g (Figure 10.22). Then it follows from (4) that the area A of the region consisting of all points in the plane having polar coordinates (r, θ) that satisfy

$$g(\theta) \le r \le f(\theta) \quad \text{and} \quad \alpha \le \theta \le \beta$$

is given by

$$A = \int_\alpha^\beta \frac{1}{2}\{[f(\theta)]^2 - [g(\theta)]^2\}\, d\theta \qquad (5)$$

If $g = 0$, then the area described in (5) is just that appearing in (4). But g need not be 0, as the next example shows.

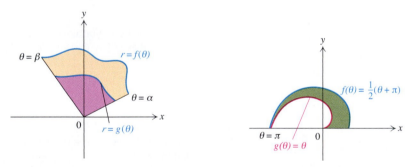

FIGURE 10.22   **FIGURE 10.23**

EXAMPLE 4   Find the area A of the region shaded in Figure 10.23.

Solution   If we let

$$f(\theta) = \frac{1}{2}(\theta + \pi) \quad \text{and} \quad g(\theta) = \theta$$

then by (5) we know that

$$A = \int_0^\pi \frac{1}{2} \{[f(\theta)]^2 - [g(\theta)]^2\} \, d\theta = \frac{1}{2} \int_0^\pi \left[\frac{1}{4}(\theta + \pi)^2 - \theta^2\right] d\theta$$

$$= \frac{1}{2}\left[\frac{1}{12}(\theta + \pi)^3 - \frac{1}{3}\theta^3\right]\Big|_0^\pi$$

$$= \frac{1}{2}\left[\left(\frac{1}{12}(8\pi^3) - \frac{1}{3}\pi^3\right) - \frac{1}{12}\pi^3\right] = \frac{\pi^3}{8} \quad \square$$

In the final example we must determine the limits of integration before we can set up the integral.

EXAMPLE 5   Find the area A of the region inside the circle $r = 2 \cos \theta$ and outside the circle $r = 1$ (Figure 10.24).

Solution   The region in question lies between the lines $\theta = \alpha$ and $\theta = \beta$ that pass through the points of intersection of the two circles (Figure 10.24). To find the points of intersection we solve the equation

$$2 \cos \theta = 1, \quad \text{or} \quad \cos \theta = \frac{1}{2}$$

We find that $\theta = -\pi/3$ or $\theta = \pi/3$, so that $\alpha = -\pi/3$ and $\beta = \pi/3$. Taking

$$f(\theta) = 2 \cos \theta \quad \text{and} \quad g(\theta) = 1$$

in (5) and using (2) of Section 5.6, we have

$$A = \int_{-\pi/3}^{\pi/3} \frac{1}{2}(4 \cos^2 \theta - 1) \, d\theta$$

$$= \int_{-\pi/3}^{\pi/3} 2 \cos^2 \theta \, d\theta - \int_{-\pi/3}^{\pi/3} \frac{1}{2} \, d\theta$$

$$= 2\left(\frac{1}{2}\theta + \frac{1}{4}\sin 2\theta\right)\Big|_{-\pi/3}^{\pi/3} - \frac{1}{2}\theta\Big|_{-\pi/3}^{\pi/3}$$

$$= \frac{\pi}{3} + \frac{\sqrt{3}}{2} \quad \square$$

The areas of some regions have been given both by a formula in rectangular coordinates ((1) of Section 5.8) and by a formula in polar coordinates ((5) above). In such cases the same area results from applying either formula (although we will not prove that here).

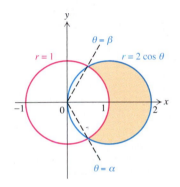

FIGURE 10.24

EXERCISES 10.2

In Exercises 1–6 find the length L of the graph of the given equation.

1. $r = 2 \cos \theta$

2. $r = \theta^2$ for $0 \leq \theta \leq 4\sqrt{2}$

3. $r = 2 - 2 \cos \theta$

4. $r = \sin^2 \dfrac{\theta}{2}$ for $0 \leq \theta \leq \pi$

5. $r = \sin^3 \dfrac{\theta}{3}$ for $0 \leq \theta \leq 2\pi$

6. $r = e^{\theta}$ for $-\ln 3 \leq \theta \leq 0$

In Exercises 7–22 find the area A of the region bounded by the graphs of the given equations.

7. $r = 4$

8. $r = a$, where $a > 0$

9. $r = 3 \sin \theta$

10. $r = 3 \sin \theta$ for $0 \leq \theta \leq \pi/3$, and the line $\theta = \pi/3$

11. $r = -2 \cos \theta$

12. $r = 9 \sin 2\theta$ (four-leaved rose)

13. $r = 9 \cos 2\theta$ for $\pi/4 \leq \theta \leq \pi/2$, and the line $\theta = \pi/2$

14. $r = -4 \sin 3\theta$

15. $r = \frac{1}{2} \cos 3\theta$

16. $r = 6 \sin 4\theta$

17. $r = 2(1 - \sin \theta)$

18. $r = 2 + 2 \cos \theta$

19. $r = 4 + 3 \cos \theta$

20. $r^2 = 9 \sin 2\theta$

21. $r^2 = 25 \cos \theta$

22. $r^2 = -\cos \theta$

In Exercises 23–29 find the area A of the region inside the first curve and outside the second curve.

23. $r = 5$ and $r = 1$

24. $r = 5$ and $r = 2(1 + \cos \theta)$

25. $r = 1$ and $r = \sin \theta$

26. $r = 1$ and $r = \cos 2\theta$

27. $r = 1$ and $r^2 = \cos 2\theta$

28. $r = 5(1 + \cos \theta)$ and $r = 2 \cos \theta$

29. $r = 2 + \cos \theta$ and $r = -\cos \theta$

In Exercises 30–34 find the area A of the indicated region.

30. The region common to the two circles $r = \cos \theta$ and $r = \sin \theta$

31. The region common to the circle $r = \cos \theta$ and the cardioid $r = 1 - \cos \theta$

32. The region inside the circle $r = \cos \theta$ and outside the cardioid $r = 1 - \cos \theta$

33. The region outside the cardioid $r = 1 + \cos \theta$ and inside the cardioid $r = 1 + \sin \theta$

*34. The region outside the small loop and inside the large loop of $r = 1 + 2 \cos \theta$

35. a. How does the length L of the polar graph of f in Figure 10.16 change if f is multiplied by a positive constant c?

b. How does the area A of the region in Figure 10.19(a) change if f is multiplied by a positive constant c?

36. The graph of

$$y^2 = \frac{x^2(1 + x)}{1 - x}$$

is called a **strophoid** (Figure 10.25).

a. Show that $r = \sec \theta - 2 \cos \theta$ for $-\pi/2 < \theta < \pi/2$ is a polar equation of this strophoid.

b. Find the area A enclosed by the loop of the strophoid.

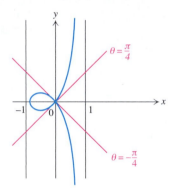

FIGURE 10.25   The strophoid for Exercise 36.

37. Suppose C is the graph of the equation $r = f(\theta)$ for $\alpha \leq \theta \leq \beta$, where $0 \leq \alpha \leq \beta \leq \pi$. Assume that f' is continuous on $[\alpha, \beta]$. Let S be the area of the surface obtained by revolving C about the x axis. Show that

$$S = \int_{\alpha}^{\beta} 2\pi f(\theta) \sin \theta \sqrt{(f'(\theta))^2 + (f(\theta))^2} \, d\theta \quad (6)$$

(*Hint:* Use (11) in Section 6.8 and (1) of this section.)

In Exercises 38–39 use (6) to determine the area S of the surface obtained by revolving the graph of the given equation about the x axis.

38. $r = 1 + \cos \theta$ for $0 \leq \theta \leq \pi$ (half a cardioid)

39. $r = \sqrt{2 \cos 2\theta}$ for $0 \leq \theta \leq \dfrac{\pi}{4}$ (one fourth of a lemniscate)

10.3 CONIC SECTIONS

**Apollonius
(ca. 262 B.C.– ca. 190 B.C.)**
Apollonius was one of the three great mathematicians of the third century B.C., along with Euclid and Archimedes. The names "parabola," "ellipse," and "hyperbola" were introduced by him in his extraordinary eight-volume treatise *Conic Sections*. Several centuries later the first historically recognized female mathematician, Hypatia (ca. 370–415), wrote commentaries on Apollonius's treatise. Interestingly, the figures of speech "parable," "ellipsis," and "hyperbole" are derived from Apollonius's conic sections.

Conic sections are among the best-known plane figures. They have been of special interest to mathematicians since the time of Apollonius, who during the third century B.C. wrote eight treatises on conic sections and became known to his contemporaries as "the Great Geometer." The conic sections arise when a double right circular cone is cut by a plane. Depending on how the plane cuts the cone, the intersection forms a curve called a parabola, an ellipse, or a hyperbola (Figure 10.26(a)–(c)) or, in "degenerate" cases, a point, a line, or two intersecting lines (Figure 10.26(d)–(f)). Since we have already studied lines in Chapter 1, we will limit our discussion in this chapter to parabolas, ellipses, and hyperbolas.

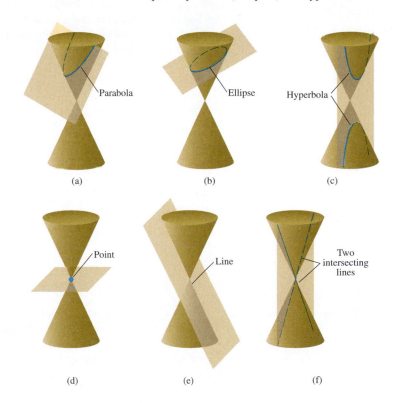

FIGURE 10.26 The conic sections.

Our definitions of these three conic sections will be given in terms of points, lines, and distances, rather than in terms of planes and cones. Using the definitions, we will derive equations of the conic sections and show that any second-degree equation $Ax^2 + Bxy + Cy^2 + Dx + Ey + F = 0$ is (except in degenerate cases) an equation of a parabola, an ellipse, or a hyperbola.

The Parabola

What do the following have in common?

1. The path of a golf ball in flight
2. The shape of the reflector in an automobile headlight

3. The shape of the mirror or reflector in certain types of telescopes

4. The shape of a cable on a suspension bridge such as the San Francisco Bay Bridge or the George Washington Bridge

The answer is that they all are related to curves called parabolas.

DEFINITION 10.1   Let *l* be a fixed line in the plane and *P* a fixed point not on *l*. The set of all points in the plane equidistant from *l* and *P* is called a **parabola**. The line *l* is called the **directrix** and the point *P* the **focus** of the parabola.

It follows from Definition 10.1 and plane geometry that a parabola is always symmetric with respect to the line through the focus perpendicular to the directrix. This line is called the **axis** of the parabola. The point midway between the focus and the directrix is the **vertex** of the parabola (Figure 10.27(a)). We say that the parabola is in **standard position** if its vertex is the origin and its axis is either the *x* axis or the *y* axis.

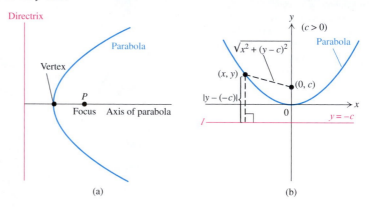

(a)   (b)

FIGURE 10.27   A parabola is symmetric with respect to its axis.

Let us now find an equation of a parabola in standard position. For the present, let us assume that the *y* axis is the axis of the parabola. Thus the focus is the point $(0, c)$ for some $c \neq 0$, and the directrix *l* is the line $y = -c$. (If $c > 0$, then the parabola is as represented in Figure 10.27(b).) By definition a point (x, y) is on the parabola if and only if the distance between (x, y) and *l* is equal to the distance between (x, y) and $(0, c)$. This means that

$$|y - (-c)| = \sqrt{(x - 0)^2 + (y - c)^2}$$

or equivalently

$$\sqrt{(y + c)^2} = \sqrt{x^2 + (y - c)^2}$$

Because the expressions within the square root signs are nonnegative, this is equivalent to the equation

$$(y + c)^2 = x^2 + (y - c)^2$$

or equivalently

$$y^2 + 2cy + c^2 = x^2 + y^2 - 2cy + c^2$$

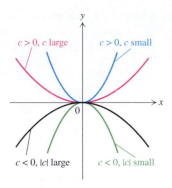

FIGURE 10.28 Parabolas with equation $x^2 = 4cy$ for various values of c.

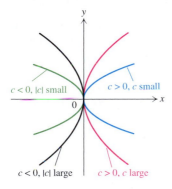

FIGURE 10.29 Parabolas with equation $y^2 = 4cx$ for various values of c.

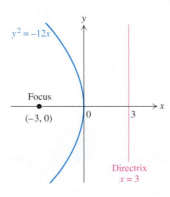

FIGURE 10.30

Simplifying, we obtain

$$x^2 = 4cy \qquad (1)$$

The graph of $x^2 = 4cy$ is symmetric wth respect to the y axis and depends on c. If $|c|$ is small, the focus and directrix are close together and the parabola looks long and slender (Figure 10.28).

Next, we assume that the axis of the parabola is the x axis, so that the focus is the point $(c, 0)$ and directrix is the line $x = -c$ for an appropriate value of c. A similar derivation (with x and y interchanged) yields the equation

$$y^2 = 4cx$$

for the parabola. In this case the parabola is symmetric with respect to the x axis, opening to the right if $c > 0$ and to the left if $c < 0$ (Figure 10.29).

For reference we list the two forms for equations of parabolas in standard position:

$$x^2 = 4cy \quad \begin{cases} \text{focus: } (0, c) \\ \text{directrix: } y = -c \\ \text{symmetry with respect to the } y \text{ axis} \end{cases} \qquad (2)$$

$$y^2 = 4cx \quad \begin{cases} \text{focus: } (c, 0) \\ \text{directrix: } x = -c \\ \text{symmetry with respect to the } x \text{ axis} \end{cases} \qquad (3)$$

Notice also that

$$2|c| = \text{distance from focus to directrix}$$

We can now analyze parabolas by investigating their equations.

EXAMPLE 1   A parabola has focus $(-3, 0)$ and directrix $x = 3$. Find an equation for the parabola, and sketch it.

Solution   The directrix is the line $x = -c$ with $-c = 3$. Thus (3) applies, and an equation of the parabola has the form $y^2 = 4(-3)x = -12x$. The parabola appears in Figure 10.30.   ❏

Parabolas that are not in standard position can be analyzed by translating coordinates by methods of Section 1.5. In particular, a parabola whose vertex is (h, k) and whose axis is parallel to the x axis or the y axis has an equation of the form

$$(y - k)^2 = 4c(x - h) \quad \text{or} \quad (x - h)^2 = 4c(y - k)$$

respectively.

EXAMPLE 2   Show that $x^2 + 2x - 6y - 23 = 0$ is an equation of a parabola. Sketch the parabola, and determine its focus, directrix, and axis.

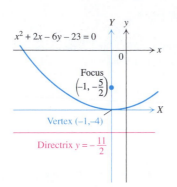

FIGURE 10.31

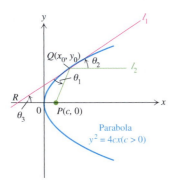

FIGURE 10.32

Very Large Array (VLA) telescopes in New Mexico. (*Gregg Hadel/Tony Stone Images*)

Solution   First we complete the square in x by adding 1 to each side of the equation. We obtain

$$x^2 + 2x + 1 - 6y - 23 = 1$$

which reduces to

$$x^2 + 2x + 1 = 6y + 24, \quad \text{or} \quad (x + 1)^2 = 6(y + 4)$$

In the translated XY system whose origin is $(-1, -4)$, we have $X = x + 1$ and $Y = y + 4$, so that the equation becomes $X^2 = 6Y$. By formula (2) the focus is $(0, \frac{3}{2})$ and the directrix is $Y = -\frac{3}{2}$ in the XY system. Therefore in the xy system the focus is $(-1, -\frac{5}{2})$, the directrix is $y = -\frac{11}{2}$, the vertex is $(-1, -4)$, and the axis is the line $x = -1$. The parabola is shown in Figure 10.31.   ❏

If either $A = 0$ or $C = 0$, but not both, then

$$Ax^2 + Cy^2 + Dx + Ey + F = 0$$

is an equation of a parabola (or a degenerate conic section). Completing the square enables us to locate the vertex, focus, directrix, and axis, as we did in Example 2.

At this point we study special properties of a line tangent to the parabola. In our discussion let us assume that the parabola is represented by the equation $y^2 = 4cx$, where $c > 0$, and let $Q(x_0, y_0)$ be an arbitrary point on the parabola, with $x_0 > 0$ and $y_0 > 0$ (Figure 10.32). We designate the line tangent to the parabola at Q by l_1. Since $y_0 > 0$ by hypothesis and $y = \sqrt{4cx}$ for all points near Q, it follows that

$$\left.\frac{dy}{dx}\right|_{x=x_0} = \left.\frac{4c}{2\sqrt{4cx}}\right|_{x=x_0} = \sqrt{\frac{c}{x_0}}$$

Thus by the point-slope formula, l_1 is given by

$$y - y_0 = \sqrt{\frac{c}{x_0}}\,(x - x_0) \tag{4}$$

We find the x intercept of l_1 by setting $y = 0$ in (4) and using the fact that $y_0 = \sqrt{4cx_0}$. We obtain

$$0 - \sqrt{4cx_0} = \sqrt{\frac{c}{x_0}}\,(x - x_0)$$

so that

$$x - x_0 = -\sqrt{\frac{x_0}{c}}\,\sqrt{4cx_0} = -2x_0$$

This means that $x = -x_0$, and thus the x intercept of l_1 is $-x_0$. With P denoting the focus $(c, 0)$ and R the point $(-x_0, 0)$, we conclude that

$$|PQ| = \sqrt{(c - x_0)^2 + y_0^2} = \sqrt{(c - x_0)^2 + 4cx_0} = \sqrt{(c + x_0)^2} = c + x_0$$
$$= |PR|$$

Consequently the triangle PQR is isosceles.

The preceding discussion relates to one of the most interesting properties of the parabola—the property that makes parabolic mirrors and telescopes so useful. From physics the **Law of Reflection** states that the angle of incidence of a light ray is equal to the angle of reflection. Thus in Figure 10.32 if l_2 is the reflected ray, then $\theta_1 = \theta_2$. However, the fact that $|PQ| = |PR|$ implies that $\theta_1 = \theta_3$. This means that $\theta_2 = \theta_3$, so that by plane geometry the reflected ray l_2 is parallel to the x axis. As a consequence, every light ray emitted from the focus of the mirror is reflected along a line parallel to the axis of the mirror. This is the reason why high-beam headlights are so blinding to drivers of oncoming cars; this is also the reason why, in a reverse way, the rays from far-away planets and stars collect at the eyepiece located at the focus of a reflecting telescope.

The Ellipse

Although ellipses had been studied by the Greeks, they were brought into prominence in the seventeenth century by Johannes Kepler's discovery that planets move around the sun in elliptical orbits. We will prove this amazing result in Section 12.7. Ellipses can be defined in the following way.

DEFINITION 10.2   Let P_1 and P_2 be two points in the plane, and let k be a number greater than the distance between P_1 and P_2. The set of all points P in the plane such that

$$|P_1P| + |P_2P| = k$$

is called an **ellipse**. The points P_1 and P_2 are called the **foci** of the ellipse.

The midpoint of the segment between the two foci is called the **center** of the ellipse (Figure 10.33). We say that an ellipse is in **standard position** if its center lies at the origin and if the foci lie on either the x axis or the y axis. In the event that the foci are the same point, the ellipse is a circle.

Let us now find an equation of an ellipse in standard position. If the foci are the points $(-c, 0)$ and $(c, 0)$, with $c \geq 0$ (Figure 10.33), then the distance between the foci is $2c$. For convenience we let $k = 2a$. It follows by hypothesis that $2a = k > 2c$, that is, $a > c$. Then by the definition of distance, a point (x, y) is on the ellipse if and only if

$$\sqrt{(x - (-c))^2 + (y - 0)^2} + \sqrt{(x - c)^2 + (y - 0)^2} = 2a$$

or

$$\sqrt{(x + c)^2 + y^2} = 2a - \sqrt{(x - c)^2 + y^2}$$

Squaring and canceling, we obtain

$$a^2 - cx = a\sqrt{(x - c)^2 + y^2}$$

Squaring again yields

$$a^4 - 2cxa^2 + c^2x^2 = a^2[(x - c)^2 + y^2]$$

which can be rewritten

$$(a^2 - c^2)x^2 + a^2y^2 = a^2(a^2 - c^2) \tag{5}$$

If we now set

$$b = \sqrt{a^2 - c^2} \tag{6}$$

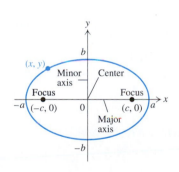

FIGURE 10.33   The ellipse
$\dfrac{x^2}{a^2} + \dfrac{y^2}{b^2} = 1$ $(a > b)$.

then $0 < b \le a$, and division by $a^2 b^2$ in (5) yields

$$\frac{x^2}{a^2} + \frac{y^2}{b^2} = 1 \qquad (7)$$

We have just seen that any point on the ellipse must satisfy equation (7). It is possible to show that any point satisfying equation (7) lies on the ellipse. Therefore (7) is an equation of the ellipse whose foci are $(-c, 0)$ and $(c, 0)$ and for which $k = 2a$.

From (7) we infer that the x intercepts of the ellipse are $-a$ and a, while the y intercepts are $-b$ and b. We call the points $(-a, 0)$ and $(a, 0)$ the **vertices** of the ellipse, the line segment between $(-a, 0)$ and $(a, 0)$ the **major axis** of the ellipse, and the line segment between $(0, -b)$ and $(0, b)$ the **minor axis** (Figure 10.33).

You can check that the upper half of the ellipse is given by

$$y = b \sqrt{1 - \frac{x^2}{a^2}} = \frac{b}{a} \sqrt{a^2 - x^2}$$

Therefore

$$\frac{dy}{dx} = \frac{-bx}{a\sqrt{a^2 - x^2}} \quad \text{and} \quad \frac{d^2y}{dx^2} = \frac{-ab}{(a^2 - x^2)^{3/2}}$$

Since the second derivative is negative, the portion of the ellipse for which $y > 0$ is concave downward. This information is consistent with the sketch in Figure 10.33.

If the foci of the ellipse are on the y axis, say at $(0, c)$ and $(0, -c)$, where $c \ge 0$, then we can find an equation of the ellipse by simply interchanging x and y in (7). We obtain

$$\frac{x^2}{b^2} + \frac{y^2}{a^2} = 1 \qquad (8)$$

The graph of this equation is the ellipse shown in Figure 10.34.

For later reference we list the two forms of equations of ellipses in standard position:

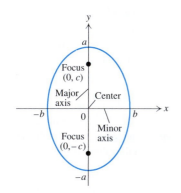

FIGURE 10.34 The ellipse $\dfrac{x^2}{b^2} + \dfrac{y^2}{a^2} = 1 \quad (a > b)$.

$$\frac{x^2}{a^2} + \frac{y^2}{b^2} = 1 \quad (a \ge b > 0) \begin{cases} \text{foci: } (-c, 0) \text{ and } (c, 0), \\ \qquad \text{where } c = \sqrt{a^2 - b^2} \\[2ex] \text{vertices: } (-a, 0) \text{ and } (a, 0) \\[2ex] \text{symmetry with respect to the} \\ \qquad x \text{ axis, } y \text{ axis, and origin} \end{cases} \qquad (9)$$

$$\frac{x^2}{b^2} + \frac{y^2}{a^2} = 1 \quad (a \ge b > 0) \begin{cases} \text{foci: } (0, -c) \text{ and } (0, c), \\ \qquad \text{where } c = \sqrt{a^2 - b^2} \\[2ex] \text{vertices: } (0, -a) \text{ and } (0, a) \\[2ex] \text{symmetry with respect to the} \\ \qquad x \text{ axis, } y \text{ axis, and origin} \end{cases} \qquad (10)$$

The numbers *a, b,* and *c* have the following geometric interpretation:

$$2a = \text{length of the major axis}$$

$$2b = \text{length of the minor axis} \qquad (11)$$

$$2c = \text{the distance between foci}$$

EXAMPLE 3   Sketch the ellipse

$$x^2 + 3y^2 = 3$$

and locate the foci and vertices.

Solution   The given equation is equivalent to

$$\frac{x^2}{3} + \frac{y^2}{1} = 1$$

This is in the form of (9), with $a = \sqrt{3}$ and $b = 1$. Hence the foci and the major axis lie on the *x* axis. Since $a = \sqrt{3}$, the vertices are $(-\sqrt{3}, 0)$ and $(\sqrt{3}, 0)$, and since

$$c = \sqrt{a^2 - b^2} = \sqrt{3 - 1} = \sqrt{2}$$

the foci are the points $(-\sqrt{2}, 0)$ and $(\sqrt{2}, 0)$. The ellipse is shown in Figure 10.35.   ❑

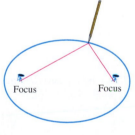

FIGURE 10.35

If the foci and the length $2a$ of the major axis are known, it is possible to draw the ellipse immediately. Simply place a tack at each focus, tie the ends of a string of length $2a$ to the tacks, and then move a pencil around with its point touching the string, keeping the string taut. The resulting figure is by definition an ellipse (Figure 10.36).

When an ellipse is not in standard position, we can analyze the ellipse by completing squares and translating the coordinate axes suitably.

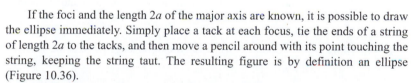

FIGURE 10.36

EXAMPLE 4   Show that

$$9x^2 - 18x + 4y^2 + 16y = 11$$

is an equation of an ellipse, and sketch the ellipse.

Solution   After squares are completed, the given equation becomes

$$9(x^2 - 2x + 1) + 4(y^2 + 4y + 4) = 11 + 9 + 16 = 36$$

so that

$$\frac{(x - 1)^2}{4} + \frac{(y + 2)^2}{9} = 1$$

In the *XY* system with origin $(1, -2)$, the equation becomes

$$\frac{X^2}{4} + \frac{Y^2}{9} = 1$$

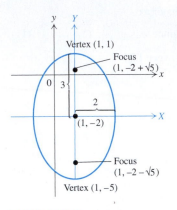

FIGURE 10.37 The ellipse $9x^2 - 18x + 4y^2 + 16y = 11$.

This is an equation of an ellipse whose vertices are $(0, -3)$ and $(0, 3)$ and whose foci are $(0, \sqrt{5})$ and $(0, -\sqrt{5})$ in the *XY* system. Therefore in the *xy* system the center of the ellipse is $(1, -2)$, the vertices are $(1, -5)$ and $(1, 1)$, and the foci are $(1, -2 - \sqrt{5})$ and $(1, -2 + \sqrt{5})$. The ellipse is shown in Figure 10.37.   ❏

Any equation of the form

$$Ax^2 + Cy^2 + Dx + Ey + F = 0$$

in which *A* and *C* are both positive or both negative (so that $AC > 0$), is an equation of an ellipse or a degenerate conic section. (See Exercise 50 for the case in which *A* and *C* are positive.) We can find the center, vertices, and foci of the ellipse by completing the squares and translating the coordinate axes.

As with parabolas, there is a reflection principle associated with ellipses. Suppose a mirror has the shape of an ellipse. If a ray is emitted from one focus, then the reflection principle says that the ray is reflected by the mirror toward the other focus (see the project for this section). The dome of a "whispering gallery" is shaped like the surface formed when an ellipse is revolved about its major axis. Any sound wave emanating from one focus bounces off the dome and is reflected to the other focus. It is said that John Quincy Adams used this attribute of the Statuary Hall of the Capitol in Washington, D.C., in order to eavesdrop on his adversaries.

Finally we note that according to Kepler's laws of planetary motion, planets move in elliptical orbits with the sun at one focus. Some comets, such as Halley's comet, also move in elliptical orbits around the sun. We also note that by classical physics, an electron in an atom moves in an elliptical orbit with the nucleus at one focus.

The Hyperbola

The last type of nondegenerate conic section is the hyperbola.

DEFINITION 10.3

Let P_1 and P_2 be two distinct points in the plane, and let *k* be a positive number less than the distance between P_1 and P_2. The set of all points *P* in the plane such that

$$\| |P_1 P| - |P_2 P| \| = k$$

is called a **hyperbola.** The points P_1 and P_2 are called the **foci** of the hyperbola.

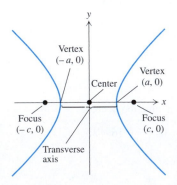

FIGURE 10.38 The hyperbola $\dfrac{x^2}{a^2} - \dfrac{y^2}{b^2} = 1$.

The point midway between the two foci is the **center** of the hyperbola, and the line through the foci is the **principal axis** (Figure 10.38). The points at which the hyperbola meets the principal axis are the **vertices** of the hyperbola, and the line segment they determine is the **transverse axis** of the hyperbola. Finally, we say that a hyperbola is in **standard position** if its center is the origin and its foci are either on the *x* axis or on the *y* axis.

Suppose the foci of a hyperbola in standard position are located at $(-c, 0)$ and $(c, 0)$, with $c > 0$ (as in Figure 10.38), so that the distance between the foci is $2c$. If we let $k = 2a$, then by hypothesis $2a = k < 2c$, so that $a < c$. By definition a point (x, y) is on the hyperbola if and only if

$$| \sqrt{(x + c)^2 + (y - 0)^2} - \sqrt{(x - c)^2 + (y - 0)^2}| = 2a$$

By algebraic manipulations similar to those made for the ellipse, we can transform this equation into the following equation of the hyperbola:

$$\frac{x^2}{a^2} - \frac{y^2}{b^2} = 1$$

where

$$b = \sqrt{c^2 - a^2}$$

For $y > 0$ we find that

$$y = \frac{b}{a} \sqrt{x^2 - a^2} \tag{12}$$

so that

$$\frac{dy}{dx} = \frac{bx}{a\sqrt{x^2 - a^2}} \quad \text{and} \quad \frac{d^2y}{dx^2} = \frac{-ab}{(x^2 - a^2)^{3/2}}$$

Therefore the portion of the hyperbola for which $y > 0$ is concave downward. For accurate graphing of the parts of the hyperbola far from the origin, we use (12), along with Example 2 of Section 4.8, to deduce that

$$\lim_{x \to \infty} \left(y - \frac{b}{a} x \right) = \frac{b}{a} \lim_{x \to \infty} (\sqrt{x^2 - a^2} - x) = 0$$

This means that the hyperbola approaches the line $y = (b/a)x$ as x approaches ∞. Because the hyperbola is symmetric with respect to the x axis, it also approaches the line $y = -(b/a)x$. We call the lines

$$y = \frac{b}{a} x \quad \text{and} \quad y = -\frac{b}{a} x$$

asymptotes of the hyperbola. This information implies that the hyperbola is as sketched in Figure 10.39(a).

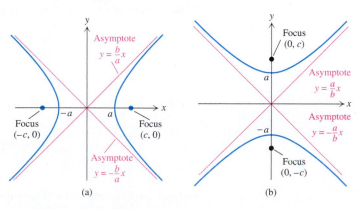

FIGURE 10.39   (a) The hyperbola $\dfrac{x^2}{a^2} - \dfrac{y^2}{b^2} = 1$. (b) The hyperbola $\dfrac{y^2}{a^2} - \dfrac{x^2}{b^2} = 1$.

Now assume that the foci of a hyperbola are located at the points $(0, -c)$ and $(0, c)$ on the y axis, and again let $k = 2a$ in Definition 10.3. Then the equation of the hyperbola becomes

$$\frac{y^2}{a^2} - \frac{x^2}{b^2} = 1$$

where, as before,

$$b = \sqrt{c^2 - a^2}$$

In this case the asymptotes are the lines $y = (a/b)x$ and $y = (-a/b)x$. The hyperbola is shown in Figure 10.39(b).

For reference we list the two forms of equations of hyperbolas in standard position:

$$\frac{x^2}{a^2} - \frac{y^2}{b^2} = 1 \begin{cases} \text{foci: } (-c, 0) \text{ and } (c, 0), \text{ where } c = \sqrt{a^2 + b^2} \\[2ex] \text{vertices: } (-a, 0) \text{ and } (a, 0) \\[2ex] \text{asymptotes: the lines } y = (b/a)x \text{ and } y = (-b/a)x \\[2ex] \text{symmetry with respect to the } x \text{ axis, } y \text{ axis, and origin} \end{cases} \quad (13)$$

$$\frac{y^2}{a^2} - \frac{x^2}{b^2} = 1 \begin{cases} \text{foci: } (0, -c) \text{ and } (0, c), \text{ where } c = \sqrt{a^2 + b^2} \\[2ex] \text{vertices: } (0, -a) \text{ and } (0, a) \\[2ex] \text{asymptotes: the lines } y = (a/b)x \text{ and } y = (-a/b)x \\[2ex] \text{symmetry with respect to the } x \text{ axis, } y \text{ axis, and origin} \end{cases} \quad (14)$$

The numbers a and c have the following geometric interpretation:

$$2a = \text{length of transverse axis}$$
$$2c = \text{distance between foci}$$

EXAMPLE 5   Find the foci and the asymptotes of the hyperbola whose equation is

$$9x^2 - 16y^2 = 144$$

and sketch the hyperbola.

Solution   Dividing both sides of the equation by 144 yields

$$\frac{x^2}{16} - \frac{y^2}{9} = 1$$

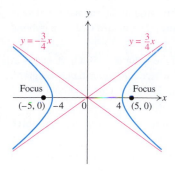

FIGURE 10.40   The hyperbola $9x^2 - 16y^2 = 144$.

This equation is in the form of (13), with $a = 4$ and $b = 3$. Therefore $c = \sqrt{4^2 + 3^2} = 5$, so that the foci are located at $(-5, 0)$ and $(5, 0)$. The asymptotes are the lines $y = \frac{3}{4}x$ and $y = -\frac{3}{4}x$. The graph is shown in Figure 10.40. ❑

If a hyperbola is situated so that a translation of the axes brings it into standard position, then we can readily sketch it.

EXAMPLE 6   Show that

$$16y^2 - x^2 - 6x - 32y = 57$$

is an equation of a hyperbola, and sketch the hyperbola.

Solution   By completing the squares we find that

$$16(y^2 - 2y + 1) - (x^2 + 6x + 9) = 57 + 16 - 9 = 64$$

so that

$$\frac{(y - 1)^2}{4} - \frac{(x + 3)^2}{64} = 1$$

If $X = x + 3$ and $Y = y - 1$, then an equation in the XY system with origin at $(-3, 1)$ is

$$\frac{Y^2}{4} - \frac{X^2}{64} = 1$$

This equation is in the form of (14), with $a = 2$ and $b = 8$. Consequently its graph is a hyperbola. Since

$$c = \sqrt{2^2 + 8^2} = 2\sqrt{17}$$

the foci are $(0, 2\sqrt{17})$ and $(0, -2\sqrt{17})$ in the XY system. Moreover, the asymptotes are $Y = \frac{1}{4}X$ and $Y = -\frac{1}{4}X$. Translating to the xy system, we find that the foci are $(-3, 1 + 2\sqrt{17})$ and $(-3, 1 - 2\sqrt{17})$, and that the asymptotes are the lines

$$y - 1 = \frac{1}{4}(x + 3) \quad \text{and} \quad y - 1 = -\frac{1}{4}(x + 3)$$

The hyperbola is sketched in Figure 10.41. ❑

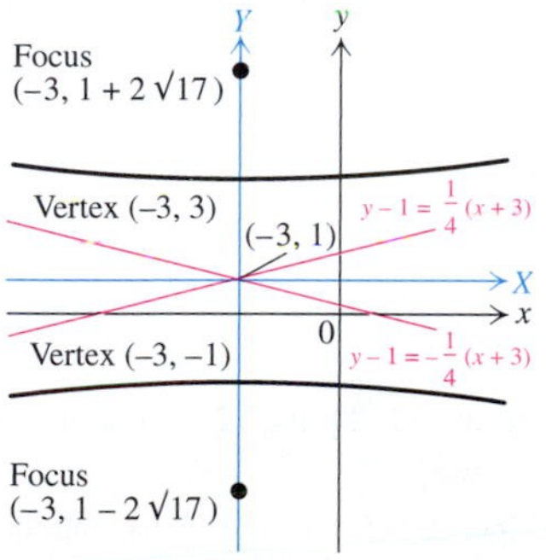

FIGURE 10.41

Any equation of the form

$$Ax^2 + Cy^2 + Dx + Ey + F = 0$$

with A and C of opposite sign (that is, $AC < 0$) is an equation of a hyperbola or a degenerate conic section (see Exercise 51). We can sketch the hyperbola after completing the squares and translating the coordinate axes. There is a reflection property for hyperbolas, just as for parabolas and ellipses. A ray aimed at one focus is reflected by the hyperbola toward the other focus (see Figure 10.42).

Next we mention a few applications of hyperbolas.

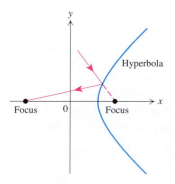

FIGURE 10.42

1. Comets that do not move in elliptical orbits around the sun almost always move in hyperbolic orbits. (In theory they can also move in parabolic orbits.)

2. Boyle's Law, relating the pressure p and the volume V of a perfect gas at constant temperature, states that $pV = c$, where c is a constant. The graph of p as a function of V is a hyperbola. In Example 1 of Section 10.4 we will demonstrate this for the case $c = 1$.

3. Hyperbolas can be used to locate the source of a sound heard at three different locations (see Exercises 58 and 59).

4. Gears in the differential of some automobiles are cut according to arcs of hyperbolas. Such cars have what is called a "hypoid" axle. (This was introduced by Packard in about 1935.)

Conclusion

The conic sections have played a vital role in mathematics for the past two thousand years. As we have noted, the three basic kinds of conic sections appear in applications of many varieties. They will appear later in the book, principally in Section 12.7 on Kepler's laws of motion of planets, and in Chapters 13 to 15 during the study of calculus in three dimensions. Finally we mention that during the past three centuries telescopes have featured parabolas, ellipses, and hyperbolas. Indeed, the early reflecting telescopes used two parabolic mirrors. The Gregory telescope combined a parabolic mirror with an elliptical mirror, and the Cassegrain telescope featured a parabolic mirror combined with a hyperbolic mirror.

EXERCISES 10.3

In Exercises 1–14 find an equation of the conic section with the given properties. Then sketch the conic section.

1. The focus of the parabola is $(-2, 0)$, and the directrix is $x = 2$.

2. The focus of the parabola is $(0, -6)$, and the directrix is $y = 6$.

3. The vertex of the parabola is $(1, 0)$, and the directrix is $x = -2$.

4. The focus of the parabola is $(3, 3)$, and the vertex is $(3, 2)$.

5. The parabola is in standard position, is symmetric with respect to the x axis, and passes through the point $(-1, 1)$.

6. The parabola is the collection of points (x, y) whose distance from $(3, 4)$ is the same as the distance from the line $y = 2$.

7. The foci of the ellipse are $(2, 0)$ and $(-2, 0)$, and the vertices are $(3, 0)$ and $(-3, 0)$.

8. The foci of the ellipse are $(2, 1)$ and $(2, -1)$, and the length of the major axis is 4.

9. The vertices of the ellipse are $(5, 0)$ and $(-5, 0)$, and the ellipse passes through $(3, -4)$.

10. The ellipse passes through $(-1, 1)$ and $(\frac{1}{2}, -2)$ and is in standard position.

11. The foci of the hyperbola are $(9, 0)$ and $(-9, 0)$, and the vertices are $(4, 0)$ and $(-4, 0)$.

12. The foci of the hyperbola are $(\sqrt{5}, 0)$ and $(-\sqrt{5}, 0)$, and the asymptotes are $y = 2x$ and $y = -2x$.

13. The foci of the hyperbola are $(0, 9)$ and $(0, -1)$, and the asymptotes are $y = \frac{4}{3}x + 4$ and $y = -\frac{4}{3}x + 4$.

14. The vertices of the hyperbola are $(4, 0)$ and $(-4, 0)$, and the asymptotes are perpendicular to one another.

In Exercises 15–26 find the foci, vertices, directrix, axis, and asymptotes, where applicable.

15. $y^2 = 3x$

16. $y^2 = -x/2$

17. $(x-1)^2 = y+2$

18. $(y+3)^2 = 4x-3$

19. $\dfrac{x^2}{9} + \dfrac{y^2}{25} = 1$

20. $3x^2 + 4y^2 = 1$

21. $4(x-1)^2 + y^2 = 1$

22. $25(x+1)^2 + (y-3)^2 = 1$

23. $\dfrac{x^2}{9} - \dfrac{y^2}{16} = 1$

24. $5y^2 - \dfrac{x^2}{4} = 1$

25. $\dfrac{(x+3)^2}{25} - \dfrac{(y+1)^2}{144} = 1$

26. $\dfrac{(y-2)^2}{121} - \dfrac{(x+2)^2}{121} = 1$

In Exercises 27–38 show that the equation represents a conic section. Sketch the conic section, and indicate all pertinent information (such as foci, directrix, asymptotes, and so on).

27. $x^2 - 6x - 2y + 1 = 0$

28. $2x^2 + 4x - 5y + 7 = 0$

29. $3y^2 - 5x + 3y = \frac{17}{4}$

30. $-4y^2 - \dfrac{x}{2} + 4y = 1$

31. $x^2 + 2y^2 - 2x - 4y = 1$

32. $9x^2 + 4y^2 - 36x + 8y + 4 = 0$

33. $x^2 - 8x + 2y^2 + 12 = 0$

34. $8x^2 + 8x + 2y^2 - 20y = 12$

35. $x^2 - y^2 + 6x + 12y = 36$

36. $x^2 - 2x - 4y^2 - 12y = -8$

37. $4x^2 - 9y^2 - 8x - 36y = 68$

38. $4x^2 - 16x - 9y^2 - 54y = 101$

39. Find an equation of the parabola that has a vertical axis, its vertex at $(-1, 3)$, and slope 2 at $x = 1$.

40. Find the point on the parabola $y^2 = 4x$ that is closest to the point $(1, 0)$.

41. The line segment that passes through the focus, is parallel to the directrix, and has its endpoints on the parabola is called the **latus rectum.** Show that if a parabola is in standard position and the focus is c units from the origin, then the length of the latus rectum is $4c$.

42. Find the number d such that the line $x + y = d$ is tangent to the parabola $x^2 = 2y$, and determine the point of tangency.

*43. Find equations for the two lines that are tangent to the parabola $x^2 = 2y$ and pass through $(-1, -4)$.

44. Find an equation of the ellipse in standard position that passes through $(0, 2)$ and has slope $1/\sqrt{2}$ at the point $(-2, y)$ with $y > 0$.

45. Find two values of d such that the line $2x + y = d$ is tangent to the ellipse $4x^2 + y^2 = 8$. Find the points of tangency.

46. Find the values of d for which the line $2y - x = d$ is tangent to the hyperbola $6y^2 - 3x^2 = 9$.

47. Find an equation for the collection of points for which the distance to $(3, 0)$ is half the distance to the line $x = -3$. Show that your equation is an equation of an ellipse.

48. Find an equation for the collection of points for which the distance to $(3, 0)$ is twice the distance to the line $x = -3$. Show that the equation represents a hyperbola.

49. Suppose a rectangle with horizontal and vertical sides is to be inscribed in the ellipse

$$\frac{x^2}{a^2} + \frac{y^2}{b^2} = 1$$

What location of the vertices of the rectangle will yield the largest area?

50. Consider the equation

$$Ax^2 + Cy^2 + Dx + Ey + F = 0 \qquad (15)$$

where A and C are positive, and let

$$r = \frac{D^2}{4A} + \frac{E^2}{4C} - F$$

By completing squares in (15), show that

a. if $r > 0$, then the graph of the equation is an ellipse.

b. if $r = 0$, then the graph of the equation is a point.

c. if $r < 0$, then the graph of the equation consists of no points.

51. Consider the equation

$$Ax^2 + Cy^2 + Dx + Ey + F = 0, \quad \text{where } AC < 0 \qquad (16)$$

and let

$$r = \frac{D^2}{4A} + \frac{E^2}{4C} - F$$

a. Show that the graph of equation (16) is a hyperbola if $r \neq 0$.

b. Show that the graph of equation (16) is a pair of intersecting lines if $r = 0$.

52. Show that an equation of the line tangent to the ellipse $x^2/a^2 + y^2/b^2 = 1$ at the point (x_0, y_0) is

$$\frac{xx_0}{a^2} + \frac{yy_0}{b^2} = 1$$

53. Let $y_0 \neq 0$. Show that an equation of the line tangent to the hyperbola $x^2/a^2 - y^2/b^2 = 1$ at the point (x_0, y_0) is

$$\frac{xx_0}{a^2} - \frac{yy_0}{b^2} = 1$$

54. Show that any line parallel to, but distinct from, an asymptote of a hyperbola intersects the hyperbola exactly once.

Applications

55. Suppose a golf ball is driven so that it travels a distance of 600 feet as measured along the ground and reaches an altitude of 200 feet. If the origin represents the tee and if the ball travels along a parabolic path over the positive x axis, find an equation for the path of the golf ball.

56. The planet Mars travels around the sun in an elliptical orbit whose equation is approximately

$$\frac{x^2}{(228)^2} + \frac{y^2}{(227)^2} = 1$$

where x and y are measured in millions of kilometers. Find the ratio of the length of the major axis to the length of the minor axis.

57. The Ellipse in Washington, D.C., has a major axis approximately 0.285 mile long and a minor axis approximately 0.241 mile long. Determine the distance from a vertex to the closest focus of the Ellipse.

The Ellipse in Washington, D.C. (*Everett C. Johnson/Folio, Inc.*)

58. Thunder is heard by Marian and Jack, who are talking to each other by telephone, 8800 feet apart. Marian hears the thunder 4 seconds before Jack does. Sketch a graph of the locations where the lightning could have struck. Take the speed of sound to be 1100 feet per second. (*Hint:* Suppose Marian is at $(4400, 0)$ and Jack is at $(-4400, 0)$. Use Definition 10.3 and (13).)

*59. In Exercise 58, suppose that Marian, Jack, and Bruce are on a conference call, with Bruce midway between Marian

and Jack. If Bruce hears the thunder 1 second after Marian does, determine where the lightning strikes in relation to the three persons involved.

In a suspension bridge the cable hangs in the shape of a parabola. The horizontal distance between the supports is called the **span,** and the vertical distance between the points where the cable is attached to the supports and the lowest point of the cable is called the **sag** of the cable (Figure 10.43). In Exercises 60–62 assume that the origin is at the lowest point on the cable.

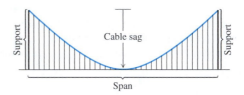

FIGURE 10.43 Suspension bridge.

60. Assume that the span and sag of a suspension bridge are a and b, respectively. Determine an equation of the parabola that represents the hanging cable. (*Hint:* Let the y axis lie along the axis of the parabola.)

61. The George Washington Bridge across the Hudson River from New York to New Jersey has a span of 3500 feet and a sag of 316 feet. Determine an equation of the parabola that represents the cable.

62. Suppose an architect wishes to design a suspension bridge for automobile traffic across the Mississippi River, and needs to make the span a half-mile long. For aesthetic reasons the architect feels that the angle the cable makes with the support should be 30°. What would be the sag?

Project

1. This project is designed to prove that an ellipse has the "reflection property": a ray emitted from one focus and reflected by the ellipse then passes through the other focus.

To that end, let

$$\frac{x^2}{a^2} + \frac{y^2}{b^2} = 1$$

be a given ellipse, and let $(-c, 0)$ and $(c, 0)$ be the foci (see Figure 10.44).

a. Let $P = (x_0, y_0)$ be a point on the ellipse, but not a vertex of the ellipse. Find the slope m of the line L

tangent to the ellipse at P.

b. Suppose that a ray of light L_1 from the focus $(c, 0)$ strikes the ellipse at P. Find the slope m_1 of L_1. Similarly, if a ray of light L_2 from the focus $(-c, 0)$ strikes the ellipse at P, find the slope m_2 of L_2.

c. Let θ_1 be the angle between L and L_1, and let θ_2 be the angle between L and L_2. Find $\tan \theta_1$ and $\tan \theta_2$.

d. Show that $\theta_1 = \theta_2$.
From physics we know that the angle of incidence in a mirror equals the angle of reflection. Therefore a ray from the focus $(c, 0)$ to P is reflected to the other focus, $(-c, 0)$.

e. Using ideas similar to those in (a)–(d), prove the reflection property for hyperbolas: A ray emitted from any point and aimed at one focus, as in Figure 10.42, is reflected by the hyperbola toward the other focus.

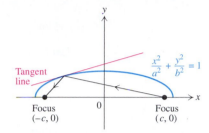

FIGURE 10.44   Figure for the Project.

10.4   ROTATION OF AXES

Let us summarize the information we have about conic sections thus far. Except in degenerate cases, the graph of the equation

$$Ax^2 + Cy^2 + Dx + Ey + F = 0 \qquad (1)$$

is

1. a parabola if $AC = 0$ but not both A and C are zero.

2. an ellipse if $AC > 0$.

3. a hyperbola if $AC < 0$.

Now we would like to analyze the graph of any second-degree equation of the form

$$Ax^2 + Bxy + Cy^2 + Dx + Ey + F = 0 \qquad (2)$$

Since (2) reduces to (1) when $B = 0$, we assume in this section that $B \neq 0$. We will show that in the XY coordinate system obtained by rotating the x and y axes through a suitable angle about the origin, (2) reduces to

$$A'X^2 + C'Y^2 + D'X + E'Y + F' = 0 \qquad (3)$$

This equation has the same form as (1) and has already been analyzed thoroughly. The purpose of rotating the x and y axes is to eliminate the xy term in (2).

To gain an understanding of the relationship between such an XY coordinate system and the given xy system, we assume that the XY coordinate system is obtained by rotating the x and y axes through an angle θ about the origin (Figure 10.45). Then any point P in the plane has coordinates (x, y) and (X, Y) in the two systems. If P is not the origin, let ϕ be the angle from the positive X axis to the line segment joining P to the origin, and let r be the length of that line segment (Figure 10.45). Then we have

$$X = r \cos \phi \quad \text{and} \quad Y = r \sin \phi$$

and

$$x = r \cos (\theta + \phi) \quad \text{and} \quad y = r \sin (\theta + \phi)$$

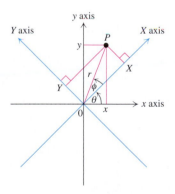

FIGURE 10.45   Rotation of axes.

Using the trigonometric identities for $\sin(\theta + \phi)$ and $\cos(\theta + \phi)$, we obtain

$$x = r\cos(\theta + \phi) = r(\cos\theta\cos\phi - \sin\theta\sin\phi)$$
$$= (\cos\theta)(r\cos\phi) - (\sin\theta)(r\sin\phi)$$

and thus

$$x = (\cos\theta)X - (\sin\theta)Y \tag{4}$$

Similarly

$$y = r\sin(\theta + \phi) = r(\sin\theta\cos\phi + \cos\theta\sin\phi)$$
$$= (\sin\theta)(r\cos\phi) + (\cos\theta)(r\sin\phi)$$

and thus

$$y = (\sin\theta)X + (\cos\theta)Y \tag{5}$$

We can now substitute (4) and (5), respectively, for x and y in (2); we conclude that

$$A[(\cos\theta)X - (\sin\theta)Y]^2 + B[(\cos\theta)X - (\sin\theta)Y][(\sin\theta)X + (\cos\theta)Y]$$
$$+ C[(\sin\theta)X + (\cos\theta)Y]^2 + D[(\cos\theta)X - (\sin\theta)Y]$$
$$+ E[(\sin\theta)X + (\cos\theta)Y] + F = 0$$

Combining like terms (those involving X^2, XY, Y^2, and so on) leads to an equation of the form

$$A'X^2 + [-2A\cos\theta\sin\theta + B(\cos^2\theta - \sin^2\theta) + 2C\cos\theta\sin\theta]XY$$
$$+ C'Y^2 + D'X + E'Y + F' = 0 \tag{6}$$

The exact expressions for A', C', D', E', and F' are irrelevant at this moment. What is important is that the equation in (6) will be in the form of (3) if the expression in brackets is 0. By the double-angle formulas, this means that

$$-A\sin 2\theta + B\cos 2\theta + C\sin 2\theta = 0$$

or

$$(A - C)\sin 2\theta = B\cos 2\theta \tag{7}$$

If $A = C$, the equation in (7) is satisfied by $\theta = \pi/4$. If $A \neq C$, then $\cos 2\theta \neq 0$, so we can divide by $(A - C)\cos 2\theta$ and deduce that

$$\tan 2\theta = \frac{\sin 2\theta}{\cos 2\theta} = \frac{B}{A - C}$$

Thus the expression in the brackets in (6) will be 0 if we choose θ so that

$$\tan 2\theta = \frac{B}{A - C} \quad \text{if } A \neq C$$
$$\theta = \frac{\pi}{4} \quad \text{if } A = C \tag{8}$$

The discussion above yields the following procedure for converting an equation in the form of (2) to an equivalent equation in the form of (3):

1. Determine θ (which may be chosen to be between 0 and $\pi/2$) by using (8) with the values of A, B, and C for the given equation.
2. Substitute the value of θ found in step 1 into (4) and (5) in order to express x and y in terms of X and Y.

3.   Substitute the expressions for x and y found in step 2 into the given equation.

4.   Simplify the equation resulting from step 3 in order to obtain an equation in the form of (3).

This procedure is illustrated in the next two examples.

EXAMPLE 1   Show that the graph of the equation

$$xy = 1$$

is a hyperbola, and find the angle θ between 0 and $\pi/2$ through which the x and y axes must be rotated for the hyperbola to be in standard position in the rotated system. Then sketch the hyperbola.

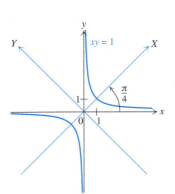

FIGURE 10.46

Solution   For the equation $xy = 1$ we have $A = 0 = C$ in (2), so that $\theta = \pi/4$ in (8). Since $\cos \pi/4 = \sin \pi/4 = \sqrt{2}/2$, formulas (4) and (5) become

$$x = \frac{\sqrt{2}}{2}(X - Y) \quad \text{and} \quad y = \frac{\sqrt{2}}{2}(X + Y)$$

Therefore the equation $xy = 1$ becomes

$$\left(\frac{\sqrt{2}}{2}(X - Y)\right)\left(\frac{\sqrt{2}}{2}(X + Y)\right) = 1 \quad \text{or} \quad \frac{X^2}{2} - \frac{Y^2}{2} = 1$$

This is an equation of a hyperbola with vertices $(-\sqrt{2}, 0)$ and $(\sqrt{2}, 0)$ in the XY coordinate system. Since the x and y axes were rotated through an angle of $\pi/4$ to obtain the XY system, the hyperbola is as shown in Figure 10.46.   ❑

EXAMPLE 2   Use rotation of axes to eliminate the xy term in the equation

$$73x^2 - 72xy + 52y^2 + 30x + 40y - 75 = 0 \qquad (9)$$

Determine the type of conic section it represents, and sketch the graph of the equation.

Solution   In this example $A = 73$, $B = -72$, and $C = 52$, so that the equation (8) becomes

$$\tan 2\theta = \frac{B}{A - C} = -\frac{72}{21} = -\frac{24}{7}$$

Since $\tan 2\theta < 0$ and $0 < \theta < \pi/2$, we have $\pi/2 < 2\theta < \pi$, so that $\cos 2\theta < 0$. From the Pythagorean Theorem we find that the hypotenuse of the triangle in Figure 10.47 is 25. Therefore

$$\cos 2\theta = -\frac{7}{25}$$

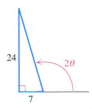

FIGURE 10.47

Since $0 < \theta < \pi/2$, both $\cos \theta$ and $\sin \theta$ are nonnegative. This implies that

$$\cos \theta = \sqrt{\frac{1 + \cos 2\theta}{2}} = \sqrt{\frac{1 - 7/25}{2}} = \frac{3}{5}$$

and

$$\sin \theta = \sqrt{\frac{1 - \cos 2\theta}{2}} = \sqrt{\frac{1 + 7/25}{2}} = \frac{4}{5}$$

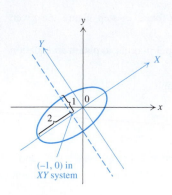

FIGURE 10.48 The ellipse $73x^2 - 72xy + 52y^2 + 30x + 40y - 75 = 0$.

From (4) and (5) we find that

$$x = \frac{3X - 4Y}{5} \quad \text{and} \quad y = \frac{4X + 3Y}{5}$$

Now we substitute for x and y in (9) to obtain

$$\frac{73}{25}(3X - 4Y)^2 - \frac{72}{25}(3X - 4Y)(4X + 3Y) + \frac{52}{25}(4X + 3Y)^2$$
$$+ \frac{30}{5}(3X - 4Y) + \frac{40}{5}(4X + 3Y) - 75 = 0$$

Expanding the squared expressions and combining like terms, we derive the equation

$$25X^2 + 100Y^2 + 50X - 75 = 0$$

Completing the square and dividing by 100 yields

$$\frac{(X + 1)^2}{4} + Y^2 = 1$$

We recognize this as an equation of an ellipse. The graph is shown in Figure 10.48, along with both sets of axes. ❑

Although the exact expressions for A' or C' in (3) are of no theoretical interest to us, it is possible to show (see Exercise 17) that

$$B^2 - 4AC = -4A'C' \tag{10}$$

The number $4A'C'$ is valuable because, as we infer from the three criteria listed at the beginning of this section, the sign of $A'C'$ (and hence of $4A'C'$) determines the type of conic section involved. Combining (10) with these three criteria, we conclude that if the graph of

$$Ax^2 + Bxy + Cy^2 + Dx + Ey + F = 0 \tag{11}$$

is not degenerate, then it is

1. a parabola if $B^2 - 4AC = 0$.
2. an ellipse if $B^2 - 4AC < 0$.
3. a hyperbola if $B^2 - 4AC > 0$.

The expression $B^2 - 4AC$ is called the **discriminant** of (11), and indeed it serves to discriminate between the different types of conic sections.

EXERCISES 10.4

In Exercises 1–10 remove the xy term by rotation of axes. Then decide what type of conic section is represented by the equation, and sketch its graph.

1. $xy = -4$
2. $x^2 + \sqrt{3}xy = 3$
3. $x^2 - xy + y^2 = 2$
4. $9x^2 - 24xy + 2y^2 - 75 = 0$
5. $145x^2 + 120xy + 180y^2 = 900$
6. $10x^2 - 12xy + 10y^2 - 16\sqrt{2}x + 16\sqrt{2}y = 16$
7. $16x^2 - 24xy + 9y^2 - 5x - 90y + 25 = 0$

8. $16x^2 + 24xy + 9y^2 + 100x - 50y = 0$

9. $2x^2 - 72xy + 23y^2 + 100x - 50y = 0$

10. $2x^2 + 4xy + 2y^2 + 28\sqrt{2}x - 12\sqrt{2}y + 16 = 0$

In Exercises 11–13 use rotation of axes to show that the graph of the given equation is a degenerate conic section.

11. $9x^2 - 24xy + 2y^2 = 0$

12. $145x^2 + 120xy + 180y^2 = 0$

13. $145x^2 + 120xy + 180y^2 = -900$

14. Show that if $B > 0$, then the graph of
$$x^2 + Bxy = F$$
is a hyperbola if $F \neq 0$, and two intersecting lines if $F = 0$.

15. Assume that $B \neq 0$. Describe the graph of
$$Bxy + Dx + Ey + F = 0$$

16. Let R be the region between the line $y = x$ and the parabola $x^2 + 2xy + y^2 - \sqrt{2}x + \sqrt{2}y = 2$. Let D be the solid region obtained by revolving R about the line $y = x$. Find the volume V of D.

*17. Determine formulas for the numbers A' and C' in (6), and show that if θ satisfies (8), then
$$B^2 - 4AC = -4A'C'$$

Project

1. In Examples 1 and 2 we were given an equation for a conic section in the xy coordinate system, and we determined its equation in the XY coordinate system obtained by rotation of the axes through an angle θ. Suppose we wish to reverse the procedure and find an equation of a conic section in the xy system from knowledge of its equation in the XY system.

   a. How would you modify the equations in (4) and (5) to obtain equations expressing X and Y in terms of x and y?

   b. Suppose the XY system is obtained by rotating the x and y axes through $\pi/4$ radians (in the counterclockwise direction). Find an equation in the xy system of the parabola with equation $Y = 4X^2$.

10.5   A UNIFIED DESCRIPTION OF CONIC SECTIONS

We defined the parabola in terms of a focus and a directrix, whereas we defined the ellipse and the hyperbola in terms of two foci. In this section we will show that ellipses and hyperbolas can also be described in terms of a focus and a directrix. Such a formulation lends itself to a description of the conic sections in polar coordinates. We will give such a description at the conclusion of the section.

Eccentricity

The key to the unified treatment of the conic sections is the concept of eccentricity. The **eccentricity** e of an ellipse or a hyperbola is defined by

$$e = \frac{\text{distance between the foci}}{\text{distance between the vertices}}$$

For an ellipse the distance between the foci is $2c$ and the distance between the vertices is $2a$. Consequently the formula for e becomes

$$e = \frac{c}{a} \tag{1}$$

From the definition of b in (6) of Section 10.3 we know that

$$c^2 = a^2 - b^2 \tag{2}$$

so that $0 \leq e < 1$. From (1) and (2) we see that

$$e^2 = 1 - \frac{b^2}{a^2} \tag{3}$$

Thus the larger e is, the smaller b^2/a^2 is, and hence the more disparate the major and minor axes are (Figure 10.49(a) and (b)). In the extreme case in which $e = 0$ we find that $a = b$, so that the ellipse is a circle.

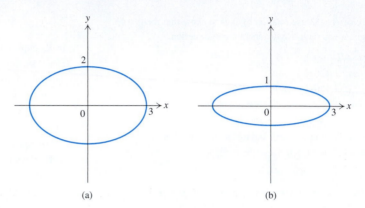

(a)   (b)

FIGURE 10.49 (a) An ellipse with eccentricity $e = \dfrac{1}{3}\sqrt{5} \approx 0.745$. (b) An ellipse with eccentricity $e = \dfrac{2}{3}\sqrt{2} \approx 0.943$.

Now let us assume that the ellipse has eccentricity $e \neq 0$ and is in standard position with one focus P_1 located at $(c, 0)$ and equation

$$\frac{x^2}{a^2} + \frac{y^2}{b^2} = 1$$

(Figure 10.50). If $P(x, y)$ is any point on the ellipse, then from (1)–(3) we deduce that

$$|PP_1| = \sqrt{(x - c)^2 + y^2} = \sqrt{x^2 - 2cx + c^2 + b^2\left(1 - \frac{x^2}{a^2}\right)}$$

$$= \sqrt{x^2\left(1 - \frac{b^2}{a^2}\right) - 2cx + c^2 + b^2}$$

$$= \sqrt{e^2x^2 - 2aex + a^2} = \sqrt{(ex - a)^2} = |ex - a| = e\left|x - \frac{a}{e}\right|$$

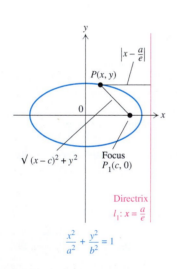

FIGURE 10.50

If we let l_1 denote the line $x = a/e$, then the distance $|Pl_1|$ from (x, y) to l_1 is $|x - a/e|$ (Figure 10.50). Consequently

$$|PP_1| = e\,|Pl_1| \tag{4}$$

We have assumed that $0 < e < 1$. But if we were to let $e = 1$ in (4), then the equation would become $|PP_1| = |Pl_1|$, which is the defining equation for a parabola with directrix l_1. With this in mind, we call the line l_1 given by $x = a/e$ a **directrix** of the ellipse. (A second directrix l_2, defined to be the line $x = -a/e$, arises in a similar way if we let P_2 be the second focus $(-c, 0)$.) Figure 10.51 indicates the relative positions of the focus and the directrix when the eccentricity varies but the major axis remains the same.

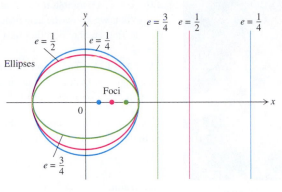

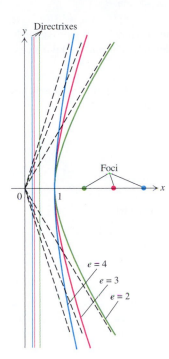

FIGURE 10.52

FIGURE 10.51

If the conic section is a hyperbola whose foci are $2c$ apart and whose transverse axis has length $2a$, the eccentricity is again given by

$$e = \frac{c}{a}$$

Since $c^2 = a^2 + b^2$, it follows that $e > 1$. In fact,

$$e^2 = 1 + \frac{b^2}{a^2}$$

For a hyperbola in standard position, if P_1 is a focus located at $(c, 0)$ and if $P(x, y)$ is any point on the hyperbola, then calculations similar to those performed above for the ellipse establish that

$$|PP_1| = \sqrt{(x - c)^2 + y^2} = e\left|x - \frac{a}{e}\right|$$

Once again we let l_1 denote the line $x = a/e$ and call l_1 a **directrix** of the hyperbola. We conclude that

$$|PP_1| = e\,|Pl_1| \tag{5}$$

This is the same formula as in (4). However, now we have $e > 1$, rather than $0 < e < 1$. By the same method you can show that if $P_2 = (-c, 0)$ and if l_2 denotes the line $x = -a/e$ (also called a directrix of the hyperbola), then any point P on the hyperbola satisfies

$$|PP_2| = e\,|Pl_2|$$

Figure 10.52 shows the relative positions of the focus and the directrix when the eccentricity varies but the transverse axis remains the same.

The formula $|PP_1| = e\,|Pl_1|$ holds for any noncircular ellipse, parabola, or hyperbola, regardless of its position in the xy plane, because translations and rotations of the coordinate axes do not change distances between fixed points and lines. It can be proved that, conversely, if l_1 is a line, P_1 a point not on l_1, and $e > 0$,

A "fish-eye" view of Mars produced from images made by the Viking Orbiter. (*U.S. Geological Survey Science Photo Library/ Photo Researchers*)

Planet	Eccentricity	Major axis ($\times 10^6$ km)
Mercury	0.206	116
Venus	0.007	216
Earth	0.017	299
Mars	0.093	456
Jupiter	0.048	1557
Saturn	0.056	2854
Uranus	0.047	5738
Neptune	0.008	8996
Pluto	0.249	11,800

Polar Equations for the Conic Sections

then the points P in the plane that satisfy the equation $|PP_1| = e|Pl_1|$ form an ellipse if $0 < e < 1$, a parabola if $e = 1$, and a hyperbola if $e > 1$.

When we know the eccentricity of a conic section and the value of a, b, or c, we can find an equation that best describes the conic section.

EXAMPLE 1   The elliptical orbit of Mars has an eccentricity of approximately 0.093, and its major axis has a length of approximately 456 million kilometers. Find an equation for the orbit.

Solution   Let us assume that the center is at the origin and that the major axis lies on the x axis and has length 456. Then $a = 228$, and therefore

$$c = ae = (228)(0.093) = 21.204$$

As a result,

$$b = \sqrt{a^2 - c^2} = \sqrt{(228)^2 - (21.204)^2} \approx 227$$

Consequently the orbit is given (approximately) by the equation

$$\frac{x^2}{(228)^2} + \frac{y^2}{(227)^2} = 1 \qquad \square$$

As you can see from Example 1, for Mars the very small eccentricity corresponds to an elliptical orbit that is nearly circular. In fact, it is apparent from the table in the sidebar that the eccentricities of most other planets are much smaller and hence their orbits are more nearly circular than is the orbit of Mars. However, the eccentricity of Mars was just large enough to lead the great German astronomer and mathematician Johannes Kepler (1571–1630) to abandon the generally accepted notion that the orbit of Mars was circular or a combination of circles, and to conclude that the orbit was actually elliptical. Kepler's conclusion came after he had made unbelievably many calculations over several years, his calculations based on incredibly accurate astronomical data of the Danish astronomer Tycho Brahe (1546–1601).

Now that we have described all noncircular conic sections in terms of distances from foci and directrixes, we will derive polar equations for these conic sections.

Let us consider a conic section having eccentricity $e > 0$, focus P_1 located at the origin, and corresponding directrix l_1 perpendicular to the x axis. Suppose l_1 lies to the right of the origin, with equation $x = k$ (Figure 10.53). If $P(x, y)$ is any point on the conic section to the left of l_1 with polar coordinates (r, θ), where $r > 0$, then

$$|PP_1| = e|Pl_1|$$

Since

$$|PP_1| = r \quad \text{and} \quad |Pl_1| = k - x = k - r\cos\theta$$

it follows that

$$r = |PP_1| = e|Pl_1| = e(k - r\cos\theta)$$

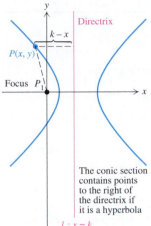

FIGURE 10.53 Conic section with eccentricity e, focus at the origin, and directrix $x = k$.

Solving for r, we obtain

$$r = \frac{ek}{1 + e \cos \theta} \tag{6}$$

Equation (6) completely describes the given conic section if it is a parabola or an ellipse, because all the points on the conic section lie to the left of l_1. However, if the conic section is a hyperbola, then it contains points to the right of l_1 as well (see Figure 10.53).

If we had chosen the coordinate system so that the directrix was the line $x = -k$, where $k > 0$, then a similar argument would demonstrate that an equation of the conic section is

$$r = \frac{ek}{1 - e \cos \theta} \tag{7}$$

Likewise, if the directrix were perpendicular to the y axis and $k > 0$, then the equation would be

$$r = \frac{ek}{1 + e \sin \theta} \quad \text{or} \quad r = \frac{ek}{1 - e \sin \theta} \tag{8}$$

depending on whether the directrix has the form $y = k$ or $y = -k$. Thus (6)–(8) are polar equations of conic sections with a focus at the origin and a directrix perpendicular to the x axis or the y axis.

EXAMPLE 2   Sketch the polar graph of the equation

$$r = \frac{6}{3 + 2 \cos \theta}$$

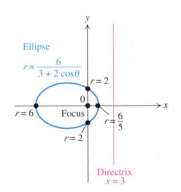

FIGURE 10.54

Solution   To put the equation into the form of one of the equations (6)–(8), we divide both numerator and denominator by 3, obtaining

$$r = \frac{2}{1 + \frac{2}{3} \cos \theta}$$

Comparing this equation with (6), we find that

$$e = \frac{2}{3} \quad \text{and} \quad k = \frac{2}{e} = \frac{2}{2/3} = 3$$

Since $e < 1$, the conic section is an ellipse with one focus at the origin; the associated directrix is the line $x = 3$ (Figure 10.54).   ❏

EXERCISES 10.5

In Exercises 1–12 find the eccentricity of the conic section with the given equation.

1. $\dfrac{x^2}{9} + \dfrac{y^2}{25} = 1$

2. $\dfrac{x^2}{64} + \dfrac{y^2}{49} = 1$

3. $\dfrac{x^2}{9} - \dfrac{y^2}{25} = 1$

4. $\dfrac{y^2}{25} - \dfrac{x^2}{9} = 1$

5. $4x^2 + y^2 = 8$

6. $6y^2 - 3x^2 = 4$

7. $4(x - 3)^2 + (y + 3)^2 = 8$

8. $x^2 = -6y$

9. $x^2 + 2y^2 - 2x - 4y = 1$

10. $x^2 + 4x - 4y^2 = 12$

11. $49x^2 - 9y^2 + 98x - 36y = 428$

12. $3y^2 - 5x + 3y - 6 = 0$

In Exercises 13–21 find an equation of the conic section possessing the given properties, and sketch the conic section.

13. The foci are $(9, 0)$ and $(-9, 0)$, and the eccentricity is $\frac{3}{5}$.

14. The foci are $(9, 0)$ and $(-9, 0)$, and the eccentricity is $\frac{5}{3}$.

15. The foci are $(0, 1)$ and $(0, -1)$, and the eccentricity is 2.

16. The vertices are $(5, 0)$ and $(-5, 0)$, and the eccentricity is $\frac{4}{5}$.

17. The vertices are $(0, 0)$ and $(0, 10)$, and the eccentricity is $\frac{13}{5}$.

18. The center is $(-2, 3)$, a vertex is $(-2, 0)$, and the eccentricity is $\frac{1}{2}$.

19. The hyperbola is in standard position, passes through $(\sqrt{20}, 8)$, and has eccentricity $\sqrt{17}$. (There are two such hyperbolas.)

20. The ellipse is in standard position, passes through $(-1, 1)$, and has eccentricity $\sqrt{3}/2$. (There are two such ellipses.)

21. The major axis of the ellipse has length 8 and is on the x axis, a focus is located at the origin, and the eccentricity is $\frac{1}{2}$. (There are two such conic sections.)

In Exercises 22–29 indicate the type of conic section represented by the given equation, and find an equation of a directrix.

22. $r = \dfrac{1}{1 + \frac{1}{2} \sin \theta}$

23. $r = \dfrac{1}{1 - 2 \cos \theta}$

24. $r = \dfrac{1}{1 - \sin \theta}$

25. $r = \dfrac{25}{5 - 3 \sin \theta}$

26. $r = \dfrac{9}{3 - 5 \cos \theta}$

27. $r = \dfrac{1}{2 - 2 \cos \theta}$

28. $r = \dfrac{12}{4 - 5 \sin \theta}$

29. $r = \dfrac{3}{1 + \sin \theta}$

30. a. Sketch the graphs of

$$r = \dfrac{1}{2 + \cos \theta} \quad \text{and} \quad r = \dfrac{1}{1 - \cos \theta}$$

b. Find the points where the graphs in (a) intersect.

Exercises 31–32 refer to the sidebar on page 700.

31. Find an approximate equation for the orbit of the earth around the sun.

*32. Assuming that the sun is located at a focus, approximate the minimum distance from the earth to the sun.

Project

1. This project analyzes what happens to the shape of a conic section and its directrix(es) when the coefficients a and b (or c in the case of a parabola) are altered.

a. Consider the ellipse

$$x^2 + \dfrac{y^2}{1/n^2} = 1$$

where n is a positive integer. Determine what happens to the ellipse's shape, the eccentricity, and the directrixes as n approaches ∞. What is the limiting shape of the ellipse as n approaches ∞?

b. Tell how your answers to (a) are altered if the ellipse has the equation

$$\dfrac{x^2}{1/n^2} + y^2 = 1$$

c. Consider the hyperbola

$$x^2 - \dfrac{y^2}{1/n^2} = 1$$

where n is a positive integer. Determine what happens to the hyperbola's shape, the eccentricity, and the directrixes as n approaches ∞. What is the limiting shape of the hyperbola?

d. Tell how your answers to (c) are altered if the hyperbola has the equation

$$\dfrac{x^2}{1/n^2} - y^2 = 1$$

e. Consider the parabola $x^2 = \frac{1}{n} y$, where n is a positive integer. Determine what happens to the parabola's shape and the directrix as n approaches ∞. What is the limiting shape of the parabola?

REVIEW

Key Terms and Expressions

Polar coordinates
Polar graph
Polar equation
Parabola
Ellipse

Hyperbola
Standard position of a conic section
Eccentricity
Rotation of axes

Key Formulas

Length of a curve in polar coordinates

$$L = \int_a^\beta \sqrt{[f'(\theta)]^2 + [f(\theta)]^2}\, d\theta$$

Area of a region in polar coordinates

$$A = \begin{cases} \int_\alpha^\beta \frac{1}{2}[f(\theta)]^2\, d\theta \\ \int_\alpha^\beta \frac{1}{2}\{[f(\theta)]^2 - [g(\theta)]^2\}\, d\theta \end{cases}$$

Parabola in standard position

$$x^2 = 4cy \quad \text{or} \quad y^2 = 4cx$$

Ellipse in standard position

$$\frac{x^2}{a^2} + \frac{y^2}{b^2} = 1 \quad \text{or} \quad \frac{x^2}{b^2} + \frac{y^2}{a^2} = 1$$

Hyperbola in standard position

$$\frac{x^2}{a^2} - \frac{y^2}{b^2} = 1 \quad \text{or} \quad \frac{y^2}{a^2} - \frac{x^2}{b^2} = 1$$

Eccentricity

$$e = \frac{c}{a}$$

Review Exercises

In Exercises 1–4 sketch the polar graph of the equation.

1. $r = \sin 5\theta$ **2.** $r = 2\cos\theta - 2$

3. $r = \sqrt{3} - 2\sin\theta$ **4.** $r^2 = \frac{1}{4}\cos 2\theta$

5. a. Sketch the polar graphs of $r = 2\sin 2\theta$ and $r = 2\sin\theta$.
  b. Find all points of intersection of the graphs found in (a).

6. Find the length L of the polar graph of the equation $r = 1 + \sin\theta$ for $0 \le \theta \le \pi$. (*Hint:* In the integral, use the fact that $\sin\theta = \cos(\pi/2 - \theta)$.)

7. Find the area A of the region common to the circles $r = 2\cos\theta$ and $r = \sin\theta + \cos\theta$.

8. Find the area A of the region inside the lemniscate $r^2 = 2\sin 2\theta$ and outside the circle $r = 1$.

In Exercises 9–16 write an equation for the conic section.

9. The parabola with focus $(-3, 4)$ and directrix $x = 5$

10. The parabola having directrix $y = -10$ and axis $x = -2$ and passing through $(6, -2)$

11. The ellipse having vertices $(0, 2\sqrt{2})$ and $(0, -2\sqrt{2})$ and passing through $(1, \sqrt{6})$

12. The ellipse with foci $(1, 1)$ and $(1, -3)$ and major axis of length 8

13. The hyperbola with vertices $(-3, 2)$ and $(1, 2)$ whose asymptotes are perpendicular to one another

14. The hyperbola with foci $(12, -3)$ and $(-8, -3)$ and asymptote $y + 3 = \frac{4}{3}(x - 2)$

15. The conic section with eccentricity 2, directrix $x = -4$, and corresponding vertex $(-2, 0)$

16. The conic section with eccentricity $\frac{1}{3}$ and foci $(0, 0)$ and $(0, -2)$

In Exercises 17–22 sketch the graph of the equation.

17. $49x^2 - 9y^2 + 98x - 36y = 428$

18. $4x^2 - 16x + y^2 - 6y = 0$

19. $9x^2 - 36x + 5y + 21 = 0$

20. $x^2 - 2x - 4y^2 - 12y = 10$

21. $4x^2 - 24xy + 11y^2 + 40x + 30y - 45 = 0$

22. $73x^2 + 72xy + 52y^2 = 25$

23. Determine an equation for the collection of points (x, y) such that the distance from (x, y) to the point $(2, 4)$ is e times the distance from (x, y) to the x axis, where

    a. $e = 3$ **b.** $e = 1$ **c.** $e = \frac{1}{2}$

In Exercises 24–25 indicate the type of conic section represented by the given equation.

24. $r = \dfrac{3}{1 + 4\sin\theta}$ **25.** $r = \dfrac{3}{4 - \cos\theta}$

26. Find all points on the parabola $y^2 = -8x$ that are closest to the point $(-10, 0)$.

27. Let $x^2 = 4cy$, with $c > 0$, and let (a, b) be outside the parabola (that is, assume that $a^2 > |4cb|$). Find equations for the two lines that are tangent to the parabola and pass through the point (a, b).

28. Show that half the length of the minor axis of an ellipse is the geometric mean of the two lengths into which a focus divides the major axis (see Exercise 85 of Section 1.1).

Topics for Discussion

1. Given a polar equation $r = f(\theta)$, how can we tell whether it is symmetric with respect to the x axis, the y axis, or the origin? Give examples of all types of symmetry. Under what conditions does the presence of two types of symmetry imply the third?

2. Derive the area of a circular sector of a circle of radius r and θ. How do sectors play a role in the derivation of the area formula for polar regions?

3. Discuss the role of eccentricity e in the analysis of conic sections. Describe how the shape of a conic section changes as e increases from 0 to 1, and from 1 toward ∞. Do you think that it would be possible to assign an eccentricity to a straight line? Explain your answer.

4. Describe the reflection properties of the parabola, the ellipse, and the hyperbola.

5. From definitions in the dictionary, explain how the common English grammar terms of "parable," "ellipsis," and "hyperbole" are derived from their conic section cousins.

Cumulative Review, Chapters 1–9

In Exercises 1–3 find the limit.

1. $\displaystyle\lim_{x\to\infty} \frac{x^2 + 1}{x\sqrt{3x^2 + 1}}$ **2.** $\displaystyle\lim_{h\to 0} \frac{\tan\left(\dfrac{\pi}{4} - h\right) - 1}{h}$

3. $\displaystyle\lim_{x\to 0} \frac{1 - \cos 4x}{3x^2}$

In Exercises 4–5 find $f'(x)$.

4. $f(x) = \cos(\ln x)$ **5.** $f(x) = \displaystyle\int_{-x}^{x^2} \sin\sqrt{t^3 + 1}\, dt$

6. Consider a line moving so that it is always tangent to the circle $x^2 + y^2 = 1$ at a point in the first quadrant. Suppose the point of tangency is moving so that its x coordinate increases at the rate of 3 units per minute. Determine how fast the x intercept of the line is decreasing when the x intercept is 2.

7. A kite is moving horizontally at a height of 30 feet above a person pulling in the string at the rate of 4 feet per second.

    a. Determine the horizontal speed of the kite when 50 feet of string is out.

    b. Would the answer to part (a) be different if only 45 feet of string were out? Explain your answer.

8. Let n be a positive integer, and consider the function f defined by $f(x) = (\ln x)/x^n$. Show that f attains a maximum value on $(0, \infty)$, and find the point at which the maximum value is attained.

9. Let c be any fixed number, and let $f(x) = \sin x \cos(x - c)$. Show that f attains its maximum value at $c/2 + \pi/4$.

10. Let $f(x) = xe^{-2x}$. Sketch the graph of f, indicating all pertinent information.

11. Let a be any positive number.

    a. Show that $\displaystyle\lim_{h\to 0} \frac{a^h - 1}{h} = \ln a$.

    b. Using the result in part (a), show that $\lim_{n\to\infty} n(\sqrt[n]{a} - 1) = \ln a$.

12. Let $f(x) = \dfrac{1}{1 + e^{1/x}}$. Show that f has an inverse, and find a formula for $f^{-1}(x)$.

In Exercises 13–16 evaluate the integral.

13. $\displaystyle\int_0^{3/4} 36x\sqrt{1 + 4x}\, dx$ **14.** $\displaystyle\int_0^{\pi/4} \frac{\tan x}{\sqrt{\sec x}}\, dx$

15. $\displaystyle\int \frac{x^2}{e^{-2x}}\, dx$ **16.** $\displaystyle\int \frac{1}{(4 - x^2)^{3/2}}\, dx$

In Exercises 17–18 determine whether the series is absolutely convergent, conditionally convergent, or divergent.

17. $\displaystyle\sum_{n=1}^{\infty} \frac{n!}{(1.4)^n}$ **18.** $\displaystyle\sum_{n=2}^{\infty} \frac{1}{n(\ln n)^{3/2}}$

19. Determine the interval of convergence of

$$\sum_{n=1}^{\infty} \frac{5n^2}{4^n(n^3 + 2)} x^n$$

Sailboats in Hawaii. (*To Nakamura/ Superstock*)

11 VECTORS, LINES AND PLANES

$\mathbf{M}$any physical and abstract quantities have only magnitude and thus can be described by numbers. Examples we have encountered are mass, cost, profit, speed, area, length, volume, and moments about an axis. Many other quantities, however, have both magnitude and direction. The most notable example is velocity, which involves not only the speed of an object but also the direction of its motion. Quantities that have both magnitude and direction are described mathematically by vectors. In this chapter we will study vectors and their applications, including the description of lines and planes in space.

Our first task will be to set up a coordinate system in space, much as we did for the plane. Using the new coordinate system, we will define the concept of vector and present various elementary properties of vectors.

705

11.1 CARTESIAN COORDINATES IN SPACE

We begin by considering three mutually perpendicular lines that pass through a point O, called the **origin** (Figure 11.1). The three lines are named the **x axis**, the **y axis**, and the **z axis**. For each of these axes we set up a correspondence between the points on the axis and the set of real numbers, letting the origin O correspond to the number 0. Often we will only draw the positive portions of the axes, as in Figure 11.1, with the positive x axis pointing toward the viewer, the positive y axis pointing toward the right, and the positive z axis pointing upward.

If P is any point in space, then the three planes through P perpendicular to the three axes intersect the x axis, the y axis, and the z axis at points corresponding to numbers x, y, and z, respectively (Figure 11.2). Therefore we associate P with the ordered triple of numbers (x, y, z) and call (x, y, z) the **rectangular coordinates** (or **Cartesian coordinates**) of P. Figure 11.3 exhibits a few points in space, along with their Cartesian coordinates.

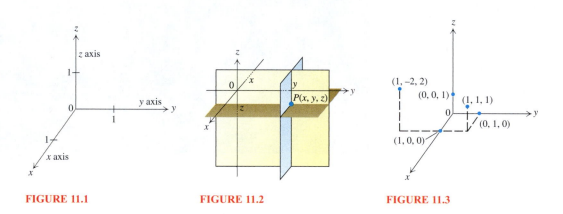

FIGURE 11.1   **FIGURE 11.2**   **FIGURE 11.3**

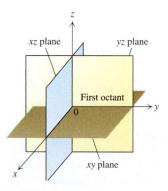

FIGURE 11.4

Under the association we have just described, each point in space is identified with an ordered triple of numbers. Conversely, each ordered triple (x, y, z) is identified with a single point whose coordinates are (x, y, z). This correspondence will enable us to describe geometric objects in space by means of equations and inequalities. Moreover, because each point in space has three coordinates, we sometimes refer to space as **three-dimensional space**.

In three-dimensional space there are three **coordinate planes**: the xy plane, which contains the x and y axes; the xz plane, which contains the x and z axes; and the yz plane, which contains the y and z axes. Since each plane divides space into two parts, the three coordinate planes together divide space into eight regions, called **octants** (Figure 11.4). The octant containing the positive x, y, and z axes is called the **first octant**.

Just as with points in the plane, the notion of distance between two points in space is fundamental. The formula for the **distance** $|PQ|$ between two points $P = (x_0, y_0, z_0)$ and $Q = (x_1, y_1, z_1)$ is

$$|PQ| = \sqrt{(x_1 - x_0)^2 + (y_1 - y_0)^2 + (z_1 - z_0)^2}\,. \qquad (1)$$

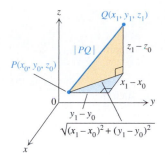

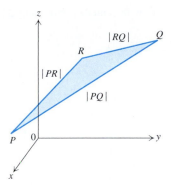

FIGURE 11.5

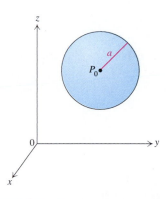

FIGURE 11.6 Triangle inequality: $|PQ| \leq |PR| + |RQ|$

as you can prove by two successive applications of the Pythagorean Theorem (Figure 11.5). If $Q = O$, then (1) reduces to

$$|PO| = |OP| = \sqrt{x_0^2 + y_0^2 + z_0^2}$$

EXAMPLE 1   Let $P = (-1, 3, 6)$ and $Q = (4, 0, 5)$. Find $|PQ|$, $|PO|$, and $|OQ|$.

Solution   From (1) we calculate that

$$|PQ| = \sqrt{(4 - (-1))^2 + (0 - 3)^2 + (5 - 6)^2} = \sqrt{35}$$
$$|PO| = \sqrt{(-1)^2 + 3^2 + 6^2} = \sqrt{46}$$
$$|OQ| = \sqrt{4^2 + 0^2 + 5^2} = \sqrt{41} \qquad \square$$

The three basic laws governing the distance between two points in space are

> $|PQ| = 0$ if and only if $P = Q$
>
> $|PQ| = |QP|$
>
> $|PQ| \leq |PR| + |RQ|$ for any third point R

The first two laws follow directly from (1). The third law, which is known as the **triangle inequality**, implies that the length of any side of a triangle does not exceed the sum of the lengths of the other two sides (Figure 11.6). (See Exercise 25 of Section 11.3 for a proof of the triangle inequality.)

The **sphere** with **center** $P_0 = (x_0, y_0, z_0)$ and **radius** a is defined to be the set of all points P whose distance from P_0 is a, that is,

$$|P_0 P| = a$$

(Figure 11.7). Thus a point $P = (x, y, z)$ lies on that sphere if and only if

$$\sqrt{(x - x_0)^2 + (y - y_0)^2 + (z - z_0)^2} = a$$

or equivalently, if and only if

$$(x - x_0)^2 + (y - y_0)^2 + (z - z_0)^2 = a^2$$

This is an equation of a sphere in space. Observe that a sphere is a surface, not a solid region.

EXAMPLE 2   Show that

$$x^2 + y^2 + z^2 = 2x + 4y - 6z$$

is an equation of a sphere. Find the center and the radius of the sphere.

Solution   We transpose terms from the right side to the left side of the equation, complete the squares, and obtain

$$(x^2 - 2x + 1) + (y^2 - 4y + 4) + (z^2 + 6z + 9) = 1 + 4 + 9$$

FIGURE 11.7 A sphere of radius a.

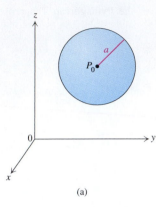

(a)

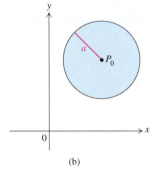

(b)

FIGURE 11.8   (a) A ball of radius a. (b) A disk of radius a.

or

$$(x-1)^2 + (y-2)^2 + (z+3)^2 = 14$$

This is an equation of the sphere with center $(1, 2, -3)$ and radius $\sqrt{14}$.   ❑

The **ball** (or **closed ball**) with center $P_0 = (x_0, y_0, z_0)$ and radius a is the collection of points $P = (x, y, z)$ such that $|P_0P| \le a$, or such that

$$\sqrt{(x-x_0)^2 + (y-y_0)^2 + (z-z_0)^2} \le a$$

(Figure 11.8(a)). This is equivalent to

$$(x-x_0)^2 + (y-y_0)^2 + (z-z_0)^2 \le a^2$$

Thus if $P_0 = (1, 2, -3)$, then $P = (x, y, z)$ is in the ball with center P_0 and radius 3 provided that $|P_0P| \le 3$, that is,

$$\sqrt{(x-1)^2 + (y-2)^2 + (z+3)^2} \le 3$$

The collection of all points (x, y, z) such that

$$(x-x_0)^2 + (y-y_0)^2 + (z-z_0)^2 < a^2$$

is the **open ball** with center (x_0, y_0, z_0) and radius a. Notice that the open ball is merely the closed ball with the outer spherical boundary removed.

A ball is the solid region in space enclosed by a sphere. The corresponding region in the plane is enclosed by a circle, and we call such a region a **disk** (or **closed disk**). The disk with center $P_0 = (x_0, y_0)$ and radius a is the collection of points $P = (x, y)$ such that $|P_0P| \le a$, or such that

$$\sqrt{(x-x_0)^2 + (y-y_0)^2} \le a$$

(Figure 11.8(b)). This is equivalent to

$$(x-x_0)^2 + (y-y_0)^2 \le a^2$$

Finally, the collection of all points (x, y) such that

$$(x-x_0)^2 + (y-y_0)^2 < a^2$$

is the **open disk** with center (x_0, y_0) and radius a.

EXERCISES 11.1

In Exercises 1–8 find the distance D between the points P and Q.

1. $P = (\sqrt{2}, 0, 0)$, $Q = (0, 1, 1)$
2. $P = (2, -1, -2)$, $Q = (3, 1, 0)$
3. $P = (-3, 4, -5)$, $Q = (0, 8, 7)$
4. $P = (4, -1, 3)$, $Q = (4, 5, 11)$

5. $P = (-1, 3, 6)$, $Q = (4, 2, 7)$
6. $P = (1, 0, -\frac{1}{2})$, $Q = (\frac{1}{2}, \frac{1}{2}\sqrt{2}, 0)$
7. $P = (2\sin x, \cos x, \tan x)$, $Q = (\sin x, 2\cos x, 0)$
8. $P = (e^x, 0, 2\sqrt{2})$, $Q = (0, e^{-x}, \sqrt{2})$
9. Show that the point $(3, 0, 2)$ is equidistant from the points $(1, -1, 5)$ and $(5, 1, -1)$.

10. Find the perimeter of the triangle with vertices $(-1, 1, 2)$, $(2, 0, 3)$, and $(3, 4, 5)$.

11. Find an equation of the sphere with radius 5 and center $(2, 1, -7)$.

12. Find an equation of the sphere with radius $\sqrt{2}$ and center $(-1, 0, 3)$.

13. Show that
$$x^2 + y^2 + z^2 - 2x - 4y + 6z = -10$$
is an equation of a sphere. Find the radius and the center of the sphere.

14. Show that
$$x^2 + y^2 + z^2 + 6x + 8y - 4z + 4 = 0$$
is an equation of a sphere. Find the radius and the center of the sphere.

15. Find an inequality satisfied by all points that belong to the closed ball with radius 6 and center $(0, -2, -3)$.

16. Find an inequality satisfied by all points that belong to the closed disk that has radius $\sqrt{3}$ and center $(\frac{1}{2}, -1)$.

17. Let $P = (1, -1, 1)$, $Q = (2, 1, -1)$, and $R = (0, 0, 0)$. By computing the lengths of the sides, show that the triangle PQR is a right triangle.

18. Let $P = (3, 0, 3)$, $Q = (2, 0, -1)$, and $R = (c, 1, 2)$. Determine the two values of c that make triangle PQR a right triangle with hypotenuse PQ.

19. Find an equation of the set of points equidistant from the points $(2, 1, 0)$ and $(4, -1, -3)$.

20. Find an equation of the set of points twice as far from the origin as from the point $(-1, 1, 1)$. Show that the set of points is a sphere and find its center.

21. Show that the midpoint of the line segment joining the points (x_0, y_0, z_0) and (x_1, y_1, z_1) is $(\frac{1}{2}(x_0 + x_1), \frac{1}{2}(y_0 + y_1), \frac{1}{2}(z_0 + z_1))$.

22. Find the midpoint of the line segment joining the points $(3, 7, 11)$ and $(-9, 8, 31)$.

23. Suppose the points $(2, -1, 3)$ and $(4, 1, 7)$ are diametrically opposite each other on a sphere. Find an equation of the sphere.

11.2   VECTORS IN SPACE

Intuitively, a vector is a directed line segment in space, often described by an arrow (Figure 11.9). Vectors appear constantly in the study of motion in space. For instance, consider a particle moving along the path shown in Figure 11.10. If at a given time t the particle is at point P, we can assign to P a vector, called the **velocity vector**, which points in the direction of the particle's motion and has length equal to the speed of the particle. Vectors are also used to describe force; the vector describing a force points in the direction in which the force acts and has length equal to the magnitude of the force.

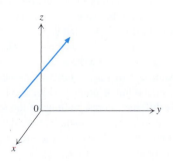

FIGURE 11.9   A vector is a directed line segment.

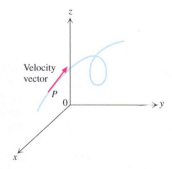

FIGURE 11.10

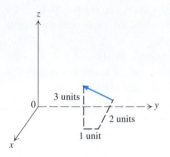

FIGURE 11.11

One obvious way to describe a vector is simply to give the coordinates of the initial and terminal points of the directed line segment associated with it. But as our comments on velocity and force might suggest, one is normally concerned more with the direction and length of a vector than with its initial and terminal points. The vector appearing in Figure 11.11 represents a change of 2 units in the positive x direction, 1 unit in the negative y direction, and 3 units in the positive z direction. These three numbers identify the direction and length of the vector, and hence we can conveniently identify the vector with the ordered triple $(2, -1, 3)$. More generally, if a directed line segment has initial point (x_0, y_0, z_0) and terminal point (x_1, y_1, z_1), then the corresponding vector represents a change of $x_1 - x_0$ units in the positive x direction, $y_1 - y_0$ units in the positive y direction, and $z_1 - z_0$ units in the positive z direction. We group these three differences into the ordered triple $(x_1 - x_0, y_1 - y_0, z_1 - z_0)$ and identify the vector with this ordered triple.

DEFINITION 11.1

A **vector** is an ordered triple (a_1, a_2, a_3) of numbers. The numbers a_1, a_2, and a_3 are called the **components** of the vector. The vector $\overrightarrow{PQ}$ associated with the directed line segment with initial point $P = (x_0, y_0, z_0)$ and terminal point $Q = (x_1, y_1, z_1)$ is $(x_1 - x_0, y_1 - y_0, z_1 - z_0)$.

Sir William Rowan Hamilton (1805–1865)
Ireland's greatest mathematician, Hamilton was a prodigy. By 13 he was fluent in more than a dozen languages, and as a 22-year-old he was appointed Royal Astronomer of Ireland. In an effort to define a multiplication for pairs of real numbers, Hamilton introduced "quaternions"—precursors of vectors. He predicted quaternions would have a profound significance in mathematical physics. It was Hamilton who introduced the term "vector."

Two vectors (a_1, a_2, a_3) and (b_1, b_2, b_3) are equal if and only if their components are equal, that is,

$$a_1 = b_1, \quad a_2 = b_2, \quad \text{and} \quad a_3 = b_3$$

EXAMPLE 1   Let $P = (1, 3, 7)$, $Q = (-1, 0, 6)$, $R = (0, -1, -2)$, and $S = (-2, -4, -3)$. Show that $\overrightarrow{PQ}$ and $\overrightarrow{RS}$ are the same vector.

Solution   Applying Definition 11.1, we find that

$$\overrightarrow{PQ} = (-2, -3, -1) = \overrightarrow{RS} \quad \square$$

In print, vectors are almost always represented by boldface letters, although the choice of the particular letters used for vectors varies widely (depending on the context). To distinguish vectors from numbers, which are also called **scalars**, we will normally denote vectors by lowercase boldface letters near the beginning of the alphabet, such as **a**, **b**, and **c**. Other letters denote special vectors; for example, the zero vector $(0, 0, 0)$ is denoted **0**. Since it is difficult to write a boldface letter by hand, vectors are usually written by placing an arrow over a symbol or expression. Thus we would write **a** as $\vec{a}$.

Each vector $\mathbf{a} = (a_1, a_2, a_3)$ can be associated with a directed line segment having an arbitrary initial point P (Figure 11.12), and different initial points in space give us different representations of the same vector. If P is the origin, then the vector $\mathbf{a} = (a_1, a_2, a_3)$ is associated with the point (a_1, a_2, a_3) in space and with the directed line segment from the origin to (a_1, a_2, a_3) (Figure 11.12). Sometimes the vector is denoted by $\langle a_1, a_2, a_3 \rangle$.

A natural way to assign a length to a vector $\mathbf{a} = (a_1, a_2, a_3)$ is to assign it the length of the directed line segment from the origin to the point (a_1, a_2, a_3).

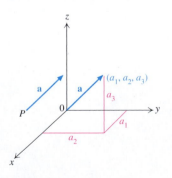

FIGURE 11.12

DEFINITION 11.2 The **length** (or **norm**) of a vector $\mathbf{a} = (a_1, a_2, a_3)$ is denoted $\| \mathbf{a} \|$ and is defined by

$$\| \mathbf{a} \| = \sqrt{a_1^2 + a_2^2 + a_3^2}$$

A **unit vector** is a vector having length 1.

For example, if $\mathbf{a} = (-1, 3, 6)$, then

$$\| \mathbf{a} \| = \sqrt{(-1)^2 + 3^2 + 6^2} = \sqrt{46}$$

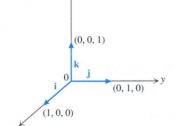

FIGURE 11.13

You can check that this is the distance between the point $(-1, 3, 6)$ and the origin. In general, the length $\| \overrightarrow{PQ} \|$ of the vector $\overrightarrow{PQ}$ is the same as the distance $| PQ |$ between the points P and Q.

There are three special unit vectors that will help in describing and operating on vectors:

$$\mathbf{i} = (1, 0, 0), \qquad \mathbf{j} = (0, 1, 0), \qquad \mathbf{k} = (0, 0, 1)$$

(see Figure 11.13). It follows from the definition of length that

$$\| \mathbf{i} \| = \| \mathbf{j} \| = \| \mathbf{k} \| = 1$$

so that $\mathbf{i}$, $\mathbf{j}$, and $\mathbf{k}$ are indeed unit vectors.

Combinations of Vectors Numbers can be added, subtracted, and multiplied; vectors can be combined in similar ways.

DEFINITION 11.3 Let $\mathbf{a} = (a_1, a_2, a_3)$ and $\mathbf{b} = (b_1, b_2, b_3)$ be vectors, and let c be a real number. Then we define the **sum $\mathbf{a} + \mathbf{b}$**, the **difference $\mathbf{a} - \mathbf{b}$**, and the **scalar multiple $c\mathbf{a}$** by

$$\mathbf{a} + \mathbf{b} = (a_1 + b_1, a_2 + b_2, a_3 + b_3)$$

$$\mathbf{a} - \mathbf{b} = (a_1 - b_1, a_2 - b_2, a_3 - b_3)$$

$$c\mathbf{a} = (ca_1, ca_2, ca_3)$$

Sometimes we write $\mathbf{a}/c$ for $(1/c)\mathbf{a}$. Thus we might express $\tfrac{1}{5}\mathbf{a}$ as $\mathbf{a}/5$.

Two types of products of vectors will be defined in Sections 11.3 and 11.4. However, we will not define any quotients of vectors.

There are many laws resulting from Definition 11.3. For example,

$$\mathbf{0} + \mathbf{a} = \mathbf{a} + \mathbf{0} = \mathbf{a} \qquad\qquad \mathbf{a} + \mathbf{b} = \mathbf{b} + \mathbf{a}$$

$$\mathbf{a} - \mathbf{b} = \mathbf{a} + (-1)\mathbf{b} \qquad\qquad c(\mathbf{a} + \mathbf{b}) = c\mathbf{a} + c\mathbf{b}$$

$$0\mathbf{a} = \mathbf{0} \qquad\qquad 1\mathbf{a} = \mathbf{a}$$

$$\mathbf{a} + (\mathbf{b} + \mathbf{c}) = (\mathbf{a} + \mathbf{b}) + \mathbf{c}$$

EXAMPLE 2   Let $\mathbf{a} = (1, -3, 2)$ and $\mathbf{b} = (-4, -1, 0)$. Find $\mathbf{a} + \mathbf{b}$, $\mathbf{a} - \mathbf{b}$, and $-\frac{1}{2}\mathbf{a}$.

Solution   From Definition 11.3 we find that

$$\mathbf{a} + \mathbf{b} = (1 + (-4), -3 + (-1), 2 + 0) = (-3, -4, 2)$$

$$\mathbf{a} - \mathbf{b} = (1 - (-4), -3 - (-1), 2 - 0) = (5, -2, 2)$$

$$-\frac{1}{2}\,\mathbf{a} = \left(-\frac{1}{2}, \frac{3}{2}, -1\right) \quad \square$$

Using addition and scalar multiplication of vectors, we can express any vector $\mathbf{a} = (a_1, a_2, a_3)$ as a combination of the special unit vectors $\mathbf{i}$, $\mathbf{j}$, and $\mathbf{k}$. Indeed,

$$\mathbf{a} = (a_1, a_2, a_3) = (a_1, 0, 0) + (0, a_2, 0) + (0, 0, a_3)$$

$$= a_1(1, 0, 0) + a_2(0, 1, 0) + a_3(0, 0, 1)$$

$$= a_1\mathbf{i} + a_2\mathbf{j} + a_3\mathbf{k}$$

In other words,

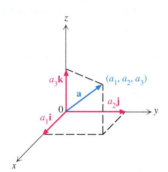

FIGURE 11.14

$$\mathbf{a} = a_1\mathbf{i} + a_2\mathbf{j} + a_3\mathbf{k} \tag{1}$$

(see Figure 11.14). For example,

$$(1, -3, 2) = \mathbf{i} - 3\mathbf{j} + 2\mathbf{k}$$

If $P = (x_0, y_0, z_0)$ and $Q = (x_1, y_1, z_1)$, then we can express $\overrightarrow{PQ}$ in the form of (1):

$$\overrightarrow{PQ} = (x_1 - x_0)\mathbf{i} + (y_1 - y_0)\mathbf{j} + (z_1 - z_0)\mathbf{k}$$

In our later study of calculus we will often find it useful to write a vector in the form of (1). Since the components of a vector $\mathbf{a}$ are the coefficients of the unit vectors $\mathbf{i}$, $\mathbf{j}$, and $\mathbf{k}$ in (1), we sometimes call a_1 the $\mathbf{i}$ **component**, a_2 the $\mathbf{j}$ **component**, and a_3 the $\mathbf{k}$ **component** of $\mathbf{a}$.

Using the notation in (1), we restate the formulas for the length, sum, difference, and scalar multiple of vectors:

$$\| a_1\mathbf{i} + a_2\mathbf{j} + a_3\mathbf{k} \| = \sqrt{a_1^2 + a_2^2 + a_3^2} \tag{2}$$

$$(a_1\mathbf{i} + a_2\mathbf{j} + a_3\mathbf{k}) + (b_1\mathbf{i} + b_2\mathbf{j} + b_3\mathbf{k}) = (a_1 + b_1)\mathbf{i} + (a_2 + b_2)\mathbf{j} + (a_3 + b_3)\mathbf{k}$$

$$(a_1\mathbf{i} + a_2\mathbf{j} + a_3\mathbf{k}) - (b_1\mathbf{i} + b_2\mathbf{j} + b_3\mathbf{k}) = (a_1 - b_1)\mathbf{i} + (a_2 - b_2)\mathbf{j} + (a_3 - b_3)\mathbf{k}$$

$$c(a_1\mathbf{i} + a_2\mathbf{j} + a_3\mathbf{k}) = ca_1\mathbf{i} + ca_2\mathbf{j} + ca_3\mathbf{k}$$

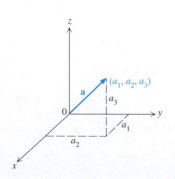

FIGURE 11.15

It follows from (2) that

$$| a_1 | \leq \| a_1\mathbf{i} + a_2\mathbf{j} + a_3\mathbf{k} \|, \qquad | a_2 | \leq \| a_1\mathbf{i} + a_2\mathbf{j} + a_3\mathbf{k} \|,$$

$$| a_3 | \leq \| a_1\mathbf{i} + a_2\mathbf{j} + a_3\mathbf{k} \|$$

(see Figure 11.15).

In summary, we list four ways of describing a vector:

1.   as (a_1, a_2, a_3), an ordered triple of numbers
2.   as (a_1, a_2, a_3), a point in space
3.   as a directed line segment with initial point (x_0, y_0, z_0) and terminal point $(x_0 + a_1, y_0 + a_2, z_0 + a_3)$
4.   as $a_1\mathbf{i} + a_2\mathbf{j} + a_3\mathbf{k}$

Geometric Interpretations of Vector Operations

The many geometric and physical meanings that can be attached to combinations of vectors make vectors very powerful tools for scientists. To interpret the sum of two vectors geometrically, we begin by letting $\mathbf{a} = a_1\mathbf{i} + a_2\mathbf{j} + a_3\mathbf{k}$ and $\mathbf{b} = b_1\mathbf{i} + b_2\mathbf{j} + b_3\mathbf{k}$. Then we can think of $\mathbf{a}$, $\mathbf{b}$, and $\mathbf{a} + \mathbf{b}$ as directed line segments from the origin to the points P, Q, and R having coordinates (a_1, a_2, a_3), (b_1, b_2, b_3), and $(a_1 + b_1, a_2 + b_2, a_3 + b_3)$, respectively (Figure 11.16). Now observe that if the directed line segment representing $\mathbf{b}$ is placed so that its initial point is P, then its terminal point will be R. Thus the vector $\mathbf{a} + \mathbf{b}$ can be obtained by placing the initial point of a representative of $\mathbf{b}$ on the terminal point of $\mathbf{a}$. Notice that the two representations of $\mathbf{a}$ and $\mathbf{b}$ shown in Figure 11.16 determine a parallelogram whose diagonal is $\mathbf{a} + \mathbf{b}$.

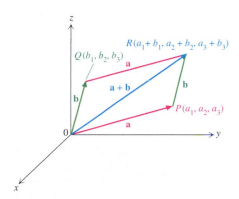

FIGURE 11.16

Next we let $\mathbf{a} = a_1\mathbf{i} + a_2\mathbf{j} + a_3\mathbf{k}$ be a vector and c be any number. It follows from the definition of length that

$$\| c\mathbf{a} \| = | c | \, \| \mathbf{a} \| \tag{3}$$

When $\mathbf{a}$ is multiplied by a positive number c, its length is multiplied by c and its direction does not change (Figure 11.17). Moreover, the vector $-\mathbf{a}$ (which denotes $(-1)\mathbf{a}$) has the same length as $\mathbf{a}$ but has the opposite direction (Figure 11.17). Thus if $\mathbf{a}$ is multiplied by a negative number c, its length is multiplied by $| c |$ and its direction is reversed (Figure 11.17). Two nonzero vectors whose initial points are the origin are considered parallel only if they lie on the same line through the origin, and in that case they are multiples of one another. Hence we make Definition 11.4, below.

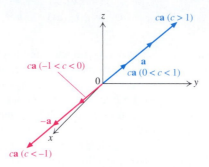

FIGURE 11.17

DEFINITION 11.4   Two nonzero vectors **a** and **b** are **parallel** if and only if there is a number c such that $\mathbf{b} = c\mathbf{a}$.

EXAMPLE 3   Let $\mathbf{a} = 6\mathbf{i} - 2\mathbf{j} + 4\mathbf{k}$ and $\mathbf{b} = -3\mathbf{i} + \mathbf{j} - 2\mathbf{k}$. Determine whether **a** and **b** are parallel.

Solution   You can check that $\mathbf{b} = -\frac{1}{2}\mathbf{a}$; consequently **a** and **b** are parallel.   ❑

The **unit vector in the direction of a nonzero vector a** is $\mathbf{a}/\|\mathbf{a}\|$. Such vectors will be particularly important in Chapter 12.

EXAMPLE 4   Find the unit vector in the direction of $4\mathbf{i} - \mathbf{j} - 3\mathbf{k}$.

Solution   Since

$$\| 4\mathbf{i} - \mathbf{j} - 3\mathbf{k} \| = \sqrt{4^2 + (-1)^2 + (-3)^2} = \sqrt{26}$$

it follows that the required vector is

$$\frac{1}{\sqrt{26}} (4\mathbf{i} - \mathbf{j} - 3\mathbf{k})   ❑$$

The difference $\mathbf{a} - \mathbf{b}$ is the sum $\mathbf{a} + (-\mathbf{b})$. Therefore our interpretations of sums and scalar multiples of vectors tell us that if the directed line segments representing **a** and **b** both have the same initial point P (such as the origin), then the directed line segment from the terminal point of **b** to the terminal point of **a** represents the vector

a − **b** (Figure 11.18(a)). From this and our interpretation of the sum of **a** and **b** we see that any nonzero vectors **a** and **b** with the same initial point determine a parallelogram whose diagonals are **a** + **b** and **a** − **b** (Figure 11.18(b)).

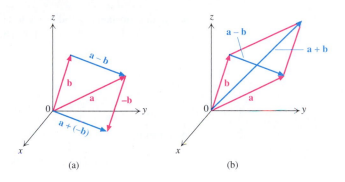

(a)   (b)

FIGURE 11.18

Applications of Vector Addition

Vector addition has been defined as in Definition 11.3 because so many physical quantities combine according to vector addition. For example, if several forces act at the same point P on an object, then the object reacts as though a single force equal to the vector sum of the several forces acts on the object.

EXAMPLE 5   Two children pull a sled by ropes 4 feet long attached to the front center of the sled. The smaller child holds the rope 2 feet above the sled and 2 feet to the side of the point of attachment, and the larger child holds the rope 3 feet above the sled and 2 feet to the opposite side of the point of attachment (Figure 11.19(a)). The smaller child exerts a force $\mathbf{F}_1$ of magnitude 5 pounds, and the taller child exerts a force $\mathbf{F}_2$ of magnitude 7 pounds. Find the resultant force $\mathbf{F}_1 + \mathbf{F}_2$ on the sled.

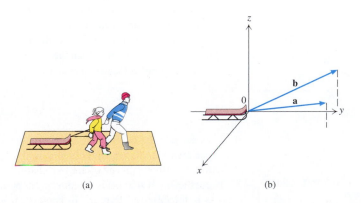

(a)   (b)

FIGURE 11.19

Solution   First we set up a coordinate system, as shown in Figure 11.19(b), with **a** representing the direction the smaller child pulls, and **b** representing the direction the larger child pulls. The forces $\mathbf{F}_1$ and $\mathbf{F}_2$ are parallel, respectively, to the vectors **a** and **b** determined by the ropes. From the given information we find that

$$\mathbf{a} = 2\mathbf{i} + a_2\mathbf{j} + 2\mathbf{k} \quad \text{and} \quad \mathbf{b} = -2\mathbf{i} + b_2\mathbf{j} + 3\mathbf{k}$$

where the constants a_2 and b_2 are to be determined. Since by hypothesis the length of each rope is 4, $\| \mathbf{a} \| = 4 = \| \mathbf{b} \|$. Thus

$$16 = \| \mathbf{a} \|^2 = 4 + a_2^2 + 4, \quad \text{so that} \quad a_2 = 2\sqrt{2}$$

and

$$16 = \| \mathbf{b} \|^2 = (-2)^2 + b_2^2 + 9, \quad \text{so that} \quad b_2 = \sqrt{3}$$

As a result,

$$\mathbf{a} = 2\mathbf{i} + 2\sqrt{2}\mathbf{j} + 2\mathbf{k} \quad \text{and} \quad \mathbf{b} = -2\mathbf{i} + \sqrt{3}\mathbf{j} + 3\mathbf{k}$$

Since $\mathbf{F}_1$ and $\mathbf{F}_2$ have magnitude 5 and 7, respectively, we have $\| \mathbf{F}_1 \| = 5$ and $\| \mathbf{F}_2 \| = 7$. Moreover, $\mathbf{F}_1$ and $\mathbf{F}_2$ are parallel to **a** and **b**, so that

$$\mathbf{F}_1 = 5\,\frac{\mathbf{a}}{\|\mathbf{a}\|} = \frac{5}{4}\,(2\mathbf{i} + 2\sqrt{2}\mathbf{j} + 2\mathbf{k})$$

and

$$\mathbf{F}_2 = 7\,\frac{\mathbf{b}}{\|\mathbf{b}\|} = \frac{7}{4}\,(-2\mathbf{i} + \sqrt{3}\mathbf{j} + 3\mathbf{k})$$

Consequently

$$\mathbf{F}_1 + \mathbf{F}_2 = \frac{5}{4}\,(2\mathbf{i} + 2\sqrt{2}\mathbf{j} + 2\mathbf{k}) + \frac{7}{4}\,(-2\mathbf{i} + \sqrt{3}\mathbf{j} + 3\mathbf{k})$$

$$= \frac{1}{4}\,[-4\mathbf{i} + (10\sqrt{2} + 7\sqrt{3})\mathbf{j} + 31\mathbf{k}] \quad \square$$

Our next illustration of vector addition is taken from electrostatics. Experiments confirm that a particle may be charged positively (as a proton is), or negatively (as an electron is). In either case the charge produces an electric force on any other charged particle. If distance is measured in meters, force in newtons, and charge in coulombs, then the electric force **F** exerted by a charge q_1 at the point P_1 on a charge q located at P is given by **Coulomb's Law**:

$$\mathbf{F} = \frac{q_1 q}{4\pi\varepsilon_0\, r^2}\,\mathbf{u} \tag{4}$$

where **u** is the unit vector in the direction of $\overrightarrow{P_1 P}$ (Figure 11.20), r is the distance between the two charges, and ε_0 is the "permittivity of empty space," equal to 8.854×10^{-12}. If q and q_1 are like charges (both positive or both negative), then the coefficient of **u** in (4) is positive and **F** points in the direction of **u**. This means that q is forced away from q_1. In contrast, if q and q_1 are opposite charges, then the coefficient is negative and **F** has the opposite direction; hence q is forced toward q_1.

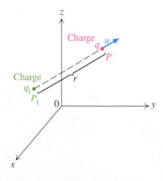

FIGURE 11.20   Coulomb's Law:
$$\mathbf{F} = \frac{q_1 q}{4\pi\varepsilon_0 r^2}\,\mathbf{u}$$

Of course, we can interchange q and q_1 and obtain similar results. Thus we have the familiar fact that "like charges repel each other and opposite charges attract each other."

If charges $q_1, q_2, \ldots, q_n$ are located at the points $P_1, P_2, \ldots, P_n$, then a **principle of superposition** states that the total electric force **F** exerted by these charges on a charge q at P is the sum of the individual forces exerted by q_1, $q_2, \ldots, q_n$, given by

$$\mathbf{F} = \sum_{j=1}^{n} \frac{q_j q}{4\pi\varepsilon_0 r_j^2}\, \mathbf{u}_j \tag{5}$$

where

$$r_j = \|\overrightarrow{P_j P}\| \quad \text{and} \quad \mathbf{u}_j = \frac{\overrightarrow{P_j P}}{\|\overrightarrow{P_j P}\|}$$

EXAMPLE 6   Suppose the charge of a proton is 1.6×10^{-19} coulombs and the charge of an electron is -1.6×10^{-19} coulombs. Find the electric force **F** exerted on a positive unit charge at the origin if the proton is located at $(3 \times 10^{-11}, 4 \times 10^{-11}, 0)$ and the electron is located at $(0, 0, 10^{-11})$.

Solution   For the unit charge at the origin let $q = 1$ and $P = (0, 0, 0)$, and for the proton and electron let

$$q_1 = 1.6 \times 10^{-19} \qquad\qquad P_1 = (3 \times 10^{-11}, 4 \times 10^{-11}, 0)$$
$$q_2 = -1.6 \times 10^{-19} \qquad\quad P_2 = (0, 0, 10^{-11})$$

Then

$$r_1 = \|\overrightarrow{P_1 P}\| = \sqrt{(3 \times 10^{-11})^2 + (4 \times 10^{-11})^2} = 10^{-11}\sqrt{3^2 + 4^2} = 5 \times 10^{-11}$$
$$r_2 = \|\overrightarrow{P_2 P}\| = 10^{-11}$$

Therefore

$$\mathbf{u}_1 = \frac{\overrightarrow{P_1 P}}{\|\overrightarrow{P_1 P}\|} = \frac{-3 \times 10^{-11}\mathbf{i} - 4 \times 10^{-11}\mathbf{j}}{5 \times 10^{-11}} = -\frac{3}{5}\mathbf{i} - \frac{4}{5}\mathbf{j}$$

$$\mathbf{u}_2 = \frac{\overrightarrow{P_2 P}}{\|\overrightarrow{P_2 P}\|} = \frac{-10^{-11}\mathbf{k}}{10^{-11}} = -\mathbf{k}$$

Consequently by (5),

$$\mathbf{F} = \frac{q_1 q}{4\pi\varepsilon_0 r_1^2}\, \mathbf{u}_1 + \frac{q_2 q}{4\pi\varepsilon_0 r_2^2}\, \mathbf{u}_2$$

$$= \frac{(1.6 \times 10^{-19})(1)}{4\pi\varepsilon_0 (5 \times 10^{-11})^2}\left(-\frac{3}{5}\mathbf{i} - \frac{4}{5}\mathbf{j}\right) + \frac{(-1.6 \times 10^{-19})(1)}{4\pi\varepsilon_0 (10^{-11})^2}(-\mathbf{k})$$

$$= \frac{1.6 \times 10^{-19}}{4\pi\varepsilon_0 10^{-22}}\left(-\frac{3}{125}\mathbf{i} - \frac{4}{125}\mathbf{j} + \mathbf{k}\right)$$

$$\approx 1.150 \times 10^{11}\,(-3\mathbf{i} - 4\mathbf{j} + 125\mathbf{k}) \quad \square$$

Vectors in the Plane

Any vector $\mathbf{a} = a_1\mathbf{i} + a_2\mathbf{j} + a_3\mathbf{k}$ for which $a_3 = 0$ can be written more easily as

$$\mathbf{a} = a_1\mathbf{i} + a_2\mathbf{j} \tag{6}$$

The point (a_1, a_2) corresponding to the vector in (6) lies in the xy plane (Figure

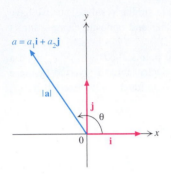

$a = a_1\mathbf{i} + a_2\mathbf{j}$

FIGURE 11.21

An iceberg in Disko Bay,
Greenland. (*H. Kanus/Superstock*)

11.21). If the vector is nonzero, then it can also be described by its length $\|\mathbf{a}\|$ and
the angle θ it makes with the positive x axis:

$$\mathbf{a} = \|\mathbf{a}\|(\cos\theta\,\mathbf{i} + \sin\theta\,\mathbf{j}) \tag{7}$$

(see Exercise 23). If $\mathbf{a}$ and $\mathbf{b}$ are vectors in the same plane and c is any number, then
$\mathbf{a} + \mathbf{b}$, $\mathbf{a} - \mathbf{b}$, and $c\mathbf{a}$ are also in that plane. In solving problems that are two-
dimensional in nature we may use vectors represented by either (6) or (7).

In theory, one way to bring water to heavily populated areas in the northeastern
part of the United States is to drag Arctic icebergs down from the Greenland coast.
Icebergs can extend several hundred feet above water level; however, because the
density of the water in an iceberg is only slightly less than that of ocean water,
approximately 90% of an iceberg resides under water. Thus the expression "the tip
of the iceberg."

EXAMPLE 7   Two tugboats are to drag a (small) iceberg. The smaller tugboat
will exert a force of 10^5 pounds in the direction 15° east of south, and the larger
tugboat will exert a force of 2×10^5 pounds. Assuming that the forces are exerted
horizontally (without any $\mathbf{k}$ component), determine the direction the larger tugboat
should be pointed in order for the iceberg to move due south.

Solution   First we set up a coordinate system with the origin at the point of
attachment of the tugboat cables to the iceberg, and with the positive x axis pointing
south and the positive y axis pointing east. Let $\mathbf{F}_1$ and $\mathbf{F}_2$ be the forces exerted by
the smaller and larger tugboats, respectively. Since 15° east of south is equivalent
to an angle of $\pi/12$ radians with respect to the positive x axis, we have

$$\mathbf{F}_1 = 10^5\left(\cos\frac{\pi}{12}\,\mathbf{i} + \sin\frac{\pi}{12}\,\mathbf{j}\right)$$

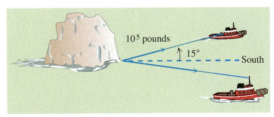

FIGURE 11.22

If θ denotes the angle in which $\mathbf{F}_2$ is directed, then

$$\mathbf{F}_2 = 2 \times 10^5\,(\cos\theta\mathbf{i} + \sin\theta\mathbf{j})$$

We need to find the value of θ such that the $\mathbf{j}$ component of $\mathbf{F}_1 + \mathbf{F}_2$ is 0 (so that the
sum of the forces is directed due south). However,

$$\mathbf{F}_1 + \mathbf{F}_2 = 10^5\left[\left(\cos\frac{\pi}{12} + 2\cos\theta\right)\mathbf{i} + \left(\sin\frac{\pi}{12} + 2\sin\theta\right)\mathbf{j}\right]$$

Therefore θ must satisfy the equation

$$\sin\frac{\pi}{12} + 2\sin\theta = 0, \quad\text{or equivalently,}\quad \sin\theta = -\frac{1}{2}\sin\frac{\pi}{12}$$

As a result,

$$\theta = \sin^{-1}\left(-\frac{1}{2}\sin\frac{\pi}{12}\right) \approx -0.1297734718$$

which is approximately $-7.4°$. Consequently the second tugboat should be directed approximately $7.4°$ west of south.   ❏

EXAMPLE 8   Show that the diagonals of a parallelogram bisect each other.

Solution   Assume that the parallelogram lies in the xy plane, as shown in Figure 11.23, and let **a** and **b** be vectors along the two sides that contain the origin. Then **a** + **b** and **a** − **b** lie along the diagonals. The vector $\frac{1}{2}(\mathbf{a} + \mathbf{b})$ joins the origin to the midpoint of the diagonal **a** + **b**, and the vector $\mathbf{b} + \frac{1}{2}(\mathbf{a} - \mathbf{b})$ joins the origin to the midpoint of the other diagonal. Since

$$\frac{1}{2}(\mathbf{a} + \mathbf{b}) = \mathbf{b} + \frac{1}{2}(\mathbf{a} - \mathbf{b})$$

it follows that the two midpoints coincide. Consequently the two diagonals bisect one another.   ❏

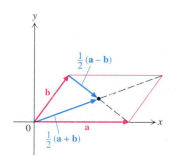

FIGURE 11.23

One exotic application of vectors involves the very precise measurement of the distance between points on the earth and the moon by means of laser beams. To understand the method, let us observe that if a ray of light is reflected in a mirror, then by the Law of Reflection the angle of incidence is equal to the angle of reflection (Figure 11.24(a)). In particular, if the mirror lies in the xy plane and the initial ray is $\mathbf{a} = a_1\mathbf{i} + a_2\mathbf{j} + a_3\mathbf{k}$, then the reflected ray is $\mathbf{b} = a_1\mathbf{i} + a_2\mathbf{j} - a_3\mathbf{k}$ (see Exercise 34a). It follows that when a ray is reflected by each of the three mutually perpendicular mirrors comprising a corner mirror, the resultant ray will be parallel to the initial ray (Figure 11.24(b)) (see Exercise 34b). In their endeavor to calculate the distance between the moon and earth, American space scientists have positioned an array of corner mirrors on the moon, and with laser beams sent from earth they have been able to calculate the distance very accurately.

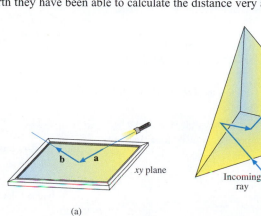

(a)

(b)

FIGURE 11.24

EXERCISES 11.2

In Exercises 1–4 write $\overrightarrow{PQ}$ as a vector in the form $a\mathbf{i} + b\mathbf{j} + c\mathbf{k}$.

1. $P = (0, 0, 0)$, $Q = (3, -4, 10)$
2. $P = (1, -3, 2)$, $Q = (3, -1, 3)$
3. $P = (0, 1, 0)$, $Q = (3, -1, \sqrt{7})$
4. $P = (2, 1, 1)$, $Q = (2, 1 + \sqrt{2}, 1 + \sqrt{3})$

In Exercises 5–8 find $\mathbf{a} + \mathbf{b}$, $\mathbf{a} - \mathbf{b}$, and $c\mathbf{a}$.

5. $\mathbf{a} = 2\mathbf{i} - 5\mathbf{j} + 10\mathbf{k}$, $\mathbf{b} = -\mathbf{i} + 2\mathbf{j} - 9\mathbf{k}$, $c = 2$
6. $\mathbf{a} = \mathbf{i} + \mathbf{j} - 3\mathbf{k}$, $\mathbf{b} = \frac{1}{2}\mathbf{i} - \frac{1}{2}\mathbf{j} - 3\mathbf{k}$, $c = -1$
7. $\mathbf{a} = 2\mathbf{i}$, $\mathbf{b} = \mathbf{j} + \mathbf{k}$, $c = \frac{1}{3}$
8. $\mathbf{a} = \mathbf{i} + 2\mathbf{j}$, $\mathbf{b} = -2\mathbf{j} + \mathbf{k}$, $c = \pi$

In Exercises 9–13 find the length of the vector.

9. $\mathbf{a} = \mathbf{i} - \mathbf{j} + \mathbf{k}$
10. $\mathbf{a} = 2\mathbf{i} + \mathbf{j} - 2\mathbf{k}$
11. $\mathbf{b} = -3\mathbf{i} + 4\mathbf{j} - 12\mathbf{k}$
12. $\mathbf{b} = 4\mathbf{i} - 8\mathbf{j} + 8\mathbf{k}$
13. $\mathbf{c} = \sqrt{2}\mathbf{i} - \mathbf{j} + \mathbf{k}$

In Exercises 14–17 find a unit vector having the same direction as the given vector.

14. $\mathbf{a} = \mathbf{i} + \mathbf{j} - \mathbf{k}$
15. $\mathbf{a} = -3\mathbf{i} + 4\mathbf{j} - 12\mathbf{k}$
16. $\mathbf{b} = 7\mathbf{i} + 12\sqrt{2}\,\mathbf{j} - 12\sqrt{2}\,\mathbf{k}$
17. $\mathbf{b} = 2\mathbf{i} - 3\mathbf{j}$

18. Let $\mathbf{a} = \mathbf{i} + 2\mathbf{j} - 3\mathbf{k}$. Find a vector $\mathbf{b}$ of length 5 whose direction is opposite to that of $\mathbf{a}$.

19. Let b_1 and b_2 be any real numbers with $|b_1| \le 1$ and $|b_2| \le 1$. Let

$$\mathbf{u} = \sqrt{1 - b_1^2}\,\mathbf{i} + b_1\sqrt{1 - b_2^2}\,\mathbf{j} + b_1 b_2 \mathbf{k}$$

Show that $\mathbf{u}$ is a unit vector.

In Exercises 20–21 find the resultant force $\mathbf{F}_1 + \mathbf{F}_2$ of the given forces $\mathbf{F}_1$ and $\mathbf{F}_2$.

20. $\mathbf{F}_1 = \sqrt{2}\mathbf{i} + \mathbf{j} - 4\mathbf{k}$, $\mathbf{F}_2 = (1 - \sqrt{2})\mathbf{i} - 5\mathbf{j} - 4\mathbf{k}$
21. $\mathbf{F}_1 = 10^{-3}\mathbf{i} + 0.12\mathbf{j} + 1.2 \times 10^4\mathbf{k}$, $\mathbf{F}_2 = 3 \times 10^{-3}\mathbf{i} + 0.39\mathbf{j} - 5 \times 10^4\mathbf{k}$

22. Let $OABCDEFG$ be a cube. Show that $\overrightarrow{OB} + \overrightarrow{OD} + \overrightarrow{OF}$ is parallel to $\overrightarrow{OG}$ (Figure 11.25).

23. **a.** Let $\mathbf{a}$ be a nonzero vector in the xy plane, and let θ be the angle from the positive x axis to $\mathbf{a}$ in the counterclockwise direction. Show that

$$\mathbf{a} = \|\mathbf{a}\|(\cos\theta\mathbf{i} + \sin\theta\mathbf{j})$$

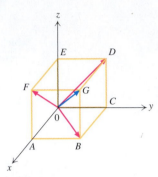

FIGURE 11.25 Figure for Exercise 22.

    b. Suppose that $\mathbf{u}$ is a unit vector in the plane. Using part (a), show that

$$\mathbf{u} = \cos\theta\mathbf{i} + \sin\theta\mathbf{j}$$

for some number θ.

24. Let $\mathbf{a}$ and $\mathbf{b}$ be two vectors with initial points at the origin and terminal points P and Q, respectively.

    a. Show that the vector $\mathbf{c}$ directed from the origin to the midpoint of PQ is $\frac{1}{2}(\mathbf{a} + \mathbf{b})$ (Figure 11.26(a)). (*Hint:* $\mathbf{c} = \mathbf{a} + \frac{1}{2}(\mathbf{b} - \mathbf{a})$.)

    b. Show that the vector $\mathbf{c}$ directed from the origin to the point on PQ two thirds of the way from P to Q is $\frac{1}{3}\mathbf{a} + \frac{2}{3}\mathbf{b}$ (Figure 11.26(b)).

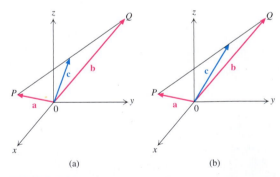

(a)   (b)

FIGURE 11.26 Figures for Exercise 24.

25. Using vectors, prove that a quadrilateral $PQRS$ is a parallelogram if the diagonals PR and QS bisect each other.

*26. Using vectors, prove that the midpoints of the sides of any quadrilateral form a parallelogram.

Applications

27. Three children tug on a ball located at O (Figure 11.27). One child pulls with a force of 20 pounds in the direction of the negative y axis. Another child pulls with a force of 100 pounds at an angle of $\pi/3$ with the positive x axis. If the total force exerted on the ball is to be **0**, find the force **F** with which the third child should pull and the tangent of the angle θ between the positive x axis and the direction in which the third child should pull.

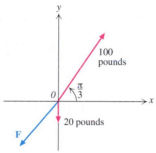

FIGURE 11.27 Figure for Exercise 27.

28. Two tugboats are pulling an ocean freighter as shown in Figure 11.28. If one tugboat exerts a force of 1000 pounds on the cable tied at A, what force must be exerted on the cable tied to the other tugboat at B if the freighter is to move along the line l?

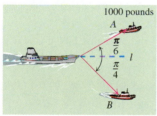

FIGURE 11.27 Figure for Exercise 28.

29. Suppose an airplane is flying in the xy plane with its body oriented at an angle of $\pi/6$ with respect to the positive x axis. If the air is moving parallel to the positive y axis at 20 miles per hour and the speed of the airplane with respect to the air is 300 miles per hour, what is the speed of the airplane with respect to the ground? (*Hint:* The velocity of the plane with respect to the ground is equal to the sum of the velocity of the plane with respect to the air and the velocity of the air with respect to the ground.)

*30. When a boat travels due north at 8 miles per hour, a vane on the boat points due northwest. When the boat anchors, the vane points due west. What is the speed of the wind? (*Hint:* See the hint for Exercise 29 and use the fact that the vane points into the wind.)

31. Suppose the electron in Example 6 were located at $(10^{-12}, 10^{-12}, 0)$ and the proton at $(0, 10^{-11}, 10^{-11})$. Find the total electric force **F** exerted by the two particles on a positive unit charge located at the origin.

32. Find the total electric force **F** exerted by the two particles in Example 6 on a positive unit charge located at the point $(0, 10^{-11}, 0)$.

*33. Suppose that point masses $m_1, m_2, \ldots, m_n$ are located at points $P_1(x_1, y_1), P_2(x_2, y_2), \ldots, P_n(x_n, y_n)$, respectively, in the xy plane. Show that the centroid of these point masses is the point $P(\overline{x}, \overline{y})$ such that

$$m_1\overrightarrow{PP_1} + m_2\overrightarrow{PP_2} + \cdots + m_n\overrightarrow{PP_n} = 0$$

(*Hint:* Write out the two components of $m_1\overrightarrow{PP_1} + m_2\overrightarrow{PP_2} + \cdots + m_n\overrightarrow{PP_n}$, and then refer to the definitions of $\overline{x}$ and $\overline{y}$ in Section 6.5.)

34. a. Suppose that the xy plane is a mirror and that a ray of light is reflected by the mirror (see Figure 11.24(a)). Let **a** and **b** be unit vectors pointing along the path of the incident and the reflected rays. Using the Law of Reflection, which states that the angle of incidence is equal to the angle of reflection, show that if $\mathbf{a} = a_1\mathbf{i} + a_2\mathbf{j} + a_3\mathbf{k}$, then $\mathbf{b} = a_1\mathbf{i} + a_2\mathbf{j} - a_3\mathbf{k}$.

*b. Suppose a ray of light is reflected in turn from each of three mutually perpendicular mirrors, which you may think of as the portions of the coordinate planes in the first octant (see Figure 11.24(b)). Show that the reflected ray is parallel to the incident ray. (*Hint:* Use (a) and the analogous results for the yz plane and the xz plane.)

11.3 THE DOT PRODUCT

Now that we have defined the sum and difference of two vectors, it is natural to ask whether a useful product of two vectors can be defined. One obvious possibility is to define multiplication of vectors componentwise, just as we did with addition and subtraction. But such a product has little physical significance and almost never arises in applications. However, two other products of vectors, known as the dot product and the cross product, have profound physical significance. First we consider the dot product, defined as follows.

DEFINITION 11.5

Let $\mathbf{a} = a_1\mathbf{i} + a_2\mathbf{j} + a_3\mathbf{k}$ and $\mathbf{b} = b_1\mathbf{i} + b_2\mathbf{j} + b_3\mathbf{k}$ be two vectors. The **dot product** (or **scalar product** or **inner product**) of $\mathbf{a}$ and $\mathbf{b}$ is the number $\mathbf{a} \cdot \mathbf{b}$ defined by

$$\mathbf{a} \cdot \mathbf{b} = a_1b_1 + a_2b_2 + a_3b_3 \qquad (1)$$

From (1) we deduce that

$$\mathbf{a} \cdot \mathbf{i} = \mathbf{i} \cdot \mathbf{a} = a_1, \quad \mathbf{a} \cdot \mathbf{j} = \mathbf{j} \cdot \mathbf{a} = a_2, \quad \text{and} \quad \mathbf{a} \cdot \mathbf{k} = \mathbf{k} \cdot \mathbf{a} = a_3$$

In particular

$$\mathbf{i} \cdot \mathbf{j} = \mathbf{j} \cdot \mathbf{k} = \mathbf{k} \cdot \mathbf{i} = 0 \quad \text{and} \quad \mathbf{i} \cdot \mathbf{i} = \mathbf{j} \cdot \mathbf{j} = \mathbf{k} \cdot \mathbf{k} = 1$$

Other dot products are almost as simple to calculate.

EXAMPLE 1   Let $\mathbf{a} = 3\mathbf{i} - \mathbf{j} - 2\mathbf{k}$ and $\mathbf{b} = 2\mathbf{i} - 3\mathbf{j} + \frac{1}{2}\mathbf{k}$. Find $\mathbf{a} \cdot \mathbf{b}$.

Solution   By (1),

$$\mathbf{a} \cdot \mathbf{b} = 3(2) + (-1)(-3) + (-2)\left(\frac{1}{2}\right) = 8 \quad \square$$

The dot product satisfies many of the laws that hold for real numbers. For example,

$\mathbf{a} \cdot \mathbf{b} = \mathbf{b} \cdot \mathbf{a}$	$(c\mathbf{a}) \cdot \mathbf{b} = c(\mathbf{a} \cdot \mathbf{b}) = \mathbf{a} \cdot (c\mathbf{b})$
$\mathbf{a} \cdot (\mathbf{b} + \mathbf{c}) = \mathbf{a} \cdot \mathbf{b} + \mathbf{a} \cdot \mathbf{c}$	$(\mathbf{a} + \mathbf{b}) \cdot \mathbf{c} = \mathbf{a} \cdot \mathbf{c} + \mathbf{b} \cdot \mathbf{c}$

The **angle** between two nonzero vectors $\mathbf{a}$ and $\mathbf{b}$ is defined to be the angle θ, where $0 \le \theta \le \pi$, formed by the corresponding directed line segments whose initial points are the origin (Figure 11.29). The relationship between the dot product and the angle between two vectors is described in the following theorem.

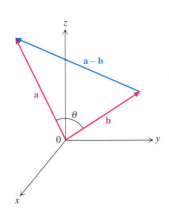

FIGURE 11.29

THEOREM 11.6

Let $\mathbf{a} = a_1\mathbf{i} + a_2\mathbf{j} + a_3\mathbf{k}$ and $\mathbf{b} = b_1\mathbf{i} + b_2\mathbf{j} + b_3\mathbf{k}$ be nonzero vectors, and let θ be the angle between $\mathbf{a}$ and $\mathbf{b}$. Then

$$\mathbf{a} \cdot \mathbf{b} = \| \mathbf{a} \| \, \| \mathbf{b} \| \cos \theta \qquad (2)$$

Proof   Formula (2) is most easily proved by applying the Law of Cosines to the triangle formed by **a**, **b**, and **a** − **b** (Figure 11.29). From this law we obtain

$$\| \mathbf{a} - \mathbf{b} \|^2 = \| \mathbf{a} \|^2 + \| \mathbf{b} \|^2 - 2 \| \mathbf{a} \| \| \mathbf{b} \| \cos \theta \tag{3}$$

where θ is the angle between **a** and **b**. Using the definition of the length of a vector and the fact that $\mathbf{a} - \mathbf{b} = (a_1 - b_1)\mathbf{i} + (a_2 - b_2)\mathbf{j} + (a_3 - b_3)\mathbf{k}$, we can rewrite (3) as

$$(a_1 - b_1)^2 + (a_2 - b_2)^2 + (a_3 - b_3)^2$$
$$= (a_1^2 + a_2^2 + a_3^2) + (b_1^2 + b_2^2 + b_3^2) - 2 \| \mathbf{a} \| \| \mathbf{b} \| \cos \theta$$

By expanding the squared terms in this equation and then canceling like terms on both sides of the equation, we obtain

$$-2a_1b_1 - 2a_2b_2 - 2a_3b_3 = -2 \| \mathbf{a} \| \| \mathbf{b} \| \cos \theta$$

from which we conclude that

$$\mathbf{a} \cdot \mathbf{b} = a_1b_1 + a_2b_2 + a_3b_3 = \| \mathbf{a} \| \| \mathbf{b} \| \cos \theta \quad \blacksquare$$

Since **a** and **b** are perpendicular if and only if $\theta = \pi/2$, or equivalently $\cos \theta = 0$, Theorem 11.6 provides us with a method for determining when two vectors are perpendicular.

COROLLARY 11.7

a.   Two nonzero vectors **a** and **b** are perpendicular if and only if $\mathbf{a} \cdot \mathbf{b} = 0$.

b.   For any vector **a**, $\mathbf{a} \cdot \mathbf{a} = \| \mathbf{a} \|^2$, or equivalently, $\| \mathbf{a} \| = \sqrt{\mathbf{a} \cdot \mathbf{a}}$.

EXAMPLE 2   Show that the vectors $\mathbf{a} = -4\mathbf{i} + 5\mathbf{j} + 7\mathbf{k}$ and $\mathbf{b} = \mathbf{i} - 2\mathbf{j} + 2\mathbf{k}$ are perpendicular.

Solution   We simply compute the dot product by using (1):

$$\mathbf{a} \cdot \mathbf{b} = (-4)(1) + (5)(-2) + (7)(2) = 0$$

It follows from Corollary 11.7(a) that **a** and **b** are perpendicular.   ❑

Solving (2) for $\cos \theta$, we obtain

$$\cos \theta = \frac{\mathbf{a} \cdot \mathbf{b}}{\| \mathbf{a} \| \| \mathbf{b} \|} \tag{4}$$

Since $0 \le \theta \le \pi$, the angle between **a** and **b** is uniquely determined by (4).

EXAMPLE 3   Find the angle between the vectors $\mathbf{a} = 2\mathbf{i} - \mathbf{j} + 2\mathbf{k}$ and $\mathbf{b} = \mathbf{i} - \mathbf{j}$.

Solution Since

$$\| \mathbf{a} \| = \sqrt{2^2 + (-1)^2 + 2^2} = 3 \quad \text{and} \quad \| \mathbf{b} \| = \sqrt{1^2 + (-1)^2 + 0^2} = \sqrt{2}$$

and since

$$\mathbf{a} \cdot \mathbf{b} = (2)(1) + (-1)(-1) + (2)(0) = 3$$

we infer from (4) that

$$\cos \theta = \frac{3}{3\sqrt{2}} = \frac{1}{\sqrt{2}} = \frac{\sqrt{2}}{2}$$

which means that $\theta = \pi/4$. ❑

The angles α, β, and γ (between 0 and π inclusive) that a nonzero vector $\mathbf{a}$ makes with the positive x, y, and z axes are called the **direction angles** of $\mathbf{a}$ (Figure 11.30). If $\mathbf{a} = a_1 \mathbf{i} + a_2 \mathbf{j} + a_3 \mathbf{k}$, then by (4) and (1),

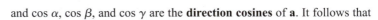

$$\cos \alpha = \frac{\mathbf{a} \cdot \mathbf{i}}{\| \mathbf{a} \| \| \mathbf{i} \|} = \frac{a_1}{\| \mathbf{a} \|}, \quad \cos \beta = \frac{\mathbf{a} \cdot \mathbf{j}}{\| \mathbf{a} \| \| \mathbf{j} \|} = \frac{a_2}{\| \mathbf{a} \|},$$

$$\cos \gamma = \frac{\mathbf{a} \cdot \mathbf{k}}{\| \mathbf{a} \| \| \mathbf{k} \|} = \frac{a_3}{\| \mathbf{a} \|}$$

and $\cos \alpha$, $\cos \beta$, and $\cos \gamma$ are the **direction cosines** of $\mathbf{a}$. It follows that

$$a_1 = \| \mathbf{a} \| \cos \alpha, \qquad a_2 = \| \mathbf{a} \| \cos \beta, \qquad a_3 = \| \mathbf{a} \| \cos \gamma$$

so that we may write $\mathbf{a}$ in the alternative form

$$\mathbf{a} = \| \mathbf{a} \| (\cos \alpha \mathbf{i} + \cos \beta \mathbf{j} + \cos \gamma \mathbf{k})$$

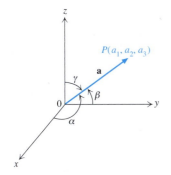

FIGURE 11.30

EXAMPLE 4 Let $\mathbf{a} = -2\mathbf{i} + 3\mathbf{j} - 5\mathbf{k}$. Find the direction cosines of $\mathbf{a}$.

Solution Since

$$\| \mathbf{a} \| = \sqrt{(-2)^2 + 3^2 + (-5)^2} = \sqrt{4 + 9 + 25} = \sqrt{38}$$

a direct calculation shows that

$$\cos \alpha = \frac{-2}{\sqrt{38}}, \qquad \cos \beta = \frac{3}{\sqrt{38}}, \qquad \cos \gamma = \frac{-5}{\sqrt{38}} \quad ❑$$

The Projection of One Vector onto Another

Suppose that two nonzero vectors $\mathbf{a}$ and $\mathbf{b}$ are positioned as in Figure 11.31(a), and that the sun casts a shadow on the line containing the vector $\mathbf{a}$. Informally we think of the shadow as determining a vector parallel to $\mathbf{a}$, which we call the projection of $\mathbf{b}$ onto $\mathbf{a}$ and denote $\mathbf{pr_a b}$. Since $\mathbf{pr_a b}$ is parallel to $\mathbf{a}$ or is $\mathbf{0}$, it must be a scalar multiple of $\mathbf{a}$. The length of $\mathbf{pr_a b}$ is evidently $\| \mathbf{b} \| | \cos \theta |$, where θ is the angle between **a** and **b** (so that $0 \le \theta \le \pi$). If $0 \le \theta \le \pi/2$, it follows that

$$\mathbf{pr_a b} = \| \mathbf{b} \| \cos \theta \, \frac{1}{\| \mathbf{a} \|} \, \mathbf{a} = \| \mathbf{b} \| \left(\frac{\mathbf{a} \cdot \mathbf{b}}{\| \mathbf{a} \| \| \mathbf{b} \|} \right) \frac{1}{\| \mathbf{a} \|} \, \mathbf{a} = \left(\frac{\mathbf{a} \cdot \mathbf{b}}{\| \mathbf{a} \|^2} \right) \mathbf{a}$$

If $\pi/2 < \theta \le \pi$, then Figure 11.31(b) applies, and in a similar way we find that

$$\mathbf{pr_a b} = \| \mathbf{b} \| \, (-\cos \theta) \, \frac{1}{\| \mathbf{a} \|} \, (-\mathbf{a}) = \| \mathbf{b} \| \, \cos \theta \, \frac{1}{\| \mathbf{a} \|} \, \mathbf{a} = \left(\frac{\mathbf{a} \cdot \mathbf{b}}{\| \mathbf{a} \|^2} \right) \mathbf{a}$$

which is the same expression for **pr_a b**. Thus we can define the projection by a single formula involving the dot product, whether $0 \le \theta \le \pi/2$ or $\pi/2 < \theta \le \pi$.

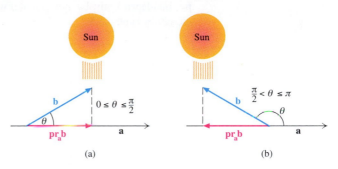

FIGURE 11.31

DEFINITION 11.8   Let **a** be a nonzero vector. The **projection** of a vector **b** onto **a** is the vector **pr_a b** defined by

$$\mathbf{pr_a b} = \left(\frac{\mathbf{a} \cdot \mathbf{b}}{\| \mathbf{a} \|^2} \right) \mathbf{a} \tag{5}$$

EXAMPLE 5   Let $\mathbf{a} = 3\mathbf{i} - \mathbf{j} - 2\mathbf{k}$ and $\mathbf{b} = 2\mathbf{i} - 3\mathbf{j} + \frac{1}{2}\mathbf{k}$, as in Example 1. Find **pr_a b**.

Solution   From Example 1 we know that $\mathbf{a} \cdot \mathbf{b} = 8$, and a simple calculation shows that $\| \mathbf{a} \|^2 = 14$. Consequently by (5) we have

$$\mathbf{pr_a b} = \frac{8}{14} \, (3\mathbf{i} - \mathbf{j} - 2\mathbf{k}) = \frac{4}{7} \, (3\mathbf{i} - \mathbf{j} - 2\mathbf{k}) \quad \square$$

For future reference we derive a formula for the length of **pr_a b** in terms of **a** and **b**. Using (5), we find that

$$\| \mathbf{pr_a b} \| = \left\| \frac{\mathbf{a} \cdot \mathbf{b}}{\| \mathbf{a} \|^2} \, \mathbf{a} \right\| = \frac{|\mathbf{a} \cdot \mathbf{b}|}{\| \mathbf{a} \|^2} \, \| \mathbf{a} \| = \frac{|\mathbf{a} \cdot \mathbf{b}|}{\| \mathbf{a} \|}$$

Thus

$$\|\mathbf{pr_a b}\| = \frac{|\mathbf{a} \cdot \mathbf{b}|}{\|\mathbf{a}\|}$$ (6)

Resolution of a Vector

If $\mathbf{a}$ and $\mathbf{a}'$ are perpendicular, then any nonzero vector $\mathbf{b}$ lying in the same plane as $\mathbf{a}$ and $\mathbf{a}'$ can be expressed as the sum of the two vectors $\mathbf{pr_a b}$ and $\mathbf{pr_{a'} b}$, which are parallel to $\mathbf{a}$ and $\mathbf{a}'$, respectively:

$$\mathbf{b} = \mathbf{pr_a b} + \mathbf{pr_{a'} b}$$

(Figure 11.32(a) and (b)). We say that $\mathbf{b}$ has been **resolved** into vectors parallel to $\mathbf{a}$ and $\mathbf{a}'$. Geometrically, $\mathbf{pr_a b}$ and $\mathbf{pr_{a'} b}$ are the legs of the right triangle that has hypotenuse $\mathbf{b}$ and has legs parallel to $\mathbf{a}$ and $\mathbf{a}'$. One can compute both $\mathbf{pr_a b}$ and $\mathbf{pr_{a'} b}$ independently, but notice that if we know $\mathbf{b}$ and $\mathbf{pr_a b}$, then we can find $\mathbf{pr_{a'} b}$ merely by subtraction:

$$\mathbf{pr_{a'} b} = \mathbf{b} - \mathbf{pr_a b}$$ (7)

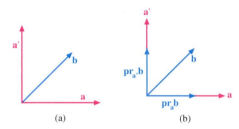

(a)   (b)

FIGURE 11.32

EXAMPLE 6   Let $\mathbf{a} = \mathbf{i} + \mathbf{j}$, $\mathbf{a}' = \mathbf{i} - \mathbf{j}$, and $\mathbf{b} = 3\mathbf{i} - 4\mathbf{j}$. Resolve $\mathbf{b}$ into vectors parallel to $\mathbf{a}$ and $\mathbf{a}'$.

Solution   Observe that all three vectors lie in the xy plane and that $\mathbf{a}$ and $\mathbf{a}'$ are perpendicular, since $\mathbf{a} \cdot \mathbf{a}' = 0$. To resolve $\mathbf{b}$, we notice that $\mathbf{a} \cdot \mathbf{b} = -1$ and $\|\mathbf{a}\|^2 = 2$, so that by (5),

$$\mathbf{pr_a b} = \frac{\mathbf{a} \cdot \mathbf{b}}{\|\mathbf{a}\|^2}\, \mathbf{a} = -\frac{1}{2}\, \mathbf{a} = -\frac{1}{2}(\mathbf{i} + \mathbf{j})$$

Then (7) yields

$$\mathbf{pr_{a'} b} = \mathbf{b} - \mathbf{pr_a b} = (3\mathbf{i} - 4\mathbf{j}) - \left(-\frac{1}{2}(\mathbf{i} + \mathbf{j})\right) = \frac{7}{2}\mathbf{i} - \frac{7}{2}\mathbf{j} = \frac{7}{2}\mathbf{a}'$$

Therefore $\mathbf{b} = -\frac{1}{2}\mathbf{a} + \frac{7}{2}\mathbf{a}'$. ☐

Work Done by a Constant Force

One application of projections of vectors arises in the definition of the work done by a force on a moving object. Suppose that a constant force $\mathbf{F}$ acts on an object

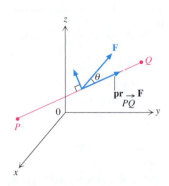

FIGURE 11.33   **F** is the sum of a vector parallel to $\overrightarrow{PQ}$ and a vector perpendicular to $\overrightarrow{PQ}$.

moving along the line from P to Q and that the direction of the force is parallel to that line. By our definition of work in Section 6.4, the work W done is either $\| \mathbf{F} \| \|\overrightarrow{PQ}\|$ or $-\| \mathbf{F} \| \|\overrightarrow{PQ}\|$, depending on whether **F** acts in the direction of $\overrightarrow{PQ}$ or in the opposite direction.

If **F** is not parallel to $\overrightarrow{PQ}$, then we resolve **F** into a vector parallel to $\overrightarrow{PQ}$ and a vector perpendicular to $\overrightarrow{PQ}$ (Figure 11.33). The vector $\mathbf{pr}_{\overrightarrow{PQ}}\,\mathbf{F}$ parallel to $\overrightarrow{PQ}$ affects the speed of the object and hence does work on the object. In contrast, the vector perpendicular to $\overrightarrow{PQ}$ does not affect the speed, so we ignore it with regard to work. Accordingly, we define the work W done by the force **F** as the work done by the force $\mathbf{pr}_{\overrightarrow{PQ}}\,\mathbf{F}$ parallel to $\overrightarrow{PQ}$. Thus if θ is the angle between **F** and $\overrightarrow{PQ}$, then

$$W = \begin{cases} \| \mathbf{pr}_{\overrightarrow{PQ}}\,\mathbf{F}\| \|\overrightarrow{PQ}\| & \text{for } 0 \le \theta \le \dfrac{\pi}{2} \\[2mm] -\| \mathbf{pr}_{\overrightarrow{PQ}}\,\mathbf{F}\| \|\overrightarrow{PQ}\| & \text{for } \dfrac{\pi}{2} < \theta \le \pi \end{cases} \tag{8}$$

But by (6),

$$\| \mathbf{pr}_{\overrightarrow{PQ}}\,\mathbf{F}\| = \frac{|\mathbf{F} \cdot \overrightarrow{PQ}|}{\|\overrightarrow{PQ}\|} = \begin{cases} \dfrac{\mathbf{F} \cdot \overrightarrow{PQ}}{\|\overrightarrow{PQ}\|} & \text{for } 0 \le \theta \le \dfrac{\pi}{2} \\[3mm] \dfrac{-\mathbf{F} \cdot \overrightarrow{PQ}}{\|\overrightarrow{PQ}\|} & \text{for } \dfrac{\pi}{2} < \theta \le \pi \end{cases} \tag{9}$$

Formulas (8) and (9) together imply that

$$W = \mathbf{F} \cdot \overrightarrow{PQ} \tag{10}$$

Using (2), we can write (10) equivalently as

$$W = \| \mathbf{F} \| \| \overrightarrow{PQ} \| \cos \theta \tag{11}$$

Hence if a force with given magnitude is applied to an object in motion, then the nearer the direction of the force is to the direction of motion, the more work the force does on the object. This is why it is easier to shovel snow with the handle held nearly horizontal than with it held up.

EXAMPLE 7   Suppose that a force **F** of 50 pounds is exerted upward on a wheelbarrow at an angle of $\pi/6$ with the horizontal and that the wheelbarrow travels 100 feet in the horizontal direction (Figure 11.34). Find the work W done by the force on the wheelbarrow.

Solution   Let us assume that the wheelbarrow travels along the x axis from 0 to 100 and that the force **F** acts in the xy plane (Figure 11.34), so that $P = (0, 0)$ and $Q = (100, 0)$. It follows that $\overrightarrow{PQ} = 100\mathbf{i}$, so that $\| \overrightarrow{PQ} \| = 100$. Since $\| \mathbf{F} \| = 50$ and the angle θ between **F** and $\overrightarrow{PQ}$ is $\pi/6$, we find from (11) that

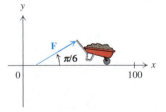

FIGURE 11.34

$$W = \| \mathbf{F} \| \|\overrightarrow{PQ}\| \cos \frac{\pi}{6} = 50 \cdot 100 \cdot \frac{\sqrt{3}}{2} = 2500\sqrt{3} \text{ (foot-pounds)} \quad \square$$

EXERCISES 11.3

In Exercises 1–4 find $\mathbf{a} \cdot \mathbf{b}$ and the cosine of the angle θ between $\mathbf{a}$ and $\mathbf{b}$.

1. $\mathbf{a} = \mathbf{i} + \mathbf{j} - \mathbf{k}$, $\mathbf{b} = 2\mathbf{i} - 3\mathbf{j} + 4\mathbf{k}$

2. $\mathbf{a} = \frac{1}{2}\mathbf{i} + \frac{1}{3}\mathbf{j} - 2\mathbf{k}$, $\mathbf{b} = 2\mathbf{i} - 2\mathbf{j} + \mathbf{k}$

3. $\mathbf{a} = \sqrt{2}\mathbf{i} + 4\mathbf{j} + \sqrt{3}\mathbf{k}$, $\mathbf{b} = -\sqrt{2}\mathbf{i} - \sqrt{3}\mathbf{j} + 2\mathbf{k}$

4. $\mathbf{a} = 4\mathbf{i} - 2\mathbf{j}$, $\mathbf{b} = -\frac{1}{2}\mathbf{i} - \mathbf{j} + \sqrt{3}\mathbf{k}$

5. Find the angle θ between the vector $\sqrt{6}\mathbf{i} + \mathbf{j} - \mathbf{k}$ and the positive x axis.

In Exercises 6–7 determine whether $\mathbf{a}$ and $\mathbf{b}$ are perpendicular.

6. $\mathbf{a} = \mathbf{i}$, $\mathbf{b} = \mathbf{j}$

7. $\mathbf{a} = \sqrt{2}\mathbf{i} + 3\mathbf{j} + \mathbf{k}$, $\mathbf{b} = -\mathbf{i} + \sqrt{2}\mathbf{j} + 5\mathbf{k}$

In Exercises 8–9 determine whether $\overrightarrow{PQ}$ and $\overrightarrow{PR}$ are perpendicular.

8. $P = (2, 1)$, $Q = (1, 4)$, and $R = (-3, 2)$

9. $P = (-1, 3, 0)$, $Q = (2, 0, 1)$, and $R = (-1, 1, -6)$

10. Show that the vectors

$$2\mathbf{i} + \mathbf{j} - \mathbf{k}, \quad 3\mathbf{i} + 7\mathbf{j} + 13\mathbf{k}, \quad \text{and} \quad 20\mathbf{i} - 29\mathbf{j} + 11\mathbf{k}$$

are mutually perpendicular.

In Exercises 11–12 find $\mathbf{pr}_{\mathbf{a}}\mathbf{b}$.

11. $\mathbf{a} = 2\mathbf{i} - \mathbf{j} + 2\mathbf{k}$, $\mathbf{b} = \mathbf{i}$

12. $\mathbf{a} = \sqrt{3}\mathbf{i} + 2\mathbf{j} - 3\mathbf{k}$, $\mathbf{b} = 4\mathbf{i} - \mathbf{j} + 2\mathbf{k}$

In Exercises 13–14 find $\mathbf{pr}_{\overrightarrow{PQ}} \overrightarrow{PR}$.

13. $P = (2, -1)$, $Q = (-1, -1)$, and $R = (-3, 3)$

14. $P = (1, 2, 3)$, $Q = (-1, 0, 1)$, and $R = (1, 1, 0)$

In Exercises 15–17 $\mathbf{a}$, $\mathbf{a}'$, and $\mathbf{b}$ lie in the same plane. Show that $\mathbf{a}$ and $\mathbf{a}'$ are perpendicular, and resolve $\mathbf{b}$ into vectors parallel to $\mathbf{a}$ and $\mathbf{a}'$.

15. $\mathbf{a} = \mathbf{i} + 2\mathbf{j} - \mathbf{k}$, $\mathbf{a}' = \mathbf{j} + 2\mathbf{k}$, $\mathbf{b} = 3\mathbf{i} + \mathbf{j} - 13\mathbf{k}$

16. $\mathbf{a} = \mathbf{i} + \mathbf{j} - \mathbf{k}$, $\mathbf{a}' = 2\mathbf{i} - 3\mathbf{j} - \mathbf{k}$, $\mathbf{b} = -5\mathbf{j} + \mathbf{k}$

17. $\mathbf{a} = 2\mathbf{i} - 4\mathbf{j} + 5\mathbf{k}$, $\mathbf{a}' = 2\mathbf{i} + 6\mathbf{j} + 4\mathbf{k}$, $\mathbf{b} = \mathbf{i} + 13\mathbf{j} + \mathbf{k}$

18. Which of the following are vertices of a right triangle?

a. $(2, 3, 4)$, $(3, 5, 5)$, and $(1, 3, 11)$

b. $(0, -1, 2)$, $(1, 4, -2)$, and $(5, 0, 1)$

19. Show by example that there exist nonzero vectors $\mathbf{a}$, $\mathbf{b}$, and $\mathbf{c}$ such that $\mathbf{a} \cdot \mathbf{b} = \mathbf{a} \cdot \mathbf{c}$, but $\mathbf{b} \neq \mathbf{c}$.

In Exercises 20–22 use the definition of the dot product to prove the statement.

20. $\mathbf{a} \cdot \mathbf{a} = \| \mathbf{a} \|^2$ for any vector $\mathbf{a}$.

21. $\mathbf{a} \cdot \mathbf{b} = \mathbf{b} \cdot \mathbf{a}$ for any vectors $\mathbf{a}$ and $\mathbf{b}$.

22. **a.** $\mathbf{a} \cdot (\mathbf{b} + \mathbf{c}) = \mathbf{a} \cdot \mathbf{b} + \mathbf{a} \cdot \mathbf{c}$ for any vectors $\mathbf{a}$, $\mathbf{b}$, and $\mathbf{c}$.

b. If $\mathbf{a}$ is perpendicular to $\mathbf{b}$ and to $\mathbf{c}$, then $\mathbf{a}$ is perpendicular to $\mathbf{b} + \mathbf{c}$.

c. Show that the vectors $\| \mathbf{b} \| \mathbf{a} + \| \mathbf{a} \| \mathbf{b}$ and $\| \mathbf{b} \| \mathbf{a} - \| \mathbf{a} \| \mathbf{b}$ are perpendicular if they are not zero.

23. Let $\mathbf{a}$ and $\mathbf{b}$ be unit vectors and let θ be the angle between $\mathbf{a}$ and $\mathbf{b}$.

a. For what value of θ in $[0, \pi]$ is $\mathbf{a} \cdot \mathbf{b}$ maximum?

b. For what value of θ in $[0, \pi]$ is $\mathbf{a} \cdot \mathbf{b}$ minimum?

c. For what value of θ in $[0, \pi]$ is $| \mathbf{a} \cdot \mathbf{b} |$ minimum?

24. a Prove the **Cauchy-Schwarz Inequality**:

$$| \mathbf{a} \cdot \mathbf{b} | \leq \| \mathbf{a} \| \ \| \mathbf{b} \|$$

b. Conclude from (a) that

$$(a_1 b_1 + a_2 b_2 + a_3 b_3)^2 \leq (a_1^2 + a_2^2 + a_3^2)(b_1^2 + b_2^2 + b_3^2)$$

for any numbers a_1, a_2, a_3, b_1, b_2, and b_3.

c. Let $b_1 = b_2 = b_3 = 1$ in (b) and conclude that

$$\left(\frac{a_1 + a_2 + a_3}{3} \right)^2 \leq \frac{a_1^2 + a_2^2 + a_3^2}{3}$$

for any numbers a_1, a_2, and a_3. (This means that the square of the average of three numbers does not exceed the average of the squares of the three numbers.)

25. a. Using the equation $\| \mathbf{a} + \mathbf{b} \|^2 = (\mathbf{a} + \mathbf{b}) \cdot (\mathbf{a} + \mathbf{b})$, prove that $\| \mathbf{a} + \mathbf{b} \|^2 = \| \mathbf{a} \|^2 + 2\mathbf{a} \cdot \mathbf{b} + \| \mathbf{b} \|^2$.

b. Using part (a), prove the **Pythagorean Theorem**:

$$\| \mathbf{a} + \mathbf{b} \|^2 = \| \mathbf{a} \|^2 + \| \mathbf{b} \|^2 \quad \text{if and only if } \mathbf{a} \text{ and } \mathbf{b} \text{ are perpendicular}$$

*c. Using part (a) and Exercise 24(a), prove the **triangle inequality**:

$$\| \mathbf{a} + \mathbf{b} \| \leq \| \mathbf{a} \| + \| \mathbf{b} \| \tag{12}$$

(The inequality in (12) is illustrated in Figure 11.35.)

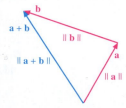

FIGURE 11.35 The triangle inequality: $\| \mathbf{a} + \mathbf{b} \| \leq \| \mathbf{a} \| + \| \mathbf{b} \|$. See Exercise 25.

26. Let P, Q, and R be three points in space. Using Exercise 25 and the fact that

$$\overrightarrow{PQ} = \overrightarrow{PR} + \overrightarrow{RQ}$$

prove the alternate form of the triangle inequality:

$$\|\overrightarrow{PQ}\| \le \|\overrightarrow{PR}\| + \|\overrightarrow{RQ}\|$$

27. a. Using the equations $\|\mathbf{a} + \mathbf{b}\|^2 = (\mathbf{a} + \mathbf{b}) \cdot (\mathbf{a} + \mathbf{b})$ and $\|\mathbf{a} - \mathbf{b}\|^2 = (\mathbf{a} - \mathbf{b}) \cdot (\mathbf{a} - \mathbf{b})$, prove the **parallelogram law**:

$$\|\mathbf{a} + \mathbf{b}\|^2 + \|\mathbf{a} - \mathbf{b}\|^2 = 2\|\mathbf{a}\|^2 + 2\|\mathbf{b}\|^2$$

(The geometric interpretation of this result is as follows: The sum of the squares of the lengths of the diagonals of the parallelogram determined by $\mathbf{a}$ and $\mathbf{b}$ is equal to the sum of the squares of the lengths of the four sides (Figure 11.36).)

b. Using the ideas of part (a), prove the **polarization identity**:

$$\|\mathbf{a} + \mathbf{b}\|^2 - \|\mathbf{a} - \mathbf{b}\|^2 = 4\mathbf{a} \cdot \mathbf{b}$$

c. Using part (b), prove that the diagonals of a parallelogram have equal length if and only if the parallelogram is a rectangle.

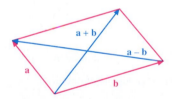

FIGURE 11.36   The parallelogram law: $\|\mathbf{a} + \mathbf{b}\|^2 + \|\mathbf{a} - \mathbf{b}\|^2 = 2\|\mathbf{a}\|^2 + 2\|\mathbf{b}\|^2$. See Exercise 27.

28. Prove that the diagonals of a rhombus (a parallelogram whose sides have equal length) are perpendicular.

29. Approximate the angle between the line segments that join the center of a cube to any two adjacent vertices of the cube.

30. Consider a rectangular parallelepiped whose base is a square of area 1 and whose height is c. Determine the value of c for which the angle formed by the lines from

the parallelepiped's center to any pair of adjacent vertices on the base is equal to $\pi/3$.

31. Let $\mathbf{a}$ and $\mathbf{b}$ be nonzero vectors, and let $\mathbf{c} = \|\mathbf{b}\|\mathbf{a} + \|\mathbf{a}\|\mathbf{b}$. Prove that if $\mathbf{c} \ne \mathbf{0}$, then $\mathbf{c}$ bisects the angle formed by $\mathbf{a}$ and $\mathbf{b}$. (*Hint:* Compute the cosine of the angle between $\mathbf{a}$ and $\mathbf{c}$ and the cosine of the angle between $\mathbf{c}$ and $\mathbf{b}$.)

Applications

32. A person pulls a sled 100 feet with a rope that makes an angle of $\pi/4$ with the horizontal ground. Find the work W done on the sled if the tension in the rope is 5 pounds.

33. A person exerts a force $\mathbf{F}$ of 100 pounds on a wheelbarrow at an angle of $\pi/6$ with respect to the ground and pushes the wheelbarrow 500 feet. Compute the work W done on the wheelbarrow.

34. A person exerts a horizontal force $\mathbf{F}$ of 20 pounds on a box and pushes it up a ramp that is 15 feet long and inclined at an angle of $\pi/6$. How much work W is done on the box?

Projects

1. Let $\mathbf{a}$ and b be nonzero vectors. Find all possible values of r such that

$$\|\mathbf{a} + \mathbf{b}\| = r(\|\mathbf{a}\| + \|\mathbf{b}\|)$$

if $\mathbf{a}$ and $\mathbf{b}$ satisfy the following conditions.

a. $\mathbf{a}$ and $\mathbf{b}$ have the same direction.

b. $\mathbf{a} = -\mathbf{b}$.

c. $\mathbf{a}$ and $\mathbf{b}$ have opposite directions but $\mathbf{a} \ne -\mathbf{b}$.

d. $\mathbf{a}$ and $\mathbf{b}$ are perpendicular to each other.

2. a. Prove that two nonzero vectors $\mathbf{a}$ and $\mathbf{b}$ are perpendicular if and only if $\|\mathbf{a}\| \le \|\mathbf{a} + c\mathbf{b}\|$ for every number c. (Hint: $\|\mathbf{a}\|^2 = \mathbf{a} \cdot \mathbf{a}$ and $\|\mathbf{a} + c\mathbf{b}\|^2 = (\mathbf{a} + c\mathbf{b}) \cdot (\mathbf{a} + c\mathbf{b})$.)

b. Let $c \ne 0$ and assume that the nonzero vectors $\mathbf{a}$ and $\mathbf{b}$ are perpendicular to one another. Draw a picture of $\mathbf{a}$, $\mathbf{b}$, and $\mathbf{a} + c\mathbf{b}$. What can you say about the comparative lengths of $\mathbf{a}$ and $\mathbf{a} + c\mathbf{b}$?

c. Draw nonperpendicular vectors $\mathbf{a}$ and $\mathbf{b}$. Find a nonzero number c such that $\|\mathbf{a} + c\mathbf{b}\|$ is less than $\|\mathbf{a}\|$, and include $\mathbf{a} + c\mathbf{b}$ in your sketch.

11.4   THE CROSS PRODUCT AND TRIPLE PRODUCTS

The second type of product of two vectors we will study is the cross product. Unlike the dot product, the cross product of two vectors is a vector.

DEFINITION 11.9

The **cross product** (or **vector product**) $\mathbf{a} \times \mathbf{b}$ of two vectors $\mathbf{a} = a_1\mathbf{i} + a_2\mathbf{j} + a_3\mathbf{k}$ and $\mathbf{b} = b_1\mathbf{i} + b_2\mathbf{j} + b_3\mathbf{k}$ is defined by

$$\mathbf{a} \times \mathbf{b} = (a_2b_3 - a_3b_2)\mathbf{i} + (a_3b_1 - a_1b_3)\mathbf{j} + (a_1b_2 - a_2b_1)\mathbf{k} \qquad (1)$$

Josiah Willard Gibbs (1839–1903)
The first great American physicist, Gibbs realized (along with Oliver Heaviside (1850–1925)) that vectors rather than Hamilton's quaternions were the proper tool for expressing many physical concepts. He introduced the notation for dot and cross products in his *Elements of Analysis,* a treatise on vectors written for students at Yale University. The formal publication of his *Vector Analysis* in 1901 strongly influenced the physics community. Amazingly, Gibbs's scientific and mathematical work was so little appreciated at Yale that for 10 years he was compelled to live on inherited income, without any salary.

The right side of (1) can be considered as the 3×3 determinant*

$$\begin{vmatrix} \mathbf{i} & \mathbf{j} & \mathbf{k} \\ a_1 & a_2 & a_3 \\ b_1 & b_2 & b_3 \end{vmatrix}$$

Such a determinant is evaluated as follows:

$$\begin{vmatrix} \mathbf{i} & \mathbf{j} & \mathbf{k} \\ a_1 & a_2 & a_3 \\ b_1 & b_2 & b_3 \end{vmatrix} = \begin{vmatrix} a_2 & a_3 \\ b_2 & b_3 \end{vmatrix}\mathbf{i} - \begin{vmatrix} a_1 & a_3 \\ b_1 & b_3 \end{vmatrix}\mathbf{j} + \begin{vmatrix} a_1 & a_2 \\ b_1 & b_2 \end{vmatrix}\mathbf{k}$$

$$= (a_2b_3 - a_3b_2)\mathbf{i} - (a_1b_3 - a_3b_1)\mathbf{j} + (a_1b_2 - a_2b_1)\mathbf{k}$$

EXAMPLE 1   Show that
$$\mathbf{i} \times \mathbf{j} = \mathbf{k}, \quad \mathbf{j} \times \mathbf{k} = \mathbf{i}, \quad \text{and} \quad \mathbf{k} \times \mathbf{i} = \mathbf{j}$$

Solution   Since $\mathbf{i} = (1)\mathbf{i} + (0)\mathbf{j} + (0)\mathbf{k}$ and $\mathbf{j} = (0)\mathbf{i} + (1)\mathbf{j} + (0)\mathbf{k}$, (1) implies that

$$\mathbf{i} \times \mathbf{j} = \begin{vmatrix} \mathbf{i} & \mathbf{j} & \mathbf{k} \\ 1 & 0 & 0 \\ 0 & 1 & 0 \end{vmatrix} = [(0)(0) - (0)(1)]\mathbf{i} + [(0)(0) - (1)(0)]\mathbf{j}$$
$$+ [(1)(1) - (0)(0)]\mathbf{k} = \mathbf{k}$$

The remaining two formulas follow in a similar fashion.   ❑

EXAMPLE 2   Let $\mathbf{a} = 2\mathbf{i} - \mathbf{j} + 3\mathbf{k}$ and $\mathbf{b} = -\mathbf{i} - 2\mathbf{j} + 4\mathbf{k}$. Calculate $\mathbf{a} \times \mathbf{b}$ and $\mathbf{b} \times \mathbf{a}$.

Solution   From the definition of the cross product, we obtain

$$\mathbf{a} \times \mathbf{b} = \begin{vmatrix} \mathbf{i} & \mathbf{j} & \mathbf{k} \\ 2 & -1 & 3 \\ -1 & -2 & 4 \end{vmatrix} = [(-1)(4) - (3)(-2)]\mathbf{i} + [(3)(-1) - (2)(4)]\mathbf{j}$$
$$+ [(2)(-2) - (-1)(-1)]\mathbf{k}$$
$$= 2\mathbf{i} - 11\mathbf{j} - 5\mathbf{k}$$

and

$$\mathbf{b} \times \mathbf{a} = \begin{vmatrix} \mathbf{i} & \mathbf{j} & \mathbf{k} \\ -1 & -2 & 4 \\ 2 & -1 & 3 \end{vmatrix} = [(-2)(3) - (4)(-1)]\mathbf{i} + [(4)(2) - (-1)(3)]\mathbf{j}$$
$$+ [(-1)(-1) - (-2)(2)]\mathbf{k}$$
$$= -2\mathbf{i} + 11\mathbf{j} + 5\mathbf{k}$$ ❑

* A general 2×2 determinant is evaluated as follows:

$$\begin{vmatrix} a & b \\ c & d \end{vmatrix} = ad - bc$$

A general 3×3 determinant is evaluated by "expanding by minors" as follows:

$$\begin{vmatrix} a_1 & a_2 & a_3 \\ b_1 & b_2 & b_3 \\ c_1 & c_2 & c_3 \end{vmatrix} = a_1\begin{vmatrix} b_2 & b_3 \\ c_2 & c_3 \end{vmatrix} - a_2\begin{vmatrix} b_1 & b_3 \\ c_1 & c_3 \end{vmatrix} + a_3\begin{vmatrix} b_1 & b_2 \\ c_1 & c_2 \end{vmatrix}$$

Notice that the vectors $\mathbf{a} \times \mathbf{b}$ and $\mathbf{b} \times \mathbf{a}$ in Example 2 are negatives of one another. This is no accident; in fact it follows directly from the definition of the cross product that if $\mathbf{a}$ and $\mathbf{b}$ are any two vectors, then

$$\mathbf{a} \times \mathbf{b} = - (\mathbf{b} \times \mathbf{a})$$

Other properties of the cross product that follow readily from the definition are

$$\mathbf{a} \times \mathbf{a} = \mathbf{0} \qquad\qquad \mathbf{a} \times (\mathbf{b} + \mathbf{c}) = (\mathbf{a} \times \mathbf{b}) + (\mathbf{a} \times \mathbf{c})$$

$$(c\mathbf{a}) \times \mathbf{b} = c(\mathbf{a} \times \mathbf{b}) = \mathbf{a} \times (c\mathbf{b}) \qquad (\mathbf{a} + \mathbf{b}) \times \mathbf{c} = (\mathbf{a} \times \mathbf{c}) + (\mathbf{b} \times \mathbf{c})$$

We also observe that

$$\mathbf{i} \times (\mathbf{i} \times \mathbf{k}) = \mathbf{i} \times (-\mathbf{j}) = -\mathbf{k} \quad \text{and} \quad (\mathbf{i} \times \mathbf{i}) \times \mathbf{k} = \mathbf{0} \times \mathbf{k} = \mathbf{0}$$

so that

$$\mathbf{i} \times (\mathbf{i} \times \mathbf{k}) \neq (\mathbf{i} \times \mathbf{i}) \times \mathbf{k}$$

Therefore $\mathbf{a} \times (\mathbf{b} \times \mathbf{c})$ and $(\mathbf{a} \times \mathbf{b}) \times \mathbf{c}$ are usually not equal (see Exercise 19(b)).

THEOREM 11.10

Let $\mathbf{a}$ and $\mathbf{b}$ be two nonzero vectors.

a.   Then

$$\mathbf{a} \cdot (\mathbf{a} \times \mathbf{b}) = 0 \quad \text{and} \quad \mathbf{b} \cdot (\mathbf{a} \times \mathbf{b}) = 0$$

Consequently if $\mathbf{a} \times \mathbf{b} \neq \mathbf{0}$, then $\mathbf{a} \times \mathbf{b}$ is perpendicular to both $\mathbf{a}$ and $\mathbf{b}$.

b.   If θ is the angle between $\mathbf{a}$ and $\mathbf{b}$ (so that $0 \leq \theta \leq \pi$), then

$$\| \mathbf{a} \times \mathbf{b} \| = \| \mathbf{a} \| \| \mathbf{b} \| \sin \theta$$

Proof   To prove (a), we simply apply the definitions of the dot product and the cross product:

$$\mathbf{a} \cdot (\mathbf{a} \times \mathbf{b}) = a_1(a_2 b_3 - a_3 b_2) + a_2(a_3 b_1 - a_1 b_3) + a_3(a_1 b_2 - a_2 b_1)$$
$$= a_1 a_2 b_3 - a_1 a_3 b_2 + a_2 a_3 b_1 - a_2 a_1 b_3 + a_3 a_1 b_2 - a_3 a_2 b_1$$
$$= 0$$

and

$$\mathbf{b} \cdot (\mathbf{a} \times \mathbf{b}) = b_1(a_2 b_3 - a_3 b_2) + b_2(a_3 b_1 - a_1 b_3) + b_3(a_1 b_2 - a_2 b_1)$$
$$= b_1 a_2 b_3 - b_1 a_3 b_2 + b_2 a_3 b_1 - b_2 a_1 b_3 + b_3 a_1 b_2 - b_3 a_2 b_1$$
$$= 0$$

For part (b) we use the definitions of the cross product and the length of a vector to deduce that

$$\| \mathbf{a} \times \mathbf{b} \|^2 = (a_2 b_3 - a_3 b_2)^2 + (a_3 b_1 - a_1 b_3)^2 + (a_1 b_2 - a_2 b_1)^2$$
$$= a_2^2 b_3^2 - 2a_2 a_3 b_2 b_3 + a_3^2 b_2^2 + a_3^2 b_1^2 - 2a_1 a_3 b_1 b_3 + a_1^2 b_3^2$$
$$\quad + a_1^2 b_2^2 - 2a_1 a_2 b_1 b_2 + a_2^2 b_1^2$$
$$= (a_1^2 + a_2^2 + a_3^2)(b_1^2 + b_2^2 + b_3^2) - (a_1 b_1 + a_2 b_2 + a_3 b_3)^2$$
$$= \| \mathbf{a} \|^2 \| \mathbf{b} \|^2 - (\mathbf{a} \cdot \mathbf{b})^2 = \| \mathbf{a} \|^2 \| \mathbf{b} \|^2 - (\| \mathbf{a} \| \| \mathbf{b} \| \cos \theta)^2$$
$$= \| \mathbf{a} \|^2 \| \mathbf{b} \|^2 (1 - \cos^2 \theta)$$

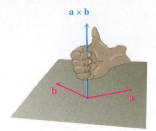

FIGURE 11.37

Therefore

$$\|\mathbf{a} \times \mathbf{b}\|^2 = \|\mathbf{a}\|^2 \|\mathbf{b}\|^2 \sin^2\theta$$

Since $\sin\theta \geq 0$ for $0 \leq \theta \leq \pi$, we can take the square root of each side of the equation and obtain

$$\|\mathbf{a} \times \mathbf{b}\| = \|\mathbf{a}\| \|\mathbf{b}\| \sin\theta \quad \blacksquare$$

From part (a) of Theorem 11.10 we see that $\mathbf{a} \times \mathbf{b}$ is perpendicular to $\mathbf{a}$ and $\mathbf{b}$, and hence to the plane determined by $\mathbf{a}$ and $\mathbf{b}$. In theory there are two possible directions for the vector $\mathbf{a} \times \mathbf{b}$. To see the direction in which $\mathbf{a} \times \mathbf{b}$ actually points, assume that θ is the angle between $\mathbf{a}$ and $\mathbf{b}$, with $0 < \theta < \pi$. It can be shown that if the fingers of one's right hand curl from $\mathbf{a}$ to $\mathbf{b}$ through θ, then the thumb will point in the direction of $\mathbf{a} \times \mathbf{b}$ (Figure 11.37). This method of determining the direction of $\mathbf{a} \times \mathbf{b}$ is called the **right-hand rule**.

Finally, the case in which $\theta = 0$ or $\theta = \pi$ is covered by Corollary 11.11.

COROLLARY 11.11   Two nonzero vectors $\mathbf{a}$ and $\mathbf{b}$ are parallel if and only if $\mathbf{a} \times \mathbf{b} = \mathbf{0}$.

As we saw in Section 11.3, the dot product is especially effective for finding the angle between two vectors and for finding the projection of one vector onto another. By Theorem 11.10(a) the cross product yields a vector perpendicular to two nonzero vectors. This will be invaluable to us in finding equations of planes in Section 11.6.

EXAMPLE 3   Let $\mathbf{a} = 4\mathbf{i} - \mathbf{j} + 3\mathbf{k}$ and $\mathbf{b} = 2\mathbf{i} + 3\mathbf{j} - \mathbf{k}$. Find a vector perpendicular to $\mathbf{a}$ and $\mathbf{b}$.

Solution   By Theorem 11.10(a) the cross product $\mathbf{a} \times \mathbf{b}$ is one such vector:

$$\mathbf{a} \times \mathbf{b} = \begin{vmatrix} \mathbf{i} & \mathbf{j} & \mathbf{k} \\ 4 & -1 & 3 \\ 2 & 3 & -1 \end{vmatrix} = (1 - 9)\mathbf{i} + (6 + 4)\mathbf{j} + (12 + 2)\mathbf{k}$$
$$= -8\mathbf{i} + 10\mathbf{j} + 14\mathbf{k} \quad \square$$

From Theorem 11.10(b) we conclude that

$\|\mathbf{a} \times \mathbf{b}\|$ = the area of the parallelogram with adjacent sides $\mathbf{a}$ and $\mathbf{b}$ (2)

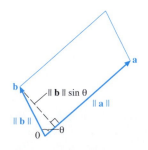

FIGURE 11.38   Area of parallelogram = $\|\mathbf{a} \times \mathbf{b}\|$.

(Figure 11.38). After all, the parallelogram has a base of length $\|\mathbf{a}\|$ and an altitude of length $\|\mathbf{b}\| \sin\theta$. Consequently the area of the parallelogram is

$$\|\mathbf{a}\| \|\mathbf{b}\| \sin\theta = \|\mathbf{a} \times \mathbf{b}\|$$

Because the triangle having $\mathbf{a}$ and $\mathbf{b}$ as two of its sides has an area one half the area of the parallelogram determined by $\mathbf{a}$ and $\mathbf{b}$, it follows from (2) that the area of the triangle equals $\frac{1}{2}\|\mathbf{a} \times \mathbf{b}\|$.

EXAMPLE 4   Let $P = (3, -2, 1)$, $Q = (7, -3, 4)$, and $R = (5, 1, 0)$. Find the area A of triangle PQR.

Solution   Let $\mathbf{a} = \overrightarrow{PQ} = 4\mathbf{i} - \mathbf{j} + 3\mathbf{k}$ and $\mathbf{b} = \overrightarrow{PR} = 2\mathbf{i} + 3\mathbf{j} - \mathbf{k}$. In Example 3 we found that

$$\mathbf{a} \times \mathbf{b} = -8\mathbf{i} + 10\mathbf{j} + 14\mathbf{k}$$

Consequently

$$A = \frac{1}{2} \, \| \, \mathbf{a} \times \mathbf{b} \, \| \; = \frac{1}{2} \sqrt{(-8)^2 + (10)^2 + (14)^2} = 3\sqrt{10} \quad \square$$

For a physical application of the cross product, suppose we wish to loosen a bolt with a wrench, as in Figure 11.39. Assume that we apply a force $\mathbf{F}$ at the point Q at the end of the wrench, which is attached to the bolt at point P. The tendency of the wrench to rotate about an axis through P perpendicular to $\overrightarrow{PQ}$ is called the **torque**, or **moment**, of $\mathbf{F}$, and is defined by

$$\mathbf{M} = \overrightarrow{PQ} \times \mathbf{F}$$

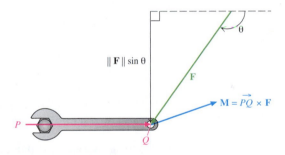

FIGURE 11.39

It follows that

$$\| \, \mathbf{M} \, \| = \| \, \overrightarrow{PQ} \, \| \, \| \, \mathbf{F} \, \| \, \sin \theta$$

where θ is the angle between $\overrightarrow{PQ}$ and $\mathbf{F}$, with $0 \leq \theta \leq \pi$ (Figure 11.39). Notice that $\| \, \overrightarrow{PQ} \, \|$ is the distance from the point P to the point Q at which the force is applied, and $\| \, \mathbf{F} \, \| \, \sin \theta$ is the magnitude of the projection of $\mathbf{F}$ onto a vector perpendicular to $\overrightarrow{PQ}$. The torque becomes larger if the force is applied farther from P, or the magnitude of $\mathbf{F}$ is increased, or the force is applied more nearly perpendicular to the vector $\overrightarrow{PQ}$. In terms of a wrench and bolt, these mean, respectively, that the wrench is larger, or one pushes harder on the wrench, or one pushes more nearly perpendicular to the wrench. Finally we mention that the line through P and parallel to $\mathbf{M}$ is called the **axis of rotation**.

Triple Products

Let $\mathbf{a} = a_1\mathbf{i} + a_2\mathbf{j} + a_3\mathbf{k}$, $\mathbf{b} = b_1\mathbf{i} + b_2\mathbf{j} + b_3\mathbf{k}$, and $\mathbf{c} = c_1\mathbf{i} + c_2\mathbf{j} + c_3\mathbf{k}$. The products $\mathbf{a} \cdot (\mathbf{b} \times \mathbf{c})$, $(\mathbf{a} \times \mathbf{b}) \cdot \mathbf{c}$, $(\mathbf{a} \times \mathbf{b}) \times \mathbf{c}$, and $\mathbf{a} \times (\mathbf{b} \times \mathbf{c})$ occasionally arise in physical applications. The first two products are called **triple scalar products**, since they are scalars (that is, numbers). The last two are called **triple vector products**, since they are vectors.

In Theorem 11.10(a) we considered two triple scalar products, $\mathbf{a} \cdot (\mathbf{a} \times \mathbf{b})$ and $\mathbf{b} \cdot (\mathbf{a} \times \mathbf{b})$. We observed that they are always 0 and consequently are equal to each other. In fact, for any $\mathbf{a}$, $\mathbf{b}$, and $\mathbf{c}$ we find that

$$\mathbf{a} \cdot (\mathbf{b} \times \mathbf{c}) = a_1(b_2c_3 - b_3c_2) + a_2(b_3c_1 - b_1c_3) + a_3(b_1c_2 - b_2c_1)$$

$$(\mathbf{a} \times \mathbf{b}) \cdot \mathbf{c} = (a_2b_3 - a_3b_2)c_1 + (a_3b_1 - a_1b_3)c_2 + (a_1b_2 - a_2b_1)c_3$$

and we see by multiplying out that the right-hand sides are equal. It follows that

$$\mathbf{a} \cdot (\mathbf{b} \times \mathbf{c}) = (\mathbf{a} \times \mathbf{b}) \cdot \mathbf{c} \tag{3}$$

The quantity $\mathbf{a} \cdot (\mathbf{b} \times \mathbf{c})$ is useful because

$$|\,\mathbf{a} \cdot (\mathbf{b} \times \mathbf{c})\,| = \text{the volume of the parallelepiped with sides } \mathbf{a}, \mathbf{b}, \text{ and } \mathbf{c} \tag{4}$$

(Figure 11.40). Indeed, $\|\,\mathbf{b} \times \mathbf{c}\,\|$ is the area of the base of this parallelepiped, and if θ is the angle between $\mathbf{a}$ and $\mathbf{b} \times \mathbf{c}$, then $\|\,\mathbf{a}\,\|\,|\cos\theta\,|$ is the height, so that the volume V is given by

$$V = (\text{height})(\text{area of the base}) = (\,\|\,\mathbf{a}\,\|\,|\cos\theta\,|\,)\,\|\,\mathbf{b} \times \mathbf{c}\,\| = |\,\mathbf{a} \cdot (\mathbf{b} \times \mathbf{c})\,|$$

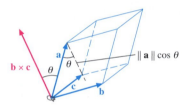

FIGURE 11.40 Volume $= |\,\mathbf{a} \cdot (\mathbf{b} \times \mathbf{c})\,|$.

Since V is the volume of the parallelepiped with sides $\mathbf{a}$, $\mathbf{b}$, and $\mathbf{c}$, the vectors can be permuted without changing V. Thus

$$V = |\,\mathbf{b} \cdot (\mathbf{c} \times \mathbf{a})\,| = |\,\mathbf{c} \cdot (\mathbf{a} \times \mathbf{b})\,|$$

We can use the determinant to facilitate computation of $\mathbf{a} \cdot (\mathbf{b} \times \mathbf{c})$:

$$\mathbf{a} \cdot (\mathbf{b} \times \mathbf{c}) = \begin{vmatrix} a_1 & a_2 & a_3 \\ b_1 & b_2 & b_3 \\ c_1 & c_2 & c_3 \end{vmatrix} = \begin{aligned} & (a_1b_2c_3 + a_2b_3c_1 + a_3b_1c_2) \\ & - (a_3b_2c_1 + a_2b_1c_3 + a_1b_3c_2) \end{aligned} \tag{5}$$

EXAMPLE 5   Let $\mathbf{a} = \mathbf{i} - \mathbf{j}$, $\mathbf{b} = 2\mathbf{i} + 3\mathbf{j} - \mathbf{k}$, and $\mathbf{c} = -\mathbf{i} + 2\mathbf{k}$. Find the volume V of the parallelepiped determined by $\mathbf{a}$, $\mathbf{b}$, and $\mathbf{c}$.

Solution   By (4), $V = |\,\mathbf{a} \cdot (\mathbf{b} \times \mathbf{c})\,|$, and by (5),

$$\mathbf{a} \cdot (\mathbf{b} \times \mathbf{c}) = \begin{vmatrix} 1 & -1 & 0 \\ 2 & 3 & -1 \\ -1 & 0 & 2 \end{vmatrix} = (6 - 1 + 0) - (0 - 4 + 0) = 9$$

Thus $V = 9$. ☐

For reference we list three additional formulas, which will occasionally be useful. The first is

$$\mathbf{a} \times (\mathbf{b} \times \mathbf{c}) = (\mathbf{a} \cdot \mathbf{c})\mathbf{b} - (\mathbf{a} \cdot \mathbf{b})\mathbf{c} \tag{6}$$

which can be verified by expanding each side (see Exercise 18). This equation is sometimes written in the equivalent form

$$\mathbf{a} \times (\mathbf{b} \times \mathbf{c}) = \mathbf{b}(\mathbf{a} \cdot \mathbf{c}) - \mathbf{c}(\mathbf{a} \cdot \mathbf{b}) \tag{7}$$

known as "the *bac* − *cab* rule." A special case of (7) occurs when $\mathbf{a} = \mathbf{b}$ and $\mathbf{c}$ is perpendicular to $\mathbf{a}$. In that case we have

$$\mathbf{a} \times (\mathbf{a} \times \mathbf{c}) = \mathbf{a}(0) - \mathbf{c}(\mathbf{a} \cdot \mathbf{a})$$

or more simply

$$\mathbf{a} \times (\mathbf{a} \times \mathbf{c}) = -\| \mathbf{a} \|^2 \mathbf{c} \tag{8}$$

EXERCISES 11.4

In Exercises 1–5 find $\mathbf{a} \times \mathbf{b}$ and $\mathbf{c} \cdot (\mathbf{a} \times \mathbf{b})$.

1. $\mathbf{a} = \mathbf{i} + \mathbf{j}$, $\mathbf{b} = \mathbf{j} + \mathbf{k}$, $\mathbf{c} = -\mathbf{i} - 3\mathbf{j} + 4\mathbf{k}$
2. $\mathbf{a} = \mathbf{i} + \mathbf{j} + \mathbf{k}$, $\mathbf{b} = \mathbf{i} - \mathbf{k}$, $\mathbf{c} = \mathbf{i} + \mathbf{j} - \mathbf{k}$
3. $\mathbf{a} = 2\mathbf{i} + 3\mathbf{j} - \mathbf{k}$, $\mathbf{b} = -\mathbf{i} + 4\mathbf{j} + 5\mathbf{k}$, $\mathbf{c} = 2\mathbf{i} + 3\mathbf{j} + 4\mathbf{k}$
4. $\mathbf{a} = 3\mathbf{i} + 4\mathbf{j} + 12\mathbf{k}$, $\mathbf{b} = 3\mathbf{i} + 4\mathbf{j} - 12\mathbf{k}$, $\mathbf{c} = \frac{1}{8}\mathbf{i} - \frac{1}{12}\mathbf{j} + \frac{1}{16}\mathbf{k}$
5. $\mathbf{a} = 3\mathbf{i} + 4\mathbf{j} + 12\mathbf{k}$, $\mathbf{b} = 3\mathbf{i} + 4\mathbf{j} + 12\mathbf{k}$, $\mathbf{c} = \mathbf{i} + \mathbf{j}$
6. Using the cross product, find the sine of the angle between the vectors $\mathbf{a}$ and $\mathbf{b}$ in Exercise 4.
7. From the definition of the cross product prove that $\mathbf{a} \times \mathbf{b} = -(\mathbf{b} \times \mathbf{a})$.
8. From the definition of the cross product prove that $\mathbf{a} \times (\mathbf{b} + \mathbf{c}) = \mathbf{a} \times \mathbf{b} + \mathbf{a} \times \mathbf{c}$.
9. Suppose that $\mathbf{a} + \mathbf{b} + \mathbf{c} = \mathbf{0}$. Show that $\mathbf{a} \times \mathbf{b} = \mathbf{b} \times \mathbf{c} = \mathbf{c} \times \mathbf{a}$. (*Hint:* Expand $\mathbf{a} \times (\mathbf{a} + \mathbf{b} + \mathbf{c})$ and $\mathbf{b} \times (\mathbf{a} + \mathbf{b} + \mathbf{c})$.)
10. Find a vector perpendicular to both $\mathbf{i} - 3\mathbf{j} + 2\mathbf{k}$ and $-2\mathbf{i} + \mathbf{j} - 5\mathbf{k}$.
11. Find the volume V of a parallelepiped with vertices $(0, 0, 0)$, $(2, -3, 4)$, $(1, 1, -1)$, and $(4, -1, -1)$, the first three of which determine one face. (*Note:* The parallelepiped is not unique, but its volume *is* unique.)
12. Find a formula for the area A of the parallelogram whose vertices, in order, are P, Q, R, and S.
13. Find a formula for the area A of the triangle whose vertices are P, Q, and R.
14. Find nonzero vectors $\mathbf{a}$, $\mathbf{b}$, and $\mathbf{c}$ in space such that $\mathbf{a} \times \mathbf{b} = \mathbf{a} \times \mathbf{c}$ but $\mathbf{b} \neq \mathbf{c}$.

15. Assume that $\mathbf{a} \neq \mathbf{0}$, $\mathbf{a} \cdot \mathbf{b} = \mathbf{a} \cdot \mathbf{c}$, and $\mathbf{a} \times \mathbf{b} = \mathbf{a} \times \mathbf{c}$. Does it follow that $\mathbf{b} = \mathbf{c}$? Support your answer.
16. Let $\mathbf{a} = 2\mathbf{i} - 3\mathbf{j} + 4\mathbf{k}$, $\mathbf{b} = \frac{1}{2}\mathbf{i} + \mathbf{j} - \mathbf{k}$, $\mathbf{c} = 4\mathbf{i} - 5\mathbf{j} + 6\mathbf{k}$. Find $\mathbf{a} \times (\mathbf{b} \times \mathbf{c})$ directly, and then use the *bac* − *cab* rule (7) to compute $\mathbf{a} \times (\mathbf{b} \times \mathbf{c})$.
17. Use (3) and (8) to show that if $\mathbf{u}$ is perpendicular to $\mathbf{a}$ or to $\mathbf{b}$, then
$$(\mathbf{u} \times \mathbf{a}) \cdot (\mathbf{u} \times \mathbf{b}) = \| \mathbf{u} \|^2 (\mathbf{a} \cdot \mathbf{b})$$
*18. Prove (6), and hence the *bac* − *cab* rule. (*Hint:* Using the definition of the cross product, find an expression for $\mathbf{b} \times \mathbf{c}$ and then one for $\mathbf{a} \times (\mathbf{b} \times \mathbf{c})$.)
19. a. Using the *bac* − *cab* rule, prove **Jacobi's identity**
$$\mathbf{a} \times (\mathbf{b} \times \mathbf{c}) + \mathbf{b} \times (\mathbf{c} \times \mathbf{a}) + \mathbf{c} \times (\mathbf{a} \times \mathbf{b}) = \mathbf{0}$$
   b. Using (a), show that $\mathbf{a} \times (\mathbf{b} \times \mathbf{c}) = (\mathbf{a} \times \mathbf{b}) \times \mathbf{c}$ if and only if $\mathbf{b} \times (\mathbf{c} \times \mathbf{a}) = \mathbf{0}$.
20. Express $(\mathbf{a} \times \mathbf{b}) \times \mathbf{c}$ as the difference of a vector parallel to $\mathbf{a}$ and a vector parallel to $\mathbf{b}$.

Applications

21. Suppose the arm of a stapler is resting at an angle of $\pi/6$ from the horizontal and is $\frac{3}{2}$ feet long (Figure 11.41). If a downward force of 32 pounds is applied to the tip of the arm, determine the moment $\mathbf{M}$ of the force about the axis of the stapler.
22. A 5-pound fish hangs from a fishing line on a 4-foot fishing pole held at an angle of $\pi/4$ with respect to the horizontal. Find the moment $\mathbf{M}$ due to the weight of the fish about the handle end of the pole.

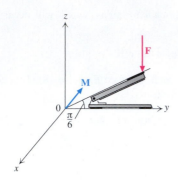

FIGURE 11.41 Figure for Exercise 21.

Project

1. Determine which of the following expressions are vectors, which are scalars, and which are undefined.

a. $(\mathbf{a} \times \mathbf{b}) \cdot \mathbf{c}$

b. $\mathbf{a} \cdot (\mathbf{b} \times \mathbf{c})$

c. $\mathbf{a} \times (\mathbf{b} \times \mathbf{c})$

d. $\mathbf{a} \cdot (\mathbf{b} \cdot \mathbf{c})$

e. $(\mathbf{a} \cdot \mathbf{b})\mathbf{c}$

f. $(\mathbf{a} \times \mathbf{b})\mathbf{c}$

g. $(\mathbf{a} \times \mathbf{b}) \times (\mathbf{c} \times \mathbf{d})$

h. $\mathbf{a} \times \mathbf{b} \times \mathbf{c} \times \mathbf{d}$

i. $\mathbf{a} \cdot (\mathbf{b} \times \mathbf{c})\mathbf{d}$

j. $\mathbf{a} \cdot (\mathbf{b} \times \mathbf{c}) \times \mathbf{d}$

11.5 LINES IN SPACE

Since we often think of vectors as directed line segments, it should not be surprising that vectors and lines are intimately related. In this section we will use vectors to describe lines.

Equations of Lines in Space

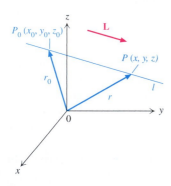

FIGURE 11.42

We say that a vector $\mathbf{L}$ and a line l are **parallel** if $\mathbf{L}$ is parallel to the vector $\overrightarrow{P_0P}$ joining any two distinct points P_0 and P on l (Figure 11.42). It follows from Euclidean geometry that a line l in space is uniquely determined by a point $P_0 = (x_0, y_0, z_0)$ on l and a vector $\mathbf{L}$ parallel to the line. Thus a point $P = (x, y, z)$ is on l if and only if $\overrightarrow{P_0P}$ is parallel to $\mathbf{L}$. By Definition 11.4 this means that

$$\overrightarrow{P_0P} = t\mathbf{L} \tag{1}$$

for a suitable number t. If $\mathbf{r}_0 = x_0\mathbf{i} + y_0\mathbf{j} + z_0\mathbf{k}$ and $\mathbf{r} = x\mathbf{i} + y\mathbf{j} + z\mathbf{k}$, then $\overrightarrow{P_0P} = \mathbf{r} - \mathbf{r}_0$, so that (1) can be rewritten

$$\mathbf{r} = \mathbf{r}_0 + t\mathbf{L} \tag{2}$$

This is a **vector equation** of l. As t varies over the real numbers, the vector $\mathbf{r}$ traces out the points on the line l. Since $\mathbf{r}_0$ can be any vector that joins the origin to a point on l, and since $\mathbf{L}$ can be any vector parallel to l, there are many different vector equations of a given line l.

EXAMPLE 1 Find a vector equation of the line that contains $(-1, 3, 0)$ and is parallel to $2\mathbf{i} - 3\mathbf{j} - \mathbf{k}$.

Solution From (2) we obtain

$$\mathbf{r} = (-\mathbf{i} + 3\mathbf{j}) + t(2\mathbf{i} - 3\mathbf{j} - \mathbf{k})$$

or alternatively,

$$\mathbf{r} = (-1 + 2t)\mathbf{i} + (3 - 3t)\mathbf{j} - t\mathbf{k} \quad \square$$

Suppose we let $\mathbf{L} = a\mathbf{i} + b\mathbf{j} + c\mathbf{k}$. Then (2) can be rewritten

$$x\mathbf{i} + y\mathbf{j} + z\mathbf{k} = (x_0\mathbf{i} + y_0\mathbf{j} + z_0\mathbf{k}) + t(a\mathbf{i} + b\mathbf{j} + c\mathbf{k})$$

or equivalently,

$$x = x_0 + at, \quad y = y_0 + bt, \quad z = z_0 + ct \tag{3}$$

These equations are called **parametric equations** of l, and t is called a **parameter**. In (3), (x_0, y_0, z_0) is a point on the line l and $a\mathbf{i} + b\mathbf{j} + c\mathbf{k}$ is a vector parallel to l. If $z_0 = 0$ and $c = 0$, then l lies in the xy plane and has parametric equations

$$x = x_0 + at, \quad y = y_0 + bt, \quad z = 0$$

which are compatible with the parametric equations for a line given in (2) of Section 6.7.

EXAMPLE 2   Find parametric equations of the line that contains $(2, -4, 1)$ and is parallel to $3\mathbf{i} + \frac{1}{2}\mathbf{j} - \mathbf{k}$.

Solution   Taking $(x_0, y_0, z_0) = (2, -4, 1)$ and $a\mathbf{i} + b\mathbf{j} + c\mathbf{k} = 3\mathbf{i} + \frac{1}{2}\mathbf{j} - \mathbf{k}$ in (3) gives us the parametric equations

$$x = 2 + 3t, \quad y = -4 + \frac{1}{2}t, \quad z = 1 - t \quad \square$$

Another set of equations for l is obtained by eliminating t from the equations in (3). If a, b, and c are all nonzero, we can solve each of the equations in (3) for t and then equate the results, obtaining

$$\frac{x - x_0}{a} = \frac{y - y_0}{b} = \frac{z - z_0}{c} \tag{4}$$

These are called **symmetric equations** of l. Notice that the coordinates x_0, y_0, z_0 of the point P_0 on l appear in the numerators in (4) and that the components a, b, c of a vector $\mathbf{L}$ parallel to l appear in the denominators in (4).

EXAMPLE 3   Find symmetric equations of the line l containing the points $P_1 = (4, -6, 5)$ and $P_2 = (2, -3, 0)$.

Solution   In order to use (4) we find a vector $\mathbf{L}$ parallel to l. Since P_1 and P_2 are distinct points lying on l, the vector $\overrightarrow{P_1P_2}$ will serve. Therefore we let

$$\mathbf{L} = \overrightarrow{P_1P_2} = -2\mathbf{i} + 3\mathbf{j} - 5\mathbf{k}$$

For this choice of $\mathbf{L}$, if we use the hypothesis that $(4, -6, 5)$ is on l, then by (4) symmetric equations of l are

$$\frac{x - 4}{-2} = \frac{y + 6}{3} = \frac{z - 5}{-5}$$

Similarly, if we use the hypothesis that $(2, -3, 0)$ is on l, then symmetric equations of l are

$$\frac{x - 2}{-2} = \frac{y + 3}{3} = \frac{z}{-5} \quad \square$$

EXAMPLE 4   Let l be the line with equations

$$\frac{x - 3}{-2} = \frac{y - 1}{-1} = z + 2$$

Find a vector parallel to l, and find two points on l.

Solution   The numbers in the denominators of the given equations (including 1 for $z + 2$ because $z + 2 = (z + 2)/1$) are the components of a vector $\mathbf{L}$ parallel to the line l:

$$\mathbf{L} = -2\mathbf{i} - \mathbf{j} + \mathbf{k}$$

The point $P_0 = (3, 1, -2)$ lies on l, as we can see by letting $x = 3$, $y = 1$, and $z = -2$ in the given symmetric equations of l. To find a second point P_1 on l, we can let $x = 1$. Then the given set of equations becomes

$$\frac{1 - 3}{-2} = \frac{y - 1}{-1} = z + 2, \text{ so that } 1 = 1 - y = z + 2$$

Therefore $y = 0$ and $z = -1$. Consequently $P_1 = (1, 0, -1)$ is also on l. $\square$

We obtained symmetric equations of a line l from the parametric equations of l under the assumption that the numbers a, b, and c in (3) were all nonzero. If one or more of these numbers is 0, then the symmetric equations of the line in (4) must be altered to avoid dividing by 0. Suppose $a = 0$ but b and c are nonzero. Then (3) becomes

$$x = x_0, \quad y = y_0 + bt, \quad z = z_0 + ct$$

This means that the x coordinate of every point on the line is x_0. By solving for t in the last two parametric equations above we obtain

$$x = x_0, \qquad \frac{y - y_0}{b} = \frac{z - z_0}{c}$$

for symmetric equations of l. For example, the line passing through $(-2, -1, 1)$ parallel to $2\mathbf{j} - 3\mathbf{k}$ is described by

$$x = -2, \qquad \frac{y + 1}{2} = \frac{z - 1}{-3}$$

Going a step further, we observe that if $a = b = 0$ and $c \neq 0$, then (3) becomes

$$x = x_0, \quad y = y_0, \quad z = z_0 + ct$$

Consequently symmetric equations of *l* are simply

$$x = x_0, \qquad y = y_0$$

This implies that each point on *l* has *x* coordinate x_0 and *y* coordinate y_0. Hence *l* is the vertical line passing through $(x_0, y_0, 0)$. Similar analyses can be made if $b = 0$ or $c = 0$.

EXAMPLE 5   Find symmetric equations of the line *l* that contains the point $(10, -1, 1)$ and is parallel to the vector $2\mathbf{i} - 3\mathbf{k}$.

Solution   In this case $b = 0$ and (3) becomes

$$x = 10 + 2t, \quad y = -1 + 0t, \quad z = 1 - 3t$$

Solving for *t* in the first and third equations yields the symmetric equations

$$y = -1, \qquad \frac{x - 10}{2} = \frac{z - 1}{-3} \quad \square$$

Distance from a Point to a Line

Many otherwise long and difficult computations can be simplified with the help of vector methods. A good example is the computation of the distance between a given line and a given point not on the line.

THEOREM 11.12

Let *l* be a line parallel to a vector $\mathbf{L}$, and let P_1 be a point *not* on *l*. Then the distance D between P_1 and *l* is given by

$$D = \frac{\| \mathbf{L} \times \overrightarrow{P_0 P_1} \|}{\| \mathbf{L} \|} \tag{5}$$

where P_0 is any point on *l*.

Proof   Let θ be the angle between $\mathbf{L}$ and $\overrightarrow{P_0 P_1}$, so that $0 \le \theta \le \pi$ (Figure 11.43). Then

$$D = \| \overrightarrow{P_0 P_1} \| \sin \theta$$

Since

$$\| \mathbf{L} \times \overrightarrow{P_0 P_1} \| = \| \mathbf{L} \| \, \| \overrightarrow{P_0 P_1} \| \, \sin \theta$$

it follows that

$$D = \frac{\| \mathbf{L} \times \overrightarrow{P_0 P_1} \|}{\| \mathbf{L} \|}$$

as we wished to prove.   ∎

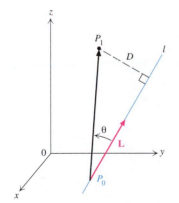

FIGURE 11.43

EXAMPLE 6   Find the distance D from the point $(2, 1, -1)$ to the line with parametric equations $x = 3t, \ y = 1 + 2t, \ z = -5 - t$.

Solution Notice that $(0, 1, -5)$ is on the line, $(2, 1, -1)$ is not on the line, and $3\mathbf{i} + 2\mathbf{j} - \mathbf{k}$ is parallel to the line. Therefore we let $P_0 = (0, 1, -5)$, $P_1 = (2, 1, -1)$, and $\mathbf{L} = 3\mathbf{i} + 2\mathbf{j} - \mathbf{k}$ in (5) and thereby obtain

$$D = \frac{\| \mathbf{L} \times \overrightarrow{P_0 P_1} \|}{\| \mathbf{L} \|} = \frac{\| (3\mathbf{i} + 2\mathbf{j} - \mathbf{k}) \times (2\mathbf{i} + 4\mathbf{k}) \|}{\| 3\mathbf{i} + 2\mathbf{j} - \mathbf{k} \|}$$

Now $\| 3\mathbf{i} + 2\mathbf{j} - \mathbf{k} \| = \sqrt{3^2 + 2^2 + (-1)^2} = \sqrt{14}$, and

$$(3\mathbf{i} + 2\mathbf{j} - \mathbf{k}) \times (2\mathbf{i} + 4\mathbf{k}) = \begin{vmatrix} \mathbf{i} & \mathbf{j} & \mathbf{k} \\ 3 & 2 & -1 \\ 2 & 0 & 4 \end{vmatrix} = 8\mathbf{i} - 14\mathbf{j} - 4\mathbf{k}$$

so that

$$\| (3\mathbf{i} + 2\mathbf{j} - \mathbf{k}) \times (2\mathbf{i} + 4\mathbf{k}) \| = \| 8\mathbf{i} - 14\mathbf{j} - 4\mathbf{k} \|$$
$$= \sqrt{8^2 + (-14)^2 + (-4)^2} = \sqrt{276}$$

Consequently

$$D = \frac{\| (3\mathbf{i} + 2\mathbf{j} - \mathbf{k}) \times (2\mathbf{i} + 4\mathbf{k}) \|}{\| 3\mathbf{i} + 2\mathbf{j} - \mathbf{k} \|} = \frac{\sqrt{276}}{\sqrt{14}} = \frac{\sqrt{138}}{\sqrt{7}} \quad \square$$

The set of points whose distances from a given line l are all the same number $a > 0$ is called a **circular cylinder** with axis l and radius a (Figure 11.44). Since the distance from any point $P = (x, y, z)$ to the z axis is $\sqrt{x^2 + y^2}$, it follows that P lies on the cylinder whose axis is the z axis and whose radius is a (Figure 11.45) if and only if $\sqrt{x^2 + y^2} = a$, or equivalently,

$$x^2 + y^2 = a^2 \tag{6}$$

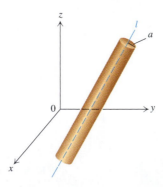

FIGURE 11.44 A circular cylinder.

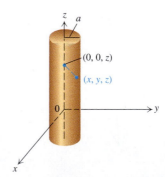

FIGURE 11.45 The circular cylinder: $x^2 + y^2 = a^2$.

Caution: Be sure to distinguish between the graph of (6) in *three-dimensional space,* which is a cylinder, and the graph of (6) in *two-dimensional space,* which is a circle.

EXERCISES 11.5

In Exercises 1–7 find a vector equation, parametric equations, and symmetric equations for the line that contains the given point and is parallel to the vector **L**.

1. $(-2, 1, 0)$; $\mathbf{L} = 3\mathbf{i} - \mathbf{j} + 5\mathbf{k}$

2. $(0, 0, 0)$; $\mathbf{L} = 11\mathbf{i} - 13\mathbf{j} - 15\mathbf{k}$

3. $(3, 4, 5)$; $\mathbf{L} = \frac{1}{2}\mathbf{i} - \frac{1}{3}\mathbf{j} + \frac{1}{6}\mathbf{k}$

4. $(-3, 6, 2)$; $\mathbf{L} = \mathbf{i} - \mathbf{j}$ **5.** $(2, 0, 5)$; $\mathbf{L} = 2\mathbf{j} + 3\mathbf{k}$

6. $(7, -1, 2)$; $\mathbf{L} = \mathbf{k}$ **7.** $(4, 2, -1)$; $\mathbf{L} = \mathbf{j}$

8. Find parametric equations for the line that contains the point $(3, -1, 2)$ and is parallel to the line with equations
$$\frac{x-1}{4} = \frac{y+3}{2} = z$$

9. Find parametric equations for the line containing the points $(-1, 1, 0)$ and $(-2, 5, 7)$.

10. Find parametric equations for the line containing the points $(-1, 1, 0)$ and $(-1, 5, 7)$.

11. Find symmetric equations for the line containing the points $(-1, 1, 0)$ and $(-1, 1, 7)$.

12. Show that the line containing the points $(1, 7, 5)$ and $(3, 2, -1)$ is parallel to the line containing $(2, -2, 5)$ and $(-2, 8, 17)$.

13. Show that the line containing the points $(2, -1, 3)$ and $(0, 7, 9)$ is perpendicular to the line containing the points $(4, -9, -3)$ and $(7, -6, -6)$.

14. Show that the line containing the points $(5, 7, 9)$ and $(4, 11, 9)$ is parallel to the line with equations
$$\frac{x-1}{-3} = \frac{y-2}{12}, \quad z = 5$$

15. Show that the line containing the points $(7, 7, -6)$ and $(6, 8, -5)$ is perpendicular to the line with equations
$$\frac{x}{7} = \frac{y-3}{4} = \frac{z+9}{3}$$

In Exercises 16–18 let l be the line that passes through $P_1 = (-1, -2, -3)$ and $P_2 = (2, -1, 0)$. Find parametric equations for l for which the given conditions are satisfied.

16. P_1 corresponds to $t = 0$ and P_2 corresponds to $t = 1$.

17. P_1 corresponds to $t = 0$ and P_2 corresponds to $t = 2$.

18. P_1 corresponds to $t = -1$ and P_2 corresponds to $t = 4$.

19. Find the distance D from the point $(5, 0, -4)$ to the line with equations
$$x - 1 = \frac{y+2}{-2} = \frac{z+1}{3}$$

20. Find the distance D from the point $(2, 1, 0)$ to the line with equations

$$x = -2, \quad y + 1 = z$$

21. Find the distance D from the origin to the line that contains the point $(-3, -3, 3)$ and is parallel to the vector $2\mathbf{i} - 3\mathbf{j} + 5\mathbf{k}$.

22. Find the distance D between the lines $x - 2y = 1$ and $2x - 4y = 3$ in the xy plane.

23. Find the distance D between the parallel lines
$$\frac{x-1}{2} = \frac{y+1}{-1} = \frac{z-2}{-2} \quad \text{and} \quad \frac{x}{2} = \frac{y-2}{-1} = \frac{z-3}{-2}$$

In Exercises 24–26 find an equation of the cylinder.

24. The cylinder with radius 3 whose axis is the x axis

25. The cylinder with radius $\sqrt{2}$ whose axis is the y axis

*26. The cylinder with radius 5 whose axis is the line with equations $x = y = z$ (*Hint:* Use (5) to find a formula for the distance from any point (x, y, z) on the cylinder to the axis of the cylinder.)

27. Are $(1, 4, 2)$, $(4, -3, -5)$, and $(-5, -10, -8)$ points on the same line? Explain your answer.

28. Find numbers x and y such that the point $(x, y, 1)$ lies on the line passing through $(2, 5, 7)$ and $(0, 3, 2)$.

29. Let **L** be a vector parallel to a given line l. Show that if **a** and **b** have initial points at the origin and terminal points on l, then $\mathbf{L} \times \mathbf{a} = \mathbf{L} \times \mathbf{b}$ (Figure 11.46).

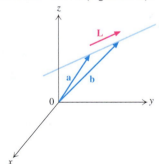

FIGURE 11.46   Figure for Exercise 29.

Project

1. Consider the line l with symmetric equations
$$\frac{x - x_0}{a} = \frac{y - y_0}{b} = \frac{z - z_0}{c}$$
and assume that (x_0, y_0, z_0) is different from the origin.

   a. Find the distance D between l and the origin.

   b. What conditions on a, b, and c are necessary in order that l does not intersect the xy plane?

c. Assuming that l does not intersect the origin, find the distance D between l and the xy plane.

d. Find a formula for the cylinder that is a given distance D from its axis l.

e. Suppose that l_1 is the line with symmetric equations

$$\frac{x-2}{1} = \frac{y+1}{-1} \quad \text{and} \quad z = 3$$

Find an equation of the cylinder whose radius is 1 and whose axis is l_1.

11.6   PLANES IN SPACE

There is only one plane that contains a given point and is perpendicular to a given line. Similarly, there is only one plane that contains a given point and is perpendicular to a given nonzero vector (Figure 11.47). In other words, a plane is determined by a point and a nonzero vector. For this reason vectors are indispensable to the analysis of planes.

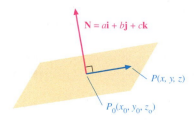

FIGURE 11.47

Equations of Planes

Let $P_0 = (x_0, y_0, z_0)$ be a given point and $\mathbf{N} = a\mathbf{i} + b\mathbf{j} + c\mathbf{k}$ a nonzero vector. Then a point $P = (x, y, z)$ lies on the plane $\mathcal{P}$ that contains P_0 and is perpendicular to $\mathbf{N}$ if and only if the vector

$$\overrightarrow{P_0P} = (x - x_0)\mathbf{i} + (y - y_0)\mathbf{j} + (z - z_0)\mathbf{k}$$

is perpendicular to $\mathbf{N}$ (Figure 11.47). This means simply that

$$\mathbf{N} \cdot \overrightarrow{P_0P} = 0$$

which is equivalent to

$$a(x - x_0) + b(y - y_0) + c(z - z_0) = 0 \tag{1}$$

In (1), (x_0, y_0, z_0) is a point on the plane, and the vector $a\mathbf{i} + b\mathbf{j} + c\mathbf{k}$ is perpendicular to the plane.

By expanding the left side of (1) and letting $d = ax_0 + by_0 + cz_0$, we obtain the equivalent equation

$$ax + by + cz = d \tag{2}$$

Both (1) and (2) are equations of the plane $\mathscr{P}$ that is perpendicular to **N** and contains P_0. The vector **N** is said to be **normal** to $\mathscr{P}$. Observe that the components of **N** appear as coefficients in (1) and (2), and the coordinates of the point P_0 in the plane appear inside the parentheses in (1).

EXAMPLE 1   Find an equation of the plane that contains the point $(-2, 4, 5)$ and has normal vector $7\mathbf{i} - 6\mathbf{k}$.

Solution   We substitute

$$x_0 = -2, \quad y_0 = 4, \quad z_0 = 5 \quad \text{and} \quad a = 7, \quad b = 0, \quad c = -6$$

into (1) and obtain the equation

$$7(x + 2) + 0(y - 4) - 6(z - 5) = 0$$

Collecting terms, we have

$$7x - 6z = -44 \quad \square$$

EXAMPLE 2   Find an equation of the plane $\mathscr{P}$ that contains the point $(\frac{1}{2}, 0, 3)$ and is perpendicular to the line l with equations

$$\frac{x + 1}{4} = \frac{y - 2}{-1} = \frac{z}{5}$$

Solution   Since l is perpendicular to $\mathscr{P}$, any vector parallel to l (such as $4\mathbf{i} - \mathbf{j} + 5\mathbf{k}$) is normal to $\mathscr{P}$. Therefore we may let $\mathbf{N} = 4\mathbf{i} - \mathbf{j} + 5\mathbf{k}$, and hence an equation of the plane is

$$4\left(x - \frac{1}{2}\right) + (-1)(y - 0) + 5(z - 3) = 0$$

or $4x - y + 5z = 17 \quad \square$

EXAMPLE 3   Find a vector normal to the plane $2x - 3y + 7z = -35$.

Solution   One such vector can be read directly from the equation: it is $2\mathbf{i} - 3\mathbf{j} + 7\mathbf{k}$. $\square$

EXAMPLE 4   Sketch the plane $3x + 2y + z = 6$.

Solution   The plane is determined by any three distinct points on it. The points we will use are the x, y, and z intercepts. To find the x intercept we set y and z equal to 0 and solve for x in the given equation:

$$3x + 2(0) + 0 = 6$$
$$x = 2$$

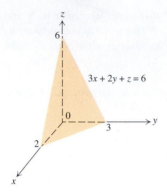

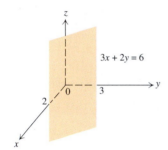

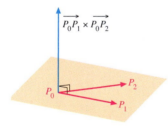

For the y intercept we set x and z equal to 0 and solve for y:

$$3(0) + 2y + 0 = 6$$
$$y = 3$$

Similarly, to find the z intercept we set x and y equal to 0 and solve for z:

$$3(0) + 2(0) + z = 6$$
$$z = 6$$

Thus the points $(2, 0, 0)$, $(0, 3, 0)$ and $(0, 0, 6)$ are on the plane. In Figure 11.48 we have drawn the portion of the plane in the first octant. ❏

FIGURE 11.48

EXAMPLE 5   Sketch the plane $3x + 2y = 6$.

Solution   Proceeding as in Example 4, we find that the x intercept is 2 and the y intercept is 3, so that the points $(2, 0, 0)$ and $(0, 3, 0)$ are on the plane. If we try to find a z intercept by setting x and y equal to 0, we obtain

$$3(0) + 2(0) \overset{?}{=} 6$$

Thus the plane contains no z intercept. It follows that the plane is perpendicular to the xy plane. In Figure 11.49 we have drawn a portion of the plane. ❏

FIGURE 11.49

Three noncollinear points P_0, P_1, and P_2 determine a plane. To find a vector normal to the plane, we first form the two vectors $\overrightarrow{P_0P_1}$ and $\overrightarrow{P_0P_2}$ and compute their cross product. Since $\overrightarrow{P_0P_1}$ and $\overrightarrow{P_0P_2}$ are parallel to the plane, their cross product is perpendicular to both vectors and hence to the plane (Figure 11.50).

EXAMPLE 6   Find an equation of the plane containing the points $P_0 = (1, 0, 2)$, $P_1 = (-1, 3, 4)$, and $P_2 = (3, 5, 7)$.

Solution   First we notice that the vectors $\overrightarrow{P_0P_1} = -2\mathbf{i} + 3\mathbf{j} + 2\mathbf{k}$ and $\overrightarrow{P_0P_2} = 2\mathbf{i} + 5\mathbf{j} + 5\mathbf{k}$ are not parallel (and thus P_0, P_1, and P_2 determine a plane). For a normal vector we can therefore take

$$\mathbf{N} = \overrightarrow{P_0P_1} \times \overrightarrow{P_0P_2} = \begin{vmatrix} \mathbf{i} & \mathbf{j} & \mathbf{k} \\ -2 & 3 & 2 \\ 2 & 5 & 5 \end{vmatrix} = 5\mathbf{i} + 14\mathbf{j} - 16\mathbf{k}$$

FIGURE 11.50

Since $\mathbf{N}$ is normal to the plane and $(1, 0, 2)$ lies on the plane, we use (1) to conclude that an equation of the plane is

$$5(x - 1) + 14(y - 0) - 16(z - 2) = 0$$

or more simply,

$$5x + 14y - 16z = -27$$ ❏

EXAMPLE 7   Consider the planes $\mathscr{P}_0$ and $\mathscr{P}_1$ that are described by

$$x - 2y + 4z = 1 \quad \text{and} \quad 2x + y - 4z = -1$$

respectively. Show that the two planes intersect, and find equations for the line l of intersection.

Solution   Let $\mathbf{N}_0 = \mathbf{i} - 2\mathbf{j} + 4\mathbf{k}$ and $\mathbf{N}_1 = 2\mathbf{i} + \mathbf{j} - 4\mathbf{k}$, so that $\mathbf{N}_0$ and $\mathbf{N}_1$ are normals of $\mathscr{P}_0$ and $\mathscr{P}_1$, respectively. Since

$$\mathbf{N}_0 \times \mathbf{N}_1 = \begin{vmatrix} \mathbf{i} & \mathbf{j} & \mathbf{k} \\ 1 & -2 & 4 \\ 2 & 1 & -4 \end{vmatrix} = 4\mathbf{i} + 12\mathbf{j} + 5\mathbf{k} \neq \mathbf{0}$$

it follows that the two planes are not parallel, and thus they intersect in a line l. Notice that l is in $\mathscr{P}_0$, so is perpendicular to $\mathbf{N}_0$. Similarly l is perpendicular to $\mathbf{N}_1$. Therefore l is parallel to $\mathbf{N}_0 \times \mathbf{N}_1 = 4\mathbf{i} + 12\mathbf{j} + 5\mathbf{k}$. To find a point on l, we select a fixed value of x and solve the two equations for the plane that result. To that end, let $x = 1$. Then from the equation of the plane $\mathscr{P}_0$,

$$1 - 2y + 4z = 1, \quad \text{so that} \quad y = 2z$$

Similarly, substituting 1 for x in the equation of the plane $\mathscr{P}_1$ yields

$$2 + y - 4z = -1, \quad \text{so that} \quad y - 4z = -3$$

Finally, if we substitute $2z$ for y in $y - 4z = -3$, we obtain

$$2z - 4z = -3, \quad \text{so that} \quad z = \frac{3}{2}$$

Thus $y = 2(\frac{3}{2}) = 3$. Consequently the point $(1, 3, \frac{3}{2})$ is on the line l. Having a vector parallel to l and a point on it, we are ready to give equations for l:

$$\frac{x - 1}{4} = \frac{y - 3}{12} = \frac{z - 3/2}{5} \quad \square$$

Distance from a Point to a Plane

Vector methods greatly simplify the calculation of distances between points and planes, just as between points and lines.

Let $\mathscr{P}$ be a plane with normal $\mathbf{N}$, and let P_1 be any point not on $\mathscr{P}$. Then the distance D between P_1 and $\mathscr{P}$ is given by

$$D = \| \, \mathbf{pr}_{\mathbf{N}} \, \overrightarrow{P_0P_1} \, \|$$

where P_0 is any point on $\mathscr{P}$. (See Figure 11.51(a) and (b).) However, by (6) of Section 11.3,

$$\| \, \mathbf{pr}_{\mathbf{N}} \, \overrightarrow{P_0P_1} \, \| = \frac{|\mathbf{N} \cdot \overrightarrow{P_0P_1}|}{\| \mathbf{N} \|}$$

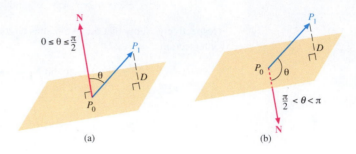

FIGURE 11.51

We summarize these observations in Theorem 11.13.

THEOREM 11.13

Let $\mathscr{P}$ be a plane with normal $\mathbf{N}$, and let P_1 be any point not on $\mathscr{P}$. Then the distance D between P_1 and $\mathscr{P}$ is given by

$$D = \frac{|\mathbf{N} \cdot \overrightarrow{P_0 P_1}|}{\|\mathbf{N}\|} \tag{3}$$

where P_0 is any point on $\mathscr{P}$.

EXAMPLE 8   Calculate the distance D between the point $P_1 = (-1, 1, 2)$ and the plane

$$3x - 2y + z = 1$$

Solution   The equation of the plane yields a vector $\mathbf{N}$ normal to the plane:

$$\mathbf{N} = 3\mathbf{i} - 2\mathbf{j} + \mathbf{k}$$

From the equation of the plane we can also find a point on the plane by letting $x = y = 0$. Then $z = 1$, so that $P_0 = (0, 0, 1)$ is on the plane. Straightforward computations yield

$$\overrightarrow{P_0 P_1} = (-1 - 0)\mathbf{i} + (1 - 0)\mathbf{j} + (2 - 1)\mathbf{k} = -\mathbf{i} + \mathbf{j} + \mathbf{k}$$

and

$$\mathbf{N} \cdot \overrightarrow{P_0 P_1} = (3)(-1) + (-2)(1) + (1)(1) = -3 - 2 + 1 = -4$$

Therefore (3) implies that

$$D = \frac{|\mathbf{N} \cdot \overrightarrow{P_0 P_1}|}{\|\mathbf{N}\|} = \frac{|-4|}{\sqrt{(3)^2 + (-2)^2 + (1)^2}} = \frac{4}{\sqrt{14}} \quad \square$$

EXERCISES 11.6

In Exercises 1–5 find an equation of the plane that contains P_0 and has normal vector $\mathbf{N}$.

1. $P_0 = (-1, 2, 3)$, $\mathbf{N} = -4\mathbf{i} + 15\mathbf{j} - \frac{1}{2}\mathbf{k}$

2. $P_0 = (\pi, 0, -\pi)$, $\mathbf{N} = 2\mathbf{i} + 3\mathbf{j} - 4\mathbf{k}$

3. $P_0 = (9, 17, -7)$, $\mathbf{N} = 2\mathbf{i} - 3\mathbf{k}$

4. $P_0 = (-1, -1, -1)$, $\mathbf{N} = \dfrac{1}{\sqrt{2}}(\mathbf{i} + \mathbf{j} - \mathbf{k})$

5. $P_0 = (2, 3, -5)$, $\mathbf{N} = \mathbf{j}$

6. Find an equation of the plane that contains the points $(2, -1, 4)$, $(5, 3, 5)$, and $(2, 4, 3)$.

7. Find an equation of the plane that contains the point $(1, -1, 2)$ and the line with symmetric equations
$$x + 2 = y + 1 = \frac{z + 5}{2}$$

8. Find an equation of the plane that contains the two parallel lines
$$\frac{x - 1}{3} = \frac{y + 1}{2} = \frac{z - 5}{4}$$
and
$$\frac{x + 3}{3} = \frac{y - 4}{2} = \frac{z}{4}$$

9. Find an equation of the plane that contains the point $(2, \frac{1}{2}, \frac{1}{3})$ and is perpendicular to the line having parametric equations
$$x = \pi + 2t, \quad y = 2\pi + 5t, \quad z = 9t$$

10. Find parametric equations for the line that passes through $(2, -1, 0)$ and is perpendicular to the plane $2x - 3y + 4z = 5$.

11. Let l be the intersection of the two planes
$$2x - 3y + 4z = 2 \quad \text{and} \quad x - z = 1$$

   a. Find a vector equation of l.

   b. Find an equation of the plane that is perpendicular to l and contains the point $(-9, 12, 14)$.

12. Let l be the line
$$\frac{x + 1}{2} = \frac{y + 3}{3} = -z$$
and let $\mathcal{P}$ be the plane
$$3x - 2y + 4z = -1$$

   a. Find the point of intersection P_0 of l and $\mathcal{P}$.

   b. Find an equation of the plane perpendicular to l at P_0.

   c. Find symmetric equations of the line perpendicular to $\mathcal{P}$ at P_0.

13. Find the distance D between the point $(3, -1, 4)$ and the plane $2x - y + z = 5$.

14. Find the distance D between the point $(2, 0, -4)$ and the plane $x + 2y + 4z - 3 = 0$.

15. Show that the distance D between the origin and the plane $ax + by + cz = d$ is $|d| / \sqrt{a^2 + b^2 + c^2}$.

16. Show that the planes
$$a_1 x + b_1 y + c_1 z = d_1 \quad \text{and} \quad a_2 x + b_2 y + c_2 z = d_2$$
are perpendicular if and only if $a_1 a_2 + b_1 b_2 + c_1 c_2 = 0$.

17. The set of all points equidistant from $(3, 1, 5)$ and $(5, -1, 3)$ is a plane. Find an equation of that plane.

In Exercises 18–21 sketch the plane.

18. $2x + y + z = 4$

19. $-\frac{1}{2}x + \frac{1}{3}y - z = 1$

20. $2x - z = 1$

21. $4y + 3z = 6$

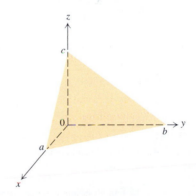

FIGURE 11.52   Figure for Exercise 22.

22. Show that an equation of the plane having x intercept a, y intercept b, and z intercept c is
$$\frac{x}{a} + \frac{y}{b} + \frac{z}{c} = 1$$
provided that a, b, and c are all nonzero (Figure 11.52).

23. Show that the points $(2, 3, 2)$, $(1, -1, -3)$, $(1, 0, -1)$, and $(5, 9, 5)$ all lie on the same plane.

24. Let l be the line
$$\frac{x - 1}{2} = \frac{y + 1}{3} = \frac{z + 5}{7}$$
and $\mathcal{P}$ be the plane
$$2(x - 1) + 2(y + 3) - z = 0$$

Find the two points on l at a distance 3 from $\mathcal{P}$.

25. Suppose planes $\mathcal{P}_1$ and $\mathcal{P}_2$ intersect, and let **a** and **b** lie on $\mathcal{P}_1$ and **c** and **d** on $\mathcal{P}_2$. Assume that **a** and **b** are not parallel, and **c** and **d** are not parallel. Show that $(\mathbf{a} \times \mathbf{b}) \times (\mathbf{c} \times \mathbf{d})$ is parallel to the intersection of $\mathcal{P}_1$ and $\mathcal{P}_2$.

26. Which of the following planes are identical, which are parallel to each other, and which are perpendicular?

    a. $x + 2y - 3z = 2$ **b.** $15x - 9y + z = 2$

    c. $-2x - 4y + 6z + 4 = 0$ **d.** $5x - 3y + \frac{1}{3}z - 1 = 0$

27. Which of the following planes are identical, which are parallel to each other, and which are perpendicular?

    a. $x + y - z + 3 = 0$ **b.** $x - y - 2 = 0$

    c. $y - z = 2$ **d.** $x + y = -5$

28. Find an equation of the plane that contains the point $(-1, 2, 3)$ and is

    a. parallel to the xy plane.

    b. perpendicular to the x axis.

    c. perpendicular to the y axis.

29. Find an equation of the plane that contains the point $(-4, -5, -3)$ and is

    a. parallel to the yz plane.

    b. parallel to the xz plane.

    c. perpendicular to the z axis.

30. Find an equation of the plane that contains the points $(-2, 1, 4)$ and $(0, 3, 1)$ and contains a line parallel to the vector $2\mathbf{i} - 4\mathbf{j} + 6\mathbf{k}$.

In Exercises 31–34 find all points of intersection of the three planes.

31. $x + y = 1; \; y + z = 2; \; x + z = 3$

32. $x + y - z = 2; \; -x + 2y - z = 3; \; x - y - z = 0$

33. $2x - 3y - z = 1; \; x + 3y + z = -2; \; y + z = 2$

34. $2x + 4y - 6z = 4; \; x - y - z - 2 = 0; \; x + 2y - 3z = 2$

In Exercises 35–37 find the distance D between the pair of parallel planes.

35. $x - y + z = 2$ and $3x - 3y + 3z = 1$

36. $y - 2z = 4$ and $-2y + 4z = 6$

37. $2x - 3y + 4z = 5$ and $4x - 6y + 8z = -1$

Project

1. **a.** The angle between two nonparallel planes $\mathcal{P}_1$ and $\mathcal{P}_2$ is the angle θ in $[0, \pi/2]$ between the vectors normal to $\mathcal{P}_1$ and $\mathcal{P}_2$.

       i. Consider the planes $\mathcal{P}_1$ and $\mathcal{P}_2$, defined by
    $$2x - y + z = 0 \quad \text{and} \quad x + y + 3z = 1$$
    respectively. Approximate the angle θ between $\mathcal{P}_1$ and $\mathcal{P}_2$.

    ii. Let $\mathcal{P}_1$ be the plane
    $$x - \frac{1}{2}y + \frac{\sqrt{11}}{2}z = 0.$$
    Find an equation of a plane $\mathcal{P}_2$ that makes an angle of $\pi/3$ with $\mathcal{P}_1$. (*Hint:* It might help if you use a unit normal for $\mathcal{P}_2$.)

    b. Suppose two nonparallel planes have normals $\mathbf{N}_1$ and $\mathbf{N}_2$, respectively. Let P_0 be a point in the intersection of the two planes. Show that a point P is in the intersection if and only if $(\mathbf{N}_1 \times \mathbf{N}_2) \times \overrightarrow{PP_0} = \mathbf{0}$. (*Hint:* First show that P lies in the intersection if and only if $\overrightarrow{PP_0}$ is perpendicular to both $\mathbf{N}_1$ and $\mathbf{N}_2$.)

    c. Suppose that the planes $\mathcal{P}_1$ and $\mathcal{P}_2$ are parallel, with equations $ax + by + cz = d_1$ and $ax + by + cz = d_2$, respectively. Find a formula for the distance D between $\mathcal{P}_1$ and $\mathcal{P}_2$.

REVIEW

Key Terms and Expressions

Three-dimensional space
Vector
  component
  length
Unit vector

Dot product; cross product
Resolution of a vector
Projection of one vector onto another
Vector, parametric, and symmetric equations of a line
Normal to a plane

Key Formulas

$$|PQ| = \sqrt{(x_1 - x_0)^2 + (y_1 - y_0)^2 + (z_1 - z_0)^2}$$

$$\overrightarrow{PQ} = (x_1 - x_0)\mathbf{i} + (y_1 - y_0)\mathbf{j} + (z_1 - z_0)\mathbf{k}$$

$$\|a_1\mathbf{i} + a_2\mathbf{j} + a_3\mathbf{k}\| = \sqrt{a_1^2 + a_2^2 + a_3^2}$$

$$\mathbf{a} \cdot \mathbf{b} = a_1b_1 + a_2b_2 + a_3b_3$$

$$\mathbf{a} \cdot \mathbf{b} = \|\mathbf{a}\| \, \|\mathbf{b}\| \, \cos\theta$$

$$\mathbf{pr_a b} = \frac{\mathbf{a} \cdot \mathbf{b}}{\|\mathbf{a}\|^2} \mathbf{a}$$

$$\mathbf{a} \times \mathbf{b} = (a_2b_3 - a_3b_2)\mathbf{i} + (a_3b_1 - a_1b_3)\mathbf{j} + (a_1b_2 - a_2b_1)\mathbf{k}$$

$$\|\mathbf{a} \times \mathbf{b}\| = \|\mathbf{a}\| \, \|\mathbf{b}\| \, \sin\theta$$

$$\mathbf{a} \times (\mathbf{b} \times \mathbf{c}) = \mathbf{b}(\mathbf{a} \cdot \mathbf{c}) - \mathbf{c}(\mathbf{a} \cdot \mathbf{b})$$

Vector equation of a line: $\mathbf{r} = \mathbf{r}_0 + t\mathbf{L}$

Parametric equations of a line:

$$x = x_0 + at, \quad y = y_0 + bt, \quad z = z_0 + ct$$

Symmetric equations of a line:

$$\frac{x - x_0}{a} = \frac{y - y_0}{b} = \frac{z - z_0}{c} \quad \text{if } a \neq 0, b \neq 0, c \neq 0$$

Distance between a point P_1 and a line l parallel to $\mathbf{L}$:

$$\frac{\|\mathbf{L} \times \overrightarrow{P_0P_1}\|}{\|\mathbf{L}\|}, \quad \text{where } P_0 \text{ is on } l$$

Equation of a plane: $ax + by + cz = d$

Distance between a point P_1 and a plane $\mathcal{P}$:

$$\frac{|\mathbf{N} \cdot \overrightarrow{P_0P_1}|}{\|\mathbf{N}\|}, \quad \text{where } P_0 \text{ is on } \mathcal{P}$$

Review Exercises

In Exercises 1–3 find $2\mathbf{a} + \mathbf{b} - 3\mathbf{c}$, $\mathbf{a} \times \mathbf{b}$, $\mathbf{c} \cdot (\mathbf{a} \times \mathbf{b})$, and $\mathbf{a} \times (\mathbf{b} \times \mathbf{c})$.

1. $\mathbf{a} = 2\mathbf{i} - 3\mathbf{j} + \mathbf{k}$, $\mathbf{b} = \mathbf{i} - \mathbf{j}$, $\mathbf{c} = \mathbf{j} - 3\mathbf{k}$

2. $\mathbf{a} = \frac{1}{2}\mathbf{i} - \mathbf{j} + 2\mathbf{k}$, $\mathbf{b} = 2\mathbf{i} - 4\mathbf{j} + 6\mathbf{k}$, $\mathbf{c} = \mathbf{i} - 5\mathbf{j} + 6\mathbf{k}$

3. $\mathbf{a} = 3\mathbf{i} - 2\mathbf{j} + \mathbf{k}$, $\mathbf{b} = 5\mathbf{i} - 2\mathbf{j} + \mathbf{k}$, $\mathbf{c} = \mathbf{j} - \mathbf{k}$

4. Find the cosine of the angle between the vectors $3\mathbf{i} - 4\mathbf{j} + 12\mathbf{k}$ and $\mathbf{i} - \mathbf{k}$.

5. Let $P = (1, -2, 3)$ and let $\mathbf{a} = 2\mathbf{i} - 2\mathbf{j} + \mathbf{k}$. Find a point Q such that $\overrightarrow{PQ}$ and $\mathbf{a}$ are the same vector.

6. Let $\mathbf{a} = 2\mathbf{i} - 3\mathbf{j} + 5\mathbf{k}$ and $\mathbf{b} = 5\mathbf{i} + 3\mathbf{j} - 7\mathbf{k}$.
   a. Find $\mathbf{a} \cdot \mathbf{b}$.   b. Find $\mathbf{a} \times \mathbf{b}$.   c. Find $\mathbf{pr_a b}$.

7. Resolve the vector $2\mathbf{i} - \mathbf{j} - \mathbf{k}$ into vectors parallel to $2\mathbf{j} + \mathbf{k}$ and $-20\mathbf{i} - 2\mathbf{j} + 4\mathbf{k}$. (Note that all three vectors lie in the same plane.)

8. Find the area A of the triangle with vertices $(1, 1, 1)$, $(2, 3, 5)$, and $(-1, 3, 1)$.

9. Show that $(\frac{1}{2}, \frac{1}{3}, 0)$, $(1, 1, -1)$, and $(-2, -3, 5)$ are collinear points, and find symmetric equations for the line containing them.

10. Let $P = (2, 5, -7)$ and $Q = (4, 3, 8)$.
    a. Find $\overrightarrow{PQ}$.   b. Find $\|\overrightarrow{PQ}\|$.
    c. For the line l through P and Q, find
      i. a vector equation of l.
      ii. parametric equations of l.
      iii. symmetric equations of l.

11. Find a vector equation for the line that contains $(-3, -3, 1)$ and is perpendicular to the plane $2x - 3y + 4z = 7$.

12. Find an equation for the plane that contains the point $(-1, 3, 2)$ and has normal $\mathbf{N} = 2\mathbf{i} + \mathbf{j} - \mathbf{k}$.

13. Show that the points $(-1, 1, 1)$, $(0, 2, 1)$, $(0, 0, \frac{3}{2})$, and $(13, -1, 5)$ lie on the same plane.

14. Find an equation of the plane that contains the points $(1, 0, -1)$, $(-5, 3, 2)$, and $(2, -1, 4)$.

15. Find an equation of the plane that is parallel to the z axis and contains the points $(3, -1, 5)$ and $(7, 9, 4)$. (A plane is parallel to a line if any vector normal to the plane is perpendicular to the line.)

16. Show that the line

$$\frac{x + 5}{7} = \frac{y - 11}{9} = \frac{z}{45}$$

is parallel to the plane with equation $9x - 2y - z = 0$.

17. Find the distance D from the point $(1, -2, 5)$ to the plane $3(x - 1) - 4(y + 2) + 12z = 0$.

18. Find the distance D from the point $(1, -2, 5)$ to the line

$$x = 1 + 3t, \quad y = -2 - 4t, \quad z = 12t$$

19. In each of parts (a)–(c) determine whether or not the given planes have either a point or a line in common. If they do, then find the point or the line.
    a. $3x - y + z = 2$, $y = 4$, $z = 1$

b. $2x + y - 2z = 1$, $3x + y - z = 2$, $x - y + z = 0$

c. $2x - 11y + 6z = -2$, $2x - 3y + 2z = 2$,
$2x - 9y + 5z = -1$

20. Suppose $a \neq 0$, $b \neq 0$, and $c \neq 0$. Find symmetric equations for the line that contains the point (a, b, c) and is parallel to the vector $a\mathbf{i} + b\mathbf{j} + c\mathbf{k}$. Show that the origin lies on the line.

21. Suppose the point (a, b, c) is not the origin. Find an equation of the plane that contains the point (a, b, c) and is perpendicular to $a\mathbf{i} + b\mathbf{j} + c\mathbf{k}$.

*22. Let $\mathcal{P}_0$ and $\mathcal{P}_1$ be the two parallel planes given respectively by $2x - 3y + 4z = 2$ and $2x - 3y + 4z = 6$. Let l be the line with symmetric equations

$$\frac{x - 5}{3} = \frac{y + 3}{2} = \frac{z}{6}$$

a. Without finding the points of intersection, show that l intersects both $\mathcal{P}_0$ and $\mathcal{P}_1$.

b. Assuming that the points of intersection are Q_0 and Q_1, determine $|\overrightarrow{Q_0Q_1}|$ without finding either Q_0 or Q_1. (*Hint:* First find the distance between the two planes.)

23. Use vectors to show that any angle inscribed in a semicircle is a right angle.

24. Let $\mathbf{a}$, $\mathbf{b}$, and $\mathbf{c}$ be the vectors pointing from the three vertices of a triangle to the midpoints of the opposite sides. Show that $\mathbf{a} + \mathbf{b} + \mathbf{c} = \mathbf{0}$.

25. Let $\mathbf{a}$ and $\mathbf{b}$ be nonzero vectors.

a. Show that if $\mathbf{a} \cdot \mathbf{c} = \mathbf{b} \cdot \mathbf{c}$ for every vector $\mathbf{c}$, then $\mathbf{a} = \mathbf{b}$. (*Hint:* Take $\mathbf{c} = \mathbf{a} - \mathbf{b}$, and show that $(\mathbf{a} - \mathbf{b}) \cdot (\mathbf{a} - \mathbf{b}) = 0$.)

b. Show that if $\mathbf{a} \times \mathbf{c} = \mathbf{b} \times \mathbf{c}$ for every vector $\mathbf{c}$, then $\mathbf{a} = \mathbf{b}$. (*Hint:* Take $\mathbf{c}$ to be a nonzero vector perpendicular to $\mathbf{a} - \mathbf{b}$.)

26. Using the dot product, prove the trigonometric identity

$$\cos(x + y) = \cos x \cos y - \sin x \sin y$$

(*Hint:* Use the unit vectors $\mathbf{a} = \cos x\mathbf{i} + \sin x\mathbf{j}$ and $\mathbf{b} = \cos y\mathbf{i} - \sin y\mathbf{j}$.)

27. a. Show that

$$\| \mathbf{a} + \mathbf{b} \|^2 \| \mathbf{a} - \mathbf{b} \|^2 = (\| \mathbf{a} \|^2 + \| \mathbf{b} \|^2)^2 - 4(\mathbf{a} \cdot \mathbf{b})^2$$

(*Hint:* $\| \mathbf{a} + \mathbf{b} \|^2 = (\mathbf{a} + \mathbf{b}) \cdot (\mathbf{a} + \mathbf{b})$ and $\| \mathbf{a} - \mathbf{b} \|^2 = (\mathbf{a} - \mathbf{b}) \cdot (\mathbf{a} - \mathbf{b})$.)

b. Use part (a) to show that

$$\| \mathbf{a} + \mathbf{b} \| \, \| \mathbf{a} - \mathbf{b} \| \leq \| \mathbf{a} \|^2 + \| \mathbf{b} \|^2$$

28. Suppose that $\mathbf{a}$, $\mathbf{b}$, and $\mathbf{c}$ are mutually perpendicular unit vectors and that $\mathbf{d} = a\mathbf{a} + b\mathbf{b} + c\mathbf{c}$. Show that

$$\| \mathbf{d} \| = \sqrt{a^2 + b^2 + c^2}$$

(*Hint:* $\| \mathbf{d} \|^2 = \mathbf{d} \cdot \mathbf{d}$.)

29. Let $\mathbf{u}$ be a unit vector, ϕ the angle between $\mathbf{u}$ and the positive z axis, and θ as in Figure 11.53. Show that

$$\mathbf{u} = \cos \theta \sin \phi\mathbf{i} + \sin \theta \sin \phi\mathbf{j} + \cos \phi\mathbf{k}$$

(The angles ϕ and θ will play a major role in Section 14.6.)

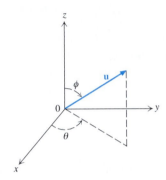

FIGURE 11.53   Figure for Exercise 29.

Applications

30. Suppose that forces of 500 pounds and 300 pounds are applied at the same point and that the angle between the two forces is $\pi/3$. Find the magnitude of the resultant force and the cosine of the angle θ the resultant force makes with respect to the 500-pound force.

31. To provide traction for a broken leg, a horizontal force of 5 pounds is to be obtained by means of a 5-pound weight as shown in Figure 11.54.

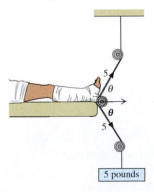

FIGURE 11.54   Figure for Exercise 31.

a. What angle θ should be chosen to accomplish this?

b. Approximate the angle θ that would provide a horizontal force of 6 pounds.

32. A river $\frac{1}{2}$ mile wide flows parallel to the shore at the rate of 5 miles per hour. If a motorboat can move at 10 miles per hour in still water, at what angle with respect to the shore should the boat be pointed in order to travel perpendicular to the shore? How long will it take to cross the river?

33. A jet pointed north travels with a motor speed of 500 miles per hour. The jet stream is clocked at 100 miles per hour and flows east. What is the ground speed of the jet?

34. Suppose particles whose charges are 3.2×10^{-19}, -6.4×10^{-19}, and 4.8×10^{-19} are located, respectively, at $(10^{-12}, 0, 0)$, $(0, 2 \times 10^{-12}, 0)$, and $(0, 0, 3 \times 10^{-12})$. Find the total electric force **F** exerted by the three charged particles on a positive unit charge located at the origin.

Topics for Discussion

1. List four ways of describing a vector. Discuss conditions under which each of the representations of a vector is the reasonable choice.

2. For what kind of results do we use the dot product? the cross product?

3. Write a formula for the area of the polygon $PQRS$ in terms of vectors in the case in which the figure is a

a. rectangle.   **b.** rhombus.   **c.** right trapezoid.

4. In your own words, derive a formula for the distance from a point P to a line l, and from P to a plane $\mathscr{P}$.

Cumulative Review, Chapters 1–10

1. Let $f(x) = \sqrt{\dfrac{1 - 2x}{1 - 3x}}$.

a. Find the domain of f.

b. Find $\lim_{x \to -\infty} f(x)$.

In Exercises 2–3 find the limit.

2. $\lim_{x \to 0^+} x(\ln x)^3$ **3.** $\lim_{x \to \infty} x^{\tan\,(1/x)}$

4. Let $f(x) = (1 - 2x)/(1 + 3x)$. Find all values of x for which $f'(x) = -20$.

5. Let $f(x) = \ln(\pi/2 + \tan^{-1} x)$.

a. Show that f^{-1} exists, and determine the domain and range of f^{-1}.

b. Find $(f^{-1})'(\ln(\pi/3))$.

c. Find a formula for f^{-1}.

6. Recall that the line normal to the graph of a function f at $(a, f(a))$ is perpendicular to the line tangent at $(a, f(a))$. Find an equation for the line in the xy plane normal to the graph of $y = \sin x$ at $(\pi/4, \sqrt{2}/2)$.

7. Consider the portion of the cycloid parametrized by $x = t - \sin t$, $y = 1 - \cos t$ for $0 < t < \pi$.

a. Use the fact that

$$\frac{dy}{dx} = \frac{dy/dt}{dx/dt}$$

to show that

$$\frac{dy}{dx} = \frac{\sin t}{1 - \cos t} = \frac{\sqrt{2y - y^2}}{y}$$

b. Using the parametric equations, show that $x = \cos^{-1}(1 - y) - \sqrt{2y - y^2}$.

c. Use implicit differentiation on the equation of part (b) to show that $dy/dx = \sqrt{2y - y^2}/y$.

8. At the moment that the sun's elevation is $30°$, a ball is dropped from the top of a pole 96 feet tall. Assuming no air friction and an unimpaired shadow from the sun, determine how fast the shadow is moving

a. after 1 second.

b. the moment the ball hits the ground.

9. Let $f(x) = x(x^2 - 5)^2$. Sketch the graph of f, indicating all pertinent information.

10. A window with a perimeter of 16 feet has the shape of a rectangle with an isosceles right triangle attached to the top. Assume that the hypotenuse of the triangle is the top side of the rectangular part of the window. Show that the maximum amount of light enters the window if the length of each leg of the triangle equals the length of the vertical sides of the rectangle.

11. Let $f(x) = x + 1/x$. Find the area A of the region between the graph of f and the line $y = \frac{5}{2}$ on $[1, 3]$.

In Exercises 12–14 evaluate the integral.

12. $\displaystyle\int \frac{1}{t^2} \csc\frac{1}{t}\,\tan^2\frac{1}{t}\,dt$ **13.** $\displaystyle\int \frac{x^3}{x^2 - x + 1}\,dx$

14. $\displaystyle\int \frac{1}{x^6 \sqrt{2x^2 - 1}}\,dx$

15. Determine whether $\int_2^\infty x e^{-3x^2}\,dx$ converges or diverges. If it converges, determine its value.

16. Let $f(x) = x^2/\sqrt{x^2 + 9}$ for $0 \leq x \leq 3$, and let R be the region between the graph of f and the x axis. Find the volume V of the solid obtained by revolving R about the y axis.

17. Let C be described parametrically by $x = e^{-t} \sin 2t$ and $y = e^{-t} \cos 2t$ for $-1 \leq t \leq 1$. Find the length L of C.

18. Find the area A of the region common to the cardioids $r = 1 + \cos \theta$ and $r = 1 - \cos \theta$.

19. Find all values for c for which $\lim_{n \to \infty} (\sqrt{n + c} - \sqrt{n})$ exists as a real number.

20. Find the numerical value of $\displaystyle\sum_{n=1}^{\infty} \frac{4}{n(n + 2)}$.

In Exercises 21–22 determine whether the series converges absolutely, converges conditionally, or diverges.

21. $\displaystyle\sum_{n=0}^{\infty} (-1)^n \frac{n^2}{\sqrt{n^4 + 2}}$

22. $\displaystyle\sum_{n=1}^{\infty} \frac{2 \cdot 4 \cdot 6 \cdots (2n - 2)(2n)}{n^n}$

23. Let $f(x) = \sum_{n=1}^{\infty} x^n/(n + 2)$. Find the Taylor series about 0 for $\int_0^x f(t)\, dt$, and determine the interval of convergence for the Taylor series.

Helical pumpkin tendrils. (*Dwight R. Kuhn*)

12 VECTOR-VALUED FUNCTIONS AND CURVES IN SPACE

Until now we have studied functions whose domains and ranges both consist of real numbers. In this chapter we will investigate a different type of function, whose domain consists of real numbers and whose range consists of vectors. Such functions, which are called vector-valued functions, have many applications. In mathematics they are an indispensable tool for studying curves in space. As for physical applications, they present a convenient way to describe the motion of a planet as well as the shape of a DNA molecule.

12.1 DEFINITIONS AND EXAMPLES

A correspondence between one set of numbers and another is a function. In the study of motion we frequently encounter a correspondence between a set of numbers and a set of vectors. Such an association determines a vector-valued function.

DEFINITION 12.1

A **vector-valued function** consists of two parts: a **domain,** which is a collection of numbers, and a **rule,** which assigns to each number in the domain one and only one vector.

The numbers in the domain of a vector-valued function will usually be denoted *t*; the reason is that in most applications the domain of such a function will represent an interval of time. The vector-valued functions themselves will normally be denoted by the boldface letters **F**, **G**, and **H**. The total collection of vectors assigned by a vector-valued function to members of its domain is called the **range** of the function. Unless otherwise indicated, when a vector-valued function is defined by a formula, the domain consists of all numbers for which the formula is meaningful. For instance, vector-valued functions are defined by the formulas

$$\mathbf{F}(t) = (2 + 3t)\mathbf{i} + (-1 + t)\mathbf{j} + 2\mathbf{k}$$

$$\mathbf{G}(t) = \mathbf{i} + \sqrt{1 + t}\,\mathbf{j} + t\mathbf{k}$$

$$\mathbf{H}(t) = a(t - \sin t)\mathbf{i} + a(1 + \cos t)\mathbf{j} \quad \text{for } 0 \leq t \leq 4\pi$$

and the assumed domains of **F**, **G**, and **H** are $(-\infty, \infty)$, $[-1, \infty)$, and $[0, 4\pi]$, respectively.

From now on we will refer to any function whose domain and range are sets of real numbers as a **real-valued function,** in order to distinguish it from a vector-valued function. Every vector-valued function **F** corresponds to three real-valued functions f_1, f_2, and f_3 in the following way: for each number *t* in the domain of **F**, let $f_1(t), f_2(t)$, and $f_3(t)$ denote the **i**, **j**, and **k** components, respectively, of the vector **F**(*t*). Then the domain of each of the functions f_1, f_2, and f_3 is the same as the domain of **F**, and

$$\mathbf{F}(t) = f_1(t)\mathbf{i} + f_2(t)\mathbf{j} + f_3(t)\mathbf{k} \quad \text{for } t \text{ in the domain of } \mathbf{F}$$

The functions f_1, f_2, and f_3 are the **component functions** of **F**.

EXAMPLE 1   Let

$$\mathbf{F}(t) = \ln t\mathbf{i} + \sqrt{1 - t}\,\mathbf{j} + t^4\mathbf{k}$$

Determine the domain of **F** and its component functions.

Solution   The domain of **F** consists of those values of *t* for which $\ln t$, $\sqrt{1 - t}$, and t^4 are all defined, which is the interval $(0, 1]$. By observation we see that

$$f_1(t) = \ln t, \quad f_2(t) = \sqrt{1 - t}, \quad \text{and} \quad f_3(t) = t^4 \quad \square$$

Usually it is impossible to draw the graph of a vector-valued function, because in theory we would need four dimensions to do so (one dimension for the domain and three dimensions for the range). However, we can depict a vector-valued function **F** by drawing only its range. If we think of **F**(*t*) as a point in space, then as *t* increases, **F**(*t*) traces out a curve *C* in space (Figure 12.1), the arrow on the curve indicating the direction in which the curve is traced out as *t* increases. The range of **F** is associated with the curve *C*. If

$$\mathbf{F}(t) = f_1(t)\mathbf{i} + f_2(t)\mathbf{j} + f_3(t)\mathbf{k} \quad \text{for } t \text{ in the domain of } \mathbf{F}$$

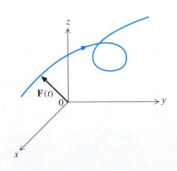

FIGURE 12.1

then we can describe C by the **parametric equations**

$$x = f_1(t), \quad y = f_2(t), \quad z = f_3(t)$$

Thus a curve in space can be represented either by a vector-valued function or by a set of parametric equations.

In case $f_3(t) = 0$ for all t in the domain of $\mathbf{F}$, the curve lies in the xy plane and can be described by the parametric equations

$$x = f_1(t) \quad \text{and} \quad y = f_2(t)$$

as discussed in Section 6.7.

EXAMPLE 2   Let

$$\mathbf{F}(t) = (2 + 3t)\mathbf{i} + (-1 + t)\mathbf{j} - 2t\mathbf{k}$$

Sketch the curve traced out by $\mathbf{F}$.

Solution   If we let (x, y, z) be the point on the curve corresponding to $\mathbf{F}(t)$, then

$$x = 2 + 3t, \quad y = -1 + t, \quad \text{and} \quad z = -2t$$

From (3) of Section 11.5 we recognize these as parametric equations of the line containing the point $(2, -1, 0)$ and parallel to the vector $3\mathbf{i} + \mathbf{j} - 2\mathbf{k}$ (Figure 12.2).   ❏

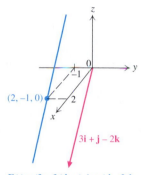

$\mathbf{F}(t) = (2 + 3t)\mathbf{i} + (-1 + t)\mathbf{j} - 2t\mathbf{k}$

FIGURE 12.2

More generally, let

$$\mathbf{F}(t) = (x_0 + at)\mathbf{i} + (y_0 + bt)\mathbf{j} + (z_0 + ct)\mathbf{k}$$

where $x_0, y_0, z_0, a, b,$ and c are constants and $a, b,$ and c are not all 0. An analysis similar to that in Example 2 shows that the curve traced out by $\mathbf{F}$ is the straight line passing through the point (x_0, y_0, z_0) and parallel to the vector $a\mathbf{i} + b\mathbf{j} + c\mathbf{k}$. For that reason $\mathbf{F}$ is called a **linear function.**

EXAMPLE 3   Let

$$\mathbf{F}(t) = \cos t\,\mathbf{i} + \sin t\,\mathbf{j}$$

Show that $\mathbf{F}$ traces out, in the counterclockwise direction, the unit circle in the xy plane with center at the origin.

Solution   The curve traced out by $\mathbf{F}$ lies in the xy plane because the $\mathbf{k}$ component of $\mathbf{F}$ is 0. Since

$$\| \mathbf{F}(t) \| = \sqrt{\cos^2 t + \sin^2 t} = 1 \quad \text{for all } t$$

it follows that the curve lies on the unit circle. Moreover, since $\mathbf{F}(t)$ makes an angle of t radians with the positive x axis and t assumes all real values (Figure 12.3), every point on the circle is $\mathbf{F}(t)$ for some value of t. Consequently the whole circle is traced out by $\mathbf{F}$. Finally, as t increases, the vector $\mathbf{F}(t)$ moves counterclockwise around the circle.   ❏

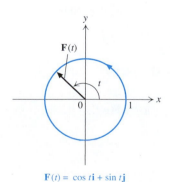

$\mathbf{F}(t) = \cos t\,\mathbf{i} + \sin t\,\mathbf{j}$

FIGURE 12.3

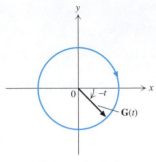

$$G(t) = \cos t\mathbf{i} - \sin t\mathbf{j}$$

FIGURE 12.4

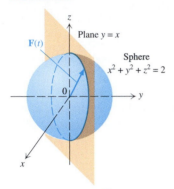

$$\mathbf{F}(t) = \cos t\mathbf{i} + \cos t\mathbf{j} + \sqrt{2}\sin t\mathbf{k}$$

FIGURE 12.5

Combinations of Vector-Valued Functions

From now on we will refer to the circle described in Example 3 as the **standard unit circle.** It will appear several times in this chapter.

You can check that the unit circle is traced out in the clockwise direction by the function **G**, where

$$\mathbf{G}(t) = \cos t\mathbf{i} - \sin t\mathbf{j} \tag{1}$$

(Figure 12.4).

EXAMPLE 4 Let

$$\mathbf{F}(t) = \cos t\mathbf{i} + \cos t\mathbf{j} + \sqrt{2}\sin t\mathbf{k}$$

Sketch the curve traced out by **F**.

Solution In order to understand the curve, we first notice that if (x, y, z) is the point on the curve corresponding to $\mathbf{F}(t)$, then

$$x = \cos t, \quad y = \cos t, \quad \text{and} \quad z = \sqrt{2}\sin t$$

Since $x = y$ for all t, every point on the curve lies on the plane $x = y$. It might not be so obvious that every point on the curve also lies on a sphere. To show this we observe that

$$x^2 + y^2 + z^2 = \cos^2 t + \cos^2 t + 2\sin^2 t = 2(\cos^2 t + \sin^2 t) = 2$$

Therefore the point (x, y, z) is on the sphere of radius $\sqrt{2}$. The intersection of the plane and the sphere is a circle. Because the plane $x = y$ passes through the center of the sphere $x^2 + y^2 + z^2 = 2$, the circle is a great circle of the sphere. Again it can be shown that every point on the circle is actually traced out by **F** (Figure 12.5). ❑

Before sketching additional curves traced out by vector-valued functions, we define several combinations of vector-valued functions.

DEFINITION 12.2   Let **F** and **G** be vector-valued functions and f and g real-valued functions. Then the functions $\mathbf{F} + \mathbf{G}$, $\mathbf{F} - \mathbf{G}$, $f\mathbf{F}$, $\mathbf{F} \cdot \mathbf{G}$, $\mathbf{F} \times \mathbf{G}$, and $\mathbf{F} \circ g$ are defined by

$$(\mathbf{F} + \mathbf{G})(t) = \mathbf{F}(t) + \mathbf{G}(t) \qquad\qquad (\mathbf{F} \cdot \mathbf{G})(t) = \mathbf{F}(t) \cdot \mathbf{G}(t)$$

$$(\mathbf{F} - \mathbf{G})(t) = \mathbf{F}(t) - \mathbf{G}(t) \qquad\qquad (\mathbf{F} \times \mathbf{G})(t) = \mathbf{F}(t) \times \mathbf{G}(t)$$

$$(f\mathbf{F})(t) = f(t)\mathbf{F}(t) \qquad\qquad (\mathbf{F} \circ g)(t) = \mathbf{F}(g(t))$$

EXAMPLE 5 Let

$$\mathbf{F}(t) = \cos t\mathbf{i} + \sin t\mathbf{j} + t\mathbf{k} \quad \text{and} \quad \mathbf{G}(t) = -\sin t\mathbf{i} + \cos t\mathbf{j} + t\mathbf{k}$$

If $g(t) = \sqrt{t}$, find $\mathbf{F} + \mathbf{G}$, $\mathbf{F} - \mathbf{G}$, $g\mathbf{F}$, $\mathbf{F} \cdot \mathbf{G}$, $\mathbf{F} \times \mathbf{G}$, and $\mathbf{F} \circ g$.

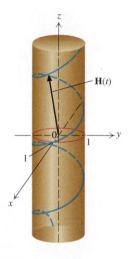

FIGURE 12.6   A circular helix:
$\mathbf{H}(t) = \cos t\mathbf{i} + \sin t\mathbf{j} + t\mathbf{k}$.

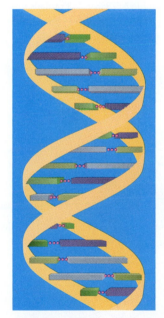

In 1962 Francis Crick, James
Watson, and Maurice Wilkins won
the Nobel Prize for their discovery
that the DNA molecule is in the
shape of a double helix (two
intertwining helical strands of
nucleic acid polymer linked
through hydrogen bonds).
(Illustration by Lili Robins)

Solution   From the definitions we find that

$$(\mathbf{F} + \mathbf{G})(t) = (\cos t - \sin t)\mathbf{i} + (\sin t + \cos t)\mathbf{j} + 2t\mathbf{k}$$

$$(\mathbf{F} - \mathbf{G})(t) = (\cos t + \sin t)\mathbf{i} + (\sin t - \cos t)\mathbf{j}$$

$$(g\mathbf{F})(t) = \sqrt{t}\cos t\mathbf{i} + \sqrt{t}\sin t\mathbf{j} + t^{3/2}\mathbf{k}$$

$$(\mathbf{F} \cdot \mathbf{G})(t) = -\cos t \sin t + \sin t \cos t + t^2 = t^2$$

$$(\mathbf{F} \times \mathbf{G})(t) = t(\sin t - \cos t)\mathbf{i} - t(\sin t + \cos t)\mathbf{j} + \mathbf{k}$$

$$(\mathbf{F} \circ g)(t) = \mathbf{F}(g(t)) = \cos \sqrt{t}\,\mathbf{i} + \sin \sqrt{t}\,\mathbf{j} + \sqrt{t}\,\mathbf{k}$$

The domain of each function except $g\mathbf{F}$ and $\mathbf{F} \circ g$ consists of all real numbers; the
domains of $g\mathbf{F}$ and $\mathbf{F} \circ g$ consist of $[0, \infty)$, since this is the domain of g.   ❑

In the next example we decompose a vector-valued function into the sum of
two other vector-valued functions in order to draw the curve traced out.

EXAMPLE 6   Let

$$\mathbf{H}(t) = \cos t\mathbf{i} + \sin t\mathbf{j} + t\mathbf{k}$$

Sketch the curve traced out by $\mathbf{H}$.

Solution   If we let

$$\mathbf{F}(t) = \cos t\mathbf{i} + \sin t\mathbf{j} \quad \text{and} \quad \mathbf{G}(t) = t\mathbf{k}$$

then we have

$$\mathbf{H}(t) = \mathbf{F}(t) + \mathbf{G}(t) = \mathbf{F}(t) + t\mathbf{k}$$

so that the point corresponding to $\mathbf{H}(t)$ lies $|t|$ units directly above or below the
point corresponding to $\mathbf{F}(t)$. Since we already know that the curve traced out by $\mathbf{F}$
is the standard unit circle (see Figure 12.3), the curve traced out by $\mathbf{H}$ is as shown
in Figure 12.6.   ❑

The curve described in Example 6 is called a **circular helix.** Notice that it lies
on the circular cylinder $x^2 + y^2 = 1$. We mention that not only is a helix found in the
makeup of a DNA molecule (in fact, a double helix!), but also the path of a charged
particle moving in a uniform magnetic field is either linear or helical.

For our final example we return to the **cycloid.** As discussed in Section 6.7, the
cycloid is the curve traced out by a point on a circle as the circle rolls along a line.
Here we will derive a formula for the cycloid.

EXAMPLE 7   A circle of radius r rolls along the x axis in the positive direction,
rotating at the rate of 1 radian per unit of time. Let P be the point on the circle that
is at the origin at time 0. Find a vector-valued function $\mathbf{F}$ that traces out the curve
described by P.

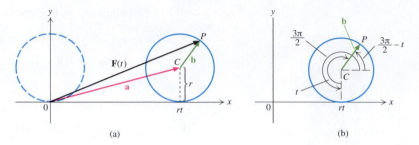

FIGURE 12.7

Solution   As can be seen from Figure 12.7(a), the vector $\mathbf{F}(t)$ from the origin to P at time t can be written as $\mathbf{a} + \mathbf{b}$, where $\mathbf{a}$ points from the origin to the center C of the circle, and $\mathbf{b}$ points from C to P. Since the circle rotates t radians in t units of time and has radius r, the circle travels a distance rt along the x axis in t units of time (see (1) of Section 1.7). Thus the $\mathbf{i}$ component of $\mathbf{a}$ is rt. Since C is r units from the x axis, the $\mathbf{j}$ component of $\mathbf{a}$ is r. Consequently $\mathbf{a} = rt\mathbf{i} + r\mathbf{j}$. In order to express $\mathbf{b}$ in terms of t, we first use the hypothesis that the circle rotates at the rate of 1 radian per unit of time, so that t is the number of radians through which $\overrightarrow{CP}$ has rotated since time 0. Therefore at time t the vector $\overrightarrow{CP}$ makes an angle of $(3\pi/2 - t)$ radians with the positive x axis (Figure 12.7(b)). Since the length of $\mathbf{b}$ is the radius r of the circle, we conclude that $\mathbf{b} = \overrightarrow{CP} = r\cos(3\pi/2 - t)\mathbf{i} + r\sin(3\pi/2 - t)\mathbf{j}$. Consequently

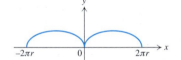

FIGURE 12.8   Two arches of a cycloid.

$$\mathbf{F}(t) = \mathbf{a} + \mathbf{b} = (rt\mathbf{i} + r\mathbf{j}) + r\cos\left(\frac{3}{2}\pi - t\right)\mathbf{i} + r\sin\left(\frac{3}{2}\pi - t\right)\mathbf{j}$$

$$= r(t - \sin t)\mathbf{i} + r(1 - \cos t)\mathbf{j} \quad \square$$

Two arches of a cycloid are shown in Figure 12.8.

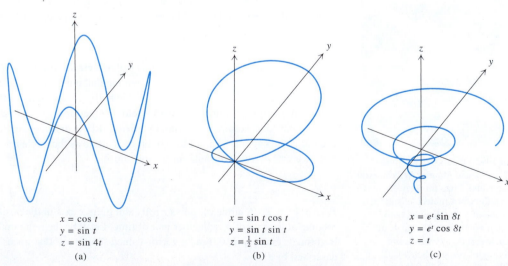

$x = \cos t$	$x = \sin t \cos t$	$x = e^t \sin 8t$
$y = \sin t$	$y = \sin t \sin t$	$y = e^t \cos 8t$
$z = \sin 4t$	$z = \frac{1}{2}\sin t$	$z = t$
(a)	(b)	(c)

FIGURE 12.9   Some curves produced by Mathematica.

So far the curves appearing in the text have been fairly easy to sketch. More complicated curves can be drawn by using computer software packages. In Figure 12.9 we display three curves produced by Mathematica.

EXERCISES 12.1

In Exercises 1–12 determine the domain and the component functions of the given function.

1. $\mathbf{F}(t) = t\mathbf{i} + t^2\mathbf{j} + t^3\mathbf{k}$

2. $\mathbf{F}(t) = \sqrt{t + 1}\,\mathbf{i} + \sqrt{1 - t}\,\mathbf{j} + \mathbf{k}$

3. $\mathbf{F}(t) = \tanh t\mathbf{i} - \dfrac{1}{t^2 - 4}\mathbf{k}$

4. $\mathbf{F}(t) = [(t^2 - 1)\mathbf{i} + \ln t\mathbf{j} + \cot t\mathbf{k}]$

$$\times \left[(4 - t^2)\mathbf{i} + e^{-5t}\mathbf{j} + \frac{1}{t}\mathbf{k} \right]$$

5. $\mathbf{F}(t) = (t\mathbf{i} + \mathbf{j}) \times (\mathbf{i} - t^2\mathbf{j} + 2\sqrt{t}\,\mathbf{k})$

6. $\mathbf{F} - \mathbf{G}$, where $\mathbf{F}(t) = 2t\mathbf{i} + t^2\mathbf{j} - \ln t\mathbf{k}$ and $\mathbf{G}(t) = e^t\mathbf{i} + e^{-t}\mathbf{j} + 2t\mathbf{k}$

7. $2\mathbf{F} - 3\mathbf{G}$, where $\mathbf{F}(t) = t\mathbf{i} + t^2\mathbf{j} + t^3\mathbf{k}$ and $\mathbf{G}(t) = \cos t\mathbf{i} + \sin t\mathbf{j} + \mathbf{k}$

8. $\mathbf{F} \times \mathbf{G}$, where $\mathbf{F}(t) = t\mathbf{i} + t^2\mathbf{j} + t^3\mathbf{k}$ and $\mathbf{G}(t) = \cos t\mathbf{i} + \sin t\mathbf{j} + \mathbf{k}$

9. $\mathbf{F} \times \mathbf{G}$, where $\mathbf{F}(t) = t\mathbf{j} - (1/\sqrt{t})\mathbf{k}$ and $\mathbf{G}(t) = (t - \sin t)\mathbf{i} + (1 - \cos t)\mathbf{j}$

10. $f\mathbf{F}$, where $\mathbf{F}(t) = \ln t\mathbf{i} - 4e^{2t}\mathbf{j} + (\sqrt{t - 1}/t)\mathbf{k}$ and $f(t) = \sqrt{t}$

11. $\mathbf{F} \circ g$, where $\mathbf{F}(t) = \cos t\mathbf{i} + \sin t\mathbf{j} + \sqrt{t + 2}\,\mathbf{k}$ and $g(t) = t^{1/3}$

12. $\mathbf{F} \circ g$, where $\mathbf{F}(t) = e^{-2t}\mathbf{i} + e^{(t^2)}\mathbf{j} + t^3\mathbf{k}$ and $g(t) = \ln t$

In Exercises 13–26 sketch the curve traced out by the vector-valued function. Indicate the direction in which the curve is traced out.

13. $\mathbf{F}(t) = t\mathbf{i}$ for $-1 \leq t \leq \frac{1}{2}$

14. $\mathbf{F}(t) = \cos \pi t\mathbf{k}$ for $-1 \leq t \leq \frac{1}{3}$

15. $\mathbf{F}(t) = t\mathbf{i} + t\mathbf{j} + t\mathbf{k}$

16. $\mathbf{F}(t) = 2t\mathbf{i} - 3t\mathbf{j} + \mathbf{k}$

17. $\mathbf{F}(t) = (2t - 1)\mathbf{i} + (t + 1)\mathbf{j} + 3t\mathbf{k}$

18. $\mathbf{F}(t) = (1 - t)\mathbf{i} + (3t - \frac{1}{2})\mathbf{j} - (-4 + t)\mathbf{k}$

19. $\mathbf{F}(t) = -16t^2\mathbf{k}$ for $t \geq 0$

20. $\mathbf{F}(t) = t\mathbf{j} + t^2\mathbf{k}$

21. $\mathbf{F}(t) = \cos 3t\mathbf{i} + \sin 3t\mathbf{j}$ for $0 \leq t \leq \pi/2$

22. $\mathbf{F}(t) = 2\cos t\mathbf{i} - \sin t\mathbf{j} - 3\mathbf{k}$ for $-\pi \leq t \leq 0$

23. $\mathbf{F}(t) = \cos t\mathbf{i} + t\mathbf{j} - \sin t\mathbf{k}$ for $0 \leq t \leq 2\pi$

24. $\mathbf{F}(t) = \cos t\mathbf{i} + \sin t\mathbf{j} + t^2\mathbf{k}$

25. $\mathbf{F}(t) = 3\sin t\mathbf{i} + 3\sin t\mathbf{j} - 3\sqrt{2}\cos t\mathbf{k}$

26. $\mathbf{F}(t) = \sqrt{2}\cos t\mathbf{i} - 2\sin t\mathbf{j} + \sqrt{2}\cos t\mathbf{k}$

In Exercises 27–30 plot the curve traced out by the vector-valued function. Indicate the direction in which the curve is traced out.

27. $\mathbf{F}(t) = \cos 12t\mathbf{i} + \sin 12t\mathbf{j} + t\mathbf{k}$ for $0 \leq t \leq 2\pi$

28. $\mathbf{F}(t) = t\cos 8t\mathbf{i} + t\sin 8t\mathbf{j} + 2t\mathbf{k}$ for $0 \leq t \leq 2\pi$

29. $\mathbf{F}(t) = t\mathbf{i} + (1 - t^2)\cos 18t\mathbf{j} + (1 - t^2)\sin 18t\mathbf{k}$ for $-1 \leq t \leq 1$

30. $\mathbf{F}(t) = \cos t\mathbf{i} + \sin 3t\mathbf{j} + \sin 4t\mathbf{k}$ for $0 \leq t \leq 2\pi$

31. Let $\mathbf{F}(t) = 2t\mathbf{i} + (t + 1)\mathbf{j} - 3t\mathbf{k}$, $\mathbf{G}(t) = (t + 1)\mathbf{i} + (3t - 2)\mathbf{j} + t\mathbf{k}$, and $g(t) = \cos t$. Show that the curves traced out by the following functions are lines or line segments by finding parametric equations for each.

    a. $\mathbf{F} - \mathbf{G}$ **b.** $\mathbf{F} + 3\mathbf{G}$ **c.** $\mathbf{F} \circ g$

32. Find the points of intersection of the cylinder $x^2 + y^2 = 4$ and the curve traced out by

$$\mathbf{F}(t) = t\cos \pi t\mathbf{i} + t\sin \pi t\mathbf{j} + t\mathbf{k}$$

33. Find the points of intersection of the sphere $x^2 + y^2 + z^2 = 10$ and the curve traced out by

$$\mathbf{F}(t) = \cos \pi t\mathbf{i} + \sin \pi t\mathbf{j} + t\mathbf{k}$$

34. Let $\mathbf{r}(t) = \cos t\mathbf{i} + \sin t\mathbf{j} + (a\cos t + b\sin t)\mathbf{k}$ for $0 \leq t \leq 2\pi$, where a and b are real numbers with $ab \neq 0$. Show that the curve C traced out by $\mathbf{r}$ lies in the intersection of a plane and a circular cylinder. (It can be shown that C is therefore an ellipse.)

35. A disk of radius r rolls at the rate of 1 radian per unit of time along a line. The curve traced out by a point P on the

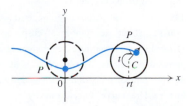

FIGURE 12.10   A trochoid is traced out by P as the circle rolls along the x axis. See Exercise 35.

disk located at a distance b from the center of the disk is called a **trochoid**. (If $b = r$, the trochoid is a cycloid.) Using the method employed in Example 7, show that the trochoid is traced out by the vector-valued function **F** defined by

$$\mathbf{F}(t) = (rt - b \sin t)\mathbf{i} + (r - b \cos t)\mathbf{j}$$

(*Hint:* Assume that at time $t = 0$, both the point P and the center of the disk lie on the positive y axis, as shown in Figure 12.10.)

36. Suppose a circle of radius r rolls at the rate of 1 radian per unit of time along the inside of a circle of radius $b > r$. The curve traced out by a point on the circumference of the rolling circle is a **hypocycloid** (Figure 12.11).

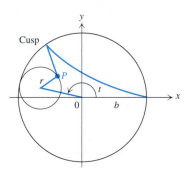

FIGURE 12.11 Graph for Exercise 36.

a. Show that the hypocycloid is traced out by the vector-valued function **F**, where

$$\mathbf{F}(t) = \left[(b - r) \cos t + r \cos\left(\frac{b - r}{r} t\right)\right]\mathbf{i}$$
$$+ \left[(b - r) \sin t - r \sin\left(\frac{b - r}{r} t\right)\right]\mathbf{j}$$

b. Show that if $b = 2r$, then the curve traced out by **F** is a straight line. (This provides a method of converting circular motion to oscillatory linear motion, which is important in engineering.)

 c. Let $b = nr$ for some positive integer $n > 2$. What can you say about the number of cusps on the graph? (*Hint:* Plot the graph for select values of n.)

*37. Suppose that a circle of radius b rolls at the rate of 1 radian per unit of time along the outside of a circle of radius r. The curve traced out by a point on the circumference of the rolling circle is called an **epicycloid** (Figure 12.12). Show that the epicycloid is traced out by

the vector-valued function **F**, where

$$\mathbf{F}(t) = \left[(r + b) \cos t - b \cos\left(\frac{r + b}{b} t\right)\right]\mathbf{i}$$
$$+ \left[(r + b) \sin t - b \sin\left(\frac{r + b}{b} t\right)\right]\mathbf{j}$$

If $b = r$, the epicycloid is called a **cardioid** (Figure 12.13).

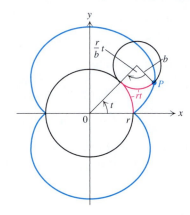

FIGURE 12.12 An epicycloid with $b = \frac{1}{2}r$. See Exercise 37.

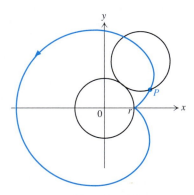

FIGURE 12.13 A cardioid. See Exercise 37.

38. A helical staircase circles once around a cylindrical water tower 200 feet tall and 100 feet in diameter. Find a formula for a vector-valued function that represents the staircase. (*Hint:* Place the coordinate system so that the axis of the cylinder is the z axis.)

12.2   LIMITS AND CONTINUITY OF VECTOR-VALUED FUNCTIONS

The definitions of limits and continuity for vector-valued functions are virtually identical with the corresponding definitions for real-valued functions.

DEFINITION 12.3

Let $\mathbf{F}$ be a vector-valued function defined at each point in some open interval containing t_0, except possibly at t_0 itself. A vector $\mathbf{L}$ is the **limit of $\mathbf{F}(t)$ as t approaches t_0** (or $\mathbf{L}$ is the **limit of $\mathbf{F}$ at t_0**) if for every $\varepsilon > 0$ there is a number $\delta > 0$ such that

$$\text{if } 0 < |\,t - t_0\,| < \delta, \quad \text{then } \|\,\mathbf{F}(t) - \mathbf{L}\,\| < \varepsilon$$

In this case we write

$$\lim_{t \to t_0} \mathbf{F}(t) = \mathbf{L}$$

and say that $\lim_{t \to t_0} \mathbf{F}(t)$ **exists.**

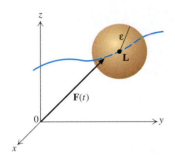

FIGURE 12.14

To say that $\lim_{t \to t_0} \mathbf{F}(t) = \mathbf{L}$ means intuitively that we can make $\mathbf{F}(t)$ as close to $\mathbf{L}$ as we wish by taking t sufficiently close to t_0. If we think of $\mathbf{F}(t)$ and $\mathbf{L}$ as points in space, then we have $\|\,\mathbf{F}(t) - \mathbf{L}\,\| < \varepsilon$ if and only if $\mathbf{F}(t)$ lies in the open ball of radius ε and center $\mathbf{L}$ (Figure 12.14). Therefore Definition 12.3 can be interpreted geometrically as follows: $\lim_{t \to t_0} \mathbf{F}(t) = \mathbf{L}$ if for any open ball with center $\mathbf{L}$ there is an open interval I about t_0 such that $\mathbf{F}$ assigns to each number in I (except possibly t_0) a point in the ball.

In the next theorem we establish the useful fact that the limit of a vector-valued function can be determined from the limits of its component functions.

THEOREM 12.4

Let $\mathbf{F}(t) = f_1(t)\mathbf{i} + f_2(t)\mathbf{j} + f_3(t)\mathbf{k}$. Then $\mathbf{F}$ has a limit at t_0 if and only if $f_1, f_2,$ and f_3 have limits at t_0. In that case

$$\lim_{t \to t_0} \mathbf{F}(t) = \left[\lim_{t \to t_0} f_1(t) \right] \mathbf{i} + \left[\lim_{t \to t_0} f_2(t) \right] \mathbf{j} + \left[\lim_{t \to t_0} f_3(t) \right] \mathbf{k}$$

Proof   First assume that $\lim_{t \to t_0} \mathbf{F}(t) = \mathbf{L} = a\mathbf{i} + b\mathbf{j} + c\mathbf{k}$. This means that for any $\varepsilon > 0$ there is a number $\delta > 0$ such that

$$\text{if } 0 < |\,t - t_0\,| < \delta, \quad \text{then } \|\,\mathbf{F}(t) - \mathbf{L}\,\| < \varepsilon$$

For such values of t we find that

$$|\,f_1(t) - a\,| = \sqrt{(f_1(t) - a)^2} \le \sqrt{(f_1(t) - a)^2 + (f_2(t) - b)^2 + (f_3(t) - c)^2}$$

$$= \|\,\mathbf{F}(t) - \mathbf{L}\,\| < \varepsilon$$

Therefore if $0 < |\,t - t_0\,| < \delta$, then $|\,f_1(t) - a\,| < \varepsilon$. This verifies that $\lim_{t \to t_0} f_1(t) = a$. Similar arguments can be used to show that $\lim_{t \to t_0} f_2(t) = b$ and $\lim_{t \to t_0} f_3(t) = c$. Conversely, assume that

$$\lim_{t \to t_0} f_1(t) = a, \quad \lim_{t \to t_0} f_2(t) = b, \quad \lim_{t \to t_0} f_3(t) = c$$

and let

$$L = a\mathbf{i} + b\mathbf{j} + c\mathbf{k}$$

This means that for any $\varepsilon > 0$ there is a number $\delta > 0$ such that if $0 < |t - t_0| < \delta$, then

$$|f_1(t) - a| < \frac{\varepsilon}{\sqrt{3}}, \quad |f_2(t) - b| < \frac{\varepsilon}{\sqrt{3}}, \quad \text{and} \quad |f_3(t) - c| < \frac{\varepsilon}{\sqrt{3}}$$

Hence if $0 < |t - t_0| < \delta$, then

$$\| \mathbf{F}(t) - \mathbf{L} \| = \sqrt{(f_1(t) - a)^2 + (f_2(t) - b)^2 + (f_3(t) - c)^2}$$

$$< \sqrt{\left(\frac{\varepsilon}{\sqrt{3}}\right)^2 + \left(\frac{\varepsilon}{\sqrt{3}}\right)^2 + \left(\frac{\varepsilon}{\sqrt{3}}\right)^2} = \varepsilon$$

It follows from Definition 12.3 that $\lim_{t \to t_0} \mathbf{F}(t)$ exists. Moreover,

$$\lim_{t \to t_0} \mathbf{F}(t) = \mathbf{L} = a\mathbf{i} + b\mathbf{j} + c\mathbf{k} = \lim_{t \to t_0} f_1(t)\mathbf{i} + \lim_{t \to t_0} f_2(t)\mathbf{j} + \lim_{t \to t_0} f_3(t)\mathbf{k} \quad ■$$

EXAMPLE 1   Find

$$\lim_{t \to 0} \left(2 \cos t\mathbf{i} + \frac{\sin t}{t}\mathbf{j} + t^2\mathbf{k} \right)$$

Solution   By Theorem 12.4 we may evaluate the limit componentwise:

$$\lim_{t \to 0} \left(2 \cos t\mathbf{i} + \frac{\sin t}{t}\mathbf{j} + t^2\mathbf{k} \right) = \left(\lim_{t \to 0} 2 \cos t \right)\mathbf{i} + \left(\lim_{t \to 0} \frac{\sin t}{t} \right)\mathbf{j} + \left(\lim_{t \to 0} t^2 \right)\mathbf{k}$$

$$= 2\mathbf{i} + \mathbf{j} \quad □$$

Theorem 12.4 also provides an easy way of finding formulas for limits of combinations of vector-valued functions.

THEOREM 12.5   Let $\mathbf{F}$ and $\mathbf{G}$ be vector-valued functions, and let f and g be real-valued functions. Assume that $\lim_{t \to t_0} \mathbf{F}(t)$ and $\lim_{t \to t_0} \mathbf{G}(t)$ exist and that $\lim_{t \to t_0} f(t)$ exists and $\lim_{s \to s_0} g(s) = t_0$. Then

a. $\lim_{t \to t_0} (\mathbf{F} + \mathbf{G})(t) = \lim_{t \to t_0} \mathbf{F}(t) + \lim_{t \to t_0} \mathbf{G}(t)$

b. $\lim_{t \to t_0} (\mathbf{F} - \mathbf{G})(t) = \lim_{t \to t_0} \mathbf{F}(t) - \lim_{t \to t_0} \mathbf{G}(t)$

c. $\lim_{t \to t_0} f\mathbf{F}(t) = \lim_{t \to t_0} f(t) \lim_{t \to t_0} \mathbf{F}(t)$

d. $\lim_{t \to t_0} (\mathbf{F} \cdot \mathbf{G})(t) = \lim_{t \to t_0} \mathbf{F}(t) \cdot \lim_{t \to t_0} \mathbf{G}(t)$

e. $\lim_{t \to t_0} (\mathbf{F} \times \mathbf{G})(t) = \lim_{t \to t_0} \mathbf{F}(t) \times \lim_{t \to t_0} \mathbf{G}(t)$

f. $\lim_{s \to s_0} (\mathbf{F} \circ g)(s) = \lim_{t \to t_0} \mathbf{F}(t)$ if $g(s) \neq t_0$ for all s in an open interval about s_0

Proof   We prove only (d); the proofs of (a), (b), (c), (e), and (f) are similar. Let

$$\mathbf{F}(t) = f_1(t)\mathbf{i} + f_2(t)\mathbf{j} + f_3(t)\mathbf{k} \quad \text{and} \quad \mathbf{G}(t) = g_1(t)\mathbf{i} + g_2(t)\mathbf{j} + g_3(t)\mathbf{k}$$

To evaluate $\lim_{t \to t_0} (\mathbf{F} \cdot \mathbf{G})(t)$, we note that by Theorem 12.4, f_1, f_2, f_3, g_1, g_2, and g_3 have limits at t_0, so we can employ the limit rules of Chapter 2:

$$\lim_{t \to t_0} (\mathbf{F} \cdot \mathbf{G})(t) = \lim_{t \to t_0} [f_1(t)g_1(t) + f_2(t)g_2(t) + f_3(t)g_3(t)]$$

$$= \lim_{t \to t_0} f_1(t) \lim_{t \to t_0} g_1(t) + \lim_{t \to t_0} f_2(t) \lim_{t \to t_0} g_2(t) + \lim_{t \to t_0} f_3(t) \lim_{t \to t_0} g_3(t)$$

To evaluate $\lim_{t \to t_0} \mathbf{F}(t) \cdot \lim_{t \to t_0} \mathbf{G}(t)$, we first apply Theorem 12.4:

$$\lim_{t \to t_0} \mathbf{F}(t) \cdot \lim_{t \to t_0} \mathbf{G}(t) = \left(\lim_{t \to t_0} f_1(t)\mathbf{i} + \lim_{t \to t_0} f_2(t)\mathbf{j} + \lim_{t \to t_0} f_3(t)\mathbf{k} \right)$$

$$\cdot \left(\lim_{t \to t_0} g_1(t)\mathbf{i} + \lim_{t \to t_0} g_2(t)\mathbf{j} + \lim_{t \to t_0} g_3(t)\mathbf{k} \right)$$

Multiplying out the dot product on the right verifies that

$$\lim_{t \to t_0} (\mathbf{F} \cdot \mathbf{G})(t) = \lim_{t \to t_0} \mathbf{F}(t) \cdot \lim_{t \to t_0} \mathbf{G}(t) \quad ■$$

EXAMPLE 2   Let

$$\mathbf{F}(t) = \cos \pi t \mathbf{i} + 2 \sin \pi t \mathbf{j} + 4t^2 \mathbf{k} \quad \text{and} \quad \mathbf{G}(t) = t\mathbf{i} + t^3\mathbf{k}$$

Find $\lim_{t \to 1} (\mathbf{F} \cdot \mathbf{G})(t)$ and $\lim_{t \to 1} (\mathbf{F} \times \mathbf{G})(t)$.

Solution   For each of the required limits we have a choice of methods. We can (1) find the product of $\mathbf{F}$ and $\mathbf{G}$ and take the limit of the product, or (2) find $\lim_{t \to 1} \mathbf{F}(t)$ and $\lim_{t \to 1} \mathbf{G}(t)$, take the product of these limits, and apply Theorem 12.5. For $\lim_{t \to 1} (\mathbf{F} \cdot \mathbf{G})(t)$ we use method (1). Since

$$(\mathbf{F} \cdot \mathbf{G})(t) = (\cos \pi t \mathbf{i} + 2 \sin \pi t \mathbf{j} + 4t^2 \mathbf{k}) \cdot (t\mathbf{i} + t^3\mathbf{k}) = t \cos \pi t + 4t^5$$

it follows that

$$\lim_{t \to 1} (\mathbf{F} \cdot \mathbf{G})(t) = \lim_{t \to 1} (t \cos \pi t + 4t^5) = \cos \pi + 4 = 3$$

For $\lim_{t \to 1} (\mathbf{F} \times \mathbf{G})(t)$ we use method (2). By Theorem 12.4,

$$\lim_{t \to 1} \mathbf{F}(t) = \cos \pi \mathbf{i} + 2 \sin \pi \mathbf{j} + 4\mathbf{k} = -\mathbf{i} + 4\mathbf{k}$$

and

$$\lim_{t \to 1} \mathbf{G}(t) = \mathbf{i} + \mathbf{k}$$

Therefore by Theorem 12.5(e) we deduce that

$$\lim_{t \to 1} (\mathbf{F} \times \mathbf{G})(t) = (-\mathbf{i} + 4\mathbf{k}) \times (\mathbf{i} + \mathbf{k}) = 5\mathbf{j} \quad □$$

Now that we have the concept of limit for vector-valued functions, we can define continuity for such functions. Again, our definition is essentially the same as for real-valued functions.

DEFINITION 12.6   A vector-valued function $\mathbf{F}$ is **continuous** at a point t_0 in its domain if

$$\lim_{t \to t_0} \mathbf{F}(t) = \mathbf{F}(t_0)$$

Since continuity is defined in terms of limits and since limits can be computed componentwise (by Theorem 12.4), you can easily prove the following theorem.

THEOREM 12.7   A vector-valued function $\mathbf{F}$ is continuous at t_0 if and only if each of its component functions is continuous at t_0.

Theorem 12.7 makes it easy to define continuity on an open or closed interval I: We say that a vector-valued function $\mathbf{F}$ is **continuous on I** if the component functions of $\mathbf{F}$ are continuous on I (see Definition 2.12).

EXERCISES 12.2

In Exercises 1–10 compute the limit or explain why it does not exist.

1. $\lim_{t \to 4} (\mathbf{i} - \mathbf{j} + \mathbf{k})$ **2.** $\lim_{t \to -1} (3\mathbf{i} + t\mathbf{j} + t^5\mathbf{k})$

3. $\lim_{t \to \pi} (\tan t\mathbf{i} + 3t\mathbf{j} - 4\mathbf{k})$

4. $\lim_{t \to 0} \left(\dfrac{\sin t}{t} \mathbf{i} + e^t \mathbf{j} + (t + \sqrt{2})\mathbf{k} \right)$

5. $\lim_{t \to 2} \mathbf{F}(t)$, where

$$\mathbf{F}(t) = \begin{cases} 5\mathbf{i} - \sqrt{2t^2 + 2t + 4}\,\mathbf{j} + e^{-(t-2)}\mathbf{k} & \text{for } t < 2 \\ (t^2 + 1)\mathbf{i} + (4 - t^3)\,\mathbf{j} + \mathbf{k} & \text{for } t > 2 \end{cases}$$

6. $\lim_{t \to 0} \mathbf{F}(t)$, where

$$\mathbf{F}(t) = \begin{cases} t\mathbf{i} + e^{-1/t^2}\mathbf{j} + t^2\mathbf{k} & \text{for } t \neq 0 \\ \mathbf{j} & \text{for } t = 0 \end{cases}$$

7. $\lim_{t \to 0} (\mathbf{F} - \mathbf{G})(t)$, where $\mathbf{F}(t) = e^{-1/t^2} \mathbf{i} + \cos t\mathbf{j} + t^3\mathbf{k}$ and

$$\mathbf{G}(t) = -\pi\mathbf{i} + \dfrac{1 + \cos t}{t} \mathbf{j}$$

8. $\lim_{t \to 1} (\mathbf{F} \cdot \mathbf{G})(t)$, where

$$\mathbf{F}(t) = \dfrac{\sin (t - 1)}{t - 1} \mathbf{i} + \dfrac{t + 3}{t - 2} \mathbf{j} + \cos \pi t\mathbf{k} \quad \text{and}$$

$$\mathbf{G}(t) = (t^2 + 1)\mathbf{i} - \dfrac{t - 2}{t + 3} \mathbf{j} - \sqrt{t^2 + 1}\mathbf{k}$$

9. $\lim_{t \to 3} \left(\dfrac{t^2 - 5t + 6}{t - 3} \mathbf{i} + \dfrac{t^2 - 2t - 3}{t - 3} \mathbf{j} + \dfrac{t^2 + 4t - 21}{t - 3} \mathbf{k} \right)$

10. $\lim_{t \to 1} \left(\dfrac{t^2 + 1}{t - 1} \mathbf{i} + \dfrac{t^2 - 1}{t + 1} \mathbf{j} + \dfrac{t^2 + 7t - 8}{t - 1} \mathbf{k} \right)$

11. a. Formulate a definition of the right-hand limit $\lim_{t \to t_0^+} \mathbf{F}(t)$ analogous to Definition 12.3, and show that the limit can be computed componentwise.

  b. Formulate a definition of the left-hand limit $\lim_{t \to t_0^-} \mathbf{F}(t)$ analogous to Definition 12.3, and show that the limit can be computed componentwise.

12. Use Exercise 11 to compute the limits that exist.

  a. $\lim_{t \to 0^+} \left(t\mathbf{i} + 2t^{1/4}\mathbf{j} - \dfrac{\ln t}{t} \mathbf{k} \right)$

  b. $\lim_{t \to 1^+} (e^{1/(1-t)}\mathbf{i} + \sqrt{t - 1}\,\mathbf{j} + \ln t\mathbf{k})$

  c. $\lim_{t \to 1^-} [\sqrt{1 - t}\mathbf{i} - (1 - t) \ln (1 - t)\mathbf{j}]$

13. a. Formulate a definition of $\lim_{t \to \infty} \mathbf{F}(t)$, and show that the limit can be computed componentwise.

  b. Evaluate

$$\lim_{t \to \infty} \left(\dfrac{1}{t} \mathbf{i} + \dfrac{t - 1}{t + 1} \mathbf{j} + \dfrac{\sin t^3}{t^2} \mathbf{k} \right)$$

14. Let $\mathbf{F}$ and $\mathbf{G}$ be continuous at t_0, and let c be a number. Prove that the following functions are continuous at t_0.
  a. $\mathbf{F} + \mathbf{G}$ **b.** $c\mathbf{F}$
  c. $\| \mathbf{F} \|$ (where $\| \mathbf{F} \| (t) = \| \mathbf{F}(t) \|$ for all t in the domain of $\mathbf{F}$)
  d. $\mathbf{F} \cdot \mathbf{G}$ **e.** $\mathbf{F} \times \mathbf{G}$

12.3   DERIVATIVES AND INTEGRALS OF VECTOR-VALUED FUNCTIONS

Our definition of the derivative of a vector-valued function, like the definitions of limits and continuity of vector-valued functions, is closely related to the corresponding definition for real-valued functions.

DEFINITION 12.8

Let t_0 be a number in the domain of a vector-valued function $\mathbf{F}$. If

$$\lim_{t \to t_0} \frac{\mathbf{F}(t) - \mathbf{F}(t_0)}{t - t_0}$$

exists, we call this limit the **derivative** of $\mathbf{F}$ at t_0 and write

$$\mathbf{F}'(t_0) = \lim_{t \to t_0} \frac{\mathbf{F}(t) - \mathbf{F}(t_0)}{t - t_0}$$

In this case we say that **F has a derivative at t_0**, that **F is differentiable at t_0**, or that **$\mathbf{F}'(t_0)$ exists.**

We define the **derivative of F** to be the vector-valued function $\mathbf{F}'$ whose domain consists of all numbers t such that $\mathbf{F}$ is differentiable at t, and whose value at any number t in its domain is $\mathbf{F}'(t)$. The Leibniz notation for the derivative of $\mathbf{F}$ is $d\mathbf{F}/dt$.

We can interpret $\mathbf{F}'(t_0)$ geometrically as follows. Denote by C the curve traced out by $\mathbf{F}$. Let P and P_0 be the points on C corresponding to $\mathbf{F}(t)$ and $\mathbf{F}(t_0)$, respectively. Then $\mathbf{F}(t) - \mathbf{F}(t_0)$ has the same direction as the "secant vector" $\overrightarrow{P_0P}$. Thus if $t > t_0$, then

$$\frac{\mathbf{F}(t) - \mathbf{F}(t_0)}{t - t_0}$$

also has the same direction as $\overrightarrow{P_0P}$ (Figure 12.15(a)), whereas if $t < t_0$, then it has the opposite direction, that is, the same direction as $\overrightarrow{PP_0}$ (Figure 12.15(b)). Therefore if $\mathbf{F}'(t_0)$ exists, it is the limit of vectors parallel to secant vectors passing through P_0 and pointing in the same general direction in which C is traced out by $\mathbf{F}$. Accordingly, it is reasonable to say that $\mathbf{F}'(t_0)$ is tangent to C at P_0 (Figure 12.15(c)).

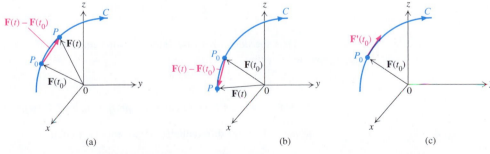

(a)   (b)   (c)

FIGURE 12.15

If $\mathbf{F}(t) = f_1(t)\mathbf{i} + f_2(t)\mathbf{j} + f_3(t)\mathbf{k}$, then

$$\frac{\mathbf{F}(t) - \mathbf{F}(t_0)}{t - t_0} = \frac{f_1(t) - f_1(t_0)}{t - t_0}\mathbf{i} + \frac{f_2(t) - f_2(t_0)}{t - t_0}\mathbf{j} + \frac{f_3(t) - f_3(t_0)}{t - t_0}\mathbf{k}$$

Using Theorem 12.4, we find that $\mathbf{F}'(t_0)$ can be obtained directly from the derivatives of the component functions.

THEOREM 12.9   Let $\mathbf{F}(t) = f_1(t)\mathbf{i} + f_2(t)\mathbf{j} + f_3(t)\mathbf{k}$. Then $\mathbf{F}$ is differentiable at t_0 if and only if f_1, f_2, and f_3 are differentiable at t_0. In that case,

$$\mathbf{F}'(t_0) = f_1'(t_0)\mathbf{i} + f_2'(t_0)\mathbf{j} + f_3'(t_0)\mathbf{k}$$

As a result, finding derivatives of vector-valued functions is as easy (or as difficult) as finding derivatives of real-valued functions. First we discuss the derivative of a constant vector-valued function.

EXAMPLE 1   Let $\mathbf{F}(t) = a\mathbf{i} + b\mathbf{j} + c\mathbf{k}$ for all t. Show that $\mathbf{F}' = \mathbf{0}$.

Solution   Since $f_1(t) = a$, $f_2(t) = b$, and $f_3(t) = c$, the component functions are all constant functions. Consequently

$$\mathbf{F}'(t) = 0\mathbf{i} + 0\mathbf{j} + 0\mathbf{k} = \mathbf{0} \quad \square$$

We can express the result of Example 1 by saying that the derivative of a constant vector-valued function is $\mathbf{0}$. Next we turn to the derivative of a linear vector-valued function.

EXAMPLE 2   Let $\mathbf{F}(t) = (x_0 + at)\mathbf{i} + (y_0 + bt)\mathbf{j} + (z_0 + ct)\mathbf{k}$. Show that $d\mathbf{F}/dt = a\mathbf{i} + b\mathbf{j} + c\mathbf{k}$.

Solution   In this case $f_1(t) = x_0 + at$, $f_2(t) = y_0 + bt$, and $f_3(t) = z_0 + ct$. Therefore Theorem 12.9 tells us immediately that

$$\frac{d\mathbf{F}}{dt} = a\mathbf{i} + b\mathbf{j} + c\mathbf{k} \quad \square$$

Thus the derivative of a linear vector-valued function is a constant vector-valued function.

EXAMPLE 3   Let $\mathbf{F}(t) = t \cos t\mathbf{i} + t \sin t\mathbf{j} + t\mathbf{k}$. Find $\mathbf{F}'(\pi)$.

Solution   First we differentiate componentwise to obtain

$$\mathbf{F}'(t) = (\cos t - t \sin t)\mathbf{i} + (\sin t + t \cos t)\mathbf{j} + \mathbf{k}$$

Then we substitute the value π for t:

$$\mathbf{F}'(\pi) = [-1 - \pi(0)]\mathbf{i} + [0 + \pi(-1)]\mathbf{j} + \mathbf{k} = -\mathbf{i} - \pi\mathbf{j} + \mathbf{k} \quad \square$$

In the remainder of this chapter we will most often encounter vector-valued functions defined on intervals. If I is an interval, then we say that $\mathbf{F}$ is **differentiable on I** if the components of $\mathbf{F}$ are differentiable on I. If I is a closed interval $[a, b]$, this implies in particular that the component functions have appropriate one-sided derivatives at a and b (see (7) of Section 3.2). For example, the function $\mathbf{F}$ defined by

$$\mathbf{F}(t) = |t|\,\mathbf{i} + |1 - t|\,\mathbf{j}$$

is differentiable on $[0, 1]$ but is not differentiable on $[-1, 1]$ since $\mathbf{F}'(0)$ does not exist, or on $[0, 2]$ since $\mathbf{F}'(1)$ does not exist.

Almost all the differentiation rules proved in Chapter 3 have counterparts for vector-valued functions. Since these rules can be proved with simple modifications of the proofs used in Chapter 3, we state them here without proof.

THEOREM 12.10   Let $\mathbf{F}$, $\mathbf{G}$, and f be differentiable at t_0, and let g be differentiable at s_0 with $g(s_0) = t_0$. Then the following hold.

a. $(\mathbf{F} + \mathbf{G})'(t_0) = \mathbf{F}'(t_0) + \mathbf{G}'(t_0)$

b. $(\mathbf{F} - \mathbf{G})'(t_0) = \mathbf{F}'(t_0) - \mathbf{G}'(t_0)$

c. $(f\mathbf{F})'(t_0) = f'(t_0)\mathbf{F}(t_0) + f(t_0)\mathbf{F}'(t_0)$

d. $(\mathbf{F} \cdot \mathbf{G})'(t_0) = \mathbf{F}'(t_0) \cdot \mathbf{G}(t_0) + \mathbf{F}(t_0) \cdot \mathbf{G}'(t_0)$

e. $(\mathbf{F} \times \mathbf{G})'(t_0) = \mathbf{F}'(t_0) \times \mathbf{G}(t_0) + \mathbf{F}(t_0) \times \mathbf{G}'(t_0)$

f. $(\mathbf{F} \circ g)'(s_0) = \mathbf{F}'(g(s_0))g'(s_0) = \mathbf{F}'(t_0)g'(s_0)$

Theorem 12.10 provides us with two methods of finding the derivative of several kinds of combinations of vector-valued functions $\mathbf{F}$ and $\mathbf{G}$. In particular, to find $(\mathbf{F} \cdot \mathbf{G})'(t)$, we can either calculate $(\mathbf{F} \cdot \mathbf{G})(t)$ and then differentiate it to find $(\mathbf{F} \cdot \mathbf{G})'(t)$, or calculate $\mathbf{F}'(t)$ and $\mathbf{G}'(t)$ and then use part (d) of Theorem 12.10 to find $(\mathbf{F} \cdot \mathbf{G})'(t)$. In Example 4 we will apply both methods, and see that the first method is the simpler for that example.

EXAMPLE 4   Let $\mathbf{F}(t) = \tan^{-1} t\mathbf{i} + 5\mathbf{k}$ and $\mathbf{G}(t) = \mathbf{i} + \ln t\mathbf{j} - 2t\mathbf{k}$. Find $(\mathbf{F} \cdot \mathbf{G})'(t)$.

Solution   First we calculate $\mathbf{F} \cdot \mathbf{G}$ and then differentiate. We find that $(\mathbf{F} \cdot \mathbf{G})(t) = \tan^{-1} t - 10t$, and conclude that

$$(\mathbf{F} \cdot \mathbf{G})'(t) = \frac{1}{t^2 + 1} - 10$$

For the second method we first notice that

$$\mathbf{F}'(t) = \frac{1}{t^2 + 1}\mathbf{i} \quad \text{and} \quad \mathbf{G}'(t) = \frac{1}{t}\mathbf{j} - 2\mathbf{k}$$

Then by part (d) of Theorem 12.10 we conclude that

$$(\mathbf{F} \cdot \mathbf{G})'(t) = \mathbf{F}'(t) \cdot \mathbf{G}(t) + \mathbf{F}(t) \cdot \mathbf{G}'(t)$$

$$= \left(\frac{1}{t^2 + 1} \mathbf{i} \right) \cdot (\mathbf{i} + \ln t \, \mathbf{j} - 2t\mathbf{k}) + (\tan^{-1} t\mathbf{i} + 5\mathbf{k}) \cdot \left(\frac{1}{t} \, \mathbf{j} - 2\mathbf{k} \right)$$

$$= \frac{1}{t^2 + 1} - 10$$

Clearly the first method is more efficient in the present case.   ❑

An elementary but significant consequence of Theorem 12.10 is the following.

COROLLARY 12.11   Let $\mathbf{F}$ be differentiable on an interval I, and assume that there is a number c such that

$$\| \mathbf{F}(t) \| = c \quad \text{for } t \text{ in } I$$

Then

$$\mathbf{F}(t) \cdot \mathbf{F}'(t) = 0 \quad \text{for } t \text{ in } I$$

Proof   Since $\| \mathbf{F}(t) \| = c$ by hypothesis, it follows that

$$(\mathbf{F} \cdot \mathbf{F})(t) = \mathbf{F}(t) \cdot \mathbf{F}(t) = \| \mathbf{F}(t) \|^2 = c^2 \quad \text{for } t \text{ in } I$$

Therefore $\mathbf{F} \cdot \mathbf{F}$ is a constant real-valued function, which by Example 1 has derivative 0. As a result, Theorem 12.10(d) implies that

$$\mathbf{F}'(t) \cdot \mathbf{F}(t) + \mathbf{F}(t) \cdot \mathbf{F}'(t) = (\mathbf{F} \cdot \mathbf{F})'(t) = 0$$

so that

$$\mathbf{F}(t) \cdot \mathbf{F}'(t) = 0 \quad \text{for } t \text{ in } I \quad ■$$

Corollary 12.11 tells us that if $\| \mathbf{F} \|$ is a constant real-valued function, then for each t in the domain of $\mathbf{F}$, one of the following is true:

1. $\mathbf{F}(t) = \mathbf{0}$
2. $\mathbf{F}'(t) = \mathbf{0}$
3. $\mathbf{F}(t)$ and $\mathbf{F}'(t)$ are perpendicular

For example, if S is a sphere centered at the origin and if $\mathbf{F}$ represents the position of an object moving on S, then $\| \mathbf{F} \|$ is a constant function, so that $\mathbf{F}(t)$ is perpendicular to $\mathbf{F}'(t)$ for all t for which $\mathbf{F}'(t) \neq \mathbf{0}$. A special case of this occurs with circular motion; we will return to this in Example 8.

The **second derivative** of $\mathbf{F}$ is defined to be the derivative of $\mathbf{F}'$, denoted $\mathbf{F}''$. If

$$\mathbf{F}(t) = f_1(t)\mathbf{i} + f_2(t)\mathbf{j} + f_3(t)\mathbf{k}$$

then two applications of Theorem 12.9 yield

$$\mathbf{F}''(t) = f_1''(t)\mathbf{i} + f_2''(t) \, \mathbf{j} + f_3''(t)\mathbf{k}$$

For example, if $\mathbf{F}(t) = (t^{1/2} + t)\mathbf{i} + (t^3 + 1)\mathbf{j} + e^{2t}\mathbf{k}$, then

$$\mathbf{F}''(t) = -\frac{1}{4} t^{-3/2}\mathbf{i} + 6t \, \mathbf{j} + 4e^{2t}\mathbf{k}$$

Velocity and Acceleration

The most important physical applications of derivatives of vector-valued functions arise in the study of motion. As an object moves through space, the coordinates x, y, and z of its location are functions of time. Let us assume that these functions are twice differentiable and define the position, velocity, speed, and acceleration as follows:

Position: $\mathbf{r}(t) = x(t)\mathbf{i} + y(t)\mathbf{j} + z(t)\mathbf{k}$

Velocity: $\mathbf{v}(t) = \dfrac{d\mathbf{r}}{dt} = \dfrac{dx}{dt}\mathbf{i} + \dfrac{dy}{dt}\mathbf{j} + \dfrac{dz}{dt}\mathbf{k}$

Speed: $\| \mathbf{v}(t) \| = \sqrt{\left(\dfrac{dx}{dt}\right)^2 + \left(\dfrac{dy}{dt}\right)^2 + \left(\dfrac{dz}{dt}\right)^2}$

Acceleration: $\mathbf{a}(t) = \dfrac{d\mathbf{v}}{dt} = \dfrac{d^2\mathbf{r}}{dt^2} = \dfrac{d^2x}{dt^2}\mathbf{i} + \dfrac{d^2y}{dt^2}\mathbf{j} + \dfrac{d^2z}{dt^2}\mathbf{k}$

The position vector $\mathbf{r}(t)$ is often called the **radial vector** or **radius vector** (this is why the symbol $\mathbf{r}$ is used). Usually $\mathbf{r}(t)$ is represented by a directed line segment from the origin to the position of the object at time t. If t_0 is fixed and $t \neq t_0$, then the vector $\mathbf{r}(t) - \mathbf{r}(t_0)$, sometimes called the **displacement vector,** is directed from the position of the object at time t_0 to the position at time t (Figure 12.16). The **average velocity** is then defined as

$$\frac{\mathbf{r}(t) - \mathbf{r}(t_0)}{t - t_0}$$

Since

$$\mathbf{v}(t_0) = \mathbf{r}'(t_0) = \lim_{t \to t_0} \frac{\mathbf{r}(t) - \mathbf{r}(t_0)}{t - t_0}$$

velocity is the limit of average velocity.

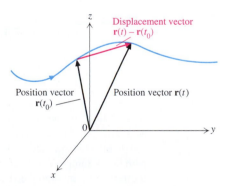

FIGURE 12.16

EXAMPLE 5   Suppose an object remains at rest at the point (a, b, c). Show that $\mathbf{v} = \mathbf{a} = \mathbf{0}$.

Solution The position is constant and therefore is given by

$$\mathbf{r}(t) = a\mathbf{i} + b\mathbf{j} + c\mathbf{k}$$

Consequently

$$\mathbf{v}(t) = \frac{d\mathbf{r}}{dt} = 0 \quad \text{and} \quad \mathbf{a}(t) = \frac{d\mathbf{v}}{dt} = 0 \quad \square$$

EXAMPLE 6 Suppose the position vector of an object is given by

$$\mathbf{r}(t) = (x_0 + at)\mathbf{i} + (y_0 + bt)\mathbf{j} + (z_0 + ct)\mathbf{k}$$

Determine the velocity, speed, and acceleration of the object.

Solution Differentiating, we find that

$$\mathbf{v}(t) = \frac{d\mathbf{r}}{dt} = a\mathbf{i} + b\mathbf{j} + c\mathbf{k}$$

and hence that

$$\mathbf{a}(t) = \frac{d\mathbf{v}}{dt} = \mathbf{0}$$

The speed is the length of the velocity vector, which in this case means that

$$\| \mathbf{v}(t) \| = \sqrt{a^2 + b^2 + c^2} \quad \square$$

In Example 6 the object moves along the line whose parametric equations are

$$x = x_0 + at, \quad y = y_0 + bt, \quad z = z_0 + ct$$

with constant velocity and zero acceleration. If we had assumed that

$$\mathbf{r}(t) = (x_0 + at^2)\mathbf{i} + (y_0 + bt^2)\mathbf{j} + (z_0 + ct^2)\mathbf{k}$$

with *a, b,* or *c* different from 0, then the object would still have moved along the same line. However, the velocity and acceleration would be given by

$$\mathbf{v}(t) = 2at\mathbf{i} + 2bt\mathbf{j} + 2ct\mathbf{k}$$

and

$$\mathbf{a}(t) = 2a\mathbf{i} + 2b\mathbf{j} + 2c\mathbf{k}$$

In this case the object would move faster as t increases with $t > 0$, and the acceleration, while constant, would not be $\mathbf{0}$.

In the next example we study the motion of an object that traces out a cycloidal path (see Example 7 of Section 12.1). You might visualize the object as a point on the rim of an automobile tire that is moving along a straight line with constant speed.

EXAMPLE 7 Let

$$\mathbf{r}(t) = r(t - \sin t)\mathbf{i} + r(1 - \cos t)\mathbf{j}$$

Find those values of t for which $\mathbf{v}(t) = \mathbf{0}$.

Solution   By differentiating **r** we find that

$$\mathbf{v}(t) = \frac{d\mathbf{r}}{dt} = r(1 - \cos t)\mathbf{i} + r \sin t\mathbf{j}$$

Therefore

$$\mathbf{v}(t) = \mathbf{0} \quad \text{if and only if} \quad r(1 - \cos t) = 0 = r \sin t$$

The last condition means that t is an integral multiple of 2π.   ❏

If, in Example 7, t is an integral multiple of 2π, then $\mathbf{r}(t)$ corresponds to one of the isolated points on the cycloid called cusps that lie on the x axis (Figure 12.17). It is at precisely such points on the cycloid that the velocity is $\mathbf{0}$. We remark that the function **r** is differentiable for all t. However, if we express the curve as the graph of $y = f(x)$, then the function f is not differentiable at any integral multiple of 2π.

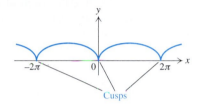

FIGURE 12.17

EXAMPLE 8   An object moves counterclockwise along a circle of radius $r_0 > 0$ with a constant speed $v_0 > 0$. Show that the position vector of the object is given by the equation

$$\mathbf{r}(t) = r_0\left(\cos \frac{v_0 t}{r_0}\mathbf{i} + \sin \frac{v_0 t}{r_0}\mathbf{j} \right)$$

Find formulas for the velocity and acceleration of the object.

Solution   We set up the coordinate system so that the circle lies in the xy plane with the origin as its center and so that the object is on the positive x axis at time 0 and moves counterclockwise around the circle. For any time t let $\theta(t)$ be the angle from the positive x axis to the vector $\mathbf{r}(t)$ (Figure 12.18). Then

$$\mathbf{r}(t) = r_0[\cos \theta(t)\mathbf{i} + \sin \theta(t)\mathbf{j}] \tag{1}$$

To obtain an expression for $\theta(t)$ in terms of t, r_0, and v_0, we differentiate both sides of (1) and obtain

$$\mathbf{v}(t) = r_0\theta'(t)[-\sin \theta(t)\mathbf{i} + \cos \theta(t)\mathbf{j}] \tag{2}$$

Because the object moves counterclockwise, θ is increasing; consequently $\theta'(t) > 0$. Therefore since $r_0 > 0$, (2) implies that

$$\| \mathbf{v}(t) \| = | r_0\theta'(t) | \sqrt{(-\sin \theta(t))^2 + (\cos \theta(t))^2} = | r_0\theta'(t) | = r_0\theta'(t) \tag{3}$$

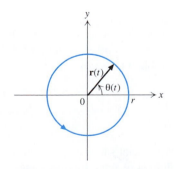

FIGURE 12.18

But $\| \mathbf{v}(t) \| = v_0$ by hypothesis, so that (3) becomes $\theta'(t) = v_0/r_0$. Since $\theta(0) = 0$ by assumption, we can integrate to obtain $\theta(t) = v_0t/r_0$. Substitution into (1) now yields

$$\mathbf{r}(t) = r_0\left(\cos \frac{v_0 t}{r_0} \mathbf{i} + \sin \frac{v_0 t}{r_0} \mathbf{j} \right)$$

By differentiating twice in succession, we find that

$$\mathbf{v}(t) = v_0\left(-\sin \frac{v_0 t}{r_0} \mathbf{i} + \cos \frac{v_0 t}{r_0} \mathbf{j} \right)$$

and

$$\mathbf{a}(t) = -\frac{v_0^2}{r_0}\left(\cos \frac{v_0 t}{r_0} \mathbf{i} + \sin \frac{v_0 t}{r_0} \mathbf{j} \right) = -\frac{v_0^2}{r_0^2} \mathbf{r}(t) \tag{4}$$

This completes the solution. ❏

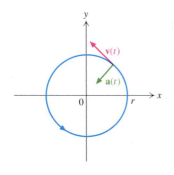

FIGURE 12.19

Because the speed of the object in Example 8 is constant, its velocity and acceleration are perpendicular (see Exercise 43). Despite the fact that the *speed* of the object is constant, the *velocity* $\mathbf{v}$ is not constant, nor is the acceleration constant. In fact, (4) implies that the acceleration is directed toward the center of the circle (Figure 12.19). Such an acceleration is called a **centripetal acceleration,** and any force producing a centripetal acceleration is called a **centripetal force.** Since $\| \mathbf{r}(t) \| = r_0$ by (1), the magnitude of the acceleration is given by

$$\| \mathbf{a}(t) \| = \frac{v_0^2}{r_0^2} \| \mathbf{r}(t) \| = \frac{v_0^2}{r_0} \tag{5}$$

Now suppose a satellite travels in a circular orbit around the earth and is influenced only by the earth's gravity. We will derive a formula relating the speed v_0 of the satellite to the radius r_0 of the orbit. To that end, recall that by Newton's Law of Gravitation, the magnitude $\| \mathbf{a} \|$ of the acceleration of an object r_0 meters from the center of the earth and influenced only by the earth's gravity is given by

$$\| \mathbf{a}(t) \| = \frac{C}{r_0^2} \tag{6}$$

where C is a constant. Combining this equation with (5), we find that

$$\frac{v_0^2}{r_0} = \frac{C}{r_0^2}, \quad \text{so that} \quad v_0^2 = \frac{C}{r_0} \tag{7}$$

Since the Law of Gravitation holds even if the object is at the earth's surface, that is, if r_0 is equal to the radius of the earth (which is approximately 6.37×10^6 meters), and since the acceleration due to gravity on the surface of the earth is approximately 9.8 meters per second per second, it follows from (6) that

$$9.8 \approx \frac{C}{(6.37 \times 10^6)^2}$$

Thus
$$C \approx (9.8)(6.37 \times 10^6)^2$$

Now suppose a weather satellite moves with constant speed v_0 in a circular orbit 6600 kilometers from the center of the earth (that is, some 140 miles above ground). Then by (7),

$$v_0 = \sqrt{\frac{C}{r_0}} \approx \sqrt{\frac{(9.8)(6.37 \times 10^6)^2}{6600 \times 10^3}} \approx 7760 \text{ (meters per second)}$$

which is approximately 17,400 miles per hour.

Finally suppose that a satellite travels in a circular orbit of radius r_0 and has speed v_0. The **period** T of the satellite is the time it takes to circle the earth once. During one period the satellite travels a distance of $2\pi r_0$, so that

$$v_0 = \frac{\text{distance}}{\text{time}} = \frac{2\pi r_0}{T}$$

Consequently

$$T^2 = \frac{4\pi^2 r_0^2}{v_0^2}$$

By (7), we infer that

$$T^2 = \frac{4\pi^2 r_0^2}{C/r_0} = \frac{4\pi^2 r_0^3}{C}$$

This is a special case of Kepler's famous Third Law of planetary motion: The square of the period is directly proportional to the cube of the radius of the orbit. We will discuss the Third Law (as well as the First and Second Laws) of Kepler in Section 12.7.

Integrals of Vector-Valued Functions

Since a vector-valued function $\mathbf{F}$ is determined by its component functions, we define the integral of $\mathbf{F}$ in terms of its component functions.

DEFINITION 12.12   Let

$$\mathbf{F}(t) = f_1(t)\mathbf{i} + f_2(t)\mathbf{j} + f_3(t)\mathbf{k}$$

where f_1, f_2, and f_3 are continuous on $[a, b]$. Then the **definite integral** $\int_a^b \mathbf{F}(t)\, dt$ and the **indefinite integral** $\int \mathbf{F}(t)\, dt$ are defined by

$$\int_a^b \mathbf{F}(t)\, dt = \left(\int_a^b f_1(t)\, dt \right)\mathbf{i} + \left(\int_a^b f_2(t)\, dt \right)\mathbf{j} + \left(\int_a^b f_3(t)\, dt \right)\mathbf{k}$$

and

$$\int \mathbf{F}(t)\, dt = \left(\int f_1(t)\, dt \right)\mathbf{i} + \left(\int f_2(t)\, dt \right)\mathbf{j} + \left(\int f_3(t)\, dt \right)\mathbf{k}$$

EXAMPLE 9   Let

$$\mathbf{F}(t) = t\mathbf{i} + t^2\mathbf{j} + \sin t\,\mathbf{k}$$

Find $\int \mathbf{F}(t)\, dt$ and $\int_0^\pi \mathbf{F}(t)\, dt$.

Solution   Integrating componentwise, we find that

$$\int \mathbf{F}(t)\, dt = \left(\int t\, dt \right)\mathbf{i} + \left(\int t^2\, dt \right)\mathbf{j} + \left(\int \sin t\, dt \right)\mathbf{k}$$

$$= \left(\frac{1}{2}t^2 + C_1 \right)\mathbf{i} + \left(\frac{1}{3}t^3 + C_2 \right)\mathbf{j} + (-\cos t + C_3)\mathbf{k}$$

Therefore

$$\int_0^\pi \mathbf{F}(t)\,dt = \left(\frac{1}{2}t^2\Big|_0^\pi\right)\mathbf{i} + \left(\frac{1}{3}t^3\Big|_0^\pi\right)\mathbf{j} + \left(-\cos t\Big|_0^\pi\right)\mathbf{k}$$

$$= \frac{1}{2}\pi^2\mathbf{i} + \frac{1}{3}\pi^3\mathbf{j} + 2\mathbf{k} \quad \square$$

If

$$\mathbf{F}(t) = f_1(t)\mathbf{i} + f_2(t)\mathbf{j} + f_3(t)\mathbf{k}$$

where f_1, f_2, and f_3 are continuously differentiable, then by Theorem 12.9 and Definition 12.12,

$$\int \mathbf{F}'(t)\,dt = \int [f_1'(t)\mathbf{i} + f_2'(t)\mathbf{j} + f_3'(t)\mathbf{k}]\,dt$$

$$= \left(\int f_1'(t)\,dt\right)\mathbf{i} + \left(\int f_2'(t)\,dt\right)\mathbf{j} + \left(\int f_3'(t)\,dt\right)\mathbf{k}$$

$$= [f_1(t) + C_1]\mathbf{i} + [f_2(t) + C_2]\mathbf{j} + [f_3(t) + C_3]\mathbf{k}$$

$$= (f_1(t)\mathbf{i} + f_2(t)\mathbf{j} + f_3(t)\mathbf{k}) + (C_1\mathbf{i} + C_2\mathbf{j} + C_3\mathbf{k})$$

$$= \mathbf{F}(t) + (C_1\mathbf{i} + C_2\mathbf{j} + C_3\mathbf{k})$$

This proves that

$$\int \mathbf{F}'(t)\,dt = \mathbf{F}(t) + \mathbf{C} \tag{8}$$

where $\mathbf{C}$ is a constant vector. Formula (8) is valuable in the study of motion. Indeed, since the velocity $\mathbf{v}$ is the derivative of the position $\mathbf{r}$ and the acceleration $\mathbf{a}$ is the derivative of the velocity $\mathbf{v}$, it follows from (8) that

$$\int \mathbf{v}(t)\,dt = \int \mathbf{r}'(t)\,dt = \mathbf{r}(t) + \mathbf{C}$$

and

$$\int \mathbf{a}(t)\,dt = \int \mathbf{v}'(t)\,dt = \mathbf{v}(t) + \mathbf{C}$$

We will use these formulas in discussing Newton's Second Law of Motion.

The Vector Form of Newton's Second Law of Motion

The most basic physical law pertaining to motion is Newton's Second Law of Motion. Suppose that at any time t an object of mass m experiences a force $\mathbf{F}(t)$ and undergoes an acceleration $\mathbf{a}(t)$. Then Newton's Second Law of Motion states that

$$\mathbf{F}(t) = m\mathbf{a}(t) \tag{9}$$

From this law we can determine the acceleration of an object once its mass and the force acting on it are known. Since the velocity is the integral of the acceleration and the position is the integral of the velocity, it is possible (at least in principle) to determine the position of an object of known mass, initial position, and initial velocity once the force acting on it is known. Normally the initial position and velocity are taken at time $t = 0$.

For an elementary application of these ideas, we consider the motion of an object that remains near some point P on the earth's surface and moves only under the influence of the earth's gravity. By Newton's Law of Gravitation the gravitational force on the object is nearly constant and is exerted downward (toward the center of the earth). We choose a coordinate system so that the positive z axis emanates from the center of the earth and points upward, passing through P. Consequently the acceleration is $-g\mathbf{k}$, where g is a constant approximately equal to 9.8 meters (or 32 feet) per second per second.

EXAMPLE 10   An object has initial position $\mathbf{r}_0$ and initial velocity $\mathbf{v}_0$ and undergoes a constant acceleration $-g\mathbf{k}$. Show that the position of the object at any time t is given by

$$\mathbf{r}(t) = -\frac{1}{2} gt^2\mathbf{k} + t\mathbf{v}_0 + \mathbf{r}_0$$

Solution   By hypothesis $\mathbf{a}(t) = -g\mathbf{k}$. Successive integrations yield

$$\mathbf{v}(t) = \int \mathbf{a}(t)\,dt = \int -g\mathbf{k}\,dt = -gt\mathbf{k} + \mathbf{C} \qquad (10)$$

and
$$\mathbf{r}(t) = \int \mathbf{v}(t)\,dt = -\frac{1}{2} gt^2\mathbf{k} + t\mathbf{C} + \mathbf{C}_1 \qquad (11)$$

where $\mathbf{C}$ and $\mathbf{C}_1$ are constant vectors. By assumption $\mathbf{v}(0) = \mathbf{v}_0$, so that by letting $t = 0$ in (10) we find that $\mathbf{C} = \mathbf{v}_0$. Similarly, $\mathbf{r}(0) = \mathbf{r}_0$, so that $\mathbf{C}_1 = \mathbf{r}_0$ by (11). Therefore (11) can be rewritten

$$\mathbf{r}(t) = -\frac{1}{2} gt^2\mathbf{k} + t\mathbf{v}_0 + \mathbf{r}_0 \qquad \square$$

Hank Aaron. (*Focus on Sports*)

EXAMPLE 11   A baseball is hit 4 feet above the ground at 100 feet per second and at an angle of $\pi/6$ with respect to the ground. How long does it take for the baseball to hit the ground, and how far away from home plate is it then?

Solution   We choose a coordinate system so that the xy plane represents the ground, the ball travels in the yz plane with the $\mathbf{j}$ component of its position vector increasing, and the initial position of the ball is $\mathbf{r}_0 = 4\mathbf{k}$. Since the initial angle is $\pi/6$ and the initial speed of the ball is 100 feet per second, so that $\|\mathbf{v}_0\| = 100$, the initial velocity is given by

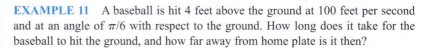

$$\mathbf{v}_0 = 100\left(\cos\frac{\pi}{6}\,\mathbf{j} + \sin\frac{\pi}{6}\,\mathbf{k}\right) = 100\left(\frac{\sqrt{3}}{2}\,\mathbf{j} + \frac{1}{2}\,\mathbf{k}\right) = 50(\sqrt{3}\,\mathbf{j} + \mathbf{k})$$

From Example 10 we know that the position of the baseball is given by

$$\mathbf{r}(t) = -\frac{1}{2}gt^2\mathbf{k} + t\mathbf{v}_0 + \mathbf{r}_0$$

Substituting for $\mathbf{r}_0$ and $\mathbf{v}_0$, we find that

$$\mathbf{r}(t) = -\frac{1}{2}gt^2\mathbf{k} + t[50(\sqrt{3}\,\mathbf{j} + \mathbf{k})] + 4\mathbf{k}$$

$$= 50\sqrt{3}\,t\mathbf{j} + \left(4 + 50t - \frac{1}{2}gt^2\right)\mathbf{k}$$

When the ball hits the ground, the $\mathbf{k}$ component of $\mathbf{r}(t)$ must be 0. This happens at the time $t > 0$ such that

$$4 + 50t - \frac{gt^2}{2} = 0$$

which means that

$$t = \frac{50 + \sqrt{2500 + 8g}}{g}$$

Taking $g = 32$, we find that $t \approx 3.2$, so the ball hits the ground after approximately 3.2 seconds. By that time the ball is $50\sqrt{3}\,(3.2) \approx 277$ feet from home plate. ◻

EXERCISES 12.3

In Exercises 1–16 find the derivative of the function.

1. $\mathbf{F}(t) = \mathbf{i} + t\mathbf{j} + t^5\mathbf{k}$

2. $\mathbf{F}(t) = 3\mathbf{i} + (t^2 + t)\mathbf{j} - t\mathbf{k}$

3. $\mathbf{F}(t) = (1 + t)^{3/2}\mathbf{i} - (1 - t)^{3/2}\mathbf{j} + \frac{3}{2}t\mathbf{k}$

4. $\mathbf{F}(t) = t^2\cos t\mathbf{i} + t^3\sin t\mathbf{j} + t^4\mathbf{k}$

5. $\mathbf{F}(t) = \tan t\mathbf{i} + \mathbf{j} + \sec t\mathbf{k}$

6. $\mathbf{F}(t) = e^t\cos t\mathbf{i} - e^t\sin t\mathbf{k}$

7. $\mathbf{F}(t) = \cosh t\mathbf{i} + \sinh t\mathbf{j} - \sqrt{t}\mathbf{k}$

8. $\mathbf{F}(t) = \sin^{-1}4t\mathbf{i} - 3\tan^{-1}(2t - 1)\mathbf{j} + 7\ln 3t^2\mathbf{k}$

9. $4\mathbf{F} - 2\mathbf{G}$, where $\mathbf{F}(t) = 2\sec t\mathbf{i} - 3\mathbf{j} + \csc t\mathbf{k}$ and $\mathbf{G}(t) = 3t\mathbf{i} - t^2\mathbf{j} - 4\csc t\mathbf{k}$

10. $\mathbf{F} \cdot \mathbf{G}$, where $\mathbf{F}(t) = 2\sec t\mathbf{i} - 3\mathbf{j} + \csc t\mathbf{k}$ and $\mathbf{G}(t) = 3t\mathbf{i} - t^2\mathbf{j} - 4\csc t\mathbf{k}$

11. $\mathbf{F} \times \mathbf{G}$, where $\mathbf{F}(t) = 2\sec t\mathbf{i} - 3\mathbf{j} + \csc t\mathbf{k}$ and $\mathbf{G}(t) = \ln t\mathbf{k}$

12. $\mathbf{F} \cdot \mathbf{G}$, where $\mathbf{F}(t) = (3/t)\mathbf{i} - \mathbf{j}$ and $\mathbf{G}(t) = t\mathbf{i} - e^{-t}\mathbf{j}$

13. $\mathbf{F} \times \mathbf{G}$, where $\mathbf{F}(t) = (3/t)\mathbf{i} - \mathbf{j}$ and $\mathbf{G}(t) = t\mathbf{i} - e^{-t}\mathbf{j}$

14. $\mathbf{F} \cdot \mathbf{G}$, where

$$\mathbf{F}(t) = \frac{-2t}{1 + t^2}\mathbf{i} + \frac{1 + 2t^2}{1 + t^2}\mathbf{j}$$

and

$$\mathbf{G}(t) = \frac{-2 - 4t^2}{(1 + t^2)^2}\mathbf{i} - \frac{4t}{(1 + t^2)^2}\mathbf{j}$$

15. $\mathbf{F} \circ g$, where $\mathbf{F}(t) = \ln t\mathbf{i} - 4e^{2t}\mathbf{j} + \dfrac{t - 1}{t}\mathbf{k}$ and $g(t) = \sqrt{t}$

16. $\mathbf{F} \circ g$, where $\mathbf{F}(t) = t^3\mathbf{i} - \sqrt{3}t\mathbf{j} + \dfrac{1}{t^2}\mathbf{k}$ and $g(t) = \cos t$

In Exercises 17–21 evaluate the integral.

17. $\displaystyle\int \left(t^2\mathbf{i} - (3t - 1)\mathbf{j} - \frac{1}{t^3}\mathbf{k}\right)dt$

18. $\displaystyle\int (t\cos t\mathbf{i} + t\sin t\mathbf{j} + 3t^4\mathbf{k})\,dt$

19. $\displaystyle\int_0^1 (e^t\mathbf{i} + e^{-t}\mathbf{j} + 2t\mathbf{k})\,dt$

20. $\displaystyle\int_0^1 (\cosh t\mathbf{i} + \sinh t\mathbf{j} + \mathbf{k})\,dt$

21. $\displaystyle\int_{-1}^1 [(1 + t)^{3/2}\mathbf{i} + (1 - t)^{3/2}\mathbf{j}]\,dt$

In Exercises 22–27 find the velocity, speed, and acceleration of an object having the given position function.

22. $\mathbf{r}(t) = 3t\mathbf{i} + 2t\mathbf{j} - 16t^2\mathbf{k}$

23. $\mathbf{r}(t) = \cos t\mathbf{i} + \sin t\mathbf{j} - 16t^2\mathbf{k}$

24. $\mathbf{r}(t) = e^{-t}\mathbf{i} + e^{-t}\mathbf{j}$

25. $\mathbf{r}(t) = 2t\mathbf{i} + t^2\mathbf{j} + \ln t\mathbf{k}$

26. $\mathbf{r}(t) = \cosh t\mathbf{i} + \sinh t\mathbf{j} + t\mathbf{k}$

27. $\mathbf{r}(t) = e^t \sin t\mathbf{i} + e^t \cos t\mathbf{j} + e^t\mathbf{k}$

In Exercises 28–32 find the position, velocity, and speed of an object having the given acceleration, initial velocity, and initial position.

28. $\mathbf{a}(t) = -32\mathbf{k}$; $\mathbf{v}_0 = \mathbf{0}$; $\mathbf{r}_0 = \mathbf{0}$

29. $\mathbf{a}(t) = -32\mathbf{k}$; $\mathbf{v}_0 = \mathbf{i} + \mathbf{j}$; $\mathbf{r}_0 = \mathbf{0}$

30. $\mathbf{a}(t) = -32\mathbf{k}$; $\mathbf{v}_0 = 3\mathbf{i} - 2\mathbf{j} + \mathbf{k}$; $\mathbf{r}_0 = 5\mathbf{j} + 2\mathbf{k}$

31. $\mathbf{a}(t) = -\cos t\mathbf{i} - \sin t\mathbf{j}$; $\mathbf{v}_0 = \mathbf{k}$; $\mathbf{r}_0 = \mathbf{i}$

32. $\mathbf{a}(t) = e^t\mathbf{i} + e^{-t}\mathbf{j}$; $\mathbf{v}_0 = \mathbf{i} - \mathbf{j} + \sqrt{2}\mathbf{k}$; $\mathbf{r}_0 = \mathbf{i} + \mathbf{j}$

33. Let $\mathbf{F}(t) = \displaystyle\int_0^t (u \tan u^3\mathbf{i} + \cos e^u\mathbf{j} + e^{(u^2)}\mathbf{k})\, du$. Find $\mathbf{F}'(t)$.

34. Let $\mathbf{G}(t) = \displaystyle\int_0^{t^2} (\cos u\mathbf{i} + e^{-(u^2)}\mathbf{j} + \tan u\mathbf{k})\, du$. Find $\mathbf{G}'(t)$.

35. Let

$$\mathbf{F}(t) = \frac{4t}{1 + 4t^2}\mathbf{i} + \frac{1 - 4t^2}{1 + 4t^2}\mathbf{j}$$

Using Corollary 12.11, show that $\mathbf{F}(t) \cdot \mathbf{F}'(t) = 0$ for all t.

36. Let $\mathbf{F}(t) = \cos t\mathbf{i} + \sin t\mathbf{j}$. Show that there is a value of t in $(0, \pi)$ such that $[\mathbf{F}(\pi) - \mathbf{F}(0)]/(\pi - 0)$ is parallel to $\mathbf{F}'(t)$, but there is no value of t in $(0, \pi)$ such that

$$\mathbf{F}'(t) = \frac{\mathbf{F}(\pi) - \mathbf{F}(0)}{\pi - 0}$$

37. Let $\mathbf{F}(t) = \sin t\mathbf{i} - \cos t\mathbf{j}$. Show that for all t the vectors $\mathbf{F}(t)$ and $\mathbf{F}''(t)$ are parallel. Determine whether there is a value of t for which $\mathbf{F}(t)$ and $\mathbf{F}''(t)$ have the same direction.

38. Let $\mathbf{F}(t) = e^{-2t}\mathbf{i} + e^{2t}\mathbf{k}$. Show that for all t the vectors $\mathbf{F}(t)$ and $\mathbf{F}''(t)$ are parallel. Determine whether there is a value of t for which $\mathbf{F}(t)$ and $\mathbf{F}''(t)$ have the same direction.

39. Show that if $\mathbf{F}''(t)$ exists for all t, then

$$\frac{d}{dt}(\mathbf{F} \times \mathbf{F}') = \mathbf{F}(t) \times \mathbf{F}''(t)$$

40. Suppose $\mathbf{F}(t)$ is parallel to $\mathbf{F}''(t)$ for all t. Show that $\mathbf{F} \times \mathbf{F}'$ is constant. (*Hint:* Use Exercise 39.)

41. Let $\mathbf{r}(t) = \cos t\mathbf{i} + \sin t\mathbf{j} + t\mathbf{k}$.
 a. For any t, find a formula for the angle $\theta(t)$ between $\mathbf{r}(t)$ and $\mathbf{v}(t)$.
 b. Find the angle $\theta(0)$.
 c. Find the value of the angle $\theta(\pi/2)$.

42. Let $\mathbf{F}$, $\mathbf{G}$, and $\mathbf{H}$ be differentiable. Using the differentiation rules for the dot and cross products of vector-valued functions, verify that

$$\frac{d}{dt}[\mathbf{F} \cdot (\mathbf{G} \times \mathbf{H})] = \frac{d\mathbf{F}}{dt} \cdot (\mathbf{G} \times \mathbf{H}) + \mathbf{F} \cdot \left(\frac{d\mathbf{G}}{dt} \times \mathbf{H}\right)$$

$$+ \mathbf{F} \cdot \left(\mathbf{G} \times \frac{d\mathbf{H}}{dt}\right)$$

43. Use Corollary 12.11 to show that if an object has a constant speed, then its acceleration vector is perpendicular to its velocity vector.

44. Show that $\| \mathbf{v} \| \dfrac{d}{dt} \| \mathbf{v} \| = \mathbf{v} \cdot \mathbf{a}$. (*Hint:* $\| \mathbf{v} \|^2 = \mathbf{v} \cdot \mathbf{v}$.)

Applications

45. If an object has mass m and position $\mathbf{r}(t)$ at time t, then its **kinetic energy** $K(t)$ at time t is defined by

$$K(t) = \frac{1}{2} m \| \mathbf{v}(t) \|^2$$

Suppose a ball is dropped from a height of 96 feet and later is thrown straight down from the same height with an initial speed of 80 feet per second. How much larger is the kinetic energy when the ball hits the ground the second time than the first time?

46. A force $\mathbf{F}$ acts on an object that moves with velocity $\mathbf{v}$. Show that the derivative of the kinetic energy is given by

$$K'(t) = \mathbf{F}(t) \cdot \mathbf{v}(t)$$

(*Hint:* Recall that $\mathbf{F} = m\mathbf{a}$ and $\| \mathbf{v}(t) \|^2 = \mathbf{v}(t) \cdot \mathbf{v}(t)$.)

47. Suppose the position of a golf ball is given by

$$\mathbf{r}(t) = 90\sqrt{2}\, t\mathbf{i} + 90\sqrt{2}\, t\mathbf{j} + (64t - 16t^2)\mathbf{k} \quad \text{for } t \geq 0$$

 a. Find the initial position and the initial velocity of the golf ball (assuming distances are in feet and time in seconds).
 b. Show that the golf ball strikes the ground at time $t = 4$, and determine the distance from its initial position.

48. An object leaves the point $(0, 0, 1)$ with initial velocity $\mathbf{v}_0 = 2\mathbf{i} + 3\mathbf{k}$. Thereafter it is subject only to the force of gravity. Find a formula for the position of the object at any time $t > 0$. Use feet and seconds.

49. A ping-pong ball rolls off a table 2.6 feet high with an initial speed of 2 feet per second.
 a. How long after it leaves the table does it hit the floor? (Note that the velocity vector is horizontal when the ball leaves the table.)
 b. At what speed does it hit the floor?

50. The *Nimbus* weather satellite orbits the earth in a circular orbit approximately 500 miles above the surface of the earth. Find the speed of the satellite.

51. A satellite can remain stationary over a fixed point on earth if it orbits the earth in a circular orbit with a period of 24 hours. This is the basic principle of a communications satellite. Find the speed and the radius of the orbit of such a satellite. (*Hint:* Use (7) and the fact that the circumference of the orbit equals both $2\pi r_0$ and $24(3600)v_0$.)

52. A bobsled moving at a constant speed of 60 miles per hour rounds a circular turn with a radius of 100 feet. Find the magnitude of the bobsled's centripetal acceleration.

53. The accelerator at the Fermi National Accelerator Laboratory in Batavia, Illinois, is circular with a radius of 1 kilometer. Find the magnitude of the centripetal acceleration of a proton moving around the accelerator with a constant speed of
   a. 2.5×10^5 kilometers per second
   b. 2.9×10^5 kilometers per second

*54. Suppose that a gun is aimed directly at a target and that a projectile is fired from the gun at the instant the target is dropped (so its initial velocity is **0**). Show that if the target is in range, then it will automatically be hit by the projectile (Figure 12.20).

Project

1. Evangelista Torricelli (1608–1647), the Italian physicist and mathematician who invented the barometer, is credited with a result called Torricelli's Theorem:

Suppose a cylindrical container stands on the floor and has an open top. Water is H feet deep in the container, and a spout is located h feet below the water level (Figure 12.21). Then the speed of the water as it flows horizontally through the spout is $\sqrt{2gh}$.

   a. Using the ideas in the solution of Example 10, find a formula for the position $r(t)$ of a droplet t seconds after it leaves the container (and before it hits the floor).
   b. Determine how long it takes for a droplet on its journey from the spout to the floor.
   c. Find the value of h that maximizes the distance R from the container to the point at which the droplet hits the floor. Then find the corresponding value of R.
   d. Show that for the maximum distance R, the speed at which the droplet hits the floor is $\sqrt{2gh}$.

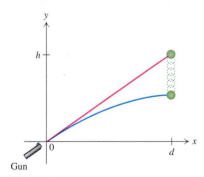

FIGURE 12.20 Figure for Exercise 54.

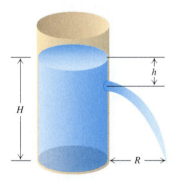

FIGURE 12.21 Figure for the Project.

12.4 SPACE CURVES AND THEIR LENGTHS

In Section 12.1 we discussed curves traced out by vector-valued functions. From now on we will use the word "curve" in a more restricted sense.

DEFINITION 12.13   A **space curve** (or simply **curve**) is the range of a continuous vector-valued function on an interval of real numbers.

Essentially all of the vector-valued functions we have encountered are continuous, and consequently their ranges are curves. In particular, any point, line, line segment, circle, parabola, cycloid, or circular helix is a curve. We will generally use C to denote a curve and $\mathbf{r}$ to denote a vector-valued function whose range is a curve C. In that case we say that C is **parametrized** by $\mathbf{r}$, or that $\mathbf{r}$ is a **parametrization** of C. However, we will occasionally refer to $\mathbf{r}$ as a curve and say, for example, "the curve $\mathbf{r}(t) = e^t\mathbf{i} + \cos t\mathbf{j} + 3\mathbf{k}$." The component functions of a vector-valued function $\mathbf{r}$ are usually denoted x, y, and z. Thus

$$\mathbf{r}(t) = x(t)\mathbf{i} + y(t)\mathbf{j} + z(t)\mathbf{k}$$

Corresponding parametric equations for the curve are

$$x = x(t), \quad y = y(t), \quad z = z(t)$$

Since we are accustomed to thinking of the graph of any continuous real-valued function as a curve, we now verify that Definition 12.13 is broad enough to include such curves.

EXAMPLE 1   Let f be a continuous real-valued function on an interval I. Show that the graph of f is a curve, and find a parametrization of that curve.

Solution   The vector-valued function $\mathbf{r}$ defined by

$$\mathbf{r}(t) = t\mathbf{i} + f(t)\mathbf{j} \quad \text{for } t \text{ in } I$$

is continuous and traces out the graph of f (Figure 12.22). Thus the graph of f is the range of $\mathbf{r}$, so by Definition 12.13 the graph of f is a curve.   ❑

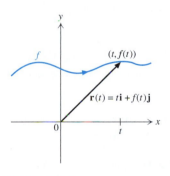

As we have just seen, it is possible to parametrize the graph of any continuous real-valued function. For instance, if $f(t) = \ln t$, then

$$\mathbf{r}(t) = t\mathbf{i} + \ln t\mathbf{j} \quad \text{for } t > 0$$

FIGURE 12.22

parametrizes the graph of f.

Properties of Curves

We will discuss three of the many properties a curve can possess: the properties of being closed, smooth, and piecewise smooth.

DEFINITION 12.14   A curve C is **closed** if it has a parametrization whose domain is a closed interval $[a, b]$ such that $\mathbf{r}(a) = \mathbf{r}(b)$, but otherwise $\mathbf{r}(t_1) \neq \mathbf{r}(t_2)$ for $t_1 \neq t_2$, with at most finitely many exceptions.

To put it informally, a curve is closed if its initial and terminal points coincide (Figure 12.23(a)), and in addition if it crosses itself only finitely many times. For example, since

$$\mathbf{r}(t) = \cos t\mathbf{i} + \sin t\mathbf{j} \quad \text{for } 0 \leq t \leq 2\pi$$

parametrizes the standard unit circle and since $\mathbf{r}(0) = \mathbf{r}(2\pi)$, the standard unit circle is closed. In fact, any circle or ellipse is closed, as is a figure 8, which crosses itself once. In contrast, the curve sketched in Figure 12.23(b) is not closed. Line segments, circular helixes, and cycloids also are not closed.

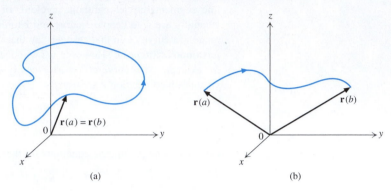

FIGURE 12.23   (a) A closed curve. (b) A curve that is not closed.

Next we consider smoothness of curves.

DEFINITION 12.15

a. A vector-valued function **r** defined on an interval I is **smooth** if **r** has a continuous derivative on I and $\mathbf{r}'(t) \neq \mathbf{0}$ for each interior point t. A curve C is **smooth** if it has a smooth parametrization.

b. A continuous vector-valued function **r** defined on an interval I is **piecewise smooth** if I is composed of a finite number of subintervals on each of which **r** is smooth. A curve C is **piecewise smooth** if it has a piecewise smooth parametrization.

Intuitively, a curve is smooth if it is composed of one piece with no sharp corners (Figure 12.24(a)). A curve is piecewise smooth if it is composed of one piece and has at most a finite number of sharp corners (Figure 12.24(b)). Of course, smooth curves are always piecewise smooth, but a piecewise smooth curve is not necessarily smooth. We will show in Example 4 that the curve composed of two arches of the cycloid is piecewise smooth (although it can be shown not to be smooth because of its cusp at 0).

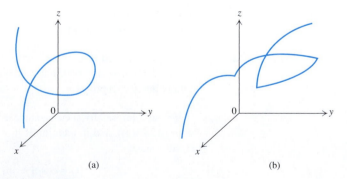

FIGURE 12.24   (a) A smooth curve. (b) A piecewise smooth curve.

EXAMPLE 2   Show that the standard unit circle is smooth.

Solution   The circle can be parametrized by

$$\mathbf{r}(t) = \cos t\mathbf{i} + \sin t\mathbf{j} \quad \text{for } 0 \leq t \leq 2\pi$$

The function $\mathbf{r}$ is differentiable on $[0, 2\pi]$, and

$$\mathbf{r}'(t) = -\sin t\mathbf{i} + \cos t\mathbf{j} \quad \text{for } 0 \leq t \leq 2\pi$$

Therefore $\mathbf{r}'$ is continuous on $[0, 2\pi]$, and

$$\| \mathbf{r}'(t) \| = \sqrt{(-\sin t)^2 + (\cos t)^2} = 1$$

From this we conclude that $\mathbf{r}'(t) \neq \mathbf{0}$ for each t in $[0, 2\pi]$. It follows that the circle is smooth.   ❑

EXAMPLE 3   Show that the helix

$$\mathbf{r}(t) = \cos t\mathbf{i} + \sin t\mathbf{j} + t\mathbf{k}$$

is smooth.

Solution   Since $\mathbf{r}'(t) = -\sin t\mathbf{i} + \cos t\mathbf{j} + \mathbf{k}$, it follows that $\mathbf{r}'$ is continuous. Moreover, $\mathbf{r}'(t) \neq \mathbf{0}$ for every t because the $\mathbf{k}$ component of $\mathbf{r}'(t)$ is 1 for all t. Therefore the helix is smooth (see Figure 12.25).   ❑

EXAMPLE 4   Show that the curve

$$\mathbf{r}(t) = r(t - \sin t)\mathbf{i} + r(1 - \cos t)\mathbf{j} \quad \text{for } -2\pi \leq t \leq 2\pi$$

which parametrizes two arches of a cycloid, is piecewise smooth.

Solution   We find that

$$\mathbf{r}'(t) = r(1 - \cos t)\mathbf{i} + r \sin t\mathbf{j}$$

Even though $\mathbf{r}'$ is continuous, $\mathbf{r}$ is not smooth because $\mathbf{r}'(0) = \mathbf{0}$. However, $\mathbf{r}'(t) \neq \mathbf{0}$ for $-2\pi < t < 0$ and for $0 < t < 2\pi$. Therefore $\mathbf{r}$ is smooth on $[-2\pi, 0]$ and on $[0, 2\pi]$, and hence $\mathbf{r}$ is piecewise smooth (Figure 12.26).   ❑

Suppose (x_0, y_0, z_0) and (x_1, y_1, z_1) are distinct points in space, and consider the parametric equations

$$x = x_0 + (x_1 - x_0)t, \quad y = y_0 + (y_1 - y_0)t, \quad z = z_0 + (z_1 - z_0)t \qquad (1)$$

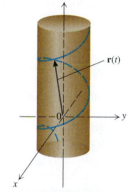

FIGURE 12.25   A circular helix: $\mathbf{r}(t) = \cos t\mathbf{i} + \sin t\mathbf{j} + t\mathbf{k}$.

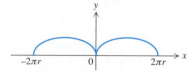

FIGURE 12.26   Two arches of a cycloid.

Since $(x, y, z) = (x_0, y_0, z_0)$ for $t = 0$ and $(x, y, z) = (x_1, y_1, z_1)$ for $t = 1$, it follows that (1) gives parametric equations for the line through (x_0, y_0, z_0) and (x_1, y_1, z_1). Moreover,

$$\mathbf{r}(t) = [x_0 + (x_1 - x_0)t]\mathbf{i} + [y_0 + (y_1 - y_0)t]\mathbf{j} + [z_0 + (z_1 - z_0)t]\mathbf{k} \quad \text{for } 0 \leq t \leq 1$$

is a smooth parametrization of the line segment from (x_0, y_0, z_0) to (x_1, y_1, z_1).

EXAMPLE 5   Find a smooth parametrization of the line segment from $(4, 3, 5)$ to $(2, 8, 5)$.

Solution   We apply the general method just described, with $x_0 = 4$, $y_0 = 3$, $z_0 = 5$, $x_1 = 2$, $y_1 = 8$, and $z_1 = 5$. We find that

$$\mathbf{r}(t) = (4 - 2t)\mathbf{i} + (3 + 5t)\mathbf{j} + 5\mathbf{k} \quad \text{for } 0 \leq t \leq 1$$

is a smooth parametrization of the line segment.   ❏

Length of a Curve

By means of a method similar to the ones we employed in Sections 6.2 and 6.7 to define length for curves in the plane, we will now assign length to curves in space.

In studying the length of curves we must be careful to avoid any parametrization that traces out a curve more than once. For example, the standard unit circle is traced out twice by the parametrization

$$\mathbf{r}(t) = \cos t\mathbf{i} + \sin t\mathbf{j} \quad \text{for } 0 \leq t \leq 4\pi$$

If we employed $\mathbf{r}$ to compute the length (circumference) of the circle, our answer would be twice the length we normally attribute to the circle. An admissible parametrization will trace out a curve only once, except that finitely many points may be traced out more than once, as happens with the parametrization

$$\mathbf{r}(t) = \cos t\mathbf{i} + \sin t\mathbf{j} \quad \text{for } 0 \leq t \leq 2\pi$$

of the standard unit circle. In this case the single point $(1, 0)$ is traced out twice, for $t = 0$ and $t = 2\pi$.

We begin by finding the length of the simplest type of curve. Let C be the line segment joining the points (x_0, y_0, z_0) and (x_1, y_1, z_1) in space (Figure 12.27). Then we define the length L of C as the distance between the two points by the formula for distance given in (1) of Section 11.1:

$$L = \sqrt{(x_1 - x_0)^2 + (y_1 - y_0)^2 + (z_1 - z_0)^2}$$

Next, let C be a smooth curve and

$$\mathbf{r}(t) = x(t)\mathbf{i} + y(t)\mathbf{j} + z(t)\mathbf{k} \quad \text{for } a \leq t \leq b$$

be a smooth parametrization of C. Let $P = \{t_0, t_1, t_2, \ldots, t_n\}$ be any partition of $[a, b]$, and consider the (black) polygonal curve shown in Figure 12.28. We denote the length of the kth portion of C by ΔL_k. If Δt_k is small, then ΔL_k is approximately equal to the length of the corresponding line segment. In other words,

$$\Delta L_k \approx \sqrt{[x(t_k) - x(t_{k-1})]^2 + [y(t_k) - y(t_{k-1})]^2 + [z(t_k) - z(t_{k-1})]^2}$$

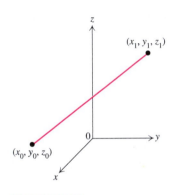

FIGURE 12.27

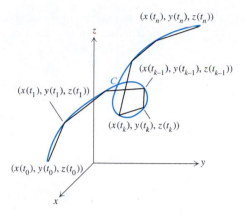

By the Mean Value Theorem there are numbers u_k, v_k, and w_k in $[t_{k-1}, t_k]$ such that

$$x(t_k) - x(t_{k-1}) = x'(u_k)\Delta t_k, \quad y(t_k) - y(t_{k-1}) = y'(v_k)\Delta t_k,$$

$$z(t_k) - z(t_{k-1}) = z'(w_k)\Delta t_k$$

where $\Delta t_k = t_k - t_{k-1}$. Consequently

$$\Delta L_k \approx \sqrt{(x'(u_k))^2(\Delta t_k)^2 + (y'(v_k))^2(\Delta t_k)^2 + (z'(w_k))^2(\Delta t_k)^2}$$

or equivalently,

$$\Delta L_k \approx \sqrt{(x'(u_k))^2 + (y'(v_k))^2 + (z'(w_k))^2}\ \Delta t_k$$

Therefore the total length L of C, which is the sum of $\Delta L_1, \Delta L_2, \ldots, \Delta L_n$, should be approximately

$$\sum_{k=1}^{n} \sqrt{(x'(u_k))^2 + (y'(v_k))^2 + (z'(w_k))^2}\ \Delta t_k$$

Since

$$\lim_{\|P\| \to 0} \sum_{k=1}^{n} \sqrt{(x'(u_k))^2 + (y'(v_k))^2 + (z'(w_k))^2}\ \Delta t_k = \int_a^b \sqrt{(x'(t))^2 + (y'(t))^2 + (z'(t))^2}\ dt$$

and since

$$\sqrt{(x'(t))^2 + (y'(t))^2 + (z'(t))^2} = \|\mathbf{r}'(t)\|$$

it follows that L should be approximately $\int_a^b \|\mathbf{r}'(t)\|\ dt$. The same approximation occurs when C is only piecewise smooth. This leads us to the definition of the length of C.

DEFINITION 12.16   Let C be a curve with a piecewise smooth parametrization $\mathbf{r}$ defined on $[a, b]$. Then the **length** L of C is defined by

$$L = \int_a^b \| \mathbf{r}'(t) \| \, dt = \int_a^b \left\| \frac{d\mathbf{r}}{dt} \right\| dt \qquad (2)$$

If

$$\mathbf{r}(t) = x(t)\mathbf{i} + y(t)\mathbf{j} + z(t)\mathbf{k} \quad \text{for } a \leq t \leq b$$

then (2) can be rewritten as

$$
\begin{aligned}
L &= \int_a^b \sqrt{(x'(t))^2 + (y'(t))^2 + (z'(t))^2} \, dt \\[2mm]
&= \int_a^b \sqrt{\left(\frac{dx}{dt}\right)^2 + \left(\frac{dy}{dt}\right)^2 + \left(\frac{dz}{dt}\right)^2} \, dt
\end{aligned}
\qquad (3)
$$

EXAMPLE 6   Find the length L of the segment of the circular helix

$$\mathbf{r}(t) = \cos t\,\mathbf{i} + \sin t\,\mathbf{j} + t\mathbf{k} \quad \text{for } 0 \leq t \leq 2\pi$$

Solution   Applying (3), we have

$$L = \int_0^{2\pi} \sqrt{(-\sin t)^2 + (\cos t)^2 + 1^2} \, dt = \int_0^{2\pi} \sqrt{2} \, dt = 2\sqrt{2}\pi \quad \square$$

EXAMPLE 7   Find the length L of the twisted cubic curve

$$\mathbf{r}(t) = t\mathbf{i} + \frac{\sqrt{6}}{2} t^2\mathbf{j} + t^3\mathbf{k} \quad \text{for } -1 \leq t \leq 1$$

Solution   Notice that

$$\mathbf{r}'(t) = \mathbf{i} + \sqrt{6}\, t\mathbf{j} + 3t^2\mathbf{k} \quad \text{and thus} \quad \| \mathbf{r}'(t) \| = \sqrt{1 + 6t^2 + 9t^4}$$

Therefore it follows from (3) that

$$L = \int_{-1}^1 \sqrt{1 + 6t^2 + 9t^4} \, dt = \int_{-1}^1 (1 + 3t^2) \, dt = (t + t^3)\Big|_{-1}^1 = 4 \quad \square$$

If the function in Example 7 were slightly different, say,

$$\mathbf{r}_1(t) = t\mathbf{i} + t^2\mathbf{j} + t^3\mathbf{k} \quad \text{for } -1 \leq t \leq 1$$

then the length of the corresponding curve would be

$$L = \int_{-1}^1 \sqrt{1 + 4t^2 + 9t^4} \, dt$$

Unfortunately, it is impossible to evaluate this integral by elementary means; its value can only be approximated. For this one might use Simpson's Rule or a computer algebra system. In fact, the lengths of most curves can only be approximated.

We close the discussion by noting that if C is composed of two curves C_1 and C_2, each of which is smooth, then the length of C is the sum of the lengths of C_1 and C_2.

Independence of Parametrization

Every curve has many parametrizations. For example, the line segment from $(\frac{1}{4}, \frac{1}{4}, \frac{1}{4})$ to $(1, 1, 1)$ is parametrized by each of the following functions:

$$\mathbf{r}_1(t) = t\mathbf{i} + t\mathbf{j} + t\mathbf{k} \qquad \text{for } \frac{1}{4} \le t \le 1$$

$$\mathbf{r}_2(t) = (t-2)\mathbf{i} + (t-2)\mathbf{j} + (t-2)\mathbf{k} \qquad \text{for } \frac{9}{4} \le t \le 3$$

$$\mathbf{r}_3(t) = t^2\mathbf{i} + t^2\mathbf{j} + t^2\mathbf{k} \qquad \text{for } \frac{1}{2} \le t \le 1$$

If we compute the length of the line segment by applying Definition 12.16 to $\mathbf{r}_1$, $\mathbf{r}_2$, and $\mathbf{r}_3$, we obtain the same length for all three parametrizations. It can be shown that we would obtain the same length for any piecewise smooth curve by using any of its piecewise smooth parametrizations. We express this by saying that the length of the curve is **independent of parametrization.**

The Arc Length Function

Let C be a smooth curve parametrized on an interval I by

$$\mathbf{r}(t) = x(t)\mathbf{i} + y(t)\mathbf{j} + z(t)\mathbf{k} \quad \text{for } t \text{ in } I$$

and let a be a fixed number in I. We define the **arc length function** s by

$$s(t) = \int_a^t \| \mathbf{r}'(u) \| \, du = \int_a^t \sqrt{(x'(u))^2 + (y'(u))^2 + (z'(u))^2} \, du \quad \text{for } t \text{ in } I \quad (4)$$

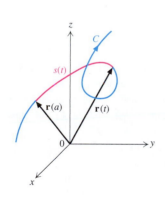

FIGURE 12.29

Notice that if $t \ge a$, then $s(t)$ is the length of the portion of the curve between $\mathbf{r}(a)$ and $\mathbf{r}(t)$ (Figure 12.29), and if $\mathbf{r}(t)$ denotes the position of an object at time $t \ge a$, then $s(t)$ is the distance traveled by the object between time a and time t.

If we differentiate the expressions in (4) with respect to t, we obtain

$$\frac{ds}{dt} = s'(t) = \| \mathbf{r}'(t) \| = \sqrt{(x'(t))^2 + (y'(t))^2 + (z'(t))^2}$$

or equivalently,

$$\frac{ds}{dt} = \left\| \frac{d\mathbf{r}}{dt} \right\| = \sqrt{\left(\frac{dx}{dt}\right)^2 + \left(\frac{dy}{dt}\right)^2 + \left(\frac{dz}{dt}\right)^2} \qquad (5)$$

For later reference we mention that if $\mathbf{r}$ denotes the position of an object in motion and $\mathbf{v}$ its velocity, then from (5) and the definition of speed in Section 12.3 we have

$$\frac{ds}{dt} = \| \mathbf{r}'(t) \| = \| \mathbf{v}(t) \| \ge 0 \qquad (6)$$

Moreover, since **r** is smooth, it follows that $ds/dt > 0$ at each interior point of I. Consequently s is an increasing function of t on I, so that t can be regarded as a function of s (see Theorem 7.3). Therefore any quantity depending on t also depends on s. In particular, if C is parametrized by $\mathbf{r}(t)$ for t in $[a, b]$ and if C has length L, then C can be parametrized by $\mathbf{r}(t(s))$ for s in $[0, L]$.

We mentioned earlier that it is usually impossible to compute the length of a curve. Consequently it is usually impossible as well to find a formula for the arc length function. However, ds/dt can often be computed by means of equation (5), without computing any lengths of curves, and in Section 12.5 we will need to do so.

EXAMPLE 8   Suppose that

$$\mathbf{r}(t) = t\mathbf{i} + t^2\mathbf{j} + t^3\mathbf{k}$$

Find ds/dt.

Solution   Using (5), we find that

$$\frac{ds}{dt} = \sqrt{\left(\frac{dx}{dt}\right)^2 + \left(\frac{dy}{dt}\right)^2 + \left(\frac{dz}{dt}\right)^2}$$

$$= \sqrt{1 + (2t)^2 + (3t^2)^2} = \sqrt{1 + 4t^2 + 9t^4} \quad \square$$

EXERCISES 12.4

In Exercises 1–10 determine which of the parametrizations are smooth, which are piecewise smooth, and which are neither.

1. $\mathbf{r}(t) = t\mathbf{i} + t^2\mathbf{j} + t^3\mathbf{k}$
2. $\mathbf{r}(t) = (t-1)\mathbf{i} + (t-1)\mathbf{j} + (t-1)\mathbf{k}$
3. $\mathbf{r}(t) = |t|\,\mathbf{i} + t\mathbf{j} + t\mathbf{k}$
4. $\mathbf{r}(t) = t^{2/3}\mathbf{i} + t\mathbf{j} + t^2\mathbf{k}$
5. $\mathbf{r}(t) = (1+t)^{3/2}\mathbf{i} + (1-t)^{3/2}\mathbf{j} + \frac{3}{2}t\mathbf{k}$
6. $\mathbf{r}(t) = \sin t\mathbf{i} + \cos t\mathbf{j} + t^2\mathbf{k}$
7. $\mathbf{r}(t) = \cos^2 t\mathbf{i} + \sin^2 t\mathbf{j} + t^2\mathbf{k}$
8. $\mathbf{r}(t) = e^t\mathbf{i} + e^{-t}\mathbf{j} + 2t\mathbf{k}$
9. $\mathbf{r}(t) = (e^t - t)\mathbf{i} + t^2\mathbf{j} + t^3\mathbf{k}$
10. $\mathbf{r}(t) = 2t\mathbf{i} + t^2\mathbf{j} + \ln t\mathbf{k}$

In Exercises 11–20 find a smooth parametrization of the curve described.

11. The straight line from $(-3, 2, 1)$ to $(4, 0, 5)$
12. The straight line from $(0, 3, -2)$ to $(6, \frac{1}{2}, -2)$
13. The circle in the xy plane centered at the origin with radius 6
14. The circle in the plane $z = -1$ centered at $(2, 4, -1)$ with radius $\frac{5}{2}$

15. The semicircle in the xy plane that passes through $(1, 0)$, $(0, 1)$, and $(-1, 0)$
16. The quarter circle in the xy plane whose endpoints are $(1, 0)$ and $(0, -1)$ and whose center is the origin
17. The quarter circle in the plane $z = 4$ whose endpoints are $(\sqrt{2}/2, \sqrt{2}/2, 4)$ and $(\sqrt{2}/2, -\sqrt{2}/2, 4)$ and whose center is $(0, 0, 4)$
18. The graph of f, where $f(x) = x^2 + 1$
19. The graph of $y = \tan x$ for $0 \le x \le \pi/4$
20. The graph of $y = x^5 - x^2 + 5$ for $-1 \le x \le 0$

In Exercises 21–28 find the length L of the curve.

21. $\mathbf{r}(t) = \cos^3 t\mathbf{i} + \sin^3 t\mathbf{j}$ for $0 \le t \le 2\pi$
22. $\mathbf{r}(t) = 2t\mathbf{i} + t^2\mathbf{j} + \ln t\mathbf{k}$ for $1 \le t \le 2$
23. $\mathbf{r}(t) = \frac{1}{3}(1+t)^{3/2}\mathbf{i} + \frac{1}{3}(1-t)^{3/2}\mathbf{j} + \frac{1}{2}t\mathbf{k}$ for $-1 \le t \le 1$
24. $\mathbf{r}(t) = \cosh t\mathbf{i} + \sinh t\mathbf{j} + t\mathbf{k}$ for $0 \le t \le 1$
25. $\mathbf{r}(t) = e^t\mathbf{i} + e^{-t}\mathbf{j} + \sqrt{2}t\mathbf{k}$ for $0 \le t \le 1$
26. $\mathbf{r}(t) = \cos t\mathbf{i} + \sin t\mathbf{j} + t^{3/2}\mathbf{k}$ for $0 \le t \le \frac{20}{3}$
27. $\mathbf{r}(t) = 2(t^2 - 1)^{3/2}\mathbf{i} + 3t^2\mathbf{j} + 3t^2\mathbf{k}$ for $1 \le t \le \sqrt{8}$
28. $\mathbf{r}(t) = (3t - t^3)\mathbf{i} + 3t^2\mathbf{j} + (3t + t^3)\mathbf{k}$ for $0 \le t \le 1$

In Exercises 29–33 find ds/dt.

29. $\mathbf{r}(t) = \sin 2t\mathbf{i} + \cos 2t\mathbf{j} + \frac{2}{3}t^{3/2}\mathbf{k}$

30. $\mathbf{r}(t) = \frac{1}{3}t^3\mathbf{i} + \frac{\sqrt{2}}{2}t^2\mathbf{j} + t\mathbf{k}$

31. $\mathbf{r}(t) = t\cos t\mathbf{i} + t\sin t\mathbf{j} + t\mathbf{k}$

32. $\mathbf{r}(t) = 2t\mathbf{i} + t^2\mathbf{j} + \frac{1}{3}t^3\mathbf{k}$

33. $\mathbf{r}(t) = (t - \sin t)\mathbf{i} + (1 - \cos t)\mathbf{j} + t\mathbf{k}$

In Exercises 34–37 use Simpson's Rule with $n = 10$ to approximate the length L of the curve.

34. $\mathbf{r}(t) = \sin t\mathbf{i} + \cos t\mathbf{j} + \frac{1}{3}t^3\mathbf{k}$ for $-1 \le t \le 1$

35. $\mathbf{r}(t) = \cos 2t\mathbf{i} + \sin 2t\mathbf{j} + \frac{4}{5}t^{5/2}\mathbf{k}$ for $0 \le t \le 2$

36. $\mathbf{r}(t) = t\mathbf{i} + t^2\mathbf{j} + t^3\mathbf{k}$ for $-1 \le t \le 1$

37. $\mathbf{r}(t) = \frac{1}{2}t^2\mathbf{i} + \frac{1}{3}t^3\mathbf{j} + \frac{1}{4}t^4\mathbf{k}$ for $0 \le t \le 1$

In Exercises 38–39 find $s(t)$. Take $a = 0$.

38. $\mathbf{r}(t) = r(t - \sin t)\mathbf{i} + r(1 - \cos t)\mathbf{j}$ for $0 \le t \le 2\pi$

39. $\mathbf{r}(t) = 2t\mathbf{i} + \sin t\mathbf{j} - \cos t\mathbf{k}$ for $0 \le t \le \pi$

*40. Let $\mathbf{r}$ be a piecewise smooth function defined on an interval I. Then $\mathbf{r}$ is said to be **parametrized by arc length** if there is a number a such that $s(t) = t - a$ for all t in I. Show that $\mathbf{r}$ is parametrized by arc length if and only if $\|\mathbf{r}'(t)\| = 1$ for all t in I.

41. The **Cauchy-Crofton formula** gives an estimate of the length of a plane curve. To present the formula, we let $\mathscr{F}$ be a family of parallel lines spaced a distance $d > 0$ apart, and consider the family $\mathscr{G}$ of lines obtained by rotating each line in $\mathscr{F}$ through angles of 0, $\pi/4$, $\pi/2$, and $3\pi/4$ radians about a fixed point in the plane. If C is the curve whose length L we wish to estimate, then let n be the total number of intersections of the lines in $\mathscr{G}$ with C, counting each time each line crosses C. Then

$$L \approx \frac{\pi n d}{8} \qquad (7)$$

a. Show that (7) yields the exact circumference of a circle if two lines tangent to the circle are rotated about the center of the circle.

*b. Using (7), approximate the length $\int_{-1}^{1} \sqrt{1 + x^4}\,dx$ of the curve $y = x^3/3$ for $-1 \le x \le 1$. Rotate the lines $x = -1$, $x = -\frac{2}{3}$, $x = -\frac{1}{3}$, $x = 0$, $x = \frac{1}{3}$, $x = \frac{2}{3}$, and $x = 1$ about the origin.

Application

42. The radius of each helix in a DNA molecule is approximately 10^{-8} centimeters, and there are about 2.9×10^8 complete turns. Over each complete turn the helix climbs approximately 3.4×10^{-8} centimeters up the axis of the helix. Determine the length of each helix in a DNA molecule.

Project

1. a. Consider the curve C_1 parametrized by

$$\mathbf{r}(t) = \frac{1 - t^2}{1 + t^2}\mathbf{i} + \frac{2t}{1 + t^2}\mathbf{j}$$

  i. Show that C_1 is a part of the unit circle centered at the origin, and find $\mathbf{r}(-1)$, $\mathbf{r}(0)$, and $\mathbf{r}(1)$.

  ii. Find $\lim_{t \to -\infty} \mathbf{r}(t)$ and $\lim_{t \to \infty} \mathbf{r}(t)$.

  iii. Show that C_1 is the entire unit circle except one point, and find that point.

b. Consider the curve C_2 parametrized by

$$\mathbf{r}(t) = \frac{t^2 + 1}{t^2 - 1}\mathbf{i} + \frac{2t}{t^2 - 1}\mathbf{j}$$

  i. Show that C_2 is a part of the hyperbola $x^2 - y^2 = 1$.

  ii. Determine which points, if any, of the hyperbola *are not* on C_2.

c. Consider the curve C_3 parametrized by

$$\mathbf{r}(t) = \frac{1 - t^2}{1 + t^2}\mathbf{i} + \frac{t}{1 + t^2}\mathbf{j}$$

Use the ideas of part (a) to show that C_3 is a portion of an ellipse, and find an equation of the ellipse in the form $x^2/a^2 + y^2/b^2 = 1$.

12.5   TANGENTS AND NORMALS TO CURVES

Tangent lines and normal lines to the graph of a real-valued function f are defined by means of the derivative of f. Now we will use the derivatives of vector-valued functions to make analogous definitions for curves in space. These tangents and normals will be vectors rather than lines.

Tangents to Curves

Because any smooth curve necessarily has a smooth parametrization, we will assume in this section that any parametrization we encounter for a smooth curve is smooth.

DEFINITION 12.17

Let C be a smooth curve and $\mathbf{r}$ a (smooth) parametrization of C defined on an interval I. Then for any interior point t of I, the **tangent vector** $\mathbf{T}(t)$ at the point $\mathbf{r}(t)$ is defined by

$$\mathbf{T}(t) = \frac{\mathbf{r}'(t)}{\|\mathbf{r}'(t)\|} = \frac{d\mathbf{r}/dt}{\|d\mathbf{r}/dt\|} \tag{1}$$

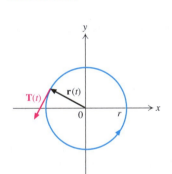

FIGURE 12.30

Since $\mathbf{T}(t)$ is a unit vector, it is sometimes called a **unit tangent vector,** and is uniquely determined by its direction. In Section 12.3 we saw that the vector $\mathbf{r}'(t)$ can be regarded as tangent to the curve traced out by $\mathbf{r}$. Since $\mathbf{r}'(t)/\|\mathbf{r}'(t)\|$ has the same direction as $\mathbf{r}'(t)$, our name for $\mathbf{T}(t)$ is reasonable (Figure 12.30).

EXAMPLE 1   Consider the circle of radius r parametrized by

$$\mathbf{r}(t) = r\cos t\mathbf{i} + r\sin t\mathbf{j} \quad \text{for } 0 \leq t \leq 2\pi$$

Find a formula for the tangent vector $\mathbf{T}(t)$, and calculate $\mathbf{T}(5\pi/6)$.

Solution   Since

$$\mathbf{r}'(t) = -r\sin t\mathbf{i} + r\cos t\mathbf{j}$$

we have

$$\|\mathbf{r}'(t)\| = \sqrt{(-r\sin t)^2 + (r\cos t)^2} = r$$

Therefore by (1),

$$\mathbf{T}(t) = \frac{\mathbf{r}'(t)}{\|\mathbf{r}'(t)\|} = -\sin t\mathbf{i} + \cos t\mathbf{j}$$

Consequently

$$\mathbf{T}\left(\frac{5\pi}{6}\right) = -\frac{1}{2}\mathbf{i} - \frac{\sqrt{3}}{2}\mathbf{j} \quad \square$$

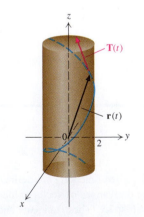

FIGURE 12.31

Notice that for the position function $\mathbf{r}(t)$ in Example 1 we have

$$\mathbf{T}(t) \cdot \mathbf{r}(t) = (-\sin t)(r\cos t) + (\cos t)(r\sin t) = 0 \quad \text{for all } t$$

This means that the tangent vector is perpendicular to the position vector for all t (Figure 12.31).

EXAMPLE 2   Find a formula for the tangent $\mathbf{T}(t)$ to the circular helix

$$\mathbf{r}(t) = 2\cos t\mathbf{i} + 2\sin t\mathbf{j} + 3t\mathbf{k}$$

Solution   First we calculate the derivative and its length:

$$\frac{d\mathbf{r}}{dt} = -2\sin t\mathbf{i} + 2\cos t\mathbf{j} + 3\mathbf{k}$$

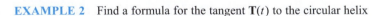

FIGURE 12.32

$$\left\| \frac{d\mathbf{r}}{dt} \right\| = \sqrt{(-2 \sin t)^2 + (2 \cos t)^2 + 3^2} = 13$$

Hence we conclude from (1) that

$$\mathbf{T}(t) = \frac{d\mathbf{r}/dt}{\| \, d\mathbf{r}/dt \, \|} = -\frac{2}{\sqrt{13}} \sin t\,\mathbf{i} + \frac{2}{\sqrt{13}} \cos t\,\mathbf{j} + \frac{3}{\sqrt{13}}\,\mathbf{k}$$

(Figure 12.32).   ❑

Normals to Curves

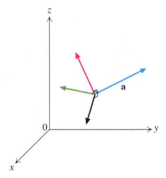

FIGURE 12.33

The line normal to the graph of a real-valued function at a given point is by definition perpendicular to the tangent at that point. Similarly, in defining a vector normal to a curve in space at a given point on the curve, we wish the normal to be perpendicular to the tangent vector. However, since there are many vectors in space that are perpendicular to any given vector (Figure 12.33), we need a way of choosing only one of them.

Now suppose a smooth curve C has a parametrization $\mathbf{r}$ that not only is smooth but also has a smooth derivative $\mathbf{r}'$. Then the tangent $\mathbf{T}$, which is defined in terms of $\mathbf{r}'$, is differentiable. Moreover, $\| \mathbf{T}(t) \| = 1$ for all t in the domain of $\mathbf{T}$. It follows from Corollary 12.11 that whenever $\mathbf{T}'(t)$ exists, it satisfies

$$\mathbf{T}'(t) \cdot \mathbf{T}(t) = 0$$

Therefore if $\mathbf{T}'(t) \neq \mathbf{0}$ then $\mathbf{T}'(t)$ is perpendicular to $\mathbf{T}(t)$, and the unit vector in the direction of $\mathbf{T}'(t)$ is our choice for the normal to C at $\mathbf{r}(t)$.

DEFINITION 12.18

Let C be a smooth curve, and let $\mathbf{r}$ be a (smooth) parametrization of C defined on an interval I such that $\mathbf{r}'$ is smooth. Then for any interior point t of I for which $\mathbf{T}'(t) \neq \mathbf{0}$, the **normal vector $\mathbf{N}\,(t)$** at the point $\mathbf{r}(t)$ is defined by

$$\mathbf{N}(t) = \frac{T'(t)}{\| \, \mathbf{T}'(t) \, \|} = \frac{d\mathbf{T}/dt}{\| \, d\mathbf{T}/dt \, \|} \tag{2}$$

Notice that the normal vector $\mathbf{N}(t)$ is a unit vector. For that reason it is sometimes called a **unit normal vector** or **principal normal vector.**

EXAMPLE 3   Find a formula for the normal $\mathbf{N}(t)$ to the circle

$$\mathbf{r}(t) = r \cos t\,\mathbf{i} + r \sin t\,\mathbf{j}$$

Solution   From Example 1 we know that

$$\mathbf{T}(t) = -\sin t\,\mathbf{i} + \cos t\,\mathbf{j}$$

Therefore

$$\mathbf{T}'(t) = -\cos t\,\mathbf{i} - \sin t\,\mathbf{j}$$

Since $\| \mathbf{T}'(t) \| = 1$ for all t, it follows from (2) that

$$\mathbf{N}(t) = \frac{\mathbf{T}'(t)}{\| \, \mathbf{T}'(t) \, \|} = -\cos t\,\mathbf{i} - \sin t\,\mathbf{j} = -\frac{1}{r}\,\mathbf{r}(t) \quad ❑$$

We can conclude from Example 3 that the normal to the path of an object moving counterclockwise about a circle points toward the center of the circle,

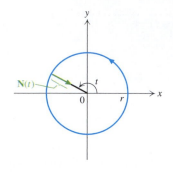

FIGURE 12.34 The normal to a circle points toward the center.

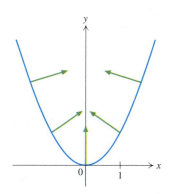

FIGURE 12.35 Normals to the parabola $\mathbf{r}(t) = t\mathbf{i} + t^2\mathbf{j}$.

opposite to the radial vector (Figure 12.34). If the circle is traversed clockwise, then the resulting normal also points toward the center of the circle. In either case the normal vector is perpendicular to the tangent vector at any point on the circle.

EXAMPLE 4 Find a formula for the normal $\mathbf{N}(t)$ for the parabola

$$\mathbf{r}(t) = t\mathbf{i} + t^2\mathbf{j}$$

Solution First we find that

$$\frac{d\mathbf{r}}{dt} = \mathbf{i} + 2t\mathbf{j} \quad \text{and} \quad \left\|\frac{d\mathbf{r}}{dt}\right\| = \sqrt{1 + 4t^2}$$

Therefore

$$\mathbf{T}(t) = \frac{1}{\sqrt{1 + 4t^2}}\mathbf{i} + \frac{2t}{\sqrt{1 + 4t^2}}\mathbf{j}$$

Then

$$\mathbf{T}'(t) = \frac{-4t}{(1 + 4t^2)^{3/2}}\mathbf{i} + \frac{2}{(1 + 4t^2)^{3/2}}\mathbf{j} \quad \text{and} \quad \|\mathbf{T}'(t)\| = \frac{2}{1 + 4t^2}$$

so we obtain the formula

$$\mathbf{N}(t) = \frac{-2t}{\sqrt{1 + 4t^2}}\mathbf{i} + \frac{1}{\sqrt{1 + 4t^2}}\mathbf{j}$$

In Figure 12.35 we have drawn a few normal vectors for the parabola. ❏

The curves we investigated in Examples 3 and 4 lie in the xy plane. Wherever either of the curves is concave upward, the normal vector has a positive $\mathbf{j}$ component (that is, the normal vector points upward), and wherever either is concave downward, the normal vector has a negative $\mathbf{j}$ component (that is, the normal vector points downward). It is possible to show that these properties are shared by any smooth curve in the xy plane. Thus the normal vectors for the graph of the sine function are as shown in Figure 12.36. You can show that $\mathbf{T}'(\pi) = \mathbf{0}$, so that $\mathbf{N}(\pi)$ does not exist. (See Exercise 27.) More generally, at inflection points of a smooth curve in the xy plane there is no normal vector.

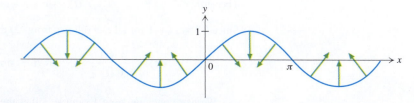

FIGURE 12.36 The sine curve: $\mathbf{r}(t) = t\mathbf{i} + \sin t\mathbf{j}$.

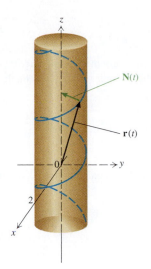

FIGURE 12.37

EXAMPLE 5   Find a formula for the normal $\mathbf{N}(t)$ to the circular helix

$$\mathbf{r}(t) = 2\cos t\mathbf{i} + 2\sin t\mathbf{j} + 3t\mathbf{k}$$

Solution   From Example 2 we have

$$\mathbf{T}(t) = -\frac{2}{\sqrt{13}}\sin t\mathbf{i} + \frac{2}{\sqrt{13}}\cos t\mathbf{j} + \frac{3}{\sqrt{13}}\mathbf{k}$$

Therefore

$$\frac{d\mathbf{T}}{dt} = -\frac{2}{\sqrt{13}}\cos t\mathbf{i} - \frac{2}{\sqrt{13}}\sin t\mathbf{j} \quad \text{and} \quad \left\|\frac{d\mathbf{T}}{dt}\right\| = \frac{2}{\sqrt{13}}$$

Consequently from (2) we deduce that

$$\mathbf{N}(t) = \frac{d\mathbf{T}/dt}{\|\,d\mathbf{T}/dt\,\|} = \frac{-(2/\sqrt{13})\cos t\mathbf{i} - (2/\sqrt{13})\sin t\mathbf{j}}{2/\sqrt{13}}$$

$$= -\cos t\mathbf{i} - \sin t\mathbf{j} \quad \square$$

We see from Example 5 that the vector normal to the helix is always perpendicular to and directed toward the z axis (Figure 12.37).

Tangential and Normal Components of Acceleration

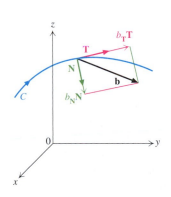

FIGURE 12.38

Since the tangent vector $\mathbf{T}$ and the normal vector $\mathbf{N}$ at any point on a smooth curve C are perpendicular, it follows from our discussion in Section 11.3 of the resolution of a vector that any vector $\mathbf{b}$ in the plane determined by $\mathbf{T}$ and $\mathbf{N}$ can be expressed in the form

$$\mathbf{b} = b_{\mathbf{T}}\mathbf{T} + b_{\mathbf{N}}\mathbf{N}$$

(Figure 12.38). We call $b_{\mathbf{T}}$ and $b_{\mathbf{N}}$ the **tangential** and **normal components** of $\mathbf{b}$, respectively. Our next tasks will be to show that the velocity and acceleration vectors of an object moving along C lie in the plane determined by $\mathbf{T}$ and $\mathbf{N}$ and to find their tangential and normal components.

To consider the velocity $\mathbf{v}$, we let $\mathbf{r}$ denote the position of the object and assume that $\mathbf{T}$ and $\mathbf{N}$ exist. Then by (1) we have

$$\mathbf{v} = \frac{d\mathbf{r}}{dt} = \left\|\frac{d\mathbf{r}}{dt}\right\|\mathbf{T} = \|\mathbf{v}\|\mathbf{T} \tag{3}$$

Thus the tangential component of $\mathbf{v}$ is the speed $\|\mathbf{v}\|$ of the object, and the normal component of the velocity is 0.

To obtain the acceleration $\mathbf{a}$, we differentiate the left-hand and right-hand expressions in (3), obtaining

$$\mathbf{a} = \frac{d\mathbf{v}}{dt} = \frac{d\|\mathbf{v}\|}{dt}\mathbf{T} + \|\mathbf{v}\|\frac{d\mathbf{T}}{dt}$$

Since $d\mathbf{T}/dt = \|d\mathbf{T}/dt\|\mathbf{N}$ from (2), the equation for $\mathbf{a}$ becomes

$$\mathbf{a} = \frac{d\|\mathbf{v}\|}{dt}\mathbf{T} + \|\mathbf{v}\|\left\|\frac{d\mathbf{T}}{dt}\right\|\mathbf{N}$$

Hence

$$\mathbf{a} = a_{\mathbf{T}}\mathbf{T} + a_{\mathbf{N}}\mathbf{N} \tag{4}$$

where
$$a_{\mathbf{T}} = \frac{d\|\mathbf{v}\|}{dt} \quad \text{and} \quad a_{\mathbf{N}} = \|\mathbf{v}\| \left\|\frac{d\mathbf{T}}{dt}\right\| \qquad (5)$$

The numbers $a_{\mathbf{T}}$ are $a_{\mathbf{N}}$ are the **tangential** and **normal components of acceleration**. Since $\mathbf{T}$ and $\mathbf{N}$ are mutually perpendicular unit vectors, we conclude from (4) that

$$\|\mathbf{a}\|^2 = \mathbf{a} \cdot \mathbf{a} = (a_{\mathbf{T}}\mathbf{T} + a_{\mathbf{N}}\mathbf{N}) \cdot (a_{\mathbf{T}}\mathbf{T} + a_{\mathbf{N}}\mathbf{N}) = a_{\mathbf{T}}^2 \|\mathbf{T}\|^2 + a_{\mathbf{N}}^2 \|\mathbf{N}\|^2$$

$$= a_{\mathbf{T}}^2 + a_{\mathbf{N}}^2$$

It follows that $a_{\mathbf{N}}$, which is often quite difficult to calculate by (5), can be calculated from $\mathbf{a}$ and $a_{\mathbf{T}}$:

$$a_{\mathbf{N}} = \sqrt{\|\mathbf{a}\|^2 - a_{\mathbf{T}}^2} \qquad (6)$$

EXAMPLE 6   Let
$$\mathbf{r}(t) = t^2\mathbf{i} + t\mathbf{j} + t^2\mathbf{k}$$

Find the tangential and normal components of acceleration.

Solution   By differentiating $\mathbf{r}$ we find that
$$\mathbf{v} = 2t\mathbf{i} + \mathbf{j} + 2t\mathbf{k}$$

and thus
$$\|\mathbf{v}\| = \sqrt{(2t)^2 + 1 + (2t)^2} = \sqrt{8t^2 + 1}$$

From this and (5) it follows that
$$a_{\mathbf{T}} = \frac{d\|\mathbf{v}\|}{dt} = \frac{8t}{\sqrt{8t^2 + 1}}$$

In order to calculate $a_{\mathbf{N}}$, we first notice that

$$\mathbf{a} = \frac{d\mathbf{v}}{dt} = 2\mathbf{i} + 2\mathbf{k}$$

so that
$$\|\mathbf{a}\| = \sqrt{2^2 + 2^2} = 2\sqrt{2}$$

Therefore we conclude from (6) that

$$a_{\mathbf{N}} = \sqrt{\|\mathbf{a}\|^2 - a_{\mathbf{T}}^2} = \sqrt{8 - \frac{64t^2}{8t^2 + 1}} = \frac{2\sqrt{2}}{\sqrt{8t^2 + 1}} \qquad \square$$

Alternative forms for $a_{\mathbf{T}}$ and $a_{\mathbf{N}}$ that can be easier to use are

$$a_{\mathbf{T}} = \frac{\mathbf{v} \cdot \mathbf{a}}{\|\mathbf{v}\|} \quad \text{and} \quad a_{\mathbf{N}} = \frac{\|\mathbf{v} \times \mathbf{a}\|}{\|\mathbf{v}\|} \qquad (7)$$

Indeed, from (4) and (3), and the fact that $\mathbf{T} \cdot \mathbf{T} = 1$ and $\mathbf{T} \cdot \mathbf{N} = 0$, we deduce that

$$\frac{\mathbf{v}}{\|\mathbf{v}\|} \cdot \mathbf{a} = a_{\mathbf{T}} \frac{\mathbf{v}}{\|\mathbf{v}\|} \cdot \mathbf{T} + a_{\mathbf{N}} \frac{\mathbf{v}}{\|\mathbf{v}\|} \cdot \mathbf{N} = a_{\mathbf{T}}\mathbf{T} \cdot \mathbf{T} + a_{\mathbf{N}}\mathbf{T} \cdot \mathbf{N} = a_{\mathbf{T}}$$

Analogously, by (4) and (3), along with the fact that $\mathbf{T} \times \mathbf{T} = \mathbf{0}$, we have

$$\frac{\mathbf{v}}{\|\mathbf{v}\|} \times \mathbf{a} = a_{\mathbf{T}} \frac{\mathbf{v}}{\|\mathbf{v}\|} \times \mathbf{T} + a_{\mathbf{N}} \frac{\mathbf{v}}{\|\mathbf{v}\|} \times \mathbf{N} = a_{\mathbf{T}} \mathbf{T} \times \mathbf{T} + a_{\mathbf{N}} \mathbf{T} \times \mathbf{N} = a_{\mathbf{N}} \mathbf{T} \times \mathbf{N}$$

Since $a_{\mathbf{N}} \geq 0$ by (5) and since $\| \mathbf{T} \times \mathbf{N} \| = 1$, we conclude that

$$\frac{\|\mathbf{v} \times \mathbf{a}\|}{\|\mathbf{v}\|} = |a_{\mathbf{N}}| \, \| \mathbf{T} \times \mathbf{N} \| = a_{\mathbf{N}}$$

Let C be a smooth curve parametrized by $\mathbf{r}(t)$ for t in an interval I. The plane containing the tangent and normal vectors $\mathbf{T}$ and $\mathbf{N}$ is called the **osculating plane** associated with C. In general the osculating plane varies as the point $\mathbf{r}(t)$ moves along C.

In addition to the unit vectors $\mathbf{T}$ and $\mathbf{N}$ we introduce a third vector, $\mathbf{B}$, defined by

$$\mathbf{B} = \mathbf{T} \times \mathbf{N}$$

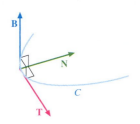

FIGURE 12.39

Then $\mathbf{B}$ is perpendicular to $\mathbf{T}$ and $\mathbf{N}$, and since $\mathbf{T}$ and $\mathbf{N}$ are unit vectors, it follows that $\mathbf{B}$ is also a unit vector. This vector is called the **binormal** vector and can be computed by means of the formula

$$\mathbf{B} = \frac{\mathbf{v} \times \mathbf{a}}{\|\mathbf{v} \times \mathbf{a}\|} \tag{8}$$

(see Exercise 31).

The vectors $\mathbf{T}$, $\mathbf{N}$, and $\mathbf{B}$ are mutually perpendicular unit vectors. As with the vectors $\mathbf{i}$, $\mathbf{j}$, and $\mathbf{k}$, if the fingers on the right hand curve around from $\mathbf{T}$ toward $\mathbf{N}$, then the thumb will point in the direction of $\mathbf{B}$ (Figure 12.39). The $\mathbf{T}$, $\mathbf{N}$, and $\mathbf{B}$ system is used in the analysis of trajectories of aircraft and spacecraft.

Orientations of Curves

Any piecewise smooth parametrization $\mathbf{r} = x\mathbf{i} + y\mathbf{j} + z\mathbf{k}$ of a curve C determines a tangent vector at all but finitely many points of C. Since each tangent vector points in the direction in which the curve is traced out by $\mathbf{r}$, we say that $\mathbf{r}$ determines an **orientation** (or **direction**) of C (Figure 12.40(a)).

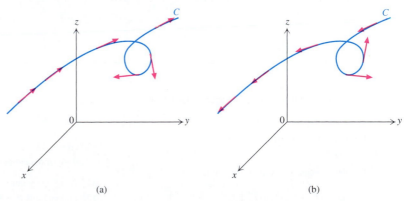

(a)   (b)

FIGURE 12.40   (a) An orientation of a piecewise smooth curve C. (b) The opposite orientation of C.

Usually there are two orientations for a given piecewise smooth curve, much as there are two directions in which one can drive on a highway. The tangent vectors for one orientation have directions opposite to the tangent vectors for the other orientation. Once a piecewise smooth curve C has a given orientation, the tangent vectors to C are uniquely defined, independent of any parametrization $\mathbf{r}$ of C.

Suppose $\mathbf{r}$ is a piecewise smooth parametrization of C on $[a, b]$, and let

$$\mathbf{r}_1(t) = \mathbf{r}(a + b - t) \quad \text{for } a \le t \le b$$

Then $\mathbf{r}_1$ is also a piecewise smooth parametrization of C and determines an orientation opposite to the orientation determined by $\mathbf{r}$ (Figure 12.40(b)). When we consider C with a given orientation, we designate the curve with the opposite orientation by $-C$. In the case of a plane curve that is closed and does not "cross itself," the two orientations are called the **counterclockwise orientation** and the **clockwise orientation** (Figure 12.41(a) and (b)).

An **oriented curve** is a piecewise smooth curve with a particular orientation associated with it. In Chapter 15 we will make use of oriented curves.

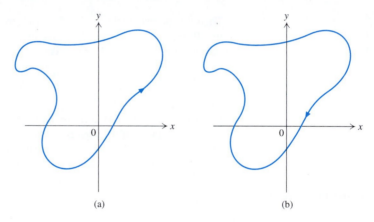

(a)   (b)

FIGURE 12.41   (a) Counterclockwise orientation of a plane curve. (b) Clockwise orientation of a plane curve.

EXERCISES 12.5

In Exercises 1–10 first show that $\| \mathbf{r}'(t) \|$ is as given. Then find the tangent and the normal of the curve parametrized by $\mathbf{r}$.

1. $\mathbf{r}(t) = (t^2 + 4)\mathbf{i} + 2t\mathbf{j}; \| \mathbf{r}'(t) \| = 2\sqrt{t^2 + 1}$

2. $\mathbf{r}(t) = \cos^3 t\mathbf{i} + \sin^3 t\mathbf{j}$ for $\pi/6 \le t \le \pi/3; \| \mathbf{r}'(t) \| = 3 \sin t \cos t$

3. $\mathbf{r}(t) = \cos t\mathbf{i} + \cos t\mathbf{j} + \sqrt{2} \sin t\mathbf{k}; \| \mathbf{r}'(t) \| = \sqrt{2}$

4. $\mathbf{r}(t) = \frac{1}{3}(1 + t)^{3/2}\mathbf{i} + \frac{1}{3}(1 - t)^{3/2}\mathbf{j} + \dfrac{\sqrt{2}}{2} t\mathbf{k}; \| \mathbf{r}'(t) \| = 1$

5. $\mathbf{r}(t) = 2t\mathbf{i} + t^2\mathbf{j} + \frac{1}{3}t^3\mathbf{k}; \| \mathbf{r}'(t) \| = 2 + t^2$

6. $\mathbf{r}(t) = \frac{4}{5} \cos t\mathbf{i} + (1 - \sin t)\mathbf{j} - \frac{3}{5} \cos t\mathbf{k}; \| \mathbf{r}'(t) \| = 1$

7. $\mathbf{r}(t) = e^t\mathbf{i} + e^{-t}\mathbf{j} + \sqrt{2} t\mathbf{k}; \| \mathbf{r}'(t) \| = e^t + e^{-t}$

8. $\mathbf{r}(t) = \cosh t\mathbf{i} + \sinh t\mathbf{j} + t\mathbf{k}; \| \mathbf{r}'(t) \| = \sqrt{2} \cosh t$

9. $\mathbf{r}(t) = 2t\mathbf{i} + t^2\mathbf{j} + \ln t\mathbf{k}; \| \mathbf{r}'(t) \| = \dfrac{2t^2 + 1}{t}$

10. $\mathbf{r}(t) = 2t^{9/2}\mathbf{i} + \frac{3}{2}\sqrt{2} t^3\mathbf{j} + \frac{3}{2}\sqrt{2} t^3\mathbf{k}; \| \mathbf{r}'(t) \| = 9t^2\sqrt{t^3 + 1}$

In Exercises 11–15 find the tangential and normal components $a_\mathbf{T}$ and $a_\mathbf{N}$ of acceleration.

11. $\mathbf{r}(t) = r(t - \sin t)\mathbf{i} + r(1 - \cos t)\mathbf{j}$

12. $\mathbf{r}(t) = 2 \cos t\mathbf{i} + 3 \sin t\mathbf{j}$

13. $\mathbf{r}(t) = 2t\mathbf{i} + t^2\mathbf{j} + \frac{1}{3}t^3\mathbf{k}$

14. $\mathbf{r}(t) = \frac{4}{5} \cos t\mathbf{i} + (1 - \sin t)\mathbf{j} - \frac{3}{5} \cos t\mathbf{k}$

15. $\mathbf{r}(t) = e^t\mathbf{i} + e^{-t}\mathbf{j} + \sqrt{2} t\mathbf{k}$

In Exercises 16–21 find a piecewise smooth (smooth if possible) parametrization with the given orientation for the curve.

16. The circle in the plane $x = -2$ centered at the point $(-2, -2, -1)$, with radius 3 and with a clockwise orientation as viewed from the yz plane

17. The semicircle in the yz plane that begins at $(0, 0, 4)$, passes through $(0, -4, 0)$, and terminates at $(0, 0, -4)$

18. The curve shown in Figure 12.42(a)

19. The curve shown in Figure 12.42(b)

20. The curve shown in Figure 12.42(c)

21. The curve shown in Figure 12.42(d)

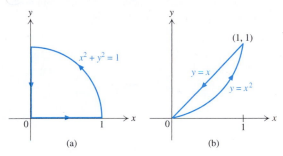

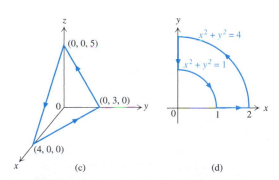

FIGURE 12.42   Figures for Exercises 18 through 21.

22. Show that the curves parametrized by

$$\mathbf{r}_1(t) = \left(\frac{1}{2}t^2 + t\right)\mathbf{i} + (t + 1)\mathbf{j} - t\mathbf{k}$$

and $\mathbf{r}_2(t) = \sin t\mathbf{i} + e^t\mathbf{j} - \tan t\mathbf{k}$

intersect at the point $(0, 1, 0)$ and that the vectors tangent to the two curves at $(0, 1, 0)$ are parallel.

23. Show that the curves parametrized by

$$\mathbf{r}_1(t) = t\mathbf{i} + 2t\mathbf{j} + t^2\mathbf{k}$$

and $\mathbf{r}_2(t) = t^2\mathbf{i} + (1 - t)\mathbf{j} + (2 - t^2)\mathbf{k}$

intersect at the point $(1, 2, 1)$ and that the vectors tangent to the two curves at $(1, 2, 1)$ are perpendicular. (*Hint:* For what values of t do $\mathbf{r}_1$ and $\mathbf{r}_2$ trace out the point $(1, 2, 1)$?)

24. Suppose an object moves in such a way that its acceleration vector is always perpendicular to its velocity

vector. Show that the speed of the object is constant.

25. Prove that the tangential component of acceleration of an object is 0 if and only if the speed is constant.

26. In Example 1 we showed that a vector tangent to a circle with center at the origin is perpendicular to the position vector. Of course, such a circle lies on the surface of a sphere centered at the origin. Using Corollary 12.11, prove more generally that the vector tangent to any smooth curve lying on the surface of a sphere centered at the origin is perpendicular to the position vector.

27. Show that the graph of the sine function has no normal vector at $(n\pi, 0)$ for any integer n (see Figure 12.36).

28. Let $\mathbf{r}(t) = x(t)\mathbf{i} + y(t)\mathbf{j}$ for $a \leq t \leq b$ be a smooth parametrization of a curve C. Find a formula for $\mathbf{T}(t)$ in terms of x and y and their derivatives.

*29. Show that $\mathbf{N} = \dfrac{(\mathbf{v} \cdot \mathbf{v})\mathbf{a} - (\mathbf{v} \cdot \mathbf{a})\mathbf{v}}{\| (\mathbf{v} \cdot \mathbf{v})\mathbf{a} - (\mathbf{v} \cdot \mathbf{a})\mathbf{v} \|}$, provided the denominator is nonzero. (*Hint:* Use (3), (4), and (7).)

30. Consider the helix parametrized by $\mathbf{r}(t) = 2 \cos t\mathbf{i} + 2 \sin t\mathbf{j} + 3t\mathbf{k}$. Use (8) to find $\mathbf{B}(t)$.

31. **a** Use (4) and the fact that $\mathbf{v}$ and $\mathbf{T}$ point in the same direction in order to show that $\mathbf{v} \times \mathbf{a}$ and $\mathbf{T} \times \mathbf{N}$ point in the same direction.

 b Using part (a) but without calculating the cross product $\mathbf{T} \times \mathbf{N}$, show that

$$\mathbf{B} = \frac{\mathbf{v} \times \mathbf{a}}{\| \mathbf{v} \times \mathbf{a} \|}$$

32. Show that $\mathbf{T}(t) = \mathbf{N}(t) \times \mathbf{B}(t)$ for all t.

Application

*33. The **rated speed** v_R of a banked curve on a road is the maximum speed a car can attain on the curve without skidding outward, under the assumption that there is no friction between the road and the tires (under icy road conditions, for example). Suppose that the curve is circular with radius ρ and that it is banked at an angle θ with respect to the horizontal (Figure 12.43). When a car of mass m traverses the curve with a constant speed v, there are two forces acting on the car: the vertical force (directed downward) due to the weight of the car, and a force $\mathbf{F_n}$ exerted by and normal to the road. At the rated speed the vertical component of $\mathbf{F_n}$ balances the weight of the car, so that

$$\| \mathbf{F_n} \| \cos \theta = mg \qquad (9)$$

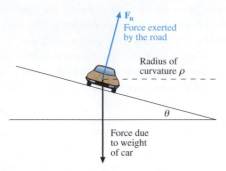

FIGURE 12.43   Figure for Exercise 33.

The horizontal component of $\mathbf{F_n}$ produces a centripetal force on the car, so that by Newton's Second Law of Motion and (5) in Section 12.3,

$$\| \mathbf{F_n} \| \sin \theta = \frac{mv_R^2}{\rho} \qquad (10)$$

a. Using (9) and (10), show that
$$v_R^2 = \rho g \tan \theta$$
b. Find the rated speed of a circular curve with radius 500 feet that is banked at an angle of $\pi/12$.
c. What radius of the curve will result in doubling the rated speed found in (b) for the same angle θ of banking?

Formula One racing car on a banked curve. (*Thomas Zimmerman/Tony Stone Images*)

Project

1. Consider a wheel of radius r rolling along the positive x axis at t radians per unit time, and assume that the wheel is tangent to the x axis at the origin at time $t = 0$. The path of a point on the rim of the wheel can be parametrized by

$$\mathbf{r}(t) = r(t - \sin t)\,\mathbf{i} + r(1 - \cos t)\,\mathbf{j} \quad \text{for } 0 \leq t \leq 2\pi$$

(Figure 12.44). Let $P(t)$ be a point on the wheel that is at the origin at time $t = 0$, and let $Q(t)$ be the point on the positive x axis that touches the wheel at time t.

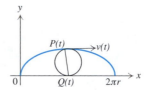

FIGURE 12.44   Figure for project.

a. Find $v(t)$.
b. Use (a) to show that $v(t)$ is perpendicular to $\overrightarrow{PQ}$ at any time t. (This means that if P is at the origin at time $t = 0$, then at any time afterwards, P moves in the direction perpendicular to the line joining P to the point of contact Q of the circle and the x axis.)
c. Calculate $\| v(t) \|$, and determine where on the wheel $P(t)$ has the greatest speed, and where it has the minimal speed.
d. Use (b) and (c) to draw $v(t)$ for $P(t)$ at several different positions on the wheel. (Note that the length of $v(t)$ is longer or shorter, depending on the location of $P(t)$ on the wheel.)
e. Use (c) to show that the speed of $P(t)$ at the top of the wheel is twice the speed of the center of the wheel.
f. Using the half-angle formulas for $\sin(t/2)$ and $\cos(t/2)$, show that $\mathbf{T}(t) = \sin(t/2)\mathbf{i} + \cos(t/2)\mathbf{j}$ for $0 < t < 2\pi$.

12.6   CURVATURE

Let C be a smooth curve. The length of the tangent vector is always 1; the direction can vary from point to point, according to the nature of the curve. For example, if the curve is a straight line, as in Figure 12.45(a), then the direction of the tangent vector is constant. If the curve undulates gently, as in Figure 12.45(b), then the tangent vector changes direction slowly along the curve. Finally, if the curve twists

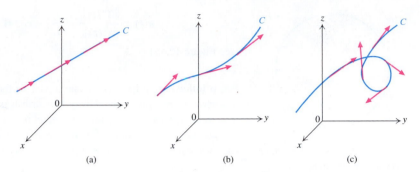

FIGURE 12.45   (a) The tangent vector is constant. (b) The tangent vector changes direction slowly. (c) The tangent vector changes direction rapidly.

as in Figure 12.45(c), then the tangent vector changes direction rapidly. Figure 12.45(a)–(c) suggests that the rate of change $d\mathbf{T}/ds$ of the tangent vector with respect to the arc length function s is closely related to the rate at which C bends in space. By the Chain Rule,

$$\frac{d\mathbf{T}}{dt} = \frac{d\mathbf{T}}{ds}\frac{ds}{dt}$$

It follows that

$$\frac{d\mathbf{T}}{ds} = \frac{d\mathbf{T}/dt}{ds/dt} = \frac{d\mathbf{T}/dt}{\|\,d\mathbf{r}/dt\,\|}$$

This suggests the following definition of the curvature κ of a curve (κ is the Greek letter kappa).

DEFINITION 12.19   Let C have a smooth parametrization $\mathbf{r}$ such that $\mathbf{r}'$ is differentiable. Then the **curvature** κ of C is defined by the formula

$$\kappa(t) = \frac{\|\,\mathbf{T}'(t)\,\|}{\|\,\mathbf{r}'(t)\,\|} = \frac{\|\,d\mathbf{T}/dt\,\|}{\|\,d\mathbf{r}/dt\,\|} \tag{1}$$

By definition $\kappa(t)$ is a nonnegative number. It is possible for $\kappa(t)$ to be 0; in fact, the curvature of a straight line is 0 at every point, since the tangent vector is constant. We can also readily determine the curvature of a circle.

EXAMPLE 1   Show that the curvature κ of a circle of radius r is $1/r$.

Solution   If the center of the circle is (x_0, y_0), then we parametrize the circle by

$$\mathbf{r}(t) = (x_0 + r\cos t)\mathbf{i} + (y_0 + r\sin t)\mathbf{j}$$

Then $\mathbf{r}'(t) = -r\sin t\mathbf{i} + r\cos t\mathbf{j}$

Moreover,

$$\|\,\mathbf{r}'(t)\,\| = r \quad \text{and} \quad \mathbf{T}(t) = -\sin t\mathbf{i} + \cos t\mathbf{j}$$

Therefore $\mathbf{T}'(t) = -\cos t\mathbf{i} - \sin t\mathbf{j}$, so that by (1),

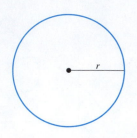

FIGURE 12.46 A circle of radius r has constant curvature $1/r$.

$$\kappa(t) = \frac{\|\mathbf{T}'(t)\|}{\|\mathbf{r}'(t)\|} = \frac{1}{r} \| - \cos t\mathbf{i} - \sin t\mathbf{j} \| = \frac{1}{r}$$

(see Figure 12.46). ☐

It follows from Example 1 that the larger the radius of a circle is, the smaller its curvature is. This explains why the earth was once thought to be flat; its very large radius gives it a curvature imperceptible to anyone on it.

Because the curvature of a circle with radius r is $1/r$, we define the **radius of curvature** $\rho(t)$ of a curve C at a point P corresponding to t to be

$$\rho(t) = \frac{1}{\kappa(t)}$$

Thus the radius of curvature of a circle is the radius. The radius of curvature is important in engineering problems.

In our next example the curve has a variable curvature.

EXAMPLE 2 Find the curvature κ of the parabola parametrized by

$$\mathbf{r}(t) = t\mathbf{i} + t^2\mathbf{j}$$

Solution First we make the necessary computations:

$$\frac{d\mathbf{r}}{dt} = \mathbf{i} + 2t\mathbf{j}$$

$$\left\| \frac{d\mathbf{r}}{dt} \right\| = \sqrt{1 + 4t^2}$$

$$\mathbf{T}(t) = \frac{1}{\sqrt{1 + 4t^2}} \mathbf{i} + \frac{2t}{\sqrt{1 + 4t^2}} \mathbf{j}$$

$$\frac{d\mathbf{T}}{dt} = \frac{-4t}{(1 + 4t^2)^{3/2}} \mathbf{i} + \frac{2}{(1 + 4t^2)^{3/2}} \mathbf{j}$$

$$\left\| \frac{d\mathbf{T}}{dt} \right\| = \frac{2}{1 + 4t^2}$$

Finally, we use (1) to obtain

$$\kappa(t) = \frac{\| d\mathbf{T}/dt \|}{\| d\mathbf{r}/dt \|} = \frac{2/(1 + 4t^2)}{(1 + 4t^2)^{1/2}} = \frac{2}{(1 + 4t^2)^{3/2}} \quad ☐$$

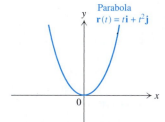

FIGURE 12.47 The maximum curvature is at the origin.

Notice that the curvature of the parabola is largest when $t = 0$ and gradually diminishes as $|t|$ increases (Figure 12.47). Since the curve is symmetric with respect to the y axis, naturally the curvature is the same for t and for $-t$.

Alternative Formulas for Curvature

By using the decomposition of the velocity and acceleration vectors into their tangential and normal components, we can obtain a formula for curvature that gives rise to simpler computations than (1) does.

Let $\mathbf{r}$ be a smooth parametrization of a smooth curve C with tangent $\mathbf{T}$ and normal $\mathbf{N}$. Recall from (3) and (4) in Section 12.5 that the velocity and acceleration of an object moving along C with position $\mathbf{r}$ are given by

$$\mathbf{v} = \|\,\mathbf{v}\,\|\,\mathbf{T} \quad \text{and} \quad \mathbf{a} = a_T\mathbf{T} + a_N\mathbf{N}$$

Then since $\mathbf{T} \times \mathbf{T} = \mathbf{0}$ we find that

$$
\begin{aligned}
\mathbf{v} \times \mathbf{a} &= (\,\|\,\mathbf{v}\,\|\,\mathbf{T}) \times (a_T\mathbf{T} + a_N\mathbf{N}) \\
&= [(\,\|\,\mathbf{v}\,\|\,\mathbf{T}) \times (a_T\mathbf{T})] + [(\,\|\,\mathbf{v}\,\|\,\mathbf{T}) \times (a_N\mathbf{N})] \\
&= (\,\|\,\mathbf{v}\,\|\,a_N)(\mathbf{T} \times \mathbf{N})
\end{aligned}
\tag{2}
$$

Since $a_N = \|\,\mathbf{v}\,\|\,\|\,d\mathbf{T}/dt\,\|$ by (5) in Section 12.5, and since $\mathbf{T}$ and $\mathbf{N}$ are perpendicular unit vectors, it follows from (2) that

$$\|\,\mathbf{v} \times \mathbf{a}\,\| = \|\,\mathbf{v}\,\|\,a_N = \|\,\mathbf{v}\,\|^2 \left\|\frac{d\mathbf{T}}{dt}\right\|$$

However,

$$\kappa = \frac{\|\,d\mathbf{T}/dt\,\|}{\|\,d\mathbf{r}/dt\,\|} = \frac{\|\,d\mathbf{T}/dt\,\|}{\|\,\mathbf{v}\,\|}$$

so that

$$\kappa = \frac{\|\,\mathbf{v} \times \mathbf{a}\,\|}{\|\,\mathbf{v}\,\|^3} \tag{3}$$

EXAMPLE 3   Using (3), find the curvature κ of the twisted cubic

$$\mathbf{r}(t) = \frac{1}{3}t^3\mathbf{i} + \frac{\sqrt{2}}{2}t^2\mathbf{j} + t\mathbf{k}$$

Solution   We have

$$\mathbf{v} = \frac{d\mathbf{r}}{dt} = t^2\mathbf{i} + \sqrt{2}\,t\mathbf{j} + \mathbf{k}$$

$$\|\,\mathbf{v}\,\| = \sqrt{t^4 + 2t^2 + 1} = t^2 + 1$$

$$\mathbf{a} = \frac{d\mathbf{v}}{dt} = 2t\mathbf{i} + \sqrt{2}\,\mathbf{j}$$

Therefore

$$\mathbf{v} \times \mathbf{a} = \begin{vmatrix} \mathbf{i} & \mathbf{j} & \mathbf{k} \\ t^2 & \sqrt{2}\,t & 1 \\ 2t & \sqrt{2} & 0 \end{vmatrix} = -\sqrt{2}\,\mathbf{i} + 2t\mathbf{j} - \sqrt{2}\,t^2\mathbf{k}$$

and thus

$$
\begin{aligned}
\|\,\mathbf{v} \times \mathbf{a}\,\| &= \sqrt{(-\sqrt{2})^2 + (2t)^2 + (-\sqrt{2}\,t^2)^2} \\
&= \sqrt{2 + 4t^2 + 2t^4} = \sqrt{2}(t^2 + 1)
\end{aligned}
$$

Consequently by (3),

$$\kappa = \frac{\|\,\mathbf{v} \times \mathbf{a}\,\|}{\|\,\mathbf{v}\,\|^3} = \frac{\sqrt{2}(t^2 + 1)}{(t^2 + 1)^3} = \frac{\sqrt{2}}{(t^2 + 1)^2} \quad \square$$

You might graph the twisted cubic curve in Example 3 on a computer to confirm that the curvature is greatest near the origin and nearly 0 far away from the origin.

In the event that $\mathbf{r}$ represents an object moving on a curve in the xy plane, we have

$$\mathbf{r} = x\mathbf{i} + y\mathbf{j}$$

$$\mathbf{v} = \frac{dx}{dt}\mathbf{i} + \frac{dy}{dt}\mathbf{j} = x'(t)\mathbf{i} + y'(t)\mathbf{j}$$

$$\mathbf{a} = \frac{d^2x}{dt^2}\mathbf{i} + \frac{d^2y}{dt^2}\mathbf{j} = x''(t)\mathbf{i} + y''(t)\mathbf{j}$$

By calculating the cross product in (3) we obtain the formula

$$\kappa = \frac{|x'(t)\,y''(t) - x''(t)\,y'(t)|}{[(x'(t))^2 + (y'(t))^2]^{3/2}} \tag{4}$$

which can be more compactly displayed if we use dots to indicate differentiation with respect to t:

$$\kappa = \frac{|\dot{x}\,\ddot{y} - \ddot{x}\,\dot{y}|}{(\dot{x}^2 + \dot{y}^2)^{3/2}}$$

If $x = t$ (so that y is also a function of x), then

$$\kappa = \frac{|y''(x)|}{[1 + (y'(x))^2]^{3/2}} \tag{5}$$

EXAMPLE 4   Let one arch of a cycloid be described by

$$\mathbf{r}(t) = r(t - \sin t)\mathbf{i} + r(1 - \cos t)\mathbf{j} \quad \text{for } 0 < t < 2\pi$$

Find the curvature κ by means of (4).

Solution   Here we have

$x(t) = r(t - \sin t)$	$y(t) = r(1 - \cos t)$
$x'(t) = r(1 - \cos t)$	$y'(t) = r \sin t$
$x''(t) = r \sin t$	$y''(t) = r \cos t$

Then by (4),

$$\kappa = \frac{|r(1 - \cos t)(r \cos t) - (r \sin t)(r \sin t)|}{[r^2(1 - \cos t)^2 + r^2 \sin^2 t]^{3/2}}$$

$$= \frac{r^2(1 - \cos t)}{[r^2(2 - 2 \cos t)]^{3/2}} = \frac{1}{2^{3/2}r\sqrt{1 - \cos t}} \quad \square$$

Notice that as t approaches 0 or 2π, the curvature becomes unbounded. In fact there is no curvature for $t = 0$ or $t = 2\pi$, which correspond to cusps on the cycloid. (See Figure 12.48.)

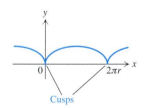

FIGURE 12.48   A cycloid.

EXERCISES 12.6

In Exercises 1–10 use (1) to find the curvature of the curve traced out by $\mathbf{r}$:

1. $\mathbf{r}(t) = (t^2 + 4)\mathbf{i} + 2t\mathbf{j}$

2. $\mathbf{r}(t) = \cos^3 t\mathbf{i} + \sin^3 t\mathbf{j}$ for $\pi/6 \le t \le \pi/3$

3. $\mathbf{r}(t) = \cos t\mathbf{i} + \cos t\mathbf{j} + \sqrt{2} \sin t\mathbf{k}$

4. $\mathbf{r}(t) = \frac{1}{3}(1 + t)^{3/2}\mathbf{i} + \frac{1}{3}(1 - t)^{3/2}\mathbf{j} + \frac{\sqrt{2}}{2}t\mathbf{k}$

5. $\mathbf{r}(t) = 2t\mathbf{i} + t^2\mathbf{j} + \frac{1}{3}t^3\mathbf{k}$

6. $\mathbf{r}(t) = \frac{4}{5}\cos t\mathbf{i} + (1 - \sin t)\mathbf{j} - \frac{3}{5}\cos t\mathbf{k}$

7. $\mathbf{r}(t) = e^t\mathbf{i} + e^{-t}\mathbf{j} + \sqrt{2}t\mathbf{k}$

8. $\mathbf{r}(t) = \cosh t\mathbf{i} + \sinh t\mathbf{j} + t\mathbf{k}$

9. $\mathbf{r}(t) = 2t\mathbf{i} + t^2\mathbf{j} + \ln t\mathbf{k}$

10. $\mathbf{r}(t) = 2t^{9/2}\mathbf{i} + \frac{3}{2}\sqrt{2}t^3\mathbf{j} + \frac{3}{2}\sqrt{2}t^3\mathbf{k}$

In Exercises 11–15 use (3) to find the curvature κ of the curve parametrized by $\mathbf{r}$.

11. $\mathbf{r}(t) = (2t - 1)\mathbf{i} + (t^2 + 1)\mathbf{j}$

12. $\mathbf{r}(t) = t\cos t\mathbf{i} + t\sin t\mathbf{j}$

13. $\mathbf{r}(t) = e^t\sin t\mathbf{i} + e^t\cos t\mathbf{j} + t\mathbf{k}$

14. $\mathbf{r}(t) = \frac{1}{3}(t^2 - 1)^{3/2}\mathbf{i} + \frac{\sqrt{2}}{4}t^2\mathbf{j} + \frac{\sqrt{2}}{4}t^2\mathbf{k}$

15. $\mathbf{r}(t) = \sin t\mathbf{i} + \cos t\mathbf{j} + \frac{2}{3}t^{3/2}\mathbf{k}$

In Exercises 16–19 use (4) to find the curvature κ of the plane curve parametrized by $\mathbf{r}$. Then find the radius of curvature at the point on the curve corresponding to the given value of t_0.

16. $\mathbf{r}(t) = 2\cos t\mathbf{i} + 3\sin t\mathbf{j}$; $t_0 = 0$

17. $\mathbf{r}(t) = 2\cos t\mathbf{i} + 3\sin t\mathbf{j}$; $t_0 = \pi/2$

18. $\mathbf{r}(t) = 2\cosh t\mathbf{i} + 3\sinh t\mathbf{j}$; $t_0 = 0$

19. $\mathbf{r}(t) = t\mathbf{i} + \frac{1}{3}t^3\mathbf{j}$; $t_0 = 1$

In Exercises 20–23 use (5) to find the curvature κ of the graph of the given equation.

20. $y = \sin x$ **21.** $y = \ln x$

22. $y = x^{1/3}$ for $x > 0$ **23.** $y = \frac{1}{x}$ for $x < 0$

24. Find all points on the ellipse $4x^2 + 9y^2 = 36$ at which the curvature is maximum and all points at which it is minimum. (*Hint:* The ellipse is parametrized by $\mathbf{r}(t) = 3\cos t\mathbf{i} + 2\sin t\mathbf{j}$ for $0 \le t \le 2\pi$.)

25. Find the point on the graph of $y = e^x$ at which the curvature is maximum.

26. Let g be a polynomial function, and define

$$f(x) = \begin{cases} -x & \text{for } x < -1 \\ g(x) & \text{for } -1 \le x \le 1 \\ x & \text{for } x > 1 \end{cases}$$

a. Show that if $g(x) = -\frac{1}{8}x^4 + \frac{3}{4}x^2 + \frac{3}{8}$, then the curvature of the graph of f is continuous.

b. Show that if g is *any* third-degree polynomial, then the curvature of the graph of f is not continuous simultaneously at $x = -1$ and $x = 1$.

27. Show that the helix $\mathbf{r}(t) = \cos t\mathbf{i} + \sin t\mathbf{j} + t\mathbf{k}$ has constant curvature.

28. Suppose the graph of f has an inflection point at $(a, f(a))$,

and assume that $f''(a)$ exists. Show that the curvature of the graph at $(a, f(a))$ is 0.

29. Using (5) in Section 12.5 and Definition 12.19, show that $a_\mathbf{N} = \kappa \| \mathbf{v} \|^2$.

30. Assume that f has a second derivative. Show that the curvature of the polar graph of the equation $r = f(\theta)$ is given by

$$\kappa(\theta) = \frac{|2[f'(\theta)]^2 - f(\theta)f''(\theta) + [f(\theta)]^2|}{[(f'(\theta))^2 + (f(\theta))^2]^{3/2}}$$

(*Hint:* Remember that $x = r\cos\theta$ and $y = r\sin\theta$. Then apply (4) to the parametrization $r(\theta) = f(\theta)\cos\theta\mathbf{i} + f(\theta)\sin\theta\mathbf{j}$.)

31. Using the result of Exercise 30, find the curvature of the single leaf of the three-leaved rose $r = \sin 3\theta$ for $0 < \theta < \pi/3$.

32. Using the result of Exercise 30, find the curvature of the cardioid $r = 1 - \cos\theta$ for $0 < \theta < 2\pi$.

*33. a. Let κ be any nonnegative function that is differentiable on an interval I, and let a be any interior point of I. Let

$$\theta(t) = \int_a^t \kappa(u)\,du$$

and

$$\mathbf{r}(t) = \left[\int_a^t \cos\theta(u)\,du\right]\mathbf{i} + \left[\int_a^t \sin\theta(u)\,du\right]\mathbf{j}$$

Show that the curvature of the curve traced out by $\mathbf{r}$ is κ. (Thus we have a way of describing a curve with any prescribed curvature.)

b. Using (a), find a parametrization on $(-1, 1)$ of a curve whose curvature is

$$\kappa(t) = \frac{1}{\sqrt{1 - t^2}} \quad \text{for } -1 < t < 1$$

c. Using (a), find a parametrization of a curve whose curvature is

$$\kappa(t) = \frac{1}{1 + t^2}$$

Applications

34. Railroad tracks are curved in such a way that the curvature exists and is continuous at each point. Which of the following functions could trace out a railroad track?

a. $\mathbf{r}(t) = t\mathbf{i}$ for $t < 0$, and $\mathbf{r}(t) = t\mathbf{i} + t^2\mathbf{j}$ for $t \ge 0$

b. $\mathbf{r}(t) = t\mathbf{i}$ for $t < 0$, and $\mathbf{r}(t) = t\mathbf{i} + t^{7/3}\mathbf{j}$ for $t \ge 0$

c. $\mathbf{r}(t) = t\mathbf{i}$ for $t < 0$, and $\mathbf{r}(t) = t\mathbf{i} + t^3\mathbf{j}$ for $t \ge 0$

35. A car travels around a circular track whose radius is 729 feet. At an instant when the speed of the car is 81 feet per second, the brakes are applied, reducing the speed of the car to 0 in 9 seconds at a constant rate (that is, $d \parallel \mathbf{v} \parallel /dt$ is constant). Find the tangential and normal components of acceleration of the car, and find the magnitude of the acceleration. (*Hint:* Take $t = 0$ at the instant when the brakes are applied, express $\parallel \mathbf{v} \parallel$ as a function of t, and use Exercise 29.)

36. A ball weighing 2 pounds swings back and forth at the end of a string 3 feet long. Assume that when the string makes an angle of $\pi/6$ with respect to the vertical, the magnitude of the force exerted on the ball by the string is 4 pounds (Figure 12.49). Determine the acceleration and speed of the ball. (*Hint:* To determine the acceleration $\mathbf{a}$, use Newton's Second Law of Motion $\mathbf{F} = m\mathbf{a}$, where $m =$ weight/$g = \frac{2}{32} = \frac{1}{16}$ is the mass of the ball and $\mathbf{F}$ is the force acting on the ball. Then $\mathbf{F} = \mathbf{F}_1 + \mathbf{F}_2$, where

$$\mathbf{F}_1 = -mg\mathbf{k} = 2\mathbf{k}$$

is the force due to gravity and

$$\mathbf{F}_2 = 4\left(\cos \frac{2\pi}{3} \mathbf{j} + \sin \frac{2\pi}{3} \mathbf{k} \right) = -2\mathbf{j} + 2\sqrt{3}\mathbf{k}$$

is the force exerted on the ball by the string. To determine the speed, find $a_{\mathbf{N}}$ by using the formula $a_{\mathbf{N}} = \mathbf{a} \cdot \mathbf{N}$ and then use Exercise 29.)

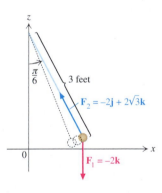

FIGURE 12.49 Figure for Exercise 36.

Project

1. Let C be a curve with a smooth parametrization $\mathbf{r}$ such that $\mathbf{r}'$ is differentiable. The purpose of this project is to prove the so-called **Frenet-Serret** formulas, which were discovered independently by the French mathematicians Frederick Frenet (1816–1900) and Alfred Serret (1819–1885) during the middle of the 19th century. The Frenet-Serret formulas are

$$\frac{d\mathbf{T}}{ds} = \kappa\mathbf{N}, \quad \frac{d\mathbf{N}}{ds} = -\kappa\mathbf{T} + \tau\mathbf{B} \quad \text{and} \quad \frac{d\mathbf{B}}{ds} = -\tau\mathbf{N}$$

where s is the arclength function, $\mathbf{T}$ is the tangent vector, $\mathbf{N}$ is the normal vector, κ is the curvature, τ is the torsion that measures how much the curve C twists outside the plane determined by $\mathbf{T}$ and $\mathbf{N}$, and $\mathbf{B} = \mathbf{T} \times \mathbf{N}$ is the binormal to C, that is, $\mathbf{B}$ is normal to the tangent and the normal vectors to C. These formulas play a large role in advanced geometry topics.

a. Prove that $\dfrac{d\mathbf{T}}{ds} = \kappa\mathbf{N}$. (*Hint:* Use the definitions of κ and $\mathbf{N}$ and the Chain Rule.)

b. Prove that $\dfrac{d\mathbf{B}}{ds}$ is perpendicular to $\mathbf{T}$. (*Hint:* Use Theorem 12.10(e) and part (a).)

c. Prove that $\dfrac{d\mathbf{B}}{ds}$ is perpendicular to $\mathbf{B}$. (*Hint:* Use the fact that $\mathbf{B}$ is a unit vector, and apply Corollary 12.11.)

d. Prove that $\mathbf{B} \times \mathbf{T} = \mathbf{N}$ and $\mathbf{B} \times \mathbf{N} = -\mathbf{T}$. (*Hint:* Use (8) in Section 11.4.)

e. Use parts (b) and (c) to prove that $\dfrac{d\mathbf{B}}{ds}$ is parallel to $\mathbf{N}$.

   Let the **torsion** τ be defined by $\dfrac{d\mathbf{B}}{ds} = -\tau\mathbf{N}$. (The minus sign is traditional.)

f. Prove that the torsion $\tau = 0$ for a plane curve.

g. Prove that $\dfrac{d\mathbf{N}}{ds} = -\kappa\mathbf{T} + \tau\mathbf{B}$. (*Hint:* Use (a), (d), and (e), and also Theorem 12.10(e).)

h. Find the torsion for the helix parametrized by

$$\mathbf{r}(t) = r \cos t\mathbf{i} + r \sin t\mathbf{j} + c\mathbf{k}$$

where r and c are constants.

12.7 KEPLER'S LAWS OF MOTION

The motion of celestial bodies has intrigued people for centuries. However, the first person to state precise and accurate laws regarding the orbits of the planets in our solar system was the German mathematician and astronomer Johannes Kepler,

Johannes Kepler
When Kepler was 30 years old, he was appointed court astronomer to Kaiser Rudolph II, succeeding the famous Danish-Swedish astronomer Tycho Brahe. With the new position, Kepler inherited Brahe's monumental detailed records of the motion of nearby planets. After studying the data diligently for several years, Kepler discovered his first two laws of planetary motion in 1609; they were substantiated by Brahe's data. It took another 10 years for Kepler to formulate his third law. These three laws serve as an accomplishment of incredible proportions.

who lived before the time of Newton. After many years of calculations based primarily on the extremely detailed observations recorded by the Danish astronomer Tycho Brahe, Kepler formulated three laws that describe planetary motion. We will derive these three laws from two fundamental laws of Newton: the Second Law of Motion and the Law of Gravitation.

Consider one object moving around another, much larger, object. You might think of a planet or comet traveling around a star, or an artificial satellite or the moon traveling around the earth. For simplicity we will call the larger object a sun and the smaller one a planet. We assume that the sun's gravitational force is so much larger than all other forces acting on the planet that such other forces produce a negligible effect on the planet's path. Thus we ignore all objects in the universe except a sun and one planet moving around it.

Let us select a coordinate system with the sun at the origin, and let $\mathbf{r}$ be the position vector from the sun to the planet. Newton's laws governing $\mathbf{r}$ are

$$\text{Newton's Second Law of Motion:} \qquad \mathbf{F} = m\frac{d^2\mathbf{r}}{dt^2} \tag{1}$$

$$\text{Newton's Law of Gravitation:} \qquad \mathbf{F} = \frac{-GMm}{r^2}\frac{\mathbf{r}}{r} = \frac{-GMm}{r^3}\mathbf{r} \tag{2}$$

where $\mathbf{F}$ is the gravitational force on the planet with mass m, M is the mass of the sun, G is the universal gravitational constant, and $r = \| \mathbf{r} \|$. Notice that $\mathbf{r}$ is a (not necessarily constant) function of t. Before deriving Kepler's laws we will show that the planet moves in a single plane.

Notice from (2) that $\mathbf{F}$ and $\mathbf{r}$ are parallel, so that $\mathbf{r} \times \mathbf{F} = \mathbf{0}$. Combining this fact with (1), we find that

$$\mathbf{r} \times \frac{d^2\mathbf{r}}{dt^2} = \mathbf{r} \times \left(\frac{1}{m}\mathbf{F}\right) = \frac{1}{m}(\mathbf{r} \times \mathbf{F}) = 0$$

Consequently

$$\frac{d}{dt}\left(\mathbf{r} \times \frac{d\mathbf{r}}{dt}\right) = \left(\frac{d\mathbf{r}}{dt} \times \frac{d\mathbf{r}}{dt}\right) + \left(\mathbf{r} \times \frac{d^2\mathbf{r}}{dt^2}\right) = 0 + 0 = 0 \tag{3}$$

This means that the vector-valued function $\mathbf{r} \times (d\mathbf{r}/dt)$ is a constant vector, a result that holds provided only that the sun is located at the origin of the coordinate system. If we now restrict our attention to a coordinate system whose positive z axis points in the same direction as $\mathbf{r} \times (d\mathbf{r}/dt)$, it follows that

$$\mathbf{r} \times \frac{d\mathbf{r}}{dt} = p\mathbf{k} \tag{4}$$

where p is a suitable positive constant. From (4) we infer that at any time t, $\mathbf{r}$ is perpendicular to $\mathbf{k}$, and consequently the object is constrained to move in a single plane perpendicular to $\mathbf{k}$. In particular:

> Planetary motion is planar.

We will henceforth assume that the planet moves in the xy plane, so that $\mathbf{r}$ lies in the xy plane.

Next we show that the orbit in which the planet moves is a conic section. Let **u** be the unit vector pointing in the direction of **r**, so that $\mathbf{r} = r\mathbf{u}$ and consequently **u** is perpendicular to **k** and $r > 0$ for all t. Differentiation of **r** with respect to t yields

$$\frac{d\mathbf{r}}{dt} = \frac{d}{dt}(r\mathbf{u}) = \frac{dr}{dt}\mathbf{u} + r\frac{d\mathbf{u}}{dt} \tag{5}$$

From (4) and (5) we deduce that

$$p\mathbf{k} = \mathbf{r} \times \frac{d\mathbf{r}}{dt} = (r\mathbf{u}) \times \left(\frac{dr}{dt}\mathbf{u} + r\frac{d\mathbf{u}}{dt}\right)$$

$$= r\frac{dr}{dt}\overbrace{(\mathbf{u} \times \mathbf{u})}^{=\,0} + r^2\left(\mathbf{u} \times \frac{d\mathbf{u}}{dt}\right)$$

Therefore

$$p\mathbf{k} = r^2\left(\mathbf{u} \times \frac{d\mathbf{u}}{dt}\right) \tag{6}$$

Next we combine (1) and (2), which gives

$$\frac{d^2\mathbf{r}}{dt^2} = \frac{\mathbf{F}}{m} = -\frac{GM}{r^2}\frac{\mathbf{r}}{r}$$

and then we substitute **u** for $\mathbf{r}/r$ to obtain

$$\frac{d^2\mathbf{r}}{dt^2} = \frac{-GM}{r^2}\mathbf{u} \tag{7}$$

Formulas (6) and (7) together yield

$$\frac{d^2\mathbf{r}}{dt^2} \times p\mathbf{k} = \frac{-GM}{r^2}\mathbf{u} \times r^2\left(\mathbf{u} \times \frac{d\mathbf{u}}{dt}\right) = -GM\mathbf{u} \times \left(\mathbf{u} \times \frac{d\mathbf{u}}{dt}\right)$$

But $\|\mathbf{u}\| = 1$, and consequently $d\mathbf{u}/dt$ is perpendicular to **u** (see Corollary 12.11). Therefore the formula $\mathbf{a} \times (\mathbf{a} \times \mathbf{c}) = -\|\mathbf{a}\|^2\mathbf{c}$ (which is (8) of Section 11.4) yields

$$\frac{d^2\mathbf{r}}{dt^2} \times p\mathbf{k} = -GM\left(-\|\mathbf{u}\|^2\frac{d\mathbf{u}}{dt}\right) = GM\frac{d\mathbf{u}}{dt} \tag{8}$$

Since $p\mathbf{k}$ is a constant function of t, the left-hand side of (8) is the derivative of $(d\mathbf{r}/dt) \times p\mathbf{k}$. Thus

$$\frac{d}{dt}\left(\frac{d\mathbf{r}}{dt} \times p\mathbf{k}\right) = \frac{d^2\mathbf{r}}{dt^2} \times p\mathbf{k} = \frac{d}{dt}(GM\mathbf{u})$$

and it follows that

$$\frac{d\mathbf{r}}{dt} \times p\mathbf{k} = GM\mathbf{u} + \mathbf{w}_1 \tag{9}$$

where $\mathbf{w}_1$ is a suitable constant vector. Since $(d\mathbf{r}/dt) \times p\mathbf{k}$ is perpendicular to **k**, as is **u**, it follows that $\mathbf{w}_1$ is perpendicular to **k** and hence lies in the xy plane. Therefore we may make the further assumption that the coordinate system is situated with the positive x axis pointing in the direction of $\mathbf{w}_1$. If we let θ be the angle from $\mathbf{w}_1$ to **u** at any given time, then (r, θ) is a set of polar coordinates of the position of the planet at that time (Figure 12.50). Now we let $w = \|\mathbf{w}_1\|$, so that

$$\mathbf{u} \cdot \mathbf{w}_1 = \|\mathbf{u}\|\,\|\mathbf{w}_1\|\cos\theta = w\cos\theta \tag{10}$$

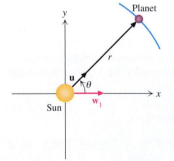

FIGURE 12.50

From (4), (9), and (10), as well as the formula $(\mathbf{a} \times \mathbf{b}) \cdot \mathbf{c} = \mathbf{a} \cdot (\mathbf{b} \times \mathbf{c})$ (which is (3) of Section 11.4), we obtain

$$p^2 = (p\mathbf{k}) \cdot (p\mathbf{k}) \stackrel{(4)}{=} \left(\mathbf{r} \times \frac{d\mathbf{r}}{dt}\right) \cdot p\mathbf{k} = \mathbf{r} \cdot \left(\frac{d\mathbf{r}}{dt} \times p\mathbf{k}\right)$$

$$\stackrel{(9)}{=} (r\mathbf{u}) \cdot (GM\mathbf{u} + \mathbf{w}_1) = rGM(\mathbf{u} \cdot \mathbf{u}) + r(\mathbf{u} \cdot \mathbf{w}_1) \stackrel{(10)}{=} GMr + wr \cos\theta$$

Thus we have

$$p^2 = GMr + wr \cos\theta \qquad (11)$$

In order to interpret (11), we change from polar coordinates to rectangular coordinates and rewrite (11) as

$$p^2 = GM\sqrt{x^2 + y^2} + wx$$

so that

$$GM\sqrt{x^2 + y^2} = p^2 - wx$$

Squaring both sides yields

$$G^2M^2(x^2 + y^2) = p^4 - 2wp^2x + w^2x^2 \qquad (12)$$

Equation (12) has the form of a conic section, regardless of the constant w, and it only remains to determine which conic sections arise for different values of w.

If $w = 0$, then (12) yields

$$x^2 + y^2 = \frac{p^4}{G^2M^2} \qquad (13)$$

which is an equation of a circle with radius p^2/GM.

If $w = GM$, then we can rewrite (12) as

$$G^2M^2y^2 = p^4 - 2GMp^2x$$

which is an equation of a parabola.

Finally, if $w \neq 0$ and $w \neq GM$, then we may rewrite (12) as

$$(G^2M^2 - w^2)x^2 + 2wp^2x + G^2M^2y^2 = p^4 \qquad (14)$$

This is an equation of an ellipse if $G^2M^2 > w^2$ and an equation of a hyperbola if $G^2M^2 < w^2$.

Therefore, regardless of the value of w, the orbit of the object around its sun is a conic section. If the orbiting object returns to its initial position periodically, as planets actually do, then the orbit is necessarily elliptical. In this case it can be shown that one of the foci is at the origin (see Exercise 7), where the sun is located. Thus we have Kepler's First Law of planetary motion, announced in 1609:

> A planet revolves around the sun in an elliptical orbit.

As we noted above, the sun is located at a focus of the elliptical orbit, not at the center of the orbit as one might conjecture. It turns out that if the orbit is described by (14), then the points on the ellipse closest to and farthest from the sun occur on the x axis (Figure 12.51), to the right and left of the origin, respectively

Saturn. (*Superstock*)

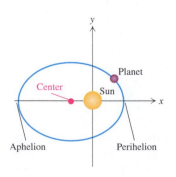

FIGURE 12.51 A planet moves around the sun in an elliptical orbit.

(see Exercise 2). The nearest point is called the **perihelion** of the orbit, and the farthest point is called the **aphelion.*** For an object moving in an elliptical orbit about the earth, the terms are **perigee** and **apogee,** respectively.†

   Theoretically, a comet can have an elliptical, parabolic, or hyperbolic orbit around the sun. If the orbit is parabolic or hyperbolic, then we can view the comet at most twice. However, if the orbit is elliptical, then the comet can return to view periodically. The most illustrious example of a comet with elliptical orbit is Halley's comet, which has a period of about 76 years. Halley's comet reappeared during 1985–1986, its perihelion occurring on February 9, 1986. Artificial satellites can have any of the three types of orbits. Certain earth satellites, such as those in the Syncom series, have been placed in circular orbits around the earth.

Kepler's Second Law

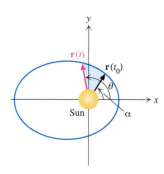

FIGURE 12.52   The position vector sweeps out area at a constant rate.

As a planet orbits the sun, the position vector **r** from the sun to the planet describes an elliptical region. Let A denote the area of the region swept out by r from an initial time t_0 to time t (see the shaded region in Figure 12.52). Let α and θ be the angles made by the vectors r(t_0) and r(t), respectively, with the positive x axis. Finally, let $r = f(\phi)$ be a polar equation of the elliptical orbit traced out by the planet. Then from (4) of Section 10.2,

$$A = \int_{\alpha}^{\theta} \frac{1}{2} [f(\phi)]^2 \, d\phi$$

Differentiating both sides of this equation with respect to θ, we find that

$$\frac{dA}{d\theta} = \frac{1}{2}(f(\theta))^2 = \frac{1}{2}r^2$$

and consequently

$$\frac{dA}{dt} = \frac{dA}{d\theta}\frac{d\theta}{dt} = \frac{1}{2}r^2\frac{d\theta}{dt} \tag{15}$$

Recall that $\mathbf{r} = r\mathbf{u}$. Since **u** is a unit vector and θ is the angle between **u** and the positive x axis, we deduce that

$$\mathbf{u} = \cos\theta\mathbf{i} + \sin\theta\mathbf{j}$$

Therefore

$$\frac{d\mathbf{u}}{dt} = \frac{d\mathbf{u}}{d\theta}\frac{d\theta}{dt} = \frac{d\theta}{dt}(-\sin\theta\mathbf{i} + \cos\theta\mathbf{j})$$

Since

$$(\cos\theta\mathbf{i} + \sin\theta\mathbf{j}) \times (-\sin\theta\mathbf{i} + \cos\theta\mathbf{j}) = \mathbf{k}$$

it follows that

$$\mathbf{u} \times \frac{d\mathbf{u}}{dt} = (\cos\theta\mathbf{i} + \sin\theta\mathbf{j}) \times \frac{d\theta}{dt}(-\sin\theta\mathbf{i} + \cos\theta\mathbf{j}) = \frac{d\theta}{dt}\mathbf{k} \tag{16}$$

Substituting (16) into (6), we obtain

$$p\mathbf{k} = r^2\left(\mathbf{u} \times \frac{d\mathbf{u}}{dt}\right) = r^2\frac{d\theta}{dt}\mathbf{k}$$

Therefore

$$p = r^2\frac{d\theta}{dt} \tag{17}$$

---

* **Perihelion:** Pronounced "per-i-*heel*-yon"; **aphelion:** pronounced "a-*feel*-yon."
† **Perigee:** Pronounced "*per*-i-jee"; **apogee:** pronounced "*ap*-o-jee."

Combining (15) and (17), we conclude that

$$\frac{dA}{dt} = \frac{1}{2} r^2 \frac{d\theta}{dt} = \frac{p}{2} \qquad (18)$$

Equation (18) brings us to Kepler's Second Law, also announced in 1609:

> The position vector from the sun to a planet sweeps out area at a constant rate.

An immediate consequence of this law is the fact that when a planet is near its sun, it moves more swiftly than when it is farther away.

Kepler's Third Law

Since the area swept out by the position vector changes at a constant rate, we find by integrating the expressions in (18) that

$$A(t_1) - A(t_0) = \int_{t_0}^{t_1} \frac{dA}{dt}\, dt = \int_{t_0}^{t_1} \frac{p}{2}\, dt = \frac{p(t_1 - t_0)}{2}$$

If T denotes the period of the planet about the sun, then

$$A(t_0 + T) - A(t_0) = \frac{pT}{2} \qquad (19)$$

But during an interval of duration T, the position vector sweeps out the complete ellipse, whose area is given by (19). If the major and minor axes of the ellipse have lengths $2a$ and $2b$, respectively, then the area of the ellipse is πab (see the comment after Example 2 of Section 8.3). Consequently

$$\pi ab = A(t_0 + T) - A(t_0) = \frac{pT}{2}$$

so that

$$T = \frac{2\pi ab}{p} \qquad (20)$$

A suitable alteration of (20) will yield Kepler's Third Law.

First we put (14) into standard form by completing the square and dividing by the right-hand side:

$$\frac{\left(x + \dfrac{wp^2}{G^2M^2 - w^2}\right)^2}{\dfrac{p^4 G^2 M^2}{(G^2M^2 - w^2)^2}} + \frac{y^2}{\dfrac{p^4}{G^2M^2 - w^2}} = 1 \qquad (21)$$

This implies that

$$a^2 = \frac{p^4 G^2 M^2}{(G^2M^2 - w^2)^2} \quad \text{and} \quad b^2 = \frac{p^4}{G^2M^2 - w^2} \qquad (22)$$

Therefore by (20),

$$T^2 = \frac{4\pi^2 a^2 b^2}{p^2} = \frac{4\pi^2 p^6 G^2 M^2}{(G^2M^2 - w^2)^3} = \frac{4\pi^2 a^3}{GM} \qquad (23)$$

Equation (23) yields Kepler's Third Law, announced in 1619:

Mary Fairfax Somerville (1780–1872)

Somerville was a Scotswoman who overcame the difficulties of being a female intellectual in nineteenth-century Britain. In 1830 she published *The Mechanisms of the Heavens,* which was a very lucid exposition of Laplace's work on celestial mechanics and for nearly a century was required reading for students in British universities. This book prompted John Adams to look for a new planet (later to be called Neptune) to explain the anomalous orbit of Uranus.

The square of the period of a planet is $4\pi^2/GM$ times the cube of half the length of the major axis of its orbit.

Observe that the number $4\pi^2/GM$ is independent of the planet orbiting the sun. Thus the period of a planet about its sun depends only on the length of the major axis of the orbit.

Kepler's Third Law can be used to determine the mass of a planet, provided there is an object (either a moon or a man-made spacecraft) in orbit around the planet. By observation, one can calculate the period T of the orbiting object, as well as the length $2a$ of the major axis of its elliptical orbit. Knowing T, a, and G, one can then determine the mass M of the planet from Kepler's Third Law.

In our derivation of Kepler's Laws we ignored all objects in the universe except one given planet and the sun. However, the planets Neptune and Uranus are sometimes close enough together to affect each other's orbits measurably. In fact, Neptune was discovered only after astronomers observed certain irregularities in the nearly elliptical orbit of Uranus and tried to explain them by the existence of a nearby planet. Working independently, both John Adams in Britain and Urbain Le Verrier in France calculated what the position of the unknown planet in the sky would have to be in order to produce the observed irregularities in the orbit of Uranus. In 1846, using Le Verrier's predictions, the German astronomer Johann Galle was able to locate Neptune with his telescope in a few hours.

EXERCISES 12.7

The exercises for this section will employ the following data.

r_0 = minimum distance from an object to the sun

v_0 = speed of an object when its distance from the sun is r_0

$a^2 = b^2 + c^2$

eccentricity $= \dfrac{c}{a}$

$GM_e = 1.237 \times 10^{12}$ (miles cubed per hour squared), where M_e is the mass of the earth

1. **a.** From (6) show that

$$p = r^2 \left\| \frac{d\mathbf{u}}{dt} \right\|$$

(*Hint:* Because $\mathbf{u}$ is a unit vector, $\mathbf{u}$ and $d\mathbf{u}/dt$ are perpendicular to one another.)

b. Whatever kind of conic section constitutes the orbit of an object about the sun, the distance r from the object to the sun is a differentiable function of time t, and has a minimum value for some special value of t. From (5) show that

$$\frac{d\mathbf{r}}{dt} = r \frac{d\mathbf{u}}{dt}$$

when r is minimum.

c. Using (a), (b), and the definitions of r_0 and v_0, show that

$$p = r_0 v_0$$

d. Use (4) to prove that $p = 0$ for straight-line motion.

2. **a.** Solve (11) for r and use Exercise 1(c) to show that

$$r = \frac{r_0^2 v_0^2}{GM + w \cos \theta}$$

b. Find a value of θ that produces the minimum value r_0 of r, and calculate r_0 in terms of v_0, G, M, and w.

c. Assuming that the orbit is elliptical, find a value of θ which produces the maximum value of r.

3. Suppose an object moves in a circular orbit. Use (13) and Exercise 1(c) to show that

$$v_0 = \sqrt{\frac{GM}{r_0}}$$

4. Find the speed of a satellite orbiting the earth in a circular orbit with radius 5000 miles. (*Hint:* Take M equal to the mass M_e of the earth.)

5. Using Kepler's Third Law with $M = M_e$, find the period of a satellite orbiting the earth in a circular orbit with radius 5000 miles.

6. The earth makes one complete revolution around its axis in approximately 23.9344 hours. If we wish a satellite to have a circular orbit with period equal to 23.9344 hours, what would be its velocity and its distance from the surface of the earth (3960 miles from the center of the earth)? (*Hint:* Use Kepler's Third Law, with a denoting the distance between the satellite and the center of the earth. Also use Exercise 3 with $M = M_e$.)

7. Show that the sun is located at one of the foci of the elliptical orbit of any of its planets. (*Hint:* Using (22), compute $c = \sqrt{a^2 - b^2}$ and compare your answer with the first term in (21). Remember that the sun is located at the origin and that by (21) the center of the ellipse is

$$(-wp^2/(G^2M^2 - w^2), 0)$$

8. Using Exercise 2, show that $w = r_0 v_0^2 - GM$.

9. **a**. Use (9) and the fact that $d\mathbf{r}/dt$ is perpendicular to $\mathbf{k}$ to show that the speed $\| d\mathbf{r}/dt \|$ of an object in orbit is given by

$$\left\| \frac{d\mathbf{r}}{dt} \right\| = \frac{1}{p} \| GM\mathbf{u} + \mathbf{w}_1 \|$$

   b. The vector $\mathbf{w}_1$ has length w and is directed along the positive x axis. Use this information, along with Exercises 1(c) and 8 and the formula derived in part (a), to show that the maximum speed (the maximum value of $\| d\mathbf{r}/dt \|$) is v_0.

10. The earth takes 365.256 days to orbit the sun. The orbit has an eccentricity of 0.016732, and the value of a is approximately 92,955,821 miles.

   a. Using this information, first find b, and then use (20) to calculate the value of p for the earth.

   b. Using the fact that $r_0 = a - c$, calculate r_0, the minimum distance from the earth to the sun.

   c. Calculate v_0, the maximum speed of the earth as it orbits the sun. (*Hint:* Use (a), (b), and Exercise 1(c).)

   d. Use (23) to calculate the mass of the sun. (*Note:* Take $G = 3.024 \times 10^{-12}$ miles cubed per hour squared per slug. Your answer will be in slugs.)

11. Suppose the maximum and minimum altitudes above the surface of the earth of a satellite moving in an elliptical orbit about the earth are 3100 and 100 miles, respectively.

   a. Find the maximum speed of the satellite. (*Hint:* Recall that $r_0 = a - c$, and note that $2a = 3100 + 100 + 2(3960) = 11,120$. Calculate a, c, b, p, and then r_0, by means of (23), (20), and Exercise 1(c).)

   b. Find the minimum speed of the satellite. (*Hint:* Use v_0, p, and r_0 from part (a); then find w from Exercise 8. Finally, use the equation in Exercise 9(a). Assume that

the minimum speed occurs at aphelion on the negative x axis.)

12. A satellite is launched from 100 miles above the earth at 20,000 miles per hour in a direction parallel to the surface of the earth that takes the satellite into an elliptical orbit whose minimum distance from the surface of the earth is 100 miles.

   a. Determine an equation for the orbit of the satellite. (*Hint:* From Exercises 1(c) and 8, find p and w, and then obtain the answer in the form of (14).)

   b. Compute the distance from the apogee to the surface of the earth.

   c. Compute the period of the satellite.

13. Suppose an object is in a parabolic orbit. Using Exercise 8 and the fact that $GM = w$, show that

$$v_0 = \sqrt{\frac{2GM}{r_0}}$$

14. If a satellite is close to the surface of the earth and its distance from the center of the earth is r_0, then the gravitational acceleration GM/r_0^2 of the satellite is approximately g. In that case the value of v_0 found in Exercise 13 is approximately given by

$$v_0 = \sqrt{2gr_0}$$

If the satellite has this velocity, it enters a parabolic orbit and disappears. Taking the radius of the earth to be 3960 miles and assuming that a satellite is to achieve a parabolic orbit at an altitude of 100 miles, what will that velocity need to be? (*Note:* Take $g = 7.855 \times 10^4$ miles per hour per hour.)

15. The moon has a mass of (approximately) 0.0123 times the mass of the earth and is (approximately) 240,000 miles from the earth. If a spacecraft is on a line between the moon and the earth, 4080 miles from the center of the moon, what is the ratio of the gravitational force of the earth on the spacecraft to the gravitational force of the moon on the spacecraft? (*Hint:* Use Newton's Law of Gravitation for the earth and for the moon.)

16. **a**. If a satellite is to orbit the moon in a circular orbit with radius 1200 miles, what must the speed of the satellite be? (*Hint:* See Exercise 15 for the mass of the moon, and apply Exercise 3.)

   b. To what value should the speed be increased if the satellite is to achieve escape velocity at a point the same distance from the center of the moon? (*Hint:* Apply Exercise 13.)

17. At its closest, Halley's comet is 5.31×10^7 kilometers from the sun. Its period is 75.6 years. What is its distance

when it is farthest from the sun? (*Hint:* For the sun, GM $\approx 1.323 \times 10^{26}$ kilometers cubed per year squared.)

18. Let **r** and **v** be the position and velocity vectors, respectively, of an object moving about the sun, and let

$$\mathbf{L}(t) = \mathbf{r}(t) \times m\mathbf{v}(t) \quad \text{for all } t$$

where m is the mass of the object. Then **L** is called the **angular momentum** of the object about its sun. Use (3) to show that $\mathbf{L}'(t) = \mathbf{0}$ for all t, and conclude that **L** is a constant function. This fact is often referred to as the **conservation of angular momentum** of the object.

19. Suppose a planet with mass m and speed v moves in a circular orbit with radius r. Then the gravitational force **F** is a centripetal force. From (5) of Section 12.3, the magnitude $\| \mathbf{F} \|$ of this force is given by

$$\| \mathbf{F} \| = m \| \mathbf{a} \| = \frac{mv^2}{r} \qquad (24)$$

a. Without using any formulas from the present section, show that the period T of the planet is given by

$$T = \frac{2\pi r}{v} \qquad (25)$$

b. Observe that Kepler's Third Law has the form

$$T^2 = cr^3 \qquad (26)$$

where c is a suitable positive constant. Using (24)–(26), show that

$$\| \mathbf{F} \| = \frac{4\pi^2}{c} \frac{m}{r^2} \qquad (27)$$

(Before Newton announced his Law of Gravitation, the validity of (27) was noticed by Robert Hooke and was mentioned about 1679 in a recently discovered letter from Hooke to Newton.)

REVIEW

Key Terms and Expressions

Vector-valued function; component function
Limit of a vector-valued function
Vector-valued function continuous at t_0; continuity on an interval
Derivative of a vector-valued function; differentiability on an interval
Definite integral of a vector-valued function
Space curve; parametrization of a curve

Smooth curve; piecewise smooth curve
Length of a curve
Tangent vector
Normal vector
Orientation of a curve; oriented curve
Curvature
Tangential and normal components of acceleration

Key Formulas

$$\mathbf{v} = \frac{d\mathbf{r}}{dt}, \quad \mathbf{a} = \frac{d\mathbf{v}}{dt}, \quad \| \mathbf{v} \| = \frac{ds}{dt}$$

$$L = \int_a^b \sqrt{\left(\frac{dx}{dt}\right)^2 + \left(\frac{dy}{dt}\right)^2 + \left(\frac{dz}{dt}\right)^2} \, dt$$

$$\left\| \frac{d\mathbf{r}}{dt} \right\| = \frac{ds}{dt} = \sqrt{\left(\frac{dx}{dt}\right)^2 + \left(\frac{dy}{dt}\right)^2 + \left(\frac{dz}{dt}\right)^2}$$

$$\mathbf{T}(t) = \frac{\mathbf{r}'(t)}{\| \mathbf{r}'(t) \|} = \frac{d\mathbf{r}/dt}{\| d\mathbf{r}/dt \|}$$

$$\mathbf{N}(t) = \frac{\mathbf{T}'(t)}{\| \mathbf{T}'(t) \|} = \frac{d\mathbf{T}/dt}{\| d\mathbf{T}/dt \|}$$

$$\kappa(t) = \frac{\| \mathbf{T}'(t) \|}{\| \mathbf{r}'(t) \|} = \frac{\| \mathbf{v} \times \mathbf{a} \|}{\| \mathbf{v} \|^3}$$

$$a_\mathbf{T} = \frac{d \| \mathbf{v} \|}{dt} \quad \text{and} \quad a_\mathbf{N} = \sqrt{\| \mathbf{a} \|^2 - a_\mathbf{T}^2}$$

Review Exercises

In Exercises 1–3 sketch the curve traced out by the function.

1. $\mathbf{F}(t) = \sin 2t\mathbf{i} - \cos 2t\mathbf{j}$ for $0 \le t \le 4\pi$

2. $\mathbf{F}(t) = 4\sqrt{2} \cos t\mathbf{i} - 4 \sin t\mathbf{j} - 4 \sin t\mathbf{k}$

3. $[(\mathbf{F} \times \mathbf{G}) \times \mathbf{H}](t)$, where $\mathbf{F}(t) = t\mathbf{i} + \mathbf{j}$, $\mathbf{G}(t) = \mathbf{j} + t\mathbf{k}$, and $\mathbf{H}(t) = t\mathbf{j}$

4. Let $\mathbf{F}(t) = e^t\mathbf{i} + t^2\mathbf{j} + e^{-t}\mathbf{k}$ and $\mathbf{G}(t) = t\mathbf{i} + e^{-t}\mathbf{j} + e^t\mathbf{k}$.

a. Find $(\mathbf{F} \cdot \mathbf{G})'(t)$.   **b.** Find $(\mathbf{F} \times \mathbf{G})'(t)$.

5. Let $\mathbf{F}(t) = \dfrac{1}{t}\mathbf{i} + t\mathbf{j}$ and $\mathbf{G}(t) = t^2\mathbf{j} - \dfrac{1}{t^2}\mathbf{k}$.

   a. Find $(\mathbf{F} \cdot \mathbf{G})'(t)$.   **b.** Find $(\mathbf{F} \times \mathbf{G})'(t)$.

6. Let $\mathbf{F}(t) = \ln t\mathbf{i} + t\ln t\mathbf{j} - \ln 6t\mathbf{k}$ and $g(t) = e^{2t}$. Find $(\mathbf{F} \circ g)'(t)$.

7. Find

$$\int \left(\tan 2\pi t\mathbf{i} + \sec^2 2\pi t\,\mathbf{j} + \frac{4}{1 + t^2}\mathbf{k} \right) dt$$

8. Let

$$\mathbf{F}(t) = \frac{2t}{1 + t^2}\mathbf{i} + \frac{1 - t^2}{1 + t^2}\mathbf{j}$$

Use Corollary 12.11 to show that $\mathbf{F}$ is perpendicular to $\mathbf{F}'$.

9. Let $\mathbf{r}(t) = e^t \cos t\mathbf{i} + e^t \sin t\mathbf{j}$. Find the length of the curve that $\mathbf{r}$ traces out from

   a. $t = 0$ to $t = 3\pi$ **b.** $t = -2\pi$ to $t = 1$

10. An object moves along the curve $y = \ln(\sec x)$ for $-\pi/2 < x < \pi/2$, in such a manner that its x coordinate at any time t is given by $x(t) = e^t$. Find the speed of the object at time 0.

11. Suppose a car traverses the curve $y = \frac{2}{3}x^{3/2}$ in such a manner that its x coordinate at any time $t > \frac{3}{2}$ is given by

$$x(t) = \left(\tfrac{3}{2}t\right)^{2/3} - 1$$

Show that the speed of the car is constant.

12. Show that the Folium of Descartes

$$\mathbf{r}(t) = \frac{3t}{1 + t^3}\mathbf{i} + \frac{3t^2}{1 + t^3}\mathbf{j}$$

passes through the origin and that the tangent to the curve at the origin is parallel to the x axis.

13. Let

$$\mathbf{r}(t) = (3t - t^3)\mathbf{i} + 3t^2\mathbf{j} + (3t + t^3)\mathbf{k}$$

Find the curvature κ of the curve that $\mathbf{r}$ traces out.

14. Show that the curvature κ of the catenary $\mathbf{r}(t) = t\mathbf{i} + \cosh t\mathbf{j}$ is given by the formula

$$\kappa(t) = \frac{1}{\cosh^2 t}$$

15. Let C be the graph of the equation $xy = 1$ for $x > 0$.

   a. Find a formula for the curvature of C.

   b. Find the maximum value of the curvature of C.

   c. Determine the radius of curvature of C at $(1, 1)$.

16. Let $\mathbf{r}(t) = t\mathbf{i} + (2\sqrt{2}/3)t^{3/2}\mathbf{j} + (1/2)t^2\mathbf{k}$ be a parametrization of a curve.

   a. Determine whether the curve is smooth, piecewise smooth, or neither.

b. Find the length L of the portion of the curve that lies between $\mathbf{r}(0)$ and $\mathbf{r}(1)$.

   c. Find $\mathbf{v}$, $\|\mathbf{v}\|$, and $\mathbf{a}$.

   d. Find $a_{\mathbf{T}}$ and $a_{\mathbf{N}}$.

   e. Find $\kappa(t)$.

17. Let the position of an object in motion be traced out by

$$\mathbf{r}(t) = e^t \cos t\mathbf{i} + e^t \sin t\mathbf{j} + e^t\mathbf{k}$$

Find the velocity and acceleration of the object, and then compute the curvature and the radius of curvature of the curve traced out.

In Exercises 18–19 find the tangent, normal, and curvature of the curve traced out by $\mathbf{r}$.

18. $\mathbf{r}(t) = e^{2t}\mathbf{i} + 2\sqrt{2}e^t\mathbf{j} + 2t\mathbf{k}$

19. $\mathbf{r}(t) = (t - \sin t)\mathbf{i} + (1 - \cos t)\mathbf{j} + 4\sin\dfrac{t}{2}\mathbf{k}$

*20. The curve

$$\mathbf{r}(t) = \sin t\mathbf{i} + \left[\cos t + \ln\left(\tan\frac{t}{2}\right)\right]\mathbf{j} \quad \text{for } 0 < t < \pi$$

is called a **tractrix**. Show that the length of the segment of the line tangent to the tractrix from any point of tangency to the point of intersection with the y axis is always 1. (*Hint:* Use the fact that $2(\sin t/2)(\cos t/2) = \sin t$.)

21. Assume that an object moves so that its velocity and acceleration vectors are always unit vectors. Prove that the curvature of the curve traversed by the object is always 1.

Topics for Discussion

1. How does a vector-valued function differ from a real-valued function? How are real-valued functions used in order to study vector-valued functions?

2. The following items are important in the study of real-valued functions: continuity, derivatives, the Chain Rule, extreme values, integrals, and length of a curve. Which of these do you think would be important in the study of vector-valued functions? Explain your answer.

3. Give a geometric interpretation of the derivative of a vector-valued function.

4. In Section 12.4 we derived the position and velocity of an object under the assumption that the acceleration is constant. Suppose that the acceleration is merely continuous, but not necessarily constant. In theory, how could we alter the results in order to derive the position and velocity of the object?

5. Give two parametrizations of the unit circle, one that is smooth and one that is not smooth. Why do you suppose smoothness requires a nonvanishing derivative of the position function?

6. How do the tangent and normal vectors relate to velocity and acceleration, respectively? For planar curves, how do these vectors relate to one another geometrically?

7. Describe Kepler's three Laws of Motion in your own words.

Cumulative Review, Chapters 1–11

In Exercises 1–2 find the limit.

1. $\displaystyle\lim_{x\to-3^+}\left(\frac{1}{x+3}-\frac{1}{|x-3|}\right)$

2. $\displaystyle\lim_{x\to0^+}(\cos x)^{1/x^2}$

3. Find $\displaystyle\lim_{y\to0}\frac{x^2-y^3}{x^2+y^2}$. (*Hint:* There are two cases: $x = 0$ and $x \neq 0$.)

4. Let $f(x) = \begin{cases} \dfrac{1}{x}\sin x^2 & \text{for } x \neq 0 \\ 0 & \text{for } x = 0 \end{cases}$

   a. Show that f is continuous at 0.
   b. Show that f is differentiable at 0 and find $f'(0)$.

5. Let $f(x) = \sin^{-1}\left(\dfrac{\sin x}{5}\right)$. Find $f'(x)$.

6. Determine the points on the ellipse $x^2/4 + y^2/16 = 1$ at which the tangents have slope 1.

7. Consider the equation $(x^3 - 2y^3)/(x^2 + y^2) = 6$. Use implicit differentiation to determine all points on the graph of the equation at which the tangent is horizontal. (*Hint:* -4 is a solution of $z^3 + 3z^2 + 16 = 0$.)

8. Use a linear approximation to estimate the value of $\sqrt{63}$.

9. Two sets of railroad tracks meet at an angle of $\pi/3$. At noon a train on one set of tracks is 100 miles from the junction and is traveling toward the junction at 60 miles per hour. On the other tracks a second train is 120 miles from the junction and is traveling toward it at 80 miles per hour. Determine the rate at which the trains are approaching each other at 1 p.m. (*Hint:* Use the Law of Cosines.)

10. Let $f(x) = 2x/\sqrt{x^2 + 4}$. Sketch the graph of f, indicating all pertinent information.

11. Suppose that there are 77 feet of fencing, to be placed along three sides of a rectangular garden with a house comprising the fourth side. Assuming that a 3-foot-wide gate is to be made out of other material, find the dimensions that will maximize the area of the garden.

(*Hint:* There are two cases, depending on the side on which the gate is placed.)

12. Among all circular sectors with a perimeter of 10 centimeters, determine the one with maximum area.

13. Suppose that after 5 years only 10% of a given radioactive substance has decayed. Find the half-life of the substance.

In Exercises 14–16 evaluate the integral.

14. $\displaystyle\int \sqrt{x}\cos\sqrt{x}\,dx$

15. $\displaystyle\int \frac{x}{(x+2)(x^2+6)}\,dx$

16. $\displaystyle\int 27\,t^2\sqrt{3t-2}\,dt$

17. Determine whether $\displaystyle\int_0^{\pi/4}\frac{\cos\sqrt{x}}{\sqrt{x}}\,dx$ converges. If it converges, determine its value.

18. The base of a solid is the semicircle $x^2 + y^2 = 1$ with $y \geq 0$. Suppose the cross sections of the solid perpendicular to the x axis are isosceles right triangles with hypotenuse on the base. Determine the volume V of the solid.

19. A conical tank 10 feet high and 5 feet in radius with its vertex at the top has water to a depth of 4 feet. Determine the work W done in raising the water up to a height 3 feet above the top of the cone.

20. Let $f(x) = \frac{1}{4}x^2 - \frac{1}{2}\ln x$ for $1 \leq x \leq 3$. Find the length L of the graph of f.

21. Determine whether $\left\{\dfrac{(-1)^n n}{n+1}\right\}_{n=1}^{\infty}$ converges or diverges.

In Exercises 22–23 determine whether the series converges absolutely, converges conditionally, or diverges.

22. $\displaystyle\sum_{n=1}^{\infty}\frac{\sinh n}{n^2}$

23. $\displaystyle\sum_{n=0}^{\infty}\frac{n^n}{3^n\,n!}$

24. Find the interval of convergence of $\displaystyle\sum_{n=1}^{\infty}\frac{3^n}{4^{2n}}x^{2n}$.

25. Let $f(x) = -2x/(1 + x^2)^2$. Find a power series representation for f, and determine the interval of convergence for the power series.

26. A unit vector $\mathbf{v}$ is perpendicular to both the line

$$\frac{x-1}{2} = \frac{y+1}{-4} = z$$

and the vector $2\mathbf{i} - 3\mathbf{j} - \mathbf{k}$, and has negative $\mathbf{k}$ component. Find $\mathbf{v}$.

27. Determine an equation of the plane that passes through the points $(1, -1, 2)$, $(2, 3, -1)$, and $(0, 2, 0)$.

A vivid rainbow. (*David Olsen/Tony Stone Images*)

13   PARTIAL DERIVATIVES

Until now we have studied functions of a single variable (both real-valued and vector-valued). Many phenomena in the physical world can be described by such functions, but most quantities actually depend on more than one variable. For example, the volume of a rectangular box depends on its length, width, and height; the temperature at a point on a metal plate depends on the coordinates of the point (and possibly on time as well). Any quantity that depends on several other quantities can be thought of as determining a function of several variables. The primary concern of this chapter is the extension of the concept of differentiation to functions of several variables. In Chapter 14 we will discuss integration of such functions.

13.1   FUNCTIONS OF SEVERAL VARIABLES

A vector-valued function assigns vectors (that is, points in the plane or in space) to real numbers. A function of several variables does the opposite; it assigns real numbers to points in the plane or in space.

813

DEFINITION 13.1   A **function of several variables** consists of two parts: a **domain,** which is a collection of points in the plane or in space, and a **rule,** which assigns to each member of the domain one and only one real number.

A function of several variables is called a **function of two variables** if its domain is a set of points in the plane and a **function of three variables** if its domain is a set of points in space. Although we will not discuss functions of more than three variables in this book, it is possible to extend the theory in this chapter to such functions. To distinguish functions of several variables from those whose domains are sets of real numbers, we will refer to the latter type as **functions of a single variable.**

The value of a function f of two variables at a point (x, y) is denoted $f(x, y)$, and the value of a function f of three variables at a point (x, y, z) is denoted $f(x, y, z)$. As with functions of a single variable, we frequently specify a function of several variables by one or more formulas for its values at the various points in the domain. Unless otherwise indicated, the domain of a function so specified consists of all points in space (or in the plane for a function of two variables) for which the formula is meaningful. For example, if

$$f(x, y) = \sqrt{9 - 4x^2 - y^2}$$

then the domain of f consists of all (x, y) such that $9 - 4x^2 - y^2 \geq 0$, that is, $4x^2 + y^2 \leq 9$. Thus the domain of f consists of all points inside or on the ellipse $4x^2 + y^2 = 9$. If

$$g(x, y, z) = \sqrt{x^2 + y^2 + z^2}$$

then the domain of g consists of all points (x, y, z) in space, because $x^2 + y^2 + z^2 \geq 0$ for all (x, y, z). In contrast, if

$$f(x, y, z) = xyz \quad \text{for } x \geq 0, y \geq 0, \text{ and } z \geq 0$$

then the domain of f is the first octant.

The following are some functions of several variables that we will encounter later in this chapter, along with geometric or physical interpretations.

1. $f(x, y) = xy$ for $x \geq 0$ and $y \geq 0$: area of a rectangle
2. $f(x, y, z) = xyz$ for $x \geq 0$, $y \geq 0$, and $z \geq 0$: volume of a rectangular parallelepiped
3. $f(x, y, z) = 2xy + 2yz + 2xz$ for $x \geq 0$, $y \geq 0$, and $z \geq 0$: surface area of a rectangular parallelepiped
4. $f(x, y, z) = \dfrac{c}{x^2 + y^2 + z^2}$, where c is a positive constant: magnitude of the gravitational force exerted by the sun, located at the origin, on a unit mass at (x, y, z)
5. $f(x, y, z) = \dfrac{c}{\sqrt{x^2 + y^2}}$, where c is a constant: strength of the electric field at (x, y, z) due to an infinitely long wire lying along the z axis

Combinations of Functions of Several Variables

The sum, product, and quotient of two functions f and g of several variables are defined exactly as you would expect. For functions of two variables the formulas are

$$(f + g)(x, y) = f(x, y) + g(x, y)$$
$$(f - g)(x, y) = f(x, y) - g(x, y)$$
$$(fg)(x, y) = f(x, y)g(x, y)$$
$$\left(\frac{f}{g}\right)(x, y) = \frac{f(x, y)}{g(x, y)}$$

The formulas for functions of three variables are analogous. The domains of $f + g$, $f - g$, and fg consist of all points simultaneously in the domain of f and the domain of g, whereas the domain of f/g consists of all points simultaneously in the domain of f and the domain of g at which g does not assume the value 0. For example, if

$$F(x, y) = axy + \frac{by}{x}$$

then F may be thought of as the sum of the functions f and g, where

$$f(x, y) = axy \quad \text{and} \quad g(x, y) = \frac{by}{x}$$

A function f of two variables x and y is a **polynomial function** if it is a sum of functions of the form $cx^m y^n$, where c is a number and m and n are nonnegative integers. A **rational function** is, as with functions of one variable, the quotient of two polynomial functions. Similar terminology is used for polynomial and rational functions of three variables. Thus if

$$f(x, y) = 3x^2 y^2 + 4x^2 - 7y^2 + 3x + \sqrt{2}$$

and
$$g(x, y, z) = 4xyz^5 - 6$$

then f and g are polynomial functions, and if

$$f(x, y) = \frac{x^3 + y^3}{x^2 + y^2}$$

and
$$g(x, y, z) = \frac{xy}{1 + x^2 + y^2 + z^2}$$

then f and g are rational functions.

It is also possible to form composites using functions of several variables. If f is a function of two variables and g is a function of a single variable, then the function $g \circ f$ is defined by

$$(g \circ f)(x, y) = g(f(x, y))$$

for all (x, y) in the domain of f such that $f(x, y)$ is in the domain of g. The definition of the composite of a function of three variables and a function of a single variable is similar.

EXAMPLE 1   Let $F(x, y) = \sqrt{\ln(4 - x^2 - y^2)}$. Find a function f of two variables and a function g of one variable such that $F = g \circ f$. Find the domain of F.

Solution   There are many ways to express F as a composite. A natural way is to let

$$f(x, y) = \ln (4 - x^2 - y^2) \quad \text{and} \quad g(t) = \sqrt{t}$$

Then $F = g \circ f$. The domain of f consists of all (x, y) for which $4 - x^2 - y^2 > 0$, and the domain of g consists of all nonnegative numbers. Therefore the domain of $g \circ f$ consists of all (x, y) such that $4 - x^2 - y^2 > 0$ and $\ln (4 - x^2 - y^2) \geq 0$. Since $\ln (4 - x^2 - y^2) \geq 0$ if and only if $4 - x^2 - y^2 \geq 1$, we conclude that the domain of $F = g \circ f$ consists of all (x, y) for which $4 - x^2 - y^2 \geq 1$, or equivalently, $x^2 + y^2 \leq 3$.   ❑

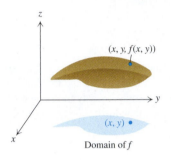

FIGURE 13.1   The graph of a function f of two variables.

The **graph** of a function f of two variables is the collection of points $(x, y, f(x, y))$ for which (x, y) is in the domain of f (Figure 13.1). It is customary to let $z = f(x, y)$; then the graph of f consists of all points (x, y, z) such that $z = f(x, y)$.

In sketching the graph of a function f of two variables, it is often helpful to determine the intersections of the graph of f with planes of the form $z = c$ (Figure 13.2). We call each such intersection the **trace** of the graph of f in the plane $z = c$. Thus the trace of the graph of f in the plane $z = c$ is the collection of points (x, y, c) such that $f(x, y) = c$. If we think of the graph of f as the surface of a mountain, then the trace in a plane $z = c$ is a curve of constant altitude. A mountain climber walking along such a trace would neither ascend nor descend.

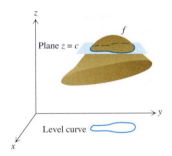

FIGURE 13.2   The trace of the graph of f in the plane $z = c$, and the level curve $f(x, y) = c$.

Closely related to the trace is the notion of level curve. The set of points (x, y) in the xy plane such that $f(x, y) = c$ is a **level curve** of f (Figure 13.2); we identify the level curve with the equation $f(x, y) = c$ and call the equation a level curve of f. Notice that the trace of f in the plane $z = c$ either lies directly above or directly below the level curve $f(x, y) = c$ or coincides with it. The idea behind level curves is to provide three-dimensional information in a two-dimensional setting. For example, level curves are employed in contour maps to indicate elevations of points on a mountain, such as Mount St. Helens (Figure 13.3(a)); a single level curve represents points of identical altitude. On a weather map a level curve represents points with identical temperature or barometric pressure (Figure 13.3(b)). Recently color-coded contour maps have been produced to depict, for example, ozone levels in the Southern Hemisphere (Figure 13.3(c)).

EXAMPLE 2   Let $f(x, y) = 8 - 2x - 4y$. Sketch the graph of f, and determine the level curves.

Solution   If we let $z = f(x, y)$, then the equation becomes

$$z = 8 - 2x - 4y$$

This is an equation of a plane with x intercept 4, y intercept 2, and z intercept 8. The portion of the plane in the first octant is sketched in Figure 13.4. For any value of c, the level curve $f(x, y) = c$ is the line in the xy plane with equation $2x + 4y = 8 - c$.   ❑

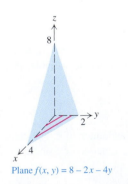

Plane $f(x, y) = 8 - 2x - 4y$

FIGURE 13.4   The level curves are parallel lines.

It is possible to show that any plane not perpendicular to the xy plane is the graph of a function f of the form $f(x, y) = ax + by + c$, and conversely, that the graph of any function of this form is a plane. The level curves of such a function are lines unless $a = b = 0$, in which case the level curve is the whole plane.

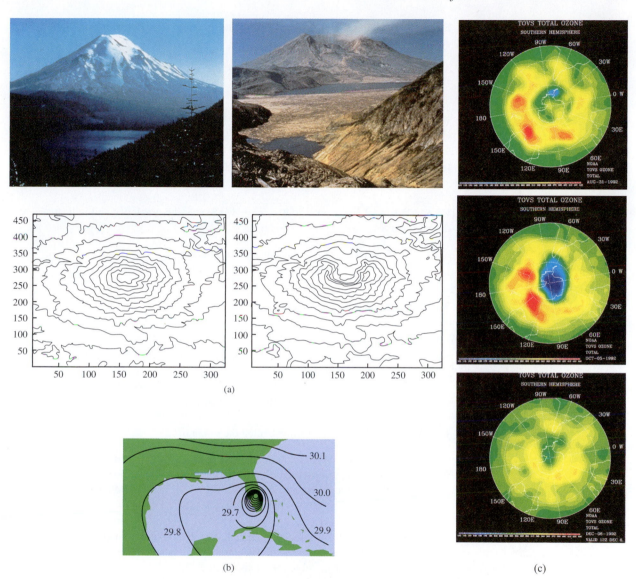

FIGURE 13.3 (a) Left, Mt. St. Helens before the eruption on May 18, 1980. Right, Mt. St. Helens after the eruption on May 18, 1980. (Photo credit: U.S.D.A./Mt. St. Helens N.V.M.) (b) A Caribbean hurricane with its isobars, as recorded in Miami, Florida. (c) The ozone hole. The color scale shows areas of lowest ozone in dark blue and highest ozone in red. Notice that the small ozone hole on August 31 developed into a large hole by October 5 and then closed up by December 6. (Climate Analysis Center/N.O.A.A.)

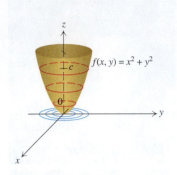

FIGURE 13.5   The level curves are circles.

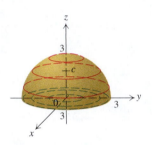

FIGURE 13.6   The level curves are circles.

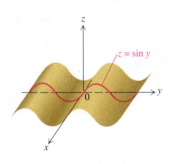

FIGURE 13.7

EXAMPLE 3   Let $f(x, y) = x^2 + y^2$. Sketch the graph of f and determine the level curves.

Solution   If $c > 0$, then the level curve $f(x, y) = c$ is given by

$$x^2 + y^2 = c$$

which is a circle with radius $\sqrt{c}$. Therefore the trace of the graph in the plane $z = c$ is also a circle with radius $\sqrt{c}$. The level curve $f(x, y) = 0$ is the point $(0, 0)$. For $c < 0$ the level curve $f(x, y) = c$ contains no points. The intersections of the graph with the planes $x = 0$ and $y = 0$ are both parabolas. The graph is sketched in Figure 13.5.   ❑

EXAMPLE 4   Let $f(x, y) = \sqrt{9 - x^2 - y^2}$. Sketch the graph of f and indicate the level curves.

Solution   If we let $z = f(x, y)$, the equation of f becomes

$$z = \sqrt{9 - x^2 - y^2} \tag{1}$$

Squaring both sides of this equation and transposing x^2 and y^2, we obtain

$$x^2 + y^2 + z^2 = 9$$

which we recognize as the equation of a sphere centered at the origin with radius 3. Since (1) holds only for $z \geq 0$, we conclude that the graph of f is the hemisphere sketched in Figure 13.6. In this case the level curve $f(x, y) = c$ is a circle if $0 \leq c < 3$ and is the point $(0, 0)$ if $c = 3$.   ❑

EXAMPLE 5   Let $f(x, y) = \sin y$. Sketch the graph of f.

Solution   Since the value of $f(x, y)$ is independent of x, it follows that if $(0, y, \sin y)$ is a point on the graph, then so is $(x, y, \sin y)$ for any x, that is, the entire line parallel to the x axis through the point $(0, y, \sin y)$ is on the graph. Thus the whole surface is determined by a sine curve in the yz plane (Figure 13.7).   ❑

We call the graph of a function like that in Example 5 a **sheet** or, more precisely, a sheet parallel to the x axis. The graph of the function whose equation is $f(x, y) = -x^3$ is a sheet parallel to the y axis (Figure 13.8).

Level curves of a function f can provide meaningful information about the graph of f. For example, Figure 13.9 displays the graph of a function, along with

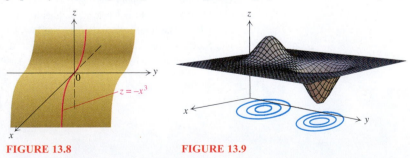

FIGURE 13.8   **FIGURE 13.9**

several of its level curves. Notice that one set of nested concentric circles is associated with a hill, whereas the other set is associated with a valley.

It is important not to be misled by the simplicity of the graphs in Figures 13.4 to 13.8. The problems inherent in graphing functions of two variables are many times as great as in graphing functions of a single variable. To see the difficulties that can arise, you might try sketching the graph of the innocuous looking function defined by

$$f(x, y) = \sin x + \sin y$$

A sketch of the graph of f appears below, along with several other graphs. All were generated on a computer by means of Mathematica.

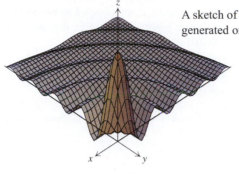

$$f(x, y) = \frac{\sin \sqrt{x^2 + y^2}}{\sqrt{x^2 + y^2}}$$

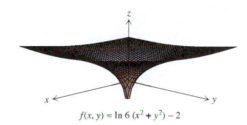

$$f(x, y) = \ln 6 \, (x^2 + y^2) - 2$$

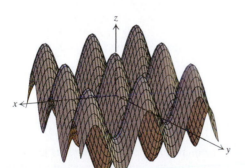

$$f(x, y) = \sin x + \sin y$$

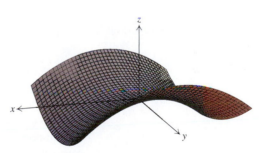

$$f(x, y) = y^2 - x^2$$

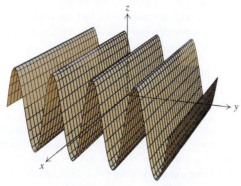

$$f(x, y) = \frac{3}{4} \sin 2y$$

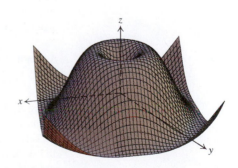

$$f(x, y) = \sin \sqrt{x^2 + y^2}$$

Level Surfaces

Although we can sketch the graphs of many functions of two variables, it is impossible to sketch the graph of any function of three variables, for that would entail four dimensions. However, we can gain information about a function f of three variables from what we call its level surfaces. For any number c the set of points (x, y, z) for which $f(x, y, z) = c$ is called a **level surface** of f, and we identify a level surface with the corresponding equation $f(x, y, z) = c$. Level surfaces of functions of three variables are analogous to level curves of functions of two variables. If $f(x, y, z)$ denotes the temperature at any point (x, y, z) in a region in space, then the level surface $f(x, y, z) = c$ is the surface on which the temperature is constantly c, and is called an **isothermal surface.** Similarly, if $V(x, y, z)$ is the voltage (or potential) at (x, y, z), then the level surface $V(x, y, z) = c$ is called an **equipotential surface.**

In Chapter 11 we encountered three kinds of level surfaces. First, spheres centered at the origin are level surfaces of the function

$$f(x, y, z) = x^2 + y^2 + z^2$$

since $x^2 + y^2 + z^2 = c^2$ is an equation of such a sphere for $c \neq 0$. Second, any cylinder whose axis is the z axis is a level surface of the function

$$f(x, y, z) = x^2 + y^2$$

because $x^2 + y^2 = c^2$ is an equation of such a cylinder if $c \neq 0$. Finally, planes are level surfaces of functions of the form

$$f(x, y, z) = ax + by + cz$$

provided that a, b, and c are not all 0.

We also observe that the graph of any function f of two variables is a level surface. We need only let

$$g(x, y, z) = z - f(x, y)$$

and notice that

$$g(x, y, z) = 0 \quad \text{if and only if} \quad z = f(x, y)$$

Thus the level surface $g(x, y, z) = 0$ is the graph of f, or equivalently, the graph of the equation $z = f(x, y)$. This is why we often call the graph of a function of, or an equation in, two variables a **surface.**

In sketching a level surface we will use the intersections of the level surface with planes of the form $x = c$ or $y = c$, as well as those of the form $z = c$. In each case the intersection of the level surface with the plane is called the **trace** of the level surface in that plane. The most important level surfaces are those called quadric surfaces, which we discuss and graph next.

Quadric Surfaces

A **quadric surface** is a level surface of a polynomial function f given by

$$f(x, y, z) = Ax^2 + By^2 + Cz^2 + Dxy + Exz + Fyz + Gx + Hy + Iz + J$$

where $A, B, \ldots, J$ are constants. Quadric surfaces are three-dimensional versions of the conic sections (parabolas, ellipses and hyperbolas) presented in Sections 10.3–10.5. Quadric surfaces fall into nine major classes. We will list these classes and sketch one surface in each class, assuming in each case that the constants a, b, and c are positive.

Ellipsoid: $\dfrac{x^2}{a^2} + \dfrac{y^2}{b^2} + \dfrac{z^2}{c^2} = 1$

The trace of the ellipsoid in any plane parallel to a coordinate plane is either an ellipse, a point, or empty (Figure 13.10). If $a = b = c$, the equation becomes $x^2 + y^2 + z^2 = a^2$, and the surface is a sphere.

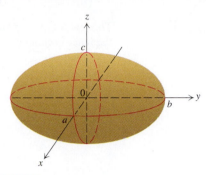

FIGURE 13.10   An ellipsoid.

Elliptic cylinder: $\dfrac{x^2}{a^2} + \dfrac{y^2}{b^2} = 1$

The trace of the elliptic cylinder in any plane parallel to the xy plane is the ellipse $x^2/a^2 + y^2/b^2 = 1$ (Figure 13.11). If $a = b$, the surface is a **circular cylinder.**

Elliptic double cone: $\dfrac{x^2}{a^2} + \dfrac{y^2}{b^2} = \dfrac{z^2}{c^2}$

The trace of the cone in any plane parallel to the xy plane is either an ellipse (a circle if $a = b$) or a point (Figure 13.12). The traces in the yz and xz planes consist of two lines through the origin. If $a = b$, the surface is called a **circular double cone.**

Elliptic paraboloid: $\dfrac{x^2}{a^2} + \dfrac{y^2}{b^2} = \dfrac{z}{c}$

The trace of the paraboloid in any plane parallel to the xy plane is either an ellipse (a circle if $a = b$), a point, or empty. The traces in the yz and xz planes are parabolas (Figure 13.13). If $a = b$, the surface is called a **circular paraboloid.** (The surface in Figure 13.5 is a circular paraboloid.)

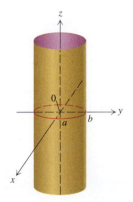

FIGURE 13.11   An elliptic cylinder.

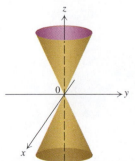

FIGURE 13.12   An elliptic double cone.

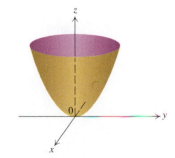

FIGURE 13.13   An elliptic paraboloid.

Parabolic sheet (or **parabolic cylinder**): $z = ax^2$

The trace of the sheet in the xz plane is the parabola $z = ax^2$ (Figure 13.14). The trace in any plane parallel to the xz plane is a translation of that parabola along the y axis.

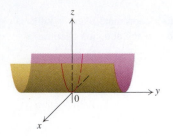

FIGURE 13.14   A parabolic sheet.

Hyperbolic paraboloid: $\dfrac{y^2}{b^2} - \dfrac{x^2}{a^2} = \dfrac{z}{c}$

The traces in the yz and xz planes are parabolas, the former opening upward and the latter opening downward (Figure 13.15). The trace in the xy plane consists of two intersecting lines. The trace in any other plane parallel to the xy plane is a hyperbola. The surface has the appearance of a saddle.

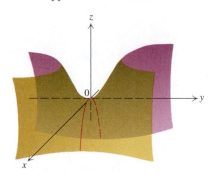

FIGURE 13.15   A hyperbolic paraboloid.

Two hyperbolic sheets (or **hyperbolic cylinder**): $\dfrac{y^2}{b^2} - \dfrac{x^2}{a^2} = 1$

The trace in any plane parallel to the xy plane is the hyperbola $y^2/b^2 - x^2/a^2 = 1$ (Figure 13.16).

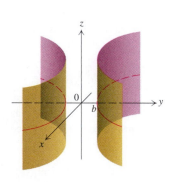

FIGURE 13.16   Two hyperbolic sheets.

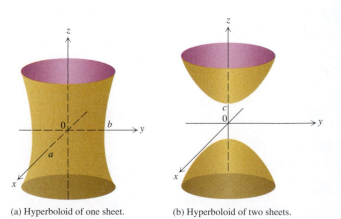

(a) Hyperboloid of one sheet.   (b) Hyperboloid of two sheets.

FIGURE 13.17

Hyperboloid of one sheet: $\dfrac{x^2}{a^2} + \dfrac{y^2}{b^2} - \dfrac{z^2}{c^2} = 1$

The traces in the yz and xz planes are hyperbolas. The trace in any plane parallel to the xy plane is an ellipse (a circle if $a = b$) (Figure 13.17(a)).

Hyperboloid of two sheets: $\dfrac{z^2}{c^2} - \dfrac{x^2}{a^2} - \dfrac{y^2}{b^2} = 1$

The trace in any plane parallel to the xz or the yz plane is a hyperbola. The trace in any plane parallel to the xy plane is an ellipse (a circle if $a = b$), a point, or empty (Figure 13.17(b)).

We end the section with the graph of a paraboloid that is positioned differently from the paraboloid in Figure 13.13.

EXAMPLE 6   Sketch the graph of $z = 4 - x^2 - y^2$.

Solution   The graph of $z = x^2 + y^2$ is a circular paraboloid opening upward with lowest point at the origin, as in Figure 13.13. Thus the graph of $z = -x^2 - y^2$ is a circular paraboloid opening downward with highest point at the origin. To obtain the graph of $z = 4 - x^2 - y^2$, we only need to translate the graph of $z = -x^2 - y^2$ upward four units. The resulting graph is shown in Figure 13.18.   ❑

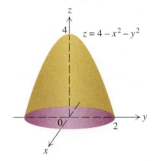

$z = 4 - x^2 - y^2$

FIGURE 13.18

EXERCISES 13.1

In Exercises 1–12 find the domain of the function.

1. $f(x, y) = \sqrt{x} + \sqrt{y}$ **2.** $f(x, y) = \sqrt{x + y}$

3. $f(x, y) = \dfrac{y}{x} - \dfrac{x}{y}$ **4.** $f(x, y) = \sin \dfrac{1}{xy}$

5. $g(x, y) = \sqrt{x^2 + y^2 - 25}$ **6.** $g(x, y) = \sqrt{25 - x^2 - y^2}$

7. $f(x, y) = \dfrac{1}{x + y}$ **8.** $f(u, v) = \ln \dfrac{u^2 + v^2}{(u^2 - v^2)^2}$

9. $f(x, y, z) = \sqrt{1 - x^2 - y^2 - z^2}$

10. $f(x, y, z) = \dfrac{1}{xyz}$ **11.** $g(x, y, z) = \dfrac{x}{y} - \dfrac{y}{z} + \dfrac{z}{x}$

12. $f(x, y, z) = \dfrac{xyz}{(x + y)^3 - (x + z)^3}$

In Exercises 13–18 sketch the level curve $f(x, y) = c$.

13. $f(x, y) = 3x - y$; $c = 2, 3$

14. $f(x, y) = 6x^2$; $c = 6, 24$

15. $f(x, y) = x^2 + 4y^2$; $c = 1, 4$

16. $f(x, y) = x^2 - y$; $c = -2, 2$

17. $f(x, y) = x^2 - y^2$; $c = -1, 0, 1$

18. $f(x, y) = 2y - \cos x$; $c = 0, 1, 2$

In Exercises 19–23 sketch the graph of f.

19. $f(x, y) = x + 2y$ **20.** $f(x, y) = 2x - 3y + 4$

21. $f(x, y) = \sqrt{4 - x^2 - y^2}$ **22.** $f(x, y) = \sqrt{4x^2 + 9y^2}$

23. $f(x, y) = x^{1/3}$

In Exercises 24–32 sketch the graph of the equation.

24. $z = 2$ **25.** $x = -3$

26. $z = y^2$ **27.** $z = x^3 + 1$

28. $z = y^3 - 1$ **29.** $x = \sqrt{1 - y^2}$

30. $x = \sqrt{4 - y^2 - z^2}$ **31.** $x = \sqrt{y^2 + 4z^2}$

32. $y = \sqrt{1 - x^2 - z^2}$

In Exercises 33–37 sketch the level surface $f(x, y, z) = c$.

33. $f(x, y, z) = 2x - 4y + z$; $c = -1$

34. $f(x, y, z) = x^2 + y^2 + z^2$; $c = 2$

35. $f(x, y, z) = 4x^2 + 4y^2 + z^2$; $c = 1$

36. $f(x, y, z) = x^2 + y^2 - z^2$; $c = 0$

37. $f(x, y, z) = z - 1 - x^2 - y^2$; $c = 2$

In Exercises 38–55 sketch the quadric surface.

38. $\dfrac{x^2}{4} + y^2 + \dfrac{z^2}{9} = 1$ **39.** $x^2 + 2y^2 + 3z^2 = 6$

40. $x^2 + z^2 = 4$ **41.** $y^2 + z^2 = 9$

42. $z = x^2 + \dfrac{y^2}{9}$ **43.** $x = y^2 + \dfrac{z^2}{4}$

44. $z^2 = x^2 + 4y^2$ **45.** $x^2 = 9y^2 + 4z^2$
46. $y = 1 - x^2$ **47.** $x = z^2 + 3$
48. $z = y^2 - 4x^2$ **49.** $x = 4z^2 - y^2$
50. $y^2 - x^2 = 4$ **51.** $z^2 - y^2 = 9$
52. $z^2 + 4y^2 - 2x^2 = 1$ **53.** $4x^2 + y^2 - z^2 = 16$
54. $z^2 - 4y^2 - x^2 = 1$ **55.** $x^2 - 9y^2 - 4z^2 = 36$

56. In each of the following, determine a function f of two variables (different from F) and a function g of one variable such that $F = g \circ f$.

  a. $F(x, y) = \sqrt{4 - x^2 - y^2}$ **b.** $F(x, y) = e^{x\sqrt{y}}$

In Exercises 57–60 match the given function with one of the given surfaces and one of the given sets of level curves.

57. $f(x, y) = |x| + |y|$ **58.** $g(x, y) = x^2 + y$
59. $h(x, y) = x\, e^{-(x^2 + y^2)}$ **60.** $k(x, y) = \sin x \sin y$

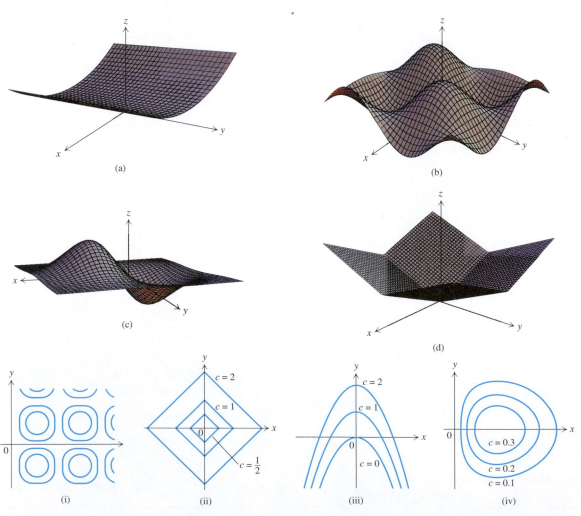

(a)

(b)

(c)

(d)

(i)   (ii)   (iii)   (iv)

61. Express the height h of a right circular cylinder as a function of the volume V and radius r.

62. Express the radius r of the base of a right circular cone as a function of the volume V and height h.

63. Express the surface area S of a rectangular box with no top as a function of the dimensions x, y, and z.

Applications

64. Express the amount A of metal required to make a storage box in the shape of a rectangular parallelepiped as a function of the length x, width y, and height z if the box is to have 12 compartments in 2 rows of 6 each and no top (Figure 13.19).

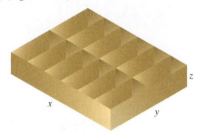

FIGURE 13.19

65. Express the cost C of painting a rectangular wall as a function of the dimensions x and y (in meters) if the cost per square meter is $0.30.

66. Express the cost C of painting a rectangular wall as a function of the dimensions x and y (in meters) if the cost per square meter is $0.30 and the wall contains a window 1 square meter in area.

67. The strength of the electric field at (x, y, z) due to an infinitely long charged wire lying along the z axis is given by

$$E(x, y, z) = \frac{c}{\sqrt{x^2 + y^2}}$$

where c is a positive constant. Describe the level surfaces of E.

68. The magnitude of the gravitational force exerted on a unit mass at (x, y, z) by a point mass located at the origin is given by

$$F(x, y, z) = \frac{c}{x^2 + y^2 + z^2}$$

where c is a positive constant. Describe the level surfaces of F.

69. Suppose a thin metal plate occupies the first quadrant of the xy plane and the temperature at (x, y) is given by

$$T(x, y) = xy$$

Describe the isothermal curves, that is, the level curves of T.

70. Let $f(x, y) = (x + 1)(y + 2)$ for $x \geq 0$ and $y \geq 0$. Sketch the level curves $f(x, y) = 3$ and $f(x, y) = 4$. (If f represents a utility function for two competing goods such as beer and wine, then the level curves are called **indifference curves**.)

Project

1. In the same way that conic sections can have different shapes—more oblong or more rounded—quadric surfaces come in various sizes and shapes. In this project we will explore relative shapes of ellipsoids, double cones and paraboloids.

a. Consider the ellipsoid

$$\frac{x^2}{a^2} + \frac{y^2}{b^2} + \frac{z^2}{c^2} = 1$$

  i. What happens to the shape of the ellipsoid if $a = b = c$ and a approaches 0? What is the limiting figure?

  ii. What happens to the shape of the ellipsoid if $a = 1 = b$ and c approaches 0? What is the limiting figure?

  iii. What happens to the shape of the ellipsoid if $a = 1 = b$ and c approaches ∞? What is the limiting figure?

b. Consider the double cone

$$\frac{x^2}{a^2} + \frac{y^2}{b^2} = \frac{z^2}{c^2}$$

  i. What happens to the shape of the double cone if $a = b = c$ and a approaches 0, or a approaches ∞? What is the limiting figure?

  ii. What happens to the shape of the double cone if $a = 1 = b$ and c approaches 0? What is the limiting figure?

  iii. What happens to the shape of the double cone if $a = 1 = b$ and c approaches ∞? What is the limiting figure?

c. Consider the elliptic paraboloid

$$z = \frac{x^2}{a^2} + \frac{y^2}{b^2}$$

  i. What happens to the shape of the paraboloid if $a = b$ and a approaches 0? What is the limiting figure?

  ii. What happens to the shape of the paraboloid if $a = b$ and a approaches ∞? What is the limiting figure?

  iii. What happens to the shape of the paraboloid if $a = 1$ and b approaches 0? What is the limiting figure?

13.2   LIMITS AND CONTINUITY

The definitions of the limit of a real-valued function in Section 2.2 and the limit of a vector-valued function in Section 12.2 provide the models for our definition of the limit of a function of several variables. Intuitively, L is the limit of $f(x, y)$ as (x, y) approaches (x_0, y_0) if $f(x, y)$ is as close to L as we wish whenever (x, y) is close enough to (x_0, y_0). Similarly, L is the limit of $f(x, y, z)$ as (x, y, z) approaches (x_0, y_0, z_0) if $f(x, y, z)$ is as close to L as we wish whenever (x, y, z) is close enough to (x_0, y_0, z_0). To formalize these ideas, we recall that the distance between the points (x, y) and (x_0, y_0) in the plane is less than δ if

$$\sqrt{(x - x_0)^2 + (y - y_0)^2} < \delta$$

and the distance between the points (x, y, z) and (x_0, y_0, z_0) in space is less than δ if

$$\sqrt{(x - x_0)^2 + (y - y_0)^2 + (z - z_0)^2} < \delta$$

DEFINITION 13.2   Let f be defined throughout an open disk centered at (x_0, y_0), except possibly at (x_0, y_0) itself, and let L be a number. Then L is the **limit of f at (x_0, y_0)** if for every $\varepsilon > 0$ there is a $\delta > 0$ such that

$$\text{if } 0 < \sqrt{(x - x_0)^2 + (y - y_0)^2} < \delta, \quad \text{then } |f(x, y) - L| < \varepsilon$$

In this case we write

$$\lim_{(x,y) \to (x_0, y_0)} f(x, y) = L$$

and say that $\lim_{(x, y) \to (x_0, y_0)} f(x, y)$ exists. Similarly, let f be defined throughout an open ball centered at (x_0, y_0, z_0), except possibly at (x_0, y_0, z_0) itself. Then L is the **limit of f at (x_0, y_0, z_0)** if for every $\varepsilon > 0$ there is a $\delta > 0$ such that

$$\text{if } 0 < \sqrt{(x - x_0)^2 + (y - y_0)^2 + (z - z_0)^2} < \delta, \quad \text{then } |f(x, y, z) - L| < \varepsilon$$

In this case we write

$$\lim_{(x, y, z) \to (x_0, y_0, z_0)} f(x, y, z) = L$$

and say that $\lim_{(x, y, z) \to (x_0, y_0, z_0)} f(x, y, z)$ exists.

It follows from the first part of Definition 13.2 that if $\lim_{(x, y) \to (x_0, y_0)} f(x, y) = L$, then for each $\varepsilon > 0$ the value assigned to every point in a sufficiently small open disk about (x_0, y_0) (except possibly (x_0, y_0)) lies in the interval $(L - \varepsilon, L - \varepsilon)$ (Figure 13.20). This is analogous to the geometric interpretation of limit given in Section 2.2. There is no analogous pictorial interpretation of $\lim_{(x_0, y_0, z_0)} f(x, y, z) = L$, because that would require four dimensions.

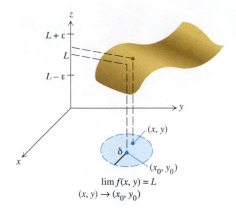

$$\lim_{(x, y) \to (x_0, y_0)} f(x, y) = L$$

FIGURE 13.20

EXAMPLE 1   Show that

$$\lim_{(x, y) \to (x_0, y_0)} x = x_0 \quad \text{and} \quad \lim_{(x, y) \to (x_0, y_0)} y = y_0$$

Solution   Let $\varepsilon > 0$. Observe that

$$\sqrt{(x - x_0)^2} \le \sqrt{(x - x_0)^2 + (y - y_0)^2}$$

Therefore if we let $\delta = \varepsilon$, it follows that

$$\text{if } 0 < \sqrt{(x - x_0)^2 + (y - y_0)^2} < \delta, \quad \text{then } |x - x_0| = \sqrt{(x - x_0)^2} < \delta = \varepsilon$$

This proves that $\lim_{(x, y) \to (x_0, y_0)} x = x_0$. The second limit is established analogously. ❑

EXAMPLE 2   Show that

$$\lim_{(x, y, z) \to (x_0, y_0, z_0)} x = x_0, \quad \lim_{(x, y, z) \to (x_0, y_0, z_0)} y = y_0, \quad \text{and} \quad \lim_{(x, y, z) \to (x_0, y_0, z_0)} z = z_0$$

Solution   Let $\varepsilon > 0$. Since

$$\sqrt{(x - x_0)^2} \le \sqrt{(x - x_0)^2 + (y - y_0)^2 + (z - z_0)^2}$$

we can let $\delta = \varepsilon$ and deduce that

$$\text{if } 0 < \sqrt{(x - x_0)^2 + (y - y_0)^2 + (z - z_0)^2} < \delta$$

$$\text{then } |x - x_0| = \sqrt{(x - x_0)^2} < \delta = \varepsilon$$

Thus the first limit is verified. The proofs for the other two limits are similar. ❑

The limit formulas for sums, products, and quotients have counterparts for functions of several variables. For functions of two variables, they are as follows:

If $\lim_{(x,\,y)\to(x_0,\,y_0)} f(x, y)$ and $\lim_{(x,\,y)\to(x_0,\,y_0)} g(x, y)$ exist, then

$$\lim_{(x,\,y)\to(x_0,\,y_0)} (f + g)(x, y) = \lim_{(x,\,y)\to(x_0,\,y_0)} f(x, y) + \lim_{(x,y)\to(x_0,\,y_0)} g(x, y)$$

$$\lim_{(x,\,y)\to(x_0,\,y_0)} (f - g)(x, y) = \lim_{(x,\,y)\to(x_0,\,y_0)} f(x, y) - \lim_{(x,\,y)\to(x_0,\,y_0)} g(x, y)$$

$$\lim_{(x,\,y)\to(x_0,\,y_0)} (fg)(x, y) = \lim_{(x,\,y)\to(x_0,\,y_0)} f(x, y) \lim_{(x,\,y)\to(x_0,\,y_0)} g(x, y)$$

$$\lim_{(x,\,y)\to(x_0,\,y_0)} \left(\frac{f}{g}\right)(x, y) = \frac{\lim_{(x,\,y)\to(x_0,\,y_0)} f(x, y)}{\lim_{(x,\,y)\to(x_0,\,y_0)} g(x, y)} \quad \text{provided} \quad \lim_{(x,\,y)\to(x_0,\,y_0)} g(x, y) \neq 0$$

The formulas for limits of functions of three variables are similar.

EXAMPLE 3   Show that

$$\lim_{(x,\,y)\to(-1,\,2)} \frac{x^3 + y^3}{x^2 + y^2} = \frac{7}{5} \quad \text{and} \quad \lim_{(x,\,y)\to(0,\,0)} \frac{x^3 + y^3}{x^2 + y^2} = 0$$

Solution   Since

$$\lim_{(x,\,y)\to(-1,\,2)} x = -1 \quad \text{and} \quad \lim_{(x,\,y)\to(-1,\,2)} y = 2$$

the product formula yields

$$\lim_{(x,\,y)\to(-1,\,2)} x^3 = -1, \qquad \lim_{(x,\,y)\to(-1,\,2)} y^3 = 8,$$

$$\lim_{(x,\,y)\to(-1,\,2)} x^2 = 1, \qquad \lim_{(x,\,y)\to(-1,\,2)} y^2 = 4$$

Then the sum and quotient formulas combine to yield

$$\lim_{(x,\,y)\to(-1,\,2)} \frac{x^3 + y^3}{x^2 + y^2} = \frac{\lim_{(x,\,y)\to(-1,2)} (x^3 + y^3)}{\lim_{(x,\,y)\to(-1,\,2)} (x^2 + y^2)}$$

$$= \frac{\lim_{(x,\,y)\to(-1,\,2)} x^3 + \lim_{(x,\,y)\to(-1,\,2)} y^3}{\lim_{(x,\,y)\to(-1,\,2)} x^2 + \lim_{(x,\,y)\to(-1,\,2)} y^2} = \frac{-1 + 8}{1 + 4} = \frac{7}{5}$$

We cannot use quite the same procedure to verify the second limit, since $\lim_{(x,\,y)\to(0,\,0)} (x^2 + y^2) = 0$, so the quotient formula does not apply. To verify the second limit, we will show first that

$$\lim_{(x,\,y)\to(0,\,0)} \frac{x^3}{x^2 + y^2} = 0 \tag{1}$$

For this limit we observe that

$$0 \leq \left| \frac{x^3}{x^2 + y^2} \right| \leq \left| \frac{x^3}{x^2} \right| = |x|$$

Since $\lim_{(x, y)\to(0, 0)} x = 0$ by Example 1, a version of the Squeezing Theorem for functions of two variables yields (1). A similar argument shows that

$$\lim_{(x, y)\to(0, 0)} \frac{y^3}{x^2 + y^2} = 0 \tag{2}$$

We can use the sum formula to combine the limits in (1) and (2):

$$\lim_{(x, y)\to(0, 0)} \frac{x^3 + y^3}{x^2 + y^2} = \lim_{(x, y)\to(0, 0)} \frac{x^3}{x^2 + y^2} + \lim_{(x, y)\to(0, 0)} \frac{y^3}{x^2 + y^2} = 0 + 0 = 0 \quad \square$$

An alternative way to show that

$$\lim_{(x, y)\to(0, 0)} \frac{x^3 + y^3}{x^2 + y^2} = 0 \tag{3}$$

involves polar coordinates. Notice that if $x = r\cos\theta$ and $y = r\sin\theta$, then (x, y) approaches $(0, 0)$ if and only if r approaches 0. Since

$$\frac{x^3 + y^3}{x^2 + y^2} = \frac{r^3\cos^3\theta + r^3\sin^3\theta}{r^2} = r(\cos^3\theta + \sin^3\theta)$$

and since $\lim_{r\to 0} r(\cos^3\theta + \sin^3\theta) = 0$, it follows that (3) is valid.

There are also versions of the Substitution Rule for functions of several variables. For the two-variable case we suppose that

$$\lim_{(x, y)\to(x_0, y_0)} f(x, y) = L$$

and that g is a function of a single variable which is continuous at L. Then

$$\lim_{(x, y)\to(x_0, y_0)} g(f(x, y)) = g(L)$$

EXAMPLE 4   Find $\displaystyle\lim_{(x, y)\to(e, 1)} \ln\frac{x}{y}$.

Solution   First we let

$$f(x, y) = \frac{x}{y} \quad \text{and} \quad g(t) = \ln t$$

Then by the quotient formula for limits,

$$\lim_{(x, y)\to(e, 1)} f(x, y) = \lim_{(x, y)\to(e, 1)} \frac{x}{y} = \frac{\displaystyle\lim_{(x, y)\to(e, 1)} x}{\displaystyle\lim_{(x, y)\to(e, 1)} y} = \frac{e}{1} = e$$

Since g is continuous, it follows from the preceding substitution formula that

$$\lim_{(x, y)\to(e, 1)} \ln\frac{x}{y} = \lim_{(x, y)\to(e, 1)} g(f(x, y)) = g(e) = \ln e = 1 \quad \square$$

From Theorem 2.5 we know that if f is a function of one variable, then in order that $\lim_{x\to a} f(x) = L$, both one-sided limits $\lim_{x\to a^-} f(x)$ and $\lim_{x\to a^+} f(x)$ must exist and equal L. Now if f is a function of two variables and

$\lim_{(x, y) \to (x_0, y_0)} f(x, y) = L$, then Definition 13.2 implies that $f(x, y)$ must approach L as (x, y) approaches (x_0, y_0) along *each line* (or curve) through (x_0, y_0). Thus to show that $\lim_{(x, y) \to (x_0, y_0)} f(x, y)$ does *not* exist, it suffices to show that $f(x, y)$ approaches two different numbers as (x, y) approaches (x_0, y_0) along two distinct lines (or curves) through (x_0, y_0).

EXAMPLE 5   Let

$$f(x, y) = \frac{y^2 - x^2}{y^2 + x^2}$$

Show that $\lim_{(x, y) \to (0, 0)} f(x, y)$ does not exist.

Solution   Notice that if $y = 0$ and $x \neq 0$ then $f(x, y) = -1$, whereas if $x = 0$ and $y \neq 0$ then $f(x, y) = 1$. Thus $f(x, y)$ approaches -1 as (x, y) approaches $(0, 0)$ along the line $y = 0$, and $f(x, y)$ approaches 1 as (x, y) approaches $(0, 0)$ along the line $x = 0$. Therefore f does not have a limit at $(0, 0)$. The computer-drawn graph of f appearing in Figure 13.21 supports the claim that f does not have a limit at $(0, 0)$.   ◻

FIGURE 13.21

Continuity

Now that limits have been defined, continuity is defined in the standard way.

DEFINITION 13.3   **a.**   Suppose a function f of two variables is defined throughout an open disk about (x_0, y_0). Then f is **continuous at (x_0, y_0)** if

$$\lim_{(x, y) \to (x_0, y_0)} f(x, y) = f(x_0, y_0)$$

b.   Suppose a function f of three variables is defined throughout an open ball about (x_0, y_0, z_0). Then f is **continuous at (x_0, y_0, z_0)** if

$$\lim_{(x, y, z) \to (x_0, y_0, z_0)} f(x, y, z) = f(x_0, y_0, z_0)$$

If a function f is continuous at each point in its domain, then we will refer to f as a **continuous function.**

Example 1 tells us that if $f(x, y) = x$ and $g(x, y) = y$, then f and g are continuous functions. Similarly, if $f(x, y, z) = x$, $g(x, y, z) = y$, and $h(x, y, z) = z$, then by Example 2, f, g, and h are continuous functions. Sums, products, and quotients of continuous functions are continuous. Thus polynomial functions and rational functions are continuous.

Finally, we observe that if f is a function of two variables and g is a function of a single variable, then the continuity of f at (x_0, y_0) and the continuity of g at $f(x_0, y_0)$ together imply that $g \circ f$ is continuous at (x_0, y_0). A similar result holds when f is a function of three variables.

EXAMPLE 6   Let

$$F(x, y) = \sin \frac{xy}{1 + x^2 + y^2}$$

Show that F is a continuous function.

Solution   First we let

$$f(x, y) = \frac{xy}{1 + x^2 + y^2} \quad \text{and} \quad g(t) = \sin t$$

Then $F = g \circ f$. Furthermore, f and g are continuous functions, and consequently so is their composite F.   ❏

Continuity on a Set

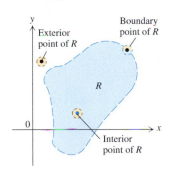

FIGURE 13.22

Let R be a set in the plane. Then for each point P in the plane, one of the following conditions holds:

1. There is an open disk centered at P and contained in R. In this case P is an **interior point** of R.

2. There is an open disk centered at P and containing *no* points of R. In this case P is an **exterior point** of R.

3. Every open disk centered at P contains a point in R and a point outside of R. In this case P is a **boundary point** of R.

Figure 13.22 displays each kind of point. The collection of boundary points of R is the **boundary** of R. The boundaries of most regions we will encounter are easily described. For example, the boundary of the disk $x^2 + y^2 \leq r^2$ is the circle $x^2 + y^2 = r^2$ (Figure 13.23(a)), and the boundary of a rectangular region is the rectangle (Figure 13.23(b)). Finally, a line segment or any other piecewise smooth curve defined on a closed interval

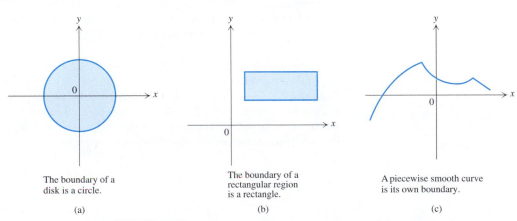

The boundary of a disk is a circle.

(a)

The boundary of a rectangular region is a rectangle.

(b)

A piecewise smooth curve is its own boundary.

(c)

FIGURE 13.23

[a, b] is its own boundary (Figure 13.23(c)). If a set R contains its boundary, then R is **closed.** Thus the regions appearing in Figure 13.23(a)–(c) are closed sets.

Suppose a plane region R is contained in the domain of a function f. We would like to define "continuity of f on R" in a way analogous to the definition of continuity of a function of a single variable on a closed interval. Observe that continuity of f at an interior point (x_0, y_0) of R has already been defined in Definition 13.3. For continuity of f at a point (x_0, y_0) on the boundary of R, we modify slightly the definition of limit in Definition 13.3.

Let (x_0, y_0) be on the boundary of R. A number L is the **limit of f restricted to R at (x_0, y_0)** if for every $\varepsilon > 0$ there is a number $\delta > 0$ such that

$$\text{if } (x, y) \text{ is in } R \text{ and } 0 < \sqrt{(x - x_0)^2 + (y - y_0)^2} < \delta \quad \text{then } |f(x, y) - L| < \varepsilon$$

In this case we say that the limit of f restricted to R exists at (x_0, y_0) and write

$$\lim_{(x, y) \underset{R}{\to} (x_0, y_0)} f(x, y) = L$$

In the event that f is continuous at every interior point of R and

$$\lim_{(x, y) \underset{R}{\to} (x_0, y_0)} f(x, y) = f(x_0, y_0)$$

for every boundary point (x_0, y_0) of R, we say that f is **continuous on R.**

When R is the domain of f, we omit R and write

$$\lim_{(x, y) \to (x_0, y_0)} f(x, y) = L$$

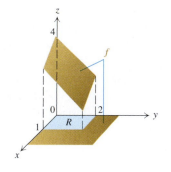

FIGURE 13.24

EXAMPLE 7   Let R be the rectangular region consisting of all points (x, y) such that $0 \le x \le 1$ and $0 \le y \le 2$. Let

$$f(x, y) = \begin{cases} 4 - x - y & \text{for } (x, y) \text{ in } R \\ 0 & \text{for } (x, y) \text{ not in } R \end{cases}$$

Show that f is continuous on R but f is not a continuous function.

Solution   Since the polynomial $4 - x - y$ is continuous, both the polynomial and f are continuous on R. Next, since $|f(x, y)| \ge 1$ for all (x, y) in R and $f(x, y) = 0$ for (x, y) not in R, it follows from Definition 13.2 that f has no limit at any boundary point of R and thus is not continuous (Figure 13.24).   ❑

Our notion of continuity on a set containing its boundary is a two-dimensional analogue of continuity of a function of a single variable on a closed interval (see Definition 2.12 in Section 2.5). The boundary of a solid region in space is defined in a way similar to the method for plane regions (but with open balls replacing open disks). For example, the boundary of a ball is a sphere, and the boundary of a rectangular parallelepiped consists of the six faces enclosing it.

EXERCISES 13.2

In Exercises 1–15 evaluate the limit.

1. $\lim_{(x, y)\to(2, 4)} (x + \frac{1}{2})$

2. $\lim_{(x, y)\to(1,-2)} (2x^3 - 4xy + 5y^2)$

3. $\lim_{(x, y)\to(1,0)} \dfrac{x^2 - xy + 1}{x^2 + y^2}$

4. $\lim_{(x, y, z)\to(-1, 2, 0)} (x^2 + 3y - 4z^2 + 2)$

5. $\lim_{(x, y, z)\to(2, 1, -1)} \dfrac{2x^2y - xz^2}{y^2 - xz}$

6. $\lim_{(x, y)\to(-1, 1)} \dfrac{x^2 + 2xy^2 + y^4}{1 + y^2}$

7. $\lim_{(x, y)\to(2, 1)} \dfrac{x^3 + 2x^2y - xy - 2y^2}{x + 2y}$

8. $\lim_{(x, y)\to(\ln 2, 0)} e^{2x + y^2}$

9. $\lim_{(x, y, z)\to(\pi/2, -\pi/2, 0)} \cos (x + y + z)$

10. $\lim_{(x, y)\to(0, 1)} \dfrac{\sin xy}{y}$

11. $\lim_{(x, y)\to(0, 0)} \dfrac{\sin (x^2 + y^2)}{x^2 + y^2}$

12. $\lim_{(x, y)\to(0, 0)} \dfrac{y(e^x - 1)}{\sqrt{x^2 + y^2}}$

13. $\lim_{(x, y)\to(0, 0)} xy \dfrac{x^2 - y^2}{x^2 + y^2}$

14. $\lim_{(x, y)\to(0, 0)} \dfrac{xy}{\sqrt{x^2 + y^2}}$

15. $\lim_{(x, y, z)\to(0, 0, 0)} \dfrac{x^3 + y^3 + z^3}{x^2 + y^2 + z^2}$

16. Determine whether $\lim_{(x, y)\to(0, 0)} \dfrac{xy}{x^2 + y^2}$ exists.

17. Determine whether $\lim_{(x, y)\to(0, 0)} \dfrac{x^2}{x^2 + y^2}$ exists.

18. Determine whether $\lim_{(x, y)\to(0, 0)} \dfrac{x^2 \sin x}{x^2 + y^2}$ exists.

In Exercises 19–22 use the definition of $\lim_{(x, y)\underset{R}{\to}(x_0, y_0)} f(x, y)$ to determine whether the given limit exists for the given region R. If the limit exists, find it.

19. $\lim_{(x, y)\underset{R}{\to}(0, 0)} \dfrac{\sin(x - y)}{x - y}$; R consists of all (x, y) such that $x \neq y$

20. $\lim_{(x, y)\underset{R}{\to}(0, 0)} \dfrac{x}{x - y}$; R consists of all (x, y) such that $x \neq y$

21. $\lim_{(x, y)\underset{R}{\to}(0, 0)} \dfrac{y}{x}$; R consists of all (x, y) such that $x \neq 0$

22. $\lim_{(x, y)\underset{R}{\to}(0, 0)} x\, e^{-1/|y|}$; R consists of all (x, y) such that $y \neq 0$

In Exercises 23–28 explain why f is continuous.

23. $f(x, y) = xy^2$

24. $f(x, y, z) = 3x^2z - \pi \dfrac{xy}{z}$

25. $f(x, y) = \dfrac{x^2 + y}{x^2 + y^2 - 1}$

26. $f(x, y, z) = \sin (xyz - 1)$

27. $f(x, y, z) = \ln (e^x + e^{yz})$

28. $f(x, y) = \begin{cases} \dfrac{x^2y^2}{x^2 + y^2} & \text{for } (x, y) \neq (0, 0) \\ 0 & \text{for } (x, y) = (0, 0) \end{cases}$

29. Let
$$f(x, y) = \begin{cases} \dfrac{xy^2}{x^4 + y^4} & \text{for } (x, y) \neq (0, 0) \\ 0 & \text{for } (x, y) = (0, 0) \end{cases}$$
a. Show that f is continuous in each variable separately at $(0, 0)$, that is, $f(x, 0)$ is a continuous function of x at 0, and $f(0, y)$ is a continuous function of y at 0.
b. Show that f is not continuous at $(0, 0)$.

*30. Let
$$f(x, y) = \begin{cases} \dfrac{\sin xy}{x^2 + y^2} & \text{for } (x, y) \neq (0, 0) \\ 0 & \text{for } (x, y) = (0, 0) \end{cases}$$
Show that f is not continuous at $(0, 0)$.

31. Describe the boundaries of the following regions.
a. The disk with center $(-3, 2)$ and radius 6
b. The rectangular region with vertices $(0, 0)$, $(2, 0)$, $(0, -3)$, and $(2, -3)$
c. The triangular region with vertices $(-1, 1)$, $(1, 1)$, and $(0, -5)$
d. The upper half of the xy plane, consisting of all (x, y) such that $y \geq 0$
e. The graph of the parabola $y = 4x^2$
f. The entire plane except the origin

32. Let R be the set of all (x, y) such that $|y| < |x^3|$, and let
$$f(x, y) = \dfrac{y}{x}$$
Determine whether $\lim_{(x, y)\underset{R}{\to}(0, 0)} f(x, y)$ exists. Compare your answer with the one to Exercise 21.

In Exercises 33–37 determine whether f is continuous on the given region R.

33. $f(x, y) = \begin{cases} 1 & \text{for } x^2 + y^2 \leq 9 \\ 0 & \text{for } x^2 + y^2 > 9 \end{cases}$
R is the disk $x^2 + y^2 \leq 9$.

34. f is as in Exercise 33; R is the disk $x^2 + y^2 \leq 4$.

35. f is as in Exercise 33; R is the disk $x^2 + y^2 \leq 16$.

36. $f(x, y) = \begin{cases} \dfrac{\sin \sqrt{1 - x^2 - y^2}}{\sqrt{1 - x^2 - y^2}} & \text{for } x^2 + y^2 < 1 \\ 1 & \text{for } x^2 + y^2 = 1 \end{cases}$

R is the disk $x^2 + y^2 \leq 1$.

37. $f(x, y) = \begin{cases} e^{-(1 + x^2)/y} & \text{for } y \neq 0 \\ 0 & \text{for } y = 0 \end{cases}$

R is the upper half plane $y \geq 0$.

38. a. Give a definition of the boundary of a set R in space.

b. Give a definition of the limit of a function at a boundary point P of a given set R in space.

c. Give a definition of continuity of a function on a set R in space.

Applications

39. When x moles of sulfuric acid are mixed with y moles of water, the heat $Q(x, y)$ produced is given by

$$Q(x, y) = \frac{17{,}860xy}{(1.798)x + y} \quad \text{for } x > 0 \text{ and } y > 0$$

Determine whether Q has a limit at $(0, 0)$, and if so, compute its value.

40. The function f defined by

$$f(x, y) = \frac{100ax}{ax + by} \quad \text{for } x > 0 \text{ and } y > 0$$

where a and b are positive constants, appears in the study of the relationship of blood flow through the right lung to the total blood flow in the system. Determine whether f has a limit at $(0, 0)$.

13.3   PARTIAL DERIVATIVES

Consider the cost C of manufacturing ski hats as a function of the cost M of the material and the cost L of the labor. The graph of C could be as in Figure 13.25(a). If we wished to analyze the dependence of C on M when the cost of labor is fixed at L_0, then we would have a function $C(M, L_0)$, which is a function of M alone. Its graph is a curve lying in the plane $L = L_0$ (Figure 13.25(b)). Moreover, as a function of M alone, the cost function C appears to have a derivative at any M. We call such a derivative a partial derivative because one variable is fixed. Similarly, we could consider C as a function of L alone when M is held fixed at a value M_0.

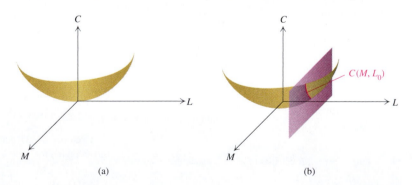

(a)   (b)

FIGURE 13.25

More generally, if f is any function of the two variables x and y, then we can define what it means for f to have a partial derivative at (x_0, y_0) when x is fixed with $x = x_0$, or when y is fixed with $y = y_0$.

DEFINITION 13.4

Let f be a function of two variables, and let (x_0, y_0) be in the domain of f. The **partial derivative of f with respect to x at (x_0, y_0)** is defined by

$$f_x(x_0, y_0) = \lim_{h \to 0} \frac{f(x_0 + h, y_0) - f(x_0, y_0)}{h}$$

provided that this limit exists. The **partial derivative of f with respect to y at (x_0, y_0)** is defined by

$$f_y(x_0, y_0) = \lim_{h \to 0} \frac{f(x_0, y_0 + h) - f(x_0, y_0)}{h}$$

provided that this limit exists.

The functions f_x and f_y that arise through partial differentiation and are defined by

$$f_x(x, y) = \lim_{h \to 0} \frac{f(x + h, y) - f(x, y)}{h}$$

and

$$f_y(x, y) = \lim_{h \to 0} \frac{f(x, y + h) - f(x, y)}{h}$$

are called the **partial derivatives** of f and are frequently denoted $\partial f/\partial x$ and $\partial f/\partial y$. We read $\partial f/\partial x$ as "the partial of f with respect to x." If we specify a function by an equation of the form $z = f(x, y)$ then we would write $\partial z/\partial x$ and $\partial z/\partial y$. Other notations in use are $f_1(x, y)$ and $f_2(x, y)$, as well as $\partial_x f$ and $\partial_y f$. Moreover, if we use other variables, such as u and v, rather than x and y, to denote points in the plane, then the partial derivatives of f would be denoted f_u and f_v, or $\partial f/\partial u$ and $\partial f/\partial v$.

Finding partial derivatives of a function of two variables is no more difficult than finding derivatives of functions of a single variable. After all, one variable is held fixed, so that the function becomes in effect a function of one variable.

The Symbol ∂

The symbol ∂, which is not a Greek letter, was introduced by the French mathematician Adrien-Marie Legendre (1752–1833). Gradually it became accepted after appearing in 1841 in a paper by Carl Gustav Jacob Jacobi (1804–1851).

EXAMPLE 1 Let $f(x, y) = 24xy - 6x^2 y$. Find f_x and f_y, and evaluate f_x and f_y at $(1, 2)$.

Solution By holding y constant and differentiating f with respect to x, we find that

$$f_x(x, y) = 24y - 12xy$$

so that $f_x(1, 2) = 48 - 24 = 24$. By holding x constant and differentiating f with respect to y, we find that

$$f_y(x, y) = 24x - 6x^2$$

so that $f_y(1, 2) = 24 - 6 = 18$. ❑

EXAMPLE 2 Let $z = x^2 \cos y$. Find $\dfrac{\partial z}{\partial x}$ and $\dfrac{\partial z}{\partial y}$.

Solution We find immediately that

$$\frac{\partial z}{\partial x} = 2x \cos y \quad \text{and} \quad \frac{\partial z}{\partial y} = -x^2 \sin y \qquad ❑$$

The sum, product, and quotient rules for derivatives have counterparts for partial derivatives. Thus if f and g have partial derivatives, then

$$(f+g)_x = f_x + g_x \quad \text{and} \quad (f+g)_y = f_y + g_y$$

$$(f-g)_x = f_x - g_x \quad \text{and} \quad (f-g)_y = f_y - g_y$$

$$(fg)_x = f_x g + f g_x \quad \text{and} \quad (fg)_y = f_y g + f g_y$$

$$\left(\frac{f}{g}\right)_x = \frac{f_x g - f g_x}{g^2} \quad \text{and} \quad \left(\frac{f}{g}\right)_y = \frac{f_y g - f g_y}{g^2}$$

It follows that a polynomial or rational function has partial derivatives at each point in its domain.

EXAMPLE 3 Let

$$f(x, y) = \frac{x^3 y - x y^3}{x^2 + y^2}$$

Find f_x and f_y.

Solution   Since f is a rational function, $f_x(x, y)$ and $f_y(x, y)$ are defined for all x and y such that the denominator $x^2 + y^2$ is not 0, that is, for all points except the origin. To find $f_x(x, y)$ we hold y constant and use the quotient rule for partial derivatives to differentiate with respect to x:

$$f_x(x, y) = \frac{(3x^2 y - y^3)(x^2 + y^2) - (x^3 y - x y^3)2x}{(x^2 + y^2)^2}$$

$$= \frac{x^4 y + 4x^2 y^3 - y^5}{(x^2 + y^2)^2}$$

For $f_y(x, y)$ we hold x constant and use the quotient rule for partial derivatives to differentiate with respect to y. You can check that

$$f_y(x, y) = \frac{x^5 - 4x^3 y^2 - x y^4}{(x^2 + y^2)^2} \quad \square$$

In the next example we must refer directly to Definition 13.4 in order to find a partial derivative.

EXAMPLE 4 Let

$$f(x, y) = \begin{cases} \dfrac{x^3 y - x y^3}{x^2 + y^2} & \text{for } (x, y) \neq (0, 0) \\ 0 & \text{for } (x, y) = (0, 0) \end{cases}$$

Show that $f_x(0, 0) = f_y(0, 0) = 0$.

Solution   Since $f(h, 0) = 0$ for all h, we have

$$f_x(0, 0) = \lim_{h \to 0} \frac{f(h, 0) - f(0, 0)}{h - 0} = \lim_{h \to 0} \frac{0 - 0}{h - 0} = 0$$

Similarly, $f(0, h) = 0$ for all h, so that

$$f_y(0, 0) = \lim_{h \to 0} \frac{f(0, h) - f(0, 0)}{h - 0} = \lim_{h \to 0} \frac{0 - 0}{h - 0} = 0$$

The computer-drawn graph of f appearing in Figure 13.26 lends credence to the fact that the partial derivatives of f at $(0, 0)$ are zero. ❑

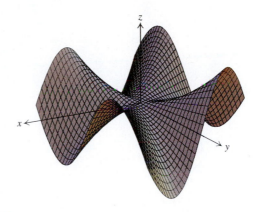

FIGURE 13.26

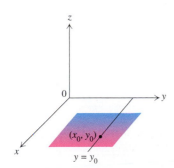

FIGURE 13.27

We can think of the partial derivative $f_x(x_0, y_0)$ as the rate of change of $f(x, y)$ at (x_0, y_0) with respect to x when y is held constant. For example, suppose $f(x, y)$ is the temperature at any point (x, y) on a flat metal plate lying on the xy plane. Then $f_x(x_0, y_0)$ is the rate at which the temperature changes at (x_0, y_0) along the line through (x_0, y_0) parallel to the x axis (Figure 13.27). If the temperature increases as x increases, then $f_x(x_0, y_0) > 0$, whereas if the temperature decreases as x increases, then $f_x(x_0, y_0) < 0$. Similarly, the partial derivative $f_y(x_0, y_0)$ is the rate at which the temperature changes at (x_0, y_0) along the line through (x_0, y_0) parallel to the y axis.

Geometrically, the partial derivatives $f_x(x_0, y_0)$ and $f_y(x_0, y_0)$ describe how the graph of f is slanted near the point $(x_0, y_0, f(x_0, y_0))$. Since $f_x(x_0, y_0)$ is obtained by holding y fixed at y_0, let us consider the curve C_1 determined by the intersection of the graph of f and the plane $y = y_0$ (Figure 13.28). If $h \neq 0$, then

$$\frac{f(x_0 + h, y_0) - f(x_0, y_0)}{h}$$

is the slope of the secant line through the points $(x_0, y_0, f(x_0, y_0))$ and $(x_0 + h, y_0, f(x_0 + h, y_0))$ on C_1 in the plane $y = y_0$. Since

$$f_x(x_0, y_0) = \lim_{h \to 0} \frac{f(x_0 + h, y_0) - f(x_0, y_0)}{h}$$

it follows that $f_x(x_0, y_0)$ is the slope of the line tangent to C_1 at $(x_0, y_0, f(x_0, y_0))$.

This implies that the vector $\mathbf{i} + f_x(x_0, y_0)\mathbf{k}$ is tangent to C_1 at $(x_0, y_0, f(x_0, y_0))$, a fact that we will use later. An analogous argument shows that $f_y(x_0, y_0)$ is the slope of the line tangent to the curve C_2 obtained by intersecting the graph of f and the plane $x = x_0$ (Figure 13.28). The corresponding tangent vector is $\mathbf{j} + f_y(x_0, y_0)\mathbf{k}$. In summary we have the following:

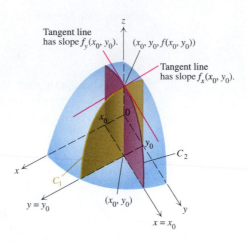

FIGURE 13.28

$\mathbf{i} + f_x(x_0, y_0)\mathbf{k}$ in the plane $y = y_0$ is tangent to C_1 at $(x_0, y_0, f(x_0, y_0))$

$\mathbf{j} + f_y(x_0, y_0)\mathbf{k}$ in the plane $x = x_0$ is tangent to C_2 at $(x_0, y_0, f(x_0, y_0))$ (1)

For functions of three variables, partial derivatives at (x_0, y_0, z_0) are defined as follows:

$$f_x(x_0, y_0, z_0) = \lim_{h \to 0} \frac{f(x_0 + h, y_0, z_0) - f(x_0, y_0, z_0)}{h}$$

$$f_y(x_0, y_0, z_0) = \lim_{h \to 0} \frac{f(x_0, y_0 + h, z_0) - f(x_0, y_0, z_0)}{h} \qquad (2)$$

$$f_z(x_0, y_0, z_0) = \lim_{h \to 0} \frac{f(x_0, y_0, z_0 + h) - f(x_0, y_0, z_0)}{h}$$

provided that these limits exist. The formulas in (2) give rise to the partial derivative functions f_x, f_y, and f_z. Alternative notations for these partial derivatives are $\partial f/\partial x$, $\partial f/\partial y$, and $\partial f/\partial z$, or $\partial w/\partial x$, $\partial w/\partial y$, and $\partial w/\partial z$ if $w = f(x, y, z)$.

EXAMPLE 5   Let $f(x, y, z) = e^{2x} \cos z + e^{3y} \sin z$. Find the partial derivatives of f.

Solution   First, we hold y and z constant and differentiate with respect to x:

$$\frac{\partial f}{\partial x} = 2e^{2x} \cos z$$

Next, we hold x and z constant and differentiate with respect to y:

$$\frac{\partial f}{\partial y} = 3e^{3y} \sin z$$

Finally, we hold x and y constant and differentiate with respect to z:

$$\frac{\partial f}{\partial z} = -e^{2x} \sin z + e^{3y} \cos z \quad \square$$

Higher-Order Partial Derivatives

A function of one variable may have second, third, and higher derivatives. There is an analogue for functions of several variables. If f is a function of the variables x and y, then the functions f_x and f_y may each have two partial derivatives. In this case the partial derivatives of f_x and f_y would generate four new partial derivatives:

$$(f_x)_x, \quad \text{usually denoted} \quad f_{xx} \quad \text{or} \quad \frac{\partial^2 f}{\partial x^2}$$

$$(f_x)_y, \quad \text{usually denoted} \quad f_{xy} \quad \text{or} \quad \frac{\partial^2 f}{\partial y \, \partial x}$$

$$(f_y)_x, \quad \text{usually denoted} \quad f_{yx} \quad \text{or} \quad \frac{\partial^2 f}{\partial x \, \partial y}$$

$$(f_y)_y, \quad \text{usually denoted} \quad f_{yy} \quad \text{or} \quad \frac{\partial^2 f}{\partial y^2}$$

The partial derivatives just described are called **second partial derivatives of** f; f_{xy} and f_{yx} are usually called **mixed partial derivatives of** f, or more briefly, **mixed partials.** To avoid confusion we will often refer to f_x, f_y, and f_z as **first partial derivatives,** or **first partials,** or **partials.**

> **Caution:**   Notice that the order in which x and y appear in the expression f_{xy} is opposite to the order in which they appear in $\partial^2 f/\partial y \partial x$:
>
> $$\frac{\partial^2 f}{\partial y \, \partial x} = \frac{\partial}{\partial y}\left(\frac{\partial f}{\partial x}\right) \quad \text{and} \quad f_{xy} = (f_x)_y$$

EXAMPLE 6   Let $f(x, y) = \sin xy^2$. Find all second partial derivatives of f.

Solution   The first partials are given by

$$f_x(x, y) = y^2 \cos xy^2 \quad \text{and} \quad f_y(x, y) = 2xy \cos xy^2$$

We obtain the second partials by computing the partial derivatives of the first partials:

$$f_{xx}(x, y) = -y^4 \sin xy^2$$

$$f_{xy}(x, y) = 2y \cos xy^2 - 2xy^3 \sin xy^2$$

$$f_{yx}(x, y) = 2y \cos xy^2 - 2xy^3 \sin xy^2$$

$$f_{yy}(x, y) = 2x \cos xy^2 - 4x^2 y^2 \sin xy^2 \quad \square$$

As with first partial derivatives, we sometimes must refer directly to Definition 13.4 to determine a second partial derivative.

EXAMPLE 7   Let

$$f(x, y) = \begin{cases} \dfrac{x^3 y - xy^3}{x^2 + y^2} & \text{for } (x, y) \neq (0, 0) \\ 0 & \text{for } (x, y) = (0, 0) \end{cases}$$

Show that $f_{xy}(0, 0) \neq f_{yx}(0, 0)$.

Solution   From Example 3 we deduce that

$$f_x(0, y) = \frac{0^4 y + 4(0^2) y^3 - y^5}{(0^2 + y^2)^2} = -y \quad \text{for } y \neq 0$$

$$f_y(x, 0) = \frac{x^5 - 4x^3(0^2) - x(0^4)}{(x^2 + 0^2)^2} = x \quad \text{for } x \neq 0$$

Furthermore, $f_x(0, 0) = 0 = f_y(0, 0)$ by Example 4. These formulas imply that

$$f_{xy}(0, 0) = \lim_{h \to 0} \frac{f_x(0, h) - f_x(0, 0)}{h - 0} = \lim_{h \to 0} \frac{-h - 0}{h - 0} = -1$$

$$f_{yx}(0, 0) = \lim_{h \to 0} \frac{f_y(h, 0) - f_y(0, 0)}{h - 0} = \lim_{h \to 0} \frac{h - 0}{h - 0} = 1$$

Consequently $f_{xy}(0, 0) \neq f_{yx}(0, 0)$.   ◻

The mixed partials computed in Example 6 are equal, whereas those in Example 7 are not equal. However, the mixed partials of the functions we usually encounter are equal, thanks to the following theorem, whose proof we omit.

THEOREM 13.5

Let f be a function of two variables, and assume that f_{xy} and f_{yx} are continuous at (x_0, y_0). Then

$$f_{xy}(x_0, y_0) = f_{yx}(x_0, y_0)$$

Second partials of a function of three variables are defined in the same way as for a function of two variables. Moreover, if f is such a function and if f_{xy} and f_{yx} are continuous at a point (x_0, y_0, z_0), then

$$f_{xy}(x_0, y_0, z_0) = f_{yx}(x_0, y_0, z_0)$$

Analogous statements hold for f_{xz} and f_{zx}, and for f_{yz} and f_{zy}.

Differentiation Under the Integral Sign

Sometimes a function F of a single variable is defined as a definite integral of a function f of two variables:

$$F(x) = \int_c^d f(x, y) \, dy \tag{3}$$

Under mild conditions on f, the derivative of F can be found by simply taking the partial derivative of f with respect to x on the right side of (3). The following theorem states the result.

THEOREM 13.6   Let R be a rectangle consisting of all points (x, y) for which $a \le x \le b$ and $c \le y \le d$. Suppose that f is a function of two variables defined on R such that f and f_x are continuous on R. Then $\int_c^d f(x, y)\, dy$ is a differentiable function of x, and

$$\frac{d}{dx} \int_c^d f(x, y)\, dy = \int_c^d \frac{\partial f}{\partial x}(x, y)\, dy \tag{4}$$

EXAMPLE 8   Find $\dfrac{d}{dx} \displaystyle\int_1^2 \dfrac{e^{xy}}{y}\, dy$.

Solution   By (4),

$$\frac{d}{dx} \int_1^2 \frac{e^{xy}}{y}\, dy = \int_1^2 \frac{\partial}{\partial x}\left(\frac{e^{xy}}{y}\right) dy = \int_1^2 e^{xy}\, dy = \frac{e^{xy}}{x}\Big|_1^2 = \frac{1}{x}\left(e^{2x} - e^{x}\right) \qquad \square$$

Formula (4) arises, for example, in certain kinds of problems involving analysis of statistical data.

Analysis of a Rainbow

Rainbows have fascinated people from ancient times until the present. Through the ages great scientists have attempted to analyze the rainbow. Aristotle considered rainbows to be produced by the sun's rays reflected in droplets. By the fourteenth century, Theodoric of Freiburg correctly and unambiguously explained that a rainbow is caused by a combination of reflections and refractions of rays in droplets. However, it was left to Descartes and Newton to discover the finer details of the nature of color, and to provide a description of the major attributes of rainbows. Other more subtle aspects of rainbows have continued to puzzle scientists until this day. A comprehensive history of the analysis of the rainbow is contained in Carl B. Boyer's book, *The Rainbow* (Princeton University Press, 1987).

Now we will use partial derivatives in order to describe the order of the colors of a rainbow. Each ray of sunlight contains all the colors in the visible spectrum, from red to violet. We will refer to a single such color in a given ray as a **monochromatic ray.** When a monochromatic ray from the sun hits a droplet of water, the ray is refracted in accordance with Snell's Law:

$$\sin i = \mu \sin r$$

where i is the angle of incidence, r is the angle of refraction, and μ is the **index of refraction** of water for the monochromatic ray (Figure 13.29(a)). The index μ depends only on the color, and is defined by the formula

$$\mu = \frac{\text{speed of the monochromatic ray in a vacuum}}{\text{speed of the monochromatic ray in pure water}}$$

For the visible spectrum μ varies from approximately 1.330 for red to 1.342 for violet. The main rainbow one sees in the sky is created by various monochromatic rays that have been refracted, then reflected, and finally refracted again by droplets of water, as in Figure 13.29(b). The **scattering angle** θ formed by the path of a monochromatic ray before the ray enters the droplet and the path after the ray leaves the droplet depends on both μ and i. The angles θ, r, and i are related by

$$\theta = 4r - 2i$$

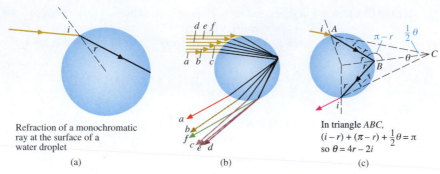

Refraction of a monochromatic
ray at the surface of a
water droplet

(a)

(b)

In triangle ABC,
$(i - r) + (\pi - r) + \frac{1}{2}\theta = \pi$
so $\theta = 4r - 2i$

(c)

FIGURE 13.29

(Figure 13.29(c)). Solving for r in the equation $\sin i = \mu \sin r$, we find that

$$r = \sin^{-1}\left(\frac{\sin i}{\mu}\right)$$

so that we obtain the formula

$$\theta(\mu, i) = 4 \sin^{-1}\left(\frac{\sin i}{\mu}\right) - 2i \tag{5}$$

for θ as a function of μ and i.

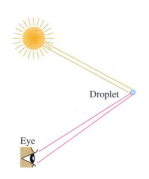

Droplet

Eye

FIGURE 13.30

Monochromatic rays emanating from the sun travel in almost, but not quite, parallel paths, and a human eye can receive rays from a single droplet through a tiny but positive angle (Figure 13.30). These two facts imply that if an observer is located so that a ray with angle of incidence i enters the eye, then other rays with nearly the same angle of incidence can still enter the eye. In 1637 Descartes observed that for rays of a single color, the less θ changes for small changes in i, the more intense is the impression of that color on the eye. This condition corresponds to $|\partial\theta/\partial i|$ being small. It follows that the strongest optical impression occurs when

$$\frac{\partial\theta}{\partial i}(\mu, i) = 0 \tag{6}$$

and this effect is so pronounced that the color with index of refraction μ is perceived only when i satisfies equation (6). Let us refer to such an angle i as i_μ. In Example 9 we show that i_μ is unique (for a given value of μ) and find a formula for i_μ in terms of μ. Then from (5) it is easy to calculate the corresponding value $\theta(\mu, i_\mu)$ of the scattering angle through which a ray passes when when it is observed by the eye.

EXAMPLE 9   Let

$$\theta(\mu, i) = 4 \sin^{-1}\left(\frac{\sin i}{\mu}\right) - 2i$$

For any given μ, find the angle i_μ for which

$$\frac{\partial\theta}{\partial i}(\mu, i_\mu) = 0$$

Solution   We find that

$$\frac{\partial \theta}{\partial i}(\mu, i) = 4 \frac{(\cos i)/\mu}{\sqrt{1 - (\sin^2 i)/\mu^2}} - 2 = 4 \frac{\cos i}{\sqrt{\mu^2 - \sin^2 i}} - 2$$

Consequently

$$\frac{\partial \theta}{\partial i}(\mu, i) = 0$$

provided that

$$4 \frac{\cos i}{\sqrt{\mu^2 - \sin^2 i}} = 2$$

or

$$\sqrt{\mu^2 - \sin^2 i} = 2 \cos i \qquad (7)$$

Squaring both sides of (7), substituting $1 - \sin^2 i$ for $\cos^2 i$, and solving for $\sin i$, we obtain

$$\sin i = \sqrt{(4 - \mu^2)/3}$$

Consequently

$$i_\mu = \sin^{-1} \sqrt{(4 - \mu^2)/3} \qquad \square$$

Using the values of μ given earlier for red and for violet along with the formula just obtained for i_μ, we calculate that, in degrees,

$$i_{\text{red}} \approx 59.6° \quad \text{and} \quad i_{\text{violet}} \approx 58.9°$$

and by (5) this means that

$$\theta(\text{red}, i_{\text{red}}) \approx 42.5° \quad \text{and} \quad \theta(\text{violet}, i_{\text{violet}}) \approx 40.8°$$

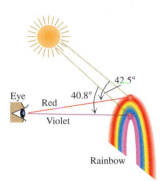

FIGURE 13.31

Thus the largest scattering angle θ is approximately 42.5°. This tells us that for a person standing on the ground to see a rainbow, the sun can be no higher than 42.5° above the horizon, because otherwise the terminal rays would point upward, rather than downward toward the observer's eye (Figure 13.31).

Next we show that $\theta(\mu, i_\mu)$ is a decreasing function of μ and then determine the order of the colors in the rainbow. Let $f(\mu) = \theta(\mu, i_\mu)$. From (5) we have

$$f(\mu) = 4 \sin^{-1}\left(\frac{\sin i_\mu}{\mu}\right) - 2i_\mu$$

Therefore

$$\frac{df}{d\mu} = \frac{4}{\sqrt{1 - (\sin^2 i_\mu)/\mu^2}} \left(\frac{\mu(\cos i_\mu)(di_\mu/d\mu) - \sin i_\mu}{\mu^2}\right) - 2\frac{di_\mu}{d\mu}$$

$$= \frac{4 \cos i_\mu (di_\mu/d\mu)}{\sqrt{\mu^2 - \sin^2 i_\mu}} - \frac{4 \sin i_\mu}{\mu\sqrt{\mu^2 - \sin^2 i_\mu}} - 2\frac{di_\mu}{d\mu}$$

$$\overset{(7)}{=} \frac{4 \cos i_\mu (di_\mu/d\mu)}{2 \cos i_\mu} - \frac{4 \sin i_\mu}{2\mu \cos i_\mu} - 2\frac{di_\mu}{d\mu}$$

$$= -\frac{2}{\mu} \tan i_\mu$$

A double rainbow. (*CNRI/Science Photo Library/Photo Researchers*)

Since i_μ is an acute angle, $df/d\mu$ is negative, so that θ is a decreasing function of μ. In other words, $\theta(\mu, i_\mu)$ is smallest for the largest value of μ, which corresponds to violet, and is largest for the smallest value of μ, which corresponds to red. It follows that the red is at the top and violet is at the bottom of the rainbow (Figure 13.31).

Sometimes one can see a second, fainter rainbow, higher in the sky than the first, created by monochromatic rays that have been reflected *twice* instead of once within the droplet of water. It turns out that the colors of the second rainbow are reversed, that is, violet is at the top and red at the bottom (see Project 1).

EXERCISES 13.3

In Exercises 1–19 find the first partial derivatives of the function.

1. $f(x, y) = \frac{2}{3}x^{3/2}$

2. $f(x, y) = 9 - x^2 - 4y^2$

3. $f(x, y) = 2x + 3x^2y^4$

4. $g(x, y) = x^3 e^{2y}$

5. $g(u, v) = \dfrac{u^3 + v^3}{u^2 + v^2}$

6. $f(x, y) = \sqrt{x^2 + y^2}$

7. $f(x, y) = \sqrt{4 - x^2 - 9y^2}$

8. $z = \sqrt{\frac{1}{4}x^2 - y^2}$

9. $z = \sqrt{(1 - x^{2/3})^3 - y^2}$

10. $w = \cos \dfrac{u}{v}$

11. $z = (\sin x^2 y)^3$

12. $z = x^y$

13. $f(x, y, z) = x^2 y^5 + xz^2$

14. $f(x, y, z) = x(\cos y)e^z$

15. $f(x, y, z) = \dfrac{x + y + z}{xy + yz + zx}$

16. $f(u, v, w) = \dfrac{1}{\sqrt{u^2 + v^2 + w^2}}$

17. $w = e^x(\cos y + \sin z)$

18. $w = \left(\dfrac{x}{y}\right)^z$

19. $w = \sin^{-1}\dfrac{1}{1 + xyz^2}$

In Exercises 20–25 find the first partial derivatives of f at the given point.

20. $f(x, y) = x^4 - 6x^2 - 3xy^2 + 17$; $(-1, 2)$

21. $f(x, y) = \sqrt{4x^2 + y^2}$; $(2, -3)$

22. $f(x, y, z) = xy^2 \sin z$; $(-1, 2, 0)$

23. $f(x, y, z) = e^{2x - 4y - z}$; $(0, -1, 1)$

24. $f(x, y) = \begin{cases} \dfrac{x^3 + y^3}{x^2 + y^2} & \text{for } (x, y) \ne (0, 0) \\ 0 & \text{for } (x, y) = (0, 0) \end{cases}$; $(0, 0)$

25. $f(x, y) = \begin{cases} \dfrac{x^2 y^3}{x^2 + 4y^3} & \text{for } (x, y) \ne (0, 0) \\ 0 & \text{for } (x, y) = (0, 0) \end{cases}$; $(0, 0)$

26. Let $f(x, y) = \displaystyle\int_1^x P(t)\, dt + \int_1^y Q(t)\, dt$, where P and Q are continuous. Find f_x and f_y.

27. Let $f(x, y) = \displaystyle\int_\pi^{x^2 + y^2} \sin t^2\, dt$. Find f_x and f_y.

28. Let
$$f(x, y) = (\tan x)(y^{xy - \sin y})e^{\cos y} + \ln(1 + x^2)\cos(x + 1)^y$$
Find $f_y(0, 1)$. (*Hint:* Use Definition 13.4.)

In Exercises 29–33 find f_{xy} and f_{yx}.

29. $f(x, y) = 3x^2 - \sqrt{2}\, xy^2 + y^5 - 2$

30. $f(x, y) = \dfrac{x^2 - y^2}{x^2 + y^2}$

31. $f(x, y) = \sqrt{x^2 + y^2}$

32. $f(x, y, z) = x^4 - 2x^2 y\sqrt{z} + 3yz^4 + 2$

33. $f(x, y, z) = z \cos xy$

In Exercises 34–38 find f_{xx}, f_{yy}, and f_{zz} (where applicable).

34. $f(x, y) = \sqrt{16 - 9x^2 - 4y^2}$

35. $f(x, y) = e^{x - 2y}$

36. $f(x, y) = \displaystyle\int_0^x \sin t^2\, dt \int_0^y \cos t^2\, dt$

37. $f(x, y, z) = \sqrt{x^2 + y^2 + z^2}$

38. $f(x, y, z) = e^{x^2} \sin yz + \ln(x^2 + y^2 + z^2)$

39. Find symmetric equations for the line that lies in the plane $y = 1$ and is tangent to the intersection of the plane and the paraboloid $z = x^2 + 16y^2$ at $(-3, 1, 25)$.

40. Find symmetric equations for the line that lies in the plane $x = 2$ and is tangent to the intersection of the plane and the cone $z = \sqrt{x^2 + y^2}$ at $(2, 2\sqrt{3}, 4)$.

In Exercises 41–44 find $\sqrt{f_x^2 + f_y^2 + 1}$.

41. $f(x, y) = 1 - x$

42. $f(x, y) = 4 - y^2$

43. $f(x, y) = \sqrt{x^2 + y^2}$

44. $f(x, y) = \sqrt{1 - x^2 - y^2}$

45. Show that the function in Exercise 29 of Section 13.2 has partial derivatives at $(0, 0)$ even though it is not continuous at $(0, 0)$.

46. Rectangular and polar coordinates in the plane are related by the equations $x = r \cos \theta$, $y = r \sin \theta$, $r = \sqrt{x^2 + y^2}$, $\theta = \tan^{-1} y/x$. Find the following partial derivatives.

  a. $\dfrac{\partial x}{\partial r}$ **b.** $\dfrac{\partial x}{\partial \theta}$ **c.** $\dfrac{\partial y}{\partial r}$ **d.** $\dfrac{\partial y}{\partial \theta}$

  e. $\dfrac{\partial r}{\partial x}$ **f.** $\dfrac{\partial r}{\partial y}$ **g.** $\dfrac{\partial \theta}{\partial x}$ **h.** $\dfrac{\partial \theta}{\partial y}$

47. If $z = e^{-ay} \cos ax$, show that

$$\frac{\partial^2 z}{\partial x^2} = a \frac{\partial z}{\partial y}$$

48. Let $z = x^c e^{-y/x}$, where c is a constant. Find the value of c such that

$$\frac{\partial z}{\partial x} = y \frac{\partial^2 z}{\partial y^2} + \frac{\partial z}{\partial y}$$

In Exercises 49–51 show that the functions u and v satisfy the **Cauchy-Riemann equations** $u_x = v_y$ and $u_y = -v_x$.

49. $u = x^2 - y^2$, $v = 2xy$

50. $u = x^3 - 3xy^2$, $v = 3x^2 y - y^3$

51. $u = e^x \cos y$, $v = e^x \sin y$

A function z satisfies **Laplace's equation** if

$$\frac{\partial^2 z}{\partial x^2} + \frac{\partial^2 z}{\partial y^2} = 0$$

In Exercises 52–53 show that the function satisfies Laplace's equation.

52. $z = x^4 - 6x^2 y^2 + y^4$ **53.** $z = \ln (x^2 + y^2)$

54. Show that if u and v satisfy the Cauchy-Riemann equations (as in Exercises 49–51) and the mixed partials of u and v are continuous, then u and v satisfy Laplace's equation.

55. Show that Example 7 does not contradict Theorem 13.5 by computing $f_{xy}(x, 0)$ for $x \neq 0$, and then verifying that f_{xy} is not continuous at $(0, 0)$.

56. Let f be a function of two variables with partials of all orders. Then f has four second partials, f_{xx}, f_{xy}, f_{yx}, and f_{yy}. If the second partials are continuous, only three of them can be distinct.

  a. How many third-order partials does f have? If these are continuous, what is the maximum number that can be distinct?

  b. Generalize (a) to nth-order partials. (*Hint:* Observe

that if all nth-order partials are continuous, then any nth-order partial is equal to another nth-order partial in which all the differentiations with respect to x are performed first and the differentiations with respect to y are done last.)

57. Let M have continuous partials on a rectangle bounded by $x = a$, $x = b$, $y = c$, and $y = d$. Show that

$$\int_a^b \frac{\partial M}{\partial x} (x, y)\, dx = M(b, y) - M(a, y) \quad \text{for } c \leq y \leq d$$

and

$$\int_c^d \frac{\partial M}{\partial y} (x, y)\, dy = M(x, d) - M(x, c) \quad \text{for } a \leq x \leq b$$

58. Using (4), find the derivatives with respect to x of the following functions.

  a. $\displaystyle\int_1^{\sqrt{3}} \ln (x^2 + y^2)\, dy$ **b.** $\displaystyle\int_1^2 \frac{1}{y} \tan^{-1} \frac{x}{y}\, dy$

Applications

59. If c is a constant, then an equation of the form

$$\frac{\partial u}{\partial t} = c \frac{\partial^2 u}{\partial x^2}$$

is called a **diffusion equation.**

  a. Show that if $u = e^{ax + bt}$, where a and b are constant, then u satisfies the diffusion equation with $c = b/a^2$.

  ***b.** Show that if

$$u = u(x, t) = \frac{1}{\sqrt{t}} e^{-x^2/at}$$

where a is a constant, then u satisfies a diffusion equation. [The number $u(x, t)$ might represent the concentration of a drug at a point x in a muscle at time t. For each value of t the graph of u (considered as a function of x) is a bell-shaped curve. As t increases, the curve becomes flatter (Figure 13.32).]

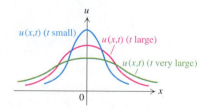

FIGURE 13.32 Figure for Exercise 59.

60. Let m_1 and m_2 be masses, with $m_1 \geq m_2$, and assume that they are connected to an apparatus called an Atwood machine (see Figure 13.33). The acceleration a of the mass m_1 downward is given by

$$a = \frac{m_1 - m_2}{m_1 + m_2} g$$

where g is the acceleration due to gravity. Show that

$$m_1 \frac{\partial a}{\partial m_1} + m_2 \frac{\partial a}{\partial m_2} = 0$$

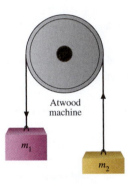

Atwood
machine

FIGURE 13.33 Figure for Exercise 60.

61. When two resistors having resistances R_1 and R_2 are connected in parallel, the combined resistance R is given by $R = R_1 R_2 / (R_1 + R_2)$. Show that

$$\frac{\partial^2 R}{\partial R_1^2} \frac{\partial^2 R}{\partial R_2^2} = \frac{4R^2}{(R_1 + R_2)^4}$$

62. The kinetic energy K of a body with mass m and velocity v is given by $K = \frac{1}{2} mv^2$. Show that

$$\frac{\partial K}{\partial m} \frac{\partial^2 K}{\partial v^2} = K$$

63. The ideal gas law states that if n moles of a gas has volume V and temperature T and is under pressure p, then $pV = nkT$, where k is the universal gas constant. Show that

$$\frac{\partial V}{\partial T} \frac{\partial T}{\partial p} \frac{\partial p}{\partial V} = -1$$

64. The index of refraction of water at 20°C for yellow light from a sodium flame is 1.333. Determine the angle $\theta(\mu, i_\mu)$ for $\mu = 1.333$.

65. Let D be a solid region whose boundary is a cylinder of radius r and height h, capped on each end by a hemisphere. The volume V and the surface area S of D are given by

$$V = \pi r^2 h + 2\left(\frac{2}{3}\pi r^3\right) = \pi r^2 \left(h + \frac{4}{3}r\right)$$

$$S = 2\pi rh + 2(2\pi r^2) = 2\pi r(h + 2r)$$

If D represents a bacterium, then the rate R at which a chemical substance can be absorbed into D is given by $R = c(S/V)$, where c is a positive constant. Show that $\partial R / \partial r < 0$ and $\partial R / \partial h < 0$. (The result implies that an increase in either the radius or height of the bacterium decreases the rate of chemical absorption.)

66. When an X-ray beam passes through an object, the intensity of the beam at a distance x from the point of entry is given by

$$I(x, \sigma) = I_0 e^{-\sigma x}$$

where I_0 is the intensity at entry and σ depends on the frequency of the X-rays (Figure 13.34). Suppose we wish to check the uniformity of thickness of an aluminum bar 1 centimeter thick by X-raying it. Then

$$\frac{\partial I}{\partial x}(1, \sigma)$$

is a measure of the variation of intensity with small variations of thickness. Thus

$$\left| \frac{\partial I}{\partial x}(1, \sigma) \right|$$

is a measure of the "contrast" of the X-ray.

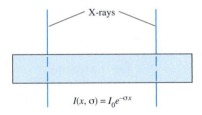

X-rays

$$I(x, \sigma) = I_0 e^{-\sigma x}$$

FIGURE 13.34 Figure for Exercise 66.

a. Find the value of σ that maximizes

$$\left| \frac{\partial I}{\partial x}(1, \sigma) \right|$$

(From that value one could determine the X-ray frequency that would provide the best contrast.)

b. Suppose a doctor X-rays a bone x_0 centimeters thick to search for a fracture. Find the value of σ that maximizes

$$\left| \frac{\partial I}{\partial x}(x_0, \sigma) \right|$$

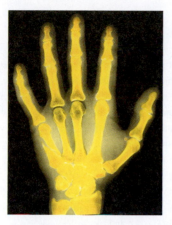

An X-ray of a human hand.

67. If $f(x, y)$ is the amount of a commodity produced from x units of capital and y units of labor, then f is called a **production function.** If

$$f(x, y) = x^{\alpha} y^{\beta} \quad \text{for } x > 0 \text{ and } y > 0$$

where α and β are positive constants less than 1, then f is called a **Cobb-Douglas production function.**

a. Show that $f(tx, ty) = t^{\alpha + \beta} f(x, y)$.

b. If $z = f(x, y)$, show that

$$\frac{1}{z} \frac{\partial z}{\partial x} = \frac{\alpha}{x} \quad \text{and} \quad \frac{1}{z} \frac{\partial z}{\partial y} = \frac{\beta}{y}$$

and that

$$x \frac{\partial z}{\partial x} + y \frac{\partial z}{\partial y} = (\alpha + \beta)z$$

68. If f is a production function (see Exercise 67), then $\partial f / \partial x$ and $\partial f / \partial y$ are called the **marginal productivity of capital** and the **marginal productivity of labor,** respectively. If

$$f(x, y) = 60x^{1/2} y^{2/3} \quad \text{for } x \geq 0 \text{ and } y \geq 0$$

find the pairs (x, y) at which the marginal productivity of capital is equal to the marginal productivity of labor.

69. It seems reasonable that an increase in taxation on a commodity would decrease the production of that commodity. The following argument supports that claim. Assume that all required derivatives exist. For any $x \geq 0$, let $P_0(x)$ be the profit before taxes on x units produced. Let $P(x, t)$ denote the profit after taxes on x units produced with tax t on each unit. Assume that at any tax rate t the company will maximize its profits by producing $f(t)$ units so that

$$\frac{\partial P}{\partial x} (f(t), t) = 0 \tag{9}$$

$$\frac{\partial^2 P}{\partial x^2} (f(t), t) < 0 \tag{10}$$

(The conditions in (9) and (10) are just those required for the Second Derivative Test.)

a. Show that $P(x, t) = P_0(x) - tx$.

b. Using (a), show that

$$\frac{\partial P}{\partial x} (x, t) = P_0'(x) - t \quad \text{and} \quad \frac{\partial^2 P}{\partial x^2} (x, t) = P_0''(x)$$

c. From (9) and (b), show that $P_0'(f(t)) - t = 0$.

d. By differentiating both sides of the equation in (c) and by using (b) and (10), show that

$$f'(t) = \frac{1}{P_0''(f(t))} = \frac{1}{\dfrac{\partial^2 P}{\partial x^2} (f(t), t)} < 0$$

(Thus the production tends to decrease as the tax rate increases.)

Projects

1. For a second, fainter rainbow (called the secondary rainbow) that sometimes we are lucky enough to see near the main, primary rainbow, a monochromatic ray is refracted as it enters the droplet, then is reflected twice (not once, as for the primary rainbow) within the droplet, and finally is refracted again as it leaves the droplet (Figure 13.35).

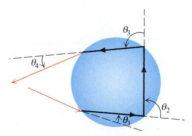

FIGURE 13.35 Figure for Project.

a. Use Figure 13.35 and the discussion in the text on the primary rainbow to show that the angle $\theta^* = \theta_1 + \theta_2 + \theta_3 + \theta_4$ through which the ray is rotated is given by

$$\theta^*(\mu, i) = 2i - 6r + 2\pi = 2i - 6 \sin^{-1}\left(\frac{\sin i}{\mu}\right) + 2\pi$$

b. Show that for any given value of μ, the angle i_μ for which

$$\frac{\partial \theta^*}{\partial i} (\mu, i_\mu) = 0$$

is given by $i_\mu = \sin^{-1} \sqrt{(9 - \mu^2)/8}$. (*Hint:* Pattern your solution after the solution of Example 9.

c. Using the values of μ for red and violet given in the text, deduce that

$$\theta_2 \text{ (red, } i_{\text{red}}) \approx 230.1° \quad \text{and} \quad \theta_2 \text{ (violet, } i_{\text{violet}}) \approx 233.2°$$

Therefore the net angles are approximately 50.1° for red and 50.3° for violet. By calculations that are similar to those used for θ, it can be shown that the net angles increase from red through orange, yellow, green, and blue to violet.

d. Convince yourself that (c) implies the following:

  i. For the secondary rainbow, red lies below violet (in contrast to the primary rainbow), so that the order of colors from top to bottom is reversed from the order in the primary rainbow.

  ii. The secondary rainbow is wider than the primary rainbow.

e. Find a formula for the angle θ_3 corresponding to the tertiary rainbow, in which monochromatic rays are reflected three times in a given droplet. Then show that θ_3(red) $\approx 317.5°$ and θ_3(violet) $\approx 321.9°$. Finally, respond to the question: Why can't we see such a rainbow?

2. Here we will study those functions $u(x, t)$ that satisfy the **wave equation**

$$\frac{\partial^2 u}{\partial x^2} = r^2 \frac{\partial^2 u}{\partial t^2}$$

where r is a nonzero constant.

a. Let $u(x, t) = \sin ax \sin bt$, where a and b are positive constants. Show that u satisfies the wave equation for all x and t, and find the corresponding value of r in terms of a and b. (If t represents time, and x represents distance along a line, then u could describe, for example, the motion of a vibrating string whose ends are fixed.)

b. Let $u(x, t) = e^{f(x, t)}$, where $f(x, t) = ax^2 + bxt + ct^2$, with positive constants a, b, and c. Show that u satisfies the wave equation with $r = 1$ for all x and t only if $a = c$ and $b = 2a$.

c. Let $u(x, t) = Ax^2 + Bxt + Ct^2 + ax + bt + c$. Determine all such polynomials that satisfy the wave equation with a given value of r, for all x and t.

d. Let $u(x, t) = e^{ax + bt}$, where a and b are nonzero constants. Determine the relationship between a and b that needs to exist for u to satisfy the wave equation with a given value of r.

e. Let $u(x, t) = e^{t - x}$. Show that u satisfies the wave equation and the diffusion equation of Exercise 59. What are the constants r and c for the wave equation and the diffusion equation, respectively?

13.4   THE CHAIN RULE

The Chain Rule for functions of one variable provides a formula for differentiating the composite of two functions f and g. If $u = f(x)$ and $y = g(u) = g(f(x))$, the Chain Rule can be expressed in the Leibniz notation as

$$\frac{dy}{dx} = \frac{dy}{du} \frac{du}{dx}$$

Composites involving functions of several variables have their own versions of the Chain Rule, which involve derivatives and partial derivatives. In this section we will introduce various versions of the Chain Rule, apply them in several examples, and then present the theoretical basis for, and proof of, the Chain Rule, which includes the notion of differentiability for functions of several variables.

Versions of the Chain Rule

First we present two versions of the Chain Rule for functions of two variables. In each statement we assume that all functions have the required derivatives.

a. Let $z = f(x, y)$, $x = g_1(t)$, and $y = g_2(t)$. Then $z = f(g_1(t), g_2(t))$, and

$$\frac{dz}{dt} = \frac{\partial z}{\partial x}\frac{dx}{dt} + \frac{\partial z}{\partial y}\frac{dy}{dt} \qquad (1)$$

b. Let $z = f(x, y)$, $x = g_1(u, v)$, and $y = g_2(u, v)$. Then $z = f(g_1(u, v), g_2(u, v))$, and

$$\frac{\partial z}{\partial u} = \frac{\partial z}{\partial x}\frac{\partial x}{\partial u} + \frac{\partial z}{\partial y}\frac{\partial y}{\partial u}$$

$$(2)$$

$$\frac{\partial z}{\partial v} = \frac{\partial z}{\partial x}\frac{\partial x}{\partial v} + \frac{\partial z}{\partial y}\frac{\partial y}{\partial v}$$

EXAMPLE 1   Let $z = x^2 e^y$, $x = \sin t$, and $y = t^3$. Find dz/dt.

Solution   Using (1), we find that

$$\frac{dz}{dt} = \frac{\partial z}{\partial x}\frac{dx}{dt} + \frac{\partial z}{\partial y}\frac{dy}{dt} = 2xe^y \cos t + x^2 e^y (3t^2)$$

$$= 2(\sin t)e^{(t^3)} \cos t + 3(\sin^2 t)e^{(t^3)}t^2 \qquad \square$$

EXAMPLE 2   Let $z = x \ln y$, $x = u^2 + v^2$, and $y = u^2 - v^2$. Find $\partial z/\partial u$ and $\partial z/\partial v$.

Solution   Using (2), we find that

$$\frac{\partial z}{\partial u} = \frac{\partial z}{\partial x}\frac{\partial x}{\partial u} + \frac{\partial z}{\partial y}\frac{\partial y}{\partial u} = (\ln y)(2u) + \left(\frac{x}{y}\right)(2u)$$

$$= 2u \ln (u^2 - v^2) + 2u\frac{u^2 + v^2}{u^2 - v^2}$$

and
$$\frac{\partial z}{\partial v} = \frac{\partial z}{\partial x}\frac{\partial x}{\partial v} + \frac{\partial z}{\partial y}\frac{\partial y}{\partial v} = (\ln y)(2v) + \left(\frac{x}{y}\right)(-2v)$$

$$= 2v \ln (u^2 - v^2) - 2v\frac{u^2 + v^2}{u^2 - v^2} \qquad \square$$

Rather than state additional versions of the Chain Rule, we will give a general procedure for finding derivatives of composite functions of several variables. First we draw a diagram that indicates how the variables are related. For example, the diagram corresponding to (2) is

The diagram is read from left to right and shows that z depends on x and y, which in turn depend on u and v. To obtain $\partial z/\partial u$ we find the product of the partial

derivatives along each individual path from z to u (that is, $(\partial z/\partial x)(\partial x/\partial u)$ and $(\partial z/\partial y)(\partial y/\partial u)$, and then add these products:

$$\frac{\partial z}{\partial u} = \frac{\partial z}{\partial x}\frac{\partial x}{\partial u} + \frac{\partial z}{\partial y}\frac{\partial y}{\partial u}$$

This is the first equation in (2). The second equation in (2), which gives $\partial z/\partial v$, is obtained by using all paths from z to v. Analogous diagrams apply to other versions of the Chain Rule, and are illustrated in the following examples.

EXAMPLE 3   Let $w = x \cos yz^2$, $x = \sin t$, $y = t^2$, and $z = e^t$. Find dw/dt.

Solution   The appropriate diagram is

Since there are three paths from w to t, we find that

$$\frac{dw}{dt} = \frac{\partial w}{\partial x}\frac{dx}{dt} + \frac{\partial w}{\partial y}\frac{dy}{dt} + \frac{\partial w}{\partial z}\frac{dz}{dt}$$

$$= \cos yz^2 \cos t - xz^2(\sin yz^2)(2t) - 2xyz(\sin yz^2)(e^t)$$

$$= \cos(t^2 e^{2t}) \cos t - 2te^{2t}\sin t \sin(t^2 e^{2t}) - 2t^2 e^{2t}\sin t \sin(t^2 e^{2t}) \qquad \square$$

EXAMPLE 4   Let $w = \sqrt{x} + y^2 z^3$, $x = 1 + u^2 + v^2$, $y = uv$, and $z = 3u$. Find $\partial w/\partial u$ and $\partial w/\partial v$.

Solution   The diagram that applies here is

It follows that

$$\frac{\partial w}{\partial u} = \frac{\partial w}{\partial x}\frac{\partial x}{\partial u} + \frac{\partial w}{\partial y}\frac{\partial y}{\partial u} + \frac{\partial w}{\partial z}\frac{\partial z}{\partial u}$$

$$= \frac{1}{2\sqrt{x}}(2u) + 2yz^3(v) + 3y^2 z^2(3) = \frac{u}{\sqrt{1 + u^2 + v^2}} + 54u^4 v^2 + 81u^4 v^2$$

$$= \frac{u}{\sqrt{1 + u^2 + v^2}} + 135u^4 v^2$$

and

$$\frac{\partial w}{\partial v} = \frac{\partial w}{\partial x}\frac{\partial x}{\partial v} + \frac{\partial w}{\partial y}\frac{\partial y}{\partial v} + \frac{\partial w}{\partial z}\frac{\partial z}{\partial v}$$

$$= \frac{1}{2\sqrt{x}}(2v) + 2yz^3(u) + 3y^2z^2(0) = \frac{v}{\sqrt{1 + u^2 + v^2}} + 54u^5v \qquad \square$$

EXAMPLE 5   Suppose that $w = f(x, y)$, $x = g(u, v)$, $y = h(u, v)$, $u = j(t)$, and $v = k(t)$. Find a formula for dw/dt.

Solution   We draw the diagram

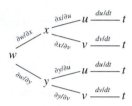

Using the four paths leading from w to t, we find that

$$\frac{dw}{dt} = \frac{\partial w}{\partial x}\frac{\partial x}{\partial u}\frac{du}{dt} + \frac{\partial w}{\partial x}\frac{\partial x}{\partial v}\frac{dv}{dt} + \frac{\partial w}{\partial y}\frac{\partial y}{\partial u}\frac{du}{dt} + \frac{\partial w}{\partial y}\frac{\partial y}{\partial v}\frac{dv}{dt} \qquad \square$$

EXAMPLE 6   Let $z = f(u - v, v - u)$. Show that

$$\frac{\partial z}{\partial u} + \frac{\partial z}{\partial v} = 0$$

Solution   If we let $x = u - v$ and $y = v - u$, then $z = f(x, y)$, and (2) applies:

$$\frac{\partial z}{\partial u} = \frac{\partial z}{\partial x}\frac{\partial x}{\partial u} + \frac{\partial z}{\partial y}\frac{\partial y}{\partial u} = \frac{\partial z}{\partial x}(1) + \frac{\partial z}{\partial y}(-1)$$

$$\frac{\partial z}{\partial v} = \frac{\partial z}{\partial x}\frac{\partial x}{\partial v} + \frac{\partial z}{\partial y}\frac{\partial y}{\partial v} = \frac{\partial z}{\partial x}(-1) + \frac{\partial z}{\partial y}(1)$$

Therefore

$$\frac{\partial z}{\partial u} + \frac{\partial z}{\partial v} = 0 \qquad \square$$

The Chain Rule for functions of several variables is particularly well adapted to related rates problems.

EXAMPLE 7   Suppose the volume V of a conical sand pile grows at a rate of 4 cubic inches per second and the radius r of the circular base grows at a rate of e^{-r} inches per second. Find the rate at which the height h of the cone is growing at the instant when the volume is 60 cubic inches and the radius is 6 inches.

Solution   We wish to find dh/dt at the instant when $V = 60$ and $r = 6$. The formula for the volume of the cone is

$$V = \frac{1}{3} \pi r^2 h$$

so the formula for the height is

$$h = \frac{3V}{\pi r^2}$$

The diagram for dh/dt is

$$h \overset{\partial h/\partial V}{\underset{\partial h/\partial r}{\diagdown}} \begin{array}{l} V \xrightarrow{dV/dt} t \\[1em] r \xrightarrow{dr/dt} t \end{array}$$

Since $dV/dt = 4$ and $dr/dt = e^{-r}$ by hypothesis, we find that

$$\frac{dh}{dt} = \frac{\partial h}{\partial V} \frac{dV}{dt} + \frac{\partial h}{\partial r} \frac{dr}{dt} = \frac{3}{\pi r^2} (4) + \frac{-6V}{\pi r^3} (e^{-r}) = \frac{6}{\pi r^2} \left(2 - \frac{V}{r} e^{-r} \right)$$

If t_0 denotes the time at which the volume is 60 cubic inches and the radius is 6 inches, then

$$\left. \frac{dh}{dt} \right|_{t=t_0} = \frac{6}{\pi 6^2} \left(2 - \frac{60}{6} e^{-6} \right) = \frac{1}{3\pi} (1 - 5e^{-6}) \text{ (inches per second)} \qquad \square$$

Implicit Differentiation

Through the Chain Rule we can reinterpret implicit differentiation, which we introduced in Section 3.6. The next example, which we will use in the discussion of implicit differentiation, is useful in its own right.

EXAMPLE 8   Suppose that $w = f(x, y)$ and $y = g(x)$. Find a formula for dw/dx.

Solution   The appropriate diagram is as follows:

$$w \overset{\partial w/\partial x}{\underset{\partial w/\partial y}{\diagdown}} \begin{array}{l} x \\[1em] y \xrightarrow{dy/dx} x \end{array}$$

It follows that

$$\frac{dw}{dx} = \frac{\partial w}{\partial x} + \frac{\partial w}{\partial y} \frac{dy}{dx} \tag{3}$$

(In this formula, $\partial w/\partial x$ refers to the partial derivative of $z = w(x, y)$ with respect to x, whereas dw/dx refers to the derivative of $z = f(x, g(x))$ with respect to x.) $\square$

Now suppose that f is a function of two variables that has partial derivatives, and assume that the equation $f(x, y) = 0$ defines a differentiable function $y = g(x)$ of x, so that $f(x, g(x)) = 0$. If $w = f(x, y)$, then by assumption,

$$\frac{dw}{dx} = \frac{d}{dx} (f(x, g(x))) = \frac{d}{dx} (0) = 0$$

Therefore by (3),

$$0 = \frac{dw}{dx} = \frac{\partial w}{\partial x} + \frac{\partial w}{\partial y}\frac{dy}{dx}$$

Finally, if $\partial w/\partial y \neq 0$, then by solving for dy/dx, we obtain

$$\frac{dy}{dx} = \frac{-\partial w/\partial x}{\partial w/\partial y} \tag{4}$$

Now we apply (4) to find a derivative that we obtained by implicit differentiation in Example 3 of Section 3.6.

EXAMPLE 9   Let $x^3 + y^3 = 2xy$. Find dy/dx.

Solution   Let $w = x^3 + y^3 - 2xy$. Then

$$\frac{\partial w}{\partial x} = 3x^2 - 2y \quad\text{and}\quad \frac{\partial w}{\partial y} = 3y^2 - 2x$$

and (4) implies that

$$\frac{dy}{dx} = \frac{-\partial w/\partial x}{\partial w/\partial y} = \frac{-(3x^2 - 2y)}{3y^2 - 2x} = \frac{2y - 3x^2}{3y^2 - 2x} \qquad \square$$

Differentiability and the Theory behind the Chain Rule

In proving one version of the Chain Rule we will use the formula that appears in the following theorem and gives rise to the concept of differentiability for functions of several variables.

THEOREM 13.7

Let f be a function having partial derivatives throughout a set containing an open disk D centered at (x_0, y_0). If f_x and f_y are continuous at (x_0, y_0) then there are functions ε_1 and ε_2 of two variables such that

$$f(x, y) - f(x_0, y_0) = f_x(x_0, y_0)(x - x_0) + f_y(x_0, y_0)(y - y_0)$$
$$+ \varepsilon_1(x, y)(x - x_0) + \varepsilon_2(x, y)(y - y_0) \text{ for } (x, y) \text{ in } D$$

where $\lim_{(x, y)\to(x_0, y_0)} \varepsilon_1(x, y) = 0$ and $\lim_{(x, y)\to(x_0, y_0)} \varepsilon_2(x, y) = 0$.

Proof   Let (x, y) be any point in D. First we write

$$f(x, y) - f(x_0, y_0) = [f(x, y) - f(x_0, y)] + [f(x_0, y) - f(x_0, y_0)]$$

and let

$$g(t) = f(t, y) \quad\text{and}\quad G(t) = f(x_0, t)$$

Applying the Mean Value Theorem to g on the interval with endpoints x and x_0 and to G on the interval with endpoints y and y_0, we obtain a number $u(x)$ between x and x_0 and a number $v(y)$ between y and y_0 such that

$$f(x, y) - f(x_0, y) = g(x) - g(x_0) = g'(u(x))(x - x_0)$$
$$= f_x(u(x), y)(x - x_0) \tag{5}$$

and

$$f(x_0, y) - f(x_0, y_0) = G(y) - G(y_0) = G'(v(y))(y - y_0)$$
$$= f_y(x_0, v(y))(y - y_0) \tag{6}$$

Equations (5) and (6) yield

$$f(x, y) - f(x_0, y_0) = [f(x, y) - f(x_0, y)] + [f(x_0, y) - f(x_0, y_0)]$$
$$= f_x(u(x), y)(x - x_0) + f_y(x_0, v(y))(y - y_0)$$

$$= f_x(x_0, y_0)(x - x_0) + \overbrace{[f_x(u(x), y) - f_x(x_0, y_0)]}^{\varepsilon_1(x, y)}(x - x_0)$$

$$+ f_y(x_0, y_0)(y - y_0) + \overbrace{[f_y(x_0, v(y)) - f_y(x_0, y_0)]}^{\varepsilon_2(x, y)}(y - y_0)$$

$$= f_x(x_0, y_0)(x - x_0) + f_y(x_0, y_0)(y - y_0) + \varepsilon_1(x, y)(x - x_0)$$

$$+ \varepsilon_2(x, y)(y - y_0)$$

Since f_x and f_y are continuous at (x_0, y_0), and since $\lim_{(x, y) \to (x_0, y_0)} u(x) = x_0$ and $\lim_{(x, y) \to (x_0, y_0)} v(y) = y_0$, both $\varepsilon_1(x, y)$ and $\varepsilon_2(x, y)$ approach 0 as (x, y) approaches (x_0, y_0). ■

We use the formula in Theorem 13.7 in the following definition.

DEFINITION 13.8

A function f of two variables is **differentiable at (x_0, y_0)** if there exist an open disk D centered at (x_0, y_0) and functions ε_1 and ε_2 of two variables such that

$$f(x, y) - f(x_0, y_0) = f_x(x_0, y_0)(x - x_0) + f_y(x_0, y_0)(y - y_0)$$
$$+ \varepsilon_1(x, y)(x - x_0) + \varepsilon_2(x, y)(y - y_0) \quad \text{for } (x, y) \text{ in } D \qquad (7)$$

where $\lim_{(x, y) \to (x_0, y_0)} \varepsilon_1(x, y) = 0$ and $\lim_{(x, y) \to (x_0, y_0)} \varepsilon_2(x, y) = 0$.

In terms of differentiability, Theorem 13.7 states that if f has partial derivatives on a disk D centered at (x_0, y_0) and if f_x and f_y are continuous at (x_0, y_0), then f is differentiable at (x_0, y_0). It is usually easier to verify that a function f has continuous partial derivatives throughout a disk centered at (x_0, y_0) than to apply Definition 13.8 to prove that f is differentiable at (x_0, y_0). Because most of the functions we will encounter have continuous partial derivatives at all or almost all points in their respective domains, they are differentiable at all (or almost all) points in their domains. In particular every rational function is differentiable at every point in its domain.

Now we will use differentiability to prove the version of the Chain Rule associated with (1).

THEOREM 13.9
Chain Rule (First Version)

Let f be a function of two variables that is differentiable at (x_0, y_0), and let g_1 and g_2 be functions of one variable that are differentiable at t_0. Suppose $x_0 = g_1(t_0)$ and $y_0 = g_2(t_0)$, and let

$$F(t) = f(g_1(t), g_2(t))$$

Then F is differentiable at t_0, and

$$F'(t_0) = f_x(x_0, y_0)g_1'(t_0) + f_y(x_0, y_0)g_2'(t_0)$$

Proof   Since g_1 and g_2 are differentiable at t_0, they are continuous at t_0, so that

$$\lim_{t \to t_0} g_1(t) = x_0 \quad \text{and} \quad \lim_{t \to t_0} g_2(t) = y_0$$

Because f is differentiable at (x_0, y_0) by hypothesis, the functions ε_1 and ε_2 appearing in (7) have the limit 0 at (x_0, y_0); hence a two-variable substitution yields

$$\lim_{t \to t_0} \varepsilon_1(g_1(t), g_2(t)) = 0 \quad \text{and} \quad \lim_{t \to t_0} \varepsilon_2(g_1(t), g_2(t)) = 0$$

Using (7) with x replaced by $g_1(t)$ and y replaced by $g_2(t)$, we find that

$$
\begin{aligned}
F'(t_0) &= \lim_{t \to t_0} \frac{F(t) - F(t_0)}{t - t_0} = \lim_{t \to t_0} \left(\frac{f(g_1(t), g_2(t)) - f(g_1(t_0), g_2(t_0))}{t - t_0} \right) \\
&= \lim_{t \to t_0} \left(f_x(x_0, y_0) \frac{g_1(t) - g_1(t_0)}{t - t_0} + f_y(x_0, y_0) \frac{g_2(t) - g_2(t_0)}{t - t_0} \right. \\
&\quad \left. + \varepsilon_1(g_1(t), g_2(t)) \frac{g_1(t) - g_1(t_0)}{t - t_0} + \varepsilon_2(g_1(t), g_2(t)) \frac{g_2(t) - g_2(t_0)}{t - t_0} \right) \\
&= f_x(x_0, y_0) g_1'(t_0) + f_y(x_0, y_0) g_2'(t_0) + 0 \cdot g_1'(t_0) + 0 \cdot g_2'(t_0) \\
&= f_x(x_0, y_0) g_1'(t_0) + f_y(x_0, y_0) g_2'(t_0) \quad \blacksquare
\end{aligned}
$$

We state the version of the Chain Rule associated with (2), but omit its proof, which follows along the same lines as that for the first version.

THEOREM 13.10
Chain Rule (Second Version)

Let f be a function of two variables that is differentiable at (x_0, y_0), and let g_1 and g_2 be functions of two variables having partial derivatives at (u_0, v_0). Suppose $x_0 = g_1(u_0, v_0)$ and $y_0 = g_2(u_0, v_0)$, and let

$$F(u, v) = f(g_1(u, v), g_2(u, v))$$

Then F has partial derivatives at (u_0, v_0), and

$$F_u(u_0, v_0) = f_x(x_0, y_0)(g_1)_u(u_0, v_0) + f_y(x_0, y_0)(g_2)_u(u_0, v_0)$$

$$F_v(u_0, v_0) = f_x(x_0, y_0)(g_1)_v(u_0, v_0) + f_y(x_0, y_0)(g_2)_v(u_0, v_0)$$

The concept of differentiability extends to functions of three variables.

DEFINITION 13.11

A function f of three variables is **differentiable at (x_0, y_0, z_0)** if there exist an open ball B centered at (x_0, y_0, z_0) and functions ε_1, ε_2, and ε_3 of three variables such that

$$
\begin{aligned}
f(x, y, z) - f(x_0, y_0, z_0) = {} & f_x(x_0, y_0, z_0)(x - x_0) + f_y(x_0, y_0, z_0)(y - y_0) \\
& + f_z(x_0, y_0, z_0)(z - z_0) + \varepsilon_1(x, y, z)(x - x_0) + \varepsilon_2(x, y, z)(y - y_0) \\
& + \varepsilon_3(x, y, z)(z - z_0) \quad \text{for all } (x, y, z) \text{ in } B
\end{aligned}
$$

where $\lim_{(x, y, z) \to (x_0, y_0, z_0)} \varepsilon_1(x, y, z) = 0$, $\lim_{(x, y, z) \to (x_0, y_0, z_0)} \varepsilon_2(x, y, z) = 0$, and $\lim_{(x, y, z) \to (x_0, y_0, z_0)} \varepsilon_3(x, y, z) = 0$.

A theorem analogous to Theorem 13.7 implies that if f has partial derivatives on a ball centered at (x_0, y_0, z_0) and if f_x, f_y, and f_z are continuous at (x_0, y_0, z_0), then f is differentiable at (x_0, y_0, z_0). Again, most of the functions of three variables we will encounter will be differentiable at all or at almost all points in their domains. Moreover, if f is differentiable at (x_0, y_0, z_0), then theorems analogous to Theorems 13.9 and 13.10 hold. Since the Chain Rule is used in the proofs of many theorems to appear in this chapter, differentiability will often occur as a hypothesis in the theorems.

EXERCISES 13.4

In Exercises 1–5 compute dz/dt.

1. $z = 2x^2 - 3y^3; x = \sqrt{t}, y = e^{2t}$
2. $z = \ln(3x^2 + y^3); x = e^{2t}, y = t^{1/3}$
3. $z = \sin x + \cos xy; x = t^2, y = 1$
4. $z = \tan^{-1}(y^2 - x^2); x = \sin t, y = \cos t$
5. $z = \sqrt{2x - 4y}; x = \ln t, y = 1 - 3t^3$

In Exercises 6–10 compute $\partial z/\partial u$ and $\partial z/\partial v$.

6. $z = \dfrac{x}{y^2}; x = u + v - 1, y = u - v - 1$

7. $z = \dfrac{4}{xy} - \dfrac{x}{y}; x = u^2, y = uv$

8. $z = 16 - 4x^2 - y^2; x = u \sin v, y = v \cos u$
9. $z = \ln(x^2 - y^2); x = u - v, y = u^2 + v^2$
10. $z = 2e^{x^2 y}; x = \sqrt{uv}, y = 1/u$

In Exercises 11–14 compute $\partial z/\partial r$ and $\partial z/\partial s$.

11. $z = \sin 2u \cos 3v; u = (r + s)^2, v = (r - s)^2$
12. $z = \ln u + \ln v; u = 4^{rs}, v = 4^{r/s}$
13. $z = ue^v + ve^{-u}; u = \ln r, v = s \ln r$
14. $z = 2^{u-v}; u = r \cos s, v = r \sin s$

In Exercises 15–20 compute dw/dt.

15. $w = \dfrac{x}{y} - \dfrac{z}{x}; x = \sin t, y = \cos t, z = \tan t$

16. $w = \dfrac{z}{xy^2} - 3; x = \dfrac{1}{t^2}, y = -5t, z = \sqrt{t}$

17. $w = \sqrt{x^2 + y^2 + z^2}; x = e^t, y = e^{-t}, z = 2t$
18. $w = \ln(x^2 + y^2 + z^2); x = \sin t, y = \cos t, z = e^{-t^2}$
19. $w = \sin xy^2 z^3; x = 3t, y = t^{1/2}, z = t^{1/3}$
20. $w = \sqrt{x^2 + y^2} - \sqrt{y^3 - z^3}; x = t^2, y = t^3, z = -t^3$

In Exercises 21–24 find $\partial w/\partial u$ and $\partial w/\partial v$.

21. $w = \dfrac{yz}{x^2 + xy}; x = u^2, y = v^2, z = u^2 - v^2$

22. $w = x^2 - 2y - 7z; x = v \cos(\pi - u)$,
    $y = u \sin(\pi - v), z = uv$
23. $w = y \ln xz; x = ve^u, y = u^2 v^4, z = ue^v$
24. $w = e^{x/y} + e^{z/x}; x = \dfrac{\ln u}{v}, y = \ln u, z = \dfrac{\ln u}{uv}$

In Exercises 25–29 find dy/dx by implicit differentiation.

25. $x^3 + 4x^2 y - 3xy^2 + 2y^3 + 5 = 0$
26. $x^{2/3} + y^{2/3} = 2$ 27. $x^2 + y^2 + \sin xy^2 = 0$
28. $e^{x/y} + \ln y/x + 15 = 0$ 29. $x^2 = \dfrac{y^2}{y^2 - 1}$

In Exercises 30–32 assume that the equation defines z implicitly as a function of x and y, and use "implicit partial differentiation" to find $\partial z/\partial x$ and $\partial z/\partial y$.

30. $x^2 z^2 - 2xyz + z^3 y^2 = 3$ 31. $x - yz + \cos xyz = 2$

32. $\dfrac{1}{z} + \dfrac{1}{y + z} + \dfrac{1}{x + y + z} = \dfrac{1}{2}$

33. Let $z = f(x - y)$. Show that $\dfrac{\partial z}{\partial x} = -\dfrac{\partial z}{\partial y}$.

34. Let $w = f(x - y, y - z, z - x)$. Show that
$$\frac{\partial w}{\partial x} + \frac{\partial w}{\partial y} + \frac{\partial w}{\partial z} = 0$$

35. Let $z = f(y + ax) + g(y - ax)$, with $a \neq 0$. Show that z satisfies the wave equation
$$\frac{\partial^2 z}{\partial x^2} = a^2 \frac{\partial^2 z}{\partial y^2}$$

*36. Let $z = f(x, y), x = r \cos \theta$, and $y = r \sin \theta$.

   a. Show that $\dfrac{\partial z}{\partial x} = \dfrac{\partial z}{\partial r} \cos \theta - \dfrac{\partial z}{\partial \theta} \dfrac{\sin \theta}{r}$

   and $\dfrac{\partial z}{\partial y} = \dfrac{\partial z}{\partial r} \sin \theta + \dfrac{\partial z}{\partial \theta} \dfrac{\cos \theta}{r}$.

b. Show that $\left(\dfrac{\partial z}{\partial x}\right)^2 + \left(\dfrac{\partial z}{\partial y}\right)^2 = \left(\dfrac{\partial z}{\partial r}\right)^2 + \dfrac{1}{r^2}\left(\dfrac{\partial z}{\partial \theta}\right)^2$.

*37. Let $w = f(x, y)$, $x = e^s \cos t$, and $y = e^s \sin t$. Assuming that the second partials of f exist, show that

$$\frac{\partial^2 w}{\partial x^2} + \frac{\partial^2 w}{\partial y^2} = e^{-2s}\left(\frac{\partial^2 w}{\partial s^2} + \frac{\partial^2 w}{\partial t^2}\right)$$

38. A function f of two variables is **homogeneous of degree** **n** if for any real number t we have

$$f(tx, ty) = t^n f(x, y) \tag{8}$$

Show that in this case

$$xf_x(x, y) + yf_y(x, y) = nf(x, y)$$

(*Hint:* Differentiate both sides of (8) with respect to t, and then set $t = 1$.)

39. Let

$$f(x, y) = \tan \frac{x^2 + y^2}{xy}$$

Use Exercise 38 to show that

$$xf_x(x, y) + yf_y(x, y) = 0$$

40. Show that if f is differentiable at (x_0, y_0), then f is continuous at (x_0, y_0). (*Hint:* Using (7), show that $\lim_{(x, y)\to(x_0, y_0)} f(x, y) = f(x_0, y_0)$.)

41. Let

$$f(x, y) = \begin{cases} \dfrac{xy}{x^2 + y^2} & \text{for } (x, y) \neq (0, 0) \\ 0 & \text{for } (x, y) = (0, 0) \end{cases}$$

Use the result of Exercise 40 to show that f is not differentiable at $(0, 0)$. (Notice that f has partial derivatives at $(0, 0)$ despite its nondifferentiability there.)

A sequoia tree in Yosemite National Park.
(*Larry Brownstein/Rainbow*)

Applications

42. A tree trunk may be considered a circular cylinder. Suppose the diameter of the trunk increases 1 inch per year and the height of the trunk increases 6 inches per year. How fast is the volume of wood in the trunk increasing when it is 100 inches high and 5 inches in diameter?

43. The time rate Q of flow of fluid through a cylindrical tube (such as a windpipe) with radius r and height l is given by

$$Q = \frac{\pi p r^4}{8 l \eta}$$

where η is the viscosity of the fluid and p is the difference in pressure at the two ends of the tube. Suppose the length of the tube remains constant, while the radius increases at the rate of $\frac{1}{10}$ and the pressure decreases at the rate of $\frac{1}{5}$. Find the rate of change of Q with respect to time.

44. A road is perpendicular to a train track. Suppose a car approaches the intersection of the road and the track at 20 miles per hour, while a train approaches at 100 miles per hour. At what rate is the distance between the car and the train changing when the car is 0.5 miles from the intersection and the train is 1.2 miles from the intersection?

45. The mass of a rocket lifting off from earth is decreasing (due to fuel consumption) at the rate of 40 kilograms per second. How fast is the magnitude F of the force of gravity decreasing when the rocket is 6400 kilometers from the center of the earth and is rising with a velocity of 100 kilometers per second? (*Hint:* By Newton's Law of Gravitation, $F = GMm/r^2$, where G is the universal gravitational constant, M is the mass of the earth, m is the mass of the rocket, and r is the distance between the rocket and the center of the earth.)

Project

1. Suppose F is a differentiable function of x, y and z, and suppose that the equation $F(x, y, z) = 0$ defines z as a differentiable function of x and y. (For example, if $F(x, y, z) = 3x^2 + 2y^2 + z^2 - 1$, then the equation $F(x, y, z) = 0$, which is equivalent to $3x^2 + 2y^2 + z^2 = 1$, defines a function z as a differentiable function of x and y (actually two such functions!).

a. Use the Chain Rule in this section to prove that

$$\frac{\partial z(x, y)}{\partial x} = \frac{-F_x(x, y, z)}{F_z(x, y, z)}$$

b. Consider the surface defined by $3x^2 + 2y^2 + z^2 = 1$. Find $\partial z/\partial x$ and $\partial z/\partial y$ by (a), and by solving for z and taking the partial derivative directly.

c. Suppose that the equation $F(x, y, z) = 0$ also defines x as a differentiable function of y and z, and defines y as a differentiable function of x and z. Use (a) to show that

$$\frac{\partial x(y, z)}{\partial y} \frac{\partial y(x, z)}{\partial z} \frac{\partial z(x, y)}{\partial x} = -1$$

One could use the method in (c) to derive the consequence of the ideal gas law that appears in Exercise 63 of Section 13.3:

$$\frac{\partial V}{\partial T} \frac{\partial T}{\partial p} \frac{\partial p}{\partial V} = -1$$

where V represents volume, T represents temperature, and p represents pressure for an ideal gas.

13.5 DIRECTIONAL DERIVATIVES

We may think of the partial derivatives of a function as describing the rate of change of the function along lines parallel to the coordinate axes. Now we discuss the rate of change of a function f at a point (x_0, y_0) along a line that is not necessarily parallel to any coordinate axis. To be specific, let l be a line in the xy plane that is parametrized by

$$x = x_0 + a_1 t \quad \text{and} \quad y = y_0 + a_2 t$$

where $\mathbf{u} = a_1 \mathbf{i} + a_2 \mathbf{j}$ is a unit vector (Figure 13.36(a)). Then for any $h \neq 0$,

$$\frac{f(x_0 + a_1 h, y_0 + a_2 h) - f(x_0, y_0)}{h}$$

is the slope of the corresponding secant line through the points P and Q in Figure 13.36(b), which lie in the vertical plane $\mathcal{P}$ containing l. It follows that if C is the curve determined by the intersection of the graph of f and the plane $\mathcal{P}$, then

$$\lim_{h \to 0} \frac{f(x_0 + a_1 h, y_0 + a_2 h) - f(x_0, y_0)}{h}$$

would be the slope of the line tangent to C at $(x_0, y_0, f(x_0, y_0))$. Because of the intimate relationship between derivatives and slopes of tangents for functions of one variable, we are led to refer to the limit above as the derivative of f in the direction of $\mathbf{u} = a_1 \mathbf{i} + a_2 \mathbf{j}$.

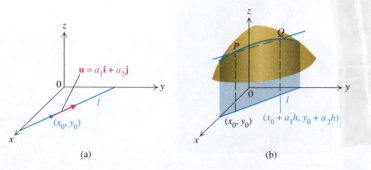

(a)   (b)

FIGURE 13.36

DEFINITION 13.12   Let f be a function defined on a set containing a disk D centered at (x_0, y_0), and let $\mathbf{u} = a_1\mathbf{i} + a_2\mathbf{j}$ be a unit vector. Then the **directional derivative** of f at (x_0, y_0) in the direction of $\mathbf{u}$, denoted $D_{\mathbf{u}}f(x_0, y_0)$, is defined by

$$D_{\mathbf{u}}f(x_0, y_0) = \lim_{h \to 0} \frac{f(x_0 + ha_1, y_0 + ha_2) - f(x_0, y_0)}{h}$$

provided that this limit exists.

Observe that if $\mathbf{u} = \mathbf{i}$, then $a_1 = 1$ and $a_2 = 0$, so that

$$D_{\mathbf{u}}f(x_0, y_0) = \lim_{h \to 0} \frac{f(x_0 + h, y_0) - f(x_0, y_0)}{h} = f_x(x_0, y_0)$$

and if $\mathbf{u} = \mathbf{j}$, then $a_1 = 0$ and $a_2 = 1$, so that

$$D_{\mathbf{u}}f(x_0, y_0) = \lim_{h \to 0} \frac{f(x_0, y_0 + h) - f(x_0, y_0)}{h} = f_y(x_0, y_0)$$

Thus the first partial derivatives of f are special cases of directional derivatives—in the directions of the positive coordinate axes. It turns out that if f is differentiable at (x_0, y_0), then any directional derivative can be evaluated by means of the two partial derivatives.

THEOREM 13.13   Let f be differentiable at (x_0, y_0). Then f has a directional derivative at (x_0, y_0) in every direction. Moreover, if $\mathbf{u} = a_1\mathbf{i} + a_2\mathbf{j}$ is a unit vector, then

$$D_{\mathbf{u}}f(x_0, y_0) = f_x(x_0, y_0)a_1 + f_y(x_0, y_0)a_2 \qquad (1)$$

Proof   Let

$$F(h) = f(x_0 + ha_1, y_0 + ha_2)$$

Then

$$\frac{F(h) - F(0)}{h - 0} = \frac{f(x_0 + ha_1, y_0 + ha_2) - f(x_0, y_0)}{h}$$

so that $D_{\mathbf{u}}f(x_0, y_0)$ exists if and only if $F'(0)$ exists. If we let

$$g_1(h) = x_0 + ha_1 \quad \text{and} \quad g_2(h) = y_0 + ha_2$$

then $F(h) = f(g_1(h), g_2(h))$, and also $g_1(0) = x_0$ and $g_2(0) = y_0$. With h replacing t and 0 replacing t_0, the hypotheses of the first version of the Chain Rule (Theorem 13.9) are satisfied. Consequently $F'(0)$ exists, and

$$D_{\mathbf{u}}f(x_0, y_0) = F'(0) = f_x(x_0, y_0)g_1'(0) + f_y(x_0, y_0)g_2'(0)$$
$$= f_x(x_0, y_0)a_1 + f_y(x_0, y_0)a_2 \quad \blacksquare$$

EXAMPLE 1   Let $f(x, y) = 6 - 3x^2 - y^2$, and let $\mathbf{u} = (1/\sqrt{2})\mathbf{i} - (1/\sqrt{2})\mathbf{j}$. Find $D_{\mathbf{u}}f(1, 2)$.

Solution   Notice that **u** is a unit vector. We will compute $D_{\mathbf{u}}f(1, 2)$ by means of (1). First we calculate the partial derivatives of f:

$$f_x(x, y) = -6x \quad \text{and} \quad f_y(x, y) = -2y$$

Therefore $f_x(1, 2) = -6$ and $f_y(1, 2) = -4$, so by (1),

$$D_{\mathbf{u}}f(1, 2) = f_x(1, 2)\left(\frac{1}{\sqrt{2}}\right) + f_y(1, 2)\left(-\frac{1}{\sqrt{2}}\right)$$

$$= (-6)\left(\frac{1}{\sqrt{2}}\right) + (-4)\left(-\frac{1}{\sqrt{2}}\right) = -\sqrt{2} \quad \square$$

In the expression $D_{\mathbf{u}}f(x_0, y_0)$, **u** represents a unit vector. The directional derivative in the direction of an arbitrary nonzero vector **a** is defined to be $D_{\mathbf{u}}f(x_0, y_0)$, where **u** is the unit vector in the direction of **a**, that is, $\mathbf{u} = \mathbf{a}/\|\mathbf{a}\|$.

EXAMPLE 2   Let $f(x, y) = xy^2$ and let $\mathbf{a} = \mathbf{i} - 2\mathbf{j}$. Find the directional derivative of f at $(-3, 1)$ in the direction of **a**.

Solution   In this case $\|\mathbf{a}\| = \sqrt{1^2 + (-2)^2} = \sqrt{5}$, so we will find $D_{\mathbf{u}}f(-3, 1)$, where

$$\mathbf{u} = \frac{1}{\|\mathbf{a}\|}\mathbf{a} = \frac{1}{\sqrt{5}}\mathbf{i} - \frac{2}{\sqrt{5}}\mathbf{j}$$

Since

$$f_x(x, y) = y^2 \quad \text{and} \quad f_y(x, y) = 2xy$$

it follows that $f_x(-3, 1) = 1$ and $f_y(-3, 1) = -6$. By (1),

$$D_{\mathbf{u}}f(-3, 1) = f_x(-3, 1)\left(\frac{1}{\sqrt{5}}\right) + f_y(-3, 1)\left(\frac{-2}{\sqrt{5}}\right)$$

$$= 1\left(\frac{1}{\sqrt{5}}\right) + (-6)\left(\frac{-2}{\sqrt{5}}\right) = \frac{13}{5}\sqrt{5} \quad \square$$

In defining the directional derivative for functions of three variables we make only the necessary changes in notation. Let $\mathbf{u} = a_1\mathbf{i} + a_2\mathbf{j} + a_3\mathbf{k}$ be a unit vector in space. The **directional derivative** $D_{\mathbf{u}}f(x_0, y_0, z_0)$ is defined by

$$D_{\mathbf{u}}f(x_0, y_0, z_0) = \lim_{h \to 0} \frac{f(x_0 + ha_1, y_0 + ha_2, z_0 + ha_3) - f(x_0, y_0, z_0)}{h}$$

provided that this limit exists. Moreover, if f is differentiable at (x_0, y_0, z_0), then the analogue of Theorem 13.13 holds:

$$D_{\mathbf{u}}f(x_0, y_0, z_0) = f_x(x_0, y_0, z_0)a_1 + f_y(x_0, y_0, z_0)a_2 + f_z(x_0, y_0, z_0)a_3 \qquad (2)$$

As before, if **a** is any nonzero vector in space, then we define the directional derivative of f at (x_0, y_0, z_0) in the direction of $\mathbf{a}$ to be $D_{\mathbf{u}}f(x_0, y_0, z_0)$, where $\mathbf{u} = \mathbf{a}/\|\mathbf{a}\|$.

EXAMPLE 3   Let $f(x, y, z) = xe^{y^2z}$, and let $\mathbf{a} = \mathbf{i} - \mathbf{j} + \sqrt{2}\mathbf{k}$. Find the directional derivative of f at $(2, 1, 0)$ in the direction of $\mathbf{a}$.

Solution   First we find the partial derivatives of f:

$$f_x(x, y, z) = e^{y^2z}, \quad f_y(x, y, z) = 2xyze^{y^2z}, \quad f_z(x, y, z) = xy^2e^{y^2z}$$

Since $\|\mathbf{a}\| = \sqrt{1^2 + (-1)^2 + (\sqrt{2})^2} = 2$, we apply (2) with

$$\mathbf{u} = \frac{1}{2}\mathbf{a} = \frac{1}{2}\mathbf{i} - \frac{1}{2}\mathbf{j} + \frac{\sqrt{2}}{2}\mathbf{k}$$

and obtain

$$D_{\mathbf{u}}f(2, 1, 0) = f_x(2, 1, 0)\left(\frac{1}{2}\right) + f_y(2, 1, 0)\left(-\frac{1}{2}\right) + f_z(2, 1, 0)\left(\frac{\sqrt{2}}{2}\right)$$

$$= 1\left(\frac{1}{2}\right) + 0\left(-\frac{1}{2}\right) + 2\left(\frac{\sqrt{2}}{2}\right) = \frac{1}{2} + \sqrt{2} \quad \square$$

EXERCISES 13.5

In Exercises 1–14 find the directional derivative of f at the point P in the direction of $\mathbf{a}$.

1. $f(x, y) = 2x^2 - 3xy + y^2 + 15$; $P = (1, 1)$;

$\mathbf{a} = \dfrac{1}{\sqrt{2}}\mathbf{i} + \dfrac{1}{\sqrt{2}}\mathbf{j}$

2. $f(x, y) = x^2 + y^2$; $P = (1, 2)$; $\mathbf{a} = \dfrac{1}{\sqrt{3}}\mathbf{i} - \dfrac{\sqrt{2}}{\sqrt{3}}\mathbf{j}$

3. $f(x, y) = \dfrac{x^2 - y^2}{x^2 + y^2}$; $P = (3, 4)$; $\mathbf{a} = \dfrac{1}{2}\mathbf{i} - \dfrac{\sqrt{3}}{2}\mathbf{j}$

4. $f(x, y) = x - y^2$; $P = (2, -3)$; $\mathbf{a} = \mathbf{i} + 2\mathbf{j}$

5. $f(x, y) = e^{4y}$; $P = (\frac{1}{2}, \frac{1}{4})$; $\mathbf{a} = 4\mathbf{i}$

6. $f(x, y) = \sin xy^2$; $P = (1/\pi, \pi)$; $\mathbf{a} = \mathbf{i} - 3\mathbf{j}$

7. $f(x, y) = \tan(x + 2y)$; $P = (0, \pi/6)$; $\mathbf{a} = -4\mathbf{i} + 5\mathbf{j}$

8. $f(x, y, z) = 3x - 2y + 4z$; $P = (1, -1, 2)$; $\mathbf{a} = \mathbf{i} + \mathbf{j} + \mathbf{k}$

9. $f(x, y, z) = x^3y^2z$; $P = (2, -1, 2)$; $\mathbf{a} = 2\mathbf{i} - \mathbf{j} - 2\mathbf{k}$

10. $f(x, y, z) = xy - yz + 3xz$; $P = (1, -1, 3)$;

$\mathbf{a} = -\mathbf{i} + 3\mathbf{j} + 2\mathbf{k}$

11. $f(x, y, z) = \dfrac{x - y - z}{x + y + z}$; $P = (2, 1, -1)$; $\mathbf{a} = -2\mathbf{i} - \mathbf{j} - \mathbf{k}$

12. $f(x, y, z) = e^{x^2 + y^2 + z^2}$; $P = (0, 0, 0)$; $\mathbf{a} = -\mathbf{i} + \mathbf{j} - \mathbf{k}$

13. $f(x, y, z) = yz2^x$; $P = (1, -1, 1)$; $\mathbf{a} = 2\mathbf{j} - \mathbf{k}$

14. $f(x, y, z) = y\sin^{-1}xz$; $P = (1/\sqrt{2}, 0, 1/\sqrt{2})$;

$\mathbf{a} = -2\mathbf{i} - 2\mathbf{j} - 2\mathbf{k}$

15. Let $f(x, y) = \sinh(y \ln x)$. Find the directional derivative of f at $(2, -1)$ in the direction away from the origin.

16. Let $f(x, y) = 2x + x^2y + y \sin y$ and let $\mathbf{u} = a\mathbf{i} + b\mathbf{j}$ be a unit vector.

a. Express $D_{\mathbf{u}}f(1, 0)$ in terms of a and b.

b. Using the result of part (a), find the values of a and b for which $D_{\mathbf{u}}f(1, 0)$ is maximum.

17. A particle travels along the parabolic path $y = x^2$ from left to right, where x and y are in meters. The temperature at any point (x, y) in the plane is given in degrees Celsius by

$$T(x, y) = \frac{40}{1 + x^2 + y^2}$$

a. What is the rate of change of the temperature experienced by the particle as it travels along the parabola at the moment the particle passes through the point $(-2, 4)$? (*Hint:* Find an appropriate directional derivative.)

b. Suppose the particle travels at a speed of 5 meters per second. How fast is the temperature changing with respect to time as the particle passes through the point $(-2, 4)$?

13.6 THE GRADIENT

By means of the first partials of a function we can define a vector called the gradient. This vector is important in the analysis of directional derivatives, plays a crucial role in the definition of the plane tangent to the graph of a function of two variables, and has special significance in physical applications.

DEFINITION 13.14

a. Let f be a function of two variables that has partial derivatives at (x_0, y_0). Then the **gradient** of f at (x_0, y_0), which is denoted $\operatorname{grad} f(x_0, y_0)$ or $\nabla f(x_0, y_0)$, is defined by

$$\operatorname{grad} f(x_0, y_0) = \nabla f(x_0, y_0) = f_x(x_0, y_0)\mathbf{i} + f_y(x_0, y_0)\mathbf{j}$$

b. Let f be a function of three variables that has partial derivatives at (x_0, y_0, z_0). Then the **gradient** of f at (x_0, y_0, z_0), which is denoted $\operatorname{grad} f(x_0, y_0, z_0)$ or $\nabla f(x_0, y_0, z_0)$, is defined by

$$\operatorname{grad} f(x_0, y_0, z_0) = \nabla f(x_0, y_0, z_0)$$
$$= f_x(x_0, y_0, z_0)\mathbf{i} + f_y(x_0, y_0, z_0)\mathbf{j} + f_z(x_0, y_0, z_0)\mathbf{k}$$

The symbol ∇, which is read "del," is an inverted capital Greek delta. Whereas the partial derivative is a number (i.e., scalar), the gradient is a vector. Thus we now have both scalar and vector derivatives of a function of several variables.

EXAMPLE 1 Let $f(x, y) = \sin xy$. Find a formula for the gradient of f, and in particular find $\operatorname{grad} f(\pi/3, 1)$.

Solution Definition 13.14 tells us that

$$\operatorname{grad} f(x, y) = f_x(x, y)\mathbf{i} + f_y(x, y)\mathbf{j} = y \cos xy\,\mathbf{i} + x \cos xy\,\mathbf{j}$$

Consequently

$$\operatorname{grad} f\left(\frac{\pi}{3}, 1\right) = \cos\frac{\pi}{3}\mathbf{i} + \frac{\pi}{3}\cos\frac{\pi}{3}\mathbf{j} = \frac{1}{2}\mathbf{i} + \frac{\pi}{6}\mathbf{j} \qquad \square$$

EXAMPLE 2 Let $f(x, y, z) = 1/\sqrt{x^2 + y^2 + z^2}$. Find a formula for the gradient of f, and in particular find $\nabla f(2\sqrt{2}, 2\sqrt{2}, -3)$.

Solution By definition,

$$\nabla f(x, y, z) = f_x(x, y, z)\mathbf{i} + f_y(x, y, z)\mathbf{j} + f_z(x, y, z)\mathbf{k}$$

$$= -\frac{x}{(x^2 + y^2 + z^2)^{3/2}}\mathbf{i} - \frac{y}{(x^2 + y^2 + z^2)^{3/2}}\mathbf{j} - \frac{z}{(x^2 + y^2 + z^2)^{3/2}}\mathbf{k}$$

$$= \frac{-1}{(x^2 + y^2 + z^2)^{3/2}}(x\mathbf{i} + y\mathbf{j} + z\mathbf{k})$$

Since $(2\sqrt{2})^2 + (2\sqrt{2})^2 + (-3)^2 = 25$, we find that

$$\nabla f(2\sqrt{2}, 2\sqrt{2}, -3) = \frac{1}{125}(-2\sqrt{2}\mathbf{i} - 2\sqrt{2}\mathbf{j} + 3\mathbf{k}) \qquad \square$$

If f is a function of two variables that is differentiable at (x_0, y_0), and if $\mathbf{u} = a_1\mathbf{i} + a_2\mathbf{j}$ is a unit vector in the xy plane, then

$$\text{grad } f(x_0, y_0) \cdot \mathbf{u} = \text{grad } f(x_0, y_0) \cdot (a_1\mathbf{i} + a_2\mathbf{j}) = f_x(x_0, y_0)a_1 + f_y(x_0, y_0)a_2$$

Thus by (1) of Section 13.5,

$$D_{\mathbf{u}}f(x_0, y_0) = [\text{grad } f(x_0, y_0)] \cdot \mathbf{u} \qquad\qquad (1)$$

This formula shows the relationship of the directional derivative to the gradient and the direction $\mathbf{u}$. If grad $f(x_0, y_0) = \mathbf{0}$, then (1) implies that $D_{\mathbf{u}}f(x_0, y_0) = 0$ for every choice of $\mathbf{u}$. If grad $f(x_0, y_0) \neq \mathbf{0}$, then we can determine the largest value of $D_{\mathbf{u}}f(x_0, y_0)$ as a function of $\mathbf{u}$; moreover, we can determine the direction $\mathbf{u}$ that produces this value of $D_{\mathbf{u}}f(x_0, y_0)$. Let $\mathbf{u}$ be any unit vector in the plane, and let ϕ be the angle between $\mathbf{u}$ and grad $f(x_0, y_0)$. From (1) we deduce that

$$D_{\mathbf{u}}f(x_0, y_0) = [\text{grad } f(x_0, y_0)] \cdot \mathbf{u} = \| \mathbf{u} \| \, \|\text{grad } f(x_0, y_0) \| \cos \phi$$
$$= \| \text{grad } f(x_0, y_0) \| \cos \phi$$

From this we see that the largest value of $D_{\mathbf{u}}f(x_0, y_0)$ is $\| \text{grad } f(x_0, y_0) \|$, and this value is assumed when $\phi = 0$, that is, when $\mathbf{u}$ points in the same direction as grad $f(x_0, y_0)$.

We summarize the results of the preceding paragraph as a theorem:

THEOREM 13.15   Let f be a function of two variables that is differentiable at (x_0, y_0).

a.   For any unit vector $\mathbf{u}$,

$$D_{\mathbf{u}}f(x_0, y_0) = [\text{grad } f(x_0, y_0)] \cdot \mathbf{u}$$

b.   The maximum value of $D_{\mathbf{u}}f(x_0, y_0)$ is $\| \text{grad } f(x_0, y_0) \|$.

c.   If grad $f(x_0, y_0) \neq \mathbf{0}$, then $D_{\mathbf{u}}f(x_0, y_0)$, regarded as a function of $\mathbf{u}$, attains its maximum value when $\mathbf{u}$ points in the same direction as grad $f(x_0, y_0)$.

Theorem 13.15(c) could also be interpreted as saying that at (x_0, y_0), f **increases most rapidly** in the direction of grad $f(x_0, y_0)$. Pictorially, Theorem 13.15(c) says that at $(x_0, y_0, f(x_0, y_0))$ the graph of f has the steepest incline in the direction of the gradient (Figure 13.37); thus if the graph represents the surface of a mountain, then at the point $(x_0, y_0, f(x_0, y_0))$ on the side of the mountain, the incline would be steepest in the direction of the gradient. A skier would descend fastest by skiing in the opposite direction.

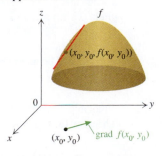

FIGURE 13.37   At (x_0, y_0), f increases most rapidly in the direction of grad $f(x_0, y_0)$.

(R. Dahlquist/Superstock)

EXAMPLE 3   Let $f(x, y) = 6 - 3x^2 - y^2$. Determine the direction in which f increases most rapidly at $(1, 2)$, and find the maximal directional derivative at $(1, 2)$.

Solution   We have

$$f_x(x, y) = -6x \quad \text{and} \quad f_y(x, y) = -2y$$

Therefore

$$\operatorname{grad} f(1, 2) = f_x(1, 2)\mathbf{i} + f_y(1, 2)\mathbf{j} = -6\mathbf{i} - 4\mathbf{j}$$

so that f increases most rapidly at $(1, 2)$ in the direction of the gradient $-6\mathbf{i} - 4\mathbf{j}$. Moreover, the maximal directional derivative at $(1, 2)$ is $\|-6\mathbf{i} - 4\mathbf{j}\| = \sqrt{36 + 16} = \sqrt{52}$, which means that the tangent plane is very steep. ❏

There is a theorem for functions of three variables analogous to Theorem 13.15. Since the proofs of both theorems are so similar, we state the three-variable version without proof.

THEOREM 13.16

Let f be a function of three variables that is differentiable at (x_0, y_0, z_0).

a. For any unit vector $\mathbf{u}$ in space,

$$D_{\mathbf{u}}f(x_0, y_0, z_0) = [\operatorname{grad} f(x_0, y_0, z_0)] \cdot \mathbf{u}$$

b. The largest value of $D_{\mathbf{u}}f(x_0, y_0, z_0)$ is $\| \operatorname{grad} f(x_0, y_0, z_0) \|$.

c. If $\operatorname{grad} f(x_0, y_0, z_0) \neq \mathbf{0}$, then $D_{\mathbf{u}}f(x_0, y_0, z_0)$, regarded as a function of $\mathbf{u}$, attains its maximum value when $\mathbf{u}$ points in the same direction as $\operatorname{grad} f(x_0, y_0, z_0)$.

The Gradient as a Normal Vector

Let C be a level curve $f(x, y) = c$ of a function f that is differentiable at a point (x_0, y_0) on C. As a point (x, y) moves along C, the value $f(x, y)$ is constant by hypothesis and hence does not change. This suggests that at (x_0, y_0) the rate of change of f in the direction of a unit vector $\mathbf{u}$ tangent to C is 0, that is, $D_{\mathbf{u}}f(x_0, y_0) = 0$ (Figure 13.38). However, (1) implies that $D_{\mathbf{u}}f(x_0, y_0) = 0$ for any unit vector $\mathbf{u}$ perpendicular to $\operatorname{grad} f(x_0, y_0)$. Combining these results, we can reasonably conjecture that $\operatorname{grad} f(x_0, y_0)$ is perpendicular, or equivalently, normal to C at (x_0, y_0) (Figure 13.38).

THEOREM 13.17

Let C be a level curve $f(x, y) = c$ of a function f. Let (x_0, y_0) be a point on C, and assume that f is differentiable at (x_0, y_0). If C is smooth and $\operatorname{grad} f(x_0, y_0) \neq \mathbf{0}$, then $\operatorname{grad} f(x_0, y_0)$ is normal to C at (x_0, y_0).

Proof   Let I be an interval, and

$$\mathbf{r}(t) = x(t)\mathbf{i} + y(t)\mathbf{j} \quad \text{for } t \text{ in } I$$

a smooth parametrization of C. Since $f(x, y) = f(x(t), y(t))$ for t in I, the Chain Rule (Theorem 13.9) implies that

$$\frac{d}{dt} f(x(t), y(t)) = \frac{\partial f}{\partial x}\frac{dx}{dt} + \frac{\partial f}{\partial y}\frac{dy}{dt} = [\operatorname{grad} f(x(t), y(t))] \cdot \frac{d\mathbf{r}}{dt}$$

FIGURE 13.38

But $f(x(t), y(t)) = c$ for t in I, so that

$$\frac{d}{dt} f(x(t), y(t)) = 0$$

and consequently

$$0 = [\text{grad } f(x(t), y(t))] \cdot \frac{d\mathbf{r}}{dt}$$

Since $\mathbf{r}$ is smooth by assumption, it follows that $d\mathbf{r}/dt$, which is tangent to C, is nonzero. Therefore grad $f(x_0, y_0)$, which is also nonzero by assumption, is perpendicular to the vector tangent to C at (x_0, y_0). Therefore grad $f(x_0, y_0)$ is normal to C at (x_0, y_0). ■

EXAMPLE 4   Assuming that the curve $x^2 - xy + 3y^2 = 5$ is smooth, find a unit vector that is perpendicular to the curve at $(1, -1)$.

Solution   Let $f(x, y) = x^2 - xy + 3y^2$, so that the given curve is the level curve $f(x, y) = 5$. Since $f(1, -1) = 5$, the point $(1, -1)$ lies on the given level curve. By Theorem 13.17, grad $f(1, -1)$ is perpendicular to the given curve at $(1, -1)$. We find that

$$\text{grad } f(x, y) = (2x - y)\mathbf{i} + (-x + 6y)\mathbf{j}$$

and thus

$$\text{grad } f(1, -1) = 3\mathbf{i} - 7\mathbf{j}$$

Therefore the unit vector

$$\frac{1}{\|3\mathbf{i} - 7\mathbf{j}\|} (3\mathbf{i} - 7\mathbf{j}) = \frac{1}{\sqrt{58}} (3\mathbf{i} - 7\mathbf{j})$$

is the desired vector perpendicular to the given curve. ❑

Now let f be a function of three variables that is differentiable at a point (x_0, y_0, z_0), and let $f(x_0, y_0, z_0) = c$. If C is any smooth curve on the level surface $f(x, y, z) = c$ and if C passes through (x_0, y_0, z_0), then an argument similar to the one used in proving Theorem 13.17 shows that if grad $f(x_0, y_0, z_0) \neq \mathbf{0}$, then grad $f(x_0, y_0, z_0)$ is perpendicular to the tangent vector of C at (x_0, y_0, z_0) (Figure 13.39). Because grad $f(x_0, y_0, z_0)$ is perpendicular to the tangent vector to any such curve C at (x_0, y_0, z_0), all such tangents lie in a single plane called a tangent plane.

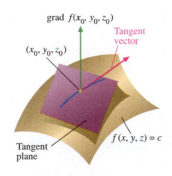

grad $f(x_0, y_0, z_0)$

Tangent vector

(x_0, y_0, z_0)

Tangent plane

$f(x, y, z) = c$

FIGURE 13.39

DEFINITION 13.18   Let f be differentiable at a point (x_0, y_0, z_0) on a level surface S of f. If grad $f(x_0, y_0, z_0) \neq \mathbf{0}$, then the plane through (x_0, y_0, z_0) whose normal is grad $f(x_0, y_0, z_0)$ is the plane **tangent** to S at (x_0, y_0, z_0). Any vector that is perpendicular to this tangent plane is said to be **normal** to S.

By Definition 13.18, grad $f(x_0, y_0, z_0)$ is normal to S at (x_0, y_0, z_0). Since

$$\text{grad } f(x_0, y_0, z_0) = f_x(x_0, y_0, z_0)\mathbf{i} + f_y(x_0, y_0, z_0)\mathbf{j} + f_z(x_0, y_0, z_0)\mathbf{k}$$

and since grad $f(x_0, y_0, z_0)$ is normal to the tangent plane at (x_0, y_0, z_0), an equation of the tangent plane is

$$f_x(x_0, y_0, z_0)(x - x_0) + f_y(x_0, y_0, z_0)(y - y_0) + f_z(x_0, y_0, z_0)(z - z_0) = 0$$

EXAMPLE 5 Find an equation of the plane tangent to the sphere $x^2 + y^2 + z^2 = 4$ at the point $(-1, 1, \sqrt{2})$.

Solution The sphere is the level surface $f(x, y, z) = 4$, where $f(x, y, z) = x^2 + y^2 + z^2$. The partials of f are given by

$$f_x(x, y, z) = 2x, \quad f_y(x, y, z) = 2y, \quad f_z(x, y, z) = 2z$$

so that $f_x(-1, 1, \sqrt{2}) = -2, \quad f_y(-1, 1, \sqrt{2}) = 2, \quad f_z(-1, 1, \sqrt{2}) = 2\sqrt{2}$

Therefore an equation of the plane tangent at $(-1, 1, \sqrt{2})$ is

$$-2(x - (-1)) + 2(y - 1) + 2\sqrt{2}(z - \sqrt{2}) = 0$$

or equivalently,

$$-x + y + \sqrt{2}z = 4 \qquad \square$$

Now suppose that f is a function of two variables that is differentiable at (x_0, y_0). In order to obtain an equation of the plane tangent to the graph of f at (x_0, y_0), we can think of the graph of f as the level surface

$$g(x, y, z) = 0, \quad \text{where } g(x, y, z) = f(x, y) - z$$

Notice that

$$\operatorname{grad} g(x_0, y_0, f(x_0, y_0)) = f_x(x_0, y_0)\mathbf{i} + f_y(x_0, y_0)\mathbf{j} - \mathbf{k} \tag{2}$$

and $\operatorname{grad} g(x_0, y_0, f(x_0, y_0))$ is normal to the plane tangent at $(x_0, y_0, f(x_0, y_0))$. This means that

$$f_x(x_0, y_0)(x - x_0) + f_y(x_0, y_0)(y - y_0) - (z - f(x_0, y_0)) = 0 \tag{3}$$

or equivalently,

$$z = f(x_0, y_0) + f_x(x_0, y_0)(x - x_0) + f_y(x_0, y_0)(y - y_0) \tag{4}$$

is an equation of the tangent plane of f at $(x_0, y_0, f(x_0, y_0))$. The vector given in (2) is said to be **normal** to the graph of f at $(x_0, y_0, f(x_0, y_0))$.

Let us observe that the formula for the normal in (2) is consistent with the results of Section 13.3 involving lines tangent to the graph of f. Specifically, recall from (1) in Section 13.3 that the vector $\mathbf{v} = \mathbf{j} + f_y(x_0, y_0)\mathbf{k}$ is tangent to the curve formed by the intersection of the graph of f and the plane $x = x_0$, and that $\mathbf{w} = \mathbf{i} + f_x(x_0, y_0)\mathbf{k}$ is tangent to the intersection of the graph of f and the plane $y = y_0$ (see Figure 13.40). Therefore $\mathbf{v}$ and $\mathbf{w}$ are parallel to the tangent plane of f at $(x_0, y_0, f(x_0, y_0))$, so that

$$\mathbf{v} \times \mathbf{w} = \begin{vmatrix} \mathbf{i} & \mathbf{j} & \mathbf{k} \\ 0 & 1 & f_y(x_0, y_0) \\ 1 & 0 & f_x(x_0, y_0) \end{vmatrix} = f_x(x_0, y_0)\mathbf{i} + f_y(x_0, y_0)\mathbf{j} - \mathbf{k}$$

is perpendicular (and hence normal) to the tangent plane of f at $(x_0, y_0, f(x_0, y_0))$—which agrees with (2).

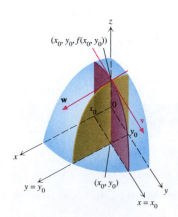

FIGURE 13.40

EXAMPLE 6   Let $f(x, y) = 6 - 3x^2 - y^2$. Find a vector normal to the graph of f at $(1, 2, -1)$, and find an equation of the plane tangent to the graph of f at $(1, 2, -1)$.

Solution   First we calculate that

$$f_x(x, y) = -6x \quad \text{and} \quad f_y(x, y) = -2y$$

so that

$$f_x(1, 2) = -6 \quad \text{and} \quad f_y(1, 2) = -4$$

Therefore by (2) a vector normal to the graph of f at $(1, 2, -1)$ is given by

$$f_x(1, 2)\mathbf{i} + f_y(1, 2)\mathbf{j} - \mathbf{k} = -6\mathbf{i} - 4\mathbf{j} - \mathbf{k}$$

Then (3) implies that an equation of the tangent plane of f at $(1, 2, -1)$ is

$$-6(x - 1) - 4(y - 2) - (z + 1) = 0$$

or

$$6x + 4y + z = 13 \quad \square$$

The gradient arises in physical situations. Let us recall from Section 13.1 that if $T(x, y, z)$ is the temperature at any point (x, y, z), then the level surfaces of T are isothermal surfaces. On an isothermal surface the temperature is constant, and no heat flows along such a surface. Instead, heat flows in a direction perpendicular to an isothermal surface; more precisely, it flows in the direction of the gradient. Similarly, if $V(x, y, z)$ represents the voltage at the point (x, y, z), then $-\text{grad } V(x, y, z)$ turns out to be the electric force that would be exerted on a positive unit charge at (x, y, z). This force is perpendicular to the equipotential surface at (x, y, z).

EXERCISES 13.6

In Exercises 1–8 find the gradient of the function.

1. $f(x, y) = 3x - 5y$ **2.** $f(x, y) = y^2 + x \sin x^2 y$

3. $g(x, y) = e^{-2x} \ln (y - 4)$ **4.** $f(x, y) = \dfrac{xy - 1}{x^2 + y^2}$

5. $f(x, y, z) = 2x^2 - y^2 - 4z^2$

6. $f(x, y, z) = (2x + y^2 + z^3)^{5/2}$

7. $g(x, y, z) = \dfrac{-x + y}{-x + z}$ **8.** $g(x, y, z) = -x^2 y^3 e^{(z^2)}$

In Exercises 9–14 find the gradient of the function at the given point.

9. $f(x, y) = \dfrac{x + 3y}{5x + 2y}; \ (-1, \frac{3}{2})$

10. $f(x, y) = x \cos xy; \ (1, -\pi)$

11. $g(x, y) = x \ln (x + y); \ (-2, 3)$

12. $f(x, y, z) = z - \sqrt{x^2 + y^2}; \ (3, -4, 7)$

13. $f(x, y, z) = ze^{-x} \tan y; \ (0, \pi, -2)$

14. $g(x, y, z) = e^x(\sin y + \sin z); \ (1, \pi/2, \pi/2)$

In Exercises 15–20 find the direction in which f increases most rapidly at the given point, and find the maximal directional derivative at that point.

15. $f(x, y) = e^x(\cos y + \sin y); \ (0, 0)$

16. $f(x, y) = e^{2x}(\cos y - \sin y); \ (\frac{1}{6}, -\pi/2)$

17. $f(x, y) = 3x^2 + 4y^2; \ (-1, 1)$

18. $f(x, y, z) = \ln (x^2 + y^2 + z^2); \ (2, 0, 1)$

19. $f(x, y, z) = e^x + e^y + e^{2z}; \ (1, 1, -1)$

20. $f(x, y, z) = \cos xyz; \ (\frac{1}{3}, \frac{1}{2}, \pi)$

From (1) it follows that the directional derivative of a function f at a point is smallest in the direction opposite to the gradient of f at that point. Thus we say that a function **decreases most rapidly** in the direction opposite the gradient. In Exercises 21–23 find the direction in which the function decreases most rapidly at the given point.

21. $f(x, y) = \sin \pi xy; \ (\frac{1}{2}, \frac{2}{3})$

22. $f(x, y) = \tan^{-1} (x - y); \ (2, -2)$

23. $f(x, y, z) = \dfrac{x - z}{y + z}; \ (-1, 1, 3)$

In Exercises 24–26 find a vector that is normal to the graph of the equation at the given point. Assume that each curve is smooth.

24. $x^3 - 3x^2y + y^2 = 5$; $(1, -1)$

25. $\sin \pi xy = \sqrt{3}/2$; $(\frac{1}{6}, 2)$ **26.** $e^{x^2y} = 2$; $(1, \ln 2)$

In Exercises 27–30 find a vector that is normal to the graph of f at the given point.

27. $f(x, y) = 3x^2 + 4y^2$; $(-2, 1, 16)$

28. $f(x, y) = \sqrt{4 - x^2 - y^2}$; $(-1, 1, \sqrt{2})$

29. $f(x, y) = 1 - x^2$; $(0, 2, 1)$ **30.** $f(x, y) = y^2 e^x$; $(0, -3, 9)$

In Exercises 31–38 find an equation of the plane tangent to the graph of the given function at the indicated point(s).

31. $f(x, y) = xy - x + y + 5$; $(0, 2, 7)$

32. $f(x, y) = \dfrac{x + 2}{y + 1}$; $(2, 3, 1)$

33. $g(x, y) = \sin \pi xy$; $(-\sqrt{2}, \sqrt{2}, 0)$ and $(-\frac{1}{2}, \frac{1}{3}, -\frac{1}{2})$

34. $f(x, y) = e^{x + y^2}$; $(-1, 0, e^{-1})$ and $(0, 1, e)$

35. $f(x, y) = (2 + x - y)^2$; $(3, -1, 36)$

36. $g(x, y) = 4x^2 + y^2 - 1$; $(2, 1, 16)$

37. $f(x, y) = \ln(x^2 + y^2)$; $(-1, 0, 0)$ and $(-1, 1, \ln 2)$

38. $f(x, y) = \begin{cases} \dfrac{x^3 - y^3}{x^2 + y^2} & \text{for } (x, y) \neq (0, 0) \\ 0 & \text{for } (x, y) = (0, 0) \end{cases}$; $(1, 0, 1)$

In Exercises 39–45 find an equation of the plane tangent to the given surface at the given point.

39. $x^2 + y^2 + z^2 = 1$; $(\frac{1}{2}, -\frac{1}{2}, -1/\sqrt{2})$

40. $\dfrac{x^2}{4} + \dfrac{y^2}{9} + \dfrac{z^2}{16} = 1$; $(0, 0, -4)$

41. $xyz = 1$; $(\frac{1}{2}, -2, -1)$ **42.** $ye^{xy} + z^2 = 0$; $(0, -1, 1)$

43. $\sin(xy) = 2 - z^2$; $(\pi, \frac{1}{2}, -1)$

44. $x^2 = y^2$; $(1, -1, 10)$ **45.** $z = \ln \sqrt{x^2 + 1}$; $(0, 2, 0)$

46. Find the point on the hyperbolic paraboloid $z = x^2 - 3y^2$ at which the tangent plane is parallel to the plane $8x + 3y - z = 4$.

47. Find the point on the paraboloid $z = 9 - 4x^2 - y^2$ at which the tangent plane is parallel to the plane $z = 4y$.

48. Show that the surfaces $z = \sqrt{x^2 + y^2}$ and $10z = 25 + x^2 + y^2$ have the same tangent plane at $(3, 4, 5)$.

49. Show that the surfaces $z = xy - 2$ and $x^2 + y^2 + z^2 = 3$ have the same tangent plane at $(1, 1, -1)$.

We say that two surfaces are **normal** at a given point if their tangent planes at that point are perpendicular to one another. In Exercises 50–51 show that the pair of surfaces are normal at the given point.

50. $x^2 + y^2 + z^2 = 16$ and $z^2 = x^2 + y^2$; $(2, 2, 2\sqrt{2})$

51. $z = x^2 + 4y^2 - 12$ and $8z = 4x + y^2 + 19$; $(-3, -1, 1)$

*52. Show that the line determined by the intersection of the plane $z = 0$ and the plane tangent to the surface $z^2(x^2 + y^2) = 4$ at a point of the form $(2 \cos \theta, 2 \sin \theta, 1)$ is tangent to the circle $x^2 + y^2 = 16$ at the point $(4 \cos \theta, 4 \sin \theta)$.

53. Show that an equation of the plane tangent to the ellipsoid

$$\frac{x^2}{a^2} + \frac{y^2}{b^2} + \frac{z^2}{c^2} = 1$$

at the point (x_0, y_0, z_0) is

$$\frac{xx_0}{a^2} + \frac{yy_0}{b^2} + \frac{zz_0}{c^2} = 1$$

54. Let $c \neq 0$. Show that an equation of the plane tangent to the paraboloid

$$cz = \frac{x^2}{a^2} + \frac{y^2}{b^2}$$

at the point (x_0, y_0, z_0) is

$$c(z + z_0) = \frac{2xx_0}{a^2} + \frac{2yy_0}{b^2}$$

55. Let g be a differentiable function of one variable and let $f(x, y) = xg(y/x)$. Show that every plane tangent to the graph of f passes through the origin.

56. Show that every line normal to the sphere $x^2 + y^2 + z^2 = 1$ passes through the origin.

57. Show that every line normal to the double cone $z^2 = x^2 + y^2$ intersects the z axis.

58. Show that there is exactly one plane tangent to the paraboloid $z = x^2 + y^2$ and parallel to any given nonvertical plane.

In Exercises 59–61 use implicit differentiation to find $\partial z/\partial x$ and $\partial z/\partial y$ at the given point. Then find an equation of the plane tangent to the level surface at that point.

59. $x^2 - y^2 - z^2 = 1$; $(\sqrt{2}, 0, 1)$

60. $xyz = 1$; $(2, -3, -\frac{1}{6})$

61. $\ln x + \ln y + \ln z = 1$; $(1, 1, e)$

62. A mountain climber's oxygen mask is leaking. If the surface of the mountain is represented by $z = 5 - x^2 - 2y^2$ and the climber is at $(\frac{1}{2}, -\frac{1}{2}, \frac{17}{4})$, in what direction should the climber turn to descend most rapidly?

63. Suppose $T(x, y, z) = x^3y + 3x^2y^2z$. Find the directional derivative of T at $(1, 1, -1)$ in the direction of the gradient.

Applications

64. A metal plate with vertices $(1, 1)$, $(5, 1)$, $(1, 3)$, and $(5, 3)$ is heated by a flame at the origin, and the temperature at a point on the plate is inversely proportional to the distance from the origin. If an ant is located at the point $(3, 2)$, in what direction should the ant crawl to cool the fastest?

65. Suppose the quadric surface $x^2 - y^2 = z$ is an equipotential surface. Show that the electric force on a positive unit charge at the origin is perpendicular to the xy plane.

Project

1. This project focuses on tangent planes and normal lines of special surfaces.

a. Consider the double cone $x^2 + y^2 = z^2$.

   i. Show that every plane tangent to the double cone passes through the origin.

   ii. Show that every line normal to the double cone passes through the z axis.

b. Consider the sphere $x^2 + y^2 + z^2 = a^2$.

   i. Let P be any plane that is tangent to the sphere, but not perpendicular to any of the coordinate axes. If the point of tangency is (x_0, y_0, z_0), show that the x, y, and z intercepts of P are a^2/x_0, a^2/y_0, and a^2/z_0, respectively.

   ii. Show that every line normal to the sphere passes through the origin. Does that happen if the sphere is replaced by a non-spherical ellipsoid? Explain why or why not.

c. Consider the three-dimensional version of the astroid:

$$x^{2/3} + y^{2/3} + z^{2/3} = a^{2/3}$$

   i. Use a computer to sketch the graph of the equation.

   ii. Let P be any plane that is tangent to the surface and is not parallel to any of the coordinate planes. Show that the sum of the squares of the intercepts of the plane is a^2.

13.7 TANGENT PLANE APPROXIMATIONS AND DIFFERENTIALS

Suppose f is differentiable at (x_0, y_0). Then by Definition 13.8,

$$f(x, y) - f(x_0, y_0) = f_x(x_0, y_0)(x - x_0) + f_y(x_0, y_0)(y - y_0)$$
$$+ \varepsilon_1(x, y)(x - x_0) + \varepsilon_2(x, y)(y - y_0) \qquad (1)$$

where
$$\lim_{(x, y) \to (x_0, y_0)} \varepsilon_1(x, y) = \lim_{(x, y) \to (x_0, y_0)} \varepsilon_2(x, y) = 0 \qquad (2)$$

Since the limits in (2) are 0, the two numbers $f(x, y) - f(x_0, y_0)$ and $f_x(x_0, y_0)(x - x_0) + f_y(x_0, y_0)(y - y_0)$ are approximately equal when (x, y) is close to (x_0, y_0):

$$f(x, y) - f(x_0, y_0) \approx f_x(x_0, y_0)(x - x_0) + f_y(x_0, y_0)(y - y_0) \qquad (3)$$

or equivalently,

$$f(x, y) \approx f(x_0, y_0) + f_x(x_0, y_0)(x - x_0) + f_y(x_0, y_0)(y - y_0) \qquad (4)$$

Now recall from (4) of Section 13.6 that any point (x, y, z) on the plane tangent to the graph of f at $(x_0, y_0, f(x_0, y_0))$ satisfies

$$z = f(x_0, y_0) + f_x(x_0, y_0)(x - x_0) + f_y(x_0, y_0)(y - y_0) \qquad (5)$$

Observe that the right sides of (4) and (5) are identical. Therefore using the right side of (4) to approximate $f(x, y)$ amounts to using the point (x, y, z) on the tangent plane to approximate the point $(x, y, f(x, y))$ on the graph of f (Figure 13.41). For this reason the approximation of $f(x, y)$ by (4) is called a **tangent plane approximation** and is reminiscent of the tangent line (or linear) approximation introduced in Section 3.8.

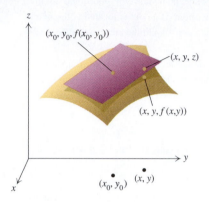

FIGURE 13.41

In order to emphasize that we will consider only points (x, y) that are close to (x_0, y_0), we replace x by $x_0 + h$ and y by $y_0 + k$, where h and k are assumed to be small. Then (4) becomes

$$f(x_0 + h, y_0 + k) \approx f(x_0, y_0) + f_x(x_0, y_0)h + f_y(x_0, y_0)k \tag{6}$$

Although we will not discuss estimates of the error introduced by using the right side of (6) to approximate $f(x_0 + h, y_0 + k)$, the following example shows how the approximation is carried out.

EXAMPLE 1 Approximate $\sqrt{(3.012)^2 + (3.997)^2}$.

Solution Since 3.012 is close to 3 and 3.997 is close to 4, and since $\sqrt{3^2 + 4^2} = 5$, we let

$$f(x, y) = \sqrt{x^2 + y^2}$$

and take $x_0 = 3$, $y_0 = 4$, $h = 0.012$, and $k = -0.003$. Next we find that

$$f_x(x, y) = \frac{x}{\sqrt{x^2 + y^2}} \quad \text{and} \quad f_y(x, y) = \frac{y}{\sqrt{x^2 + y^2}}$$

so that $f_x(3, 4) = \frac{3}{5}$ and $f_y(3, 4) = \frac{4}{5}$. Then by (6) we have

$$\sqrt{(3.012)^2 + (3.997)^2} = f(3.012, 3.997)$$

$$\approx f(3, 4) + f_x(3, 4)(0.012) + f_y(3, 4)(-0.003)$$

$$= 5 + \frac{3}{5}(0.012) + \frac{4}{5}(-0.003) = 5.0048 \qquad \square$$

The actual value of the given square root is 5.00481, accurate to 6 places. The accuracy of the estimate is excellent, and finding the estimate involved little computation.

There is a similar approximation for a function f of three variables that is differentiable at a point (x_0, y_0, z_0). By Definition 13.11,

$$f(x, y, z) - f(x_0, y_0, z_0) = f_x(x_0, y_0, z_0)(x - x_0) + f_y(x_0, y_0, z_0)(y - y_0)$$
$$+ f_z(x_0, y_0, z_0)(z - z_0) + \varepsilon_1(x, y, z)(x - x_0)$$
$$+ \varepsilon_2(x, y, z)(y - y_0) + \varepsilon_3(x, y, z)(z - z_0)$$

where

$$\lim_{(x, y, z) \to (x_0, y_0, z_0)} \varepsilon_1(x, y, z) = \lim_{(x, y, z) \to (x_0, y_0, z_0)} \varepsilon_2(x, y, z)$$

$$= \lim_{(x, y, z) \to (x_0, y_0, z_0)} \varepsilon_3(x, y, z) = 0$$

This leads to the approximation

$$f(x_0 + h, y_0 + k, z_0 + l) \approx f(x_0, y_0, z_0) + f_x(x_0, y_0, z_0)h$$
$$+ f_y(x_0, y_0, z_0)k + f_z(x_0, y_0, z_0)l \qquad (7)$$

where h, k and l are assumed to be small.

EXAMPLE 2 Suppose a cardboard box in the shape of a rectangular parallelepiped has outer dimensions 14, 14, and 28 inches. If the cardboard is $\frac{1}{8}$ inch thick, approximate the volume of cardboard.

Solution Let $V(x, y, z) = xyz$. Then $V(14, 14, 28)$ is the volume of a box with outer dimensions 14, 14, and 28. If we let $x_0 = y_0 = 14$, $z_0 = 28$, and $h = k = l = -\frac{1}{4}$, then we seek

$$V(14, 14, 28) - V\left(14 - \frac{1}{4}, 14 - \frac{1}{4}, 28 - \frac{1}{4}\right)$$

But by (7) we find that

$$V(14, 14, 28) - V\left(14 - \frac{1}{4}, 14 - \frac{1}{4}, 28 - \frac{1}{4}\right)$$

$$\approx -V_x(14, 14, 28)h - V_y(14, 14, 28)k - V_z(14, 14, 28)l$$

$$= -14 \cdot 28\left(-\frac{1}{4}\right) - 14 \cdot 28\left(-\frac{1}{4}\right) - 14 \cdot 14\left(-\frac{1}{4}\right) = 245$$

Thus the volume of the cardboard is approximately 245 cubic inches. ❑

The actual volume of cardboard is 241.515625 cubic inches, so the estimate is in error by less than 4 cubic inches.

Differentials

If f is a function of two variables, we can replace (x_0, y_0) by any point (x, y) in the domain of f at which f is differentiable and transform (6) into

$$f(x + h, y + k) - f(x, y) \approx f_x(x, y)h + f_y(x, y)k \tag{8}$$

The number $f_x(x, y)h + f_y(x, y)k$ on the right side of (8) is usually called the **differential** (or **total differential**) of f (at (x, y) with increments h and k) and is denoted df. Thus

$$df = f_x(x, y)h + f_y(x, y)k \tag{9}$$

Of course, df depends on $x, y, h,$ and k, even though they are not indicated in the notation df.

If $g_1(x, y) = x$ and $g_2(x, y) = y$, then the differential dg_1 is denoted dx, and the differential dg_2 is denoted dy. Since

$$(g_1)_x(x, y) = 1, \quad (g_1)_y(x, y) = 0, \quad (g_2)_x(x, y) = 0, \quad (g_2)_y(x, y) = 1$$

we have

$$dx = dg_1 = 1 \cdot h + 0 \cdot k = h \quad \text{and} \quad dy = dg_2 = 0 \cdot h + 1 \cdot k = k$$

Therefore we can rewrite (9) as

$$df = f_x(x, y)\,dx + f_y(x, y)\,dy \quad \text{or} \quad df = \frac{\partial f}{\partial x}dx + \frac{\partial f}{\partial y}dy \tag{10}$$

EXAMPLE 3   Let $f(x, y) = xy^2 + y \sin x$. Find df.

Solution   Since

$$\frac{\partial f}{\partial x} = y^2 + y \cos x \quad \text{and} \quad \frac{\partial f}{\partial y} = 2xy + \sin x$$

we deduce from (10) that

$$df = (y^2 + y \cos x)\,dx + (2xy + \sin x)\,dy \qquad \square$$

If f is a function of three variables that is differentiable at (x_0, y_0, z_0), then the **differential** df is defined by

$$df = f_x(x, y, z)h + f_y(x, y, z)k + f_z(x, y, z)l \tag{11}$$

Other forms for df are

$$df = f_x(x, y, z)\,dx + f_y(x, y, z)\,dy + f_z(x, y, z)\,dz$$
$$df = \frac{\partial f}{\partial x}dx + \frac{\partial f}{\partial y}dy + \frac{\partial f}{\partial z}dz \tag{12}$$

EXAMPLE 4   Let $f(x, y, z) = x^2 \ln(y - z)$. Find df.

Solution   Since

$$\frac{\partial f}{\partial x} = 2x \ln(y - z), \quad \frac{\partial f}{\partial y} = \frac{x^2}{y - z}, \quad \frac{\partial f}{\partial z} = \frac{-x^2}{y - z} = \frac{x^2}{z - y}$$

(12) tells us that

$$df = 2x \ln(y - z)\,dx + \frac{x^2}{y - z}\,dy + \frac{x^2}{z - y}\,dz \qquad \square$$

EXERCISES 13.7

In Exercises 1–8 approximate the value of f at the given point.

1. $f(x, y) = \sqrt{x^2 + y^2}$; $(3.01, 4.03)$

2. $f(x, y) = \sqrt{x^2 + y}$; $(3.02, -4.98)$

3. $f(x, y) = \ln(x^2 + y^2)$; $(-0.03, 0.98)$

4. $f(x, y) = \sin \pi xy$; $(-1.97, 2.005)$

5. $f(x, y) = \tan xy$; $(0.99\pi, 0.24)$

6. $f(x, y) = \sqrt{6 - x^2 - y^2}$; $(0.987, 1.013)$

7. $f(x, y, z) = \sqrt{x^2 + y^2 + z^2}$; $(3.01, 4.02, 11.98)$

8. $f(x, y, z) = xyz^2$; $(-2.1, 1.01, 0.989)$

In Exercises 9–12 approximate the number.

9. $\sqrt[4]{(1.9)^3 + (2.1)^3}$ **10.** $(16.05)^{1/4}(7.95)^{2/3}$

11. $e^{0.1} \ln 0.9$ **12.** $\sin \dfrac{9\pi}{20} \cos \dfrac{9\pi}{30}$

In Exercises 13–22 determine df.

13. $f(x, y) = 3x^3 - x^2 y + y + 17$

14. $f(x, y) = y \ln \dfrac{1 + x}{1 - x}$

15. $f(x, y) = x^2 + y^2$

16. $f(x, y) = \cos(x + y) + \cos(x - y)$

17. $f(x, y) = x \tan y + y \cot x$

18. $f(x, y, z) = x^2 + y^2 + z^2$

19. $f(x, y, z) = z^2 \sqrt{1 + x^2 + y^2}$

20. $f(x, y, z) = \ln \sqrt{x^2 + y^2 + z^2}$

21. $f(x, y, z) = xe^{y^2 - z^2}$

22. $f(x, y, z) = \dfrac{x}{x^2 + y^2 + z^2}$

Applications

 23. When two resistors having resistances R_1 and R_2 are connected in parallel, the resistance of the combination is given by

$$R = \frac{R_1 R_2}{R_1 + R_2}$$

Suppose R_1 and R_2 are measured as 2 and 6 ohms, respectively, so that the corresponding value of R is 1.5. If the measurement error in R_1 is at most 0.01 ohms and the measurement error in R_2 is at most 0.02, estimate the maximum error in R.

24. Suppose a gas station has an underground tank in the shape of a rectangular parallelepiped that measures 10 feet by 8 feet by 6 feet, with error of at most 0.005 feet in each measurement. If the tank is filled with gasoline costing \$10 per cubic foot, estimate by how much the cost of the gasoline can vary from \$4800.

13.8   EXTREME VALUES

Like functions of one variable, functions of several variables can have maximum and minimum values on a given set. As in the one-variable case, the notion of relative extreme values will facilitate the study of extreme values of functions of several variables. The problem of recognizing and classifying relative extreme values is much more difficult for functions of several variables than for functions of one variable, so in this section we will limit our discussion to the two-variable case. Applications of extreme values of functions of several variables are many and varied, because functions in the physical world often have more than one variable. Among other applications, we will use the results of extreme values to determine the largest rectangular parallelepiped that can be mailed parcel post (see Example 6).

DEFINITION 13.19

Let f be a function of two variables, R a set contained in the domain of f, and (x_0, y_0) a point in R. Then f **has a maximum value** (respectively, a **minimum value**) **on R at (x_0, y_0)** if $f(x, y) \leq f(x_0, y_0)$ (respectively, $f(x, y) \geq f(x_0, y_0)$) for all (x, y) in R. If R is the domain of f, we say that f has a maximum value (respectively, a minimum value) at (x_0, y_0). Furthermore, f **has a relative maximum value** (respectively, a **relative minimum value**) **at (x_0, y_0)** if there is an open disk D centered at (x_0, y_0) and contained in the domain of f such that $f(x, y) \leq f(x_0, y_0)$ (respectively, $f(x, y) \geq f(x_0, y_0)$) for all (x, y) in D.

We combine maximum and minimum values under the heading **extreme values,** and relative maximum and relative minimum values under the heading **relative extreme values.** If there is a disk centered at (x_0, y_0) that is contained in the domain of f, then an extreme value of f at (x_0, y_0) is also a relative extreme value.

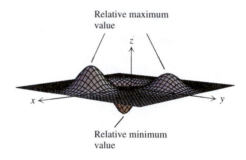

Relative maximum value

Relative minimum value

FIGURE 13.42

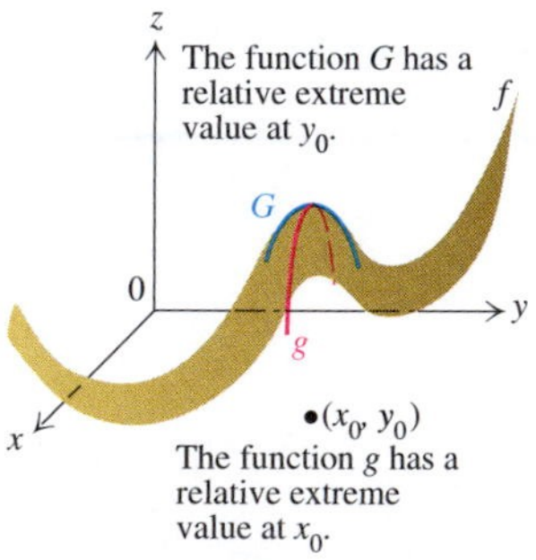

Relative extreme values correspond to the hilltops and valley bottoms on the graph of a function (Figure 13.42). Recall that a relative extreme value of a function f of a single variable occurs at a critical number—a number x_0 such that either $f'(x_0) = 0$ or $f'(x_0)$ does not exist. As a first step in identifying relative extreme values of functions of two variables, we assume that f has a relative extreme value at (x_0, y_0) and let

$$g(x) = f(x, y_0) \quad \text{and} \quad G(y) = f(x_0, y)$$

Then g has a relative extreme value at x_0, and G has a relative extreme value at y_0 (Figure 13.43). Therefore if $f_x(x_0, y_0)$ and $f_y(x_0, y_0)$ exist, then

$$f_x(x_0, y_0) = g'(x_0) = 0 \quad \text{and} \quad f_y(x_0, y_0) = G'(y_0) = 0$$

This proves the following theorem.

The function G has a relative extreme value at y_0.

G

g

$\bullet (x_0, y_0)$

The function g has a relative extreme value at x_0.

FIGURE 13.43

THEOREM 13.20

Let f have a relative extreme value at (x_0, y_0). If f has partial derivatives at (x_0, y_0), then

$$f_x(x_0, y_0) = f_y(x_0, y_0) = 0$$

or equivalently, grad $f(x_0, y_0) = \mathbf{0}$.

From Theorem 13.20 we see that relative extreme values of f occur only at those points at which the partial derivatives of f exist and are 0, or at which one or both of the partial derivatives does not exist. We say that f has a **critical point** at a point (x_0, y_0) in the domain of f if $f_x(x_0, y_0) = f_y(x_0, y_0) = 0$, or if one of the partial derivatives does not exist. Then we can interpret Theorem 13.20 as saying that f has relative extreme values only at critical points in its domain.

EXAMPLE 1   Let $f(x, y) = 3 - x^2 + 2x - y^2 - 4y$. Find all critical points of f.

Solution   The partial derivatives of f exist at every point in the domain of f, so relative extreme values can occur only at points at which both partial derivatives are 0. The partial derivatives are

$$f_x(x, y) = -2x + 2 \quad \text{and} \quad f_y(x, y) = -2y - 4$$

Therefore $f_x(x, y) = 0$ only if $x = 1$, and $f_y(x, y) = 0$ only if $y = -2$. This means that $f_x(x, y) = f_y(x, y) = 0$ only if $(x, y) = (1, -2)$, and thus $(1, -2)$ is the only critical point of f.   ❏

We can determine whether the function in Example 1 has a relative extreme value at $(1, -2)$ by completing squares in the formula for $f(x, y)$. Thus

$$3 - x^2 + 2x - y^2 - 4y = 3 - (x^2 - 2x + 1) - (y^2 + 4y + 4) + 1 + 4$$
$$= 8 - (x - 1)^2 - (y + 2)^2$$

Since $(x - 1)^2 \geq 0$ and $(y + 2)^2 \geq 0$ for all x and y, it is apparent that $f(1, -2) = 8$ is a relative maximum value of f; in fact, we can even conclude that 8 is the maximum value of f.

EXAMPLE 2   Let $f(x, y) = \sqrt{x^2 + y^2}$. Determine all critical points and all relative extreme values of f.

Solution   We find that

$$f_x(x, y) = \frac{x}{\sqrt{x^2 + y^2}} \quad \text{and} \quad f_y(x, y) = \frac{y}{\sqrt{x^2 + y^2}}$$

Since the partial derivatives exist at all points except the origin, and since the origin is in the domain of f, the origin is a critical point of f. Because $f_x(x, y) = 0$ only if $x = 0$ and $f_y(x, y) = 0$ only if $y = 0$, there is no point (x, y) such that $f_x(x, y) = f_y(x, y) = 0$. Consequently $(0, 0)$ is the only critical point of f. Since $f(0, 0) = 0$ and $f(x, y) \geq 0$ for all (x, y), it follows that 0 is the only (relative) minimum value of f, and there is no relative maximum value.   ❏

The next example illustrates the fact that a function need not have a relative extreme value at a critical point.

EXAMPLE 3   Let $f(x, y) = y^2 - x^2$. Show that the origin is the only critical point but that $f(0, 0)$ is not a relative extreme value of f.

Solution   In this case

$$f_x(x, y) = -2x \quad \text{and} \quad f_y(x, y) = 2y$$

so that $(0, 0)$ is the unique critical point of f. However, $f(0, 0) = 0$ is not a relative extreme value of f, because $f(x, 0) = -x^2 < 0$ for $x \neq 0$, and $f(0, y) = y^2 > 0$ for $y \neq 0$. Consequently f has no relative extreme values (Figure 13.44). ❑

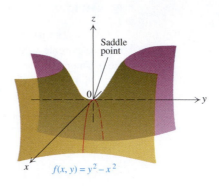

$$f(x, y) = y^2 - x^2$$

FIGURE 13.44

If we consider the values of the function f in Example 3 at points along the x axis, we obtain a function of x that has its maximum value at the origin. However, if we consider the values of f at points along the y axis, we obtain a function of y that has its minimum value at the origin. Thus the graph of f resembles a saddle (see Figure 13.44), and that is why f does not have a relative extreme value at $(0, 0)$, despite the fact that $f_x(0, 0) = f_y(0, 0) = 0$. More generally, if f is a function for which $f_x(x_0, y_0) = f_y(x_0, y_0) = 0$, we say that $(x_0, y_0, f(x_0, y_0))$ is a **saddle point of f,** or that f **has a saddle point at (x_0, y_0),** if there is a disk centered at (x_0, y_0) such that the following condition holds:

f assumes its maximum value on one diameter of the disk only at (x_0, y_0), and assumes its minimum value on another diameter of the disk only at (x_0, y_0)

From this definition and our preceding comments we conclude that if $f(x, y) = y^2 - x^2$, then f has a saddle point at the origin.

The Second Partials Test

Suppose (x_0, y_0) is a critical point of a function f. How can we determine whether f has a relative extreme value or a saddle point at (x_0, y_0)?

In order to see what is involved in answering such a question, let

$$f(x, y) = Ax^2 + 2Bxy + Cy^2 \tag{1}$$

and suppose that $A \neq 0$. Then we can rewrite $f(x, y)$ in the following way:

$$f(x, y) = Ax^2 + 2Bxy + Cy^2 = A\left(x^2 + \frac{2B}{A}xy + \frac{B^2}{A^2}y^2 + \frac{C}{A}y^2 - \frac{B^2}{A^2}y^2\right)$$

$$= A\left[\left(x + \frac{B}{A}y\right)^2 + \frac{1}{A^2}(AC - B^2)y^2\right]$$

Thus

$$f(x, y) = A\, g(x, y) \tag{2}$$

where

$$g(x, y) = \left(x + \frac{B}{A}\, y\right)^2 + \frac{1}{A^2}\,(AC - B^2)\, y^2 \tag{3}$$

Now assume that $AC - B^2 > 0$. Then $g(x, y) \geq 0$ for all (x, y), and $g(x, y) = 0$ only if

$$x + \frac{B}{A}\, y = 0 \quad \text{and} \quad y = 0$$

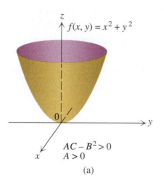

$f(x, y) = x^2 + y^2$

$AC - B^2 > 0$
$A > 0$

(a)

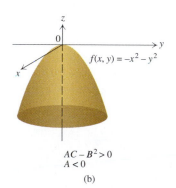

$f(x, y) = -x^2 - y^2$

$AC - B^2 > 0$
$A < 0$

(b)

FIGURE 13.45

This means that $g(x, y) > 0$ unless $(x, y) = (0, 0)$. Returning to (2), we see that if $A > 0$, then $f(x, y) \geq 0 = f(0, 0)$ for all (x, y), so that f has a minimum value and it occurs at $(0, 0)$. Similarly, if $A < 0$, then $f(x, y) \leq 0 = f(0, 0)$ for all (x, y), so that f has a maximum value and it occurs at $(0, 0)$. Thus when $AC - B^2 > 0$, f has either a minimum or maximum value at $(0, 0)$, depending on whether $A > 0$ or $A < 0$. The first of these cases occurs, for example, if $f(x, y) = x^2 + y^2$, and the second occurs if $f(x, y) = -x^2 - y^2$ (Figure 13.45(a) and (b)).

Next, suppose $AC - B^2 < 0$. Then the first expression on the right side of (3) is nonnegative, and the second expression is nonpositive. It can be shown that $(0, 0)$ is a saddle point (see Exercise 38). Therefore we conclude that if f has the form in (1), then f has a relative extreme value at $(0, 0)$ if $AC - B^2 > 0$ and has a saddle point at $(0, 0)$ if $AC - B^2 < 0$.

In order to extend the preceding results to more general functions, let us observe that if f is defined as in (1), then

$$f_{xx}(x, y) = 2A, \quad f_{yy}(x, y) = 2C, \quad \text{and} \quad f_{xy}(x, y) = 2B$$

Thus

$$f_{xx}(x, y)\, f_{yy}(x, y) - [f_{xy}(x, y)]^2 = 4(AC - B^2) \tag{4}$$

It follows that the left-hand side of (4) is positive (or negative) precisely when $AC - B^2$ is positive (or negative). For many functions f (not just one of the form in (1)), the left side of (4) can be used in order to determine whether a critical point of f yields a relative extreme value or a saddle point. The next theorem, aptly called the Second Partials Test because of the appearance of second partial derivatives, provides the criterion. The proof is omitted.

THEOREM 13.21
Second Partials Test

Assume that f has a critical point at (x_0, y_0) and that f has continuous second partial derivatives in a disk centered at (x_0, y_0). Let

$$D(x_0, y_0) = f_{xx}(x_0, y_0)\, f_{yy}(x_0, y_0) - [f_{xy}(x_0, y_0)]^2$$

a. If $D(x_0, y_0) > 0$ and $f_{xx}(x_0, y_0) < 0$ (or $f_{yy}(x_0, y_0) < 0$), then f has a relative maximum value at (x_0, y_0).

b. If $D(x_0, y_0) > 0$ and $f_{xx}(x_0, y_0) > 0$ (or $f_{yy}(x_0, y_0) > 0$), then f has a relative minimum value at (x_0, y_0).

c. If $D(x_0, y_0) < 0$, then f has a saddle point at (x_0, y_0).

Finally, if $D(x_0, y_0) = 0$, then f may or may not have a relative extreme value, or a saddle point, at (x_0, y_0).

The expression $D(x_0, y_0)$ in the Second Partials Test is called the **discriminant of f at (x_0, y_0).** It can also be given in determinant form:

$$D(x_0, y_0) = \begin{vmatrix} f_{xx}(x_0, y_0) & f_{xy}(x_0, y_0) \\ f_{xy}(x_0, y_0) & f_{yy}(x_0, y_0) \end{vmatrix}$$

Notice that if $D(x_0, y_0) > 0$, then $f_{xx}(x_0, y_0)$ and $f_{yy}(x_0, y_0)$ have the same sign, that is, $f_{xx}(x_0, y_0) > 0$ and $f_{yy}(x_0, y_0) > 0$, or $f_{xx}(x_0, y_0) < 0$ and $f_{yy}(x_0, y_0) < 0$. Thus if $D(x_0, y_0) > 0$, then the sign of either $f_{xx}(x_0, y_0)$ or $f_{yy}(x_0, y_0)$ determines the nature of the relative extreme value.

A critical point (x_0, y_0) is said to be **degenerate** if $D(x_0, y_0) = 0$ and **nondegenerate** otherwise. If a critical point is nondegenerate, the Second Partials Test determines the nature of the critical point. In contrast, the test implies nothing about a degenerate critical point. Indeed, f may have a relative extreme value or a saddle point or neither at a degenerate critical point. For example, let

$$f(x, y) = y^4 - x^4, \quad g(x, y) = x^4 + y^4, \quad \text{and} \quad h(x, y) = x^5 + x^3 + y^3$$

Then the point $(0, 0)$ is a critical point of f, g, and h, and in each case,

$$D(0, 0) = 0$$

However, a simple calculation shows that f has a saddle point at $(0, 0)$, whereas $g(x, y) \geq 0$ for all (x, y) and $g(0, 0) = 0$, so that g has a minimum value at $(0, 0)$. Finally, it turns out that on no line through the origin in the xy plane does h have a relative extreme value at $(0, 0)$; thus h has neither a relative extreme value nor a saddle point at $(0, 0)$.

EXAMPLE 4   Let

$$f(x, y) = x^2 - 2xy + \frac{1}{3}y^3 - 3y$$

Using the Second Partials Test, determine at which points f has relative extreme values and at which points f has saddle points.

Solution   We find that

$$f_x(x, y) = 2x - 2y \quad \text{and} \quad f_y(x, y) = -2x + y^2 - 3$$

Observe that $f_x(x, y) = 0$ if $x = y$ and $f_y(x, y) = 0$ if $-2x + y^2 - 3 = 0$. Thus (x, y) is a critical point if

$$x = y \quad \text{and} \quad -2x + y^2 - 3 = 0$$

By substituting y for x we can transform the second equation into $y^2 - 2y - 3 = 0$. The two solutions of this equation are $y = 3$ and $y = -1$. Thus the critical points of f are $(3, 3)$ and $(-1, -1)$. For the second partials of f we find that

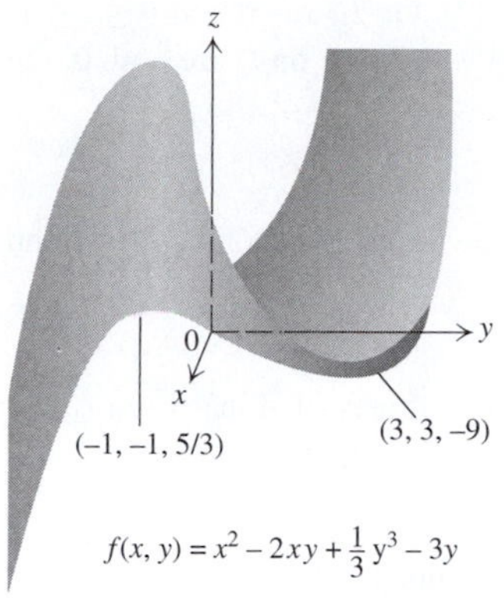

$$f_{xx}(x, y) = 2, \quad f_{yy}(x, y) = 2y, \quad \text{and} \quad f_{xy}(x, y) = -2$$

Therefore

$$D(3, 3) = f_{xx}(3, 3) f_{yy}(3, 3) - [f_{xy}(3, 3)]^2 = (2)(6) - (-2)^2 = 8 > 0$$

and

$$D(-1, -1) = f_{xx}(-1, -1) f_{yy}(-1, -1) - [f_{xy}(-1, -1)]^2$$
$$= (2)(-2) - (-2)^2 = -8 < 0$$

Since $D(3, 3) > 0$ and $f_{xx}(3, 3) = 2 > 0$, f has a relative minimum value at $(3, 3)$. Since $D(-1, -1) < 0$, f has a saddle point at $(-1, -1)$ (Figure 13.46). ❑

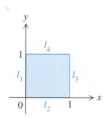

$$f(x, y) = x^2 - 2xy + \frac{1}{3} y^3 - 3y$$

(−1, −1, 5/3)

(3, 3, −9)

FIGURE 13.46

Extreme Values on a Set

Let R be a set in the plane. Assume that R is bounded (that is, R is contained in some rectangle) and closed (that is, R contains its boundary). Under these conditions there is a theorem (which we state without proof) for functions of two variables analogous to the Maximum-Minimum Theorem (Theorem 4.2).

THEOREM 13.22
Maximum-Minimum
Theorem for Two Variables

Let R be a closed, bounded set in the plane, and let f be continuous on R. Then f has both a maximum value and a minimum value on R.

By Theorem 13.20 and the comments that follow it, if R is closed and bounded, and if f has an extreme value on R at (x_0, y_0), then (x_0, y_0) is either a critical point of f or a boundary point of R. This observation, along with the Maximum-Minimum Theorem for Two Variables, provides us with a method of finding extreme values:

1. Find the critical points of f in R, and compute the values of f at these points.
2. Find the extreme values of f on the boundary of R.
3. The maximum value of f on R will be the largest of the values computed in steps 1 and 2, and the minimum value of f on R will be the smallest of those values.

EXAMPLE 5 Let $f(x, y) = xy - x^2$, and let R be the square region shown in Figure 13.47. Find the extreme values of f on R.

Solution By the Maximum-Minimum Theorem f has extreme values on R. Since

$$f_x(x, y) = y - 2x \quad \text{and} \quad f_y(x, y) = x$$

it follows that $f_x(x, y) = 0$ if $y = 2x$ and $f_y(x, y) = 0$ if $x = 0$. Thus the only critical point of f is $(0, 0)$, which happens to be a boundary point of R. Therefore the extreme values of f on R must occur on the boundary of R, which is composed of

FIGURE 13.47

the four line segments l_1, l_2, l_3, and l_4 (Figure 13.47). On l_1, $x = 0$ and $0 \leq y \leq 1$, and since $f(0, y) = 0$, the maximum and minimum values of f on l_1 are both 0. On l_2, $y = 0$ and $0 \leq x \leq 1$, and since

$$f(x, 0) = -x^2$$

the maximum value of f on l_2 is 0 and the minimum value is -1. On l_3, $x = 1$ and $0 \leq y \leq 1$, and since

$$f(1, y) = y - 1$$

the maximum value of f on l_3 is 0 and the minimum value is -1. Finally, on l_4, $y = 1$ and $0 \leq x \leq 1$, and since

$$f(x, 1) = x - x^2$$

the maximum value of f on l_4 is $\frac{1}{4}$ and the minimum value is 0, as you can verify by the methods of Chapter 4. By comparing the extreme values of f on l_1, l_2, l_3, and l_4, we conclude that the maximum value of f on R is $\frac{1}{4}$ and the minimum value is -1. ❏

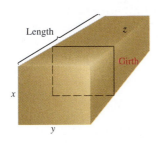

FIGURE 13.48

EXAMPLE 6   Under present Post Office regulations a package in the shape of a rectangular parallelepiped can be mailed parcel post only if the sum of the length of the longest side and corresponding girth (Figure 13.48) of the package is not more than 108 inches. Find the largest volume V of such a package.

Solution   Let x, y, and z represent the dimensions of the package, with z denoting the length of the largest side. Then $2x + 2y$ is the girth (Figure 13.48), and

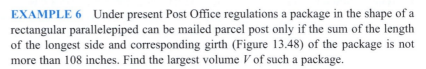

$$V = xyz \quad \text{and} \quad \overbrace{2x + 2y}^{\text{girth}} + \overbrace{z}^{\text{length}} \leq 108$$

We will assume that $z \geq x \geq 0$ and $z \geq y \geq 0$. Since we seek the largest possible volume, we may assume that

$$2x + 2y + z = 108, \quad \text{or equivalently,} \quad z = 108 - 2x - 2y \qquad (5)$$

Substituting for z in the equation $V = xyz$, we obtain

$$V = xy(108 - 2x - 2y) = 108xy - 2x^2y - 2xy^2 \qquad (6)$$

Now we wish to find the maximum value of V as a function of x and y. Since $z \geq x$ and $z \geq y$ by hypothesis, it follows that $108 - 2x - 2y = z \geq x$ and $108 - 2x - 2y = z \geq y$. Therefore x and y must satisfy the following four inequalities:

$$108 - 3x - 2y \geq 0, \quad 108 - 2x - 3y \geq 0, \quad x \geq 0 \text{ and } y \geq 0$$

The collection of all (x, y) satisfying the four inequalities forms a closed region R whose boundary consists of the four line segments

$$108 - 3x - 2y = 0, \quad 108 - 2x - 3y = 0, \quad x = 0 \text{ and } y = 0$$

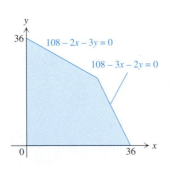

FIGURE 13.49

(Figure 13.49). Our goal is to find the maximum value of V for (x, y) in R. Such a maximum value exists by the Maximum-Minimum Theorem. It can occur only at a boundary point of R or at a critical point in the interior of R.

To find the critical points in the interior of R, we take partial derivatives of V:

$$\frac{\partial V}{\partial x} = 108y - 4xy - 2y^2 \quad \text{and} \quad \frac{\partial V}{\partial y} = 108x - 2x^2 - 4xy$$

Then $\partial V/\partial x = 0$ if $108y - 4xy - 2y^2 = 0$, that is, if $y = 0$ or if $108 - 4x - 2y = 0$. Similarly, $\partial V/\partial y = 0$ if $108x - 2x^2 - 4xy = 0$, that is, if $x = 0$ or if $108 - 2x - 4y = 0$. Since (x, y) can be a critical point in the interior of R only if $x \neq 0$ and $y \neq 0$, it follows that (x, y) is such a critical point only if

$$108 - 4x - 2y = 0 \quad \text{and} \quad 108 - 2x - 4y = 0 \qquad (7)$$

Solving for y in the first equation, we obtain $y = 54 - 2x$. Substituting for y in the second equation gives us

$$0 = 108 - 2x - 4(54 - 2x) = -108 + 6x$$

so that $x = 18$. Then

$$y = 54 - 2(18) = 18$$

Thus $(18, 18)$ is the only critical point in the interior of R. By (6) it follows that $V(18, 18) = 108(18)(18) - 2(18)^2(18) - 2(18)(18)^2 = 11{,}664$.

In order to complete the solution we will find the maximum value of V on the boundary of R, and compare it with the value $11{,}664$ just obtained. The boundary of R consists of four line segments (see Figure 13.49 again). On the horizontal line segment, $y = 0$, so that $V = 0$. Similarly, on the vertical line segment, $x = 0$, so that again $V = 0$. Next, on the lower diagonal line segment, $108 - 3x - 2y = 0$, so that $y = 54 - \frac{3}{2}x$. In addition,

$$\text{if } 108 - 3x - 2y = 0 \quad \text{then } 108 - 2x - 2y = x$$

By (5), this means that $z = x$. Substituting for y and z, we find that

$$V = xyz = x\left(54 - \frac{3}{2}x\right)x = 54x^2 - \frac{3}{2}x^3$$

Thus $dV/dx = 0$ if $108x - \frac{9}{2}x^2 = 0$, which implies that $x = 0$ or $x = 24$. If $x = 0$ then $V = 0$; if $x = 24$, then

$$y = 54 - \frac{3}{2}(24) = 18 \quad \text{and} \quad z = x = 24$$

It follows that $V = (24)(18)(24) = 10{,}368$. Consequently the maximum value of V on the lower diagonal line segment is $10{,}368$. Similar calculations (or the fact that x and y play the same role) show that the maximum value of V on the upper diagonal line segment is also $10{,}368$. Thus the maximum value of V on the entire boundary of R is $10{,}368$. Comparing this with the volume of $11{,}664$ corresponding to the critical point, we conclude that the maximum volume of a mailable package that is in the form of a rectangular parallelepiped is $11{,}664$ (cubic inches). ❑

When we solved the equations $\partial V/\partial x = 0$ and $\partial V/\partial y = 0$ in Example 6, we obtained two equations involving x and y (see (7)). Then we solved for y in one equation and substituted for it in the other equation to find the possible values for x. This is the general method of attack: When you obtain two equations in two unknowns, try to solve for one variable in one of the equations; then substitute for that variable in the other equation and find the required value of the remaining variable.

EXERCISES 13.8

In Exercises 1–23 find all critical points. Determine whether each critical point yields a relative maximum value, a relative minimum value, or a saddle point.

1. $f(x, y) = x^2 + 2y^2 - 6x + 8y - 1$

2. $f(x, y) = x^2 - 2y^2 - 6x + 8y + 3$

3. $f(x, y) = x^2 + 6xy + 2y^2 - 6x + 10y - 2$

4. $g(x, y) = x^2 - xy - 2y^2 + 7x - 8y + 3$

5. $k(x, y) = -x^2 - 2xy - 2y^2 + 6x - 10y + 5$

6. $f(x, y) = -x^2 + 4xy + y^2 - 2x + 9$

7. $f(x, y) = x^2y - 2xy + 2y^2 - 15y$

8. $f(x, y) = x^3 - 6x^2 - 3y^2$

9. $f(x, y) = 3x^2 - 3xy^2 + y^3 + 3y^2$

10. $f(u, v) = u^3 + v^3 - 6uv$

11. $f(x, y) = 4xy + 2x^2y - xy^2$

12. $f(x, y) = \dfrac{1}{x} + \dfrac{1}{y} + xy$ **13.** $f(x, y) = x^2 - e^{y^2}$

14. $f(x, y) = (y - 2) \ln xy$ **15.** $k(x, y) = e^x \sin y$

16. $f(x, y) = e^x(\sin y - 1)$ **17.** $f(u, v) = |u| + |v|$

18. $g(u, v) = 3 - |u - 2| + |v + 1|$

19. $f(x, y) = e^{xy}$

20. $f(x, y) = \sin x + \sin y$ for $0 < x < \pi/2$, $0 < y < \pi/2$

*21. $f(x, y) = \sin x + \sin y$ **22.** $k(u, v) = (u + v)^2$

23. $f(x, y) = (y + ax + b)^2$, where a and b are constants

*24. Let a and b be nonzero and $f(x, y) = (ax^2 + by^2)e^{-x^2 - y^2}$. Show that if $a \neq b$, then there are five critical points of f, whereas if $a = b$, then the critical points consist of a circle and its center.

In Exercises 25–28 find the extreme values of f on R.

25. $f(x, y) = x^2 - y^2$; R is the disk $x^2 + y^2 \leq 1$.

26. $f(x, y) = ye^{-x}$; R is the rectangular region with vertices $(0, 0)$, $(\ln 2, 0)$, $(\ln 2, 3)$, $(0, 3)$.

27. $f(x, y) = 2 \sin x + 3 \cos y$; R is the square region with vertices $(0, -\pi/2)$, $(\pi, -\pi/2)$, $(\pi, \pi/2)$, $(0, \pi/2)$.

28. $f(x, y) = e^{x^2 - y^2}$; R is the ring bounded by the circles $x^2 + y^2 = \frac{1}{2}$ and $x^2 + y^2 = 2$.

In the remaining exercises in this section, assume that the required extreme values exist.

29. Find the three positive numbers whose sum is 48 and whose product is as large as possible. Calculate the product.

30. Find the three positive numbers whose product is 48 and whose sum is as small as possible. Calculate the sum.

31. Show that the box in the shape of a rectangular parallelepiped whose volume is the largest of any inscribed in a given sphere is a cube.

32. A rectangular box without top is to have a volume of 32 cubic meters. Find the dimensions of such a box having the smallest possible surface area.

33. Find the point in space the sum of whose coordinates is 48 and whose distance from the origin is minimum.

34. Find a vector in space whose length is 16 and whose components have the largest possible sum.

35. Let $f(x, y) = e^{-y^2}(2x^3 - 3x^2 + 1) + e^{-y}(2x^3 - 3x^2)$ (see Figure 13.50).

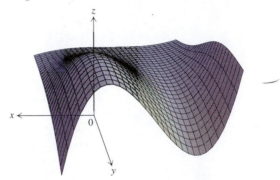

FIGURE 13.50   Graph for Exercise 35.

a. Show that f has exactly one critical point (x_0, y_0).

b. Show that $f(x_0, y_0)$ is a relative maximum value.

c. Show that f has no maximum value. (Contrast this result with (1) in Section 4.6.)

One measure of the closeness of a line l with equation $y = mx + b$ to the point (x_1, y_1) is the square $[y_1 - (mx_1 + b)]^2$ of the distance between (x_1, y_1) and l measured along the vertical line $x = x_1$ (Figure 13.51). For n fixed points (x_1, y_1), (x_2, y_2), ... , (x_n, y_n), the corresponding measure of the closeness of the line l to these points is

$$f(m, b) = \sum_{k=1}^{n} (y_k - (mx_k + b))^2$$

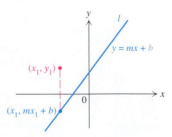

FIGURE 13.51

It can be shown that f has a minimum value $f(m_0, b_0)$, which can be determined by solving the equations $f_m(m, b) = 0$ and $f_b(m, b) = 0$ for m and b. The line $y = m_0x + b_0$ is called the **line of best fit** for the n given points (Figure 13.52). (Statisticians also call this line the **line of regression.** This terminology originated from a statistical study of the tendency of the height of the offspring of tall parents to regress toward the average height.) The method of determining the line of best fit to n given points is called the **method of least squares.** In Exercises 36–37 use the method of least squares to determine the line of best fit for the given collection of points.

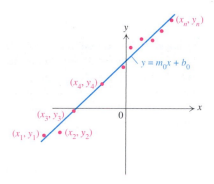

FIGURE 13.52

36. $(1, 1), (2, 3), (3, 4)$ **37.** $(0, 0), (1, -1), (-2, 1)$

38. Let $f(x, y) = Ax^2 + 2Bxy + Cy^2$, as in (1). Assume that $A \neq 0$ and $AC - B^2 < 0$. Verify that there are points (x, y) as close to $(0, 0)$ as one wishes such that $f(x, y) > 0$, and other points (x, y) as close to $(0, 0)$ as one wishes such that $f(x, y) < 0$. Conclude that f has a saddle point at $(0, 0)$. (*Hint:* Consider points of the form $(x, 0)$ and $(-By/A, y)$.)

Applications

39. A rectangular box is tied once each way around with a string of fixed length l, and without knots (Figure 13.53). Find the maximum possible volume of the package.

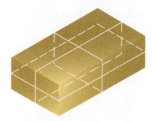

FIGURE 13.53   Figure for Exercise 39.

40. A textbook warehouse in the shape of a rectangular parallelepiped with volume 960,000 cubic feet is to be erected. Assume that because of decorations, the front wall will cost twice as much per square foot as the side and back walls and the floor, and the roof will cost $\frac{3}{2}$ as much as the side walls. Find the dimensions of the warehouse that will minimize the cost.

*41. A forest ranger must walk from a certain spot in a thicket back to the ranger station, first through the thicket and then through marshland, and finally along a road, as in Figure 13.54. Suppose the ranger can proceed through the thicket at 3 kilometers per hour, through the marshland at 4 kilometers per hour, and on the road at 5 kilometers per hour. What is the most expeditious route?

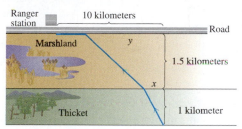

FIGURE 13.54   Figure for Exercise 41.

*42. A rectangular piece of tin with width l is to be bent as shown in Figure 13.55. Show that the maximal cross-sectional area is obtained if $x = \frac{1}{3}l$ and $\theta = \pi/3$. (*Hint:* After taking partial derivatives, eliminate l from the equations you must solve.)

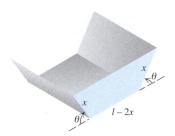

FIGURE 13.55   Figure for Exercise 42.

Project

1. Consider the function $f(x, y) = x^2 + cxy + y^2$, where c is a constant. We want to see what effect various values of c have on the nature of critical points of f.
   a. Let $c = 2$. Show that the critical points of f have the form $(x, -x)$. First determine whether the Second

Partials Test applies, and then determine whether the critical points yield relative maximum values, relative minimum values, or saddle points.

b. Let $c = -2$. First find the critical points of f. Then determine whether the Second Partials Test applies, and finally determine whether the critical points yield relative maximum values, relative minimum values, or saddle points.

c. Let $|c| \neq 2$. Find the critical points of f. Then determine whether the Second Partials Test applies, and finally determine whether the critical points yield relative maximum values, relative minimum values, or saddle points.

13.9   LAGRANGE MULTIPLIERS

Suppose that two new firms want to be hooked up to an electric supply route. Assume that for purposes of economy, the electric company decides to run electric lines to the two firms from a single terminal on the supply route. The problem is to find the point on the supply route that minimizes the total length of electric lines. In the present section we will describe a method, called Lagrange multipliers, that could assist us in finding a solution to this problem.

Lagrange Multipliers for Functions of Two Variables

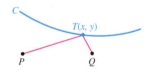

FIGURE 13.56

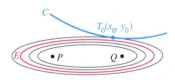

FIGURE 13.57

In preparation for the method of Lagrange multipliers, we will consider the supply route problem in more detail. Let the two firms be located at the points P and Q. Suppose the electric supply route is denoted by C, with $T = T(x, y)$ any point on C (Figure 13.56). Finally, let $f(x, y)$ be the sum of the distances from P to T and from T to Q, so that $f(x, y) = |PT| + |TQ|$.

The points S in the xy plane for which $|PS| + |SQ|$ is constant form an ellipse with foci at P and Q. Thus in order to find the point $T_0(x_0, y_0)$ on C that minimizes the distance f, we need only consider all the ellipses that have foci at P and Q and also contact C. Of these ellipses, we must find the one, which we will denote by E, for which $|PS| + |SQ|$ is smallest (see Figure 13.57). Of course this procedure only gives us an idea of where the point $T_0(x_0, y_0)$ is, but does not tell us how to find its coordinates.

To devise a method of determining coordinates of the point $T_0(x_0, y_0)$, suppose C is the level curve $g(x, y) = c$ for a function g whose gradient exists and is never **0**. The key observation is that the smallest admissible ellipse E is tangent to C. (If it were not tangent to C, then it would contact C in more than one point, and a smaller admissible ellipse could be found. See Figure 13.57 again.) This means that at $T_0(x_0, y_0)$, the normals to E and to C are parallel. Since E and C are level curves for f and g, respectively, it follows from Theorem 13.17 that grad $f(x_0, y_0)$ and grad $g(x_0, y_0)$ are normal to E and C. Consequently there is a scalar λ such that

$$\text{grad } f(x_0, y_0) = \lambda \text{ grad } g(x_0, y_0)$$

We conclude that if $T(x_0, y_0)$ is to yield the minimum value of f on C, then we need only determine the points (x, y) on C such that

$$\text{grad } f(x, y) = \lambda \text{ grad } g(x, y) \tag{1}$$

for some scalar λ.

The next theorem implies that under very general conditions on the functions f and g, the equation in (1) can assist in finding extreme values of f subject to a condition of the form $g(x, y) = c$.

THEOREM 13.23

Let f and g be differentiable at (x_0, y_0). Let C be the level curve $g(x, y) = c$ that contains (x_0, y_0). Assume that C is smooth, and that (x_0, y_0) is not an endpoint of the curve. If grad $g(x_0, y_0) \neq \mathbf{0}$ and if f has an extreme value on C at (x_0, y_0), then there is a number λ such that

$$\text{grad } f(x_0, y_0) = \lambda \text{ grad } g(x_0, y_0) \tag{2}$$

Proof   If grad $f(x_0, y_0) = \mathbf{0}$, then (2) is satisfied with $\lambda = 0$. Thus for the rest of the proof we will assume that grad $f(x_0, y_0) \neq \mathbf{0}$. Let I be an interval and

$$\mathbf{r}(t) = x(t)\mathbf{i} + y(t)\mathbf{j} \quad \text{for } t \text{ in } I$$

a smooth parametrization of C. Let t_0 be such that $\mathbf{r}(t_0)$ corresponds to the point (x_0, y_0). Then t_0 is not an endpoint of I since (x_0, y_0) is not an endpoint of C. Finally, let F be defined by

$$F(t) = f(x(t), y(t)) \quad \text{for } t \text{ in } I$$

As in the proof of Theorem 13.17, the Chain Rule yields

$$F'(t) = \frac{dF}{dt} = \frac{\partial f}{\partial x}\frac{dx}{dt} + \frac{\partial f}{\partial y}\frac{dy}{dt} = \text{grad } f(x(t), y(t)) \cdot \mathbf{r}'(t)$$

Since f has an extreme value on C at $(x_0, y_0) = (x(t_0), y(t_0))$, it follows that F has an extreme value on I at t_0. Since F is differentiable on I and t_0 is not an endpoint of I it follows that $F'(t_0) = 0$. Therefore

$$0 = F'(t_0) = \text{grad } f(x_0, y_0) \cdot \mathbf{r}'(t_0)$$

But $\mathbf{r}'(t_0) \neq \mathbf{0}$ since $\mathbf{r}$ is a smooth parametrization of C, and grad $f(x_0, y_0) \neq \mathbf{0}$ by assumption. Thus grad $f(x_0, y_0)$ is perpendicular to $\mathbf{r}'(t_0)$, which itself is tangent to C. Therefore grad $f(x_0, y_0)$ is normal to C. But grad $g(x_0, y_0)$ is normal to C by Theorem 13.17. Consequently grad $f(x_0, y_0)$ and grad $g(x_0, y_0)$ are parallel; this yields (2).   ■

The number λ in (2) is a "multiplier," called a **Lagrange multiplier** after the French-Italian mathematician Joseph Lagrange, whom we have encountered before. The condition $g(x, y) = c$ appearing in the theorem is called a **constraint** (or **side condition**) for the function f.

Observe that the equation grad $f(x_0, y_0) = \lambda$ grad $g(x_0, y_0)$ in (2) is equivalent to the pair of equations

$$f_x(x_0, y_0) = \lambda g_x(x_0, y_0) \quad \text{and} \quad f_y(x_0, y_0) = \lambda g_y(x_0, y_0)$$

The method of determining extreme values by means of Lagrange multipliers proceeds as follows:

1. Assume that f has an extreme value on the level curve $g(x, y) = c$.
2. Solve the equations

constraint $\qquad g(x, y) = c$

grad $f(x, y) = \lambda$ grad $g(x, y)$ $\quad \begin{cases} f_x(x, y) = \lambda g_x(x, y) \\ f_y(x, y) = \lambda g_y(x, y) \end{cases}$

3. Calculate the value of f at each point (x, y) that arises in step 2, and at each endpoint (if any) of the curve. If f has a maximum value on the level curve $g(x, y) = c$, it will be the largest of the values computed; if f has a minimum value on the level curve, it will be the smallest of the values computed.

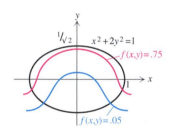

FIGURE 13.58

EXAMPLE 1 Let $f(x, y) = x^2 + 4y^3$. Find the extreme values of f on the ellipse $x^2 + 2y^2 = 1$, and the points at which they occur.

Solution The ellipse and two level curves for f are shown in Figure 13.58. To find the extreme values, we let

$$g(x, y) = x^2 + 2y^2$$

so that the constraint is $g(x, y) = x^2 + 2y^2 = 1$. Since

$$\text{grad } f(x, y) = 2x\mathbf{i} + 12y^2\mathbf{j} \quad \text{and} \quad \text{grad } g(x, y) = 2x\mathbf{i} + 4y\mathbf{j}$$

the equations we will use to find x and y are

constraint $\qquad\qquad x^2 + 2y^2 = 1 \qquad (3)$

grad $f(x, y) = \lambda$ grad $g(x, y)$ $\quad \begin{cases} 2x = 2x\lambda & (4) \\ 12y^2 = 4y\lambda & (5) \end{cases}$

By (4), either $x = 0$ or $\lambda = 1$. If $x = 0$, then (3) implies that $2y^2 = 1$, so that either $y = 1/\sqrt{2}$ or $y = -1/\sqrt{2}$. If $\lambda = 1$, then (5) becomes $12y^2 = 4y$, which means that $y = 0$ or $y = \frac{1}{3}$. By (3),

$$\text{if } y = 0, \quad \text{then } x^2 + 2(0)^2 = 1, \quad \text{so } x = 1 \text{ or } x = -1$$

$$\text{if } y = \frac{1}{3}, \quad \text{then } x^2 + 2\left(\frac{1}{3}\right)^2 = 1, \quad \text{so } x = \frac{\sqrt{7}}{3} \text{ or } x = -\frac{\sqrt{7}}{3}$$

Thus the only possible extreme values of f occur at $(0, 1/\sqrt{2})$, $(0, -1/\sqrt{2})$, $(1, 0)$, $(-1, 0)$, $(\sqrt{7}/3, \frac{1}{3})$, and $(-\sqrt{7}/3, \frac{1}{3})$. Since

$$f\left(0, \frac{1}{\sqrt{2}}\right) = \sqrt{2} \qquad\qquad f(1, 0) = 1 = f(-1, 0)$$

$$f\left(0, -\frac{1}{\sqrt{2}}\right) = -\sqrt{2} \qquad f\left(\frac{\sqrt{7}}{3}, \frac{1}{3}\right) = \frac{25}{27} = f\left(-\frac{\sqrt{7}}{3}, \frac{1}{3}\right)$$

we conclude that the maximum value $\sqrt{2}$ of f occurs at $(0, 1/\sqrt{2})$ and the minimum value $-\sqrt{2}$ of f occurs at $(0, -1/\sqrt{2})$. ❑

EXAMPLE 2   Let $f(x, y) = 3x^2 + 2y^2 - 4y + 1$. Find the extreme values of f on the disk $x^2 + y^2 \leq 16$.

Solution   By the Maximum-Minimum Theorem f has extreme values on the disk, and it can have them only on the boundary $x^2 + y^2 = 16$ or at critical points in the interior.

First, we use Lagrange multipliers to find the possible extreme values of f on the circle $x^2 + y^2 = 16$. Let $g(x, y) = x^2 + y^2$, so that the constraint is $g(x, y) = x^2 + y^2 = 16$. Since

$$\operatorname{grad} f(x, y) = 6x\mathbf{i} + (4y - 4)\mathbf{j} \quad \text{and} \quad \operatorname{grad} g(x, y) = 2x\mathbf{i} + 2y\mathbf{j}$$

the equations we will use to find x and y are

$$\begin{array}{lll} \text{constraint} & x^2 + y^2 = 16 & (6) \\[2mm] \operatorname{grad} f(x, y) = \lambda \operatorname{grad} g(x, y) & \begin{cases} 6x = 2x\lambda & (7) \\ 4y - 4 = 2y\lambda & (8) \end{cases} \end{array}$$

By (7), either $x = 0$ or $\lambda = 3$. If $x = 0$, then it follows from (6) that $y = 4$ or $y = -4$. If $\lambda = 3$, then (8) becomes

$$4y - 4 = 6y, \quad \text{so that} \quad y = -2$$

Then (6) implies that

$$x^2 + (-2)^2 = 16, \quad \text{so that} \quad x = \sqrt{12} \quad \text{or} \quad x = -\sqrt{12}$$

Thus f can have its extreme values on the circle $x^2 + y^2 = 16$ only at $(0, 4)$, $(0, -4)$, $(\sqrt{12}, -2)$, or $(-\sqrt{12}, -2)$.

Turning to the interior of the disk, we find that

$$f_x(x, y) = 6x \quad \text{and} \quad f_y(x, y) = 4y - 4$$

so that $f_x(x, y) = 0 = f_y(x, y)$ only if $x = 0$ and $y = 1$. Thus f can also have an extreme value on the disk at $(0, 1)$. Finally, we calculate that

$$\begin{array}{ll} f(0, 4) = 17 & f(\sqrt{12}, -2) = 53 = f(-\sqrt{12}, -2) \\[2mm] f(0, -4) = 49 & f(0, 1) = -1 \end{array}$$

Our conclusion is that the maximum value of f on the disk $x^2 + y^2 \leq 16$ is 53 and the minimum value is -1.   ❑

The Lagrange Method for Functions of Three Variables

Next we consider the problem of finding extreme values of a function of three variables subject to a constraint of the form $g(x, y, z) = c$. By an argument similar to that used for functions of two variables, it is possible to show that if f has such an extreme value at (x_0, y_0, z_0), then $\operatorname{grad} f(x_0, y_0, z_0)$ and $\operatorname{grad} g(x_0, y_0, z_0)$, if not $\mathbf{0}$, are both normal to the level surface $g(x, y, z) = c$ at (x_0, y_0, z_0), and hence are parallel to each other. Thus there is a number λ, again called a Lagrange multiplier, such that

$$\operatorname{grad} f(x_0, y_0, z_0) = \lambda \operatorname{grad} g(x_0, y_0, z_0)$$

To find the extreme values of f subject to the constraint $g(x, y, z) = c$, we follow the same approach as in steps 1–3 for functions of two variables:

1. Assume that f has an extreme value on the level surface $g(x, y, z) = c$.
2. Solve the equations

constraint $\qquad\qquad\qquad\qquad\qquad$ $g(x, y, z) = c$

$\text{grad } f(x, y, z) = \lambda \text{ grad } g(x, y)$ $\quad\begin{cases} f_x(x, y, z) = \lambda g_x(x, y, z) \\ f_y(x, y, z) = \lambda g_y(x, y, z) \\ f_z(x, y, z) = \lambda g_z(x, y, z) \end{cases}$

3. Calculate $f(x, y, z)$ for each point (x, y, z) that arises from step 2. If f has a maximum (minimum) value on the level surface, it will be the largest (smallest) of the values computed.

EXAMPLE 3   Suppose heavy-duty tape is to be applied on the bottom and side edges of a rectangular carton (Figure 13.59). If 96 inches of tape are available, find the maximum volume of the carton.

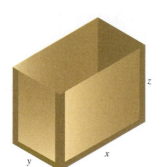

FIGURE 13.59

Solution   Let x denote the length, y the width, and z the height of the carton. If V represents the volume, then $V(x, y, z) = xyz$. Letting $g(x, y, z)$ denote the length of the tape used, we find that $g(x, y, z) = 2x + 2y + 4z$. Because

$$\text{grad } V(x, y, z) = yz\mathbf{i} + xz\mathbf{j} + xy\mathbf{k} \quad \text{and} \quad \text{grad } g(x, y, z) = 2\mathbf{i} + 2\mathbf{j} + 4\mathbf{k}$$

the equations we will use to find x, y, and z are

constraint $\qquad\qquad\qquad\qquad\qquad$ $2x + 2y + 4z = 96 \qquad (9)$

$\text{grad } V(x, y, z) = \lambda \text{ grad } g(x, y, z)$ $\quad\begin{cases} yz = 2\lambda & (10) \\ xz = 2\lambda & (11) \\ xy = 4\lambda & (12) \end{cases}$

First we solve for λ in terms of x, y, and z in (10)–(12), obtaining

$$\lambda = \frac{yz}{2} = \frac{xz}{2} = \frac{xy}{4} \qquad (13)$$

Since $V(x, y, z) = 0$ if x, y, or z is 0, and since 0 is obviously not the maximum value of V subject to (9), we can assume that x, y, and z are different from 0. Then (13) tells us that $x = y$ and $z = \frac{1}{2}y$. Substituting for x and z in (9) yields

$$96 = 2y + 2y + 4\left(\frac{1}{2}y\right) = 6y, \quad \text{so that} \quad y = 16$$

Thus $x = 16$ and $z = \frac{1}{2}(16) = 8$, and therefore $V(16, 16, 8)$ is the only possible extreme value of V subject to the constraint. Since we are assuming that V has a maximum value subject to the constraint, we conclude that $V(16, 16, 8) = 2048$ is that value. Consequently the maximum possible volume of the carton is 2048 cubic inches. ❏

EXAMPLE 4   Find the minimum distance from a point on the surface $xy + 2xz = 5\sqrt{5}$ to the origin.

Solution   We could let

$$f_1(x, y, z) = \sqrt{x^2 + y^2 + z^2}$$

which represents the distance from (x, y, z) to the origin, and seek the minimum value of f_1 on the surface. However, the computations involved in using Lagrange multipliers will be simplified if we minimize f_1^2 on the surface. Thus we let

$$f(x, y, z) = x^2 + y^2 + z^2$$

and minimize f subject to the constraint

$$g(x, y, z) = xy + 2xz = 5\sqrt{5}$$

Notice that f and f_1 have extreme values at identical points, so using f instead of f_1 will not alter the point we find whose distance from the origin is minimum. Because

$$\text{grad } f(x, y, z) = 2x\mathbf{i} + 2y\mathbf{j} + 2z\mathbf{k}$$

and $$\text{grad } g(x, y, z) = (y + 2z)\mathbf{i} + x\mathbf{j} + 2x\mathbf{k}$$

the equations we will use to find x, y, and z are

$$\text{constraint} \qquad\qquad xy + 2xz = 5\sqrt{5} \qquad\qquad (14)$$

$$\text{grad } f(x, y, z) = \lambda \text{ grad } g(x, y, z) \quad \begin{cases} 2x = (y + 2z)\lambda & (15) \\ 2y = x\lambda & (16) \\ 2z = 2x\lambda & (17) \end{cases}$$

If λ were 0, then $x = y = z = 0$ by (15)–(17), so (14) would not hold. Thus $\lambda \neq 0$, so that by (16) and (17),

$$x \overset{(16)}{=} \frac{2y}{\lambda} \quad \text{and} \quad z \overset{(17)}{=} x\lambda \overset{(16)}{=} 2y \qquad\qquad (18)$$

If y were 0, then by (18) we would have $x = 0$ and $z = 0$, which would mean that (14) would not hold. Thus $y \neq 0$. Using (18) to substitute for x and z in (15), we find that

$$2\left(\frac{2y}{\lambda}\right) = [y + 2(2y)]\lambda$$

so that since $y \neq 0$,

$$\frac{4}{\lambda} = 5\lambda, \quad \text{or} \quad \lambda^2 = \frac{4}{5} \qquad\qquad (19)$$

Using (18) to substitute for x and z in (14), we obtain

$$5\sqrt{5} = \left(\frac{2y}{\lambda}\right)y + 2\left(\frac{2y}{\lambda}\right)(2y) = \frac{10}{\lambda}y^2 \qquad\qquad (20)$$

From (20) we see that λ is positive. Therefore (19) implies that $\lambda = 2/\sqrt{5}$, and then (20) implies that $y = 1$ or $y = -1$. By (18),

$$\text{if } y = 1, \qquad \text{then } x = \frac{2y}{\lambda} = \sqrt{5} \qquad \text{and } z = 2y = 2$$

$$\text{if } y = -1, \qquad \text{then } x = \frac{2y}{\lambda} = -\sqrt{5} \qquad \text{and } z = 2y = -2$$

Consequently the only points on the surface $xy + 2xz = 5\sqrt{5}$ that can have the minimum distance from the origin are $(\sqrt{5}, 1, 2)$ and $(-\sqrt{5}, -1, -2)$. Since by assumption the minimum distance exists and is the minimum value of f_1 on the surface, and since

$$f_1(\sqrt{5}, 1, 2) = \sqrt{10} = f_1(-\sqrt{5}, -1, -2)$$

the minimum distance from a point on the surface to the origin is $\sqrt{10}$. ❑

EXERCISES 13.9

In Exercises 1–6 find the extreme values of f subject to the given constraint. In each case assume that the extreme values exist.

1. $f(x, y) = x + y^2$; $x^2 + y^2 = 4$

2. $f(x, y) = xy$; $(x + 1)^2 + y^2 = 1$

3. $f(x, y) = x^3 + 2y^3$; $x^2 + y^2 = 1$

4. $f(x, y, z) = y^3 + xz^2$; $x^2 + y^2 + z^2 = 1$

5. $f(x, y, z) = xyz$; $x^2 + y^2 + 4z^2 = 6$

6. $f(x, y, z) = xy + yz$; $x^2 + y^2 + z^2 = 8$

In Exercises 7–10 find the minimum value of f subject to the given constraint. In each case assume that the minimum value exists.

7. $f(x, y) = 4x^2 + y^3 + 3y + 7$; $2x^2 + \frac{3}{2}y^2 = \frac{3}{2}$

8. $f(x, y, z) = x^2 + 2y^2 + z^2$; $x + y + z = 4$

9. $f(x, y, z) = x^4 + 8y^4 + 27z^4$; $x + y + z = \frac{11}{12}$

10. $f(x, y, z) = 3z - x - 2y$; $z = x^2 + 4y^2$

In Exercises 11–14 find the extreme values of f in the region described by the given inequalities. In each case assume that the extreme values exist.

11. $f(x, y) = 2x^2 + y^2 + 2y - 3$; $x^2 + y^2 \leq 4$

12. $f(x, y) = x^3 + x^2 + \frac{y^2}{3}$; $x^2 + y^2 \leq 36$

13. $f(x, y) = xy$; $2x^2 + y^2 \leq 4$

14. $f(x, y) = 16 - x^2 - 4y^2$; $x^4 + 2y^4 \leq 1$

In the remaining exercises in this section, assume that the required extreme values exist.

15. Find the points on the surface $x^2 - yz = 1$ that are closest to the origin.

16. Find the points on the sphere $x^2 + y^2 + z^2 = 1$ that are closest to or farthest from the point $(4, 2, 1)$.

17. A triangle is to be inscribed in the ellipse $\frac{1}{4}x^2 + y^2 = 1$ with one vertex of the triangle at $(-2, 0)$ and the opposite side perpendicular to the x axis. Find the largest possible area of the triangle.

18. Let x and y denote the acute angles of a right triangle. Find the maximum value of $\sin x \sin y$.

19. Let x, y, and z denote the angles of an arbitrary triangle. Find the maximum value of $\sin x \sin y \sin z$.

20. Find the minimum volume of a tetrahedron in the first octant bounded by the planes $x = 0$, $y = 0$, $z = 0$, and a plane tangent to the sphere $x^2 + y^2 + z^2 = 1$. (*Hint:* If the plane is tangent to the sphere at the point (x_0, y_0, z_0), then the volume of the tetrahedron is $1/(6x_0 y_0 z_0)$.)

21. A rectangular parallelepiped lies in the first octant, with three sides on the coordinate planes and one vertex on the plane $2x + y + 4z = 12$. Find the maximum possible volume of the parallelepiped.

Use Lagrange multipliers to solve Exercises 22–25, which also appear as maximum-minimum problems in Chapter 4.

22. A cylindrical can with bottom and no top has volume V. Find the radius of the can with the smallest possible surface area. (This is Exercise 32 in Section 4.6.)

23. Find the points on the parabola $y = x^2 + 2x$ that are closest to the point $(-1, 0)$. (This is Exercise 25 in Section 4.6.)

24. A rectangular printed page is to have margins 2 inches wide at the top and the bottom and margins 1 inch wide on each of the two sides. If the page is to have 35 square inches of printing, determine the minimum possible area of the page itself. (This is Exercise 31 in Section 4.6.)

25. An isosceles triangle is inscribed in a circle of radius r. Find the maximum possible area of the triangle. (This is Exercise 60 in Section 4.1.)

Applications

26. A construction company needs a type of funnel to reduce spillage when trucks are loaded with sand. The funnel is to consist of a circular cylinder with radius 3 feet on top of a cone with the same radius (Figure 13.60).

a. If the entire funnel is to have a capacity (volume) of 300 cubic feet, find the heights H and h of the cylinder and cone that minimize the amount of material (surface area) needed.

b. How, if at all, would the answer to part (a) be altered if the capacity were to be 400 cubic feet?

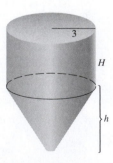

FIGURE 13.60 Figure for Exercise 26.

27. A rectangular metal storage box with volume 12 cubic inches is to be made in the form shown in Figure 13.61. Find the dimensions that will minimize the amount of metal required for the box. (*Hint:* Solve for λ in each equation you obtain.)

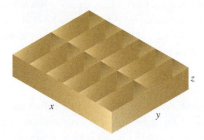

FIGURE 13.61   Figure for Exercise 27.

28. A rectangular box, open at the top, is to have a volume of 1728 cubic inches. Find the dimensions that will minimize the cost of the box if

   a. the material for the bottom costs 16 times as much per unit area as the material for the sides.

   b. the material for the bottom costs twice as much per unit area as the material for the sides.

29. In this exercise you will solve a realistic version of Exercise 35 in Section 4.6. A cylindrical pipe of radius r and length l must be slid on the floor from one corridor to a perpendicular corridor, each 3 meters wide (Figure 13.62). Find the dimensions of the pipe that maximize its volume V. (*Hint:* As in Figure 13.62, assume that the opposite ends of the pipe touch the walls when the angle between the pipe and the wall is $\pi/4$.)

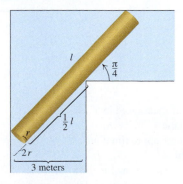

FIGURE 13.62   Figure for Exercise 29.

30. Suppose that on your vacation you plan to spend x days in San Francisco, y days in your home town, and z days in New York. You calculate that your total enjoyment $f(x, y, z)$ will be given by

$$f(x, y, z) = 2x + y + 2z$$

If plans and financial limitations dictate that

$$x^2 + y^2 + z^2 = 225$$

how long should each stay be to maximize your enjoyment?

31. The ground state energy $E(x, y, z)$ of a particle of mass m in a rectangular box with dimensions x, y, and z is given by

$$E(x, y, z) = \frac{h^2}{8m}\left(\frac{1}{x^2} + \frac{1}{y^2} + \frac{1}{z^2}\right)$$

where h is a constant. Assuming that the volume V of the box is fixed, find the values of x, y, and z that minimize the value of E.

32. The object distance p, image distance q, and focal length f of a simple lens satisfy the equation

$$\frac{1}{p} + \frac{1}{q} = \frac{1}{f}$$

Determine the minimum distance $p + q$ between the object and the image for a given focal length.

33. Fermat's Principle states that light always travels the path between points requiring the least time. Suppose light travels from a point A in one medium in which it has velocity v to a point B in a second medium in which it has velocity u. Using the fact that in a single medium, light travels in a straight line, and using Figure 13.63, show that the light is bent according to Snell's Law:

$$\frac{\sin\theta}{v} = \frac{\sin\phi}{u}$$

(*Hint:* The constraint is $x + y = l$, where x and y are the distances indicated in Figure 13.63.)

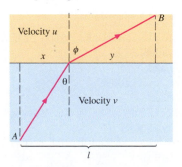

FIGURE 13.63   Figure for Exercise 33.

34. A pharmaceutical company plans to make capsules containing a given volume V of medicine. One executive would like to have the capsules in the form of a right circular cylinder having length h and base radius r with a hemisphere at each end (see Figure 13.64). A second executive objects to the wastefulness of materials, and contends that the same volume could be contained in a spherical capsule having a smaller surface area. Which executive should get the next promotion?

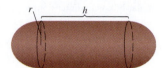

FIGURE 13.64 Figure for Exercise 34.

35. Let x represent capital and y labor in the manufacture of $f(x, y)$ units of a given product. Assume that capital costs a dollars per unit and labor costs b dollars per unit and that there are c dollars available, so that $ax + by = c$.

a. Using Lagrange multipliers, show that production is maximum at the point (x_0, y_0) such that

$$\frac{f_x(x_0, y_0)}{a} = \frac{f_y(x_0, y_0)}{b} = \lambda$$

where λ is the Lagrange multiplier, called the **equimarginal productivity** of the production function f.

b. Let f be the Cobb-Douglas production function given by

$$f(x, y) = x^\alpha y^\beta \quad \text{for } x > 0 \text{ and } y > 0$$

where α and β are positive constants less than 1. Using (a), show that at maximum production $f(x_0, y_0)$

we have

$$\frac{y_0}{x_0} = \frac{\beta a}{\alpha b}$$

which is independent of the money available.

*36. Let x, y, and z be positive numbers. Show that the geometric mean $(xyz)^{1/3}$ of x, y, and z is less than or equal to their arithmetic mean $(x + y + z)/3$. (*Hint:* Maximize $(xyz)^{1/3}$ subject to the constraint $x + y + z = c$, where c is a fixed number.)

Project

1. Lagrange multipliers can be used to find the extreme values of functions subject to more than one constraint. Suppose we wish to find the extreme values of a function f of three variables satisfying the two constraints

$$g_1(x, y, z) = c_1 \quad \text{and} \quad g_2(x, y, z) = c_2 \qquad (21)$$

where c_1 and c_2 are constants. The method is to solve the equation

$$\text{grad } f(x, y, z) = \lambda \text{ grad } g_1(x, y, z) + \mu \text{ grad } g_2(x, y, z)$$

along with the constraints in (21) for (x, y, z) and for λ and μ if necessary and then to determine the largest and smallest values of f. Both λ and μ are called Lagrange multipliers. Use this method to solve the following.

a. Find the minimum distance between the origin and a point on the intersection of the paraboloid $z = \frac{3}{2} - x^2 - y^2$ and the plane $x + 2y = 1$.

b. Find the distance from the point $(2, -2, 3)$ to the intersection of the planes

$$2x - y + 3z = 1 \quad \text{and} \quad -x + 3y + z = -3$$

REVIEW

Key Terms and Expressions

Function of several variables; graph of a function of several variables
Trace in a plane
Level curve; level surface
Quadric surface
Limits and continuity of a function of several variables; continuity on a set
Partial derivative; second partial derivative; mixed partial derivative
Differentiability at a point
Directional derivative

Gradient
Direction of most rapid increase
Tangent plane
Tangent plane approximation
Differential
Extreme value of a function of several variables; extreme value on a set; relative extreme value
Critical point of a function of several variables
Saddle point
Lagrange multiplier

Key Formulas

$$f_x(x_0, y_0) = \lim_{h \to 0} \frac{f(x_0 + h, y_0) - f(x_0, y_0)}{h}$$

$$f_y(x_0, y_0) = \lim_{h \to 0} \frac{f(x_0, y_0 + h) - f(x_0, y_0)}{h}$$

$$f_x(x_0, y_0, z_0) = \lim_{h \to 0} \frac{f(x_0 + h, y_0, z_0) - f(x_0, y_0, z_0)}{h}$$

$$f_y(x_0, y_0, z_0) = \lim_{h \to 0} \frac{f(x_0, y_0 + h, z_0) - f(x_0, y_0, z_0)}{h}$$

$$f_z(x_0, y_0, z_0) = \lim_{h \to 0} \frac{f(x_0, y_0, z_0 + h) - f(x_0, y_0, z_0)}{h}$$

$$\text{grad } f(x_0, y_0) = \nabla f(x_0, y_0) = f_x(x_0, y_0)\mathbf{i} + f_y(x_0, y_0)\mathbf{j}$$

$$\text{grad } f(x_0, y_0, z_0) = \nabla f(x_0, y_0, z_0) = f_x(x_0, y_0, z_0)\mathbf{i}$$
$$+ f_y(x_0, y_0, z_0)\mathbf{j} + f_z(x_0, y_0, z_0)\mathbf{k}$$

$$D_{\mathbf{u}}f(x_0, y_0) = [\text{grad } f(x_0, y_0)] \cdot \mathbf{u}$$

$$D_{\mathbf{u}}f(x_0, y_0, z_0) = [\text{grad } f(x_0, y_0, z_0)] \cdot \mathbf{u}$$

$$z = f(x_0, y_0) + f_x(x_0, y_0)(x - x_0) + f_y(x_0, y_0)(y - y_0)$$

Key Theorems

Chain Rule
Second Partials Test
Maximum-Minimum Theorem for Two Variables

Review Exercises

In Exercises 1–3 find the domain of the function.

1. $f(x, y) = \sqrt{1 - \frac{1}{4}x^2 - \frac{1}{25}y^2}$

2. $g(u, v) = \sin^{-1}(u - v)$

3. $k(x, y, z) = \ln(x - y + z)$

In Exercises 4–6 find the level curve $f(x, y) = c$.

4. $f(x, y) = y - x + 4$; $c = -1, 6$

5. $f(x, y) = xy$; $c = 1, -1$ **6.** $f(x, y) = \frac{x^2 - y^2}{x^2 + y^2}$; $c = \frac{1}{2}$

In Exercises 7–8 sketch the graph of the level surface $f(x, y, z) = c$.

7. $f(x, y, z) = \sqrt{4 - x^2 - y^2} - z$; $c = 1, -1$

8. $f(x, y, z) = 4z - x^2 - y^2$; $c = 2, 0$

In Exercises 9–13 sketch the quadric surface.

9. $x^2 = 4y^2 + z^2$ **10.** $z = 2x^2 - y^2$

11. $z = 2x^2 + y^2$ **12.** $x = y^2$

13. $y^2 - x^2 - z^2 = 8$

In Exercises 14–16 evaluate the limit.

14. $\displaystyle\lim_{(x, y) \to (0, 0)} \frac{y^2}{\sqrt{x^2 + y^2}}$

15. $\displaystyle\lim_{(x, y) \to (-2, \sqrt{2})} \frac{x^4 + x^2y^2 - 6y^4}{x^2 - 2y^2}$

16. $\displaystyle\lim_{(x, y, z) \to (-1, 1, 2)} \frac{2x^2 + 4xy - 6x^3z^2}{xyz^2}$

In Exercises 17–21 find the first partial derivatives.

17. $f(x, y) = 4x^3 - 3y^2$ **18.** $g(x, y) = \dfrac{x - y}{x^2 + 2y^2}$

19. $f(x, y, z) = e^{x^2}\ln(y^2 - 3z)$ **20.** $f(x, y, z) = \dfrac{\cos z^4}{xy^2}$

21. $k(x, y, z) = [\sqrt{z} \tan(x^2 + y)]^{5/2}$

In Exercises 22–23 show that $f_y = g_x$.

22. $f(x, y) = x - \cos y$, $g(x, y) = x \sin y$

23. $f(x, y) = y + 2xe^y$, $g(x, y) = x + x^2e^y$

In Exercises 24–26 find all second partials.

24. $f(x, y) = \sin^{-1}(x^2 - y^2)$ **25.** $g(u, v) = \ln \dfrac{u^2}{e^v}$

26. $f(x, y, z) = x \sin yz^2$

27. Let $z = e^{-ay}\cos bx$. Show that

$$a^2 \frac{\partial^2 z}{\partial x^2} + b^2 \frac{\partial^2 z}{\partial y^2} = 0$$

28. Let $z = \sqrt{x^2 + y^2}$. Show that

$$\frac{\partial^2 z}{\partial x^2} \frac{\partial^2 z}{\partial y^2} = \left(\frac{\partial^2 z}{\partial x \, \partial y}\right)^2$$

29. Compute dz/dt if $z = e^{x^2 - e^{y/2}}$, $x = t^2$, $y = t^3 - t$.

30. Compute $\partial z/\partial u$ and $\partial z/\partial v$ if $z = \sin xy - y^2\cos x$, $x = u^2v$, $y = 1/v$.

31. Compute dw/dt if $w = \sqrt{x^2 + y^2 z^4}$, $x = 2t$, $y = t^3$, $z = 1/t$.

32. Compute $\partial w/\partial u$ and $\partial w/\partial v$ if $w = x^2 + y \sin yz$, $x = u^2 + v^2$, $y = uv$, $z = u^2 - v^2$.

33. Let $z = f(x^2 + y^2)$. Show that

$$y \frac{\partial z}{\partial x} - x \frac{\partial z}{\partial y} = 0$$

34. Let $\tan^{-1} xy + \sin^{-1} xy = \pi/2$. Find dy/dx by implicit differentiation.

35. Let $xyz + 1/xyz = z^3$. Use implicit partial differentiation to find $\partial z/\partial x$ and $\partial z/\partial y$.

In Exercises 36–38 find the directional derivative of f at the given point P in the direction of **a**.

36. $f(x, y) = 4 - x^2 + 3y^2 + y$; $P = (-1, 0)$;
$\mathbf{a} = (-1/\sqrt{2})\mathbf{i} + (1/\sqrt{2})\mathbf{j}$

37. $f(x, y, z) = 1/(x^2 + y^2 + z^2)$; $P = (-1, 0, 2)$;
$\mathbf{a} = \mathbf{i} - \mathbf{j} - \mathbf{k}$

38. $f(x, y, z) = \csc (yz + x)$; $P = (\pi, -\pi/4, 1)$;
$\mathbf{a} = -3\mathbf{i} - \mathbf{j} - 2\mathbf{k}$

In Exercises 39–40 find the gradient of the function at the given point.

39. $f(x, y) = e^{2x} \ln y$; $(0, 1)$

40. $f(x, y, z) = x^2 \cos y \sin \left(\dfrac{\pi}{2} \sin z \right)$; $(1, -\pi/6, \pi/6)$

41. Let $f(x, y) = \cos xy$. Find the direction in which f is increasing most rapidly at $(\frac{1}{2}, \pi)$.

42. Let $f(x, y, z) = xye^z$. Find the direction in which f is increasing most rapidly at $(2, 3, 0)$.

In Exercises 43–44 find a vector normal to the graph of the equation at the given point. Assume in each case that the graph is a smooth curve.

43. $\tan^{-1} (x^2 + y) = \pi/4$; $(\frac{1}{2}, \frac{3}{4})$

44. $x^4 - 3x^2 y + 2y^3 = 11$; $(-1, 2)$

In Exercises 45–47 find a vector normal to the graph of f at the given point and an equation of the plane tangent to the graph of f at the given point.

45. $f(x, y) = 1/x - 1/y$; $(-\frac{1}{2}, \frac{1}{3}, -5)$

46. $f(x, y) = ye^x$; $(0, 1, 1)$

47. $f(x, y) = \sqrt{3x^2 + 2y^2 + 2}$; $(4, -5, 10)$

In Exercises 48–50 find an equation of the plane tangent to the level surface at the given point.

48. $x^2 - y^2 - z^2 = 1$; $(3, -2, 2)$

49. $xe^{yz} - 2y = -1$; $(1, 1, 0)$

50. $\sin xy + \cos yz = 0$; $(0, \pi/2, 1)$

51. Find the points on the surface $2x^3 + y - z^2 = 5$ at which the tangent plane is parallel to the plane $24x + y - 6z = 3$.

52. Show that there are exactly two planes tangent to the sphere $x^2 + y^2 + z^2 = 1$ and parallel to any given plane.

53. Show that the surfaces $z = x^2 + 4y^2$ and $z = 4x + y^2 - 4$ have the same tangent plane at $(2, 0, 4)$.

54. Show that the plane tangent to the elliptic hyperboloid

$$\frac{x^2}{a^2} - \frac{y^2}{b^2} + \frac{z^2}{c^2} = 1$$

at (x_0, y_0, z_0) has equation

$$\frac{xx_0}{a^2} - \frac{yy_0}{b^2} + \frac{zz_0}{c^2} = 1$$

55. Let $f(x, y) = \tan^{-1} \dfrac{x}{1 + y}$. Approximate $f(0.97, 0.05)$.

56. Let $f(x, y, z) = \sqrt{3x^2 + y^2 + 5z^2}$. Find an approximation for $f(2.9, 2.1, 0.9)$.

57. Let $f(x, y) = \ln (x/y)$. Find df.

58. Let $f(x, y, z) = x^2 y - e^{(z^2)} \cos yz$. Find df.

In Exercises 59–62 find all critical points. Specify which yield relative maximum values, which yield relative minimum values, and which yield saddle points.

59. $f(x, y) = x^2 + y^2 - 2x - 4y + 5$

60. $f(x, y) = x^2 + y - \dfrac{1}{2y^2}$ **61.** $f(x, y) = xy + \dfrac{8}{x^2} + \dfrac{8}{y^2}$

62. $f(x, y) = -2x^2 + xy + y^2 - 4x + 3y - 1$

In Exercises 63–66 find all extreme values of f subject to the given constraint. Assume that the extreme values exist.

63. $f(x, y) = 3x^2 - xy + y^2$; $3x^2 + y^2 = 3$

64. $f(x, y) = x^2 + y^2 - 2x - 4y - 6$; $x^2 + y^2 \leq 16$

65. $f(x, y, z) = \dfrac{x^2 + y^2}{z^2 + 5}$; $x^2 + y^2 - z = 2$

66. $f(x, y, z) = \dfrac{x^2 + y^2}{z^2 + 5}$; $x^2 + y^2 - 2 \leq z \leq 6$

67. Find the dimensions of the rectangular parallelepiped with faces parallel to the coordinate planes whose volume is the largest of any inscribed in the ellipsoid $x^2 + 4y^2 + 9z^2 = 36$. Assume that there exists a largest volume.

Applications

68. The image distance q from a lens is related to the object distance p and the focal length f of the lens by

$$q = \frac{pf}{p - f}$$

Show that $q_f(p, f) > 0 > q_p(p, f)$ for all (p, f) in the domain of q.

69. Consider the pendulum shown in Figure 13.65. The angle θ of deflection depends on time; for convenience we denote $d\theta/dt$ by $\dot{\theta}$ and $d^2\theta/dt^2$ by $\ddot{\theta}$ in this exercise. If the pendulum has small vibrations, then the kinetic energy T and the potential energy V of the pendulum are given (approximately) by

$$T = \frac{ml^2\,\dot{\theta}^2}{2} \quad \text{and} \quad V = \frac{mgl\theta^2}{2} \tag{1}$$

where g is the acceleration due to gravity, l is the length of the pendulum cord, and m is the mass of the object at the bottom of the pendulum. It can be shown that T and V satisfy **Lagrange's equation**

$$\frac{d}{dt}\frac{\partial T}{\partial \dot{\theta}} + \frac{\partial V}{\partial} = 0$$

Using (1), show that Lagrange's equation reduces to

$$\ddot{\theta} + \frac{g}{l}\theta = 0$$

Lagrange's equation is important in the branch of physics known as mechanics.

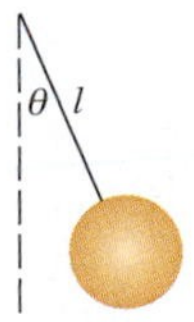

Pendulum

FIGURE 13.65   Figure for Exercise 69.

*70. A farmer wishes to employ tomato pickers to harvest tomatoes. Each picker can harvest 625 tomatoes per hour and is paid \$6 per hour. In addition, the farmer must pay a supervisor \$10 per hour and the union \$10 for each picker employed. Finally, if V tomatoes are picked, then a service charge of $\$50{,}000/\sqrt{V}$ is levied against the farmer. Show that the total cost to the farmer is minimum if five pickers are employed, and determine the number V that would be picked at this minimum cost.

Topics for Discussion

1. Discuss the similarities and differences between

   a. the derivative of a real-valued function.

   b. a partial derivative.

   c. a directional derivative.

2. Suppose the first partial derivatives of a function exist at the origin. Geometrically, what does this mean about f? Find a function whose first partial derivatives exist at the origin but whose other directional derivatives fail to exist. What additional condition on f implies the existence of all directional derivatives of f at the origin?

3. Discuss the method of using the Chain Rule for functions of several variables.

4. What is a gradient? List at least two ways in which the gradient is used in the text.

5. Describe how the following are related to one another: gradient, directional derivative, and direction of maximum increase.

6. Suppose a function f of two variables has a critical point at the origin. What are the four possible kinds of behavior for f at the origin? Give an example for each.

7. Describe the procedure for determining the extreme values of a function f of two variables whose partial derivatives exist throughout the closed unit disk centered at the origin.

Cumulative Review, Chapters 1–12

In Exercises 1–2 find the limit.

1. $\displaystyle\lim_{h\to 0}\frac{\sqrt{x - h} - \sqrt{x}}{h}$ **2.** $\displaystyle\lim_{x\to 0^+}(\tan^{-1} x)^x$

3. Consider the parabola $y = 4x^2$. Find an equation for each tangent to the parabola that passes through the point $(0, -2)$.

4. Let $f(x) = \ln(x + \sqrt{1 + x^2})$. Show that $\displaystyle\lim_{x\to\infty} xf'(x) = 1$.

5. Let n be a positive integer, and $f(x) = (\sin nx)(\sin^n x)$. Show that $f'(x) = n(\sin^{n-1} x)(\sin(n + 1)x)$.

6. Let $f(x) = \int_{x^2}^{\pi/2}\sin\sqrt{t}\,dt$. Find $f'(x)$.

7. Suppose that $\sqrt{x} + \sqrt{y} = 9$. Use implicit differentiation to find d^2y/dx^2.

8. The sides of an equilateral triangle grow at the rate of 2 centimeters per minute. An inscribed circle is attached to the triangle and grows with the triangle. Determine the rate at which the area of the region between the circle and the triangle grows when the sides of the triangle are 12 centimeters long. (*Hint:* The radius of the circle is $\sqrt{3}/6$ times as long as a side of the triangle.)

9. Let

$$f(x) = \frac{2 - x^2}{1 - x^2}$$

Sketch the graph of f, indicating all pertinent information.

10. A manufacturer plans to create a rectangular box with square bottom and top out of 960 square inches of cardboard. If the bottom will have 3 layers of cardboard and the top 2 layers, determine the dimensions that will maximize the volume V. (Assume that the cardboard has negligible thickness.)

11. Find the area A of the region between the curves $y = 2x^3 + 2x^2 + 10x$ and $y = x^3 - 3x^2 + 4x$.

In Exercises 12–14 evaluate the integral.

12. $\displaystyle\int x^2 \tan^{-1} x \, dx$ 13. $\displaystyle\int (\cos^2 \theta - \cos^3 \theta) \, d\theta$

14. $\displaystyle\int \frac{x^5}{\sqrt{1 - x^2}} \, dx$

15. Determine whether

$$\int_1^\infty \frac{x^3}{(3 + x^4)^{3/2}} \, dx$$

converges or diverges. If it converges, determine its value.

16. A football-like solid is obtained by revolving around the x axis the graph of $y = x \cos x$ for $0 \le x \le \pi/2$. Determine the volume V of the solid.

17. Let R be the region in the xy plane between the graphs of $y = 2 - x^2$ and $y = 1$. Determine the center of gravity of R.

18. Sketch the polar graph of $r = 1 - 2 \sin \theta$.

19. Find the area A of the region inside the circle $r = \sin \theta$ and outside the circle $r = \sqrt{3} \cos \theta$.

20. Evaluate $\displaystyle\lim_{n \to \infty} \frac{(\sin n!) \ln n}{n}$.

21. Find the numerical value of $\displaystyle\sum_{n=1}^{\infty} \left(\frac{1}{\sqrt{n}} - \frac{1}{\sqrt{n+1}} \right)$.

22. Determine the interval of convergence of $\displaystyle\sum_{n=1}^{\infty} \frac{x^n}{n4^n}$.

23. Let $f(x) = 2x^3 e^{-(x^3)}$. Find the Taylor series of f and f', and determine their radii of convergence.

24. Determine which, if either, is farther from the plane $x - 2y + 3z = 6$: the point $(-1, 3, 4)$ or the point $(1, 2, 0)$.

25. A curve C is parametrized by $\mathbf{r}(t) = t\mathbf{i} + t^2\mathbf{j} + \frac{2}{3}t^3\mathbf{k}$.

    a. Find the value of t for which the tangent vector $\mathbf{T}(t)$ is parallel to the vector $\mathbf{i} - \mathbf{j} + \frac{1}{2}\mathbf{k}$.

    b. Find the curvature $\kappa(t)$ for an arbitrary value of t, and show that it is maximum for $t = 0$.

26. Solve the differential equation

$$\frac{dy}{dx} = \frac{ay + b}{cy + d}$$

where a, b, c, and d are constants with $a \ne 0$.

An iceberg. (*H. Kanus/Superstock*)

14 MULTIPLE INTEGRALS

Our definition in Chapter 5 of the integral $\int_a^b f(x)\, dx$ of a function that is continuous on an interval $[a, b]$ was motivated by means of area. Integrals of functions of two or three variables, which we will define in this chapter, will enable us to compute areas of more complicated regions, volumes of many types of solid regions, and masses and centers of gravity of two- and three-dimensional objects. The proofs of the theorems stated in this chapter are complicated and normally are given only in advanced calculus books; thus we will omit them.

14.1 DOUBLE INTEGRALS

From now on we will refer to the definite integral $\int_a^b f(x)\, dx$ of a function f that is continuous on an interval $[a, b]$ as a **single integral.** In this section we will define the double integral of a function that is continuous on a certain type of plane region. The definition of the double integral will be motivated by means of certain elementary properties we expect volume to possess.

897

Volume and the Double Integral

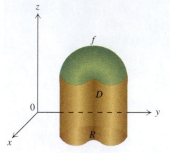

FIGURE 14.1

FIGURE 14.2

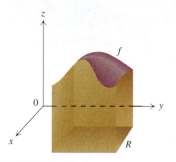

FIGURE 14.3

Consider a region R in the xy plane, a function f that is nonnegative and continuous on R, and the solid region D shown in Figure 14.1, bounded below by R, above by the graph of f, and on the sides by the vertical surface passing through the boundary of R. We call D the **solid region between the graph of f and R.** Our goal is to define the volume of D.

We make the basic assumption that if a rectangular parallelepiped has height c and base area A, then its volume V is given by

$$V = cA$$

(Figure 14.2). This is a three-dimensional version of the Rectangle Property of single variables.

To determine the volume of the more general region D, first we consider the case in which f is nonnegative and continuous on a rectangle R. The solid region D between the graph of f and R is shown in Figure 14.3. We partition R into subrectangles $R_1, R_2, \ldots, R_n$. For $1 \leq k \leq n$, we erect a rectangular parallelepiped whose base is the subrectangle R_k and whose height is $f(x_k, y_k)$, where (x_k, y_k) is an arbitrarily chosen point in R_k (see Figure 14.4). Denoting the area of R_k by ΔA_k, we find that the volume V_k of the kth solid is $f(x_k, y_k)\, \Delta A_k$. If the dimensions of the rectangles are small, then the n parallelepipeds together have a volume that is approximately the volume V of D. In mathematical terms this means that

$$V \approx \sum_{k=1}^{n} f(x_k, y_k)\, \Delta A_k \qquad (1)$$

Moreover, it seems reasonable that by making the dimensions of all the subrectangles sufficiently small, we could ensure that the sum in (1) is as close to V as we like. It turns out that if f is continuous on the rectangle R (as we have assumed), then even if f assumes negative values, the sum on the right side of (1) approaches a unique limit I as the dimensions of the subrectangles shrink to 0, no matter how the points (x_k, y_k) for $1 \leq k \leq n$ are chosen in their respective subrectangles. If f is nonnegative and continuous on R, we then define the volume V of D to be the number I, as the preceding discussion suggests.

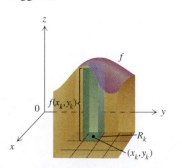

FIGURE 14.4

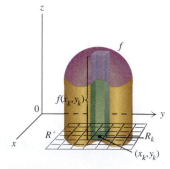

FIGURE 14.5

Next we assume once again that f is continuous and nonnegative on R, but we relax the assumption that R is a rectangle, and assume only that R is closed and bounded, as in Figure 14.1. This means that R contains its boundary and is contained in a rectangle R'. We partition R' into a collection P of subrectangles.

In general, some of the subrectangles in P will be entirely contained in R, some only partially contained in R, and some will contain no points of R (Figure 14.5). We number the rectangles in P so that $R_1, R_2, \ldots, R_n$ are those that are entirely contained in R and, for the moment, we ignore the other subrectangles of R. As before, let (x_k, y_k) be an arbitrary point in R_k, ΔA_k the area of R_k, and V_k the volume of the parallelepiped with base R_k and height $f(x_k, y_k)$. Following the idea in Chapter 5, we let $\| P \|$ be the **norm** of P, that is, the largest dimension of the subrectangles comprising P. If the norm of P is small enough, then, as before, the collection of parallelepipeds have a volume that we should expect to be approximately the volume V of D, that is, $V \approx \sum_{k=1}^{n} f(x_k, y_k) \Delta A_k$. However, because we have ignored the subrectangles in P that are only partially contained in R, it is possible, depending on how complicated the boundary of R is, that the approximation might not be so good. More precisely, it could be that the boundary of R is so complicated that $\sum_{k=1}^{n} f(x_k, y_k) \Delta A_k$ does not approach a limit as the norm of P tends to 0. This dichotomy leads us to the following definition, which applies to functions that are not necessarily nonnegative.

DEFINITION 14.1

Let f be continuous on a closed, bounded region R in the xy plane. We say that f is **integrable** on R if there is a number I with the following property:

For every $\varepsilon > 0$ there is a number $\delta > 0$ such that for any rectangle R' containing R and any partition P of R' into subrectangles, of which $R_1, R_2, \ldots, R_n$ are those that are entirely contained in R,

$$\text{if } \| P \| < \delta, \quad \text{then} \quad \left| I - \sum_{k=1}^{n} f(x_k, y_k) \Delta A_k \right| < \varepsilon$$

for any choice of (x_k, y_k) in R_k for $1 \leq k \leq n$.

The property in Definition 14.1 is usually expressed as

$$I = \lim_{\|P\| \to 0} \sum_{k=1}^{n} f(x_k, y_k) \Delta A_k \tag{2}$$

The sum

$$\sum_{k=1}^{n} f(x_k, y_k) \Delta A_k$$

is called a **Riemann sum** for f on R. It is entirely analogous to the corresponding concept in Chapter 5 for functions of a single variable: The number I in (2) is the limit of the Riemann sums of f on R as the norm of the partitions approaches 0. For functions of two variables we will call the limit the double integral of f over R.

DEFINITION 14.2

Let R be a closed, bounded region in the xy plane, and let f be continuous on R.
a. If f is integrable on R, we call the number I in Definition 14.1 the **double integral** of f over R, and denote it by

$$\iint\limits_{R} f(x, y) \, dA$$

b. If f is nonnegative and integrable on R, then the volume V of the solid region between the graph of f and the xy plane is given by

$$V = \iint\limits_{R} f(x, y)\, dA$$

It is practically impossible, except in very special cases, to evaluate a double integral by computing Riemann sums. The next example illustrates how one computes a Riemann sum in a simple case.

EXAMPLE 1 Let R be the triangular region between the graph of $y = 1 - x$ and the x axis on $[-2, 1]$ (Figure 14.6), and suppose that

$$f(x, y) = 4 - y \quad \text{for } (x, y) \text{ in } R$$

Let R' and the partition $P = \{R_1, R_2, \ldots, R_9\}$ of R' be as shown in Figure 14.6. Find the Riemann sum for f that results from choosing (x_k, y_k) to be the midpoint of R_k for $1 \leq k \leq 3$.

Solution Notice that each of the rectangles has area 1, so that $\Delta A_k = 1$ for all k. Since only the three rectangles R_1, R_2, and R_3 in the partition are entirely contained in R, the required Riemann sum is

$$\sum_{k=1}^{3} f(x_k, y_k)\, \Delta A_k = f\left(-\frac{3}{2}, \frac{1}{2}\right) \Delta A_1 + f\left(-\frac{3}{2}, \frac{3}{2}\right) \Delta A_2 + f\left(-\frac{1}{2}, \frac{1}{2}\right) \Delta A_3$$

$$= \frac{7}{2} \cdot 1 + \frac{5}{2} \cdot 1 + \frac{7}{2} \cdot 1 = \frac{19}{2} \qquad \square$$

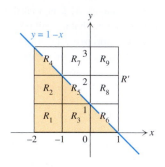

FIGURE 14.6

Vertically and Horizontally Simple Regions

As we have mentioned, it is difficult to compute double integrals by Riemann sums; it also happens that there are regions on which not every continuous function is integrable. Hence we restrict our attention to two particular types of regions, on which every continuous function is integrable and for which we have a straightforward method of computing double integrals.

DEFINITION 14.3

a. A plane region R is **vertically simple** if there are two continuous functions g_1 and g_2 on an interval $[a, b]$ such that $g_1(x) \leq g_2(x)$ for $a \leq x \leq b$ and such that R is the region between the graphs of g_1 and g_2 on $[a, b]$ (Figure 14.7). In this case we say that R is **the vertically simple region between the graphs of g_1 and g_2 on $[a, b]$.**

b. A plane region R is **horizontally simple** if there are two continuous functions h_1 and h_2 on an interval $[c, d]$ such that $h_1(y) \leq h_2(y)$ for $c \leq y \leq d$ and such that R is the region between the graphs of h_1 and h_2 on $[c, d]$ (Figure 14.8). In this case we say that R is **the horizontally simple region between the graphs of h_1 and h_2 on $[c, d]$.**

c. A plane region R is **simple** if it is both vertically simple and horizontally simple (Figure 14.9).

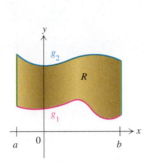

FIGURE 14.7   A vertically simple region.

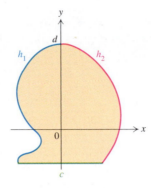

FIGURE 14.8   A horizontally simple region.

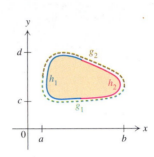

FIGURE 14.9   A simple region.

Vertical lines intersect the boundary of a vertically simple region R at most twice, except for those vertical lines composing part of the boundary of R. Rectangles, triangles, and circles are vertically simple, as are the regions depicted in Figure 14.10(a) and (b); however, the regions in Figure 14.10(c) and (d) are not vertically simple. Similarly, horizontal lines intersect the boundary of a horizontally simple region R at most twice, except for those horizontal lines comprising part of the boundary of R. Again rectangles, triangles, and circles are horizontally simple, as are the regions drawn in Figure 14.10(a) and (c). However, the regions in Figure 14.10(b) and (d) are not horizontally simple. Finally, the region in Figure 14.10(a) is simple, and those in Figure 14.10(b)–(d) are not simple.

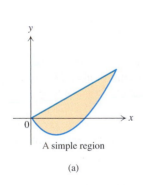

A simple region

(a)

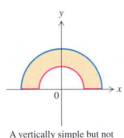

A vertically simple but not horizontally simple region

(b)

A horizontally simple but not vertically simple region

(c)

A region that is neither vertically nor horizontally simple

(d)

FIGURE 14.10

EXAMPLE 2   Let R be the region between the graphs of $y = x^2$ and $y = x + 6$. Show that R is simple.

Solution   First we determine where the two graphs intersect. Observe that

$$y = x^2 \quad \text{and} \quad y = x + 6$$

only if

$$x^2 = y = x + 6$$

which means that $x = -2$ or $x = 3$. Thus the graphs intersect at the points $(-2, 4)$ and

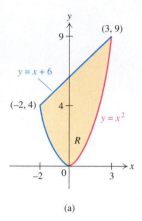
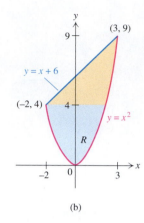

(a)   (b)

FIGURE 14.11

$(3, 9)$ (Figure 14.11(a)). Therefore if $g_1(x) = x^2$ and $g_2(x) = x + 6$, then $g_1 \leq g_2$ on $[-2, 3]$, so R is the vertically simple region between the graphs of g_1 and g_2 on $[-2, 3]$. To prove that R is horizontally simple, we notice that R is composed of two portions, one for which $0 \leq y \leq 4$ and the other for which $4 \leq y \leq 9$ (Figure 14.11(b)). Thus (x, y) is in R provided that either

$$0 \leq y \leq 4 \quad \text{and} \quad -\sqrt{y} \leq x \leq \sqrt{y}$$

or

$$4 \leq y \leq 9 \quad \text{and} \quad y - 6 \leq x \leq \sqrt{y}$$

Consequently if we let

$$h_1(y) = \begin{cases} -\sqrt{y} & \text{for } 0 \leq y \leq 4 \\ y - 6 & \text{for } 4 \leq y \leq 9 \end{cases} \quad \text{and} \quad h_2(y) = \sqrt{y}$$

then $h_1 \leq h_2$ on $[0, 9]$, so R is the horizontally simple region between the graphs of h_1 and h_2 on $[0, 9]$. Since R is both vertically and horizontally simple, R is simple by definition. ❑

Although the region in Example 2 is both vertically and horizontally simple, it is more easily described as a vertically simple region than as a horizontally simple region.

Evaluation of Double Integrals

To explain how we can evaluate double integrals over vertically simple regions, let f be nonnegative and continuous on a vertically simple region R between the graphs of g_1 and g_2 on $[a, b]$. Let D be the solid region between the graph of f and R. Since f is continuous in each variable separately, the cross-sectional area $A(x)$ of D at any x in $[a, b]$ is given by

$$A(x) = \int_{g_1(x)}^{g_2(x)} f(x, y) \, dy \tag{3}$$

(Figure 14.12). It can be shown that A is continuous on $[a, b]$; hence by the cross-sectional definition of volume in Section 6.1 and (3), the volume V of D is given by

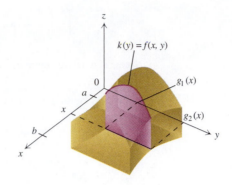

FIGURE 14.12

$$V = \int_a^b A(x)\, dx = \int_a^b \left[\int_{g_1(x)}^{g_2(x)} f(x, y)\, dy \right] dx$$

But by Definition 14.2, V is also given by

$$V = \iint_R f(x, y)\, dA$$

Therefore

$$\iint_R f(x, y)\, dA = \int_a^b \left[\int_{g_1(x)}^{g_2(x)} f(x, y)\, dy \right] dx$$

Similarly, if R is the horizontally simple region between the graphs of h_1 and h_2 on $[c, d]$, then

$$\iint_R f(x, y)\, dA = \int_c^d \left[\int_{h_1(y)}^{h_2(y)} f(x, y)\, dx \right] dy$$

The integrals

$$\int_a^b \left[\int_{g_1(x)}^{g_2(x)} f(x, y)\, dy \right] dx \quad \text{and} \quad \int_c^d \left[\int_{h_1(y)}^{h_2(y)} f(x, y)\, dx \right] dy \qquad (4)$$

Iterated Integrals

All early double integration involved iterated integrals. Although several mathematicians had used them, it was Euler who presented the first systematic account of iterated integrals in a paper in 1769.

are called **iterated integrals,** because they are performed iteratively. That is, to evaluate $\int_a^b \left[\int_{g_1(x)}^{g_2(x)} f(x, y)\, dy \right] dx$, we first evaluate $\int_{g_1(x)}^{g_2(x)} f(x, y)\, dy$ with x fixed and then integrate the resulting function with respect to x. Normally we omit the brackets appearing in (4) and write

$$\int_a^b \int_{g_1(x)}^{g_2(x)} f(x, y)\, dy\, dx \quad \text{for} \quad \int_a^b \left[\int_{g_1(x)}^{g_2(x)} f(x, y)\, dy \right] dx$$

and

$$\int_c^d \int_{h_1(y)}^{h_2(y)} f(x, y) \, dx \, dy \quad \text{for} \quad \int_c^d \left[\int_{h_1(y)}^{h_2(y)} f(x, y) \, dx \right] dy$$

Our results are summarized in the following theorem.

THEOREM 14.4

Let f be continuous on a region R in the xy plane.

a. If R is the vertically simple region between the graphs of g_1 and g_2 on $[a, b]$, then f is integrable on R, and

$$\iint_R f(x, y) \, dA = \int_a^b \int_{g_1(x)}^{g_2(x)} f(x, y) \, dy \, dx$$

b. If R is the horizontally simple region between the graphs of h_1 and h_2 on $[c, d]$, then f is integrable on R, and

$$\iint_R f(x, y) \, dA = \int_c^d \int_{h_1(y)}^{h_2(y)} f(x, y) \, dx \, dy$$

Observe that if R is simple, then by Theorem 14.4(a) and (b),

$$\iint_R f(x, y) \, dA = \int_a^b \int_{g_1(x)}^{g_2(x)} f(x, y) \, dy \, dx = \int_c^d \int_{h_1(y)}^{h_2(y)} f(x, y) \, dx \, dy$$

In $\int_a^b \int_{g_1(x)}^{g_2(x)} f(x, y) \, dy \, dx$ the limits of integration for y must be functions of x alone (appearances of y are not allowed), and the limits of integration for x must be constants (appearances of x or y are not allowed). Analogously, in $\int_c^d \int_{h_1(y)}^{h_2(y)} f(x, y) \, dx \, dy$ the limits of integration for x must be functions of y alone, and the limits of integration for y must be constants.

EXAMPLE 3 Let R be the rectangular region bounded by the lines $x = -1$, $x = 2$, $y = 0$, and $y = 2$. Find $\iint_R x^2 y \, dA$.

Solution The region R is the vertically simple region between the graphs of

$$y = 0 \quad \text{and} \quad y = 2 \quad \text{for } -1 \le x \le 2$$

Therefore

$$\iint_R f(x, y) \, dA = \int_{-1}^2 \int_0^2 x^2 y \, dy \, dx$$

To evaluate the iterated integral, we first compute $\int_0^2 x^2 y \, dy$ for each x in $[-1, 2]$. We obtain

$$\int_0^2 x^2 y \, dy = x^2 \int_0^2 y \, dy = x^2 \left(\frac{1}{2} y^2 \right) \Big|_0^2 = 2x^2$$

because x is held constant when we integrate with respect to y. We conclude that

$$\iint\limits_{R} f(x, y)\, dA = \int_{-1}^{2}\int_{0}^{2} x^2 y\, dy\, dx = \int_{-1}^{2} 2x^2\, dx = \frac{2}{3} x^3 \Big|_{-1}^{2} = 6 \qquad \square$$

We can also evaluate the double integral in Example 3 by regarding R as the horizontally simple region between the graphs of

$$x = -1 \quad \text{and} \quad x = 2 \quad \text{for } 0 \leq y \leq 2$$

Using this approach, we obtain

$$\iint\limits_{R} f(x, y)\, dA = \int_{0}^{2}\int_{-1}^{2} x^2 y\, dx\, dy = \int_{0}^{2} y\left(\frac{1}{3} x^3 \Big|_{-1}^{2}\right) dy$$

$$= \int_{0}^{2} y\left(\frac{8}{3} - \left(-\frac{1}{3}\right)\right) dy = \int_{0}^{2} 3y\, dy = \frac{3}{2} y^2 \Big|_{0}^{2} = 6$$

EXAMPLE 4   Evaluate $\iint_{R}(4-y)\, dA$, where R is the triangular region in Figure 14.13.

Solution   Again the region is simple, so that we can calculate the double integral by means of two different iterated integrals. Let us regard R as the vertically simple region between the graphs of

$$y = 0 \quad \text{and} \quad y = 1 - x \quad \text{for } -2 \leq x \leq 1$$

It follows that

$$\iint\limits_{R}(4-y)\, dA = \int_{-2}^{1}\int_{0}^{1-x}(4-y)\, dy\, dx = \int_{-2}^{1}\left(4y - \frac{y^2}{2}\right)\Big|_{0}^{1-x} dx$$

$$= \int_{-2}^{1}\left(4(1-x) - \frac{(1-x)^2}{2}\right) dx = \int_{-2}^{1}\left(\frac{7}{2} - 3x - \frac{x^2}{2}\right) dx$$

$$= \left(\frac{7}{2} x - \frac{3}{2} x^2 - \frac{x^3}{6}\right)\Big|_{-2}^{1} = \frac{27}{2} \qquad \square$$

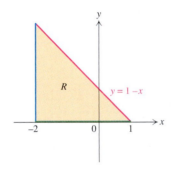

FIGURE 14.13

When the region R in Example 4 is regarded as the horizontally simple region between the graphs of

$$x = -2 \quad \text{and} \quad x = 1 - y \quad \text{for } 0 \leq y \leq 3$$

we find that

$$\iint\limits_{R}(4-y)\, dA = \int_{0}^{3}\int_{-2}^{1-y}(4-y)\, dx\, dy = \int_{0}^{3}(4-y)x \Big|_{-2}^{1-y} dy$$

$$= \int_{0}^{3}(4-y)[(1-y) - (-2)]\, dy = \int_{0}^{3}(12 - 7y + y^2)\, dy$$

$$= \left(12y - \frac{7}{2} y^2 + \frac{y^3}{3}\right)\Big|_{0}^{3} = \frac{27}{2}$$

Thus we obtain the same value for $\iint_{R} f(x, y)\, dA$, whether we consider R as a vertically simple region or a horizontally simple region.

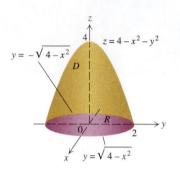

FIGURE 14.14

EXAMPLE 5 Find the volume V of the solid region D bounded above by the paraboloid $z = 4 - x^2 - y^2$ and below by the xy plane.

Solution The intersection of D and the xy plane is the region R bounded by the circle $x^2 + y^2 = 4$ (Figure 14.14). Since $z = 4 - x^2 - y^2 \geq 0$ for all (x, y) in R, it follows from Definition 14.2(b) that

$$V = \iint_R (4 - x^2 - y^2) \, dA$$

Since R is the vertically simple region between the graphs of

$$y = -\sqrt{4 - x^2} \quad \text{and} \quad y = \sqrt{4 - x^2} \quad \text{for } -2 \leq x \leq 2$$

we conclude that

$$V = \int_{-2}^{2} \int_{-\sqrt{4-x^2}}^{\sqrt{4-x^2}} (4 - x^2 - y^2) \, dy \, dx$$

$$= \int_{-2}^{2} \left((4 - x^2)y - \frac{y^3}{3} \right) \Big|_{-\sqrt{4-x^2}}^{\sqrt{4-x^2}} dx$$

$$= 2 \int_{-2}^{2} \left((4 - x^2)\sqrt{4 - x^2} - \frac{(4 - x^2)^{3/2}}{3} \right) dx$$

$$= \frac{4}{3} \int_{-2}^{2} (4 - x^2)^{3/2} \, dx \overset{x = 2 \sin \theta}{=} \frac{64}{3} \int_{-\pi/2}^{\pi/2} \cos^4 \theta \, d\theta$$

$$\overset{\substack{(13) \text{ of} \\ \text{Section 8.1}}}{=} \frac{64}{3} \left(\frac{1}{4} \cos^3 \theta \sin \theta + \frac{3}{8} \cos \theta \sin \theta + \frac{3}{8} \theta \right) \Big|_{-\pi/2}^{\pi/2} = 8\pi \quad \square$$

In Section 14.2 we will discuss the use of polar coordinates to evaluate double integrals. With polar coordinates the calculations needed in order to find the volume of the region in Example 5 will be greatly reduced.

We define the area A of a plane region R by

$$A = \iint_R 1 \, dA \tag{5}$$

When R is the region between the graphs of two continuous functions g_1 and g_2 on $[a, b]$ such that $g_1 \leq g_2$, we have two definitions of the area A of R: (5) above and Definition 5.22 in Section 5.8. However, since Theorem 14.4 implies that

$$A = \iint_R 1 \, dA = \int_a^b \int_{g_1(x)}^{g_2(x)} 1 \, dy \, dx = \int_a^b [g_2(x) - g_1(x)] \, dx$$

both definitions yield the same value for A. For instance, if R is the circle of radius

3 centered at the origin, then

$$A = \iint\limits_{R} 1 \, dA = \int_{-3}^{3} \int_{-\sqrt{9-x^2}}^{\sqrt{9-x^2}} 1 \, dy \, dx$$

$$= 2 \int_{-3}^{3} \sqrt{9-x^2} \, dx = 18 \int_{-\pi/2}^{\pi/2} \cos^2 \theta \, d\theta$$

$$= 18 \int_{-\pi/2}^{\pi/2} \left(\frac{1}{2} + \frac{1}{2} \cos 2\theta \right) d\theta$$

$$= 9 \left(\theta + \frac{1}{2} \sin 2\theta \right) \Bigg|_{-\pi/2}^{\pi/2} = 9\pi$$

Reversing the Order of Integration

In case R is a simple region, $\iint_R f(x, y) \, dA$ can be evaluated as either $\int_a^b \int_{g_1(x)}^{g_2(x)} f(x, y) \, dy \, dx$ or $\int_c^d \int_{h_1(y)}^{h_2(y)} f(x, y) \, dx \, dy$. Which iterated integral we use depends on the situation—the integrand, the limits of integration, and our convenience. It sometimes happens that one iterated integral is either difficult or impossible to evaluate, whereas the other iterated integral can be evaluated easily. The change from one iterated integral to the other is called **reversing the order of integration,** since it involves changing from $dy \, dx$ to $dx \, dy$, or vice versa.

EXAMPLE 6   By reversing the order of integration, evaluate

$$\int_{0}^{9} \int_{\sqrt{y}}^{3} \sin \pi x^3 \, dx \, dy$$

Solution   Notice that there is no simple antiderivative of $\sin \pi x^3$, so we cannot evaluate the given iterated integral. This suggests reversing the order of integration. To reverse the order of integration we must determine the region R over which the integration is performed. From the limits on the given iterated integral we infer that R is the horizontally simple region between the graphs of

$$x = \sqrt{y} \quad \text{and} \quad x = 3 \quad \text{for } 0 \le y \le 9$$

(Figure 14.15). Since R is also the vertically simple region between the graphs of

$$y = 0 \quad \text{and} \quad y = x^2 \quad \text{for } 0 \le x \le 3$$

we find that

$$\int_{0}^{9} \int_{\sqrt{y}}^{3} \sin \pi x^3 \, dx \, dy = \iint\limits_{R} \sin \pi x^3 \, dA = \int_{0}^{3} \int_{0}^{x^2} \sin \pi x^3 \, dy \, dx$$

$$= \int_{0}^{3} y \sin \pi x^3 \Bigg|_{0}^{x^2} dx = \int_{0}^{3} x^2 \sin \pi x^3 \, dx$$

$$= \frac{-1}{3\pi} \cos \pi x^3 \Bigg|_{0}^{3} = \frac{2}{3\pi} \qquad \square$$

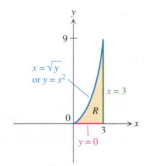

FIGURE 14.15

Reversing the order of integration applies only to simple regions. If such a region R is given as a vertically simple (respectively, horizontally simple) region, then reversing the order of integration amounts to formulating R as a horizontally simple (respectively, vertically simple) region.

Double Integration over More General Regions

If R is composed of two or more vertically or horizontally simple subregions R_1, $R_2, \ldots, R_n$, with the property that any two subregions have only boundaries in common (see Figure 14.16 for an example), then any function f that is continuous on R is integrable on R, and

$$\iint_R f(x, y)\, dA = \iint_{R_1} f(x, y)\, dA + \iint_{R_2} f(x, y)\, dA + \cdots + \iint_{R_n} f(x, y)\, dA \quad (6)$$

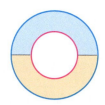

FIGURE 14.16

EXAMPLE 7 Let R be the region between the graphs of $y = x$ and $y = x^3$. Evaluate $\iint_R (x - 1)\, dA$.

Solution   The graphs of the two equations intersect at (x, y) if

$$x = y = x^3$$

which happens if $x = -1$, $x = 0$, or $x = 1$. Therefore R is composed of the two vertically simple regions R_1 and R_2, as depicted in Figure 14.17. Consequently

$$\iint_R (x - 1)\, dA = \iint_{R_1} (x - 1)\, dA + \iint_{R_2} (x - 1)\, dA$$

$$= \int_{-1}^0 \int_x^{x^3} (x - 1)\, dy\, dx + \int_0^1 \int_{x^3}^x (x - 1)\, dy\, dx$$

$$= \int_{-1}^0 (x - 1)y \Big|_x^{x^3}\, dx + \int_0^1 (x - 1)y \Big|_{x^3}^x\, dx$$

$$= \int_{-1}^0 (x - 1)(x^3 - x)\, dx + \int_0^1 (x - 1)(x - x^3)\, dx$$

$$= \int_{-1}^0 (x^4 - x^3 - x^2 + x)\, dx - \int_0^1 (x^4 - x^3 - x^2 + x)\, dx$$

$$= \left(\frac{x^5}{5} - \frac{x^4}{4} - \frac{x^3}{3} + \frac{x^2}{2} \right)\Big|_{-1}^0 - \left(\frac{x^5}{5} - \frac{x^4}{4} - \frac{x^3}{3} + \frac{x^2}{2} \right)\Big|_0^1$$

$$= -\frac{1}{2} \quad \square$$

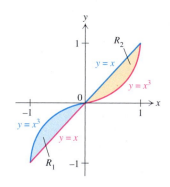

FIGURE 14.17

Symmetry in Double Integrals

Let us reconsider the solid region D whose volume we calculated in Example 5 by evaluating the double integral

$$\iint_R (4 - x^2 - y^2)\, dA, \quad \text{where } R \text{ is the disk } x^2 + y^2 \leq 4$$

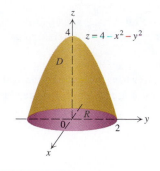

FIGURE 14.18

(Figure 14.18). Notice that R is symmetric with respect to the y axis and $4 - x^2 - y^2$ is even as a function of x alone. Therefore if R_1 denotes the front half of the disk R (for which $x \geq 0$) and R_2 the rear half of R (for which $x \leq 0$), then

$$\iint\limits_{R_1} (4 - x^2 - y^2)\, dA = \iint\limits_{R_2} (4 - x^2 - y^2)\, dA$$

Along with (6), this means that

$$\iint\limits_{R} (4 - x^2 - y^2)\, dA = \iint\limits_{R_1} (4 - x^2 - y^2)\, dA + \iint\limits_{R_2} (4 - x^2 - y^2)\, dA$$

$$= 2 \iint\limits_{R_1} (4 - x^2 - y^2)\, dx$$

Similarly, since R is symmetric with respect to the x axis and $4 - x^2 - y^2$ is even as a function of y alone,

$$\iint\limits_{R} (4 - x^2 - y^2)\, dA = 2 \iint\limits_{R_3} (4 - x^2 - y^2)\, dA$$

where R_3 is the right half of R (for which $y \geq 0$). Combining both symmetries together, we conclude that

$$\iint\limits_{R} (4 - x^2 - y^2)\, dA = 4 \iint\limits_{R_4} (4 - x^2 - y^2)\, dA$$

where R_4 is the portion of R in the first quadrant. Consequently

$$\iint\limits_{R} (4 - x^2 - y^2)\, dA = 4 \int_0^2 \int_0^{\sqrt{4-x^2}} (4 - x^2 - y^2)\, dy\, dx$$

This iterated integral is somewhat easier to evaluate than the one used in the solution of Example 5.

More generally, if R is symmetric with respect to the y axis and f is even as a function of x alone, then

$$\iint\limits_{R} f(x, y)\, dA = 2 \iint\limits_{R_1} f(x, y)\, dA$$

where R_1 is the half of R on one side of the yz plane. Examples are

$$f(x, y) = x^2 y^3, \quad f(x, y) = \cos x \sin y, \quad \text{and} \quad f(x, y) = e^y \ln(x^4 + 1)$$

Analogously, if R is symmetric with respect to the x axis and f is even as a function of y alone, then

$$\iint\limits_{R} f(x, y)\, dA = 2 \iint\limits_{R_2} f(x, y)\, dA$$

where R_2 is the half of R on one side of the xz plane.

By contrast, consider the integral $\iint_R y^3\, dA$, where R again is the disk $x^2 + y^2 \le 4$. Let R_1 and R_2 be the left and right halves of R, respectively. Since y^3 is an odd function, we deduce that

$$\iint\limits_{R_2} y^3\, dA = -\iint\limits_{R_1} y^3\, dA \qquad (7)$$

Combining (6) and (7) yields

$$\iint\limits_{R} y^3\, dA = \iint\limits_{R_1} y^3\, dA + \iint\limits_{R_2} y^3\, dA = \iint\limits_{R_1} y^3\, dA - \iint\limits_{R_1} y^3\, dA = 0$$

More generally, if R is symmetric with respect to the x axis and if f is an odd function when considered as a function of y alone, then $\iint_R f(x, y)\, dA = 0$. Analogously, if R is symmetric with respect to the y axis and if f is odd when considered as a function of x alone, then $\iint_R f(x, y)\, dA = 0$. For example, suppose that R is the semielliptical region $x^2 + 4y^2 \le 4$ for which $x \ge 0$ (Figure 14.19). If $f(x, y) = \sin(x^2 y^3)$, then since $\sin(x^2 y^3)$ is odd as a function of y and R is symmetric with respect to the x axis, we can conclude without trying to evaluate the double integral that

$$\iint\limits_{R} \sin(x^2 y^3)\, dA = 0$$

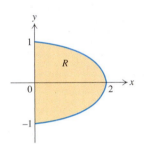

FIGURE 14.19   *R is the semi-elliptical region $x^2 + 4y^2 \le 4,\ x \ge 0$.*

If R were a different region that did not possess the appropriate symmetry property, then there would be virtually no hope of evaluating $\iint_R \sin(x^2 y^3)\, dA$.

As the preceding observations suggest, it is good to be watchful for opportunities to use symmetry in order to facilitate evaluation of double integrals.

EXERCISES 14.1

1. Let R, f, R', and P be as in Example 1. Find the Riemann sum that employs
   a. the lower left vertices of R_1, R_2, and R_3.
   b. the upper right vertices of R_1, R_2, and R_3.

2. Let R be the triangular region bounded by the lines $y = 2x$, $x = 0$, and $y = 4$, and let R' be the rectangle whose sides are the lines $x = 0$, $x = 2$, $y = 0$, and $y = 4$. Suppose the partition P of R' consists of the squares whose sides are 1 unit long, and consider those that are entirely contained in R'. If $f(x, y) = x + y$ for (x, y) in R, find the Riemann sum of f that

a. uses the lower left vertices of the rectangles of P.
b. uses the midpoints of the rectangles of P.

In Exercises 3–19 evaluate the iterated integral.

3. $\displaystyle\int_0^1 \int_{-1}^1 x\, dy\, dx$

4. $\displaystyle\int_{-5}^{-3} \int_{-2}^3 y\, dy\, dx$

5. $\displaystyle\int_0^1 \int_0^1 e^{x+y}\, dy\, dx$

6. $\displaystyle\int_0^1 \int_1^5 \frac{1}{r}\, dr\, ds$

7. $\displaystyle\int_0^1 \int_x^{x^2} 1\, dy\, dx$

8. $\displaystyle\int_0^1 \int_0^3 x\sqrt{x^2 + y}\, dy\, dx$

9. $\int_0^1 \int_0^y x\sqrt{y^2 - x^2}\, dx\, dy$ **10.** $\int_0^2 \int_0^{\sqrt{4-y^2}} y\, dx\, dy$

11. $\int_0^2 \int_0^{\sqrt{4-y^2}} x\, dx\, dy$ **12.** $\int_0^{2\pi} \int_0^1 r\sin\theta\, dr\, d\theta$

13. $\int_0^{2\pi} \int_0^1 r\sqrt{1-r^2}\, dr\, d\theta$ **14.** $\int_0^{2\pi} \int_0^{1+\cos\theta} r\, dr\, d\theta$

15. $\int_1^3 \int_0^x \frac{2}{x^2+y^2}\, dy\, dx$ **16.** $\int_1^e \int_1^{\ln y} e^x\, dx\, dy$

17. $\int_0^1 \int_0^x e^{(x^2)}\, dy\, dx$ **18.** $\int_{\ln(\pi/6)}^{\ln(\pi/2)} \int_0^{e^y} \cos e^y\, dx\, dy$

19. $\int_{-1}^1 \int_0^2 x^{15} e^{x^2 y^2}\, dy\, dx$

In Exercises 20–33 express the double integral as an iterated integral and evaluate it.

20. $\iint_R (x+y)\, dA$; R is the rectangular region bounded by the lines $x=2$, $x=3$, $y=4$, and $y=6$.

21. $\iint_R (x+y)\, dA$; R is the triangular region bounded by the lines $y=2x$, $x=0$, and $y=4$.

22. $\iint_R xy^2\, dA$; R is the region between the parabola $x = 4-y^2$ and the y axis on $[-1, 1]$.

23. $\iint_R x\, dA$; R is the trapezoidal region bounded by the lines $x=3$, $x=5$, $y=1$, and $y=x$.

24. $\iint_R x(x-1)e^{xy}\, dA$; R is the triangular region bounded by the lines $x=0$, $y=0$, and $x+y=2$.

25. $\iint_R (3x-5)\, dA$; R is the triangular region bounded by the lines $y=5+x$, $y=-x+7$, and $x=10$.

26. $\iint_R xy\, dA$; R is the region bounded by the graphs of $y=x$ and $y=x^2$.

27. $\iint_R 1\, dA$; R is the region between the graphs of $y=1+x$ and $y=\sin x$ on $[\pi, 2\pi]$.

28. $\iint_R (4+x^2)\, dA$; R is the region bounded by the graphs of $y=1+x^2$ and $y=9-x^2$.

29. $\iint_R (1-y)\, dA$; R is the region bounded by the graphs of $x=y^2$ and $x=2-y$.

30. $\iint_R x\, dA$; R is the portion of the disk $x^2 + y^2 \le 16$ in the second quadrant.

31. $\iint_R x\sqrt{y^2+1}\, dA$; R is the region above the line $y = \frac{1}{2}$ and inside the circle $x^2+y^2=1$.

32. $\iint_R 2y\, dA$; R is the region between the graph of $y=\sin x$ and the x axis on $[0, 3\pi/2]$.

33. $\iint_R x^2\, dA$; R is the region between the graphs of $y = x^3 + x^2 + 1$ and $y = x^3 + x + 1$ on $[-1, 1]$.

In Exercises 34–44 find the volume V of the region, using the methods of this section.

34. The solid region bounded above by the parabolic sheet $z = 1 - x^2$, below by the xy plane, and on the sides by the planes $y = -1$ and $y = 2$.

35. The solid region bounded by the plane $x + 2y + 3z = 6$ and the coordinate planes

36. The solid region bounded by the planes $z = 1 + x + y$, $x = 2$, $y = 1$, and the coordinate planes

37. The solid region in the first octant bounded by the paraboloid $z = x^2 + y^2$, the plane $x + y = 1$, and the coordinate planes

38. The solid region in the first octant bounded by the parabolic sheet $z = x^2$ and the planes $x = 2y$, $y = 0$, $z = 0$, and $x = 2$

39. The solid region in the first octant bounded by the cylinder $x^2 + y^2 = 4$, the plane $z = y$, the xy plane, and the yz plane

40. The solid region bounded above by the parabolic sheet $z = y^2$, on the sides by the sheet $x = 1 - (y-1)^2$ and the plane $y = x$, and on the bottom by the xy plane

41. The solid region in the first octant that is common to the cylinders $x^2 + y^2 = 1$ and $x^2 + z^2 = 1$ (*Hint:* Over what region in the xy plane does the region lie? Integrate first with respect to y.)

42. The region bounded above by the sphere $x^2 + y^2 + z^2 = 1$ and below by the xy plane

43. The solid region bounded above by the surface $z = xy$, below by the xy plane, and on the sides by the plane $y = x$ and the surface $y = x^3$ (*Hint:* There are two parts to the region in the xy plane.)

44. The region bounded above by the plane $z = 10 + 2x + 3y$, below by the xy plane, and on the sides by the surfaces $y = x^2$ and $y = x$

In Exercises 45–50 the iterated integral represents the volume of a solid region D. Sketch the region D.

45. $\int_{-2}^1 \int_1^4 3\, dy\, dx$ **46.** $\int_{-3}^3 \int_{-\sqrt{9-x^2}}^{\sqrt{9-x^2}} 5\, dy\, dx$

47. $\int_{-5}^5 \int_{-\sqrt{25-x^2}}^{\sqrt{25-x^2}} \sqrt{25 - x^2 - y^2}\, dy\, dx$

48. $\int_0^5 \int_0^{\sqrt{25-x^2}} \sqrt{25 - x^2 - y^2}\, dy\, dx$

49. $\int_{-2}^2 \int_{-\sqrt{4-y^2}}^{\sqrt{4-y^2}} (16 - 4x^2 - 4y^2)\, dx\, dy$

50. $\int_{-8}^{8} \int_{-\sqrt{64-x^2}}^{\sqrt{64-x^2}} (16 - 2\sqrt{x^2 + y^2})\, dy\, dx$

In Exercises 51–56 find the area A of the region in the xy plane by the methods of this section.

51. The region bounded by the parabolas $y = x^2$ and $y = \sqrt{x}$ on $[1, 4]$

52. The region bounded by the graphs of the equations $y = \cosh x$ and $y = \sinh x$ on $[-1, 1]$ (Leave your answer in terms of cosh or sinh.)

53. The region between the parabolas $x = y^2$ and $x = 32 - y^2$

54. The region bounded by the parabolas $x = y^2$ and $x = 4y^2 - 3$

55. The region in the first quadrant that is bounded by the graph of $x = 2 - y^2$ and by the lines $x = 0$ and $x = y$

56. The region in the first quadrant that is bounded by the graphs of $x + y = 4$ and $y = 3/x$

In Exercises 57–64 reverse the order of integration and evaluate the resulting integral.

57. $\int_{0}^{1} \int_{y}^{1} e^{(x^2)}\, dx\, dy$ **58.** $\int_{1}^{4} \int_{\sqrt{y}}^{2} \sin\left(\frac{x^3}{3} - x\right) dx\, dy$

59. $\int_{0}^{2} \int_{1+y^2}^{5} ye^{(x-1)^2}\, dx\, dy$

60. $\int_{0}^{1} \int_{\sin^{-1}y}^{\pi/2} \sec^2(\cos x)\, dx\, dy$

61. $\int_{1}^{e} \int_{0}^{\ln x} y\, dy\, dx$

62. $\int_{1}^{e} \int_{1/e}^{1/y} \cos (x - \ln x)\, dx\, dy$

63. $\int_{0}^{\pi^{1/3}} \int_{y^2}^{\pi^{2/3}} \sin x^{3/2}\, dx\, dy$

64. $\int_{0}^{\sqrt{\pi/2}} \int_{y}^{\sqrt{\pi/2}} y^2 \sin x^2\, dx\, dy$ (*Hint:* After changing the order of integration, make the substitution $u = x^2$.)

In Exercises 65–67 approximate the integral with the help of a computer algebra system.

65. $\int_{0}^{1} \int_{0}^{1} 4e^{-(x^2 + y^2)}\, dy\, dx$ **66.** $\int_{0}^{4} \int_{0}^{4} 4e^{-(x^2 + y^2)}\, dy\, dx$

67. $\int_{0}^{1} \int_{0}^{x} \sqrt{x^4 + y^4}\, dy\, dx$ (*Hint:* How does the integral $\int_{0}^{1} \int_{0}^{x} \sqrt{x^4 + y^4}\, dy\, dx$ compare with $\int_{0}^{1} \int_{0}^{1} \sqrt{x^4 + y^4}\, dy\, dx$?)

Let R be a vertically or horizontally simple region in the xy plane, and let f be continuous on R. The **average value** (or **mean value**) f_{av} of f on R is given by

$$f_{av} = \frac{1}{\text{area of } R} \iint_R f(x, y)\, dA$$

In Exercises 68–69 find the average value of f on R.

68. $f(x, y) = x^2 \sin (xy)$; R is the region bounded by the right triangle with vertices $(0, 0)$, $(\sqrt{\pi/2}, 0)$, and $(\sqrt{\pi/2}, \sqrt{\pi/2})$.

69. $f(x, y) = x^{77} e^{x^2 y^4}$; R is the disk $x^2 + y^2 \le 4$.

70. Suppose f is continuous on a disk R with center (x_0, y_0). If $\iint_{R_0} f(x, y)\, dA = 0$ for every rectangle R_0 contained in R, show that $f(x_0, y_0) = 0$. (*Hint:* If $f(x_0, y_0) > 0$, then $f(x, y) \ge \frac{1}{2}f(x_0, y_0)$ for all (x, y) in some rectangle R_0 about (x_0, y_0).)

Applications

71. The table below includes heights (in meters above sea level) of an iceberg. The iceberg is contained in a rectangular region R that is 50 meters by 40 meters. The region R is subdivided into squares 10 meters on a side, and in each square a measurement of the height is taken. The results are the numbers in the table. Approximate the volume V of the portion of the iceberg above sea level.

6	8	7	8	5
7	9	9	5	3
6	8	6	3	0
8	7	4	2	1

An iceberg. (H. Kanus/Superstock)

72. A terraced hill is in the shape of an ellipse whose major axis is 80 meters long and minor axis is 40 meters long. Measurements of the height at each of the terraces are taken, and appear in Figure 14.20. Approximate the volume V of the hill.

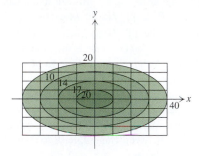

FIGURE 14.20   The terraced hill for Exercise 72.

73. A square glass surface 1 meter on a side has a variable color and, as a result, the fraction of incident light that it absorbs varies over the surface. Suppose the fraction it absorbs per unit area at (x, y) is $\frac{1}{3}(1 + y + x^2)$. Find the fraction of light absorbed by the entire surface.

Projects

1. In Section 14.1 we defined the double integral $\iint_R f(x, y)\, dA$, where R is a vertically or horizontally simple region in the xy plane. The definition involved circumscribing a rectangle R' about R, partitioning R' into subrectangles and considering the collection R_1, $R_2, \ldots, R_n$, which we will call P_e, that are entirely contained in R. We also mentioned the collection R_1, $R_2, \ldots, R_n, \ldots, R_m$, which we will call P_p, of subrectangles in R' that are partially contained in R. Let S_e and S_p be the corresponding sums defined by

$$S_e = \sum_{k=1}^{n} f(x_k, y_k)\, \Delta A_k \quad \text{and} \quad S_p = \sum_{k=1}^{m} f(x_k, y_k)\, \Delta A_k$$

Then S_e is the ordinary Riemann sum associated with the partition P_e, and S_p is an alternative "Riemann sum" that involves all the subrectangles of R' that have any point in common with R.

In this project we will return to Example 1, and derive results that suggest that the sums S_e and S_p should approach the same value as the norm of the partition of R' tends to 0. Recall from Example 1 that $f(x, y) = 4 - y$, and R is the triangular region in Figure 14.6. Next, let (x_k, y_k) be the lower left vertex of the kth subrectangle, and let ΔA_k the area of the kth subrectangle.

a. Let the side length of R_k be 1, for all k. Find S_e and S_p, and then compute $S_e - S_p$ (which is the difference in the sums).

b. Let the side length of R_k be 1/2, for all k. Compute $S_e - S_p$.

c. Let the side length of R_k be 1/10, for all k. Compute $S_e - S_p$.

d. Suppose that j is a positive integer, and let the side length of R_k be $1/j$, for all k. Find a formula for $S_e - S_p$. (Your formula will show that S_e and S_p get arbitrarily close together as j increases without bound.)

2. Suppose that f is continuous for $0 \le x \le 1$ and $0 \le y \le 1$, and that $0 < a < 1$. For each of the following iterated integrals, sketch the region R of integration. Then reverse the order of integration.

a. $\int_0^1 \int_0^{\min\{a, y\}} f(x, y)\, dx\, dy$, where $\min\{a, y\}$ denotes the smaller of the two values a and y (and depends on the value of y in the interval $[0, 1]$).

b. $\int_0^1 \int_{\max\{a, y\}}^{1} f(x, y)\, dx\, dy$

c. $\int_0^1 \int_0^{\min\{a, y^2\}} f(x, y)\, dx\, dy$

d. $\int_0^1 \int_{\max\{a, y^2\}}^{1} f(x, y)\, dx\, dy$

14.2   DOUBLE INTEGRALS IN POLAR COORDINATES

Certain curves in the xy plane, such as circles and cardioids, can be described more easily in polar coordinates than in rectangular coordinates. Thus it is reasonable to expect double integrals over regions enclosed by such curves to be more easily evaluated by means of polar coordinates.

The type of region over which we will integrate in polar coordinates can be described as follows. Suppose that h_1 and h_2 are continuous on an interval $[\alpha, \beta]$, where $0 \le \beta - \alpha \le 2\pi$, and that

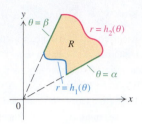

FIGURE 14.21 The region between the polar graphs of h_1 and h_2 on $[\alpha, \beta]$.

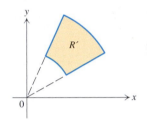

FIGURE 14.22 A circular arch.

$$0 \leq h_1(\theta) \leq h_2(\theta) \quad \text{for } \alpha \leq \theta \leq \beta$$

Let R be the closed region in the xy plane bounded by the lines $\theta = \alpha$ and $\theta = \beta$ and by the polar graphs of $r = h_1(\theta)$ and $r = h_2(\theta)$ (Figure 14.21). We say that R is **the region between the polar graphs of h_1 and h_2 on $[\alpha, \beta]$.** Such a region R need not be vertically or horizontally simple. Nevertheless, it turns out that every function that is continuous on R is integrable on R, so $\iint_R f(x, y)\, dA$ exists.

To explain intuitively how to evaluate $\iint_R f(x, y)\, dA$, we assume that f is continuous and nonnegative on R, so that $\iint_R f(x, y)\, dA$ is the volume of the solid region D between the graph of f and R. Although the definition of $\iint_R f(x, y)\, dA$ was based on partitions of R by rectangles, it is also possible to partition R by other types of regions. When working in polar coordinates, we first circumscribe R by a circular arch R', which is the region between two circular arcs (Figure 14.22), and then partition R' into circular subarches, numbered so that $R_1, R_2, \ldots, R_n$ are the ones entirely contained in R. For each k between 1 and n, let R_k be bounded by the lines $\theta = \alpha_k$ and $\theta = \beta_k$ and by the circular arcs $r = s_k$ and $r = t_k$ (Figure 14.23). Furthermore, let

$$r_k = \frac{1}{2}(s_k + t_k) \quad \text{and} \quad \theta_k = \frac{1}{2}(\alpha_k + \beta_k)$$

so that (r_k, θ_k) are polar coordinates of the "center" of R_k. From (3) of Section 10.2, we find that the area of R_k is given by

$$\frac{1}{2}t_k^2(\beta_k - \alpha_k) - \frac{1}{2}s_k^2(\beta_k - \alpha_k) = \frac{1}{2}(t_k + s_k)(t_k - s_k)(\beta_k - \alpha_k)$$

$$= r_k(t_k - s_k)(\beta_k - \alpha_k)$$

The volume of the solid region with height $f(r_k \cos \theta_k, r_k \sin \theta_k)$ and base R_k is approximately the volume of the portion of D that lies over R_k (Figure 14.23).

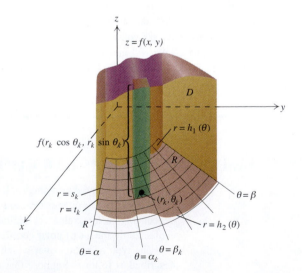

FIGURE 14.23

Therefore

$$\iint\limits_R f(x, y)\, dA \approx \sum_{k=1}^{n} f(r_k \cos \theta_k, r_k \sin \theta_k) r_k (t_k - s_k)(\beta_k - \alpha_k) \quad (1)$$

However, R corresponds to the horizontally simple region S in the $r\theta$ plane between the graphs of $r = h_1(\theta)$ and $r = h_2(\theta)$ on $[\alpha, \beta]$ (Figure 14.24), and each circular arch R_k into which we partitioned R' corresponds to the rectangle S_k in the $r\theta$ plane bounded by the lines $\theta = \alpha_k$, $\theta = \beta_k$, $r = s_k$, and $r = t_k$. The area ΔA_k of S_k

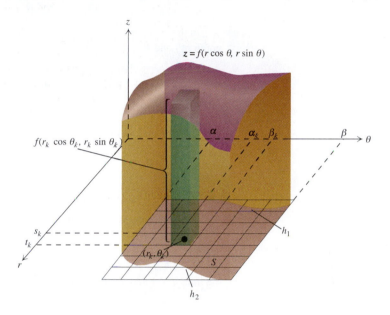

FIGURE 14.24

is $(t_k - s_k)(\beta_k - \alpha_k)$. Consequently

$$\sum_{k=1}^{n} f(r_k \cos \theta_k, r_k \sin \theta_k) r_k (t_k - s_k)(\beta_k - \alpha_k) = \sum_{k=1}^{n} f(r_k \cos \theta_k, r_k \sin \theta_k) r_k\, \Delta A_k \quad (2)$$

and the right side of (2) is a Riemann sum associated with the double integral $\iint_S f(r \cos \theta, r \sin \theta) r\, dA$, so that

$$\sum_{k=1}^{n} f(r_k \cos \theta, r_k \sin \theta) r_k\, \Delta A_k \approx \iint\limits_S f(r \cos \theta, r \sin \theta) r\, dA \quad (3)$$

Since S is horizontally simple, we know from Theorem 14.4(b) that

$$\iint\limits_S f(r \cos \theta, r \sin \theta) r\, dA = \int_{\alpha}^{\beta} \int_{h_1(\theta)}^{h_2(\theta)} f(r \cos \theta, r \sin \theta) r\, dr\, d\theta \quad (4)$$

Equations (1)–(4) suggest the following theorem.

THEOREM 14.5

Suppose that h_1 and h_2 are continuous on $[\alpha, \beta]$, where $0 \le \beta - \alpha \le 2\pi$, and that $0 \le h_1(\theta) \le h_2(\theta)$ for $\alpha \le \theta \le \beta$. Let R be the region between the polar graphs of

$$r = h_1(\theta) \quad \text{and} \quad r = h_2(\theta) \quad \text{for } \alpha \le \theta \le \beta$$

If f is continuous on R, then

$$\iint\limits_{R} f(x, y)\, dA = \int_{\alpha}^{\beta} \int_{h_1(\theta)}^{h_2(\theta)} f(r \cos\theta, r \sin\theta) r\, dr\, d\theta \qquad (5)$$

In the event that f is nonnegative on R, the volume V of the region between the graph of f and R is given by

$$V = \int_{\alpha}^{\beta} \int_{h_1(\theta)}^{h_2(\theta)} f(r \cos\theta, r \sin\theta) r\, dr\, d\theta$$

and the area A of R is given by

$$A = \int_{\alpha}^{\beta} \int_{h_1(\theta)}^{h_2(\theta)} r\, dr\, d\theta$$

These two formulas follow from (5) and the formulas for volume and area presented in Section 14.1.

EXAMPLE 1 Suppose R is the region bounded by the circles $r = 1$ and $r = 2$ and the lines $\theta = \alpha$ and $\theta = \beta$, where $0 \le \beta - \alpha \le 2\pi$. Express

$$\iint\limits_{R} (3x + 8y^2)\, dA$$

as an iterated integral in polar coordinates. Then evaluate the iterated integral for

   a. $\alpha = 0$ and $\beta = \pi/2$ (R is a quarter-ring)

   b. $\alpha = 0$ and $\beta = \pi$ (R is a half-ring)

   c. $\alpha = 0$ and $\beta = 2\pi$ (R is a ring)

Solution By Theorem 14.5 we have

$$\iint\limits_{R} (3x + 8y^2)\, dA = \int_{\alpha}^{\beta} \int_{1}^{2} (3r \cos\theta + 8r^2 \sin^2\theta) r\, dr\, d\theta$$

If R is the quarter-ring (Figure 14.25(a)), we find that

$$\iint\limits_{R} (3x + 8y^2)\, dA = \int_0^{\pi/2} \int_1^2 (3r\cos\theta + 8r^2\sin^2\theta)r\, dr\, d\theta$$

$$= \int_0^{\pi/2} \int_1^2 (3r^2\cos\theta + 8r^3\sin^2\theta)\, dr\, d\theta$$

$$= \int_0^{\pi/2} (r^3\cos\theta + 2r^4\sin^2\theta)\bigg|_1^2\, d\theta$$

$$= \int_0^{\pi/2} (7\cos\theta + 30\sin^2\theta)\, d\theta$$

$$= \int_0^{\pi/2} (7\cos\theta + 15 - 15\cos 2\theta)\, d\theta$$

$$= \left(7\sin\theta + 15\theta - \frac{15}{2}\sin 2\theta\right)\bigg|_0^{\pi/2} = 7 + \frac{15}{2}\pi$$

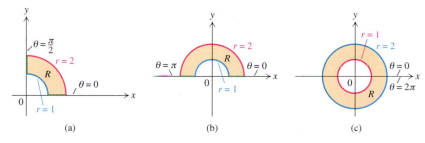

(a)   (b)   (c)

FIGURE 14.25

For the half-ring (Figure 14.25(b)) the outer limit $\pi/2$ of integration is replaced by π, so we obtain

$$\iint\limits_{R} (3x + 8y^2)\, dA = \int_0^{\pi} \int_1^2 (3r\cos\theta + 8r^2\sin^2\theta)r\, dr\, d\theta$$

$$= \left(7\sin\theta + 15\theta - \frac{15}{2}\sin 2\theta\right)\bigg|_0^{\pi} = 15\pi$$

For the ring (Figure 14.25(c)), we substitute 2π for π in the outer limits of integration, obtaining

$$\iint\limits_{R} (3x + 8y^2)\, dA = \int_0^{2\pi} \int_1^2 (3r\cos\theta + 8r^2\sin^2\theta)r\, dr\, d\theta$$

$$= \left(7\sin\theta + 15\theta - \frac{15}{2}\sin 2\theta\right)\bigg|_0^{2\pi} = 30\pi \qquad \square$$

EXAMPLE 2   Let D be the solid region bounded above by the paraboloid $z = 4 - x^2 - y^2$ and below by the xy plane. Find the volume V of D.

Solution   From Example 5 of Section 14.1, the region R over which the integral is to be taken is bounded by the circle $x^2 + y^2 = 4$, whose equation in polar coordinates is $r = 2$; therefore D can be described as the region between the paraboloid and the disk $x^2 + y^2 \leq 4$ (Figure 14.26). As a result,

$$V = \iint\limits_R (4 - x^2 - y^2)\, dA = \int_0^{2\pi} \int_0^2 (4 - r^2) r\, dr\, d\theta$$

$$= \int_0^{2\pi} \int_0^2 (4r - r^3)\, dr\, d\theta = \int_0^{2\pi} \left(2r^2 - \frac{r^4}{4} \right) \Bigg|_0^2 d\theta$$

$$= \int_0^{2\pi} 4\, d\theta = 8\pi \quad \square$$

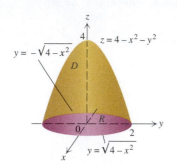

FIGURE 14.26

Although the volume we just calculated is the same as the one calculated in Example 5 of Section 14.1, the use of polar coordinates simplified the solution considerably.

EXAMPLE 3   Suppose D is the solid region bounded on the sides by the cylinder $r = \cos\theta$, above by the cone $z = 16 - \sqrt{x^2 + y^2}$, and below by the xy plane (Figure 14.27). Find the volume V of D.

Solution   Observe that R is the region between the polar graphs of

$$r = 0 \quad \text{and} \quad r = \cos\theta \quad \text{for} \quad -\frac{\pi}{2} \leq \theta \leq \frac{\pi}{2}.$$

Therefore

$$V = \iint\limits_R (16 - \sqrt{x^2 + y^2})\, dA = \int_{-\pi/2}^{\pi/2} \int_0^{\cos\theta} (16 - r) r\, dr\, d\theta$$

$$= \int_{-\pi/2}^{\pi/2} \int_0^{\cos\theta} (16r - r^2)\, dr\, d\theta = \int_{-\pi/2}^{\pi/2} \left(8r^2 - \frac{r^3}{3} \right) \Bigg|_0^{\cos\theta} d\theta$$

$$= \int_{-\pi/2}^{\pi/2} \left(8\cos^2\theta - \frac{\cos^3\theta}{3} \right) d\theta$$

$$= \int_{-\pi/2}^{\pi/2} \left(4 + 4\cos 2\theta - \frac{\cos\theta}{3} + \frac{\cos\theta \sin^2\theta}{3} \right) d\theta$$

$$= \left(4\theta + 2\sin 2\theta - \frac{\sin\theta}{3} + \frac{\sin^3\theta}{9} \right) \Bigg|_{-\pi/2}^{\pi/2}$$

$$= 4\pi - \frac{4}{9} \quad \square$$

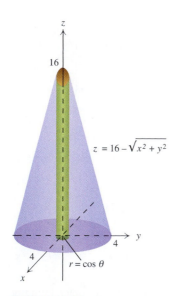

FIGURE 14.27

Caution: When a double integral is evaluated by means of formula (5), h_1 and h_2 must be nonnegative on $[\alpha, \beta]$. Notice that the circle that bounds the disk in Example 3 can also be described as the graph of $r = \cos\theta$ for $0 \leq \theta \leq \pi$. However, $\cos\theta$ is negative for $\pi/2 \leq \theta \leq \pi$, and you can show that

$$\int_0^\pi \int_0^{\cos\theta} (16 - r)r\,dr\,d\theta \neq \int_{-\pi/2}^{\pi/2} \int_0^{\cos\theta} (16 - r)r\,dr\,d\theta$$

We also remark that the reason for our stipulation that $0 \leq \beta - \alpha \leq 2\pi$ at the beginning of this section was to ensure that each point in R corresponds to exactly one point in the $r\theta$ plane. However, certain other regions, such as the region between two appropriately chosen spirals, have this property, even if we take $\alpha = 0$ and $\beta = 3\pi$ (in which case $\beta - \alpha > 2\pi$).

EXAMPLE 4   Let R be the region between the polar graphs of $r = \theta$ and $r = 2\theta$ for $0 \leq \theta \leq 3\pi$ (Figure 14.28). Evaluate $\iint_R (x^2 + y^2)\,dA$.

Solution   We find that

$$\iint_R (x^2 + y^2)\,dA = \int_0^{3\pi} \int_\theta^{2\theta} (r^2)r\,dr\,d\theta = \int_0^{3\pi} \int_\theta^{2\theta} r^3\,dr\,d\theta$$

$$= \int_0^{3\pi} \frac{r^4}{4}\Big|_\theta^{2\theta}\,d\theta = \frac{15}{4}\int_0^{3\pi} \theta^4\,d\theta$$

$$= \frac{15}{4}\frac{\theta^5}{5}\Big|_0^{3\pi} = \frac{729}{4}\pi^5 \qquad \square$$

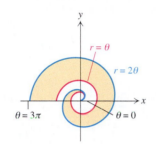

FIGURE 14.28

Evaluating the double integral in Example 4 with the iterated integrals of Section 14.1 would be very difficult.

In Sections 7.2 and 8.7 we discussed the standard normal density, $y = e^{-x^2/2}/\sqrt{2\pi}$. We promised that in this section we would prove that

$$\int_{-\infty}^{\infty} \frac{1}{\sqrt{2\pi}} e^{-x^2/2}\,dx = 1$$

that is, that the area under the graph of the integrand is 1 (Figure 14.29). Now, with double integration in polar coordinates available, we are able to fulfill this promise.

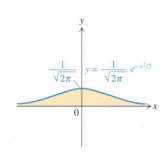

FIGURE 14.29

EXAMPLE 5   Show that $\displaystyle\int_{-\infty}^{\infty} \frac{1}{\sqrt{2\pi}} e^{-x^2/2}\,dx = 1$.

Solution   Using the fact that the integrand is symmetric with respect to the y axis, we find that

$$\int_{-\infty}^{\infty} \frac{1}{\sqrt{2\pi}} e^{-x^2/2}\,dx = 2\int_0^{\infty} \frac{1}{\sqrt{2\pi}} e^{-x^2/2}\,dx = \sqrt{\frac{2}{\pi}}\int_0^{\infty} e^{-x^2/2}\,dx$$

Therefore it suffices to show that

$$\sqrt{\frac{2}{\pi}}\int_0^{\infty} e^{-x^2/2}\,dx = 1, \quad \text{or equivalently,} \quad \int_0^{\infty} e^{-x^2/2}\,dx = \sqrt{\frac{\pi}{2}}$$

Although we cannot evaluate the last integral directly, it is possible by means of

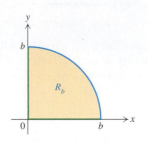

FIGURE 14.30

polar coordinates to evaluate the related iterated integral $\int_0^\infty \int_0^\infty e^{-(x^2+y^2)}\,dy\,dx$, which we regard as an improper integral as follows:

$$\int_0^\infty \int_0^\infty e^{-(x^2+y^2)}\,dy\,dx = \lim_{b\to\infty}\iint\limits_{R_b} e^{-(x^2+y^2)}\,dA$$

where R_b is the quarter of the disk $x^2 + y^2 \le b^2$ that lies in the first quadrant (Figure 14.30). Using this observation and changing to polar coordinates, we find that

$$\int_0^\infty \int_0^\infty e^{-(x^2+y^2)}\,dy\,dx = \lim_{b\to\infty}\iint\limits_{R_b} e^{-(x^2+y^2)}\,dA = \lim_{b\to\infty}\int_0^{\pi/2}\int_0^b e^{-r^2}\,r\,dr\,d\theta$$

$$= \lim_{b\to\infty}\int_0^{\pi/2}\left(-\frac{1}{2}e^{-r^2}\right)\Big|_0^b\,d\theta = \lim_{b\to\infty}\int_0^{\pi/2}\frac{1}{2}\left(1 - e^{-b^2}\right)d\theta$$

$$= \lim_{b\to\infty}\frac{1}{2}\left(1 - e^{-b^2}\right)\theta\Big|_0^{\pi/2} = \lim_{b\to\infty}\frac{\pi}{4}\left(1 - e^{-b^2}\right) = \frac{1}{4}\pi$$

Thus

$$\frac{1}{4}\pi = \int_0^\infty \int_0^\infty e^{-(x^2+y^2)}\,dy\,dx = \int_0^\infty \int_0^\infty e^{-x^2}e^{-y^2}\,dy\,dx$$

$$= \int_0^\infty e^{-x^2}\left(\int_0^\infty e^{-y^2}\,dy\right)dx$$

$$= \left(\int_0^\infty e^{-x^2}\,dx\right)\left(\int_0^\infty e^{-y^2}\,dy\right) = \left(\int_0^\infty e^{-x^2}\,dx\right)^2$$

Consequently

$$\int_0^\infty e^{-x^2}\,dx = \frac{1}{2}\sqrt{\pi} \qquad (6)$$

Finally, we make the substitution

$$u = \frac{x}{\sqrt{2}}, \quad \text{so that} \quad du = \frac{1}{\sqrt{2}}\,dx$$

We obtain

$$\int_0^\infty e^{-x^2/2}\,dx = \int_0^\infty e^{-u^2}\sqrt{2}\,du \stackrel{(6)}{=} \sqrt{2}\,\frac{1}{2}\sqrt{\pi} = \sqrt{\frac{\pi}{2}}$$

which is what we desired to show.   ❑

EXERCISES 14.2

In Exercises 1–6 express the integral as an iterated integral in polar coordinates, and then evaluate it.

1. $\iint_R xy \, dA$, where R is the region bounded by the circle $r = 5$

2. $\iint_R y \, dA$, where R is the region bounded by the circle $r = \cos \theta$

3. $\iint_R (x + y) \, dA$, where R is the region in the first quadrant bounded by the lines $y = 0$ and $y = \sqrt{3}\, x$ and the circle $r = 2$

4. $\iint_R (x^2 + y^2)^{1/2} \, dA$, where R is the region bounded by the limaçon $r = 2 + \cos \theta$

5. $\iint_R (x^2 + y^2) \, dA$, where R is the region bounded by the cardioid $r = 2(1 + \sin \theta)$

6. $\iint_R x^2 \, dA$, where R is the region bounded by the circle $r = 4 \sin \theta$

In Exercises 7–13 find the volume V of the region by using iterated integrals in polar coordinates.

7. The solid region bounded by a hemisphere with radius 3

8. The solid region bounded by the planes $z = 0$ and $z = 4$ and the cylinder $r = 2 \cos \theta$

9. The solid region bounded above by the plane $z = 4 + x + 2y$, on the sides by the cylinder $x^2 + y^2 = 1$, and below by the xy plane

10. The solid region bounded above by the plane $z = x$, on the sides by the cylinder $x^2 + y^2 = x$, and below by the xy plane

11. The solid region inside the sphere $x^2 + y^2 + z^2 = 4$, outside the cylinder $x^2 + y^2 = 1$, and above the xy plane

12. The solid region bounded on the sides by the paraboloid $z = 4 - x^2 - y^2$, above by the plane $z = 3$, and below by the xy plane.

13. The solid region above the xy plane bounded on the sides by the cylinder $x^2 + y^2 - 4x = 0$ and above by the cone $z^2 = x^2 + y^2$

In Exercises 14–21 find the area A of the region in the xy plane by means of iterated integrals in polar coordinates.

14. The circular sector bounded by the graph of $r = 1$ on $[0, \alpha]$, where $0 \le \alpha \le 2\pi$

15. The region bounded by the limaçon $r = 2 + \sin \theta$

16. One leaf of the three-leaved rose bounded by the graph of $r = 2 \sin 3\theta$

17. The region bounded by the lemniscate $r^2 = 4 \cos 2\theta$ (*Hint:* First calculate the area of the portion for which $-\pi/4 \le \theta \le \pi/4$.)

18. The region inside the cardioid $r = 1 + \cos \theta$ and outside the circle $r = \frac{1}{2}$

19. The region inside the large loop and outside the small loop of $r = 1 + 2 \sin \theta$

*20. The region bounded by the graph of $r = \sin \theta - \cos \theta$ (*Hint:* Be careful with the limits of integration.)

21. The region between the spirals $r = e^\theta$ and $r = e^{2\theta}$ on $[0, 3\pi]$

In Exercises 22–29 change the integral to an iterated integral in polar coordinates, and then evaluate it.

22. $\displaystyle\int_0^1 \int_0^{\sqrt{1-x^2}} 1 \, dy \, dx$ 23. $\displaystyle\int_0^1 \int_y^{\sqrt{2-y^2}} 1 \, dx \, dy$

24. $\displaystyle\int_{3/\sqrt{2}}^3 \int_0^{\sqrt{9-x^2}} \frac{1}{\sqrt{x^2 + y^2}} \, dy \, dx$

25. $\displaystyle\int_0^1 \int_0^{\sqrt{1-y^2}} \sin (x^2 + y^2) \, dx \, dy$

26. $\displaystyle\int_0^1 \int_0^{\sqrt{1-x^2}} e^{\sqrt{x^2 + y^2}} \, dy \, dx$

27. $\displaystyle\int_0^1 \int_0^{\sqrt{1-x^2}} e^{-(x^2 + y^2)} \, dy \, dx$

28. $\displaystyle\int_0^1 \int_{\sqrt{x-x^2}}^{\sqrt{1-x^2}} 1 \, dy \, dx$ 29. $\displaystyle\int_0^1 \int_{-\sqrt{x-x^2}}^{\sqrt{x-x^2}} (x^2 + y^2) \, dy \, dx$

30. Let S be the region in the $r\theta$ plane bounded by the straight lines $r = s$ and $r = t$ with $s < t$, and $\theta = \alpha$ and $\theta = \beta$ with $0 \le \beta - \alpha \le 2\pi$. Let R be the region in the xy plane bounded by the polar graphs $r = s$, $r = t$, $\theta = \alpha$, and $\theta = \beta$. By what factor must the area of S be multiplied to yield the area of R?

31. **a.** Let f be defined by the formula $f(x) = \int_x^1 e^{-t^2} \, dt$, for $0 \le x \le 1$. Find the average value f_{av} of f on $[0, 1]$, by reversing the order of integration on an appropriate iterated integral. (Even though we cannot integrate $\int_x^1 e^{-t^2} \, dt$ to obtain a numerical value for $f(x)$, we can nevertheless find the average of the function f on $[0, 1]$.)

    b. Suppose g is defined on $[0, 1]$ by the formula $g(x) = \int_0^x e^{-t^2} \, dt$, for $0 \le x \le 1$. Is it possible to find the average value g_{av} of g on $[0, 1]$ by integration? Explain your answer.

14.3 SURFACE AREA

In Section 6.3 we presented formulas for the surface area of the surface generated by revolving a curve in the *xy* plane about the *x* axis. The double integral will play a prominent role in this section as we define the surface area *S* of more general kinds of surfaces.

Our basic approach is as follows. We know from (2) in Section 11.4 how to compute the area of any parallelogram in space: it is the norm $\| \mathbf{a} \times \mathbf{b} \|$ of the two vectors $\mathbf{a}$ and $\mathbf{b}$ representing adjacent sides of the parallelogram. For a more complicated surface Σ, we will just break Σ up into small pieces and approximate the area of each piece by calculating the area of an appropriate parallelogram in a plane that is tangent to that piece of Σ.

To carry out this approach, let Σ be the graph in space of a function f that is continuous on a vertically or horizontally simple region R in the *xy* plane, and assume that f has continuous partial derivatives on R. If R' is a rectangle containing R, then let us consider a partition P of R' into subrectangles, numbered so that $R_1, R_2, \ldots, R_n$ are the ones entirely contained in R. For any integer k between 1 and n let (x_k, y_k) be the corner of R_k with the smallest values of x_k and y_k, as in Figure 14.31. Let Δx_k and Δy_k be the lengths of two adjacent sides of R_k, and let ΔA_k be the area of R_k. The projection of R_k onto the plane tangent to Σ at the point $(x_k, y_k, f(x_k, y_k))$ yields a parallelogram R'_k (Figure 14.31). We think of the area $\Delta A'_k$ of R'_k as an approximation to the area of the portion of the surface Σ that corresponds to R_k. The vectors $\mathbf{i} + f_x(x_k, y_k)\mathbf{k}$ and $\mathbf{j} + f_y(x_k, y_k)\mathbf{k}$ are parallel to two adjacent sides of R'_k (see Section 13.3). This implies that

$$\mathbf{B}_k = \Delta x_k \mathbf{i} + f_x(x_k, y_k)\, \Delta x_k \mathbf{k} \quad \text{and} \quad \mathbf{C}_k = \Delta y_k \mathbf{j} + f_y(x_k, y_k)\, \Delta y_k \mathbf{k}$$

form two adjacent sides of R'_k. By the results of Section 11.4, the area $\Delta A'_k$ of R'_k is the length of the cross product of the sides $\mathbf{B}_k$ and $\mathbf{C}_k$. As a result,

$$
\begin{aligned}
\Delta A'_k &= \| \mathbf{B}_k \times \mathbf{C}_k \| \\
&= \| -f_x(x_k, y_k)\, \Delta x_k\, \Delta y_k \mathbf{i} - f_y(x_k, y_k)\, \Delta x_k\, \Delta y_k \mathbf{j} + \Delta x_k\, \Delta y_k \mathbf{k} \| \\
&= \sqrt{[f_x(x_k, y_k)]^2 + [f_y(x_k, y_k)]^2 + 1}\; \Delta x_k\, \Delta y_k \\
&= \sqrt{[f_x(x_k, y_k)]^2 + [f_y(x_k, y_k)]^2 + 1}\; \Delta A_k
\end{aligned}
$$

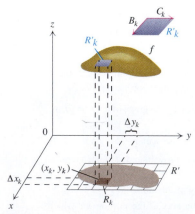

FIGURE 14.31

A geodesic dome. (*Stephen R. Swinburne/Stock Boston*)

Therefore

$$\sum_{k=1}^{n} \Delta A_k' = \sum_{k=1}^{n} \sqrt{[f_x(x_k, y_k)]^2 + [f_y(x, y)]^2 + 1} \, dA$$

The sum $\sum_{k=1}^{n} \Delta A_k'$ should approximate the surface area S of Σ, while the sum on the right-hand side of the equation is a Riemann sum approximating

$$\iint_R \sqrt{[f_x(x, y)]^2 + [f_y(x, y)]^2 + 1} \, dA$$

Geometrically the Riemann sum corresponds to replacement of small portions of the surface Σ by corresponding portions of tangent planes. The resulting surface looks like a geodesic dome, whose area approximates the area S of Σ. This suggests the following definition of surface area.

DEFINITION 14.6   Let R be a vertically or horizontally simple region, and let f have continuous partial derivatives on R. If Σ is the graph of f on R, then the **surface area** S of Σ is defined by

$$S = \iint_R \sqrt{[f_x(x, y)]^2 + [f_y(x, y)]^2 + 1} \, dA \tag{1}$$

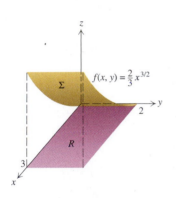

FIGURE 14.32

It is usually difficult or impossible to compute surface area by this formula. However, in some cases it is possible.

EXAMPLE 1   Let R be the rectangular region bounded by the lines

$$x = 0, \quad x = 3, \quad y = 0, \quad y = 2$$

and let $f(x, y) = \frac{2}{3}x^{3/2}$. Find the surface area S of the portion of the graph of f that lies over R (Figure 14.32).

Solution   Notice that

$$f_x(x, y) = x^{1/2} \quad \text{and} \quad f_y(x, y) = 0 \quad \text{for } (x, y) \text{ in } R$$

Consequently (1) implies that

$$S = \iint_R \sqrt{(x^{1/2})^2 + 0 + 1} \, dA = \int_0^3 \int_0^2 \sqrt{x + 1} \, dy \, dx$$

$$= 2 \int_0^3 \sqrt{x + 1} \, dx = \frac{4}{3}(x + 1)^{3/2} \Big|_0^3 = \frac{28}{3} \quad \square$$

EXAMPLE 2   Find the surface area S of the portion of the paraboloid

$$z = 4 - x^2 - y^2$$

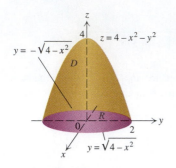

FIGURE 14.33

that lies above the *xy* plane (see Figure 14.33).

Solution   The given surface lies over the region *R* in the *xy* plane bounded by the circle $x^2 + y^2 = 4$. If $f(x, y) = 4 - x^2 - y^2$, then

$$f_x(x, y) = -2x \quad \text{and} \quad f_y(x, y) = -2y$$

By (1),

$$S = \iint_R \sqrt{4x^2 + 4y^2 + 1} \, dA$$

This double integral is suitable for evaluation by polar coordinates:

$$S = \int_0^{2\pi} \int_0^2 \sqrt{4r^2 + 1} \, r \, dr \, d\theta = \frac{1}{12} \int_0^{2\pi} (4r^2 + 1)^{3/2} \Big|_0^2 \, d\theta$$

$$= \frac{1}{12} \int_0^{2\pi} (17^{3/2} - 1) \, d\theta = \frac{1}{6} \pi (17^{3/2} - 1) \qquad \square$$

EXAMPLE 3   Let $a > 0$. Find the surface area *S* of the frustum of the cone

$$z = a\sqrt{x^2 + y^2}$$

with minimum and maximum radii r_1 and r_2, respectively (Figure 14.34).

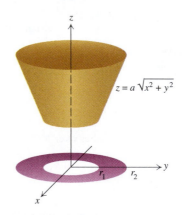

FIGURE 14.34

Solution   The given surface lies over the region *R* in the *xy* plane bounded by the annulus

$$r_1^2 \leq x^2 + y^2 \leq r_2^2$$

If $f(x, y) = a\sqrt{x^2 + y^2}$, then

$$f_x(x, y) = \frac{ax}{\sqrt{x^2 + y^2}} \quad \text{and} \quad f_y(x, y) = \frac{ay}{\sqrt{x^2 + y^2}}$$

so that by (1),

$$S = \iint\limits_{R} \sqrt{\left(\frac{ax}{\sqrt{x^2+y^2}}\right)^2 + \left(\frac{ay}{\sqrt{x^2+y^2}}\right)^2 + 1}\, dA$$

$$= \iint\limits_{R} \sqrt{\frac{a^2 x^2}{x^2+y^2} + \frac{a^2 y^2}{x^2+y^2} + 1}\, dA = \iint\limits_{R} \sqrt{a^2+1}\, dA$$

$$= \sqrt{a^2+1} \iint\limits_{R} 1\, dA = \sqrt{a^2+1} \quad (\text{area of } R)$$

Since the area of the annulus R is $\pi r_2^2 - \pi r_1^2 = \pi(r_2^2 - r_1^2)$, we find that

$$S = \pi \sqrt{a^2+1}\,(r_2^2 - r_1^2) \qquad \square$$

The formula for the surface area of the frustum obtained in Example 3 looks rather different from the one given in (1) of Section 6.3, which involved the slant height l of the frustum. In order to show that the formulas are consistent with one another, we now determine the slant height of the frustum in Example 3.

The intersection of the frustum with the first quadrant of the yz plane is the line segment in the yz plane having equation $z = ay$ for $r_1 \le y \le r_2$. The slant height l, which is the length of this line segment, is given by

$$l = \sqrt{(r_2 - r_1)^2 + (ar_2 - ar_1)^2} = \sqrt{(1+a^2)(r_2-r_1)^2}$$

$$= \sqrt{1+a^2}\,(r_2 - r_1) = \sqrt{a^2+1}\,(r_2 - r_1)$$

Thus the surface area S can be written as follows:

$$S = \pi \underbrace{\sqrt{a^2+1}\,(r_2 - r_1)}_{l} (r_2 + r_1) = \pi(r_1 + r_2)l$$

This is consistent with the formula appearing in (1) of Section 6.3.

The formula for surface area in this section can be shown to be compatible with the formula in (4) of Section 6.3 when the surface is obtained by revolving the graph of a continuous, nonnegative function about the x axis. (See Section 14.9 and in particular Exercise 26 in Section 14.9.)

EXERCISES 14.3

In Exercises 1–10 find the surface area of the given surface.

1. The portion of the plane $x + 2y + 3z = 6$ in the first octant
2. The portion of the paraboloid $z = 9 - x^2 - y^2$ above the xy plane

3. The portion of the paraboloid $z = 9 - x^2 - y^2$ above the plane $z = 5$
4. The portion of the sphere $x^2 + y^2 + z^2 = 4$ that is inside the cylinder $x^2 + y^2 = 1$

5. The portion of the sphere $x^2 + y^2 + z^2 = 16$ that is inside the cylinder $x^2 - 4x + y^2 = 0$

6. The portion of the cylinder $x^2 + z^2 = 9$ that is directly above the triangle with vertices $(0, 0, 0)$, $(1, 0, 0)$, and $(1, 1, 0)$

7. The portion of the parabolic sheet $z = x^2$ directly above the triangle with vertices $(0, 0, 0)$, $(1, 0, 0)$, and $(1, 1, 0)$

8. The portion of the sphere $x^2 + y^2 + z^2 = 9$ that is inside the paraboloid $x^2 + y^2 = 8z$

9. The portion of the sphere $x^2 + y^2 + z^2 = 14z$ that is inside the paraboloid $x^2 + y^2 = 5z$

10. The portion of the graph of $z = \frac{2}{3}\sqrt{2}\, x^{3/2} + \frac{2}{3} y^{3/2}$ directly over the region in the xy plane between the graph of $y = x^2$ and the x axis on $[0, 1]$

14.4 TRIPLE INTEGRALS

An analogue of single and double integrals, called the triple integral, can be defined for functions of three variables.

Definition of the Triple Integral

Let R be a vertically or horizontally simple region in the xy plane, and let F_1 and F_2 be continuous on R and satisfy

$$F_1(x, y) \leq F_2(x, y) \quad \text{for } (x, y) \text{ in } R$$

Let D denote the solid region consisting of all points (x, y, z) such that

$$(x, y) \text{ is in } R \quad \text{and} \quad F_1(x, y) \leq z \leq F_2(x, y)$$

(Figure 14.35). We refer to D as the **solid region between the graphs of F_1 and F_2 on R**. In examples it will be convenient to refer to D as the solid region between the graphs of $z = F_1(x, y)$ and $z = F_2(x, y)$ for (x, y) in R.

In order to define integrals of continuous functions on D, it is necessary to make further assumptions on the boundary of D that normally appear only in advanced calculus texts. We will assume that every solid region that appears in the remainder of this book satisfies these assumptions.

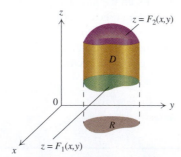

FIGURE 14.35 The solid region between the graphs of F_1 and F_2 on R.

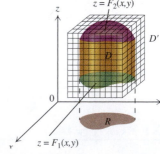

FIGURE 14.36

In preparation for the definition of the triple integral, let D' be a rectangular parallelepiped containing D, and let P partition D' into smaller rectangular parallelepipeds (Figure 14.36), numbered so that $D_1, D_2, \ldots, D_n$ are the ones entirely contained in D. In addition, we assume that f is continuous on D, and we let (x_k, y_k, z_k) be an arbitrary point in D_k for $1 \leq k \leq n$. If ΔV_k denotes the volume of D_k then the sum

$$\sum_{k=1}^{n} f(x_k, y_k, z_k) \, \Delta V_k$$

is a **Riemann sum** for f on D. It can be shown that the Riemann sums converge to a unique number I in the following sense.

THEOREM 14.7

Let f be continuous on the solid region D between the graphs of two continuous functions F_1 and F_2 on a vertically or horizontally simple region R in the xy plane. Then there is a unique number I with the following property:

Let D' be any parallelepiped containing D. For any number $\varepsilon > 0$ there is a number $\delta > 0$ that has the following property: For any partition P of D' into subparallelepipeds whose dimensions are all less than δ, and with $D_1, D_2, \ldots, D_n$ the subparallelepipeds in D' that are entirely contained in D,

$$\left| I - \sum_{k=1}^{n} f(x_k, y_k, z_k) \, \Delta V_k \right| < \varepsilon$$

where (x_k, y_k, z_k) is an arbitrary point in D_k and ΔV_k denotes the volume of D_k, for $1 \leq k \leq n$.

We call the number I the triple integral of f on D.

DEFINITION 14.8

Let D be the solid region between the graphs of two continuous functions F_1 and F_2 on a vertically or horizontally simple region R in the xy plane. If f is continuous on D, then we write

$$\iiint_D f(x, y, z) \, dV$$

for the unique number I to which the Riemann sums converge by Theorem 14.7. The number $\iiint_D f(x, y, z) \, dV$ is called the **triple integral** of f on D.

The conclusion of Theorem 14.7 is usually expressed as

$$\iiint_D f(x, y, z) \, dV = \lim_{\|P\| \to 0} \sum_{k=1}^{n} f(x_k, y_k, z_k) \, \Delta V_k$$

where $\| P \|$ is the largest of the dimensions of the subparallelepipeds in P and is called the **norm of the partition P.**

Evaluation of Triple Integrals

In general, Riemann sums are not very effective in evaluating a triple integral $\iiint_D f(x, y, z) \, dV$. Once again, iterated integrals (this time three integrals in succession) provide a method of evaluating triple integrals.

THEOREM 14.9   Let D be the solid region between the graphs of two continuous functions F_1 and F_2 on a vertically or horizontally simple region R in the xy plane, and let f be continuous on D. Then

$$\iiint\limits_{D} f(x, y, z)\, dV = \iint\limits_{R} \left(\int_{F_1(x, y)}^{F_2(x, y)} f(x, y, z)\, dz \right) dA$$

We evaluate $\int_{F_1(x,y)}^{F_2(x,y)} f(x, y, z)\, dz$ by integrating with respect to z while both x and y are held fixed, thus obtaining a number depending on x and y. If R is the vertically simple region between the graphs of g_1 and g_2 on $[a, b]$, we evaluate the double integral over R by using Theorem 14.4(a), obtaining

$$\iiint\limits_{D} f(x, y, z)\, dV = \int_a^b \left[\int_{g_1(x)}^{g_2(x)} \left(\int_{F_1(x, y)}^{F_2(x, y)} f(x, y, z)\, dz \right) dy \right] dx \quad (1)$$

Similarly, if R is the horizontally simple region between the graphs of h_1 and h_2 on $[c, d]$, then from Theorem 14.4(b) we obtain

$$\iiint\limits_{D} f(x, y, z)\, dV = \int_c^d \left[\int_{h_1(y)}^{h_2(y)} \left(\int_{F_1(x, y)}^{F_2(x, y)} f(x, y, z)\, dz \right) dx \right] dy \quad (2)$$

The integrals on the right sides of (1) and (2) are called (triple) **iterated integrals.** Normally we omit the parentheses and brackets and write

$$\int_a^b \int_{g_1(x)}^{g_2(x)} \int_{F_1(x, y)}^{F_2(x, y)} f(x, y, z)\, dz\, dy\, dx$$

for

$$\int_a^b \left[\int_{g_1(x)}^{g_2(x)} \left(\int_{F_1(x, y)}^{F_2(x, y)} f(x, y, z)\, dz \right) dy \right] dx$$

and

$$\int_c^d \int_{h_1(y)}^{h_2(y)} \int_{F_1(x, y)}^{F_2(x, y)} f(x, y, z)\, dz\, dx\, dy$$

for

$$\int_c^d \left[\int_{h_1(y)}^{h_2(y)} \left(\int_{F_1(x, y)}^{F_2(x, y)} f(x, y, z)\, dz \right) dx \right] dy$$

In

$$\int_a^b \int_{g_1(x)}^{g_2(x)} \int_{F_1(x, y)}^{F_2(x, y)} f(x, y, z)\, dz\, dy\, dx$$

the limits of integration for z must be functions of x and y alone, the limits of integration for y must be functions of x alone, and the limits of integration for x must be constants. Analogous comments apply to the other iterated integrals.

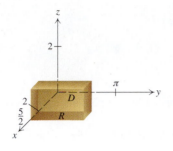

FIGURE 14.37

EXAMPLE 1   Let R be the rectangular region in the xy plane bounded by the lines $x = 2$, $x = 5/2$, $y = 0$, and $y = \pi$, and let D be the parallelepiped between the graphs of

$$z = 0 \quad \text{and} \quad z = 2 \quad \text{on } R$$

(Figure 14.37). Evaluate $\iiint_D zx \sin xy \, dV$.

Solution   By (1) we have

$$\iiint_D zx \sin xy \, dV = \int_2^{5/2} \int_0^{\pi} \int_0^2 zx \sin xy \, dz \, dy \, dx$$

$$= \int_2^{5/2} \int_0^{\pi} \left(\frac{z^2 x}{2} \sin xy \right) \Big|_0^2 \, dy \, dx$$

$$= \int_2^{5/2} \int_0^{\pi} 2x \sin xy \, dy \, dx = \int_2^{5/2} -2 \cos xy \Big|_0^{\pi} \, dx$$

$$= \int_2^{5/2} (2 - 2 \cos \pi x) \, dx = \left(2x - \frac{2}{\pi} \sin \pi x \right) \Big|_2^{5/2}$$

$$= \left(5 - \frac{2}{\pi} \sin \frac{5\pi}{2} \right) - \left(4 - \frac{2}{\pi} \sin 2\pi \right) = 1 - \frac{2}{\pi} \qquad \square$$

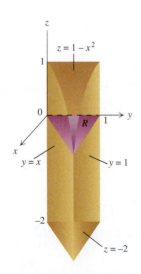

FIGURE 14.38

EXAMPLE 2   Let R be the triangular region in the xy plane between the graphs of $y = x$ and $y = 1$ for $0 \le x \le 1$, and let D be the solid region between the graphs of the surfaces

$$z = -2 \quad \text{and} \quad z = 1 - x^2 \quad \text{for } (x, y) \text{ in } R$$

(Figure 14.38). Evaluate $\iiint_D (x + 1) \, dV$.

Solution   By (1) we have

$$\iiint_D (x + 1) \, dV = \int_0^1 \int_x^1 \int_{-2}^{1 - x^2} (x + 1) \, dz \, dy \, dx$$

$$= \int_0^1 \int_x^1 (x + 1)z \Big|_{-2}^{1 - x^2} \, dy \, dx$$

$$= \int_0^1 \int_x^1 (x + 1)[(1 - x^2) - (-2)] \, dy \, dx$$

$$= \int_0^1 \int_x^1 (-x^3 - x^2 + 3x + 3) \, dy \, dx$$

$$= \int_0^1 (-x^3 - x^2 + 3x + 3)y \Big|_x^1 \, dx$$

$$= \int_0^1 (x^4 - 4x^2 + 3) \, dx$$

$$= \left(\frac{1}{5}x^5 - \frac{4}{3}x^3 + 3x \right)\bigg|_0^1 = \frac{28}{15} \quad \square$$

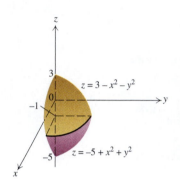

FIGURE 14.39

EXAMPLE 3 Let D be the solid region bounded by the portions of the two circular paraboloids

$$z = 3 - x^2 - y^2 \quad \text{and} \quad z = -5 + x^2 + y^2$$

for which $x \geq 0$ and $y \geq 0$ (Figure 14.39). Evaluate $\iiint_D y \, dV$.

Solution   To be able to use (2), we must determine the region R in the xy plane such that D is the solid region between the two paraboloids on R. For this purpose we first determine where the two paraboloids intersect. At any point (x, y, z) of intersection, x and y must satisfy

$$3 - x^2 - y^2 = z = -5 + x^2 + y^2$$

which is equivalent to $x^2 + y^2 = 4$. But if $x^2 + y^2 = 4$, then

$$z = 3 - x^2 - y^2 = 3 - 4 = -1$$

so the intersection lies in the plane $z = -1$. We find that the corresponding region R in the xy plane is the horizontally simple region in the first quadrant that lies <u>inside</u> the circle $x^2 + y^2 = 4$, and hence between the graphs of $x = 0$ and $x = \sqrt{4 - y^2}$ for $0 \leq y \leq 2$. Since

$$3 - x^2 - y^2 \geq -5 + x^2 + y^2 \quad \text{for } (x, y) \text{ in } R$$

we have

$$\iiint_D y \, dV = \int_0^2 \int_0^{\sqrt{4-y^2}} \int_{-5+x^2+y^2}^{3-x^2-y^2} y \, dz \, dx \, dy$$

$$= \int_0^2 \int_0^{\sqrt{4-y^2}} yz \bigg|_{-5+x^2+y^2}^{3-x^2-y^2} dx \, dy$$

$$= \int_0^2 \int_0^{\sqrt{4-y^2}} y(8 - 2x^2 - 2y^2) \, dx \, dy$$

$$= \int_0^2 \left[y(8 - 2y^2)x - \frac{2}{3}yx^3 \right]\bigg|_0^{\sqrt{4-y^2}} dy$$

$$= \int_0^2 \left[y(8 - 2y^2)\sqrt{4 - y^2} - \frac{2}{3}y(4 - y^2)^{3/2} \right] dy$$

$$= \frac{4}{3}\int_0^2 y(4 - y^2)^{3/2} \, dy = \frac{-4}{15}(4 - y^2)^{5/2}\bigg|_0^2 = \frac{128}{15} \quad \square$$

Volume by Triple Integration

Let D be the solid region between the graphs of two continuous functions F_1 and F_2 on a vertically or horizontally simple region R in the xy plane. The volume V of D is defined by

$$V = \iiint_D 1 \, dV \qquad (3)$$

By Theorem 14.9 we can rewrite the integral for the volume as a double integral over R:

$$V = \iiint_D 1 \, dV = \iint_R \left[\int_{F_1(x, y)}^{F_2(x, y)} 1 \, dz \right] dA = \iint_R [F_2(x, y) - F_1(x, y)] \, dA$$

In case F_1 is the constant function 0, this double integral is just the double integral appearing in Definition 14.2(b) with F_2 replacing f. Consequently (3) defines volume for a larger class of solid regions than Definition 14.2(b) does (because F_1 need not be 0).

EXAMPLE 4   Let D be the solid region in the first octant between the graphs of

$$z = x^2 + 2y + 1 \quad \text{and} \quad z = y + 2$$

Find the volume V of D.

Solution   First we must determine the region R. At any point (x, y, z) of intersection of the surfaces $z = x^2 + 2y + 1$ and $z = y + 2$ we have

$$x^2 + 2y + 1 = y + 2$$

or equivalently,

$$y = 1 - x^2$$

Since D is in the first octant, R must be in the first quadrant. Therefore R is the vertically simple region between the graphs of $y = 0$ and $y = 1 - x^2$ on $[0, 1]$ (Figure 14.40). Since $x^2 + 2y + 1 \le y + 2$ for (x, y) in R, it follows from (3) that

$$V = \iiint_D 1 \, dV = \int_0^1 \int_0^{1-x^2} \int_{x^2+2y+1}^{y+2} 1 \, dz \, dy \, dx$$

$$= \int_0^1 \int_0^{1-x^2} z \Big|_{x^2+2y+1}^{y+2} \, dy \, dx = \int_0^1 \int_0^{1-x^2} (1 - x^2 - y) \, dy \, dx$$

$$= \int_0^1 \left[(1 - x^2)y - \frac{1}{2} y^2 \right] \Big|_0^{1-x^2} dx = \int_0^1 \left[(1 - x^2)^2 - \frac{1}{2}(1 - x^2)^2 \right] dx$$

$$= \int_0^1 \frac{1}{2}(1 - x^2)^2 \, dx = \frac{1}{2} \left(x - \frac{2x^3}{3} + \frac{x^5}{5} \right) \Big|_0^1 = \frac{4}{15} \qquad \Box$$

EXAMPLE 5   Suppose that f is a nonnegative continuous function on $[a, b]$ and R is the region in the xy plane between the graph of f and the x axis on $[a, b]$. Let D be the solid region generated by revolving R about the x axis (Figure 14.41). Show that the formula for the volume V of D given in (3) yields formula (3) of Section 6.1.

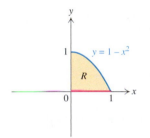

FIGURE 14.40

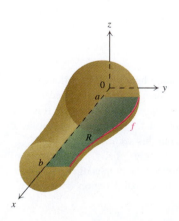

FIGURE 14.41

Solution Let R_1 be the vertically simple region in the xy plane between the graphs of $-f$ and f on $[a, b]$. Then D is the solid region between the surfaces

$$z = -\sqrt{[f(x)]^2 - y^2} \quad \text{and} \quad z = \sqrt{[f(x)]^2 - y^2} \quad \text{on } R_1$$

Thus (3) implies that

$$V = \iiint_D 1 \, dV = \int_a^b \int_{-f(x)}^{f(x)} \int_{-\sqrt{[f(x)]^2 - y^2}}^{\sqrt{[f(x)]^2 - y^2}} 1 \, dz \, dy \, dx$$

$$= \int_a^b \int_{-f(x)}^{f(x)} z \Big|_{-\sqrt{[f(x)]^2 - y^2}}^{\sqrt{[f(x)]^2 - y^2}} \, dy \, dx$$

$$= \int_a^b \int_{-f(x)}^{f(x)} 2\sqrt{[f(x)]^2 - y^2} \, dy \, dx$$

Making the substitution

$$y = f(x) \sin u, \quad \text{so that} \quad dy = f(x) \cos u \, du$$

we obtain

$$\int_a^b \int_{-f(x)}^{f(x)} 2\sqrt{[f(x)]^2 - y^2} \, dy \, dx = \int_a^b \int_{-\pi/2}^{\pi/2} [2f(x) \cos u] f(x) \cos u \, du \, dx$$

$$= \int_a^b \int_{-\pi/2}^{\pi/2} 2[f(x)]^2 \cos^2 u \, du \, dx$$

$$= \int_a^b 2[f(x)]^2 \left(\frac{1}{2} u + \frac{1}{4} \sin 2u\right) \Big|_{-\pi/2}^{\pi/2} \, dx$$

$$= \int_a^b \pi [f(x)]^2 \, dx$$

Thus

$$V = \int_a^b \pi [f(x)]^2 \, dx$$

which is (3) of Section 6.1. ❏

Example 5 shows that our formulas for volume here and in Section 6.1 are compatible. Of course, the one presented here is much more general, because the solid region need not be generated by revolving a region about the x axis. It is also possible to show that the formula in (7) in Section 6.1 for the volume of a solid of revolution about the y axis is compatible with (3) in this section.

It is sometimes advantageous to interchange the roles of x, y, and z; instead of regarding a given region D as the solid region between the graphs of two functions of x and y, we occasionally wish to regard D as the solid region between two functions of x and z or two functions of y and z.

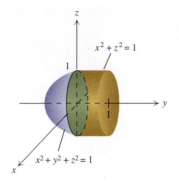

FIGURE 14.42

EXAMPLE 6   Evaluate $\iiint_D y\sqrt{1-x^2}\, dV$, where D is the region depicted in Figure 14.42.

Solution   It is convenient to think of D as the region between the graphs of $y = -\sqrt{1-x^2-z^2}$ and $y = 1$ on the region in the xz plane bounded by the circle $x^2 + z^2 = 1$. Using this approach, we find that

$$\iiint_D y\sqrt{1-x^2}\, dV = \int_{-1}^{1} \int_{-\sqrt{1-x^2}}^{\sqrt{1-x^2}} \int_{-\sqrt{1-x^2-z^2}}^{1} y\sqrt{1-x^2}\, dy\, dz\, dx$$

$$= \int_{-1}^{1} \int_{-\sqrt{1-x^2}}^{\sqrt{1-x^2}} \left(\frac{1}{2}\sqrt{1-x^2}\, y^2 \Big|_{-\sqrt{1-x^2-z^2}}^{1} \right) dz\, dx$$

$$= \int_{-1}^{1} \int_{-\sqrt{1-x^2}}^{\sqrt{1-x^2}} \frac{1}{2}\sqrt{1-x^2}\,(x^2 + z^2)\, dz\, dx$$

$$= \int_{-1}^{1} \frac{1}{2}\sqrt{1-x^2}\left(x^2 z + \frac{1}{3}z^3 \right)\Big|_{-\sqrt{1-x^2}}^{\sqrt{1-x^2}} dx$$

$$= \int_{-1}^{1} \left[x^2(1-x^2) + \frac{1}{3}(1-x^2)^2 \right] dx$$

$$= \int_{-1}^{1} \left(-\frac{2}{3}x^4 + \frac{1}{3}x^2 + \frac{1}{3} \right) dx$$

$$= \left(-\frac{2}{15}x^5 + \frac{1}{9}x^3 + \frac{1}{3}x \right)\Big|_{-1}^{1} = \frac{28}{45} \qquad \square$$

Suppose D is composed of two or more subregions $D_1, D_2, \ldots, D_n$ of the type appearing in Definition 14.8 and having at most boundaries in common. Then we define $\iiint_D f(x, y, z)\, dV$ by the formula

$$\iiint_D f(x, y, z)\, dV = \iiint_{D_1} f(x, y, z)\, dV + \iiint_{D_2} f(x, y, z)\, dV + \cdots$$

$$+ \iiint_{D_n} f(x, y, z)\, dV$$

(4)

Mass and Charge

According to the molecular theory of matter, any piece of matter is just a collection of molecules and consequently has a mass equal to the sum of the masses of its molecules. However, the molecules are so numerous that finding such a sum would defy even modern computers. In order to calculate the mass of an object like a cube of ice or a steel beam, one idealizes the mass and thinks of it as being "spread everywhere" throughout the object (not only at the locations of the molecules). If the mass m of the object is also distributed uniformly, or homogeneously, throughout the object and the volume V of the object is not 0, then we can define the **mass density** δ of the object by the equation

$$\delta = \frac{m}{V} \tag{5}$$

This equation applies to any such object, which might be a quart of oil, a cube of ice, or a steel beam.

In contrast, consider a can of frozen orange juice. Due to settling, the bottom portion of the frozen orange juice would probably be heavier than the top, in which case the mass would not be distributed uniformly throughout the can. In order to describe the mass in the different parts of the can, we define the **mass density** $\delta(x, y, z)$ at the point (x, y, z) by

$$\delta(x, y, z) = \lim_{\|\Delta V\| \to 0} \frac{\Delta m}{\Delta V} \tag{6}$$

provided that this limit exists, where Δm is the mass and ΔV the volume of a parallelepiped centered at (x, y, z), and where $\|\Delta V\|$ is the largest dimension of the parallelepiped (Figure 14.43). Formula (6) applies to any object for which the limit exists. Our assumption that the mass is "spread everywhere" throughout the object is formalized as the assumption that δ is defined and continuous on D. Under this hypothesis, scientists are able to describe many physical phenomena with remarkable accuracy. Our immediate goal is to show that under this hypothesis the total mass m of the object is given by

$$m = \iiint_D \delta(x, y, z)\, dV$$

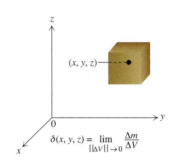

FIGURE 14.43

Let D' be any parallelepiped containing D, and let P be a partition of D' into subparallelepipeds, numbered so that $D_1, D_2, \ldots, D_n$ are the ones entirely contained in D. For each k between 1 and n, choose an arbitrary point (x_k, y_k, z_k) in D_k, and let Δm_k be the mass of D_k and ΔV_k the volume of D_k. Formula (6) suggests that $\delta(x_k, y_k, z_k)\, \Delta V_k$ approximates Δm_k if the dimensions of D_k are small. Since $m = \Delta m_1 + \Delta m_2 + \cdots + \Delta m_n$, m is approximated by

$$\sum_{k=1}^{n} \delta(x_k, y_k, z_k)\, \Delta V_k$$

which is a Riemann sum for δ on D. From Definition 14.8 we conclude that the **mass** of D is given by

$$m = \iiint_D \delta(x, y, z)\, dV \tag{7}$$

EXAMPLE 7   A cubical object occupying the solid region D shown in Figure 14.44 has a density given by $\delta(x, y, z) = 1 + xyz$ for (x, y, z) in D. Find the mass m of the object.

Solution   By (7) we find that

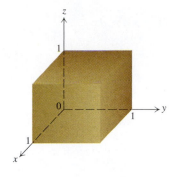

FIGURE 14.44

$$m = \iiint\limits_D \delta(x, y, z)\, dV = \iiint\limits_D (1 + xyz)\, dV$$

$$= \int_0^1 \int_0^1 \int_0^1 (1 + xyz)\, dz\, dy\, dx = \int_0^1 \int_0^1 \left(z + \frac{1}{2} xyz^2\right)\Big|_0^1 dy\, dx$$

$$= \int_0^1 \int_0^1 \left(1 + \frac{1}{2} xy\right) dy\, dx = \int_0^1 \left(y + \frac{1}{4} xy^2\right)\Big|_0^1 dx$$

$$= \int_0^1 \left(1 + \frac{1}{4} x\right) dx = \left(x + \frac{1}{8} x^2\right)\Big|_0^1 = \frac{9}{8} \qquad \square$$

If an object such as a copper ball has a homogeneous charge distribution (that is, if any two portions of the object having the same volume have the same charge), then we define the **charge density** ρ of the object by

$$\rho = \frac{q}{V}$$

where q is the charge in the object and V is its volume. If an object does not have a homogeneous charge distribution, the **charge density** $\rho(x, y, z)$ at any point (x, y, z) in the object is defined by

$$\rho(x, y, z) = \lim_{\|\Delta V\| \to 0} \frac{\Delta q}{\Delta V}$$

provided that this limit exists, where Δq is the charge in a parallelepiped centered at (x, y, z), ΔV is the volume of the parallelepiped, and $\| \Delta V \|$ is the largest dimension of the parallelepiped. The same argument we used to derive (7) shows that the **total charge** q in a charged object is given by

$$q = \iiint\limits_D \rho(x, y, z)\, dV \qquad (8)$$

where D is the region occupied by the object.

EXAMPLE 8   Suppose the charge density $\rho(x, y, z)$ at any point (x, y, z) occupying the ball $x^2 + y^2 + z^2 \leq 1$ is equal to the distance from (x, y, z) to the xy plane. Find the total charge q in the ball.

Solution   Let D be the ball. We know that $\rho(x, y, z) = |z|$. Since D is symmetric with respects to the x, y, and z axes, and since $|z|$ is an even function, an application of (8) yields

$$q = \iiint\limits_D \rho(x, y, z)\, dV = 8 \int_0^1 \int_0^{\sqrt{1-x^2}} \int_0^{\sqrt{1-x^2-y^2}} |z|\, dz\, dy\, dx$$

$$= 8 \int_0^1 \int_0^{\sqrt{1-x^2}} \frac{1}{2} z^2 \Big|_0^{\sqrt{1-x^2-y^2}} dy\, dx = 4 \int_0^1 \int_0^{\sqrt{1-x^2}} (1 - x^2 - y^2)\, dy\, dx$$

$$= 4 \int_0^1 \left((1 - x^2)y - \frac{y^3}{3} \right) \Big|_0^{\sqrt{1-x^2}} .dx = \frac{8}{3} \int_0^1 (1 - x^2)^{3/2} \, dx$$

$$\stackrel{x = \sin \theta}{=} \frac{8}{3} \int_0^{\pi/2} \cos^4 \theta \, d\theta$$

$$\stackrel{\substack{(13) \text{ of} \\ \text{Section 8.1}}}{=} \frac{8}{3} \left(\frac{1}{4} \cos^3 \theta \sin \theta + \frac{3}{8} \cos \theta \sin \theta + \frac{3}{8} \theta \right) \Big|_0^{\pi/2}$$

$$= \frac{\pi}{2} \quad \square$$

EXERCISES 14.4

In Exercises 1–10 evaluate the iterated integral.

1. $\int_0^3 \int_{-1}^1 \int_2^4 (y - xz) \, dz \, dy \, dx$

2. $\int_0^3 \int_{-1}^1 \int_2^4 (y - xz) \, dy \, dx \, dz$

3. $\int_{-1}^1 \int_0^x \int_{x-y}^{x+y} (z - 2x - y) \, dz \, dy \, dx$

4. $\int_0^{\pi/2} \int_0^1 \int_0^{\sqrt{1-x^2}} x \cos z \, dy \, dx \, dz$

5. $\int_0^{\ln 3} \int_0^1 \int_0^y (z^2 + 1) e^{(y^2)} \, dx \, dz \, dy$

6. $\int_0^{\sqrt{\pi/6}} \int_0^y \int_0^y (1 + y^2 z \cos xz) \, dx \, dz \, dy$

7. $\int_{-13}^{13} \int_1^e \int_0^{1/\sqrt{x}} z(\ln x)^2 \, dz \, dx \, dy$

8. $\int_{-1}^1 \int_{-\sqrt{1-y^2}}^{\sqrt{1-y^2}} \int_{-\sqrt{1-x^2-y^2}}^{\sqrt{1-x^2-y^2}} x^2 y^2 z \, dz \, dx \, dy$

9. $\int_0^{\pi/2} \int_0^{\pi/2} \int_0^{\sin z} x^2 \sin y \, dx \, dy \, dz$

10. $\int_{-\pi/2}^{\pi/2} \int_{-\cos z}^{\cos z} \int_{-\cos zy}^{\cos zy} x \cos zy \, dx \, dy \, dz$

In Exercises 11–19 evaluate the integral.

11. $\iiint_D e^y \, dV$, where D is the solid region bounded by the planes $y = 1$, $z = 0$, $y = x$, $y = -x$, and $z = y$

12. $\iiint_D 1/x \, dV$, where D is the prism bounded by the planes $x + y + z = 4$, $y = x$, $x = 1$, $x = 2$, $z = 0$, and $y = 0$

13. $\iiint_D y e^{xy} \, dV$, where D is the cube bounded by the planes $x = 1$, $x = 3$, $y = 0$, $y = 2$, $z = -2$, and $z = 0$

14. $\iiint_D xy \, dV$, where D is the solid region in the first octant bounded above by the hemisphere $z = \sqrt{4 - x^2 - y^2}$ and on the sides and bottom by the coordinate planes

15. $\iiint_D zy \, dV$, where D is the solid region in the first octant bounded above by the plane $z = 1$ and below by the cone $z = \sqrt{x^2 + y^2}$

16. $\iiint_D (x + z) \, dV$, where D is the solid region bounded by the cylinder $x^2 + z^2 = 1$ and by the planes $y = -4$, $y = 5$, $x = 0$, and $x = 1$.

17. $\iiint_D z \, dV$, where D is the solid region bounded above by the sphere $x^2 + y^2 + z^2 = 9$, below by the plane $z = 0$, and on the sides by the planes $x = -1$, $x = 1$, $y = -1$, and $y = 1$

18. $\iiint_D xz \, dV$, where D is the solid region in the first octant bounded above by the sphere $x^2 + y^2 + z^2 = 4$, below by the plane $z = 0$, and on the sides by the planes $x = 0$ and $y = 0$ and the cylinder $x^2 + y^2 = 1$

19. $\iiint_D 3xy \, dV$, where D is the solid region bounded below by the cone $z = \sqrt{x^2 + y^2}$ and above by the cylinder $x^2 + z^2 = 1$

In Exercises 20–26 find the volume V of the region.

20. The solid region bounded by the plane $x + 3y + 6z = 1$ and the three coordinate planes

21. The solid region in the first octant bounded by the planes $z = 10 + x + y$, $y = 2 - x$, $y = x$, $z = 0$, and $x = 0$

22. The solid region in the first octant bounded above by the plane $z = 2x$, below by the xy plane, and on the sides by the elliptic cylinder $2x^2 + y^2 = 1$ and the plane $y = 0$

23. The solid region bounded above by the circular paraboloid $z = 4(x^2 + y^2)$, below by the plane $z = -2$, and on the sides by the parabolic sheet $y = x^2$ and the plane $y = x$

24. The solid region bounded above by the elliptic paraboloid $z = 2x^2 + 3y^2$, on the sides by the parabolic sheet $y^2 = 1 - x$ and the plane $x = 0$, and below by the plane $z = 0$ (*Hint:* Integrate with respect to x before y.)

*25. The solid region bounded above by the plane $z = h$ ($h > 0$) and below by the cone $z = h\sqrt{x^2 + y^2}$ (The answer is the volume of a cone whose height is h and whose base has radius 1.)

*26. A right pyramid with height h and a square base having sides of length s (*Hint:* Place the base on the xy plane with center at the origin and vertices on the axes. Compute the volume of the portion in the first octant and multiply by 4.)

27. An object occupies the tetrahedron in the first octant bounded by the coordinate planes and the plane $2x + y + z = 1$. Find the total mass m of the object if the mass density at a given point in the object is

  a. equal to the distance from the xy plane.
  b. equal to twice the distance from the xz plane.
  c. equal to the distance from the yz plane.

28. An object occupies a cube of volume 8 with three of its faces on the coordinate planes. If the mass density at any point in the cube is equal to the square of its distance from the origin, find the total mass m of the object.

29. An object occupies the solid region bounded above by the sphere $x^2 + y^2 + z^2 = 9$ and below by the xy plane. If the charge density of the object at any point is equal to its distance from the xy plane, find the total charge q in the object.

30. An object occupies the solid region in the first octant bounded by the coordinate planes and the two cylinders $x^2 + y^2 = 4$ and $y^2 + z^2 = 4$. If the charge density at any point (x, y, z) is x, find the total charge q in the object.

31. Suppose D is the solid region between the graphs of two continuous functions and has nonzero volume V. If f is continuous on D, then the **average value f_{av} of f on D** is defined to be $(1/V) \iiint_D f(x, y, z) \, dV$. Under the following conditions, find the average value of f on D.

  a. $f(x, y, z) = x + y + z$; D is the solid region bounded by the sheet $z = x^2$ and by the planes $z = x$, $y = 0$, and $y = 1$.
  b. $f(x, y, z) = xy$; D is the solid region in the first octant bounded by the circular paraboloids $z = 2 - x^2 - y^2$ and $z = x^2 + y^2$ and the planes $x = 0$ and $y = 0$.

14.5 TRIPLE INTEGRALS IN CYLINDRICAL COORDINATES

Just as certain double integrals are easier to evaluate by means of polar coordinates than by rectangular coordinates, certain triple integrals are easier to evaluate by coordinates other than rectangular coordinates. In Sections 14.5 and 14.6 we introduce two additional types of coordinates—cylindrical and spherical—and explain how to evaluate triple integrals by means of these coordinates.

Cylindrical Coordinates

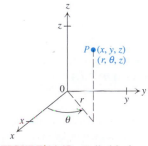

FIGURE 14.45 Cylindrical coordinates.

Let (x, y, z) be the rectangular coordinates of a point P in space. If (r, θ) is a set of polar coordinates for the point (x, y), then we call (r, θ, z) a set of **cylindrical coordinates** for P (Figure 14.45). Given the rectangular coordinates (x, y, z) of a point P, we can determine a set of cylindrical coordinates for P with the aid of the formulas

$$x^2 + y^2 = r^2 \quad \text{and} \quad \tan \theta = \frac{y}{x} \quad (\text{if } x \neq 0)$$

Conversely, from any set (r, θ, z) of cylindrical coordinates of a point P we can determine the rectangular coordinates (x, y, z) of P by the formulas

$$x = r \cos \theta \quad \text{and} \quad y = r \sin \theta$$

Let us compare equations in rectangular coordinates and in cylindrical coordinates for several surfaces (Table 14.1). These surfaces, which are displayed in Figure 14.46(a)–(d), will recur frequently in this and the next chapter. The simplicity of the formula $r = a$ for the cylinder suggests a reason for the name "cylindrical coordinates." We usually refer to the double cone simply as a cone, and to its upper and lower portions as its **upper nappe** and **lower nappe.**

TABLE 14.1

Surface	Rectangular	Cylindrical
Cylinder	$x^2 + y^2 = a^2$	$r = a$
Sphere	$x^2 + y^2 + z^2 = a^2$	$r^2 + z^2 = a^2$
Double circular cone	$x^2 + y^2 = a^2 z^2$	$r = az$ or $z = r \cot \phi_0$
Circular paraboloid	$x^2 + y^2 = az$	$r^2 = az$

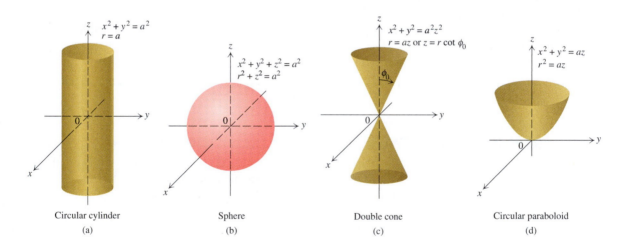

Circular cylinder
(a)

Sphere
(b)

Double cone
(c)

Circular paraboloid
(d)

FIGURE 14.46

Triple Integrals in Cylindrical Coordinates

Theorem 14.9 asserts that under suitable conditions

$$\iiint_D f(x, y, z)\, dV = \iint_R \left(\int_{F_1(x, y)}^{F_2(x, y)} f(x, y, z)\, dz \right) dA$$

We also know how to evaluate double integrals by means of polar coordinates, thanks to Theorem 14.5. Together these two results give us a method of evaluating triple integrals by means of cylindrical coordinates.

THEOREM 14.10

Let D be the solid region between the graphs of F_1 and F_2 on R, where R is the plane region between the polar graphs of h_1 and h_2 on $[\alpha, \beta]$, with $0 \le \beta - \alpha \le 2\pi$ and $0 \le h_1(\theta) \le h_2(\theta)$ for $\alpha \le \theta \le \beta$. If f is continuous on D, then

$$\iiint_D f(x, y, z)\, dV = \int_\alpha^\beta \int_{h_1(\theta)}^{h_2(\theta)} \int_{F_1(r\cos\theta, r\sin\theta)}^{F_2(r\cos\theta, r\sin\theta)} f(r\cos\theta, r\sin\theta, z) r\, dz\, dr\, d\theta$$

Integration by means of cylindrical coordinates is especially effective when expressions containing $x^2 + y^2$ appear in the integrand or in the limits of integration and the region over which the integration is taken is easily described by polar coordinates.

EXAMPLE 1 Let D be the solid region bounded above by the plane $y + z = 4$, below by the xy plane, and on the sides by the cylinder $x^2 + y^2 = 16$ (Figure 14.47). Evaluate $\iiint_D \sqrt{x^2 + y^2}\, dV$.

Solution Observe that D is the solid region between the graphs of $z = 0$ and $z = 4 - y$ on R, where R is the disk $x^2 + y^2 \le 16$. In polar coordinates R is the region between the polar graphs of

$$r = 0 \quad \text{and} \quad r = 4 \quad \text{for } 0 \le \theta \le 2\pi$$

Consequently in cylindrical coordinates D is the solid region between the graphs of

$$z = 0 \quad \text{and} \quad z = 4 - r\sin\theta \quad \text{for } (r, \theta) \text{ in } R$$

Then Theorem 14.10 implies that

$$\iiint_D \sqrt{x^2 + y^2}\, dV = \int_0^{2\pi} \int_0^4 \int_0^{4 - r\sin\theta} r \cdot r\, dz\, dr\, d\theta$$

$$= \int_0^{2\pi} \int_0^4 r^2 z \Big|_0^{4 - r\sin\theta}\, dr\, d\theta$$

$$= \int_0^{2\pi} \int_0^4 (4r^2 - r^3 \sin\theta)\, dr\, d\theta$$

$$= \int_0^{2\pi} \left(\frac{4}{3} r^3 - \frac{r^4}{4} \sin\theta \right) \Big|_0^4\, d\theta$$

$$= \int_0^{2\pi} \left(\frac{256}{3} - 64\sin\theta \right) d\theta$$

$$= \left(\frac{256}{3}\theta + 64\cos\theta \right) \Big|_0^{2\pi} = \frac{512}{3}\pi \qquad \square$$

EXAMPLE 2 Find the volume V of the solid region that the cylinder $r = a\cos\theta$ cuts out of the sphere of radius a centered at the origin (Figure 14.48).

Solution In this case R is bounded by the polar graphs of

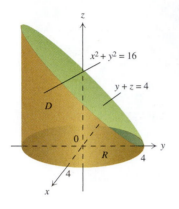

$x^2 + y^2 = 16$

$y + z = 4$

D

R

FIGURE 14.47

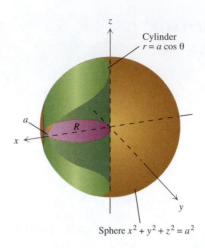

FIGURE 14.48

$$r = 0 \quad \text{and} \quad r = a\cos\theta \quad \text{for} \quad -\frac{\pi}{2} \le \theta \le \frac{\pi}{2}$$

and D is the solid region determined by the sphere between the graphs of

$$z = -\sqrt{a^2 - r^2} \quad \text{and} \quad z = \sqrt{a^2 - r^2} \quad \text{for } (r, \theta) \text{ in } R$$

Using symmetry, we find that

$$V = \iiint\limits_{D} 1 \, dV = \int_{-\pi/2}^{\pi/2} \int_{0}^{a\cos\theta} \int_{-\sqrt{a^2-r^2}}^{\sqrt{a^2-r^2}} 1 \cdot r \, dz \, dr \, d\theta$$

$$= 4 \int_{0}^{\pi/2} \int_{0}^{a\cos\theta} \int_{0}^{\sqrt{a^2-r^2}} r \, dz \, dr \, d\theta = 4 \int_{0}^{\pi/2} \int_{0}^{a\cos\theta} r\sqrt{a^2 - r^2} \, dr \, d\theta$$

$$= 4 \int_{0}^{\pi/2} -\frac{1}{3} (a^2 - r^2)^{3/2} \Big|_{0}^{a\cos\theta} d\theta = \frac{4}{3} \int_{0}^{\pi/2} \left[a^3 - (a^2 - a^2\cos^2\theta)^{3/2} \right] d\theta$$

$$= \frac{4}{3} a^3 \int_{0}^{\pi/2} (1 - \sin^3\theta) \, d\theta = \frac{4}{3} a^3 \int_{0}^{\pi/2} \left[1 - (1 - \cos^2\theta)\sin\theta \right] d\theta$$

$$= \frac{4}{3} a^3 \left(\theta + \cos\theta - \frac{1}{3}\cos^3\theta \right) \Big|_{0}^{\pi/2} = \frac{2}{9} a^3 (3\pi - 4) \qquad \square$$

EXAMPLE 3 Let D be the solid region bounded above by the sphere $r^2 + z^2 = a^2$ and below by the nappe of the cone $z = r \cot\phi$ that makes an angle of ϕ with the positive z axis. Show that the volume V of D is given by

$$V = \frac{2\pi a^3}{3} (1 - \cos\phi)$$

Solution First we consider the case $0 \le \phi \le \pi/2$. Then the nappe in question is the upper nappe (Figure 14.49). To determine where the sphere and the cone intersect, we substitute $r \cot \phi$ for z in the equation $r^2 + z^2 = a^2$ of the sphere and obtain

$$r^2(1 + \cot^2 \phi) = a^2$$

This implies that

$$r \csc \phi = a, \quad \text{or} \quad r = a \sin \phi$$

It follows that D is the solid region between the cone and the sphere on the plane region R bounded by the circle $r = a \sin \phi$ (see Figure 14.49).

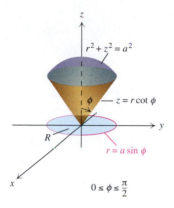

FIGURE 14.49

As a result,

$$
\begin{aligned}
V = \iiint_D 1 \, dV &= \int_0^{2\pi} \int_0^{a \sin \phi} \int_{r \cot \phi}^{\sqrt{a^2 - r^2}} 1 \cdot r \, dz \, dr \, d\theta \\
&= \int_0^{2\pi} \int_0^{a \sin \phi} rz \Big|_{r \cot \phi}^{\sqrt{a^2 - r^2}} dr \, d\theta \\
&= \int_0^{2\pi} \int_0^{a \sin \phi} \left(r\sqrt{a^2 - r^2} - r^2 \cot \phi \right) dr \, d\theta \\
&= \int_0^{2\pi} \left(-\frac{1}{3}(a^2 - r^2)^{3/2} - \frac{r^3}{3} \cot \phi \right) \Big|_0^{a \sin \phi} d\theta \\
&= \int_0^{2\pi} \left(-\frac{1}{3}(a^2 - a^2 \sin^2 \phi)^{3/2} - \frac{a^3 \sin^3 \phi}{3} \cot \phi + \frac{a^3}{3} \right) d\theta \\
&= \frac{a^3}{3}(-\cos^3 \phi - \sin^2 \phi \cos \phi + 1) \int_0^{2\pi} 1 \, d\theta \\
&= \frac{2\pi a^3}{3}[-(\cos^2 \phi + \sin^2 \phi)\cos \phi + 1] = \frac{2\pi a^3}{3}(1 - \cos \phi)
\end{aligned}
$$

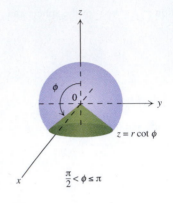

$z = r \cot \phi$

ϕ

0

$\dfrac{\pi}{2} < \phi \leq \pi$

FIGURE 14.50

If $\pi/2 < \phi \leq \pi$, then the nappe under consideration is the lower nappe (Figure 14.50). The volume V we seek is the difference between the volume $4\pi a^3/3$ of the complete sphere and the volume of the solid region bounded below by the sphere and above by the lower nappe of the cone. It follows from the first part of the solution that the latter volume is

$$\frac{2\pi a^3}{3}\left[1 - \cos\left(\pi - \phi\right)\right]$$

Thus

$$V = \frac{4\pi a^3}{3} - \frac{2\pi a^3}{3}\left[1 - \cos\left(\pi - \phi\right)\right]$$

$$= \frac{4\pi a^3}{3} - \frac{2\pi a^3}{3}\left(1 + \cos\phi\right)$$

$$= \frac{2\pi a^3}{3}\left(1 - \cos\phi\right) \qquad \square$$

It is not uncommon to encounter a triple iterated integral in rectangular coordinates that could be evaluated more easily by changing to cylindrical coordinates. The procedure is first to describe in cylindrical coordinates the region over which the integration is to be performed and then to evaluate the corresponding iterated integral in cylindrical coordinates.

EXAMPLE 4   Evaluate $\displaystyle\int_{-2}^{2}\int_{-\sqrt{4-x^2}}^{\sqrt{4-x^2}}\int_{(x^2+y^2)^2}^{1} x^2\,dz\,dy\,dx.$

Solution   The limits -2 and 2 on the first integral and $-\sqrt{4-x^2}$ and $\sqrt{4-x^2}$ on the second integral tell us that those two integrals are taken over the region bounded by the circle $x^2 + y^2 = 4$, or $r = 2$, in the xy plane. It follows that

$$\int_{-2}^{2}\int_{-\sqrt{4-x^2}}^{\sqrt{4-x^2}}\int_{(x^2+y^2)^2}^{1} x^2\,dz\,dy\,dx = \int_{0}^{2\pi}\int_{0}^{2}\int_{r^4}^{1} (r\cos\theta)^2 r\,dz\,dr\,d\theta$$

$$= \int_{0}^{2\pi}\int_{0}^{2} \left[(r^3\cos^2\theta)z\right]\Big|_{r^4}^{1}\,dr\,d\theta$$

$$= \int_{0}^{2\pi}\int_{0}^{2} (r^3 - r^7)\cos^2\theta\,dr\,d\theta$$

$$= \int_{0}^{2\pi} \left(\frac{r^4}{4} - \frac{r^8}{8}\right)\cos^2\theta\,\Big|_{0}^{2}\,d\theta$$

$$= -28\int_{0}^{2\pi} \cos^2\theta\,d\theta$$

$$= -28\int_{0}^{2\pi} \left(\frac{1}{2} + \frac{1}{2}\cos 2\theta\right)d\theta$$

$$= -14\left(\theta + \frac{1}{2}\sin 2\theta\right)\Big|_{0}^{2\pi} = -28\pi \qquad \square$$

EXERCISES 14.5

In Exercises 1–8 write the equation in cylindrical coordinates, and sketch its graph.

1. $y = -4$ 2. $x = 5z$ 3. $x + y + z = 3$
4. $x^2 + y^2 + z^2 = 16$ 5. $x^2 + y^2 + z = 1$ 6. $x^2 + y^2 + z^2 = 0$
7. $4x^2 + 4y^2 - z^2 = 0$ 8. $x^2 + y^2 + 3z^2 = 9$

In Exercises 9–13 evaluate the iterated integral.

9. $\displaystyle\int_0^{2\pi}\int_1^2\int_0^5 e^z r\, dz\, dr\, d\theta$

10. $\displaystyle\int_0^{\pi/2}\int_0^1\int_0^{\sqrt{1-r^2}} r\sin\theta\, dz\, dr\, d\theta$

11. $\displaystyle\int_{-\pi/2}^0\int_0^{2\sin\theta}\int_0^{r^2} r^2\cos\theta\, dz\, dr\, d\theta$

12. $\displaystyle\int_0^{\pi/4}\int_0^{1+\cos\theta}\int_0^r 1\, dz\, dr\, d\theta$

13. $\displaystyle\int_{-\pi/4}^{\pi/4}\int_0^{1-2\cos^2\theta}\int_0^1 r\sin\theta\, dz\, dr\, d\theta$

In Exercises 14–18 express the triple integral as an iterated integral in cylindrical coordinates. Then evaluate it.

14. $\iiint_D (x^2 + y^2)\, dV$, where D is the solid region bounded by the cylinder $x^2 + y^2 = 1$ and the planes $z = 0$ and $z = 4$

15. $\iiint_D z\, dV$, where D is the portion of the ball $x^2 + y^2 + z^2 \le 1$ that lies in the first octant

16. $\iiint_D y^2\, dV$, where D is the solid region common to the cylinder $x^2 + y^2 = 1$ and the sphere $x^2 + y^2 + z^2 = 4$

17. $\iiint_D xz\, dV$, where D is the portion of the ball $x^2 + y^2 + z^2 \le 4$ in the first octant

18. $\iiint_D yz\, dV$, where D is the solid region in the first octant bounded by the sphere $x^2 + y^2 + z^2 = 1$, the circular cylinder $r = \cos\theta$, and the planes $y = 0$ and $z = 0$

In Exercises 19–31 find the volume V of the region.

19. The solid region bounded below by the surface $z = \sqrt{r}$ and above by the plane $z = 1$

20. The solid region bounded above by the sphere $x^2 + y^2 + z^2 = 2$ and below by the circular paraboloid $z = x^2 + y^2$

21. The solid region bounded above by the surface $z = e^{-x^2 - y^2}$, below by the xy plane, and on the sides by the cylinder $x^2 + y^2 = 1$

22. The solid region above the plane $z = 1$ and inside the sphere $x^2 + y^2 + z^2 = 2$

23. The solid region inside the sphere $x^2 + y^2 + z^2 = 4$ and above the upper nappe of the cone $z^2 = 3x^2 + 3y^2$

24. The solid region bounded above by the paraboloid $z = 1 - x^2 - y^2$ and below by the plane $z = -3$

25. The solid region inside the cone $z = r$ and between the planes $z = 1$ and $z = 2$

26. The solid region bounded above by the cone $z = 8 - \sqrt{x^2 + y^2}$, on the sides by the cylinder $x^2 + y^2 = 2x$, and below by the xy plane

27. The solid region bounded above by the plane $z = y$ and below by the paraboloid $z = x^2 + y^2$

28. The solid region inside the sphere $x^2 + y^2 + z^2 = 4a^2$ and the cylinder $x^2 + y^2 = a^2$

29. The solid region inside the sphere $r^2 + z^2 = a^2$ and the cylinder $r = a\sin\theta$

30. The solid region bounded on the sides by the surface $r = 1 + \cos\theta$, above by the upper nappe of the cone $z = r$, and below by the xy plane

31. The solid region in the first octant bounded by the coordinate planes, the circular paraboloid $z = r^2$, and the surface $r^2 = 4\cos\theta$

32. An object occupies the solid region bounded by the cylinder $x^2 + y^2 = 9$ and the planes $z = 0$ and $z = 5$. If the mass density at any point is equal to the distance from the point to the axis of the cylinder, find the total mass m of the object.

33. An object occupies the solid region bounded by the upper nappe of the cone $z^2 = 9x^2 + 9y^2$ and the plane $z = 9$. Find the total mass m of the object if the mass density at (x, y, z) is equal to the distance from (x, y, z) to the top.

34. A cylindrical hole of radius b is bored out of the center of a spherical ball of radius a, where $0 < b < a$. How much volume is retained? (This exercise is related to Review Exercise 8 of Chapter 6.)

35. Find the volume V of the solid region bounded above by the plane $z = h$ ($h > 0$) and below by the upper nappe of the cone $z^2 = h^2(x^2 + y^2)$.

36. Let $f(x, y, z) = 6 - x^2 - y^2$. Find the average value f_{av} of f on the cylindrical region D bounded below by the xy plane, above by the plane $z = 3$, and on the sides by the cylinder $x^2 + y^2 = 4$.

Application

37. When pineapple juice is left standing, it settles so that it thickens toward the bottom. Suppose a hemispherical glass 8 centimeters tall is full of pineapple juice, and the density of the juice, in grams per cubic centimeter, is given by $\delta(x, y, z) = 1 - \frac{1}{16}z$ for $-8 \le z \le 0$, where $z = 0$ corresponds to the top of the glass. Find the total mass m of the juice.

14.6   TRIPLE INTEGRALS IN SPHERICAL COORDINATES

The final coordinate system we discuss is the spherical coordinate system. This system simplifies evaluation of triple integrals over solid regions bounded by surfaces such as spheres and cones.

Spherical Coordinates

Let (x, y, z) and (r, θ, z) be, respectively, sets of rectangular and cylindrical coordinates (see Figure 14.51) for a point P in space, with $r \geq 0$. We let

$$\rho = \text{the length of the line segment } OP = \sqrt{x^2 + y^2 + z^2}$$

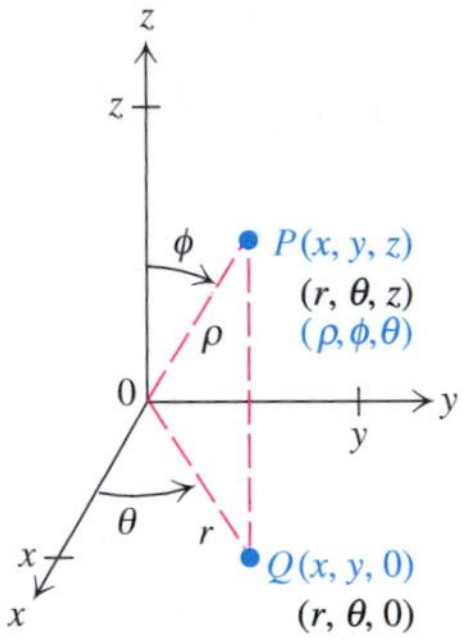

If $\rho \neq 0$, then we let

$$\phi = \text{the angle } OP \text{ makes with the positive } z \text{ axis, with } 0 \leq \phi \leq \pi$$

If $\rho = 0$, then ϕ may be chosen arbitrarily. Finally, we let

$$\theta = \text{the angle } OQ \text{ makes with the positive } x \text{ axis}$$

(Figure 14.51). The point P is specified by the three quantities ρ, ϕ, and θ, and we call the triple (ρ, ϕ, θ) a set of **spherical coordinates** for P. Figure 14.52(a) and (b) depicts the points A and B with spherical coordinates $(1, \pi/6, \pi/4)$ and $(1, 2\pi/3, \pi/4)$, respectively.

From trigonometry we find that

$$r = \rho \sin \phi \quad \text{and} \quad z = \rho \cos \phi$$

These equations, along with the polar coordinate formulas

$$x = r \cos \theta \text{ and } y = r \sin \theta$$

yield the following formulas for converting from spherical coordinates to rectangular coordinates:

FIGURE 14.51   Spherical coordinates.

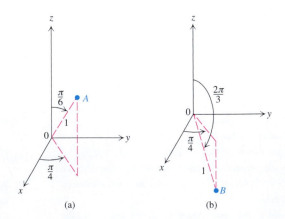

(a)   (b)

FIGURE 14.52

$$x = r \cos \theta = \rho \sin \phi \cos \theta$$
$$y = r \sin \theta = \rho \sin \phi \sin \theta$$
$$z = \rho \cos \phi$$
$$x^2 + y^2 = \rho^2 \sin^2 \phi$$

EXAMPLE 1   A point P has spherical coordinates $(8, 2\pi/3, -\pi/6)$. Find the rectangular coordinates for P.

Solution   The rectangular coordinates for P are given by

$$x = 8 \sin\left(\frac{2\pi}{3}\right) \cos\left(-\frac{\pi}{6}\right) = 8\left(\frac{\sqrt{3}}{2}\right)\left(\frac{\sqrt{3}}{2}\right) = 6$$

$$y = 8 \sin\left(\frac{2\pi}{3}\right) \sin\left(-\frac{\pi}{6}\right) = 8\left(\frac{\sqrt{3}}{2}\right)\left(-\frac{1}{2}\right) = -2\sqrt{3}$$

$$z = 8 \cos\left(\frac{2\pi}{3}\right) = 8\left(-\frac{1}{2}\right) = -4 \quad \square$$

Spheres centered at the origin have particularly simple equations in spherical coordinates. Indeed, from the definition of ρ we find that an equation in spherical coordinates of the sphere centered at the origin with radius a is $\rho = a$.

Analogously, if $0 < \alpha < \pi$ and $\alpha \neq \pi/2$, then the graph of the equation $\phi = \alpha$ is a nappe of a cone whose angle with respect to the positive z axis is α. Notice that if $0 < \alpha < \pi/2$, then the nappe of the cone opens upward, whereas if $\pi/2 < \alpha < \pi$, then the nappe opens downward (Figure 14.53(a) and (b)). The equations $\phi = 0$, $\phi = \pi$, and $\phi = \pi/2$ yield the positive z axis, negative z axis, and xy plane, respectively.

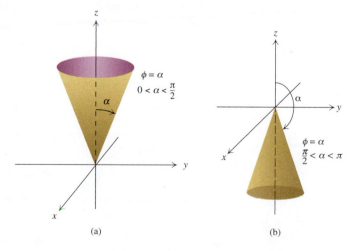

(a)   (b)

FIGURE 14.53

TABLE 14.2

Surface	Equation
Sphere	$\rho = a$
Cone	$\phi = \alpha$
Vertical half-plane	$\theta = \alpha$

In Table 14.2 we list the surfaces on which the spherical coordinates ρ, ϕ, and θ are constant.

Formulas in spherical coordinates for horizontal planes are a little less simple than those for vertical half-planes.

EXAMPLE 2   Find an equation in spherical coordinates for the plane $z = 2$.

Solution   Since $z = \rho \cos \phi$, the equation $z = 2$ becomes

$$\rho \cos \phi = 2$$

Dividing by $\cos \phi$, which cannot be 0 if $\rho \cos \phi = 2$, we find that

$$\rho = \frac{2}{\cos \phi} = 2 \sec \phi$$

Thus an equation in spherical coordinates for the plane $z = 2$ is $\rho = 2 \sec \phi$.   ❑

Triple Integrals in Spherical Coordinates

Just as parallelepipeds form the basic three-dimensional solids for integration in rectangular coordinates, spherical wedges form the basic solids for integration in spherical coordinates. A **spherical wedge** D is a solid region bounded by the surfaces

$$\rho = \rho_0 \quad \text{and} \quad \rho = \rho_1, \quad \phi = \phi_0 \quad \text{and} \quad \phi = \phi_1, \quad \theta = \theta_0 \quad \text{and} \quad \theta = \theta_1$$

where

$$0 \leq \rho_0 \leq \rho_1, \quad 0 \leq \phi_0 \leq \phi_1 \leq \pi, \quad \text{and} \quad 0 \leq \theta_1 - \theta_0 \leq 2\pi$$

(Figure 14.54).

Our next goal is to estimate the volume V of D, under the conditions that $\rho_1 - \rho_0$, $\phi_1 - \phi_0$, and $\theta_1 - \theta_0$ are each small. In that case adjacent sides of D are nearly perpendicular to each other, so that if the sides are labeled as in Figure 14.54, then

$$V \approx abc$$

The formula for a is easy:

$$a = \rho_1 - \rho_0$$

Using the fact that b is the length of an arc of angle $\phi_1 - \phi_0$ in a circle of radius ρ_1, we deduce that

$$b = \rho_1(\phi_1 - \phi_0)$$

Finally, c is approximately the length of an arc of angle $\theta_1 - \theta_0$ in a circle of radius $r_1 = \rho_1 \sin \phi_1$, so that

$$c \approx r_1(\theta_1 - \theta_0) = (\rho_1 \sin \phi_1)(\theta_1 - \theta_0)$$

Consequently

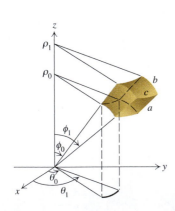

FIGURE 14.54

$$V \approx abc \approx \overbrace{(\rho_1 - \rho_0)}^{a}\overbrace{[\rho_1(\phi_1 - \phi_0)]}^{b}\overbrace{[(\rho_1 \sin \phi_1)(\theta_1 - \theta_0)]}^{\approx c}$$

$$= (\rho_1^2 \sin \phi_1)(\rho_1 - \rho_0)(\phi_1 - \phi_0)(\theta_1 - \theta_0)$$

It is possible to show that if ρ_1^* and ϕ_1^* are chosen appropriately, with $\rho_0 \leq \rho_1^* \leq \rho_1$ and $\phi_0 \leq \phi_1^* \leq \phi_1$, then

$$V = (\rho_1^*)^2 (\sin \phi_1^*)(\rho_1 - \rho_0)(\phi_1 - \phi_0)(\theta_1 - \theta_0) \tag{1}$$

This formula will be of fundamental use in deriving a formula for evaluating triple integrals by means of spherical coordinates.

Let $h_1, h_2, F_1,$ and F_2 be continuous, and let D be the collection of all points in space with spherical coordinates (ρ, ϕ, θ) such that

$$\alpha \leq \theta \leq \beta, \quad 0 \leq h_1(\theta) \leq \phi \leq h_2(\theta) \leq \pi, \quad \text{and} \quad F_1(\phi, \theta) \leq \rho \leq F_2(\phi, \theta)$$

with $0 \leq \beta - \alpha \leq 2\pi$. We wish to find a way of evaluating $\iiint_D f(x, y, z)\, dV$, where f is continuous on D.

First circumscribe D with a spherical wedge D', and then partition D' into smaller spherical wedges, of which $D_1, D_2, \ldots, D_n$ are entirely contained in D. For each k between 1 and n, let D_k have dimensions $\Delta\rho_k, \Delta\phi_k,$ and $\Delta\theta_k$ in spherical coordinates. By (1) the volume V_k of the wedge D_k is given by

$$V_k = (\rho_k^*)^2 \sin \phi_k^* \, \Delta\rho_k \, \Delta\phi_k \, \Delta\theta_k$$

where $(\rho_k^*, \phi_k^*, \theta_k^*)$ are spherical coordinates of a point in D_k. This suggests that the triple integral $\iiint_D f(x, y, z)\, dV$ is approximately

$$\sum_{k=1}^{n} f(\rho_k^* \sin \phi_k^* \cos \theta_k^*, \rho_k^* \sin \phi_k^* \sin \theta_k^*, \rho_k^* \cos \phi_k^*)(\rho_k^*)^2 \sin \phi_k^* \, \Delta\rho_k \, \Delta\phi_k \, \Delta\theta_k$$

By an argument similar to the one used in Section 14.2, it can be shown that this sum is a Riemann sum for the triple integral

$$\int_{\alpha}^{\beta} \int_{h_1(\theta)}^{h_2(\theta)} \int_{F_1(\phi, \theta)}^{F_2(\phi, \theta)} f(\rho \sin \phi \cos \theta, \rho \sin \phi \sin \theta, \rho \cos \phi)\rho^2 \sin \phi \, d\rho \, d\phi \, d\theta$$

This leads us to the following theorem.

THEOREM 14.11

Let α and β be real numbers with $\alpha \leq \beta \leq \alpha + 2\pi$. Let $h_1, h_2, F_1,$ and F_2 be continuous functions with $0 \leq h_1 \leq h_2 \leq \pi$ and $0 \leq F_1 \leq F_2$. Let D be the solid region consisting of all points in space whose spherical coordinates (ρ, ϕ, θ) satisfy

$$\alpha \leq \theta \leq \beta$$

$$h_1(\theta) \leq \phi \leq h_2(\theta)$$

$$F_1(\phi, \theta) \leq \rho \leq F_2(\phi, \theta)$$

If f is continuous on D, then

$$\iiint\limits_{D} f(x, y, z) \, dV$$

$$= \int_{\alpha}^{\beta} \int_{h_1(\theta)}^{h_2(\theta)} \int_{F_1(\phi, \theta)}^{F_2(\phi, \theta)} f(\rho \sin \phi \cos \theta, \rho \sin \phi \sin \theta, \rho \cos \phi) \rho^2 \sin \phi \, d\rho \, d\phi \, d\theta$$

When you find the limits of integration for a triple integral in spherical coordinates, keep in mind that

ρ measures distance from the origin, so $\rho \geq 0$

ϕ measures the angle from the positive z axis, and $0 \leq \phi \leq \pi$

θ measures the horizontal angle from the positive x axis

EXAMPLE 3   Let D be the solid region between the spheres $\rho = 1$ and $\rho = 2$. Evaluate $\iiint_D z^2 \, dV$.

Solution   Observe that D is the collection of points with spherical coordinates (ρ, ϕ, θ) such that

$$0 \leq \theta \leq 2\pi, \quad 0 \leq \phi \leq \pi, \quad \text{and} \quad 1 \leq \rho \leq 2$$

Since $z = \rho \cos \phi$, Theorem 14.11 tells us that

$$\iiint\limits_{D} z^2 \, dV = \int_0^{2\pi} \int_0^{\pi} \int_1^2 (\rho^2 \cos^2 \phi) \rho^2 \sin \phi \, d\rho \, d\phi \, d\theta$$

$$= \int_0^{2\pi} \int_0^{\pi} \int_1^2 \rho^4 \cos^2 \phi \sin \phi \, d\rho \, d\phi \, d\theta$$

$$= \int_0^{2\pi} \int_0^{\pi} \frac{1}{5} \rho^5 \cos^2 \phi \sin \phi \Big|_1^2 \, d\phi \, d\theta$$

$$= \int_0^{2\pi} \int_0^{\pi} \frac{31}{5} \cos^2 \phi \sin \phi \, d\phi \, d\theta$$

$$= \int_0^{2\pi} \frac{31}{5} \left[\frac{1}{3} (-\cos^3 \phi) \Big|_0^{\pi} \right] d\theta$$

$$= \frac{62}{15} \int_0^{2\pi} 1 \, d\theta = \frac{124}{15} \pi \quad \square$$

EXAMPLE 4   Find the volume V of the solid region D between the spheres $x^2 + y^2 + z^2 = 1$ and $x^2 + y^2 + z^2 = 9$, and above the upper nappe of the cone $z^2 = 3(x^2 + y^2)$ (Figure 14.55).

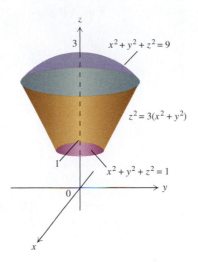

Solution In spherical coordinates the equations of the given spheres are $\rho = 1$ and $\rho = 3$. Next, recall that $z^2 = \rho^2 \cos^2 \phi$ and $x^2 + y^2 = r^2 = \rho^2 \sin^2 \phi$. Therefore the equation $z^2 = 3(x^2 + y^2)$ becomes

$$\rho^2 \cos^2 \phi = 3\rho^2 \sin^2 \phi$$

It follows that $\rho = 0$ or $\cos^2 \phi = 3 \sin^2 \phi$, so that if $\rho \neq 0$, then

$$\tan^2 \phi = \frac{1}{3}$$

Since D lies above the upper nappe of the cone $z^2 = 3(x^2 + y^2)$, we have $0 \leq \phi \leq \pi/2$, so $\tan \phi = 1/\sqrt{3}$. Thus $\phi = \pi/6$. Consequently D is the collection of points with spherical coordinates (ρ, ϕ, θ) such that

$$1 \leq \rho \leq 3, \quad 0 \leq \phi \leq \frac{\pi}{6}, \quad \text{and} \quad 0 \leq \theta \leq 2\pi$$

Therefore

$$V = \iiint_D 1 \, dV = \int_0^{2\pi} \int_0^{\pi/6} \int_1^3 \rho^2 \sin \phi \, d\rho \, d\phi \, d\theta$$

$$= \int_0^{2\pi} \int_0^{\pi/6} \left[\frac{\rho^3}{3} \Big|_1^3 \right] \sin \phi \, d\phi \, d\theta = \int_0^{2\pi} \int_0^{\pi/6} \frac{26}{3} \sin \phi \, d\phi \, d\theta$$

$$= \frac{26}{3} \int_0^{2\pi} (-\cos \phi) \Big|_0^{\pi/6} d\theta = \frac{26}{3} \left(1 - \frac{\sqrt{3}}{2} \right) \int_0^{2\pi} 1 \, d\theta$$

$$= \frac{26}{3} \pi (2 - \sqrt{3}) \qquad \square$$

EXAMPLE 5   Find the volume V of the solid region enclosed by the torus $\rho = 3 \sin \phi$ (Figure 14.56).

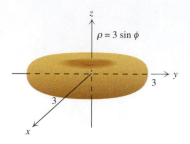

Solution   In this case D is the collection of points with spherical coordinates (ρ, ϕ, θ) such that

$$0 \le \theta \le 2\pi, \quad 0 \le \phi \le \pi, \quad \text{and} \quad 0 \le \rho \le 3 \sin \phi$$

Consequently

$$V = \iiint\limits_{D} 1 \, dV = \int_0^{2\pi} \int_0^{\pi} \int_0^{3 \sin \phi} \rho^2 \sin \phi \, d\rho \, d\phi \, d\theta$$

$$= \int_0^{2\pi} \int_0^{\pi} \frac{\rho^3}{3} \sin \phi \Big|_0^{3 \sin \phi} \, d\phi \, d\theta = \int_0^{2\pi} \int_0^{\pi} 9 \sin^4 \phi \, d\phi \, d\theta$$

$$\overset{\underset{\text{(12) of}}{\text{Section 8.1}}}{=} \int_0^{2\pi} 9 \left(-\frac{1}{4} \sin^3 \phi \cos \phi - \frac{3}{8} \sin \phi \cos \phi + \frac{3}{8} \phi \right) \Big|_0^{\pi} \, d\theta$$

$$= \int_0^{2\pi} \frac{27\pi}{8} \, d\theta = \frac{27\pi^2}{4} \quad \square$$

Finding the volume of the region in Example 5 by rectangular or cylindrical coordinates would be very complex if at all feasible.

EXAMPLE 6   A star occupies a spherical region D centered at the origin with radius a. Its density is given by

$$\delta(x, y, z) = ce^{-[(x^2 + y^2 + z^2)/a^2]^{3/2}}$$

where c is a positive constant. Determine the total mass m of the star.

Solution   Using (7) of Section 14.4 and using spherical coordinates, we find that

$$m = \iiint\limits_{D} \delta(x, y, z) \, dV = \iiint\limits_{D} ce^{-[(x^2 + y^2 + z^2)/a^2]^{3/2}} \, dV$$

$$= \int_0^{2\pi} \int_0^{\pi} \int_0^{a} ce^{-\rho^3/a^3} \rho^2 \sin \phi \, d\rho \, d\phi \, d\theta$$

$$= \int_0^{2\pi} \int_0^{\pi} -\frac{a^3c}{3} e^{-\rho^3/a^3} \sin \phi \Big|_0^a d\phi \, d\theta$$

$$= \int_0^{2\pi} \int_0^{\pi} \left[\frac{a^3c}{3} \left(1 - \frac{1}{e} \right) \right] \sin \phi \, d\phi \, d\theta$$

$$= \frac{a^3c}{3} \left(1 - \frac{1}{e} \right) \int_0^{2\pi} (-\cos \phi) \Big|_0^{\pi} d\theta$$

$$= \frac{a^3c}{3} \left(1 - \frac{1}{e} \right) \int_0^{2\pi} 2 \, d\theta = \frac{4}{3} \pi a^3 c \left(1 - \frac{1}{e} \right) \qquad \square$$

EXERCISES 14.6

1. Plot the points having the following spherical coordinates. Then give their rectangular coordinates.

   a. $\left(1, \frac{\pi}{2}, \frac{\pi}{6} \right)$ **b.** $\left(2, \pi, \frac{\pi}{2} \right)$

   c. $\left(3, \frac{\pi}{4}, \frac{4\pi}{3} \right)$ **d.** $\left(\frac{1}{2}, \frac{\pi}{3}, \frac{5\pi}{4} \right)$

   e. $\left(1, 0, \frac{7\pi}{6} \right)$ **f.** $\left(5, \frac{\pi}{2}, 0 \right)$

2. Give a set of spherical coordinates of the points having the following rectangular coordinates.

   a. $(1, 0, 1)$ **b.** $(3, 0, 0)$

   c. $(2, 2, 2\sqrt{2}/\sqrt{3})$ **d.** $(2\sqrt{2}, -2\sqrt{2}, -4\sqrt{3})$

In Exercises 3–7 evaluate the iterated integral.

3. $\displaystyle\int_0^{2\pi} \int_0^{\pi/4} \int_0^1 \rho^2 \sin \phi \, d\rho \, d\phi \, d\theta$

4. $\displaystyle\int_0^{2\pi} \int_0^{\pi/2} \int_1^3 \rho^3 \cos \phi \sin \phi \, d\rho \, d\phi \, d\theta$

5. $\displaystyle\int_0^{\pi} \int_{\pi/2}^{\pi} \int_1^2 \rho^4 \sin^2 \phi \cos^2 \theta \, d\rho \, d\phi \, d\theta$

6. $\displaystyle\int_0^{\pi} \int_0^{\pi/2} \int_0^{\sin \phi} \rho^2 \sin \phi \, d\rho \, d\phi \, d\theta$

7. $\displaystyle\int_{\pi/4}^{\pi/3} \int_0^{\theta} \int_0^{9 \sec \phi} \rho \cos^2 \phi \cos \theta \, d\rho \, d\phi \, d\theta$

In Exercises 8–14 express the integral as an iterated integral in spherical coordinates. Then evaluate it.

8. $\iiint_D (x^2 + y^2) \, dV$, where D is the ball $x^2 + y^2 + z^2 \le 1$

9. $\iiint_D x^2 \, dV$, where D is the solid region between the spheres $x^2 + y^2 + z^2 = 4$ and $x^2 + y^2 + z^2 = 9$

10. $\iiint_D (x^2 + y^2 + z^2)^2 \, dV$, where D is the solid region bounded above by the sphere $x^2 + y^2 + z^2 = 1$ and below by the upper nappe of the cone $z^2 = x^2 + y^2$

11. $\displaystyle\iiint_D \frac{1}{x^2 + y^2 + z^2} \, dV$, where D is the solid region above the xy plane bounded by the cone $z = \sqrt{3x^2 + 3y^2}$ and the spheres $x^2 + y^2 + z^2 = 9$ and $x^2 + y^2 + z^2 = 81$

12. $\iiint_D (z^2 + 1) \, dV$, where D is the solid region in the first octant bounded by the spheres $x^2 + y^2 + z^2 = 1$ and $x^2 + y^2 + z^2 = 2$ and by the coordinate planes

13. $\iiint_D \sqrt{z} \, dV$, where D is the solid region in the first octant bounded by the sphere $x^2 + y^2 + z^2 = 16$ and the planes $z = 0$, $x = \sqrt{3}y$, and $x = y$

14. $\iiint_D (x^2 + y^2 + z^2) \, dV$, where D is the solid region inside the sphere $\rho = 2 \cos \phi$ and outside the sphere $\rho = \cos \phi$

In Exercises 15–21 find the volume V of the region.

15. The solid region bounded above by the sphere $\rho = 2$ and below by the cone $\phi = \pi/4$

16. The solid region bounded above by the sphere $x^2 + y^2 + z^2 = 4$ and below by the upper nappe of the cone $z^2 = 3(x^2 + y^2)$

17. The solid region between the spheres $x^2 + y^2 + z^2 = 1$ and $x^2 + y^2 + z^2 = 4$ and below the upper nappe of the cone $3z^2 = x^2 + y^2$

18. The solid region bounded above by the sphere $x^2 + y^2 + z^2 = 4z$ and below by the upper nappe of the cone $z^2 = x^2 + y^2$

19. The solid region bounded above by the upper nappe of the cone $x^2 + y^2 = z^2$, on the sides by the cylinder $x^2 + y^2 = 4$, and below by the xy plane

20. The smaller of the two solid regions bounded by the sphere $\rho = 5$ and the half-planes $\theta = \pi/6$ and $\theta = \pi/3$

*21. The solid region bounded above by the sphere $x^2 + y^2 + z^2 = 64$ and below by the plane $z = -4\sqrt{3}$

22. Find the volume V of the solid region bounded below by the xy plane and above by the surface $\rho = 1 + \cos \phi$.

23. Find the volume V of the solid region bounded by the surface $\rho = \cos \phi$.

24. Find the total mass m of an object occupying the solid region bounded above by the sphere $x^2 + y^2 + z^2 = 4$ and below by the upper nappe of the cone $x^2 + y^2 = z^2$. Assume that the mass density at the point (x, y, z) is equal to the distance from (x, y, z) to the origin.

25. Find the total mass m of an object occupying the solid region bounded by the spheres $x^2 + y^2 + z^2 = 4$ and $x^2 + y^2 + z^2 = 16$, with mass density at (x, y, z) equal to the reciprocal of the distance from (x, y, z) to the origin.

26. By changing to spherical coordinates, evaluate the triple integral

$$\int_{-\infty}^{\infty} \int_{-\infty}^{\infty} \int_{-\infty}^{\infty} \sqrt{x^2 + y^2 + z^2}\, e^{-(x^2 + y^2 + z^2)}\, dz\, dy\, dx$$

as $\lim_{b \to \infty} \iiint_{D_b} \sqrt{x^2 + y^2 + z^2}\, e^{-(x^2 + y^2 + z^2)}\, dV$ where D_b is a ball of radius b centered at the origin.

27. Let $f(x, y, z) = \sin (x^2 + y^2 + z^2)^{3/2}$. Find the average value f_{av} of f on the unit ball centered at the origin.

14.7 MOMENTS AND CENTERS OF GRAVITY

In Section 6.5 we defined the moments and center of gravity of a plane region. In this section we will use double integrals to define moments and centers of gravity of more general plane regions. Then we will turn to solid regions and define their moments (about the three coordinate planes) and centers of gravity.

Moments and Centers of Gravity of Plane Regions

Let R be a rectangle with center (x, y) and area A. Recall from Section 6.5 that the moments M_x and M_y of R about the x axis and y axis are given by

$$M_x = yA \quad \text{and} \quad M_y = xA$$

More generally, if R is a vertically or horizontally simple region, let R' be a rectangle that circumscribes R. Partition R' into subrectangles, numbered so that $R_1, R_2, \ldots, R_n$ are the ones entirely contained in R. For any k between 1 and n, let (x_k, y_k) be the center of R_k and ΔA_k the area of R_k. It follows that the moment ΔM_k of R_k about the x axis is $y_k \Delta A_k$. Moreover, if all the rectangles $R_1, R_2, \ldots, R_n$ are sufficiently small, the moment M_x of R about the x axis should be approximately the moment of the region comprised of the rectangles $R_1, R_2, \ldots, R_n$, which means the sum of the moments $\Delta M_1, \Delta M_2, \ldots, \Delta M_n$. Consequently M_x should be approximately $\sum_{k=1}^{n} y_k \Delta A_k$, which is a Riemann sum for the function y on R. Accordingly we should define M_x by the formula

$$M_x = \iint_R y\, dA$$

Similar reasoning applies to M_y. Thus we make the following definition.

DEFINITION 14.12

Let R be a vertically or horizontally simple region. The **moment M_x of R about the x axis** and the **moment M_y of R about the y axis** are defined by

$$M_x = \iint_R y\, dA \quad \text{and} \quad M_y = \iint_R x\, dA$$

If R has positive area A, then the **center of gravity** (or **center of mass,** or **centroid**) of R is the point $(\overline{x}, \overline{y})$ defined by

$$\overline{x} = \frac{M_y}{A} = \frac{\iint_R x\, dA}{\iint_R 1\, dA} \quad \text{and} \quad \overline{y} = \frac{M_x}{A} = \frac{\iint_R y\, dA}{\iint_R 1\, dA}$$

In the event that R is vertically simple, we can obtain our original formulas for M_x and M_y (see Definition 6.5) from those in Definition 14.12. Indeed, if R is the region between the graphs of f and g on $[a, b]$ and if $g(x) \le f(x)$ for $a \le x \le b$, then

$$M_x = \iint_R y\, dA = \int_a^b \int_{g(x)}^{f(x)} y\, dy\, dx = \int_a^b \frac{1}{2} y^2 \Big|_{g(x)}^{f(x)} dx$$

$$= \frac{1}{2} \int_a^b \{ [f(x)]^2 - [g(x)]^2 \}\, dx$$

and

$$M_y = \iint_R x\, dA = \int_a^b \int_{g(x)}^{f(x)} x\, dy\, dx = \int_a^b xy \Big|_{g(x)}^{f(x)} dx = \int_a^b x[f(x) - g(x)]\, dx$$

EXAMPLE 1   Let R be the plane region bounded by the line $y = x$ and the parabola $y = 2 - x^2$ (Figure 14.57). Find the moments of R about the x and y axes, and determine the center of gravity of R.

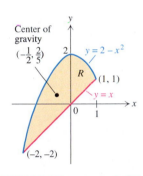

FIGURE 14.57

Solution   The line and the parabola intersect at the points (x, y) satisfying

$$x = y = 2 - x^2 \quad \text{that is,} \quad x^2 + x - 2 = 0$$

This means that $x = -2$ or $x = 1$, and thus the points of intersection are $(-2, -2)$ and $(1, 1)$. Consequently by Definition 14.12,

$$M_x = \iint_R y\, dA = \int_{-2}^1 \int_x^{2-x^2} y\, dy\, dx = \int_{-2}^1 \frac{y^2}{2} \Big|_x^{2-x^2} dx$$

$$= \frac{1}{2} \int_{-2}^1 (x^4 - 5x^2 + 4)\, dx = \frac{1}{2} \left(\frac{x^5}{5} - \frac{5}{3} x^3 + 4x \right) \Big|_{-2}^1 = \frac{9}{5}$$

and

$$M_y = \iint_R x\, dA = \int_{-2}^1 \int_x^{2-x^2} x\, dy\, dx = \int_{-2}^1 xy \Big|_x^{2-x^2} dx$$

$$= \int_{-2}^1 x(2 - x^2 - x)\, dx = \left(x^2 - \frac{x^4}{4} - \frac{x^3}{3} \right) \Big|_{-2}^1 = -\frac{9}{4}$$

Since the area A of R is given by

$$A = \iint_R 1\, dA = \int_{-2}^1 \int_x^{2-x^2} 1\, dy\, dx = \int_{-2}^1 (2 - x^2 - x)\, dx$$

$$= \left(2x - \frac{x^3}{3} - \frac{x^2}{2} \right) \Big|_{-2}^1 = \frac{9}{2}$$

it follows that

$$\bar{x} = \frac{M_y}{A} = \frac{-9/4}{9/2} = -\frac{1}{2} \quad \text{and} \quad \bar{y} = \frac{M_x}{A} = \frac{9/5}{9/2} = \frac{2}{5}$$

Consequently the center of gravity of R is $\left(-\frac{1}{2}, \frac{2}{5}\right)$ (Figure 14.57).   ❑

Moments and Centers of Gravity of Solid Regions

If a point mass having mass m is located at (x, y, z), we define its moments M_{xy}, M_{xz}, and M_{yz} about the coordinate planes as follows:

$$\text{moment about the } xy \text{ plane} = M_{xy} = zm$$
$$\text{moment about the } xz \text{ plane} = M_{xz} = ym \qquad (1)$$
$$\text{moment about the } yz \text{ plane} = M_{yz} = xm$$

Now consider an object occupying a solid region D and having a continuous mass density δ. To define the moments and center of gravity of the object, we first circumscribe D with a rectangular parallelepiped D' and then partition D' into rectangular parallelepipeds, of which $D_1, D_2, \ldots, D_n$ are entirely contained in D. For each k between 1 and n, let ΔV_k be the volume of D_k, and choose any point (x_k, y_k, z_k) in D_k. If all the dimensions of D_k are small, then since δ is by assumption continuous on D_k, (5) of Section 14.4 implies that the mass in D_k is approximately $\delta(x_k, y_k, z_k) \Delta V_k$, and the distance from any point in D_k to the xy plane is approximately z_k. Accordingly, the moment of D_k about the xy plane should be approximately equal to the moment $z_k \delta(x_k, y_k, z_k) \Delta V_k$ of a point mass located at (x_k, y_k, z_k) and having mass $\delta(x_k, y_k, z_k) \Delta V_k$. It follows that the moment of the whole solid region D about the xy plane should be approximately equal to

$$\sum_{k=1}^{n} z_k \delta(x_k, y_k, z_k) \Delta V_k$$

which is a Riemann sum for the function $z\delta$ on D. Consequently we are led to define the moment M_{xy} of D about the xy plane by the formula

$$M_{xy} = \iiint_D z\delta(x, y, z) \, dV$$

Similar considerations apply to the moments about the xz and yz planes.

DEFINITION 14.13

Suppose an object with continuous mass density δ occupies a solid region D. Then the object's **moment M_{xy} about the xy plane**, its **moment M_{xz} about the xz plane**, and its **moment M_{yz} about the yz plane** are defined by

$$M_{xy} = \iiint_D z\delta(x, y, z) \, dV$$

$$M_{xz} = \iiint_D y\delta(x, y, z) \, dV$$

$$M_{yz} = \iiint_D x\delta(x, y, z) \, dV$$

If the mass m of the object is positive, then the **center of gravity** (or **center of mass**) (x, y, z) of the object is defined by

$$\bar{x} = \frac{M_{yz}}{m} = \frac{\iiint_D x\delta(x, y, z)\, dV}{\iiint_D \delta(x, y, z)\, dV}$$

$$\bar{y} = \frac{M_{xz}}{m} = \frac{\iiint_D y\delta(x, y, z)\, dV}{\iiint_D \delta(x, y, z)\, dV}$$

$$\bar{z} = \frac{M_{xy}}{m} = \frac{\iiint_D z\delta(x, y, z)\, dV}{\iiint_D \delta(x, y, z)\, dV}$$

Notice that a point mass with mass m located at the center of gravity of a solid region D has the same moments about the coordinate planes as D has. When the mass density is constant, the object's center of gravity is often referred to as its **centroid**; in that case the center of gravity (or centroid) is independent of the mass density. As a result one often speaks of the centroid of a solid region.

If the mass density δ and the region D an object occupies are both symmetric with respect to a plane, then the center of gravity lies on that plane. In particular, if the mass density is constant and D is symmetric with respect to all the coordinate planes (as is a sphere centered at the origin), then the center of gravity of D lies on each of the coordinate planes and consequently is the origin.

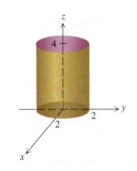

FIGURE 14.58

EXAMPLE 2   Suppose an object occupying the solid region D bounded by the circular cylinder shown in Figure 14.58 has a mass density δ given by

$$\delta(x, y, z) = 20 - z^2$$

Compute the moments M_{xy}, M_{xz}, and M_{yz}, and then determine the center of gravity of the object.

Solution   By definition,

$$M_{xz} = \iiint_D y\delta(x, y, z)\, dV = \iiint_D y(20 - z^2)\, dV$$

and you can verify that the triple integral is 0 either by carrying out the integration or by noticing that D is symmetric with respect to the xz plane and that the integrand is an odd function of y. By a similar argument,

$$M_{yz} = \iiint_D x\delta(x, y, z)\, dV = \iiint_D x(20 - z^2)\, dV = 0$$

For the computation of M_{xy} it is convenient to use cylindrical coordinates. By doing so we obtain

$$M_{xy} = \iiint_D z\delta(x, y, z)\, dV = \iiint_D z(20 - z^2)\, dV$$

$$= \int_0^{2\pi} \int_0^2 \int_0^4 (20z - z^3)r\, dz\, dr\, d\theta = \int_0^{2\pi} \int_0^2 \left(10z^2 r - \frac{z^4 r}{4}\right)\Big|_0^4 dr\, d\theta$$

$$= \int_0^{2\pi} \int_0^2 96r\, dr\, d\theta = \int_0^{2\pi} 48r^2 \Big|_0^2 d\theta = \int_0^{2\pi} 192\, d\theta = 384\pi$$

To determine the center of gravity, we must still compute the mass m of the object and then gather our information:

$$m = \iiint_D \delta(x, y, z)\, dV = \iiint_D (20 - z^2)\, dV$$

$$= \int_0^{2\pi} \int_0^2 \int_0^4 (20 - z^2)r\, dz\, dr\, d\theta = \int_0^{2\pi} \int_0^2 \left(20z - \frac{z^3}{3}\right)\Big|_0^4 r\, dr\, d\theta$$

$$= \int_0^{2\pi} \int_0^2 \frac{176}{3} r\, dr\, d\theta = \frac{176}{3} \int_0^{2\pi} \frac{r^2}{2}\Big|_0^2 d\theta = \frac{352}{3} \int_0^{2\pi} 1\, d\theta = \frac{704}{3}\pi$$

Therefore the center of gravity is $(\overline{x}, \overline{y}, \overline{z})$, where

$$\overline{x} = \frac{M_{yz}}{m} = 0, \quad \overline{y} = \frac{M_{xz}}{m} = 0, \quad \overline{z} = \frac{M_{xy}}{m} = \frac{384\pi}{704\pi/3} = \frac{18}{11} \qquad \square$$

Notice that the center of gravity of the object occupying the cylinder is $(0, 0, \frac{18}{11})$, whereas the centroid of the solid region enclosed by the cylinder is the center $(0, 0, 2)$ of the region.

Moments of Inertia

Assume that a body occupies a solid region D, and has mass density $\delta(x, y, z)$ at the point (x, y, z), as defined in (6) of Section 14.4. A quantity called the **moment of inertia** measures the extent to which the body resists rotation when the body is at rest, or continues rotation when rotating.

The moments of inertia about the three coordinate axes are:

$$\text{moment of inertia about the } x \text{ axis} = I_x = \iiint_D (y^2 + z^2)\, \delta(x, y, z)\, dV$$

$$\text{moment of inertia about the } y \text{ axis} = I_y = \iiint_D (x^2 + z^2)\, \delta(x, y, z)\, dV$$

$$\text{moment of inertia about the } z \text{ axis} = I_z = \iiint_D (x^2 + y^2)\, \delta(x, y, z)\, dV$$

For I_x the expression $y^2 + z^2$ in the integrand represents the square of the distance of the point (x, y, z) from the x axis. We see that the larger the mass density is or the farther the body is from the x axis, the larger the moment of inertia about the x axis will be. If mass and length are in kilograms and meters, respectively, then the units for the moment of inertia are kilogram-meters squared.

EXAMPLE 3   Consider a rectangular aluminum plate 2 meters wide, 3 meters long, and 0.1 meter thick, located in the first octant as in Figure 14.59. Assume that the plate is homogeneous, with mass density 2700 kilograms per cubic meter. Find the moment of inertia of the plate about the z axis.

Solution   Letting D denote the region occupied by the plate, we find that

$$I_z = \iiint_D (x^2 + y^2)(2700)\, dV = (2700) \int_0^3 \int_0^2 \int_0^{0.1} (x^2 + y^2)\, dz\, dy\, dx$$

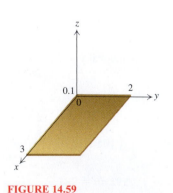

FIGURE 14.59

$$= (270) \int_0^3 \int_0^2 (x^2 + y^2)\, dy\, dx = (270) \int_0^3 \left(x^2 y + \frac{1}{3} y^3 \right) \Big|_0^2 \, dx$$

$$= (270) \int_0^3 \left(2x^2 + \frac{8}{3} \right) dx = (270) \left(\frac{2}{3} x^3 + \frac{8}{3} x \right) \Big|_0^3$$

$$= 7020 \text{ (kilogram-meters squared)} \quad \square$$

Moments of inertia are important in the study of rotation about an axis (such as a rotating propeller). As we mentioned above, the moment of inertia I about an axis measures the extent to which a body resists changes in rotation about the axis. Another quantity, called torque (see Section 11.4) and normally represented by the Greek letter τ, measures the tendency of an exterior force to produce changes in rotation about the axis. These two quantities, along with the angular acceleration α of the body rotating about the axis, are related by the equation

$$\tau = I\,\alpha$$

This equation resembles Newton's famous Second Law of Motion, $F = ma$. Thus torque, moment of inertia, and angular acceleration of a body rotating about an axis are the rotational analogues of force, mass, and acceleration.

EXERCISES 14.7

In Exercises 1–6 determine the center of gravity of the plane region. Use symmetry where applicable.

1. The region between the graphs of $y = 5$ and $y = 1 + x^2$
2. The region between the graphs of $y = x^2$ and $y = x^4$
3. The region inside the circles $(x - 1)^2 + y^2 = 1$ and $x^2 + (y - 1)^2 = 1$
4. The leaf of the four-leaved rose $r = \sin 2\theta$ that lies in the first quadrant
5. The region bounded by the cardioid $r = 1 + \cos\theta$
6. The region inside the circle $r = 2\sin\theta$ and outside the circle $r = 1$

In Exercises 7–13 find the centroid of the region. Use symmetry wherever possible to reduce calculations.

7. The solid region bounded above by the sphere $x^2 + y^2 + z^2 = a^2$ and below by the xy plane
8. The solid region bounded below by the paraboloid $z = 4x^2 + 4y^2$ and above by the plane $z = 2$
9. The solid region bounded above by the plane $z = 1$ and below by the upper nappe of the cone $z^2 = 9x^2 + 9y^2$
10. The solid region bounded above by the sphere $x^2 + y^2 + z^2 = 1$ and below by the cone $z = \sqrt{x^2 + y^2}$
11. The solid region in the first octant bounded by the planes $x = 0$ and $y = 0$ and the paraboloids $z = 1 - x^2 - y^2$ and $z = x^2 + y^2$

12. The solid region in the first octant bounded by the coordinate planes and the planes $z = 1 + x + 2y$, $x = 1$, and $y = 1$
13. The pyramid with vertices $(1, 0, 0)$, $(0, 1, 0)$, $(-1, 0, 0)$, $(0, -1, 0)$, and $(0, 0, 2)$

In Exercises 14–20 find the center of gravity of an object that occupies the given region and has the given mass density.

14. The ball $x^2 + y^2 + z^2 \le 4$; $\delta(x, y, z)$ is equal to
    a. the distance from (x, y, z) to the origin.
    b. $1 + z^2$.
15. The solid region inside the sphere $x^2 + y^2 + z^2 = 4$ and inside the cylinder $x^2 + y^2 = 2$; $\delta(x, y, z) = z^2 + 1$
16. The solid region bounded by the paraboloids $z = 1 - x^2 - y^2$ and $z = x^2 + y^2$; $\delta(x, y, z) = 2 - z$
17. The cube in the first octant with sides of length 2 and one vertex at the origin; $\delta(x, y, z) = 1 + x$
18. The solid region bounded by the sheet $z = 1 - x^2$ and the planes $z = 0$, $y = -1$, and $y = 1$; $\delta(x, y, z) = z(y + 2)$
19. The solid region bounded above by the sphere $x^2 + y^2 + z^2 = 9$ and below by the xy plane; $\delta(x, y, z)$ is equal to the distance from (x, y, z) to the z axis.
20. The solid region bounded above by the sphere $x^2 + y^2 + z^2 = 4$ and below by the upper nappe of the cone $z^2 = x^2 + y^2$; $\delta(x, y, z) = z^2(x^2 + y^2 + z^2)$

21. A cylindrical can containing pineapple juice is 20 centimeters tall and 8 centimeters in diameter. Set up a coordinate system with the xy plane at the bottom of the can, and assume that due to settling, the density of the pineapple juice is given by

$$\delta(x, y, z) = a(40 - z)$$

where a is a positive constant. Determine the center of gravity of the pineapple juice.

22. A cube with side length 1 and constant density 1 is located on top of another cube with side length 1 and constant density 10. Determine the center of gravity of the combination of cubes.

In Exercises 23–26 find the moments of inertia about the coordinate axes for the given region and mass density.

23. The ball $x^2 + y^2 + z^2 \leq 25$; $\delta(x, y, z) = 5$
24. The ball $x^2 + y^2 + z^2 \leq 4$; $\delta(x, y\, z) = x^2 + y^2 + z^2$
25. The solid region bounded by the cylinder $x^2 + y^2 = 4$ and the planes $z = 0$ and $z = 6$; $\delta(x, y, z) = 2$
26. The solid region bounded above by the plane $z = 4$ and

below by the upper nappe of the cone $x^2 + y^2 = z^2$; $\delta(x, y, z) = x^2 + y^2$

Applications

27. A cubical block of ice 10 centimeters on a side is spun around an edge. Compute the moment of inertia I about an edge, assuming that the density of the ice is 0.917 gram per cubic centimeter.

28. One component of a steel forging is a rectangular bar 12 centimeters long, 6 centimeters wide, and 3 centimeters tall. Assuming that the density of the bar is 7.83 grams per cubic centimeter, find the moment of inertia I about

  a. the vertical axis through the center of the bar.
  b. a vertical axis through a vertical edge.
  c. one of the longest horizontal edges.

29. A turntable of a record player is in the shape of a solid cylinder that has a radius of 15 centimeters, thickness of 3 centimeters, and mass of 2 kilograms. Assume that the density of the turntable is constant, denoted by δ. Find the moment of inertia I about the central axis.

14.8   CHANGE OF VARIABLES IN MULTIPLE INTEGRALS

In Section 14.2 we used the formula

$$\iint_R f(x, y)\, dA = \int_\alpha^\beta \int_{h_1(\theta)}^{h_2(\theta)} f(r \cos \theta, r \sin \theta) r\, dr\, d\theta$$

to evaluate integrals by means of polar coordinates rather than rectangular coordinates. Other formulas arose in Sections 14.5 and 14.6 for evaluating integrals by means of cylindrical or spherical coordinates instead of rectangular coordinates. Such formulas are called **change of variables formulas,** because they allow us to evaluate a multiple integral in a coordinate system other than the given coordinate system.

The present section is devoted to more general change of variables formulas—formulas that will allow us to integrate with respect to a variety of other coordinates rather than with respect to rectangular coordinates. The usefulness of such a formula becomes apparent when we seek to evaluate the integral $\iint_R x^3 y\, dA$, where R is the region in Figure 14.60(a). Evaluating the double integral by means of rectangular coordinates could be achieved, but it would necessitate evaluating three separate iterated integrals, one over each of the portions labeled I, II, and III in Figure 14.60(b). However, it is possible (as you will see in Example 2), to make a change of variables that converts $\iint_R x^3 y\, dA$ to an integral over the region S

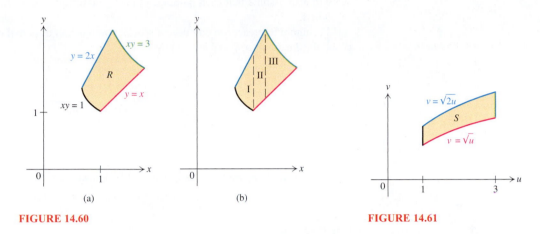

FIGURE 14.60   **FIGURE 14.61**

shown in Figure 14.61; this new double integral is easily evaluated.

Change of Variables for Double Integrals

Before we derive the change of variables formula for double integrals, let us return to formula (5) of Section 14.2:

$$\iint_R f(x, y)\, dA = \int_\alpha^\beta \int_{h_1(\theta)}^{h_2(\theta)} f(r \cos \theta, r \sin \theta) r\, dr\, d\theta \tag{1}$$

In order to explain the origin of the extra r on the right side, recall that x, y, r, and θ are related by the formulas

$$x = r \cos \theta \quad \text{and} \quad y = r \sin \theta$$

Now the area of a rectangle in the $r\theta$ plane of dimensions Δr and $\Delta \theta$ is $\Delta r\, \Delta \theta$ (Figure 14.62(a)), whereas the area of the associated arched region in the xy plane (Figure 14.62(b)) is $r\, \Delta r\, \Delta \theta$, where r is the average radius of the arched region. One can think of r as the "magnification factor" by which the area of a sufficiently small region in the $r\theta$ plane must be multiplied to obtain the area of an associated region in the xy plane.

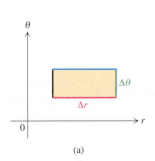

(a)

More generally, let R be a given region in the xy plane, and consider a new set of variables u and v. Suppose that for (u, v) in a set S in the uv plane, u and v are related to x and y by the formulas

$$x = g_1(u, v) \quad \text{and} \quad y = g_2(u, v) \tag{2}$$

where g_1 and g_2 are continuously differentiable (and (x, y) is in R). Our goal is to find an appropriate function $J(u, v)$ such that

$$\iint_R f(x, y)\, dA = \iint_S f(g_1(u, v), g_2(u\, v))\, J(u, v)\, dA \tag{3}$$

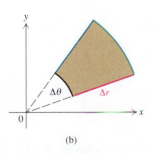

(b)

FIGURE 14.62

irrespective of the continuous function f. The function defined by (2) is called a **transformation**; we will denote it by T. Thus T is defined by

$$T(u, v) = (x, y) \quad \text{for } (u, v) \text{ in } S$$

We will also assume that each point (x, y) in R is assigned to exactly one point (u, v) in S.

For future reference we will need a special combination of partial derivatives called the **Jacobian** of T and defined by

$$\frac{\partial(x, y)}{\partial(u, v)} = \begin{vmatrix} \dfrac{\partial x}{\partial u} & \dfrac{\partial x}{\partial v} \\ \dfrac{\partial y}{\partial u} & \dfrac{\partial y}{\partial v} \end{vmatrix} = \frac{\partial x}{\partial u}\frac{\partial y}{\partial v} - \frac{\partial x}{\partial v}\frac{\partial y}{\partial u}$$

EXAMPLE 1   Suppose the transformation T is defined by

$$x = \frac{u}{v} \quad \text{and} \quad y = v$$

Find the Jacobian of T.

Solution   By definition,

$$\frac{\partial(x, y)}{\partial(u, v)} = \begin{vmatrix} \dfrac{\partial x}{\partial u} & \dfrac{\partial x}{\partial v} \\ \dfrac{\partial y}{\partial u} & \dfrac{\partial y}{\partial v} \end{vmatrix} = \begin{vmatrix} \dfrac{1}{v} & -\dfrac{u}{v^2} \\ 0 & 1 \end{vmatrix} = \frac{1}{v} \qquad \square$$

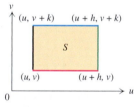

A rectangular region S in the uv plane

(a)

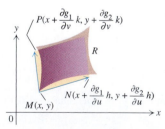

The image R of S in the xy plane

(b)

FIGURE 14.63

Now let T be a transformation that maps points in the uv plane into points in the xy plane and is given by (2). Assume that S is a rectangle with vertices (u, v), $(u + h, v)$, $(u + h, v + k)$, and $(u, v + k)$ (Figure 14.63(a)), so that the area A_S of S is hk. If h and k are small, the continuity of g_1 and g_2 implies that the image R of S is approximately a parallelogram in the xy plane three of whose vertices are $(x, y) = (g_1(u, v), g_2(u, v))$, $(g_1(u + h, v), g_2(u + h, v))$, and $(g_1(u, v + k), g_2(u, v + k))$. But since v is constant on the line joining (u, v) and $(u + h, v)$, it follows from the definition of the partial derivatives $\partial g_1/\partial u$ and $\partial g_2/\partial u$ that if h is small, then

$$g_1(u + h, v) \approx g_1(u, v) + \frac{\partial g_1}{\partial u}h = x + \frac{\partial g_1}{\partial u}h \tag{4}$$

$$g_2(u + h, v) \approx g_2(u, v) + \frac{\partial g_2}{\partial u}h = y + \frac{\partial g_2}{\partial u}h \tag{5}$$

Similarly, u is constant on the line joining (u, v) and $(u, v + k)$, so that if k is small, we have

$$g_1(u, v + k) \approx g_1(u, v) + \frac{\partial g_1}{\partial v}k = x + \frac{\partial g_1}{\partial v}k \tag{6}$$

$$g_2(u, v + k) \approx g_2(u, v) + \frac{\partial g_2}{\partial v}k = y + \frac{\partial g_2}{\partial v}k \tag{7}$$

Using (4)–(7) we find that if h and k are small, the parallelogram is very close to the parallelogram determined by the points

$$M = (x, y), \quad N = \left(x + \frac{\partial g_1}{\partial u}h, \; y + \frac{\partial g_2}{\partial u}h\right), \quad P = \left(x + \frac{\partial g_1}{\partial v}k, \; y + \frac{\partial g_2}{\partial v}k\right)$$

(Figure 14.63(b)). The area A of the latter parallelogram is the length of the cross product $\overrightarrow{MN} \times \overrightarrow{MP}$ (see (2) in Section 11.4). Since

$$\overrightarrow{MN} \times \overrightarrow{MP} = \begin{vmatrix} \mathbf{i} & \mathbf{j} & \mathbf{k} \\ \dfrac{\partial g_1}{\partial u}h & \dfrac{\partial g_2}{\partial u}h & 0 \\ \dfrac{\partial g_1}{\partial v}k & \dfrac{\partial g_2}{\partial v}k & 0 \end{vmatrix} = \left[\left(\frac{\partial g_1}{\partial u}h\right)\left(\frac{\partial g_2}{\partial v}k\right) - \left(\frac{\partial g_2}{\partial u}h\right)\left(\frac{\partial g_1}{\partial v}k\right)\right]\mathbf{k}$$

it follows that

$$||\overrightarrow{MN} \times \overrightarrow{MP}|| = \left|\frac{\partial g_1}{\partial u}\frac{\partial g_2}{\partial v} - \frac{\partial g_2}{\partial u}\frac{\partial g_1}{\partial v}\right| hk = \left|\frac{\partial(x, y)}{\partial(u, v)}\right| hk = \left|\frac{\partial(x, y)}{\partial(u, v)}\right| A_S$$

We conclude that

$$A \approx \left|\frac{\partial(x, y)}{\partial(u, v)}\right| A_S \tag{8}$$

Therefore the area of the image R of the rectangle S is approximately the product of the area of S and the absolute value of the Jacobian of T.

Now we no longer assume that R is rectangular but do assume that R has a piecewise smooth boundary and that f is continuous on R. Since we can approximate the double integral $\iint_R f(x, y)\, dA$ as accurately as we wish by Riemann sums of the form $\sum_{k=1}^{n} f(x_k, y_k)\, \Delta A_k$, where ΔA_k is the area of a suitable small rectangle R_k contained in R for $k = 1, \ldots, n$, one can use (8) to prove the following theorem.

THEOREM 14.14   Suppose S and R are sets in the uv and xy planes, respectively, each with a boundary that is a piecewise smooth curve. Let T be a transformation from S to R, defined by

$$x = g_1(u, v) \quad \text{and} \quad y = g_2(u, v)$$

where g_1 and g_2 have continuous partial derivatives on S. Suppose also that each point (x, y) in R is the image of a unique point (u, v) in S and that $\partial(x, y)/\partial(u, v) \neq 0$ throughout S except possibly at finitely many points. Finally, assume that f is continuous and integrable on R. Then

$$\iint_R f(x, y)\, dA = \iint_S f(g_1(u, v), g_2(u, v)) \left|\frac{\partial(x, y)}{\partial(u, v)}\right| dA \tag{9}$$

Thus the function $J(u, v)$ we sought in (3) is given by

$$J(u, v) = \left|\frac{\partial(x, y)}{\partial(u, v)}\right|$$

To change from rectangular to polar coordinates, we have

$$x = r \cos \theta \quad \text{and} \quad y = r \sin \theta \qquad (10)$$

For the transformation defined by (10) we find that

$$\frac{\partial(x, y)}{\partial(r, \theta)} = \begin{vmatrix} \dfrac{\partial x}{\partial r} & \dfrac{\partial x}{\partial \theta} \\[2mm] \dfrac{\partial y}{\partial r} & \dfrac{\partial y}{\partial \theta} \end{vmatrix} = \begin{vmatrix} \cos \theta & -r \sin \theta \\ \sin \theta & r \cos \theta \end{vmatrix} = r \cos^2 \theta + r \sin^2 \theta = r$$

Thus

$$\left| \frac{\partial(x, y)}{\partial(r, \theta)} \right| = |r| = r$$

since $r \geq 0$. Therefore if a region S in the $r\theta$ plane corresponds to a region R under the transformation given in (10), then (9) yields

$$\iint\limits_R f(x, y)\, dA = \iint\limits_S f(r \cos \theta, r \sin \theta) r\, dA$$

which is equivalent to the polar coordinate change of variable formula in (1).

To implement (9) when the transformation T is specified, we need first to calculate $\partial(x, y)/\partial(u, v)$, which is generally easy, and to find S, which may not be so easy. It is usually simplest to find S by determining the curves (or lines) in the uv plane corresponding to the boundary of R; then S will be the region bounded by these curves and lines.

EXAMPLE 2   Evaluate $\iint_R x^3 y\, dA$, where R is the region in the first quadrant bounded by the lines $y = x$ and $y = 2x$, and by the hyperbolas $xy = 1$ and $xy = 3$ (Figure 14.64). Let T be defined by

$$x = \frac{u}{v} \quad \text{and} \quad y = v \qquad (11)$$

Solution   As we computed in Example 1,

$$\frac{\partial(x, y)}{\partial(u, v)} = \frac{1}{v}$$

Next we determine the region S in the uv plane that is mapped onto R by T. For points on the hyperbola $xy = 1$ we substitute from (11) to obtain

$$1 = xy = \left(\frac{u}{v} \right) v = u$$

so that T maps the line $u = 1$ onto the hyperbola $xy = 1$. Similarly, for the hyperbola $xy = 3$ we obtain

$$3 = xy = \left(\frac{u}{v} \right) v = u$$

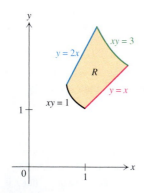

FIGURE 14.64

so that T maps the line $u = 3$ onto the hyperbola $xy = 3$. Next, for points on the line $y = x$, we substitute from (11) to obtain

$$v = y = x = \frac{u}{v}$$

so that $v^2 = u$, or equivalently, $v = \sqrt{u}$. Similarly, for points on the line $y = 2x$ we obtain

$$v = y = 2x = 2\left(\frac{u}{v}\right)$$

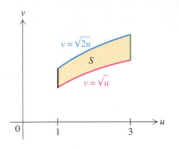

FIGURE 14.65

so that $v^2 = 2u$, or equivalently, $v = \sqrt{2u}$. Consequently S is the region in the uv plane bounded by the lines $u = 1$ and $u = 3$, and by the parabolas $v = \sqrt{u}$ and $v = \sqrt{2u}$ (Figure 14.65). Now we can apply (9) to conclude that

$$\iint_R x^3 y \, dA \stackrel{(9)}{=} \iint_S \left(\frac{u}{v}\right)^3 v \left|\frac{\partial(x, y)}{\partial(u, v)}\right| dA = \int_1^3 \int_{\sqrt{u}}^{\sqrt{2u}} \frac{u^3}{v^3} v \frac{1}{v} \, dv \, du$$

$$= \int_1^3 \int_{\sqrt{u}}^{\sqrt{2u}} \frac{u^3}{v^3} \, dv \, du = \int_1^3 -\frac{1}{2} \frac{u^3}{v^2} \Big|_{\sqrt{u}}^{\sqrt{2u}} du$$

$$= \int_1^3 \frac{1}{4} u^2 \, du = \frac{1}{12} u^3 \Big|_1^3 = \frac{13}{6} \qquad \square$$

Suppose we have a double integral $\iint_R f(x, y) \, dA$ in which either f or R is complicated, so that the integral is hard to evaluate, and suppose no transformation is prescribed. Then we proceed in the following way:

1. Define a transformation T_R from R into the uv plane. For this you will need to use your knowledge of R and perhaps the integrand.
2. Determine the image S of R under the transformation.
3. Solve for x and y from the equations defining T_R in order to obtain a transformation T from S to R.
4. Compute $\dfrac{\partial(x, y)}{\partial(u, v)}$.
5. Evaluate $\iint_R f(x, y) \, dA$ by means of (9).

The examples below use the procedure just outlined.

EXAMPLE 3   Evaluate $\iint_R 7xy \, dA$, where the boundary of R is the parallelogram determined by the lines $2x + 3y = 1$, $2x + 3y = 3$, $x - 2y = 2$, and $x - 2y = -2$ (Figure 14.66).

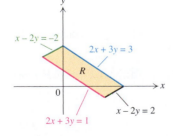

FIGURE 14.66

Solution   As you can see from the figure, the region R is not so easy to integrate over. Since $2x + 3y$ and $x - 2y$ appear in the description of R, we let T_R be defined by

$$u = 2x + 3y \quad \text{and} \quad v = x - 2y \tag{12}$$

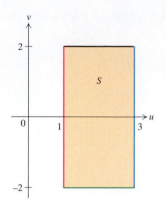

FIGURE 14.67

To find the image S of R under T_R, we notice that by our definition of u the images of the lines $2x + 3y = 1$ and $2x + 3y = 3$ are the lines $u = 1$ and $u = 3$ in the uv plane. Similarly, the images of the lines $x - 2y = 2$ and $x - 2y = -2$ are the lines $v = 2$ and $v = -2$, respectively. Therefore S is the rectangular region bounded by the lines $u = 1$, $u = 3$, $v = 2$, and $v = -2$ (Figure 14.67). Next we solve for x and y in (12). We find that

$$u - 2v = (2x + 3y) - 2(x - 2y) = 7y$$

and

$$2u + 3v = 2(2x + 3y) + 3(x - 2y) = 7x$$

Therefore

$$x = \frac{1}{7}(2u + 3v) \quad \text{and} \quad y = \frac{1}{7}(u - 2v)$$

which defines a transformation T from S onto R. Since

$$\frac{\partial(x, y)}{\partial(u, v)} = \begin{vmatrix} \dfrac{\partial x}{\partial u} & \dfrac{\partial x}{\partial v} \\ \dfrac{\partial y}{\partial u} & \dfrac{\partial y}{\partial v} \end{vmatrix} = \begin{vmatrix} \dfrac{2}{7} & \dfrac{3}{7} \\ \dfrac{1}{7} & -\dfrac{2}{7} \end{vmatrix} = \left(\frac{2}{7}\right)\left(-\frac{2}{7}\right) - \left(\frac{3}{7}\right)\left(\frac{1}{7}\right) = -\frac{1}{7}$$

it follows from (9) that

$$\iint\limits_R 7xy \, dA = \iint\limits_S \left[(2u + 3v)\right]\left[\frac{1}{7}(u - 2v)\right]\left|-\frac{1}{7}\right| dA$$

$$= \iint\limits_S \frac{1}{49}(2u^2 - uv - 6v^2) \, dA = \int_1^3 \int_{-2}^2 \frac{1}{49}(2u^2 - uv - 6v^2) \, dv \, du$$

$$= \frac{1}{49} \int_1^3 \left(2u^2 v - \frac{1}{2} uv^2 - 2v^3\right)\Big|_{-2}^2 du = \frac{1}{49} \int_1^3 (8u^2 - 32) \, du$$

$$= \frac{1}{49}\left(\frac{8}{3}u^3 - 32u\right)\Big|_1^3 = \frac{1}{49}\left(72 - 96 - \frac{8}{3} + 32\right) = \frac{16}{147} \qquad \square$$

Integration over an elliptical region is most often facilitated by transforming the region into a circular one and then proceeding by polar coordinates (when feasible). The following example illustrates this procedure.

EXAMPLE 4 Evaluate $\iint_R (x^2 + y^2) \, dA$, where R is the region bounded by the ellipse $x^2 + y^2/4 = 1$ (Figure 14.68).

Solution We will transform the elliptical region R into a circular region. For this, let T_R be the transformation defined by

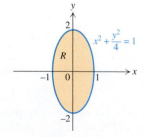

FIGURE 14.68

$$u = x \quad \text{and} \quad v = \frac{1}{2}y \tag{13}$$

This transformation maps the ellipse in the xy plane into the unit circle in the uv plane, because

$$1 = x^2 + \frac{y^2}{4} = u^2 + v^2$$

Thus S is the unit disk $u^2 + v^2 \le 1$ (Figure 14.69). Solving for x and y in (13) is trivial, and yields

$$x = u \quad \text{and} \quad y = 2v$$

We find that

$$\frac{\partial(x, y)}{\partial(u, v)} = \begin{vmatrix} \dfrac{\partial x}{\partial u} & \dfrac{\partial x}{\partial v} \\[2mm] \dfrac{\partial y}{\partial u} & \dfrac{\partial y}{\partial v} \end{vmatrix} = \begin{vmatrix} 1 & 0 \\ 0 & 2 \end{vmatrix} = 2$$

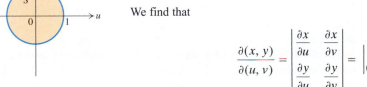

FIGURE 14.69

Therefore by (9),

$$\iint\limits_R (x^2 + y^2)\, dA = \iint\limits_S (u^2 + (2v)^2) \left| \frac{\partial(x, y)}{\partial(u, v)} \right| dA = \iint\limits_S (u^2 + 4v^2) 2\, dA$$

Converting to polar coordinates with $u = r \cos \theta$ and $v = r \sin \theta$, we obtain

$$\iint\limits_S (u^2 + 4v^2) 2\, dA = 2 \iint\limits_S (u^2 + v^2 + 3v^2)\, dA$$

$$= 2 \int_0^{2\pi} \int_0^1 (r^2 + 3r^2 \sin^2 \theta) r\, dr\, d\theta$$

$$= 2 \int_0^{2\pi} \int_0^1 (1 + 3 \sin^2 \theta) r^3\, dr\, d\theta$$

$$= 2 \int_0^{2\pi} (1 + 3 \sin^2 \theta) \left. \frac{r^4}{4} \right|_0^1 d\theta$$

$$= \frac{1}{2} \int_0^{2\pi} (1 + 3 \sin^2 \theta)\, d\theta$$

$$= \frac{1}{2} \left[\theta + 3 \left(\frac{1}{2} \theta - \frac{1}{4} \sin 2\theta \right) \right] \Big|_0^{2\pi} = \frac{5}{2} \pi \qquad \square$$

As we will see in the next example, sometimes it is not necessary to solve for x and y directly in the integrand. Since it is possible to prove that

$$\frac{\partial(x, y)}{\partial(u, v)} = \frac{1}{\dfrac{\partial(u, v)}{\partial(x, y)}} \tag{14}$$

we can dispense with actually computing T from T_R in the evaluation of the integral.

EXAMPLE 5   Evaluate $\displaystyle\iint_R \frac{\sin(x-y)}{\cos(x+y)}\,dA$, where R is the triangular region bounded by the lines $y=0$, $y=x$, and $x+y=\pi/4$ (Figure 14.70).

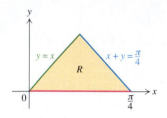

FIGURE 14.70

Solution   This time the integrand is complicated, so that direct integration appears to be impossible. In an effort to simplify the integrand we let T_R be defined by

$$u = x - y \quad \text{and} \quad v = x + y$$

Now we compute the region S. We notice that if $y=0$, then $u=x=v$, so the image of the line $y=0$ is the line $u=v$. Next, if $y=x$, then $u=x-y=0$, so the image of the line $y=x$ is the line $u=0$. Finally, if $x+y=\pi/4$, then $v=x+y=\pi/4$, so the image of the line $x+y=\pi/4$ is the line $v=\pi/4$. Therefore the image S of R is the triangular region bounded by the lines $u=v$, $u=0$, and $v=\pi/4$ (Figure 14.71). By definition,

$$\frac{\partial(u,v)}{\partial(x,y)} = \begin{vmatrix} \dfrac{\partial u}{\partial x} & \dfrac{\partial u}{\partial y} \\[2mm] \dfrac{\partial v}{\partial x} & \dfrac{\partial v}{\partial y} \end{vmatrix} = \begin{vmatrix} 1 & -1 \\ 1 & 1 \end{vmatrix} = (1)(1) - (-1)(1) = 2$$

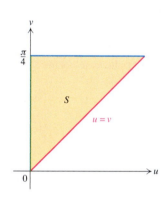

FIGURE 14.71

Therefore it follows from (9) and (14) that

$$\iint_R \frac{\sin(x-y)}{\cos(x+y)}\,dA \overset{(9)}{=} \iint_S \frac{\sin u}{\cos v}\left|\frac{\partial(x,y)}{\partial(u,v)}\right|\,dA \overset{(14)}{=} \int_0^{\pi/4}\int_0^{v} \frac{\sin u}{\cos v}\frac{1}{2}\,du\,dv$$

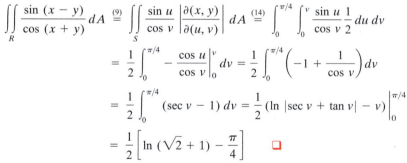

$$= \frac{1}{2}\int_0^{\pi/4} \left.-\frac{\cos u}{\cos v}\right|_0^{v}\,dv = \frac{1}{2}\int_0^{\pi/4}\left(-1+\frac{1}{\cos v}\right)\,dv$$

$$= \frac{1}{2}\int_0^{\pi/4} (\sec v - 1)\,dv = \frac{1}{2}\left(\ln|\sec v + \tan v| - v\right)\Big|_0^{\pi/4}$$

$$= \frac{1}{2}\left[\ln(\sqrt{2}+1) - \frac{\pi}{4}\right] \qquad \square$$

Change of Variables for Triple Integrals

The change of variables formula for triple integrals is similar to the formula for double integrals. If a transformation T is defined by

$$x = g_1(u,v,w), \quad y = g_2(u,v,w), \quad \text{and} \quad z = g_3(u,v,w)$$

and if T maps a region E in uvw space onto a region D in xyz space, then the formula for triple integrals corresponding to (9) is

$$\iiint_D f(x,y,z)\,dV = \iiint_E f(g_1(u,v,w), g_2(u,v,w), g_3(u,v,w))\left|\frac{\partial(x,y,z)}{\partial(u,v,w)}\right|\,dV \quad (15)$$

where

$$\frac{\partial(x,y,z)}{\partial(u,v,w)} = \frac{\partial x}{\partial u}\left(\frac{\partial y}{\partial v}\frac{\partial z}{\partial w} - \frac{\partial y}{\partial w}\frac{\partial z}{\partial v}\right) + \frac{\partial x}{\partial v}\left(\frac{\partial y}{\partial w}\frac{\partial z}{\partial u} - \frac{\partial y}{\partial u}\frac{\partial z}{\partial w}\right) + \frac{\partial x}{\partial w}\left(\frac{\partial y}{\partial u}\frac{\partial z}{\partial v} - \frac{\partial y}{\partial v}\frac{\partial z}{\partial u}\right)$$

We abbreviate the preceding equation by

$$\frac{\partial(x, y, z)}{\partial(u, v, w)} = \begin{vmatrix} \dfrac{\partial x}{\partial u} & \dfrac{\partial x}{\partial v} & \dfrac{\partial x}{\partial w} \\[2mm] \dfrac{\partial y}{\partial u} & \dfrac{\partial y}{\partial v} & \dfrac{\partial y}{\partial w} \\[2mm] \dfrac{\partial z}{\partial u} & \dfrac{\partial z}{\partial v} & \dfrac{\partial z}{\partial w} \end{vmatrix} \tag{16}$$

The expression in (16) is called the **Jacobian** of the transformation T. By change of variables we will derive the formula for integrals in spherical coordinates.

EXAMPLE 6   Using (15), derive the formula for integrals in spherical coordinates that appeared in Theorem 14.11.

Solution   In this case we have

$$x = \rho \sin \phi \cos \theta, \quad y = \rho \sin \phi \sin \theta, \quad \text{and} \quad z = \rho \cos \phi$$

Therefore by (16),

$$\frac{\partial(x, y, z)}{\partial(\rho, \phi, \theta)} = \begin{vmatrix} \sin \phi \cos \theta & \rho \cos \phi \cos \theta & -\rho \sin \phi \sin \theta \\ \sin \phi \sin \theta & \rho \cos \phi \sin \theta & \rho \sin \phi \cos \theta \\ \cos \phi & -\rho \sin \phi & 0 \end{vmatrix}$$

$$= \sin \phi \cos \theta \, (0 + \rho^2 \sin^2 \phi \cos \theta) + \rho \cos \phi \cos \theta \, (\rho \sin \phi \cos \phi \cos \theta)$$
$$\quad - \rho \sin \phi \sin \theta \, (-\rho \sin^2 \phi \sin \theta - \rho \cos^2 \phi \sin \theta)$$

$$= \rho^2 \sin \phi \, (\sin^2 \phi \cos^2 \theta + \cos^2 \phi \cos^2 \theta + \sin^2 \phi \sin^2 \theta + \cos^2 \phi \sin^2 \theta)$$

$$= \rho^2 \sin \phi$$

Since $0 \le \phi \le \pi$ and thus $\rho^2 \sin \phi \ge 0$, it follows from (15) that

$$\iiint\limits_{D} f(x, y, z) \, dV = \iiint\limits_{E} f(\rho \sin \phi \cos \theta, \rho \sin \phi \sin \theta, \rho \cos \phi) \rho^2 \sin \phi \, dV$$

If the region E in $\rho\phi\theta$ space is defined by

$$\alpha \le \theta \le \beta, \quad h_1(\theta) \le \phi \le h_2(\theta), \quad \text{and} \quad F_1(\phi, \theta) \le \rho \le F_2(\phi, \theta)$$

where $h_1, h_2, F_1,$ and F_2 are continuous and where $0 \le \beta - \alpha \le 2\pi$, $0 \le h_1 \le h_2 \le \pi$, and $0 \le F_1 \le F_2$, then (15) becomes

$$\iiint\limits_{D} f(x, y, z) \, dV$$

$$= \int_{\alpha}^{\beta} \int_{h_1(\theta)}^{h_2(\theta)} \int_{F_1(\phi,\theta)}^{F_2(\phi,\theta)} f(\rho \sin \phi \cos \theta, \rho \sin \phi \sin \theta, \rho \cos \phi) \rho^2 \sin \phi \, d\rho \, d\phi \, d\theta$$

which is the formula appearing in Theorem 14.11.   ❑

By this formula we can see that the expression $\rho^2 \sin \phi$ appearing in the spherical coordinate integral theorem (Theorem 14.11) is the Jacobian $\frac{\partial(x, y, z)}{\partial(\rho, \phi, \theta)}$ of the map that transforms spherical coordinates to rectangular coordinates.

EXERCISES 14.8

In Exercises 1–8 find the Jacobian of the transformation.

1. $x = 3u - 4v, \ y = \dfrac{1}{2}u + \dfrac{1}{6}v$

2. $x = 3u - 6v, \ y = -2u + 4v$

3. $x = uv, \ y = u^2 + v^2$

4. $x = \cos u + \sin v, \ y = -\sin u + \cos v$

5. $x = e^v, \ y = ue^v$

6. $x = u - \ln v, \ y = v + \ln u$

7. $x = au, \ y = bv, \ z = w$

8. $x = \dfrac{u}{v^2}, \ y = \dfrac{v}{w^2}, \ z = \dfrac{w}{u^2}$

In Exercises 9–17 evaluate the integral by using the given transformation.

9. $\displaystyle\iint_R \frac{y}{x - 3y} \, dA$, where R is the region bounded by the lines $y = 1$, $y = \frac{1}{4}x$, and $x - 3y = e$; let $x = 3u + v$, $y = u$

10. $\iint_R y^2 \, dA$, where R is the region bounded by the ellipse $4x^2 + 9y^2 = 1$; let $x = \frac{1}{2}u$, $y = \frac{1}{3}v$

11. $\iint_R xy^2 \, dA$, where R is the region bounded by the lines $x - y = 2$, $x - y = -1$, $2x + 3y = 1$, and $2x + 3y = 0$; let $x = \frac{1}{5}(3u + v)$, $y = \frac{1}{5}(v - 2u)$

12. $\iint_R e^{(y-x)^2} \, dA$, where R is the region bounded by the lines $y = \frac{3}{2}x$, $y = 2x$, and $y = x + 1$; let $x = u + v$, $y = u + 2v$

13. $\iint_R \dfrac{y}{x} e^{x^2 - y^2} \, dA$, where R is the region in the first quadrant bounded by the hyperbolas $x^2 - y^2 = 1$ and $x^2 - y^2 = 4$ and by the lines $x = 2y$ and $x = \sqrt{2}y$; let $x = u \sec v$ and $y = u \tan v$ for $u > 0$ and $0 < v < \pi/2$

14. $\iint_R y \cos xy \, dA$, where R is bounded by the curves $xy = \pi/2$, $xy = \pi$, $y(2 - x) = 2$, and $y(2 - x) = 4$; let $x = \dfrac{2v}{u + v}$, $y = u + v$

15. $\iint_R e^{x^2 - y^2} \, dA$, where R is the region in the first quadrant bounded by the curves $x^2 - y^2 = 1$, $x^2 - y^2 = 4$, $y = 0$, and $y = \frac{3}{5}x$; let $x = u \cosh v$, $y = u \sinh v$ for $u > 0$

16. $\iiint_D x \, dV$, where D is the solid region in the first octant bounded by the ellipsoid

$$\frac{x^2}{4} + \frac{y^2}{9} + z^2 = 1.$$

Let $x = 2u$, $y = 3v$, $z = w$.

17. $\iiint_D (x^2 + y^2) \, dV$, where D is the solid region in the first octant bounded by the xz and yz planes, the paraboloids $z = x^2 + y^2$ and $z = 4(x^2 + y^2)$, and the planes $z = 1$ and $z = 4$; let $x = (v/u) \cos w$, $y = (v/u) \sin w$, $z = v^2$ for $u > 0$, $v \geq 0$, and $0 \leq w \leq \pi/2$

In Exercises 18–25 evaluate the integral by a suitable change of variables.

18. $\iint_R 49x^2 y \, dA$, where R is the region bounded by the lines $2x - y = 1$, $2x - y = -2$, $x + 3y = 0$, and $x + 3y = 1$

19. $\displaystyle\iint_R \left(\frac{x - 2y}{x + 2y}\right)^3 \, dA$, where R is the region bounded by the lines $x - 2y = 1$, $x - 2y = 2$, $x + 2y = 1$, and $x + 2y = 3$

20. $\displaystyle\iint_R \left(1 + \frac{x^2}{16} + \frac{y^2}{25}\right)^{3/2} \, dA$, where R is the region bounded by the ellipse $\dfrac{x^2}{16} + \dfrac{y^2}{25} = 1$

21. $\iint_R x^2 \, dA$, where R is the region bounded by the ellipse $\dfrac{x^2}{a^2} + \dfrac{y^2}{b^2} = 1$

22. $\iint_R \cos(x^2 + 4y^2 + \pi - 1) \, dA$, where R is the region above the x axis bounded by the ellipse $x^2 + 4y^2 = 1$ and the x axis

23. $\displaystyle\iint_R \sin\left[\pi\left(\frac{y - x}{y + x}\right)\right] \, dA$, where R is the region bounded by the lines $x + y = 1$, $x + y = 2$, $x = 0$, and $y = 0$

24. $\iint_R (x - y)^2 \sin^2(x + y) \, dA$, where R is the region bounded by the parallelogram with vertices $(\pi, 0)$, $(2\pi, \pi)$, $(\pi, 2\pi)$, and $(0, \pi)$

25. $\iint_R e^{(2x - y)/(x + y)} \, dA$, where R is the region bounded by the lines $x = 2y$, $y = 2x$, $x + y = 1$, and $x + y = 2$

In Exercises 26–27 find the area A of R by using the given transformation.

26. R is the region in the first quadrant bounded by the curves $xy = 1$, $xy = 2$, $y = x$, and $y = 4x$. Let $x = u/v$ and $y = v$.

27. R is the region in the first quadrant bounded by $x^2 - y^2 = a^2$, $x^2 - y^2 = b^2$, $y = 0$, and $y = \frac{1}{2}x$, where $b \geq a > 0$. Let $x = u \cosh v$ and $y = u \sinh v$.

28. Using a change of variables, determine the volume V of the solid region D bounded by the ellipsoid $x^2 + 2y^2 + 4z^2 = 1$.

29. Let $a > 0$ and $b > 0$. By using a suitable change of variables, show that the area A of the ellipse $x^2/a^2 + y^2/b^2 = 1$ is πab.

30. The transformation

$$x = \frac{1}{\sqrt{2}}(u - v) \quad \text{and} \quad y = \frac{1}{\sqrt{2}}(u + v)$$

is equivalent to a rotation of axes through $\pi/4$ radians.

a. Show that the Folium of Descartes $x^3 - \sqrt{2}\,xy + y^3 = 0$ corresponds to the curve C in the uv plane that has equation

$$u^3 + 3uv^2 - u^2 + v^2 = 0$$

b. Solve the equation in part (a) for v^2 as a function of u. Using that result, show that the area A of the loop of the folium is given by

$$A = 2\int_0^1 u\sqrt{\frac{1 - u}{3u + 1}}\, du$$

(See Figure 14.72.)

 c. Use a computer algebra system to assist in evaluating the integral in part (b).

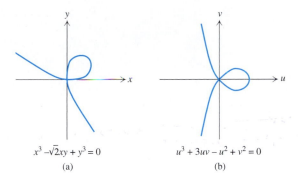

$$x^3 - \sqrt{2}xy + y^3 = 0 \qquad u^3 + 3uv - u^2 + v^2 = 0$$
(a)   (b)

FIGURE 14.72   The Folium of Descartes is rotated through $\pi/4$. See Exercise 30.

31. Find the area A of the ellipse $x^2 - xy + y^2 = 2$. (*Hint*: Use the transformation in Exercise 30, along with the result of Exercise 29.)

Project

1. Suppose that D is the unit ball $x^2 + y^2 + z^2 \leq 1$ and let E be the half unit ball for which $z \geq 0$. Suppose that $m = \iiint_E f(x, y, z)\, dV$. In terms of m, what can you say about $\iiint_D f(x, y, z)\, dV$ if
   a. $f(x, y, -z) = f(x, y, z)$ for all (x, y, z) in D?
   b. $f(-x, -y, -z) = -f(x, y, z)$ for all (x, y, z) in D?
   c. $f(-x, -y, -z) = f(x, y, z)$ for all (x, y, z) in D?

14.9   PARAMETRIZED SURFACES

In Section 6.7 we introduced parametrized curves, because some very common curves such as the cycloid (Figure 14.73) are difficult or impossible to describe in simple rectangular coordinates. Similarly, there are surfaces that are difficult to describe in rectangular, cylindrical or spherical coordinates, but which can be described by other sets of parameters. For example, consider the helicoid in Figure 14.74. If it represents a ramp in a parking garage, we might want to know its surface area. Although there is no simple way to describe the helicoid in terms of rectangular, cylindrical or spherical coordinates, it does have a reasonable parametric description, and using it we will be able to determine its surface area.

Before introducing parametrized surfaces, let us recall two basic ways of representing parametrically a curve in the plane. Either we can express the coordinates (x, y) of an arbitrary point on the curve in terms of t using parametric

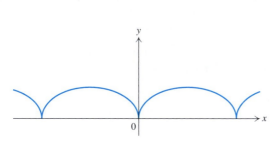

FIGURE 14.73   A cycloid.

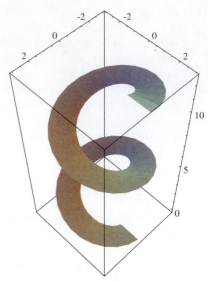

FIGURE 14.74   A helicoid drawn in Mathematica.

equations of the form

$$x = x(t) \quad \text{and} \quad y = y(t) \quad \text{for} \quad a \le t \le b$$

or we can consider the curve as traced out by the corresponding vector-valued function

$$\mathbf{r}(t) = x(t)\mathbf{i} + y(t)\mathbf{j} \quad \text{for} \quad a \le t \le b$$

For example, the circle of radius r centered at the origin has the parametrization

$$x = r \cos t \quad \text{and} \quad y = r \sin t \quad \text{for} \quad 0 \le t \le 2\pi$$

and is traced out by the vector-valued function $\mathbf{r} = r \cos t\mathbf{i} + r \sin t\mathbf{j}$.

For curves in space (with three coordinates), the corresponding formulas are

$$x = x(t), \quad y = y(t), \quad \text{and} \quad z = z(t) \quad \text{for} \quad a \le t \le b$$

or

$$\mathbf{r}(t) = x(t)\mathbf{i} + y(t)\mathbf{j} + z(t)\mathbf{k} \quad \text{for} \quad a \le t \le b$$

Thus a standard vertically oriented helix has parametrization

$$x = r \cos t, \quad y = r \sin t, \quad \text{and} \quad z = t \quad \text{for all} \quad t$$

(refer to Figure 12.6).

Now we return to surfaces. Suppose that a surface Σ can be expressed in terms of two variables u and v as

$$x = x(u, v), \quad y = y(u, v), \quad \text{and} \quad z = z(u, v) \tag{1}$$

for all (u, v) lying in a planar region R. Then Σ is said to be a **parametrized surface** parametrized by u and v. In vector form (1) is written

$$\mathbf{r}(u, v) = x(u, v)\mathbf{i} + y(u, v)\mathbf{j} + z(u, v)\mathbf{k} \qquad (2)$$

We may think of Σ as the image of R under the map $\mathbf{r}$ (Figure 14.75). We will usually employ the vector form appearing in (2) for parametrized surfaces. Also, we will generally assume that R is the entire uv plane, a connected subset of the uv plane that is bounded by finitely many smooth curves (such as lines and circles), or a finite union of vertically or horizontally simple regions. Moreover, we will assume that $\mathbf{r}(u_1, v_1) \neq \mathbf{r}(u_2, v_2)$ for any two interior points (u_1, v_1) and (u_2, v_2) of the domain of $\mathbf{r}$. This will ensure that each point of Σ is traced out only once by the various values of (u, v), except possibly for points on the boundary of the domain of $\mathbf{r}$.

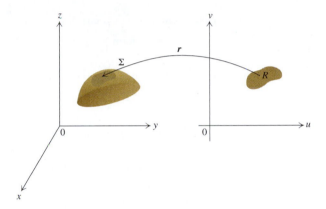

FIGURE 14.75

The parameters need not be denoted by u and v. In Examples 1 and 2 we will parametrize a sphere and a cone in terms of pairs of spherical coordinates. Before we give these examples, let us recall from Section 14.6 that the spherical coordinates ρ, ϕ, and θ are related to the rectangular coordinates x, y, and z by the equations

$$x = \rho \sin \phi \cos \theta, \quad y = \rho \sin \phi \sin \theta, \quad z = \rho \cos \phi \qquad (3)$$

EXAMPLE 1 Find a parametrization for the sphere Σ whose equation is $x^2 + y^2 + z^2 = a^2$ (Figure 14.76).

Solution Using (3) and the fact that $\rho = \sqrt{x^2 + y^2 + z^2} = a$ at every point on Σ, we obtain

$$x = a \sin \phi \cos \theta, y = a \sin \phi \sin \theta, z = a \cos \phi \text{ for } 0 \leq \phi \leq \pi \text{ and } 0 \leq \theta \leq 2\pi$$

as the desired parametrization. In vector notation the parametrization is

$$\mathbf{r}(\phi, \theta) = a \sin \phi \cos \theta \mathbf{i} + a \sin \phi \sin \theta \mathbf{j} + a \cos \phi \mathbf{k}$$

for $\qquad\qquad 0 \leq \phi \leq \pi \text{ and } 0 \leq \theta \leq 2\pi \qquad$ ❑

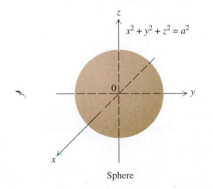

Sphere

FIGURE 14.76

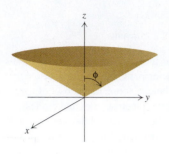

FIGURE 14.77

EXAMPLE 2   Find a parametrization for the part of the double cone $x^2 + y^2 = 3z^2$ for which $z \geq 0$ (that is, the upper nappe of the cone) in terms of ρ and θ (Figure 14.77).

Solution   Notice first that the angle ϕ at which the cone opens up is constant. Using the fact that $r^2 = x^2 + y^2 = 3z^2$, along with Table 1 in Section 14.5, we find that at every point (except the origin) of Σ,

$$\tan \phi = \frac{r}{z} = \frac{\sqrt{3z^2}}{z} = \sqrt{3}$$

This means that $\phi = \pi/3$. Thus the equations in (3) become

$$x = \rho \sin \frac{\pi}{3} \cos \theta = \frac{\sqrt{3}}{2} \rho \cos \theta, \ y = \rho \sin \frac{\pi}{3} \sin \theta = \frac{\sqrt{3}}{2} \rho \sin \theta, \ z = \rho \cos \frac{\pi}{3} = \frac{1}{2} \rho$$

or in vector form,

$$\mathbf{r}(\rho, \theta) = \frac{\sqrt{3}}{2} \rho \cos \theta \mathbf{i} + \frac{\sqrt{3}}{2} \rho \sin \theta \mathbf{j} + \frac{1}{2} \rho \mathbf{k} \quad \text{for} \quad \rho \geq 0 \quad \text{and} \quad 0 \leq \theta \leq 2\pi \qquad \square$$

We remark that in the parametrization for the cone given in Example 2, the origin is represented, or "traced out," infinitely often by the parametrization because

$$\mathbf{r}(0, \theta) = \frac{\sqrt{3}}{2}(0) \cos \theta \mathbf{i} + \frac{\sqrt{3}}{2}(0) \sin \theta \mathbf{j} + \frac{1}{2}(0)\mathbf{k} = 0$$

for all values of θ in $[0, 2\pi]$. For our discussion there is no problem, since the origin is a single point of the cone and since the points $(0, \theta)$ lie on the boundary of the domain of $\mathbf{r}$.

In the next example, we will parametrize a cylinder.

EXAMPLE 3   Find a parametrization for the cylinder $x^2 + y^2 = 4$ (Figure 14.78).

Solution   Notice that z can assume any value since it does not appear in the formula for the cylinder, and that the intersection of the cylinder with the xy-plane is a circle with radius 2. In polar coordinates that circle has the parametric representation

$$x = 2 \cos \theta \quad \text{and} \quad y = 2 \sin \theta \quad \text{for} \quad 0 \leq \theta \leq 2\pi$$

Thus the parametrization we seek is

$$\mathbf{r}(\theta, z) = 2 \cos \theta \mathbf{i} + 2 \sin \theta \mathbf{j} + z\mathbf{k} \quad \text{for} \quad 0 \leq \theta \leq 2\pi \quad \text{and any} \quad z \qquad \square$$

Any surface that is the graph of a function f on a region R in the xy-plane can be (trivially) considered as a parametrized surface by giving it the parametrization

$$\mathbf{r}(x, y) = x\mathbf{i} + y\mathbf{j} + f(x, y)\mathbf{k} \tag{4}$$

or more simply,

$$x = x, \quad y = y, \quad z = f(x, y) \text{ for } (x, y) \quad \text{in} \quad R$$

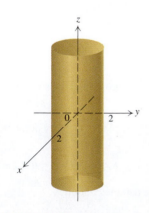

FIGURE 14.78

EXAMPLE 4   Sketch the surface Σ parametrized by

$$\mathbf{r}(x, y) = x\mathbf{i} + y\mathbf{j} + (4 - x^2 - y^2)\mathbf{k}$$

for (x, y) in the circular region defined by $x^2 + y^2 \leq 4$.

Solution   By Example 6 in Section 13.1, the graph of $z = f(x, y) = 4 - x^2 - y^2$ is a circular paraboloid. Since $4 - x^2 - y^2 \geq 0$ precisely when $x^2 + y^2 \leq 4$, it follows that Σ is the portion of the paraboloid that lies above the xy-plane (Figure 14.79).   ❑

Planes are examples of surfaces that are graphs of functions, and they are easy to describe parametrically. Indeed, the plane $3x - 2y + z = 1$ can be described parametrically by first solving for z to obtain $z = 1 - 3x + 2y$, and then converting to

$$\mathbf{r}(x, y) = x\mathbf{i} + y\mathbf{j} + (1 - 3x + 2y)\mathbf{k}$$

However, when we encounter only a portion Σ of a plane, it may be more convenient to describe it by using other representations, or parametrizations.

EXAMPLE 5   Find a parametric representation for the portion Σ of the plane $z = 1 - 3x + 2y$ that lies inside the cylinder $x^2 + y^2 = 1$ (Figure 14.80).

Solution   The cylinder intersects the xy plane in a circle of radius 1 that bounds the region R defined by

$$x = r \cos \theta \quad \text{and} \quad y = r \sin \theta, \quad \text{where} \quad 0 \leq r \leq 1 \quad \text{and} \quad 0 \leq \theta \leq 2\pi$$

Here r and θ represent polar coordinates in the xy plane. We find that points on the plane $z = 1 - 3x + 2y$ are given by the equation $z = 1 - 3\,r \cos \theta + 2\,r \sin \theta$. Our information thus yields the parametrization

$$\mathbf{r}(r, \theta) = r \cos \theta\mathbf{i} + r \sin \theta\mathbf{j} + (1 - 3\,r \cos \theta + 2\,r \sin \theta)\mathbf{k}$$

for

$$0 \leq r \leq 1 \quad \text{and} \quad 0 \leq \theta \leq 2\pi \quad ❑$$

We complete this subsection with a parametrization of a helicoid, or ramp, Σ, pictured in Figure 14.81:

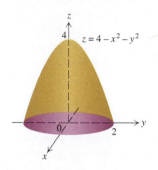

FIGURE 14.79

FIGURE 14.80   The part of a plane inside a cylinder.

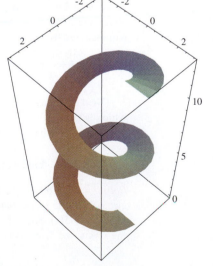

FIGURE 14.81   A helicoid.

$$\mathbf{r}(u, v) = u \cos v\mathbf{i} + u \sin v\mathbf{j} + v\mathbf{k} \quad \text{for} \quad 1 \le u \le 2 \quad \text{and} \quad 0 \le v \le 4\pi \quad (5)$$

We observe that for any given value of v, $\mathbf{r}(u, v)$ for $1 \le u \le 2$ traces out a line segment of length 1 in the plane $z = v$, and for any value of u, $\mathbf{r}(u, v)$ traces out two complete revolutions of a helix with radius u. You might think of the helicoid as a ribbon one unit wide, which makes two complete revolutions about the z axis as the height of the ribbon increases from 0 to 2 from the horizontal coordinate plane.

Surfaces of Revolution

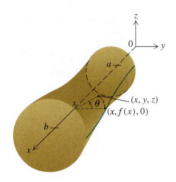

FIGURE 14.82

Another way of producing a surface is to rotate a plane curve about an axis. Such a surface is called a **surface of revolution**. Examples are easy to find: a sphere is obtained by revolving a semicircle about an axis through its center, and cones and cylinders are obtained in a similar manner.

Provided that we have enough information about the curve, we can find a formula for a parametrization of the resulting surface of revolution. Suppose, for example, that a surface Σ is produced by revolving the graph of a nonnegative function f on an interval $[a, b]$ around the x axis, as we discussed in Section 6.3 (Figure 14.82). For each x between a and b the set of all points (x, y, z) on Σ with first coordinate x lies on a circle with radius $f(x)$ (Figure 14.82). It follows that $y = f(x) \cos \theta$ and $z = f(x) \sin \theta$, where θ is the angle shown in Figure 14.82. Therefore Σ can be parametrized by

$$x = x, \quad y = f(x) \cos \theta, \quad z = f(x) \sin \theta \quad \text{for} \quad a \le x \le b \quad \text{and} \quad 0 \le \theta \le 2\pi$$

or in vector form,

$$\mathbf{r}(x, \theta) = x\mathbf{i} + f(x) \cos \theta\mathbf{j} + f(x) \sin \theta\mathbf{k} \quad \text{for} \quad a \le x \le b \quad \text{and} \quad 0 \le \theta \le 2\pi \quad (6)$$

Notice that the parameters are x and θ. If the curve were revolved around the y axis, then the parameters would be y and θ.

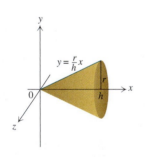

FIGURE 14.83   A cone as a surface of revolution.

EXAMPLE 6   Find a parametrization of a cone whose axis is the x axis, height is h, and base radius is r.

Solution.   The cone is produced by revolving the graph of $y = \frac{r}{h}x$ for $0 \le x \le h$ about the x axis, as you can check (Figure 14.83). Thus the cone can be parametrized by

$$\mathbf{r}(x, \theta) = x\mathbf{i} + \frac{r}{h}x \cos \theta\mathbf{j} + \frac{r}{h}x \sin \theta\mathbf{k} \quad \text{for} \quad 0 \le x \le h \quad \text{and} \quad 0 \le \theta \le 2\pi \quad \square$$

Next we determine a parametrization for a torus (doughnut-shaped surface), which is the surface obtained by revolving a circle about an axis that lies in the same plane as the circle but lies outside the circle.

EXAMPLE 7   Find a parametrization for the torus Σ shown in Figure 14.84.

Solution   Let b and r be as in Figure 14.85 (a). Consider any point $(p, 0, q)$ on the circle in the xz plane that is rotated to obtain Σ, and let θ be the angle shown in Figure 14.85(a). Since the center of the circle is b units from the origin and lies on the x axis, we deduce that

$$p = b + r \cos \theta \quad \text{and} \quad q = r \sin \theta \quad (7)$$

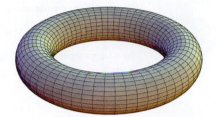

FIGURE 14.84   A torus drawn in Mathematica.

Now consider any point (x, y, z) on Σ that arises when the point $(p, 0, q)$ is revolved around the z axis (Figure 14.85(a)). Then (x, y, z) lies on the plane $z = q$, which means that $z = q = r \sin \theta$. Let ϕ denote the angle through which the point travels as it proceeds from the location $(p, 0, q)$ to the location (x, y, z), as in Figure 14.85(b).

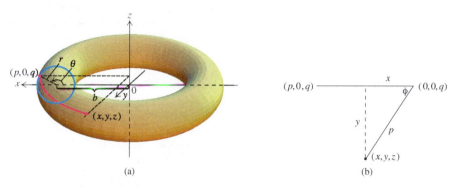

(a)   (b)

FIGURE 14.85

Since the circular arc through which $(p, 0, q)$ travels to get to the point (x, y, z) has radius $p = b + r \cos \theta$, we deduce that

$$x = p \cos \phi \quad \text{and} \quad y = p \sin \phi \tag{8}$$

We conclude from (7) and (8) that Σ is parametrized by

$$\mathbf{r}(\phi, \theta) = (b + r \cos \theta) \cos \phi \mathbf{i} + (b + r \cos \theta) \sin \phi \mathbf{j} + r \sin \theta \mathbf{k}$$

for $0 \leq \phi \leq 2\pi$ and $0 \leq \theta \leq 2\pi$ ☐

Figure 14.84 was drawn in Mathematica. The curves appearing on the surface result from keeping either the angle ϕ constant or the angle θ constant. Such curves are called **grid curves** of the surface.

Smooth Surfaces

In order for us to be able to study tangent planes and surface area of parametrized surfaces, we now will focus on parametrized surfaces that are smooth, in the sense that they don't have any creases or sharp points. Thus a sphere or ellipsoid fits the bill, whereas a cone (which has a sharp point) and a cube (which has creases) do not (Figure 14.86).

FIGURE 14.86

Let Σ be a parametrized surface, and let

$$\mathbf{r}(u, v) = x(u, v)\mathbf{i} + y(u, v)\mathbf{j} + z(u, v)\mathbf{k}$$

be a parametrization of Σ on a region R in the uv plane. If the functions x, y, and z have partial derivatives at a point (u, v) in R, then the **partial derivatives** of the vector-valued function $\mathbf{r}$ at (u, v) are defined by

$$\mathbf{r}_u(u, v) = \frac{\partial \mathbf{r}}{\partial u}(u, v) = \frac{\partial x}{\partial u}(u, v)\mathbf{i} + \frac{\partial y}{\partial u}(u, v)\mathbf{j} + \frac{\partial z}{\partial u}(u, v)\mathbf{k} \qquad (9)$$

and

$$\mathbf{r}_v(u, v) = \frac{\partial \mathbf{r}}{\partial v}(u, v) = \frac{\partial x}{\partial v}(u, v)\mathbf{i} + \frac{\partial y}{\partial v}(u, v)\mathbf{j} + \frac{\partial z}{\partial v}(u, v)\mathbf{k} \qquad (10)$$

If we don't place restrictions on the partial derivatives, then the "surface" could be merely a set of points or curves and not really a surface. For example, if at each point in the region R the partial derivative of $\mathbf{r}$ with respect to u is 0, then $\mathbf{r}$ depends only on v and hence the "surface" is just a curve (or a point). A more subtle example, where the partial derivatives are not 0, is given by any function $\mathbf{r}$ of the form

$$\mathbf{r}(u, v) = g(u, v)(\mathbf{i} + \mathbf{j} + \mathbf{k}) \qquad (11)$$

where g is a real-valued function of two variables. It follows that irrespective of the values of u and v, the vector $\mathbf{r}(u, v)$ is parallel to the vector $\mathbf{i} + \mathbf{j} + \mathbf{k}$, so that the image of $\mathbf{r}$ on any domain R in the plane is not a surface but is (part of) a line.

You might wonder how we can tell whether the image of a vector-valued function is really a surface, or whether it is "degenerate," that is, a curve or an even smaller set of points. We can get a hint from analyzing the partial derivatives of the function in (11). If g has partial derivatives with respect to u and v, then so does $\mathbf{r}$, and

$$\mathbf{r}_u(u, v) = g_u(u, v)(\mathbf{i} + \mathbf{j} + \mathbf{k}) \text{ and } \mathbf{r}_v(u, v) = g_v(u, v)(\mathbf{i} + \mathbf{j} + \mathbf{k})$$

Thus although both $\mathbf{r}_u(u, v)$ and $\mathbf{r}_v(u, v)$ may be nonzero, they are nevertheless parallel and hence their cross product $\mathbf{r}_u(u, v) \times \mathbf{r}_v(u, v) = \mathbf{0}$.

We can avoid such "degenerate surfaces" by assuming that $\mathbf{r}_u(u, v) \times \mathbf{r}_v(u, v) \neq \mathbf{0}$. This implies that neither $\mathbf{r}_u(u, v)$ nor $\mathbf{r}_v(u, v)$ is $\mathbf{0}$, and that $\mathbf{r}_u(u, v)$ and $\mathbf{r}_v(u, v)$ are not parallel and hence determine a plane. Thus we are led to the following definition.

DEFINITION 14.15

Let Σ be a surface. We say that Σ is **smooth at a point** (x_0, y_0, z_0) on Σ if there is a vector-valued function $\mathbf{r}$ defined on an open subset U of the uv plane, such that $\mathbf{r}(u, v)$ lies on Σ for every (u, v) in U and $\mathbf{r}(u_0, v_0) = (x_0, y_0, z_0)$ for some (u_0, v_0) in U, and such that $\mathbf{r}$ has continuous partial derivatives on U with $\mathbf{r}_u(u_0, v_0) \times \mathbf{r}_v(u_0, v_0) \neq \mathbf{0}$. We say that Σ is **smooth** if Σ is smooth at each of its points. Finally, Σ is **piecewise smooth** if it consists of finitely many smooth subsurfaces $\Sigma_1, \Sigma_2, \ldots, \Sigma_n$, and finitely many smooth curves or points that serve as common boundaries to these subsurfaces.

It is easy to check that a sphere, paraboloid, plane, and torus are smooth (see Examples 1, 4, 5, and 7, respectively). By contrast, a cube is not smooth but is piecewise smooth because it is composed of six smooth faces all of which are joined together by line segments (see Figure 14.86 again). Finally, we mention that a cone such as the one in Example 2 is *not* smooth because of the point at the origin. However, if we limit the domain and let Σ_1 be the entire cone except the origin, then we can provide Σ_1 with a smooth parametrization:

$$\mathbf{r}(\rho, \theta) = \frac{\sqrt{3}}{2} \rho \cos \theta \mathbf{i} + \frac{\sqrt{3}}{2} \rho \sin \theta\, j + \frac{1}{2} \rho \mathbf{k} \quad \text{for } \rho > 0 \text{ and } 0 \leq \theta \leq 2\pi$$

Tangent Planes

As before, consider the surface Σ and a point (x_0, y_0, z_0) on Σ at which Σ is smooth. Let

$$\mathbf{r}(u, v) = x(u, v)\mathbf{i} + y(u, v)\mathbf{j} + z(u, v)\mathbf{k} \quad \text{for } (u, v) \text{ in the planar region } R$$

be a parametrization of Σ as guaranteed by Definition 14.15, with $\mathbf{r}(u_0, v_0) = (x_0, y_0, z_0)$. Holding $v = v_0$ and letting u vary, we obtain a parametrized curve C_1 on the surface Σ that passes through the point (x_0, y_0, z_0) and is parametrized by

$$\rho_1(u) = x(u, v_0)\mathbf{i} + y(u, v_0)\,\mathbf{j} + z(u, v_0)\mathbf{k} \quad \text{for } u \text{ near } u_0$$

Then the partial derivative $\mathbf{r}_u(u_0, v_0)$ is precisely the derivative $\rho_1'(u_0)$, and hence is a vector that is tangent to C_1 at the point (x_0, y_0, z_0) (Figure 14.87).

Similarly, by holding $u = u_0$ and letting v vary, we obtain a parametrized curve C_2 on the surface Σ that passes through the point (x_0, y_0, z_0) and is parametrized by

$$\rho_2(u) = x(u_0, v)\mathbf{i} + y(u_0, v)\,\mathbf{j} + z(u_0, v)\mathbf{k} \quad \text{for } v \text{ near } v_0$$

Then the partial derivative $\mathbf{r}_v(u_0, v_0)$ is precisely the derivative $\rho_2'(v_0)$, and hence is a vector that is tangent to C_2 at the point (x_0, y_0, z_0) (Figure 14.87).

Because $\mathbf{r}_u(u_0, v_0) \times \mathbf{r}_v(u_0, v_0) \neq \mathbf{0}$ by hypothesis, the vectors $\mathbf{r}_u(u_0, v_0)$ and $\mathbf{r}_v(u_0, v_0)$ are nonzero and nonparallel. Therefore they determine a plane containing the point (x_0, y_0, z_0). Using the Chain Rule for functions of two variables, one can

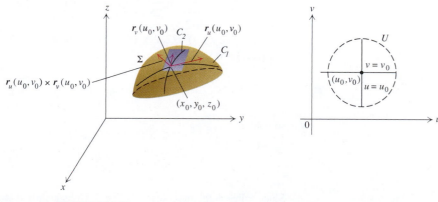

FIGURE 14.87

prove more generally that the plane contains the tangent vector at (x_0, y_0, z_0) to any smooth curve on Σ passing through (x_0, y_0, z_0). Thus the plane is called the **tangent plane** to the surface Σ at the point (x_0, y_0, z_0). By definition of the tangent plane, the cross product

$$\mathbf{r}_u(u_0, v_0) \times \mathbf{r}_v(u_0, v_0)$$

is normal to the tangent plane at $\mathbf{r}(u_0, v_0)$ (Figure 14.87). In addition, the vector

$$\mathbf{n} = \frac{\mathbf{r}_u(u, v) \times \mathbf{r}_v(u, v)}{\|\mathbf{r}_u(u, v) \times \mathbf{r}_v(u, v)\|} \tag{12}$$

is a unit vector normal to the surface Σ. This will be useful in Section 15.6.

If Σ is smooth, then the vectors $\mathbf{r}_u(u, v)$ and $\mathbf{r}_v(u, v)$ and hence the normal vector $\mathbf{r}_u(u, v) \times \mathbf{r}_v(u, v)$ vary continuously as the point (x_0, y_0, z_0) moves continuously over Σ. Thus the tangent plane also changes continuously. This is the rationale for the term "smooth surface."

EXAMPLE 8 Let Σ be the cone whose axis is the x axis, height is 5, and base radius is 2. Find an equation of the plane tangent to Σ at the point $P = (2, 2/5, 2\sqrt{3}/5)$.

Solution   By Example 6 with $h = 5$ and $r = 2$, a parametrization is given by

$$\mathbf{r}(x, \theta) = x\mathbf{i} + \frac{2}{5}x \cos \theta \, \mathbf{j} + \frac{2}{5}x \sin \theta \, \mathbf{k} \quad \text{for } 0 \le x \le 5 \text{ and } 0 \le \theta \le 2\pi$$

Therefore

$$\mathbf{r}_x(x, \theta) \times \mathbf{r}_\theta(x, \theta) = \begin{vmatrix} \mathbf{i} & \mathbf{j} & \mathbf{k} \\ 1 & \frac{2}{5} \cos \theta & \frac{2}{5} \sin \theta \\ 0 & -\frac{2}{5}x \sin \theta & \frac{2}{5}x \cos \theta \end{vmatrix}$$

$$= \frac{4}{25}x(\cos^2 \theta + \sin^2 \theta)\mathbf{i} - \frac{2}{5}x \cos \theta \, \mathbf{j} - \frac{2}{5}x \sin \theta \mathbf{k}$$

$$= \frac{4}{25}x\mathbf{i} - \frac{2}{5}x \cos \theta \, \mathbf{j} - \frac{2}{5}x \sin \theta \mathbf{k}$$

Since $P = (x_0, y_0, z_0) = (2, 2/5, 2\sqrt{3}/5)$, we have $x_0 = 2$. Moreover, if θ_0 is such that $(x_0, y_0, z_0) = \mathbf{r}(x_0, \theta_0)$, then

$$y_0 = \frac{2}{5}x_0 \cos \theta_0 \quad \text{and} \quad z_0 = \frac{2}{5}x_0 \sin \theta_0$$

so that

$$\tan \theta_0 = \frac{z_0}{y_0} = \frac{2\sqrt{3}/5}{2/5} = \sqrt{3}$$

Since P lies in the first octant, $\theta_0 = \pi/3$. Consquently if we let $x_0 = 2$ and $\theta_0 = \pi/3$, we find that

$$\mathbf{r}_x(x_0, \theta_0) \times \mathbf{r}_\theta(x_0, \theta_0) = \mathbf{r}_x(2, \pi/3) \times r_\theta(2, \pi/3) = \frac{8}{25}\mathbf{i} - \frac{2}{5}\mathbf{j} - \frac{2\sqrt{3}}{5}\mathbf{k}$$

Simplifying the fractions by multiplying the vector by 25/2, we deduce that the vector $4\mathbf{i} - 5\mathbf{j} - 5\sqrt{3}\mathbf{k}$ is normal to the tangent plane at the point $(2, 2/5, 2\sqrt{3}/5)$. Thus an equation of the tangent plane is

$$4(x - 2) - 5\left(y - \frac{2}{5}\right) - 5\sqrt{3}\left(z - \frac{2\sqrt{3}}{5}\right) = 0 \qquad \square$$

Area of a Parametrized Surface

In Section 14.3 we defined the area of a surface that is the graph of a continuous function f with continuous partial derivatives on a vertically or horizontally simple region R in the plane. The derivation of the formula presented in that section can be modified to apply to smooth parametrized surfaces.

We begin with the necessary notation. Let Σ be a smooth parametrized surface with parametrization

$$\mathbf{r}(u, v) = x(u, v)\mathbf{i} + y(u, v)\mathbf{j} + z(u, v)\mathbf{k}, \quad \text{with} \quad (u, v) \quad \text{in} \quad R$$

where R is a vertically or horizontally simple region in the uv plane. Let a rectangle R' contain R, and partition R' into subrectangles, of which $R_1, R_2, \ldots, R_n$ are the subrectangles that lie entirely in R. For any k between 1 and n, let (u_k, v_k) be the vertex of R_k with the smallest values of u and v (Figure 14.88). In addition, let Δu_k and Δv_k be the lengths of adjacent sides of the rectangle R_k appearing in Figure 14.88, so that the area ΔA_k of R_k is $\Delta u_k \Delta v_k$.

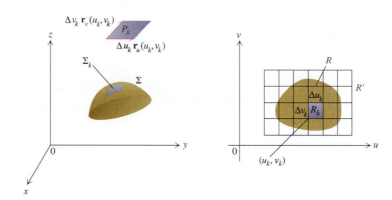

FIGURE 14.88

For simplicity we will denote

$$x_k = x(u_k, v_k), \quad y_k = y(u_k, v_k), \quad z_k = z(u_k, v_k)$$

and consider the plane tangent to Σ at the point (x_k, y_k, z_k). By the definition of the tangent plane, the vectors $\mathbf{r}_u(u_k, v_k)$ and $\mathbf{r}_v(u_k, v_k)$ emanating from (x_k, y_k, z_k) lie in that tangent plane.

Because we already know how to find areas of parallelograms, we first let Σ_k be the part of Σ corresponding to R_k, and then find the area of a suitable small parallelogram in the plane tangent to Σ_k at the point (x_k, y_k, z_k). That area will approximate the area ΔS_k of Σ_k. Proceeding as we did in the introductory discussion of Section 14.3, we consider the parallelogram P_k in the tangent plane with vertex at (x_k, y_k, z_k) and with vectors $\Delta u_k\, \mathbf{r}_u\,(u_k, v_k)$ and $\Delta v_k\, \mathbf{r}_v\,(u_k, v_k)$ for adjacent sides. We note that the area ΔP_k of P_k is the norm of the cross product of the two vectors (see (2) in Section 11.4). Therefore

$$\Delta S_k \approx \Delta P_k = \|\Delta u_k\, \mathbf{r}_u\,(u_k, v_k) \times \Delta v_k\, \mathbf{r}_v\,(u_k, v_k)\| = \|\,\mathbf{r}_u\,(u_k, v_k) \times \mathbf{r}_v\,(u_k, v_k)\|\, \Delta u_k\, \Delta v_k$$

Since $\Delta A_k = \Delta u_k\, \Delta v_k$, we find that

$$\Delta S_k \approx \|\,\mathbf{r}_u\,(u_k, v_k) \times \mathbf{r}_v\,(u_k, v_k)\,\|\, \Delta A_k \tag{13}$$

It follows that the area of Σ_k is approximately the area of R_k, multiplied by the factor $\|\,\mathbf{r}_u\,(u_k, v_k) \times \mathbf{r}_v\,(u_k, v_k)\|$, which is called the "local magnification factor."

Consequently the area of the entire surface Σ, which is the sum of the areas of $\Sigma_1, \Sigma_2, \ldots, \Sigma_n$, is approximately equal to

$$\sum_{k=1}^{n} \|\,\mathbf{r}_u\,(u_k, v_k) \times \mathbf{r}_v\,(u_k, v_k)\,\|\, \Delta A_k$$

which is a Riemann sum for the double integral

$$\iint_R \|\,\mathbf{r}_u\,(u, v) \times \mathbf{r}_v\,(u, v)\,\|\, dA$$

Since it is reasonable that the approximation by Riemann sums can be made arbitrarily close to the area of Σ by taking fine enough partitions of R', we are led to the following definition.

DEFINITION 14.16   Let Σ be a smooth surface parametrized by

$$\mathbf{r}\,(u, v) = x(u, v)\mathbf{i} + y(u, v)\mathbf{j} + z(u, v)\mathbf{k}$$

on a vertically or horizontally simple region R in the uv plane. Then the **surface area** S of Σ is defined by

$$S = \iint_R \|\,\mathbf{r}_u\,(u, v) \times \mathbf{r}_v\,(u, v)\,\|\, dA \tag{14}$$

The **surface area** of a piecewise smooth surface is the sum of the surface areas of the component smooth surfaces.

It can be proved that the surface area of a smooth surface Σ is independent of which parametrization $\mathbf{r}$ is used in (14).

The formula in (14) is compact. By evaluating the cross product we can derive an alternate formula. Specifically, notice that

$$\mathbf{r}_u\,(u, v) \times \mathbf{r}_v\,(u, v) = \begin{vmatrix} \mathbf{i} & \mathbf{j} & \mathbf{k} \\ \dfrac{\partial x}{\partial u} & \dfrac{\partial y}{\partial u} & \dfrac{\partial z}{\partial u} \\ \dfrac{\partial x}{\partial v} & \dfrac{\partial y}{\partial v} & \dfrac{\partial z}{\partial v} \end{vmatrix}$$

$$= \left(\frac{\partial y}{\partial u}\frac{\partial z}{\partial v} - \frac{\partial z}{\partial u}\frac{\partial y}{\partial v}\right)\mathbf{i} + \left(\frac{\partial z}{\partial u}\frac{\partial x}{\partial v} - \frac{\partial x}{\partial u}\frac{\partial z}{\partial v}\right)\mathbf{j} + \left(\frac{\partial x}{\partial u}\frac{\partial y}{\partial v} - \frac{\partial y}{\partial u}\frac{\partial x}{\partial v}\right)\mathbf{k}$$

Using the notation for the Jacobian from Section 14.8, we can rewrite the equation as

$$\mathbf{r}_u(u, v) \times \mathbf{r}_v(u, v) = \frac{\partial(y, z)}{\partial(u, v)}\mathbf{i} + \frac{\partial(z, x)}{\partial(u, v)}\mathbf{j} + \frac{\partial(x, y)}{\partial(u, v)}\mathbf{k} \qquad (15)$$

Thus we have two additional equivalent forms for the surface area:

$$S = \iint\limits_{R} \sqrt{\left(\frac{\partial y}{\partial u}\frac{\partial z}{\partial v} - \frac{\partial z}{\partial u}\frac{\partial y}{\partial v}\right)^2 + \left(\frac{\partial z}{\partial u}\frac{\partial x}{\partial v} - \frac{\partial x}{\partial u}\frac{\partial z}{\partial v}\right)^2 + \left(\frac{\partial x}{\partial u}\frac{\partial y}{\partial v} - \frac{\partial y}{\partial u}\frac{\partial x}{\partial v}\right)^2}\, dA \quad (16)$$

and

$$S = \iint\limits_{R} \sqrt{\left(\frac{\partial(y, z)}{\partial(u, v)}\right)^2 + \left(\frac{\partial(z, x)}{\partial(u, v)}\right)^2 + \left(\frac{\partial(x, y)}{\partial(u, v)}\right)^2}\, dA \qquad (17)$$

Equation (17) is reminiscent of (3) in Section 12.4 for the length of a vector-valued curve.

In Example 9 we will compute the surface area S of a sphere of radius a, which will corroborate the well-known formula $S = 4\pi\, r^2$ for the surface area of a sphere of radius r.

EXAMPLE 9   Use (14) and the parametrization in Example 1 to find the area S of a sphere of radius a.

Solution   Recall from Example 1 that the sphere is parametrized by

$$x = a\sin\phi\cos\theta, \quad y = a\sin\phi\sin\theta, \quad z = a\cos\phi \quad \text{for } 0 \le \phi \le \pi \text{ and } 0 \le \theta \le 2\pi$$

Therefore

$$\mathbf{r}_\phi(\phi, \theta) \times \mathbf{r}_\theta(\phi, \theta) = \begin{vmatrix} \mathbf{i} & \mathbf{j} & \mathbf{k} \\ \dfrac{\partial x}{\partial \phi} & \dfrac{\partial y}{\partial \phi} & \dfrac{\partial z}{\partial \phi} \\ \dfrac{\partial x}{\partial \theta} & \dfrac{\partial y}{\partial \theta} & \dfrac{\partial z}{\partial \theta} \end{vmatrix} = \begin{vmatrix} \mathbf{i} & \mathbf{j} & \mathbf{k} \\ a\cos\phi\cos\theta & a\cos\phi\sin\theta & -a\sin\phi \\ -a\sin\phi\sin\theta & a\sin\phi\cos\theta & 0 \end{vmatrix}$$

$$= a^2\sin^2\phi\cos\theta\,\mathbf{i} + a^2\sin^2\phi\sin\theta\,\mathbf{j} + a^2\cos\phi\sin\phi\,\mathbf{k}$$

Consequently

$$||\mathbf{r}_\phi(\phi, \theta) \times \mathbf{r}_\theta(\phi, \theta)|| = \sqrt{(a^4\sin^4\phi(\cos^2\theta + \sin^2\theta) + a^4\cos^2\phi\sin^2\phi}$$

$$= \sqrt{a^4\sin^2\phi\,(\sin^2\phi + \cos^2\phi)} = a^2\sqrt{\sin^2\phi}$$

Notice that if $0 \leq \phi \leq \pi$ then $\sin \phi \geq 0$, so that $a^2 \sqrt{\sin^2 \phi} = a^2 \sin \phi$. Thus by (14) and the preceding equations, the surface area S is given by

$$S = \iint\limits_{R} a^2 \sin \phi \, dA = \int_0^{\pi} \int_0^{2\pi} a^2 \sin \phi \, d\theta \, d\phi$$

$$= \int_0^{\pi} 2a^2 \pi \sin \phi \, d\phi = -2a^2 \pi \cos \phi \, \Big|_0^{\pi} = 4 \pi a^2 \qquad \square$$

This result is a well-known result that would have required improper integrals to evaluate by methods of Section 6.3 or 14.3.

Our next example would be considerably more difficult to evaluate with rectangular coordinates.

EXAMPLE 10   Let Σ be the portion of the cylinder $x^2 + y^2 = 4$ that is bounded below by the xy plane and above by the slanted plane $z = y + 1$ (see Figure 14.89). Find the area S of Σ.

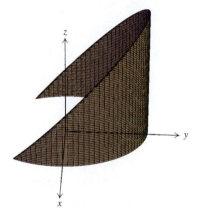

FIGURE 14.89   A Mathematica plot of part of a cylinder.

Solution   Since the cylinder has radius 2, it can be parametrized in the standard way by

$$\mathbf{r}(z, \theta) = 2 \cos \theta \mathbf{i} + 2 \sin \theta \mathbf{j} + z\mathbf{k} \quad \text{for} \quad 0 \leq \theta \leq 2\pi \quad \text{and all real} \quad z$$

However, the cylinder can equally well be parametrized with any interval of θ of length 2π. For the present example it will be convenient to use the interval $-\pi/2 \leq \theta \leq 3\pi/2$. The portion Σ of the cylinder whose area we want to find lies between the xy plane $z = 0$ and the slanted plane $z = y + 1$. If a point $P = (x, y, z)$ lies on Σ (which is part of the cylinder), then $0 \leq z \leq y + 1 = 2 \sin \theta + 1$, so that $0 \leq 2 \sin \theta + 1$, or equivalently, $\sin \theta \geq -1/2$. It follows that $-\pi/6 \leq \theta \leq 7\pi/6$. Therefore Σ is parametrized by

$$\mathbf{r}(z, \theta) = 2 \cos \theta \mathbf{i} + 2 \sin \theta \mathbf{j} + z\mathbf{k}, \quad \text{for} \quad 0 \leq z \leq 2 \sin \theta + 1 \quad \text{and} \quad -\pi/6 \leq \theta \leq 7\pi/6$$

A calculation shows that

$$\mathbf{r}_z \times \mathbf{r}_\theta = \begin{vmatrix} \mathbf{i} & \mathbf{j} & \mathbf{k} \\ 0 & 0 & 1 \\ -2\sin\theta & 2\cos\theta & 0 \end{vmatrix} = -2\cos\theta\,\mathbf{i} - 2\sin\theta\,\mathbf{j}$$

Thus

$$\|\mathbf{r}_z \times \mathbf{r}_\theta\| = \sqrt{(-2\cos\theta)^2 + (-2\sin\theta)^2} = 2$$

Therefore by (14),

$$S = \int_{-\pi/6}^{7\pi/6} \int_0^{2\sin\theta+1} \|\mathbf{r}_z \times \mathbf{r}_\theta\| \, dz \, d\theta = \int_{-\pi/6}^{7\pi/6} \int_0^{2\sin\theta+1} 2 \, dz \, d\theta$$

$$= \int_{-\pi/6}^{7\pi/6} (4\sin\theta + 2) \, d\theta = (-4\cos\theta + 2\theta)\Big|_{-\pi/6}^{7\pi/6} = 4\sqrt{3} + 8\pi/3 \qquad \square$$

Now we return to the helicoid, or ramp, and determine its surface area.

EXAMPLE 11   Consider the helicoid Σ parametrized by

$$\mathbf{r} = u\cos v\,\mathbf{i} + u\sin v\,\mathbf{j} + v\,\mathbf{k} \quad \text{for} \quad 1 \le u \le \sqrt{3} \quad \text{and} \quad 0 \le v \le 4\pi$$

Find the surface area S of Σ.

Solution   First we find that

$$\mathbf{r}_u \times \mathbf{r}_v = \begin{vmatrix} \mathbf{i} & \mathbf{j} & \mathbf{k} \\ \cos v & \sin v & 0 \\ -u\sin v & u\cos v & 1 \end{vmatrix} = \sin v\,\mathbf{i} - \cos v\,\mathbf{j} + u\,\mathbf{k}$$

Thus

$$\|\mathbf{r}_u \times \mathbf{r}_v\| = \sqrt{\sin^2 v + (-\cos v)^2 + u^2} = \sqrt{1 + u^2}$$

Therefore by (14),

$$S = \int_1^{\sqrt{3}} \int_0^{4\pi} \|\mathbf{r}_u \times \mathbf{r}_v\| \, dv \, du = \int_1^{\sqrt{3}} \int_0^{4\pi} \sqrt{1 + u^2} \, dv \, du$$

$$= \int_1^{\sqrt{3}} 4\pi \sqrt{1 + u^2} \, du \overset{u=\tan\theta}{=} \int_{\pi/4}^{\pi/3} 4\pi\sec^3\theta \, d\theta$$

$$= 4\pi\left(\frac{1}{2}\sec\theta\tan\theta + \frac{1}{2}\ln|\sec\theta + \tan\theta|\right)\Bigg|_{\pi4}^{\pi/3}$$

$$= 4\pi\left(\sqrt{3} - \frac{1}{2}\sqrt{2} + \frac{1}{2}\ln\frac{2 + \sqrt{3}}{\sqrt{2} + 1}\right)$$

$$\approx 15.61668343 \qquad \square$$

Suppose that Σ is the graph of a function f with continuous partial derivatives f_x and f_y on its domain R, which is a vertically or horizontally simple region in the xy plane. Then (4) tells us that Σ is parametrized by

$$\mathbf{r} = x\mathbf{i} + y\mathbf{j} + f(x, y)\mathbf{k} \quad \text{for} \quad (x, y) \quad \text{in} \quad R$$

This means that

$$\mathbf{r}_x\,(x,\,y) \times \mathbf{r}_y\,(x,\,y) = \begin{vmatrix} \mathbf{i} & \mathbf{j} & \mathbf{k} \\ 1 & 0 & f_x\,(x,\,y) \\ 0 & 1 & f_y\,(x,\,y) \end{vmatrix} = -f_x\,(x,\,y)\mathbf{i} - f_y\,(x,\,y)\mathbf{j} + \mathbf{k} \qquad (18)$$

so that

$$\|\mathbf{r}_x\,(x,\,y) \times \mathbf{r}_y(x,\,y)\| = \sqrt{[f_x(x,\,y)]^2 + [f_y(x,\,y)]^2 + 1} \qquad (19)$$

Using (19) together with (14), we deduce that

$$S = \iint\limits_{R} \sqrt{[f_x(x,\,y)]^2 + [f_y(x,\,y)]^2 + 1}\; dA$$

This formula is the same as (1) of Section 14.3. It is also possible to prove (see Exercise 26) that if Σ is generated by revolving the graph of a function f of a single variable about the x axis, then the formula in (14) applied to the natural parametrization given in (6) of the present section reduces to formula (4) of Section 6.3. Thus the definition of surface area in Definition 14.16 agrees with, and extends, all of our earlier definitions of surface area.

EXERCISES 14.9

1. Show that $\mathbf{r}(u,\,v) = (u + v)\mathbf{i} + (u - v)\mathbf{j} + 4uv\,\mathbf{k}$ is a parametrization of the hyperbolic paraboloid $z = x^2 - y^2$.

2. Show that the hyperboloid of one sheet

$$\frac{x^2}{a^2} + \frac{y^2}{b^2} - \frac{z^2}{c^2} = 1$$

is parametrized by

$x = a\cos u\cosh v,\quad y = b\sin u\cosh v,\quad z = c\sinh v$

for $0 \le u \le 2\pi$ and arbitrary v

3. Show that the hyperboloid of two sheets

$$\frac{z^2}{c^2} - \frac{x^2}{a^2} - \frac{y^2}{b^2} = 1$$

is parametrized by

$x = a\cos u\sinh v,\quad y = b\sin u\sinh v,\quad z = \pm\,c\cosh v$

for $0 \le u \le 2\pi$ and arbitrary v

(The upper sheet occurs with the $+ c$ and the lower sheet occurs with the $-c$.)

4. Show that $\mathbf{r}(u,\,v) = \cos u\cos v\,\mathbf{i} + \cos u\sin v\,\mathbf{j} + (1 + \sin u)\mathbf{k}$ parametrizes a sphere Σ. Find the center and radius of Σ.

In Exercises 5–10 find a parametrization of the surface Σ.

5. Σ is the part of the cylinder $x^2 + y^2 = 1$ that lies between the planes $z = -1$ and $z = 1$.

6. Σ is the plane $2x + 3y + 2z = 5$.

7. Σ is the portion of the plane $x - y + z = 3$ that lies inside the cylinder $x^2 + z^2 = 4$.

8. Σ is the elliptic cylinder $\dfrac{y^2}{4} + \dfrac{z^2}{9} = 16$.

9. Σ is the cone $y^2 + z^2 = x^2$.

10. Σ is the part of the paraboloid $x = y^2 + z^2$ for which $x \le 4$.

In Exercises 11–15 find a parametrization for the surface of revolution Σ.

11. Σ is obtained by revolving about the x axis the graph of $y = \cos x$ for $-\pi/2 \le x \le \pi/2$.

12. Σ is obtained by revolving about the y axis the graph of $z = \sqrt{y}\ \ $ for $1 \le y \le 4$.

13. Σ is obtained by revolving about the z axis the graph of $x = e^z$ for $0 \le z \le 1$.

14. Σ is the cone obtained by rotating the line $y = x$ around the x axis, for $x \ge 0$.

15. Σ is Gabriel's horn, obtained by revolving about the x axis the graph of $y = 1/x$ for $x \ge 1$.

In Exercises 16–21 find an equation of the plane tangent to the parametrized surface Σ at the point corresponding to $(u_0,\,v_0)$.

16. Σ is parametrized by
$\mathbf{r}(u,\,v) = u^2\,\mathbf{i} + uv\,\mathbf{j} + v^2\,\mathbf{k};\ (u_0,\,v_0) = (1,\,2)$.

17. Σ is the ellipsoid parametrized by
$\mathbf{r}(u,\,v) = 2\sin u\cos v\,\mathbf{i} + 3\sin u\sin v\,\mathbf{j} + 4\cos u\,\mathbf{k};$
$(u_0,\,v_0) = (\pi/4,\,-\,\pi/4)$.

18. Σ is the elliptic cone parametrized by

$\mathbf{r}(u, v) = 4u \cos v\,\mathbf{i} + 6u \sin v\,\mathbf{j} + u\,\mathbf{k}; \quad (u_0, v_0) = (1, \pi/6)$.

19. Σ is the elliptic paraboloid parametrized by

$x = u^2, \; y = 3u \cos v, \; z = 5\,u \sin v; \quad (u_0, v_0) = (2, \pi)$.

20. Σ is the cylinder parametrized by

$$\mathbf{r}(u, v) = \cos u\,\mathbf{i} + \frac{1}{\sqrt{2}}(v + \sin u)\mathbf{j} + \frac{1}{\sqrt{2}}(v - \sin u)\mathbf{k};$$

$(u_0, v_0) = (\pi/2, 1)$.

21. Σ is the torus parametrized by $\mathbf{r}(u, v) =$
$(3 + 2 \cos u) \sin v\,\mathbf{i} + (3 + 2 \cos u) \cos v\,\mathbf{j} + 2 \sin u\,\mathbf{k}$;
$(u_0, v_0) = (-\pi/2, 0)$.

In Exercises 22–24 find the area S of the surface Σ.

22. Σ is the portion of the sphere $\mathbf{r} = 2 \sin \phi \cos \theta\,\mathbf{i} +$
$2 \sin \phi \sin \theta\,\mathbf{j} + 2 \cos \phi\,\mathbf{k}$ that lies above the plane $z = 1$.
(*Hint:* What are the ranges of ϕ and θ?)

23. Σ is the torus obtained by rotating the circle $(x - b)^2 + z^2 = r^2$ about the z axis, where $0 < r < b$. (*Hint:* See Example 7.)

24. Σ is the portion of the hyperboloid of one sheet parametrized by

$x = \cos u \cosh v, \quad y = \sin u \cosh v, \quad z = c \sinh v$
for $\;0 \le u \le 2\pi \;$ and $0 \le v \le 1$

25. Let Σ be a more general helicoid than that described in (5), parametrized by

$$\mathbf{r}(u, v) = u \cos \omega v\,\mathbf{i} + u \sin \omega v\,\mathbf{j} + cv\,\mathbf{k}$$

with $0 \le u \le r$ and $0 \le v \le \alpha$, where the width is r units, v represents the angular measure, ω represents the rate of turning, and α is the maximum angular measure. Find the area S of the part of the helicoid represented by one complete revolution.

26. Let f be a function with continuous derivatives on $[a, b]$, and let Σ be the surface obtained by revolving the graph of f about the x axis. Using the parametrization in (6), show that the area S of Σ is given by

$$S = \int_a^b 2\pi f(x) \sqrt{1 + (f'(x))^2}\, dx$$

27. Suppose a surface Σ is given in cylindrical coordinates by $z = f(r, \theta)$ for (r, θ) in a region R in the $r\theta$ plane.

a. Show that the area S of Σ is given by

$$S = \iint_R \sqrt{r^2[f_r(r, \theta)]^2 + [f_\theta(r, \theta)]^2 + r^2}\, dA$$

b. Use the formula in (a) to find the area S of the part of the paraboloid $z = 5 - r^2$ inside the cylinder $r = 3$.

28. Show that the torus as parametrized in Example 7 (with $b = 1 = r$) is the graph of each of the following equations.

a. $(\sqrt{x^2 + y^2} - 1)^2 + z^2 = 1$

b. $\rho = 2 \sin \phi$ (where ρ and ϕ represent spherical coordinates)

29. Consider the torus parametrized by

$\mathbf{r}(\phi, \theta) = (1 + \cos \theta) \cos \phi\,\mathbf{i} + (1 + \cos \theta) \sin \phi\,\mathbf{j} + \sin \theta\,\mathbf{k}$
for $0 \le \phi \le 2\pi$ and $0 \le \theta \le 2\pi$ as in Example 7 (with $b = 1 = r$).

a. For all t, let $\mathbf{s}(t) = \mathbf{r}(2t, 3t)$ parametrize the curve C on the torus. Show that $\mathbf{s}$ is periodic, and find the period.

b. Use a computer to draw the torus and the curve C on it.

c. Suppose that b is an irrational number, and let $\mathbf{w}(t) = \mathbf{r}(t, bt)$ for all t. Prove that $\mathbf{w}$ traces out a curve C_b on the torus, but that $\mathbf{w}$ is *not* periodic, and that C_b does not intersect itself.

Applications

30. a. Show that the parametrization

$$x = \frac{2r^2 u}{u^2 + v^2 + r^2}, \quad y = \frac{2r^2 v}{u^2 + v^2 + r^2},$$
$$z = \frac{r(u^2 + v^2 - r^2)}{u^2 + v^2 + r^2}$$

where (u, v) ranges over the entire uv plane, parametrizes the entire sphere $x^2 + y^2 + z^2 = r^2$ except for the "north pole." This parametrization arises from the so-called **stereographic projection** (see Figure 14.90).

b. Which points on the sphere correspond to points on the circle $u^2 + v^2 = r^2$ in the uv plane, and which correspond to points inside the circle, and which correspond to points outside the circle?

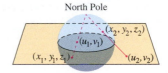

North Pole

FIGURE 14.90   Figure for Exercise 30.

31. a. In Example 1 we gave one parametrization of a sphere Σ of radius a centered at the origin. Now show that Σ can alternatively be parametrized by

$\mathbf{r}(u, v) = a \cos v \cos u\,\mathbf{i} + a \cos v \sin u\,\mathbf{j} + a \sin v\,\mathbf{k}$
for $0 \le u \le 2\pi \;$ and $- \pi/2 \le v \le \pi/2$

b. When applied to the earth (which is assumed to be spherical), the parameter u in (a) refers to longitude, and v to latitude. To see how we can make this identification, first position the earth on the xyz coordinate system with the north pole and south pole on the z axis, the equator on the xy plane, and Greenwich corresponding to a point $(x, 0, z)$, where

$x > 0$ and $z > 0$. Taking Greenwich to be at $0° = 360°$ longitude, with longitude positive and decreasing as one travels east from Greenwich, we find that u radians corresponds to a longitude of $(360 - u \frac{180}{\pi})$ degrees on earth. Similarly, taking the equator to be $0°$ latitude, with latitude positive north of the equator and negative south of the equator, we find that v radians corresponds to the latitude of $(90 - v \frac{180}{\pi})$ degrees on earth.

 i. Sketch the earth with axes as described above, and note the equator and Greenwich on the sketch.

 ii. Washington, D.C., has a latitude of approximately $39°$ north (of the equator) and a longitude of approximately $77°$ west of Greenwich. What values of u and v yield the position of Washington, D.C.?

c. What percentage of the area of the earth lies between the latitudes $-\pi/4$ and $\pi/4$ (that is, between $45°$ in the southern hemisphere and $45°$ in the northern hemisphere)?

32. a. Consider a sphere of radius a. Let Σ be a "spherical rectangle," with boundary composed of four curves: the latitudes $\phi = \phi_1$ and $\phi = \phi_2$ and the longitudes $\theta = \theta_1$ and $\theta = \theta_2$. Find a formula for the area S of Σ.

b. Wyoming is nearly a rectangle on our earth (that is, a spherical rectangle). Its latitude ranges between approximately $104.1°$ and $110.9°$ north, and its longitude ranges between approximately $41°$ and $45°$ west of Greenwich. Using 3960 miles for the radius of the earth, approximate the area S of Wyoming. (Its real area is approximately 97,818 square miles.)

c. Suppose that a "rectangular" region Σ has its bottom boundary on the equator, and has the same change in longitude and the same change in latitude as Wyoming has. Find the area S of Σ, and compare it with the area of Wyoming.

Projects

1. A smooth surface Σ with a smooth boundary B is called a **minimal surface** if Σ has the minimum surface area of all smooth surfaces with boundary B. It is known that soap bubbles that stretch between a closed boundary curve form minimal surfaces. In this project we will look at three minimal surfaces.

If Σ is defined by $z = f(x, y)$, where f has continuous second partials, then the criterion for Σ to be a minimal surface is that the following formula be satisfied:

$$(1 + z_y^2) z_{xx} + (1 + z_x^2) z_{yy} = 2z_x z_y z_{xy}$$

a. Let R be a region in the xy plane with smooth boundary. Show that R is a minimal surface for the boundary of R.

b. Consider the helicoid described by formula (5), parametrized by

$$\mathbf{r}(u, v) = u \cos v \, \mathbf{i} + u \sin v \, \mathbf{j} + v\mathbf{k}$$

for $1 \leq u \leq 2$ and $0 \leq v \leq 4\pi$. Show that in rectangular coordinates the helicoid satisfies the equation $y = x \tan z$, or equivalently, $z = \tan^{-1} (y/x)$. Then show that the helicoid is a minimal surface.

c. **Scherk's surface** is defined by

$$\mathbf{r}(u, v) = u \, \mathbf{i} + v \, \mathbf{j} + \ln \frac{\cos u}{\cos v} \, \mathbf{k}, \quad \text{or equivalently,}$$

$$e^z \cos y = \cos x.$$

 i. Show that the two formulas for Scherk's surface are equivalent.

 ii. Show that Scherk's surface is a minimal surface.

2. Let $\mathbf{r}(u, v) = x(u, v) \, \mathbf{i} + y(u, v) \, \mathbf{j} + z(u, v) \, \mathbf{k}$ parametrize a smooth surface Σ. The curves on Σ that result from keeping either u constant or v constant are called **grid curves** for Σ (with respect to the given parametrization). The grid curves are reminiscent of level curves discussed in Section 13.1.

a. Consider the sphere of radius a, parametrized by

$$\mathbf{r}(\phi, \theta) = a \sin \phi \cos \theta \, \mathbf{i} + a \sin \phi \sin \theta \, \mathbf{j} + a \cos \phi \, \mathbf{k}$$

for $0 \leq \phi \leq \pi$ and $0 \leq \theta \leq 2\pi$, as in Examples 1 and 9. Show that $\mathbf{r}_\phi(\phi, \theta)$ and $\mathbf{r}_\theta(\phi, \theta)$ are perpendicular to one another, for any pair (ϕ, θ) in the domain of $\mathbf{r}$. An immediate consequence is that the grid curves corrsponding to fixed ϕ and those for fixed θ are perpendicular to one another. We say that the grid curves of Σ form an **orthogonal family**. The word "orthogonal" is derived from the Greek expression meaning right-angled.

b. Show that the grid curves of the cone parametrized as in Example 8 form an orthogonal family.

c. Show that the grid curves of the torus parametrized as in Example 7 form an orthogonal family.

d. Let Σ have the "Cartesian" parametrization in (4):

$$\mathbf{r} = x \, \mathbf{i} + y \, \mathbf{j} + f(x, y) \, \mathbf{k}$$

Under what very special conditions would this parametrization of Σ have the property that the grid curves form an orthogonal family? Find an example of a surface that is not planar whose Cartesian parametrization yields an orthogonal family of level curves.

REVIEW

Key Terms and Expressions

Double integral

Volume; area

Vertically simple region; horizontally simple region; simple region

Iterated integral

Surface area

Solid region between the graphs of F_1 and F_2 on R

Triple integral

Mass; mass density

Charge; charge density

Cylindrical coordinates

Spherical coordinates

Moment about an axis; center of gravity of a plane region

Moment about a plane; center of gravity of a solid region

Moment of inertia

Jacobian

Parametrized surface

Surface of revolution

Smooth surface

Piecewise smooth surface

Tangent plane

Area of a parametrized surface

Key Formulas

$$A = \iint\limits_R 1 \, dA$$

$$V = \iiint\limits_D 1 \, dV$$

$$S = \iint\limits_R \sqrt{[f_x(x, y)]^2 + [f_y(x, y)]^2 + 1} \, dA$$

$$\iint\limits_R f(x, y) \, dA = \int_a^b \int_{g_1(x)}^{g_2(x)} f(x, y) \, dy \, dx$$

$$\iint\limits_R f(x, y) \, dA = \int_c^d \int_{h_1(y)}^{h_2(y)} f(x, y) \, dx \, dy$$

$$\iint\limits_R f(x, y) \, dA = \int_\alpha^\beta \int_{h_1(\theta)}^{h_2(\theta)} f(r \cos \theta, r \sin \theta) r \, dr \, d\theta$$

$$\iiint\limits_D f(x, y, z) \, dV = \int_a^b \int_{g_1(x)}^{g_2(x)} \int_{F_1(x,y)}^{F_2(x,y)} f(x, y, z) \, dz \, dy \, dx$$

$$\iiint\limits_D f(x, y, z) \, dV = \int_c^d \int_{h_1(y)}^{h_2(y)} \int_{F_1(x,y)}^{F_2(x,y)} f(x, y, z) \, dz \, dx \, dy$$

$$\iiint\limits_D f(x, y, z) \, dV = \int_\alpha^\beta \int_{h_1(\theta)}^{h_2(\theta)} \int_{F_1(r\cos\theta, r\sin\theta)}^{F_2(r\cos\theta, r\sin\theta)} f(r \cos \theta, r \sin \theta, z) r \, dz \, dr \, d\theta$$

$$\iiint\limits_D f(x, y, z) \, dV = \int_\alpha^\beta \int_{h_1(\theta)}^{h_2(\theta)} \int_{F_1(\phi,\theta)}^{F_2(\phi,\theta)} f(\rho \sin \phi \cos \theta, \rho \sin \phi \sin \theta, \rho \cos \phi) \rho^2 \sin \phi \, d\rho \, d\phi \, d\theta$$

$$\iint\limits_R f(x, y) \, dA = \iint\limits_S f(g_1(u, v), g_2(u, v)) \left| \frac{\partial(x, y)}{\partial(u, v)} \right| \, dA$$

$$n = \frac{\mathbf{r}_u(u, v) \times \mathbf{r}_v(u, v)}{||\mathbf{r}_u(u, v) \times \mathbf{r}_v(u, v)||}$$

$$S = \iint\limits_R ||\mathbf{r}_u(u, v) \times \mathbf{r}_v(u, v)|| \, dA$$

$$S = \iint\limits_R \sqrt{\left(\frac{\partial(y, z)}{\partial(u, v)} \right)^2 + \left(\frac{\partial(z, x)}{\partial(u, v)} \right)^2 + \left(\frac{\partial(x, y)}{\partial(u, v)} \right)^2} \, dA$$

$$S = \iint\limits_R \sqrt{(f_x(x, y))^2 + (f_y(x, y))^2 + 1} \, dA$$

Review Exercises

In Exercises 1–4 evaluate the interated integrals.

1. $\displaystyle\int_0^1 \int_x^{3x} ye^{(x^3)}\,dy\,dx$ **2.** $\displaystyle\int_0^{\sqrt{\pi}} \int_0^x \sin x^2\,dy\,dx$

3. $\displaystyle\int_{-1}^1 \int_0^2 \int_{2x}^{5x} e^{xy}\,dz\,dy\,dx$

4. $\displaystyle\int_1^e \int_0^x \int_0^{1/(x+y)} \ln(x+y)\,dz\,dy\,dx$

In Exercises 5–7 reverse the order of integration, and then evaluate.

5. $\displaystyle\int_0^1 \int_{\sqrt{x}}^1 e^{(y^3)}\,dy\,dx$ **6.** $\displaystyle\int_1^9 \int_{\sqrt{y}}^3 \frac{e^{(x^2-2x)}}{x+1}\,dx\,dy$

7. $\displaystyle\int_1^{\sqrt{3}} \int_x^{\sqrt{3}} \frac{x}{(x^2+y^2)^{3/2}}\,dy\,dx$

In Exercises 8–11 find the area A of the region in the xy plane by means of double integrals.

8. The region bounded by the graphs of $y^2 = x$ and $y = x^3$

9. The region bounded by the graphs of $\sqrt{x} + \sqrt{y} = \sqrt{a}$ and $x + y = a$

10. The region inside the cardioid $r = 1 + \sin\theta$ and outside the cardioid $r = 1 + \cos\theta$

11. The region outside the limaçon $r = 3 - \sin\theta$ and inside the circle $r = 5\sin\theta$

In Exercises 12–17 evaluate the multiple integral.

12. $\iint_R \sin(x+y)\,dA$, where R is the region in the xy plane bounded by the lines $y = x$, $y = 0$, and $x = \pi/2$

13. $\iint_R (3x-5)\,dA$, where R is the region bounded by the lines $y = 5 + x$, $y = -x + 7$, $x = 0$, and $x = 1$

14. $\iint_R (4+x^2)\,dA$, where R is the region between the parabolas $y = 1 + x^2$ and $y = 3 - x^2$

15. $\iiint_D (z^2+1)\,dV$, where D is the solid region bounded below by the upper nappe of the cone $z^2 = 3x^2 + 3y^2$ and above by the sphere $x^2 + y^2 + z^2 = 4$

16. $\iiint_D xy\,dV$, where D is the solid region in the first octant bounded by the coordinate planes and the cylinders $x^2 + y^2 = 1$ and $x^2 + z^2 = 1$

17. $\iiint_D xyz\,dV$, where D is the solid region bounded below by the hemisphere $z = -\sqrt{9 - x^2 - y^2}$ and above by the xy plane

In Exercises 18–20 find the surface area S of the surface.

18. The portion of the parabolic sheet $z = \frac{1}{2}y^2$ cut out by the planes $y = x$, $y = 2\sqrt{2}$, and $x = 0$

19. The portion of the surface $z = xy$ that is inside the cylinder $x^2 + y^2 = 1$

20. The portion of the surface $z = \frac{2}{3}(x^{3/2} + y^{3/2})$ cut out by the planes $x = 0$, $y = 1$, and $x = 7y$

In Exercises 21–30 find the volume V of the region.

21. The solid region bounded by the paraboloid $z = x^2 + y^2$ and the upper nappe of the cone $x^2 + y^2 = z^2$

22. The solid region bounded by the surface $z = e^x$ and the planes $x = y$, $y = 0$, $x = 1$, and $z = 0$

23. The solid region bounded on the sides by the cylinder $r = 4\sin\theta$, above by the cone $r = z$, and below by the xy plane

24. The solid region in the first octant bounded by the coordinate planes, the cylinder $x^2 + y^2 = 5y$, and the sphere $x^2 + y^2 + z^2 = 25$

25. The solid region bounded below by the paraboloid $z = 4x^2 + y^2$ and above by the parabolic sheet $z = 16 - 3y^2$

26. The solid region bounded by the paraboloids $\frac{1}{2}z = -9 + x^2 + y^2$ and $x^2 + y^2 + z = 9$

27. The solid region in the first octant bounded by the planes $z = x + 2y$ and $6 = x + 3y$, and by the coordinate planes

28. The solid region in the first octant bounded above by the plane $z = 3$, below by the upper nappe of the cone $x^2 + y^2 = 3z^2$, and on the sides by the planes $y = 0$ and $x = \sqrt{3}y$

29. The solid region bounded above by the sphere $x^2 + y^2 + z^2 = 49$ and below by the paraboloid $x^2 + y^2 = 3z + 21$

30. The solid region bounded on the sides by the cylinder $x^2 + y^2 = 6x$, above by the paraboloid $x^2 + y^2 = 4z$, and below by the plane $z = -2$

31. Find the center of mass of an object occupying the region bounded by the cone $z = \sqrt{x^2 + y^2}$ and the plane $z = 3$, if the mass density at any point in the object is equal to the distance from the point to the xy plane.

32. Find the total mass and the center of mass of a body that is bounded on the sides by the cylinder $x^2 + y^2 = 4$, above by the cone $z = \sqrt{x^2 + y^2}$, and below by the xy plane, if the mass density at (x, y, z) is given by $\delta(x, y, z) = z + 3$.

33. Find the centroid of the solid region bounded below by the paraboloid $z = x^2 + y^2$ and above by the upper nappe of the cone $z^2 = x^2 + y^2$.

34. Find the mass of an object occupying a ball of radius a if the mass density at any point is equal to its distance from the outer boundary of the ball.

In Exercises 35–36 evaluate the integral by using the given transformation.

35. $\iint_R x \, dA$, where R is the region bounded by $xy = 1$, $xy = 2$, $x(1 - y) = 1$, and $x(1 - y) = 2$; let $x = u + v$ and $y = v/(u + v)$

36. $\iint_R (x - y^2) \, dA$, where R is the region bounded by $x = y^2 - y$, $x = 2y + y^2$, and $y = 2$; let $x = 2u - v + (u + v)^2$ and $y = u + v$

In Exercises 37–38 evaluate the integral by using a suitable change of variables.

37. $\iint_R (2x - y^2) \, dA$, where R is the region bounded by the lines $x - y = 1$, $x - y = 3$, $x + y = 2$, and $x + y = 4$

38. $\displaystyle\iint_R \cos\left(\frac{y - x}{y + x}\right) dA$, where R is the region bounded by the lines $x + y = 2$, $x + y = 4$, $x = 0$, and $y = 0$

39. Find a parametrization involving polar coordinates in the xz plane of the part of the plane $x + y + z = 2$ that lies inside the cylinder $y^2 + z^2 = 4$.

40. Let C be a curve in the first quadrant of the xy plane parametrized by $x = f(u)$ and $y = g(u)$ for $a \le u \le b$. Let Σ be the surface obtained by revolving C about the x axis. Find a parametrization for Σ.

41. Let Σ be the surface parametrized by $x = u^2 + v^2$, $y = u^2 - v^2$, $z = u - v$.

a. Show that the point $(2, 0, -2)$ is on Σ.

b. Find an equation of the plane that is tangent to Σ at the point $(2, 0, -2)$.

42. Let Σ be parametrized by
$$\mathbf{r}(u, v) = au \cos v \, \mathbf{i} + au \sin v \, \mathbf{j} + u^2 \, \mathbf{k}$$
for $0 \le u \le 2$ and $0 \le v \le 2\pi$.

a. Describe the surface Σ in rectangular coordinates.

b. Find the area S of Σ.

43. Consider the portion Σ of the hyperboloid of two sheets parametrized by
$$x = \cos u \sinh v, \quad y = \sin u \sinh v, \quad z = \cosh v$$
for $0 \le u \le 2\pi$ and $0 \le v \le 1$. Find the area S of Σ.

44. Consider the ellipsoid
$$\frac{x^2}{a^2} + \frac{y^2}{a^2} + \frac{z^2}{c^2} = 1$$

a. Find a parametrization for the ellipsoid in the style of the parametrization for a sphere in Example 1 of Section 14.9.

b. Use the parametrization you obtain in (a) to find a formula for the area Σ of the ellipsoid.

c. Approximate the value of the area S of the ellipsoid.

d. Suppose $a = 2$ and $c = 3$ in the formula for the ellipsoid. Find an equation for the plane tangent to the ellipsoid at the point $(1, -1, 3\sqrt{2}/2)$.

45. Find the area S of the part of the sphere $x^2 + y^2 + z^2 = 1$ that lies above the xy plane and inside the (vertical) cylinder $r = \sin \theta$ (where r and θ represent polar coordinates in the xy plane).

Application

46. One of the hills that is presently being filled at a landfill on Staten Island is roughly in the shape of an elliptical cone with an elliptical base whose major axis and minor axis are approximately 2600 feet long and 1300 feet long, respectively. When completed in a decade or two, the landfill is expected to be approximately 435 feet tall.

a. Find an equation for such an idealized cone.

b. Find the volume V of the cone given in (a).

Topics for Discussion

1. What is the difference between a vertically simple region and a horizontally simple region? Give an example of a vertically simple region that is not horizontally simple, and vice versa. How do the vertically and horizontally simple regions aid in calculating the values of double integrals?

2. Describe conditions on the integrand of a double integral in order that symmetry of the plane region R of integration can be used in order to simplify the calculation of the integral. How can symmetry of a solid region be used in calculation of triple integrals?

3. Explain the process of reversing the order of integration. Give examples of double integrals which can be more easily evaluated by reversing the order of integration.

4. List the three coordinate systems for three-dimensional space that are used in this chapter for triple integration. For what types of regions would each tend to be most successful?

5. What role does the Jacobian play in integration by change of variables? Describe the change of variables that converts from spherical coordinates to rectangular coordinates.

Cumulative Review, Chapters 1–13

In Exercises 1–2 find the limit.

1. $\displaystyle\lim_{h \to 0} \frac{2x^2h - 5xh^2 - 6h^3}{3xh + 2h^2}$

2. $\displaystyle\lim_{x \to \infty} \left(1 - \frac{2}{x^2}\right)^{4x^2}$

3. Let $f(x) = \sqrt{x + \sqrt{x^2 - 4}}$.

a. Find the domain of f.

b. Find $f'(x)$.

4. Let $f(x) = x^{(\sin x)/x}$. Find $f'(\pi)$.

5. Find the two points on the graph of the equation $x^2 - xy + y^2 = 4$ at which the slope of the tangent line is 1.

6. A pyramid whose base is a square 4 inches on a side and whose height is 6 inches is being filled with water at a constant rate. Determine the rate that will make the water level rise $\frac{1}{2}$ inch per minute when the water level is 2 inches above the base. (*Hint:* The volume V of a pyramid of base side a and height h is given by $V = \frac{1}{3}a^2h$.)

7. A rectangular toolshed 7 feet tall is to be made from cedar, and is to have a volume of 1512 cubic feet. If the front and back are one and a half times as expensive per square foot as the two sides, and if the top is twice as expensive per square foot as the two sides, determine the dimensions that will minimize the cost of the toolshed.

8. Let l denote the portion of a line tangent to the curve $y = 1/(2x)$ that lies in the first quadrant. Determine the shortest possible length for l.

9. Let

$$f(x) = \frac{x}{\ln x}$$

Sketch the graph of f, indicating all pertinent information.

In Exercises 10–12 evaluate the integral.

10. $\displaystyle\int (2x + 9x^3)\sqrt{1 + 4x^2 + 9x^4}\, dx$

11. $\displaystyle\int \tan x \sin^2 x \cos^5 x\, dx$

12. $\displaystyle\int \frac{2x^3 + 3}{x^3 - 2x^2 + x}\, dx$

13. Consider

$$\int_{1/2}^{1} \frac{1}{\sqrt{1 - x^2}}\, dx$$

Determine whether the region between the graph of the integrand and the x axis on $[\frac{1}{2}, 1]$ has finite area. If it does, calculate the area.

14. The base of a solid is a square of side length 1, and the cross sections perpendicular to a fixed diagonal of the square are semicircular. Find the volume V of the solid.

15. The curve C is parametrized by $x = \frac{2}{3}\sin^{3/2} t$, $y = \sin t$ for $0 \le t \le \pi/2$. Determine the length L of C.

16. Sketch the graph of $r = \cos 5\theta$.

17. Show that $\left\{ \dfrac{(k!)^3 27^k}{(3k)!} \right\}_{k=1}^{\infty}$ is an increasing sequence.

18. Determine whether

$$\left\{ 1 + 4 + \frac{4^2}{2!} + \frac{4^3}{3!} + \cdots + \frac{4^n}{n!} \right\}_{n=1}^{\infty}$$

converges or diverges.

19. Find the interval of convergence of $\displaystyle\sum_{n=1}^{\infty} \frac{(-1)^n}{n + 1} x^{2n}$.

20. Use Taylor series to approximate $\int_0^1 \sin(x^4)\, dx$ to within 10^{-5} of the exact value.

21. A person exerts a force of 100 pounds at a 45° angle from the vertical in raising a sack of concrete mix 6 feet vertically. Determine the work W done on the sack.

22. A projectile is fired from the back of a truck 4 feet above the ground. If the projectile has an initial speed of 1500 feet per second and the angle of elevation is 30°, find the position and velocity of the projectile at any time t until it hits the ground.

23. Let $\mathbf{r}(t) = t^3\mathbf{i} + 6t\mathbf{j} + 3t^2\mathbf{k}$.

a. Find the tangent vector $\mathbf{T}(t)$, and determine any value(s) of t for which $\mathbf{T}(t)$ is parallel to the y axis.

b. Find the normal vector $\mathbf{N}(t)$ for any value of t.

c. Find the unit vector perpendicular to both $\mathbf{T}(1)$ and $\mathbf{N}(1)$ and having a negative $\mathbf{j}$-component.

24. Let $g(x, y) = \begin{cases} \dfrac{\sin(x^2(y^2 + 1))}{x} & \text{for } x \neq 0 \\ 0 & \text{for } x = 0 \end{cases}$

a. Find $g_x(0, 0)$ and $g_y(0, 0)$.

b. Determine whether g is continuous at $(0, 0)$.

25. Consider the paraboloid $z = 1 - x^2 - y^2$.

a. Sketch the paraboloid.

b. Find symmetric equations for the line l that is normal to the paraboloid at $(-1, 1, -1)$.

26. Determine the dimensions of the parallelepiped with maximum volume that lies inside the first octant and inside the ellipsoid $x^2 + 4y^2 + 8z^2 = 24$, and three of whose sides lie along the coordinate planes. Assume the desired parallelepiped exists.

White water rafting. (*Superstock*)

15 CALCULUS OF VECTOR FIELDS

In this chapter we study calculus of a type of function called a vector field, which assigns vectors to points in space. The gravitational field of the earth is an example of a vector field.

The chapter opens with an introduction to vector fields. In Section 15.2 we define the line integral and employ it to give a formula for the work done by a force on an object as it travels through space. The following section is devoted to the Fundamental Theorem of Line Integrals, which in spirit is similar to the Fundamental Theorem of Calculus.

The surface integral is defined in Section 15.5 and is used to calculate the amount of fluid flowing through a surface, such as a membrane. The chapter culminates in three important higher-dimensional analogues of the Fundamental Theorem of Calculus: Green's Theorem, Stokes's Theorem, and the Divergence Theorem.

15.1 VECTOR FIELDS

The gravitational field of the earth associates with the point (x, y, z) in space the force that the earth would exert on a unit mass located at (x, y, z). Analogously, the electric field due to a given charge associates with the point (x, y, z) in space the electric force that the given charge would exert on a positive unit charge located at (x, y, z). In both cases we associate a vector with each point of a given region in space. Such associations are called vector fields.

DEFINITION 15.1   A **vector field F** consists of two parts: a collection D of points in space, called the **domain,** and a **rule,** which assigns to each point (x, y, z) in D one and only one vector $\mathbf{F}(x, y, z)$.

Of course, it is impossible to graph a vector field, because this would require six dimensions (three for the domain and three for the range). However, it is possible to represent a vector field **F** graphically by drawing the vector $\mathbf{F}(x, y, z)$ as an arrow emanating from (x, y, z). Although we cannot show $\mathbf{F}(x, y, z)$ for every point (x, y, z), we can usually obtain a good impression of the vector field by graphing $\mathbf{F}(x, y, z)$ for several choices of (x, y, z). For example, if $\mathbf{F}(x, y, z) = \mathbf{j}$, then **F** could be represented as in Figure 15.1(a), and if $\mathbf{F}(x, y, z) = z\mathbf{j}$, then the graph in Figure 15.1(b) would represent **F**. In the special case in which the domain and range of **F** are contained in the xy plane, we let (x, y) and $\mathbf{F}(x, y)$ denote the points in the domain and range, respectively, and we represent **F** by vectors in the xy plane. For example, if $\mathbf{F}(x, y) = x\mathbf{i} + y\mathbf{j}$, then **F** could be represented as in Figure 15.1(c).

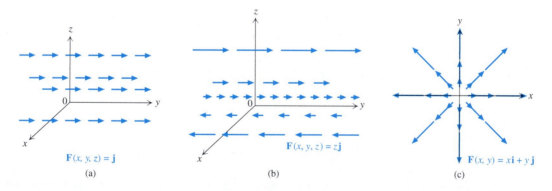

FIGURE 15.1   The graphical representation of three vector fields.

Next we obtain formulas for the gravitational and electric fields. According to Newton's Law of Gravitation, the gravitational force $\mathbf{F}(x, y, z)$ exerted by a given point mass m at the origin on a unit point mass located at a point (x, y, z) other than the origin is given by

$$\mathbf{F}(x, y, z) = \frac{Gm}{x^2 + y^2 + z^2}\, \mathbf{u}(x, y, z)$$

where G is the universal gravitational constant and $\mathbf{u}(x, y, z)$ is the unit vector emanating from (x, y, z) and directed toward the origin. The vector field $\mathbf{F}$ is called the **gravitational field** of the point mass. Because $\mathbf{u}(x, y, z)$ points from (x, y, z) toward the origin, it has the same direction as $-x\mathbf{i} - y\mathbf{j} - z\mathbf{k}$. Since $\mathbf{u}(x, y, z)$ is a unit vector, it can be written in the form

$$\mathbf{u}(x, y, z) = \frac{-1}{\sqrt{x^2 + y^2 + z^2}}\,(x\mathbf{i} + y\mathbf{j} + z\mathbf{k})$$

and consequently

$$\mathbf{F}(x, y, z) = \frac{-Gm}{(x^2 + y^2 + z^2)^{3/2}}\,(x\mathbf{i} + y\mathbf{i} + z\mathbf{k}) \qquad (1)$$

The gravitational field $\mathbf{F}$ has the properties that $\mathbf{F}(x, y, z)$ always points toward the origin and, moreover, that the magnitude of $\mathbf{F}(x, y, z)$ is the same for all points (x, y, z) located the same distance from the origin (Figure 15.2). A vector field that represents force and has these properties is called a **central force field.**

In physics a point (x, y, z) in space is often represented by the vector

$$\mathbf{r} = x\mathbf{i} + y\mathbf{j} + z\mathbf{k}$$

Thus in terms of the vector $\mathbf{r}$, the gravitational field of a single point mass at the origin can be written

$$\mathbf{F}(\mathbf{r}) = \frac{-Gm}{\|\mathbf{r}\|^3}\,\mathbf{r}$$

However, the gravitational field of a collection of masses can be more complicated. For example, consider a binary star system consisting of two nearby stars, A and B. In the vicinity of A the gravitational force from B is small in comparison with the gravitational force from A, so at such points the gravitational field of the system is nearly the same as the gravitational field from A. By contrast, at points approximately midway between A and B the gravitational forces from the two stars can nearly cancel each other, so that gravitational force is small. Figure 15.3 portrays the part of the gravitational field in a plane containing the centers of the stars.

FIGURE 15.2   The gravitational field of the earth.

The binary star system ARLac in the constellation Lacerta. (Max-Planck-Institut für Physik and Astrophysik/Science Photo Library/Photo Researchers)

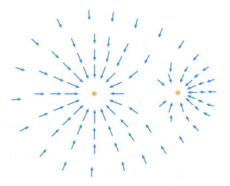

FIGURE 15.3   The gravitational field of a binary star system.

If a charge q is located at the origin, then according to Coulomb's Law the electric force $\mathbf{E}(x, y, z)$ exerted by the charge on a positive unit charge located at a point (x, y, z) other than the origin is given by the equation

$$\mathbf{E}(x, y, z) = \frac{q}{4\pi\varepsilon_0(x^2 + y^2 + z^2)}\,\mathbf{u}(x, y, z)$$

or

$$\mathbf{E}(x, y, z) = \frac{q}{4\pi\varepsilon_0(x^2 + y^2 + z^2)^{3/2}}\,(x\mathbf{i} + y\mathbf{j} + z\mathbf{k})$$

where $\mathbf{u}$ is the unit vector emanating from the origin directed toward (x, y, z) and ε_0 is a constant called the permittivity of empty space (see (4) of Section 11.2). We also assume that distance is measured in meters, charge in coulombs, and force in newtons. The vector field $\mathbf{E}$ is called the **electric field** of the point charge. Like the gravitational field, the electric field of a point charge is a central force field.

The **electric field** at any point (x, y, z) due to a finite collection of charges is defined to be the total force the charges would exert on a positive unit charge at (x, y, z). Since the total force exerted is the vector sum of the forces exerted by the individual charges, it follows that the electric field of the collection of charges is equal to the sum of the electric fields of the individual charges. This result is known as a **superposition principle.**

Vector fields can also describe the motion of a fluid. If at each point (x, y, z) in a given region we let $\mathbf{v}(x, y, z)$ denote the velocity of the fluid at (x, y, z), then $\mathbf{v}$ is a vector field, called the **velocity field** of the fluid. For example, Figure 15.4 might represent the velocity of water in a sink south of the equator.

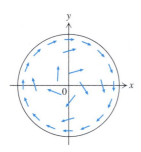

FIGURE 15.4 Velocity of water at a few points in a sink south of the equator.

Just as a vector can be expressed in terms of its three components, a vector field $\mathbf{F}$ can be expressed in terms of three **component functions** M, N, and P:

$$\mathbf{F}(x, y, z) = M(x, y, z)\mathbf{i} + N(x, y, z)\mathbf{j} + P(x, y, z)\mathbf{k}$$

or in condensed form,

$$\mathbf{F} = M\mathbf{i} + N\mathbf{j} + P\mathbf{k}$$

Using components, we can rewrite formula (1) for the gravitational field as

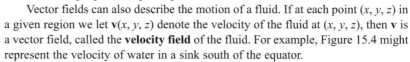

$$\mathbf{F}(x, y, z) = \overbrace{\frac{-Gmx}{(x^2 + y^2 + z^2)^{3/2}}}^{M(x,\,y,\,z)}\mathbf{i} + \overbrace{\frac{-Gmy}{(x^2 + y^2 + z^2)^{3/2}}}^{N(x,\,y,\,z)}\mathbf{j} + \overbrace{\frac{-Gmz}{(x^2 + y^2 + z^2)^{3/2}}}^{P(x,\,y,\,z)}\mathbf{k} \quad (2)$$

Now let $\mathbf{F} = M\mathbf{i} + N\mathbf{j} + P\mathbf{k}$ be a vector field. We say $\mathbf{F}$ is **continuous** at (x, y, z) if and only if M, N, and P are continuous at (x, y, z). The gravitational field $\mathbf{F}$ and the electric field $\mathbf{E}$ defined above are continuous at every point in their domains. In fact every vector field we will consider is continuous at every point in its domain.

The Gradient as a Vector Field

Suppose f is a differentiable function of three variables. Then the gradient of f, defined in Section 13.6, is actually a vector field, denoted grad f or ∇f and given by

$$\text{grad } f(x, y, z) = \nabla f(x, y, z) = \frac{\partial f}{\partial x}(x, y, z)\mathbf{i} + \frac{\partial f}{\partial y}(x, y, z)\mathbf{j} + \frac{\partial f}{\partial z}(x, y, z)\mathbf{k}$$

If a vector field $\mathbf{F}$ is equal to grad f for some differentiable function f of several variables, then $\mathbf{F}$ is called a **conservative vector field,** and f is a **potential function*** for $\mathbf{F}$. Many vector fields that arise in physics are conservative. For example, we can show that the gravitational field $\mathbf{F}$ of a point mass is conservative. Indeed, we know from Example 2 of Section 13.6 that if

$$f_1(x, y, z) = \frac{1}{(x^2 + y^2 + z^2)^{1/2}}$$

then

$$\text{grad } f_1(x, y, z) = \frac{-1}{(x^2 + y^2 + z^2)^{3/2}} (x\mathbf{i} + y\mathbf{j} + z\mathbf{k})$$

Consequently if

$$f(x, y, z) = \frac{Gm}{(x^2 + y^2 + z^2)^{1/2}}$$

then the formula in (1) for $\mathbf{F}$ implies that

$$\text{grad } f(x, y, z) = \frac{-Gm}{(x^2 + y^2 + z^2)^{3/2}} (x\mathbf{i} + y\mathbf{j} + z\mathbf{k}) = \mathbf{F}(x, y, z) \qquad (3)$$

Therefore f is a potential function for $\mathbf{F}$, and thus $\mathbf{F}$ is conservative. Because the electric field of a point charge differs from the gravitational field of a point mass by only a constant factor, the electric field is also conservative.

The Divergence of a Vector Field

There are two types of derivatives of a vector field, one that is a real-valued function and one that is a vector field. We begin with the real-valued derivative.

DEFINITION 15.2

Let $\mathbf{F} = M\mathbf{i} + N\mathbf{j} + P\mathbf{k}$ be a vector field such that $\partial M/\partial x$, $\partial N/\partial y$, and $\partial P/\partial z$ exist. Then the **divergence** of $\mathbf{F}$, denoted div $\mathbf{F}$ or $\nabla \cdot \mathbf{F}$, is the function defined by

$$\text{div } \mathbf{F}(x, y, z) = \nabla \cdot \mathbf{F}(x, y, z)$$

$$= \frac{\partial M}{\partial x}(x, y, z) + \frac{\partial N}{\partial y}(x, y, z) + \frac{\partial P}{\partial z}(x, y, z)$$

EXAMPLE 1   Let $\mathbf{F}$ be the gravitational field given by (2). Show that div $\mathbf{F} = 0$.

Solution   We find that

$$\frac{\partial M}{\partial x}(x, y, z) = \frac{-Gm(x^2 + y^2 + z^2)^{3/2} - (-Gmx)(\frac{3}{2})(2x)(x^2 + y^2 + z^2)^{1/2}}{(x^2 + y^2 + z^2)^3}$$

$$= \frac{Gm(2x^2 - y^2 - z^2)}{(x^2 + y^2 + z^2)^{5/2}}$$

---

* In physics the potential function for $\mathbf{F}$ is taken to be the function f such that $\mathbf{F} = -\text{grad } f$.

In a similar fashion we find that

$$\frac{\partial N}{\partial y}(x, y, z) = \frac{Gm(2y^2 - z^2 - x^2)}{(x^2 + y^2 + z^2)^{5/2}}$$

and

$$\frac{\partial P}{\partial z}(x, y, z) = \frac{Gm(2z^2 - x^2 - y^2)}{(x^2 + y^2 + z^2)^{5/2}}$$

Therefore

$$\text{div } \mathbf{F} = \frac{\partial M}{\partial x} + \frac{\partial N}{\partial y} + \frac{\partial P}{\partial z} = 0 \qquad \square$$

If div $\mathbf{F} = 0$, then $\mathbf{F}$ is said to be **divergence free** or **solenoidal.** Example 1 shows that the gravitational field is divergence free.

A revealing interpretation of div $\mathbf{F}$ arises from the study of fluid flowing through a region. Suppose $\mathbf{v}$ represents the velocity field of a fluid, such as air, flowing through a surface, such as a screen. Then in physical terms, div $\mathbf{v}(x, y, z)$ represents the rate (with respect to time) of mass flow per unit volume of the fluid from the point (x, y, z). (We will examine this interpretation of div $\mathbf{v}$ further in Section 15.8.) A point (x, y, z) is a **source** if div $\mathbf{v}(x, y, z) > 0$; this means that there is a positive mass flow *from* the point (x, y, z). By contrast, (x, y, z) is a **sink** if div $\mathbf{v}(x, y, z) < 0$; this means that there is a positive flow *to* the point (x, y, z). Finally, if div $\mathbf{v}(x, y, z) = 0$ for all (x, y, z) in a region, then there are neither sources nor sinks in the region. A fluid whose velocity field is divergence free is called **incompressible.**

EXAMPLE 2   Let $\mathbf{v}(x, y, z) = x^3yz^2\mathbf{i} + x^2y^2z^2\mathbf{j} + x^2yz^3\mathbf{k}$. Determine which points in space are sources and which are sinks.

Solution   A straightforward calculation shows that

$$\text{div } \mathbf{v}(x, y, z) = 3x^2yz^2 + 2x^2yz^2 + 3x^2yz^2 = 8x^2yz^2$$

Thus div $\mathbf{v}(x, y, z) = 0$ if (x, y, z) lies on any of the coordinate planes. Moreover, div $\mathbf{v}(x, y, z) > 0$ if $y > 0$ and x and z are not 0, whereas div $\mathbf{v}(x, y, z) < 0$ if $y < 0$ and x and z are not 0. Consequently the sources lie to the right of the xz plane and the sinks to the left. $\square$

The Curl of a Vector Field

The second type of derivative of a vector field is a vector field.

DEFINITION 15.3

Let $\mathbf{F} = M\mathbf{i} + N\mathbf{j} + P\mathbf{k}$ be a vector field such that the first partial derivatives of M, N, and P all exist. Then the **curl** of $\mathbf{F}$, which is denoted curl $\mathbf{F}$ or $\nabla \times \mathbf{F}$, is defined by

$$\text{curl } \mathbf{F}(x, y, z) = \nabla \times \mathbf{F}(x, y, z)$$

$$= \left(\frac{\partial P}{\partial y} - \frac{\partial N}{\partial z}\right)\mathbf{i} + \left(\frac{\partial M}{\partial z} - \frac{\partial P}{\partial x}\right)\mathbf{j} + \left(\frac{\partial N}{\partial x} - \frac{\partial M}{\partial y}\right)\mathbf{k}$$

We often express curl **F** symbolically in determinant form, which is evaluated in the same way we evaluate determinants:

$$\begin{vmatrix} \mathbf{i} & \mathbf{j} & \mathbf{k} \\ \dfrac{\partial}{\partial x} & \dfrac{\partial}{\partial y} & \dfrac{\partial}{\partial z} \\ M & N & P \end{vmatrix} = \left(\frac{\partial P}{\partial y} - \frac{\partial N}{\partial z}\right)\mathbf{i} + \left(\frac{\partial M}{\partial z} - \frac{\partial P}{\partial x}\right)\mathbf{j} + \left(\frac{\partial N}{\partial x} - \frac{\partial M}{\partial y}\right)\mathbf{k}$$

EXAMPLE 3   Let $\mathbf{F}(x, y, z) = xz\mathbf{i} + xy^2z\mathbf{j} - e^{2y}\mathbf{k}$. Find curl **F**.

Solution   By definition

$$\text{curl } \mathbf{F}(x, y, z) = \begin{vmatrix} \mathbf{i} & \mathbf{j} & \mathbf{k} \\ \dfrac{\partial}{\partial x} & \dfrac{\partial}{\partial y} & \dfrac{\partial}{\partial z} \\ xz & xy^2z & -e^{2y} \end{vmatrix}$$

$$= (-2e^{2y} - xy^2)\mathbf{i} + (x - 0)\mathbf{j} + (y^2z - 0)\mathbf{k}$$

$$= (-2e^{2y} - xy^2)\mathbf{i} + x\mathbf{j} + y^2z\mathbf{k} \qquad \square$$

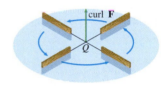

FIGURE 15.5

Suppose that **v** represents the velocity field of a fluid flowing through a solid region. Then it turns out that curl **v** measures the tendency of the fluid to curl, or rotate, about an axis. More specifically, particles in the fluid tend to rotate about the axis that points in the direction of curl $\mathbf{v}(x, y, z)$, and the length of curl $\mathbf{v}(x, y, z)$ measures the swiftness of the motion of the particles around the axis (Figure 15.5). (We will return to this interpretation of curl **v** in Section 15.7.) Formerly the term for curl **F** was "rot **F**," where "rot" is an abbreviation for "rotation." If curl $\mathbf{F} = \mathbf{0}$, then **F** is said to be **irrotational,** whether or not **F** represents a velocity field.

Among the several useful relations between the gradient, divergence, and curl, the two that occur most often are

$$\text{div (curl } \mathbf{F}) = 0 \tag{4}$$

and

$$\text{curl (grad } f) = \mathbf{0} \tag{5}$$

The proofs of (4) and (5) involve performing the required differentiations, and they depend on the equality of the mixed partials that arise (see Exercises 27 and 28).

Another important formula is

$$\text{div(grad } f) = \frac{\partial^2 f}{\partial x^2} + \frac{\partial^2 f}{\partial y^2} + \frac{\partial^2 f}{\partial z^2}$$

The right side of this formula is the **Laplacian** of f, usually denoted $\nabla^2 f$. A function that satisfies the equation

$$\nabla^2 f = 0$$

which is known as **Laplace's equation,** is said to be **harmonic.** Harmonic functions are important in physics.

Let f, M, and N be functions of two variables, and let $\mathbf{F} = M\mathbf{i} + N\mathbf{j}$. Then the two-dimensional versions of the gradient, divergence, curl, and Laplacian, which we will employ later, are

$$\operatorname{grad} f(x, y) = \frac{\partial f}{\partial x}\mathbf{i} + \frac{\partial f}{\partial y}\mathbf{j} \qquad \operatorname{curl} \mathbf{F}(x, y) = \left(\frac{\partial N}{\partial x} - \frac{\partial M}{\partial y}\right)\mathbf{k}$$

$$\operatorname{div} \mathbf{F}(x, y) = \frac{\partial M}{\partial x} + \frac{\partial N}{\partial y} \qquad \nabla^2 f(x, y) = \frac{\partial^2 f}{\partial x^2} + \frac{\partial^2 f}{\partial y^2}$$

Recovering a Function from Its Gradient

Just as a function of a single variable can be recovered from its derivative by integration, a function of several variables can sometimes be recovered from its gradient by successive integrations. We illustrate the procedure in the next two examples.

EXAMPLE 4   Find a function f of two variables such that
$$\operatorname{grad} f(x, y) = y^3\mathbf{i} + 3xy^2\mathbf{j}$$

Solution   Since

$$\frac{\partial f}{\partial x}\mathbf{i} + \frac{\partial f}{\partial y}\mathbf{j} = \operatorname{grad} f(x, y) = y^3\mathbf{i} + 3xy^2\mathbf{j}$$

we have
$$\frac{\partial f}{\partial x} = y^3 \quad \text{and} \quad \frac{\partial f}{\partial y} = 3xy^2 \tag{6}$$

By integrating both sides of the first equation in (6) with respect to x we obtain
$$f(x, y) = xy^3 + g(y)$$

where $g(y)$ is constant with respect to x. Taking partial derivatives of both sides with respect to y, we find that

$$\frac{\partial f}{\partial y} = 3xy^2 + \frac{dg}{dy} \tag{7}$$

Comparison of (7) and the second equation in (6) reveals that

$$\frac{dg}{dy} = 0 \quad \text{so that} \quad g(y) = C$$

where C is a constant. Therefore

$$f(x, y) = xy^3 + C \qquad \square$$

The constant C in the solution of Example 4 corresponds to the constant of integration in the indefinite integral of a function of a single variable.

EXAMPLE 5   Find a function f of three variables such that
$$\operatorname{grad} f(x, y, z) = (2xy + z^2)\mathbf{i} + x^2\mathbf{j} + (2xz + \pi \cos \pi z)\mathbf{k}$$

Solution   Since

$$\frac{\partial f}{\partial x}\mathbf{i} + \frac{\partial f}{\partial y}\mathbf{j} + \frac{\partial f}{\partial z}\mathbf{k} = \operatorname{grad} f(x, y, z) = (2xy + z^2)\mathbf{i} + x^2\mathbf{j} + (2xz + \pi \cos \pi z)\mathbf{k}$$

we have

$$\frac{\partial f}{\partial x} = 2xy + z^2, \quad \frac{\partial f}{\partial y} = x^2, \quad \text{and} \quad \frac{\partial f}{\partial z} = 2xz + \pi \cos \pi z \qquad (8)$$

Integrating both sides of the first equation in (8) with respect to x, we obtain

$$f(x, y, z) = x^2y + xz^2 + g(y, z) \qquad (9)$$

where g is constant with respect to x. Differentiation of both sides of (9) with respect to y yields

$$\frac{\partial f}{\partial y} = x^2 + \frac{\partial g}{\partial y} \qquad (10)$$

Comparing (10) with the second equation in (8), we find that

$$\frac{\partial g}{\partial y} = 0$$

so that g is constant with respect to y. Thus (9) can be rewritten as

$$f(x, y, z) = x^2y + xz^2 + h(z) \qquad (11)$$

for an appropriate function h of z. Next we differentiate both sides of (11) with respect to z and obtain

$$\frac{\partial f}{\partial z} = 2xz + \frac{dh}{dz} \qquad (12)$$

Comparing (12) with the third equation in (8), we find that

$$\frac{dh}{dz} = \pi \cos \pi z$$

Thus $h(z) = \sin \pi z + C$ for some constant C. Consequently (11) becomes

$$f(x, y, z) = x^2y + xz^2 + \sin \pi z + C \qquad \square$$

If $\mathbf{F} = M\mathbf{i} + N\mathbf{j} + P\mathbf{k}$ is a vector field such that M, N, and P have continuous partial derivatives, and if there is a function f such that $\mathbf{F} = \text{grad } f$, then (5) implies that

$$\text{curl } \mathbf{F} = \text{curl } (\text{grad } f) = \mathbf{0}$$

But curl $\mathbf{F} = \mathbf{0}$ is equivalent to

$$\frac{\partial P}{\partial y} = \frac{\partial N}{\partial z}, \quad \frac{\partial M}{\partial z} = \frac{\partial P}{\partial x}, \quad \text{and} \quad \frac{\partial N}{\partial x} = \frac{\partial M}{\partial y} \qquad (13)$$

This argument is not reversible. If the equations in (13) hold for a vector field $\mathbf{F} = M\mathbf{i} + N\mathbf{j} + P\mathbf{k}$, then $\mathbf{F}$ is not necessarily the gradient of a function, that is, $\mathbf{F}$ is not necessarily conservative (see Exercise 10 of Section 15.3). However, if the domain D of $\mathbf{F}$ is all of three-dimensional space or a ball or a parallelepiped (or more generally, if D contains no "holes"), then the argument is reversible. We summarize these results in the next theorem.

THEOREM 15.4   Let $\mathbf{F} = M\mathbf{i} + N\mathbf{j} + P\mathbf{k}$ be a vector field. If there is a function f having continuous mixed partials whose gradient is $\mathbf{F}$, then curl $\mathbf{F} = \mathbf{0}$, that is,

$$\frac{\partial P}{\partial y} = \frac{\partial N}{\partial z}, \quad \frac{\partial M}{\partial z} = \frac{\partial P}{\partial x}, \quad \text{and} \quad \frac{\partial N}{\partial x} = \frac{\partial M}{\partial y} \qquad (14)$$

If the domain of $\mathbf{F}$ is all of three-dimensional space and if the equations in (14) are satisfied, then there is a function f such that $\mathbf{F} = \text{grad } f$.

EXAMPLE 6   Let

$$\mathbf{F}(x, y, z) = 2xyz\mathbf{i} + x^2 z\mathbf{j} + (x^2 y + 1)\mathbf{k}$$

and

$$\mathbf{G}(x, y, z) = yz \cos xy\mathbf{i} + xz \cos xy\mathbf{j} + \cos xy\mathbf{k}$$

Show that $\mathbf{F}$ is the gradient of some function but that $\mathbf{G}$ is not the gradient of any function.

Solution   It is routine to verify that for $\mathbf{F}$ we have

$$\frac{\partial P}{\partial y} = x^2 = \frac{\partial N}{\partial z}, \quad \frac{\partial M}{\partial z} = 2xy = \frac{\partial P}{\partial x}, \quad \text{and} \quad \frac{\partial N}{\partial x} = 2xz = \frac{\partial M}{\partial y}$$

Since the domain of $\mathbf{F}$ is all of three-dimensional space, Theorem 15.4 implies that $\mathbf{F}$ is the gradient of some function. However, for $\mathbf{G}$ we have

$$\frac{\partial P}{\partial y} = -x \sin xy \quad \text{and} \quad \frac{\partial N}{\partial z} = x \cos xy$$

so that the first equation in (14) is not satisfied. By Theorem 15.4, $\mathbf{G}$ cannot be the gradient of any function.   ❑

In case a vector field $\mathbf{F}$ is given by

$$\mathbf{F}(x, y) = M(x, y)\mathbf{i} + N(x, y)\mathbf{j}$$

the conditions in (14) reduce to

$$\frac{\partial N}{\partial x} = \frac{\partial M}{\partial y}$$

and the corresponding statement in Theorem 15.4 holds for such vector fields.

EXAMPLE 7   Let

$$\mathbf{F}(x, y) = y^2 e^{xy}\mathbf{i} + (1 + xy)e^{xy}\mathbf{j}$$

and

$$\mathbf{G}(x, y) = \frac{x}{y}\mathbf{i} + \frac{y}{x}$$

Show that $\mathbf{F}$ is the gradient of some function but that $\mathbf{G}$ is not the gradient of any function.

Solution   For **F** we have

$$\frac{\partial N}{\partial x} = ye^{xy} + (1 + xy)ye^{xy} = (2y + xy^2)e^{xy}$$

and

$$\frac{\partial M}{\partial y} = 2ye^{xy} + y^2xe^{xy} = (2y + xy^2)e^{xy}$$

Since $\partial N/\partial x = \partial M/\partial y$ and the domain of **F** is the xy plane, we conclude from the two-dimensional version of Theorem 15.4 that **F** is the gradient of some function. For **G** we find that

$$\frac{\partial N}{\partial x} = \frac{-y}{x^2} \quad \text{and} \quad \frac{\partial M}{\partial y} = \frac{-x}{y^2}$$

so that **G** is not the gradient of any function.   ❏

Fractals

Early in the section we observed that, in general, it is impossible to graph vector fields because that would entail six dimensions. Despite this fact, with the help of computer graphics it is possible to produce fascinating patterns that arise from vector fields mapping the plane into itself.

For simplicity let R^2 denote the plane, that is, the set of all points in the xy plane. A function **F** whose domain is R^2 and range is contained in R^2 is a **contraction** if there is a number s with $0 \le s < 1$ such that

$$\| \mathbf{F}(\mathbf{u}) - \mathbf{F}(\mathbf{v}) \| \le s \| \mathbf{u} - \mathbf{v} \| \quad \text{for all } \mathbf{u} \text{ and } \mathbf{v} \text{ in } R^2$$

The smallest such number s is the **contraction constant** of **F**. In effect a contraction reduces, or contracts, distances between pairs of points in R^2. For example, if

$$\mathbf{F}(x, y) = \left(\frac{1}{2} x + 1, \frac{1}{3} y + 2 \right) \tag{15}$$

then for any pair of points (x, y) and (u, v),

$$\| \mathbf{F}(x, y) - \mathbf{F}(u, v) \| = \left\| \left(\frac{1}{2} x + 1, \frac{1}{3} y + 2 \right) - \left(\frac{1}{2} u + 1, \frac{1}{3} v + 2 \right) \right\|$$

$$= \left\| \left(\frac{1}{2} (x - u), \frac{1}{3} (y - v) \right) \right\|$$

$$= \sqrt{\frac{1}{4} (x - u)^2 + \frac{1}{9} (y - v)^2}$$

$$< \frac{1}{2} \sqrt{(x - u)^2 + (y - v)^2}$$

$$= \frac{1}{2} \| (x, y) - (u, v) \|$$

Thus **F** is a contraction whose contraction constant is less than or equal to $\frac{1}{2}$. You can check that $\| \mathbf{F}(x, 0) - \mathbf{F}(u, 0) \| = \| (\frac{1}{2}x - \frac{1}{2}u, 0) \| = \frac{1}{2} \| (x, 0) - (u, 0) \|$, so that the contraction constant of **F** is $\frac{1}{2}$.

Generally it is not simple to determine if a given vector field on R^2 is a contraction. However, it can be shown that **F** is a contraction if it has the form

$$\mathbf{F}(x, y) = (ax + by + s, cx + dy + t) \tag{16}$$

where $a, b, c, d, s,$ and t are constants with $|a| \le \frac{1}{2}, |b| \le \frac{1}{2}, |c| \le \frac{1}{2},$ and $|d| \le \frac{1}{2}$. The function **F** in (16) is called **affine.** It maps (0, 0) into (s, t), and maps each line in the *xy* plane into a line (or a point). Affine functions are easy for computers to process.

Let **F** be a contraction of R^2. Consider the sequence $(\mathbf{v}_n)_{n=0}^{\infty}$ of points in the *xy* plane defined by selecting $\mathbf{v}_0$ arbitrarily and then recursively letting

$$\mathbf{v}_1 = \mathbf{F}(\mathbf{v}_0), \quad \mathbf{v}_2 = \mathbf{F}(\mathbf{v}_1), \quad \text{and in general,} \quad \mathbf{v}_{n+1} = \mathbf{F}(\mathbf{v}_n)$$

The points $\mathbf{v}_1, \mathbf{v}_2, \ldots$ are the iterates of $\mathbf{v}_0$ for **F**. For the contraction **F** defined in (15), the first few iterates of $\mathbf{v}_0 = (-1, 2)$ are

$$\left(\frac{1}{2}, \frac{8}{3}\right), \left(\frac{5}{4}, \frac{26}{9}\right), \left(\frac{13}{8}, \frac{80}{27}\right), \left(\frac{29}{16}, \frac{242}{81}\right), \quad \text{and} \quad \left(\frac{61}{32}, \frac{728}{243}\right)$$

Additional calculations would show that as *n* increases without bound, $\mathbf{v}_n$ converges to (2, 3). Moreover, it is possible to show that if $\mathbf{v}_0$ is *any* point in the plane, then its iterates converge to (2, 3).

A famous and powerful theorem called the **Contraction Mapping Theorem** states that if **F** is *any* contraction defined on R^2, then there is a point $\mathbf{v}^*$ such that iterates of any point $\mathbf{v}_0$ in the plane converge to $\mathbf{v}^*$, and moreover, $\mathbf{F}(\mathbf{v}^*) = \mathbf{v}^*$. Thus the iterates of any point $\mathbf{v}_0$ converge to a fixed point of **F**. Now if $\mathbf{v}^*$ and $\mathbf{w}^*$ were two fixed points of **F**, then $\mathbf{F}(\mathbf{v}^*) = \mathbf{v}^*$ and $\mathbf{F}(\mathbf{w}^*) = \mathbf{w}^*$, so that

$$\| \mathbf{v}^* - \mathbf{w}^* \| = \| \mathbf{F}(\mathbf{v}^*) - \mathbf{F}(\mathbf{w}^*) \| \leq s \| \mathbf{v}^* - \mathbf{w}^* \|$$

Since $0 \leq s < 1$, it follows that $\mathbf{v}^* - \mathbf{w}^* = \mathbf{0}$, and hence $\mathbf{v} = \mathbf{w}$. Therefore **F** has only one fixed point. We conclude from the Contraction Mapping Theorem that if **F** is a contraction, then the iterates of every point in the plane converge to the lone fixed point of **F**.

The iterates of a point with respect to a single contraction **F** are interesting only in that they converge to the fixed point of **F**. A much more complex and interesting pattern can unfold if we evaluate the iterates of not just one contraction but rather several contractions $\mathbf{F}_1, \mathbf{F}_2, \ldots, \mathbf{F}_k$. Let $\mathbf{v}_0$ be any element of R^2. Then let $\mathbf{v}_1 = \mathbf{F}_{i_1}(\mathbf{v}_0)$, where i_1 is selected at random from the integers $1, 2, \ldots, k$. Next let $\mathbf{v}_2 = \mathbf{F}_{i_2}(\mathbf{v}_1)$, where i_2 is also selected at random. Continuing the same procedure indefinitely yields the sequence $(\mathbf{v}_n)_{n=0}^{\infty}$ of elements of R^2. Recently it has been shown that, except possibly for the first few elements of the sequence, the sequence approximates a

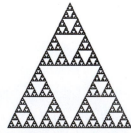

(a)

$F_1(x, y) = (\frac{1}{2}x, \frac{1}{2}y)$

$F_2(x, y) = (\frac{1}{2}x + \frac{1}{2}, \frac{1}{2}y)$

$F_3(x, y) = (\frac{1}{2}x + \frac{1}{4}, \frac{1}{2}y + \frac{1}{2})$

(b)

$F_1(x, y) = (\frac{1}{2}x, \frac{1}{2}y)$

$F_2(x, y) = (\frac{1}{2}x + \frac{1}{2}, \frac{1}{2}y)$

$F_3(x, y) = (-\frac{1}{2}y + 1, \frac{1}{2}x + \frac{1}{2})$

(c)

$F_1(x, y) = (0.25x - 0.25, 0.25y + 0.25)$

$F_2(x, y) = (0.288x - 0.167y - 0.5,$
$\qquad 0.167x + 0.288y - 0.5)$

$F_3(x, y) = (0.25x - 0.746, 0.25y + 0.746)$

$F_4(x, y) = (0.5x + 0.5, 0.5y - 0.5)$

$F_5(x, y) = (0.18x + 0.174y + 0.5,$
$\qquad - 0.174x + 0.18y + 0.5)$

(d)

$F_1(x, y) = (.33x - .66, .33y + .66)$

$F_2(x, y) = (.325x + .325y, -.325x$
$\qquad + .325y)$

$F_3(x, y) = (.33x - .66, .33y - .66)$

$F_4(x, y) = (.33x + .66, .33y - .66)$

$F_5(x, y) = (.33x + .66, .33y + .66)$

FIGURE 15.6

closed and bounded subset A of R^2 that is the same, independent of our choice of $\mathbf{v}_0$. Generally the set A, which is called a **fractal,** is impossible to draw by hand. Figure 15.6(a)–(d) displays a number of fractals, along with the corresponding contractions. Notice that each of the fractals is comprised of miniature copies of itself, and that each miniature copy is made up of even smaller copies, and so forth. This attribute of the fractals in Figure 15.6(a)–(d) is called **self-similarity.** It is the self-similarity of a fractal that makes it impossible to draw accurately by hand, but more amenable to produce on the computer.

EXERCISES 15.1

In Exercises 1–2 represent $\mathbf{F}$ graphically.

1. $\mathbf{F}(x, y) = \mathbf{i} + y\mathbf{j}$ **2.** $\mathbf{F}(x, y) = y\mathbf{i} - x\mathbf{j}$

In Exercises 3–10 find the curl and the divergence of the given vector field.

3. $\mathbf{F}(x, y) = x\mathbf{i} + y\mathbf{j}$

4. $\mathbf{F}(x, y) = \dfrac{x}{x^2 + y^2}\mathbf{i} + \dfrac{y}{x^2 + y^2}\mathbf{j}$

5. $\mathbf{F}(x, y, z) = x^2\mathbf{i} + y^2\mathbf{j} + z^2\mathbf{k}$

6. $\mathbf{F}(x, y, z) = \cos x\mathbf{i} + \sin y\mathbf{j} + e^{xy}\mathbf{k}$

7. $\mathbf{F}(x, y, z) = \dfrac{-x}{z}\mathbf{i} - \dfrac{y}{z}\mathbf{j} + \dfrac{1}{z}\mathbf{k}$

8. $\mathbf{F}(x, y, z) = (y + z)\mathbf{i} + (z + x)\mathbf{j} + (x + y)\mathbf{k}$

9. $\mathbf{F}(x, y, z) = e^x \cos y\mathbf{i} + e^x \sin y\mathbf{j} + z\mathbf{k}$

10. $\mathbf{F}(x, y, z) = \dfrac{x}{(x^2 + y^2)^{3/2}}\mathbf{i} + \dfrac{y}{(x^2 + y^2)^{3/2}}\mathbf{j} + \mathbf{k}$

11. Is the divergence of the vector field represented in Figure 15.7 positive or negative?

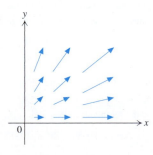

FIGURE 15.7   Figure for Exercise 11.

12. Does the curl of the vector field represented in Figure 15.8 point in the direction of $\mathbf{k}$ or in the direction of $-\mathbf{k}$?

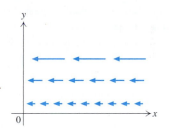

FIGURE 15.8   Figure for Exercise 12.

In Exercises 13–16 show that f satisfies Laplace's equation.

13. $f(x, y) = x^2 - y^2$

14. $f(x, y) = \tan^{-1}\dfrac{y}{x}$

15. $f(x, y, z) = x^2 + y^2 - 2z^2$

16. $f(x, y, z) = \dfrac{1}{\sqrt{x^2 + y^2 + z^2}}$

In Exercises 17–24 determine whether $\mathbf{F}$ is the gradient of some function f. If it is, find such a function f.

17. $\mathbf{F}(x, y) = e^y\mathbf{i} + (xe^y + y)\mathbf{j}$

18. $\mathbf{F}(x, y) = y^2 e^{xy}\mathbf{i} + (1 + xy)e^{xy}\mathbf{j}$

19. $\mathbf{F}(x, y, z) = 2xyz\mathbf{i} + x^2 z\mathbf{j} + (x^2 y + 1)\mathbf{k}$

20. $\mathbf{F}(x, y, z) = yz\mathbf{i} + xz\mathbf{j} + xy\mathbf{k}$

21. $\mathbf{F}(x, y, z) = xz\mathbf{i} + yz\mathbf{j} + xz\mathbf{k}$

22. $\mathbf{F}(x, y, z) = (2xz + 1)\mathbf{i} + 2y(z + 1)\mathbf{j} + (x^2 + y^2 + 3z^2)\mathbf{k}$

23. $\mathbf{F}(x, y, z) = (y^2 + x^2)\mathbf{i} + (z^2 + y^2)\mathbf{j} + (x^2 + z^2)\mathbf{k}$

24. $\mathbf{F}(x, y, z) = yze^{xy}\mathbf{i} + xze^{xy}\mathbf{j} + (e^{xy} + \cos z)\mathbf{k}$

25. Let f and g be functions of several variables, and let $\mathbf{F}$ and $\mathbf{G}$ be vector fields. Decide which of the following expressions represent vector fields, which represent functions of several variables, and which are meaningless.

a. grad (fg) **b.** grad $\mathbf{F}$ **c.** curl $(\text{grad } f)$
d. grad $(\text{div } \mathbf{F})$ **e.** curl $(\text{curl } \mathbf{F})$ **f.** div $(\text{grad } f)$
g. $(\text{grad } f) \times (\text{curl } \mathbf{F})$ **h.** div $(\text{curl } (\text{grad } f))$
i. curl $(\text{div } (\text{grad } f))$

26. Let $\mathbf{F}$ and $\mathbf{G}$ be vector fields, and let f be a function of three variables. Then $f\mathbf{F}$, $\mathbf{F} \cdot \mathbf{G}$, and $\mathbf{F} \times \mathbf{G}$ are defined by the following formulas:

$$(f\mathbf{F})(x, y, z) = f(x, y, z)\mathbf{F}(x, y, z)$$

$$(\mathbf{F} \cdot \mathbf{G})(x, y, z) = \mathbf{F}(x, y, z) \cdot \mathbf{G}(x, y, z)$$

$$(\mathbf{F} \times \mathbf{G})(x, y, z) = \mathbf{F}(x, y, z) \times \mathbf{G}(x, y, z)$$

  a. Use components of the vector field $\mathbf{F}$ to prove that $f\mathbf{F}$ is a continuous vector field if f and $\mathbf{F}$ are continuous.

  b. Use components of the vector fields $\mathbf{F}$ and $\mathbf{G}$ to prove that $\mathbf{F} \cdot \mathbf{G}$ is a continuous function of several variables if $\mathbf{F}$ and $\mathbf{G}$ are continuous.

  c. Use components of the vector fields $\mathbf{F}$ and $\mathbf{G}$ to prove that $\mathbf{F} \times \mathbf{G}$ is a continuous vector field if $\mathbf{F}$ and $\mathbf{G}$ are continuous.

In Exercises 27–31 verify the identity. Use the formulas in Exercise 26 for $f\mathbf{F}$, $\mathbf{F} \cdot \mathbf{G}$, and $\mathbf{F} \times \mathbf{G}$, and assume that the required partial derivatives exist and are continuous.

27. div $(\text{curl } \mathbf{F}) = 0$ (Thus the curl of a vector field is solenoidal.)

28. curl $(\text{grad } f) = \mathbf{0}$ (Thus the gradient of a function is irrotational.)

29. div $(f\mathbf{F}) = f \text{ div } \mathbf{F} + (\text{grad } f) \cdot \mathbf{F}$

30. div $(\mathbf{F} \times \mathbf{G}) = (\text{curl } \mathbf{F}) \cdot \mathbf{G} - \mathbf{F} \cdot (\text{curl } \mathbf{G})$ (Thus the cross product of two irrotational vector fields is solenoidal.)

31. curl $(f\mathbf{F}) = f(\text{curl } \mathbf{F}) + (\text{grad } f) \times \mathbf{F}$

32. **a.** Suppose f is continuous on a simple region R in the xy plane. If grad $f(x, y) = \mathbf{0}$ for all (x, y) in R, show that f is constant.

  b. Suppose f is continuous on a simple solid region D. If grad $f(x, y, z) = \mathbf{0}$ for all (x, y, z) in D, show that f is constant.

In Exercises 33–36 assume that all functions and all components of vector fields have the required continuous partial derivatives.

33. Let $\mathbf{F}$ be a constant vector, and let $\mathbf{G}(x, y, z) = x\mathbf{i} + y\mathbf{j} + z\mathbf{k}$. Show that curl $(\mathbf{F} \times \mathbf{G}) = 2\mathbf{F}$.

34. Let $\mathbf{F}$ be a vector field, f a function of several variables, and $\mathbf{G} = \mathbf{F} + \text{grad } f$. Prove that curl $\mathbf{G} = \text{curl } \mathbf{F}$. (*Hint:* First show that if $\mathbf{F}_1$ and $\mathbf{F}_2$ are vector fields, then curl $(\mathbf{F}_1 + \mathbf{F}_2) = \text{curl } \mathbf{F}_1 + \text{curl } \mathbf{F}_2$.)

35. Show that if $\mathbf{F}$ and $\mathbf{G}$ are conservative, then $\mathbf{F} + \mathbf{G}$ is also conservative.

36. Let $\mathbf{F}(x, y, z) = M(y, z)\mathbf{i} + N(x, z)\mathbf{j} + P(x, y)\mathbf{k}$. Show that $\mathbf{F}$ is solenoidal.

Applications

37. An object having mass m is spun around in a circular orbit with angular velocity ω and is subject to a centrifugal force $\mathbf{F}$ given by

$$\mathbf{F}(x, y, z) = m\omega^2(x\mathbf{i} + y\mathbf{j} + z\mathbf{k})$$

Show that the function f defined by

$$f(x, y, z) = \frac{m\omega^2}{2}(x^2 + y^2 + z^2)$$

is a potential function for $\mathbf{F}$.

38. Suppose that the potential function f for an electric field $\mathbf{E}$ produced by an electric dipole at the origin is given by

$$f(x, y, z) = \frac{ax + by + cz}{(x^2 + y^2 + z^2)^{3/2}}$$

where a, b, and c are constants. Find the electric field $\mathbf{E}$.

39. A vector field $\mathbf{G}$ is a **vector potential** of a vector field $\mathbf{F}$ if $\mathbf{F} = \text{curl } \mathbf{G}$. Use Exercise 27 to show that if $\mathbf{F}$ has a vector potential, then $\mathbf{F}$ is solenoidal.

40. Suppose the magnetic induction field $\mathbf{B}$ associated with a current in a wire is given by

$$\mathbf{B}(x, y, z) = I\left(\frac{-y}{x^2 + y^2}\mathbf{i} + \frac{x}{x^2 + y^2}\mathbf{j}\right)$$

where I is a constant. Let

$$\mathbf{G}(x, y, z) = \frac{-I}{2}\ln(x^2 + y^2)\mathbf{k}$$

Show that $\mathbf{G}$ is a vector potential for $\mathbf{B}$.

41. Let $\mathbf{F}(x, y, z) = 8\mathbf{i}$. Find a vector potential for $\mathbf{F}$.

42. If $\mathbf{v}$ represents the velocity field of a homogeneous fluid that rotates at a constant angular velocity ω about the z axis, then

$$\mathbf{v}(x, y, z) = -\omega y\mathbf{i} + \omega x\mathbf{j}$$

Show that curl $\mathbf{v}(x, y, z)$ depends only on ω and not on (x, y, z).

43. Suppose the temperature in a region is given by

$$T(x, y, z) = 30 - 2x^2 - y^2 - 4z^2$$

  a. Show that grad T (called the **temperature gradient**) is continuous.

  b. Determine whether grad T is a central force field.

15.2   LINE INTEGRALS

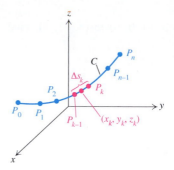

FIGURE 15.9

In this section we define a single integral that is more general than the single integral introduced in Chapter 5. Instead of integrating a function over an interval $[a, b]$, we will integrate over a curve C in space. Such an integral is called a line integral, although the term "curve integral" would be more appropriate. Line integrals have many physical applications, and we will motivate the definition of the line integral by one such application: the mass of a thin wire with a known mass density.

Suppose that a wire lies along a piecewise smooth curve C of finite length and that the mass density (mass per unit length) of the wire at any point (x, y, z) on C is $f(x, y, z)$. Consider any partition P of C obtained by choosing points P_0, P_1, $P_2, \ldots, P_n$ of subdivision along C (Figure 15.9). For each integer k between 1 and n let (x_k, y_k, z_k) be any point on C between P_{k-1} and P_k, and let Δs_k be the length of the portion of C between P_{k-1} and P_k. If Δs_k is small, then the mass of the portion of the wire between P_{k-1} and P_k should be approximately $f(x_k, y_k, z_k)\Delta s_k$, which is the product of the mass density at (x_k, y_k, z_k) and the length Δs_k. Therefore the mass m of the whole wire should be approximately $\sum_{k=1}^{n} f(x_k, y_k, z_k)\,\Delta s_k$. It seems reasonable to expect the sum

$$\sum_{k=1}^{n} f(x_k, y_k, z_k)\,\Delta s_k \tag{1}$$

to approach m as the largest of the lengths $\Delta s_1, \Delta s_2, \ldots, \Delta s_n$, which we denote $\| P \|$, approaches 0. It can be proved that if f is continuous at every point on C, then the sum in (1) does in fact approach a limit, and we call that limit the line integral of f over C.

DEFINITION 15.5

Let f be continuous on a piecewise smooth curve C with finite length. Then the **line integral** $\int_C f(x, y, z)\,ds$ of f over C is defined by

$$\int_C f(x, y, z)\,ds = \lim_{\|P\|\to 0} \sum_{k=1}^{n} f(x_k, y_k, z_k)\,\Delta s_k \tag{2}$$

If C is a closed curve, then the notation $\oint_C f(x, y, z)\,ds$ is sometimes used in place of $\int_C f(x, y, z)\,ds$.

In case f is the mass density of a wire, we compute the mass m of the wire by the formula

$$m = \int_C f(x, y, z)\,ds \tag{3}$$

Evaluation of the line integral $\int_C f(x, y, z)\,ds$ by means of the sum in (2) is an unwieldy process. It is usually simpler to employ a parametrization of the curve C. For the present let us assume that C is parametrized by a smooth vector-valued function $\mathbf{r} = x\mathbf{i} + y\mathbf{j} + z\mathbf{k}$ on an interval $[a, b]$. By using the fact that $\| d\mathbf{r}/dt \| = ds/dt$, one can show that

$$\int_C f(x, y, z)\,ds = \int_a^b f(x(t), y(t), z(t)) \left\| \frac{d\mathbf{r}}{dt} \right\| dt \tag{4}$$

and that any smooth parametrization **r** of the curve C yields the same value of $\int_C f(x, y, z)\, ds$.

EXAMPLE 1 Let C be the line segment from $(0, 0, 0)$ to $(1, -3, 2)$. Find $\int_C (x + y^2 - 2z)\, ds$.

Solution We parametrize C by

$$\mathbf{r}(t) = t\mathbf{i} - 3t\mathbf{j} + 2t\mathbf{k} \quad \text{for } 0 \le t \le 1$$

Then since $x(t) = t$, $y(t) = -3t$, $z(t) = 2t$, and

$$\left\| \frac{d\mathbf{r}}{dt} \right\| = \sqrt{1^2 + (-3)^2 + 2^2} = \sqrt{14}$$

we deduce from (4) that

$$\int_C (x + y^2 - 2z)\, ds = \int_0^1 [t + (-3t)^2 - 2(2t)]\sqrt{14}\, dt$$

$$= \sqrt{14} \int_0^1 (-3t + 9t^2)\, dt$$

$$= \sqrt{14}\left(\frac{-3}{2} t^2 + 3t^3 \right)\Big|_0^1 = \frac{3}{2}\sqrt{14} \qquad \square$$

EXAMPLE 2 Let C be the twisted cubic curve parametrized by

$$\mathbf{r}(t) = t\mathbf{i} + t^2\mathbf{j} + t^3\mathbf{k} \quad \text{for } 0 \le t \le \frac{1}{2}$$

Find

$$\int_C (8x + 36z)\, ds$$

Solution First we notice that $x(t) = t$, $y(t) = t^2$, $z(t) = t^3$, and

$$\frac{d\mathbf{r}}{dt} = \mathbf{i} + 2t\mathbf{j} + 3t^2\mathbf{k}$$

Then

$$\left\| \frac{d\mathbf{r}}{dt} \right\| = \sqrt{1^2 + (2t)^2 + (3t^2)^2} = \sqrt{1 + 4t^2 + 9t^4}$$

It follows from (4) that

$$\int_C (8x + 36z)\, ds = \int_0^{1/2} (8t + 36t^3)\sqrt{1 + 4t^2 + 9t^4}\, dt$$

$$= \frac{2}{3}(1 + 4t^2 + 9t^4)^{3/2}\Big|_0^{1/2}$$

$$= \frac{2}{3}\left(\frac{41\sqrt{41}}{64} - 1 \right) \qquad \square$$

You might imagine in Example 2 that we are computing the mass of a wire bent in the shape of a twisted cubic curve.

The line segment C from $(a, 0, 0)$ to $(b, 0, 0)$ on the x axis is parametrized by

$$\mathbf{r}(t) = t\mathbf{i} \quad \text{for } a \leq t \leq b$$

Consequently if f is continuous on C, formula (4) becomes

$$\int_C f(x, y, z)\, ds = \int_a^b f(x(t), y(t), z(t)) \left\| \frac{d\mathbf{r}}{dt} \right\| dt$$

$$= \int_a^b f(t, 0, 0)(1)\, dt = \int_a^b f(t, 0, 0)\, dt$$

By identifying f with f_0, where

$$f_0(t) = f(t, 0, 0) \quad \text{for } a \leq t \leq b$$

we find that

$$\int_C f(x, y, z)\, ds = \int_a^b f(t, 0, 0)\, dt = \int_a^b f_0(t)\, dt$$

Thus the line integral of a function f that is continuous on a closed interval on the x axis is the same as the single integral of the associated function f_0 of a single variable.

We observe also that if $\mathbf{r}$ parametrizes C on $[a, b]$, then

$$\int_C 1\, ds = \int_a^b \left\| \frac{d\mathbf{r}}{dt} \right\| dt$$

and the expression on the right is the length of the curve C, by Definition 12.16 of Section 12.4. Consequently the length of a smooth curve in space may be given in terms of a line integral.

It follows from the definition of line integral that if f is continuous on a piecewise smooth curve C composed of smooth curves $C_1, C_2, \ldots, C_n$, then

$$\int_C f(x, y, z)\, ds = \int_{C_1} f(x, y, z)\, ds + \int_{C_2} f(x, y, z)\, ds + \cdots + \int_{C_n} f(x, y, z)\, ds$$

EXAMPLE 3   Let C be composed of the two curves C_1 and C_2 shown in Figure 15.10. Evaluate $\int_C (1 + xy)\, ds$.

Solution   The curves C_1 and C_2 can be parametrized as follows:

$$C_1: \quad \mathbf{r}_1(t) = 2\cos t\mathbf{i} + 2\sin t\mathbf{j} \quad \text{for } -\frac{\pi}{2} \leq t \leq \frac{\pi}{2}$$

$$C_2: \quad \mathbf{r}_2(t) = -2t\mathbf{j} \quad \text{for } -1 \leq t \leq 1$$

Consequently

$$\left\| \frac{d\mathbf{r}_1}{dt} \right\| = \sqrt{(-2\sin t)^2 + (2\cos t)^2} = 2 \quad \text{and} \quad \left\| \frac{d\mathbf{r}_2}{dt} \right\| = 2$$

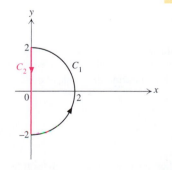

FIGURE 15.10

so that

$$\int_{C_1} (1 + xy)\, ds = \int_{-\pi/2}^{\pi/2} [1 + (2 \cos t)(2 \sin t)]2\, dt = 2(t + 2 \sin^2 t)\Big|_{-\pi/2}^{\pi/2} = 2\pi$$

and

$$\int_{C_2} (1 + xy)\, ds = \int_{-1}^{1} [1 + 0(-2t)]2\, dt = 2t\Big|_{-1}^{1} = 4$$

Therefore

$$\int_{C} (1 + xy)\, ds = \int_{C_1} (1 + xy)\, ds + \int_{C_2} (1 + xy)\, ds = 2\pi + 4 \qquad \square$$

Line Integrals of Vector Fields

Recall from Section 11.3 that if a constant force $\mathbf{F}$ is applied to an object moving along a straight line from a point P to a point Q in space, then the work W done on the object by the force is given by

$$W = \mathbf{F} \cdot \overrightarrow{PQ}$$

Now assume that the object moves along a smooth curve C of finite length in space and that the vector field $\mathbf{F}$ represents a continuous force on the object. Our task is to define the work done by the force on the object as it traverses C.

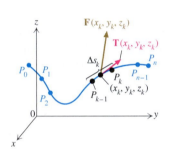

FIGURE 15.11

We orient C in the direction in which the object moves and consider any partition P of C obtained by choosing points $P_0, P_1, P_2, \ldots, P_n$ of subdivision along C (Figure 15.11). For each integer k between 1 and n, let Δs_k be the length of the portion of the curve between P_{k-1} and P_k, and let (x_k, y_k, z_k) be a point on that portion of the curve. If Δs_k is small, then as the object moves along C from P_{k-1} to P_k, it proceeds very nearly in the direction of the tangent vector $\mathbf{T}(x_k, y_k, z_k)$ at (x_k, y_k, z_k). Since by hypothesis $\mathbf{F}$ is continuous on C, the values of $\mathbf{F}$ along the curve from P_{k-1} to P_k are close to $\mathbf{F}(x_k, y_k, z_k)$. Consequently the amount of work done on the object on this portion of the curve C should be approximately

$$\mathbf{F}(x_k, y_k, z_k) \cdot [\Delta s_k \mathbf{T}(x_k, y_k, z_k)]$$

which is the same as $[\mathbf{F}(x_k, y_k, z_k) \cdot \mathbf{T}(x_k, y_k, z_k)]\, \Delta s_k$. It follows that the total amount of work done on the object as it moves along the entire curve C should be approximately

$$\sum_{k=1}^{n} \left[\mathbf{F}(x_k, y_k, z_k) \cdot \mathbf{T}(x_k, y_k, z_k) \right] \Delta s_k$$

It seems reasonable that by making all the lengths $\Delta s_1, \Delta s_2, \ldots, \Delta s_n$ sufficiently small, we could ensure that this sum is as close to the total work as we like. Therefore we replace f in Definition 15.5 by $\mathbf{F} \cdot \mathbf{T}$ and define the work W done by the force to be

$$W = \int_{C} \mathbf{F}(x, y, z) \cdot \mathbf{T}(x, y, z)\, ds \qquad (5)$$

Integrals in the form of (5) will appear in other contexts. For that reason we introduce an abbreviated notation for the integral.

DEFINITION 15.6

Let **F** be a continuous vector field defined on a smooth oriented curve C. Then the **line integral** of **F** over C, denoted $\int_C \mathbf{F} \cdot d\mathbf{r}$, is defined by

$$\int_C \mathbf{F} \cdot d\mathbf{r} = \int_C \mathbf{F}(x, y, z) \cdot \mathbf{T}(x, y, z)\, ds$$

where $\mathbf{T}(x, y, z)$ is the tangent vector at (x, y, z) for the given orientation of C.

Caution: The line integral $\int_C f(x, y, z)\, ds$ in Definition 15.5 does not require C to be oriented. However, for the line integral $\int_C \mathbf{F} \cdot d\mathbf{r}$ of the vector field **F**, the curve C must be oriented.

If **F** represents the force on an object moving along a curve C that is oriented in the direction of motion, then the **work** W done by the force on the object as it traverses C is given by

$$W = \int_C \mathbf{F} \cdot d\mathbf{r} \tag{6}$$

The integral $\int_C \mathbf{F} \cdot d\mathbf{r}$ depends on the orientation of C, but not on any particular smooth parametrization **r** that induces the orientation of C. To prepare $\int_C \mathbf{F} \cdot d\mathbf{r}$ for evaluation, let $\mathbf{r} = x\mathbf{i} + y\mathbf{j} + z\mathbf{k}$ be a parametrization of C with domain $[a, b]$, and assume that the parametrization induces the given orientation on C. Then we know from (1) of Section 12.5 that

$$\mathbf{T}(x(t), y(t), z(t)) = \frac{d\mathbf{r}/dt}{\|d\mathbf{r}/dt\|}$$

so it follows from (4) and Definition 15.6 that

$$\int_C \mathbf{F} \cdot d\mathbf{r} = \int_a^b \left[\mathbf{F}(x(t), y(t), z(t)) \cdot \frac{d\mathbf{r}/dt}{\|d\mathbf{r}/dt\|} \right] \left\| \frac{d\mathbf{r}}{dt} \right\| dt$$

$$= \int_a^b \mathbf{F}(x(t), y(t), z(t)) \cdot \frac{d\mathbf{r}}{dt}\, dt$$

Thus

$$\int_C \mathbf{F} \cdot d\mathbf{r} = \int_a^b \mathbf{F}(x(t), y(t), z(t)) \cdot \frac{d\mathbf{r}}{dt}\, dt \tag{7}$$

The right-hand integral in (7) is the integral we use to evaluate $\int_C \mathbf{F} \cdot d\mathbf{r}$.

EXAMPLE 4   A particle moves upward along the circular helix C_1 parametrized by

$$\mathbf{r}(t) = \cos t\,\mathbf{i} + \sin t\,\mathbf{j} + t\,\mathbf{k} \quad \text{for } 0 \le t \le 2\pi$$

(Figure 15.12), under a force given by

$$\mathbf{F}(x, y, z) = -zy\mathbf{i} + zx\mathbf{j} + xy\mathbf{k}$$

Find the work W done on the particle by the force.

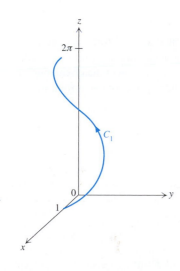

FIGURE 15.12   Part of a circular helix.

Solution   We have $x(t) = \cos t$, $y(t) = \sin t$, $z(t) = t$, and

$$\frac{d\mathbf{r}}{dt} = -\sin t\mathbf{i} + \cos t\mathbf{j} + \mathbf{k}$$

Then we use (6) and (7) to conclude that

$$W = \int_{C_1} \mathbf{F} \cdot d\mathbf{r} = \int_0^{2\pi} \mathbf{F}(x(t), y(t), z(t)) \cdot \frac{d\mathbf{r}}{dt}\, dt$$

$$= \int_0^{2\pi} (-t \sin t\mathbf{i} + t \cos t\mathbf{j} + \cos t \sin t\mathbf{k}) \cdot (-\sin t\mathbf{i} + \cos t\mathbf{j} + \mathbf{k})\, dt$$

$$= \int_0^{2\pi} (t \sin^2 t + t \cos^2 t + \cos t \sin t)\, dt$$

$$= \int_0^{2\pi} (t + \cos t \sin t)\, dt = \left(\frac{1}{2}t^2 + \frac{1}{2}\sin^2 t\right)\Bigg|_0^{2\pi} = 2\pi^2 \qquad \square$$

EXAMPLE 5   Assume that the particle in Example 4 moves under the same force and with the same initial and terminal points, but along the line segment C_2 parametrized by

$$\mathbf{r}(t) = \mathbf{i} + t\mathbf{k} \quad \text{for } 0 \le t \le 2\pi$$

(Figure 15.13). Find the work W done on the particle by the force.

Solution   Here $x(t) = 1$, $y(t) = 0$, $z(t) = t$, and $d\mathbf{r}/dt = \mathbf{k}$. Consequently

$$W = \int_{C_2} \mathbf{F} \cdot d\mathbf{r} = \int_0^{2\pi} \mathbf{F}(x(t), y(t), z(t)) \cdot \frac{d\mathbf{r}}{dt}\, dt$$

$$= \int_0^{2\pi} (-(t)(0)\mathbf{i} + (t)(1)\mathbf{j} + (1)(0)\mathbf{k}) \cdot \mathbf{k}\, dt = \int_0^{2\pi} 0\, dt = 0 \qquad \square$$

One implication of the last two examples is that a given force may perform different amounts of work if the paths along which it acts are different.

The orientation of the curve $-C$ is by definition opposite to the orientation of C. Since the tangent vector at any point on $-C$ is the negative of the tangent vector at the same point on C, Definition 15.6 tells us that

$$\int_{-C} \mathbf{F} \cdot d\mathbf{r} = -\int_{C} \mathbf{F} \cdot d\mathbf{r}$$

For example, if $\mathbf{F}$ and C_1 are as in Example 4, then

$$\int_{-C_1} \mathbf{F} \cdot d\mathbf{r} = -\int_{C_1} \mathbf{F} \cdot d\mathbf{r} = -2\pi^2$$

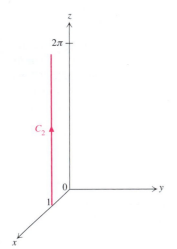

FIGURE 15.13

Alternative Form of the Line Integral

Let $\mathbf{F} = M\mathbf{i} + N\mathbf{j} + P\mathbf{k}$ be a continuous vector field defined on a smooth oriented curve C parametrized by

$$\mathbf{r}(t) = x(t)\mathbf{i} + y(t)\mathbf{j} + z(t)\mathbf{k} \quad \text{for } a \le t \le b$$

Then by (7),

$$\int_C \mathbf{F} \cdot d\mathbf{r} = \int_a^b \mathbf{F}(x(t), y(t), z(t)) \cdot \frac{d\mathbf{r}}{dt} \, dt$$

$$= \int_a^b [M(x(t), y(t), z(t))\mathbf{i} + N(x(t), y(t), z(t))\mathbf{j}$$

$$+ \, P(x(t), y(t), z(t))\mathbf{k}] \cdot \left[\frac{dx}{dt}\mathbf{i} + \frac{dy}{dt}\mathbf{j} + \frac{dz}{dt}\mathbf{k}\right] dt$$

$$= \int_a^b \left[M(x(t), y(t), z(t)) \frac{dx}{dt} + N(x(t), y(t), z(t)) \frac{dy}{dt}\right.$$

$$\left. + \, P(x(t), y(t), z(t)) \frac{dz}{dt}\right] dt$$

It is common to write the final integral above in the abbreviated form

$$\int_C M(x, y, z) \, dx + N(x, y, z) \, dy + P(x, y, z) \, dz \tag{8}$$

or even more briefly as

$$\int_C M \, dx + N \, dy + P \, dz$$

Thus the integral in (8) is just another notation for $\int_C \mathbf{F} \cdot d\mathbf{r}$ and is usually evaluated by means of the formula

$$\int_C M(x, y, z) \, dx + N(x, y, z) \, dy + P(x, y, z) \, dz$$

$$= \int_a^b \left[M(x(t), y(t), z(t)) \frac{dx}{dt} + N(x(t), y(t), z(t)) \frac{dy}{dt}\right. \tag{9}$$

$$\left. + \, P(x(t), y(t), z(t)) \frac{dz}{dt}\right] dt$$

EXAMPLE 6   Let C be the twisted cubic curve parametrized by

$$\mathbf{r}(t) = t\mathbf{i} + t^2\mathbf{j} + t^3\mathbf{k} \quad \text{for } 0 \le t \le 1$$

Evaluate $\int_C xy \, dx + 3zx \, dy - 5x^2 yz \, dz$.

Solution   Notice that

$$x(t) = t, \quad \frac{dx}{dt} = 1; \quad y(t) = t^2, \quad \frac{dy}{dt} = 2t; \quad z(t) = t^3, \quad \frac{dz}{dt} = 3t^2$$

Consequently (9) implies that

$$\int_C xy \, dx + 3zx \, dy - 5x^2 yz \, dz = \int_0^1 \left[(t)(t^2)(1) + 3(t^3)(t)(2t) - 5(t)^2(t^2)(t^3)(3t^2) \right] dt$$

$$= \int_0^1 (t^3 + 6t^5 - 15t^9) \, dt$$

$$= \left(\frac{t^4}{4} + t^6 - \frac{3t^{10}}{2} \right) \Bigg|_0^1 = -\frac{1}{4} \qquad \square$$

In the event that $\mathbf{F} = M\mathbf{i}$, so that $N = P = 0$, we write the expression in (8) as

$$\int_C M(x, y, z) \, dx, \quad \text{or} \quad \int_C M \, dx$$

The integrals $\int_C N(x, y, z) \, dy$ and $\int_C P(x, y, z) \, dz$ have analogous interpretations. When we evaluate these reduced line integrals, we do so by the formulas

$$
\begin{aligned}
\int_C M(x, y, z) \, dx &= \int_a^b \left[M(x(t), y(t), z(t)) \frac{dx}{dt} \right] dt \\
\int_C N(x, y, z) \, dy &= \int_a^b \left[N(x(t), y(t), z(t)) \frac{dy}{dt} \right] dt \qquad (10) \\
\int_C P(x, y, z) \, dz &= \int_a^b \left[P(x(t), y(t), z(t)) \frac{dz}{dt} \right] dt
\end{aligned}
$$

EXAMPLE 7 Let C be the unit circle in the xz plane, oriented by the parametrization

$$\mathbf{r}(t) = \cos t\mathbf{i} + \sin t\mathbf{k} \quad \text{for } 0 \le t \le 2\pi$$

Find $\int_C (y + z) \, dx$.

Solution Here $x(t) = \cos t$, $y(t) = 0$, $z(t) = \sin t$, and $dx/dt = -\sin t$. The first formula in (10) tells us that

$$
\begin{aligned}
\int_C (y + z) \, dx &= \int_0^{2\pi} (0 + \sin t)(-\sin t) \, dt \\
&= -\int_0^{2\pi} \sin^2 t \, dt = -\int_0^{2\pi} \left(\frac{1}{2} - \frac{1}{2} \cos 2t \right) dt \\
&= -\left(\frac{1}{2}t - \frac{1}{4} \sin 2t \right) \Bigg|_0^{2\pi} = -\pi \qquad \square
\end{aligned}
$$

If an oriented curve C is not smooth but is piecewise smooth, composed of smooth curves $C_1, C_2, \ldots, C_n$, then the line integral $\int_C M(x, y, z) \, dx + N(x, y, z) \, dy + P(x, y, z) \, dz$ is defined by

$$
\begin{aligned}
&\int_C M(x, y, z) \, dx + N(x, y, z) \, dy + P(x, y, z) \, dz \\
&= \sum_{k=1}^n \int_{C_k} M(x, y, z) \, dx + N(x, y, z) \, dy + P(x, y, z) \, dz
\end{aligned} \qquad (11)
$$

Each of the line integrals on the right side of (11) is evaluated by means of (9). In our original notation for line integrals of vector fields, (11) becomes

$$\int_C \mathbf{F} \cdot d\mathbf{r} = \sum_{k=1}^{n} \int_{C_k} \mathbf{F} \cdot d\mathbf{r} \tag{12}$$

EXAMPLE 8   Suppose C_0, C_1, C_2, and C_3 are as shown in Figure 15.14, and let C be composed of C_1, C_2, and C_3. Show that

$$\int_C yz\, dx + xz\, dy + xy\, dz = \int_{C_0} yz\, dx + xz\, dy + xy\, dz$$

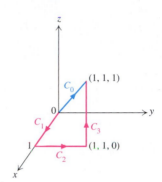

FIGURE 15.14   Two paths joining the origin and the point $(1, 1, 1)$.

Solution   The curves are parametrized as follows:

$$C_0: \quad \mathbf{r}_0(t) = t\mathbf{i} + t\mathbf{j} + t\mathbf{k} \quad \text{for } 0 \leq t \leq 1 \quad \left(\frac{dx}{dt} = \frac{dy}{dt} = \frac{dz}{dt} = 1 \right)$$

$$C_1: \quad \mathbf{r}_1(t) = t\mathbf{i} \quad \text{for } 0 \leq t \leq 1 \quad \left(\frac{dx}{dt} = 1, \frac{dy}{dt} = \frac{dz}{dt} = 0 \right)$$

$$C_2: \quad \mathbf{r}_2(t) = \mathbf{i} + t\mathbf{j} \quad \text{for } 0 \leq t \leq 1 \quad \left(\frac{dx}{dt} = \frac{dz}{dt} = 0, \frac{dy}{dt} = 1 \right)$$

$$C_3: \quad \mathbf{r}_3(t) = \mathbf{i} + \mathbf{j} + t\mathbf{k} \quad \text{for } 0 \leq t \leq 1 \quad \left(\frac{dx}{dt} = \frac{dy}{dt} = 0, \frac{dz}{dt} = 1 \right)$$

Consequently we infer from (9) that

$$\int_{C_0} yz\, dx + xz\, dy + xy\, dz = \int_0^1 [(t \cdot t)1 + (t \cdot t)1 + (t \cdot t)1]\, dt = \int_0^1 3t^2\, dt = 1$$

$$\int_{C_1} yz\, dx + xz\, dy + xy\, dz = \int_0^1 [(0 \cdot 0)1 + (t \cdot 0)0 + (t \cdot 0)0]\, dt = \int_0^1 0\, dt = 0$$

$$\int_{C_2} yz\, dx + xz\, dy + xy\, dz = \int_0^1 [(t \cdot 0)0 + (1 \cdot 0)1 + (1 \cdot t)0]\, dt = \int_0^1 0\, dt = 0$$

$$\int_{C_3} yz\, dx + xz\, dy + xy\, dz = \int_0^1 [(1 \cdot t)0 + (1 \cdot t)0 + (1 \cdot 1)1]\, dt = \int_0^1 1\, dt = 1$$

Combining these equations according to (11), we find that

$$\int_C yz\, dx + xz\, dy + xy\, dz = 1 = \int_{C_0} yz\, dx + xz\, dy + xy\, dz \quad \square$$

In Section 15.3 we will see why the integrals in Example 8 are equal, despite the fact that C and C_0 are very different from each other.

In many physical applications one wishes to integrate a function of two variables x and y over a curve that lies in the xy plane. Of course, such a plane curve is a space curve, and consequently the procedures we have discussed for evaluating line integrals still apply. In particular, the two-dimensional versions of

(7) and (9) are

$$\int_C \mathbf{F} \cdot d\mathbf{r} = \int_a^b \mathbf{F}(x(t), y(t)) \cdot \frac{d\mathbf{r}}{dt} \, dt$$

and

$$\int_C M(x, y)dx + N(x, y)dy = \int_a^b \left[M(x(t), y(t)) \frac{dx}{dt} + N(x(t), y(t)) \frac{dy}{dt} \right] dt$$

EXAMPLE 9   Evaluate $\int_C y \, dx + xy \, dy$, where C is the circle $x^2 + y^2 = 4$, oriented counterclockwise.

Solution   The circle is parametrized by

$$\mathbf{r}(t) = 2 \cos t\mathbf{i} + 2 \sin t\mathbf{j} \quad \text{for } 0 \le t \le 2\pi$$

Consequently

$$\int_C y \, dx + xy \, dy = \int_0^{2\pi} \left[(2 \sin t)(-2 \sin t) + (2 \cos t)(2 \sin t)(2 \cos t) \right] dt$$

$$= 4 \int_0^{2\pi} (-\sin^2 t + 2 \cos^2 t \sin t) \, dt$$

$$= 4 \int_0^{2\pi} \left(-\frac{1}{2} + \frac{1}{2} \cos 2t + 2 \cos^2 t \sin t \right) dt$$

$$= 4 \left(-\frac{1}{2}t + \frac{1}{4} \sin 2t - \frac{2}{3} \cos^3 t \right) \Big|_0^{2\pi} = -4\pi \quad \square$$

EXERCISES 15.2

In Exercises 1–10 evaluate the line integral.

1. $\int_C (9 + 8y^{1/2}) \, ds$, where C is parametrized by $\mathbf{r}(t) = 2t^{3/2}\mathbf{i} + t^2\mathbf{j}$ for $0 \le t \le 1$

2. $\int_C xy \, ds$, where C is the circle $x^2 + y^2 = 4$

3. $\int_C y \, ds$, where C is parametrized by $\mathbf{r}(t) = t\mathbf{i} + t^3\mathbf{j}$ for $-1 \le t \le 0$

4. $\int_C (x^3 + y^3) \, ds$, where C is parametrized by $\mathbf{r}(t) = \cos^3 t\mathbf{i} + \sin^3 t\mathbf{j}$ for $0 \le t \le \pi/2$

5. $\int_C 2xyz \, ds$, where C is parametrized by $\mathbf{r}(t) = e^t\mathbf{i} + e^{-t}\mathbf{j} + \sqrt{2} \, t\mathbf{k}$ for $0 \le t \le 1$

6. $\int_C (x - z^2) \, ds$, where C is parametrized by $\mathbf{r}(t) = \ln t\mathbf{i} - t^2\mathbf{j} + 2t\mathbf{k}$ for $1 \le t \le 2$

7. $\int_C (1 + \frac{9}{4}z^{2/3})^{1/4} \, ds$, where C is parametrized by $\mathbf{r}(t) = \cos t\mathbf{i} + \sin t\mathbf{j} + t^{3/2}\mathbf{k}$ for $0 \le t \le \frac{20}{3}$

8. $\int_C (2xy - 5yz) \, ds$, where C is the line segment from $(1, 0, 1)$ to $(0, 3, 2)$

9. $\int_C (y + 2z) \, ds$, where C is the triangular path from $(-1, 0, 0)$ to $(0, 1, 0)$, from $(0, 1, 0)$ to $(0, 0, 1)$, and from $(0, 0, 1)$ to $(-1, 0, 0)$

10. $\int_C (3x - 2y + z) \, ds$, where C is as shown in Figure 15.15

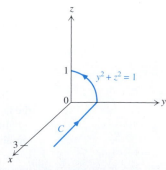

FIGURE 15.15   Graph for Exercise 10.

In Exercises 11–17 evaluate $\int_C \mathbf{F} \cdot d\mathbf{r}$, where C is parametrized by $\mathbf{r}(t)$.

11. $\mathbf{F}(x, y, z) = z\mathbf{i} - y\mathbf{j} - x\mathbf{k}$;
    $\mathbf{r}(t) = 5\mathbf{i} - \sin t\mathbf{j} - \cos t\mathbf{k}$ for $0 \leq t \leq \pi/4$

12. $\mathbf{F}(x, y, z) = y\mathbf{i} + x\mathbf{j} + z^3\mathbf{k}$;
    $\mathbf{r}(t) = (1 - t)\mathbf{i} + t\mathbf{j} + \pi t\mathbf{k}$ for $0 \leq t \leq 1$

13. $\mathbf{F}(x, y, z) = y\mathbf{i} + xy\mathbf{j} + z^3\mathbf{k}$;
    $\mathbf{r}(t) = \cos t\mathbf{i} + \sin t\mathbf{j} + 2t\mathbf{k}$ for $0 \leq t \leq \pi/2$

14. $\mathbf{F}(x, y, z) = -z\mathbf{i} + x\mathbf{k}$;
    $\mathbf{r}(t) = \cos t\mathbf{i} + \sin t\mathbf{k}$ for $0 \leq t \leq \pi$

15. $\mathbf{F}(x, y, z) = -z\mathbf{i} + x\mathbf{k}$; $\mathbf{r}$ parametrizes the curve in Exercise 14 in the opposite direction

16. $\mathbf{F}(x, y, z) = ye^{(x^2)}\mathbf{i} + xe^{(y^2)}\mathbf{j} + \cosh xy^3\mathbf{k}$;
    $\mathbf{r}(t) = e^t\mathbf{i} + e^t\mathbf{j} + 3\mathbf{k}$ for $0 \leq t \leq 1$

17. $\mathbf{F}(x, y, z) = 5e^{\sin \pi x}\mathbf{i} - 4e^{\cos \pi x}\mathbf{j}$;
    $\mathbf{r}(t) = \frac{1}{2}\mathbf{i} + 2\mathbf{j} - \ln(\cosh t)\mathbf{k}$ for $0 \leq t \leq \pi/6$

18. Let $\mathbf{F}(x, y, z) = xy\mathbf{i} + 2z\mathbf{j} + (y + z)\mathbf{k}$. Find $\int_C \mathbf{F} \cdot d\mathbf{r}$, where
    a. C is composed of the line segments from $(0, 0, 0)$ to $(0, -1, 0)$ and from $(0, -1, 0)$ to $(1, 1, 2)$.
    b. C is composed of the line segments from $(0, 0, 0)$ to $(0, 1, 1)$ and from $(0, 1, 1)$ to $(1, 1, 2)$.
    c. C is the parabolic curve parametrized by $\mathbf{r}(t) = t\mathbf{i} + t\mathbf{j} + 2t^2\mathbf{k}$ for $0 \leq t \leq 1$.
    d. In each case the initial and terminal points of C are the same. Are the values of the line integrals equal?

In Exercises 19–28 evaluate the line integral.

19. $\int_C y\, dx - x\, dy + xyz^2\, dz$, where C is parametrized by $\mathbf{r}(t) = e^{-t}\mathbf{i} + e^t\mathbf{j} + t\mathbf{k}$ for $0 \leq t \leq 1$

20. $\int_C e^x\, dx + xy\, dy + xyz\, dz$, where C is parametrized by $\mathbf{r}(t) = t\mathbf{i} + t\mathbf{j} + 2t\mathbf{k}$ for $-1 \leq t \leq 1$

21. $\int_C e^x\, dx + xy\, dy + xyz\, dz$, where C is the curve opposite to the one in Exercise 20

22. $\int_C y(x^2 + y^2)\, dx - x(x^2+y^2)\, dy + xy\, dz$, where C is parametrized by $\mathbf{r}(t) = \cos t\mathbf{i} + \sin t\mathbf{j} + t\mathbf{k}$ for $-\pi \leq t \leq \pi$

23. $\int_C xy\, dx + (x + z)\, dy + z^2\, dz$, where C is parametrized by $\mathbf{r}(t) = (t + 1)\mathbf{i} + (t - 1)\mathbf{j} + t^2\mathbf{k}$ for $-1 \leq t \leq 2$

24. $\int_C x\, dx + y\, dy + xy\, dz$, where C is parametrized by $\mathbf{r}(t) = \cos t\mathbf{i} + \sin t\mathbf{j} + \cos t\mathbf{k}$ for $-\pi/2 \leq t \leq 0$

25. $\int_C \dfrac{1}{1 + x^2}\, dx + \dfrac{2}{1 + y^2}\, dy$, where C is the quarter unit circle from $(1, 0)$ to $(0, 1)$

26. $\int_C z\, dx + xy\, dy$, where C is the line segment from $(-1, 1, 0)$ to $(2, 1, 0)$

27. $\int_C x \ln(xz/y)\, dx + \cos(\pi xy/z)\, dy$, where C is parametrized by $\mathbf{r}(t) = t\mathbf{i} + t^2\mathbf{j} + t^3\mathbf{k}$ for $1 \leq t \leq 2$

28. $\int_C x \ln(xz/y)\, dx + \cos(\pi xy/z)\, dy$, where C is composed

of the line segments from $(1, 1, 1)$ to $(2, 2, 2)$ and from $(2, 2, 2)$ to $(4, 2, 2)$

29. Evaluate $\int_C y\, dx + z\, dy + x\, dz$, where C is composed of
    a. the line segments from $(0, 0, 0)$ to $(0, -5, 0)$ and from $(0, -5, 0)$ to $(0, 1, 1)$.
    b. the line segments from $(0, 0, 0)$ to $(1, 0, 0)$ and from $(1, 0, 0)$ to $(0, 1, 1)$.

30. Let C_1 be parametrized by
    $$\mathbf{r}_1(t) = t\mathbf{i} + t\mathbf{j} + t\mathbf{k} \quad \text{for } 0 \leq t \leq \frac{1}{2}$$
    and C_2 by
    $$\mathbf{r}_2(t) = \sin t\mathbf{i} + \sin t\mathbf{j} + \sin t\mathbf{k} \quad \text{for } 0 \leq t \leq \frac{\pi}{6}$$
    Evaluate $\int_{C_1} (xy + z)\, ds$ and $\int_{C_2} (xy + z)\, ds$. Are the answers the same? Explain why or why not.

31. Suppose the first component of a parametrization of a piecewise smooth curve C is constant, and let M be continuous on C. Show that $\int_C M(x, y, z)\, dx = 0$.

32. Suppose the mass density at a given point on a thin wire is equal to the square of the distance from that point to the x axis. If the wire is helical and is parametrized by
    $$\mathbf{r}(t) = \sin t\mathbf{i} - \cos t\mathbf{j} + 4t\mathbf{k} \quad \text{for } \pi \leq t \leq 2\pi$$
    find the mass m of the wire.

33. Suppose the mass density at a given point on the wire in Exercise 32 is equal to the square of the distance from that point to the origin. Find the mass m of the wire.

34. Let $\mathbf{F}(x, y, z) = xy\mathbf{i} + yz\mathbf{j} + xz\mathbf{k}$. Find the work done by the force $\mathbf{F}$ on an object as it traverses the twisted quartic curve C parametrized by
    $$\mathbf{r}(t) = t\mathbf{i} + t^2\mathbf{j} + t^4\mathbf{k} \quad \text{for } 0 \leq t \leq 1$$

35. Let $\mathbf{F}(x, y, z) = (2x - y)\mathbf{i} + 2z\mathbf{j} + (y - z)\mathbf{k}$.
    a. Find the work W done by the force $\mathbf{F}$ on an object moving from $(0, 0, 0)$ to $(1, 1, 1)$ along a straight line.
    b. Find the work W done as the object moves along the curve parametrized by
    $$\mathbf{r}(t) = \sin \frac{\pi t}{2}\mathbf{i} + \sin \frac{\pi t}{2}\mathbf{j} + t\mathbf{k} \quad \text{for } 0 \leq t \leq 1$$

Applications

36. A painter weighing 120 pounds carries a pail of paint weighing 30 pounds up a helical staircase surrounding a circular cylindrical water tower. If the tower is 200 feet tall and 100 feet in diameter and the painter makes exactly four revolutions during the ascent to the top, how much work is done by gravity on the painter and the pail during the ascent?

37. If the pail in Exercise 36 steadily leaks 1 pound of paint for every 10 feet the painter rises, how much less work would

be done by gravity on the painter and the pail during the ascent than would be done if there were no leak?

38. A satellite weighing 6000 pounds orbits the earth in a circular orbit 4400 miles from the center of the earth. Find the work W done on the satellite by gravity during

a. half a revolution.   **b.** one complete revolution.

39. a. Suppose a continuous force acts on an object in a direction normal to the path of the object. Show that the work done by the force on the object is 0.

   b. Redo Exercise 38 using part (a) of this exercise.

15.3   THE FUNDAMENTAL THEOREM OF LINE INTEGRALS

This section is devoted to a generalization of the Fundamental Theorem of Calculus that is applicable to line integrals of the form $\int_C \mathbf{F} \cdot d\mathbf{r}$ for certain vector fields $\mathbf{F}$ and for certain curves C. Recall from Section 12.5 that an oriented curve is by definition piecewise smooth.

THEOREM 15.7
Fundamental Theorem of Line Integrals

Let C be an oriented curve with initial point (x_0, y_0, z_0) and terminal point (x_1, y_1, z_1). Let f be a function of three variables that is differentiable at every point on C, and assume that grad f is continuous on C. Then

$$\int_C \text{grad } f \cdot d\mathbf{r} = f(x_1, y_1, z_1) - f(x_0, y_0, z_0)$$

Proof   First we consider the case in which C is smooth. Let $\mathbf{r}$ be a smooth parametrization of C defined on an interval $[a, b]$. Then by (7) of Section 15.2, the Chain Rule in Section 13.4, and the Fundamental Theorem of Calculus,

$$\int_C \text{grad } f \cdot d\mathbf{r} \overset{\underset{\text{(7) of Section 15.2}}{}}{=} \int_a^b \left[\text{grad } f(x(t), y(t), z(t)) \right] \cdot \frac{d\mathbf{r}}{dt}\, dt$$

$$= \int_a^b \left[\frac{\partial f}{\partial x}(x(t), y(t), z(t))\frac{dx}{dt} + \frac{\partial f}{\partial y}(x(t), y(t), z(t))\frac{dy}{dt} \right.$$

$$\left. + \frac{\partial f}{\partial z}(x(t), y(t), z(t))\frac{dz}{dt} \right] dt$$

$$\overset{\underset{\text{Rule}}{\text{Chain}}}{=} \int_a^b \frac{d}{dt}\left[f(x(t), y(t), z(t)) \right] dt$$

$$= f(x(b), y(b), z(b)) - f(x(a), y(a), z(a))$$

$$= f(x_1, y_1, z_1) - f(x_0, y_0, z_0)$$

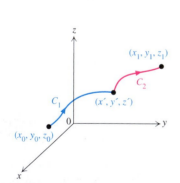

FIGURE 15.16   *C* is composed of C_1 and C_2.

If C is only piecewise smooth, we can prove the result by combining line integrals over the smooth portions of the curve. We will carry out the proof for the case in which C is composed of two smooth curves C_1 and C_2, where C_1 joins (x_0, y_0, z_0) to (x', y', z'), and C_2 joins (x', y', z') to (x_1, y_1, z_1) (Figure 15.16). Using the result we just proved for smooth curves, we have

$$\int_C \text{grad } f \cdot d\mathbf{r} = \int_{C_1} \text{grad } f \cdot d\mathbf{r} + \int_{C_2} \text{grad } f \cdot d\mathbf{r}$$

$$= [f(x', y', z') - f(x_0, y_0, z_0)] + [f(x_1, y_1, z_1) - f(x', y', z')]$$

$$= f(x_1, y_1, z_1) - f(x_0, y_0, z_0) \quad \blacksquare$$

If a continuous vector field **F** is the gradient of a function f, then

$$\int_C \mathbf{F} \cdot d\mathbf{r} = \int_C \operatorname{grad} f \cdot d\mathbf{r}$$

so that the formula in the Fundamental Theorem of Line Integrals becomes

$$\int_C \mathbf{F} \cdot d\mathbf{r} = f(x_1, y_1, z_1) - f(x_0, y_0, z_0) \tag{1}$$

This is reminiscent of the formula

$$\int_a^b f'(x)\, dx = f(b) - f(a)$$

for functions of one variable, which follows from the Fundamental Theorem of Calculus. If C is piecewise smooth but we do not have a parametrization of C, equation (1) enables us to evaluate $\int_C \mathbf{F} \cdot d\mathbf{r}$ provided that we know the end points of C and can find a potential function for **F**. Finding a potential function f for **F**, as described in Section 15.1, thus becomes the central problem in the evaluation of $\int_C \mathbf{F} \cdot d\mathbf{r}$ by means of (1), as the following examples indicate.

EXAMPLE 1   Let C be the curve from $(1, -1, -\tfrac{1}{2})$ to $(1, 1, \tfrac{1}{2})$ parametrized by

$$\mathbf{r}(t) = -\cos \pi t^4 \mathbf{i} + t^{5/3} \mathbf{j} + \frac{t}{t^2 + 1} \mathbf{k} \quad \text{for} \quad -1 \le t \le 1$$

and let $\mathbf{F}(x, y, z) = (2xy + z^2)\mathbf{i} + x^2\mathbf{j} + (2xz + \pi \cos \pi z)\mathbf{k}$. Find $\int_C \mathbf{F} \cdot d\mathbf{r}$.

Solution   Notice that $\mathbf{r}(t)$ is very complicated, so it is difficult to imagine evaluating the integral by methods of the preceding section. By Example 5 of Section 15.1, $\mathbf{F} = \operatorname{grad} f$, where

$$f(x, y, z) = x^2y + xz^2 + \sin \pi z$$

As a result, the Fundamental Theorem of Line Integrals implies that

$$\int_C \mathbf{F} \cdot d\mathbf{r} = f(1, 1, \tfrac{1}{2}) - f(1, -1, -\tfrac{1}{2})$$

$$= (1 + \tfrac{1}{4} + 1) - (-1 + \tfrac{1}{4} - 1) = 4 \qquad \square$$

EXAMPLE 2   Let C be a piecewise smooth curve from $(1, -2, 3)$ to $(1, 0, 0)$ and **F** the gravitational field of a point mass m at the origin. Find the work W done by **F** on a unit point mass that traverses C.

Solution   The force on the unit mass is equal to the gravitational field **F**, and by (3) of Section 15.1 we know that the gravitational field **F** is the gradient of f, where

$$f(x, y, z) = \frac{Gm}{(x^2 + y^2 + z^2)^{1/2}}$$

Consequently

$$W = \int_C \mathbf{F} \cdot d\mathbf{r} = f(1, 0, 0) - f(1, -2, 3)$$

$$= Gm - \frac{Gm}{(1 + 4 + 9)^{1/2}} = Gm\left(1 - \frac{1}{\sqrt{14}}\right) \qquad \square$$

Naturally, the Fundamental Theorem of Line Integrals can be formulated in two dimensions: Let f be a function of two variables that is differentiable at every point on an oriented curve C with initial point (x_0, y_0) and terminal point (x_1, y_1), and let grad f be continuous on C. Then

$$\int_C \text{grad } f \cdot d\mathbf{r} = f(x_1, y_1) - f(x_0, y_0)$$

Moreover, if $\mathbf{F}$ is a continuous vector field such that $\mathbf{F} = \text{grad } f$, then

$$\int_C \mathbf{F} \cdot d\mathbf{r} = f(x_1, y_1) - f(x_0, y_0)$$

EXAMPLE 3   Let C be the portion of the parabola $y = x^2$ with initial point $(0, 0)$ and terminal point $(2, 4)$, and let

$$\mathbf{F}(x, y) = y^3\mathbf{i} + 3xy^2\mathbf{j}$$

Find $\int_C \mathbf{F} \cdot d\mathbf{r}$.

Solution   By Example 4 of Section 15.1, $\mathbf{F} = \text{grad } f$ if $f(x, y) = xy^3$. Consequently

$$\int_C \mathbf{F} \cdot d\mathbf{r} = f(2, 4) - f(0, 0) = 2 \cdot 4^3 - 0 = 128 \qquad \square$$

Independence of Path

The Fundamental Theorem of Line Integrals says that if $\mathbf{F}$ is the gradient of a differentiable function f and has domain D, then the value of a line integral $\int_C \mathbf{F} \cdot d\mathbf{r}$ depends only on the initial and terminal points of the oriented curve C in D. Thus if C_1 is another oriented curve lying in D and having the same initial and terminal points as C, then

$$\int_{C_1} \mathbf{F} \cdot d\mathbf{r} = \int_C \mathbf{F} \cdot d\mathbf{r}$$

More generally, if $\mathbf{F}$ is a continuous vector field with domain D and $\int_{C_1} \mathbf{F} \cdot d\mathbf{r} = \int_{C_2} \mathbf{F} \cdot d\mathbf{r}$ for any two oriented curves C_1 and C_2 lying in D and having the same initial and terminal points, we say that $\int_C \mathbf{F} \cdot d\mathbf{r}$ is **independent of path.** Rephrased in this terminology, the Fundamental Theorem of Line Integrals says that if $\mathbf{F}$ is continuous and is the gradient of some differentiable function, then $\int_C \mathbf{F} \cdot d\mathbf{r}$ is independent of path.

The condition that $\int_C \mathbf{F} \cdot d\mathbf{r}$ is independent of path for oriented curves in a region D implies that if C is any closed, oriented curve in D, then $\int_C \mathbf{F} \cdot d\mathbf{r} = 0$. The reason is that C can be regarded as composed of two oriented curves C_1 and C_2, where the terminal point of C_1 is the initial point of C_2 and the initial point of C_1 is the terminal point of C_2 (Figure 15.17). Since C_1 and $-C_2$ have the same initial and terminal points, the independence of path implies that

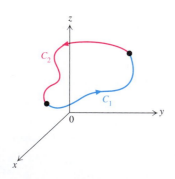

FIGURE 15.17   *C* is composed of C_1 and C_2.

$$\int_C \mathbf{F} \cdot d\mathbf{r} = \int_{C_1} \mathbf{F} \cdot d\mathbf{r} + \int_{C_2} \mathbf{F} \cdot d\mathbf{r} = \int_{C_1} \mathbf{F} \cdot d\mathbf{r} - \int_{-C_2} \mathbf{F} \cdot d\mathbf{r} = 0$$

A somewhat more technical argument shows that if $\int_C \mathbf{F} \cdot d\mathbf{r} = 0$ for every closed, oriented curve in D, then $\mathbf{F} = \operatorname{grad} f$ for some function f. Hence we have the following three equivalent conditions:

1. $\mathbf{F} = \operatorname{grad} f$ for some function f; that is, $\mathbf{F}$ is conservative.
2. $\int_C \mathbf{F} \cdot d\mathbf{r}$ is independent of path.
3. $\int_C \mathbf{F} \cdot d\mathbf{r} = 0$ for every closed, oriented curve C lying in the domain of $\mathbf{F}$.

By (5) of Section 15.1, condition 1 implies a fourth condition:

4. curl $\mathbf{F} = \mathbf{0}$.

If the domain of $\mathbf{F}$ is all of three-dimensional space, or any region with no "holes," then condition 4 implies condition 1, so that conditions 1–4 are equivalent.

Because condition 1 implies condition 3, the work done by a conservative, continuous force on an object that traverses a closed, piecewise smooth curve is 0. In particular, this is true of the gravitational and electric fields described in Section 15.1, since both are conservative.

Conservation of Energy

Now we use Newton's Second Law of Motion to discuss further the work done by a force on an object. We will derive the Law of Conservation of Energy for conservative force fields (a special case of which we discussed in Section 6.4).

Suppose an object moves under the influence of a continuous force $\mathbf{F}$ along a smooth curve C parametrized by

$$\mathbf{r}(t) = x(t)\mathbf{i} + y(t)\mathbf{j} + z(t)\mathbf{k} \quad \text{for} \quad a \le t \le b$$

For the sake of brevity we write

$$\mathbf{F}(\mathbf{r}(t)) \quad \text{for} \quad \mathbf{F}(x(t), y(t), z(t))$$

as is often done. Newton's Second Law of Motion states that the force $\mathbf{F}$ and the acceleration $\mathbf{a}$ of the object are related by

$$\mathbf{F}(\mathbf{r}(t)) = m\mathbf{a}(t) = m\mathbf{r}''(t) \quad \text{for } a \le t \le b \tag{2}$$

The work W done by the force $\mathbf{F}$ on the object is given by

$$W = \int_C \mathbf{F} \cdot d\mathbf{r} = \int_a^b \mathbf{F}(\mathbf{r}(t)) \cdot \mathbf{r}'(t)\, dt = \int_a^b m\mathbf{r}''(t) \cdot \mathbf{r}'(t)\, dt$$

$$= \frac{m}{2} \int_a^b \frac{d}{dt}[\mathbf{r}'(t) \cdot \mathbf{r}'(t)]\, dt = \frac{m}{2} \int_a^b \frac{d}{dt} \|\mathbf{r}'(t)\|^2\, dt$$

By the Fundamental Theorem of Calculus,

$$\frac{m}{2} \int_a^b \frac{d}{dt} \|\mathbf{r}'(t)\|^2\, dt = \frac{m}{2} \|\mathbf{r}'(t)\|^2 \Big|_a^b = \frac{m}{2} \|\mathbf{r}'(b)\|^2 - \frac{m}{2} \|\mathbf{r}'(a)\|^2$$

Thus

$$W = \frac{m}{2} \|\mathbf{r}'(b)\|^2 - \frac{m}{2} \|\mathbf{r}'(a)\|^2 \tag{3}$$

For any number t in $[a, b]$, the quantity

$$\frac{m}{2} \|\mathbf{r}'(t)\|^2$$

is called the **kinetic energy** of the object. (Notice that $\|\mathbf{r}'(t)\|$ is the speed of the object, so that the kinetic energy is one half the product of the mass and the square of the speed.) Therefore (3) expresses the fact that the work done by the force on the object from the beginning to the end of the curve is equal to the change in the kinetic energy of the object from the initial point to the terminal point of the curve. It also follows from (3) that the work depends only on the speeds at the two endpoints of the curve and on the mass.

Now let us assume that $\mathbf{F}$ is a conservative force. Then there is a function f such that $\mathbf{F} = -\text{grad } f$ (here the minus sign is included, as is done in physics). The function f is called the **potential energy function** of the object corresponding to $\mathbf{F}$. It follows from the Chain Rule that

$$\frac{d}{dt} f(\mathbf{r}(t)) = \frac{d}{dt} f(x(t), y(t), z(t)) = \frac{\partial f}{\partial x} \frac{dx}{dt} + \frac{\partial f}{\partial y} \frac{dy}{dt} + \frac{\partial f}{\partial z} \frac{dz}{dt}$$

$$= \left(\frac{\partial f}{\partial x} \mathbf{i} + \frac{\partial f}{\partial y} \mathbf{j} + \frac{\partial f}{\partial z} \mathbf{k} \right) \cdot \left(\frac{dx}{dt} \mathbf{i} + \frac{dy}{dt} \mathbf{j} + \frac{dz}{dt} \mathbf{k} \right)$$

$$= [\text{grad } f(\mathbf{r}(t))] \cdot \mathbf{r}'(t)$$

Therefore

$$\frac{d}{dt} f(\mathbf{r}(t)) = [\text{grad } f(\mathbf{r}(t))] \cdot \mathbf{r}'(t) \tag{4}$$

Using (2) and (4) and the assumption that $\mathbf{F} = -\text{grad } f$, we find that

$$\frac{d}{dt} \left(\frac{m}{2} \|\mathbf{r}'(t)\|^2 + f(\mathbf{r}(t)) \right) = \frac{d}{dt} \left(\frac{m}{2} \mathbf{r}'(t) \cdot \mathbf{r}'(t) + f(\mathbf{r}(t)) \right)$$

$$\overset{(4)}{=} m\mathbf{r}''(t) \cdot \mathbf{r}'(t) + [\text{grad } f(\mathbf{r}(t))] \cdot \mathbf{r}'(t)$$

$$= [m\mathbf{r}''(t) + \text{grad } f(\mathbf{r}(t))] \cdot \mathbf{r}'(t)$$

$$= [m\mathbf{r}''(t) - \mathbf{F}(\mathbf{r}(t))] \cdot \mathbf{r}'(t)$$

$$\overset{(2)}{=} 0 \cdot \mathbf{r}'(t) = 0$$

Thus

$$\frac{d}{dt} \left(\frac{m}{2} \|\mathbf{r}'(t)\|^2 + f(\mathbf{r}(t)) \right) = 0$$

When we integrate both sides, we obtain

$$\frac{m}{2} \|\mathbf{r}'(t)\|^2 + f(\mathbf{r}(t)) = c \tag{5}$$

for some constant c. Correctly interpreted, (5) expresses the **Law of Conservation of Energy** of Newtonian mechanics:

> The sum of the kinetic energy and the potential energy of an object due to a conservative force is constant.

It can be shown that any central force field is conservative (see Exercise 11). By contrast, frictional forces are not conservative because more work is required to overcome friction over long paths than over short paths joining the same points. For an example of another kind of nonconservative force, see Exercise 16.

EXERCISES 15.3

In Exercises 1–7 show that the line integral is independent of path, and evaluate the integral.

1. $\int_C (e^x + y)\,dx + (x + 2y)\,dy$; C is any piecewise smooth curve in the xy plane from $(0, 1)$ to $(2, 3)$.

2. $\int_C (2xy^2 + 1)\,dx + 2x^2 y\,dy$; C is any piecewise smooth curve in the xy plane from $(-1, 2)$ to $(2, 3)$.

3. $\int_C y\,dx + (x + z)\,dy + y\,dz$; C is parametrized by

$$\mathbf{r}(t) = \frac{t^2 + 1}{t^2 - 1}\mathbf{i} + \cos \pi t\mathbf{j} + 2t \sin \pi t\mathbf{k} \quad \text{for } 0 \le t \le \tfrac{1}{2}.$$

4. $\int_C (\cos x + 2yz)\,dx + (\sin y + 2xz)\,dy + (z + 2xy)\,dz$; C is any piecewise smooth curve from $(0, 0, 0)$ to $(\pi, \pi, 1/\pi)$.

5.
$$\int_C \frac{x}{1 + x^2 + y^2 + z^2}\,dx + \frac{y}{1 + x^2 + y^2 + z^2}\,dy$$
$$+ \frac{z}{1 + x^2 + y^2 + z^2}\,dz;$$

C is parametrized by $\mathbf{r}(t) = t\mathbf{i} + t^2\mathbf{j} + t^4\mathbf{k}$ for $0 \le t \le 1$.

6. $\int_C (y + 2xe^y)\,dx + (x + x^2 e^y)\,dy$; C is parametrized by $\mathbf{r}(t) = t^{1/2}\mathbf{i} + \ln t\mathbf{j} + t\mathbf{k}$ for $1 \le t \le 4$.

7.
$$\int_C e^{-x} \ln y\,dx - \frac{e^{-x}}{y}\,dy + z\,dz; \quad C \text{ is parametrized by}$$

$\mathbf{r}(t) = (t - 1)\mathbf{i} + e^t\mathbf{j} + (t^2 + 1)\mathbf{k}$ for $0 \le t \le 1$.

8. Evaluate $\int_C (x + \cos \pi y)\,dx - \pi x \sin \pi y\,dy$, where C is the curve parametrized by $\mathbf{r}(t) = e^t\mathbf{i} + t\mathbf{j} + (\tan \pi t/4)\mathbf{k}$ for $0 \le t \le 1$.

9. Suppose that f, g, and h are continuous functions. Prove that $\int_C f(x)\,dx + g(y)\,dy + h(z)\,dz$ is independent of path.

10. Let
$$\mathbf{F}(x, y) = \frac{y}{x^2 + y^2}\mathbf{i} - \frac{x}{x^2 + y^2}\mathbf{j}$$

a. Show that curl $\mathbf{F} = \mathbf{0}$.

b. Let R be any region in the plane that contains the circle $x^2 + y^2 = 1$ but does not contain the origin. Show that $\int_C \mathbf{F} \cdot d\mathbf{r}$ is not independent of path in R, and hence $\mathbf{F}$ is not conservative. (*Hint:* Calculate first $\int_{C_1} \mathbf{F} \cdot d\mathbf{r}$ and then $\int_{C_2} \mathbf{F} \cdot d\mathbf{r}$, where C_1 is the semicircle parametrized by $\mathbf{r}_1(t) = \cos t\mathbf{i} + \sin t\mathbf{j}$ for $0 \le t \le \pi$ and C_2 is the semicircle parametrized by $\mathbf{r}_2(t) = \cos t\mathbf{i} - \sin t\mathbf{j}$ for $0 \le t \le \pi$.)

11. Let g be a continuous function of one variable, and let
$$\mathbf{F}(x, y, z) = [g(x^2 + y^2 + z^2)](x\mathbf{i} + y\mathbf{j} + z\mathbf{k}) \quad (6)$$

a. Show that $\mathbf{F}$ is conservative. (*Hint:* Show that $\mathbf{F} = \text{grad } f$, where $f(x, y, z) = \tfrac{1}{2}h(x^2 + y^2 + z^2)$ and $h(u) = \int g(u)\,du$.)

b. Show that $\mathbf{F}$ is irrotational.

(Because central force fields are in the form of (6), parts (a) and (b) show that they are both conservative and irrotational.)

Applications

12. Find the work W done on a rocket weighing 5000 pounds by the earth's gravitational field when the rocket descends to earth from a distance of 7460 miles from the center of the earth. (*Hint:* Take the radius of the earth to be 3960 miles, and recall that the gravitational field is conservative. The weight of an object is the magnitude of the gravitational force on it at the earth's surface.)

13. Assume that an electron is located at the origin and has a charge of $(-1.6) \times 10^{-19}$ coulombs. Find the work W done by the electric field on a positive unit charge that moves from a distance of 10^{-11} meters to a distance of 10^{-12} meters from the electron. (*Hint:* The electric field is conservative. Take $4\pi\varepsilon_0 = 1.113 \times 10^{-10}$. Your answer will be in joules.)

14. The force $\mathbf{F}(x, y)$ exerted at the point (x, y) by a two-dimensional linear oscillator at the origin is given by

$$\mathbf{F}(x, y) = -ax\mathbf{i} - ay\mathbf{j}$$

where a is a positive constant. Find the work W done by the force on an object that moves from $(3, -6)$ to $(1, -2)$.

15. Suppose the speed of an object having a mass of 5 kilograms and moving in a conservative force field decreases from 50 meters per second to 10 meters per second. Find the increase in potential energy of the object. (Your answer will be in joules.)

16. The speed of water is greater in the middle of a river than along the banks. The velocity field of the river creates a force field $\mathbf{F}$ for a canoe traveling in the river. By considering the path in Figure 15.18, explain why $\mathbf{F}$ should not be conservative. (*Hint:* Why are $\int_{C_2} \mathbf{F} \cdot d\mathbf{r} = 0$ and $\int_{C_4} \mathbf{F} \cdot d\mathbf{r} = 0$?)

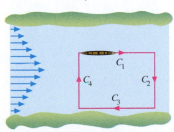

FIGURE 15.18   Figure for Exercise 16.

15.4   GREEN'S THEOREM

If f is continuous on a closed interval $[a, b]$ and if F is an antiderivative of f on $[a, b]$, then the Fundamental Theorem of Calculus tells us that

$$\int_a^b f(x)\, dx = F(b) - F(a)$$

An alternative way of interpreting the Fundamental Theorem is that we can evaluate $\int_a^b f(x)\, dx$ by evaluating an antiderivative F at the boundary points a and b of the interval $[a, b]$. In the preceding section we discovered a similar result for certain line integrals. Now we will present an analogue of the Fundamental Theorem that applies to regions in the plane.

Although the result we will derive applies to almost any plane region you are likely to encounter, the proof is complicated when very general regions are involved. Hence we will state the theorem, which is usually called Green's Theorem, only for simple regions in the plane.

THEOREM 15.8
Green's Theorem

Let R be a simple region in the xy plane with a piecewise smooth boundary C oriented counterclockwise. Let M and N be functions of two variables having continuous partial derivatives on R. Then

$$\int_C M(x, y)\, dx + N(x, y)\, dy = \iint_R \left(\frac{\partial N}{\partial x} - \frac{\partial M}{\partial y} \right) dA$$

George Green (1793–1841)
By profession a baker, Green was a self-taught mathematician and physicist who privately published the book *An Essay on the Application of Mathematical Analysis to the Theories of Electricity and Magnetism.* This treatise contained results related to the theorem that now bears his name. Some years later he entered Cambridge University as a fellow.

Proof   It is sufficient to prove that

$$\int_C N(x, y)\, dy = \iint_R \frac{\partial N}{\partial x}\, dA \tag{1}$$

and

$$\int_C M(x, y)\, dx = -\iint_R \frac{\partial M}{\partial y}\, dA \tag{2}$$

since the result follows from adding these two equations. We will prove only (2), because (1) follows by a similar argument. To prove (2), we assume that R is the region between the graphs of g_1 and g_2 on $[a, b]$. On the one hand, this means that

$$-\iint_R \frac{\partial M}{\partial y}\, dA = -\int_a^b \int_{g_1(x)}^{g_2(x)} \frac{\partial M}{\partial y}\, dy\, dx = -\int_a^b M(x, y) \Big|_{g_1(x)}^{g_2(x)} dx$$

$$= -\int_a^b \left[M(x, g_2(x)) - M(x, g_1(x)) \right] dx$$

$$= \int_a^b \left[M(x, g_1(x)) - M(x, g_2(x)) \right] dx$$

On the other hand, the boundary C of R is composed of the curves C_1, C_2, C_3, and C_4 (Figure 15.19), parametrized as follows:

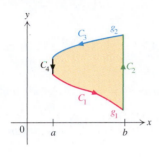

FIGURE 15.19

C_1: $\mathbf{r}_1(t) = t\mathbf{i} + g_1(t)\mathbf{j}$ for $a \le t \le b$

C_2: $\mathbf{r}_2(t) = b\mathbf{i} + t\mathbf{j}$ for $g_1(b) \le t \le g_2(b)$

C_3: $\mathbf{r}_3(t) = (a + b - t)\mathbf{i} + g_2(a + b - t)\mathbf{j}$ for $a \le t \le b$

C_4: $\mathbf{r}_4(t) = a\mathbf{i} + [g_1(a) + g_2(a) - t]\mathbf{j}$ for $g_1(a) \le t \le g_2(a)$

(C_2 or C_4 or both may contain only one point.) We will use (10) of Section 15.2 to evaluate $\int_{C_i} M(x, y)\, dx$ for $i = 1, 2, 3, 4$. First we notice that x is constant on C_2 and on C_4, so that $dx/dt = 0$ and thus

$$\int_{C_2} M(x, y)\, dx = 0 \quad \text{and} \quad \int_{C_4} M(x, y)\, dx = 0$$

Therefore

$$\int_C M(x, y)\, dx = \int_{C_1} M(x, y)\, dx + \int_{C_2} M(x, y)\, dx + \int_{C_3} M(x, y)\, dx + \int_{C_4} M(x, y)\, dx$$

$$= \int_{C_1} M(x, y)\, dx + \int_{C_3} M(x, y)\, dx$$

$$= \int_a^b M(t, g_1(t))(1)\, dt + \int_a^b M(a + b - t, g_2(a + b - t))(-1)\, dt$$

$$\overset{u = a + b - t}{=} \int_a^b M(t, g_1(t))\, dt + \int_b^a M(u, g_2(u))\, du$$

$$= \int_a^b [M(x, g_1(x)) - M(x, g_2(x))]\, dx$$

Notice that the last integral appears at the end of our calculation of the double integral above. Thus

$$\int_C M(x, y)\, dx = -\iint_R \frac{\partial M}{\partial y}\, dA \qquad \blacksquare$$

The assumption in Green's Theorem that C is the boundary of R and is oriented counterclockwise implies that if we traverse C in the direction defined by the orientation of C, the region R will always lie to our left (as in Figure 15.19).

By using Green's Theorem we can sometimes evaluate a line integral $\int_C M(x, y)\, dx + N(x, y)\, dy$ without using a parametrization of C. Example 1 illustrates this consequence.

EXAMPLE 1   Find $\int_C - x^2 y\, dx + x^3\, dy$, where C is the circle $x^2 + y^2 = 4$, oriented counterclockwise.

Solution   We can find the line integral directly, but it is simpler to use Green's Theorem with $M(x, y) = -x^2 y$, $N(x, y) = x^3$, and R equal to the disk $x^2 + y^2 \le 4$. Since

$$\frac{\partial N}{\partial x} = 3x^2 \quad \text{and} \quad \frac{\partial M}{\partial y} = -x^2$$

we have

$$\iint\limits_R \left(\frac{\partial N}{\partial x} - \frac{\partial M}{\partial y}\right) dA = \iint\limits_R (3x^2 + x^2)\, dA = \int_0^{2\pi} \int_0^2 4(r\cos\theta)^2 r\, dr\, d\theta$$

$$= \int_0^{2\pi} \int_0^2 4r^3 \cos^2\theta\, dr\, d\theta = \int_0^{2\pi} (\cos^2\theta) r^4 \Big|_0^2 \, d\theta$$

$$= 16 \int_0^{2\pi} \cos^2\theta\, d\theta = 16 \int_0^{2\pi} \left(\frac{1}{2} + \frac{1}{2}\cos 2\theta\right) d\theta$$

$$= 16 \left(\frac{\theta}{2} + \frac{1}{4}\sin 2\theta\right) \Big|_0^{2\pi} = 16\pi$$

Thus by Green's Theorem,

$$\int_C - x^2 y\, dx + x^3\, dy = 16\pi \qquad \square$$

EXAMPLE 2 Let C be the closed curve described in Figure 15.20, oriented counterclockwise. Evaluate

$$\int_C 2y^3\, dx + (x^4 + 6y^2 x)\, dy$$

Solution Again, we can either evaluate the line integral or use Green's Theorem and evaluate a double integral. As before, the latter is more efficient:

$$\int_C 2y^3\, dx + (x^4 + 6y^2 x)\, dy = \iint\limits_R \left[\frac{\partial}{\partial x}(x^4 + 6y^2 x) - \frac{\partial}{\partial y}(2y^3)\right] dA$$

$$= \int_0^1 \int_0^{(1-x^4)^{1/4}} 4x^3\, dy\, dx$$

$$= \int_0^1 4x^3 (1 - x^4)^{1/4}\, dx$$

$$= -\frac{4}{5}(1 - x^4)^{5/4} \Big|_0^1 = \frac{4}{5} \qquad \square$$

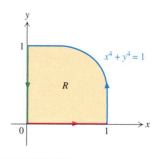

FIGURE 15.20

Green's Theorem can also be used in the reverse way, that is, to evaluate a double integral by evaluating a line integral. In particular, suppose we wish to evaluate $\iint_R 1\, dA$, which represents the area of R. If R is a simple region with a piecewise smooth boundary C and if M and N are chosen so that

$$\frac{\partial N}{\partial x} - \frac{\partial M}{\partial y} = 1 \tag{3}$$

then by Green's Theorem,

$$\iint\limits_R 1\, dA = \iint\limits_R \left(\frac{\partial N}{\partial x} - \frac{\partial M}{\partial y}\right) dA = \int_C M(x, y)\, dx + N(x, y)\, dy$$

where C is the boundary of R, oriented counterclockwise. Of the many possible choices of M and N satisfying equation (3), the three most commonly used are

$$M(x, y) = 0 \qquad \text{and} \quad N(x, y) = x$$

$$M(x, y) = -y \qquad \text{and} \quad N(x, y) = 0$$

$$M(x, y) = -\frac{1}{2}y \quad \text{and} \quad N(x, y) = \frac{1}{2}x$$

These three sets of equations for M and N yield the following formulas for the area A of R:

$$A = \int_C x\, dy = -\int_C y\, dx = \frac{1}{2}\int_C x\, dy - y\, dx \qquad (4)$$

Of course, the first two integrals appear to be simpler than the third, but there are occasions when the third is the easiest to evaluate.

EXAMPLE 3 Use (4) to compute the area A of the region R enclosed by the ellipse

$$\frac{x^2}{a^2} + \frac{y^2}{b^2} = 1$$

Solution   The ellipse is oriented counterclockwise by the parametrization

$$\mathbf{r}(t) = a \cos t\mathbf{i} + b \sin t\mathbf{j} \quad \text{for } 0 \le t \le 2\pi$$

Therefore by (4),

$$A = \frac{1}{2}\int_C x\, dy - y\, dx = \frac{1}{2}\int_0^{2\pi} \left[(a \cos t)(b \cos t) - (b \sin t)(-a \sin t)\right] dt$$

$$= \frac{1}{2}\int_0^{2\pi} ab\, dt = \pi ab \qquad \square$$

The solution of Example 3 is our third method of finding the area of an ellipse: by parametric equations (Section 6.8), by trigonometric substitution (Section 8.3), and by line integrals.

Green's Theorem applies to many regions that are not simple but can be broken up into collections of simple regions.

EXAMPLE 4 Show that Green's Theorem holds for the semiannular region R shown in Figure 15.21(a).

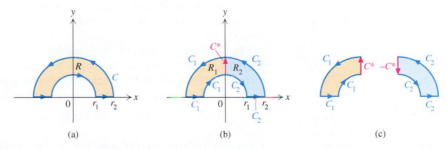

(a)   (b)   (c)

FIGURE 15.21 A nonsimple region subdivided into simple regions.

Solution   The region R is vertically simple but not horizontally simple, so Green's Theorem as stated does not apply. However, let us insert the line C^* indicated in Figure 15.21(b). Then R is divided into two simple subregions R_1 and R_2, with C_1 and C^* composing the boundary of R_1 and C_2 and $-C^*$ composing the boundary of R_2 (Figure 15.21(c)). Since each boundary is oriented counterclockwise, we can use Green's Theorem on R_1 and R_2:

$$\iint_R \left(\frac{\partial N}{\partial x} - \frac{\partial M}{\partial y} \right) dA = \iint_{R_1} \left(\frac{\partial N}{\partial x} - \frac{\partial M}{\partial y} \right) dA + \iint_{R_2} \left(\frac{\partial N}{\partial x} - \frac{\partial M}{\partial y} \right) dA$$

$$= \int_{C_1} M(x, y)\, dx + N(x, y)\, dy + \int_{C^*} M(x, y)\, dx + N(x, y)\, dy$$

$$+ \int_{C_2} M(x, y)\, dx + N(x, y)\, dy$$

$$+ \int_{-C^*} M(x, y)\, dx + N(x, y)\, dy$$

$$= \int_{C_1} M(x, y)\, dx + N(x, y)\, dy + \int_{C_2} M(x, y)\, dx + N(x, y)\, dy$$

$$= \int_C M(x, y)\, dx + N(x, y)\, dy \qquad \square$$

EXAMPLE 5   Let R be the semiannular region shown in Figure 15.21(a), where $r_1 = 1$ and $r_2 = 2$. Find

$$\int_C y^3\, dx - x^3\, dy$$

Solution   By Example 4 we may apply Green's Theorem to obtain

$$\int_C y^3\, dx - x^3\, dy = \iint_R \left[\frac{\partial}{\partial x}(-x^3) - \frac{\partial}{\partial y}(y^3) \right] dA = -\iint_R 3(x^2 + y^2)\, dA$$

$$= -\int_0^\pi \int_1^2 (3r^2) r\, dr\, d\theta = -\int_0^\pi \frac{3r^4}{4} \Big|_1^2 d\theta$$

$$= -\int_0^\pi \frac{45}{4}\, d\theta = -\frac{45}{4}\pi \qquad \square$$

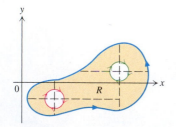

FIGURE 15.22   A nonsimple region subdivided into simple regions.

To evaluate the integral in Example 5 by parametrizing the boundary of R and then evaluating the line integral would be tedious, because there would be four integrals to evaluate.

The procedure followed in Example 4—dividing a region into simple subregions—can be applied to almost any plane region that one ordinarily encounters (Figure 15.22). Thus Green's Theorem applies to very general regions.

We now have three ways of evaluating line integrals of the form $\int_C \mathbf{F} \cdot d\mathbf{r}$:

1.   Parametrize C and then use the formula

$$\int_C \mathbf{F} \cdot d\mathbf{r} = \int_C M\, dx + N\, dy$$

2. Use the Fundamental Theorem of Line Integrals (*provided* that **F** is a gradient).

3. Use Green's Theorem (*provided* that C is closed).

Alternative Forms of Green's Theorem

To conclude, we discuss two other forms of Green's Theorem. Let $\mathbf{F} = M\mathbf{i} + N\mathbf{j}$ be a continuous vector field defined on a simple region R in the xy plane, and assume that the range of **F** is contained in the xy plane. Furthermore, let the boundary C of R have its counterclockwise orientation. By Green's Theorem,

$$\int_C \mathbf{F} \cdot d\mathbf{r} = \int_C M(x, y)\, dx + N(x, y)\, dy = \iint_R \left(\frac{\partial N}{\partial x} - \frac{\partial M}{\partial y} \right) dA$$

Since

$$\operatorname{curl} \mathbf{F} = \left(\frac{\partial N}{\partial x} - \frac{\partial M}{\partial y} \right) \mathbf{k}$$

it follows that

$$(\operatorname{curl} \mathbf{F}) \cdot \mathbf{k} = \left[\left(\frac{\partial N}{\partial x} - \frac{\partial M}{\partial y} \right) \mathbf{k} \right] \cdot \mathbf{k} = \frac{\partial N}{\partial x} - \frac{\partial M}{\partial y}$$

so that

$$\int_C \mathbf{F} \cdot d\mathbf{r} = \iint_R (\operatorname{curl} \mathbf{F}) \cdot \mathbf{k}\, dA \tag{5}$$

To obtain the second alternative form of Green's Theorem, we assume that the boundary C of a simple region R is oriented counterclockwise by a smooth parametrization

$$\mathbf{r}(t) = x(t)\mathbf{i} + y(t)\mathbf{j} \quad \text{for } a \le t \le b$$

Then the tangent **T** of C is given by

$$\mathbf{T}(t) = \frac{x'(t)}{\|\mathbf{r}'(t)\|}\mathbf{i} + \frac{y'(t)}{\|\mathbf{r}'(t)\|}\mathbf{j}$$

Let

$$\mathbf{n}(t) = \frac{y'(t)}{\|\mathbf{r}'(t)\|}\mathbf{i} - \frac{x'(t)}{\|\mathbf{r}'(t)\|}\mathbf{j}$$

Then **n** lies in the xy plane and is perpendicular to **T**, so **n** is parallel to the normal vector **N** of C. (It can be shown that **n** points "out of" the region R.) If $\mathbf{F} = M\mathbf{i} + N\mathbf{j}$ is a continuous vector field, then by (4) of Section 15.2,

$$\int_C \mathbf{F} \cdot \mathbf{n}\, ds = \int_a^b (\mathbf{F} \cdot \mathbf{n})(t)\, \|\mathbf{r}'(t)\|\, dt$$

$$= \int_a^b [M(x(t), y(t))\mathbf{i} + N(x(t), y(t))\mathbf{j}] \cdot \left(\frac{y'(t)}{\|\mathbf{r}'(t)\|}\mathbf{i} - \frac{x'(t)}{\|\mathbf{r}'(t)\|}\mathbf{j} \right) \|\mathbf{r}'(t)\|\, dt$$

$$= \int_a^b [M(x(t), y(t))y'(t) - N(x(t), y(t))x'(t)]\, dt$$

$$= \int_C M(x, y)\, dy - N(x, y)\, dx$$

Then using Green's Theorem and the fact that div $\mathbf{F}(x, y) = \partial M/\partial x + \partial N/\partial y$, we find that

$$\int_C \mathbf{F} \cdot \mathbf{n} \, ds = \int_C M(x, y) \, dy - N(x, y) \, dx = \iint_R \left(\frac{\partial M}{\partial x} + \frac{\partial N}{\partial y} \right) dA$$

$$= \iint_R \operatorname{div} \mathbf{F}(x, y) \, dA$$

Thus
$$\int_C \mathbf{F} \cdot \mathbf{n} \, ds = \iint_R \operatorname{div} \mathbf{F}(x, y) \, dA \tag{6}$$

Formulas (5) and (6) are related to formulas appearing in two other generalizations of the Fundamental Theorem, which we will present in Sections 15.7 and 15.8.

EXERCISES 15.4

In Exercises 1–5 find $\int_C M(x, y) \, dx + N(x, y) \, dy$, where C is oriented counterclockwise.

1. $M(x, y) = y$, $N(x, y) = 0$; C is composed of the portion of the quarter circle $x^2 + y^2 = 4$ in the first quadrant and of the intervals $[0, 2]$ on the x and y axes.

2. $M(x, y) = 0$, $N(x, y) = x$; C is the circle $x^2 + y^2 = 9$.

3. $M(x, y) = xy$, $N(x, y) = x^{3/2} + y^{3/2}$; C is the square with vertices $(0, 0)$, $(1, 0)$, $(1, 1)$, and $(0, 1)$.

4. $M(x, y) = y \cos x$, $N(x, y) = x \sin y$; C is the triangle with vertices $(0, 0)$, $(\pi/2, 0)$, and $(0, \pi/2)$.

5. $M(x, y) = (x^2 + y^2)^{3/2} = N(x, y)$; C is the circle $x^2 + y^2 = 1$.

In Exercises 6–14 use Green's Theorem to evaluate the line integral. Assume that each curve is oriented counterclockwise.

6. $\int_C (y^3 + y) \, dx + 3y^2 x \, dy$; C is the circle $x^2 + y^2 = 100$.

7. $\int_C y \, dx - x \, dy$; C is the cardioid $r = 1 - \cos \theta$.

8. $\int_C (2xy^3 + \cos x) \, dx + (3x^2 y^2 + 5x) \, dy$; C is the circle $x^2 + y^2 = 64$.

9. $\int_C e^x \sin y \, dx + e^x \cos y \, dy$; C is composed of the graph of $\sqrt{x} + \sqrt{y} = 5$ and the intervals $[0, 25]$ on the x and y axes.

10. $\int_C (x \cos^2 x - y) \, dx + (x - e^y) \, dy$; C is the square with vertices $(-1, 1)$, $(-1, 0)$, $(0, 0)$, and $(0, 1)$.

11. $\int_C xy \, dx + (\frac{1}{2}x^2 + xy) \, dy$; C is composed of the interval $[-1, 1]$ on the x axis and the top half of the ellipse $x^2 + 4y^2 = 1$.

12. $\int_C xy^2 \, dx + (x^4 + x^2 y) \, dy$; C is the graph of $x^4 + y^4 = 1$.

13. $\int_C (\cos^3 x + e^x) \, dx + e^y \, dy$; C is the graph of $x^6 + y^8 = 1$.

14. $\int_C \ln (x^2 + y^2) \, dx + \ln (x^2 + y^2) \, dy$; C is the boundary of the semiannular region in Figure 15.21(a), with $r_1 = 1$ and $r_2 = 2$.

In Exercises 15–18 use Green's Theorem to evaluate $\int_C \mathbf{F} \cdot d\mathbf{r}$, where C is oriented counterclockwise.

15. $\mathbf{F}(x, y) = y\mathbf{i} + 3x\mathbf{j}$; C is the circle $x^2 + y^2 = 4$.

16. $\mathbf{F}(x, y) = y^4\mathbf{i} + x^3\mathbf{j}$; C is the square with vertices $(-2, -2)$, $(-2, 2)$, $(2, -2)$, and $(2, 2)$.

17. $\mathbf{F}(x, y) = y \sin x\mathbf{i} - \cos x\mathbf{j}$; C is composed of the semicircle $x^2 + y^2 = 9$ for $y \geq 0$, and the line $y = 0$ for $-3 \leq x \leq 3$.

18. $\mathbf{F}(x, y) = y(x^2 + y^2)\mathbf{i} - x(x^2 + y^2)\mathbf{j}$; C is the unit circle $x^2 + y^2 = 1$.

In Exercises 19–21 use Green's Theorem to find the area A of the given region.

19. The region bounded below by the x axis and above by one arch of the cycloid parametrized by

$$\mathbf{r}(t) = (t - \sin t)\mathbf{i} + (1 - \cos t)\mathbf{j} \quad \text{for } 0 \leq t \leq 2\pi$$

(*Hint:* To obtain the area of the region you must use the correct orientation of the boundary.)

20. The region bounded by the hypocycloid parametrized by

$$\mathbf{r}(t) = \cos^3 t\mathbf{i} + \sin^3 t\mathbf{j} \quad \text{for } 0 \leq t \leq 2\pi$$

21. The region bounded by the y axis, the line $y = \frac{1}{4}$, and the curve parametrized by

$$\mathbf{r}(t) = \sin \pi t\mathbf{i} + t(1 - t)\mathbf{j} \quad \text{for } 0 \leq t \leq \frac{1}{2}$$

22. Suppose R is a simple region in the xy plane, with a piece-

wise smooth boundary C and area A. Use Green's Theorem to show that the center of gravity $(\overline{x}, \overline{y})$ of R is given by

$$\overline{x} = \frac{1}{2A}\int_C x^2\, dy \quad \text{and} \quad \overline{y} = -\frac{1}{2A}\int_C y^2\, dx$$

23. Use (4) and the result of Exercise 22 to find the center of gravity of the region R, where R is bounded by the curve C parametrized by $\mathbf{r}(t) = (2\cos t - \sin 2t)\mathbf{i} + 2\sin t\,\mathbf{j}$ for $0 \le t \le 2\pi$ (Figure 15.23).

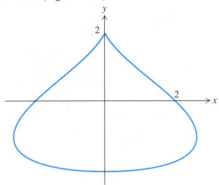

FIGURE 15.23   Graph for Exercise 23.

*24. Let

$$M(x, y) = \frac{-y}{x^2 + y^2} \quad \text{and} \quad N(x, y) = \frac{x}{x^2 + y^2}$$

a. Verify that

$$\int_C M(x, y)\, dx + N(x, y)\, dy = \iint_R \left(\frac{\partial N}{\partial x} - \frac{\partial M}{\partial y}\right) dA$$

if R is the annulus whose boundary C is composed of the circle $x^2 + y^2 = 4$ oriented counterclockwise and the circle $x^2 + y^2 = 1$ oriented clockwise.

b. Show that

$$\int_C M(x, y)\, dx + N(x, y)\, dy \ne \iint_R \left(\frac{\partial N}{\partial x} - \frac{\partial M}{\partial y}\right) dA$$

if R is the disk whose boundary C is the circle $x^2 + y^2 = 1$.

c. Why does the result of part (b) not contradict Green's Theorem?

25. Assume that R is a simple region and that C_1 and C_2 are piecewise smooth closed curves in R, both oriented counterclockwise. Suppose $\partial N/\partial x = \partial M/\partial y$ on R. Use Green's Theorem to prove that

$$\int_{C_1} M(x, y)\, dx + N(x, y)\, dy = \int_{C_2} M(x, y)\, dx + N(x, y)\, dy$$

*26. Prove the formula in (1), which yields the second half of the proof of Green's Theorem.

27. Let g be a function and $\mathbf{F}$ a vector field, and assume that g and $\mathbf{F}$ are defined on a simple region R in the xy plane with a piecewise smooth boundary C oriented counterclockwise. Assume that the partial derivatives of g and of the component functions of $\mathbf{F}$ exist and are continuous throughout R.

a. Using the identity

$$\text{div}\,(g\mathbf{F}) = g\,\text{div}\,\mathbf{F} + (\text{grad}\,g)\cdot\mathbf{F}$$

and the alternative form of Green's Theorem in (6), show that

$$\int_C g\mathbf{F}\cdot\mathbf{n}\, ds = \iint_R \left[g\,\text{div}\,\mathbf{F} + (\text{grad}\,g)\cdot\mathbf{F}\right] dA \quad (7)$$

b. Suppose f is a function of two variables having continuous partial derivatives throughout R. Show that

$$\int_C g(\text{grad}\,f)\cdot\mathbf{n}\, ds$$

$$= \iint_R \left[g\nabla^2 f + (\text{grad}\,g)\cdot(\text{grad}\,f)\right] dA \quad (8)$$

Equation (8) is known as **Green's Theorem in the first form.**

c. Interchange the roles of f and g in (8), subtract the two sides of the resulting equation from (8), and deduce that

$$\int_C (g\,\text{grad}\,f - f\,\text{grad}\,g)\cdot\mathbf{n}\, ds$$

$$= \iint_R (g\nabla^2 f - f\nabla^2 g)\, dA \quad (9)$$

Equation (9) is known as **Green's Theorem in the second form.** (Formulas (8) and (9), which appear in a three-dimensional form in Green's book, are two of the formulas usually known as "Green's identities." But they are the closest of the results in his book to the theorem called Green's Theorem.)

28. Let f be defined on a simple region R with piecewise smooth boundary C, and assume that $f(x, y) = 0$ for (x, y) on C. Assume also that $\nabla^2 f = 0$ on R.

a. Taking $f = g$ in (8), prove that

$$\iint_{R_0} \|\text{grad}\,f\|^2\, dA = 0$$

for every rectangle R_0 contained in R.

b. Use (a) and Exercise 71 of Section 14.1 to deduce that $\text{grad}\,f(x, y) = \mathbf{0}$ for (x, y) in R.

c. Use (b) and Exercise 32(a) of Section 15.1 to conclude that f is constant on R.

29. Suppose f is a function of two variables that is harmonic throughout a simple region R. Use Green's Theorem to show that $\int_C -f_y\, dx + f_x\, dy$ is independent of path in R.

Application

*30. In this exercise you will use techniques of this section to rederive Kepler's Second Law, which describes the rate at which the position vector from the sun to a planet sweeps out area.

    a. Let $\mathbf{r} = x\mathbf{i} + y\mathbf{j}$. Using (4) of Section 12.7, with τ replacing t, show that

$$x\frac{dy}{d\tau} - y\frac{dx}{d\tau} = p$$

    b. Let (x_0, y_0) be the position of a planet at a fixed time t_0, and (x, y) the position at an arbitrary later time t. Let $A(t)$ be the area of the region bounded by the curve C that is composed of the curves C_1, C_2, and C_3 in Figure 15.24. Show that

$$A(t) = \frac{1}{2}\int_C x\,dy - y\,dx = \frac{1}{2}\int_{t_0}^{t}\left(x\frac{dy}{d\tau} - y\frac{dx}{d\tau}\right)d\tau$$

    (*Hint:* Use (7) in Section 15.2 to demonstrate that $\frac{1}{2}\int_{C_1} x\,dy - y\,dx = \frac{1}{2}\int_{C_2} x\,dy - y\,dx = 0$.)

    c. Using (a) and (b), show that $dA/dt = p/2$. (This yields the same result as (18) of Section 12.7, which led us to Kepler's Second Law.)

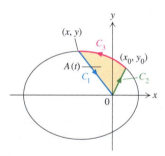

FIGURE 15.24 Graph for Exercise 30(b).

Project

1. a. Let C be the line segment from a point (x_1, y_1) to a point (x_2, y_2) in the plane. Show that

$$\frac{1}{2}\int_C x\,dy - y\,dx = \frac{1}{2}(x_1 y_2 - x_2 y_1)$$

    b. Suppose the vertices of a polygon, labeled counterclockwise, are (x_1, y_1), (x_2, y_2), . . . , (x_n, y_n). Using part (a), show that the area A of the polygon is given by

$$A = \frac{1}{2}(x_1 y_2 - x_2 y_1) + \frac{1}{2}(x_2 y_3 - x_3 y_2) + \cdots$$
$$+ \frac{1}{2}(x_{n-1} y_n - x_n y_{n-1}) + \frac{1}{2}(x_n y_1 - x_1 y_n)$$

    c. Find the area of the quadrilateral with vertices $(0, 0)$, $(1, 0)$, $(2, 3)$, and $(-1, 1)$.

    d. Consider the regular polygon with n sides, inscribed in the unit circle $x^2 + y^2 = 1$, with one vertex $(x_1, y_1) = (1, 0)$, and the next vertex counterclockwise labeled (x_2, y_2). Find a formula for (x_2, y_2), and then show that

$$\frac{1}{2}(x_1 y_2 - x_2 y_1) = \frac{1}{n} \text{ (area of the polygon)}$$

    e. Use the result of (d) to find the area A_n of the regular polygon that is inscribed in the unit circle and has n sides, where

      i. $n = 4$

      ii. $n = 8$

      iii. $n = 15$

15.5   SURFACE INTEGRALS

Now we turn to a higher-dimensional analogue of line integrals, called surface integrals. These too help to describe certain physical phenomena, and they appear in the statements of two more generalizations of the Fundamental Theorem of Calculus: Stokes's Theorem and the Divergence Theorem, which we will present in the last two sections of this chapter.

    Section 15.2 concerned integrals over piecewise smooth curves. In this section the surface integrals will be defined over piecewise smooth surfaces. Initially let us assume that Σ is a smooth surface, parametrized by

$$\mathbf{r}(u, v) = x(u, v)\mathbf{i} + y(u, v)\mathbf{j} + z(u, v)\mathbf{k}$$

where (u, v) ranges through the points in a region R in the uv plane that is a finite union of vertically or horizontally simple regions. You might think of Σ as a thin metal surface such as a shield against x-rays, which is denser in the middle than at the sides. Let g be a function that is continuous at every point of Σ. For example, $g(x, y, z)$ could represent the mass density (mass per unit area) of Σ at the point (x, y, z), or $g(x, y, z)$ could represent electric charge or heat per unit area. To motivate the discussion leading up to the definition of surface integral, we will assume that g represents mass density and then define the mass of the plate Σ.

Proceeding as we did when we introduced the plane tangent to a parametrized surface in Section 14.9, let us select a rectangle R' that contains R, and partition R' into subrectangles with sides parallel to the u and v axes, with $R_1, R_2, \ldots, R_n$ denoting those subrectangles that are entirely contained in R. For $1 \le k \le n$, let Σ_k be the set of points in Σ that correspond to points in R_k, and let (u_k, v_k) be the vertex of R_k with the smallest values of u_k and v_k (Figure 15.25). Furthermore, for

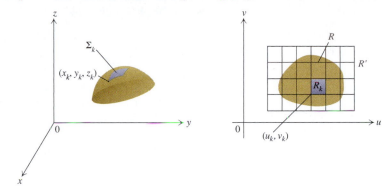

FIGURE 15.25

simplicity we let $x_k = x(u_k, v_k)$, $y_k = y(u_k, v_k)$, and $z_k = z(u_k, v_k)$. Then (x_k, y_k, z_k) is a point on Σ_k. Recalling that mass equals the product of area and density, we multiply the area ΔS_k of Σ_k by the density $g(x_k, y_k, z_k) = g(x(u_k, v_k), y(u_k, v_k), z(u_k, v_k))$ at the point (x_k, y_k, z_k) to obtain an approximation for the mass of Σ_k. Since

$$\Delta S_k \approx \| \mathbf{r}_u(u_k, v_k) \times \mathbf{r}_v(u_k, v_k) \| \, \Delta A_k$$

by (13) in Section 14.9, the mass of Σ_k is therefore approximately equal to

$$g(x(u_k, v_k), y(u_k, v_k), z(u_k, v_k)) \| \mathbf{r}_u(u_k, v_k) \times \mathbf{r}_v(u_k, v_k) \| \, \Delta A_k$$

Consequently the mass of Σ, which equals the sum of the masses $\Sigma_1, \Sigma_2, \ldots, \Sigma_n$, is approximately equal to

$$\sum_{k=1}^{n} g(x(u_k, v_k), y(u_k, v_k), z(u_k, v_k)) \| \mathbf{r}_u(u_k, v_k) \times \mathbf{r}_v(u_k, v_k) \| \, \Delta A_k$$

Since the preceding sum is a Riemann sum for

$$\iint_R g(x(u, v), y(u, v), z(u, v)) \| \mathbf{r}_u(u, v) \times \mathbf{r}_v(u, v) \| \, dA$$

and since it seems reasonable that the approximation can be made as accurate as we wish by taking a fine enough partition of R', we make the following definition.

DEFINITION 15.9

Let Σ be a smooth surface with parametrization

$$\mathbf{r}(u, v) = x(u, v)\mathbf{i} + y(u, v)\mathbf{j} + z(u, v)\mathbf{k} \tag{1}$$

and assume that g is continuous on Σ. Then the **surface integral** $\iint_\Sigma g(x, y, z)\, dS$ is defined by

$$\iint_\Sigma g(x, y, z)\, dS = \iint_R g(x(u, v), y(u, v), z(u, v))\, \|\mathbf{r}_u(u, v) \times \mathbf{r}_v(u, v)\|\, dA \tag{2}$$

If Σ represents a thin metal surface and $g(x, y, z)$ the mass density at any point (x, y, z) of Σ, then according to the discussion preceding the definition above, the mass m of the plate is given by

$$m = \iint_\Sigma g(x, y, z)\, dS$$

It is possible to prove that the value of the surface integral in Definition 15.9 is independent of which parametrization $\mathbf{r}$ is used in (2). However, the proof is technical, and we omit it.

EXAMPLE 1 Evaluate $\iint_\Sigma e^z\, dS$, where Σ is the sphere $x^2 + y^2 + z^2 = 9$.

Solution We use a standard parametrization of the sphere of radius a (see Example 1 in Section 14.9):

$$\mathbf{r}(\phi, \theta) = a \sin \phi \cos \theta\, \mathbf{i} + a \sin \phi \sin \theta\, \mathbf{j} + a \cos \phi\, \mathbf{k}$$

where (ϕ, θ) ranges through the rectangle R defined by $0 \leq \phi \leq \pi$ and $0 \leq \theta \leq 2\pi$. Below we will use the fact, derived in the solution of Example 9 in Section 14.9, that

$$\|\mathbf{r}_\phi(\phi, \theta) \times \mathbf{r}_\theta(\phi, \theta)\| = a^2 \sin \phi \tag{3}$$

Continuing with our solution, we apply Definition 15.9 as well as (3) with $a = 3$, and deduce that

$$\iint_\Sigma e^z\, dS = \iint_R e^{3 \cos \phi} \|\mathbf{r}_\phi(\phi, \theta) \times \mathbf{r}_\theta(\phi, \theta)\|\, dA = \int_0^\pi \int_0^{2\pi} e^{3 \cos \phi}\, (9 \sin \phi)\, dA$$

$$= \int_0^\pi 18\pi e^{3 \cos \phi} \sin \phi\, d\phi = -6\pi e^{3 \cos \phi} \Big|_0^\pi = 6\pi(e^3 - e^{-3}) \qquad \square$$

EXAMPLE 2 Evaluate $\iint_\Sigma (1 + z)\, dS$, where Σ is the hemisphere $z = \sqrt{1 - x^2 - y^2}$.

Solution We will use the same parametrization for the hemisphere as we have used for a sphere (but with restricted domain):

$\mathbf{r}(\phi, \theta) = \sin \phi \cos \theta \, \mathbf{i} + \sin \phi \sin \theta \, \mathbf{j} + \cos \phi \, \mathbf{k}, \quad \text{for } 0 \le \phi \le \pi/2 \text{ and } 0 \le \theta \le 2\pi$

In this example the hemisphere has radius 1, so that (3) becomes

$$\| \mathbf{r}_\phi \times \mathbf{r}_\theta \| = \sin \phi$$

Therefore

$$\iint_\Sigma (1 + z) \, dS = \int_0^{2\pi} \int_0^{\pi/2} (1 + \cos \phi) \sin \phi \, d\phi \, d\theta$$

$$= 2\pi \left(-\cos \phi - \frac{1}{2} \cos^2 \phi \right) \bigg|_0^{\pi/2} = 3\pi \quad \square$$

When Σ is the graph of a function f of two variables x and y on a region R in the xy plane, then Σ has the simplified parametrization

$$\mathbf{r}(x, y) = x \, \mathbf{i} + y \, \mathbf{j} + f(x, y) \, \mathbf{k}$$

We can use (19) in Section 14.9 and rewrite (2) as

$$\iint_\Sigma g(x, y, z) dS = \iint_R g(x, y, f(x, y)) \sqrt{[f_x(x, y)]^2 + [f_y(x, y)]^2 + 1} \, dA \quad (4)$$

EXAMPLE 3   Evaluate $\iint_\Sigma y \, dS$, where Σ is the surface $z = x + y^2$ for $0 \le x \le 1$ and $0 \le y \le 2$ (Figure 15.26).

Solution   We use (4) with $g(x, y, z) = y$ and $f(x, y) = x + y^2$, and R equal to the rectangle defined by $0 \le x \le 1$ and $0 \le y \le 2$. Since $f_x(x, y) = 1$ and $f_y(x, y) = 2y$, (4) yields

$$\iint_\Sigma y \, dS = \iint_R y \sqrt{1^2 + (2y)^2 + 1} \, dA = \int_0^2 \int_0^1 y \sqrt{2 + 4y^2} \, dx \, dy$$

$$= \int_0^2 y \sqrt{2 + 4y^2} \, dy$$

$$= \frac{1}{12} (2 + 4y^2)^{3/2} \bigg|_0^2 = \frac{1}{12} (18 \sqrt{18} - 2\sqrt{2}) = \frac{13 \sqrt{2}}{3} \quad \square$$

FIGURE 15.26

Some surfaces cannot be expressed as graphs of functions of x and y but can be expressed as graphs of functions of x and z or of y and z. In such cases we can still apply (4), with the roles of x, y, and z interchanged.

EXAMPLE 4   Evaluate $\iint_\Sigma (x + y + z) \, dS$, where Σ is the portion of the plane $x + y = 1$ in the first octant for which $0 \le z \le 1$ (Figure 15.27).

Solution   Since z does not appear in the equation $x + y = 1$ that defines Σ, there is no function f of x and y whose graph is Σ. However, we can solve the equation for y in terms of x, obtaining $y = 1 - x$. Thus if we let R be the square in the xz plane consisting of all (x, z) for which $0 \le x \le 1$ and $0 \le z \le 1$, and if

$$y = f(x, z) = 1 - x$$

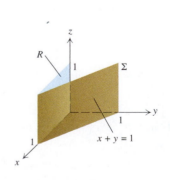

FIGURE 15.27

then Σ is the graph of $y = f(x, z)$ on R. Interchanging the roles of y and z in (4), we obtain

$$\sqrt{[f_x(x, y)]^2 + [f_z(x, y)]^2 + 1} = \sqrt{(-1)^2 + 0 + 1} = \sqrt{2}$$

and finally

$$\iint_\Sigma (x + y + z)\, dS = \iint_R (x + (1 - x) + z) \sqrt{2}\, dA$$

$$= \sqrt{2} \int_0^1 \int_0^1 (1 + z)\, dx\, dz = \sqrt{2} \int_0^1 (1 + z)\, dz$$

$$= \sqrt{2}\left(z + \frac{z^2}{2} \right)\Big|_0^1 = \frac{3\sqrt{2}}{2} \qquad \square$$

We could have solved Example 4 by solving for x as a function of y and z, and integrating over a region in the yz plane.

If Σ is a piecewise smooth surface, that is, a finite union of smooth surfaces $\Sigma_1, \Sigma_2, \ldots, \Sigma_n$, then we define

$$\iint_\Sigma g(x, y, z)\, dS = \iint_{\Sigma_1} g(x, y, z)\, dS + \iint_{\Sigma_2} g(x, y, z)\, dS + \cdots + \iint_{\Sigma_n} g(x, y, z)\, dS$$

Now we will solve an example for which the piecewise smooth surface has three smooth components.

EXAMPLE 5   Let Σ be the portion of the cylinder $x^2 + y^2 = 1$ between the xy plane and the plane $z = 1$, together with the portions of those planes inside the cylinder. Evaluate $\iint_\Sigma z\, dS$.

Solution   The side surface consists of the cylinder parametrized by

$$\mathbf{r}_1(\theta, z) = \cos\theta\,\mathbf{i} + \sin\theta\,\mathbf{j} + z\mathbf{k} \quad \text{for } 0 \le z \le 1 \text{ and } 0 \le \theta \le 2\pi$$

and the lower and upper plane surfaces are parametrized by

$$\mathbf{r}_2(r, \theta) = r\cos\theta\,\mathbf{i} + r\sin\theta\,\mathbf{j} + 0\mathbf{k} \quad \text{for } 0 \le r \le 1 \text{ and } 0 \le \theta \le 2\pi$$

and

$$\mathbf{r}_3(r, \theta) = r\cos\theta\,\mathbf{i} + r\sin\theta\,\mathbf{j} + \mathbf{k} \quad \text{for } 0 \le r \le 1 \text{ and } 0 \le \theta \le 2\pi$$

respectively. Therefore

$$(\mathbf{r}_1)_\theta \times (\mathbf{r}_1)_z = \begin{vmatrix} \mathbf{i} & \mathbf{j} & \mathbf{k} \\ -\sin\theta & \cos\theta & 0 \\ 0 & 0 & 1 \end{vmatrix} = \cos\theta\,\mathbf{i} + \sin\theta\,\mathbf{j}$$

$$(\mathbf{r}_2)_r \times (\mathbf{r}_2)_\theta = \begin{vmatrix} \mathbf{i} & \mathbf{j} & \mathbf{k} \\ \cos\theta & \sin\theta & 0 \\ -r\sin\theta & r\cos\theta & 0 \end{vmatrix} = r\mathbf{k}$$

$$(\mathbf{r}_3)_r \times (\mathbf{r}_3)_\theta = \begin{vmatrix} \mathbf{i} & \mathbf{j} & \mathbf{k} \\ \cos\theta & \sin\theta & 0 \\ -r\sin\theta & r\cos\theta & 0 \end{vmatrix} = r\mathbf{k}$$

It follows that

$$\| (\mathbf{r}_1)_\theta \times (\mathbf{r}_1)_z \| = \sqrt{\cos^2\theta + \sin^2\theta} = 1$$

$$\| (\mathbf{r}_2)_r \times (\mathbf{r}_2)_\theta \| = r$$

$$\| (\mathbf{r}_3)_r \times (\mathbf{r}_3)_\theta \| = r$$

Since

$$\iint\limits_\Sigma z\, dS = \iint\limits_{\Sigma_1} z\, dS + \iint\limits_{\Sigma_2} z\, dS + \iint\limits_{\Sigma_3} z\, dS$$

we calculate that

$$\iint\limits_{\Sigma_1} z\, dS = \int_0^1 \int_0^{2\pi} z \| (\mathbf{r}_1)_\theta \times (\mathbf{r}_1)_z \|\, d\theta\, dz = \int_0^1 \int_0^{2\pi} z\, d\theta\, dz = \int_0^1 2\pi z\, dz = \pi$$

$$\iint\limits_{\Sigma_2} z\, dS = \int_0^{2\pi} \int_0^1 0 \| (\mathbf{r}_2)_r \times (\mathbf{r}_2)_\theta \|\, dr\, d\theta = 0$$

$$\iint\limits_{\Sigma_3} dS = \int_0^{2\pi} \int_0^1 1 \| (\mathbf{r}_3)_r \times (\mathbf{r}_3)_\theta \|\, dr\, d\theta = \int_0^{2\pi} \int_0^1 r\, dr\, d\theta = \int_0^{2\pi} \frac{1}{2}\, d\theta = \pi$$

Adding these three surface integrals together, we conclude that

$$\iint\limits_\Sigma z\, dS = \pi + 0 + \pi = 2\pi \qquad \square$$

We observe that without a parametrization for the cylindrical portion of the surface, the integral would have been more difficult to evaluate (involving the evaluation of improper integrals).

In conclusion we remark that if the function g has the property that $g(x, y, z) = 1$ for all (x, y, z) on the surface Σ, then the integral on the right side of (2) agrees with the integral in Definition 14.16, which yields the area S of Σ. Therefore the area S of Σ is also given by

$$S = \iint\limits_\Sigma 1\, dS$$

EXERCISES 15.5

In Exercises 1–12 evaluate $\iint_\Sigma g(x, y, z)\, dS$.

1. $g(x, y, z) = x$; Σ is the part of the plane $2x + 3y + z = 6$ in the first octant.

2. $g(x, y, z) = z^2$; Σ is the part of the cone $z = \sqrt{x^2 + y^2}$ between the planes $z = 1$ and $z = 3$.

3. $g(x, y, z) = 2x^2 + 1$; Σ is the part of the plane $z = 3x - 2$ inside the cylinder $x^2 + y^2 = 4$.

4. $g(x, y, z) = z^2$; Σ is the part of the sphere $x^2 + y^2 + z^2 = 9$ in the first octant.

5. $g(x, y, z) = \sqrt{4x^2 + 4y^2 + 1}$; Σ is the part of the paraboloid $z = x^2 + y^2$ below the plane $y = z$.

6. $g(x, y, z) = xy$; Σ is the part of the paraboloid $z = 4 - x^2 - y^2$ that lies above the xy plane. (*Hint:* Use rectangular coordinates.)

7. $g(x, y, z) = y$; Σ is the part of the parabolic sheet $z = 4 - y^2$ for which $0 \le x \le 3$ and $0 \le y \le 2$.

8. $g(x, y, z) = x^2 z$; Σ is the part of the cylinder $x^2 + z^2 = 1$ between the planes $y = -1$ and $y = 2$ and above the xy plane.

9. $g(x, y, z) = z(x^2 + y^2)$; Σ is the hemisphere given by $z = \sqrt{4 - x^2 - y^2}$.

10. $g(x, y, z) = x^2 + y^2$; Σ is composed of the part of the paraboloid $z = 1 - x^2 - y^2$ above the xy plane, and the part of the xy plane that lies inside the circle $x^2 + y^2 = 1$.

11. $g(x, y, z) = x + y$; Σ is the cube with vertices $(0, 0, 0)$, $(1, 0, 0)$, $(1, 1, 0)$, $(0, 1, 0)$, $(0, 0, 1)$, $(1, 0, 1)$, $(1, 1, 1)$, and $(0, 1, 1)$.

12. $g(x, y, z) = x + 1$; Σ is the tetrahedron with vertices $(0, 0, 0)$, $(1, 0, 0)$, $(0, 1, 0)$, $(0, 0, 1)$.

13. Suppose that a thin metal funnel has the shape of the conical surface $z = 2\sqrt{x^2 + y^2}$ for $\frac{1}{2} \le z \le 4$ and that the mass density of the funnel is given by $\delta(x, y, z) = 6 - z$. Compute the mass m of the funnel.

14. Suppose the mass density at any point (x, y, z) of a thin spherical metal shell with radius 5 is given by $\delta(x, y, z) = 1 + z^2$. Compute the mass m of the shell.

Applications

Let D be a solid region filled with water and Σ a surface in contact with the water (Figure 15.28). The hydrostatic pressure $p(x, y, z)$ at any point (x, y, z) on Σ is defined to be the force per unit surface area of Σ, that is,

$$p(x, y, z) = \lim_{\varepsilon \to 0} \frac{\Delta F_\varepsilon}{\Delta S_\varepsilon}$$

where ΔF_ε is the hydrostatic force exerted by the water on the portion Σ_ε of Σ that is at a distance less than ε from (x, y, z), and where ΔS_ε is the surface area of Σ_ε. It can be shown that

$$p(x, y, z) = 62.5(z_0 - z)$$

where z_0 is the z coordinate of the highest point(s) in D, and that the hydrostatic force F exerted by the water on the entire surface Σ is given by

$$F = \iint_\Sigma p(x, y, z) \, dS = \iint_\Sigma 62.5(z_0 - z) \, dS$$

(We assume that length is measured in feet and force in pounds. Water weighs approximately 62.5 pounds per cubic foot.)

In Exercises 15–17 assume that the given water tank is full of water, and compute the hydrostatic force **F** on the tank.

15. A hemispherical tank having a radius of 10 feet and a flat bottom

16. A semicylindrical tank having a radius of 4 feet, a length of 12 feet, a flat top, and a round bottom

17. A spherical tank having a radius of 10 feet

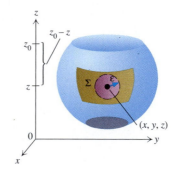

FIGURE 15.28

15.6 INTEGRALS OVER ORIENTED SURFACES

In Section 15.2 we first defined the line integral $\int_C f(x, y, z) \, ds$ of a function over a curve C and then used it to define the line integral $\int_C \mathbf{F} \cdot d\mathbf{r}$ of a vector field over an oriented curve C. Now that we have defined the surface integral $\iint_\Sigma g(x, y, z) \, dS$ of a function over a surface, we will introduce the notion of orientation of a surface and then define the surface integral of a vector field over an oriented surface.

For surface integrals of vector fields we must further limit the surfaces over which we integrate to those having two sides—which for many surfaces (such as a sphere) you might think of as the outside and the inside. In order to identify two sides of a surface Σ, we assume that Σ has a tangent plane at each of its nonboundary points. At such a point on the surface two (unit) normal vectors exist, and they have opposite directions (Figure 15.29(a)). If it is possible to select one normal at each nonboundary point in such a way that the chosen normal varies continuously on Σ, then the surface Σ is said to be **orientable,** or **two-sided,** and the selection of the normal gives an **orientation** to Σ and thus makes Σ an **oriented surface.** In such a case there are two possible orientations (Figure 15.29(b) and (c)). Spheres, paraboloids, and all the other surfaces we have encountered thus far are orientable (Figure 15.29(d) and (e)). When Σ is the boundary of a solid region D in space, we customarily choose the normal to Σ that is directed outward from D (Figure 15.29(f)).

Some surfaces are not orientable; the most celebrated one is the Möbius band*, which can be made with a strip of paper, curved and given a half turn before attaching the ends (Figure 15.30). If you think of the normal at a given point of the

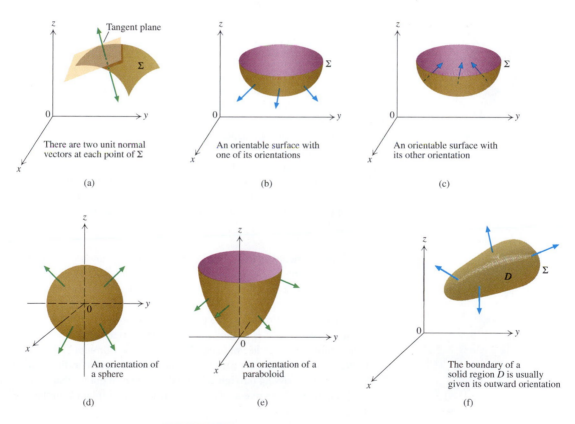

(a)   There are two unit normal vectors at each point of Σ

(b)   An orientable surface with one of its orientations

(c)   An orientable surface with its other orientation

(d)   An orientation of a sphere

(e)   An orientation of a paraboloid

(f)   The boundary of a solid region D is usually given its outward orientation

FIGURE 15.29

* **Möbius:** This name is German and is pronounced, approximately, "*Mer*-bius," but without the "r" sound.

Augustus Ferdinand Möbius (1790–1868)
One of several prominent nineteenth-century geometers, Möbius was professor at Leipzig University. About 1865 he wrote a paper in which he discussed a surface that has only one side and one edge, and which became known as a Möbius band or Möbius strip.

Möbius band as a toy soldier, and if you lead the soldier around the band along a curve whose initial and terminal points are the same, you will find that the soldier's head points in one direction at the outset and in the opposite direction at the end (Figure 15.30). This suggests (but does not prove) that the band has no orientation. It is one-sided, not two-sided.

FIGURE 15.30   A Möbius band is not orientable.

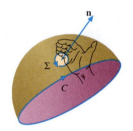

FIGURE 15.31

In the remainder of this chapter we will assume that all surfaces under consideration are orientable. If Σ is an oriented surface bounded by a curve C, then the orientation of Σ induces an orientation on C. To see how this is done, imagine placing the side of your right hand at any point on Σ, with the thumb pointed in the direction of the normal that gives Σ its orientation. If you move your right hand toward C, always keeping your thumb pointed in the direction of the normal, then the remaining fingers naturally curl in a way that defines a direction, or orientation, on C (Figure 15.31). That orientation is called the **induced orientation** on C. Notice that if the orientation of Σ is reversed, then the orientation induced on C is also reversed.

Flux Integrals

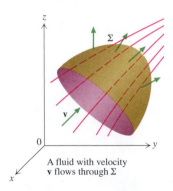

A fluid with velocity **v** flows through Σ

FIGURE 15.32

Suppose Σ is an oriented surface, which means that we can choose a unit normal vector **n** at each nonboundary point of Σ so that **n** varies continuously over Σ. For the present, think of Σ as a membrane through which a fluid of constant density δ is flowing with a constant velocity **v**. We wish to determine the rate with respect to time at which mass is flowing in the direction of **n** through the surface Σ (Figure 15.32). In the event that Σ is a plane surface having surface area S, with **n** pointing in the same direction as **v**, the volume of fluid that passes through Σ between a time t and a later time $t + h$ is $h\|\mathbf{v}\|S$, the volume of the cylinder in Figure 15.33(a). Therefore the mass passing through Σ in the direction of **n** during that time interval is $\delta h\|\mathbf{v}\|S$, and hence the rate at which mass is flowing through Σ in the direction of **n** is

$$\frac{\delta h\|\mathbf{v}\|S}{h} = \delta\|\mathbf{v}\|S$$

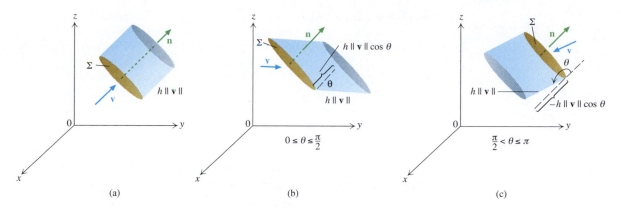

FIGURE 15.33

If Σ is a plane surface but **n** does not necessarily have the same direction as **v**, let θ be the angle between **n** and **v**, so that by convention $0 \le \theta \le \pi$. Then the volume of fluid that passes through Σ is $h\|\mathbf{v}\|(\cos\theta)S$ if $0 \le \theta \le \pi/2$ and is $-h\|\mathbf{v}\|(\cos\theta)S$ if $\pi/2 < \theta \le \pi$ (Figure 15.33(b) and (c)), for each of these numbers is the volume of the solid in the respective figure. If $0 \le \theta < \pi/2$, then the mass flows through the surface in the general direction of **n**; if $\pi/2 < \theta \le \pi$, then the mass flows in the opposite direction. Consequently in either case the amount of mass flowing through Σ *in the general direction of* **n** is $\delta h\|\mathbf{v}\|(\cos\theta)S = \delta h(\mathbf{v}\cdot\mathbf{n})S$. (If this number is negative, mass is actually flowing in the general direction opposite to **n**.) It follows that the rate of mass flow through Σ in the general direction of **n** is

$$\frac{\delta h(\mathbf{v}\cdot\mathbf{n})S}{h} = \delta(\mathbf{v}\cdot\mathbf{n})S$$

It is not necessary for the surface Σ to be a plane region or for the velocity of the fluid to be constant. Let us dispense with these hypotheses and assume only that Σ is a surface with a given orientation and surface area S. Assume also that the velocity **v** is a continuous vector field on Σ. We partition Σ into subsurfaces $\Sigma_1, \Sigma_2, \ldots, \Sigma_n$, each of which is so small that it is nearly a plane surface. If ΔS_k is the surface area of Σ_k and (x_k, y_k, z_k) is an arbitrary point on Σ_k, then the discussion in the preceding paragraph shows that the amount of mass flowing through Σ_k in the general direction of **n** between a time t and a later time $t + h$ should be approximately

$$\delta h[\mathbf{v}(x_k, y_k, z_k)\cdot\mathbf{n}(x_k, y_k, z_k)]\,\Delta S_k$$

This means that the amount of mass flowing through the entire surface Σ in the general direction of **n** during that time should be approximately

$$\sum_{k=1}^{n}\delta h[\mathbf{v}(x_k, y_k, z_k)\cdot\mathbf{n}(x_k, y_k, z_k)]\,\Delta S_k$$

Consequently the rate of mass flow should be approximately

$$\sum_{k=1}^{n}\delta[\mathbf{v}(x_k, y_k, z_k)\cdot\mathbf{n}(x_k, y_k, z_k)]\,\Delta S_k$$

We conclude that the rate of mass flow of the liquid through the whole surface Σ

should be

$$\iint_{\Sigma} \delta[\mathbf{v}(x, y, z) \cdot \mathbf{n}(x, y, z)] \, dS$$

or in abbreviated form,

$$\iint_{\Sigma} \delta \mathbf{v} \cdot \mathbf{n} \, dS \tag{1}$$

Because (1) is derived in terms of fluid motion, the integral in (1) is called a **flux integral** after the Latin word *fluxus,* meaning "a flow." In fact, any integral of the form

$$\iint_{\Sigma} \mathbf{F} \cdot \mathbf{n} \, dS \tag{2}$$

is called a **flux integral,** whether or not $\mathbf{F} = \delta \mathbf{v}$.

> **Caution:** Notice that the surface integral $\iint_{\Sigma} g(x, y, z) \, dS$ does not require Σ to be oriented. However, for the flux integral $\iint_{\Sigma} \mathbf{F} \cdot \mathbf{n} \, dS$, Σ *must* be oriented.

Flux integrals may look very difficult to evaluate. However, when Σ has a parametrization

$$\mathbf{r}(u, v) = x(u, v)\mathbf{i} + y(u, v)\mathbf{j} + z(u, v)\,\mathbf{k}$$

for u and v in a region R in the uv plane, we can use previous results to simplify the flux integral in (2).

To begin, we recall from (12) in Section 14.9 that the vector

$$\frac{\mathbf{r}_u(u, v) \times \mathbf{r}_v(u, v)}{\|\mathbf{r}_u(u, v) \times \mathbf{r}_v(u, v)\|}$$

is a unit vector normal to the surface Σ at the point $(x(u, v), y(u, v), z(u, v))$. The unit vector $\mathbf{n}$ that we will choose at the point $(x(u, v), y(u, v), z(u, v))$ is given by

$$\mathbf{n} = \frac{\mathbf{r}_u(u, v) \times \mathbf{r}_v(u, v)}{\|\mathbf{r}_u(u, v) \times \mathbf{r}_v(u, v)\|} \quad \text{or} \quad \mathbf{n} = -\frac{\mathbf{r}_u(u, v) \times \mathbf{r}_v(u, v)}{\|\mathbf{r}_u(u, v) \times \mathbf{r}_v(u, v)\|} \tag{3}$$

where we choose the first formula if $\mathbf{r}_u \times \mathbf{r}_v$ happens to be in the direction of the normal that is compatible with the given orientation of Σ, and where we choose the second formula if $\mathbf{r}_u \times \mathbf{r}_v$ has direction opposite to the given orientation of Σ. Now observe from (2) in Section 15.5 that

$$\iint_{\Sigma} \mathbf{F} \cdot \mathbf{n} \, dS = \iint_{R} [\mathbf{F}(x(u, v), y(u, v), z(u, v)) \cdot \mathbf{n}] \, \|\mathbf{r}_u(u, v) \times \mathbf{r}_v(u, v)\| \, dA$$

From the last two equations we deduce that

$$\iint\limits_{\Sigma} \mathbf{F} \cdot \mathbf{n}\, dS = \pm \iint\limits_{R} \mathbf{F}(x(u, v), y(u, v), z(u, v)) \cdot [\mathbf{r}_u(u, v) \times \mathbf{r}_v(u, v)]\, dA \quad (4)$$

where the plus (minus) sign is chosen if $\mathbf{r}_u \times \mathbf{r}_v$ is in the direction (opposite direction) of the normal for the given orientation of Σ.

Next, suppose that $\mathbf{F} = M\mathbf{i} + N\mathbf{j} + P\mathbf{k}$. Then we can use (15) of Section 14.9 to rewrite (4) above as

$$\iint\limits_{\Sigma} \mathbf{F} \cdot \mathbf{n}\, dS = \pm \iint\limits_{R} \left[M(x(u, v), y(u, v), z(u, v)) \frac{\partial(y, z)}{\partial(u, v)} \right.$$

$$+ N(x(u, v), y(u, v), z(u, v)) \frac{\partial(z, x)}{\partial(u, v)}$$

$$\left. + P(x(u, v), y(u, v), z(u, v)) \frac{\partial(x, y)}{\partial(u, v)} \right] dA \quad (5)$$

where as before the plus, or minus, sign results from the orientation of Σ.

Either (4) or (5) can be used to evaluate a flux integral over a parametrized surface.

EXAMPLE 1   Let Σ be the sphere $x^2 + y^2 + z^2 = 9$, oriented with the normal that is directed outward. Evaluate $\iint_{\Sigma} \mathbf{F} \cdot \mathbf{n}\, dS$, where $\mathbf{F}(x, y, z) = z^2\mathbf{k}$.

Solution   For the given $\mathbf{F}$, we use (4) with $M = N = 0$ and $P = z^2$. From Example 1 in Section 14.9 we know that the sphere is parametrized by

$$x = 3 \sin \phi \cos \theta, \quad y = 3 \sin \phi \sin \theta, \quad z = 3 \cos \phi$$

for

$$0 \le \phi \le \pi \quad \text{and } 0 \le \theta \le 2\pi$$

Therefore

$$\mathbf{F}(x(\phi, \theta), y(\phi, \theta), z(\phi, \theta)) = z^2\mathbf{k} = 9 \cos^2 \phi\, \mathbf{k}$$

Moreover, from the solution of Example 9 in Section 14.9, we have

$$\mathbf{r}_\phi(\phi, \theta) \times \mathbf{r}_\theta(\phi, \theta) = 9 \sin^2 \phi \cos \theta\, \mathbf{i} + 9 \sin^2 \phi \sin \theta\, \mathbf{j} + 9 \cos \phi \sin \theta\, \mathbf{k}$$

so that

$$\mathbf{F} \cdot (\mathbf{r}_u \times \mathbf{r}_v) = 81 \cos^3 \phi \sin \phi$$

Since the sphere Σ is by assumption oriented with normal pointed outward, and since $9 \cos \phi \sin \theta \ge 0$ on the upper part of the sphere (where $0 \le \phi \le \pi/2$) and $9 \cos \phi \sin \theta \le 0$ on the lower part of the sphere (where $\pi/2 \le \phi \le \pi$), it follows that $\mathbf{r}_\phi(\phi, \theta) \times \mathbf{r}_\theta(\phi, \theta)$ points outward from the sphere, in the direction of the given orientation of Σ. Therefore in applying (4) we use the plus sign and obtain

$$\iint\limits_{\Sigma} \mathbf{F} \cdot \mathbf{n}\, dS = \int_0^\pi \int_0^{2\pi} 81 \cos^3 \phi \sin \phi\, d\theta\, d\phi = (2\pi) \left(-\frac{81}{4} \cos^4 \phi \right) \Bigg|_0^\pi = 81\pi \quad \square$$

For the case in which Σ is the graph of a function f with continuous partial derivatives on a region R in the xy plane that is composed of vertically and horizontally simple regions, the formula in (4) can be made more explicit. Indeed, in this case Σ can be parametrized by $\mathbf{r}(x, y) = x\mathbf{i} + y\mathbf{j} + f(x, y)\mathbf{k}$, and by (18) in Section 14.9,

$$\mathbf{r}_x(x, y) \times \mathbf{r}_y(x, y) = -f_x(x, y)\mathbf{i} - f_y(x, y)\mathbf{j} + \mathbf{k}$$

Next, let $\mathbf{F}(x, y, z) = M(x, y, z)\mathbf{i} + N(x, y, z)\mathbf{j} + P(x, y, z)\,\mathbf{k}$. Then (4) reduces to

$$\iint_{\Sigma} \mathbf{F} \cdot \mathbf{n}\, dS = \pm \iint_{R} [(-M(x, y, f(x, y))\, f_x(x, y) - N(x, y, f(x, y))\, f_y(x, y) + P(x, y, f(x, y))]\, dA \qquad (6)$$

where the plus sign in (6) is chosen if Σ is oriented with normals pointing upward (so that the normals have positive $\mathbf{k}$ component); analogously, the minus sign in (6) is chosen if Σ is oriented with normals pointing downward (so that the normals have negative $\mathbf{k}$ component).

EXAMPLE 2   Suppose Σ is the part of the paraboloid $z = 1 - x^2 - y^2$ that lies above the xy plane and is oriented by the normal directed upward (Figure 15.34). Assume that the velocity of a fluid with constant density δ is given by $\mathbf{v}(x, y, z) = x\mathbf{i} + y\mathbf{j} + 2z\mathbf{k}$. Determine the rate of mass flow through Σ in the direction of $\mathbf{n}$.

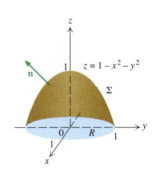

FIGURE 15.34

Solution   According to (1), the rate of mass flow is equal to the surface integral

$$\iint_{\Sigma} \delta\mathbf{v} \cdot \mathbf{n}\, dS$$

Since the normal is directed upward, we will evaluate this integral by using (6) with the plus sign and with

$$\mathbf{F}(x, y, z) = \delta\mathbf{v}(x, y, z) = \delta x\mathbf{i} + \delta y\mathbf{j} + 2\delta z\mathbf{k}$$

so that

$$M(x, y, z) = \delta x, \quad N(x, y, z) = \delta y, \quad \text{and} \quad P(x, y, z) = 2\delta z$$

Let R denote the disk with boundary $x^2 + y^2 = 1$. If

$$f(x, y) = 1 - x^2 - y^2$$

then Σ is the graph of $z = f(x, y)$ on R. Since

$$f_x(x, y) = -2x \quad \text{and} \quad f_y(x, y) = -2y$$

(6) implies that

$$\iint_{\Sigma} \delta\mathbf{v} \cdot \mathbf{n}\, dS = \iint_{R} [-\delta x(-2x) - \delta y(-2y) + 2\delta(1 - x^2 - y^2)]\, dA$$

$$= \delta \iint_{R} 2\, dA = \delta \int_{0}^{2\pi} \int_{0}^{1} 2r\, dr\, d\theta = \delta \int_{0}^{2\pi} r^2 \Big|_{0}^{1} d\theta$$

$$= \delta \int_{0}^{2\pi} 1\, d\theta = 2\pi\delta \qquad \square$$

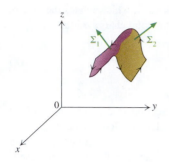

FIGURE 15.35

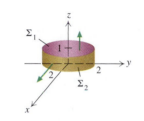

FIGURE 15.36

Flux integrals can be defined for a surface Σ composed of several oriented surfaces $\Sigma_1, \Sigma_2, \ldots, \Sigma_n$, provided that the surfaces induce opposite orientations on the common curves that bind them together (Figure 15.35). We simply let

$$\iint_{\Sigma} \mathbf{F} \cdot \mathbf{n}\, dS = \iint_{\Sigma_1} \mathbf{F} \cdot \mathbf{n}\, dS + \iint_{\Sigma_2} \mathbf{F} \cdot \mathbf{n}\, dS + \cdots + \iint_{\Sigma_n} \mathbf{F} \cdot \mathbf{n}\, dS \quad (7)$$

EXAMPLE 3   Let Σ be the surface in Figure 15.36 consisting of a top Σ_1 and cylindrical sides Σ_2, but no bottom. Assume that Σ has the orientation indicated in the figure (upward on top and outward on the sides). Evaluate $\iint_{\Sigma} \mathbf{F} \cdot \mathbf{n}\, dS$, where $\mathbf{F}(x, y, z) = x\mathbf{i} + z\mathbf{k}$.

Solution   By (7),

$$\iint_{\Sigma} \mathbf{F} \cdot \mathbf{n}\, dS = \iint_{\Sigma_1} \mathbf{F} \cdot \mathbf{n}_1\, dS + \iint_{\Sigma_2} \mathbf{F} \cdot \mathbf{n}_2\, dS$$

For the top we notice that $\mathbf{F}(x, y, z) = x\mathbf{i} + z\mathbf{k}$ and $\mathbf{n}_1 = \mathbf{k}$ (since the normal is pointed upward). Thus $\mathbf{F}(x, y, z) \cdot \mathbf{n}_1 = (x\mathbf{i} + z\mathbf{k}) \cdot \mathbf{k} = z = 1$ for all (x, y, z) on Σ_1. Therefore

$$\iint_{\Sigma_1} \mathbf{F} \cdot \mathbf{n}_1\, dS = \iint_{\Sigma_1} 1\, dS = \text{area of } \Sigma_1 = 4\pi$$

For the flux integral over the cylinder Σ_2 we use the parametrization

$$\mathbf{r}(\theta, z) = 2 \cos \theta \mathbf{i} + 2 \sin \theta \mathbf{j} + z\mathbf{k} \quad \text{for } 0 \le \theta \le 2\pi \quad \text{and } 0 \le z \le 1$$

(as in the solution of Example 3 in Section 14.9). In order to apply (4), we first calculate that

$$\mathbf{r}_\theta(\theta, z) \times \mathbf{r}_z(\theta, z) = \begin{vmatrix} \mathbf{i} & \mathbf{j} & \mathbf{k} \\ -2 \sin \theta & 2 \cos \theta & 0 \\ 0 & 0 & 1 \end{vmatrix} = 2 \cos \theta \mathbf{i} + 2 \sin \theta \mathbf{j}$$

Notice that for $0 < \theta < 2\pi$ the cross product points outward from the cylindrical sides, in agreement with the orientation of Σ. Therefore in (4) we will use the plus sign. Next we observe that $\mathbf{F}(x, y, z) = x\mathbf{i} + z\mathbf{k} = 2 \cos \theta \mathbf{i} + z\mathbf{k}$. Together with (4) this means that

$$\iint_{\Sigma_2} \mathbf{F} \cdot \mathbf{n}_2\, dS = \int_0^{2\pi} \int_0^1 (2 \cos \theta \mathbf{i} + z\mathbf{k}) \cdot (2 \cos \theta \mathbf{i} + 2 \sin \theta \mathbf{j})\, dz\, d\theta$$

$$= \int_0^{2\pi} \int_0^1 4 \cos^2 \theta\, dz\, d\theta = \int_0^{2\pi} (2 + 2 \cos 2\theta)\, d\theta = (2\theta + \sin 2\theta) \Big|_0^{2\pi} = 4\pi$$

We conclude that

$$\iint_{\Sigma} \mathbf{F} \cdot \mathbf{n}\, dS = \iint_{\Sigma_1} \mathbf{F} \cdot \mathbf{n}_1\, dS + \iint_{\Sigma_2} \mathbf{F} \cdot \mathbf{n}_2\, dS = 4\pi + 4\pi = 8\pi \quad \square$$

Gauss's Law

One of the basic laws in electrostatics is **Gauss's Law,** which relates the electric field **E** on the boundary Σ of a solid region D to the total charge q in D:

$$\iint_\Sigma \mathbf{E} \cdot \mathbf{n} \, dS = \frac{q}{\varepsilon_0} \qquad (8)$$

where ε_0 is the permittivity of empty space (see Section 11.2 for the value of ε_0). Physicists call this flux integral the **flux of the electric field through Σ.** From Gauss's Law, if one knows the electric field **E** on Σ, one can compute the total charge q in D. For example, if the vector field **F** appearing in Example 1 represents the electric field over the sphere of radius 3, we would be able to deduce directly from the solution to Example 1 and Gauss's Law that

$$81\pi = \frac{q}{\varepsilon_0}$$

and hence the total charge q inside the sphere of radius 3 is

$$q = 81\pi\varepsilon_0$$

Conversely, it is possible to use Gauss's Law to find the electric field produced by a charged object whose charge distribution possesses sufficient symmetry. We illustrate the method by finding the electric field of a uniformly charged, infinitely long wire (an idealized telephone wire) whose charge density λ (charge per unit length) is constant. For simplicity we select a coordinate system so that the wire lies on the z axis.

Let (x, y, z) be any point not on the wire. There is an electric field at (x, y, z) due to the portion of the wire below (x, y, z), and there is an electric field due to the portion of the wire above (x, y, z) (Figure 15.37). Since the charge on the wire is uniformly distributed, these two electric fields have the same magnitude and make the same acute angle θ with the wire. It follows from the superposition principle that the electric field at (x, y, z) due to the entire wire is the sum of these two electric fields and, by vector addition, is perpendicular to the wire (Figure 15.37). We deduce from the symmetry of the charge distribution that the magnitude of the electric field at any two points having the same distance from the wire must be equal. Consequently the magnitude of the electric field at any point depends only

Karl Friedrich Gauss (1777–1855)
Already acknowledged in Chapter 8 in connection with the normal, or Gaussian, distribution, Gauss ranks with the handful of towering mathematical geniuses of all time. A child prodigy, he is said to have noticed an error in his father's bookkeeping while only 3 years old. Gauss made fundamental contributions to electricity and magnetism, astronomy, and many branches of mathematics.

Nighttime lightning. (*Tony Stone Images*)

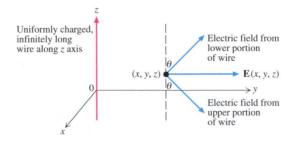

FIGURE 15.37

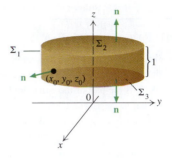

FIGURE 15.38

on the distance from the point to the wire. As a result there exists a real-valued function f of a single variable (the distance from the z axis) such that for any point (x, y, z) not on the z axis, the electric field at (x, y, z) is given by

$$\mathbf{E}(x, y, z) = f(\sqrt{x^2 + y^2}) \frac{x\mathbf{i} + y\mathbf{j}}{\sqrt{x^2 + y^2}}$$

Next we determine the value of $f(\sqrt{x_0^2 + y_0^2})$ for any specific point (x_0, y_0, z_0) not on the z axis. Let Σ be the surface shown in Figure 15.38, composed of a portion Σ_1 of a circular cylinder with height 1, a top Σ_2, and a bottom Σ_3. Then $\mathbf{E} \cdot \mathbf{n} = 0$ on Σ_2 and Σ_3, since $\mathbf{E}$ is perpendicular to the z axis and since on Σ_2 and Σ_3 the normal $\mathbf{n}$ is parallel to the z axis. Consequently

$$\iint_{\Sigma} \mathbf{E} \cdot \mathbf{n} \, dS = \iint_{\Sigma_1} \mathbf{E} \cdot \mathbf{n} \, dS + \iint_{\Sigma_2} \mathbf{E} \cdot \mathbf{n} \, dS + \iint_{\Sigma_3} \mathbf{E} \cdot \mathbf{n} \, dS = \iint_{\Sigma_1} \mathbf{E} \cdot \mathbf{n} \, dS$$

On Σ_1 we have

$$\mathbf{E}(x, y, z) = f(\sqrt{x^2 + y^2}) \frac{x\mathbf{i} + y\mathbf{j}}{\sqrt{x^2 + y^2}} = f(\sqrt{x^2 + y^2}) \, \mathbf{n}$$

Using the fact that $f(\sqrt{x^2 + y^2})$ is constant on Σ_1, we deduce that

$$\iint_{\Sigma} \mathbf{E} \cdot \mathbf{n} \, dS = \iint_{\Sigma_1} \mathbf{E} \cdot \mathbf{n} \, dS = \iint_{\Sigma_1} [f(\sqrt{x^2 + y^2}) \, \mathbf{n}] \cdot \mathbf{n} \, dS$$

$$= \iint_{\Sigma_1} f(\sqrt{x^2 + y^2}) \, dS = \iint_{\Sigma_1} f(\sqrt{x_0^2 + y_0^2}) \, dS$$

$$= f(\sqrt{x_0^2 + y_0^2}) \cdot \text{surface area of } \Sigma_1$$

$$= f(\sqrt{x_0^2 + y_0^2})(2\pi \sqrt{x_0^2 + y_0^2})$$

Thus if q is the total charge inside Σ, then by Gauss's Law,

$$f(\sqrt{x_0^2 + y_0^2}) 2\pi \sqrt{x_0^2 + y_0^2} = \iint_{\Sigma} \mathbf{E} \cdot \mathbf{n} \, dS = \frac{q}{\varepsilon_0}$$

However, the total charge q inside Σ is simply the charge on the portion of the wire inside Σ. By hypothesis the length of the wire inside Σ is 1 and the charge per unit length of wire is λ; it follows that $q = \lambda \cdot 1 = \lambda$. Consequently

$$f(\sqrt{x_0^2 + y_0^2}) = \frac{\lambda}{2\pi\varepsilon_0} \frac{1}{\sqrt{x_0^2 + y_0^2}}$$

We conclude that the electric field at (x_0, y_0, z_0) is given by

$$\mathbf{E}(x_0, y_0, z_0) = \frac{\lambda}{2\pi\varepsilon_0} \frac{1}{\sqrt{x_0^2 + y_0^2}} \frac{x_0\mathbf{i} + y_0\mathbf{j}}{\sqrt{x_0^2 + y_0^2}} = \frac{\lambda}{2\pi\varepsilon_0} \frac{x_0\mathbf{i} + y_0\mathbf{j}}{x_0^2 + y_0^2}$$

From the formula we see that the electric field is completely determined once we know the charge density λ along the wire.

EXERCISES 15.6

In Exercises 1–4 sketch the boundary of the surface and indicate the induced orientation on the boundary.

1. The surface shown in Figure 15.39(a)
2. The surface shown in Figure 15.39(b)
3. The surface shown in Figure 15.39(c)
4. The surface shown in Figure 15.39(d)

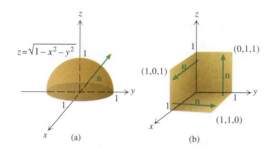

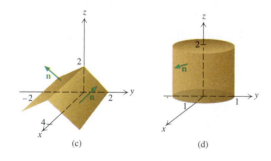

FIGURE 15.39   Figures for Exercises 1–4.

In Exercises 5–9 use a parametrization and (4) or (5) to evaluate $\iint_\Sigma F \cdot n \, dS$.

5. $\mathbf{F}(x, y, z) = -y\mathbf{i} + x\mathbf{j} + z^4\mathbf{k}$; Σ is the sphere $x^2 + y^2 + z^2 = 4$; $\mathbf{n}$ is directed outward from the sphere.

6. $\mathbf{F}(x, y, z) = x\mathbf{i} + y\mathbf{j} + z\mathbf{k}$; Σ is the part of the cylinder $x^2 + z^2 = 1$ between the planes $y = -2$ and $y = 1$; $\mathbf{n}$ is directed away from the y axis.

7. $\mathbf{F}(x, y, z) = \mathbf{i} + \mathbf{j} + 2\mathbf{k}$; Σ is the lower half $z = -2\sqrt{1 - x^2 - y^2}$ of an ellipsoid; $\mathbf{n}$ is directed upward.

8. $\mathbf{F}(x, y, z) = y\mathbf{i} - x\mathbf{j} + z^2\mathbf{k}$; Σ is the part of the cone $z = \sqrt{x^2 + y^2}$ above the quarter disk of radius 1 in the first quadrant of the xy plane; $\mathbf{n}$ is directed upward.

9. $\mathbf{F}(x, y, z) = \sin(x^2 + y^2)(\mathbf{i} + y\mathbf{j})$; Σ is composed of the part of the cylinder $x^2 + y^2 = \pi/2$ between the planes $z = 0$ and $z = 1$, including both the bottom ($z = 0$) and the top ($z = 1$); $\mathbf{n}$ is directed outward from the cylinder.

In Exercises 10–12 use (4) or (6) to evaluate $\iint_\Sigma F \cdot n \, dS$.

10. $\mathbf{F}(x, y, z) = y\mathbf{i} - x\mathbf{j} + 8\mathbf{k}$; Σ is the part of the paraboloid $z = 9 - x^2 - y^2$ above the xy plane; $\mathbf{n}$ is directed upward.

11. $\mathbf{F}(x, y, z) = x\mathbf{i} - \mathbf{j} + 2x^2\mathbf{k}$; Σ is the part of the paraboloid $z = x^2 + y^2$ above the region in the xy plane bounded by the parabolas $x = 1 - y^2$ and $x = y^2 - 1$; $\mathbf{n}$ is directed downward.

12. $\mathbf{F}(x, y, z) = x\mathbf{i} + y\mathbf{j} + z\mathbf{k}$; Σ is the cube with vertices $(0, 0, 0)$, $(0, 0, 1)$, $(0, 1, 0)$, $(0, 1, 1)$, $(1, 0, 0)$, $(1, 0, 1)$, $(1, 1, 0)$, and $(1, 1, 1)$; $\mathbf{n}$ is directed outward from the cube.

Applications

13. Suppose a fluid having constant density 50 flows with velocity $\mathbf{v} = x\mathbf{i} + y\mathbf{j} + z\mathbf{k}$. Determine the rate of mass flow through the sphere $x^2 + y^2 + z^2 = 10$ in the direction of the outward normal.

14. The cross sections of a straight canal are rectangles 40 feet wide and 10 feet high. Set up a coordinate system with the z axis vertical and the y axis running down the center of the canal bottom. Suppose the canal is full of water flowing with velocity

$$\mathbf{v}(x, y, z) = z(400 - x^2)\mathbf{j}$$

where time is measured in minutes. Calculate the rate of flow of mass through a cross section perpendicular to the y axis. The mass density of water is approximately 1.95 slugs per cubic foot.

15. Suppose an electric field is given by $\mathbf{E}(x, y, z) = x\mathbf{i} + y\mathbf{j}$. Let Σ be the part of the cone $x^2 + z^2 = y^2$ that lies above the xy plane and between the planes $y = 0$ and $y = 1$. Find the flux of $\mathbf{E}$ through Σ in the direction of the normal that points upward.

16. Suppose an electric field is given by

$$\mathbf{E}(x, y, z) = 2x\mathbf{i} + 2y\mathbf{j} + 4z\mathbf{k}$$

Use Gauss's Law to find the total charge q inside the cube whose center is the origin, whose volume is 8, and whose sides are parallel to the coordinate planes.

17. Suppose Σ is the unit sphere centered at the origin, and consider the electric field $\mathbf{E}(x, y, z) = \dfrac{q}{4\pi\varepsilon_0}(x\mathbf{i} + y\mathbf{j} + z\mathbf{k})$ due to a point charge q located at the origin (see Section 15.1). Prove that Gauss's Law holds for Σ; that is, prove that $\iint_\Sigma \mathbf{E} \cdot \mathbf{n} \, dS = q/\varepsilon_0$.

18. Suppose the xy plane is uniformly charged with charge density σ (charge per unit area). Use Gauss's Law and symmetry to show that the electric field is given by

$$\mathbf{E}(x, y, z) = \begin{cases} \dfrac{\sigma}{2\varepsilon_0}\mathbf{k} & \text{for } z > 0 \\[2ex] \dfrac{-\sigma}{2\varepsilon_0}\mathbf{k} & \text{for } z < 0 \end{cases}$$

(*Hint:* Integrate over the surface of a rectangular parallelepiped, as shown in Figure 15.40.)

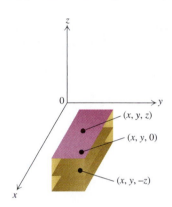

FIGURE 15.40   Rectangular parallelepiped with cross-sectional area 1. See Exercise 18.

*19. Suppose a sphere with radius a and center at the origin is uniformly charged with charge density σ (charge per unit area). Use Gauss's Law and symmetry to show that the electric field is given by

$$\mathbf{E}(x, y, z) = \begin{cases} \mathbf{0} & \text{for } 0 \le x^2 + y^2 + z^2 < a^2 \\[2ex] \dfrac{a^2\sigma}{\varepsilon_0(x^2 + y^2 + z^2)^{3/2}}(x\mathbf{i} + y\mathbf{j} + z\mathbf{k}) \\[1ex] & \text{for } x^2 + y^2 + z^2 \ge a^2 \end{cases}$$

*20. Suppose charge is distributed uniformly throughout the interior of a sphere of radius a, with charge density ρ_0 (charge per unit volume). Use Gauss's Law and symmetry to show that the electric field is given by

$$\mathbf{E}(x, y, z) = \begin{cases} \dfrac{\rho_0}{3\varepsilon_0}(x\mathbf{i} + y\mathbf{j} + z\mathbf{k}) \\[1ex] \quad\quad\quad \text{for } 0 \le x^2 + y^2 + z^2 \le a^2 \\[2ex] \dfrac{\rho_0}{3\varepsilon_0}\dfrac{a^3}{(x^2 + y^2 + z^2)^{3/2}}(x\mathbf{i} + y\mathbf{j} + z\mathbf{k}) \\[1ex] \quad\quad\quad \text{for } x^2 + y^2 + z^2 > a^2 \end{cases}$$

15.7   STOKES'S THEOREM

Stokes's Theorem is a three-dimensional version of Green's Theorem, involving three-dimensional surfaces and their boundaries rather than plane regions and their boundaries.

THEOREM 15.10
Stokes's Theorem

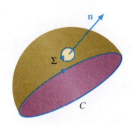

FIGURE 15.41

Let Σ be an oriented surface with normal $\mathbf{n}$ and finite surface area. Assume that Σ is bounded by a closed, piecewise smooth curve C whose orientation is induced by Σ (Figure 15.41). Let $\mathbf{F}$ be a continuous vector field defined on Σ, and assume that the component functions of $\mathbf{F}$ have continuous partial derivatives at each nonboundary point of Σ. Then

$$\int_C \mathbf{F} \cdot d\mathbf{r} = \iint_\Sigma (\text{curl } \mathbf{F}) \cdot \mathbf{n} \, dS \tag{1}$$

If $\mathbf{F} = M\mathbf{i} + N\mathbf{j} + P\mathbf{k}$, then

$$\int_C M(x, y, z) \, dx + N(x, y, z) \, dy + P(x, y, z) \, dz = \iint_\Sigma (\text{curl } \mathbf{F}) \cdot \mathbf{n} \, dS \tag{2}$$

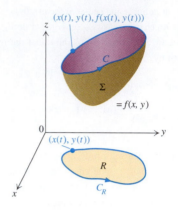

FIGURE 15.42

That Stokes's Theorem is a genuine extension of Green's Theorem follows directly from the alternative form of Green's Theorem given in (5) of Section 15.4, because the normal vector **k** induces the counterclockwise orientation on the boundary of any surface in the xy plane.

The proof of Stokes's Theorem is too complicated for this book, but we will now give an idea of how to prove (2) in the special case that Σ is the graph of a function f with continuous partials on a simple region R in the xy plane whose boundary C_R is piecewise smooth, and for which the image of C_R is C (Figure 15.42). We will also assume that on Σ the normal **n** points upward.

Since

$$\operatorname{curl} \mathbf{F} = \left(\frac{\partial P}{\partial y} - \frac{\partial N}{\partial z}\right)\mathbf{i} + \left(\frac{\partial M}{\partial z} - \frac{\partial P}{\partial x}\right)\mathbf{j} + \left(\frac{\partial N}{\partial x} - \frac{\partial M}{\partial y}\right)\mathbf{k}$$

it follows from (6) of Section 15.6 that

$$\iint_{\Sigma} (\operatorname{curl} \mathbf{F}) \cdot \mathbf{n} \, dS$$

$$= \iint_{R} \left[-\left(\frac{\partial P}{\partial y} - \frac{\partial N}{\partial z}\right)f_x - \left(\frac{\partial M}{\partial z} - \frac{\partial P}{\partial x}\right)f_y + \left(\frac{\partial N}{\partial x} - \frac{\partial M}{\partial y}\right) \right] dA$$

where all partial derivatives are to be evaluated at $(x, y, f(x, y))$. Thus (2) can be rewritten

$$\int_C M \, dx + N \, dy + P \, dz$$

$$= \iint_{R} \left[-\left(\frac{\partial P}{\partial y} - \frac{\partial N}{\partial z}\right)f_x - \left(\frac{\partial M}{\partial z} - \frac{\partial P}{\partial x}\right)f_y + \left(\frac{\partial N}{\partial x} - \frac{\partial M}{\partial y}\right) \right] dA \quad (3)$$

Now we will show that

$$\int_C M \, dx = -\iint_{R} \left(\frac{\partial M}{\partial y} + \frac{\partial M}{\partial z} f_y\right) dA \quad (4)$$

If we let $M_R(x, y) = M(x, y, f(x, y))$ for (x, y) in R, then by the Chain Rule,

$$\frac{\partial M_R}{\partial y} = \frac{\partial M}{\partial y} + \frac{\partial M}{\partial z} f_y \quad (5)$$

Next we let $\mathbf{r}(t) = x(t)\mathbf{i} + y(t)\mathbf{j}$ for $a \leq t \leq b$ be a parametrization of C_R. Then a parametrization of C is given by

$$\mathbf{r}_1(t) = x(t)\mathbf{i} + y(t)\mathbf{j} + f(x(t), y(t))\mathbf{k} \quad \text{for } a \leq t \leq b$$

Therefore

$$\int_C M(x, y, z)\, dx = \int_a^b M(x(t), y(t), f(x(t), y(t)))x'(t)\, dt$$

$$= \int_a^b M_R(x(t), y(t))x'(t)\, dt$$

$$= \int_{C_R} M_R(x, y)\, dx$$

so that more succinctly,

$$\int_C M\, dx = \int_{C_R} M_R\, dx \tag{6}$$

Using (2) in the proof of Green's Theorem to relate integrals involving R and C_R, we find that

$$\int_C M\, dx \overset{(6)}{=} \int_{C_R} M_R\, dx \overset{\underset{\text{Theorem}}{\text{Green's}}}{=} -\iint_R \frac{\partial M_R}{\partial y}\, dA \overset{(5)}{=} -\iint_R \left(\frac{\partial M}{\partial y} + \frac{\partial M}{\partial z} f_y \right) dA$$

which yields (4). Using the same ideas, one can show that

$$\int_C N\, dy = \iint_R \left(\frac{\partial N}{\partial x} + \frac{\partial N}{\partial z} f_x \right) dA \tag{7}$$

and with a little more work that

$$\int_C P\, dz = \iint_R \left(\frac{\partial P}{\partial x} f_y - \frac{\partial P}{\partial y} f_x \right) dA \tag{8}$$

The sum of the left sides of (4), (7), and (8) is the left side of (2), and similarly for the right sides. This completes the proof of Stokes's Theorem for the special case.

EXAMPLE 1   Let C be the oriented triangle described in Figure 15.43, which lies in the plane $z = y/2$. If

$$\mathbf{F}(x, y, z) = -3y^2\mathbf{i} + 4z\mathbf{j} + 6x\mathbf{k}$$

calculate $\int_C \mathbf{F} \cdot d\mathbf{r}$.

Solution   Direct calculation of the line integral would require evaluation of three separate line integrals, one over each of the line segments C_1, C_2, and C_3 composing C. However, if we apply (1) then we need only evaluate $\iint_\Sigma (\text{curl } \mathbf{F}) \cdot \mathbf{n}\, dS$, where Σ is the triangular surface having boundary C and is oriented by the normal directed upward. First we find that

$$\text{curl } \mathbf{F}(x, y, z) = \begin{vmatrix} \mathbf{i} & \mathbf{j} & \mathbf{k} \\ \dfrac{\partial}{\partial x} & \dfrac{\partial}{\partial y} & \dfrac{\partial}{\partial z} \\ -3y^2 & 4z & 6x \end{vmatrix} = -4\mathbf{i} - 6\mathbf{j} + 6y\mathbf{k}$$

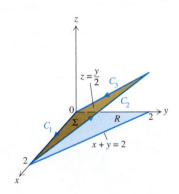

FIGURE 15.43

If $f(x, y) = y/2$, then Σ is the graph of f on the triangular region R in the first quadrant bounded by the coordinate axes and the line $x + y = 2$. Now we use (6) of Section 15.6, with curl $\mathbf{F}$ replacing $\mathbf{F}$, to obtain

$$\iint\limits_{\Sigma} (\text{curl } \mathbf{F}) \cdot \mathbf{n} \, dS = \int_0^2 \int_0^{2-x} \left[-(-4)0 - (-6)\frac{1}{2} + 6y \right] dy \, dx$$

$$= \int_0^2 \int_0^{2-x} (3 + 6y) \, dy \, dx$$

$$= 3 \int_0^2 (y + y^2) \Big|_0^{2-x} dx$$

$$= 3 \int_0^2 (6 - 5x + x^2) \, dx$$

$$= 3 \left(6x - \frac{5}{2} x^2 + \frac{x^3}{3} \right) \Big|_0^2 = 14$$

Thus $\int_C \mathbf{F} \cdot d\mathbf{r} = 14$. ❑

EXAMPLE 2   Let C be the intersection of the paraboloid $z = x^2 + y^2$ and the plane $z = y$, and give C its counterclockwise orientation as viewed from the positive z axis (Figure 15.44). Evaluate $\int_C xy \, dx + x^2 \, dy + z^2 \, dz$.

Solution   Let

$$\mathbf{F}(x, y, z) = xy\mathbf{i} + x^2\mathbf{j} + z^2\mathbf{k}$$

and let Σ be the portion of the plane $z = y$ that lies inside the paraboloid. By (2) we can evaluate the given line integral by evaluating

$$\iint\limits_{\Sigma} (\text{curl } \mathbf{F}) \cdot \mathbf{n} \, dS$$

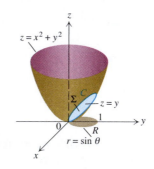

FIGURE 15.44

Notice that if (x, y, z) is on C, then $x^2 + y^2 = y$, and this is an equation of the circular cylinder having equation $r = \sin \theta$ in cylindrical coordinates. Therefore if R is the region in the xy plane bounded by the circle $r = \sin \theta$, then Σ is the graph of $z = y$ on R. When we orient Σ by the normal directed upward, the induced orientation on C is counterclockwise, as prescribed. Since

$$\text{curl } \mathbf{F}(x, y, z) = \begin{vmatrix} \mathbf{i} & \mathbf{j} & \mathbf{k} \\ \dfrac{\partial}{\partial x} & \dfrac{\partial}{\partial y} & \dfrac{\partial}{\partial z} \\ xy & x^2 & z^2 \end{vmatrix} = x\mathbf{k}$$

and $f(x, y) = y$, we conclude from (2) above and (6) of Section 15.6 that

$$\int_C xy\,dx + x^2\,dy + z^2\,dz = \iint_{\Sigma}(\operatorname{curl}\mathbf{F})\cdot\mathbf{n}\,dS = \iint_R [-0(0) - 0(1) + x]\,dA$$

$$= \iint_R x\,dA = \int_0^{\pi}\int_0^{\sin\theta}(r\cos\theta)r\,dr\,d\theta$$

$$= \int_0^{\pi}\frac{r^3}{3}\cos\theta\,\Big|_0^{\sin\theta}\,d\theta = \frac{1}{3}\int_0^{\pi}\sin^3\theta\cos\theta\,d\theta$$

$$= \frac{1}{12}\sin^4\theta\,\Big|_0^{\pi} = 0 \qquad \square$$

Suppose two oriented surfaces Σ_1 and Σ_2 are bounded by the same curve C and induce the same orientation on C. If $\mathbf{n}_1$ and $\mathbf{n}_2$ denote the normals of Σ_1 and Σ_2, respectively, then by Stokes's Theorem we infer that

$$\iint_{\Sigma_1}(\operatorname{curl}\mathbf{F})\cdot\mathbf{n}_1\,dS = \int_C \mathbf{F}\cdot d\mathbf{r} = \iint_{\Sigma_2}(\operatorname{curl}\mathbf{F})\cdot\mathbf{n}_2\,dS \qquad (9)$$

for any vector field $\mathbf{F}$ whose components have continuous partial derivatives on both Σ_1 and Σ_2. In cases where it is difficult to integrate over Σ_1, it may be easier to integrate over Σ_2.

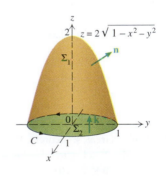

FIGURE 15.45

EXAMPLE 3   Let Σ_1 be the semiellipsoid $z = 2\sqrt{1 - x^2 - y^2}$, oriented so that the normal $\mathbf{n}$ is directed upward (Figure 15.45), and let

$$\mathbf{F}(x, y, z) = x^2\mathbf{i} + y^2\mathbf{j} + z^2\tan xy\,\mathbf{k}$$

Evaluate $\iint_{\Sigma_1}(\operatorname{curl}\mathbf{F})\cdot\mathbf{n}\,dS$.

Solution   First we compute curl $\mathbf{F}$:

$$\operatorname{curl}\mathbf{F}(x, y, z) = \begin{vmatrix} \mathbf{i} & \mathbf{j} & \mathbf{k} \\ \dfrac{\partial}{\partial x} & \dfrac{\partial}{\partial y} & \dfrac{\partial}{\partial z} \\ x^2 & y^2 & z^2\tan xy \end{vmatrix} = xz^2\sec^2 xy\,\mathbf{i} - yz^2\sec^2 xy\,\mathbf{j}$$

We could determine the normal $\mathbf{n}$ of Σ_1 and then evaluate $\iint_{\Sigma_1}(\operatorname{curl}\mathbf{F})\cdot\mathbf{n}\,dS$. But notice from Figure 15.45 that the unit disk Σ_2 in the xy plane has the same boundary as Σ_1 has, and when Σ_2 is oriented with normal $\mathbf{n}$ directed upward, the induced orientations on the common boundary of Σ_1 and Σ_2 are identical. Thus by (9),

$$\iint_{\Sigma_1}(\operatorname{curl}\mathbf{F})\cdot\mathbf{n}\,dS = \iint_{\Sigma_2}(\operatorname{curl}\mathbf{F})\cdot\mathbf{n}\,dS$$

The integral over Σ_2 is easily evaluated, since the normal to Σ_2 is $\mathbf{k}$, which is perpendicular to curl **F**. We obtain

$$\iint\limits_{\Sigma_1} (\text{curl } \mathbf{F}) \cdot \mathbf{n} \, dS = \iint\limits_{\Sigma_2} (\text{curl } \mathbf{F}) \cdot \mathbf{n} \, dS = \iint\limits_{\Sigma_2} 0 \, dS = 0 \qquad \square$$

Imagine that a fluid flows through a cylinder capped by the surface $z = 2\sqrt{1 - x^2 - y^2}$ and that the vector field curl **F** represents the velocity of the fluid. Then $\iint_{\Sigma_1} \delta(\text{curl } \mathbf{F}) \cdot \mathbf{n} \, dS$ would represent the rate of mass flow through Σ_1 (see (1) in Section 15.6.). Stokes's Theorem tells us that the rate of mass flow through Σ_1 is the same as the rate of mass flow through the disk Σ_2 with the same boundary.

Now suppose two oriented surfaces Σ_1 and Σ_2 are bounded by the same curve C but induce opposite orientations on C. Then Stokes's Theorem implies that

$$\iint\limits_{\Sigma_1} (\text{curl } \mathbf{F}) \cdot \mathbf{n} \, dS = -\iint\limits_{\Sigma_2} (\text{curl } \mathbf{F}) \cdot \mathbf{n} \, dS \tag{10}$$

EXAMPLE 4   Let Σ be the unit sphere $x^2 + y^2 + z^2 = 1$, oriented by the normal **n** that is directed outward. Let **F** be any vector field whose component functions have continuous partial derivatives on Σ. Show that $\iint_{\Sigma}(\text{curl } \mathbf{F}) \cdot \mathbf{n} \, dS = 0$.

Solution   Let Σ_1 and Σ_2 be the upper and lower hemispheres of Σ (Figure 15.46). Then the normals that are directed outward from Σ_1 and Σ_2 induce opposite orientations on the circle $x^2 + y^2 = 1$ in the xy plane, which is the boundary of both Σ_1 and Σ_2. By (10) we conclude that

$$\iint\limits_{\Sigma} (\text{curl } \mathbf{F}) \cdot \mathbf{n} \, dS = \iint\limits_{\Sigma_1} (\text{curl } \mathbf{F}) \cdot \mathbf{n} \, dS + \iint\limits_{\Sigma_2} (\text{curl } \mathbf{F}) \cdot \mathbf{n} \, dS = 0 \qquad \square$$

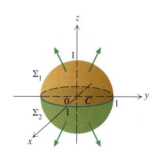

FIGURE 15.46

Stokes's Theorem can be extended to many surfaces that are not covered by our original version, in much the same way that we extended Green's Theorem. Let us show that Stokes's Theorem holds for the circular cylinder Σ in Figure 15.47(a), with normal directed outward. Stokes's Theorem does not apply directly, because the boundary of Σ is composed of two disjoint circles C_1 and C_2 and hence is not piecewise smooth. But suppose we "cut" the cylinder along two lines, as shown in Figure 15.47(b). Then we can apply Stokes's Theorem to each of the resulting half-cylinders Σ_1 and Σ_2 and the corresponding curves C_3 and C_4, which bound them (Figure 15.47(c)). Since the lines forming the cut each receive opposite orientations from the two half-cylinders, the line integrals over them cancel each other. Therefore, by two applications of Stokes's Theorem, we obtain

$$\iint\limits_{\Sigma} (\text{curl } \mathbf{F}) \cdot \mathbf{n} \, dS = \iint\limits_{\Sigma_1} (\text{curl } \mathbf{F}) \cdot \mathbf{n}_1 \, dS + \iint\limits_{\Sigma_2} (\text{curl } \mathbf{F}) \cdot \mathbf{n}_2 \, dS$$

$$= \int_{C_3} \mathbf{F} \cdot d\mathbf{r} + \int_{C_4} \mathbf{F} \cdot d\mathbf{r} = \int_{C_1} \mathbf{F} \cdot d\mathbf{r} + \int_{C_2} \mathbf{F} \cdot d\mathbf{r}$$

Thus Stokes's Theorem applies if we add the line integrals over both parts C_1 and

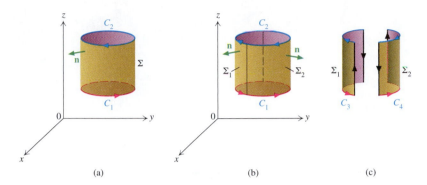

FIGURE 15.47

C_2 of the boundary, each with its induced orientation.

Stokes's Theorem enables us to interpret curl **v** when **v** represents the velocity of a fluid in motion. Let Σ be an oriented surface whose boundary C has an induced counterclockwise orientation, and let **v** denote the velocity of a fluid. If **r** parametrizes C on $[a, b]$, then the closer the directions of **v** and $d\mathbf{r}/dt$ are to one another, the larger $\mathbf{v} \cdot d\mathbf{r}/dt$ is (Figure 15.48). Since $d\mathbf{r}/dt$ is always parallel to the tangent of C and since

$$\int_C \mathbf{v} \cdot d\mathbf{r} = \int_a^b \mathbf{v} \cdot \frac{d\mathbf{r}}{dt}\, dt$$

it follows that the closer the directions of **v** and the tangent are, the larger $\int_C \mathbf{v} \cdot d\mathbf{r}$ is. Consequently $\int_C \mathbf{v} \cdot d\mathbf{r}$ measures the tendency of the fluid to move counterclockwise around C. As a result, $\int_C \mathbf{v} \cdot d\mathbf{r}$ is often called the **circulation of the fluid around C.**

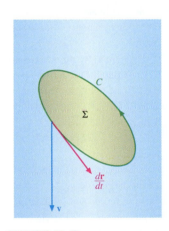

FIGURE 15.48

If (x_0, y_0, z_0) is on Σ and if Σ is very small and has area S, then $\iint_\Sigma (\text{curl } \mathbf{v}) \cdot \mathbf{n}\, dS$ is very nearly $[\text{curl } \mathbf{v}(x_0, y_0, z_0)] \cdot \mathbf{n}(x_0, y_0, z_0)S$. Because Stokes's Theorem tells us that

$$\int_C \mathbf{v} \cdot d\mathbf{r} = \iint_\Sigma (\text{curl } \mathbf{v}) \cdot \mathbf{n}\, dS$$

we conclude that for small Σ, the circulation is approximately $[\text{curl } \mathbf{v}(x_0, y_0, z_0)] \cdot \mathbf{n}(x_0, y_0, z_0)S$. Thus $[\text{curl } \mathbf{v}(x_0, y_0, z_0)] \cdot \mathbf{n}(x_0, y_0, z_0)$ is approximately the rate of circulation per unit area in the counterclockwise direction at (x_0, y_0, z_0). The larger the circulation is per unit area, the larger the rotation, or curling, around (x_0, y_0, z_0). Moreover, the closer to one another the directions of $\mathbf{n}(x_0, y_0, z_0)$ and curl $\mathbf{v}(x_0, y_0, z_0)$ are, the larger $[\text{curl } \mathbf{v}(x_0, y_0, z_0)] \cdot \mathbf{n}(x_0, y_0, z_0)$ is. Consequently the curling effect is greatest about the axis parallel to curl $\mathbf{v}(x_0, y_0, z_0)$. If we imagine a small paddle wheel placed in the fluid at (x_0, y_0, z_0), as in Figure 15.49, then the paddle wheel will spin fastest if its axis is parallel to the curl $\mathbf{v}(x_0, y_0, z_0)$.

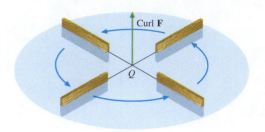

FIGURE 15.49

EXERCISES 15.7

In Exercises 1–5 use Stokes's Theorem to compute $\int_C \mathbf{F} \cdot d\mathbf{r}$, where C is the curve that bounds Σ and that has the induced orientation from Σ.

1. $\mathbf{F}(x, y, z) = z\mathbf{i} + x\mathbf{j} + y\mathbf{k}$; Σ is the part of the paraboloid $z = 1 - x^2 - y^2$ in the first octant; $\mathbf{n}$ is directed downward.

2. $\mathbf{F}(x, y, z) = z^2\mathbf{i} - y^2\mathbf{j}$; Σ is composed of the three squares shown with their normals in Figure 15.50.

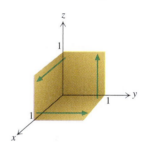

FIGURE 15.50

3. $\mathbf{F}(x, y, z) = y\mathbf{i} - x\mathbf{j} + z\mathbf{k}$; Σ is composed of the part of the cylinder $x^2 + y^2 = 1$ between the planes $z = 0$ and $z = 1$ and the part of the plane $z = 1$ inside the cylinder $x^2 + y^2 = 1$; $\mathbf{n}$ is directed away from the z axis on the cylinder and upward on the plane.

4. $\mathbf{F}(x, y, z) = y^2\mathbf{i} + xy\mathbf{j} - 2xz\mathbf{k}$; Σ is the hemisphere $z = \sqrt{4 - x^2 - y^2}$; $\mathbf{n}$ is directed upward.

5. $\mathbf{F}(x, y, z) = 2y\mathbf{i} + 3z\mathbf{j} - 2x\mathbf{k}$; Σ is the part of the sphere $x^2 + y^2 + z^2 = 1$ in the first octant; $\mathbf{n}$ is directed upward.

In Exercises 6–11 use Stokes's Theorem to evaluate $\int_C \mathbf{F} \cdot d\mathbf{r}$. In each case assume that C has its counterclockwise orientation as viewed from above.

6. $\mathbf{F}(x, y, z) = \dfrac{1}{\sqrt{x^2 + y^2 + z^2 + 1}}(x\mathbf{i} + y\mathbf{j} + z\mathbf{k})$;

C is the intersection of the paraboloid $2z = x^2 + y^2$ and the cylinder $x^2 + y^2 = 2x$.

7. $\mathbf{F}(x, y, z) = xz\mathbf{i} + y^2\mathbf{j} + x^2\mathbf{k}$; C is the intersection of the plane $x + y + z = 5$ and elliptic cylinder $x^2 + y^2/4 = 1$.

8. $\mathbf{F}(x, y, z) = 3y\mathbf{i} + 2z\mathbf{j} - x\mathbf{k}$; C is the triangle with vertices $(1, 0, 0)$, $(0, 1, 0)$, and $(0, 0, 1)$.

9. $\mathbf{F}(x, y, z) = y(x^2 + y^2)\mathbf{i} - x(x^2 + y^2)\mathbf{j}$; C is the rectangle with vertices $(0, 0, 0)$, $(1, 0, 0)$, $(1, 1, 1)$, and $(0, 1, 1)$.

10. $\mathbf{F}(x, y, z) = \ln(y^2 + 1)\mathbf{i} + x\mathbf{j} + (x + y)\mathbf{k}$; C is the rectangle in the xy plane with vertices $(0, 0)$, $(3, 0)$, $(3, 1)$, and $(0, 1)$.

11. $\mathbf{F}(x, y, z) = (z - y)\mathbf{i} + y\mathbf{j} + x\mathbf{k}$; C is the intersection of the circular cylinder $r = \cos\theta$ and the part of the sphere $x^2 + y^2 + z^2 = 1$ above the xy plane.

12. Let $\mathbf{F}(x, y, z) = yz\mathbf{k}$, and let C be the boundary of the part of the cone $z = \sqrt{x^2 + y^2}$ in the first octant between the planes $z = 2$ and $z = 3$. Give the counterclockwise orientation to the part of C in the plane $z = 2$ and the clockwise orientation to the part of C in the plane $z = 3$, as seen from above. Use Stokes's Theorem to evaluate $\int_C \mathbf{F} \cdot d\mathbf{r}$.

13. Let $\mathbf{F}(x, y, z) = (x^2 + z)\mathbf{i} + (y^2 + x)\mathbf{j} + (z^2 + y)\mathbf{k}$, and let C be the intersection of the sphere $x^2 + y^2 + z^2 = 1$ and the cone $z = \sqrt{x^2 + y^2}$. Use Stokes's Theorem and (9) to evaluate $\int_C \mathbf{F} \cdot d\mathbf{r}$, where C has its counterclockwise orientation as seen from above.

In Exercises 14–20 use Stokes's Theorem, and (9) where appropriate, to evaluate $\iint_\Sigma (\text{curl } \mathbf{F}) \cdot \mathbf{n}\, dS$.

14. $\mathbf{F}(x, y, z) = xz\mathbf{i} + (y^2 + 2x)\mathbf{j} + x\mathbf{k}$; Σ is the part of the paraboloid $z = 9 - x^2 - y^2$ above the xy plane; $\mathbf{n}$ is directed upward.

15. $\mathbf{F}(x, y, z) = x\mathbf{i} + (x^2 + y^2 + z^2)\mathbf{j} + z(y^4 - 1)\mathbf{k}$; Σ is composed of the four upper sides of the pyramid whose apex is $(0, 0, 6)$ and whose base is the square in the xy plane with vertices $(-1, -1)$, $(1, -1)$, $(1, 1)$, and $(-1, 1)$; $\mathbf{n}$ is directed upward.

16. $\mathbf{F}(x, y, z) = z^2 y\mathbf{i} - x\mathbf{j} + z \sin x^2 y^2 z\mathbf{k}$; Σ is the part of the cylinder $x^2 + y^2 = 15$ between the planes $z = -1$ and $z = 2$; $\mathbf{n}$ is directed away from the z axis.

17. $\mathbf{F}(x, y, z) = x \sin z\mathbf{i} + xy\mathbf{j} + yz\mathbf{k}$; Σ is composed of all faces of the cube with vertices $(0, 0, 0)$, $(1, 0, 0)$, $(1, 1, 0)$, $(0, 1, 0)$, $(0, 0, 1)$, $(1, 0, 1)$, $(1, 1, 1)$, and $(0, 1, 1)$ except the face in the plane $z = 1$; $\mathbf{n}$ is directed outward from the cube.

18. $\mathbf{F}(x, y, z) = (3y^2 z + y)\mathbf{i} + y\mathbf{j} + 3xy^2\mathbf{k}$; Σ is the upper half of the torus obtained by rotating the circle $r = \sin \theta$ about the x axis; $\mathbf{n}$ is directed outward from the torus.

19. $\mathbf{F}(x, y, z) = xz^2\mathbf{i} + x^3\mathbf{j} + \cos xz\mathbf{k}$; Σ is the part of the ellipsoid $x^2 + y^2 + 3z^2 = 1$ below the xy plane; $\mathbf{n}$ is directed outward from the ellipsoid.

20. $\mathbf{F}(x, y, z) = (z + 1)y^3\mathbf{i} - (z + 1)xy\mathbf{j} + e^{x^2 y^2}\mathbf{k}$; Σ is the part of the surface $x^2 + y^2 + z^6 = 1$ above the xy plane; $\mathbf{n}$ is directed upward.

21. Suppose Σ is an oriented surface with oriented boundary C and $\mathbf{F}$ is a constant vector field defined on Σ. Show that $\int_C \mathbf{F} \cdot d\mathbf{r} = 0$.

22. Suppose Σ is an ellipsoid with normal directed outward

and $\mathbf{F}$ a vector field whose component functions have continuous partial derivatives on Σ. Show that $\iint_\Sigma (\text{curl } \mathbf{F}) \cdot \mathbf{n} \, dS = 0$.

Applications

23. Let Σ be the portion of the sphere $x^2 + y^2 + z^2 = 1$ above the plane $z = y$, and orient Σ by the normal directed outward from the sphere. Assume that the velocity $\mathbf{v}$ of a fluid passing through Σ is given by
$$\mathbf{v}(x, y, z) = x^3\mathbf{i} - zy\mathbf{j} + x\mathbf{k}$$
Find the circulation of the fluid around the boundary of Σ.

24. Let Σ consist of the portion of the paraboloid
$$2z = x^2 + y^2$$
below the plane $z = x$, and orient Σ by the normal directed downward. Suppose a fluid is passing through Σ with a velocity $\mathbf{v}$ given by
$$\mathbf{v}(x, y, z) = (\sin xz + 2yz)\mathbf{i}$$
$$+ (\cosh y + 2xz)\mathbf{j} + (e^{x^2 z^2} + 5y)\mathbf{k}$$
Find the circulation of the fluid around the boundary of Σ.

15.8   THE DIVERGENCE THEOREM

Simple solid region

FIGURE 15.51

The Divergence Theorem is our second and final higher-dimensional analogue of Green's Theorem. To simplify the statement of the Divergence Theorem, we will call a solid region D a **simple solid region** if D is the solid region between the graphs of two functions F_1 and F_2 on a simple region R in the xy plane and if D has the corresponding properties with respect to the xz plane and the yz plane. Regions bounded by spheres, hemispheres, ellipsoids, cubes, and tetrahedrons are simple solid regions. A simple solid region has an interior and an exterior, separated by a boundary surface (Figure 15.51). We assume from now on that any simple solid region D we consider has an orientable boundary surface, and we will orient it by the normal directed outward from D (Figure 15.51). This brings us to the Divergence Theorem, also called Gauss's Theorem.

THEOREM 15.11
The Divergence Theorem

Let D be a simple solid region whose boundary surface Σ is oriented by the normal $\mathbf{n}$ directed outward from D, and let $\mathbf{F}$ be a vector field whose component functions have continuous partial derivatives on D. Then
$$\iint_\Sigma \mathbf{F} \cdot \mathbf{n} \, dS = \iiint_D \text{div } \mathbf{F}(x, y, z) \, dV$$

Proof   Let $\mathbf{F} = M\mathbf{i} + N\mathbf{j} + P\mathbf{k}$. Then the formula $\iint_\Sigma \mathbf{F} \cdot \mathbf{n} \, dS =$

$\iiint_D$ div $\mathbf{F}(x, y, z)\, dV$ is equivalent to

$$\iint_\Sigma M\mathbf{i} \cdot \mathbf{n}\, dS + \iint_\Sigma N\mathbf{j} \cdot \mathbf{n}\, dS + \iint_\Sigma P\mathbf{k} \cdot \mathbf{n}\, dS$$

$$= \iiint_D \frac{\partial M}{\partial x}\, dV + \iiint_D \frac{\partial N}{\partial y}\, dV + \iiint_D \frac{\partial P}{\partial z}\, dV$$

Therefore it suffices to show that

$$\iint_\Sigma M\mathbf{i} \cdot \mathbf{n}\, dS = \iiint_D \frac{\partial M}{\partial x}\, dV, \quad \iint_\Sigma N\mathbf{j} \cdot \mathbf{n}\, dS = \iiint_D \frac{\partial N}{\partial y}\, dV$$

$$\iint_\Sigma P\mathbf{k} \cdot \mathbf{n}\, dS = \iiint_D \frac{\partial P}{\partial z}\, dV$$

Since all three formulas are proved in analogous ways, we prove only the third:

$$\iint_\Sigma P\mathbf{k} \cdot \mathbf{n}\, dS = \iiint_D \frac{\partial P}{\partial z}\, dV \tag{1}$$

By hypothesis D is a simple solid region; hence there exist a simple region R in the xy plane and continuous functions F_1 and F_2 on R such that D consists of all points (x, y, z) satisfying

$$(x, y) \text{ is in } R \quad \text{and} \quad F_1(x, y) \le z \le F_2(x, y)$$

For the triple integral in (1) we obtain

$$\iiint_D \frac{\partial P}{\partial z}\, dV = \iint_R \left[\int_{F_1(x,y)}^{F_2(x,y)} \frac{\partial P}{\partial z}(x, y, z)\, dz \right] dA$$

$$= \iint_R [P(x, y, F_2(x, y)) - P(x, y, F_1(x, y))]\, dA \tag{2}$$

For the surface integral in (1) we first observe that Σ is composed of three subsurfaces shown in Figure 15.52:

the bottom Σ_1, consisting of all points $(x, y, F_1(x, y))$ for (x, y) in R
the top Σ_2, consisting of all points $(x, y, F_2(x, y))$ for (x, y) in R
the sides Σ_3, consisting of all points (x, y, z) for which (x, y) is on the boundary of R and

$$F_1(x, y) \le z \le F_2(x, y)$$

(In some cases, such as when Σ is a sphere, Σ_3 will not appear.) Now on Σ_1 the normal $\mathbf{n}$ is directed downward. Hence by (6) of Section 15.6,

$$\iint_{\Sigma_1} P\mathbf{k} \cdot \mathbf{n}\, dS = -\iint_R P(x, y, F_1(x, y))\, dA$$

Next, on Σ_2 the normal $\mathbf{n}$ is directed upward. Hence by (6) of Section 15.6,

$$\iint_{\Sigma_2} P\mathbf{k} \cdot \mathbf{n}\, dS = \iint_R P(x, y, F_2(x, y))\, dA$$

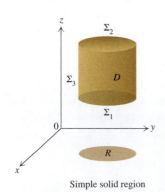

Simple solid region

FIGURE 15.52

Finally, on Σ_3 the normal **n** is horizontal and hence is perpendicular to **k**. Consequently

$$\iint\limits_{\Sigma_3} P\mathbf{k} \cdot \mathbf{n} \, dS = \iint\limits_{\Sigma_3} 0 \, dS = 0$$

Putting our information about the three surface integrals together, we conclude that

$$\iint\limits_{\Sigma} P\mathbf{k} \cdot \mathbf{n} \, dS = -\iint\limits_{R} P(x, y, F_1(x, y)) \, dA + \iint\limits_{R} P(x, y, F_2(x, y)) \, dA + 0$$

$$= \iint\limits_{R} [P(x, y, F_2(x, y)) - P(x, y, F_1(x, y))] \, dA$$

This is the same expression we found in (2) for

$$\iiint\limits_{D} \frac{\partial P}{\partial z} (x, y, z) \, dV \quad \blacksquare$$

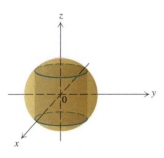

FIGURE 15.53

Again we have a theorem that relates an integral over a region to another integral over its boundary; in that sense it is quite similar to the Fundamental Theorem of Calculus. Moreover, observe that the Divergence Theorem is a higher-dimensional analogue of the second alternative form of Green's Theorem, given in (6) of Section 15.4. In addition, the Divergence Theorem holds more generally for any solid region that can be decomposed into finitely many simple solid regions of the type appearing in Theorem 15.11. Thus it holds for a donut or for the region outside a cylinder and inside a sphere (Figure 15.53).

EXAMPLE 1   Suppose Σ is the sphere $x^2 + y^2 + z^2 = 4$, and let $\mathbf{F}(x, y, z) = 3x\mathbf{i} + 4y\mathbf{j} + 5z\mathbf{k}$. Evaluate $\iint_{\Sigma} \mathbf{F} \cdot \mathbf{n} \, dS$.

Solution   Evaluation of $\iint_{\Sigma} \mathbf{F} \cdot \mathbf{n} \, dS$ itself would involve calculating two surface integrals, one each over the upper and the lower hemispheres of Σ. However, if we use the Divergence Theorem, we need only evaluate $\iiint_{D} \text{div } \mathbf{F}(x, y, z) \, dV$, where D is the ball $x^2 + y^2 + z^2 \leq 4$. Since

$$\text{div } \mathbf{F}(x, y, z) = 3 + 4 + 5 = 12$$

it follows that

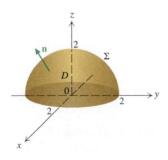

FIGURE 15.54

$$\iint\limits_{\Sigma} \mathbf{F} \cdot \mathbf{n} \, dS = \iiint\limits_{D} \text{div } \mathbf{F}(x, y, z) \, dV = \iiint\limits_{D} 12 \, dV = 12\left(\frac{4}{3}\pi(2^3)\right) = 128\pi \quad \square$$

EXAMPLE 2   Let D be the region bounded by the xy plane and the hemisphere shown in Figure 15.54, and let $\mathbf{F}(x, y, z) = x^3\mathbf{i} + y^3\mathbf{j} + z^3\mathbf{k}$. Evaluate $\iint_{\Sigma} \mathbf{F} \cdot \mathbf{n} \, dS$, where Σ is the boundary of D.

Solution   Since a direct calculation of the surface integral would involve an improper integral, we turn to the Divergence Theorem. In order to use it, we calculate that

$$\text{div } \mathbf{F}(x, y, z) = 3x^2 + 3y^2 + 3z^2 = 3(x^2 + y^2 + z^2)$$

and conclude that

$$\iint_{\Sigma} \mathbf{F} \cdot \mathbf{n} \, dS = \iiint_{D} \text{div } \mathbf{F}(x, y, z) \, dV = \iiint_{D} 3(x^2 + y^2 + z^2) \, dV$$

$$= \int_{0}^{2\pi} \int_{0}^{\pi/2} \int_{0}^{2} (3\rho^2)\rho^2 \sin \phi \, d\rho \, d\phi \, d\theta$$

$$= 3 \int_{0}^{2\pi} \int_{0}^{\pi/2} \frac{\rho^5}{5} \sin \phi \bigg|_{0}^{2} d\phi \, d\theta = \frac{96}{5} \int_{0}^{2\pi} \int_{0}^{\pi/2} \sin \phi \, d\phi \, d\theta$$

$$= \frac{96}{5} \int_{0}^{2\pi} - \cos \phi \bigg|_{0}^{\pi/2} d\theta = \frac{96}{5} \int_{0}^{2\pi} 1 \, d\theta = \frac{192}{5} \pi \qquad \square$$

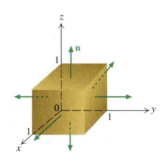

FIGURE 15.55

EXAMPLE 3   Let Σ be the cube shown in Figure 15.55, and let $\mathbf{F}(x, y, z) = x\mathbf{i} + y\mathbf{j} + z^2\mathbf{k}$. Find $\iint_{\Sigma} \mathbf{F} \cdot \mathbf{n} \, dS$.

Solution   Direct evaluation of the surface integral would involve six integrals, one over each of the six faces of the cube. However, if we apply the Divergence Theorem, we need evaluate only one triple integral. To calculate the triple integral, we first find that

$$\text{div } \mathbf{F}(x, y, z) = 1 + 1 + 2z = 2(1 + z)$$

Letting D denote the solid region whose boundary is Σ and applying the Divergence Theorem, we compute that

$$\iint_{\Sigma} \mathbf{F} \cdot \mathbf{n} \, dS = \iiint_{D} 2(1 + z) \, dV = \int_{0}^{1} \int_{0}^{1} \int_{0}^{1} 2(1 + z) \, dz \, dy \, dx$$

$$= \int_{0}^{1} \int_{0}^{1} (1 + z)^2 \bigg|_{0}^{1} dy \, dx = \int_{0}^{1} \int_{0}^{1} 3 \, dy \, dx = 3 \qquad \square$$

By means of the Divergence Theorem it is possible to reinterpret the divergence of a vector field in terms of fluid motion. Suppose (x_0, y_0, z_0) is a fixed point in space and D is a ball centered at (x_0, y_0, z_0) with boundary Σ and volume V. Assume that the vector field $\mathbf{v}$ represents the velocity of a fluid with density 1. If D is very small, then $\iiint_{D} \text{div } \mathbf{v}(x, y, z) \, dV$ is very nearly $[\text{div } \mathbf{v}(x_0, y_0, z_0)]V$. Therefore the Divergence Theorem implies that

$$\iint_{\Sigma} \mathbf{v} \cdot \mathbf{n} \, dS = \iiint_{D} \text{div } \mathbf{v}(x, y, z) \, dV \approx [\text{div } \mathbf{v}(x_0, y_0, z_0)]V$$

so that

$$\text{div } \mathbf{v}(x_0, y_0, z_0) \approx \frac{1}{V} \iint_{\Sigma} \mathbf{v} \cdot \mathbf{n} \, dS \tag{3}$$

Now recall that $\iint_{\Sigma} \mathbf{v} \cdot \mathbf{n} \, dS$ equals the rate (with respect to time) of mass flow from D outward through Σ. Thus (3) suggests that $\text{div } \mathbf{v}(x_0, y_0, z_0)$ should be the rate of mass flow per unit volume at (x_0, y_0, z_0) in the direction of the outward normal. Hence the name "divergence" of $\mathbf{v}$ at a point.

Maxwell's Equations

**James Clerk Maxwell
(1831–1879)**
Born in Edinburgh, Scotland,
Maxwell is considered to have
been the greatest theoretical
physicist of the nineteenth century.
In addition to his fundamental
laws of electromagnetism, he
developed the theory of kinetic
gases, made profound discoveries
concerning the rings of Saturn,
and analyzed color vision. He also
introduced the curl and
divergence.

One of the reasons that Stokes's Theorem and the Divergence Theorem are so important is that they play a central role in the study of electricity and magnetism. At every point (x, y, z) in space there is an electric field $\mathbf{E}(x, y, z)$ and a magnetic field $\mathbf{B}(x, y, z)$. During the eighteenth century and early in the nineteenth century, mathematicians and physicists studied electricity and magnetism. Their work culminated in the appearance of four equations now called the **Maxwell equations,** formulated by James Clerk Maxwell and published in 1865. The four equations are:

$$\text{div } \mathbf{E} = \frac{1}{\varepsilon_0}\, \rho \tag{4}$$

$$\text{div } \mathbf{B} = 0 \tag{5}$$

$$\text{curl } \mathbf{E} = -\frac{\partial \mathbf{B}}{\partial t} \tag{6}$$

$$\text{curl } \mathbf{B} = \mu_0 \left(\varepsilon_0 \frac{\partial \mathbf{E}}{\partial t} + \mathbf{J} \right) \tag{7}$$

In the equations, ρ denotes the charge density, $\mathbf{J}$ the current density, and μ_0 the permeability constant of space (as distinguished from the permittivity of space ε_0).

These four equations, which connect the seemingly unrelated phenomena of electricity and magnetism, are fundamental to the study of electromagnetism. Because of (4)–(7), Maxwell was led to conjecture that energy is transmitted by electromagnetic waves, which move with the speed of light. From this he concluded that light itself is an electromagnetic wave.

Equation (4) is often called the electric flux equation because it is a consequence of Gauss's Law involving the electric flux over the boundary of a region:

$$\iint_\Sigma \mathbf{E} \cdot n\, dS = \frac{q}{\varepsilon_0} \tag{8}$$

(see (8) in Section 15.6). Indeed, assume that div $\mathbf{E}$ is continuous in a small ball ΔD centered at the point (x_0, y_0, z_0) with boundary $\Delta\Sigma$. If the volume of ΔD is ΔV, then by (3) (with $\mathbf{E}$ substituted for $\mathbf{v}$) and (8),

$$\text{div } \mathbf{E}\,(x_0, y_0, z_0) \approx \frac{1}{\Delta V} \iint_{\Delta\Sigma} \mathbf{E} \cdot \mathbf{n}\, dS = \frac{1}{\Delta V} \frac{q}{\varepsilon_0}$$

Since ρ is charge per unit volume, we conclude that

$$\text{div } \mathbf{E}\,(x_0, y_0, z_0) = \lim_{\|\Delta V\| \to 0} \frac{1}{\varepsilon_0} \frac{q}{\Delta V} = \frac{1}{\varepsilon_0} \rho(x_0, y_0, z_0)$$

Thus Gauss's Law implies (4), which is the first of Maxwell's equations listed above.

Next we turn to Maxwell's second equation, appearing as (5). Experiment has shown the magnetic flux through the boundary Σ of any simple solid region D is 0, so that

$$\iint_\Sigma \mathbf{B} \cdot \mathbf{n}\, dS = 0$$

(Physicists explain this by saying that the number of magnetic field lines that enter a solid region is the same as the number that leave that region.) By the Divergence

Theorem, we deduce that

$$0 = \iint_{\Sigma} \mathbf{B} \cdot \mathbf{n} \, dS = \iiint_{D} \operatorname{div} \mathbf{B} \, dV \tag{9}$$

Assuming that div $\mathbf{B}$ is continuous, this means that div $\mathbf{B} = 0$ (see Exercise 30). This yields (5). It is often called the magnetic flux equation because of the appearance of the magnetic flux in (9).

Maxwell's third equation, which is (6), is a reformulation of the famous Law of Induction discovered by Michael Faraday in 1832. Notice that it relates the electric and magnetic fields, and in fact implies that a changing magnetic field creates an electric field.

It was the fourth equation of Maxwell, appearing in (7), that led to the discovery of electromagnetic waves. Equation (7) is an analogue of (6), since it implies that a changing electric field creates a magnetic field. Derivations of (6) and (7) are beyond the scope of this book.

The prediction of electromagnetic waves by Maxwell was later confirmed experimentally by Heinrich Hertz (1857–1894), discoverer of radio waves and the physicist for whom "hertz," the measure of frequency, is named. The Very Large Array (VLA) of 27 radio telescopes in Augustine Plaines, New Mexico, are designed to intercept electromagnetic waves from outer space.

Very Large Array (VLA) telescopes in New Mexico. (*Gregg Hadel/Tony Stone Images*)

To give an indication of the immense importance of Maxwell's four equations, we end with a quotation of Hertz: "One cannot escape the feeling that these mathematical formulas have an independent existence and an intelligence of their own, that they are wiser than we are, wiser even than their discoverers, that we get more out of them than we put into them."

EXERCISES 15.8

In Exercises 1–8 determine whether the given region is a simple solid region.

1. The solid region inside the cylinder $x^2 + y^2 = 1$ and between the planes $z = 0$ and $z = 1$.
2. The solid region inside the ellipsoid $x^2 + y^2 + 2z^2 = 1$
3. The solid region bounded by the planes $x = 0$, $y = 0$, $z = 0$, and $x + y + z = 1$
4. The solid region bounded by the paraboloids $z = 1 - x^2 - y^2$ and $z = x^2 + y^2 - 1$
5. The solid region inside the double cone $z^2 = x^2 + y^2$ and between the planes $z = -1$ and $z = 1$
6. The solid region bounded by the parabolic sheet $y = z^2$ and the planes $x = 0$, $x = 1$, and $y = 1$
7. The solid region bounded by the surface $y = \sin x$ and the planes $x = -2\pi$, $x = 2\pi$, $y = 1$, $z = 0$, and $z = 3$
8. The solid region bounded by the surface $z = y^3 - y$ and the planes $z = 2$, $x = 0$, $x = 1$, $y = -2$, and $y = 1$

In Exercises 9–23 use the Divergence Theorem to compute $\iint_{\Sigma} \mathbf{F} \cdot \mathbf{n} \, dS$, where $\mathbf{n}$ is the normal to Σ that is directed outward.

9. $\mathbf{F}(x, y, z) = x^2\mathbf{i} + xy\mathbf{j} - 2xz\mathbf{k}$; Σ is the tetrahedron with vertices $(0, 0, 0)$, $(1, 0, 0)$, $(0, 1, 0)$, and $(0, 0, 1)$.
10. $\mathbf{F}(x, y, z) = x^2\mathbf{i} + y^2\mathbf{j} + z^2\mathbf{k}$; Σ is the parallelepiped with vertices $(0, 0, 0)$, $(1, 0, 0)$, $(1, 2, 0)$, $(0, 2, 0)$, $(0, 0, 3)$, $(1, 0, 3)$, $(1, 2, 3)$, and $(0, 2, 3)$.
11. $\mathbf{F}(x, y, z) = x\mathbf{i} + y\mathbf{j} + z\mathbf{k}$; Σ is the boundary of the solid region in the first octant that is inside the cylinder $x^2 + y^2 = 1$ and between the planes $z = 0$ and $z = 1$.
12. $\mathbf{F}(x, y, z) = 2x\mathbf{i} + xy\mathbf{j} + xz\mathbf{k}$; Σ is the sphere $x^2 + y^2 + z^2 = 1$.
13. $\mathbf{F}(x, y, z) = x\mathbf{i} + y\mathbf{j} + z\mathbf{k}$; Σ is composed of the hemisphere $z = \sqrt{1 - x^2 - y^2}$ and the disk in the xy plane bounded by the circle $x^2 + y^2 = 1$.
14. $\mathbf{F}(x, y, z) = (x - \cos x)\mathbf{i} + (y - y \sin x)\mathbf{j} + 2z\mathbf{k}$; Σ is the tetrahedron described in Exercise 9.

15. $\mathbf{F}(x, y, z) = x^2\mathbf{i} + y^2\mathbf{j} + z^2\mathbf{k}$; Σ is the boundary of the solid region inside the cylinder $x^2 + y^2 = 4$ and between the planes $z = 0$ and $z = 2$.

16. $\mathbf{F}(x, y, z) = 3x\mathbf{i} - 2y\mathbf{j} + z\mathbf{k}$; Σ is the sphere $x^2 + y^2 + z^2 = 4$.

17. $\mathbf{F}(x, y, z) = y(x^2 + y^2)^{3/2}\mathbf{i} - x(x^2 + y^2)^{3/2}\mathbf{j} + (z + 1)\mathbf{k}$; Σ is the boundary of the solid region bounded above by the plane $z = 2x$ and below by the paraboloid $z = x^2 + y^2$.

18. $\mathbf{F}(x, y, z) = yz\mathbf{i} + xy\mathbf{j} + xz\mathbf{k}$; Σ is the boundary of the solid region inside the cylinder $x^2 + z^2 = 1$ and between the planes $y = -1$ and $y = 1$.

19. $\mathbf{F}(x, y, z) = -2x\mathbf{i} + 4y\mathbf{j} - 7z\mathbf{k}$; Σ is the boundary of the solid region inside the sphere $x^2 + y^2 + z^2 = 4$ and outside the cylinder $x^2 + y^2 = 1$.

20. $\mathbf{F}(x, y, z) = y\mathbf{i} + x\mathbf{j} + 8\mathbf{k}$; Σ is the boundary of the solid region bounded above by the paraboloid $z = 1 - x^2 - y^2$ and below by the xy plane.

21. $\mathbf{F}(x, y, z) = x^2\mathbf{i} + y\mathbf{j} - 2z^2\mathbf{k}$; Σ is the boundary of the solid region bounded below by the xy plane, above by the plane $z = x$, and on the sides by the parabolic sheet $y^2 = 2 - x$.

22. $\mathbf{F}(x, y, z) = x^2y\mathbf{i} + yz\mathbf{j} + z^2\mathbf{k}$; Σ is the boundary of the solid region bounded by the planes $z = 1$ and $x + y = 1$ and the coordinate planes.

23. $\mathbf{F}(x, y, z) = (x^2 + y^2 + z^2)(x\mathbf{i} + y\mathbf{j})$; Σ is the sphere $x^2 + y^2 + z^2 = 9$.

24. Let
$$\mathbf{F}(x, y, z) = x^3y^2z^2\mathbf{i} - x^4yz^2\mathbf{j}$$
Evaluate $\iiint_D \operatorname{div}\mathbf{F}(x, y, z)\, dV$ by using the Divergence Theorem, where D is the ball $x^2 + y^2 + z^2 \le 1$.

25. Let $\mathbf{F}$ and $\mathbf{G}$ be vector fields defined on a simple solid region D with boundary Σ. Assume that $\mathbf{F} = \operatorname{curl}\mathbf{G}$ on D and that the components of $\mathbf{F}$ have continuous partial derivatives on D. Show that $\iint_\Sigma \mathbf{F}\cdot\mathbf{n}\, dS = 0$.

26. Let $\mathbf{F}(x, y, z) = x\mathbf{i} + y\mathbf{j} + z\mathbf{k}$, and let D be a simple solid region with boundary Σ and normal $\mathbf{n}$ directed outward. Show that the volume V of D is given by the formula
$$V = \frac{1}{3}\iint_\Sigma \mathbf{F}\cdot\mathbf{n}\, dS$$

27. Use Exercise 26 to find the volume of a circular cone with height h and radius a.

28. Let f and g be functions of three variables whose partial derivatives are continuous on a simple solid region D having boundary Σ.
   a. Prove that
$$\iint_\Sigma (g\operatorname{grad} f)\cdot\mathbf{n}\, dS$$

$$= \iiint_D [g\nabla^2 f + (\operatorname{grad} g)\cdot(\operatorname{grad} f)]\, dV$$

(*Hint*: Use the identity
$$\operatorname{div}(g\mathbf{F}) = g\operatorname{div}\mathbf{F} + (\operatorname{grad} g)\cdot\mathbf{F}$$

   b. Using part (a), prove that
$$\iint_\Sigma (g\operatorname{grad} f - f\operatorname{grad} g)\cdot\mathbf{n}\, dS$$

$$= \iiint_D (g\nabla^2 f - f\nabla^2 g)\, dV$$

(The formulas in (a) and (b) are those Green actually presented in his book on electricity and magnetism. See also Exercise 27 of Section 15.4.)

29. Suppose $\mathbf{F}$ is a constant vector field defined on a simple solid region D having boundary Σ. Prove that $\iint_\Sigma \mathbf{F}\cdot\mathbf{n}\, dS = 0$.

30. Assume that $\operatorname{div}\mathbf{B}$ is continuous and $\iiint_D \operatorname{div}\mathbf{B}\, dV = 0$ for every simple solid region D. Prove that $\operatorname{div}\mathbf{B}(x, y, z) = 0$ for all (x, y, z). (*Hint*: Assume that $\operatorname{div}\mathbf{B}(x_0, y_0, z_0) \neq 0$ for some point (x_0, y_0, z_0), and obtain a contradiction.)

Applications

31. Let D be a simple solid region, with boundary Σ oriented by the normal $\mathbf{n}$ that is directed outward. Suppose there is a continuous distribution of charge in D, and let q be the total charge in D. Use the Divergence Theorem and (4) to prove that $\iint_\Sigma \mathbf{E}\cdot\mathbf{n}\, dS = q/\varepsilon_0$.

32. Under certain circumstances an electromagnetic wave consists of an electric field $\mathbf{E}$ and a magnetic field $\mathbf{B}$ that are perpendicular to each other and to the direction of propagation of the wave (Figure 15.56). Such a wave is called a "plane wave." Assume that the wave moves in the direction of the positive y axis, and that
$$\mathbf{E}(x, y, z, t) = E(y, t)\mathbf{k}\quad\text{and}\quad \mathbf{B}(x, y, z, t) = B(y, t)\mathbf{i}$$
where E and B are real-valued functions of two variables having continuous second partial derivatives. Assume also that $\mathbf{J} = \mathbf{0}$ in (7) (as would be the case in empty space).
   a. Show that
$$\operatorname{curl}\mathbf{E} = \frac{\partial E}{\partial y}\mathbf{i}\quad\text{and}\quad \operatorname{curl}\mathbf{B} = -\frac{\partial B}{\partial y}\mathbf{k}$$
   b. From (6), (7), and part (a), show that
$$\frac{\partial E}{\partial y} = -\frac{\partial B}{\partial t}\quad\text{and}\quad \frac{\partial B}{\partial y} = -\frac{1}{c^2}\frac{\partial E}{\partial t}\tag{10}$$
where $c = 1/\sqrt{\mu_0\varepsilon_0}$, which turns out to be the speed of light (approximately 2.99792×10^8 meters per second).

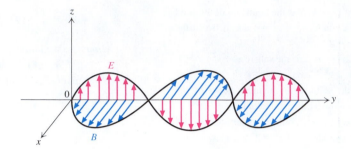

FIGURE 15.56 A plane electromagnetic wave moving in the positive y direction. See Exercise 32.

c. From part (b) show that E and B satisfy the wave equations

$$\frac{\partial^2 E}{\partial y^2} = \frac{1}{c^2}\frac{\partial^2 E}{\partial t^2} \quad \text{and} \quad \frac{\partial^2 B}{\partial y^2} = \frac{1}{c^2}\frac{\partial^2 B}{\partial t^2} \quad (11)$$

33. Let a, b, and ω be positive constants, and let c be the speed of light. Also let

$$E(y, t) = a\cos\omega(y/c - t), \quad B(y, t) = b\cos\omega(y/c - t)$$

so that the amplitudes of the electric and magnetic fields are a and b, respectively. Show that (10) in Exercise 32 is satisfied if and only if $a = bc$. (This would mean that the ratio of the amplitude of the electric field to the amplitude of the magnetic field equals the speed of light. That would indicate why it might be very hard to detect electromagnetic waves.)

34. Let a and b be arbitrary constants, and let c be the speed of light. Show that if

$$E(y, t) = ac\, e^{b(y - ct)} \quad \text{and} \quad B(y, t) = ae^{b(y - ct)}$$

then (10) is satisfied.

35. One of Maxwell's equations implies that $\mathbf{B} = \operatorname{curl}\mathbf{F}$ for a suitable vector field $\mathbf{F}$. Let D be a simple solid region having boundary Σ. Find the fallacy in the following steps, which "prove" that $\mathbf{B} = \mathbf{0}$ on D (that is, that magnetic fields do not exist).

1. $\operatorname{div}\mathbf{B} = 0$

2. $\iint_\Sigma \mathbf{B}\cdot\mathbf{n}\, dS = \iiint_D \operatorname{div}\mathbf{B}\, dV = 0$ by the Divergence Theorem

3. Let C be any closed, oriented curve on Σ. Then by Stokes's Theorem,

$$\int_C \mathbf{F}\cdot d\mathbf{r} = \iint_\Sigma (\operatorname{curl}\mathbf{F})\cdot\mathbf{n}\, dS = \iint_\Sigma \mathbf{B}\cdot\mathbf{n}\, dS = 0$$

4. By step 3 of this exercise and by the equivalence of conditions (1) and (3) of Section 15.3 for independence of path, $\mathbf{F} = \operatorname{grad} f$ for a suitable function f of several variables.

5. $\mathbf{B} = \operatorname{curl}\mathbf{F} = \operatorname{curl}(\operatorname{grad} f) = \mathbf{0}$

(Adapted from George Arfken, "Magnetic Fields Are Not Real," *American Journal of Physics* **27** (1959), p. 526.)

REVIEW

Key Terms and Expressions

Vector field; conservative vector field
Gradient
Divergence
Curl
Potential function
Line integral

Independence of path
Surface integral
Flux integral
Orientation of a surface; oriented surface; induced orientation
Simple solid region

Key Formulas

$$\int_C f(x, y, z)\, ds \;=\; \int_a^b f(x(t), y(t), z(t)) \left\| \frac{d\mathbf{r}}{dt} \right\| dt$$

$$\int_C \mathbf{F} \cdot d\mathbf{r} \;=\; \int_a^b \mathbf{F}(x(t), y(t), z(t)) \cdot \frac{d\mathbf{r}}{dt}\, dt$$

$$\int_C M\, dx + N\, dy + P\, dz = \int_a^b \left[M(x(t), y(t), z(t)) \frac{dx}{dt} \right.$$
$$\left. + N(x(t), y(t), z(t)) \frac{dy}{dt} + P(x(t), y(t), z(t)) \frac{dz}{dt} \right] dt$$

$$\int_C \operatorname{grad} f \cdot d\mathbf{r} = f(x_1, y_1, z_1) - f(x_0, y_0, z_0)$$

$$\int_C M(x, y)\, dx + N(x, y)\, dy = \iint_R \left(\frac{\partial N}{\partial x} - \frac{\partial M}{\partial y} \right) dA$$

$$\iint_\Sigma g(x, y, z)\, dS \;=\;$$

$$\iint_R g(x, y, f(x, y)) \sqrt{[f_x(x, y)]^2 + [f_y(x, y)]^2 + 1}\, dA$$

$$\iint_\Sigma \mathbf{F} \cdot \mathbf{n}\, dS = \;\pm \iint_R \big[M(x, y, f(x, y)) f_x(x, y) \big.$$
$$\big. + N(x, y, f(x, y)) f_y(x, y) - P(x, y, f(x, y)) \big]\, dA$$

$$\int_C \mathbf{F} \cdot d\mathbf{r} = \iint_\Sigma (\operatorname{curl} \mathbf{F}) \cdot \mathbf{n}\, dS$$

$$\iint_\Sigma \mathbf{F} \cdot \mathbf{n}\, dS = \iiint_D \operatorname{div} \mathbf{F}(x, y, z)\, dV$$

Key Theorems

Fundamental Theorem of Line Integrals
Green's Theorem

Stokes's Theorem
Divergence Theorem

Evaluation of Integrals

Let $\mathbf{F} = M\mathbf{i} + N\mathbf{j} + P\mathbf{k}$ (where M, N, and P are functions of x, y, z), and $\mathbf{r}(t) = x(t)\mathbf{i} + y(t)\mathbf{j} + z(t)\mathbf{k}$ for $a \le t \le b$ parametrize C.

$$\int_C \mathbf{F} \cdot d\mathbf{r} \quad \text{or} \quad \int_C M\, dx + N\, dy + P\, dz$$

In theory one can always integrate it as a line integral:

$$\int_C \mathbf{F} \cdot d\mathbf{r} \;=\; \int_a^b \left(M \frac{dx}{dt} + N \frac{dy}{dt} + P \frac{dz}{dt} \right) dt, \quad \text{by (7) or (9) of 15.2}$$

If $\mathbf{F} = \operatorname{grad} f$, then one can use the Fundamental Theorem of Line Integrals:

$$\int_C \mathbf{F} \cdot d\mathbf{r} = f(\text{terminal point of } C) - f(\text{initial point of } C)$$

If C is the boundary of plane region R, and if N_x and M_y are continuous, then one can use Green's Theorem:

$$\int_C \mathbf{F} \cdot d\mathbf{r} = \iint_R \left(\frac{\partial N}{\partial x} - \frac{\partial M}{\partial y} \right) dA$$

If C is the boundary of surface Σ, and if curl $\mathbf{F}$ is continuous, then one can use Stokes's Theorem:

$$\int_C \mathbf{F} \cdot d\mathbf{r} = \iint_\Sigma (\operatorname{curl} \mathbf{F}) \cdot \mathbf{n}\, dS$$

In theory one can always evaluate it as a surface integral:

$$\iint_\Sigma \mathbf{F} \cdot \mathbf{n} \, dS = \iint_R \pm (-Mf_x - Nf_y + P) \, dA, \quad \text{by (6) of 15.6}$$

where Σ is the graph of f on R

If Σ has boundary C and $\mathbf{F} = \text{curl } \mathbf{G}$, then one can use Stokes's Theorem:

$$\iint_\Sigma \mathbf{F} \cdot \mathbf{n} \, dS = \int_C \mathbf{G} \cdot d\mathbf{r}$$

If Σ is the boundary of a simple solid region D, div $\mathbf{F}$ is continuous on D, and $\mathbf{n}$ is pointed outward from D, then one can use the Divergence Theorem:

$$\iint_\Sigma \mathbf{F} \cdot \mathbf{n} \, dS = \iiint_D (\text{div } \mathbf{F}) \, dV$$

$$\iint_\Sigma \mathbf{F} \cdot \mathbf{n} \, dS$$

Review Exercises

1. Let $\mathbf{a}$ and $\mathbf{b}$ be constant vectors, and let $\mathbf{r}(x, y, z) = x\mathbf{i} + y\mathbf{j} + z\mathbf{k}$. Show that
$$\mathbf{a} \times \mathbf{b} = \text{grad } [\mathbf{a} \cdot (\mathbf{b} \times \mathbf{r})]$$

2. Let f be a function of several variables whose component functions have continuous second derivatives. Show that
$$\text{curl}(f \text{ grad } f) = \mathbf{0}$$
(*Hint*: Use Exercises 28 and 31 of Section 15.1.)

In Exercises 3–4 find a function f having the given gradient.

3. $\text{grad } f(x, y) = (y^2 - y \sin xy)\mathbf{i} + (2xy - x \sin xy)\mathbf{j}$

4. $\text{grad } f(x, y, z) = ye^z\mathbf{i} + (xe^z + e^y)\mathbf{j} + (xy + 1)e^z\mathbf{k}$

In Exercises 5–26 evaluate the integral.

5. $\int_C (xy + z^2) \, ds$, where C is parametrized by $\mathbf{r}(t) = \cos t\mathbf{i} + \sin t\mathbf{j} + t\mathbf{k}$ for $\pi/4 \le t \le 3\pi/4$

6. $\int_C (2xy - 3yz) \, ds$, where C is composed of the line segments from $(1, 0, 4)$ to $(0, 3, 2)$ and from $(0, 3, 2)$ to $(0, 0, 0)$

7. $\int_C xy \, dx + z \cos x \, dy + z \, dz$, where C is parametrized by $\mathbf{r}(t) = t\mathbf{i} + \cos t\mathbf{j} + \sin t\mathbf{k}$ for $0 \le t \le \pi/2$

8. $\int_C y \sin 2x \, dx + \sin^2 x \, dy$, where C is parametrized by $\mathbf{r}(t) = t\mathbf{i} + \sin t\mathbf{j}$ for $0 \le t \le \pi/3$

9. $\int_C x \, ds$, where C is the graph of $y = x^2$ for $0 \le x \le 1$

10. $\int_C (1 + y \sin z) \, dx + (1 + x \sin z) \, dy + xy \cos z \, dz$, where C is the curve parametrized by $\mathbf{r}(t) = \tan^5 t\mathbf{i} + \cos^4 t\mathbf{j} + t\mathbf{k}$ for $0 \le t \le \pi/4$

11. $\int_C e^x \cos z \, dx + y \, dy - e^x \sin z \, dz$, where C is parametrized by $\mathbf{r}(t) = e^{t^2(t^2-1)}\mathbf{i} + t\mathbf{j} + te^{t^{17}-1}\mathbf{k}$ for $0 \le t \le 1$

12. $\iint_\Sigma \mathbf{F} \cdot \mathbf{n} \, dS$, where
$$\mathbf{F}(x, y, z) = x\sqrt{x^2 + y^2 + z^2}\mathbf{i} + y\sqrt{x^2 + y^2 + z^2}\mathbf{j} + z\sqrt{x^2 + y^2 + z^2}\mathbf{k}$$
and where Σ is the torus $\rho = \sin \phi$ oriented by the normal directed outward

13. $\iint_\Sigma \mathbf{F} \cdot \mathbf{n} \, dS$, where $\mathbf{F}(x, y, z) = x\mathbf{i} + xy\mathbf{j} + (1 + z)\mathbf{k}$ and where Σ is the boundary of the solid region in the first octant bounded by the coordinate planes, the plane $z = 1 + x$, and the parabolic sheet $x = 1 - y^2$ (Σ oriented by the normal directed outward)

14. $\int_C (e^x - 4y \sin^2 x) \, dx + (2x + \sin 2x) \, dy$, where C is the square with vertices $(2, 2)$, $(2, -2)$, $(-2, 2)$, and $(-2, -2)$, oriented counterclockwise

15. $\iint_\Sigma \mathbf{F} \cdot \mathbf{n} \, dS$, where $\mathbf{F}(x, y, z) = xyz\mathbf{i} + 2yz\mathbf{j} + x^3y\mathbf{k}$, and where Σ is the boundary of the part of the cylindrical region $x^2 + y^2 \le 4$ between the planes $z = -2$ and $z = 3$ (Σ oriented by the normal directed outward)

16. Evaluate $\iint_\Sigma \mathbf{F} \cdot \mathbf{n} \, dS$, where $\mathbf{F} = y\mathbf{i} - x\mathbf{j} + z\mathbf{k}$ and where Σ is the portion of the hyperbolic paraboloid $x = u + v$, $y = u - v$, $z = uv$ that is cut out by the planes $y = x$, $y = -x$, and $x = 1$, oriented by the normal pointed upward)

17. $\iint_\Sigma \mathbf{F} \cdot \mathbf{n} \, dS$, where $\mathbf{F}(x, y, z) = y\mathbf{i} - x\mathbf{j} + z\mathbf{k}$ and where Σ is the portion of the paraboloid $z = -1 + x^2 + y^2$ below the plane $z = 1$ (Σ oriented by the normal directed downward)

18. $\int_C \mathbf{F} \cdot d\mathbf{r}$, where $\mathbf{F}(x, y, z) = xyz\mathbf{i} + 2x^2z\mathbf{j} + y^6\mathbf{k}$ and where C is the boundary of the rectangle in the plane $z = y$ for which $-1 \leq x \leq 1$ and $0 \leq y \leq 2$, oriented counterclockwise as viewed from above

19. $\iint_\Sigma (\text{curl } \mathbf{F}) \cdot \mathbf{n} \, dS$, where $\mathbf{F}(x, y, z) = x^3 y \mathbf{i} - y^3 \mathbf{j} + z \sec xyz\mathbf{k}$ and where Σ is the portion of the sphere $x^2 + y^2 + z^2 = 2$ above the plane $z = 1$ (Σ oriented by the normal directed downward)

20. $\int_C \sin x \sin y \, dx - \cos x \cos y \, dy$, where C is the circle $x^2 + y^2 = 4$, oriented clockwise

21. $\iint_\Sigma \mathbf{F} \cdot \mathbf{n} \, dS$, where

$$\mathbf{F}(x, y, z) = \tan x\mathbf{i} - y(1 + \tan^2 x)\mathbf{j} - 6z\mathbf{k}$$

and where Σ is composed of the hemisphere $z = \sqrt{4 - x^2 - y^2}$ and the disk $x^2 + y^2 \leq 4$ in the xy plane (Σ oriented by the normal directed outward)

22. $\int_C (y^4 + x^3 y^2) \, dx + x^2 y^3 \, dy$, where C is the boundary of the region between the graphs of $y = x^2$ and $y = 1$, oriented counterclockwise

23. $\int_C y \, dx + y \, dy + x^2 \, dz$, where C is the intersection of the surfaces $z = x^2 + y^2$ and $z = 1 - y^2$, oriented counterclockwise as viewed from above

24. $\iint_\Sigma \mathbf{F} \cdot \mathbf{n} \, dS$, where $\mathbf{F}(x, y, z) = x^2\mathbf{i} + 2yz\mathbf{j} - z^2\mathbf{k}$ and where Σ is the boundary of the solid region in the first octant bounded by the xy and yz coordinate planes, by the surface $z = y^2$, and by the cylinder $r = \sin \theta$ (Σ oriented by the normal directed outward)

25. $\int_C \mathbf{F} \cdot d\mathbf{r}$, where $\mathbf{F}(x, y, z) = x^4\mathbf{i} - y^2\mathbf{j} + z\mathbf{k}$ and where C is parametrized by $\mathbf{r}(t) = t\mathbf{i} - t^3\mathbf{k}$ for $-1 \leq t \leq 1$

26. $\iint_\Sigma \mathbf{F} \cdot \mathbf{n} \, dS$, where $\mathbf{F}(x, y, z) = yz\mathbf{k}$ and where Σ is the boundary of the solid region in the first octant bounded by the cone $z^2 = x^2 + y^2$ and the planes $x = 0$, $y = 0$, $z = 2$, and $z = 3$ (Σ oriented by the normal directed outward)

27. Let a force $\mathbf{F}$ be given by

$$\mathbf{F}(x, y, z) = y\mathbf{i} + x\mathbf{j} + z^3\mathbf{k}$$

Find the work W done by $\mathbf{F}$ on an object that moves from $(1, 0, 0)$ to $(0, 1, \pi)$

a. along a straight line.

b. along a helix parametrized by $\mathbf{r}(t) = \cos t\mathbf{i} + \sin t\mathbf{j} + 2t\mathbf{k}$ for $0 \leq t \leq \pi/2$.

28. Find the area A of the region enclosed by the curve parametrized by $\mathbf{r}(t) = \cos t\mathbf{i} + \sin 2t\mathbf{j}$ for $-\pi/2 \leq t \leq \pi/2$.

Topics for Discussion

1. How does a vector field differ from a function of one variable, a vector-valued function, and a function of several variables? Describe phenomena that can be described by vector fields.

2. How are the curl and divergence of a vector field defined, and what are their physical interpretations?

3. How does a line integral differ from the original integral as defined in Chapter 5? Write down three different forms of line integrals. Under what conditions is it possible to evaluate a line integral by evaluating instead a double integral?

4. Describe the Fundamental Theorem of Line Integrals in your own words. Write down an example of a line integral that cannot be evaluated by means of the Fundamental Theorem of Line Integrals.

5. What is a conservative field? Give examples of conservative forces and of nonconservative forces.

6. How does a surface integral differ from the double integral as defined in Chapter 14? Write down two different forms of surface integrals. Under what conditions is it possible to evaluate a surface integral by evaluating instead a triple integral? Is it ever possible to evaluate a surface integral both by evaluating a triple integral and by evaluating a line integral? Explain.

7. What is the difference between Green's Theorem and Stokes's Theorem? If a line integral can be evaluated by means of Green's Theorem, can it automatically be evaluated by means of Stokes's Theorem? Explain.

8. At the outset of the chapter we said that Green's Theorem, Stokes's Theorem, and the Divergence Theorem are higher-dimensional analogues of the Fundamental Theorem of Calculus. Justify this statement.

Cumulative Review, Chapters 1–14

In Exercises 1–2 find the limit.

1. $\displaystyle\lim_{x \to -\infty} \frac{1}{x^2 + 2} \sqrt{\frac{x^5 - 1}{2x + 1}}$

2. $\displaystyle\lim_{x \to \pi/2} \frac{\ln (\sin x)}{(\pi - 2x)^2}$

3. Let $f(x) = \sin (x^3)$. Find the smallest positive integer n such that $f^{(n)}(0) \neq 0$.

4. Determine whether there is a nonzero value of c for which the parabola $y = cx^2$ and the circle $x^2 + y^2 = 4$ intersect at right angles.

5. A carpenter drills downward through the center of the top of a 2×4 inch plank at the rate of $\frac{1}{12}$ inch per second. If the drill bit is conical with a radius of $\frac{1}{2}$ inch and a height of 1 inch, determine the rate at which the volume of the

drilled hole increases when the point of the drill is $\frac{2}{3}$ inch into the wood.

6. A right circular cone is inscribed in a sphere of radius r. Find the height h of the cone with the largest possible volume.

7. Let
$$f(x) = \frac{x}{x^2 + 2x + 2}$$
Sketch the graph of f, indicating all pertinent information.

8. Determine the area A of the region bounded by the curves $y = 2x^2$ and $16x = y^4$.

In Exercises 9–10 evaluate the integral.

9. $\int x^2 \sec^2 x \tan x \, dx$

10. $\int \frac{3x + 4}{x^3 + 4x} dx$

11. Determine whether
$$\int_0^2 \frac{x^2}{\sqrt{4 - x^2}} dx$$
converges or diverges. If it converges, find its value.

12. Let R be the region bounded by the curve $y = x^2 + 1$ and the line $y = 2x + 1$. Find the volume V of the solid generated by revolving R about the
   a. x axis.   b. y axis.

13. Let $(x^2 + y^2)^3 = 3x^2 y^2$.
   a. Convert the equation to polar coordinates, and then sketch its graph.
   b. Determine the maximum distance between a point on the graph of the equation and the origin.

14. Find the sum of the series $\sum_{n=2}^{\infty} 3(.3)^{2n}$.

15. Determine whether
$$\sum_{n=1}^{\infty} (-1)^n \left(1 + \frac{1}{n}\right)^{-n^2}$$
converges absolutely, converges conditionally, or diverges.

16. Find the interval of convergence of $\sum_{n=0}^{\infty} (\cosh n) x^n$.

17. Show that
$$\frac{5}{6} \leq \int_0^1 e^{-t^2/2} \, dt \leq \frac{103}{120}$$
(*Hint:* Use the Taylor series expansion for $e^{-t^2/2}$.)

18. Consider the point $P = (0, 1, 2)$ and the plane $4x - y + z = 12$.
   a. Find symmetric equations for the line l that passes through P and is perpendicular to the plane.
   b. Find the distance D between the plane and P.

19. Let $\mathbf{r}(t) = t\mathbf{i} + \frac{4}{3} t^{3/2}\mathbf{j} + t^2\mathbf{k}$ represent the motion of a particle traversing a curve C.
   a. Find the positive value of b that makes the length L of the curve C from $t = 0$ to $t = b$ exactly 30 units.
   b. Show that the tangential component of acceleration is constant, and find that constant.
   c. Find the normal component of acceleration, and show that it is a decreasing function on $(0, \infty)$.

20. Let $f(x, y) = \sqrt{2x^2 + y^2}$.
   a. Find $f_{yy}(x, y)$.
   b. Find the unit vector $\mathbf{u}$ such that at the point $(1, \sqrt{2})$, f decreases most rapidly in the direction of $\mathbf{u}$.

21. Consider the two surfaces $x^2 - 2y^2 + z^2 = 0$ and $xyz = 1$. Show that at each point of their intersection their tangent planes are perpendicular to one another. (*Hint:* Do not find any points of intersection.)

22. Assuming that the minimum distance exists, find the minimum distance from the origin $(0, 0, 0)$ to the surface $xy^2 z^4 = 32$.

23. Evaluate $\int_1^8 \int_{y^{1/3}}^2 \cos\left(\frac{x^4}{4} - x\right) dx \, dy$.

24. Determine the surface area S of that portion of the surface $z = x^2 - y^2$ inside the cylinder $x^2 + y^2 = 4$.

25. Compute $\iiint_D z \, dV$, where D is the solid region bounded below by the paraboloid $2z = x^2 + y^2$ and above by the sphere $x^2 + y^2 + z^2 = 8$.

26. Evaluate $\iint_R (y - x) \sin (y + x)^3 \, dA$, where R is the triangular region with vertices $(0, 0)$, $(2, 2)$, and $(0, 4)$.

Appendix

PROOFS OF SELECTED THEOREMS

In the Appendix we present the proofs of all theorems that were not proved in Chapters 1 to 9. In addition, we prove one version of l'Hôpital's Rule. The first collection of theorems will be proved from the definition of limit and other definitions and results already appearing in the text.

First we need an elementary observation about the limits of a pair of functions. In finding the limits of two functions at a point by use of ε and δ, we may choose the same δ for both functions. We will prove this in the following lemma.

LEMMA A.1

Assume that $\lim_{x \to a} f(x) = L$ and $\lim_{x \to a} g(x) = M$. For any $\varepsilon > 0$, there is a common $\delta > 0$ such that if $0 < |x - a| < \delta$, then both

$$|f(x) - L| < \varepsilon \quad \text{and} \quad |g(x) - M| < \varepsilon$$

Proof For any $\varepsilon > 0$, there is a $\delta_1 > 0$ such that

$$\text{if } 0 < |x - a| < \delta_1, \quad \text{then } |f(x) - L| < \varepsilon$$

and a $\delta_2 > 0$ such that

$$\text{if } 0 < |x - a| < \delta_2, \quad \text{then } |g(x) - M| < \varepsilon$$

Let δ be the minimum of δ_1 and δ_2; then $\delta > 0$. Moreover, if $0 < |x - a| < \delta$, then we certainly have both $0 < |x - a| < \delta_1$ and $0 < |x - a| < \delta_2$. Therefore

$$|f(x) - L| < \varepsilon \quad \text{and} \quad |g(x) - M| < \varepsilon \quad \blacksquare$$

We now turn to the limit theorems. First we prove that if f has a limit at a, then f has *only one* limit at a.

A-1

THEOREM A.2
Uniqueness of Limits

If the limit of a function f at a exists, then this limit is unique. Equivalently, if L and M are both limits of f at a, then $L = M$.

Proof Let $\varepsilon > 0$, and let $f = g$ in Lemma A.1. The lemma says that there is a common $\delta > 0$ such that if $0 < |x - a| < \delta$, then

$$|f(x) - L| < \varepsilon \quad \text{and} \quad |f(x) - M| < \varepsilon$$

If we choose an x such that $0 < |x - a| < \delta$, we can conclude that

$$|L - M| = |L - f(x) + f(x) - M|$$
$$\leq |L - f(x)| + |f(x) - M| < \varepsilon + \varepsilon = 2\varepsilon$$

Consequently for *any* positive ε, the distance from L to M is less than 2ε. This implies that $L = M$. ■

THEOREM A.3
Squeezing Theorem
(Theorem 2.3)

Assume that $f(x) \leq g(x) \leq h(x)$ for all x in some open interval I about a, except possibly a itself. If $\lim_{x \to a} f(x) = \lim_{x \to a} h(x) = L$, then $\lim_{x \to a} g(x)$ exists and $\lim_{x \to a} g(x) = L$.

Proof Let $\varepsilon > 0$. By Lemma A.1 there is a common $\delta > 0$ such that if $0 < |x - a| < \delta$, then x is in I and

$$|f(x) - L| < \varepsilon \quad \text{and} \quad |h(x) - L| < \varepsilon$$

Since $f(x) \leq g(x)$, the first inequality above implies that

$$L - \varepsilon < f(x) \leq g(x) \tag{1}$$

Similarly, since $|h(x) - L| < \varepsilon$ and $g(x) \leq h(x)$,

$$g(x) \leq h(x) < L + \varepsilon \tag{2}$$

Together (1) and (2) imply that

$$L - \varepsilon < g(x) < L + \varepsilon$$

and thus

$$|g(x) - L| < \varepsilon \quad ■$$

THEOREM A.4
Sum Rule for Limits
(Theorem 2.2)

Assume that $\lim_{x \to a} f(x)$ and $\lim_{x \to a} g(x)$ both exist. Then $\lim_{x \to a} [f(x) + g(x)]$ exists, and

$$\lim_{x \to a}[f(x) + g(x)] = \lim_{x \to a} f(x) + \lim_{x \to a} g(x)$$

Proof Let $\lim_{x \to a} f(x) = L$, $\lim_{x \to a} g(x) = M$, and $\varepsilon > 0$. We will show that there is a $\delta > 0$ such that if $0 < |x - a| < \delta$, then $|[f(x) + g(x)] - (L + M)| < \varepsilon$. Suppose we replace ε by $\varepsilon/2$ in Lemma A.1. The lemma implies that there is a common

$\delta > 0$ such that if $0 < |x - a| < \delta$, then

$$|f(x) - L| < \frac{\varepsilon}{2} \quad \text{and} \quad |g(x) - M| < \frac{\varepsilon}{2}$$

Therefore

$$|[f(x) + g(x)] - (L + M)| = |[f(x) - L] + [g(x) - M]|$$

$$\leq |f(x) - L| + |g(x) - M|$$

$$< \frac{\varepsilon}{2} + \frac{\varepsilon}{2} = \varepsilon \quad \blacksquare$$

THEOREM A.5
Constant Multiple Rule for
Limits
(Theorem 2.2)

Assume that $\lim_{x \to a} f(x)$ exists and that c is any real number. Then $\lim_{x \to a} cf(x)$ exists, and

$$\lim_{x \to a} cf(x) = c \lim_{x \to a} f(x)$$

Proof If $c = 0$, the proof is trivial, because both sides of the equation are 0. If $c \neq 0$, we let $\lim_{x \to a} f(x) = L$ and choose an arbitrary $\varepsilon > 0$. We will show that for some $\delta > 0$, if $0 < |x - a| < \delta$, then $|cf(x) - cL| < \varepsilon$. Note that $\varepsilon/|c| > 0$, so that corresponding to this positive quantity there is a $\delta > 0$ such that if $0 < |x - a| < \delta$, we have

$$|f(x) - L| < \varepsilon/|c|$$

Consequently

$$|cf(x) - cL| = |c||f(x) - L| < |c|\frac{\varepsilon}{|c|} = \varepsilon$$

and thus

$$\lim_{x \to a} cf(x) = cL = c \lim_{x \to a} f(x) \quad \blacksquare$$

To prove that the limit of a product is the product of the limits, we first consider the special case in which the limits are 0.

LEMMA A.6

Assume that $\lim_{x \to a} f(x) = 0$ and $\lim_{x \to a} g(x) = 0$. Then $\lim_{x \to a} f(x)g(x)$ exists, and

$$\lim_{x \to a} f(x)g(x) = 0$$

Proof Let $\varepsilon > 0$. We may assume that $0 < \varepsilon < 1$. Let 0 replace both L and M in Lemma A.1. The lemma says that we can find a common $\delta > 0$ such that if $0 < |x - a| < \delta$, then

$$|f(x) - 0| < \varepsilon \quad \text{and} \quad |g(x) - 0| < \varepsilon$$

Thus

$$|f(x)g(x) - 0| = |f(x) - 0||g(x) - 0| < \varepsilon \cdot \varepsilon = \varepsilon^2 < \varepsilon \qquad ■$$

THEOREM A.7
Product Rule for Limits
(Theorem 2.2)

Assume that $\lim_{x \to a} f(x)$ and $\lim_{x \to a} g(x)$ both exist. Then $\lim_{x \to a} f(x)g(x)$ exists, and

$$\lim_{x \to a} f(x)g(x) = \lim_{x \to a} f(x) \lim_{x \to a} g(x)$$

Proof Let $\lim_{x \to a} f(x) = L$ and $\lim_{x \to a} g(x) = M$. Then from the Sum Rule we deduce that

$$\lim_{x \to a} [f(x) - L] = \lim_{x \to a} [f(x) + (-L)] = \lim_{x \to a} f(x) + \lim_{x \to a} (-L)$$

$$= L + (-L) = 0$$

Similarly,

$$\lim_{x \to a} [g(x) - M] = 0$$

Next, the product $f(x)g(x)$ can be written in the form

$$f(x)g(x) = [f(x) - L][g(x) - M] + [Lg(x) + Mf(x)] - LM$$

as can be verified by multiplying out $[f(x) - L][g(x) - M]$. Now by Lemma A.6,

$$\lim_{x \to a} [f(x) - L][g(x) - M] = 0$$

Moreover, the Constant Multiple Rule yields

$$\lim_{x \to a} Lg(x) = LM, \quad \lim_{x \to a} f(x)M = M, \quad \text{and} \quad \lim_{x \to a} (-LM) = -LM$$

By the Sum Rule we obtain

$$\lim_{x \to a} f(x)g(x) = \lim_{x \to a} [f(x) - L][g(x) - M] + \lim_{x \to a} Lg(x) + \lim_{x \to a} f(x)M + \lim_{x \to a} (-LM)$$

$$= 0 + LM + LM - LM$$

$$= LM \qquad ■$$

Before proving the Quotient Rule for Limits, we prove a preliminary result.

LEMMA A.8

If $\lim_{x \to a} g(x) = M > 0$, then there is a $\delta > 0$ such that

$$\text{if } 0 < |x - a| < \delta, \quad \text{then } g(x) > \frac{1}{2}M$$

If $\lim_{x \to a} g(x) = M < 0$, then there is a $\delta > 0$ such that

$$\text{if } 0 < |x - a| < \delta, \quad \text{then } g(x) < \frac{1}{2}M$$

Proof Suppose that $\lim_{x \to a} g(x) = M > 0$. Then by Definition 2.1 with $\varepsilon = \frac{1}{2}M$, there is a $\delta > 0$ such that

$$\text{if } 0 < |x - a| < \delta, \quad \text{then } |g(x) - M| < \frac{1}{2}M$$

Consequently if $0 < |x - a| < \delta$, then

$$g(x) = M - (M - g(x)) \geq M - |M - g(x)| > M - \frac{1}{2}M = \frac{1}{2}M$$

Similarly, if $\lim_{x \to a} g(x) = M < 0$, then there is a $\delta > 0$ such that

$$\text{if } 0 < |x - a| < \delta, \quad \text{then } |g(x) - M| < -\frac{1}{2}M$$

Consequently if $0 < |x - a| < \delta$, then

$$g(x) = (g(x) - M) + M \leq |g(x) - M| + M < -\frac{1}{2}M + M = \frac{1}{2}M \quad \blacksquare$$

It follows directly from Lemma A.8 that

$$\text{if } \lim_{x \to a} g(x) > 0, \quad \text{then } g(x) > 0 \text{ for all } x \text{ sufficiently close to } a$$

and

$$\text{if } \lim_{x \to a} g(x) < 0, \quad \text{then } g(x) < 0 \text{ for all } x \text{ sufficiently close to } a$$

Now we are prepared to prove the Quotient Rule for Limits.

THEOREM A.9
Quotient Rule for Limits
(Theorem 2.2)

Suppose that $\lim_{x \to a} f(x)$ and $\lim_{x \to a} g(x)$ exist and that $\lim_{x \to a} g(x) \neq 0$. Then $\lim_{x \to a} f(x)/g(x)$ exists, and

$$\lim_{x \to a} \frac{f(x)}{g(x)} = \frac{\lim_{x \to a} f(x)}{\lim_{x \to a} g(x)}$$

Proof Let $\lim_{x \to a} f(x) = L$ and $\lim_{x \to a} g(x) = M \neq 0$. From Lemma A.8 it follows that there is a $\delta > 0$ such that if $0 < |x - a| < \delta$, then $|g(x)| > |M|/2$. For such x we have

$$\left| \frac{1}{Mg(x)} \right| < \frac{2}{M^2}$$

Consequently

$$\left| \frac{f(x)}{g(x)} - \frac{L}{M} \right| = \left| \frac{Mf(x) - Lg(x)}{Mg(x)} \right| = |Mf(x) - Lg(x)| \left| \frac{1}{Mg(x)} \right|$$

$$\leq |Mf(x) - Lg(x)| \left(\frac{2}{M^2} \right)$$

As a result, for $0 < |x - a| < \delta$, the function

$$\left| \frac{f(x)}{g(x)} - \frac{L}{M} \right|$$

is squeezed between the function 0 and the function

$$|Mf(x) - Lg(x)| \left(\frac{2}{M^2} \right)$$

But by the Sum and Constant Multiple Rules,

$$\lim_{x \to a} [Mf(x) - Lg(x)] = M \lim_{x \to a} f(x) - L \lim_{x \to a} g(x)$$

$$= ML - LM = 0$$

Thus

$$\lim_{x \to a} |Mf(x) - Lg(x)| \left(\frac{2}{M^2} \right) = 0$$

Therefore by the Squeezing Theorem we have

$$\lim_{x \to a} \left| \frac{f(x)}{g(x)} - \frac{L}{M} \right| = 0$$

which is equivalent to

$$\lim_{x \to a} \frac{f(x)}{g(x)} = \frac{L}{M} \qquad \blacksquare$$

THEOREM A.10
Substitution Rule for Limits

Suppose $\lim_{x \to a} f(x) = c$ and $f(x) \neq c$ for all x in some open interval about a, with the possible exception of a itself. Suppose also that $\lim_{y \to c} g(y)$ exists. Then

$$\lim_{x \to a} g(f(x)) = \lim_{y \to c} g(y)$$

Proof Let $\varepsilon > 0$. Suppose $\lim_{y \to c} g(y) = L$. Then there is a $\delta_1 > 0$ such that

if $\qquad\qquad 0 < |y - c| < \delta_1, \quad$ then $|g(y) - L| < \varepsilon$ $\qquad\qquad$ (3)

Since $\lim_{x \to a} f(x) = c$ and $\delta_1 > 0$, there is a $\delta > 0$ such that

$$\text{if } 0 < |x - a| < \delta, \quad \text{then } |f(x) - c| < \delta_1$$

By hypothesis, δ may be chosen so small that if $0 < |x - a| < \delta$, then $f(x) \neq c$. Thus

$$\text{if } 0 < |x - a| < \delta, \quad \text{then } 0 < |f(x) - c| < \delta_1$$

and hence by (3),

$$|g(f(x)) - L| < \varepsilon$$

Consequently

$$\lim_{x \to a} g(f(x)) = L = \lim_{y \to c} g(y) \qquad \blacksquare$$

Our final limit theorem shows how one-sided limits and two-sided limits are related.

THEOREM A.11
(Theorem 2.5)

Let f be defined in an open interval about a, except possibly at a itself. Then $\lim_{x \to a} f(x)$ exists if and only if both one-sided limits, $\lim_{x \to a^+} f(x)$ and $\lim_{x \to a^-} f(x)$ exist, and

$$\lim_{x \to a^+} f(x) = \lim_{x \to a^-} f(x)$$

In that case

$$\lim_{x \to a} f(x) = \lim_{x \to a^+} f(x) = \lim_{x \to a^-} f(x)$$

Proof   From the definitions, if $\lim_{x \to a} f(x) = L$, then $\lim_{x \to a^+} f(x) = L$ and $\lim_{x \to a^-} f(x) = L$. Conversely, assume that $\lim_{x \to a^+} f(x) = L$ and $\lim_{x \to a^-} f(x) = L$, and let $\varepsilon > 0$. It follows from Definition 2.4 that there exist $\delta_1 > 0$ and $\delta_2 > 0$ such that

$$\text{if } 0 < x - a < \delta_1, \quad \text{then } |f(x) - L| < \varepsilon$$

and

$$\text{if } -\delta_2 < x - a < 0, \quad \text{then } |f(x) - L| < \varepsilon$$

Let $\delta > 0$ be the minimum of δ_1 and δ_2. It follows directly that

$$\text{if } 0 < |x - a| < \delta, \quad \text{then } |f(x) - L| < \varepsilon$$

This implies that $\lim_{x \to a} f(x) = L$.   ■

Next we prove the Chain Rule for derivatives.

THEOREM A.12
Chain Rule
(Theorem 3.8)

Let f be differentiable at a, and let g be differentiable at $f(a)$. Then $g \circ f$ is differentiable at a, and

$$(g \circ f)'(a) = g'(f(a))f'(a)$$

Proof   Let

$$G(x) = \begin{cases} \dfrac{g(x) - g(f(a))}{x - f(a)} & \text{for } x \neq f(a) \\ g'(f(a)) & \text{for } x = f(a) \end{cases}$$

By the definition of the derivative of g at $f(a)$ and by the Substitution Rule,

$$\lim_{x \to a} G(f(x)) = \lim_{y \to f(a)} G(y) = \lim_{y \to f(a)} \frac{g(y) - g(f(a))}{y - f(a)} = g'(f(a))$$

By considering the cases $f(x) = f(a)$ and $f(x) \neq f(a)$ separately, you can show that

$$\frac{g(f(x)) - g(f(a))}{x - a} = G(f(x))\left(\frac{f(x) - f(a)}{x - a}\right)$$

By the Product Rule we conclude that

$$\lim_{x \to a} \frac{g(f(x)) - g(f(a))}{x - a} = \lim_{x \to a} G(f(x)) \lim_{x \to a} \left(\frac{f(x) - f(a)}{x - a} \right)$$

$$= g'(f(a))f'(a)$$

Thus $(g \circ f)'(a) = g'(f(a))f'(a)$. ∎

The Least Upper Bound Axiom and Its Consequences

The proofs of the theorems in this subsection rely on a property of the real numbers that has not yet been discussed in this book. Before introducing this property, we extend the notion of boundedness that was introduced for intervals in Section 1.1. We make the blanket assumption that any set discussed in this Appendix contains at least one number.

DEFINITION A.13

a. A set S of numbers is **bounded above** if there is a number M such that $x \leq M$ for every x in S. Any such number M is called an **upper bound** of S.

b. A set S of numbers is **bounded below** if there is a number m such that $m \leq x$ for every x in S. Any such number m is called a **lower bound** of S.

c. A set of numbers is **bounded** if it is bounded above and below.

Any interval of the form (a, b) $[a, b]$, $(a, b]$, or $[a, b)$ is bounded according to Definition A.13. In each case b is an upper bound and a is a lower bound of the interval. Any interval of the form $(-\infty, a)$ or $(-\infty, a]$ is bounded above but not below. Analogously, any interval of the form (a, ∞) or $[a, \infty)$ is bounded below but not above. Hence Definition A.13 implies that intervals of the form (a, ∞), $[a, \infty)$, $(-\infty, a)$, or $(-\infty, a]$ are not bounded.

If M_1 is an upper bound of a set S and if $M_2 > M_1$, then M_2 is also an upper bound of S. Thus if a set has an upper bound, then it has infinitely many upper bounds, and an upper bound of such a set may be chosen arbitrarily large. Our interest will center on the possibility that there exists a *smallest* upper bound of a set.

DEFINITION A.14

a. A number M is a **least upper bound** of a set S if M is an upper bound of S and if no upper bound of S is less than M.

b. A number m is a **greatest lower bound** of a set S if m is a lower bound of S and if no lower bound of S is greater than m.

If M_1 and M_2 are both least upper bounds of S, then by definition neither can be less than the other, and thus $M_1 = M_2$. We conclude that a set can have at most one least upper bound. Similarly, a set can have at most one greatest lower bound. The least upper bound of an interval of the form (a, b), $[a, b]$, $(a, b]$, or $[a, b)$ is b, and the greatest lower bound is a. From these examples we see that the least upper bound, and also the greatest lower bound, may or may not be in a given set. For an example not involving an interval, let S be the set of all rational numbers in the open interval $(0, 1)$ that have decimal expansions containing only finitely many zeros and nines and no other integers. Some examples of numbers in S are 0.9999, 0.900009, and 0.90900909. The greatest lower bound of S is 0, and the least upper bound of S is 1. However, neither 0 nor 1 belongs to S. In contrast, an unbounded interval like (a, ∞) and $[a, \infty)$ has no least upper bound, and neither $(-\infty, a)$ nor $(-\infty, a]$ has a greatest lower bound.

Now we state the property of real numbers mentioned earlier.

> **Least Upper Bound Axiom**
> Every set of real numbers that is bounded above has a least upper bound.

It is possible to prove that the Least Upper Bound Axiom is equivalent to the property that every set of real numbers that is bounded below has a greatest lower bound (see Exercise 11). It follows that any set that is contained in a bounded interval has a least upper bound and a greatest lower bound, each of which may or may not be contained in the set itself.

We are now in a position to prove the theorems that rely on the Least Upper Bound Axiom.

THEOREM A.15
Intermediate Value Theorem
(Theorem 2.12)

Suppose f is continuous on a closed interval $[a, b]$. Let p be any number between $f(a)$ and $f(b)$, so that $f(a) \leq p \leq f(b)$ or $f(b) \leq p \leq f(a)$. Then there exists a number c in $[a, b]$ such that $f(c) = p$.

Proof If $f(a) = p$ or $f(b) = p$, then let $c = a$ or $c = b$. Otherwise, either $f(a) < p < f(b)$ or $f(b) < p < f(a)$. The proof of the theorem is similar for the two cases, so we will only prove the theorem under the assumption that $f(a) < p < f(b)$. Let S be the collection of all x in $[a, b]$ such that $f(x) < p$. Then a is in S and b is an upper bound of S. By the Least Upper Bound Axiom, S has a least upper bound, which we will call c; notice that $a \leq c \leq b$. Either $f(c) < p$, $f(c) > p$, or $f(c) = p$. If $f(c) < p$, it follows that $c < b$, since $p < f(b)$ by prior assumption, and that $f(c) + \varepsilon < p$ for sufficiently small ε. Since f is continuous at c, there is a $\delta > 0$ such that if $|x - c| < \delta$, then x is in $[a, b]$ and $|f(c) - f(x)| < \varepsilon$. In particular if we take $x = c + \delta/2$, then we have $|f(c) - f(c + \delta/2)| < \varepsilon$, which implies that

$$f\left(c + \frac{\delta}{2}\right) < f(c) + \varepsilon < p$$

But then $c + \delta/2$ is in S, so c is not the least upper bound of S. Consequently the assumption that $f(c) < p$ is false. Analogously, the assumption that $f(c) > p$ is false. Therefore $f(c) = p$. ■

One implication of the Least Upper Bound Axiom that is particularly useful in proofs involves covers of sets. A collection $\mathscr{C}$ of open intervals is a **cover** of a set S, or **covers** S, if every point in S is in at least one of the intervals comprising $\mathscr{C}$. The following theorem, due to Eduard Heine (1821–1881) and Emile Borel (1871–1956), is derived from the Least Upper Bound Axiom.

THEOREM A.16 **Heine-Borel Theorem**	Let $\mathscr{C}$ be a collection of open intervals that cover a closed interval $[a, b]$. Then there is a finite subcollection $\mathscr{C}_0$ of $\mathscr{C}$ that covers $[a, b]$.

Proof Let S consist of all x in $[a, b]$ such that some finite subcollection of $\mathscr{C}$ covers $[a, x]$. Then a is in S and b is an upper bound of S. By the Least Upper Bound Axiom, S has a least upper bound c, and $a \leq c \leq b$. Since $\mathscr{C}$ covers $[a, b]$, one member of $\mathscr{C}$, say (d, e), contains c. Since $d < c$ and c is the least upper bound of S, d is not an upper bound of S. Consequently there is a number x in S such that $d < x \leq c$. Since x is in S, a finite subcollection $\mathscr{C}_1$ of $\mathscr{C}$ covers $[a, x]$. Let $\mathscr{C}_0$ consist of $\mathscr{C}_1$ and (d, e). Since (d, e) contains $[x, c]$, $\mathscr{C}_0$ covers $[a, c]$. It cannot be the case that $c < b$, because then there would be a number y in $[a, b]$ such that $c < y < e$, and thus $\mathscr{C}_0$ would cover $[a, y]$. That would mean that y is in S, so that c would not be an upper bound of S, a contradiction to our assumption. Thus $c = b$; since $c < e$, $\mathscr{C}_0$ covers $[a, b]$. ■

We now use the Heine-Borel Theorem to prove a preliminary result on boundedness of functions. We say that a function f is **bounded on a set** S if there are numbers m and M such that

$$m \leq f(x) \leq M \quad \text{for all } x \text{ in } S$$

that is, if the values assumed by f on S form a bounded set.

THEOREM A.17	If f is continuous on a closed interval $[a, b]$, then f is bounded on $[a, b]$.

Proof Since f is continuous on $[a, b]$, it follows that for each t in $[a, b]$ there is a number $\delta(t) > 0$ such that if x is in $[a, b]$ and $t - \delta(t) < x < t + \delta(t)$, then $f(t) - 1 < f(x) < f(t) + 1$. The collection of open intervals $(t - \delta(t), t + \delta(t))$, for t in $[a, b]$, forms a cover $\mathscr{C}$ of $[a, b]$. By the Heine-Borel Theorem there is a finite subcollection $\mathscr{C}_0$ of $\mathscr{C}$ that covers $[a, b]$. Let the intervals comprising $\mathscr{C}_0$ be $(t_1 - \delta(t_1), t_1 + \delta(t_1)), (t_2 - \delta(t_2), t_2 + \delta(t_2)), \ldots, (t_n - \delta(t_n), t_n + \delta(t_n))$, and let m be the smallest of the numbers $f(t_1) - 1, f(t_2) - 1, \ldots, f(t_n) - 1$, and M the largest of the numbers $f(t_1) + 1, f(t_2) + 1, \ldots, f(t_n) + 1$. Each x in $[a, b]$ lies in one of the subintervals $(t_k - \delta(t_k), t_k + \delta(t_k))$ comprising $\mathscr{C}_0$, so $m \leq f(t_k) - 1 < f(x) < f(t_k) + 1 \leq M$. Therefore f is bounded on $[a, b]$. ■

THEOREM A.18 **Maximum-Minimum** **Theorem** (Theorem 4.2)	Let f be continuous on a closed interval $[a, b]$. Then f has a maximum and a minimum value on $[a, b]$.

Proof Let S be the set of values assumed by f on $[a, b]$. By Theorem A.17, f is bounded, and thus S is bounded. Applying the Least Upper Bound Axiom and the remark following it, we conclude that S has both a least upper bound M and a greatest lower bound m. If $f(x) \neq M$ for all x in $[a, b]$, then let g be defined by

$$g(x) = \frac{1}{M - f(x)} \quad \text{for } a \leq x \leq b$$

By construction g is positive and continuous on $[a, b]$, so Theorem A.17 implies that g is bounded. Thus there is a positive number L such that

$$\frac{1}{M - f(x)} \leq L \quad \text{for } a \leq x \leq b$$

Solving for $f(x)$, we obtain

$$f(x) \leq M - \frac{1}{L} \quad \text{for } a \leq x \leq b$$

Therefore $M - 1/L$ is an upper bound of S, and this contradicts the assumption that M is the *least* upper bound of S. Consequently the assumption that $f(x) \neq M$ for all x in $[a, b]$ is false, so $f(c) = M$ for some number c in $[a, b]$. In other words f has a maximum value on $[a, b]$. A similar argument proves that for an appropriate number d in $[a, b]$, we have $f(d) = m$, so that f has a minimum value on $[a, b]$. ■

Recall that a function f is continuous on an interval I if for every x in I and every $\varepsilon > 0$ there is a $\delta > 0$, depending on ε and x, such that

$$\text{if } y \text{ is in } I \text{ and } |x - y| < \delta, \quad \text{then } |f(x) - f(y)| < \varepsilon$$

We emphasize that δ depends on both ε and x. There are functions f and intervals I for which δ need only depend on ε, and not on x.

DEFINITION A.19   A function f is **uniformly continuous on an interval** I if for every $\varepsilon > 0$ there is a $\delta > 0$ such that

$$\text{if } x \text{ and } y \text{ are in } I \text{ and } |x - y| < \delta, \quad \text{then } |f(x) - f(y)| < \varepsilon$$

Notice that any uniformly continuous function is continuous. The converse is false, as Exercise 13 shows. However, it turns out that any function that is continuous on a *closed, bounded* interval is also uniformly continuous on the interval. That is the content of the following theorem, which provides the key to our proof of the existence of the definite integral.

THEOREM A.20   If f is continuous on a closed interval $[a, b]$, then f is uniformly continuous on $[a, b]$.

Proof   Let $\varepsilon > 0$. Since f is continuous on $[a, b]$, for each t in $[a, b]$ there is a number $\delta(t) > 0$ such that

$$\text{if } x \text{ is in } [a, b] \text{ and } |t - x| < \delta(t), \quad \text{then } |f(x) - f(t)| < \frac{\varepsilon}{2}$$

The collection $\mathscr{C}$ of open intervals $(t - \frac{1}{2}\delta(t), t + \frac{1}{2}\delta(t))$, for t in $[a, b]$, covers $[a, b]$. By the Heine-Borel Theorem a finite subcollection $\mathscr{C}_0$ covers $[a, b]$. Let the intervals comprising $\mathscr{C}_0$ be $(t_1 - \frac{1}{2}\delta(t_1), t_1 + \frac{1}{2}\delta(t_1))$, $(t_2 - \frac{1}{2}\delta(t_2), t_2 + \frac{1}{2}\delta(t_2))$, . . . , $(t_n - \frac{1}{2}\delta(t_n), t_n + \frac{1}{2}\delta(t_n))$, and let δ be the smallest of the numbers $\frac{1}{2}\delta(t_1)$,

$\frac{1}{2}\delta(t_2), \ldots, \frac{1}{2}\delta(t_n)$. If x is in $[a, b]$, then x must lie in one such interval, say $(t_k - \frac{1}{2}\delta(t_k), t_k + \frac{1}{2}\delta(t_k))$. Then $|x - t_k| < \frac{1}{2}\delta(t_k)$. Therefore if y is in $[a, b]$ and $|y - x| < \delta$, then

$$|y - t_k| = |y - x + x - t_k| \le |y - x| + |x - t_k| < \delta + \tfrac{1}{2}\delta(t_k) \le \delta(t_k)$$

By definition of $\delta(t_k)$ it follows that

$$|f(x) - f(t_k)| < \frac{\varepsilon}{2} \quad \text{and} \quad |f(y) - f(t_k)| < \frac{\varepsilon}{2}$$

Therefore

$$|f(x) - f(y)| = |f(x) - f(t_k) + f(t_k) - f(y)|$$
$$\le |f(x) - f(t_k)| + |f(t_k) - f(y)|$$
$$< \frac{\varepsilon}{2} + \frac{\varepsilon}{2} = \varepsilon$$

Thus f is uniformly continuous on $[a, b]$ by Definition A.19. ■

Our next goal is to prove Theorem 5.3, which concerns the existence of a unique number I to which the Riemann sums for a continuous function on a closed interval $[a, b]$ converge. We begin with a preliminary result about lower and upper sums.

THEOREM A.21

> Let f be continuous on $[a, b]$. Then there is a unique number I satisfying
>
> $$L_f(P) \le I \le U_f(P)$$
>
> for every partition P of $[a, b]$.

Proof We observed in Sections 5.1 and 5.2 that every lower sum of f on $[a, b]$ is less than or equal to every upper sum. Thus the collection $\mathcal{C}_1$ of all lower sums is bounded above (by any upper sum) and the collection $\mathcal{C}_2$ of all upper sums is bounded below (by any lower sum). By the Least Upper Bound Axiom, $\mathcal{C}_1$ has a least upper bound L and $\mathcal{C}_2$ has a greatest lower bound G. From our preceding remarks it follows that

$$L_f(P) \le L \le G \le U_f(P)$$

for each partition P of $[a, b]$. Moreover, any number I satisfying

$$L_f(P) \le I \le U_f(P)$$

for each partition P of $[a, b]$ must satisfy

$$L \le I \le G$$

since L is the least upper bound of the lower sums and G is the greatest lower bound of the upper sums. Hence to complete the proof of the theorem it is enough to prove that $L = G$. Let $\varepsilon > 0$. Since f is continuous on $[a, b]$, it follows from Theorem A.20

that f is uniformly continuous on $[a, b]$. Thus there is a $\delta > 0$ such that

$$\text{if } x \text{ and } y \text{ are in } [a, b] \text{ and } |x - y| < \delta, \quad \text{then } |f(x) - f(y)| < \frac{\varepsilon}{b - a}$$

Let $P = \{x_0, x_1, x_2, \ldots, x_n\}$ be a partition of $[a, b]$ such that $\Delta x_k < \delta$ for $1 \leq k \leq n$, and let M_k and m_k be, respectively, the largest and smallest values of f on $[x_{k-1}, x_k]$. Then

$$\begin{aligned}
U_f(P) - L_f(P) &= (M_1 \, \Delta x_1 + M_2 \, \Delta x_2 + \cdots + M_n \, \Delta x_n) \\
&\quad - (m_1 \, \Delta x_1 + m_2 \, \Delta x_2 + \cdots + m_n \, \Delta x_n) \\
&= (M_1 - m_1) \, \Delta x_1 + (M_2 - m_2) \, \Delta x_2 + \cdots + (M_n - m_n) \, \Delta x_n \\
&< \frac{\varepsilon}{b - a} (\Delta x_1 + \Delta x_2 + \cdots + \Delta x_n) \\
&= \frac{\varepsilon}{b - a} (b - a) = \varepsilon
\end{aligned}$$

Since $L_f(P) \leq L \leq G \leq U_f(P)$, it follows that

$$0 \leq G - L \leq U_f(P) - L_f(P) < \varepsilon$$

Because ε was arbitrary, we conclude that $L = G$.   ■

We now incorporate the proof of Theorem A.21 into a proof of Theorem 5.3.

THEOREM A.22
(Theorem 5.3)

Let f be continuous on $[a, b]$. Then there is a unique number I with the following property: For any $\varepsilon > 0$ there is a number $\delta > 0$ such that the following statement holds: If $P = \{x_0, x_1, \ldots, x_n\}$ is any partition of $[a, b]$ each of whose subintervals has length less than δ, and if $x_{k-1} \leq t_k \leq x_k$ for each k between 1 and n, then the associated Riemann sum $\sum_{k=1}^{n} f(t_k) \, \Delta x_k$ satisfies

$$\left| I - \sum_{k=1}^{n} f(t_k) \, \Delta x_k \right| < \varepsilon$$

Proof   Let I be the number guaranteed by Theorem A.21. For any $\varepsilon > 0$ choose $\delta > 0$ such that

$$\text{if } x \text{ and } y \text{ are in } [a, b] \text{ and } |x - y| < \delta, \quad \text{then } |f(x) - f(y)| < \frac{\varepsilon}{b - a}$$

If P is chosen so that $\Delta x_k < \delta$ for each k, then by the proof of Theorem A.21,

$$U_f(P) - L_f(P) < \varepsilon$$

Moreover, if $x_{k-1} \leq t_k \leq x_k$ for $1 \leq k \leq n$, then

$$m_k \leq f(t_k) \leq M_k \quad \text{for } 1 \leq k \leq n$$

It follows that

$$L_f(P) = \sum_{k=1}^{n} m_k \, \Delta x_k \leq \sum_{k=1}^{n} f(t_k) \, \Delta x_k \leq \sum_{k=1}^{n} M_k \, \Delta x_k = U_f(P)$$

Since

$$L_f(P) \le I \le U_f(P)$$

we conclude that

$$\left| I - \sum_{k=1}^{n} f(t_k)\, \Delta x_k \right| \le U_f(P) - L_f(P) < \varepsilon$$

It is possible to prove that any number I satisfying the condition in the theorem must satisfy the condition in Theorem A.21 and hence is unique. ■

THEOREM A.23
Addition Property
(Theorem 5.8)

Let f be continuous on an interval containing a, b, and c. Then

$$\int_a^b f(x)\, dx = \int_a^c f(x)\, dx + \int_c^b f(x)\, dx$$

Proof   First we assume that $a < c < b$. By Definition 5.4 and Theorems A.21 and A.22, $\int_a^b f(x)\, dx$ is the *unique* number such that

$$L_f(P) \le \int_a^b f(x)\, dx \le U_f(P)$$

for any partition P of $[a, b]$. Therefore it suffices to show that

$$L_f(P) \le \int_a^c f(x)\, dx + \int_c^b f(x)\, dx \le U_f(P) \quad \text{for any } P \qquad (4)$$

To that end, let P be a partition of $[a, b]$, and consider the case in which the point c is in the partition P. Let P_1 be the collection of points in P that are in $[a, c]$, and let P_2 be the collection of points in P that are in $[c, b]$. Then

$$L_f(P) = L_f(P_1) + L_f(P_2) \quad \text{and} \quad U_f(P) = U_f(P_1) + U_f(P_2) \qquad (5)$$

Moreover, by the properties of lower and upper sums, we have

$$L_f(P_1) \le \int_a^c f(x)\, dx \le U_f(P_1) \quad \text{and} \quad L_f(P_2) \le \int_c^b f(x)\, dx \le U_f(P_2) \qquad (6)$$

Adding the two sets of inequalities in (6) and applying the equations in (5), we obtain

$$L_f(P) = L_f(P_1) + L_f(P_2) \le \int_a^c f(x)\, dx + \int_c^b f(x)\, dx \le U_f(P_1) + U_f(P_2) = U_f(P)$$

Therefore (4) is satisfied in the case that $a < c < b$ and c is in the partition P.

Next, consider the case in which $P = \{x_0, x_1, \ldots, x_n\}$ is a given partition and c is not in P. Then c must be in one of the subintervals associated with P, so there is an integer k such that $x_{k-1} < c < x_k$. Let P' be the partition $\{x_0, x_1, \ldots, x_{k-1}, c, x_k, \ldots, x_n\}$, which is obtained from P by inserting the point c. Since c belongs to P', we know from the preceding case that

$$L_f(P') \leq \int_a^c f(x)\, dx + \int_a^b f(x)\, dx \leq U_f(P')$$

Therefore in order to prove (4) it suffices to prove that

$$L_f(P) \leq L_f(P') \quad \text{and} \quad U_f(P') \leq U_f(P) \tag{7}$$

For this we observe that P and P' differ only in the number c, which is in $[x_{k-1}, x_k]$. Moreover, the minimum value of f on $[x_{k-1}, x_k]$ is less than or equal to the minimum values of f on the separate subintervals $[x_{k-1}, c]$ and $[c, x_k]$, and similarly, the maximum value of f on $[x_{k-1}, x_k]$ is greater than or equal to the maximum values of f on the separate subintervals $[x_{k-1}, c]$ and $[c, x_k]$. Therefore

$$L(P) \leq L(P') \quad \text{and} \quad U(P') \leq U(P)$$

Consequently (7) is proved

For any other ordering of a, b, and c the result follows from the first part of the proof and from Definition 5.6. We carry out the proof for the case $b < c < a$; the others are completely analogous. If $b < c < a$, then

$$\int_a^b f(x)\, dx = -\int_b^a f(x)\, dx = -\left(\int_b^c f(x)\, dx + \int_c^a f(x)\, dx \right)$$

$$= -\int_b^c f(x)\, dx - \int_c^a f(x)\, dx = -\int_c^a f(x)\, dx - \int_b^c f(x)\, dx$$

$$= \int_a^c f(x)\, dx + \int_c^b f(x)\, dx \quad \blacksquare$$

By Theorem 7.3, if f is increasing or decreasing, then f^{-1} exists. There is a converse to that result which applies to functions that are continuous on intervals, and whose proof uses the Intermediate Value Theorem.

LEMMA A.24   Let f be continuous on an interval I. If f has an inverse, then f is either increasing or decreasing on I.

Proof   Since f^{-1} exists, if a, b, and c are distinct numbers in I, then $f(a), f(b)$, and $f(c)$ are also distinct numbers. Suppose f were neither increasing nor decreasing. Then there would necessarily be numbers a, b, and c in I such that $a < b < c$ and such that $f(b)$ does not lie between $f(a)$ and $f(c)$. This means that either

$$f(c) \quad \text{lies between} \quad f(a) \text{ and } f(b)$$

or

$$f(a) \quad \text{lies between} \quad f(b) \text{ and } f(c)$$

(Figure A.1(a) and (b)). If the first condition holds, then by the Intermediate Value Theorem there is a number d between a and b such that $f(d) = f(c)$ (Figure A.1(a)).

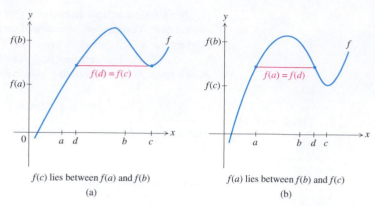

$f(c)$ lies between $f(a)$ and $f(b)$

(a)

$f(a)$ lies between $f(b)$ and $f(c)$

(b)

FIGURE A.1

If the second condition holds, then by the Intermediate Value Theorem there is a number d between b and c such that $f(d) = f(a)$ (Figure A.1(b)). In either case, f assigns the same value to two distinct points (d and c, or d and a). Consequently f cannot have an inverse, which contradicts the hypothesis. Thus f must be either increasing or decreasing. ■

If f is continuous on an interval I, then by the Intermediate Value Theorem, the values assigned by f to the points in I form an interval J. We now prove that if in addition f has an inverse, then f^{-1} is continuous.

THEOREM A.25
(Theorem 7.4)

Let f be continuous on an interval I, and let the values assigned by f to the points in I form the interval J. If f has an inverse, then f^{-1} is continuous on J.

Proof Let c be any interior point of J, and let $\varepsilon > 0$. We must find a $\delta > 0$ such that

$$\text{if } |y - c| < \delta, \quad \text{then } |f^{-1}(y) - f^{-1}(c)| < \varepsilon$$

Because c is an interior point of J, and because f and hence f^{-1} are either increasing or decreasing by Lemma A.24, it follows that $f^{-1}(c)$ is an interior point of I. Therefore there exists an $\varepsilon_1 > 0$ such that $\varepsilon_1 < \varepsilon$ and such that the interval K described by

$$K = (f^{-1}(c) - \varepsilon_1, f^{-1}(c) + \varepsilon_1)$$

is contained in I. Since f is continuous and either increasing or decreasing on the interval K, the values f assigns to the points of K form an interval J_0 in J by the Intermediate Value Theorem. In particular J_0 contains an interval of the form $(c - \delta, c + \delta)$ about c. But then by the definition of K we conclude that

$$\text{if } |y - c| < \delta, \quad \text{then } |f^{-1}(y) - f^{-1}(c)| < \varepsilon_1 < \varepsilon$$

Consequently f^{-1} is continuous at c, when c is an interior point of J. An analogous argument proves that f is continuous at any endpoint of J (if there are any). ■

Next we will prove a version of l'Hôpital's Rule. As preparation we first prove a generalization of the Mean Value Theorem.

THEOREM A.26
Generalized Mean Value Theorem

Let f and g be continuous on $[a, b]$ and differentiable on (a, b). If $g'(x) \neq 0$ for $a < x < b$, then there is a number c in (a, b) that

$$\frac{f(b) - f(a)}{g(b) - g(a)} = \frac{f'(c)}{g'(c)} \tag{8}$$

Proof   We introduce a special function much like the one appearing in the proof of the Mean Value Theorem:

$$h(x) = [f(b) - f(a)]g(x) - [g(b) - g(a)]f(x) \quad \text{for } a \leq x \leq b$$

Being a combination of f and g, the function h is continuous on $[a, b]$ and differentiable on (a, b). Thus by the Mean Value Theorem there is a number c in (a, b) such that

$$\frac{h(b) - h(a)}{b - a} = h'(c)$$

But a simple calculation reveals that $h(a) = h(b)$, and thus $h'(c) = 0$. Another calculation shows that

$$h'(x) = [f(b) - f(a)]g'(x) - [g(b) - g(a)]f'(x) \quad \text{for } a < x < b$$

Since $h'(c) = 0$, it follows that

$$[f(b) - f(a)]g'(c) - [g(b) - g(a)]f'(c) = 0 \tag{9}$$

By assumption, $g'(x) \neq 0$ for $a < x < b$. Consequently $g'(c) \neq 0$. It also follows that $g(a) \neq g(b)$ (because if $g(a) = g(b)$, then the Mean Value Theorem would imply that $g'(x) = 0$ for some x in (a, b), contradicting the hypothesis). Since $g'(c) \neq 0$ and $g(a) \neq g(b)$, we can divide both sides of (9) by $[g(b) - g(a)]g'(c)$ to obtain

$$\frac{f(b) - f(a)}{g(b) - g(a)} - \frac{f'(c)}{g'(c)} = 0$$

which is equivalent to (8). ■

If we let $g(x) = x$ for $a \leq x \leq b$, then (8) reduces to the equation of the Mean Value Theorem, because in this case

$$\frac{f(b) - f(a)}{b - a} = \frac{f(b) - f(a)}{g(b) - g(a)} = \frac{f'(c)}{g'(c)} = \frac{f'(c)}{1} = f'(c)$$

This justifies the name of the Generalized Mean Value Theorem.

THEOREM A.27
L'Hôpital's Rule

Let L be a real number or ∞ or $-\infty$.

Suppose f and g are differentiable on (a, b) and $g'(x) \neq 0$ for $a < x < b$. If

$$\lim_{x \to a^+} f(x) = 0 = \lim_{x \to a^+} g(x) \quad \text{and} \quad \lim_{x \to a^+} \frac{f'(x)}{g'(x)} = L$$

then

$$\lim_{x \to a^+} \frac{f(x)}{g(x)} = L = \lim_{x \to a^+} \frac{f'(x)}{g'(x)}$$

An analogous result holds if $\lim_{x \to a^+}$ is replaced by $\lim_{x \to b^-}$, or by $\lim_{x \to c}$, where c is any number in (a, b). (In the latter case f and g need not be differentiable at c.)

Proof   We establish the formula involving the right-hand limits. Define F and G on $[a, b)$ by

$$F(x) = \begin{cases} f(x) & \text{for } a < x < b \\ 0 & \text{for } x = a \end{cases}$$

$$G(x) = \begin{cases} g(x) & \text{for } a < x < b \\ 0 & \text{for } x = a \end{cases}$$

Then

$$\lim_{x \to a^+} F(x) = \lim_{x \to a^+} f(x) = 0 = F(a)$$

so that F is continuous on $[a, b)$. The same is true of G. Moreover, F and G are differentiable on (a, b), since they agree with f and g, respectively, on (a, b). Consequently if x is any number in (a, b), then F and G are continuous on $[a, x]$ and differentiable on (a, x). By the Generalized Mean Value Theorem, this means that there is a number $c(x)$ in (a, x) such that

$$\frac{F(x)}{G(x)} = \frac{F(x) - F(a)}{G(x) - G(a)} = \frac{F'(c(x))}{G'(c(x))}$$

Because $F = f$ and $G = g$ on (a, b), this mean that

$$\frac{f(x)}{g(x)} = \frac{f'(c(x))}{g'(c(x))}$$

Since $a < c(x) < x$, we know that

$$\lim_{x \to a^+} c(x) = a$$

so we can use the Substitution Rule with $y = c(x)$ to conclude that

$$\lim_{x \to a^+} \frac{f(x)}{g(x)} = \lim_{x \to a^+} \frac{f'(c(x))}{g'(c(x))} = \lim_{y \to a^+} \frac{f'(y)}{g'(y)} = \lim_{x \to a^+} \frac{f'(x)}{g'(x)} = L$$

This proves the equation involving right-hand limits. The results involving left-hand and two-sided limits are proved analogously. ■

The final theorem we prove in this subsection concerns sequences.

THEOREM A.28
(Theorem 9.6)

If a bounded sequence is either increasing or decreasing, then it converges.

Proof Let $\{a_n\}_{n=1}^\infty$ be a bounded, increasing sequence. By the Least Upper Bound Axiom there is a least upper bound L of the set S consisting of the numbers $a_1, a_2, a_3, \ldots$. We will show that $\lim_{n\to\infty} a_n = L$. Let $\varepsilon > 0$. Since L is the least upper bound of S, $L - \varepsilon$ is not an upper bound of S. Thus there is a number a_N in S such that $L - \varepsilon < a_N \leq L$. Since the given sequence is increasing, we have $a_N \leq a_n$ and hence $L - \varepsilon < a_N \leq a_n \leq L$ for any $n \geq N$. Thus

$$|a_n - L| < \varepsilon \quad \text{for } n \geq N$$

Therefore

$$\lim_{n\to\infty} a_n = L$$

A similar proof shows that every bounded, decreasing sequence converges. ∎

Power Series Theorems

We now prepare to prove the Differentiation Theorem for power series.

LEMMA A.29

The power series $\sum_{n=0}^\infty c_n x^n$ and $\sum_{n=1}^\infty n c_n x^{n-1}$ have the same radius of convergence.

Proof Let s be a fixed nonzero number and let $0 < |x| < |s|$. Assuming that $\sum_{n=0}^\infty c_n s^n$ converges, we will prove that $\sum_{n=1}^\infty n c_n x^{n-1}$ converges. First we notice that since $\sum_{n=0}^\infty c_n s^n$ converges, it follows that

$$\lim_{n\to\infty} \left| \frac{c_n s^n}{x} \right| = \frac{1}{x} \lim_{n\to\infty} |c_n s^n| = 0$$

Consequently there is a positive integer N such that

$$\left| \frac{c_n s^n}{x} \right| \leq 1 \quad \text{for } n \geq N$$

This means that

$$|n c_n x^{n-1}| = \left| n c_n \frac{x^n}{x} \frac{s^n}{s^n} \right| = \left| \frac{c_n s^n}{x} \right| n \left| \frac{x}{s} \right|^n \leq n \left| \frac{x}{s} \right|^n \quad \text{for } n \geq N$$

Since $|x| < |s|$ and thus $|x/s| < 1$, and since $\sum_{n=1}^\infty n x^n$ converges for $|x| < 1$ by Example 4 of Section 9.8, we know that $\sum_{n=N}^\infty n\,|x/s|^n$ converges. From the Comparison Test we conclude that $\sum_{n=N}^\infty n c_n x^{n-1}$ converges, and hence $\sum_{n=1}^\infty n c_n x^{n-1}$ also converges. Since x was arbitrary with $0 < |x| < |s|$, it follows that the radius of convergence of the series $\sum_{n=1}^\infty n c_n x^{n-1}$ is at least as large as the radius of convergence of the series $\sum_{n=0}^{\infty} c_n x^n$. Conversely, if $\sum_{n=1}^{\infty} nc_n s^{n-1}$ converges and $|x| < |s|$, then

$$\sum_{n=1}^{\infty} xnc_n x^{n-1} = \sum_{n=1}^{\infty} nc_n x^n$$

and hence $\sum_{n=1}^{\infty} nc_n x^n$ converges absolutely. Since

$$|c_n| \leq |nc_n| \quad \text{for all } n \geq 1$$

it follows from the Comparison Test that $\sum_{n=1}^{\infty} c_n x^n$ converges. Thus the radius of convergence of $\sum_{n=0}^{\infty} c_n x^n$ is at least as large as the radius of convergence of $\sum_{n=1}^{\infty} nc_n x^{n-1}$. This completes the proof of the converse. ■

COROLLARY A.30

The power series $\sum_{n=0}^{\infty} c_n x^n$, $\sum_{n=1}^{\infty} nc_n x^{n-1}$, and $\sum_{n=2}^{\infty} n(n-1)c_n x^{n-2}$ have the same radius of convergence.

Proof The proof consists merely of replacing $\sum_{n=0}^{\infty} c_n x^n$ and $\sum_{n=1}^{\infty} nc_n x^{n-1}$ in Lemma A.29 by $\sum_{n=1}^{\infty} nc_n x^{n-1}$ and $\sum_{n=2}^{\infty} n(n-1)c_n x^{n-2}$, respectively. The conclusion is that all three series have the same radius of convergence. ■

As a final preparation for proving the Differentiation Theorem, we present two applications of the Mean Value Theorem, both of which will appear in the proof. First, if $f(z) = z^n$ and if t and x are two distinct numbers, then there is a number s_n between t and x such that

$$\frac{f(t) - f(x)}{t - x} = f'(s_n)$$

This implies that

$$\frac{t^n - x^n}{t - x} = ns_n^{n-1} \tag{10}$$

Second, if $g(z) = z^{n-1}$ and if x and s_n are two distinct numbers, then there is a number r_n between x and s_n such that

$$\frac{g(x) - g(s_n)}{x - s_n} = g'(r_n)$$

Substitution for values of g and g' yields

$$x^{n-1} - s_n^{n-1} = (x - s_n)(n - 1)r_n^{n-2} \tag{11}$$

Now we are ready to prove the Differentiation Theorem.

THEOREM A.31
Differentiation Theorem for Power Series
(Theorem 9.25)

Let $\sum_{n=0}^{\infty} c_n x^n$ be a power series with radius of convergence $R > 0$. Then $\sum_{n=1}^{\infty} nc_n x^{n-1}$ has the same radius of convergence, and

$$\frac{d}{dx}\left(\sum_{n=0}^{\infty} c_n x^n\right) = \sum_{n=1}^{\infty} nc_n x^{n-1} = \sum_{n=1}^{\infty} \frac{d}{dx}(c_n x^n) \quad \text{for } |x| < R \quad (12)$$

Proof Let x be any number in $(-R, R)$, and let $f(t) = \sum_{n=0}^{\infty} c_n t^n$, for $|t| < R$. We will show that

$$\lim_{t \to x}\left(\sum_{n=1}^{\infty} nc_n x^{n-1} - \frac{f(t) - f(x)}{t - x}\right) = 0$$

that is,

$$\lim_{t \to x}\left(\sum_{n=1}^{\infty} nc_n x^{n-1} - \frac{\sum_{n=0}^{\infty} c_n t^n - \sum_{n=0}^{\infty} c_n x^n}{t - x}\right) = 0$$

This will verify the first equality in (12). To that end, we first choose a number b such that $|x| < b < R$. Then $\sum_{n=0}^{\infty} |c_n| b^n$ converges, and thus by Corollary A.30 the series $\sum_{n=2}^{\infty} n(n-1)|c_n| b^{n-2}$ converges. For later use we let

$$c = \sum_{n=2}^{\infty} n(n-1)|c_n| b^{n-2}$$

Next, observe that for any t in $(-b, b)$ distinct from x we have

$$\sum_{n=1}^{\infty} nc_n x^{n-1} - \frac{\sum_{n=0}^{\infty} c_n t^n - \sum_{n=0}^{\infty} c_n x^n}{t - x} = \sum_{n=1}^{\infty} nc_n x^{n-1} - \sum_{n=0}^{\infty} c_n \frac{t^n - x^n}{t - x}$$

$$\overset{(10)}{=} \sum_{n=1}^{\infty} nc_n x^{n-1} - \sum_{n=1}^{\infty} c_n n s_n^{n-1}$$

$$= \sum_{n=1}^{\infty} nc_n(x^{n-1} - s_n^{n-1})$$

$$\overset{(11)}{=} \sum_{n=2}^{\infty} (x - s_n)n(n-1)c_n r_n^{n-2}$$

By the definition of s_n we know that

$$|x - s_n| \leq |x - t| \tag{13}$$

and by the definition of r_n we know that $|r_n| < b < R$. Consequently from our calculations above and from (13) we obtain

$$\lim_{t \to x} \left| \sum_{n=1}^{\infty} nc_n x^{n-1} - \frac{\sum_{n=0}^{\infty} c_n t^n - \sum_{n=0}^{\infty} c_n x^n}{t - x} \right| = \lim_{t \to x} \left| \sum_{n=2}^{\infty} (x - s_n) n(n-1) c_n r_n^{n-2} \right|$$

$$\leq \lim_{t \to x} |x - t| \sum_{n=2}^{\infty} n(n-1)|c_n| b^{n-2}$$

$$= \lim_{t \to x} |x - t| c = 0$$

This completes the proof that the first equality in (12) holds. Since

$$n c_n x^{n-1} = \frac{d}{dx}(c_n x^n)$$

the second equality in (12) also holds. ∎

We now use the Differentiation Theorem to prove the Integration Theorem for Power Series.

THEOREM A.32
Integration Theorem for
Power Series
(Theorem 9.28)

Let $\sum_{n=0}^{\infty} c_n x^n$ be a power series with radius of convergence $R > 0$. Then for any x in $(-R, R)$ we have

$$\int_0^x \left(\sum_{n=0}^{\infty} c_n t^n \right) dt = \sum_{n=0}^{\infty} \frac{c_n}{n+1} x^{n+1} = \sum_{n=0}^{\infty} \left(\int_0^x c_n t^n \, dt \right) \qquad (14)$$

Proof Since $\sum_{n=0}^{\infty} c_n x^n$ converges absolutely for $-R < x < R$ and since

$$\left| \frac{c_n x^n}{n+1} \right| \leq |c_n x^n|$$

we know by the Comparison Test that $\sum_{n=0}^{\infty} c_n x^n / (n+1)$ also converges for $-R < x < R$, and hence

$$\sum_{n=0}^{\infty} \frac{c_n}{n+1} x^{n+1}$$

converges for $-R < x < R$. Since $\sum_{n=0}^{\infty} c_n t^n$ is continuous on $[0, x]$, it follows from the Fundamental Theorem of Calculus that

$$\frac{d}{dx} \left(\int_0^x \sum_{n=0}^{\infty} c_n t^n \, dt \right) = \sum_{n=0}^{\infty} c_n x^n$$

Moreover, from the Differentiation Theorem,

$$\frac{d}{dx} \left(\sum_{n=0}^{\infty} \frac{c_n}{n+1} x^{n+1} \right) = \sum_{n=0}^{\infty} \frac{(n+1)c_n}{n+1} x^n = \sum_{n=0}^{\infty} c_n x^n$$

As a result,

$$\int_0^x \left(\sum_{n=0}^\infty c_n t^n \right) dt = \sum_{n=0}^\infty \frac{c_n}{n+1} x^{n+1} + C$$

for an appropriate constant C. However, taking $x = 0$, we find that

$$0 = \int_0^0 \left(\sum_{n=0}^\infty c_n t^n \right) dt = \sum_{n=0}^\infty \frac{c_n}{n+1} 0^{n+1} + C = C$$

so that $C = 0$. This completes the proof of the first part of (14). The second part of (14) follows from the fact that

$$\int_0^x t^n \, dt = \frac{x^{n+1}}{n+1} \qquad \blacksquare$$

Finally, we prove Taylor's Theorem.

THEOREM A.33
Taylor's Theorem
(Theorem 9.31)

Let n be a nonnegative integer, and suppose $f^{(n+1)}(x)$ exists for each x in an open interval I containing a. For each $x \neq a$ in I, there is a number t_x strictly between a and x such that

$$f(x) = f(a) + f'(a)(x - a) + \cdots + \frac{f^{(n)}(a)}{n!}(x - a)^n +$$

$$\frac{f^{(n+1)}(t_x)}{(n+1)!}(x - a)^{n+1} \qquad (15)$$

and $$r_n(x) = \frac{f^{(n+1)}(t_x)}{(n+1)!}(x - a)^{n+1} \qquad (16)$$

Proof In the proof of the Mean Value Theorem we introduced an auxiliary function g, which allowed us to apply Rolle's Theorem. Here we also introduce an auxiliary function g. For any fixed $x \neq a$ in I and any t in I, let

$$g(t) = f(x) - f(t) - f'(t)(x - t) - \frac{f''(t)}{2!}(x - t)^2 - \frac{f^{(3)}(t)}{3!}(x - t)^3 - \cdots$$

$$- \frac{f^{(n)}(t)}{n!}(x - t)^n - r_n(x) \frac{(x - t)^{n+1}}{(x - a)^{n+1}}$$

Then $g(x) = g(a) = 0$, as you can check by substituting and using the definition of $r_n(x)$. Next we find $g'(t)$ (remembering that x is fixed):

$$g'(t) = 0 - f'(t) + [f'(t) - f''(t)(x - t)]$$

$$+ \left(\frac{2f''(t)}{2!}(x - t) - \frac{f^{(3)}(t)}{2!}(x - t)^2 \right)$$

$$+ \left(\frac{3f^{(3)}(t)}{3!}(x - t)^2 - \frac{f^{(4)}(t)}{3!}(x - t)^3 \right) + \cdots$$

$$+ \left(\frac{nf^{(n)}(t)}{n!}(x - t)^{n-1} - \frac{f^{(n+1)}(t)}{n!}(x - t)^n \right)$$

$$+ (n + 1)r_n(x)\frac{(x - t)^n}{(x - a)^{n+1}}$$

As you can see, adjacent pairs of terms cancel each other, leaving

$$g'(t) = \frac{-f^{(n+1)}(t)}{n!}(x - t)^n + (n + 1)r_n(x)\frac{(x - t)^n}{(x - a)^{n+1}}$$

By Rolle's Theorem, there is a number t_x strictly between a and x such that $g'(t_x) = 0$. This means that

$$0 = \frac{-f^{(n+1)}(t_x)}{n!}(x - t_x)^n + (n + 1)r_n(x)\frac{(x - t_x)^n}{(x - a)^{n+1}}$$

Solving for $r_n(x)$, we obtain

$$r_n(x) = \frac{f^{(n+1)}(t_x)}{(n + 1)!}(x - a)^{n+1}$$

This proves (16), and (15) now follows from the definition of $r_n(x)$.   ∎

EXERCISES

In Exercises 1–8 find the least upper bound and the greatest lower bound of the given set.

1. $(-1, 1)$ 2. $[\frac{1}{2}, 100]$ 3. $[0, \pi)$ 4. $(-9.9, \sqrt{2})$

5. The set consisting of $(0, 1)$ and $(2, 5]$

6. The set consisting of 4 and the interval $(-1, 3)$

7. The set consisting of 0 and all numbers of the form $1/n$, where n is a positive integer

8. The set S consisting of all numbers in $(\frac{3}{10}, \frac{1}{3})$ that have decimal expansions containing only finitely many zeros and threes after the decimal point and no other integers

9. Show that the set of positive integers has no upper bound. (*Hint:* If there were an upper bound, then there would be a least upper bound M. Using the fact that $n + 1$ is a positive integer if n is a positive integer, show that it would follow that $n \leq M - 1$ for every positive integer n, and obtain a contradiction.)

10. **a.** Show that if $a > 0$, then the set S consisting of the numbers of the form na, where n is a positive integer, has no upper bound. (*Hint:* Using the idea of the hint for Exercise 9, show that if M were a least upper bound of S, then $M - a$ would be an upper bound of S, which is a contradiction.)

   b. Using (a), prove **Archimedes' Principle:** If b is any number and $a > 0$, there is a positive integer n such that $na > b$.

11. Using the Least Upper Bound Axiom, prove that every set that is bounded below has a greatest lower bound.

12. Using Exercise 9, prove that $\lim_{n \to \infty} 1/n = 0$. (*Hint:* Show that if the given result did not hold, then the set of positive integers would be bounded above.)

13. Show that the continuous function $1/x$ is not uniformly continuous on $(0, 1)$. (*Hint:* Let $\varepsilon = 1$, and show that for any positive $\delta < 1$, if $x = \delta$ and $y = \frac{1}{2}\delta$, then $|x - y| < \delta$ but $|1/x - 1/y| \geq \varepsilon$.)

ANSWERS TO ODD-NUMBERED EXERCISES

CHAPTER 1

Section 1.1

1. $a > b$ **3.** $a > b$ **5.** $\sqrt{2} > 1.41$

7. closed, bounded   **9.** open, unbounded

11. closed, unbounded   **13.** closed, unbounded

15. $(-3, 4)$ **17.** $(1, \infty)$

19. $\left(-\infty, -\frac{7}{6}\right)$ **21.** $\left[1, \frac{7}{2}\right)$

23. union of $\left(-\infty, -\frac{1}{2}\right]$ and $[1, \infty)$

25. union of $\left(-\infty, -\frac{1}{3}\right)$ and $\left(0, \frac{2}{3}\right)$

27. union of $(-\infty, -3)$ and $(-1, \infty)$

29. $\left(-\infty, \frac{3}{2}\right]$ **31.** $\left(\frac{1}{2}, \infty\right)$

33. union of $\left(-\infty, -\sqrt{6}\right)$, $(-2, 0)$, and $(2, \sqrt{6})$

35. union of $(-\infty, -2]$, $(-1, 1)$, and $(1, \infty)$

37. $\left(-\infty, \frac{3}{2}\right)$ **39.** $\left(-1, -\frac{1}{3}\right)$ **41.** $[-3, 1)$

43. -3 **45.** 10 **47.** $-1, 1$

49. $-1, 3$ **51.** $-\frac{5}{6}$ **53.** $-1, 0, 1$

55. $0, -2$ **57.** 0 **59.** $(1, 3)$

61. $(-1.01, -0.99)$

63. union of $(-\infty, -6]$ and $[0, \infty)$

65. union of $(-\infty, -1]$ and $[0, \infty)$

67. union of $\left(-\infty, -\frac{1}{6}\right)$ and $\left(\frac{1}{2}, \infty\right)$

69. $\left(\frac{3}{2}, \frac{5}{2}\right)$

71. approximately $2.498407507 \times 10^{54}$

73. approximately $9.034120564 \times 10^{-64}$

75. union of $(-\infty, 10.5)$ and $(14.5, \infty)$

77. yes   **79.** no

Section 1.2

1.

3. 5

5. $4\sqrt{2}$

7. $9\sqrt{2}$

9. $\sqrt{6 - 2\sqrt{6}}$

11. $\sqrt{2}\,|b - a|$

13. $y = 3x - 7$

15. $y = -x + 1$

17. $y = -x$

19. $y = 3x - 3$

21. $m = -1, b = 0$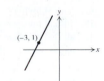

23. $m = 2, b = 7$

A-25

25. $m = -2$, $b = 4$

27. perpendicular; $\left(-\frac{8}{5}, -\frac{1}{5}\right)$

29. parallel   **31.** perpendicular; $(1, -1)$

33. parallel   **35.** parallel

37. $y = 3x - 7$ **39.** $y = -x$

41. $y = \frac{2}{3}x - \frac{1}{3}$ **43.** $y = -\frac{1}{2}x - \frac{7}{2}$

45. $y = \frac{3}{2}x$ **47.** $y = -\frac{1}{2}x - 3$

51. $y - 4 = -(x + 2)$

53.   **55.**   **57.**

59.   **61.**

63. a. the x axis   **b.** the y axis

65. $(-3, 7)$, $(-7, 7)$, and $(-7, 3)$

71. approximately 17.9 feet

Section 1.3

1. $\sqrt{3}$, $\sqrt{3}$ **3.** 1, 1   **5.** $\frac{1}{4}$

7. $3, -\frac{1}{2}$ **9.** $\frac{1}{8}$

11. approximately 3.863431953, approximately 3.150887574

13. approximately -1.572145546, approximately 417.6780725

15. all real numbers   **17.** $[-2, 8]$

19. $[-2, \infty)$

21. union of $(-\infty, 0]$ and $[1, \infty)$

23. union of $(-\infty, -\sqrt{3}/3]$ and $[\sqrt{3}/3, \infty)$

25. all real numbers

27. all real numbers except 1

29. all real numbers except -4 and 4

31. all real numbers

33. union of $[-4, -1]$ and $(0, 6)$

35. union of $[-3, -2\sqrt{2}]$ and $[2\sqrt{2}, 3]$

37. the set consisting of -1

39. $(-\infty, 10)$

41. all real numbers except 0

43. a, e, f, g, and i define functions.

45. $f_1 = f_5$ and $f_2 = f_3$

47. a. union of $(-\infty, -1]$ and $[1, \infty)$

49. $f(x) = \frac{1}{2}x^2 \sqrt[3]{x/5}$ for $x \geq 0$

51. $A(x) = \frac{1}{4}\sqrt{3}x^2$ for $x \geq 0$

53. a. $h(t) = -4.9t^2 - 43.9t + 254$

55. 5.5125 meters per second

57. a. 26.275 meters after $\frac{1}{2}$ second, and 20.1 meters after 1 second

b. approximately 1.6 seconds

59. a. $L = gT^2/4\pi^2$

b. approximately 9.4 seconds

61. $D(t) = \begin{cases} 400t & \text{for } 0 \leq t < 2 \\ |1600 - 400t| & \text{for } 2 \leq t \leq 5 \end{cases}$

63. a. 0; $\dfrac{2}{c + 2d}$; no response   **b.** $x = \dfrac{cR(x)}{1 - dR(x)}$

Section 1.4

1.   **3.**   **5.**

7.   **9.**   **11.**

13.   **15.**   **17.**

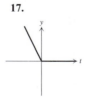

19.   **21.**   **23.**

25.   **27.**

29. approximately -0.1232

31. approximately -1.82494

33. approximately -0.0175242

35. not the graph of a function

37. the graph of a function

39. not the graph of a function

41. not the graph of a function

43. the x and y axes; not the graph of a function

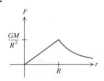

45. not the graph of a function

49. a. $-2, -1, 1, 3$

b. union of $[-4, -2), (-1, 1)$, and $(3, 4]$

c. union of $(-2, -1)$ and $(1, 3)$

d. none

e. $[-4, 4]$

51. $p = 2, q = 1, r = -2$

53. f: revenue function; g: cost function; h: profit function

55.

57. a.

59.

Section 1.5

1. y intercepts: $-\sqrt{\frac{2}{3}}, \sqrt{\frac{2}{3}}$; x intercept: -2; symmetry with respect to the x axis

3. no y intercepts; x intercepts: $-1, 1$; symmetry with respect to the x axis, y axis, origin

5. y intercept: 0; x intercept: 0; symmetry with respect to the y axis

7. no intercepts; symmetry with respect to the x axis, y axis, origin

9. no y intercepts; x intercepts: $-1, 1$; symmetry with respect to the origin

11. y intercept: 0; x intercepts: $[0, 1)$; no symmetry

13. y intercept: 3; x intercepts: $-3, 3$; symmetry with respect to the y axis

15. y intercept: 1; x intercept: 1; no symmetry

17. y intercept: 0; x intercept: 0; symmetry with respect to the origin

19. y intercept: -3; x intercepts: $-\sqrt{3}, \sqrt{3}$; symmetry with respect to the y axis

21. y intercepts: $-1, 1$; no x intercepts; symmetry with respect to the x axis, y axis, origin

23. y intercepts: $-2, 2$; x intercept: 2; symmetry with respect to the x axis

25. y intercept: 0; x intercept: 0; symmetry with respect to the x axis, y axis, origin

27.

29.

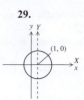

31.

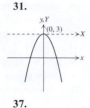

33.

35.

37.

39. a. odd **b.** even **c.** neither **d.** neither **e.** even
f. odd **g.** odd **h.** even **i.** odd
41. $\frac{3}{2} + \frac{1}{2}\sqrt{5}$ and $\frac{3}{2} - \frac{1}{2}\sqrt{5}$
43. no real zero
45. approximately -0.20 and 1.24
49. $(-1, 0)$ and approximately $(1.54, 4.67)$
51. approximately $(-1.21, \infty)$
53. a.

b.

55. The graph of g is d units above the graph of f if $d \geq 0$, and is $-d$ units below the graph of f if $d < 0$.
59. symmetry with respect to the line $x = c$
61.

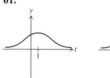

The curve moves to the right.

Section 1.6

1. -1 **3.** $-\frac{33}{4}$ **5.** 6
7. $2x + 5$ **9.** $\frac{529}{257}$ **11.** $\frac{4}{41}\sqrt{3}$
13. 0
15. $(f + g)(x) = \dfrac{x^2 - 2x + 3}{x - 1}$ for $x \neq 1$
   $(fg)(x) = 2$ for $x \neq 1$
   $(f/g)(x) = 2/(x - 1)^2$ for $x \neq 1$
17. $(f + g)(t) = t^{3/4} + t^2 + 3$ for $t \geq 0$
   $(fg)(t) = t^{11/4} + 3t^{3/4}$ for $t \geq 0$
   $(f/g)(t) = t^{3/4}/(t^2 + 3)$ for $t \geq 0$
19. $(g \circ f)(x) = -2x + 7$ for all x
   $(f \circ g)(x) = -2x - 4$ for all x

21. $(g \circ f)(x) = \sqrt{x^2} = |x|$ for all x
   $(f \circ g)(x) = x$ for $x \geq 0$
23. $(g \circ f)(x) = x - 5\sqrt{x} + 6$ for $x \geq 0$
   $(f \circ g)(x) = \sqrt{x^2 - 5x + 6}$ for $x \leq 2$ or $x \geq 3$
25. $(g \circ f)(x) = (x - 1)/x$ for $x \neq 0, 1$
   $(f \circ g)(x) = -(x + 1)/x$ for $x \neq -1, 0$
27. $f(x) = x - 3$; $g(x) = \sqrt{x}$
29. $f(x) = 3x^2 - 5\sqrt{x}$; $g(x) = x^{1/3}$
31. $f(x) = x + 3$; $g(x) = 1/(x^2 + 1)$ or $f(x) = (x + 3)^2 + 1$;
   $g(x) = 1/x$)
33. $f(x) = \sqrt{x} - 1$; $g(x) = \sqrt{x}$ (or $f(x) = \sqrt{x}$; $g(x) = \sqrt{x - 1}$)
35. $g(x) = -|x - 2|$ **37.** $[-3, 1]$
39. f is red, and g is green; $f + g$ is blue; fg is black.
41. a. $c = -1$ **b.** $c \approx -0.6$
43. The domain of h is $[-1, 2]$.
45. all functions
47. $f(x) \geq 0$ for all x in the domain of f.
57. approaches 0
59. oscillates between approximately 0.80 and 0.51
61. oscillates between approximately 0.16, 0.50, and 0.96
63. $P(x) = -\frac{1}{10}x^4 + x^2 + 24x - 38$; $P(1) = -13.1$; $P(2) = 12.4$
65. a. $V(r(s)) = \dfrac{1}{6\sqrt{\pi}} s^{3/2}$ for $s \geq 0$ **b.** $\sqrt{6/\pi}$

Section 1.7

1. a. $7\pi/6$ **b.** $-9\pi/4$ **c.** $\pi/180$
3. a. $-\frac{1}{2}$ **b.** $-\frac{1}{2}\sqrt{3}$ **c.** $-\frac{1}{2}\sqrt{2}$ **d.** $-\frac{1}{2}\sqrt{3}$ **e.** $\sqrt{3}$
   f. -1 **g.** $\sqrt{3}$ **h.** $\frac{1}{3}\sqrt{3}$ **i.** -1 **j.** 2 **k.** 1 **l.** $\frac{2}{3}\sqrt{3}$
5. $\tan x = -\frac{4}{3}$, $\cot x = -\frac{3}{4}$, $\sec x = -\frac{5}{3}$, $\csc x = \frac{5}{4}$
7. $7\pi/6$, $11\pi/6$ **9.** 0, $\pi/3$, π, $5\pi/3$
11. union of $[0, 7\pi/6)$ and $(11\pi/6, 2\pi)$
13. union of $[\pi/4, \pi/2)$ and $[5\pi/4, 3\pi/2)$
15. union of $(0, \pi/4]$, $(\pi/2, 3\pi/4]$, $(\pi, 5\pi/4]$, and $(3\pi/2, 7\pi/4]$
17. y intercept: -1; x intercepts: $\pi/2 + n\pi$ for any integer n;
   symmetric with respect to the y axis; even function

19. no y intercept; x intercepts: $\pi/2 + n\pi$ for any integer n;
   symmetric with respect to the origin; odd function

21. y intercept: 1; no x intercepts; symmetric with respect to the y axis; even function. The graph is the one in Figure 1.70(c).

23. y intercept: 0; x intercepts: $n\pi/2$ for any integer n; symmetric with respect to the origin; odd function

25. $\frac{1}{4}\sqrt{2}(\sqrt{3}+1)$

27. $\pi/6 + 2n\pi$, $5\pi/6 + 2n\pi$, and $3\pi/2 + 2n\pi$ for any integer n

33. $\frac{14}{5}$

35. a. π **b.** $2\pi/3$ **c.** π **d.** π

39. They approach a number that is approximately 0.739.

41. The distance d satisfies $d = 2 \sec \theta$.

43. $\pi/4$; the vision becomes blocked.

45. b. approximately 783.4 feet

47. b. T is greatest when $\theta = \pi$, and is smallest when $\theta = 0$.

49. a. 5 seconds   **b.** 12

c.

d. positive values: flow into the lungs; negative values: flow out of the lungs

e. approximately -0.29

Section 1.8

1. 3   **3.** $3x$ **5.** 1

7. $2x$ **9.** $\frac{1}{2}$ **11.** $-\frac{1}{2}x$

13. 0.2893842179   **15.** 11.31370850

17. $\dfrac{\ln 5}{\ln 3} \approx 1.464973521$ **19.** $\dfrac{1}{\ln \pi} \approx 0.8735685268$

21. $(\ln x)^2 \neq \ln (x^2)$ **23.** even

25.

27.

29. $(\frac{1}{2}, e^{1/2})$ **31.** $(\frac{1}{2} \ln 3, 3^{3/2})$

33. $(1, 0)$ **35.** $\frac{1}{6}(1 + \sqrt{13})$

45. 0.2591711018   **47.** 0.1 millimeters

49. approximately 0.7943282347

51. approximately 7

53. a. approximately 40 decibels

b. approximately 50 decibels

c. approximately 140 decibels

d. approximately 170 decibels

55. 10^{-10}

57. a. approximately 29.92 inches of mercury

b. approximately 11.01 inches of mercury

c. approximately 4.05 inches of mercury

59. approximately 1200

Chapter 1 Review Exercises

1. union of $(-\infty, -\frac{3}{2})$ and $(4, \infty)$

3. $[\frac{5}{2}, 3)$ **5.** $(\frac{7}{12}, \frac{3}{4})$ **7. a.** a **b.** 0

9. $y - 1 = -4(x - 0)$ **11.** $y = -\frac{1}{3}x + 6$

13. perpendicular

15. a. $y = -\frac{1}{2}x - \frac{3}{2}$ **b.** $y = 2x - 4$

19. all real numbers except -3, -1, and 0

21. $(0, \infty)$

23.   **25.**

27. y intercept: 1; x intercept: 1; no symmetry

29. y intercepts: $-\sqrt{3}, \sqrt{3}$; x intercepts: $-\sqrt{3}, \sqrt{3}$; symmetry with respect to the x axis, y axis, and origin

31. y intercept: 1; x intercepts: $\pi/2 + 2n\pi$ for any integer n; not symmetric with respect to either axis or origin

33.

35. domain: $(-1, 1)$; x and y intercepts: 0; symmetric with respect to the y axis; even function

37. even   **39.** even

41.   **43.**

45. a. $f(x) = \begin{cases} -2x + 5 & \text{for } x \le 2 \\ 1 & \text{for } 2 < x < 3 \\ 2x - 5 & \text{for } x \ge 3 \end{cases}$

$g(x) = \begin{cases} -1 & \text{for } x \le 2 \\ 2x - 5 & \text{for } 2 < x < 3 \\ 1 & \text{for } x \ge 3 \end{cases}$

b.

47. $(f - g)(x) = \dfrac{3}{(x - 3)(x - 1)}$ for $x \ne -1$, 1, and 3

$\left(\dfrac{f}{g}\right)(x) = \dfrac{x + 2}{x - 1}$ for $x \ne -1$, 1, and 3

49. a. domain of f: $[-1, \infty)$; domain of g: $[-2, \infty)$; domain of h: union of $(-\infty, -2]$ and $[-1, \infty)$

b. domain of fg: $[-1, \infty)$. The domain of fg has fewer numbers than does the domain of h.

55. $\cos x = \frac{1}{3}\sqrt{5}$; $\cot x = -\frac{1}{2}\sqrt{5}$; $\sec x = \frac{3}{5}\sqrt{5}$; $\csc x = -\frac{3}{2}$

57. union of $[\pi/4, 3\pi/4]$ and $[5\pi/4, 7\pi/4]$

59. a. $a = 2$, $b = \pi/3 + 2n\pi$ for any integer n; $a = -2$, $b = 4\pi/3 + 2n\pi$ for any integer n

b. $a = 2$, $b = -\pi/6 + 2n\pi$ for any integer n; $a = -2$, $b = 5\pi/6 + 2n\pi$ for any integer n

61. $(-\infty, \ln 2]$

63. a. red curve in Fig. 1.86a   **b.** blue curve in Fig. 1.86b
c. blue curve in Fig. 1.86a   **d.** red curve in Fig. 1.86b
e. green curve in Fig. 1.86a   **f.** green curve in Fig. 1.86b

67. no

69. $(1, 3)$ (opposite $(2, 0)$), $(5, 1)$ (opposite $(0, 1)$), or $(-1, -1)$ (opposite $(3, 2)$)

71. You pass through B.   **73.** $600 \sin \frac{4}{3} \approx 583$ feet

75. a. 64,049   **b.** approximately 73,098

CHAPTER 2

Section 2.1

1. 3   **3.** 0   **5.** $-\frac{1}{11}$ **7.** $\frac{2}{5}$

9. -4 **11.** -6 **13.** 3   **15.** 3   **17.** 1

19. 1.003346721, 1.000033335, 1.000000333, 1.003346721, 1.000033335, 1.000000333; $\lim_{x \to 0} f(x) = 1$

21. 0.0499583472, 0.0049999583, 0.0005, -0.0499583472, -0.0049999583, -0.0005; $\lim_{x \to 0} f(x) = 0$

23. 0.6164048071, 0.6001600395, 0.6000016, 0.6164048071, 0.6001600395, 0.6000016; $\lim_{x \to 0} \dfrac{\sin 3x}{\sin 5x} = 0.6$

25. 0.0500417084, 0.0050000417, 0.0004999999, -0.0500417084, -0.0050000417, -0.0004999999; $\lim_{x \to 0} (\csc x - \cot x) = 0$

27. 3.498588076, 3.045453395, 3.004504503, 2.591817793, 2.955446645, 2.995504497; $\lim_{x \to 0} \dfrac{e^{3x} - 1}{x} = 3$

29. 1.6   **31.** 0

33. The limit does not exist.   **35. b.** 1

37. $(f(2) - f(1))/(2 - 1)$ is larger.

39. 6   **41.** 16

43. 0.5129329439, 0.5012541824, 0.5001250418, 0.4879016417, 0.4987541511, 0.4998750416; slope: 0.5

45. 1.583844403, 1.570925532, 1.570797619, 1.583844403, 1.570925532, 1.570797619; slope: approximately 1.57

47. average velocity at time $t = t + \frac{1}{2}$; velocity at time $\frac{1}{2} = 1$

49. a. Use data for $t = 50$ and either $t = 49$ or 51; ignore data for $t = 48$ and 50.

b. 0.6 miles per hour

Section 2.2

1. -5 **3.** $\frac{1}{2}$ **5.** 3

7. $\frac{3}{2}$ **9.** 0   **15.** $y = \pi$

17. $y = 4x - \frac{1}{2}$ **19.** $y = 5x - 1$ **21.** 2

23. -128 **25.** -56

27. yes; $f(x) = mx + b$ **29.** A tangent line exists.

31. a. 0.1   **b.** $-\frac{1}{2} + \frac{1}{2}\sqrt{1 + 4\varepsilon}$

35. -9.8 meters per second

37. a. -25.6 meters per second   **b.** 25.6 meters per second

39. a. 0 meters per second   **b.** yes

Section 2.3

1. 14
3. 144
5. $1/e$
7. 1
9. $-\frac{3}{8}\sqrt{3}$
11. $\frac{1}{32}$
13. $e^{1/2}$
15. 0
17. 0
19. 0
21. $\frac{1}{2}$
23. -2
25. -2
27. -8
29. $\frac{1}{2}$
31. $\sqrt{2}/8$
33. -4
35. 2
37. 0
39. 0
41. 0
43. $y = 6(x - 1)$
45. $y - \frac{1}{2} = -\frac{1}{4}(x + 1)$
47. $y = 1$
49. $y - \sqrt{3} = -\frac{\sqrt{3}}{3}(x - 1)$
51. yes
53. yes
55. yes
57. a. 3.5   **b.** 3   **c.** -4 **61.** -2
65. a. $f(x) = -1/x$ for $-1 \le x < 0$; $f(0) = 0$; $f(x) = 1/x$ for $0 < x \le 1$
  b. $f(x) = 100/x$ for $0 < |x| \le 1$; $f(0) = 0$;
  c. $\lim_{x \to 0} f(x)$ exists.
67. 0   **69. a.** $L \le M$ **b.** $L \le M$
73. a. Product Rule not applicable   **b.** 1

Section 2.4

1. -19
3. 4
5. $\frac{5}{2}$
7. -1
9. $-\infty$
11. $-\infty$
13. $-\infty$
15. ∞
17. $-\infty$
19. $-\infty$
21. ∞
23. 0
25. ∞
27. ∞
29. $-\infty$
31. $-\infty$
33. The limit does not exist.   **35.** $\frac{1}{2}$ **37.** $\frac{1}{6}$
39. 0   **41.** The limit does not exist.   **43.** $-\infty$
45. -4 **47.** $x = 3$ **49.** $x = -1$ and $x = 1$
51. $x = -2$ and $x = 0$ **53.** $x = 0$
55. none   **57.** $x = \pi/2 + n\pi$ for every integer n
59. $x = 0$ **61.** $x = 0$
63.

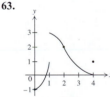

69. a. $-2\sqrt{197}$ meters per second
  b. approximately 30.2 meters
71. no

Section 2.5

1. yes   **3.** yes   **5.** the graphs in (b) and (f)
7. continuous from the right
9. continuous from the left at 0
11. continuous at 0; continuous from the right at 4
13. continuous at 0; continuous from the left at 1
15. a. redefine $f(3) = 3$ to make continuous at 3
  b. cannot redefine to make continuous at 1
43. union of $(-\infty, 2)$ and $(2, 5)$
45. union of $(-\infty, -3]$ and $[-1, 0]$

47. union of $(-\infty, -\frac{1}{2})$, $(\frac{1}{2}, 1]$, and $[3, \infty)$
49. $(1, 3)$ **51.** $\frac{23}{16}$ **53.** $\frac{23}{16}$ **55.** $\frac{35}{16}$ **57.** $\frac{13}{16}$
59. $0.738\ldots$ **61.** $0.567\ldots$
67. odd integers greater than 1
69. a. continuous   **b.** continuous   **c.** discontinuous
  d. continuous   **e.** discontinuous
77. yes

Chapter 2 Review Exercises

1. $-\frac{5}{3}$ **3.** 2   **5.** -1
7. $\frac{15}{2}$ **9.** 0   **11.** $-\infty$ **13.** 3
15. $\frac{3}{2}\sqrt{2}$ **17.** The limit does not exist.
19. 0   **21.** $x = 4$
23. $x = -2$ and $x = 7$ **25.** $x = -2$
27. $x = 0$ **29.** $x = 0$
31. continuous from the left at $-\sqrt{13}$
33. continuous from the right at 2
41. 10
43. a. 0   **b.** 1   **c.** ∞ **d.** $-\infty$
45. union of $(-\infty, -1)$, $(-1, \frac{3}{2})$, and $(4, \infty)$
47. $\frac{13}{16}$ **49.** 0.456   **51.** $b - a$ **59.** $(4, 2)$
61. a. 0.30625 meters   **b.** 2.45 meters per second

CHAPTER 3

Section 3.1

1. 0   **3.** 2   **5.** 0
7. 5   **9.** $-\frac{1}{16}$ **11.** 4
13. $f'(x) = 0$ **15.** $f'(x) = -10x$
17. $g'(x) = 3x^2$ **19.** $k'(x) = -2/x^3$
21. $dy/dx = 0$ **23.** $dy/dx = 6x$
25. 0   **27.** 4
29. no derivative   **31.** no derivative
33. 1   **35.** no derivative
37. $y = -4x - 4$ **39.** $y = \frac{1}{4}x + 1$
41. $f'(0) \approx -2, f'(1) \approx -1.5, f'(2) \approx -0.5, f'(3) \approx 0.7$
43. decreasing on $[-10, 2]$; increasing on $[2, 10]$
45. decreasing on $[-10, -1]$ and on $[0, 2]$; increasing on $[-1, 0]$ and on $[2, 10]$
47. increasing on $[-10, 10]$
49. $m_2 = 4.1; f'(2) = 4; |f'(2) - m_2| = 0.1$
51. x **53.** $-\sin x$
55. approximately 6.9999848
57. approximately 1.648710434
59. a. $f(x) = x^4 + c$, for any constant c; 32   **b.** $f(x) = x^4 + c$, for any constant c; 32
61. $\lim_{h \to 0} \dfrac{f(a - h) - f(a)}{h} = -f'(a)$
71. 2 meters per second   **73.** $-\frac{3}{2500}$
75. a. 8 thousand dollars per thousand gallons   **b.** $\frac{11}{8}$

77. 8 thousand dollars per thousand gallons; the same
81. a. 2 meters per minute   **b.** 3 meters per minute

Section 3.2

1. 0   **3.** 3; 0   **5.** 8
7. 10   **9.** $0; \frac{1}{2}\sqrt{3}$ **11.** -2
13. $5x^4$ **15.** $1/(x+1)^2$
17. $4x/(x^2+1)^2$ **19.** $3 \sin x$
21. $\frac{2}{3}x^{-1/3}$ **23.** $1/(2\sqrt{x}-1)$ **25.** $2e^{2x}$
33. a. -3 **b.** 5   **c.** 3 or -3
   d. $-\pi/6 + 2n\pi$ or $\pi/6 + 2n\pi$ for any integer n
35. because $\frac{1}{2}x^{-1/2}$ is not continuous on $[0, 1]$, whereas the other function is
37. the graph in Figure 3.13(b)
39. $h = f'$ and $k = g'$
41. a. $y - 1 = \frac{1}{2}(x+1)$ **b.** $y + 5 = -\frac{1}{2}(x+1)$
   c. $y = -x$
43. $v(t) = -3 \cos t; v(\pi/6) = -\frac{3}{2}\sqrt{3}$
45. a. no   **b.** no   **49.** $2\pi r$
51. a. $\frac{1}{2}\sqrt{3}\, x$, where x is the length of a side   **b.** 2
53. a. 7 minutes
   b. 2 minutes (from 5 to 7 minutes after leaving home)
   c. from 0 to 5 minutes after leaving home

Section 3.3

1. $-12x^2$ **3.** $16x^3 + 9x^2 + 4x + 1$
5. $36/t^{10}$ **7.** $4x + 7$
9. $-1/x^2 + 2/x^3$ **11.** $12x^{-4} - 2 \sin x$
13. $-6z^2 + 4 \sec z \tan z$ **15.** $2z \sin z + z^2 \cos z$
17. $2 \sin x \cos x = \sin 2x$ **19.** $-14/(4x-1)^2$
21. $-1/(t^2 + 4t + 4)$
23. $-4(t^2 + 12t + 26)/(t^2 + t - 20)^2$
25. $-\csc^2 x$
27. $\frac{1}{2}(1/\sqrt{y}) \sec y + \sqrt{y} \sec y \tan y$
29. $e^x + 1/e^x$ **31.** $(2 - 2t)/e^t$
33. $18x^2 - 26x - 5$ **35.** $2x - 2/x^3$
37. $(-x^6 + 4x^3 + 3x^2)/(x^4 + 1)^2$
39. $-\csc^2 x + \sec^2 x$
41. $\dfrac{(x^3 + x) \cos x + (1 - x^2) \sin x}{(x^2 + 1)^2}$

43. $\dfrac{2 \sin x + 2x \cos x - 2x \sin x}{e^x}$

45. 63   **47.** $-16/\pi$ **49.** $2e$
51. $y = x - 8$ **53.** $y = x + 1 - \pi/2$
55. $(1, 6)$ and $(2, 5)$
57. a. Two lines: the points of tangency of the first line are $(1, 2)$ and $(-1, -2)$; the points of tangency of the second line are $(-1, 2)$ and $(1, -2)$
   b. Two lines: the points of tangency of the first line are $(\frac{1}{2}(a + b), a + \frac{1}{4}(a + b)^2)$ and $(-\frac{1}{2}(a + b),$

$-b - \frac{1}{4}(a + b)^2)$; the points of tangency of the second line are $(-\frac{1}{2}(a + b), a + \frac{1}{4}(a + b)^2)$ and $(\frac{1}{2}(a + b), -b - \frac{1}{4}(a + b)^2)$

Section 3.4

1. $\frac{9}{4}x^{5/4}$ **3.** $-\frac{9}{2}(1 - 3x)^{1/2}$
5. $(2 - 14x^2)/\sqrt{2 - 7x^2}$ **7.** $5 \cos 5t$
9. $4 \sin^3 t \cos t - 4 \cos^3 t \sin t$
11. $(-\frac{1}{3} \cos x)(1 - \sin x)^{-2/3}$
13. $-[\sin (\sin x)] \cos x$ **15.** $\dfrac{6(x-1)^2}{(x+1)^4}$
17. $\dfrac{3x - 5}{x^2(5 - 2x)^{3/2}}$ **19.** $\cos \dfrac{1}{x} + \dfrac{1}{x} \sin \dfrac{1}{x}$
21. $[2z - (2z)^{1/3}]^{-1/2}[1 - \frac{1}{3}(2z)^{-2/3}]$
23. $-36z^5 \cos (3z^6) \sin (3z^6)$
25. $-2 [\sin (1 + \tan 2x)] \sec^2 2x$
27. $5x^4 e^{(x^5)}$ **29.** $3e^{3t}(\sec^2(e^{3t}))$
31. $\dfrac{2t}{t^2 + 1}$ **33.** $\dfrac{2 \ln t}{t}$
35. $5(\ln 3)3^{5x - 7}$ **37.** $\dfrac{2x}{(\ln 3)(x^2 + 4)}$
39. $-2x^{-5/3}$ **41.** $-(1 + 6x^2)/\sqrt{1 + 3x^2}$
43. $-\dfrac{2}{3}\dfrac{\sin x + x \cos x}{(x \sin x)^{5/3}}$ **45.** $\frac{3}{2} \tan^2 (\frac{1}{2}x) \sec^2 (\frac{1}{2}x)$
47. $-(\sin x)e^{\cos x}$
49. $a^x[(\ln a)(\cos bx) - b \sin bx]$
51. $5y^4 \, dy/dx$ **53.** $-\dfrac{2}{y^2}\dfrac{dy}{dx}$
55. $\left(\dfrac{1}{2\sqrt{y}} \cos \sqrt{y}\right)\dfrac{dy}{dx}$ **57.** $3x^2y^2 + 2x^3y\dfrac{dy}{dx}$
59. $\left(x + y\dfrac{dy}{dx}\right)/\sqrt{x^2 + y^2}$ **61.** $1/\sqrt{x^2 - 1}$
63. $y = -2x + 1$ **65.** $y = 2$
67. $y = -6x + 2$ **69.** yes
71. $s'(x) = \dfrac{\pi}{180}c(x)$
   $c'(x) = -\dfrac{\pi}{180}s(x)$
73. $g'(x) = -f'(-x)$
75. $F'(t) = 14{,}000 \sin \dfrac{\pi t}{24}$
77. a. $-\dfrac{40}{V}\left(1 - \dfrac{v}{V}\right)^{-0.6}$
   b. $\dfrac{40v}{V^2}\left(1 - \dfrac{v}{V}\right)^{-0.6}$

79. a. $v'(r) = -\dfrac{96,000}{r^2}\left(\dfrac{192,000}{r} + v_0^2 - 48\right)^{-1/2}$

    b. $-\sqrt{6}/72,000$ miles per second per mile

81. $dV/dt = 40\pi r^2$

83. $\dfrac{dA}{dh} = \tfrac{2}{3}\sqrt{3}h; \dfrac{dA}{dh}\Big|_{h=\sqrt{3}} = 2$

Section 3.5

1. 0 **3.** $-240x^3 + 6x^2 + \tfrac{1}{2}(1-x)^{-3/2}$

5. $192(1-4x)^{-4}$ **7.** $an(n+1)x^{-n-2}$

9. $6x(2x^3+1)/(x^3-1)^3$

11. $2\sec^2 x \tan x$

13. $3(x^2+\sin x)[2(2x+\cos x)^2 + (x^2+\sin x)(2-\sin x)]$

15. $-\tfrac{1}{4}(1+\sin x)^{-3/2}\cos^2 x - \tfrac{1}{2}(1+\sin x)^{-1/2}\sin x$

17. $\dfrac{2}{x^3}e^{1/x} + \dfrac{1}{x^4}e^{1/x}$ **19.** $\dfrac{2-2x^2}{(1+x^2)^2}$ **21.** $\tfrac{3}{4}x^{-1/2}$

23. $6(x^4 - \tan x)(4x^3 - \sec^2 x)^2 + $
              $3(x^4 - \tan x)^2(12x^2 - 2\sec^2 x \tan x)$

25. $2a$ **27.** $2/(3-x)^3$

29. $\csc x \cot^2 x + \csc^3 x$ **31.** $2e^x \cos x$

33. $2\ln x + 3$ **35.** 0 **37.** $-12x\sin x^2 - 8x^3\cos x^2$

39. $-6/x^4$ **41.** $1440/(4x+5)^4$

43. $12e^{x^2} + 8x^3 e^{x^2}$ **45.** $\tfrac{105}{8}x^{1/2} - \tfrac{15}{4}x^{-1/2}$

47. $-\csc x \cot^3 x - 5\csc^3 x \cot x$

49. $6a$

51. $5040x^4 + 270x^2 + \tfrac{135}{64}x^{-13/4} + 48x^{-5}$

53. $\pi^4 \sin \pi x$ **55.** $4e^{-\sqrt{2}x}$

57. $v(t) = -32t + 3; a(t) = -32$

59. $v(t) = 2\cos t + 3\sin t; a(t) = -2\sin t + 3\cos t$

61. 0 **63.** e^x **65.** $(-1)^n e^{-x}(x-n)$

67. $\dfrac{n!}{(1-x)^{n+1}}$ **69.** $h = f$ and $g = f'$

71. a. $(2x + x^4)f(x)$ **b.** 87π

73. $f''(x)\,g(x) + 2f'(x)g'(x) + f(x)\,g''(x)$

75. $-4\pi^2\omega^2 a \cos(2\pi\omega t)$

79. b. $c = (\ln b)/ka$ **c.** $a/2$

Section 3.6

1. $\dfrac{4x^3}{3y}$ **3.** $\dfrac{2}{(2y+1)(1-x)^2}$

5. $\dfrac{\sec^2 x}{\sec y \tan y}$ **7.** $\dfrac{3(y^2+1)^2}{(y^2+1)\cos y - 2y\sin y}$

9. $\dfrac{-2x - 2xy^2}{2x^2y + 3y^2}$ **11.** $\dfrac{x^4 + y^2}{xy - x^3y}$

13. $\dfrac{-y(xy)^{-1/2} - (x+2y)^{-1/2}}{x(xy)^{-1/2} + (x+2y)^{-1/2}}$ **15.** $\dfrac{2x - e^y}{xe^y - 1}$

17. $\dfrac{2x - \sqrt{y}/x}{1 + ((\ln x)/2\sqrt{y}) - 2y}$ **19.** 0

21. $-\tfrac{1}{2}$ **23.** $-\tfrac{7}{2}$ **25.** $-\tfrac{1}{9}$

27. -4 **29.** $\tfrac{1}{8}$

31. $y = \tfrac{3}{4}x - \tfrac{9}{2}$ **33.** $y = -3x + \pi$

35. $\dfrac{2y^4 - 3x^2}{4y^7}$ **37.** $\dfrac{(\tan 2y)(2\sec^2 2y + 1)}{x^2}$

39. $\dfrac{x}{y}\dfrac{dx}{dt}$ **41.** $-\dfrac{\tan y}{x}\dfrac{dx}{dt}$

43. $-\dfrac{y^2\sin(xy^2)}{1 + 2xy\sin(xy^2)}\dfrac{dx}{dt}$

45. a. $\dfrac{dy}{dx} = \dfrac{2x}{3y^2}; y = x^{2/3}$ **b.** $\dfrac{dy}{dx} = \dfrac{-xy^2}{4}; y = \dfrac{8}{x^2+4}$

    c. $\dfrac{dy}{dx} = \dfrac{-2x}{3y^2(x^2-1)^2}; y = \left(\dfrac{x^2}{x^2-1}\right)^{1/3}$

47. $(1,1)$ **49.** 16

51. $(\sqrt{6}/4, \sqrt{2}/4), (\sqrt{6}/4, -\sqrt{2}/4),$
    $(-\sqrt{6}/4, \sqrt{2}/4), (-\sqrt{6}/4, -\sqrt{2}/4)$

Section 3.7

1. decreasing at 32π cubic centimeters per minute

3. decreasing at $12/\pi^2$ inches per hour

5. increasing at $1/(64\pi^{1/3}3^{2/3})$ centimeters per second

7. increasing at $3/(20\pi)$ centimeters per minute

9. $\tfrac{4}{3}\sqrt{10}$ feet per second

11. a. sliding at $\tfrac{3}{2}$ feet per second

    b. decreasing at $\tfrac{7}{4}$ square feet per second

13. rising at $\tfrac{1}{4}\sqrt{3}$ feet per minute

15. increasing at 50 pounds per square centimeter per second

17. $\sqrt{3}$

19. rising at $\tfrac{3}{4}$ foot per second

21. pulled in at $\tfrac{24}{13}$ feet per second

23. moving at 10 feet per second

25. $\tfrac{6}{25}$ radians per second

27. $\tfrac{3}{16}$ mile

29. $5\sqrt{3}$ feet per second

31. increasing at $\tfrac{1500}{29}$ radians per hour

33. 24 feet per second

35. $\tfrac{3}{250}$

37. 160π feet per minute

39. a. decreasing at $\tfrac{9}{7}$ feet per second

    b. decreasing at $\tfrac{3}{4}$ feet per second

Section 3.8

1. 10.05

3. $3 + \frac{2}{27} \approx 3.074074074$

5. 85

7. $\frac{1}{2}\sqrt{3} + \pi/156 \approx 0.8861638182$

9. $\sqrt{2}(1 - \pi/68) \approx 1.348877049$

11. 1.1

13. 0.1

15. 0.05

17. 0.02

19. $15x^2\, dx$

21. $-\sin x \cos(\cos x)\, dx$

23. $2x^3(1 + x^4)^{-1/2}\, dx$

25. $\pi^2/12{,}168$

27. $1/2187$

29. approximately 3.072245085

31. approximately 0.8847589319

33. approximately 0.095

35. 1.879385242

37. 2.094551482

39. -0.8228756555

41. 4.493409458

43. 0.5671432904

45. 3.872983346

47. 2.080083823

49. -0.1823735451 and 0.6001766211

51. Get approximate zero 0.1230777986, which is outside $[-2, 0]$.

53. The successive approximations oscillate between $-1/\sqrt{5}$ and $1/\sqrt{5}$.

57. approximately 2.094395102 cubic feet

59. **a.** $2\sqrt{3}/3 \approx 1.154700538$
   b. approximately 1.00000004

61. **a.** 505   **b.** 510   **c.** 520, 540, 570, 570, 560, respectively

Chapter 3 Review Exercises

1. $-12x^2 - 4/x^3$

3. $-(2x + 1)/(2x - 1)^3$

5. $-\sin t \sin 2t + 2\cos t \cos 2t$

7. $5\tan t + 5t\sec^2 t + 9\sec 3t \tan 3t$

9. $2e^{2t}\ln(3 + e^t) + \dfrac{e^{3t}}{3 + e^t}$

11. $12x^2 - \sqrt{3} - 2/(5x^2)$

13. $\dfrac{\cos x - 1 + \tan^2 x}{(1 - \sec x)^2}$

15. $3x^2\sqrt{x^2 - 4} + x^4/\sqrt{x^2 - 4}$

17. $e^x + xe^x + 5e^{-x}$

19. $y = 5x$

21. $y = x$

23. $y = \frac{13}{4}x - \frac{25}{4}$

25. $y = 2x$

27. $33x^{10} - 180x^4$

29. $3(t^2 + 9)^{1/2} + 3t^2(t^2 + 9)^{-1/2}$

31. $2 + 1/x$

33. $\dfrac{8xy - y}{9y^2 - 4x^2 + x}$

35. $\dfrac{2\sqrt{x} - y}{2\sqrt{x}(\sqrt{x} + 1)}$

37. $-\dfrac{y\cos xy^2}{3y + 2x\cos xy^2}$

39. $\frac{6}{5}$

41. $-\dfrac{y}{x}\dfrac{dx}{dt}$

43. $(2x\cos x - x^2\sin x)\, dx$

45. $5(x - e^x)^4(1 - e^x)\, dx$

47. linear approximation: 4.166666667; parabolic approximation: 4.162037037

49. 1.895494267, -1.895494267, and 0

51. 1.763222834

55. $2a^{1/2}f'(a)$

59. 2

61. the graph in Figure 3.61(c)

65. $\frac{16}{9}$

Cumulative Review Exercises (Chapters 1–2)

1. union of $(-\sqrt{3}, 0)$ and $(1, \sqrt{3})$

2. $(0, 1)$

3. union of $(-1, -\frac{1}{5})$ and $(\frac{1}{5}, 1)$

4. union of $[0, \pi/6]$ and $[5\pi/6, 2\pi)$

5. union of $(-\infty, -\frac{1}{2})$ and $(\frac{2}{3}, \infty)$

7. $(-\infty, -1]$

8. **a.** all real numbers except 3 and 4   **b.** $\dfrac{1}{-12 + 7x - x^2}$

9. -1

10. 1

11. -1

12. ∞

13. 2

14. It is continuous at -1.

15.

16. $\frac{31}{16}$

CHAPTER 4

Section 4.1

1. -2

3. $-2, 0, 1$

5. $-1, 1$

7. no critical numbers

9. $\pi/2 + n\pi$ for any integer n

11. $(2n + 1)\pi$ for any integer n

13. 2

15. $-2, 0$

17. $1/e$

19. minimum value: $f(\frac{1}{2}) = -\frac{1}{4}$; maximum value: $f(2) = 2$

21. no extreme values

23. minimum value: $k(0) = 1$; maximum value: $k(3) = \sqrt{10}$

25. minimum value: $f(0) = 0$; no maximum value

27. minimum value: $f(\pi^3/8) = -1$; maximum value: $f(-\pi^3/27) = \frac{1}{2}\sqrt{3}$

29. no extreme values

31. minimum value: $f(-1) = -e^{-1}$; maximum value: $f(0) = 0$

33. minimum value: $f(2) = 2 - 2\ln 2$; maximum value: $f(\frac{1}{2}) = \frac{1}{2} + 2\ln 2$

35. approximately -2.879385242 and 0.5320888862

37. critical number: approximately -0.3517337112; minimum value: approximately $f(-0.3517337112) \approx 0.8271840261$; maximum value: $f(1) = 1 + e$

45. a. 10 feet long (parallel to the house), 5 feet long (perpendicular to the house)

    b. 800 square feet

47. 2

49. b. approximately 0.287863324 radians (or 16.5°)

51. should not switch

53. base side: $\sqrt[3]{4}$ feet; height: $6/\sqrt[3]{16}$ feet

55. all sides of length 1

57. a. $20\sqrt{2}$ centimeters **b.** 800 cubic centimeters

59. 18

61. $\frac{4}{27}\pi R^2 H$

Section 4.2

1. 2   **3.** $-\frac{2}{3}\sqrt{3}$ **5.** -1

7. $1 - \frac{1}{3}\sqrt{21}$ **9.** $(\frac{7}{3})^{3/2}$

11. $m = 4,\ c \approx 1.618033989$

17. $9 < 28^{2/3} \le \frac{83}{9}$ **19.** $|\sqrt{3} - 1.7| \le \frac{11}{340}$

31. yes   **33.** $(1, 0)$

Section 4.3

1. C **3.** $\frac{3}{2}x^2 + C$ **5.** $-\frac{1}{3}x^3 + C$

7. $-\cos x + C$ **9.** $\frac{1}{2}\sin^2 x + C$ **11.** $e^x + C$

13. $-2x$ **15.** $\frac{1}{3}x^3 - 5$

17. $\sin x + 1 - \sqrt{3}/2$

19. $e^x + 9$ **21.** $C_1 x + C_2$ **23.** $-x + 2$

25. $-\sin x - 3x + 4$

27. $C_1 x^3 + C_2 x^2 + C_3 x + C_4$

29. decreasing on $(-\infty, -\frac{1}{2})$; increasing on $[-\frac{1}{2}, \infty)$

31. increasing on $(-\infty, \infty)$

33. decreasing on $(-\infty, \frac{3}{2}]$; increasing on $[\frac{3}{2}, \infty)$

35. decreasing on $[-\sqrt{2/5}, \sqrt{2/5}]$; increasing on $(-\infty, -\sqrt{2/5}]$ and $[\sqrt{2/5}, \infty)$

37. decreasing on $[0, 4]$; increasing on $[-4, 0]$

39. decreasing on $(-\infty, -3)$ and $(-3, \infty)$

41. decreasing on $[0, \infty)$; increasing on $(-\infty, 0]$

43. increasing on $(\pi/2 + n\pi, \pi/2 + (n+1)\pi)$ for any integer n

45. decreasing on $[-\pi/6 + 2n\pi, 7\pi/6 + 2n\pi]$ for any integer n; increasing on $[7\pi/6 + 2n\pi, 11\pi/6 + 2n\pi]$ for any integer n

47. decreasing on $(-\infty, -1]$; increasing on $[-1, \infty)$

49. decreasing on $(0, 1]$; increasing on $[1, \infty)$

51. decreasing on $[-1, 1]$

53. decreasing on $[0, 1]$; increasing on $[-1, 0]$

67. $f(x) = \begin{cases} x^3 & \text{for } -1 < x < 0 \\ 0 & \text{for } x = 0 \\ 2e^x - 2 & \text{for } 0 < x < 1 \end{cases}$

69.

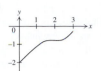

71. c. no   **d.** no

75. v is increasing on $[0, a/2]$ and decreasing on $[a/2, a]$.

77. increasing on $(0, \sqrt{1/(CL)}\,]$

79. $T = -\frac{1}{2}m\omega^2 r^2/l$

Section 4.4

1. a. 4 days **b.** $2\dfrac{\ln 3}{\ln 2} \approx 3.17$ (days)

3. $\dfrac{41 \ln 1.25}{6 \ln 2} \approx 2.2$ (days ago)

5. $\dfrac{20 \ln 2.5}{\ln 2} \approx 26.44$ (years)

9. approximately $-\dfrac{1590}{\ln 2}\ln 0.9 \approx 241.7$ (years)

11. approximately 1.44 percent

13. $100e^{(\ln 2)/4.07} \approx 119$ (milligrams)

15. $e^{-0.2} \approx 0.819$

17. a. approximately 29.92 (inches of mercury)

    b. approximately $29.92e^{-0.1} \approx 11.01$ (inches of mercury)

    c. approximately $29.92e^{-0.2} \approx 4.049$ (inches of mercury)

19. $200e^{(\ln 2)/10} \approx 214$ (milligrams)

21. $10 \ln 2 \approx 6.93$ (percent)

23. after $\dfrac{\ln \frac{1}{2}}{\ln \frac{3}{4}} \approx 2.41$ minutes

27. a. approximately 1,150,000 years

    b. approximately 0.908

Section 4.5

1. negative to positive at -3

3. negative to positive at -1 and 1; positive to negative at 0

5. negative to positive at 1; positive to negative at -1

7. negative to positive at $4\pi/3 + 2n\pi$ for any integer n; positive to negative at $2\pi/3 + 2n\pi$ for any integer n

9. relative maximum value: $f(\frac{1}{2}) = \frac{31}{4}$

11. relative minimum value: $f(0) = 4$; relative maximum value: $f(-2) = 8$

13. relative minimum value: $g(-\frac{1}{2}) = 3$

15. relative maximum value: $f(2) = \frac{1}{12}$

17. relative minimum value: $f(-\frac{1}{2}\sqrt{2}) = -\frac{1}{2}$; relative maximum value: $f(\frac{1}{2}\sqrt{2}) = \frac{1}{2}$

19. relative minimum values: $k(5\pi/6 + 2n\pi) = -\frac{1}{2}\sqrt{3} + 5\pi/12 + n\pi$ for any integer n; relative maximum values: $k(\pi/6 + 2n\pi) = \frac{1}{2}\sqrt{3} + \pi/12 + n\pi$ for any integer n

21. relative minimum value: $k(0) = 0$

23. relative minimum value: $f(0) = 0$; relative maximum value: $f(2) = 4e^{-2}$

25. relative maximum value: $f(\frac{3}{8}) = -\frac{7}{16}$

27. relative minimum value: $f(4) = -79$; relative maximum value: $f(-2) = 29$

29. relative minimum values: $f(-\frac{1}{2}) = \frac{1}{16}$ and $f(\frac{3}{2}) = -\frac{127}{16}$; relative maximum value: $f(0) = \frac{1}{2}$

31. relative minimum value: $f\left(\dfrac{1}{\sqrt[3]{2}}\right) = \dfrac{1}{2^{2/3}} + \sqrt[3]{2} + 1$

33. relative minimum value: $f(\pi/4 + n\pi) = -\sqrt{2}$ for any odd integer n; relative maximum value: $f(\pi/4 + n\pi) = \sqrt{2}$ for any even integer n

35. no relative extreme values

37. relative minimum value: $f(-4) = -4$

39. no relative extreme values

41. relative minimum value: $f(-1) = -3$

43. relative minimum value: $f(-1) = f(1) = 0$; relative maximum value: $f(0) = 1$

45. relative minimum value: $f(-1) = f(2) = 0$; relative maximum value: $f(\frac{1}{2}) = \frac{81}{16}$

47. relative minimum value: $f(0) = 1$

49. 0   **51.** 0   **53.** 0

55. -0.5 (relative maximum value), 0 (relative minimum value)

57. -5.0 (relative maximum value), -1.2 (relative minimum value), 1.2 (relative maximum value), 5.0 (relative minimum value)

61. a. $f(x) = x^{4/3}$ **b.** $f(x) = x^{5/3}$

Section 4.6

1. $a/2$ **5.** $4a$ **7.** $e^{-1/2}$ **9.** 9 and 9

11. base side length: 2 meters; height: 1 meter

15. width of rectangle: $12/(6 - \sqrt{3}) \approx 2.8$ feet; height of rectangle: $(18 - 6\sqrt{3})/(6 - \sqrt{3}) \approx 1.8$ feet

17. length of side along highway: $\frac{1}{2}\sqrt{2}$ kilometers; length of other side: $\sqrt{2}$ kilometers

23. b. no

25. $(-1 + \sqrt{2}/2, -\frac{1}{2})$ and $(-1 - \sqrt{2}/2, -\frac{1}{2})$

27. a. $p \le \frac{1}{2}$ **b.** $p > \frac{1}{2}$

29. a. the point $\frac{6}{7}\sqrt{7}$ miles down the road

b. the point at which the car is located

c. If $c > \frac{6}{7}\sqrt{7}$, then walk toward the point $\frac{6}{7}\sqrt{7}$ down the road. If $c \le \frac{6}{7}\sqrt{7}$, then walk toward the car.

33. $\frac{2}{3}\sqrt{3}\,R$ **35.** $20\sqrt{2}$ feet

37. 10%   **39.** $20\sqrt{5c/b}$

Section 4.7

1. concave downward on $(-\infty, \infty)$

3. concave upward on $(2, \infty)$; concave downward on $(-\infty, 2)$

5. concave upward on $(-\sqrt{3}, 0)$ and $(\sqrt{3}, \infty)$; concave downward on $(-\infty, -\sqrt{3})$ and $(0, \sqrt{3})$

7. concave upward on $(-2, \infty)$; concave downward on $(-\infty, -2)$

9. concave upward on $(0, \infty)$

11. concave upward on $(0, \infty)$; concave downward on $(-\infty, 0)$

13. concave upward on $(-\infty, \infty)$

15. concave upward on $(0, \infty)$; concave downward on $(-\infty, 0)$

17. concave upward on $(n\pi + \pi/2, (n+1)\pi)$ for any integer n; concave downward on $(n\pi, n\pi + \pi/2)$ for any integer n

19. concave upward on
$(-\pi/2 + 2n\pi, \pi/2 + 2n\pi)$ for any integer
n; concave downward
on $(\pi/2 + 2n\pi, 3\pi/2 + 2n\pi)$ for any
integer n

21. $(-2, 0)$

23. $(-1, 9)$

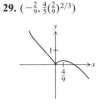

25. $(0, 0)$ and $(-\frac{2}{3}, -\frac{16}{27})$

27. $(-1/\sqrt[3]{2}, \frac{11}{8})$, $(0, 0)$, and
$(1/\sqrt[3]{2}, -\frac{11}{8})$

29. $(-\frac{2}{9}, \frac{4}{5}(\frac{2}{9})^{2/3})$

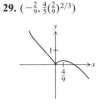

31. $(n\pi, 0)$ for any integer n

33. -0.7492409172 **35.** -0.588532744

37. a. c   **b.** a or d   **c.** b   **d.** d   **e.** e

39. a. f'' is negative on $(-3, -1)$ and on $(1, 3)$; it is positive
on $(-1, 1)$.
  b. $x = -1$ and $x = 1$ **c.** $-2, 0, 2$

41. The graph of f has greater concavity than the graph of g.

43. The concavity of the graph of $-f$ is opposite to that of f.

47. There are one or three inflection points.

49. There are at most $n - 2$ inflection points.

Section 4.8

1. 0	**3.** $\frac{1}{3}$
5. 2	**7.** ∞
9. 0	**11.** 0
13. $-\infty$	**15.** 0
17. 0	**19.** 1 **21.** 0

23. horizontal asymptote:
  $y = 0$

25. horizontal asymptote:
  $y = \frac{3}{2}$

27. horizontal asymptote:
  $y = -2$

29. horizontal asymptote:
  $y = 1$

31. horizontal asymptote:
  $y = 1$

33. horizontal asymptote:
  $y = 0$

35. horizontal asymptote: $y = 0$

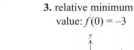

37. horizontal asymptote: $y = 1/\sqrt{2}$; vertical asymptotes:
  $x = -4$ and $x = 4$

39. vertical asymptotes: $x = 1$ and $x = -1$

41. vertical asymptote: $x = 0$

43. $\frac{2}{3}$ **45.** 1

47. a. undefined   **b.** 0   **c.** undefined   **d.** $-\infty$
  e. undefined   **f.** $-\infty$ **g.** $-\infty$ **h.** $-\infty$
  i. $x = -2$ and $x = 0$ **j.** $y = 0$

53. c. yes

55. a. v^* **b.** v^* is the "terminal velocity."

57. b

Section 4.9

1. inflection point: $(0, 2)$

3. relative minimum
  value: $f(0) = -3$

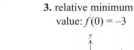

5. relative maximum value: $g(-2) = -4$; relative minimum value: $g(2) = 4$

7. relative maximum value: $g(2) = \frac{1}{4}$; inflection point: $(3, \frac{2}{9})$

25. relative minimum value: $f(0) = 0$

27. relative maximum value: $f(1) = 1$

9.

11. relative maximum value: $k(\frac{1}{2}) = 1$; relative minimum value: $k(-\frac{1}{2}) = -1$; inflection points: $(-\frac{1}{2}\sqrt{3}, -\frac{1}{2}\sqrt{3})$, $(0, 0)$, and $(\frac{1}{2}\sqrt{3}, \frac{1}{2}\sqrt{3})$

29. relative maximum value: $g(\pi/2 + n\pi) = 1$ for any integer n; relative minimum value: $g(n\pi) = 0$ for any integer n

31. relative maximum value: $g(\pi/3 + 2n\pi) = 2$ for any integer n; relative minimum value: $g(-2\pi/3 + 2n\pi) = -2$ for any integer n; inflection points: $(-\pi/6 + 2n\pi, 0)$ and $(5\pi/6 + 2n\pi, 0)$ for any integer n

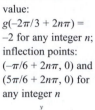

13. relative maximum value: $f(0) = -1$

15. The graph is that of Exercise 13 shifted up 1 unit.

33. relative maximum value: $g(\pi/2 + n\pi) = 1$ for any integer n; relative minimum value: $g(n\pi) = 0$ for any integer n; inflection points: $(\pi/4 + n\pi/2, \frac{1}{2})$ for any integer n

35. inflection point: $(0, \frac{1}{2})$

17. relative minimum value: $f(0) = 0$; inflection point: $(\frac{1}{2}, \frac{2}{9})$

19. relative maximum value: $f(\sqrt{3}) = -\frac{3}{2}\sqrt{3}$; relative minimum value: $f(-\sqrt{3}) = \frac{3}{2}\sqrt{3}$; inflection point: $(0, 0)$

37. relative minimum value: $f(0) = \ln 2$

39.

21. inflection points: $(-\sqrt[3]{\frac{1}{2}}, -\frac{2}{3})$ and $(0, -1)$

23.

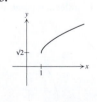

41. inflection points:
$(-\sqrt{3}/2, -\sqrt{3}/4)$ and
$(\sqrt{3}/2, \sqrt{3}/4)$

43.

45. x intercepts: approx.
$-2.55, 0.55$; relative
maximum value:
$f(-1) = -1$; relative
minimum value:
$f(-2) = -2 = f(0)$;
inflection points:
$(-1 - \frac{1}{3}\sqrt{3}, -\frac{14}{9})$,
$(-1 + \frac{1}{3}\sqrt{3}, -\frac{14}{9})$

47. relative maximum
value:
$$f(1 + \sqrt{2}) = \frac{\sqrt{2}}{4 + 3\sqrt{2}};$$
relative minimum
value:
$$f(1 - \sqrt{2}) = \frac{\sqrt{2}}{3\sqrt{2} - 4};$$
inflection point: ap-
prox. $(3.85, 0.15)$

51.

53.

55.

Chapter 4 Review Exercises

1. $0, \frac{8}{5}, 2$
3. minimum value: $f(-\frac{1}{2}) = \frac{3}{4}$; maximum value: $f(2) = 7$
5. minimum value: $f(-\frac{1}{2}\sqrt{2}) = -\sqrt{2}$; maximum value:
   $f(1) = 1$
7. **b.** because f is not continuous at 0
9. $f(x) = \frac{1}{3}x^3 + \cos x + C$
11. $f(x) = \frac{1}{12}x^4 - 2x^2 + C_1 x + C_2$
13. $(-\infty, \infty)$

15. increasing on $[2n\pi - \pi/3, 2n\pi + \pi/3]$; decreasing on
   $[2n\pi + \pi/3, 2n\pi + \pi/2)$, $(2n\pi + \pi/2, 2n\pi + 3\pi/2)$, and
   $(2n\pi + 3\pi/2, 2n\pi + 5\pi/3]$ for any integer n
19. relative minimum values: $f(0) = 3$ and $f(2) = -5$; relative
   maximum value: $f(\frac{1}{2}) = \frac{55}{16}$
21. relative minimum value: $f(-1) = f(2) = 0$; relative
   maximum value: $f(0) = 16$
23. concave upward on $(-\infty, -2)$ and $(1, \infty)$; concave
   downward on $(-2, 1)$
25. concave upward on $(-\infty, -\sqrt[4]{\frac{3}{5}})$ and $(\sqrt[4]{\frac{3}{5}}, \infty)$;
   concave downward on $(-\sqrt[4]{\frac{3}{5}}, \sqrt[4]{\frac{3}{5}})$
27. increasing on $(-\infty, -1]$ and $[1, \infty)$; decreasing on $[-1, 1]$;
   concave upward on $(-\frac{1}{2}, \infty)$; concave downward on
   $(-\infty, -\frac{1}{2})$
29. relative maximum value: $f(-\sqrt{2}) = 4\sqrt{2} - 1$; relative
   minimum value: $f(\sqrt{2}) = -4\sqrt{2} - 1$; inflection
   point: $(0, -1)$

31. relative maximum value: $f(-\sqrt{3}) = \frac{2}{9}\sqrt{3}$; relative
   minimum value: $f(\sqrt{3}) = -\frac{2}{9}\sqrt{3}$; inflection points:
   $(-\sqrt{6}, \frac{5}{36}\sqrt{6})$ and $(\sqrt{6}, -\frac{5}{36}\sqrt{6})$

33. relative maximum
   value: $k(0) = -\frac{1}{4}$

35. relative maximum
   value: $f(\ln \frac{3}{2}) = -\frac{9}{8}$;
   inflection point:
   $(\ln \frac{9}{4}, \frac{80}{729})$

37.

39.

41. m even, $n = 3m$ **43.** e^{-1}

45. a. approximately 4.5 billion years
   b. approximately 99.8%

47. 48 miles per hour

49. side length of base: 10 feet; height: 8 feet

51. $125,000

53.

55. 160 feet

59. a. $T = \dfrac{a \sec \theta}{v} + \dfrac{k - a \tan \theta}{w}$ **c.** Jo

61. radius: $\frac{1}{6}\sqrt{6l}$; length: $\frac{2}{3}\sqrt{3l}$

63. 60 feet from the house closest to the street

65. no

Cumulative Review Exercises (Chapters 1–3)

1. union of $(-1/\sqrt[3]{2}, 0)$ and $(1/\sqrt[3]{2}, \infty)$

2. $[-7, 7]$

3. union of $(-\sqrt[4]{15}, -1]$ and $[1, \sqrt[4]{15})$

4. $3\pi/2 + 2n\pi$ and $[\pi/6 + 2n\pi, 5\pi/6 + 2n\pi]$ for any integer n

5. a. union of $(-\infty, \frac{1}{3})$ and $(\frac{1}{2}, \infty)$ **b.** $\sqrt{\dfrac{2x - 1}{6x - 2}}$

6. $f(x) = 1 + x$ and $g(x) = \sin(3x^2 - 2)$

7. $-\infty$ **8.** ∞ **9.** 2 **10.** ∞

11. $\left(\frac{15}{4}, \frac{17}{16}\right)$

12. a. It is. **b.** It is not.

13. $-\frac{1}{8}$

14. $-(3x^2 \sin x^3) \cos (\cos x^3)$

15. $\dfrac{(\sec^2 x) e^{\tan x} - 1}{(x - e^{\tan x})^2}$

16. a. $(0, \infty)$ **b.** $\dfrac{e^x}{1 - e^x}$ **c.** ∞

18. a. $t = 1$ **b.** $-\frac{1}{2}$

19. $\dfrac{4 - y^2}{2xy + 3}$

20. $2\sqrt{3}$ units per second

21. $\dfrac{\pi}{16}$ square feet

22. 1.308571201

23. approximately 4.5×10^8 kilometers

CHAPTER 5
Section 5.1

1. $L_f(P) = 3/2$; $U_f(P) = 9/2$

3. $L_f(P) = -\pi/2$; $U_f(P) = \pi/2$

5. left sum: 11/6; right sum: 13/12

7. $L_f(P) = 27/4$; $U_f(P) = 33/4$

9. $L_f(P) = 13/12$; $U_f(P) = 11/6$

11. $L_f(P) = \pi/2$; $U_f(P) = \sqrt{2}\pi/2$

13. $L_f(P) \approx 0.6039259454$; $U_f(P) \approx 0.7150370565$

15. $L_f(P) \approx 0.0518487986$; $U_f(P) \approx 0.3984223889$

17. left sum: -15; right sum: -9; midpoint sum: -12

19. left sum: approximately 0.927222222; right sum: approximately 0.482777778; midpoint sum: approximately 0.6481464407

21. left sum: approximately 4.163897657; right sum: approximately 4.163897657; midpoint sum: approximately 4.15942373

23. left sum: approximately 2.12401742; right sum: approximately 2.193332139; midpoint sum: approximately 2.158987223

25. left sum: approximately 0.927222222

27. upper sum: approximately 0.7113948156

29. We cannot find $U_f(P)$ since f has no maximum value on $[0, x_1]$. No problem for $L_f(P)$.

33. a. Approximately 0.9352941176×10^4, that is, $\$9,352.94 \times 10^4$
   b. Approximately 1.202905554×10^4, that is, $\$12,029.06$. Adding partition points increases the lower sum.

Section 5.2

1. $L_f(P) = 4$; $U_f(P) = 12$

3. $L_f(P) = -3\sqrt{2}\pi/8$; $U_f(P) = 3\sqrt{2}\pi/8$

5. left sum: 2; right sum: 8; midpoint sum: 9/2

7. left sum: $-\pi/3$; right sum: $\pi/3$; midpoint sum: 0

9. approximately 2.008248408

11. approximately 0.1427055747

13. approximately 2.762521203

15. 0 **17.** -125 **19.** approximately 2/3

21. 25/2 **23.** 15/2

33. a. 16 **b.** -8 **c.** 0 **d.** $-1/(2\pi)$

37. $\displaystyle\int_a^b x^n dx = \dfrac{1}{n + 1}(b^{n+1} - a^{n+1})$

39. a. $1/(12n^2)$ **b.** $9/(8n^2)$ **c.** $1/(12n^2)$

43. a. $A \approx 18.85539331$ **b.** $A = \pi ab$

45. 69,800 square feet

Section 5.3

1. 14 **3.** 30

5. $\frac{1}{2} + \frac{3}{2} = 2$ **7.** $-\frac{1}{3} + \frac{8}{3} = \frac{7}{3}$

9. $a = 3, b = 2$　　**11.** $a = 5, b = 1$

13. $m = \dfrac{1}{3}, M = \dfrac{1}{2}; \dfrac{1}{3} \le \displaystyle\int_2^3 \dfrac{1}{x}\, dx \le \dfrac{1}{2}$

15. $m = \dfrac{1}{2}, M = \dfrac{\sqrt{2}}{2}; \dfrac{\pi}{24} \le \displaystyle\int_{\pi/4}^{\pi/3} \cos x\, dx \le \dfrac{\sqrt{2}\pi}{24}$

17. $\frac{1}{2}$　　　　**19.** $\frac{1}{3}$　　　　**23.** $\frac{5}{6}$

25. b. approximately 1.65

29. approximately 54.4 degrees Fahrenheit

Section 5.4

1. $x(1 + x^3)^{29}$　　**3.** $-1/y^3$　　**5.** $2x^3 \sin x^2$

7. $-(1 + y^2)^{1/2} + 2y(1 + y^4)^{1/2}$

9. $\frac{512}{5}x(1 + 16x^2)^{-1/5}$　　**11.** 4

13. -4　　　　　　　　**15.** 12

17. $\frac{1}{101}$　　　　　　　　**19.** 0

21. $\frac{9}{2}(4^{2/9} - 1)$　　　　**23.** $10\pi - 2\pi^2 + 8.625$

25. $\frac{9}{2}$　　**27.** $\frac{1}{2}\sqrt{3}$　　**29.** $\frac{1}{2} - \frac{1}{2}\sqrt{2}$

31. $\frac{7}{24}$　　**33.** $\ln 2$　　**35.** $e^2 - 1$

37. $\sqrt{3}$　　**39.** 1　　**41.** $\frac{2}{5}$

43. $\frac{3}{2}$　　**45.** $\frac{14}{3}$　　**47.** 1

49. 1　　**51.** $\frac{1}{4}\pi$　　**57.** $\frac{25}{3}$

59. a. $f(t) = 5t^2 - \frac{1}{3}t^3$　　**b.** $f(5) = \frac{250}{3}$

61. $672{,}000/\pi$ tons

63. a. \$111.01 dollars　　**b.** \$400.01 dollars

65. a. 8 feet per second　　**b.** 8 seconds

67. $\int_{t_1}^{t_2} R(t)\, dt$

71. $781{,}250{,}000\pi$ foot-pounds

73. 1800 feet

75. a. $p(t_2) - p(t_1) = \int_{t_1}^{t_2} F\, dt$

　　b. The impulse is 0.95 kilogram meter per second. The average force is 950 Newtons.

Section 5.5

1. $x^2 - 7x + C$

3. $\frac{3}{2}x^{4/3} - \frac{12}{7}x^{7/4} + \frac{5}{7}x^{7/5} + C$

5. $\dfrac{t^6}{6} + \dfrac{1}{3t^3} + C$　　　**7.** $2 \sin x - \frac{5}{2}x^2 + C$

9. $-3 \cot x - \frac{1}{2}x^2 + C$　　**11.** $\frac{4}{3}t^3 + 2t^2 + t + C$

13. $x + 2 \ln x - 1/x + C$　　**15.** $-\frac{15}{2}$

17. $3 - 5\sqrt{2}$　　　　　　**19.** $\dfrac{9\pi^2}{32} + \dfrac{2}{\pi} + \dfrac{\sqrt{2}}{2}$

21. $-1 - \frac{5}{3}\sqrt{3}$　　　　**23.** $-\frac{136}{3}$

25. $\pi + \dfrac{5}{\pi} + \dfrac{\pi^2}{4}$　　　**27.** $-\frac{142}{3}$

29. $\frac{5}{2}$　　　**31.** $4 - 2e^\pi$　　　**33.** 18

35. $\int 20x(1 + x^2)^9\, dx = (1 + x^2)^{10} + C$

37. $\int (x \cos x + 2 \sin x)\, dx = x \sin x - \cos x + C$

39. $\int 21 \sin^6 x \cos x\, dx = 3 \sin^7 x + C$

41. $e^{(x^2)} - e^{-x} + C$　　　**43.** 10

45. 12　　　　　　　**47.** $3 - \sqrt{2}/2$

49. $12 - 4 \ln 2$　　　**55.** $\frac{1}{12}\sqrt{3}\pi$

65. a. $2\sqrt{2} - 1$　　**b.** 1

Section 5.6

1. $\frac{1}{6}(4x - 5)^{3/2} + C$　　**3.** $(\sin \pi x)/\pi + C$

5. $\frac{1}{2}\sin x^2 + C$　　　**7.** $\frac{1}{3}\cos^{-3} t + C$

9. $-\dfrac{2}{5} \cdot \dfrac{1}{(t^2 - 3t + 1)^{5/2}} + C$

11. $\frac{2}{5}(x + 1)^{5/2} - \frac{4}{3}(x + 1)^{3/2} + C$

13. $\frac{2}{3}(3 + \sec x)^{3/2} + C$　　**15.** $\frac{1}{2}e^{(x^2)} + C$

17. $\frac{1}{2}\ln(x^2 + 1) + C$　　**19.** $\frac{1}{13}(x^3 + 1)^{13} + C$

21. $\frac{1}{3}(4 + x^{3/2})^2 + C$　　**23.** $\frac{2}{9}(3x + 7)^{3/2} + C$

25. $\frac{1}{3}(1 + 2x + 4x^2)^{3/2} + C$

27. 0　　　　　　　**29.** $\frac{1}{7}\sin^7 t + C$

31. $\frac{1}{3}(\sin 2z)^{3/2} + C$　　**33.** $\sqrt{2} - 1$

35. $2 \tan \sqrt{z} + C$

37. $\frac{1}{3}(w^2 + 1)^{3/2} + (w^2 + 1)^{1/2} + C$

39. $\frac{1}{2}(27 - 5\sqrt{5})$

41. $-\frac{1}{2}\cos(1 + e^{2x}) + C$

43. $\frac{1}{4}\ln(1 + x^4) + C$

45. $\frac{2}{5}(x + 2)^{5/2} - \frac{4}{3}(x + 2)^{3/2} + C$

47. $\frac{96}{5}$

49. $-\frac{1}{256}\left[\frac{1}{3}(1 - 8t)^{3/2} - \frac{2}{5}(1 - 8t)^{5/2} + \frac{1}{7}(1 - 8t)^{7/2}\right] + C$

51. $\frac{26}{15}$　　**53.** $\frac{14}{3}$　　**55.** $\frac{3}{20}$　　**57.** $\frac{38}{3}$

61. b. $1/858 \approx 0.0011655012$

65. $1/x$ is not continuous (or even defined) on $[-1, 1]$.

67. a. $1/4$

　　b. The electron is (essentially) certain to be in the interval $[0, 1]$.

Section 5.7

1. $\ln 4$　　　　　　**3.** $2 \ln 3$

5. domain: $(-1, \infty)$; $f'(x) = \dfrac{1}{x + 1}$

7. domain: union of $(-\infty, 2)$ and $(3, \infty)$;

　　$f'(x) = \dfrac{1}{2(x - 3)(x - 2)}$

9. domain: $(0, \infty)$; $f'(t) = [\cos (\ln t)] \dfrac{1}{t}$

11. domain: $(1, \infty)$; $f'(x) = \dfrac{1}{x \ln x}$

13. $\dfrac{dy}{dx} = \dfrac{(y^2 + x) \ln (y^2 + x) + x}{5(y^2 + x) - 2xy}$

15.

17. relative minimum value: $f(0) = 0$; inflection points: $(-1, \ln 2)$ and $(1, \ln 2)$

19. 1.531584394
21. $\ln |x - 1| + C$
23. $\frac{1}{2} \ln (x^2 + 4) + C$
25. $-\frac{2}{3} \ln 2$
27. $-\ln 2$
29. $\frac{1}{2} (\ln z)^2 + C$
31. $\frac{1}{2} (\ln (\ln t))^2 + C$
33. $\ln | \sin t | + C$
35. $\ln | x \sin x + \cos x | + C$
37. 1
39. $\frac{1}{2} \ln 2 - \frac{1}{8}$
41. $\left[(x + 1)^{1/5}(2x + 3)^2(7 - 4x)^{-1/2} \right] \cdot$
$\left[\dfrac{1}{5(x + 1)} + \dfrac{4}{2x + 3} + \dfrac{2}{7 - 4x} \right]$
43. $\sqrt[3]{(x + 3)^2(2x - 1)/(4x + 5)^4} \cdot$
$\left(\dfrac{2}{3x + 9} + \dfrac{2}{6x - 3} - \dfrac{16}{12x + 15} \right)$
45. $\dfrac{x^{3/2}e^{-x^2}}{1 - e^x} \left(\dfrac{3}{2x} - 2x + \dfrac{e^x}{1 - e^x} \right)$
53. $y = -x + 2$
55. b. $1/e$ **c.**

57. $W = c \ln(V_2/V_1)$
59. a. approximately 89.7834 kilograms
   b. approximately 10.2204 kilograms

Section 5.8

1. $\frac{4}{3}$ **3.** $\frac{2}{3}$ **5.** $\frac{1}{3}$

7. $\frac{56}{3}$ **9.** $3\sqrt{2} - 2$ **11.** $(\ln 2)^2$
13. $\frac{27}{4}$ **15.** $\frac{11}{6}$ **17.** $\frac{2}{3}\sqrt{3} + 2$
19. $\ln 2$ **21.** $\frac{6}{5}\sqrt{3} + \frac{26}{15}$
23. $1 - e^{-1} + \frac{1}{2}e^{-2} + \frac{1}{2}e^2 - e$
25. 1   **27.** 36   **29.** $\frac{37}{12}$
31. 12   **33.** $\frac{16}{3}$ **35.** 4
37. $\frac{1}{3}$ **39.** $\frac{343}{24}$ **43.** It is not.

Chapter 5 Review Exercises

1. left sum: $\frac{1}{6}\pi^2\sqrt{3}$; right sum: $\frac{1}{6}\pi^2\sqrt{3} - \frac{3}{4}\pi^2$;
   midpoint sum: $\frac{1}{3}\pi^2 - \frac{5}{16}\sqrt{2}\pi^2$
3. $\frac{5}{8}x^{8/5} - 3x^{8/3} + C$
5. $\frac{1}{4}x^4 - \frac{3}{2}x^2 + 2x - 2 \ln |x| + C$
7. $(1 + \sqrt{x + 1})^2 + C$
9. $-\frac{1}{12} \cos^4 3t + C$
11. $\frac{4}{5}(1 + \sqrt{x})^{5/2} - \frac{4}{3}(1 + \sqrt{x})^{3/2} + C$
13. $-\frac{1}{2} - 2^{5/3}$ **15.** 20
17. $\frac{2}{5} \ln 2$ **19.** $\ln(1 + e)$
21. $2 \ln 2 - \ln (2 - \sqrt{2})$ **23.** $\frac{195}{4}$
25. $68 - 6\sqrt{2}$
27. $3 - e + \ln \dfrac{(1 + e)^2}{16} \approx 0.1356528243$
29. $\frac{128}{3}$ **31.** $\frac{1}{2}$
33.   **35.** $x\sqrt{1 + x^5}$

37. $1/(x \ln x)$ **39.** 0
41. $\dfrac{(4 - \cos x)^{1/3}\sqrt{2x - 5}}{\sqrt[3]{x + 5}} \cdot$
$\left(\dfrac{\sin x}{12 - 3 \cos x} + \dfrac{1}{2x - 5} - \dfrac{1}{3x + 15} \right)$
43. b. $\frac{1}{3}(x^2 + 6)^{3/2} + C$
45. c. $\frac{1}{2}[\ln (x + 1)]^2 + C$
49. c. lower bound: 0; upper bound: $\frac{5}{6}$
53. a. $f(x) > 0$ for x in I
   b. b such that $f(b) = 0$
   c. f increasing on I
55. 242 feet   **57.** 6050 R **59. b.** $e^{-(10^4)}$
61. approximately 28 joules per mole

Cumulative Review Exercises (Chapters 1–4)

1. union of $(-\infty, 2)$ and $(4, \infty)$
2. **a.** domain of $f \circ g$: all numbers except $\frac{2}{3}$; domain of $g \circ f$: all numbers except $-\frac{2}{3}$
   c. no
3. $-\infty$ 4. $-\frac{2}{5}$ 5. 2
6. **a.** It is.   **b.** It is not.
7. $e^{1/(x^2+1)}\left[\dfrac{-2x}{(x^2+1)^2}\right]$
8. $-\frac{1}{2}$ 11. $-\frac{3}{16}$ 12. $f''(x) = 4f(x)$
13.

14. relative maximum value: $f(-2) = 0$; relative minimum value: $f(4) = 12$

15. relative minimum value: $f(2) = \frac{3}{4}$; inflection point: $\left(3, \frac{7}{9}\right)$

16. $-1/(24\pi)$ inch per second
17. **a.** -64 feet per second   **b.** 2 seconds
18. $\frac{5}{2}$ 19. \$3.50
20. 1 and 3 inches
21. approximately 0.2766633735 radian (or 15.9°) per second

CHAPTER 6
Section 6.1

1. $\pi/4$ 3. $\pi/2$ 5. π
7. $\frac{2}{9}\pi (27 - 2\sqrt{2})$ 9. $\frac{19}{4}\pi$ 11. 4π
13. π 15. 540π 17. $\frac{37}{15}\pi$
19. $\frac{4}{3}\pi$ 21. π 23. $\pi (\ln 2)^2$
25. $\frac{20}{3}\pi$ 27. $\frac{1}{18}\pi(35 - 16\sqrt{2})$
29. $\frac{8}{3}\pi$ 31. $\frac{8}{3}\pi$ 33. $\frac{16}{3}\pi$
35. $\frac{500}{3}\sqrt{3}$ 37. $V = \int_a^b \pi[f(x) - c]^2\, dx$

39. $\frac{\pi}{4}(3 - 4e^{-2} + e^{-4})$ 41. $\frac{7}{30}\pi$
43. approximately 43.3%
45. **a.** π **b.** $\pi/3$ **c.** infinite volume
47. $V = \int_a^b 2\pi(x - c)f(x)\, dx$
49. $43\pi/6$ 51. π
53. $\frac{32,000}{3}$ cubic feet   55. $\frac{2}{3}\pi$ cubic centimeters
57. $2/\pi$ inches per second   59. $\frac{191}{480}\pi$ cubic centimeters
61. 2000π cubic centimeters

Section 6.2

1. $4\sqrt{5}$ 3. $5 + \frac{1}{8}\ln\frac{3}{2}$
5. $\frac{1923}{128}$ 7. $\ln 5 - \frac{7}{8}\ln 2 + \frac{3}{16}$
9. $\frac{1}{8}(\sqrt{3} - \sqrt{2}) + \ln\dfrac{2 + \sqrt{3}}{\sqrt{2} + 1}$
11. $\frac{17}{16} + \frac{1}{32}\pi$ 13. $\ln(\sqrt{2} + 1)$
15. $\dfrac{2}{n + 2}(4^{(n+2)/2} - 2^{(n+2)/2})$
17. $\dfrac{2n}{2n + 1}(\sqrt{2}\, 2^n - 1)$
19. **a.** $y = (r^{2/3} - x^{2/3})^{2/3}$ for $0 \le x \le r$
    b. $L_\varepsilon = \frac{3}{2}r - \frac{3}{2}\varepsilon^{2/3}r^{1/3}$; $L = \frac{3}{2}r$

Section 6.3

1. 8π 3. $\frac{49}{3}\pi$
5. $\frac{\pi}{4}(e^2 - 2 + e^{-2}) = \frac{\pi}{4}(e - e^{-1})^2$ 7. $\frac{291}{256}\pi$
9. **a.** $S = 4\pi ra$ **b.** yes
11. $S_\varepsilon = \frac{12}{5}\pi(1 - \varepsilon^{2/3})^{5/2}$; $S = \frac{12}{5}\pi$
13. $\frac{800}{3}\pi(2\sqrt{2} - 1) \approx 1531.78$ square centimeters

Section 6.4

1. 3384 joules   3. 80 foot-pounds
5. 2×10^7 joules   7. 2.5×10^7 ergs
9. 10^8 ergs   11. 9.6×10^6 ergs
13. 1,850,000 foot-pounds   15. 6750π foot-pounds
17. 17,920 foot-pounds   19. 0
21. 60.28 ergs
23. **a.** 1.4 foot-pounds   **b.** 0.2 foot-pounds
25. **a.** approximately 16,000 foot-pounds
    b. 19,200 foot-pounds
27. 880 foot-pounds
29. approximately 1.651×10^{12} foot-pounds
31. 16 foot-pounds

Section 6.5

1. same side as the 15-kilogram child, 1 meter from the axis of revolution

3. inside

5. $(\frac{10}{9}, -\frac{4}{9})$

7. $(\frac{2}{3}, 0)$

9. $(\frac{14}{9}, \frac{7}{2})$

11. $\left(\dfrac{10 - 3\sqrt{6}}{3(1 - \sqrt{6} + \sqrt{3})}, \dfrac{\ln 2}{4(1 - \sqrt{6} + \sqrt{3})} \right)$

13. $(0, \frac{38}{35})$ 15. $(\frac{2}{3}, \frac{4}{3})$ 17. $(0, 3)$ 19. $(0, 0)$

21. $(0, h/3)$ 23. $(3a/4, 3h/10)$ 27. $8\pi\sqrt{2}$

Section 6.6

1. 140.625 pounds

3. $\dfrac{125}{18}\sqrt{3} \approx 12.0281$ pounds

5. $\frac{125}{3}\sqrt{3}(\frac{9}{2} - \sqrt{3}) \approx 199.760$ pounds for each triangle pointing downward, and $\frac{125}{3}\sqrt{3}(\frac{9}{2} - \frac{1}{2}\sqrt{3}) \approx 262.260$ pounds for those pointing upward

11. **a.** 10,000 pounds   **b.** 40,000 pounds
    c. 70,000 pounds
    d. $30,000\sqrt{226} \approx 451,000$ pounds

13. $937,500\pi \approx 2,945,000$ pounds

Section 6.7

1. $x^2 + y^2 = 4$

3. $(x - 2)^2 + (y + 1)^2 = 1$

5. $x = -y$

7. $x = 3$ for $-2 \le y \le -1$

9. $y = x^{2/3}$

11. $y = 1/x^3$ for $x > 0$

13. **a.** $\pi/4, 5\pi/4$ **b.** no such value   **c.** no such value

15. **a.** $x^3 + y^3 = 3xy$

b. The portion above the x axis is traced out as t increases without bound. The portion below the x axis is traced out as t increases toward -1.

17. **a.** The number of loops increases.
    b. dy/dx is larger for t near $\pi/2$ or $3\pi/2$ than for t near 0.

19. **b.** $x^2 + y^2 = 1$, except $(0, -1)$

21. $x = \begin{cases} 0 & \text{for } -\frac{1}{2}\pi \le t \le \frac{1}{2}\pi \\ -\frac{1}{4}\pi \cos t & \text{for } \frac{1}{2}\pi < t \le \frac{3}{2}\pi \\ \frac{1}{4}\pi \cos t & \text{for } \frac{3}{2}\pi < t \le \frac{5}{2}\pi \end{cases}$

$y = \begin{cases} t & \text{for } -\frac{1}{2}\pi \le t \le \frac{1}{2}\pi \\ \frac{1}{4}\pi(1 + \sin t) & \text{for } \frac{1}{2}\pi < t \le \frac{3}{2}\pi \\ -\frac{1}{4}\pi(1 + \sin t) & \text{for } \frac{3}{2}\pi < t \le \frac{5}{2}\pi \end{cases}$

Section 6.8

1. $\frac{1}{27}(13\sqrt{13} - 8)$ 3. $\frac{1}{8}\pi^2$ 5. $\frac{31}{60}\sqrt{2}\pi$ 7. **b.** no

9. **a.** $\int_0^{2\pi} \sqrt{a^2 \sin^2 t + b^2 \cos^2 t}\, dt$ **b.** $2\pi a$ **c.** $4a$

11. $n = 2$ or 3

Chapter 6 Review Exercises

1. $\pi(\frac{7}{3} + \frac{1}{2}e^{-2})$ 3. $\pi(e + e^{-1} - 2)$

5. **a.** $\frac{64}{5}\pi$ **b.** $(\frac{8}{5}, \frac{16}{7})$ **c.** $\frac{64}{5}\pi$

7. **a.** $4\pi c$ **b.** $4\pi c + \frac{2}{3}\pi c^2$

9. $20\sqrt{3}$ cubic feet   11. 50

13. $e - \frac{1}{4}e^{-1} - \frac{3}{4}$ 15. $\pi(e^2 - \frac{1}{16}e^{-2} + \frac{1}{16})$

17. 20 foot-pounds   19. 243,000π foot-pounds

21. $(\frac{1}{2}, \frac{2}{5})$ 23. **a.** $\left(\dfrac{2}{\ln 3}, \dfrac{6 + c}{3 \ln 3} \right)$

25.

27. $\frac{1}{4}(e^2 + e^{-2} - 2)$ 29. $\frac{1}{15}\pi(56 - 2\sqrt{2})$ 31. **b.** $c \ln 2$

33. **a.** 171.9π pounds   **b.** 57.3π pounds

35. **b.** $f(h) = \sqrt{\dfrac{cA}{\pi k}}\, h^{1/4}$ for $0 \le h \le b$

Cumulative Review Exercises (Chapters 1–5)

1. ∞ 2. 0   3. 0   4. $-\frac{2}{3}$

5. **a.** all numbers except -1 and 1
   c. yes, the lines $x = -1$ and $x = 1$

7. $f^{(24)}(x) = 24 \sin x + (x - 1)\cos x$

8. $y = 2$

9. $\frac{9}{2}$ inches per minute

10. relative maximum value: $f(-\sqrt{\tfrac{3}{2}}) = f(\sqrt{\tfrac{3}{2}}) = \tfrac{1}{4}$; relative minimum value: $f(0) = -2$; inflection points: $(-\sqrt{\tfrac{1}{2}}, -\tfrac{3}{4})$ and $(\sqrt{\tfrac{1}{2}}, -\tfrac{3}{4})$

11. inflection points: $(-1, -\tfrac{8}{3})$, $(0, 0)$, $(1, \tfrac{8}{3})$

12. $\tfrac{53}{55}$ miles per minute   **13.** $\pi/6$

14. radius: 2 meters; height: 10 meters

15. $(-1, 1)$ and $(1, 1)$ **16.** $5 + \ln \tfrac{3}{2}$

18. $\sqrt{2x - 3} + C$ **19.** $\ln 3$

20. $\tfrac{6}{13}(1 + \sqrt{x})^{13/3} - \tfrac{9}{5}(1 + \sqrt{x})^{10/3} + \tfrac{18}{7}(1 + \sqrt{x})^{7/3} - \tfrac{3}{2}(1 + \sqrt{x})^{4/3} + C$

21. $\tfrac{3}{8}(\ln t)^{8/3} + C$

22. a. $(101)^{10} + 1$ **b.** not possible

CHAPTER 7

Section 7.1

1. inverse exists; domain: $(-\infty, \infty)$; range: $(-\infty, \infty)$

3. inverse does not exist

5. inverse exists; domain: $[0, \infty)$; range: $(-\infty, 4]$

7. inverse does not exist

9. inverse exists; domain: $(-\infty, \infty)$; range: $(-\infty, \infty)$

11. inverse does not exist

13. inverse exists: domain: $(-\infty, \infty)$; range: $(-\infty, 3)$

19. k is the inverse of f; h is the inverse of g

25. $f^{-1}(x) = -\sqrt[3]{\dfrac{x + 1}{4}}$

27. $g^{-1}(x) = x^2 - 1$ for $x \geq 0$

29. $k^{-1}(t) = \dfrac{t + 1}{1 - t}$ **31.** $(-\infty, 0]$ or $[0, \infty)$

33. $[-\sqrt{5/3}, \sqrt{5/3}]$ (or $(-\infty, -\sqrt{5/3}]$ or $[\sqrt{5/3}, \infty)$)

35. $(-\infty, 0]$ or $[0, \infty)$

37. any interval of the form $[n\pi, (n + 1)\pi]$, where n is an integer

39. any interval of the form $[n\pi/2, (n + 1)\pi/2]$, where n is an integer

41. $\tfrac{1}{3}$ **43.** $\tfrac{1}{2}$ **45.** $\tfrac{1}{4}$ **47.** $\tfrac{1}{6}$

49. $1/(9x^8 + 7)$ **51.** $(x^3 + 1)/3x^2$

53. $1/\cos x$ for $-\pi/2 < x < \pi/2$

55. $[1, 4]$, $[4, 6]$, $[6, 8]$, $[8, 9]$

59. a. no   **b.** yes

71. $\tfrac{16}{35}$ **75.** $\tfrac{8}{27}(10^{3/2} - 1)$

Section 7.2

1. $\dfrac{e^x - e^{-x}}{e^x + e^{-x}}$ **3.** $\dfrac{-2e^x}{(e^x - 1)^2}$

5. $\dfrac{y - 2x^2 y e^y}{x^3 y e^y - x}$

7. a. $16\, e^x \sin x$ **b.** $16^{10}\, e^x \sin x$

11. relative minimum value: $f(0) = 2$

13. $e^{e^x - 1} + C$

15. $e^{\sqrt{3}} - 1$

17. $\sqrt{e^{2t} - 4} + C$

19. $-\frac{1}{2}[\ln (1 + e^{-t})]^2 + C$

21. $\ln (e^x + 1) + C$

23. $-e^{-(e^x)} + C$

25. $(-2, 1)$

27. $f^{-1}(x) = \ln \left(\dfrac{1 + x}{1 - x} \right)$

29. approximately 1.84140566

33. $\frac{1}{2}(e + e^{-1}) - 1$

35. $\pi(e + e^{-1} - 2)$ **37.** $\sqrt{2}\,(e^{\pi} - 1)$ **39.** 1

45. **a.** $f(t) = C_1 e^{2t}$ for some constant C_1, $g(t) = C_2 e^{6t}$ for some constant C_2; $y = \dfrac{C_2}{C_1^3} x^3$

   b. C is the graph of $y = \frac{3}{8} x^3$

47. **a.** $-VC$ **b.**   **c.** $RC \ln 10$

Section 7.3

1. $(-\ln 2)2^{-x}$

3. $t^t(\ln t + 1)$

5. $t^{2/t} \left(\dfrac{2 - 2 \ln t}{t^2} \right)$

7. $(-\sin x)(\cos x)^{\cos x} [\ln (\cos x) + 1]$

9. $2\sqrt{2}(2x)^{\sqrt{2} - 1}$

11. inflection point: $(0, 0)$

13. $\dfrac{1}{\ln 2} 2^x + C$

15. $\dfrac{8}{\ln 3}$

17. $\dfrac{-1}{2 \ln 5} 5^{-x^2} + C$

19. $\dfrac{1}{2\pi + 1} x^{2\pi + 1} + C$

21. $\dfrac{7}{\ln 2}$

23. $\dfrac{8}{\ln 3} - \dfrac{3}{\ln 2}$

25. approximately 1.464973521

27. approximately 0.8735685268

29. $\frac{1}{3}\pi \ln 3$

33. concave upward if $a > 1$; concave downward if $0 < a < 1$

35. $2\sqrt{2}$ **37.** **b.** $\log_4 x = \log_4 7 \log_7 x$

39. $a^{2/(a+1)}$ **41.** $3 \log_3 2 \approx 1.89$

43. **a.** $dy/dt = k(\ln a)(\ln b)b^t a^{(b^t)}$

   b. The level of diffusion is increasing.

   c. k

Section 7.4

1. 0

3. 0

5. $\dfrac{1 + e^2}{1 - e^2}$

7. $\frac{4}{3}$

9. $\frac{17}{15}$

11. $\frac{2}{3}\sqrt{2}$

13. $\dfrac{x^2 - 1}{2x}$

15. $\dfrac{x^2 - 1}{x^2 + 1}$

17. $\mathrm{sech}^2 x$

19. $-\mathrm{sech}\, x \tanh x$

21. $-\dfrac{1}{2\sqrt{x}} \mathrm{sech} \sqrt{x} \tanh \sqrt{x}$

23. $-\dfrac{2x}{\sqrt{1 - x^2}} \sinh \sqrt{1 - x^2} \cosh \sqrt{1 - x^2}$

25. $2e^{2x}(\sec^2 e^{2x}) \sinh (\tan e^{2x})$

29. $\tanh x + C$ **31.** $(\cosh 2^x)/\ln 2 + C$

33. $\ln \dfrac{10 + \sqrt{101}}{5 + \sqrt{26}}$ **35.** $16(e - e^{-1})$

37. The identity $\cosh^2 x = 1 + \sinh^2 x$ does not help.

39. $\left(\dfrac{\sqrt{2} - 1}{\ln (1 + \sqrt{2})}, \dfrac{\pi}{8 \ln (1 + \sqrt{2})} \right)$

47. **a.** 625 **b.** approximately 600

Section 7.5

1. $\pi/3$ **3.** $\pi/4$ **5.** $-\pi/6$

7. $\pi/6$ **9.** $5\pi/4$ **11.** $-\frac{1}{2}$

13. 1 **15.** 2 **17.** $\pi/3$

19. $\sqrt{1 - x^2}$ **21.** $\sqrt{x^2 + 1}$ **23.** $x^2/\sqrt{x^4 + 1}$

25. $1 - 2x^2$ **27.** $\dfrac{3}{\sqrt{1 - 9x^2}}$ **29.** $\dfrac{1}{2(t + 1)\sqrt{t}}$

31. $\dfrac{x}{(2 - x^2)\sqrt{1 - x^2}}$ **33.** $\frac{1}{4} \tan^{-1} \frac{1}{4}x + C$

35. $\frac{1}{12} \tan^{-1} \frac{3}{4}x + C$

37. $\dfrac{\sqrt{2}}{4} \tan^{-1}[(x + 1)/\sqrt{2}] + C$

39. $\frac{1}{2} \sin^{-1} \frac{2}{3}x + C$ **41.** $\frac{1}{5} \sec^{-1} \frac{1}{5}x + C$

43. $\frac{1}{4} \sin^{-1} x^4 + C$ **45.** $-\tan^{-1} e^{-x} + C$

47. $\frac{1}{4}(\tan^{-1} 2x)^2 + C$ **49.** $\frac{1}{3} \tan^{-1}(\frac{1}{3} \sin t) + C$

51. $\frac{1}{8} \sec^{-1}(2 \sin 4x) + C$ **53.** $\frac{1}{6}\pi$

55. $\frac{1}{12}\pi$ **57.** $\frac{1}{3} - \frac{1}{18}\pi\sqrt{3}$

59. $\frac{1}{6}\pi^2$ **61.** $\frac{1}{2}\pi - 1$

69. On $(-\infty, 0)$, $f(x) = -\pi/2$; on $(0, \infty)$, $f(x) = \pi/2$.

73. **b.** $\frac{1}{2}\pi$

75. 2.5 feet above the floor; yes, the same

77. **a.** approximately 15.4 meters

   b. approximately 5 hours and 28 minutes

Section 7.6

1. $16a^{15}$ **3.** 0 **5.** $\frac{8}{5}$

7. 1 **9.** 1 **11.** ∞

13. 1 **15.** 0 **17.** 2

19. $\frac{4}{9}$ **21.** 1 **23.** 0

25. $\ln 5 - \ln 3$ **27.** 1 **29.** 6

31. 0 **33.** 0 **35.** 0

37. 1 **39.** 0 **41.** 1

43. 2 **45.** ∞ **47.** 1

49. 0 **51.** $2a$

53. relative maximum value: $f(1) = e^{-1}$; inflection point: $(2, 2e^{-2})$

57. a. because $\lim_{x \to \pi/2} x = \pi/2$ (so is not 0, ∞, or $-\infty$)

　b. $2/\pi$

63. $\frac{1}{12}$

Section 7.7

13.　　**15.**　　**17.**

21. $\frac{1}{4}$

23. a. The graph rises from left to right.

　b. The graph rises from left to right and is concave upward.

25. $c \geq 0$ **27.** b/a

Section 7.8

1. $y^2 - x^2 = C$ **3.** $\tan^{-1} y = \tan^{-1} x + C$

5. $|1 - e^{-y}| = C/(1 + e^x)$

7. $y = -\frac{1}{2} \ln (-x^2 + 4x + 1)$

9. $\frac{1}{3} y^3 = x - \ln |x| + 8$

11. $y = Ce^{1/x}$ **13.** $y = 2 + Ce^{-2x}$

15. $y = e^x(\ln |1 - e^{-x}| + C)$

17. $y = 1 + C \cos x$ **19.** $y = -2e^{-3x} - 2e^{-5x}$

21. $y = 1 + (\sqrt{2} + 1)/(\sec x + \tan x)$

25. $xy^2 = C$ **29. a.** $s = \frac{1}{2} gt^2$

31. $C \approx 0.0068$; $k \approx 1.2956$

Chapter 7 Review Exercises

1. yes **3.** yes **5.** no

7. $f^{-1}(x) = (x + 2)/(x + 3)$

11. $(\ln 2)2^x/(1 + 2^x)$ **13.** $1/\cosh x$

15. $-\dfrac{2x}{3(1 - x^2)^{2/3} \sqrt{1 - (1 - x^2)^{2/3}}}$

17. $\dfrac{2x}{(\ln 4)(1 + x^4) \tan^{-1} x^2}$ **19.** $2\sqrt{1 + e^x} + C$

21. $\sinh^{-1}(e^x) + C$ **23.** $\frac{1}{2} \ln (e^{2x} + 1) + C$

25. $\dfrac{4}{15 \ln 5}$ **27.** $\frac{3}{2} \tan^{-1} 2t + C$

29. $\pi/6$ **31.** $2 \tan^{-1} e^x + C$

33. $\frac{1}{12} \sqrt{6} \tan^{-1} \left(\dfrac{x^2 + 2}{\sqrt{6}} \right) + C$

35. a **37.** 1 **39.** 0 **41.** e^2

43. $-\dfrac{1}{x} - \dfrac{1}{2} e^{-(y^2)} = C$

45. $y = \frac{1}{5} x^3 + 2x + C/x^2$

47. $x^2 = y + \ln y - 1$

49. $y = \frac{1}{2}(7e^{2x} - 3)$

51. relative minimum value: $f(\frac{1}{2} e^{-1}) = 1/e^{1/2e}$

57. b. 1

61. $2(e^3 - e^2)$

65. a. $\dfrac{2GmM}{RL} \tan^{-1} \dfrac{L}{2R}$

Cumulative Review Exercises (Chapters 1–6)

1. $\frac{1}{8} \sqrt{2}$ **2.** 0

3. $-\dfrac{e^{-x} - 2 - e^x}{(1 + e^x)^2}$ **4.** $2e^{(2x+1)^2} - e^{(x^2)}$

8. inflection points: $(\frac{3}{2}\pi + 2n\pi, 0)$ for any integer n; vertical asymptotes: $x = \frac{1}{2}\pi + 2n\pi$ for any integer n

9. inflection point: $(0, 1)$; horizontal asymptotes: $y = e$ and $y = 0$

10. $c = \frac{1}{16}$; point of tangency: $(1, 4)$

11. $25\sqrt{3}$ feet per second   **12.** $288\sqrt{3}$ square inches

13. $\frac{1}{2} \ln (4 + x^2) + C$ **14.** $\dfrac{\pi}{18}\sqrt{3}$

15. $80 + 5\pi \approx 95.7$ miles   **16.** $\frac{7}{3} - 4 \ln 5 + 8 \ln 2$

17. $\frac{74}{9}$ **18.** $\frac{1}{4}$

19. a. 43π **b.** $(0, 2)$

20. $\frac{1,228,250}{27}$ foot-pounds

CHAPTER 8

Section 8.1

1. $-x \cos x + \sin x + C$ **3.** $\frac{1}{2}x^2 \ln x - \frac{1}{4}x^2 + C$

5. $x(\ln x)^2 - 2x \ln x + 2x + C$

7. $\frac{1}{4}x^4 \ln x - \frac{1}{16}x^4 + C$

9. $\frac{1}{4}x^2 e^{4x} - \frac{1}{8}xe^{4x} + \frac{1}{32}e^{4x} + C$

11. $x^3 \sin x + 3x^2 \cos x - 6x \sin x - 6 \cos x + C$

13. $\frac{1}{6}e^{3x} \cos 3x + \frac{1}{6}e^{3x} \sin 3x + C$

15. $\dfrac{t}{\ln 2} 2^t - \dfrac{1}{(\ln 2)^2} 2^t + C$

17. $\left(\dfrac{t^2}{\ln 4} - \dfrac{2t}{(\ln 4)^2} + \dfrac{2}{(\ln 4)^3} \right) 4^t + C$

19. $t \cosh t - \sinh t + C$

21. $x \tan^{-1} x - \frac{1}{2} \ln (1 + x^2) + C$

23. $x \cos^{-1}(-7x) + \frac{1}{7}\sqrt{1 - 49x^2} + C$

25. $\dfrac{m}{n + 1} x^{n+1} \ln x - \dfrac{m}{(n + 1)^2} x^{n+1} + C$

27. $\frac{1}{2}[x \cos (\ln x) + x \sin (\ln x)] + C$

29. $\frac{4}{25}e^5 + \frac{1}{25}$ **31.** -2π

33. $\pi/4 - \pi\sqrt{3}/3 + \frac{1}{2} \ln 2$ **35.** $2 \ln 2 - 1$

37. $-\dfrac{x}{a} \cos ax + \dfrac{1}{a^2} \sin ax + C$

39. $-\cos x \tan^{-1}(\cos x) + \frac{1}{2} \ln (1 + \cos^2 x) + C$

41. $2\sqrt{t} \sin \sqrt{t} + 2 \cos \sqrt{t} + C$

43. $\dfrac{5\sqrt{2}}{6}$

51. $x (\ln x)^3 - 3x (\ln x)^2 + 6x \ln x - 6x + C$

55. $2 \ln 2 - 1$ **57.** 1

59. $\frac{9}{2} \ln 3 - \frac{28}{9}$ **61.** $\frac{4}{3}\pi$

63. $\left(\dfrac{\pi\sqrt{2} - 4}{4\sqrt{2} - 4}, \dfrac{1}{4\sqrt{2} - 4} \right)$

65. a. $\frac{\pi}{2} - 1$ **b.** 0 **c.** $2 \ln 2 - \frac{3}{4}$

67. a. $f(x) = \begin{cases} 2Hx/L + H & \text{for } -L/2 \le x \le 0 \\ -2Hx/L + H & \text{for } 0 < x \le L/2 \end{cases}$

  b. $4HL/\pi^2$

Section 8.2

1. $-\frac{1}{3} \cos^3 x + \frac{1}{5} \cos^5 x + C$

3. $\frac{1}{12} \sin^4 3x + C$

5. $\dfrac{1}{3} \cos^3 \dfrac{1}{x} - \dfrac{2}{5} \cos^5 \dfrac{1}{x} + \dfrac{1}{7} \cos^7 \dfrac{1}{x} + C$

7. $\frac{1}{8}y - \frac{1}{32} \sin 4y + C$

9. $-\frac{1}{128} \sin^3 2x \cos 2x - \frac{3}{256} \sin 2x \cos 2x + \frac{3}{128}x + C$

11. $-\frac{1}{9} \sin^{-9} x + \frac{1}{7} \sin^{-7} x + C$

13. $\frac{17}{8}x - \frac{1}{32} \sin 4x + C$ **15.** $\frac{3}{2}\sqrt{2} - 2$

17. $\frac{1}{6} \tan^6 x + C$ **19.** $\frac{7}{24}$

21. $-\frac{2}{3} - \frac{1}{3}\sqrt{2}$

23. $\frac{2}{5} \sec^5 \sqrt{x} - \frac{2}{3} \sec^3 \sqrt{x} + C$

25. $\frac{1}{6} \tan^6 x + \frac{1}{4} \tan^4 x + C$ **27.** $\frac{1}{5} \sec^5 x + C$

29. $-\frac{1}{4} \cot^4 x + C$ **31.** $\frac{2}{15}(\sqrt{2} + 1)$

33. $\frac{1}{2} \sin^2 x + C$ **35.** $\dfrac{1}{3 \cos^3 x} + C$

37. $\frac{1}{3} \sin^3 x - \frac{1}{5} \sin^5 x + C$ **39.** $\frac{1}{2} \sin^2 x + C$

41. $\tan x - x + C$

43. $\frac{1}{3} \tan^3 x - \tan x + x + C$

45. $\frac{1}{2} \cos x - \frac{1}{10} \cos 5x + C$

47. $\frac{1}{4} \cos 2x + \frac{1}{12} \cos 6x + C$

49. $3 \cos \frac{1}{6}x - \frac{3}{7} \cos \frac{7}{6}x + C$

51. $-\frac{1}{10} \sin 5x + \frac{1}{2} \sin x + C$

53. $\frac{1}{4} \sin 2x + \frac{1}{16} \sin 8x + C$

55. $2 - \sqrt{2}$ **57.** $\ln | 1 - \cos x | + C$

61. $\frac{2}{15}$ **63.** $\frac{1}{3} (2 + \sqrt{2})$

65. $\pi[\sqrt{2} + \ln (\sqrt{2} + 1)]$

Section 8.3

1. $\pi/8$ **3.** π

5. $\dfrac{1}{3} \dfrac{x}{\sqrt{3 - x^2}} + C$ **7.** $\dfrac{x}{\sqrt{x^2 + 1}} + C$

9. $\frac{1}{25}\sqrt{5}$ **11.** $2 - \pi/2$

13. $\frac{1}{54}\tan^{-1}\frac{t}{3} + \frac{t}{18(9+t^2)} + C$

15. $\pi/12$

17. $\frac{1}{2}\ln|\sqrt{4x^2+4x+2}+(2x+1)| + C$

19. $\frac{2w^3}{3(1-2w^2)^{3/2}} + \frac{w}{(1-2w^2)^{1/2}} + C$

21. $\pi/2$

23. $\frac{1}{8}\sin^{-1}(2x-1) + \frac{1}{4}(2x-1)\sqrt{x-x^2} + C$

25. $\frac{1}{18}x\sqrt{9x^2-1} + \frac{1}{54}\ln|3x+\sqrt{9x^2-1}| + C$

27. $-\frac{1}{2}\ln\left|\frac{\sqrt{x^2+4}}{x}+\frac{2}{x}\right| + C$

29. $\frac{\sqrt{4x^2-9}}{9x} + C$

31. $\frac{1}{2}x\sqrt{1+x^2} - \frac{1}{2}\ln|\sqrt{1+x^2}+x| + C$

33. $\frac{-3x^3}{16(9x^2-4)^{3/2}} + \frac{x}{16(9x^2-4)^{1/2}} + C$

35. $\frac{x\sqrt{4+x^2}}{2} + 2\ln\left|\frac{\sqrt{4+x^2}}{2}+\frac{x}{2}\right| + C$

37. $\frac{1}{81}\left(\frac{3\sqrt{3}}{8} - \frac{5\sqrt{2}}{12}\right)$

39. $\frac{1}{2}\sqrt{2x^2+12x+19} -$
$\frac{3}{2}\sqrt{2}\ln|\sqrt{2x^2+12x+19}+\sqrt{2}(x+3)| + C$

41. $\frac{1}{2}(x+3)\sqrt{x^2+6x+5} -$
$2\ln\left|\frac{x+3}{2}+\frac{\sqrt{x^2+6x+5}}{2}\right| + C$

43. $\frac{1}{2}e^w\sqrt{1+e^{2w}} + \frac{1}{2}\ln|\sqrt{1+e^{2w}}+e^w| + C$

45. $\frac{1}{2}x^2\sin^{-1}x - \frac{1}{4}\sin^{-1}x + \frac{1}{4}x\sqrt{1-x^2} + C$

47. $\pi/4$ 49. $\frac{9}{2}\sqrt{2}+\frac{9}{2}\ln(\sqrt{2}+1)$

51. $4\sqrt{2}\pi[\sqrt{2}+\frac{1}{2}\ln(\sqrt{2}+1) - \frac{1}{2}\ln(\sqrt{2}-1)]$

53. $\frac{1}{2}\sqrt{2}+\frac{1}{2}\ln(\sqrt{2}+1)$ 55. $\frac{17\pi}{12}+\frac{\pi^2}{2}$

57. b. $\pi(750)(640) = 480,000\pi$ square feet

Section 8.4

1. $x - \ln|x+1| + C$

3. $x + \frac{1}{2}\ln\left|\frac{x-1}{x+1}\right| + C$

5. $\ln\left|\frac{x^4}{(x-1)^3}\right| - \frac{5}{x-1} + C$

7. $\ln\frac{12}{7}$

9. $\frac{15}{2}\ln|t-5| - \frac{9}{2}\ln|t-3| + C$

11. $1 - \frac{1}{2}\ln 2$

13. $x - \frac{1}{2}\ln|x+1| + \frac{3}{2}\ln|x-1| + C$

15. $\frac{1}{2}\ln 3 - \frac{1}{6}\sqrt{3}\pi$

17. $3\ln|x-2| - \frac{6}{x-2} + C$

19. $\ln\left|\frac{x-1}{x-2}\right| + C$

21. $\frac{1}{2}u^2 - 2u + 3\ln|u+1| + \frac{1}{u+1} + C$

23. $\frac{-1}{4(x+1)} + \frac{1}{4}\ln\left|\frac{x+1}{x-1}\right| - \frac{1}{4(x-1)} + C$

25. $\frac{2}{9}\ln\left|\frac{x-2}{x+1}\right| - \frac{1}{3(x+1)} + C$

27. $\ln|x+1| - \frac{1}{2}\ln(x^2+1) + \tan^{-1}x + \frac{x}{x^2+1} + C$

29. $\frac{1}{2}\ln|x^2-x+4| - \frac{1}{\sqrt{15}}\tan^{-1}\frac{\sqrt{15}(2x-1)}{15} + C$

31. $\ln\left|\frac{\sqrt{x+1}-1}{\sqrt{x+1}+1}\right| + C$

33. $\frac{6}{7}x^{7/6} - \frac{6}{5}x^{5/6} + 2x^{1/2} - 6x^{1/6} + 6\tan^{-1}x^{1/6} + C$

35. $\frac{4}{3} - \ln 3$

37. $\sin x - \tan^{-1}(\sin x) + C$

39. $-\frac{1}{3}\ln|1-e^x| + \frac{1}{6}\ln(1+e^x+e^{2x}) +$
$\frac{1}{\sqrt{3}}\tan^{-1}\frac{2e^x+1}{\sqrt{3}} + C$

41. $\frac{1}{2}x^2\tan^{-1}x - \frac{1}{2}x + \frac{1}{2}\tan^{-1}x + C$

43. $x\ln(x^2+1) - 2x + 2\tan^{-1}x + C$

45. $-\frac{\sqrt{x+4}}{x} + \frac{1}{2}\ln\left|\frac{\sqrt{x+4}-2}{\sqrt{x+4}+2}\right| + C$

47. $\frac{1}{10}\ln\left|\frac{x^{10}}{1-x^{10}}\right| + C$

49. $\frac{9}{2} - \frac{1}{2}\ln 10$

51. $-\frac{5}{2}\ln 2 + \frac{9}{10}\ln 5 + \frac{\pi}{10} + \frac{2}{5}\tan^{-1}2$

53. $2\pi(\frac{5}{2} - 3\ln 2)$ 55. $3 + \ln 2$

57. $\frac{1}{a}\ln\frac{y}{1-y} + C$ 59. b. $a^{c-b}b^{a-c}c^{b-a}$

Section 8.5

1. $\frac{x}{2}\sqrt{x^2+9} + \frac{9}{2}\ln|x+\sqrt{x^2+9}| + C$

3. $\frac{4}{101}e^5[(5\sin\frac{1}{2} - \frac{1}{2}\cos\frac{1}{2}) + \frac{1}{2}]$

5. $\frac{1}{12}\ln\left|\frac{2x-3}{2x+3}\right| + C$

7. $\sqrt{10x - \dfrac{1}{4}x^2} + 10\cos^{-1}\left(1 - \dfrac{x}{20}\right) + C$

9. $\frac{2}{3}e^{\sqrt{x}}(2\cosh 2\sqrt{x} - \sinh 2\sqrt{x}) + C$

11. $e\sin e + \cos e - \sin 1 - \cos 1$

13. $\frac{1}{2}x[\sin(\ln x) - \cos(\ln x)] + C$

15. $\frac{1}{3}(2x - \sqrt{x} - 3)\sqrt{2\sqrt{x} - x} + \cos^{-1}(1 - \sqrt{x}) + C$

17. $-\frac{1}{2}(4 + x)\sqrt{4 - x^2} + 2\sin^{-1}\dfrac{x}{2} + C$

19. $\text{Sqrt}[a^2 + x^2] \cdot$

$$\left(\dfrac{-a^4x}{16} + \dfrac{a^2x^3}{24} + \dfrac{x^5}{6}\right) + \dfrac{a^6\,\text{Log}[x + \text{Sqrt}[a^2 + x^2]]}{16}$$

21. Mathematica cannot evaluate the integral.

Section 8.6

1. $T \approx 1.106746032$
   $S \approx 1.098941799$

3. $T \approx 0.7462107961$
   $S \approx 0.7468249483$

5. $E_{10}^T \leq \dfrac{1}{75}$

7. $E_{10}^S \leq \dfrac{162}{5 \times 10^4} = 3.24 \times 10^{-3}$

9. $E_{10}^S \leq \dfrac{0.8}{1.8 \times 10^6} \approx 4.444444444 \times 10^{-7}$

11. 1.154724372

13. a. 1.32 **b.** 1.286666667

15. a. i. 3.139925989 **ii.** 3.141592614
   b. $0.001666665; 4 \times 10^{-8}$

17. a. 70 **b.** 8

21. b. It will decrease.   **c.** Yes. It would increase.

25. approximately 20.9 miles per hour

27. approximately 6620 cases

29. $\dfrac{\pi}{12}(20 + 8\sqrt{26}) \approx 15.9153$

31. approximately 12.1 square miles

33. approximately \$1,250,000,000

Section 8.7

1. converges; 10

3. diverges

5. converges; $\frac{3}{2}$

7. diverges

9. converges; $2\sqrt{2}$

11. diverges

13. converges; 0

15. diverges

17. diverges

19. diverges

21. converges; 0

23. converges; $\pi/2$

25. diverges

27. converges; $2^{-\pi+1}/(\pi - 1)$

29. diverges

31. converges; $\frac{1}{2}$

33. diverges

35. diverges

37. converges; $1/[2(\ln 2)^2]$

39. diverges

41. diverges

43. diverges

45. converges; 1

47. diverges

49. converges; $\pi/2$

51. converges; $\pi/4$

53. diverges

55. diverges

57. converges; 0

59. converges; π

61. b. no

63. finite area: $\frac{1}{3}$

65. infinite area

67. infinite area

69. converges

71. converges

73. diverges

83. $1/c$

85. a. approximately 8060 years
   b. approximately 2090 years

87. $g(x) = 1.5cx^{-2.5}$ for $x \geq s$

Chapter 8 Review Exercises

1. $x\ln(x^2 + 9) - 2x + 6\tan^{-1}x/3 + C$

3. $-x\cot x + \ln|\sin x| + C$

5. $x\sinh x - \cosh x + C$

7. $\frac{1}{4}x^2 + \frac{1}{4}x\sin 2x + \frac{1}{8}\cos 2x + C$

9. $\frac{1}{6}\sin^2 x^3 + C$

11. $\frac{1}{4}\tan^4 x - \frac{1}{2}\tan^2 x - \ln|\cos x| + C$

13. $\frac{1}{2}x^2\sin x^2 + \frac{1}{2}\cos x^2 + C$

15. $-\frac{2}{81}(1 - 3t)^{3/2} + \frac{4}{135}(1 - 3t)^{5/2} - \frac{2}{189}(1 - 3t)^{7/2} + C$

17. $-\csc x + \cot x + x + C$

19. $\dfrac{8}{243} \cdot \dfrac{(x + 1)^3}{x^2 + 2x + 10)^{3/2}} + \dfrac{2}{3} \cdot \dfrac{1}{(x^2 + 2x + 10)^{3/2}} +$

$$\dfrac{1}{81} \cdot \dfrac{x + 1}{(x^2 + 2x + 10)^{1/2}} + C$$

21. $x - \dfrac{3}{2}\tan^{-1}x + \dfrac{1}{2} \cdot \dfrac{x}{x^2 + 1} + C$

23. $\frac{2}{3}\ln|x + 6| + \frac{1}{3}\ln|x - 3| + C$

25. improper; diverges

27. proper; converges; $\sqrt{2}\pi/4 - \ln(\sqrt{2} + 1)$

29. proper; $2 + \ln\frac{4}{9}$ **31.** proper; $\frac{9}{560}\sqrt{2}$

33. proper; $\frac{1}{4}$

35. proper; $\ln(\sqrt{2} + 1) - \frac{1}{2}\sqrt{2}$

37. proper; $\pi/9$

39. proper; $\sqrt{3} + \frac{1}{2}\ln(2 + \sqrt{3})$

41. improper; diverges

43. improper; diverges

45. improper; converges; $-\frac{1}{4}$

47. improper; diverges

49. improper; converges; $\frac{1}{2}$

51. improper; converges; $\frac{1}{2}\ln 2$

53. $\frac{1}{2}(\ln x)^2 + C$

57. $\frac{1}{2} \ln |\sec^2 x + \sec x \tan x| - \frac{1}{2} \cos x + \frac{1}{2}x + C$

59. a. 1.518524414   **b.** 1.55008698

61. a. 1.007085359   **b.** 1.007133034

63. $9\pi/4$

65. infinite area

67. infinite area

69. finite area: $\frac{1}{4}$

Cumulative Review Exercises (Chapters 1–7)

1. $\frac{1}{6}$ **2.** e^{-3} **3.** 1

4. $\dfrac{-5}{2(3x-2)^{3/2}(x+1)^{1/2}}$ **5.** $\dfrac{e^{3x} + 2e^{2x}}{(e^x + 1)^2}$

6. $f^{(5)}(x) = \dfrac{120}{(1-x)^6}$ **8.** $\frac{3}{4}$ radian per second

9. approximately 1.146193221

10. relative maximum value: $f(0) = 0$; relative minimum value: $f(\frac{4}{3}) = (-\frac{2}{3})2^{2/3} \approx -1.06$; inflection point: (2, 0)

11. relative minimum value: $f(-3) = -\frac{1}{12}$; inflection point: $(-6, -2/27)$

12. $\frac{3}{4}\sqrt{3}a^2$ **13.** isosceles triangle

14. $\dfrac{-24 \ln 0.01}{\ln 2} \approx 159.453$ hours

15. $\ln |x^2 + 4x - 5| + C$

16. $2\sqrt{\sin x} + \sin x + C$ **17.** $\frac{1}{2}e + e^{-1/2} - \frac{3}{2}$

18. $\frac{1}{6}\pi$ **19.** $10\sqrt{10} - 2\sqrt{2}$

20. $\frac{2}{75}\pi (\sqrt{2} + 1)$

CHAPTER 9
Section 9.1

1. $p_0(x) = 0, p_1(x) = x, p_2(x) = x,$ and $p_3(x) = x - \frac{1}{6}x^3$

3. $p_0(x) = 1, p_1(x) = 1 - 2x, p_2(x) = 1 - 2x + 2x^2,$ and $p_3(x) = 1 - 2x + 2x^2 - \frac{4}{3}x^3$

5. $p_0(x) = 0, p_1(x) = 2x, p_2(x) = 2x,$ and $p_3(x) = 2x - \frac{4}{3}x^3$

7. $-2 - x + x^2$

9. $1 - x + x^2 - x^3 + \cdots + (-1)^n x^n$

11. $p_n(x) = 1 - x + \dfrac{x^2}{2!} - \dfrac{x^3}{3!} + \cdots + \dfrac{(-1)^n}{n!} x^n$

13. $p_{2n+1}(x) = p_{2n}(x) = 1 + \dfrac{x^2}{2!} + \dfrac{x^4}{4!} + \dfrac{x^6}{6!} + \cdots + \dfrac{x^{2n}}{(2n)!}$

15. $x - \dfrac{x^3}{3!} + \dfrac{x^5}{5!} - \dfrac{x^7}{7!} + \cdots + \dfrac{(-1)^n}{(2n+1)!} x^{2n+1}$

17. $p_2(x) = 0$ for all x

19. $p_3(x) = 1 - x^2$

21. $p_2(x) = -\frac{1}{2}x^2$

23. $p_3(x) = p_2(x) = 1 + \frac{1}{2}x^2$

25. $p_2(x) = 0$

27. a. $p_2(x) = 1 + \frac{1}{2}x - \frac{1}{8}x^2$

     b. 1.375   **c.** 1.04875

Section 9.2

1. $\frac{1}{3}, \frac{1}{4}, \frac{1}{5}, \frac{1}{6}$ **3.** $0, \frac{1}{3}, \frac{1}{2}, \frac{3}{5}$ **13.** π

15. 0   **17.** 0   **19.** 0

21. $-\infty$ **23.** π **25.** $e^{0.05}$

27. 1   **29.** $\pi/4$ **31.** 0

33. diverges   **35.** converges; 0

37. converges; 0   **39.** 9

41. 183   **43.** 0   **45.** $\frac{1}{2}$

49. approximately 112 million; better with the Verhulst sequence

51. $a_n = 1000[1 + (.01)r]^n$

53. b. $Pe^{(0.01)r}$ **c.** approximately 32¢

Section 9.3

1. bounded   **3.** unbounded   **5.** unbounded

7. 2   **9.** 0   **11.** 0

13. 1   **15.** $\frac{2}{3}$ **17.** 1

19. 1   **21.** $\frac{1}{2}$ **23.** 0

25. 0   **27.** 0   **29.** 1

33. 0.7390851343   **35.** 1

Section 9.4

1. 4   **3.** $\frac{40}{27}$

5. $\frac{13}{60}$ **7.** diverges

9. could converge   **11.** diverges

13. diverges   **15.** $s_j = j$; diverges

17. $s_{2j} = 0, s_{2j+1} = 1$; diverges

19. $s_j = \dfrac{1}{2} - \dfrac{1}{j+2}; \displaystyle\sum_{n=1}^{\infty}\left(\dfrac{1}{n+1} - \dfrac{1}{n+2}\right) = \dfrac{1}{2}$

21. $s_j = 1 - (j+1)^3$; diverges

25. $\frac{20}{3}$

27. $\frac{10}{13}$

29. $\frac{5}{2}$

31. $\frac{405}{2}$

33. diverges

35. $\frac{57}{343} + \sum_{n=4}^{\infty} \left(\frac{1}{7}\right)^n$

37. $\frac{4}{5} + \sum_{n=4}^{\infty} \frac{1}{n^2 + 1}$

39. $\frac{49}{36} + \sum_{n=4}^{\infty} \frac{1}{n^2}$

41. $\frac{8}{11}$

43. $\frac{232}{999}$

45. $\frac{2756}{100} + \frac{123}{99,900} = \frac{917,789}{33,300}$

47. $\frac{864}{1000} = \frac{108}{125}$

49. $\ln 2 - \frac{5}{6}$

51. a. 11   **b.** 31

53. One of the series does not converge.

55. $c_0/(1 - e^{-bt_0})$

57. approximately 0.846 second

59. $w_n = \sum_{k=1}^{n} 2500(1 - p)^k$ **61. a.** $\frac{2}{5}\sqrt{3}$

63. $\frac{1000}{9}$

65. a. yes; between 10 and 11 minutes

   b. between 272,400,599 and 272,400,600 minutes

Section 9.5

1. converges

3. converges

5. diverges

7. converges

9. converges

11. diverges

13. converges

15. converges

17. diverges

19. converges

21. converges

23. converges

25. diverges

27. 5

29. 6.44×10^{17}

33. $p > 1$

37. 1

39. a. infinite   **b.** finite

Section 9.6

1. diverges

3. diverges

5. converges

7. diverges

9. converges

11. converges

13. diverges

15. converges

17. diverges

19. converges

21. converges

23. converges

27. b. $\frac{2}{19}$

29. b. $\sum_{n=1}^{\infty} n^n$

Section 9.7

1. converges

3. diverges

5. converges

7. converges

9. converges

11. diverges

13. diverges

15. $p > 0$

17. $\frac{1}{121}$

19. $1/\sqrt{15}$

21. 0.9011164764

23. converges conditionally

25. diverges

27. diverges

29. converges absolutely

31. converges absolutely

33. converges absolutely

39. $\sum_{n=1}^{\infty} (-1)^n (1/\sqrt{n})$

Section 9.8

1. $[-1, 1]$

3. $[-3, 3)$

5. $[-1, 1]$

7. $(-1, 1]$

9. $(-\infty, \infty)$

11. $(-\infty, \infty)$

13. $\left[-\frac{3}{2}, \frac{3}{2}\right)$

15. $(-1, 1)$

17. $[-1, 1]$

19. $(-1, 1)$

21. e

23. 1

25. 3

27. $f'(x) = \sum_{n=1}^{\infty} n(n + 1)x^{n-1};\ \int_0^x f(t)\,dt = \sum_{n=1}^{\infty} x^{n+1}$

29. $f'(x) = \sum_{n=1}^{\infty} 5nx^{n^2-1}$;

$$\int_0^x f(t)\,dt = \sum_{n=1}^{\infty} \frac{5}{n(n^2 + 1)} x^{n^2 + 1}$$

31. 0.8131655739

33. 0.7968055556

35. 0.0308035714

37. 1.4626369

41. c. $f(x) = \sin x,\ g(x) = \cos x$ **43.** 1

45. c. 0.6885416667

47. 2.521×10^{-7}

49. b. $\sum_{n=0}^{\infty} \frac{(-1)^n}{4n + 3}\left(\frac{1}{2}\right)^{4n+3}$

51. a. $\sum_{n=0}^{\infty} \frac{1}{n!} x^{n+1}$

53. 3.1415926536

57. 4.5

Section 9.9

1. $43 - 26(x + 3) + 4(x + 3)^2$

3. $\sum_{n=0}^{\infty} \frac{e^2}{n!} (x - 2)^n$

5. $\ln 2 + \sum_{n=1}^{\infty} \frac{(-1)^{n+1}}{n2^n} (x - 2)^n$

7. $\sum_{n=0}^{\infty} \frac{(-1)^n 2^{2n+1}}{(2n + 1)!} x^{2n+1}$

9. $\ln 3 + \sum_{n=0}^{\infty} \frac{(-1)^n}{n + 1} (x - 1)^{n+1}$

11. $\sum_{n=0}^{\infty} \frac{(-1)^n}{n + 1} x^{2n+3}$ **13.** $\sum_{n=0}^{\infty} \frac{(\ln 2)^n}{n!} x^n$

15. $\sum_{n=0}^{\infty} \frac{(-1)^n}{2^{n+1}} (x - 1)^{n+1}$

17. $1 + \sum_{n=1}^{\infty} \frac{(-1)^n 2^{2n-1}}{(2n)!} x^{2n}$

19. $\sum_{n=0}^{\infty} \frac{(-1)^n}{(2n + 1)!} x^{2n}$

21. $\frac{\sqrt{3}}{2} \sum_{n=0}^{\infty} (-1)^n \frac{(x - \pi/3)^{2n}}{(2n)!} +$

$\frac{1}{2} \sum_{n=0}^{\infty} (-1)^n \frac{(x - \pi/3)^{2n+1}}{(2n + 1)!}$

23. 1.6484375

25. 0.0953103333

27. 0.9009801767

29. 0.6913580247

31. 4

33. 10

37. $\sum_{n=0}^{\infty} (2^n + 1)x^n$ **39. b.** $\sum_{n=0}^{\infty} \frac{4^n n!}{(2n + 1)!} x^{2n+1}$

Section 9.10

1. $1 + \frac{1}{2}(\frac{1}{20}) - \frac{1}{8}(\frac{1}{20})^2 \approx 1.0246875$

3. $3(1 + \frac{1}{4}(\frac{2}{81})) \approx 3.018518519$

5. $2(1 + \frac{1}{6}(\frac{1}{64})) \approx 2.005208333$

7. $\sum_{n=0}^{\infty} \binom{-\frac{1}{2}}{n} x^n; \ 1 - \frac{1}{2}x + \frac{3}{8}x^2 - \frac{5}{16}x^3$

9. $\sum_{n=0}^{\infty} \binom{-\frac{8}{5}}{n} x^n; \ 1 - \frac{8}{5}x + \frac{52}{25}x^2 - \frac{312}{125}x^3$

11. $\sum_{n=0}^{\infty} (-1)^n \binom{-\frac{1}{2}}{n} x^{2n+1}; \ x + \frac{1}{2}x^3 + \frac{3}{8}x^5 + \frac{5}{16}x^7$

13. $\sum_{n=0}^{\infty} (-1)^n \binom{\frac{1}{2}}{n} (x+1)^{2n};$

$1 - \frac{1}{2}(x+1)^2 - \frac{1}{8}(x+1)^4 - \frac{1}{16}(x+1)^6$

17. a. $\sum_{n=0}^{\infty} \binom{\frac{1}{2}}{n}\left(\frac{1}{a}\right)^{2n-1} \frac{1}{2n+1} x^{2n+1};$ radius of convergence: a

b. $\sum_{n=0}^{\infty} (-1)^n \binom{\frac{1}{2}}{n}\left(\frac{1}{a}\right)^{2n-1} \frac{1}{2n+1} x^{2n+1};$ radius of convergence: a

21. b. 5; 1.089923093   **23.** 1.25 miles

25. 508.402 feet

Chapter 9 Review Exercises

1. -1 **3.** e^e **5.** 1

7. converges   **9.** converges   **11.** diverges

13. converges   **15.** converges   **17.** $\frac{90,349}{3330}$

19. a. yes **b.** not possible

21. e. 1.45   **23.** -0.0495224715

29. $(-\sqrt[3]{25/3}, \sqrt[3]{25/3})$ **31.** $4/e^2$

33. $p_2(x) = 1$

35. $\sum_{n=0}^{\infty} (-1)^n \frac{\sqrt{2}}{2} \left[\frac{(x - \pi/4)^{2n}}{(2n)!} + \frac{(x - \pi/4)^{2n+1}}{(2n+1)!} \right]$

37. $1 - \sum_{n=0}^{\infty} 2(-1)^n x^n$

39. 2.030517578   **41. a.** $-\frac{1}{3}$

43. a. $\frac{48}{25}$ **b.** converges; 2   **c.** 0   **d.** $2/[n(n-1)]$

47. $\frac{19}{80} = 0.2375$

Cumulative Review Exercises (Chapters 1–8)

1. $-\infty$ **2.** 0   **3.** ∞ **4.** $1/(2x\sqrt{x-1})$

5. a. $(\frac{5}{3}, \infty)$ **b.** $3/[(3x-4) \ln (3x-4)]$

6. 0   **7.** $\ln \frac{3}{2}$

8. relative maximum value: $f(2) = 4e^{-2}$; relative minimum value: $f(0) = 0$; inflection points: $(2 - \sqrt{2}, 16 - 4\sqrt{2}e^{\sqrt{2}-2})$ and $(2 + \sqrt{2}, 16 + 4\sqrt{2}e^{-2-\sqrt{2}})$

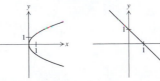

9. $1/(4\pi)$ foot per second   **10.** $\pi p^3/54$

11. $\frac{1}{10}e^{-3x}(-3 \cos x + \sin x) + C$

12. $\frac{5}{16}$ **13.** $\frac{1}{8}(\sqrt{2} - \ln (\sqrt{2} + 1))$

14. $6 \ln 3 - 4 \ln 5 - \frac{3}{20}$ **15.** converges; 1

16. diverges   **17.** $(3 \ln 2 - 1, \frac{33}{8})$

18. 4   **19.** $\frac{112,000}{3}$ foot-pounds

CHAPTER 10

Section 10.1

1. a. $(\frac{3}{2}\sqrt{2}, \frac{3}{2}\sqrt{2})$ **b.** $(-\sqrt{3}, 1)$

c. $(\frac{3}{2}, \frac{3}{2}\sqrt{3})$ **d.** $(5, 0)$

e. $(0, -2)$ **f.** $(0, 2)$

g. $(-2\sqrt{2}, 2\sqrt{2})$ **h.** $(0, 0)$

i. $(-\frac{1}{2}, \frac{1}{2}\sqrt{3})$ **j.** $(-\frac{1}{2}, -\frac{1}{2}\sqrt{3})$

k. $(0, -1)$ **l.** $(-\frac{3}{2}\sqrt{3}, -\frac{3}{2})$

3. $r = \dfrac{4}{2 \cos \theta + 3 \sin \theta}$ **5.** $r = \dfrac{1}{\sqrt{8 \sin^2 \theta + 1}}$

7. $r^2 = \cos 2\theta$ **9.** $r = 4 \sin \theta$

11. $r = \dfrac{3 \cos^2 \theta - \sin^2 \theta}{\cos \theta}$ **13.** $x^2 + y^2 = 3x$

15. $x = 3y$ **17.** $(x^2 + y^2)^3 = 4x^2y^2$

21.   **23.**   **25.**

27.   **29.**   **31.**

33.   **35.**   **37.** symmetry with respect to the x axis

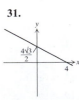

39. symmetry with respect to the y axis

41. symmetry with respect to the y axis

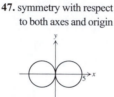

43. symmetry with respect to both axes and origin

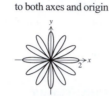

45. symmetry with respect to both axes and origin

47. symmetry with respect to both axes and origin

49. symmetry with respect to the y axis

51.

53. a. symmetry with respect to both axes and origin

    b. symmetry with respect to both axes and origin

    c. symmetry with respect to the x axis

55. b. a rose with n leaves

59. $r^2 = 2 \cos 2\theta$

Section 10.2

1. 2π **3.** 16 **5.** $\pi + 3\sqrt{3}/8$

7. 16π **9.** $9\pi/4$ **11.** π

13. $81\pi/16$ **15.** $\pi/16$ **17.** 6π

19. $41\pi/2$ **21.** 50 **23.** 24π

25. $3\pi/4$ **27.** $\pi - 1$ **29.** $17\pi/4$

31. $7\pi/12 - \sqrt{3}$ **33.** $2\sqrt{2}$

35. a. L is multiplied by c.   **b.** A is multiplied by c^2.

39. $2\pi(2 - \sqrt{2})$

Section 10.3

1. $y^2 = -8x$

3. $y^2 = 12(x - 1)$

5. $y^2 = -x$

7. $\dfrac{x^2}{9} + \dfrac{y^2}{5} = 1$

9. $x^2 + y^2 = 25$

11. $\dfrac{x^2}{16} - \dfrac{y^2}{65} = 1$

13. $\dfrac{(y - 4)^2}{16} - \dfrac{x^2}{9} = 1$

15. focus: $(\frac{3}{4}, 0)$; vertex: $(0, 0)$; directrix: $x = -\frac{3}{4}$; axis: $y = 0$

17. focus: $(1, -\frac{7}{4})$; vertex: $(1, -2)$; directrix: $y = -\frac{9}{4}$; axis: $x = 1$

19. foci: $(0, -4)$ and $(0, 4)$; vertices: $(0, -5)$ and $(0, 5)$

21. foci: $(1, -\sqrt{3}/2)$ and $(1, \sqrt{3}/2)$; vertices: $(1, -1)$ and $(1, 1)$

23. foci: $(-5, 0)$ and $(5, 0)$; vertices: $(-3, 0)$ and $(3, 0)$; asymptotes: $y = \frac{4}{3}x$ and $y = -\frac{4}{3}x$

25. foci: $(-16, -1)$ and $(10, -1)$; vertices: $(-8, -1)$ and $(2, -1)$; asymptotes: $y + 1 = \frac{12}{5}(x + 3)$ and $y + 1 = -\frac{12}{5}(x + 3)$

27. focus: $(3, -3.5)$; directrix: $y = -4.5$

29. focus: $(-\frac{7}{12}, -\frac{1}{2})$; directrix: $x = -\frac{19}{12}$

31. foci: $(1 - \sqrt{2}, 1)$ and $(1 + \sqrt{2}, 1)$; vertices: $(-1, 1)$ and $(3, 1)$

33. foci: $(4 - \sqrt{2}, 0)$ and $(4 + \sqrt{2}, 0)$; vertices: $(2, 0)$ and $(6, 0)$

35. vertices: $(-6, 6)$ and $(0, 6)$; asymptotes: $y - 6 = x + 3$ and $y - 6 = -(x + 3)$

37. vertices: $(-2, -2)$ and $(4, -2)$; asymptotes: $y + 2 = \frac{2}{3}(x - 1)$ and $y + 2 = -\frac{2}{3}(x - 1)$

39. $(x + 1)^2 = 2(y - 3)$

43. $y - 8 = -4(x + 4)$ and $y - 2 = 2(x - 2)$

45. d: -4 and 4; points of tangency: $(-1, -2)$ and $(1, 2)$

47. $(x - 5)^2/16 + y^2/12 = 1$

49. $(\pm a\sqrt{2}/2, \pm b\sqrt{2}/2)$

55. $(x - 300)^2 = -450(y - 200)$

57. approximately 0.066434 mile

59. at $(3375, 825\sqrt{35})$ or $(3575, -825\sqrt{35})$ if Marian is at $(4400, 0)$ and Jack is at $(-4400, 0)$.

61. $x^2 = \dfrac{765{,}625}{79}\, y$

Section 10.4

1. $Y^2/8 - X^2/8 = 1$; hyperbola

3. $X^2/4 + 3Y^2/4 = 1$; ellipse

5. $X^2/4 + Y^2/9 = 1$; ellipse

7. $(Y - 1)^2 = 3X$; parabola

9. $2(Y - 1)^2 - (X - 1)^2 = 1$; hyperbola

15. a hyperbola or two intersecting lines

17. $A' = A \cos^2 \theta + B \sin \theta \cos \theta + C \sin^2 \theta$; $C' = A \sin^2 \theta - B \sin \theta \cos \theta + C \cos^2 \theta$

Section 10.5

1. $\frac{4}{5}$ **3.** $\sqrt{34}/3$ **5.** $\sqrt{3}/2$

7. $\sqrt{3}/2$ **9.** $\sqrt{2}/2$ **11.** $\sqrt{58}/3$

13. $x^2/225 + y^2/144 = 1$ **15.** $4y^2 - 4x^2/3 = 1$

17. $(y - 5)^2/25 - x^2/144 = 1$

19. $x^2/16 - y^2/256 = 1$ or $4y^2/251 - x^2/1004 = 1$

21. $(x - 2)^2/16 + y^2/12 = 1$ or $(x + 2)^2/16 + y^2/12 = 1$

23. hyperbola; directrix: $x = -\frac{1}{2}$

25. ellipse; directrix: $y = -\frac{25}{3}$

27. parabola; directrix: $x = -\frac{1}{2}$

29. parabola; directrix: $y = 3$

31. $x^2/(149.5)^2 + y^2/(149.48)^2 = 1$

Chapter 10 Review

1.

3.

5. a.   **b.** $(0, 0)$, $(\sqrt{3}, \pi/3)$, and $(-\sqrt{3}, 5\pi/3)$

7. $\pi/2 - \frac{1}{2}$

9. $(y - 4)^2 = -16(x - 1)$

11. $x^2/4 + y^2/8 = 1$

13. $(x + 1)^2/4 - (y - 2)^2/4 = 1$

15. $(x + 6)^2/16 - y^2/48 = 1$

17.

19.   **21.**

23. a. $4(y + \frac{1}{2})^2/9 - (x - 2)^2/18 = 1$
   b. $(x - 2)^2 = 8(y - 2)$
   c. $3(x - 2)^2/16 + 9(y - \frac{16}{3})^2/64 = 1$

25. ellipse

27. $y - \dfrac{(a + \sqrt{a^2 - 4bc})^2}{4c} =$

$\dfrac{a + \sqrt{a^2 - 4bc}}{2c}(x - a - \sqrt{a^2 - 4bc})$

and $y - \dfrac{(a - \sqrt{a^2 - 4bc})^2}{4c} =$

$\dfrac{a - \sqrt{a^2 - 4bc}}{2c}(x - a + \sqrt{a^2 - 4bc})$

Cumulative Review Exercises (Chapters 1–9)

1. $1/\sqrt{3}$ **2.** -2 **3.** $\frac{8}{3}$

4. $-(1/x) \sin (\ln x)$

5. $\sin \sqrt{-x^3 + 1} + 2x \sin \sqrt{x^6 + 1}$

6. 12 units per minute

7. a. -5 feet per second   **b.** yes

8. maximum occurs at $e^{1/n}$

10. relative maximum value: $f(\frac{1}{2}) = \frac{1}{2}e^{-1} = 1/(2e)$;
   inflection point: $(1, e^{-2})$

12. $f^{-1}(x) = 1/\ln (1/x - 1)$ for $0 < x < 1$ and $x \neq \frac{1}{2}$

13. $\frac{87}{5}$ **14.** $2 - 2^{3/4}$

15. $\frac{1}{2}x^2e^{2x} - \frac{1}{2}xe^{2x} + \frac{1}{4}e^{2x} + C$

16. $\dfrac{x}{4\sqrt{4 - x^2}} + C$ **17.** diverges

18. converges absolutely   **19.** $[-4, 4)$

CHAPTER 11
Section 11.1

1. 2   **3.** 13

5. $3\sqrt{3}$ **7.** $|\sec x|$

11. $(x - 2)^2 + (y - 1)^2 + (z + 7)^2 = 25$

13. radius: 2; center: $(1, 2, -3)$

15. $x^2 + (y + 2)^2 + (z + 3)^2 \leq 36$

19. $4x - 4y - 6z = 21$

23. $(x - 3)^2 + y^2 + (z - 5)^2 = 6$

Section 11.2

1. $3\mathbf{i} - 4\mathbf{j} + 10\mathbf{k}$ **3.** $3\mathbf{i} - 2\mathbf{j} + \sqrt{7}\,\mathbf{k}$

5. $\mathbf{a} + \mathbf{b} = \mathbf{i} - 3\mathbf{j} + \mathbf{k}$; $\mathbf{a} - \mathbf{b} = 3\mathbf{i} - 7\mathbf{j} + 19\mathbf{k}$;
   $c\mathbf{a} = 4\mathbf{i} - 10\mathbf{j} + 20\mathbf{k}$

7. $\mathbf{a} + \mathbf{b} = 2\mathbf{i} + \mathbf{j} + \mathbf{k}$; $\mathbf{a} - \mathbf{b} = 2\mathbf{i} - \mathbf{j} - \mathbf{k}$; $c\mathbf{a} = \frac{2}{3}\mathbf{i}$

9. $\sqrt{3}$ **11.** 13

13. 2   **15.** $-\frac{3}{13}\mathbf{i} + \frac{4}{13}\mathbf{j} - \frac{12}{13}\mathbf{k}$

17. $\dfrac{2}{\sqrt{13}}\mathbf{i} - \dfrac{3}{\sqrt{13}}\mathbf{j}$

21. $4 \times 10^{-3}\mathbf{i} + 0.51\mathbf{j} - 3.8 \times 10^4\mathbf{k}$

27. $\mathbf{F} = -50\mathbf{i} + (20 - 50\sqrt{3})\mathbf{j}$; $\sqrt{3} - \frac{2}{5}$

29. $\sqrt{96{,}400} \approx 310$ miles per hour

31. $\dfrac{100\sqrt{2}}{\pi\varepsilon_0}(100\mathbf{i} + 99\mathbf{j} - \mathbf{k})$

Section 11.3

1. $\mathbf{a} \cdot \mathbf{b} = -5$; $\cos \theta = -5/\sqrt{87}$
3. $\mathbf{a} \cdot \mathbf{b} = -2 - 2\sqrt{3}$; $\cos \theta = -\frac{2}{3}(1 + \sqrt{3})/\sqrt{21}$
5. $\pi/6$　　　　　　　　7. not perpendicular
9. perpendicular　　　　11. $\frac{4}{9}\mathbf{i} - \frac{2}{9}\mathbf{j} + \frac{4}{9}\mathbf{k}$
13. $-5\mathbf{i}$　　　　　　15. $\mathbf{b} = 3\mathbf{a} - 5\mathbf{a}'$
17. $\mathbf{b} = -\mathbf{a} + \frac{3}{2}\mathbf{a}'$　　23. **a.** 0　**b.** π　**c.** $\pi/2$
29. approximately 1.231 radians ($\approx 70.53°$)
33. $25{,}000\sqrt{3}$ foot-pounds

Section 11.4

1. $\mathbf{i} - \mathbf{j} + \mathbf{k}$; 6　　　　3. $19\mathbf{i} - 9\mathbf{j} + 11\mathbf{k}$; 55
5. $\mathbf{0}$; 0　　　　　　11. 15
13. $A = \frac{1}{2}\|\overrightarrow{PQ} \times \overrightarrow{QR}\| = \frac{1}{2}\|\overrightarrow{PQ} \times \overrightarrow{PR}\| = \frac{1}{2}\|\overrightarrow{PR} \times \overrightarrow{QR}\|$
15. yes　　　　　　　21. $-24\sqrt{3}\mathbf{i}$

Section 11.5

1. $\mathbf{r} = (-2 + 3t)\mathbf{i} + (1 - t)\mathbf{j} + 5t\mathbf{k}$;
   $x = -2 + 3t, y = 1 - t, z = 5t$
   $$\frac{x + 2}{3} = \frac{y - 1}{-1} = \frac{z}{5}$$
3. $\mathbf{r} = (3 + \frac{1}{2}t)\mathbf{i} + (4 - \frac{1}{3}t)\mathbf{j} + (5 + \frac{1}{6}t)\mathbf{k}$;
   $x = 3 + \frac{1}{2}t, y = 4 - \frac{1}{3}t, z = 5 + \frac{1}{6}t$;
   $$\frac{x - 3}{\frac{1}{2}} = \frac{y - 4}{-\frac{1}{3}} = \frac{z - 5}{\frac{1}{6}}$$
5. $\mathbf{r} = 2\mathbf{i} + 2t\mathbf{j} + (5 + 3t)\mathbf{k}$; $x = 2, y = 2t, z = 5 + 3t$;
   $x = 2, y/2 = (z - 5)/3$
7. $\mathbf{r} = 4\mathbf{i} + (2 + t)\mathbf{j} - \mathbf{k}$; $x = 4, y = 2 + t, z = -1$;
   $x = 4, z = -1$
9. $x = -1 - t, y = 1 + 4t, z = 7t$
11. $x = -1, y = 1$
17. $x = -1 + \frac{3}{2}t, y = -2 + \frac{1}{2}t, z = -3 + \frac{3}{2}t$
19. $5\sqrt{13}/14$　　21. $3\sqrt{39}/19$　　23. $5\sqrt{2}/3$
25. $x^2 + z^2 = 2$　　　27. no

Section 11.6

1. $8x - 30y + z = -65$　　3. $2x - 3z = 39$
5. $y = 3$　　　　　　　7. $7x - y - 3z = 2$
9. $4x + 10y + 18z = 19$
11. **a.** $\mathbf{r} = (1 + t)\mathbf{i} + 2t\mathbf{j} + t\mathbf{k}$
    b. $x + 2y + z = 29$
13. $\sqrt{6}$　　　　　　17. $x - y - z = 0$
19. 　　　　　　　　　21.

27. The planes in (a) and (b) are perpendicular, as are the planes in (b) and (d).
29. **a.** $x = -4$　　**b.** $y = -5$　　**c.** $z = -3$
31. $(1, 0, 2)$　　　　33. $(-\frac{1}{3}, -\frac{11}{6}, \frac{23}{6})$
35. $5\sqrt{3}/9$　　　　37. $\frac{11}{58}\sqrt{29}$

Chapter 11 Review Exercises

1. $2\mathbf{a} + \mathbf{b} - 3\mathbf{c} = 5\mathbf{i} - 10\mathbf{j} + 11\mathbf{k}$; $\mathbf{a} \times \mathbf{b} = \mathbf{i} + \mathbf{j} + \mathbf{k}$;
   $\mathbf{c} \cdot (\mathbf{a} \times \mathbf{b}) = -2$; $\mathbf{a} \times (\mathbf{b} \times \mathbf{c}) = -6\mathbf{i} + \mathbf{j} + 15\mathbf{k}$
3. $2\mathbf{a} + \mathbf{b} - 3\mathbf{c} = 11\mathbf{i} - 9\mathbf{j} + 6\mathbf{k}$; $\mathbf{a} \times \mathbf{b} = 2\mathbf{j} + 4\mathbf{k}$;
   $\mathbf{c} \cdot (\mathbf{a} \times \mathbf{b}) = -2$; $\mathbf{a} \times (\mathbf{b} \times \mathbf{c}) = -15\mathbf{i} - 14\mathbf{j} + 17\mathbf{k}$
5. $(3, -4, 4)$
7. $-\frac{3}{5}\mathbf{a} - \frac{1}{10}\mathbf{a}'$
9. $$\frac{x - 1}{-3} = \frac{y - 1}{-4} = \frac{z + 1}{6}$$
11. $\mathbf{r} = (-3 + 2t)\mathbf{i} + (-3 - 3t)\mathbf{j} + (1 + 4t)\mathbf{k}$
15. $5x - 2y = 17$　　　　17. $\frac{60}{13}$
19. **a.** point $(\frac{5}{3}, 4, 1)$　　**b.** point $(\frac{1}{2}, 1, \frac{1}{2})$
    c. line $x = 2 - \frac{1}{2}t, y = t, z = -1 + 2t$
21. $ax + by + cz = a^2 + b^2 + c^2$
31. **a.** $\pi/3$
    b. approximately 0.927295218 radian (≈ 53 degrees)
33. $100\sqrt{26} \approx 509.902$ miles per hour

Cumulative Review Exercises (Chapters 1–10)

1. **a.** union of $(-\infty, \frac{1}{3})$ and $[\frac{1}{2}, \infty)$　　**b.** $\sqrt{\frac{2}{3}}$
2. 0　　　　　3. 1　　　　4. $-\frac{1}{2}$ and $-\frac{1}{6}$
5. **a.** domain: $(-\infty, \ln \pi)$; range: $(-\infty, \infty)$
   b. $\frac{4}{9}\pi$　　**c.** $f^{-1}(x) = \tan(e^x - \pi/2)$ for $x < \ln \pi$
6. $y - \frac{1}{2}\sqrt{2} = -\sqrt{2}(x - \frac{1}{4}\pi)$
8. **a.** $32\sqrt{3}$ feet per second　　**b.** $96\sqrt{2}$ feet per second
9. relative maximum values: $f(-\sqrt{5}) = 0$ and $f(1) = 16$;
   relative minimum values: $f(-1) = -16$ and $f(\sqrt{5}) = 0$;
   inflection points: $(-\sqrt{3}, -4\sqrt{3})$, $(0, 0)$, and $(\sqrt{3}, 4\sqrt{3})$

11. $1 + \ln \frac{3}{4}$　　　　12. $-\sec(1/t) + C$
13. $\frac{1}{2}x^2 + x - \dfrac{2}{\sqrt{3}} \tan^{-1} \dfrac{2x - 1}{\sqrt{3}} + C$
14. $4\sqrt{2}\left(1 - \dfrac{1}{2x^2}\right)^{1/2} - \dfrac{8}{3}\sqrt{2}\left(1 - \dfrac{1}{2x^2}\right)^{3/2} +$
    $\dfrac{4}{5}\sqrt{2}\left(1 - \dfrac{1}{2x^2}\right)^{5/2} + C$

15. converges; $\frac{1}{6}e^{-12}$

16. $18\pi(2 - \sqrt{2})$

17. $\sqrt{5}\,(e - e^{-1})$

18. $\frac{3}{2}\pi - 4$

19. all real numbers

20. 3

21. diverges

22. converges absolutely

23. $\displaystyle\sum_{n=1}^{\infty} \frac{x^{n+1}}{(n+1)(n+2)}$; interval of convergence: $[-1, 1]$

CHAPTER 12

Section 12.1

1. domain: $(-\infty, \infty)$; $f_1(t) = t$; $f_2(t) = t^2$; $f_3(t) = t^3$

3. domain: union of $(-\infty, -2)$, $(-2, 2)$, and $(2, \infty)$;
   $f_1(t) = \tanh t$; $f_2(t) = 0$; $f_3(t) = -1/(t^2 - 4)$

5. domain: $[0, \infty)$; $f_1(t) = 2\sqrt{t}$; $f_2(t) = -2t^{3/2}$; $f_3(t) = -(t^3 + 1)$

7. domain: $(-\infty, \infty)$; $f_1(t) = 2t - 3\cos t$;
   $f_2(t) = 2t^2 - 3\sin t$; $f_3(t) = 2t^3 - 3$

9. domain: $(0, \infty)$; $f_1(t) = (1/\sqrt{t})(1 - \cos t)$;
   $f_2(t) = -(1/\sqrt{t})(t - \sin t)$; $f_3(t) = -t(t - \sin t)$

11. domain: $[-8, \infty)$; $f_1(t) = \cos t^{1/3}$; $f_2(t) = \sin t^{1/3}$;
   $f_3(t) = \sqrt{t^{1/3} + 2}$

13.

15.

17.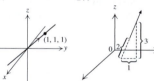

19.

21.

23.

25.

31. **a.** $x = t - 1$; $y = -2t + 3$; $z = -4t$
   b. $x = 5t + 3$; $y = 10t - 5$; $z = 0$
   c. $x = 2\cos t$; $y = \cos t + 1$; $z = -3\cos t$

33. $(-1, 0, 3)$ and $(-1, 0, -3)$

Section 12.2

1. $\mathbf{i} - \mathbf{j} + \mathbf{k}$

3. $3\pi\mathbf{j} - 4\mathbf{k}$

5. $5\mathbf{i} - 4\mathbf{j} + \mathbf{k}$

7. does not exist because $\lim_{t\to 0} (1 + \cos t)/t$ does not exist

9. $\mathbf{i} + 4\mathbf{j} + 10\mathbf{k}$

13. **b.** $\mathbf{j}$

Section 12.3

1. $\mathbf{j} + 5t^4\mathbf{k}$

3. $\frac{3}{2}(1 + t)^{1/2}\mathbf{i} + \frac{3}{2}(1 - t)^{1/2}\mathbf{j} + \frac{3}{2}\mathbf{k}$

5. $\sec^2 t\mathbf{i} + \sec t\tan t\mathbf{k}$

7. $\sinh t\mathbf{i} + \cosh t\mathbf{j} - \dfrac{1}{2\sqrt{t}}\mathbf{k}$

9. $(8\sec t\tan t - 6)\mathbf{i} + 4t\mathbf{j} - 12\csc t\cot t\mathbf{k}$

11. $-\dfrac{3}{t}\mathbf{i} - \left(2\sec t\tan t\ln t + \dfrac{2}{t}\sec t\right)\mathbf{j}$

13. $\left(\dfrac{3}{t^2}e^{-t} + \dfrac{3}{t}e^{-t} + 1\right)\mathbf{k}$

15. $\dfrac{1}{2t}\mathbf{i} - \dfrac{4e^{2\sqrt{t}}}{\sqrt{t}}\mathbf{j} + \dfrac{1}{2}t^{-3/2}\mathbf{k}$

17. $\dfrac{t^3}{3}\mathbf{i} - \left(\dfrac{3}{2}t^2 - t\right)\mathbf{j} + \dfrac{1}{2t^2}\mathbf{k} + \mathbf{C}$

19. $(e - 1)\mathbf{i} + (1 - e^{-1})\mathbf{j} + \mathbf{k}$

21. $\frac{8}{5}\sqrt{2}(\mathbf{i} + \mathbf{j})$

23. $\mathbf{v}(t) = -\sin t\mathbf{i} + \cos t\mathbf{j} - 32 t\mathbf{k}$; $\|\mathbf{v}(t)\| = \sqrt{1 + 1024t^2}$;
   $\mathbf{a}(t) = -\cos t\mathbf{i} - \sin t\mathbf{j} - 32\mathbf{k}$

25. $\mathbf{v}(t) = 2\mathbf{i} + 2t\mathbf{j} + \dfrac{1}{t}\mathbf{k}$; $\|\mathbf{v}(t)\| = \dfrac{2t^2 + 1}{t}$;
   $\mathbf{a}(t) = 2\mathbf{j} - \dfrac{1}{t^2}\mathbf{k}$

27. $\mathbf{v}(t) = (e^t\sin t + e^t\cos t)\mathbf{i} + (e^t\cos t - e^t\sin t)\mathbf{j} + e^t\mathbf{k}$;
   $\|\mathbf{v}(t)\| = e^t\sqrt{3}$; $\mathbf{a}(t) = 2e^t\cos t\mathbf{i} - 2e^t\sin t\mathbf{j} + e^t\mathbf{k}$

29. $\mathbf{r}(t) = t\mathbf{i} + t\mathbf{j} - 16t^2\mathbf{k}$; $\mathbf{v}(t) = \mathbf{i} + \mathbf{j} - 32t\mathbf{k}$;
   $\|\mathbf{v}(t)\| = \sqrt{2 + 1024t^2}$

31. $\mathbf{r}(t) = \cos t\mathbf{i} + (\sin t - t)\mathbf{j} + t\mathbf{k}$;
   $\mathbf{v}(t) = -\sin t\mathbf{i} + (\cos t - 1)\mathbf{j} + \mathbf{k}$;
   $\|\mathbf{v}(t)\| = \sqrt{3 - 2\cos t}$

33. $t\tan t^3\mathbf{i} + \cos e^t\mathbf{j} + e^{(t^2)}\mathbf{k}$

37. There is no such value of t.

41. **a.** $\theta(t) = \cos^{-1}\dfrac{t}{2(1 + t^2)}$ **b.** $\dfrac{\pi}{2}$
   c. approximately 0.9316761004 radian

45. 3200 m

47. **a.** $\mathbf{r}_0 = \mathbf{0}$; $\mathbf{v}_0 = 90\sqrt{2}\mathbf{i} + 90\sqrt{2}\mathbf{j} + 64\mathbf{k}$ **b.** 720 feet

49. **a.** $\sqrt{2.6}/4 \approx 0.40$ seconds
   b. $\sqrt{170.4} \approx 13$ feet per second

51. speed: approximately 6857.41 miles per hour; radius: approximately 22,233 miles

53. a. 6.25×10^{10} kilometers per second per second
   b. 8.41×10^{10} kilometers per second per second

Section 12.4

1. smooth   **3.** piecewise smooth

5. smooth   **7.** piecewise smooth

9. piecewise smooth

11. $\mathbf{r}(t) = (-3 + 7t)\mathbf{i} + (2 - 2t)\mathbf{j} + (1 + 4t)\mathbf{k}$ for $0 \leq t \leq 1$

13. $\mathbf{r}(t) = 6 \cos t\mathbf{i} + 6 \sin t\mathbf{j}$ for $0 \leq t \leq 2\pi$

15. $\mathbf{r}(t) = \cos t\mathbf{i} + \sin t\mathbf{j}$ for $0 \leq t \leq \pi$

17. $\mathbf{r}(t) = \cos t\mathbf{i} - \sin t\mathbf{j} + 4\mathbf{k}$ for $-\pi/4 \leq t \leq \pi/4$

19. $\mathbf{r}(t) = t\mathbf{i} + \tan t\mathbf{j}$ for $0 \leq t \leq \pi/4$

21. 6   **23.** $\sqrt{3}$ **25.** $e - e^{-1}$

27. $54 - 4\sqrt{2}$ **29.** $\sqrt{4 + t}$ **31.** $\sqrt{2 + t^2}$

33. $\sqrt{3 - 2\cos t}$ **35.** $L \approx 6.482559337$

37. $L \approx 0.668455969$ **39.** $\sqrt{5}\,t$

41. b. approximately $\frac{3}{4}\pi$

Section 12.5

1. $\mathbf{T} = \dfrac{t}{\sqrt{t^2 + 1}}\mathbf{i} + \dfrac{1}{\sqrt{t^2 + 1}}\mathbf{j};$

$\mathbf{N} = \dfrac{1}{\sqrt{1 + t^2}}\mathbf{i} - \dfrac{t}{\sqrt{1 + t^2}}\mathbf{j}$

3. $\mathbf{T} = -\dfrac{\sin t}{\sqrt{2}}\mathbf{i} - \dfrac{\sin t}{\sqrt{2}}\mathbf{j} + \cos t\mathbf{k};$

$\mathbf{N} = -\dfrac{\cos t}{\sqrt{2}}\mathbf{i} - \dfrac{\cos t}{\sqrt{2}}\mathbf{j} - \sin t\mathbf{k}$

5. $\mathbf{T} = \dfrac{2}{2 + t^2}\mathbf{i} + \dfrac{2t}{2 + t^2}\mathbf{j} + \dfrac{t^2}{2 + t^2}\mathbf{k};$

$\mathbf{N} = -\dfrac{2t}{2 + t^2}\mathbf{i} + \dfrac{2 - t^2}{2 + t^2}\mathbf{j} + \dfrac{2t}{2 + t^2}\mathbf{k}$

7. $\mathbf{T} = \dfrac{e^t}{e^t + e^{-t}}\mathbf{i} - \dfrac{e^{-t}}{e^t + e^{-t}}\mathbf{j} + \dfrac{\sqrt{2}}{e^t + e^{-t}}\mathbf{k};$

$\mathbf{N} = \dfrac{\sqrt{2}}{e^t + e^{-t}}\mathbf{i} + \dfrac{\sqrt{2}}{e^t + e^{-t}}\mathbf{j} - \dfrac{e^t - e^{-t}}{e^t + e^{-t}}\mathbf{k}$

9. $\mathbf{T} = \dfrac{2t}{2t^2 + 1}\mathbf{i} + \dfrac{2t^2}{2t^2 + 1}\mathbf{j} + \dfrac{1}{2t^2 + 1}\mathbf{k};$

$\mathbf{N} = \dfrac{1 - 2t^2}{2t^2 + 1}\mathbf{i} + \dfrac{2t}{2t^2 + 1}\mathbf{j} - \dfrac{2t}{2t^2 + 1}\mathbf{k}$

11. $a_{\mathbf{T}} = \dfrac{r^2 \sin t}{\sqrt{2r^2(1 - \cos t)}};\ a_{\mathbf{N}} = \dfrac{|r|}{2}\sqrt{1 - \cos t}$

13. $a_{\mathbf{T}} = 2t;\ a_{\mathbf{N}} = 2$

15. $a_{\mathbf{T}} = e^t - e^{-t};\ a_{\mathbf{N}} = \sqrt{2}$

17. $\mathbf{r}(t) = -4 \sin t\mathbf{j} + 4 \cos t\mathbf{k}$ for $0 \leq t \leq \pi$

19. $\mathbf{r}(t) = \begin{cases} t\mathbf{i} + t^2\mathbf{j} & \text{for } 0 \leq t \leq 1 \\ (2 - t)\mathbf{i} + (2 - t)\mathbf{j} & \text{for } 1 \leq t \leq 2 \end{cases}$

21. $\mathbf{r}(t) = \begin{cases} (t + 1)\mathbf{i} \text{ for } 0 \leq t \leq 1 \\ 2 \cos \dfrac{\pi}{2}(t - 1)\mathbf{i} + 2 \sin \dfrac{\pi}{2}(t - 1)\mathbf{j} \\ \qquad\qquad\qquad \text{for } 1 \leq t \leq 2 \\ (4 - t)\mathbf{j} \text{ for } 2 \leq t \leq 3 \\ \sin \dfrac{\pi}{2}(t - 3)\mathbf{i} + \cos \dfrac{\pi}{2}(t - 3)\mathbf{j} \\ \qquad\qquad\qquad \text{for } 3 \leq t \leq 4 \end{cases}$

33. b. approximately 65.5 feet per second (44.6 miles per hour)   **c.** 2000 feet

Section 12.6

1. $\dfrac{1}{2(t^2 + 1)^{3/2}}$ **3.** $\dfrac{1}{\sqrt{2}}$ **5.** $\dfrac{2}{(2 + t^2)^2}$

7. $\dfrac{\sqrt{2}}{(e^t + e^{-t})^2}$ **9.** $\dfrac{2t}{(2t^2 + 1)^2}$ **11.** $\dfrac{1}{2(1 + t^2)^{3/2}}$

13. $\dfrac{2e^t(1 + e^{2t})^{1/2}}{(2e^{2t} + 1)^{3/2}}$ **15.** $\dfrac{2t + 1}{2\sqrt{t}\,(1 + t)^{3/2}}$

17. $\dfrac{6}{(4 + 5 \cos^2 t)^{3/2}}; \dfrac{4}{3}$ **19.** $\dfrac{2|t|}{(1 + t^4)^{3/2}}; \sqrt{2}$

21. $\dfrac{x}{(1 + x^2)^{3/2}}$ **23.** $\dfrac{-2x^3}{(1 + x^4)^{3/2}}$

25. $(-\frac{1}{2}\ln 2, \frac{1}{2}\sqrt{2})$ **31.** $\dfrac{8 \cos^2 3\theta + 10}{(8 \cos^2 3\theta + 1)^{3/2}}$

33. b. $\mathbf{r}(t) = \left(\dfrac{t}{2}\sqrt{1 - t^2} + \dfrac{1}{2}\sin^{-1} t\right)\mathbf{i} + \dfrac{t^2}{2}\mathbf{j}$

   c. $\mathbf{r}(t) = \ln(t + \sqrt{1 + t^2})\mathbf{i} + (\sqrt{1 + t^2} - 1)\mathbf{j}$

35. $a_{\mathbf{T}} = -9;\ a_{\mathbf{N}} = \dfrac{(9 - t)^2}{9}; \|\mathbf{a}\| = \sqrt{81 + \dfrac{(9 - t)^4}{81}}$

Section 12.7

5. $\sqrt{\dfrac{4\pi^2(5000)^3}{1.237 \times 10^{12}}} \approx 1.99733$ hours

11. a. approximately 19,669.2 miles per hour
   b. approximately 11,311.2 miles per hour

15. approximately 0.024316

17. approximately 5.31×10^9 kilometers

Chapter 12 Review Exercises

1.

3.

5. a. $3t^2$ **b.** $\frac{1}{t^2}\mathbf{i} - \frac{3}{t^4}\mathbf{j} + \mathbf{k}$

7. $-\frac{1}{2\pi}\ln|\cos 2\pi t|\mathbf{i} + \frac{1}{2\pi}\tan 2\pi t\mathbf{j} + 4\tan^{-1}t\mathbf{k} + \mathbf{C} \approx$
10.4337 feet per second

9. a. $\sqrt{2}(e^{3\pi} - 1)$ **b.** $\sqrt{2}(e - e^{-2\pi})$

13. $\dfrac{1}{3(1 + t^2)^2}$

15. a. $\kappa = 2x^3/(x^4 + 1)^{3/2}$ **b.** $\kappa(1) = \frac{1}{2}\sqrt{2}$ **c.** $\sqrt{2}$

17. $\mathbf{v} = e^t(\cos t - \sin t)\mathbf{i} + e^t(\sin t + \cos t)\mathbf{j} + e^t\mathbf{k}$; $\mathbf{a} = -2e^t\sin t\mathbf{i} + 2e^t\cos t\mathbf{j} + e^t\mathbf{k}$; $\kappa = \frac{1}{3}\sqrt{2}e^{-t}$; $\rho = \frac{3}{2}\sqrt{2}e^t$

19. $\mathbf{T} = \frac{1}{2}(1 - \cos t)\mathbf{i} + \frac{1}{2}\sin t\,\mathbf{j} + \cos\frac{t}{2}\mathbf{k}$

$\mathbf{N} = \dfrac{\sqrt{2}\sin t}{\sqrt{3 - \cos t}}\mathbf{i} + \dfrac{\sqrt{2}\cos t}{\sqrt{3 - \cos t}}\mathbf{j} - \dfrac{\sqrt{2}\sin t/2}{\sqrt{3 - \cos t}}\mathbf{k}$;

$\kappa = \dfrac{\sqrt{2}}{8}\sqrt{3 - \cos t}$

Cumulative Review Exercises (Chapters 1–11)

1. ∞ **2.** $e^{-1/2}$

3. 0 for $x = 0$; 1 for $x \neq 0$ **5.** $\dfrac{\cos x}{\sqrt{25 - \sin^2 x}}$

6. $(-2\sqrt{5}/5, 8\sqrt{5}/5)$ and $(2\sqrt{5}/5, -8\sqrt{5}/5)$

7. $(0, -3), (4, -4)$ **8.** $\frac{127}{16}$

9. 70 miles per hour **10.** inflection point: $(0, 0)$

11. 20 feet (perpendicular to house) and 40 feet (parallel to house)

12. radius: $\frac{5}{2}$; angle: 2 radians

13. approximately 32.8941 years

14. $2x\sin\sqrt{x} + 4\sqrt{x}\cos\sqrt{x} - 4\sin\sqrt{x} + C$

15. $-\frac{1}{5}\ln|x + 2| + \frac{1}{10}\ln(x^2 + 6) + \dfrac{3}{5\sqrt{6}}\tan^{-1}\dfrac{x}{\sqrt{6}} + C$

16. $\frac{2}{7}(3t - 2)^{7/2} + \frac{8}{5}(3t - 2)^{5/2} + \frac{8}{3}(3t - 2)^{3/2} + C$

17. converges; 2 **18.** $\frac{1}{3}$

19. $46{,}250\pi$ foot-pounds **20.** $2 + \frac{1}{2}\ln 3$

21. diverges **22.** diverges

23. converges absolutely **24.** $(-4/\sqrt{3}, 4/\sqrt{3})$

25. $\displaystyle\sum_{n=1}^{\infty}(-1)^n(2n)x^{2n-1}$; interval of convergence: $(-1, 1)$

26. $-(7/\sqrt{69})\mathbf{i} - (4/\sqrt{69})\mathbf{j} - (2/\sqrt{69})\mathbf{k}$

27. $x + 5y + 7z = 10$

CHAPTER 13

Section 13.1

1. all (x, y) such that $x \geq 0$ and $y \geq 0$

3. all (x, y) such that $x \neq 0$ and $y \neq 0$

5. all (x, y) such that $x^2 + y^2 \geq 25$

7. all (x, y) such that $x + y \neq 0$

9. all (x, y, z) such that $x^2 + y^2 + z^2 \leq 1$

11. all (x, y, z) such that $x \neq 0$, $y \neq 0$, and $z \neq 0$

13.
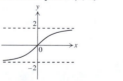

15.

17.

19.

21.

23.

25.

27.

29.
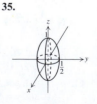

31.
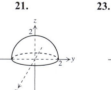

33.

35.

37.

39.

41.

43.

45.

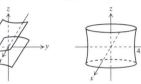

47.

49.

51.

53.

55.

57. surface (d), level curves (ii)

59. surface (c), level curves (iv)

61. $h = V/\pi r^2$ for $r > 0$ and $V > 0$

63. $S = xy + 2xz + 2yz$ for $x > 0$, $y > 0$, and $z > 0$

65. $C = \frac{3}{10}xy$ for $x > 0$ and $y > 0$

67. circular cylinders

69. hyperbolas, or two intersecting lines

Section 13.2

1. $\frac{5}{2}$ **3.** 2   **5.** 2

7. 3   **9.** 1   **11.** 1

13. 0   **15.** 0

17. The limit does not exist.

19. The limit exists and equals 1.

21. The limit does not exist.

31. a. $(x + 3)^2 + (y - 2)^2 = 36$

b. $y = 0$ and $0 \le x \le 2$, $x = 2$ and $-3 \le y \le 0$, $y = -3$
and $0 \le x \le 2$, and $x = 0$ and $-3 \le y \le 0$

c. $y = 1$ and $-1 \le x \le 1$, $y = -6x - 5$ and $-1 \le x \le$
0, and $y = 6x - 5$ and $0 \le x \le 1$

d. the x axis   **e.** $y = 4x^2$ **f.** the origin

33. f is continuous on R.

35. f is not continuous on R.

37. f is continuous on R.

39. The limit exists and equals 0.

Section 13.3

1. $f_x(x, y) = x^{1/2}$; $f_y(x, y) = 0$

3. $f_x(x, y) = 2 + 6xy^4$; $f_y(x, y) = 12x^2y^3$

5. $g_u(u, v) = \dfrac{u^4 + 3u^2v^2 - 2uv^3}{(u^2 + v^2)^2}$;

$g_v(u, v) = \dfrac{v^4 + 3u^2v^2 - 2u^3v}{(u^2 + v^2)^2}$

7. $f_x(x, y) = \dfrac{-x}{\sqrt{4 - x^2 - 9y^2}}$; $f_y(x, y) = \dfrac{-9y}{\sqrt{4 - x^2 - 9y^2}}$

9. $\dfrac{\partial z}{\partial x} = \dfrac{-(1 - x^{2/3})^2}{x^{1/3}\sqrt{(1 - x^{2/3})^3 - y^2}}$;

$\dfrac{\partial z}{\partial y} = \dfrac{-y}{\sqrt{(1 - x^{2/3})^3 - y^2}}$

11. $\dfrac{\partial z}{\partial x} = 6xy \sin^2 x^2y \cos x^2y$;

$\dfrac{\partial z}{\partial y} = 3x^2 \sin^2 x^2y \cos x^2y$

13. $f_x(x, y, z) = 2xy^5 + z^2$; $f_y(x, y, z) = 5x^2y^4$; $f_z(x, y, z) = 2xz$

15. $f_x(x, y, z) = \dfrac{-(y^2 + yz + z^2)}{(xy + yz + zx)^2}$;

$f_y(x, y, z) = \dfrac{-(z^2 + zx + x^2)}{(xy + yz + zx)^2}$;

$f_z(x, y, z) = \dfrac{-(x^2 + xy + y^2)}{(xy + yz + zx)^2}$

17. $\dfrac{\partial w}{\partial x} = e^x(\cos y + \sin z)$; $\dfrac{\partial w}{\partial y} = -e^x \sin y$; $\dfrac{\partial w}{\partial z} = e^x \cos z$

19. $\dfrac{\partial w}{\partial x} = \dfrac{-yz^2}{(1 + xyz^2)^2 \sqrt{1 - \left(\dfrac{1}{1 + xyz^2}\right)^2}}$;

$\dfrac{\partial w}{\partial y} = \dfrac{-xz^2}{(1 + xyz^2)^2 \sqrt{1 - \left(\dfrac{1}{1 + xyz^2}\right)^2}}$;

$\dfrac{\partial w}{\partial z} = \dfrac{-2xyz}{(1 + xyz^2)^2 \sqrt{1 - \left(\dfrac{1}{1 + xyz^2}\right)^2}}$

21. $f_x(2, -3) = \frac{8}{5}$; $f_y(2, -3) = -\frac{3}{5}$

23. $f_x(0, -1, 1) = 2e^3$; $f_y(0, -1, 1) = -4e^3$; $f_z(0, -1, 1) = -e^3$

25. $f_x(0, 0) = 0 = f_y(0, 0)$

27. $f_x(x, y) = 2x \sin (x^2 + y^2)^2$; $f_y(x, y) = 2y \sin (x^2 + y^2)^2$

29. $f_{xy}(x, y) = -2\sqrt{2}y = f_{yx}(x, y)$

31. $f_{xy}(x, y) = \dfrac{-xy}{(x^2 + y^2)^{3/2}} = f_{yx}(x, y)$

33. $f_{xy}(x, y, z) = -z \sin xy - xyz \cos xy = f_{yx}(x, y, z)$

35. $f_{xx}(x, y) = e^{x - 2y}$; $f_{yy}(x, y) = 4e^{x - 2y}$

37. $f_{xx}(x, y, z) = \dfrac{y^2 + z^2}{(x^2 + y^2 + z^2)^{3/2}}$;

$f_{yy}(x, y, z) = \dfrac{x^2 + z^2}{(x^2 + y^2 + z^2)^{3/2}}$;

$f_{zz}(x, y, z) = \dfrac{x^2 + y^2}{(x^2 + y^2 + z^2)^{3/2}}$

39. $y = 1, z - 25 = -6(x + 3)$

41. $\sqrt{2}$ **43.** $\sqrt{2}$

65. $\theta(\mu, i_\mu) \approx 0.734402$ (approximately 42.1°)

Section 13.4

1. $2 - 18e^{6t}$ **3.** $2t(\cos t^2 - \sin t^2)$

5. $\dfrac{1 + 18t^3}{t\sqrt{2 \ln t - 4(1 - 3t^3)}}$

7. $\dfrac{\partial z}{\partial u} = \dfrac{-12 - u^4}{u^4 v}$; $\dfrac{\partial z}{\partial v} = \dfrac{-4 + u^4}{u^3 v^2}$

9. $\dfrac{\partial z}{\partial u} = \dfrac{2(u - v) - 4u(u^2 + v^2)}{(u - v)^2 - (u^2 + v^2)^2}$;

$\dfrac{\partial z}{\partial v} = \dfrac{-2(u - v) - 4v(u^2 + v^2)}{(u - v)^2 - (u^2 + v^2)^2}$

11. $\dfrac{\partial z}{\partial r} = 4(r + s) \cos [2(r + s)^2] \cos [3(r - s)^2]$

$\qquad - 6(r - s) \sin [2(r + s)^2] \sin [3(r - s)^2]$;

$\dfrac{\partial z}{\partial s} = 4(r + s) \cos [2(r + s)^2] \cos [3(r - s)^2]$

$\qquad + 6(r - s) \sin [2(r + s)^2] \sin [3(r - s)^2]$

13. $\dfrac{\partial z}{\partial r} = \dfrac{1}{r}\left(r^s - \dfrac{s}{r} \ln r\right) + \dfrac{s}{r}\left(r^s \ln r + \dfrac{1}{r}\right)$;

$\dfrac{\partial z}{\partial s} = \left(r^s \ln r + \dfrac{1}{r}\right) \ln r$

15. $1 + \csc t + \tan^2 t - \csc t \sec^2 t$

17. $\dfrac{e^{2t} - e^{-2t} + 4t}{\sqrt{e^{2t} + e^{-2t} + 4t^2}}$ **19.** $9t^2 \cos 3t^3$

21. $\dfrac{\partial w}{\partial u} = \dfrac{2v^2}{u^3(u^2 + v^2)^2} (-u^4 + 2u^2v^2 + v^4)$;

$\dfrac{\partial w}{\partial v} = \dfrac{2v}{u^2(u^2 + v^2)^2} (u^4 - 2u^2v^2 - v^4)$

23. $\dfrac{\partial w}{\partial u} = uv^4[1 + u + 2 \ln (uve^u e^v)]$;

$\dfrac{\partial w}{\partial v} = u^2v^3[1 + v + 4 \ln (uve^u e^v)]$

25. $\dfrac{-3x^2 - 8xy + 3y^2}{4x^2 - 6xy + 6y^2}$ **27.** $\dfrac{-2x - y^2 \cos xy^2}{2y + 2xy \cos xy^2}$

29. $-\dfrac{x(y^2 - 1)^2}{y}$

31. $\dfrac{\partial z}{\partial x} = \dfrac{1 - yz \sin xyz}{y + xy \sin xyz}$; $\dfrac{\partial z}{\partial y} = \dfrac{-z - xz \sin xyz}{y + xy \sin xyz}$

43. $\dfrac{dQ}{dt} = \dfrac{\pi r^3}{40l\eta} (2p - r)$

45. $\dfrac{dF}{dt} = -\dfrac{GM}{6400^2}\left(40 + \dfrac{m}{32}\right)$

Section 13.5

1. 0   **3.** $(96 + 72\sqrt{3})/625$

5. 0   **7.** $24/\sqrt{41}$

9. $\frac{64}{3}$ **11.** $\sqrt{6}/3$

13. $\frac{6}{5}\sqrt{5}$ **15.** $-\frac{1}{4}\sqrt{5}(1 + \ln 2)$

17. a. $\dfrac{160}{49\sqrt{17}}$

b. $\dfrac{800}{49\sqrt{17}} \approx 3.96$ degrees Celsius per second

Section 13.6

1. $3\mathbf{i} - 5\mathbf{j}$

3. $-2e^{-2x} \ln (y - 4)\mathbf{i} + \dfrac{e^{-2x}}{y - 4}\mathbf{j}$

5. $4x\mathbf{i} - 2y\mathbf{j} - 8z\mathbf{k}$

7. $\dfrac{y - z}{(-x + z)^2}\mathbf{i} + \dfrac{1}{-x + z}\mathbf{j} + \dfrac{x - y}{(-x + z)^2}\mathbf{k}$

9. $-\frac{39}{8}\mathbf{i} - \frac{13}{4}\mathbf{j}$ **11.** $-2\mathbf{i} - 2\mathbf{j}$

13. $-2\mathbf{j}$ **15.** $\mathbf{i} + \mathbf{j}$; $\sqrt{2}$

17. $-6\mathbf{i} + 8\mathbf{j}$; 10

19. $e\mathbf{i} + e\mathbf{j} + 2e^{-2}\mathbf{k}$; $e\sqrt{2 + 4e^{-6}}$

21. $(\pi/12)(-4\mathbf{i} - 3\mathbf{j})$ **23.** $\frac{1}{4}(-\mathbf{i} - \mathbf{j})$

25. $\pi\mathbf{i} + \dfrac{\pi}{12}\mathbf{j}$ **27.** $-12\mathbf{i} + 8\mathbf{j} - \mathbf{k}$

29. $-\mathbf{k}$ **31.** $x + y - z = -5$

33. $\pi\sqrt{2}x - \pi\sqrt{2}y - z = -4\pi$ and

$$\frac{\pi\sqrt{3}}{6}x - \frac{\pi\sqrt{3}}{4}y - z = -\frac{\pi\sqrt{3}}{6} + \frac{1}{2}$$

35. $12x - 12y - z = 12$

37. $2x + z = -2$ and $x - y + z = \ln 2 - 2$

39. $x - y - \sqrt{2}z = 2$ **41.** $2x - \frac{1}{2}y - z = 3$

43. $z = -1$ **45.** $z = 0$

47. $(0, -2, 5)$

59. $\dfrac{\partial z}{\partial x}\bigg|_{(\sqrt{2},\,0,\,1)} = \sqrt{2}; \dfrac{\partial z}{\partial y}\bigg|_{(\sqrt{2},\,0,\,1)} = 0; \sqrt{2}x - z = 1$

61. $\dfrac{\partial z}{\partial x}\bigg|_{(1,\,1,\,e)} = -e = \dfrac{\partial z}{\partial y}\bigg|_{(1,\,1,\,e)}; ex + ey + z = 3e$

63. $\sqrt{43}$

Section 13.7

1. 5.03   **3.** -0.04

5. $1 - (0.025)\pi \approx 0.921460$

7. 12.99   **9.** 2   **11.** -0.1

13. $(9x^2 - 2xy)dx + (-x^2 + 1)dy$

15. $2x\,dx + 2y\,dy$

17. $(\tan y - y\csc^2 x)dx + (x\sec^2 y + \cot x)\,dy$

19. $\dfrac{xz^2}{\sqrt{1 + x^2 + y^2}}\,dx + \dfrac{yz^2}{\sqrt{1 + x^2 + y^2}}\,dy$
$+ 2z\sqrt{1 + x^2 + y^2}\,dz$

21. $e^{y^2 - z^2}\,dx + 2yxe^{y^2 - z^2}\,dy - 2zxe^{y^2 - z^2}\,dz$

23. approximately $\frac{11}{1600} = 0.006875$ ohms

Section 13.8

1. relative minimum value at $(3, -2)$

3. saddle point at $(-3, 2)$

5. relative maximum value at $(11, -8)$

7. relative minimum value at $(1, 4)$; saddle points at $(-3, 0)$ and $(5, 0)$

9. relative minimum value at $(0, 0)$; saddle points at $(\frac{1}{2}, -1)$ and $(2, 2)$

11. relative minimum value at $(-\frac{2}{3}, \frac{4}{3})$; saddle points at $(0, 0)$, $(0, 4)$, and $(-2, 0)$

13. saddle point at $(0, 0)$

15. no critical points

17. Any point of the form $(u, 0)$ or $(0, v)$ is a critical point; relative minimum value at $(0, 0)$

19. saddle point at $(0, 0)$

21. critical point at $(\pi/2 + m\pi, \pi/2 + n\pi)$, for any integers m, n; relative maximum value if m, n are even; relative minimum value if m, n are odd; otherwise a saddle point

23. relative minimum value at any point on the line $y = -ax - b$

25. maximum value: 1; minimum value: -1

27. maximum value: 5; minimum value: 0

29. All three numbers are 16; the product is 4096.

33. $(16, 16, 16)$ **37.** $y = -\frac{9}{14}x - \frac{3}{14}$

39. $l^3/1728$

41. the route with $x = \frac{3}{4}$ (kilometers) and $y = 2$ (kilometers)

Section 13.9

1. maximum value: 17/4; minimum value: -2

3. maximum value: 2; minimum value: -2

5. maximum value: $\sqrt{2}$; minimum value: $-\sqrt{2}$

7. 3   **9.** $\frac{11}{96}$

11. maximum value: 6; minimum value: -4

13. maximum value: $\sqrt{2}$; minimum value: $-\sqrt{2}$

15. $(1, 0, 0)$ and $(-1, 0, 0)$

17. $\frac{3}{2}\sqrt{3}$ **19.** $\frac{3}{8}\sqrt{3}$ **21.** 8

23. $(-1 - \frac{1}{2}\sqrt{2}, -\frac{1}{2})$ and $(-1 + \frac{1}{2}\sqrt{2}, -\frac{1}{2})$

25. $\frac{3}{4}\sqrt{3}r^2$

27. $x = (196)^{1/3}$, $y = (\frac{108}{7})^{1/3}$, and $z = (\frac{4}{7})^{1/3}$

29. length: $2\sqrt{2}$ meters; radius: $\sqrt{2}$ meters

31. $x = y = z = V^{1/3}$

Chapter 13 Review Exercises

1. all (x, y) such that $x^2/4 + y^2/25 \le 1$

3. all (x, y, z) such that $x - y + z > 0$

5. the hyperbolas $xy = 1$ and $xy = -1$, respectively

7.

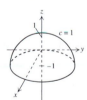 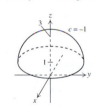

9.   **11.**   **13.**

15. 10

17. $f_x(x, y) = 12x^2; f_y(x, y) = -6y$

19. $f_x(x, y, z) = 2xe^{x^2}\ln(y^2 - 3z); f_y(x, y, z) = \dfrac{2ye^{x^2}}{y^2 - 3z};$

$f_z(x, y, z) = \dfrac{-3e^{x^2}}{y^2 - 3z}$

21. $k_x(x, y, z) = 5xz^{5/4}[\tan^{3/2}(x^2 + y)] \sec^2(x^2 + y);$
  $k_y(x, y, z) = \frac{5}{2}z^{5/4}[\tan^{3/2}(x^2 + y)] \sec^2(x^2 + y);$
  $k_z(x, y, z) = \frac{5}{4}\sqrt[4]{z}\tan^{5/2}(x^2 + y)$

25. $g_{uu}(u, v) = -2/u^2;\ g_{uv}(u, v) = g_{vu}(u, v) = g_{vv}(u, v) = 0$

29. $4t^3 e^{t^4} - \frac{1}{2}(3t^2 - 1)e^{(t^3 - t)/2}$

31. $\sqrt{5}t/|t|$

35. $\dfrac{\partial z}{\partial x} = \dfrac{z(x^2y^2z^2 - 1)}{x(3xyz^4 + 1 - x^2y^2z^2)};$

  $\dfrac{\partial z}{\partial y} = \dfrac{z(x^2y^2z^2 - 1)}{y(3xyz^4 + 1 - x^2y^2z^2)}$

37. $2\sqrt{3}/25$ **39.** $\mathbf{j}$

41. $-2\pi\mathbf{i} - \mathbf{j}$ **43.** $\frac{1}{2}\mathbf{i} + \frac{1}{2}\mathbf{j}$

45. normal: $-4\mathbf{i} + 9\mathbf{j} - \mathbf{k}$;
  equation of plane: $-4x + 9y - z = 10$

47. normal: $\frac{6}{5}\mathbf{i} - \mathbf{j} - \mathbf{k}$;
  equation of plane: $6x - 5y - 5z = -1$

49. $x - 2y + z = -1$

51. $(2, -2, 3)$ and $(-2, 30, 3)$

55. $\pi/4 - 0.04 \approx 0.745398$

57. $\dfrac{1}{x}\,dx - \dfrac{1}{y}\,dy$

59. relative minimum value at $(1, 2)$

61. relative minimum values at $(2, 2)$ and $(-2, -2)$

63. maximum value: $3 + \sqrt{3}/2$; minimum value: $3 - \sqrt{3}/2$

65. maximum value: $\frac{1}{2}$; minimum value: 0

67. $4\sqrt{3}, 2\sqrt{3},$ and $\frac{4}{3}\sqrt{3}$

Cumulative Review Exercises (Chapters 1–12)

1. $-1/(2\sqrt{x})$ for $x > 0$ **2.** 1

3. $y = 4\sqrt{2}x - 2$ and $y = -4\sqrt{2}x - 2$

6. $-2x \sin|x|$

7. $\frac{9}{2}x^{-3/2}$

8. $12\sqrt{3} - 4\pi$ square centimeters per minute

9. relative minimum value: $f(0) = 2$

10. height: 20 inches; side length of bottom: 8 inches

11. $\frac{37}{12}$

12. $\frac{1}{3}x^3 \tan^{-1} x - \frac{1}{6}x^2 + \frac{1}{6}\ln(1 + x^2) + C$

13. $\frac{1}{2}\theta + \frac{1}{4}\sin 2\theta - \sin\theta + \frac{1}{3}\sin^3\theta + C$

14. $-\frac{2}{5}(4 - x^2)^{5/2} - \frac{1}{6}x(4 - x^2)^{3/2} - x\sqrt{4 - x^2} -$
  $4 \sin^{-1}\dfrac{x}{2} + C$

15. converges; $\frac{1}{4}$ **16.** $\pi^4/48 - \pi^2/8$

17. $(0, 7/5)$ **18.**

19. $\frac{1}{24}\pi + \frac{1}{4}\sqrt{3}$ **20.** 0

21. 1 **22.** $[-4, 4)$

23. of f: $\displaystyle\sum_{n=0}^{\infty} \dfrac{2(-1)^n}{n!}x^{3n+3}$; of f': $\displaystyle\sum_{n=0}^{\infty} \dfrac{2(-1)^n(3n+3)}{n!}x^{3n+2}$;
  radius of convergence: ∞

24. $(1, 2, 0)$

25. a. $t = -\frac{1}{2}$ **b.** $2/(2t^2 + 1)^2$

26. $x + C = \dfrac{c}{a}y + \dfrac{ad - bc}{a^2}\ln|ay + b|$

CHAPTER 14
Section 14.1

1. a. 11 **b.** 8 **3.** 1 **5.** $e^2 - 2e + 1$

7. $-\frac{1}{6}$ **9.** $\frac{1}{12}$ **11.** $\frac{8}{3}$

13. $2\pi/3$ **15.** $(\pi/2)\ln 3$ **17.** $\frac{1}{2}(e - 1)$

19. 0 **21.** $\frac{40}{3}$ **23.** $\frac{74}{3}$

25. 1296 **27.** $\frac{3}{2}\pi^2 + \pi + 2$ **29.** $\frac{27}{4}$

31. 0 **33.** $\frac{1}{2}$ **35.** 6

37. $\frac{1}{6}$ **39.** $\frac{8}{3}$ **41.** $\frac{2}{3}$

43. $\frac{1}{8}$ **45.** **47.**

 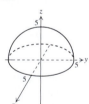

49. **51.** $\frac{49}{3}$ **53.** $\frac{512}{3}$

55. $\frac{4}{3}\sqrt{2} - \frac{7}{6}$ **57.** $\frac{1}{2}(e - 1)$ **59.** $\frac{1}{4}(e^{16} - 1)$

61. $e/2 - 1$ **63.** $\frac{4}{3}$ **65.** 2.23099

67. 0.272357　　　　**69.** 0

71. approximately 11,200 cubic meters　　**73.** $\frac{11}{18}$

Section 14.2

1. 0　　　　　　　**3.** $\frac{4}{3}(\sqrt{3}+1)$　　　**5.** 35π

7. 18π　　　　　**9.** 4π　　　　　**11.** $2\sqrt{3}\pi$

13. $\frac{256}{9}$　　　　　**15.** $9\pi/2$　　　　**17.** 4

19. $\pi+3\sqrt{3}$　　　　　**21.** $\frac{1}{8}(e^{12\pi}-2e^{6\pi}+1)$

23. $\pi/4$　　　　　　**25.** $(\pi/4)(1-\cos 1)$

27. $(\pi/4)(1-e^{-1})$　　　**29.** $3\pi/32$

31. a. $\frac{1}{2}(1-e^{-1})$　　**b.** no

Section 14.3

1. $3\sqrt{14}$　　　　　　　**3.** $(\pi/6)(17^{3/2}-1)$

5. $32(\pi-2)$　　　**7.** $\frac{1}{12}(5^{3/2}-1)$　　**9.** 70π

Section 14.4

1. -54　　　　　**3.** 0　　　　　**5.** $\frac{2}{3}e^{(\ln 3)^2}-\frac{2}{3}$

7. $\frac{13}{3}$　　　　　**9.** $\frac{2}{9}$　　　　　**11.** $2e-4$

13. $\frac{2}{3}e^6-2e^2+\frac{4}{3}$　　**15.** $\frac{1}{15}$　　　　**17.** $\frac{50}{3}$

19. 0　　　　　　**21.** $\frac{34}{3}$　　　　**23.** $\frac{71}{105}$

25. $\frac{1}{3}\pi h$　　　　**27. a.** $\frac{1}{48}$　**b.** $\frac{1}{24}$　**c.** $\frac{1}{96}$

29. $81\pi/4$　　　　**31. a.** $\frac{7}{5}$　**b.** $1/(3\pi)$

Section 14.5

1. $r\sin\theta=-4$

3. $r(\cos\theta+\sin\theta)+z=3$

5. $r^2+z=1$　　　　　**7.** $4r^2=z^2$; or $z=2r$

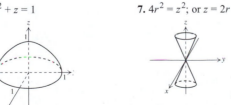

9. $3\pi(e^5-1)$　　**11.** $-\frac{16}{15}$　　　**13.** 0

15. $\pi/16$　　　　**17.** $\frac{32}{15}$　　　**19.** $\pi/5$

21. $\pi(1-e^{-1})$　　**23.** $\frac{8}{3}\pi(2-\sqrt{3})$　**25.** $\frac{7}{3}\pi$

27. $\pi/32$　　　　　　**29.** $\dfrac{4a^3}{3}\left(\dfrac{\pi}{2}-\dfrac{2}{3}\right)$

31. π　　　　　**33.** $\frac{243}{4}\pi$　　　**35.** $\pi h/3$

37. $\frac{1216}{3}\pi\approx 1273$ (grams)

Section 14.6

1. a. $(\sqrt{3}/2,\frac{1}{2},0)$

　b. $(0,0,-2)$

　c. $(-\frac{3}{4}\sqrt{2},-\frac{3}{4}\sqrt{6},\frac{3}{2}\sqrt{2})$

　d. $(-\sqrt{6}/8,-\sqrt{6}/8,\frac{1}{4})$

　e. $(0,0,1)$

　f. $(5,0,0)$

3. $(\pi/3)(2-\sqrt{2})$　　　　**5.** $\frac{31}{40}\pi^2$

7. $\frac{27}{4}\pi(\sqrt{3}-\frac{3}{4}\sqrt{2})+\frac{81}{4}(1-\sqrt{2})$

9. $\frac{844}{15}\pi$　　　**11.** $6\pi(2-\sqrt{3})$　**13.** $\frac{128}{63}\pi$

15. $\frac{8}{3}\pi(2-\sqrt{2})$　**17.** 7π　　　**19.** $16\pi/3$

21. $64\pi(16+9\sqrt{3})/3$　　**23.** $\pi/6$

25. 24π　　　　　　**27.** $1-\cos 1$

Section 14.7

1. $(0,\frac{17}{5})$　　　　　**3.** $(\frac{1}{2},\frac{1}{2})$

5. $(\frac{5}{6},0)$　　　　　**7.** $(0,0,\frac{3}{8}a)$

9. $(0,0,\frac{3}{4})$

11. $(8\sqrt{2}/(15\pi),8\sqrt{2}/(15\pi),\frac{1}{2})$

13. $(0,0,\frac{1}{2})$　　　　**15.** $(0,0,0)$

17. $(\frac{7}{6},1,1)$　　　　**19.** $(0,0,16/(5\pi))$

21. $(0,0,\frac{80}{9})$

23. $I_x=I_y=I_z=\frac{25{,}000}{3}\pi$

25. $I_x=I_y=624\pi$; $I_z=96\pi$

27. approximately 6.113×10^4 gram centimeters squared

29. 225 kilogram centimeters squared (or 2.25×10^5 gram centimeters squared)

Section 14.8

1. $\frac{5}{2}$　　　　**3.** $2v^2-2u^2$　　　**5.** $-e^{2v}$

7. ab　　　　**9.** $\frac{1}{4}(e^2-3)$　　**11.** $\frac{133}{2500}$

13. $\frac{1}{4}(e^4-e)\ln\frac{3}{2}$　　　**15.** $\frac{1}{2}(e^4-e)\ln 2$

17. $\frac{315}{128}\pi$　　　　　　**19.** $\frac{5}{12}$

21. $\frac{1}{4}a^3 b\pi$　　　　　**23.** 0

25. $\frac{1}{2}(e-1)$　　　　　**27.** $\frac{1}{4}(b^2-a^2)\ln 3$

31. $4\pi/\sqrt{3}$

Section 14.9

5. $\mathbf{r}(\theta, z) = \cos\theta\mathbf{i} + \sin\theta\mathbf{j} + z\mathbf{k}$ for $0 \le \theta \le 2\pi$ and $-1 \le z \le 1$

7. $\mathbf{r}(r, \theta) = r\cos\theta\,\mathbf{i} + (r\cos\theta + r\sin\theta - 3)\,\mathbf{j} + r\sin\theta\,\mathbf{k}$, for $0 \le r \le 2$ and $0 \le \theta \le 2\pi$

9. $\mathbf{r}(\rho, \theta) = \dfrac{1}{\sqrt{2}}\,\rho\,\mathbf{i} + \dfrac{1}{\sqrt{2}}\,\rho\sin\theta\mathbf{j} + \dfrac{1}{\sqrt{2}}\,\rho\cos\theta\mathbf{k}$ for $0 \le \rho$ and $0 \le \theta \le 2\pi$

11. $\mathbf{r}(x, \theta) = x\mathbf{i} + \cos x\cos\theta\,\mathbf{j} + \cos x\sin\theta\,\mathbf{k}$ for $-\pi/2 \le x \le 2\pi$ and $0 \le \theta \le 2\pi$

13. $\mathbf{r}(\theta, z) = e^z\cos\theta\mathbf{i} + e^z\sin\theta\mathbf{j} + z\mathbf{k}$ for $0 \le z \le 1$ and $0 \le \theta \le 2\pi$

15. $\mathbf{r}(x, \theta) = x\mathbf{i} + \dfrac{1}{x}\cos\theta\mathbf{j} + \dfrac{1}{x}\sin\theta\mathbf{k}$ for $x \ge 1$

17. $3\sqrt{2}(x - 1) - 2\sqrt{2}(y + 3/2) + 3(z - 2\sqrt{2}) = 0$

19. $30(x - 4) + 40(y + 6) = 0$ or $3x + 4y + 12 = 0$

21. $z = -2$

23. $4\pi^2 br$

25. $\dfrac{\alpha}{\omega}\left[\dfrac{\omega r}{2}\sqrt{c^2 + \omega^2 r^2} + \right.$
$\left. \dfrac{c}{2}\ln\left|\omega r + \sqrt{c^2 + \omega^2 r^2}\right| - \dfrac{\alpha c^2}{2\omega}\ln|c| \right]$

27. b. $\dfrac{\pi}{6}(37\sqrt{37} - 1)$

31. b. ii. 39° north (latitude) is approximately .89 radians 77° west (longitude) is approximately 1.34 radians
c. approximately 70.7%

Chapter 14 Review Exercises

1. $\frac{4}{3}(e - 1)$ 　　　　　　　**3.** $\frac{3}{2}(e^2 - e^{-2}) - 6$

5. $\frac{1}{3}(e - 1)$

7. $\ln\dfrac{2 + \sqrt{3}}{1 + \sqrt{2}} - \dfrac{\sqrt{2}}{4}\ln 3$

9. $\frac{1}{3}a^2$ 　　　　　**11.** $\pi + 6\sqrt{3}$

13. -4 　　　　　**15.** $\frac{16}{15}\pi(9 - 4\sqrt{3})$

17. 0 　　　　　**19.** $\frac{2}{3}\pi(2\sqrt{2} - 1)$

21. $\pi/6$ 　　　**23.** $\frac{256}{9}$ 　　　**25.** 32π

27. 20 　　　**29.** $\frac{471}{2}\pi$ 　　　**31.** $(0, 0, \frac{12}{5})$

33. $(0, 0, \frac{1}{2})$ 　　　**35.** 1 　　　**37.** $\frac{55}{6}$

39. $\mathbf{r} = r\cos\theta\,\mathbf{i} + (2 - r\cos\theta - r\sin\theta)\mathbf{j} + r\sin\theta\,\mathbf{k}$ for $0 \le r \le 2$ and $0 \le \theta \le 2\pi$

41. b. $4(x - 2) + 8(z + 2) = 0$, 　or　 $x + 2z = -2$

43. $2\pi\left(\cosh 1 + \dfrac{1}{3}\cosh^3 1\right) - \dfrac{8\pi}{3}$ 　　**45.** $\pi - 2$

Cumulative Review Exercises (Chapters 1–13)

1. $\frac{2}{3}x$ 　　　　　　　**2.** e^{-8}

3. a. $[2, \infty)$ 　　**b.** $\dfrac{1}{2\sqrt{x + \sqrt{x^2 - 4}}}\left(1 + \dfrac{x}{\sqrt{x^2 - 4}}\right)$

4. $-(\ln\pi)/\pi$

5. $(\frac{2}{3}\sqrt{3}, -\frac{2}{3}\sqrt{3})$ and $(-\frac{2}{3}\sqrt{3}, \frac{2}{3}\sqrt{3})$

6. $\frac{32}{9}$ cubic inches per minute

7. front: 12 feet; side: 18 feet 　　　　**8.** 2

9. relative minimum value: $f(e) = e$; inflection point: $(e^2, \frac{1}{2}e^2)$

10. $\frac{1}{6}(1 + 4x^2 + 9x^4)^{3/2} + C$

11. $-\frac{1}{5}\cos^5 x + \frac{1}{7}\cos^7 x + C$

12. $2x + 3\ln|x| + \ln|x - 1| - \dfrac{5}{x - 1} + C$

13. finite area; $\dfrac{\pi}{3}$ 　　　　　**14.** $\sqrt{2}\,\pi/12$

15. $\frac{2}{3}(2^{3/2} - 1)$ 　　　　　**16.**

18. converges 　　　　　**19.** $[-1, 1]$

20. approximately 0.1875763126

21. $300\sqrt{2}$ foot-pounds

22. $\mathbf{v}(t) = 750\sqrt{3}\mathbf{i} + (750 - 32t)\mathbf{k}$

23. a. $\mathbf{T}(t) = \dfrac{t^2}{t^2 + 2}\mathbf{i} + \dfrac{2}{t^2 + 2}\mathbf{j} + \dfrac{2t}{t^2 + 2}\mathbf{k}$; 0

b. $\dfrac{2t}{t^2 + 2}\mathbf{i} - \dfrac{2t}{t^2 + 2}\mathbf{j} + \dfrac{2 - t^2}{t^2 + 2}\mathbf{k}$

c. $-\frac{2}{3}\mathbf{i} - \frac{1}{3}\mathbf{j} + \frac{2}{3}\mathbf{k}$

24. a. $g_x(0, 0) = 1; g_y(0, 0) = 0$ 　**b.** It is continuous.

25. a.

b. $\dfrac{x + 1}{2} = \dfrac{y - 1}{-2} = \dfrac{z + 1}{-1}$

26. $2\sqrt{2}$, $\sqrt{2}$, and 1

CHAPTER 15

Section 15.1

1.

3. curl $\mathbf{F}(x, y) = \mathbf{0}$; div $\mathbf{F}(x, y) = 2$

5. curl $\mathbf{F}(x, y, z) = \mathbf{0}$; div $\mathbf{F}(x, y, z) = 2(x + y + z)$

7. curl $\mathbf{F}(x, y, z) = -\dfrac{y}{z^2}\mathbf{i} + \dfrac{x}{z^2}\mathbf{j}$;

div $\mathbf{F}(x, y, z) = -\dfrac{2}{z} - \dfrac{1}{z^2}$

9. curl $\mathbf{F}(x, y, z) = 2e^x \sin y\mathbf{k}$;

div $\mathbf{F}(x, y, z) = 2e^x \cos y + 1$

11. positive

17. $f(x, y) = xe^y + y^2/2 + C$

19. $f(x, y, z) = x^2yz + z + C$

21. not a gradient   **23.** not a gradient

25. a. vector field   **b.** meaningless

    c. vector field

    d. vector field

    e. vector field

    f. function of several variables

    g. vector field

    h. function of several variables

    i. meaningless

41. $\mathbf{G}(x, y, z) = 8y\mathbf{k}$ (Other solutions are possible.)

43. b. not a central force field

Section 15.2

1. $\frac{26}{3}\sqrt{13}$ **3.** $\frac{1}{54}(1 - 10^{3/2})$

5. $4\sqrt{2}(1 - e^{-1})$ **7.** $\frac{2032}{63}$

9. $3\sqrt{2}$ **11.** $\frac{5}{2}\sqrt{2} - \frac{21}{4}$

13. $-\pi/4 + \pi^4/4 + \frac{1}{3}$ **15.** $-\pi$

17. 0 **19.** $-\frac{5}{3}$

21. $e^{-1} - e - \frac{2}{3}$ **23.** $\frac{57}{2}$

25. $\pi/4$ **27.** $4\ln 2 - \frac{9}{2}$

29. a. 3 **b.** $\frac{1}{2}$ **33.** $\sqrt{17}(\pi + \frac{112}{3}\pi^3)$

35. a. $\frac{3}{2}$ **b.** $2 - 2/\pi$ **37.** $-28{,}000$ foot-pounds

Section 15.3

1. $e^2 + 13$ **3.** 1

5. $\ln 2$ **7.** $\frac{1}{2}$

13. approximately 1294 joules

15. 6000 joules

Section 15.4

1. $-\pi$ **3.** $\frac{1}{2}$ **5.** 0

7. -3π **9.** 0 **11.** $\frac{1}{6}$

13. 0 **15.** 8π **17.** 0

19. 3π **21.** $1/\pi - 2/\pi^2$ **23.** $(0, -\frac{1}{2})$

Section 15.5

1. $3\sqrt{14}$ **3.** $12\pi\sqrt{10}$

5. $\frac{5}{8}\pi$ **7.** $\frac{1}{4}(17^{3/2} - 1)$

9. 16π **11.** 6

13. $\frac{623}{48}\sqrt{5}\pi$ **15.** $125{,}000\pi$ (pounds)

17. $250{,}000\pi$ (pounds)

Section 15.6

1.   **3.**

5. 0 **7.** 2π

9. $\frac{1}{2}\pi^2$ **11.** 0

13. $2000\pi\sqrt{10}$ **15.** $-\pi/6$

Section 15.7

1. $-\frac{4}{3} - \pi/4$ **3.** -2π **5.** $-3\pi/4$

7. 0 **9.** $-\frac{8}{3}$ **11.** $\pi/4$

13. $\pi/2$ **15.** 0 **17.** $-\frac{1}{2}$

19. $-3\pi/4$ **23.** $\pi/\sqrt{2}$

Section 15.8

1. yes   **3.** yes   **5.** no

7. no   **9.** $\frac{1}{24}$ **11.** $3\pi/4$

13. 2π **15.** 16π **17.** $\pi/2$

19. $-20\pi\sqrt{3}$ **21.** $\frac{32}{15}\sqrt{2}$ **23.** 648π

27. $\frac{1}{3}\pi a^2 h$

35. Σ is the boundary of a simple solid region, so has no boundary curve C.

Chapter 15 Review Exercises

3. $f(x, y) = xy^2 + \cos xy + C$

5. $\frac{13}{96}\sqrt{2}\pi^3$ **7.** $\pi/2 - \frac{5}{6}$

9. $\frac{1}{12}(5^{3/2} - 1)$ **11.** $\frac{1}{2} + e(\cos 1 - 1)$

13. $\frac{16}{7}$

15. 20π

17. 0

19. 0

21. -32π

23. $-\frac{1}{2}\sqrt{2}\,\pi$

25. $\frac{2}{5}$

27. a. $\frac{1}{4}\pi^4$ **b.** $\frac{1}{4}\pi^4$

Cumulative Review Exercises (Chapters 1–14)

1. $\frac{1}{2}\sqrt{2}$

2. $-\frac{1}{8}$

3. 3

4. There are none.

5. $\pi/108$ cubic inches per second

6. $\frac{4}{3}r$

7. relative minimum value: $f(-\sqrt{2}) = \dfrac{-1 - \sqrt{2}}{2}$;

relative maximum value: $f(\sqrt{2}) = \dfrac{-1 + \sqrt{2}}{2}$;

inflection points: $(-2, -1), \left(1 - \sqrt{3}, \dfrac{-1 - \sqrt{3}}{4}\right)$,

$\left(1 + \sqrt{3}, \dfrac{-1 + \sqrt{3}}{4}\right)$

8. $\frac{14}{15}$

9. $\frac{1}{2}x^2 \tan^2 x - x \tan x - \ln |\cos x| + \frac{1}{2}x^2 + C$

10. $\ln |x| - \frac{1}{2}\ln (x^2 + 4) + \frac{3}{2}\tan^{-1}\dfrac{x}{2} + C$

11. converges; π

12. a. $\frac{104}{15}$ **b.** $\frac{8}{3}\pi$

13. a. $r = \frac{1}{2}\sqrt{3}\,\sin 2\theta$ **b.** $\frac{1}{2}\sqrt{3}$

14. $\frac{243}{9100}$

15. converges absolutely

16. $(-1/e, 1/e)$

18. a. $x/4 = (y - 1)/(-1) = (z - 2)/1$ **b.** $\frac{11}{6}\sqrt{2}$

19. a. 5 **b.** 2 **c.** $1/\sqrt{t}$

20. a. $\dfrac{2x^2}{(2x^2 + y^2)^{3/2}}$ **b.** $-\sqrt{\frac{2}{3}}\mathbf{i} - \sqrt{\frac{1}{3}}\mathbf{j}$

22. $\sqrt{7}$

23. $\sin 2 + \sin \frac{3}{4}$

24. $\dfrac{\pi}{6}(17^{3/2} - 1)$

25. $\frac{28}{3}\pi$

26. $\frac{1}{12}(1 - \cos 64)$

APPENDIX

1. least upper bound: 1; greatest lower bound: -1

3. least upper bound: π; greatest lower bound: 0

5. least upper bound: 5; greatest lower bound: 0

7. least upper bound: 1; greatest lower bound: 0

INDEX OF SYMBOLS

INDEX